Animal Physiology

From Genes to Organisms second edition

Lauralee Sherwood
Department of Physiology and Pharmacology
School of Medicine
West Virginia University

Hillar Klandorf
Division of Animal and Nutritional Sciences
West Virginia University

Paul H. Yancey
Department of Biology
Whitman College

BROOKS/COLE
CENGAGE Learning

Australia • Brazil • Japan • Korea • Mexico • Singapore • Spain • United Kingdom • United States

BROOKS/COLE
CENGAGE Learning™

Animal Physiology: From Genes to Organisms, Second Edition
Lauralee Sherwood, Hillar Klandorf, and Paul H. Yancey

Publisher: Yolanda Cossio

Developmental Editor: Suzannah Alexander

Assistant Editor: Alexis Glubka

Editorial Assistant: Lauren Crosby

Media Editor: Lauren Oliveira

Marketing Manager: Tom Ziolkowski

Marketing Coordinator: Jing Hu

Marketing Communications Manager: Linda Yip

Content Project Manager: Hal Humphrey

Design Director: Rob Hugel

Art Director: John Walker

Print Buyer: Karen Hunt

Rights Acquisitions Specialist: Roberta Broyer

Production Service: Graphic World Inc.

Text Designer: Jeanne Calabrese

Photo Researcher: Jeremy Glover, Bill Smith Group

Text Researcher: Sue C. Howard

Copy Editor: Graphic World Inc.

Illustrators: Precision Graphics, Thompson Type, Graphic World Inc.

Cover Designer: John Walker

Cover Image: Northern hawk owl; Ben Cranke/Getty Images

Compositor: Graphic World Inc.

Library of Congress Control Number: 2011935353

ISBN-13: 978-0-8400-6865-1

ISBN-10: 0-8400-6865-4

Brooks/Cole
20 Davis Drive
Belmont, CA 94002-3098
USA

Cengage Learning is a leading provider of customized learning solutions with office locations around the globe, including Singapore, the United Kingdom, Australia, Mexico, Brazil, and Japan. Locate your local office at **www.cengage.com/global.**

Cengage Learning products are represented in Canada by Nelson Education, Ltd.

To learn more about Brooks/Cole, visit **www.cengage.com/brookscole.**

Purchase any of our products at your local college store or at our preferred online store **www.CengageBrain.com.**

Printed in the United States of America
1 2 3 4 5 6 7 15 14 13 12 11

Dedication

To my family, for all they have done for me in the past, all they mean to me in the present, and all I hope will yet be in the future.

And to Hillar and Paul for taking the ball and running with it. Their expertise and dedication made this book a reality.

Lauralee Sherwood

To my family, Britt, Alex, and Emma.

To the many individuals who have contributed to this edition, particularly our editors, Yolanda Cossio and Suzannah Alexander.

A special thank you to Paul and Laurie for making this work possible. It remains a privilege to share this undertaking with you. Your insights and clarity of thought are amazing!

And finally to one of the best scientists I know, Gerald Lincoln, whose contribution to the subject of seasonal breeding in animals was brilliant. His enthusiasm for life is infectious.

Hillar Klandorf

To my wife, son, and parents for all their support, and Hillar and Laurie for their hard work and dedication.

Paul H. Yancey

Brief Contents

Contents

Section I: Foundations

Section II: Whole-Body Regulation and Integration

Section III: Support and Movement

8   Muscle Physiology   335

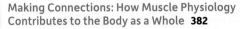
Section IV: Self-Maintenance

Section V: Reproduction

Preface

"There is grandeur in this view of life, with its several powers, having been originally breathed into a few forms or into one; and that, whilst this planet has gone cycling on according to the fixed law of gravity, from so simple a beginning endless forms most beautiful and most wonderful have been, and are being, evolved." —Charles Darwin, last line of *The Origin of Species.*

Even though the authors of this text have been teaching biology for many years, we continue to be awed by the grandeur of life as so eloquently expressed by Charles Darwin. As animal physiologists, we are continually amazed by the wondrously intricate functions that mesh together to sustain an animal's body. As biologists, we are also fascinated by the powerful and yet quirky mechanisms of evolution, which have allowed life to adapt from the bitter cold of Antarctica to the scalding heat of hydrothermal vents. Moreover, the fact that we humans have structures, processes, and genes in our cells that connect us to every life form on this planet is both humbling and inspiring. It is the grand view of life that we try to convey to our students.

This second edition of *Animal Physiology: From Genes to Organisms* focuses on the mechanisms and evolutionary histories of animal body functions from genes to organ systems to the whole organism interacting with its environment. This text is designed for mid- to upper-level undergraduate courses. Because this book may be the only exposure many undergraduate students will have to physiology beyond the introductory level, all aspects of physiology and many types of animals are covered. This text serves for general biology and zoology programs, preprofessional programs in animal husbandry, veterinary medicine and science, marine biology, and wildlife and fisheries biology, as well as for human health sciences.

We have three major goals in writing this second edition. First, to better understand scientific knowledge, students should realize that our knowledge is incomplete and yet continually advancing, and they should know how that knowledge is obtained. Therefore, pertinent information based on recent discoveries, as well as new research methodologies, has been included in all chapters, with new controversial ideas and hypotheses presented to illustrate that physiology is a dynamic, changing discipline. Students can be assured of the timeliness and accuracy of the material presented. Our second goal has been to clarify and modify figures and text as needed to strengthen our coverage of physiological features and to ensure that the text and art work well together. Third, we have sought to strengthen our comparative, evolutionary approach with greater emphasis on some nonmammalian systems as some of the more remarkable recent discoveries have been made in this arena.

Even the most tantalizing subject can be difficult to comprehend if not effectively presented. Thus, this book has a logical, understandable format with an emphasis on how each concept is an integral part of the entire subject. Too often, students view the components and functions of an animal as isolated entities; by understanding how each component depends on the others, students can appreciate the integrated functioning of the whole organism. Moreover, this text is designed to promote understanding of the basic principles and concepts of physiology rather than memorization of details. The text is written in simple, straightforward language, and every effort has been made to ensure smooth reading through good transitions, logical reasoning, and integration of ideas throughout the text.

Major Organizing Principles

This text follows an "integrative systems" approach, which involves both vertical and horizontal integration. *Vertical integration* concerns the hierarchical nature of physiology, in which organ systems and their functions arise from cells and their molecular components, and from the evolutionary forces that gave rise to those components. After an introduction to the major themes of physiology, the chapters proceed from molecular and cellular physiology to classical organ systems. Organ systems are grouped according to universal functions of whole-body regulation and integration, support and movement, maintenance, and reproduction. For each organ system, both the evolutionary forces and the cellular and molecular underpinnings are discussed, including new research in genomics, which is having a major impact on all areas of biology. Where necessary, sufficient relevant anatomy is included to make the inseparable relation between structure and function meaningful.

Horizontal integration concerns the interactions of all organ systems to yield a whole functioning organism. As we noted earlier, it is all too easy to view each system as an isolated entity. This text works to overcome this in two ways. First, this text's central themes are homeostasis and other integrated forms of regulation—how processes from the genetic to the behavioral level allow an animal's body to meet changing demands while maintaining the internal consistency necessary for all cells and organs to function. Each chapter not only emphasizes these themes but also indicates interactions with other systems. Second, special integrative chapters—Fluid and Acid–Base Balance and Energy Balance and Thermal Physiology—focus on crucial whole-body phenomena that are dependent on more than one organ system.

In addition to an integrated systems organization, this text also incorporates a *comparative approach* throughout that illustrates respective solutions animals have evolved to cope with the environment. Mechanisms used by major and exemplary groups of vertebrates, insects, and other illustrative groups are included for three purposes. First, comparisons among types of animals are selected to illuminate important universal functions and principles, emphasizing the unity of life. Second, unique or striking adaptations that reveal the diversity that can result from evolutionary adaptation are featured. Finally, this text incorporates detailed coverage of animal species of relevance to those preparing for animal-related careers, especially vertebrates.

The scope of this text has been limited by judicious selection of pertinent content that students can reasonably be expected to assimilate in a one-semester course. To keep pace with today's rapid advances in the biological sciences, students interested in careers related to animal physiology must be able to draw on their overall understanding of concepts instead of merely recalling isolated facts. For this reason, this text is designed to promote understanding of the basic principles and concepts of physiology rather than the memorization of details. Yet depth, where needed, is not sacrificed. All aspects of physiology receive coverage, including separate chapters on immune and reproductive systems and detailed coverage of pregastric and postgastric digestion.

Learning Aids and High-Interest Features

Pedagogical Illustrations Anatomic illustrations, schematic representations, photographs, tables, and graphs complement and reinforce the written material. Widespread use of integrated descriptions within figures, including numerous process-oriented figures with step-by-step descriptions, allows visually oriented students to review processes through figures. Flow diagrams are used extensively to help students integrate the written information. In the flow diagrams, lighter and darker shades of the same color denote a decrease or increase in a controlled variable, such as blood pressure or the concentration of blood glucose. Physical entities, such as body structures and chemicals, are distinguished visually from actions.

Feedforward Statements as Subsection Titles Instead of traditional short topic titles for each major subsection (for example, "Heart valves"), feedforward statements alert students to the main point of the subsection to come (for example, "Heart valves ensure that the blood flows in the proper direction through the heart"). These headings also break up large concepts into smaller, more manageable pieces for the student, and as an added bonus, the listing of these headings in the Contents at the beginning of the book serves as a set of objectives for each chapter.

Cross-Referencing Many chapters build on material presented in preceding chapters, yet each chapter is designed to stand on its own to allow the instructor flexibility in curriculum design. With extensive cross-references provided, the sequence of presentation can be varied at the instructor's discretion.

Unanswered Questions The physiological sciences are changing constantly due to new discoveries and challenges of old ideas. In most chapters, *Unanswered Questions* present short overviews and questions on controversial ideas and new hypotheses to illustrate the dynamic nature of physiology research and to stimulate students to propose their own ideas. Many of these have been updated for the second edition.

Boxed Features More detailed boxed features are inserted where appropriate within chapters. *Beyond the Basics* exposes students to high-interest, tangentially relevant information on diverse topics, including historical perspectives, environmental issues, and common diseases. *Challenges and Controversies* boxes introduce the frontiers of physiological research with new hypotheses and unsolved questions, going in more depth than in the *Unanswered Questions*. *A Closer Look at Adaptation* illustrates the scope and limits of evolutionary adaptation with selected illustrative examples. *Molecular Biology and Genomics* provides examples of cutting-edge research resulting from the animal genome projects and related research to uncover the physiological functions of newly discovered genes. Most of these boxes are new or updated in this second edition.

Evolutionary Foundations Most chapters have an overview of major evolutionary events and selective forces, typically at the chapter start. All topics are placed in a true biological context that melds both the proximate (mechanistic) approach of traditional physiology with the evolutionary explanations behind them; this helps explain not only the power of natural selection but also the often quirky and even illogical features of life.

Cellular, Molecular, and Genomic Foundations Even though this text is primarily organized according to body systems, it provides coverage of cellular, molecular, and genomic topics in early chapters and incorporates these throughout as a basis for understanding organ function. The current tools used to explore molecular physiology are also outlined in most chapters. This is the cutting edge of physiology research.

Pathophysiology This text is also designed to help students realize that they are learning worthwhile and applicable material. Because many students using this text will have a clinical component to their careers, reference to important pathophysiological issues demonstrates the content's relevance to their professional goals. For example, this text includes the effects of environmental toxins on some body systems, an issue of increasing concern.

Key Terms, Word Derivations, and Glossary with Phonetic Pronunciations Key terms are defined as they appear in the text. A glossary at the end of the book, which includes phonetic pronunciations of the entries, enables students to quickly review key terms. Because physiology is laden with new vocabulary words, many of which are rather intimidating at first glance, word derivations are often provided to enhance understanding.

Integration Following our integrated systems approach, each chapter includes a *Making Connections* section that helps students put into perspective how the system just discussed contributes to the whole animal and important regulatory functions.

Chapter Summaries Each chapter now ends with a *Chapter Summary* of the key concepts as a bulleted list.

Review Questions Two new features help the student retain information. First, *Check Your Understanding* questions have been added after each major section to help review the main ideas just discussed. Answers to these questions appear in an appendix at the end of the book. Second, at each chapter's end are *Review, Synthesize, and Analyze* essay questions that integrate the chapter material or delve deeper into a topic and challenge the student on a higher level.

Suggested Readings Each chapter concludes with a list of both recent and classic articles as well as books that give important details on the chapter material.

Online Review The following online materials are designed to help students review foundation concepts presumably covered in prerequisite courses:

- **A Review of Chemical Principles:** This review is offered online at CengageBrain.com as a reference for students who need to go over basic chemistry concepts that are essential to understanding physiology.
- **Storage, Replication, and Expression of Genetic Information:** This review includes a discussion of DNA and chromosomes, protein synthesis, cell division, and mutations and is also offered online at CengageBrain.com.

New to the Second Edition

This edition has undergone extensive revision. Some of the changes are as follows:

New and Enhanced Artwork More than 90% of the art has been upgraded in this edition, with more three-dimensional art; many conceptually redesigned figures to enhance student understanding; brighter, more contemporary, and more visually appealing colors; and more consistent style throughout. New pictures of various animals that illustrate key physiological adaptations have been added. Also, icons of physical entities are incorporated into many of the flow diagrams to aid students in learning what structures are involved in specific actions.

In addition, many new figures have been added, such as those on the following topics: a cell's response to oxidative stress under normal and pathological states and the effects of chronic stress on cellular aging; the mechanism by which new synaptic connections are formed in association with long-term memory; cephalopod eye formation; pineal function in seasonally breeding animals; new insights into the mechanism by which the "catch state" is generated in bivalves; lipid transport for insects and mammals; a possible acquired immune response in insects; crosscurrent flow in a bird lung; protein and carbohydrate metabolism in the rumen; and global warming and animal metabolism.

New and Updated Content We have incorporated appropriate, timely material throughout, as the following examples illustrate: Chapter 2 has a new section on generating stem cells and new findings on microRNAs and small interfering RNAs; Chapter 5 contains updated material on learning and a new section on sleep; Chapter 6 includes updates on TRP-type receptor channels in most types of sensory cells; Chapter 7 contains new material on TSH in seasonally breeding animals, updates on nonvertebrate hormones, and new coverage of epigenetics; Chapter 8 includes updates on the role of titin and new concepts in muscle fatigue; Chapter 9 contains new information on genes in the evolution of the four-chambered heart; Chapter 10 has updates on dendritic cells, regulatory T cells, nanotubes, lamprey immunity, and insect immunity; Chapter 11 includes recent findings on insect respiration, birdlike crocodile lungs, and mammoth hemoglobin; Chapter 12 includes recent findings on insect (anti)diuretic hormones and exciting new hypotheses on the mammalian renal countercurrent mechanism; Chapter 13 has a new section on the effects of ocean acidification on animals; Chapter 14 presents updated material on methane production in ruminants and climate change; Chapter 15 includes a major new section on the effects of climate change on thermal physiology and updates on hypotheses about metabolic scaling and on gastrointestinal hormones in energy regulation; and Chapter 16 includes novel material on seasonal breeding in mammals, the female sexual act, and more depth on the topic of reproduction in nonvertebrates as well as new information on the rut in deer and the multiple roles of prolactin.

Increased Evolutionary and Comparative Aspects We have added additional detail on physiological adaptations, including the following examples: Chapter 1 has added information on vestigial features, cost–benefit trade-offs in evolution, and the Krogh principle; Chapter 6 contains new material on the use of infrasound and ultrasound in various animals, Pax6 genes in animal eye evolution, electroreception and magnetoreception mechanisms and electric organs, and color and UV vision in fish, reptiles, birds, mammals (including whales), and mantis shrimp; Chapter 7 has incorporated additional comparative aspects of thyroid function; Chapter 9 contains new material on hemolymph proteins in arthropods, the evolution of clotting in mammals, and cardiac output in various endotherms and ectotherms; Chapter 10 has been extensively rewritten with an evolutionary overview of universal and nonvertebrate immune functions; Chapter 11 has been extensively rewritten with an evolutionary overview of respiratory functions and more comparative and evolutionary details such as dinosaur and bird connections, nonhemoglobin respiratory pigments, hemoglobin oxygen affinities in various species, and comparisons of high flyers and deep divers; Chapter 12 includes more comparative examples on urea excretion and vertebrate renal development and evolution, and the entire chapter has been reorganized to reflect a more explicit evolutionary framework; Chapter 13 includes a table comparing water balance in humans and kangaroo rats and a new box on the evolution of osmoregulation in vertebrates; and Chapter 15 includes a new table comparing energy metabolism in humans and hummingbirds and new boxes on the evolution of endothermy and the thermal maximum for animal life.

Instructor and Student Ancillaries

PowerLecture This convenient tool makes it easy for instructors to create customized lectures. Each chapter includes the following features, all organized by chapter:

- Lecture slides
- All chapter art and photos
- Animations and videos
- *Test Bank,* which includes more than 900 questions in multiple-choice, true/false, matching, fill-in-the-blank, and essay formats

This single disc places all the media resources at your fingertips.

CourseMate Cengage Learning's Biology CourseMate brings course concepts to life with interactive learning, study, and exam preparation tools that support the printed textbook or the included eBook. With CourseMate, professors can use the included Engagement Tracker to assess student preparation and engagement either for the class as a whole or for individual students. Students can access an interactive eBook, chapter-specific interactive learning tools, including flash cards, quizzes, videos, and more in their Biology CourseMate, accessed through **CengageBrain.com.**

Acknowledgments

It takes the effort of many individuals to create a textbook. The authors would like to thank the following reviewers who provided excellent feedback during the various stages of the writing process.

Joe Bastian, *The University of Oklahoma*
Frank Blecha, *Kansas State University*
Ken Blemings, *West Virginia University*
Charles E. Booth, *Eastern Connecticut State University*
Randal K. Buddington, *Mississippi State University*
David Bunick, *University of Illinois at Urbana*
Mark L. Burleson, *University of Texas at Arlington*
Craig Cady, *Bradley University*
Heather K. Caldwell, *Kent State University*
Maria Carro, *Universidad de Leon*
Randy Cohen, *California State University, Northridge*
Victoria Connaughton, *American University*
Robin L. Cooper, *University of Kentucky*
J. Crivello, *University of Connecticut*
Dane Alan Crossley II, *University of North Dakota*
Bob Dailey, *West Virginia University*
Gerrit de Boer, *University of Kansas*
William L. Diehl-Jones, *University of Waterloo*
Michael E. Dorcas, *Davidson College*
Tina Dow, *West Virginia University*
Lisa M. Duich-Perry, *Chaminade University of Honolulu*
Gene Felton, *West Virginia University*
David P. Froman, *Oregon State University*
Joey Gigilotti, *West Virginia University*
David S. Gonzales, *Arizona State University*
Will Hoover, *West Virginia University*
Keith Inskeep, *West Virginia University*
Gabriel Kass-Simon, *University of Rhode Island*
Phil Keeting, *West Virginia University*
Merideth Kamradt Krevosky, *Bridgewater State College*
James Larimer, *University of Texas at Austin*
Jack R. Layne, Jr., *Slippery Rock University*
John C. Lee, *Virginia-Maryland Regional College of Veterinary Medicine (Virginia Tech)*

Gerald Lincoln, *University of Edinburgh*
James A. Long, *Boise State University*
Catherine Loudon, *University of Kansas*
David A. Lovejoy, *University of Toronto*
Duncan MacKenzie, *Texas A&M University*
Margaret MacNeil, *York College, City University of New York*
David S. Mallory, *Marshall University*
Bill Martin, *West Virginia University*
Sarah L. Milton, *Florida Atlantic University*
Robert J. Omeljaniuk, *Lakehead University*
Sanford E. Ostroy, *Purdue University*
John W. Parrish, Jr., *Georgia Southern University*
Valerie M. Pasztor, *McGill University*
Gilbert R. Pitts, *Austin Peay State University*
Donald R. Powers, *George Fox University*
Bill Radke, *University of Central Oklahoma*
Chris R. Ross, *Kansas State University*
Dean D. Schwartz, *Auburn University*
Peter Sharp, *Roslin Research Institute, Scotland*
Donald E. Spiers, *University of Missouri*
Rebekah J. Thomas, *Texas A&M University*
Andrea R. Tilden, *Macalester College*
Janet Tou, *West Virginia University*
Alexa Tullis, *University of Puget Sound*
Norma Venable, *West Virginia University*
Carol Vleck, *Iowa State University*
Linda Vona-Davis, *West Virginia University*
Melvin Weisbart, *University of Regina*
Matt Wilson, *West Virginia University*
Jianbo Yao, *West Virginia University*
Yong Zhu, *East Carolina University*

About the Authors

L. Sherwood

Lauralee Sherwood is a Professor Emerita of Physiology at West Virginia University. Having earned a D.V.M. degree from Michigan State University, she continued to have an interest in animals even though her faculty appointment for 43 years was in the School of Medicine. She participated as a guest lecturer in Animal Science classes regularly. Dr. Sherwood has authored two human physiology textbooks: *Human Physiology: From Cells to Systems,* Eighth Edition, Brooks/Cole Publishing, and *Fundamentals of Human Physiology,* Fourth Edition, Brooks/Cole Publishing. She brings to this animal physiology textbook her knowledge of veterinary medicine, her years of classroom teaching experience, and her text-writing expertise. She has received numerous teaching awards, including an Amoco Foundation Outstanding Teacher Award, a Golden Key National Honor Society Outstanding Faculty Award, several listings in Who's Who Among America's Teachers, and the Dean's Award for Excellence in Education.

H. Klandorf

Hillar Klandorf is a Professor of Animal and Nutritional Sciences at West Virginia University. He completed his B.Sc. at UCLA and his Ph.D. degree at the Roslin Research Institute in Edinburgh, Scotland, in Avian Physiology. After completing two postdoctoral research programs, he returned to UCLA to investigate agents that affect the onset of diabetes in animal models. Currently, he is at West Virginia University, investigating the role of uric acid in limiting oxidative stress and tissue aging in birds. He instructs animal physiology and animal behavior classes.

Paul H. Yancey is a Professor of Biology at Whitman College. After earning a Ph.D. in Marine Biology (Animal Physiology and Biochemistry emphasis) from UCSD, he has conducted research at the University of St. Andrews (Scotland), the National Institutes of Health, the Mt. Desert Island Biological Laboratory, the Monterey Bay Aquarium Research Institute, the University of Otago (New Zealand), and the University of Hawai'i, as well as at Whitman. His research is on biochemical and physiological adaptations to water stress, temperature, and pressure in the deep sea. His research includes coral, mollusks, crustacea, hydrothermal vent clams and worms, hagfish, bony fish, cartilaginous fish, and mammalian kidneys in health and disease (including diabetes). He teaches animal physiology, human anatomy and physiology, marine biology, and bioethics, and he has won several teaching awards. He brings to this text a broad evolutionary perspective, and 30 years of teaching animal physiology using texts supplemented with a wide variety of external materials.

P. Yancey

Size Matters. A tiny mouse, sitting on an elephant's trunk, has a much greater surface-area:volume ratio than the elephant. This difference profoundly affects their physiologies.

Bob Elsdale/Getty Images

1

Homeostasis and Integration: The Foundations of Physiology

1.1 Introduction

Life. The word *life* is the essence of the biological sciences, and yet has no clear definition. We seem to recognize instinctively what is alive and what is not alive on Earth, but we founder when contemplating how alien life on another planet could be recognized. How might we define life? Most attempts to define it focus on the **functions** of living systems, that is, on the dynamic processes that life forms have and that nonliving things do not. Living things organize themselves using energy and raw materials from their surroundings (a set of processes called *metabolism*), maintain integrity in the face of disturbances (a process called *homeostasis*), and reproduce. Such abilities are what physiology is all about—the study of the functions of organisms, or how life works.

Biological adaptations have two levels of explanation: the mechanistic (or proximate) and the evolutionary (or ultimate)

Physiologists often view organisms as organic machines whose mechanisms of action can be explained in terms of cause-and-effect sequences of physical and chemical processes—the same types of processes that occur in other components of the universe. However, biological traits, such as pelage coloration or visual acuity, differ from the rest of the universe in one crucial way: They are the result of millions of years of evolution through random variation and **natural selection**: the process in which members of a species having genetic characteristics that enhance survival are able to produce more surviving offspring than other members not expressing those characteristics. Therefore, scientists must recognize that biological phenomena always

have two different levels of explanation or description:

- The *mechanistic or proximate explanation:* This is the answer to "How does it work?" This focus emphasizes the mechanism or composition of a function or structure, and is the traditional core of the physiological and medical sciences.
- The *evolutionary or ultimate explanation:* This is the answer to "How did it get to be this way?" This emphasis recognizes that biological features are the result of variation and natural selection, that is, evolutionary history. A species must cope with a variety of *selective pressures* (environmental stresses, mating, and so on). Many genes within a species differ among individuals, and under a particular selective pressure some gene variants better equip an animal for survival, which allows it to reproduce better than do other variants.

Those gene variants get passed on to more offspring and thus the species gradually changes genetically over time. This process never ends, because selective pressures (such as climate and predator–prey interactions) change over time. As you'll see, to fully understand why a certain feature is the way it is, we often need to know a species' history.

Because of this process of natural selection, physiologists assume that most organismal traits are **adaptations**, that is, beneficial features that enhance overall survival of the species. When conducting an evolutionary analysis, biologists distinguish between two kinds of adaptations. **Homologous** traits in two different organisms are those that are related by common ancestry, such as bird wings and human arms: Both have the same basic bony components because both evolved from limbs of four-legged reptiles. In contrast, **analogous** traits are those with similar structures and/or functions but which evolved independently; for example, bird and insect wings serve similar functions but share no common ancestral structure.

Natural selection is a powerful force, which often results in exquisite adaptations such as wings. However, it is important to recognize that selection is also constrained by the past. That is, evolution works primarily through modifications of previously evolved features and variations of these features and so does not always result in the most logical or optimal design. We'll look at examples shortly, but first warn you about another kind of explanation.

Adaptations reflect evolutionary history including cost–benefit trade-offs

In addition to mechanistic and evolutionary explanations, there are purely teleological approaches to explaining the various features of organisms. In a **teleological approach**, phenomena that occur in organisms are explained in terms of their particular **purpose** (a function that is useful or beneficial) in fulfilling an organismal need, without necessarily considering how this outcome is accomplished (mechanism) or how it evolved. Teleology is commonly used because most biological mechanisms do serve a useful purpose (having been selected through evolutionary time to do so). If you try to find the thread of "purpose" in what you are studying, you can avoid a good deal of pure memorization!

However, be aware of a pitfall: Teleology can be dangerous thinking because it assumes that features are always logical, having evolved for idealized purposes. But there are at least two reasons this may not be so. First, adaptations are often a record of historical compromises; that is, they are not necessarily the most logical solution to a problem. Second, some structures may be **vestigial**, that is, having *no* (or very minimal) useful function. They likely evolved from useful features in ancestors but have not been eliminated by selection. These constraints on natural selection have been often overlooked in the physiological sciences, especially since the modern physiological sciences started in the 1600s, long before Charles Darwin presented his theory of evolution (in 1859, the year he published *On the Origin of Species*). However, a recent movement called *evolutionary physiology,* and its applied counterpart *Darwinian medicine,* is re-evaluating many teleological explanations in traditional physiology. As we shall see in later chapters, this has often profound implications for the understanding of physiology.

Simple examples can help you to distinguish among these approaches (mechanistic/proximate, evolutionary, teleological) and the role of historical constraints. First, let's take a reasonably logical, useful adaptation: shivering in mammals. The *mechanistic* or *proximate* explanation of why a mammal shivers is that, when temperature-sensitive nerve cells detect a fall in body temperature, they signal the *hypothalamus*, the part of the brain responsible for temperature regulation. In turn, the hypothalamus activates nerve pathways that trigger rapid oscillating muscle contractions (that is, shivering) to produce heat (in a process called *negative feedback homeostasis*, a topic we explain in detail shortly). A *teleological* explanation of why a mammal shivers when it is cold is "to keep warm" to maintain a relatively constant, warm interior. But this does not answer the question of why it is useful for most mammals to keep constantly warm. It is not necessary for the vast majority of life in variable temperate habitats, or even for a hibernating mammal in winter. To explain this, we must analyze the course of evolution among mammals to try to determine why regulation of body temperature enhanced survival in this group, and why 37° to 39°C (about 99° to 102°F) was the ideal temperature selected for most placental (but not other) mammals.

Now let's take a less logical example, the human spinal column and its series of bony vertebrae joined by cushioning pads. The mechanistic description is that the spine is a movable support device for the upper body and a protective conduit for the major nerve cords. Teleologically, we might say that vertebrae are a logical way to provide both flexibility for movement and rigidity for support and protection of the nerves. But this ignores the fact that the spine in humans is a suboptimal design! Back problems—pinched nerves, damaged cushioning pads, and so on—are among the most common ailments of humankind. To explain this, we must look at the course of evolution. The spine originally evolved as a *horizontal* flexible swimming device in fishes, animals that are not concerned with gravity because they float. Later the spine became adapted into a (still horizontal) support system against gravity in four-legged land vertebrates. But it only recently began evolving into a *vertical* support in our hominid ancestors, and thus has not been optimized. Indeed, it may never be, because the spine is fundamentally a horizontal system in its inception.

Adaptations that are not "perfect" show another aspect of evolutionary history called *cost–benefit trade-offs.* Essentially, adaptations have costs that may negatively impact other adaptations. The most basic trade-off involves energy, which is typically limited. It takes energy to make structures such as large brains, skeletons, and embryos, and an organism cannot "do it all." For example, animals like seastars that produce thousands or millions of embryos do not protect nor provide much yolk (energy) for them, while animals such as birds or primates that spend considerable energy nourishing and protecting their young consequently do not have many offspring. Large structures may also have other costs such as reduced speed and maneuverability; for example, a large thick shell on a snail's body clearly slows that animal down. You are familiar with similar cost–benefit trade-offs in human engineering. Consider that small, lightweight automobiles get high mileage but are more likely to suffer damage in an accident than a large well-armored (but low-mileage) vehicle.

Finally, consider vestigial features. Their origins are usually revealed by useful homologous features in other species. An example is the *dewclaw*, a tiny digit on the side of the foot in many mammals and reptiles (including birds). Dewclaws evolved from useful fifth digits, and may still provide a gripping function in some species, but in many dogs, rear dewclaws hang loosely with little muscle or bone and have no discernable function. There can even be vestigial genes called *pseudogenes* (although see Chapter 2 for a possible remnant function of these). For example, humans cannot synthesize vitamin C (needed for survival; see p. 65) and so must have it in their diets, and yet their genetic codes contain a nonfunctional gene for a protein to make the vitamin. Presumably, our ancestors ate enough dietary vitamin C (e.g., in fruit) so that when this gene mutated and shut off, it did not impair survival. As another example, short-lived mammals express the gene for uricase, which inactivates uric acid (a potent antioxidant; Chapter 12), while long-lived primates, birds, and reptiles do not express this gene. In this case, we can only speculate on what events led to the suppression of this gene product.

As we examine animal adaptations throughout this text, keep in mind these lessons of the human spine, the slow snail's shell, and the dog dewclaw: *Evolution builds on the past* and therefore is often constrained by it, and *selection often involves trade-offs and compromises.*

Physiology is an integrative discipline

As the example of the spine shows, physiology is closely interrelated with *anatomy,* the study of the structure of organisms. Just as the functioning of an automobile depends on the shapes, organization, and interactions of its various parts, the structure and function of an animal's body are inseparable. But physiology also integrates many other scientific fields. *Physics* is necessary to understand processes in organisms such as electrical conduction, fluid dynamics, and leverage exerted by musculoskeletal systems. Of course, all life processes depend on chemical reactions, so a solid understanding of *chemistry* is crucial. Also, biological functions depend on large biological molecules such as DNA and proteins (see Chapter 2), and therefore *biochemistry* and *molecular biology* provide a major foundation for physiological understanding. At the other end of the spectrum, we need to know something about an organism's *ecology* to explain, for example, why certain functions evolved.

Molecular biology is the most rapidly growing biological science, and its integration with physiology is now the dominant force in many physiological studies. Driving much of this growth are organismal **genome projects** (the complete analysis of genetic codes of key species, including humans). The genetic code ultimately gives rise to crucial parts of the functioning organism, and those functions in turn affect how that code is used (see Chapter 2 for details). Many physiologists are dedicating their research to understanding how genes are regulated and how they are used in physiological processes.

Genome studies are also having a major impact on human understanding of evolution. In particular, these studies are revealing the stunningly complex and often illogical nature of evolution at the most fundamental level. The human genome, for example, appears riddled with formerly functional genes that have mutated into pseudogenes and contains a large percentage of code that seems to have no function at all now or in the past (although certain recent studies hint at some functions for these codes, as we'll describe later). Furthermore, at least some genes seem to have arisen from bacteria and viruses!

Physiology is also a comparative discipline, often following the Krogh principle

As we have just seen, complete physiological understanding must integrate in a "vertical" sense: from the atomic level to the whole organism to its environmental interactions to the evolutionary forces that have acted on that organism's ancestors. But another major approach in physiology—*comparative physiology*—works in a "horizontal" sense, by comparing physiological features in different types of organisms. Such research is important in at least three ways. First, we learn about the uniqueness and diversity of life on Earth by discovering the amazing features that one or a few types of organism have evolved but that others have not. For example, one of the pioneers of modern comparative physiology in the mid-20th century, Per Scholander (1905–1980; a naturalized U.S. citizen born in Sweden) analyzed (among many other things) physiological adaptations of diving mammals, the filling mechanism of fish swim bladders, and thermal adaptations of Arctic mammals and insects.

Such basic knowledge is fascinating in its own right but can sometimes have practical applications. See the box, *A Closer Look at Adaptation: Geckos on Glass.*

Second, comparative studies help us understand the trade-offs and constraints found in many biological features, as you've seen. Thirdly, comparative studies help us find out what physiological functions are universal rather than unique. This broadens our understanding of the basic nature of life. For example, from studying the adaptations of other animals to environmental temperature we learn that the mammalian need to maintain a nearly constant body temperature is not universal, but we also learn (as we will see in Chapter 15) that there are universal effects of temperature on biological molecules to which all life must adapt.

Comparative physiology research on both the diversity and universality of adaptation is often motivated by the **August Krogh principle**, formulated by German physiologist Hans Kreb in 1975: "For a large number of problems there will be some animal of choice, or a few such animals, on which it can be most conveniently studied." This principle was based on remarks made by the great Danish physiologist August Krogh (1874–1949), founder of Denmark's first animal physiology laboratory, who championed the integrative and comparative approaches to physiology. He believed that the diversity of adaptations on Earth is such that, for any particular physiological process, some species will have adapted in such a way that it provides an ideal model system for studying that process. In turn, this knowledge can provide insight into that process in a much wider variety of organisms. For example, to understand how animal neurons (nerve cells) produce electrical signals, the English physiologist John Zachary Young in the 1930s chose the giant squid axon because it was much easier to pierce with electrodes than the much smaller mammalian neurons. As you'll learn in Chapter 4, Young and his followers using squid axons discovered basic principles that apply to all animal neurons, including our own.

A Closer Look at Adaptation
Geckos on Glass

Very few animals can climb up glass windows and walk upside down on smooth ceilings. You have undoubtedly noticed flies and spiders doing this. If you've lived in or visited tropical habitats, you may have seen much larger animals—geckos—performing the same remarkable feat. How do they do it? Only recently has the mechanism for these lizards' phenomenal abilities been uncovered. The feet of geckos have small pads studded with millions of hairlike bristles called *setae*. In turn, setae tips sprout hundreds of tiny branches, called *spatulae*, each only 200 nanometers wide. Flies and spiders have similar structures. Kellar Autumn and colleagues have found that spatulae provide a huge surface area for weak molecular attractions called *van der Waals* forces (about which you probably learned in a chemistry class). Not only do these forces allow a gecko to cling against gravity, but they also allow rapid detachment of the spatulae as the animal runs up the window. Here we see how basic chemistry directly explains whole-animal structure and function. Moreover, the unique ability of gecko feet to attach and detach repeatedly has inspired scientists to design reusable adhesive tape based on this mechanism. This is an example of a growing movement called **biomimicry**—the emulation of organismal adaptations by humans to create useful products.

 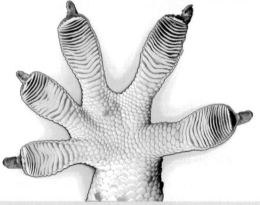

(a) A tokay gecko.

(b) The foot of the gecko. Each pad consists of many lamellae (folds), which in turn have millions of tiny bristles that adhere to surfaces through van der Waals interactions.

© Mark Moffett/Minden Pictures

Working on a variety of animals, Krogh made major discoveries about (among other things) the use of radioactive isotopes in metabolic research, oxygen movement from the lungs to the blood by diffusion (work aided by his wife, Marie), metabolism in insect flight, and the regulation of capillary blood flow (for which he was awarded the Nobel Prize in Physiology or Medicine in 1920). He inspired generations of comparative physiologists, for example his daughter, Bodil Schmidt-Nielsen (1918–) and her husband Knut (1915–2007), a Norwegian who came to work in Krogh's laboratory. The Schmidt-Nielsens came to America at Per Scholander's invitation, where they pioneered studies of (among other things) animals in extreme environments such as deserts (we'll see some of their work in later chapters). In concert with the Krogh principle, broadly used mechanisms are often best revealed by studying them at the extremes of adaptation.

The inherently broad nature of physiology presents a major challenge to learning. To help handle this challenge, as we tell the story of how animals work, we integrate the physical, chemical, molecular/genetic, anatomic, ecological, and evolutionary features necessary for understanding function. Also, to provide breadth in the comparative sense we discuss a variety of animal types. To fully understand these compari-

sons, you should develop a grasp of the basic evolutionary relationships of animal phyla. If you have not learned about these, or need a review, look at Figure 1-1. In particular, be aware that the dichotomy of "vertebrates" and "invertebrates," widely used for the animal kingdom, is evolutionarily meaningless. In fact, there are over 30 animal phyla (groups distinguished by major body plans), and vertebrates are only a subphylum within one of those, the phylum Chordata (unified by all members having a dorsal *notochord* at some stage in their life cycles). Also, the phylum Arthropoda ("jointed legs"), which includes insects, is by far the most successful in terms of numbers of species and sheer mass of animal life. We will use instead the terms vertebrate and nonvertebrate in this text, and we also name the pertinent phyla.

check your understanding 1.1

Explain the terms "proximate" and "evolutionary" explanations and give an example of each.

What are *adaptations* and why aren't they always logical, optimal features?

Describe the Krogh principle and give an example.

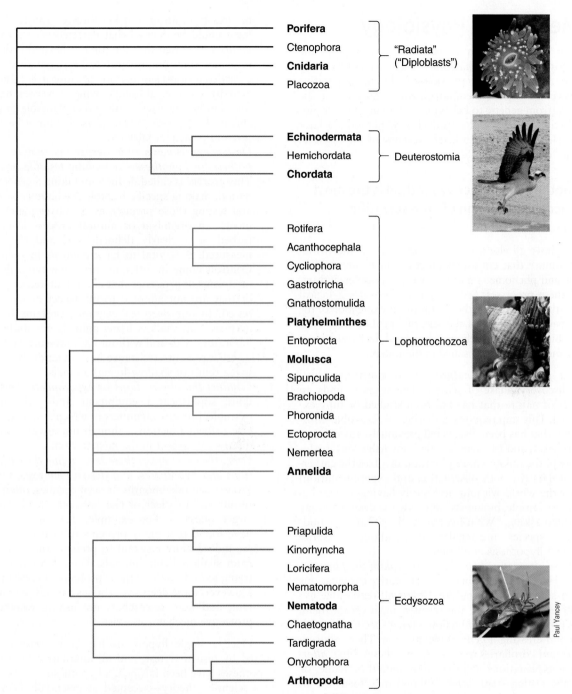

FIGURE 1-1 Animal evolutionary tree, showing the relationship among phyla as deduced from morphology coupled with molecular analyses. The three major branches are the Radiata, the Deuterostomia, and the Protostomia (not labeled, but containing the Ecdysozoa—groups that molt—and Lophotrochozoa). The nine most successful phyla, to which we refer most often in this text, are **bold**; these are the **Porifera** (sponges), **Cnidaria** (jellies, hydras, corals, anemones, etc.), **Echinodermata** (seastars, sea urchins, sea cucumbers, etc.), **Chordata** (tunicates, vertebrates, etc.), **Platyhelminthes** (flatworms), **Mollusca** (snails, octopods, bivalves, etc.), **Annelida** (segmented worms), **Nematoda** (round worms), and **Arthropoda** (crustacea, insects, arachnids, etc.).

Source: http://www.mhhe.com/biosci/pae/zoology/animalphylogenetics/section02.mhtml, from *Teaching Animal Molecular Phylogenetics* by C. Leon Harris. Used by permission of The McGraw-Hill Companies.

1.2 Methods in Physiology

Physiology is, of course, a science and thus employs the same philosophical logic that other scientists use to study the world. But because of physiology's broad, integrative nature, we can fairly say it employs a wider range of specific techniques (from genetic to behavioral) than most other biological disciplines. We will introduce key methods in later chapters. Here we provide a brief overview of the essential aspects of "doing science."

The hypothetico-deductive method is the most widely accepted version of "the scientific method"

Science is above all else a *way of thinking* about the world. First, it assumes that consistent processes in nature explain structures and phenomena and that we can uncover and understand these processes. Second, science provides a logical approach (the "scientific method") for this investigation of the universe. Several versions of the scientific method have been put forth, but the most widely used is termed the **hypothetico-deductive method**. It can be stated in these steps:

1. *Ask a question about nature.* This is sometimes called the **discovery phase** of science. The scientist finds an aspect of nature that has not been studied or fully explained. This step requires gathering all accessible information that has been discovered previously, any existing hypotheses, and (if there is little knowledge yet) *exploration* of the unknown. Exploration involves the gathering of data (by both observation and experimentation) about the world without necessarily having a hypothesis. For example, biologists exploring the deep sea today are often asking, "What lives in the deep?" and are finding new species on a regular basis, without necessarily testing a hypothesis at all times.

2. *Propose alternative hypotheses to explain the phenomenon.* **Hypotheses** are, in essence, tentative explanations about some aspect of nature. Unless all reasonable hypotheses already exist, the scientist must create them. This process is called **induction**: taking specific information and creating a general explanation. There are two important corollaries to making hypotheses. First, *alternative* explanations or predictions should be made, or else the testing steps (which follow) may focus on an erroneous explanation and miss a correct one. For example, a deep-sea physiologist might ask how life survives the crushing pressures of the deep, which are known to inhibit the functions of biomolecules such as proteins. He or she might propose that deep-sea organisms (1) have evolved proteins that are resistant to pressure, or (2) suffer from pressure inhibition and thus have sluggish metabolisms. (Can you think of other hypotheses?)

   Second, scientific hypotheses must be **testable** (amenable to experiments or observation), and, in principle, **falsifiable** (capable in principle of being disproved by experiment or observation), or they are of little use in science (whether they are true or not!). For example, *vitalism,* a view of life prominent before the 20th century, hypothesized that living entities get their dynamic natures from an unseen life force or spirit. No experiment could be devised that detected such a force, and vitalists proposed that the force is beyond ordinary nature and thus unmeasurable. Thus, vitalism is today considered nonscientific, not because it is false (it may actually be true, although living processes have been found to be consistent with and explainable by physicochemical processes), but because it cannot be tested or (even in principle) falsified.

3. *Design experiments or observations that test the hypothesis or hypotheses by making testable predictions.* This process is called **deduction**: taking a general explanation, making specific testable predictions based on it, and testing those predictions. In physiological experiments, an organism or natural process is often perturbed in a clearly defined way, and the outcome measured. It is vital to have *controls* in such testing. Controls come in different forms but typically involve organisms or processes that are not altered or perturbed or that are not adapted to an aspect of nature being tested. In our deep-sea example, proteins from both deep-sea and shallow-living animals are studied in the laboratory with and without high pressure to determine whether pressure resistance has in fact evolved in deep-living (but not shallow-living) animals.

4. *Conduct the observations or experiments.* This can require sophisticated equipment and techniques (for example, pressure chambers in which scientists can analyze protein functions), though not always. We briefly discuss this aspect in the next section.

5. *Using the outcome of these tests, refine the earlier questions and hypotheses, and design new tests.* Often this process can take months or even decades, often because results are not clear or test only one small aspect of a larger question. For example, deep-sea physiologists have found that many proteins from deep-sea animals are indeed more resistant to pressure than are proteins from shallow-living animals. The researchers are now trying to find out how those proteins can resist pressure. However, some deep-sea proteins are *not* pressure resistant, and now researchers are asking whether other protective mechanisms exist.

Once a single hypothesis has been consistently supported by test after test, and all alternative (and testable) hypotheses have been falsified, a hypothesis may be elevated to a scientific **theory**—accepted as essentially correct, but with details yet to discover.

Physiological observations and experiments use techniques from the molecular to the behavioral level, for both discovery and hypothesis testing. As we noted earlier, the integrative nature of physiology presents a challenge to the researcher. Understanding the full scope of physiology requires that scientists conduct observations and experiments that range from the functions of specific molecules to the behaviors that animals use with internal processes. Understanding the methods can be important to understanding the knowledge they yield. Although this text does not focus heavily on such methods, we cover selected ones in the appropriate chapters. For example, Chapter 2 discusses a number of important techniques scientists use in genomic and other molecular studies.

It is worth repeating that scientific experiments and observations are used for both discovery (before there are detailed hypotheses) and hypothesis testing. The analysis of organismal genomes, for example, was largely a discovery process in the initial phases.

check your understanding 1.2

Describe the hypothetico-deductive method.

1.3 Levels of Organization in Organisms

One triumph of scientific study in the 19th century was the development of the cell theory of life. Regardless of the difficulties in defining "life," the foundation of all functions of life, and thus of all living organisms, is the **cell**—the smallest unit capable of carrying out the processes associated with life on Earth. A cell consists of an outer barrier called a **membrane** and an internal (semigelatinous) fluid called the **cytosol**, which contains water, salts, small organic molecules, and **macromolecules**. Macromolecules include **deoxyribonucleic** and **ribonucleic acid** (**DNA** and **RNA**), which carry the genetic instructions for making most cell components, and **proteins,** molecules with complex shapes, which carry out those instructions by assembling into many cell structures, regulating most cell functions, and catalyzing most cell reactions (see Chapter 2).

Life has the properties of self-organization, self-regulation, self-support and movement, and self-reproduction

Cells are classified traditionally in two distinct forms: **prokaryotic cells** (in two groups, the Bacteria and Archaea), which for the most part lack complex internal membranous structures such as nuclei; and **eukaryotic cells**, which do have these structures, found in the kingdoms Protista (an awkward kingdom that includes protozoa and unicellular and multicellular algae), Fungi, Plantae (higher plants), and Animalia. However, because the two prokaryotic groups are fundamentally different from each other, biologists now classify all life in three *domains*: archaea, bacteria, and eukarya. The simplest life forms are **unicellular** (single-celled) or colonial (loose collections of similar cells), typically prokaryotes and many protists (eukaryotes) such as amoebae. More complex organisms such as animals are **multicellular** (many-celled)—an aggregate of hundreds (in certain worms) to trillions (in mammals) of cooperating eukaryotic cells with distinct specializations in structure and function. As you will see later (p. 385), physical constraints such as *diffusion* limit the size of a single cell, so that large body size depended on the evolution of multicellularity.

All cells, whether they exist as solitary cells or as part of a multicellular organism, perform certain basic functions essential for survival of the cell, which in turn aid in survival of the organism. These basic cell functions include the following:

1. *Self-organization:* Using resources from the environment to create the cell. This function includes many steps, including:
   - Obtaining energy and raw materials from the environment surrounding the cell (for plants, these are light and inorganic nutrients; for most animals, these come from food and oxygen). Initially, this uptake occurs through the cell's membrane.
   - Performing various chemical reactions that use energy and raw materials to provide energy stores and building blocks for the cell's other needs. Most cell reactions are catalyzed by large molecules called **enzymes**, which in most cases are complex proteins that increase the rate of specific chemical reactions. In most animal cells, the overall process can be roughly summarized thus:

$$\text{Food} + O_2 \rightarrow CO_2 + H_2O + \text{energy stores} + \text{buildings-block molecules}$$

   - Eliminating to the cell's surrounding environment (through the membrane) any waste products of metabolism, such as carbon dioxide and ammonia or other nitrogenous waste.
   - Synthesizing proteins and other components needed for cell structure, for growth, and for carrying out particular cell functions (using energy stores and building blocks).
2. *Self-regulation:* Maintaining self-integrity in the face of disturbances, including:
   - Being sensitive and responsive to changes in the environment surrounding the cell (such as temperature, osmotic status).
   - Controlling the exchange of materials between the cell and its surrounding environment.
   - Repairing damage to the cell.
   - Correcting deviations in internal conditions, which threaten other functions (*homeostasis*).
3. *Self-support and movement:* Having structures that give specific form to the cell, the ability to move materials within the cell, and (in some cases) the ability to move the whole cell or organism through the surrounding environment.
4. *Self-replication:* Reproducing to carry on the species, and to repair damage. Some animal cells, such as nerve cells and muscle cells in mammals, are unable or have greatly restricted abilities to replicate. When trauma or disease processes destroy these cells, they are generally not replaced.

Cells are remarkable in the similarity with which they carry out these functions. Thus, all cells share many common characteristics. In multicellular organisms, each cell also performs a specialized function, which is usually a modification or elaboration of a basic cell function. Here are a few examples in animals:

- As a specialization of basic cell protein-synthesizing ability, *gland cells* of *digestive systems* secrete digestive enzymes.
- By enhancing the basic ability of cells to respond to changes in their environments, *neurons* generate and transmit to other regions electrical impulses that relay information about stimuli such as light, temperature, nutrient concentrations, and so on.

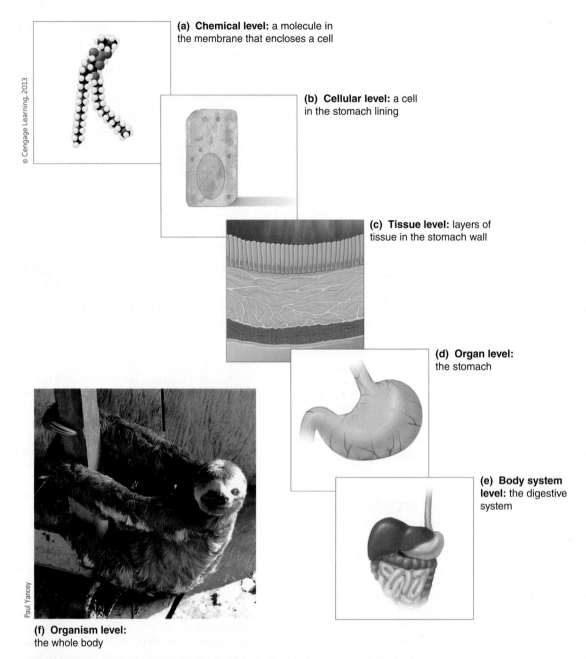

(a) **Chemical level:** a molecule in the membrane that encloses a cell

(b) **Cellular level:** a cell in the stomach lining

(c) **Tissue level:** layers of tissue in the stomach wall

(d) **Organ level:** the stomach

(e) **Body system level:** the digestive system

(f) **Organism level:** the whole body

FIGURE 1-2 Levels of organization in an animal body, showing an example for each level.

- Through specializations in the basic ability of cells to transport molecules through membranes, *kidney cells and tubules* selectively retain substances needed by animals while eliminating unwanted substances in the urine.
- Based on the inherent capability of eukaryotic cells to produce intracellular movement, *muscles* use movement of internal protein filaments to bring about shortening of the whole organ.

It is important to recognize that cells perform these specialized activities in addition to carrying on the unceasing, fundamental activities required of all cells. The fundamental cell activities are essential for the survival of each individual cell, whereas the specialized contributions and interactions among the cells of a multicellular organism contribute to the survival of the whole organism.

Cells are progressively organized into tissues, organs, systems, and finally the whole body

Just as a machine does not function unless all its various parts are properly assembled, the cells of organisms must be specifically organized to carry out the life-sustaining processes of the whole, such as (in animals) digestion, respiration, and circulation. Multicellular organisms have four levels of organization: cells, tissues, organs, and systems (Figure 1-2).

In animals, cells of similar structure and function are organized into **tissues** (groups of cells with similar structures and functions), of which there are four primary types: *epithelial, connective, muscular,* and *nervous.* Each tissue consists of specialized cells along with varying amounts of extracellular ("outside the cell") material (Figure 1-3).

Organ: Body structure that integrates different tissues and carries out a specific function

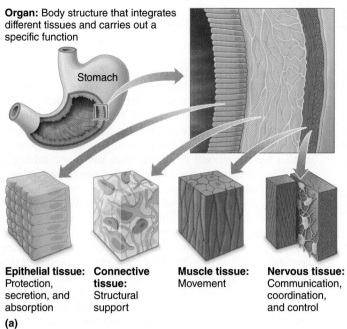

Stomach

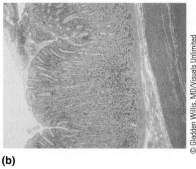

(b)

Epithelial tissue: Protection, secretion, and absorption

Connective tissue: Structural support

Muscle tissue: Movement

Nervous tissue: Communication, coordination, and control

(a)

FIGURE 1-3 **The stomach as an organ made up of all four primary tissue types.** (a) The four tissue components; (b) Low-power microscope view of the mammalian stomach, showing three of the primary tissue layers (left folded layer = epithelial, middle layer = connective, right darker layer = muscle).

1. **Epithelial tissue** is made up of cells specialized in the exchange of materials. This tissue is organized into two general types of structures: *sheets* and secretory *glands*. Epithelial cells are joined together very tightly to form **epithelial sheets** that cover and line various organs. For example, the outer layer of the skin is epithelial tissue, as is the inner lining of the digestive tract. In general, these epithelial sheets serve as boundaries that separate animals from the external environment and from the contents of cavities that communicate with the external environment, such as the digestive tract lumen. (A **lumen** is the cavity within a hollow organ or tube, and in the case of digestive tracts, is technically outside of the body!). Only selected transfer of materials is permitted between the regions separated by an epithelial barrier. The type and extent of controlled exchange varies, depending on the location and function of the epithelial tissue. For example, very little can be exchanged across the skin and the external environment in most animals, whereas the epithelial cells lining the digestive tract are specialized for absorbing water and nutrients.

   • **Glands** are epithelial tissue derivatives that are specialized for secretion. Secretion is the release from a cell, in response to appropriate stimulation, of specific products that have in large part been synthesized by the cell. Glands are formed during embryonic development by pockets of epithelial tissue that dip inward from the surface. There are two categories of glands: *exocrine* and *endocrine*. If during development the connecting cells between the epithelial surface cells and the secretory gland cells within the depths of the invagination remain intact as a duct between the gland and the surface, an exocrine gland is formed.

   • **Exocrine glands** (*exo* means "external"; *crine* means "secretion") secrete through ducts to the outside of organisms (or into a cavity that communicates with the outside). Examples are sweat and digestive glands. If, in contrast, the connecting cells disappear during development and the secretory gland cells are isolated from the surface, an endocrine gland is formed. **Endocrine glands** (*endo,* "internal") lack ducts and release their secretory products, known as **hormones,** internally into the blood within the animal (see Chapter 7). For example, the vertebrate parathyroid gland secretes parathyroid hormone into the blood, which transports this hormone to its sites of action at the bones and kidneys.

2. **Connective tissue** is distinguished by having relatively few cells dispersed within an abundance of extracellular material that they secrete. As its name implies, connective tissue connects, supports, and anchors various body parts. It includes such diverse structures as the **loose connective tissue** that attaches epithelial tissue to underlying structures; **tendons**, which attach skeletal muscles to bones; **bone**, which gives vertebrates shape, support, and protection; and **blood** (or **hemolymph**), which transports materials from one part of a body to another, and in a sense, connects all cells of the body. Except for blood, the cells within connective tissue produce specific molecules that they release into the extracellular spaces between the cells. One such molecule is the rubber-band-like protein fiber **elastin**, whose elastic properties facilitate the stretching and recoiling of structures such as lungs, which alternately inflate and deflate during breathing (see p. 520).

3. **Muscular tissue** consists of cells specialized for contraction and force generation. Vertebrates have three types of muscle tissue (Chapter 8): **skeletal muscle**, which causes movement of the skeleton; **cardiac muscle**, which is responsible for pumping blood out of the heart; and **smooth muscle**, which encloses and controls movement of contents through hollow tubes and organs, such as the digestive tract.

4. **Nervous tissue** consists of cells specialized for initiation and transmission of electrical impulses, sometimes over long distances (Chapters 4, 5). These electrical impulses act as signals that relay information from one part of an organism to another. Nervous tissue in vertebrates is found in (1) the brain; (2) the spinal cord; (3) epithelial linings that monitor the external environment as well as the status of various internal factors that are subject to regulation, such as temperature, blood pressure, and gut contents; and (4) muscles, glands, and other *effector* organs (see p. 14).

Epithelial, connective, muscular, and nervous tissue are the primary tissues in a classical sense. The term *tissue* is also frequently used to refer to the aggregate of various cellular and extracellular components that make up a particular organ (for example, lung tissue or liver tissue).

Organs consist of two or more types of primary tissue organized to perform a particular function or functions. The stomach is an example of an organ made up of all four primary tissue types (Figure 1-3), which function collectively to store ingested food and begin digestion as well as to move it forward into the rest of the digestive tract. Vertebrate stomachs are lined with epithelial tissue that restricts the transfer of harsh digestive chemicals and undigested food from the stomach lumen into the blood. Gland cells, derived from epithelia in the stomach, include exocrine cells, which secrete digesting juices into the lumen, and endocrine cells, which secrete hormones that help regulate the stomach's exocrine secretion and muscle contraction. The walls of the stomach contain smooth muscle tissue whose contraction mixes ingested food with the digestive juices and propels the mixture forward into the intestine. Also within the walls is nervous tissue, which, along with hormones, controls muscle contraction and gland secretion. These various tissues are all bound together by connective tissue.

Organs are further organized into **organ systems**, each of which is a collection of organs that perform related functions and interact to accomplish a common activity that is essential for survival of the whole body. For example, the digestive system consists (in vertebrates) of the mouth, salivary glands, pharynx (throat), esophagus, stomach, pancreas, liver, gallbladder, small intestine, and large intestine (and sometimes other organs). These organs collectively break down food into small nutrient molecules that can be absorbed into the blood. We end this chapter with a closer look at these systems.

The total animal body—a single, independently living individual—consists of the various organ systems structurally and functionally linked together as an entity that is separate from the external (outside organisms) environment. Thus, an animal is made up of living cells organized into life-sustaining systems.

check your understanding 1.3

Describe the basic functions of life.

What are the four major tissue types? For two of them, describe the major subtypes.

Describe the construction of organs from tissues.

1.4 Size and Scale among Organisms

Because life on Earth ranges from unicellular prokaryotes to large multicellular eukaryotes such as whales, organisms range in mass (using the metric weights in grams) over a scale of 10^{20} (the number 10 followed by 20 zeroes)! The table inside the back cover gives examples of the size of some organisms to give you a sense of this enormous range. Independent of other factors, the size of an organism has important implications for its structures and functions. For example, why are there huge mammals, but no huge insects? The study of the effects of size on anatomy and physiology is called **scaling**. An example is lifespan: Larger animals generally live longer than smaller ones. Even more interestingly, a bird lives about 2.5 times longer than a mammal of comparable body size. Scaling phenomena such as this one show up in many chapters of this book (the lifespan of birds, for example, appears in Chapters 2 and 12), so it is important to examine this principle now. The most important size effect may be a simple mathematical concept known as the **surface-area-to-volume ratio**.

The larger the organism, the smaller the surface-area-to-volume ratio

Consider a small spherical cell of minute size. It has a certain volume and mass, and a certain surface area of its outer boundary (Figure 1-4). From simple mathematics, we know the volume is related to the *cube of the radius* of the object. The surface area is related to the *square of the radius*. Now consider a cell (or organism) 10 times bigger (see Figure 1-4). Because of the cubed factor, the volume of it is 1,000 (10^3) times bigger than the first object, but because of the squared factor, the surface area is only 100 times (10^2) bigger. Its surface-area-to-volume ratio is one tenth that of the smaller sphere. This has tremendous implications for a variety of living functions. For instance, the entire mass of both cells needs to obtain nutrients and get rid of wastes through the surface boundary, but the larger object is at a 10-fold disadvantage because of its relatively smaller ratio (that is, relatively smaller surface area to support the volume). This is one reason why larger animals have evolved highly branched (that is, increased surface area) circulatory and respiratory systems. We examine this development in Chapters 9 and 11.

However, if there is something that each cell or organism needs to retain, such as heat, the larger object is at a 10-fold advantage because heat is typically produced by the volume of the cell and lost through its surface area. In Chapter 15 we examine this aspect of physiology in detail.

Scaling also affects skeletal mechanics. The ability of a skeleton (or a tree trunk, for that matter) to support mass against gravity increases with the cross-sectional area of the skeletal element (such as a leg bone or insect limb exoskeleton). But the weight being supported increases as the cube of the animal size. Thus, a large animal such as an elephant must have massive limbs in comparison to its body, whereas a tiny spider can support itself with limbs that are much thinner in comparison to its body.

It is important to remember this basic phenomenon-Note that larger objects have larger surface areas on an absolute scale, but it is the ratio that matters. Put simply, all other things being equal, *larger organisms have smaller surface-area-to-volume ratios*.

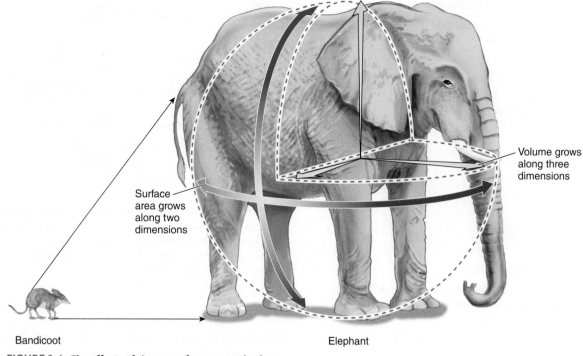

FIGURE 1-4 **The effects of size on surface area and volume.**

© Cengage Learning, 2013

check your understanding 1.4

How does surface area change with size and why is it important?

1.5 Homeostasis: Basic Mechanisms and Enhancements

Earlier we discussed several functions of cells (and thus life): self-organization, self-regulation, support and movement, and self-replication. To many biologists, the ability to self-regulate is the most crucial of these functions for distinguishing living from nonliving things. For example, a fire can arise spontaneously (given an initial energy source such as lightning) and can maintain itself by using combustible materials in the environment, can move by spreading to nearby fuel sources, and can replicate through sparks. But it takes no corrective action when water is dumped on it. Living systems, in contrast, can correct for at least some disturbances. Indeed, one recent definition of life that has been proposed is *"a system that tries to regulate itself to preserve its identity."*

Maintenance of a desired state in the face of disturbances has been given a term that is central to physiology: **homeostasis** (*homeo*, "similar"; *stasis*, "to stand or stay," although used here to mean "state"). The concept began with the French scientist Claude Bernard (1813–1878). His ambition was to be a playwright, but he turned to medicine and then to physiological research, which he termed "experimental medicine." Indeed, he pioneered the application of the scientific method to medicine and so is often called "the father of modern physiology." He documented the abilities of mammals to maintain a relatively constant state of the internal environment (which he called *le milieu*

interieur). The concept languished until the American physiologist Walter B. Cannon (1871–1945) took an integrative approach to physiological function, further developing Bernard's ideas and coining the term *homeostasis*. It is important to note that Cannon used *homeo* for "similar," not *homo* for "same". We'll see why shortly.

Unicellular organisms have some homeostatic abilities, such as the ability to maintain energy levels, acid–base balance, and volume. In animals, although individual cells also have some homeostatic abilities, many regulatory processes occur at the level of the whole organism. Let's now examine this crucial concept in detail.

Body cells are in contact with a privately maintained internal environment instead of with the external environment that surrounds organisms

If each cell has basic survival skills, why must cells perform specialized tasks and be organized according to specialization into systems dedicated to the whole body's survival? The vast majority of cells are not in direct contact with the external environment and cannot function without contributions from the other cells. In contrast, a unicellular organism such as an amoeba can directly obtain nutrients and O_2 from its immediate external surroundings and eliminate wastes back into those surroundings. In a multicellular animal, a muscle cell has the same need for life-supporting nutrients and O_2 uptake and waste elimination, yet the muscle cannot directly make these exchanges with the external environment, because the cells are physically isolated from it.

How can a muscle cell make vital exchanges with the external environment with which it has no contact? The key is the presence of an aqueous internal environment with

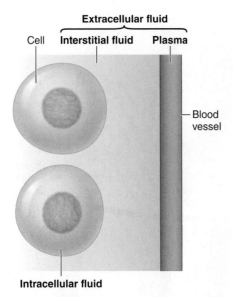

FIGURE 1-5 Components of the extracellular fluid (ECF; internal environment), and location of the intracellular fluid (ICF).

© Cengage Learning, 2013

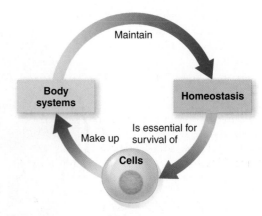

FIGURE 1-6 **Interdependent relationship of cells, body systems, and homeostasis.** The depicted interdependent relationship serves as a key foundation for modern-day physiology: *Homeostasis is essential for the survival of cells, body systems maintain homeostasis, and cells make up body systems.*

© Cengage Learning, 2013

which body cells are in direct contact. This internal environment is outside the cells but inside bodies. It consists of the **extracellular** (*extra* means "outside of") **fluid,** or **ECF,** which in vertebrates is made up of *plasma,* the fluid portion of the blood, and *interstitial fluid,* which surrounds and bathes the cells (Figure 1-5). Various epithelial layers accomplish exchanges between the external environment and the internal environment. For example, a digestive system transfers the nutrients required by all body cells from the external environment into the ECF (such as blood, which is part of the internal body environment). Likewise, a respiratory system transfers O_2 from the external environment into the ECF. A circulatory system, if present, distributes these nutrients and O_2 throughout the body. As a result, the nutrients and O_2 originally obtained from the external environment are delivered to the interstitial fluid that surrounds the cells. Membranes of cells, in turn, make life-sustaining exchanges between the ECF and their interiors, the **ICF** (**intracellular fluid,** see Figure 1-5). (Note that although the term *intracellular fluid* is widely used, the fluid inside cells is actually somewhat gelatinous.) No matter how remote a cell is from the external surface, it can take in from the internal environment the nutrients and O_2 needed to support its own existence. Similarly, metabolic end products produced by the cells (such as CO_2 and ammonia) are extruded into the interstitial fluid, where (in more complex animals) they are picked up by the circulation and transported to organs that specialize in eliminating these wastes from the internal environment to the external environment. For example, lungs (in air breathers) remove CO_2 from the blood, and kidneys remove many other wastes for elimination in urine.

Thus, a body cell takes in essential nutrients from and eliminates wastes into its watery surroundings, just as an amoeba does. The major difference is that body systems must help maintain the composition of the internal environment so that this fluid continuously remains suitable to support the existence of all body cells. In contrast, an amoeba does nothing to regulate its surroundings, which are essentially infinite.

Homeostasis is essential for proper cell function, and most cells, as part of an organized system, contribute to homeostasis

Animal cells can live and function only when they are bathed by an ECF that is compatible with their survival; thus, many aspects of the chemical composition and physical state of the internal environment can be allowed to deviate only within narrow limits. As cells remove nutrients and O_2 from the internal environment, these essential materials must constantly be replenished for each cell's ongoing maintenance of life processes to continue. Likewise, wastes must be removed from the internal environment so they do not reach toxic levels. Other elements in the internal environment that are important for maintaining life also must be kept relatively constant.

The functions performed by the various body systems contribute to homeostasis, thereby maintaining within animals the environment required for the survival and function of all the cells of which organisms consist. This is one of the most important themes of physiology and of this book: *Homeostasis is essential for the survival of each cell, and each cell, through its specialized activities, contributes as part of a body system to the maintenance of the internal environment shared by all cells* (Figure 1-6).

The factors of the internal environment that are often homeostatically regulated include the following:

1. *Concentration of energy-rich molecules.* Nondormant cells need a consistent supply of such molecules (both externally and internally) to serve as a metabolic fuel for life-sustaining and specialized cell activities.

2. *Concentration of O_2 and CO_2.* Cells need O_2 to perform chemical reactions that extract from nutrient molecules the most energy possible for use by the cell. The CO_2 produced during these chemical reactions must be balanced by removal of CO_2 from the body (for example, by lungs or gills) so that CO_2 does not increase the acidity (H^+ concentration) of the internal environment. This balance is accomplished through these reactions:

$$CO_2 + H_2O \leftrightarrow H_2CO_3 \leftrightarrow H^+ + HCO_3^-$$

3. *Concentration of waste products.* Various chemical reactions produce end products that can be toxic to cells if these wastes are allowed to accumulate beyond a certain limit.

4. *pH.* Among the most pronounced effects of changes in the pH (acidity) of the internal fluid environment are alterations in the electrical signaling mechanism of nerve cells and in the enzyme activities of all cells.

5. *Concentration of water, salt, and other electrolytes.* Because the relative concentrations of salt ions (mainly Na^+ and Cl^-), organic solutes, and water in the ECF influence how much water enters or leaves the cells, these concentrations may be regulated to maintain the proper volume of the cells. Cells do not function normally when they are swollen or shrunken. Other ions perform numerous vital functions. For example, the rhythmic beating of a vertebrate heart depends on a relatively constant concentration of potassium (K^+) in the ECF.

In addition, some animals regulate the following:

6. *Volume and pressure.* The circulating component of the internal environment, the plasma in animals with circulatory systems, must be maintained at an adequate volume and pressure to ensure bodywide distribution of this link between the external environment and the cells.

7. *Temperature.* Cells function optimally within a narrow temperature range. Chemical reactions slow down if they are too cold, and, worse yet, membrane and protein structures are disrupted if the environment gets too hot.

8. *Social parameters.* Sometimes homeostasis can extend beyond the individual to a social level. This has been documented in social insects—in termite mounds, for example. In what has been termed *social homeostasis,* the colony of termites acts as a single "superorganism," regulating its population density and number of different termite types (such as workers, soldiers) to nearly constant levels (moreover, temperature of the mound is also regulated).

The fact that the internal environment has some stable features does not mean that its composition, volume, and other characteristics are absolutely unchanging. First, external and internal factors continuously threaten to disrupt homeostasis. When this occurs, appropriate opposing reactions (or behaviors) are initiated to restore these conditions. For example, exposure to a cold environmental temperature tends to reduce a mammal's internal temperature. To counter this, compensatory shivering may be initiated, which internally generates heat that restores body temperature to normal. (Alternatively or in conjunction with shivering, the animal may seek a warmer location.) Likewise, addition of CO_2 into the internal environment as a result of energy-generating chemical reactions tends to raise the concentration of this gas (and the acid level) within organisms. This triggers an increase in breathing rate in many animals. As a result, the extra CO_2 (or HCO_3^-) is released to the external environment, restoring the CO_2 concentration in the ECF to normal. Thus, *homeostasis should be viewed not as a fixed state but as a dynamic steady state* in which the changes that do occur are minimized by compensatory physiological responses. This is why Cannon chose the prefix *homeo* ("similar") instead of *homo* ("same"). For each factor in the internal environment, the small fluctuations around the optimal level are normally kept within the narrow limits compatible with life by carefully regulated mechanisms.

Animals vary in their homeostatic abilities

The term *homeostasis* was developed primarily from research on mammals, which do exhibit a remarkably consistent (though nevertheless modestly varying) internal environment. However, many kinds of multicellular organisms exhibit far more variability in their internal environment than do mammals, yet they survive. Physiologists use specific terms to describe these abilities.

Regulators, Conformers, Avoiders There are two strategies that can be utilized by organisms from the viewpoint of internal environments: *regulators,* which use internal mechanisms to defend a relatively constant state, and *conformers,* whose internal state varies with that of the environment. Some physiologists recognize a third category, *avoiders,* which may not be capable of internal regulation but which nevertheless can minimize internal variations by avoiding environmental disturbances (Figure 1-7). For example, most organisms (including nonanimals) are thermoconformers (body temperatures equal to that of the environment), whereas birds and mammals are generally thermoregulators, maintaining a stable internal temperature regardless of the environmental temperature. But some thermoconformers (such as some fish and insects) can avoid large changes in body temperature by changing their locations in the environment (see Chapter 15).

Enantiostasis Some animals have variable internal components that achieve a kind of homeostasis termed **enantiostasis** (*enantio,* "opposing"). This term refers to consistency of function achieved by changing one physiological variable to counteract a change in another. The term was first used for blue crabs, which suffer a decrease in internal salt composition when they leave the ocean and enter a lower-salinity environment (an estuary). Researchers have found that reduced salt-ion levels inhibit oxygen binding by *hemocyanin* (Chapter 9), a large biomolecule (a protein similar to hemoglobin in your red blood cells), which transports oxygen in the crabs' circulatory fluid. Yet the crabs do not suffer from this, because they make their internal fluids more alkaline (less acidic) by increasing production of ammonia, a strong base. Higher alkalinity in turn increases oxygen binding by hemocyanin and thus offsets the ion effects. Thus, a functional homeostasis of hemocyanin is achieved.

Negative feedback is the main regulatory mechanism for homeostasis

In order to maintain homeostasis, organisms must be able to detect any deviations in internal factors that need to be held within narrow limits, and must be able to control the various body systems that respond to these deviations. For example, to maintain body temperature at an optimal value, mammals must be able to detect a change in internal temperature and then appropriately alter heat production or loss so that internal temperature returns to the desirable level.

Homeostatic control mechanisms primarily operate on the principle of negative feedback. **Negative feedback** occurs when a change in a controlled variable triggers a response that opposes the change, driving the variable in the opposite direction of the initial change. Feedback systems can be very simple *unreferenced* ones, with imprecise control, or *refer-*

Challenges and Controversies
Can a Planet Have Physiology?

As we have discussed, one of the main features that seems to define life itself is the ability to react to, and adjust for, environmental disturbances. In recent years, a number of scientists headed by James Lovelock have proposed that planet Earth itself exhibits this key property of life. Called the Gaia hypothesis, this concept is based on the observation that the conditions of weather and nutrient cycling that favor life have been relatively constant for hundreds of millions, perhaps billions of years. Although periods of prolonged cold (ice ages) have been interspersed with eons of global warmth, at no time have conditions deviated from the range necessary for life. Lovelock and his supporters propose

that life itself, in conjunction with larger geologic and atmospheric processes, maintains this favorable state in homeostatic feedback processes some collectively call "geophysiology." As an example of such a feedback process, researchers have found that cloud formation can be triggered by an atmospheric gas called *dimethyl sulfide (DMS)*. DMS in turn is a breakdown product of an *osmolyte* (a molecule used to regulate cell water content; see Chapter 13) in marine phytoplankton (single-cell algae such as diatoms that are Earth's most common photosynthesizers). The hypothesis is that, as Earth's climate enters any warming trend, plankton growth increases and so,

therefore, does DMS production. In turn, more clouds form that, by blocking sunlight, oppose the heating trend. Thus, the temperature of the planet's atmosphere is prevented from extreme changes:

Increasing atmospheric temperature → more DMS release from phytoplankton → more clouds form → less sunlight reaches Earth → atmospheric cooling

If this is correct, would it be fair to think of Earth itself as a giant living organism (called Gaia)? The other properties of life such as reproduction and evolution may influence your answer to this intriguing question.

(a)   (b)   (c)

FIGURE 1-7 Examples of regulators, avoiders, and conformers. (a) Regulator: a capuchin monkey's internal body temperature is relatively constant regardless of environment; (b) Avoider: monarch butterflies migrate in the autumn from freezing northern habitats to warm southern habitats to avoid extreme body temperatures. (c) Conformer: a marine snail's body temperature matches that of the external aquatic environment. (The monarch also can regulate its body temperature daily by basking in the sun, but at night it becomes a conformer.)

enced with a built-in **set point** that gives the ideal state. An example of an unreferenced system has been proposed for Earth's climate and global warming As Earth's atmosphere begins to heat up from excess carbon dioxide, certain processes may be triggered that produce more clouds (see box, *Challenges and Controversies: Can a Planet Have Physiology?*). Clouds may reduce input from the sun and so oppose the warming trend. But there is no precise regulation, because there is no set point.

In contrast, a referenced negative-feedback system typically has these components: a **sensor** to measure the variable being regulated, an **integrator** that compares the sensed information with a set point, and an **effector**, the device or process that actually makes the corrective response (Figure

1-8a). A common example of a negative feedback is control of room temperature (the controlled variable) by a system that includes a sensing and integrating thermostatic device, a furnace and an air conditioner, and all their electrical connections (Figure 1-8b). The room temperature is altered by the activity of the effectors—the furnace (a heat source) and air conditioner (a heat remover). Suppose someone opens the room door, and too much cold air enters. To switch the effectors on or off appropriately, the control system must "know" (measure) what the actual room temperature is, "compare" it with the desired room temperature (set point), and adjust the output of the effectors to bring the actual temperature to the desired level. A thermosensing device in the thermostat, which monitors the magnitude of the con-

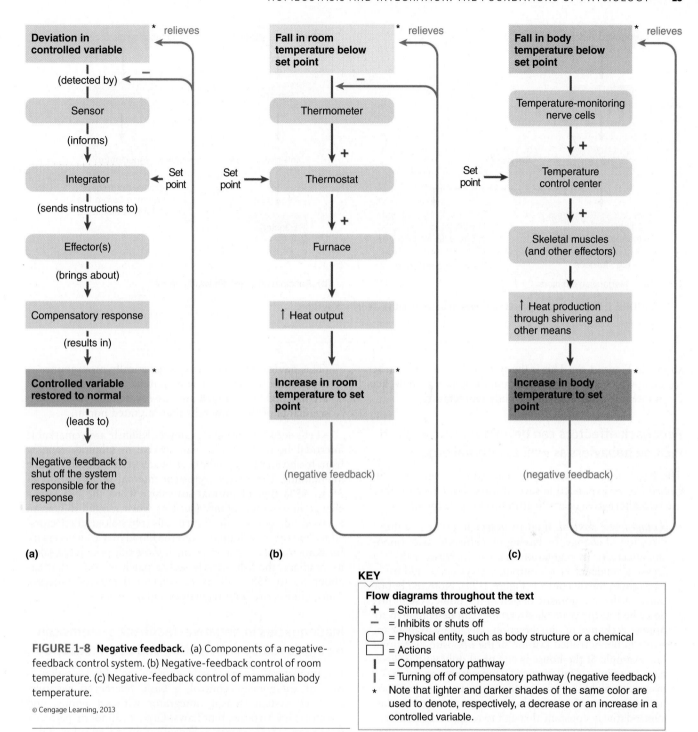

FIGURE 1-8 Negative feedback. (a) Components of a negative-feedback control system. (b) Negative-feedback control of room temperature. (c) Negative-feedback control of mammalian body temperature.

© Cengage Learning, 2013

trolled variable, provides information about the actual room temperature. The thermostat setting provides the desired temperature level, or *set point*. The thermostat acts as an integrator: It compares the sensor's input with the set point and, in this case, adjusts the heat output of the furnace to oppose the deviation from the set point.

When cold is being compensated for, the heat produced by the furnace counteracts or is "negative" to the original fall in the temperature. Note that this is still termed "negative" feedback even though heat is being added. Once the room temperature reaches the set point, the activating mechanism in the thermostat and consequently the effector is switched off. If heat production were to continue un-

abated, the room temperature would be increased above the set point. Overshooting far beyond the set point does not occur because the air temperature "feeds back" to shut off the thermostat that triggered heat output, thereby limiting its own production by altering the signal that initiated the heat production. Thus, the control system takes corrective actions to prevent the controlled variable from drifting too far below or too far above the set point.

Homeostatic systems in organisms operate very similarly with negative feedback to maintain a controlled factor in a relatively steady state. For an example, you can see the similarities between the thermostat system of Figure 1-8b and one aspect of the mammalian thermoregulatory system,

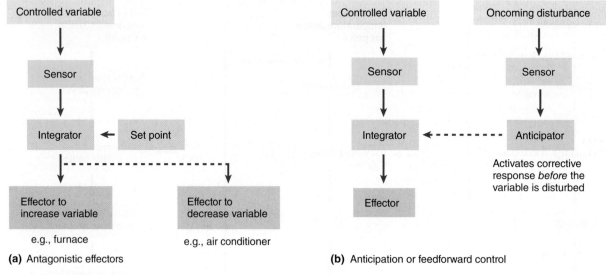

(a) Antagonistic effectors

(b) Anticipation or feedforward control

FIGURE 1-9 Expansion and enhancement of basic negative-feedback system.
© Cengage Learning, 2013

shown in Figure 1-8c. This is the system we described earlier (p. 2), in which a region of the brain, the hypothalamus, acts as an integrator for controlling body temperature.

Feedback effectors can be antagonistic and can include behaviors as well as internal organs

The basic negative-feedback mechanism we have just discussed can be expanded in several ways. For effectors, there are two alternative design features to keep in mind:

- *Antagonistic control.* If corrections can occur in either direction of change, by having two effectors with opposite effects, or by regulation that can increase or decrease a single effector's output, the system is said to have **antagonistic control** (Figure 1-9a). Note the importance of the antagonistic design in the home system just described: If the controlled variable can be regulated to oppose a change in one direction only, the variable can move in uncontrolled fashion in the opposite direction. For example, if the house is equipped only with a furnace to oppose a fall in room temperature, no mechanism is available to prevent the house from getting too hot in summer. However, the room temperature can be kept relatively constant through two opposing mechanisms, one that heats and one that cools the room (that is, an air conditioner), despite wide variations in the temperature of the external environment.

- *Behaviors as effectors.* An effector for a disturbance is typically an organ that directly adjusts some internal state (such as a muscle that shivers to produce heat). However, **behaviors** (specific animal movement patterns used in specific situations, typically requiring coordinated actions of many neural and muscular components) can also serve as corrective processes, and are thus also effectors. Recall that some animals are *avoiders;* these typically use behavioral effectors. For example, monarch butterflies (see Figure 1-7) migrate from the northern United States and Canada to California and Mexico for the winter, thus avoiding extreme cold.

(This navigational behavior is genetically programmed to an amazing degree: The monarchs that migrate south are hatched in the north and are several-generation descendants of the monarchs that migrated north.)

Let's look at another example. Killifish are remarkable intertidal bony fishes that can survive in salinities ranging from freshwater to well above seawater concentrations, while maintaining a relatively consistent internal salinity at about 35 to 45% that of normal seawater. When the fish face a change in external salinity (such as when rainfall dilutes an exposed tide pool), cells in their gills (physiological effectors) adjust the rate of salt intake into the blood to a level necessary for homeostasis. At the same time, if the tide pool has a salinity gradient, the fish actively seek a patch of water that has about 35 to 45% salinity (a behavioral effector process). Thus, killifish are both regulators and avoiders.

Inadequacies in negative-feedback systems can be improved by anticipation and acclimatization

It is important to note that, even with the added sophistication of antagonistic control, a basic referenced negative-feedback system (sensor, integrator with set points, and effectors) has two inherent flaws. First, it cannot be perfectly *homo*static ("same state"), because it must first suffer a disturbing change before it can react and make corrections, resulting in a *delayed response*. Similarly, there is another delay in shutting off an effector, and so basic systems tend to overshoot the set point somewhat. Thus, many regulated parameters—body temperature, hormone levels, water balance, for example—fluctuate continuously (hence the prefix *homeo*-). Second, some components may not work well in a new environment or situation, such as after an animal migrates from low to high altitude. Evolution (and human engineers) have come up with two ways to enhance regulation to overcome these problems (which exist to varying degrees among systems and animals).

- *Anticipation or feedforward system.* An anticipation or feedforward system reduces the delay phenomenon by detecting or predicting an oncoming disturbance *before* a regulated state is changed (for example, perturbed beyond the set point) (Figure 1-9b). It may also predict the results of an effector's output before reaching the set point. The anticipator then activates the appropriate response *in advance* of the change or suppresses the effector before it overshoots the set point. For example, modern electronic thermostats have anticipator circuits that shut off the furnace just before the set point is actually reached, so that the hot air already on its way from the furnace to the thermostat brings the room to the set point without overshooting. As an organismal example, temperature sensors in a mammal's skin detect external heat or cold, and can trigger corrective responses *before* the internal "core" temperature is disturbed. Similarly, when a meal is still in the digestive tract, a feedforward mechanism increases secretion of a hormone (insulin) that promotes cellular uptake and storage of ingested nutrients after they have been absorbed from the digestive tract. This anticipatory response helps limit the rise in blood nutrient concentration that follows nutrient absorption.

- *Acclimatization systems.* Acclimatization systems are mechanisms that alter existing feedback and other components, usually over many days, to work better in a new situation. Examples of new situations include migrating to high altitude or entering a different season. To continue the temperature theme, mammals may increase their hair or fat layers as winter approaches and shed them in the spring, thus augmenting the basic internal negative-feedback system.

Some processes that are technically forms of acclimatization are not always designated as such, because they do not involve habitat features. For example, the muscle mass of an animal increases over time if locomotory activity undergoes a prolonged increase, as during migration. This form of acclimatization is often called **upregulation**. The opposite effect, **downregulation**, can also occur, as when muscle mass declines during more sedentary periods.

A note on terminology: The term **acclimation** refers to acclimatization processes that take place in a controlled situation, such as a laboratory. Acclimation and acclimatization processes, as well as up- and downregulation, are often called *adaptations* by many people. For example, a mountain climber may say he or she has "adapted" to high altitude. However, this term can be misleading, because evolutionary biologists use "adaptation" to refer to genetic changes that occur over many generations through variation and natural selection (p. 1). You will encounter both terms—*acclimatization, adaptation*—frequently, and for the latter term, you should always note whether it is being used in the between-generations (evolutionary) or the within-organism (physiological) sense.

check your understanding 1.5

Define *homeostasis* and give an example of why it is important.

Describe a negative-feedback system with antagonistic effectors.

Discuss how anticipation and acclimatization improve such a system.

Kevin Johnson, U.S. Geological Survey

Newly hatched brine shrimp. Before hatching, they were in an embryonic cyst for several years in a dormant state.

1.6 Regulated Change

Not everything in organisms is homeostatic by any means, not even some of those traditionally considered to be classic examples. Mammalian body temperature is again a good example: Although normally quite narrowly regulated in most mammals most of the time, body temperature varies considerably with fever, between wake and sleep, and in hibernation.

Some internal processes are not always homeostatic but may be changed by reset and positive-feedback systems

There are many nonhomeostatic outcomes of acclimatization to changing environmental stresses. For many animals, negative-feedback regulation cannot be modified to keep working at an optimum, and acclimatization involves switching to a state of **dormancy** (a state of greatly reduced metabolism), such as *hibernation* by a small mammal in which body temperature falls dramatically. Or consider the embryos of a brine shrimp (sold commercially as "sea monkeys") in the Great Salt Lake in Utah. When the embryos are washed up on a dry shoreline, they can lose over 98% of their internal water, yet remain viable by entering a resistant cyst state (which can hatch out years later if put in water). These animals maintain homeostasis only in the broadest sense of "survival." (Compare this to mammals, which die if they lose more than 10 to 15% body water!)

Furthermore, many functions such as locomotion, growth and development, and neural signaling are not homeostatic at all (though they may contribute to homeostasis in some fashion). Some regulatory systems are simply switched on when stimulated for a particular need, and are then switched off. Examples of such "on-demand" regula-

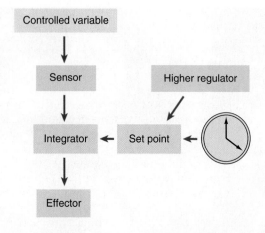

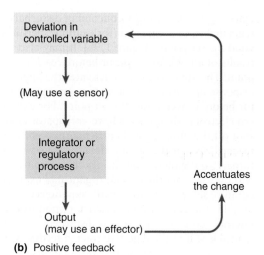

(a) Reset control of negative feedback by a higher system or clock

(b) Positive feedback

Signal from mature fetus

↓

Uterus begins contractions ←

↓   Contractions
                           enhanced
Stretch
sensors

↓

Mother's hypothalamus

↓

Pituitary gland

↓

Oxytocin secreted ─────────

(c) Example of positive feedback: birth of a mammal

FIGURE 1-10 **Mechanisms for regulated change.**

© Cengage Learning, 2013

tion are muscles and glands in the digestive tract, which are locally activated to churn and digest the food when present and shut off when not. This can contribute to other homeostatic needs such as nutrient concentrations in the blood, but this regulation is not homeostatic in itself. Some physiologists have termed nonhomeostatic regulation **rheostasis** ("variable state").

Evolution has also come up with two other mechanisms for regulating useful nonhomeostatic change, which we mention frequently in later chapters:

- **Reset system:** one that changes the set point of a negative-feedback system, in a *temporary*, *permanent*, or *cyclic* fashion (Figure 1-10a). Fever is an example of a temporary reset of body temperature. At sexual maturity in a mammal, sex hormone concentrations are permanently reset to higher levels. Annual hibernation (such as by a squirrel) and many reproductive cycles are additional examples of cyclical resets regulated by internal biological clocks (see Chapter 7). You can probably think of both temporary and cyclical mechanisms for the room thermostat system of Figure 1-8b.

- **Positive-feedback systems** create rapid change when conditions demand a rapid change from a set point. In negative feedback, a control system's output is regulated to resist change, so that the regulated variable is maintained at a relatively steady set point. With positive feedback, however, the output is continually enhanced so that the controlled variable continues to move in the direction of the initial change (Figure 1-10b). Instead of bringing about a response that counteracts the initial change, positive feedback reinforces the change in the same direction. Such action is comparable to the heat generated by a furnace triggering the thermostat to call for even more heat output from the furnace so that the room temperature continuously rises. Because positive feedback moves the controlled variable even farther from a steady state, it does not occur purposefully very often in organisms, where the major goal is maintenance of stable, homeostatic conditions. Indeed, unintended positive feedback can lead to rapid death. For example, in congestive heart failure a weak mammalian heart results in a drop in blood pressure, triggering negative-feedback systems that retain fluid in the body in an attempt to increase blood pressure (more fluid in blood vessels increases pressure). However, more fluid increases stress on the heart, weakening it further, dropping pressure again, and so the cycle continues until the heart fails completely.

Nevertheless, positive feedback is often very useful. Examples include neuron action potentials, lactation, blood clotting, bile release, many mating behaviors and orgasms, ovulation, and some reactions of immune systems are all positive feedback. These all have one thing in common: A relatively rapid change is needed because of a new situation in which the homeostatic set point is no longer appropriate. As a detailed example, consider the birth of a mammalian infant. The hormone *oxytocin* causes powerful contractions of the uterus. As uterine contractions push the baby against the cervix (the exit from the uterus), the resultant stretching of the cervix triggers a sequence of events that releases even more oxytocin, which causes even stronger uterine contractions, triggering the release of more oxytocin, and so on (Figure 1-10c). Positive feedback here is useful to help the infant be born relatively quickly once it reaches a key stage of developmental maturity.

Except for permanent and cyclical reset systems, non-homeostatic "change" mechanisms (on-demand, temporary reset, and positive feedback) need to be shut off once they are no longer needed. The mechanisms for this vary but typically involve the loss of the initiating signal. For example, the oxytocin positive-feedback loop stops once the fetus is no longer pushing on the cervix. Like negative-feedback systems, some of these change systems can be activated in feedforward fashion and improved or adjusted up or down by acclimatization. For example, the activation of digestive muscles and glands (described earlier) can be feedforward-activated by the taste and smell of food. And the internal clock that cyclically resets a human's body temperature each night and day can be acclimatized to a new time zone by local sunlight exposure.

Disruptions in regulation can lead to illness and death

When one or more of an animal's systems fail to function properly, homeostasis (or the ability to regulate change) is disrupted, and all the cells may suffer because they no longer have a suitable environment in which to live and function. Various pathophysiological states ensue, depending on the type and extent of homeostatic disruption. The term *pathophysiology* refers to the abnormal functioning of organisms (altered physiology) associated with disease. When a homeostatic disruption becomes so severe that it is no longer compatible with survival, death results. An example is that of congestive heart failure, mentioned earlier.

check your understanding 1.6

Describe how reset and positive-feedback systems create useful change.

1.7 Organization of Regulatory and Organ Systems

Control of an animal's functions is often—erroneously—portrayed as a centralized command system with the brain and endocrine systems regulating everything. In fact, such a design would be needlessly cumbersome and slow because every tiny response needed by each cell or tissue for homeostasis would require these events: a signal to a central command organ; a decision; a return signal; and finally a local response. For many simple responses, the time delay could be harmful; for example, you would not want to wait for some central endocrine gland to signal the need for a blood clot when your foot is injured and bleeding. Therefore, control in an animal is actually **hierarchical**, much like governmental systems in many countries!

Homeostasis (and other regulation) is hierarchically distributed

The United States and Canada, for example, have at least four levels in this hierarchy in order of increasing priority: city, county, state/province, and federal (or national) governments. Basic local needs are delegated to city governments because local knowledge of the situation is better, and it would be unnecessarily slow to wait for interactions with the national level (considerable delays occur if a city resident requires new federal permission for construction work on his or her home). But needs that involve the whole country (such as distributing food supplies and defending the country) often require the highest level of control. Similarly, many simple, common homeostatic responses in animals are controlled locally, at the level of individual cells, tissues, or organs. Only when more complex responses are needed is a whole-body ("federal" or "national") system needed.

Let's examine this hierarchy in more detail. At the most basic level, an individual cell can regulate its internal energy supplies, ion concentrations, and cell volume, at least to some extent. Beyond the cellular level, control systems that operate to maintain homeostasis are often grouped into two broad classes in this hierarchical design—intrinsic and extrinsic controls. **Intrinsic controls** are those regulated by a single tissue or organ on its own (akin to a county or state/province government). For example, as an exercising muscle rapidly uses up O_2 and produces CO_2 to generate energy to support its contractile activity, the O_2 concentration falls and the CO_2 concentration increases within the muscle. These local chemical changes are directly detected by the smooth muscle in the walls of the blood vessels that supply the exercising muscle and result in the relaxation of the smooth muscle and dilation (open widely) of the vessels, which increases blood flow into the exercising muscle. This local mechanism contributes to the maintenance of an optimal level of O_2 and CO_2 in the internal fluid environment surrounding the exercising muscle's cells. As we discussed earlier, this response is generated at the local level, which avoids unnecessary delays and possibly avoids the reduced precision of generalized whole-body responses.

However, many factors in the internal environment are maintained by **extrinsic controls**, which are regulatory mechanisms initiated outside an organ to alter its activity

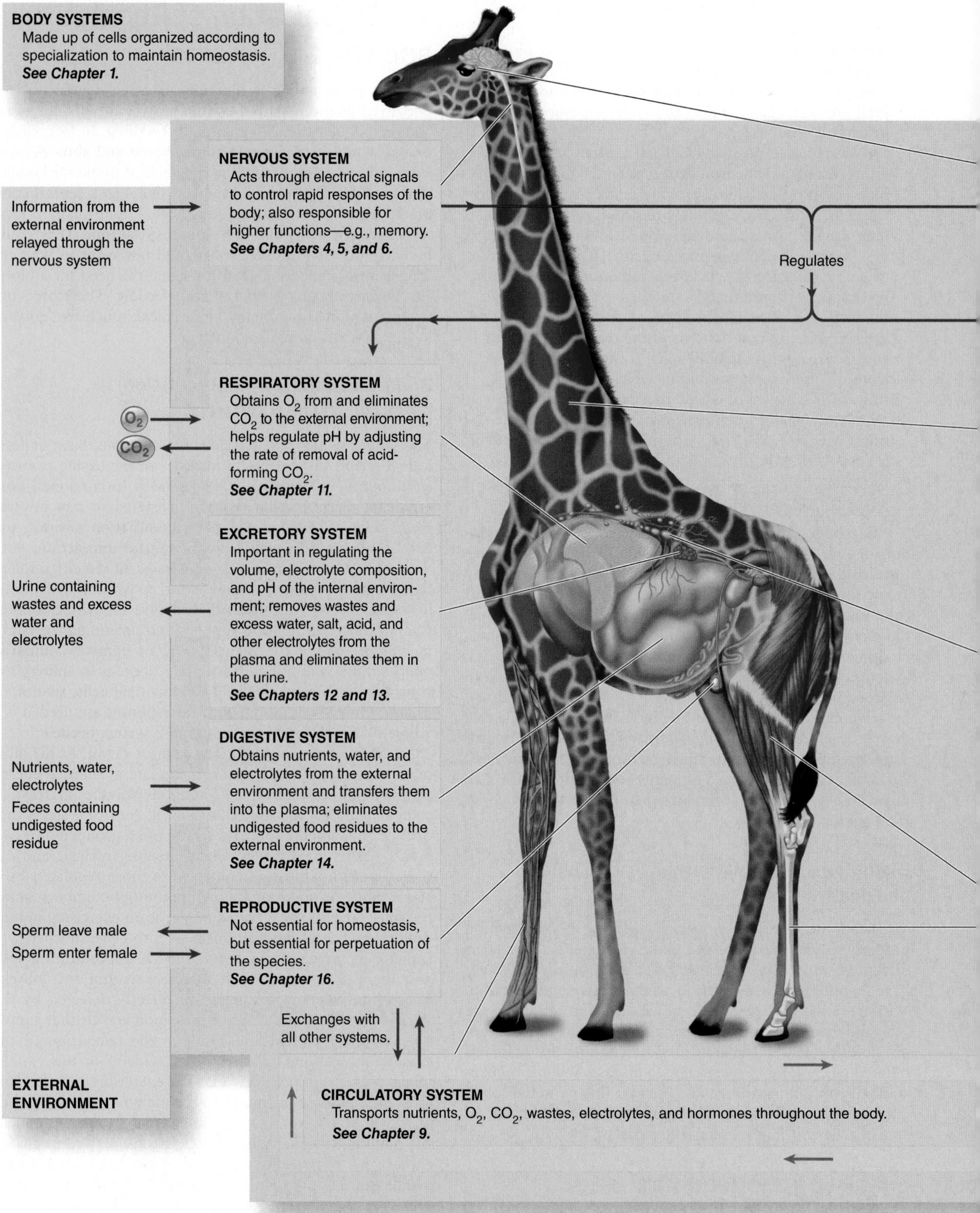

BODY SYSTEMS
Made up of cells organized according to specialization to maintain homeostasis.
See Chapter 1.

NERVOUS SYSTEM
Acts through electrical signals to control rapid responses of the body; also responsible for higher functions—e.g., memory.
See Chapters 4, 5, and 6.

Information from the external environment relayed through the nervous system

Regulates

RESPIRATORY SYSTEM
Obtains O_2 from and eliminates CO_2 to the external environment; helps regulate pH by adjusting the rate of removal of acid-forming CO_2.
See Chapter 11.

O_2

CO_2

EXCRETORY SYSTEM
Important in regulating the volume, electrolyte composition, and pH of the internal environment; removes wastes and excess water, salt, acid, and other electrolytes from the plasma and eliminates them in the urine.
See Chapters 12 and 13.

Urine containing wastes and excess water and electrolytes

DIGESTIVE SYSTEM
Obtains nutrients, water, and electrolytes from the external environment and transfers them into the plasma; eliminates undigested food residues to the external environment.
See Chapter 14.

Nutrients, water, electrolytes

Feces containing undigested food residue

REPRODUCTIVE SYSTEM
Not essential for homeostasis, but essential for perpetuation of the species.
See Chapter 16.

Sperm leave male

Sperm enter female

Exchanges with all other systems.

EXTERNAL ENVIRONMENT

CIRCULATORY SYSTEM
Transports nutrients, O_2, CO_2, wastes, electrolytes, and hormones throughout the body.
See Chapter 9.

FIGURE 1-11 The major vertebrate body systems and their roles.

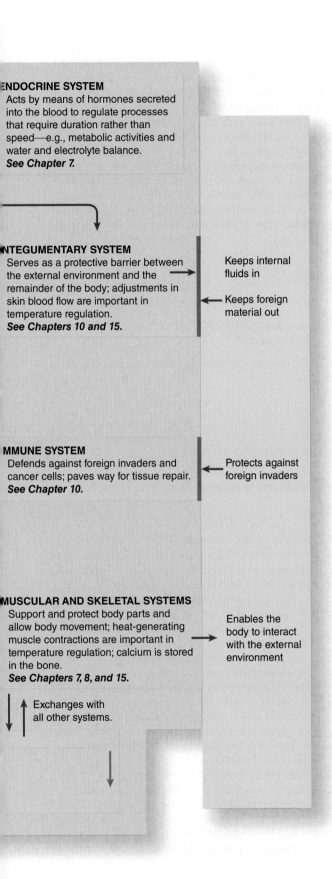

ENDOCRINE SYSTEM

Acts by means of hormones secreted into the blood to regulate processes that require duration rather than speed—e.g., metabolic activities and water and electrolyte balance. *See Chapter 7.*

INTEGUMENTARY SYSTEM

Serves as a protective barrier between the external environment and the remainder of the body; adjustments in skin blood flow are important in temperature regulation. *See Chapters 10 and 15.*

Keeps internal fluids in

Keeps foreign material out

IMMUNE SYSTEM

Defends against foreign invaders and cancer cells; paves way for tissue repair. *See Chapter 10.*

Protects against foreign invaders

MUSCULAR AND SKELETAL SYSTEMS

Support and protect body parts and allow body movement; heat-generating muscle contractions are important in temperature regulation; calcium is stored in the bone. *See Chapters 7, 8, and 15.*

Enables the body to interact with the external environment

Exchanges with all other systems.

(the "national government" level). Often this involves coordination of two or more organs. Extrinsic control of the various organs and systems is accomplished by the nervous, endocrine, and (to some extent) immune systems. Extrinsic control permits coordinated regulation of several organs toward a common goal; in contrast, intrinsic controls are self-serving for the organ in which they occur. Coordinated, overall regulatory mechanisms are critical for maintaining the needs of the internal environment as a whole. For example, the nervous system can take priority over an individual organ's regulation of blood flow during exercise (diverting blood from digestive to muscular systems; just as the national government can sometimes override state/province and city governments).

Organ systems can be grouped according to whole-body contributions

Now that you have considered the basic features of living systems and the importance of homeostasis, let's briefly examine the actual working components of animals that carry out these functions. These components govern the organization of most of this book. As we said earlier, physiologists typically categorize organs in animals into functional **systems** based on their contributions to the whole organism. In turn, these systems can be grouped into four categories based on the broad view of life itself: whole-body control systems; maintenance systems; support and movement systems; and reproductive systems. There are about 11 major body systems (see Figure 1-11 for vertebrates), as follows.

Whole-Body Control Systems Whole-body control systems regulate and coordinate homeostatic and other functions of the other systems for the good of the whole body. Keep in mind that these do not control all activities in the hierarchical control design.

1. **Nervous systems** control and coordinate bodily activities that require swift responses. They are especially important in detecting and initiating reactions to changes in the external environment. Furthermore, they are responsible for more complex "higher" functions such as consciousness, learning and memory, and creativity. In animals with central brains, this system is also hierarchical, for example, with short **reflex** responses (automatic, unlearned reactions to a stimulus) and longer more complex responses.
2. The hormone-secreting glands of **endocrine systems** regulate activities that require duration rather than speed. These systems are especially important in controlling reproductive cycles, the concentration of nutrients, and the internal environment's volume and electrolyte composition.

Support and Movement Systems From a purely homeostatic view, support and movement are not necessarily directed toward maintaining homeostasis, although holding position ("posture") is a type of homeostasis. And the ability to protect from harm and obtain food contributes to the broad homeostasis of body integrity and survival.

3. **Skeletal systems** provide support and protection for the soft tissues and organs. In vertebrates, the skeleton also

functions in homeostatic regulation of calcium (Ca^{2+}), an electrolyte whose plasma concentration must be maintained within very narrow limits. Together with the muscular system, the skeletal system also enables movement of animals and their parts.

4. **Muscular systems** move the skeletal components to which the skeletal muscles are attached. Furthermore, the heat generated by muscle contraction is important in temperature regulation in some animals, especially birds, mammals, and some insects and fishes.

Maintenance Systems These are organ systems that do most of the actual work of maintaining the internal environment. The systems and their most important contributions to homeostasis are as follows:

5. **Circulatory systems** are the transport systems that carry materials such as nutrients, O_2, CO_2, nitrogenous wastes, electrolytes, heat, and hormones from one part of an animal to another.

6. **Defense** or **immune systems** defend against foreign invaders (viruses, bacteria, parasites) and body cells that have become cancerous. They also pave the way for repair or replacement of injured or worn-out cells. Because some defensive responses can coordinate activities of organs throughout the body (at least in mammals), in recent years many researchers have come to regard the mammalian immune system as the third whole-body control system (along with nervous and endocrine systems). For now, we will leave it under maintenance systems.

7. **Respiratory systems** obtain O_2 from and eliminate CO_2 to the external environment. By adjusting the rate of removal of acid-forming CO_2, a respiratory system is also important in maintaining the proper pH (acid–base balance) of the internal environment.

8. **Excretory systems** remove excess water, salt, acid, and other electrolyte from the plasma and eliminate them, along with waste products other than CO_2.

9. **Digestive systems** break down dietary food into small nutrient molecules that can be absorbed into the circulatory system for distribution to all cells. They also transfer water and electrolytes from the external environment into the internal environment. A digestive system eliminates undigested food residues to the external environment in the feces. (The vertebrate liver also excretes some wastes—see Chapter 12.)

10. **Integumentary systems** serve as outer, protective physical barriers that prevent internal fluid from being lost from organisms and foreign microorganisms from entering them. These organs are also important for regulating body temperature in some animals.

Reproductive Systems Reproductive systems are not essential for homeostasis of the individual and therefore are not essential for survival of the individual. They are essential, however, for perpetuation of the species.

11. **Reproductive systems** consist of organs that produce gametes (eggs, sperm), organs to deliver them and, in some animals, support systems for offspring (such as yolk production, and embryo and fetal development).

As we examine each of these systems in greater detail, always keep in mind that a body is a coordinated whole even though each system provides its own special contributions. It is easy to forget that all the parts actually fit together into a functioning, interdependent whole body. Accordingly, each chapter concludes with a brief overview of the integrative, whole-body contributions of the system.

Keep another point in mind as you read through the book: The functioning whole is greater than the sum of its separate parts. Through specialization, cooperation, and interdependence, cells combine to form a coordinated, unique, single living organism with more diverse and complex capabilities than is possessed by any of the cells that make it up.

check your understanding 1.7

What are intrinsic and extrinsic controls? Illustrate.

Why is it important that the brain not exclusively control all of an animal's functions?

Why is it advantageous for physiologists to organize organs into functional systems?

making connections

How Homeostasis and Integration Contribute to the Body as a Whole

In this chapter, you have learned that homeostasis is a dynamic steady state of the constituents in the internal fluid environment (the extracellular fluid) that surrounds and exchanges materials with the cells. You have also learned how homeostasis can be enhanced with anticipation and acclimatization, and how nonhomeostatic regulated changes are controlled by resets and positive feedback. Homeostasis and regulated changes are regulatory features essential for the survival and normal functioning of cells, and each cell, through its specialized activities, contributes as part of a body system to these regulatory features.

This integrated relationship among cells, organ systems, and the whole organism is the foundation of physiology and the central theme of this book. We have already described how cells are organized according to specialization into body systems. How the key functions of life and body systems maintain this internal consistency and necessary changes are the topics covered in the remainder of this book. Each chapter concludes with this capstone feature to facilitate your understanding of (1) how the systems and functions under discussion contribute to whole-body homeostasis and regulated change and (2) the integrated interactions and interdependency of body systems.

Chapter Summary

- Biological adaptations have two levels of explanation: the mechanistic (or proximate) and the evolutionary (or ultimate). Evolution by natural selection occurs when offspring exhibit differential survival and reproduction due to variations in their genes. Adaptations reflect evolutionary history including constraints and cost–benefit trade-offs, so are not always logically optimal.

- Physiology is an integrative discipline, using physics, genetics, ecology, and so on; and it is also comparative, using animals of all types to reveal both the diversity and the universality of processes. This often follows the Krogh principle, which states that, for every adaptation, there will be a particular species in which it is most conveniently studied.

- The hypothetico-deductive method is the most widely accepted version of "the scientific method." Data are gathered about an aspect of nature in the "discovery" phase, then alternative hypotheses are made to explain them. Testable deductions must be made from the hypotheses, with experiments and observations designed to test them, leading to support, refinement, or rejection of hypotheses.

- Life has the properties of self-organization, self-regulation, self-support and movement, and self-reproduction. Cells are progressively organized into tissues, organs, systems, and finally the whole body. Tissues are epithelial for linings, connective for support, neural for electrical communication, and muscular for contraction.

- Size matters: Many anatomical and physiological features are scaled to (that is, are dependent on) body size. For example, the larger the organism, the smaller the surface-area-to-volume ratio; and larger animals usually live longer than smaller ones.

- Body cells are in contact with a privately maintained internal environment instead of with the external environment that surrounds organisms. Homeostasis is the maintenance of a consistent internal state and is essential for proper cell function; each cell in turn, as part of an organized system, contributes to homeostasis. Negative feedback, such as a home thermostat, is the main mechanism for homeostasis, with sensors, integrators with set points, and effectors. Some animals are good internal regulators, while others behaviorally avoid disturbances, and others do not avoid or internally regulate but conform to the environment. Some use enantiostasis: varying one body parameter to achieve homeostasis in another.

- Feedback effectors can be antagonistic (e.g., shivering and sweating devices) and can include behaviors as well as internal organs. Inadequacies in negative-feedback systems can be improved by anticipation (initiating corrections before a disturbance) and acclimatization (improving response of a feedback component in a new situation).

- Some internal processes are not always homeostatic but may be changed by reset systems, which raise or lower a negative-feedback system set point, and positive-feedback systems, which rapidly enhance changes. Disruptions in regulation can lead to illness and death.

- Homeostasis (and other regulation) is hierarchically distributed, with intrinsic regulation occurring locally in cells and tissues for rapid control of basic needs, and extrinsic (mainly neural and hormonal) for coordinating multiple body parts, often overriding intrinsic controls for the good of the whole.

- Organ systems can be grouped according to whole-body contributions, following universal life properties, into regulatory (neural, hormonal), maintenance (digestive, respiratory, circulatory, immune, excretory, integumentary), support and movement (skeletal and muscular), and reproductive systems.

Review, Synthesize, and Analyze

1. Some people dismiss the concept of evolution as "just a theory." Using the hypothetico-deductive method, explain why most scientists accept evolution as fully established.

2. Explain why the term *homeostasis* (similar state) was chosen rather than *homostasis* (same state).

3. Why is positive feedback sometimes useful, but also often dangerous, even lethal?

4. What are intrinsic and extrinsic controls, why are they analogous to the U.S. government system of city, county, state, and federal levels, and why is this hierarchy useful?

5. Why is the Krogh principle useful in the field of animal physiology?

Suggested Readings

Boyd, C. A. R., & D. Noble, eds. (1993). *The Logic of Life: The Challenge of Integrative Physiology.* Oxford, UK: Oxford University Press. Discusses ideas on the evolution and regulation of physiological processes.

Burggren, W. W. (2000). Developmental physiology, animal models, and the August Krogh principle. *Zoology* 102:148–156. On using the Krogh principle.

Calder, W. A. (1996). *Size, Function and Life History.* Cambridge, MA: Harvard University Press. A classic book on the scaling of organisms.

Kelly, K. (1995). *Out of Control: The New Biology of Machines, Social Systems and the Economic World.* Reading, MA: Addison-Wesley. Discusses how biological concepts of feedback and hierarchical regulation are being applied to human society and industry.

McGowan, C. (1994). *Diatoms to Dinosaurs: The Size and Scale of Living Things*. Washington, DC: Island Press. Elegantly explains the concept of scaling.

Mangum, C., & D. Towle. (1977). Physiological adaptation to unstable environments. *American Scientist* 65:67–75. Introduces the concept of enantiostasis.

Moore, J. A. (1999). *Science as a Way of Knowing*. Cambridge, MA: Harvard University Press. A detailed, highly readable look at the scientific method and scientific thinking.

Mrosovsky, N. (1997). *Rheostasis: The Physiology of Change*. Oxford, UK: Oxford University Press. Discusses how the concept of homeostasis is overemphasized at the expense of regulated change or rheostasis.

Nesse, R. M., & G. C. Williams. (1996). *Why We Get Sick: The New Science of Darwinian Medicine*. New York: Vintage Books. A classic book introducing this new field arising out of evolutionary physiology.

Patrick, J. (1999). *Homeostasis*. physioweb.med.uvm.edu/homeostasis/. Includes links to online interactive simulations of simple and complex negative feedback.

Schmidt-Nielsen, B. (1984). August and Marie Krogh and respiratory physiology. *Journal of Applied Physiology* 57: 293–303. A history of the Kroghs' contributions by a protégé.

Schmidt-Nielsen, K. (1997). *Animal Physiology: Adaptation and Environment*. Cambridge & New York: Cambridge University Press. A classic comparative textbook.

Schulkin, J. (2003). *Rethinking Homeostasis: Allostatic Regulation in Physiology and Pathophysiology*. Cambridge, MA: MIT Press. Discusses important nonhomeostatic and enantiostatic processes of mammalian bodies.

Volk, T. (1997). *Gaia's Body: Toward a Physiology of Earth*. New York: Copernicus Books. Discusses feedback and other regulatory processes in the planetary geo- and biosphere.

CourseMate

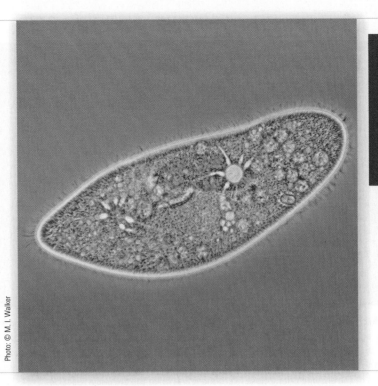

2

Cellular and Molecular Physiology

2.1 Introduction

As we noted in Chapter 1, cells are distinct from their inanimate constituents because they have the ability to grow, replicate, perform complex metabolic reactions, and respond to environmental stimuli. Modern cellular physiologists are unraveling many of the broader mysteries of how cells work, both as living systems individually and in their specialized contributions to multicellular organisms. Comparative physiologists study how an individual species is adapted to a specific environment in part by analyzing the molecular structure and organization of its cells. Analyses of the structures of living cells, and of the molecular components of which they are made, show the deep interconnectedness of all life through evolution. In this chapter, we'll review those structures and molecules.

It is important to realize that, although researchers have analyzed structures and functions of most of the molecular constituents of cells, no one has thus far been able to assemble building-block chemicals artificially into a living system in a laboratory environment (although attempts are under way). Scientists have created a new bacterium with an artificial genome, but a previously living cell was still needed in

which to insert the synthesized genome. (**Genome** refers to the entire genetic code of an organism.) It is not merely the presence of various molecules that confers the unique characteristics of life; it is the complex organization and interaction of these molecules within the cell that is critical. Much remains to be learned about the molecular organization necessary for life, and tools that allow us to do so are continuously increasing in sophistication and power.

Water, other inorganic chemicals, and four types of organic molecules are the universal components of cells

Living systems are constructed from two broad classes of chemicals called *inorganic* and *organic molecules.* Scientists originally designated organic chemicals as those from living or once-living sources, but today the term *organic* refers to almost all molecules made from the element carbon (and invariably hydrogen, with the smallest organic molecule being methane or natural gas, CH_4). Based on this defini-

tion, organic material has now been confirmed on other planets and a moon (Titan) in our solar system, in nebulae in space, and even on a distant planet, Osiris, by the Hubble Space Telescope. Located 150 light-years from Earth (about 900 trillion miles), Osiris also has oxygen in its atmosphere. However, the development of life on Osiris is unlikely because its close proximity to its sun (4.3 million miles) probably makes it too hot.

All other chemicals are termed *inorganic,* even small carbon-containing molecules such as carbon dioxide (CO_2). Although cells contain a large variety of inorganic and organic components, there are some fundamental, universal types. We assume you are familiar with the basic inorganic and organic molecules of life; for those who are not, details are summarized here.

Water (H_2O) is the universal inorganic molecule of life on Earth. Its properties such as its *polarity* (unequal sharing of electrons between hydrogen and oxygen) and hydrogen-bond formation make it unique among molecules in the universe for serving as a

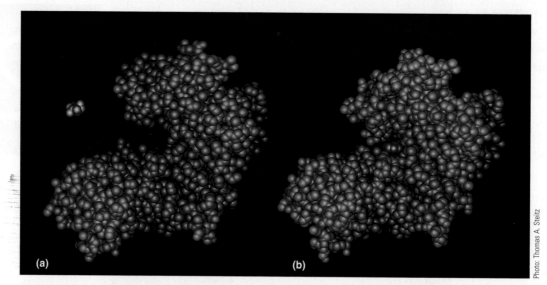

FIGURE 2-1 **The complex structures of proteins are mainly determined by weak bonds between amino-acid side groups, allowing proteins to be dynamic and flexible.** Here, a model of an enzyme, hexokinase, is at work. Hexokinase catalyzes phosphorylation of glucose. (a) A glucose molecule (color-coded *red*) is heading toward the active site, a cleft in the enzyme (*green*). (b) When the glucose molecule makes contact with the site, parts of the enzyme briefly close in around it and drive the chemical reaction.

medium for the reactions of life. In particular, polarity allows water molecules to bind to any molecule that is charged or polar. Indeed, water is sometimes called "the universal solvent." In addition, other small inorganic ions such as sodium (Na^+), potassium (K^+), chloride (Cl^-) and phosphate (PO_4^{3-}) appear to be universal solutes (dissolved molecules) in cells; we explore the roles of these and other inorganic chemicals later.

Because of the unique bonding ability of carbon, millions of different organic molecules are possible. However, it is possible to simplify this tremendous diversity into a few categories. The building blocks of organic molecules are called **monomers**, and chains of monomers linked together are called **polymers**. The largest polymers, called **macromolecules**, are the most important organic components of life. The basic categories of organic molecules are as follows:

1. **Carbohydrates** are made of carbon, hydrogen, and oxygen. The smallest building-block (monomer) units are simple sugars or **monosaccharides** such as **glucose**, the dominant energy molecule in most animals. In turn, these can form polymers called **polysaccharides**, such as **glycogen** (a branching chain of glucose molecules and an important energy storage molecule in animals). In addition to energy, carbohydrates can be used to form important structures such as *cellulose*, one of the constituents of plant cell walls, and *chitin*, a component of the exoskeleton of arthropods. Sugars are also often found attached to proteins to form **glycoproteins**, which are important components of cell membranes and signaling systems (see Chapter 3).

2. **Lipids** are made of carbon and hydrogen (and sometimes oxygen) and are *hydrophobic* ("water fearing"); that is, they do not easily dissolve in water. This arises from the fact that C–H bonds are nonpolar because they share electrons equally and cannot form hydrogen

bonds with water. You probably know that oil, a hydrophobic hydrocarbon, floats on water. This is because water molecules make hydrogen bonds with each other but not with oil molecules, which are in effect squeezed out as water molecules attract each other. (Lipids then rise to the top because they are less dense than water.) This lipid–water interaction is one of the most important chemical forces in living systems. Key examples of lipids are **fatty acid** monomers, which provide energy and serve as signal molecules; **triglycerides** (polymers of three fatty acids and glycerol), which are energy stores in animals; *phospholipids*, used to make membranes; and *cholesterol*, which is used to make membranes, steroid hormones, and bile.

3. **Amino acids** and **proteins** are made of carbon, hydrogen, oxygen, and nitrogen in a specific arrangement. Carboxylic amino acids, the most biologically common type, have an amino group (NH_2), an acid group (carboxyl, or COOH), and a side group that gives each amino acid unique properties. Amino acids may be used for energy, as *osmolytes* (see Chapter 13), or as signal molecules (Chapter 4), but their most crucial role is to form polymers called *peptides* and *proteins*. Peptides and proteins are formed from 20 different carboxylic amino acids (*glycine, alanine,* etc.), arranged in any order and to almost any length, providing an endless list of possible types of proteins. These polymers are the principal dynamic molecular agents of life, forming regulating compounds, such as signal molecules and receptors; transporters that move molecules through membranes; structural and functional components (such as those in muscle filaments); and enzymes, the primary catalysts of life's chemical reactions (Figure 2-1a).

Ultimately a protein's structure and function is as a result of the properties of its amino acid composition and the respective chemical properties of its constituent amino acids (especially their side groups), and how they interact with each other and the surrounding medium. The highly complex three-dimensional structures that result from these molecular interactions determine a protein's function; that is, shape determines function. There are up to four levels of structure involving peptide chains and proteins.

4. **Nucleotides** and **nucleic acids** are made of carbon, nitrogen, oxygen, hydrogen, and phosphate in a particular arrangement. The monomers are called **nucleotides**, which have three components: a sugar (ribose or deoxyribose), phosphate, and a **base** that is one of five major types designated **A, G, C, T,** and **U** (adenine, guanine, cytosine, thymine, and uridine). These can form two types of polymers differing in the sugar component and one of the bases:
   - **Deoxyribonucleic acid,** or **DNA,** made of A, G, C, and T, is the basis for storage of genetic information and the process of inheritance.
   - **Ribonucleic acid,** or **RNA,** made of A, G, C, and U, helps convert that information into cell structures and working machinery.

We'll examine the functions of DNA and RNA later after reviewing basic features of cells.

All cells have two major subdivisions: the plasma membrane and the cytoplasm

Recall that life falls into three domains of *archaea, bacteria,* and *eukarya,* but in terms of basic cell features, biologists have traditionally designated two basic types—*prokaryotic* and *eukaryotic* (Chapter 1). Regardless of classification, all cells share many common features, having two major subdivisions—the *plasma membrane* and the *cytoplasm*—and having DNA-based genetic information systems. Eukaryotic cells have a third subdivision, the *nucleus,* to house the DNA, along with other membrane-bound structures called *organelles* in the cytoplasm.

The **plasma** or **cell membrane** is a very thin oily barrier that encloses each cell, separating the cell's contents from the surrounding environment. Its composition, further detailed in Chapter 3, is primarily composed of phospholipids, proteins and glycoproteins, and often cholesterol. The fluid contained within all the cells of the body is known collectively as **intracellular fluid (ICF),** and the fluid outside the cells is referred to as **extracellular fluid (ECF).** The plasma membrane does not merely serve as a mechanical barrier to hold in the contents of the cells; it also can selectively control movement of many molecules between the ICF and ECF. The plasma membrane can be likened to the gated walls that enclosed ancient cities. Through this structure, the cell can regulate the entry of nutrients and other needed supplies and the export of products manufactured within, while at the same time guarding against unwanted traffic into or out of the cell. However, not all molecules are selectively regulated; for example, water and CO_2 molecules (because they are so small) and hydrophobic molecules (except very large ones) can typically diffuse through most membranes.

The **cytoplasm** is that portion of the cell interior not occupied by the nucleus. It contains a number of discrete, specialized *organelles* (the cell's "little organs") and the *cyto-*

skeleton (a scaffolding of proteins that serve as the cell's "bone and muscle") dispersed within the *cytosol* (a complex, gel-like liquid). **Organelles** are distinct, highly organized structures that perform specialized functions within the cell.

Many macromolecular structures need to be flexible to function and to be regulated, and protecting those structures is the basis for many forms of homeostasis

Large molecular complexes, the framework of living cells, can be very vulnerable because their higher-level structures are often held together with weak bonds such as hydrogen bonds, as opposed to the strong or covalent bonds that join atoms within single molecules. Reliance on weak bonds is important because they give macromolecular structures dynamic features that could not occur with stronger, more rigid covalent bonds. Membranes, for example, are held together primarily by the result of hydrophobic interactions (Chapter 3) and thus can be flexible. Flexibility is necessary for cell shape changes and for many transport proteins to move molecules through membranes. The DNA double helix relies on hydrogen bonding so it can be readily disassembled for RNA formation and DNA replication. In addition, the complex three-dimensional structures of proteins are mainly determined by weak bonds between amino-acid side groups, allowing many proteins to be dynamic and flexible. Dynamic changes in proteins are crucial for everything from transport to enzyme catalysis. An example of the latter is shown in Figure 2-1b.

Another reason why many protein molecules are inherently flexible is that this characteristic allows their shapes to alter in order to regulate their activities, that is, be activated or inhibited. Useful conformational changes are induced in at least four ways: (1) from the binding of a regulatory molecule to a special binding site, a process called **allosteric** ("other site") **modulation;** (2) from a change in an electric field; (3) from physical deformation; and (4) by **phosphorylation**—the addition of a phosphate group by an enzyme called a **kinase** or by removal of phosphate by a **phosphatase.** We present examples of all four in later chapters.

But there is a price to be paid for having flexible macromolecules. The cost is that the shapes of such molecules can be disturbed by changes in common environmental factors. Temperature, for example, can disrupt weak bonds so that proteins lose their three-dimensional conformation, which is essential for their function. Similarly, inorganic ions can bind to charged side chains of amino acids, disrupting their weak bond interactions. *Thus, one of the major forces in the evolution of homeostasis*—the process of maintaining a consistent interior environment (for example, in pH level or salt content)—*has been the provision of an optimal environment for the functioning of macromolecules.*

Archaeal and bacterial cells have a simpler organization than eukaryotic cells

The two prokaryotic cell types, Archaea and Bacteria, have an outer plasma membrane but do not contain a separate nucleus or (for the most part) other organelles per se. The prokaryotic cytoplasm is the site where the cell converts energy and synthesizes the molecules necessary for cell

TABLE 2-1 Summary of Cytoplasm Components

Cytoplasm Component	Structure	Function
MEMBRANOUS ORGANELLES		
Endoplasmic Reticulum	Extensive, continuous membranous network of fluid-filled tubules and flattened sacs, partially studded with ribosomes	Forms new cell membrane and other cell components and manufactures products for secretion
Golgi Complex	Sets of stacked, flattened membranous sacs	Modifies, packages, and distributes newly synthesized proteins
Lysosomes	Membranous sacs containing hydrolytic enzymes	Serve as cell's digestive system, destroying foreign substances and cellular debris
Peroxisomes	Membranous sacs containing oxidative enzymes	Perform detoxification activities
Mitochondria	Rod- or oval-shaped bodies enclosed by two membranes, with the inner membrane folded into cristae that project into an interior matrix	Act as energy organelles; major site of ATP production; contain enzymes for citric acid cycle, proteins of electron transport system, and ATP synthase
NONMEMBRANOUS ORGANELLES		
Ribosomes	Granules of RNA and proteins—some attached to rough ER, some free in cytosol	Serve as workbenches for protein synthesis
Proteosomes	Tunnel-like protein complexes	Serve to break down unwanted proteins
Vaults	Shaped like hollow octagonal barrels	Serve as cellular trucks for transport from nucleus to cytoplasm
Centrosome/ Centrioles	A pair of cylindrical structures at right angles to each other (centrioles) surrounded by an amorphous mass	Form and organize the microtubule cytoskeleton
CYTOSOL		
Intermediary Metabolism Enzymes	Dispersed within the cytosol	Facilitate intracellular reactions involving degradation, synthesis, and transformation of small organic molecules
Transport, Secretory, and Endocytic Vesicles	Transiently formed, membrane-enclosed products synthesized within or engulfed by the cell	Transport and/or store products being moved within, out of, or into the cell, respectively
Inclusions	Glycogen granules, fat droplets	Store excess nutrients
Cytoskeleton		As an integrated whole, serves as the cell's "bone and muscle"
Microtubules	Long, slender, hollow tubes composed of tubulin molecules	Maintain asymmetric cell shapes and coordinate complex cell movements, specifically serving as highways for transport of secretory vesicles within cell, serving as main structural and functional component of cilia and flagella, and forming mitotic spindle during cell division
Microfilaments	Intertwined helical chains of actin molecules; microfilaments composed of myosin molecules also present in muscle cells	Play a vital role in various cellular contractile systems, including muscle contraction and amoeboid movement; serve as a mechanical stiffener for microvilli
Intermediate Filaments	Irregular, threadlike proteins	Help resist mechanical stress

growth and function. Even though there is no distinct nucleus, prokaryotes have a distinct nuclear area with a single, circular chromosome. Given these similarities, why are Archaea and Bacteria now separated into their own domains, with some biologists even calling for the abolishment of the term *prokaryote*?

In fact, though looking similar in the microscope, Archaea and Bacteria differ significantly in many fundamental

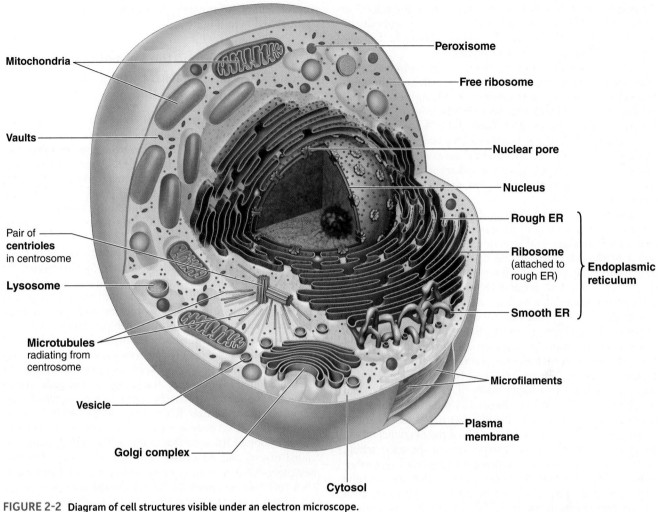

FIGURE 2-2 Diagram of cell structures visible under an electron microscope.

ways, such as in their genetic sequences, mechanisms of gene regulation, and cell-wall components, with some archaeal features being more closely related to eukaryotic ones than to bacterial ones. In terms of importance for animal physiology, true bacteria include many of the familiar pathogens that cause animal diseases (but note there are even more species that are not pathogenic). Archaea, though they have now been found in most habitats on Earth, were originally noted for their ability to colonize extreme environments, such as hot springs and highly saline lakes, and to exhibit unusual metabolisms as well. Though Archaea are not known to cause animal diseases, one group of particular importance is the *methanogens*, methane-producing Archaea that populate digestive tracts of many animals (see Chapter 14) as well as low-oxygen sediments of aquatic habitats.

Interestingly, some bacteria (not Archaea) do possess one membrane-bound organelle—called the *acidocalcisome*—that resembles a lysosome (p. 44) and plays an important role in osmoregulation, and which has been found in many eukaryotes as well. Its ancestral role is believed to be linked to its ability to store cations and phosphate, with cation and pH homeostasis derived following the divergence of prokaryotes and eukaryotes.

Eukaryotic cells are subdivided into the plasma membrane, nucleus, and cytoplasm

Eukaryotes are all based on a complex cell type with a plasma membrane and a distinct *nucleus* and other organelles in the cytoplasm's **cytosol**, a complex, semigelatinous watery mass filling the cell interior. Table 2-1 summarizes the main division of a eukaryotic cell and the functions of the major components. Nearly all eukaryotic cells contain five main types of **membranous organelles**—the *endoplasmic reticulum, Golgi complex, lysosomes, peroxisomes,* and *mitochondria* (Figure 2-2; Table 2-1). These organelles are similar in all eukaryotic cells, although there are some variations depending on the specialized capabilities of each cell type. Organelles increase the efficiency of metabolism (especially in cells larger than bacteria) by acting as scaffolds for macromolecules (such as enzymes) participating together in a series of reactions, and by serving as intracellular "specialty shops." Each contains a specific set of chemicals for carrying out a particular cell function. This compartmentalization is evolutionarily advantageous because it permits chemical activities that would not be compatible with each other to occur simultaneously within the cell. For example, the enzymes that destroy unwanted proteins in the cell do so within the protective confines of the lysosomes without the risk of destroying essential cell proteins. About half of the total cell volume is occupied by organelles.

The cytosol also contains many small molecules, proteins such as enzymes, and critical large **nonmembranous organelles**: *ribosomes, proteasomes, vaults,* and several types of *cytoskeleton*—filamentous proteins that control cell shape and movements (Figure 2-2; Table 2-1). We discuss these later. Many fundamental metabolic chemical reactions take place in the cytosol. The cytoskeletal network, which elaborately laces the cytosol, gives the eukaryotic cell its shape, provides for its internal organization, and regulates its movements (as you'll see).

The nucleus, which is typically the largest single organized cell component, can be seen under magnification as a distinct spherical or oval structure, usually located near the center of the cell. Surrounding it is a double-layered membrane with many pores, which separates the nucleus from the remainder of the cell. Sequestered within the nucleus is the cell's genetic material, which we discuss in the next section.

Evidence shows that the eukaryotic cell evolved from a *symbiosis* (two different species living intimately together) between two or more prokaryotic cell types. Mitochondria and chloroplasts (organelles for photosynthesis in plants and algae) are descended from free-living bacterial ancestors that took up residence in a larger host cell, possibly an archaeal cell. This idea was first hypothesized by the Russian biologist Konstantin Mereschcowsky in 1905, but it was the American biologist Lynn Margulis who assembled the overwhelmingly convincing evidence in the 1960s. Among the most telling pieces of evidence are the occurrence of bacterial-type ribosomes (complexes that synthesize proteins that we examine later) and genes (the units of inheritance to which we turn our attention next) in bacterial-like chromosomes (strands of genetic material) inside chloroplasts and mitochondria.

Let's now examine each of the eukaryotic components in more detail.

check your understanding 2.1

What are a eukaryotic cell's three major subdivisions?

Why is it important for macromolecular cellular structures to be flexible? Provide an example.

2.2 Nucleus, Chromosomes, and Genes

The **nucleus** of eukaryotic cells (as well as the cytosol of prokaryotic cells) contains the materials for genetic instructions and inheritance. The key molecule is the double helix formed by DNA, which is packaged with proteins called *histones* in the nucleus to form complexes called **chromosomes**. DNA has two important functions: providing a code of information for RNA and protein synthesis and serving as a genetic blueprint during cell replication (for genetic inheritance). By coding for the kinds and amounts of various enzymes and other proteins that are produced, the nucleus indirectly governs most cell activities and serves as the cell's

control center. Information critical to animal physiology is summarized here (Figure 2-3).

DNA contains codes in the form of genes for making RNAs and proteins through the processes of transcription and translation

Recall that DNA is constructed from the four nucleotide bases A, G, T, and C. Because these bases exhibit complementary binding—that is, they bind to each other in pairs of A to T, G to C—the double helix of DNA serves as a genetic blueprint during cell replication to ensure that the cell produces additional cells just like itself, thus continuing the identical type of cell line within the body. Furthermore, in the reproductive cells, the DNA blueprint passes on genetic characteristics to future generations.

As with amino acids in proteins, the sequence of A, G, T, and C can vary to form countless codes, or "instructions." An individual coding sequence, along with regulatory codes that we discuss shortly, is called a *gene*. Within any single species, there may be tens of thousands of genes; and the variations that occur among all species on Earth are limitless. This is analogous to using the relatively few letters of an alphabet to make unlimited numbers of sentences. Gene codes ultimately direct the synthesis of specific RNA and proteins within the cell. First, in the process of **transcription**, the gene (DNA strand that codes for a particular protein) is *transcribed* (copied) into a **pre-messenger RNA** molecule (pre-mRNA) by an enzyme called *RNA polymerase*. This pre-mRNA is a complementary copy of the gene that contains the coding sequences (called *exons*) and noncoding sequences (*introns*). The pre-messenger RNA in turn is processed into a final **messenger RNA** (mRNA) by removal of the dispersed introns and ligation of the exons, a process called **mRNA splicing**. This step is followed by the addition of noncoding signal sequences to the leading and trailing ends of the molecule. The mature mRNA can now exit the nucleus through its membrane pores. Within the cytoplasm in the process called **translation**, mRNA delivers the coded message to a *ribosome* (a "factory" made of protein and **ribosomal RNA, or rRNA;** p. 38), which "reads" the mRNA code in order to translate it into the appropriate amino acid sequence for the designated protein being synthesized. Finally, **transfer RNA** (tRNA) transfers the appropriate amino acids from the cytoplasm to the ribosome to be added to the protein under construction (Figure 2-3). (Note that rRNA and tRNA are themselves the products of gene transcription; thus, not all genes code for mRNAs.)

Different genes are expressed in different tissues and organs

Because each cell in a multicellular eukaryote usually has identical DNA sequences, you might assume that all cells would produce the same proteins. This is not the case, however, because—in addition to common or "housekeeping" functions that all cells require—different cell types transcribe different sets of genes and thus synthesize different proteins for their specialized roles. For example, all mammalian cells contain the genes for hemoglobin proteins—which serve to bind and carry oxygen—but these genes are only *expressed* (that is, transcribed and translated to produce proteins) in erythrocytes (red blood cells). In addition,

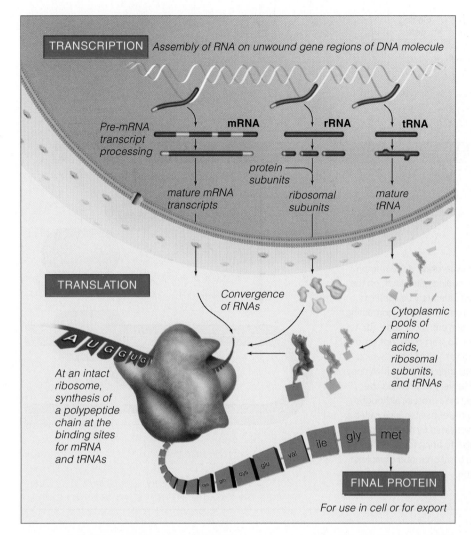

TRANSCRIPTION *Assembly of RNA on unwound gene regions of DNA molecule*

Pre-mRNA transcript processing   **mRNA**   **rRNA**   **tRNA**

protein subunits

mature mRNA transcripts   *ribosomal subunits*   *mature tRNA*

TRANSLATION

Convergence of RNAs

Cytoplasmic pools of amino acids, ribosomal subunits, and tRNAs

At an intact ribosome, synthesis of a polypeptide chain at the binding sites for mRNA and tRNAs

A U G G U G

cys gln cys glu val ile gly met

FINAL PROTEIN

For use in cell or for export

FIGURE 2-3 Summary of the flow of genetic information from DNA to proteins in eukaryotic cells. The DNA is transcribed into RNA in the nucleus; RNA is translated in the cytoplasm. Prokaryotic cells do not have a nucleus; transcription and translation proceed in their cytoplasm.

© Cengage Learning, 2013

many genes are active only under certain conditions or life stages. For example, many genes involved in animal sexual maturation are inactive in pre-adults.

How is this **differential gene expression** accomplished? There appear to be two levels of control:

1. *Regulation of individual genes with promoters and transcription factors.* A complete gene does not just consist of a code for an RNA strand (also known as the open reading frame), but also has regulatory DNA sequences called **promoters** that help regulate the gene's expression (Figure 2-4a). Special regulatory proteins—themselves the products of other genes—control the process by regulating the packaging of the DNA and the action of RNA polymerase. One DNA sequence, the *TATA box* (so called because it contains the nucleotide sequence TATAA), must be bound by proteins forming a *basal transcription* complex (Figure 2-4b), which recognizes the promoter code and then activates RNA polymerase to make the pre-mRNA copy of the nearby protein-coding sequences. However, this does not provide specificity for different cell types, because these general factors are often the same for numerous genes in numerous cell types. Other promoter sequences, often called **enhancers**, typically must be activated first. In one

common process, specific transcription factors that are different in different cell types must bind to a particular DNA code in an enhancer. This bound complex of proteins then activates the basal complex and ultimately RNA polymerase (Figure 2-4b). In recent years, transcription factors acting as "pause buttons" have been found that do the opposite, that is, they cause polymerases to stop in place temporarily.

2. *Regulation of transcription factors in different tissues and at different stages.* Through the process of cell differentiation in embryonic development, each specialized cell type expresses its own set of genes for specific transcription factors. The transcription factors in turn activate genes for organ-specific proteins and thus yield specialized cell functions. However, this process does not work just in embryos. Gene expression is constantly being altered in adult animals by various metabolic factors and signaling molecules. Enhancers are also called **response elements** if they respond (via specific transcription factors) to metabolic factors and signal molecules such as hormones. As we discuss later, many hormones exert their effects by controlling specific transcription factors that control gene expression through response elements (Figure 2-4b).

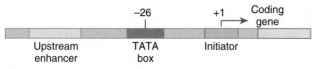

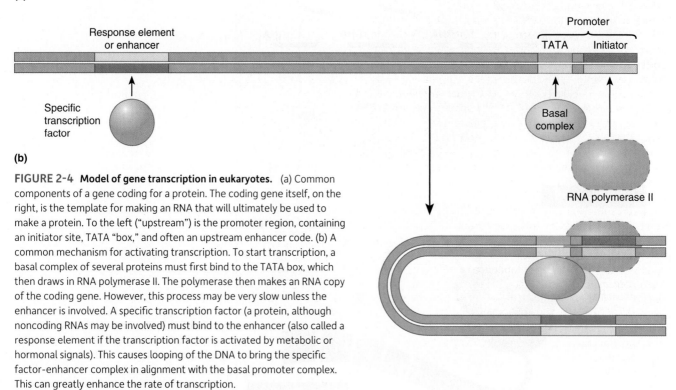

(a)

(b)

FIGURE 2-4 Model of gene transcription in eukaryotes.   (a) Common components of a gene coding for a protein. The coding gene itself, on the right, is the template for making an RNA that will ultimately be used to make a protein. To the left ("upstream") is the promoter region, containing an initiator site, TATA "box," and often an upstream enhancer code. (b) A common mechanism for activating transcription. To start transcription, a basal complex of several proteins must first bind to the TATA box, which then draws in RNA polymerase II. The polymerase then makes an RNA copy of the coding gene. However, this process may be very slow unless the enhancer is involved. A specific transcription factor (a protein, although noncoding RNAs may be involved) must bind to the enhancer (also called a response element if the transcription factor is activated by metabolic or hormonal signals). This causes looping of the DNA to bring the specific factor-enhancer complex in alignment with the basal promoter complex. This can greatly enhance the rate of transcription.

© Cengage Learning, 2013

Alternative splicing of RNA and some "junk" DNA are important in gene expression and evolution

These steps above constitute the basic processes of transcription and translation. However, there is a very important modification of these steps called **alternative splicing,** which is performed by a special RNA–protein complex called a **spliceosome.** By using different combinations of exons, spliceosomes can make more than one mRNA code and so lead to more than one type of protein. For example, flight muscles of dragonflies (which spend much of their lives flying to hunt, fight, and mate) can make six different forms from one gene of a contraction-regulating protein called *troponin* (Chapter 8). Each form may help them acclimatize their muscles to a different behavioral use, developmental stage, and/or environmental condition. Alternative splicing is critical for generating many protein forms since most organisms have a surprisingly low number of protein-coding genes. Consider that the

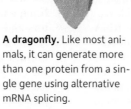

A dragonfly. Like most animals, it can generate more than one protein from a single gene using alternative mRNA splicing.

Photo: Paul Yancey

nematode worm *Caenorhabditis elegans* (which has only around 1,000 total cells in its body) has about 19,000 protein-coding genes, while humans (with trillions of cells) have only a modest amount more, about 25,000. However, humans generate a far greater number of mRNAs and thus proteins than do the worms.

These relatively low numbers of protein-coding genes are even more surprising when we consider that they make up only a small percentage of the total DNA, for example, only about 1.5% in humans. The rest has long been called "junk" DNA because it had been thought to be largely useless. Some probably is; however, unlike protein-coding genes, the amount of non-protein-coding DNA does roughly correlate with organismal complexity, suggesting important roles that are now being discovered. As an example, consider two species of North American voles that differ dramatically in their behaviors in part due to "junk" DNA differences. Prairie voles are socially monogamous while meadow voles are promiscuous. The former have many more *receptors* (cell membrane proteins that bind external signal molecules, p. 93) for a major hormone in brain areas regulating social interactions than do the latter. Remarkably, the difference is in part due to a longer stretch of noncoding DNA in the promoter region of the prairie vole's gene, which somehow increases the gene's expression. The "junk" DNA's role was tested by engineering (p. 34) prairie voles with a shorter stretch; they behaved monogamously like meadow voles! This suggests major behavioral differences could evolve from very simple genetic changes.

In addition, some introns and other noncoding RNAs become *microRNAs* (miRNAs) and *small interfering RNAs* (siRNA), classes of RNAs that regulate gene expression by guiding mRNA splicing, or—in a process called **RNA interference** or **RNAi**—by binding to mRNAs to inhibit their translation (e.g., during development or viral invasion). As an example, researchers have found that one particular miRNA decreases rapidly when a zebra finch learns a new song. Presumably, this reduction allows certain genes involved in song memory to be expressed.

Finally, a striking study in 2010 of "junk" DNA usefulness reported that *pseudogenes* (nonfunctioning protein-coding genes that were probably active in an organism's ancestors) can make untranslated mRNAs that bind with interfering miRNAs, which would otherwise have inhibited translation of mRNA from a "good" gene similar to the pseudogene. This may be a new type of gene regulation, though its importance is not yet known.

Telomeres protect chromosome ends, and their loss is associated with aging

Eukaryotic chromosomes are linear, and researchers have found that the ends lose segments of DNA every time a chromosome is replicated during cell division. To prevent loss of coding genes, the chromosomes have stretches of repeated, noncoding DNA sequences called *telomeres*. Each time a cell divides, a piece of telomere is lost, but presumably no harm is done because the lost DNA nucleotide sequences contain no genes. However, that changes once the telomeres are gone. At that point, useful genes begin to be lost during cell replication, causing cell malfunctions, and in fact, most cells cease dividing. Scientists have found that this feature of chromosome structure, which sets an upper limit on the number of times a cell can divide, correlates with the aging of individuals. Indeed, researchers have found a linear correlation between the length of telomere lost per year and the lifespan of five species of birds and eight species of mammals. For example, zebra finches lose about 500 nucleotides of telomere each year and live for about 5 years. Common terns, in contrast, lose only about 50 nucleotides per year and live for about 26 years. These rates vary among individuals within a species, and another study on tree swallows found that 1-year-olds with short telomeres had lower survival than 1-year-olds with longer telomeres.

The rate of telomere loss is not random or uncontrollable: Telomeres can be repaired by an enzyme called *telomerase*. For reasons not fully understood, the rate of telomere repair, and thus presumably the longevity of species, varies widely among organisms and is subject to natural selection. One possibility that limits the evolution of higher repair rates in short-lived species is cancer. Researchers have found that high telomerase activity does increase the number of times a cell can divide but also correlates with a greater chance of that cell becoming cancerous, that is, repeatedly dividing without stopping and thus forming a tumor.

check your understanding 2.2

What are the basic processes of transcription?

How is differential gene expression accomplished?

How is gene expression regulated?

2.3 Methodology

Many molecular processes we have just discussed are used throughout this text. Because this is the most rapidly growing area of biology, it is useful to understand how molecular knowledge is acquired. Here we introduce some of the major molecular techniques mentioned later in the text.

Driving much of the growth in knowledge about molecular biology are the *genome projects*. In the last years of the 20th century, scientists undertook an international effort to decipher the complete human genetic sequence—the arrangement of the A, G, T, and C nucleotides. Dubbed the Human Genome Project, its goal was to yield insight into the codes responsible for human genetics, development, physiological functions, evolution, and diseases. At about the same time, researchers initiated genome projects for various microbes (some of which were the first to be completed), some plants, and several other animal species. Since then, many species including protistan and fungal ones have been added to the list. The deciphering of genomes is often an example of the discovery phase of science (p. 6). For example, recently the first amphibian genome (in the Western clawed frog) was deciphered, and a large (150 million nucleotide) sequence was found that is virtually identical in chickens and humans. Genome analysis also can result in, and be used to test, new hypotheses about evolution and genes. See the box, *Molecular Biology and Genomics: Genomics and Evolution.*

Although the wealth of genome data available is enormous and growing daily, that does not mean we are close to achieving a genetic understanding of physiology. The role of any individual gene in an organism's physiology is sometimes readily apparent—think of hemoglobin genes, for example, which code for the proteins that carry oxygen in blood—but often it is not. Many genes being discovered by genome analysis have functions that are either speculative or fully unknown, and many proteins have been found that are similarly mysterious. Because of the enormous complexity and intricacy of cellular pathways, it clearly will be many decades before scientists know what each and every gene does, how each evolved, how each is regulated throughout an organism's life cycle, and how each and its products interact with other genes. Uncovering gene functions is the goal of the new field of *genomics,* which uses a variety of new technologies to reveal the connections between genes and cell (or organismal) function. A companion field, *proteomics* (also called *proteonomics*), is focused on discovering the functions of proteins.

In later chapters, we describe some of the recent findings of genomics and proteomics. Next we briefly describe some of the key methods being used.

Molecular Biology and Genomics
Genomics and Evolution

Genomic data have yielded many exciting findings important for understanding animal physiology and its evolution. Evolutionary change is primarily based on changes—mutations—in the DNA code that increase the fitness of the organism. Mutations have been the subject of genetic analysis for more than 100 years, yielding many important discoveries about the evolution of new features via changes in genetic codes. The new technologies have enormously increased the pace of discovery. For example, the traditional view holds that DNA mutates randomly. But genome analysis is revealing that some organisms have stretches of DNA that mutate at extraordinarily high rates, resulting in apparently useful evolutionary diversity. One example has been found in an interesting group of tropical mollusks, the predatory cone snails. Not only do these snails have colorful shells prized by collectors around the world, they also possess venomous proteins—*conotoxins*—that are used to paralyze prey fish, shrimp, and other mollusks (the venom blocks ion channels, described in Chapter 4). A conotoxin is being used medically, for example, to deaden nerves in people suffering from chronic pain. Each of the 500 species of cone snails has a unique set of genes coding for unique conotoxins. Analysis reveals that sections of the venom genes have mutated at rates far faster than the rest of the snail genome. The causes of this rapid mutation are uncertain but have played a role in the evolutionary diversity of cone snail species and their predatory physiologies.

This type of "deliberate" rapid mutation in a selective section of DNA is called **focused variation**. It may be a widespread phenomenon, thought to occur in the evolution of snake and scorpion venom, for example. It probably occurs in the evolution of parasite antigens (proteins on the surface that an animal's immune system recognizes as foreign), allowing some parasites to evade host immune defenses previously induced against them.

Genomics is also revealing the importance of **horizontal gene transfer** in evolution. This so-called *jumping gene* phenomenon—which can result from viral infection and other mechanisms we'll not describe—is the permanent insertion of a gene from one species into the genome of another.

The cone shell, *Conus textile*.

This occurs frequently in microbes; for example, some bacteria can gain antibiotic-resistance genes from other bacteria (even other species) that evolved them. But evidence for this phenomenon has also been found in eukaryotic genomes as well (as you will see later for mitochondrial genes). Perhaps the most dramatic is an animal genome with photosynthetic genes! Many cells of the seaslug *Elysia chlorotica* have photosynthetic organelles called *chloroplasts*, which the slug obtains from eating algae and which produce sugars for the animal using solar energy. Recently, researchers found that the slug's genome has permanently obtained the genes for synthesizing *chlorophyll*, the algal and plant compound responsible for capturing the sun's energy.

Immunofluorescence An important technique that predates modern genomics and proteomics, but that is used widely in conjunction with them, is **immunofluorescence**. This method relies on the ability of mammals and birds to generate specific antibodies (immune proteins; Chapter 10) that bind to foreign proteins entering their bodies. A target protein of interest in species A is purified and injected into species B, typically a laboratory rabbit. The rabbit's immune system makes antibodies to the protein. The antibodies are purified, and a fluorescent molecule is covalently bound to each. The new fluorescent antibody is then injected into species A, or added to a solution bathing a tissue slice from species A. The antibody binds only to places where the target protein occurs, and the fluorescent molecule allows the researchers to visualize its locations through a microscope (Figure 2-5). In this way, proteins of known functions, and "new" proteins discovered by proteomics, can be located in specific tissues and at specific developmental stages of a particular species.

Gene Knockout, Knock-In, and RNA Interference One way to elucidate gene functions is to block the function of a gene and then study resulting defects (if any) in the animal's structures and/or physiology to gain clues about gene (and protein) function. A commonly used approach is the **knock-out method**, in which a laboratory animal is raised from a

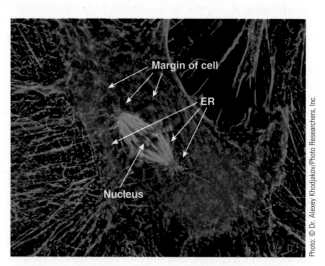

FIGURE 2-5 Immunofluorescence. Immunofluorescent light micrograph of a section through a kangaroo rat kidney epithelial cell during mitosis (cell division). Antibodies have been used to attach fluorescent dyes to specific cell structures. The nuclear membrane (not seen) that encloses the chromosomes (*blue*) is breaking down. The actin microfilaments (*red*) and tubulin microtubules (*green*) of the cytoskeleton maintain the structure of the cell. The two centrioles (*pink dots*, upper left and lower right) are microtubule-organizing centers.

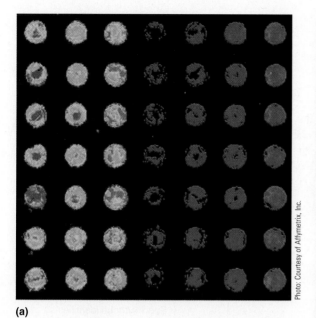

(a)

Photo: Courtesy of Affymetrix, Inc.

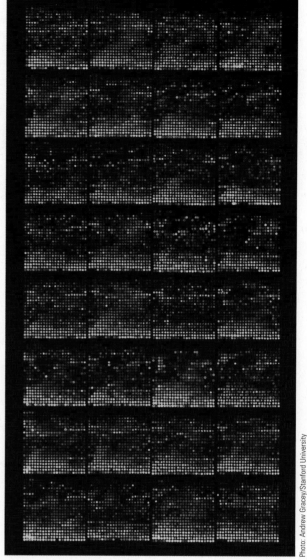

Photo: Andrew Gracey/Stanford University

(b)

FIGURE 2-6 DNA Microarrays. (a) A microarray chip. Thousands of DNA samples are placed on a 1-cm-square slide by robotics. These slides are then bound by complementary DNA or RNA and visualized using fluorescent markers. The location and intensity of the fluorescent signal is analyzed by computer to establish which genes are currently being transcribed and to what extent. (b) A DNA microarray from DNA of a carp, hybridized with liver cDNAs from control animals (labeled green) and from animals subjected to cold temperatures (red). Yellow spots indicate genes with expressions equal in the two groups; green spots are genes expressed more strongly in control animals; and red spots are those expressed more strongly in cooled animals.

fertilized egg that has had a specific gene disrupted. In the opposite approach, **knock-in**, genes are inserted into specific locations. Results of such experiments can be surprising. For example, researchers in Korea knocked out in mice the gene for an enzyme that adds the sugar fucose (a hexose sugar) to proteins. The result was female mice that act like males! Apparently a key protein that regulates masculinizing hormones in fetuses could not work without fucose attached. These methods are not always useful, however, if a knockout yields aborted development or widespread aberrations. An analogy is studying an automobile's internal workings by selectively removing instructions in a robotic factory for assembling key components—components you know little about as you begin. If you mutate an unknown instruction yielding a car that will not start and leaks fuel from a loose tube between fuel tank and engine, you would conclude the instructions were for attaching the tube. But if another mutated instruction leads to no car but just a pile of its parts, conclusions would be trickier. Was the instruction a master command to activate the assembly robots? Or just for setting up the initial framework of the car? Or perhaps for how to use bolts, rivets, and welds? Similarly, knockout changes in cells that are lethal in early development can be hard to interpret. Nevertheless, there have been other useful knockouts that you will see in later chapters.

Another gene-suppression method uses RNA interference (p. 33). Artificially synthesized interfering RNAs are used as a tool to inhibit specific mRNA translation in cells, both to investigate the roles of that messenger and its protein and to inhibit disease processes.

DNA Microarrays Another powerful new technology is called the **DNA microarray**, or colloquially, the "gene chip." In its basic form, a gene chip is a small plate (membrane or glass slide) onto which thousands of different genes from an organism are bound. This technique takes advantage of the fact that DNA and RNA sequences can bind in complementary fashion to one another. RNA from a particular organ is isolated and converted into corresponding *complementary DNA* (called *cDNAs*). The cDNA is labeled with a fluorescent molecule or radioactive atoms and used as a "probe": The cDNAs are exposed to the chip, and any genes active in that organ are revealed by the binding of the cDNA to the genes bound to the plate. Different levels of fluorescent intensities indicate the relative activities (or expression) of these genes. In this way, researchers can determine what genes are active at any given time in a life cycle, and during stress conditions and disease (Figure 2-6). Much of microarray research is "discovery science" (p. 6) in that gene

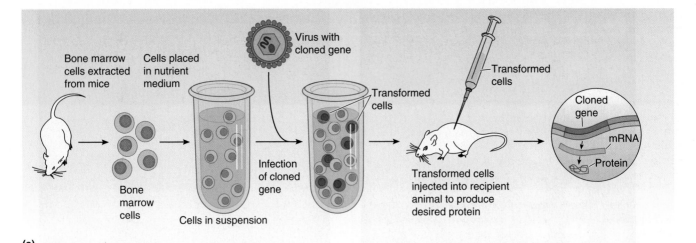

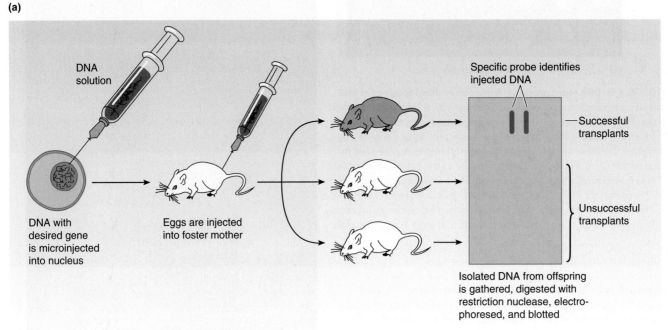

FIGURE 2-7 Gene therapy and transgenics. (a) Gene therapy in bone marrow cells. A cloned gene that directs the synthesis of a missing protein is introduced into bone marrow cells from the mouse. The transformed cells are replaced in the mouse's body, where they produce the desired protein. (b) The procedure for producing transgenic mammals. The two mice on the bottom have not received new genes. The mouse on the top has received a new gene, which it can pass on to its offspring. The mouse on the top is a transgenic organism, and the altered gene is a transgene.

Source: (a) Adapted from Dealing with Genes: The Language of Heredity, by Paul Berg and Maxine Singer, © 1992 by University Science Books.
(b) Campbell, Biochemistry, p. 381.

expression patterns are often conducted without a detailed hypothesis, but data for devising new hypotheses can arise from this exploration.

Databases Genomics also relies heavily on the fact that all life is related via evolution. Thus, the nucleotide sequences of related genes in different organisms are similar. Analysis based on this principle makes extensive use of computers and the Internet, which provide access to databases containing most sequenced genes to date. There are at least two important ways to use these data. First, if a gene is discovered that has an unknown function, its sequence can be compared to all related genes in all analyzed species. This

should give a clue to the "new" gene's function because some of the related genes in the database are likely to have known functions. Second, the sequence of a gene with a known function can be used to look for similar (undiscovered) genes in the genome databases. In this way, previously unsuspected genes (and possibly proteins) can be found in other tissues in the same organisms, or in other species.

Gene Therapy and Transgenics The ultimate goal of genomics is to understand genes so that they can be used in practical ways. An increasingly important application is that of **gene therapy**, in which a defect arising from a mutated gene is corrected by inserting a normal gene (Figure 2-7a). This can be done in a fertilized egg, or in an adult tissue. For example, researchers found that one form of canine blindness—*Leber congenital amaurosis*—can arise from a single defective gene that codes for a protein called RPE65. This protein helps synthesize a pigment used by the retina to receive light. Researchers created a modified nonpathological virus carrying a normal RPE65 gene and injected it into the right retina of blind dogs (the left retina received a control injection). Amazingly, the dogs regained some visual function and could navigate around obstacles that they had formerly bumped into repeatedly in a dimly lit room. Testing of this gene therapy on blind humans began in 2008, with some success. In 2010, the vision of mice with *retinitis pigmentosa* blindness was improved with a new nonviral form of DNA delivery called *compacted DNA nanoparticles,* in which a single gene of therapeutic value is compacted into a small polymer rod small enough to enter nuclear pores.

The technology used in gene therapy can also be used to create **transgenic** animals (Figure 2-7b). Here, a gene from one species is inserted into a fertilized egg or adult tissue of another (though unlike the knock-in method, insertion is not designed to be at a specific site). This may confer commercially useful properties into an agricultural animal; for example, sheep have been created that carry genes for human proteins with medical uses. The sheep produce milk containing these human proteins. Also, mosquitoes have been engineered to make proteins that block malarial parasites from entering their circulatory fluids, which may help slow the spread of malaria to humans. The process may also be used to create markers or tracers of gene activity for basic research. In 2000, the first transgenic primate—a rhesus monkey named ANDi—was created, with a jellyfish gene that produces *green fluorescent protein (GFP),* a protein for bioluminescence. This gene, when attached to a host species' gene, provides a visual marker of that gene's activity in different organs and stages (see Immunofluorescence, p. 34).

Cloning Finally, another technology based on the nucleus is that of **whole-animal cloning.** Because most adult cells contain the same DNA sequences on chromosomes throughout an organism's life cycle, the adult DNA, in theory, can code for a genetically identical offspring—a clone. In one common method of cloning called *nuclear transplantation,* the nucleus of a fertilized egg is removed or destroyed, and a carefully selected nucleus from an adult cell is injected in its place. Poorly understood factors synthesized within the egg somehow rewind the internal clock of the adult genetic material and restore it to the *pluripotent* state. When the adult DNA is exposed to the proper signals (such as specific transcription factors), the genetic program initiates the development of a new animal (Figure 2-8). Frogs were cloned in this manner in the 1960s, but only in the late 1990s was a mammal—a sheep named Dolly, in Scotland—produced by cloning. Since then, many other mammals have been cloned, including rodents, cows, goats, cats, a mule, and a horse. Strains of medically and agriculturally useful animals might be maintained by cloning. In addition, this technology might be used to save endangered species. However, it is not clear how successful the process will be in the long run. Many of the cloned mammals have shown signs of premature aging, probably because they developed from adult DNA that had suffered damage (e.g., to telomeres) during the parent's life. In addition to the physiological problems, the use of this and other genetic technologies remains fraught with ethical dilemmas. See *Unanswered Questions: Are cloning and genetic engineering ethical?*

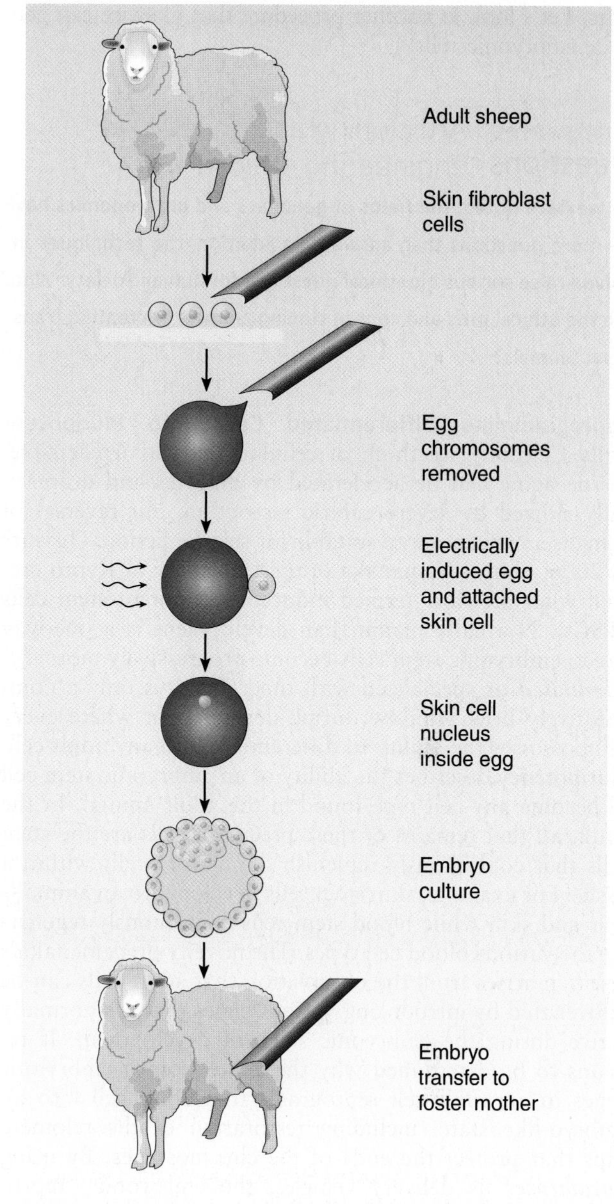

FIGURE 2-8 Nuclear-transplantation techniques in mammals. The genetic material is removed from the recipient cell (an egg), and then replaced by a nucleus from a donor cell. The resulting embryo is then transferred to a surrogate mother. The clones are genetically identical to the donor.

Source: J. B. Gurdon, Alan Colman, The future of cloning, 'Nature', December 16, 1999, p. 743. Used by permission from Macmillan Publishers Ltd, copyright 1999.

Note that cloning is not simply an artificial technique. Basic cell division is a form of cloning. Many multicellular organisms, including many "primitive" animals, reproduce by cloning (although not by nuclear transplantation). Adult sea anemones and corals (Cnidarians), for example, can split in two to produce two genetically identical individuals (see Chapter 16, Reproduction).

Cloning is not the only procedure that is capable of reprogramming the DNA of an adult animal into an embryonic state. Let's look at another procedure that likewise can generate embryonic cells.

unanswered | Are cloning and genetic
Questions | engineering ethical?

As we have noted, the fields of genomics and proteonomics have far more questions than answers. In addition, the techniques involved raise serious bioethical questions for human society. What are the ethical pros and cons in cloning animals or creating transgenic animals? <<

Reprogramming Differentiated Cells into Pluripotent Cells Certainly we think of cellular aging as irreversible: Tissue aging can be accelerated by diabetes and dramatically slowed by severe caloric restriction, but reversal of aging is a concept more suitable for science fiction. Or is it? In 2006, Shinya Yamanaka of the University of Kyoto created what are now termed **induced pluripotent stem cells (iPSCs)**. Normally mammalian development is a one-way street; embryonic stem cells become progressively more *differentiated* or specialized with time. There is only a comparatively brief window during development where every cell possesses the ability to differentiate into any adult cell. **Pluripotency** describes the ability of an embryonic stem cell to become any cell type found in the adult animal. In the adult, all that remains of these precursor cells are the stem cells that continuously replenish the mature cells within a tissue. For example, skin stem cells develop into an animal's hair and skin while blood stem cells continuously regenerate the various blood cell types. The novelty of Yamanaka's research arises from the observation that adult cells can be rejuvenated by introducing various genes that are normally active during the embryonic stage of development. It remains to be determined why the expression of embryonic genes in an adult cell reprograms that adult cell into an embryo-like state, including restoration of the telomere caps that protect the ends of the chromosomes. By using *retroviruses* as delivery vehicles, the embryonic "reprogramming" genes can be transferred into the host cell genome. More recent studies have demonstrated that the production of iPSCs requires the retrovirus carrier to transport genetic material from as few as two of the original two dozen reprogramming embryonic genes or more simply by introducing the proteins encoded by four of the reprogramming genes directly into the cell.

At the present time, about a dozen different adult cell types from four different species (mouse, rat, monkey, and human) have been successfully reprogrammed into iPSCs. Ultimately, the objective of researchers is to produce iPSCs without using viruses as carriers, which reduces the cancer risk, and instead expose adult cells to a cocktail of drugs that mimic the effect of the reprogramming genes. Consider the remarkable implications of this research: employing a comparatively simple procedure to use an animal's own iPSCs to produce replacement parts for cells and organs damaged by disease. In 2011, Roberto Bolli reported the ability of heart stem cells to repair heart tissue damaged after a heart attack. Within seven days after the infarct, the stem cells had infiltrated the entire region of the infarct resulting in a gradual improvement in cardiovascular function.

Now that we have examined some key techniques of molecular investigation, we will turn our attention to organelles, other molecular complexes, and metabolism. Recall that in addition to the nucleus, eukaryotic cells contain several other large structures, including other membrane-bound structures or organelles.

check your understanding 2.3

What are the potential uses of gene therapy?

How has the old dogma that the specialty of cells is irreversibly set once they have differentiated been challenged by the discovery of iPSCs?

How are the technological approaches of cloning and iPSCs similar to each other?

2.4 Ribosomes and Endoplasmic Reticulum

Ribosomes are ribosomal RNA–protein complexes that synthesize proteins, indirectly under the direction of nuclear DNA. Recall that mRNA carries the genetic message from the nucleus to the ribosome "workbench," where protein synthesis takes place (see Figure 2-2). Note that prokaryotes also have ribosomes, with a slightly different structure.

Ribosomes are found in two locations. Unattached or "free" ribosomes are dispersed throughout the cytosol. Bound ribosomes are found on membranes of a major organelle, the **endoplasmic reticulum (ER)**—an elaborate, fluid-filled membranous system distributed extensively throughout the cytosol, where it is primarily a protein-manufacturing factory. Two distinct types of ER—rough and smooth—can be distinguished. The **rough ER** consists of stacks of relatively flattened interconnected sacs, whereas the **smooth ER** is a meshwork of tiny interconnected tubules (Figure 2-9). Even though these two regions differ considerably in appearance and function, the ER is thought to be one continuous organelle with many interconnected channels. The relative amount of smooth and rough ER varies between cells, depending on the activity of the cell.

The rough endoplasmic reticulum and its ribosomes synthesize proteins for secretion and membrane construction

The outer surface of the rough ER is studded with ribosomes, giving it a "rough" or granular appearance. These ribosomes synthesize and release a variety of new proteins into the ER **lumen**, the fluid-filled space enclosed by the ER membrane. These proteins serve one of two purposes: (1) Some are destined for export to the cell's exterior as secretory products, such as proteinaceous hormones or enzymes. (2) Others are transported to sites within the cell for use in constructing new cell membrane (either new plasma or organelle membrane) or other protein components of organelles. Cell membranes consist mostly of lipids (fats) and proteins. The membranous wall of the ER also contains enzymes essential for the synthesis of nearly all the lipids

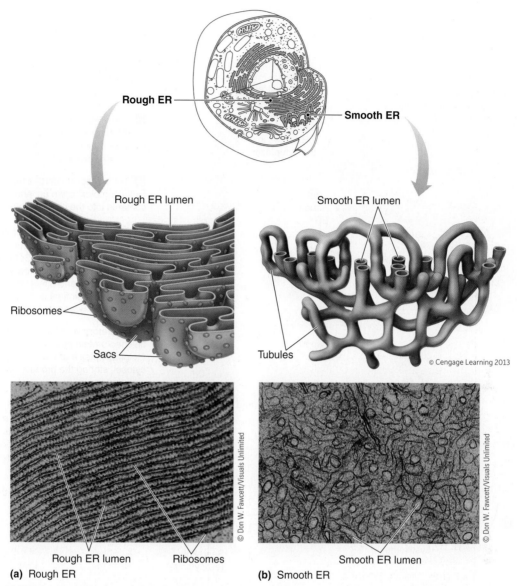

Rough ER

Smooth ER

Rough ER lumen

Smooth ER lumen

Ribosomes

Sacs

Tubules

© Cengage Learning 2013

© Don W. Fawcett/Visuals Unlimited

Rough ER lumen   Ribosomes

(a) Rough ER

© Don W. Fawcett/Visuals Unlimited

Smooth ER lumen

(b) Smooth ER

FIGURE 2-9 Endoplasmic reticulum (ER). (a) Diagram and electron micrograph of the rough ER, which consists of stacks of relatively flattened interconnected sacs studded with ribosomes. (b) Diagram and electron micrograph of the smooth ER, which is a meshwork of tiny interconnected tubules. The rough ER and smooth ER are connected, making one continuous organelle.

needed to produce new membranes. These newly synthesized lipids enter the ER lumen along with the proteins. Predictably, the rough ER is most abundant in cells specialized for protein secretion (for example, cells that secrete digestive enzymes) or in cells that require extensive membrane synthesis (for example, rapidly growing cells such as immature egg cells).

Within the ER lumen, a newly synthesized protein is folded into its final conformation and may also be modified in other ways, such as being pruned or having sugar molecules attached to it (*glycosylation*). After this processing, a new protein cannot pass out through the lumen of the ER membrane and therefore becomes permanently separated from the cytosol as soon as it has been synthesized. In contrast to the rough ER ribosomes, free ribosomes synthesize proteins that are used within the cytosol. In this way, newly

produced molecules that are destined for export out of the cell or for synthesis of new cell components (those synthesized by the ER) are physically separated from those that belong in the cytosol (those produced by the free ribosomes).

How do the newly synthesized molecules within the ER lumen get to their destinations at other sites inside the cell or to the outside of the cell? They do so through the action of the smooth endoplasmic reticulum.

The smooth endoplasmic reticulum packages new proteins in transport vesicles

The smooth ER does not synthesize proteins because it does not contain ribosomes; hence it is "smooth." Instead, it serves a variety of other purposes that vary among cell types.

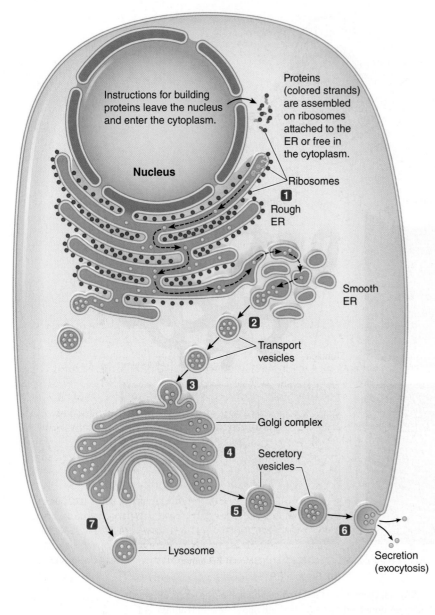

FIGURE 2-10 **Overview of the secretion process for proteins synthesized by the endoplasmic reticulum.**
Note that the secretory product never comes into contact with the cytosol.

© Cengage Learning, 2013

In most cells, the smooth ER is rather sparse and serves primarily as a central packaging and discharge site for molecules that are to be transported from the ER. Newly synthesized proteins and lipids pass from the rough ER to gather in the smooth ER. Portions of the smooth ER then "bud off" (that is, balloon outward, then are pinched off), forming **transport vesicles** that enclose the new molecules in a spherical capsule derived from the smooth ER membrane (Figure 2-10). (A **vesicle** is a fluid-filled, membrane-enclosed intracellular cargo container.) Newly synthesized membrane components are rapidly incorporated into the ER membrane itself to replace the membrane that was used to "wrap" the transport vesicle. Transport vesicles move to the Golgi complex for further processing of their cargo (discussed shortly).

In contrast to the sparseness of the smooth ER in most cells, some specialized types of cells have an extensive smooth ER, which confers additional cell functions as follows:

• The smooth ER is abundant in cells that specialize in lipid metabolism—for example, cells that secrete *steroid hormones* in arthropods and vertebrates (see Chapter 7). (A steroid is a special type of lipid derived from cholesterol.) The membranous wall of the smooth ER contains enzymes for synthesis of lipids. By themselves, lipid-producing enzymes in the rough ER membrane cannot carry out enough lipid synthesis to maintain adequate steroid-hormone secretion levels, so these cells have an expanded smooth-ER compartment to house the additional enzymes necessary to keep pace with demands for hormone secretion.

- In vertebrate liver cells, the smooth ER contains enzymes specialized for detoxifying toxic compounds produced within the body by the metabolism of substances that enter from the outside (such as drugs and plant toxins). These detoxification enzymes alter toxic substances so that the latter can be eliminated more readily via excretory organs. The amount of smooth ER available in liver cells for the task of detoxification can vary dramatically, depending on the need. For example, if phenobarbital (a barbiturate drug used as a sedative) is administered in large quantities to a mammal, the amount of smooth ER with its associated detoxification enzymes doubles within a few days (a form of *acclimatization* or *upregulation*, p. 17), only to return to normal within five days after drug administration ceases. The mechanisms involved in regulating these changes are not well understood.
- Muscle cells have evolved another specialized use for the smooth ER. They have an elaborate, modified smooth ER known as the *sarcoplasmic reticulum,* which stores calcium and plays an important role in muscle contraction (see Chapter 8).

check your understanding 2.4

Describe the structure of the endoplasmic reticulum, distinguishing between the rough and the smooth ER. What is the function of each?

2.5 Golgi Complex

Closely associated with the endoplasmic reticulum is the **Golgi complex,** which consists of sets of flattened, slightly curved, membrane-enclosed sacs, or *cisternae,* stacked in layers (see Figure 2-10). The sacs within each Golgi stack do not come into physical contact with one another. Note that the flattened sacs are thin in the middle but have dilated, or bulging, edges. The number of Golgi stacks varies, depending on the cell type. Some cells have only one stack, whereas cells highly specialized for protein secretion may have hundreds of stacks.

Transport vesicles carry their cargo to the Golgi complex for further processing

The majority of the newly synthesized molecules that have just budded off from the smooth ER enter a Golgi stack. When a transport vesicle carrying its newly synthesized cargo reaches a Golgi stack, the vesicle membrane fuses with the membrane of the sac closest to the center of the cell. The vesicle membrane opens up and becomes integrated into the Golgi membrane, and the contents of the vesicle are released to the interior of the sac (see Figure 2-10). (Note that vesicles are typically associated with cytoskeletal filaments, which are discussed later in this chapter.)

These newly synthesized raw materials continue to travel via vesicles through the layers of the Golgi stack, from the innermost sac closest to the ER to the outermost sac near the plasma membrane. During this transit, two important interrelated functions take place:

1. *Processing the raw materials into finished products.* Within the Golgi complex, the "raw" proteins from the ER are modified into their final form, such as attaching carbohydrates. The biochemical pathways that the proteins undergo during their passage through the Golgi complex are elaborate, precisely programmed, and specific for each final product.
2. *Sorting and directing the finished products to their final destinations.* The Golgi complex is responsible for sorting and segregating different types of products according to their function and destination, namely, products (1) to be secreted to the cell's exterior, (2) to be used for construction of new plasma membrane, or (3) to be incorporated into other organelles, especially lysosomes.

The Golgi complex packages secretory vesicles for release by exocytosis

How does the Golgi complex sort and direct finished proteins to the proper destinations inside and outside the cell? Finished products collect within the dilated edges of the Golgi complex's sacs. The dilated edge of the outermost sac then pinches off to form a membrane-enclosed vesicle containing a selected product. For each type of product to read its appropriate site of function, each distinct type of vesicle takes up a specific product before budding off. Vesicles with their selected cargo destined for different sites are wrapped in membranes containing distinctly different surface proteins. Each type of surface protein serves as a specific **docking marker** (like an address on an envelope). A vesicle can "dock" and "unload" its selected cargo only at the appropriate **docking-marker acceptor**, a protein located only at the proper destination within the cell (like a house address). Thus, Golgi products reach their appropriate site of function because they are sorted and delivered like addressed envelopes containing particular pieces of mail being delivered only to the appropriate house addresses.

Specialized secretory cells include endocrine cells that secrete protein hormones and digestive gland cells that secrete digestive enzymes. In secretory cells, numerous large **secretory vesicles** (or granules), which contain proteins to be secreted, bud off from the Golgi stacks. Secretory vesicles, which are about 200 times larger than transport vesicles, store the secretory proteins until the cell is stimulated by a specific signal that indicates a need for release of that particular secretory product. On the appropriate signal, a vesicle moves to the cell's periphery, fuses with the plasma membrane, and empties its contents to the outside (Figures 2-10 and 2-12a). This mechanism—releasing substances originating within the cell to the exterior—is called **exocytosis** (*exo,* "out of "; *cyto,* "cell"). Exocytosis is the primary mechanism for accomplishing secretion. Secretory vesicles fuse only with the plasma membrane and not with any of the internal membranes that enclose organelles, thereby preventing fruitless or even dangerous discharge of secretory products into the organelles.

Exocytosis can be an explosive event because vesicle contents are often ejected. The most extreme example may be in Cnidaria (anemones, coral, hydras, jellies, and so on). All Cnidaria sting prey by using epithelial cells with special

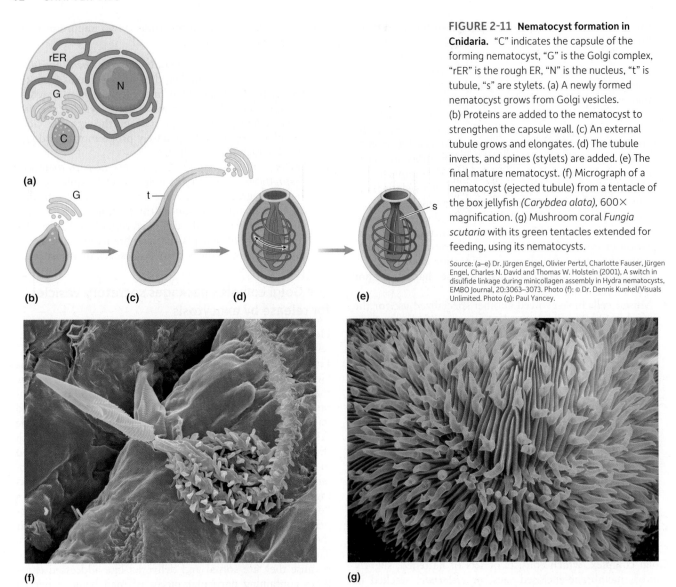

FIGURE 2-11 Nematocyst formation in Cnidaria. "C" indicates the capsule of the forming nematocyst, "G" is the Golgi complex, "rER" is the rough ER, "N" is the nucleus, "t" is tubule, "s" are stylets. (a) A newly formed nematocyst grows from Golgi vesicles. (b) Proteins are added to the nematocyst to strengthen the capsule wall. (c) An external tubule grows and elongates. (d) The tubule inverts, and spines (stylets) are added. (e) The final mature nematocyst. (f) Micrograph of a nematocyst (ejected tubule) from a tentacle of the box jellyfish *(Carybdea alata)*, 600× magnification. (g) Mushroom coral *Fungia scutaria* with its green tentacles extended for feeding, using its nematocysts.

Source: (a–e) Dr. Jürgen Engel, Olivier Pertzl, Charlotte Fauser, Jürgen Engel, Charles N. David and Thomas W. Holstein (2001), A switch in disulfide linkage during minicollagen assembly in Hydra nematocysts, EMBO Journal, 20:3063–3073. Photo (f): © Dr. Dennis Kunkel/Visuals Unlimited. Photo (g): Paul Yancey.

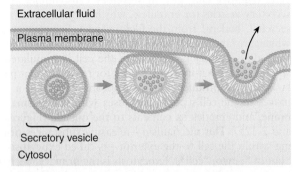

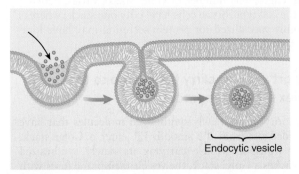

(a) Exocytosis: A secretory vesicle fuses with the plasma membrane, releasing the vesicle contents to the cell exterior. The vesicle membrane becomes part of the plasma membrane.

(b) Endocytosis: Materials from the cell exterior are enclosed in a segment of the plasma membrane that pockets inward and pinches off as an endocytic vesicle.

FIGURE 2-12 Exocytosis and endocytosis.

© Cengage Learning, 2013

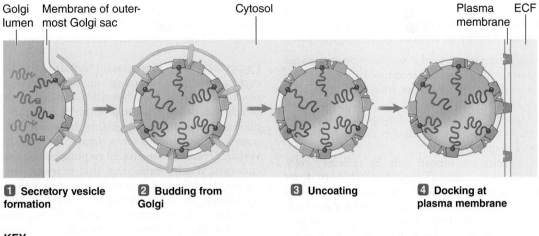

Golgi lumen | Membrane of outer-most Golgi sac | Cytosol | Plasma membrane | ECF

1 Secretory vesicle formation

2 Budding from Golgi

3 Uncoating

4 Docking at plasma membrane

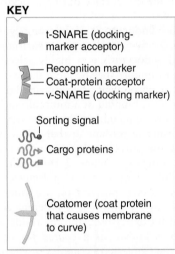

KEY

▶ t-SNARE (docking-marker acceptor)

Recognition marker
Coat-protein acceptor
v-SNARE (docking marker)

Sorting signal

Cargo proteins

Coatomer (coat protein that causes membrane to curve)

1 Recognition markers in the membrane of the outermost Golgi sac capture the appropriate cargo from the Golgi lumen by binding only with the sorting signals of the protein molecules to be secreted. The membrane that will wrap the vesicle is coated with coatomer, which causes the membrane to curve, forming a bud.

2 The membrane closes beneath the bud, pinching off the secretory vesicle.

3 The vesicle loses its coating, exposing v-SNARE docking markers on the vesicle surface.

4 The v-SNAREs bind only with the t-SNARE docking-marker acceptors of the targeted plasma membrane, ensuring that secretory vesicles empty their contents to the cell's exterior.

FIGURE 2-13 Packaging, docking, and release of secretory vesicles. The diagram series illustrates secretory vesicle formation and budding with the aid of a coat protein and docking with the plasma membrane by means of v-SNAREs and t-SNARES. Secretion can then occur by exocytosis.

Golgi-derived vesicles called **nematocysts**, capsules each having a double-layered wall and an inverted tube studded with barbs (often with a toxin). On contact with prey, the capsule exocytoses the barbed tube, which inverts, exposing the spines and jabbing the prey (Figure 2-11). This exocytosis is one of the fastest biological actions known. It appears driven in part by *osmosis,* which we cover in the next chapter. (Briefly, water tends to move from dilute to concentrated solutions.) The capsule is full of cations and long amino-acid chains called *polyglutamate,* serving as *osmolytes* (see Chapter 13) that create a solution much more concentrated than the external environment. The rigid capsule prevents premature discharge. Prey contact triggers a lid on the capsule to open, and water rushes in by osmosis, inverting the tube.

Let's now look in more detail at how secretory vesicles take up specific products in the Golgi stacks for release into the ECF and why they are able to dock only at the plasma membrane (Figure 2-13).

- The newly finished proteins destined for secretion contain a unique sequence of amino acids known as a *sorting signal,* and the interior surface of the Golgi membrane contains *recognition markers,* proteins that recognize and attract specific sorting signals. Recognition of the right protein's sorting signal by the complementary membrane marker ensures that the proper cargo is captured and packaged into the secretory vesicle.

- *Coat proteins* called *coatomer* from the cytosol bind with another specific protein facing the outer surface of the membrane. The linking of these coat proteins causes the surface membrane of the Golgi sac to curve and form a dome-shaped bud around the captured cargo. Eventually, the surface membrane closes and pinches off the vesicle.

- After budding off, the vesicle sheds its coat proteins and exposes the docking markers, known as *v-SNAREs,* which face the outer surface of the vesicle membrane.

- A v-SNARE can only bind lock-and-key fashion with its docking marker acceptor, called a *t-SNARE,* on the targeted membrane. In the case of secretory vesicles, the targeted membrane is the plasma membrane, the designated site for secretion to take place. Thus, the v-SNAREs of secretory vesicles fuse only with the t-SNAREs of the plasma membrane. Once a vesicle has docked at the appropriate membrane by means of matching SNAREs, the two membranes completely fuse; then the vesicle empties its contents at the targeted site.

By manufacturing its particular secretory protein ahead of time and storing this product in secretory vesicles, a secretory cell has a readily available reserve from which to secrete large amounts of this product on demand. If a secretory cell had to synthesize all its product on the spot as needed for export, the cell would be more limited in its ability to meet varying levels of demand.

Secretory vesicles are formed only by secretory cells. However, the Golgi complex of these and other cell types sorts and packages newly synthesized products for different destinations within the cell. In each case, a particular vesicle captures a specific kind of cargo from among the many proteins in the Golgi lumen, and then addresses each shipping container for a distinct destination.

check your understanding 2.5

How does the Golgi complex sort and direct finished proteins to the proper destinations inside and outside the cell?

How are the coat proteins different from docking markers?

2.6 Lysosomes and Proteasomes

Two major structures in the cytosol are important for breaking down unwanted materials: the *lysosome* organelle and the *proteasome* complex.

Lysosomes digest extracellular material brought into the cell by phagocytosis

Lysosomes are small organelles that break down organic molecules (*lys* means "breakdown"; *some* means "body"). Instead of having a uniform structure, as is characteristic of all other organelles, lysosomes vary in size and shape, depending on the contents they are digesting. Most commonly, lysosomes are small (0.2 to 0.5 mm in diameter) oval or spherical membrane-bound bodies. On average, a cell contains about 300 lysosomes. They were first described in 1955 by Christian De Duve, who noted that lysosomes were acidic organelles containing an array of powerful **hydrolytic enzymes** or **hydrolases**, which catalyze **hydrolysis**, reactions that break down organic molecules by the addition of water at a bond site (*hydrolysis* means "splitting with water"). In lysosomes the organic molecules are cell debris and foreign material, such as bacteria, that have been brought into the cell. Each lysosome contains more than 30 different hydrolases synthesized in the ER and transported to the Golgi complex for packaging into the budding lysosome. Lysosomal enzymes are similar to the hydrolytic enzymes that the digestive system secretes to digest food. Thus, lysosomes serve as the intracellular "digestive system."

Extracellular material to be attacked by lysosomal enzymes is brought into the cell through the process of phagocytosis, a type of endocytosis. **Endocytosis** (Figure 2-12b), the reverse of exocytosis, refers to the internalization of extracellular material within a cell (*endo* means "within"). Endocytosis can be accomplished in three ways—*pinocytosis, receptor-mediated endocytosis,* and *phagocytosis*—depending on the contents of the internalized material and the cell type.

Pinocytosis With **pinocytosis** ("cell drinking"), a droplet of extracellular fluid is taken up nonselectively. First, the plasma membrane dips inward (or invaginates), forming a pouch that contains a small bit of ECF (Figure 2-14a). The plasma membrane then seals at the surface of the pouch, trapping the contents in a small, intracellular **endocytic vesicle.** *Dynamin,* the protein responsible for pinching off an endocytic vesicle, forms rings that wrap around and "wring the neck" of the pouch, severing the vesicle from the surface membrane. Besides bringing ECF into a cell, pinocytosis provides a means to retrieve extra plasma membrane that has been added to the cell surface during exocytosis.

Receptor-Mediated Endocytosis Unlike pinocytosis, which involves the nonselective uptake of the surrounding fluid, **receptor-mediated endocytosis** is a highly selective process that enables cells to import specific large molecules that the cell needs from its environment. Receptor-mediated endocytosis is triggered by the binding of a molecule such as a protein to a specific surface membrane receptor site. This binding causes the plasma membrane at that site to invaginate, and then seal at the surface where it now contains the protein/receptor complex (Figure 2-14b). The pouch is formed by the linkage of *clathrin* molecules, which are membrane-deforming coat proteins on the inner surface of the plasma membrane. Clathrin is a different coat protein than the one used for exocytosis, and the resulting clathrin-containing pouch is known as a *coated pit* because it is coated with clathrin. Cholesterol complexes, vitamin B_{12}, the hormone insulin, and iron are examples of substances selectively taken into cells by receptor-mediated endocytosis.

Unfortunately, some viruses can sneak into cells by exploiting this mechanism. For instance, flu viruses and HIV/FIV, the viruses that cause human and feline AIDS (see p. 482), respectively, gain entry to cells via receptor-mediated endocytosis. They do so by binding with membrane receptor sites normally designed to trigger internalization of a needed molecule.

Phagocytosis During **phagocytosis** ("cell eating"), large multimolecular particles are internalized. Most body cells perform pinocytosis, many carry out receptor-mediated endocytosis, but only a few specialized cells are capable of phagocytosis. The latter are the "professional" phagocytes, the most notable being certain types of protists and immune cells that are crucial defense mechanisms in all animals (see Chapter 10). When an immune cell encounters a large multimolecular particle, such as a bacterium or tissue debris, it extends surface projections known as *pseudopodia* ("false feet"), or *pseudopods,* that completely surround or engulf the particle and trap it within an internalized vesicle (see Figure 2-14c). A lysosome fuses with the membrane of the internalized vesicle and releases its hydrolytic enzymes into the vesicle, where they attack the trapped material without damaging the rest of the cell. The enzymes largely break down the material into reusable raw ingredients, such as amino acids, glucose, and fatty acids. These small products pass through the lysosomal membrane and enter the cytosol.

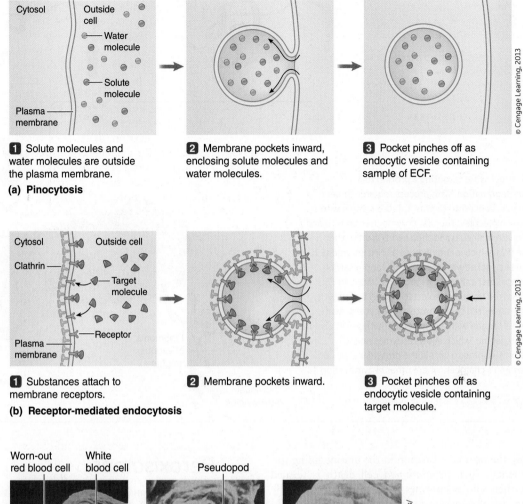

1 Solute molecules and water molecules are outside the plasma membrane.

(a) Pinocytosis

2 Membrane pockets inward, enclosing solute molecules and water molecules.

3 Pocket pinches off as endocytic vesicle containing sample of ECF.

1 Substances attach to membrane receptors.

(b) Receptor-mediated endocytosis

2 Membrane pockets inward.

3 Pocket pinches off as endocytic vesicle containing target molecule.

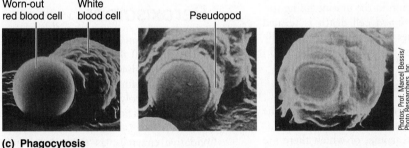

(c) Phagocytosis

FIGURE 2-14 Forms of endocytosis. (a) Diagram of pinocytosis. The surface membrane dips inward to form a pouch, then seals the surface, forming an intracellular endocytic vesicle that nonselectively internalizes a bit of ECF. (b) Diagram of receptor-mediated endocytosis. When a large molecule such as a protein attaches to a specific surface receptor, the membrane pockets inward with the aid of a coat protein, forming a coated pit, then pinches off to selectively internalize the molecule in an endocytic vesicle. (c) Scanning electron micrograph series of a white blood cell phagocytizing an old, worn-out red blood cell by extending surface projections known as pseudopods that wrap around and seal in the targeted material.

Lysosomes remove worn out organelles

Lysosomes can also fuse with aged or damaged organelles to remove parts of the cell. This selective self-digestion makes way for new replacement parts. All organelles are renewable. If the whole cell is severely damaged or dies, the lysosomes rupture and release their destructive enzymes into the cytosol so that the cell digests itself entirely. In most tissues, elimination of a nonfunctional cell clears the way for its replacement with a healthy new one through cell division. However, in tissues in which cell reproduction rarely occurs, such as mammalian heart and brain tissue, scar tissue replaces the self-destructed dead cells.

In specific instances, lysosomes cause intentional self-destruction of healthy cells. This happens as a normal part of embryonic development when certain unwanted tissues that form are programmed for destruction. For example, embryonic ducts in mammals that are capable of forming a male reproductive tract are deliberately destroyed during the development of a female fetus. Similarly, many larval (caterpillar) tissues must be destroyed in a butterfly's pupal stage. Lysosomes also play an important role in tissue regression,

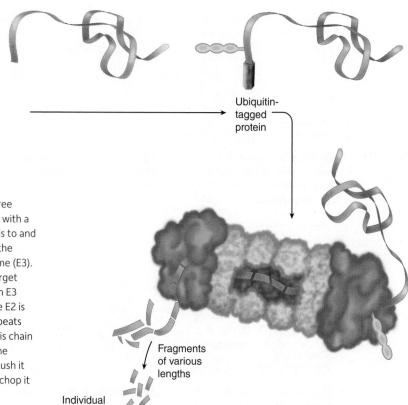

Ubiquitin-
tagged
protein

FIGURE 2-15 Process for targeting a protein to a proteasome for degradation. This process requires three enzymes working in concert to tag the ill-fated protein with a chain of ubiquitin molecules. The first enzyme (E1) binds to and activates a ubiquitin molecule and then hands it off to the second enzyme (E2), which in turn joins to a third enzyme (E3). E3 enzymes are like socket wrenches that fit various target proteins using "sockets" called F-box proteins. When an E3 binds to a protein, the ubiquitin molecule carried by the E2 is broken off and transferred to the protein. The cycle repeats until the protein is tagged with a chain of ubiquitins. This chain binds to the proteasome, which allows enzymes near the opening of the proteasome to unfold the protein and push it into the proteasome's chamber, where other enzymes chop it to pieces.

Source: From Scientific American, January, 2001. Used by permission.

Fragments
of various
lengths

Individual
amino acids

such as during the normal reduction in the uterine lining following pregnancy. Such *programmed* cell death is termed **apoptosis.** (Other cellular enzymes and organelles, especially mitochondria, are also involved in apoptosis; see Chapter 3.)

Proteasomes destroy internal proteins

The other important intracellular digestive apparatus is the **proteasome,** a large tunnel-like structure made of numerous proteins (Figure 2-15). The proteasome, of which there are many thousands per cell, has the job of taking in internal cell proteins (rather than external ones) and chopping them up into reusable amino acids. Why would a cell want to destroy its own proteins? Although cell proteins are made for specific purposes, they outlive their usefulness if that purpose is only needed temporarily. For example, during cell division, many regulatory proteins have to be made, then destroyed, for the cell to move through the various stages of mitosis. Also, in extreme starvation proteins must be broken down to provide emergency energy supplies. Some cell proteins are not useful at all due to errors in their production (such as misfolding) or even genetic mutations. Without proteasomes, unneeded proteins could accumulate in cells to harmful levels.

Proteins are not randomly destroyed by proteasomes. Instead, special enzymes recognize unwanted proteins (in ways not completely understood) and "tag" them with a tiny protein called *ubiquitin.* The ubiquitin-tagged protein is then recognized by the proteasome and drawn in to its disassembly tunnel. However, as in the case of lysosomes, this destruction process can go awry and cause harm. For example, some viruses have evolved their own proteins that can tag useful cellular defense proteins for proteasome destruction.

2.7 Peroxisomes

Typically, several hundred small **peroxisomes** that are about one third to one half the average size of lysosomes are present in a cell. Peroxisomes, like lysosomes, are membrane-enclosed sacs containing enzymes, but unlike lysosomes, which contain hydrolytic enzymes, peroxisomes house several powerful *oxidative enzymes* and contain most of the cell's *catalase.*

Oxidative enzymes, as the name implies, use oxygen (O_2), to strip hydrogen from certain organic molecules. This reaction helps detoxify various wastes produced within the cell or foreign toxic compounds that have entered the cell, such as ethanol that is consumed in alcoholic beverages.

The major product generated in the peroxisome, *hydrogen peroxide* (H_2O_2), is formed by molecular oxygen and the hydrogen atoms stripped from the toxic molecule. Hydrogen peroxide is potentially destructive if allowed to accumulate or escape from the confines of the peroxisome. However, peroxisomes also contain an abundance of **catalase,** an enzyme that decomposes potent H_2O_2 into harmless H_2O and O_2. This latter reaction is an important safety mechanism that destroys the potentially deadly peroxide at the site of its production, thereby preventing its possible devastating escape into the cytosol.

check your understanding 2.6, 2.7

Compare the function of lysosomes, proteosomes, and peroxisomes.

Why is it important that a cell have a source of antioxidants?

2.8 Mitochondria and Energy Metabolism

Mitochondria are the energy organelles or "power plants" of the cell; they extract energy from the nutrients in food and transform it into a usable form to support cell activities. As you will see in Chapter 8, mitochondria can also take up Ca^{2+} under physiological conditions, often at sufficient magnitude to influence cytoplasmic Ca^{2+} homeostasis. About 90% of the energy that cells—and, accordingly, that tissues, organs, and the whole body—need to survive and function is generated by mitochondria. The mammalian organ most dependent on mitochondrial energy production is the brain, followed by the heart and some forms of skeletal muscle, and the kidneys and hormone-synthesizing tissues.

Mitochondria are generally rod- or oval-shaped structures about the size of a bacterium. However, they can vary in size and shape depending on the tissue type, even within a given animal. The mitochondria of arthropod motor neurons are rodlike, with branches extending 1 μm or more. The similarity in basic size of mitochondria and bacteria is not a coincidence; as we noted earlier, there is considerable evidence that mitochondria are descendants of bacteria that invaded or were engulfed by primitive cells early in evolutionary history. Mitochondrial DNA (*mtDNA*), exclusively inherited from the maternal side, contains genes for producing many of the molecules mitochondria need to transduce energy. (Mitochondrial ancestors probably had all the necessary genes, but some have transferred horizontally [p. 34] into nuclear DNA.) Recent research suggests that flaws gradually accumulate in mitochondrial DNA over an animal's lifetime, and the resultant progressive impairment in mitochondrial energy transduction has been implicated in a wide variety of disorders, including muscle weakness, seizures, blindness, and degenerative illnesses associated with aging. In addition, mtDNA can be used to date ancient biological samples. By measuring the mutations that accumulate in the mitochondrial DNA, researchers can assign an approximate age to the sequence. In this fashion mtDNA acts like a molecular clock. mtDNA is used because the cell has only two copies of the nuclear DNA, whereas thousands of copies of mtDNA are found in each cell.

Mitochondria are enclosed by two membranes

Like nuclei, mitochondria are enclosed by a double membrane—a smooth outer membrane that surrounds the mitochondrion itself and an inner membrane that forms a series of infoldings or shelves called **cristae**, which project into an inner cavity filled with a gel-like solution known as the **matrix** (Figure 2-16). Cristae contain crucial proteins that ultimately are responsible for converting much of the energy in food into a usable form (the electron transport proteins, described shortly). The generous folds of the inner membrane greatly increase the surface area available for housing these important proteins. (This is a common theme in physiology: Increasing the surface area of various body structures by means of surface projections or infoldings makes more surface area available to participate in the structure's functions.) The matrix consists of a concentrated mixture of hundreds of different dissolved enzymes (the cit-

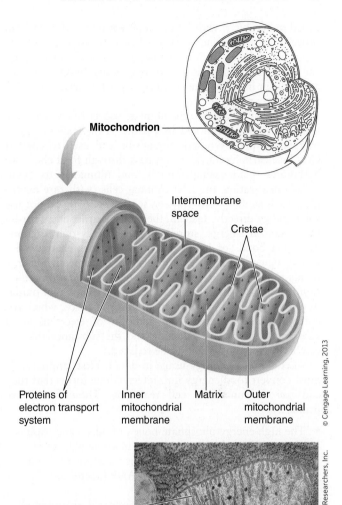

Mitochondrion

Intermembrane space

Cristae

Proteins of electron transport system

Inner mitochondrial membrane

Matrix

Outer mitochondrial membrane

Cristae

© Cengage Learning, 2013

© Bill Longcore/Photo Researchers, Inc.

FIGURE 2-16 Mitochondrion. Diagram and electron micrograph of a mitochondrion. Note that the outer membrane is smooth while the inner membrane forms folds known as cristae that extend into the matrix. An intermembrane space separates the outer and inner membranes. The electron transport proteins embedded in the cristae are ultimately responsible for converting much of the energy of food into a usable form.

ric acid cycle enzymes, soon to be described) that are important in preparing nutrient molecules for the final extraction of usable energy by the cristae proteins.

Aerobic metabolism in mitochondria relies on O_2 to convert energy in food into ATP

Every cell of an animal relies on specific biochemical pathways to generate energy from the food it consumes. One unifying principle of biochemistry is the similarity in the major pathways of metabolism that organisms use. Some of these pathways are termed **aerobic** ("with air" or "with O_2") and require the consumption of O_2, whereas others are called **anaerobic** ("without air" or "without O_2"), and can proceed in the absence of O_2. Most cell types can switch from one pathway to the other depending on the physiological state of the organism and availability of O_2. Aerobic metabolism in eukaryotes requires mitochondria; anaerobic may or may not.

For most organisms, the ultimate source of energy is the sun (exceptions are noted in Chapter 14). Solar energy is trapped by photosynthetic organisms and used to convert CO_2 into carbohydrates, which pass through food chains to herbivores (plant-eating animals) and ultimately to carnivores (meat-eating animals). Animal cells (with rare exceptions such as the photosynthetic seaslug, p. 690) cannot use solar energy directly. Instead, they must extract energy from the chemical energy stored in the energy-rich carbon–hydrogen bonds of ingested food. Note that energy is released when electrons are transferred from high-energy bonds to *electron acceptors* like oxygen in *oxidation-reduction* reactions. (Recall from your chemistry class that **oxidation** is the removal of electrons from a substance, which become **oxidized**, whereas the addition of electrons to a substance is termed a **reduction** and the substance that receives the electrons is said to be **reduced**.)

Let's look at energy usage in detail. First, animal cells must convert food energy into intermediate forms that they can use to run other cell reactions. These forms are primarily:

- The high-energy phosphate bonds of **adenosine triphosphate (ATP)**, which consists of adenosine with three phosphate groups attached, and is made by adding a phosphate (with an electron) to **ADP (adenosine diphosphate)**.
- The electrons (with hydrogen) in reduced **nicotinamide adenine dinucleotide (NADH)**, made by transferring two electrons and a hydrogen to the oxidized form **NAD⁺**, an electron acceptor derived from the B vitamin *niacin*.

ATP and NADH are universal energy carriers—the common "currency" of life on Earth—because these energized biomolecules contain chemically useful forms of stored energy. NADH carries energy-rich electrons that can be used to reduce other organic molecules. Each NADH molecule is actually worth almost three ATPs! As you will see, the hydrogen and electrons of NADH may be transferred ultimately to *oxygen*, the final electron acceptor, in order to make more ATP. Similarly, ATP carries a high-energy bond in the terminal phosphate; when this is split, a substantial amount of energy is released. The energy released can drive the energy-requiring activities within the cell, such as contraction of a muscle, transport of ions across a membrane, or synthesis of a new protein. To "cash in" the energy currency of ATP, enzymes split the terminal phosphate bond of ATP, which yields ADP plus inorganic phosphate (P_i) plus energy:

$$\text{ATP} \xrightarrow{\text{splitting}} \text{ADP} + P_i + \text{energy for use by the cell} + \text{heat}$$

In this energy scheme, food might be thought of as the "crude fuel," whereas ATP is the "refined fuel" for operating the cell's machinery. ATP is constantly being used by cell processes and must constantly be replenished.

Let's elaborate on this fuel conversion process (Figure 2-17). In the vast majority of animals, the cell has several options available to it when its supply of ATP molecules becomes depleted:

- Energy-rich nutrients stored within the cell can be broken down to generate ATP.
- Similar nutrients can be mobilized from specialized storage cells in the animal and transferred, via a circulatory system or diffusion, to energy-depleted cells.
- Dietary energy nutrients (i.e., food) can be digested, or broken down, by the digestive system into smaller absorbable units that can be transferred (again, by circulation or diffusion) from the lumen of the digestive tract to cells whose energy supplies are depleted (see Chapter 14).

The main endogenous fuels stored by most animal cells are *glycogen* and *triglycerides* (defined earlier, p. 26), and—in some marine animals—**waxes** (a fatty acid linked to a fatty alcohol). Vertebrates store glycogen in the liver, muscles, and glia, whereas triglycerides are stored in *adipose* (fat) tissue. Insects store huge reserves of carbohydrate and triglycerides in their *fat body*, the insect equivalent of a liver. Generally, periods of short-term strenuous activity rely on the combustion of glycogen whereas during prolonged activity there is a transition to the combustion of fat. For example, during foraging flight, locusts rely on carbohydrate sources, but during long-distance migratory flight, they rely on the combustion of fat. Dragonflies use fats to power their prolonged hovering behaviors. In salmon (fish that migrate between rivers and the ocean), so-called expendable proteins can be converted into amino acids and used as fuel during migration. Similarly, amino acids can be used to fuel energy needs in some species of marine nonvertebrates as well as in insects. In many vertebrates, the use of protein stores as a fuel to generate ATP represents a "last ditch" effort for survival. For example, Emperor penguins, who fast for prolonged periods in Antarctica during the winter months when near-shore seas are frozen over, rely on stored triglycerides for fuel generation. When these supplies are exhausted and the body is forced to burn its protein stores, the penguin returns to the sea to replenish its fuel.

Glycolysis The most common energy pathway begins with glucose, which can be derived from most dietary carbohydrates (such as starch, a plant storage polymer composed of glucose units). Stored carbohydrates and some amino acids can also be converted into glucose. When delivered to the cells by a circulatory system, the nutrient molecules are transported across the plasma membrane into the cytosol. Among the thousands of enzymes within the cytosol are those responsible for **glycolysis**, a chemical process that breaks down the simple 6-carbon sugar molecule, glucose, into two **pyruvate** molecules, each of which contains three carbons. There are actually 10 separate sequential reactions in glycolysis, each catalyzed by a separate enzyme. Glycolysis is common to almost every cell (prokaryotic and eukaryotic) and is believed to be the most ancient of the metabolic pathways, having arisen before oxygen accumulated in the atmosphere. During this process, two hydrogens and two electrons are released and (except for one H^+) transferred to NAD⁺ molecules (becoming NADH) for later use. In addition, some of the energy stored in the chemical bonds of glucose is used to convert ADP into ATP (Figure 2-18). Importantly, although glycolysis is widespread, it is not very

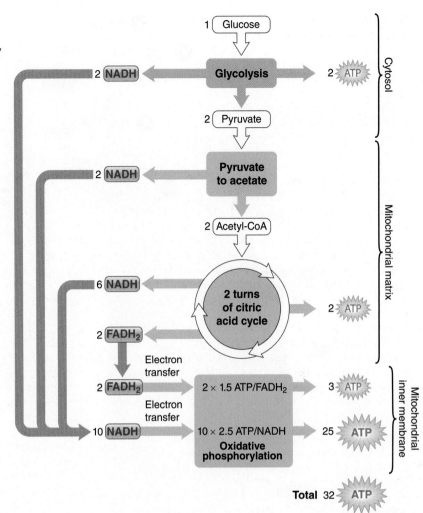

FIGURE 2-17 Summary of ATP production from the complete oxidation of one molecule of glucose. The total of 32 ATP assumes that electrons carried by each NADH yield 2.5 ATP and those carried by each $FADH_2$ yield 1.5 ATP during oxidative phosphorylation.

© Cengage Learning, 2013

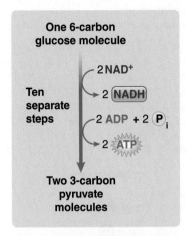

FIGURE 2-18 Glycolysis in the cytosol. Glycolysis splits glucose (six carbons) into two pyruvate molecules (three carbons each), with a net yield of 2 ATP plus 2 NADH (available for further energy extraction by the electron transport system).

© Cengage Learning, 2013

efficient in terms of energy extraction: *One molecule of glucose yields only two molecules of ATP in glycolysis.* Much of the energy originally contained in the glucose molecule is still locked in the chemical bonds of the pyruvate molecules. The low-energy yield of glycolysis is sufficient to sustain the needs of many single-celled anaerobic organisms and temporarily supports some aerobic cells through periods of reduced oxygen supply, but is grossly insufficient to sustain the long-term energy requirements of large, active animals. This is where the mitochondria come into play.

Citric Acid Cycle The pyruvate produced by glycolysis in the cytosol can be selectively transported into the mitochondrial matrix. Here it is further broken down into a 2-carbon molecule, acetic acid, by enzymatic removal of one of the carbons in the form of carbon dioxide (CO_2), which eventually is eliminated from the body as an end product, or waste (Figure 2-19). During this breakdown process, a carbon–hydrogen bond is disrupted, so a hydrogen atom and two electrons are also released and transferred to make another NADH from NAD^+. The acetic acid thus formed combines with *coenzyme A*, a derivative of pantothenic acid (a B vitamin), producing the compound acetyl coenzyme A or *acetyl-CoA.*

Acetyl-CoA then enters the **citric acid cycle,** a cyclical series of eight biochemical reactions catalyzed by the enzymes of the mitochondrial matrix. This cycle is also known as the *Krebs cycle,* in honor of its principal discoverer, or the *tricarboxylic acid (TCA) cycle,* because citric acid contains three carboxylic acid groups. Notably, the pyruvate formed in glycolysis can be further oxidized to generate more ATP *only* when oxygen is available to the animal. This cycle of reactions can be compared to one revolution around a Ferris wheel. (Keep in mind that Figure 2-19 is highly schematic. It depicts a cyclical series of biochemical reactions. The molecules themselves are not physically moved around in a cycle.) On the top of the Ferris wheel, acetyl-CoA, a 2-carbon molecule, enters a seat already occupied by *oxaloacetic acid,* a 4-carbon molecule. These two molecules link together to form a 6-carbon *citric acid* molecule and the trip around the citric acid cycle begins. As the seat moves around the cycle, at each new position, matrix enzymes modify the passenger molecule to form a slightly different molecule. These molecular alterations have the following important consequences:

1. Two carbons are "kicked off the ride"—released one at a time from 6-carbon citrate, converting it back into 4-carbon oxaloacetate, which is now available at the

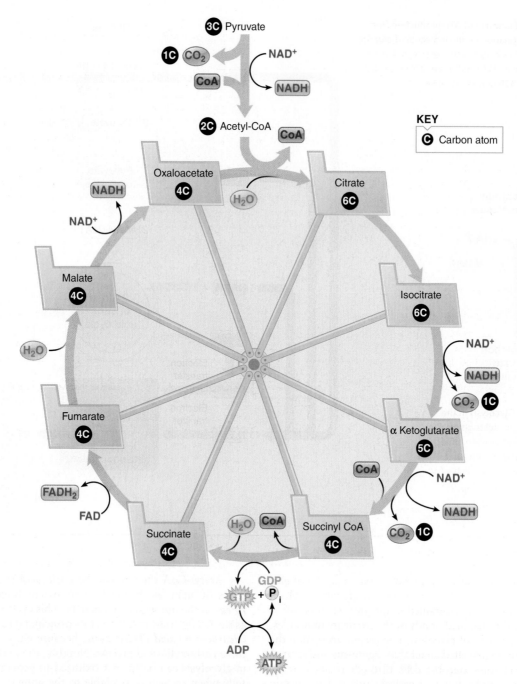

FIGURE 2-19 Citric acid cycle in the mitochondrial matrix. The two carbons entering the cycle by means of acetyl-CoA are eventually converted to CO_2, with oxaloacetate, which accepts acetyl-CoA, being regenerated at the end of the cyclical pathway. The hydrogens released at specific points along the pathway bind to the hydrogen carrier molecules NAD^+ and FAD for further processing by the electron transport system. One molecule of ATP is generated for each molecule of acetyl-CoA that enters the citric acid cycle, for a total of two molecules of ATP for each molecule of processed glucose.

© Cengage Learning, 2013

top of the cycle to pick up another acetyl-CoA for another revolution through the cycle. CoA is recycled too; it is released at the end, making it available to bind with a new acetate to form another acetyl-CoA.

2. The released carbon atoms become two molecules of CO_2. This CO_2, as well as the CO_2 produced during the formation of acetic acid from pyruvate, diffuses out of the

mitochondrial matrix and subsequently out of the cell to enter the ECF. In animals with circulatory systems such as blood, the gas (and its derivative, bicarbonate; p. 540) is carried to the lungs or gills, where it is eliminated. (Note that the oxygen in the CO_2 molecule is derived from the molecules that were involved in the reactions, not from free molecular oxygen supplied by breathing.)

3. Hydrogen atoms and their electrons are also "bumped off" during the cycle at four of the chemical conversion steps. The key purpose of the citric acid cycle is to produce these hydrogens for entry into the electron transport system in the inner mitochondrial membrane. These hydrogens and electrons are "captured" by two acceptors—NAD$^+$ and **flavine adenine dinucleotide (FAD)**, a derivative of the B vitamin *riboflavin*, which becomes FADH$_2$. Three NADH and one FADH$_2$ are produced for each turn of the citric acid cycle.

4. One more molecule of ATP is produced for each molecule of acetyl-CoA processed. Actually, ATP is not directly produced by the citric acid cycle. The released energy is used to directly link inorganic phosphate to **guanosine diphosphate (GDP)** to form **guanosine triphosphate (GTP)**, a high-energy molecule similar to ATP. The energy from GTP can then be transferred to ATP as follows:

$$ADP + GTP \rightarrow ATP + GDP$$

Because each glucose molecule is converted into two acetic acid molecules, fueling two turns of the citric acid cycle, two more ATP molecules are produced from each glucose molecule.

These two additional ATPs are still not much of an energy profit. However, the citric acid cycle is important in preparing the hydrogen carrier molecules for their entry into the final stage, *oxidative phosphorylation*, which produces far more energy than the sparse amount of ATP produced by the cycle itself.

Electron Transport Chain and Oxidative Phosphorylation

Considerable untapped energy is still stored in the released hydrogen atoms, which contain electrons at high-energy levels. So far we have not mentioned any role for free oxygen in these reactions. That is because free (or molecular) oxygen—O$_2$—is not a reactant in the pathways of either glycolysis or the citric acid cycle. However, oxygen is essential as the final electron acceptor, as you will see, in the disposition of the reduced intermediate acceptors, NADH and FADH$_2$. The "big payoff" comes when NADH and FADH$_2$ enter the electron transport chain (step 1), which consists of electron carrier molecules located in the inner mitochondrial membrane lining the cristae (Figure 2-20). The high-energy electrons are extracted from the hydrogens held in NADH and FADH$_2$ and transferred sequentially to the electron carrier molecules, freeing NAD and FAD to pick up more hydrogen atoms (step 2). The electron transport molecules are arranged in a specific order on the inner membrane so that the high-energy electrons are progressively transferred through a chain of reactions, with the electrons falling to successively lower energy levels at each step (step 3).

This electron transport chain is also called the **respiratory chain** because it relates to use of O$_2$. A fundamental question now arises: Why is oxygen essential for the survival of most organisms? Ultimately, it is the passage of these electrons generated from foodstuffs to available O$_2$ derived from the environment (air or water) that ensures survival. Electrons that become bound to O$_2$ are in their lowest energy state. Oxygen thus enters the mitochondria to serve as the final electron acceptor of the electron transport chain. This negatively charged oxygen (negative because it has acquired additional electrons) then combines with the positively charged hydrogen ions (positive because they have donated the electrons at the beginning of the electron transport chain) to form two waters (H$_2$O) (step 4).

As the electrons move through this chain of reactions to ever-lower energy levels, they release energy. Part of the released energy is lost as heat, but some is harnessed by the mitochondrion to synthesize ATP through the following steps. At three sites in the electron transport system (Complexes I, III, and IV), the *energy released during the transfer of electrons is used to transport hydrogen ions* (H$^+$) across the inner mitochondrial membrane from the matrix to the space between the inner and outer mitochondrial membranes, the *intermembrane space* (step 5). As a result, hydrogen ions are more heavily concentrated in the intermembrane space than in the matrix. Consider the amazing fact that the mitochondrion has created a powerful force, that is, a gradient available to do work. This H$^+$ gradient generated by the electron transport system (step 6) can now supply the energy that drives ATP synthesis by the membrane-bound mitochondrial enzyme ATP synthase.

ATP synthase consists of a *basal unit* embedded in the inner membrane, connected by a stalk to a *headpiece* located in the matrix, with the *stator* bridging the basal unit and headpiece. Because H$^+$ ions are more heavily concentrated in the intermembrane space than in the matrix, they have a strong tendency to flow back into the matrix through the inner membrane via channels formed between the basal units and stators of the ATP synthase complexes (step 7). This flow of H$^+$ ions activates ATP synthase and powers ATP synthesis by the headpiece, a process called **chemiosmosis**. Passage of H$^+$ ions through the channel makes the headpiece and stalk spin like a top (step 8), similar to the flow of water making a waterwheel turn. As a result of the changes in its shape and position as it turns, the headpiece is able to sequentially pick up ADP and P$_i$, combine them, and release the ATP product (step 9).

Oxidative phosphorylation encompasses the entire process by which ATP synthase synthesizes ATP by phosphorylating (adding a phosphate to) ADP using the energy released by electrons as they are transferred to O$_2$ by the electron transport system. The harnessing of energy into a useful form as the electrons tumble from a high-energy state to a low-energy state can be likened to a power plant converting the energy of water tumbling down a waterfall (while turning a turbine) into electricity.

The series of steps that lead to oxidative phosphorylation may at first seem unnecessarily complicated. Why not just directly oxidize, or "burn," food molecules to release their energy? Oxidation was originally used to describe the reaction that occurs when fuel substances are burned in air, in which oxygen directly accepts the electrons derived from the fuels and all the energy stored in the food molecule is released explosively as heat (Figure 2-21). However, in the animal, oxidation of food molecules actually occurs in many small, controlled steps so that the food molecule's chemical energy is gradually made available for convenient packaging in a storage form useful to the cell. (Analogously, the fuel in an automobile's tank is not normally burned all at once—that would produce an explosion. Rather, fuel is fed a little at a time into the engine combustion chambers.) Approximately 70% of the chemical energy released in the oxidation of glucose to CO$_2$ and H$_2$O is recovered as energy in the form

1 The high-energy electrons extracted from the hydrogens in NADH and FADH$_2$ are transferred from one electron-carrier molecule to another.

2 The NADH and FADH$_2$ are converted to NAD$^+$ and FAD, which frees them to pick up more hydrogen atoms released during glycolysis and the citric acid cycle.

3 The high-energy electrons fall to successively lower energy levels as they are transferred from carrier to carrier through the electron transport system.

4 The electrons are passed to O$_2$, the final electron acceptor of the electron transport system. This oxygen, now negatively charged because it has acquired additional electrons, combines with H$^+$ ions, which are positively charged because they donated electrons at the beginning of the electron transport system, to form H$_2$O.

5 As electrons move through the electron transport system, they release free energy. Part of the released energy is lost as heat, but some is harnessed by the mitochondrion to transport H$^+$ across the inner mitochondrial membrane from the matrix to the intermembrane space at Complexes I, III, and IV.

6 As a result, H$^+$ ions are more heavily concentrated in the intermembrane space than in the matrix. This H$^+$ gradient supplies the energy that drives ATP synthesis by ATP synthase.

7 Because of this gradient, H$^+$ ions have a strong tendency to flow into the matrix across the inner membrane via channels between the basal units and stators of the ATP synthase complexes.

8 This flow of H$^+$ ions activates ATP synthase and powers ATP synthesis by the headpiece, a process called **chemiosmosis**. Passage of H$^+$ ions through the channel makes the headpiece and stalk spin like a top.

9 As a result of changes in its shape and position as it turns, the headpiece picks up ADP and P$_i$, combines them, and releases the ATP product.

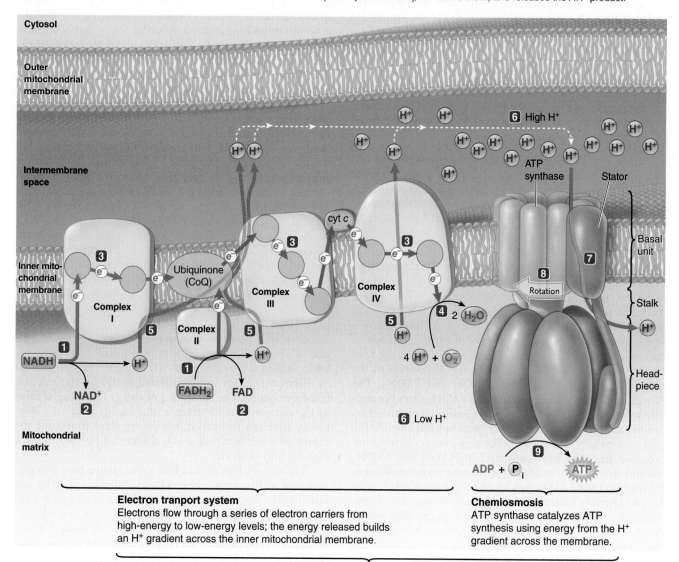

FIGURE 2-20 Oxidative phosphorylation at the mitochondrial inner membrane. Oxidative phosphorylation involves the electron transport system (steps 1–6) and chemiosmosis by ATP synthase (steps 7–9). The pink circles in the electron transport system represent specific electron carriers.

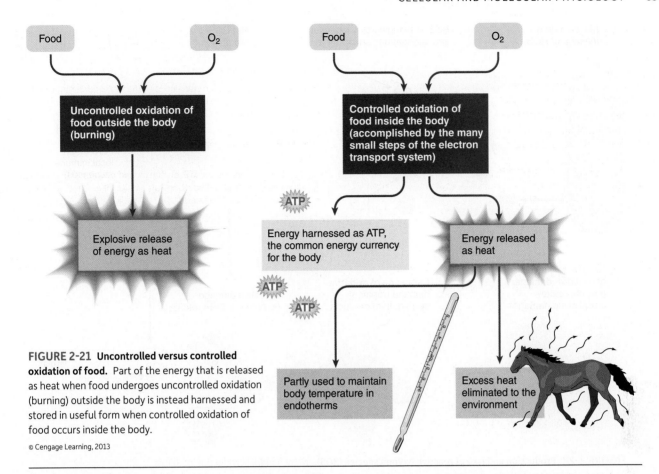

FIGURE 2-21 Uncontrolled versus controlled oxidation of food. Part of the energy that is released as heat when food undergoes uncontrolled oxidation (burning) outside the body is instead harnessed and stored in useful form when controlled oxidation of food occurs inside the body.

© Cengage Learning, 2013

of ATP, and much less is converted to heat. The heat produced is not completely wasted energy in endotherms such as birds and mammals; it is used to help maintain body temperature, with any excess heat being lost to the environment. (Endotherms are organisms that generate internal heat for thermoregulation; see Chapter 15.)

When enough O_2 is present, mitochondrial processing (both the citric acid cycle in the matrix and the electron transport chain on the cristae) harnesses sufficient energy to generate 26 more molecules of ATP, for a total net yield of 30 ATPs per molecule of glucose processed after glycolysis. The overall reaction for oxidation of food molecules to yield energy is as follows:

$$Food + O_2 \rightarrow CO_2 + H_2O + ATP$$

| (necessary for oxidative phosphorylation) | (produced primarily by the citric acid cycle) | (produced primarily by the electron transport chain) |

Glucose, the principal nutrient derived from dietary carbohydrates, is the fuel preference of most cells. The overall equation for glucose including the ATP of glycolysis is

$$C_6H_{12}O_6 + 6\,O_2 \rightarrow 6\,CO_2 + 6\,H_2O + 32\,ATP$$

However, nutrient molecules derived from fats (fatty acids) and, if necessary, from protein (amino acids) can also participate at specific points in this overall chemical reaction to eventually produce energy. Amino acids are usually used for protein synthesis instead of energy production, but they can be used as fuel if insufficient glucose and fat are available.

Mitochondrial metabolism can create oxidative stress

Unfortunately, leakage of some of the electrons associated with mitochondrial metabolism can potentially damage the tissues of the host animal and contribute to the aging process. This requires cellular strategies to detoxify and limit the production of certain metabolites of molecular oxygen known as *reactive oxygen species (ROS)*. ROS encompass a variety of diverse chemicals, which includes superoxide anions, hydroxyl radicals, peroxynitrite, and hydrogen peroxide (Figure 2-22). Some ROS are **free radicals** (molecules containing one or more unpaired electrons and thus react readily with other molecules, acquiring or giving up an electron to achieve stability). Radicals, such as hydroxyl radicals and superoxide, are extremely unstable. In contrast, other forms such as hydrogen peroxide are comparatively long-lived. ROS production is countered by an elaborate antioxidant system that includes the enzymatic scavengers *superoxide dismutase (SOD)*, *catalase,* and *glutathione peroxidase*. SOD increases the conversion of superoxide to hydrogen peroxide, while catalase and glutathione peroxidase convert hydrogen peroxide to water. Levels of catalase tend to correlate positively with lifespan in mammals. A variety of other nonenzymatic, low-molecular-weight molecules are also important in scavenging ROS and include *ascorbate* (vitamin C), *tocopherols* (vitamin E derivatives), *flavonoids, carotenoids* (vitamin A derivatives), and perhaps most important in long-lived species, *uric acid,* which is present in high concentrations within cells. Dietary intake of carotenoids, which vary seasonally, are also suggested to increase antioxidant protection in some species (see Chapter 10, p. 458).

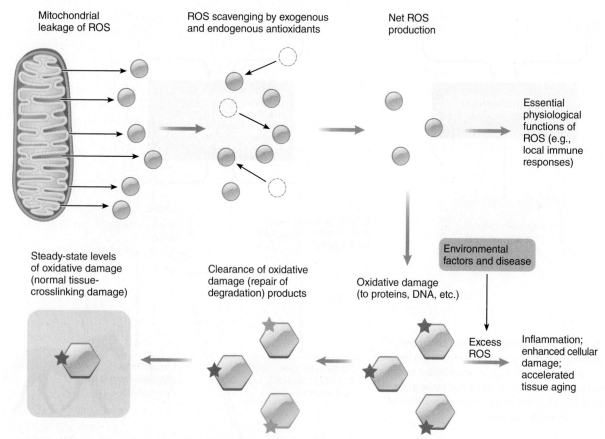

FIGURE 2-22 Production and fate of reactive oxygen species (ROS). Mitochondrial production of ROS is counteracted by both endogenous and exogenous antioxidants. Net ROS production can be harnessed by the immune system for local immune responses. Excess leakage of ROS leads to oxidative damage of cells and their constituents. Under normal physiological conditions, some damage to cellular components is reversible; however, there is also a steady-state accumulation of tissue crosslinks in a species-specific pattern of aging. Excessive ROS production results from inflammation associated with disease as well as environmental factors such as dietary toxins and radiation, which can lead to accelerated tissue aging and/or enhanced cellular damage leading to specific complications.

© Cengage Learning, 2013

Oxidative stress occurs when reactive oxygen species production overwhelms the antioxidant defenses, as with inflammation. In such coupled oxidation-reduction reactions, also termed **redox reactions**, electrons release some of their energy as they pass from a donor to an acceptor molecule. An organism is in redox balance if the electrons that are generated through metabolism are removed at an equivalent rate. Problems arise when these reactive species are not scavenged by antioxidants but instead react with other components of the cell, generating an imbalance in the redox reaction favoring oxidation, which results in oxidative stress. Longevity of a species as well as the ability of individual organisms to withstand oxidative stress and inflammation can thus be critically linked to the efficacy of its antioxidant system.

ROS are not always useless by-products. ROS generation by phagocytic cells is a critical host defense mechanism used to combat infection (see Chapter 10). Further, cytosolic ROS production under conditions of metabolic stress may trigger specific signaling pathways within the cell, activating mechanisms to cope with stress.

Mitochondrial densities vary among tissue and organ types such as muscles

Concentrations of ATP normally remain fairly constant in any particular tissue, even during high metabolic activity. How, then, can cells maintain ATP supplies during the transition from rest to work, particularly in skeletal muscles? Consider the different rates of ATP use in muscles at maximum energy output. For example, the flight muscle of the locust uses ATP at about 5,400 mole/g/min. In comparison, ATP use is about 600 mole/g/min in hummingbird flight muscle and only 30 mole/g/min in an exercising human's limb muscles. One solution is to alter the total number of mitochondria per unit volume—that is, the *density* of mitochondria—within each cell. Indeed, muscle cells generally have mitochondrial densities corresponding to their energy loads (see Chapter 8). Skeletal muscles of terrestrial vertebrates have 1 to 10% of their volume as mitochondria (in comparison to 30 to 50% in vertebrate hearts). As befits its enormous energy demand, locust and dragonfly flight muscle has up to 50% of their volume as mitochondria. As

A Carolina locust. The flight muscles of this animal are densely packed with mitochondria, which in turn have an unusually high packing density of cristae for ATP production.

an extreme example, specialized modified muscles called *heater cells*, located at the base of the brain in billfish (such as swordfish), keep temperatures in the brain and eye above ambient temperature (see Chapter 15). These cells contain one of the highest mitochondrial densities of any animal cell, ranging from 55 to 70% of total cell volume. Mitochondrial density is not necessarily fixed; it can be upregulated in many skeletal muscles after a prolonged increase in muscular activity (see Chapter 8). Acclimation to cold temperatures (see Chapter 15) also involves change in mitochondrial numbers. For example, the mitochondrial cell volume increases from 2.9 to 4.5% in the aerobic (red) muscle fibers in striped bass following acclimation from 25 to 5°C.

The second solution to high energy demands is to increase the *packing of cristae* within mitochondria. In particular, the packing is two to three times denser in mitochondria of insect flight muscle, hummingbird flight muscle, skipjack tuna swimming (red) muscle, and leg muscles of mammalian endurance "athletes" such as pronghorn antelopes.

Phosphagens provide a rapid source for ATP production

A third solution to meeting the cells' ATP requirements involves the mobilization of tissue-specific *phosphagens* (see Chapter 8). Phosphagens are organic phosphate compounds present in high concentrations in the muscle cell that can transfer a high-energy phosphate group to ADP in order to regenerate ATP. They are made from ATP in times of plenty (when the cell is resting and has plenty of energy intake). In

mammals, phosphagens are in sufficient supply to produce ATP for only a short time, such as up to about 10 seconds in humans. Thereafter the concentration of phosphagens must also be replenished by oxidation of sugars and fatty acids. The highest concentrations of phosphagens are in white skeletal muscle fibers, cells that rely heavily on anaerobic energy processes (see Chapter 8). There are somewhat lower concentrations in red skeletal muscle fibers (those that rely on aerobic metabolism), the heart, and the brain. The phosphagen of vertebrates is *creatine phosphate*, whereas nonvertebrates use a variety of organic phosphate compounds, the best investigated being *arginine phosphate*. Insect flight muscle contains a surprisingly low amount of arginine phosphate, suggesting that the efficiency of oxygen delivery to the muscle tissue is adequate to meet the requirements of aerobic metabolism.

Oxygen deficiency forces cells to rely on glycolysis and other anaerobic reactions, producing lactate, propionic acid, octopine, or other end products

Phosphagens can produce ATP in the absence of oxygen, but for a very limited time. How can animals meet their metabolic requirements in the absence of molecular oxygen for prolonged periods? Consider that in the absence of O_2, the electron transport chain cannot accept electrons from NADH and $FADH_2$ and so "shuts down." This occurs during bursts of high activity (such as a cheetah running) and when animals must cease breathing (as in an oyster closing its shell tightly during low tide). In such cases, glycolysis is usually the primary pathway for ATP replenishment. Recall that glycolysis takes place in the cytosol and involves the breakdown of glucose into pyruvate, producing a low yield of two molecules of ATP per molecule of glucose. Oxidative phosphorylation, the source of 26 out of every 30 ATP molecules, ceases, and the untapped energy of the glucose molecule may remain locked in the bonds of the pyruvate molecules. In some groups such as most vertebrates, pyruvate is converted into lactate (Figure 2-23) by an enzyme called *lactate dehydrogenase (LDH)*, in a reaction that regenerates NAD+, the hydrogen-electron acceptor needed for earlier steps in glycolysis to proceed (p. 48):

$$2 \text{ Pyruvate} + 2 \text{ NADH} \rightarrow 2 \text{ lactate} + 2 \text{ NAD}^+$$

Lactate thus steadily accumulates in the tissues and gradually reduces the pH. Lactate production is necessary because the electron transport chain can no longer regenerate the acceptors NAD+. We will encounter this reaction and LDH in later chapters.

Significant modifications to anaerobic end products have evolved at two points in the glycolytic pathway.

1. *Higher ATP yield.* The first modification is at the *phosphoenolpyruvate (PEP)* step, the net ATP-yielding step just before pyruvate formation in glycolysis:

$$2 \text{ PEP} + 2 \text{ ADP} \rightarrow 2 \text{ pyruvate} + 2 \text{ ATP}$$

Many animals (bivalve mollusks such as oysters, nematodes, parasitic helminths, and the swamp-dwelling annelid *Alma emeni*) gain more than two ATPs per glucose by channeling PEP into a partial set of reactions in the

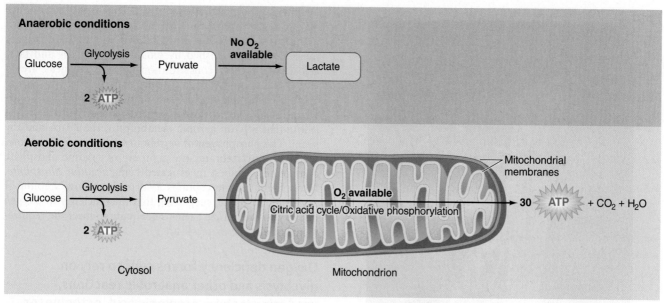

Anaerobic conditions

FIGURE 2-23 Comparison of energy yield and products under anaerobic and aerobic conditions. In anaerobic conditions, only 2 ATP are produced for every glucose molecule processed, but in aerobic conditions a total of 32 ATP are produced per glucose molecule.

© Cengage Learning, 2013

citric acid cycle, in reverse direction. The end product is usually either succinic acid, yielding four ATPs per glucose, or propionic acid, yielding six ATPs per glucose (compare the following with Figure 2-18):

> **2 PEP** → 2 oxaloacetic acid (+ 2 ATP) → 2 malic acid → 2 fumaric acid → 2 succinic acid (+ 2 ATP) → 2 succinyl CoA → 2 methylmalonyl CoA → 2 propionyl CoA (+ 2 ATP) → 2 propionic acid

Some nonvertebrates (such as bivalve mollusks) can use **aspartate** (an amino acid) as an anaerobic fuel. Aspartate is particularly high in oyster and mussel hearts. It is converted to oxaloacetate (by removal of ammonia), which enters the pathway just described.

2. *Alternative end products.* The second modification point is the conversion of pyruvate for the purposes of regenerating NAD^+. The acid problem created by lactate production is avoided in some organisms, in which pyruvate is converted to a variety of nonacidic organic products, including

- *Strombine, tauropine, lysopine, octopine, alanine,* and *alanopine* in mollusks. The first four compounds are produced by joining pyruvate to an amino acid: glycine, taurine, lysine, and arginine, respectively.
- *Ethanol* (alcohol) and *acetate.* Yeasts metabolize glucose anaerobically to two ethanol and two CO_2 molecules, a process used in manufacturing beer, for example. Goldfish wintering at the bottom of ice-covered ponds and lakes produce ethanol as the oxygen levels drop. This is the only vertebrate known to use this pathway.

Ultimately, end products such as lactate that accumulate in body fluids are converted back to pyruvate using NAD^+ as the electron acceptor. Pyruvate can then either enter into the citric acid cycle or be reconverted back into glucose or glycogen. Regardless of the disposition of pyruvate, oxygen is required. A functional electron transport system is essential to capture the energy from the NADH and $FADH_2$ during these reactions. The conversion of lactate to glucose is an example of *gluconeogenesis* ("new formation of glucose"). Basically, it is a reversal of glycolysis, because the principal intermediates of the pathway are the same. However, the gluconeogenic pathway involves several different enzymes and requires the expenditure of six molecules of ATP for each molecule of glucose generated. The ATP is necessary because glucose synthesis is energetically "uphill," that is, an energy-requiring process (in contrast to glycolysis, which is "downhill," in terms of energy expenditure).

Tolerance of oxygen deficiency varies widely among organisms

Many animals require oxygen continuously as the final electron acceptor for survival, and are classified as **obligate aerobes**. In mammals, most cells cannot survive more than a few minutes without oxygen. Anaerobic pathways are, however, important during intense bursts of activity. Skeletal muscle cells in particular take advantage of this ability during short bursts of strenuous exercise, when energy demands for contractile activity outstrip the animal's ability to bring adequate O_2 to the exercising muscles to support oxidative phosphorylation. For example, cheetahs are noted for their remarkable sprint speeds. However, the metabolic costs incurred to the animal include an enormous *oxygen debt* and a reduction in blood and tissue pH (acidosis; see Chapter 13) caused by the accumulation of lactate. Cheetahs can sustain this intense activity for only brief periods of time and must rest for lengthy periods before resuming any activity.

Three species of *loriciferans,* tiny sediment-dwelling marine animals that belong to their own phylum first described in 1983, have recently been found in sediment cores obtained from the bottom of a briny (hyper-salty), acidic and sulfurous basin of anoxic (oxygen-starved) water 3.5 kilometers below the Mediterranean Sea's surface. These creatures are almost a millimeter long and appear like jellyfish sprouting from a conical shell. They may be the first known animal example of an obligate anaerobe. Adaptations to their anoxic environment include a lack of mitochondria and the presence of **hydrogenosomes**, respiratory organelles that generate ATP and molecular hydrogen found previously only in some obligate anaerobic fungi and protists. Similar to mitochondria in structure but lacking a genome, they are bounded by a double membrane, the inner one with some cristaelike projections. Discoveries such as these offer the possibility of identifying additional metazoan life in other anoxic regions of the planet.

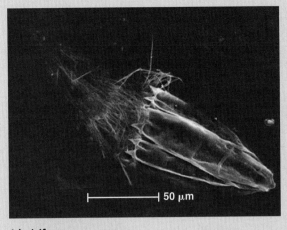

A loriciferan.

Source: Danovaro et al. BMC Biology 2010 8:30

In some species of fish, anaerobic pathways can fuel intermediate swimming speeds (between cruising and sprint), for sustained periods of time (tens of minutes). Lactate accumulates in large quantities in their white muscle fibers, and, unlike in mammals, removing these end products requires many hours of recovery. Because of the depletion of glucose stores and the poor return in energy yield on glucose, animals can rely on anaerobic metabolism for only relatively brief periods. When energy stores become depleted, vertebrate muscles fatigue and lose their ability to contract.

Some species of organisms—the **facultative anaerobes**—can adapt to anaerobic conditions for periods ranging from days to months by substituting other electron acceptors for oxygen, or by entering dormant states. The latter include embryos of brine shrimp (p. 17) and cysts of tardigrades (p. 614), which can survive inactive without oxygen indefinitely. Active facultative anaerobes include some parasitic worms such as tapeworms that can survive without oxygen indefinitely, and bivalve mollusks that live in the intertidal zone. For extended periods of time each day, these animals close their shells when exposed to air. Limiting desiccation by the sun in this way also renders them unable to exchange respiratory gases. Thus, they rely on the alternative pathways described earlier. Snapping turtles, the longest hibernating reptile in North America, bury themselves in the mud for up to eight months each year. Anaerobic glycolysis supplies the energy needs of these animals, at the expense of a buildup of lactate (particularly in the shell; see Chapter 13). However, in these and most vertebrates the brain and the heart normally continue their reliance on oxygen, albeit at a reduced rate.

A third group of organisms, called **obligate anaerobes,** are inhibited or killed in the presence of oxygen and thrive in anaerobic environments. These are primarily some archaea, bacteria (such as *Clostridium,* the source of botulism in improperly canned foods), and protozoa such as *Entamoeba,* which causes intestinal diseases such as dysentery in many mammals. Until recently, no animal was thought to be an obligate anaerobe, but that changed in 2010: see the box, *A Closer Look at Adaptation: Anoxic Animals.*

Animal cells that rely on anaerobic metabolism must have available to them an immediately available reservoir of stored energy supplies. For example, in vertebrate white skeletal muscle fibers, fuel is stored within the cell in the form of glycogen. Tissue glycogen can then be readily mobilized to glucose in times of energy expenditure. In many nonvertebrates, glycogen stores can form a significant proportion of the actual muscle tissue weight (see Chapter 8), and some have alternative fuels such as aspartate (p. 56). Phosphagens (p. 55) can also be used anaerobically, though generally for only a short time before they are exhausted.

check your understanding 2.8

How does the density of mitochondria in muscle tissue correlate with the relative activity of an organism?

What end products arise from mitochondrial metabolism? What are their fates?

Describe the structure of mitochondria, and explain their role in respiration.

What mechanisms permit an organism to survive periods without oxygen?

2.9 Vaults

Vaults, which are three times as large as ribosomes, are shaped like octagonal barrels (Figure 2-24). They have been found in many (but not all) animal groups and some protists. Their name comes from their multiple arches, which reminded their discoverers of vaulted or cathedral ceilings. A cell may contain thousands of vaults, but they have been elusive until recently because they do not show up with ordinary staining techniques. Two clues to the function of vaults may be their octagonal shape and their hollow interior. Intriguingly, the nuclear pores are also octagonal and the same size as vaults, leading to speculation that vaults may be cellular "trucks." According to this proposal, vaults would dock at nuclear pores, pick up molecules such as mRNAs or ribosome subunits synthesized in the nucleus, and deliver their cargo elsewhere in the cell such as the

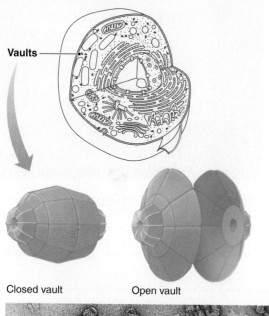

Vaults

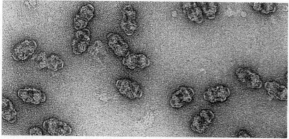

Closed vault   Open vault

FIGURE 2-24   Vaults. Diagram of closed and open vaults and electron micrograph of vaults, which are octagonal barrel-shaped nonmembranous organelles believed to transport either messenger RNA or the ribosomal subunits from the nucleus to cytoplasmic ribosomes.

© Dr. Leonard H. Rome/UCLA School of Medicine

rough ER. Ongoing research supports vaults' role in nucleus-to-cytoplasm transport, but what cargo they are carrying is uncertain.

2.10 Cytosol

Recall that the cytosol is the semiliquid portion of the cytoplasm that surrounds the organelles. It occupies about 55% of the total cell volume. Its amorphous appearance under an electron microscope belies the fact that the cytosol is not a uniform liquid mixture but is actually more like a highly organized, gelatinous mass with differences in composition and gelatinous consistency between various regions and states of the cell.

The cytosol is important in intermediary metabolism, ribosomal protein synthesis, and storage of fat and glycogen

Four general categories of activities are associated with the cytosol: (1) enzymatic regulation of intermediary metabolism; (2) ribosomal protein synthesis; (3) storage of fat and

carbohydrate; and (4) temporary storage of vesicles. Dispersed throughout the cytosol is a *cytoskeleton* that gives shape to the cell, provides an intracellular organizational framework, and is responsible for various cell movements.

Enzymatic Regulation of Intermediary Metabolism The term *intermediary metabolism* refers collectively to the large set of intracellular chemical reactions that involve the degradation, synthesis, and transformation of small organic molecules such as simple sugars, amino acids, and fatty acids. These reactions, regulated by feedback loops, are critical for ultimately capturing energy to be used for cellular activities and for providing the raw materials needed for maintaining cell structure and function and for cell growth. All intermediary metabolism occurs in the cytoplasm and mostly in the cytosol. Thousands of enzymes involved in glycolysis and other intermediary biochemical reactions are found in the cytosol.

Ribosome Protein Synthesis Also dispersed throughout the cytosol are the free ribosomes, as we noted earlier. Often, cytosolic ribosomes that are synthesizing identical proteins are clustered together in "assembly lines" known as polyribosomes.

Storage of Fat and Glycogen Excess nutrients not immediately used for ATP production are converted in the cytosol into storage forms that are readily visible even under a light microscope. Such nonpermanent masses of stored material are known as *inclusions*. The main forms, as we noted earlier, are glycogen and triglycerides. Storage of energy in the form of triglyceride lipids represents the largest and most important form of storing extra energy because lipids have almost twice the energy density of carbohydrates and proteins. Intermediates of lipid metabolism are channeled directly into the tricarboxylic acid cycle of mitochondria, yielding proportionately greater amounts of ATP. (Curiously, mitochondria of cartilaginous fishes cannot use fatty acids, but those of all other vertebrates can.) Through a microscope, small fat droplets can be seen within the cytosol in various cells. In insect fat bodies and vertebrate adipose tissue, the stored triglycerides can occupy almost the entire cytosol, coalescing to form one large fat droplet (Figure 2-25a). The other visible storage product is glycogen, which appears as aggregates or clusters dispersed throughout the cell (Figure 2-25b). Cells vary in their ability to store glycogen, with liver and muscle cells having the greatest stores. When food is not available to provide fuel for the citric acid cycle and electron transport chain, stored glycogen and fat are broken down to release glucose and fatty acids, respectively, which can feed into the mitochondrial energy-producing machinery.

Storage of Vesicles Secretory vesicles that have been processed and packaged by the endoplasmic reticulum and Golgi complex also remain in the cytosol, where they are stored until signaled to empty their contents extracellularly. In addition, transport vesicles are transiently present in the cytosol as they carry their selected cargo from one membrane-enclosed organelle compartment to another or to the plasma membrane. Likewise, endocytotic vesicles can be found transiently in the cytosol until the cell disposes of these packages it has internalized.

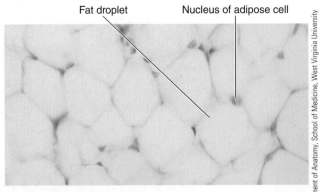

Fat droplet   Nucleus of adipose cell

(a) Fat storage in adipose cells

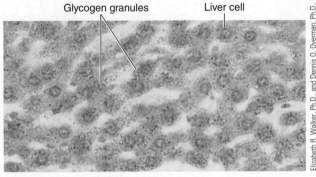

Glycogen granules   Liver cell

(b) Glycogen storage in liver cells

FIGURE 2-25 **Inclusions.** (a) Light micrograph showing fat storage in adipose cells. A fat droplet occupies almost the entire cytosol of each cell. (b) Light micrograph showing glycogen storage in liver cells. The red-stained granules throughout the cytosol of each liver cell are glycogen deposits.

Elizabeth R. Walker, Ph.D., and Dennis O. Overman, Ph.D., Department of Anatomy, School of Medicine, West Virginia University

check your understanding 2.9, 2.10

How does the shape of a vault provide a clue as to one of its possible functions?

How is the relative fat content of an adipose cell different from that of other cells? In what biological form are fats and sugars stored?

2.11 Centrosome, Centrioles, and Microtubule Organization

The **centrosome,** or **cell center,** located near the nucleus, consists of the centrioles surrounded by an amorphous mass of proteins. The **centrioles** are a pair of short cylindrical structures that lie at right angles to each other at the centrosome's center. The centrosome is the cell's main **microtubule organizing center** (MTOC). Microtubules are one of the components of the cytoskeleton. When a cell is not dividing, microtubules are formed from the amorphous mass and radiate outward in all directions from the centrosome (Figure 2-2). These microtubules anchor many of the membranous organelles and also serve as "highways" along which vesicles are transported within the cell by "molecular motors." In some cells, the centrioles form cilia and flagella, which are elongated, slender motile structures made up of bundles of microtubules. During

cell division, the centrioles form a mitotic spindle out of microtubules to direct movement of chromosomes. You will learn about the microtubules in the next section.

2.12 Cytoskeleton: Cell "Bone and Muscle"

Permeating the cytosol is the **cytoskeleton,** a complex protein network that acts as the "bone and muscle" of the cell. The distinct shape, size, complexity, and intracellular specialization of the various body cells necessitate intracellular scaffolding to support and organize the cell components into an appropriate arrangement and to control their movements. These functions are performed by the cytoskeleton. This elaborate network has at least three distinct elements (Figure 2-26): (1) *microtubules,* (2) *microfilaments,* and (3) *intermediate filaments* (Table 2-1). The different parts of the cytoskeleton are structurally linked and functionally coordinated to provide certain integrated functions for the cell. Because of the complexity of this network and the variety of functions it serves, we discuss its elements separately.

Microtubules are essential for maintaining asymmetric cell shapes and are important in complex cell movements

The **microtubules** are the largest of the cytoskeletal elements. They are very slender (22 nanometer, or nm, in diameter; 1 nm = 1 billionth of a meter), long, hollow, unbranched tubes composed primarily of **tubulin,** a small, globular protein molecule (6 nm in diameter) (Figure 2-26a).

Microtubules position many of the cytoplasmic organelles, such as the ER, Golgi complex, lysosomes, and mitochondria. They are also essential for maintaining an asymmetric cell shape, such as that of a nerve cell, whose elongated *axon* may extend a meter in length or more in a large vertebrate, from the origin of the cell body in the spinal cord to the termination of the axon at a muscle (see Chapter 4). Microtubules, along with specialized intermediate filaments, stabilize this asymmetric axonal extension.

Microtubules also play an important role in coordinating numerous complex cell movements, including (1) transport of secretory vesicles from one region of the cell to another, (2) movement of specialized cell projections such as cilia and flagella, and (3) distribution of chromosomes during cell division through formation of a mitotic spindle. Let's examine each of these roles.

Transport of Secretory Vesicles Secretory vesicles produced by the Golgi complex are too big to move quickly by simple diffusion, particularly in large and long cells like neurons (Chapter 4) and tall thin epithelial cells. Microtubules solve this problem by providing a "highway" for vesicular traffic (and for mitochondria as well), with the driving force dependent on ATP. **Molecular motors** are the transporters. A molecular motor is a protein that attaches to the particle to be transported, then uses energy harnessed from ATP to "walk" along the microtubule with the particle riding in "piggyback" fashion (*motor* means "movement"). **Kinesin,** one such motor protein, consists of two "feet," a stalk, and

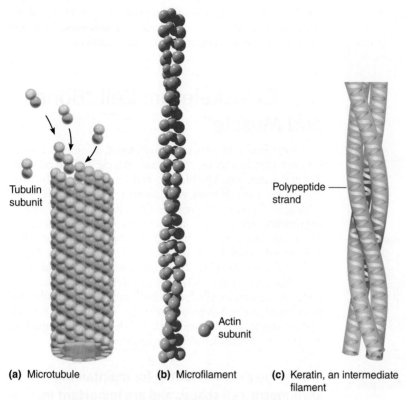

Tubulin subunit

Actin subunit

Polypeptide strand

(a) Microtubule   **(b)** Microfilament   **(c)** Keratin, an intermediate filament

FIGURE 2-26 **Components of the cytoskeleton.** (a) Microtubules, the largest of the cytoskeletal elements, are long, hollow tubes formed by two slightly different variants of globular-shaped tubulin molecules. (b) Most microfilaments, the smallest of the cytoskeletal elements, consist of two chains of actin molecules wrapped around each other. (c) The intermediate filament keratin, found in skin, is made of three polypeptide strands wound around one another. The composition of intermediate filaments, which are intermediate in size between the microtubules and microfilaments, varies among different cell types.

© Cengage Learning, 2013

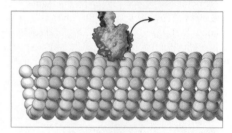

a fanlike tail (Figure 2-27). The tail binds to the secretory vesicle to be moved, and the feet move one at a time, as if walking. They alternately attach to one tubulin molecule on the microtubule, bend and push forward, then let go. During this process, the back foot swings ahead of what was the front foot and then attaches to the next tubulin molecule farther down the microtubule. The process is repeated over and over as kinesin moves its cargo to the plasma membrane by using each of the tubulin molecules as a stepping-stone. You'll see a detailed example for neurons in Chapter 4.

Movement of Cilia and Flagella Microtubules are also the dominant structural and functional components of cilia and flagella. These specialized protrusions from the cell surface allow a cell to move materials across its surface (in the case of a stationary cell) or to propel itself through its environment (in the case of a motile cell or animal). **Cilia** (meaning "eyelashes"; singular, *cilium*) are numerous tiny, hairlike protrusions found in large numbers on the surface of a ciliated cell. **Flagella** (meaning "whips"; singular, *flagellum*) are whiplike appendages; typically, a cell has one or a few flagella at most. Even though they project from the surface of the cell, cilia and flagella are both intracellular structures covered by the plasma membrane.

FIGURE 2-27 **How a kinesin molecule "walks."** A kinesin molecule walks along the surface of a microtubule by alternately attaching and releasing its "feet" as it cyclically swings the rear foot ahead of the front foot.

© Cengage Learning, 2013

Cilia beat or stroke in unison, much like the coordinated efforts of a rowing team. Each cilium exerts a rapid active stroke, which moves material on the cell surface forward. This stroke is followed by a recovery phase in which the cilium more slowly returns to its original position with a kind of unrolling, backward movement that does not exert much force, so there is little pushing in the reverse direction. On a ciliated swimming animal like a ctenophore (comb jelly), hundreds of coordinated cilia push on the external seawater to create propulsion. Mammals have ciliated cells that line the respiratory tract (to help remove foreign particles from the lungs, Figure 2-28) and the oviduct of the female repro-

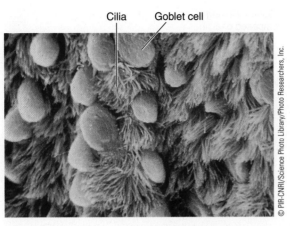

FIGURE 2-28 Cilia in the respiratory tract. Scanning electron micrograph of cilia on cells lining the human respiratory tract. The respiratory airways are lined by goblet cells, which secrete a sticky mucus that traps inspired particles, and epithelial cells that bear numerous hairlike cilia. The cilia all beat in the same direction to sweep inspired particles up and out of the airways.

Cilia   Goblet cell

© PIR-CNRI/Science Photo Library/Photo Researchers, Inc.

FIGURE 2-29 Internal structure of a cilium or flagellum. (a) The relationship between the microtubules and the centriole turned basal body of a cilium or flagellum. (b) Diagram of a cilium or flagellum in cross section showing the characteristic "9 + 2" arrangement of microtubules along with the dynein arms and other accessory proteins that hold the system together. (c) Electron micrograph of a flagellum in cross section; individual tubulin molecules are visible in the microtubule walls. (d) Depiction of bending of a cilium or flagellum caused by microtubule sliding brought about by dynein "walking."

Don Fawcett/Photo Researchers, Inc.

ductive tract. You'll learn more about these in Chapter 11 and about the reproductive tract in Chapter 16.

In addition to the multiple motile cilia found in cells in these specific locations, almost all vertebrate cells possess a single nonmotile *primary cilium*. Until recently, primary cilia were considered useless vestiges, but growing evidence suggests that they act as microscopic sensory organs that sample the extracellular environment. For example, the primary cilium of olfactory neurons contains receptors that detect odors, while the primary cilium of the epithelial cells lining the kidney tubules extends from the apical surface to monitor flow rate through the tubule. They may also be critical for receiving regulatory signals involved in controlling growth, cell differentiation, and cell proliferation (expansion of a given cell type).

The only vertebrate cells that bear flagella are sperm. The whiplike motion of the flagellum or "tail" enables a sperm to move through its environment. This ability is particularly useful when the sperm maneuvers for final penetration of the ovum during fertilization. Flagella are found in a few other animal cell types, such as the *choanocytes* (cells that create feeding current) of sponges that we discuss in Chapter 9.

Cilia and flagella arise from centrioles and consequently have the same basic internal structure. During cellular division they form after a duplicated centriole moves to a position just under the plasma membrane, where microtubules grow outward from the centriole in an orderly pattern to form the motile appendage. The centriole remains at the base of the developed cilium or flagellum and becomes the so-called **basal body**, a short cylinder composed of a parallel microtubular symmetry similar to that of the cilium or flagellum. Those consist of nine fused pairs of microtubules (doublets) arranged in an outer ring around two single unfused microtubules in the center (Figure 2-29). This characteristic

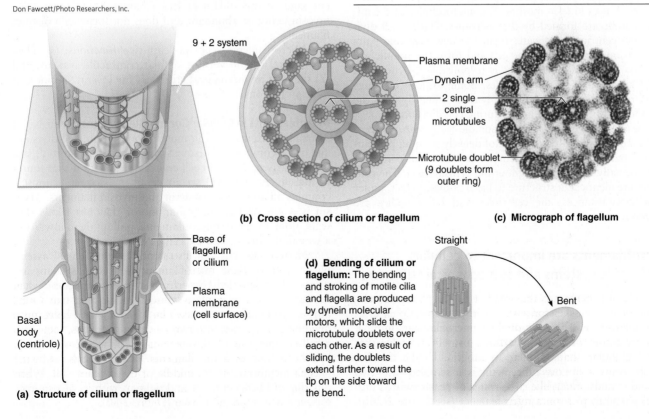

"9 + 2" array of microtubules extends throughout the length of the motile appendage. (In contrast, a primary cilium lacks the central pair of microtubules giving rise to a "9 + 0" array, which does not beat.)

Associated with these microtubules are accessory proteins that maintain the microtubules' organization and play an essential part in the microtubular movement that causes the entire structure to bend. The most important of these accessory proteins is **dynein**, which forms a set of armlike projections from each doublet of microtubules (see Figure 2-29b). The sliding of adjacent microtubule doublets past each other produces the bending movements of cilia and flagella. Sliding is accomplished by the dynein arms, which are motor proteins such as kinesin. The dynein arms have the capability of splitting ATP and then using the released energy to "crawl" along the neighboring microtubule doublet (similar to kinesin movement) to cause relative displacement of the doublets. Groups of cilia working together are oriented to beat in the same direction and contract in a synchronized manner through controlling mechanisms that are poorly understood. These mechanisms appear to involve the single microtubules at the cilium's center and their surrounding accessory proteins, the inner sheath (see Figure 2-29).

Formation of the Mitotic Spindle Cell division involves two discrete but related activities: **mitosis** (nuclear division) and **cytokinesis** (cytoplasmic division). Mitosis involves replication of the DNA-containing chromosomes, which are then evenly distributed in the two halves of the cell. In cytokinesis, the plasma membrane is constricted in the middle of the cell, and the two halves separate into two new daughter cells, each with a full complement of chromosomes.

The replicated chromosomes are pulled apart by a cellular apparatus called the **mitotic spindle**, which is transiently assembled from microtubules only during cell division and which uses kinesin motors. The microtubules of the mitotic spindle are formed by the centrioles. During cell division, the centrioles replicate; then, the new centriole pairs move to opposite ends of the cell and form the spindle apparatus. The spindle apparatus is responsible for the organized assemblage of microtubules. Some anticancer drugs interfere with microtubule assembly that ordinarily pulls the chromosomes to opposite poles during cell division.

Besides their role in mitotic-spindle formation, the centrioles and surrounding complex of densely staining proteinaceous material together assemble the many microtubules that normally radiate throughout the cytoskeleton. The centrioles are identical in structure to basal bodies. In fact, under some circumstances, the centrioles and basal bodies are interconvertible.

Microfilaments are important to cellular contractile systems and as mechanical stiffeners

The **microfilaments** are the smallest (6 nm diameter) elements of the cytoskeleton visible with a conventional electron microscope. The most obvious microfilaments in most cells are those composed of **actin**, a protein molecule that has a globular shape similar to tubulin. Unlike tubulin, which forms a hollow tube, actin is assembled into two twisted strands, much like two strings of pearls twisted into a helix (spiral) to form a microfilament (see Figure 2-26b).

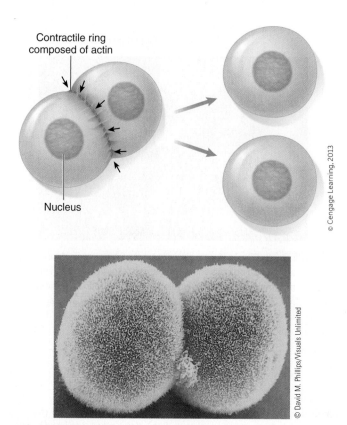

Contractile ring composed of actin

Nucleus

© Cengage Learning, 2013

© David M. Phillips/Visuals Unlimited

FIGURE 2-30 Cytokinesis. Diagram and micrograph of a cell undergoing cytokinesis, in which a contractile ring composed of actin filaments tightens, squeezing apart the two duplicate cell halves formed by mitosis.

In muscle cells, another protein called *myosin* forms a different kind of microfilament (see Chapter 8). In most cells, myosin is not as abundant and does not form such distinct filaments.

Microfilaments serve at least two functions: (1) They play a vital role in various cellular contractile systems, and (2) they act as mechanical stiffeners for several specific cellular projections.

Microfilaments in Cellular Contractile Systems Actin-based assemblies are involved in muscle contraction, cell division, and cell locomotion. The most obvious, best organized, and most clearly understood cell contractile system is that found in muscle. As you will see in detail in Chapter 8, muscle contains an abundance of actin and myosin filaments. **Myosin** is a kinesin-like molecular motor that has heads that walk along the actin microfilaments, pulling them inward in a way that causes a muscle cell to contract.

Nonmuscle cells can also contain "musclelike" assemblies. Some of these microfilament contractile systems are transiently assembled to perform a specific function when needed. A good example is the contractile ring that forms during **cytokinesis,** the process by which the two halves of a dividing cell separate into two new daughter cells, each with a full complement of chromosomes. The ring consists of a beltlike bundle of actin filaments located just beneath the plasma membrane in the middle of the dividing cell. When this ring of fibers contracts and tightens with its myosin motors, it pinches the cell in two (Figure 2-30).

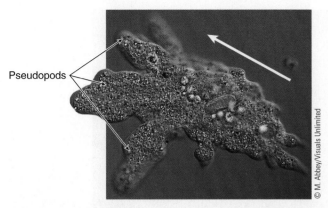

FIGURE 2-31 **An amoeba undergoing amoeboid movement.**

Pseudopods

© M. Abbey/Visuals Unlimited

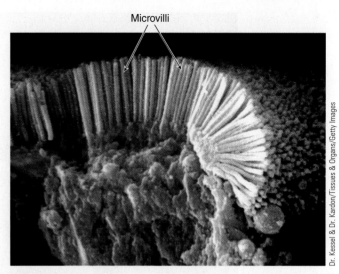

Microvilli

Dr. Kessel & Dr. Kardon/Tissues & Organs/Getty Images

FIGURE 2-32 **Microvilli in the small intestine.** Scanning electron micrograph showing microvilli on the surface of a small-intestine epithelial cell.

Complex actin-based assemblies are also responsible for most cell locomotion. Four types of adult mammalian cells are capable of moving on their own—sperm, white blood cells, fibroblasts, and skin cells. In nonvertebrates, amoebocytes—immune cells—crawl similarly. Sperm move by the flagellar mechanism already described. Motility for the other cells is accomplished by **amoeboid movement,** a cell-crawling process that depends on the activity of their actin filaments, in a mechanism similar to that used by amoebae to maneuver through their environment. When crawling, the motile cell forms *pseudopods* at the "front" or leading edge of the cell in the direction of the target (Figure 2-31). For example, the target that triggers amoeboid movement might be the proximity of food in the case of an amoeba, or a bacterium in the case of a white blood cell or amoebocyte. Pseudopods are formed as a result of the organized assembly and disassembly of branching actin networks. During amoeboid movement, actin filaments continuously grow at the cell's leading edge through the addition of actin molecules at the front of the actin chain. This filament growth protrudes that portion of the cell forward as a pseudopod. Simultaneously, actin molecules at the rear of the filament are being disassembled and transferred to the front of the line. Thus, the filament does not get any longer; it stays the same length but moves forward through the continuous transfer of actin molecules from the rear to the front of the filament in what is termed *treadmilling* fashion. The cell attaches the pseudopod to the surrounding connective tissue and at the same time detaches from its older adhesion site at the rear. The cell uses the new adhesion site at the leading edge as an anchor, then pulls the bulk of its body forward through cytoskeletal contraction using myosin motors.

Microfilaments as Mechanical Stiffeners Besides their role in cell contractile systems, the actin filaments' second major function is to serve as mechanical supports or stiffeners for several cellular extensions, of which the most common are microvilli. **Microvilli** are microscopic, nonmotile, hairlike projections from the surface of many epithelial cells linings, such as the vertebrate small intestine and kidney tubules. Within each microvillus, a core consisting of parallel actin filaments linked together forms a rigid mechanical stiffener that keeps these valuable surface projections intact. A single small-intestinal cell may have several thousand microvilli, which are packed together like the bristles of a brush, pro-

jecting from the cell's free surface (Figure 2-32). Their presence greatly increases the surface area available for transferring material across the plasma membrane, such as for absorbing digested nutrients in an intestine or recovering useful substances passing through the kidney. You will see more important examples in the sensory systems of fish and the mammalian inner ear in Chapter 6.

Both microtubules and microfilaments form both stable structures, such as microvilli, and transient structures, such as mitotic spindles and contractile rings, as the need arises. Pools of unassembled tubulin and actin subunits in the cytosol can be rapidly assembled into organized structures to perform specific activities, then disassembled when they are no longer needed.

Intermediate filaments are important in regions of the cell subject to mechanical stress

The **intermediate filaments** are intermediate in size between the microtubules and the microfilaments (7 to 11 nm in diameter)—hence their name. The proteins that compose the intermediate filaments vary between cell types, but in general they appear as irregular, threadlike molecules. These proteins form tough, durable fibers that play a central role in maintaining the structural integrity of a cell and in resisting mechanical stresses externally applied to a cell. In contrast to the other cytoskeletal elements, the intermediate filaments are highly stable structures. No evidence exists for a reversible pool between unassembled and assembled intermediate-filament proteins. The different types of intermediate filaments are tailored to suit their structural or tension-bearing role in specific cell types. In general, only one class of intermediate filament is found in a particular cell type. Several important examples follow:

- Neurofilaments are intermediate filaments in nerve cell axons (see Chapter 4). Together with microtubules, neurofilaments strengthen and stabilize these elongated cellular extensions.

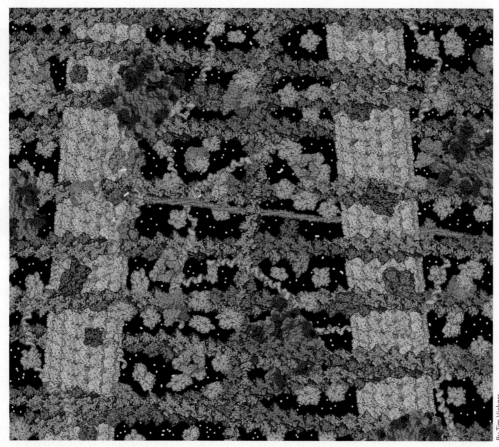

By Tim Vickers

FIGURE 2-33 The crowded cytosol. Diagram of cell interior with microtubules (*light blue*), actin filaments (*dark blue*), ribosomes (*orange* and *purple*), soluble proteins (*light blue*), kinesin (*red*), small molecules (*white*) and RNA (*pink*).

- Intermediate filaments in skeletal muscle cells hold the actin–myosin contractile units in proper alignment (Chapter 8).
- Skin cells contain irregular networks of intermediate filaments made of the protein **keratin** (see Figure 2-26c). These intracellular filaments interconnect with extracellular filaments that tie adjacent cells together, creating a continuous filamentous network that extends throughout the skin and gives it strength. When the surface skin cells die, their tough keratin skeletons persist to form a protective waterproof outer layer. Hair and nails are also keratin-based structures.

Emphasizing the importance of intermediate filaments in some specialized cell types, intermediate filaments account for up to 85% of the total protein in nerve cells and keratin-producing skin cells, whereas these filaments constitute only about 1% of other cells' total protein on average.

The cytoskeleton functions as an integrated whole, links other parts of the cell, and contributes to cytosolic "crowding"

Collectively, the cytoskeletal elements and their interconnections support the plasma membrane and are responsible for the particular shape, rigidity, and spatial geometry of each different cell type. Furthermore, growing evidence suggests that the cytoskeleton serves as a lattice to organize groups of enzymes for many cellular activities. This internal framework thus acts as the cell's "skeleton" (Figure 2-33; see also Figure 2-5). New studies hint that the cytoskeleton as a whole is not merely a supporting structure that maintains the tensional integrity of the cell but may serve as a mechanical communications system as well. Various components of the cytoskeleton behave as if they were structurally connected or "hard-wired" to each other as well as to the surface plasma membrane and to the nucleus. This force-carrying network may serve as a mechanism by which mechanical forces acting on the cell surface reach all the way from the plasma membrane through the cytoskeleton to ultimately influence gene regulation in the nucleus. Furthermore, as you have learned, the coordinated action of the cytoskeletal elements directs intracellular transport and regulates numerous cellular movements and thereby also serves as the cell's "muscle."

As we finish this look at the major components of eukaryotic cells, you may wonder how all these structures fit within a cell. In fact, the total number of organelles along with macromolecular complexes do fit, but they create a crowded cytosol. These structures are estimated to take up to 40% of the physical space inside a cell (Figure 2-33). Such crowding is predicted to slow down diffusion rates considerably, much as a person will take longer to move through a crowd than across an empty field. Conversely, crowding is predicted to increase the rates of chemical reactions between large macromolecules in the same region, such as transcription regulating proteins binding to DNA in the nucleus. The overall importance of these effects is currently being investigated and is likely to be of even greater importance in dehydration stresses, such as the cells of an animal losing too much water would suffer.

check your understanding 2.11, 2.12

List and describe the functions of microtubules and microfilaments.

Which of these is responsible for maintaining the cell in an asymmetric shape, such as that of a nerve cell?

2.13 Cell-to-Cell Adhesions

In multicellular organisms, plasma membranes not only serve as the outer boundaries of all cells but also participate in cell-to-cell adhesions, allowing groups of cells to bind together into tissues and to be packaged further into organs. Organization of cells into appropriate groupings may be at least partially attributable to the carbohydrate chains on the membrane surface. Once arranged, cells are held together by three different means: (1) *cell adhesion molecules (CAMs)* in the cells' plasma membranes, (2) the *extracellular matrix*, and (3) specialized *cell junctions*.

The extracellular matrix serves as the biological "glue"

Many cells within a tissue are not in direct physical contact with neighboring cells. Instead, they are held together by the **extracellular matrix (ECM)**, an intricate meshwork of fibrous proteins embedded in a watery, gel-like substance composed of complex carbohydrates. The watery gel, an ECF component usually called the interstitial fluid, provides a pathway for diffusion of nutrients, wastes, and other water-soluble traffic between the blood and tissue cells. Interwoven within this gel are three major types of protein fibers:

1. **Collagen** forms cablelike fibers or sheets that provide tensile strength (resistance to longitudinal stress). Collagen is the most abundant protein in mammals, making up nearly half of the total body protein by weight. In *scurvy*, a human condition caused by vitamin C deficiency (due to a human pseudogene, p. 33), these fibers are not properly formed. As a result, the tissues, especially those of the skin and blood vessels, become very fragile. This leads to bleeding in the skin and mucous membranes, which is especially noticeable in the gums.
2. **Elastin** is a rubberlike protein fiber most abundant in tissues that must be capable of easily stretching and then recoiling after the stretching force is removed. It is found, for example, in mammalian lungs, which stretch and recoil as air moves in and out.
3. **Fibronectin** promotes cell adhesion and holds cells in position. Researchers have found reduced amounts of

this protein within certain types of cancerous tissue, possibly explaining why cancer cells do not adhere well to each other but tend to break loose and metastasize (spread elsewhere in the body).

Local cells, most commonly **fibroblasts** ("fiber formers") present in the matrix, secrete the extracellular matrix. Often the matrix and the cells within it are known collectively as **connective tissue** because they connect cells together into tissues and tissues into organs. The exact composition of extracellular matrix components varies for different tissues, providing distinct local environments for the various cell types in the body. In some tissues, the matrix becomes highly specialized to form such structures as cartilage or tendons or, on appropriate calcification, the hardened structures of bones and teeth.

Contrary to long-held belief, the ECM is not just a passive scaffolding for cellular attachment but also helps regulate the behavior and functions of the cells with which it interacts. Cells are able to function normally and indeed even to survive only when associated with their normal matrix components. The matrix is especially influential in cell growth and differentiation. In a vertebrate's body, only circulating blood cells are designed to survive and function without attaching to the ECM.

Some cells are directly linked together by specialized cell junctions

In tissues where the cells lie in close proximity to each other, CAMs provide some tissue cohesion as they "Velcro" adjacent cells to each other. In addition, some cells within given types of tissues are directly linked by one of three types of specialized cell junctions: (1) *desmosomes* (adhering junctions), (2) *tight junctions* (impermeable junctions), or (3) *gap junctions* (communicating junctions).

Desmosomes Acting like "spot welds" that anchor together two closely adjacent but nontouching cells, **desmosomes** consists of two components: (1) a pair of dense, buttonlike cytoplasmic thickenings known as *plaque* located on the inner surface of each of the two adjacent cells; and (2) strong glycoprotein filaments containing *cadherins* (a type of CAM) that extend across the space between the two cells and attach to the plaque on both sides (Figure 2-34). These intercellular filaments bind adjacent plasma membranes together so they resist being pulled apart. Thus, desmosomes are adhering junctions. They are the strongest cell-to-cell connections.

Desmosomes are distributed widely throughout animal bodies. Desmosomes are most abundant in tissues that are subject to considerable stretching, such as the skin, heart, and the uterus. In these tissues, functional groups of cells are riveted together by desmosomes. Furthermore, intermediate cytoskeletal filaments, such as tough keratin filaments in the skin, stretch across the interior of these cells and attach to the desmosome plaques located on opposite sides of the cell's inner surface. This arrangement forms a continuous network of strong fibers throughout the tissue, both through and between the cells, much like a continuous line of people firmly holding hands. This interlinking fibrous network provides tensile strength, reducing the chances of the tissue being torn when stretched.

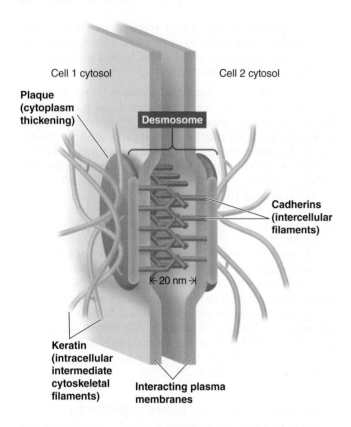

FIGURE 2-34 Desmosome. Desmosomes are adhering junctions that spot-rivet cells, anchoring them together in tissues subject to considerable stretching.

© Cengage Learning, 2013

Tight Junctions Tight junctions join sheets of epithelial tissue. Epithelial tissue covers or produces the surface of animal bodies and lines all their internal cavities. Usually these sheets serve as highly selective barriers between two compartments that have considerably different chemical compositions. For example, the epithelial sheet lining a digestive tract separates the food and potent digestive juices within the inner cavity (lumen) from the ECF that lies on the other side. To accomplish this, the lateral (side) edges of the adjacent cells in the epithelial sheet are joined together in a tight seal near their luminal border by "kiss sites," which are sites at which membrane *junctional proteins* from two adjacent cells fuse directly (Figure 2-35). These tight junctions are impermeable and thus prevent materials from passing between the cells. Passage across the epithelial barrier, therefore, must take place through the cells, not between them. This *transcellular* ("through the cell") traffic is regulated by means of the channels and carrier proteins present in the plasma membrane. If the cells were not joined by tight junctions, uncontrolled exchange of molecules could take place between the compartments by unregulated traffic through the spaces between adjacent cells (*paracellular* pathway). Thus, tight junctions prevent undesirable leaks within epithelial sheets. (See Chapters 12 and 14 for specific details on epithelial transport in kidneys and digestive tracts.)

Gap Junctions At a **gap junction,** a gap exists between adjacent cells that are linked by small connecting tunnels known as *connexons* made of *connexin* proteins in vertebrates and

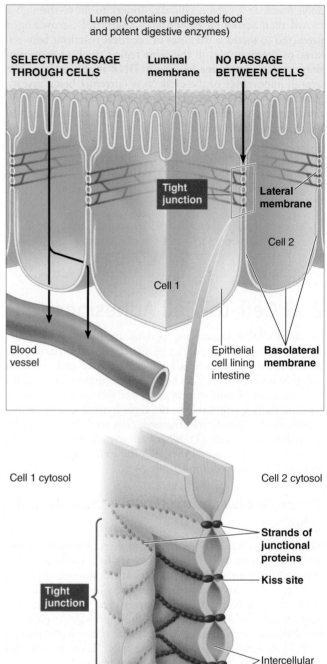

FIGURE 2-35 Tight junction. Tight junctions are impermeable junctions that join the lateral edges of epithelial cells near their luminal borders, thus preventing movement of materials between the cells. Only regulated passage of materials can occur through these cells, which form highly selective barriers that separate two compartments of highly different chemical composition.

© Cengage Learning, 2013

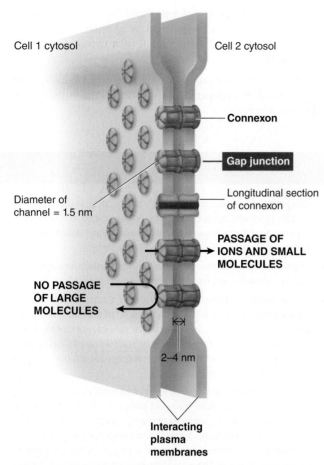

Cell 1 cytosol   Cell 2 cytosol

Connexon

Gap junction

Longitudinal section
of connexon

Diameter of
channel = 1.5 nm

**PASSAGE OF
IONS AND SMALL
MOLECULES**

**NO PASSAGE
OF LARGE
MOLECULES**

2–4 nm

**Interacting
plasma
membranes**

FIGURE 2-36 Gap junction. Gap junctions are communicating junctions made up of connexons, which form tunnels that permit movement of charge-carrying ions and other small molecules between two adjacent cells.

© Cengage Learning, 2013

innexin proteins in other animals (e.g., arthropods, nematodes). A **connexon** is made up of six protein subunits arranged in a hollow tubelike structure. Two connexons, one from each of the plasma membranes of two adjacent cells, extend outward and join end-to-end to form a connecting tunnel between the two cells (Figure 2-36). The small diameter of the tunnels permits small water-soluble particles such as ions (electrically charged particles) to pass between the connected cells but precludes passage of large molecules such as proteins. Gap junctions are found in most organs, but are perhaps best studied in heart muscle and smooth muscle. Movement of ions between cells through gap junctions plays an important role in transmitting electrical activity throughout an entire muscle mass. Because this electrical activity brings about contraction, the presence of gap junctions enables synchronized contraction of a whole muscle mass, such as the heart (see Chapter 9). Some nerve cells are also connected in this way. For example, crayfish escape attack by rapidly flipping their tails. Gap junctions allow for rapid conduction of the electrical signal (see Chapter 4).

Many tissues that are not electrically active also are connected by gap junctions, where they permit unrestricted passage of small nutrient molecules between cells. For example,

glucose, amino acids, and other nutrients pass through gap junctions to a developing egg cell from surrounding cells within the mammalian ovary, thus helping the egg stockpile these essential nutrients. Furthermore, these gap junctions also serve as avenues for the direct transfer of small signaling molecules from one cell to the next. Such transfer permits the cells connected by gap junctions to directly communicate with each other. This communication provides one possible mechanism by which cooperative cell activity may be coordinated. In the next chapter, we'll describe another recently discovered tunnel-like communication device called a *nanotube*.

The cytosol's surrounding barrier, the plasma membrane, is intimately involved in desmosome, tight and gap junctions, and nanotubes. The properties of this major structure are explored in the next chapter, including other ways by which cells "talk to each other."

check your understanding 2.13

How do cells that are subject to considerable mechanical stress resist being pulled apart?

Through what structure can the electrical activity in one cell be transmitted to another? Explain.

making connections

How Cellular and Molecular Physiology Contribute to the Body as a Whole

The ability of cells to perform functions essential for their own survival as well as specialized tasks that contribute to homeostasis and regulated changes within an animal ultimately depends on the successful and cooperative operation of the intracellular components. For example, genes must be regulated at the right time and place to construct cell structures. Also, to support life-sustaining activities, all cells must convert energy into a usable form from nutrient molecules. Energy is converted intracellularly by chemical reactions that take place within the cytosol and mitochondria.

In addition to being essential for basic cell survival, the organelles and cytoskeleton also participate in many cells' specialized tasks that contribute to homeostasis. Here are several examples:

- Nerve and endocrine cells both release chemical messengers that are important in regulatory activities aimed at maintaining homeostasis—for example, chemical messengers released from nerve cells stimulate the respiratory muscles, which accomplish life-sustaining exchanges of O_2 and CO_2 between the body and environment through breathing. These chemical messengers (neurotransmitters in nerve cells and hormones in endocrine cells) are all produced by the endoplasmic reticulum and Golgi complex and released by exocytosis from cells when needed.

- The ability of muscle cells to contract depends on their highly developed cytoskeletal microfilaments sliding past each other. Muscle contraction is responsible for many homeostatic activities, including (1) contraction of heart muscle; (2) contraction of the muscles attached to skeletal elements, which enables behaviors, such as procuring food, fleeing predators, and mating; and (3) contraction of the muscle in the walls of digestive tracts, which moves the food along the tracts.

- White blood cells and amoebocytes help animal bodies resist infection by making extensive use of lysosomal destruction of engulfed particles as they police the body for microbial invaders. These cells are able to roam the

body fluids by means of amoeboid movement, a cell-crawling process accomplished by alternate assembly and disassembly of actin, one of their cytoskeletal components.

As we begin to examine the various organs and systems, keep in mind that proper cell functioning is the foundation of all organ activities.

Chapter Summary

- Water, other inorganic chemicals, carbohydrates, lipids, proteins, and nucleic acids are the universal components of cells.

- Many macromolecular structures need to be flexible to function and to be regulated, and protecting those structures is the basis for many forms of homeostasis.

- Prokaryotic (archaeal, bacterial) cells have a simpler organization than eukaryotes as exemplified by the lack of a separate nucleus and most organelles. Eukaryotic cells are subdivided into the plasma membrane, nucleus, and cytoplasm with organelles.

- DNA contains codes in the form of genes for making proteins through the processes of transcription and translation with different genes expressed in different tissues and organs. Individual genes are regulated by promoters and transcription factors, whereas in different tissues transcription is regulated by transcription factors expressed at different stages of development. Driving much of the growth in knowledge about molecular biology are the genome projects, which utilize a variety of experimental procedures to elucidate the connections between the gene and the cell.

- The rough endoplasmic reticulum synthesizes proteins for secretion and membrane construction, whereas the smooth endoplasmic reticulum packages new proteins in transport vesicles. Transport vesicles carry their cargo to the Golgi complex for further processing. Once there, the Golgi complex packages secretory vesicles for release by exocytosis.

- Lysosomes digest extracellular material brought into the cell by phagocytosis, whereas proteasomes destroy internal proteins, that is, intracellular proteins that have outlived their usefulness.

- Mitochondria are the energy organelles or "power plants" of the cell; using oxygen, they extract energy from the nutrients in food and transform it into a usable form—primarily ATP—to support cell activities. However, leakage of some of the electrons associated with mitochondrial metabolism can potentially damage the tissues of the host animal and contribute to the aging process.

- Oxygen deficiency forces cells to rely on glycolysis and other anaerobic reactions, producing lactate, propionic acid, octopine, or other end products. However, tolerance of oxygen deficiency varies widely among tissues as well as different organisms.

- The cytosol is important in intermediary metabolism, ribosomal protein synthesis, and storage of fat and glycogen.

- Permeating the cytosol is the cytoskeleton, a complex protein network that acts as the "bone and muscle" of the cell. The distinct shape, size, complexity, and intracellular specialization of the various body cells necessitate intracellular scaffolding to support and organize the cell components into an appropriate arrangement and to control their movements.

- The microtubules, the largest of the cytoskeletal elements, are important for intracellular transport and cilia and flagella. Microfilaments are the smallest element of the cytoskeleton and are important to cellular contractile systems and as mechanical stiffeners. Intermediate filaments are important in regions of the cell subject to mechanical stress. The cytoskeleton functions as an integrated whole and links other parts of the cell, possibly as a lattice to organize groups of enzymes for many cellular activities.

- The extracellular matrix serves as the biological "glue." It is an intricate meshwork of fibrous proteins embedded in the interstitial fluid. Interwoven within this gel are three major types of protein fibers: collagen, elastin, and fibronectin.

- Some cells are directly linked together by specialized cell junctions. Desmosomes act like "spot welds" that anchor together two closely adjacent but nontouching cells; tight junctions join sheets of epithelial tissue and prevent undesirable leaks from one compartment to another; and gap junctions allow small molecules to pass from one cell to another.

Review, Synthesize, and Analyze

1. What organelles serve as the intracellular digestive system? What type of enzymes do they contain? What functions do these organelles serve?

2. How are free radicals important in the aging process? For species with comparative longevity, what is the mechanism by which they are able to overcome the degenerative processes associated with free radicals?

3. Distinguish among *cellular respiration, oxidative phosphorylation,* and *chemiosmosis.*

4. On an expedition to survey life in deep ocean waters in the Mediterranean Sea, you recover a Loriciferan from 350 km below the surface in the hostile L'Atalante basin. As a biologist onboard the vessel, how would you determine if the animals were really surviving in such an anaerobic environment rather than simply drifting down from elsewhere and dying?

5. Is iPSC technology the long sought after "fountain of youth"? Speculate on a mechanism by which the expression of embryonic genes can reprogram a differentiated cell.

6. You have been provided a strand of hair from a wooly mammoth. How would you determine its approximate age? How might you try to resurrect a living mammoth?

7. Oxygen can be considered a double-edged sword. Although essential for life, it also generates free radicals that contribute to the aging process. What other elements could alternatively be substituted for oxygen as a final electron acceptor? Specify any advantages/disadvantages associated with its use.

Suggested Readings

Alberts, B., A. Johnson, J. Lewis, M. Raff, K. Roberts, & P. Walter. (2007). *Molecular Biology of the Cell*. New York: Garland. A best-selling textbook on this topic.

Ellis, R. J., & A. P. Minton. (2003). Join the crowd. *Nature* 425:27–28. A science news story on crowding in the cell cytosol.

Hochedlinger, K. (2010). Your inner healers. *Scientific American* 302: 46–53. A review of the methodologies by which stem cells are generated.

Marshall, J.M., & C.E. Taylor. (2009). Malaria control with transgenic mosquitoes. *PLoS Medicine* 6: e1000020, found at www.plosmedicine .org/article/info%3Adoi%2F10.1371%2Fjournal.pmed.1000020

Mattick, J. S. (2003). Challenging the dogma: The hidden layer of non–protein-coding RNAs in complex organisms. *BioEssays* 25:930–939. An update on noncoding RNAs.

Moyes, C. D., & D. A. Hood (2003). Origins and consequences of mitochondrial variation in vertebrate muscle. *Annual Review of Physiology* 65:177–201.

Wells, W. A. (2001). Containing the bullet. *Journal of Cell Biology* 154:13. Available online at www.jcb.org/cgi/content/full/154/1/13. A review of nematocyst physiology.

Wilmut, I. (1998). Cloning for medicine. *Scientific American* 279:59–63. The process by which clones are generated.

CourseMate

3

Mark Conlin / Alamy

Membrane Physiology

3.1 Membrane Structure and Composition

The survival of every animal cell depends on the maintenance of an intracellular content unique for that cell type despite the remarkably different composition of the extracellular fluid surrounding it. This difference in fluid composition inside and outside a cell is maintained by the **plasma membrane,** an extremely thin layer of lipids and proteins that forms the outer boundary of every cell and encloses the intracellular contents. In addition to serving as a mechanical barrier that traps needed molecules within the cell, the plasma membrane plays an active role in determining the composition of the cell by selectively permitting specific substances to pass between the cell and its environment. Besides controlling the entry of nutrient molecules and the exit of waste products, the plasma membrane maintains differences in ion concentrations between the cell's interior and exterior. These ionic differences, as you will learn, are important in the electrical

activity of cells. Also, the plasma membrane plays a key role in the ability of a cell to respond to changes, or signals, in the cell's environment. No matter what the cell type, these common functions accomplished by the plasma membrane are crucial to the cell's survival and to the cell's ability to perform specialized homeostatic activities.

In this chapter, we examine the common structural and functional patterns shared by plasma membranes of all cells. We also explore how many of the functional differences between cell types are due to subtle variations in the composition of their plasma membranes. For example, modifications in specific protein components of the plasma membranes of various cell types enable different cells to interact in different ways with essentially the same extracellular fluid environment. To illustrate, thyroid gland cells are the only cells in a vertebrate to use iodine. Appropriately, a membrane protein unique to the plasma membranes of thyroid gland cells permits these cells (and no others) to take up iodine from the blood.

The plasma membrane is a fluid lipid bilayer embedded with proteins

The plasma membrane is too thin to be seen under an ordinary light microscope, but with an electron microscope it appears as a **trilaminar structure** (three-layered; *lamina* means "layer") consisting of two dark layers separated by a light middle layer (Figure 3-1). The specific arrangement of the molecules that make up the plasma membrane is responsible for this "sandwich" appearance.

All plasma membranes consist mostly of lipids (fats) and proteins plus small amounts of carbohydrates. The most abundant membrane lipids are phospholipids, with lesser amounts of cholesterol. **Phospholipids** have a *polar* (electrically charged) head containing a negatively charged phosphate group and two *nonpolar* (electrically neutral) fatty acid tails (Figure 3-2a). The polar end is *hydrophilic* ("water loving")

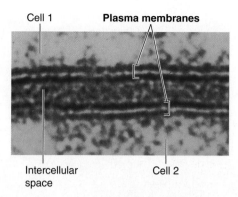

FIGURE 3-1 **Trilaminar appearance of a plasma membrane in an electron micrograph.** Depicted are the plasma membranes of two adjacent cells. Note that each membrane appears as two dark layers separated by a light middle layer.

© Don W. Fawcett/Visuals Unlimited

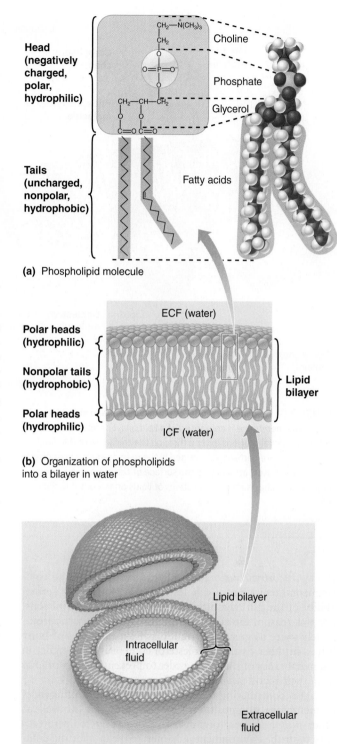

(a) Phospholipid molecule

(b) Organization of phospholipids into a bilayer in water

(c) Separation of ECF and ICF by the lipid bilayer

FIGURE 3-2 **Structure and organization of phospholipid molecules in a lipid bilayer.** (a) Phospholipid molecule. (b) In water, phospholipid molecules organize themselves into a lipid bilayer with the polar heads interacting with the polar water molecules at each surface and the nonpolar tails all facing the interior of the bilayer. (c) An exaggerated view of the plasma membrane enclosing a cell, separating the ICF from the ECF.

© Cengage Learning, 2013

because it can interact with water molecules, which are also polar; the nonpolar end is *hydrophobic* ("water fearing") and will not mix with water. Such two-sided molecules self-assemble into a **lipid bilayer**, a double layer of lipid molecules, when in contact with water (Figure 3-2b). The hydrophobic tails bury themselves in the center away from the water, whereas the hydrophilic heads line up on both sides in contact with the water. The outer surface of the bilayer is exposed to extracellular fluid (ECF), whereas the inner surface is in contact with the intracellular fluid (ICF) (Figure 3-2c). (Do not confuse the bilayer with the trilaminar appearance it gives!)

This lipid bilayer is not a rigid structure at normal body temperatures but instead is fluid, with a consistency more like liquid cooking oil than like solid shortening. The phospholipids, which are not held together by chemical bonds, are constantly moving. They can twirl, vibrate, and move around within their own half of the bilayer, exchanging places millions of times a second. This phospholipid movement accounts in large part for membrane *fluidity*. A certain degree of fluidity is required, for example, for the proper action of transport and channel proteins (p. 73), and for cell shape changes. Conversely, an overly fluid membrane is too leaky and can harm or kill cells. We'll explore fluidity in more detail shortly.

Cholesterol contributes to both the fluidity and the stability of the membrane (Figure 3-3). By being tucked in between the phospholipid molecules, cholesterol molecules prevent the fatty acid chains from packing together and crystallizing, a process that would drastically reduce membrane fluidity. Through their spatial relationship with phospholipid molecules, cholesterol molecules also help stabilize the phospholipids' position. Cholesterol also affects the functional attributes of cell membranes such as the activities of various membrane proteins. In animals exposed to varied environmental temperatures, cholesterol provides some rigidity to the membrane and can counter some of the temperature-induced perturbations in membrane fluidity (which can be detrimental at high temperatures). For example, concentrations of cholesterol in the plasma membranes of rainbow

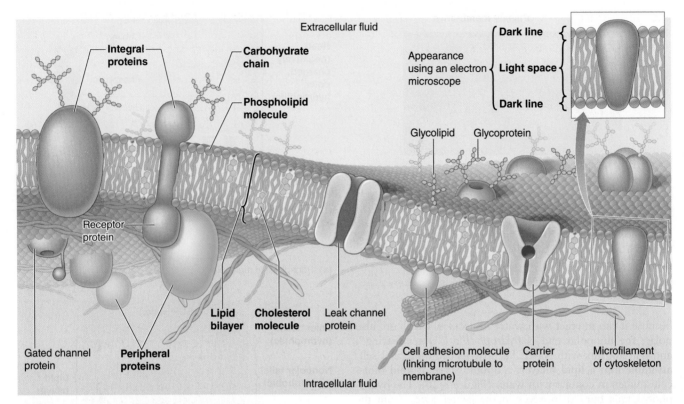

Extracellular fluid

Integral
proteins

Carbohydrate
chain

Phospholipid
molecule

Appearance
using an electron
microscope

Dark line

Light space

Dark line

Glycolipid   Glycoprotein

Receptor
protein

Gated channel
protein

Peripheral
proteins

Lipid
bilayer

Cholesterol
molecule

Leak channel
protein

Cell adhesion molecule
(linking microtubule to
membrane)

Carrier
protein

Microfilament
of cytoskeleton

Intracellular fluid

FIGURE 3-3 Fluid mosaic model of plasma membrane structure.
The plasma membrane is composed of a lipid bilayer embedded with
proteins. Integral proteins extend through the thickness of the
membrane or are partially submerged in the membrane, and peripheral
proteins are loosely attached to the surface of the membrane. Short
carbohydrate chains attach to proteins or lipids on the outer surface
only.

© Cengage Learning, 2013

trout (*Oncorhynchus mykiss*) increase with increasing environmental temperature. Cholesterol also reduces the permeability of the gill membrane to water and the flow of ions. For this reason, the concentration of cholesterol is approximately twice as great in the gill plasma membranes of bony fish as in other tissues. Cholesterol content is even higher in gill membranes of sharks, in order to prevent the loss of *urea* from their blood (see Chapter 13).

Also contributing to membrane fluidity are differences in membrane composition. For example, fish, as well as other *poikilotherms* (organisms whose body temperatures vary with the environment; Chapter 15), can alter the composition of their lipids to synthesize membranes with physical properties that let them adapt and function at low ambient temperatures. Cold temperatures cause the membrane lipids to become more rigid (like butter placed in a refrigerator). For this reason, membrane lipids of animals adapted to cold environments are enriched in **polyunsaturated fatty acids (PUFAs)** at the expense of a reduction in **saturated fatty acids (SFAs)**. The addition of double bonds (points of unsaturation) into the fatty acid molecule results in chains that do not stack together well and so do not become too rigid in the cold, thus offsetting the effects of low temperatures on fluidity. There are two distinctly different groups of PUFAs: the omega-6 series, in which the first double bond is

found between the sixth and seventh carbon atoms in the fatty acid chain, and the omega-3 series, in which the first double bond is located between the third and fourth carbon atoms. For every 20°C decrease in ambient temperature, there is a 1.3- to 1.4-fold increase in the ratio of UFAs to SFAs. Characteristics of lipids containing PUFAs include a reduction in membrane order because of their more expanded conformation. Under frigid conditions these lipids pack less efficiently. For example, brain fatty acids of Antarctic fish are enriched in 24-carbon PUFAs. The proportion of PUFAs in membranes of cold-adapted marine fish may be as high as 42%, compared to 33% in species inhabiting more temperate waters. (See Chapter 15 for more on temperature and membranes.)

The most prominent long-chain omega-3 PUFAs in fish oils are **eicosapentaenoic acid (EPA)**, which is 22 carbons long, with five double bonds (20:5, omega-3) and **docosahexaenoic acid (DHA)**, which is 22 carbons long, with six double bonds (22:6, omega-3). DHA is also important for mammalian brain development, perhaps because it increases membrane fluidity at *synapses* (neural junctions), aiding neurotransmitter release (see pp. 131–132). The highest amounts of EPA and DHA are in fatty fish such as mackerel, herring, and salmon; white fish such as cod and flounder contain only small amounts of omega-3 PUFAs.

Attached to or inserted within the lipid bilayer are the **membrane proteins** (Figure 3-3). **Integral proteins** are embedded in the lipid bilayer; if they extend through the entire thickness of the membrane, they are called **transmembrane proteins** (*trans* means "across"). Like phospholipids, integral proteins also have hydrophilic and hydrophobic regions. **Peripheral proteins** are usually polar molecules that do not penetrate the membrane. They stud only the outer or inner surface, anchored by relatively weak chemical bonds with the polar parts of integral membrane proteins or membrane lipids. The plasma membrane has about 50 times more lipid molecules than protein molecules. However, proteins account for nearly half of the membrane's mass because they are much larger than the lipids. The fluidity of the lipid bilayer enables many membrane proteins to float freely like "icebergs" in a moving "sea" of lipid, although the cytoskeleton restricts the mobility of proteins that perform a specialized function in a specific area of the cell. This view of membrane structure is known as the **fluid mosaic model**, in reference to the membrane fluidity and the ever-changing mosaic pattern of the proteins embedded within the lipid bilayer.

A small amount of **membrane carbohydrate** is located only at the outer surface. Short-chain carbohydrates protrude like tiny antennas from the outer surface, bound primarily to membrane proteins and to a lesser extent to lipids. These sugary combinations are known as *glycoproteins* and *glycolipids,* respectively (Figure 3-3), and the "sugar coating" they form is called the *glycocalyx* (*glyco* means "sweet"; *calyx* means "husk").

The different components of the plasma membrane carry out a variety of functions. As a quick preview, the lipid bilayer forms the primary barrier to diffusion, the proteins perform most of the specific membrane functions, and the carbohydrates play an important role in "self"-recognition processes and cell-to-cell interactions.

The mosaic model and membrane-skeleton fence model describe membrane structure and function

The fluid mosaic model can account for the trilaminar appearance of the plasma membrane. When researchers use stains to help them visualize the plasma membrane under an electron microscope, the two dark lines represent the hydrophilic polar regions of the lipid and protein molecules that have taken up the stain. The light space between corresponds to the poorly stained hydrophobic core formed by the nonpolar regions of these molecules.

Even though the outer and inner layers have the same appearance when viewed with an electron microscope, our description makes clear that the plasma membrane actually is asymmetric; that is, the surface of the membrane facing the extracellular fluid is strikingly different from the surface facing the cytoplasm. Carbohydrate is located only on the outer surface; the outer and inner surfaces bear different amounts and types of protein; and even the lipid composition of the outer half of the bilayer varies somewhat from the inner half. This distinct sidedness of the membrane is related to the different functions carried out at the outer and inner surfaces.

Furthermore, considerable heterogeneity exists within each half of the membrane. For example, recent information suggests that not all membrane proteins are as freely mobile and uniformly distributed as proposed by the fluid mosaic model. To explain this phenomenon, researchers proposed the **membrane-skeleton fence model**, in which the mobility of proteins that perform a specialized function in a specific area of the membrane is restricted by several means. Whereas some proteins can drift randomly within the membrane (similar to a dog that freely roams the neighborhood), scientists think other proteins are tethered to the cytoskeleton (see p. 64) (much like a dog restricted by a leash). These so-called *nondiffusing* or *immobile* proteins constitute a sizable proportion of the total membrane fraction. Still other proteins appear restricted to **confinement zones** by fine cytoskeletal *meshwork fences* within the membrane (similar to a dog confined by a fence to its own backyard). Sometimes these barriers to mobility temporarily open so that proteins can move between confinement zones. Finally, a few proteins undergo rapid, highly directed transport toward a specific region of the membrane, perhaps being carried by motor proteins similar to kinesin (see p. 60). (This is analogous to a dog being carried by its owner to a particular destination.) Thus, the emerging view is that the plasma membrane is a highly complex, dynamic, regionally differentiated structure.

The lipid bilayer forms the basic structural barrier that encloses the cell

In addition to providing fluidity, the lipid bilayer serves at least two other important functions:

1. It forms the basic structure of the membrane. The phospholipids can be visualized as the "pickets" that form the "fence" around the cell.
2. Its hydrophobic interior serves as a barrier to passage of water-soluble substances between the ICF and ECF. Most water-soluble substances cannot dissolve in and pass through the lipid bilayer. (However, water molecules themselves and some other small uncharged molecules such as CO_2 are small enough to pass between the molecules that form this barrier.) By means of this barrier, the cell can maintain different mixtures and concentrations of solutes (dissolved substances) inside and outside the cell.

The membrane proteins perform a variety of specific membrane functions

A variety of different proteins within the plasma membrane serve the following specialized functions (see Figure 3-3), some of which we will discuss in detail later:

1. *Channels.* Some transmembrane proteins form water-filled pathways, or channels, across the lipid bilayer. Their presence enables water-soluble ions that are small enough to enter a channel to pass through the membrane without coming into direct contact with the hydrophobic lipid interior.
2. *Carriers.* Another group of proteins that span the membrane serve as **carrier proteins**, which transfer larger, specific substances across the membrane that cannot cross on their own. Both channels and carriers help determine what substances are transferred between the ICF and ECF.
3. *Receptors.* Many proteins on the outer surface serve as **receptors**, proteins with sites that "recognize" and bind with specific molecules in the cell's environment. This binding initiates a series of membrane and intracellular events (to be described later) that alter the activity of the particular cell. In this way, chemical messengers in the blood, such as hormones, can influence only the specific cells that have receptors for the messenger while having no effect on other cells, even though every cell is exposed

to the same messenger via its widespread distribution by the blood. To illustrate, the mammalian anterior pituitary gland secretes into the blood thyroid-stimulating hormone (TSH), which can attach to the surface receptors of thyroid gland cells to stimulate secretion of thyroid hormone.

4. *Docking-marker acceptors.* Still other proteins on the inner surface serve as **docking-marker acceptors** that bind lock-and-key fashion with the docking markers of secretory vesicles (see p. 41). Secretion is initiated as stimulatory signals trigger the fusion of the secretory vesicle membrane with the inner surface of the plasma membrane through interactions between their SNARE proteins (see Chapter 2, p. 43). The secretory vesicle subsequently opens up and empties its contents to the outside by exocytosis.

5. *Enzymes.* Another group of proteins function as **membrane-bound enzymes** that control specific chemical reactions at either the inner or the outer cell surface. These also vary among cell types. For example, the outer layer of the plasma membrane of skeletal muscle cells contains an enzyme that destroys the neurotransmitter that triggers muscle contraction, enabling the muscle to relax.

6. *CAMs.* Still other proteins serve as **cell adhesion molecules (CAMs)**. Many CAMs protrude from the outer membrane surface and form loops or other appendages that the cells use to grip each other or grasp the connective tissue fibers between cells. For example, *cadherins*, a type of CAM found on the surface of adjacent cells, interlock in zipper fashion to help hold the cells within tissues and organs together. Other CAMs, such as the *integrins*, span the plasma membrane where they serve as a structural link between the outer membrane and its extracellular surroundings and also connect the inner membrane surface to the intracellular cytoskeletal scaffolding. Besides mechanically linking the cell's external environment and intracellular components, integrins also relay regulatory signals through the plasma membrane in either direction. Some CAMs also function as "signaling molecules" in cell growth and in inflammatory responses, among other things.

7. *Self-identity markers.* Finally, still other proteins on the outer membrane surface, especially in conjunction with carbohydrates, are important in cells' ability to recognize "self" (that is, cells of the same type) and in cell-to-cell interactions, as described next.

The membrane carbohydrates serve as self-identity markers

The short sugar chains on the outer-membrane surface serve as self-identity markers that enable cells to identify and interact with each other in the following ways.

1. Different cell types have different markers. The unique combination of sugar chains projecting from the surface membrane serves as the "trademark" of a particular cell type, enabling a cell to recognize others of its own kind. These carbohydrate chains play an important role in recognition of "self" and in cell-to-cell interactions. Cells can recognize other cells of the same type and join together to form tissues. This is especially important

during embryonic development. If cultures of embryonic cells of two different types such as nerve cells and muscle cells, are mixed, the cells sort themselves into separate aggregates of nerve cells and muscle cells. "Self"-recognition is also crucial in all animal immune responses (e.g., *major histocompatibility complex* in mammals, Chapter 10, p. 483).

2. Carbohydrate-containing surface markers are also involved in tissue growth, which is normally held within certain limits of cell density. Cells do not "trespass" across the boundaries of neighboring tissues; that is, they do not overgrow their own territory. The exception is the uncontrolled spread of cancer cells, which have been shown to bear abnormal surface carbohydrate markers.

We now turn our attention to the topic of membrane transport, focusing on how the plasma membrane selectively controls what enters and exits the cell.

check your understanding 3.1

How do lipids organize themselves to form a plasma membrane, and what purpose does the membrane's hydrophobic interior serve?

What are the functions of and ways of altering membrane fluidity? What structural changes in membrane composition help maintain fluidity at elevated environmental temperatures? At low environmental temperatures?

What functions do membrane proteins have?

What functions do membrane carbohydrates have?

3.2 Unassisted Membrane Transport

Anything that passes between a cell and the surrounding extracellular fluid must be able to penetrate the plasma membrane. If a substance can cross the membrane, the membrane is said to be **permeable** to that substance; if a substance cannot pass, the membrane is **impermeable** to it. The plasma membrane is **selectively permeable** in that it lets some particles pass through while excluding others. In some instances permeability can be controlled, as exemplified by the opening or closing of specific channels.

Lipid-soluble substances and small polar molecules can passively diffuse through the plasma membrane down their electrochemical gradient

Two properties of particles influence whether they can permeate the plasma membrane without any assistance: (1) the relative solubility of the particle in lipid and (2) the size of the particle. Highly lipid-soluble particles can dissolve in the lipid bilayer and pass through the membrane. Uncharged or nonpolar molecules (such as O_2, CO_2, alcohol, steroids, and fatty acids) are very lipid-soluble and readily permeate the membrane. Charged particles (ions such as Na^+ and K^+) and polar molecules (such as glucose and proteins) have low

FIGURE 3-4 Diffusion.
(a) Diffusion down a
concentration gradient.
(b) Dynamic equilibrium, with
no net diffusion occurring.

© Cengage Learning, 2013

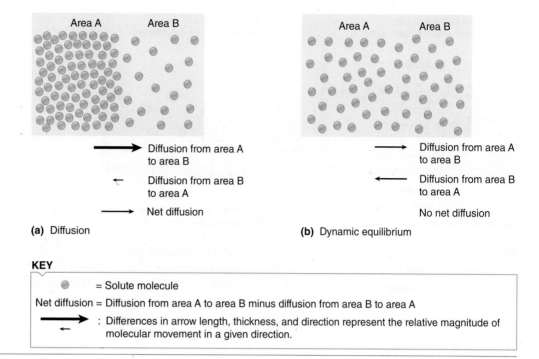

(a) Diffusion   **(b)** Dynamic equilibrium

KEY

⬤ = Solute molecule

Net diffusion = Diffusion from area A to area B minus diffusion from area B to area A

➡ : Differences in arrow length, thickness, and direction represent the relative magnitude of
←   molecular movement in a given direction.

---

lipid solubility but are very soluble in water. The lipid bi-layer serves as an impermeable barrier to such particles.

Yet some of these particles, for example glucose and Na⁺ and K⁺, must cross the membrane for the cell to survive and function. (Most cells use glucose as their fuel of choice to produce ATP.) Cells have several means of assisted transport—the channels and carriers noted earlier—to move particles across the membrane that must enter or leave the cell but cannot do so unaided. We will look at these in more detail later.

Even though a particle might be capable of permeating the membrane by virtue of its lipid solubility or its ability to fit through a channel, some force is needed to produce its movement across the membrane. Two general types of forces are involved: (1) **passive forces**, which do not require the cell to expend energy to produce movement; and (2) **active forces**, which require cellular energy (ATP) to be expended to transport a substance across the membrane. Let's now examine the various methods of passive and active membrane transport.

In unassisted membrane transport, particles that can permeate the membrane diffuse passively down their concentration gradient

Molecules that can penetrate the plasma membrane on their own—that is, unassisted—are passively driven across the membrane by two forces: *diffusion* down a concentration gradient and/or *conduction* along an electrical gradient. We will first examine diffusion.

All molecules are in continuous random motion at temperatures above absolute zero as a result of heat (thermal) energy. Because of this Brownian motion (named after the Scottish botanist Robert Brown, who in 1827 described random motions of floating pollen), each molecule moves separately and randomly in any direction. This motion is most evident in liquids and gases, where the individual molecules

have more room to move before colliding with another molecule, bouncing off each other in different directions like billiard balls striking each other. However, the greater the molecular concentration of a substance in a solution, the greater the likelihood of collisions. As a result, if a substance can move between regions with different concentrations, a nonrandom net movement called **diffusion** (from *diffusus*, spread out) occurs. To see how, note the example in Figure 3-4a where the concentration differs between area A and area B in a solution. Such a difference in between two adjacent areas is called a **concentration** (or **chemical**) **gradient.** Random molecular collisions occur more frequently in area A than in B. As a result, more molecules bounce from area A into area B than in the opposite direction. In both areas, the *individual molecules move randomly and in all directions*, but the *net movement of molecules by diffusion is not random*, but rather from the area of higher concentration to the area of lower concentration.

The term **net diffusion** refers to the difference between two opposing movements. If 10 molecules move from A to B while 2 molecules simultaneously move from B to A, the net diffusion is 8 molecules moving from A to B. Molecules spread in this way until the substance is uniformly distributed between the two areas and a concentration gradient no longer exists (Figure 3-4b). At this point, even though movement is still taking place, no *net* diffusion is occurring, because the opposing movements exactly counterbalance each other. Movement of molecules from A to B is exactly matched by movement of molecules from B to A. This situation is known as a **steady state** or **equilibrium.**

What happens if a plasma membrane separates different concentrations of a substance? If the substance can permeate the membrane, net diffusion of the substance occurs through the membrane down its concentration gradient from the area of high concentration to the area of low concentration (Figure 3-5a). No energy is required for this movement, so it is a passive mechanism of membrane transport. For example, O₂ is

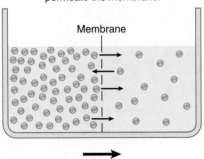

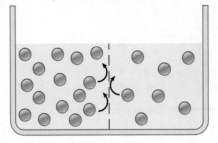

(a) Diffusion occurs

(b) No diffusion occurs

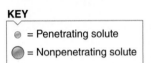

KEY

= Penetrating solute

= Nonpenetrating solute

FIGURE 3-5 **Diffusion through a membrane.** (a) Net diffusion of a penetrating solute across the membrane down a concentration gradient. (b) No diffusion of a nonpenetrating solute through the membrane despite the presence of a concentration gradient.

© Cengage Learning, 2013

transferred across the lung membrane by this means. The blood carried to the lungs is low in O_2, having given up O_2 to the body tissues for cellular metabolism. The air in the lungs, in contrast, is high in O_2 because it is continuously exchanged with fresh air by the process of breathing. Because of this concentration gradient, net diffusion of O_2 occurs from the lungs into the blood as blood flows through the lungs. Thus, as blood leaves the lungs for delivery to the tissues, it is high in O_2. (Recall from Chapter 1 that August and Marie Krogh were the first to demonstrate that oxygen moves by diffusion in lungs.) Let's look at membrane diffusion more closely.

Fick's Law of Diffusion Several factors in addition to the concentration gradient influence the rate of net diffusion across a membrane. In 1855 Adolf Fick in Germany derived an expression that gives the rate, termed Q, at which molecules diffuse from one area to another (**Fick's law of diffusion,** Table 3-1), with the following factors:

1. *The magnitude (or steepness) of the concentration gradient.* The greater the difference in concentration, the greater the rate of diffusion. For example, let's follow CO_2 in a running lizard. The working muscles produce CO_2 more rapidly as a result of burning additional fuel to produce the extra ATP they need to power the stepped-up, energy-demanding contractile activity. The resultant increase in CO_2 level in the muscles creates a greater-than-normal CO_2 difference between the muscles and the blood. Because of this larger gradient, more CO_2 than usual enters the blood. As this blood reaches the lizard's lungs, a greater-than-normal CO_2 difference exists between the blood and air in the lungs. Accordingly, more CO_2 than normal diffuses from the blood into the air before equilibrium is achieved. This extra CO_2 is subsequently breathed out to the environment. Thus, any additional CO_2 produced by exercising muscles is eliminated from the body simply as a result of the increase in CO_2 concentration gradient.

2. *The permeability of the membrane to the substance.* The more permeable the membrane is to a substance, the

TABLE 3-1 Factors Influencing the Rate of Net Diffusion of a Substance across a Membrane (Fick's Law of Diffusion)

Factor	Effect on Rate of Net Diffusion
↑ Concentration gradient of substance (ΔC)	↑
↑ Surface area of membrane (A)	↑
↑ Lipid solubility (β)	↑
↑ Molecular weight of substance (MW)	↓
↑ Distance (thickness) (ΔX)	↓

Modified Fick's equation:

$$\text{Net rate of diffusion } (Q) = \frac{\Delta C \cdot A \cdot \beta}{\sqrt{MW} \cdot \Delta X}$$

$$\left[\text{diffusion coefficient } (D) \propto \frac{1}{\sqrt{MW}} \right]$$

$$\left[\text{permeability } (P) = \frac{D\beta}{\Delta X} \right]$$

Restated $Q \propto \Delta C \cdot A \cdot P$

© Cengage Learning, 2013

more rapidly the substance can diffuse down its concentration gradient. Note that if the membrane is impermeable to the substance, no diffusion can take place across the membrane, even though a concentration gradient may exist (Figure 3-5b). For example, because the plasma membrane is impermeable to the vital intracellular proteins, they cannot escape from the cell, even though they are in much greater concentration in the ICF than in the ECF.

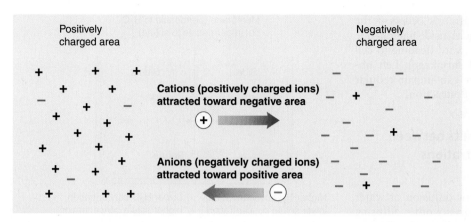

FIGURE 3-6 Movement along an electrical gradient.

© Cengage Learning, 2013

3. *The surface area of the membrane across which diffusion is taking place.* Obviously, the larger the surface area available, the greater the rate of diffusion it can accommodate. For example, absorption of nutrients across the membranes in the small intestine is enhanced by surface foldings (*villi*) and projections (*microvilli*) that greatly increase the available absorptive surface area (Chapter 14).

4. *The molecular weight of the substance.* Lighter molecules such as O_2 and CO_2 bounce farther on collision than do heavier molecules. Consequently, O_2 and CO_2 diffuse rapidly, permitting rapid exchanges of these gases across the respiratory membranes.

5. *The distance through which diffusion must take place.* The greater the diffusion barrier thickness, the slower the rate of diffusion. Accordingly, membranes across which diffusing particles must travel are normally relatively thin, such as the membranes separating the air and blood in vertebrate lungs.

6. *Temperature.* Generally, as temperatures increase, the higher kinetic energy of molecules will increase diffusion rates. Permeability may also be affected, generally increasing as temperature rises.

One of the most important aspects of diffusion to keep in mind is how extraordinarily *slow* the process is because of its inherent randomness and molecular scale. For example, oxygen moving only by diffusion through a nostril would take many years to reach equilibrium within a large air-breathing animal. Thus, an animal that needs to transport molecules such as oxygen and hormones over large distances cannot rely on diffusion. As you will see, evolution resulted in alternative transport systems, termed *bulk transport* mechanisms, that move the entire medium carrying molecules of interest. *The slowness of diffusion, and thus its inadequacy for movement over large distances, has been the primary force in the evolution of circulatory systems* and other bulk transport processes in multicellular animals (Chapter 9).

Ions that can permeate the membrane also conduct passively along their electrical gradient

Movement of **ions** (electrically charged particles that have either lost or gained an electron) is also affected by their electrical charge. Like charges (those with the same kind of charge) repel each other, whereas opposite charges attract each other. If a relative difference in charge exists between two adjacent areas (Figure 3-6), the positively charged ions

(**cations**) tend to move toward the more negatively charged area, whereas the negatively charged ions (**anions**) tend to move toward the more positively charged area. A difference in charge between two adjacent areas thus produces an **electrical gradient** that passively induces ion movement—a process called *conduction*. When an electrical gradient exists between the ICF and ECF, only ions that can permeate the plasma membrane can conduct down this gradient. The simultaneous existence of an electrical gradient and concentration (chemical) gradient for a particular ion is called an **electrochemical gradient**. Later in this chapter, you will learn how electrochemical gradients contribute to the electrical properties of the plasma membrane.

You will frequently find that ion movement is called "diffusion," but this is improper terminology when charge differences are causing movement. Recall that diffusion is based on random molecular motions, and is very slow. Ion conduction, in contrast, is much faster because of charge interaction, which is a strong directional force (think about how fast two magnets can move toward each other, an analogous situation). If the electrical systems in your house relied on diffusion (rather than conductance) of electrons, it would take many decades for the electrons to reach your house from a power plant!

Osmosis is the net movement of water through a membrane

Water molecules can readily permeate most plasma membranes. Even though water molecules are strongly polar, they are small enough to slip through momentary spaces created between the phospholipid molecules' tails as they sway and move within the lipid bilayer. This simple fact has enormous implications for cell physiology and evolution. At some early point in the evolution of the first cells, there would have been an unequal distribution of molecules: a high concentration of organic molecules on the inside and a low concentration on the outside. Why might such differences be a potential problem for cells? The answer lies with the difference in total **solute** (dissolved substance) concentrations on either side of the membrane, coupled with water's ability to penetrate most membranes. As a membrane-enclosed system, a cell must contend with a powerful force, **osmosis**—the result of movement of water down its own concentration gradient through a membrane (note that diffusion of water in an open system without a membrane is not osmosis, but simply diffusion). Osmosis occurs if the total concentration of solutes is not equal inside and outside the cell. Osmotic imbalances can be inward (leading to cell swelling) or outward (leading to cell shrinkage). Left unchecked, osmotic movement of water can impair cellular function and, if extreme, result in cell destruction.

Osmosis across membranes results between solutions with different concentrations of a nonpenetrating solute

In simplest terms, the driving force for diffusion of water across the membrane is the same as for any other diffusing molecule, namely, its concentration gradient. Usually, the term **concentration** refers to the density of the solute in a given volume of water. It is important to recognize, however, that the addition of a solute to pure water in essence decreases the water concentration.

Compare the water and solute concentrations in two compartments separated by a membrane in Figure 3-7. The container labeled side 1 has a low solute concentration and thus a high water concentration. In side 2, more of the water molecules have been replaced by solute, and the water concentration decreases correspondingly. In this situation, the solution in side 2 with higher solute concentration is said to be **hyperosmotic** with respect to the other. Conversely, side 1, with higher water concentration, is **hypo-osmotic**. Note that in Figure 3-7, the solutions are separated by a membrane that permits passage of water but not solute. Thus, water will move down its own concentration gradient from side 1 (lower solute/higher water concentrations) to side 2 (higher solute/ lower water concentrations). Because solutions are always referred to in terms of concentration of solute, *water moves by osmosis to the area of higher solute concentration.* The solute can be thought of as "drawing," or attracting, water, but osmosis is almost always portrayed as a special case of diffusion, in which water molecules move by this inherently random process. But when we actually measure the rate of osmosis across membranes, we find that for many solutes the rate is much faster than that predicted by Fick's law. In fact, the physical chemistry of osmosis is not fully understood (see *Unanswered Questions: Is osmosis really the simple diffusion of water down its concentration gradient?* on p. 80). However, we will treat it as a form of diffusion.

As a result of water movement alone, side 2's volume increases while the volume of side 1 correspondingly decreases. Loss of water from side 1 increases the solute concentration on side 1, whereas addition of water to side 2 reduces the solute concentration on that side. Eventually, the concentrations of water and solute on the two sides of the membrane become equal, and net diffusion of water ceases. The side originally containing the greater solute concentration has a larger volume, having gained water. With all concentration gradients abolished, an **isosmotic** state (*isos,* "equal"), net osmosis ceases.

What happens if a nonpenetrating solute is present on side 2 and pure water is present on side 1 (Figure 3-8)? Osmosis again occurs from side 1 to side 2, but the concentrations between the two compartments can never become equal. No matter how dilute side 2 becomes because of water diffusing into it, it can never become pure water, nor can side 1 ever acquire any solute. Because equilibrium is impossible

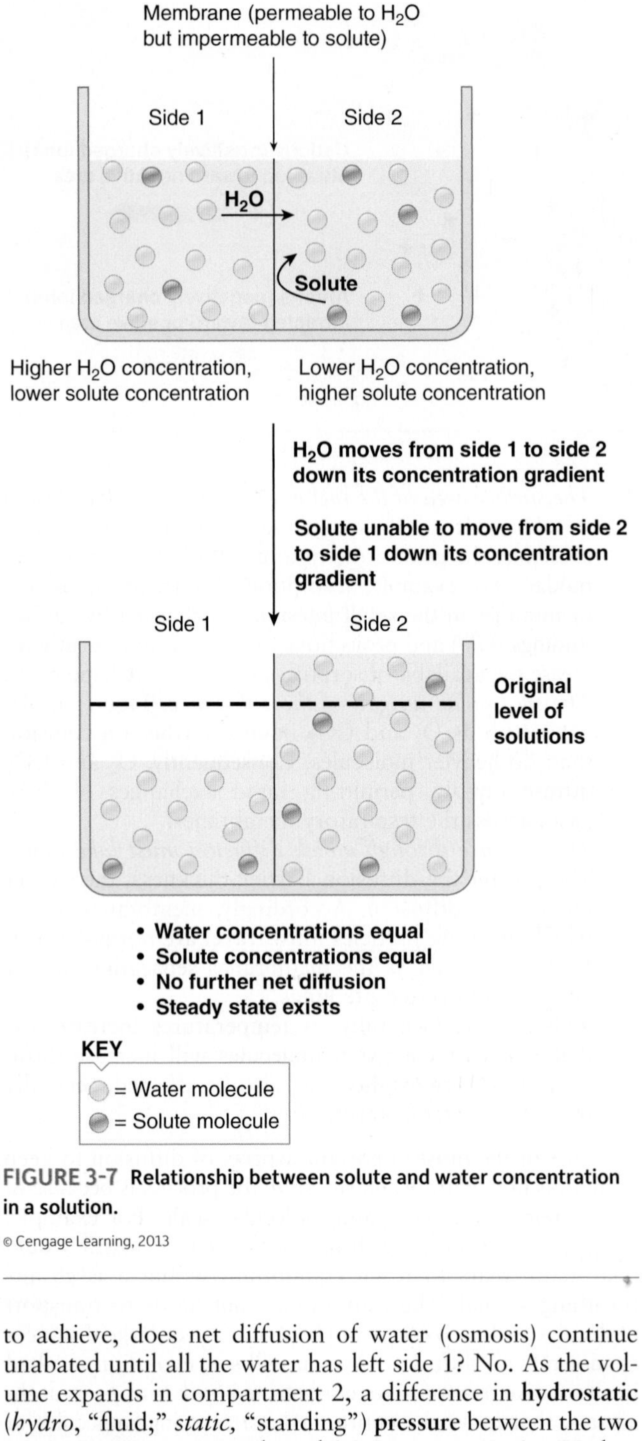

Membrane (permeable to H_2O but impermeable to solute)

Higher H_2O concentration, lower solute concentration

Lower H_2O concentration, higher solute concentration

H_2O moves from side 1 to side 2 down its concentration gradient

Solute unable to move from side 2 to side 1 down its concentration gradient

Original level of solutions

- **Water concentrations equal**
- **Solute concentrations equal**
- **No further net diffusion**
- **Steady state exists**

KEY

⬤ = Water molecule

⬤ = Solute molecule

FIGURE 3-7 Relationship between solute and water concentration in a solution.

© Cengage Learning, 2013

to achieve, does net diffusion of water (osmosis) continue unabated until all the water has left side 1? No. As the volume expands in compartment 2, a difference in **hydrostatic** (*hydro,* "fluid;" *static,* "standing") **pressure** between the two compartments is created, and it opposes osmosis. (Hydrostatic pressure is the pressure exerted by a standing, or stationary, fluid on an object—in this case the plasma membrane.) The hydrostatic pressure exerted by the larger volume of fluid on side 2 is greater than the pressure exerted by water moving from side 1. This hydrostatic pressure difference tends to push fluid from side 2 to side 1. The magnitude of opposing pressure necessary to completely stop osmosis is defined as the **osmotic pressure** of the solution on side 2. The osmotic pressure can be related directly to the concentration of nonpenetrating solute. The greater the concentration of nonpenetrating solute → the lower the concentration of water → the greater the drive for water to move by osmosis from pure water into the solution → the greater the opposing

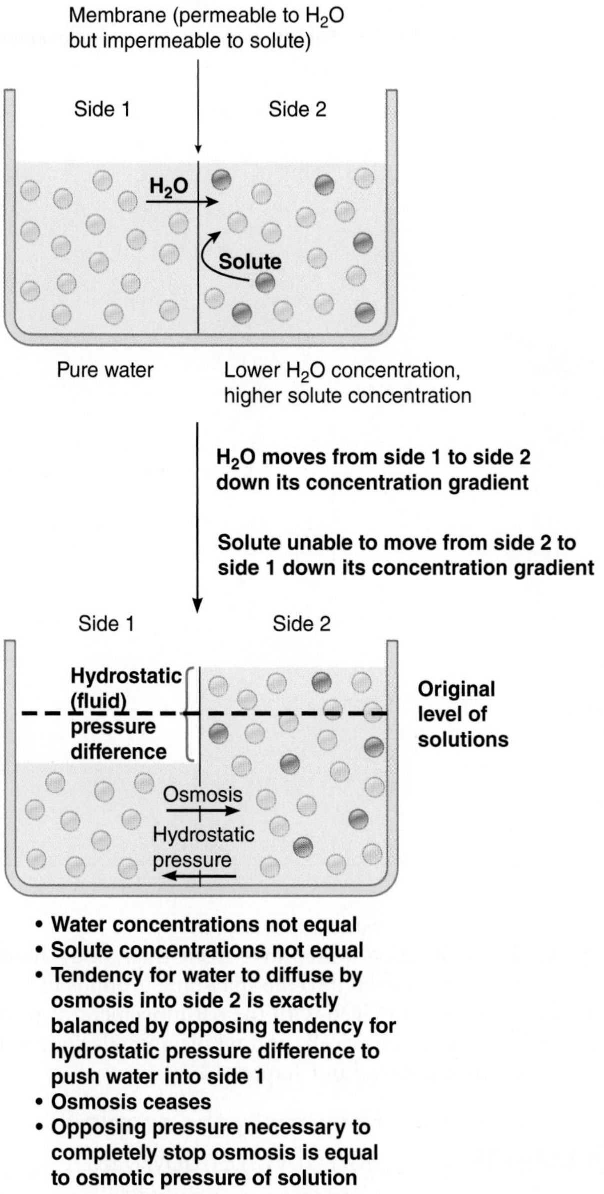

FIGURE 3-8 Osmosis when pure water is separated from a solution containing a nonpenetrating solute.

© Cengage Learning, 2013

pressure required to stop the osmotic flow → the greater the osmotic pressure of the solution. Therefore, *a solution with a high solute concentration exerts greater osmotic pressure than does a solution with a lower solute concentration.*

Colligative properties of solutions are dependent solely on the number of dissolved particles

Osmotic pressure is one of the so-called **colligative properties** of solutes, which are those that depend solely on the number of dissolved particles in a given volume of solution, not on the chemical properties of the solute. There are four such properties for an idealized solute (one that fits this description) at 1 mole in 1 kg of water (defined as 1 *molal*): (1) it has an osmotic pressure of 22.4 atmospheres (atm), (2) it raises the boiling point by 0.54°C, (3) it depresses the

freezing point by 1.86°C, and (4) it reduces vapor pressure. The effect on freezing point is important in the context of animals adapting to freezing temperatures (Chapter 15).

Osmotic pressure can be measured in atmospheres, but is more conveniently expressed as **osmolality**, as osmoles/kg water (osmol/kg) or milliosmoles/kg water (mosmol/kg), because for an ideal solute this osmotic value is equal to its *molal* concentration. Note that a 1-molal solution (1 mole/kg water) is not quite equal to a 1-*molar* solution (1 mole/liter solution), but in the range of concentrations for most biological fluids, the two are nearly equal. Since biologists tend to use molar (M) rather than molal because it is easier to measure a liter than to weigh a kg of water, **osmolarity** with corresponding abbreviations **Osm** (osmoles/liter) and **mOsm** (milliosmoles/liter) are commonly used. An ideal solute dissolved at a concentration of 1 M yields about 1 Osm or 1,000 mOsm. For example, because a molecule of glucose remains as an intact molecule when in solution, 1 M glucose equals 1 Osm. By contrast, because a molecule of NaCl dissociates (separates) into 2 ions—Na^+ and Cl^-—when in solution, 1 M NaCl equals about 2 Osm (actually 1.86 Osm, because Na^+ and Cl^- ions in solution occasionally bind together temporarily to act as one particle). Because osmolarity depends on the number, not the nature, of particles, any mixture of particles can contribute to the osmolality of a solution. Two useful examples of osmotic pressures in the biological world are about 1,000 mOsm for average seawater, and 300 mOsm (approximately) for most mammalian internal fluids. In fact, the basic dissolved constituents of cells—inorganic ions plus metabolites, proteins, and so forth—typically yield an osmotic pressure of roughly 250–400 mOsm in most organisms.

Tonicity refers to the effect the concentration of nonpenetrating solutes in a solution has on cell volume

Whereas the osmotic pressure of a solution refers to a colligative property, the **tonicity** of a solution refers to the effect on cell volume of the concentration of nonpenetrating solutes in the solution surrounding the cell. Solutes that can penetrate the plasma membrane quickly become equally distributed between the ECF and ICF (intracellular fluid) and thus do not contribute to long-term osmotic or tonicity differences. An **isotonic solution** (*iso* means "equal") has the same concentration of nonpenetrating solutes as do normal body cells. When a cell is bathed in an isotonic solution, no water enters or leaves the cell by osmosis, so cell volume remains constant. Thus, the ECF in animals is normally isotonic to cells so that no net diffusion of water occurs across the plasma membranes. This is important because cells, especially neural cells, do not function properly if they are swollen or shrunken. Any change in the concentration of nonpenetrating solutes in the ECF produces a corresponding change in the water concentration difference across the plasma membrane. The resultant osmotic movement of water brings about changes in cell volume.

The easiest way to demonstrate this phenomenon is to place mammalian red blood cells in solutions with varying concentrations of nonpenetrating solutes (Figure 3-9). Normally the plasma in which red blood cells are suspended has the same osmotic activity as the fluid inside these cells, en-

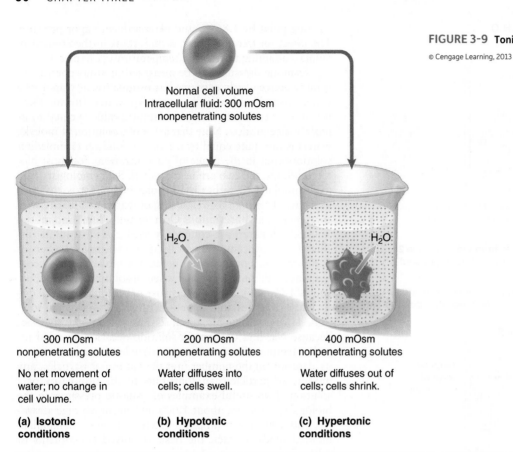

Normal cell volume
Intracellular fluid: 300 mOsm
nonpenetrating solutes

H_2O

H_2O

300 mOsm nonpenetrating solutes	200 mOsm nonpenetrating solutes	400 mOsm nonpenetrating solutes
No net movement of water; no change in cell volume.	Water diffuses into cells; cells swell.	Water diffuses out of cells; cells shrink.
(a) Isotonic conditions	**(b) Hypotonic conditions**	**(c) Hypertonic conditions**

FIGURE 3-9 **Tonicity and osmotic water movement.**
© Cengage Learning, 2013

abling the cells to maintain a constant volume. If red blood cells are placed in a hypo-osmotic solution, water enters the cells by osmosis. Net gain of water by the cells causes them to swell, perhaps to the point of rupturing, or lysing. In this situation causing swelling, the solution is called **hypotonic** (*hypo* means "below"). If, in contrast, red blood cells are placed in a concentrated or hyperosmotic solution, the cells shrink as they lose water by osmosis. In this situation, the solution is called **hypertonic** (*hyper* means "above"). When a red blood cell decreases in volume, its surface area does not decrease correspondingly, so the cell assumes a *crenated*, or spiky, shape. Thus, it is critical that the concentration of nonpenetrating solutes in the ECF quickly be restored to normal should the ECF become hypotonic (as with ingesting too much water) or hypertonic (as with losing too much water through severe diarrhea). (See Chapters 12 and 13 for further details about the important homeostatic mechanisms that regulate ECF osmolarity.)

Given that water penetrates membranes readily, whereas many solutes do not, a top priority of every animal cell is **osmotic homeostasis**, or **cell volume regulation**. This is the ability of a cell to adapt to an inequality of total solutes between its internal composition and the surrounding fluid. Inequalities can readily arise because a cell has a minimal requirement for internal solutes necessary for life, but the total concentration of these solutes may not match that of the external environment. Imbalances can occur from selective transport of solutes in and out of the cell as well (see next section). And imbalances may change in magnitude and direction within an animal's life cycle. For example, in the ocean the epithelial cells covering the gills of a salmon are

responsible for limiting water loss from their body fluids, whereas during migration into the spawning grounds in fresh water, the cells are involved with the net movement of water into the body fluids. How cells and whole animals cope with these demands is covered in Chapter 13.

unanswered
Questions | Is osmosis really the simple diffusion of water down its concentration gradient?

Although osmosis is generally characterized as water diffusing down its concentration gradient, created by differences in dissolved solutes that reduce water concentration, considerable evidence shows this is an oversimplification. In particular, some dissolved salts actually *increase* the effective concentration of water, and yet water in a lower-salt solution always osmoses through a membrane into a higher-salt solution. G. Hulett in 1902 first proposed an alternative explanation for osmosis. He hypothesized that the *pressure of the dissolved molecules colliding with the membrane* causes water molecules to move. The solution with the higher concentration has more collisions and thus exerts more pressure. This appears to alter the bonding forces between water molecules such that these molecules are pulled through the membrane into the side with higher salt. In recent decades, H. T. Hammel has provided considerable evidence that Hulett was correct. For this reason, the phenomenon of osmosis still remains incompletely understood. **<<**

Distinguish between diffusion and osmosis.

Why is the movement of ions down an electrical gradient not an example of diffusion?

What happens to the concentration of water inside a cell if ECF solute concentrations increase? Decrease? Explain why for both cases.

Explain the component terms of Fick's law with biological examples.

3.3 Assisted Membrane Transport

All kinds of transport we have discussed thus far—diffusion down concentration gradients, conduction along electrical gradients, and osmosis—produce net movement of molecules capable of permeating the plasma membrane by virtue of their lipid solubility or small size. Large, poorly lipid-soluble molecules such as proteins, glucose, and amino acids cannot cross the plasma membrane on their own no matter what forces are acting on them. Most small, charged molecules (*ions*) have the same property. This impermeability ensures that the large intracellular proteins and critical ions cannot escape from the cell. It is important that these molecules remain in the cell where they belong and can carry out their functions.

However, because of this impermeability, the cell must provide mechanisms for deliberately transporting some of these types of molecules into or out of the cell when it is necessary. For example, the cell must usher into the cell essential nutrients, such as glucose for energy and amino acids for the synthesis of proteins, and transport out of the cell metabolic wastes and secretory products, such as water-soluble protein hormones and digestive enzymes. Some ions must enter or leave the cell in, for example, signaling processes such as those of neurons. Cells use two classes of mechanisms to accomplish selective transport processes for poorly permeable molecules: (1) *channel and carrier-mediated transport* for transfer of small water-soluble substances across the membrane, and (2) *vesicular transport* for movement of large molecules and even multimolecular particles. Let's now examine these processes.

Channel transport is accomplished by protein pores, while carrier-mediated transport generally requires proteins that "flip-flop"

As you saw earlier, both channels and carriers are transmembrane proteins that serve as selective avenues for the movement of water-soluble substances across the membrane, but there are notable differences between them. (1) Only ions (and water, in the case of *aquaporins*; see next column) fit through the narrow channels, whereas moderately sized polar molecules such as glucose and amino acids are transported across the membrane by carriers. (2) Channels called **gated channels** can be open or closed as a result of changes in channel shape in response to a controlling mechanism; others called **leak channels** are open all the time, thus permitting unregulated leakage of their chosen ion across the membrane through the channels down their gradients. Similar to leak channels, carriers are almost always "open for business," but they work by very different mechanisms. (3) Movement through channels is considerably faster than carrier-mediated transport. When open for traffic, channels are open at both sides of the membrane at the same time, permitting continuous, rapid movement of ions between the ECF and ICF through these nonstop passageways. By contrast, carriers are never open to both the ECF and ICF simultaneously. They must change shape to alternately pick up passenger molecules on one side of the membrane and then drop them off on the other side, a time-consuming process. Whereas a carrier may move up to 5,000 particles per second across the membrane, 5 million ions may pass through a channel in a second.

Channels are highly selective. Not only does their small diameter preclude passage of particles greater than 0.8 nanometer (nm) in diameter (1 nm = 1 billionth of a meter), but a given channel can also selectively attract or repel particular ions. For example, sodium (Na^+) channels can accommodate the passage of only Na^+, whereas only K^+ can pass through potassium (K^+) channels. This selectivity is due to specific arrangements of various chemical groups on the interior surfaces of the proteins that form the channel walls. Cells vary in the number, kind, and activity of channels they have. As an example of a channel, we will examine water channels, leaving additional details on ion channels for the next chapter.

Aquaporins Osmotic water movement across a basic lipid bilayer membrane is relatively slow. For many membranes, the rate of osmosis is far too great to be explained by this process. Investigation of membrane proteins led to the discovery of special water channels called **aquaporins** (Figure 3-10). They are found in all membranes where rapid water movement is necessary: 1 to 10 billion water molecules per second can pass through an aquaporin in single file. Their discovery began with bovine lens in the 1980s: Michael Gorin, Barbara Yancey, and colleagues analyzed the gene for a major lens protein whose function was not known and deduced from the amino acid code that it was likely to be an aqueous channel. The name "aquaporin" was subsequently coined in the 1990s following elucidation of the complete structure and function of the aquaporin in mammalian red blood cells by Peter Agre's research team (resulting in a 2003 Nobel Prize in Chemistry for Agre). Since those discoveries, aquaporins in all kingdoms of life have been found; in mammals alone, there are at least 13 different aquaporin-type proteins, designated (in order of discovery) AQP0 (lens) to AQP12 (pancreas). Different cell types vary in their density of aquaporins and thus in their water permeability. You'll see a major role for aquaporins in water regulation by the mammalian kidney in Chapter 12.

Carrier-Mediated Transport Carrier proteins span the thickness of the plasma membrane and can undergo reversible changes in shape so that specific binding sites can alternately be exposed at either side of the membrane. That is, the carrier "flip-flops" so that binding sites located in the interior of the carrier are alternately exposed to the ECF and the ICF. Figure 3-11 is a schematic representation of how

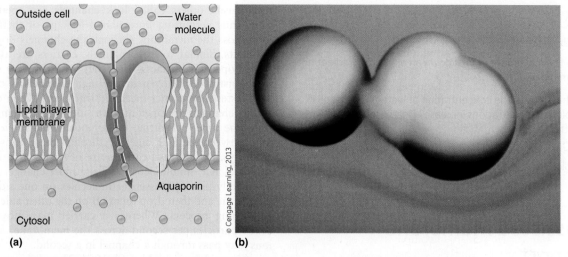

(a) **(b)**

FIGURE 3-10 Aquaporins. (a) Structure of an aquaporin forming a channel in a membrane. The channel is lined with hydrophilic amino acids that attract water molecules in single file. (b) One of the Nobel Prize–winning experiments from Peter Agre's laboratory. Incubation in hypotonic buffer fails to cause swelling of a control frog oocyte (*left*), due to low water permeability. In contrast, an oocyte injected with mammalian aquaporin RNA (*right*) exhibits high water permeability due to production of aquaporin proteins, and the cell has exploded.

Source: (b) Reprinted from Trends in Biochemical Sciences, 19/10, Peter Agre and Maarten J. Chrispeels, Aquaporins: water channel proteins of plant and animal cells, 421-425, 1994, with permission from Elsevier.

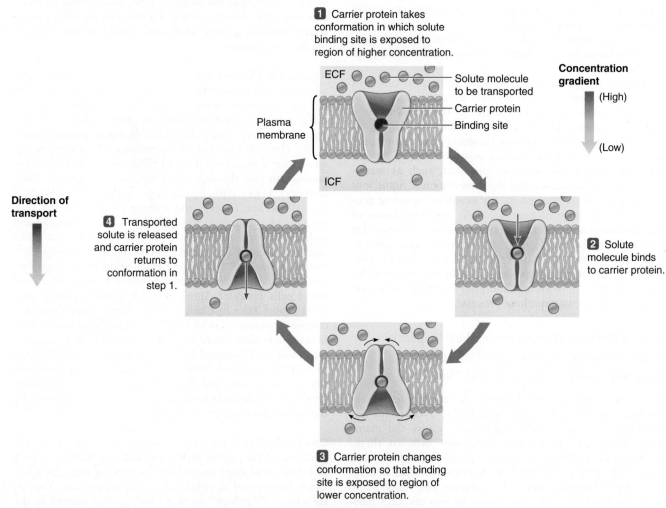

1 Carrier protein takes conformation in which solute binding site is exposed to region of higher concentration.

Concentration gradient

(High)

(Low)

ECF

Solute molecule to be transported

Carrier protein

Binding site

Plasma membrane

ICF

Direction of transport

4 Transported solute is released and carrier protein returns to conformation in step 1.

2 Solute molecule binds to carrier protein.

3 Carrier protein changes conformation so that binding site is exposed to region of lower concentration.

FIGURE 3-11 Model for facilitated diffusion, a passive form of carrier-mediated transport.

this **carrier-mediated transport** process takes place. As the molecule to be transported attaches to a binding site on the carrier on one side of the membrane (step 1), it triggers a change in the carrier's shape that causes the same site to be exposed to the other side of the membrane (step 2). Then, having been moved in this way from one side of the membrane to the other, the bound molecule detaches from the carrier (step 3). After the passenger detaches, the carrier reverts back to its original shape (step 4).

Carrier-mediated transport systems display three important characteristics that determine the kind and amount of material that can be transferred across the membrane: *specificity, saturation,* and *competition.*

1. *Specificity.* As with channels, each carrier protein is specialized to transport a specific substance, or at most a few closely related chemical compounds. For example, amino acids cannot bind to glucose carriers, although several structurally similar amino acids may be able to use the same carrier. Cells vary in the types of carriers they have, thus permitting transport selectivity among cells. For example, as we noted earlier, the plasma membranes of vertebrate thyroid gland cells uniquely have carriers for iodine. A number of inherited diseases involve defects in transport systems for a particular substance. *Cysteinuria* (cysteine in the urine) is such a disease, involving defective cysteine carriers in human and dog kidney membranes. This transport system normally removes cysteine from the fluid destined to become urine and returns this essential amino acid to the blood. When this carrier is malfunctional, large quantities of cysteine remain in the urine, where it is relatively insoluble and tends to precipitate. This is one cause of urinary or kidney stones.

2. *Saturation.* A limited number of carrier binding sites are available within a particular plasma membrane for a specific substance. Therefore, there is a limit to the amount of a substance that a carrier can transport across the membrane in a given time. This limit is known as the **transport maximum** (T_m). Until the T_m is reached, the number of carrier binding sites occupied by a substance and, accordingly, the substance's rate of transport across the membrane are directly related to its concentration. The more of a substance available to be transported, the more will be transported. When the T_m is reached, the carrier is saturated (all binding sites are occupied), and the rate of the substance's transport across the membrane is maximal. Further increases in the substance's concentration are not accompanied by corresponding increases in the rate of transport (Figure 3-12).

As an analogy, consider a ferry boat that can maximally carry 100 people across a river during one trip in an hour. If 25 people are on hand to board the ferry, 25 will be transported that hour. Doubling the number of people on hand to 50 doubles the rate of transport to 50 people that hour. Such a direct relationship exists between the number of people waiting to board (the concentration) and the rate of transport until the ferry is fully occupied (its T_m is reached). The ferry can maximally transport 100 people per hour. Even if 150 people are waiting to board, still only 100 will be transported per hour.

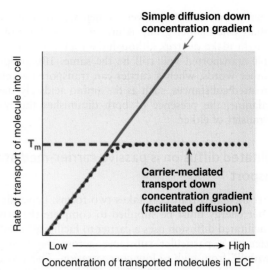

FIGURE 3-12 **Comparison of carrier-mediated transport and simple diffusion down a concentration gradient.** With simple diffusion of a molecule down its concentration gradient, the rate of transport of the molecule into the cell is directly proportional to the extracellular concentration of the molecule. With carrier-mediated transport of a molecule down its concentration gradient, the rate of transport of the molecule into the cell is directly proportional to the extracellular concentration of the molecule until the carrier is saturated, at which time the rate of transport reaches the transport maximum (T_m). After T_m is reached, the rate of transport levels off despite further increases in the ECF concentration of the molecule.

© Cengage Learning, 2013

Saturation of carriers is a critical rate-limiting factor in the transport of selected substances across vertebrate kidney membranes during urine formation and across the intestinal membranes during absorption of digested foods. Furthermore, it is sometimes possible to regulate (for example, by hormones) the rate of carrier-mediated transport by varying the affinity (attraction) of the binding site for its passenger or by varying the number of carriers ("ferry boats"). For example, the hormone insulin in vertebrates greatly increases the carrier-mediated transport of glucose into most cells by promoting an increase in the number of glucose carriers in the cell's plasma membranes. Deficiency of insulin secretion (diabetes mellitus) drastically impairs a cell's ability to take up and use glucose as the primary energy source.

3. *Competition.* Several closely related compounds may compete for a ride across the membrane on the same carrier. If a given binding site can be occupied by more than one type of molecule, the rate of transport of each substance is lower when both molecules are present than when either is present by itself. To illustrate, assume the ferry has 100 seats (binding sites) that can be occupied by either men or women. If only men are waiting to board, up to 100 men can be transported during each trip; the same holds true if only women are waiting to board. If, however, both men and women are waiting to board, they will compete for the available seats so

that fewer men and fewer women will be transported than when either group is present alone. Fifty of each might make the trip, although the total number of people transported will still be the same, 100 people. In other words, when a carrier can transport two closely related substances, such as the amino acids glycine and alanine, the presence of both diminishes the rate of transfer of either.

Facilitated diffusion is passive carrier-mediated transport

Carrier-mediated transport takes two forms, depending on whether energy must be supplied to complete the process. (1) **Facilitated diffusion** uses a carrier to facilitate (assist) the transfer of a particular substance across the membrane "downhill" from high to low concentration. This process is passive and does not require energy because movement occurs naturally down a concentration gradient. (2) **Active transport**, in contrast, requires the carrier to expend energy to transfer its passenger "uphill" *against* a concentration gradient, from an area of lower concentration to an area of higher concentration. An analogous situation is a ball on a hill with a barrier in front of it. If a way is provided to move through the barrier—for example, a partition that folds down when the ball bumps it—no energy is required to move the ball downhill (once rolling because of gravity, the ball will penetrate the barrier and continue to the bottom of the hill). Getting the ball uphill, however, requires the use of energy; for example, the barrier may have a powered catapult to flip the ball to the top.

The most notable example of facilitated diffusion is the transport of glucose into cells. Glucose can be at a higher concentration in the blood than in the tissues. Fresh supplies of this nutrient are regularly added to the blood by eating and from reserve energy stores in the body. Simultaneously, the cells may metabolize glucose almost as rapidly as it enters the cells from the blood. As a result, there can be a continuous gradient for net diffusion of glucose into the cells. Being a polar molecule, however, glucose cannot cross cell membranes on its own and is too large to fit through a channel. Without glucose carrier molecules (called *glucose transporters,* or *GLUT*) to facilitate membrane transport of glucose, the cells would be deprived of their preferred source of fuel.

The binding sites on facilitated diffusion carriers can bind with their passenger molecules when exposed to either side of the membrane (see Figure 3-11). As a result of thermal energy, these carriers undergo spontaneous changes in shape, alternately exposing their binding sites to the ECF or ICF. Passenger binding triggers the carrier to flip its conformation and drop off the passenger on the opposite side of the membrane. Because passengers are more likely to bind with the carrier on the high-concentration side than on the low-concentration side, the net movement always proceeds down the concentration gradient from higher to lower concentration. As is characteristic of mediated transport, the rate of facilitated diffusion is limited by saturation of the carrier binding sites, unlike the rate of simple diffusion, which is always directly proportional to the concentration gradient (Figure 3-12).

Active transport is carrier-mediated transport that uses energy and moves a substance against its concentration gradient

Active transport also involves the use of a protein transporter (channel or carrier) to transfer a specific substance across the membrane, but in this case the transporter moves the substance uphill against its concentration gradient. Active transport comes in two forms. In **primary active transport**, energy is *directly* required to move a substance against its concentration gradient; the transporter (a modified channel in the case of ions) splits ATP to power the transport process. In **secondary active transport**, energy is required in the entire process, but it is *not directly* used to produce uphill movement. That is, the carrier does not split ATP; instead, it moves a molecule uphill by using "secondhand" energy stored in the form of an **ion concentration gradient** (most commonly an Na^+ gradient). This ion gradient is built up by primary active transport of the ion by a different transporter. Let's examine each process in more detail.

Primary Active Transport With active transport, the binding site has a greater affinity for its passenger on the low-concentration side as a result of *phosphorylation* (p. 27) of the transporter on this side (Figure 3-13, step 1). The transporter acts as an enzyme that has ATPase activity, which means it splits the terminal phosphate from an ATP molecule to yield ADP and inorganic phosphate plus free energy (step 1) (Do not confuse *ATPase,* which splits ATP, with *ATP synthase,* which synthesizes ATP.) The phosphate group then attaches to the transporter, increasing the affinity of its binding site for the ion. As a result the ion to be transported binds to the transporter on the low-concentration side (step 2). In response to this binding, the transporter changes its conformation so that the ion is now exposed to the high-concentration side of the membrane (step 3). The change in transporter shape reduces the affinity of the binding site for the passenger, so the ion is released on the high-concentration side. Simultaneously the change in shape is also accompanied by *dephosphorylation* (step 4). The transporter then returns to its original conformation (step 5). Thus, ATP energy is used in the phosphorylation–dephosphorylation cycle of the transporter. It alters the affinity of the transporter's binding sites on opposite sides of the membrane so that transported ions are moved uphill from an area of low concentration to an area of higher concentration.

Primary active-transport transporters all move charged ions, namely Na^+, K^+, H^+, Cl^- or Ca^{2+}, across the membrane. Analysis of the structures of these transporters has been difficult, but recent studies on the *Na^+/K^+ ATPase* transporter (which we will describe shortly) show that it is probably a modified channel, not a carrier. This was discovered when a coral toxin was applied to the transporter, resulting in it being locked "open" so that the slow pump became a fast leak channel for Na^+. Rather than calling them channels, however, these active transporters are usually called "*pumps,*" analogous to water pumps that require energy to lift water against the downward pull of gravity. There are numerous pump proteins, especially those that move protons and other ions essential to life.

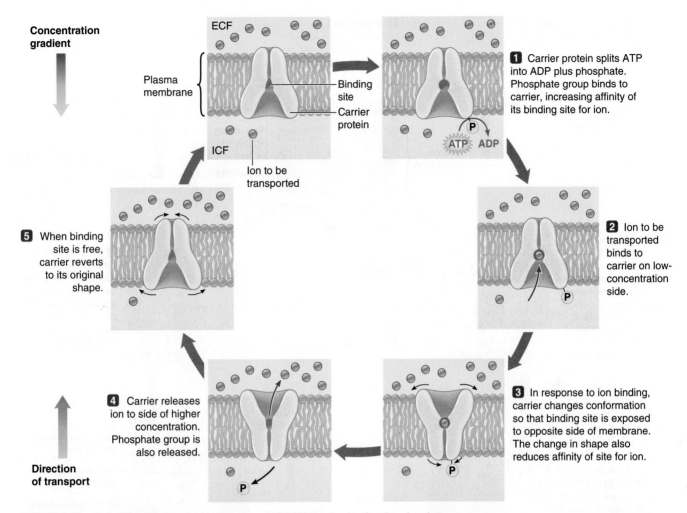

Concentration gradient

Plasma membrane

ECF

Binding site

Carrier protein

ICF

Ion to be transported

1 Carrier protein splits ATP into ADP plus phosphate. Phosphate group binds to carrier, increasing affinity of its binding site for ion.

ATP → ADP   P

5 When binding site is free, carrier reverts to its original shape.

2 Ion to be transported binds to carrier on low-concentration side.

4 Carrier releases ion to side of higher concentration. Phosphate group is also released.

Direction of transport

3 In response to ion binding, carrier changes conformation so that binding site is exposed to opposite side of membrane. The change in shape also reduces affinity of site for ion.

FIGURE 3-13 **Model for active transport.** The energy of ATP is required in the phosphorylation–dephosphorylation cycle of the carrier to transport the molecule uphill from a region of low concentration to a region of high concentration.

© Cengage Learning, 2013

The simplest active-transport systems pump a single type of passenger. An example is the **hydrogen-ion** (proton or H^+) **pump.** One type of this pump—**the F-ATPase**—may have been the first ATPase pump to evolve; as we saw in Chapter 2, they are used to generate energy by creating a hydrogen gradient. Another widespread type, the **V-ATPase** pump, is used to remove H^+ from body fluids, such as by the mammalian kidney, insect Malpighian tubules, and crustacean gills for excretion (see Chapter 12).

Other more complicated active transport mechanisms involve the transfer of two different passengers, either simultaneously in the same direction (called **symporting**) or sequentially in opposite directions (**antiporting**). For example, specialized vertebrate stomach cells use another type of proton pump to transport H^+ into the stomach lumen in exchange for a K^+ ion. This pump moves H^+ against a tremendous gradient: The concentration of H^+ in the stomach lumen is three to four million times greater than in the blood.

Primary Active Transport: The Na^+/K^+ ATPase Pump The most important example of an antiport pump is undoubtedly the **Na^+/K^+ ATPase pump (Na^+/K^+ pump** for short), found in the plasma membrane of all cells. This transporter transports Na^+ out of the cell, concentrating it in the ECF of multicellular organisms, and picks up K^+ from the outside, concentrating it in the ICF (Figure 3-14). The pump has high affinity for Na^+ on the ICF side. Binding of Na^+ to the transporter triggers the splitting of ATP through ATPase activity and subsequent phosphorylation of the transporter on the intracellular side, inducing a change in transporter shape, which moves Na^+ to the exterior. Simultaneously the change in shape increases the transporter's affinity for K^+ on the ECF side. Binding of K^+ leads to dephosphorylation of the transporter, inducing a second change in transporter shape, which transfers K^+ into the cytoplasm. There is not a direct exchange of Na^+ for K^+, however. The Na^+/K^+ pump moves three Na^+ out of the cell for every two K^+ it pumps in. (To appreciate the magnitude of active Na^+/K^+ pumping that

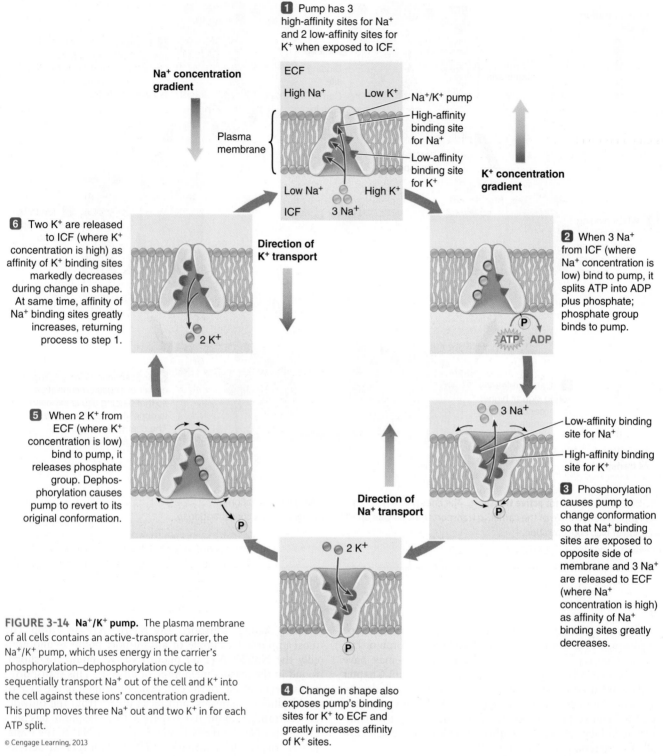

Na+ concentration gradient

1 Pump has 3 high-affinity sites for Na+ and 2 low-affinity sites for K+ when exposed to ICF.

ECF

High Na+   Low K+   — Na+/K+ pump

                             — High-affinity binding site for Na+

Plasma membrane

                             — Low-affinity binding site for K+

Low Na+   High K+

ICF   3 Na+

Direction of K+ transport

K+ concentration gradient

6 Two K+ are released to ICF (where K+ concentration is high) as affinity of K+ binding sites markedly decreases during change in shape. At same time, affinity of Na+ binding sites greatly increases, returning process to step 1.

2 K+

2 When 3 Na+ from ICF (where Na+ concentration is low) bind to pump, it splits ATP into ADP plus phosphate; phosphate group binds to pump.

P
ATP ADP

5 When 2 K+ from ECF (where K+ concentration is low) bind to pump, it releases phosphate group. Dephosphorylation causes pump to revert to its original conformation.

P

3 Na+

— Low-affinity binding site for Na+

— High-affinity binding site for K+

P

3 Phosphorylation causes pump to change conformation so that Na+ binding sites are exposed to opposite side of membrane and 3 Na+ are released to ECF (where Na+ concentration is high) as affinity of Na+ binding sites greatly decreases.

Direction of Na+ transport

2 K+

P

4 Change in shape also exposes pump's binding sites for K+ to ECF and greatly increases affinity of K+ sites.

FIGURE 3-14 **Na+/K+ pump.** The plasma membrane of all cells contains an active-transport carrier, the Na+/K+ pump, which uses energy in the carrier's phosphorylation–dephosphorylation cycle to sequentially transport Na+ out of the cell and K+ into the cell against these ions' concentration gradient. This pump moves three Na+ out and two K+ in for each ATP split.

© Cengage Learning, 2013

takes place, consider that a single nerve cell membrane contains approximately 1 million Na+/K+ pumps capable of transporting about 200 million ions per second.)

The Na+/K+ pump (Figure 3-14) plays two important roles:

1. It maintains Na+ and K+ concentration gradients across the plasma membrane of all cells. These gradients are

critically important for secondary active transport (discussed in detail shortly) as well as for the ability of nerve and muscle cells to generate electrical signals essential to their functioning (a topic discussed in the next chapter).

2. It helps regulate cell volume by controlling the concentrations of solutes inside the cell and thus minimizing osmotic effects that would induce swelling or shrinking of the cell.

The activity of the Na^+/K^+ pump is inhibited by the highly toxic metabolic inhibitor **ouabain**. Ouabain is isolated from the bark and root of the ouabaio tree and has been used by tribes in Africa since antiquity as an arrow poison. Oxygen deprivation can also reduce activity of the Na^+/K^+ pump because of the decline in ATP production. In vertebrates the number of Na^+/K^+ pump units in the cell membrane is regulated, in part, by thyroid hormone (see p. 301). Salt water acclimation in trout and salmonids (such as chum and pink salmon) involves an increase in gill basolateral Na^+/K^+ pump units as the freshwater acclimated epithelial cells transition to a salt secreting epithelium. The activity of these pump units in maintaining the electrochemical gradients across the plasma membrane accounts for approximately 25% of the standard metabolic rate in these animals (see Chapter 15).

Secondary Active Transport With primary active transport, energy is *directly* required to move a substance uphill. In addition, many cells can also actively transport certain solutes by moving them uphill from low to high concentration using *symporters*—also called **cotransport carriers**—that have two binding sites, one for an ion and one for the nutrient molecule. This secondary active transport may at first seem confusing because it uses "secondhand" energy stored in the form of an established **ion electrochemical gradient** (for example, an Na^+ gradient) to move the cotransported molecule uphill. This is a very efficient interaction. The cotransported molecule is essentially getting a free ride because Na^+ must be pumped out anyway to maintain the electrical and osmotic integrity of the cell. Cell energy (in the form of ATP) is not directly used by the carrier protein, but rather the energy stored in the Na^+ gradient (maintained with ATP) is used. Let us examine this more closely in mammals, in which (unlike most cells of the body) the intestinal and kidney cells transport glucose and amino acids by cotransport. Intestinal cells transport these nutrients from inside the intestinal lumen into the blood, concentrating them in the blood until none of these molecules are left in the lumen to be lost in the feces. The kidney cells save these nutrient molecules for the body by transporting them out of the fluid that is to become urine, moving them against a concentration gradient into the blood. These carriers that transport glucose against its concentration gradient from the lumen in the intestine and kidneys are distinct from the glucose facilitated-diffusion carriers found in all cells.

Here we focus specifically on symport in intestinal epithelial cells. This carrier, known as the **sodium and glucose cotransporter** or **SGLT**, is located in the luminal membrane (the membrane facing the intestinal lumen). The Na^+/K^+ pumps lie in the basolateral membrane (the membrane at the base of the cell opposite the lumen and along the lateral edge of the cell below the tight junction; see Figure 2-35). More Na^+ is present in the lumen than inside the cells because the basolateral energy-requiring Na^+/K^+ pump keeps the intracellular Na^+ concentration low. Because of this Na^+ concentration difference, more Na^+ binds to the SGLT when it is exposed to the lumen than when it is exposed to the ICF (Figure 3-15, step 1). Binding of Na^+ to SGLT then increases the carrier's affinity for its other passenger (for example, glucose), so the carrier has a high affinity for glucose when exposed to the outside (step 2a). When both Na^+ and glucose are bound to the carrier, it undergoes a conformational change and opens to the inside of the cell (step 2b). Both Na^+ and glucose are released to the interior, Na^+ because of the lower intracellular Na^+ concentration, and glucose because of the reduced affinity of the binding site on release of Na^+ (step 2c).

Note that the movement of Na^+ into the cell by SGLT carrier is downhill, but the movement of glucose is uphill, because glucose becomes concentrated in the cell. The released Na^+ is quickly pumped out by the Na^+/K^+ pumps, keeping the level of intracellular Na^+ low. The glucose that is carried into the cell by this process then passively moves out of the cell across the basolateral border down its concentration gradient and enters the blood (step 3). This movement is accomplished by facilitated diffusion, mediated by another carrier in the plasma membrane. This passive carrier is GLUT (p. 84), identical to the one that transports glucose into other cells, but in intestinal and mammalian kidney cells it transports glucose out of the cell. The difference depends on the direction of the glucose concentration gradient. In the case of intestinal and kidney cells, because of secondary active transport, glucose is in higher concentration inside the cells.

In the previous example, the symporter is responsible for the inward movement of both Na^+ and the substrate molecule. Molecules can also be moved in *opposing* directions by transport carriers as long as there is an established concentration or electrical (or electrochemical) gradient for one of the molecules. In this *antiport* (also known as *countertransport*) exchange, the solute and Na^+ move through the membrane in opposite directions, as you saw for the Na^+/K^+ pump, but here powered by the Na^+ gradient rather than ATP. For example, the uptake of Na^+ through the skin of frogs uses a carrier that extrudes H^+. This carrier plays an important role in maintaining the appropriate pH inside the cells (a fluid becomes more acidic as its H^+ concentration rises). Adding *amiloride* to the water inhibits both the uptake of Na^+ and the excretion of H^+. Similarly, the Na^+/H^+ and Cl^-/HCO_3^- antiporters in the gill epithelium of freshwater teleosts (higher bony fishes) are responsible for the uptake of Na^+ and Cl^- from the environment in exchange for the loss of hydrogen and bicarbonate (HCO_3^-) ions. Chloride ion can be pumped from solutions as low as 0.02 mM in fresh water (FW) across the gill surface to approximately 100 mM in the ECF. In these examples, the source of energy for the transport of Na^+ and Cl^- is derived from gradients of H^+ and HCO_3^- produced as by-products of metabolism in the animal's body. This mechanism permits these animals to restore valuable solute lost from the body through the gills and urine in exchange for removing these waste by-products.

Before leaving the topic of pump- and carrier-mediated transport, think about all the activities that rely on carrier assistance. All cells depend on carriers for the uptake of glucose and amino acids, which serve as the major source of energy and structural building blocks, respectively. Na^+/K^+ pumps are essential for generating cellular electrical activity and for ensuring that cells have an appropriate intracellular concentration of osmotically active ions. Active transport, both primary or secondary, is used extensively to accomplish the specialized functions of the nervous and digestive systems as well as excretory organs and all types of muscle.

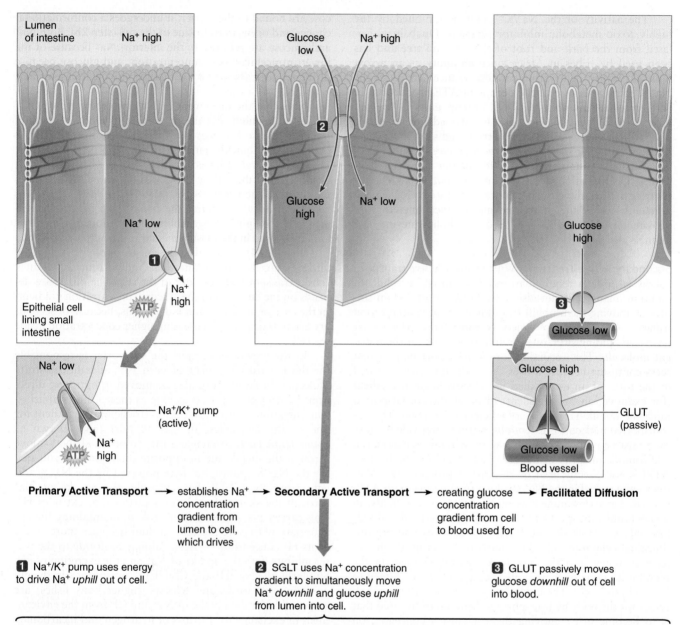

Primary Active Transport → establishes Na⁺ concentration gradient from lumen to cell, which drives → **Secondary Active Transport** → creating glucose concentration gradient from cell to blood used for → **Facilitated Diffusion**

1 Na⁺/K⁺ pump uses energy to drive Na⁺ *uphill* out of cell.

2 SGLT uses Na⁺ concentration gradient to simultaneously move Na⁺ *downhill* and glucose *uphill* from lumen into cell.

3 GLUT passively moves glucose *downhill* out of cell into blood.

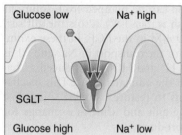

2a Binding of Na⁺ on luminal side, where Na⁺ concentration is higher, increases affinity of SGLT for glucose. Therefore, glucose also binds to SGLT on luminal side, where glucose concentration is lower.

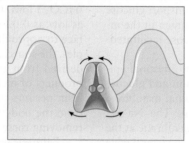

2b When both Na⁺ and glucose are bound, SGLT changes shape, opening to cell interior.

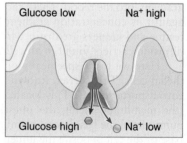

2c SGLT releases Na⁺ to cell interior, where Na⁺ concentration is lower. Because affinity of SGLT for glucose decreases on release of Na⁺, SGLT also releases glucose to cell interior, where glucose concentration is higher.

FIGURE 3-15 Symport of glucose. Glucose is transported across intestinal and kidney cells against its concentration gradient by means of secondary active transport mediated by the sodium and glucose cotransporter (SGLT) at the cells' luminal membrane.

With vesicular transport, material is moved into or out of the cell wrapped in membrane

The special carrier-mediated transport systems embedded in the plasma membrane can selectively transport ions and small polar molecules. But how do large polar molecules, such as the protein hormones secreted by endocrine cells, or even multimolecular materials, such as the bacteria ingested by white blood cells, leave or enter the cell? These materials cannot cross the plasma membrane, even with assistance of single membrane proteins. These large particles are transferred between the ICF and ECF by being wrapped in a membrane-enclosed vesicle, a process known as **vesicular transport**. Vesicular transport requires energy expenditure by the cell, so this is an *active* method of membrane transport. Energy is needed to accomplish vesicle formation and vesicle movement within the cell. Recall from Chapter 2 that transport into the cell in this manner is termed *endocytosis*, while transport out of the cell is *exocytosis*. An important feature of vesicular transport is that the materials sequestered within the vesicles do not mix with the cytosol. The vesicles are designed to recognize and fuse only with a targeted membrane, ensuring a directed transfer of designated large polar molecules or multimolecular materials between the cell's interior and exterior.

Endocytosis In endocytosis, the plasma membrane surrounds the substance to be ingested, then fuses over the surface, pinching off a membrane-enclosed vesicle so that the engulfed material is trapped within the cell (see Figure 2-12b). Recall that there are three forms of endocytosis (p. 44), depending on the nature of the material internalized: *pinocytosis* (nonselective uptake of ECF), *receptor-mediated endocytosis* (selective uptake of a large molecule), and *phagocytosis* (selective uptake of a multimolecular particle or cell). Once inside the cell, an engulfed vesicle has two possible destinies:

1. In many instances, lysosomes fuse with the vesicle to degrade and release its contents into the intracellular fluid (p. 45).
2. In some cells, the endocytotic vesicle bypasses the lysosomes and travels to the opposite side of the cell, where it releases its contents by exocytosis. This provides a pathway to shuttle intact particles through the cell. Such vesicular traffic is one means by which materials are transferred through the thin cells lining the capillaries of vertebrates, across which exchanges are made between the blood and surrounding tissues.

Exocytosis In exocytosis, almost the reverse of endocytosis occurs: A membrane-enclosed vesicle formed within the cell fuses with the plasma membrane, then opens up and releases its contents to the exterior (p. 41). This serves two different purposes:

1. It provides a mechanism for secreting large polar molecules, such as proteins that cannot cross the plasma membrane. In this case, the vesicular contents are highly specific and are released only on receipt of appropriate signals.
2. It enables the cell to add specific components to the membrane, such as selected carriers, channels, or receptors, depending on the cell's needs.

Balance between Endocytosis and Exocytosis The rate of endocytosis and exocytosis must be kept in balance to maintain a constant membrane surface area and cell volume. More than 100% of the plasma membrane may be used in an hour to wrap internalized vesicles in a cell actively involved in endocytosis, necessitating rapid replacement of surface membrane by exocytosis. In contrast, when a secretory cell is stimulated to secrete, it may temporarily insert up to 30 times its surface membrane through exocytosis. This added membrane must be specifically retrieved by an equivalent level of endocytotic activity.

Caveolae There may be a fourth vesicular mechanism by which cells can exchange materials with their environment, called **caveolae** ("tiny caves"). These are tiny, flask-shaped indentations dimpling the plasma membrane of many epithelial and smooth muscle cells. Researchers have observed these small, flask-shaped pits for more than 45 years through electron microscopy. However, it was not until the mid-1990s that evidence accumulated indicating that caveolae may have two functions:

1. *Another route for transport into cells:* An abundance of receptors cluster in these tiny chambers, some of which appear to play a role in cellular uptake of small molecules and ions. The best-studied example is the transport of the B vitamin *folic acid* into the cell. When folic acid binds with its receptors, which are concentrated in the portions of the plasma membrane that form the caveolae, the extracellular openings of these tiny caves close off. The high concentration of folic acid within a closed caveolar compartment encourages the movement of this vitamin across the caveolar membrane into the cytoplasm. Cellular uptake through the cyclic opening and closing of caveolae has been termed **potocytosis**.
2. *A "switchboard" for relaying signals from many extracellular chemical messengers into the cell's interior:* Many membrane receptors important in signal mechanisms are concentrated in the caveolae, leading one investigator to call these tiny caves "signaling organelles." Signaling is a critical topic we cover in detail in the next section.

Our discussion of membrane transport is now complete; Table 3-2 summarizes the established pathways by which materials can pass between the ECF and ICF.

check your understanding 3.3

Are aquaporins channels or carriers? Explain.

How is symporting by facilitated diffusion similar to and yet different from symporting by secondary active transport? Illustrate with examples.

How does the Na^+/K^+ ATPase antiport pump work, and why is it so important?

Does the process of exocytosis require energy to work? Explain.

TABLE 3-2 Methods of Membrane Transport and Their Characteristics

Method of Transport	Substances Involved	Energy Requirements and Force-Producing Movement	Limit to Transport
		SIMPLE DIFFUSION	
Diffusion through Lipid Bilayer	Nonpolar molecules of any size (e.g., O_2, CO_2, fatty acids)	Passive; molecules move down concentration gradient (from high to low concentration)	Continues until gradient is abolished (dynamic equilibrium with no net diffusion)
Diffusion through Protein Channel	Specific small ions (e.g., Na^+, K^+, Ca^{2+}, Cl^-)	Passive; ions move down electrochemical gradient through open channels (from high to low concentration and by attraction of ion to area of opposite charge)	Continues until there is no net movement and dynamic equilibrium is established
Osmosis	Water only	Passive; water moves down its own concentration gradient (to area of lower water concentration, i.e., higher solute concentration)	Continues until concentration difference is abolished or until stopped by opposing hydrostatic pressure or until cell is destroyed
		CARRIER-MEDIATED TRANSPORT	
Facilitated Diffusion	Specific polar molecules for which carrier is available (e.g., glucose)	Passive; molecules move down concentration gradient (from high to low concentration)	Displays a transport maximum (T_m); carrier can become saturated
Primary Active Transport	Specific cations for which carriers are available (e.g., Na^+, K^+, H^+, Ca^{2+})	Active; ions move against concentration gradient (from low to high concentration); requires ATP	Displays a transport maximum; carrier can become saturated
Secondary Active Transport (Symport or Antiport)	Specific polar molecules and ions for which coupled transport carriers are available (e.g., glucose, amino acids for symport; some ions for antiport)	Active; substance moves against concentration gradient (from low to high concentration); driven directly by ion gradient (usually Na^+) established by ATP-requiring primary pump. In symport, cotransported molecule and driving ion move in same direction; in antiport, transported solute and driving ion move in opposite directions	Displays a transport maximum; coupled transport carrier can become saturated
		VESICULAR TRANSPORT	
Endocytosis			
Pinocytosis	Small volume of ECF fluid; also important in membrane recycling	Active; plasma membrane dips inward and pinches off at surface, forming internalized vesicle	Control poorly understood
Receptor-Mediated Endocytosis	Specific large polar molecule (e.g., protein)	Active; plasma membrane dips inward and pinches off at surface, forming internalized vesicle	Necessitates binding to specific receptor on membrane surface
Phagocytosis	Multimolecular particles (e.g., bacteria and cellular debris)	Active; cell extends pseudopods that surround particle, forming internalized vesicle	Necessitates binding to specific receptor on membrane surface
Exocytosis	Secretory products (e.g., hormones and enzymes) as well as large molecules that pass through cell intact; also important in membrane recycling	Active; increase in cytosolic Ca^{2+} induces fusion of secretory vesicle with plasma membrane; vesicle opens up and releases contents to outside	Secretion triggered by specific neural or hormonal stimuli; other controls involved in transcellular traffic and membrane recycling not known

3.4 Intercellular Communication and Signal Transduction

Communication is critical for the survival of the society of cells that collectively compose the animal. The ability of cells to communicate with each other is essential for coordination of their diverse activities to maintain homeostasis as well as to control growth and development of the animal as a whole.

Communication between cells is orchestrated by direct contact and by extracellular chemical messengers

Intercellular communication can take place either directly or indirectly (Figure 3-16).

Direct Intercellular Communication This involves physical contact between the interacting cells by three mechanisms (Figure 3-16 a to c):

1. *Through gap junctions* (Figure 3-16a): The most intimate means of intercellular communication is through gap junctions, the minute tunnels that bridge the cytoplasm of neighboring cells in some types of tissues. Through these specialized anatomic arrangements (see Figure 2-34), small molecules and ions are directly exchanged between interacting cells without ever entering the ECF. Gap junctions are especially important in permitting electrical signals to spread from one cell to the next in cardiac and smooth muscle. However, in response to an increase in intracellular Ca^{2+} or H^+ ion concentrations, gap junctions rapidly close.

2. *Through transient direct linkup of surface markers* (Figure 3-16b): Some cells, such as those of the immune system, have specialized markers on the surface membrane that allow them to directly link up transiently and interact with certain other cells that have compatible markers for transient interactions. This is the means by which the phagocytes of the body's defense system specifically recognize and selectively destroy only undesirable cells, such as microbial invaders, while leaving the body's own healthy cells alone (Chapter 10).

3. *Through nanotubes* (Figure 3-16c): These are recently discovered long tubes with an internal actin-filament support surrounded by a plasma membrane that permits contiguous contact between two cells. They allow transfer of not only signal molecules but also whole organelles such as mitochondria. They have been identified between some developing insect and mammalian cells and between some mammalian immune cells.

Indirect Intercellular Communication The most common means by which cells communicate with each other is through intercellular chemical messengers, of which there are six categories based on modes of action: *paracrines, neurotransmitters, hormones, neurohormones, pheromones,* and *cytokines.* In each case, a specific chemical messenger is synthesized by specialized cells to serve a designated purpose. On being released into the ECF by appropriate stimulation, these signaling agents act on other particular cells, the messenger's **target cells,** in a prescribed manner. To exert its effect, an intercellular messenger, also known as a **ligand,** must bind with target-cell receptors specific for it. Receptors are typically transmembrane glycoproteins, that is, integral proteins that extend through the thickness of the plasma membrane and have carbohydrate chains attached on their outer surface (see Figure 3-3). A given cell may have thousands to as many as a few million receptors, of which hundreds to as many as 100,000 may be for the same chemical messenger. Different cell types have distinct combinations of receptors, allowing them to react individually to various regulatory extracellular chemical messengers. The modes of action of the six categories of chemical messenger are as follows (Figure 3-16 d to h):

- **Paracrines** (Figure 3-16d) are local chemical messengers whose effect is exerted only on neighboring cells in the immediate environment of their site of secretion. (If their action affects the same cell that secreted them, they are called **autocrines.**) Because paracrines are distributed by simple diffusion, their action is restricted to short distances. They do not gain entry to the blood in any significant quantity, because they are rapidly inactivated by locally existing enzymes. One example of a paracrine is *histamine,* which is released from a specific type of mammalian connective tissue cell during an inflammatory response within an invaded or injured tissue. Among other things, histamine dilates (opens more widely) the blood vessels in the vicinity to increase blood flow to the tissue. This action brings additional blood-borne immune-response supplies into the affected area (Chapter 10). Paracrines must be distinguished from chemicals that influence neighboring cells after being released nonspecifically. For example, an increased local concentration of CO_2 in an exercising muscle promotes local dilation of the blood vessels supplying the muscle. The resultant increased blood flow helps meet the more active tissue's increased metabolic demands. However, CO_2 is produced by all cells and is not specifically released to accomplish this particular response, so it and similar molecules are not paracrines.

- **Neurotransmitters** (Figure 3-16e) are used by neurons (nerve cells), which communicate directly with the cells they innervate (their target cells) by releasing these signal molecules in response to electrical signals (see p. 132). Like paracrines, neurotransmitters are very short-range chemical messengers, which diffuse from their site of release across a narrow extracellular space to act locally on only an adjoining target cell, which is usually another neuron, a muscle, or a gland. (Note that neurons may also release true paracrines, called *neuromodulators,* as discussed in Chapter 4).

- **Hormones** (Figure 3-16f) are long-range chemical messengers that are specifically secreted into the circulation by endocrine (ductless) glands (see p. 9) in response to an appropriate signal. Circulatory fluid (blood or hemolymph) carries the messengers to other sites in the body, where they exert their effects on their target cells some distance away from their site of release. Only the target cells of a particular hormone bear receptors to bind with this hormone. Nontarget cells are not influenced by any blood-borne hormones that reach them.

- **Neurohormones** (Figure 3-16g) are hormones released by **neurosecretory neurons.** These respond to and con-

DIRECT INTERCELLULAR COMMUNICATION

INDIRECT INTERCELLULAR COMMUNICATION VIA EXTRACELLULAR CHEMICAL MESSENGERS

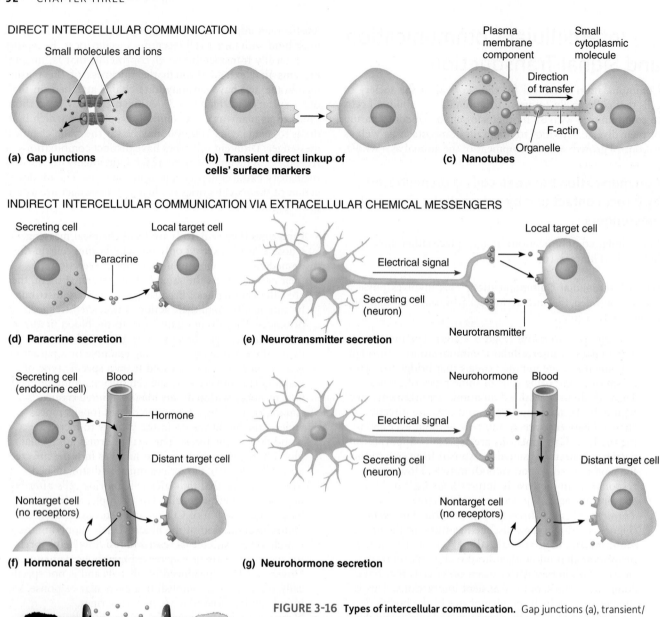

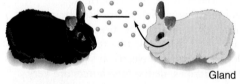

(h) Pheromones

FIGURE 3-16 Types of intercellular communication. Gap junctions (a), transient/direct linkup of cells (b), and nanotubes (c) are means of direct communication between cells. Paracrines (d), neurotransmitters (e), hormones (f), neurohormones (g), and pheromones (h) are all extracellular chemical messengers that accomplish indirect communication between cells. These chemical messengers differ in their source and the distance they travel to reach their target cells.

© Cengage Learning, 2013

duct electrical signals like neurons, but instead of directly innervating target cells, they release neurohormones into the circulatory fluid. The neurohormone is then distributed to the target cells. Thus, like endocrine cells, neurosecretory neurons release blood- or hemolymph-borne chemical messengers, whereas ordinary neurons secrete short-range neurotransmitters into a confined space.

• **Pheromones** (Figure 3-16h) are chemical signals released into the environment, usually by glands, and travel through the air or water to sensory cells in another animal. Such signals are used for sexual activity (such as signaling of readiness to mate), marking of territories,

and other behaviors related to interactions among individuals of a species.

• **Cytokines** (not shown in Figure 3-16) are another class of signal molecules, which scientists historically discovered in the mammalian immune system and initially studied outside the framework of the traditional chemical messengers. Cytokines are regulatory peptides with a variety of effects that can be local, like paracrines, or at a distance, like hormones. But they are not produced by specialized gland or paracrine cells. Instead, they can be made by almost any body cell with another (nonsignaling) primary function, and are generally involved in development and immunity.

Structures of Signal Molecules Molecules used as messengers make up a long list that is still growing. The list can be simplified into categories of chemical structures, as follows:

- *Eicosanoids:* Derivatives of fatty acids, **eicosanoids** are used primarily as paracrines. Examples include **prostaglandins**, which regulate vertebrate smooth-muscle action among other effects (Chapter 16), and **anandimide** (also spelled *anandamide*), a paracrine neuromodulator in the mammalian brain involved in appetite and memory (p. 196).
- *Gases:* Three inorganic gases, though normally considered toxic pollutants, have now been found to be produced in animals as natural messengers. **Nitric oxide (NO)**, a small, highly reactive, short-lived gas molecule once known primarily as a toxic air pollutant, has been found as a paracrine in many animal phyla. Also, researchers have found evidence that both **carbon monoxide (CO)** and **hydrogen sulfide (H_2S)** are paracrines. We'll examine NO and H_2S functions later.
- *Purines:* The primary messengers in this chemical group are **adenosine** and **ATP**. They are used as paracrines and neurotransmitters in addition to their roles in nucleic acids and cellular energy. Adenosine is one major target of **caffeine** (see later discussion).
- *Amines:* This large category consists of a few amino acids and numerous derivatives. It includes many paracrines, such as histamine and serotonin, and many hormones, such as epinephrine (Chapter 7).
- *Peptides and proteins:* Numerous amino-acid sequences are produced as messengers, as paracrines, cytokines, neurohormones, and hormones (Chapter 7).
- *Steroids:* These hydrophobic molecules are derived from cholesterol. They are primarily hormones and sometimes pheromones (Chapter 7). Their lipid nature permits them to diffuse directly through cell membranes.
- *Retinoids:* These messengers are derived from vitamin A and are mainly paracrines involved in development and differentiation. These will not be considered further.

Chemical messengers can have multiple, unrelated effects because of the "same key, different locks" principle

As we have noted, chemical messengers generally exert their effects by binding to receptor proteins. (We discuss the details of this shortly.) It is important to realize that one cannot deduce what the cellular response to any particular messenger will be from the messenger's chemical structure. This is because the *response is determined by the receptor*. Thus, a particular hormone or other signaling molecule can bind to a receptor on target cell A and trigger response 1, and also (at the same time or different time) bind to a receptor on target cell B and trigger a different response 2. Responses 1 and 2 may be completely unrelated. This may be called the *"same key, different locks"* principle. Think of a messenger as something like a key that can fit a variety of locks (receptors). Lock A may start your car, whereas lock B may open your house, and in principle, you could have one key that does both. (We don't usually have single keys that do both, but then again, evolution is often more economical than we are.)

TABLE 3-3 Some Functions of Nitric Oxide (NO)

- Causes relaxation of vertebrate arteriolar smooth muscle, helping to control blood flow through the tissues and in maintaining mean arterial blood pressure (Chapter 9).
- Dilates the arterioles of the mammalian penis and clitoris, thus serving as the direct mediator of erection of these reproductive organs. Erection is accomplished by rapid engorgement of these organs with blood (Chapter 16).
- Used as chemical warfare against bacteria and cancer cells by phagocytic cells of various animal immune systems (Chapter 10).
- Interferes with mammalian platelet function and blood clotting at sites of vessel damage.
- Serves as a novel type of neurotransmitter in brains and elsewhere (Chapter 5).
- Plays a role in the changes underlying memory (Chapter 5).
- By promoting relaxation of vertebrate digestive-tract smooth muscle, helps regulate peristalsis, a type of contraction that pushes digestive tract contents forward.
- Relaxes the smooth muscle cells in the airways of mammalian lungs, helping keep these passages open to facilitate movement of air in and out of the lungs (Chapter 11).
- Modulates the filtering process involved in mammalian urine formation.
- Directs blood flow to O_2-starved tissues in vertebrates.
- May play a role in relaxation of vertebrate skeletal muscle.
- In concert with CO, is involved in the cellular resistance to stress.
- Modulates light production in fireflies.

© Cengage Learning, 2013

Nitric oxide (NO) illustrates this principle nicely. Studies have revealed an astonishing number of biological roles for NO, many unrelated (Table 3-3). NO is derived from the amino acid arginine in an oxidative reaction that consumes oxygen and NADPH (an electron-hydrogen donor similar to NADH, p. 48). Because NO is a gas, and so cannot be stored and released as needed, the synthesis of NO is tightly regulated. Three nitric oxide synthase (NOS) isoenzymes, each the product of a unique gene, have been characterized. Nitric oxide has a variety of paracrine actions in almost all animal phyla that have been studied, ranging from memory in neurons to control of blood flow (for an example, see the box, *A Closer Look at Adaptation: Lighting Up the Night* on p. 96). In 1998, the Nobel Prize was awarded to Robert Furchgott, Louis Ignarro, and Ferid Murad for their pioneering work in this complex field.

This principle explains why so many drugs have so-called *side effects*: For example, the paracrine histamine has multiple unrelated roles. When released by certain immune cells (Chapter 10), it binds to H1 receptors, triggering parts of the inflammation process. In the stomach, histamine binds to H2 receptors that regulate acid production; and in the brain, it binds to H3 receptors that induce alertness. These functions appear to be independent. They explain why many antihistamine drugs, designed to block inflammation, also make you drowsy: Those drugs bind to and block histamine from binding both H1 and H3 receptors.

Extracellular chemical messengers bring about cell responses primarily by signal transduction

Extracellular chemical messengers, such as hormones delivered by the blood or neurotransmitters released from nerve endings, usually do not trigger changes in target cells directly. Some cannot even enter their target cells to bring about the desired intracellular response. Instead, most of these messengers issue their orders by binding with receptors (with specific binding sites for one type of messenger) that trigger a biochemical chain of events inside the target cell. This process is called **signal transduction,** in which incoming signals (instructions from extracellular chemical messengers) are conveyed to the target cell's interior for execution. (A *transducer* is a device that receives energy or information from one system and transmits it in a different form to another system. For example, your radio receives radio waves sent out from the broadcast station and transmits these signals in the form of sound waves that can be detected by your ears.) The combination of extracellular messenger with receptor triggers a sequence of intracellular events that ultimately controls a particular cellular activity important in maintaining homeostasis or activating a regulated change, such as membrane transport, secretion, metabolism, contraction, or division and differentiation.

Signal transduction occurs by two broad types of mechanisms, depending on the messenger and receptor type. (1) **Lipophilic** (lipid-soluble) extracellular messengers gain entry into the cell by dissolving in and passing through the lipid bilayer of the target cell's plasma membrane. Small gases also penetrate the membrane. These messengers usually bind to receptors inside the target cell to initiate the desired intracellular response themselves, usually by changing gene activity. (2) In contrast, most **lipophobic** (water-soluble) extracellular messengers cannot gain entry to the target cell because they are poorly soluble in lipid and cannot dissolve in the plasma membrane. These messengers signal the cell to perform a given response by first binding with surface membrane receptors specific for that given messenger. Let's examine these two types in more depth.

Internal receptors bind lipophilic and gaseous external messengers that enter the cell

Dispersed within the cytoplasm and nucleus of most cells are special protein receptors that bind small and hydrophobic chemical messengers that can penetrate the cell membrane. These messengers include steroid hormones, eicosanoids, the hydrophobic amine hormones of the thyroid gland, and gases. (In recent years, some lipophilic messengers have been found to act on receptors within the same cell that produced them, a process called **intracrine** signaling.) Following the binding of messenger to the receptor, at least two different events can ensue, depending on the system:

- *Nitric oxide*: The internal receptor for NO is also an enzyme, *guanylyl cyclase*, which produces a second messenger, *cyclic GMP*. This in turn activates other cell proteins.
- *Thyroid and steroid hormones*: These have what might be viewed as more profound effects, because they alter gene transcription. Here the internal receptor is a **transcription factor,** a protein that, when bound with its messenger, can recognize a particular promoter sequence of genes (the *enhancer* or *response element* sequence, p. 31). The adjoining coding gene is then transcribed, an mRNA is translated, and new proteins are produced in the cell. Details of this mode of action are covered at the gene level in Chapter 2, and at the hormonal level in Chapter 7.

Some lipophobic external messengers bind to plasma membrane receptors, which alter ion channels or trigger phosphorylating enzymes (kinases)

The second broad mechanism of transduction involves the lipophobic messengers that bind to membrane receptors. Despite the wide range of possible responses, binding of the receptor with the extracellular messenger (the **first messenger**) brings about the desired intracellular response by only three general means: (1) by opening or closing specific channels in the membrane to regulate the movement of particular ions into or out of the cell, (2) by activating an enzyme that phosphorylates a cell protein, or (3) by transferring the signal to an intracellular chemical messenger (the **second messenger**), which in turn triggers a preprogrammed series of biochemical events within the cell. Because of the universal nature of these events, let's examine each more closely.

Chemically Gated Receptor-Channels Some first messengers bind to receptors that open or close *gated-receptor-channels* that regulate movement of particular ions across the membrane. In many cases, the *receptor itself serves as an ion channel* (Figure 3-17a). When the appropriate extracellular messenger binds to the **receptor-channel,** the channel opens or closes, depending on the signal. There are three different types of gated channels, depending on which of the following mechanisms bring about channel opening and closing: (1) **chemically gated** or **ligand gated,** by binding of an extracellular chemical messenger to a specific membrane receptor that is in close association with, or an integral part of, the channel; (2) **voltage gated,** by changes in electrical current in the plasma membrane; and (3) **mechanically gated,** by stretching or other mechanical deformation of the channel. For now, we will concentrate on chemical gating and defer the other two methods of controlling channels until later (for example, see p. 117).

For other channels, the receptor is a separate protein located near the channel. In this case, binding of the extracellular messenger with its receptor activates membrane-bound intermediaries known as **G proteins,** which in turn open (or in some instances close) the appropriate adjacent channel. (The mechanism of action of G proteins is discussed shortly.)

By opening or closing specific channels, extracellular messengers can regulate the flow of particular ions across

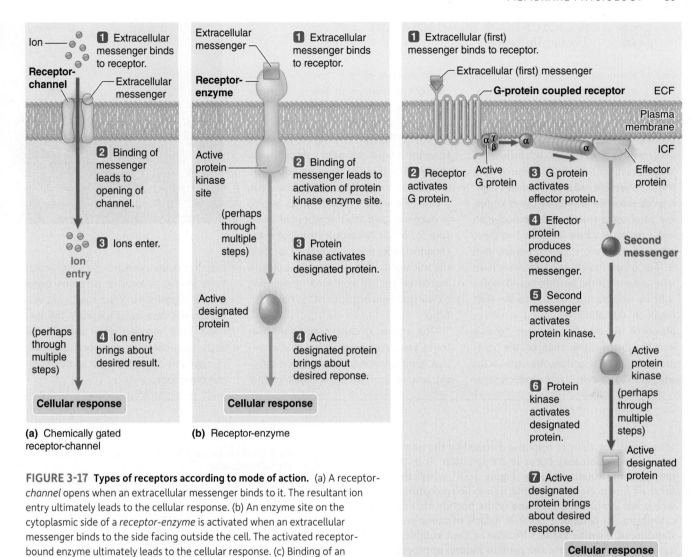

FIGURE 3-17 Types of receptors according to mode of action. (a) A receptor-*channel* opens when an extracellular messenger binds to it. The resultant ion entry ultimately leads to the cellular response. (b) An enzyme site on the cytoplasmic side of a *receptor-enzyme* is activated when an extracellular messenger binds to the side facing outside the cell. The activated receptor-bound enzyme ultimately leads to the cellular response. (c) Binding of an extracellular (first) messenger to the extracellular side of a *G-protein coupled receptor* activates a membrane-bound effector protein by means of a G-protein intermediary. The effector protein produces an intracellular second messenger, which ultimately leads to the cellular response.

© Cengage Learning, 2013

the membrane. This ionic movement can be responsible for two different cellular events:

1. A small, short-lived movement of Na^+, K^+, or both across the membrane alters the electrical activity of cells that are capable of generating electrical signals (or impulses), such as nerve and muscle cells.
2. A transient flow of calcium (Ca^{2+}) into the cell through opened Ca^{2+} channels triggers an alteration in shape and function of specific intracellular proteins, which leads to the cell's response. Illustrative is the increase in cytosolic Ca^{2+} responsible for triggering the release of secretory product from many gland cells.

On completion of the response, the extracellular messenger is removed from the receptor site, and the chemically gated channels close once again. The ions that moved across the membrane through opened channels to trigger the response are returned to their original location by spe-

cial membrane carriers. Cytosolic Ca^{2+} may be increased in another way besides entry from the ECF through opened membrane channels: Intracellular Ca^{2+} stores can be released in a second-messenger pathway, as described in the next section.

Phosphorylating Enzymes and the Tyrosine Kinase Pathway Some receptors have an associated enzyme, called a **protein kinase**, that phosphorylates a target cell protein (Figure 3-17b). This activity is activated when a first messenger binds to the receptor. Recall that *phosphorylation* refers to the transfer of a phosphate group from ATP to the protein at the expense of degrading ATP to ADP. Attachment of a phosphate group to a protein induces the protein to change its shape and function (either activating or inhibiting it) to bring about the desired response.

In response to phosphorylation, these proteins alter their shape and function, that is, are activated, to ultimately

A Closer Look at Adaptation
Lighting Up the Night

One of the most memorable natural displays that humans encounter are the summertime light shows of lightning bugs or fireflies. Male fireflies (actually a type of beetle) produce yellowish light in their abdomens as a way to attract mates; the longer and brighter a male produces light, the more likely he will attract females. Light is produced when an enzyme called *luciferase* catalyses a reaction between oxygen and a molecule called *luciferin.* This takes place in organelles called *peroxisomes* (p. 46) in light-organ cells. Scientists have long known that the beetle's brain triggers light by neurons that extend to the light organ in the abdomen. But the nerves attach to the air tubes (trachea) leading into the organ rather than on the light-producing cells. How does the signal actually get to those cells? Light production re-

quires oxygen, which is normally too low for light production in the light organs because mitochondria consume oxygen in the course of normal aerobic metabolism (pp. 51–53). The secret to light regulation is to let the oxygen get to the light cells. New studies show that the nerve signal triggers an enzyme to produce *nitric oxide (NO),* a paracrine signal researchers discovered in mammals as a local regulator of blood flow. In fireflies, NO appears to inhibit mitochondrial processes, allowing oxygen from the trachea to diffuse to the light cells. A crucial experiment consisted of exposing a dissected light organ to NO in the lab. It lit up!

The control of light production illustrates two major mechanisms of intercellular communication. First, a neuron carrying a signal from a central regulator (the

Photo: © Michael Jeffords

brain) releases a neurotransmitter molecule; second, a local or paracrine signal (NO) is triggered to affect the target system (in this case, mitochondria in the light cells). We'll mention many examples of such long-distance and local signaling mechanisms throughout our study of animal physiology.

---

accomplish the cellular response dictated by the first messenger. Transduction may occur in a single step, as in the common **tyrosine kinase** pathway (Figure 3-18), where the *receptor itself functions as an enzyme,* a so-called **receptor-enzyme**, which has a protein kinase site on its portion that faces the cytoplasm. In actuality, there is a lack of complete understanding of tyrosine kinases because of their complexity and multiplicity. To activate the protein kinase, appropriate extracellular messengers must bind with two of these receptors, which assemble into a pair. On activation, the receptor's protein kinase adds phosphate groups to tyrosine residues incorporated in the receptor itself (**autophosphorylation**) (Figure 3-18). Designated proteins inside the cell recognize and bind to the phosphorylated receptor. Then the receptor's protein kinase adds phosphate groups to the bound proteins. As a result of phosphorylation, the proteins change shape and function (are activated), enabling them to bring about the desired cellular response. In this pathway, the portions of the proteins that are specifically phosphorylated in both the receptor and designated proteins contain the amino acid tyrosine. Because of this specificity, such receptor-enzymes are called **tyrosine kinases.**

The hormone insulin, which plays a major role in maintaining glucose homeostasis, exerts its effects via tyrosine kinases. Also, many growth factors that help regulate cell growth and division, such as *nerve growth factor* and *epidermal growth factor,* act via this pathway. Receptor tyrosine kinases also play crucial roles in normal physiological processes including embryogenesis, cell proliferation and cell death. This network can also detect, amplify, filter, and process a variety of environmental and intercellular cues.

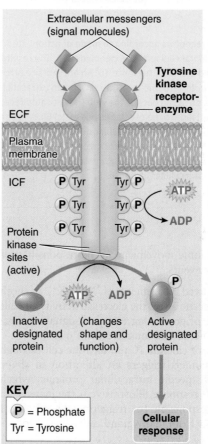

1. Two extracellular messengers bind to two receptors and receptors pair, activating receptor's protein kinase site.

2. Protein kinase site self-phosphorylates receptor's tyrosines.

3. Inactive designated protein binds to receptor, which phosphorylates protein, activating it.

4. Active designated protein brings about desired response.

KEY

P = Phosphate
Tyr = Tyrosine

FIGURE 3-18 Tyrosine kinase pathway.

© Cengage Learning, 2013

Other lipophobic external messengers bind to plasma membrane receptors, which stimulate second-messenger production, often via G proteins

Most lipophobic hormones work more indirectly than the mechanisms we've just described by binding to **G-protein–coupled membrane receptors (GPCRs)**, which in turn exert effects in target cells by acting through a **second-messenger system** to alter the activity of preexisting proteins. There are two major second-messenger pathways: One uses **cyclic adenosine monophosphate (cyclic AMP, or cAMP)** as a second messenger, while the other employs **diacylglyerol-inositol trisphosphate** and Ca^{2+} in this role.

The G-protein system is found on the inner surface of the plasma membrane, as an intermediary between the receptor and effector protein (Figure 3-17c). G proteins are so named because they are bound to guanine nucleotides—*guanosine triphosphate (GTP)* when active or *guanosine diphosphate (GDP)* when inactive. An inactive G protein consists of a complex of alpha (α), beta (β), and gamma (γ) subunits, with a GDP molecule bound to the α subunit. A number of different G proteins with varying α subunits have been identified. The different G proteins are activated in response to binding of various first messengers to surface receptors. When an appropriate extracellular messenger (a first messenger) binds with its receptor, the receptor attaches to the appropriate G protein, resulting in release of GDP from the G-protein complex. GTP then attaches to the α subunit, an action that activates the G protein. Once activated, the G protein breaks away from the G-protein complex and moves along the inner surface of the plasma membrane until it reaches an effector protein. An effector protein is either an enzyme or an ion channel within the membrane. The α subunit links up with the effector protein and alters its activity. Researchers have identified more than 300 different receptors that convey instructions of extracellular messengers through the membrane to effector proteins by means of G proteins.

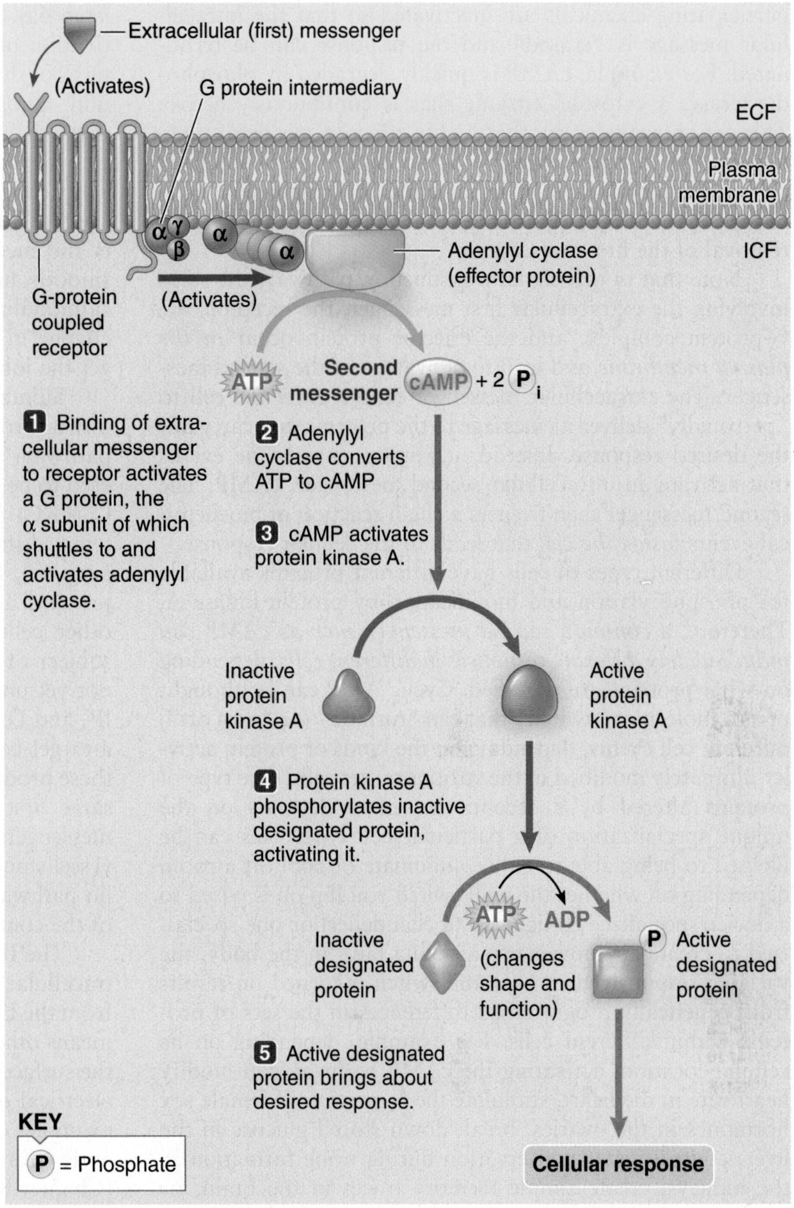

FIGURE 3-19 Mechanism of action of hydrophilic hormones via activation of the cyclic AMP second-messenger pathway.

© Cengage Learning, 2013

Cyclic AMP Second-Messenger GPCR Pathway Cyclic AMP is the most widely used second messenger. In the following description of the cAMP pathway, the numbered steps correlate to the numbered steps in Figure 3-19.

Binding of the first messenger to its surface membrane receptor (GPCR) activates the associated G protein, which in turn activates the effector protein, in this case the enzyme **adenylyl cyclase** (step 1), located on the cytoplasmic side of the plasma membrane. Adenylyl cyclase induces the conversion of intracellular ATP to cAMP by cleaving off two of the phosphates (step 2). (This is the same ATP used as the common energy currency in the body.) Acting as the intracellular second messenger, cAMP triggers a preprogrammed series of biochemical steps within the cell to bring about the response

dictated by the first messenger. To begin, cyclic AMP activates a specific intracellular enzyme, **protein kinase A** (step 3). Protein kinase A, in turn, phosphorylates a designated preexisting intracellular protein, such as an enzyme important in a particular metabolic pathway. Phosphorylation causes the protein to change its shape and function, thereby activating it (step 4). This activated protein brings about the target cell's ultimate response to the first messenger (step 5). For example, the activity of a particular enzymatic protein that regulates a specific metabolic event may be increased or decreased. After the response is completed, the subunit cleaves off a phosphate, converting GTP to GDP, in essence shutting itself off, then rejoins the β and γ subunits to restore the inactive G-protein complex. Cyclic AMP and the other participating chemicals are inactivated so that the intracellular message is "erased" and the response can be terminated. For example, cAMP is quickly degraded by **phosphodiesterase**, a cytosolic enzyme that is continuously active. This action provides another highly effective means of turning off the response when it is no longer needed. Other complementary means of terminating the response are removal of the added phosphates by protein phosphatase or removal of the first messenger.

Note that in this signal transduction pathway, the steps involving the extracellular first messenger, the receptor, the G-protein complex, and the effector protein occur *in the plasma membrane* and lead to activation of the second messenger. The extracellular messenger cannot enter the cell to "personally" deliver its message to the proteins that carry out the desired response. Instead, it initiates membrane events that activate an intracellular second messenger, cAMP. The second messenger then triggers a chain reaction of biochemical events *inside the cell* that leads to the cellular response.

Different types of cells have different proteins available for phosphorylation and modification by protein kinase A. Therefore, a *common second messenger such as cAMP can induce widely differing responses in different cells*, depending on what proteins are modified. Cyclic AMP can be thought of as a molecular "switch" that can "turn on" (or "turn off") different cell events, depending on the kinds of protein activity ultimately modified in the various target cells. The type of proteins altered by a second messenger depends on the unique specialization of a particular cell type. This can be likened to being able to either illuminate or cool off a room depending on whether the wall switch you flip on is wired to a device specialized to light up (a chandelier) or one specialized to create air movement (a ceiling fan). In the body, the variable responsiveness once the switch is turned on results from genetically programmed differences in the sets of proteins within different cells. For example, depending on its cellular location, activating the cAMP pathway can modify heart rate in the heart, stimulate the formation of female sex hormones in the ovaries, break down stored glucose in the liver, control water conservation during urine formation in the kidneys, create simple memory traces in the brain, or cause "perception" of a sweet taste by a taste bud.

Diacylglycerol-Inositol Trisphosphate-Ca^{2+} Second-Messenger Pathway Some cells use Ca^{2+} instead of cAMP as a second messenger. In such cases, binding of the first messenger to a GPCR eventually leads by means of G proteins to activation of the enzyme **phospholipase C**, an effector protein bound to the inner side of the membrane (step 1 in Figure 3-20). This enzyme breaks down **phosphatidylinositol bisphosphate** (abbreviated **PIP_2**), a component of the tails of the phospholipid molecules within the membrane itself. The products of PIP_2 breakdown are **diacylglycerol (DAG)** and **inositol trisphosphate (IP_3)** (step 1). Lipid-soluble DAG remains in the lipid bilayer of the plasma membrane, but water-soluble IP_3 diffuses into the cytosol. IP_3 mobilizes intracellular Ca^{2+} stored in the endoplasmic reticulum to increase cytosolic Ca^{2+} by binding with IP_3-gated receptor-channels in the ER membrane (step 3a). Calcium then takes on the role of second messenger, ultimately bringing about the response commanded by the first messenger. Many of the Ca^{2+}-dependent cellular events are triggered by activation of **calmodulin**, an intracellular Ca^{2+}-binding protein

(step 4a). Activation of calmodulin by Ca^{2+} is similar to activation of protein kinase A by cAMP. From here, the patterns of the two pathways are similar. The activated calmodulin phosphorylates the designated proteins (perhaps through multiple steps), thereby causing these proteins to change their shape and function, activating them (step 5a). The active designated proteins bring about the ultimate desired cellular response (step 6a). For example, this pathway is the means by which chemical messengers can activate smooth muscle contraction. Note that in this pathway, calmodulin, not protein kinase, brings about the necessary change in form and function of the designated proteins to get the job done.

Simultaneous to the IP_3 pathway, the other PIP_2 breakdown product, DAG, sets off another second-messenger pathway. (IP_3 and DAG themselves are sometimes considered to be second messengers.) DAG activates **protein kinase C (PKC)** (step 3b), which phosphorylates designated proteins, different from those phosphorylated by calmodulin (step 4b). The resultant change in shape and function of these proteins activates them. These active proteins produce another cellular response (step 5b). Although currently the subject of considerable investigation, the DAG pathway is not yet understood as well as the other signaling pathways. IP_3 and DAG typically trigger complementary actions inside a target cell to accomplish a common goal because both of these products are formed at the same time in response to the same first messenger. For example, extracellular chemical messengers promote increased contractile activity of blood-vessel smooth muscle via the IP_3/intracellular Ca^{2+}/calmodulin pathway, and the DAG pathway enhances the sensitivity of the contractile apparatus to Ca^{2+}.

The IP_3 pathway is not the only means of increasing intracellular Ca^{2+}. Intracellular Ca^{2+} can be increased by entry from the ECF or by release from Ca^{2+} stores inside the cell via means other than the IP_3 pathway. Calcium channels in both the surface membrane and in the ER may be opened by either electrical or chemical means. In later chapters, you will see examples of voltage changes regulating Ca^{2+} channels in synaptic transmission (Chapter 4) and muscle contraction (Chapter 8). The pathways can get even more complex than this. Calcium entering from the ECF can serve as a second messenger to trigger an even larger release of Ca^{2+} from intracellular stores, as it does to bring about contraction in cardiac muscle (Chapter 9). This all sounds confusing, but these examples are meant to illustrate the complexity of signaling, not to overwhelm.

Furthermore, the cAMP and Ca^{2+} pathways frequently overlap in bringing about a particular cellular activity. For example, cAMP and Ca^{2+} can influence each other. Calcium-activated calmodulin can regulate adenylyl cyclase and thus influence cAMP, whereas protein kinase A may phosphorylate and thereby change the activity of Ca^{2+} channels or carriers. Therefore, the cell-signaling pathways may communicate with one another to integrate their responses. This interpathway interaction is called **cross talk**.

Although the cAMP and Ca^{2+} pathways are the most prevalent second-messenger systems, they are not the only ones. For example, in a few cells **cyclic guanosine monophosphate (cyclic GMP)** serves as a second messenger in a system analogous to the cAMP system. In other cells, the second messenger is still unknown. Remember that activation of second messengers is a universal mechanism employed by a

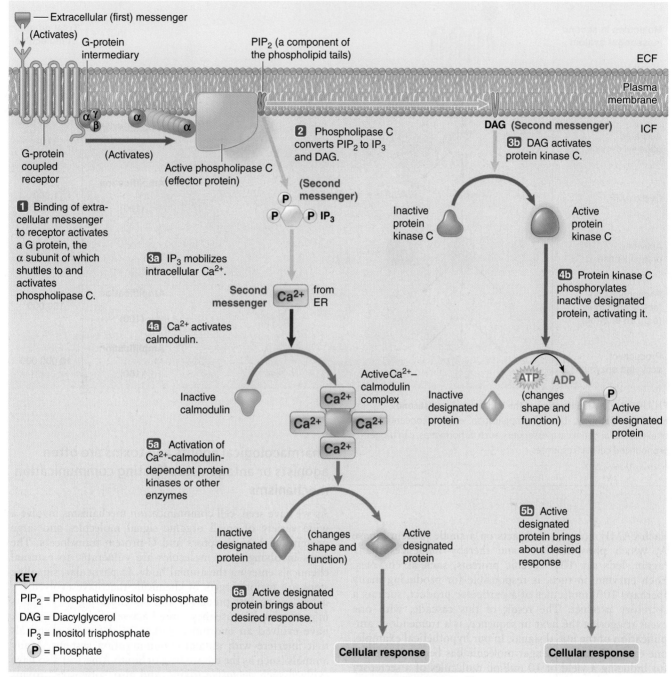

FIGURE 3-20 Mechanism of action of hydrophilic hormones via concurrent activation of the IP_3/Ca^{2+} second-messenger pathway and the DAG pathway.

© Cengage Learning, 2013

variety of extracellular messengers in addition to hydrophilic hormones. (See the box, *Molecular Biology and Genomics: Programmed Cell Suicide* on p. 101, for a surprising signal-transduction pathway—one that causes a cell to kill itself.)

Amplification by a Second-Messenger Pathway Several remaining points about receptor activation and the ensuing events merit attention. First, considering the number of steps in a second-messenger relay chain, you might wonder why so many cell types use the same complex system to

accomplish such a wide range of functions. The multiple steps of a second-messenger pathway are actually advantageous because the cascading (multiplying) effect of these pathways greatly amplifies the initial signal (Figure 3-21). *Amplification* means that the output of a system is much greater than the input. Using the cAMP pathway as an example, binding of one extracellular messenger molecule to a receptor activates a number of adenylyl cyclase molecules (let's arbitrarily say 10), each of which activates many (in our hypothetical example, let's say 100) cAMP molecules.

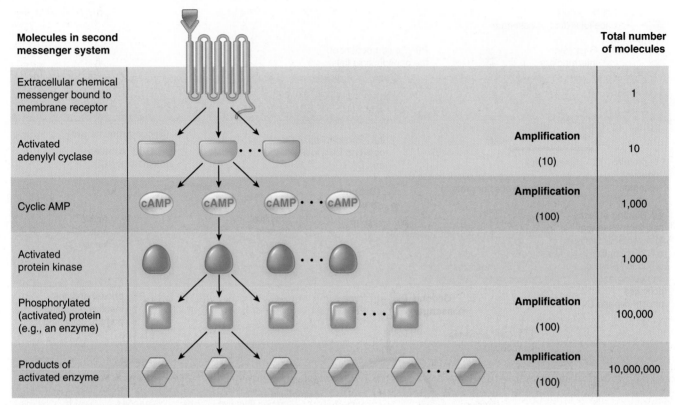

Molecules in second
messenger system

Total number
of molecules

	Total number of molecules
Extracellular chemical messenger bound to membrane receptor	1
Activated adenylyl cyclase	**Amplification** (10) — 10
Cyclic AMP	**Amplification** (100) — 1,000
Activated protein kinase	1,000
Phosphorylated (activated) protein (e.g., an enzyme)	**Amplification** (100) — 100,000
Products of activated enzyme	**Amplification** (100) — 10,000,000

FIGURE 3-21 Amplification of the initial signal by a second-messenger pathway. Through amplification, very low concentrations of extracellular chemical messengers, such as hormones, can trigger pronounced cellular responses.

© Cengage Learning, 2013

Each cAMP molecule then acts on a single protein kinase A, which phosphorylates and thereby influences many (again, let's say 100) specific proteins, such as enzymes. Each enzyme, in turn, is responsible for producing many (perhaps 100) molecules of a particular product, such as a secretory product. The result of this cascade, with one event triggering the next in sequence, is a tremendous amplification of the initial signal. In our hypothetical example, one extracellular messenger molecule has been responsible for inducing a yield of 10 million molecules of a secretory product. In this way, very low concentrations of hormones and other chemical messengers can trigger pronounced cell responses.

Regulation of Receptors Although membrane receptors serve as links between extracellular first messengers and intracellular second messengers in the regulation of specific cellular activities, the receptors themselves are also frequently subject to regulation. In many instances, receptor number and affinity (attraction of a receptor for its chemical messenger) can be altered, depending on the circumstances. For example, the number of receptors for the hormone insulin can be decreased in response to a chronic elevation of insulin in mammalian blood. This is usually called *downregulation* if receptors are decreased in number, or *upregulation* if increased (akin to acclimatization, p. 16).

Pharmacological agents and toxins are often agonists or antagonists affecting communication mechanisms

As we have seen, cell communication mechanisms involve a wide variety of small organic signal molecules and large proteins such as receptors and G-protein transducers. The steps involving these molecules are vulnerable to external chemicals entering the animal body. In particular, signaling steps are major targets of chemicals that have evolved for defending against predators and for capturing prey. Plants in particular, because they cannot actively fight or run away, have evolved an enormous variety of defensive chemicals that interfere with animal communication systems. Many animals (such as the curare frog of South America) also have evolved such defensive toxins, and also "offensive" toxins such as snake venom that are used to capture prey. Chemicals that affect a communication mechanism fall into two broad categories:

- **Antagonists** block a step in a communication pathway. Often these are chemicals that are similar in molecular structure to a signal molecule, but different enough to cause interference. Frequently, an antagonist binds to a receptor without activating that receptor, while at the same time blocking the binding of the normal signal molecule. For example, the active agent of the curare frog, *D-tubocurarine*, binds to and blocks the acetylcholine receptor at nerve–muscle synapses. As we shall see later (p. 134), acetylcholine normally triggers receptors that initiate muscle contraction; by blocking the receptor, the toxin causes paralysis. Another example is *antihistamine*, discussed earlier (p. 94).

Molecular Biology and Genomics
Programmed Cell Suicide

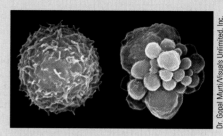

A normal cell (*left*) and a cell undergoing apoptosis (*right*).

In the vast majority of cases, the signal transduction pathways triggered by the binding of an extracellular first messenger to a membrane receptor are aimed at promoting useful functions, growth, survival, or reproduction of the cell. By contrast, most animal cells have a built-in pathway that, if triggered, causes the cell to commit suicide by activating intracellular protein-snipping enzymes, which slice the cell into small, disposable pieces. Such deliberate programmed cell death is termed **apoptosis** (meaning "dropping off" of cells that are no longer useful, much as autumn leaves drop off trees; pronounced "app-oh-TOE-sis" with a silent "p" as in "pterodactyl"). Apoptosis differs from the other form of cell death, **necrosis** (meaning "make dead"). Necrosis is uncontrolled, accidental, messy murder of useful cells that have been severely injured by, for example, a physical blow, O_2 deprivation, or disease.

Apoptosis is a normal part of life: Individual cells that have become superfluous or disordered are triggered to self-destruct for the greater good of maintaining the whole body's health or to aid development. Genomic analysis shows that even the most primitive animals, sponges, have apoptosis genes, but protists do not. Scientists believe that cell suicide, along with cell–cell adhesion and growth-control mechanisms, were key innovations in the evolution of multicellularity. Here are some examples of apoptotic functions:

1. *Development.* Certain unwanted cells produced during development are programmed to kill themselves as a body is sculpted into its final form. For example, apoptosis deliberately prunes the embryonic ducts capable of forming a mammalian male reproductive tract during the development of a female. Likewise, apoptosis destroys many larval tissues when a caterpillar pupates and becomes a butterfly.

2. *Turnover.* Optimal functioning of most tissues depends on a balance between controlled production of new cells and regulated cellular self-destruction. This balance maintains the proper number of cells in a given tissue while ensuring a controlled supply of fresh cells that are at their peak of performance.

3. *Defense.* Apoptosis provides a means to remove cells infected with harmful viruses. Furthermore, infection-fighting white blood cells (in vertebrates) that have finished their prescribed function and are no longer needed, execute themselves. Also, apoptosis often removes cells that have suffered irreparable damage by exposure to radiation or other poisons, and cells that are becoming cancerous.

A cell signaled to commit suicide detaches itself from its neighbors, then shrinks instead of swelling and bursting. As its lethal weapon, the suicidal cell activates a cascade of normally inactive intracellular protein-cutting enzymes, the **caspases**, which kill the cell from within. When a cell has been signaled to undergo apoptosis, the mitochondria become leaky, permitting *cytochrome c* to leak out into the cytosol, where it activates the caspase cascade. (Cytochrome c, a component of the electron transport chain, usually participates in oxidative phosphorylation to produce ATP; see Figure 2-20, p. 52). Once unleashed, the caspases act like molecular scissors to systematically dismantle the cell. Snipping protein after protein, they chop up the nucleus, then break down the internal shape-holding cytoskeleton, and finally fragment the cell itself into disposable membrane-enclosed packets (see accompanying photo). Importantly, the contents of the dying cell remain wrapped by plasma membrane throughout the entire self-execution process, thus avoiding the spewing of potentially harmful intracellular contents characteristic of necrosis. Cells in the vicinity swiftly engulf and destroy the apoptotic cell fragments by phagocytosis. The breakdown products are then recycled for other purposes as needed.

Evolutionary differences in apoptosis genes may explain a curious difference between humans and chimpanzees: The former are subjected to a variety of deadly cancers while the latter very rarely get cancer. Genomic analysis shows that, while human–chimp genes differ by an average of two amino acid mutations per protein, genes for apoptotic and DNA-repair proteins differ by eight amino acid mutations. This has resulted in cancerous cells avoiding apoptosis more often in humans than in chimps. Why did these differences evolve? One hint is that the same genes that reduce apoptosis in humans allow for sperm—which undergo apoptosis as they age—to survive longer and thus increase human fertility.

- **Agonists** activate a step in a communication pathway. Again, often these are chemicals that are similar in molecular structure to a signal molecule, but in this case they cause the same general effect; for example, an agonist may bind to a receptor and activate it, just as the normal signal molecule would. However, the agonist is often structurally somewhat different so that it may bind for an abnormally long time, or it may not be broken down effectively to shut off the stimulus. For example, *nicotine* (found in tobacco) is a long-lasting agonist of one class of acetylcholine receptors ("nicotinic receptors"; see p. 164).

A common class of antagonistic molecule is the **methylxanthines**, *caffeine* and *theophylline*, molecules that have structures similar to those of cAMP and adenosine. Caffeine is found in coffee and cacao beans (chocolate). In tea leaves, the concentration of theophylline can be as high as 3.5%. The activity of PDE (phosphodiesterase, which shuts off the cAMP cascade) is decreased by these antagonistic chemicals, which bind to the enzyme. These compounds also block the receptors for **adenosine**, a paracrine that usually dampens neural activity by opening potassium channels (Chapter 4). Methyl xanthines are noted for their powerful stimulatory effects on the central nervous system and (in vertebrates) heart function. These compounds probably evolved to disrupt the feeding behavior of herbivorous insects. When administered artificially to animals, methyl xanthines exert unpredictable effects on their behavior. For example, caffeine administration prevents spiders from spinning a functional web. In many cases, scientists do not know whether inhibition of PDE or adenosine receptors or both are the basis of these effects.

Much medical and physiological research is based on using natural agonists and antagonists (or laboratory-designed ones) as pharmacologic agents (or drugs). Medicine, of course, uses these for treating various diseases, whereas physiologists use them as probes of signal functions.

check your understanding 3.4

Describe the major mechanisms of signal transduction by lipophilic external messengers using internal receptors, and by lipophobic messengers and membrane receptors using ion channels, kinases, and G proteins.

By what process does a cell commit suicide? What cellular organelles are utilized in the process?

3.5 Membrane Potential

Earlier we noted that solutes are unequally distributed between the interior and exterior of cells. We have also discussed how some cells can exploit these differences in solute concentration on either side of the membrane and can selectively change the permeability of the membrane to particular solutes. In most cells, inorganic solutes are unevenly distributed as well as organic ones, with **sodium (Na^+)** and chloride (Cl^-) dominating outside and potassium (K^+) (along with organic molecules) dominating inside. Put another way, Na^+ and Cl^- are usually the major **osmolytes** (solutes involved in osmotic pressure regulation) in the **ECF (extracellular** fluid) in animals, whereas K^+ is the dominant inorganic osmolyte in the **ICF (intracellular fluid)**. Indeed, the occurrence of K^+ as the dominant intracellular cation appears universal from prokaryotes on up and suggests that this feature originated with the earliest life forms. (In Chapter 13, we'll consider why higher K^+ than Na^+ in cells is adaptive for all life.) **Excitable cells** such as neurons regulate permeability of these ions (especially Na^+ and K^+) as a signaling mechanism, as we discuss in the next chapter.

Membrane potential is a separation of opposite charges across the plasma membrane

All plasma membranes have a membrane potential, or are polarized electrically. The term **membrane potential** refers to a separation of charges across the membrane, or to a difference in the relative number of cations and anions in the ICF and ECF. Recall that opposite charges tend to attract each other and like charges tend to repel each other. Work must be performed (energy expended) to separate opposite charges after they have come together. Conversely, when oppositely charged particles have been separated, the electrical force of attraction between them can be harnessed to perform work when the charges are permitted to come together again. This is the basic principle underlying electrically powered devices. We use the term *membrane potential* because separated charges now have the potential to do work. Potential is measured in units of volts (the same unit used for the voltage in electrical devices), but because the membrane potential is relatively low the unit used is **millivolts** (mV) (1 mV = 1/1,000 volt).

Because the concept of potential is fundamental to understanding nerve and muscle physiology, it is important to understand clearly what this term means. The membrane in Figure 3-22a is electrically neutral. An equal number of positive (+) and negative (−) charges are on each side of the membrane, so no membrane potential exists. In Figure 3-22b, some of the + charges from the right side have been moved to the left. Now the left has an excess of + charges, leaving an excess of − charges on the right. In other words, there is a separation of opposite charges across the membrane, or a difference in the relative number of + and − charges between the two sides (that is, a membrane potential exists). The attractive force between these separated charges causes them to accumulate in a thin layer along the outer and inner surfaces of the plasma membrane (Figure 3-22c). It is important to realize that these separated charges represent only a small fraction of the total number of charged particles (ions) present in the ICF and ECF. The vast majority of the fluid inside and outside the cells is electrically neutral (Figure 3-22d). The electrically balanced ions can be ignored because they do not contribute to membrane potential. Thus, an almost insignificant fraction of the total number of charged particles present in the body fluids is responsible for the membrane potential.

Note that the membrane itself is not charged. The term *membrane potential* refers to the difference in charge between the wafer-thin regions of ICF and ECF lying next to the inside and outside of the membrane, respectively. The magnitude of the potential depends on the degree of separation of the opposite charges: The greater the number of charges separated, the larger the potential.

Membrane potential is primarily due to differences in the distribution and permeability of key ions

All living cells have a membrane potential characterized by a slight excess of positive charges outside and a corresponding slight excess of negative charges on the inside. The cells of *excitable tissues*—namely nerve cells and muscle cells—have the ability to exploit this relationship and produce rapid, transient changes in their membrane potential when excited. These brief fluctuations in potential serve as electrical signals. The constant membrane potential present in nonexcitable tissues and in excitable tissues when they are at rest—that is, when they are not producing electrical signals—is known as the **resting membrane potential**. We concentrate now on the generation and maintenance of the resting potential, and in later chap-

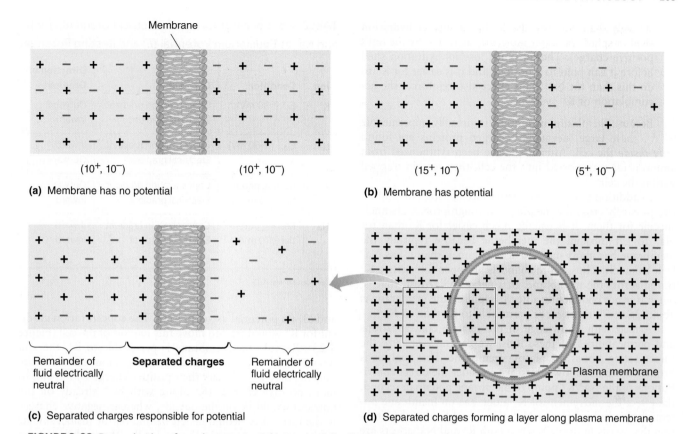

FIGURE 3-22 Determination of membrane potential by unequal distribution of positive and negative charges across the membrane. (a) When the positive and negative charges are equally balanced on each side of the membrane, no membrane potential exists. (b) When opposite charges are separated across the membrane, membrane potential exists. (c) The unbalanced charges responsible for the potential accumulate in a thin layer along opposite surfaces of the membrane. (d) The vast majority of the fluid in the ECF and ICF is electrically neutral. The unbalanced charges accumulate along the plasma membrane.

© Cengage Learning, 2013

ters, we examine the changes that take place in excitable tissues during signaling.

The unequal distribution of a few key ions between the ICF and ECF and their selective movement through the plasma membrane govern the electrical properties of the membrane. In an animal body, electrical charges are carried by ions. The ions primarily responsible for generating the resting membrane potential are Na^+, K^+, and A^-. The last refers to the large, negatively charged (anionic) intracellular proteins. Other ions (calcium, magnesium, chloride, bicarbonate, and phosphate, to name a few) do not make a direct contribution to the electrical properties of the plasma membrane in most cells, even though they play other important roles in the body. The concentrations of these ions differ from one animal to another, but the factors that regulate their distribution are the same.

The concentrations and relative permeabilities of the ions critical to membrane electrical activity are compared in Table 3-4. Note that *Na^+ is in greater concentration in the extracellular fluid and K+ is in much higher concentration in the intracellular fluid.* These concentration differences are maintained in two very distinct ways:

1. *The Na^+/K^+ pump* (Figure 3-14) *at the expense of energy (ATP).*

TABLE 3-4 Concentration and Permeability of Ions Responsible for Membrane Potential in a Resting Nerve Cell

Ion	Concentration (millimoles/liter)		Permeability Relative
	Extracellular	Intracellular	
Na^+	150	15	1
K^+	5	150	25–30
A^-	0	65	0

© Cengage Learning, 2013

2. *Different solubilities in cell water and affinity for cell proteins.* Due to attraction of water molecules by proteins, water inside cells appears to form a hydrogen-bonded network that is different from the network of water by itself. Experimentally, researchers have found that K^+ is more soluble in this internal water than is Na^+ and that this leads to K^+ preferentially entering a cell. Also, the negative charges of proteins attract K^+ more

strongly than Na^+ because K^+ has a smaller **hydration shell** (a sphere of water molecules attracted by the ion's positive charge). The ion must lose its hydration shell before it can bind to a protein, and it is easier for K^+ to do this than for Na^+. This effect also reinforces the accumulation of K^+ over Na^+.

Because the plasma membrane is virtually impermeable to A^-, these large, negatively charged proteins are found *only inside* the cell. After they have been synthesized from amino acids transported into the cell, they remain trapped within the cell.

In addition to the active carrier mechanism, Na^+ and K^+ can passively cross the membrane through protein channels specific for them. It is usually much easier for K^+ than for Na^+ to get through the membrane because typically the membrane has many more K^+ leak channels than it has Na^+ leak channels. Thus, more channels are open for passive K^+ traffic across the membrane. In a nerve cell at rest (that is, when it is not conducting a nerve impulse), the membrane is about 25 to 30 times more permeable to K^+ than to Na^+.

Armed with knowledge of the relative concentrations and permeabilities of these ions, we can now analyze the forces acting across the plasma membrane. This analysis is broken down as follows: First we consider the direct contributions of the Na^+/K^+ pump to membrane potential; second, the effect the movement of K^+ alone has on membrane potential; third, the effect of Na^+ alone; and finally, the situation that exists in the cells when both K^+ and Na^+ effects are taking place concurrently. Table 3-5 summarizes the concentration and electrical gradients that exist for K^+ and for Na^+ under various conditions. Remember throughout this discussion that the concentration gradient for K^+ is always outward and the concentration gradient for Na^+ is always inward because the Na^+/K^+ pump maintains a higher concentration of K^+ inside the cell and a higher concentration of Na^+ outside the cell. Also, note that because K^+ and Na^+ are both cations (positively charged), *the electrical gradient for both is always toward the negatively charged side of the membrane.*

Effect of Sodium/Potassium Pump on Membrane Potential

Recall that Na^+/K^+ pump transports three Na^+ out for every two K^+ it transports in. Because this unequal transport moves positive ions only, it generates a membrane potential, with the outside becoming relatively more positive than the inside as more ions are transported out than in. However, this active transport mechanism separates only enough charges to generate an almost negligible membrane potential of 1 mV to 3 mV, with the interior negative to the exterior of the cell. The vast majority of the membrane potential results from the passive diffusion of K^+ (and to a lesser extent, Na^+) down concentration gradients. Thus, most of the Na^+/K^+ pump's role in producing membrane potential is indirect, through its critical contribution to maintaining the concentration gradients directly responsible for the ion movements that generate most of the potential.

Effect of the Movement of Potassium Alone on Membrane Potential: Equilibrium Potential for K^+

Let's consider a hypothetical situation characterized by (1) the concentrations that exist for K^+ and A^- across the plasma membrane, (2) free permeability of the membrane to K^+ but not to A^-,

TABLE 3-5 Concentration and Electrical Gradients for K^+ and Na^+ at Equilibrium Potential (E) and Resting Potential

Ion	Condition	Gradient	Direction of Gradient
K^+	E_{K^+} (−90 mV)	Concentration gradient Electrical gradient	Outward Inward
Na^+	E_{Na^+} (+60 mV)	Concentration gradient Electrical gradient	Inward Outward
K^+	Resting potential (−70 mV)	Concentration gradient Electrical gradient	Outward Inward
Na^+	Resting potential (−70 mV)	Concentration gradient Electrical gradient	Inward Inward

© Cengage Learning, 2013

and (3) no potential as yet present. The concentration gradient for K^+ would tend to move this ion out of the cell (Figure 3-23). Because the membrane is permeable to K^+, this ion would readily pass through. As potassium ions moved to the outside, they would carry their positive charge with them, so more positive charges (K^+ along with Na^+ already on the outside) would be on the outside, whereas negative charges in the form of A^- would be left behind on the inside, similar to the situation shown in Figure 3-22b. (Remember that the large protein anions cannot diffuse out, despite a tremendous concentration gradient.) A membrane potential would now exist. Because an electrical gradient would also be present, K^+, being a positively charged ion, would be attracted toward the negatively charged interior and repelled by the positively charged exterior. Thus, two opposing forces would now be acting on K^+: the concentration gradient tending to move K^+ out of the cell and the electrical conduction gradient tending to move the ions into the cell.

Initially, the concentration gradient would be stronger than the electrical gradient, so net diffusion of K^+ out of the cell would continue and the membrane potential would increase. As more and more K^+ moved down its concentration gradient and out of the cell, however, the opposing electrical gradient would also become greater as the outside became increasingly more positive and the inside more negative. You might think that the outward concentration gradient for K^+ would gradually decrease as K^+ leaves the cell down this gradient. Surprisingly, however, the K^+ concentration gradient would remain essentially constant despite the outward movement of K^+. The reason is that only infinitesimal movement of K^+ out of the cell would bring about rather large changes in membrane potential. Accordingly, such an extremely small number of K^+ ions present in the cell would have to leave the cell to establish an opposing electrical gradient, that the K^+ concentration inside and outside the cell would remain essentially unaltered. As K^+ continued to move out down its unchanging concentration gradient, the inward electrical gradient would continue to increase in strength. Net outward movement would gradually be reduced as the strength of the electrical gradient approached that of the concentration gradient. Finally, when these two forces exactly balanced each other (that is, when they were in equilibrium), no further net movement of K^+ would occur. The po-

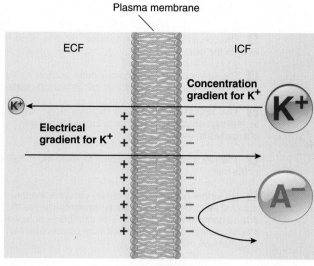

Plasma membrane

ECF | ICF

Concentration gradient for K$^+$

Electrical gradient for K$^+$

E_{K^+} = −90 mV

1 The concentration gradient for K$^+$ tends to move this ion out of the cell.

2 The outside of the cell becomes more positive as K$^+$ ions move to the outside down their concentration gradient.

3 The membrane is impermeable to the large intracellular protein anion (A$^-$). The inside of the cell becomes more negative as K$^+$ ions move out, leaving behind A$^-$.

4 The resulting electrical gradient tends to move K$^+$ into the cell.

5 No further net movement of K$^+$ occurs when the inward electrical gradient exactly counterbalances the outward concentration gradient. The membrane potential at this equilibrium point is the equilibrium potential for K$^+$ (E_{K^+}) at −90 mV.

FIGURE 3-23 Equilibrium potential for K$^+$.

© Cengage Learning, 2013

tential that would exist at this equilibrium is known as the **equilibrium potential** for K$^+$ (E_{K^+}). At this point, a large concentration gradient for K$^+$ would still exist, but no more net movement of K$^+$ would occur out of the cell down this concentration gradient because of the exactly equal opposing electrical gradient (Figure 3-23).

The membrane potential at E_{K^+} is −90 mV. It is not really a negative potential. By convention, *the sign always designates the polarity of the excess charge on the inside of the membrane.* A membrane potential of −90 mV means that the potential is of a magnitude of 90 mV, with the inside being negative relative to the outside. A potential of +90 mV would have the same strength, but in this case the inside would be more positive than the outside.

The equilibrium potential for a given ion of differing concentrations across a membrane can be calculated by means of the **Nernst equation** as follows:

$$E_{ion} = \frac{61}{z} \log \frac{C_0}{C_i}$$

where

E_{ion} = equilibrium potential for ion in mV

61 = a constant that incorporates the universal gas constant (R), absolute temperature (T), and an electrical constant known as Faraday (F), along with the conversion of the natural logarithm (*ln*) to the logarithm to base 10 (*log*); 61 = RT/F.

z = the ion's valence; z = 1 for K$^+$ and Na$^+$, the ions that contribute to membrane potential (for divalent ions such as Ca^{2+}, z = 2)

C_0 = concentration of the ion outside the cell (ECF) in millimoles/liter (millimolar, mM)

C_i = concentration of the ion inside the cell (ICF) in mM

Given the mammalian ECF concentration of K$^+$ at 5 mM and the ICF concentration at 150 mM:

$$E_{K^+} = 61 \log \frac{5 \text{ mM}}{150 \text{ mM}}$$

$$= 61 \log \frac{1}{30}$$

Because the log of 1/30 = −1.477,

$$E_{K^+} = 61(-1.477) = -90 \text{ mV}$$

Because 61 is a constant (for a mammal with a 37°C body temperature), the equilibrium potential is essentially a measure of the membrane potential (that is, the magnitude of the electrical gradient) that exactly counterbalances the concentration gradient that exists for the ion (that is, the ratio between the ion's concentration outside and inside the cell). For a poikilotherm with a lower body temperature of 18°C, the constant is reduced to 58. Note that the larger the concentration gradient for an ion, the greater the ion's equilibrium potential. A comparably greater opposing electrical gradient would be required to counterbalance the larger concentration gradient.

Effect of Movement of Sodium Alone on Membrane Potential: Equilibrium Potential for Na$^+$ A similar hypothetical situation could be developed for Na$^+$ alone (Figure 3-24). The concentration gradient for Na$^+$ would move this ion into the cell, building up positive charges on the interior of the membrane and leaving negative charges unbalanced outside (primarily in the form of chloride, Cl$^-$; Na$^+$ and Cl$^-$— that is, salt—are the predominant ECF ions). Net diffusion inward would continue until equilibrium was established by the development of an opposing electrical gradient that exactly counterbalanced the concentration gradient. At this point, given the concentrations for Na$^+$, the **Na$^+$ equilibrium potential** (E_{Na^+}) would be +61 mV. Given that the ECF

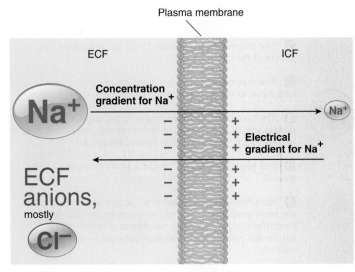

Plasma membrane

ECF

ICF

Na⁺

Concentration gradient for Na⁺

Na⁺

Electrical gradient for Na⁺

ECF anions,

mostly

Cl⁻

$E_{Na^+} = +60 \text{ mV}$

1 The concentration gradient for Na⁺ tends to move this ion into the cell.

2 The inside of the cell becomes more positive as Na⁺ ions move to the inside down their concentration gradient.

3 The outside becomes more negative as Na⁺ ions move in, leaving behind in the ECF unbalanced negatively charged ions, mostly Cl⁻.

4 The resulting electrical gradient tends to move Na⁺ out of the cell.

5 No further net movement of Na⁺ occurs when the outward electrical gradient exactly counterbalances the inward concentration gradient. The membrane potential at this equilibrium point is the equilibrium potential for Na⁺ (E_{Na^+}) at +60 mV.

FIGURE 3-24 Equilibrium potential for Na⁺.

© Cengage Learning, 2013

concentration of Na⁺ is 150 mM and the ICF concentration is 15 mM,

$$E_{Na^+} = 61 \log \frac{150 \text{ mM}}{15 \text{ mM}}$$

$$= 61 \log 10$$

Because the log of 10 = 1,

$$E_{Na^+} = 61(1) = 61 \text{ mV}$$

In this case, the inside of the cell would be positive, in contrast to the equilibrium potential for K⁺. The magnitude of E_{Na^+} is somewhat less than for E_{K^+} (60 mV compared to 90 mV) because the concentration gradient for Na⁺ is not as large (Table 3-4); thus, the opposing electrical gradient (membrane potential) is not as great at equilibrium.

Concurrent Potassium and Sodium Effects on Membrane Potential Neither K⁺ nor Na⁺ are alone in the body fluids, so equilibrium potentials exist only in hypothetical or experimental conditions. In a living cell, the effects of both K⁺ and Na⁺ must be taken into account. *The greater the permeability of the plasma membrane for a given ion, the greater the tendency for that ion to drive the membrane potential toward the ion's own equilibrium potential.* Because the membrane at rest is 25 to 30 times more permeable to K⁺ than to Na⁺, K⁺ passes through more readily than Na⁺; thus, K⁺ influences the resting membrane potential to a much greater extent than does Na⁺. Recall that K⁺ acting alone would establish an equilibrium potential of −90 mV. The membrane is somewhat permeable to Na⁺, however, so some Na⁺ enters the cell in a limited attempt to reach its equilibrium potential. This Na⁺ influx neutralizes, or cancels, some of the potential produced by K⁺ alone if Na⁺ were not present.

To facilitate an understanding of this concept, assume that each separated pair of charges in Figure 3-25 represents 10 mV of potential. (This is not technically correct, because in reality, many separated charges must be present to account

for a potential of 10 mV.) In this simplified example, nine separated pluses and minuses, with the minuses on the inside, would represent the E_{K^+} of −90 mV. Superimposing the slight influence of Na⁺ on this K⁺-dominated membrane, assume that two sodium ions enter the cell down the Na⁺ concentration and electrical gradients. (Note that the electrical gradient for Na⁺ is now inward in contrast to the outward electrical gradient for Na⁺ at E_{Na^+}. At E_{Na^+}, the inside of the cell is positive as a result of the inward movement of Na⁺ down its concentration and electrical gradients. In a resting nerve cell, however, the inside is negative because of the dominant influence of K⁺ on membrane potential. Thus, both the concentration and electrical gradients now favor the inward movement of Na⁺). The inward movement of these two positively charged sodium ions neutralizes some of the potential established by K⁺, so now only seven pairs of charges are separated, and the potential is −70 mV. This is the **resting membrane potential** of a typical nerve cell. All cells have a negative resting membrane potential with values that range from −30 to −100 mV. The resting potential of a cell is much closer to E_{K^+} than to E_{Na^+} because of the greater permeability of the membrane to K⁺, but it is slightly less than E_{K^+} (−70 mV is a lower potential than −90 mV) because of the weak influence of Na⁺.

Membrane potential can be measured directly in experimental conditions by recording the voltage difference between the inside and outside of the cell, or it can be calculated using the **Goldman-Hodgkin-Katz equation (GHK equation)**, which takes into account the relative permeabilities and concentration gradients of all permeable ions. The stable, resting membrane is permeable to K⁺, Na⁺, and Cl⁻, but for reasons to be described later, Cl⁻ does not directly contribute to potential in most cells. Therefore, we can ignore it when calculating membrane potential, making the simplified GHK equation:

$$V_m = 61 \log \frac{P_{K^+}[K^+]_o + P_{Na^+}[Na^+]_o}{P_{K^+}[K^+]_i + P_{Na^+}[Na^+]_i}$$

1 The Na⁺/K⁺ pump actively transports Na⁺ out of and K⁺ into the cell, keeping the concentration of Na⁺ high in the ECF and the concentration of K⁺ high in the ICF.

2 Given the concentration gradients that exist across the plasma membrane, K⁺ tends to drive membrane potential to the equilibrium potential for K⁺ (−90 mV), whereas Na⁺ tends to drive membrane potential to the equilibrium potential for Na⁺ (+60 mV).

3 However, K⁺ exerts the dominant effect on resting membrane potential because the membrane is more permeable to K⁺. As a result, resting potential (−70 mV) is much closer to E_{K^+} than to E_{Na^+}.

4 During the establishment of resting potential, the relatively large net diffusion of K⁺ outward does not produce a potential of −90 mV because the resting membrane is slightly permeable to Na⁺ and the relatively small net diffusion of Na⁺ inward neutralizes (in *gray* shading) some of the potential that would be created by K⁺ alone, bringing resting potential to 70 mV, slightly less than E_{K^+}.

5 The negatively charged intracellular proteins (A⁻) that cannot cross the membrane remain unbalanced inside the cell during the net outward movement of the positively charged ions, so the inside of the cell is more negative than the outside.

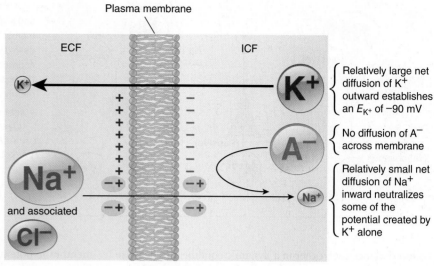

Resting membrane potential = −70 mV

FIGURE 3-25 **Effect of concurrent K⁺ and Na⁺ movement on establishing the resting membrane potential.**

© Cengage Learning, 2013

where

$$V_m = \text{membrane potential in mV}$$

$$61 = \text{a constant representing RT/zF, when z = 1, as it does for K}^+ \text{ and Na}^+$$

$$P_{K^+}, P_{Na^+} = \text{permeabilities for K}^+ \text{ and Na}^+, \text{ respectively}$$

$$[K^+]o, [Na^+] = \text{concentration of K}^+ \text{ and Na}^+ \text{ outside the cell in mM, respectively}$$

$$[K^+]_i, [Na^+]_i = \text{concentration of K}^+ \text{ and Na}^+ \text{ inside the cell in mM, respectively.}$$

The GHK equation is basically an expanded version of the Nernst equation. The Nernst equation can only be used to calculate the potential generated by a specific ion, but the GHK equation takes into account the combined contributions to potential of all ions moving across the membrane. Assuming the resting membrane is 25 times more permeable to K⁺ than to Na⁺, then the relative permeabilities are $P_{K^+} = 1.0$ and $P_{Na^+} = 0.04$ (1/25 of 1.0). Given these permeabilities and the concentrations for K⁺ and Na⁺ in the ECF and ICF listed in Table 3-4,

$$V_m = 61 \log \frac{(1)(5) + (0.04)(150)}{(1)(150) + (0.04)(15)}$$

$$= \log \frac{5 + 6}{150 + 0.6}$$

$$= 61 \log 0.073$$

Because the log of 0.073 is −1.137,

$$V_m = 61 \, (-1.137) = -69 \text{ mV}$$

Adding −1 mV of potential generated directly by the Na⁺/K⁺ pump to this value totals −70 mV for the resting membrane potential.

Balance of Passive Leaks and Active Pumping at Resting Membrane Potential At resting potential, neither K⁺ nor Na⁺ is at equilibrium. A potential of −70 mV does not exactly counterbalance the concentration gradient for K⁺; it takes a potential of −90 mV to do that. Thus, there is a continual tendency for K⁺ to passively *leak* out through its channels. In the case of Na⁺, the concentration and electrical gradients do not even oppose each other; they both favor the inward movement of Na⁺. Therefore, Na⁺ continually leaks inward down its electrochemical gradient, but only slowly because of its low permeability; that is, because of the scarcity of Na⁺ leak channels.

Because such leaking goes on all the time, why does the intracellular concentration of K⁺ not continue to fall and the concentration of Na⁺ inside the cell progressively increase? This does not happen, because of the Na⁺/K⁺ pump. This active transport mechanism counterbalances the rate of leakage (Figure 3-26). At resting potential, the pump transports back into the cell essentially the same number of potassium ions that have leaked out and simultaneously transports to the outside the sodium ions that have leaked in. At this point, a steady state exists: No net movement of any ions takes place, because all passive leaks are exactly balanced by active pumping. No net change takes place in either a steady state or a dynamic equilibrium, but in a *steady state* energy must be used to maintain the constancy, whereas in a *dynamic equilibrium* no energy is needed to maintain the constancy. That is, opposing passive and active forces counterbalance each other in a steady state and opposing passive forces

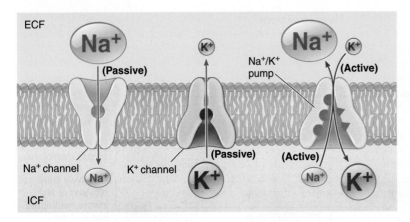

FIGURE 3-26 Counterbalance between passive Na⁺ and K⁺ leaks and the active Na⁺/K⁺ pump. At resting membrane potential, the passive leaks of Na⁺ and K⁺ down their electrochemical gradients are counterbalanced by the active Na⁺/K⁺ pump, so that there is no net movement of Na⁺ and K⁺, and the membrane potential remains constant.

© Cengage Learning, 2013

counterbalance each other in a dynamic equilibrium. Because in the steady state across the membrane the active pump offsets the passive leaks, the concentration gradients for K⁺ and Na⁺ remain constant across the membrane. Thus, not only is the Na⁺/K⁺ pump initially responsible for the Na⁺ and K⁺ concentration differences across the membrane, but it also maintains these differences.

As just discussed, it is the presence of these concentration gradients, together with the difference in permeability of the membrane to these ions, that accounts for the magnitude of the resting membrane potential. In this resting state, the potential remains constant. There is no net movement of any ions. All passive forces are exactly balanced by active forces. A steady state exists, even though there is still a strong concentration gradient for both K⁺ and Na⁺ in opposite directions, as well as a slight excess of positive charges in the ECF accompanied by a corresponding slight excess of negative charges in the ICF (enough to account for a potential of the magnitude of 70 mV).

Chloride Movement at Resting Membrane Potential Thus far, we have largely ignored one other ion present in high concentration in the ECF: Cl⁻. Chloride is the principal ECF anion. Its equilibrium potential is −70 mV to −80 mV, similar to the resting membrane potential. Movement alone of negatively charged Cl⁻ into the cell down its concentration gradient would produce an opposing electrical gradient, with the inside negative compared to the outside. When physiologists were first examining the ionic effects that could account for the membrane potential, it was tempting to think that Cl⁻ movements and establishment of the Cl⁻ equilibrium potential could be solely responsible for producing the identical resting membrane potential. Actually, the reverse is true. The membrane potential is responsible for driving the distribution of Cl⁻ across the membrane.

Most cells are highly permeable to Cl⁻ but have no active transport mechanisms for Cl⁻. With no active forces acting on it, Cl⁻ passively distributes itself to achieve an individual state of equilibrium. In this case, Cl⁻ is driven out of the cell, establishing an inward concentration gradient that exactly counterbalances the outward electrical gradient (that is, the resting membrane potential) produced by K⁺ and Na⁺ movement. Thus, the concentration difference for Cl⁻ between the ECF and ICF is brought about passively by the presence of the membrane potential, rather than being maintained by an active pump, as is the case for K⁺ and Na⁺. Therefore, in most cells Cl⁻ does not influence membrane potential; instead, membrane potential passively influences the Cl⁻ distribution. (Some specialized cells have an active Cl⁻ pump, with subsequent movement of Cl⁻ accounting for part of the potential.)

Specialized Use of Membrane Potential in Nerve and Muscle Cells Nerve and muscle and cells have developed a specialized use for membrane potential. They can rapidly and transiently alter their membrane permeabilities to the involved ions in response to appropriate stimulation, thereby bringing about fluctuations in membrane potential. The rapid fluctuations in potential are responsible for producing nerve impulses in sensory and nerve cells and for triggering contraction in muscle cells. These activities are the focus of most of the next chapters. Even though all cells display a membrane potential, its significance in other cells is uncertain.

check your understanding 3.5

What is an electrochemical gradient? What purpose does it serve?

Explain the forces that keep K⁺ high inside cells.

Explain why Na⁺ is low inside cells despite a large concentration gradient inwards.

making connections

How Membrane Physiology Contributes to the Body as a Whole

All cells of an animal's body must obtain vital materials, such as nutrients and O_2, from the surrounding ECF; they must also eliminate wastes to the ECF and release secretory products, such as chemical messengers and digestive enzymes. Thus, transport of materials across the plasma membrane between the ECF and ICF is essential for cell survival, and the constituents of the ECF must be homeostatically maintained to support these life-sustaining exchanges.

Many cell types use membrane transport to carry out their specialized activities geared toward maintaining homeostasis. Here are several examples:

1. Absorption of nutrients from the digestive tract lumen involves the transport of these energy-giving molecules across the membranes of the cells lining the tract.

2. Exchange of O_2 and CO_2 between the air and blood in the lungs involves the diffusion across the membranes of the cells lining the air sacs and blood vessels of the lungs.

3. Urine is formed by the selective transfer of materials between the blood and the fluid within the kidney tubules across the membranes of the cells lining the tubules.

4. The beating of the heart is triggered by cyclic changes in the transport of Na^+, K^+, and Ca^{2+} across the heart cells' membranes.

5. Secretion of chemical messengers such as neurotransmitters from nerve cells and hormones from endocrine cells involves the transport of these regulatory products to the ECF on appropriate stimulation.

In addition to providing selective transport of materials between the ECF and ICF, the plasma membrane contains receptors for binding with specific chemical messengers that regulate various cell activities, many of which are specialized activities aimed toward maintaining homeostasis. For example, the hormone vasopressin, which is secreted in response to a water deficit in the body, binds with receptors in the plasma membrane of a specific type of kidney cell. This binding triggers the cells to conserve water during urine formation by promoting the insertion of additional aquaporins (water channels) in the plasma membrane of these cells, thus helping alleviate the water deficit that initiated the response.

All living cells have a membrane potential; the specialized activities of nerve and muscle cells depend on these cells' ability to change their membrane potential rapidly on appropriate stimulation. The transient, rapid changes in potential in nerve cells serve as electrical signals, which provide a means to transmit information along nerve pathways. This information is used to accomplish regulatory adjustments, such as restoring blood pressure to normal when signaled that it has fallen too low. Rapid changes in membrane potential also trigger muscle contraction.

Chapter Summary

- The lipid bilayer forms the basic structural barrier that encloses the cell. The membrane proteins perform a variety of specific membrane functions, which include the transport of substances in and out of the cell. Membrane carbohydrates serve as self-identity markers.

- The fluid mosaic and membrane-skeleton fence models describe membrane structure and function.

- Lipid-soluble substances and small polar molecules that can permeate the membrane diffuse passively down their concentration gradient. Ions that can permeate the membrane also move passively along their electrical (conduction) gradient.

- The slowness of diffusion, and thus its inadequacy for movement over large distances, has been the primary force in the evolution of circulatory systems and other bulk transport processes in multicellular animals. Fick's law of diffusion states that concentration gradient, temperature, molecular weight of the diffusing substance, thickness, and surface area of the membrane affect the net rate of diffusion across a membrane.

- Osmosis is the net movement of water through a membrane. Because solutions are always referred to in terms of concentration of solute, water moves by osmosis to the area of higher solute concentration. Osmotic pressure is one of the so-called colligative properties of solutes. These properties depend solely on the number of dissolved particles in a given volume of solution, not on the chemical properties of the solute. Tonicity refers to the effect the concentration of nonpenetrating solutes in a solution has on cell volume.

- Special mechanisms are used to transport selected molecules unable to cross the plasma membrane on their own. Channel proteins form selective pores for specific ions or, in the case of aquaporins, for water. Carrier- and pump-mediated transport systems display three important characteristics that determine the kind and amount of material that can be transferred across the membrane: specificity, saturation, and competition. Active transport is carrier-mediated transport that uses energy and moves a substance against its concentration gradient. For example, the Na^+/K^+ pump moves three Na^+ out of the cell for every two K^+ it pumps in. With secondary active transport, energy is required in the entire process, but it is not directly required to run the pump; rather it uses the energy of an Na^+ gradient to move substances such as glucose against a gradient.

- Large particles are transferred between the ICF and ECF not by crossing the membrane but by being wrapped in a membrane-enclosed vesicle, a process known as vesicular transport. Caveolae may play roles in membrane transport and signal transduction.

- Communication between cells is largely orchestrated by extracellular chemical messengers, of which there are six types: paracrines, neurotransmitters, hormones, neurohormones, pheromones, and cytokines.

- Binding of the membrane receptor with the extracellular chemical messenger (the first messenger) brings about the desired intracellular response by only three general means: (1) by opening or closing specific ion channels in the membrane to regulate the movement of particular ions into or out of the cell, (2) by activating an enzyme that phosphorylates a cell protein, or (3) by transferring the signal to an intracellular chemical messenger (the second messenger), which in turn triggers a preprogrammed series of biochemical events within the cell.

- Some extracellular chemical messengers activate receptor-enzymes via a tyrosine kinase receptor, an enzyme that transfers a phosphate group from ATP to a particular intracellular protein.

- Some extracellular messengers bind with receptors that activate membrane-bound intermediaries known as G proteins, which in turn open (or in some instances close) the appropriate adjacent channel or amplify enzymes that activate second messengers such as cyclic AMP, which in turn activates additional steps. Each step in the signal transduction pathway amplifies the initial signal.

- Membrane potential is a separation of opposite charges across the plasma membrane, due primarily to differences in the distribution and permeability of key ions, with K^+ and large anions dominating in the ICF and Na^+ and Cl^- dominating in the ECF. The Na^+/K^+ pump's role in producing membrane potential is indirect, through its critical contribution to maintaining the concentration gradients directly responsible for the ion movements that generate most of the potential. The resting membrane potential is near the equilibrium for K^+, whose outward diffusion gradient is opposed by negative charges in the cell. Membrane potential can be measured directly by recording the voltage difference between the inside and outside of the cell, or it can be calculated using the Goldman-Hodgkin-Katz equation, which takes into account the relative permeabilities and concentration gradients of all permeable ions.

Review, Synthesize, and Analyze

1. What advantages are there to using diffusion as a transport mechanism versus active transport of a substance?

2. A trout exposed to acid mine drainage experiences an increase in mucus production along the gills. According to Fick's law of diffusion, how would that affect the movement of oxygen through the gill capillaries?

3. Why would hyponatremia (low plasma sodium concentrations) in an animal be better treated with an Na^+/glucose solution rather than the ingestion of salt tablets alone? Normal plasma sodium concentrations are between 135 and 145 mEq/L, while severe diarrhea or vomiting can lower concentrations to less than 125 mEq/L.

4. What advantages are there for cells to use different second messenger systems?

5. How does the Na^+/K^+ pump contribute to the development of a membrane potential? How does the movement of ICF K^+ ions down its concentration gradient affect membrane potential?

6. How is primary active transport different from that of secondary active transport? Which form of transport does the Na^+/K^+ pump use? Explain.

7. Explain how the cascading effect of hormonal pathways amplifies the response.

8. Consider the effect of a twofold increase in the ECF concentration of Na^+ or K^+ ions. Which increase in ion concentration is the more deadly and why? Hint: Consider if the calculated decrease in E_{K^+} or the increase in E_{Na^+} bring the resting membrane potential of the nerve or muscle cell closer to threshold potential? Why would this change in ion concentration stop the heart from beating?

Suggested Readings

Hammel, H. T. (1999). Evolving ideas about osmosis and capillary fluid exchange. *FASEB Journal* 13:213–231. Describes an alternative explanation for osmosis.

Kew, J., & C. Davies, eds. (2009). *Ion Channels: From Structure to Function.* Oxford, U.K.: University Press. A comprehensive overview of ion channel functions.

Ling, G. N., & G. Bohr. (1970). Studies on ion distribution in living cells. Part 2. Cooperative interaction between intracellular K^+ and Na^+. *Biophysical Journal* 10:519. Discusses why K^+ universally accumulates more than Na^+ in cells.

Lucky, M. (2008). *Membrane Structural Biology.* Cambridge, U.K.: Cambridge University Press.

Singer, S. J., & G. Nicholson. (1972). The fluid mosaic model of the structure of cell membranes. *Science* 175:720–731. The classic paper establishing the modern membrane model.

CourseMate

Access an interactive eBook, chapter-specific interactive learning tools, including flashcards, quizzes, videos and more in your Biology CourseMate, accessed through **CengageBrain.com.**

The squid *Loligo* has giant nerve axons for rapid locomotion. The comparatively large size of these neurons permitted the intracellular placement of recording electrodes for measurement of voltage changes during the course of an action potential. These studies, by A. Hodgkin and A. Huxley (in Britain) and by K. Cole and H. Curtis (in America) in the 1930s, yielded the first detailed analysis of ion movements and voltage changes during nerve signaling.

photolibrary.com

Neuronal Physiology

4.1 Introduction

Every cell displays a membrane potential, which refers to a separation of positive and negative charges across the membrane, as discussed in the preceding chapter. This negative potential is related to the uneven distribution of Na$^+$, K$^+$, and large intracellular protein anions between the intracellular fluid (ICF) and extracellular fluid (ECF), and to the differential permeability of the plasma membrane to these ions (see pp. 102–104).

Excitable cells such as neurons and muscles evolved for rapid signaling, coordination, and movement

A number of cell types in eukaryotes have evolved a specialized use for this membrane potential. They can undergo transient, rapid changes in their membrane potentials that serve as either rapid electrical signals or stimulators of rapid cellular action. These cells, called **excitable cells**, are best known in animals. Here we explore the two main types of excitable animal cells, *neurons* (or *nerve cells*) and *muscle cells*. (Some endocrine cells and ova are also excitable.) In Chapter 3, you learned that the membrane potential that exists when a neuron or muscle cell is not excited is called the *resting potential* (although the membrane is far from "resting" because of the balanced leak–pump activity constantly going on). In contrast, when these cells are excited, a rapid, self-propagating electrical signal along the membrane called an *action potential* is produced. Evolution of neurons and muscles enabled rapid long-distance communication, coordination of activities between cells, and rapid movement. These abilities are especially important for survival in organisms that require quick responses often involving more than one organ.

We can see the evolutionary roots of these abilities in the protozoan *Paramecium* (see photo p. 25). When a *Paramecium* bumps into an obstacle, an action potential is initiated that rapidly and without decay sweeps around the entire cell. This triggers a nearly simultaneous reversal in the direction of beating of its cilia, and the cell quickly backs away. This response may have evolved because these unicellular protists are large, and movement of signals by simple diffusion would be much too slow (a topic we will explore in Chapters 9 and 11).

Neurons use similar electrical signals to receive, process, initiate, and transmit messages. In this chapter, we consider how neurons undergo changes in potential to accomplish their function. In muscle cells, these electrical signals initiate contraction, which will be discussed later (Chapters 8 and 9).

Terminology and Methodology Before understanding what these electrical signals are and how they are created, you must become familiar with the following terms used to describe changes in potential, as pictured in Figure 4-1:

1. **Polarization:** Charges are separated across the plasma membrane, so that the membrane has potential to do work. Any time the value of the membrane potential is not 0 mV, in either the positive or negative direction, the membrane

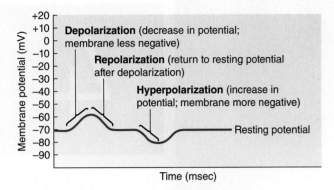

FIGURE 4-1 Types of changes in membrane potential.

© Cengage Learning, 2013

is in a state of polarization. Recall that the magnitude of the potential is directly proportional to the number of positive and negative charges separated by the membrane and that the sign of the potential (+ or −) always designates whether excess positive or excess negative charges are present, respectively, on the inside of the membrane.

2. **Depolarization:** A change in potential that makes the membrane less polarized than at resting potential. Since excitable cells have negative resting potentials (at about −70 mV in a typical neuron, p. 106), depolarization decreases membrane potential in the positive direction. A cell's potential can become positive as it does during an action potential when the membrane potential reverses itself (for example, becoming +30 mV).

3. **Repolarization:** The membrane returns to resting potential after having been depolarized.

4. **Hyperpolarization:** A change in potential that makes the membrane more polarized (more negative) than at resting potential. Hyperpolarization increases membrane potential, moving it even farther from 0 mV (for instance, a change from −70 mV to −80 mV); more charges are separated than at resting potential.

One possibly confusing point should be clarified. On the device used for recording rapid changes in potential, during a depolarization when the inside becomes *less* negative than at resting, this *decrease* in the magnitude of the potential is represented as an *upward* deflection. By contrast, during a hyperpolarization when the inside becomes *more* negative than at resting, this *increase* in the magnitude of the potential is represented by a *downward* deflection.

It is also important to understand how researchers obtained much of the information in this chapter (and the end of Chapter 3). There are two major types of techniques. The first involves the use of various **microelectrodes** that can be inserted into a neuron with little damage. With these electrodes (in conjunction with electrodes outside the cell), local potentials across membranes can be measured accurately (see the caption to the chapter opening photograph). In addition, such electrodes can be used to produce a **voltage clamp**, in which the potential across a membrane is held at a constant value by an electronic circuit. This allows the researcher to determine, for example, what kinds of ions

move across the membrane at any given voltage. The second group of methodologies involves **patch clamping**, in which a tiny pipette is attached to a plasma membrane with gentle suction. This forms a tight seal around one patch of a membrane. With a fine-enough pipette, a researcher can isolate a *single* ion channel or receptor protein and measure its properties. Through the pipette, various substances can be added (such as signal molecules to bind to receptors); and electrodes within and external to the pipette can be used to measure or control (for example, clamp) potentials.

In addition, much research on neuron function, including all the pioneering work, has been done on nonvertebrate preparations (see the opening photograph of the chapter). Over the years, researchers have found that neurons work by nearly identical membrane and molecular mechanisms in all animal phyla that have been studied. Thus, there has been a high degree of conservation since the first animal neurons evolved.

Electrical signals are produced by changes in ion movement through ion channels across the plasma membrane

Changes in membrane potential (depolarization or hyperpolarization) are brought about by changes in ion movement across the membrane. Changes in ion movement in turn are brought about by changes in membrane permeability in response to *triggering events*. Depending on the type of electrical signal, a triggering event might be (1) a stimulus, such as sound waves stimulating specialized neural endings in an animal's ear; (2) a change in the electrical field in the vicinity of an ion channel within the membrane of an excitable cell; (3) an interaction of a chemical messenger with a surface receptor on a neuron or muscle cell membrane: or (4) a spontaneous change of potential caused by inherent imbalances in the leak–pump cycle. (You will learn more about the nature of these various triggering events as our discussion of electrical signals continues.)

Because ions responsible for carrying charge cannot penetrate the plasma membrane's lipid bilayer, these charges can cross the membrane only through channels specific for them. As described in Chapter 3, there are two types of channels: **leak channels** or nongated channels, which are open all the time, and **gated channels**, which can be opened or closed in response to specific triggering events. It is the "gates" guarding particular ionic channels that triggering events open or close to alter membrane permeability. The subsequent ionic movements redistribute charge across the membrane and thus change membrane potential.

Like many proteins, gated channels are inherently flexible proteins whose conformations can be altered in response to external factors (Chapter 3). In this case, the "gates" alternately permit or block ion passage through the channel. There are at least four kinds of gated channels, depending on the factor that induces the change in channel conformation; we'll explore the first two in this chapter:

1. **Voltage-gated ion channels,** which open or close in response to changes in membrane potential; these are crucial to action potentials.

2. **Chemically gated channels** (also called **ligand gated**), which change conformation *allosterically* (p. 27) in

response to the binding of a specific chemical messenger with a membrane receptor that is in close association with the channel. You have seen such channels in signal transduction (Figure 3-17) and will see them in *synapses* in this chapter.

3. **Mechanically gated channels**, which respond to stretching or other mechanical deformation such as touch; these are important in sensory transduction (Chapter 6).

4. **Thermally gated channels** respond to local changes in temperature (heat or cold), also important in sensory transduction (Chapter 6).

Channel opening results in one of two basic forms of electrical signals: (1) the more primitive *graded potentials*, which serve as short-distance signals that decay over distance, and (2) *action potentials*, which, as we noted earlier, signal over longer distances without decay. Let's examine these signals in more detail and then explore how neurons use these channels and signals to convey messages.

check your understanding 4.1

What are the two major types of excitable tissue in animals, and their evolutionary roots?

Define the following terms: *polarization, depolarization, hyperpolarization, repolarization,* and *resting potential.*

How do ions cross the plasma membrane?

4.2 Graded Potentials

Graded potentials are local changes in membrane potential that occur in varying grades or degrees of magnitude or strength. For example, membrane potential could change from −70 mV to −60 mV (a 10 mV graded potential depolarization) or from −70 mV to −90 mV (a 20mV graded potential hyperpolarization).

The stronger a triggering event, the larger the resultant graded potential

Graded potentials are usually produced by a specific triggering event that causes gated ion channels to open in a specialized region of the excitable cell membrane. Most commonly, gated Na+ channels open in response to the event, leading to the inward movement of Na+ down its concentration and electrical gradients. The resultant depolarization—the graded potential—is confined to this relatively small specialized region of the plasma membrane.

The magnitude of this initial graded potential (that is, the difference between the new potential and the resting potential) is related to the magnitude of the triggering event: *The stronger the triggering event, the more gated channels that open, the greater the charge entering the cell, and the larger the graded potential at the point of origin.* Also, the longer the duration of the triggering event, the longer the duration of the graded potential (Figure 4-2). Graded potentials can either be depolarizing or hyperpolarizing. We'll examine depolarizating ones here, but later you will see how hyperpolarizing ones are important.

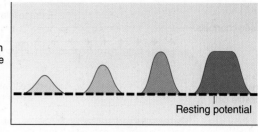
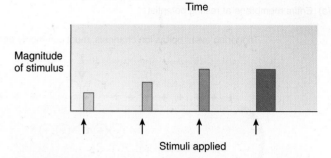

FIGURE 4-2 The magnitude and duration of a graded potential. The magnitude and duration of a graded potential depend directly on the strength and duration of the triggering event, such as a stimulus.

© Cengage Learning, 2013

Graded potentials spread by passive current flow and are impeded by resistances

When a graded potential occurs locally in a nerve or muscle cell membrane, the rest of the membrane remains at resting potential. The temporarily depolarized region is called an *active area*. Note from Figure 4-3b that inside the cell, the active area is relatively more positive than the neighboring *inactive areas* that are still at resting potential. Outside the cell, the active area is relatively less positive than these adjacent areas. Because of this difference in potential, electrical charges, in this case carried by ions, passively flow between the active and adjacent resting regions on both the inside and outside of the membrane.

Any flow of electrical charges is called a **current**. By convention, the direction of current flow is always designated by the direction in which the positive charges are moving (Figure 4-3c). Inside the cell, positive charges flow through the ICF away from the relatively more positive depolarized active region toward the more negative adjacent resting regions. Similarly, outside the cell, positive charges flow through the ECF from the more positive adjacent inactive regions toward the relatively more negative active region. Ion movement (that is, current) is occurring *along* the membrane between regions next to each other on the same side of the membrane. This flow is in contrast to ionic movement *across* the membrane through ionic channels with which you are more familiar.

As a result of local current flow between an active depolarized area and an adjacent inactive area, potential in the previously inactive area alters. Positive charges have flowed into this adjacent area on the inside, while simultaneously positive charges have flowed out of this area on the outside. Thus, at this adjacent site, the inside is more positive (or less negative) than before (Figure 4-3c). Stated differently, the

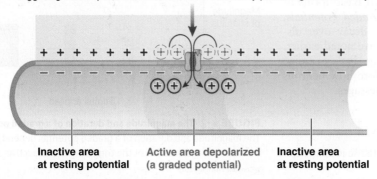

Extracellular fluid

Closed channels

$+ + + + + + + + + + + + + + + + + +$

$- - - - - - - - - - - - - - - - - -$

Unbalanced charges distributed across the plasma membrane that are responsible for membrane potential

Portion of an excitable cell

Intracellular fluid

(a) Entire membrane at resting potential

Triggering event opens ion channels, most commonly permitting net Na⁺ entry

Inactive area at resting potential **Active area depolarized (a graded potential)** **Inactive area at resting potential**

(b) Inward movement of Na⁺ depolarizes membrane, producing a graded potential

Current flows between the active and adjacent inactive areas

Inactive area **Previously inactive area being depolarized** **Original active area** **Previously inactive area being depolarized** **Inactive area**

Spread of depolarization

(c) Depolarization spreads by local current flow to adjacent inactive areas, away from point of origin

FIGURE 4-3 Current flow during a graded potential. (a) The membrane of an excitable cell at resting potential. (b) A triggering event opens ion channels, usually leading to net Na⁺ entry that depolarizes the membrane at this site. The adjacent inactive areas are still at resting potential. (c) Local current flows between the active and adjacent inactive areas, resulting in depolarization of the previously inactive areas. In this way, the depolarization spreads away from its point of origin.

© Cengage Learning, 2013

previously inactive adjacent region has been depolarized; thus the graded potential has spread. This area's potential now differs from that of the inactive region immediately next to it on the other side, inducing further current flow at this new site, and so on. In this manner, current spreads in both directions away from the initial site of the potential change.

The amount of current that flows between two areas depends on the potential difference between the areas and on the resistance of the material through which the charges are moving. **Resistance** is the hindrance to electrical charge movement. The greater the difference in potential, the greater the current flow. Similarly, the lower the resistance, the greater the current flow. *Conductors* have low resistance and provide little hindrance to current flow. Electrical wires and the intracellular and extracellular fluid are all good conductors, so current readily flows through them. *Insulators* have high resistance and greatly hinder movement of charge. The rubber or plastic surrounding electrical wires has high resis-

tance, as do lipids. Thus, current does not flow across the plasma membrane's lipid bilayer. Current, carried by ions, can move across the membrane only through ion channels.

Graded potentials die out over short distances

The passive current flow between active and adjacent inactive areas is similar to the means by which current is carried through electrical wires. You know from experience that current can leak out of an electrical wire unless the wire is covered with an insulating material such as plastic. Likewise, current is lost across the cell membrane as charge-carrying ions leak out through the "uninsulated" parts of the membrane, that is, by moving outward down their electrochemical gradient through open channels. Because of this current loss, the magnitude of the local current progressively diminishes with increasing distance from the initial site of origin (Figure 4-4). Thus, the magnitude of the graded potential continues

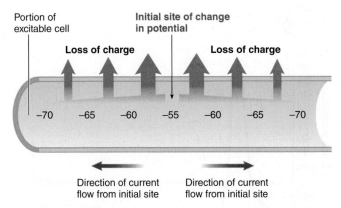

Portion of
excitable cell

**Initial site of change
in potential**

Loss of charge   **Loss of charge**

−70 −65 −60 −55 −60 −65 −70

Direction of current
flow from initial site

Direction of current
flow from initial site

* Numbers refer to the local potential in mV
at various points along the membrane.

(a) Current loss across the membrane

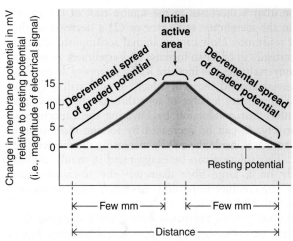

Change in membrane potential in mV
relative to resting potential
(i.e., magnitude of electrical signal)

**Initial
active
area**

Decremental spread
of graded potential

Decremental spread
of graded potential

15

10

5

0

Resting potential

Few mm   Few mm

Distance

(b) Decremental spread of graded potentials

FIGURE 4-4 **Current loss across the plasma membrane leading to
decremental spread of a graded potential.** (a) Leakage of charge-
carrying ions across the plasma membrane results in progressive loss
of current with increasing distance from the initial site of the change in
potential. (b) Because of leaks in current, the magnitude of a graded
potential continues to decrease as it passively spreads from the initial
active area. The potential dies out altogether within a few millimeters
of its site of initiation.

© Cengage Learning, 2013

to decrease the farther it moves away from the initial active
area. Another way of saying this is that the spread of a
graded potential is **decremental** (gradually decreases). Note
that in Figure 4-4, the magnitude of the initial potential
change is 15 mV (a change from −70 mV to −55 mV), then
decreases as it moves along the membrane, continuing to
diminish the farther it moves away from the initial active
area until there is no longer a potential change. In this way,
these local currents die out within a few millimeters from the
initial site of potential change and consequently can function
as signals only for very short distances.

For this reason, this form of electrical depolarization is
termed **passive conduction**. An analogy is sound waves. The
magnitude of the sound waves is greatest (loudest) at the
point of origin, then progressively decreases and eventually
dies off as the sound waves spread out from this initial site.
The passive movement of the electrical depolarization can be
described by **Ohm's law**, where ΔV_m is the actual voltage
change across the membrane, ΔI the amount of current (amperes), and R the resistance encountered by the current passing along the membrane (ohms).

$$\Delta V_m = \Delta I \times R \text{ or } \Delta I = \Delta V_m \times g$$

Ohm's law states that the *reduction* in voltage across a
membrane in response to a current flowing through the cell
is directly related to the amount of current multiplied by the
resistance of the membrane. As current flows inside the cell,
a constant proportion leaks through the membrane at each
point. Membrane resistance is also the reciprocal of membrane *conductance* (g), where conductance depends on both
the permeability and driving force for the movement of ions
(the unit of conductance is the *siemen*). The passive movement of current across the cell membrane is termed **electro-**

tonic conduction. Comparable to the flow of electricity
through wires, the speed of electrotonic spread is extremely
rapid.

The extent of electrotonic spread is determined in part
by membrane resistance and membrane capacitance. **Capacitance** is a measure of the amount of charge that can be maintained across an insulating gap. The lipid bilayer is an effective capacitor because it can maintain the separation of
charged ions across a relatively narrow space. Because of
these two properties of a cell membrane, there is, for every
excitable cell, only a finite distance that current can passively
flow from the activated region. The fraction that leaks
through the membrane flows back in a path to the source, so
less current is available to influence the membrane potential
at a more distant site. The distance from the active area that
a signal can be passively transmitted is best described by the
following relationship:

$$V_x = V_0 e^{-x/\lambda}$$

In this equation, V_x is the potential measured at a distance x
from the active area whereas V_0 is a measurement of the
original amount of potential generated at the active area.
The length constant (λ) is related to the resistance of the
membrane, cytoplasm, and the ECF by the following
equation:

$$\lambda = \sqrt{\frac{R_m}{R_i}}$$

In this equation R_m is the resistance of the membrane and
R_i is a measurement of the internal and external resistances.
Under normal circumstances the contribution of the external
resistance to R_i is comparatively small. The length constant
can also be defined as the distance from V_0 that the potential
is reduced 63% from its initial value. Thus, when comparing
the rate of signal propagation between neurons, the one with
the higher value of λ indicates the higher velocity. Neurons
that rely on the electrotonic flow of current for communication purposes are rarely more than a few millimeters in length
and are noted for their high membrane resistance. These neurons are termed **nonspiking** neurons; examples include stretch
sensory neurons in arthropods (p. 217). Determination of the
length constant also provides some insight into structural
adaptations that species have evolved to increase λ.

Note that λ increases as the square root of (1) an increase in the membrane resistance or (2) a decrease in the internal resistance. How can an animal accomplish either of these strategies in order to increase λ? Vertebrates as well as some nonvertebrates can increase R_m by wrapping an insulating material around the cell, which, as you will see later (p. 126), minimizes leakage of ions through the membrane and confers fast conduction by lowering the capacitance. Alternatively, R_i can be decreased by forcing more of the current to travel inside the cell by increasing the diameter of the cell. This solution can be exaggerated in many animals that rely on a large fiber diameter—the so-called giant neurons—to conduct important messages, such as escape responses (p. 129).

Although graded potentials have limited signaling distance, they are critically important to neural function, as we explain in later chapters. The following are all graded potentials: *postsynaptic potentials, receptor potentials, end-plate potentials, pacemaker potentials,* and *slow-wave potentials.* These terms may be unfamiliar to you now, but you will become well acquainted with them as we continue discussing neural, sensory, and muscle physiology. We are including this list here because it is the only place in this book that we group all these graded potentials together. For now, it is enough to say that for the most part, excitable cells produce one of these types of graded potentials in response to a triggering event. In turn, graded potentials can initiate *action potentials,* the long-distance signals, in an excitable cell.

check your understanding 4.2

What properties are associated with a graded potential?

How does Ohm's law explain the passive movement of current down an axon?

4.3 Action Potentials

Action potentials are brief, rapid, large (100 mV) changes in membrane potential during which the potential actually reverses so that the inside of the excitable cell transiently becomes more positive than the outside. As with a graded potential, a single action potential involves only a small portion of the total excitable cell membrane. Unlike graded potentials, however, action potentials are conducted, or propagated, throughout the membrane in *nondecremental* fashion; that is, they do not diminish in strength as they travel from their site of initiation throughout the remainder of the cell membrane. This is why action potentials can serve as faithful long-distance signals.

Think about the nerve cell that brings about contraction of muscle cells in a mammal's leg for flexing the toes. To bend the toes, action potentials are sent via a neuron from the brain to the spinal cord to another neuron, where an action potential is initiated. This action potential travels in undiminishing fashion quickly all the way down the neuron's long axon, which runs through the leg to terminate on foot muscle cells. The magnitude of the action potential at the end of the muscle axon is identical to the magnitude of the action potential first initiated in the spinal cord; it has not weakened or died off.

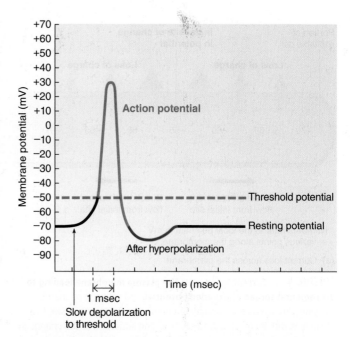

FIGURE 4-5 Changes in membrane potential during an action potential.

© Cengage Learning, 2013

As we noted earlier, action potentials may have first evolved in animal-like protists, and in Chapter 5, we'll explore their use in different animal groups and their respective nervous systems. In this chapter, we will focus on basic neural physiology. Let's begin with the changes in permeability and ionic movements responsible for generating an action potential before turning our attention to the means by which action potentials spread throughout the cell membrane without diminishing.

During an action potential, the membrane potential rapidly and transiently reverses

If it is of sufficient magnitude, a graded potential can initiate an action potential before the graded potential dies off. You will discover the means by which this initiation is accomplished for the various types of graded potentials in more detail later. Typically, the portion of the excitable membrane where graded potentials are produced in response to a triggering event does not undergo an action potential. Instead, passive current flow from the region where the graded potential is taking place depolarizes adjacent portions of the membrane where action potentials can occur.

Depolarization from the resting potential of −70 mV (Figure 4-5) proceeds slowly at first until it reaches a critical level known as **threshold potential,** typically between −50 and −55 mV or +10 to +15 mV above resting potential. At threshold potential, an explosive depolarization takes place. A recording of the potential at this time shows a sharp upward deflection as the potential rapidly reverses itself so that the inside of the cell becomes positive compared to the outside. Peak potential is usually +30 mV to +40 mV, depending on the excitable cell. Just as rapidly, the membrane repolarizes, dropping back to resting potential. Often the forces responsible for driving the membrane back to resting poten-

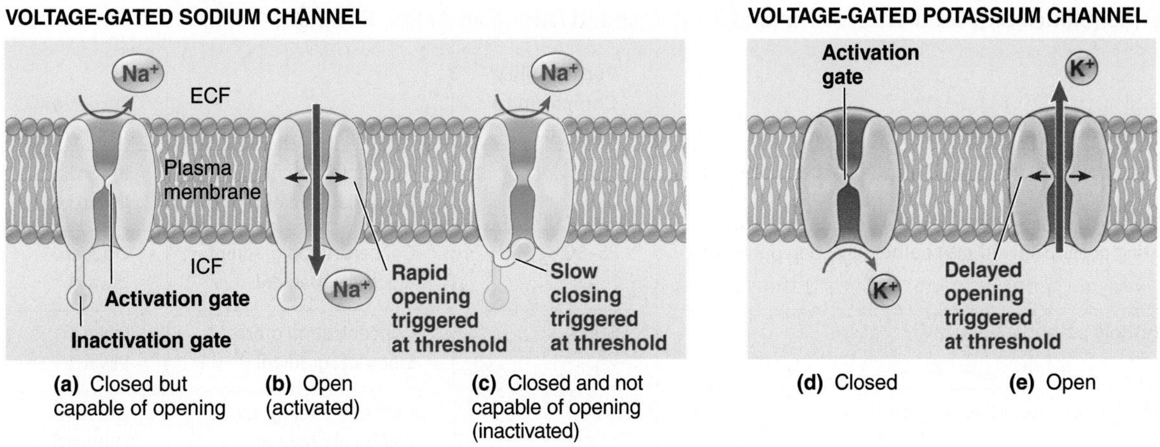

VOLTAGE-GATED SODIUM CHANNEL

VOLTAGE-GATED POTASSIUM CHANNEL

(a) Closed but capable of opening

(b) Open (activated)

(c) Closed and not capable of opening (inactivated)

(d) Closed

(e) Open

FIGURE 4-6 **Conformations of voltage-gated sodium and potassium channels.**

© Cengage Learning, 2013

tial push it too far, causing a transient **after hyperpolarization**, during which the inside of the membrane briefly becomes even more negative than normal (for example, −80 mV) before the resting potential is restored.

The action potential is the entire rapid change in potential from threshold to peak and then back to resting. Unlike the variable duration of a graded potential, the duration of an action potential is always the same in a given excitable cell. In a neuron, an action potential lasts for only 1 msec (0.001 sec). It lasts longer in muscle, with the duration varying depending on the muscle type. The portion of the action potential during which the potential is reversed (between 0 mV and +30 mV) is called the **overshoot**. Often an action potential is called a **spike** because of its spiky appearance on an oscilloscope. Alternatively, when an excitable membrane is triggered to undergo an action potential, it is said to **fire**. Thus, the terms *action potential, spike,* and *firing* all refer to the same phenomenon of rapid potential reversal. If threshold potential is not reached by the initial triggering depolarization, no action potential takes place. Thus, threshold is a critical *all-or-none* event. Either the membrane is depolarized to threshold and an action potential takes place, or threshold is not reached in response to the depolarizing event and no action potential occurs.

Marked changes in membrane permeability and ion movement enhanced by positive feedback lead to an action potential

How is the membrane potential, which is usually maintained at a constant resting level, thrown out of balance to such an extent as to produce an action potential? Recall that K^+ contributes the most to establishing the resting potential because the membrane at rest is much more permeable to K^+ than to Na^+. During an action potential, marked changes in membrane permeability to Na^+ and K^+ take place because of voltage-gated channels, permitting rapid fluxes of these ions down their electrochemical gradients. As you will see, the most rapid changes are accelerated by *positive feedback*. These ion movements carry the current responsible for the potential changes that occur during an action potential.

Voltage-Gated Na^+ and K^+ Channels Two specific types of channels are of major importance in developing an action potential: **voltage-gated Na^+ channels** and **voltage-gated K^+ channels**. Voltage-gated channels consist of proteins that have a number of charged groups. The electric field (potential) surrounding the channels can distort the channel structure as charged portions of the channel proteins are electrically attracted or repelled by charges in the fluids surrounding the membrane. Unlike the majority of membrane proteins, which remain stable despite fluctuations in membrane potential, the voltage-gated channel proteins are exquisitely sensitive to voltage changes. Small distortions in channel shape induced by potential changes can cause them to alternate to another conformation. Here is another example of how subtle changes in structure can have a profound influence on function. For example, all vertebrates, as well as many other animals evolved for speed, have comparatively fast-opening Na^+ channels; in contrast, sluggish animals such as sea slugs have Na^+ channels that open at rates approximately 10 times slower. Fast-opening channels increase survival rates because electrical impulses are more rapidly transmitted to muscles used in catching prey or avoiding predators (note that sluggish animals have other defenses).

The voltage-gated Na^+ channel has two gates: an activation gate and an inactivation gate (Figure 4-6). The **activation gate** guards the channel by opening and closing like a hinged door. The **inactivation gate** consists of a ball-and-chain-like sequence of amino acids at the channel opening facing the ICF. This gate is open when the ball is dangling free on its chain and closed when the ball binds to its receptor located at the channel opening, thus blocking the opening. Both gates must be open to permit passage of Na^+ through the channel, and closure of either gate prevents passage. This voltage-gated Na^+ channel can exist in three different conformations: (1) closed but capable of opening (activation gate closed, inactivation gate open, Figure 4-6a; (2) open, or activated (both gates open, Figure 4-6b); and (3) closed and not capable of opening (activation gate open, inactivation gate closed, Figure 4-6c). These channels are widespread in animals, with mammals having at least nine genes for different variants that evolved by gene duplication. However, while Ca^{2+} and K^+ channels are ubiquitous among organisms, gated Na^+

TABLE 4-1 Permeabilities and Gradients for Na⁺ and K⁺ at Rest and During an Action Potential

Ion	Condition	Permeability Compared to Resting P_{Na^+}	Gradient	Direction of Gradient
Na⁺	Resting potential (-70 mV) before an action potential	1	Concentration gradient Electrical gradient	Inward Inward
K⁺	Resting potential (-70 mV) before an action potential	25–30×	Concentration gradient Electrical gradient	Outward Inward
Na⁺	Threshold potential (-50 mV)	600×	Concentration gradient Electrical gradient	Inward Inward
K⁺	Peak of action potential ($+30$ mV)	300×	Concentration gradient Electrical gradient	Outward Outward
Na⁺	Resting potential (-70 mV) after an action potential	1	Concentration gradient Electrical gradient	Inward Inward
K⁺	Resting potential (-70 mV) after an action potential	50–75×	Concentration gradient Electrical gradient	Outward Inward

© Cengage Learning, 2013

channels are found almost exclusively in animals (although few protists, including certain marine diatoms, have signal-generating Na⁺ channels). Most, if not all, animal voltage-gated channels probably evolved from a common ancestral protein. Cnidaria (such as sea anemones) are a possible exception: They have Na⁺ channels that are insensitive to tetrodotoxin (see p. 143), suggesting that their structure differs from that of other animals' channels.

The basic voltage-gated K⁺ channel of excitable tissues is simpler. It has only one gate, which can be either open or closed (Figure 4-6d and e). This is an ancient type of channel; related ones have been found in most organisms, including bacteria and plants. Neurons have a specific subclass of gated K-channels called **shaker K⁺-channels** (because mutant forms in fruit flies cause uncontrollable leg shaking). They represent an ancient family found in all animal neurons. For example, the simple nervous systems of jellies (Cnidaria) express at least two shaker-type genes; and the genome of the electric fish *Apteronotus* has at least 10 such genes (electric fish are discussed in Chapter 6).

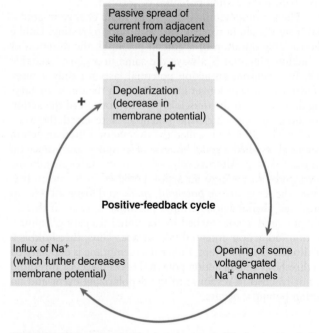

FIGURE 4-7 Positive-feedback cycle responsible for opening Na⁺ channels at threshold.

© Cengage Learning, 2013

unanswered Questions How do voltage gates actually open and close?

It is still not fully understood how the so-called gates on neural voltage-gated channels actually work. However, recent studies on voltage-gated K⁺ channels are revealing.

Researchers used X-ray crystallography to get detailed pictures of these channels from bacteria and mammals. Both channels contain a central pore for K⁺ ions, surrounded by what the researcher's term "voltage-sensor paddles." These are folds on the outside of the protein with hydrophobic and cationic (positively charged) amino acids, and with flexible hinge regions. These features suggest that paddles react to voltage changes in the nearby membrane and move on their hinges in response. The paddle's

structure appears to be ancient and highly conserved: When the gene for a rat shaker K⁺ channel was modified by replacing its paddle code with that from a deep-sea archaean from a hydrothermal vent, the resulting protein still worked normally! <<

Changes in Permeability and Ion Movement during an Action Potential As we noted earlier, the action potential is rapid due to positive feedback. Figure 4-7 shows the overall feedback cycle, while Figure 4-8 summarizes the key channel

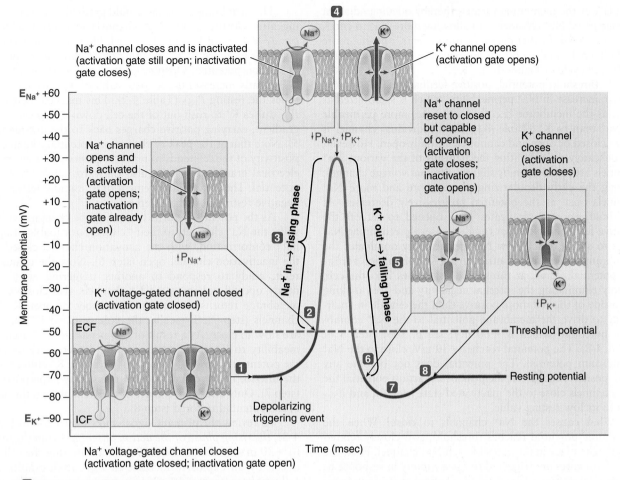

FIGURE 4-8 **Permeability changes and ion fluxes during an action potential.**

1 Resting potential: all voltage-gated channels closed.

2 At threshold, Na⁺ activation gate opens and P_{Na^+} rises.

3 Na⁺ enters cell, causing explosive depolarization to +30 mV, which generates rising phase of action potential.

4 At peak of action potential, Na⁺ inactivation gate closes and P_{Na^+} falls, ending net movement of Na⁺ into cell. At the same time, K⁺ activation gate opens and P_{K^+} rises.

5 K⁺ leaves cell, causing its repolarization to resting potential, which generates falling phase of action potential.

6 On return to resting potential, Na⁺ activation gate closes and inactivation gate opens, resetting channel to respond to another depolarizing triggering event.

7 Further outward movement of K⁺ through still-open K⁺ channel briefly hyperpolarizes membrane, which generates after hyperpolarization.

8 K⁺ activation gate closes, and membrane returns to resting potential.

© Cengage Learning, 2013

and potential changes. At resting potential (−70 mV), all the voltage-gated Na⁺ and K⁺ channels are closed (step 1 of Figure 4-8), with the Na⁺ channels' activation gates closed and their inactivation gates open; that is, the *"closed but capable of opening"* conformation. Therefore, passage of Na⁺ and K⁺ is prevented through these channels at resting potential. However, because of the presence of many K⁺ leak channels, the resting membrane is 25 to 30 times more permeable to K⁺ than to Na⁺ (Table 4-1).

When current spreads passively from an adjacent site already depolarized into a new region at resting potential, the new region of membrane starts to depolarize toward threshold (top of Figure 4-7), causing the activation gates of some of its voltage-gated Na⁺ channels to open, so that both gates of these activated channels are now open (step 2 of Figure 4-8). Because both the concentration and electrical gradients for Na⁺ favor its movement into the cell, Na⁺ starts to move in (step 3). The inward movement of positively charged Na⁺

depolarizes the membrane further, thereby opening adjacent voltage-gated Na^+ channels and allowing more Na^+ to enter. As a result, still further depolarization occurs, opening more Na^+ channels, and so on, in an explosive positive-feedback cycle (full cycle of Figure 4-7).

At threshold potential, positive feedback ensures an explosive increase in Na^+ permeability, which is symbolized as P_{Na^+}, as the membrane becomes 600 times more permeable to Na^+ than to K^+ (Table 4-1). Each individual channel is either closed or open and cannot be partially open. However, the delicately poised gating mechanisms of the various Na^+ channels are jolted open by slightly different voltage changes. During the early depolarizing phase, more and more Na^+ channels open as the potential progressively decreases. At threshold, enough Na^+ gates have opened to set off the positive-feedback cycle that rapidly causes the remaining Na^+ gates to swing open. Now Na^+ permeability dominates the membrane potential, in contrast to the K^+ domination at resting potential. Thus, at threshold Na^+ rushes into the cell, rapidly eliminating the internal negativity and even making the inside of the cell more positive than the outside in an attempt to drive the membrane potential to the Na^+ equilibrium potential (which is +60 mV; see p. 106) (step 3 of Figure 4-8). The potential reaches +30 mV, close to the Na^+ equilibrium potential. The potential does not become any more positive, because at the peak of the action potential the Na^+ channels close to the inactivated state (step 4), and P_{Na^+} falls to its low resting value.

What causes the Na^+ channels to close? When the membrane potential reaches threshold, two closely related events take place in the gates of each Na^+ channel. First, the activation gates are triggered to *open rapidly* in response to the depolarization, converting the channel to its open (activated) conformation (Figure 4-6b). Surprisingly, the conformational change that opens the channel also allows the inactivation gate's ball to bind to the channel opening, thereby physically blocking the mouth of the channel. This might seem counterproductive, but the closure process takes time, so the inactivation gate *closes slowly* compared to the rapidity of channel opening (Figure 4-6c). Meanwhile, during the 0.5-msec delay after the activation gate opens and before the inactivation gate closes, both gates are open and Na^+ rushes into the cell through these open channels, bringing the action potential to its peak. Then the inactivation gate closes, membrane permeability to Na^+ plummets to its low resting value, and further Na^+ entry is prevented. The channel remains in this *"closed and not capable of opening"* conformation until the membrane potential has been restored to its resting value.

Simultaneous with inactivation of the Na^+ channels at the peak of the action potential, the voltage-gated K^+ channels are opened (step 4, Figure 4-8). This opening of the K^+ channel gate is a delayed voltage-gated response triggered by the initial depolarization to threshold. Thus, three action potential–related events occur at threshold: (1) the rapid opening of the Na^+ activation gates, which permits Na^+ to enter, moving the potential from threshold to its positive peak; (2) the slow closing of the Na^+ inactivation gates, which halts further Na^+ entry after a brief time delay, thus keeping the potential from rising any further and thus stopping the positive feedback; and (3) the slow opening of the K^+ gates, which, is in large part responsible for the potential plummeting from its peak back to resting.

The membrane potential would gradually return to resting after closure of the Na^+ channels in concert with the steady leakage of K^+ out of the cell. However, the return to resting is hastened by the opening of K^+ gates at the peak of the action potential. Opening of the voltage-gated K^+ channels greatly increases the K^+ permeability (P_{K^+}) to about 300 times the resting P_{Na^+} (Table 4-1). This marked increase in P_{K^+} causes K^+ to rush out of the cell down its concentration gradient, carrying positive charges back to the outside (step 5). Note that at the peak of the action potential, the internal positivity of the cell tends to repel the positive K^+ ions, so the electrical gradient for K^+ is also outward, unlike at resting potential. The outward movement of K^+ rapidly restores the negative resting potential (step 6).

As the potential returns to resting, the changing voltage shifts the Na^+ channels to their "closed but capable of opening" conformation, with the activation channel closed and the inactivation channel open (step 6). Now the channel is reset, ready to respond to another triggering event. The newly opened voltage-gated K^+ channels also close, so the membrane returns to the resting number of open K^+ leak channels (step 7). Typically, the voltage-gated K^+ channels are slow to close. As a result of this persistent increased permeability to K^+, more K^+ leaves than is necessary to bring the potential to resting. This slight excessive K^+ efflux causes the interior of the cell to become transiently hyperpolarized (step 7). Only after final closure of these channels is the membrane potential restored (step 8).

To review the primary signal spike (steps 2–6 of Figure 4-8), *the rising phase of the action potential* (from threshold to +30 mV) *is due to Na^+ influx* (Na^+ entering the cell) induced by an explosive increase in P_{Na^+} at threshold. *The falling phase* (from +30 mV to resting potential) *is brought about by K^+ efflux* (K^+ leaving the cell) caused by the marked increase in P_{K^+} occurring simultaneously with the inactivation of the Na^+ channels at the peak of the action potential.

The Na^+/K^+ ATPase pump gradually restores the concentration gradients disrupted by action potentials

At the completion of an action potential, the membrane potential has been restored to its resting condition, but the ion distribution has been altered slightly. Sodium has entered the cell during the rising phase, and a comparable amount of K^+ has left during the falling phase. Recall that the bulk of the cell is electrically neutral; the potential difference between the ECF and ICF is measurable only nm (nanometers) deep along the surface of the membrane. It is the task of the Na^+/K^+ pump (a slow modified channel requiring ATP, Figure 3-14) to restore these ions to their original locations in the long run, but not after each action potential.

The active pumping process takes much longer to restore Na^+ and K^+ to their original locations than it takes for the passive fluxes of these ions during an action potential. However, the membrane does not need to wait until the Na^+/K^+ pump slowly restores the concentration gradients before it can undergo another action potential. Actually, the movement of only relatively few of the total number of Na^+ and K^+ ions present is responsible for the dramatic

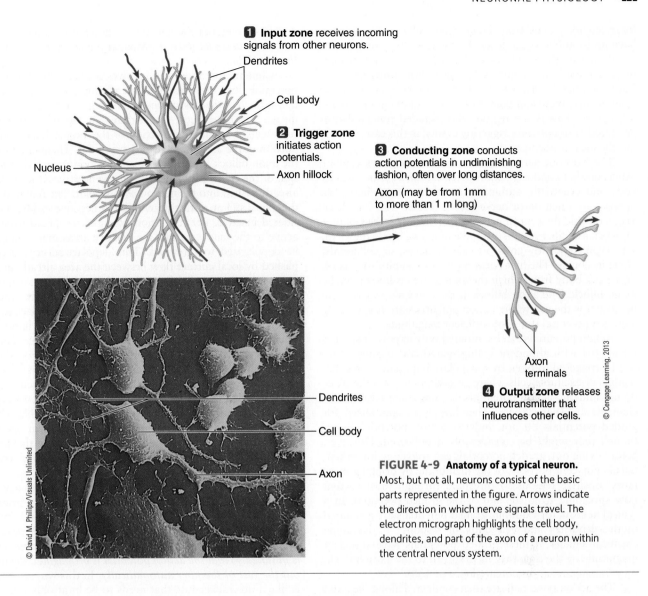

1 Input zone receives incoming signals from other neurons.

Dendrites

Cell body

Nucleus

2 Trigger zone initiates action potentials.

Axon hillock

3 Conducting zone conducts action potentials in undiminishing fashion, often over long distances.

Axon (may be from 1mm to more than 1 m long)

Axon terminals

4 Output zone releases neurotransmitter that influences other cells.

Dendrites

Cell body

Axon

© David M. Philips/Visuals Unlimited

© Cengage Learning, 2013

FIGURE 4-9 Anatomy of a typical neuron. Most, but not all, neurons consist of the basic parts represented in the figure. Arrows indicate the direction in which nerve signals travel. The electron micrograph highlights the cell body, dendrites, and part of the axon of a neuron within the central nervous system.

swings in potential that occur during an action potential. Only about 1 out of every 100,000 K^+ ions present in the cell leaves during an action potential, while a comparable number of Na^+ ions enter from the ECF. The movement of this extremely small proportion of the total Na^+ and K^+ during a single action potential produces dramatic 100 mV changes in potential (between -70 and $+30$ mV) but only infinitesimal changes in the ICF and ECF concentrations of these ions. Much more K^+ is still inside the cell than outside, and Na^+ is still predominantly an extracellular cation. Consequently, the Na^+ and K^+ concentration gradients still exist, so repeated action potentials can occur without the pump having to keep pace to restore the gradients.

Were it not for the pump, of course, after a finite number of fluxes, equilibrium would be reached and generation of action potentials would be impossible. If the concentrations of Na^+ and K^+ were equal between the ECF and ICF, changes in permeability to these ions would not bring about ionic fluxes, so no change in potential would occur. Thus, the Na^+/K^+ pump is critical to maintaining the concentration gradients in the long run. However, it is not directly involved in the ion fluxes or potential changes that occur during an action potential.

Action potentials are propagated from the axon hillock to the axon terminals

A single action potential involves only a small patch of the total surface membrane of an excitable cell. But if action potentials are to serve as long-distance signals, they cannot be merely isolated events occurring in a limited area of a neuron or muscle cell membrane. Mechanisms must exist to conduct or spread the action potential throughout the entire cell membrane. Furthermore, the signal must somehow be transmitted from one cell to the next cell (for example, along specific nerve pathways). Let's first examine how an action potential (nerve impulse) is conducted throughout a neuron before turning our attention to how the impulse is passed to another cell.

A single nerve cell, or **neuron**, consists of three basic parts: the *cell body*, the *dendrites*, and the *axon*, although there are variations in structure, depending on the location and function of the neuron. (The distinctions of other specialized neurons are described later.) The nucleus and organelles are housed in the **cell body** (Figure 4-9), from which numerous extensions known as **dendrites** typically project like antennae to increase the surface area available for receiving signals from other nerve cells. Some neurons have up to 400,000 dendrites. In most neurons, the plasma membrane of the dendrites and cell body contain protein receptors that bind chemical messengers from other neurons. Therefore, the dendrites and cell body are the neuron's *input zone* because these components receive and integrate incoming signals. This is the region where graded potentials are produced in response to triggering events, in this case, incoming chemical messengers.

The **axon,** or **nerve fiber,** is a single, elongated, tubular extension that conducts action potentials *away from* the cell body and eventually terminates at other cells. A **nerve** (as opposed to a neuron or nerve cell) is defined as a bundle of axons outside the central nervous system (CNS; see Chapter 5), whereas a **fiber tract** is a bundle of axons inside the CNS. The axon frequently gives off side branches, or **collaterals,** along its course. The first portion of the axon plus the region of the cell body from which the axon leaves is known as the **axon hillock.** The axon hillock is the neuron's *trigger zone* because it is the site where action potentials are triggered by a graded potential if it is of sufficient magnitude.

Action potentials can be initiated only in portions of the membrane with abundant voltage-gated Na^+ channels that can be triggered to open by a depolarizing event. Typically, regions of excitable cells where graded potentials take place do not undergo action potentials because voltage-gated Na^+ channels are sparse there. Therefore, sites specialized for graded potentials do not undergo action potentials, even though they might be considerably depolarized. However, before dying out, graded potentials can trigger action potentials in adjacent portions of the membrane by bringing these more sensitive regions to threshold through local current flow spreading from the site of the graded potential. In a typical neuron, for example, graded potentials are generated in the dendrites and cell body in response to incoming chemical signals. If these graded potentials have sufficient magnitude by the time they have spread to the axon hillock, they initiate an action potential (see p. 138).

The action potentials are then conducted along the axon from the axon hillock to the typically highly branching endings at the **axon terminals.** These terminals release chemical messengers that simultaneously affect target cells with which they come into close association. Functionally, therefore, the axon is the *conducting zone* of the neuron, whereas the axon terminals constitute the *output zone.* (The major exception to this typical neuronal structure and functional organization are neurons specialized to carry sensory information, as described in Chapter 6.)

Axons vary in length from less than a millimeter in neurons that communicate only with neighboring cells to longer than a meter in neurons that communicate with distant parts of a nervous system or with peripheral organs. For example, the axon of the nerve cell innervating a mammalian foot must travel the distance between the origin of its cell body within the posterior spinal cord all the way down the leg to the foot.

Once initiated, action potentials are conducted over the surface of an axon

Once an action potential is initiated at the axon hillock, no further triggering event is necessary to activate the remainder of the axon. The action potential is automatically conducted throughout the neuron, without further stimulation, by one of two methods of propagation: *contiguous conduction* or *saltatory conduction.*

Contiguous conduction involves the spread of the action potential along every patch of membrane down the length of the axon (*contiguous* means "touching" or "next to in sequence"). This process is illustrated in Figure 4-10, which represents a longitudinal section of the axon hillock and the portion of the axon immediately beyond it. The membrane at the axon hillock is at the peak of an action potential. The inside of the cell is positive in this active area because Na^+ has already rushed into the neuron at this point. The remainder of the axon, still at resting potential and negative inside, is considered inactive. For the action potential to spread from the active to the inactive areas, the inactive areas must somehow be depolarized to threshold. This depolarization is accomplished by local current flow between the area already undergoing an action potential and the adjacent inactive area, similar to the spread of graded potentials. Recall that the rate at which the region ahead of the active area is depolarized depends on the length constant λ. Because opposite charges attract, current can flow locally between the active area and the neighboring inactive area on both the inside and the outside of the membrane. This local current flow in effect neutralizes or eliminates some of the unbalanced charges in the inactive area, a depolarizing effect that quickly brings the involved inactive area to threshold, at which time the voltage-gated Na^+ channels in this region of the membrane are all thrown open by positive feedback, leading to an action potential in this previously inactive area. Meanwhile, the original active area returns to resting potential as a result of K^+ efflux.

In turn, beyond the new active area is another inactive area, so the same thing happens again. This cycle repeats itself in a chain reaction until the action potential has spread to the end of the axon. *Once an action potential is initiated in one part of a neuron membrane, a self-perpetuating cycle is initiated so that the action potential is propagated throughout the rest of the fiber automatically.* In this way, the axon is like a firecracker fuse that needs to be lit at only one end. Once ignited, the fire spreads down the fuse; it is not necessary to hold a match to every separate section of the fuse. Therefore, a new action potential can be initiated in two ways, in both cases involving passive spread of current from an adjacent site already depolarized. An action potential is initiated in the axon hillock in the first place by depolarizing current spreading from a graded potential in the cell body and dendrites. During propagation of the action potential down the axon, each new action potential is initiated by depolarizing local current flow spreading from the preceding site undergoing an action potential.

Note that the original action potential waveform does not travel along the membrane. Instead, it triggers an identical new action potential in the adjacent area of the membrane, with this process being repeated along the axon's length. An analogy is the spectator "wave" often seen at a sports stadium. Each section of spectators stands up (the rising phase of an action potential), then sits down (the falling phase) in sequence one after another as the wave moves around the stadium. The wave, not individual spectators (individual action potentials), travels around the stadium (conduction of the action potential along the axon). Because each new action potential in the conduction process is a fresh event dependent on the induced permeability changes and electrochemical

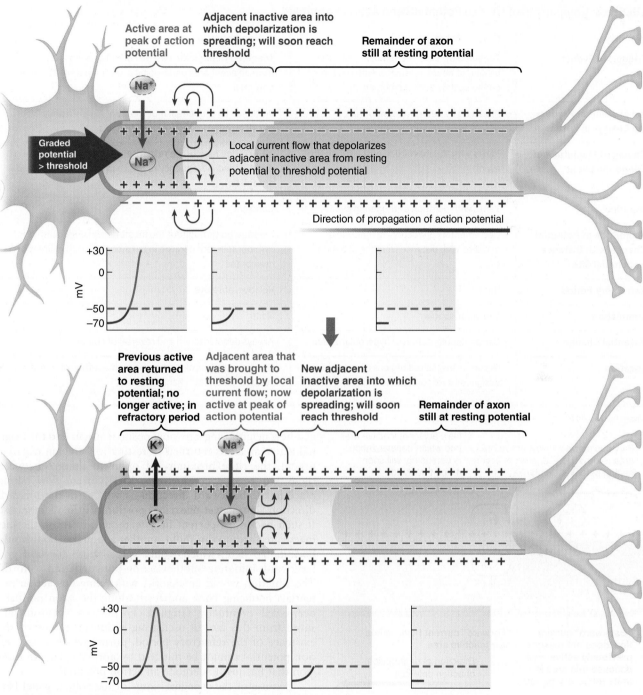

FIGURE 4-10 Contiguous conduction. Local current flow between the active area at the peak of an action potential and the adjacent inactive area still at resting potential reduces the potential in this contiguous inactive area to threshold, which triggers an action potential in the previously inactive area. The original active area returns to resting potential, and the new active area induces an action potential in the next adjacent inactive area by local current flow as the cycle repeats itself down the length of the axon.

© Cengage Learning, 2013

gradients, which are virtually identical down the length of the axon, the action potential arriving at the end of the axon is identical to the original one, no matter how long the axon. Thus, an action potential is spread throughout the axon in undiminished fashion. In this way, action potentials can serve as long-distance signals without attenuation or distortion.

Again, this nondecremental propagation of an action potential is in contrast to the decremental spread of a graded potential, which dies out over a very short distance because it is not able to regenerate itself. Table 4-2 summarizes the differences between graded potentials and action potentials, some of which are yet to be discussed.

TABLE 4-2 Comparison of Graded Potentials and Action Potentials

Property	Graded Potentials	Action Potentials
Triggering Events	Triggered by a specific stimulus, or by the binding of neurotransmitter with receptor on the postsynaptic membrane	Triggered through the passive spread of depolarization from adjacent area undergoing a graded potential or action potential
Ion Movement Producing Change in Potential	Produced by net movement of Na^+, K^+, Cl^-, or Ca^{2+} across plasma membrane	Produced by sequential movement of Na^+ into and K^+ out of cell through voltage-gated channels
Coding of Magnitude of Triggering Event	Varies with the magnitude of the triggering event	All-or-none membrane response; magnitude of triggering event coded in frequency rather than amplitude of action potentials
Duration	Varies with duration of triggering event	Constant
Magnitude of Potential Change with Distance from Initial Site	Decremental conduction; magnitude diminishes with distance from the initiation site	Propagated throughout the membrane in undiminishing fashion; self-regenerated in neighboring inactive areas of membrane
Refractory Period	None	Relative, absolute
Summation	Temporal, spatial	None
Potential Change	Can be depolarization or hyperpolarization	Always depolarization and reversal of charges
Location	Occurs in specialized regions of membrane designed to respond to the triggering event	Occurs in regions of membrane with a sufficient number of voltage-gated Na^+ channels

© Cengage Learning, 2013

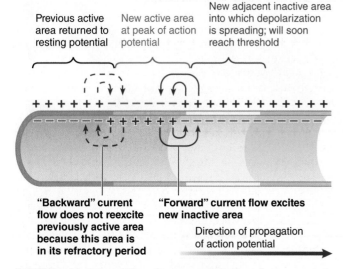

"Backward" current flow does not reexcite previously active area because this area is in its refractory period

"Forward" current flow excites new inactive area

Direction of propagation of action potential ⟶

FIGURE 4-11 Value of the refractory period. The refractory period prevents "backward" current flow. During an action potential and slightly afterward, an area cannot be restimulated by normal events to undergo another action potential. Thus, the refractory period ensures that an action potential can be propagated only in the forward direction along the axon.

© Cengage Learning, 2013

The refractory period ensures unidirectional propagation of the action potential

What ensures the one-way propagation of an action potential away from the initial site of activation? Note from Figure 4-11 that once the action potential has been regenerated

at a new neighboring site (now positive inside) and the original active area has returned to resting (once again negative inside), the close proximity of opposite charges between these two areas is conducive to local current flow taking place in the backward direction as well as in the forward direction (into as yet unexcited portions of the membrane). If such backward current flow were able to bring the just inactivated area to threshold, another action potential would be initiated here, which would spread both forward and backward, initiating another action potential, and so forth. The situation would be chaotic, with numerous action potentials bouncing back and forth along the axon until the nerve cell eventually fatigued. Fortunately, neurons are saved from this fate of oscillating action potentials by the existence of the **refractory period**, during which a new action potential cannot be initiated by normal events in a region that has just undergone an action potential.

Because of the changing status of the voltage-gated Na^+ and K^+ channels during and after an action potential, the refractory period has two components: the *absolute refractory period* and the *relative refractory period*. During the time that a particular patch of axonal membrane is undergoing an action potential, it is incapable of initiating another action potential, no matter how strongly it is stimulated by a triggering event. This time period when a recently activated patch of membrane is completely refractory (meaning "stubborn" or unresponsive) to further stimulation is known as the **absolute refractory period** (Figure 4-12). Once the voltage-gated Na^+ channels have flipped to their open, or activated, state, they cannot be triggered to open again in response to another depolarizing triggering event, no matter how strong, until resting potential is restored and the channels are reset to their original positions. Accordingly, the absolute refractory period lasts the entire time from opening of the voltage-gated Na^+ channels'

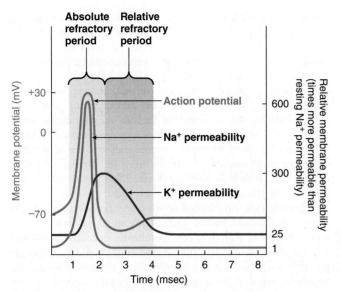

Absolute refractory period **Relative refractory period**

Action potential

Na⁺ permeability

K⁺ permeability

FIGURE 4-12 Absolute and relative refractory periods. During the absolute refractory period, the portion of the membrane that has just undergone an action potential cannot be restimulated. This period corresponds to the time during which the Na⁺ gates are not in their resting conformation. During the relative refractory period, the membrane can be restimulated only by a stronger stimulus than is usually necessary. This period corresponds to the time during which the K⁺ gates opened during the action potential have not yet closed, coupled with lingering inactivation of the voltage-gated Na⁺ channels.

© Cengage Learning, 2013

activation gates at threshold, through closure of their inactivation gates at the peak of the action potential ("closed and not capable of opening" state), until the return to resting potential when the channels' activation gates close and inactivation gates open once again; that is, until the channels are in their "closed but capable of opening" conformation. Only then can they respond to another depolarization to initiate another action potential. Because of this absolute refractory period, one action potential must be over before another can be initiated at the same site. Action potentials cannot overlap or be added one on top of another "piggyback fashion."

Following the absolute refractory period is a **relative refractory period**, during which a second action potential can be produced only by a triggering event considerably stronger than usual. The relative refractory period occurs after the action potential is complete because of a twofold effect: lingering inactivation of the voltage-gated Na⁺ channels and slowness to close of the voltage-gated K⁺ channels that opened at the peak of the action potential. During this time, fewer than normal voltage-gated Na⁺ channels are in a position to be jolted open by a depolarizing triggering event. Simultaneously, K⁺ is still leaving through its slow-to-close channels during the *after hyperpolarization phase*. The less-than-normal Na⁺ entry in response to another triggering event is opposed by a persistent hyperpolarizing outward leak of K⁺ through its not-yet-closed channels, and thus a greater-than-normal depolarizing triggering event is needed to bring the membrane to threshold during the relative refractory period.

By the time the original site has recovered from its refractory period and is capable of being restimulated by normal

current flow, the action potential is so far away that it can no longer influence the original site. Thus, *the refractory period ensures the unidirectional propagation of the action potential down the axon away from the initial site of activation.*

The refractory period also limits the frequency of action potentials

The refractory period also sets an upper limit on the *frequency* of action potentials; that is, it determines the maximum number of new action potentials that can be initiated and propagated along the fiber in a given period of time. As you will see shortly, frequency is the major method of carrying information by action potentials. The length of the refractory period varies for different types of neurons. The longer the refractory period, the greater the delay before the new action potential can be initiated and the lower the frequency with which a neuron can respond to repeated or ongoing stimulation.

Action potentials occur in all-or-none fashion

If any portion of the neuronal membrane is depolarized to threshold, an action potential is initiated and relayed throughout the membrane in undiminished fashion. Furthermore, once threshold has been reached, the resultant action potential always goes to maximal height. The reason for this effect is the positive-feedback aspect we discussed earlier (Figure 4-7), whose magnitude is not altered by the strength of the depolarizing triggering event. However, a triggering event that fails to depolarize the membrane to threshold does not trigger an action potential at all. Thus, *an excitable membrane either responds to a triggering event with a maximal action potential that spreads nondecrementally throughout the membrane, or it does not respond with an action potential at all.* This property is called the **all-or-none law**.

This all-or-none concept is like lighting a fuse with applied heat. Either a heated fuse does not reach its ignition temperature (threshold is not reached), or it is heated enough to start combustion (threshold is reached), which then self-propagates down the fuse. A higher amount of applied heat does not produce a greater or faster fuse fire. Just as it is not possible to ignite a fuse halfway, it is normally not possible to have a halfway action potential.

The threshold phenomenon allows some discrimination between important and unimportant stimuli or other triggering events. Stimuli too weak to bring the membrane to threshold do not initiate an action potential and therefore do not clutter up the nervous system by transmitting insignificant signals.

The strength of a stimulus is coded primarily by the frequency of action potentials

How is it possible to differentiate between two stimuli of varying intensities if both bring the membrane to threshold and generate action potentials of the same magnitude? For example, how can an animal distinguish between a warm and a hot surface if both trigger identical action potentials in a nerve fiber relaying information about skin temperature to the central nervous system? The answer lies primarily in the *frequency* with which the action potentials are generated. A stronger stimulus does not produce a larger action potential, but it does trigger a greater *number* of action potentials per second to be propagated along the fiber. This occurs because graded potentials that depolarize neurons typically last much longer than one action potential. A strong graded potential will rapidly push the neuron back to threshold repeatedly after each refractory period. In contrast, a weak graded potential will only gradually do so, resulting in many fewer action potentials per unit of time.

In addition, a stronger stimulus in a region will result in more neurons reaching threshold, thus increasing the total information sent to the central nervous system. For example, lightly touch this page with your finger and note the area of skin in contact with the page. Now, press down more firmly and note that a larger surface area of skin is in contact with the page. Therefore, more neurons are brought to threshold with this stronger touch stimulus.

Recent research suggests that more complex coding mechanisms also occur in some neural systems. For example, the *temporal (timing) pattern* of action potentials can carry information. Suppose two neurons are both transmitting 100 action potentials per second. If one does so in an even pattern, but the other does so in five dense groups of 20 action potentials separated by long time gaps, the target neuron reacts differently to each. The degree to which neurons use frequency versus temporal coding, particularly inside brains, is still uncertain.

Once initiated, the velocity, or speed, with which an action potential travels down the axon depends on two factors: (1) whether the fiber is *myelinated* and (2) the diameter of the fiber. Contiguous conduction occurs in unmyelinated fibers. In this case, as we just said, each individual action potential initiates an identical new action potential in the next contiguous (bordering) segment of the axon membrane so that every portion of the membrane undergoes an action potential as this electrical signal is conducted from the beginning to the end of the axon. A faster method of propagation, *saltatory conduction,* takes place in myelinated fibers. Next let's see how a myelinated fiber compares with an unmyelinated fiber, then see how saltatory conduction compares with contiguous conduction.

Myelination increases the speed of conduction of action potentials and conserves energy in the process

Myelinated fibers, found in vertebrates (excluding lampreys and hagfishes) and a few other animals, are axons covered with myelin at regular intervals along the length of the axon (Figure 4-13a). **Myelin** consists primarily of lipids. Because the water-soluble ions responsible for carrying current across the membrane cannot permeate this thick lipid barrier, the myelin coating acts as an insulator, just like rubber around an electrical wire, to prevent current leakage across the myelinated portion of the membrane. Myelin is not actually a part of the nerve cell but consists of separate myelin-forming cells that tightly wrap themselves around the axon in concentric rings. These myelin-forming cells are **oligodendrocytes** in the central nervous system (the brain and spinal cord) and **Schwann cells** in the peripheral nervous system (the nerves running between the central nervous system and the various regions of the body) (Figure 4-13b and c). Each

patch of lipid-rich myelin consists of multiple layers of the myelin-forming cell's plasma membrane (predominantly the lipid bilayer) as the cell wraps itself around and around the axon. A patch of vertebrate myelin might be made up of as many as 300 layers of wrapped lipid bilayers.

Myelin appears to be rare outside of vertebrates (see *Unanswered Questions: How many animal groups have myelin?*). Disadvantages of myelin include the significant costs in metabolic and biosynthetic resources required to produce the many layers of lipid-rich membrane; and for some species, key components of myelin may be limiting such as cholesterol, which is not synthesized by protostomes and hence is an essential nutrient in their diet.

Between the myelinated regions, the axonal membrane is bare and exposed to the ECF. Only at these bare spaces, called **nodes of Ranvier,** can membrane potential exist and current flow across the membrane (Figure 4-13). Voltage-gated Na^+ channels are concentrated at the nodal areas; the myelin-covered regions are almost devoid of these special passageways. By contrast, an unmyelinated fiber has a uniform density of voltage-gated Na^+ channels throughout its entire length. As you know, action potentials can be generated only at portions of the membrane furnished with an abundance of these channels. Spread of action potentials in myelinated fibers is called *saltatory conduction,* which we will now examine.

The nodes are usually about 1 mm apart, close enough that local current from an active node can reach an adjacent node before dying off. When an action potential occurs at one of the nodes, opposite charges attract from the adjacent inactive node, reducing its potential to threshold so that it undergoes an action potential, and so on. Consequently, in a myelinated fiber, the impulse "jumps" from node to node, skipping over the myelinated sections of the axon (Figure 4-14); this process is called **saltatory conduction** (the Latin word *saltere* means "to leap"). Saltatory conduction propagates action potentials more rapidly than does conduction by local current flow because the action potential leaps over myelinated sections but must be regenerated within every section of an unmyelinated axonal membrane from beginning to end. Myelinated fibers conduct impulses about 50 times faster than unmyelinated fibers of comparable size. You can think of myelinated fibers as the "superhighways" and unmyelinated fibers as the "back roads" of the nervous system when it comes to the speed with which information can be transmitted. Thus, the most urgent types of information are transmitted via myelinated fibers, whereas nerve pathways carrying less urgent information are unmyelinaed.

Besides permitting action potentials to travel faster, myelination also conserves energy. Because the ion fluxes associated with action potentials are confined to the nodal regions, the energy-consuming Na^+/K^+ pump must restore fewer ions to their respective sides of the membrane after propagation of an action potential.

Insect neurons are coated with a sheath that regulates ion balance

Insects do not have a myelin sheath per se but, rather, loosely coat the axon with a cellular *nerve sheath*. There are no nodes for enhancing conduction speed. The purpose of the sheath becomes evident when you examine the concen-

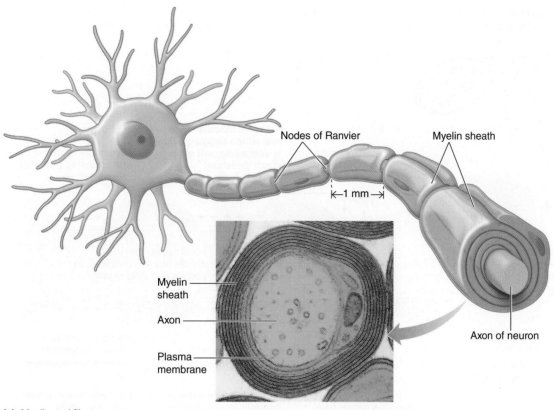

Nodes of Ranvier

Myelin sheath

←1 mm→

Myelin sheath

Axon

Plasma membrane

Axon of neuron

(a) Myelinated fiber

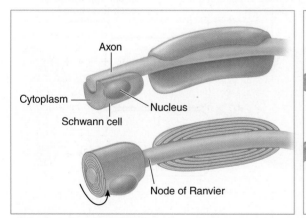

Axon

Cytoplasm

Nucleus

Schwann cell

Node of Ranvier

(b) Schwann cells in peripheral nervous system

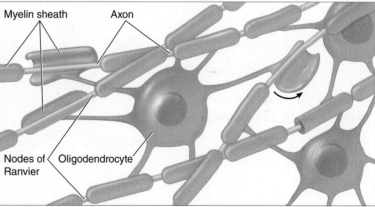

Myelin sheath

Axon

Nodes of Ranvier

Oligodendrocyte

(c) Oligodendrocytes in central nervous system

FIGURE 4-13 Myelinated fibers. (a) A myelinated fiber is surrounded by myelin at regular intervals. The intervening bare, unmyelinated regions are known as nodes of Ranvier. The electron micrograph shows a myelinated fiber in cross section at a myelinated region. (b) In the PNS, each patch of myelin is formed by a separate Schwann cell that wraps itself tightly around the nerve fiber. (c) In the CNS, each of the several processes ("arms") of a myelin-forming oligodendrocyte forms a patch of myelin around a separate nerve fiber.

© C. Raines/Visuals Unlimited

tration of ions in the hemolymph of herbivorous (plant-eating) insects. Plants and fruits are high in K⁺ but low in Na⁺ content, so animals that feed exclusively on plant material potentially may consume insufficient Na⁺. In herbivorous insects, concentrations of Na⁺ in the hemolymph range from 2 to 15 mM lower than that measured in the ICF—too low to sustain propagation of an action potential. Com-

pared to vertebrates, concentrations of K⁺ are elevated in the hemolymph (20 to 60 mM), although still lower than in the ICF. These insects have solved the dietary ion problem by evolving the nerve fat-body sheath, which aids in propagation of an action potential. Because the nerve sheath is only loosely wrapped around the axon, it can actively regulate the local fluid bathing the axon. Measurement of Na⁺

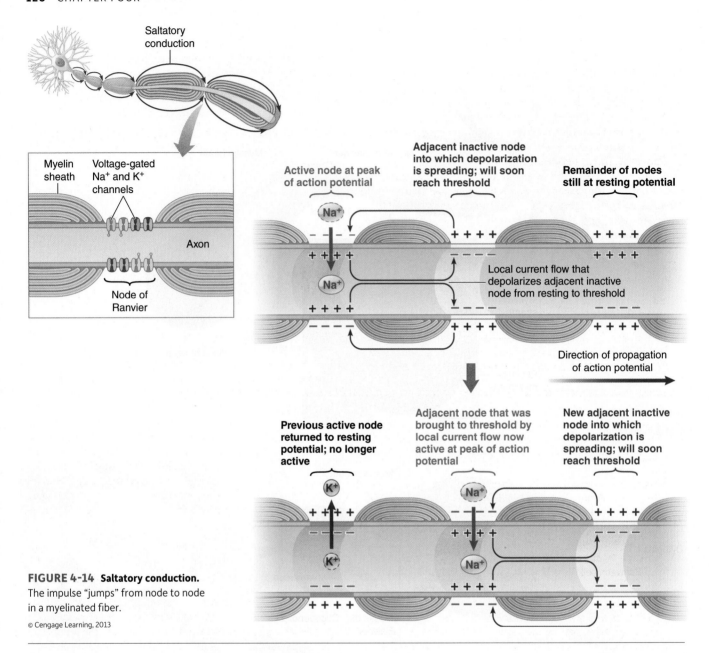

FIGURE 4-14 Saltatory conduction.
The impulse "jumps" from node to node
in a myelinated fiber.

© Cengage Learning, 2013

concentrations in this fluid reveals concentrations about 10-fold higher than in the hemolymph, concentrations now high enough to sustain propagation of an action potential. In contrast, herbivorous mammals such as deer meet this Na⁺ need by locating natural salty minerals ("licks"), while farmers provide dietary mineral supplements to domesticated species or provide access to artificial salt licks.

unanswered Questions How many animal groups have myelin?

For decades, scientists thought only vertebrates accelerate action potentials with insulation. For example, most protostomes, which include mollusks and insects, completely lack myelin. In recent years, however, researchers have found such insulation in several nonvertebrates. For example, neurons of Oligochaetes (annelids

such as earthworms) have wrappings of 20 to 200 bilayers. Myelin structures have also been found in shrimp and, unexpectedly, in copepods. Copepods are tiny (almost microscopic) planktonic crustaceans that inhabit all oceans and, as the primary ocean herbivores, are thought to be the most numerous animals on Earth. But scientists always assumed their tiny size meant unmyelinated axons would work fast enough. In 1998, a chance microscopic observation by researchers at the University of Hawaii led to the discovery that copepods have myelin-like insulation with spiral wrapping. What purpose might this serve in the survival of this important but tiny animal? And how many other types of animals might have myelin that we have not yet discovered? See Suggested Readings by Lenz, Hartline, and Davis (2000) and Hartline (2008). ◀◀

Fiber diameter also influences the velocity of action potential propagation

Besides the effect of myelination, the diameter of the fiber also influences the speed with which an axon can conduct action potentials. The magnitude of current flow (that is, the amount of charge that moves) depends not only on the difference in potential between two adjacent electrically charged regions but also on the resistance to electrical charge movement between the two regions. When fiber diameter increases, the resistance to local current decreases. Thus, the larger the diameter of the nerve fiber, the faster it can propagate action potentials. Fast conduction of an impulse increases the possibility of an animal escaping from a predator. Other uses include prey capture or a response to stress. Compared to slow conduction pathways, fast conduction requires considerably greater energy expenditure.

In some species, fiber diameters have become so large that they are called **giant nerve axons**. They are characterized as axons with diameters that are markedly larger than the diameters of other axons in the same animal and by their role in the coordination of abrupt withdrawal or escape movements. Giant axons may arise from the fusion of smaller fibers (up to 1,500 cell bodies). Such multicellular forms occur in numerous groups, including annelids, decapod crustaceans, and cephalopods and in many vertebrates. Giant axons were first identified in the squid *Loligo;* indeed, researchers first used these axons to uncover the mechanisms of resting and action potentials (see opening picture of this chapter).

There are also unicellular giant axons, found in cockroaches, crayfish, and some annelids. They are important for escaping predators or catching prey. For example, in a crayfish, the perception of danger is relayed by a giant axon to muscles for the tail, which rapidly flips, propelling the animal away from the danger. Some vertebrates also have giant axons called **Mauthner neurons** that, in fish and larval amphibians, arise from the brain stem and run to tail muscles. These too are used for tail-flip escape behaviors.

Even with increased size, unmyelinated nerves are slower than many myelinated ones. In the former, an action potential in a giant axon is conducted at velocities from 5 to 25 meters/sec, compared with a velocity of 0.7 meters/sec in small unmyelinated fibers such as those supplying a vertebrate digestive tract. In contrast, large myelinated fibers, such as those supplying skeletal muscles, can conduct action potentials at a speed of up to 120 meters/sec. This variability in speed of propagation of action potentials is related to the urgency of the information being conveyed. A signal to skeletal muscles to execute a particular movement (for example, to prevent you from falling as you trip on something) must be transmitted more rapidly than a signal to modify a slow-acting digestive process. Were it not for myelination, axon diameters within urgent nerve pathways would have to be very large to achieve the necessary conduction velocities, as you have seen. Vertebrates (and at least some annelids and crustacea) have escaped the necessity of very large fibers by wrapping the axons in myelin to permit economical, rapid, long-distance signaling.

check your understanding 4.3

Define the following terms: *threshold potential, action potential, refractory period,* and *all-or-none law.*

How does each of the three components of a neuron contribute to its function?

Compare contiguous conduction with saltatory conduction, and the ways by which various animal neurons increase the speed of both.

4.4 Electrical and Chemical Synapses

What happens once an action potential reaches the end of an axon? Where the action potential waveform must somehow transmit its message to another cell—called the *target* cell—is at the region termed the **synapse** (*synapsis* means "juncture"). Charles Sherrington in 1897 first recognized that a different form of communication occurred at this juncture.

The signaling cell is called the *presynaptic* neuron, and its target, also called *postsynaptic* cell, is typically another neuron or effector such as a gland or muscle. When the action potential waveform reaches the presynaptic axon terminal, it alters the activity of the target cells on which the neuron terminates. There are only two basic kinds of transmission of signals at this point: electrical (direct) or chemical (indirect), which differ in signal transfer mechanisms. Although electrical transmission is not as common as chemical, it is much simpler. Initially, let's consider electrical transmission, to set the stage for the more complex chemical transmission process.

Electrical synapses transfer action potential waveforms through gap junctions

Action potential waveforms are essentially transmitted across **electrical synapses** unperturbed, moving along the axon as if the synapse was not present. This is so because the cytoplasm of the presynaptic neuron is in direct contact with the postsynaptic neuron via *gap junctions* (Chapter 2, pp. 66–67). Because the resistance of the gap junction membranes to the flow of current is low, this permits a nearly full-strength signal to be induced in the postsynaptic cell. Also, compared to a chemical synapse, electrical transmission occurs with negligible time delay. This feature makes electrical synapses useful as a mechanism for escape responses in nonvertebrates. Electrical synapses were first discovered in the abdominal nerve cord of crayfish in 1959 by E. Furshpan and D. Potter and have since been identified in numerous other animals, including some vertebrate excitable cells such as retinal neurons and smooth and cardiac muscle fibers. The rapidity of electrical communication is a feature exploited by these vertebrate tissues to coordinate the activity of a group of cells (for cardiac cells, see Chapter 9). In addition, in some neural pathways, signal transmission is both electrical and chemical. This is found in the escape response of some fishes, which flip their tails quickly when startled.

Chemical synapses convert action potentials into organic chemical messengers that are exocytosed into the synaptic gap

Nerves and their target cells do not actually make direct contact at the second type of synapse, the **chemical synapse**. The gap, or **synaptic cleft**, between these two structures is too large (approximately 20 to 40 nm wide) to permit electrical transmission of an impulse between them (that is, the action potential cannot "jump" that far). Instead, an organic *chemical messenger*—generically called a **neurotransmitter**—is used to carry the signal between the neuron terminal and the target cell. Briefly, the messenger is exocytosed from the neuron and diffuses to the target, where it binds to **receptor** proteins. The receptors in turn alter the activity of the target cell. (Again, as with ions in action potentials, the giant neurons of squid provided the first evidence that transmission of a neural signal across the synapse required the release of a transmitter substance.) Because chemical synapses rely on diffusion of the neurotransmitter, they are much slower (by several orders of magnitude) than electrical synapses; but they have two advantages over electrical synapses. (1) They typically operate in one direction only, so that false messages do not inadvertently travel from effectors to integrators; that is, the presynaptic neuron brings about changes in membrane potential of the postsynaptic (target) membrane, but not vice versa. However, we later discuss *retrograde* messengers in which a signal is sent "backward." (2) They allow for various kinds of signaling events other than simply triggering action potentials in the target. How these features arise and how they are used will shortly become apparent when we examine the events that occur at a synapse.

Neurotransmitters come in several chemical classes of signal molecules (Chapter 3, p. 93), but the most common ones are *amines* (Table 4-3, listing classical neurotransmitters). Examples you will encounter later in this chapter include glutamate, norepinephrine, dopamine, and serotonin. Some of these are highly conserved in evolution, serving as neurotransmitters in many different animal phyla. For example, *serotonin* is found in vertebrate nervous systems; it is used in synapses involved with sleep, pain, aggression, sexual behavior, and food intake. The transmitter *dopamine* is found in mammalian neurons involved with control of locomotion and reward. In snails (a mollusk), serotonin has been found to stimulate foot locomotion, whereas dopamine inhibits it. Serotonin has even been found (using immunofluorescence, p. 34) in synaptic vessels of a sea anemone (a cnidarian).

TABLE 4-3 Some Common Neurotransmitters

Acetylcholine	Histamine
Dopamine	Glycine
Norepinephrine	Glutamate
Epinephrine	Adenosine
Serotonin	Gamma-aminobutyric acid (GABA)

Depending on where a neuron terminates, a chemical synapse can cause another neuron to convey an electrical message along a nerve pathway, a muscle cell to contract, a gland cell to secrete, or some other function. When a neuron terminates on an effector, the neuron is said to **innervate** the structure. The junctions between nerves and glands that they innervate are described later. For now, we will concentrate on the junction between two neurons, and between a neuron and a skeletal muscle. Let's begin with two neurons.

check your understanding 4.4

List some characteristics of neurotransmitters.

Why do synapses normally operate in one direction only?

Why is transmission across a chemical synapse slower than an electrical synapse?

4.5 Neuron-to-Neuron Synapses

Typically, a neuron-to-neuron synapse involves a junction between an axon terminal of a presynaptic neuron and the dendrites or cell body of a postsynaptic neuron. Less frequently, axon-to-axon and dendrite-to-dendrite connections occur. Most neuronal cell bodies and associated dendrites receive thousands of synaptic inputs, which are axon terminals from many other neurons. Neurologists have estimated that some neurons within the central nervous system receive as many as 100,000 synaptic inputs (Figure 4-15).

The anatomy of one of these synapses is shown in Figure 4-16. The axon terminal of the presynaptic neuron, which conducts its action potentials *toward* the synapse, ends in a slight swelling, the **synaptic knob**. The synaptic knob contains synaptic vesicles, which store neurotransmitters synthesized and packaged by the Golgi apparatus (p. 41) of the presynaptic neuron. The synaptic knob comes into close proximity to, but (because of the cleft or gap) does not actually directly contact, the postsynaptic neuron. The part of the postsynaptic membrane immediately underlying the synaptic knob is the **subsynaptic membrane** (*sub*, "under").

A neurotransmitter carries the signal across a fast synapse and opens a chemically gated channel

Most, but not all, neurotransmitters function by changing the conformation of *chemically gated channels,* thereby altering membrane permeability and ionic fluxes across the postsynaptic membrane. Synapses involving these rapid responses are considered **"fast" synapses**. The events that occur at these common types of synapse are summarized next and illustrated in Figure 4-16 (we investigate **"slow"** synapses later):

1. When an action potential in a presynaptic neuron has been propagated to the axon terminal (step 1 in Figure 4-16), this change in potential triggers the opening of voltage-gated Ca^{2+} channels in the synaptic knob.
2. Because Ca^{2+} is in much higher concentration in the ECF, and because the cell has a negative charge, this ion flows into the synaptic knob through the opened channels (step 2).

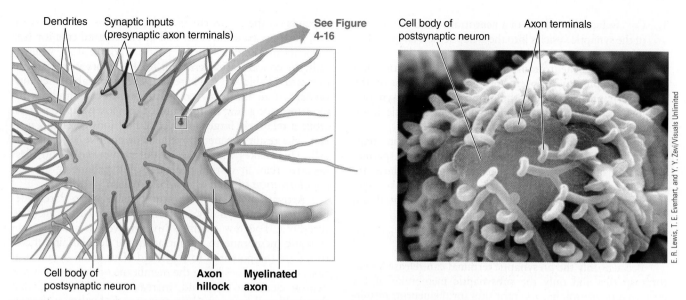

FIGURE 4-15 **Synaptic inputs (presynaptic axon terminals) to the cell body and dendrites of a single postsynaptic neuron.** Note that the drying process used to prepare the neuron for the electron micrograph has toppled the presynaptic axon terminals and pulled them away from the postsynaptic cell body.

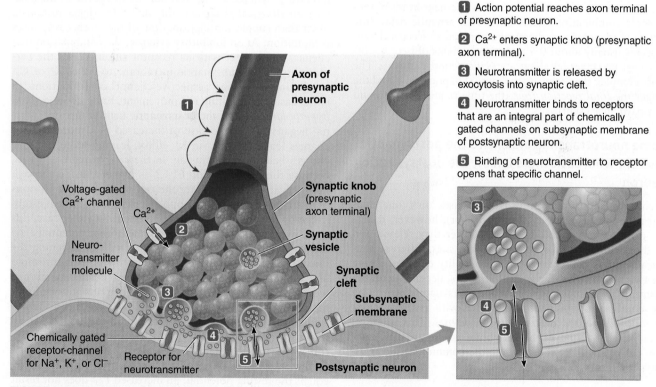

1 Action potential reaches axon terminal of presynaptic neuron.

2 Ca^{2+} enters synaptic knob (presynaptic axon terminal).

3 Neurotransmitter is released by exocytosis into synaptic cleft.

4 Neurotransmitter binds to receptors that are an integral part of chemically gated channels on subsynaptic membrane of postsynaptic neuron.

5 Binding of neurotransmitter to receptor opens that specific channel.

FIGURE 4-16 **Structure and function of a single synapse.** The numbered steps designate the sequence of events that take place at a synapse. The blowup depicts the release by exocytosis of neurotransmitter from the presynaptic axon terminal and its subsequent binding with receptors specific for it on the subsynaptic membrane of the postsynaptic neuron.

3. Ca^{2+} induces the release of a neurotransmitter from some of the synaptic vesicles into the synaptic cleft (step 3). The release is accomplished by exocytosis (see p. 41).
4. The released neurotransmitter diffuses across the cleft and combines with specific protein receptor sites on the **subsynaptic membrane**, the portion of the postsynaptic membrane immediately underlying the synaptic knob (*sub* means "under") (step 4).
5. This binding triggers the opening of specific ion channels in the subsynaptic membrane, changing the permeability of the postsynaptic neuron (step 5). These are chemically gated channels, in contrast to the voltage-gated channels responsible for the action potential and for the Ca^{2+} influx into the synaptic knob.

Finally, a mechanism is needed to stop the signal. We will look at mechanisms for this later.

Because only the presynaptic terminal can release a neurotransmitter and only the subsynaptic membrane of the postsynaptic neuron has receptor sites for the neurotransmitter, the synapse can operate only in the direction from presynaptic to postsynaptic neuron. This advantage over electrical synapses prevents signals from moving backward, which would interfere with the ability of neurons to control their targets.

Conversion of the electrical signal in the presynaptic neuron (an action potential) to an electrical signal in the postsynaptic neuron by chemical means (via the neurotransmitter–receptor combination) takes time. This **synaptic delay** (the major disadvantage of chemical compared to electrical synapses) is usually about 0.5 to 1 msec. Chains of neurons often must be traversed along a specific neural pathway. The more complex the pathway, the more synaptic delays and the longer the *total reaction time* (the time required to respond to a particular event).

Some neurotransmitters excite the postsynaptic neuron, whereas others inhibit the postsynaptic neuron

Each presynaptic neuron usually releases only one neurotransmitter (exceptions are noted later); however, different neurons vary in the neurotransmitter they release. On binding with their subsynaptic receptor sites, different neurotransmitters cause different permeability changes. There are two broad types of chemical synapses, depending on the permeability changes induced in the postsynaptic neuron by the combination of neurotransmitter with receptor sites: *excitatory synapses* and *inhibitory synapses*. This situation is unlike that found at the vertebrate neuromuscular junction (as you will see later) where the communication is always excitatory.

Excitatory Synapses At an **excitatory synapse**, the response to the binding of neurotransmitter to the receptor-channel is the opening of nonspecific cation channels in the subsynaptic membrane that permit simultaneous passage of Na^+ and K^+ through them. (These are a different type of channel from those you have encountered before.) Thus, permeability to both these ions is increased at the same time. How much of each ion conducts and/or diffuses through an open channel depends on the electrochemical gradients. At resting potential, both the concentration and electrical gradients for Na^+ favor its movement into the postsynaptic neuron, whereas only the concentration gradient for K^+ favors its movement outward (Table 4-1). Therefore, the permeability change induced at an excitatory synapse results in the simultaneous movement of a few K^+ ions out of the postsynaptic neuron while a relatively larger number of Na^+ ions enter this neuron. The result is a net movement of positive ions into the cell. This makes the inside of the membrane slightly less negative than at resting potential, thus producing a *small depolarization* of the postsynaptic neuron.

Activation of one excitatory synapse can rarely depolarize the postsynaptic neuron sufficiently to bring it to threshold. Too few channels are involved at a single subsynaptic membrane to permit adequate depolarizing fluxes to reduce the potential to threshold. This small depolarization, however, does bring the membrane of the postsynaptic neuron closer to threshold, increasing the likelihood that threshold will be reached (in response to further excitatory input) and an action potential will occur. That is, the membrane is more excitable (easier to bring to threshold) than when at rest. Accordingly, such a postsynaptic potential change occurring at an excitatory synapse is called an **excitatory postsynaptic potential**, or EPSP (Figure 4-17a).

Inhibitory Synapses The second advantage of a chemical over an electrical synapse is the ability to trigger responses other than simple stimulation. One of these other responses is inhibition. At an **inhibitory synapse**, the combination of a chemical messenger with its receptor site increases the permeability of the subsynaptic membrane to either K^+ or Cl^- by altering these ions' respective channel conformations. In either case, the resulting ion movements bring about a *small hyperpolarization* of the postsynaptic neuron (greater internal negativity). In the case of increased P_{K^+}, more positive charges leave the cell via K^+ efflux, leaving more negative charges behind on the inside. In the case of increased P_{Cl^-}, because the concentration of Cl^- is higher outside the cell, more negative charges enter the cell in the form of Cl^- ions than are driven out by the opposing electrical gradient established by the resting membrane potential. In either case, this small hyperpolarization moves the membrane potential even farther away from threshold (Figure 4-17b), lessening the likelihood that the postsynaptic neuron will reach threshold and undergo an action potential. That is, the membrane would be harder to bring to threshold by excitatory input than when it is at resting potential. Under these circumstances the membrane is said to be inhibited, and the small hyperpolarization of the postsynaptic cell is called an **inhibitory postsynaptic potential**, or IPSP.

In cells where the equilibrium potential for Cl^- exactly equals the resting potential, an increased P_{Cl^-} does not result in a hyperpolarization because there is no driving force to produce Cl^- movement. Nevertheless, opening of Cl^- channels in these cells is still inhibitory because it tends to hold the membrane at resting potential, thus reducing the likelihood that threshold will be reached. That is, if a small amount of Na^+ enters from another channel, the cell becomes slightly less negative, which in turn allows some Cl^- to enter the cell, offsetting the effect of Na^+.

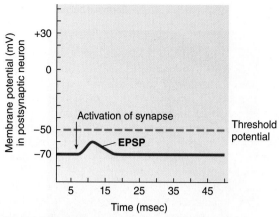

(a) **Excitatory synapse**

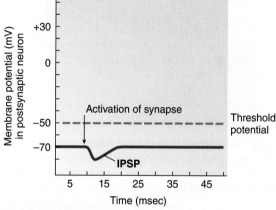

(b) **Inhibitory synapse**

FIGURE 4-17 Postsynaptic potentials. (a) An excitatory postsynaptic potential (EPSP) brought about by activation of an excitatory presynaptic input brings the postsynaptic neuron closer to threshold potential. (b) An inhibitory postsynaptic potential (IPSP) brought about by activation of an inhibitory presynaptic input moves the postsynaptic neuron farther from threshold potential.

© Cengage Learning, 2013

Each neurotransmitter–receptor combination always produces the same response

The binding of a specific neurotransmitter at fast synapses with the appropriate subsynaptic receptors always leads to the same change in permeability and resultant change in potential of a particular postsynaptic membrane. That is, the response to a given neurotransmitter–receptor combination is always constant. Some neurotransmitters (for example, glutamate, the most common excitatory neurotransmitter in the brain) typically bring about EPSPs, whereas others (for example, *gamma-aminobutyric acid,* or *GABA,* the vertebrate brain's main inhibitory neurotransmitter) always induce IPSPs. Moreover, a single fast neuron only makes one class of neurotransmitter (excitatory or inhibitory). That is, *a given synapse is either excitatory or inhibitory.* However, this does not necessarily mean that synapses are invariant over the long term. At least in developing spinal cord in frog embryos, artificially varying the rate of electrical stimulation can change the type of neurotransmitter an individual neuron makes from excitatory to inhibitory and vice versa.

How frequently this type of switching occurs in nature—if at all—is unknown.

Although each neurotransmitter–receptor combination yields the same response, some neurotransmitters (for example, norepinephrine) may produce an EPSP at one synapse and an IPSP at another synapse. This is a result of having different types of receptors (an example of the "same key, different locks" principle, p. 93).

Recent evidence suggests that in some instances two different neurotransmitters can be released simultaneously from a single axon terminal; however, they are not in conflict. For example, glycine and GABA, both of which produce inhibitory responses, can be packaged and released from the same synaptic vesicles. Scientists speculate that the fast-acting glycine and the more slowly acting GABA may complement each other in the control of activities that depend on precise timing, for example, coordination of complex movements. The brain can also synthesize neurosteroids, which enhance GABA receptor function, and so fine-tune neuronal activity.

Neurotransmitters are quickly removed from the synaptic cleft

As long as the neurotransmitter remains bound to the receptor sites, the alteration in membrane permeability responsible for the EPSP or IPSP continues. It is important that the neurotransmitter be removed after producing the appropriate response in the postsynaptic neuron so that the postsynaptic "slate" is "wiped clean," ready to receive additional messages from the same or other presynaptic inputs. Thus, after combining with the postsynaptic receptor chemical, transmitters are removed and the response is terminated. This occurs in three general ways. The neurotransmitter may (1) *diffuse* away from the synaptic cleft, (2) be *inactivated by a specific enzyme* in a subsynaptic membrane, or (3) be transported back into the axon terminal by *reuptake* carriers in the presynaptic membrane. Once within the synaptic knob, the transmitter can be stored and released another time (recycled) in response to a subsequent action potential or destroyed by enzymes within the synaptic knob. The method employed depends on the particular synapse. We describe a detailed example at neuromuscular synapses shortly.

Neurotransmitters in slow synapses function through intracellular second-messenger systems

As we have seen, neurotransmitters usually function by changing the conformation of chemically gated channels, in so-called fast synapses. Another mode of synaptic transmission used by some neurotransmitters, such as *serotonin,* involves the activation of intracellular second messengers, such as cAMP (see p. 97), within the postsynaptic neuron (these neurotransmitters are sometimes called *neuromodulators,* as are *neuropeptides*; we will discuss these later). Synapses that lead to responses mediated by second messengers are known as **"slow"** synapses because these responses take longer and often last longer (because of second-messenger cascades, Figure 3-17c) than those accomplished by fast synapses. Activation of the intracellular messenger, cyclic AMP, can induce both short- and long-term effects. In the short term, cAMP can lead to a prolonged opening of specific ionic gates. The gating effects can be either excitatory or inhibitory. In addition, cAMP may trigger more long-term changes in the postsynaptic cell, even to the extent of altering the cell's genetic expression by (indirectly) activating transcription factors. Such long-term cellular changes are linked to neuronal growth and development and play a role in learning and memory, which we will discuss in detail in the next chapter (p. 194).

check your understanding 4.5

Describe the events that occur at a chemical synapse.

How are so-called fast synapses different from slow synapses?

Compare the events that occur at excitatory and inhibitory synapses.

4.6 Neuromuscular Synapses

Now let's turn to a second example of a chemical synapse. Action potentials are conducted from the cell body in the CNS to skeletal muscle through large, myelinated **motor neurons**. As the axon approaches a muscle, it divides into many terminal branches and loses its myelin sheath. Each of these axon terminals forms a special synaptic junction, a **neuromuscular junction**, with one of the many muscle cells that compose the whole muscle (Figure 4-18). Each branch innervates only one muscle cell; therefore, each muscle cell has only one neuromuscular junction. A single muscle cell, called a **muscle fiber**, is long and cylindrical in shape. The axon terminal is enlarged into a knoblike structure, the **terminal button**, or **bouton**, which fits into a shallow depression, or groove, in the underlying muscle fiber (Figure 4-19). Some scientists alternatively call the neuromuscular junction a *motor end plate*. However, we reserve the term **motor end plate** for the specialized portion of the muscle cell membrane immediately under the terminal button.

Normally, each vertebrate muscle fiber is innervated at one or two end plates, although in amphibians and lizards specialized "slow" muscle fibers receive multiterminal innervation; that is, they are innervated by numerous chemical synapses along the length of each fiber. Interestingly, these fibers lack all-or-none action potentials and rather depend on the copious distribution of neuromuscular junctions to generate graded contractions. Tension produced in these muscles thus depends on the frequency of motor neuron output and is associated with slow, sustained contractions. (See Chapter 8 for details on typical neuromuscular control.)

Acetylcholine, a fast excitatory neurotransmitter, links electrical signals in motor neurons with electrical signals in skeletal muscle cells

Here let's take a look at the events at a muscle synapse in a vertebrate, following Figure 4-19. We will focus on one terminal button, but the same events take place concurrently at all the terminal buttons of a given neuromuscular junction. Each terminal button contains thousands of vesicles that store many molecules of the neurotransmitter **acetylcholine (ACh)**. Arrival of an action potential at the axon terminal triggers events similar to steps at the neuron-to-neuron synapse (Figure 4-16), with steps 1–2 being identical. These are followed by (Figure 4-19):

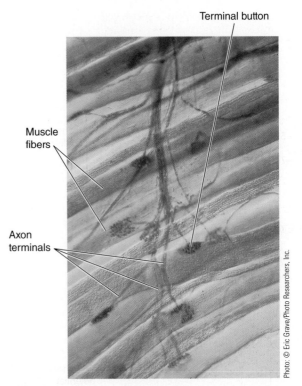

Terminal button

Muscle fibers

Axon terminals

Photo: © Eric Grave/Photo Researchers, Inc.

FIGURE 4-18 Motor neuron innervating skeletal muscle cells. When a motor neuron reaches a skeletal muscle, it divides into many terminal branches, each of which forms a neuromuscular junction with a single muscle cell (muscle fiber).

3. Release of ACh by exocytosis from several hundred of the vesicles into the cleft, and diffusion of ACh across the cleft (step 3).

4. Binding of ACh molecules to **receptors** in the postsynaptic membrane (step 4) (these **cholinergic receptors** are of the *nicotinic* type; see p. 164).

5. Opening of chemical-messenger–gated channels in the motor end plate, which depolarizes the motor end plate (step 5).

Steps 6–8 involve changes in the muscle potential, a graded one known as the **end-plate potential (EPP)**. The motor end-plate region itself does not have a threshold potential, so an action potential cannot be initiated at this site. However, an EPP brings about an action potential in the rest of the muscle fiber, as follows (continuing with Figure 4-19):

6. The neuromuscular junction is usually located in the middle of the long cylindrical muscle fiber. When an EPP takes place at the motor end plate, local current flow occurs between the depolarized end plate and the adjacent, resting cell membrane in both directions, reducing the potential to threshold in the adjacent areas (step 6).

7. The subsequent action potential initiated at these sites is propagated throughout the muscle fiber membrane by contiguous conduction (opening voltage-gated Na^+ channels) (step 7) and thus conduction (see Figure 4-10).

8. The spread runs in both directions, away from the motor end plate toward both ends of the fiber. This electrical activity initiates contraction of the muscle fiber (step 8).

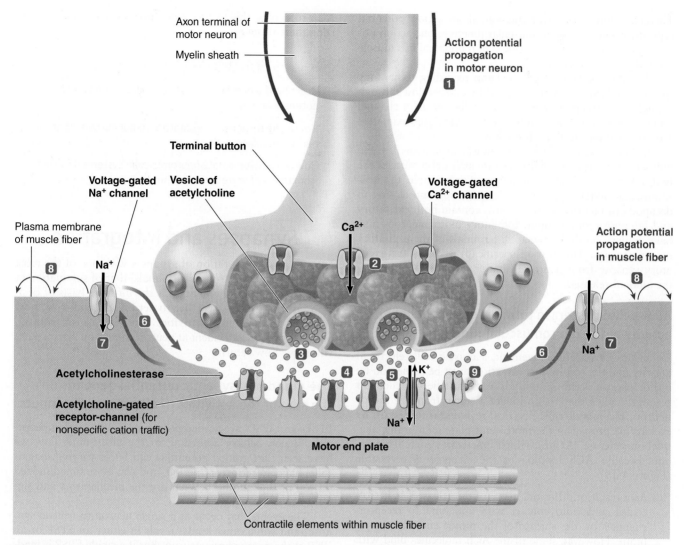

FIGURE 4-19 Events at a neuromuscular junction.

© Cengage Learning, 2013

Labels in figure:

Axon terminal of motor neuron

Myelin sheath

Action potential propagation in motor neuron ❶

Terminal button

Voltage-gated Na⁺ channel

Vesicle of acetylcholine

Ca^{2+}

Voltage-gated Ca²⁺ channel

Plasma membrane of muscle fiber

Action potential propagation in muscle fiber

❽ Na⁺ ❷

❻ ❼ ❸ ❹ ❺ K⁺ ❾

Acetylcholinesterase

Acetylcholine-gated receptor-channel (for nonspecific cation traffic)

Na⁺

Motor end plate

❻ Na⁺ ❼ ❽

Contractile elements within muscle fiber

❶ An action potential in a motor neuron is propagated to the axon terminal (terminal button).

❷ This local action potential triggers the opening of voltage-gated Ca²⁺ channels and the subsequent entry of Ca²⁺ into the terminal button.

❸ Ca²⁺ triggers the release of acetylcholine (ACh) by exocytosis from a portion of the vesicles.

❹ ACh diffuses across the space separating the nerve and muscle cells and binds with receptor-channels specific for it on the motor end plate of the muscle cell membrane.

❺ This binding brings about the opening of these nonspecific cation channels, leading to a relatively large movement of Na⁺ into the muscle cell compared to a smaller movement of K⁺ outward.

❻ The result is an end-plate potential. Local current flow occurs between the depolarized end plate and the adjacent membrane.

❼ This local current flow opens voltage-gated Na⁺ channels in the adjacent membrane.

❽ The resultant Na⁺ entry reduces the potential to threshold, initiating an action potential, which is propagated throughout the muscle fiber.

❾ ACh is subsequently destroyed by acetylcholinesterase, an enzyme located on the motor end-plate membrane, terminating the muscle cell's response.

The graded EPP is similar to an EPSP (p. 132) except that the magnitude of an EPP is much larger for the following reasons: (1) a neuromuscular junction consists of multiple terminal buttons, each of which simultaneously releases ACh on activation of the axon terminal; (2) more transmitter is released from a terminal button than from a *presynaptic knob* in response to an action potential; (3) the motor end plate has a larger surface area bearing a higher density of transmitter receptor sites, and so, accordingly, has more sites for binding with transmitter than does a subsynaptic membrane; and (4) accordingly, many more receptor-channels are opened in response to the transmitter–receptor complex in the motor end plate. This permits a greater net influx of Na⁺ ions and a larger depolarization.

Thus by means of ACh, an action potential in a motor neuron brings about an action potential and subsequent contraction in the muscle fiber. Unlike synaptic transmission among neurons, the magnitude of a single EPP is normally sufficient to cause an action potential in the muscle cell.

Therefore, one-to-one transmission of an action potential typically occurs at a neuromuscular junction; one action potential in a nerve cell triggers one action potential in a muscle cell that it innervates.

The development of patch clamp recording technology (p. 112) enabled researchers to measure current flow through a single ACh receptor channel. Generally, the ACh channel remains open for 1 to 3 msec and permits approximately 1 to 5×10^4 ions to flow through it.

Unlike most vertebrate neuromuscular systems (as you will see later), neurons that innervate arthropod muscles are both inhibitory as well as excitatory. For example, the neuromuscular systems regulating claw opening and closing in decapod crustaceans (such as crabs) contain both excitatory and inhibitory motor input. Inhibitory motor neurons also have endings on the presynaptic terminals of the excitatory motor neurons. This feature specifically protects the excitatory synapse from neurotransmitter depletion by limiting the amount released.

Acetylcholinesterase terminates acetylcholine activity at the neuromuscular junction

To ensure purposeful movement, a muscle cell's electrical activity and the resultant contraction must be turned on by motor neuron action potentials and must be switched off promptly when there is no longer a signal from the motor neuron. The muscle cells electrical response is turned off by an enzyme present in the motor end-plate membrane, **acetylcholinesterase (AChE)**, which inactivates ACh. This occurs because ACh binding to the receptor is reversible (Figure 4-19):

9. As a result of diffusion, many of the released ACh molecules come into contact with and bind to receptor-channels on the surface of the motor end-plate membrane. However, some of the ACh molecules bind with AChE, which is also at the end-plate surface. Being quickly inactivated, this ACh never contributes to the end-plate potential. The acetylcholine that does bind with receptor-channels does so very briefly (for about 1 millionth of a second), then detaches. Some of the ACh molecules quickly rebind with receptor-channels, keeping the end-plate channels open, but some diffuse deeper into the folds of the motor end plate, where AChE is located. AChE cleaves ACh into fragments, deactivating it (step 9). As this process is repeated, more and more ACh is inactivated until it has been virtually removed from the cleft within a few milliseconds after its release.

Removal of ACh terminates the EPP so the remainder of the muscle cell membrane returns to resting potential. Now the muscle can relax. Or, if sustained contraction is essential for the desired movement, another motor neuron action potential leads to the release of more ACh, which keeps the contractile process going (see p. 352). By removing contraction-inducing ACh from the motor end plate, AChE permits the choice of allowing relaxation to take place (no more ACh released) or keeping the contraction going (more ACh released), depending on the body's momentary needs.

ACh receptors were first isolated from the electric organs of marine electric fish (see p. 203). Because the electric organs of electric rays, for example, consist of high concentrations of modified neuromuscular junctions with a high density of ACh receptors, this feature aided investigators in characterizing these proteins.

check your understanding 4.6

Describe the events at a muscle synapse when an action potential arrives.

How is the response to excitatory stimuli turned off at the neuromuscular junction?

How are nonvertebrate neuromuscular systems different from those of vertebrate systems?

4.7 Synapses and Integration

Let's end this exploration of synapses with one of the most important functions found in animals: that of **summation** at chemical synapses. Summation is the ultimate basis of decision making (that is, whether to act or not, and if so, how strongly) in all animal activities involving neural control. Summation is most evident at neuron-to-neuron synapses, to which we now return.

The grand postsynaptic potential depends on the sum of the activities of all presynaptic inputs

EPSPs and IPSPs are graded potentials. Unlike action potentials, which behave according to the all-or-none law, we have seen that graded potentials can be of varying magnitude, have no refractory period, and can be summed (added on top of one another). What are the mechanisms and significance of summation?

The events that occur at a single neuron-to-neuron synapse (and some other kinds) result in either an EPSP or an IPSP at the postsynaptic neuron. But if a single EPSP is inadequate to bring the postsynaptic neuron to threshold and an IPSP moves it even farther from threshold, how is it possible to initiate an action potential in the postsynaptic neuron? Recall that a typical neuronal cell body receives thousands of presynaptic inputs from many other neurons. Some of these presynaptic inputs may be carrying sensory information brought from the environment; some may be signaling internal changes in homeostatic balance; others may be transmitting signals from control centers in the brain; and still others may arrive carrying other bits of information. At any given time, any number of these presynaptic neurons (probably hundreds) may be firing and thus influencing the postsynaptic neuron's level of activity. The total potential in the postsynaptic neuron, the **grand postsynaptic potential (GPSP)**, is a composite of all EPSPs and IPSPs occurring at approximately the same time.

The postsynaptic neuron can be brought to threshold in two ways: (1) *temporal summation* and (2) *spatial summation*. To illustrate these methods of summation, we will examine the possible interactions of three presynaptic inputs—two excitatory (Ex1 and Ex2) and one inhibitory (In1)—on a hypothetical postsynaptic neuron (Figure 4-20). The recording represents the potential in the postsynaptic cell. Bear in mind during our discussion of this simplified version that many thousands of synapses are actually interacting in the same way on a single cell body and its dendrites.

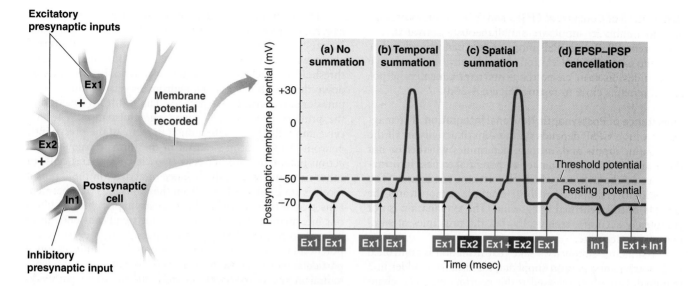

Excitatory presynaptic inputs

Ex1

Ex2

Membrane potential recorded

Postsynaptic cell

In1

Inhibitory presynaptic input

(a) If an excitatory presynaptic input (Ex1) is stimulated a second time after the first EPSP in the postsynaptic cell has died off, a second EPSP of the same magnitude will occur.

(b) If, however, Ex1 is stimulated a second time before the first EPSP has died off, the second EPSP will add onto, or sum with, the first EPSP, resulting in *temporal summation*, which may bring the postsynaptic cell to threshold.

(c) The postsynaptic cell may also be brought to threshold by *spatial summation* of EPSPs that are initiated by simultaneous activation of two (Ex1 and Ex2) or more excitatory presynaptic inputs.

(d) Simultaneous activation of an excitatory (Ex1) and inhibitory (In1) presynaptic input does not change the postsynaptic potential, because the resultant EPSP and IPSP cancel each other out.

FIGURE 4-20 **Determination of the grand postsynaptic potential by the sum of activity in the presynaptic inputs.** Two excitatory (Ex1 and Ex2) and one inhibitory (In1) presynaptic inputs terminate on this hypothetical postsynaptic neuron. The potential of the postsynaptic neuron is being recorded. For simplicity in the figure, summation of two EPSPs brings the postsynaptic neuron to threshold, but in reality, many EPSPs must sum to reach threshold.

© Cengage Learning, 2013

Temporal Summation Suppose Ex1 has an action potential that causes an EPSP in the postsynaptic neuron. After this EPSP has died off, if another action potential occurs in Ex1, an EPSP of the same magnitude takes place before dying off (Figure 4-20a). Next suppose that Ex1 has two action potentials in close succession to each other (Figure 4-20b). The first action potential in Ex1 produces an EPSP in the postsynaptic neuron. Although the postsynaptic membrane is still partially depolarized from this first EPSP (before it has returned to resting), the second presynaptic action potential produces a second EPSP in the postsynaptic neuron. Because graded potentials do not have a refractory period, the second EPSP adds on to the first EPSP, bringing the membrane to threshold, so that an action potential can occur in the postsynaptic neuron.

The summing of several EPSPs occurring very close together in time because of successive firing of a single presynaptic neuron is known as **temporal summation** (the Latin word *tempus* means "time"). In reality, up to 50 EPSPs might be needed to bring the postsynaptic membrane to threshold. Each action potential in a presynaptic neuron triggers the emptying of a certain number of synaptic vesicles. The amount of neurotransmitter released and the resultant magnitude of the change in postsynaptic potential are thus directly related to the frequency of presynaptic action poten-

tials. One way, then, in which the postsynaptic membrane can be brought to threshold is through rapid, repetitive excitation from a single, persistent input.

Spatial Summation Let's now see what will happen in the postsynaptic neuron if both excitatory inputs are stimulated simultaneously (Figure 4-20c). An action potential in either Ex1 or Ex2 will produce an EPSP in the postsynaptic neuron; however, neither of these alone will bring the membrane to threshold to elicit a postsynaptic action potential. But simultaneous action potentials in Ex1 and Ex2 produce EPSPs that add to each other, bringing the postsynaptic membrane to threshold, so that an action potential does occur. Such summation of EPSPs originating simultaneously from several different presynaptic inputs (that is, from different points in "space") is known as **spatial summation**. A second way, therefore, to elicit an action potential in a postsynaptic cell is through concurrent activation of several excitatory inputs. Again, in reality, up to 50 EPSPs arriving simultaneously on the postsynaptic membrane are required to bring it to threshold.

Similarly, IPSPs can undergo temporal and spatial summation. As IPSPs add together, however, they progressively move the potential farther from threshold.

Cancelation of Concurrent EPSPs and IPSPs If an excitatory and an inhibitory input are simultaneously activated, the concurrent EPSP and IPSP more or less cancel each other out. The extent of cancelation depends on their respective magnitudes. In most cases, the postsynaptic membrane potential remains close to resting (Figure 4-20d).

Importance of Postsynaptic Neuronal Integration The magnitude of the GPSP depends on the sum of activity in all the presynaptic inputs and, in turn, determines whether or not the neuron will undergo an action potential to pass information on to the cells at which the neuron terminates. The following oversimplified real-life example demonstrates the benefits of this neuronal integration. The explanation is not completely accurate, technically, but the principles of summation are accurate.

Assume for simplicity's sake that urination is controlled by a postsynaptic neuron supplying the urinary bladder in a mammal. Glutamate released at this neuron opens Na^+ channels, while GABA opens Cl^- channels. When this neuron fires due to the former, its own neurotransmitter release induces the bladder to contract. (Actually, the control of urination also involves urethral sphincter muscles, but we will simplify the process here.) As the bladder starts to fill with urine and becomes stretched, a reflex is initiated that ultimately produces EPSPs in the postsynaptic neuron responsible for causing bladder contraction. Partial filling of the bladder does not cause sufficient excitation to bring the neuron to threshold, so urination does not take place (Figure 4-20a). As the bladder progressively fills, the frequency of action potentials progressively increases in the presynaptic neuron Ex1, leading to more rapid formation of EPSPs in the postsynaptic neuron (Ex1 in Figure 4-20b). Thus, the frequency of EPSP formation arising from Ex1 activity signals the postsynaptic neuron of the extent of bladder filling. When the bladder becomes sufficiently stretched that the Ex1-generated EPSPs are temporally summed to threshold, the postsynaptic neuron undergoes an action potential that stimulates bladder contraction.

What if the time is inopportune for urination? Consider a pet dog with its eye set on a favorite fire hydrant or a bobcat moving to the boundary of its territory to mark it, each needing to stop urination before the animal reaches its goal. Presynaptic inputs originating in higher levels of the brain responsible for voluntary control can produce IPSPs (using GABA) at the bladder postsynaptic neuron (In1 in Figure 4-20d). These "voluntary" IPSPs in effect cancel out the "reflex" EPSPs triggered by stretching of the bladder. Thus, the postsynaptic neuron remains at resting potential and does not have an action potential, so the bladder is prevented from contracting and emptying even though it is full.

What if the bladder is only partially filled, so that the presynaptic input originating from this source (Ex1) is insufficient to bring the postsynaptic neuron to threshold to cause bladder contraction, yet what if the dog has found another inviting fire hydrant? The dog can voluntarily activate an excitatory presynaptic neuron (Ex2 in Figure 4-20c), which spatially summates with the reflex-activated presynaptic neuron (Ex1) to bring the postsynaptic neuron to threshold. This achieves the action potential necessary to stimulate bladder contraction, even though the bladder is not full, so the animal can leave its mark.

This example illustrates the importance of postsynaptic neuronal integration. Each postsynaptic neuron in a sense "computes" all the input it receives and makes a "decision" about whether to pass the information on (that is, whether threshold is reached and an action potential is transmitted down the axon). In this way, neurons serve as complex computational devices, or integrators. The dendrites function as the primary processors of incoming information. They receive and tally the signals coming in from all the presynaptic neurons. Each neuron's output in the form of frequency of action potentials to other cells (muscle cells, gland cells, or other neurons) reflects the balance of activity in the inputs it receives via EPSPs or IPSPs from the thousands of other neurons that terminate on it. Each postsynaptic neuron filters out information it receives that is not significant enough to bring it to threshold and does not pass it on. If every action potential in every presynaptic neuron that impinges on a particular postsynaptic neuron were to cause an action potential in the postsynaptic neuron, the neuronal pathways would be overwhelmed with trivia. Only if an EPSP is reinforced by other supporting signals through summation is the information passed on. Furthermore, interaction of EPSPs and IPSPs provide a way for one set of signals to offset another allowing a fine degree of discrimination and control in determining what information will be passed on. Thus, unlike an electrical synapse, a chemical synapse is more than a simple on–off switch because many factors can influence the generation of a new action potential in the postsynaptic cell. Whether or not the postsynaptic neuron has an action potential or not depends on the relative balance of information coming in via presynaptic neurons at all of its excitatory and inhibitory synapses.

Let's now see why action potentials are initiated at the axon hillock.

Action potentials are initiated at the axon hillock because it has the lowest threshold

Threshold potential is not uniform throughout the postsynaptic neuron. The lowest threshold is present at the axon hillock because this region has a much greater density of voltage-gated Na^+ channels than anywhere else in the neuron. For this reason, the axon hillock is considerably more responsive than the dendrites or remainder of the cell body to changes in potential. Because of local current flow, changes in membrane potential (EPSPs or IPSPs) occurring anywhere on the cell body or dendrites spread throughout the cell body, dendrites, and axon hillock. When summation of EPSPs takes place, the lower threshold of the axon hillock is reached first, whereas the cell body and dendrites at the same potential are still considerably below their own much higher thresholds. Therefore, an action potential originates in the axon hillock and is propagated from there to the end of the axon.

Neuropeptides act primarily as neuromodulators

Researchers recently discovered that in addition to the classical neurotransmitters just described, some neurons also release *neuropeptides* (Table 4-4). Neuropeptides differ from classical neurotransmitters in several important ways. Classical neurotransmitters at fast synapses are small mol-

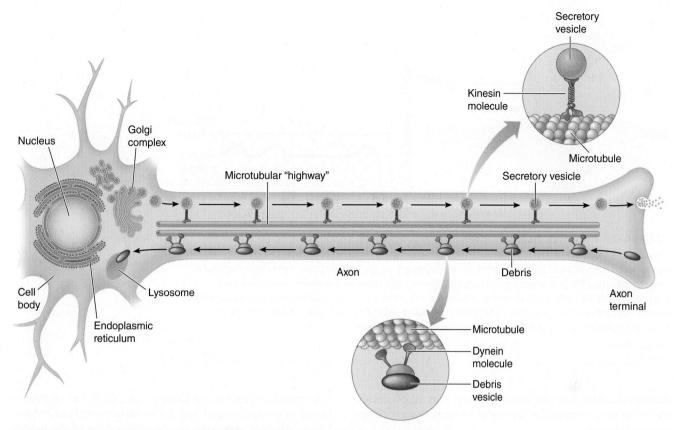

FIGURE 4-21 Two-way vesicular axonal transport facilitated by the microtubular "highway" in a neuron. Secretory vesicles are transported from the site of production in the cell body along a microtubule "highway" to the terminal end for secretion. Vesicles containing debris are transported in the opposite direction for degradation in the cell body. The top enlargement depicts kinesin, a molecular motor, carrying a secretory vesicle down the microtubule by using its "feet" to "step" on one tubulin molecule after another. The bottom enlargement depicts another molecular motor, dynein, transporting the debris up the microtubule.

© Cengage Learning, 2013

ecules that typically trigger the opening of specific ion channels to bring about a change in potential in the postsynaptic neuron (an EPSP or IPSP) within a few milliseconds or less. Most classical neurotransmitters are primarily amino acids or closely related compounds synthesized and packaged locally in small synaptic vesicles in the cytosol of the axon terminal. Neurotransmitters at slow synapses activate slower second-messenger systems, but they are made and released in the same way and have similar structures as fast neurotransmitters.

Neuropeptides are larger molecules made up of anywhere from 2 to about 40 amino acids. They are synthesized in the neuronal cell body in the endoplasmic reticulum and Golgi complex (see p. 41) and are subsequently moved by axonal transport along the microtubular highways to the axon terminal (Figure 4-21). Neuropeptides are not stored within small synaptic vesicles but instead are packaged in large **dense-core vesicles,** which are also present in the axon terminal. The dense-core vesicles undergo Ca²⁺-induced exocytosis and release neuropeptides at the same time that the neurotransmitter is released from the synaptic vesicles. An axon terminal typically releases only a single classical

TABLE 4-4 Some Known or Suspected Neuropeptides

β-endorphin	Motilin
Adrenocorticotropic hormone (ACTH)	Insulin
α-melanocyte-stimulating hormone (MSH)	Glucagon
Thyrotropin-releasing hormone (TRH)	Angiotensin II
Gonadotropin-releasing hormone (GnRH)	Bradykinin
Somatostatin	Vasopressin
Vasoactive intestinal polypeptide (VIP)	Oxytocin
Cholecystokinin (CCK)	Carnosine
Leucine enkephalin	Bombesin
Substance P	Gastrin
Methionine enkephalin	Neurotensin

© Cengage Learning, 2013

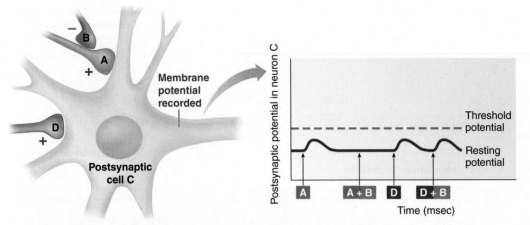

FIGURE 4-22 Presynaptic inhibition. A, an excitatory terminal ending on postsynaptic cell C, is itself innervated by inhibitory terminal B. Stimulation of terminal A alone produces an EPSP in cell C, but simultaneous stimulation of terminal B prevents the release of excitatory neurotransmitter from terminal A. Consequently, no EPSP is produced in cell C despite the fact that terminal A has been stimulated. Such presynaptic inhibition selectively depresses activity from terminal A without suppressing any other excitatory input to cell C. Stimulation of excitatory terminal D produces an EPSP in cell C even though inhibitory terminal B is simultaneously stimulated because terminal B only inhibits terminal A.

© Cengage Learning, 2013

neurotransmitter, but the same terminal may also contain one or more neuropeptides that are cosecreted along with the neurotransmitter.

Even though neuropeptides are currently the subject of intense investigation, knowledge about their functions and control is still sketchy. For example, proteomic and genomic analysis of the honeybee has revealed 36 genes that code for 100 neuropeptides in the bees' brains, but the functions of these proteins are largely unknown. Some neuropeptides may function as true neurotransmitters, but most are believed to function as neuromodulators. **Neuromodulators** are chemical messengers that do not cause the formation of EPSPs and IPSPs but rather bring about long-term changes that subtly *modulate*—depress or enhance—the action of the synapse. (Neurotransmitters like serotonin at slow synapses also activate slower second-messenger systems and are called neuromodulators by some investigators.) Neuromodulators can bind to neuronal receptors at nonsynaptic sites and they often activate second-messenger systems. These in turn may, for example, alter the sensitivity of the postsynaptic neuron to a particular neurotransmitter by causing long-term changes in the number of subsynaptic receptor sites for the neurotransmitter. Neuromodulators may also act on presynaptic sites. For example, a neuromodulator may influence the level of an enzyme critical in the synthesis of a neurotransmitter by a presynaptic neuron. Thus, neuromodulators delicately fine-tune the synaptic response. The effect may last for days or even months or years. Like some slow neurotransmitters, neuromodulators are involved with more long-lasting events such as learning and motivation.

Interestingly, in "same-key-different-locks" fashion (p. 93), vertebrate neuromodulators include many substances that also have distinctly different roles in other body regions as hormones released into the blood from neural and nonneural tissues. For example, *cholecystokinin (CCK)* is a well-known hormone released from the small intestine that causes the gallbladder to contract and release

bile into the intestine (see Chapter 14). CCK has also been found in axon terminal vesicles in various areas of the rat brain, where it mediates satiety (the feeling of no longer being hungry; Chapter 15) and perhaps other unknown activities. As is the case for many neuropeptides, CCK was named for its first-discovered role as a hormone (*chole*, "bile"; *cysto*, "bladder"; *kinin*, "contraction").

Presynaptic inhibition or facilitation can selectively alter the effectiveness of a given presynaptic input

Besides neuromodulation, presynaptic inhibition or facilitation is another means of depressing or enhancing synaptic effectiveness. Sometimes a third neuron influences activity between a presynaptic ending and a postsynaptic neuron. A presynaptic axon terminal (labeled A in Figure 4-22) may itself be innervated by another axon terminal (labeled B). The neurotransmitter released from modulatory terminal B binds with receptor sites on terminal A. This binding alters the amount of transmitter released from terminal A in response to action potentials. If the amount of transmitter released from A is reduced, the phenomenon is known as **presynaptic inhibition.** If the release of transmitter is enhanced, the effect is **presynaptic facilitation.**

Let's look more closely at how this process works. You know that Ca^{2+} entry into an axon terminal induces the release of neurotransmitter by exocytosis of synaptic vesicles. The amount of neurotransmitter released from terminal A depends on how much cytosolic Ca^{2+} enters this terminal in response to an action potential. Ca^{2+} entry into terminal A, in turn, can be influenced by activity in modulatory terminal B. Let's use presynaptic inhibition to illustrate (Figure 4-22). The amount of transmitter released from presynaptic terminal A, an excitatory input in our example, influences the potential in the postsynaptic neuron on which it terminates (labeled C

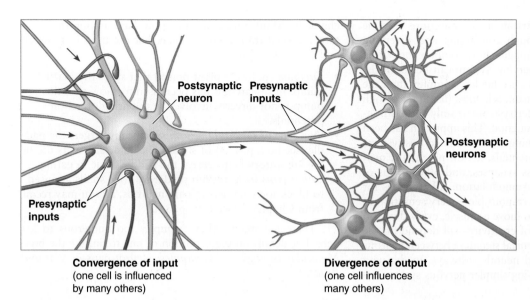

FIGURE 4-23 Convergence and divergence. Arrows indicate the direction in which information is being conveyed.

© Cengage Learning, 2013

Convergence of input
(one cell is influenced
by many others)

Divergence of output
(one cell influences
many others)

in the figure). Firing of A alone will bring about an EPSP in postsynaptic neuron C. Now consider that B is stimulated simultaneously with A, reducing Ca^{2+} entry into terminal A. Less Ca^{2+} entry means less neurotransmitter release from A. Note that modulatory neuron B can suppress neurotransmitter release from A only when A is firing. If this presynaptic inhibition by B prevents A from releasing its neurotransmitter, the formation of EPSPs on postsynaptic membrane C from input A is specifically prevented. As a result, no change in the potential of the postsynaptic neuron occurs despite action potentials in A.

Could the same thing be accomplished by a simultaneous production of an IPSP through activation of an inhibitory input to negate an EPSP produced by activation of A? Not quite. Activation of an inhibitory input to cell C in Figure 4-22 would produce an IPSP in cell C, but this IPSP could cancel out not only an EPSP from excitatory presynaptic input A, but also any EPSPs produced by other excitatory terminals, such as terminal D in the figure. The entire postsynaptic membrane is hyperpolarized by IPSPs, negating (canceling) excitatory information fed into any part of the cell from any presynaptic input. By contrast, presynaptic inhibition (or presynaptic facilitation) works in a much more specific way than does the action of inhibitory inputs to the postsynaptic cell.

Presynaptic inhibition provides a means by which certain inputs to the postsynaptic neuron can be *selectively* inhibited without affecting contributions of any other inputs. For example, firing of B specifically prevents formation of an EPSP in the postsynaptic neuron from excitatory presynaptic neuron A but does not influence other excitatory presynaptic inputs. Excitatory input D can still produce an EPSP in the postsynaptic neuron even when B is firing. This type of neuronal integration is another means by which electrical signaling between neurons can be carefully fine-tuned.

Retrograde messengers travel "backward" and also modulate synaptic function

Although the primary message being relayed through a synapse travels in one direction in most animals, researchers are finding an increasing number of examples of **retrograde** signals. In such cases, a stimulated postsynaptic cell may produce and release a signal molecule that diffuses backward, that is, to the presynaptic terminal. Once there, it does not trigger a message to travel up the axon; rather, it modifies the sensitivity of the terminal in a prolonged or longterm way (a type of neuromodulation). One example is in the neuromuscular synapse of the laboratory fruit fly (*Drosophila*). The primary neurotransmitter from the neuron to the muscle is glutamate (not ACh as in vertebrates). The (postsynaptic) muscle cell also produces a peptide that travels to the (presynaptic) neuron. There it enhances release of glutamate. The system appears to work in negative-feedback fashion for "synaptic homeostasis." The muscle produces *more* peptide when it is unstimulated and *less* when stimulated. This tends to keep the amount of glutamate released at a consistent level. In the next chapter, we describe how *nitric oxide* in a vertebrate can be used in a positive-feedback retrograde system (p. 199).

Neurons are linked to each other through convergence and divergence to form complex nerve pathways

Two important relationships exist between neurons: convergence and divergence. A given neuron may have many other neurons synapsing on it. Such a relationship is known as **convergence** (Figure 4-23). Through this converging input, a single cell is influenced by thousands of other cells. This single cell, in turn, influences the level of activity in many other cells by divergence of output. The term **divergence** refers to the branching of axon terminals so that a single cell synapses with and influences many other cells.

Note that a particular neuron is postsynaptic to the neurons converging on it but presynaptic to the other cells on which it terminates. Thus, the terms *presynaptic* and *postsynaptic* refer only to a single synapse. Most neurons are presynaptic to one group of neurons and postsynaptic to another group.

There are an estimated 100 billion neurons and 10^{14} (100 quadrillion) synapses in the human brain alone. When you consider the vast and intricate interconnections possible between these neurons through converging and diverging pathways, you can begin to imagine how complex the wiring mechanism of an animal's nervous system can be. Even the most sophisticated computers are far less complex than the vertebrate brain. For this reason, scientists (using the Krogh principle, p. 3) look at simpler systems (mainly certain non-vertebrates) to study neural function. This approach has been very successful because the "language" of *all* nervous systems is in the form of graded potentials, action potentials, neurotransmitter signaling across synapses, and other forms of chemical chatter such as neuromodulation. All activities for which the nervous system is responsible—every sensation in an animal, every command to move a muscle, every thought, every memory, every spark of creativity—all depend on the patterns of electrical and chemical signaling between neurons along these complexly wired neural pathways. In the next chapter, you will see how using simpler nervous systems reveals universal principles.

check your understanding 4.7

Why are action potentials initiated at the axon hillock?

How does presynaptic inhibition or facilitation depress or enhance synaptic activity?

Through what process does a single neuron influence the activity of others? How is a single neuron influenced by other neurons?

4.8 Neural Signaling and External Agents

Neural signaling mechanisms are the targets of many external chemical and physical agents. Their effects play an important role in evolutionary adaptation, research, and medicine. Some are accidentally acquired as poisons or by disease processes, and some are toxins evolved in prey organisms for defense and in predators for paralyzing or killing prey. Many plants and fungi, for example, have evolved defensive neurotoxins presumably because there is no chance these can poison their own tissues. In addition, some of those agents as well as purely human-made ones can be used to help researchers probe the mechanisms of neural signaling, and may also be used as pharmacologic agents (drugs). Virtually every step of signal transmission is vulnerable. We end this chapter with illustrative examples.

Synthetic and natural toxins can alter resting potentials and action potentials

Let's start with the electrical signaling processes. External chemicals have long been used for research and medicine. For example, researchers have studied the role of K^+ channels in resting and action potentials using the synthetic compound *tetraethylammonium (TEA)*. TEA selectively blocks most types of gated K^+ channels when placed at either the ICF or ECF. However, it does not block the K^+ leak channels. Many predators and prey have toxins that also target axonal transmission; some that are used in research and medicine include:

- *Ouabain* (p. 87), a plant poison that stops the Na^+/K^+ pump. In research, it has been used to show that this pump is not directly involved in the firing of action potentials.
- One type of *tarantula venom* specifically targets the paddle (see box, p. 118) of shaker K-channels, and its use in the laboratory helps reveal the paddle's role in gating.
- *Conotoxins* from certain predatory cone snails are used to block ion channels in pain neurons, as you may recall from Chapter 2 (see box, p. 34).

There are many other examples too numerous to list here. For details on the evolution of one toxin, see the box, *Molecular Biology and Genomics: Neurotoxins in War and Peace.*

Drugs, natural toxins, and pollutants can modify synaptic transmission between neurons

Most drugs that influence the nervous system perform their function by altering synaptic mechanisms. Synaptic drugs may block an undesirable effect (antagonist) or enhance a desirable effect (agonist). Possible actions include (1) altering the synthesis, axonal transport, storage, or release of a neurotransmitter; (2) modifying neurotransmitter interaction with the postsynaptic receptor; (3) influencing neurotransmitter reuptake or destruction; and (4) replacing a deficient neurotransmitter with a substitute transmitter.

Consider the reuptake step. The widely prescribed drug *Prozac* (fluoxetine), used for treatment of some abnormal moods and behaviors, alters serotonin reuptake (see the box, *Challenges and Controversies: Synaptic Solutions to Pet Problems*). Similarly, **cocaine** (an *alkaloid* from the coca plant where it is thought to poison insect nerves) blocks reuptake of the neurotransmitter **dopamine** at presynaptic terminals by competitively binding with the dopamine reuptake transporter, a protein molecule that picks up released dopamine from the synaptic cleft and shuttles it back to the axon terminal. With cocaine occupying the dopamine transporter, dopamine remains in the synaptic cleft longer than usual and continues to interact with its postsynaptic receptor sites. The result is prolonged activation of neural pathways that use this chemical as a neurotransmitter. Among these pathways are those that play a role in emotional responses, especially feelings of reward (at least in humans). In essence, when cocaine is present, the neural switches in the reward pathway are locked in the "on" position.

Cocaine is addictive to humans and laboratory rodents (who will choose cocaine over food and water even to the point of death) because the involved neurons become *desensitized* to the drug. Permanent molecular changes of the involved neurons result from cocaine usage such that they cannot transmit normally across synapses without increasingly higher doses of the drug. After the neurons have been incessantly stimulated for an extended time by cocaine, a transcription factor (p. 31) called δ-*FosB* is activated by unknown means. This factor, in turn, activates the gene for an enzyme called *cyclin-dependent kinase 5* or *Cdk5*. Cdk5 greatly reduces sensitivity for cocaine (again, by unknown

Molecular Biology and Genomics
Neurotoxins in War and Peace

Neurotoxins in nature are widespread "warfare" mechanisms for both attack and defense; their targets are animal nerves, but they are made not only in some animals but also in some bacteria, protists, fungi, and plants. A well-studied example is **tetrodotoxin (TTX)**, best known in the tropical marine Japanese pufferfish *(Fugu rubripes)* and related species. Because of the toxicity of this compound, chefs in Japan require years of training before they are permitted to serve pufferfish to their patrons. TTX is specific for voltage-gated Na^+ channels and binds to the extracellular region of the channel. However, if injected inside the neuron, TTX has no effect on action potential propagation, demonstrating that TTX is specific in blocking the entry of Na^+.

TTX is widespread in the animal kingdom, found not only in pufferfish but also in other not closely related fish, the blue-ringed octopus, some seastars, crabs, snails, tunicates, flatworms, ribbon worms, some terrestrial frogs and newts, and even algae. Have these diverse organisms evolved the genes for this toxin independently, or do they all have something in common? Recent studies make it clear that most, or perhaps all of these organisms, lack the ability to synthesize TTX. Pufferfish, one of the animals whose genome has been sequenced, have no genes related to TTX synthesis, and they cannot make it if grown in isolated aquaria. It is now clear that *symbiotic bacteria* manufacture the toxin for their hosts. The blue-ringed octopus of Australia, for example, has been found to harbor these bacteria. Thus, it now seems that a peaceful, beneficial coexistence has independently evolved between these bacteria and a number of animal hosts. Presumably, the bacteria are given nourishment and protection in the symbiosis. In turn, the TTX made by the bacteria is used by some of these organisms as defense (such as algae and pufferfish) and by others to kill prey (such as carnivorous snails).

TTX is one of the most potent blockers of animal sodium channels known. How do the host animals themselves survive their bacterial partners? Genetic analysis reveals that the pufferfish has a single point mutation in its sodium-channel gene, rendering the channel protein resistant to TTX binding. Presumably similar mutations have occurred in other TTX hosts. But resistant channels are found elsewhere: Garter

Photo: © Masa Ushioda/Imagequestmarine.com

snakes that eat TTX-bearing newts have recently been found to have these as well.

The TTX story illustrates a general principle of biology called **coevolution**, in which adaptations in one species select for new adaptations in other species—those that are prey, predator, competitor, or symbiote. In turn, these changes in the second species may lead to yet newer adaptations in the first species . . . and so on.

---

mechanisms). Whatever the mechanism, the synapses require more and more cocaine to achieve a stimulus and thus a reward feeling. When cocaine is no longer available, the sense of reward cannot be achieved, because the normal level of dopamine release is no longer sufficient to stimulate the postsynaptic neuron. Note that this is a form of *acclimatization* or *downregulation* (p. 17).

Other drugs and toxins may cause nervous system disorders by acting at either presynaptic or postsynaptic sites agonistically or antagonistically. For example, *caffeine*, another compound in certain plants that disrupts neural function in herbivorous insects, blocks the receptors for the inhibitory neuromodulator *adenosine*. Adenosine is used in the sleep circuits of mammalian brains, explaining caffeine's stimulatory effects. *Ethyl alcohol*, in contrast, enhances the action of inhibitory GABA receptors (p. 133), one of the reasons it lowers reaction speeds.

Two other neural poisons, *strychnine* and *tetanus toxin*, act at different synaptic sites to block inhibitory impulses, yet have similar outcomes. One (strychnine) blocks specific postsynaptic inhibitory receptors, whereas the other (tetanus toxin) prevents the presynaptic release of a specific inhibitory neurotransmitter:

- *Strychnine*, a poison used to kill rodents that infest buildings, competes with an inhibitory neurotransmitter, glycine, at the postsynaptic receptor site. This antagonist combines with the receptor but does not directly alter the potential of the postsynaptic cell. Instead, strychnine blocks the receptor so that glycine cannot bind when it is released from the inhibitory presynaptic ending. Thus, postsynaptic inhibition (formation of IPSPs) is abolished in nerve pathways that use glycine as an inhibitory transmitter. Unchecked excitatory pathways lead to convulsions, muscle spasticity, and death.

- *Tetanus toxin* from *Clostridium tetani* bacteria, in contrast, prevents release of another inhibitory transmitter, gamma-aminobutyric acid (GABA), from inhibitory presynaptic inputs terminating on neurons that supply skeletal muscles. Unchecked excitatory inputs to these neurons result in uncontrolled muscle spasms. These spasms occur especially in the jaw muscles early in the disease, giving rise to the common name of **lockjaw** for this condition. Later, they progress to the muscles responsible for breathing, at which point the disease proves fatal.

Some pollutants can inhibit neural signaling. Lead (Pb^{2+}), for example, is an important environmental toxicant because it competes with Ca^{2+} for entry into the terminal button. Once ingested by an animal, Pb^{2+} at high doses inhibits neural transmission, which can kill the animal.

Ever since the identification of the synapse as the functional junction between two neurons, the synapse has been regarded as the most vulnerable physiological site for manipulation and artificial regulation. Behavioral pharmacotherapy, the use of drugs to treat behavioral problems, is an example of a science that has evolved from this original precept. Interestingly, scientists observed that during the course of treatment with antihistamines for congestion, some human patients noted a significant improvement in emotional state. In fact, many psychoactive drugs are chemically similar to antihistamines. Psychoactive drugs exert their behavioral effects by acting on neurotransmitters and neuromodulators in the brain (see p. 189). For example, treatment with drugs that selectively inhibit reuptake of *serotonin* into the neuron (such as **Prozac**, an **SSRI**—selective serotonin reuptake inhibitor) results in a higher level of activity in any part of the nervous system that uses serotonin as a chemical signal.

Serotonin is the common name for **5-hydroxytryptamine (5-HT)**. It is synthesized in the brain from the amino acid tryptophan. Because there are at least nine different serotonin receptor subtypes, each localized separately in the mammalian brain and responsible for different behavioral responses, regulation of the serotonin receptor system is particularly complex. In mammals, serotonin plays an important role in the control of sleep, pain, aggression, sexual behavior, and food intake. This diverse range of responses is not surprising, considering that serotonin neurons project fibers to virtually all parts of the central nervous system. Documented effects of SSRIs in dogs include a reduction in the intensity of obsessive–compulsive disorders such as excessive licking behavior, tail mutilation, separation anxieties, dominance-related aggression, and so forth. In cats, SSRIs are used for the treatment of psychogenic alopecia (unnatural loss of hair), offensive aggression, and urine spraying. Prozac has been used to treat cribbing (force-swallowing gulps of air) in horses and bulimia (episodes of binge eating that continue until terminated by abdominal pain or vomiting) in pigs.

Tricyclic antidepressants (so called because of their three-ringed chemical structure) also block the reuptake of serotonin and catecholamines from synapses, which promotes a state of calmness. Tricyclics can be used to treat urine spraying in cats, urine marking in dogs, and separation anxiety in both cats and dogs. These drugs are also useful in the treatment of stereotypies. **Stereotypies** are repetitive behaviors that perform no function (such as pacing in zoo animals). They arise from chronic confinement, chronic conflict, or sensory deprivation, and researchers suggest they provide some relief. Chemical intervention is normally warranted to eliminate stereotypies because, once incorporated into an animal's behavioral repertoire, the stereotypy continues to be performed even in the absence of the initiating stressor.

Neuroleptics represent a family of drugs whose mode of action is the antagonism of dopamine (p. 189) and are effective in treating certain phobias, fears, and anxieties. Phobias are extreme and apparently irrational fears whose origins can normally be identified. For example, animals may be fearful of people or of other animals, sensory stimuli such as sounds (thunderstorms), odors, or touch, or exhibit phobias under specific circumstances such as visiting the veterinarian's office. These compounds are also effective in preventing anxiety-related inappropriate urine marking and spraying in cats. Neuroleptics are used in the treatment of stereotypic disorders such as flank biting in horses, fur sucking in cats, and feather picking in birds. Although many of these and other drugs show promise in the treatment of fear-based conditions in pets, veterinarians recommend that chemical therapy be combined with behavioral modification in order to achieve long-lasting results.

Other drugs and diseases that influence synaptic transmission are too numerous to mention, but as these examples illustrate, any site along the synaptic pathway is vulnerable to interference, either pharmacologic (drug induced) or pathologic (disease induced).

The neuromuscular junction is vulnerable to several chemical agents and diseases

Several chemical agents and diseases affect the neuromuscular synapse by acting at different sites in the transmission process. Two well-known toxins—*black widow spider venom* and *curare*—alter the function of ACh, but in different ways.

The venom of black widow spiders exerts its deadly effect on small prey animals (not humans) by causing an explosive release of ACh from the storage vesicles, not only at neuromuscular junctions but at all cholinergic sites. Thus, the toxin is an agonist (p. 101). All cholinergic sites undergo prolonged depolarization, the most detrimental consequence of which is respiratory failure in a bitten mammal. Breathing is accomplished by alternate contraction and relaxation of skeletal muscles, particularly the diaphragm. Respiratory paralysis occurs as a result of prolonged depolarization of the mammalian diaphragm. During this so-called **depolarization block**, the voltage-gated Na^+ channels are trapped in their inactivated state (that is, their activation gate remains open and their inactivation gate remains closed; they are in their "closed and not capable" of opening conformation; see p. 120). This depolarization block prohibits the initiation of new action potentials and resultant contraction of the diaphragm. As a consequence, the spider's victim cannot breathe.

Other chemicals are antagonists (p. 100) that interfere with neuromuscular junction activity by blocking the effect of released ACh. The best-known example is **curare** (curare is derived from *wurari* meaning "poison," a word from the Carib language of the Macusi Indians of Guyana), which is obtained from tropical plants of the family Loganiaceae in South America. Curare reversibly binds to the ACh receptor sites on the motor end plate. Unlike ACh, however, curare does not alter membrane permeability, nor is it inactivated by AChE. When ACh receptor sites are occupied by curare, ACh cannot combine with these sites to open the channels that would permit the ionic movement responsible for an EPP. Consequently, because muscle action potentials cannot occur in response to nerve impulses to these muscles, paralysis ensues. The toxin presumably evolved to protect these frogs from predators: When sufficient curare is present to effectively block a significant number of ACh receptor sites, the predator dies from respiratory paralysis caused by an inability to contract the diaphragm. Some indigenous people in South America use curare as a deadly arrowhead poison. Curare and related drugs have also been used medically (at very low doses!) during surgery to help achieve more complete skeletal muscle relaxation with less anesthetic.

An example of a disease that affects the junction is *myasthenia gravis*, characterized by extreme muscle weakness. It is caused by antibodies of an animal's own immune system that attack ACh receptors of the junction. This *autoimmune* (p. 486) disease has been documented in cats, dogs, and humans. The drug **neostigmine**, which inhibits AChE, is used to treat this because its action prolongs the activity of whatever amount of ACh is released at the synapse.

Temperature and pressure also influence action potentials and synapses

Finally, evolutionary adaptation of animals has been greatly affected by physical factors that alter neural functions. In particular, temperature as well as pressure exerts pronounced effects on membrane processes. Recall from Chapter 3 (p. 71) that cold temperatures can make membranes too rigid, and high temperatures can make them too fluid. Temperature also directly affects the rates of all reactions and protein structures (p. 13), including channel opening and closing. The effects of temperature are discussed in detail in Chapter 15.

Effects of an increase in pressure on cells are more complex, and most are not discussed here. Briefly, high pressure can inhibit the proper folding of proteins, reduce movement of proteins during shape changes, and reduce membrane fluidity. Such effects presumably explain the effects of pressure on animals that are unadapted to pressure, which include the following:

- A reduction in the activity of the Na^+/K^+ pump
- A decrease in the rate at which an action potential is propagated
- A decrease in the rate of axonal repolarization and an increase in axon potential duration
- Inhibition of excitatory synaptic transmission, because of reduced neurotransmitter release from the presynaptic terminal
- Reduced binding of ACh to its receptor on the muscle membrane

All these effects are potentially detrimental to animals living permanently in the deep sea. However, some of these effects are minimized or not apparent in such animals (e.g., deep-sea fish and shrimp). Adaptations include membranes with higher fluidity, achieved by a decrease in the ratio of saturated to unsaturated fatty acids (as you saw in Chapter 3 for temperature, p. 72). We discuss another form of adaptation in Chapter 13.

Such adaptations have not occurred in shallow-living animals that occasionally experience high pressure—in particular, deep-diving marine mammals such as Weddell seals and sperm whales. However, these effects may be useful to these divers, contributing to an energy-saving reduction in heart rate when the animals descend to depth. For example, at high pressures (150 atm), the conduction velocity of an impulse in a seal's heart is reduced by 40%. We explore these and other aspects of deep diving in Chapter 11.

check your understanding 4.8

Give two examples of toxins that modify action potentials and two that modify synaptic transmission, and describe how they work.

How do temperature and pressure affect neural signaling?

making connections

How Neuronal Physiology Contributes to the Body as a Whole

Neurons are specialized to receive, process, encode, and rapidly transmit information from one part of the body to another. The information is transmitted over intricate nerve pathways by propagation of action potentials along the neuron's length as well as by chemical transmission of the signal from neuron to neuron and from neuron to muscles and glands through neurotransmitter–receptor interactions at synapses and from neuron to muscles and glands through other neurotransmitter–receptor interactions at these junctions.

Collectively, neurons make up the nervous system, one of the two major control systems of the body. To maintain homeostasis, cells must communicate so that they work together to accomplish life-sustaining activities. Many of the activities controlled by the nervous system are thus geared toward maintaining homeostasis. Some neuronal electrical signals convey information about changes to which the body must respond in order to maintain homeostasis—for example, these signals convey information about a reduction in blood pressure. Other neuronal electrical signals convey messages to muscles and glands to stimulate appropriate responses to counteract these changes—for example, the nervous system through its electrical signals initiates adjustments in heart and blood vessel activity to restore blood pressure to normal when it starts to fall, or to initiate behaviors that aid survival such as feeding when low internal energy supplies are sensed.

The specialization of muscle cells, contraction, also depends on these cells' ability to undergo action potentials. Action potentials trigger muscle contractions, many of which are important in maintaining homeostasis. For example, beating of the heart, mixing of ingested food with digestive enzymes, and shivering to generate heat when a mammal is cold are all accomplished by muscle contractions.

Chapter Summary

- Neurons and muscle cells are excitable tissues; that is, they can produce electrical signals when stimulated.

- Electrical signals are produced by changes in ion movement through ion channels across the plasma membrane. There are two types of channels: leak channels, or nongated channels, which are open all the time; and gated channels, which can be opened or closed in response to specific triggering events.

- There are at least three kinds of gated channels: (1) voltage-gated ion channels, which open or close in response to changes in membrane potential; (2) chemically gated channels, which change in response to the binding of a specific chemical messenger with a membrane receptor that is in close association with the channel; and (3) mechanically gated channels, which respond to stretching or other mechanical deformation.

- There are two basic forms of electrical signals: (1) the more primitive graded potentials, which serve as short-distance decaying signals, and (2) action potentials, which signal over longer distances without decaying.

- The stronger a triggering event, the larger the resultant graded potential.

- The passive movement of the electrical depolarization in a graded potential can be described by Ohm's law, where ΔV_m is the actual voltage change across the membrane, ΔI the amount of current (amperes), and R the resistance encountered by the current passing along the membrane (ohms).

- Action potentials are brief, rapid, large (100 mV) changes in membrane potential during which the potential actually reverses so that the inside of the excitable cell transiently becomes more positive than the outside. Action potentials are propagated from the axon hillock to the axon terminals. Action potentials occur in all-or-none fashion.

- The action potential is an electrical waveform composed of depolarization, repolarization, and hyperpolarization. The rising phase of the action potential (from threshold to +30 mV) is due to Na^+ influx (Na^+ entering the cell) induced by an explosive increase in P_{Na^+} at threshold. The falling phase (from +30 mV to resting potential) is brought about by K^+ efflux (K^+ leaving the cell) caused by the marked increase in P_{K^+} occurring simultaneously with the inactivation of the Na^+ channels at the peak of the action potential.

- The Na^+/K^+ ATPase pump gradually restores the concentration gradients disrupted by action potentials.

- A single nerve cell, or neuron, consists of three basic parts: the cell body, the dendrites, and the axon, although there are variations in structure, depending on the location and function of the neuron.

- An excitable membrane either responds to a triggering event with a maximal action potential that spreads nondecrementally throughout the membrane, or it does not respond with an action potential at all. Once initiated, action potentials are conducted over the surface of an axon.

- The refractory period ensures unidirectional propagation of the action potential and limits the frequency of action potentials. The strength of a stimulus is coded by the frequency of action potentials.

- Myelination increases the speed of conduction of action potentials and conserves energy in the process. In a myelinated fiber, the impulse "jumps" from node to node, skipping over the myelinated sections of the axon; this process is called saltatory conduction. Fiber diameter also influences the velocity of action potential propagation.

- Electrical synapses transfer action potential waveforms through gap junctions, whereas chemical synapses convert action potentials into organic chemical messengers that are exocytosed into the synaptic gap.

- Most, but not all, neurotransmitters function by changing the conformation of chemically gated channels, thereby altering membrane permeability and ionic fluxes across the postsynaptic membrane. Some neurotransmitters excite the postsynaptic neuron, whereas others inhibit the postsynaptic neuron.

- Acetylcholine, a fast excitatory neurotransmitter, links electrical signals in motor neurons with electrical signals in skeletal muscle cells. Acetylcholinesterase terminates acetylcholine activity at the neuromuscular junction.

- The grand postsynaptic potential depends on the sum of the activities of all presynaptic inputs. The postsynaptic neuron can be brought to threshold in two ways: (1) temporal summation and (2) spatial summation.

- Presynaptic inhibition or facilitation can selectively alter the effectiveness of a given presynaptic input. Retrograde messengers travel "backward" and also modulate synaptic function.

- Neurons are linked to each other through convergence and divergence to form complex nerve pathways.

- Drugs and natural toxins, such as ouabain, tetrodotoxin, caffeine, cocaine, and Prozac, as well as pollutants such as lead, can modify action potentials and synaptic transmission between neurons and the neuromuscular junction.

Review, Synthesize, and Analyze

1. Why is the electrotonic flow of current along a membrane faster than current flow generated by action potentials?

2. What is the relative importance of the Na^+/K^+ pump in action potential propagation?

3. How does a neuron ensure that action potential propagation is one-way?

4. Compare the role of myelin in vertebrates with that in nonvertebrates. Consider that the rate of nerve conduction observed in the mylinated axons of the penaeid shrimp is comparable to that observed in vertebrates (200 m/sec).

5. Why does that portion of the excitable membrane where graded potentials are initiated not undergo an action potential?

6. How does the delayed opening of K⁺ channels lead to the repolarization stage of impulse propagation?

7. Can the movement of charges along an axon truly be considered a "flow"?

8. Through what process can the postsynaptic membrane influence the activity of the presynaptic membrane? Under what circumstances is this communication important?

9. Why are there multiple neurotransmitters? Would it not be simpler to have fewer neurotransmitters associated with postsynaptic receptors specific to a particular function?

10. The synthetic compound *tetraethylammonium (TEA)* selectively blocks most types of gated K⁺ channels, but it does not block the K⁺ leak channels. Explain what happens to membrane potential after an action potential spike with TEA present, and why.

Suggested Readings

Cowan, W. M., T. C. Südhoff, & C. F. Stevens, eds. (2001). *Synapses.* Baltimore: Johns Hopkins University Press. A compendium of reviews on synaptic physiology.

Doyle, D. A., J. Morais Cabral, R. A. Pfuetzner, et al. (1998). The structure of the potassium channel: Molecular basis of K⁺ conduction and selectivity. *Science* 280:69–77. The first detailed structural analysis of K-channels.

Hartline, D. K. (2008). What is myelin? *Neuron Glia Biology* 4:153–163. A review of myelins in the animal kingdom.

Hodgkin, A. L. (1994). *Chance and Design, Reminiscences of Science in Peace and War.* New York: Cambridge University Press. A personal view of the pioneering work on neurons.

Kandel, E. R., J. H. Schwartz, & T. M. Jessell, eds. (2000). *Principles of Neural Science,* 4th ed. New York: McGraw-Hill/Appleton & Lange. An excellent textbook on neuroscience.

Katz, P. S. (1999). *Beyond Neurotransmission: Neuromodulation and Its Importance for Information Processing.* New York: Oxford University Press. A review of neuromodulation and how it differs from classic neurotransmission.

Lenz, P. H., D. K. Hartline, & A. D. Davis. (2000). The need for speed. Part 1. Fast reactions and myelinated axons in copepods. *Journal of Comparative Physiology* 186:337–345. The discovery of myelin in copepods.

Levitan, I. B., & L. K. Kaczmarek. (2001). *The Neuron: Cell and Molecular Biology,* 3rd ed. Oxford, UK: Oxford University Press. A textbook on neuroscience at the molecular level.

Matthews, G. G. (2001). *Neurobiology: Molecules, Cells and Systems.* Palo Alto, CA: Blackwell Science. An excellent textbook on neurobiology.

Neher, E. & B. Sakmann. (1992). The patch clamp technique. *Scientific American* 266:44–51. Excellent review of the development and use of this classic physiological tool.

Salinas, P. C. (2003). Backchat at the synapse. *Nature* 425:464–466. On retrograde transmission and a new discovery in *Drosophila.*

Unwin, N. (1995). Acetylcholine receptor channel imaged in the open state. *Nature* 373:37–43. A detailed structural analysis of receptor channel function.

CourseMate

Access an interactive eBook, chapter-specific interactive learning tools, including flashcards, quizzes, videos and more in your Biology CourseMate, accessed through **CengageBrain.com.**

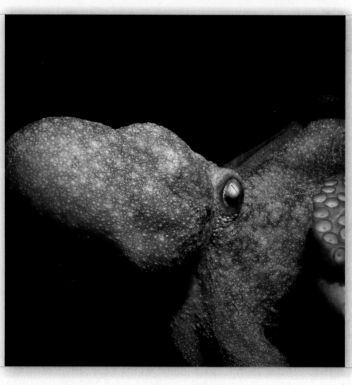

Photo: © Stephen Frink/CORBIS

An octopus has the largest brain:body ratio outside the vertebrates.

5

Nervous Systems

5.1 Evolution of Nervous Systems

With this chapter, we begin examining the organ systems of animals, starting with those involved in whole-body regulation. The nervous system is one of two such control systems of an animal body, the other being the endocrine system (although some researchers include the immune system as a third). In general, the nervous system coordinates rapid responses, whereas the endocrine (hormonal) system (Chapter 7) regulates activities that require duration rather than speed. Consider that electrical conduction is faster and more precise than

© Thomas D. Mangelsen/Images of Nature

Primates (along with dolphins) have the highest brain:body ratio in the animal kingdom.

the diffusion of chemicals, something that enables more rapid detection and a coordinated response to either perceived threats or available opportunities. With increasing complexity, a nervous system can recall past experiences and anticipate the future.

Nervous systems evolved from simple reflex arcs to centralized brains with distributed, hierarchic regulation

In all animals the ultimate function of the nervous system is to rapidly translate sensory information into action potentials (APs), which can be processed (integrated) into an adaptive response via activation of effector organs. The adaptive response is often one that, in negative-feedback fashion, corrects in some way for a disturbance that initiated the sensory information (see Figure 1-8). Although feedback regulation can occur locally within cells and tissues, nervous systems have

the advantages of (1) providing rapid regulation over greater distances, and (2) coordinating multiple organs. If the regulatory response involves movement via one or more effector skeletal muscles, the response is called a **behavior**. The most basic pathway for accomplishing this is known as a **reflex arc**, with the simplest of these having a sensory neuron that also controls an effector cell directly, as shown in Figure 5-1a.

However, most reflex arcs have two or more neurons involved, with information being transmitted through one or more intermediate synapses (Figure 5-1b, c). These intermediate synapses constitute the simplest form of a **central nervous system (CNS)**, where signals from more than one neuron can be integrated. The ability to have multiple synaptic connections allowed for the evolution of complex control of responses. For example, an integrator neuron can have inputs from a "higher" integrator such as a brain (Figure 5-1d). The higher regulatory center modifies the basic reflex response: It may

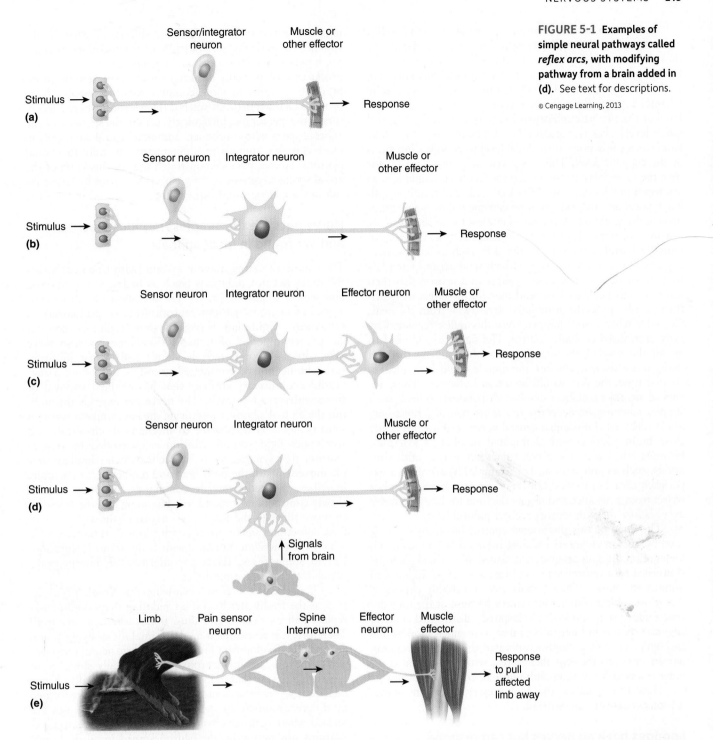

FIGURE 5-1 Examples of simple neural pathways called *reflex arcs*, with modifying pathway from a brain added in (d). See text for descriptions.

© Cengage Learning, 2013

suppress the reflex, alter its sensitivity, or activate it in advance of a disturbance in *anticipation* fashion. The "higher" input may modify the reflex based on past experience, that is, on *memory*. Integrating neurons in a CNS also allows for sophisticated behaviors, which involve multiple sensory inputs and multiple controls of effector muscles.

Because of these advantages, the evolution of a CNS is a major trend in animal evolution. This has been generally associated with two other significant trends: (1) the evolution of longitudinal *centralization*, the association of diffuse neuronal cell bodies into a distinct longitudinal *nerve cord*, as well as (2) *cephalization*, the concentration of neurons in the head, the leading part of an animal's body that typically deals with more environmental information than do other parts. That is, over evolutionary time in many animal groups, the interneuronal component progressively expanded, formed more complex interconnections especially at the head end of the nervous system, forming a brain. From the brain, many neural control pathways became routed through the centralized nerve cords. Associated with this evolution was the acquisition of more specialized receptors for sensing the external environment and the routing of their information through the nerve cord to the brain for processing.

Let's examine the relationship between simple reflex circuits and a CNS with cephalization in more detail. Although more advanced animals do indeed have centralized brains, it is important to note that no animal has fully centralized cephalic control of all its functions. As you saw in Chapter 1, many functions in advanced animals remain regulated at the *intrinsic* cell/organ levels and at "lower" neural reflex levels, that is, regulation is *distributed* in a hierarchy from simple functions at the local level to complex functions at the cephalic level. There is a very important reason for this: the time-delay phenomenon for feedback regulation, as discussed in Chapter 1 (p. 16). A centralized brain is not always necessary and may even be detrimental for some kinds of neural regulation. A simple example—the *withdrawal reflex* in a mammal (Figure 5-1e)—illustrates this. Here, a potentially harmful stimulus to the skin such as intense heat triggers heat and pain sensors, which send action potentials to a spinal (CNS) integrating synapse and neuron that then sends signals back to a skeletal muscle, the effector, which then rapidly pulls the stimulated area away from the heat. Consider what would happen were this reflex regulated by fully centralized cephalic control. The signal would need to go all the way to the brain, presumably passing through many more synapses, before the muscle could be activated. By that time, the skin would be seriously burned. Thus, advanced neural regulation involves distributed control, with simple, *common homeostatic functions typically remaining at the reflex level* through a central nerve cord (or physically close brain regions) with a minimal number of synapses. However, the brain's usefulness comes into play in other situations, such as anticipation. For example, based on past experience of a burn, the brain may detect a potentially hot object from a distance and trigger the limb (or body) to move away *before* the skin sensors register painful heat.

The logic of this distributed control hierarchy is being used by human engineers building robots that mimic animal locomotion. Initially, engineers designed all movements to be controlled by a centralized computer using sensor inputs and outputs to motors. These robots were invariably clumsy at even the simplest walking movement because of the time delays. Later, using "biologically inspired" distributed control, engineers began to build robots that were much more stable and agile. To do so, they incorporated the simplest walking movements into the legs themselves, with the central computer reserved for dealing with unusual, complex motions.

Now let's examine the evolutionary trends in animal nervous systems in more detail.

Sponges have no nerves but can respond to stimuli using electrical signals

Sponges are interesting because they are the only animals without neurons. Yet some sponges react to external stimuli in reflex fashion. For example, the deep-sea sponge *Rhabdocalyptus dawsoni* stops its feeding currents if sediment-laden water enters its pores (threatening to clog them) or if the body is touched. Action potentials have been measured spreading from the outer cells to the rest of the sponge body, including the flagellate cells that create feeding currents (see Chapter 9, p. 386). Unlike action potentials that last for 1 msec in more complex animals, the action potential in the sponge lasts for about 5 seconds. The electrical pathway is thought to be the *trabecular reticulum*, a **syncytium** (a group of joined cells acting as one functional unit) that connects all parts of the sponge. The reticulum is not made of true neurons. Sponges do, however, contain some of the genes associated with the production of proteins localized on the postsynaptic membrane of a nerve. In more complex nervous systems, these proteins characteristically anchor the neurotransmitter receptors in the membrane. Intriguingly, cells in the sponge's larvae express genes whose products stimulate neural stem cells in more complex animals to differentiate into fully functional neurons. Studies have demonstrated that the insertion of the larval sponges' genes into frog embryos or fruit fly larvae result in the generation of extra neurons.

Nerve nets are the simplest nervous systems and are found in most animals

The simplest existing nervous system (with true neurons) is the nerve net in Cnidarians (such as hydra, sea anemones, and jellies) (Figure 5-2a, b). These consist of diffuse networks of neurons without centralization or cephalization. Of course, Cnidarian nervous system evolution has occurred over a period of perhaps 700 million years, so nerve nets, although comparatively simple and diffuse in structure, should not be considered primitive but rather a successful evolutionary strategy that has enabled survival of these widespread animals. The nerve net extends throughout the animal's body, monitoring the environment through sensory neurons responsive to mechanical, chemical, and sometimes light stimuli. Motion is controlled by *motorneurons* that synapse on muscle cells, which stimulate simple movements of the body wall and tentacles; for example, in jellies, the net primarily commands the body wall to slowly contract and expand for swimming and the tentacles to move through the water. The axons of the sensory neurons sometimes make direct connections to motorneurons, which then activate muscles with little or no integration. But in some cases, there are intermediate interneurons, which integrate information.

Of particular interest (following the Krogh principle, p. 3) is the freshwater hydra, a Cnidarian that continuously replaces all its tissues, and thus, unless eaten, is immortal! Therefore, it is an excellent model animal for studying neural growth and development and the most basic neural functions and behaviors. Hydra neurons are continually sloughed at the extremities, so that individual neurons continually change their relative position to each other. New neurons are formed by differentiation from *stem cells* (p. 3), and synapses are formed where neurons cross each other. For this reason, excitations originating in one neuron spread in all directions along multiple paths. Sensory cells are situated in the epidermis such that the animal can respond to external stimuli from any direction. Branches of the nerve net are connected to these sensors and to epithelial cells that have contractile properties.

Neuron density in the nerve net is higher in the hydra's head and foot region compared to its body column, but it is insufficient to label either of these regions as an integrating center in the sense of a CNS. However, not all hydra neurons are in diffuse nets. Some neurons are concentrated in well-defined bundles, which are believed important in simple reflex behavioral responses. Feeding behavior, for example, involves interconnections between sensory neurons in the tentacles and contractile epithelial cells lining the tentacles

(a)

FIGURE 5-2 **Nerve net of a sea anemone, a Cnidarian.** (a) A living anemone. (b) Diagram of nerve net with sensory and contractile cells. The cell types are embedded in epithelium between the outer epidermis and a jellylike midlayer (*mesoglea*) of the body wall. (c) Model of possible origin of the nervous system in a Cnidarian. A *multifunctional* cell (as found in some larvae jellies) with a sensory body that triggered its own contractile region (left panel) may have split into separate sensory and muscle cells with a connecting interneuron.

Sources: (a) Paul Yancey; (b) C. Starr & R. Taggart (2004), *Biology: The Unity and Diversity of Life*, 10th ed. (Belmont, CA: Brooks/Cole), Figure 34.13; (c) G. Miller (2009), On the origin of the nervous system. *Science* 325:24–26.

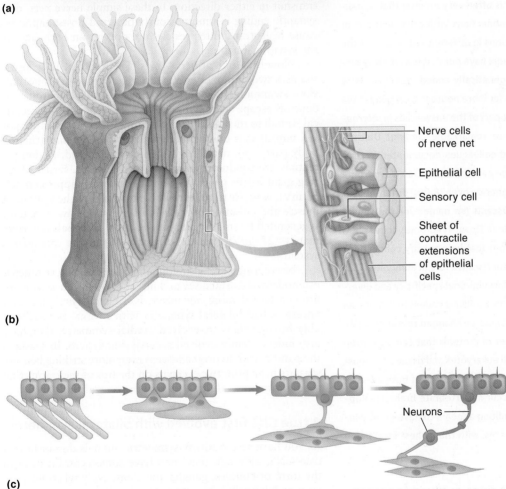

Nerve cells of nerve net

Epithelial cell

Sensory cell

Sheet of contractile extensions of epithelial cells

(b)

Neurons

(c)

and surrounding the mouth. When the sensory receptors come into contact with a potential food item, the ensuing signals generate a reflex response that ultimately draws the food item into the mouth. These bundles may represent the evolutionary precursors of central nervous systems.

Nerve nets are found not just in Cnidarians. Every phylum, including the chordates, has diffuse nerve nets retained from early evolutionary stages. In vertebrates, nets take the form of a **nerve plexus** (network of interwoven nerves),

located in the intestinal wall, which is responsible for propelling digesta (food being processed) along the intestine (p. 661). In some animals, the nerve net may have a sensory function, whereas in others, nerve nets are used to modulate the activity of an organ by releasing neurotransmitters or neuromodulators (p. 138). And in some instances, nerve nets may be used to coordinate behavior such as locomotion. In mollusks, for example, motor nerve nets are associated with local reflex pathways.

| How did nervous systems evolve?

We do not know how the first nervous system evolved, but a clue comes from primitive excitable cells in the bells of some jelly (Cnidarian) larvae (Figure 5-2c). These *multifunctional* cells are both sensors and effectors: If the outer sensory part of the cell is touched, action potentials are initiated that travel to the inner musclelike part of the cell, which contracts to cause the jelly to swim. This cell may have evolved into separate sensors and muscles with interconnecting neurons (also found in jellies) as a way to coordinate distant regions of the body. See the Suggested Reading by Miller (2009).

As more complex animals evolved, nervous systems made of tens of thousands to billions of neurons arose. These cells form networks in which the "wiring" is often very precise; that is, individual neurons make specific connections with other neurons or target organs in the same locations in different individuals of the same species. As some researchers have put it, it is as if every single neuron has its own unique genetically coded "address." How can this happen when there are far more neurons than genes? We don't yet fully know, but at least part of the answer lies in *alternative RNA splicing* (p. 32). In a remarkable finding, Dietmar Schmucker, James Clemens, and colleagues documented in *Drosophila* (fruit flies) that one gene called *Dscam* (Down syndrome cell adhesion molecule) could generate over 38,000 different proteins via alternative splicing! Dscams (so named for the similar gene in humans that plays a role in Down syndrome) act as CAM receptors (p. 74) on growing axons that help specify connections to target cells. (See also p. 490 for Dscam's role in immunity.)

Another mechanism to explain neuronal specificity and diversity is *RNA editing*. Here, an existing mRNA's code is altered by an enzyme; for example, cytidine (C) can be changed to a uridine (U). This further amplifies the number of proteins that can arise from one gene, and occurs widely in all eukaryotes and tissues, but most prominently in neural tissue. Considerable levels of both alternative splicing and RNA editing occur in vertebrate brain development. Intriguingly, neural RNA editing appears to have greatly increased in the evolution of primates, with the highest level being found in humans. <<

Simple ganglia and nerve rings evolved for more complex behavior

More complicated neural circuitry can be found in **medusae** (free-swimming forms such as scyphozoan or true jellies, cubomedusae or box jellies, and hydromedusae—jellylike larvae of hydroids). In these radially symmetric animals, a distribution of sensory structures over a full 360 degrees is advantageous because there is no optimal direction from which sensory information is received. Unlike sessile hydra, however, swimming Cnidarians must coordinate more complex behaviors, especially swimming and orienting in the

open water. In these animals, radial symmetry is often associated with (1) a **nerve ring** or "ring brain," and (2) (in some of these animals) selected neurons are condensed into simple *ganglia*. A **ganglion** (singular of *ganglia*) is a group or cluster of nerve cell bodies, often with related functions. Like hydra bundles, ganglia and rings represent steps of centralization between nerve nets and true brains.

An example of ganglia is found in the **rhopalium**, a complex of primitive sensory organs and integrating ganglia in medusae. It has **ocelli** (simple eye structures, p. 240) and **statocysts** (fluid-filled sacs that help indicate position as the animal moves, p. 218). The rhopalium ganglia in scyphomedusae (true jellies) also function as **pacemakers** (excitable cells that fire spontaneously), which generate the swimming rhythm by continually stimulating waves of contraction. Transmission of action potentials in either direction down an axon is achieved by the presence of gap junctions as well as *bidirectional chemical synapses,* which allow neurons to transmit in either direction. In these simple nerve nets, each synaptic ending is simultaneously pre- and postsynaptic because neurotransmitter vesicles are present on both sides of the synaptic junction.

Now let's look at nerve rings. The hydromedusa *Aglantha* uses two: an inner pacemaker ring that controls normal, slow swimming, and an outer ring in the bell margin that controls escape swimming behavior in response to mechanical stimuli to the body. The outer ring has a single giant axon (35 mm) that is unique: It forms a perfect *annulus* (ringlike body part) with neither a beginning nor an end. Action potentials are conducted at rates up to 4 m/sec. Firing of the ring giant axons generates action potentials (approximately 95 mV), which lead to forceful contractions of the swimming muscle and pulsation of the bell. In contrast, slow swimming is generated by smaller-amplitude action potentials (approximately 27 mV), which originate from the pacemaker system.

Nerve rings are also found in the radially symmetric echinoderms (seastars, sea urchins, and so on). These animals are considered more advanced than Cnidarians, and their ancestors had bilateral symmetry with a CNS. But presumably because they "re-evolved" radial symmetry, they have also independently evolved a nerve ring system. In seastars, the parallel with hydromedusae is even more striking because seastars have primitive eyespots on the tips of their arms that send information about light into the central ring.

A true CNS first evolved with bilateral symmetry

As you have seen, radially symmetric animals do not have a true CNS, although they may have some centralization in the form of bundles, ganglia and rings. Only when we progress to bilaterally symmetric animals such as platyhelminths (flatworms, tapeworms, and so on) do we find the beginnings of a true CNS—longitudinal nerve cords that coordinate nervous activity—in conjunction with a **peripheral nervous system (PNS)**, a communication network that extends between the CNS and all parts of the animal's body. Flatworms represent an early stage in the evolution of a CNS (Figure 5-3a). They have two anterior ganglia from which branch two principal nerve cords with lateral neurons connecting the two cords and projecting to the various regions of the body. Collectively, the ganglia can be thought of as a rudimentary brain because they serve as an integrat-

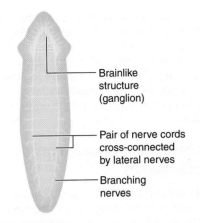

- Brainlike structure (ganglion)
- Pair of nerve cords cross-connected by lateral nerves
- Branching nerves

(a) Flatworm

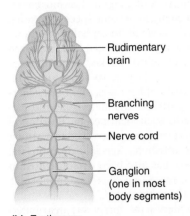

- Rudimentary brain
- Branching nerves
- Nerve cord
- Ganglion (one in most body segments)

(b) Earthworm

FIGURE 5-3 **Bilateral nervous systems of a few nonvertebrates.** (a) Flatworm, (b) earthworm, (c) crayfish, (d) grasshopper, (e) cuttle (or cuttlefish), a cephalopod mollusk.

Sources: (a)-(d) C. Starr & R. Taggart (2004), *Biology: The Unity and Diversity of Life*, 10th ed. Belmont, (CA: Brooks/Cole), Figure 34.14; (e) © Richard E. Young, Michael Vecchione, Katharina M. Mangold, 1999, www.fortunecity.com/emachines/e11/86/graphics/cephpod/CEPHPOD6.gif.

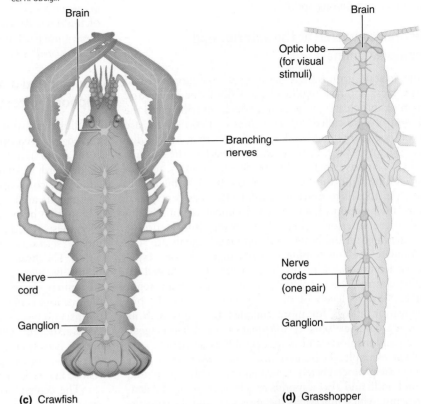

- Brain
- Branching nerves
- Nerve cord
- Ganglion

(c) Crawfish

- Brain
- Optic lobe (for visual stimuli)
- Branching nerves
- Nerve cords (one pair)
- Ganglion

(d) Grasshopper

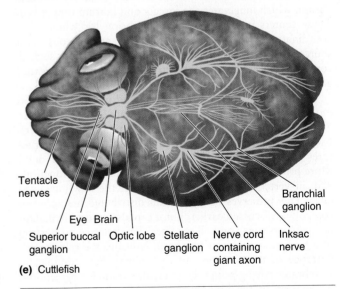

- Tentacle nerves
- Eye Brain
- Superior buccal ganglion
- Optic lobe
- Stellate ganglion
- Nerve cord containing giant axon
- Inksac nerve
- Branchial ganglion

(e) Cuttlefish

ing center for particular nerve pathways. Ganglia within the head of a flatworm coordinate signals arising from sensory organs such as the eyespots. The longitudinal nerve cord evolved for transmitting motor signals to effectors located bilaterally along the animal's elongated body.

The next evolutionary advance is found in a more complex ganglionic CNS, a characteristic of more advanced protostomes such as annelids, arthropods, and some mollusks (Figure 5-3b, c, d, e). In segmented animals, the CNS consists of a chain of *segmental ganglia*. Paired bundles of axons, termed **connectives**, link one ganglia to its neighbors. For example, the CNS of annelids (Figure 5-3b) consists of a bilobed *brain* with a double nerve cord with ganglia (one in most body segments), connecting with separate sensory and motor neurons (in the PNS). These networks rely on one-way conduction along the nerve axons. Ganglia are larger in arthropods (Figure 5-3c, d) and are associated with more highly developed sense organs. Generally, there is one ganglion for each thoracic and abdominal body segment, although some arthropods show secondary fusion of some of these ganglia. Ganglia in each segment are specialized in coordinating the regional function of each body segment, that is, legs, wings, tail or abdomen, and structures on the head. *Decentralized* brain function best describes the neural circuitry of these animals. For example, the ability to copulate is not in the least impaired in a decapitated male praying mantis. The female can be steadily consuming her mate while

being inseminated at the same time! (Nutrients provided by the male in this manner ultimately help ensure the development of the ova within the female.) You will soon see that elements of segmental organization are also found in vertebrates and govern the regulation of many reflex actions.

Arthropod ganglia consist of an outer rind and an inner core. In some ganglia, virtually all the neuron cell bodies are located in the rind, with the central region containing the synaptic contacts and bundles of nerve processes (called the neuropil). The packing of pre- and postsynaptic neuronal branches into a dense neuropil reduces transmission delays that are characteristic of animals with diffuse nerve nets. The degree to which the nervous system is organized into a cellular rind varies among species.

True brains evolved at the anterior end of complex animals

Finally, in some groups there has been an evolutionary trend for the most anterior region of the CNS to be larger and contain sizable clusters of neurons, which contributed to the evolution of a so-called *superganglion* or **brain**. These are found in the more complex nonvertebrates that we just discussed (Figure 5-3b, c, d, e). As we noted earlier, the anterior position of brains is believed to be an evolutionary adaptation resulting from the tendency of the head of bilaterally symmetric animals to move forward, so that newly acquired sensory information about the environment is initially detected in the anterior region. Some sensory information is then integrated in the brain, and command signals are sent to modulate the functioning of particular effectors, especially skeletal muscles. Again, in the logic of distributed hierarchic control, the simplest basic responses remain regulated at "lower" levels of the CNS in most animals, but cephalization allows for more complex functions such as processing complex sensory information (such as images), learning and memory, and ultimately self-awareness.

As an example, the brain of insects consists of three pairs of lobes, the **protocerebrum**, which innervates the compound eyes and ocelli and also responds to olfactory input, the **deutocerebrum**, which innervates the antennae, and the **tritocerebrum**, which innervates the foregut and labrum (upper lip). These centers process sensory information for whole-body coordination, such as moving towards an odorous food source. However, as we noted earlier, some functions such as copulation in praying mantises are located in "downstream" ganglia.

The most advanced brains are found in the cephalopod mollusks (squids, cuttles, nautiloids, and octopods; Figure 5-3e) and in vertebrates (which we examine in detail later). In general, evolution has favored the addition of new structures to brains rather than modification of existing ones. More primitive structures have retained their basic functions and, in vertebrates, are layered beneath the newly evolved ones. This preserves the important distributed hierarchy of control we discussed earlier, but it can also lead to maladaptive behaviors (as you will see later). The relative proportion of a particular region of the brain indicates the relative importance of that function to the animal. For example, in vertebrates relying primarily on tactile stimuli for perception, the regions of the brain that process this information are significantly larger. The large *optic lobe* of birds and *visual cortex* of primates are examples of the importance of visual information for survival in these animals. In contrast, weakly electrical fish that use electrical currents for intraspecies communications exhibit relative hypertrophy of the *cerebellum* and *rhombencephalon*, those areas of the brain involved in processing electrosensory information.

It has generally been thought that the brains of protostomes, such as annelids and arthropods, evolved independently of deuterostome (e.g., chordate) brains (see Figure 1-1). However, recent studies by Detlev Arendt and colleagues may have overthrown this idea. They examined the expression of microRNAs (regulators of gene expression; p. 33) and also 42 genes essential to brain development in the brains of a marine annelid, an insect, and a vertebrate. To everyone's surprise, they found that the timing and location of the microRNAs and genes were amazingly similar in all three animals. Thus, cephalization may have begun only once in the history of evolution.

Cephalopod brains have complex structures supporting complex behaviors

An illustrative trend of cephalization and centralization occurred in cephalopods. Mollusks in general have two nerve cords running down each side of the body (in contrast to the single cord in many other animal types; see Figure 5-3). Along the cords are several pairs of ganglia. The most anterior ganglia are joined and enlarged in some species, forming a brain that in most mollusks is small. In cephalopods, however, these joined ganglia are greatly enlarged and organized into a series of specialized, tightly packed lobes (Figure 5-3e). The brain has many similarities to those of vertebrates: Both have lobes with complex folds, and both generate similar electrical wave patterns. Distinct visual and tactile memory centers have also been identified in cephalopod brains, and one area, the vertical lobe, is involved in learning tasks. Another vertebrate attribute, specialization in the brain's hemispheres (p. 194), has also been demonstrated in octopods. For example, octopuses, which rely on monocular vision, favor one eye over the other.

The cephalopod CNS shows the hierarchic distributed arrangement that we have discussed earlier. Indeed, one recent study revealed that the octopus's arms contain simple motor programs that are completely independent of the brain. That is, certain stereotypical movements used daily, such as reaching out and then curling up around prey animals, occur even in a stimulated arm removed from the animal. Thus, the brain probably sees the prey, sends a "go" command to the arms, and the local "reach out and grab" program in the arm itself does the rest. Again, this avoids feedback delays to the brain that might allow the prey to escape. (As you will see, many motor programs in vertebrates are similarly arranged.)

The large brain of cephalopods is involved more directly in other, more complex processes. Controversial evidence suggests that octopods can learn as effectively as some mammals and may even be able to learn by observing other octopods. Moreover, individual octopods and cuttles seem to "invent" unique behaviors not found in others of their species, suggesting that such behaviors are probably not innate. Squid, cuttle, and octopod brains also control an elaborate array of **chromatophores** in their skins. These pigment cells of varying colors can be neuronally (and therefore rapidly) expanded or contracted to display different color patterns. In some cuttles, around 40 distinct color patterns have been observed being used in various behaviors such as mating, territorial displays, and possibly other social interactions such as alarm warnings. Cuttlefish are the only nonvertebrates known to exhibit specific reactions to different kinds of predators. Some researchers have even proposed that these patterns form a kind of language. Color patterns are also used for camouflage, often in complex ways. In some octopods, camouflage patterns are coupled with sophisticated mimicry behavior; for example, a Malaysian octopod species has been seen using its arms and colors to simulate sea snakes and molding its entire body to mimic a flounder or a seastar (p. 357).

Despite similarities between cephalopod and vertebrate brains, they clearly evolved independently, because detailed structures are quite different. For example, the cephalopod brain is wrapped around the esophagus instead of lying in a cranium. An example of the often-illogical nature of evolution (pp. 2–3), this is not a very good "design" because spines of prey animals can actually pierce the cephalopod brain if they get caught in the esophagus!

Vertebrate brain size varies up to 30-fold for a given body size

The total number of neurons in nervous systems ranges from only several hundred in simple animals, such as nematodes (approximately 300 neurons), to approximately 10^8 in some cephalopods and 10^{11} in large mammals. Because of these numbers as well as the complexity of their neural circuitry, brain function in cephalopods rivals that of some vertebrates, as evidenced by the intricacy of their behavior compared to other nonvertebrates. In the human brain, the number of neurons is approximately 8 to 9×10^{10}, whereas larger mammals such as elephants and whales show a further twofold increase in number. This example illustrates that the total number of neurons does not correspond to intellect! The exhibition of specific behavioral patterns is far more dependent on the internal wiring of the brain rather than on its overall size and depends on the ratio of the brain size to body size.

The logarithm of brain weight as a function of body weight is typically linear for a variety of vertebrates, with a positive slope indicating that the overall size of an animal is related to its relative brain size (Figure 5-4). (The slope is less than 1.0, meaning that brain size does not increase as much as body weight does. This *scaling* phenomenon is explored for metabolism in Chapter 15.) However, as you have just seen, the various regions of the brain do not necessarily vary directly with body weight. For example, the forebrain shows a disproportionately large size relative to body weight in primates, whereas there is a relatively large cerebellum and rhombencephalon in many bony fishes. There are also significant exceptions to the basic scaling of total brain size with some species exhibiting five times the predicted size, whereas others have approximately one-fifth the predicted size. Brain size varies approximately 30-fold for a given body size when considered across all vertebrate radiations. Some interesting trends include:

- Agnathans (hagfish, lampreys) generally have the smallest brains for their body size, but hagfishes have brains two to three times larger than lampreys of the same body size.
- Many cartilaginous fishes (sharks and rays) have brains as large for their body size as those of many birds and mammals.
- Among amphibians, frogs generally have larger brains than salamanders, whereas the brains of reptiles are two to three times larger than the brains of most amphibians of the same body size.
- The most complex vertebrate classes, birds and mammals, have brains that are up to 10-fold larger than the brains of reptiles of the same body size. Interestingly, the

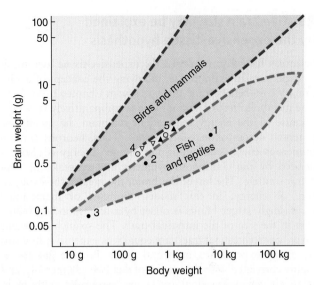

FIGURE 5-4 Weights of adult cephalopod brains compared to those of various vertebrates. Upward triangle *Sepia* squid; downward triangle *Loligo* squid; 1, *Octopus vulgaris*; 2, *O. salutii*; 3, *O. defillipi*; 4 and 5, squids *Illex* and *Todarodes*.

Sources: Modified from H. J. Jerison (1976), Paleoneurology and the evolution of mind, *Scientific American* 234:90–100; and K. Mangold-Wirz (1966), Cerebralisation und Ontogenesemodus bei Eutherien, *Acta Anatomica* 63:449–508

brains of octopods and squids are larger on average (relative to body size) than most animals, except some cartilaginous fishes, birds, and mammals (Figure 5-4).

- The largest brains relative to body size among birds are found in the perching birds (woodpeckers and parrots), whereas the smallest brains are in the granivores (quail and pigeons).
- The brains of migratory birds and bats are generally smaller than those of similar-sized nonmigrators, suggesting that the weight of a large brain is too costly for long-range migration.
- Brain size of primates, in general, is much greater than the average for mammals of the same body size, although the second largest relative brain size (after humans) is found in dolphins (cetaceans). In contrast, the insectivores, rodents, and nonplacental mammals (marsupials, monotremes) have the smallest brains.

For primates, numerous theories have been put forward to explain differences in relative brain size. For example, primates that consume leaves generally have smaller brains than those that eat fruits. This observation leads to the idea that a larger brain is required of animals that must use their intellect to locate ripened fruits, compared to those that simply find leaves. Carnivores (who must hunt wary prey) also tend to have larger brains than herbivores. Also, selection for an increase in the size of a single brain center leads to an overall increase in brain size. For example, dolphins with their relatively large brains are carnivores that use echolocation, a process requiring considerable neural power, corresponding to a relative large auditory cortex. In contrast, baleen whales, which do not echolocate and do not hunt wary prey, have brains that are relatively small for their body size. A relationship between relative size of the brain and the complexity of social organization (which is quite advanced in dolphins, for example) has also been suggested.

Relative brain size may be explained by the expensive-tissue hypothesis

Scientists have also proposed the **expensive-tissue hypothesis** to explain relative brain size, following the concept of evolutionary cost–benefit trade-offs you saw in Chapter 1 (pp. 2–3). The metabolic cost of neural tissue is comparatively very high because of the energy required to maintain the membrane potential across axonal membranes. Thus, neurons contain numerous energy-consuming ion pumps. Energy is also expended by the continuous synthesis and reuptake of neurotransmitters. The larger the brain, the greater the expenditure of energy. In this model, the cost of maintaining increasingly larger brains is offset by a corresponding reduction in the size of the intestinal tract. The splanchnic organs (liver, gastrointestinal tract) are equally as expensive, in terms of energy expenditure, as neural tissue. Because gut size is highly correlated with the quality of diet (see Chapter 14), and relatively small intestinal tracts are associated with high-quality foodstuffs, a reduction in the size of one of these metabolically expensive organs is "traded off" against a greater degree of cephalization. Thus, the increase in brain size could not have been achieved without a concomitant increase in quality of the diet, and high-quality diets are associated with reduced size (and energy cost) of the intestinal tract. This pattern is seen in a general comparison between carnivores and herbivores: The former are large-brained, with high-quality diets, whereas the latter are small-brained and often eat low-quality fibrous plants that require large digestive tracts to process.

This largely *proximate* (p. 1) hypothesis does not explain which evolved first: a higher-quality diet (which then favored greater intelligence), or nondietary selective forces favoring a larger brain (which then gave the ability to find higher-quality food). Also, an exception to this relationship has been found in bats: Insect-eating bats have smaller brains than their fruit-eating relatives. In this example, researchers found no correlations either with social complexity or with the relative size of the digestive tract. Finally, an additional explanation for the relative brain size of mammals takes into account the relative brain size of the mother and the brain size of the developing offspring. The **maternal-energy hypothesis** states that ultimate brain size depends on the allocation of maternal resources to fetal and postnatal development. Because development of the brain begins at an early stage of development, resources provided by the mother can be directly linked to brain development in her offspring, particularly in placental mammals. Note, though, that this is also a proximate hypothesis that does not explain what selective forces would be at work in maternal allocation.

Some nervous systems exhibit plasticity

Ultimately, the way animals act and react depends on organized, discrete neuronal processing. In addition to brain–body size ratios, another key factor in complexity of this processing is change during development and aging. Many nonvertebrate nervous systems appear to have "hard-wired" processing; that is, the neural cell structures underlying behavior are relatively fixed once they have formed. However, cephalopods and vertebrates have nervous systems with some degree of **plasticity** (modifiable cell structures, particularly the number of synaptic connections, axonal and dendritic branches, and neurons). When the immature nervous system develops according to its genetic plan, an overabundance of neurons and synapses is formed in most vertebrates. Depending on external stimuli and the extent that these pathways are used, some are retained, firmly established, and even enhanced with new branches and connections, whereas others are eliminated. Once the nervous system has matured, ongoing modifications still occur as an animal continues to learn from its unique set of experiences, though less readily. The ability of neurons to respond structurally to usage and incoming stimuli is the basis for certain forms of learning as well as for nervous system regeneration. The nervous system in complex animals thus represents a network of living cells that are modifiable throughout a lifetime.

check your understanding 5.1

What are the components of a reflex arc?

What features led to the evolution of centralization and cephalization?

Why are complex ganglionic nervous systems characteristic of advanced nonvertebrates?

What factors are important in determining relative brain size in an animal?

5.2 The Vertebrate Nervous System: Overview and Peripheral System

As with bilateral animals in general, the nervous system of vertebrates is functionally organized into a *central nervous system (CNS)* and a *peripheral nervous system (PNS)*. In vertebrates, the CNS consists of the brain and spinal cord, and the PNS consists of nerve fibers that carry information between the CNS and other parts of the body (the periphery) (Figure 5-5). In contrast to the organization of the arthropod CNS, which consists of a brain and discontinuous ganglia linked by connectives, the vertebrate CNS arises from a single neural tube and is organized into a continuous column of neuron cell bodies and bundles of diverse axons. There is some similarity, however, in that branches of the vertebrate PNS have a segmented pattern (Table 5-1). Most of the neuronal cell bodies are contained within the CNS. However, the PNS also includes ganglia (also in a segmented pattern) that contain the cell bodies of most sensory neurons and some nonmotor effector neurons. The PNS is further subdivided into *afferent* and *efferent* divisions. The afferent division (*a* is from *ad*, "toward," as in *advance*; and *ferent*, "carrying") carries information from *sensors* to the CNS, appraising it of the external environment and providing status reports on internal activities being regulated by the nervous system. In the next chapter, on sensory physiology, we focus on this afferent division. The **efferent division** (*e* is from *ex*, "from," as in *exit*) transmits instructions from the CNS to **effector organs**—the muscles, glands, and other organs that carry out the orders to bring about the desired action. The efferent nervous system in turn has two or three components (Figure 5-5):

1. The **autonomic nervous system** fibers, which innervate smooth muscle, cardiac muscle, glands, and other non-

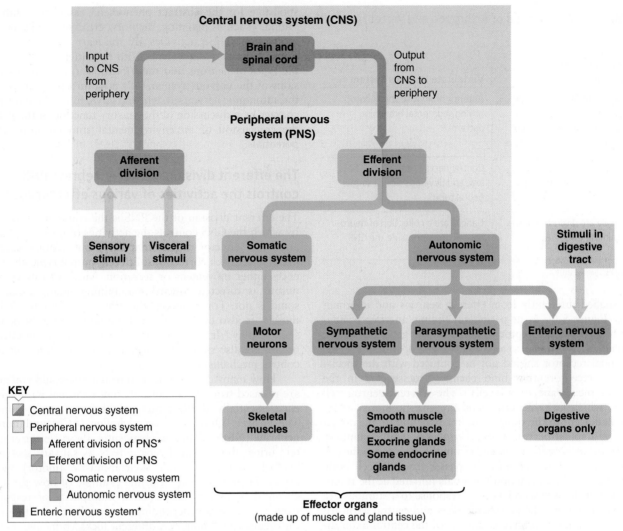

FIGURE 5-5 Organization of the vertebrate nervous system. Afferent fibers travel within the same nerves as efferent fibers but in the opposite direction. The enteric nervous system lies entirely within the wall of the digestive tract.

© Cengage Learning, 2013

motor organs (such as brown adipose tissue, p. 746, and some immune organs, p. 488). In most vertebrates, this system is further subdivided into the **sympathetic nervous system** and the **parasympathetic nervous system**, both of which innervate most of the visceral organs. Not all vertebrates have such clear subdivisions, as evidenced in lampreys where the sympathetic and parasympathetic divisions are difficult to distinguish. In hagfish, the autonomic nervous system is poorly developed.

2. The **somatic nervous system**, which consists of the fibers of the **motor neurons** that supply the skeletal muscles.

3. In addition to the CNS and PNS, the **enteric nervous system** is an extensive nerve network in the wall of the digestive tract. Digestive activities are controlled by the autonomic nervous system and the enteric nervous system as well as by hormones. The enteric nervous system can act independently of the rest of the nervous system but is also influenced by autonomic fibers that terminate on the enteric neurons. Sometimes the enteric nervous system is considered a third component of the autonomic nervous system, one that supplies the digestive organs only.

The somatic system is sometimes called the *voluntary* system, whereas the autonomic and enteric systems are often termed *involuntary*. However, these terms are based on human biology. For other vertebrates, it is more accurate to say simply that the somatic (motor) system controls reflex and more complex nonreflex *skeletal muscle* actions, whereas the other systems operate all other innervated effector organs, generally independently of higher brain regions involved in locomotion, learning, and memory. It is also important to recognize that all these "nervous systems" are really subdivisions of a single, integrated nervous system. They are arbitrary divisions based on differences in the structure, location, and functions of the various diverse parts of the whole nervous system.

The three classes of neurons are afferent neurons, efferent neurons, and interneurons

Three classes of neurons make up the nervous system: *afferent neurons, efferent neurons,* and *interneurons.* The afferent division of the PNS consists of **afferent neurons,** which

TABLE 5-1 Comparison of Arthropod and Vertebrate Nervous Systems

Arthropod Nervous System	Vertebrate Nervous System
Anterior brain or large ganglion, with ganglia in body segments linked by paired bundles of connective fibers	Anterior brain with spinal cord having segmental branching pattern
Nerve cords are ventral, solid, and double in origin	Central nerve cord is dorsal, hollow, and develops from a single neural tube
Ganglia are discrete entities within the CNS	Ganglia are a collection of neuronal cell bodies outside the CNS

© Cengage Learning, 2013

are shaped differently from efferent neurons and interneurons (Figure 5-6). At its peripheral ending, an afferent neuron has a **sensory receptor** (Chapter 6) that generates action potentials in response to a particular type of stimulus. (This neuronal receptor should not be confused with the special protein receptors that bind chemical messengers in the plasma membrane of all cells.) The afferent-neuron cell body, which lacks dendrites and presynaptic inputs, is located adjacent to the spinal cord in a ganglion (in the *dorsal root* of a spinal nerve). A long *peripheral axon*, commonly called the *afferent fiber*, extends from the receptor to the cell body, and a short *central axon* passes from the cell body into the spinal cord. Action potentials initiated at the receptor end of the peripheral axon in response to a stimulus are propagated along the peripheral axon and central axon toward the spinal cord. The terminals of the central axon diverge and synapse with other neurons within the spinal cord, disseminating information about the stimulus. Afferent neurons lie primarily within the PNS: only a small portion of their central axon endings project into the spinal cord to relay signals from the periphery to the CNS.

Efferent neurons also lie primarily in the PNS (Figure 5-6). The cell bodies of efferent neurons originate in the CNS, where many centrally located presynaptic inputs converge on them to influence their outputs to the effector organs. Efferent axons (*efferent fibers*) leave the CNS to course their way to the muscles or glands they innervate, conveying their integrated output for the effector organs to put into effect. (An autonomic pathway actually consists of a two-neuron chain between the CNS and the effector organ.)

Interneurons lie entirely within the CNS. In mammals, about 99% of all neurons belong to this category. The human CNS is estimated to have 100 billion interneurons! These neurons, which are the integrator neurons we saw at the beginning of this chapter, serve two main roles. First, as their name implies, they lie between (*inter*, "between") the afferent and efferent neurons and are important in the integration of peripheral responses to peripheral information, as you saw with the withdrawal reflex earlier (Figure 5-1e). The more complex the required action, the greater the number of interneurons interposed between the afferent message and efferent response. Second, in the most advanced vertebrates, interconnections between interneurons themselves are re-

sponsible for the abstract phenomena associated with the "mind," such as planning, memory, creativity, intellect, and motivation. These activities are the least understood functions of the nervous system. With this brief introduction to the types of neurons and their location in the various divisions of the nervous system, let's now turn our attention to the autonomic nervous system and CNS, followed in the next chapter by a discussion of the sensory function of the PNS—the conversion of an environmental stimulus into action potentials.

The efferent division of the vertebrate PNS controls the activities of various effector organs

The efferent division of the PNS is the communication link by which the CNS controls the activities of muscles, glands, and other effector organs that carry out actions, such as making feedback corrections. Efferent output typically influences either movement or secretion. Much of this efferent output is directed toward maintaining homeostasis, but some is not. For example, the efferent output to skeletal muscles is also directed toward nonhomeostatic behaviors, such as play. (It is important to realize that many effector organs are also subject to hormonal control and/or intrinsic control mechanisms.)

How many different neurotransmitters would you guess are released from the various efferent neuronal terminals to elicit essentially all the controlled effector organ responses? The answer: Only two—**acetylcholine** (**ACh**) and **norepinephrine** (**NE**)![1] Acting independently, these neurotransmitters bring about such diverse effects as salivary secretion, bladder contraction, and skeletal-muscle (motor) movements. These effects are a prime example of how the same chemical messenger may elicit a multiplicity of responses from various tissues, depending on specialization of the effector organs (the "same key, different locks" principle, p. 93).

The sympathetic and parasympathetic branches of the autonomic system innervate most visceral organs to control essential visceral functions

Functionally, the autonomic nervous system is involved in many basic, physiological functions that are critical to the survival of a vertebrate. Specifically, it regulates visceral activities normally outside the realm of higher brain control, such as circulation, digestion, thermoregulation, and pupil size (to name a few), and afferent information arising from the viscera (internal organs) usually does not reach higher brain levels (conscious levels in humans). Examples of visceral afferent information include input from the baroreceptors that monitor blood pressure and input from the chemo-

---

[1] *Noradrenaline (norepinephrine)* is chemically very similar to *adrenaline (epinephrine)*, the primary hormone product secreted by the adrenal medulla gland. Because a U.S. pharmaceutical company marketed this product for use as a drug under the trade name Adrenalin, the scientific community in the United States prefers the alternative name "epinephrine" as a generic term for this chemical messenger, and accordingly, "noradrenaline" is known as "norepinephrine." In most other English-speaking countries, however, "adrenaline" and "noradrenaline" are the terms of choice.

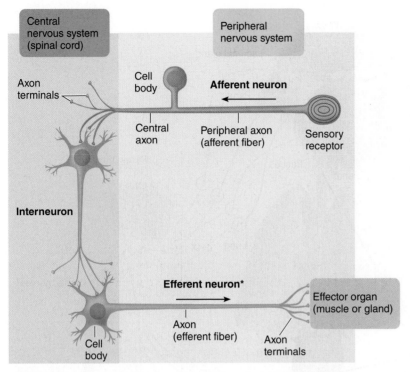

FIGURE 5-6 **Structure and location of the three functional classes of neurons.**
*Efferent autonomic nerve pathways consist of a two-neuron chain between the CNS and the effector organ.

© Cengage Learning, 2013

receptors that monitor the protein or fat content of ingested food. This input is used to direct the activity of the autonomic efferent neurons.

Most visceral organs are innervated by both sympathetic and parasympathetic nerve fibers (Figure 5-7). Innervation of a single organ by both branches of the autonomic nervous system is known as **dual innervation**. Table 5-2 summarizes the major effects of these autonomic branches. Although the details of this wide array of autonomic responses are described more fully in later chapters that discuss the individual organs involved, several general concepts can be derived now. As can be seen from the table, the sympathetic and parasympathetic nervous systems generally exert opposite effects in a particular organ. For example, sympathetic stimulation increases the heart rate, whereas parasympathetic stimulation decreases it; sympathetic stimulation slows down movement within the digestive tract, whereas parasympathetic stimulation enhances digestive motility. Note that both systems increase the activity of some organs and reduce the activity of others.

Rather than memorizing a list such as that presented in Table 5-2, you will do better to logically deduce the actions of the two systems based on an understanding of the circumstances under which each system dominates. Usually, both systems are partially active; that is, normally some level of action potential activity exists in both the sympathetic and the parasympathetic fibers supplying a particular organ. This ongoing activity is called **sympathetic** or **parasympathetic tone** or **tonic activity**. Under given circumstances, activity of one division can dominate the other. *Sympathetic dominance* to a particular organ exists when the sympathetic fibers' rate of firing to that organ increases above tonic level, coupled with a simultaneous decrease below tonic level in the parasympathetic fibers' frequency of action potentials to the same organ. The reverse situation is true for *parasympathetic dominance.* Shifts in balance between sympathetic and parasympathetic activity can be accomplished discretely for individual organs to meet specific demands (for example, sympathetically induced dilation of the pupil in dim light; see p. 242), or a more generalized, widespread discharge of one autonomic system in favor of the other can be elicited to control bodywide functions. Massive widespread discharges take place more frequently in the sympathetic system. The value of this potential for massive sympathetic discharge is evident, considering the circumstances during which this system usually dominates.

The sympathetic system dominates in times of "fight or flight"

The sympathetic system promotes responses that prepare a vertebrate for strenuous physical activity in the face of emergency or other stressful situations, such as a physical threat from the environment. This response is typically referred to as a **fight-or-flight response** because the sympathetic system readies the body to fight against or flee from the threat. The changes are not homeostatic in every case; rather, many physiological systems are *reset* (p. 18) to different levels. In other cases, the changes are anticipatory, to maintain homeostasis with little or no delays. Think about the body resources needed in such circumstances. The heart beats more rapidly and more forcefully; the respiratory airways open wide to permit maximal air flow; glycogen (stored sugar) and fat stores are broken down to release extra fuel into the blood; and blood vessels supplying skeletal muscles dilate (open more widely). All these responses are aimed at providing increased flow of oxygenated, nutrient-rich blood to the skeletal muscles in anticipation of strenuous physical activity. Furthermore, the pupils dilate and the eyes adjust for far vision, enabling the animal to make a quick visual assessment of the entire threatening scene. In some species, sweating or panting is promoted, in anticipation of excess heat production by the physical exertion. Because digestive and urinary activities are inessential in meeting the threat, the sympathetic system inhibits (i.e., resets lower) these activities.

The parasympathetic system dominates in times of "rest and digest"

The parasympathetic system, in contrast, dominates in quiet, relaxed situations. Under such nonthreatening circumstances, the body can be concerned with its own "general housekeeping" activities, such as digestion and emptying of the urinary bladder. The parasympathetic system promotes these types of bodily functions while slowing down those activities that are enhanced by the sympathetic system. In a tranquil setting, for example, the animal does not need to have its heart beating rapidly and forcefully.

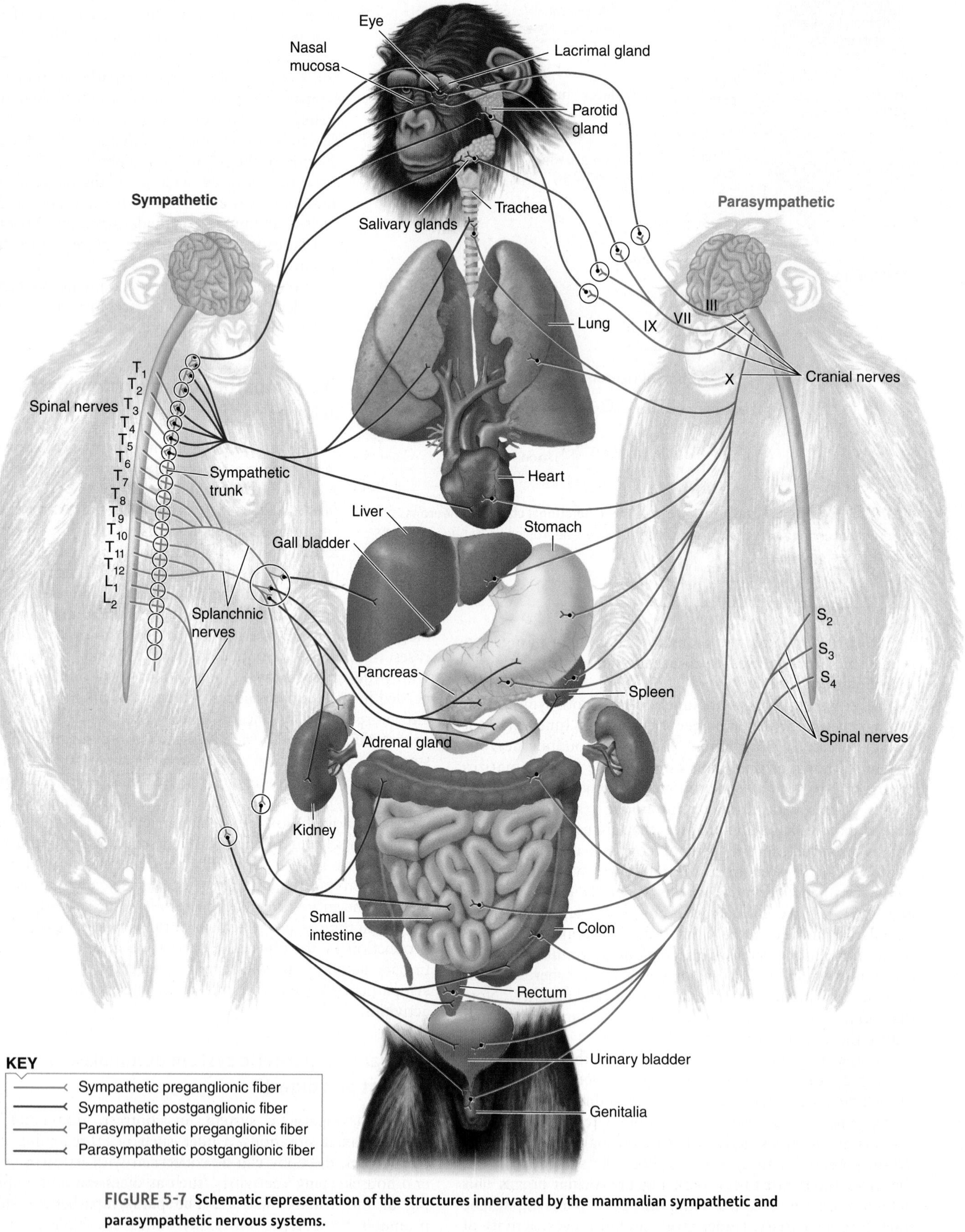

Sympathetic

Parasympathetic

Eye

Nasal mucosa

Lacrimal gland

Parotid gland

Trachea

Salivary glands

Lung

III

VII

IX

X

Cranial nerves

T₁
T₂
T₃
T₄
T₅
T₆
T₇
T₈
T₉
T₁₀
T₁₁
T₁₂
L₁
L₂

Spinal nerves

Sympathetic trunk

Heart

Liver

Stomach

Gall bladder

Splanchnic nerves

Pancreas

Spleen

S₂
S₃
S₄

Spinal nerves

Adrenal gland

Kidney

Small intestine

Colon

Rectum

Urinary bladder

Genitalia

KEY

⊰	Sympathetic preganglionic fiber
⊰	Sympathetic postganglionic fiber
⊰	Parasympathetic preganglionic fiber
⊰	Parasympathetic postganglionic fiber

FIGURE 5-7 Schematic representation of the structures innervated by the mammalian sympathetic and parasympathetic nervous systems.

© Cengage Learning, 2013

TABLE 5-2 Effects of the Autonomic Nervous System on Various Organs

Organ	Effect of Sympathetic Stimulation (and Types of Adrenergic Receptors)	Effect of Parasympathetic Stimulation
Heart	Increased rate, increased force of contraction (of whole heart) (β_1)	Decreased rate, decreased force of contraction (of atria only)
Blood Vessels	Constriction (α_1)	Dilation of vessels supplying the penis and clitoris only
Lungs	Dilation of bronchioles (β_2)	Constriction of bronchioles
	Inhibition of mucus secretion (α)	Stimulation of mucus secretion
Digestive Tract	Decreased motility (α_2, β_2)	Increased motility
	Contraction of sphincters (to prevent forward movement of contents) (α_1)	Relaxation of sphincters (to permit forward movement of contents)
	Inhibition of digestive secretions (α_2)	Stimulation of digestive secretions
Gallbladder	Relaxation	Contraction (emptying)
Urinary Bladder	Relaxation (β_2)	Contraction (emptying)
Eye	Dilation of pupil (contraction of radial muscle) (α_1)	Constriction of pupil (contraction of circular muscle)
	Adjustment for far vision (β_2)	Adjustment for near vision
Liver (Glycogen Stores)	Glycogenolysis (glucose released) (β_2)	None
Adipose Cells (Fat Stores)	Lipolysis (fatty acids released) (β_2)	None
Exocrine Glands		
Exocrine pancreas	Inhibition of pancreatic exocrine secretion (α_2)	Stimulation of pancreatic exocrine secretion (important for digestion)
Sweat glands	Stimulation of secretion by most sweat glands (α_1)	Stimulation of secretion by some sweat glands
Salivary glands	Stimulation of small volume of thick saliva rich in mucus (α_1)	Stimulation of large volume of watery saliva rich in enzymes
Endocrine Glands		
Adrenal medulla	Stimulation of epinephrine and norepinephrine secretion (cholinergic)	None
Endocrine pancreas	Inhibition of insulin secretion; stimulation of glucagon secretion (α_2)	Stimulation of insulin and glucagon secretion
Genitals	Ejaculation and orgasmic contractions (males); orgasmic contractions (females) (α_1)	Erection (caused by dilation of blood vessels in penis [male] and clitoris [female])
Brain Activity	Increased alertness (receptors unknown)	None

© Cengage Learning, 2013

Dual innervation gives precise, antagonistic control

What is the advantage of dual innervation of organs with nerve fibers whose actions oppose each other? It enables precise control over an organ's activity, like having both an accelerator and a brake to control the speed of a car. If an animal suddenly darts across the road as you are driving, you could eventually stop if you simply took your foot off the accelerator, but you might stop too slowly to avoid hitting the animal. If you simultaneously apply the brake as you lift up on the accelerator, however, you can come to a more rapid, controlled stop. In a similar manner, a sympathetically accelerated heart rate could gradually be slowed to normal after a stressful situation by decreasing the rate of firing in the cardiac sympathetic nerve (letting up on the accelerator), but the heart rate can be reduced more rapidly by simultaneously increasing activity in the parasympathetic

FIGURE 5-8 Autonomic nerve pathway.

© Cengage Learning, 2013

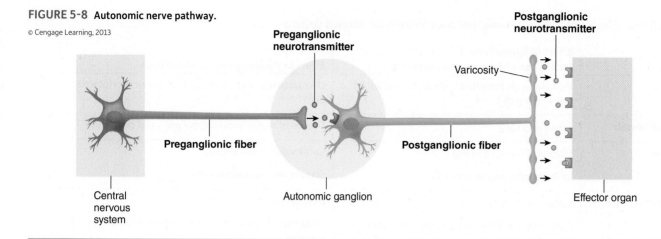

supply to the heart (applying the brake). Indeed, the two divisions of the autonomic nervous system are usually reciprocally controlled; increased activity in one division is accompanied by a corresponding decrease in the other. Thus, dual innervation of an organ by the two branches of the autonomic system permits more precise control over the organ. This is the *antagonistic negative-feedback* design we saw in Chapter 1 (p. 16).

There are several exceptions to the general rule of dual reciprocal innervation by the two branches of the autonomic nervous system (Table 5-2); the most notable are the following:

- *Innervated blood vessels* (most arterioles and veins are innervated; arteries and capillaries are not) for the most part receive only sympathetic nerve fibers. Regulation is accomplished by increasing or decreasing the firing rate above or below the tonic level in these sympathetic fibers. The only blood vessels to receive both sympathetic and parasympathetic fibers are those supplying the penis and clitoris. The precise vascular control this dual innervation affords these organs is important in accomplishing erection (see Chapter 16).
- *Most sweat glands* are innervated only by sympathetic nerves. The postganglionic fibers of these nerves are unusual because they secrete ACh rather than NE.
- *Salivary glands* are innervated by both autonomic divisions, but unlike elsewhere, sympathetic and parasympathetic activity is not antagonistic. Both stimulate salivary secretion, but the saliva's volume and composition differ, depending on which autonomic branch is dominant.

An autonomic nerve pathway consists of a two-neuron chain

Each autonomic nerve pathway extending from the CNS to an innervated organ consists of a two-neuron chain (Figure 5-8) (except to the adrenal medulla; you will learn more about this exception later). The cell body of the first neuron in the series is located in the CNS. Its axon, the **preganglionic fiber**, synapses with the cell body of the second neuron, which lies within a ganglion outside of the CNS. The axon of the second neuron, the **postganglionic fiber**, innervates the effector organ.

Sympathetic nerve fibers originate in the thoracic and lumbar regions of the spinal cord (see p. 176). Most sympa-

thetic preganglionic fibers in birds and mammals are very short, synapsing with cell bodies of postganglionic neurons within paravertebral (*para*, "near") ganglia that lie in a **sympathetic ganglion chain** (the **sympathetic trunk**) located along either side of the spinal cord (Figures 5-7 and 5-9). Long, postganglionic fibers originating in the ganglion chain terminate on the effector organs. Some preganglionic fibers pass through the ganglion chain without synapsing and terminate later in sympathetic *collateral ganglia* located about halfway between the CNS and the innervated organs, with postganglionic fibers traveling the remainder of the distance. An unusual group of ganglia found in birds, referred to as the *intestinal* or *Remak's* nerve, consists of both sympathetic and parasympathetic components. This nerve is located along the entire length of the alimentary canal from the duodenum to the cloaca.

Parasympathetic preganglionic fibers arise from the cranial (brain) and sacral areas (lower spinal cord) of the CNS. (Some cranial nerves contain parasympathetic fibers.) These fibers are long in comparison to sympathetic preganglionic fibers because they do not end until they reach **terminal ganglia** that lie in or near the effector organs. Very short postganglionic fibers terminate on the cells of the organ itself.

Parasympathetic postganglionic fibers release acetylcholine; sympathetic ones release norepinephrine

Sympathetic and parasympathetic preganglionic fibers release the same neurotransmitter, ACh (acetylcholine), but the postganglionic endings of these two systems release different neurotransmitters that influence the effector organs. Parasympathetic postganglionic fibers release acetylcholine. Accordingly, they, along with all autonomic preganglionic fibers, are called **cholinergic fibers**. Most sympathetic postganglionic fibers, in contrast, are called **adrenergic fibers** because they release NE (norepinephrine or noradrenaline). Both acetylcholine and norepinephrine also serve as chemical messengers elsewhere in the body (Table 5-3).

Postganglionic autonomic fibers do not end in a single terminal swelling like a synaptic knob. Instead, the terminal branches of autonomic fibers contain numerous swellings, or **varicosities**, that simultaneously release neurotransmitter over a large area of the innervated organ rather than on single cells (Figures 5-8 and 8-31). This diffuse release of neurotransmitter, coupled with the fact that any resulting change in electrical

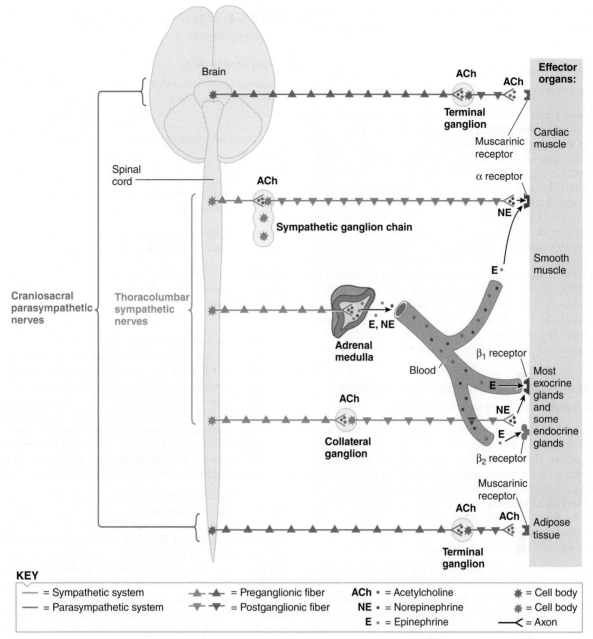

FIGURE 5-9 Autonomic nervous system. The sympathetic nervous system, which originates in the thoracolumbar regions of the spinal cord, has short cholinergic (acetylcholine-releasing) preganglionic fibers and long adrenergic (norepinephrine-releasing) postganglionic fibers. The parasympathetic nervous system, which originates in the brain and sacral region of the spinal cord, has long cholinergic preganglionic fibers and short cholinergic postganglionic fibers. In most instances, sympathetic and parasympathetic postganglionic fibers both innervate the same effector organs. The adrenal medulla is a modified sympathetic ganglion, which releases epinephrine and norepinephrine into the blood. Nicotinic cholinergic receptors are located in the autonomic ganglia and adrenal medulla and respond to ACh released by all autonomic preganglionic fibers. Muscarinic cholinergic receptors are located at the autonomic effectors and respond to ACh released by parasympathetic postganglionic fibers. α_1, α_2, β_1, β_2 adrenergic receptors are variably located at the autonomic effectors and differentially respond to norepinephrine released by sympathetic postganglionic fibers and to epinephrine released by the adrenal medulla.

© Cengage Learning, 2013

activity is spread throughout a smooth or cardiac muscle mass via gap junctions (see p. 66), means that whole organs instead of discrete cells are typically influenced by autonomic activity.

The adrenal medulla, an endocrine gland, is a modified part of the sympathetic nervous system

The *adrenal gland,* which lies above the kidney on each side (*ad,* "next to"; *renal,* "kidney"), is an endocrine gland consisting of an outer portion, the *adrenal cortex,* and an inner portion, the *adrenal medulla.* The **adrenal medulla** is a modified sympathetic ganglion that does not give rise to postganglionic fibers. Instead, on stimulation by the preganglionic fiber that originates in the CNS, it secretes hormones directly into the blood (Figures 5-9 and 5-10). Not surprisingly, the hormones are identical or similar to postganglionic sympathetic neurotransmitters. In mammals, NE ranges from about 2% of the adrenal medullary hormone output in rabbits to almost 50% in pigs, while the output of norepinephrine is as high as 80% in chickens and whales. In all species studied, fetal adrenal tissue contains predominantly NE. The remaining secretion is the closely related substance **epinephrine (adrenaline)** (see footnote 1, p. 158). These hormones, in general, reinforce activity of the sympathetic nervous system.

Several different receptor types are available for each autonomic neurotransmitter

Because each autonomic neurotransmitter and medullary hormone stimulates activity in some tissues but inhibits activity in others, the particular responses must depend on the specialization of the tissue cells rather than on properties of the chemicals themselves. Responsive tissue cells have one or more of several different types of plasma membrane receptor proteins for these chemical messengers ("same key, different locks"). Binding of a neurotransmitter to a receptor induces the tissue-specific response, as follows:

1. Cholinergic **nicotinic receptors.** Two types of acetylcholine (*cholinergic*) receptors—*nicotinic* and *muscarinic*—have been identified on the basis of their response to particular drugs (Table 5-4). **Nicotinic receptors** (activated by the tobacco plant derivative nicotine) are found on the postganglionic cell bodies in all autonomic ganglia. Nicotinic receptors respond to acetylcholine released from both sympathetic and parasympathetic preganglionic fibers. Binding of ACh to these receptors brings about opening of nonspecific cation channels in the postganglionic cell that permit passage of both Na^+ and K^+. Because of the greater electrochemical gradient for Na^+ than for K^+, more Na^+ enters the cell than K^+ leaves, depolarizing the postganglionic cell.
2. Cholinergic **muscarinic receptors** (activated by the mushroom poison muscarine) are found on effector cell membranes (smooth muscle, cardiac muscle, and glands). They bind with ACh released from parasympathetic postganglionic fibers. There are five subtypes of muscarinic receptors, all of which are linked to G proteins that activate second-messenger systems that lead to the target cell response.
3. **Adrenergic receptors.** There are two major classes of adrenergic receptors for norepinephrine and epinephrine based on the ability of various drugs to either initi-

TABLE 5-3 Sites of Release for Acetylcholine and Norepinephrine

Acetylcholine (ACh)	Norepinephrine (NE)
All preganglionic terminals of the autonomic nervous system	Most sympathetic autonomic nervous system postganglionic terminals
All parasympathetic postganglionic terminals	Adrenal medulla
Sympathetic postganglionic terminals at sweat glands and some blood vessels in skeletal muscle	Central nervous system
Terminals of efferent neurons supplying skeletal muscle (motor neurons)	
Central nervous system	

© Cengage Learning, 2013

ate or prevent responses in the effector organ. These receptors are designated as **alpha (α)** and **beta (β) receptors,** with further subclassifications of α_1 and α_2 and β_1 and β_2 receptors. These various receptor types are distinctly distributed among the effector organs, as shown in Table 5-2. Receptors of the β_2 type bind primarily with epinephrine, whereas β_1 receptors have about equal affinities for norepinephrine and epinephrine, and α receptors of both subtypes have a greater sensitivity to norepinephrine than to epinephrine (Figure 5-10).

All adrenergic receptors are coupled to G proteins, but the ensuing pathway differs for the various receptor types. Activation of both β_1 and β_2 receptors brings about the target cell response by means of the cyclic AMP second-messenger system (see p. 97). Stimulation of α_1 receptors elicits the desired response via the IP_3/Ca^{2+} second-messenger system (see p. 98). By contrast, binding of a neurotransmitter to an α_2 receptor blocks cyclic AMP production in the target cell. Accordingly, activation of α_2 receptors brings about an inhibitory response in the effector organ, such as decreased smooth muscle contraction in the digestive tract. Activation of α_1 receptors, in contrast, usually brings about an excitatory response in the effector organ—for example, arteriolar constriction caused by increased contraction of the smooth muscle in the walls of these blood vessels. The α_1 receptors are present in most sympathetic target tissues. Stimulation of β_1 receptors, which are found primarily in the heart, also causes an excitatory response, namely, increased rate and force of cardiac contraction. The response to β_2 receptor activation is generally inhibitory, such as arteriolar or bronchiolar (respiratory airway) dilation caused by relaxation of the smooth muscle in the walls of these tubular structures. As a quick rule of thumb, activation of the subscript "1" versions of adrenergic receptors leads to excitatory responses and activation of the subscript "2" versions leads to inhibitory responses (Table 5-2).

Autonomic Agonists and Antagonists Because activation of various receptor types brings about different responses to the same autonomic messenger, these receptors can be manipulated fairly selectively by drugs. Drugs are available that

TABLE 5-4 Location of Nicotinic and Muscarinic Cholinergic Receptors

Type of Receptor	Site of Receptor	Respond to Acetylcholine Released from:
Nicotinic Receptors	All autonomic ganglia	Sympathetic and parasympathetic preganglionic fibers
	Motor end plates of skeletal muscle fibers	Motor neurons
	Some CNS cell bodies and dendrites	Some CNS presynaptic terminals
Muscarinic Receptors	Effector cells (cardiac muscle, smooth muscle, glands)	Parasympathetic postganglionic fibers
	Some CNS cell bodies and dendrites	Some CNS presynaptic terminals

© Cengage Learning, 2013

selectively enhance or mimic (agonists) or block (antagonists) (p. 100) autonomic responses at each of the receptor types. Some are only of experimental interest, but others are very important therapeutically. For example, *atropine* blocks the effect of ACh at muscarinic receptors but does not affect nicotinic receptors. Because the ACh released at both parasympathetic and sympathetic preganglionic fibers combines with nicotinic receptors, blockage at nicotinic synapses would knock out both of these autonomic branches. By acting selectively to interfere with ACh action only at muscarinic junctions, which are the sites of parasympathetic postganglionic action, atropine effectively blocks parasympathetic effects but does not influence sympathetic activity at all. This principle is used to suppress salivary and bronchial secretions before surgery to reduce the risk of a patient (human or veterinary) inhaling these secretions into the lungs.

Many regions of the central nervous system are involved in the control of autonomic activities

Messages from the CNS are delivered to cardiac muscle, smooth muscle, glands, and other nonskeletal-muscle effectors via the autonomic nerves, but what regions of the CNS regulate autonomic output? Autonomic control of these effectors is mediated by reflexes and through centrally located control centers. Going back one step further, ultimately information carried to the CNS via the visceral afferents is used to determine the appropriate output via the autonomic efferents to the effectors to maintain homeostasis.

1. Some autonomic reflexes, such as urination, defecation, and erection, are integrated at the *spinal cord* (reflex) level, but all these spinal reflexes are subject to control by higher levels (following the basic design of Figure 5-1e).

2. The *medulla* within the brainstem is the region most directly responsible for autonomic output. Centers for

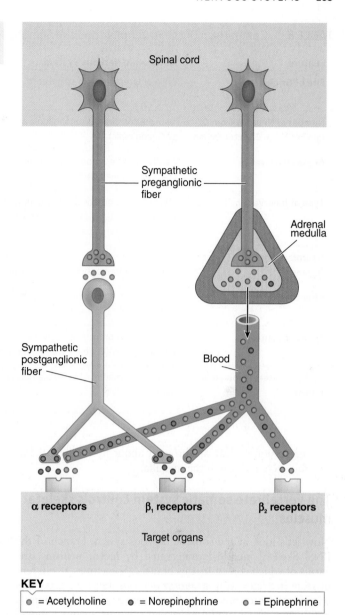

KEY

● = Acetylcholine	● = Norepinephrine	● = Epinephrine

FIGURE 5-10 Comparison of the release and binding to receptors of epinephrine and norepinephrine. Norepinephrine is released both as a neurotransmitter from sympathetic postganglionic fibers and as a hormone from the adrenal medulla. Beta$_1$ (β_1) receptors bind equally with both norepinephrine and epinephrine, whereas beta$_2$ (β_2) receptors bind primarily with epinephrine and alpha (α) receptors of both subtypes have a greater affinity for norepinephrine than for epinephrine.

© Cengage Learning, 2013

controlling cardiovascular, respiratory, and digestive activity via the autonomic system are located there. Some of these are also subject to higher control (such as breath holding).

3. The *hypothalamus* plays an important role in maintaining stability of the internal environment (e.g., temperature, osmotic, energy homeostasis), and in integrating the autonomic, somatic, and endocrine responses that automatically accompany various emotional and behavioral states. For example, the increased heart rate, blood pressure, and respiratory activity associated with displays of

TABLE 5-5 Comparison of the Autonomic and the Somatic Nervous Systems

Feature	Autonomic Nervous System	Somatic Nervous System
Site of Origin	Brain or lateral horn of spinal cord	Ventral horn of spinal cord for most; those supplying muscles in head originate in brain
Number of Neurons from Origin in CNS to Effector Organ	Two-neuron chain (preganglionic and postganglionic)	Single neuron (motor neuron)
Organs Innervated	Cardiac muscle, smooth muscle, exocrine and some endocrine glands	Skeletal muscle
Type of Innervation	Most effector organs dually innervated by the two antagonistic branches of this system (sympathetic and parasympathetic)	Effector organs innervated only by motor neurons
Neurotransmitter at Effector Organs	May be acetylcholine (parasympathetic terminals) or norepinephrine (sympathetic terminals)	Only acetylcholine
Effects on Effector Organs	Either stimulation or inhibition (antagonistic actions of two branches)	Stimulation only (inhibition possible only centrally through IPSPs on cell body of motor neuron)
Types of Control	Under reflex control; may be voluntarily controlled with biofeedback techniques and training	Subject to both reflex and nonreflex control
Higher Centers Involved in Control	Spinal cord, medulla, hypothalamus, prefrontal association cortex	Spinal cord, motor cortex, basal nuclei, cerebellum, brainstem

aggression or fear are brought about by the hypothalamus acting through the medulla.

The vertebrate somatic system controls skeletal muscles

Now let's look at the somatic nervous system branch of the PNS. Skeletal muscle is innervated by **motor neurons,** the axons of which constitute this branch. (Sometimes all efferent neurons are referred to as *motor neurons,* but we reserve this term for the efferent somatic fibers that supply skeletal muscle and bring about body movement; *motor* means "movement.") The cell bodies of these motor neurons are located within the ventral horn of the spinal cord. The only exception is that the cell bodies of motor neurons supplying muscles in the head are in the brainstem. Unlike the two-neuron chain of autonomic nerve fibers, the axon of a motor neuron is continuous from its origin in the spinal cord to its termination on skeletal muscle. Motor-neuron axon terminals release ACh, which brings about excitation and contraction of the innervated muscle fibers. Motor neurons can only stimulate skeletal muscles, in contrast to autonomic fibers, which may either stimulate or inhibit their effector organs. Skeletal muscle activity can be inhibited only within the CNS through activation of inhibitory synaptic input to the cell bodies and dendrites of the motor neurons supplying that particular muscle.

Motor neurons are the final common pathway

Motor neuron dendrites and cell bodies are influenced by many converging presynaptic inputs, both excitatory and inhibitory. Some of these inputs are part of spinal-reflex pathways originating with peripheral sensory receptors. Others are part of descending pathways originating within the brain. Areas of the brain that exert control over skeletal muscle movements include the motor regions of the cortex, the basal nuclei, the cerebellum, and the brainstem (see p. 174).

Motor neurons are considered the **final common pathway** because the only way any other parts of the nervous system can influence skeletal muscle activity is by acting on these motor neurons. The level of activity in a motor neuron and its subsequent output to the skeletal muscle fibers it innervates depend on the relative balance of EPSPs and IPSPs (see p. 132) brought about by its presynaptic inputs originating from these diverse sites in the brain.

The somatic system is considered under cerebral control, but much of the skeletal muscle activity involving posture, balance, and stereotypic movement is controlled at lower levels, following the logic of a hierarchical system. A vertebrate animal's cerebrum may make the decision to start walking, but the cerebrum itself does not usually control the alternate contraction and relaxation of the involved muscles, because these movements are coordinated by lower brain centers such as the cerebellum, brainstem, or spine (hence the infamous tales of a chicken running around even after its head has been severed. Similarly, a shark's body will continue to swim even if its brain is destroyed).

Table 5-5 summarizes the features of the two divisions of the efferent nervous system discussed in this chapter.

check your understanding 5.2

What features characterize the peripheral nervous system, and how is it different from the central nervous system?

How do receptor types determine neurotransmitter function?

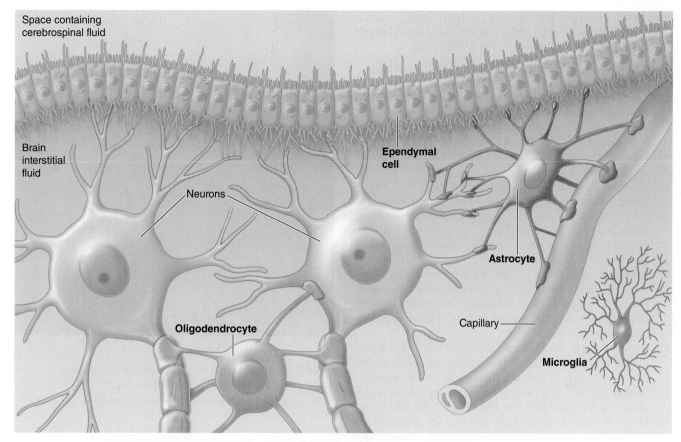

FIGURE 5-11 Glial cells of the mammalian central nervous system. The glial cells include the astrocytes, oligodendrocytes, microglia, and ependymal cells.

© Cengage Learning, 2013

5.3 The Vertebrate Nervous System: Central System

As we've discussed, the vertebrate central nervous system (CNS), which consists of the brain and spinal cord, receives input about the external and internal environment from the afferent neurons. The CNS sorts and processes this input, then initiates appropriate directions in the efferent PNS, which carry the instructions to glands or muscles to bring about the desired response—some type of secretion or movement. In general, the nervous system acts by means of its action potentials to control the rapid responses of the body, which are directed toward maintaining homeostasis.

Many cells within the CNS are not neurons but **glial cells** or **neuroglia**, which serve as the connective tissue of the CNS and as such help support the neurons. For unknown reasons, their densities correlate with brain size, making up only a few percent of a nematode worm brain, 25% of a fruit fly brain, 65% of a mouse brain, 90% of a human brain, and 97% of an elephant brain. Despite their large numbers in the human brain, the glial cells occupy only about half the volume because they do not branch as extensively as neurons do.

Glial cells support the interneurons physically, metabolically, and functionally

Unlike neurons, glial cells generally do not initiate or conduct nerve impulses. They are important in the viability of the CNS, however. For much of the time since discovering them in the 19th century, scientists thought the glial cells were passive "mortar" that physically supported the functionally important neurons (*glia* is Greek for *glue*). In the last decade, however, the varied and important roles of these dynamic cells have become apparent. The glial cells serve as the connective tissue of the CNS and as such help support the neurons both physically and metabolically. They homeostatically maintain the composition of the specialized extracellular environment surrounding the neurons within the narrow limits optimal for normal neuronal function. Furthermore, they actively modulate synaptic function and are considered nearly as important as neurons to learning and memory. Certain glial cells can also take up and destroy neurotransmitters released from neighboring neurons. The four major types of glial cells in the CNS are *astrocytes, oligodendrocytes, ependymal cells,* and *microglia* (Figure 5-11). Let's examine each in turn.

Astrocytes Named for their starlike shape (*astro,* "star"; *cyte,* "cell") (Figure 5-12), **astrocytes** are the most abundant glial cells. They fill a number of critical functions:

1. As the main "glue" of the CNS, they hold the neurons together in proper spatial relationships.
2. They serve as a scaffold to guide neurons to their proper final destination during fetal brain development.
3. They induce the small blood vessels of the brain to establishment the blood–brain barrier, a tight barricade between the blood and brain (soon to be described in greater detail).
4. Through their close association with both local capillaries and neurons, astrocytes help transfer nutrients from the blood to the neurons and store some glycogen.
5. They are important in the repair of brain injuries and in neural scar formation.
6. Astrocytes play a role in neurotransmitter activity. They take up glutamate and gamma-aminobutyric acid (GABA), excitatory and inhibitory neurotransmitters, respectively, thus bringing the actions of these chemical messengers to a halt.
7. Astrocytes take up excess K^+ from the brain ECF when high action-potential activity outpaces the ability of the Na^+/K^+ pump to return the effluxed K^+ to the neurons. (Recall that K^+ leaves a neuron during the falling phase of an action potential; see Figure 4-8.) By taking up excess K^+, the astrocytes help maintain the proper brain ECF ion concentration to sustain normal neural excitability. If brain ECF K^+ levels were allowed to rise, the resultant lower K^+ concentration gradient between the neuronal ICF and surrounding ECF would reduce the neuronal membrane closer to threshold, even at rest. This would increase the excitability of the brain. In fact, an elevation in brain ECF K^+ concentration may be one of the factors responsible for the brain cells' explosive convulsive discharge that occurs during epileptic seizures, which occur in some humans and other mammals including some dogs.
8. Astrocytes and other glial cells enhance synapse formation and modify synaptic transmission. Elongated, fine starlike extensions from an astrocyte's cell body are often wedged between presynaptic and postsynaptic portions of adjacent neurons. Recently scientists learned that an astrocyte can selectively retract these fine processes through actin activity, in a sort of reverse of pseudopod formation in amoeboid movement (see p. 44). Withdrawal of these processes may allow new synapses to form between neurons that were otherwise physically separated by the intervening process. Astrocytes also appear to influence synapse formation and function by chemical means.
9. Astrocytes communicate with neurons and with one another by means of chemical signals in two ways. First, gap junctions (see p. 66) have been identified between astrocytes themselves and between astrocytes and neurons. Chemical signals pass directly between cells through the gap junctions without entering the ECF. Second, astrocytes have receptors for the common neurotransmitter glutamate. Furthermore, the firing of neurons in the brain in some instances triggers the release of ATP along with the classical neurotransmitter from the axon terminal. Binding of glutamate to an astro-

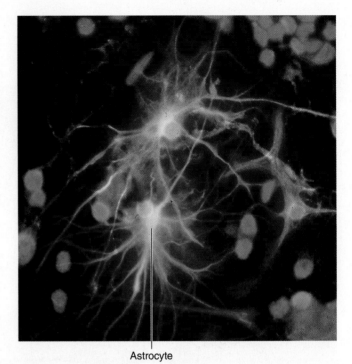

Astrocyte

FIGURE 5-12 Astrocytes. Note the starlike shape of these astrocytes, which have been grown in tissue culture.

Source: Nancy Kedersha, Ph.D., Research Scientist, Cell Biology, ImmunoGen, Inc. (Harvard Medical School).

cyte's receptors and/or detection of extracellular ATP by the astrocyte leads to calcium influx into this glial cell. The resultant rise in intracellular calcium prompts the astrocyte itself to release ATP, thereby activating adjacent glial cells. In this way, astrocytes share information about action potential activity in a nearby neuron. In addition, astrocytes and other glial cells can also release the same neurotransmitters as neurons do, as well as other chemical signals. These extracellular chemical signals from the glia can affect neuronal excitability and strengthen synaptic activity, such as by increasing neuronal release of neurotransmitter or promoting the formation of new synapses. Moreover, astrocytes secrete *thrombospondin,* a large protein that can trigger synapse formation. Such glial modulation of synaptic activity is likely important in memory and learning.

Scientists are trying to sort out the two-directional chatter that takes place between and among the glial cells and neurons because this dialogue plays an important role in the processing of information in the brain. In fact, some neuroscientists suggest that synapses should be considered "three-party" junctures involving the glial cells as well as the presynaptic and postsynaptic neurons. This point of view is indicative of the increasingly important role being placed on astrocytes in synapse function. Thus, astrocytes have come a long way from their earlier reputation as "support staff" for neurons; these glial cells might turn out to be the "board members" coordinating and/or regulating the neurons.

Oligodendrocytes The insulating myelin sheaths around axons in the CNS are formed by **oligodendrocytes.** An oligodendrocyte has several elongated projections, each of which is wrapped spirally around a section of an interneuronal axon to form a patch of myelin, similar to that made by *Schwann cells* in the PNS (see Figure 4-13, p. 127, and Figure 5-11).

Ependymal Cells Lining the internal cavities of the vertebrate CNS are **ependymal cells**. As the nervous system develops in an embryo from a hollow neural tube, the original central cavity of this tube is maintained and modified to form the *ventricles* of the brain (Figure 5-13) and the central canal of the spinal cord. The ependymal cells lining the ventricles contribute to the formation of *cerebrospinal fluid*, to be discussed shortly. Ependymal cells are one of the few cell types to bear cilia, whose beating action contributes to the flow of cerebrospinal fluid throughout the ventricles.

Ependymal cells also serve as neural stem cells with the potential of forming not only other glial cells but new neurons as well. Traditional view has long held that new neurons are not produced anywhere in the mature mammalian brain. Then, in the late 1990s, scientists discovered that new neurons are produced in one restricted site, namely in a specific part of the hippocampus, a structure important for learning and memory. The discovery that ependymal cells are precursors for new neurons indicates that the adult brain has more potential for repairing damaged regions than previously assumed.

Microglia The immune defense of the CNS depends on **microglia**, scavengers derived from the same tissue that gives rise to *monocytes*, a type of white blood cell that exits the blood and sets up residence as frontline defense agents in various tissues throughout the body (see Chapter 10). Microglia migrate during embryonic development to the CNS. There they remain stationary until activated by an infection or injury.

In the resting state, microglia are wispy cells with many long branches that radiate outward. Recent evidence suggests that resting microglia are not just waiting watchfully. They release low levels of growth factors, such as *nerve growth factor,* which help neurons and other glial cells survive and thrive. When trouble occurs in the CNS, microglia retract their branches, round up, and become highly mobile, moving toward the affected area to remove any foreign invaders or tissue debris. Activated microglia release destructive chemicals for assault against their target.

Researchers increasingly suspect that excessive release of these chemicals from overzealous microglia may damage the neurons they are meant to protect, thus contributing to the insidious neuronal damage associated with *neurodegenerative diseases.*

The delicate central nervous tissue is well protected

CNS tissue is very delicate. This characteristic, coupled with the fact that damaged nerve cells cannot be replaced because most neurons cannot divide, makes it imperative that this fragile, irreplaceable tissue be well protected. Four major features help protect the vertebrate CNS from injury; we will examine the last two of these in detail:

1. It is enclosed by hard, bony structures: The **cranium (skull)** encases the brain, and the vertebral column surrounds the spinal cord.

2. Three protective and nourishing membranes, the **meninges**, lie between the bony covering and the nervous tissue. From the outermost to the innermost layer, these are the *dura mater*, the *arachnoid mater*, and the *pia mater* (*mater,* "mother").
3. The brain "floats" in a special cushioning fluid, the *cerebrospinal fluid* (*CSF*).
4. A highly selective *blood–brain barrier* limits access of blood-borne materials into the vulnerable brain tissue.

The brain floats in its own special cerebrospinal fluid

Cerebrospinal fluid (**CSF**) surrounds and cushions the brain and spinal cord. The CSF has about the same density as the brain itself, so the brain essentially floats or is suspended in its special fluid environment. A major function of CSF is to serve as a shock-absorbing fluid to minimize damage of the brain when subjected to sudden, jarring movements. Consider the impact on the brain when two combative rams charge into each other.

In addition to protecting the delicate brain from mechanical trauma, the CSF plays an important role in the exchange of materials between the neural cells and the interstitial fluid surrounding the brain. Only the brain interstitial fluid—not the blood or CSF—comes into direct contact with the neurons and glial cells. Because the brain interstitial fluid directly bathes the neural cells, its composition is influenced more by changes in the composition of the CSF than by alterations in the blood. Materials are exchanged fairly freely between the CSF and brain interstitial fluid, whereas only limited exchange occurs between the blood and brain interstitial fluid. Thus, the composition of the CSF must be carefully regulated.

Cerebrospinal fluid is formed primarily as a result of selective transport mechanisms across the membranes of **choroid plexuses** found in particular regions of the ventricle cavities of the brain. Choroid plexuses consist of richly vascularized **pia mater** tissue, which supplies blood to the outer cortex and which dips into pockets formed by ependymal cells. The composition of CSF differs from that of blood. For example, CSF is lower in K^+ and slightly higher in Na^+, making the brain interstitial fluid an ideal environment for movement of these ions down concentration gradients, a process essential for conduction of nerve impulses. The biggest difference is the near absence of plasma proteins in the CSF. Plasma proteins cannot exit the brain capillaries to leave the blood during formation of CSF.

A highly selective blood–brain barrier carefully regulates exchanges between the blood and brain

The vertebrate brain is carefully shielded from harmful changes in the blood by a highly selective **blood–brain barrier (BBB)**. Throughout a mammal, exchange of materials between the blood and the surrounding interstitial fluid can take place only across the walls of capillaries, the smallest of blood vessels (Chapter 9). Unlike the rather free exchange across most capillaries in the body, permissible exchanges across brain capillaries are strictly limited. Changes in most

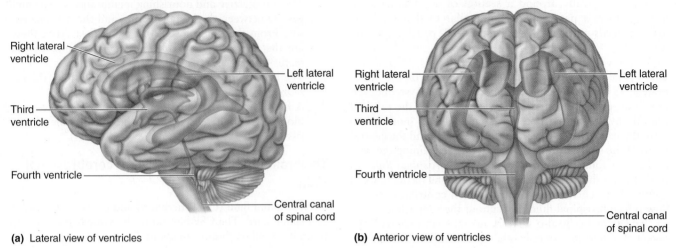

(a) Lateral view of ventricles

(b) Anterior view of ventricles

FIGURE 5-13 The ventricles of the human brain.

© Cengage Learning, 2013

plasma constituents do not easily influence the composition of brain interstitial fluid because only selected exchanges can be made. For example, even if the K^+ level in the blood is doubled, little change occurs in the K^+ concentration of the fluid bathing the central neurons. This is beneficial because alterations in interstitial fluid K^+ would be detrimental to neuronal function.

The BBB consists of both anatomic and physiologic features. A single layer of endothelial cells forms capillary walls throughout the body. Usually, holes or pores between the cells making up the capillary wall (Figure 9-50) permit free exchange of all plasma components, except the large plasma proteins, with the surrounding interstitial fluid. In the brain capillaries, however, the cells are joined by *tight junctions,* which completely seal the capillary wall so that nothing can be exchanged across the wall by passing between the cells (Figure 5-14). The only possible exchanges are through the capillary cells themselves. Lipid-soluble substances such as O_2, CO_2, alcohol, and steroid hormones penetrate these cells easily by dissolving in the lipid plasma membrane. Small water molecules also diffuse through readily, by passing between the phospholipid molecules of the plasma membrane or through aquaporins (water channels, p. 81). All other substances exchanged between the blood and brain interstitial fluid, including such essential materials as glucose, amino acids, and ions, are transported by highly selective membrane-bound carriers. For example, the uptake of glucose is accomplished by primary active transport by a specific glucose transporter. Thus, transport across brain capillary walls *between* the wall-forming cells is prevented *anatomically,* and transport *through* the cells is restricted *physiologically.* Together, these mechanisms constitute the BBB.

By strictly limiting exchange between the blood and brain, the BBB protects the delicate brain from chemical fluctuations in the blood and minimizes the possibility that potentially harmful blood-borne substances might reach the central neural tissue. On the negative side, the BBB limits the use of drugs for the treatment of brain and spinal cord disorders because many drugs are unable to penetrate this barrier. To "fool" the brain, some drugs can be attached to specific delivery vehicles, such as proteins or fatty acids that are nor-

mally transported across the barrier. Once in the brain the protein or fatty acid is metabolized, leaving the drug to pursue its therapeutic task.

Certain areas of the brain are not subject to the BBB, most notably a portion of the hypothalamus. Functioning of the hypothalamus depends on its "sampling" the blood and adjusting its controlling output accordingly to maintain homeostasis. Part of this output is in the form of hormones that must enter hypothalamic capillaries to be transported to their sites of action. Appropriately, these hypothalamic capillaries are not sealed by tight junctions.

The brain depends on delivery of oxygen and glucose by the blood

Even though many substances in the blood never actually come in contact with the brain tissue, the brain, more than any other tissue, is highly dependent on a constant blood supply. Unlike most tissues, which can resort to anaerobic metabolism to produce ATP in the absence of O_2 for at least short periods (see p. 55), most vertebrate brains cannot produce ATP in the absence of O_2. Scientists recently discovered an O_2-binding protein, *neuroglobin,* in the brain. This molecule, which is similar to hemoglobin, the O_2-carrying protein in red blood cells, is thought to play a key role in O_2 diffusion and storage in neurons (see p. 533 for details). Also, in contrast to most tissues, which can use other sources of fuel for energy production in lieu of glucose, the mammalian brain uses only glucose under normal physiological conditions, and glucose plus ketone bodies during starvation. However, except for small amounts of glycogen (a glucose polymer, p. 26) stored in some astrocytes, the brain does not store these nutrients. Therefore, the brain is absolutely dependent on a continuous, adequate blood supply of O_2 and glucose. Accordingly, brain damage results if this organ (particularly in birds and mammals) is deprived of its critical O_2 supply for more than a few minutes or if its glucose supply is cut off for more than 10 to 15 minutes. However, some vertebrates can cope without O_2 for extended periods; for example, some fish and turtles spend the winter living under the ice without O_2 (see Chapter 11). At

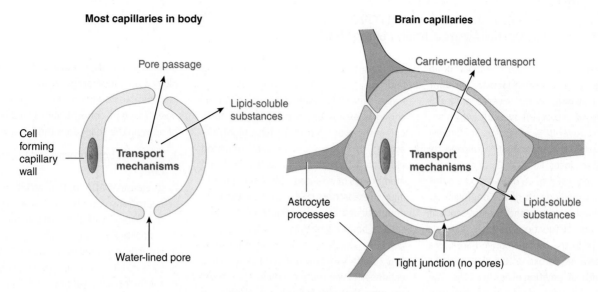

Most capillaries in body

Pore passage

Lipid-soluble substances

Cell forming capillary wall

Transport mechanisms

Water-lined pore

Brain capillaries

Carrier-mediated transport

Transport mechanisms

Astrocyte processes

Lipid-soluble substances

Tight junction (no pores)

Capillaries in cross section

FIGURE 5-14 Blood–brain barrier. Unlike most capillaries in the body, the cells forming the walls of brain capillaries are joined by tight junctions that prevent materials from passing between the cells. The only passage across brain capillaries is through the cells that form the capillary walls. With the exception of lipid-soluble substances and water, passage of all other materials through these cells is physiologically regulated by carrier-mediated systems, which are not present in capillaries elsewhere.

© Cengage Learning, 2013

low temperatures their cells consume ATP at a substantially lower rate.

In contrast to birds and mammals, brains in insects and carp store considerably more glycogen, particularly in glial cells, from which it can be transported to nerve cells in the form of alanine (after breakdown of glycogen to glucose to pyruvate and transamination to alanine). In contrast to neurons of honeybees, which lack enzymes to metabolize fatty acids, the brains of moths have a high capacity to oxidize fatty acids as fuel. The extraordinary energy-requiring demands of the moth's long-distance migrations are supported by mobilization of lipid fuel from the fat body and delivered to the brain via the hemolymph.

Brains of virtually all vertebrates display a degree of plasticity

Normally, damaged axons cannot regenerate within the CNS as they can in the peripheral nervous system, because the myelin-forming oligodendrocytes of the CNS release nerve growth–inhibiting proteins. However, the brains of virtually all vertebrates, and many nonvertebrates, display a degree of **plasticity**, that is, an ability to change or be functionally remodeled in response to the demands placed on it. This ability is more pronounced in early development, but even adult animals retain some plasticity. The term *plasticity* is used to describe this ability because plastics can be manipulated into any desired shape to serve a particular purpose. When an area of the brain associated with a particular activity is destroyed, other areas of the brain may gradually assume some or all of the responsibilities of the damaged region. The underlying molecular mechanisms responsible for the brain's plasticity are only beginning to be unraveled. Current evidence suggests that the formation of new neural

pathways between existing neurons in response to changes in experience are mediated in part by alterations in dendritic shape resulting from modifications in certain cytoskeletal elements (see p. 59). Its dendrites undergo **arborization** (*arbor* is Latin for "tree"): more elongated and branched dendrites with small connecting outgrowths called **dendritic spines,** so a neuron can receive and integrate more signals from other neurons. Thus, the precise synaptic connections between neurons are not fixed but can be modified by experience. Conversely, synaptic connections to a particular neuron may be reduced or eliminated if there is a sustained level of reduced input. This mechanism may also be important in seasonally breeding animals, where synaptic connections between differing regions of the brain are believed to change cyclically during the breeding cycle.

Recently, Gerhard Schratt and colleagues found that a microRNA (p. 33) called *miR-134* is expressed only in dendrites of rat hippocampal neurons, and that it stops the growth of dendritic spines. Like most microRNAs, miR-134 works by blocking translation of specific mRNAs, in this case one for a protein kinase (p. 95) called *Limk1* that enhances spine growth. Stimulated neurons somehow reduce miR-134 inhibition of *Limk1* translation and thus grow more connections.

Neurogenesis is a normal function of several regions of the brain

Another manifestation of plasticity is **neurogenesis**—the reproduction of neurons. Once thought to be absent in the entire mammalian CNS, cell proliferation and track tracing studies in laboratory rodents demonstrate that there are two main sites of neurogenesis in the adult brain. The first is the *hippocampus* (a key memory center we will encounter later)

The most common cause of human brain damage is a **cerebrovascular accident (CVA** or **stroke)**. When a cerebral (brain) blood vessel is blocked by a clot (which accounts for over 80% of strokes in humans) or ruptures, the brain tissue supplied by that vessel loses its vital O_2 and glucose supply. The result of O_2 deprivation is damage and usually death of the deprived tissue. Neural damage (and the subsequent loss of neural function) normally extends well beyond the blood-deprived area as a result of a neurotoxic effect that leads to the death of additional nearby cells. The initial O_2-deprived cells die by *necrosis* (unintentional cell death), but the doomed neighbors undergo *apoptosis* (deliberate cell suicide; see p. 101). In a process known as **excitotoxicity**, affected cells release excessive amounts of glutamate, the common excitatory neurotransmitter. The excitatory overdose of glutamate from the damaged brain cells binds with and overexcites surrounding neurons. Specifically, glutamate binds with excitatory receptors known as *NMDA receptors* (p. 199), which function as Ca^{2+} channels. As a result of excessive glutamate binding, they remain open for too long, permitting too much Ca^{2+} to rush into the affected neighboring neurons. This elevated intracellular Ca^{2+} triggers the cells to self-destruct, the mitochondria explosively producing *free radicals* in the process (p. 53). These highly reactive particles contain an unpaired electron, which can damage the DNA and membranes of affected cells. Adding to the injury, researchers speculate that the Ca^{2+} apoptotic signal may spread from the dying cells to abutting healthy cells through gap junctions, which allow Ca^{2+} and other small ions to diffuse freely between cells. These actions kill even more neurons. Thus, most neurons that die following a stroke are originally unharmed cells that commit suicide in response to the chain of reactions unleashed by the toxic release of glutamate from the initial site of O_2 deprivation.

Following the Krogh principle (p. 3) Gillian Renshaw and her colleagues wondered how the epaulette shark, a reef inhabitant in the waters of eastern Australia, could survive lengthy periods of anoxia. At low tides these animals can become trapped in the shallow waters and occasionally are left completely exposed to the air, a situation that would kill nonadapted fish in a matter of minutes. Unlike the situation in mammals, the epaulette shark employs a unique strategy to survive in these inhospitable conditions. In response to hypoxia (oxygen concentrations below 30% of normal) the shark does the following:

1. It reduces both its heart and respiratory rates.
2. It releases GABA (p. 133), an inhibitory neurotransmitter that counteracts the excitotoxicity of glutamate in nonessential brain areas.
3. It becomes completely inactive but retains its electroception sense to monitor the presence of any predators or prey.
4. Most importantly, it somehow inhibits the neuronal mitochondria, resulting in a savings in energy. Turning off the mitochondria effectively prevents the production of free radicals once oxygen is restored to the animal.

Unfortunately what works for the shark—GABA effectively blocking the toxic effects of glutamate—does not work in mammals. However, these studies again illustrate how animals can devise unique solutions to survive seemingly inhospitable conditions.

---

with regard to the formation of spatial memory and the development of cognitive skills. The hippocampus is a highly layered tissue derived by the infolding of the ventricles. These neurons are organized into interconnecting layers, which have a cellular appearance in histological sections and appear *granular*. Adjacent tissue is less organized and termed *agranular*—it is here where most cell proliferation occurs. Throughout the brain ventricular system there are residual stem/progenitor cells capable of regeneration. Like a conveyor belt, cells are generated in the agranular zone of the hippocampus and migrate into the granular zone where they differentiate into functional neurons. These newly differentiated cells undergo arborization to link with existing hippocampal cells, an event associated with the ability to form a memory, as illustrated when test animals are trained to more quickly find a reward in a maze. These inputs eventually result in a local increase in size of a particular region as evidenced by larger hippocampuses in animals that use food caches to store food and must recall their locations. Similarly, the hippocampus of experienced migratory birds is larger than that of comparably sized inexperienced individuals.

These newly derived cells become those most important in memory formation, and are accompanied by cell loss through apoptosis in the hippocampus, which appears to be required for the subsequent formation of new lineages. If this process is blocked, as occurs, for example, from excessive secretion of adrenal glucocorticoids in chronic stress or adrenal disease, depressive behavior and amnesia result. This can also be exemplified in dominant–subordinate relationships where reduced cognitive abilities, as exemplified by social depression, are associated with the subordinate animal. It is likely that older cells are triggered into apoptosis to provoke neurogenesis from the tissue stem/progenitor pool to maintain normal function. This implies that local signals communicate between cells; thus, there is a negative-feedback control between the differentiated cells and the stem/progenitor pool that ensures regeneration.

The second site of neurogenesis is in the *subventricular zone (SVZ)* associated with the lateral ventricles of the brain where major cell proliferation occurs. These cells migrate in a *rostral* direction (moving forwards in the brain) along defined axonal tracts to reach the olfactory epithelium where they form a regenerative population of sensory cells required for the detection of odors. Besides the hippocampus and SVZ, there is also the *hypothalamus*, where cell division occurs in relation to season and other environmental signals for the generation of long-term cycles, and the *spinal canal*,

Songbirds are noteworthy in that there is a constant turnover of neurons in certain regions of their brains. The canary's forebrain (which contains about 7 million neurons), adds approximately 20,000 new neurons each day to replace the ones that die. Within the forebrain the pathway for song production is anatomically distinct from the rest of the brain. Discrete song-responsive nuclei in the high vocal center (HVc) connect to the robust nucleus of the archistriatum (RA), which then connects to the motor neurons controlling the musculature of the syrinx and vocal tract muscles, in conjunction with the respiratory and postural systems. Male canaries sing only during the spring breeding season and then again in the fall as they begin learning a new song repertoire for the subsequent breeding season. During singing, the pattern of firing of neurons in the HVc can be associated with the production of syllables, that is, a group of notes. Each pattern of notes is associated with a stable and unique pattern of neuronal activity. In contrast, the RA exhibits precisely timed bursts of activity that are uniquely associated with note identity.

Canaries are sexually dimorphic, that is, the HVc and the RA are anatomically distinct between males and females. However, females can be induced to sing if they are administered the hormone testosterone. Associated with song production, both the HVc and RA in the treated females increase in size. Male songbirds with the largest HVc and RA are also the best singers and so are the most likely to attract a mate successfully during the breeding season. In another species, the marsh wren, there is a 40% greater size of the HVc and 30% greater size of the

A woodthrush, one of the most melodious of avian singers.

RA nucleus of males living along the west coast of North America compared to males living on the east coast. Associated with this, west coast wrens have song repertoires three times greater than their east coast cousins, most likely related to the greater competetion of west coast males for mates.

The neural circuitry associated with song learning is initially formed in response to both listening and practicing of the adult song. The HVc receives auditory input from the primary forebrain auditory nucleus. During this developmental period, neurogenesis is restricted to the song nuclei. Song development normally begins one month after hatch. At this time, the size of the HVc is approximately 10 to 15% of the adult HVc. In the fall, the number of RA-projecting neurons increase as the birds modify their song in preparation for the next breeding season. Even in old birds, new neurons have been shown to project an axon exactly the distance between the HVc and the RA, successfully linking these two brain regions. At the same time, some of the RA-projecting neurons die. This continual turnover of neurons indicates the remarkable plasticity of the songbird brain.

where cell division can sometimes repair neural damage due to injury. A remarkable regeneration of neurons has also been associated with the regulation of the *song control system* of songbirds (see *Challenges and Controversies: Neural Plasticity: A Song for All Seasons*).

Neurogenesis is an area of exciting new research in regenerative medicine. Long gone is the old dogma that there is no neural regeneration in the adult brain. Regeneration is required for normal function—remember, use it or lose it!

Mammalian neural tissue is susceptible to neurodegenerative disorders

Although the mammalian brain and other parts of the CNS are well protected, they can suffer from a number of damaging disorders, including degenerative diseases (in which neural tissue dies) such as Alzheimer's in humans. Among the most perplexing of these diseases are the *transmissible spongiform encephalopathies (TSE)*. The TSE family of disease includes *bovine spongiform encephalopathy (BSE),* commonly referred to as "mad cow disease"; *scrapie,* which affects sheep and goats; *transmissible mink encephalopathy* in minks; *feline spongiform encephalopathy* in cats; *chronic wasting disease* in deer and elk; and *kuru, classical,* and a *variant Creutzfeldt-Jacob disease* in humans (*vCJD*) that appears to be caused by the BSE agent following ingestion of infected

bovine neural tissue. BSE was first diagnosed in 1986 in Great Britain and first appeared in America in 2004. Chronic wasting disease has been found mainly in central western United States and Canada, especially in Colorado and Wyoming.

The agent responsible for BSE and other TSEs is smaller than the smallest known viruses. It is a highly stable protein called a **prion**, short for *proteinaceous infectious particle.* Like a clicker toy, prions can exist in two stable conformations, one of which is active or dominant while the other is recessive. The neuronal genes that code for prions give rise to the recessive form, which appears to have roles in normal brain functions. For example, mice with their prion genes knocked out (p. 34) cannot make proper myelin sheaths. Also, prionlike proteins may be important to long-term memory formation (as we will discuss later).

However, the recessive form of prions can be converted into the dominant form in the event of an animal consuming food that contains the active form. Once in the dominant form prions initiate a self-perpetuating conversion process that eventually is lethal for the animal. Neurons eventually become clogged with prions, which severely impair their function. Ultimately the assault leads to **spongiform damage**— a stipple of microscopic holes that accumulate as the brain reacts to the infection.

Unfortunately, the BSE agent is extremely resistant to heat and is partially resistant to proteases (enzymes that degrade other proteins), and does not evoke any detectable immune responses or inflammatory reactions in the infected animal. The incubation period (the time from when an animal becomes infected until it first shows disease signs) ranges from two to eight years. Cattle affected by BSE experience progressive degeneration of the nervous system, which leads to nervousness or aggression, abnormal posture, lack of coordination, and low milk production. Following the onset of clinical signs, the animal's condition steadily declines until it either dies or is destroyed. This period normally takes from two to six months. At present there is no treatment for the disease or vaccine to prevent it.

The BSE agent has been identified only in brain tissue, spinal cord, and retina. In experimentally infected cattle, BSE has been localized in the dorsal root ganglion, trigeminal ganglion, distal ileum, and the bone marrow, although there is no evidence of infection in milk or in muscle tissue.

check your understanding 5.3

What structures play a role in protecting vertebrate central nervous tissue?

How are brain capillaries structurally different from systemic capillaries?

Define *neural plasticity*. Other than neurons lining the nasal cavity and those in taste buds, can adult vertebrate neurons divide? Explain.

5.4 Brain Evolution in Vertebrates

At the beginning of this chapter, we discussed brain evolution across the animal kingdom. Here we examine the specific details for vertebrates.

Newer, more sophisticated regions of the vertebrate brain are piled on top of older, more primitive regions

Even though the vertebrate brain is a functional whole, it is organized into several different regions. The parts of the brain can be arbitrarily grouped in various ways based on anatomic distinctions, functional specialization, and evolutionary origin. A vertebrate brain is classically organized into three distinct regions—the hindbrain, midbrain, and forebrain—which form from three successive portions of the embryonic neural tube (Figure 5-15a). But these do not readily follow functional aspects of the brain. Therefore, we use the following regions (Table 5-6):

1. Brainstem
   a. Medulla oblongata
   b. Pons
   c. Midbrain
2. Cerebellum
3. Forebrain (with Limbic System)
   a. Diencephalon: i. Hypothalamus; ii. Thalamus
   b. Cerebrum: i. Basal nuclei; ii. Cerebral cortex

The order in which these components are listed generally represents both their anatomic location (from bottom to top in upright animals, or back to front in horizontal animals) and their complexity and sophistication of function (from the least specialized to the most specialized level). As we noted earlier in this chapter, the more sophisticated layers of the brain were added on to the more primitive layers, so the order in the list also represents an evolutionary sequence of oldest to newest. The fundamental regional structure of the brain remains similar between all modern vertebrates, with the exception of jawless vertebrates. The relatively conservative nature of these major brain divisions suggests that much of the organization of these structures must have arisen with the origin of vertebrates or shortly thereafter. Let's examine the brain regions and their evolution in more detail.

Brainstem The **brainstem**, the smallest and least changed region of the vertebrate brain, is continuous with the spinal cord (Table 5-6). It is the most ancient part of the brain, having evolved approximately 500 million years ago. It encompasses the **medulla oblongata** (also called just the *medulla*) and **pons**: brain centers that control many life-sustaining processes, such as breathing and circulation, as well as certain baseline activities of skeletal muscle that are common to all vertebrates. In teleosts (which are the majority of bony fishes), one notable specialization is a region devoted to taste, whereas in electric fishes there is an area that governs activation of the electric organs. The anterior brainstem is the **midbrain**, which helps coordinate reflex responses to sight and sound.

Cerebellum Attached at the top rear portion of the brainstem is the **cerebellum**, another structure that has changed comparatively little with evolution. Only the jawless vertebrates, hagfishes and lampreys, lack a cerebellum. A cerebellum-like structure composed of a central body and paired lobes arose with the origin of the jawed fishes. The cerebellum is concerned with maintaining proper position of the body in space and coordination of motor activity (movement). For example, signals from the balance organs and the auditory and visual systems are integrated in the cerebellum and the information used to orient an animal relative to its surroundings. Flying mammals (such as bats) and birds have a comparatively large cerebellum because of the complexity associated with flight.

Forebrain Using an ice cream cone as an analogy with the brainstem and cerebellum as the "cone," the **forebrain** sits anteriorly like layered scoops of ice cream. The forebrain has changed more than other regions of the brain during vertebrate evolution (Figure 5-15b). It has two major subdivisions—an inner **diencephalon** and an outer **cerebrum** "scoops." The diencephalon houses two regions of specific nuclei: the **hypothalamus**, which controls many homeostatic functions important in maintaining stability of the internal environment, and the **thalamus**, which performs some primitive sensory processing and is a relay station for almost all sensory input and for some outgoing commands. In fishes and amphibians, the forebrain is primarily devoted to olfactory input.

The cerebrum, the outermost "scoop," becomes progressively larger and more highly convoluted (that is, with tortuous ridges delineated by deep grooves or folds) the more

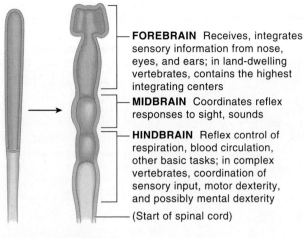

FOREBRAIN Receives, integrates sensory information from nose, eyes, and ears; in land-dwelling vertebrates, contains the highest integrating centers

MIDBRAIN Coordinates reflex responses to sight, sounds

HINDBRAIN Reflex control of respiration, blood circulation, other basic tasks; in complex vertebrates, coordination of sensory input, motor dexterity, and possibly mental dexterity

(Start of spinal cord)

(a) Expansion of the dorsal, hollow nerve cord into more complex, functionally distinct regions in certain lineages

FIGURE 5-15 Evolutionary trend toward an expanded, more complex vertebrate brain. The trend became apparent after morphological comparisons were made of the brains of some vertebrates. These dorsal views are not to the same scale.

Source: C. Starr & R. Taggart (2004), *Biology: The Unity and Diversity of Life*, 10th ed. (Belmont, CA: Brooks/Cole), Figure 34.15.

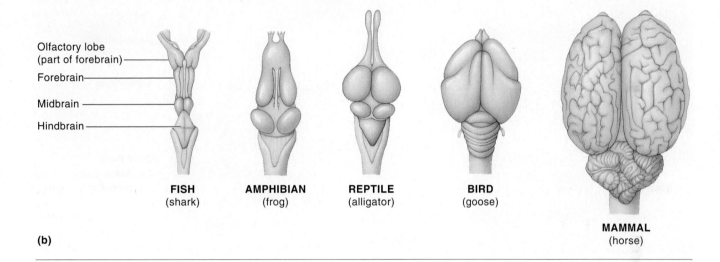

Olfactory lobe
(part of forebrain)

Forebrain

Midbrain

Hindbrain

FISH
(shark)

AMPHIBIAN
(frog)

REPTILE
(alligator)

BIRD
(goose)

MAMMAL
(horse)

(b)

advanced the vertebrate species is. The number of interconnections also increases; for example, in the cortex of the primate brain, each neuron can receive inputs from as many as 10,000 other neurons. The cerebrum is most highly developed in humans, where it constitutes about 80% of the total brain weight. The outer layer of the cerebrum is the highly convoluted **cerebral cortex**, which caps an inner core called the **basal nuclei**. The myriad convolutions of the cerebral cortex in higher mammals give it the appearance of a much-folded walnut (Figure 5-15b). The cortex is perfectly smooth in many lower mammals. Without these surface wrinkles, the primate cortex would take up to three times the area it does, and accordingly, would not fit like a cover over the underlying structures. The **hyperpallium** (formerly referred to as the hyperstriatum) of birds is comparable with the mammalian cerebral cortex and is most highly developed in parrots, crows and passerines. The cortex plays a key role in the most sophisticated neural functions, with three key regions we will explore in depth later: (1) the **motor cortex**, which initiates nonreflex movements; (2) five **sensory cortexes**, which deal with sensory perception; and (3) **association cortexes**, the

sites of complex memory, integration, and planning, as well as (in some species) self-awareness, language, and personality traits. It is the highest, most complex integrating area of the brain. The forebrain also houses the **limbic system**, an ancient memory and emotional complex consisting of parts of the cerebral cortex, basal ganglia, and diencephalon.

Each of these regions of the vertebrate brain is discussed in turn, following evolution by starting with the lowest level, the spinal cord, and moving to the highest, the cerebral cortex.

check your understanding 5.4

Which vertebrate brain structure has changed the least over evolutionary time? Why do you think this is so?

Which region of the brain has changed the most over evolutionary time in certain mammals? Why do you think this is so?

TABLE 5-6 Overview of Structures and Functions of the Major Components of the Mammalian Brain (Shown for a Human)

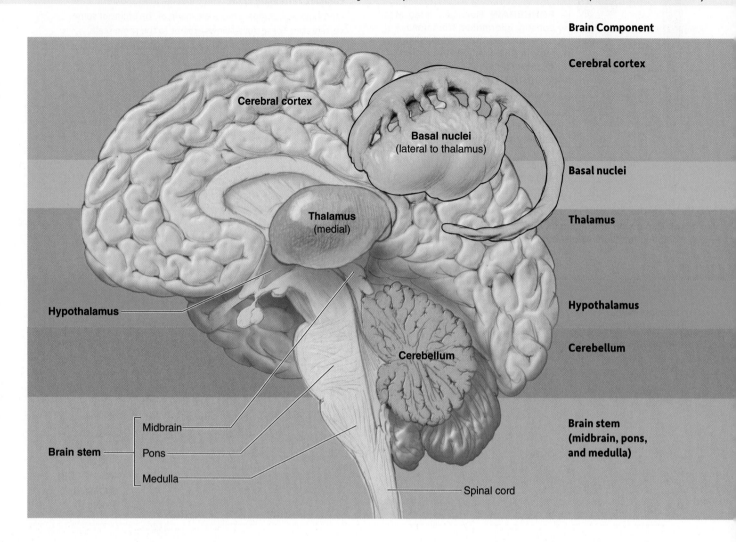

Brain Component

Cerebral cortex

Basal nuclei

Thalamus

Hypothalamus

Cerebellum

Brain stem (midbrain, pons, and medulla)

5.5 Spinal Cord

We begin with the oldest component of the CNS, the spinal cord.

The spinal cord retains an inherent segmental organization characteristic of many nonvertebrates

The **spinal cord** is a long, slender cylinder of nerve tissue that extends from the brainstem. Comparable to the organization of segmented ganglia in some nonvertebrates, the spinal cord and its associated nerves have retained a fundamental segmental organization. In humans, it is about 45 cm long (18 inches) and 2 cm in diameter (about the size of your thumb). Exiting through a large hole in the base of the skull, the spinal cord is enclosed by the protective vertebral column as it extends posteriorly through the vertebral canal. Paired *spinal nerves* emerge from the spinal cord through spaces formed between the bony, winglike arches of adjacent vertebrae (Figure 5-16). The actual number of spinal cord segments differs between species, with elongated animals having a greater number of spinal cord segments, whereas short animals have fewer.

The spinal nerves are named according to the region of the vertebral column from which they emerge: Humans have 8 pairs of **cervical** (neck) nerves (that is, C1–C8), 12 **thoracic** (chest) nerves, 5 **lumbar** (abdominal) nerves, 5 **sacral** (pelvic) nerves, and 1 **coccygeal** (tailbone) nerve. The spinal cord extends only to the level of the first or second lumbar vertebra (about waist level in humans), so the nerve roots of the remaining nerves are greatly elongated in order to exit the vertebral column at their appropriate space. The thick bundle of elongated nerve roots within the lower vertebral canal is known as the **cauda equina** ("horse's tail") because of its appearance. Thoracic and lumbar regions of the avian spinal cord are highly fused because of their adaptation for flight. For this reason individual vertebrae and their associated ganglia are not numbered.

The white matter of the spinal cord is organized into tracts

Although there are some slight regional variations within a particular species, the cross-sectional anatomy of the spinal cord is generally the same throughout its length (Figure 5-17). However, among different types of vertebrates the

Major Functions

1. Sensory perception
2. Voluntary control of muscle
3. Language
4. Personality traits
5. Sophisticated mental events, such as thinking, memory decision making, creativity, and self-consciousness

1. Inhibition of muscle tone
2. Coordination of slow, sustained movement
3. Supression of useless patterns of movement

1. Relay station for all synaptic input
2. Crude awareness of sensation
3. Some degree of consciousness
4. Role in motor control

1. Regulation of many homeostatic functions, such as temperature control, thirst, urine output, and food intake
2. Important link between nervous and endocrine systems
3. Extensive involvement with emotion and basic behavioral patterns

1. Maintenance of balance
2. Enhancement of muscle tone
3. Coordination and planning of skilled voluntary muscle activity

1. Origin of majority of peripheral cranial nerves
2. Cardiovascular, respiratory, and digestive control centers
3. Regulation of muscle reflexes involved with equilibrium and posture
4. Reception and integration of all synaptic input from spinal cord; arousal and activation of cerebral cortex
5. Role in sleep–wake cycle

© Cengage Learning, 2013

size and number of cord segments is related to the particular locomotory strategies used. For example, species with well-developed limbs show distinct enlargements of the thoracic and lumbar regions, reflecting the greater number of neurons within the spinal cord at these levels. In contrast, snakes (which have no limbs) lack any enlargement in this region of the spinal cord.

Throughout the entire CNS, **gray matter** consists predominantly of densely packaged cell bodies and their dendrites as well as glial cells. Bundles or **tracts** of myelinated nerve fibers (axons) constitute the **white matter**; the white appearance is due to the lipid (fat) composition of the myelin. The gray matter can be viewed as the "computers" of the CNS and the white matter as the "wires" that connect the computers to each other. Integration of neural input and initiation of neural output take place at synapses within the gray matter. The gray matter in the spinal cord forms a butterfly-shaped region on the inside and is surrounded by the outer white matter (Figure 5-17). The tracts are grouped into columns that extend the length of the cord. Each of these begins or ends within a particular area of the brain, and each is specific in the type of information that it transmits. Some are **ascending** (cord to brain) **tracts** that transmit to the brain signals derived from afferent input. Others are **descending**

(brain to cord) **tracts** that relay messages from the brain to efferent neurons (Figure 5-18).

The tracts are generally named for their origin and termination. For example, the **ventral spinocerebellar tract** is an ascending pathway that originates in the spinal cord and runs up the ventral (toward the front) margin of the cord with several synapses along the way until it eventually terminates in the cerebellum (Figure 5-18a). This tract carries information derived from muscle stretch receptors that has been delivered to the spinal cord by afferent fibers for use by the spinocerebellum. In contrast, the **ventral corticospinal tract** is a descending pathway that originates in the motor region of the cerebral cortex, then travels down the ventral portion of the spinal cord, and terminates in the spinal cord on the cell bodies of efferent motor neurons supplying skeletal muscles (Figure 5-18b). Because various types of signals are carried in different tracts within the spinal cord, damage to particular areas of the cord can interfere with some functions while other functions remain intact.

Each horn of the spinal cord gray matter houses a different type of neuronal cell body

The centrally located gray matter is also functionally organized (Figure 5-19). The central canal, which is filled with CSF, lies in the center of the gray matter. Each half of the gray matter is divided into a **dorsal** (**posterior**) (toward the back) horn, a **ventral** (**anterior**) **horn**, and a **lateral horn**. The dorsal horn contains cell bodies of interneurons on which afferent neurons terminate. The ventral horn contains cell bodies of the efferent motor neurons supplying skeletal muscles. Autonomic nerve fibers supplying cardiac and smooth muscle and exocrine glands originate at cell bodies found in the lateral horn.

Spinal nerves contain both afferent and efferent fibers

Spinal nerves (bundles of axons) connect with each side of the spinal cord by a **dorsal root** and a **ventral root** (Figure 5-17). Afferent fibers carrying incoming signals enter the spinal cord through the dorsal root; efferent fibers carrying outgoing signals leave through the ventral root. The cell bodies for the afferent neurons at each level are clustered together in a **dorsal root ganglion**. (A collection of neuronal cell bodies or a group of interneurons in the PNS is called a *ganglion*, whereas a functional collection of cell bodies within the CNS is referred to as a *center* or a *nucleus*.) The cell bodies for the efferent neurons originate in the gray matter and send axons out through the ventral root.

The dorsal and ventral roots at each level join to form a **spinal nerve** that emerges from the vertebral column (Figure 5-17). A spinal nerve contains both afferent and efferent fibers traversing between a particular region of the body and the spinal cord. Note the difference between a *nerve* and a *neuron*. A nerve is a bundle of peripheral neuronal axons, some afferent and some efferent, enclosed by a connective tissue covering and following the same pathway (Figure 5-20). A nerve does not contain a complete nerve cell, only the axonal portions of many neurons. (By this definition, there are no nerves in the CNS! Bundles of axons in the CNS are called *tracts*.) The individual fibers within a nerve generally do not have any direct influence on each other. They travel together for convenience, just as many individual tele-

FIGURE 5-16 Location of the spinal cord relative to the vertebral column.

Source: Adapted from C. Starr & R. Taggart, *Biology: The Unity and Diversity of Life*, 8th ed., Figure 35.9a, p. 577. Copyright 1998 Wadsworth Publishing Company.

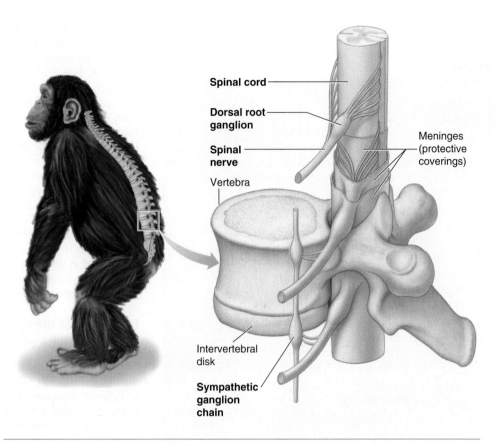

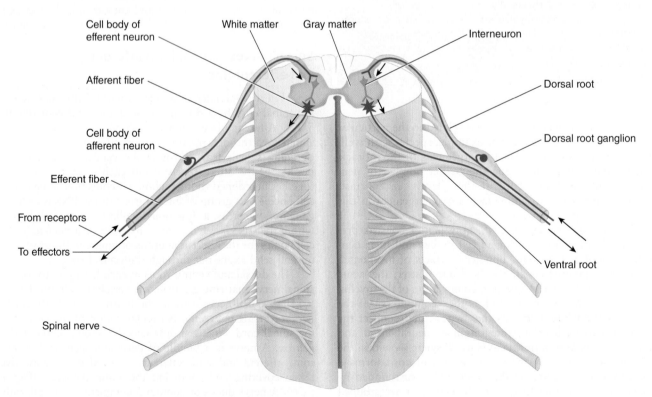

FIGURE 5-17 Spinal cord in cross section. Schematic representation of the spinal cord in cross section showing the relationship between the spinal cord and spinal nerves. The afferent fibers exit through the dorsal root, and the efferent fibers exit through the ventral root. Afferent and efferent fibers are enclosed together within a spinal nerve.

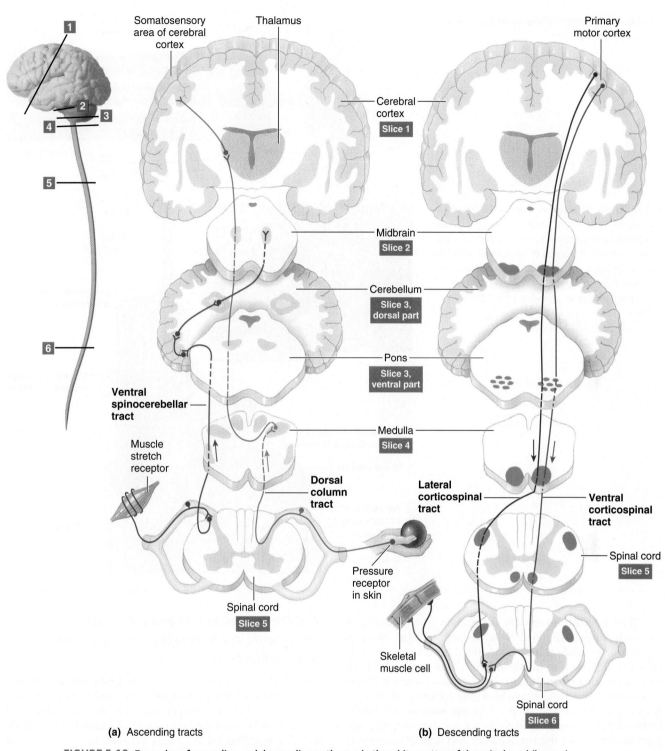

(a) Ascending tracts **(b)** Descending tracts

FIGURE 5-18 Examples of ascending and descending pathways in the white matter of the spinal cord (human).
(a) Cord-to-brain pathways of several ascending tracts (a dorsal column tract and ventral spinocerebellar tract).
(b) Brain-to-cord pathways of several descending tracts (lateral corticospinal and ventral corticospinal tracts).
© Cengage Learning, 2013

phone lines are carried within a telephone cable, yet any particular connection can be private without interference or influence from other lines in the cable.

The spinal nerves, along with the cranial nerves that arise from the brain, constitute the peripheral nervous system (PNS) that we discussed earlier. It is important to reiterate that the PNS is not a separate nervous system: Although the PNS entails axons from sensory and CNS neurons, it is structurally continuous with the CNS. After they emerge, these spinal nerves progressively branch to form a vast network of peripheral nerves that supply the tissues. Each segment of the spinal cord gives rise to a pair of spinal nerves that ultimately supply a particular region of the body with both afferent and efferent fibers.

FIGURE 5-19 **Regions of the spinal gray matter.**

© Cengage Learning, 2013

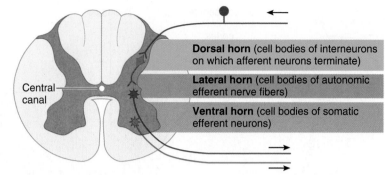

Central canal

Dorsal horn (cell bodies of interneurons on which afferent neurons terminate)

Lateral horn (cell bodies of autonomic efferent nerve fibers)

Ventral horn (cell bodies of somatic efferent neurons)

FIGURE 5-20 **Structure of a nerve.** Neuronal axons (both afferent and efferent fibers) are bundled together into connective-tissue-wrapped fascicles. A nerve consists of a group of fascicles enclosed by a connective tissue covering and following the same pathway. The photograph is a scanning electron micrograph of several nerve fascicles in cross section.

Dr. R. G. Kessel and Dr. R. H. Kardon/ Visuals Unlimited.

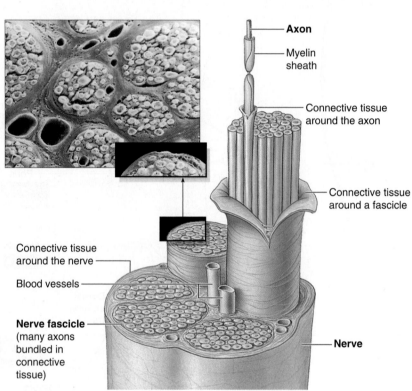

Axon

Myelin sheath

Connective tissue around the axon

Connective tissue around a fascicle

Connective tissue around the nerve

Blood vessels

Nerve fascicle (many axons bundled in connective tissue)

Nerve

With reference to sensory input, each specific region of the body surface supplied by a particular spinal nerve is called a **dermatome**. These same spinal nerves also carry fibers that branch off to supply internal organs, and sometimes pain originating from one of these organs is "referred" (that is, perceived as arising from the wrong location) to the corresponding dermatome supplied by the same spinal nerve. **Referred pain** originating in the heart, for example, may appear to come from the left shoulder and arm in humans or from the withers region (on the back at the base of the neck) in a ruminant. Pain appears to arise from the withers in response to the penetration of a nail, for example, through the ruminant gut into the pericardium (p. 403), resulting in pericardial inflammation called **traumatic pericarditis**. This is relatively common in cattle allowed to graze in old pastures. The mechanism responsible for referred pain is not completely understood. Presumably, inputs arising from the heart share a pathway to the brain in common with inputs from the upper extremities. The higher perception levels, being more accustomed to receiving sensory input from the withers than from the heart, may interpret the pain input from the heart as having arisen from the withers.

Many reflex responses and patterned movements in vertebrates are integrated in the spinal cord

The spinal cord's location between the brain and afferent and efferent fibers of the PNS enables the cord to fulfill its two primary functions: (1) serving as a link for transmission of information between the brain and the remainder of the body and (2) integrating reflex activity between afferent input and efferent output without involving the brain. This type of reflex activity is known as a **spinal reflex**. Often neuronal connections that produce patterned or stereotyped movements find their origins in the spinal cord. For example, as we noted earlier, a rudimentary walking rhythm is retained in chickens in the absence of any input from the brain. In fishes, the spinal cord can mediate near-normal swimming and visceral functions. Although the brain can override specific behavioral patterns that originate from the spinal cord, the complexity of the spinal neuron circuitry is sufficient to generate these complex and coordinated behaviors.

As you saw at the beginning of this chapter, a **reflex** is an innate response that occurs without higher neural func-

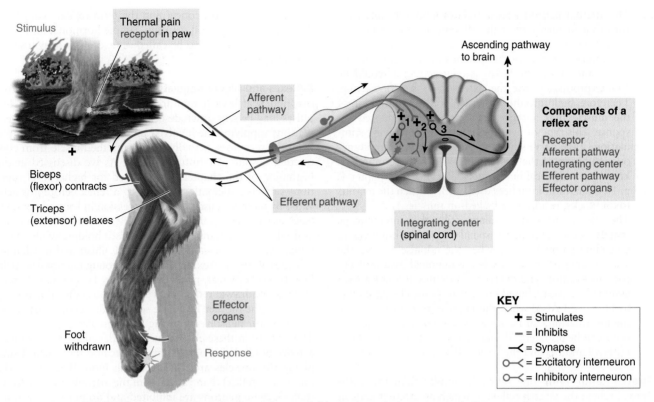

FIGURE 5-21 The withdrawal reflex. When a painful stimulus activates a receptor in the paw, action potentials are generated in the corresponding afferent pathway, which propagates the electrical signals to the CNS. Once the afferent neuron enters the spinal cord, it diverges and terminates on three different types of interneurons (only one of each type is depicted): (1) excitatory interneurons, which in turn stimulate the efferent motor neurons to the biceps, causing the leg to flex and pull the foot away from the painful stimulus; (2) inhibitory interneurons, which inhibit the efferent motor neurons to the triceps, thus preventing counterproductive contraction of this antagonistic muscle; and (3) interneurons that carry the signal up the spinal cord via an ascending pathway to the brain for awareness of pain, memory storage, and so on.

tions (anticipation and learning), such as pulling a limb away from a hot object (Figure 5-1e) or the scratch reflex so elegantly performed by a pet dog. Some physiologists suggest that some responses that are acquired, such as hunting or flying skills, should be called conditioned reflexes, as long as they become "automatic." However, this is misleading, because these learned reactions usually involve higher brain regions such as the motor cortex and cerebellum, and this suggestion ignores the importance of hierarchic control in neural evolution. In some cases, though, some true reflexes through the spine, brainstem, or hypothalamus may improve through *potentiated* synapses, in which case these can be considered true acquired reflexes. We explore potentiation later in the chapter.

Recall that the neural pathway involved in accomplishing reflex activity is the **reflex arc** (Figure 5-1). The spinal cord and brainstem are responsible for integrating basic reflexes, whereas higher brain levels usually process so-called acquired reflexes. The integrating center processes all information available to it from the receptor as well as from all other inputs, then "makes a decision" about the appropriate response. Unlike higher-level behaviors, in which any one of a number of responses is possible, a reflex response is predictable because the pathway between the receptor and effector is always the same.

Withdrawal Reflex Now that you have learned more about vertebrate nervous systems, let us revisit the withdrawal spinal reflex of Figure 5-1e in more detail in Figure 5-21. When an animal touches a hot object (such as an ember following a forest fire), a reflex is initiated to pull the limb away from the object (to withdraw from the painful stimulus). The skin has different receptors for warmth, cold, light touch, pressure, and pain. Even though all information is sent to the CNS by way of action potentials, the CNS can distinguish between various stimuli because specific receptors and consequently different afferent pathways are activated by different stimuli. When a receptor is stimulated sufficiently to reach threshold, an action potential is generated in the afferent neuron. The stronger the stimulus, the greater the frequency of action potentials generated and propagated to the CNS. Once the afferent neuron enters the spinal cord, it diverges to synapse with the following different interneurons (the numbers correspond to those on Figure 5-21).

1. An excited afferent neuron stimulates excitatory interneurons that in turn stimulate the efferent motor neurons supplying the biceps, the muscle in the limb that bends the knee joint. The resultant contraction of this flexor pulls the foot away from the hot ember.

2. The afferent neuron also stimulates inhibitory interneurons that in turn inhibit the efferent neurons supplying the triceps to prevent it from contracting. This extensor is the muscle in the leg that extends (straightens out) the knee joint. When the flexor is contracting, it would be counterproductive for the extensor to be contracting. Therefore, built into the withdrawal reflex is inhibition of the muscle that antagonizes (opposes) the desired response. This type of neuronal connection involving stimulation of the nerve supply to one muscle and simultaneous inhibition of the nerves to its antagonistic muscle is known as **reciprocal innervation**. (Recall from Chapter 4, p. 143, that inhibition here uses GABA, which the tetanus toxin blocks, resulting in locked-up muscles.)

3. The afferent neuron stimulates still other interneurons that carry the signal up the spinal cord to the brain via an ascending pathway. Only when the impulse reaches the sensory area of the cortex is the mammal aware of the pain, its location, and the type of stimulus. (Or at least we assume it is aware, based on their responses being similar to humans'.) Also, when the impulse reaches the brain, the information can be stored as memory, and other actions can be initiated. All this activity at the cortical level is above and beyond the basic reflex.

Stretch Reflex Only one reflex is simpler than the withdrawal reflex: the **stretch reflex**, in which an afferent neuron originating at a stretch-detecting receptor in a skeletal muscle terminates directly on the efferent neuron supplying the same skeletal muscle to cause it to contract and counteract the stretch. This is the **monosynaptic** (*mono,* "one") reflex of Figure 5-1b, so named because the only synapse in the reflex arc is the one between the afferent neuron and the efferent neuron. The withdrawal reflex and all other reflexes are **polysynaptic** (*poly,* "many"), because interneurons are interposed in the reflex pathway, and, therefore, a number of synapses are involved. The withdrawal reflex is an example of a polysynaptic basic spinal reflex.

Other Reflex Activity Spinal reflex action is not necessarily limited to motor responses on the side of the body to which the stimulus is applied. Assume an animal steps on a sharp rock fragment. The ensuing reflex response includes the same neurons as those used for the withdrawal reflex but also recruits neurons that control extension of the opposite limb. A reflex arc is initiated to withdraw the injured foot from the painful stimulus, while the opposite leg simultaneously prepares to suddenly bear the extra weight so that the animal does not lose balance (Figure 5-22). Unimpeded bending of the injured extremity's knee is accomplished by concurrent reflex stimulation of the muscles that flex the knee and inhibition of the muscles that extend the knee. At the same time, unimpeded extension of the opposite limb's knee is accomplished by activation of pathways that cross over to the opposite side of the spinal cord to reflexly stimulate this knee's extensors and inhibit its flexors. This **crossed extensor reflex** ensures that the opposite limb will be in a position to bear the weight of the body as the injured limb is withdrawn from the stimulus.

Besides protective reflexes such as the withdrawal reflex and simple postural reflexes such as the crossed extensor reflex, basic spinal reflexes also mediate emptying of pelvic organs (for example, urination, defecation, and expulsion of semen).

One final point, concerning the term *reflex:* Not all reflex activity is based on neurons. Simple hormonal and paracrine responses can also be called *reflexes,* as you will see in later chapters.

Reflexes and Higher Regulation Most spinal reflexes can be overridden at least temporarily by higher brain centers. Impulses may be sent down descending pathways to the efferent neurons supplying the involved muscles to override the input from the receptors, actually preventing the muscle from contracting despite a harmful stimulus. As we discussed at the beginning of the chapter, this evolved for higher-level functions such as anticipation (activating a reflex before it is activated by its own sensors) and suppression in learned contexts (such as suppressing the urination reflex when urination is not desired, or, in humans, conscious breath holding while swimming). The mechanisms involved were illustrated in Chapter 4 (Figure 4-20). For example, when a young mammal is suckling its mother's teat, pain receptors may be stimulated, initiating a withdrawal reflex. However, because the animal must be fed, in the mother IPSPs are sent via descending pathways to the motor neurons supplying the musculature of the limb. The activity in these efferent neurons depends on the sum of activity of all their synaptic inputs. Because the neurons supplying the muscles are now receiving more IPSPs from the brain (non-reflex) than EPSPs from the afferent pain pathway (reflex), these neurons are inhibited and do not reach threshold. Therefore, the muscles are not stimulated to contract and the mother remains stationary. In this way, the withdrawal reflex has been overridden.

Activity of command fibers elicits a fixed action pattern

Finally, there is another key example of higher regulation controlling lower neural circuits. Activation of certain **command fibers** located within the CNS of many animals (not just vertebrates) can innately generate *coordinated* behaviors of varying degrees of complexity, that is, **fixed action patterns**. Fixed action patterns are stereotyped behaviors that are generated in response to specific stimuli and, once activated, carry out a complete sequence of movements. They are often characteristic for a particular species or group of animals, and they involve more motions than the simple reflexes we have been considering. Because they are automatic and innate, these patterns are considered reflexes. Examples include mating dances and escape responses. For example, the giant Mauthner neurons (p. 129) in the brains of teleost fish activate muscles in a set sequence via the spine to generate the *startle response,* characterized by a vigorous muscle contraction on one side of the body that move the body quickly out of harm's way. There are two Mauthner cells located on each side of the brain. Tapping the glass of an aquarium creates a mechanical disturbance in the water, which generates sensory input into these cells. The neuron that first receives the signal sends inhibitory input to its counterpart Mauthner cell while at the same time activating muscles on one side of the body, permitting the escape response to occur.

Fixed action patterns are widespread among animals. For example, arthropods and mollusks have a variety of such patterns for feeding, flight, walking and gill ventilation. For example, the crayfish tail flip (in which the tail is repeatedly flipped for rapid swimming) is illustrative of a rapid escape

FIGURE 5-22 The crossed extensor reflex coupled with the withdrawal reflex. (a) The withdrawal reflex, which causes flexion of the injured extremity to withdraw from a painful stimulus. (b) The crossed extensor reflex, which extends the opposite limb to support the full weight of the body.

© Cengage Learning, 2013

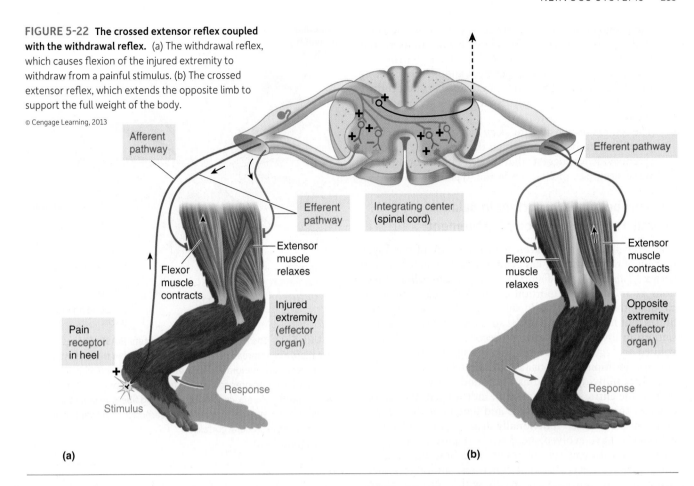

behavior. Because the tail flip is generated in specific giant neurons (medial giant and lateral giant neurons), these neurons are considered to function as command fibers. More complex behaviors such as the patterned movements insects use in molting their old exoskeletons are also driven by command fibers. *Ecdysis behavior* of grasshoppers and crickets involves a series of synchronous muscular contractions in each of the body segments, which loosens the old exoskeleton. Subsequently, the large muscles of the body wall contract in a peristaltic fashion from posterior to anterior, which then forces the animal out of the exoskeleton.

check your understanding 5.5

What is the difference between a nerve and a neuron?

Compare spinal reflexes and fixed action patterns.

Why is reciprocal innervation important in the withdrawal reflex?

5.6 Brainstem and Cerebellum

Now that we've examined the spine, let's turn to its connections to the brain.

The Brainstem Is a Critical Connecting Link between the Brain and the Spinal Cord All incoming and outgoing fibers traversing between the periphery and higher brain centers must pass through the brainstem (with exception of the olfactory and optic nerves), with incoming fibers relaying sensory information to the brain and outgoing fibers carrying command signals from the brain for efferent output. A few fibers merely pass through, but most synapse within the brainstem for important processing. The functions of the brainstem include the following:

1. *Sensation input and motor output in the head and neck via cranial nerves.* Twelve pairs of **cranial nerves** arise from the brainstem; most supply structures in the head and neck with both sensory and motor fibers (Figure 5-7). They are important in sight, hearing, taste, smell, sensation of the face and scalp, eye movement, chewing, swallowing, facial expressions, and salivation. An exception is cranial nerve X, the **vagus nerve** (in Figure 5-7), which innervates organs in the thoracic and abdominal cavities. The vagus is the major nerve of the parasympathetic nervous system.
2. *Reflex control of heart, blood vessels, respiration, and digestion.* Collected within the *pons* and *medulla* are integrating neuronal clusters that control these functions primarily for homeostasis (see Chapters 9 and 11).
3. *Modulating the sense of pain* (see p. 260).
4. *Regulation of muscle reflexes involved in equilibrium and posture.*
5. *Receiving and integrating all synaptic input via the reticular formation.* A widespread network of interconnected neurons called the **reticular formation** runs throughout the entire brainstem and into the thalamus. Ascending fibers originating in the reticular formation

carry signals upward to arouse and activate the cerebral cortex (Figure 5-23). These fibers compose the **reticular activating system (RAS)**, which controls the overall degree of cortical alertness and is important in the ability to direct attention. In turn, fibers descending from the cortex, especially its motor areas, can activate the RAS.

6. The centers responsible for *sleep* traditionally have been considered to be housed within the brainstem, although recent evidence suggests that the center promoting slow-wave sleep (see p. 204) lies in the hypothalamus.

The cerebellum is important in balance as well as in coordination of movement

The cerebellum, which is attached to the back of the upper portion of the brainstem, lies behind the occipital lobe of the cortex (Table 5-6 and Figure 5-24). The cerebellum functions as an integrative portion of the brain that consists of circuitry that is functionally similar in all classes of vertebrates that have this region. The surface of the cerebellum is smooth in lower vertebrates but becomes increasingly convoluted in higher vertebrates, increasing available surface area for additional neurons (as in the cerebrum).

More individual neurons are found in the cerebellum than in the entire rest of the brain, indicative of the importance of this structure. In birds and mammals the cerebellum consists of three functionally distinct parts, which are believed to have evolved successively (Figure 5-24). These parts have different sets of inputs and outputs, and, accordingly, each has different functions, although collectively they are concerned primarily with control of motor activity.

*   The **vestibulocerebellum** is important for the maintenance of balance and controls eye movement.
*   The **cerebrocerebellum** plays a role in planning nonreflex muscle activity by providing input to the cortical motor areas. This is also the cerebellar region that stores *procedural memories* (which we discuss later, p. 195).
*   The **spinocerebellum** enhances muscle tone and coordinates skilled, nonreflex movements. This brain region is especially important in ensuring the accurate timing of various muscle contractions to coordinate movements involving multiple joints. This region also receives input from peripheral receptors that appraise it of the body movements and positions that are actually taking place. The spinocerebellum essentially acts as "middle management," comparing the "intentions" or "orders" of the higher centers (especially the **motor cortex**) with the "performance" of the muscles and then correcting any "errors" or deviations from the intended movement (Figure 5-25). The spinocerebellum even appears able to predict the position of a body part in the next fraction of a second and to make adjustments accordingly.

The motor command for a particular nonreflex activity arises from the motor cortex, but coordination of the actual execution of that activity is accomplished by the cerebellum and certain subcortical regions, which we will examine next. You will learn more about motor control when we cover muscle physiology in Chapter 8.

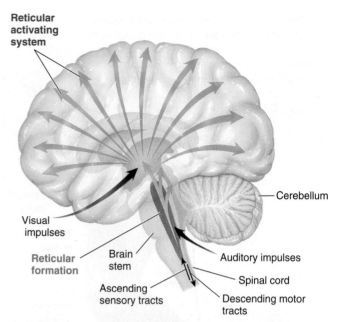

FIGURE 5-23 The reticular activating system (RAS). The reticular formation, a widespread network of neurons within the brain stem (in *red*), receives and integrates all synaptic input. The reticular activating system, which promotes cortical alertness and helps direct attention toward specific events, consists of ascending fibers (in *blue*) that originate in the reticular formation and carry signals upward to arouse and activate the cerebral cortex.

© Cengage Learning, 2013

check your understanding 5.6

What are the key functions of the brainstem?

What are the key functions of the cerebellum as a coordinator and predictor?

5.7 Basal Nuclei, Thalamus, Hypothalamus, and Limbic System

We now move up to the **subcortical** ("under the cortex") **regions** of the brain, which interact extensively with the cortex in the performance of their functions. These regions include the *basal nuclei,* located in the cerebrum, and the *thalamus* and *hypothalamus,* located in the diencephalon. The *limbic system* includes portions of these structures, as well as cortical regions.

The basal nuclei play an important inhibitory role in motor control

The **basal nuclei** (also known as **basal ganglia**) consist of several masses of gray matter located deep within the cerebral white matter (Table 5-6). In the nervous system, a **nucleus** (plural, **nuclei**) refers to a functional aggregation of neuronal cell bodies (thus the term *ganglion* is probably more appropriate, but *nucleus* is unfortunately more widely

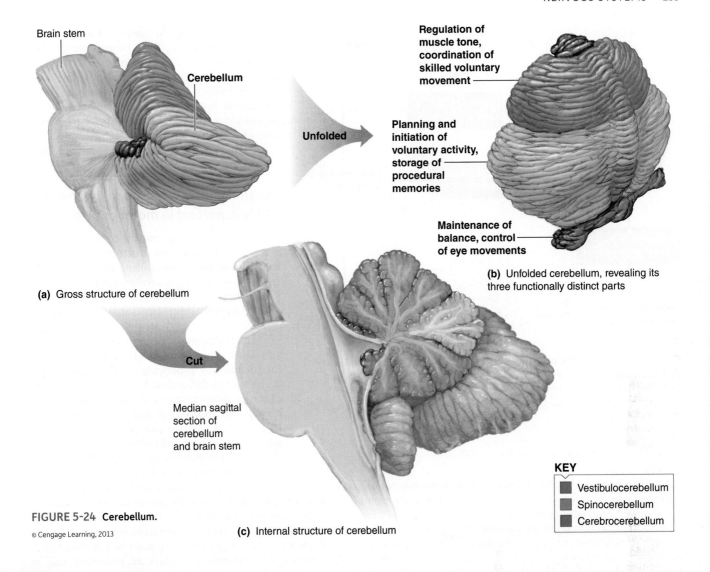

Brain stem

Cerebellum

Unfolded

Regulation of
muscle tone,
coordination of
skilled voluntary
movement

Planning and
initiation of
voluntary activity,
storage of
procedural
memories

Maintenance of
balance, control
of eye movements

(b) Unfolded cerebellum, revealing its
three functionally distinct parts

(a) Gross structure of cerebellum

Cut

Median sagittal
section of
cerebellum
and brain stem

FIGURE 5-24 Cerebellum.

© Cengage Learning, 2013

(c) Internal structure of cerebellum

KEY

Vestibulocerebellum
Spinocerebellum
Cerebrocerebellum

Challenges and Controversies
Is the Cerebellum a "Smith Predictor"?

The cerebellum has long been known to be important for smooth, coordinated muscle movements, but scientists still do not entirely understand how it does this. One hypothesis that explains its actions well is that this brain region is a biological version of a Smith predictor. Smith predictors originated in industry for situations with feedback delays, such as a grasping robotic arm in a factory that crushes a soft object before feedback sensors can inform the controller that the object is soft. The Smith predictor has an "internal model" of the action it is about to take

that allows it to initiate certain motions *before* feedback information can inform whether it is correct or not. The cerebellum seems to act this way, particularly once it has learned a skilled motor task. Consider a child throwing a ball. Her higher (prefrontal) center triggers her arm to move forward then waits for feedback to tell it when the arm is in the correct position to release the ball. But by the time the release command arrives at the hand, the arm has moved too far, so the ball goes down at her feet. As the action is learned, though, her spinocerebellum will send the

release command just *before* her hand arrives at the proper angle to prevent this overshooting. Neural recordings suggest that the cerebellum makes such smoothing, coordinating commands even in the absence of feedback from sensors. These ongoing *anticipatory* adjustments are especially important for rapidly changing activities such as climbing, swinging through trees, and running, in which delays in sensory feedback would make the action clumsy. See the Suggested Reading by Miall, Weir, Wolpert, and Stein (1993) at the end of this chapter.

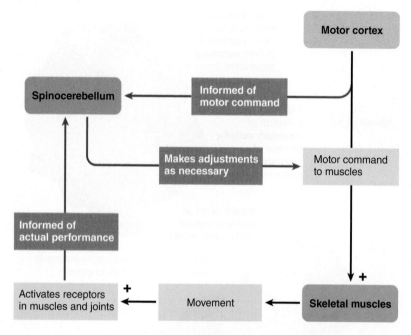

FIGURE 5-25 Role of the spinocerebellum in subconscious control of voluntary motor activity. In coordinating rapid, phasic motor activity, the spinocerebellum compares the "intentions" of higher motor centers with the "performance" of the muscles and corrects any "errors" by making the necessary adjustments to accomplish the intended movement.

© Cengage Learning, 2013

used). The basal nuclei play a complex role in the control of movement in addition to having nonmotor functions that are less well understood. In particular, the basal nuclei are important in (1) inhibiting muscle tone throughout the body (proper muscle tone is normally maintained by a balance of excitatory and inhibitory inputs to the neurons that innervate skeletal muscles); (2) selecting and maintaining purposeful motor activity while suppressing useless or unwanted patterns of movement; and (3) helping monitor and coordinate slow, sustained contractions, especially those related to posture and support. The basal nuclei do not directly influence the efferent motor neurons that bring about muscle contraction but act instead by modifying ongoing activity in motor pathways.

To accomplish these complex integrative roles, the basal nuclei receive and send out much information, as is indicated by the tremendous number of fibers linking them to other regions of the brain. One important pathway consists of strategic interconnections that form a complex feedback loop linking the cerebral cortex (especially its motor regions), the basal nuclei, and the *thalamus* (Table 5-6). The thalamus (which we'll explore further shortly) positively reinforces nonreflex motor behavior initiated by the cortex, whereas the basal nuclei modulate this activity by exerting an inhibitory effect on the thalamus to eliminate antagonistic or unnecessary movements. The basal nuclei also exert an inhibitory effect on motor activity by acting through neurons in the brainstem.

The cerebellum and basal nuclei both monitor and adjust motor activity commanded from the motor cortex, and like the cerebellum, the basal nuclei do not have any direct influence on the efferent motor neurons. Both function indi-

rectly by modifying the output of major motor systems of the brain. Even though the cerebellum and basal nuclei both help to coordinate skeletal motor activity, they play different roles. The cerebellum helps maintain balance; helps coordinate fast, phasic motor activity; and enhances muscle tone. The basal nuclei are important in coordinating slow, sustained movement related to posture and support, and they also function in inhibiting muscle tone.

The thalamus is a sensory relay station and is important in motor control

Deep within the brain near the basal nuclei is the diencephalon, a midline structure that forms the walls of the third ventricular cavity, one of the spaces through which cerebrospinal fluid flows. Recall that its two main parts are the thalamus and the hypothalamus (see Table 5-6 and Figure 5-26).

The **thalamus**, as we noted earlier, serves as a "relay station" and synaptic integrating center for preliminary processing of all sensory input on its way to the cortex. It screens out insignificant signals and routes the important sensory impulses to appropriate areas of the higher sensory cortexes that we will examine later. Along with the brainstem and higher cortical areas, the thalamus is important in an animal's ability to direct attention to stimuli of interest. For example, mammalian parents can sleep soundly through the noise of wind or rain but be instantly aware of their offspring's slightest whimper. The thalamus is also capable of crude awareness of various types of sensation but cannot distinguish their location or intensity. As described in the preceding section, the thalamus also plays an important role in motor control by positively reinforcing nonreflex motor behavior initiated by the cortex.

The hypothalamus regulates many homeostatic functions

The **hypothalamus** is a collection of specific nuclei and associated fibers that lie beneath the thalamus. It is an integrating center for many important homeostatic functions that we will explore in later chapters and serves as an important link between the autonomic nervous system and the endocrine system. Specifically, the hypothalamus (1) controls body temperature (although this ability varies considerably among vertebrates, as you will see in Chapter 15); (2) controls thirst and urine output (Chapters 12 and 13); (3) controls food intake (Chapter 15); (4) controls anterior pituitary hormone secretion; (5) produces posterior pituitary hormones (Chapter 7 and others); (6) controls uterine contractions and milk ejection (Chapter 16); (7) serves as a major autonomic nervous system coordinating center, which in turn affects all smooth muscle, cardiac muscle, and exocrine glands; (8) plays a role in emotional and behavioral patterns; and (9) participates in the sleep–wake cycle.

As this list indicates, the hypothalamus (along with the brainstem) is the area of the brain most notably involved in regulating the direct homeostasis of the internal environ-

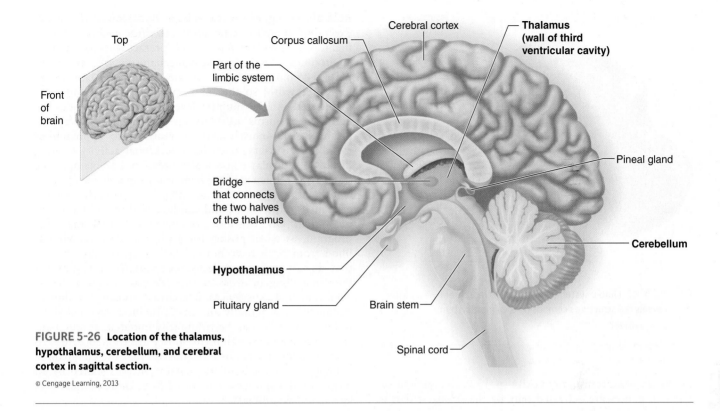

FIGURE 5-26 **Location of the thalamus, hypothalamus, cerebellum, and cerebral cortex in sagittal section.**

© Cengage Learning, 2013

ment. For example, when a mammalian body is cold, its hypothalamus initiates internal responses to increase heat production (such as shivering) and to decrease heat loss (such as constricting the skin blood vessels to reduce the flow of warm blood to the body surface, where heat could be lost to the external environment). Other areas of the brain, such as the cerebral cortex, act more indirectly to regulate the internal environment. For example, a cat that feels cold is motivated to seek a warm, sunny place to nap. Even these nonreflex behavioral activities are strongly influenced by the hypothalamus, which, as a part of the *limbic system* (see the next section), functions in conjunction with the cortex in controlling emotions and motivated behavior.

The hypothalamus represents an intermediate level in the distributed hierarchy of neural control and is considered a reflex integrator. That is, it receives input from various internal sensors and sends out effector commands with fewer synapses than would be required if the cerebrum were involved. The hypothalamus is also physically closer to body effectors than is the cerebrum.

The limbic system plays a key role in animal motivation and memory

The **limbic system** is not a separate structure but refers to a ring of forebrain structures that surround the brainstem and are interconnected by intricate neuronal pathways (Figure 5-27). The limbic system includes portions of each of the following: ancient parts of the cerebral cortex, the basal nuclei, the thalamus, and the hypothalamus (although in birds the limbic system remains poorly defined). This complex interacting network is associated with emotions, basic

survival and sociosexual behavioral patterns, motivation, and learning. Two ancient cortical regions we will explore in detail are the *hippocampus,* a vital center for memory processing (see Section 5.9), and the *amygdala.*

The amygdala processes and learns inputs that give rise to the sensation of fear

The **amygdala** of mammalian limbic system is located on the interior underside of the temporal lobe. It is believed to be the homologue of the reptilian and avian **paleostriatum.** It is an especially important region for processing inputs that give rise to the sensation of fear, a sense that normally helps animals avoid danger. Research studies have shown that the amygdala is the region where the link between an *unconditioned* stimulus and a *conditioned* stimulus is formed (see p. 198). For example, when a rat hears a sound and receives a shock, there is a pronounced increase in the firing rate of neurons in the amygdala. This leads to a strengthening of synaptic connections in this region, which, once laid down, are retained for the lifetime of the animal. The amygdala, in turn, activates the animal's fight-or-flight stress system that is controlled by the hypothalamus (p. 310), in anticipation of strenuous escape (or fight) activity. Thus, the next time the rat's ear receives that sound, the amygdala gears up the body for fight-or-flight even before the higher sensory centers "perceive" the sound. After the higher centers process the information, actual fight-or-flight behavior can either be initiated at a high rate (because the body is flooded with oxygen and glucose), or it can suppress the response if the information is judged to be nondangerous.

Normally, in the amygdala the circuits that learn these *emotional memories* are held in check by the release of the

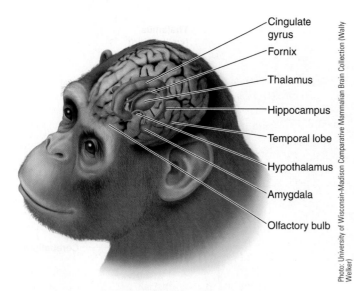

Cingulate gyrus
Fornix
Thalamus
Hippocampus
Temporal lobe
Hypothalamus
Amygdala
Olfactory bulb

Photo: University of Wisconsin-Madison Comparative Mammalian Brain Collection (Wally Welker)

FIGURE 5-27 Limbic system. This partially transparent view of the brain reveals the structures composing the limbic system.

Source: brainmuseum.org.

inhibitory neurotransmitter GABA (p. 133), so that fight-or-flight responses are activated only by the strongest danger signals. GABA is believed to permit the amygdala to filter out unthreatening stimuli. If, however, the amygdala fails to adequately filter enough incoming information, and, like a leaky valve, permits the entry of too much information, even mildly frightening stimuli can be perceived as terrifying. This syndrome is seen in overly anxious humans and domestic mammals.

The limbic system and higher cortex participate in the control of basic behavioral patterns

Basic innate (inborn) behavioral patterns controlled at least in part by the limbic system include those aimed at survival of the animal (attack, searching for food) and those directed toward perpetuation of the species (sociosexual behaviors conducive to mating). In experimental mammals, stimulation of the limbic system brings about complex and even bizarre behaviors. For example, stimulation in one area can elicit responses of anger and rage in a normally docile animal, whereas stimulation in another area results in placidity and tameness, even in an otherwise vicious animal. Stimulation in yet another limbic area can induce sexual behaviors such as copulatory movements in rodents.

The relationships among the hypothalamus, limbic system, and higher cortical regions regarding emotions and behavior are still not well understood. Apparently the extensive involvement of the hypothalamus in the limbic system governs the internal responses of various body systems in preparation for appropriate action to accompany a particular emotional state. For example, the hypothalamus controls the increased heart rate and respiratory rate, elevation of blood pressure, and diversion of blood to skeletal muscles that occur in anticipation of attack or when angered. These preparatory changes in internal state require no cerebral control (conscious in humans), and as you have seen, emotional memories in the amygdala can trigger them via the hypothalamus.

Role of the Higher Cortex in Basic Behavioral Patterns In executing complex behavioral activities such as attack, flight, or mating, an animal must interact with the external environment. Higher cortical mechanisms are called into play to connect the limbic system and hypothalamus with the outer world so that appropriate overt behaviors are manifested. At the simplest level, the cortex provides the neural mechanisms necessary for implementing the appropriate skeletal muscle activity required to approach or avoid an adversary, participate in sexual activity (mating display), or display emotional expression. For example, the stereotypic sequence of movement for the universal human emotional expression of smiling is apparently preprogrammed in the cortex and can be called forth by the limbic system. One can also voluntarily call forth the smile program, as when posing for a picture. Even individuals blind from birth have normal facial expressions; that is, they do not learn to smile by observation. Smiling means the same thing in every culture, despite widely differing environmental experiences. In contrast, so-called smiling in chimpanzees is actually an innate, limbic-controlled show of fear; but it too can be voluntarily initiated, as chimpanzees in human entertainment shows are trained to do, to make human viewers believe (falsely) that the chimpanzee is happy or amused. Such behavior patterns shared by all members of a species are believed to be more abundant in less advanced animals.

Motivated behaviors are goal directed

An animal tends to reinforce behaviors that have proved gratifying and to suppress behaviors that have been associated with unpleasant experiences. Certain regions of the limbic system have been designated as "reward" and "punishment" centers because stimulation in these respective areas gives rise to pleasant or unpleasant sensations in humans and responses that appear to evoke similar sensations in other mammals. When a self-stimulating device is implanted in a reward center, an experimental rodent will self-deliver up to 5,000 stimulations per hour and will even shun food when starving, in preference for the pleasure derived from self-stimulation. In contrast, when the device is implanted in a punishment center, animals will avoid stimulation at all costs. Reward centers are found most abundantly in regions mediating the highly motivated activities of eating, drinking, and sexual activity.

Motivation is the ability to direct behavior toward specific goals. The concept of animal motivation encompasses many factors, which include subjective emotional feelings and moods (such as rage and fear) that only human subjects can report, plus the overt physical responses that occur in association with these feelings. Animals must continually evaluate the value or worth of a particular stimulus and make decisions weighing the beneficial versus the painful consequences of their choices. These responses include specific behavioral patterns, for example, preparation for attack or defense when approached by an adversary. Evidence (from stimulating or inactivating various brain regions) points to a central role for the brainstem and the limbic system in all aspects of emotion. Some goal-directed behaviors are aimed at satisfying specific identifiable physical needs related to homeostasis. **Homeostatic drives** represent the subjective urges associated with specific bodily needs that motivate appropriate behavior to satisfy those needs. As an example, the sensation of thirst accompanying a water deficit in the body drives an animal to drink to satisfy the homeostatic need for water. Animal behavior is influenced by experience, learning, and habit, shaped in a complex framework of unique personal gratifications.

Norepinephrine, dopamine, and serotonin are neurotransmitters in pathways for emotion and behavior

The underlying neurophysiological mechanisms responsible for the psychological observations of motivated behavior and emotions largely remain a mystery, although the neurotransmitters norepinephrine (NE), dopamine, and serotonin all have been implicated. NE and dopamine, both chemically classified as *catecholamines,* are known to be transmitters in the regions that elicit the highest rates of self-stimulation in animals equipped with self-administering devices. Imbalances in these are associated with various mental disorders in humans, and possibly other mammals. For example, a functional deficiency of serotonin or NE or both is implicated in **depression**, a disorder characterized by a pervasive unpleasant mood accompanied by a generalized loss of interests and inability to experience pleasure. All effective antidepressant drugs increase the available concentration of these neurotransmitters in the CNS. *Prozac,* the most widely prescribed drug in American psychotherapy, is illustrative. It blocks the reuptake transporters (p. 133) for released serotonin, thus prolonging serotonin activity at synapses. It is often prescribed for humans and companion animals suffering from certain forms of depression and other behavior disorders (see the box, *Concepts and Controversies: Synaptic Solutions to Pet Problems* in Chapter 4, p. 144). Researchers are optimistic that as our understanding of the molecular mechanisms of mental disorders is expanded in the future, many psychological problems can be corrected or managed through drug intervention.

check your understanding 5.7

What role do the basal nuclei play in motor control?

Why is the hypothalamus important in the body's homeostatic functions?

Explain the roles that the limbic system has in reactions to fear and sociosexual behaviors related to mating.

5.8 Mammalian Cerebral Cortex

We now turn to the top of the ice cream cone. The cerebrum, by far the largest region of mammalian brains, is divided into two halves, the right and left **cerebral hemispheres**. They are connected to each other by the **corpus callosum**, a thick band consisting of an estimated 300 million neuronal axons traversing between the two hemispheres (Figure 5-26). The corpus callosum is the body's "information highway." The two hemispheres communicate and cooperate with each other by means of constant information exchange through this neural connection.

The cerebral cortex is an outer shell of gray matter covering an inner core of white matter

Each hemisphere consists of a thin outer shell of gray matter, the **cerebral cortex**, covering a thick central core of white matter, the cerebral medulla. The fiber tracts in the white matter transmit signals from one part of the cerebral cortex to another or between the cortex and other regions of the CNS. Such communication between different areas of the cortex and elsewhere facilitates integration of their activity. This integration of neural input is essential for even a relatively simple task such as selecting a food item to eat. Vision and smell of the food is received by one area of the cortex, reception of its palatability takes place in another area, and movement is initiated by still another area. More complex functions may also come into play; for example, a particular food item may have made an animal sick, and its memory therefore may make it reject this particular item even though it might be nutritious. How such complex processing actually works is still poorly understood.

The cerebral cortex is organized into layers and functional columns

The cerebral cortex is organized into six well-defined layers based on varying distributions of the cell bodies and locally associated fibers of several distinctive cell types. These layers are organized into functional vertical columns that extend perpendicularly about 2 mm from the cortical surface down through the thickness of the cortex to the underlying white matter. The neurons within a given column are believed to function as a "team," with each cell being involved in different aspects of the same specific activity—for example, perceptual processing of the same stimulus from the same location.

The functional differences between various areas of the cortex result from different layering patterns within the columns and from different input–output connections, not from the presence of unique cell types or different neuronal mechanisms. For example, those regions of the cortex responsible for perception of senses have an expanded layer 4, a layer rich in **stellate cells**, which are responsible for initial processing of sensory input to the cortex. In contrast, the cortical areas that control output to skeletal muscles have a thickened layer 5, which contains an abundance of large **pyramidal cells**. These cells send fibers down the spinal cord from the cortex to terminate on the efferent motor neurons that innervate the skeletal muscles. Pyramidal cells are thought to be the major neurons responsible for higher "decision" events, such as integrating multiple sensory inputs with memory and then outputting appropriate commands.

The four pairs of lobes in the cerebral cortex are specialized for different activities

We now consider the locations of the major functional areas of the mammalian cerebral cortex. Throughout this discussion, keep in mind that even though a discrete activity is ultimately attributed to a particular region of the brain, no part of the brain functions in isolation. Each part depends on complex interplay among numerous other regions for both incoming and outgoing messages.

FIGURE 5-28 **Major functional divisions of the mammalian cerebrum, shown for a rat, cat, chimpanzee, and human.** Regions devoted to *motor* control are colored red. Regions devoted to processing *sensory* information are colored as indicated in the legend boxes. Regions dedicated to "higher" *associative* functions are shaded in cream. Associative functions (those involved in complex memory, planning, and so on) have expanded in certain mammalian groups such as primates.

© Cengage Learning, 2013

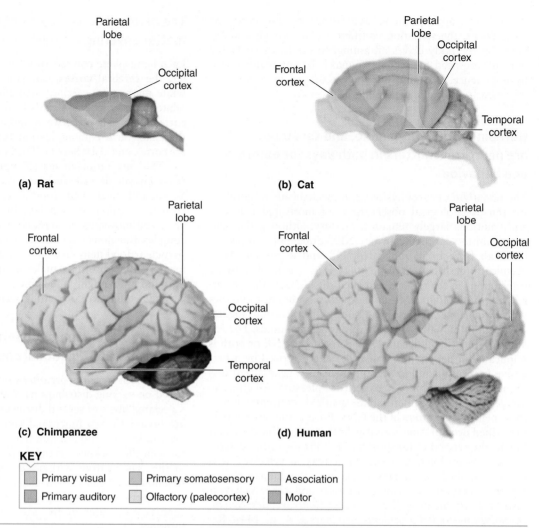

(a) Rat

(b) Cat

(c) Chimpanzee

(d) Human

KEY

▢ Primary visual	▢ Primary somatosensory	▢ Association
▢ Primary auditory	▢ Olfactory (paleocortex)	▢ Motor

The anatomic landmarks used in cortical mapping are certain deep folds that divide each half of the cortex into four major lobes: the *occipital, temporal, parietal,* and *frontal* lobes (Figure 5-28). Recall that there are three functional types of cortexes: *motor, sensory* and *association* (p. 175). Note in Figure 5-28 the relative sizes of these types in various mammals.

During the following discussion of the major activities attributed to various regions of these lobes, refer to the basic functional map of the human cortex in Figure 5-29a. Also think about the typical flow of information, as illustrated in Figure 5-29b for human language. First, as seen in the "Seeing" and "Hearing" images, sensory input arrives at a *sensory* cortex (from the thalamus) for initial perception, then the information goes to a nearby *association* area for processing (e.g., comparing new information with memory, or correlating two or more senses). Wernicke's area (Figure 5-29a) does this for language. Then information goes to the frontal association area for planning a response ("Thinking" image). Broca's area (Figure 5-29a) does this for speech. Finally, if action is decided upon, commands go to the *motor* cortex, which initiates muscle activity ("Speaking" image).

Now let's look at the sensory, motor and associative functions of the lobes. The **occipital lobes,** which are located posteriorly (at the back of the head), are a sensory and association area that carries out the initial perception and processing of visual input in the cortex. Similarly, auditory (sound) sensation is initially received and processed by the **temporal lobes,** located laterally (on the sides of the head) (Figure 5-29a and b). You will learn more about the functions of these regions in Chapter 6 when we discuss vision and hearing.

The *parietal* lobes and *frontal* lobes, located on the top of the head, are separated by the **central sulcus,** a deep infolding that runs roughly down the middle of the lateral surface of each hemisphere. The parietal lobes lie to the rear of the central sulcus on each side, and the frontal lobes lie in front of it. The **parietal lobes** are primarily responsible for receiving and processing body sensory input (touch, temperature, pain, etc.). The **frontal lobes** contain motor and higher association areas and are responsible for four main functions: (1) nonreflex motor activity, (2) vocal ability (in mammals with this capability), (3) some types of *long-term* memory, and (4) higher mental functions such as planning involving multiple sensory inputs and memory.

Now return to Figure 5-28 and note these lobes and their motor, sensory, and association components in different mammals. Compare, for example, the yellow-colored olfactory cortexes of rat and human (Figure 5-28a versus d), and think about why the differences occur. Note also that the cream-colored association cortexes have expanded more than other regions in primate evolution.

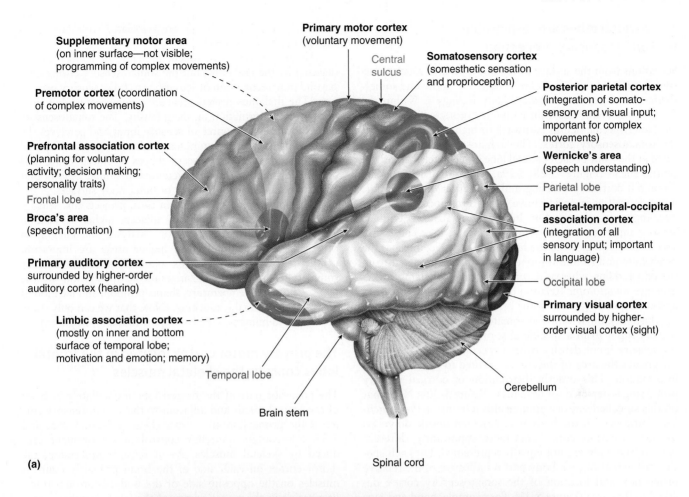

Supplementary motor area (on inner surface—not visible; programming of complex movements)

Premotor cortex (coordination of complex movements)

Prefrontal association cortex (planning for voluntary activity; decision making; personality traits)

Frontal lobe

Broca's area (speech formation)

Primary auditory cortex surrounded by higher-order auditory cortex (hearing)

Limbic association cortex (mostly on inner and bottom surface of temporal lobe; motivation and emotion; memory)

Temporal lobe

Brain stem

Primary motor cortex (voluntary movement)

Central sulcus

Somatosensory cortex (somesthetic sensation and proprioception)

Posterior parietal cortex (integration of somatosensory and visual input; important for complex movements)

Wernicke's area (speech understanding)

Parietal lobe

Parietal-temporal-occipital association cortex (integration of all sensory input; important in language)

Occipital lobe

Primary visual cortex surrounded by higher-order visual cortex (sight)

Cerebellum

Spinal cord

(a)

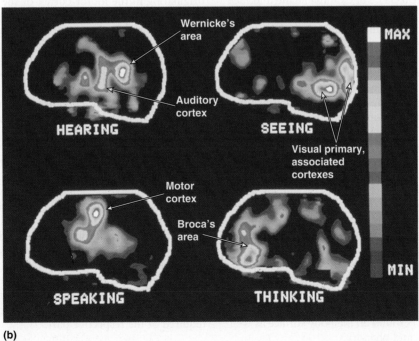

(b)

FIGURE 5-29 Functional areas of the human cerebral cortex. (a) Various regions of the cerebral cortex are primarily responsible for various aspects of neural processing, as indicated in this lateral view of the brain. (b) Different areas of the brain "light up" on positron emission tomography (PET) scans as a person performs different tasks. PET scans detect the magnitude of blood flow in various regions of the brain. Because more blood flows into a particular region of the brain when it is more active, neuroscientists can use PET scans to "take pictures" of the brain at work on various tasks.

Courtesy of Washington University School of Medicine, St. Louis.

The parietal lobes are responsible for somatosensory processing

Sensations from the surface of the body, such as touch, pressure, heat, cold, and pain, are collectively known as **somesthetic sensations** (*somesthetic*, "body feelings"). Within the CNS of mammals, this information is "projected" (transmitted along specific neural pathways to higher brain levels) to the **somatosensory cortex**. The somatosensory cortex is located at the front of each parietal lobe immediately behind the central sulcus (Figures 5-28, 5-29a, and 5-30a). It is the site for initial cortical processing and perception of somesthetic input as well as proprioceptive input. **Proprioception** is the awareness of body position. In nonmammalian vertebrates, sensory input generated from proprioceptive receptors are sent to several regions of the brain. For example, in fishes the cerebellum and tectum respond to stimulation initiated from the skin and fins, whereas in turtles and frogs, thalamic nuclei generate *somatotopic* (relation between specific body regions and corresponding areas of the brain) responses.

Each region within the somatosensory cortex receives sensory input from a specific area of the body. The greater the sensory input density from a specific area of the body, the greater the area of the cortex devoted to processing this information. This orderly distribution of cortical sensory processing is depicted for humans in Figure 5-30b. Note that on this so-called **sensory homunculus** in humans (*homunculus*, "little man"), the body is represented upside down on the somatosensory cortex and, more importantly, different parts of the body are not equally represented. For each species, the extent of each body part is indicative of the relative proportion and location of the somatosensory cortex devoted to that area (Figure 5-31). For example, hand and face areas are large in monkeys, mouthparts in rabbits, claws and forelimbs in cats, face in dogs. In cats this cortical representation is referred to a *felunculus* and in dogs it is a *canunculus*. In humans the exaggerated size of the face, tongue, hands, and genitalia is indicative of the high degree of sensory perception and density of nerves associated with these body parts. These cortical maps of the body surface are subject of constant modification in response to changing input from the sensory pathways. Learning, as you will see later, encourages both stronger synaptic connections as well as new synaptic growth.

These maps of the sensory world of an animal were put together by experiments on several species of animals. Electrodes were placed in the cerebral cortex of anesthetized animals, and recordings of clusters of neurons were made in response to visual, auditory, or somatosensory stimulation. For example, tactile stimulation of the animal's body activated neurons in the rostral regions of the cortex. These studies established that there are maps of the sensory surfaces in the cerebral cortices of all vertebrates. The rhombencephalon of the catfish, for example, contains extensive sensory maps of the taste receptors found on the body surface as well as those from the mouth and pharynx.

The somatosensory cortex on each side of the brain for the most part receives sensory input from the opposite side of the body because most of the ascending pathways carrying sensory information up the spinal cord cross over to the opposite side before eventually terminating in the cortex. Thus, damage to the somatosensory cortex in the left hemisphere produces sensory deficits on the right side of the body, whereas sensory losses on the left side are associated with damage to the right half of the cortex.

Simple awareness of touch, pressure, or temperature is detected by the thalamus, but the somatosensory cortex goes beyond pure recognition of sensations to fuller sensory perception. The thalamus detects changes in temperature but does not provide information on the intensity. The somatosensory cortex localizes the source of sensory input and perceives the level of intensity of the stimulus. It also is capable of spatial discrimination, so it can discern shapes of objects being held and can distinguish subtle differences in similar objects that come into contact with the skin or outer body surfaces.

The somatosensory cortex, in turn, projects this sensory input via white matter fibers to adjacent higher association areas for even further elaboration, analysis, and integration of sensory information. These higher areas are important in the perception of complex patterns of somatosensory stimulation—for example, simultaneous perception of the texture, firmness, temperature, shape, position, and location of an object the animal is touching. (Note that we can only study perception in humans, who can report to the investigator.)

The primary motor cortex located in the frontal lobes controls the skeletal muscles

The posterior part of the frontal lobe immediately in front of the central sulcus and adjacent to the somatosensory cortex is the **primary motor cortex** (Figures 5-28, 5-29a, and 5-30c). It confers nonreflex control over movement produced by skeletal muscles. As in sensory processing, the motor cortex on each side of the brain primarily controls muscles on the opposite side of the body. Neuronal tracts originating in the motor cortex of the left hemisphere cross over in the medulla oblongata before passing down the spinal cord to terminate on efferent motor neurons that trigger skeletal muscle contraction on the right side of the body. Accordingly, damage to the motor cortex on the left side of the brain produces paralysis on the right side of the body, and the converse is also true.

Stimulation of different areas of the primary motor cortex brings about movement in different regions of the body. Like the sensory homunculus for the somatosensory cortex, the **motor homunculus**, which depicts the location and relative amount of motor cortex devoted to output to the muscles of each body part, is upside down and distorted (Figure 5-30c). For humans, the fingers, thumbs, and muscles important in vocalization, especially those of the lips and tongue, are grossly exaggerated in size, indicative of the fine degree of motor control with which these body parts are endowed. Compare this to how little brain tissue is devoted to the trunk, arms, and lower extremities, which are not capable of such complex movements. Thus, the extent of representation in the motor cortex, as seen in the somatosensory cortex, is proportional to the precision and complexity of motor skills required of the respective part.

The higher motor areas are important in motor control

Even though signals from the primary motor cortex terminate on the efferent neurons that trigger nonreflex skeletal muscle contraction, the motor cortex is not the only region of the brain involved with motor control. First, as you have

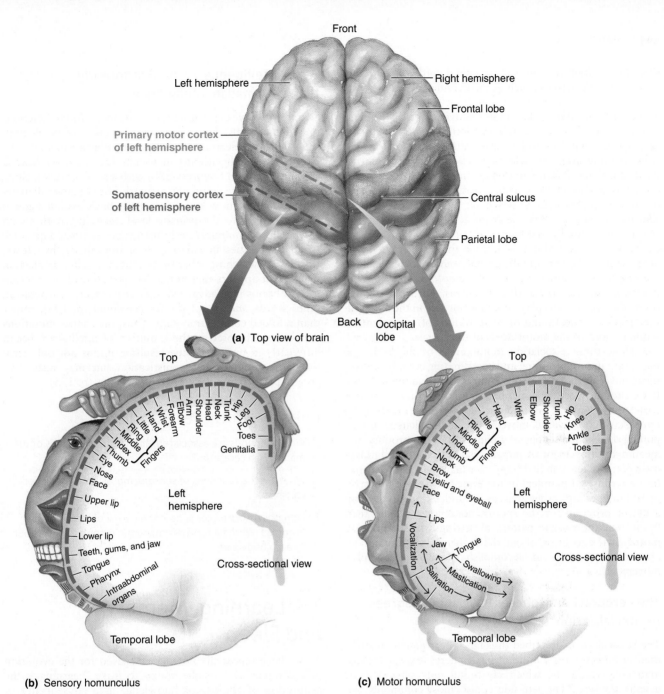

FIGURE 5-30 Somatotopic maps of the human somatosensory and primary motor cortexes. (a) Top view of cerebral hemispheres showing somatosensory cortex and primary motor cortex. (b) Sensory homunculus showing the distribution of sensory input to the somatosensory cortex from different parts of the body. The distorted graphic representation of the body parts indicates the relative proportion of the somatosensory cortex devoted to reception of sensory input from each area. (c) Motor homunculus showing the distribution of motor output from the primary motor cortex to different parts of the body. The distorted graphic representation of the body parts indicates the relative proportion of the primary motor cortex devoted to controlling skeletal muscles in each area.

© Cengage Learning, 2013

Rabbit   Cat   Monkey   Human

FIGURE 5-31 Somatosensory projections of four mammals. Each is distorted to represent the extent of innervation to different body parts. Disproportionately large projection areas in the somatosensory cortex are associated with densely innervated tissue.

© Cengage Learning, 2013

seen, lower brain regions and the spinal cord control reflex skeletal muscle activity, such as the maintenance of posture. The cerebellum also plays an important role in monitoring and coordinating nonreflex motor activity that the primary motor cortex has set in motion. Second, although fibers originating from the motor cortex can activate motor neurons to bring about muscle contraction, the motor cortex itself does not *initiate* nonreflex movement. The motor cortex is activated by a widespread pattern of neuronal discharge, the **readiness potential**, which occurs about 750 msec before specific electrical activity is detectable in the motor cortex. Three higher motor association areas of the brain are involved in this decision-making period. These areas, which all command the primary motor cortex, include the *supplementary motor area,* the *premotor cortex,* and *the posterior parietal cortex* (Figure 5-29a). Furthermore, the cerebellum appears to play an important role in the anticipatory planning and timing of certain kinds of movement by sending input to the motor areas of the cortex.

These three association motor areas of the brain are important in programming and coordinating complex movements involving simultaneous contraction of many muscles. Even though electrical stimulation of the primary motor cortex brings about contraction of particular muscles, no purposeful coordinated movement can be elicited, just as pulling on isolated strings of a puppet does not produce any meaningful movement. A puppet displays purposeful movements only when a skilled puppeteer manipulates the strings in a coordinated manner. In the same way, these other regions (and perhaps other areas as yet undetermined) develop a **motor program** for the specific nonreflex task and then "pull" the appropriate pattern of "strings" in the primary motor cortex to bring about the sequenced contraction of appropriate muscles to accomplish the desired complex movement.

The cerebral hemispheres have some degree of specialization

The cortical areas described thus far appear equally distributed in both the right and left hemispheres (except for human language areas, which are found only on one side, usually the left). The left side is also most commonly the dominant hemisphere for fine motor control, at least in primates. Thus, most humans are right-handed because the left side of the brain controls the right side of the body. In contrast, most parrots are left-footed, with the right side of the brain controlling the left side of the body. Each hemisphere is somewhat specialized in the types of mental activities it carries out best. In humans, the **left cerebral hemisphere** excels in the performance of logical, analytical, sequential, and verbal tasks, such as math, language forms, and philosophy. In contrast, the **right cerebral hemisphere** excels in nonlanguage skills, especially spatial perception and artistic and musical endeavors. Whereas the left hemisphere tends to process information in a fragmentary way, the right hemisphere views the world in a big-picture, holistic way. Normally, much sharing of information occurs between the two hemispheres so that they complement each other, but in many individuals the skills associated with one hemisphere are more strongly developed. Left cerebral-hemisphere dominance tends to be associated with "thinkers," whereas the right-hemispheric skills dominate in "creators."

Avian intelligence evolved differently than mammalian intelligence

So far, we have been discussing advanced brain functions only in mammals. The story for birds illustrates the flexibility of natural selection. Birds have long been thought to be inferior in intelligence to mammals because they have a smooth cerebrum (Figure 5-28b) with a small cortex lacking the computer-like columnar network of the mammalian cerebrum. But we now know that many birds exhibit signs of intelligence up to the primate level, including both sound and visual communication in many species (including some language abilities in parrots), tool use (sticks) by crows, learning by watching other birds, and even self-awareness in magpies. It appears that limbic-like structures have evolved in bird brains into advanced associative neural clusters—in other words, *nuclei*—of the *hyperpallium* (p. 175), rather than a folded columnar system. Thus, the chance mutations that became favored in the evolution of intelligence began differently in birds, and the resulting brains not only rival those of mammals in sophistication but may also have greater plasticity and regenerative abilities (p. 173).

check your understanding 5.8

What functions are associated with the four divisions of the mammalian cerebrum?

What are the functions of sensory, motor, and association cortexes?

Somatosensory projections in the cortex of mammals are distorted. What factors determine the extent of innervation to each body part?

5.9 Learning, Memory, and Sleep

As we have repeatedly noted, one reason for the evolution of a complex CNS is the ability to *learn*. **Learning** is the acquisition of abilities or knowledge as a consequence of experience, instruction, or both. It is widely believed that rewards and punishments are integral parts of many types of learning, at least in higher vertebrates (and perhaps cephalopods). If an animal is rewarded on responding in a particular way to a stimulus, the likelihood increases that it will respond in the same way again to the same stimulus as a consequence of this experience. Conversely, if a particular response is accompanied by punishment, the animal is less likely to repeat the same response to the same stimulus. When behavioral responses that give rise to pleasure are reinforced or those accompanied by punishment are avoided, learning has taken place. Housebreaking a puppy is an example. If the puppy is praised when it urinates outdoors but scolded when it wets the carpet, it soon learns the acceptable place to empty its bladder. Wild mammals, except those specifically marking territorial boundaries, have no such training experience, but they can learn about other aspects of their environments through reward (such as finding and remembering a source of tasty food) and punishment (such as encountering and remembering the den of a predator).

TABLE 5-7 Comparison of Short-Term and Long-Term Memory

Characteristic	Short-Term Memory	Long-Term Memory
Time of Storage after Acquisition of New Information	Immediate	Later; must be transferred from short-term to long-term memory through consolidation; enhanced by practice or recycling of information through short-term mode
Duration	Lasts for seconds to hours	Retained for days to years
Capacity of Storage	Limited	Very large
Retrieval Time (Remembering)	Rapid retrieval	Slower retrieval, except for thoroughly ingrained memories, which are rapidly retrieved
Inability to Retrieve (Forgetting)	Permanently forgotten; memory fades quickly unless consolidated into long-term memory	Usually only transiently unable to access; relatively stable memory trace
Mechanism of Storage	Involves transient modifications in functions of preexisting synapses, such as altering amount of neurotransmitter released	Involves relatively permanent functional or structural changes between existing neurons, such as formation of new synapses; synthesis of new proteins plays a key role

© Cengage Learning, 2013

(Note again that the perception of "reward" and "punishment" are derived solely from human subjects; we infer similar feelings in some animals because of similar reactions.) Thus, learning is a change in behavior that occurs as a result of experiences. It is highly dependent on the animal's interaction with its environment. The only limits to the effects that environmental influences can have on learning are the biological constraints imposed by species-specific and individual genetic endowments.

Memory comes in two forms—declarative and procedural—and is laid down in stages

Memory is the storage of acquired knowledge or abilities for later recall. It is widely recognized that there are at least two categories of memory: (1) **declarative** or **explicit** memory—the learning of events, places, and so on [sometimes subdivided into **semantic memories** (memories of facts) and **episodic memories** (memories of events)]; and (2) **procedural** or **implicit** memory—the learning of skilled motor movements. Learning and both types of memory form the basis by which animals adapt their behaviors to their particular external circumstances. Without these mechanisms, planning for successful interactions and intentional avoidance of predictably disagreeable circumstances would be impossible.

Not all forms of learning provide well for this flexibility, however, as evidenced by the phenomenon of **imprinting**, in which newborns are programmed to learn that situations and objects encountered early in life are both normal and important for life. For example, many newly hatched birds are programmed to learn that the first moving object they encounter is a parent. Konrad Lorenz first demonstrated this programmed learning by exposing newly hatched goslings to a substitute mother. Rather than follow their biological parent, the goslings could be induced to follow a variety of inanimate objects or another animal, including a human. In birds and some species of mammals there is a narrowly defined period in which the nervous system is responsive to this form of learning. For example, foals exposed to a variety of handling and grooming procedures within 24 hours after birth are less fearful of these procedures later in life (such as the vibration and sound of clippers, and rhythmic palpation of both the inside and outside of the ears, nostrils, mouth, anus, genitals). If properly desensitized, these animals are more easily examined and treated by veterinarians and other professionals dealing with horses. Conversely, infant chimpanzees raised in highly stressful situations have hair-trigger fight-or-flight responses for the rest of their lives.

The neural change responsible for retention or storage of knowledge is known as the **memory trace**. Generally concepts, not *verbatim* information, seem to be stored. As you read this page, you are storing the concept discussed, not the specific words. Later, when you retrieve the concept from memory, you will convert it into your own words. It is possible, however, to memorize bits of information word by word.

Storage of acquired information is accomplished in at least two stages: short-term memory and long-term memory (Table 5-7 and Figure 5-32). **Short-term memory** lasts for seconds to hours, whereas **long-term memory** is retained for days to years. The process of transferring and fixing short-term memory traces into long-term memory stores is known as **consolidation**. Stored knowledge is of no use unless it can be retrieved and used to influence current or future behavior.

A recently developed concept is that of a **working memory**, or what has been called "the erasable blackboard of the mind" (Figure 5-32). Working memory involves comparing current sensory data with relevant stored knowledge and manipulating that information, as in being able to plan a future action based on a current situation coupled with past experience. Newly acquired information is initially deposited in short-term memory, which has a limited capacity for storage. Information in short-term memory has one of two eventual fates. Either it is soon forgotten (for example, forgetting a telephone number after you have looked it up and finished dialing), or it is transferred into the more permanent long-term memory mode through *active practice*. The recycling of newly acquired information through short-term memory increases the likelihood of long-term memory consolidation.

FIGURE 5-32 Memory storage.
© Cengage Learning, 2013

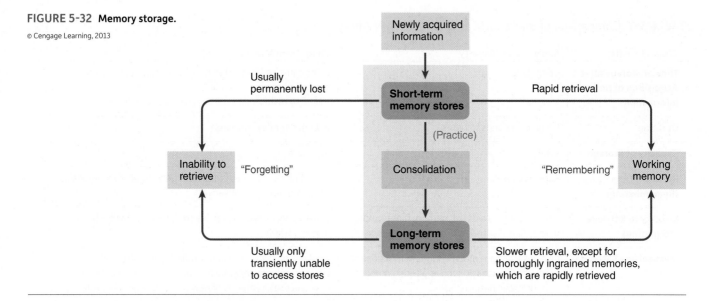

(Therefore, when you cram for an exam, your long-term retention of the information is poor!)

Short-term memory has very limited capacity; indeed, there is evidence that all vertebrates have only about 7 to 10 distinct short-term "registers." That is, a vertebrate brain cannot deal with more than 7 to 10 distinct items on a short-term basis (think about how many numbers you can remember briefly, as in a telephone number). Prey fish that school are thought to take advantage of this. A predator such as a barracuda approaching a school of prey herring may initially focus on one individual to pursue. But by initiating certain stereotypical maneuvers (such as the flash expansion, in which the fish move outward radially with each individual moving off at a unique angle from the predator), the barracuda's brain seems to become confused by having too many moving targets for its short-term memory to cope with.

The storage capacity of the long-term declarative memory bank in mammals is much larger than the capacity of short-term memory. Different informational aspects of long-term declarative memory traces seem to be processed and codified, then stored in conjunction with other memories of the same type; for example, visual memories are stored separately from auditory memories. This organization facilitates future searching of memory stores to retrieve desired information.

Because long-term memory stores are larger, it often takes longer to retrieve information from long-term declarative memory than from short-term memory. *Remembering* is the process of retrieving specific information from memory stores; *forgetting* is the inability to retrieve stored information or to erase it. Information lost from short-term memory is permanently forgotten, but information in long-term storage is frequently forgotten only transiently. Forgetting short-term memory is important, to make room for new information. There is evidence that short-term memories are in fact actively erased: There are neuromodulators, especially **anandamide** (also spelled *anandimide*), that reset short-term memory neurons to their unpotentiated state, particularly after some stressful events. (The active ingredient in marijuana, *tetrahydrocannabinol*, is an anandamide agonist, explaining why this drug interferes with short-term memory.)

Memory traces are present in multiple regions of the brain

What parts of the CNS govern memory? We here focus primarily on vertebrates but consider other animals for some aspects of memory later. There is no single "memory center" in the vertebrate brain. Instead, the neurons involved in memory traces are widely distributed throughout the subcortical and cortical regions of the brain. The regions of the brain implicated in memory include the hippocampus and other parts of the limbic system, and associated structures of the medial (inner) temporal lobes, the prefrontal cortex, the cerebellum and other regions of the cerebral cortex.

The Hippocampus and Declarative Memories The **hippocampus** ("horse-headed sea monster"), the elongated, seahorse-shaped medial portion of the temporal lobe that is part of the limbic system (see Figure 5-27), plays a vital role (as we noted on p. 172) in short-term declarative memory involving the integration of various related stimuli, and is also crucial for consolidation into long-term memory. The hippocampus is believed to store new long-term memories only temporarily and then transfer them to other cortical sites for more permanent storage. The sites for long-term storage of various types of memories are only beginning to be identified by neuroscientists, but they include the prefrontal cortex and medial temporal lobes.

Use of declarative memory requires active recall. Accessing and manipulating these long-term stores, as well as comparing them to current information and short-term memories, appear to be carried out by the *working memory* of the *prefrontal cortex* (see Figure 5-29). Consider what happens, for example, when your prefrontal cortex is planning to tell your motor cortex to initiate walking across a street. As you look right and then left down the street, your working memory uses (1) the current state of the street to the left, (2) the short-term hippocampal memory of the street a second ago when you looked to the right, and (3) long-term memories about the danger of moving vehicles. A monkey planning to descend from a tree might go through a similar sequence as it scans the ground for predators.

The Cerebellum and Procedural Memories The cerebellum and some motor cortical regions play an essential role in the "how to" *procedural memories* involving motor skills gained through repetitive training. In contrast to declarative memories, which are recollected from previous experiences by the "highest" brain centers (that is, consciously in humans), procedural memories can be brought forth without the involvement of those centers (that is, subconsciously in humans). As you saw earlier, learned skilled motor behaviors may be initiated by higher brain centers (that is, the prefrontal or premotor cortex), but after that they are taken over by the lower centers that have learned the detailed movements. This generally makes the behavioral movements much smoother (see the box on the cerebellum as a Smith predictor, p. 185).

Short-term and long-term memory involve different molecular mechanisms

Another question besides the "where" of memory is the "how" of memory. Obviously, some change must take place within the neural circuitry of the brain to account for the altered behavior that follows learning. Recent research shows that simple short-term memories can reside in a single neuron, but more complex memories are thought to reside in changes in the pattern of signals transmitted across synapses within a neuronal *network*. However, different mechanisms are responsible for short- and long-term memory. Short-term memory involves transient modifications in the function of preexisting synapses, such as a temporary alteration in the amount of neurotransmitter released in response to stimulation within affected nerve pathways. Long-term memory, in contrast, involves relatively permanent functional or structural changes between existing neurons in the brain. Let's look at each of these types of memory in more detail.

Short-term memory involves transient changes in synaptic activity

To investigate short-term memory, we depart from vertebrates for the moment. Following the Krogh principle (p. 3), ingenious experiments were undertaken by Eric Kandel using the giant marine snail *Aplysia* (Figure 5-33), for which he was awarded the Nobel Prize in Physiology or Medicine in 2000. *Aplysia* were selected as a model for the study of memory for several reasons: (1) they have comparatively large neurons which can be easily identified and recorded, (2) their nervous systems are relatively small (only about 20,000 neurons), and (3) they exhibit simple behaviors with rudimentary learning. The experiments have shown that two forms of short-term memory—*habituation* and *sensitization*—are due to modification of different membrane channel proteins in presynaptic terminals of specific afferent neurons. This modification in turn brings about changes in transmitter release.

These studies have exploited the defensive withdrawal reflexes of the external organs of the mantle cavity. This reflex is analogous to vertebrate withdrawal responses. In mollusks, the mantle cavity is a respiratory chamber housing the gill (Figure 5-33). Surrounding the cavity is a protective covering, the *mantle shelf*, which terminates in a fleshy spout, the *siphon*. When the siphon or mantle shelf is stimulated by

touch, the siphon and gill are vigorously withdrawn into the mantle cavity under the shelf. Afferent neurons responding to touch of the siphon or mantle shelf directly synapse on efferent motor neurons controlling gill withdrawal. This simple reflex can be modified by experience in two ways:

- **Habituation** represents a decreased responsiveness to repetitive presentations of an indifferent stimulus—that is, one that is neither rewarding nor punishing. The snail becomes habituated when its siphon is repeatedly touched; that is, it learns to ignore an indifferent stimulus and no longer withdraws its gill in response. (A natural example would be an ocean current.)
- **Sensitization** refers to increased responsiveness to mild stimuli following a strong or noxious stimulus; after having been subjected to a threatening stimulus an animal responds more attentively to almost any other stimulus. Sensitization, a more complex form of learning, takes place in *Aplysia* when it is given an electric shock to the tail followed immediately by a touch on the siphon, or repeated shocks to the tail (a natural analogue might be a fish nibbling on the tail). Subsequently, the snail withdraws its gill more vigorously in response to even mild stimuli it once failed to react to. The animal in effect has learned that something truly dangerous such as a biting fish is in the vicinity.

Interestingly, these different forms of learning affect the same site—the synapse between a siphon afferent and a gill efferent—in opposite ways. Habituation depresses this synaptic activity, whereas sensitization enhances it. These transient modifications persist for as long as the time course of the memory.

Mechanism of Habituation Habituation in *Aplysia* develops after about 10–15 nonharmful stimulations. Scientists have found that this is based on *prolonged closing of Ca^{2+} channels*, reducing Ca^{2+} entry into the presynaptic terminal, which leads to a decrease in quantity of neurotransmitter released. As a consequence, the postsynaptic potential is reduced compared to normal, resulting in a decrease or absence of the gill withdrawal controlled by the postsynaptic efferent neuron (Figure 5-33, left-hand chart). Thus, the memory for habituation in *Aplysia* is stored in the form of modification of specific Ca^{2+} channels. With no further training, this reduced responsiveness lasts an hour or more. The synaptic events responsible for habituation are also observed in other species studied. This suggests that Ca^{2+} channel modification may be a general mechanism of habituation, although in advanced species the involvement of intervening interneurons makes the process somewhat more complicated. Habituation is probably the most common form of learning; by learning to ignore indifferent stimuli, the animal is free to focus on other more important stimuli.

Mechanism of Sensitization Sensitization in *Aplysia* has also been shown to involve channel modification, although a different channel and mechanism are involved. In contrast to what happens in habituation, *Ca^{2+} entry into the presynaptic terminal is enhanced* in sensitization. The *nociceptors* (pain receptors) activated by the electric shocks to the tail connect with the presynaptic terminal of the sensory touch neuron in the gill-withdrawal reflex pathway. The subsequent increase in the release of excitatory neurotransmitter

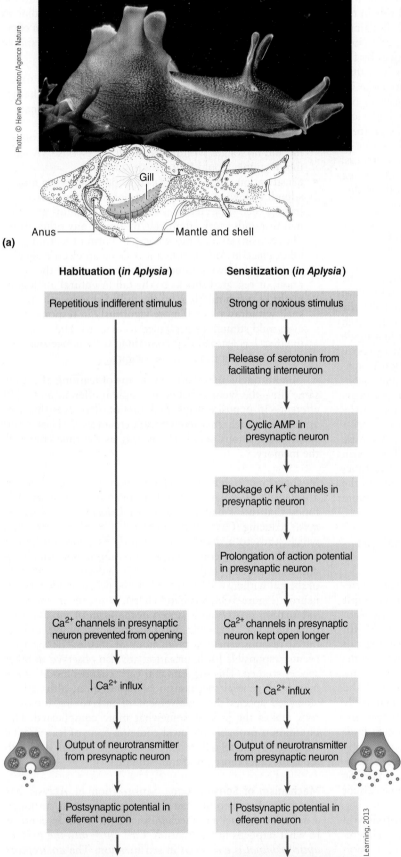

(a)

Anus

Gill

Mantle and shell

(b)

Habituation (*in Aplysia*)

Repetitious indifferent stimulus

↓

Ca²⁺ channels in presynaptic neuron prevented from opening

↓

↓ Ca²⁺ influx

↓

↓ Output of neurotransmitter from presynaptic neuron

↓

↓ Postsynaptic potential in efferent neuron

↓

Reduced behavioral response to indifferent stimuli

Sensitization (*in Aplysia*)

Strong or noxious stimulus

↓

Release of serotonin from facilitating interneuron

↓

↑ Cyclic AMP in presynaptic neuron

↓

Blockage of K⁺ channels in presynaptic neuron

↓

Prolongation of action potential in presynaptic neuron

↓

Ca²⁺ channels in presynaptic neuron kept open longer

↓

↑ Ca²⁺ influx

↓

↑ Output of neurotransmitter from presynaptic neuron

↓

↑ Postsynaptic potential in efferent neuron

↓

Enhanced behavioral response to mild stimuli

© Cengage Learning, 2013

Photo: © Herve Chaumeton/Agence Nature

FIGURE 5-33 Gill of a sea hare (*Aplysia*), a gastropod, and habituation and sensitization. (a) The gill lies within the mantle cavity, into which it is withdrawn in response to a gentle tap on the siphon or mantle. (b) Researchers have shown that in the sea snail *Aplysia* two forms of short-term memory—habituation and sensitization—result from opposite changes in neurotransmitter release from the same presynaptic neuron, caused by different transient channel modifications. See text for details.

Source: (a) C. Starr & R. Taggart (2004), *Biology: The Unity and Diversity of Life*, 10th ed. (Belmont, CA: Brooks/Cole), Figure 40.4.

from the touch neuron strengthens the synaptic potential in the motor cell. Sensitization thus elicits a larger postsynaptic potential (EPSP) in the motor neuron, which results in a more vigorous gill-withdrawal response. Sensitization does not have a direct effect on the presynaptic Ca²⁺ channels. Instead, it indirectly enhances Ca²⁺ entry via presynaptic facilitation (see p. 140), that is, enhanced neurotransmitter release from the terminals of the sensory touch neurons. Sensitization thus involves a more complex response at the molecular level. For this form of learning to occur, noxious stimulation of the tail (unconditioned stimulus) must be either repeated or paired with stimulation of the touch neurons in the siphon or mantle shelf (conditioned stimulus). The conditioned stimulus normally produces only a small response, whereas a noxious or unconditioned stimulus generates a powerful response. After pairing of these two stimuli, the gill withdrawal response to stimulation of the siphon is enhanced. The conditioned stimulus must *precede* the unconditioned stimulus by a narrow critical period of 0.5 sec. Biochemical events initiated in the sensory touch neuron are then exaggerated by neurotransmitter (serotonin) released from the pain pathway. This results in an increase in the release of the excitatory neurotransmitter glutamate from the touch neuron onto the motor neuron innervating the gill. Here is a summary of the events associated with sensitization (Figure 5-33, right-hand chart):

1. Arrival of an action potential in the touch neuron within 0.5 sec of activation of the nociceptor (pain) pathway. Release of *serotonin* by an interneuron in the pain pathway.

2. Binding of serotonin to a specific receptor in the sensory touch neuron and activation of adenylyl cyclase and generation of *cAMP* from ATP (see p. 97).

3. Activation of a *cAMP-dependent protein kinase, protein kinase A* (see p. 97).

4. Protein kinase phosphorylation of a specific class of K⁺ channels, which reduces the efflux of K⁺ from the touch neuron,

prolonging the action potential and permitting the Ca^{2+} channels to remain open for an extended period of time.

5. A greater quantity of glutamate release from the sensory touch neuron.

6. Ca^{2+} activation of the protein *calmodulin,* priming adenylyl cyclase to generate additional cAMP.

7. Duration of the action potential increased by 20%, which leads to a more vigorous withdrawal of the gill for a prolonged period.

Thus, existing synaptic pathways may be functionally interrupted (habituated) or enhanced (sensitized) during simple learning. Habituation and sensitization persist for about an hour in *Aplysia,* the time it takes for cAMP concentrations to return to normal. However, if noxious stimuli (four to five tail shocks) are repeatedly administered to *Aplysia,* behavioral sensitization can last from one to several days. This phenomenon is known as **long-term potentiation (LTP)**—a prolonged increase in the strength of existing synaptic connections in activated pathways following brief periods of repetitive stimulation. That is, this connection gets stronger the more often it is used. Sensitization alerts animals to predators and other potentially harmful stimuli and so is critical for their survival. Researchers speculate that much of short-term memory in vertebrates is similarly a temporary modification of already existing processes. The cyclic AMP cascade, particularly the activation of protein kinase, plays an important role in elementary forms of learning and memory, to which we now turn our attention.

Mechanism of Long-Term Potentiation in Mammals LTP has been shown to last for days or even weeks in vertebrates—long enough for this short-term memory to be consolidated into more permanent long-term memory, which you may recall is the role of the hippocampus (p. 172), where LTP is especially prevalent. When LTP occurs, simultaneous activation of both the presynaptic and the postsynaptic neurons at a given excitatory synapse results in long-lasting modifications that enhance the ability of the presynaptic neuron to excite the postsynaptic neuron (Figure 5-34). Because both the presynaptic neuron and the postsynaptic neurons must be activated at the same time, the development of LTP is restricted to the pathway that is stimulated. With LTP, signals from a given presynaptic cell to a postsynaptic neuron become stronger with repeated use. Keep in mind that strengthening of synaptic activity results in the formation of more EPSPs in the postsynaptic neuron, ultimately resulting in more action potentials being sent along this postsynaptic cell to other neurons.

Enhanced synaptic transmission could theoretically result from either changes in the postsynaptic neuron (such as increased responsiveness to the neurotransmitter) or in the presynaptic neuron (such as increased release of neurotransmitter). The underlying mechanisms for LTP are still the subject of much research and debate. Most likely, multiple mechanisms are involved in this complex phenomenon. It appears that there are several forms of LTP, some arising from changes only in the postsynaptic neuron and others also having a presynaptic component. Based on current scientific evidence, the following is a plausible mechanism for LTP involving both a postsynaptic change and a presynaptic modification.

LTP begins when a presynaptic neuron releases the common excitatory neurotransmitter glutamate in response to an action potential (step 1). Glutamate binds to two types of receptors on the postsynaptic neuron: *AMPA receptors* and *NMDA receptors* (step 2). An **AMPA receptor** is a chemically mediated receptor-channel that opens on binding of glutamate (step 3) and permits net entry of Na^+ ions, leading to formation of an EPSP at the postsynaptic neuron (step 4; see p. 132). This is the ordinary receptor at excitatory synapses about which you already learned. An **NMDA receptor** is a receptor-channel that permits Ca^{2+} entry when it is open. This receptor-channel is unusual because it is both chemically gated and voltage dependent. It is closed by both a gate and by a magnesium ion (Mg^{2+}) that physically blocks the channel opening at resting potential. Two events must happen almost simultaneously to open an NMDA receptor-channel: presynaptic glutamate release and postsynaptic depolarization by other inputs. The gate opens on binding of glutamate, but this action alone does not permit Ca^{2+} entry. Additional depolarization of the postsynaptic neuron beyond that produced by the EPSP resulting from glutamate binding to the AMPA receptor is needed to depolarize the postsynaptic neuron enough to force Mg^{2+} out of the channel (step 5). The postsynaptic cell can be sufficiently depolarized to expel Mg^{2+} in two ways: by repeated input from this single excitatory presynaptic neuron, resulting in temporal summation of EPSPs from this source (see p. 136), or by additional excitatory input from another presynaptic neuron at about the same time. When the NMDA receptor channel opens as a result of simultaneous gate opening and Mg^{2+} expulsion, Ca^{2+} enters the postsynaptic cell (step 6). The entering Ca^{2+} activates a Ca^{2+} second-messenger pathway in this neuron. This second-messenger pathway leads to the physical insertion of additional AMPA receptors in the postsynaptic membrane (step 7). Because of the increased availability of AMPA receptors, the postsynaptic cell exhibits a greater EPSP response to subsequent release of glutamate from the presynaptic cell. This heightened sensitivity of the postsynaptic neuron to glutamate from the presynaptic cell helps maintain LTP.

Furthermore, at some synapses activation of the Ca^{2+} second-messenger pathway in the postsynaptic neuron causes this cell to release a **retrograde** ("going backward") paracrine that diffuses to the presynaptic neuron (step 8). Here, the retrograde paracrine activates a second-messenger pathway in the presynaptic neuron, ultimately enhancing the release of glutamate from the presynaptic neuron (step 9). This positive feedback strengthens the signaling process at this synapse, also helping sustain LTP. Note that in this mechanism, a chemical factor from the postsynaptic neuron influences the presynaptic neuron, just the opposite direction of neurotransmitter activity at a synapse. The retrograde paracrine is distinct from classical neurotransmitters or neuropeptides. Most investigators believe that the retrograde messenger is *nitric oxide,* a chemical that performs a wide array of other functions (see p. 93).

The modifications that take place during the development of LTP are sustained long after the activity that led to these changes has ceased. Therefore, information can be transmitted along this same synaptic pathway more efficiently when activated in the future. That is, the synapse "remembers." Note that LTP develops in response to frequent activity across a synapse as a result of repetitive, intense firing of a given input (as with repeatedly practicing a

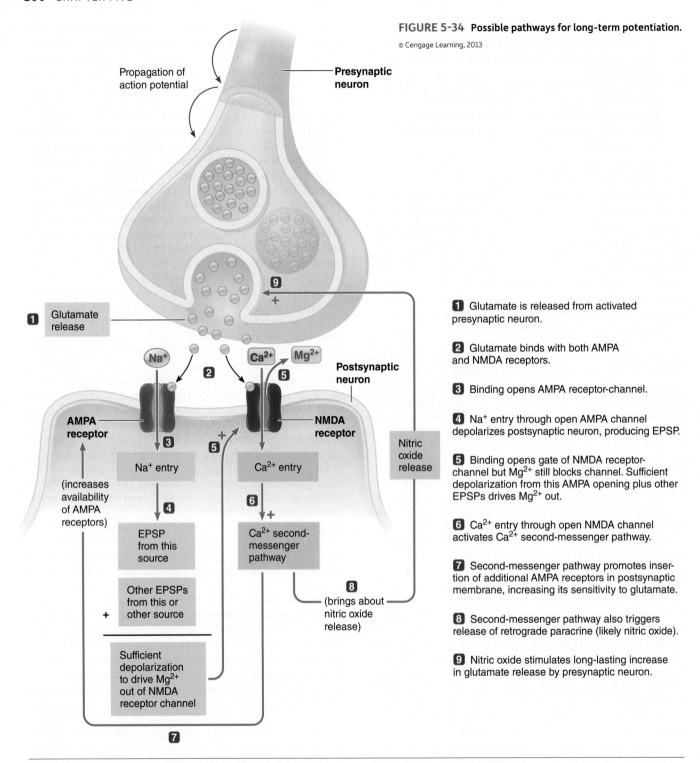

FIGURE 5-34 Possible pathways for long-term potentiation.
© Cengage Learning, 2013

1 Glutamate is released from activated presynaptic neuron.

2 Glutamate binds with both AMPA and NMDA receptors.

3 Binding opens AMPA receptor-channel.

4 Na^+ entry through open AMPA channel depolarizes postsynaptic neuron, producing EPSP.

5 Binding opens gate of NMDA receptor-channel but Mg^{2+} still blocks channel. Sufficient depolarization from this AMPA opening plus other EPSPs drives Mg^{2+} out.

6 Ca^{2+} entry through open NMDA channel activates Ca^{2+} second-messenger pathway.

7 Second-messenger pathway promotes insertion of additional AMPA receptors in postsynaptic membrane, increasing its sensitivity to glutamate.

8 Second-messenger pathway also triggers release of retrograde paracrine (likely nitric oxide).

9 Nitric oxide stimulates long-lasting increase in glutamate release by presynaptic neuron.

particular display) or to the linking of one input with another input firing at the same time. For example, consider a zoo animal whose mouth waters in anticipation of the imminent arrival of dinner when hearing a food cart. Earlier in its life, only the taste and feel of food in the mouth could trigger salivation. However, through experience, neurons in the pathway that control salivation link input arising from the sound of the cart with input from food in its mouth. After the sound-input pathway is strengthened through development of LTP and ultimate consolidation into long-term storage, the sound alone can cause salivation.

The *ethanol* in alcoholic beverages blocks the NMDA receptors, which is likely the reason people have difficulty remembering what happened during a time of heavy drinking. Ethanol also enhances the actions of GABA, the major inhibitory neurotransmitter in the brain, thus depressing the overall activity of the CNS (dampening reaction speed, for example).

Studies suggest a regulatory role for the cAMP second-messenger pathway in the development and maintenance of LTP in addition to the Ca^{2+} second-messenger pathway. Participation of cAMP may hold a key to linking short-term memory to long-term memory consolidation.

Can intelligence and memory be changed by single genes? We have discussed a number of proteins and genes (as well as neuropeptides) associated with learning and memory. Thus, many researchers have believed that complex higher brain functions such as intelligence are the result of many interacting genes (as well as environmental influences) and thus cannot be readily manipulated. But others believe that single genes may have profound influences on memory and perhaps intelligence. Many studies are being done to test these ideas. Some focus on changes in gene expression during learning. For example, one research team put rats to a standard learning task, and used a gene chip (p. 25) to analyze gene expression in the hippocampus at various times before, during, and after the task. They found that learning was associated with significant changes in expression of about 30 genes. The roles of these genes are mostly still uncertain, but one of them, *Fibroblast Growth Factor (FGF) 18*, was tested further because it increased at every stage of learning. Significantly, injections of FGF-18 into rats improved the ability of rats to learn a new task.

Other studies approach the problem by genetically engineering rodents with ex-cess genes suspected of being involved in learning, or by knocking out such genes. One study engineered mice to overexpress growth-associated protein *GAP-43*, a target of the second-messenger transducer protein kinase C, and which is suspected of being crucial to the growth of new synapses during learning. Another study engineered mice that had enhanced synaptic release of glutamate. In both cases, mice exhibited significantly greater long-term learning abilities.

Yet another study knocked out a gene called *NR1,* which codes for part of the NMDA receptor associated with memory formation. The gene was knocked out in the hippocampus of mice, and the mice indeed could not learn very well. The next study engineered overexpression with another NMDA gene, called *NR2B,* that is active only in the forebrain. The mice's forebrains then had an NMDA receptor with a major difference: The channel of the receptor stayed open (during learning) about 2.5 times longer than normal (250 msec instead of 100 msec). These mice were smarter—they solved problems such as mazes about 40% faster than normal mice—and their memories were better.

Other such studies have focused on drugs that target genes and proteins involved in learning. For example, a drug that enhances *CREB* activity (see main text below) led to dramatic improvement in learning in mice with cognitive impairments. Another study involved hamsters with learning impairments caused by a lack of circadian rhythm (p. 283); their mutation resulted in constant release of the inhibitory neurotransmitter GABA (p. 133) in hippocampus memory neurons. A drug that blocked GABA binding restored their learning abilities. Such studies are aimed at the development of drugs that might help with certain human mental disorders.

The idea of drugs or genetic manipulations to increase human learning may be appealing, but there may harmful trade-offs (p. 2). Humans who are born with "photographic memories" (people who, for example, can read a long text just once and then can recite it verbatim) often have other cognitive problems, for example, problems with *understanding* a text because of an overwhelming focus on details. Simple pharmaceutical or genetic manipulations of human learning may thus not work as planned.

Long-term memory involves formation of new, permanent synaptic connections

Whereas short-term memory involves transient strengthening of preexisting synapses, long-term memory storage requires the activation of specific genes that control protein synthesis. These proteins, in turn, are needed for the formation of new synaptic connections. Thus, long-term memory storage involves permanent physical changes in the brain.

Studies comparing the brains of experimental rodents reared in a sensory-deprived environment with those raised in a sensory-rich environment demonstrate readily observable microscopic differences. The animals afforded more environmental interactions—and, therefore, more opportunity to learn—have greater branching, "spine" formation and elongation of dendrites in nerve cells in regions of the brain involved with memory storage. Greater dendritic surface area presumably provides more sites for synapses. Thus, long-term memory may be stored at least in part by a particular pattern of dendritic branching and synaptic contacts.

We have seen that sensitization in *Aplysia* increases concentrations of cAMP and protein kinase A in synapses. But what happens after repeated bouts of noxious stimuli? Repeated learning trials causes protein kinase A to move to the nucleus, where it activates a regulatory protein, **cAMP response-element binding protein (CREB-1)**, which binds to a promoter (the cAMP response element) (Figure 5-35a). This is the molecular switch that activates (turns on) a group of genes, the **immediate early genes (IEGs)**, which play a critical role in memory consolidation. These genes govern the synthesis of the proteins that encode long-term memory. If the action of CREB-1 is somehow blocked (e.g., by the use of specific repressors), then the long term strengthening of these synaptic connections is prevented. Consolidation of procedural as well as declarative memories depends on CREB-1. CREB-1 serves as a common molecular switch for converting both these categories of memory from the transient short-term to the permanent long-term mode. Recently, researchers have found another protein kinase *PKM*, which is somehow activated during sensitization. Blocking PKM with an injected inhibitor prevents long-term memory formation. PKM is unique in that, once activated, it cannot be shut off, thus insuring memory persistence.

Interestingly, protein kinase A recruits another kinase, called **MAP kinase**, which also migrates to the nucleus and *inactivates* a second form of the CREB protein. In contrast to CREB-1, which activates gene expression, **CREB-2** suppresses gene expression. The shifting balance between positive and repressive factors is believed to ensure that only information relevant to the animal, not everything encountered,

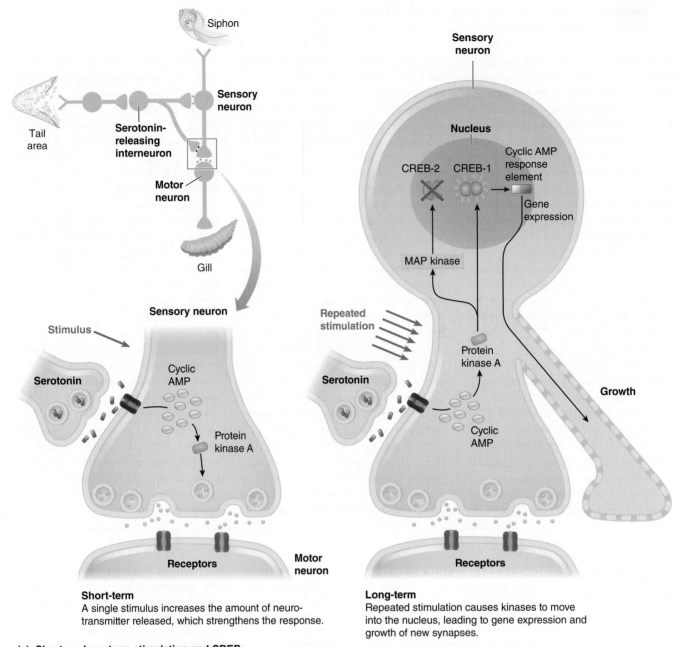

Short-term
A single stimulus increases the amount of neuro-transmitter released, which strengthens the response.

Long-term
Repeated stimulation causes kinases to move into the nucleus, leading to gene expression and growth of new synapses.

(a) Short- vs long-term stimulation and CREB

FIGURE 5-35 The role of CREB and CPEB in long-term memory in _Aplysia_. (a) Left panel: a summary of short-term sensitization (see Figure 5-33 and main text). Right panel: prolonged stimulation leads to gene activation via _cAMP response-element binding protein 1 (CREB-1),_ with the resulting growth changes shown.

is put into long-term storage. Thus, formation of enduring memories involves not only the activation of positive regulatory factors (CREB-1, PKM) that favor memory storage but also the turning off of inhibitory, constraining factors (CREB-2) that prevent memory storage. In life-threatening situations, sufficient MAP kinase molecules are presumably sent to the nucleus rapidly enough to inactivate the CREB-2 molecules, so that protein kinase A activates sufficient CREB-1 molecules to ensure that the animal recalls the details of this experience for future reference.

Initially, proteins synthesized in the nucleus are shipped to the activated synapses where they initiate synaptic growth. However, for the growth of new synapses to be sustained,

locally synthesized proteins are now required. For this to occur dormant mRNA synthesized in the nucleus must be activated at these particular synapses by a novel protein, **CPEB (cytoplasmic polyadenylation element-binding protein)**, which regulates local protein synthesis (Figure 5-35b). Of note, CPEB is a _prion_-like protein (p. 173), that is, it self-perpetuates: Once activated at a particular synapse, it remains active for as long as the memory persists. And it is serotonin that is responsible for the activation of CPEB.

Returning to mammals, most investigation of learning and memory has focused on changes in synaptic connections within the brain's gray matter. To complicate the issue even further, scientists now have evidence that white matter also

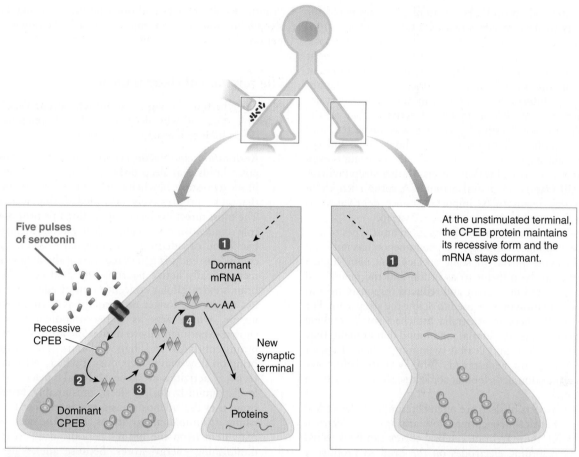

(b) Long-term memory and the prion-like CPEB protein. As a result of a prior stimulus, the sensory cell's nucleus has sent dormant messenger RNA (mRNA) to all axon terminals (1). Five pulses of serotonin at one terminal convert a prion-like protein (CPEB) that is present at all synapses into a dominant, self-perpetuating form (2). Dominant CPEB can convert recessive CPEBs to the dominant form (3). Dominant CPEB activates dormant messenger RNA (4). The activated messenger RNA regulates protein synthesis at the new synaptic terminal, stabilizes the synapse, and perpetuates the memory.

FIGURE 5-35, cont'd The role of CREB and CPEB in long-term memory in *Aplysia*. (b) The role of *CPEB (cytoplasmic polyadenylation element-binding protein)* in long-term memory.

Source: (a) Eric Kandal (2006), *In Search of Memory: The Emergence of a New Science of Mind* (New York: Norton). (b) 'Figure 19-4' from IN SEARCH OF MEMORY: THE EMERGENCE OF A NEW SCIENCE OF MIND by Eric Kandel. Copyright © 2006 by Eric R. Kandel. Used by permission of W.W. Norton & Company, Inc.

changes during learning and memory formation as more myelin surrounds axons, especially during adolescence, speeding up transmission between connected neurons. Apparently neurons produce chemical signals such as **neuregulin** that regulate the extent to which myelin-forming cells wrap themselves around the axon. The amount of neuregulin produced is correlated with the extent of action potential propagation within the axon. Accordingly, researchers propose that conduction velocity can be increased by means of further myelination in more active pathways, and that these changes support learning and memory. In addition to a probable role of white matter, numerous hormones and neuropeptides have also been shown to affect learning and memory processes. We discussed one of these, anandamide, earlier (p. 196).

Complex memories and learning may reside in neuronal networks

Although we have seen how the beginnings of learning and memory probably begin with changes in synaptic functions and connections, we have not addressed the larger question of how such changes actually encode a retrievable complex memory such as an image of an object or the location of a home den or lair. This question about neuronal function remains largely unsolved. Working hypotheses and models that have made some headway in explaining complex learning are based on **neuronal networks** (not to be confused with nerve nets, p. 150). These are largely computer models that simulate a number of neurons with synaptic connections that excite, inhibit, modulate, and change with usage. Crude forms of image recognition and other simple learning have been demonstrated in these models. Whether these represent the way a real nervous system works is still under investigation.

Now let's turn to another incompletely understood phenomenon that may be intimately tied to memory and learning: *sleep*.

Sleep is a nearly universal phenomenon, and has two states in birds and mammals

Sleep is a widespread phenomenon characterized by several features: (1) periods of minimal movement; (2) reduced responsiveness to external stimuli; (3) rapid reversibility (as opposed to hibernation, which is not); and often (4) a characteristic body posture. For example, bats are noted for their ability to sleep upside down with talons locked onto a cave roof or tree branch (this allows them to drop off to take flight, since bats have trouble taking off from an upright roosting position like birds). However, despite its apparent quiescent nature, sleep is an *active* process, not just the absence of wakefulness. Sleep, at least in mammals, is not accompanied by a reduction in neural activity (that is, the brain cells are not "resting"), as once was suspected, and overall metabolic rate drops by only about 10%. Indeed, during certain stages of sleep, O_2 uptake by the brain is even increased above normal waking levels. Rather sleep involves a profound *change* in neural activity. A certain length of sleep per day seems to be important, since deprivation is normally followed by longer "rebound" sleep.

Some form of sleep probably occurs in most types of animals except sponges, although the phenomenon has not been studied in many phyla. In some species true neurological sleep states can be difficult to ascertain and the behavioral indices of minimal movement and reduced responsiveness to external stimulation may be used to define the sleep state. For example, researchers in Australia attached small tracking devices to box jellies (Cnidaria), and found that at night, they sank to the seafloor and ceased swimming. Thus, sleep, or sleeplike states, are widespread. Why is that? Before we tackle that question, let's look at some key aspects of sleep in more detail.

Neural aspects of sleep are often determined by electrical recordings. Extracellular current flow arising from electrical activity within the vertebrate cerebral cortex can be detected by placing recording electrodes on the head to produce a graphic record known as an **electroencephalogram**, or **EEG**. These "brain waves" for the most part are not due to action potentials but instead represent the momentary collective postsynaptic potential activity (that is, EPSPs and IPSPs; see p. 132) in the cell bodies and dendrites located in the cortical layers under the recording electrode. Electrical activity can always be recorded from the living brain, even during sleep and unconscious states, but the waveforms vary, depending on the degree of activity in the cerebral cortex. Interestingly, some cetaceans and some birds can sleep with one half of the brain while the other half is awake, so-called **unihemispheriuc slow-wave sleep**. This may be so that they keep a lookout for predators. In the case of cetaceans, this may also allow them to regularly surface to breathe, while for migrating birds, this might let them keep flying on long transoceanic migrations without stopping.

In birds and mammals two types of sleep can be characterized by different EEG patterns as well as different behaviors: *slow-wave sleep* and *paradoxical,* or *REM, sleep.* **Slow-wave sleep** is characterized by slow, large EEG waves and by considerable muscle tone with frequent body shifts. **Paradoxical** or **REM sleep** has rapid small EEG waves similar to the wake state, but with muscle tone absent and little movement except for breathing and rapid eye movements (hence the term *REM*). Reptiles such as the Mexican giant tortoise and spiny-tailed iguana exhibit two distinct sleep phases with similar characteristics. Although still speculative, recent studies suggest that these forms of sleep serve different purposes. Birds and nonhuman mammals have cycles of non-REM and REM sleep comparable to humans, though birds' cycles are much shorter. Short bouts of REM sleep concomitant with unihemispheric sleep enable some species of birds to sleep "on the wing." Horses and other herbivorous ungulates can sleep while standing, but must necessarily lie down for brief periods when shifting from slow-wave to REM sleep.

The function of sleep is unclear

Now let's turn to the big question: why sleep? Despite considerable research, why sleep occurs largely remains a mystery. Let's look at the major hypotheses.

1. *Restoration and recovery:* One widely accepted proposal holds that sleep provides "catch-up" time for the brain to restore biochemical or physiological processes that have progressively degraded during wakefulness. The most direct evidence supporting this proposal is the role of *adenosine* as a neural sleep factor. Adenosine, the backbone of adenosine triphosphate (ATP), the cell's energy currency, is generated during the awake state by metabolically active neurons and glial cells. Thus, the brain's extracellular concentration of adenosine continues to rise the longer an animal has been awake. Adenosine, which acts as a neuromodulator, has been shown experimentally to inhibit the arousal centers (the brainstem RAS and hypothalamus. This action can bring on slow-wave sleep, during which restoration and recovery activities are believed to take place. Adenosine levels diminish during sleep, presumably because the brain uses this adenosine as a raw ingredient for replenishing its limited energy stores. Thus, the need for sleep may stem from the brain's periodic need to replenish diminishing energy stores. Because adenosine reflects the level of brain cell activity, the concentration of this chemical in the brain may serve as a gauge of how much energy has been depleted.

 Another "restoration and recovery" proposal suggests that slow-wave sleep provides time for the brain to repair the damage caused by the toxic free radicals produced as by-products of the stepped-up metabolism during the waking state. A possible "restoration and recovery" function of REM sleep is to let some of the neural pathways regain full sensitivity. When an animal is awake, the brain neurons that release the neurotransmitters norepinephrine and serotonin are maximally and continuously active. Release of these neurotransmitters ceases during REM sleep. Studies suggest that constant release of norepinephrine and serotonin can desensitize their receptors. Perhaps REM sleep is needed to restore receptor sensitivity for optimal functioning during the next period of wakefulness.

 In support of these ideas is the observation that smaller, high-metabolic-rate mammals need much more sleep than large, low-metabolic-rate ones (a *scaling* phenomenon; see p. 10). For example, in hours per day, a brown bat sleeps 20, a house cat 13.5, a human 8, and an Asiatic elephant 3.1. The correlation is not perfect—a lion, for example, may sleep as much as a house cat—but overall the trend is quite striking.

2. *Memory processing:* Another leading theory has nothing to do with restoration and recovery. Instead, other researchers believe that sleep is necessary to allow the brain to "shift gears" to accomplish the long-term structural and chemical adjustments necessary for learning and memory. This theory might explain the observation

that young mammals tend to sleep for a larger proportion of each 24-hour day than adults (although some dolphins and orcas do not sleep at all during the first weeks of life, likely because of their vulnerability to predators). Young mammals' highly plastic brains are rapidly undergoing profound synaptic modifications in response to environmental stimulation. In contrast, neural changes in older animals are less dramatic. Some evidence suggests that the different types of sleep might be involved in consolidation, editing and even elimination of different kinds of memories, with declarative memories being processed during slow-wave sleep and procedural memories during REM sleep. A recent memory-related theory is that sleep, especially slow-wave sleep, is a time for replaying the events of the day, not only to help consolidate memories but perhaps to make recent experiences more meaningful by catching information missed on first pass and by "connecting the dots" between new pieces of information. Also, some evidence shows that recent memories deemed irrelevant are deleted (forgotten), an essential action to prevent what might become an overwhelming amount of stored "data." Recall that the neuromodulator *anandamide* is involved in this process in some memory regions (p. 196).

Even the nematode worm *Caenorhabditis elegans* has a sleeplike period during its early development. Importantly, this occurs when its nervous system is undergoing its most rapid changes in synaptic growth and pruning.

3. *Energy conservation* is another theory of why animals need sleep. This may indeed be the case for less complex animals such as box jellies; but this proposal is not as widely accepted for vertebrates because of the level of neural activity during sleep.

These various hypotheses are not mutually exclusive. Sleep might serve multiple purposes.

The term *consciousness* refers to a subjective awareness of the world and self

This ends our coverage of nervous system evolution and mechanisms. There is one final topic of complex brain function we have not analyzed: *consciousness*. This term refers to subjective awareness of the external world and self. Even though it is clear that the final level of awareness resides in the cerebral cortex, and although there is evidence that a crude sense of awareness depends on the thalamus, conscious experience depends on the integrated functioning of many parts of the nervous system. The cellular and molecular basis underlying consciousness remains one of the greatest unanswered questions in neuroscience. Although consciousness is a thorny topic, some researchers believe that many primates (in addition to humans), African parrots, elephants, and dolphins have a sense of self and personal experience over time. For example, these animals and humans (but not other animals that have been tested) recognize themselves in mirrors. Perhaps other animals have a personal self-image as well.

The following states of consciousness are listed in decreasing order of arousal level, based on the extent of interaction between peripheral stimuli and the brain:

- maximum alertness
- wakefulness
- sleep (several different types)
- coma

Maximum alertness depends on attention-getting sensory input that "energizes" the RAS and subsequently the activity level of the CNS as a whole. At the other extreme, coma is the total unresponsiveness of a mammal to external stimuli, caused either by brainstem damage that interferes with the RAS or by widespread depression of the cerebral cortex, such as accompanies O_2 deprivation.

check your understanding 5.9

What is the difference between habituation and sensitization? Which is the more common?

For long-term memory to occur, what must happen?

What factors affect long-term memory storage?

making connections

How Nervous Systems Contribute to the Body as a Whole

To interact in appropriate ways with the external environment to sustain an animal's viability, such as in the acquisition of food, and to make the internal adjustments needed to maintain homeostasis, the animal's body must be informed about any changes taking place in the external and internal environment and must be able to process this information and send messages to various effectors to accomplish the desired results. The nervous system, one of the body's two major control systems, plays a central role in this life-sustaining communication. The central nervous system (CNS) receives information about the external and internal environment by means of afferent peripheral nerves. After sorting, processing, and integrating this input, the CNS sends commands by means of efferent peripheral nerves to bring about appropriate muscular contractions, glandular secretions, and other effector functions. Electrical signaling allows for rapid responses.

Much neural regulation is aimed toward maintaining homeostasis, whereas some is for regulated changes. For example, when informed by the afferent nervous system that blood pressure has fallen in a mammal, the CNS sends appropriate commands to the heart and blood vessels to increase the blood pressure to normal. When a predator is detected through the eyes, and recognized in the brain, commands may be sent to initiate rapid running. Were it not for this processing and integrative ability of the CNS, coordinated regulation of multiple organs in complex animals would be impossible.

The autonomic nervous system in vertebrates plays a major role in the following range of homeostatic activities:

- Regulation of blood pressure.
- Control of digestive juice secretion and of digestive tract contractions that mix ingested food with the digestive juices.
- Control of sweating to help maintain body temperature.

The somatic nervous system of vertebrates, the efferent branch that innervates skeletal muscle, contributes to homeostasis by stimulating the following activities:

- Skeletal muscle contractions that enable the body to move in relation to the external environment contribute to homeostasis by, for example, moving the body toward food or away from harm.
- Skeletal muscle contraction also accomplishes breathing to maintain appropriate levels of O_2 and CO_2 in the body.
- Shivering is a skeletal muscle activity important in the maintenance of body temperature.

Nervous systems have become progressively more complex during evolution, with more sophisticated control of homeostasis and regulated change. However, more primitive regions retain many controls necessary for survival. For example, in vertebrates, the spinal cord integrates many basic protective and evacuative reflexes that do not require higher levels, such as withdrawal from a painful stimulus and emptying of the urinary bladder. The newer levels piled on top of the older ones progressively modify, enhance, or nullify actions coordinated by lower levels in a hierarchy of command, using new capabilities such as complex perceptual awareness of the external environment, initiation of nonreflex skeletal muscle movements, memory and anticipation, and consciousness. Learning and both types of memory form the basis by which animals adapt their behaviors to their particular external circumstances. Without these mechanisms, planning for successful interactions and intentional avoidance of predictably disagreeable circumstances would be impossible.

Chapter Summary

- Nervous systems evolved from simple reflex arcs to centralized brains with distributed, hierarchic regulation.

- Sponges have no nerves but can respond to stimuli using electrical signals.

- Nerve nets are the simplest nervous systems and are found in most animals. Simple ganglia and nerve rings evolved for more complex behavior.

- There has been an evolutionary trend for the most anterior region of the CNS to be larger and contain sizable clusters of neurons, which contributed to the evolution of a so-called superganglion or brain.

- The logarithm of brain weight as a function of body weight is typically linear for a variety of vertebrates, with a positive slope indicating that the overall size of an animal is related to its relative brain size.

- As with bilateral animals in general, the nervous system of vertebrates is organized into a central nervous system (CNS) and a peripheral nervous system (PNS).

- Three classes of neurons make up the nervous system: afferent neurons, efferent neurons, and interneurons.

- The autonomic efferent division of the PNS is the communication link by which the CNS controls the activities of muscles, glands, and other effector organs that carry out actions, such as making feedback correction. Each autonomic nerve pathway extending from the CNS to an innervated organ consists of a two-neuron chain.

- The sympathetic system promotes responses that prepare a vertebrate animal for strenuous physical activity in the face of emergency or other stressful situations, such as a physical threat from the environment. The parasympathetic system, in contrast, dominates in quiet, relaxed situations. Sympathetic and parasympathetic preganglionic fibers release the same neurotransmitter, ACh (acetylcholine), but the postganglionic endings of these two systems release different neurotransmitters that influence the effector organs. Dual innervation of organs with nerve fibers whose actions oppose each other enables precise control over an organ's activity.

- Responsive tissue cells have one or more of several different types of plasma membrane receptor proteins for the respective neurotransmitters released from the presynaptic terminal. Binding of a neurotransmitter to a receptor induces the tissue-specific response.

- About 90% of the cells within the CNS are not neurons but glial cells, which serve as the connective tissue of the CNS and as such help support the neurons both physically and metabolically.

- A highly selective blood–brain barrier carefully regulates exchanges between the blood and brain. The brain floats in a cushioning cerebrospinal fluid.

- Neurogenesis is a normal function of several regions of the brain. Brains of virtually all vertebrates display a degree of plasticity.

- Newer, more sophisticated regions of the vertebrate brain are piled on top of older, more primitive regions. A vertebrate brain is classically organized into three distinct regions—the hindbrain, midbrain, and forebrain. The forebrain has changed more than other regions of the brain during vertebrate evolution.

- The spinal cord is strategically located between the brain and afferent and efferent fibers of the PNS; this location enables the spinal cord to fulfill its two primary functions: (1) serving as a link for transmission of information between the brain and the remainder of the body and (2) integrating reflex activity between afferent input and efferent output without involving the brain.

- The brainstem is a critical connecting link between the remainder of the brain and the spinal cord. The cerebellum is important in balance as well as in planning and execution of nonreflex movements, coordinating by anticipating needed motions.

- The basal nuclei play a complex role in the control of movement in addition to having nonmotor functions that are less well understood.

- The thalamus serves as a "relay station" and synaptic integrating center for preliminary processing of all sensory input on its way to the cortex.

- The hypothalamus is a collection of specific nuclei and associated fibers that lie beneath the thalamus. It is an integrating center for many important homeostatic functions and serves as an important link between the autonomic nervous system and the endocrine system.

- The cerebrum, by far the largest region of mammalian brains, is divided into two halves, the right and left hemispheres. Functionally, it has sensory, motor, and associative cortexes.

- Sensations from the surface of the body, such as touch, pressure, heat, cold, and pain, are collectively known as somesthetic sensations. Within the CNS of mammals, this information is transmitted along specific neural pathways to higher brain levels to the somatosensory cortex.

- The primary motor cortex located in the frontal lobes controls the skeletal muscles. The motor cortex is activated by a widespread pattern of neuronal discharge, the readiness potential.

- The limbic system controls basic emotions, responses to fear, and primitive memory.

- Learning is the acquisition of abilities or knowledge as a consequence of experience, instruction, or both. Memory comes in two forms—declarative and procedural—and is laid down in stages.

- Two forms of short-term memory—habituation and sensitization—are due to modification of different membrane channel proteins in presynaptic terminals of specific afferent neurons, which in turn brings about changes in transmitter release. In habituation, repeated stimulation leads to closing of the Ca^{2+} channels, reducing

Ca^{2+} entry into the presynaptic terminal, which leads to a decrease in quantity of neurotransmitter released. In contrast, Ca^{2+} entry into the presynaptic terminal is enhanced in sensitization.

- Whereas short-term memory involves transient strengthening of preexisting synapses, long-term memory storage requires the activation of specific genes that control protein synthesis. These proteins, in turn, are needed for the formation of new synaptic connections.

- In birds and mammals, two types of sleep can be characterized by different EEG patterns as well as different behaviors: slow-wave sleep and paradoxical, or REM, sleep. Sleep's role is unclear, but may serve for restoration and recovery and/or to process memory.

Review, Synthesize, and Analyze

1. What factors determine the complexity of a nervous system?

2. Would you expect the self-perpetuating form of CPEB (which maintains protein synthesis in a particular neuron) to be capable of reverting back to its inactive state? Support your reasoning.

3. If tight junctions limit transport of substances through brain capillaries, how can veterinarians treat disorders of the brain?

4. Why is plasticity an important feature of neural function?

5. What is the relationship between the basal nuclei and the thalamus?

6. Describe the process by which a permanent memory is made.

7. Are genes that encode suppressor proteins for converting short-term to long-term memory set at a low or high threshold? Explain your reasoning.

8. Compare the evolution of intelligence in birds and mammals.

Suggested Readings

Beer, R. D., H. J. Chiel, R. D. Quinn, K. S. Espenschied, & P. Larsson. (1992). A distributed neural network architecture for hexapod robot locomotion. *Neural Computation* 4:356–365. Shows how "biological inspiration" is aiding robot engineering.

Delcomyn, F., R. Gillette, C. L. Prosser, W. T. Greenough, & A. Spencer. (1991). In C. L. Prosser, ed., *Neural and Integrative Animal Physiology,* 4th ed. New York: Wiley-Liss. A classic encyclopedic textbook of comparative physiology.

Greenspan, R. J. (2007). *An Introduction to Nervous Systems.* Cold Spring Harbor, NY: Cold Spring Harbor Laboratory Press. Introduces students to major concepts in neuroscience with a focus on invertebrate behavior.

Kandal, Eric. (2006). *In Search of Memory. The Emergence of a New Science of Mind.* New York: Norton. An autobiography of a remarkable individual. His insights and explanations into the complex arena of memory assist the student in comprehending this demanding discipline.

Matthews, G. G. (2001). *Neurobiology: Molecules, Cells and Systems.* Palo Alto, CA: Blackwell Science. An excellent textbook on neurobiology at all levels.

Matynia, A., S. A. Kushner, & A. J. Silva. (2002). Genetic approaches to molecular and cellular cognition: A focus on LTP and learning and memory. *Annual Review of Genetics* 2002: 687–724. A review of molecular research on learning.

Miall, R. C., D. J. Weir, D. M. Wolpert, & J. F. Stein. (1993). Is the cerebellum a Smith Predictor? *Journal of Motor Behavior* 25:203–216. A thoughtful discussion of the role of the industry tool, the Smith Predictor, in explaining the function of the cerebellum.

Miller, G. (2009). On the origin of the nervous system. *Science* 325: 24–26. A synopsis of the available information on the origins of the nervous system.

Northcutt, R. G. (2002). Understanding vertebrate brain evolution. *Integrative and Comparative Biology* 42:743–756. A concise review of this topic.

Randall, D., W. Burggren, & K. French. (2002). *Eckert Animal Physiology. Mechanisms and Adaptation,* 5th ed. New York: W. H. Freeman. A classic animal physiology text with particularly comprehensive coverage of neural functions and systems.

Satterlie, R. A. (2002). Neuronal control of swimming in jellyfish: A comparative story. *Canadian Journal of Zoology* 80:1654–1669. A review of evolution and physiology of jelly nervous systems and swimming.

CourseMate

The great white shark, *Carcharodon carcharius.* Like other sharks, it has electrical sensors in its snout that allow it to find prey even in murky waters and when its eyes are covered by protective flaps during an attack.

Sensory Physiology

6.1 Evolution and Roles of Senses

In Chapter 1, you learned that organisms must gather information about their environments and internal states if they are to successfully achieve regulated changes and homeostasis. The shark above, for example, finds its prey by detecting sound waves, light patterns, chemicals in the water, the flow of water, and electrical fields. This multifaceted information allows its nervous system to regulate muscles for swimming toward, and subsequently locating and consuming, the prey. In turn, prey consumption aids in internal energy homeostasis. In animals, the ability to gather such information is typically done by **sensory cells**, each of which responds to a specific **stimulus**—defined as any agent detectable by an organism—such as light, vibration, and so on. In this chapter we will examine how these cells work.

Sensory cells have ion channels and receptor proteins with specific modalities

Animal nervous systems transmit information by electrical signals (graded and action potentials, Chapter 5). Thus, sensory cells must be specialized to perform **transduction**: the conversion of one form of energy—an external stimulus—into electrical energy that can be interpreted by a nervous system. The process of transduction is inextricably linked to the opening and closing of *ion channels,* often due to the actions of *receptor proteins* in plasma membranes, which bind specific chemicals. As you saw in Chapters 3 and 4, ion channels and receptor proteins are universal components of living systems, found even in prokaryotes. Recall that channels fall into two broad types: **leak** channels (which are always open) and **gated** channels, that is, ones that open and/or close in response specific stimuli (p. 81). The

latter come in four forms (p. 94): (1) *mechanically gated,* (2) *chemically gated,* (3) *voltage-gated,* and (4) *thermally gated* channels. Let's begin with a look at the evolution of receptor proteins and ion channels in sensory functions.

1. *Mechanically gated* were probably the first gated ion channels to evolve. Sensitive to stretch, they may have served as osmotic stress sensors in prokaryotes, opening when a cell was deformed by changes in cell volume. By opening, such channels *transduce* mechanical deformation into an ion current. In turn, the current triggers compensatory processes.

2. *Chemically gated* channels are also ancient, probably arising initially for sensing chemicals in the environment. Such sensing by single cells often involves *chemoreceptor proteins* rather than, or in addition to, gated channels. A key example is found in **chemotaxis**: movement

of an organism toward or away from a chemical substance. For example, bacteria and protozoa residing in ruminant stomachs are attracted to "desirable" molecules such as sugars, amino acids, and small peptides, which are associated with the arrival of a fresh bolus of food. They also can move away from toxins. Both responses follow the binding of the molecules to surface-membrane receptor proteins, which transduce the signal to the cell interior to alter flagellar motions. In many bacteria, these chemoreceptor proteins are strategically localized at the forward-moving pole of the cell.

Receptor-based chemoreception is at the heart of the evolution of multicellularity, as evidenced by *intercellular communication* in the slime mold (*Dictyostelium discoideum*). Starvation induces the single-celled amoeba to secrete *cAMP* (p. 97) into the environment, which serves as a chemoattractant for neighboring amoeba. Chemotaxis leads the amoebas to aggregate into large-bodied slugs, which can move to a new food source or turn into a reproductive stage to spread spores to new habitats. Activation of the membrane-bound cAMP receptor protein triggers a G-protein-linked biochemical cascade (p. 97), which leads to the developmental changes. As you will see, this mechanism was retained by animals in various forms of chemosensing.

3. *Voltage-gated channel*s are also ancient, but their ancestral functions are unclear. For example, in the common gut and laboratory bacterium *Escherichia coli,* two types of membrane channels have been identified, one of which responds to deformation, and another that is voltage gated (and with an uncertain function, though it may be involved in sugar transport). Voltage-gated K^+ channels are also found in the plasma membranes of yeast cells and various plant cells. For example, these channels are essential to the opening of stomata, the pores of leaves that allow for gas exchange. Moreover, K^+- and Ca^{2+}-based action potentials are generated in response to touch in fast-moving plants such the insectivorous Venus flytrap. However, Na^+ channels are found only in animals and a few protists, where they are essential to long-range signaling involved in both sensory input and effector control (Chapters 4 and 5).

4. *Thermally gated channels* are widespread in animals and are also found in some animal-like protists including *Paramecium* (protozoa with swimming cilia that cover much of the cell; p. 25). However, the evolution of these channels is poorly known outside of animals.

Paramecium are of great interest to animal physiologists because they can actually be considered free-swimming sensory cells with most of the sensory capabilities of animals. That is, they can respond to a variety of environmental stimuli—not only thermal but also mechanical, chemical, *photic* (light), and ionic—through changes in their membrane potentials. They have multiple types of ion-selective membrane channels that are voltage, chemically, mechanically, and thermally gated, and they have receptor proteins that react to certain chemicals that indirectly affect membrane potential via second messenger. They respond behaviorally by changing either their cell shape or the pattern of ciliary beating. Thus, much of the machinery associated with the function of sensory neurons arose long before the evolution of nervous systems.

The four types of gated channels, some leak channels, and various receptor proteins provide the basic transduction mechanisms for animal sensory cells. Animal physiologists recognize six to eight types of such sensors based on their **modalities**—the kind of stimulus they react to. (Note again that the term *receptor* can refer both to a protein and to a whole sensory cell; in this chapter, we will refer to the proteins as *receptor proteins* or *sites,* while calling the cells *receptors, receptor cells* or *sensors* interchangeably.)

- **Mechanoreceptors** are sensitive to mechanical energy through various mechanically gated channels. Examples include touch and pressure sensors in the skin; skeletal muscle sensors sensitive to stretch; sound, motion, and balance sensors in the mammalian ear; and osmosensors and baroreceptors in vertebrate bloodstreams.
- **Chemoreceptors** are sensitive to specific chemicals due primarily to chemical receptor proteins and chemically gated channels. Chemoreceptors include those for smell and taste, as well as those that detect O_2 and CO_2 concentrations in the blood and the chemical content of the digestive tract. Leak and voltage-gated channels are involved in sensing ions, for example, in salty food or as the primary component of the osmotic concentration of blood.
- **Thermoreceptors** are sensitive to heat and cold by having thermally gated channels.
- **Photoreceptors** are sensitive to photic energy, including ultraviolet and infrared wavelengths. They have light-activated receptor proteins which in turn control chemically gated channels.
- **Electroreceptors** are responsive to electric fields by poorly known mechanisms, most likely involving leak channels or voltage-gate channels.
- **Magnetoreceptors** are responsive to magnetic fields. These may use a special type of mechanical gating.

In addition, one other type of stimulus is sometimes classified separately:

- **Nociceptors**, or **pain receptors**, are sensitive to tissue damage such as pinching or burning, to distortion of tissue, or to harmful chemicals. These are thought to be types of chemo- and mechanoreceptors. Intense stimulation of almost any sensor is also perceived as painful, at least to mammals.

Animal senses can be classified into three roles: sensing the external environment, the internal environment, and body motion and position

In complex animals, the number and types of sensory cells are quite extensive, far more than the classic "five senses" frequently taught. In addition to categorizing sensors by modality (above), physiologists recognize three primary *roles* of receptor cells:

- **Interoreceptors** detect information about the internal body fluids usually crucial to homeostasis, such as blood pressure and the concentration of O_2.
- **Proprioceptors** send information about movement and the position of an animal's body or certain parts such as limbs.
- **Exteroreceptors** are the "classic" senses that detect external stimuli such as light, chemicals (for taste, smell), touch, temperature, electromagnetic fields, and sound.

In vertebrates, interoreceptors are located at numerous sites, mainly in association with blood vessels and gut fluids. Proprioceptors are located mainly in the muscles, tendons, and joints, and also in the inner ear. Finally, exteroreception information is sometimes categorized as (1) **somesthetic sensation** arising mainly from the body surface (touch, pressure, temperature, pain, and electrical fields), and (2) **special senses**, including *vision, hearing, taste,* and *smell*. Somesthetic sensors, found diffusely in many organs but most notably in the integument, provide information about an animal's interaction with the environment across its entire surface and about some internal physical stresses. In contrast, each of the special senses has highly localized, extensively specialized sensors (usually in a distinct, dedicated sensory organ) that respond to unique environmental stimuli. Each animal group has evolved specific senses that are essential for survival in its particular environment.

The peripheral nervous system (PNS) carries information about body position and motions and about internal and external environments from the sensory cells to the central nervous system (CNS: spine and brain in vertebrates). For example, the incoming pathway for information derived from visceral interoreceptors is called a **visceral afferent**. Processing of sensory input by the CNS, followed by efferent activation of appropriate effectors, is essential for interaction with the environment for basic survival (for example, food procurement and defense from predators).

Perception is not reality

If the sensory information from the PNS reaches the brain, it may be *perceived* by the animal. **Perception** is an animal's interpretation of the external world as created by the brain from a pattern of nerve impulses delivered to it from sensory cells. This is primarily a human concept because we cannot know what another animal truly perceives.

unanswered Questions | Do fish perceive pain?

Recent studies with fish illustrate the difficulties in demonstrating perception in nonhuman animals. Bony fish (teleosts) have skin sensors that are similar to pain sensors in mammals, whereas earlier evolved fish (cartilaginous fishes, lampreys) do not. To investigate the possible perception of pain in fish, researchers injected acetic acid or bee venom (irritants) or saline (a control) into the lips of trout. The trout with acetic acid exhibited complex nonreflex behaviors, such as rubbing their lips on the aquarium walls, which indicated a perception of pain. Similar experiments in more primitive fish did not induce such behaviors. Other researchers attached tiny heaters to goldfish after some were injected with pain-suppressing morphine and others with saline. After heat exposure, the fish without painkillers retreated to one spot in their tank where they remained, as if they were fearful or wary of their environments. This suggested some perception of pain occurred. However, other researchers note that fish lack the neocortical centers associated with pain perception in mammals, and their responses in these studies could be avoidance behaviors triggered as reflexes in the absence of

true perception. Thus, it is still not certain whether any fish actually perceive pain. See the Suggested Readings by Rose (2002), Sneddon et al. (2003), and Nordgreen et al. (2009). <<

Is the environment, as we humans and other animals perceive it, reality? The answer is a resounding no. Perception is different from what is really "out there" for several reasons. Humans, for example, have sensors that detect only a limited number of existing energy forms. We perceive sounds, colors, shapes, textures, smells, tastes, and temperature but are not informed of magnetic and electric forces and polarized and ultraviolet light waves (stimuli that some animals can sense), because we lack sensors for the latter energy forms. Our response range is limited even for the energy forms for which we do have sensors. For example, dogs can hear a dog whistle pitched above our level of detection. Second, the information channels to our brains are not high-fidelity recorders. During precortical processing of sensory input, many features of stimuli are *filtered*: Some features are accentuated, and others are suppressed or ignored. For example, the retina of mammals, horseshoe crabs, and many other animals actively enhances *contrast*, making fuzzy boundaries appear as sharp edges. Furthermore, the mammalian retina also enhances light signals from any object that is moving relative to the background light.

Third, the cerebral cortex further manipulates the data, comparing the sensory input with other incoming information as well as with memories of past experiences to extract the significant features—for example, sifting out a friend's words from the hubbub of sound in a school cafeteria. In the process, the cortex often fills in or distorts the information to abstract a logical perception; that is, it "completes the picture." Human brains, for example, are primed to detect faces even with limited input; as a result, we easily "see" faces where there are none, such as in natural stone formations and a mountain on Mars (or Figure 6-1a). Our brains are also primed to recognize that an object will appear smaller the further away it is. Many optical illusions, used by artists, architects, and photographers, can use this expectation to make objects look smaller or larger than they are (Figure 6-1b). Chimpanzees are also subjected to such optical illusions, and at least one other animal may actually exploit one. Consider the male great bowerbird of Australia, which builds tunnel-like nests to attract females for mating. Remarkably, he selects stones of various sizes and lays them out from the backside of the tunnel in increasing size, to form a "courtyard" (Figure 6-1c). This array of stones may trick the female brain (as it does the human brain) into perceiving that the courtyard is much smaller than it is, since normally a courtyard would have paving stones of equal size that should appear to get smaller with distance. The illusion may cause the male sitting in the small-appearing courtyard to look much larger than he is. When John Endler and colleagues, who discovered the illusion, reversed the stone array, the male bird dutiful restored it to the original pattern, showing that it was deliberate.

Optical illusions illustrate how the brain interprets reality according to its own rules. Thus, perceptions do not replicate reality. Species equipped with different types and sensitivities of sensors and numbers of sensors, with different neural processing, presumably would perceive a markedly different, but nevertheless filtered, world from ours.

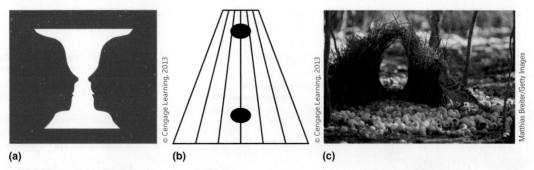

(a)   **(b)**   **(c)**

FIGURE 6-1 Optical Illusions. (a) Do you see two faces in profile, or a wineglass? Can you see both at once? (b) Forced perspective illusion. The two ovals are identical in size, but the upper one appears larger. The drawing simulates a path receding into the distance, and our brains expect that "distant" objects should appear small. (c) Forced perspective illusion. A male great bower bird builds a tunnel-like nest and "courtyard" of small to large stones that possibly make him look bigger to a female entering the nest (see main text).

check your understanding 6.1

List and describe the receptor types according to their modalities and roles.

Describe the ancient roles of receptors and ion channels in sensing.

6.2 Receptor Cell Physiology

As we've discussed, stimuli exist in a variety modalities, such as heat, sound, and chemical changes. Sensory cells must detect these different forms of energy using their ion channels and/or receptor proteins.

Receptors have differential sensitivities to various stimuli

Each type of receptor (sensory cell) is specialized to respond more readily to one type of stimulus, called its **adequate stimulus,** than to other stimuli. For example, receptors in the eye are most sensitive to light, certain receptors in the ear to sound waves, and warmth receptors in the skin to heat energy. An animal cannot "see" with its ears or "hear" with its eyes because of this differential sensitivity of receptors, a principle Johannes Müller formulated in 1835 and entitled the **doctrine of specific nerve energies.** Some receptors can respond weakly to stimuli other than their adequate stimulus, but even when activated by a different stimulus, a receptor still gives rise to the sensation usually detected by that receptor type. As an example, the adequate stimulus for eye receptors (photoreceptors) is light, to which they are exquisitely sensitive, but these receptors can also be activated to a lesser degree by mechanical stimulation. When hit in the eye, a person often "sees stars" because the mechanical pressure stimulates the photoreceptors, and the brain interprets the action potentials as light. Thus, the sensation perceived depends on the type of receptor stimulated rather than on the type of stimulus.

Some sensations in humans are compound sensations in that their perception arises from central integration of several, simultaneously activated, primary sensory inputs. For example, the perception of wetness comes from touch, pressure, and thermal receptor input; there appears to be no such thing as a "wet receptor."

Uses for Information Detected by Receptors The information detected by receptors is conveyed via afferent neurons to the CNS to be used for a variety of purposes:

- Afferent input is essential for control of efferent output, both for regulating motor behavior in accordance with external circumstances and for coordinating internal activities to maintain homeostasis.
- Central processing of sensory information gives rise to an animal's perceptions of the environment in the higher brain centers. In humans at least, it can also evoke emotional responses.
- Processing of sensory input by the reticular activating system in the vertebrate brain stem is critical for cortical arousal and consciousness (see p. 184).
- Finally, selected information delivered to the CNS may be learned for future reference.

Next let's examine how adequate stimuli initiate action potentials that ultimately are used for these purposes.

A stimulus alters the receptor's permeability, leading to a graded receptor potential

A receptor may be located either (1) on an afferent neuron with a specialized dendrite or (2) a separate cell closely associated with a dendrite of an afferent neuron (Figure 6-2). As we noted earlier, stimulation of a receptor alters its permeability to selective ion(s). The means by which this permeability change takes place is individualized for each receptor type. Because the electrochemical driving force is greater for Na^+ than for other small ions at resting potential, the predominant channel is an Na^+ one, leading to an inward flux of Na^+ that depolarizes the receptor membrane. (There are exceptions; for example, vertebrate photoreceptors are hyperpolarized on stimulation, as you will see later.)

This local depolarizing change in potential is known as a **receptor potential** in the case of a separate receptor cell or as a **generator potential** if the receptor is a specialized afferent

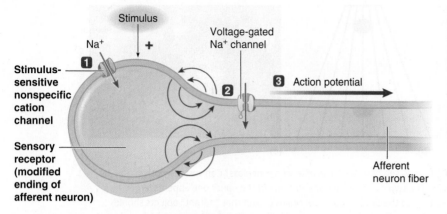

(a) Receptor potential in specialized afferent ending

1 In sensory receptors that are specialized afferent neuron endings, stimulus opens stimulus-sensitive channels, permitting net Na⁺ entry that produces receptor potential.

2 Local current flow between depolarized receptor ending and adjacent region opens voltage-gated Na⁺ channels.

3 Na⁺ entry initiates action potential in afferent fiber that self-propagates to CNS.

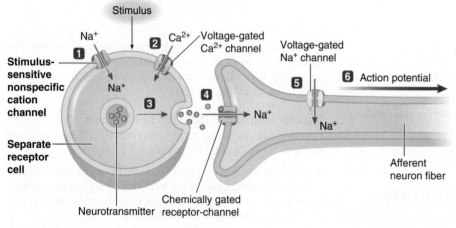

(b) Receptor potential in separate receptor cell

1 In sensory receptors that are separate cells, stimulus opens stimulus-sensitive channels, permitting net Na⁺ entry that produces receptor potential.

2 This local depolarization opens voltage-gated Ca²⁺ channels.

3 Ca²⁺ entry triggers exocytosis of neurotransmitter.

4 Neurotransmitter binding opens chemically gated receptor-channels at afferent ending, permitting net Na⁺ entry.

5 Resultant depolarization opens voltage-gated Na⁺ channels in adjacent region.

6 Na⁺ entry initiates action potential in afferent fiber that self-propagates to CNS.

FIGURE 6-2 Conversion of receptor and generator potentials into action potentials. (a) Specialized afferent ending as sensory receptor. Local current flow between a depolarized receptor ending undergoing a receptor potential and the adjacent region initiates an action potential in the afferent fiber by opening voltage-gated Na⁺ channels. (b) Separate receptor cell as sensory receptor. The depolarized receptor cell undergoing a receptor potential releases a neurotransmitter that binds with chemically gated channels in the afferent fiber ending. This binding leads to a depolarization that opens voltage-gated Na⁺ channels, initiating an action potential in the afferent fiber.

© Cengage Learning, 2013

neuron. For convenience, from here on we refer to both receptor and generator potentials as *receptor potentials*. The receptor potential is graded (see p. 113) with an amplitude and duration that can vary, depending on the strength and the rate of application or removal of the stimulus. That is, *the stronger the stimulus, the greater the permeability change and the larger the receptor potential.* As is true of all graded potentials, receptor potentials have no refractory period (p. 124), so summation in response to rapidly successive stimuli is possible. Because the receptor region lacks voltage-gated action-potential channels, the receptor potential must be converted into action potentials for long-distance transmission.

Receptor potentials may initiate action potentials in the afferent neuron

If of sufficient magnitude, a receptor potential can initiate an action potential in the afferent neuron membrane adjacent to the receptor cell or dendritic region by triggering the opening of voltage-gated Na⁺ channels in this region. The means by which the Na⁺ channels are opened differ depending on whether the receptor is a separate cell or a specialized afferent ending.

- In the case of a specialized afferent ending (Figure 6-2a), local current (from the receptor potential) flows between the activated receptor and the nearby *trigger zone* (p. 121), opening voltage-gated Na⁺ channels.
- In the case of a separate receptor, such as in the retina, a receptor potential triggers the release of a chemical messenger—a *neurotransmitter*—that diffuses across a synapse between the receptor and an afferent neuron

(Figure 6-2b). Binding of the neurotransmitter with specific protein receptors on the afferent neuron opens chemically gated Na^+ channels (see pp. 130–132). The subsequent graded current travels to the trigger zone of the afferent and opens voltage-gated Na^+ channels.

In either case, if the magnitude of the resulting ionic flux is sufficient to bring the trigger-zone membrane to threshold, action potentials are initiated that self-propagate along the afferent fiber to the CNS. Of course, not all stimuli result in action potentials. For example, mosquitoes can land so lightly on the skin of a mammal that the receptor potentials generated in touch sensors do not reach threshold.

Note that in an afferent neuron the trigger zone differs from that in an efferent neuron or interneuron. In the latter two types of neurons, the trigger zones are located in the axon hillock at the beginning of the axon adjacent to the cell body (see p. 121). By contrast, action potentials are initiated at the peripheral end of an afferent nerve fiber adjacent to the receptor region, often a long distance from the cell body (Figure 6-3).

A larger receptor potential does not bring about a larger action potential (recall the all-or-none law, p. 125), but will induce a more rapid firing of action potentials. This is one way stimulus strength is coded: The *stronger* the stimulus, the greater the *frequency* of action potentials (Figure 6-4). Stimulus strength is also reflected by the *size of the area* stimulated. Stronger stimuli usually affect larger areas, so a correspondingly greater population of receptors responds. For example, a light touch does not activate as many pressure receptors in the skin as a more forceful touch applied to the same area. Stimulus intensity is therefore distinguished both by the frequency of action potentials generated in the afferent neuron (**frequency code**) and by the number of receptors activated within the area (**population code**).

Receptors may fire continuously or adapt slowly or rapidly to sustained stimulation

Stimuli of the same intensity do not always bring about receptor potentials of the same magnitude from the same receptor. Some receptors can diminish the extent of their depolarization despite sustained stimulus strength, a phenomenon called **adaptation** (see p. 17 for caveats regarding this term). That is, the receptor "adapts" to the stimulus by decreasing its response to it or not responding at all.

Tonic and Phasic Receptors The two types of receptors—*tonic* and *phasic* receptors—differ in speed of adaptation. **Tonic receptors** do not adapt at all, or adapt slowly (Figure 6-5a). These receptors are important in situations where maintained information about a stimulus is valuable. Examples are muscle stretch proprioceptors, which monitor muscle length, and joint proprioceptors, which measure the degree of joint flexion. To maintain posture and balance, the CNS must continually be informed about the degree of muscle length and joint position. It is important, therefore, that these receptors do not adapt to a stimulus but continue to generate action potentials to relay this information to the CNS.

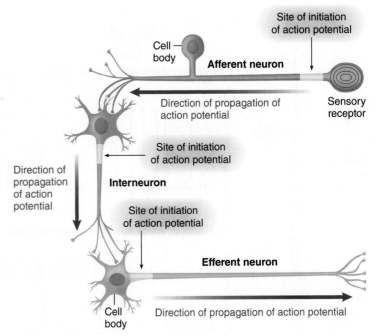

FIGURE 6-3 Comparison of the site of initiation of an action potential in the three types of neurons.

© Cengage Learning, 2013

Phasic receptors, in contrast, adapt rapidly; that is, they quickly stop responding to a maintained stimulus. Some of these receptors respond with a slight depolarization when the stimulus is removed, the **off response** (Figure 6-5b). Rapidly adapting receptors include *tactile (touch)* receptors in the skin that signal changes in pressure on the skin surface. Because these receptors adapt rapidly, a horse does not remain conscious of a saddle on its back or a bit in its mouth. You, for example, are not continually conscious of wearing your watch, rings, and clothing. When you put something on, you soon become accustomed to it because of these receptors' adaptation. When you take the item off, you are aware of its removal because of the off response.

Thus, phasic receptors essentially ignore continuous information. Why is this useful to survival? There are two interrelated reasons. (1) Often it is important to pay attention to a *change* in stimulus rather than status quo information. A sudden change in an environmental stimulus may signify danger, the arrival of prey, and so on, whereas, after a short time, if the stimulus continues and is of no threat or use, it can be ignored. (2) Brains have limited sensory processing capabilities. Ignoring an unimportant continuous stimulus prevents the brain from being overloaded with too much information. While this is useful, it can also be a problem in some situations. Many predators make use of this phenomenon in, for example, stalking or ambushing behaviors. Consider a lioness stalking a gazelle as she moves at a slow even pace, or freezes in position for an extended time. Until she makes her final running attack, the gazelle may stop paying attention to her as she gradually gets closer. Also, brain scans reveal that when a person is doing a visual task akin to driving a car while talking (even hands-free) on a mobile phone, his or her brain significantly reduces attention to the visual task of driving to focus on the conversation and so increases the odds of an accident.

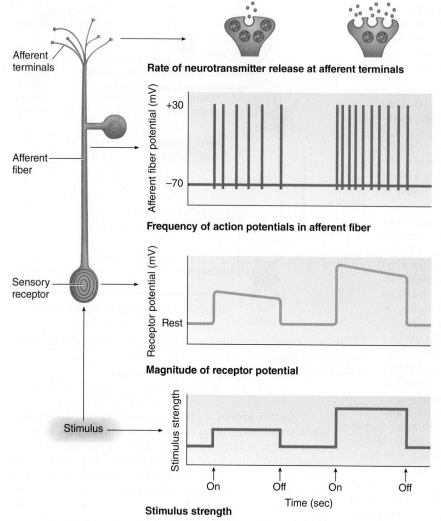

Afferent terminals

Afferent fiber

Sensory receptor

Stimulus

Rate of neurotransmitter release at afferent terminals

Afferent fiber potential (mV)

+30

−70

Frequency of action potentials in afferent fiber

Receptor potential (mV)

Rest

Magnitude of receptor potential

Stimulus strength

On   Off   On   Off

Time (sec)

Stimulus strength

FIGURE 6-4 Magnitude of receptor potential, frequency of action potentials in afferent fiber, and rate of neurotransmitter release at afferent terminals as a function of stimulus strength.

© Cengage Learning, 2013

receptor potential of a magnitude that reflects the intensity of the stimulus. As the stimulus continues, the pressure energy is dissipated because it causes the layers to slip (just as steady pressure on a peeled onion causes its layers to slip). Because this physical effect removes the steady component of the applied pressure, the underlying neuronal ending no longer responds with a receptor potential. Also contributing to adaptation is the electrochemical component, which involves changes in ionic movement across the receptor membrane. In the corpuscle, for reasons unknown, the Na^+ channels that opened in response to the stimulus are slowly inactivated, thus reducing the inward flow of Na^+ ions that was responsible for the receptor potential. With prolonged depolarization, the voltage-gated Na^+ channels cannot open, regardless of stimulus intensity.

Intrinsic Mechanism of Adaptation in Olfactory Receptors Olfactory (smell) receptors can also adapt, as you might recall from the last time you entered a musty room—at first you notice the musty odor, but after a few minutes you do not. As you will see later, binding of an odorant molecule to a receptor leads to the opening of Ca^{2+} channels. Evidence suggests that ions entering these open channels slowly activate a protein that in turn closes the ion channels. Importantly, some olfactory cells—particularly those that detect odors of dangerous substances—do not adapt. Rotting food, for example, produces sulfurous odors that do not lead to adaptation. (Similar odors are added by gas companies to natural gas—which is odorless—so that people will not ignore gas leaks.) The skunk takes advantage of this by producing a defensive cocktail of seven compounds, six of them based on sulfur. The cocktail slowly reacts with water to release more of the strong-smelling substances. Potential predators sprayed with this cocktail are continually bothered by a noxious smell until the odor molecules themselves decay.

The proximate mechanism by which adaptation is accomplished varies for different receptors and has not been fully elucidated for all receptor types. The known mechanisms fall into the two regulatory categories of *intrinsic* (local) and *extrinsic* (p. 19). Intrinsic mechanisms occur in the receptor cell itself and are not controlled by the brain. Extrinsic ones, on the other hand, are accomplished by the brain and so can be turned on or off rapidly. Let's look at examples.

Intrinsic Mechanisms of Adaptation in the Pacinian Corpuscle One receptor type that has been extensively studied is the mammalian **Pacinian corpuscle**, a rapidly adapting skin receptor that detects pressure and vibration. A comparable sensory receptor, the **Herbst corpuscle**, is the most widely distributed receptor in the skin of birds and can be found in the bills of some aquatic species. Adaptation in a Pacinian corpuscle is believed to involve both mechanical and electrochemical intrinsic components. The mechanical component depends on the physical properties of this receptor. A Pacinian corpuscle is a specialized receptor ending that consists of concentric layers of connective tissue resembling layers of an onion wrapped around the peripheral terminal of an afferent neuron (Figure 6-6). When pressure is first applied to the Pacinian corpuscle, the underlying terminal responds with a

Extrinsic Mechanism of Adaptation in Hearing In contrast to the preceding examples, reception of sound can be dampened by the brain in mammals (i.e., extrinsically). Sound vibrations are detected by *hair cells* in the inner ear (which we will discuss later). In turn, these hair cells release neurotransmitter onto an afferent neuron. The brain has an efferent neuron connecting to this synapse that releases GABA, thus suppressing the signal. This is what your brain does when it chooses to ignore so-called white noise like the hum of a constant fan motor. However, unlike the intrinsic

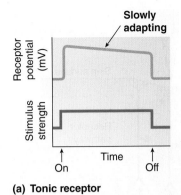

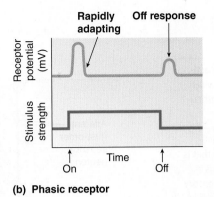

(a) Tonic receptor

(b) Phasic receptor

FIGURE 6-5 Tonic and phasic receptors. (a) A tonic receptor does not adapt at all or adapts slowly to a sustained stimulus and thus provides continuous information about the stimulus. (b) A phasic receptor adapts rapidly to a sustained stimulus and frequently exhibits an off response when the stimulus is removed. Thus, the receptor signals changes in stimulus intensity rather than relaying status quo information.

© Cengage Learning, 2013

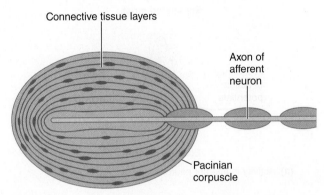

FIGURE 6-6 Pacinian corpuscle. The Pacinian corpuscle consists of concentric layers of connective tissue wrapped around the peripheral terminal of an afferent neuron.

© Cengage Learning, 2013

mechanisms of smell (which you cannot consciously reactivate), you can instantly hear the fan noise if someone calls your attention to it (as your brain shuts off its GABA neuron). This important mechanism allows, for example, an animal to ignore the sound of running water when that information is useless (freeing up the brain to listen for prey or predators), or to instantly hear it if it is thirsty.

Adaptation should not be confused with *habituation* (see p. 197). Although both these phenomena involve decreased neural responsiveness to repetitive stimuli, they operate at different points in the neural pathway. Adaptation is a receptor adjustment in the PNS, whereas habituation involves a modification in synaptic effectiveness in the CNS.

Each somatosensory pathway is "labeled" according to modality and location

On reaching the spinal cord of vertebrates, afferent information has two possible destinations: (1) It may become part of a reflex arc, bringing about an appropriate effector response, and/or (2) it may be relayed upward to the brain via ascending pathways for further processing (and possible conscious awareness). Pathways conveying sensory information consist of discrete chains of neurons, or **labeled lines,** synaptically interconnected in a particular sequence to accomplish progressively more sophisticated processing of the sensory information.

Labeled Lines Sensory signals begin in the periphery and are carried by the PNS either to the spinal cord or the medulla, depending on which sensory pathway is involved. There a secondary neuron has an axon that synapses in the *thalamus* "relay station" (p. 186), which in turn sends an axon to the appropriate *sensory cortex* (see Figure 5-26a). With each step, the input may be enhanced or filtered. In sum, a particular sensory modality detected by a specialized receptor type is sent over a specific afferent and ascending pathway to excite a defined area in the cortex. Thus, different types of incoming information are kept separated within specific labeled lines between the periphery and the cortex. In this way, even though all information is propagated to the CNS via the same type of signal (action potentials), the brain can decode the type and location of the stimulus, following Müller's doctrine of specific nerve energies (p. 211). The mechanism of perception in the specific sensory cortexes is not understood, but if one experimentally stimulates one of them (for example, by an electrode in the optical cortex), a human subject will report the perception of that cortex's speciality (e.g., light). Table 6-1 summarizes how the CNS is informed of the type (what?), location (where?), and intensity (how much?) of a stimulus.

Acuity is influenced by receptive field size and filtered by lateral inhibition

Let's look at sensory enhancement and filtering more closely, using skin senses as an example. Each sensory neuron responds to stimulus information only within a circumscribed region of the skin surface surrounding it; this region is known as its **receptive field.** The size of a receptive field varies inversely with the density of receptors in the region; the more closely receptors of a particular type are spaced, the smaller the area of skin each monitors. The smaller the receptive field in a region, the greater its **acuity** or **discriminative ability.** Compare the tactile (touch) discrimination in your fingertips with that in your elbow by "feeling" the same object with both. For example, you can discern more precise information about the object with your richly innervated fingertips because the receptive fields there are small; as a result, each neuron signals information about small, discrete portions of the object's surface. An estimated 17,000 tactile mechanoreceptors are present in the fingertips and palm of each human hand, whereas the skin over the elbow is served by relatively few sensory endings with larger receptive fields. Thus, within each large receptive field, subtle differences in pressure cannot be discriminated. The

TABLE 6-1 Coding of Sensory Information

Stimulus Property	Mechanism of Coding
Type of Stimulus (Stimulus Modality)	Distinguished by the type of receptor activated and the specific pathway over which this information is transmitted to a particular area of the cerebral cortex
Location of Stimulus	Distinguished by the location of the activated receptor field and the pathway that is subsequently activated to transmit this information to the area of the somatosensory cortex representing that particular location
Intensity of Stimulus (Stimulus Strength)	Distinguished by the frequency of action potentials initiated in an activated afferent neuron and the number of receptors (and afferent neurons) activated

© Cengage Learning, 2013

distorted cortical representation of various body parts in the *sensory homunculus* (see Figure 5-30b) corresponds precisely with the innervation density; more cortical space is allotted for sensory reception from areas with smaller receptive fields and, accordingly, greater tactile discriminative ability.

Besides receptor density, a second factor influencing acuity is **lateral inhibition**, which enhances *contrast*. You can appreciate the importance of this phenomenon by slightly indenting the surface of your skin with the point of a pencil (Figure 6-7a). The receptive field is excited immediately under the center of the pencil point where the stimulus is most intense, but the surrounding receptive fields are also stimulated, only to a lesser extent because they are less distorted. If information from these marginally excited afferent fibers in the fringe of the stimulus area were to reach the cortex, localization of the pencil point would be blurred. To facilitate localization and sharpen contrast, lateral inhibition occurs within the CNS (Figure 6-7b). The most strongly activated signal pathway originating from the center of the stimulus area inhibits the less excited pathways from the fringe areas. This occurs via inhibitory interneurons that pass laterally between ascending fibers serving neighboring receptive fields. Blocking further transmission in the weaker inputs increases the contrast between wanted and unwanted information so that the pencil point can be precisely localized.

As we noted earlier, perception is not reality. The reality is that the pencil distorts more of your skin than just at the point. The perception, however, is just of the point itself, which is presumably more useful information. The extent of lateral inhibition within sensory pathways varies for different modalities. Those with the most lateral inhibition—touch and vision—bring about the most accurate localization. As we mentioned earlier, retinas enhance the contrast of edges, making blurry boundaries appear sharp, by using lateral inhibition.

Now we turn to the major organs and modalities of senses, primarily the extero- and proprioceptors. The roles of most interoceptors are discussed in later chapters.

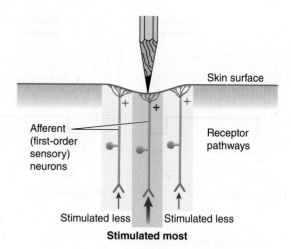

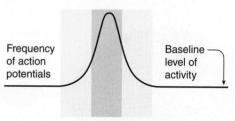

(a) Activity in afferent neurons

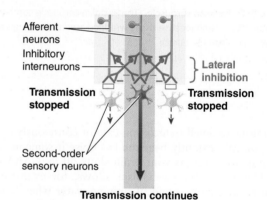

(b) Lateral inhibition

FIGURE 6-7 Lateral inhibition. (a) The receptor at the site of most intense stimulation is activated to the greatest extent. Surrounding receptors are also stimulated but to a lesser degree. (b) The most intensely activated receptor pathway halts transmission of impulses in the less intensely stimulated pathways through lateral inhibition. This process facilitates localization of the site of stimulation.

© Cengage Learning, 2013

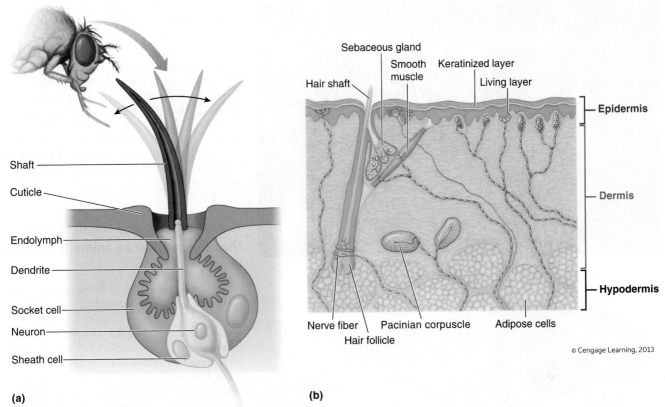

Shaft
Cuticle
Endolymph
Dendrite
Socket cell
Neuron
Sheath cell

(a)

Sebaceous gland
Smooth muscle
Hair shaft
Keratinized layer
Living layer
Epidermis
Dermis
Hypodermis
Nerve fiber
Hair follicle
Pacinian corpuscle
Adipose cells

© Cengage Learning, 2013

(b)

FIGURE 6-8 Mechanoreception. (a) Diagram of a bristle mechanoreceptor in *Drosophila,* consisting of a bristle and three cells: a socket cell, a sheath cell, and a neuron. When the bristle moves, it distorts the membrane of the neuron's dendrites, opening up mechanogated channels. (b) Vertebrate skin mechanoreceptors include free nerve endings that wrap around the base of the hair shaft.

Source: (a) R. C. Walker, A. T. Willingham, & C. S. Zucker, 2000, "A *Drosophila* mechanosensory transduction channel," *Science* 187:2229–2234, Figure 1A. Copyright © 2000 AAAS. Reprinted by permission.

check your understanding 6.2

Compare tonic and phasic receptors and their uses.

Explain how acuity is influenced by receptive field size and by lateral inhibition.

Define and compare *adequate stimulus, receptor potential, graded potential,* and *action potential* as well as their roles in the sensory process.

6.3 Mechanoreception: Touch, Pressure, and Proprioception

Detection of physical forces in the environment is possibly the most ancient sense, as we noted earlier. Virtually all single-celled motile organisms react to surface contact with an external object. In animals, this mode of sensing is found in integuments and some internal organs for detecting the somesthetic senses of touch and pressure. Mechanoreception is also used in proprioception and hearing. Though all of these senses are presumed to use mechanically gated ion channels, only a few of the channels involved have been identified.

Animal integuments may have both touch and pressure receptors

Touch and pressure receptors in integuments are typically in sensory dendrites (Figure 6-8a). Researchers have investigated these channels in the exoskeleton of insects, specifically the sensors located at the hinge of leg bristles. The channel proteins have external protein fibers attached to them, and when these fibers are stretched or distorted (as when vertebrate skin is touched, or when the insect bristle moves), they essentially pull open the gate of the ion channel. Such external fibers attached to channels may also act as levers, amplifying very small movements into large ones at the channel. Cations entering the sensory dendrite then create a receptor potential and, if the signal is strong enough, action potentials.

The fruit fly channel is called *TRP-N1,* a mechanically gated Ca^{2+} channel. Of note is that TRP-N1 belongs to the superfamily of *transient receptor-potential (TRP)* proteins involved in sensory transduction in eukaryotes from yeast to mammals (as you will see later).

The skin of vertebrates has three or more types of touch and pressure mechanoreceptors. Earlier we saw the Pacinian corpuscle, which detects deep pressure. Closer to the skin surface are highly sensitive touch sensors consisting of

late very precise movements of the small limbs of arthropods. However, as you recall, these signals cannot travel long distances.

The stretch receptors in vertebrates are **muscle spindles, Golgi tendon organs,** and other such sensors, discussed in Chapter 8. In humans, these sensors provide feedback to the brain in the so-called **kinesthetic sense,** a conscious "sixth sense" that lets you know where your limbs are even when you have your eyes closed. These sensors also provide feedback to the cerebellum while it is learning motor skills to produce smooth coordination (p. 184).

The statocyst is the simplest organ that can monitor an animal's position in space

Numerous groups of animals have gravity receptors known as **statocysts** (Greek *statos,* "*standing*"; *kystis,* "bag") (Figure 6-9), which are considered the simplest organs of equilibrium. This sense organ is particularly important in animals that are essentially neutrally buoyant, such as fish, because they do not get information about gravity from other sensory sources. (Notably, insects lack this special sense organ and instead rely on visual and joint proprioceptors to orient themselves.) A statocyst is essentially a hollow chamber lined with ciliated mechanoreceptors that contain dense, movable objects, the **statoliths.** For example, the statoliths of lobsters are sand grains glued together by mucus. In other animals, the statoliths consist of calcium carbonate fused together to resemble tiny marbles. When a body movement tilts the statocyst, the statoliths contained inside it move in the same direction, bending the sensory hairs, opening mechanically gated channels and thus generating action potentials. On reception and integration of the neural signals in the brain, motor neurons are activated, which can restore the animal to its equilibrium position.

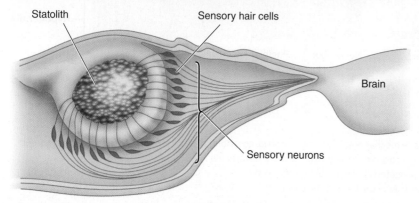

FIGURE 6-9 A statocyst, an organ of equilibrium. Shown here is the organ in a lobster's antennule.

© Cengage Learning, 2013

numerous dendrites with a large surface area, which can act as levers when pushed. There are also touch mechanoreceptors at the base of hairs, much like those associated with insect bristles (Figure 6-8b).

Proprioceptors in the muscles, tendons, and joints give information on limb position and motion

Another major use of mechanoreception is the detection of motion and position. This sense is critical to most animals' ability to navigate and orient properly in the environment and to move in a coordinated manner. Examples of these mechanoreceptors include the stretch and related sensors associated with muscle movement. In fruit flies and the nematode *C. elegans,* the TRP-N1 channel has been identified (by fluorescent labeling, p. 34) in proprioceptor neurons. Mutant flies and worms with nonfunctional TRP-N1 genes had uncoordinated muscle control, and the flies could neither walk nor fly. In arthropod limbs, many of these receptors are **nonspiking** neurons, that is, ones that do not generate action potentials. Rather they transmit information in strictly *graded* (rather than frequency) fashion. This may allow for a continuous range of corrective signals (as opposed to the quantized nature of frequency coding) to regu-

The lateral line system of amphibians and fish detects motion in the surrounding water

In fish, a specialized somesthetic and proprioceptive mechanoreceptor system extends the length of the animal's body. Information about the fish's orientation with respect to gravity, its swimming velocity and details about water currents and vibrations are collected by a string of **neuromast cells** (Figure 6-10a and b), the basic sensory unit of the **lateral line system.** Fish behavior linked to lateral line function includes object and predator avoidance and the ability to school. Cave and deep-sea fishes, as well as species active at night, rely on highly sensitive lateral line systems for detecting both prey and predators.

Neuromasts evolved in early vertebrate ancestors and persist in bony fish, lampreys, sharks, and aquatic amphibians. As you will see shortly, derivatives are also found in the inner ear of reptiles, birds, and mammals for proprioception and hearing. Neuromasts may occur free on the skin's surface, in small pits called *pit organs*, or may be lined up in rows within fluid-filled grooves or canals. The lateral line system in fish runs along the sides of the body onto the head, where it divides into three branches, two to the snout and one to the lower jaw. In many species of fish the lateral, often-pigmented rows of canal pores give rise to the name of the organ. Each neuromast is dome shaped, with up to several hundred mechanoreceptor sensory **hair cells** clustered at its base. Microvillar processes called **stereocilia** (Figure 6-10b) are the actual sensory transducers that protrude from the sensory hair cells into a jellylike substance (at the base of the dome). The cilia of the hair cells may actually extend up to 50 microns into the surrounding water, usually embedded in an overlying caplike, gelatinous layer, the **cupula**. Stereocilia are arranged together into ciliary bundles and are oriented according to size. The tallest row of stereocilia lies closest to an elongated cilium, the **kinocilium** (Figure 6-10b). Signals are sent to the brain in response to the bending of the kinocilium and stereocilia, using a mechanism identical to that of the mammalian inner ear, which we describe shortly.

A swimming fish sets up a pressure wave in the water that can be detected by the lateral line systems of other fishes. It also sets up a bow wave in front of itself, the pressure of which is higher than that of the wave flow along its sides. These minute differences in pressure are detected by its own lateral line system. As the fish approaches a solid object, such as a rock or the glass wall of an aquarium, the pressure waves around its body are distorted enough that the lateral line system registers these changes and results in the fish swerving away. Schooling fish also coordinate their motions by detecting pressure and flow changes, allowing precise movements.

Hair cells transduce movement using externally gated mechanical channels (thought to be TRP channels, p. 217). The stereocilia of each hair cell are organized into rows of graded heights ranging from short to tall in a staircase pattern (Figure 6-11a). Importantly, external **tip links**, which are CAMs (cell adhesion molecules, p. 65), link the tips of stereocilia in adjacent rows (Figure 6-11a). In neuromasts, as well as all the inner-ear senses of reptiles, birds, and mammals, hair cells normally send out continuous bursts of nerve impulses when they are at rest (that is, the cilia are unbent; Figure 6-11b left). This is because the mechanically gated channels in the ciliary tips are partly open. When a force causes the cilia to bend (by pressure waves, sound waves, or gravity) towards the tallest stereocilium (sometimes called the kinocilium), the tip links stretch, tugging open the cation channels to which they are attached (Figure 6-11b middle). When more cation channels are pulled open, more positive ions enter the hair cell, depolarizing it. When a force moves the cilia in the opposite direction, the hair bundle bends away from the tallest stereocilium, slackening the tip links and closing all the channels. As a result, ion entry ceases, hyperpolarizing the hair cell (Figure 6-11b right).

In fish neuromasts, the ion channels are for Ca^{2+}. In the mammalian hearing organ (the *cochlea*, which we will describe shortly), the stimulating ion is unusual because of the unique composition of the special ECF (*endolymph*) that bathes the stereocilia. In sharp contrast to ECF elsewhere, endolymph has a higher concentration of K^+ than found inside the hair cell. Some cation channels are open in a resting

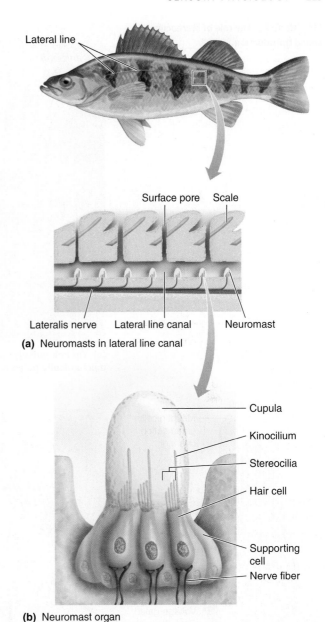

(a) Neuromasts in lateral line canal

(b) Neuromast organ

FIGURE 6-10 The lateral line system in fishes. (a) The lateral line system consists of a row of *neuromasts* set in pits and canals along each side of the body. (b) Vertical section through the skin and lateral line canal.

Source: Modified from K. Liem, W. Bemis, W. Walker, & L. Grande, 2001, *Functional Anatomy of Vertebrates: An Evolutionary Perspective*, 3rd ed. (Belmont, CA: Brooks/Cole), Figure 12-8, p. 406.

hair cell, allowing low-level K^+ entry down its concentration gradient.

The vestibular apparatus of vertebrates detects position and motion of the head and is important for equilibrium and coordination of head, eye, and body movements

The advanced vertebrate inner ear has two specialized sensory components using hair cells: the **cochlea** (from the Greek *kochlias*, "snail"), a tightly coiled structure in mammals and

FIGURE 6-11 The role of stereocilia in sound transduction.

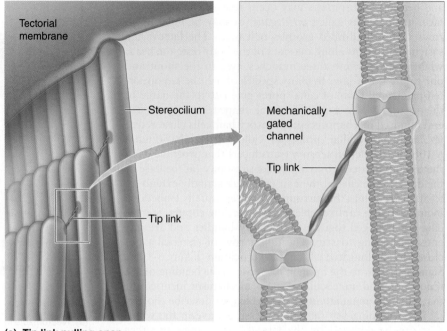

(a) Tip link pulling open mechanically gated channel

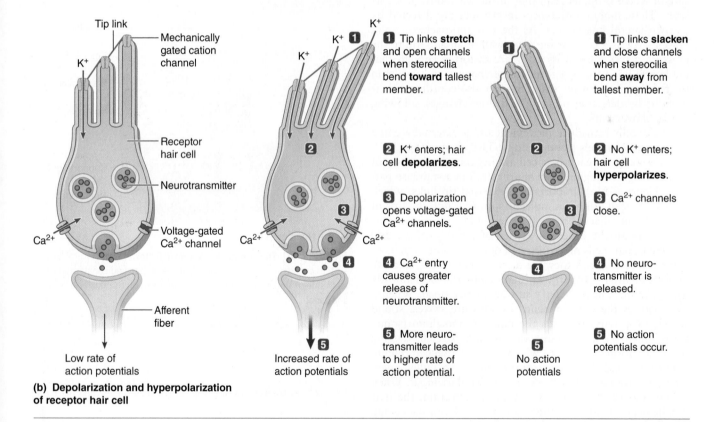

(b) Depolarization and hyperpolarization of receptor hair cell

a curved tube in birds and reptiles, which is used in hearing; and the **vestibular apparatus,** which provides proprioceptive information essential for the sense of equilibrium and for coordinating head movements with eye and postural movements. The vestibular apparatus consists of two sets of structures lying within a tunneled-out region of the temporal bone near the cochlea—the *semicircular canals* and the *otolith or-*

gans (the *utricle* and *saccule*) (Figure 6-12a). Vertebrates without a cochlea (fish and amphibians) also use the vestibular apparatus for hearing. The vestibular hair cells have synapses with afferent neurons whose axons join with those of the other vestibular structures to form the **vestibular nerve.** This nerve in turn unites with the **auditory nerve** from the cochlea to form the **vestibulocochlear nerve.**

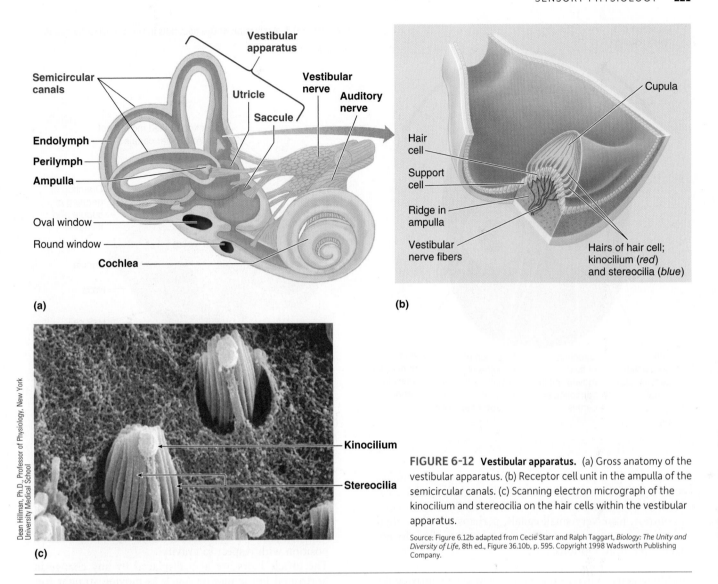

(a)

(b)

(c)

Dean Hillman, Ph.D., Professor of Physiology, New York University Medical School

—— **Kinocilium**

—— **Stereocilia**

FIGURE 6-12 Vestibular apparatus. (a) Gross anatomy of the vestibular apparatus. (b) Receptor cell unit in the ampulla of the semicircular canals. (c) Scanning electron micrograph of the kinocilium and stereocilia on the hair cells within the vestibular apparatus.

Source: Figure 6.12b adapted from Cecie Starr and Ralph Taggart, *Biology: The Unity and Diversity of Life*, 8th ed., Figure 36.10b, p. 595. Copyright 1998 Wadsworth Publishing Company.

Semicircular Canals Let's first examine the vestibular senses. The **semicircular canals** (Figure 6-12a) detect *rotational or angular acceleration or deceleration of the head,* such as when starting or stopping, spinning, diving, or turning the head. Each ear contains three semicircular canals arranged three-dimensionally in planes at right angles to each other. The receptive hair cells of each semicircular canal lie on top of a ridge located in the **ampulla,** a swelling at the base of the canal (Figures 6-13a and b). The hairs are embedded in a cupula (as in a neuromast), which protrudes into the endolymph within the ampulla. The cupula sways in the direction of fluid movement, much like seaweed leaning in the direction of the prevailing tide.

Acceleration or deceleration during rotation of the head in any direction causes endolymph movement in at least one of the semicircular canals because of their three-dimensional arrangement (Figure 6-13a and b). As the head starts to move, the bony canal and the ridge of hair cells embedded in the cupula move with the head. Initially, however, the fluid within the canal, not being attached to the skull, does not move in the direction of the rotation but lags behind because of its inertia. (Because of inertia, a resting object remains at rest, and a moving object continues to move in the same di-

rection unless acted on by some external force.) Thus, the endolymph that is in the same plane as the head movement is shifted in the opposite direction from the movement (like your body tilting rightward as the car in which you are riding suddenly turns left). This relative fluid movement causes the cupula to lean in the opposite direction from the head movement, resulting in the bending of its embedded sensory hair. If the head movement continues at the same rate in the same direction, the endolymph catches up and moves in unison with the head so that the hairs return to their unbent position. When the head slows down and stops, the reverse occurs. The endolymph briefly continues to move in the direction of the rotation while the head decelerates to a stop. When the endolymph gradually comes to a halt, the hairs straighten again. Thus, the semicircular canals detect *changes in the rate of rotational movement of the head.* They do not respond when the head is motionless or during circular motion at a constant speed. (However, think about why you get dizzy if you spin around repeatedly.)

The size of the semicircular canals has changed greatly with evolutionary adaptation. In primates that swing through trees and flying vertebrates, the canals are very large, evolved for processing complex motion in three

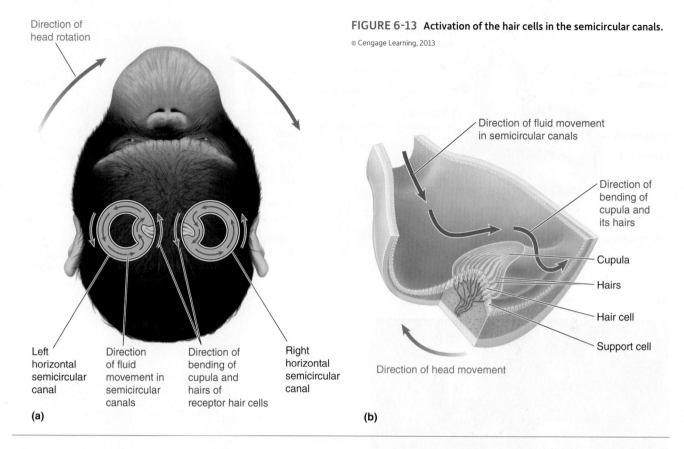

Direction of head rotation

Left horizontal semicircular canal

Direction of fluid movement in semicircular canals

Direction of bending of cupula and hairs of receptor hair cells

Right horizontal semicircular canal

(a)

FIGURE 6-13 Activation of the hair cells in the semicircular canals.
© Cengage Learning, 2013

Direction of fluid movement in semicircular canals

Direction of bending of cupula and its hairs

Cupula

Hairs

Hair cell

Support cell

Direction of head movement

(b)

dimensions. They are of moderate size in terrestrial animals, including extinct terrestrial ancestors of whales (whose canals can be reconstructed from fossil skulls). Modern whales, in contrast, have very small canals, perhaps to reduce their sensitivity, to prevent seasickness, or because they do not make rapid head motions.

Otolith Organs Whereas the semicircular canals provide the CNS with information about rotational changes in head movement, the **otolith organs** provide information about the position of the head relative to gravity and changes in rate of linear motion. The otolith organs, the **utricle** and **saccule**, are saclike structures housed within a bony chamber situated between the semicircular canals and the cochlea (see Figure 6-12a). The cilia of the hair cells in these organs also protrude into an overlying gelatinous sheet, whose movement displaces the cilia and results in changes in hair cell potential. Many tiny crystals of calcium carbonate—the **otoliths** ("ear stones")—similar to the statoliths associated with statocysts in many nonvertebrates—are suspended within the gelatinous layer, making it heavier and giving it more inertia than the surrounding fluid (Figure 6-14). When an animal is in an upright position, the hairs within the utricles are oriented vertically and the saccule hairs are lined up horizontally.

Let's look at the utricle as an example. Its otolith-embedded, gelatinous mass shifts positions and bends the hairs in two ways, described as follows.

1. When the head is tilted in any direction other than vertical (that is, other than straight up and down), the hairs are bent in the direction of the tilt because of the gravitational force exerted on the top-heavy gelatinous layer (Figure 6-14b). Within the utricles on each side of the head, some of the hair cell bundles are oriented to depolarize and others to hyperpolarize when the head is in any position other than upright. The CNS thus receives different patterns of neural activity, depending on head position with respect to gravity.

2. The utricle hairs are also displaced by any change in horizontal linear motion (such as moving straight forward, backward, or to the side). As an animal starts to walk forward (Figure 6-14c), the top-heavy otolith membrane at first lags behind the endolymph and hair cells because of its greater inertia. The hairs are thus bent to the rear, in the opposite direction of the forward head movement. If the walking pace is maintained, the gelatinous layer soon catches up and moves at the same rate as the head so that the hairs are no longer bent. When the animal stops walking, the otolith sheet continues to move forward briefly as the head slows and stops, bending the hairs toward the front. Thus, the hair cells of the utricle detect horizontally directed linear acceleration and deceleration, but they do not provide information about movement in a straight line at constant speed.

The saccule functions similarly to the utricle, except that it responds selectively to the tilting of the head away from a horizontal position (such as diving) and to vertically directed linear acceleration and deceleration (such as leaping up and down). An interesting variation occurs in flatfish such as flounders (*Pleuronectiformes*): Fry (immature fish), which have body forms and orientation similar to most fish, metamorphose into adults that lie and swim on one side. Although one eye migrates across the top of the head, leav-

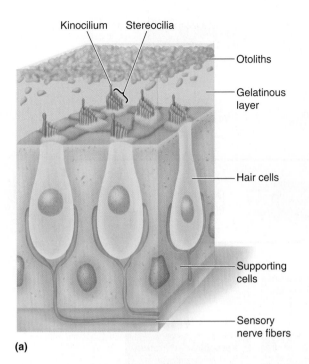

Kinocilium   Stereocilia

Otoliths

Gelatinous layer

Hair cells

Supporting cells

Sensory nerve fibers

(a)

FIGURE 6-14 Utricle. (a) Receptor unit in utricle. (b) Activation of the utricle by a change in head position. (c) Activation of the utricle by horizontal linear acceleration.

© Cengage Learning, 2013

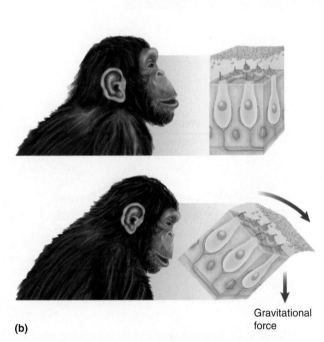

(b)

Gravitational force

(c)

ing the bottom side eyeless, the vestibular system does not change in orientation, and so functions at 90 degrees relative to that of other fish. Ichthyologists believe the saccule replaces the utricle in posture regulation, suggesting that central nervous pathways have reorganized.

Signals arising from the various components of the vestibular apparatus are carried through the vestibulocochlear nerve to the cerebellum and the **vestibular nuclei,** a cluster of neuronal cell bodies in the brain stem. Here, the vestibular information is integrated with input from the skin surface, eyes, joints, and muscles for (1) maintaining balance and desired posture; (2) controlling the external eye muscles so that the eyes remain fixed on the same point, despite head

movement; and (3) perceiving motion and orientation (Figure 6-15).

check your understanding 6.3

Why are proprioceptors important for locomotion?

Describe how hair cells work in general, and in the fish lateral line system.

Compare the functions of statocysts, semicircular canals, utricle, and saccule.

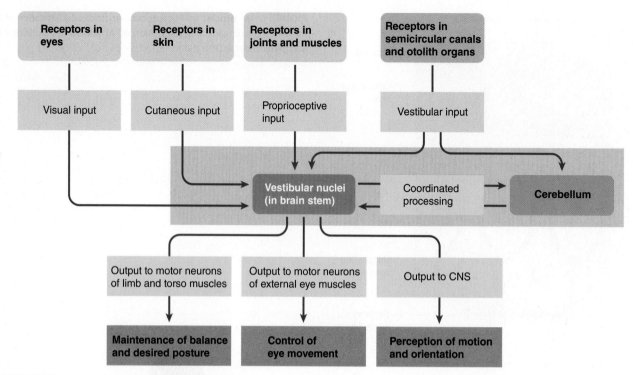

FIGURE 6-15 Input and output of the vestibular nuclei.

© Cengage Learning, 2013

6.4 Mechanoreception: Ears and Hearing

Hearing is the neural perception of sound energy, which travels as **sound waves** through a medium—liquid, gas, or solid (but unlike light, not in a vacuum). Waves are traveling vibrations of air or liquid that consist of regions of high pressure caused by compression of molecules, alternating with regions of low pressure caused by rarefaction of the molecules. Detection of sound waves is less common than photoreception and touch; for example, there is no evidence for sound sensing in single-celled organisms. But many animals do detect sound, especially vertebrates and arthropods, using some form of mechanoreceptor, often organized into complex organs called *ears*. Like light and smell, but unlike taste and touch, sound is a long-range "early warning" signal used to detect predators, competitors, physical threats, potential prey, and mating calls.

Fishes hear with lateral lines and otolith organs, aided by the gas bladder in some species

Fishes can hear noises for some of these purposes, such as the intense "whistle" of the male toadfish used as a mating call during the breeding season. Because sound waves are waves of pressure, the lateral lines in some species are used to detect very-low-frequency sounds of 100 Hz or less. Researchers think fish and amphibians also detect sounds that penetrate the body using hair cells of otolith organs in their inner ears. Some fish (such as minnows and catfish) have a **Weberian apparatus** (Figure 6-16), consisting of bone and fluid sacs that transfer sound from the *gas (or swim) bladder* (p. 539) to the inner ear. This is advantageous because the gas bladder is much more sensitive to sound vibrations than are the otolith organs. Whether directly stimulated by sound waves or indirectly by the gas bladder, otoliths in motion result in a shearing action on the cilia of the sensory hair cells and thus in generation of action potentials. Many fishes (such as cod, *Gadus morhua*) also can localize sound, although the precise mechanism is not known.

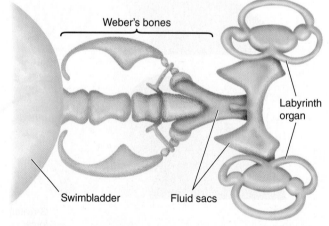

FIGURE 6-16 Weberian apparatus for fish hearing, which includes Weber's bones and two fluid sacs connecting the swim bladder with the otolith organs.

© Cengage Learning, 2013

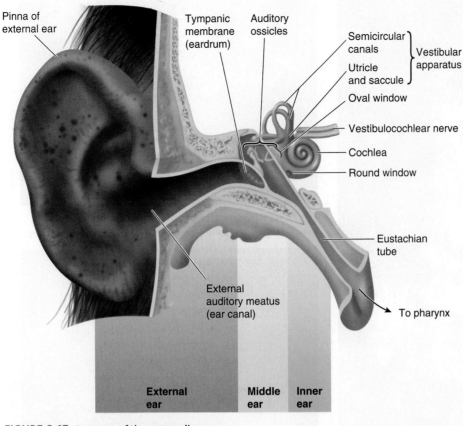

Pinna of external ear

Tympanic membrane (eardrum)

Auditory ossicles

Semicircular canals
Utricle and saccule
} Vestibular apparatus

Oval window

Vestibulocochlear nerve

Cochlea

Round window

Eustachian tube

To pharynx

External auditory meatus (ear canal)

External ear   **Middle ear**   **Inner ear**

FIGURE 6-17   Anatomy of the mammalian ear.

© Cengage Learning, 2013

Many terrestrial vertebrates and insects receive sound initially with a tympanic membrane, aided by other external features in some groups

Compared to animals in aquatic habitats, animals living in air evolved different structures capable of receiving and processing sounds traveling through the air. In terrestrial vertebrates and many insects, initial sound reception involves a **tympanic membrane** (eardrum) or comparable structure that vibrates as sound hits it. As we will explore later, many insects have such structures on their abdominal segments or legs. Among vertebrates, amphibians and some reptiles have a tympanic membrane located at the back of the animal's head on the skull surface. In other reptiles, including birds and in mammals, there is a canal leading from the external environment to the tympanic membrane. Snakes lack external ears, but recent research shows that their *jaws* can pick up sound vibrations in the ground and transmit them to their inner ears.

Vertebrates typically have two ears, which allows for localization of sounds. Whether a sound is approaching from the right or left is determined by two cues. First, the sound wave reaches the ear closer to the sound source slightly before it arrives at the farther ear. Second, the sound is less intense as it reaches the farther ear, because the head acts as a sound barrier that partially disrupts the propagation of sound waves. The auditory cortex integrates all these cues to determine the location of the sound source.

Let's look at mammals in more detail. In mammals the **external ear** (Figure 6-17) consists of the *pinna* (ear), **exter-**nal auditory meatus (ear canal), and tympanic membrane. Unique to mammals, the **pinna**, a prominent skin-covered flap of cartilage, collects sound waves and channels them down the external ear canal, improving hearing acuity. Most mammals (dogs, for example) can cock their ears in the direction of sound to collect more sound waves, but human ears are relatively immobile. Because of its shape, the pinna partially shields sound waves that approach the ear from the rear and thus helps an animal distinguish whether a sound is coming from directly in front or behind.

Among nonmammalian vertebrates, birds have the most highly evolved auditory system. The avian external ear is inconspicuous, although ear-flaps or specialized feathers may reflect or amplify sound, as in the barn owl (*Tyto alba*). In some species of owls, the external openings in their left and right ears differ in height and direction on the head in order to more accurately locate a particular sound source. Remarkably, barn owls can successfully hunt prey (such as a field mouse making a faint noise) in total darkness. In the owl's midbrain, certain cells fire only when sounds are received from a particular spot in the external environment. In principle, this could give the owl the ability to "see" (that is, visually perceive) its prey's location using sound!

Recall that the inner ear of terrestrial vertebrates evolved a separate sound-sensing organ, the *cochlea*. We explore this and *middle ear* structures in more detail below.

Sound is characterized by its intensity (loudness), pitch (tone), and timbre (quality)

Let us look first at the transmission and properties of sound in the environment. Sound waves can travel through air, solids and liquids. Because water is less compressible than air, sound attenuation is not as great. Thus, in water, sound travels at an average of 1,500 m/sec, compared to 340 m/sec in air. However, both the temperature of the water and the hydrostatic pressure can affect the measured speed. The greater the depth, the heavier the column of water above, and the faster sound waves are propagated.

Sound waves have three features—*intensity, pitch,* and *timbre*—all of which are important to the sense of hearing (see Figure 6-18).

Intensity   The intensity, or loudness, of a sound depends on the *amplitude* of the sound waves, or the pressure differences between a high-pressure region of compression and a low-pressure region of rarefaction. Loudness is expressed in decibels (dB), which are a logarithmic measure of intensity compared with the faintest sound that can be heard—the

Collared scops owl (*Otus lettia*). This bird hunts at night and has enhanced two major sensory systems to aid in this behavior. First, it has a specialized auditory system that amplifies sounds and localizes prey in three dimensions (see text). Second, it has large eyes with a reflective layer, the tapetum lucidum, behind the retina. This layer reflects photons back into the retina that were not detected initially (see p. 245).

Tympanic membranes (the large round "eardrum" seen on the head) vibrate with sound waves in the frog's initial step in hearing.

A dolphin head's dome-shaped melon focuses sound waves for the process of echolocation.

hearing threshold. Because of the logarithmic relationship, every 10 decibels indicates a 10-fold increase in loudness. Within the hearing range, the greater the amplitude, the louder the sound. An animal's ear may detect a wide range of sound intensities, from the slightest rustle of a leaf (about 10 dB) to the comparatively loud roar of a waterfall (perhaps 100 dB, a 100 million times louder than the leaf rustling).

Pitch The pitch or tone of a sound (for example, whether it is a C or a G note) is determined by the *frequency* of vibrations. The greater the frequency of vibration, the higher the pitch. Human ears can detect sound waves with frequencies from 20 to 20,000 cycles per second (Hertz, Hz) but are most sensitive to frequencies between 1,000 and 4,000 Hz (or 1 to 4 kilohertz, or kHz). In contrast, goldfish (*Carassius auratus*) detect frequencies between 50 to 3,000 Hz with the

best sensitivity between 200 to 1,000 Hz. Dogs, on the other hand, can hear sound frequencies up to 40 kHz. These higher frequencies, or **ultrasound,** consist of frequencies greater than those audible by the human ear, whereas **infrasound** consists of sounds below the range of frequencies audible by humans (20 Hz).

Infrasound is subject to very little attenuation in the environment and so can function in long-distance communication, and it travels well in the ground as well as in air and water. Salts in seawater more rapidly absorb high-frequency sounds, whereas low-frequency acoustic waves are not affected by the molecular structure of the salts and may travel for thousands of kilometers. Infrasound communication is particularly useful for animals separated by great distances. For example, whales and elephants use such infrasound for such communication. Baleen whales such as humpbacks may broadcasting mating calls or the location of feeding

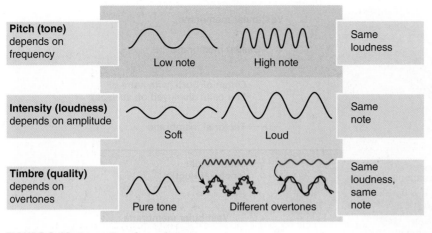

Pitch (tone) depends on frequency	Low note	High note	Same loudness
Intensity (loudness) depends on amplitude	Soft	Loud	Same note
Timbre (quality) depends on overtones	Pure tone	Different overtones	Same loudness, same note

FIGURE 6-18 Properties of sound waves.

© Cengage Learning, 2013

grounds over hundreds of kilometers using frequencies less than 60 Hz. Elephants use frequencies in the range of 14 to 25 Hz, which can travel in the ground over distances as great as several kilometers. This is valuable because once every four years and for only four days, a female elephant makes a special call that signals mature elephant bulls in the area that she is in estrus. Movement of separate elephant herds can also be coordinated by this form of communication. (Interestingly, research suggests that elephants can pick up these sound wave vibrations in the ground through their feet!)

Ultrasound, in contrast to infrasound, is useful because its numerous small (i.e., high frequency) waves can reflect well off very small features, making it ideal for **echolocation:** the use of sound echoes to detect objects in the environment. Echolocation (also called **biosonar** in reference to the human invention SONAR for SOund Navigation And Ranging) is particularly useful for animals active at night or living in dark waters. Consider insectivorous bats, which locate prey at night by emitting successive pulses of ultrasound; the duration of each varies from about 10 to 15 ms (5 pulses per second) and consists of a descending frequency spectrum that varies from approximately 100 kHz to 20 kHz. It has been shown that varying the frequency spectrum this way aids in discriminating details of echo patterns. Faint reflections (echoes) from the insect are then detected by enlarged pinnae on the bat (see p. 400), and its brain uses the information to compute location, size, and speed of the insects. Like the owl, bats may actually visualize their environments using sound.

Other vertebrates that can obtain information from echoes of the sounds they produce include toothed whales, shrews, and several species of birds such as cave swiftlets. Toothed whales (Odontoceti) such as dolphins generate patterned arrangements of clicks with sound frequencies between 200 Hz to 32 kHz for echolocating prey items (such as squid) and other objects in the environment. (They also use sounds of many frequencies for communication.) In dolphin echolocation, physiologists think click sounds (1) are produced by *lips* near the blowhole; (2) become focused by the *melon,* a fat-filled organ in the forehead, much as a lens focuses light; (3) travel until they hit a sound-reflecting object (such as another animal); (4) reflect back ("echo") toward the dolphin; and (5) are received in the *lower jaw* where a *fat body* focuses the waves into (6) the inner ear. The dolphin's brain then computes the distance of the object by the time delay of the echo, and by making repeated clicks (like bats) can judge the size and speed of the object. Again, actual visualization through sound is thought to occur in dolphin brains.

Timbre The **timbre,** or **quality,** of a sound depends on its *overtones,* which are additional frequencies superimposed on top of the fundamental pitch or tone. A tuning fork has a pure tone, but most sounds lack purity. For example, complex mixtures of overtones impart different sounds to different instruments playing the same note (a C note sounds different on a trumpet from the way it sounds on a piano). Overtones are likewise responsible for the characteristic differences in voices. Timbre enables the listener to distinguish the source of sound waves because each source produces a different pattern of overtones. Moreover, an echolocating dolphin can learn about an object's composition, because the timbre of reflected sounds differs depending on that composition.

The external ear and middle ear convert airborne sound waves into fluid vibrations in the inner ear

Now we turn to the fate of sound waves that reach an animal, focusing primarily on mammals. The specialized receptors for sound are located in the fluid-filled inner ear in terrestrial vertebrates. Airborne sound waves must therefore be channeled toward and transferred into the inner ear while at the same time compensating for the loss in sound energy that naturally occurs as sound waves pass from air into water. This function is performed by the external ear (described earlier) and the *middle ear.*

Recall the **tympanic membrane (eardrum),** stretched across the entrance to the middle ear, vibrates when struck by sound waves. The alternating higher- and lower-pressure regions of a sound wave cause the exquisitely sensitive eardrum to bow inward and outward in unison with the wave's frequency. For the membrane to be free to move as sound waves strike it, the resting air pressure on both sides of the tympanic membrane must be equal. The outside of the eardrum is exposed to atmospheric pressure through the ear canal. The inside of the eardrum facing the middle-ear cavity is also exposed to atmospheric pressure via the **Eustachian (auditory) tube,** which connects the middle ear to the **pharynx** (back of the throat) (see Figure 6-17). The Eustachian tube is normally closed, but it can be pulled open by yawning, chewing, and swallowing, permitting air pressure within the middle ear to equilibrate with atmospheric pressure. Infections originating in the throat sometimes spread through this tube to the middle ear.

Middle Ear The **middle ear** of mammals transfers the vibratory movements of the tympanic membrane to the fluid of the inner ear. This transfer is facilitated by a movable chain

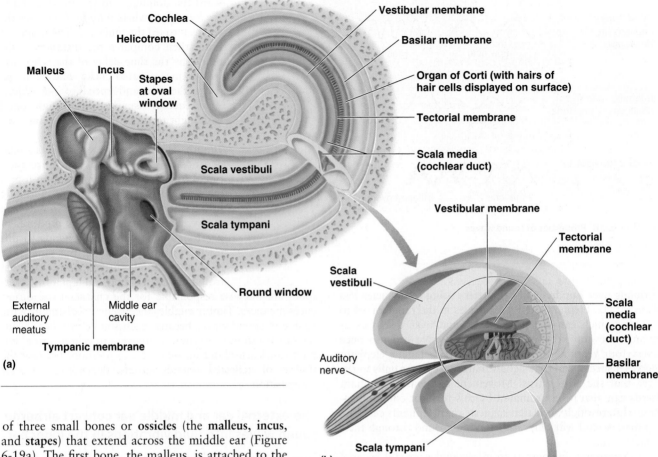

(a)

(b)

The stereocilia (hairs) from the hair cells of the basilar membrane contact the overlying tectorial membrane. These hairs are bent when the basilar membrane is deflected in relation to the stationary tectorial membrane. This bending of the inner hair cells' hairs opens mechanically gated channels, leading to ion movements that result in a receptor potential.

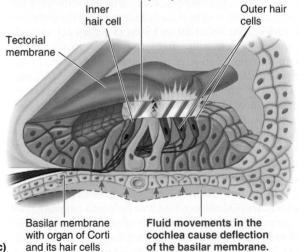

(c)

FIGURE 6-19 Middle ear and cochlea. (a) Gross anatomy of the middle ear and cochlea, with the cochlea "unrolled." (b) Cross section of the cochlea. (c) Enlargement of the organ of Corti. The stereocilia (hairs) from the hair cells of the basilar membrane are embedded in the overlying tectorial membrane. These hairs are bent when the basilar membrane is deflected in relation to the stationary tectorial membrane. This bending opens channels, leading to ion movements that result in a receptor potential.

© Cengage Learning, 2013

of three small bones or **ossicles** (the **malleus, incus,** and **stapes**) that extend across the middle ear (Figure 6-19a). The first bone, the malleus, is attached to the tympanic membrane, and the last bone, the stapes, is attached to the **oval window,** the entrance into the fluid-filled cochlea. As the tympanic membrane vibrates in response to sound waves, the chain of bones is set into motion at the same frequency, transmitting this frequency of movement from the tympanic membrane to the oval window. The resulting pressure on the oval window with each vibration produces wavelike movements in the inner ear fluid at the same frequency as the original sound waves. To create the greater pressure required to set cochlear fluid in motion, two mechanisms related to the ossicular system come into play. First, because the surface area of the tympanic membrane is much larger than that of the oval window, pressure is increased as force exerted on the tympanic membrane is conveyed to the oval window (pressure = force/unit area). Second, the lever action of the ossicles provides an additional mechanical advantage. Together, these mechanisms increase the force exerted on the oval window by 20 times what it would be if the sound wave struck the oval window directly. This additional pressure is sufficient to set the cochlear fluid in motion.

The tympanic membrane of anuran amphibians and reptiles (including birds) is connected to the inner ear by a single, bony ossicle, the **columella,** homologous to the stapes of mammals. Both embryological and fossil evidence show that the malleus, incus, and stapes evolved from jaw structures. Indeed, in 2011, a reptilian-like mammalian fossil was found in which there was a connection between the middle ear bones and a support bone arising from the jaw.

Several tiny muscles in the middle ear contract reflexively in response to loud sounds (over 70 dB in humans), tightening the tympanic membrane and limiting movement of the ossicular chain. This reduced movement of middle ear structures diminishes the transmission of loud sound waves to the inner ear, to protect the delicate sensory apparatus from damage. This reflex response is relatively slow, however, happening at least 40 msec after exposure to a loud sound. It thus provides protection only from prolonged loud sounds, not from sudden sounds like an explosion. Taking advantage of this reflex, World War II antiaircraft guns were designed to make a loud prefiring sound, to protect the gunner's ears from the much louder boom of the actual firing.

The cochlea of the inner ear contains the organ of Corti, the sense organ for hearing

The cochlear portion of the mammalian inner ear is a coiled tubular system lying deep within the temporal bone (see Figure 6-17). It is easier to understand the functional components of the mammalian cochlea by "uncoiling" it somewhat, as shown in Figures 6-19a and 6-20a. The cochlea is divided throughout most of its length into three fluid-filled longitudinal compartments. A blind-ended **cochlear duct**, or **scala media**, constitutes the middle compartment. It tunnels lengthwise through the center of the cochlea, almost but not quite reaching its end. The upper compartment, the **scala vestibuli**, follows the inner contours of the spiral, while the **scala tympani**, the lower compartment, follows the outer contours (Figure 6-19a and b). The fluid within the scala vestibuli and scala tympani is called **perilymph**. The cochlear duct contains a slightly different fluid, the **endolymph** (Figure 6-20a). The region beyond the tip of the cochlear duct where the fluid in the upper and lower compartments is continuous is called the **helicotrema**. The scala vestibuli is sealed from the middle ear cavity by the oval window, to which the stapes is attached. Another small membrane-covered opening, the **round window**, seals the scala tympani from the middle ear. The thin **vestibular membrane** forms the ceiling of the cochlear duct and separates it from the scala vestibuli. The **basilar membrane** forms the floor of the cochlear duct, separating it from the scala tympani. The basilar membrane is especially important because it bears the **organ of Corti**, the sense organ for hearing (Figure 6-19c).

Hair cells in the organ of Corti transduce fluid movements into neural signals

The organ of Corti, which rests on top of the basilar membrane throughout its full length, contains hair cells that are the receptors for sound. The hair cells (15,000 within each human cochlea) are arranged in four parallel rows along the length of the basilar membrane: one row of **inner hair cells** and three rows of **outer hair cells** (Figure 6-19c). Protruding from the surface of each hair cell are about 100 stereocilia, which you saw earlier (p. 219). These stereocilia are mechanically embedded in the **tectorial membrane**, an awning-like projection overhanging the organ of Corti throughout its length (Figure 6-19b and c).

The cochlea of other terrestrial vertebrates is a curved tube. In birds, the organ has short and tall (rather than inner and outer) hair cells embedded in the tectorial membrane. In contrast, the hair cells of non-avian reptiles project into the endolymph without an overlying tectorial membrane, and amphibians lack both tectorial and basilar membranes.

Returning to mammals, the pistonlike action of the stapes against the oval window sets up pressure waves in the upper compartment. Because fluid is incompressible, pressure is dissipated in two ways as the stapes causes the oval window to bulge inward: (1) displacement of the round window and (2) deflection of the basilar membrane (Figure 6-20a). In the first of these pathways, the pressure wave pushes the perilymph forward in the upper compartment, then around the helicotrema, and into the lower compartment, where it causes the round window to bulge outward into the middle-ear cavity to compensate for the pressure increase. As the stapes rocks backward and pulls the oval window outward toward the middle ear, the perilymph shifts in the opposite direction, displacing the round window inward. This pathway does not result in sound reception; it just dissipates pressure.

Pressure waves of frequencies associated with sound reception take a "shortcut." Pressure waves in the upper compartment are transferred through the thin vestibular membrane, into the cochlear duct, and then through the basilar membrane into the lower compartment, where they cause the round window to alternately bulge outward and inward. The main difference in this pathway is that *transmission of pressure waves through the basilar membrane causes this membrane to vibrate in synchrony with the pressure wave*. Because the organ of Corti rides on the basilar membrane, the hair cells also move up and down as that membrane oscillates.

Role of the Inner Hair Cells The inner and outer hair cells differ in function. The inner hair cells are the ones that "hear": They transform the mechanical forces of sound (cochlear fluid vibration) into the electrical impulses of hearing (action potentials propagating auditory messages to the cerebral cortex). Because the stereocilia of these receptor cells contact the stiff, stationary tectorial membrane, they are bent back and forth when the oscillating basilar membrane shifts their position in relationship to the tectorial membrane (Figure 6-19c). This back-and-forth mechanical deformation of the hairs alternately opens and closes mechanically gated cation channels (see Figure 6-11) in the hair cell, resulting in alternating depolarizing and hyperpolarizing potential changes—the receptor potential—at the same frequency as the original sound stimulus. The inner hair cells communicate, via a chemical synapse, with the terminals of afferent nerve fibers making up the **auditory (cochlear) nerve**. Thus, sound waves are translated into neural signals that the brain can perceive as sound sensations (Figure 6-21).

Role of the Outer Hair Cells Whereas the inner hair cells send auditory signals to the brain over afferent fibers, the outer hair cells do not signal the brain about sounds. Instead, the outer hair cells actively and rapidly elongate in response to changes in membrane potential, a function known as *electromotility*. These changes in length are believed to amplify or accentuate the motion of the basilar membrane. An analogy is a person deliberately pushing the pendulum of a grandfather clock in time with its swing, to accentuate its motion. Researchers speculate that such modi-

Cochlear duct
Vestibular membrane
Scala vestibuli
Incus
Malleus
Oval window
Tectorial membrane
Helicotrema
Cochlea
Perilymph
Hairs
Endolymph
Organ of Corti
Stapes
Perilymph
Basilar membrane
Scala tympani
Tympanic membrane
Round window

Fluid movement within the perilymph set up by vibration of the oval window follows two pathways:

Pathway 1: Through the scala vestibuli, around the heliocotrema, and through the scala tympani, causing the round window to vibrate. This pathway just dissipates sound energy.

(a)

Pathway 2: A "shortcut" from the scala vestibuli through the basilar membrane to the scala tympani. This pathway triggers activation of the receptors for sound by bending the hairs of hair cells as the organ of Corti on top of the vibrating basilar membrane is displaced in relation to the overlying tectorial membrane.

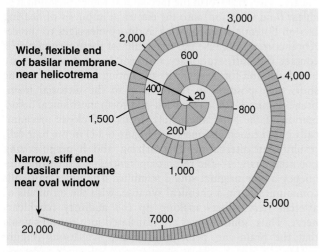

Wide, flexible end of basilar membrane near helicotrema

Narrow, stiff end of basilar membrane near oval window

3,000
2,000
600
400
20
4,000
800
1,500
200
1,000
5,000
7,000
20,000

(b) The numbers indicate the frequencies in cycles per second with which different regions of the basilar membrane maximally vibrate.

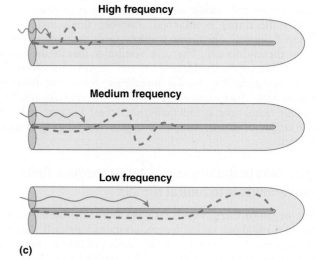

High frequency

Medium frequency

Low frequency

(c)

FIGURE 6-20 Transmission of sound waves in the mammalian ear. (a) Fluid movement within the perilymph set up by vibration of the oval window follows two pathways, one dissipating sound energy and the other initiating the receptor potential. (b) Different regions of the basilar membrane vibrate maximally at different frequencies. (c) The narrow, stiff end of the basilar membrane nearest the oval window vibrates best with high-frequency pitches. The wide, flexible end of the basilar membrane near the helicotrema vibrates best with low-frequency pitches.

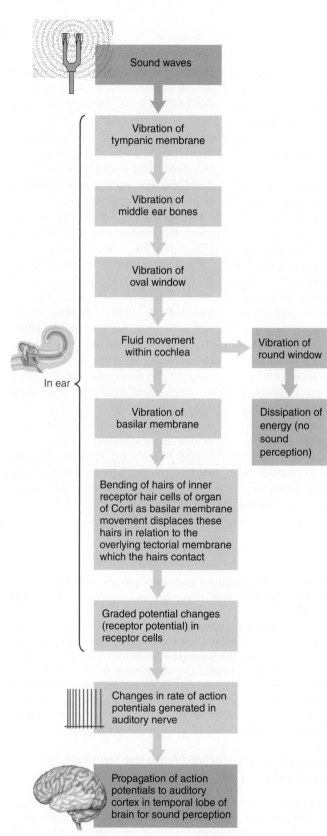

FIGURE 6-21 Pathway for sound transduction.

© Cengage Learning, 2013

fication of basilar membrane movement improves and tunes the stimulation of the inner hair cells. Thus, the outer hair cells enhance the response of the inner hair cells, making them exquisitely sensitive to sound intensity and highly discriminatory between various pitches of sound.

Pitch discrimination depends on the region of the basilar membrane that vibrates

Pitch discrimination (that is, the ability to distinguish between various frequencies of incoming sound waves) depends on the shape and properties of the basilar membrane. Different regions of the basilar membrane naturally vibrate maximally at different frequencies; that is, each frequency displays peak vibration at a different position along the membrane. The *narrow* end nearest the oval window, which is narrow and stiff, vibrates best with *high-frequency* pitches, whereas the *wide* end nearest the helicotrema, which is wide and flexible, vibrates maximally with *low-frequency* tones (Figure 6-20c; note again that the cochlea has been diagrammatically "unrolled" in this figure to aid understanding, but the real cochlea does not do this!). The pitches in between are sorted out along the length of the membrane from higher to lower frequency. As a sound wave of a particular frequency is set up in the cochlea by oscillation of the stapes, the wave travels to the region of the basilar membrane that naturally responds maximally to that frequency. The energy of the pressure wave is dissipated with this vigorous membrane oscillation, so the wave dies out at the region of maximal displacement.

The hair cells in the region of peak vibration of the basilar membrane undergo the most mechanical deformation and accordingly are the most excited. You can think of the organ of Corti as a piano with 15,000 strings (represented by the 15,000 hair cells). In a piano, the resonant frequency of each string is determined by its width and stiffness; similarly, each part of the basilar membrane has a resonant frequency based on its width and stiffness of its connective protein fibers such as collagen. Each hair cell is also "tuned" to an optimal sound frequency, determined by its location on the organ of Corti. Since different sound waves promote maximal movement of different regions of the basilar membrane, they activate differently tuned hair cells (that is, different sound waves vibrate different "piano strings"). This information is propagated to the CNS, which interprets the pattern of hair cell stimulation as a sound of a particular frequency. Modern techniques have determined that the basilar membrane is so fine-tuned that the peak membrane response to a single pitch probably extends no more than the width of a few hair cells, possibly due to *lateral inhibition* (see Figure 6-7).

Overtones of varying frequencies cause many points along the basilar membrane to vibrate simultaneously but less intensely than the fundamental tone, enabling the CNS to distinguish the timbre of the sound (**timbre discrimination**).

Loudness discrimination depends on the amplitude of vibration

Intensity (loudness) discrimination depends on the amplitude of vibration. As sound waves originating from louder sound sources strike the eardrum, they cause it to vibrate more vigorously (that is, bulge in and out to a greater extent) but at

Molecular Biology and Genomics
The Smell and Taste of Evolution

Until the advent of modern genetic techniques, the receptor proteins involved in many senses were unknown. But with these new methods, genes can be found once a single receptor's structure has been found by searching the genome for similar codes and then seeing where (if anywhere) those genes are expressed (see Chapter 2). In several studies that searched animal genomes for codes related to a known taste receptor, researchers have found that there are many (not one) genes for variants of bitter receptors, ranging from about 50 in a frog, 30 to 40 in various mammals including humans, to fewer than 10 in zebra fish and pufferfish. Multiple sweet receptors have also been found. Furthermore, the researchers found that these genes were being expressed in the mouse tongue. The sense of taste may be more sophisticated than originally thought, particularly for bitter substances, which can be dangerous and which have evolved into an enormous array of defensive molecules in plants.

The most recent studies on taste involve "knocking out" genes (p. 34) for taste receptors. In one study, mice were engineered by knocking out a sweet-receptor gene (called *T1R2*) and replacing it with the human equivalent. The mice were highly attracted to the artificial sweetener *aspartame*, which humans taste as sweet but which mice normally do not detect.

Genomic database searches have revealed odorant receptors of humans, dogs, mice, catfish, salamanders, many bird species, and insects. A surprisingly large family of vertebrate odor receptors (OR) has been found, with approximately 50 OR genes in fish to about 1,000 in mammals (though not all are functional). OR genes in birds are quite variable, estimated to range from 107 in galah cockatoos to 600 in kiwis. In mice, animals highly reliant on their sense of smell, 80% of the 1,000 or so OR genes are functional; the other 20% have mutated into unusable pseudogenes. In comparison, the proportion of OR pseudogenes in chimpanzees is about 30%, and in humans it is 60%! This is an excellent example of *vestigial structures* (p. 2), which, like remnant hindlimbs in snakes and whales, provide some of the strongest evidence for the theory of evolution. The loss of 60% of functional OR genes in humans accords with what people have long known about human sense of smell: It has clearly degenerated in comparison to other mammals. Yoav Gilad and colleagues have found that the percentage of OR pseudogenes correlates among humans and Old World and New World primates with the degree of reliance on color vision.

Finally, the first insect taste-receptor genes have been recently isolated in *Drosophila* (fruit flies). Here again, the fly genome was scanned for codes of known taste receptors, and the results were about 75 potential taste genes. Many of these were found to be expressed in the fly's taste organ, the proboscis **labellum**. The researchers in this study hope the information will allow the design of bad-tasting (to an insect) chemicals for use in repelling pests that eat crops and damage trees.

Clearly, genomic analysis continues to produce many surprising and exciting results for physiologists. Many long-standing mysteries such as sensory mechanisms are now becoming amenable to further physiological analysis.

---

the same frequency as a softer sound of the same pitch. The greater tympanic membrane deflection is converted into a greater amplitude of basilar membrane movement in the region of peak responsiveness. The CNS interprets greater basilar membrane oscillation as a louder sound.

The auditory cortex is mapped according to tone

Just as various regions of the basilar membrane are associated with particular tones, the **primary auditory cortex** (see Figure 5-17a) in the temporal lobe is also *tonotopically* organized. Each region of the basilar membrane is linked to a specific region of the primary auditory cortex. Accordingly, specific cortical neurons are activated only by particular tones; that is, each region of the auditory cortex becomes excited only in response to a specific tone detected by a selected portion of the basilar membrane.

The neural pathway between the organ of Corti and the auditory cortex, beginning with the auditory nerve, involves several synapses en route, including the *brainstem* (which uses the auditory input for alertness and arousal) and the *thalamus,* which sorts and relays the signals upward. Unlike signals in the visual pathways, auditory signals from each ear are transmitted to both temporal lobes because the fibers partially cross over in the brainstem. For this reason, a disruption of the auditory pathways on one side beyond the brain stem does not affect hearing in either ear to any extent.

The primary auditory cortex appears to perceive discrete sounds, whereas the surrounding *associative* auditory cortex integrates the separate sounds into a coherent, meaningful pattern. Consider the complexity of the task accomplished by the auditory system in sorting out a myriad of noises in the environment. An animal's ability to discriminate the sound of an approaching predator from that of background noises can be a matter of survival.

Insect ears are derived from the respiratory tracheal system

Let's conclude our examination of hearing with the most successful group of all animals. Insects use auditory information for a variety of behaviors, including social dominance as well as communication with the opposite sex. In crickets the prothoracic walking leg contains a specialized auditory organ, which is derived from a modified trachea. In contrast, locusts' ears are located on the first abdominal segment and consist of modified spiracles (see Figure 11-3e, p. 496). Sensory cells are aggregated into clusters directly coupled to the **tympanum,** a structure similar in appearance to the frog tympanic membrane. The tympanum is exposed to air on both sides of the membrane. Thus, as in mammals, any changes in air pressure induced by sound waves causes the tympanum to vibrate. Vibration of the tympanum activates specific receptors, generating action potentials, which are sent to the central nervous system. Depending on the source of the sound, differences in the time it takes the sound waves to reach the right or left tympani can be used to establish the source. In addition, because the tracheal system is interconnected, sounds arriving at one tympanum are conducted throughout the animal and cause the opposite tympanum to move from the inside. Interestingly, rather than the ability to discriminate pitch, as in the case of vertebrates, insects' hearing organs are specialized for determining patterns, duration, and intensity of sound waves.

check your understanding 6.4

What is responsible for the pitch, intensity, and timbre of a sound? How are these different sound components used by various animals?

Describe the function of each of the following parts of the mammalian ear: pinna, ear canal, tympanic membrane, ossicles, oval window, and the various parts of the cochlea. Include a discussion of how sound waves are transduced into action potentials.

6.5 Chemoreception: Taste and Smell

As we noted earlier, detection of chemicals in the environment is a very ancient sense. Animal sensors for taste and smell, plus numerous interoreceptors, are chemoreceptors, detecting chemicals to generate neural signals, often (but not always) following binding of particular chemicals in their environment to specific receptor proteins. The functions of interoreceptors, such as those monitoring blood chemistry, are discussed in other chapters. Here we focus on the exteroreceptors for **gustation** (**taste,** or detection of molecules in objects in solids or liquids in contact with the body) and **olfaction** (**smell** or detection of individual molecules released from a distant object) exteroreceptors. Chemoreception can take place on the skin or in special organs such as antennae of insects and the tongue in the mouth and olfactory bulbs in nasal cavities of air-breathing vertebrates.

Taste and smell have numerous roles such as sensing food, kin, mates, and direction

Animal chemical senses have many functions. Clearly, taste and smell provide a "quality control" checkpoint for substances available for ingestion. The sensations of taste and smell in association with food intake in some species can influence the flow of digestive juices and affect appetite. Furthermore, stimulation of taste or smell receptors induces pleasurable or objectionable sensations (at least in humans). In many species of animals these senses also permit finding direction, seeking prey or avoiding predators, detection of damaged and diseased tissue, social recognition of kin, and sexual attraction to a mate. For example, carnivorous cone

snails (p. 34) detect prey by their odor molecules; mosquitoes find their human "prey" by smelling CO_2 and octenol, a volatile chemical in sweat; barnacle larvae settle on intertidal rocks chosen in part by smelling chemicals released by adult barnacles; mice sniff for potential mates while avoiding those that are infected with parasites; and hermit crabs find buried snail shells in part by smelling calcium leaching from them. Crabs can detect virtually all classes of biologically available molecules, including petroleum hydrocarbons and the odors of decaying flesh.

Taste and smell senses are highly variable in sensitivities among species. For example, cats cannot taste sweets because their sweet-receptor gene (presumably not useful in these meat-eaters) has become a pseudogene (p. 3). Smell is a relatively poor sense in humans (see *Molecular Biology and Genomics: The Smell and Taste of Evolution*) and is much less important in influencing our behavior compared to most mammals. Bird species are also variable. Some species of pelagic birds, which until recently ornithologists thought had no sense of smell at all, can detect the odor of *dimethyl sulfide* (DMS) from up to 4 kilometers away. White-chinned petrels, bloodhounds of the Antarctic skies, rely on the odor of DMS to locate zooplankton (such as krill, shrimplike crustaceans) on which to feed. DMS is produced when zooplankton feed on phytoplankton (single-celled algae), which contain an *osmolyte* (p. 617) called *dimethylsulfoniopropionate* that breaks down into the globally important gas DMS (see Chapter 1, p. 14). Vultures and kiwis also locate food sources by smell and have unusually large proportions of their brains devoted to the sense of smell.

The intraspecies roles of olfaction are largely mediated by *pheromones* (p. 92). In the social insects, including ants, honeybees, and termites, olfactory communication maintains the very fabric of their complex societies. Exocrine glands located on the bodies of these insects release volatile pheromones into the environment. These lightweight chemical compounds made of 5 to 20 carbon atoms are passed from one animal to another and trigger a characteristic behavioral or physiological response. Among the specific behaviors attributed to pheromones in insects include alarm reactions, orientation, swarming, food source tracking, fighting, and recognizing members of a colony. (We will examine pheromones in vertebrates shortly.) The first pheromone to be identified was **bombykol,** a pheromone released from sexually receptive female moths to attract the male. Chemoreceptors of insects can be exquisitely sensitive to their ligands: The binding of only one pheromone molecule to a receptor on the antennae of the male gypsy moth (*Lymantria dispar*) is sufficient to trigger an action potential. Relying on this remarkable degree of sensitivity, the male flies "upwind" in pursuit of the flightless, sexually receptive female.

Next, we examine the mechanism of taste (gustation) in detail and then turn our attention to smell (olfaction).

Taste sensation is coded by patterns of activity in various taste receptors

In insects, **sensilla**—specialized projections from the cuticle—permit molecules to reach their internal sensory endings through minute pores. In flies and moths, for example, these chemoreceptors are located in sensory hairlike projections

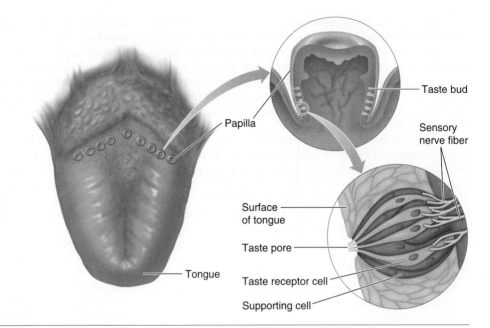

FIGURE 6-22 Location and structure of human taste buds. Taste buds are located primarily along the edges of moundlike papillae on the upper surface of the tongue. The receptor cells and supporting cells of a taste bud are arranged like slices of an orange.

© Cengage Learning, 2013

on the terminal segment of the leg as well as in the tip of the proboscis, which is used for feeding and drinking. Appropriate stimuli result in extension of the proboscis and elicitation of feeding or drinking behavior. Dendrites, containing the sensory receptors, project to the apex of each sensillum. In some species of insects (such as ants, bees, and wasps), taste organs are also located on the antennae. Insects rely on their chemical senses for feeding and mating behavior, habitat selection, and parasite–host relationships.

In higher vertebrates the chemoreceptors for taste (**gustatory**) sensation are packaged in **taste buds** (Figure 6-22), about 10,000 of which are present in the human oral cavity and throat, with the greatest percentage on the upper surface of the tongue. This compares to over 15,000 in pigs and rabbits, 550 in lizards, and only 24 in chickens. Recent studies on mammals have found that the digestive tract also contains taste receptors involved in regulation of digestion (see Chapter 14), and the lung contains bitter taste receptors that trigger relaxation of the airways, perhaps to aid in coughing out toxins. In other vertebrates, such as catfish, taste buds may be located over the entire surface of the animal's body and number over 100,000. In most species of bony fish, taste buds are located in the mouth, pharyngeal region, and gill arches. The extraoral taste receptors help in detecting and capturing the foodstuff, whereas the function of the intraoral taste system is similar to that of air-breathing animals.

A typical mammalian taste bud consists of about 50 long, spindle-shaped receptor cells packaged with supporting cells in an arrangement like slices of an orange. Each taste bud has a small opening, the **taste pore**, through which fluids in the mouth come into contact with the surface of its receptor cells.

Vertebrate taste receptor cells are modified epithelial cells with many surface folds—*microvilli*—that protrude slightly through the taste pore, greatly increasing the surface area exposed to the oral contents. The plasma membrane of the microvilli contains receptor sites that bind selectively with chemical molecules in the environment. Only chemicals

in solution—either ingested liquids or solids that have been dissolved in saliva—can attach to receptor cells and evoke the sensation of taste. Binding of a taste-provoking chemical, a **tastant,** with a receptor cell ultimately alters the cell's ionic channels to produce a depolarizing receptor potential. This receptor potential, in turn, initiates action potentials within terminal endings of afferent nerve fibers with which the receptor cell synapses.

Most receptors are carefully sheltered from direct exposure to the environment, but the taste receptor cells, by virtue of their task, frequently come into contact with potent chemicals. Unlike the eye or ear receptors, which are irreplaceable, taste receptors have a limited life span—about 10 days in humans. Epithelial cells surrounding the taste bud differentiate first into supporting cells and then into receptor cells to constantly renew the taste bud components.

Terminal afferent endings of several cranial nerves synapse with taste buds in various regions of the mouth. Signals in these sensory inputs are conveyed via synaptic stops in the brain stem and thalamus to the **cortical gustatory area,** a region in the parietal lobe adjacent to the "tongue" area of the somatosensory cortex. Unlike most sensory input, the gustatory pathways are primarily uncrossed. The brain stem also projects fibers to the hypothalamus and limbic system, presumably to add affective dimensions, such as whether the taste is pleasant or unpleasant, and to process behavioral aspects associated with taste and smell.

Mammalian taste is based on five (and perhaps more) primary tastes

Mammals can discriminate among thousands of different taste sensations, yet most tastes are thought to be varying combinations of just five tastant categories, the **primary tastes:** *salty (sodium), sour* (or acidic), *sweet, bitter,* and *umami,* a meaty or savory taste. Each receptor cell responds in varying degrees to all the primary tastes but is generally preferentially responsive to one of the taste modalities. (You may have learned that specific areas of your tongue respond to specific tastes; for example, sweet taste is sometimes shown as located at the tip of your tongue. Recent studies show that these common taste maps are erroneous. Human taste types are scattered over most of the tongue, and the map varies from individual to individual.)

As with all biological features, understanding these tastes requires both proximate and evolutionary explanations. In terms of evolution, each uniquely aids survival. In terms of mechanisms, receptor cells use different pathways to bring about a receptor potential in response to each of the five primary tastant categories (Figure 6-23):

- **Salty taste** is stimulated by chemical salts, especially Na⁺ (as in table salt, NaCl). This is important to survival because obtaining sufficient dietary NaCl is critical for osmotic balance (Chapter 13) and electrical signals (Chapter 4). Salty taste is particularly prominent in herbivores, whose plant diet is low in sodium. Transduction is direct, due to entry of positively charged Na⁺ ions through specialized Na⁺ channels in the receptor cell membrane of rats, hamsters, and primates. There are at least two types of channels, including a very specific one called *ENaC* (for epithelial Na⁺ channel; Figure 6-23). It may be a simple leak channel, though there is some evidence it might be gated. This ion movement reduces the cell's internal negativity and is believed to be responsible for receptor potential. Salty taste transduction is different among species; e.g., frog taste cells have nonspecific cation channels whereas mammalian taste cells are highly specific for Na⁺.

- **Sour taste** is caused by acids, which contain a free hydrogen ion, H⁺. The citric acid content of lemons, for example, accounts for their distinctly sour taste. Moderate sourness taste is pleasant, perhaps because it favors ingestion of acidic vitamin C. Strong acid taste may warn against spoiled food or unripe fruit. Depolarization of the receptor cell by sour tastants may occur directly by H⁺ entry (see Figure 6-23), or when H⁺ blocks K⁺ channels in the receptor cell membrane. The resultant decrease in the passive movement of positively charged K⁺ ions out of the cell reduces the internal negativity.

- **Sweet taste** is evoked by the particular configuration of small sugar molecules (Figure 6-23). Clearly this pleasant taste favors intake of this primary energy form (glucose, sucrose and so forth) in most animals. Depending on the species of animal, other organic molecules, such as saccharin, aspartame, and other artificial sweeteners, can also interact with sweet receptor binding sites. Signaling begins with the binding of glucose or another chemical to a **G-protein coupled receptor**, which leads to either (1) the cAMP second-messenger pathway, or (2) (in some species) the IP₃ pathway (see pp. 97–98). The second-messenger pathway results in either (1) phosphorylation and blockage of K⁺ channels in the receptor cell membrane, leading to a depolarizing receptor potential; or (2) IP₃-induced release of Ca²⁺ from the endoplasmic reticulum, leading to neurotransmitter release.

- **Bitter taste** is elicited by a more chemically diverse group of tastants than the other taste sensations because there are many different bitter receptors (see the box, *Molecular Biology and Genomics: The Smell and Taste of Evolution*, p. 232). This is largely a negative taste

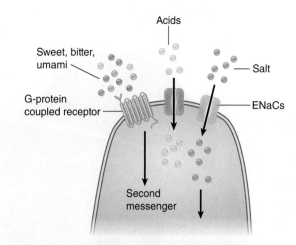

FIGURE 6-23 Transduction pathways in taste bud cells. For convenience, pathways for salt, acid, sweet, bitter, and umami are shown in one cell, but a real taste bud receptor would have only one type.

© Cengage Learning, 2013

that helps avoid noxious or toxic molecules. For example, alkaloids (such as caffeine, nicotine, strychnine, morphine, and other toxic plant derivatives) as well as many poisonous plant compounds all taste bitter. Each bitter-taste receptor cell has multiple receptors and possibly multiple signaling pathways for detecting this wide array of potentially harmful chemicals. Most or all involve G proteins (Figure 6-23). Indeed, the first G protein in taste—**gustducin**—was identified in one of the bitter-signaling pathways. Interestingly, this G protein is structurally very similar to the visual G protein, *transducin* (see p. 248).

Interestingly, humans and laboratory animals can learn to like certain bitter tastes, a phenomenon called **acquired taste**. For example, most wines and beer contain bitter compounds that are distasteful upon first drinking to humans and laboratory mice. However, as the brain learns to associate the pleasure induced by alcohol with these drinks, the associative sensory cortex somehow reverses its perception of the bitter taste.

- **Umami** (Japanese for "yummy") **taste** is triggered by amino acids such as glutamate. The presence of amino acids serves as a marker for a nutritionally protein-rich food. First named by a Japanese researcher in 1908, umami was not widely accepted as a fifth taste until the 1990s, and the actual receptor was not confirmed until 2009. Glutamate binds to this G-protein coupled receptor and activates a second-messenger system (Figure 6-23).

This list of five tastes may soon be outmoded. A new sixth taste sensation has also been proposed—*fatty*. Scientists have recently identified a potential sensor in the mouth for long-chain fatty acids in humans and rodents. This makes sense for survival, because all three major food groups—carbohydrates (sweet), protein (umami), and fats—are essential. Mice have recently been found to have a seventh taste modality with a unique receptor for *calcium* (humans may also have it). Calcium is essential for so many functions—bone formation, exocytosis, muscle contraction—that a taste for it makes evolutionary sense as well.

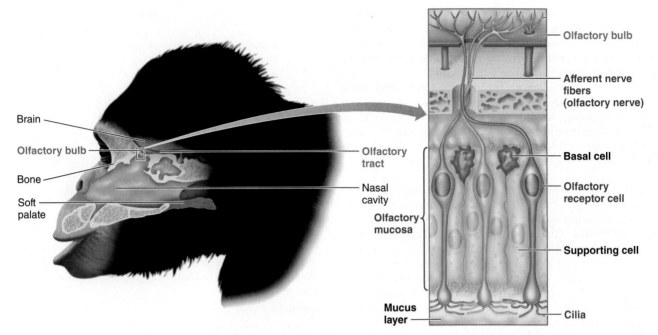

FIGURE 6-24 Location and structure of the mammalian olfactory receptors.

© Cengage Learning, 2013

The richness of fine taste discrimination beyond these five (or more) primary tastes depends on subtle differences in the stimulation patterns of all the taste buds in response to various substances, as well as to the complementary sense of smell. (As you know if you have tasted food while your nose was plugged up due to a cold, much of what humans call food's "taste" is actual a smell sensation, in part from chemicals released during chewing that enter the nasal cavity from the nostrils or the back of the mouth.) Other factors affecting taste include temperature and texture of the food as well as psychological factors associated with past experiences with the food. How the cortex accomplishes the complex perceptual processing of taste sensation is currently not known.

We now turn to the sensation of smell.

The olfactory receptors in the nose are specialized endings of renewable afferent neurons

The **olfactory mucosa** of vertebrates, located in the **nasal fossae** (upper tract of the respiratory airways), contains three cell types: *olfactory receptors, supporting cells,* and *basal cells* (Figure 6-24). The **supporting** cells (also called **Bowman's glands**) secrete mucus, which coats the nasal passages. The **basal** cells are precursors for new olfactory receptor cells, which are replaced about every two months in humans. This is remarkable because, unlike the other special sense receptors, **olfactory receptors** are specialized endings of afferent neurons, not separate cells. The entire neuron, including its afferent axon projecting into the brain, is replaced. These are the only mammalian neurons that undergo cell division (although recent evidence suggests that new neurons can be produced in the hippocampus, a part of the brain important in forming long-term memory; see p. 171).

The receptor portion of the olfactory receptor cells consists of an enlarged knob bearing several long cilia that extend like a tassel to the surface of the mucosa. These cilia contain the binding sites for attaching **odorants,** molecules that can be smelled. During quiet breathing in air breathers, odorants typically reach the sensitive receptors only by diffusion because the olfactory mucosa is above the normal path of airflow. The act of sniffing enhances this process by drawing the air currents upward within the nasal cavity so that a greater percentage of the odoriferous molecules in the air contact the olfactory mucosa. As we noted earlier for taste, odorants also reach the olfactory mucosa during eating by wafting up to the nose from the mouth through the nostrils or pharynx (back of the throat).

The nasal cavity in both fish and aquatic amphibians is generally a pit or saclike structure on either the dorsal or ventral surface of the snout. In aquatic vertebrates water is drawn across the olfactory epithelium and odor signals become absorbed in the mucous layer, where they bind to specific receptor sites on the surface membranes of olfactory cilia in the outer mucous layer.

To be smelled, a substance must be (1) sufficiently volatile (easily vaporized or dissolved) that some of its molecules can enter the nose in the inspired air or water; and (2) sufficiently soluble that it can dissolve in the mucus layer coating the olfactory mucosa. Many thousands of odors can be distinguished by the sense of smell, although it has been difficult to categorize particular odors based on their chemical structure. Some odors quickly dissipate in the atmosphere, such as alarm substances released by ants, whereas others are intended to persist for long periods of times. For example, lipid added to the territorial scent markings of lions and tigers extends the release time of the pheromonal message released by the anal sacs.

Various parts of an odor are detected by different olfactory receptors and sorted into "smell files"

The human nose contains 5 million olfactory receptors, about 10-fold less than in rodents and dogs. Genomics reveal that most vertebrates have many more different types of odor receptor (OR) genes than genes for other senses. Although chemoreception is important in all animals, mammals can sample a larger repertoire of the environment compared to other vertebrates because they have more OR proteins (though some ORs are pseudogenes; see the box, *Molecular Biology and Genomics: The Smell and Taste of Evolution.* p. 232). Also, number of olfactory neurons in mammals is greater, varying from 60 million in rats, to 100 million in rabbits, to almost a billion in the German shepherd dog.

Like many tastants, sensing of an appropriate scent molecule begins with binding to a G-protein coupled receptor, triggering a cascade of intracellular reactions that leads to opening of Ca^{2+} or Na^+ channels. The resultant ion movement brings about a depolarizing receptor potential that generates action potentials in the afferent fiber. For unknown reasons, some members of the olfactory receptor family are also expressed in the mammalian tongue, sperm cells (p. 799), and developing rat heart, as well as in the avian notochord (non-neuronal embryonic tissue which induces formation of the neural tube).

Smell detection "dissects" an odor into various components. Each receptor responds to only one discrete component of an odor rather than to the whole odorant molecule. Accordingly, different receptors detect the various parts of an odor, and a given receptor can respond to a particular odor component shared in common by different scents, ultimately permitting spatial summation. Basically the gustatory sense determines whether the food is good or not (that is, is it safe?), whereas the olfactory sense is more discriminatory and picks up the selected differences in food items.

The afferent fibers arising from the receptor endings in the nose pass through tiny holes in the flat bone plate separating the olfactory mucosa from the overlying brain tissue (Figure 6-25). They immediately synapse in the **olfactory bulb,** a complex neural structure containing several different layers of cells that are functionally similar to the retinal layers of the eye. In humans, the twin olfactory bulbs, one on each side, are about the size of small grapes. Each olfactory bulb is lined by small, ball-like neural junctions known as **glomeruli** ("little balls") (Figure 6-25). Comparatively, fish are noted for having few glomeruli. Within each glomerulus, the terminals of receptor cells carrying information about a particular scent component synapse with the next cells in the olfactory pathway, the **mitral cells.** Because each glomerulus

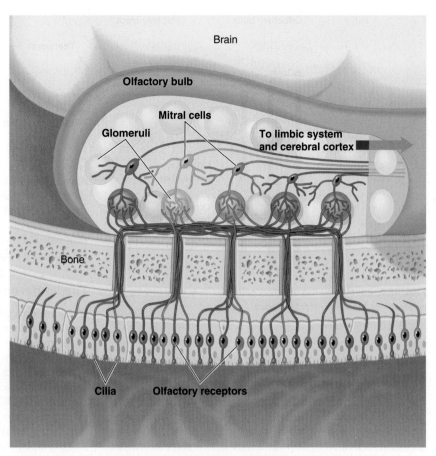

FIGURE 6-25 Processing of scents in the mammalian olfactory bulb. Each of the glomeruli lining the olfactory bulb receives synaptic input from only one type of olfactory receptor, which, in turn, responds to only one discrete component of an odorant. Thus, the glomeruli sort and file the various components of an odiferous molecule before relaying the smell signal to the mitral cells and higher brain levels for further processing.

© Cengage Learning, 2013

receives signals only from receptors that detect a particular odor component, the glomeruli serve as "smell files." The separate components of an odor are sorted into different glomeruli, one component per file. Thus, the glomeruli, which serve as the first relay station for processing olfactory information, play a key role in organizing scent perception.

The mitral cells on which the olfactory receptors terminate in the glomeruli refine the smell signals and relay them to the brain for further processing. Fibers leaving the olfactory bulb travel in two different routes (Figure 6-26):

1. A *subcortical route* going primarily to regions of the limbic system, especially the lower medial sides of the temporal lobes (considered to be the **primary olfactory cortex**). This route is for close coordination between smell and primitive memory, as well as behavioral reactions associated with feeding, mating, and direction orienting.

2. A *thalamic-cortical route.* This route, which includes hypothalamic involvement, permits conscious perception and fine discrimination of smell.

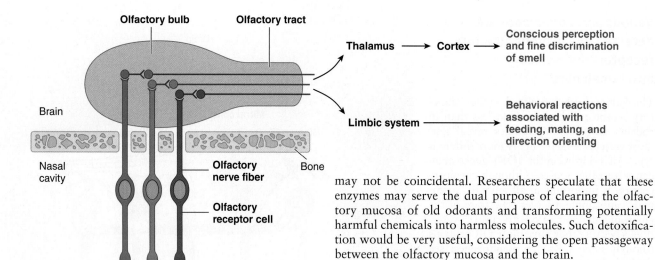

FIGURE 6-26 Mammalian olfactory pathways. See Figures 6-24 and 6-25 for a location reference point.

© Cengage Learning, 2013

Odor discrimination is coded by patterns of activity in the olfactory bulb glomeruli

Because each given odorant activates multiple receptor proteins and glomeruli in response to its various odor components, odor discrimination is based on different patterns of glomeruli activated by various scents. In this way, the cortex can distinguish about 20,000 different scents with a 1,000 or fewer different receptor proteins. This mechanism for distinguishing different odors is very effective, even in humans, who have a poor sense of smell compared to other species. (By comparison, dogs' sense of smell is hundreds of times more sensitive than that of humans). A noteworthy example is our ability to detect methyl mercaptan at a concentration of 1 molecule per 50 billion molecules in the air! This is the substance is added to odorless natural gas to enable us to detect potentially lethal gas leaks (p. 214).

The olfactory system adapts quickly, and odorants are rapidly cleared

Although the olfactory system is sensitive and highly discriminating, it is also quickly adaptive. Sensitivity to most new odors diminishes rapidly after a short period of exposure to it, even though the odor source continues to be present. This reduced sensitivity or habituation involves both intrinsic receptor adaptation (as we noted earlier, p. 214) and some sort of extrinsic adaptation in the CNS. Adaptation is specific for a particular odor, and responsiveness to other odors remains unchanged.

What clears the odorants away from their binding sites on the olfactory receptors so that the smell does not "linger" after the source is removed? Several "odor-eating" enzymes have recently been discovered in the olfactory mucosa that may serve as molecular janitors, clearing away odoriferous molecules so they do not continue to stimulate olfactory receptors. Interestingly, these odorant-clearing enzymes are very similar chemically to detoxification enzymes in the liver. (These liver enzymes inactivate potential toxins absorbed from the digestive tract; p. 685). This resemblance may not be coincidental. Researchers speculate that these enzymes may serve the dual purpose of clearing the olfactory mucosa of old odorants and transforming potentially harmful chemicals into harmless molecules. Such detoxification would be very useful, considering the open passageway between the olfactory mucosa and the brain.

The vomeronasal organ detects pheromones

In addition to the olfactory mucosa, the vertebrate nose may contain another sense organ: the **vomeronasal organ** (VNO) in mammals, including humans, and non-avian reptiles, where it is sometimes called **Jacobson's organ** (note most aquatic vertebrates and birds lack this organ). The VNO is located just inside the nose next to the vomer bone, hence its name. The VNO is called the "sexual nose" for its role in governing reproductive and social behaviors, such as identifying and attracting a mate, by reception of pheromones. For example, male ungulates, such as stallions, put their lips into the urine of an estrous mare and curl their upper lip in the **Flehman position** ("horse's laugh"), which partially blocks the nostril opening. Breathing deeply carries the urine into the VNO, where the stallion determines sexual receptivity of the mare by the concentration of pheromone. Sex pheromones are also produced in the urine of female elephants about to ovulate, which initiates the Flehman response in reproductively active males. Males first identify the urine through the sense of smell and then detect the pheromone by placing urine from the tip of their trunk to the opening of ducts leading to the VNO. Interestingly, this pheromone, *(Z)-7-dodecenyl acetate*, is identical to a substance used by more than a hundred species of moths as one of their major sex pheromones.

Males also make pheromones. For example, male mice produce an unusual protein called *darcin*, which, in their urine, makes them attractive to female mice. One of the few other mammalian pheromones that has been characterized is *androstenone*, a steroid in the saliva of male pigs and human male sweat. Release of this potent pheromone in pigs causes sows to assume the **lordosis** (mating) position. Androstenone elicits this behavior only in estrous sows, which suggests that in some species, the response to a *releaser* pheromone is context dependent and not necessarily automatic.

Although the role of the VNO in human behavior has not been validated, researchers suspect it governs spontaneous "feelings" between people, either "good chemistry," such as "love at first sight," or "bad chemistry," such as "getting bad vibes" from someone you just met. Evidence from individuals whose VNO has been surgically removed supports the existence of a subtle role for the VNO in group behavior, sexual activity, and compatibility with others, similar to the role they play in other animals. Because messages conveyed by the VNO are believed to bypass the cortex (using the limbic path of Figure 6-26), the response to the largely odorless pheromones, if real, would be largely subconscious.

Recent studies in mice have found that the VNO also has receptor proteins for the detection of damaged or infected tissues. These presumably help animals assess the health status of potential mates, and perhaps even to avoid potential disease carriers in general.

check your understanding 6.5

Describe the structure and activation of the taste receptors and why each different primary taste aids survival.

Discuss the location and activation of olfactory receptors (including the VNO) and how the brain distinguishes tens of thousands of odors from 1,000 or fewer receptor proteins.

6.6 Photoreception: Eyes and Vision

Light is one of the most useful signals in the environment, and the ability to detect it can be critical for survival. Indeed, over 90% of known animal species have some type of light sensors. In some animals (such as birds) vision may be the dominant sense in day-to-day life (and in some species vision is considered responsible for an evolutionary decline in acuity in the olfactory system, as we noted earlier).

Light detection uses universal photopigments in ciliary or rhabdomeric receptor cells

Even though organisms may be sensitive to light, their sensitivities and acuities can vary tremendously. At the most primitive level, light detection is based on structures not dedicated solely to this purpose. For example, *chromatophores* (pigment-containing skin cells, p. 154) in many phyla of nonvertebrates and classes of lower vertebrates are directly sensitive to light, responding with color and shade changes in the skin. Even a single-celled *Amoeba* responds to a flash of light by ceasing to move. However, special organs dedicated to light detection have evolved numerous times in a variety of structures. Yet even though the light-capturing organ of an animal may be structurally very different (as you will see shortly), the initial chemical pathways that transduce the light energy are similar, beginning with a highly conserved group of *photopigment molecules* (see *Molecular Biology and Genomics: The Eyes Have It*, p. 241). Furthermore, most organisms use light energies in a narrow spectral band between 400 and 700 nm (nanometers, or billionths of a meter between wave peaks), matching the major output frequencies of the sun as well as the biochemical limits of the pigments, because too little energy would not excite the pigments, whereas energy frequencies that are too high would potentially break down the constituents necessary for phototransduction. However, as you will see, a number of animals have expanded the range into the ultraviolet (UV) region of the spectrum below 400 nm.

Only two types of photoreceptor cells have evolved:

1. **Ciliary** photoreceptors have a large single cilium (see p. 60) with a highly folded membrane, often internal, that contains the photopigments. A key example is the *rod cell* of vertebrates, to be discussed later; ciliary types are also found in Cnidaria, flatworms, and some mollusks.
2. **Rhabdomeric** photoreceptors increase their surface area with numerous parallel microvilli (p. 63). These are found in arthropod compound eyes, also to be discussed later, and in most other bilateral animals with the notable exception of vertebrates. Note that this means that some phyla, including flatworms and mollusks, have both ciliary and rhabdomeric types. Why did both of these evolve? One hint comes from recent studies on the marine annelid, the ragworm *Platynereis dumerilii*. Considered a "living fossil" for its ancient features, this worm's head has two large rhabdomeric eyes for image formation and two simple ciliary *eyespots* (see below) dedicated to detecting light for regulation of *circadian rhythms* (p. 276). Thus, the two types of photoreceptor cells may have first evolved for independent functions.

Dedicated light-sensing organs range from simple eyespots to complex eyes

Considerable evolutionary inventiveness exists in the light-detecting organs that house the photoreceptors (Figure 6-27), though remarkably, most if not all share common genetic features (see the box, *Molecular Biology and Genomics: The Eyes Have It*). Scientists have discovered at least 10 different categories of eyelike structures. **Eyespots**, the simplest form, consist of a small number of photoreceptor cells (less than 100) lining an open cup or pit (Figure 6-27a). Each pit samples a relatively large area of the visual world. Representatives with these simple photoreceptor plates that have received study include planarian flatworms (Platyheminthes), jellies (Cnidarians), and seastars (Echinoderms); but similar less-studied structures are found in many animal phyla. These photoreceptors are lined by photopigment on one side and receive light from only one direction. The pigment cells enclose the dendritic endings of the sensory neurons; the neuronal axons are bundled into nerves that travel from each cup, usually to integrating nerves such as a ganglion or brain. This particular arrangement permits the animal to locate a light source and, in some cases, to orient movement. For example, butterflies sense light with their genitalia. These extraocular photoreceptors determine whether the ovipositor (external appendage that transmits the egg) in females is sufficiently pushed out from the abdominal tip on oviposition (process of egg laying) or the deposited eggs will not be properly attached to the leaf surface. In males the photoreceptors confirm correct coupling with the female during mating.

However, recognition of predator/prey movement requires a more complex optical system that permits formation of an image—a true **eye**. By reducing the size of the cup aperture to produce a **pinhole eye**, an eye can actually form an image, even though it may not be very discriminatory. The evolution of a clear focusing orb called a **lens** in some animals enhances the light-gathering power of the eye, forming what is called a **camera eye** (Figure 6-27b, c). This familiar arrangement appears in many phyla; the most ancient version is probably that evolved by the Cubozoa (box jellies), a surprisingly advanced structure for a Cnidarian (Figure 6-27d). More complex camera eyes are the optical solution evolved by vertebrates and cephalopods (the most intelligent nonvertebrates). The major alternative evolutionary solution is the **compound** eye, with densely packed units called *ommatidia,* each having its own lens and pigment-shielded photoreceptors (Figure 6-27e). Fossil evidence of ommatidia is very ancient, found in arthropods such as trilobites living during the mid-Cambrian period. Although compound eyes provide comparatively poor image formation, they are superior for detecting movement and have a wider field of view than do camera eyes.

A final improvement in eye evolution was the origin of two or more different photoreceptors sensitive to different wavelengths, a fundamental condition for color vision. Many vertebrates, for example, have this adaptation to at least some degree. Let's now examine the vertebrate eye in detail, and then take a look at some nonvertebrate eyes.

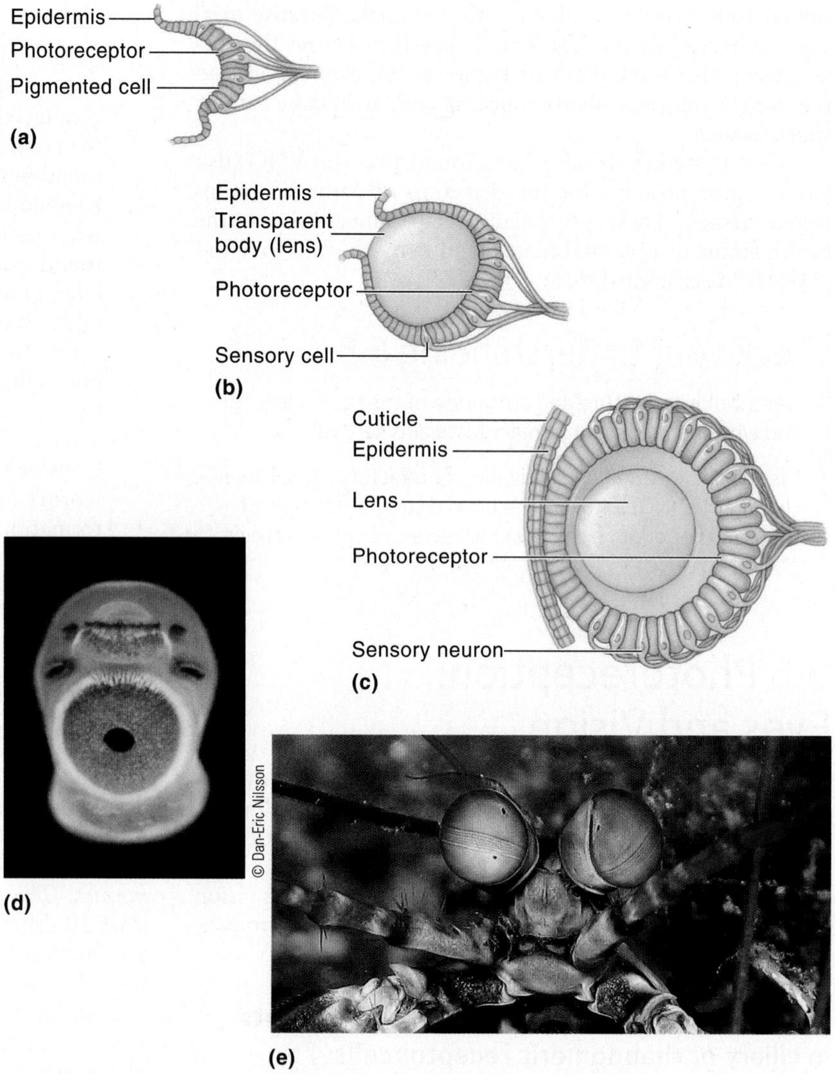

The vertebrate eye is a fluid-filled sphere enclosed by three specialized tissue layers

Camera eyes capture patterns of illumination in the environment as an "optical picture" on a layer of light-sensitive cells, the **retina,** much as a camera captures an image on film. Just as film can be developed into a visual likeness of the original image, the coded image on the retina is transmitted through a series of progressively more complex steps of visual processing until it is finally perceived in the visual associative cortex as a likeness of the original image. The retina of all vertebrates is structurally and functionally similar.

Each eye (Figure 6-28a) is a spherical, fluid-filled structure enclosed by three layers. From outermost to innermost, these layers are (1) the *sclera* and *cornea*; (2) the *choroid, ciliary body*, and *iris*; and (3) the *retina*. Most of the eyeball is covered by a tough outer layer of connective tissue, the **sclera,** which forms the visible white part of the eye (Figure 6-28b). Anteriorly (toward the front), the outer layer consists of the transparent **cornea** through which light rays pass into the interior of the eye. The middle layer underneath the sclera is the highly pigmented **choroid,** which contains many blood vessels that nourish the retina. The choroid layer becomes specialized anteriorly to form the **ciliary body** and **iris,** which are described shortly. The innermost coat under the choroid is the retina, which consists of an outer pigmented layer and an inner nervous tissue layer. The latter contains the **rods** and **cones,** the photoreceptors that convert light energy into nerve impulses. Like the black walls of a photographic stu-

(a)

(b)

(c)

(d)

(e)

FIGURE 6-27 Examples of nonvertebrate eyes. There are far more photoreceptors than can be shown in these diagrams. (a) Limpet ocellus, a shallow depression in the epidermis that incorporates light-sensitive receptors. (b) Abalone camera eye, with its spherical, transparent lens. (c) Eye of a land snail. (d) The camera eye of a Cubozoan. (e) Compound eye of a mantis shrimp, which has the most sophisticated light discrimination of any known animal.

Source: (a–c) C. Starr & R. Taggart (2004). *Biology: The Unity and Diversity of Life*, 10th ed. (Belmont, CA: Brooks/Cole).

dio, the pigment in the choroid absorbs light after it strikes the retina to prevent reflection or scattering of light within the eye.

The interior of the eye consists of two fluid-filled cavities, separated by the lens, all of which are transparent to permit light to pass through the eye from the cornea to the retina. The anterior (front) cavity between the cornea and lens contains a clear, watery fluid, the **aqueous humor,** and the larger posterior (rear) cavity between the lens and retina contains a semifluid, jellylike substance, the **vitreous humor** (Figure 6-28a). The vitreous humor is important in maintaining the spherical shape of the eyeball. The aqueous humor carries nutrients for the cornea and lens, both of which lack a blood supply. Blood vessels in these structures would impede the passage of light to the photoreceptors.

Molecular Biology and Genomics
The Eyes Have It

Genomic studies have revealed remarkable commonalities among animal eyes. Photo-pigments are almost universally the protein *opsin* and the vitamin-A derivative *retinene*. We will examine these in detail later, but it is worth noting that opsins are ancient. David Plachetzki and colleagues have recently used genomic analysis to find an opsin gene in hydras, simple freshwater Cnidarians. Though lacking eyes, hydras will react when exposed to a sudden bright light by withdrawing. Blocking the phototransduction pathway (which is active in some hydra neurons) abolishes this reaction. Remarkably, genetic sequences show that human and hydra opsins evolved from a common ancestor.

Moreover, despite the diversity of eyes at the organ level, scientists have found a striking similarity in their development. The formation of all these eyes (at least all that have been tested) is initiated by a common "master" regulatory gene, called *Pax6* in vertebrates, first discovered in mutated form in humans born either without eyes or severely deformed ones. Nearly identical genes have been found in a variety of eyes across the animal kingdom. These genes are active in embryonic stages just before eye formation, which does not occur if the gene is nonfunctional. Remarkably, these related genes are interchangeable. A fruit fly with its Pax6 homolog knocked out will not form eyes, but if a mouse or a squid Pax6 gene is engineered into it, it will form normal fly eyes! Apparently early in the evolution of animals, this gene became the master switch that turns on a cascade of other genes to construct eyes. However, this does not mean that eyes arose only once. Pax6 also regulates other developmental pathways in other body regions, such as mammalian pancreas and squid olfactory organ, so perhaps it is a general organ-development switch that was selected independently in different groups for initiation of eye development.

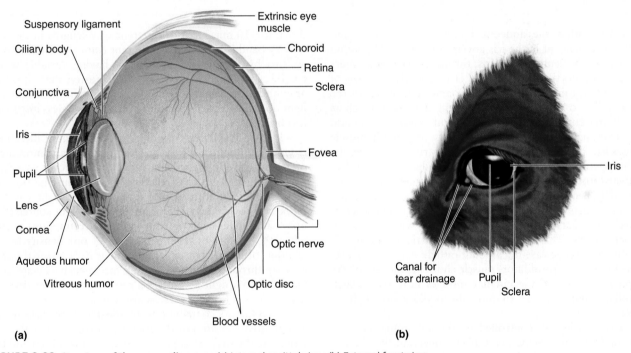

FIGURE 6-28 Structure of the mammalian eye. (a) Internal sagittal view. (b) External front view.

© Cengage Learning, 2013

The mammalian retina is comparatively well vascularized, but the presence of blood vessels in the retina interferes with visual acuity. In contrast, reptiles have retinas that completely lack blood vessels. These groups rely on a *pecten* (birds) or a very similar *conus papillaris* (non-avian reptiles) to provide oxygen and nutrients to the retina by diffusion through the vitreous body. The avian **pectens** are thin, comblike projections from the choroid into the vitreous humor. Their comblike folds of endothelial cells enhance surface area available for nutrient exchange. In nocturnal birds the pecten tends to be simplified in structure, whereas in diurnal birds it is elongated and more elaborate. Because these vessels do not lie within the retina, the eyes of birds are capable of greater visual resolution than a mammal's due to more space for photoreceptors ($1,000,000/mm^2$ in some hawks, versus 200,000 in humans). Thus, a sharp-eyed hawk can spot a field mouse from hundreds of meters away, something a mammal cannot do. Birds in general have the largest eyes relative to body size among all animals.

The amount of light entering many vertebrate eyes is controlled by the iris

Not all the light passing through the cornea reaches the light-sensitive photoreceptors, because of the **iris,** a thin, pigmented, smooth muscle that forms a visible ringlike

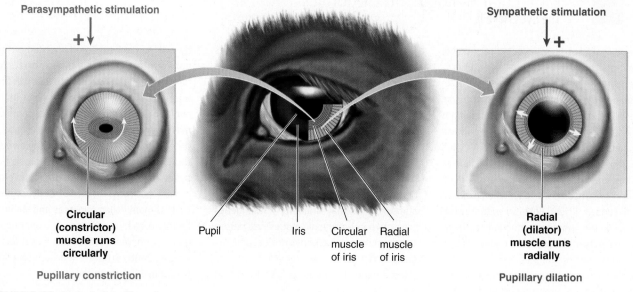

Parasympathetic stimulation

Sympathetic stimulation

Circular (constrictor) muscle runs circularly

Pupil   Iris   Circular muscle of iris   Radial muscle of iris

Radial (dilator) muscle runs radially

Pupillary constriction

Pupillary dilation

FIGURE 6-29 Control of pupillary size.

© Cengage Learning, 2013

structure within the aqueous humor (Figures 6-28 and 6-29). The pigment in the iris governs eye color. The round opening in the center of the iris, through which light enters the interior portions of the eye, is the **pupil.** The size of this opening can be adjusted by variable contraction of the iris muscles to admit more or less light as needed, much as the shutter controls the amount of light entering a camera. The mammalian iris contains two sets of smooth muscle networks, one **circular** (muscle fibers run in a ringlike fashion within the iris) and the other **radial** (fibers project outward from the pupillary margin like bicycle spokes) (Figure 6-29). Because muscle fibers shorten when they contract, the pupil gets smaller when the **circular** (or **constrictor**) **muscle** contracts and forms a smaller ring while the radial muscles relax. This reflex pupillary constriction occurs in bright light to decrease the amount of light entering the eye. When the **radial** (or **dilator**) **muscle** shortens, the size of the pupil increases and the circular muscles now relax. Such pupillary dilation occurs in dim light to allow more light to enter.

Iris muscles are controlled by the autonomic nervous system. Parasympathetic nerve fibers innervate the circular muscle, and sympathetic fibers supply the radial muscle. Acting via the autonomic nervous system, conditions other than light can induce changes in pupillary size (Figure 6-29). For example, dilation of the pupils accompanies generalized discharge of the sympathetic nervous system in response to actual or perceived danger.

The cornea and lens refract the entering light to focus the image on the retina

Light is a form of electromagnetic radiation consisting of particle-like individual packets of energy, called **photons,** that travel in a wavelike fashion. The distance between two wave peaks is known as the **wavelength.** So-called visible light (a term based on human vision) lies between 400 (violet) and 700 nm (red), which most animal eyes can sense, with some species also detecting ultraviolet wavelengths

down to 310 nm. Light of different wavelengths in this frequency band is perceived as different **color** (hue) sensations, at least by humans. In addition to having variable wavelengths, light energy also varies in **intensity;** that is, in the amplitude, or height, of the wave. Dimming a bright red light does not change its color; it just becomes less intense or less bright.

Light waves *diverge* (radiate outward) in all directions from every point of a light source. The forward movement of a light wave in a particular direction is known as a *light ray.* Divergent light rays reaching the eye must be bent inward to be focused back into a point on the light-sensitive retina to provide an accurate image of the light source (Figure 6-30a). The bending of a light ray (**refraction**) occurs when the ray passes from a medium of one density into a medium of a different density. Light travels faster through air than through other transparent media such as water and glass. When a light ray enters a medium of greater density, it slows down (the converse is also true).

Two factors contribute to the degree of refraction: the comparative *densities* of the two media (the greater the difference in density, the greater the degree of bending) and the *angle* at which the light strikes the second medium (the greater the angle, the greater the refraction). *The two structures most important in the eye's refractive ability are the cornea and the lens.* The curved corneal surface, the first structure light passes through as it enters the eye, contributes extensively to the refractive ability of a terrestrial animal's eye because the difference in density at the air/corneal interface is much greater than the differences in density between the lens and the fluids surrounding it. In contrast, the refractive power of the fish cornea is negligible because the surrounding medium of the fish cornea is water and not air. Thus, in fish, most of the refractive power is present in the lens.

With a curved surface such as a lens, the greater the curvature, the greater the degree of bending and the stronger the lens. When a light ray strikes the curved surface of any object of greater density, the direction of refraction depends

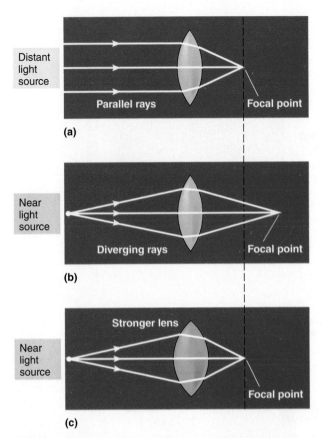

FIGURE 6-30 Focusing with a convex lens. Diverging light rays must be bent inward to be focused. (a) The rays from a distant (far) light source (more than 6 meters from the eye) are parallel by the time the rays reach the eye. (b) The rays from a near light source (less than 6 meters from the eye) are still diverging when they reach the eye. A longer distance is required for a lens of a given strength to bend the diverging rays from a near light source into focus compared to the parallel rays from a distant light source. (c) To focus both a distant and a near light source in the same distance (the distance between the lens and retina), a stronger lens must be used for the near source.

© Cengage Learning, 2013

on the angle of the curvature (Figure 6-30b, c). A lens with **convex** surfaces converges light rays, bringing them closer together, a requirement for bringing an image to a focal point. Refractive surfaces of the eye are therefore convex. (A lens with *concave* surfaces diverges light rays.)

The transparent state of the lens is due to proteins called **crystallins,** a term originating from the ancient Greeks who considered the lens to be icelike in appearance (*krusallinos,* "ice crystal"). Young lens cells contain organelles when they first form from stem cells in a fetus, but the structures are destroyed during early development, leaving a cytoplasm consisting of an usually thick solution of these crystallins. Strangely enough, the genes for crystallins evolved from the duplicated genes of various small enzymes such as *lactate dehydrogenase* of anaerobic glycolysis (p. 55). Vertebrates have multiple families of crystallins, whereas cephalopods have only a single family, the S-crystallins, which descended from a liver enzyme. For a lens to be transparent, crystallins must stay folded and evenly dispersed in order to create a glassy state.

Many vertebrate lenses in color-sensing eyes are *multifocal,* having concentric rings of different refractive indices such that the more peripheral region has a lower refractive index (a decreased ability to bend light) than does the more central region. Seeing clearly underwater also requires a special spherical lens with this property. Blue light refracts more than red light does, so having different zones of refractive index allows the lens to focus all color wavelengths properly on the retina. This gradation, from the lens's center outwards, is accomplished with progressively lower concentrations of crystallins. Animals such as humans and pigs lack multifocal lenses and so suffer from *chromatic aberration*; that is, they cannot focus on, for example, a blue and a red object at the same distance simultaneously. This problem is minimal in bright light when narrowed pupils allow light only into the lens center, where refraction is minimal; so multifocal lenses are mainly found in color-seeing nocturnal species and those fish residing in dim waters or with non-closing pupils.

In primates, there may be yellow pigment in the lens that acts as a filter, absorbing wavelengths below 400 nm to eliminate ultraviolet light. In contrast, the lens of animals that perceive ultraviolet wavelengths is clearer and permits the transmission of wavelengths as low as 310 nm. The lens thus absorbs only the more destructive region of the spectrum (nucleic acids and proteins begin to absorb light strongly at wavelengths shorter than 310 nm). In some species of fish the cornea is pigmented, which can also limit the penetration of short-wavelength light. In some species of fishes, reptiles including birds, and amphibians, colored oil droplets effectively limit the passage of short wavelengths to the photoreceptors. These oil droplets lie in the inner segments of the cone receptors.

Occasionally the crystallin fibers of the lens in older animals are progressively broken down and become so opaque that light rays cannot pass through, a condition known as a **cataract.** Cataracts often appear to have a white or crushed ice appearance and are the most common cause of blindness in dogs, in which either a genetic predisposition or diabetes is the cause. However, in the vast majority of dogs over six years of age the lens develops the more normal condition of **nuclear sclerosis,** which appears as a slight graying of the lens. It usually appears in both eyes at the same time and occurs because of compression of the crystallin fibers of the lens.

If an image is focused before it reaches the retina, or is not yet focused when it reaches the retina, it will be blurred. Light rays originating from near objects are more divergent when they reach the eye than are rays from distant sources. By the time they reach the eye, rays from light sources more than 6 meters away are considered parallel. For a given refractive ability of the eye, a near source of light requires a greater distance behind the lens for focusing than does a distant source because the near-source rays are still diverging when they reach the eye (Figure 6-30a, b).

In species where the distance between the lens and the retina always remains the same, in order to bring both near and distant light sources into focus on the retina (that is, in the same distance) a stronger lens must be used for the near source (Figure 6-30c). The strength of the lens can be adjusted through the process of *accommodation,* to which we now turn.

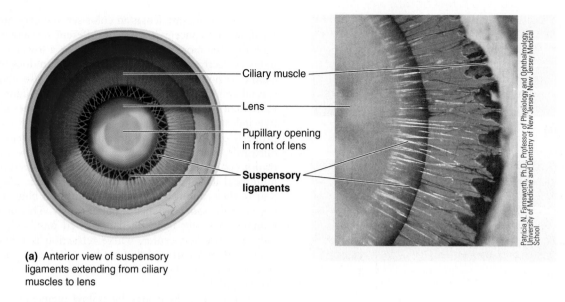

(a) Anterior view of suspensory ligaments extending from ciliary muscles to lens

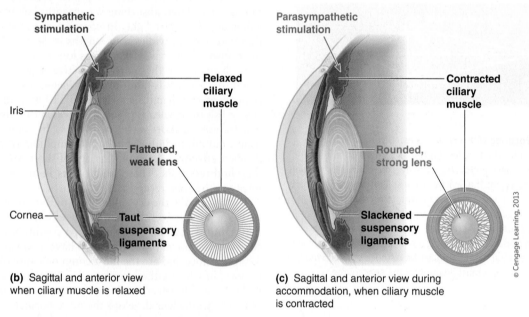

(b) Sagittal and anterior view when ciliary muscle is relaxed

(c) Sagittal and anterior view during accommodation, when ciliary muscle is contracted

FIGURE 6-31 **Mechanism of accommodation.** (a) Suspensory ligaments extend from the ciliary muscle to the outer edge of the lens. (b) When the ciliary muscle is relaxed, the suspensory ligaments are taut, putting tension on the lens so that it is flat and weak. (c) When the ciliary muscle is contracted, the suspensory ligaments become slack, reducing the tension on the lens, allowing it to assume a stronger, rounder shape because of its elasticity.

Accommodation increases the strength of the lens for near vision

The ability to adjust lens strength to focus both near and distant sources of light on the retina is called **accommodation**. This occurs in three ways. (1) The strength of the lens depends on its shape, which in turn is regulated in mammals and some species of reptiles (including all birds) by the *ciliary muscle*. (2) However, the fish lens is extremely dense (overall a high refractive index), so it has a fixed focal length, and focusing is accomplished by physical movement of the lens back and forth. (3) Moving the photoreceptor layer also works if the lens shape cannot be changed. For example, some species of annelids can increase or decrease

the fluid volume within the optic chamber to alter distance between the lens and the photoreceptors.

In mammals, the **ciliary muscle** is part of the ciliary body, an anterior specialization of the choroid layer. The ciliary body has two major components: the ciliary muscle and the capillary network that produces the aqueous humor (Figure 6-28). The ciliary muscle is a ring of smooth muscle attached to the lens by suspensory ligaments (Figure 6-31), and it works somewhat counter intuitively. When the ciliary muscle is *relaxed,* the suspensory ligaments are taut, and they *pull* the lens into a flattened, weakly refractive shape. As the muscle *contracts,* its circumference decreases, *slackening* the tension in the suspensory ligaments (Figure 6-31). When the lens is subjected to less tension by the suspensory ligaments, it assumes a more spherical shape because of its inherent elasticity. The greater curvature of the more rounded lens increases its strength, bending light rays more strongly.

In the normal mammalian eye, the ciliary muscle is relaxed and the lens is flat for distant vision, but the muscle contracts to let the lens become more convex and stronger for near vision. Sympathetic nerve fibers induce relaxation of the ciliary muscle for distant vision, whereas the parasympathetic nervous system causes the muscle's contraction for near vision. The lens of birds is softer than that of mammals and so exhibits a more rapid accommodation rate. Of note, the lens of the chameleon is remarkable for its ability to cause light to diverge rather than focus it. Studies have shown that the process of accommodation in chameleons is much greater than that of other vertebrates. First, the chameleon cornea overfocuses the image, which the lens corrects for by diverging the image just sufficiently enough to cast a magnified image on the retina.

Light must pass through several retinal layers in vertebrates before reaching the photoreceptors

The major function of the eye is to focus light rays from the environment on the *rods* and *cones,* the photoreceptor cells of the retina. The photoreceptors then transform the light energy into electrical signals for transmission to the CNS.

The receptor-containing portion of the vertebrate retina is actually an extension of the CNS and not a separate peripheral organ. During embryonic development, the retinal cells "back out" of the nervous system, so the retinal layers, surprisingly, are facing backward! The neural portion of the retina consists of three layers of excitable cells (Figure 6-32): (1) the outermost layer (closest to the choroid) containing the *rods* and/or *cones,* whose light-sensitive ends face the choroid (away from the incoming light); (2) a middle layer of *bipolar cells;* and (3) an inner layer of *ganglion cells.* **Rods** and **cones** were named for their appearance under a light microscope. In lower vertebrates, rods and cones converge via chemical synapses onto the same **bipolar neuron,** in contrast to the separate rod bipolar and cone bipolar pathways of higher vertebrates. Axons of the **ganglion cells** join together to form the **optic nerve,** which leaves the retina slightly off center. The point on the retina at which the optic nerve leaves and through which blood vessels pass is the **optic disc** (Figures 6-28a and 6-33). This region is often called the **blind spot;** no image can be detected in this area because it lacks rods and cones. We are normally not aware of the blind spot, because a combination of information from two eyes, rapid eye movements, and central processing "fills in" the missing spot. You can discover your own blind spot by a simple exercise (Figure 6-34).

The backwards nature of vertebrate retinas may be an accident of evolution that, while seeming to be illogical (see the cephalopod eye, p. 255), works well enough. However, it does create at least three problems: (1) the blind spot that required the evolution of corrective mechanisms; (2) susceptibility to *retinal detachment* (a serious problem in many mammals including dogs and humans, in which the neural layer separates from the pigment layer); and (3) light having to pass through nonsensory cell layers first, with potential distortion. Recent research has found that certain glial cells in the overlying nonsensory layer have *waveguide* properties

that direct light into the photoreceptors, compensating for problem 3. Whether this ability favored the evolution of the "backwards" layout of the retina, or evolved later to compensate for an accidentally flawed layout, is not known. The glial cells do not solve problems 1 and 2, however.

Light must pass through the ganglion and bipolar layers before reaching the photoreceptors in all areas of the retina except the **fovea.** In the primate fovea, which is a pinhead-sized depression located in the exact center of the retina (Figure 6-28a), the bipolar and ganglion cell layers are pulled aside so that light strikes the photoreceptors directly. This feature, coupled with the fact that *only* cones (which have greater acuity or discriminative ability than do rods) are found here, makes the fovea the point of greatest visual acuity (most distinct vision). Thus, you turn your eyes so that the object at which you are looking is focused on the fovea. The area immediately surrounding the fovea, the **macula lutea,** also has a high concentration of cones and fairly high acuity (Figure 6-34). Fovea are present in teleost fishes, some snakes, and birds. Dogs (and wolves) lack a fovea per se and instead have a **visual streak,** which serves as the region of highest visual acuity. In addition to a dense central area, the visual streak extends into the temporal and nasal portions of the retina. Physiologists believe these extensions enable animals to scan the horizon with enhanced visual acuity. The visual streak also contains a higher density of ganglion cells. For example, the density of ganglion cells is generally greater in wolves than in domesticated canines (12,000 to 14,000/mm^2 versus 6,400 to 14,000/mm^2). Survival pressures that maintain the elevated density of ganglion cells are likely evolutionary factors in wolves, whereas breeding programs in dogs place little selective pressure on maximizing visual acuity.

Some species of birds may have as many as two or three foveas. The **central fovea** of birds lies on the nasal side of the optic nerve and, in species that need high stereoscopic vision (hawks and eagles), produces (as we noted earlier) a resolving power approximately eight times greater than that of humans. The density of ganglion cells in this region reaches approximately 65,000 mm^2, far surpassing foveal values from mammals with high acuity (human, 38,000 mm^2; macaque, 33,000 mm^2). A second **lateral fovea** is located above the optic nerve and a third area, the so-called **linear area,** extends horizontally across the central portion of the retina. Nocturnal predators, such as owls, must summate light from both eyes under dim conditions and are characterized by a single temporal fovea.

In conditions of dim light, a layer of reflecting material, the superiorly located **tapetum lucidum,** enhances the ability of many animal species to locate and detect distant objects. In dogs the tapetum lucidum is a highly cellular structure enriched in zinc and cysteine that is between 9 and 20 layers thick at its center. Physiologists believe the tapetum reflects light that has previously passed through the retina back through it a second time, giving photoreceptors a second opportunity to capture the light stimuli. For example, the feline eye reflects approximately 130 times more light than does the human eye. However, there is a price for reflecting this light: The ability of the eye to accurately resolve the details of an image is somewhat compromised by the scattering of light during this process. This reflecting layer accounts for the eye shine you can see in animals staring back into a source of light.

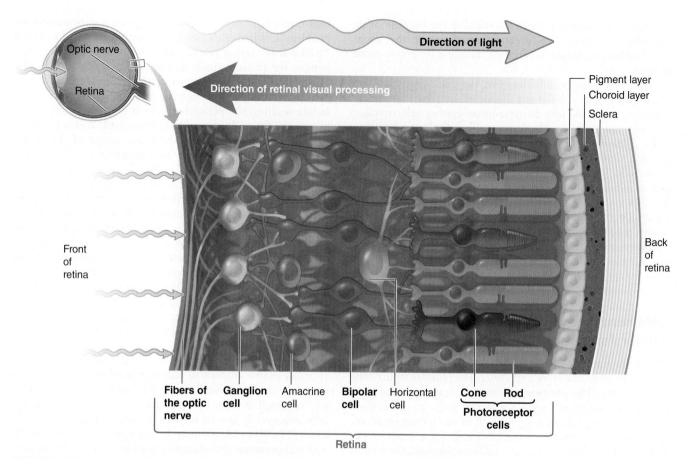

FIGURE 6-32 Retinal layers (mammalian eye). The retinal visual pathway extends from the photoreceptor cells (rods and cones, whose light-sensitive ends face the choroid *away from* the incoming light) to the bipolar cells to the ganglion cells. The horizontal and amacrine cells act locally for retinal processing of visual input.

© Cengage Learning, 2013

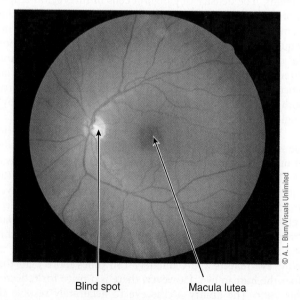

Blind spot   Macula lutea

© A. L. Blum/Visuals Unlimited

FIGURE 6-33 View of the human retina seen through an ophthalmoscope. With an ophthalmoscope, a lighted viewing instrument, it is possible to view the optic disc (blind spot) and macula lutea within the retina of the rear of the eye.

FIGURE 6-34 Demonstration of the blind spot. Find the blind spot in your left eye by closing your right eye and holding the book about 4 inches from your face. While focusing on the cross, gradually move the book away from you until the circle vanishes from view. At this time, the image of the circle is striking the blind spot of your left eye. You can similarly locate the blind spot in your right eye by closing your left eye and focusing on the circle. The cross will disappear when its image strikes the blind spot of your right eye.

© Cengage Learning, 2013

The tapetum of cats and lemurs is rich in riboflavin, which can absorb light in the shorter wavelengths (blue, 450 nm) and then emit fluorescent light at a longer wavelength (520 nm) more closely approximating the maximal sensitivity of *rhodopsin* in the rod photoreceptors. This shift in wavelength would brighten the appearance of a blue-black background and enhance the contrast between objects in the environment and the background.

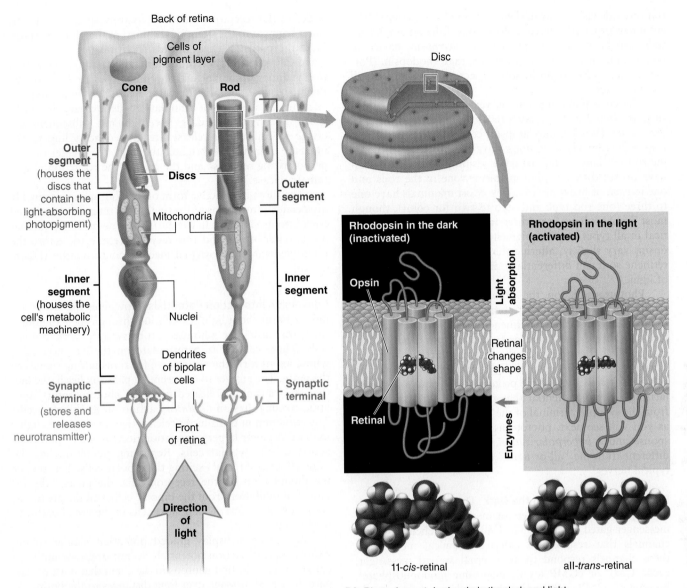

(a) Structure of rods and cones

(b) Photopigment rhodopsin in the dark and light

FIGURE 6-35 Photoreceptors. (a) The three parts of the rods and cones, the eye's photoreceptors. Note in the outer segment of the rod and cone the stacked, flattened, membranous discs, which contain an abundance of photopigment molecules. (b) A photopigment, such as rhodopsin, depicted here and found in rods, consists of opsin, a plasma-membrane protein, and retinal, a vitamin-A derivative. In the dark, 11-*cis*-retinal is bound within the interior of opsin and the photopigment is inactive. In the light, retinal changes to all-*trans*-retinal, activating the photopigment.

© Cengage Learning, 2013

Phototransduction by opsin and retinene in retinal cells converts light stimuli into neural signals

Photoreceptors consist of three parts (Figure 6-35a): (1) an *outer segment,* which lies closest to the eye's exterior, facing the choroid, and detects the light stimulus; (2) an *inner segment,* which lies in the middle of the photoreceptor's length and contains the metabolic machinery of the cell; and (3) a *synaptic terminal,* which lies closest to the eye's interior, facing the bipolar cells, and transmits the signal generated in the photoreceptor on light stimulation to these next cells in the visual pathway.

The outer segment, which is rod-shaped in rods and cone-shaped in cones (Figure 6-35a), consists of stacked, flattened, membranous discs containing an abundance of photopigment molecules. Each retina has about 150 million photoreceptors, and over a billion photopigment molecules may be packed into the outer segment of each photoreceptor.

Photopigments within the photoreceptor undergo chemical alterations when activated by light. A photopigment consists of an enzymatic protein called **opsin** combined with **retinene,** a derivative of vitamin A. **Rhodopsin,** the vertebrate rod photopigment, cannot discriminate between various wavelengths in the visible spectrum; it absorbs all visible wavelengths, though with a peak around 500 nm. Therefore, rods provide vision only in shades of gray by detecting different intensities, not different colors. Some fish and amphibian rods also have the photopigment **porphyropsin**, named in part because it is more purple in hue than rhodopsin. (Porphyropsin is also found in some insect eyes, which we will consider later.)

The cone photopigments in vertebrates, called **scotopsins**, are found in five types—**red**, **green**, **yellow**, **blue**, and **ultraviolet**. (We will look at insect color vision later.) These respond selectively to various wavelengths of light according to their names, making color vision possible. Birds can have up six different photopigments, one in the rods and one in each of five types of cones. Most mammals have one to three (one rod type and zero to two for color), though most primates have four (three for color). Retinene is identical in all types of photopigments, but the photoreceptors' opsins vary slightly, altering the electron resonance of the retinene so it can differentially absorb various wavelengths of light.

Phototransduction, the mechanism of excitation, is basically the same for all vertebrate photoreceptors (Figure 6-35b). It begins when a retinene molecule absorbs a photon, causing retinene to shape from a *cis* to a *trans* conformation. This change triggers the enzymatic activity of opsin. Through a series of steps, this light-induced breakdown and subsequent activation of the protein bring about a hyperpolarizing receptor potential that *reduces* transmitter release from the synaptic terminal of the photoreceptor. Ultimately, as you will see next, phototransduction both a *graded* response and a *hyperpolarization*, and as such, is distinctly different from the "all-or-none" depolarization response of typical nerves.

Photoreceptor Activity in the Dark The plasma membrane of a photoreceptor's outer segment contains chemical-messenger-gated Na$^+$ channels. Unlike other chemical-gated channels that respond to external chemical messengers, these channels respond to an internal second messenger, **cyclic GMP** or **cGMP** (cyclic guanosine monophosphate). Binding of cGMP to these Na$^+$ channels keeps them open. In the absence of light, the concentration of cGMP is high (Figure 6-36a). Unlike most receptors, therefore, the Na$^+$ channels of a vertebrate photoreceptor are open in the absence of stimulation, that is, in the dark. The resultant passive inward Na$^+$ leak depolarizes the photoreceptor. The passive spread of this depolarization from the outer segment (where the Na$^+$ channels are located) to the synaptic terminal (where the photoreceptor's neurotransmitter is stored) keeps the synaptic terminal's voltage-gated Ca^{2+} channels open. Calcium entry triggers the release of the neurotransmitter glutamate from the synaptic terminal while in the dark.

Photoreceptor Activity in the Light On exposure to light, the concentration of cyclic GMP is decreased through a series of biochemical steps triggered by retinene-opsin activation (Figure 6-36b). Rod and cone cells contain a G protein (see p. 97) called **transducin**. The activated opsin activates transducin, which in turn activates the enzyme **phosphodiesterase**. This enzyme degrades cyclic GMP, thus decreasing the concentration of this second messenger in the cell. This reduction in cyclic GMP permits the chemically gated Na$^+$ channels to close, which stops the depolarizing Na$^+$ leak and hyperpolarizes the membrane. This hyperpolarization,

which is the receptor potential, passively spreads from the outer segment to the synaptic terminal of the photoreceptor. Here the potential change leads to *closure* of the voltage-gated Ca^{2+} channels and reduction in glutamate release from the synaptic terminal. Thus, photoreceptors are inhibited by their adequate stimulus (hyperpolarized by light) and excited in the absence of stimulation (depolarized by darkness). The hyperpolarizing potential and subsequent decrease in transmitter release are graded according to the intensity of light. The brighter the light, the greater the hyperpolarizing response and the greater the reduction in glutamate release.

The short-lived active form of the photopigment quickly dissociates into opsin and retinal. The retinal is converted back into its *cis* form. In the dark, enzyme-mediated mechanisms rejoin opsin and this recycled retinal to restore the photopigment to its original inactive conformation (Figure 6-35b).

Conversion into Action Potentials How does the retina signal the brain about light stimulation through such an inhibitory response? The photoreceptors synapse with bipolar cells. These cells in turn terminate on the ganglion cells, whose axons form the optic nerve for transmitting signals to the brain. The answer to our seeming paradox lies in the fact that the transmitter released from the photoreceptors' synaptic terminal has an *inhibitory* action on the bipolar cells. The reduction in transmitter release that accompanies light-induced receptor hyperpolarization decreases this inhibitory action on the bipolar cells. Removing inhibition has the same effect as directly exciting the bipolar cells. The greater the illumination on the receptor cells, the greater the removal of inhibition from the bipolar cells and the greater in effect the excitation of these next cells in the visual pathway to the brain.

Bipolar cells display graded potentials similar to the photoreceptors. Action potentials do not originate until the ganglion cells, the first neurons in the chain that must propagate the visual message over long distances to the brain.

Now let us turn to the differences between rods and cones.

Rods provide indistinct gray vision at night, whereas cones provide sharp color vision during the day

The earliest mammals are believed to have been nocturnal and thus characterized by a pure rod retina, comparable to that found in living nocturnal animals such as the American flying squirrel (*Glaucomys volans*). In the human retina there are 20 times more rods than cones (120 million rods compared to 6 million cones per eye). In species with a mixture of both rods and cones, the rods are most abundant in the periphery (away from the macula/fovea). Recall that cones, if present, are most abundant in the macula/fovea regions, although the proportion of cones in this region varies between species. For example, in this region of the canine retina there are fewer than 10% cones, whereas diurnal squirrels have an all-cone region. Diurnal birds have a greater number of cones, as well as a greater cone density, than do humans.

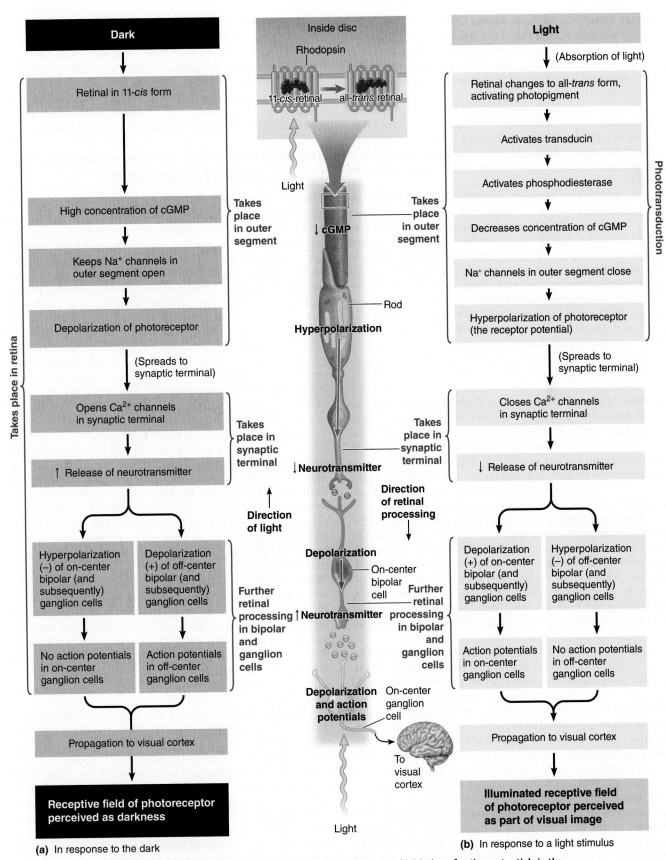

FIGURE 6-36 Phototransduction, further retinal processing, and initiation of action potentials in the vertebrate visual pathway. (a) Events occurring in the retina and visual pathway in response to the dark. (b) Events occurring in the retina and visual pathway in response to a light stimulus.

Greater Sensitivity of Rods versus Cones The outer segments are longer in rods than in cones, so they contain more photopigments and thus can absorb light more readily. Also, as you will see shortly, the way in which rods connect with other neurons in their processing pathway further increases the sensitivity of rod vision. Because rods have high sensitivity, they can respond to the dim light of night. Cones by contrast have lower sensitivity to light, being activated only by bright daylight. Thus, as we noted earlier, rods are specialized for night vision and cones for day vision. Diurnal animals use cones for day vision, which is in color and distinct.

An astonishing discovery in 2009 revealed a totally unexpected mechanism to enhance rod sensitivity. In most cells, including rods of diurnal mammals, active unwound DNA is in the center of the nuclei, with dense-packed inactive DNA in the periphery. Irina Solovei, Jochen Guck, and colleagues found that in animals that prefer low-light conditions—including cats, rats, mice, opossums and deer—the situation is reversed. Amazingly, the nuclei appear to be lenses! With loose DNA on the outside and dense DNA in the center, the refractive index of these nuclei are predicted by biophysical simulations to focus light in the rods like a tiny lens. (Whether this actually works to improve night vision remains to be tested.)

Greater Acuity of Cones versus Rods The pathways by which cones are "wired" to the other retinal neuronal layers confer high acuity (sharpness; ability to distinguish between two nearby points). Thus, cones provide sharp vision with high resolution for fine detail. By contrast, the wiring pathways of rods provide low acuity, so you can see at night with your rods but at the expense of distinctness. Let us see how the wiring patterns influence sensitivity and acuity.

Little convergence of neurons takes place in the retinal pathways for cone output (see p. 141). Each cone generally has a private line connecting it to a particular ganglion cell. In contrast, much convergence occurs in rod pathways. Each human retina has about 1 million ganglion cells, which receive the output from around 125 million photoreceptors, 120 million of which are rods. Thus, output from more than 100 rods may converge via bipolar cells on a single ganglion cell.

Before a ganglion cell can have an action potential, the cell must be brought to threshold through influence of the graded potentials in the photoreceptors to which it is wired. Because a single-cone ganglion cell is influenced by only one cone, only bright daylight is intense enough to induce a sufficient receptor potential in the cone to ultimately bring the ganglion cell to threshold. The abundant convergence in the rod visual pathways, in contrast, offers good opportunities for summation of subthreshold events in a rod ganglion cell (see p. 136). Whereas a small receptor potential induced by dim light in a single cone would not be sufficient to bring its ganglion cell to threshold, similar small receptor potentials induced by the same dim light in multiple rods converging on a single ganglion cell would have an additive effect to bring the rod ganglion cell to threshold. Because rods can bring about action potentials in response to small amounts of light, they are much more sensitive than cones. However, because cones have private lines into the optic nerve, each cone transmits information about an extremely small receptive field on the retinal surface. Cones are thus able to provide highly detailed vision at the expense of sensitivity. With rod vision, acuity is sacrificed for sensitivity. Because many rods share a single ganglion cell, once an action potential is initiated, it is impossible to discern which of the multiple rod inputs were activated to bring the ganglion cell to threshold. Objects appear fuzzy when rod vision is used because of this poor ability to distinguish between two nearby points.

The sensitivity of eyes can vary markedly through dark and light adaptation

The eyes' sensitivity to light depends on the amount of photopigment present in the rods and cones. When you go from bright sunlight into darkened surroundings, you cannot see anything at first, but gradually you begin to distinguish objects as a result of both pupil dilation and the process of **dark adaptation**. Breakdown of photopigments during exposure to sunlight tremendously decreases photoreceptor sensitivity. For example, a reduction in rhodopsin content of only 0.6% from its maximum value decreases rod sensitivity about 3,000 times. In the dark, the photopigments broken down during light exposure are gradually regenerated. (An hour is required for canine rhodopsin to regenerate itself after exposure to light, whereas in humans it regenerates much more quickly.) As a result, the sensitivity of your eyes gradually increases so that you can begin to see in the darkened surroundings.

Conversely, when you move from the dark to the light (for example, leaving a movie theater and entering the bright sunlight), your eyes are very sensitive to the dazzling light at first. With little contrast between lighter and darker parts, the entire image appears bleached. In addition to pupil constriction, as the intense light rapidly breaks some of the photopigments down, the sensitivity of the eyes decreases and normal contrasts can once again be detected, a process known as **light adaptation**. The rods are so sensitive to light that sufficient rhodopsin is broken down to essentially "burn out" the rods in bright light; that is, the rod photopigments, having already been broken down by the bright light, are no longer able to respond to the light. Furthermore, a central neural adaptive mechanism switches the eye from the rod system to the cone system on exposure to bright light. Researchers estimate that human eyes' sensitivity can change as much as 1 million times with each change event as they adjust to various levels of illumination through dark and light adaptation.

Because retinene, one of the two key photopigment components, is a derivative of vitamin A, adequate amounts of this nutrient must be available for the ongoing resynthesis of photopigments. **Night blindness** occurs because of dietary deficiencies of vitamin A. This can occurs for example in cattle lacking green food, such as those being fed old hay or grazing on very dry pastures. Although photopigment concentrations in both rods and cones are reduced in this condition, there is still enough cone photopigment to respond to the intense stimulation of bright light, except in the most severe cases. However, even modest reductions in rhodopsin content can decrease the sensitivity of rods so much that they cannot respond to dim light. Such an animal can see in the day using cones but cannot see at night because the rods are no longer functional.

Color vision depends on the ratios of stimulation of the various cone types

Certain objects in the environment such as the sun, fire, and bioluminescent organisms emit light. But how does an animal see most objects (e.g., trees, rocks), which do not emit light? Molecules in various objects selectively absorb particular wavelengths of light transmitted to them from light-emitting sources, and the unabsorbed wavelengths are *reflected* from the objects' surfaces. It is these reflected light rays that enable animals to see the objects and their color. In the open ocean, for example, water molecules absorb the longer red wavelengths of light and scatter the shorter blue wavelengths back to the observer, which can be absorbed by the photopigment in the eyes' blue cones, thereby activating them.

Color vision occurs in most vertebrates and insects. We will explore species differences later; for now, recall that each cone type is most effectively activated by a particular wavelength of light in the range of color indicated by its name. Many primates including humans have **trichromatic**, or three-color, vision, as illustrated in Figure 6-37. Cones that are most sensitive to wavelengths in the region of 560 nm are termed **long (L) cones** and contribute to the perception of red. Cones that respond best to more intermediate wavelengths of 530 nm are termed **middle (M) cones**, and contribute to the perception of green; whereas cones that respond to short wavelengths of light in the region of 420 nm are termed **short (S) cones** and contribute to the perception of blue. Animals that can perceive ultraviolet waves have a fourth type of cone, the **ultraviolet (UV) cone**, with a peak response in the region of 360–380 nm. However, cones also respond in varying degrees to other wavelengths (Figure 6-37). An animal's perception of the many colors of the world depends on the cone types' various *ratios of stimulation* in response to different wavelengths. A wavelength perceived as blue does not stimulate red or green cones at all but excites blue cones maximally (the percentages of maximal stimulation for red, green, and blue cones, respectively, are 0:0:100). The sensation of yellow, in comparison, arises from a stimulation ratio of 83:83:0, red and green cones each being stimulated 83% of maximum, whereas blue cones are not excited at all. The ratio for green is 31:67:36, and so on, with various combinations giving rise to the sensation of all the different colors. (Televisions and other video equipment make use of this phenomenon with so-called *RGB* technology, e.g., red, green, and blue pixels on a computer monitor, and RGB output cables from some video players.) *White* is a mixture of all wavelengths of light, whereas *black* is the absence of light.

The extent that each of the cone types is excited is coded and transmitted in separate parallel pathways to the brain. A distinct color vision center in the primary visual cortex has recently been identified. This center combines and processes these inputs to generate the perception of color, taking into consideration the object in comparison with its background.

Color discrimination abilities vary greatly among types of animals

Now let's compare animal color vision in more detail. Generally, the level of color discrimination correlates well with an animal's lifestyle and habitat light features. As previously

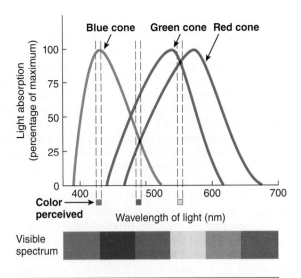

Color perceived	Percentage of maximum stimulation		
	Red cones	Green cones	Blue cones
■	0	0	100
■	31	67	36
■	83	83	0

FIGURE 6-37   Sensitivity of the different types of cones to different wavelengths. Graph shows the trichromatic eye for Old World primates (including humans), with blue (S), green (M), and red (L) cones. The ratios of stimulation of the three cone types are shown for three sample colors (see main text).

© Cengage Learning, 2013

discussed, many strictly nocturnal animals such as some species of bats have only rods. Other illustrative patterns include the following:

- *Fish*: Many shallow-water bony fish such as trout are trichromatic, with L, M, and S cones as in humans. However, most deep-sea fish (and other animals) have lost the L (red) cones, because red light from the sun does not penetrate into the depths. Remarkably, one species of deep-sea fish, the predatory dragonfish *Malacosteus niger*, has re-evolved L cones with a novel phototransduction using bacterial *chlorophyll* instead of retinene to absorb red light! This is adaptive because these fish have bioluminescent organs—*photophores*—that produce red light, which they use to illuminate prey (in contrast, most bioluminescent marine organisms make blue or green light). This allows the dragonfish's light to be virtually invisible to all but itself (a "sniper scope"). Some fish such as salmon have UV cones (we will explore the use of UV vision shortly). In at least one fish, the salmon *Oncorhynchus gorbuscha*, the UV cones switch to S (blue) types with age. This is likely because young salmon live in shallow water where UV penetrates, whereas adults go deeper where blue dominates but UV is absent. This discovery by Christiana Cheng and Iñigo Flamarique was the first to document cone-switching as a way to modulate color vision as an animal changes habitat.

- *Non-avian reptiles*: Most diurnal snakes and lizards have cones but no rods; many can see UV. Curiously, nocturnal geckos (p. 4), which evolved from diurnal ancestors, have only UV, L, and M cones (again, no rods). It is not certain how these work at night, but of note is that the cones have evolved rodlike morphologies and have been behaviorally shown to work very well in dim light.

- *Birds*: Most birds except some nocturnal species have excellent color vision, some having four (**tetrachromatic**) or five (**pentachromatic**) cone types! For example, the pigeon *Columba livia* has UV and yellow cones in addition to L, M, and S cones.
- *Mammals*: Most orders of mammals have been shown to have color vision in that they can discriminate based on color, with most (such as mice, dogs and cats) being *dichromatic* (with M, or medium-long "LM," cones and either S or UV cones). In contrast, Old World primates including humans, and some female New World monkeys, are trichromatic; for example, baboons have primarily green M (63%) and the recently evolved red L (33%) cones, whereas only 4% are blue S. Genomic analysis shows that the L opsin evolved from a few simple mutations from the M opsin. This may have been favored to help pick out ripe red fruit from among green leaves and/or to see bright colors on the faces or genitals of potential mates (as in baboons). Remarkably, mice engineered with the human L-opsin gene gained the ability to distinguish new colors at the red end of the spectrum!
- *Arthropods*: Many species of insects can distinguish color, including UV and polarized light (see the following section). Most are thought to be dichromatic, but flies and honeybees are trichromatic, while the Japanese swallowtail butterfly is tetrachromatic. The mechanism by which insects perceive color is less well understood than in vertebrates but is believed to result from expressing retinal receptors with differing sensitivities to light of different wavelengths. Flies and honeybees have UV, blue, and yellow receptors but cannot see red. The world champion in terms of color vision appears to be the mantis shrimp (see Figure 6-27e). Some species have up to *16* different photoreceptors, 12 for color (**dodedachromatic!**), and 4 for polarized light. Such vision may be important in mating but also in feeding. These highly colorful animals hide in burrows or reef crevices looking for specific prey that often have unique color patterns, such as fish and snail shells; these must be discriminated from the complex colorful reef background. The shrimp attack using powerful, extremely fast claws as stunning and shell-cracking weapons. (The claws are fast due to storage and release of spring energy; see Chapter 8, p. 350).

unanswered Questions | Why did whales lose the blues?

Whales and seals are **monochromatic**, with rods and only a medium-long "LM" cone; their blue S cones have been lost due to mutation of the S-opsin gene into a pseudogene. This is counterintuitive, since blue light is so dominant in the oceans, yet it seems to have occurred independently in whales and seals. One reason suggested by Leo Peichl and colleagues (who discovered the widespread occurrence of this phenomenon) is that the mutation occurred in a coastal semiterrestrial ancestor that foraged in murky shallow waters, which are browner (redder) and much less blue than the open sea. However, we do not know this for certain. <<

Some animals can see ultraviolent and/or polarized light

Many arthropods and some species of birds, fish, marsupial mammals, bats, and rodents have UV cones (although there are not necessarily all four types of cones in any one species). The detection of UV wavelengths may give birds and fish spatial cues that they can use in orientation and navigation during migration. Short-wavelength gradients vary depending on the sun's angle in the sky. These gradients increase in saturation as the sun moves from directly overhead to an angle of 90 degrees (that is, UV light is a greater proportion of the incoming sunlight at twilight). Some species of birds as well as some salmon migrate over thousands of kilometers and appear to rely on UV-color gradients in the sky to locate the position of the sun on days when clouds obscure it.

Territorial markings and food detection are other uses of ultraviolet sensing. Mice and voles, for example, have M and UV cones, with the latter possibly used to mark trails and see territorial scent markings (urine and feces), which are visible in UV light. In turn, kestrels—raptorial birds with UV cones—can use this information to locate areas of prey abundance. Because fresh vole urine and feces absorb UV sunlight, to a kestrel, the marks would look like dark streaks on a field. Recently, researchers have shown that a flower bat (*Glossophaga soricina*) senses ultraviolet; although it cannot distinguish colors, it may be able to see flowers better at twilight, when ultraviolet radiation is relatively more prevalent. Some pollinating insects also find desirable flowers base on their UV reflectance.

Among arthropods, honeybees have been best studied. They can detect the pattern of not only UV itself but also **polarized** light, which has electromagnetic waves all oriented at the same angle. That angle changes with respect to the movement of the sun in the sky, and the bees can use this information to communicate the location of a food source to the other bees in the hive. (This form of communication can be eliminated experimentally by using filters that block the passage of specific light forms to determine the role of those forms in animal behavior.)

Polarized light can also be used to find potential mates, prey, and predators by the pattern of polarized light reflected from the body surfaces. For example, researchers have recently shown that one species of butterfly can detect patterns of polarized light reflecting off the wings of others of its kind, a signal used in mating. And the African dung beetle uses the polarization of moonlight to orient itself so that it can move along a straight line. Once the beetle has located a food source, it quickly forms a ball of dung and rolls it rapidly away in order to avoid any aggressive encounters for food in the dung pile.

Finally, let us return to the visual champions, mantis shrimp. Recently it was found that at least two mantis shrimp species—as well as one shiny colorful scarab beetle—can see **circularly polarized** light (CPL), in which the angle of the waves rotates continuously as the light travels. CPL can be produced when light reflects off of certain surfaces (e.g., reflections off helical proteins in scarab beetle carapaces are CPL). It is not known what these arthropods use this ability for, but it may give them private communication signals that no other animal can perceive.

Visual information is separated and modified within the visual pathway before it is integrated into a perceptual image of the visual field by the vertebrate cortex

The field of view that can be seen without moving the head is known as the **visual field.** In vertebrates, the information that reaches the visual cortex in the occipital lobe is not a replica of the visual field for several reasons.

1. The image detected on the retina at the onset of visual processing is upside down and backward because the light rays bend (Figure 6-38). Once projected to the brain, the animal interprets the inverted image as being correctly oriented.

2. The information transmitted from the retina to the brain is not merely a point-to-point record of photoreceptor activation. Before the information reaches the brain, the retinal neuronal layers beyond the rods and cones reinforce selected information and suppress other information to enhance contrast. One mechanism of retinal processing is *lateral inhibition* (see Figure 6-7, p. 216), which increases the dark–bright contrast to enhance the sharpness of boundaries.

    Another mechanism of retinal processing involves differential activation of two types of ganglion cells, **on-center** and **off-center ganglion cells** (Figure 6-39). The receptive field of a cone ganglion cell is determined by the field of light detection by the cone with which it is linked. On-center and off-center ganglion cells respond in opposite ways, depending on the relative comparison of illumination between the center and periphery (surroundings) of their receptive fields. Think of the receptive field as a doughnut. An on-center ganglion cell increases its rate of firing when light is most intense at the center of its receptive field (that is, when the doughnut hole is lit up). In contrast, an off-center cell increases its firing rate when the periphery of its receptive field is most intensely illuminated (that is, when the doughnut itself is lit up). This is useful for enhancing the difference in light level between one small area at the center of a receptive field and the illumination immediately around it. By emphasizing differences in relative brightness, this mechanism helps define contours of images, but in so doing, information about absolute brightness is sacrificed.

3. Various aspects of visual information such as form, color, depth, and movement are separated and projected in parallel pathways to different regions of the cortex. Only when these separate bits of processed information are integrated by higher visual regions is a reassembled picture of the visual scene perceived. This is similar to the blobs of paint on an artist's palette versus the finished portrait; the separate pigments do not represent a portrait of a face until they are appropriately integrated on a canvas.

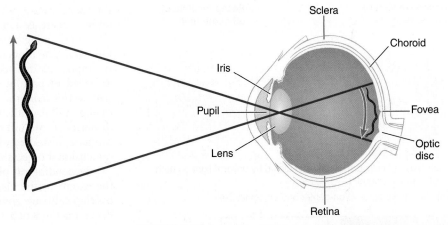

FIGURE 6-38 Inversion of the image on the retina.

© Cengage Learning, 2013

4. Because of the pattern of wiring between the eyes and the visual cortex, the left half of the cortex receives information only from the right half of the visual field as detected by both eyes, and the right half receives input only from the left half of the visual field of both eyes.

As a result of refraction, light rays from the left half of the visual field fall on the right half of the retina of both eyes (the medial or inner half of the left retina and the lateral or outer half of the right retina) (Figure 6-40). Similarly, rays from the right half of the visual field reach the left half of each retina (the lateral half of the left retina and the medial half of the right retina). Each optic nerve exiting the retina carries information from both halves of the retina it serves. The human optic nerve contains 1.2 million nerve fibers; compare this to 167,000 in canines and 116,000 to 165,000 in cats. This information is separated as the optic nerves meet at the **optic chiasm** (*chiasm* means "cross") located underneath the hypothalamus. Within the optic chiasm, the fibers from the medial half of each retina cross to the opposite side, but those from the lateral half remain on the original side. The reorganized bundles of fibers leaving the optic chiasm are known as **optic tracts**. Each optic tract carries information from the lateral half of one retina and the medial half of the other retina. Therefore, this partial crossover brings together from the two eyes fibers that carry information from the same half of the visual field. Each optic tract, in turn, delivers to the half of the brain on its same side information about the opposite half of the visual field.

The thalamus and visual cortices elaborate the visual message

The first stop in the mammalian brain for information in the visual pathway is the **lateral geniculate nucleus** in the thalamus (Figure 6-40) (**optic tectum** in fish and amphibians). It separates information received from the eyes and relays it via fiber bundles known as **optic radiations** to different zones in the cortex, each of which processes different aspects of the visual stimulus (for example, color, form, depth, movement). This sorting process is no small task, because each human optic nerve contains more than 1 million fibers carrying information from the photoreceptors in one retina. This is more than all the afferent fibers carrying

Receptive field of
on-center cell

Receptive field of
off-center cell

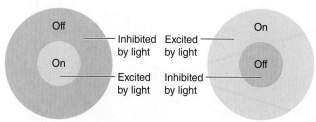

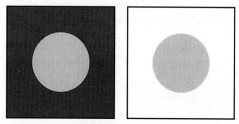

Both types of cells are weakly stimulated by uniform light on both
center and surround.

(a) Receptive fields of on-center and off-center cells

(b) Outcome of retinal processing by on-center
and off-center cells

FIGURE 6-39 Example of the outcome of retinal processing by on-center and off-center ganglion cells. (a) On-center cells are excited and off-center cells are inhibited by bright light in the centers of their receptive fields. (b) Retinal processing by on-center and off-center ganglion cells is largely responsible for enhancing differences in relative (rather than absolute) brightness, which helps define contours. Note that the gray circle surrounded by black appears brighter than the one surrounded by white, even though the two circles are identical (same shade and size).

© Cengage Learning, 2013

somatosensory input from all the regions of the body! Researchers estimate that hundreds of millions of neurons, occupying about 30% of the human cortex, participate in visual processing, compared to 8% devoted to touch perception and 3% to hearing. Yet the connections in the visual pathways are precise. The lateral geniculate nucleus and each of the zones in the cortex that process visual information have a topographic map representing the retina point for point. As with the somatosensory cortex, the neural maps of the retina are distorted. The fovea, the retinal region capable of greatest acuity, has much greater representation in the neural map than do the more peripheral regions of the retina. Interestingly, in animals born with immature visual systems such as the ferret and cats, during the first three to four weeks after birth while their eyes remain closed, the retinal ganglion cells are still developing connections among themselves as well as to other retinal cells. The long axons of the ganglion cells grow along the optic tract to each lateral geniculate nucleus of the thalamus. These ganglion cells eventually grow into layers in the lateral geniculate nucleus, although axons from each eye remain separate from each other.

Depth Perception Although each half of the visual cortex receives information simultaneously from the same part of the *visual field* as received by both eyes, the messages from

the two eyes are not identical. Depending on the species of animal, there is a potential area of overlap (Figure 6-41). The overlapping area seen by both eyes at the same time is known as the **binocular** ("two-eyed") **field** of vision, which is important for **depth perception.** Binocular vision is achieved in part by routing axons from one eye together with axons from the other eye to synapse in the same layers of the lateral geniculate nucleus. The brain uses the slight disparity in the information received from the two eyes to estimate distance, allowing an animal to perceive three-dimensional objects in spatial depth.

Depending on placement of the eyes in the skull, both the extent of the visual field and the amount of binocular overlap can vary considerably. For example, there is no evidence for binocular vision in teleost fish, where eyes are located on opposite sides of the head. Dog eyes are laterally directed approximately 20 degrees from the midline, and human and owl eyes look straight ahead. Thus, with each eye, the average dog has a monocular field of view from 135 to 150 degrees, approximately 60 to 70 degrees greater than that of humans. Although this gives dogs a greater ability to scan the horizon, the degree of binocular overlap is much less (about 30 to 60 degrees compared to 130 in cats and 140 degrees in humans). Interestingly, in some species of birds the effect of two or more foveas enables each eye to have completely separate visual fields. Such an arrangement increases the total visual field and may also permit some binocular vision.

Hierarchy of Visual Cortical Processing Within the cortex, visual information is first processed in the primary visual cortex, then is projected to *associative* visual areas for even more complex processing and abstraction. The cortex contains a hierarchy of visual cells that respond to increasingly complex stimuli. The visual cortex is precisely organized into vertical columns extending from its outer surface to the white matter. Each column is made up of cells that process the same bit of visual input. There are three types of columns: One type is devoted to input from the left or the right eye for binocular interaction and depth perception; a second type perceives form and movement; and a third types processes color. Unlike a retinal cell, which responds to the amount of light, a cortical cell fires only when it receives a particular pattern of illumination for which it is programmed. For example, the types that process form and movement come in three subtypes called **simple, complex,** and **hypercomplex cells.** Simple and complex cells are found within the columns of the primary visual cortex, whereas hypercomplex cells are found in the higher visual associative areas. Some simple cells fire only when a bar is viewed vertically in a specific location, others when a bar is horizontal, and others at various oblique orientations. Movement of a critical axis of orientation is important for response by some of the complex cells. Hypercomplex cells add a new dimension by responding only to particular edges, corners, and curves. Other aspects of visual input, such as color, are processed simultaneously through a similar hierarchical organization. In this way, the dotlike pattern of photoreceptors stimulated in the retinal image is transformed in the cortex into information about depth, color, position, orientation, movement, contour, and length. How and where the entire image is finally put together is still unresolved.

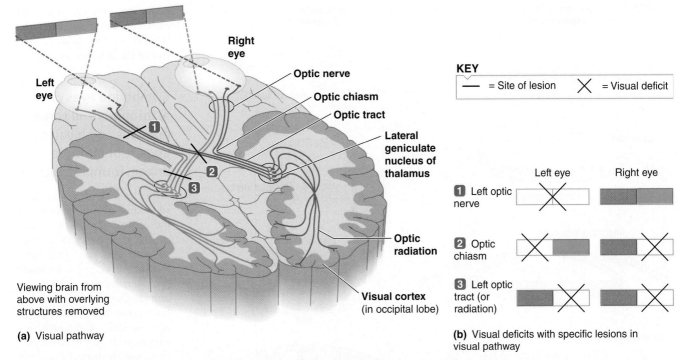

FIGURE 6-40 **The mammalian visual pathway and visual deficits associated with lesions in the pathway.** (a) Note that the left half of the visual cortex in the occipital lobe receives information from the right half of the visual field of both eyes (in *green*), and the right half of the cortex receives information from the left half of the visual field of both eyes (in *red*). (b) Each visual deficit illustrated is associated with a lesion at the corresponding numbered point of the visual pathway in part (a).

© Cengage Learning, 2013

Visual input goes to other areas of the vertebrate brain not involved in vision perception

Not all fibers in the visual pathway terminate in the visual cortexes. Some are projected to other regions of the brain for purposes other than direct vision perception. Examples of nonsight activities dependent on input from the rods and cones include (1) contribution to cortical alertness and attention, (2) control of pupil size, (3) setting the biological "clock," and (4) control of eye movements. For setting the clock, 3% of the ganglion cells are not involved in visual processing. Instead, they make **melanopsin,** a light-sensitive pigment that leads to signals through a non-optic nerve that connects the retina to the clock center of the hypothalamus (p. 282). For eye movement, each eye is equipped with a set of six **external eye muscles** that position and move the eye so that it can better locate, see, and track objects in response to retinal signals.

Cephalopod camera eyes have light-sensing cells on top of neural cells

Let's now contrast the vertebrate eye with advanced ones in other animals. The cephalopod eye is particularly instructive in terms of **convergent evolution** (that is, the independent evolution of similar structures in different species without a common ancestral structure). The backward structure of the vertebrate retina, as you saw earlier (p. 247), causes problems. Comparative studies suggest that there is no inherent advantage for this design, because the sophisticated cephalo-

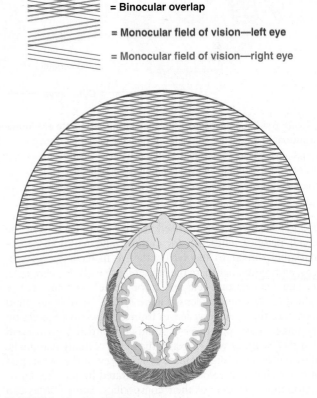

FIGURE 6-41 **Visual fields.**

© Cengage Learning, 2013

**Eye formation in
cephalopods**

**Eye formation in
vertebrates**

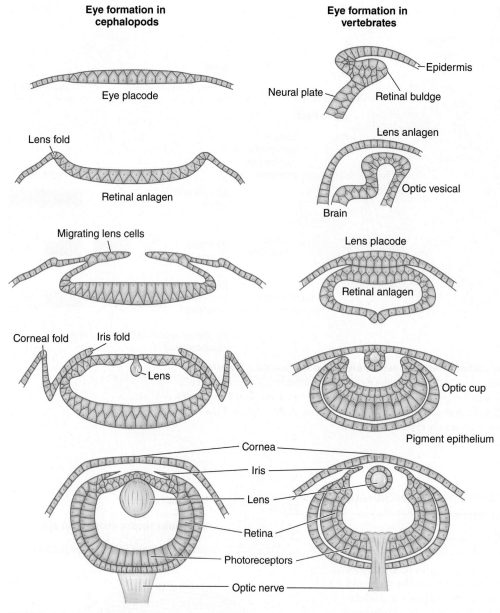

FIGURE 6-42 Comparison of cephalopod and vertebrate eye formation. Development proceeds from top to bottom. The cephalopod eye forms from the epidermis through a series of invaginations, while the vertebrate eye emerges from the internal brain and induces the overlying epidermis to form the lens. Thus, their respective retinas end up in opposite arrangements.

Source: W.A. Harris (1997). *Pax-6: Where to be conserved is not conservative. Proceedings of the National Academy of Sciences* 94: 2098–2100.

evolution, arising from the different ways in which this sensory system first evolved out of the nervous system.

Compound eyes of arthropods and some annelids and bivalves consist of multiple image-forming units

Now let's turn to the most successful group of animals, the arthropods. As we noted earlier, they have *compound*, or *faceted*, eyes. The functional unit of the compound eye is termed the rhabdomeric **ommatidium** (Figure 6-43). Each ommatidium consists of an optical, light-gathering part as well as a sensory portion, which transduces light into an action potential. The number of ommatidia can vary from one, as in the worker ant *Ponera punctatissima*, to over 10,000 in the eye of a dragonfly. Generally, in animals with only a few ommatidia the **facets** (external surface of an individual compound eye unit or ommatidium) are circular in appearance, whereas with increased numbers the facets are more densely packed together and assume a hexagonal shape for optimal packing (Figure 6-44a). Each of these optical units focuses on a separate portion of the visual field, with the conglomerate image appearing as a mosaic rather than the sharp visual image perceived by vertebrates (Figure 6-44b), much like ink dots

pod eye has evolved a reversed arrangement. The eye of the squid, octopus, cuttlefish (or cuttles), and nautiloids is remarkably similar to the vertebrate eye in having a cornea, lens, and retina, but there the resemblance ends. In these eyes, the pigment (sensory) cells are on top of the retina and receive light directly from the lens (Figure 6-42). There is no blind spot since the optic nerve leaves from the backside. Studies on the image-resolving power of squid and cuttlefish eyes reveal that they work as well or better than eyes of many mammals, allowing them to discriminate distant prey in the ocean and intricate signaling color patterns generated by the skin chromatophores (p. 154) of their conspecifics. Some cuttles can also see polarized light (p. 252). The difference between cephalopod and vertebrate retinas seems to be an accident of

that make up a newspaper photograph. Due to these separate units, compound eyes form an upright retinal image, in contrast to the inverted image displayed on the vertebrate retina. Compound eyes have lower visual acuity than do vertebrate eyes because while vertebrate photoreceptors sample approximately 0.02 degrees of the visual field, each ommatidium views as much as 2 to 3 degrees. Consequently, the visual fields of adjacent ommatidia overlap. However, because most adult insects have a pair of compound eyes that bulge out of each side of the head, the *field of vision* can be markedly greater compared to vertebrates. For example, in the water bug *Notonecta,* the field of view for the two eyes provides binocular vision in front (almost 250 degrees in the horizontal plane), above and below the head.

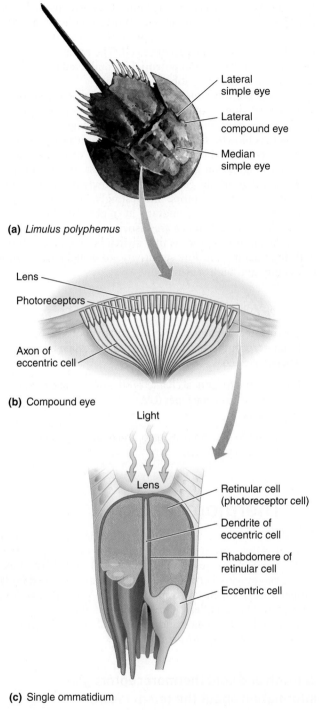

(a) *Limulus polyphemus*

(b) Compound eye

(c) Single ommatidium

FIGURE 6-43 **Anatomy of the compound eyes of the horseshoe crab,** *Limulus polyphemus.*

© Cengage Learning, 2013

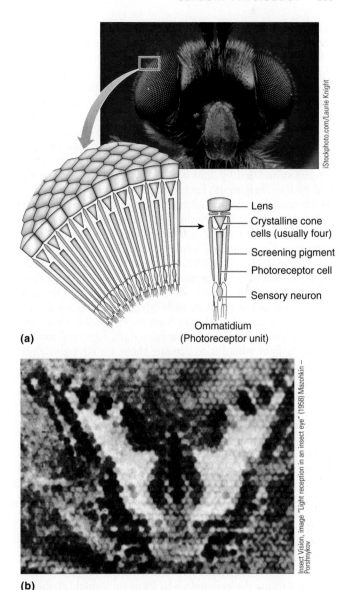

(a)

(b)

FIGURE 6-44 **Insect compound eyes.** (a) Eyes of a soldier fly. The lens of each photosensitive unit (ommatidium) directs light onto a crystalline cone, which focuses light on photoreceptor cells below it. (b) An approximation of light reception in an insect eye. This image of a butterfly formed after a researcher took a photograph through the outer surface of a compound eye that had been detached from an insect. It might not be what the insect "sees," because integration of signals sent to the brain from photoreceptors might produce a sharper image. But this representation is useful insofar as it suggests how the separate ommatidia sample the overall visual field.

Source: C. Starr & R. Taggart (2004), *Biology: The Unity and Diversity of Life*, 10th ed. (Belmont, CA: Brooks/Cole). (a) Figure 35.14, (b) Figure 35.15.

The lens of arthropod compound eyes forms from the cuticle as a biconvex corneal lens at the outer end of each ommatidium. Normally this cuticle forming the lens is transparent and colorless. In addition to having its own lens, each ommatidia contains a cluster of at least eight **retinular** or **photoreceptor cells** (Figures 6-43b and 6-44a). The retinular cells are arranged in a circle with the periphery lined with pigment to shield against leakage of light from neighboring ommatidia. At the center of the circle lies a dendrite from a

single, modified retinular cell, the so-called **eccentric cell.** The photopigment *rhodopsin* is localized in a specialized region of the retinular cell called the **rhabdomere** (Figure 6-43c). Normally the rhabdomeres of each retinular cell are tightly packed along this central region of the ommatidium. Structurally, rhabdomeres appear as microvilli that increase the surface area available for light absorption. This elongated region of photosensitive microvilli, with contributions from each of the eight or more rhabdomeres, is termed the rhabdom of the ommatidium (Figure 6-43c). The body of the eccentric cell lies on the periphery of the ommatidium, with the dendrite making direct connections with the rhabdom.

To enhance visual contrast at a light–dark edge, the eccentric cells interact with each other by the process of lateral inhibition that we noted earlier for touch (see Figure 6-7) and the vertebrate eye. Lateral inhibition also diminishes the apparent difference in brightness between areas that are uniformly illuminated, whereas in darkness the ommatidia are less stimulated by light but are comparatively less inhibited by their neighbors. Each eccentric cell sends collaterals, which make inhibitory synaptic contacts with its neighbor cells lying approximately three to five ommatidia away. In this manner, a strongly activated cell directly receiving a light stimulus generates impulses, which *decrease* impulse generation in its comparatively weakly stimulated neighbors. At the boundary between light and dark illumination, the net effect is to enhance discrimination between the two zones because the eccentric cells lying in this region receive different degrees of inhibition from the two sides. Thus, ommatidia lining the border of a light–dark edge and not receiving a light stimulus to activate them are strongly inhibited by their neighboring cells, which are better illuminated. The net effect is that the activity in these cells is comparatively low compared to ommatidia lying in either the light or dark zones.

Two basic types arthropod compound eyes have evolved. Most diurnal insects have **apposition eyes**, characterized by ommatidium that are optically isolated from each other. Thus, each ommatidium receives light from a comparatively narrow region of the visual field. However, a wide, optically clear zone between the rhabdoms and the lens characterizes **superposition eyes**, which nocturnal arthropods have. Light from different ommatidia can mix together so that when the light finally reaches the rhabdom it represents a composite from many ommatidia. The adaptive value of this arrangement lies with its ability to receive light energy striking the eye surface to focus on a single rhabdom at the expense of acuity.

In addition to arthropods, compound eyes occur in some annelids such as featherduster worms (see Figure 11-5) and bivalve mollusks called ark clams. These have very poor resolution and seem to serve mainly as sentinels for danger, warning the animal of a potential predator approaching so that it can withdraw into its tube (worm) or close up (clam).

Phototransduction in rhabdomeric eyes uses rhodopsin but otherwise differs from that in vertebrate eyes

Although compound eyes use rhodopsin and G proteins as do camera eyes, the rest of the process of phototransduction in differs remarkably from that found in vertebrates. Recall that vertebrate Na^+ channels *close* in response to an increase in light, leading to hyperpolarization in rods and cones. In contrast, rhabdomeric photoreceptors *depolarize* in response to light. Rhodopsin, via a G protein, activates phospholipase C, which in turn leads to an increased production of inositol triphosphate (IP_3) and diacylglycerol (DAG) (see p. 98). The increase in IP_3 triggers release of Ca^{2+} ions from intracellular stores. Together these two second messengers, intracellular

Ca^{2+} and DAG, govern the opening of cation channels and thus generate a depolarizing potential. Graded potentials generated in the retinular cells spread through gap junctions into the dendrite of the eccentric cell. The eccentric cell thus depolarizes and an action potential is conducted through the optic nerve to the optic lobe of the brain.

Insects use the same set of photoreceptors for vision at all light intensities. Comparative studies of photoreceptor K^+ conductance have shown that rapidly moving diurnal species such as blowflies have fast photoreceptor responses that consume considerable amounts of ATP in order to restore rapidly changing ion gradients. In contrast, species such as crane-flies that are nocturnal and fly slowly have slow (and less costly) photoreceptors. Interestingly, locusts (*Schistocerca*) are diurnal but migrate at night. In this species changes in K^+ conductance are associated with the functioning of the photoreceptors as they switch between a day state, with high acuity and low sensitivity, to a night state, with low acuity and high sensitivity.

check your understanding 6.6

Describe the steps involved in vertebrate phototransduction by photoreceptors and further retinal processing by bipolar and ganglion cells.

Compare the functional characteristics of rods and cones, and how various cone types (UV, S, M, L) are used in various vertebrates.

Compare eye structure, imaging, and phototransduction in vertebrates and insects.

6.7 Thermoreception

The ability to detect environmental temperature has clear survival value, given that biological processes are often highly temperature sensitive (Chapter 15) and thermal extremes can quickly become lethal, particularly elevated body temperatures, which can rapidly impair protein function. Vertebrates have at least three kinds of thermoreceptors, best studied in mammals and the rattlesnake: *cold* sensors, *warmth* sensors, and *infrared* sensors.

Warmth and cold thermoreceptors give information about the temperature of the environment near the body

Mammalian skin and tongue (technically an extension of the skin) have two distinct kinds of sensors for temperature. One responds to increased temperature (warmth sensors), primarily above the skin's temperature, whereas the other responds to decreasing temperature (cold sensors) below skin temperature. (Skin temperatures tend to be lower than body temperatures, around 30 to 33°C in humans.) These are used primarily for thermoregulation, informing the brain about possible disturbances to core temperature (there are also pain sensors that respond to extreme heat or cold; see the section on nociception that follows). These sensors are *anticipatory* feedback components: They inform the brain of an oncoming disturbance *before* the core temperature is ac-

(a)

(b)

FIGURE 6-45 Thermoreception in pit vipers. (a) A close-up of the Western diamond-backed rattlesnake (*Crotalus atrox*) showing a thermoreceptor pit. (b) Appearance of a live mouse's body in infrared wavelengths, converted to colors for human vision.

© Cengage Learning, 2013

tually altered, allowing corrective measures to be activated with little or no time delay (Chapter 1, pp. 16–17, and Chapter 15, p. 748).

The molecular receptors for these senses have proven remarkably elusive until recently. However, distinct mammalian heat-gated and cold-gated ion channels have now been discovered, and their genes cloned. They are of the *TRP* Ca-channel family (p. 217) and thus related to mechanoreceptors. At least three heat-gated channels are known, called TRPV1, -2, and -3. TRPV3 starts firing at about 33°C, likely providing the signal for thermoregulation. But TRPV1 only fires above 42°C, a point at which heat starts to become painful. As you will see later, this receptor also reacts to spicy food! And TRPV2 fires at 52°C and higher, also clearly for pain response rather than thermoregulation. These channels appear to be directly gated by thermal changes. (In contrast, thermosensing neurons in the nematode *C. elegans* have a thermally activated receptor that activates cGMP as a second messenger, as in vertebrate phototransduction.)

Two cold-gated channels are known, called **TRPA1** and **TRPM8** (which is also activated by menthol, the "cooling" chemical of mint). TRPA1 only opens at ice-cold temperatures and is associated with pain. TRPA1 also appears to signal temperature changes in fruit flies and nematodes. In contrast, TRPM8 opens between 8° and 28°C in mammals and seems to be the receptor for thermoregulation. Interestingly, the amphibian TRPM8 channel operates over a lower temperature range, reflecting these animals' lower basal body temperatures than mammals'.

Recently, researchers have discovered thermal sensors in sharks. The sensory cells are those that have long been known as electroreceptors, the topic of the next section, where we return to this discovery.

Infrared thermoreceptors give pit vipers the ability to detect prey and desirable thermal habitats

A very different use of thermal stimuli has evolved in at least two groups of animals, the pit vipers (Viperidae, which includes rattlesnakes), and some pythons and boas. These have extraordinarily sensitive receptors that respond to **infrared radiation,** the low-energy radiation that carries heat energy. The receptor cells are simply branched dendrites of neurons, located in small pits in the skin (Figure 6-45a), on either side of the head in pit vipers (anterior to and below the eyes), and along the jaws of pythons. Behavioral studies (in which a rattlesnake's eyes are covered with black tape) show that these pits allow detection of warm mammalian prey (such as rodents; Figure 6-45b). The snake's brain may construct a visual image from the thermal data, although measurements and calculations of sensory resolution show that the snake would see only blurry images. A mouse that is 10°C warmer than the background environment can be detected at about half a meter away. The sensors are particularly important for use in the first rapid strike to capture the prey. Because there are two pits on vipers, a type of binocular infrared vision is possible, allowing a rattlesnake to determine prey distance.

Physiologists are not sure how these receptor cells detect infrared radiation, but recent work shows they have a very heat-sensitive TRPA1 channel (76% identical to the TRPA1 that signals "ice-cold" pain in mammals; see above). Remarkably, neurophysiological recordings suggest that different sensors respond to different infrared wavelengths. Thus, the snakes may be able to see infrared light in colors! Recent behavioral studies suggest that pit vipers can also use thermal information to find desirable thermal habitats and to avoid undesirable ones. These animals, like all ectotherms, cannot regulate body temperature by internal mechanisms, and instead must rely on the environment (see Chapter 15).

check your understanding 6.7

Compare uses and mechanisms of cold, warmth, and infrared receptors.

Are warmth or cold receptors more important to an animal? Why?

6.8 Nociception: Pain

Pain is primarily a protective mechanism meant to bring to awareness (or the higher brain centers) the fact that tissue damage is occurring or is about to occur. Unlike other somatosensory modalities, the perception of pain can be influenced by other past or present experiences, for example, heightened pain perception in a dog that has been abused, or lowered pain signaling induced by stress responses in an injured animal. (Recall, though, that perception is only reportable by humans; we infer perception in other animals by their reactions, but cannot know what, if anything, they actually perceive.)

Stimulation of mammalian nociceptors elicits two types of pain signals

There are three categories of pain receptors or *nocireceptors*:

- **Mechanical nociceptors** respond to mechanical damage such as cutting, crushing, or pinching.
- **Thermal nociceptors** respond to temperature extremes, especially heat.
- **Polymodal nociceptors** respond equally to all kinds of damaging stimuli, including irritating chemicals released from injured tissues.

None of the nociceptors have specialized receptor organs; they are all naked nerve endings. At least in mammals, to some extent pain occurs in two phases with two signaling pathways: (1) a quick response to direct damage (the **fast pain pathway**, which may subside), followed by (2) a slower, prolonged chemical response to inflammation and cell contents released by damage (the **slow pain pathway**). Think about the last time you cut or burned your finger. You undoubtedly felt a sharp twinge of pain at first that stopped, followed by a more diffuse, disagreeable, persistent pain shortly thereafter. Pain impulses originating at nociceptors are transmitted to the mammalian CNS via one of two types of afferent fibers.

Fast Pain The fast initial pain has the following features:

- Arises from specific mechanical or heat nociceptors.
- Transmitted over large, myelinated **A-delta fibers** at rates of up to 30 meters/sec.
- Easily localized, but may be fast adapting giving only a brief signal

Slow Pain The subsequent slow, dull, aching sensation that persists has these properties:

- Arises from mechanical and thermal nociceptors, activated by chemicals such as **bradykinin,** a normally inactive substance that is activated by enzymes released into the ECF from damaged tissue. (It is also involved in inflammation; see p. 469).
- Transmitted by small, unmyelinated **C fibers** at a slower rate of 12 meters/sec.
- Poorly localized, but because of value to survival, does not quickly adapt to sustained stimulation. The persistence of bradykinin might explain the long-lasting, aching pain that continues after removal of the mechanical or thermal stimulus that caused the damage.

Researchers have isolated some mammalian pain receptor proteins. As you saw earlier in the section on thermoreception, one of the fast pain receptors is the thermally gated TRPV2, which fires at 52°C and higher. This channel appears to be directly gated by the "heat" stimulus. One nociceptor protein responsible for the slow burning pain associated with inflammation is called *TRPV1* (p. 259), also known as "the chili pepper receptor" because it is also activated by **capsaicin,** the chemical of seed-bearing fruits of pepper plants that causes burning sensation in the mouth (see *A Closer Look at Adaptation: Why Are Hot Spices Hot?*). Interestingly, the TRPA1 channel that we noted earlier as a "cold pain" sensor has also been found to be activated by chemicals of damaged tissues as well as by wasabi, horseradish, garlic, and mustard oil.

All nociceptors in mammals can be sensitized by **prostaglandins,** which greatly enhance the receptor response to noxious stimuli (that is, it hurts more when prostaglandins are present). Prostaglandins are fatty acid derivatives released as paracrines in tissue injury (see p. 468). Aspirin-like drugs inhibit the synthesis of these paracrines, accounting at least in part for the analgesic (pain-relieving) properties of these drugs (p. 470).

Higher-Level Processing of Pain Input Multiple structures are involved in mammalian pain processing. The primary afferent pain fibers synapse with specific second-order excitatory interneurons in the dorsal horn of the spinal cord. The two best-known neurotransmitters here are *substance P* and *glutamate.* **Substance P** activates ascending pathways that transmit nociceptive signals to higher levels for further processing (Figure 6-46a). Ascending pain pathways go first to the brainstem's *reticular formation,* which increases the level of alertness associated with the noxious encounter. The next stop is the *thalamus* relay station, and then to cortical *somatosensory* areas, which localize the pain. (Pain can still be perceived in the absence of the cortex, presumably at the level of the thalamus.) Next, associative cortical areas participate in other conscious components of the pain experience, such as deliberation (at least in humans) about the incident. Interconnections from the thalamus and reticular formation to the *hypothalamus* and *limbic system* elicit the behavioral and emotional responses accompanying the painful experience. The limbic system appears to be especially important in perceiving the unpleasant aspects of pain.

Recall that glutamate is a major excitatory neurotransmitter (p. 172). In pain afferents, it acts on two different plasma membrane receptors on the dorsal horn excitatory interneurons, with two different outcomes. First, binding of glutamate with its *AMPA receptors* (see p. 199) ultimately results in the generation of action potentials in the dorsal horn cells, transmitting the pain message to higher centers. Second, binding of glutamate with its *NMDA receptors* leads to Ca^{2+} entry into these neurons. Ca^{2+} initiates second-messenger systems that make the dorsal horn cells more excitable than usual (see p. 199). This hyperexcitability contributes in part to the exaggerated sensitivity of an injured area to subsequent exposure to painful or even normally nonpainful stimuli, such as a light touch. Think about how highly sensitive your sunburned skin is, even to clothing. This exaggerated sensitivity is presumably useful to discourage activities that could cause further damage or interfere with healing of the injured area.

The mammalian brain has a built-in analgesic system

In addition to the chain of neurons from peripheral nociceptors to higher CNS structures for pain perception, the mammalian CNS also contains a built-in pain-suppressing or **analgesic system** that suppresses transmission in the pain pathways as they enter the spinal cord. One major brainstem region involved is the **periaqueductal gray matter** (gray matter surrounding the cerebral aqueduct, a narrow canal that connects the third and fourth ventricular cavities), which stimulates particular neurons whose cell bodies lie in nuclei in the *medulla* and *reticular formation*. Electrical stimulation of any of these parts of the brain produces profound analgesia. From the nuclei, neurons terminate on inhibitory interneurons in the dorsal horn of the spinal cord (Figure 6-46b). These inhibitory interneurons release *enkephalin*, which binds with **opiate receptors** at the afferent pain-fiber terminal.

People have long known that **morphine**, a component of the opium poppy, is a powerful analgesic. Researchers considered it very unlikely that mammals have been endowed with opiate receptors only to interact with chemicals derived from a flower! They therefore began to search for the substances that normally bind with these opiate receptors. The result was the discovery of **endogenous opiates** (morphine-like substances)—the *endorphins, enkephalins,* and *dynorphin*—which are important in the body's natural analgesic system. These endogenous opiates serve as analgesic neurotransmitters. Binding of enkephalin from the dorsal-horn inhibitory interneuron with the afferent pain-fiber terminal suppresses the release of substance P via presynaptic inhibition, thereby blocking further transmission of the pain signal (see Figure 4-22, p. 140). Morphine binds to these same opiate receptors, which accounts in large part for its analgesic properties. Furthermore, injection of morphine into the periaqueductal gray matter and medulla causes profound analgesia, suggesting that endogenous opiates also are released centrally to block the descending pain-suppressing pathway.

check your understanding 6.8

Compare the fast and slow pain pathways in mammals.

Describe the built-in analgesic system of the mammalian brain.

6.9 Electroreception and Magnetoreception

Many non-visual systems have evolved for environmental navigation and object sensing in situations where light is not adequate. Examples include predators tracking prey scent trails, a harbor seal's sensitive whiskers tracking the turbulent wake of a fish, and an echolocating bat. An alternative sensory mechanism in light-limited areas evolved in a number of bony and cartilaginous fishes and at least two mammals: **electroreception** which allows for **electrolocation**—the directional detection of external electric fields.

Electroreception can be passive or active, and can be used for navigation, prey detection, and communication

Passive electroreception is the ability to detect extraneous electric fields, not an animal's own electrical currents, and is used primarily for electrolocation. **Ampullary electroreceptors** (found in almost all nonteleost fishes and in some teleosts, as well as in several amphibian species) respond to low-frequency electric signals that are typical of electrical output from animal nerves and hearts as well as the electrical discharge used in electric fish courtship. (Most animals emit a direct current or DC field in seawater because of ion currents in their excitable tissues, and a wound or even a scratch can markedly alter these electrical fields). In elasmobranchs (sharks, skates, and rays) and certain teleosts, the **ampullae of Lorenzini**, a modified group of neuromasts (p. 218), act as **electroreceptors** and are used to locate prey even if it is located below the sediment surface (Figure 6-47a). For example, a shark or skate can easily detect the faint potentials associated with the gill muscles (from minute leakage of ions) of

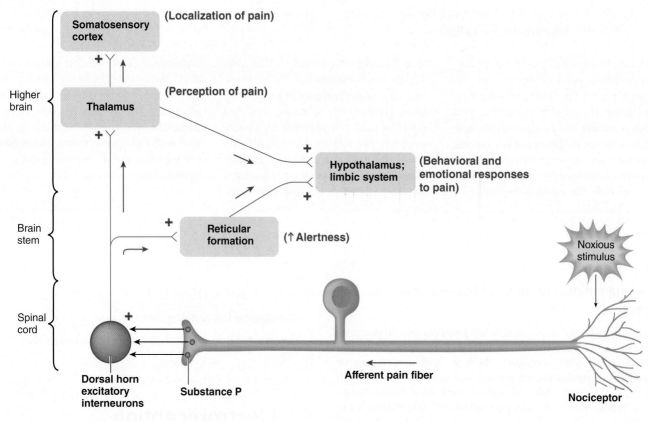

(a) Substance P pain pathway

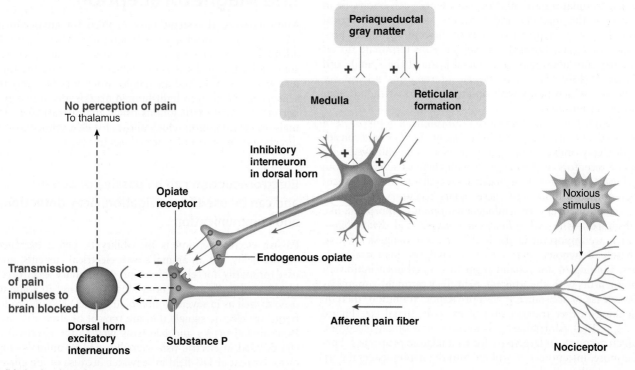

(b) Analgesic pathway

FIGURE 6-46 Substance P pain pathway and analgesic pathway. (a) When activated by a noxious stimulus, some afferent pain pathways release substance P, which activates ascending pain pathways that provide various brain regions with input for processing different aspects of the painful experience. (b) Endogenous opiates released from descending analgesic (pain-relieving) pathways bind with opiate receptors at the synaptic knob of the afferent pain fiber. This binding inhibits the release of substance P, thereby blocking transmission of pain impulses along the ascending pain pathways.

a buried flounder. The ampullae are filled with a clear, salty gel (a glycoprotein complex) that readily transmits minute electrical currents. New studies show that very small temperature changes induce voltage changes in the gel, suggesting that it serves in both electrosensing and thermosensing.

Passive electroreception has been also demonstrated in the platypus, which uses sensors in its "duckbill" to find electrical currents of prey hidden at the bottom of streams. Also, the aptly named star-nosed mole, a blind animal living in burrows, is suspected of having electroreception in its odd nose, which sprouts numerous fingerlike tentacles.

In contrast, **active electroreception** resembles echolocation in that the animal assesses its environment by actively emitting signals and receiving the feedback signal. Found in some freshwater fishes in the murky Amazon and some (also murky) African rivers, this system uses an **electric organ** in the tail section of the fish's body that discharges a current field, which emanates from its anterior body and then converges on the tip of the tail (Figure 6-47b), a transepidermal current flow. (We will describe these organs in more detail shortly.) Momentarily, the tip of the tail becomes negatively charged so that an electric current is produced in the surrounding water. This momentary change is induced repeatedly, at a high frequency. **Tuberous electroreceptors**, more numerous than ampullary electroreceptors, lie on the anterior body surface in the lateral line system and monitor changes in the local transepidermal current flow. These receptors respond to the high-frequency signals from the fish's *electric organ discharge* (which we will describe later in detail). Their distribution over the body surface provides a somatotopically organized view of the electrosensory world that resembles what might be presented to the visual system by an extended retina without a lens. These electroreceptors terminate in somatotopically organized maps in an area of the brain known as the *electrosensory lateral line lobe* (ELL), with much of the ELL circuitry dedicated to detecting small changes in amplitude associated with prey.

The reception is used in both *electrolocation* and *electrocommunication* with conspecifics. Active electroreceptive fish test the conductivity of the water around them and locate perturbations by detecting distortions produced in their own electric field. Any object that differs in impedance from the surrounding water distorts the electric field and alters the pattern of transepidermal current intensities in the area of body surface closest to the object (Figure 6-47c). The South American electric fish *Eigenmannia* generates electric organ discharges at frequencies from 250 to 600 Hz. When two fish with similar discharges meet, they shift their frequencies away from each other in order to create a larger frequency difference and to avoid "jamming" or interfering with their conspecific neighbor's own discharge pattern.

In marine fish, the electroreceptors (mainly ampullary) are connected to the body surface via long, low-resistance channels, whereas the ducts of freshwater fish receptors (mainly tuberous) are comparatively short. Because the skin resistance of freshwater fish is much higher than that of marine fish, transepidermal voltage differences are greater, which means that current flow can be detected with the much shorter ducts. When current enters an electroreceptor cell of a fish, the resulting depolarization opens voltage-gated Ca^{2+} channels in the membrane, which triggers the efflux of neurotransmitter from the receptor cell. Conse-

quently, the frequency of action potentials in the sensory fiber that innervates the receptor is increased. Although one important disadvantage of having an electric sense is its very limited range (only several meters), electrolocation permits an animal to actively function in darkness or in murky waters.

Electric organs use specialized electrocytes to generate electric organ discharges

Electric organs, located within active electroreceptive animals, produce **electric organ discharges (EODs),** currents outside the animal's body. The organs are derived from muscle tissue and consist of thin, waferlike cells called **electrocytes** stacked in columns of several thousand surrounded by a gelatinous insulating material (one side of the electrocyte is "smooth" and highly innervated, whereas the other is "rough" and consists of deep folds). The increased surface area due to the folding markedly increases the capacitance of each cell. These modified muscle cells have lost the ability to contract and are instead specialized for generating an ion current flow. When the organ is at rest, the *opposite* faces of the generating cells are at the same positive potential (that is, the extracellular fluid is positive on opposite surfaces of the electrocyte at approximately $+80$ mV). However, when the potential on one side of the electrocyte is reversed, current flows in a circuit that involves the two membranes, the cell cytoplasm and the external medium (Figure 6-47d). During a discharge, the potential on the innervated side is briefly reversed. The electrocytes are excited in synchrony by spinal nerves that generate small individual voltage gradients around each electrocyte (approximately 150 mV). Because the electrocytes are stacked in series and surrounded by insulating material, the voltage adds arithmetically, in a similar manner to batteries connected in series. A pacemaker potential whose command center is in the medulla drives the electric organ. Gap junctions assure precision in the firing of the electrocytes, maximizing signal intensity to achieve close electrotonic coupling of the medullary and spinal neurons. An additional factor that determines signal intensity is the number of columns of electrocytes capable of generating current.

In most electric fish species, EODs are limited in range to millivolts to volts and may be emitted continually for electrolocation and social communication. In contrast, a few species such as the electric eel (*Electrophorus*) can produce EODs up to several hundred volts *outside* their bodies. Such strong EODs are emitted to stun or kill prey. Because the conductivity of seawater is so high, the electric field produced by a strong discharge is essentially short-circuited and thus limited in range.

Production of EODs is energetically costly, and the South American longtail knifefish (*Sternopygus macrurus*) has evolved a clever solution. It uses its electrical sense at night, while during the day, it turns down signal generation. Neural and hormonal signals have been found to regulate the rapid insertion and removal of Na^+ channels in electrocyte membranes in a night-day (circadian) rhythm. Reducing Na^+ currents in this way saves considerable energy. Moreover, during unexpected social encounters, a surge of neural-hormonal signals can also trigger rapid insertion of channels. This system may be the fastest in the animal kingdom for regulating the number of ion channels in a membrane.

The Pacific electric ray, *Torpedo californica*, which can stun its prey with 45 volts of electricity generated by electrocytes in its wings.

Paul Yancey

Some animals can detect magnetic fields for long-range navigation, possibly by magnetic induction, magnetic minerals in receptors, or magnetochemical reactions

A final sensory mode provides a form of internal "compass." A variety of organisms, including some bacteria, mollusks, crustaceans, insects, fish, amphibians, reptiles, birds, and mammals (mole rats, at least), can detect *magnetic fields* and are capable of using this information for orientation. The lines of Earth's field are vertical at the poles and horizontal at the equator. Migratory birds are known to be able to use this magnetic field to help locate established nesting sites thousands of miles away from the wintering grounds. Similarly, sea turtles have been shown in the laboratory to respond to artificial magnetic fields, orienting their swimming directions in response.

How do these animals detect the field? Since magnetic fields penetrate bodies, no discrete sense organ is necessary, and the mechanisms have remained elusive. Three possible mechanisms have been identified (of which an animal might have more than one); but the receptors remain unknown.

1. *Magnetic induction:* The highly sensitive electroreceptors of elasmobranchs are suspected to detect Earth's magnetic field. Movement of any conductive object such as a fish body across magnetic field lines produces electric currents (a process called *magnetic induction*); these may be monitored by the electroreceptors in the ampullae of Lorenzini. This mechanism only works for animals with such sensitive receptors.

2. *Magnetic minerals* (magnetite crystals, Fe_3O_4) have been identified in various magnetically sensitive organisms including bacteria, honeybees (thorax), homing pigeons (beak), salmon (brain), rainbow trout (noses at terminal end the *trigeminal* nerve, which responds to magnetic fields), robins (nostrils), and mole rats (cornea). For magnetic sensing, magnetic crystals are arranged in a chain called a **magnetosome** within a bacterium or an animal cell. Single crystals of magnetite (50 nm in size) do not interact strongly enough with Earth's magnetic field to overcome the randomizing effects of thermal buffeting. However, once arranged in chains, their individual movements add together such that they can align with a magnetic field and interact with the receptor. How this leads to cellular signals is unknown, although moving magnetic particles could be linked to the opening of mechanically gated channels.

3. *Magnetochemical reactions:* Some chemical reactions involve free-radical formation that can be affected by magnetic fields. One class of biological candidates involve **cryptochromes**, ancient blue-light receptors used in nonvisual and nonphotosynthetic light responses as diverse as light-oriented plant growth, moonlight-induced coral spawning, and setting of circadian rhythms in animals with eyes (where the cryptochromes are located; see p. 284). Light absorption by cryptochrome does indeed cause magnetically sensitive free-radical reactions. Fruit flies lacking their cryptochrome gene lost the ability to react to magnetic fields. Some experiments with birds suggest that their cryptochrome-containing retinas can detect magnetic fields, leading to the possibility that they can visualize the Earth's fields!

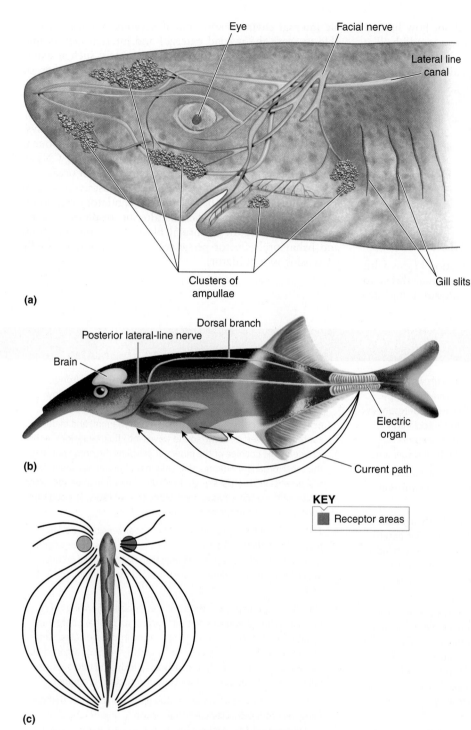

(a)

(b)

(c)

KEY

Receptor areas

FIGURE 6-47 **Electroreception and electrical discharge.** (a) Ampullary electroreceptors in a shark. Sensory nerve endings in the ampullae detect electrical currents produced by other animals; they also appear to be sensitive to changes in temperature and salinity. (b) Active electroreception in a weakly electric fish (*Gymnotus*). An electric field is set up by an organ in the tail and received by sensors in the head. (c) An external object (round circle) in the electric field distorts the field (in this case, a nonconductive object diverts the field around it), allowing the fish to detect it (a conductive object such as another animal would concentrate the field lines). (d) Transmembrane potentials in modified muscle cells to produce an EOD (electric organ discharge). The cells have smooth sections (left parts) and rough projections (right parts). At rest (*left*), the cells are negatively charged internally. Only the smooth part is innervated. When stimulated (*right*), ion channels permit Na$^+$ to flow into the smooth part, making it positive inside, but the rough part remains negative. This allows for a large external current to flow. By having cells in series, the currents in some electric fish can add up to very large voltages, which can stun prey.

© Cengage Learning, 2013

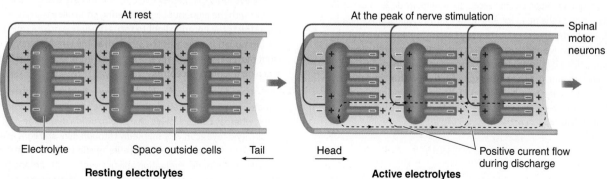

(d)

As we conclude this chapter, think about how limited our human senses are. Try to imagine what it would be like to see the world through sound echoes, find your way home following a scent trail, or navigate by visualizing the Earth's magnetic fields.

check your understanding 6.9

Discuss mechanisms and uses of electroreception and of magnetoreception.

making connections

How Senses Contribute to the Body as a Whole

As you saw in Chapter 1, biological regulation begins with the sensing of a regulated state or of a potential threat to that state. In other words, to maintain a life-sustaining sta-ble internal environment, animal nervous systems must be able to sense the myriad external and internal factors that continuously threaten to disrupt homeostasis, such as external exposure to cold, the impending attack of a predator, or internal acidity (pH) changes. Regulated (nonhomeostatic) changes such as reproduction also involve senses such as in finding a mate.

Sensory information is sent to the central nervous system (CNS), the integrating and decision-making component of the nervous system in "advanced" animals. The CNS in turn commands appropriate effector responses in organ systems to maintain the body's viability. Almost every system in an animal body can be involved in regulatory responses, including physiological (such as altering respiration in response to sensing an abnormal blood CO_2 concentration) and behavioral effector processes (such as running rapidly on sensing of a predator).

Chapter Summary

- Sensory cells have ion channels and receptor proteins with specific modalities: mechano-, chemo-, photo-, thermo-, electro-, and magnetoreception; and nociception (pain). Each receptor responds best to a single adequate stimulus (e.g., light). Receptors fall into three roles: interoceptors for internal states, proprioceptors for body position and motion, and exteroceptors for the environment.

- Perception is not reality because sensory systems enhance (as in lateral inhibition for contrast enhancement) or filter information before it reaches the brain.

- A stimulus alters the receptor's permeability, leading to a graded receptor potential. Receptor potentials may initiate action potentials in an afferent neuron. Receptors may fire continuously or adapt slowly or rapidly to sustained stimulation. Each sensory pathway is "labeled" according to modality and location, so the brain "knows" what and where it is.

- Animal integuments may have both touch and pressure receptors. Proprioceptors in the muscles, tendons, and joints give information on limb position and motion. The statocyst is the simplest organ that can monitor an animal's position in space, with mineral crystals that move and open ion channels. The lateral line system of amphibians and fish detects motion in the surrounding water with hair cells, which transduce movement using externally gated mechanical channels on cilia.

- The vestibular apparatus of vertebrate inner ears detects position and motion of the head for equilibrium and coordination of head, eye, and body movements. These use otolith minerals like statocyts, which move and push on hair cells. Fishes hear with otolith organs, aided by the gas bladder in some species. Many terrestrial vertebrates and insects receive sound initially with a tympanic membrane, aided by other external features in some groups such as the mammalian pinna. Insect ears are derived from the respiratory tracheal system.

- Sound is characterized by its intensity (loudness), pitch (tone), and timbre (quality). Pitch, for example, can be varied, with low frequencies for long-distance signals and high frequencies for echolocation. The vertebrate external ear (typanum) and middle ear bones convert airborne sound waves into fluid vibrations in the inner ear. The cochlea of the inner ear contains the organ of Corti, the sense organ for hearing. Hair cells in that organ transduce fluid movements into neural signals. Pitch discrimination depends on the region of the organ's basilar membrane that vibrates. The auditory cortex is mapped according to tone.

- Taste and smell have numerous roles such as sensing food, kin, mates, and direction. Taste sensation is coded by patterns of activity in various taste receptors. Mammalian taste is based on five (and perhaps more) primary tastes, labeled for discrimination in the brain.

- The olfactory receptors in the nose are specialized endings of renewable afferent neurons. Various parts of an odor are detected by different olfactory receptors and sorted into "smell files." Odor discrimination is coded by patterns of activity in the olfactory bulb glomeruli. The olfactory system adapts quickly, and odorants are rapidly cleared. The vomeronasal organ detects pheromones.

- Light detection uses universal photopigments in ciliary or rhabdomeric receptor cells. Dedicated light-sensing organs range from simple eyespots to complex eyes. The vertebrate eye is a fluid-filled sphere with the amount of light entering controlled by the iris. The eye refracts the entering light to focus the image on the retina. Accommodation increases the strength of the lens for near vision. Light must pass through several retinal layers in vertebrates before reaching the photoreceptors.

- Phototransduction by retinal cells converts light stimuli into neural signals via hyperpolarization. Rods provide indistinct gray vision at night, whereas cones provide sharp color vision during the day. The sensitivity of eyes can vary markedly through dark and light adaptation. Color vision depends on the ratios of stimulation of the various cone types. Color discrimination abilities vary greatly among types of animals. Some animals can see ultraviolet and/or polarized light.

- Visual information is separated and modified within the visual pathway before it is integrated into a perceptual image of the visual field by the cortex. The thalamus and visual cortices elaborate the visual message. Visual input goes to other areas of the vertebrate brain not involved in vision perception.

- Cephalopod camera eyes have light-sensing cells on top of neural cells. Compound eyes of arthropods and some annelids and bivalves consist of multiple image-forming units. Phototransduction in rhabdomeric eyes uses rhodopsin but otherwise differs from that in vertebrate eyes, causing depolarization.

- Warmth and cold thermoreceptors give information about the temperature of the environment near the body. Infrared thermoreceptors give pit vipers the ability to detect prey and desirable thermal habitats.

- Stimulation of mammalian nociceptors elicits two types of pain signals: fast signals for quick warning and slow, persisting signals to maintain awareness of an injury. The mammalian brain has a built-in analgesic system (e.g., endorphins) to shut off excessive pain signals.

- Electroreception can be passive or active and can be used for navigation, prey detection, and communication. Electric organs use specialized electrocytes to generate signals.

- Some animals can detect magnetic fields for long-range navigation, possibly by magnetic induction, magnetic minerals in receptors, or magnetochemical reactions.

Review, Synthesize, and Analyze

1. Compare the transduction mechanisms of touch, cold, vision, hearing, taste, smell, and temperature.

2. In what ways is perception not reality, and is that harmful, beneficial, or both?

3. Humans have circular pupils, whereas cats' pupils are more slitlike. For simplicity in calculation, assume the cat pupil is rectangular.
   (a) Calculate by what percentage the amount of light allowed into the eye decreases if a human's round pupil contracts by half, and if a cat's rectangular pupil contracts by half along one axis only.
   (b) Comparing these values, do humans or cats have more precise control over the amount of light falling on the retina?

4. Humans with certain nerve disorders are unable to feel pain. Why is this a problem?

5. Receptor cells that are neurons with sensory dendrites are faster than receptor cells that are separate from neurons. Why is that, and why might that be important in, for example, the sense of touch versus vision?

6. Compare the evolution and development steps for vertebrate and cephalopod eyes and the implications for their functions.

7. In what ways are vertebrate eyes superior to compound eyes? And vice versa?

8. What factors might favor the evolution of UV vision? Of pentachromatic vision?

9. Explain how humans and some other mammals can learn to like certain bitter-tasting substances and painful chemicals in food.

10. Electric fish detect time differences between different parts of their body surface within 0.5 microseconds. Provide an example of how mammals and birds similarly exploit time differences with one of their senses.

Suggested Readings

Axel, R. (1995). The molecular logic of smell. *Scientific American* 273:154–159. Discusses how 20,000 different odorants can be discriminated.

Bachmanov, A. A., & G. K. Beauchamp. (2007). Taste receptor genes. *Annual Review of Nutrition* 27: 389–414. A review of taste genes discovered through genomics.

Brown, B. R. (2003). Sensing temperature without ion channels. *Nature* 421:495. Discusses how shark's electrosensing ampullae can also sense temperature.

Lamb, T. D., S. P. Collin & E. N. Pugh, Jr. (2007). Evolution of the vertebrate eye: opsins, photoreceptors, retina and eye cup. *Nature Reviews Neuroscience* 8:960–976. Describes how a complex eye can readily evolve from simple beginnings.

Lohmann, K. J. (2010). Magnetic-field perception. *Nature* 464:1140–1142. Reviews the hypotheses for magnetic field sensing.

Nordgreen, J., J. P. Garner, A. M. Janczak, et al. (2009). Thermonociception in fish: Effects of two different doses of morphine on thermal threshold and post-test behaviour in goldfish (*Carassius auratus*). *Applied Animal Behaviour Science* 119:101–107. Discusses and tests whether fish can feel pain.

Rose, J. D. (2002). The neurobehavioral nature of fishes and the question of awareness and pain. *Reviews in Fisheries Science* 10:1–38. Also examines pain perception in fish.

Sneddon, L. U, V. A. Braithwaite, & M. J. Gentle. (2003). Do fish have nociceptors: Evidence for the evolution of a vertebrate sensory system. *Proceedings of the Royal Society—Biological Sciences* 270:1115–1121. Also examines pain perception in fish.

Walker, R. G., A. T. Willingham, & C. S. Zucker. (2000). A *Drosophila* mechanosensory transduction channel. *Science* 287:2229–2234. The first analysis of mechanoreceptor mechanism.

Young, J. M., & B. J. Trask. (2002). The sense of smell: Genomics of vertebrate odorant receptors. *Human Molecular Genetics* 11:1153–1160. Covers the evolution of olfactory receptor proteins among vertebrates.

CourseMate

Access an interactive eBook, chapter-specific interactive learning tools, including flashcards, quizzes, videos and more in your Biology CourseMate, accessed through **CengageBrain.com**.

The northern gannet (*Sula bassanus*). Diving into the cold waters of the North Sea precluded the development of a brood patch in this species. Rather, the females cradle their eggs in the webbing of their feet during incubation, achieving temperatures close to 37°C.

Endocrine Systems

7.1 Introduction: Principles of Endocrinology

The ability of cells to "communicate" with each other is ancient, preceding the evolution of multicellularity. Protozoan cells, for example, communicate through the release of chemical signals into the environment in order to mate. Recall that chemicals that elicit specific behavioral responses in other individuals of the same species are termed **pheromones** (p. 92). In animals, pheromones and internal signal molecules used locally within a tissue (that is, paracrine action, p. 91) appear in all phyla, as do signal molecules used by neurons at synapses except in the sponges. With the evolution of internal body cavities and circulatory systems, which provided an effective route for delivering signal molecules to distant tissues, there arose **endocrine systems**, which consist of ductless glands, often scattered throughout an animal's body as both discrete endocrine organs and as subsets of other organs having nonendocrine functions (as shown in Figure 7-1 for a mammal). The glands secrete **hormones** (signal molecules delivered by circulatory fluids), providing a second mechanism (in addition to nervous systems) for regulating and coordinating distant organs. The endocrine and nervous systems are specialized for controlling different types of activities. In general, the nervous system coordinates rapid, precise responses and is especially important in mediating interactions with the external environment. The endocrine system, by contrast, primarily controls activities that require duration rather than speed as well as coordinating diverse tissues (e.g. liver, muscle, and gut). In the course of animal evolution, there has been an increase in the number and complexity of processes that have come under hormonal regulation. Moreover, the two systems are often intimately linked in *neuroendocrine* interactions, in which neurons secrete neurohormones into a body fluid for transport. In vertebrates, this is reflected in a classification of glands into **central** ones linked to the central nervous system, and **peripheral** ones that are not.

Endocrinology is the study of the evolution and physiological function of hormones. We begin this chapter by examining general principles of endocrinology, and then we examine specific hormones in nonvertebrates (primarily insects) and vertebrates.

Hormones are chemically classified into three categories: peptides and proteins, amines, steroids

Hormones are not all similar chemically, but instead fall into three distinct classes according to their biochemical structure (Table 7-1; compare this table to the list of all chemical messengers on p. 93):

1. The **peptide** and **protein hormones** consist of specific amino acids arranged in a chain of varying length; the shorter chains are peptides and the longer ones are categorized as proteins. For convenience, we refer to this entire category as *peptides*. The majority of animal hormones fall into this class; an example is *insulin*, which regulates blood glucose in vertebrates.

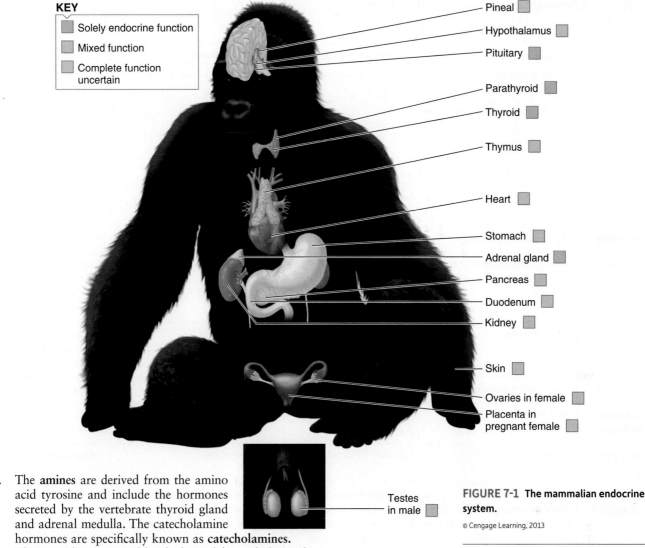

KEY

- Solely endocrine function
- Mixed function
- Complete function uncertain

- Pineal
- Hypothalamus
- Pituitary
- Parathyroid
- Thyroid
- Thymus
- Heart
- Stomach
- Adrenal gland
- Pancreas
- Duodenum
- Kidney
- Skin
- Ovaries in female
- Placenta in pregnant female
- Testes in male

FIGURE 7-1 **The mammalian endocrine system.**

© Cengage Learning, 2013

2. The **amines** are derived from the amino acid tyrosine and include the hormones secreted by the vertebrate thyroid gland and adrenal medulla. The catecholamine hormones are specifically known as **catecholamines.**

3. The **steroids** are neutral lipids derived from cholesterol. These include the hormones secreted by the molting glands of arthropods and the vertebrate adrenal cortex and gonads (sex steroids such as *estradiol* and *testosterone*), as well as some mammalian placental hormones. Insects cannot synthesize the core steroid molecule and so must ingest suitable precursors in order to synthesize their primary steroid hormone *ecdysone*.

These signal types are apparently quite ancient. Several amines and peptides identical to signal molecules found in animals have been detected as pheromones in protozoa. A peptide closely related to human insulin (with a corresponding receptor) has been found in some Porifera (sponges), although its function is unclear. In recent years, all three signal types have been found in Cnidaria. For example, estradiol and testosterone have been found in some corals, in which estradiol levels surge during mass spawning events, suggesting a regulatory role in reproduction.

Minor differences in chemical structure between hormones within each category often result in profound differences in biological response. For example, note the subtle difference between testosterone, the male sex hormone responsible for inducing the development of masculine characteristics, and estradiol, a form of estrogen, which is the feminizing female sex hormone (Figure 7-2).

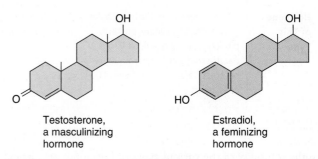

Testosterone, a masculinizing hormone

Estradiol, a feminizing hormone

FIGURE 7-2 **Comparison of two steroid hormones, testosterone and estradiol.**

© Cengage Learning, 2013

Solubility Characteristics of Hormone Classes The structural classification of hormones is of more than biochemical interest. The chemical properties of a hormone, most notably its solubility, determine the means by which the hormone is synthesized, stored, and secreted; the way it is transported in the blood; and the mechanism by which it exerts its effects at the target cell, and typically, the type of processes regulated (rapid versus long-term) and their half-lives (turnover) in the circulation. The following differences

TABLE 7-1 Chemical Classification of Vertebrate Hormones

		Amines		
Properties	**Peptides**	*Catecholamines*	*Thyroid Hormone*	**Steroids**
Structure	Chains of specific amino acids, for example: (Vasopressin)	Tyrosine derivative, for example: (Epinephrine)	Iodinated tyrosine derivative, for example: (Thyroxine, T_4)	Cholesterol derivative, for example: (Cortisol)
Solubility	Hydrophilic (lipophobic)	Hydrophilic (lipophobic)	Lipophilic (hydrophobic)	Lipophilic (hydrophobic)
Synthesis	In rough endoplasmic reticulum; packaged in Golgi complex	In cytosol	In colloid, an inland extra-cellular site	Stepwise modification of cholesterol molecule in various intracellular compartments
Storage	Large amounts in secretory granules	In chromaffin granules	In colloid	Not stored; cholesterol precursor stored in lipid droplets
Secretion	Exocytosis of granules	Exocytosis of granules	Endocytosis of colloid	Simple diffusion
Transport in Blood	As free hormone	Half bound to plasma proteins	Mostly bound to plasma proteins	Mostly bound to plasma proteins
Receptor Site	Surface of target cell	Surface of target cell	Inside target cell	Inside target cell
Mechanism of Action	Channel changes or activation of second-messenger system to alter activity of preexisting proteins that produce the effect	Activation of second messenger system to alter activity of preexisting proteins that produce the effect	Activation of specific genes to produce new proteins that produce the effect	Activation of specific genes to produce new proteins that produce the effect
Hormones of This Type	All hormones from the hypothalamus, anterior pituitary, posterior pituitary, pineal gland, pancreas, parathyroid gland, gastrointestinal tract, kidneys, liver, thyroid C cells, heart	Only hormones from the adrenal medulla	Only hormones from the thyroid follicular cells	Hormones from the adrenal cortex and gonads plus most placental hormones (vitamin D is steroid-like)

© Cengage Learning, 2013

in the solubility of the various types of hormones are critical to their function (Table 7-1):

- All peptides and catecholamines are *hydrophilic* (water loving); that is, they are highly H_2O soluble and have low lipid solubility.
- All steroid and thyroid hormones are *lipophilic* (lipid loving); that is, they have high lipid solubility and are poorly soluble in H_2O.

We are first going to consider the different ways in which these hormone types are processed at their site of origin, the endocrine cell, before comparing their means of transport and mechanisms of action.

The mechanisms of hormone synthesis, storage, and secretion vary according to the class of hormone

Because of their chemical differences, the means by which the various classes of hormones are synthesized, stored, and secreted differ as follows.

Peptide Hormones Peptide hormones are synthesized by the same method used for the manufacture of any protein that is to be exported (see Figure 2-10, p. 40). Because they are destined to be released from the endocrine cell, the synthesized hormones must be segregated from intracellular proteins by being sequestered in a membrane-enclosed compartment until they are secreted. Briefly, synthesis of peptide hormones requires the following steps:

1. Large precursor proteins, or **preprohormones,** are synthesized by ribosomes on the rough endoplasmic reticulum. The preprohormones then migrate to the Golgi complex in membrane-enclosed vesicles that pinch off from the smooth endoplasmic reticulum.

2. During their journey through the endoplasmic reticulum and Golgi complex, the large preprohormone precursor molecules are pruned first to **prohormones** and finally to **active hormones.** The peptide "scraps" that remain after the large preprohormone molecule is cleaved to form the classic hormone are often stored and cosecreted along with the hormone. This raises the possibility that these other peptides may also exert biological effects that differ from the traditional hormonal product; that is, the cell may actually be secreting multiple hormones, but the functions of the other peptide products are for the most part unknown. A known example involves the large precursor molecule **pro-opiomelanocortin (POMC)** in vertebrates. Before secretion, several diverse cell types produce POMC and slice it in unique ways, depending on the processing enzymes they possess, to yield different active products. In particular, POMC can be cleaved into *adrenocorticotropic hormone (ACTH), melanocyte-stimulating hormones (MSH),* and a morphinelike substance, β-*endorphin,* along with peptide "scraps" that have no known function. We'll examine the roles of each active product later.

3. The Golgi complex concentrates the finished hormones, then packages them into secretory vesicles that are pinched off and stored in the cytoplasm until an appropriate signal triggers their secretion. By storing peptide hormones in a readily releasable form, the gland can respond rapidly to any demands for increased secretion without first needing to increase hormone synthesis.

4. On appropriate stimulation, the secretory vesicles fuse with the plasma membrane and release their contents to the outside by the process of exocytosis (see p. 42). Such secretion usually does not go on continuously; it is triggered only by specific stimuli. The blood subsequently picks up the secreted hormone for distribution.

Steroid Hormones All steroidogenic (steroid-producing) cells perform the following steps to produce and release their hormonal product:

1. Cholesterol is the common precursor for all steroid hormones; however, numerous species cannot synthesize cholesterol and must obtain it from their diet. Vertebrate steroidogenic cells can synthesize some cholesterol on their own, although most is derived from low-density lipoproteins (LDLs) from the blood (see p. 389). Unused cholesterol may be chemically modified and stored in large amounts as lipid droplets within steroidogenic cells. Utilization of LDLs and conversion of stored cholesterol into free cholesterol for use in steroid hormone production can be closely coordinated with the animal's overall need for the hormone product.

2. Synthesis of the various steroid hormones from cholesterol requires a series of enzymatic reactions that modify the type and position of side groups attached to the cholesterol framework or the degree of saturation within the rings (Figure 7-3). Each conversion from cholesterol to a specific steroid hormone requires enzymes found in the mitochondria or endoplasmic reticulum. Accordingly, each steroidogenic organ can produce only the steroid hormones for which it has a complete set of appropriate enzymes. For example, a key enzyme necessary for the production of *cortisol* is found only in the adrenal cortex. The steroid molecule is therefore shuttled back and forth between different compartments within the steroidogenic cell for step-by-step modification until the final secretory product is formed.

3. Unlike peptide hormones, steroid hormones cannot be stored after their formation. Once formed, the lipid-soluble steroid hormones immediately diffuse through the steroidogenic cell's lipid plasma membrane to enter the circulatory system. Only the hormone precursor cholesterol is stored in significant quantities within steroidogenic cells. Accordingly, the rate of steroid hormone secretion is controlled entirely by the rate of hormone synthesis. In contrast, peptide hormone secretion is controlled primarily by regulating the release of presynthesized, stored hormone.

4. After their secretion into the circulation, some steroid hormones undergo further interconversions within the blood or other organs, where they are converted into more potent or different hormones.

Amines The amine hormones of vertebrates—thyroid hormone and catecholamine hormone—have unique synthetic and secretory pathways that are thoroughly described when each of these hormones is specifically addressed. However, in brief, the amines share the following features:

- They are derived from the naturally occurring amino acid tyrosine.
- Both types of amines are stored until they are secreted.
- Thyroid hormone also undergoes further processing once in the peripheral circulation.

Hydrophilic hormones are transported dissolved in the plasma, whereas lipid-soluble hormones are almost always transported bound to plasma proteins

All hormones are carried by the blood, but they are not all transported in the same manner:

1. The hydrophilic peptide hormones are transported simply dissolved in the plasma or, in some cases, bound to a specific carrier protein.

2. Lipophilic steroids and thyroid hormone (in vertebrates), which are poorly soluble in water, cannot dissolve in the aqueous plasma in sufficient quantities to account for their known plasma concentrations. Instead, most of these hormones circulate to their target cells reversibly bound to plasma proteins. Some plasma proteins are designed to carry only one type of hormone, whereas other plasma proteins, such as albumin, indiscriminately pick up any "hitchhiking" hormone.

Only the small, unbound, freely dissolved fraction of a lipophilic hormone is biologically active (that is, free to

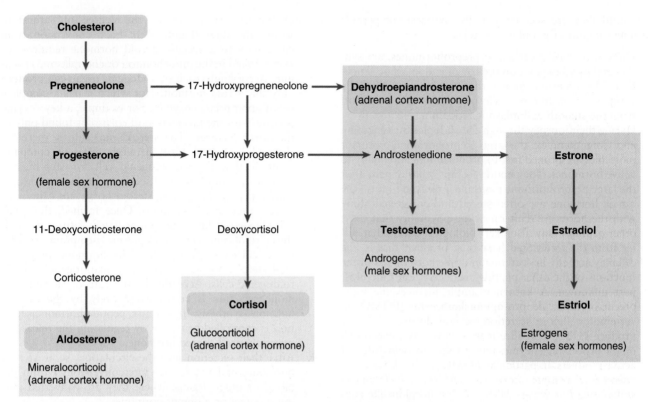

FIGURE 7-3   Steroidogenic pathways for the major steroid hormones. All steroid hormones are produced through a series of enzymatic reactions that modify cholesterol molecules, such as by varying the side groups attached to them. Each steroidogenic organ can produce only those steroid hormones for which it has a complete set of the enzymes needed to appropriately modify cholesterol. For example, the testes have the enzymes necessary to convert cholesterol into testosterone (male sex hormone), whereas the ovaries have the enzymes needed to yield progesterone and the various estrogens (female sex hormones).

© Cengage Learning, 2013

diffuse and bind with target cell receptors to exert an effect). Once a hormone has interacted with a target cell, it is rapidly inactivated and excreted so that it is no longer available to interact with another target cell. Because the *carrier-bound* hormone is in dynamic equilibrium with the *free hormone* pool, the bound form of lipophilic hormone provides a large reserve that can be called on to replenish the active free pool. To maintain normal endocrine function, the magnitude of the small, free, effective pool, rather than the total plasma concentration of a particular lipophilic hormone, is monitored and adjusted through feedback mechanisms.

Catecholamines in vertebrates are unusual in that only about 50% of these hydrophilic hormones circulate as free hormone, whereas the other 50% are loosely bound to the plasma protein albumin. Because catecholamines are water soluble, the importance of this protein binding is unclear.

Hormones exert a variety of regulatory effects throughout the body

Endocrine systems have regulatory designs similar to that of nervous systems. These are cells that sense (usually with receptor proteins) some state in the body that is being regulated, integrator cells (which may be neurons or glands), and effector organs and behaviors that carry out responses. Hormones are the messengers that transmit commands, similar to the roles of action potentials and neurotransmitters in nervous systems.

Even though hormones are distributed throughout an animal body by blood or hemolymph, only specific target cells can respond to each hormone because only the target cells have receptors (intracellular or integral plasma membrane proteins) for binding with the particular hormone. Binding of a hormone with its specific target cell receptors initiates a chain of events within the target cells to bring about the hormone's final effect. Some hormones have a single target cell type; others have many.

Tropic Hormones   The sole function of some hormones is regulating the production and secretion of another hormone. A hormone that has as its primary function the regulation of hormone secretion by another endocrine gland is classified functionally as a **tropic hormone** (*tropic* means "causing to turn" or "attracting," here meaning "stimulating"). (Do not confuse these with *trophic hormones,* because "trophic" means "nourishing." Trophic hormones are involved in triggering cell growth and development, and are usually nontropic!) Tropic hormones stimulate and maintain their endocrine target tissues. For example, the tropic hormone thyroid-stimulating hormone (TSH) from the anterior pituitary stimulates thyroid hormone secretion by the thyroid gland and also maintains the structural integrity of this gland. In the absence of TSH, the thyroid gland atrophies (shrinks) and produces very low levels of its hormone. A **nontropic hormone,** in contrast, primarily exerts its effects on nonendocrine target tissues. Thyroid hormone, which increases the rate of O_2 consumption and the metabolic activity of almost every cell of mammals and birds, is an example of a nontropic hormone.

Complexity of Endocrine Function The following factors add to the complexity of an endocrine system:

- A single endocrine gland may produce multiple hormones. The mammalian anterior pituitary, for example, secretes six different hormones, usually made by different specialized cells in that gland; each hormone is under different control mechanisms and has different functions, some being tropic and others nontropic.
- A single hormone may be secreted by more than one endocrine gland. For example, the vertebrate hypothalamus and pancreas both secrete the hormone *somatostatin.* In amphibians *thyrotropin-releasing hormone* is synthesized in the hypothalamus and the skin as well as other organs. The functional significance of this arrangement has not been established.
- Frequently, a single hormone has more than one type of target cell and therefore can induce more than one type of effect. This is the "same key, different locks" concept that we discussed in Chapter 3 (p. 93). As an example, *vasopressin* from the posterior pituitary promotes H_2O reabsorption by mammalian kidney tubules by binding with V_2 receptors on the distal and collecting tubular cells and vasoconstriction of arterioles throughout the body by binding with V_1 receptors on arteriolar smooth muscle. Sometimes hormones that have multiple target-cell types can coordinate and integrate the activities of various tissues toward a common end. For example, the effects of *insulin* from the pancreas on muscle, liver, and fat all act in concert to store nutrients after absorption of a meal.
- The secretion rate of some hormones varies considerably over the course of time in a cyclic pattern. Therefore, endocrine systems also provide temporal (time) coordination of function. This is particularly evident in the endocrine control of reproductive cycles, such as the estrous cycle, in which normal function requires highly specific patterns of change in the secretion of various hormones.
- A single target cell may be influenced by more than one hormone. Some cells contain an array of receptors for responding in different ways to different hormones. To illustrate, *insulin* promotes the conversion of glucose into *glycogen* (a storage polymer) within vertebrate liver cells by stimulating one particular hepatic enzyme, whereas another hormone, *glucagon,* by activating a different set of hepatic enzymes, enhances the degradation of glycogen into glucose within liver cells.
- The same chemical messenger may be either a hormone or a neurotransmitter, depending on its source and mode of delivery to the target cell. A prime example is norepinephrine, which is secreted as a hormone by the vertebrate adrenal medulla and released as a neurotransmitter from sympathetic postganglionic nerve fibers. The effects of the identical hormone and neurotransmitter are generally different, again illustrating the "same key, different locks" principle (p. 93).
- Some organs are exclusively endocrine in function (they specialize in hormonal secretion alone, the anterior pituitary and thyroid glands being examples), whereas other organs of the endocrine system perform nonendocrine

functions in addition to secreting hormones. For example, vertebrate testes produce sperm and also secrete the male sex-hormone *testosterone.*
- Nonnative or hormonelike substances termed **endocrine-disrupting chemicals (EDCs)**, whose structures are similar to hormones, can sometimes disrupt endocrine communication by altering hormone secretion, transport or action. These compounds are derived from the by-products of manufactured organic compounds.

This has been a brief overview of the general functions of the endocrine system. Certain hormones are introduced elsewhere and are not discussed in this chapter; these (in vertebrates) are the *gastrointestinal hormones* (Chapter 14), the renal hormones (*erythropoietin* in Chapter 9, and *renin* in Chapter 13), *atrial natriuretic peptide* from the heart (Chapter 12), and *thymosin* (Chapter 10). Most of the remainder of the major hormones are described in greater detail in this chapter.

Hormones produce their effects by altering intracellular proteins through ion fluxes, second messengers, and transcription factors

To induce their effects, hormones must bind with target cell receptors specific for them. Each interaction between a particular hormone and a target-cell receptor produces a highly characteristic response that differs among hormones and among different target cells influenced by the same hormone.

General Mechanisms of Hydrophilic and Lipophilic Hormone Action The location of the receptors within the target cell, and the mechanism by which binding of the hormone with the receptors induces a response, both vary, depending on the hormone's solubility characteristics. Receptor–hormone interactions can be grouped into two broad categories based on the location of their receptors (Table 7-1):

1. *Membrane receptors.* The hydrophilic peptides and catecholamines, which are poorly soluble in lipid, cannot pass through the lipid membrane barriers of their target cells. Instead, they bind with specific receptors located on the outer plasma membrane surface of the target cell. In turn, these receptors either alter the conformation (shape) of adjacent ion channels already present in the membrane (p. 94), or activate second-messenger systems within the target cell. Second messengers directly alter the activity of preexisting intracellular proteins, usually enzymes, to produce the desired effect.
2. *Internal receptors.* The lipophilic steroids and thyroid hormone (in the free form, not bound with a plasma protein carrier) easily pass through the surface membrane to bind with specific receptors located *inside* the target cell. The receptors inside the cell are typically *transcription factors* that regulate specific genes in the target cell that code for the formation of new intracellular proteins (pp. 31 and 94). Some lipophilic hormones also have specific receptors in the plasma membrane and cytosol of target tissues.

Membrane receptors were covered in Chapters 2 and 3; however, the lipophilic mechanism of hormonal action warrants further examination.

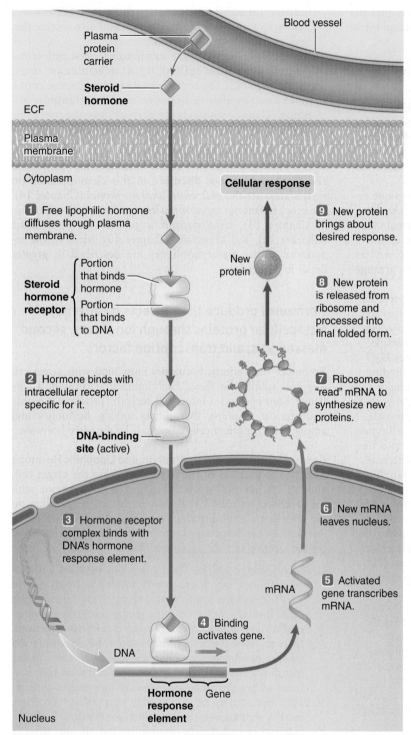

FIGURE 7-4 Mechanism of action of lipophilic hormones.

© Cengage Learning, 2013

brane of the target cell (step 1 in Figure 7-4) and binds with its specific receptor inside the cell, either in the cytoplasm or in the nucleus (step 2). Each receptor has a specific region for binding with its hormone and another region for binding with DNA. The receptor cannot bind with DNA unless it first binds with the hormone. Once the hormone is bound to the receptor, the hormone receptor complex binds with DNA at a specific attachment site on the DNA known as the **hormone response element (HRE)** (step 3). Different steroid hormones and thyroid hormone, once bound with their respective receptors, attach at different HREs on DNA. For example, the estrogen receptor complex binds at DNA's estrogen response element.

Binding of the hormone receptor complex with DNA "turns on" or activates a specific gene within the target cell (step 4). This gene contains a code for synthesizing a given protein. The code of the activated gene is transcribed into complementary messenger RNA (mRNA) (step 5). The new mRNA leaves the nucleus and enters the cytoplasm (step 6), where it binds to a ribosome, the "workbench" that mediates the assembly of new proteins. Here, mRNA directs the synthesis of the designated new proteins according to the DNA code in the activated genes (step 7). The newly synthesized protein, either enzymatic or structural, is released from the ribosome (step 8) and produces the target cell's ultimate response to the hormone (step 9). By means of this mechanism, different genes are activated by different lipophilic hormones, resulting in different biological effects.

By means of this mechanism of nuclear receptors, different lipophilic hormones activate different genes, resulting in hormone-specific biological effects. Genomic and evolutionary analysis of nuclear receptors has revealed that they share an extensive homology and so are grouped into a large superfamily divided into six subfamilies. For example, one subfamily places the vertebrate thyroid-hormone receptor in the same class as the vertebrate vitamin D receptor and the arthropod ecdysone receptor. Another subfamily contains the vertebrate steroid receptors; vertebrate genomes have six nuclear steroid receptors: two for estrogens, and one each for testosterone/androgens, glucocorticoids, mineralocorticoids (such as aldosterone), and progesterone (hormones we discuss later). Recently, researchers have found that the genome of a mollusk, the seaslug *Aplysia* (p. 198), has an estrogen-type receptor. Although this receptor is not activated by steroids, its existence suggests that steroid receptors evolved before chordates arose.

Even though most steroid actions are accomplished by hormonal binding with intracellular receptors that leads to gene activation, recent studies have unveiled another mechanism by which steroid hormones induce effects that occur too rapidly to be mediated by gene transcription. Some ste-

By stimulating genes, lipophilic hormones promote synthesis of new proteins

All lipophilic hormones (steroids and thyroid hormone) bind with intracellular receptors in their target cells by activating specific genes that cause the synthesis of new proteins, as summarized in Figure 7-4.

Free lipophilic hormone (hormone not bound with its plasma-protein carrier) diffuses through the plasma mem-

roid hormones, most notably some of the sex hormones, bind with unique steroid receptors in the plasma membrane, in addition to binding with the traditional steroid receptors in the nucleus. This membrane binding leads to *nongenomic steroid receptor actions,* that is, actions accomplished by something other than altering gene activity, such as by inducing changes in ionic flux across the membrane or by altering activity of cellular enzymes.

Onset and Duration of Hormonal Responses Compared to neural responses that are brought about within milliseconds, hormone action is relatively slow and prolonged, taking minutes to hours for the response to take place after the hormone binds to its receptor. The variability in time of onset for hormonal responses depends on the mechanism employed. Hormones that act through a second-messenger system to alter a preexisting enzyme's activity elicit full action within a few minutes. In contrast, hormonal responses that require the synthesis of new protein may take up to several hours before any response can be measured.

Also, in contrast to neural responses that are quickly terminated once the triggering signal ceases, hormonal responses persist for a period of time after the hormone is no longer bound to its receptor. Once an enzyme is activated in response to hydrophilic hormonal input, it no longer depends on the presence of the hormone. Thus, the response lasts until the enzyme is inactivated. As a result, a hormone's effect usually lasts for some time after its withdrawal. Predictably, the responses that depend on protein synthesis last longer than do those stemming from enzyme activation.

Furthermore, as we explain shortly, compared to hydrophilic hormones, lipophilic hormones usually persist longer after secretion before being inactivated.

Hormone actions are greatly amplified at the target cell

The actions of hormones are greatly amplified at the target cell. As we noted earlier, hormones exert their effect at incredibly low concentrations—as low as 1 picogram (10^{-12} gram; 1 millionth of a millionth of a gram) per mL—as opposed to the much higher localized concentration of neurotransmitter at the target cell during neural communication. Interaction of one hormonal molecule with its receptor can result in the formation of many active protein products that ultimately carry out the physiological effect. For example, one peptide hormone results in the production of many cAMP messengers, each in turn activating many latent enzymes (see p. 97). Similarly, one steroid-hormone-activated gene induces formation of many messenger RNA molecules, each of which is used to make numerous new proteins.

Endocrine-disrupting chemicals can mimic the effects of native hormones

As we noted earlier, endocrine-disrupting chemicals (EDCs) are human-made substances that are released into the environment and interfere with the endocrine modulation of neural and behavioral maturation of animals. EDCs range across all continents and oceans, are found in animal populations from the poles to the tropics, and can be passed from generation to generation. The discovery of hormonally me-

diated toxic effects in fish downstream of sewage discharge points led to the realization that wastewater contains EDCs. Even so, some scientists have suggested that environmental EDCs, except in extreme cases of a discrete spill or a point source, rarely occur in high enough concentrations to impact endocrine function. Further, natural endocrine disruption has been happening for millions of years as exemplified by some plants, which evolved these secondary products to serve as plant regulators or as endocrine-disrupting defenses against herbivores, especially insects.

Sexual development of the vertebrate brain is influenced by estrogenic (female) and androgenic (male) hormones. Researchers have found that many EDCs either mimic (as agonists) the actions of estrogens or oppose (as antagonists) their actions (anti-estrogenic); in some situations EDCs may even function as anti-androgens. For example, DDE (dichlorodiphenyldichloroethylene), a breakdown product of the pesticide DDT (dichlorodiphenyltrichloroethane), which was used globally from the 1940s to 1960s to kill mosquitoes and other insect pests, can be found in almost all living tissue on Earth and acts as an anti-androgen in mammals. More recent discoveries have demonstrated that common household products (for example, heavy-duty laundry powders and liquid detergents, personal care products, and household cleaners) contain nonionic surfactants that break down in the environment to form estrogen agonists (which interact with estrogen receptors). EDCs that interfere with sex hormone activity and production have the potential to disturb normal brain sexual development, as has been demonstrated in wildlife studies of birds, fishes, whales, porpoises, alligators, and turtles. These effects have been linked with direct exposure to sewage and industrial effluents and pesticides, and indirectly to accumulation through aquatic food webs. Similarly, EDCs that impair thyroid function are believed to contribute to learning difficulties in animals, in addition to other neurological abnormalities.

The effective plasma concentration of a hormone is normally regulated by changes in its rate of secretion

The primary function of most hormones is the regulation of various homeostatic activities. Because hormones' effects are proportional to their concentrations in the blood, it follows that these concentrations must be subject to control according to homeostatic need (Figure 7-5). The plasma concentration of free, biologically active hormone—and thus the hormone's availability to its receptors—depends on several factors: (1) the hormone's rate of secretion into the blood by the endocrine gland; (2) for a few hormones, rate of metabolic activation; (3) for lipophilic hormones, extent of binding to plasma proteins; and (4) rate of removal from the circulation by metabolic inactivation and excretion in urine. Furthermore, the magnitude of the hormonal response depends on the availability and sensitivity of the target cell's receptors for the hormone. Let's first examine the factors that influence the plasma concentration of the hormone before turning our attention to the target cells' responsiveness to the hormone.

Normally, the effective plasma concentration of a hormone is regulated by appropriate adjustments in the rate of its secretion. Secretion rates of all hormones are subject to

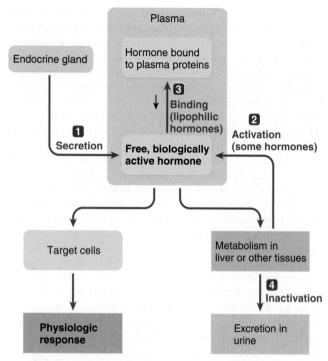

FIGURE 7-5 Factors affecting the plasma concentration of free, biologically active hormone. The plasma concentration of free, biologically active hormone, which can interact with its target cells to produce a physiological response, depends on (1) the hormone's rate of secretion by the endocrine gland, (2) its rate of metabolic activation (for a few hormones), (3) its extent of binding to plasma proteins (for lipophilic hormones), and (4) its rate of metabolic inactivation and excretion.

© Cengage Learning, 2013

control, often by a combination of several complex mechanisms. The regulatory system for each hormone is considered in detail in later sections. For now, we address these general mechanisms, which are common to many different hormones: negative-feedback control, neuroendocrine reflexes, and diurnal (circadian) rhythms.

Negative-Feedback Control Negative feedback is a prominent feature of most biological control systems (see p. 13), including hormonal. Recall that *negative feedback exists when the output of a system counteracts a change in input,* thus maintaining a controlled variable within a narrow range around a set level. Control of hormonal secretion to maintain a hormone's plasma concentration provides some classic physiological examples of negative feedback. For example, when the plasma concentration (in a mammal) of free circulating thyroid hormone falls below a given "set point," the anterior pituitary secretes thyroid-stimulating hormone (TSH), which stimulates the thyroid to increase its secretion of thyroid hormone. Thyroid hormone in turn inhibits further secretion of TSH by the anterior pituitary. Negative feedback ensures that once thyroid gland secretion has been "turned on" by TSH, it will not continue unabated but instead will be "turned off" when the appropriate level of free circulating thyroid hormone has been achieved. Thus, the effect of a particular hormone's actions can inhibit its own secretion.

Neuroendocrine Reflexes Many endocrine control systems involve **neuroendocrine reflexes,** which include neural as well as hormonal components. The purpose of such reflexes is to produce a sudden increase in hormone secretion (that is, a *reset* mechanism, p. 18, that "turns up the thermostat setting") in response to a specific stimulus, frequently a stimulus external to the body. In some instances, neural input to the endocrine gland is the only factor regulating secretion of the hormone. For example, secretion of epinephrine by the adrenal medulla is solely controlled by the sympathetic nervous system. Some endocrine control systems, in contrast, include both feedback control, which maintains a constant basal level of the hormone, and neuroendocrine resetting reflexes (which cause sudden bursts in secretion in response to a sudden increased need for the hormone). An example is the increased secretion of cortisol by the adrenal cortex during a stress response (see p. 305).

Diurnal (Circadian) and Other Biological Rhythms Although some form of negative feedback usually regulates hormone secretion rates, this does not imply that hormones are always maintained at a constant level. Instead, the secretion rates of most hormones rhythmically fluctuate up and down as a function of time. The most common endocrine rhythm is the **diurnal** ("day–night"), or **circadian** ("around a day"), **rhythm,** which is characterized by repetitive oscillations in gene expression that correspond with changes in the physiology, metabolism, and behavior in nearly all organisms on Earth, from bacteria to mammals. As we'll see, fluctuations in certain hormone levels are very regular and have a frequency of one cycle every 24 hours. This rhythmicity is caused by endogenous oscillators, called **biological clocks,** similar to the self-paced respiratory neurons in the brainstem that govern the rhythmic motions of breathing, except the time-keeping oscillators cycle on a much longer time scale. Furthermore, unlike the rhythmicity of breathing, endocrine rhythms are locked on, or **entrained,** to external cues, called **zeitgebers** ("time givers"), such as the light–dark cycle; that is, the inherent 24-hour cycles of peak and ebb of hormone secretion are set to "march in step" with cycles of light and dark. Such biological rhythms are a classic example of *anticipatory regulation* (Chapter 1): Once synchronized with a regular environmental cycle such as day–night or summer–winter, clocks allow organisms to prepare for these cyclical changes in advance of their actual occurrence. For example, *cortisol* secretion (p. 286) in a diurnal mammal rises during the late night, reaching its peak secretion in the early morning just before waking time, and then falls throughout the day to its lowest level at dusk (Figure 7-6). This prepares the animal for the stresses associated with waking up and initiating activity such as finding food.

Inherent hormonal rhythmicity and the entrainment of zeitgebers are not accomplished by the endocrine organs themselves but instead result from the central nervous system changing the set point of these organs. Indeed, neural clocks regulate many neuromuscular functions such as feeding and other behaviors, as well as hormonal secretion. In turn, the neurons themselves exhibit clock cycles due to *clock genes,* which are found in all organisms. (We examine these neural and molecular biological mechanisms in detail later; p. 284) Negative-feedback control mechanisms maintain whatever set point is established for that time of day.

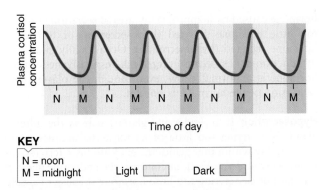

FIGURE 7-6 Diurnal rhythm of cortisol secretion.

Source: Adapted from George A. Hedge, Howard D. Colby, & Robert L. Goodman. (1987). *Clinical Endocrine Physiology*. Philadelphia: W. D. Saunders Company, Figure 1-13, p. 28.

Some endocrine cycles operate on time scales other than a circadian rhythm—some much shorter than a day and some much longer. A well-known example of the latter is the monthly menstrual cycle in humans, and a tidal rhythm called **circalunadian** ("about a lunar day") in many shoreline animals. Another common cycle is the **circannual** (yearly) rhythm in seasonally reproducing organisms and hibernating animals.

The effective plasma concentration of a hormone can be influenced by the hormone's transport, metabolism, and excretion

Even though the effective plasma concentration of a hormone is normally regulated by appropriate adjustments in the rate of secretion, alterations in the transport, metabolism, or excretion of a hormone can also influence the size of its effective pool, sometimes inappropriately. For example, the vertebrate liver synthesizes plasma proteins, so liver disease may result in abnormal endocrine activity because of a change in the balance between free and bound pools of certain hormones.

Eventually, all hormones are metabolized by enzyme-mediated reactions that modify the hormonal structure in some way. In most cases, this inactivates the hormone. Hormone metabolism is not always a mechanism for removal of used hormones, however. In some cases, a hormone is activated by metabolism; that is, the hormone's product has greater activity than the original hormone. For example, after the thyroid hormone thyroxine is secreted, it is converted to a more powerful hormone by enzymatic removal of one of the iodine atoms it contains. Usually the rate of such hormone activation is itself under hormonal or metabolic control.

Metabolic Inactivation and Urinary Excretion of Hormones The liver is the most common site for metabolic hormonal inactivation in vertebrates, but some hormones are also metabolized in the kidneys, blood, or target cells. In contrast to hormone activation, which is typically regulated, hormonal inactivation and excretion are not subject to control. The primary means of eliminating hormones and their metabolites from vertebrate blood is by urinary excretion. When liver and kidney function are normal, measuring urinary concentrations of hormones and their metabolites provides a useful, noninvasive way to assess endocrine function because the rate of excretion of these products in urine directly reflects their rate of secretion by the endocrine glands. This type of information may provide the only clue that an animal has come into breeding condition and can participate in a captive breeding program. Because the liver and kidney are important in removing hormones from the blood by means of metabolic inactivation and urinary excretion, animals with liver or kidney disease may suffer from excess activity of certain hormones solely as a result of reduced hormone elimination.

The amount of time after a hormone is secreted before it is inactivated and the means by which this is accomplished differ for different classes of hormones. In general, the hydrophilic peptides and catecholamines are easy targets for blood and tissue enzymes, so they remain in the blood only briefly (a few minutes to a few hours) before being enzymatically inactivated. In the case of some peptide hormones, the target cell actually engulfs the bound hormone by endocytosis and degrades it intracellularly. In contrast, binding of lipophilic hormones to plasma proteins renders them less vulnerable to metabolic inactivation and prevents them from escaping into the urine. Therefore, lipophilic hormones are removed from the plasma much more slowly. They may persist in the blood for hours (steroids) or up to a week in humans (thyroid hormone). In general, lipophilic hormones must undergo a series of reactions to reduce their biological activity and enhances their H_2O solubility so that they can be freed of their plasma protein carriers and be eliminated in the urine.

The responsiveness of a target cell to its hormone can be varied by regulating the number of its hormone-specific receptors

A target cell's response to a hormone is correlated with the number of the cell's receptors occupied by molecules of that hormone, which in turn depends on the number of receptors present in the target cell and on the plasma concentration of the hormone. Thus, the response of a target cell to a given plasma concentration can be fine-tuned up or down by varying the number of receptors available for hormone binding.

Down-Regulation As an illustration, when the plasma concentration of insulin is chronically elevated, the total number of target cell receptors for insulin is reduced as a direct result of the effect that an elevated level of insulin has on the insulin receptors. This phenomenon, known as **down-regulation**, constitutes an important locally acting negative-feedback mechanism that prevents the target cells from overreacting to the high concentration of insulin; that is, the target cells are desensitized to insulin, helping blunt the effect of insulin hypersecretion. This is a form of *acclimatization,* one of the key regulatory mechanisms we discussed in Chapter 1 (p. 17). Down-regulation of insulin is accomplished by the following mechanism. The binding of insulin to its surface receptors induces endocytosis of the hormone–receptor complex, which is subsequently attacked by intracellular lysosomal enzymes. This internalization serves a twofold purpose: It provides a pathway for degradation of the hormone, and it also plays a role in regulating the number of receptors available for binding on the target cell's surface. At high plasma-insulin concentrations, the number of surface receptors for insulin is gradually reduced as a result of the accelerated rate of receptor internalization and degradation brought about by increased hormonal binding. The rate of synthesis of new receptors within the endoplasmic reticulum and their insertion in the plasma membrane do not keep pace with their rate of destruction. Over time, this self-induced loss of target cell receptors for insulin reduces the target cell's sensitivity to the elevated hormone concentration.

Permissiveness, Synergism, and Antagonism A given hormone's effects are influenced not only by the concentration of the hormone itself but also by the concentrations of other hormones that interact with it. Because hormones are widely distributed through the blood, target cells may be exposed simultaneously to many different hormones, giving rise to numerous complex hormonal interactions on target cells. Hormones frequently alter the receptors for other kinds of hormones as part of their normal physiological activity. A hormone can influence the activity of another hormone at a given target cell in one of three ways: permissiveness, synergism, and antagonism.

- With **permissiveness**, one hormone must be present in adequate amounts for the full exertion of another hormone's effect. In essence, the first hormone, by enhancing a target cell's responsiveness to another hormone, "permits" this other hormone to exert its full effect. For example, thyroid hormone increases the number of receptors for epinephrine in epinephrine's target cells, increasing the effectiveness of epinephrine. Epinephrine is only marginally effective in the absence of thyroid hormone.
- **Synergism** occurs when the actions of several hormones are complementary and their combined effect is greater than the sum of their separate effects. An example is the synergistic action of follicle-stimulating hormone and testosterone, both of which are required to maintain the normal rate of sperm production. Synergism probably results from each hormone's influence on the number or affinity of receptors for the other hormone.
- **Antagonism** occurs when one hormone causes the loss of another hormone's receptors, reducing the effectiveness of the second hormone. To illustrate, progesterone (a mammalian hormone secreted during pregnancy that decreases contractions of the uterus) inhibits uterine responsiveness to estrogen (another hormone secreted during pregnancy that increases uterine contractions). By causing loss of estrogen receptors on uterine smooth muscle, progesterone prevents estrogen from exerting its excitatory effects during pregnancy and thus keeps the uterus in a quiet (noncontracting) environment suitable for the developing fetus.

Endocrine disorders are attributable to hormonal excess, hormonal deficiency, or decreased responsiveness of the target cells

From the preceding discussion, you can see that abnormalities in a hormone's effective plasma concentration can arise from a variety of factors. Endocrine disorders most commonly result from abnormal plasma concentrations of a hormone caused by inappropriate rates of secretion—that is, too little hormone secreted (**hyposecretion**) or too much hormone secreted (**hypersecretion**). Occasionally endocrine dysfunction arises because target cell responsiveness to the hormone is abnormally low, even though the plasma concentration of the hormone is normal.

Hyposecretion If an endocrine gland (such as the adrenal gland) is secreting too little of its hormone because of an abnormality within that gland, the condition is referred to as *primary hyposecretion*. If, in contrast, the endocrine gland is normal but is secreting too little hormone because of a deficiency of its tropic hormone, the condition is known as *secondary hyposecretion*. The following are among the many different factors (each listed with an example) that may be responsible for hormone deficiency: (1) *genetic* (inborn absence of an enzyme that catalyzes synthesis of the hormone); (2) *dietary* (lack of iodine, which is necessary for synthesis of thyroid hormone); (3) *chemical or toxic* (certain insecticide residues may destroy the adrenal cortex); (4) *immunologic* (auto-immune antibodies may cause self-destruction of the animal's own thyroid tissue); (5) *other disease processes* (cancer or tuberculosis may coincidentally destroy endocrine glands); (6) *iatrogenic* (physician or veterinarian-induced, such as surgical removal of a cancerous thyroid gland); and (7) *idiopathic* (meaning the cause is not known).

The most common method of treating hormone hyposecretion is to administer a hormone that is the same as (or similar to, such as from another species) the one that is deficient or missing. The sources of hormone preparation for clinical use include (1) endocrine tissues from domestic livestock; (2) placental tissue and urine of pregnant women; (3) laboratory synthesis of hormones; and (4) genetically engineered "hormone factories": bacteria into which genes coding for the production of mammalian hormones have been introduced. The method of choice for a given hormone is determined largely by its structural complexity and degree of species specificity.

Hypersecretion Like hyposecretion, hypersecretion by a particular endocrine gland is designated as primary or secondary depending on whether the defect lies in that gland or is due to excessive stimulation from the outside, respectively. Hypersecretion may be caused by (1) tumors that ignore the normal regulatory input and continuously secrete excess hormone and (2) immunologic factors, such as excessive stimulation of the thyroid gland by an abnormal antibody that mimics the action of TSH, the thyroid tropic hormone. Excessive levels of a particular hormone may also arise from substance abuse, such as the banned practice among athletes of using certain steroids that increase muscle mass by promoting protein synthesis in muscle cells, and the injections of these and other hormones into livestock to enhance production (such as GH, p. 291). Note however that such injections lead to feedback inhibition of natural hormone production (see p. 296).

There are several ways of treating hormonal hypersecretion. If a tumor is the culprit, it may be surgically removed or destroyed with radiation treatment. In some instances, drugs that block hormone synthesis or inhibit hormone secretion can limit hypersecretion. Sometimes giving drugs that inhibit the action of the hormone without actually reducing the excess hormone secretion may treat the condition.

The quantity of JH released determines the quality of the molt

JH, secreted from the corpora allata, partners the release of ecdysone and ultimately determines the "quality" of the molt. JH has two broad actions in insects: It has a "priming" or activating action whereby it prepares a tissue for hormonal response by directing the synthesis and assembly of the cell machinery necessary to carry out the responses elicited by the binding of JH to its receptor. It also performs a regulatory role by controlling the rate of functioning of the primed tissue. For example, the fat body of virtually all insects cannot synthesize the protein **vitellogenin** (protein taken up by the developing oocyte as **vitellin,** the principal yolk protein) until it has been primed by exposure to JH. Thus, in addition to controlling metamorphosis, JH regulates many aspects of reproduction as well. The fat body is most highly developed during the larval instars and normally regresses at the end of metamorphosis, except in some adult species of insects that do not actively feed.

The quantity of JH released at each molt determines whether molting will ultimately produce a larva or an adult, because when released, JH ensures that larval characteristics are retained. For example, in insects such as moths and beetles JH is released at progressively decreasing concentrations during each larval stage (or instar) until the final stage, when its release is inhibited and negligible concentrations of JH are detectable. JH release is regulated by the cerebral neuropeptide **allostatin,** which suppresses the release of JH, as well as by **allotropin,** which stimulates the release of JH. During the time of molt the tissues are exposed to a pulse of ecdysteroids in the absence of JH, which enables metamorphosis of the tissues into the adult form. However, if the corpora allata are prematurely removed from an insect, metamorphosis occurs, producing a small adult. Conversely, if JH is administered at each molt the juvenile fails to develop into the adult form.

There is also evidence that JH induces the production and release of pheromones in several species of insects (such as cockroaches, some coleopterans, and lepidopterans) and in the boll weevil. JH enhances the production of sex pheromone in the male fat body. Concentrations of JH are regulated by the proteins **juvenile hormone–binding protein (JHBP)** and **juvenile hormone esterase (JHE).** JHBP is produced by the fat body and released into the hemolymph, where it binds JH for transport and protection from degrading enzymes, including JHE. Fat bodies make JHE at increasing levels before the final larval molt to degrade JH; at high concentrations, JHE can degrade the JH bound to the transporter JHBP.

With the prevalence of insects in terrestrial ecosystems, their endocrine system is a prominent target of evolutionary adaptation in plants that are eaten by insects. Substances closely related to ecdysone occur in certain varieties of plants and are believed to protect the plants from feeding by insects. Plant substances called **precocenes** block JH production by the corpora allata, triggering a premature metamorphosis in species in which the larval stage (such as caterpillars) and not the adult is the herbivore. Similarly, to protect crops humans have targeted insect endocrine and pheromone systems. EcR, for example, is now an important target in the design of novel, environmentally safe insecticides. Also, the genes for JHBP and JHE have recently been cloned and are being studied as a mechanism for controlling development of insect pests.

Pheromones are used in mating and colonial interactions

Let's end our examination of insect endocrinology with a brief look at pheromones. Although, strictly speaking, pheromones are neither hormones nor endocrine in origin, they are used in very similar ways, traveling from one body to another and binding to receptors to stimulate specific behaviors. The most widespread use of pheromones in all animals including insects is in mating. This was first discovered in the 1870s by the French naturalist Jean-Henri Fabré. A female great peacock moth that hatched from a cocoon in his laboratory was soon surrounded by dozens of male great peacocks coming in through the windows. He postulated the existence of an attractive odorant, which we now know is the case. Males, using receptors on their antennae, are incredibly sensitive to these female **sex pheromones,** and some species can detect just a single molecule that has traveled a great distance from the female. Pheromone synthesis in insects is regulated by the neurohormone **pheromone biosynthesis activation peptide (PBAN), which** controls the enzymes involved in pheromone synthesis via G-protein coupled receptors. The release of PBAN is regulated by environmental factors such as day length and host-plant odors.

Pheromone use beyond reproduction has probably diversified more in insects than in any other animal group. Social insects, such as ants, termites. and bees, rely heavily on pheromones to control and coordinate colony functions and development of castes such as workers and soldiers. For example, **alarm pheromones** are released when a colony is attacked; these signal molecules trigger rapid and violent defense behaviors. Many biologists consider insect colonies "superorganisms," with individual insects acting like cells and organs and with pheromones acting like true hormones.

check your understanding 7.2

Describe the process of molting in insects. What factor(s) determine the onset of metamorphosis?

7.3 Vertebrate Endocrinology: Central Endocrine Glands

The central endocrine glands of vertebrates include the **hypothalamus,** the **pituitary gland,** and **pineal gland.** The hypothalamus, a part of the brain, and the posterior pituitary gland act as a unit to release hormones essential for maintaining water balance and reproduction. The hypothalamus also secretes regulatory hormones that control the hormonal output of the **anterior pituitary gland,** which secretes seven hormones that in turn largely control the hormonal output of several peripheral endocrine glands.

Earlier, we discussed the concept of biological clocks, zeitgebers, and rhythmicity in hormone production. The pineal gland is a part of the brain that secretes a hormone

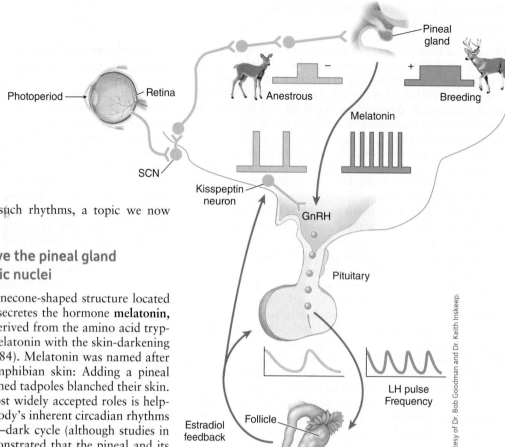

FIGURE 7-8 Pineal synchronization of reproductive activity in a seasonally breeding animal. In this example, a short-day breeder responds to a declining photoperiod by increasing melatonin secretion. In turn, GnRH pulse frequency is increased, which increases LH pulse frequency and stimulates gonadal development. Low concentrations of gonadal estrogen inhibit GnRH release during anestrus by binding to estrogen receptors on neurons in the periventricular and arcuate nuclei, limiting the release of kisspeptin. During the breeding season, this estrogen negative feedback decreases and kisspeptin stimulates GnRH pulse frequency, increasing follicular development and estrogen secretion and leading to positive feedback of estrogen, triggering an LH surge to cause ovulation.

© Cengage Learning, 2013

important in establishing such rhythms, a topic we now discuss in detail.

Biological clocks involve the pineal gland and the suprachiasmatic nuclei

The **pineal gland,** a tiny, pinecone-shaped structure located in the center of the brain, secretes the hormone **melatonin,** an indoleamine hormone derived from the amino acid tryptophan. (Do not confuse melatonin with the skin-darkening pigment, melanin—see p. 284). Melatonin was named after its lightening action on amphibian skin: Adding a pineal extract to water that contained tadpoles blanched their skin.

One of melatonin's most widely accepted roles is helping to keep the vertebrate body's inherent circadian rhythms in synchrony with the light–dark cycle (although studies in the quail and chicken demonstrated that the pineal and its melatonin rhythm are not essential for circadian function). Melatonin is the hormone of darkness: Its secretion increases up to 10-fold during the darkness of night and then falls to low levels during the light of day. Short-day breeders (such as sheep and deer) rely on steadily increasing concentrations of melatonin (Figure 7-8), whereas long-day breeders (such as many birds) rely on the progressively decreasing concentrations of melatonin to trigger reproduction. In fish, reptiles, birds, and amphibians, the pineal is photosensitive and is not directly controlled by nervous input, but rather by sunlight penetrating the skull. Pineal cells in these vertebrates are modified rod cells with reduced outer segments, containing photosensitive pigments structurally related to rhodopsin. As in mammals, there is a diurnal rhythm of melatonin, which is independent of cues from the eye. Some studies also show that melatonin controls color changes in certain ectothermic vertebrates independently of the control pathway of *melanocyte-stimulating hormone* regulation, which we will discuss later.

The master biological clock that serves as the pacemaker for circadian rhythms in mammals is the **suprachiasmatic nucleus (SCN)** (*mediobasal hypothalamus* in birds) a cluster of nerve cell bodies in the hypothalamus above the optic chiasm (the point at which part of the nerve fibers from each eye cross to the opposite half of the brain; see p. 255) (*supra,* "above"; *chiasm,* "cross"). The SCN is strategically located to receive light–dark information from the eyes. The release of glutamate by the neural input to the SCN acts to entrain the individual cells to the SCN to produce a coordinated 24-hour signal that is conveyed to multiple sites in the brain to control rhythmic patterns of body temp, feeding behavior, and so forth.

In mammals, pineal and melatonin output are under the control of the SCN. Daily changes in light intensity become the major environmental cue used to adjust this master clock. Special photoreceptors in the retina pick up light signals and transmit them directly to the SCN. Intermingled among the visually oriented retinal ganglion cells, about 1 to 2% of the retinal ganglion cells form an entirely independent light-detection system that responds to levels of illumination, like a light meter on a camera, rather than the contrasts, colors, and contours detected by the image-forming visual system. These photoreceptors are distinct from the rods and cones used to perceive, or see, light (see p. 248) and are located in the preganglionic region of the retina. These cells express **melanopsin,** a photoreceptor protein originally isolated in frogs and act to entrain the circadian clock (see *Molecular Biology and Genomics: Clocks* *and Genes*). The melanopsin-containing, illumination-detecting retinal ganglion cells cue the pineal gland about the absence or presence of light by sending their signals along the **retino-hypothalamic tract** to the SCN. This pathway is distinct from the neural systems that result in vision perception. This is the major way the internal clock is coordinated to a 24-hour day. In rodent models in which the rods and cones of the retina are absent due to genetic modification and the animals are visually blind, they are still able to adjust their circadian system normally to light due to persistent regulation by the melanopsin cells. In many blind people it is the melanopsin cells that allow a relatively normal lifestyle in spite of the lack of normal visual input. However, in a few individuals that totally lack a retina or have no eyes, the circadian system free-runs and these individuals suffer severely due to the lack of synchrony to the natural environment. These subjects can be treated successfully by the administration of melatonin at the normal sleep time to restore normal timing.

Moreover the SCN signals the adrenal gland through the sympathetic nervous system to control the production of adrenal hormones including *cortisol* or *corticosterone*. The consequent adrenal steroid hormone rhythm also plays a role in coordinating the daily biological rhythmicity. These peripheral mechanisms are particularly important in synchronizing the different body organs, which have their own biological rhythms to ensure optimal timing.

All eukaryotic cells, including all those within an animal, express clock genes

The molecular biology of the circadian clock was first elucidated in *Drosophila* and other simple models. These studies determined that clock function is generated by the interplay of a small number of genes called **clock genes,** which have now been found in all eukaryotes examined (as well as some bacteria and archaea). Clock genes interact with each other through both transcription and negative-feedback loops to produce a cycle that lasts approximately a day, that is, a circadian rhythm. Remarkably the clock genes expressed in insects and vertebrates have a major degree of homology (see *Molecular Biology and Genomics: Clocks and Genes*), which allowed the rapid discovery of the clock gene mechanisms in birds and mammals. Initially the expectation was that the clock genes would be expressed solely in the SCN, which was recognized as the central pacemaker for the circadian system. However, it was discovered that clock genes were expressed in virtually all cells in the body, consistent with their early evolution, and which endows all organ systems their own circadian rhythms. Thus, the SCN acts as the conductor, which synchronizes all the multiple organs in the body to have the appropriate timing across the day. For example, the liver produces enzymes in anticipation of mealtimes. If these rhythms are disrupted, such as travel across time zones as occurs when horses are transferred between Europe and Asia for horseracing, these internal rhythms become desynchronized. It has been demonstrated in rodents that it takes about three days for the brain to readjust to an eight-hour phase (time zone) shift. However it takes as long as seven days for the gut and liver to adjust to such a phase change.

The biological clock must be synchronized with environmental cues

On its own, the circadian biological clock generally cycles a bit slower or faster than the 24-hour environmental cycle. In many insects and vertebrates (including humans), this rhythm is slightly longer than 24 hours and so must be adjusted each day by light from the environment to reset the natural daylight cycle to a period of 24 hours. (Blind cave fish and salamanders may exhibit rhythmic behaviors, but they generally do not follow a 24-hour cycle.) If this master clock were not continuously adjusted to keep pace with the world outside, the body's circadian rhythm would lag progressively out of synchrony with the daily cycles of light and dark. The existence of internal clocks was revealed by the discovery that circadian rhythms persist even if the light–dark cycle is artificially reverted to constant light (LL) or constant dark (DD). Under these conditions the rhythm is allowed to **free-run,** and animals may entrain on other zeitgebers, including environmental temperature, or the noises associated with individuals providing food at particular time in the day. Thus, the SCN must be reset daily by external cues so that the body's biological rhythms synchronize with the activity levels driven by the surrounding environment. The SCN works in conjunction with the pineal gland and its hormonal product melatonin to synchronize the various circadian rhythms with the 24-hour day–night cycle.

Biological clocks are ultimately regulated by genes and proteins that must somehow respond to sunlight, but until recently these were almost completely unknown. See *Molecular Biology and Genomics: Clocks and Genes* for these findings.

The pituitary gland descends from the hypothalamus and consists of two or three lobes

The **pituitary gland,** or **hypophysis,** is a small endocrine gland in a bony cavity at the base of the vertebrate brain just below the hypothalamus (Figure 7-9). The pituitary is connected to the hypothalamus by a thin stalk, the **infundibulum,** which contains nerve fibers and small blood vessels.

The pituitary has two anatomically and functionally distinct lobes, the **posterior pituitary** and the **anterior pituitary.** The posterior pituitary, being derived embryonically from an outgrowth of the brain, consists of nervous tissue from the hypothalamus and thus is also termed the **neurohypophysis.** The anterior pituitary, in contrast, consists of glandular epithelial tissue derived embryonically from an outpouching of ectoderm that buds off from the roof of the mouth. Accordingly, the anterior pituitary is also known as the **adenohypophysis** (*adeno*, "glandular"). The anterior pituitary consists of the **pars distalis (PD)** (main body of the pituitary gland) and the **pars tuberalis (PT),** which forms the stalk of the pituitary gland immediately adjacent to the base of the brain. The posterior pituitary is connected to the hypothalamus by a neural pathway, whereas the anterior pituitary is connected to the hypothalamus by a vascular link.

The Intermediate Lobe and Melanocyte-Stimulating Hormone In lampreys and hagfish (jawless fishes), amphibians, reptiles, and most mammals, the adenohypophysis includes a third, well-defined intermediate lobe (pars intermedia). Although absent in birds and cetaceans (whales and dolphins),

Molecular Biology and Genomics:
Clocks and Genes

Biological clocks have been known for over a century, but the molecular mechanisms have only been uncovered in the last two decades, with the fullest details uncovered in the fungus *Neurospora,* certain plants, fruit flies, and mice. Numerous clock genes have been discovered and their expression activities tracked. In mice, transcription of clock genes named *BMAL1* and *CLOCK* within the nuclei of SCN neurons start the cycle via synthesis of the corresponding proteins *Bmal1* and *Clock* in the cytosol. *Bmal1* and *Clock* proteins are transported back into the nucleus, where they bind to and activate specific promoter sequences for genes that ultimately increase their own production. They also bind to and activate promoters for genes called *PERIOD* and *CRYPTOCHROME (CRY).* Thus, there is increasing production of *Bmal1, Clock, Period,* and *CRY* proteins. However, *Period* and *CRY* slowly but increasingly block the actions of *Bmal1* and *Clock* proteins, thus reducing the transcription of all four genes. The level of clock proteins gradually dwindles as they degrade. This removes the inhibitory influence of *Period* and *CRY.* No longer being blocked, the *BMAL1* and *CLOCK* genes once again rev up the production of more clock proteins, as the cycle repeats itself. Each cycle takes about a day. The fluctuating levels of clock proteins bring about cyclical changes in neural output from the SCN that in turn lead to cyclic changes in effector organs throughout the day. An example is the diurnal variation in cortisol secretion (see Figure 7-6). Circadian rhythms are thus linked to fluctuations in clock proteins, which use a feedback loop to control their own production at the transcriptional (gene) level. In this way, internal timekeeping is a self-sustaining mechanism built into the genetic makeup of the SCN neurons.

These clock protein cycles allow the clock mechanism to run without any input from the environment. But at some point, all clocks need to be synchronized with external signals (see text). This appears to be the role of one or more light-sensitive proteins, such as *CRY.* *CRY* belongs to the larger class of *cryptochromes* found first in plants and later in the retinas and SCN of fruit flies, mice, and humans. Cryptochromes in the retina react to blue light and are probably activated at dawn, and appear (based on knockout experiments) to be crucial to setting clocks in *Drosophila.* In mammals, **melanopsin** is thought to have that function. This protein is found in retinas of fishes, frogs, and mammals. It is a type of *opsin,* a family of genes used to make proteins for light absorption in vision (see Figure 6-35, p. 247). Melanopsin's gene, though related to visual opsins, is distinctly different enough to suggest a function other than vision. Knockout mice unable to make melanopsin were found to have 50 to 80% less sensitivity to light in setting their SCN clocks.

---

an intermediate lobe exists in the human fetus but becomes rudimentary after birth. Compared to the other regions of the pituitary gland and endocrine tissues in general, the intermediate lobe is poorly vascularized. Generally the size of the intermediate lobe is correlated with the ability of the animal to adapt to the coloration of its environment. For example, the lobe is enlarged in the chameleon, a champion of color change in lizards.

The intermediate lobe secretes several **melanocyte-stimulating hormones,** or **MSHs.** MSH is under primarily inhibitory control, as evidenced from experiments where ectopic transplantation of the entire pituitary gland lead to hypertrophy of the intermediate lobe and an increase in MSH secretion. Inhibition of MSH secretion by the hypothalamus is by the catecholamine dopamine. The two forms of melanocyte-stimulating hormone, **α-MSH** and **β-MSH,** are derived from the larger precursor protein *POMC* (p. 271). β-MSH has no known physiological role and may only be a structural component of POMC, whereas α-MSH controls skin coloration via the dispersion of storage granules containing the pigment **melanin.** These granules are in **melanocytes** (melanin-containing cells), found in the skin of most vertebrate species.

Melanocytes in the epidermis synthesize the colored substance melanin using the precursor amino acid tyrosine. Melanins that appear brown or black are referred to as **eumelanins,** whereas red or lighter colored melanins are termed **phaeomelanins.** Melanin is stored in melanin granules, or **melanosomes,** where it is released into the skin cells in response to environmental photic cues. By causing variable skin darkening in certain amphibians, reptiles, and fishes, MSHs play a vital role in the camouflage of these species. In some lower vertebrates the action of α-MSH is opposed by an antagonistic pituitary hormone **melanin-concentrating hormone (MCH).** Teleost MCH induced blanching of the skin when injected into dark-adapted fishes by stimulating melanosome pigment aggregation.

Some species of mammals and birds also rely on seasonal changes in melanin deposition in the pelage or feathers to minimize detection by predators or prey. MSH and steroid hormones can also differentially affect melanin deposition in the hairs and feathers of mammals and birds, although scientists do not understand the mechanism of melanin pigmentation of feathers. Nonetheless some mammals (such as the hare *Lepus americanus*) and some birds (such as the ptarmigan *Lagopus lagopus*) can change from a brown summer coat to a white winter coat in response to the dramatic changes in photoperiod.

In adult humans, the skin secretes small amounts of MSH. However, MSH is not involved in differences in the amount of melanin deposited in the skin in response to sunlight, nor with the process of skin tanning, although excessive MSH secretion does darken skin. Rather it is the activated form of vitamin D (alone or synergistically with other hormonal factors) that has been suggested to simulate melanogenesis (sun tanning). In animals without a pars intermedia (and so lacking a source of MSH), it is the intrinsic activity of the **melanocortin-1 receptor (MC1R)** that accounts for differences in skin color in humans, pelage (hair, fur, wool covering in a mammal), and feather pigmentation in birds. Another role for α-MSH lies in the appetite-suppressing neurons in the hypothalamus, which secrete to this hormone

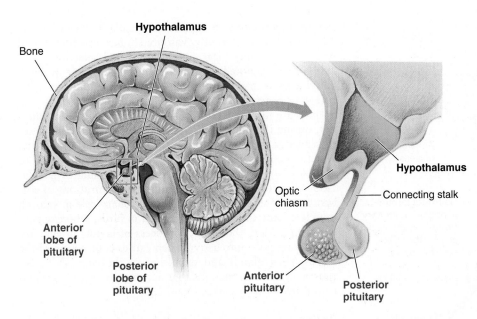

(a) Relation of pituitary gland to hypothalamus and rest of brain

(b) Enlargement of pituitary gland and its connection to hypothalamus

FIGURE 7-9 Anatomy of the pituitary gland.

© Cengage Learning, 2013

Dispersion of pigment granules in the melanocytes by β-MSH permits the skin coloration of the California lizard, *Sceloporus*, to approximate that of the background.

control food intake (see p. 725). MSH has also been shown to suppress the immune system, possibly serving in a check-and-balance way to prevent excessive immune responses.

The hypothalamus and posterior pituitary form a neurosecretory system that secretes vasopressin and oxytocin

The release of hormones from both the posterior and the anterior pituitary is directly controlled by the hypothalamus, but the nature of the relationship is entirely different. The posterior pituitary connects to the hypothalamus by a neural pathway, whereas the anterior pituitary connects to the hypothalamus by a unique vascular link.

Let's look first at the posterior pituitary. The hypothalamus and posterior pituitary form a neuroendocrine system that consists of a population of neurosecretory neurons whose cell bodies lie in two well-defined clusters in the hypothalamus, the **supraoptic** and **paraventricular nuclei**). The axons of these neurons pass down through the thin connecting stalk to terminate on capillaries in the posterior pituitary (Figure 7-10). The posterior pituitary consists of these neuronal terminals plus glial-like supporting cells called **pituicytes**. Functionally as well as anatomically, the posterior pituitary is simply an extension of the hypothalamus. This is similar to the relationship between the X-organ and sinus gland in crustaceans (p. 279).

The posterior pituitary stores and, on appropriate stimulation, releases into the blood two small peptide neurohormones, **vasopressin** and **oxytocin,** which are synthesized by the neuronal cell bodies in the hypothalamus. The form of vasopressin in mammals is **arginine vasopressin (AVP)**. Of note is the ancient origin of these two messengers—both hormones or closely related peptides appear in all animal species, from hydra to worms and snails, to fishes, birds, and mammals. In vertebrates the ancestor is thought to be **arginine vasotocin (AVT)**, a peptide found in hagfishes and lampreys, primitive jawless fishes (agnathans) that originated over 500 million years ago. AVT also found in other vertebrates (see p. 574).

The synthesized hormones are packaged in secretory granules that are transported by axoplasmic flow down the cytoplasm of the axon (see Figure 4-21) to be stored in the neuronal terminals within the posterior pituitary. Each terminal stores either vasopressin or oxytocin but not both. Thus, these hormones can be released independently as needed. On stimulatory input to the hypothalamus, either vasopressin or oxytocin is released into the systemic blood from the posterior pituitary by exocytosis of the appropriate secretory granules. This hormonal release is triggered in response to action potentials that originate in the hypothalamic cell body and sweep down the axon to the neuronal terminal in the posterior pituitary. As in any other neuron, action potentials are generated in these neurosecretory neurons in response to synaptic input to their cell bodies.

The actions of vasopressin and oxytocin are only briefly summarized here, but they are described more thoroughly in later chapters.

Vasopressin In most mammals, vasopressin (also called **antidiuretic hormone, ADH**) has two major effects that correspond to its two names: (1) it enhances retention of H_2O by kidneys (an antidiuretic effect), and (2) it causes contraction of arteriolar smooth muscle (a vasopressor effect).

Photo: Hillar Klandorf

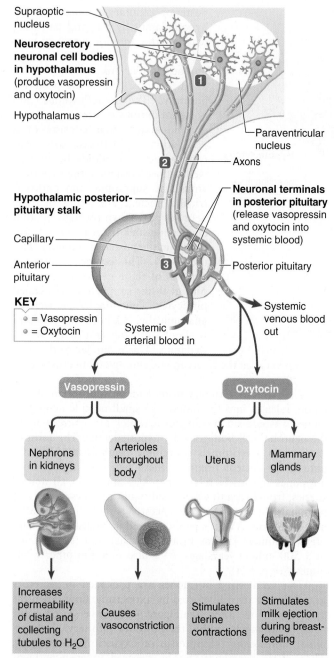

Supraoptic nucleus

Neurosecretory neuronal cell bodies in hypothalamus (produce vasopressin and oxytocin)

Hypothalamus

Paraventricular nucleus

Axons

Hypothalamic posterior-pituitary stalk

Neuronal terminals in posterior pituitary (release vasopressin and oxytocin into systemic blood)

Capillary

Anterior pituitary

Posterior pituitary

KEY

- = Vasopressin
- = Oxytocin

Systemic arterial blood in

Systemic venous blood out

Vasopressin

Oxytocin

Nephrons in kidneys

Arterioles throughout body

Uterus

Mammary glands

Increases permeability of distal and collecting tubules to H_2O

Causes vasoconstriction

Stimulates uterine contractions

Stimulates milk ejection during breast-feeding

1 The paraventricular and supraoptic nuclei both contain neurons that produce vasopressin and oxytocin. The hormone, either vasopressin or oxytocin depending on the neuron, is synthesized in the neuronal cell body in the hypothalamus.

2 The hormone travels down the axon to be stored in the neuronal terminals within the posterior pituitary.

3 When the neuron is excited, the stored hormone is released from the terminals into the systemic blood for distribution throughout the body.

FIGURE 7-10 **Relationship of the hypothalamus and posterior pituitary.**

© Cengage Learning, 2013

Under normal conditions, vasopressin is the primary endocrine factor that regulates urinary H_2O loss and overall H_2O balance. The major control for hypothalamic-induced release of vasopressin from the posterior pituitary is input from hypothalamic *osmoreceptors*, which increase vasopressin secre-

tion in response to a rise in plasma osmolarity (see Chapters 12 and 13). More recently, vasopressin has also been found to play roles in fever, learning, memory, and behavior.

Earlier we noted that the equivalent hormone in lower vertebrates is *arginine vasotocin* (AVT). In addition to its role in osmoregulation, AVT produces vasoconstriction and an increase in blood pressure in most ectotherms. In flounder, during the initial phase of adaptation from fresh water to seawater there is a transitory rise in plasma osmolarity, which increases plasma AVT. During metamorphosis of tadpoles into frogs, the increase in AVT stimulates ACTH-induced steroidogenesis, which raises concentrations of corticosterone (see p. 305). AVT's evolutionary role in smooth muscle contraction is also evidenced in birds, where it governs *oviposition*, the physical process of laying an egg; administering AVT into a laying hen causes contraction of the uterus (shell gland) and expulsion of the egg. AVT also causes uterine contractions that accompany birth in viviparous ("live birth") snakes (*Nerodia*).

Oxytocin Oxytocin in mammals has been described as the "hormone of love" for a variety of social bonding influences; for example, its release in lactating rat mothers provides the urge to nurse their pups. Behaviorally it keeps male prairie voles monogamous and induces trust in people (see Chapter 16 for details on these and other examples). Physiologically, through positive-feedback regulation, it stimulates contraction of the uterine smooth muscle to help expel the fetus during birth and promotes ejection of the milk from the mammary glands (see Chapter 16).

The nonmammalian homologue, **mesotocin (MT)**, does not affect the uterus but rather influences the blood flow to some organs (for example, in poultry an increased concentration of MT reduces the blood flow to the lower limbs and comb) in addition to reducing the circulating concentrations of aldosterone.

The anterior pituitary secretes six established hormones, many of which are tropic

Unlike the posterior pituitary, which releases hormones that are synthesized by the hypothalamus, the anterior pituitary itself synthesizes the hormones that it releases into the blood. Different cell populations within the anterior pituitary produce and secrete six established peptide hormones. The action of each of these hormones is discussed in detail in subsequent sections and chapters. For now, we briefly state their primary effects to explain their names (Figure 7-11):

1. The type of anterior pituitary cells known as **somatotropes** secrete **growth hormone (GH, somatotropin),** the primary hormone responsible for regulating overall body growth (*somato* means "body"). GH also exerts important metabolic actions.

2. **Thyrotropes** secrete **thyroid-stimulating hormone (TSH, thyrotropin),** which stimulates secretion of thyroid hormone and growth of the thyroid gland.

3. **Corticotropes** produce and release **adrenocorticotropic hormone (ACTH, or corticotropin),** the hormone that stimulates cortisol secretion by the adrenal cortex and promotes growth of the adrenal cortex. In advanced vertebrates ACTH is derived from the large precursor molecule *POMC* (p. 271) produced within the endoplas-

FIGURE 7-11 Functions of the anterior pituitary hormones. Five different endocrine cell types produce the six anterior pituitary hormones—TSH, ACTH, growth hormone, LH and FSH (produced by the same cell type), and prolactin—which exert a wide range of effects throughout the body.

© Cengage Learning, 2013

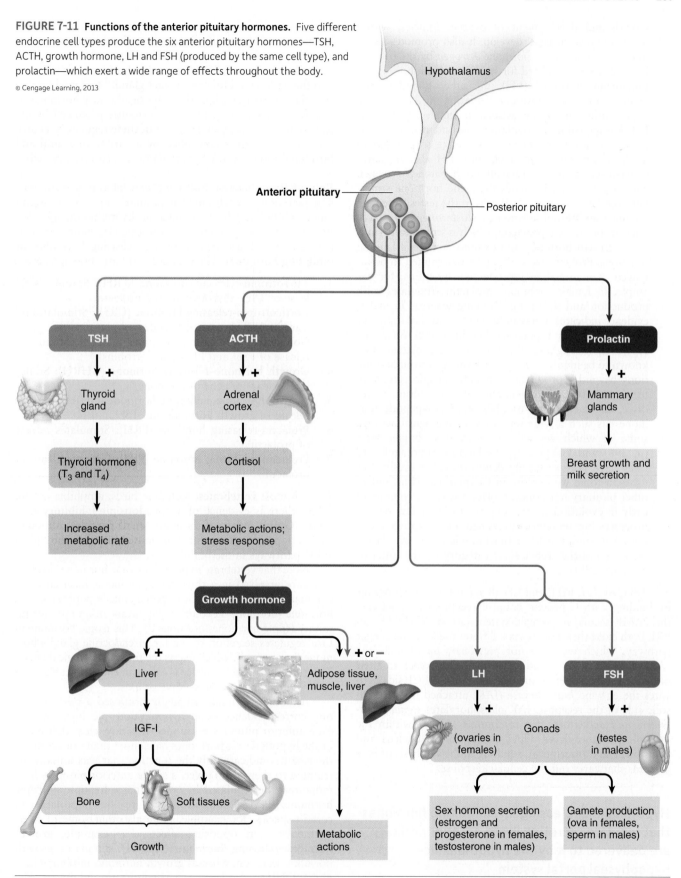

mic reticulum of the anterior pituitary's ACTH-secreting cells.

4. **Gonadotropes** secrete two **gonadotropins,** hormones that act on the gonads (reproductive organs, namely the ova-

ries and testes)—follicle-stimulating hormone and luteinizing hormone. **Follicle-stimulating hormone (FSH)** helps regulate gamete (reproductive cells, namely ova and sperm) production in both sexes. In females it stimulates growth and development of ovarian follicles, within which the ova, or eggs, develop. It also promotes secretion of the sex steroid hormone *estrogen* by the ovaries. In males FSH is required for sperm production.

5. **Luteinizing hormone (LH)**, the other gonadotropin, helps control sex hormone secretion in both sexes, among other important actions in females. In females LH is responsible for ovulation and luteinization (that is, the formation of a hormone-secreting corpus luteum in the ovary following ovulation). LH also regulates ovarian secretion of the female sex hormones, estrogen and progesterone. In males the same hormone stimulates the Leydig (interstitial) cells in the testes to secrete the male sex steroid hormone, *testosterone.*

6. **Lactotropes** secrete **prolactin**, which is structurally similar to growth hormone and exhibits some overlapping functions. Prolactin was originally determined to be involved in mammalian reproductive functions; for example, in female mammals prolactin stimulates milk production and nurturing of young whereas in males, evidence indicates it may induce production of testicular LH receptors and paternal bonding with offspring. However, throughout the vertebrates, prolactin now known to be involved in a remarkable spectrum of functions beyond reproduction, including osmoregulation, promotion of growth, support of metabolism, water drive, and metamorphosis. No other polypeptide hormone has such a wide repertoire of biological actions, some of which we will cover in this chapter later (p. 296) and Chapter 16. Prolactin's osmoregulatory roles have been primarily demonstrated in freshwater fishes, a topic we explore in Chapter 13. Thus, unlike other pituitary hormones, prolactin was not committed early in evolution to the control of one or few related processes but remained diversified and adaptive in nature. This multiplicity of functions is likely why lactotropes constitute the largest category of the anterior pituitary cell types.

TSH, ACTH, FSH, and LH all act at their target organs by binding with G-protein coupled receptors that activate the cAMP second-messenger system (see p. 97). GH and PRL both exert their effects via a different second-messenger pathway, which we have not previously mentioned—the **JAK/STAT** pathway. Binding of these hormones to their target-cell surface membrane receptors on the ECF side activates the enzyme *Janus kinase (JAK)* attached to the cytosolic side of the receptor. JAK phosphorylates *signal transducers and activators of transcription (STAT)* within the cytosol. Phosphorylated STAT moves to the nucleus and turns on gene transcription, resulting in the synthesis of new proteins that carry out the cellular response.

Hypothalamic releasing and inhibiting hormones that regulate secretion in the anterior pituitary are delivered to it by the hypothalamic–hypophyseal portal system

None of the anterior pituitary hormones are secreted at a constant rate. Even though each of these hormones has a unique control system, there are some common regulatory patterns. The two most important factors that regulate secretion of anterior pituitary hormone are (1) hypothalamic hormones and (2) feedback by target gland hormones.

Because the anterior pituitary secretes hormones that control the secretion of various other hormones, it has long had the undeserved title of "master gland." It is now known that the release of each of the anterior pituitary hormones is largely controlled by still other hormones produced by the hypothalamus, and that secretion of these regulatory neurohormones in turn is controlled by a variety of neural and hormonal inputs to the hypothalamic neurosecretory cells.

Role of Hypothalamic Releasing and Inhibiting Hormones The secretion of each anterior pituitary hormone is stimulated or inhibited by hypothalamic **hypophysiotropic** (*hypophysis* means "pituitary") **hormones**. Depending on their actions, these hormones are called **releasing hormones** or **inhibiting hormones**. Their names and basic functions are:

1. **Thyrotropin-releasing hormone (TRH):** Stimulates release of TSH (thyrotropin) and prolactin.
2. **Corticotropin-releasing hormone (CRH):** Stimulates release of ACTH (corticotropin).
3. **Gonadotropin-releasing hormone (GnRH):** Stimulates release of FSH and LH (gonadotropins).
4. **Growth hormone–releasing hormone (GHRH):** Stimulates release of growth hormone.
5. **Growth hormone–inhibiting hormone (GHIH):** Inhibits release of growth hormone and TSH.
6. **Prolactin-releasing hormone (PRH):** Stimulates release of prolactin.
7. **Prolactin-inhibiting hormone (PIH):** Inhibits release of prolactin.

Also, in most vertebrates, including birds, amphibians, and fishes, there is evidence of a **gonadotropin-inhibitory hormone (GnIH)**. Most of these are small peptides, although PIH is **dopamine** (the neurotransmitter found in the "pleasure" pathways in the brain).

Note that vertebrate hypophysiotropic hormones in most cases are involved in a three-hormone hierarchical chain of command (Figure 7-12): The hypothalamic hypophysiotropic hormone (hormone 1) controls the output of an anterior pituitary tropic hormone (hormone 2). This tropic hormone in turn regulates secretion of the target endocrine gland's hormone (hormone 3), which exerts the final physiological effect. This three-hormone sequence is called an **endocrine axis**, as in the hypothalamus–pituitary–thyroid axis.

Although scientists originally proposed a neat one-to-one correspondence—one hypophysiotropic hormone for each anterior pituitary hormone—it is now clear that many of the hypothalamic hormones have more than one effect, so their names indicate only the function that was initially attributed to them. Moreover, a single anterior pituitary hormone may be regulated by two or more hypophysiotropic hormones. Several substances elicit release of prolactin in certain species, for example. Note also that some of these hormones exert opposing effects. For example, **growth hormone–releasing hormone (GHRH)** stimulates growth hormone secretion, whereas **growth hormone–inhibiting hormone (GHIH)**, also known as **somatostatin**, inhibits it. The output of the anterior pituitary growth hormone secreting cells (that is, the rate of growth hormone secretion) in response to two such opposing inputs depends on the relative concentrations of these hypothalamic hormones as well as on the intensity of other regulatory inputs. This is analogous to the output of a nerve cell (that is, the rate of action potential propagation) depending on the relative magnitude of excitatory and inhibitory synaptic inputs (EPSPs and IPSPs) to it (see p. 137).

Role of the Hypothalamic–Hypophyseal Portal System The hypothalamic regulatory hormones reach the anterior pituitary by means of a unique vascular link. In contrast to the direct neural connection between the hypothalamus and posterior pituitary, the link between the hypothalamus and anterior pituitary is an unusual capillary-to-capillary connection, the **hypothalamic–hypophyseal portal system**. A portal system is a vascular arrangement in which venous blood flows directly from one capillary bed through a connecting vessel to another capillary bed before going back to the heart. The largest and best-known portal system is the *hepatic portal* system (see p. 685). Although much smaller, the hypothalamic–hypophyseal portal system is no less important, because it provides a critical link between the brain and much of the endocrine system. It begins in the base of the hypothalamus with a group of capillaries that recombine into small portal vessels, which pass down through the connecting stalk into the anterior pituitary. Here they branch to form most of the anterior pituitary capillaries, which in turn drain into the venous system (Figure 7-13).

As a result, almost all the blood supply to the anterior pituitary must first pass through the hypothalamus. Because materials can be exchanged between the blood and surrounding tissue only at the capillary level, the hypothalamic–hypophyseal portal system provides a route where releasing and inhibiting hormones can be picked up at the hypothalamus and delivered immediately and directly to the anterior pituitary at relatively high concentrations, completely bypassing the general circulation.

The axons of the neurosecretory neurons that produce the hypothalamic regulatory hormones terminate on the capillaries at the origin of the portal system. These hypothalamic neurons secrete their (neuro) hormones in the same way as the hypothalamic neurons that produce vasopressin and oxytocin, but into the portal vessels rather than into the general circulation.

Control of Hypothalamic Releasing and Inhibiting Hormones What regulates the secretion of these hypophysiotropic hormones? Like other neurons, the neurons secreting these regulatory hormones receive abundant input of information (both neural and hormonal and both excitatory and inhibi-

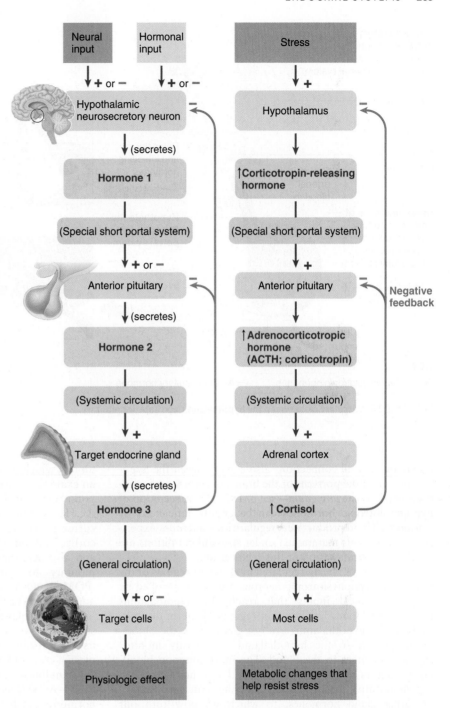

FIGURE 7-12 Hierarchic chain of command in endocrine control. The general pathway involved in the hierarchic chain of command among the hypothalamus, anterior pituitary, and peripheral target endocrine gland is depicted on the left. The pathway on the right leading to cortisol secretion provides a specific example of this endocrine chain of command.

© Cengage Learning, 2013

tory) that they must integrate. Studies are still in progress to unravel the complex neural input from many diverse areas of the brain to the hypophysiotropic secretory neurons. Some of these inputs carry information about a variety of environmental conditions. One example is the marked increase in the secretion of corticotropin-releasing hormone (CRH) in response to stressful situations (Figure 7-12).

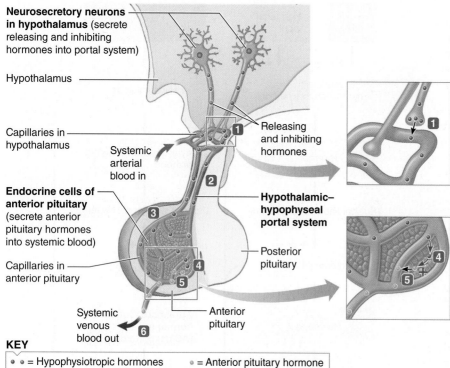

Neurosecretory neurons in hypothalamus (secrete releasing and inhibiting hormones into portal system)

Hypothalamus

Capillaries in hypothalamus

Systemic arterial blood in

Endocrine cells of anterior pituitary (secrete anterior pituitary hormones into systemic blood)

Capillaries in anterior pituitary

Systemic venous blood out

Releasing and inhibiting hormones

Hypothalamic–hypophyseal portal system

Posterior pituitary

Anterior pituitary

KEY

● ● = Hypophysiotropic hormones　　● = Anterior pituitary hormone

1 Hypophysiotropic hormones (releasing hormones and inhibiting hormones) produced by neurosecretory neurons in the hypothalamus enter the hypothalamic capillaries.

2 These hypothalamic capillaries rejoin to form the hypothalamic–hypophyseal portal system, a vascular link to the anterior pituitary.

3 The portal system branches into the capillaries of the anterior pituitary.

4 The hypophysiotropic hormones, which leave the blood across the anterior pituitary capillaries, control the release of anterior pituitary hormones.

5 When stimulated by the appropriate hypothalamic releasing hormone, the anterior pituitary secretes a given hormone into these capillaries.

6 The anterior pituitary capillaries rejoin to form a vein, through which the anterior pituitary hormones leave for ultimate distribution throughout the body by the systemic circulation.

FIGURE 7-13 Vascular link between the hypothalamus and anterior pituitary.
© Cengage Learning, 2013

Numerous neural connections also exist between the hypothalamus and the portions of the brain concerned with emotions (such as the *amygdala*—see p. 187). Thus, secretion of hypophysiotropic hormones can be greatly influenced by emotions. The reproductive irregularities sometimes experienced by animals maintained under stressful conditions are a common manifestation of this relationship.

In addition to being regulated by different regions of the brain, the hypophysiotropic neurons are also controlled by various chemical inputs that reach the hypothalamus through the blood. Unlike other regions of the brain, portions of the hypothalamus are not guarded by the blood–brain barrier, so the hypothalamus can easily monitor chemical changes in the blood. The most common blood-borne factors that influence hypothalamic neurosecretion are the negative-feedback effects of either anterior pituitary or target gland hormones, to which we now turn our attention.

Target gland hormones inhibit hypothalamic and anterior pituitary hormone secretion via negative feedback

In most cases, hypophysiotropic hormones initiate a three-hormone sequence: (1) hypophysiotropic hormone, (2) anterior pituitary tropic hormone, and (3) peripheral target–endocrine gland hormone (Figure 7-12). Typically, in addition to producing its physiologic effects, the target gland hormone also acts to suppress secretion of the tropic hormone that is driving it. This negative feedback is accomplished by the target gland hormone acting either directly on the pituitary itself or on the release of hypothalamic hormones, which in turn regulate anterior pituitary function. As an example, consider the CRH–ACTH–cortisol system. Hypothalamic CRH stimulates the anterior pituitary to secrete ACTH (corticotropin), which in turn stimulates the adrenal cortex to secrete cortisol. The final hormone in the system, cortisol, inhibits the hypothalamus to reduce CRH secretion and also acts directly on the corticotropes in the anterior pituitary to reduce synthesis of pro-opiomelanocortin (POMC) and ACTH secretion. Through this double-barreled approach, cortisol exerts negative-feedback control to stabilize its own plasma concentration. If plasma cortisol levels start to rise above a prescribed set level, cortisol suppresses its own further secretion by these inhibitory actions. If plasma cortisol levels fall below the desired set point, cortisol's inhibitory actions are reduced, so cortisol secretion (CRH–ACTH) increases accordingly. The other target-gland hormones act by similar negative-feedback loops to maintain their plasma levels relatively constant at a set point. For example, feedback of testosterone to the hypothalamus maintains levels of LH, FSH, and sex steroids via GnRH production. Male mammals injected with testosteronelike steroids can become sterile because of this feedback, which causes FSH levels to decline. Birth control pills with female sex steroids work similarly.

The one exception to the negative-feedback relationship just described is the preovulatory positive-feedback effect of estrogen on LH secretion in female mammals. This causes an extremely dramatic rise in LH secretion that triggers ovulation. This relationship is discussed further in Chapter 16.

Diurnal rhythms are superimposed on this type of stabilizing negative-feedback regulation; that is, the set point changes as a function of the time of day. Furthermore, other controlling inputs may break through the negative-feedback control to alter hormone secretion (that is, change the set point level) at times of special need. For example, stress raises the set point for cortisol secretion.

In addition, other hormones outside a particular sequence may also exert important influences, either stimulatory or inhibitory, on the secretion of hypothalamic or anterior pituitary hormones within a given sequence. For example, even though estrogen is not in the direct chain of command for prolactin secretion, this sex steroid notably enhances prolactin secretion by the anterior pituitary. This is but one example of the common phenomenon in the endocrine system that one seemingly unrelated hormone can have pronounced effects on the secretion or actions of another hormone.

check your understanding 7.3

What is the role of melatonin in determining the onset of the breeding season?

What is the importance of biological clocks in an animal, and how do they work?

How is regulation of anterior hormone release different from that of the posterior pituitary?

Under what conditons is MSH released from the pars intermedia?

7.4 Endocrine Control of Growth and Development in Vertebrates

Next we turn to growth and postnatal development, body-wide processes that are coordinated by central glands, and sometimes involve peripheral ones. Both are examples of a regulated change (p. 17) as opposed to a homeostatic process. Development involves *differentiation* of cells into specific functions, whereas growth requires net synthesis of proteins and includes lengthening of the long bones (the bones of the extremities) as well as increases in the size and number of cells in the soft tissues throughout the body. Weight gain alone is not synonymous with growth as it is meant here, because weight gain may occur as a result of retention of excess H_2O or fat without true structural growth of tissues. If the pituitary of a young vertebrate animal is surgically removed, the animal ceases growth; only by the administration of extracts of the pituitary gland can this defect can be partially remedied.

Growth depends on growth hormone but is influenced by other factors as well

Although, as the name implies, **growth hormone (GH)** is absolutely essential for growth, it alone is not wholly responsible for determining the rate and final magnitude of growth in a given animal. The following factors also affect an animal's growth potential:

- *An adequate diet,* including sufficient total protein and ample essential amino acids to accomplish the protein synthesis necessary for growth. Malnourished animals never achieve their full growth potential. The growth-stunting effects of inadequate nutrition are most profound when they occur in infancy. In severe cases, the animal may be locked into irreversible stunting of body growth and neural development. For example, about 70% of the total growth of the human brain occurs in the first two years of life. In contrast, mammals cannot exceed a *genetically determined maximum* by eating more than an adequate diet. The excess food intake produces obesity instead of growth in stature (this is not true of fish, as you will see).

- *Freedom from chronic disease and stressful environmental conditions.* Stunting of growth under adverse circumstances is due in large part to the prolonged stress-induced secretion of glucocorticoid (e.g., cortisol) secretion from the adrenal cortex. Glucocorticoids exert several potent antigrowth effects, such as promoting protein breakdown, inhibiting growth in the long bones in land vertebrates, and blocking the secretion of GH. This applies to aquatic vertebrates also; fish grown in modern aquaculture facilities do not maximize their growth potential, because of the metabolic costs associated with crowding. Even though sickly or stressed animals do not grow well, if the condition is rectified before adult size is achieved, they can rapidly catch up to their normal growth curve through a remarkable spurt in growth.

- *Normal concentrations of growth-influencing hormones.* In growing animals, numerous genes code for particular proteins and protein regulators of growth. In addition to GH, peripheral hormones, including thyroid hormone, insulin, and the sex hormones, play secondary roles in promoting growth.

Rapid-Growth Periods The rate of growth is not continuous in animals, nor are the factors responsible for promoting growth the same throughout the growth period. In mammals, fetal growth is promoted largely by certain hormones from the placenta (see p. 804), with the size at birth being determined principally by genetic and environmental factors. GH and other nonplacental hormonal factors begin to play an important role in regulating growth after birth. Genetic and nutritional factors also strongly affect growth during this period. In addition, day length affects growth in juvenile animals because day length influences energy intake and energy expenditure. For example, exposure of an animal to short day lengths, as occurs seasonally in northern and southern latitudes, results in long nocturnal periods of food deprivation. During the comparatively short light period, large amounts of food are consumed, which makes available to the tissues increased concentrations of amino acids. In a growing animal, these dietary amino acids are particularly important for protein synthesis (repair, growth, and maintenance) as well as storage for use in subsequent periods of food deprivation. Normally, when the supply of amino acids exceeds the protein synthesis capacity of the animal, the excess amino acids are oxidized and thus lost to the animal. However, during periods of food unavailability such as overnight fasting, protein synthesis rates drop significantly, which may severely limit growth rates.

Unlike birds and mammals, which reach a maximum body size and then cease growing, most cultured fish species continue to grow if food supplies remain available. However, species such as trout show periodic growth spurts, which depend on season and stage of development. For example, growth is reduced during the winter months as well as during the reproductive period. This is due in part to the lowered energy requirements of fishes because they are ectotherms (Chapter 15). Superimposed on the seasonal variations in growth rate in many fishes is a 14-day semilunar cycle of growth rate, which correlates with a rhythm in food consumption. Slow growth rates are associated with either the full or new moon, which are associated with reduced food requirements.

In contrast, most mammals display two periods of rapid growth—a postnatal growth spurt during their first years of life and a pubertal growth spurt during adolescence. Much of the growth observed in young animals is due to the increase in body protein, with skeletal muscle showing the greatest percentage change in body composition. Growth is achieved when protein synthesis rates are higher than protein breakdown rates. The productive efficiency of growing farm animals is determined by the proportion of nutrients partitioned to fat relative to muscle, and by the rate at which growth occurs.

Before puberty there is little sexual difference in height or weight in most mammals. During puberty, a marked acceleration in linear growth takes place because of the lengthening of the long bones, with males outgrowing females in most species (with some notable exceptions such as spotted hyenas and blue whales). The mechanisms responsible for the pubertal growth spurt are not clearly understood. Apparently both genetic and hormonal factors are involved. During rapid mammalian growth in puberty, there is a marked increase in the magnitude and frequency of GH release, which induces **insulin-like growth factor (IGF-I)** expression in the liver and other tissues, including the skeleton. In turn, systemic and local IGF-I contributes to the acceleration of growth during this time. Furthermore, **androgens** ("male" sex hormones), whose secretion increases dramatically at puberty, also contribute to the pubertal growth spurt by promoting protein synthesis and bone growth. The potent androgen from the male testes, testosterone, is of greatest importance in promoting a sharp increase in body size in adolescent males in most species, as well as development of male secondary sexual characteristics. In contrast, the less potent adrenal androgens from the adrenal gland, which also show a sizable increase in secretion during adolescence, are most likely important in the female pubertal growth spurt. Although estrogen secretion by the ovaries also begins during puberty, inducing development of female secondary sex characteristics, it is unclear what role this "female" sex hormone may play in the pubertal growth spurt in females. There is no doubt that testosterone and estrogen both ultimately act on bone to halt its further growth so that full adult height is attained by the end of adolescence.

Growth hormone exerts metabolic effects not related to growth

In addition to promoting growth, GH has important metabolic effects and enhances the immune system. In fishes GH has partially retained ancestral osmoregulatory features independent of its growth-promoting functions. We briefly describe GH's metabolic actions before turning our attention to its growth-promoting actions.

Metabolic Actions To exert its metabolic effects, GH binds directly with receptors on its target organs, namely adipose tissue, skeletal muscles, and liver. GH increases fatty acid levels in the blood by enhancing the breakdown of triglyceride fat stored in adipose tissue, and it increases blood glucose levels by decreasing glucose uptake by muscles and increasing glucose output by the liver. Muscles use the mobilized fatty acids instead of glucose as a metabolic fuel. Thus, the overall metabolic effect of GH is to mobilize fat stores as a major energy source while conserving glucose for glucose-dependent tissues such as the brain. The brain can use only glucose as its metabolic fuel, yet nervous tissue cannot store much glycogen (stored glucose polymer). This metabolic pattern is suitable for maintaining the body during prolonged fasting or other situations when the body's energy needs exceed available glucose stores. GH also stimulates amino acid uptake and protein synthesis, but, in most tissues, in does so *indirectly* to accomplish these growth-promoting actions. In some cases, however, it does so directly, such as protein synthesis in the liver, diaphragm, and heart. Before examining the means by which GH indirectly promotes growth, let's summarize its metabolic effects: It increases blood fatty acids, increases blood glucose and spares glucose for the brain, and stimulates protein synthesis (decreasing blood amino acids in the process).

Indirect effects of growth hormone are exerted through insulin-like growth factors

GH's indirect growth-promoting actions are mediated by IGFs (as we noted earlier), which act on the target cells to cause growth of both soft tissues and bones. IGFs are produced in many tissues and have endocrine, paracrine, and autocrine actions. Originally called **somatomedins**, these peptide mediators are now preferentially called insulin-like growth factors because they are structurally and functionally similar to insulin. Like insulin, IGFs exert their effects largely by binding with receptor-enzymes that activate designated effector proteins within the target cell by phosphorylating the tyrosines in the protein (the tyrosine kinase pathway; see p. 96). There are two IGFs—**IGF-I** and **IGF-II**. Fish are the first vertebrate group that has a complete system of hormones and receptors for the insulin, IGF-I and IGF-II molecules.

IGF-I Synthesis of IGF-I in mammals is stimulated by GH and mediates most of this hormone's growth-promoting actions. The major source of circulating IGF-I is the liver, which (acting as a peripheral endocrine gland) releases this peptide product into the blood in response to GH stimulation. However, IGF-I is produced in many if not most tissues, although it is not released into the blood from these other sites. Scientists have proposed that IGF-I produced locally in target tissues may act through paracrine means (see p. 91) for at least some of the growth-hormone-induced effects. Local production of IGF-I in target tissues may possibly be more important than delivery of blood-borne IGF-I during this time.

Production of IGF-I is controlled by a number of factors other than GH in humans, including nutritional status, age, and tissue-specific factors as follows:

- IGF-I production depends on adequate nutrition. Inadequate food intake results in reduced IGF-I production, apparently brought about by decreased sensitivity to GH of the tissues that produce IGF-I. As a result, changes in circulating IGF-I levels do not always coincide with changes in GH secretion. For example, fasting decreases IGF-I levels even though it increases GH secretion.

- Age-related factors also influence IGF-I production. A dramatic increase in circulating IGF-I levels accompanies the moderate increase in GH at puberty, which may, of course, be an important factor in the pubertal growth spurt.

- Finally, various tissue-specific stimulatory factors can increase IGF-I production in particular tissues. To illustrate, the gonadotropins and sex hormones stimulate IGF-I production within reproductive organs such as the testes in males and the ovaries and uterus in females.

Thus, control of IGF-I production is complex and subject to a variety of systemic and local factors.

IGF-II In contrast to IGF-I, **IGF-II** production does not depend on GH. IGF-II is known to be important during fetal development. Unlike IGF-I, IGF-II does not increase during the pubertal growth spurt. IGF-II continues to be produced during adulthood and may be involved in muscle growth. Recently, a study found that a simple mutation that increased IGF-II gene activity in pig muscle resulted in a 3 to 4% growth in meat (muscle) mass.

From the preceding discussion of the factors involved in controlling growth and development, you can see that many gaps still exist in the state of knowledge in this important area.

Growth hormone/IGF-I promote growth of soft tissues by stimulating hyperplasia and hypertrophy

GH stimulates growth of both soft tissues and the skeleton. GH (acting through IGF-I) is thus *trophic* (p. 272), promoting growth of soft tissues by (1) increasing the number of cells (**hyperplasia**) and (2) increasing the size of cells (**hypertrophy**). GH increases the number of cells by stimulating cell division and by preventing apoptosis (programmed cell death). GH increases the size of cells by favoring synthesis of proteins, the main structural component of cells. GH stimulates almost all aspects of protein synthesis while it simultaneously inhibits protein degradation. It promotes the uptake of amino acids (the raw materials for protein synthesis) by cells, decreasing blood amino acid levels in the process. Furthermore, it stimulates the cell machinery responsible for accomplishing protein synthesis according to the cell's genetic code. It increases DNA and RNA synthesis and increases incorporation of amino acids into the new proteins at the ribosomal level.

Growth-Promoting Actions on Bones Growth of the long bones, resulting in increased height, is the most dramatic effect of GH. **Bone** is a living tissue. Being a form of connective tissue, it consists of cells and an extracellular organic matrix known as **osteoid** that is produced by the cells. The bone cells that produce the organic matrix are known as **osteoblasts** ("bone-formers"). Osteoid is composed of col-

lagen fibers (see p. 65) in a mucopolysaccharide-rich semisolid gel. This matrix has a rubbery consistency and is responsible for the tensile strength of bone (the resilience of bone to breakage when tension is applied). Bone is hardened by impregnation with **hydroxyapatite crystals,** which consist primarily of precipitated $Ca_3(PO_4)_2$ (calcium phosphate) salts. These inorganic crystals provide the bone with compressional strength (the ability of bone to hold its shape when squeezed or compressed). Bones have structural strength approaching that of reinforced concrete, yet they are not brittle and are much lighter in weight, as a result of the structural blending of an organic scaffolding hardened by inorganic crystals. **Cartilage** is similar to bone, except that living cartilage is not calcified.

A long bone basically consists of a fairly uniform cylindrical shaft, the **diaphysis,** with a flared articulating knob at either end, an **epiphysis.** In a growing bone, the diaphysis is separated at each end from the epiphysis by a layer of cartilage known as the **epiphyseal plate** (Figure 7-14a). The central cavity of the bone is filled with bone marrow, which is the site of blood cell production (see p. 392).

Growth in *thickness* of bone is achieved by the addition of new bone on top of the already existing bone on the outer surface. This growth occurs through activity of **osteoblasts** within the **periosteum,** a connective tissue sheath that covers the outer bone surface. As new bone is being deposited by osteoblast activity on the external surface, other cells within the bone, the **osteoclasts** ("bone breakers"), dissolve the bony tissue on the inner surface adjacent to the marrow cavity (see p. 325 for more on these cells). In this way, the marrow cavity is enlarged to keep pace with the increase in the circumference of the bone shaft.

Growth in *length* of long bones is accomplished by a different mechanism from growth in thickness. Bones grow in length as a result of proliferation of the cartilage cells, or **chondrocytes,** in the epiphyseal plates (Figure 7-14b). During growth, cartilage cells on the outer edge of the plate next to the epiphysis divide and multiply, temporarily widening the epiphyseal plate. As new chondrocytes are formed on the epiphyseal border, the older cartilage cells toward the diaphyseal border are enlarging. This combination of proliferation of new cartilage cells and hypertrophy of maturing chondrocytes temporarily widens the epiphyseal plate. This thickening of the intervening cartilaginous plate pushes the bony epiphysis farther away from the diaphysis. Soon the matrix surrounding the oldest hypertrophied cartilage becomes calcified. Because cartilage lacks its own capillary network, the survival of cartilage cells depends on diffusion of nutrients and O_2 through the matrix, a process prevented by the deposition of calcium salts. As a result, the old nutrient-deprived cartilage cells on the diaphyseal border die. As osteoclasts clear away dead chondrocytes and the calcified matrix that imprisoned them, the area is invaded by osteoblasts, which swarm upward from the diaphysis, trailing their capillary supply with them. These new tenants lay down bone around the persisting remnants of disintegrating cartilage until bone entirely replaces the inner region of cartilage on the diaphyseal side of the plate. When this **ossification** ("bone formation") is complete, the bone on the diaphyseal side has lengthened and the epiphyseal plate has returned to its original thickness. The cartilage that bone has replaced on the diaphyseal end of the plate is as thick as the new cartilage on the epiphyseal end of the plate. Thus, bone

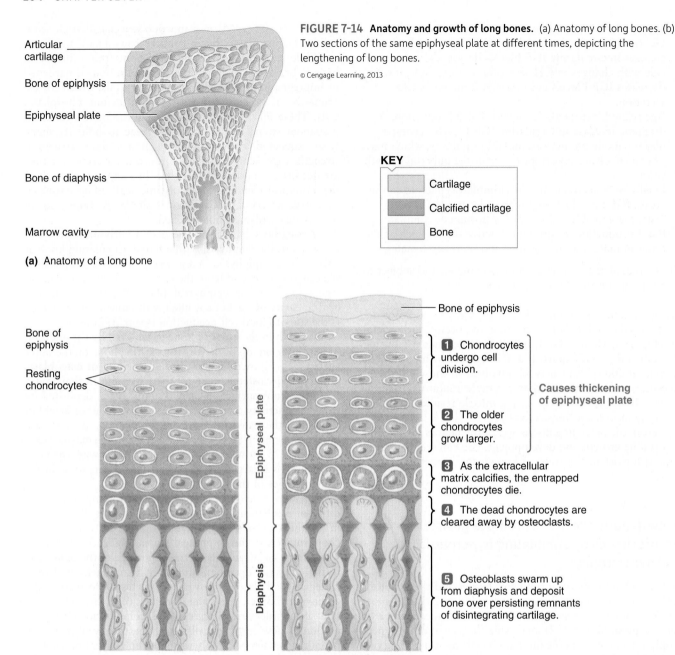

FIGURE 7-14 Anatomy and growth of long bones. (a) Anatomy of long bones. (b) Two sections of the same epiphyseal plate at different times, depicting the lengthening of long bones.

© Cengage Learning, 2013

Articular cartilage

Bone of epiphysis

Epiphyseal plate

Bone of diaphysis

Marrow cavity

(a) Anatomy of a long bone

KEY

Cartilage

Calcified cartilage

Bone

Bone of epiphysis

Bone of epiphysis

Resting chondrocytes

Epiphyseal plate

Diaphysis

1 Chondrocytes undergo cell division.

2 The older chondrocytes grow larger.

Causes thickening of epiphyseal plate

3 As the extracellular matrix calcifies, the entrapped chondrocytes die.

4 The dead chondrocytes are cleared away by osteoclasts.

5 Osteoblasts swarm up from diaphysis and deposit bone over persisting remnants of disintegrating cartilage.

(b) Two sections of the same epiphyseal plate at different times, depicting the lengthening of long bones

growth is made possible by the growth and death of cartilage, which acts like a "spacer" to push the epiphysis farther out while it provides a framework for future bone formation on the end of the diaphysis.

GH promotes growth of bone in both thickness and length. It stimulates the proliferation of epiphyseal cartilage, thereby making space for more bone formation, and also stimulates osteoblast activity. GH can promote lengthening of long bones as long as the epiphyseal plate remains cartilaginous, or is "open." At the end of adolescence, under the influence of the sex hormones, these plates completely ossify, or "close," so that the bones can grow no further in length despite the presence of GH. Thus, after the plates are closed, the individual animal does not grow any taller.

Growth hormone secretion is regulated by two hypophysiotropic hormones and influenced by a variety of other factors

The control of GH secretion is complex, with two hypothalamic hypophysiotropic hormones playing a key role.

Growth Hormone–Releasing Hormone and Growth Hormone–Inhibiting Hormone Two antagonistic regulatory hormones from the hypothalamus are involved in controlling GH secretion: growth hormone–releasing hormone (GHRH), which is stimulatory, and growth hormone–inhibiting hormone (GHIH or somatostatin), which is inhibitory (Figure 7-15). Both GHRH and somatostatin act on the anterior pituitary somatotropes by binding with G-protein coupled receptors linked to the cAMP second-messenger pathway, with GHRH increasing cAMP and somatostatin decreasing cAMP.

As with the other hypothalamus–anterior pituitary axes, negative-feedback loops participate in regulating GH secretion. Complicating the feedback loops for the hypothalamus–pituitary–liver axis is the fact that GH secretion is directly regulated by both stimulatory and inhibitory factors. Therefore negative-feedback loops involve both inhibition of stimulatory factors and stimulation of inhibitory factors. GH stimulates IGF-I secretion by the liver, and IGF-I in turn is the primary inhibitor of GH secretion by the anterior pituitary. IGF-I inhibits the somatotropes in the pituitary directly and further decreases GH secretion by inhibiting GHRH-secreting cells and stimulating the somatostatin-secreting cells in the hypothalamus, thus decreasing hypothalamic stimulation of the somatotropes. Furthermore, GH itself inhibits hypothalamic GHRH secretion and stimulates somatostatin release.

Other Factors That Influence GH Secretion A number of factors influence GH secretion by acting on the hypothalamus. GH secretion displays a well-characterized circadian rhythm. In a diurnal mammal, through most of the day GH levels tend to be low and fairly constant. About one hour after the onset of deep sleep, however, GH secretion markedly increases, then rapidly drops over the next several hours.

Superimposed on this circadian fluctuation in GH secretion are further bursts in secretion, in response to exercise, stress, and **hypoglycemia** (low blood glucose), the major stimuli for increased secretion. The benefit of increased GH secretion during these situations when energy demands outstrip the body's glucose reserves is presumably to conserve glucose for the brain and to provide fatty acids as an alternative energy source for muscle. Because GH uses up fat stores and promotes synthesis of body proteins, it encourages a change in body composition away from adipose deposition toward increased muscle protein. Accordingly, the increased GH secretion that accompanies exercise may at least in part mediate exercise's effects in reducing percentage of body fat while increasing lean body mass.

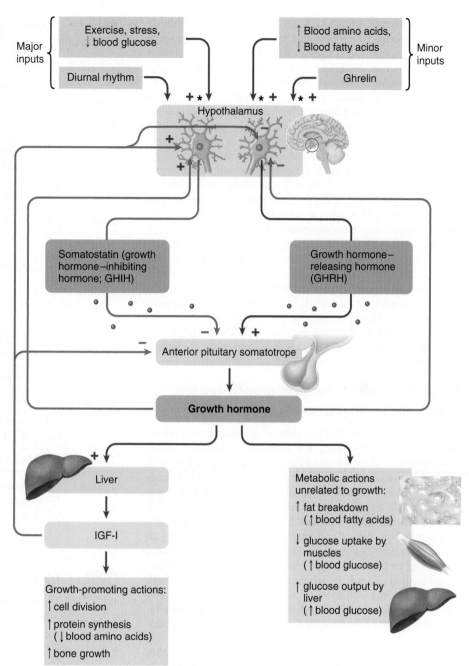

*These factors all increase growth hormone secretion, but it is unclear whether they do so by stimulating GHRH or inhibiting GHIH somatostatin, or both.

FIGURE 7-15 Control of growth hormone secretion.

© Cengage Learning, 2013

A rise in blood amino acids after a high-protein meal also enhances GH secretion. In turn, GH promotes use of these amino acids for protein synthesis. GH release is also stimulated by a decline in level of blood fatty acids. Because of the fat-mobilizing actions of GH, such regulation helps maintain fairly constant blood fatty-acid levels. Finally, *ghrelin*, a potent appetite stimulator released from the stomach, also stimulates GH secretion (see p. 712). This "hunger hormone" may play a role in coordinating growth with nutrient acquisition.

Growth Hormone Administration GH is readily available for administering to humans and to domestic and laboratory animals. In humans, it is used to treat dwarfism (and abused by some athletes). Administering GH to swine dramatically increases protein deposition and muscle mass in addition to reducing adipose tissue growth. However, deposition of greater lean tissue requires an increased amount of amino acids in the diet of treated animals. Uptake of glucose into target tissues is also reduced, which results in increased concentrations of both plasma glucose and insulin. In swine, concentrations of GH peak between four and seven hours after subcutaneous injection and are associated with increased plasma IGF-I concentrations as well. However, in poultry, unless GH is administered in pulses, there is no effect of exogenously administered GH on growth.

Administering GH to dairy cows results in an exquisite coordination of metabolism to meet the nutrient needs associated with increased synthesis of milk components (Table 7-2). Glucose production and oxidation are modified to supply the increased glucose needs of the mammary glands with such precision that plasma glucose concentrations remain unaltered. When feed intake does not meet the energy needs required for milk production, treated animals must capitalize on alternate energy sources such as fatty acids.

Prolactin has a wide range of growth-related effects, including lactogenesis

Prolactin (p. 288) is another central hormone involved in growth and development, for example, mammary gland development and lactation (milk production) to aid growth of newborn mammals. The first suggestion that a hormone of the anterior pituitary might affect lactation in mammals arose from the experiments of Stricker and Grueter in 1928, who found that lactation could be elicited in the pseudopregnant rabbit, before and after spaying, by injections of an anterior pituitary extract. Prolactin was eventually isolated and found to be essential for initiating and maintaining lactation (only in mammals, whose very name is based on this). **Lactogenesis,** the initiation of lactation, involves mammary cell differentiation and increased activity of the enzymes responsible for the production of milk components (see Chapter 16).

As we noted earlier (p. 288), prolactin has remarkably diverse roles. In addition to influencing parental behavior, some roles bear no relationship to reproduction. Recent studies suggest that prolactin in mammals may enhance the immune system and support the development of new blood vessels at the tissue level in both sexes, another growth function. But what about other vertebrates that lack milk? Lactogenic ("milk generating") effects have been found (by administering prolactin to laboratory mammals) in the pituitary extracts of mammals, fish, amphibians, reptiles, and birds, showing that all vertebrates have this hormone. Obviously, it is not involved in lactogenesis in nonmammals! For example, Riddle and coworkers in 1931 found that administering pituitary extract to pigeons resulted in development of the crop sac (see p. 668). Subsequent work demonstrated that this response in birds is caused by the same hormone that induces lactation in mammals. Prolactin in birds is also important for brood patch formation (see the accompanying box, *A Closer Look at Adaptation: Brood Patch Development: Some Have It, Others Don't!*), as well as migratory

TABLE 7-2 Effect of Bovine Somatotropin Supplement of Lactating Cows on Specific Tissues and Physiological Processes

Tissue		Process Affected during First Few Days and Weeks of Supplement
Mammary	↑	Synthesis of milk with normal composition
	↑	Uptake of all nutrients used for milk synthesis
	↑	Activity per secretory cell
	↓	Loss of secretory cells (i.e., enhanced persistence)
	↑	Blood flow consistent with increase in milk yield
Liver	↑	Basal rates of gluconeogenesis
	↓	Ability of insulin to inhibit gluconeogenesis
	φ	Glucagon effects on gluconeogenesis and/or glycogenolysis
Adipose	↓	Basal lipogenesis if in positive energy balance
	↑	Basal lipolysis if in negative energy balance
	↓	Ability of insulin to stimulate lipogenesis
	↑	Ability of insulin to inhibit lipolysis
	↑	Ability of catecholamines to stimulate lipolysis
Muscle	↓	Uptake of glucose
Pancreas	φ	Basal or glucose-stimulated secretion of insulin
	φ	Basal or insulin/glucose-stimulated secretion of glucagon
Kidney	↑	Production of 1,25 vitamin D_3
Intestine	↑	Absorption of Ca, P, and other minerals required for milk
	↑	Ability of 1,25 vitamin D_3 to stimulate calcium-binding protein
	↑	Calcium-binding protein
Whole Body	↓	Oxidation of glucose
	↑	Fatty-acid oxidation if in negative energy balance
	φ	Insulin and glucagon clearance rates
	φ	Energy expenditure for maintenance
	↑	Energy expenditure consistent with increase in milk yield (i.e., heat per unit of milk not changed)
	↑	Cardiac output consistent with increases in milk yield
	↑	Productive efficiency (milk per unit of energy intake)

Note: Changes (↑ = increased, ↓ = decreased, φ = no change).

© Cengage Learning, 2013

behavior in birds and mammals. Prolactin's effects are explored in more detail in Chapter 16.

Further evidence of the complex role of prolactin is suggested by the finding that the gene-encoding prolactin receptor is widely expressed in avian peripheral tissues, including kidneys, skin, brood patch, gut, gonads, adrenal glands, liver, adipose tissue, and spleen, whereas in the CNS the

The vast majority of birds incubate their eggs by transfer of heat between parts of their body, usually but not always their ventral surface, and the clutch of eggs. That region of the incubating bird in contact with the egg is commonly termed the incubation or **brood patch.** Hormonal regulation of this patch of skin differs between sexes as well as species of birds.

Although the evolution of this structure is open to conjecture, one possibility is that the appearance of a brood patch was a definitive step in the evolution of birds from reptiles, and it has been thus suggested to share a common ancestry with the mammalian mammary gland. The nature, formation, and structure of this patch varies enormously between species, reflecting the diverse mechanisms and methods adopted by different species in the incubation of eggs. In some birds such as the domestic pigeon and ring dove, there is a form of pocket or apterium on the ventral surface of both sexes, which is devoid of feathers throughout adult life. In other species, patch formation occurs periodically and is closely related to the period of egg laying and incubation. Furthermore, although both sexes may have the potential to develop a brood patch, it is a general observation that the patch only develops in the sex actually involved in incubation. There is also a relationship between the gender that incubates, and the form of steroid that is most effective, in combination with prolactin, to produce a full brood patch. In those species in which the female alone incu-

The brood patch of a female bluebird (*Sialia sialis*). Its function is to transfer heat to the developing chicks. Brood patch development is hormonally controlled, primarily by prolactin secretion from the adenohypophysis. In some species, only the female develops a brood patch, while in others, both the male and female develop patches.

Courtesy of Dr. Robert Whitmore

bates, it is estrogen, whereas in those species in which the male incubates, it is androgen. Consistent with this, species that might be considered as intermediate, such as the California quail and laughing gull, are sensitive, in terms of brood patch development, to both estrogen plus prolactin and androgen plus prolactin.

In most species, brood patch development involves a number of morphological changes to the ventral body area: defeathering (mainly down feathers); a significant increase in the folding of the skin, together with an infiltration of leukocytes (edema formation); a thickening of the cornified

layer of the ventral skin surface (epidermal hyperplasia); and an increase in both size and number of local blood vessels (vascularization). The structure is surprisingly sophisticated in that the musculature of arterioles supplying blood to the patch also increases and thus can shut down blood flow to this region when the parent is off the nest. In some species, such as the house sparrow (*Passer domesticus*), there is also an increase in the amount of underlying defatted tissue.

Taken together, these dramatic changes are superbly evolved to facilitate a closer contact and more effective heat transfer between the parenting bird and the surface of the eggs, while minimizing any possible damage caused to the skin by the sustained period of contact time. The function of the brood patch during the incubatory period of the breeding cycle is complex and multifactorial. Much still remains to be understood, and one early suggestion for a possible purpose of the patch, that the bird finds relief from the peripheral irritation of the developing brood patch by sitting on eggs, remains plausible.

Not all birds have such patches. For example, the northern gannet (see chapter opening picture) does not develop one and instead incubates its egg by increasing blood flow (and heat exchange) through the webbing in its feet. Because these birds dive into the cold waters of the Atlantic Ocean, this adaptation reduces potential heat loss to the environment.

gene is highly expressed in the anterior pituitary gland and in the hypothalamus.

In the previous sections, we discussed the major central endocrine glands of vertebrates, and noting the peripheral endocrine role of the liver. We now turn our attention to other peripheral glands.

check your understanding 7.4

How do the direct functions of GH differ from the indirect?

How does GH usage change body composition?

What endocrine factors influence brood patch development?

7.5 Thyroid Gland

The **thyroid gland** is a major regulator of metabolism and other metabolic and developmental processes. Thyroid hormones are produced by endocrine cells arranged in a very characteristic anatomical structure called a *thyroid follicle*. In mammals the thyroid gland consists of two lobes of thyroid follicles joined in the middle by a narrow portion of the gland, giving it a bowtie shape (Figure 7-16a). The gland is even located in the appropriate place for a bow tie in a human, lying over the trachea just below the larynx. In contrast, the thyroid gland of most nonmammalian vertebrates consists of discrete clusters of thyroid follicles, sometimes central, sometimes paired, lying at varying distances lateral to the esophagus; in agnathans and most teleosts the thyroid

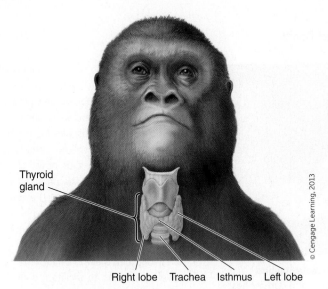

Thyroid gland

Right lobe   Trachea   Isthmus   Left lobe

(a) Gross anatomy of thyroid gland

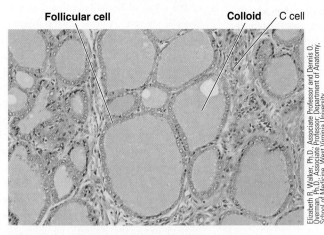

Follicular cell   Colloid   C cell

(b) Light-microscopic appearance of thyroid gland

FIGURE 7-16 Anatomy of the thyroid gland. (a) Gross anatomy of the thyroid gland, anterior view. The thyroid gland lies over the trachea just below the larynx and consists of two lobes connected by a thin strip called the *isthmus.* (b) Light-microscope appearance of the thyroid gland. The thyroid gland consists primarily of colloid-filled spheres enclosed by a single layer of follicular cells.

consists of diffuse clusters of follicles distributed throughout the ventral region of the head. In spite of the varied gross morphology of the thyroid gland, the follicular structure is highly conserved and essential for the unique method of extracellular hormone synthesis utilized by this gland.

The major thyroid hormone secretory cells are organized into colloid-filled follicles

The major thyroid secretory cells, known as **follicular cells,** are arranged into fluid-filled spheres, each of which forms a functional unit called a **follicle.** On a microscopic section (Figure 7-16b), the follicles appear as rings of follicular cells enclosing an inner lumen filled with **colloid,** a substance that serves as an extracellular storage site for thyroid hormones.

The chief constituent of the colloid is a large, complex glycoprotein known as **thyroglobulin (Tg),** within which are incorporated the thyroid hormones in their various stages of synthesis. Extracellular hormone storage is unique to the thyroid gland and probably exists because the thyroid evolved from an exocrine-secreting digestive organ called the endostyle. As we'll see, the colloid indirectly stores the scarce trace element iodine, which is an essential component of the thyroid gland, and prevents diffusion of the steroid-like thyroid hormone out of the colloid.

Thyroid hormone is synthesized and stored on the thyroglobulin molecule

The follicular cells of vertebrates produce two iodine-containing hormones derived from the amino acid tyrosine: **tetraiodothyronine (T_4, or thyroxine)** and **tri-iodothyronine (T_3).** The prefixes *tetra* and *tri* and the subscripts 4 and 3 denote the number of iodine atoms incorporated into each of these hormones. These two hormones, together referred to as **thyroid hormone,** are important regulators of development and overall basal metabolic rate. The basic ingredients for thyroid hormone synthesis are tyrosine and iodine, both of which must be taken up from the blood by the follicular cells. Tyrosine, an amino acid, is synthesized in sufficient amounts by the body, so it is not an essential dietary requirement. By contrast, the iodine needed for thyroid hormone synthesis must be obtained from dietary intake. Dietary iodine (I) is reduced to iodide (I⁻) prior to absorption by the small intestine. The synthesis, storage, and secretion of thyroid hormone involve the following steps:

1. All steps of thyroid hormone synthesis take place on the thyroglobulin molecules within the colloid. Thyroglobulin itself is produced by the endoplasmic reticulum/Golgi complex of the thyroid follicular cells. The amino acid tyrosine becomes incorporated in the much larger thyroglobulin molecules as the latter are being produced. Once produced, tyrosine-containing thyroglobulin is exported from the follicular cells into the colloid by exocytosis (step 1 in Figure 7-17).

2. The thyroid captures I⁻ from the blood and transfers it into the colloid by means of an *iodide pump,* energy-requiring transport proteins located in the outer membranes of the follicular cells (step 2). The iodide pump is a symporter driven by the Na⁺ concentration gradient established by the Na⁺/K⁺ pump at the basolateral membrane (the outer membrane of the follicular cell in contact with the interstitial fluid). The iodide pump transports Na⁺ into the follicular cell down its concentration gradient and I⁻ into the cell against its concentration gradient. Almost all the I⁻ in the body is transported against its concentration gradient to become trapped in the thyroid for the purpose of thyroid hormone synthesis. Iodine serves no other known purpose in the vertebrate body.

3. Within the membrane–colloid interface, iodide is oxidized to "active" iodide by a membrane-bound enzyme, **thyroperoxidase (TPO),** located at the luminal membrane, the membrane of the follicular cell in contact with the colloid (step 3). This active iodide exits through a channel in the luminal membrane to enter the colloid (step 4). Attachment of one iodine to tyrosine yields

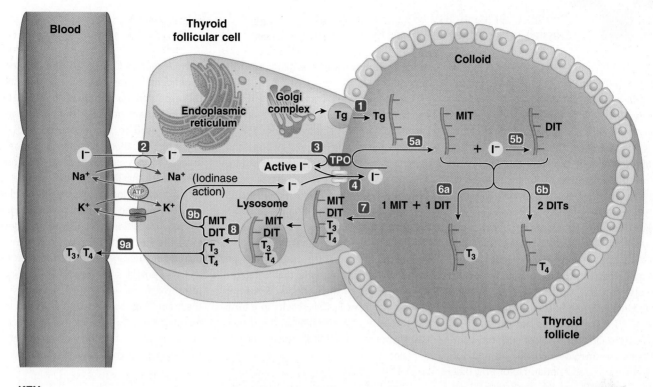

KEY

= Primary active transport	**Tg** = Thyroglobulin	**DIT** = Di-iodotyrosine
	I⁻ = Iodide	**T₃** = Tri-iodothyronine
= Secondary active transport (symporter)	**TPO** = Thyroperoxidase	**T₄** = Tetraiodothyronine (thyroxine)
	MIT = Monoiodotyrosine	

1 Tyrosine-containing Tg produced within the thyroid follicular cells by the endoplasmic reticulum/Golgi complex is transported by exocytosis into the colloid.

2 Iodide is carried by secondary active transport from the blood into the colloid by symporters in the basolateral membrane of the follicular cells.

3 In the follicular cell, the iodide is oxidized to active form by TPO at the luminal membrane.

4 The active iodide exits the cell through a luminal channel to enter the colloid.

5a Catalyzed by TPO, attachment of one iodide to tyrosine within the Tg molecule yields MIT.

5b Attachment of two iodides to tyrosine yields DIT.

6a Coupling of one MIT and one DIT yields T_3.

6b Coupling of two DITs yields T_4.

7 On appropriate stimulation, the thyroid follicular cells engulf a portion of Tg-containing colloid by phagocytosis.

8 Lysosomes attack the engulfed vesicle and split the iodinated products from Tg.

9a T_3 and T_4 diffuse into the blood (secretion).

9b MIT and DIT are deiodinated, and the freed iodide is recycled for synthesizing more hormone.

FIGURE 7-17 Synthesis, storage, and secretion of thyroid hormone. Note that the organelles are not drawn to scale. The endoplasmic reticulum/Golgi complex are proportionally too small.

monoiodotyrosine (MIT) (step 5a). Attachment of two iodines to tyrosine yields di-iodotyrosine (DIT) (step 5b).
4. Next, a coupling process occurs between the iodinated tyrosine molecules to form the thyroid hormones. Coupling of two DITs (each bearing two iodine atoms) yields tetraiodothyronine (T_4 or thyroxine), the four-iodine form of thyroid hormone (step 6b). Coupling of one MIT (with one iodine) and one DIT (with two iodines) yields tri-iodothyronine or T_3 (with three iodines) (step 6a). Coupling does not occur between two MIT molecules. All these products remain attached to thyro-

globulin by peptide bonds. Thyroid hormones remain stored in this form in the colloid until they are split off and secreted. Sufficient thyroid hormone is normally stored to supply the body's needs for several months.

Because these reactions occur within the thyroglobulin molecule, all the products remain attached to this large protein. Immobilized in this fashion they, contrary to their steroid-hormone-like behavior, cannot freely diffuse into the circulation after synthesis. Thyroid hormones remain stored in this form in the colloid until they are split off and

The anterior pituitary consists of the pars distalis (PD) (main body of the pituitary gland) and the pars tuberalis (PT), which forms the stalk of the pituitary gland immediately adjacent to the base of the brain. As we have seen, the PD receives signals from the hypothalamus via the portal system to regulate the multiple cell types in this tissue. The PT forms a specialized tissue that expresses melatonin receptors and thus responds to the daily light–dark cycle that regulates seasonal changes in pituitary function. In turn, *retrograde signaling* from the PT directs the hypothalamus to govern neuroendocrine control of reproduction, food intake, and other seasonal characteristics. PT thyrotrophs do not express TRH receptors and are not under inhibitory control by the thyroid gland, and also do not produce the transcription factor *PIT1* that drives normal PD cell differentiation. PIT1 in PD cells also regulates the genes for TSH and prolactin.

The primary cell type in the PT is a thyrotroph-like cell that is distinct from the thyrotrophs in the PD that secretes TSH to regulate the thyroid glands. In birds and mammals, the PT secretes TSH only locally, which then diffuses into the basal hypothalamus and binds to TSH receptors on the *tanycytes* (specialized cells in the ependymal layer) that line the third ventricle. In this fashion, these nonneural tanycytes can be induced to express the outer-ring (type 2) *deiodinase,* the enzyme that catalyzes the conversion of thyroxine to its biologically active form T_3. This results in a local increase in T_3 concentration in the medial basal hypothalamus (MBH), which activates a major cascade of new secretory responses that drive reproduction and other centrally related functions, which have long been recognized to be thyroid-hormone dependent. In seasonally breeding animals, the experimental introduction of TSH locally into the ventricle of the brain in animals exposed to winter photoperiod induces all the biological responses normally associated with summer photoperiod. Thus, it is the PT that initiates the response to spring photoperiods, relays the information to the brain, and activates the associated biology. In addition, it is believed that these local changes in thyroid hormone are required for neurogenesis within the tissue and this provides the basis for the long-term transformation in the cell biology of the brain. Thus, over a period of weeks and months, the animal transforms its physiology from winter to summer, readjusting a functional set point for homeostasis.

---

secreted. As a result of this unique extracellular synthetic process, researchers estimate that sufficient thyroid hormone to supply a mammal's needs for several months is normally stored in the colloid. Should a predator consume the thyroid gland of its prey, it would receive a physiologically significant dose upon consumption.

To secrete thyroid hormone, the follicular cells phagocytize thyroglobulin-laden colloid

The release of the thyroid hormones into the systemic circulation is a rather complex process, for two reasons. First, before their release T_4 and T_3 are still bound within the thyroglobulin molecule. Second, these hormones are stored in the follicular lumen, so they must be transported completely across the follicular cells to reach capillaries that course through the interstitial spaces between the follicles.

The process of thyroid hormone secretion essentially involves the follicular cells "biting off" a piece of colloid, breaking the thyroglobulin molecule down into its component parts, and "spitting out" the freed T_4 and T_3 into the blood. On appropriate stimulation for thyroid hormone secretion, the follicular cells internalize a portion of the thyroglobulin–hormone complex by phagocytizing a piece of colloid (step 7 of Figure 7-17). Within the cells, the membrane-enclosed droplets of colloid coalesce with lysosomes, whose enzymes split off the biologically active thyroid hormones, T_4 and T_3, as well as the inactive iodotyrosines, MIT and DIT (step 8). The thyroid hormones, being lipophilic, pass through the outer membranes of the follicular cells and into the blood (step 9a).

The MIT and DIT have no known endocrine function. The follicular cells contain an enzyme that swiftly removes the iodine from MIT and DIT, allowing the freed iodine to be recycled for synthesis of more hormone (step 9b). Some intact thyroglobulin is also released into the general circulation, although this protein has no known physiological function.

For the most part, both T_4 and T_3 are transported bound to specific plasma proteins

Once released into the blood, the highly lipophilic thyroid hormone molecules very quickly bind with several plasma proteins. Less than 1% of the T_3 and less than 0.1% of the T_4 remain in the **unbound (free) form**. This is remarkable, considering that only the free portion of the total thyroid hormone pool has access to the target cell receptors and thus can exert a biological effect.

Three different plasma proteins synthesized within the liver are important in thyroid hormone binding: In mammals (excluding cats), **thyroxine-binding globulin (TBG)** selectively binds all thyroid hormones even though its name specifies only "thyroxine" (T_4); binding proteins in other species include **albumin,** which nonselectively binds many lipophilic hormones, including T_4 and T_3; and **transthyretin (TTR)**, which binds the remaining T_4. TTR appears to be produced in the liver of amphibians and fish during transient periods of hyperthyroidism and more constantly in endotherms. TTR appears to be selectively secreted into the CSF by the choroid plexus of animals with a neocortex, supporting a role for this hormone in CNS development and function. The turnover of T_4 is comparatively slow in species that utilize TBG as a binding protein.

Most of the secreted T₄ is converted into T₃ outside the thyroid

In general, the proportion of T_4 and T_3 present in the thyroid gland of vertebrates is variable, although most thyroids secrete primarily T_4, most likely because more DIT is made than MIT. Regardless, in the peripheral tissues (e.g., liver and kidney), most of the secreted T_4 is converted into T_3, or *activated,* by an outer-ring deiodinase enzyme (Type II) that strips off one of its iodines. *T_3 is the major biologically active form of thyroid hormone at the cellular level.* T_4 can also be *inactivated* by inner-ring deiodinase (Type I) and converted into the metabolically inactive **reverse triiodothyronine,** or rT_3. Peripherally produced T_3 can bind to receptors with 10 times the affinity of T_4, giving peripheral cells the ability to activate their own hormone stimulation (or inactivate it by conversion to rT_3). It is now becoming apparent that regulation of these deiodinase enzymes is a major determinant of thyroid hormone stimulation of targets. For example, the stimulation of neuron growth during development of the brain is dependent on the deiodination of T_4 by astrocytes. Cells in the chick embryo produce only rT_3 until the time of piping (the bill piercing the air sack), whereon the enzymatic machinery begins to synthesize the metabolically active T_3. These inactivation pathways are also important in preserving energy stores during periods of limited food availability and are associated with reduced activity of the enzymatic machinery that generates T_3.

Thyroid hormone is the primary determinant of overall metabolic rate and exerts other effects as well

Thyroid hormone does not have any discrete target organs. It affects virtually every tissue in the body. Like all lipophilic hormones, thyroid hormone crosses the plasma membrane and binds with an intracellular receptor, in this case a nuclear receptor bound to the **thyroid-response element** of DNA. This binding alters the transcription of specific mRNAs and thus synthesis of specific new proteins, typically enzymes that carry out the cellular response.

Compared to other hormones, the action of thyroid hormone is "sluggish." The response to an increase in thyroid hormone is detectable only after a delay of several hours, and the maximal response is not evident for several days. The duration of the response is also quite long, partially because thyroid hormone is not rapidly degraded but also because the response to an increase in secretion continues to be expressed for days or even weeks after the plasma thyroid hormone concentrations have declined.

The effects of T_3 and T_4 can be grouped into several overlapping categories.

Effect on Metabolic Rate and Heat Production The evolution from ectothermy to endothermy necessitated development of a mechanism to regulate metabolic heat production. The solution provided by thyroid hormones is to increase a bird's or mammal's overall basal metabolic rate (BMR), or "idling speed" (see p. 320). This occurs by regulating mitochondrial function as well as certain mitochondrial proteins. For example, the number of active Na^+/K^+ ATPase pump units in the membrane is closely regulated in endotherms.

The transport of Na^+ relies on the hydrolysis of ATP, which yields heat as a by-product. Considering that as much as 20 to 40% of the total cell energy supply is required to maintain the pump activity, a considerable amount of heat is liberated in the process. Thyroid hormone is thus the most important regulator of the rate of O_2 consumption and energy expenditure under resting conditions. Inhibition of the Na^+/K^+ ATPase pump activity by ouabain (p. 87) markedly reduces the effect of thyroid hormone on heat production and oxygen consumption. Whereas thyroid hormones are generally elevated in endotherms, they are selectively elevated in ectotherms during periods of metabolically demanding activity (e.g., mating, migration, active feeding), although much less is known about their regulation in ectotherms.

Effect on Intermediary Metabolism In addition to increasing the general metabolic rate, thyroid hormone modulates the rates of many specific reactions involved in fuel metabolism. The effects of thyroid hormone on the metabolic fuels are multifaceted; not only can it influence both the synthesis and degradation of carbohydrate, fat, and protein, but small or large amounts of the hormone may induce opposite effects. For example, conversion of glucose to glycogen, the storage form of glucose, is facilitated by small amounts of thyroid hormone, but the reverse—breakdown of glycogen into glucose—occurs with large amounts of the hormone. Similarly, adequate amounts of thyroid hormone are essential for the protein synthesis needed for normal bodily growth, yet protein degradation effects predominate at high doses. In general, at abnormally high plasma levels of thyroid hormone, as in thyroid hypersecretion, the overall effect is to favor consumption rather than storage of fuel, as shown by depleting liver glycogen stores, depleting fat stores, and muscle wasting from protein degradation.

In many birds and mammals thyroid hormone levels vary on daily and/or seasonal bases. For example, in birds, daily changes in circulating levels of thyroid hormone are driven in part by food intake, T_3 increasing during the day associated with a reciprocal decline in T_4. In the American black bear, serum levels of T_3 and T_4 decrease during winter sleep (often loosely called *hibernation;* see p. 751) in association with reduced metabolism. An increase in thyroid hormone is also associated with the molt process in mammals and birds as well as the formation of new feathers and growth of horns or hair. For other seasonal changes, see the box *Molecular Biology and Genomics: Crosstalk between the Pituitary and Hypothalamus.*

Sympathomimetic Effect Any action similar to one produced by the sympathetic nervous system is known as a **sympathomimetic** ("sympathetic-mimicking") **effect.** Thyroid hormone increases target cell responsiveness to catecholamines (epinephrine and norepinephrine), the chemical messengers used by the sympathetic nervous system and its hormonal reinforcements from the adrenal medulla. Thyroid hormone presumably accomplishes this permissive action by causing a proliferation of specific catecholamine target-cell receptors (see p. 309). Because of this action, many of the effects observed when thyroid hormone secretion is elevated are similar to those that accompany activation of the sympathetic nervous system (a sympathomimetic effect).

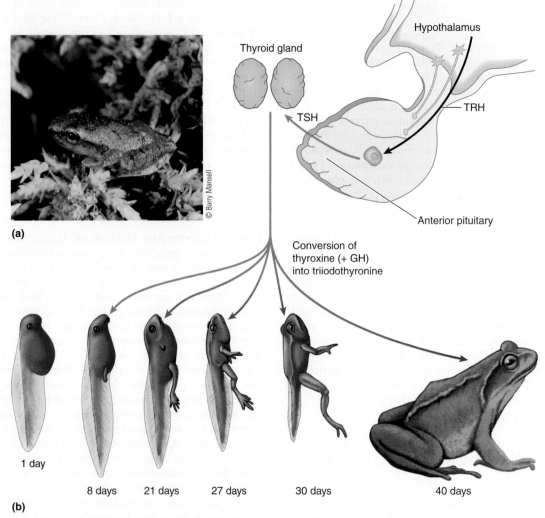

(a)

(b)

FIGURE 7-18 Thyroid hormones and amphibian metamorphosis. (a) A grass frog. The frog spends the first weeks of its life as an aquatic herbivore—a tadpole—in which the thyroid gland and other organs gradually mature. During this period, its skeletal muscles and other tissues are relatively insensitive to thyroxine, which is being synthesized in the thyroid, although little is released. At about 21 days (b), a spurt in thyroxine secretion begins to trigger major remodeling of the amphibian body. These developmental changes only occur if growth hormone also is present. The entire process may be triggered by changing environmental conditions. As metamorphosis gets under way, shifts in gene expression in cells of various tissues lead to major changes in body structures and physiological functions. Among other alterations, the tail regresses, digested away by newly synthesized lysosomal enzymes, and limbs grow. Lungs develop and gills degenerate, and the digestive tract becomes more suited to processing animal foods such as insects. The kidneys shift from excreting nitrogenous wastes as ammonia—an adaptation typical of aquatic animals—to eliminating urea.

© Cengage Learning, 2013

Effect on the Cardiovascular System Through its effect of increasing the heart's responsiveness to circulating catecholamines, thyroid hormone increases heart rate and force of contraction, thus increasing cardiac output. In addition, in response to the heat load generated by the calorigenic effect of thyroid hormone, peripheral vasodilation occurs to carry the extra heat to the body surface for elimination to the environment.

Effect on Growth and the Nervous System Thyroid hormone is essential for normal growth. Thyroid hormones act *permissively* (indirectly) in concert with other hormones in stimulating the growth process. Thyroid hormone is required for GH secretion in mammals and also promotes the effects of GH (or IGFs) on the synthesis of new structural proteins and on skeletal growth. Thyroid-deficient animals have stunted growth that is reversible with thyroid replacement therapy. Unlike excess GH, however, excess thyroid hormone does not result in excessive growth.

Thyroid hormone plays a crucial role in normal development of the nervous system, especially the CNS, an effect impeded in animals with thyroid deficiency from birth (or hatch). Thyroid hormone is also essential for normal CNS activity in adult animals. Furthermore, the conduction velocity of peripheral nerves varies directly with the availability of thyroid hormone.

Developmental Effects of Thyroid Hormone Thyroid hormone also controls *metamorphosis* in amphibians (Figure 7-18). **Metamorphosis** is associated with radical developmental changes in body structure that permit transition of an amphibian from an aquatic to a terrestrial habitat. The enzyme *hyaluronidase* is normally induced when tissues mature and thus serves as a signal for differentiation. Concentrations of hyaluronic acid are elevated in proliferating tissue and decline as the cells become increasingly differentiated (specialized). Researchers have recognized that at the time of tadpole differentiation, generation of T_3 increases, paralleled by a decrease in hyaluronic acid. Administering thyroid hormone to a tadpole also accelerates metamorphosis of the juvenile stage into the adult stage. Conversely, if the thyroid gland is prematurely removed from a tadpole, metamorphosis does not occur and the animals eventually develop into giant juveniles.

In at least one species of fish, an increase in thyroid hormone is associated with metamorphosis. In flounder, migration of the eye from one side of the head to the other is triggered by increased thyroid hormone secretion.

Thyroid hormone is regulated by the hypothalamus–pituitary–thyroid axis

Thyroid-stimulating hormone (TSH), the thyroid tropic hormone from the anterior pituitary, is the most important physiological regulator of thyroid hormone secretion (Figure 7-19). TSH stimulates almost every step of thyroid hormone synthesis and release.

In addition to enhancing thyroid hormone secretion, TSH maintains structural integrity of the thyroid gland. In the absence of TSH, the thyroid atrophies (decreases in size) and secretes its hormones at a very low rate. Conversely, it undergoes hypertrophy (increase in the size of each follicular cell) and hyperplasia (increase in number of follicular cells) in response to excess TSH stimulation, which results in thyroid enlargement or **goiter.**

In mammals, hypothalamic **thyrotropin-releasing hormone (TRH),** in tropic fashion, "turns on" TSH secretion by the anterior pituitary, whereas thyroid hormone, in negative-feedback fashion, "turns off" TSH secretion by inhibiting the anterior pituitary and hypothalamus. TRH functions via the IP_3/DAG/Ca^{2+} second-messenger pathway. Like other negative-feedback loops, the one between thyroid hormone and TSH tends to maintain a stable thyroid hormone output. However, in some nonmammalian vertebrates CRH has been found to be more important in the regulation of TSH secretion. Various types of stress inhibit TSH and thyroid hormone secretion, presumably through neural influences on the hypothalamus, although the adaptive importance of this inhibition is unclear.

Unlike most other hormonal systems, hormones in the thyroid axis in an adult mammal normally do not undergo sudden, wide swings in secretion. The relatively steady rate of thyroid hormone secretion is in keeping with the sluggish, long-lasting responses that this hormone induces; there would be no adaptive value in suddenly increasing or decreasing plasma thyroid-hormone levels. Seasonal changes in TRH are known to occur in some mammals; however, the

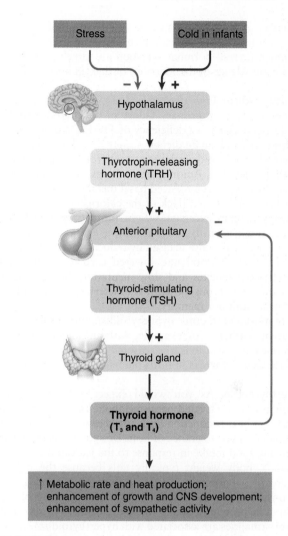

FIGURE 7-19 **Regulation of thyroid hormone secretion.**
© Cengage Learning, 2013

only known consistent factor that increases TRH secretion in mammals (and, accordingly, TSH and thyroid hormone secretion) is exposure to cold. This mechanism is highly adaptive in newborn mammals. The dramatic increase in heat-producing thyroid hormone secretion is thought to help maintain body temperature in the abrupt drop in surrounding temperature at birth, as the infant passes from the mother's warm body to the cooler environmental air. A similar TSH response to cold exposure does not occur in adult mammals. Some evidence suggests that on a longer-term basis during acclimatization to a cold environment, the concentration of hormones in this axis does increase as a means to increase the BMR and heat production.

Abnormalities of thyroid function include both hypothyroidism and hyperthyroidism

Normal thyroid function is called **euthyroidism.** Abnormalities of thyroid function are among the most common of all endocrine disorders in humans and domestic (and some wild) birds and mammals. Disorders fall into two major categories—*hypothyroidism* and *hyperthyroidism*—reflecting deficient and excess thyroid-hormone secretion, respectively.

A number of specific causes can give rise to each of these conditions. Whatever the cause, the consequences of too little or too much thyroid hormone secretion are largely predictable, given a knowledge of the functions of thyroid hormone.

Hypothyroidism Low thyroid activity, or **hypothyroidism**, can result (1) from primary failure of the thyroid gland itself; (2) secondary to a deficiency of TRH, TSH, or both; or (3) from an inadequate dietary supply of iodine. The symptoms of hypothyroidism largely stem from reduced overall metabolic activity. Among other things, a mammal with hypothyroidism has a reduced basal metabolic rate; displays poor tolerance of cold (lack of the calorigenic effect); has a tendency to gain excessive weight (not burning fuels at a normal rate); is easily fatigued (lower energy production); has a slow, weak pulse (caused by reduced rate and strength of cardiac contraction and lowered cardiac output); and exhibits slow reflexes and slow mentation (because of the effect on the nervous system), characterized by diminished alertness, and poor memory.

In most endotherms, hypothyroidism diminishes quality of fur or feathers. For example, without adequate thyroid hormone concentrations, the seasonal growth in their pelage fails to occur, whereas in birds plumage fails to develop.

Hyperthyroidism As you would expect, the hyperthyroid mammal has an elevated basal metabolic rate. The resulting increase in heat production leads to excessive perspiration or panting and poor tolerance of heat. Despite the increased appetite and food intake in response to the increased metabolic demands, body weight typically falls because the body is burning fuel abnormally fast. Net degradation of endogenous carbohydrate, fat, and protein stores occurs. The loss of skeletal muscle protein results in weakness. Various cardiovascular abnormalities are associated with hyperthyroidism, caused both by the direct effects of thyroid hormone and by its interactions with catecholamines. Heart rate and strength of contraction may rise to dangerous levels. In severe cases, the heart may fail to meet the body's metabolic demands in spite of increased cardiac output. Nervous system involvement is manifested by excessive mental alertness to the point where the animal is irritable, tense, and anxious.

Three general methods of treatment are available for suppressing excess thyroid-hormone secretion: surgical removal of a portion of the oversecreting thyroid gland; administration of radioactive iodine, which, after being concentrated in the thyroid gland by the iodine pump, selectively destroys thyroid glandular tissue; and antithyroid drugs that specifically interfere with thyroid hormone synthesis and generation of T_3.

check your understanding 7.5

How is the concentration of T_3 regulated in the circulation?

What is the role of thyroid hormone in endotherms and ectotherms?

What is the role of thyroglobulin in the synthesis of active thyroid hormone?

7.6 Adrenal Glands

The **adrenal gland** (*adrenal,* "next to the kidney") (Figure 7-20a) of higher vertebrates consists of two distinct cell types: **chromaffin** (*chroma,* "color"; *affinis,* "affinity") cells, which are derived from the neural crest, and **steroidogenic cells,** which are of mesodermal origin.

In most vertebrates, the adrenal gland consists of a steroid-secreting cortex intermingled with chromaffin tissue

The term *chromaffin* arises from the observation that the tissue stains the color brown when reacted with oxidizing agents such as chromate. In mammals there are two adrenal glands, one embedded above each kidney in a capsule of fat. The shape of the adrenal varies considerably and in some species actually fuses with the kidney. In most mammals the adrenal gland consists of an outer, steroid-secreting **adrenal cortex** and an inner, catecholamine-secreting **adrenal medulla.** For this reason the chromaffin tissue is referred to as the adrenal medulla. However, in most nonmammalian species the chromaffin tissue is not associated with any surrounding cortex and in many instances, rather than forming distinct zones, the two tissue types are intermingled in the adrenal gland. In elasmobranchs, **interrenal tissue,** which is homologous with the adrenal cortex of higher vertebrates, is organized into glands situated between the posterior regions of the kidneys.

The steroid-secreting adrenal cortex and catecholamine-secreting medulla produce hormones belonging to different chemical categories, whose functions, mechanisms of action, and regulation are entirely different.

The adrenal cortex secretes mineralocorticoids, glucocorticoids, and sex hormones

About 80% of the adrenal gland of most mammals is composed of the cortex, which consists of three different layers or zones: the **zona glomerulosa,** the outermost layer; the **zona fasciculate,** the middle and largest portion; and the **zona reticularis,** the innermost zone (Figure 7-20b). The adrenal cortex produces a number of different adrenocortical hormones, all of which are steroids derived from the common precursor molecule, cholesterol. All steroidogenic ("steroid-producing") cells are filled with lipid droplets (**liposomes**) containing cholesterol. Cholesterol is first converted to *pregnenolone,* then modified by stepwise enzymatic reactions to produce active steroid hormones (Figure 7-3). Each steroidogenic tissue has a complement of enzymes to produce one or several but not all steroid hormones. Slight variations in structure confer different functional capabilities on the various adrenocortical hormones. On the basis of their primary actions, the adrenal steroids can be divided into three categories:

1. **Mineralocorticoids,** mainly *aldosterone,* which influence mineral (electrolyte) balance, specifically Na^+ and K^+ balance (produced exclusively in the zona glomerulosa). The actions and regulation of the primary adrenocortical mineralocorticoid, **aldosterone,** are described thoroughly elsewhere (Chapter 12).
2. **Glucocorticoids,** primarily *cortisol* and *corticosterone,* which play a major role in glucose metabolism as well

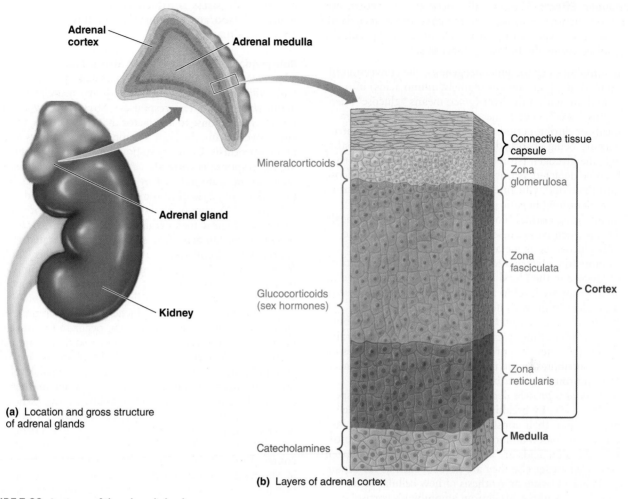

Mineralcorticoids

**Glucocorticoids
(sex hormones)**

Catecholamines

Connective tissue
capsule

Zona
glomerulosa

Zona
fasciculata

Zona
reticularis

Medulla

Cortex

Adrenal
cortex — Adrenal medulla

Adrenal gland

Kidney

(a) Location and gross structure
of adrenal glands

(b) Layers of adrenal cortex

FIGURE 7-20 **Anatomy of the adrenal glands.**
© Cengage Learning, 2013

as in protein and lipid metabolism (synthesized in the two inner layers, with the zona fasciculata being the major source of this glucocorticoid).

3. **Sex steroids** identical or similar to those produced by the gonads (testes in males, ovaries in females). The most abundant and physiologically important of the adrenocortical sex hormones is *dehydroepiandrosterone (DHEA),* a "male" sex hormone (produced by the two inner zones).

The production of adrenocortical and other steroid hormones requires a series of steps in which the cholesterol molecule undergoes various enzymatic modifications (see p. 272). The enzymes that carry out the synthesis of the major steroids are highly conserved among vertebrates and homologous to those characterized in mammals. The different functional types of adrenal steroids are produced in anatomically distinct portions of the adrenal cortex, as noted in the preceding list.

Being lipophilic, the adrenocortical hormones are all carried in the blood extensively bound to plasma proteins. About 60% of circulating aldosterone is protein bound, primarily to nonspecific albumin, whereas approximately 90% of the glucocorticoids are bound, mostly to a plasma protein specific for it called **corticosteroid-binding globulin**

(transcortin). Likewise, 98% of dehydroepiandrosterone is bound, in this case, exclusively to albumin.

Each of the adrenocortical steroid hormones binds with a receptor specific for it within the cytoplasm of the hormone's target cells: Mineralocorticoids bind to the **mineralocorticoid receptor (MR)**, glucocorticoids to the **glucocorticoid receptor (GR)**, and dehydroepiandrosterone to the **androgen receptor (AR)**. As is true of all steroid hormones, each hormone-receptor complex moves to the nucleus and binds with a complementary hormone-response element in DNA, namely the *mineralocorticoid response element, glucocorticoid response element,* and *androgen response element.* This binding initiates specific gene transcription leading to synthesis of new proteins that carry out the effects of the hormone.

Glucocorticoids exert metabolic effects and have an important role in adaptation to stress

Cortisol and **corticosterone** play an important role in carbohydrate, protein, and fat metabolism; execute significant permissive actions for other hormonal activities; and help animals cope with stress.

Metabolic Effects The overall effect of glucocorticoids' metabolic actions is to *increase the concentration of blood glucose* at the expense of protein and fat stores. Specifically, cortisol performs the following functions:

- It stimulates hepatic **gluconeogenesis,** the conversion of noncarbohydrate sources (namely, amino acids) into carbohydrate within the liver (*gluco* means "glucose"; *neo* means "new"; *genesis* means "production"). Between meals or during periods of fasting, when no new nutrients are being absorbed into the blood for use and storage, the glycogen (stored glucose) in the liver tends to become depleted as it is broken down to release glucose into the blood. Gluconeogenesis is an important factor in replenishing hepatic glycogen stores and thus in maintaining normal blood-glucose levels between meals. This is essential because the vertebrate brain can use only glucose as its metabolic fuel, yet nervous tissue cannot store glycogen to any extent. The concentration of glucose in the blood must therefore be maintained at an appropriate level to adequately supply the glucose-dependent brain with nutrients.
- It *inhibits glucose uptake* and use by many tissues, but not the brain, thus sparing glucose for use by the brain, which absolutely requires it as a metabolic fuel. This action contributes to the increase in blood glucose concentration brought about by gluconeogenesis.
- It *stimulates protein degradation* in many tissues, especially muscle. By breaking down a portion of muscle proteins into their constituent amino acids, cortisol increases the blood amino acid concentration. These mobilized amino acids are available for use in gluconeogenesis or wherever else they are needed, such as for repair of damaged tissue or synthesis of new cellular structures. For example, an anuran amphibian's cortisol is markedly elevated during metamorphosis of tadpole into frog (metamorphic climax). This is associated with increased mRNA levels of *POMC* (see p. 271) during this developmental stage.
- It *facilitates lipolysis* (*lysis,* "breakdown"), the breakdown of lipid (fat) stores in adipose tissue, releasing free fatty acids into the blood. The mobilized fatty acids are available as an alternative metabolic fuel for tissues that can use this energy source in lieu of glucose, conserving glucose for the brain.
- It stimulates *acclimatization to seawater* in euryhaline fishes (see pp. 626–627).

Permissive Actions Glucocorticoids are extremely important for their permissiveness (see p. 278). For example, in mammals glucocorticoids must be present in adequate amounts to permit the catecholamines to induce vasoconstriction. An animal lacking cortisol, if untreated, may go into circulatory shock in a stressful situation that demands immediate widespread vasoconstriction.

Effects on the Brain Glucocorticoids also affect neural functions such as memory. For example, elevated concentrations of plasma cortisol in Pacific salmon during home-stream migration aids in recalling the imprinted memory of the home-stream chemical aroma. Mammalian memory centers (such as the hippocampus) have cortisol receptors of uncertain function. In contrast to salmon, prolonged elevation in cortisol from stress is associated with memory loss in humans and laboratory mammals, but this is considered an abnormal situation.

Role in Adaptation to Long-Term Stress Glucocorticoids play a key role in adaptation to stress, particularly *long-term* stress (you will learn in the section on adrenal medulla hormones about *short-term* stress responses). **Stress** is the generalized, nonspecific response to any factor that overwhelms, or threatens to overwhelm, the body's compensatory abilities to maintain homeostasis. Contrary to popular usage, the agent inducing the response is correctly called a *stressor,* whereas *stress* refers to the state induced by the stressor. Stressors include those that are *physical* (trauma, intense heat or cold); *chemical* (reduced O_2 supply, acid–base imbalance, nutritional deficit); *physiological* (hemorrhagic shock, pain); *psychological* or *emotional* (anxiety, fear such as that generated from an approaching thunderstorm); and *social* (conflict, isolation from a group). The following types of noxious stimuli illustrate the range of factors that can induce a stress response in fish: poor water quality, handling, and confinement.

Dramatic increases in glucocorticoid secretion, mediated by the central nervous system, occur in response to all kinds of stressful situations. For this reason, measurement of cortisol/corticosterone in an animal's blood is one of the best indicators of stress. For example, cortisol levels skyrocket in infant chimps separated from their mothers, and conversely, cortisol levels decline below average when their mothers are grooming these chimps. Similarly, a horse that is separated from its herd exhibits increased cortisol and other stress symptoms, showing the importance of social contact in this herd animal. The magnitude of the increase in plasma glucocorticoid concentration is generally proportional to the intensity of the stressful stimulation; a greater increase in glucocorticoid concentrations is evoked in response to severe stress than to mild stress.

Although the precise role of glucocorticoids in adapting to stress is not known, a speculative but plausible explanation might be as follows. An animal wounded or faced with a life-threatening situation must forgo eating. A glucocortico-induced enhancement of breakdown of carbohydrate stores (which increases the availability of blood glucose) would help protect the brain from malnutrition during the imposed fasting period. Also, the amino acids liberated by protein degradation would provide a readily available supply of building blocks for tissue repair should physical injury occur. Thus, an increased pool of glucose, amino acids, and fatty acids is available for use as needed.

Anti-Inflammatory and Immunosuppressive Effects When stress is accompanied by tissue injury, inflammatory and immune responses accompany the stress response. Cortisol exerts *anti-inflammatory* and *immunosuppressive* effects to help hold these immune system responses in check and balance. An exaggerated inflammatory response has the potential of causing harm. Cortisol interferes with almost every step of inflammation. For example, among other anti-inflammatory actions, cortisol partially blocks production of inflammatory chemical mediators, such as prostaglandins and leukotrienes (see p. 470); it suppresses migration of neutrophils to the injured site and interferes with their phagocytic activity (see Figure 10-7, p. 468); and it inhibits proliferation of fibroblasts in wound repair (see p. 469). Cortisol also inhibits immune responses by interfering with antibody production by lymphocytes. Blurring the line between endocrine and immune control, lymphocytes have been shown to secrete ACTH, and some of the cytokines (such as IL-1, IL-2, and IL-6; see p. 460) released from immune cells can stimulate the hypothalamus–pituitary–adrenal axis. In feedback fashion, cortisol in turn has a profound dampening (turning-down) impact on the immune system. These interactions between the immune system and cortisol secretion help maintain immune homeostasis, an area only beginning to be explored.

Administering large amounts of glucocorticoid inhibits almost every step of the inflammatory response, making these steroids effective drugs in treating conditions in which the inflammatory response itself has become destructive. Glucocorticoids used in this manner do not affect the underlying disease process; they merely suppress the body's response to the disease. Because glucocorticoids also exert multiple inhibitory effects on the overall immune process, such as "knocking out of commission" the white blood cells responsible for antibody production as well as those that directly destroy foreign cells, these agents have also proved useful in managing various allergic disorders and in preventing organ transplant rejections.

When glucocorticoids are administered at pharmacological levels (that is, at higher-than-physiologic concentrations), not only are their anti-inflammatory and immunosuppressive effects increased but their metabolic effects are also magnified. Therefore, synthetic glucocorticoids have been developed that maximize the anti-inflammatory and immunosuppressive effects of these steroids while minimizing the metabolic effects.

Glucocorticoid secretion is directly regulated by the hypothalamic–pituitary–adrenal axis

Glucocorticoid secretion by the adrenal cortex is regulated by a negative-feedback system involving the hypothalamus and anterior pituitary (Figure 7-21). ACTH from the anterior pituitary corticotropes, acting through the cAMP pathway, stimulates the adrenal cortex to secrete cortisol. Hormonal steroid output by the adrenal cortex is seen within two minutes of exposure to ACTH, with the rate-limiting step in the biosynthesis of glucocorticoid being the conversion of cholesterol into pregnenolone (see Figure 7-3). Being tropic to the zona fasciculata and zona reticularis, ACTH stimulates both the growth and the secretory output of these two inner layers of the cortex, which shrink considerably in the absence of adequate amounts of ACTH.

The ACTH-producing cells in turn secrete only at the command of CRH from the hypothalamus. CRH stimulates the corticotropes via the cAMP pathway. The feedback control loop is completed by glucocorticoid's inhibitory actions on CRH and ACTH secretion by the hypothalamus and anterior pituitary, respectively. Superimposed on the basic negative-feedback control system are two additional factors that influence plasma glucocorticoid concentrations by changing the set points: These are *stress* and *circadian rhythms*, both of which act on the hypothalamus to vary the secretion rate of CRH.

Influence of Circadian Rhythm on Cortisol Secretion Recall that there is a characteristic circadian rhythm in plasma cortisol concentration, with the highest level occurring in

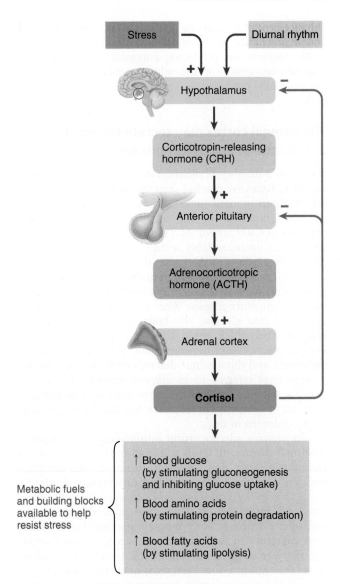

FIGURE 7-21 Control of cortisol secretion.
© Cengage Learning, 2013

diurnal animals in the morning and the lowest level at night (see Figure 7-6). This rhythm, which is governed by the SCN (the master biological clock we described earlier), is related primarily to the sleep–wake cycle. The peak and low levels are thus reversed in animals that are nocturnal.

The adrenal cortex secretes both male and female sex hormones in both sexes

In both sexes, the adrenal cortex produces both *androgens,* or "male" sex hormones, and *estrogens,* or "female" sex hormones. Under normal circumstances in mammals, the adrenal androgens and estrogens are not sufficiently abundant or powerful to induce masculinizing or feminizing effects, respectively. The only adrenal sex hormone that has any biological importance is the androgen **dehydroepiandrosterone (DHEA)**. Adrenal DHEA is overpowered by testicular testosterone in males but is of physiological significance in female humans, who otherwise lack androgens. This adrenal androgen is responsible for androgen-dependent processes in the female such as growth of pubic and axillary (armpit) hair, enhancement of the pubertal growth spurt, and development and maintenance of the female sex drive. Because the enzymes required for producing estrogens are found in very low concentrations in the adrenocortical cells, estrogens are normally produced in very small quantities from this source.

The adrenal cortex may secrete too much or too little of any of its hormones

There are a number of different disorders of adrenocortical function. Excessive cortisol secretion (**Cushing's syndrome**), which occurs in many older dogs, is most commonly caused by overstimulation of the adrenal cortex by excessive amounts of CRH and/or ACTH. A rarer cause is an adrenal tumor that uncontrollably secretes cortisol independent of ACTH. Regardless of the cause, the prominent characteristics of this syndrome are related to the exaggerated effects of glucocorticoid, with the main symptoms being reflections of excessive gluconeogenesis. When too many amino acids are converted into glucose, the body suffers from combined glucose excess (high blood glucose) and protein shortage. Because the resultant hyperglycemia, glucosuria (glucose in the urine), and thirst mimic diabetes mellitus, the condition is sometimes referred to as *adrenal diabetes*. For reasons that are unclear, some of the extra glucose is deposited as body fat in locations characteristic for this disease, typically in the abdomen in dogs.

In **primary adrenocortical insufficiency,** also known as **Addison's disease,** all layers of the adrenal cortex are undersecreting. Certain dog breeds are prone to this condition, which is most commonly caused by autoimmune destruction of the cortex by erroneous production of adrenal cortex-attacking antibodies, in which case both aldosterone and cortisol are deficient. The symptoms associated with aldosterone deficiency in Addison's disease are the most threatening. If severe enough, the condition is fatal because aldosterone is essential for life. Symptoms of cortisol deficiency are as would be expected: poor response to stress, hypoglycemia caused by reduced gluconeogenic activity, and lack of permissive action for many metabolic activities.

Having completed discussion of the adrenal cortex, we now shift attention to the adrenal medulla.

The catecholamine-secreting adrenal medulla is a modified sympathetic ganglion

The adrenal medulla is actually a modified part of the sympathetic nervous system. A sympathetic pathway consists of two neurons in sequence—a *preganglionic neuron* originating in the CNS, whose axonal fiber terminates on a second peripherally located *postganglionic neuron*, which in turn terminates on the effector organ (see Figure 5-8, p. 162). The neurotransmitter released by sympathetic postganglionic fibers is norepinephrine (NE), which interacts locally with the innervated organ by binding with specific target receptors known as *adrenergic receptors*.

The adrenal medulla consists of modified postganglionic sympathetic neurons called **chromaffin cells.** Unlike ordinary postganglionic sympathetic neurons, chromaffin cells do not have axonal fibers that terminate on effector organs. Instead, on stimulation by the preganglionic fiber the chromaffin cells release their chemical transmitter directly into the circulation (see Figure 5-9, p. 163) as a neurohormone. Like sympathetic fibers, the adrenal medulla does release NE, but in mammals its most abundant secretory output is the similar chemical messenger **epinephrine.** Both epinephrine and NE belong to the chemical class of *catecholamines,* which are derived from the amino acid tyrosine (see Table 7-1). Epinephrine is similar in structure to NE except that it has a methyl group added to it. In species where the chromaffin tissue is separated from the steroidogenic tissue, such as the dogfish shark, NE is the only catecholamine produced. However, in frogs, where chromaffin tissue is loosely intermingled with steroidogenic tissue, NE is reduced to about 55 to 70% of the total catecholamine content.

Storage of Catecholamines in Chromaffin Granules Catecholamine synthesis in vertebrates is accomplished almost entirely within the cytosol of the adrenomedullary secretory cells, with only one step taking place within the hormonal storage granules. Whereas epinephrine is produced exclusively by the adrenal medulla, the bulk of NE is produced by sympathetic postganglionic fibers. Once produced, epinephrine and NE are stored in **chromaffin granules,** which are similar to the transmitter storage vesicles in sympathetic nerve endings. Catecholamines are ultimately secreted into the circulation by exocytosis of chromaffin granules; their release is analogous to the release mechanism for secretory vesicles that contain stored peptide hormones or the release of NE at sympathetic postganglionic terminals. Once in the circulation catecholamines are rapidly degraded resulting in a fast turnover: they are specialized for short-term actions associated with the stress response.

Sympathetic stimulation of the adrenal medulla is solely responsible for epinephrine release

Catecholamine secretion by the adrenal medulla is controlled entirely by sympathetic input to the gland. When the sympathetic system is activated under conditions of fear or stress, it simultaneously triggers a surge of adrenomedullary catecholamine release, flooding the circulation with up to 300 times the normal concentration of epinephrine. Although a number of different factors influence adrenal catecholamine secretion, they all act by increasing preganglionic sympathetic impulses to the adrenal medulla. Among the major factors that stimulate increased adrenomedullary output are a variety of stressful conditions such as physical or environmental disturbances, hemorrhage, illness, exercise, hypoxia (low arterial O_2), cold exposure, and hypoglycemia.

Epinephrine reinforces the sympathetic nervous system in "fight-or-flight" short-term stress, and exerts additional metabolic effects

Catecholamine hormones are not essential for life, but virtually all organs in the body are affected by them. They play important roles in stress responses, regulation of arterial blood pressure, and control of fuel metabolism. Together, the sympathetic nervous system and adrenomedullary epinephrine mobilize the body's resources to support peak physical exertion in the face of impending danger ("short-term" stress). The sympathetic and epinephrine actions constitute a "fight-or-flight" response that prepares the animal to deal with a range of physical and environmental disturbances. An arthropod and molluscan equivalent of NE is **octopamine** (first discovered in octopus), a biogenic amine derived from the amino acid tyrosine. In many of these animals, octopamine is a neurotransmitter, neuromodulator, and neurohormone. In response to stressful situations in insects, such as handling or during the early stage of flight, octopamine mobilizes stores from the fat body for release into the hemolymph.

The following sections discuss epinephrine's major effects, which it accomplishes in collaboration with NE or performs alone to complement direct sympathetic response.

Effects on Organ Systems Together, the sympathetic nervous system and adrenomedullary epinephrine mobilize the body's resources that are ideally suited for fight-or-flight responses (see p. 159). Specifically, the sympathetic system and epinephrine increase the rate and strength of cardiac contraction, increasing cardiac output, and their generalized vasoconstrictor effects increase total peripheral resistance. These effects cause an increase in arterial blood pressure, thus ensuring an appropriate driving pressure to force blood to the organs that are most vital for meeting the emergency. Meanwhile, vasodilation of coronary and skeletal-muscle blood vessels induced by epinephrine and local metabolic factors shifts blood to the heart and skeletal muscles from other vasoconstricted regions of the body. In fishes, the physiological effects of catecholamines are also used to maintain levels of energy and oxygen supply under conditions such as hypoxia. However, in teleosts, adrenergic control of the heart is mainly neuronal, and it is only under severe stress that circulating levels of epinephrine increase sufficiently to affect cardiac function. Catecholamines also increase oxygen delivery to the tissues, change gill diffusion capacity, increase erythrocyte release from the spleen, increase the blood oxygen capacity by elevating intracellular pH, and increase blood flow and ventilation rate.

Epinephrine (but not NE) dilates the respiratory airways to reduce the resistance encountered in moving air in and out of the lungs. Epinephrine also reduces digestive activity and inhibits bladder emptying, both activities that can be "put on hold" during a fight-or-flight situation.

Metabolic Effects Epinephrine exerts some important metabolic effects, even at blood hormone concentrations lower than required for eliciting the cardiovascular responses. In general, epinephrine prompts mobilization of stored carbohydrate and fat to provide immediately available energy to fuel muscular work. Specifically, epinephrine increases the blood glucose level by several different mechanisms. First, it stimulates both hepatic (liver) gluconeogenesis and **glycogenolysis,** the latter being the breakdown of stored glycogen into glucose, which is released into the blood. Epinephrine also stimulates glycogenolysis in skeletal muscles. Because of the difference in enzyme content between liver and muscle, however, muscle glycogen is not converted to glucose for the blood. Instead, muscle glycogen is broken down via glycolysis into lactate, providing a burst of ATP for muscles in the emergency situation. In addition to raising blood glucose levels, epinephrine also raises the blood fatty-acids level by promoting lipolysis.

Epinephrine's metabolic effects are appropriate for transient fight-or-flight situations. The elevated levels of glucose and fatty acids provide additional fuel to power the muscular movement required by the situation and also ensure adequate nourishment for the brain during the crisis when no new nutrients are being consumed. Muscles can use fatty acids for energy production, but the brain cannot.

Because of its other widespread actions, epinephrine also increases overall metabolic rate. Under the influence of epinephrine, many tissues metabolize faster. For example, the work of the heart and respiratory muscles is increased, and the pace of liver metabolism is stepped up. Thus, epinephrine as well as thyroid hormone can increase metabolic rate.

Other Effects Epinephrine affects the central nervous system to promote a state of arousal and increased CNS alertness, also useful to fight-or-flight situations.

Both epinephrine and NE cause sweating in mammals that have sweat glands, which helps the body rid itself of extra heat generated by increased muscular activity. Also, epinephrine acts on smooth muscles within the eyes to dilate the pupil and flatten the lens. These actions adjust the eyes for more encompassing vision so that the whole threatening scene can be quickly viewed.

Epinephrine and norepinephrine vary in their affinities for the different adrenergic receptor types

Epinephrine and NE have differing affinities for four distinctive receptor types: α_1, α_2, β_1, and β_2 adrenergic receptors (see p. 165). NE binds predominantly with α and β_1 receptors located near postganglionic sympathetic-fiber terminals. Hormonal epinephrine, which can reach all α and β_1 receptors via its circulatory distribution, interacts with these same receptors. NE has a slightly greater affinity than epinephrine for the α receptors, and the two hormones have approximately the same potency at the β_1 receptors. Thus, epinephrine and NE exert similar effects in many tissues, with epinephrine generally reinforcing sympathetic nervous activity. In addition, epinephrine activates β_2 receptors, over which the sympathetic nervous system exerts little influence. Many of the essentially epinephrine-exclusive β_2 receptors are located at tissues not even supplied by the sympathetic nervous system but reached by epinephrine through the blood. Examples include skeletal muscle and its blood vessels and bronchiolar smooth muscle, with the effects we just discussed.

Sometimes epinephrine, through its exclusive β_2-receptor activation, brings about a different action from that elicited by NE and epinephrine action through their mutual activation of other adrenergic receptors. For example, NE and epinephrine bring about a generalized vasoconstrictor effect mediated by α_1-receptor stimulation. By contrast, epinephrine promotes vasodilation of the blood vessels that supply skeletal muscles and the heart through β_2-receptor activation.

Realize, however, that epinephrine functions only at the bidding of the sympathetic nervous system, which is solely responsible for stimulating its secretion from the adrenal medulla. Epinephrine secretion always accompanies a generalized sympathetic nervous system discharge, so sympathetic activity indirectly controls actions of epinephrine. By having the more versatile circulating epinephrine at its call, the sympathetic nervous system has a means of reinforcing its own neurotransmitter effects plus a way of executing additional actions on tissues that it does not directly innervate.

The stress response is a generalized, nonspecific pattern of neural and hormonal reactions to any situation that threatens homeostasis

Because both components of the adrenal gland play an extensive role in responding to stress, this is an appropriate place to pull together the various major factors involved in the stress response. Recall that a variety of noxious physical, chemical, physiological, and psychosocial stimuli that threaten to overwhelm the body's compensatory ability to maintain homeostasis all can elicit a stress response. Different stressors may produce some specific responses characteristic of that stressor; for example, in some mammals the specific response to cold exposure is shivering and skin vasoconstriction, whereas the specific response to bacterial invasion includes increased phagocytic activity and antibody production. In addition to their specific response, however, all stressors also produce a similar nonspecific, generalized response regardless of the type of stressor.

In the 1930s, Hans Selye was the first to recognize this commonality of responses to noxious stimuli in what he called the **general adaptation syndrome.** When a stressor is recognized, both nervous and hormonal responses are called into play to bring about defensive measures to cope with the emergency. Across the entire vertebrate lineage, the result is a state of intense readiness and mobilization of biochemical resources.

To appreciate the value of the multifaceted stress response, imagine a gazelle that has just seen a lion lurking in the grass. The major neural response to such a stressful stimulus is generalized activation of the sympathetic nervous system. The resultant increase in cardiac output and ventilation as well as diversion of blood from vasoconstricted regions of suppressed activity, such as the digestive tract and kidneys, to the more active vasodilated skeletal muscles and heart prepare the body for a fight-or-flight response. Simultaneously, the sympathetic system calls forth hormonal reinforcements in the form of a massive outpouring of epinephrine from the adrenal medulla. Epinephrine strengthens sympathetic responses and reaches places not innervated by the sympathetic system to perform additional functions, such as mobilizing carbohydrate and fat stores.

Besides epinephrine, a number of other hormones are involved in the overall stress response (Table 7-3). As you have seen, the other predominant hormonal response is activation of the CRH–ACTH–glucocorticoid system, usually for longer-term stress. In our gazelle, this system would dominate if it escapes the lion but receives a serious wound (such as a bite in its leg). Note that a major difference between epinephrine and cortisol is that the former does not promote muscle protein breakdown, whereas the latter does. It would be maladaptive to cannibalize muscles during short-term fight-or-flight, but during long-term trauma it is useful to mobilize amino acids for tissue repair in the event of an injury.

In addition to the effects of cortisol in the hypothalamus–pituitary–adrenal cortex axis, there is much evidence that

TABLE 7-3 Major Hormonal Changes during the Stress Response

Hormone	Change	Purpose Served
Epinephrine	↑	Reinforces the sympathetic nervous system to prepare the body for "fight-or-flight"
	↑	Mobilizes carbohydrate and fat energy stores; increases blood glucose and blood fatty acids
CRH–ACTH–Cortisol	↑	Mobilizes energy stores and metabolic building blocks for use as needed; increases blood glucose, blood amino acids, and blood fatty acids
	↑	ACTH facilitates learning and behavior
Glucagon	↑	Act in concert to increase blood glucose and blood fatty acids
Insulin	↓	
Renin–Angiotensin–Aldosterone; Vasopressin	↑	Conserve salt and H_2O to expand the plasma volume; help sustain blood pressure when acute loss of plasma volume occurs
		Angiotensin II and vasopressin cause arteriolar vasoconstriction to increase blood pressure
		Vasopressin facilitates learning

© Cengage Learning, 2013

ACTH may play a role in resisting stress. ACTH suppresses release of GH, TSH, and gonadotropins, suppressing growth, metabolism, and reproduction, respectively. This helps divert energy toward stress needs. ACTH is one of several peptides that facilitate learning and behavior. Thus, it is possible that an increase in ACTH during psychosocial stress might help the body cope more readily with similar stressors in the future by facilitating the learning of appropriate behavioral responses. Furthermore, ACTH is not released alone from its anterior pituitary storage vesicles. Pruning of the large POMC precursor molecule yields not only ACTH but also morphine-like **β-endorphin** and similar compounds. These compounds are cosecreted with ACTH on stimulation by CRH during stress. As a potent endogenous opiate, β-endorphin may exert a role in mediating **analgesia** (reduced pain perception) should physical injury be inflicted during stress (see p. 261), allowing an animal to ignore pain.

The multifaceted stress response is coordinated by the hypothalamus

All the individual responses to stress just described are either directly or indirectly influenced by the hypothalamus (Figure 7-22). The hypothalamus receives input concerning physical and emotional stressors from many areas of the brain and from many receptors throughout the body. In response, the

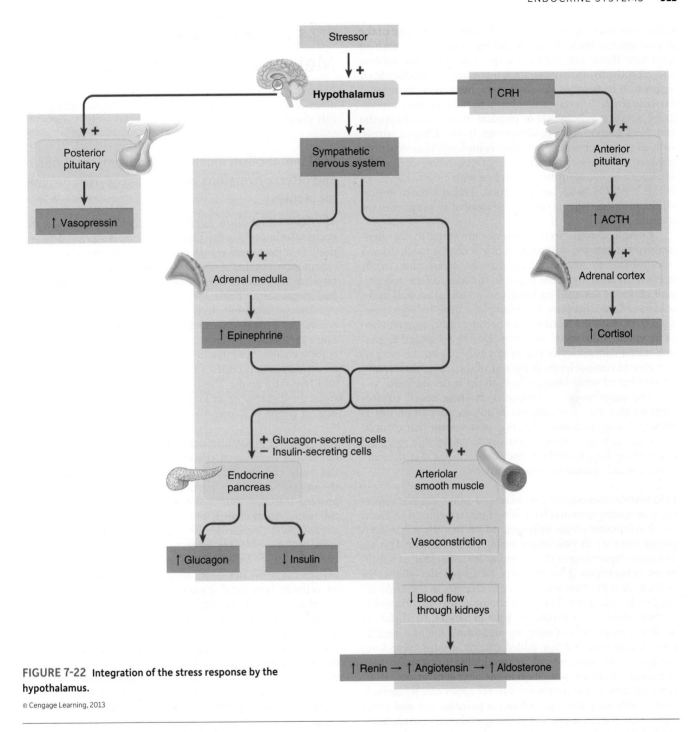

FIGURE 7-22 Integration of the stress response by the hypothalamus.

© Cengage Learning, 2013

hypothalamus directly activates the sympathetic nervous system, secretes CRH to stimulate ACTH and cortisol release, and triggers the release of vasopressin. Sympathetic stimulation in turn brings about the secretion of epinephrine, with which it has a conjoined effect on the pancreatic secretion of insulin and glucagon. Furthermore, vasoconstriction of the renal afferent arterioles by the catecholamines indirectly triggers the secretion of renin by reducing the flow of oxygenated blood through the kidneys. Renin in turn sets in motion the renin–angiotensin–aldosterone mechanism. In this way, the hypothalamus integrates the responses of both the sympathetic nervous system and the endocrine system during stress.

Activation of the stress response by chronic psychosocial stressors may be harmful

Acceleration of cardiovascular and respiratory activity, retention of salt and H_2O, and mobilization of metabolic fuels and building blocks can be of benefit in response to a physical stressor. Many stressors in an animal's life are psychosocial in nature, yet they induce these same magnified responses. Although the mobilization of body resources is appropriate in the face of real or threatened physical injury, it is often inappropriate in response to nonphysical stress. If no extra energy is demanded, no tissue damaged, and no blood lost, body stores are being broken down and fluid retained needlessly.

Other responses can also be harmful. In the acute phase of the stress response, useful behaviors necessary for dealing with an imminent threat (avoidance, escape) emerge, while feeding and reproductive behaviors are temporarily suspended. Such suspension is useful in the short term, but can be harmful in extreme stress. In every animal in which it has been investigated, severe psychological or physical stress reduces appetite and food intake often to dangerous levels. Chronic stress, such as social subordination, may completely shut down reproductive behavior in many vertebrates, where diminished sex-hormone concentrations correlate with elevated levels of POMC, ACTH, and corticosterone. Perhaps most importantly, prolonged glucocorticoid elevation also suppresses the immune system (p. 470).

Overall, chronic stress responses are often to the detriment of the stressed animal, which will have a much greater susceptibility to diseases. You may be aware that overly stressed humans are more prone to diseases; so are zoo animals in poorly designed enclosures, as are social animals in the wild in certain situations. For example, a male baboon entering a troop for the first time must become highly aggressive to be accepted in the social hierarchy. During prolonged aggressive states, cortisol levels rise in this baboon, but his actions also elevate cortisol levels in the rest of the troop. As a result, the number of white blood cells in all the males declines.

The suppression of immunity in these social stresses suggests that this phenomenon is not an adaptation but is rather an inappropriate activation of a system that evolved for more ancient, nonsocial stresses. Considerable work remains to be done to evaluate the contributions that stressors make toward disease in these situations.

Epigenetics Inadequate nutrition and other stresses affecting a pregnant mammal have long been recognized to harm the development of the offspring. More recently it has been recognized that chronic stress during pregnancy can also affect gene expression in the offspring for a number of subsequent generations. This is a type of **epigenetics**—heritable changes in genes that are not due to changes in the underlying DNA code itself. The most common way this occurs is by *methylation* of cytosine (C base of DNA). Methylation occurs in many genes during an organism's life as a regulatory mechanism, but the added methyl groups were long thought to be stripped off in developing gametes so that "original" genes are passed on to the next generation. However, we now know that some gamete genes can be methylated in different ways according to parental sex and environmental effects, a process called **genetic imprinting.** Such methylation patterns may be passed on to more than one generation. In terms of stress, several studies appear to show such effects. For example, studies in the 1980s on birth and death records of humans in an isolated Swedish town show that a severe famine may genetically affect the longevity of grandchildren. More recently, Isabelle Mansuy and colleagues reported that male mice stressed by lack of maternal care not only developed anxiety behaviors but also passed this behavior on to their offspring, which were not stressed.

check your understanding 7.6

What is stress? How is the response to short-term stress different to that of long-term stress?

What is the role of the catecholamines in the stress response?

7.7 Endocrine Control of Fuel Metabolism in Vertebrates

Next let's turn to the metabolic patterns that occur in the absence of stress, including the hormonal factors that govern this normal metabolism.

Fuel metabolism includes anabolism, catabolism, and interconversions among energy-rich organic molecules

The term **metabolism** refers to all chemical reactions that occur within body cells. Those reactions involving degradation, synthesis, and transformation of the three classes of energy-rich organic molecules—protein, carbohydrate, and fat—are collectively known as **intermediary metabolism** or **fuel metabolism** (Table 7-4).

During digestion, large nutrient molecules (**macromolecules**) are broken down into their smaller, absorbable subunits as follows: Proteins are converted into amino acids, complex carbohydrates into monosaccharides (mainly glucose), and triglycerides (dietary fats) into monoglycerides and free fatty acids. These absorbable units are transferred from the digestive tract lumen into the blood directly or by way of lymph (Chapter 14).

Anabolism and Catabolism These organic molecules are constantly exchanged between the blood and the cells of the body (Figure 7-23). The chemical reactions in which the organic molecules participate within the cells are categorized into two metabolic processes: anabolism and catabolism. **Anabolism** is the buildup or synthesis of larger organic macromolecules from the small organic molecular subunits. Anabolic reactions generally require the input of energy in the form of ATP. These reactions result in either (1) the manufacture of materials needed by the cell, such as cellular structural proteins or secretory products, or (2) storage of excess ingested nutrients not immediately needed for energy production or needed as cellular building blocks. Storage is in the form of glycogen (the storage form of glucose) or fat reservoirs. **Catabolism** is the breakdown, or degradation, of large, energy-rich organic molecules within cells. Catabolism encompasses two levels of breakdown: (1) *hydrolysis* (see p. 656) of large, cellular, organic macromolecules into their smaller subunits intracellularly as in release of glucose from glycogen stores (also occurring extracellularly in the digestive tract), and (2) oxidation of smaller subunits, such as glucose, to release energy for ATP production (see Chapter 2).

As an alternative to energy production, the smaller, multipotential organic subunits derived from intracellular hydrolysis may be released into the blood. These mobilized glucose, fatty acid, and amino acid molecules can then be used as needed for ATP production or cellular synthesis elsewhere in the body.

In an adult, the rates of anabolism and catabolism are generally in balance, so an adult body remains in a dynamic steady state and appears unchanged even though the organic molecules that determine its structure and function are continuously being turned over. During growth, anabolism exceeds intracellular catabolism.

TABLE 7-4 Summary of Reactions in Fuel Metabolism

Metabolic Process	Reaction	Consequence
Glycogenesis	Glucose → glycogen	↓ Blood glucose
Glycogenolysis	Glycogen → glucose	↑ Blood glucose
Gluconeogenesis	Amino acids → glucose	↑ Blood glucose
Protein Synthesis	Amino acids → protein	↓ Blood amino acids
Protein Degradation	Protein → amino acids	↑ Blood amino acids
Fat Synthesis (Lipogenesis or Triglyceride Synthesis)	Fatty acids and glycerol → triglycerides	↓ Blood fatty acids
Fat Breakdown (Lypolysis or Triglyceride Degradation)	Triglycerides → fatty acids and glycerol	↑ Blood fatty acids

© Cengage Learning, 2013

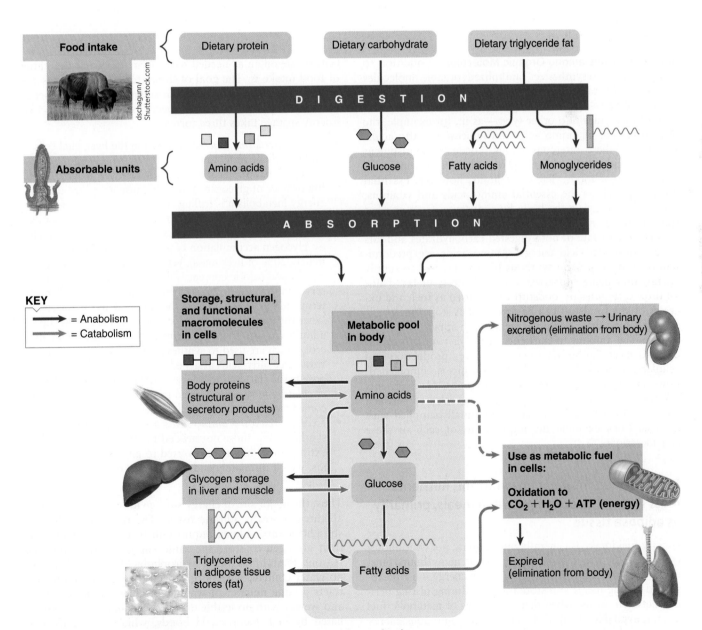

FIGURE 7-23 Summary of the major pathways involving organic nutrient molecules.

© Cengage Learning, 2013

TABLE 7-5 Stored Metabolic Fuel in the Human Body

Metabolic Fuel	Circulating Form	Storage Form	Major Storage Site	Percentage of Total Body Energy Content (and Calories*)	Reservoir Capacity	Role
Carbohydrate	Glucose	Glycogen	Liver, muscle	1% (1,500 calories)	Less than a day's worth of energy	First energy source; essential for the brain
Fat	Free fatty acids	Triglycerides	Adipose tissue	77% (143,000 calories)	About two months' worth of energy	Primary energy reservoir; energy source during a fast
Protein	Amino acids	Body proteins	Muscle	22% (41,000 calories)	Death results long before capacity is used because of structural and functional impairment	Source of glucose for the brain during a fast; last resort to meet other energy needs

*Actually refers to kilocalories; see p. 717.

Interconversions among Organic Molecules In addition to being able to resynthesize catabolized organic molecules back into the same type of molecules, many cells of the body, especially liver cells, can convert most types of small organic molecules into other types—as in, for example, the transformation of amino acids into glucose or fatty acids. Because of these interconversions, adequate nourishment can be provided by a wide range of molecules present in different types of foods. There are limits, however. **Essential nutrients,** such as the essential amino acids and vitamins, cannot be formed in the body by conversion from another type of organic molecule.

The major fate of both ingested carbohydrates and fats is catabolism to yield energy. Amino acids are predominantly used for protein synthesis but can be used to supply energy after being converted to carbohydrate or fat. Thus, all three categories of foodstuff can be used as fuel, and excesses of any foodstuff can be deposited as stored fuel.

At a superficial level, fuel metabolism appears relatively simple: The amount of nutrients in the diet must be sufficient to meet the body's needs for energy production and cellular synthesis. This simple relationship is complicated, however, by two important considerations: (1) Nutrients taken in at meals must be stored and then released between meals, and (2) the brain must be continually supplied with glucose. Let's examine the implications of each of these considerations.

Because food intake is intermittent, nutrients must be stored for use between meals, primarily as adipose tissue

Dietary fuel intake is generally intermittent, not continuous; even an animal that grazes on grass for many hours must walk, rest, and sleep. As a result, excess energy must be absorbed during meals and stored for use during fasting periods between meals, when dietary sources of metabolic fuel are not available (Table 7-5). Despite discontinuous energy intake, the body cells' demand for energy is ever-present and fluctuating. That is, energy must constantly be available for

cells to use on an as-needed basis no matter what the status of food intake is. The goal of these storage processes is *energy homeostasis:* maintaining an adequate supply of energy for both short- and long-term needs (see also Chapter 15). Energy storage takes three forms:

- *Excess circulating glucose* is stored in the liver, glial cells, and muscle of vertebrates and in the fat body in insects as *glycogen* (p. 26). For many species, researchers ignore the importance of glycogen storage because its contribution to energy metabolism is trifling compared to the energy value of lipid deposits or bulk increases in body protein. For example, the crucian carp (*Carassius carassius*) has the highest glycogen accumulation of any vertebrate (approximately 20% of liver mass), yet its contribution to body weight and caloric content is minimal. Carbohydrate reserves in the liver and muscle are useful as a limited short-term energy reserve because they are easy to mobilize. Once the liver and muscle glycogen stores are "filled up," additional glucose is transformed into fatty acids and glycerol, which are used to synthesize *triglycerides* (glycerol with three fatty acids attached; p. 26), primarily in adipose tissue (fat) and to a lesser extent in muscle. *Excess circulating fatty acids* derived from dietary intake also become incorporated into triglycerides.
- *Excess circulating amino acids* are used first for protein synthesis, but those not needed for this are not stored as extra protein but are converted to glucose and fatty acids, which ultimately end up being stored as triglycerides.

Thus, the major site of energy storage for excess nutrients of all three classes is adipose tissue. The reasons for this are straightforward. Lipids require much less space and mass than glycogen to store the same energy content, for two reasons. First, lipids (being hydrophobic) form droplets with no water content, whereas glycogen (being hydrophilic) attracts a large number of water molecules, taking up space and weight with no usable energy. Second, lipids are dominated by high-energy C–H bonds, which can be used to make NADH (see p. 48), whereas glycogen has a large number of C–O bonds that are already oxidized.

Sufficient triglyceride is stored to provide energy for prolonged periods of fasting (such as the emperor penguin, which voluntarily fasts throughout the Antarctic winter); the fatty acids released from triglyceride catabolism serve as the primary source of energy for most tissues. The catabolism of stored triglycerides frees glycerol as well as fatty acids, but quantitatively speaking, the fatty acids are far more important. Catabolism of stored fat yields 90% fatty acids and 10% glycerol by weight. Glycerol can be converted to glucose by the liver and contributes in a small way to maintaining blood glucose during periods of prolonged food withdrawal.

As a third energy reservoir, a substantial amount of energy is stored as *structural protein*, primarily in muscle, the most abundant protein mass in the body. For example, once the triglyceride reserve is exhausted in the wintering emperor penguins and they are forced to catabolize proteins, it is time to begin the long march to the sea in order to replenish body reserves. In most animals protein is not the first choice to tap as an energy source, because it serves other essential functions; in contrast, the glycogen and triglyceride reservoirs serve primarily as energy depots.

Glucose is homeostatically regulated to supply the brain and to prevent damaging processes at high concentrations

The second factor complicating fuel metabolism (in addition to intermittent intake) is that the vertebrate brain normally depends on the delivery of adequate blood glucose as its sole source of energy. Consequently, it is essential that the blood glucose concentration be maintained above a critical level, that is, that hypoglycemia be prevented. The blood glucose concentration of humans is typically 100 mg glucose/100 mL plasma and is normally maintained within the narrow limits of 70 to 110 mg/100 mL in a process of *glucose homeostasis* (you will see later how this is regulated with insulin and glucagon). Blood glucose is typically lower in large mammals and two- to sixfold higher in birds, but all have some general set-point regulation. Liver glycogen is an important reservoir for maintaining blood glucose levels during a short fast, and is the primary reservoir in daily life. However, liver glycogen is depleted relatively rapidly, so during a longer fast than that between daily meals, other mechanisms must ensure that the energy requirements of the glucose-dependent brain are met. First, when new dietary glucose is not entering the blood, tissues not obligated to use glucose shift their metabolic gears to burn fatty acids instead, thus sparing glucose for the brain. Fatty acids are made available by catabolism of triglyceride stores as an alternative energy source for non-glucose-dependent tissues. Second, amino acids can be converted to glucose by gluconeogenesis, whereas fatty acids cannot. Thus, once glycogen stores are depleted despite glucose sparing, new glucose supplies for the brain are provided by catabolism of body proteins and conversion of the freed amino acids into glucose. How then can a crucian carp survive the months of anoxia in frozen ponds? In order to master this feat, glycogen content in the crucian carp's brain increases 15-fold as a reservoir for the neural tissue, concomitant with a 10-fold decline in the activity of the Na^+/K^+ ATPase pump units.

TABLE 7-6 Comparison of Absorptive and Postabsorptive States

Metabolic Fuel	Absorptive State	Postabsorptive State
Carbohydrate	Glucose providing major energy source Glycogen synthesis and storage Excess converted to and stored as triglyceride fat	Glycogen degradation and depletion Glucose sparing to conserve glucose for the brain Production of new glucose through gluconeogenesis
Fat	Triglyceride synthesis and storage	Triglyceride catabolism Fatty acids providing the major energy source for non-glucose-dependent tissues
Protein	Protein synthesis Excess converted to and stored as triglyceride fat	Protein catabolism Amino acids used for gluconeogenesis

© Cengage Learning, 2013

An alternative fuel for the brain that is derived from fats are ketones, normally generated during starvation. We investigate ketones shortly.

A third complication of fuel metabolism is the potential damage of excessively high blood glucose (**hyperglycemia**). At levels significantly higher than an animal's set point, three harmful processes occur: (1) the osmotic effect of the glucose leads to cell dehydration; (2) glucose can spontaneously react—in a process called **glycation**—with proteins such as hemoglobin and collagen, forming a covalent bond that alters the structure and function of the proteins (see p. 564); and (3) some tissues such as lens convert excess glucose into *sorbitol*, which can cause cataracts. These effects are relatively slow, so that temporary rises in blood glucose are not necessarily dangerous (if not too extreme). Thus, glucose homeostasis is not as precise as other regulation such as mammalian body-temperature homeostasis. In summary, glucose homeostasis works to prevent short-term hypoglycemia and long-term hyperglycemia.

Metabolic fuels are stored during the absorptive state and are mobilized during the postabsorptive state

From the preceding discussion, you can see that the disposition of organic molecules depends on the body's metabolic state. Two functional metabolic states are related to eating and fasting cycles—the absorptive state and the postabsorptive state, respectively (Table 7-6).

Absorptive State After a meal, ingested nutrients are being absorbed and entering the blood during the **absorptive, or fed, state**. During this time, glucose is plentiful (assuming the meal includes carbohydrates) and serves as the major energy source. Very little of the absorbed fat and amino acids is used for energy during the absorptive state because most cells are programmed to use glucose first, when available. Extra nutrients not immediately used for energy or structural repairs are channeled into storage as glycogen or triglycerides.

Postabsorptive State In humans the average meal is completely absorbed in about four hours. Therefore, on a typical three-meals-a-day diet, no nutrients are being absorbed from the digestive tract during late morning and late afternoon and throughout the night. These times constitute the **postabsorptive**, or **fasting, state**. There are few daily fasting periods in nonruminant herbivores, which may graze most of the day, while there are many in carnivores (such as cheetahs), which may gorge on only a few large prey animals per month.

During the fasting state, endogenous energy stores are mobilized to provide energy, whereas gluconeogenesis and glucose sparing are used to maintain the blood glucose at an adequate level to nourish the brain. The synthesis of protein and fat is curtailed. Instead, stores of these organic molecules are catabolized for glucose formation and energy production, respectively. Carbohydrate synthesis does occur through gluconeogenesis, but the use of glucose for energy is greatly reduced.

Note that blood concentration of nutrients does not fluctuate markedly between absorptive and postabsorptive states. During the absorptive state, the glut of absorbed nutrients is swiftly removed from the blood and placed into storage; during the postabsorptive state, these stores are catabolized to maintain blood concentrations at levels necessary to sustain tissue energy demands.

Roles of Key Tissues in Metabolic States During these alternating metabolic states, various tissues play different roles as summarized here:

- The *liver* plays the primary role in maintaining normal blood-glucose levels. It stores glycogen when excess glucose is available, releases glucose into the blood when needed, and is the principal site for metabolic interconversions such as gluconeogenesis.
- *Adipose* tissue serves as the primary energy-storage site and is important in regulating fatty acid levels in the blood.
- *Muscle* is the primary site of amino acid storage and is the major energy user. It also stores glycogen.
- The *brain* normally needs glucose as its main energy source, yet in most species only minor glycogen storage occurs in glial cells,

Lesser energy sources are tapped as needed

Several other organic intermediates play a lesser role as energy sources—namely, glycerol, lactate, and ketones (or ketone bodies):

- As mentioned earlier, *glycerol* derived from triglyceride hydrolysis (it is the backbone to which the fatty acid chains are attached) can be converted to glucose by the liver.
- Similarly, *lactate,* which is produced by the incomplete catabolism of glucose via glycolysis in muscle (see p. 55), can also be converted to glucose in the liver.

- **Ketone bodies** (namely acetone, acetoacidic acid, and b-hydroxybutyric acid) are a group of compounds produced by the liver during glucose sparing. Unlike other tissues, when the liver uses fatty acids as an energy source, it oxidizes them only to acetyl coenzyme A (acetyl CoA), which it cannot process through the citric acid cycle for further energy extraction. Thus, the liver does not degrade fatty acids all the way to CO_2 and H_2O for maximum energy release. Instead, it partially extracts the available energy and converts the remaining energy-bearing acetyl CoA molecules into ketone bodies, which it releases into the blood. Ketone bodies serve as an alternative energy source for tissues capable of oxidizing them further by means of the citric acid cycle.

During long-term starvation, the vertebrate brain starts using ketones instead of glucose as a major energy source because glucose homeostasis cannot be maintained. Because death from starvation is usually due to protein wasting rather than to hypoglycemia, prolonged survival without any caloric intake requires that gluconeogenesis be kept to a minimum as long as the needs of the brain are not compromised. A sizable portion of cell protein can be catabolized without serious cell malfunction, but a point is finally reached at which a cannibalized cell can no longer function adequately. To ward off the fatal point of failure so long as possible during prolonged starvation, the brain starts using ketones as a major energy source, correspondingly decreasing its use of glucose. The brain's use of ketones from the liver limits the necessity of mobilizing body proteins for glucose production to nourish the brain.

Locusts and cockroaches also rely on ketone bodies during periods of starvation. The activities of ketone-body-oxidizing enzymes are also elevated in brain tissue of insects that make extensive use of fatty acids during flight and of insects that do not feed as adults.

The pancreatic hormones, insulin and glucagon, are most important in regulating fuel metabolism

How does the body "know" when to shift its metabolic gears from one of net anabolism and nutrient storage to one of net catabolism and glucose sparing? The flow of organic nutrients along metabolic pathways is influenced by a variety of hormones, including *insulin, glucagon, epinephrine, glucocorticoids* (cortisol or corticosterone), and *GH*. Under most circumstances, the pancreatic hormones, insulin and glucagon, are the dominant hormonal regulators responsible for glucose homeostasis; that is, they shift the metabolic pathways back and forth from net anabolism to net catabolism and glucose sparing, depending on whether the body is feasting or fasting, respectively.

The vertebrate **pancreas** is an organ composed of both exocrine and endocrine tissues. The exocrine portion of the pancreas secretes a watery alkaline solution and digestive enzymes through the pancreatic duct into the digestive tract lumen (Chapter 14). Scattered throughout the pancreas between the exocrine cells are clusters, or "islands," of endocrine cells known as the **islets of Langerhans** (see Figure 14-15, p. 683). These cells are the integrators of endocrine regulatory responses controlled by the pancreas. The most abundant pancreatic-endocrine cell type in mammals is the **β (beta) cell**, the site of *insulin* synthesis and secretion. The

Factors that increase blood glucose

Factors that decrease blood glucose

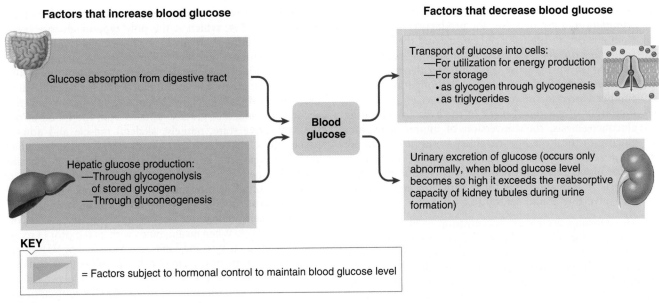

KEY

= Factors subject to hormonal control to maintain blood glucose level

FIGURE 7-24 **Factors affecting blood glucose concentration.**

© Cengage Learning, 2013

β cells are concentrated centrally in the islets, with the other cells clustered around the periphery. Insulin has been isolated from representative species in all classes of vertebrates including the agnathans. Next most important are the **α (alpha) cells**, which produce *glucagon*. The **D (delta) cells** are the pancreatic site of *somatostatin* synthesis, whereas the least common islet cells, the **F cells**, secrete *pancreatic polypeptide*, which plays a possible role in reducing appetite and food intake, is poorly understood, and will not be discussed any further.

Somatostatins Somatostatin is released from the pancreatic D cells in direct response to an increase in blood glucose and blood amino acids during absorption of a meal. By exerting inhibitory effects on digestion of nutrients and nutrient absorption, pancreatic somatostatin acts as a hormone in negative-feedback fashion to prevent excessive plasma levels of nutrients. Pancreatic somatostatin also acts in a paracrine fashion, decreasing the secretion of insulin, glucagon, and somatostatin itself, but the physiologic importance of such paracrine function has not been determined.

Somatostatin is also produced by cells lining the digestive tract, where it acts locally as a paracrine to inhibit most digestive processes. Furthermore, somatostatin (alias GHIH) is produced by the hypothalamus, where it inhibits the secretion of growth hormone and TSH.

We next consider insulin and then glucagon, followed by a discussion of how both function as an endocrine unit to shift metabolic gears between the absorptive and postabsorptive states.

Insulin lowers blood levels of glucose, amino acids, and fatty acids, and promotes their storage

In vertebrates **insulin** has important effects on carbohydrate, fat, and protein metabolism. It lowers blood levels of glucose, fatty acids, and amino acids and promotes their storage. As these nutrient molecules enter the blood during the absorptive state, insulin promotes their cellular uptake and conversion into glycogen, triglycerides, and protein, respectively. Insulin exerts its many effects either by altering transport of specific bloodborne nutrients into cells or by altering the activity of the enzymes involved in specific metabolic pathways. To accomplish its effects, in some instances insulin increases the activity of an enzyme, for example *glycogen synthase,* the key regulated enzyme that synthesizes glycogen from glucose molecules, a process known as **glycogenesis.** In other cases, however, insulin decreases the activity of an enzyme, for example, by inhibiting *hormone-sensitive lipase,* the enzyme that catalyzes the breakdown of stored triglycerides back to free fatty acids and glycerol.

Vertebrate insulin belongs to a superfamily consisting of a number of peptide hormones including **insulin-like peptides (ILPs)** in some nonvertebrates (mollusks, nematodes, insects). Nonvertebrate versions also consist of two peptide chains cross-linked by highly conserved insulin-type disulfide bonds. Some insect ILPs are neuroendocrine hormones as exemplified by locust insulin-related peptide (LIRP), which was isolated from the neurohemal lobes in the corpora cardiaca (p. 280). Functionally, these nonvertebrate insulin-like peptides act as growth factors and developmental regulators, much like vertebrate IGF. For example, in fruit flies and *C. elegans* (nematode) mutations in the insulin-like receptor genes result in defects in growth and development.

Actions on Carbohydrates The maintenance of blood glucose homeostasis is a particularly important function of the pancreas. Circulating glucose concentrations are determined by the balance among the following processes (Figure 7-24): glucose absorption from the digestive tract; transport of glucose into cells; cellular (primarily hepatic) glucose production; and (abnormally) urinary excretion of glucose. Somatostatins regulate digestive absorption, while insulin and glucagon act on cellular transport and hepatic production.

Insulin exerts four effects to lower bloodglucose levels and promote carbohydrate storage:

1. *Facilitation of glucose transport* into most cells.
2. *Stimulation of glycogenesis,* the production of glycogen from glucose, in both skeletal muscle and the liver.
3. *Inhibition of glycogenolysis,* by inhibiting breakdown of glycogen into glucose. By inhibiting degradation of glycogen, insulin likewise favors carbohydrate storage and decreases glucose output by the liver.
4. *Inhibition of gluconeogenesis* in the liver. By inhibiting gluconeogenesis, the conversion of amino acids into glucose in the liver, hepatic glucose output is reduced. Insulin does so by decreasing the amount of amino acids in the blood that are available to the liver for gluconeogenesis, and by inhibiting the hepatic enzymes required for converting amino acids into glucose.

Thus, insulin decreases concentration of blood glucose by promoting the cells' uptake of glucose from the blood for use and storage, while simultaneously blocking the two mechanisms by which the liver releases glucose into the blood (glycogenolysis and gluconeogenesis). Note that the cells stimulated by glucose are the effectors in a negative-feedback sense, because they make the actual corrective changes, whereas insulin is a signal molecule (akin to the electrical signals running from a thermostat to an air conditioner). Insulin is the *only* hormone capable of lowering glucose concentrations in the blood.

These functions have been primarily explored in depth in mammals, but in nonmammalian vertebrates experiments suggest that insulin has similar roles. For example, removing islet cells from either the teleost *Gillichthys mirabilis* (mudsucker) or the southern lamprey *Geotria australis* dramatically increases blood glucose concentrations. In general, the hypoglycemic action of insulin in ectotherms develops more slowly compared to endotherms and can require days instead of hours to achieve a full effect.

Glucose Transporter Recruitment Glucose transport between the blood and cells is accomplished by means of plasma membrane carriers known as **glucose transporters (GLUT)**. Fourteen forms of glucose transporters have been identified, named in the order they were discovered—GLUT-1, GLUT-2, and so on. The amino acid structure of each GLUT isoform is nearly identical when the structure is compared with different species, which suggests that all glucose transporters originated from a common ancestor. For example, GLUT-1 is over 95% identical between different mammalian species. These glucose transporters all accomplish passive facilitated diffusion of glucose across the plasma membrane. Once GLUT transports glucose into a cell, an enzyme within the cell immediately phosphorylates glucose to **glucose-6-phosphate**, which has no means out of the cell, unlike "plain" glucose, which could exit through the bidirectional glucose transporter. Therefore, glucose is trapped inside the cell. Furthermore, the phosphorylation of glucose as it enters the cell keeps the intracellular concentration of plain glucose low so that a gradient favoring the facilitated diffusion of glucose into the cell is maintained.

Each member of the GLUT family performs slightly different functions. For example, *GLUT-1* transports glucose across the blood-brain barrier, *GLUT-2* transfers into glucose from kidney (Chapter 12) and intestinal cells (Chapter 14) into the blood, and *GLUT-3* is the main transporter of glucose into neurons. The glucose transporter responsible for the majority of glucose uptake by most cells of the body is *GLUT-4*, which is the only type of glucose transporter that responds to insulin. Glucose molecules cannot readily penetrate most cell membranes in the absence of insulin, making most tissues highly dependent on insulin for uptake of glucose from the blood and for its subsequent use. GLUT-4 is especially abundant in the tissues that account for much of glucose uptake from the blood during the absorptive state, namely, skeletal muscle and adipose tissue cells. Unlike the other types of GLUT molecules, which are always present in the plasma membranes at the sites where they perform their functions, GLUT-4 in the absence of insulin is excluded from the plasma membrane. Insulin promotes glucose uptake by **transporter recruitment.** Insulin-dependent cells maintain a pool of intracellular vesicles containing GLUT-4. When insulin binds with its receptor (a receptor that acts as a tyrosine kinase enzyme; see p. 95) on the surface membrane of the target cell, the subsequent signaling pathway induces these vesicles to move to the plasma membrane and fuse with it, thus inserting GLUT-4 molecules into the plasma membrane. In this way, increased insulin secretion promotes a rapid 10- to 30-fold increase in glucose uptake by insulin-dependent cells. When insulin secretion decreases, these glucose transporters are retrieved from the membrane by endocytosis and returned to the intracellular pool.

Several tissues do not depend on insulin for their glucose uptake—namely, the brain, working muscles, and liver. The brain, which requires a constant supply of glucose for its minute-to-minute energy needs, is freely permeable to glucose at all times by means of GLUT-1 and GLUT-3 transporters. Skeletal muscle cells do not depend on insulin for their glucose uptake during exercise, even though they are dependent at rest. Muscle contraction triggers the insertion of GLUT-4 into the plasma membranes of exercising muscle cells in the absence of insulin. The liver also does not depend on insulin for glucose uptake, because it does not use GLUT-4. However, insulin does enhance the metabolism of glucose by the liver by stimulating the first step in glucose metabolism, the phosphorylation of glucose to form glucose-6-phosphate.

Insulin also exerts important actions on fat and protein.

Actions on Fat Insulin exerts multiple effects to lower blood fatty acids and promote triglyceride storage:

1. *Increase in the transport of glucose into adipose tissue cells* by means of GLUT-4 recruitment. Glucose serves as a precursor for the formation of fatty acids and glycerol, which are the raw materials for triglyceride synthesis.
2. *Activation of enzymes* that catalyze the production of *fatty acids* from glucose derivatives.
3. *Promotion of the uptake of fatty acids* from the blood into adipose tissue cells.
4. *Inhibition of lipolysis* (fat breakdown), by inhibiting *hormone-sensitive lipase,* the enzyme that catalyzes the breakdown of stored triglycerides back to free fatty acids and glycerol, thus reducing release of fatty acids from adipose tissue into the blood. In birds, insulin is not antilipolytic.

Collectively, these actions favor removal of glucose and fatty acids from the blood and promote their storage as triglycerides. Adipose tissue here serves as an effector.

Actions on Protein Insulin lowers blood amino acid levels and enhances protein synthesis through several effects:

1. *Promotion of the active transport of amino acids* from the blood into muscles and other tissues. This effect lowers the circulating amino acid level and provides the building blocks for protein synthesis within the cells.
2. *Enhancing the rate of amino acid incorporation into protein* by stimulating the cells' protein-synthesizing machinery.
3. *Inhibition of protein degradation.*

The collective result of these actions is a protein anabolic effect. For this reason, insulin is essential for normal growth.

Summary of Insulin's Actions In short, insulin stimulates biosynthetic pathways that increase glucose use, carbohydrate and fat storage, and protein synthesis. In so doing, this hormone lowers the blood glucose, fatty acid, and amino acid levels. This metabolic pattern is characteristic of the absorptive state. Indeed, insulin secretion rises during this state and is responsible for shifting metabolic pathways to net anabolism. Consider as well that the number of insulin receptors on a target cell reflects the carbohydrate content of the natural diet and the differential involvement of insulin in the regulation of protein or carbohydrate metabolism in different species. For example, in salmonids, a carnivorous group of fish whose diet consists mainly of protein, the number of insulin receptors is lower compared to either herbivorous (such as the African fish tilapia) or omnivorous (such as carp) species. In general, the number of insulin receptors in ectothermic vertebrates is higher than that of IGF-I receptors, whereas in endotherms the opposite holds true. The increase in ratio between insulin and IGF-I receptors during vertebrate evolution is probably related to the higher physiological response of skeletal muscles to insulin in birds and mammals.

When insulin secretion is low, the opposite effects occur. The rate of glucose entry into cells is reduced and net catabolism rather than net synthesis of glycogen, triglycerides, and protein occurs. This pattern is reminiscent of the postabsorptive state; indeed, insulin secretion falls during the postabsorptive state. However, the other major pancreatic hormone, glucagon, also plays an important role in shifting from absorptive to postabsorptive metabolic patterns, as we describe shortly.

The primary stimulus for increased insulin secretion is an increase in blood glucose concentration

The primary control of insulin secretion is a direct negative-feedback system between the pancreatic β cells and the concentration of glucose in the blood flowing to them. An elevated blood-glucose level, such as during absorption of a meal, directly stimulates synthesis and release of insulin by the β cells. The increased insulin in turn reduces blood glu-

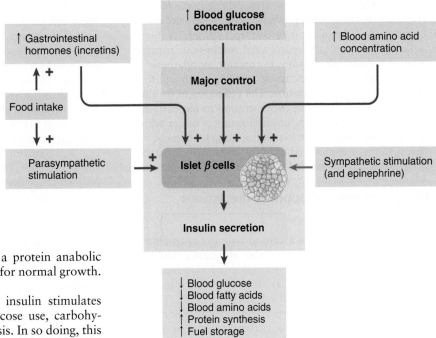

FIGURE 7-25 Factors controlling insulin secretion.

© Cengage Learning, 2013

cose to normal while it promotes use and storage of this nutrient. Conversely, a fall in blood glucose below normal, such as during fasting, directly inhibits insulin secretion. Lowering the rate of insulin secretion shifts metabolism from the absorptive to the postabsorptive pattern. Thus, this simple negative-feedback system can maintain a relatively constant supply of glucose to the tissues without requiring the participation of nerves or other hormones.

In addition to plasma glucose concentration, other inputs are involved in regulating insulin secretion, as follows (Figure 7-25):

- An elevated blood amino acid level, such as after eating a high-protein meal, directly stimulates β cells to increase insulin secretion. In negative-feedback fashion, the increased insulin enhances entry of these amino acids into the cells, lowering blood amino acid level while promoting protein synthesis.
- The major gastrointestinal hormones secreted by the digestive tract in response to food, especially *glucose-dependent insulinotropic peptide (GIP)* (see p. 711) and a similar candidate hormone *glucagon-like peptide (GLP)*, stimulate pancreatic insulin secretion in addition to having direct regulatory effects on the digestive system. Through this control, insulin secretion is increased in feedforward or anticipatory fashion (p. 16) even before nutrient absorption increases the concentration of glucose and amino acids in the blood. Hormones released from the digestive tract that "notify" the β pancreatic cell of the impending rise in blood nutrients (primarily blood glucose) are termed **incretins.** Incretins increase insulin secretion by increasing cAMP, which enhances Ca^{2+}-induced release of insulin.
- The autonomic nervous system also directly influences insulin secretion. The islets are richly innervated by both

parasympathetic (vagal) and sympathetic nerve fibers. The increase in parasympathetic activity that occurs in response to food in the digestive tract stimulates insulin release. This, too, is a feedforward response in anticipation of nutrient absorption. In contrast, sympathetic stimulation and the concurrent increase in epinephrine both inhibit insulin secretion. The reduction in insulin allows the blood glucose concentration to increase, an appropriate response to the circumstances under which generalized sympathetic activation occurs—namely, stress (fight-or-flight) and exercise. In both these situations, extra fuel is needed for increased muscle activity.

Glucagon in general opposes the actions of insulin

Even though insulin plays a central role in controlling the metabolic adjustments between the absorptive and postabsorptive states, the secretory product of vertebrate pancreatic-islet α cells, **glucagon,** is also very important. Many physiologists view the insulin-secreting β cells and the glucagon-secreting α cells as a coupled antagonistic endocrine system whose combined secretory output is a major factor in regulating fuel metabolism.

Glucagon affects many of the same metabolic processes that are influenced by insulin, but in most cases glucagon's actions are opposite to those of insulin. The major site of action of glucagon is the liver (serving as an effector), where it exerts a variety of effects on carbohydrate, fat, and protein metabolism.

Actions on Carbohydrate
The overall effects of glucagon on carbohydrate metabolism are to increase hepatic glucose production and release, and thus raise blood glucose levels. Glucagon exerts its hyperglycemic effects by decreasing glycogen synthesis, promoting glycogenolysis, and stimulating gluconeogenesis.

Actions on Fat
Glucagon also antagonizes the actions of insulin with regard to fat metabolism by promoting fat breakdown and inhibiting triglyceride synthesis. Glucagon enhances hepatic ketone production (**ketogenesis**) by promoting conversion of fatty acids to ketone bodies. Thus, blood levels of fatty acids and ketones rise under glucagon's influence.

Actions on Protein
Glucagon inhibits hepatic protein synthesis and promotes degradation of hepatic protein. Stimulation of gluconeogenesis further contributes to glucagon's catabolic effect on hepatic protein metabolism. Glucagon promotes protein catabolism in the liver, but it does not have any significant effect on blood amino acid levels because it does not affect muscle protein, the major protein store in the body.

Glucagon secretion is increased during the postabsorptive state

Considering the catabolic effects of glucagon on energy stores, you would be correct in assuming that glucagon secretion is increased during the postabsorptive state and decreased during the absorptive state, just the opposite of insulin secretion. In fact, insulin is sometimes called a "hormone of feasting" and glucagon a "hormone of fasting." Insulin tends to put nutrients in storage when their blood levels are high, such as following a meal, whereas glucagon promotes catabolism of nutrient stores between meals to keep up the blood nutrient levels, especially blood glucose. The levels of glucagon vary not only within an individual but also among species; for example, the average concentration of glucagon in birds is approximately twice as great as in mammals, presumably because of the higher energy requirements of avian tissues.

As in insulin secretion, the major factor regulating glucagon secretion is a direct effect of the blood glucose concentration on the endocrine pancreas. In this case, the pancreatic α cells increase glucagon secretion in response to a fall in blood glucose. The hyperglycemic actions of this hormone tend to restore the blood glucose concentration to normal. Conversely, an increase in blood glucose concentration, such as after a meal, inhibits glucagon secretion, which likewise tends to restore the blood glucose concentration to normal.

Insulin and glucagon work as a team to maintain blood glucose and fatty acid levels

Thus, there is a direct negative-feedback relationship between blood glucose concentration and the α cells' rate of secretion, but it is in the opposite direction of the effect of blood glucose on the β cells (Figure 7-26). Because glucagon raises blood glucose and insulin decreases blood glucose, the changes in secretion of these pancreatic hormones in response to deviations in blood glucose work together homeostatically to restore blood glucose levels to normal.

Similarly, a fall in blood fatty-acid concentration directly stimulates glucagon output and inhibits insulin output by the pancreas, both of which are negative-feedback control mechanisms to restore the blood fatty-acid level to normal.

The effect of blood amino acid concentration on the secretion of these two hormones is a different story. A rise in blood amino acid concentration stimulates *both* glucagon and insulin secretion (Figure 7-27). Why this seeming paradox, because glucagon does not affect blood amino acid concentration? Because little carbohydrate is available for absorption following consumption of a high-protein meal, the amino-acid-induced increase in insulin secretion would drive too much glucose into the cells, causing a sudden, inappropriate drop in the blood glucose level. The hyperglycemic effects of glucagon counteract the hypoglycemic actions of insulin, with the net result maintenance of normal blood-glucose levels (and prevention of hypoglycemic starvation of the brain) during absorption of a meal that is high in protein but low in carbohydrates.

Epinephrine, cortisol, growth hormone, and thyroid hormone also exert direct metabolic effects

The pancreatic hormones are the most important regulators of normal fuel metabolism. However, several other hormones exert direct metabolic effects, even though control of their secretion is keyed to factors other than transitions in metabolism between feasting and fasting states (Table 7-7).

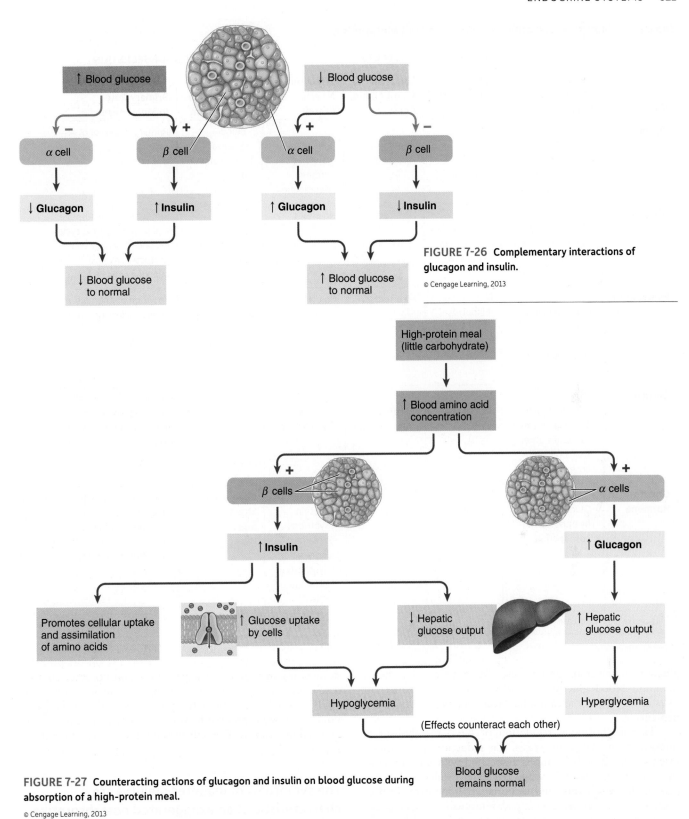

FIGURE 7-26 Complementary interactions of glucagon and insulin.

© Cengage Learning, 2013

FIGURE 7-27 Counteracting actions of glucagon and insulin on blood glucose during absorption of a high-protein meal.

© Cengage Learning, 2013

As we discussed earlier, the stress hormones, epinephrine and cortisol, both increase blood levels of glucose and fatty acids through a variety of metabolic effects, primarily in stress situations. In addition, cortisol mobilizes amino acids by promoting protein catabolism in long-term stress, including long-term starvation, in which the hormone appears to be important to maintaining blood glucose concentrations.

GH (acting through IGF-I) has protein anabolic effects in muscle. In fact, this is one of its growth-promoting features. Although GH can elevate the blood levels of glucose and fatty acids, it is normally of little importance to the overall regulation of fuel metabolism. Deep sleep, stress, exercise, and severe hypoglycemia stimulate GH secretion, possibly to provide fatty acids as an energy source and spare

TABLE 7-7 Summary of Hormonal Control of Fuel Metabolism

Hormone	MAJOR METABOLIC EFFECTS				CONTROL OF SECRETION	
	Effect on Blood Glucose	Effect on Blood Fatty Acids	Effect on Blood Amino Acids	Effect on Muscle Protein	Major Stimuli for Secretion	Primary Role in Metabolism
Insulin	↓ +Glucose uptake +Glycogenesis −Glycogenolysis −Gluconeogenesis	↓ +Triglyceride synthesis −Lipolysis	↓ +Amino acid uptake	↑ +Protein synthesis −Protein degradation	↑ Blood glucose ↑ Blood amino acids	Primary regulator of absorptive and postabsorptive cycles
Glucagon	↑ +Glycogenolysis +Gluconeogenesis −Glycogenesis	↑ +Lipolysis −Triglyceride synthesis	No effect	No effect	↓ Blood glucose ↑ Blood amino acids	Regulation of absorptive and postabsorptive cycles in concert with insulin; protection against hypoglycemia
Epinephrine	↑ +Glycogenolysis +Gluconeogenesis −Insulin secretion +Glucagon secretion	↑ +Lipolysis	No effect	No effect	Sympathetic stimulation during stress and exercise	Provision of energy for emergencies and exercise
Cortisol	↑ +Gluconeogenesis −Glucose uptake by tissues other than brain; glucose sparing	↑ +Lipolysis	↑ +Protein degradation	↓ +Protein degradation	Stress	Mobilization of metabolic fuels and building blocks during adaptation to stress
Growth Hormone	↑ −Glucose uptake by muscles; glucose sparing	↑ +Lipolysis	↓ +Amino acid uptake	↑ +Protein synthesis −Protein degradation +Synthesis of DNA and RNA	Deep sleep Stress Exercise Hypoglycemia	Promotion of growth; normally little role in metabolism; mobilization of fuels plus glucose sparing in extenuating circumstances

↑ = Increase, ↓ = Decrease

glucose for the brain under these circumstances. GH helps maintain blood glucose concentrations during short-term fasts as it opposes the catabolic effects of glucocorticoids on proteins.

Thyroid hormone regulation of basal metabolism is important for fuel homeostasis. The reduction in metabolic rate associated with fasting is associated with a decrease in the concentrations of T_3 while refeeding results in T_3 neogenesis. In ectotherms (and birds as well), increases in thyroid hormone may serve to impact intermediary metabolism by matching utilization of nutrients to nutrient intake.

Note that, with the exception of the anabolic effects of GH on protein metabolism, all the metabolic actions of these other hormones are opposite to those of insulin. Insulin alone can reduce blood glucose and blood fatty-acid levels, whereas glucagon, epinephrine, cortisol, and GH all increase blood levels of these nutrients. Because of this, these other hormones are considered **insulin antagonists.** Thus, the main reason diabetes mellitus (discussed next) has such

devastating metabolic consequences is that no other control mechanism is available to pick up the slack to promote anabolism when insulin activity is insufficient, so the catabolic reactions promoted by other hormones are allowed to proceed unchecked. The only exception is protein anabolism stimulated by GH.

The symptoms of diabetes mellitus are characteristic of an exaggerated postabsorptive state

A common and major form of failed glucose homeostasis is **diabetes mellitus** (*mellitus* means "sweet"). The acute symptoms of diabetes mellitus are attributable to inadequate insulin action coupled with glucagon excess. Because insulin is the only hormone capable of lowering blood glucose levels, one of the most prominent features of diabetes mellitus is elevated blood glucose levels, or *hyperglycemia*. Diabetes literally means "siphon" or "running through," a reference to the large urine volume accompanying the condition, caused by osmotic "attraction" from high glucose levels in the urine. Prolonged high-glucose concentrations have other long-term effects, including blindness from lens cataracts (although inappropriate voiding of the bladder is the first symptom a pet owner notices) (see the box, *Avian Longevity: Unraveling the Mystery*, p. 564).

There are two distinct types of diabetes mellitus. **Type I (insulin-dependent) diabetes mellitus** is characterized by a total or near-total lack of insulin secretion by pancreatic β cells, and occurs in many mammalian species. For example, in dogs the incidence of Type I diabetes is greater in smaller breeds, with about 1 in 400 domestic dogs and cats developing diabetes in their lifetime. Type I diabetes requires exogenous insulin for survival. It arises from an autoimmune reaction involving the erroneous, selective destruction of the pancreatic β cells by inappropriately activated T lymphocytes. The precise cause of this self-destructive immune attack remains unclear.

In humans with **Type II (non-insulin-dependent** or **maturity-onset) diabetes mellitus,** insulin secretion may be normal or even increased, but insulin's target cells are less sensitive than normal to the hormone. Various genetic and lifestyle factors, most notably obesity, are important in the development of Type II diabetes, although the exact cellular mechanisms underlying the condition have not been completely unraveled. It is thought that reduced sensitivity arises from down-regulation (see p. 277) of insulin receptors in association with obesity. In response to chronic *hyperinsulinemia* (elevated blood insulin, which may result from a diet high in rapidly digesting carbohydrates), the number of insulin receptors gradually decreases over time.

check your understanding 7.7

How do insulin and glucagon regulate fuel metabolism?

What hormone(s) predominate in the fed state versus the fasting state? What is their role?

7.8 Endocrine Control of Calcium Metabolism in Vertebrates

Besides regulating the concentration of organic nutrient molecules in the blood by manipulation of anabolic and catabolic pathways, the endocrine system also regulates the plasma concentration of a number of inorganic electrolytes. For example, aldosterone controls Na^+ and K^+ concentrations in the ECF (Chapters 12 and 13). Three other hormones—*parathyroid hormone, calcitonin,* and *vitamin D*—control calcium (Ca^{2+}) and phosphate (PO_4^{3-}) metabolism after their ingestion from the diet and/or uptake from the gill. These hormonal agents concern themselves with regulating plasma Ca^{2+}, and in the process, plasma PO_4^{3-} is also maintained. Plasma Ca^{2+} concentration is one of the most tightly controlled variables in vertebrates. The need for the precise regulation of plasma Ca^{2+} stems from its critical influence on so many general physiological processes as well as from the

additional demands imposed by puberty, lactation, or oviposition (egg laying).

Calcium metabolism in water breathers is different from that in terrestrial vertebrates because, in addition to dietary Ca^{2+}, the ion is also available from the aqueous environment. For example, in seawater the concentration of Ca^{2+} is approximately 10 mmol/L (considerably greater than in cells), whereas in freshwater the concentration of Ca^{2+} varies from 0.1 to 4 mmol/L (comparable to intracellular concentrations). Fish can transport Ca^{2+} from either environment and maintain a positive calcium balance unless environmental concentrations of Ca^{2+} are extremely low, or environmental pollutants inhibit the functioning of the transporter.

Plasma calcium must be closely regulated to prevent changes in neuromuscular excitability

In mammals, about 99% of the Ca^{2+} in the body is in crystalline form within the skeleton and teeth. Of the remaining 1%, about 0.9% is found intracellularly within the soft tissues; less than 0.1% is present in the ECF. Approximately half of the plasma Ca^{2+} is either bound to plasma proteins and therefore restricted to the plasma or is complexed with PO_4^{3-} and not free to participate in chemical reactions. The other half of the plasma Ca^{2+} is freely diffusible and can readily pass into the interstitial fluid and interact with the cells. The free Ca^{2+} in the plasma and interstitial fluid is considered a single pool. Only this free Ca^{2+} is biologically active and subject to regulation; it constitutes less than one thousandth of the total Ca^{2+} in the body.

This small, freely diffusible fraction of ECF Ca^{2+} plays a vital role in a number of essential activities, including the following:

1. *Neuromuscular excitability.* Even minor variations in the concentration of free ECF Ca^{2+} can profoundly and immediately affect the sensitivity of excitable tissues. A fall in free Ca^{2+} results in overexcitability of nerves and muscles; conversely, a rise in free Ca^{2+} depresses neuromuscular excitability. These effects result from the influence of Ca^{2+} on membrane permeability to Na^+. A decrease in free Ca^{2+} increases Na^+ permeability, with the resultant influx of Na^+ moving the resting potential closer to threshold. Consequently, in the presence of **hypocalcemia** (low blood Ca^{2+}) excitable tissues may be brought to threshold by normally ineffective physiological stimuli so that skeletal muscles discharge and contract (go into spasm) "spontaneously" (in the absence of normal stimulation). If severe enough, spastic contraction of the respiratory muscles results in death by asphyxiation. Hypercalcemia (elevated blood Ca^{2+}), in contrast, is also life threatening because it causes cardiac arrhythmias accompanied by generalized depression of neuromuscular excitability.

2. *Excitation–contraction coupling in cardiac and smooth muscle.* Ca^{2+} is the primary trigger of the contractile mechanism in muscles and arises from the ECF in cardiac and smooth muscle (Chapter 8). Note that a *rise in cytosolic Ca^{2+}* within a muscle cell causes contraction, whereas an *increase in free ECF Ca^{2+}* decreases neuromuscular excitability and reduces the likelihood of contraction occurring. Unless this point is kept in mind, it would be difficult to understand why low plasma Ca^{2+}

levels induce muscle hyperactivity when Ca^{2+} is necessary to switch on the contractile apparatus. We are talking about two different Ca^{2+} pools, which exert different effects.

3. *Stimulus–secretion coupling.* The entry of Ca^{2+} into secretory cells, which results from increased permeability to Ca^{2+} in response to appropriate stimulation, triggers the release of the secretory product by exocytosis. This process is important for the secretion of neurotransmitters by nerve cells (p. 131) and for peptide and catecholamine hormone secretion by endocrine cells.

4. *Maintenance of tight junctions between cells.* Calcium forms part of the intercellular cement that holds particular cells tightly together (p. 66).

5. *Clotting of blood.* Calcium serves as a cofactor in several steps of the cascade of reactions that lead to clot formation (p. 397).

In addition to these functions of free Ca^{2+}, intracellular Ca^{2+} serves as a second messenger in many cells and is involved in cell motility and cilia action. Here, calcium levels cannot be allowed to rise too greatly, because elevated Ca^{2+} precipitates readily with PO_4^{3-} as calcium phosphate (free PO_4^{3-} is a major anion in cells, necessary for ATP production, among other uses). Calcium phosphate is essential for bone and teeth formation but is harmful inside a cell.

Because of the profound effects of deviations in free Ca^{2+}, especially on neuromuscular excitability, the plasma concentration of this electrolyte is regulated with extraordinary precision in vertebrates. Let's see how.

Control of Ca^{2+} metabolism includes regulation of both calcium homeostasis and Ca^{2+} balance

Maintaining the proper plasma concentration of free Ca^{2+} differs from regulation of Na^+ and K^+ in two important regards. Na^+ and K^+ homeostasis is maintained primarily by regulating the urinary excretion of these electrolytes so that controlled output matches uncontrolled input (Chapters 12 and 13). In the case of Ca^{2+}, in contrast, not all the ingested Ca^{2+} is absorbed from the digestive tract; instead, the extent of absorption is hormonally controlled and is dependent on the Ca^{2+} status of the body. In addition, bone serves as a large Ca^{2+} reservoir that can be drawn on to maintain the free plasma Ca^{2+} concentration within the narrow limits compatible with life should dietary intake become too low. Exchange of Ca^{2+} between the ECF and bone is also subject to control. Similar in-house stores are not available for Na^+ and K^+.

Regulation of Ca^{2+} metabolism depends on hormonal control of exchanges between the ECF and three other compartments serving as effectors: bone, kidneys, and intestine. Accordingly, control of Ca^{2+} metabolism encompasses two aspects:

1. Regulation of **calcium homeostasis** involves the immediate adjustments required to maintain a *constant free plasma Ca^{2+} concentration* on a minute-to-minute basis. This is largely accomplished by rapid exchanges between the bone and ECF and to a lesser extent by modifications in urinary excretion of Ca^{2+}.

2. Regulation of **calcium balance** involves the more slowly responding adjustments required to maintain a *constant total amount of Ca^{2+} in the body.* Control of Ca^{2+} balance ensures that Ca^{2+} intake is equivalent to Ca^{2+} excretion over the long term (weeks to months). Calcium balance is maintained by adjusting the extent of intestinal Ca^{2+} absorption and urinary Ca^{2+} excretion.

Parathyroid hormone (PTH), the principal regulator of Ca^{2+} metabolism, acts directly or indirectly on all three effector sites. It is the primary hormone responsible for maintaining Ca^{2+} homeostasis and is essential for maintaining Ca^{2+} balance, although vitamin D also contributes in important ways to Ca^{2+} balance. The third Ca^{2+}-influencing hormone, calcitonin, is not essential for routine maintenance of Ca^{2+} homeostasis in birds and mammals, though has roles in other vertebrates. We examine the specific effects of each of these hormonal systems in more detail.

Parathyroid hormone raises free plasma calcium levels by its effects on bone, kidneys, and intestine

Parathyroid hormone (PTH), the predominant hypercalcemic hormone in some amphibians, and reptiles, birds, and mammals, is a peptide hormone secreted by the **parathyroid glands,** four small glands located on the back surface of the mammalian thyroid gland, one in each corner. In birds the number of parathyroids varies between two and four and lie caudal to the thyroid. Like aldosterone, PTH is *essential for life.* The overall effect of PTH is to increase the Ca^{2+} concentration of plasma (and, accordingly, of the entire ECF), thereby preventing hypocalcemia. In the complete absence of PTH, death ensues within a few days, usually because of asphyxiation caused by hypocalcemic spasm of respiratory muscles. By its combined actions on bone, kidneys, and intestine (effectors), PTH raises the plasma Ca^{2+} level when it starts to fall so that hypocalcemia and its effects are normally avoided. This hormone also acts to lower plasma PO_4^{3-} concentration. We consider each of these mechanisms next, beginning with an overview of bone remodeling.

Bone continuously undergoes remodeling

Recall that 99% of the body's Ca^{2+} is found in the skeleton as **hydroxyapatite,** $Ca_3(PO_4)_2$ (p. 293). Bone has numerous functions:

- Support
- Protection of vital internal organs
- Assistance in body movement by giving attachment to muscles and providing leverage
- Manufacture of blood cells (bone marrow)
- Storage depot for Ca^{2+} and PO_4^{3-}, which can be exchanged with the plasma to maintain plasma concentrations of these electrolytes

Normally, $Ca_3(PO_4)_2$ salts are in solution in the ECF, but the conditions within the bone are suitable for these salts to precipitate (crystallize) around the collagen fibers in the matrix. By mobilizing some of these Ca^{2+} stores in the bone, PTH raises the plasma Ca^{2+} concentration when it starts to fall. Egg-laying birds (and dinosaurs as well) have a highly labile reservoir of Ca^{2+} in the form of **medullary bone,** which develops within the long bones in response to an increase in gonadal steroids at puberty. Medullary bone is the most exquisitely estrogen-sensitive form of vertebrate bone. The amount of Ca^{2+} contained in each egg amounts to about

10% of the total body stores of Ca^{2+}, and mobilized medullary bone accounts for approximately 30 to 40% of eggshell calcium.

Bone Remodeling Despite the apparent inanimate nature of bone, bone constituents are continually being turned over. **Bone deposition** and **bone resorption** normally go on concurrently, so bone is constantly being remodeled, much as people remodel buildings by tearing down walls and replacing them. Bone remodeling serves two purposes: (1) It keeps the skeleton appropriately "engineered" for maximum effectiveness in its mechanical uses, and (2) it helps maintain the plasma Ca^{2+} level. Let's examine in more detail the underlying mechanisms and controlling factors for each of these purposes.

Recall that two key types of bone cells are important for bone growth, maintenance, and repair. The *osteoblasts* secrete the extracellular organic matrix within which the $Ca_3(PO_4)_2$ crystals precipitate. The *osteoclasts* resorb bone in their vicinity. (A third type, **osteocytes,** are "retired" osteoblasts imprisoned within the bony wall they have deposited around themselves.) The large, multinucleated osteoclasts attach to the organic matrix and form a "ruffled membrane" that increases its surface area in contact with the bone. Thus attached, the osteoclast actively secretes hydrochloric acid that dissolves the $Ca_3(PO_4)_2$ crystals and enzymes that break down the organic matrix. After it has created a cavity, the osteoclast moves on to an adjacent site to burrow another hole or dies by apoptosis (cell suicide), depending on the regulatory signals it receives. Osteoblasts move into the cavity and secrete osteoid to fill in the hole. Subsequent mineralization of this organic matrix results in new bone to replace the bone dissolved by the osteoclast. Thus, a constant cellular tug-of-war goes on in bone, with bone-forming osteoblasts countering the efforts of the bone-destroying osteoclasts. These construction and demolition crews, working side by side, continuously remodel bone. Throughout most of adult life, the rates of bone formation and bone resorption are about equal, so total bone mass remains fairly constant during this period.

Osteoblasts and osteoclasts both trace their origins to the bone marrow. Osteoblasts are derived from *stromal cells,* a type of connective tissue cell in the bone marrow, whereas osteoclasts differentiate from *macrophages,* a type of white blood cell (p. 465). In a unique communication system, osteoblasts and their immature precursors produce two chemical signals that govern osteoclast development and activity in opposite ways—*RANK ligand* and *osteoprotegerin*—as follows (Figure 7-28):

- **RANK ligand (RANKL)** revs up osteoclast action. (A **ligand** is a small molecule that binds with a larger protein molecule, such as an extracellular chemical messenger binding with a plasma membrane receptor.) As its name implies, RANK ligand binds to **RANK** (for *receptor activator of NF-kB*), a protein receptor on the membrane surface of nearby macrophages. This binding induces the macrophages to differentiate into osteoclasts and helps them live longer by suppressing apoptosis. As a result, bone resorption is stepped up and bone mass decreases.
- Alternatively, neighboring osteoblasts can secrete **osteoprotegerin (OPG),** which by contrast suppresses osteoclast activity. OPG secreted into the matrix serves as a

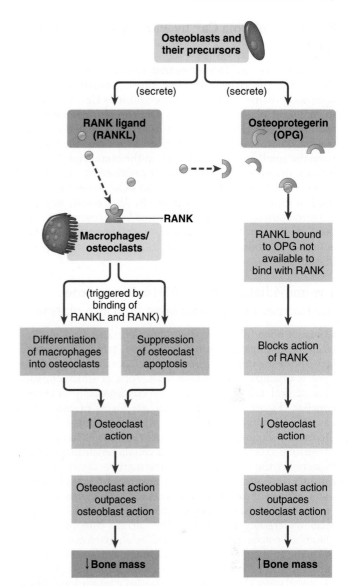

FIGURE 7-28 Role of osteoblasts in governing osteoclast development and activity.

© Cengage Learning, 2013

freestanding decoy receptor that binds with RANKL. By taking RANKL out of action so that it cannot bind with its intended RANK receptors, OPG prevents RANKL from revving up osteoclasts' bone-resorbing activity. As a result, the matrix-making osteoblasts are able to outpace the matrix-removing osteoclasts, so bone mass increases.

The balance between RANKL and OPG thus is an important determinant of bone density. If osteoblasts produce more RANKL, the more osteoclast action, the lower the bone mass. If osteoblasts produce more OPG, the less osteoclast action, the greater the bone mass. Importantly, scientists are currently unraveling the influence of various factors on this balance. For example, the female sex hormone estrogen stimulates activity of the OPG-producing gene in osteoblasts and also promotes apoptosis of osteoclasts, both mechanisms by which estrogen preserves bone mass. It has this role in males as well as females.

PTH raises plasma Ca²⁺ by withdrawing Ca²⁺ from the bone

In addition to the factors geared toward controlling the mechanical effectiveness of bone, throughout life PTH uses bone as a "bank" from which it withdraws Ca^{2+} as needed to maintain plasma Ca^{2+} level. Parathyroid hormone has two major effects on bone that raise plasma Ca^{2+} concentration. First, it induces a fast Ca^{2+} efflux into the plasma from the small *labile pool* of Ca^{2+} in the bone fluid. Second, by stimulating bone dissolution, it promotes a slow transfer into the plasma of both Ca^{2+} and PO_4^{3-} from the *stable pool* of bone minerals in bone itself. As a result, ongoing bone remodeling is tipped in favor of bone resorption over bone deposition. Let's examine more thoroughly PTH's actions in mobilizing Ca^{2+} from its labile and stable pools in bone.

The immediate effect of PTH is to promote the transfer of Ca²⁺ from bone fluid into plasma

Compact bone forms the dense outer portion of a bone. Interconnecting spicules of **trabecular bone** make up the more lacy-appearing inner core of a bone (Figure 7-29a). Compact bone is organized into **osteon** units, each of which consists of a **central canal** surrounded by concentrically arranged **lamellae** (Figure 7-29b). Lamellae are layers of osteocytes entombed within the bone they have deposited around themselves (Figure 7-29c). The osteons typically run parallel to the long axis of the bone. Blood vessels penetrate the bone from either the outer surface or the marrow cavity and run through the central canals. Osteoblasts are present along the outer surface of the bone and along the inner surfaces lining the central canals. Osteoclasts are also located on bone surfaces undergoing resorption. The surface osteoblasts and entombed osteocytes are connected by an extensive network of small, fluid-containing canals, the **canaliculi,** which allow substances to be exchanged between trapped osteocytes and the circulation. These small canals also contain long, filmy cytoplasmic extensions, or "arms," of osteocytes and osteoblasts that are connected to one another, much as if these cells were "holding hands." The "hands" of adjacent cells are connected by gap junctions, which permit communication and exchange of materials among these bone cells. The interconnecting cell network, which is called the **osteocytic–osteoblastic bone membrane,** separates the mineralized bone itself from the blood vessels within the central canals (Figure 7-30a). The small, labile pool of Ca^{2+} is in the **bone fluid** that lies between this bone membrane and the adjacent bone, both within the canaliculi and along the surface of the central canal.

PTH exerts its effects via cAMP. The earliest effect of PTH is to activate membrane-bound Ca^{2+} pumps located in the plasma membranes of the osteocytes and osteoblasts. These pumps promote movement of Ca^{2+}, without the accompaniment of PO_4^{3-}, from the bone fluid into these cells, which in turn transfer the Ca^{2+} into the plasma within the central canal. Thus, PTH stimulates the transfer of Ca^{2+} from the bone fluid across the osteocytic–osteoblastic bone membrane into the plasma. Movement of Ca^{2+} out of the labile pool across the bone membrane accounts for the fast exchange between bone and plasma (Figure 7-30b). Because of the large surface area of the osteocytic-osteoblastic membrane, small movements of Ca^{2+} across individual cells are amplified into large Ca^{2+} fluxes between the bone fluid and plasma.

After Ca^{2+} is pumped out, the bone fluid is replenished with Ca^{2+} from the partially mineralized bone along the adjacent bone surface. Thus, the fast exchange of Ca^{2+} does not involve resorption of completely mineralized bone, and bone mass is not decreased. Through this means, PTH draws Ca^{2+} out of the "quick-cash branch" of the bone bank and rapidly increases the plasma Ca^{2+} level without actually entering the bank (that is, without breaking down mineralized bone itself). Normally, this exchange is much more important for maintaining plasma Ca^{2+} concentration than is the slow exchange.

PTH's chronic effect is to promote localized dissolution of bone to release Ca²⁺ into plasma

Under conditions of chronic hypocalcemia, such as may occur with dietary Ca^{2+} deficiency, PTH influences the slow exchange of Ca^{2+} between bone itself and the ECF by promoting actual localized dissolution of bone. It does so by acting on osteoblasts, causing them to secrete RANKL, thereby indirectly stimulating osteoclasts to gobble up bone and increasing the formation of more osteoclasts while transiently inhibiting the bone-forming activity of osteoblasts. Bone contains so much Ca^{2+} compared to the plasma (more than 1,000 times as much) that even when PTH promotes increased bone resorption, there are no immediate discernible effects on the skeleton because such a tiny amount of bone is affected. Yet the negligible amount of Ca^{2+} "borrowed" from the bone bank can be lifesaving in terms of restoring free plasma Ca^{2+} level to normal. The borrowed Ca^{2+} is then redeposited in the bone at another time when Ca^{2+} supplies are more abundant. Meanwhile, the plasma Ca^{2+} level has been maintained without sacrificing bone integrity. However, prolonged excess PTH secretion over months or years eventually leads to the formation of cavities throughout the skeleton that are filled with very large, overstuffed osteoclasts.

When PTH promotes dissolution of the $Ca_3(PO_4)_2$ crystals in bone to harvest their Ca^{2+} content, both Ca^{2+} and PO_4^{3-} are released into the plasma. An elevation in plasma PO_4^{3-} is undesirable, but PTH deals with this dilemma by its actions on the kidneys.

PTH acts on the kidneys to conserve Ca²⁺ and to eliminate PO₄³⁻

PTH promotes Ca^{2+} conservation and PO_4^{3-} elimination by the kidneys during urine formation. Under the influence of PTH, the kidneys can reabsorb more of the filtered Ca^{2+}, so less Ca^{2+} escapes into urine. This effect increases the plasma Ca^{2+} level and decreases urinary Ca^{2+} losses. (It would be counterproductive to dissolve bone to obtain more Ca^{2+} only to lose it in urine.) By contrast, PTH decreases PO_4^{3-} reabsorption, thus increasing urinary PO_4^{3-} excretion. As a result, PTH reduces plasma PO_4^{3-} levels at the same time it increases plasma Ca^{2+} concentrations.

An additional function of PTH on the kidneys (besides increasing Ca^{2+} reabsorption and decreasing PO_4^{3-} reabsorption) is to enhance the activation of vitamin D by the kidneys.

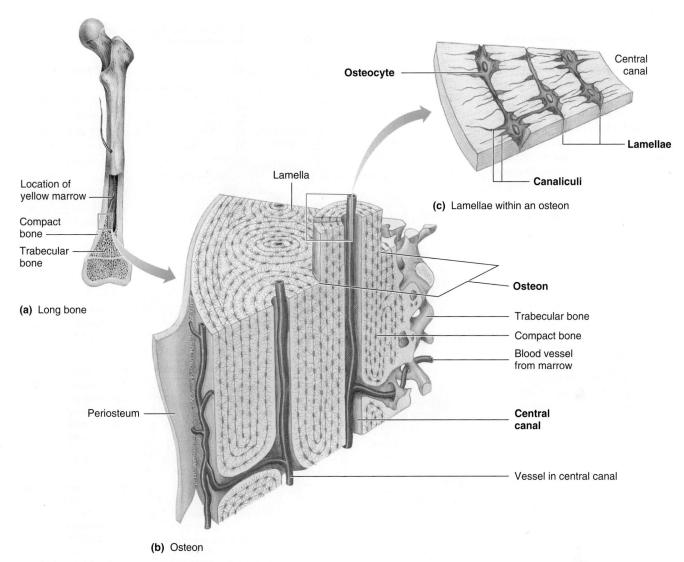

(a) Long bone

Location of yellow marrow

Compact bone

Trabecular bone

Lamella

Osteocyte

Central canal

Lamellae

Canaliculi

(c) Lamellae within an osteon

Osteon

Trabecular bone

Compact bone

Blood vessel from marrow

Central canal

Vessel in central canal

Periosteum

(b) Osteon

FIGURE 7-29 Organization of compact bone into osteons. (a) Structure of a long bone showing location of compact bone and trabecular bone. (b) An osteon, the structural unit of compact bone, consists of concentric lamellae (layers of osteocytes entombed by the bone they have deposited around themselves) surrounding a central canal. A small blood vessel branch traverses the central canal. (c) A magnification of the lamellae showing the entombed osteocytes.

Source: SPENCE, HUMAN ANATOMY & PHYSIOLOGY, 2nd Edition, © 1983. Reprinted by permission of Pearson Education, Inc., Upper Saddle River, NJ.

PTH indirectly promotes absorption of Ca^{2+} and PO_4^{3-} by the intestine

Although PTH has no direct effect on the intestine, it indirectly increases both Ca^{2+} and PO_4^{3-} absorption from the small intestine by means of its role in vitamin D activation. This vitamin, in turn, directly increases intestinal absorption of Ca^{2+} and PO_4^{3-}.

The primary signal for regulating PTH secretion is the plasma concentration of free Ca^{2+}

All the effects of PTH are aimed at raising the plasma Ca^{2+} levels. Appropriately, PTH secretion is increased in response to a fall in plasma Ca^{2+} concentration and decreased by a

rise in plasma Ca^{2+} levels. The secretory cells of the parathyroid glands are directly and exquisitely sensitive to changes in free plasma Ca^{2+}. Because PTH regulates plasma Ca^{2+} concentration, this relationship forms a simple negative-feedback loop for controlling PTH secretion without involving any nervous or other hormonal intervention, with the glands serving as sensing integrators much like a house thermostat (Figure 7-31).

Calcitonin lowers the plasma Ca^{2+} concentration, but may not be essential

Calcitonin (CT), the hormone produced by the C cells of the mammalian thyroid gland, is synthesized in anatomically distinct glands in most vertebrates. In birds the paired

FIGURE 7-30 Fast and slow exchanges of Ca²⁺ across the osteocytic–osteoblastic bone membrane. (a) Entombed osteocytes and surface osteoblasts are interconnected by long cytoplasmic processes that extend from these cells and connect to one another within the canaliculi. This interconnecting cell network, the osteocytic–osteoblastic bone membrane, separates the mineralized bone from the plasma in the central canal. Bone fluid lies between the membrane and the mineralized bone. (b) Fast exchange of Ca²⁺ between the bone and plasma is accomplished by Ca²⁺ pumps in the osteocytic–osteoblastic bone membrane that transport Ca²⁺ from the bone fluid into these bone cells, which transfer the Ca²⁺ into the plasma. Slow exchange of Ca²⁺ between the bone and plasma is accomplished by osteoclast dissolution of bone.

© Cengage Learning, 2013

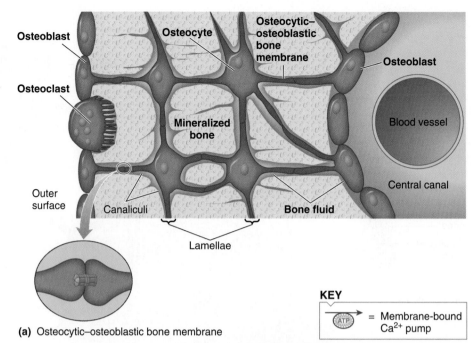

(a) Osteocytic–osteoblastic bone membrane

KEY

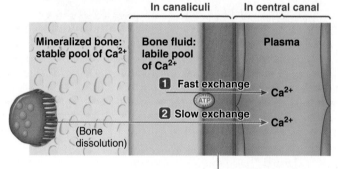

(b) Fast and slow exchange of Ca²⁺ between bone and plasma

1 In a fast exchange, Ca²⁺ is moved from the labile pool in the bone fluid into the plasma by PTH-activated Ca²⁺ pumps located in the osteocytic–osteoblastic bone membrane.

2 In a slow exchange, Ca²⁺ is moved from the stable pool in the mineralized bone into the plasma through PTH-induced dissolution of the bone by osteoclasts.

ultimobranchial glands lie posterior to the parathyroids, whereas in fish calcitonin secreting cells are located in connective tissue sheets around the heart. In fish, calcitonin has been suggested to promote bone formation by osteoblasts. Salmon CT is the most powerful form of calcitonin known in regulating calcium, and is administered to humans suffering from osteoporosis. However, its role in fish is still uncertain, and the most important hormone regulating calcium homeostasis in bony fish appears to be the hypocalcemic hormone, **stanniocalcin** (formerly called *hypocalcin*), which is synthesized in small organs termed the **corporacles of Stannius** located within the kidney. The ability to lower blood calcium is of greatest importance in seawater fishes, which are constantly exposed to the high levels of calcium present in seawater; PTH does not occur in these fishes. In response to the release of stanniocalcin, calcium uptake at the gills and intestine is rapidly reduced. In mammals, calcitonin exerts a comparatively weak effect on plasma Ca²⁺ levels.

Like PTH, CT has two effects on bone, but in this case both effects *decrease* plasma Ca²⁺ levels. First, on a short-term basis calcitonin decreases Ca²⁺ movement from the bone fluid into the plasma. Second, on a long-term basis CT decreases bone resorption by inhibiting the activity of osteoclasts. The suppression of bone resorption results in decreased plasma PO₄³⁻ levels as well as a reduced plasma Ca²⁺ concentration. CT only affects bone; it has no effect on the kidneys or intestine.

As with PTH, the primary regulator of calcitonin release is the free plasma Ca²⁺ concentration; an increase in plasma Ca²⁺ stimulates and a fall in plasma Ca²⁺ inhibits CT secretion (Figure 7-31). Because CT reduces plasma Ca²⁺

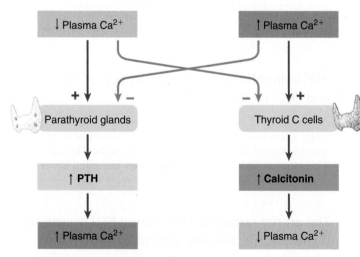

FIGURE 7-31 Negative-feedback loops controlling parathyroid hormone (PTH) and calcitonin secretion.

© Cengage Learning, 2013

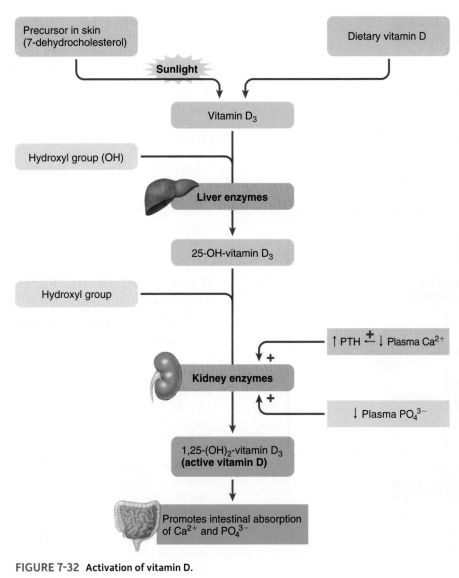

FIGURE 7-32 Activation of vitamin D.

© Cengage Learning, 2013

distant target site, the intestine. The skin, therefore, is actually an endocrine gland and vitamin D is a prohormone. In 1931, the researcher H. C. Hou found that removing the *preen gland* from chicks caused rickets to develop even if the chicks were exposed to sunlight. From this observation he deduced that birds secrete a provitamin D from the preen gland onto their feathers, where it is converted into vitamin D. Traditionally, however, this chemical messenger has been named a vitamin—that is, a dietary supplement—for two reasons. First, scientists originally discovered and isolated it from a dietary source and tagged it as a vitamin. Second, even though the skin would be an adequate source of vitamin D if it were exposed to sufficient sunlight, indoor housing of production animals requires that it be supplemented in their diets.

Activation of Vitamin D Regardless of its source, vitamin D is biologically inactive when it first enters the blood from either the skin or the digestive tract. It must be activated by two sequential biochemical alterations that involve the addition of two hydroxyl (–OH) groups (Figure 7-32). The first of these reactions occurs in the liver and the second in the kidneys. The end result is production of the active form of vitamin D, *1,25-(OH)$_2$-vitamin D$_3$*, also known as **calcitriol**. In response to a fall in plasma Ca^{2+}, PTH stimulates the kidney enzyme 1 α-hydroxylase that is involved in the second step of vitamin D activation. To a lesser extent, a fall in plasma PO_4^{3-} also enhances the activation process.

levels, this system constitutes a second simple negative-feedback control over plasma Ca^{2+} concentration, one that is antagonistic to the PTH system. The effects of CT in birds and mammals are weak, however, and do not appear to be important in normal Ca^{2+} homeostasis. Most evidence suggests, that CT plays a critical role in protecting skeletal integrity when there is a large Ca^{2+} demand, such as during pregnancy or mammary-gland feeding.

Vitamin D is actually a hormone that increases calcium absorption in the intestine

The final factor involved in the regulation of Ca^{2+} metabolism in vertebrates is **cholecalciferol**, or vitamin D, a steroid-like compound that is essential for Ca^{2+} absorption in the intestine. The livers of marine teleosts are well known to contain high concentrations of vitamin D, which has traditionally been used as a nutritional supplement. Strictly speaking, vitamin D should be considered a hormone because it can be produced in the skin from a precursor related to cholesterol (7–dehydrocholesterol) on exposure to sunlight. It is subsequently released into the blood to act at a

Function of Vitamin D The most dramatic and biologically important effect of activated vitamin D is to increase Ca^{2+} absorption in the intestine. Unlike most dietary constituents, dietary Ca^{2+} is not indiscriminately absorbed by the digestive system. In fact, the majority of ingested Ca^{2+} is typically not absorbed but is lost instead into the feces. When needed, more dietary Ca^{2+} is absorbed into the plasma under the influence of calcitriol. Calcitriol induces RNA transcription and synthesis of proteins including the calcium-binding protein, **calbindin D$_{28k}$**. For example, the onset of egg lay in birds coincides with increased 1,25-(OH)$_2$-vitamin D$_3$ concentrations, which elevates concentrations of calbindin and increases intestinal calcium transport. Independent of its effects on Ca^{2+} transport, calcitriol also increases intestinal PO_4^{3-} absorption. Furthermore, calcitriol increases the responsiveness of bone to PTH. Thus, vitamin D and PTH have a closely interdependent relationship (Figure 7-33). Vitamin D may be used in some nonvertebrates; a role for steroids in Ca^{2+} absorption in the snail *Helix aspersa* has been demonstrated, with vitamin D shown to be more effective than cholesterol.

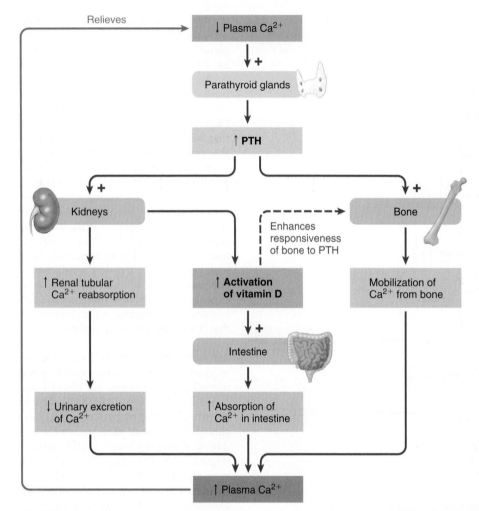

FIGURE 7-33 Interactions between PTH and vitamin D in controlling plasma calcium.

© Cengage Learning, 2013

PTH is principally responsible for controlling Ca^{2+} homeostasis, because the actions of calcitriol are too sluggish for it to contribute substantially to the minute-to-minute regulation of plasma Ca^{2+} concentration. However, both PTH and calcitriol are essential to Ca^{2+} balance, the process that ensures that, over the long term, Ca^{2+} input into the body is equivalent to Ca^{2+} output. When dietary Ca^{2+} intake is reduced, the resultant increase in PTH activates calcitriol, which increases the efficiency of uptake of ingested Ca^{2+}.

Phosphate metabolism is controlled by the same mechanisms that regulate Ca^{2+} metabolism

In the ECF, plasma PO_4^{3-} concentration is not as tightly controlled as plasma Ca^{2+} concentration. Phosphate is regulated directly by calcitriol and indirectly by the plasma Ca^{2+}–PTH feedback loop. To illustrate, a fall in plasma PO_4^{3-} concentration exerts a twofold effect to help restore the circulating PO_4^{3-} level back to normal (Figure 7-34). First, a fall in plasma PO_4^{3-} causes an increase in plasma Ca^{2+}, which directly suppresses PTH secretion. In the presence of reduced PTH, PO_4^{3-} reabsorption by the kidneys increases, returning plasma PO_4^{3-} concentration toward

normal. Second, a fall in plasma PO_4^{3-} also results in increased activation of vitamin D, which then promotes PO_4^{3-} absorption in the intestine. This further helps to alleviate the initial hypophosphatemia. Note that these changes do not compromise Ca^{2+} balance. Although the increase in activated vitamin D stimulates Ca^{2+} absorption, the concurrent fall in PTH produces a compensatory increase in urinary Ca^{2+} excretion because less of the filtered Ca^{2+} is reabsorbed. Therefore plasma Ca^{2+} remains unchanged while plasma PO_4^{3-} is being increased to normal.

Disorders in Ca^{2+} metabolism may arise from abnormal levels of parathyroid hormone or vitamin D

The primary disorders that affect Ca^{2+} metabolism are too much or too little PTH, a deficiency of vitamin D, or excessive demands on blood calcium concentrations. Let's look at three of these.

PTH Hypersecretion Excess PTH secretion, or hyperparathyroidism, occurs in domestic animals in response to diets low in calcium or vitamin D content. These may be all-meat diets for carnivores, which are low in calcium, or inappropriate diets, such as those for horses that are high in phosphate relative to calcium. Less frequently there may be a hypersecreting tumor in one of the parathyroid glands. The affected animal can be asymptomatic or symptoms can be severe, depending on the magnitude of the problem. The following are among the possible consequences of the ensuing hypercalcemia and hypophosphatemia:

- *Reduced excitability of muscle and nervous tissue* leads to muscle weakness and neurological disorders. Cardiac disorders may also occur.
- *Excessive mobilization of Ca^{2+} and PO_4^{3-} from skeletal stores* leads to thinning of bone, which may result in skeletal deformities and increased incidence of fractures.
- *An increased incidence of Ca^{2+}-containing kidney stones* occurs because the excess quantity of Ca^{2+} being filtered through the kidneys may precipitate and form stones. These stones may impair renal function. Passage of the stones through the ureters causes extreme pain.

Because of these potential multiple consequences, hyperparathyroidism has been called a disease of "bones, stones, and abdominal groans."

Vitamin D Deficiency The major consequence associated with vitamin D deficiency is impaired intestinal absorption of Ca^{2+}. In the face of reduced Ca^{2+} uptake, PTH maintains the plasma Ca^{2+} level at the expense of the bones. As a re-

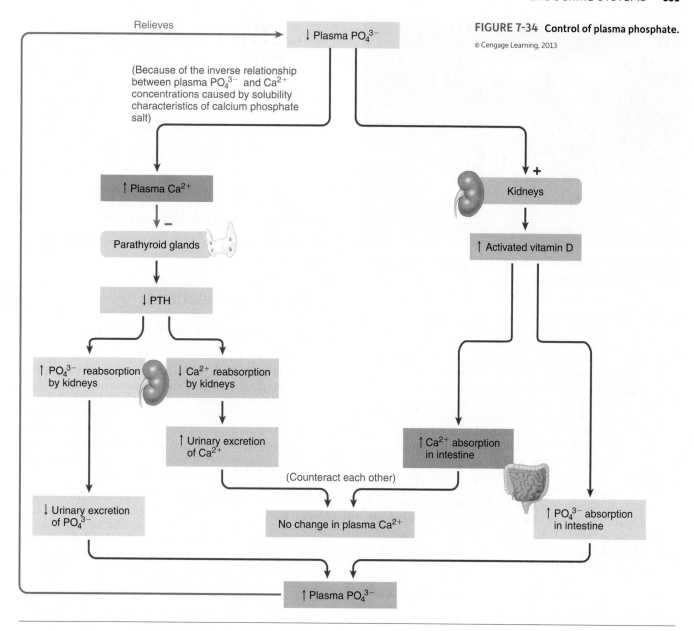

FIGURE 7-34 **Control of plasma phosphate.**
© Cengage Learning, 2013

sult, the bone matrix is not properly mineralized because Ca^{2+} salts are not available for deposition. The demineralized bones become soft and deformed, bowing to the pressures of weight bearing. This condition is known as **rickets** in prepubertal mammals and **osteomalacia** in adults. In the absence of vitamin D, amphibians (such as South African clawed toads) and reptiles are particularly susceptible to the development of rickets.

Excessive Demands for Calcium We end our discussion of disorders of calcium metabolism with a look at the effects of excessive demands. These occur most notably in farm animals bred for two specialized processes: milk production by dairy cows and egg laying by hens.

- *Parturient paresis (milk fever, or parturient hypocalcemia).* This is the most common metabolic disorder affecting high-producing dairy cattle and provides a dynamic example of calcium metabolism during the immediate postpartum period. Because colostrum con-

tains high concentrations of calcium (2 g/L) and approximately 3 g of calcium are required per hour to produce colostrum, the inability of the dairy cow to adequately mobilize calcium reserves results in the characteristic symptoms of milk fever, which include restlessness, anxiety, anorexia, incoordination, and a lack of interest in their calves. The incidence of milk fever increases with age and yield of the animals. Affected cattle usually demonstrate symptoms within 72 hours following parturition. Without intervention the cow's progress to a second stage, which is manifested by recumbency: The cows lie on their sternums with their heads turned back toward their flanks, breathing is slow and labored, and body temperatures are below normal. In the final stage of parturient paresis the cows are stretched out on their sides as they progress from dullness to coma.

If the cows are fed an *increased level* of calcium in the diet during the *prepartum* period, then which hormone is predominant during the early lactation period?

Photo: Courtesy of USDA/Agricultural Research Service

Dairy cows can develop milk fever from the excessive demands for calcium in milk production.

The answer is *calcitonin,* which functions to block the resorption of calcium by PTH at the bone. Consider the profound impact on plasma calcium concentrations as colostrum begins to be produced. The sudden demand for calcium—more than all the calcium present in the blood—results in a dramatic decline in calcium concentrations in the blood, which triggers an enormous output of PTH from the parathyroid gland. Because of the elevated plasma-calcitonin concentrations, PTH can exert its effect primarily at the intestinal level, which cannot fill the increased demands placed on it. (Note that this does not necessarily reveal an important natural role for calcitonin, because this is an artificial situation of feeding calcium to a cow bred for high milk production.) However, if the cow is placed on a *low-calcium diet* (supplemented with vitamin D) during the prepartum period, then concentrations of PTH predominate during parturition and they are eminently more successful in mobilizing calcium from the bone reservoirs. Acting in concert with vitamin D to increase concentrations of calcium-binding protein in the intestinal epithelia and calcium transport across the intestine, PTH now acts uninhibitedly on bone to effectively combat the decline in blood calcium concentrations, and thus dramatically reduces the incidence of milk fever.

- *Egg lay in female birds.* Comparable to these events, which drain calcium from the blood of lactating dairy cows, is the situation in egg-laying birds. In the laying hen, an amount of calcium equivalent to 8 to 10% of the total calcium in her body is secreted into the shell each day on which an egg is laid (**oviposition**). Researchers believe the shell, composed primarily of calcium carbonate, evolved from the soft cover of eggs of ancestral reptiles into the calcareous shell to give a measure of protection from attacks by predators and pathogens (including soil organisms). PTH can rapidly mobilize calcium from the medullary bone whenever the rate at which calcium deposition on the shell exceeds the rate at which it is absorbed from the intestinal tract. Thus, as expected, there is a reciprocal relationship between the concentrations of PTH and Ca^{2+} during the egg cycle. During the two-week period prior to the onset of lay, a whole new system of secondary bone, medul-

lary bone (p. 324), is laid down in the marrow cavities of most bones under the combined influence of estrogens and androgens secreted from the developing follicles. Interestingly, calcitriol does not stimulate the synthesis of a calcium transporter (Ca^{2+}-ATPase) in the shell gland of a laying bird (unlike the situation in the intestinal epithelial cells) but rather its production is regulated by estrogen. When a bird goes "out of lay," concentrations of estrogen decline in concert with the reduction in shell gland Ca^{2+}-ATPase activity.

check your understanding 7.8

Why must ECF Ca^{2+} be kept within such narrow limits?

How do PTH and vitamin D regulate calcium concentrations?

making connections

How the Endocrine System Contributes to the Body as a Whole

The endocrine system is one of an animal's two major control systems, the other being the nervous system (although, as you will see in Chapter 10, some physiologists consider the immune system a third major control system). Through its relatively slow-acting hormonal messengers, the endocrine system generally regulates activities that require duration rather than speed. Some endocrine systems are aimed at regulated changes such as these in vertebrates:

- Most aspects of postnatal development and growth are regulated by hormones such as GH and IGFs.
- The major hormone secreted by the adrenal medulla, epinephrine, generally reinforces activities of the sympathetic nervous system. It elevates blood pressure and also helps prepare the body for peak physical responsiveness in fight-or-flight situations. This includes increasing the plasma concentrations of glucose and fatty acids above normal, which provides additional energy sources for increased physical activity.
- The thyroid hormones, tetraiodothyronine (T_4) and triiodothyronine (T_3), increase the overall metabolic rate, influencing the rate at which cells use nutrient molecules and O_2 within the internal environment.

Other endocrine actions are directed toward maintaining homeostasis, as exemplified by the following in vertebrates:

- The actions of the thyroid hormones also produce heat, which helps maintain body temperature in endotherms.
- The adrenal cortex secretes cortisol, the primary glucocorticoid secreted by the adrenal cortex, which increases the plasma concentrations of glucose, fatty acids, and amino acids above normal. Although these actions change the concentrations of these molecules in the internal environment, they indirectly contribute to homeostasis by making the molecules readily available as energy sources or building blocks for tissue repair to help the body adapt to stressful situations.
- The two major hormones secreted by the endocrine pancreas, insulin and glucagon, are important in shifting metabolic pathways between the absorptive and postab-

sorptive states, which maintains the appropriate plasma levels of nutrient molecules, by regulating cellular uptake, storage, and release of these molecules (the crustacean hyperglycemic hormone is another example).

- Parathyroid hormone from the parathyroid glands is critical to maintaining plasma concentration of Ca^{2+}. PTH is essential for life because of Ca^{2+}'s effect on neu-

romuscular excitability. In the absence of PTH, death rapidly occurs from asphyxiation caused by pronounced spasms of the respiratory muscles.

Hormones often act in negative-feedback fashion to resist the change that induced their secretion, thus maintaining stability in the internal environment.

Chapter Summary

- The ability of cells to "communicate" with each other is ancient, preceding the evolution of multicellularity in the form of pheromones among unicellular organisms. In animals, pheromones gave rise to internal signal molecules used locally within a tissue. With the evolution of circulatory systems, which provided an effective route for delivering signal molecules to distant tissues, there arose endocrine systems, which consist of ductless glands that secrete hormones (signal molecules delivered by circulatory fluids), providing a second mechanism (in addition to nervous systems) for regulating and coordinating distant organs.

- Hormones are chemically classified into three categories: peptides and proteins, amines, and steroids. The mechanisms of hormone synthesis, storage, and secretion vary according to the class of hormone.

- Hydrophilic hormones are transported dissolved in the plasma, whereas lipid-soluble hormones are almost always transported bound to plasma proteins.

- Hormones exert a variety of regulatory effects throughout the body. Compared to neural responses that are brought about within milliseconds, hormone action is relatively slow and prolonged, taking minutes to hours for the response to take place after the hormone binds to its receptor.

- Nonnative or hormonelike substances termed endocrine-disrupting chemicals (EDCs), whose structures are similar to hormones, can sometimes disrupt endocrine communication by altering hormone secretion, transport, or action. These compounds are derived from the by-products of manufactured organic compounds.

- Hormones produce their effects by altering intracellular proteins through ion fluxes, second messengers, and transcription factors.

- The effective plasma concentration of a hormone is normally regulated by changes in its rate of secretion.

- The most common endocrine rhythm is the circadian ("around a day") rhythm, which is characterized by repetitive oscillations in gene expression that correspond with changes in the physiology, metabolism, and behavior in nearly all organisms on Earth.

- The effective plasma concentration of a hormone can be influenced by the hormone's transport, metabolism, and excretion. The responsiveness of a target cell to its hormone can be varied by regulating the number of its hormone-specific receptors.

- Endocrine disorders are attributable to hormonal excess, hormonal deficiency, or decreased responsiveness of the target cells.

- Nonvertebrate hormones are produced either by neurosecretory cells or glands in neural haemal organs where the axonal endings secrete the hormone into the haemolymph, which then can bind to specific receptors on peripheral organs.

- The pituitary has two anatomically and functionally distinct lobes, the posterior pituitary and the anterior pituitary. The hypothalamus and posterior pituitary form a neurosecretory system that secretes

vasopressin and oxytocin. The anterior pituitary secretes six established hormones, many of which are tropic. In agnathans, amphibians, reptiles, and most mammals, the adenohypophysis includes a third, well-defined intermediate lobe (pars intermedia).

- Hypothalamic releasing and inhibiting hormones are delivered to the anterior pituitary by the hypothalamic–hypophyseal portal system to control anterior pituitary hormone secretion.

- Target gland hormones inhibit hypothalamic and anterior pituitary hormone secretion via negative feedback.

- To exert its metabolic effects, growth hormone (GH) binds directly with receptors on adipose tissue, skeletal muscles, and liver. GH's indirect growth-promoting actions are mediated by Insulin-like growth factors (IGFs), which act on the target cells to cause growth of both soft tissues and bones. GH secretion is regulated by two hypophysiotropic hormones and influenced by a variety of other factors.

- Prolactin has a wide range of effects, including lactogenesis, reproductive behaviors, and ion regulation.

- Thyroid hormone is the primary determinant of overall metabolic rate and exerts other effects as well, such as amphibian development. Thyroid hormones are produced by a thyroid follicle in the thyroid gland. Most of the secreted thyroxine (T_4) is converted into triiodothyronine (T_3), or activated, by deiodinase enzyme (Type II) that strips off one of its iodines. T_3 is the major biologically active form of thyroid hormone at the cellular level.

- In most vertebrates, the adrenal gland consists of a steroid-secreting cortex intermingled with chromaffin tissue. The adrenal cortex secretes mineralocorticoids, glucocorticoids, and sex hormones.

- Glucocorticoids produced from the adrenal cortex exert metabolic effects and have an important role in adaptation to stress. Glucocorticoid secretion is directly regulated by the hypothalamic–pituitary–adrenal axis.

- Ganglionic cell bodies within the adrenal medulla release their chemical transmitter directly into the circulation on stimulation by the preganglionic fiber. Sympathetic stimulation of the adrenal medulla is solely responsible for epinephrine release.

- Epinephrine reinforces the sympathetic nervous system in "fight-or-flight" short-term stress, and exerts additional metabolic effects. The stress response is a generalized, nonspecific pattern of neural and hormonal reactions to any situation that threatens homeostasis.

- Fuel metabolism includes anabolism, catabolism, and interconversions among energy-rich organic molecules. Metabolic fuels are stored during the absorptive state and are mobilized during the postabsorptive state.

- The pancreatic hormones, insulin and glucagon, are most important in regulating fuel metabolism. Insulin lowers blood glucose, amino acid, and fatty acid levels and promotes their storage. Glucagon in general opposes the actions of insulin. Glucagon promotes catabolism of nutrient stores between meals to keep up the blood nutrient levels, especially blood glucose.

- Plasma calcium must be closely regulated to prevent changes in neuromuscular excitability. Bone is the primary location of calcium; it is constantly remodeled by actions of osteoblasts (bone builders) and osteoclasts (bone destroyers). Three hormones—parathyroid hormone, calcitonin, and vitamin D—control calcium (Ca^{2+}) and

phosphate (PO_4^{3-}) metabolism after their ingestion from the diet and/or uptake from the gill. Parathyroid hormone raises free plasma calcium levels by its effects on bone, kidneys, and intestine. PTH acts on the kidneys to conserve Ca^{2+} and to eliminate PO_4^{3-}. On a long-term basis, calcitonin decreases bone resorption by inhibiting the activity of osteoclasts. Vitamin D is a hormone that increases calcium absorption in the intestine. Phosphate metabolism is controlled by the same mechanisms that regulate Ca^{2+} metabolism.

Review, Synthesize, and Analyze

1. Would you expect the concentration of hypothalamic releasing and inhibiting hormones in a systemic venous blood sample to be higher, lower, or the same as the concentration of these hormones in a sample of hypothalamic–hypophyseal portal blood?

2. Thinking about the feedback control loop among TRH, TSH, and thyroid hormone, would you expect the concentration of TSH to be normal, above normal, or below normal in an animal whose diet is deficient in iodine (an element essential for synthesizing thyroid hormone)?

3. An animal displays symptoms of excess cortisol secretion. What factors could be measured in a blood sample to determine whether the condition is caused by a defect at the hypothalamic/anterior pituitary level or the adrenal cortex level?

4. Discuss the release of hypophysiotropic hormones from the hypothalamus.

5. How is the multifaceted general stress response manifested in a mammal?

6. How is intermediary metabolism altered in the fed versus the fasting state?

7. Glucose homeostasis is altered in diabetes. Elaborate on the changes that result in chronic hyperglycemia?

8. How does parathyroid hormone adjust calcium and phosphate levels in the blood?

9. Explain the effect of a lack of activated vitamin D (due to inadequate exposure to sunlight) on an animal. How then are animals maintained indoors for the duration of their lifetime without developing these abnormalities?

Suggested Readings

Barton, B. A. (2002). Stress in fishes: A diversity of responses with particular reference to changes in circulating corticosteroids. *Integrative and Comparative Biology* 42:517–525.

Cerami, A., H. Vlassara, & M. Brownlee. (1987). Glucose and aging. *Scientific American* 256: 90–96. An enlightening description of how glucose contributes to the aging process.

Franken, P. & D.-J. Dijk. (2009). Circadian clock genes and sleep. *European Journal of Neuroscience* 29:1820–1829. A review of clock genes and their cyclical regulation.

Hadley, M. E. (2007). *Endocrinology,* 6th ed. Englewood Cliffs, NJ: Pearson Prentice Hall. For those with an interest in comparative endocrinology.

Hirsch, P. F., G. E. Lester, & R. V. Talmage. (2001). Calcitonin, an enigmatic hormone: Does it have a function? *Journal of Musculoskeletal Neuronal Interactions* 1:299–305.

Lea, R. B., & H. Klandorf. (2002). The brood patch. In C. Deeming, ed., *Avian Eggs and Incubation.* Oxford, UK: Oxford University Press. An extensive coverage of the brood patch.

Lehman, M. N., Ladha, Z., Coolen, L. M., Hileman, S. M., Connors, J. M. & R. L. Goodman. (2010). Neuronal plasticity and seasonal reproduction in sheep. *European Journal of Neuroscience*, 32:2152–2164.

Thornton, J. W., E. Need, & D. Crews. (2003). Resurrecting the ancestral steroid receptor: Ancient origin of estrogen signaling. *Science* 301:1714–1718.

Whitmore, D., N. S. Foulkes, & P. Sassone-Corsi. (2000). Light acts directly on organs and cells in culture to set the vertebrate circadian clock. *Nature* 404:87–91.

A kangaroo has powerful muscles in its hind legs, coupled with efficient tendons that store and release energy during hopping. After reading this chapter, make a hypothesis on what type of muscle fibers you would expect to find in its hind limbs.

Muscle Physiology

8.1 Introduction

Almost all living cells have rudimentary intracellular machinery for producing movement, including that associated with the redistribution of various cell components during cell division, other types of intracellular trafficking, and movement of whole cells (Chapter 2). For example, white blood cells and amoebas use intracellular contractile *microfilaments* (p. 62) to propel themselves through their environment with pseudopodia. Other cells such as sperm and paramecia move with *microtubule*-based propulsive structures called *cilia* and *flagella* (p. 60). As the first animals evolved, larger scale movements evolved from these intracellular ones. For example, Porifera (sponges) create large-scale water currents using numerous flagellated cells. Many aquatic larvae, flatworms, and even 1.5-meter comb jellies (ctenophores) move slowly through the environment by the beating of dozens to thousands of coordinated cilia. However, other (especially faster) locomotory movements of animals required the evolution of a highly organized microfilament-based structure called a **muscle**: a specialized cell that *contracts* to pull an external structure or squeeze a fluid. With the use of skeletons or fluids, the force produced by muscle can be transmitted over a considerable distance and can generate large and often rapid movements (as you will see). In addition, the energy of muscle contraction can be stored temporarily in specialized elastic structures, and then released rapidly to create even faster movements. For example, even at near-freezing temperatures, ectothermic ("cold-blooded") chameleons can capture prey by rapidly uncoiling tongues powered not by muscles that only weakly contract in the cold environment to retract the tongue but by energy stored during that contraction in a rubber-band-like sheath of collagen inside their tongue.

These combined innovations—contractile muscle cells, force-transmitting skeletons or fluids, and elastic storage structures—are very different from flagellar and ciliary mechanisms. Moreover, though muscles use microfilaments like pseudopodia, they power locomotion outside the cell rather than within it. All animal phyla above the Porifera (that is, sponges, which have no true organs) contain clearly identifiable muscle cells, often organized into muscle organs. By far, the most important effectors for animal *behavior* are muscle organs used in coordinated patterns.

Three types of muscle create a wide range of movements

Recall from Chapter 1 (p. 9) that there are three types of muscle: *skeletal muscle, cardiac muscle,* and *smooth muscle.* Through their highly developed microfilament system, muscle cells can shorten and develop tension, which enables them to produce force and do work. In contrast to sensory systems, which transform other forms of energy in the environment into electrical signals, muscles, in response to electrical signals, convert the chemical energy of ATP into mechanical energy that can act on the environment. Controlled contraction of muscles permits the following:

1. Purposeful *locomotory* movement of the whole body or parts of the body in relation to the environment (such as walking or flying)
2. Manipulation of external objects (such as a monkey poking a straw or stick into a termite mound, up which the termites then climb)
3. Propulsion of contents through various hollow internal organs (such as circulation of blood or movement of materials through the digestive tract)

4. Emptying the contents of certain organs to the external environment (such as urination or parturition)

5. Production of heat as a metabolic by-product, which may help to maintain internal body temperatures above ambient

6. Production of sound

All muscles share numerous underlying characteristics, although there is considerable diversity among muscle cells from different animals requiring very different outputs. Examples of this diversity include fish sound-producing muscles that can contract at frequencies greater than 100 Hz (cycles per second), contraction frequencies unobtainable by their locomotory muscle counterparts. Hamlet, a species of coral-reef fish, broadcast a series of beeps and pulses ranging from 350 to 650 Hz at the critical moment when sperm and eggs are released into the environment. In contrast, the highest frequency for vertebrate locomotory muscles is an order of magnitude lower, 25 to 30 Hz for mouse and lizard so-called fast-twitch muscles at body temperatures of 35°C. You will discover that the ability to increase the speed of contraction is accomplished, in part, by reapportionment of key intracellular structures in addition to shifts in certain muscle isoforms (proteins with the same basic function but which differ slightly in their structures). Trade-offs that maximize performance of one function are usually at the expense of another. For this reason, muscles that operate at high frequencies are not effective at low frequencies and generate less force.

In most vertebrates, muscle is the largest group of tissues in the body, accounting, for example, for approximately half of the body's weight in humans. Skeletal muscle alone makes up about 40% of body weight in human males and 32% in females, with smooth and cardiac muscle making up another 10% of the total weight. Comparably, in most fishes, skeletal muscle mass varies from 20 to 50%, whereas in species noted for high rates of acceleration (such as barracudas), muscle mass ranges from 55 to 65%. In hummingbirds, flight muscles alone account for 25% of the animal's body weight.

Although the three muscle types are structurally and functionally distinct, they can be classified in two different ways according to their common characteristics (Figure 8-1). Muscles are categorized as **striated** (skeletal and cardiac muscle) or **unstriated** (smooth muscle), depending on whether alternating dark and light bands, or striations (which you will learn about), can be seen when the muscle is viewed under a light microscope.

Most of this chapter is devoted to a detailed examination of the most abundant and best understood muscle, skeletal muscle. Skeletal muscles make up the **muscular system.** Interestingly, arthropods and vertebrates share a common ancestor of striated muscle, although they diverged more than 700 million years ago. We begin by discussing skeletal muscle structure, then examine how it works from the molecular through the cellular level, and finally look at functions of the whole muscle. The chapter concludes with a discussion of the unique properties of smooth and cardiac muscle in comparison to skeletal muscle.

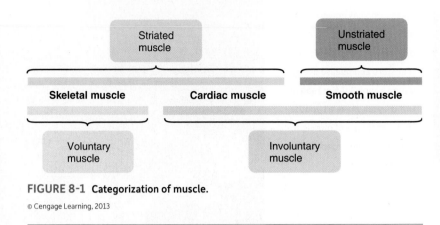

FIGURE 8-1 Categorization of muscle.
© Cengage Learning, 2013

check your understanding 8.1

Compare the different strategies used to produce movement.

8.2 Skeletal Muscle

A basic understanding of the structural components of a skeletal muscle fiber is essential to understanding how muscles contract. In the 1830s William Bowman studied the basic structure and properties of skeletal muscle from more than 40 animal species and correctly noted that muscle shortening involved specific changes in the striation pattern during nerve-stimulated contractions. We now know that the striations arise from specialized protein arrays within each cell or fiber.

Skeletal muscle consists of muscle fibers lying parallel to each other

A single skeletal muscle cell, known as a **muscle fiber,** is relatively large, elongated, and cylindrical, measuring from 10 to 100 micrometers (1 μm = 1 millionth of a meter) in diameter; they can be tens of centimeters in length in a large animal. A skeletal muscle (organ) consists of a number of muscle fibers lying parallel to each other and bundled together by connective tissue (Figure 8-2a). The fibers usually extend the entire length of the muscle. During embryonic development of vertebrates, the huge skeletal-muscle fibers are formed by the fusion of many smaller cells called **myoblasts** (*myo* means "muscle"; *blast* refers to a primitive cell that forms more specialized cells); thus, one striking feature is the presence of multiple nuclei in a single muscle cell. This multinucleate muscle cell, produced by a merging of cells, is termed a functional **syncytium** (*syn*, "together"; *cyt*, "cell").

The most predominant structural feature of most skeletal muscle fibers is the presence of numerous **myofibrils.** These specialized contractile elements, which constitute up to 90% of the volume of the muscle fiber, are cylindrical intracellular organelles, typically 1 μm in diameter, that extend the entire length of the muscle fiber (Figure 8-2b). The greater the volume of the fiber taken up by myofibrils, the greater the force per cross-sectional area that can be generated. Muscle fibers with a low percentage of myofibrils cannot generate high degrees of tension but are generally associated with the ability to turn muscles on and off quickly or to generate prolonged activity. For example, the aerobic sound-producing muscle of

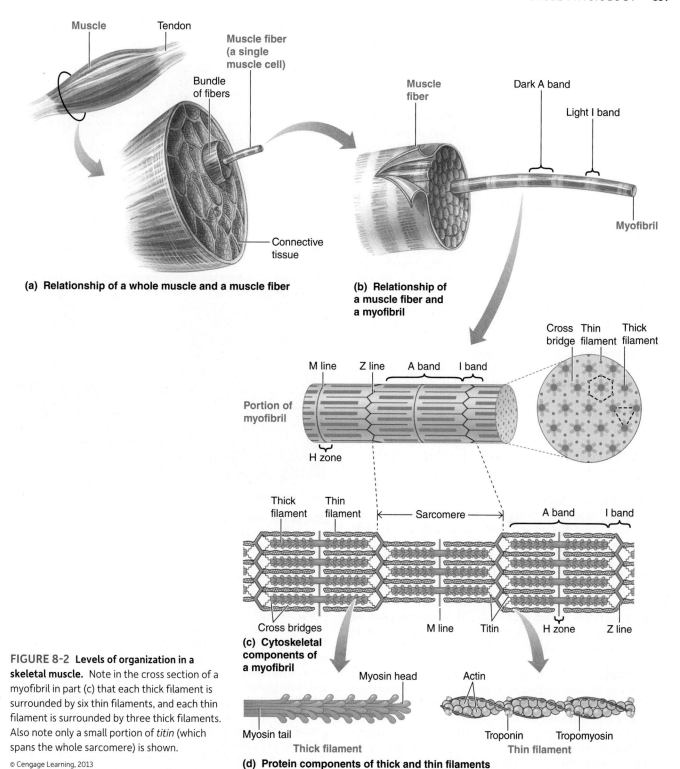

(a) Relationship of a whole muscle and a muscle fiber

(b) Relationship of a muscle fiber and a myofibril

Portion of myofibril

(c) Cytoskeletal components of a myofibril

Thick filament

Thin filament

(d) Protein components of thick and thin filaments

FIGURE 8-2 Levels of organization in a skeletal muscle. Note in the cross section of a myofibril in part (c) that each thick filament is surrounded by six thin filaments, and each thin filament is surrounded by three thick filaments. Also note only a small portion of *titin* (which spans the whole sarcomere) is shown.

© Cengage Learning, 2013

male cicadas (insects that make loud mating "songs") has a myofibrillar volume of only 22%; similarly, the high-frequency tail-shaker muscle of the rattlesnake contains only 31% myofibrils (the rest being *glycogen,* an energy store we will explore later). Conservation of function in this sound-producing muscle with those in cicadas is suggestive of convergent evolution, supporting the contention that these modifications are essential for high-frequency performance.

Each myofibril consists of a regular arrangement of highly organized cytoskeletal elements—the thick and thin filaments (Figure 8-2c). The **thick filaments,** which are 0.012 to 0.018 μm in diameter and 1.6 μm in length, are special assemblies of the protein **myosin,** whereas the **thin filaments,** which are 0.005 to 0.008 μm in diameter and 1.0 μm long, are made up primarily of the protein **actin** (with other key proteins) (Figure 8-2d). These same proteins are found in all other cells but in smaller quantities and in a less organized fashion. The levels of organization in a skeletal muscle can be summarized as follows:

Whole muscle	→	muscle fiber	→	myofibril	→	thick and thin filaments	→	myosin and actin
(an organ)		(a cell)		(a specialized intracellular structure)		(cytoskeletal elements)		(protein molecules)

Structural Organization of the Sarcomere Viewed with a light microscope, a relaxed myofibril (Figure 8-3a) displays *striations:* alternating dark bands (the A bands) and light bands (the I bands). The bands of all the myofibrils lined up parallel to each other collectively lead to the striated appearance of a skeletal muscle fiber (Figure 8-3b). Alternate stacked sets of thick and thin filaments, the **myofilaments**, which slightly overlap each other, are responsible for the A and I bands (Figure 8-2c). An **A band** consists of a stacked set of thick filaments along with the portions of the thin filaments that overlap on both ends of the thick filaments. The thick filaments lie only within the A band and extend its entire width; that is, the two ends of the thick filaments within a stack define the outer limits of a given A band. The lighter area within the middle of the A band, where the thin filaments do not reach, is known as the **H zone.** Only the central portions of the thick filaments are found in this region. The **I band** consists of the remaining portion of the thin filaments that do not project into the A band.

Visible in the middle of each I band is a dense, vertical **Z line.** The area between two Z lines is called a **sarcomere**, which is the functional unit of skeletal muscle. A **functional unit** of any organ is the smallest component that can perform all the functions of that organ. Accordingly, a sarcomere is the smallest component of a muscle fiber that is capable of contraction. The Z line is a flat cytoskeletal disc (in three dimensions) made from a cytoskeletal protein complex that connects the thin filaments of two adjoining sarcomeres. Each relaxed sarcomere is about 2.5 μm in width and consists of one whole A band and half of each of the two I bands located on either side. During growth, a muscle increases in length by adding new sarcomeres, not by increasing the size of each sarcomere. Just as the Z line or discs hold the sarcomeres together in a chain along the myofibril's length, another system of supporting proteins holds the thick filaments together vertically within each stack. These proteins can be seen as the **M line**, which extends vertically down the middle of the A band within the center of the H zone.

Cross Bridges With an electron microscope, fine **cross bridges** can be seen extending from each thick filament toward the surrounding thin filaments in the regions where the thick and thin filaments overlap (Figure 8-2c). The thin filaments are arranged hexagonally around the thick filaments: Cross bridges project from each thick filament in all six directions toward the surrounding thin filaments. Each thin filament, in turn, is surrounded by three thick filaments (Figure 8-2c). To give you an idea of the magnitude of these filaments, a single muscle fiber may contain an estimated 16 billion thick and 32 billion thin filaments, all arranged in this very precise pattern within the myofibrils.

Myosin forms the thick filaments

Each thick filament consists of several hundred myosin molecules packed together in a specific arrangement. A **myosin molecule** is a protein consisting of two identical subunits, each shaped somewhat like a golf club with two heads (Figure

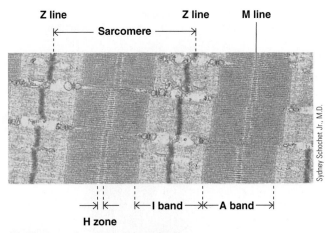

(a) Electron micrograph of a myofibril

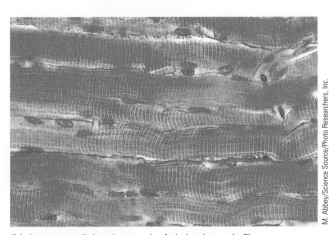

(b) Low-power light micrograph of skeletal muscle fibers

FIGURE 8-3 **Microscope view of skeletal muscle components.** (a) Note the A and I bands. (b) Note striated appearance.

Source: Reprinted with permission from Sydney Schochet Jr., M.D., Professor, Department of Pathology, School of Medicine, West Virginia University, *Diagnostic Pathology of Skeletal Muscle and Nerve*, Fig. 1–13 (Stamford, CT: Appleton & Lange, 1986).

8-4a). The protein's *tail* ends are intertwined around each other, with the two globular *heads* projecting out at one end. The two halves of each thick filament are mirror images made up of myosin molecules lying lengthwise in a regular, staggered array, with their tails oriented toward the center of the filament and their globular heads protruding outward at regular intervals (Figure 8-4b). These heads form the cross bridges between the thick and thin filaments. Each head has two important sites crucial to the contractile process: an *actin binding site* and an *ATPase (ATP-splitting) site.* Myosin filaments are linked to the Z lines by the gigantic, elastic protein **titin** (Figure 8-2c), whose function is still being investigated (see the box, *Molecular Biology and Genomics: What Are the Functions of Titin?*).

Actin, along with tropomyosin and troponin, forms the thin filaments

Thin filaments consist of three proteins: *actin, tropomyosin,* and *troponin* (Figure 8-5). **Actin** molecules, the primary structural proteins of the thin filament, are spherical. The backbone of a thin filament is formed by actin molecules joined into two strands and twisted together, like two chains of pearls wrapped around each other. Each actin molecule

Molecular Biology and Genomics:
What Are the Functions of Titin?

Titin, also known as **connectin**, is the largest protein known in any organism: The mouse homologue contains 35,213 amino acids and spans half the length of a sarcomere! A single gene for titin yields numerous different tissue-specific proteins through what is thought to be the most prolific example of *alternative RNA splicing* (p. 30). Discovered only in 1979, the roles titin plays in muscles are still not fully understood. It has been implicated in muscle development, perhaps acting as a scaffold for sarcomere assembly. Titin is also believed to keep the thick filaments centered in the sarcomere during the contraction–relaxation cycle as it binds to the Z line and M line of the sarcomere. And there is evidence of springlike mechanical roles: It may act as a locomotory spring assisting in the return of the sarcomere to its relaxed conformation or in preventing harmful overextension of the sarcomere. Recent studies shed more light on this role: Differences in the elasticity of different titin isoforms lead to high compliance (flexibility) in elephant muscle and to great stiffness in small mammals such as the shrew. Titin also has a mechanically activated *titin kinase* domain, which regulates nuclear transcription factors; these in turn control genes that may help remodel muscles to adapt to different levels of mechanical stresses. Deletion of these genes in turn leads to severe skeletal muscle *hypoplasia* (underdevelopment) during the perinatal period.

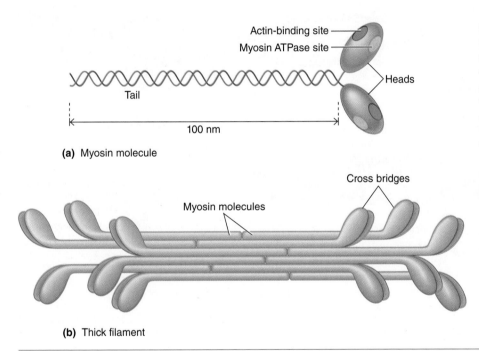

(a) Myosin molecule

(b) Thick filament

FIGURE 8-4 Structure of myosin molecules and their organization within a thick filament. (a) Each myosin molecule consists of two identical, golf-club-shaped subunits with their tails intertwined and with globular heads each containing an actin-binding site and a myosin ATPase site, projecting out at one end. (b) A thick filament is made up of myosin molecules lying lengthwise parallel to one another. Half are oriented in one direction and half in the opposite direction. The globular heads, which protrude at regular intervals along the thick filament, form the cross bridges.

© Cengage Learning, 2013

has sites for both *weak* (primarily electrostatic, that is, the interaction between positively and negatively charged moieties) and *strong* myosin attachment with a myosin cross bridge. By a mechanism to be described shortly, binding of actin and myosin molecules at the cross bridges results in energy-consuming contraction of the muscle fiber. Accordingly, actin and myosin are often called **contractile proteins**, even though, as you will see, neither myosin nor actin actually contracts. Again, myosin and actin are not unique to muscle cells, but these proteins are more abundant and more highly organized in muscle cells.

In a relaxed muscle fiber, contraction does not take place; actin cannot bind with cross bridges because of the position of the two other types of protein within the thin filament—tropomyosin and troponin. **Tropomyosin** molecules are threadlike proteins that lie end-to-end alongside the groove of the actin spiral. In this position, tropomyosin covers the actin sites that bind with the cross bridges, thus blocking the interaction that leads to muscle contraction. The other thin-filament component, **troponin**, is a protein complex consisting of three polypeptide units: one that binds to tropomyosin, one that binds to actin, and a third that can bind with Ca^{2+}.

When troponin is not bound to Ca^{2+}, this protein complex stabilizes tropomyosin in its blocking position over actin's cross-bridge binding sites (Figure 8-6a). When Ca^{2+} binds to troponin, the shape of this complex is changed in such a way that tropomyosin is allowed to slide away from its blocking position (Figure 8-6b). With tropomyosin out of the way, the myosin head can bind to actin to form cross bridges, resulting (as you will see) in muscle contraction. Tropomyosin is a flexible molecule, and its positioning on the actin filament should be considered dynamic. Tropomyosin and troponin are often called **regulatory proteins** because of their role in covering (preventing contraction) or exposing (permitting contraction) the binding sites for cross-bridge interaction between actin and myosin.

FIGURE 8-5 Composition of a thin filament. The main structural component of a thin filament is two chains of spherical actin molecules that are twisted together. Troponin molecules (which consist of three small, spherical subunits) and threadlike tropomyosin molecules are arranged to form a ribbon that lies alongside the groove of the actin helix and physically covers the binding sites on actin molecules for attachment with myosin cross bridges. (The thin filaments shown here are not drawn in proportion to the thick filaments in Figure 8-4. Thick filaments are two to three times larger in diameter than thin filaments.)

© Cengage Learning, 2013

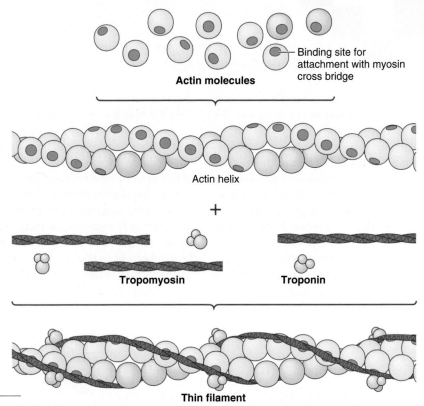

Actin molecules

Binding site for attachment with myosin cross bridge

Actin helix

+

Tropomyosin   Troponin

Thin filament

FIGURE 8-6 Role of calcium in turning on cross bridges.

© Cengage Learning, 2013

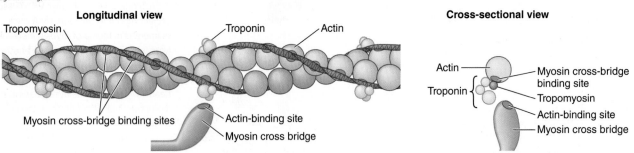

Longitudinal view

Cross-sectional view

Tropomyosin   Troponin   Actin

Actin

Myosin cross-bridge binding site

Troponin

Tropomyosin

Actin-binding site

Myosin cross bridge

Myosin cross-bridge binding sites

Actin-binding site

Myosin cross bridge

(a) Relaxed

1 No excitation.

2 No cross-bridge binding because cross-bridge binding site on actin is physically covered by troponin–tropomyosin complex.

3 Muscle fiber is relaxed.

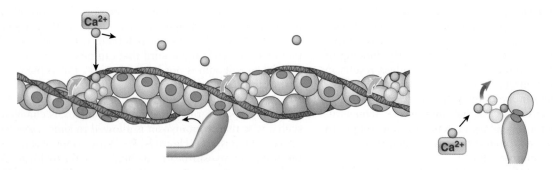

Ca²⁺

Ca²⁺

(b) Excited

1 Muscle fiber is excited and Ca²⁺ is released.

2 Released Ca²⁺ binds with troponin, pulling troponin–tropomyosin complex aside to expose cross-bridge binding site.

3 Cross-bridge binding occurs.

4 Binding of actin and myosin cross bridge triggers power stroke that pulls thin filament inward during contraction.

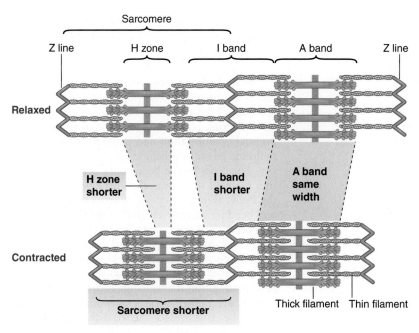

FIGURE 8-7 Changes in banding pattern during shortening. During muscle contraction, each sarcomere shortens as the thin filaments slide closer together between the thick filaments so that the Z lines are pulled closer together. The width of the A bands does not change as a muscle fiber shortens, but the I bands and H zones become shorter.

© Cengage Learning, 2013

check your understanding 8.2

What is the smallest component of a muscle fiber that is capable of contraction?

How is muscle contraction regulated?

8.3 Molecular Basis of Skeletal Muscle Contraction

How does cross-bridge interaction between actin and myosin bring about muscle contraction? How does a muscle action potential trigger this contractile process? What is the source of the Ca^{2+} that physically repositions troponin and tropomyosin to permit cross-bridge binding? We turn our attention to these topics in this section.

During contraction, cycles of cross-bridge binding and bending pull the thin filaments closer together between the stationary thick filaments, causing shortening of the sarcomeres

During contraction in mammalian striated muscle, the thin filaments on each side of a sarcomere were demonstrated to slide inward toward the A band's center (Figure 8-7). As they slide inward, the thin filaments pull closer together the Z discs to which they are attached, so the sarcomere shortens. Because all the sarcomeres throughout the muscle fiber's length shorten simultaneously, the entire fiber becomes shorter. This is known as the **sliding-filament mechanism** of muscle contraction, which was first proposed in 1954 by two separate teams, Hugh Huxley and Jean Hanson, and Andrew Huxley and R. Niedergerke. Nonvertebrate striated muscle, although considerably smaller in length, was also demonstrated to operate on this same principle. The H zone, the region in the center of the A band where the thin filaments do not reach, becomes smaller as the thin filaments approach each other when they slide more deeply inward. The H zone may even disappear if the thin filaments meet in the middle of the A band. The I band, which consists of the portions of the thin filaments that do not overlap with the thick filaments, narrows as the thin filaments further overlap the thick filaments during their inward slide. The thin filaments themselves do not change length during muscle fiber shortening. The width of the A band remains unchanged during contraction because its width is determined by the length of the thick filaments, and the thick filaments do not change length during the shortening process. Note that neither the thick nor thin filaments decrease in length to shorten the sarcomere. Instead, contraction is accomplished by the thin filaments within each sarcomere sliding closer together between the thick filaments.

Power Stroke With the tropomyosin "blockers" pulled out of the way by the binding of Ca^{2+}, thin filaments are pulled inward relative to the stationary thick filaments by myosin cross-bridge activity. Myosin is a molecular motor, similar to kinesin and dynein. Recall that kinesin and dynein have little feet that "walk" along microtubules to transport specific products from one part of the cell to another (as in protein transport within a neuronal axon; p. 139) and to move microtubules in relation to one another (as in the beating of cilia or flagella; p. 60). In the same way, the myosin heads or cross bridges "walk" along an actin filament to pull it inward relative to the stationary thick filament. Let's concentrate on a single cross-bridge interaction (Figure 8-8a). The two myosin heads of each myosin molecule act independently, with only one head attaching to actin at a given time. When myosin and actin make contact at a cross bridge, the bridge changes shape, bending 45° inward as if it were on a hinge, "stroking" toward the center of the sarcomere, like

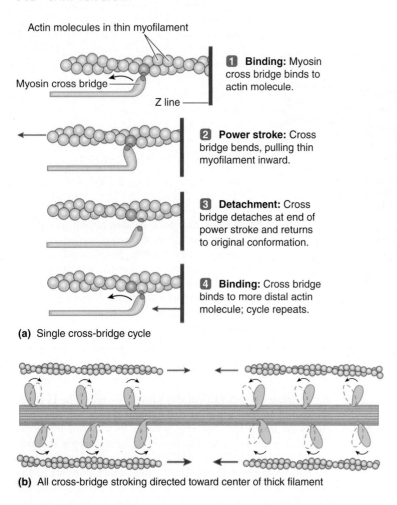

Actin molecules in thin myofilament

Myosin cross bridge

Z line

1 Binding: Myosin cross bridge binds to actin molecule.

2 Power stroke: Cross bridge bends, pulling thin myofilament inward.

3 Detachment: Cross bridge detaches at end of power stroke and returns to original conformation.

4 Binding: Cross bridge binds to more distal actin molecule; cycle repeats.

(a) Single cross-bridge cycle

(b) All cross-bridge stroking directed toward center of thick filament

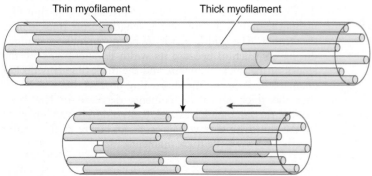

Thin myofilament   Thick myofilament

(c) Simultaneous pulling inward of all six thin filaments surrounding a thick filament

FIGURE 8-8 Cross-bridge activity. (a) During each cross-bridge cycle, the cross bridge binds with an actin molecule, bends to pull the thin filament inward during the power stroke, then detaches and returns to its resting conformation, ready to repeat the cycle. (b) The power strokes of all cross bridges extending from a thick filament are directed toward the center of the thick filament. (c) All six thin filaments surrounding each thick filament are pulled inward simultaneously through cross-bridge cycling during muscle contraction.

© Cengage Learning, 2013

Complete shortening is accomplished by repeated cycles of cross-bridge binding and bending

At the end of one cross-bridge cycle, the link between the myosin cross-bridge and actin molecule is broken. Like a boat oar being lifted and returned to its starting point, the cross bridge returns to its original conformation and binds to the next actin molecule positioned behind its previous actin partner. The cross bridge bends once again to pull the thin filament in further, then detaches and repeats the cycle. Repeated cycles of cross-bridge binding and bending successively pull in the thin filaments.

Because of the orientation of the myosin molecules within a thick filament (Figure 8-8b), all the cross bridges' power strokes are directed toward the center of the sarcomere, so that all six of the surrounding thin filaments on each end of the sarcomere are pulled inward simultaneously (Figure 8-8c). The cross bridges aligned with given thin filaments do not all stroke in unison, however. At any time during contraction, some of the cross bridges are attached to the thin filaments and are stroking, while others are returning to their original conformation in preparation for binding with another actin molecule. Thus, some cross bridges are "holding on" to the thin filaments, whereas others "let go" to bind with new actin. Were it not for this asynchronous cycling of the cross bridges, the thin filaments would slip back toward their resting position between strokes.

How is this cross-bridge cycling switched on by muscle excitation? The term **excitation–contraction coupling** refers to the series of events linking muscle excitation (the presence of an action potential in a muscle fiber) to muscle contraction (cross-bridge activity that causes the thin filaments to slide closer together to produce sarcomere shortening). Now let's turn our attention to the mechanisms of this coupling.

Calcium is the link between excitation and contraction

Skeletal muscles are stimulated to contract by release of acetylcholine (**ACh**) at neuromuscular junctions between motor neuron terminals and muscle fibers (see Figure 4-19). Recall that the binding of ACh with the motor end plate of a muscle fiber brings about permeability changes in the muscle fiber that result in an action potential that is conducted over the entire surface of the muscle cell membrane. Recall also that the enzyme **acetylcholinesterase** (**AChE**) destroys ACh to shut off the signal. Two membranous structures within the muscle fiber play an important role in linking this excitation to contraction—**transverse tubules** and the **sarcoplasmic reticulum.** Let's examine the structure and function of each.

Spread of the Action Potential Down the T Tubules At each end of the A band (mammalian and reptilian muscles) or Z disc (amphibian muscle), the surface membrane dips deeply

the stroking of a boat oar. This so-called power stroke of a cross bridge pulls inward the thin filament to which it is attached. A single power stroke pulls the thin filament inward (approximately a micron), only a minute percentage of the total shortening distance, thus the need for repeated power strokes to achieve the desired extent of muscle shortening.

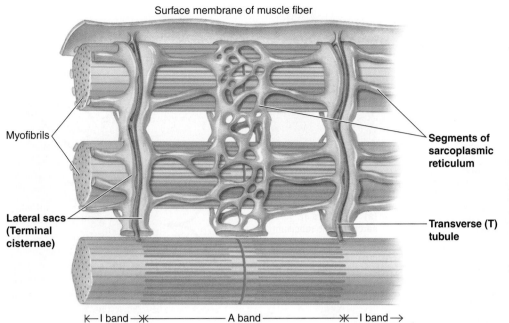

Surface membrane of muscle fiber

Myofibrils

Lateral sacs
(Terminal
cisternae)

**Segments of
sarcoplasmic
reticulum**

**Transverse (T)
tubule**

⊩—I band—⊬⊬————— A band —————⊬⊬— I band —→

FIGURE 8-9 **The T tubules and
sarcoplasmic reticulum in
relationship to the myofibrils.**
The transverse (T) tubules are
membranous, perpendicular
extensions of the surface
membrane that dip deep into the
muscle fiber at the junctions
between the A and I bands of the
myofibrils. The sarcoplasmic
reticulum is a fine, membranous
network that runs longitudinally
and surrounds each myofibril,
with separate segments
encircling each A band and I
band. The ends of each segment
are expanded to form lateral sacs
that lie next to the adjacent T
tubules.

© Cengage Learning, 2013

into the large muscle fiber to form a transverse tubule (or **T
tubule**), which runs perpendicularly from the surface of the
muscle cell membrane into the central portions of the muscle
fiber (Figure 8-9). Because the T tubule membrane is continuous with the surface membrane, an action potential on
the surface membrane also spreads down into the T tubule,
providing a means of rapidly transmitting the surface electric activity into the central portions of the fiber. The presence of a local action potential in the T tubules induces
permeability changes in a separate membranous network
within the muscle fiber, the sarcoplasmic reticulum.

Release of Calcium from the Sarcoplasmic Reticulum The
sarcoplasmic reticulum is a modified endoplasmic reticulum
(see p. 38) that is important for turning the muscle on and
off. It consists of a fine network of interconnected tubules
surrounding each myofibril like a mesh sleeve (Figure 8-9).
This membranous network runs longitudinally down the
myofibril (that is, encircles the myofibril throughout its
length), but is not continuous. Separate segments of sarcoplasmic reticulum are wrapped around each A band and
each I band. The ends of each segment expand to form
saclike regions, the lateral sacs (alternatively known as terminal cisternae), which are separated from the adjacent T
tubules by a slight gap (Figures 8-9 and 8-10). The sarcoplasmic reticulum's lateral sacs store Ca^{2+} with the aid of
Ca^{2+}-ATPase pumps first discovered by Wilhelm Hasselbach
in 1961. *Spread of an action potential down a T tubule triggers release of Ca^{2+} from the sarcoplasmic reticulum into the
cytosol to trigger contraction, and the pumps later reuptake
the Ca^{2+} for relaxation.* Performance of this pumping is affected by the number of pump units, that is, the density of
the sarcoplasmic reticulum in the muscle fiber, as well as
kinetic rates at which the pump isoforms operate. The time
course of Ca^{2+} release and reuptake during contractions is
related to the function of the muscle. Sarcoplasmic reticulum
volume in a muscle fiber varies from 3 to 30% between

muscle fiber types; thus, the greater the speed of contraction,
the greater the volume of the sarcoplasmic reticulum. Increases in cellular sarcoplasmic reticulum and mitochondrial
volumes are generally associated with reduced myofibrillar
volume, reducing force generation but permitting the muscle
to operate at higher contraction frequencies. For example,
sarcoplasmic volume in the shaker muscle fibers of rattlesnakes is approximately 26% (with mitochondrial volume
also 26% and glycogen volume at 17%), thus leaving only
31% for the myofibrillar contents (as we noted earlier).

How is a change in T tubule potential linked with the
release of Ca^{2+} from the sarcoplasmic reticulum's lateral
sacs? An orderly arrangement of **foot proteins** extends from
the sarcoplasmic reticulum and spans the gap between the
lateral sac and T tubule. Each foot protein contains four
subunits arranged in a specific pattern (Figure 8-10a). These
foot proteins not only bridge the gap but also serve as *Ca^{2+}-
release channels*. These foot protein Ca^{2+} channels are
known as **L-type calcium channels** or **ryanodine receptors**
because experimentally they can be locked in the open position by the plant chemical *ryanodine*.

Half of the sarcoplasmic reticulum's foot proteins are
"zipped together" with complementary receptors or Ca^{2+}
channels on the T tubule side of the junction. These T tubule
channels, which are made up of four subunits in exactly the
same pattern as the foot proteins, are located like mirror
images in contact with every other foot protein protruding
from the sarcoplasmic reticulum (Figure 8-10b and c). These
T tubule channels are known as **dihydropyridine receptors**
because they are blocked by the drug *dihydropyridine*. These
Ca^{2+} channels are voltage-gated sensors. When an action
potential is propagated down the T tubule, the local depolarization activates the voltage-gated dihydropyridine receptors. These activated T-tubule channels in turn trigger the
opening of the directly abutting foot proteins in the adjacent
lateral sacs of the sarcoplasmic reticulum. Opening of the
half of these Ca^{2+}-release channels in direct contact with the

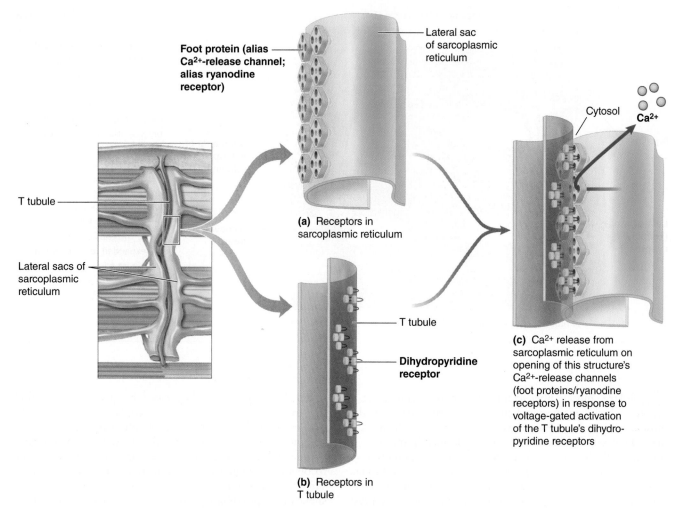

Foot protein (alias Ca²⁺-release channel; alias ryanodine receptor)

Lateral sac of sarcoplasmic reticulum

(a) Receptors in sarcoplasmic reticulum

T tubule

Lateral sacs of sarcoplasmic reticulum

T tubule

Dihydropyridine receptor

(b) Receptors in T tubule

Cytosol

Ca²⁺

(c) Ca²⁺ release from sarcoplasmic reticulum on opening of this structure's Ca²⁺-release channels (foot proteins/ryanodine receptors) in response to voltage-gated activation of the T tubule's dihydropyridine receptors

FIGURE 8-10 Relationship between dihydropyridine receptors on the T tubule and ryanodine receptors (Ca²⁺-release channels) on the adjacent lateral sacs of the sarcoplasmic reticulum.

© Cengage Learning, 2013

dihydropyridine channels triggers the opening of the other half that are not in direct contact.

Ca²⁺ moves down its electrochemical gradient into the surrounding cytosol from the sarcoplasmic reticulum's lateral sacs through all these open Ca²⁺-release channels. By slightly repositioning the troponin and tropomyosin molecules, this released Ca²⁺ exposes the binding sites on the actin molecules so that they can link with the myosin cross bridges at their complementary binding sites. The full coupling cycle (including relaxation, which we will discuss shortly) is summarized in Figure 8-11.

ATP-Powered Cross-Bridge Cycling Recall that a myosin cross bridge has both an actin-binding site and an ATPase site (Figure 8-4a). The latter is an enzymatic site that can bind the energy carrier *adenosine triphosphate (ATP)* and split it into *adenosine diphosphate (ADP)* and *inorganic phosphate (Pᵢ)*, releasing energy in the process. The breakdown of ATP occurs on the myosin cross bridge before the bridge ever links with an actin molecule (Figure 8-12, step 1). The ADP and Pᵢ remain tightly bound to the myosin, and the generated energy is stored within the cross bridge to produce a high-energy form of myosin. To use an analogy, the cross bridge is

"cocked" like a gun, ready to be fired when the trigger is pulled. Recall that when the troponin–tropomyosin complex moves out of its blocking position, the energized (cocked) myosin cross bridge can bind with an actin molecule (step 2a). This contact between myosin and actin "pulls the trigger," causing the cross-bridge bending that produces the *power stroke* (step 3). Researchers have not found the mechanism by which the chemical energy released from ATP is stored within the myosin cross bridge and then translated into the mechanical energy of the power stroke. Inorganic phosphate is released from the cross bridge during the power stroke. After the power stroke is complete, ADP is released.

When the muscle is not excited and Ca²⁺ is not released, troponin and tropomyosin remain in their blocking position so that actin and the myosin cross bridges do not bind and no power stroking takes place (step 2b).

When Pᵢ and ADP are released from myosin following contact with actin and the subsequent power stroke, the myosin ATPase site is free for attachment of another ATP molecule. The actin and myosin remain linked at the cross bridge until a fresh molecule of ATP attaches to myosin at the end of the power stroke. Attachment of the new ATP molecule permits detachment of the cross bridge, which re-

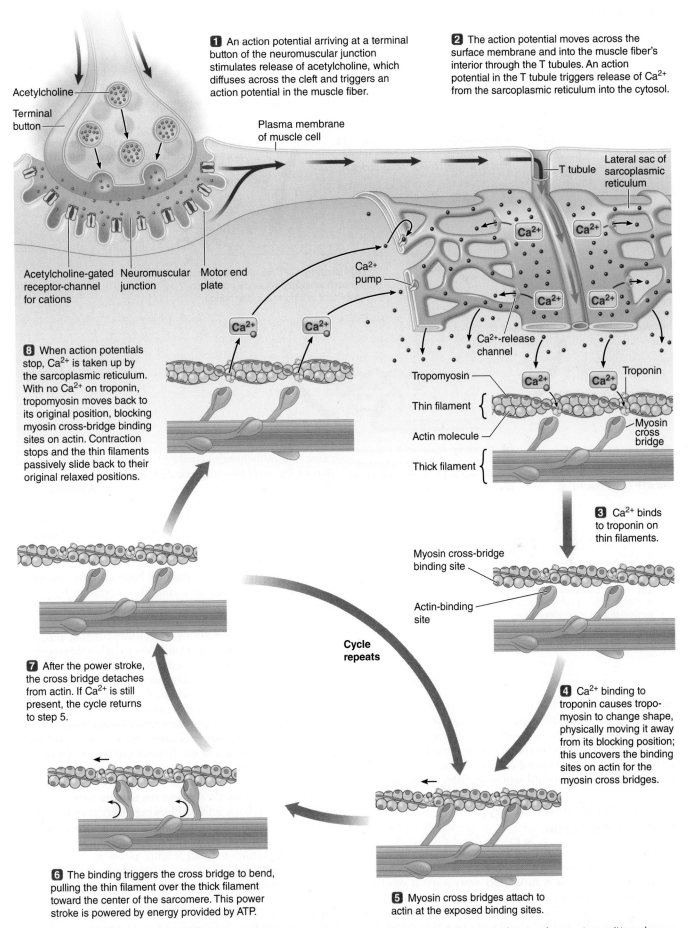

1 An action potential arriving at a terminal button of the neuromuscular junction stimulates release of acetylcholine, which diffuses across the cleft and triggers an action potential in the muscle fiber.

2 The action potential moves across the surface membrane and into the muscle fiber's interior through the T tubules. An action potential in the T tubule triggers release of Ca^{2+} from the sarcoplasmic reticulum into the cytosol.

Acetylcholine

Terminal button

Plasma membrane of muscle cell

Lateral sac of sarcoplasmic reticulum

T tubule

Ca^{2+}

Ca^{2+}

Ca^{2+}

Ca^{2+}

Acetylcholine-gated receptor-channel for cations

Neuromuscular junction

Motor end plate

Ca^{2+} pump

Ca^{2+}-release channel

Ca^{2+}

Ca^{2+}

Ca^{2+}

Ca^{2+}

Tropomyosin

Troponin

Thin filament

Actin molecule

Myosin cross bridge

Thick filament

8 When action potentials stop, Ca^{2+} is taken up by the sarcoplasmic reticulum. With no Ca^{2+} on troponin, tropomyosin moves back to its original position, blocking myosin cross-bridge binding sites on actin. Contraction stops and the thin filaments passively slide back to their original relaxed positions.

3 Ca^{2+} binds to troponin on thin filaments.

Myosin cross-bridge binding site

Actin-binding site

Cycle repeats

4 Ca^{2+} binding to troponin causes tropo-myosin to change shape, physically moving it away from its blocking position; this uncovers the binding sites on actin for the myosin cross bridges.

7 After the power stroke, the cross bridge detaches from actin. If Ca^{2+} is still present, the cycle returns to step 5.

6 The binding triggers the cross bridge to bend, pulling the thin filament over the thick filament toward the center of the sarcomere. This power stroke is powered by energy provided by ATP.

5 Myosin cross bridges attach to actin at the exposed binding sites.

FIGURE 8-11 Excitation–contraction coupling and muscle relaxation. Steps 1 through 7 show the events that couple neurotransmitter release and subsequent electrical excitation of the muscle cell with muscle contraction. At step 7, if Ca^{2+} is still present, the cross-bridge cycle returns to step 5 for another power stroke. If Ca^{2+} is no longer present as a consequence of step 8, relaxation occurs.

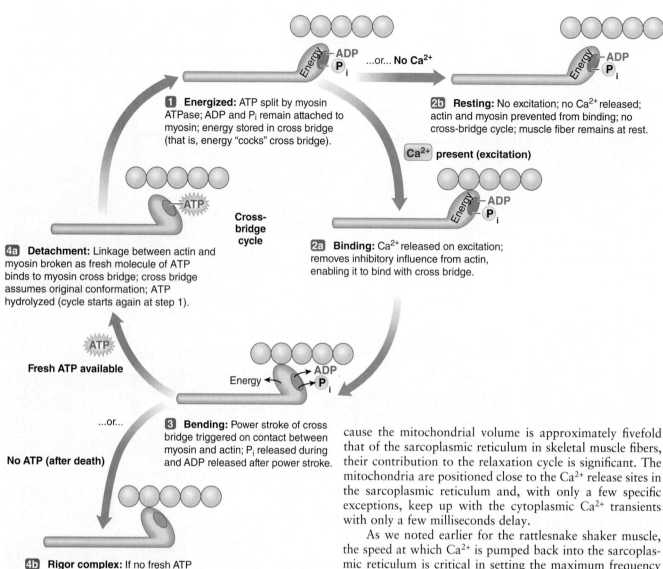

1 **Energized:** ATP split by myosin ATPase; ADP and P_i remain attached to myosin; energy stored in cross bridge (that is, energy "cocks" cross bridge).

2b **Resting:** No excitation; no Ca^{2+} released; actin and myosin prevented from binding; no cross-bridge cycle; muscle fiber remains at rest.

...or... **No Ca^{2+}**

Ca^{2+} **present (excitation)**

Cross-bridge cycle

4a **Detachment:** Linkage between actin and myosin broken as fresh molecule of ATP binds to myosin cross bridge; cross bridge assumes original conformation; ATP hydrolyzed (cycle starts again at step 1).

2a **Binding:** Ca^{2+} released on excitation; removes inhibitory influence from actin, enabling it to bind with cross bridge.

Fresh ATP available

No ATP (after death)

...or...

Energy

3 **Bending:** Power stroke of cross bridge triggered on contact between myosin and actin; P_i released during and ADP released after power stroke.

4b **Rigor complex:** If no fresh ATP available (after death), actin and myosin remain bound in rigor complex.

FIGURE 8-12 Cross-bridge cycle.

© Cengage Learning, 2013

turns to its unbent form, ready to start another cycle (step 4a). The newly attached ATP is then split by myosin ATPase, energizing the myosin cross bridge once again (step 1). On binding with another actin molecule, the energized cross bridge again bends, and so on, successively pulling the thin filament inward to accomplish contraction.

Removal of calcium is the key to muscle relaxation

As we noted earlier, **relaxation** is normally accomplished by the sarcoplasmic reticulum's Ca^{2+}–ATPase pumps. That is, the contractile process is turned off when Ca^{2+} is returned to the lateral sacs at cessation of local electrical activity (Figure 8-11). Recent research has demonstrated that the mitochondria also take up Ca^{2+} during a contraction-relaxation cycle. Each mitochondrion has a comparatively large surface area and is tightly wrapped around many myofibrils. Be-

cause the mitochondrial volume is approximately fivefold that of the sarcoplasmic reticulum in skeletal muscle fibers, their contribution to the relaxation cycle is significant. The mitochondria are positioned close to the Ca^{2+} release sites in the sarcoplasmic reticulum and, with only a few specific exceptions, keep up with the cytoplasmic Ca^{2+} transients with only a few milliseconds delay.

As we noted earlier for the rattlesnake shaker muscle, the speed at which Ca^{2+} is pumped back into the sarcoplasmic reticulum is critical in setting the maximum frequency at which a muscle can operate. In addition to increased density of Ca^{2+}–ATPase pump units, high-speed muscles also have elevated concentrations of calcium-binding proteins, such as **calsequestrin**, which help the sarcoplasmic reticulum store calcium. These calcium-binding proteins also play a critical role in reducing the duration of the twitch by rapidly lowering free (unbound) cytosolic Ca^{2+}.

Let's look at the steps more closely. When acetylcholinesterase removes ACh from the neuromuscular junction, the muscle fiber action potential ceases. When there is no longer a local action potential in the T tubules to trigger the release of Ca^{2+}, the ongoing activity of the sarcoplasmic reticulum's pumps returns the released Ca^{2+} back into its lateral sacs. Removal of cytosolic Ca^{2+} allows the troponin–tropomyosin complex to slip back into its blocking position so that actin and myosin can no longer bind at the cross bridges. The thin filaments, freed from cycles of cross-bridge attachment and pulling, can return to their resting position. The "elastic band" properties of titin may also help return the sarcomere to its unstimulated conformation. Relaxation has occurred.

Rigor Mortis Note that fresh ATP is needed not only for Ca^{2+} pumping but also for myosin to break the cross bridge link to actin at the end of a cycle, even though the ATP is not split during this dissociation process. The need for ATP

The term *meat* is generally defined as those animal tissues suitable for use as food. However, it is the musculature that undergoes postmortem changes that determine palatability and therefore consumer acceptability. At the time of death, cessation of blood flow and thus the delivery of O_2 to muscle initiate a remarkable sequence of biochemical changes that explain the conversion of muscle to meat. As the oxygen supply is depleted, movement of electrons through the citric acid cycle (p. 49) and the electron transport chain ceases. The generation of ATP is shifted from the aerobic pathway to the anaerobic pathway, comparable to the situation that occurs in living muscle when oxygen debt arises in response to periods of intense activity. For a period of time after death, ATP is produced in sufficient quantities to maintain the structural integrity of the fibers. Recall that a major feature of anaerobic metabolism is the accumulation of lactate. When the oxygen supply is cut off, NADH generated in glycolysis cannot cycle through the TCA cycle and begins to accumulate in the muscle. NADH is then converted back to NAD^+ in conjunction with lactate generation from pyruvic acid, which permits glycolysis to proceed at a rapid rate (see p. 48). Because the circulatory system is no longer functioning, lactate remains in muscle fibers and increases in concentration as postmortem metabolism proceeds. Lactate accumulation in the muscle lowers its pH, and at levels less than pH 6.0 the rate of glycolysis is reduced, along with ATP synthesis that had resulted from substrate-level phosphorylation. Mechanisms to regenerate ATP are now insufficient to prevent permanent cross-bridge formation between actin and myosin. Under these conditions, the postmortem onset of rigor mortis gets under way. As the stores of creatine phosphate are depleted, phosphorylation of ADP is insufficient to maintain the tissue in a relaxed state. Actin–myosin bridges begin to form, and this phase continues until all the creatine phosphate is exhausted and ATP can no longer be formed from ADP. This signals the completion of rigor mortis, which in chicken and fish requires less than an hour and as long as 6 to 12 hours in lamb and beef. Because the muscle glycogen content of fish is much lower than in mammals, the pH declines to only 6.8. "Resolution," or the softening of rigor mortis, results from changes in the ultrastructure of the myofilaments. These structural changes result from the action of acid and alkaline proteases, enzymes that originate in lysosomes. Normally these acid proteases, called *cathepsins,* are maintained in an inactive state; but with the reduction in pH, loss of membrane integrity, and increase in ICF Ca^{2+}, these enzymes are activated and are responsible for much of the postmortem structural changes and consequent improvement in meat tenderness.

With the advent of modern animal agriculture, generally only a few days elapse between the time when meat animals attain market weight and when they are slaughtered. Nonetheless, during this period certain factors can affect meat quality. For example, if an animal is stressed before slaughter, this may result in undesirable changes in postmortem metabolism. Glycogen deficiency in the muscle can occur in association with transport stress, fasting, or restraint. Consider also a serious injury to a deer sustained from a misdirected shot from a hunter that permitted the animal to run a considerable distance before blood loss sufficiently incapacited its ability to run. If an injured or stressed animal dies or is slaughtered before it has replenished the glycogen, a *dark, firm, dry* muscle condition can result. Muscle glycogen deficiency results in a limited generation of lactate after death because the substrate of glycolysis, glycogen, was depleted before harvest. The resultant high pH (approximately 6.8) changes the muscle color development when these muscles are exposed to O_2. This muscle tissue is dry and sticky because of its ability to bind water. Its unattractive appearance and decreased palatibility result in a considerable monetary loss for the producer. In addition, its high pH makes it more susceptible to microbial degradation.

---

for both of these processes is amply demonstrated by the phenomenon of *rigor mortis*. This "stiffness of death" is a generalized locking-in-place of the skeletal muscles that begins sometime after death (3 to 4 hours in humans, becoming complete in about 12 hours). After death, cells stop making ATP, and the cytosolic concentration of Ca^{2+} begins to rise, most likely because the muscle-cell membrane cannot keep out extracellular Ca^{2+} and perhaps also because Ca^{2+} leaks out of the lateral sacs. This Ca^{2+} moves the regulatory proteins aside, permitting actin to bind with the myosin cross bridges, which were already charged with ATP before death. Without fresh ATP, actin and myosin, once bound, cannot detach. The thick and thin filaments thus remain linked together by the immobilized cross bridges, resulting in the stiffness of dead muscles (Figure 8-12, step 4b). During the next several days, rigor mortis gradually subsides as the proteins involved in the rigor complex begin to degrade (see *Beyond the Basics: Conversion of Muscle into Meat*).

Now let's compare the duration of contractile activity to the duration of excitation, before we shift gears to discuss skeletal muscle mechanics.

Contractile activity far outlasts the electrical activity that initiated it

A single action potential in a skeletal muscle fiber lasts only 1 to 2 msec, resulting in a brief, weak contraction known as a **twitch**, which is too short and too weak to be useful (Figure 8-13). The onset of the resultant contractile response lags behind the action potential, because the entire excitation–contraction coupling process must take place before cross-bridge activity begins. In fact, the action potential is completed before the contractile apparatus even becomes operational. This time delay of a few milliseconds between stimulation and the onset of contraction is known as the **latent period.**

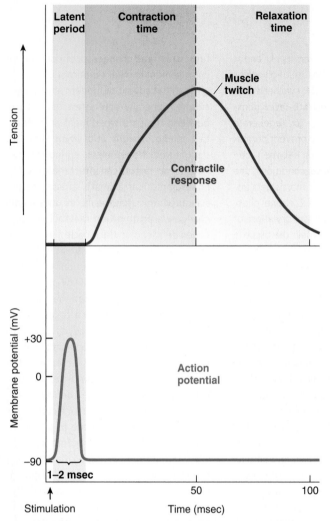

FIGURE 8-13 Relationship of an action potential to the resultant muscle twitch. The duration of the action potential is not drawn to scale but is exaggerated. Note that the resting potential of a skeletal muscle fiber is –90 mV, compared to a resting potential of –70 mV in a neuron.

© Cengage Learning, 2013

Time is also required for the generation of tension within the muscle fiber produced by means of the sliding interactions between the thick and thin filaments through cross-bridge activity. The time from the onset of contraction until peak tension is developed—the **contraction time**—averages about 50 msec in vertebrate locomotory muscles, although this time varies, depending on the type of muscle fiber. The contractile response does not cease until the lateral sacs have taken up all the Ca^{2+} released in response to the action potential. This reuptake of Ca^{2+} is also time consuming. Even after Ca^{2+} is removed, it takes time for the filaments to return to their resting positions. The time from peak tension until relaxation is complete, the **relaxation time**, usually lasts slightly longer than contraction time, another 50 msec or more. Consequently, the entire contractile response to a single action potential may last up to 100 msec or more; this is considerably longer than the duration of the action potential that initiated it (100 msec compared to 1 to 2 msec). These fundamental concepts are important in an animal's ability to produce muscle contractions of variable strength, as you will discover in the next section.

check your understanding 8.3

How does the structure of the microfilaments allow for linear contraction of a whole muscle cell?

How does an event (action potential) on the outside of a muscle fiber trigger the contractile process on the inside?

Ryanodine receptors and dihydropyridine receptors are named for the specific drugs that bind to them, but physiologically, what is their function?

Defend the statement "Mitochondria may participate in calcium uptake but cannot be relied upon to deliver calcium when needed by the cell." Under what circumstances are mitochondria unable to keep up with the Ca^{2+} transient?

8.4 Skeletal Muscle Mechanics

Thus far, we have described the contractile response in a single muscle fiber. In the body, groups of muscle fibers are organized into whole muscle organs. We now turn our attention to contraction of whole muscles. There are four major problems to be solved if muscle organs are to be useful to an animal.

First, each contraction must be precisely controlled. Tension must be developed and maintained at a level appropriate for the physiological response. There are essentially two evolutionary strategies for the gradation of tension generated in a muscle, one typified by vertebrates and the other by arthropods. Other nonvertebrates normally use variations of the solution found in arthropods.

Second, muscles typically contract linearly, but locomotory motions need more diverse motions. This is a major reason that most muscles exert force through external elements, for example, various forms of *skeletons*. For example, as we'll see later, hard skeletons with *joints* can act as levers to convert linear contraction into angular motion.

Third, muscles have a fundamental limit—the inability to expand actively due to the one-way nature of the myosin "rowing" mechanism. For a contracted muscle to expand, there are three basic mechanisms: (1) an *antagonistic* muscle typically found on the opposite side of a skeletal joint, whose contraction stretches out the other muscle (p. 350); (2) *fluid pressure* created by distant muscles that push on and re-extend a contracted muscle; and (3) **series elastic elements:** springlike structures that get stretched or compressed by a contracting muscle and that then "rebound" to release their stored energy and thereby stretch out that muscle. Titin may be one of these, but external springs are typically stronger and faster.

Fourthly, the sarcomere mechanism is too slow in strong muscles, due to trade-offs you saw earlier, to power some of the incredibly fast locomotory movements seen in the animal kingdom, like jumping fleas. This problem can also be solved by elastic mechanisms because a spring can move much faster than muscle can contract. This is the principle used in the bow-and-arrow invention of humans: Muscles are used to pull back the elastic bow cord (relatively

A Closer Look at Adaptation
Muscles Evolved for Speed

Recent studies have investigated Ca^{2+} flux in one of the fastest known vertebrate muscle, the sound-producing swim-bladder muscle of the toadfish (*Opsanus tau*). To attract egg-laying females to his nest, the male toadfish sings a "boat whistle" mating call 10 to 12 times per minute. Vocalizations are generated by repetitive contractions (at about 200 Hz) of the muscles encircling the fish's gas-filled swimbladder (see p. 539). High-frequency contractions of this muscle yield high-frequency sound production. If a muscle is to contract and relax rapidly, two conditions must apply:

1. Ca^{2+}, the trigger for muscle contraction, must both enter and leave the myoplasm rapidly (Ca^{2+} *transient*). To achieve this remarkable feat, researchers determined, the Ca^{2+} transient in the sonic muscle is the fastest ever measured for any fiber type, approximately 50-fold greater than the rate measured in comparable locomotory muscle. The importance of this observation is that the time course of Ca^{2+} uptake into the sarcoplasmic reticulum, comparable to that in a synapse, is so rapid that it has completely returned to baseline by the time of the next stimulus! In these superfast muscle fibers, this feat is achieved by (1) a maximal density of Ca^{2+} pumps in the sarcoplasmic reticulum, (2) different Ca^{2+} pump isoforms (which pump Ca^{2+} at different rates), (3) a high volume of sarcoplasmic reticulum, approaching 30% of the muscle volume, and (4) exclusion of mitochondria from the intermyofibrillar area where the Ca^{2+} release sites are located. Because mitochondria take up and release Ca^{2+} at rates different from the sarcoplasmic reticulum, their activity would hamper the rapid oscillations characteristic of the sarcoplasmic reticulum.

2. Myosin cross bridges must attach to actin and generate force soon after the Ca^{2+} concentrations rise and then must detach and stop generating force soon after the Ca^{2+} concentrations decline. A faster relaxing time also requires that troponin rapidly release its bound Ca^{2+}. In actuality, the release of Ca^{2+} from its troponin binding site in the sonic muscles is three times faster than in locomotory muscle. To permit faster cross-bridge cycling, modification in the molecular structure of troponin is required. Selection for isoforms of troponin with a lower affinity (attraction) for Ca^{2+}

helped increase beat frequency. In addition, the detachment rate of myosin–actin cross bridges in high-frequency muscles is approximately 70-fold faster than that measured in locomotory muscles, fast enough to account for the rapid relaxation rate. Because the detachment rate of cross bridges in swimbladder muscle is so fast, a low proportion of cross bridges are actually attached at any given instant. Although this represents a solution for speed, the consequences are that the muscles are weak. Fortunately, it does not require extraordinary power or force to produce sound.

Gulf Toadfish, *Opsanus beta*

---

slowly) to store energy in it; when released, the cord recoils and shoots the arrow at speeds much greater than a contracting muscle could. This is also the mechanism used in the chameleon tongue, mentioned at the beginning of the chapter. You'll see locomotory examples later.

We will encounter each solution to these four problems after concluding our discussion on the structure of muscles as organs.

Muscle organs are groups of muscle fibers bundled together by connective tissue, often attached to skeletal elements in antagonistic pairs

Each muscle (organ) is covered by a sheath of connective tissue that penetrates from the surface into the muscle to envelop each individual fiber and divide the muscle into columns or bundles. In vertebrates, the connective tissue extends beyond the ends of the muscle to form tough, collagenous **tendons** that normally attach the muscle to bones (Figure 8-2). A tendon may be quite long, attaching to a

TABLE 8-1 Determinants of Whole-Muscle Tension in Skeletal Muscle

Number of Fibers Contracting	Tension Developed by Each Contracting Fiber
Number of motor units recruited*	Frequency of stimulation (twitch summation and tetanus)*
Number of muscle fibers per motor unit	Length of fiber at onset of contraction (length–tension relationship)

Number of muscle fibers available to contract (size of muscle)	Extent of fatigue
	    Duration of activity
	    Type of fiber (fatigue-resistant oxidative or fatigue-prone glycolytic)
	
	Thickness of fiber
	    Pattern of neural activity (hypertrophy, atrophy)
	    Amount of testosterone (larger fibers in males than females)

*Factors controlled to accomplish gradation of contraction.

© Cengage Learning, 2013

bone some distance from the fleshy portion of the muscle. For example, some of the muscles involved in digit movement are found in the forelimb, with long tendons extending down to attach to the bones of the digits. (You can readily observe the movement of these tendons on the top of your hand when you wiggle your fingers.) This arrangement permits greater dexterity; for example, in humans the fingers would be much thicker and more awkward if all the muscles involved in finger movement were actually located in the fingers.

In arthropods, muscles attach to **apodemes**, ridges that project from the inner face of the exoskeleton. The apodeme is considered the arthropod equivalent of a vertebrate tendon, although it is 50 times stiffer than a typical tendon. As you will see later, both apodemes and tendons can store and release elastic energy, acting as springs.

As we noted earlier and in Chapter 5, muscles are often arranged in **antagonistic pairs** to move a body part in two opposing directions. A common pairing is that of **flexors**, which bend a limb, and **extensors**, which typically straighten the limb out. Familiar examples are the *biceps* flexor and *triceps* extensor of a mammal's limb (see Figure 5-21, p. 181). Similarly, a crab claw (a modified limb) is opened with an extensor and closed with a flexor.

Contractions of a whole muscle can be of varying strength

Recall that a **twitch** is too short and too weak to be useful (Figure 8-13). Muscle fibers are arranged into whole muscles where they can function cooperatively to produce contractions of variable grades of strength stronger than a twitch. The force produced by a single muscle can be made to vary and depends on the force required to move (or to carry an object). Most vertebrate muscle fibers contract in "all-or-none" fashion, such that a single fiber cannot produced graded contractions in a simple way. Thus, two primary factors are adjusted to accomplish gradation of whole-muscle tension: (1) *the number of muscle fibers contracting within a muscle* and (2) *the tension developed by each contracting fiber* (Table 8-1) by other than simple graded means. We discuss each of these factors in turn.

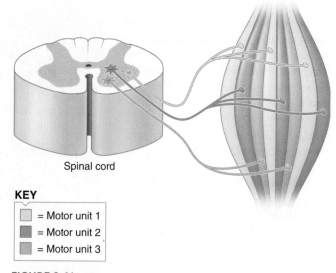

Spinal cord

KEY

☐ = Motor unit 1
■ = Motor unit 2
■ = Motor unit 3

FIGURE 8-14 Motor units in a skeletal muscle.

© Cengage Learning, 2013

The number of fibers contracting within a vertebrate muscle depends on the extent of motor unit recruitment

Because the greater the number of fibers contracting, the greater the total muscle tension, larger muscles consisting of more muscle fibers can obviously generate more tension than can smaller muscles with fewer fibers. Each whole muscle is innervated by a number of different motor neurons. Typically, hundreds to thousands of motor neurons innervate a vertebrate muscle, whereas only 1 to 10 motor neurons innervate an arthropod muscle. When a vertebrate motor neuron enters a muscle, it branches, with each axon terminal supplying a single muscle fiber (Figure 8-14). *One motor neuron innervates a number of muscle fibers, but each vertebrate muscle fiber is supplied by only one motor neuron.* When a motor neuron is activated, all the muscle fibers it supplies are stimulated to contract simultaneously.

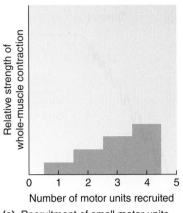

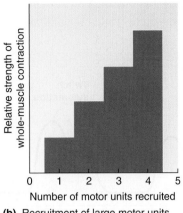

FIGURE 8-15 **Comparison of motor unit recruitment in skeletal muscles with small motor units and muscles with large motor units.** (a) Small incremental increases in strength of contraction occur during motor unit recruitment in muscles with small motor units because only a few additional fibers are called into play as each motor unit is recruited. (b) Large incremental increases in strength of contraction occur during motor unit recruitment in muscles with large motor units, because so many additional fibers are stimulated with recruitment of each additional motor unit.

© Cengage Learning, 2013

(a) Recruitment of small motor units   **(b)** Recruitment of large motor units

This team of concurrently activated components—one motor neuron plus all the muscle fibers it innervates—is called a **motor unit**. (We will examine a different arrangement in arthropods later.)

The muscle fibers that compose a vertebrate motor unit are dispersed throughout the whole muscle; thus, their simultaneous contraction results in an evenly distributed, although weak, contraction of the whole muscle. Each muscle consists of a number of intermingled motor units. For a weak contraction of the whole muscle, only one or a few of its motor units are activated. For stronger and stronger contractions, more and more motor units are recruited, or stimulated to contract, a phenomenon known as **motor unit recruitment**.

How much stronger the contraction will be with the recruitment of each additional motor unit depends on the size of the motor units (that is, the number of muscle fibers controlled by a single motor neuron) (Figure 8-15). The number of muscle fibers per motor unit and the number of motor units per muscle vary widely, depending on the specific function of the muscle. For muscles that produce precise, delicate movements, such as the external eye muscles, and the hand muscles in humans, a single motor unit may contain as few as a dozen muscle fibers. Because so few muscle fibers are involved with each motor unit, recruitment of each additional motor unit results in only a small additional increment in the whole muscle's strength of contraction. These small motor units allow a very fine degree of control over muscle tension. In contrast, in muscles used for powerful, coarsely controlled movement, such as those of most mammalian legs, a single motor unit may contain 1,500 to 2,000 muscle fibers. Recruitment of motor units in these muscles results in large incremental increases in whole-muscle tension. More powerful contractions occur at the expense of less precisely controlled gradations. Thus, the number of muscle fibers participating in the whole muscle's total contractile effort depends on the number of motor units recruited and the number of muscle fibers per motor unit in that muscle.

To delay or prevent **fatigue** (inability to maintain muscle tension at a given level) during a *sustained* contraction involving only a portion of a muscle's motor units, as is necessary in muscles supporting the weight of the body against the force of gravity, **asynchronous recruitment of motor units** takes place. The brain alternates motor unit activity, like shifts at a factory, to give motor units that have been active a chance to rest while others take over. Changing of the shifts is carefully coordinated, so that the sustained contraction is smooth rather than jerky. Asynchronous motor unit recruitment is possible only for submaximal contractions, during which only some of the motor units must maintain the desired level of tension. During maximal contractions, when all the muscle fibers must participate, it is impossible to alternate motor unit activity to prevent fatigue. This is one reason why you cannot support a heavy object as long as one that is light.

Furthermore, the type of muscle fiber that is activated varies with the magnitude of contraction. Most muscles consist of a mixture of fiber types that differ metabolically, some being more resistant to fatigue than others (as we'll discuss later). During weak or moderate endurance-type activities (aerobic activity), the motor units most resistant to fatigue are recruited first. The last fibers called into play in the face of demands for further increases in tension are those that fatigue rapidly. An animal can therefore engage in endurance activities for prolonged periods of time but can only briefly maintain bursts of all-out, powerful effort. Of course, even the muscle fibers most resistant to fatigue do eventually fatigue if required to maintain a certain level of sustained tension.

The frequency of stimulation can influence the tension developed by each vertebrate skeletal muscle fiber

Whole-muscle tension depends not only on the number of muscle fibers contracting but also on the tension developed by each contracting fiber. Various factors influence the extent to which tension can be developed. These factors, summarized in Table 8-1, include (1) frequency of stimulation; (2) length of the fiber at the onset of contraction; (3) extent of fatigue; and (4) thickness of the fiber. The other factors in Table 8-2 will be discussed later.

First let's examine the effect of frequency of stimulation.

Twitch Summation and Tetanus Even though a single action potential in a muscle fiber produces only an "all-or-none" twitch, contractions with longer duration and greater tension can be achieved by repetitive stimulation of the fiber. Let's see what happens when a second action potential occurs in a muscle fiber. If the muscle fiber has completely relaxed before the next action potential takes

FIGURE 8-16 Twitch summation and tetanus.

© Cengage Learning, 2013

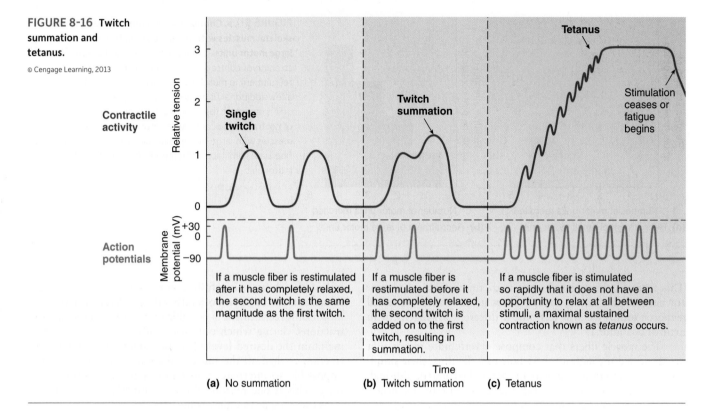

Contractile activity

Single twitch

Twitch summation

Tetanus

Stimulation ceases or fatigue begins

Action potentials

If a muscle fiber is restimulated after it has completely relaxed, the second twitch is the same magnitude as the first twitch.

If a muscle fiber is restimulated before it has completely relaxed, the second twitch is added on to the first twitch, resulting in summation.

If a muscle fiber is stimulated so rapidly that it does not have an opportunity to relax at all between stimuli, a maximal sustained contraction known as *tetanus* occurs.

Time

(a) No summation   **(b)** Twitch summation   **(c)** Tetanus

place, a second twitch of the same magnitude as the first occurs (Figure 8-16a). The same excitation–contraction events take place each time, resulting in identical twitch responses. If, however, the muscle fiber is stimulated a second time before it has completely relaxed from the first twitch, a second action potential occurs that causes a second contractile response, which is added "piggyback" on top of the first twitch (Figure 8-16b). The two twitches resulting from the two action potentials add together, or summate, to produce greater tension in the fiber than that produced by a single action potential. This twitch summation is similar to temporal summation of EPSPs at the postsynaptic neuron (see p. 137).

Twitch summation is possible only because the duration of the action potential (1 to 2 msec) is much shorter than the duration of the resultant twitch (100 msec). Remember that once an action potential has been initiated, a brief refractory period occurs during which another action potential cannot be initiated (see p. 124). It is therefore impossible to achieve summation of action potentials. The membrane must return to resting potential and recover from its refractory period before another action potential can occur. However, because the action potential and refractory period are over long before the resultant muscle twitch is completed, the muscle fiber may be restimulated while some contractile activity still exists to produce summation of the mechanical response.

If the muscle fiber is stimulated so rapidly that it does not have a chance to relax at all between stimuli, a smooth, sustained contraction of maximal strength known as **tetanus** occurs (Figure 8-16c). (Not that a tetanus is a normal process; the disease with the same name causes uncontrollable tetanus, as we described on p. 143.) A tetanic contraction is usually three to four times stronger than a single twitch.

Twitch summation results from a sustained elevation in cytosolic calcium

What is the mechanism of twitch summation and tetanus at the cellular level? The tension produced by a contracting muscle fiber increases as a result of greater cross-bridge cycling. Connective tissue, as well as other components of the muscle, such as intracellular elastic proteins, exhibits a certain degree of passive elasticity. These noncontractile tissues are part of the *series-elastic component* we noted earlier; they behave like a stretchy spring placed between the internal tension-generating elements and the bone that is to be moved against an external load. The series-elastic component (connective tissue, titin, and tendons) must be stretched to transmit the tension generated in the muscle fiber to the bone, and it takes time to stretch these elastic elements. Accordingly, two factors contribute to twitch summation: (1) sustained elevation in cytosolic Ca^{2+} permitting greater cross-bridge cycling, and (2) more time to stretch the series-elastic component.

The most important factor in the development of twitch summation is sustained elevation in cytosolic Ca^{2+} as the frequency of action potentials increases. Sufficient Ca^{2+} is released in response to a single action potential to interact with all the troponin within the cell. As a result, all the cross bridges are free to participate in the contractile response. How then, can repetitive action potentials bring about a greater contractile response? The difference depends on how long sufficient Ca^{2+} is available.

The cross bridges remain active and continue to cycle as long as sufficient Ca^{2+} is present to keep the troponin-tropomyosin complex away from the cross-bridge binding sites on actin. Each troponin–tropomyosin complex spans a distance of seven actin molecules. Thus, binding of one Ca^{2+} ion to one troponin molecule leads to the uncovering of only seven cross-bridge binding sites on the thin filament.

As soon as Ca^{2+} is released in response to an action potential, the sarcoplasmic reticulum starts pumping Ca^{2+} back into the lateral sacs. As the cytosolic Ca^{2+} concentration declines with the reuptake of Ca^{2+} by the lateral sacs, less Ca^{2+} is present to bind with troponin, so some of the troponin-tropomyosin complexes slip back into their blocking positions. Consequently, not all the cross-bridge binding sites remain available to participate in the cycling process during a single twitch induced by a single action potential. Because not all the cross bridges find a binding site, the resulting contraction during a single twitch is not of maximal strength.

If action potentials and twitches occur far enough apart in time for all the released Ca^{2+} from the first contractile response to be pumped back into the lateral sacs between the action potentials, an identical twitch response occurs as a result of the second action potential. If, however, a second action potential occurs and more Ca^{2+} is released while the Ca^{2+} that was released in response to the first action potential is being taken back up, the cytosolic Ca^{2+} concentration remains elevated. This prolonged availability of Ca^{2+} in the cytosol permits more of the cross bridges to continue participating in the cycling process for a longer time. As a result, tension development increases correspondingly. As the frequency of action potentials increases, the duration of elevated cytosolic Ca^{2+} concentration increases, and contractile activity likewise increases until a maximum tetanic contraction is reached. With tetanus, the maximum number of cross-bridge binding sites remain uncovered so that cross-bridge cycling, and consequently tension development, are at their peak.

An additional factor contributing to twitch summation is related to the elastic element. During a single twitch, the contraction does not last long enough to completely stretch the series-elastic component and allow the full sarcomere-generated tension to be transmitted to the bone. At the end of the twitch, the elastic elements slowly relax, or recoil to their initial nonstretched state. If another twitch occurs before the elastic elements have completely relaxed, the tension from the second twitch adds to the residual tension in the series-elastic component remaining from the first twitch. With greater frequencies of action potentials and more frequent twitches, less time is available for the elastic elements to recoil between twitches. Consequently, despite the all-or-none nature of a twitch, as the frequency of action potentials increases, the tension in the series-elastic component transmitted to the bone progressively increases until it reaches its maximum during tetanus.

Because skeletal muscle must be stimulated by motor neurons to contract, the nervous system plays a key role in regulating the strength of contraction. The two main factors subject to control to accomplish gradation of contraction of a given muscle are the *number of motor units stimulated* and the *frequency of their stimulation*. The areas of the vertebrate brain responsible for directing motor activity use a combination of tetanic contractions and precisely timed shifts of asynchronous motor unit recruitment to execute smooth rather than jerky contractions.

Arthropod muscle tension is controlled by gradation of contraction within a motor unit

The mechanism by which tension is developed in arthropod muscle is somewhat different from that in vertebrate twitch muscle. Most nonvertebrate muscle fibers are innervated by

more than one motor neuron. Thus, an individual muscle fiber forms a part of several motor units! Because the nervous system of nonvertebrates consists of comparatively fewer neurons, a smaller number of such overlapping motor units must generate the complete range of tension necessary for the animal to move. In some arthropod muscles, one motor neuron innervates most, if not all, of the muscle fibers with numerous axons, such that the whole muscle acts much like a single unit. To examine the mechanism by which arthropods grade individual contractions, let's look at the example of crustacean skeletal muscle fibers.

The limb muscles of crayfish are innervated by very few efferent neurons. In crayfish, crabs, and other decapod crustaceans, two muscles of the terminal joints, the "opener" and "stretcher" (extensor) muscles share one single, common excitatory neuron. Recall that most vertebrate striated muscle fibers are innervated by only one or two end plates (see p. 134), but in many nonvertebrates motor control is achieved by **multiterminal innervation** of each muscle fiber. Synaptic terminals that run the entire length of the fiber, for example, repeatedly innervate most crustacean skeletal muscle fibers. Unlike the "all-or-none" contraction of most vertebrate fibers, contraction of the crustacean fiber depends on *graded depolarization* of the muscle fiber: The greater the frequency of action potentials arriving at the end plate region, the stronger the resulting contraction. In addition, unlike the situation for vertebrate muscle fibers, they also receive *separate* presynaptic inhibitory input on the excitatory nerve endings that permit the action of the two muscles to perform separately. Modulation of contraction and a graded response is thus achieved through *inhibitory* impulses, mostly at the presynaptic level. The release of the inhibitory neurotransmitter GABA (p. 138) onto presynaptic receptors increases Cl^- conductance. An excitatory nerve impulse arriving in the terminal button is then reduced in amplitude because a repolarizing leakage of Cl^- into the nerve terminal takes place when its membrane is depolarized by the nerve impulse. An impulse of reduced amplitude consequently releases less neurotransmitter at the excitatory synapse that results in a weaker contraction.

In addition to presynaptic inhibition of arthropod muscle, a second form of inhibitory control is common. In postsynaptic inhibition, the inhibitory nerve terminals form multiterminal neuromuscular synapses along the muscle fiber. Impulses generated in the inhibitory neuron produce IPSPs in the postsynaptic muscle fiber that oppose the membrane depolarization by EPSPs and reduce the force of the resultant contraction. Of these two mechanisms of inhibition, postsynaptic inhibition is the more common. Both mechanisms are important in controlling muscle tension and, in some situations, may speed the rate of relaxation of a contracted muscle.

There is an optimal muscle length at which maximal tension can be developed on a subsequent contraction

Additional factors not directly under nervous control also influence the tension developed during contraction (Table 8-1). Among these is the length of the fiber at the onset of contraction, to which we now turn.

A relationship exists between the length of the muscle before the onset of contraction, and the tetanic tension that

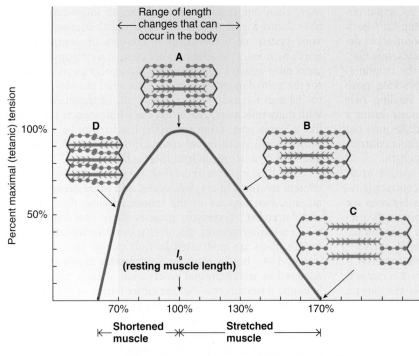

FIGURE 8-17 Length–tension relationship. Maximal tetanic contraction can be achieved when a muscle fiber is at its optimal length (l_o) before the onset of contraction, because this is the point of optimal overlap of thick-filament cross bridges and thin-filament cross-bridge binding sites (point A). The percentage of maximal tetanic contraction that can be achieved decreases when the muscle fiber is longer or shorter than l_o before contraction. When it is longer, fewer thin-filament binding sites are accessible for binding with thick-filament cross bridges because the thin filaments are pulled out from between the thick filaments (points B and C). When the fiber is shorter, fewer thin-filament binding sites are exposed to thick-filament cross bridges because the thin filaments overlap (point D). Also, further shortening and tension development are impeded as the thick filaments become forced against the Z lines (point D). In the body, the resting muscle length is at l_o. Furthermore, because of restrictions imposed by skeletal attachments, muscles cannot vary beyond 30% of their l_o in either direction (the range screened in *light green*). At the outer limits of this range, muscles still can achieve about 50% of their maximal tetanic contraction.

© Cengage Learning, 2013

the thin filaments do not overlap at l_o, lacks cross bridges; only myosin tails are here.

At Lengths Greater Than l_o At greater lengths, as when a muscle is passively stretched in our experiment (point B), the thin filaments have been pulled out from between the thick filaments, decreasing the number of actin sites available for cross-bridge binding; that is, some of the actin sites and cross bridges no longer "match up," so they "go unused." Thus, less cross-bridge activity can occur after stimulation, so less tension can be developed. In fact, when the muscle is stretched to about 70% longer than its l_o (point C), the thin filaments are completely pulled out from between the thick filaments so that no cross-bridge activity and consequently no contraction can occur.

At Lengths Less Than l_o If a muscle is compressed so that it is shorter than l_o prior to stimulation (point D), less tension can be developed, for three reasons:

1. The thin filaments from the opposite sides of the sarcomere become overlapped, decreasing the number of actin sites exposed to the cross bridges.
2. The thick filaments become forced against the Z lines, so further shortening is impeded.
3. Besides these two mechanical factors, at muscle lengths less than 80% of l_o, not as much Ca^{2+} is released during excitation–contraction coupling for reasons unknown. Furthermore, by an unknown mechanism the ability of Ca^{2+} to bind to troponin and pull the troponin–tropomyosin complex aside is reduced at shorter muscle lengths. Consequently, fewer actin sites are uncovered for participation in cross-bridge activity.

each contracting fiber can subsequently develop at that length. Every muscle has an **optimal length** (l_o) at which maximal force can be achieved on a subsequent tetanic contraction. More tension can be achieved during tetanus when beginning at the optimal muscle length than can be achieved when the contraction begins with the muscle less than or greater than its optimal length. This **length–tension relationship** can be explained by the sliding-filament mechanism of muscle contraction, which was first demonstrated on frog skeletal muscle fibers. Figure 8-16 shows the results of such a study: The tension generated by a muscle is measured after it was set at different lengths prior to stimulation (Points B, C, and D).

At Optimal Length (l_o) At l_o when maximum tension can be developed (point A in Figure 8-17), the thin filaments optimally overlap the regions of the thick filaments from which the cross bridges project. At this length, a maximal number of cross-bridge sites are accessible to the actin molecules for binding and bending. The central region of thick filaments, where

Limitations on Muscle Length The extremes in muscle length that prevent the development of tension occur only under experimental conditions, when a muscle is removed and stimulated at various lengths. *In vivo*, muscles are usually positioned such that their relaxed length is approximately their optimal length; thus, they can achieve near-maximal tetanic contraction most of the time. Because of limitations imposed by attachment to the skeleton, a muscle cannot be stretched or shortened more than 30% of its resting optimal length, and usually it deviates much less than 30% from normal length. Even at the outer limits (130 and 70% of l_o), the muscles can still generate half their maximum tension.

The factors discussed thus far that influence how much tension can be developed by a contracting muscle fiber—the frequency of stimulation and the muscle length at the onset of contraction—can vary from contraction to contraction. Other determinants of muscle fiber tension—the metabolic capability of the fiber relative to resistance to

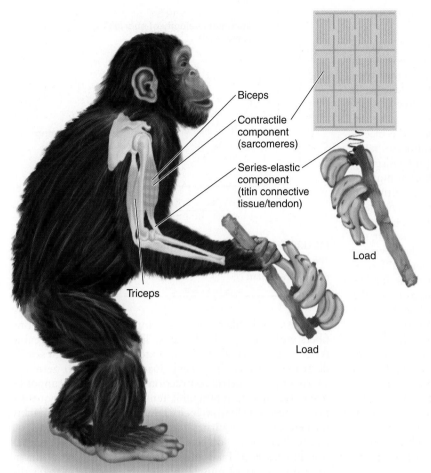

Biceps

Contractile
component
(sarcomeres)

Series-elastic
component
(titin connective
tissue/tendon)

Triceps

Load

Load

FIGURE 8-18 Relationship between the contractile component and the series-elastic component in transmitting muscle tension to bone. Muscle tension is transmitted to the bone by means of the stretching and tightening of the muscle's elastic connective tissue and tendon (similar to the stretching spring) as a result of sarcomere shortening brought about by cross-bridge cycling.

© Cengage Learning, 2013

fatigue and the thickness of the fiber—do not vary from contraction to contraction but depend on the fiber type and can be modified over a period of time. After we finish discussing skeletal muscle mechanics, we will consider these other factors in the next section, on skeletal muscle metabolism and fiber types.

Muscle tension is transmitted to skeletal elements as the contractile component tightens the series-elastic component

Tension is produced internally within the sarcomeres, the contractile component of the muscle, as a result of cross-bridge activity and the resultant sliding of filaments. However, the sarcomeres are not attached directly to the skeleton. Instead, the tension generated by these contractile elements must be transmitted to the bone or exoskeleton via the connective tissue (such as tendons) before the skeletal element can be moved (Figure 8-18). This shortening of the sarco-

meres stretches the series-elastic component. Muscle tension is transmitted to the skeleton by this tightening of the series-elastic component. This externally applied tension is responsible for moving the skeleton against a load.

A vertebrate skeletal muscle is typically attached to at least two different bones across a joint by means of tendons that extend from each end of the muscle. When the muscle shortens during contraction, the position of the joint is changed as one bone is moved in relation to the other—for example, flexion of the elbow joint by contraction of the biceps (flexor) muscle and extension of the elbow by contraction of the triceps (extensor). The end of the muscle attached to the more stationary part of the skeleton is called the **origin**, and the end attached to the skeletal part that moves is the **insertion**.

For a muscle to shorten during contraction, the tension developed in the muscle must exceed the forces that oppose movement of the bone to which the muscle's insertion is attached. In the case of elbow flexion, the opposing force, or **load**, is the weight of an object being lifted. The force exerted by the muscle on the load is equivalent to the tension generated. If you flex your elbow without lifting any external object, there is still a load, albeit a minimal one—the weight of your forearm being moved against the force of gravity.

The two primary types of contraction are isotonic and isometric

Muscle accomplishes work in a physical sense only when an object is moved. **Work** is defined as force multiplied by distance. **Force** can be equated to the muscle tension required to overcome the load (the weight of the object being moved, such as a bone). The amount of work accomplished by a contracting muscle therefore depends on how much an object weighs and how far it is moved. There are two primary types of contraction, depending on whether the muscle changes length during contraction. In an **isotonic contraction**, muscle tension remains constant as the muscle changes length, doing work. In an **isometric contraction**, the muscle is prevented from shortening, so tension develops at constant muscle length; thus no work is done. The same internal events occur in both isotonic and isometric contractions: The tension-generating contractile process is turned on by muscle excitation; the cross bridges start cycling; and filament sliding shortens the sarcomeres, which stretches the series-elastic component to exert tension on the bone at the site of the muscle's insertion.

Considering the biceps as an example, assume a chimpanzee is going to lift an object (as in Figure 8-18). When the tension developing in the biceps becomes great enough to overcome the weight of the object, the animal can lift the object, with the whole muscle shortening in the process. Because the weight of the object does not change as it is lifted, the muscle tension remains constant throughout the period of shortening. This is an *isotonic* (literally, "constant tension") *contraction*. Isotonic contractions are used for body movements and for moving external objects. This type of isotonic contraction, in which the muscle shortens, is sometimes called a **concentric** one. However, another type called **eccentric** occurs when the *muscle lengthens* because it is being stretched by an external force, even while it is attempting to contract. An example is lowering a load to the ground. During this action, the muscle fibers in the biceps are lengthening but are still working in opposition to being stretched. This tension supports the weight of the object.

Not all vertebrate muscle contractions shorten muscles and move bones, however. What happens when lifting an object too heavy to move upwards (that is, if the tension in the muscles is just equal to that required to hold the load in place)? In this case, the muscle cannot shorten and lift the object but remains at constant length despite the development of tension, so an *isometric* ("constant length") *contraction* occurs. In addition to occurring when the load is too great, isometric contractions also take place when the tension developed in the muscle is deliberately less than that needed to move the load. In this case, the goal is to keep the muscle at fixed length although it is capable of developing more tension. These submaximal isometric contractions are important for maintaining posture (such as keeping the legs stiff while standing) and for supporting objects in a fixed position. During a given movement, a muscle may shift between isotonic and isometric contractions. For example, when the animal picks up a banana to eat it, the biceps undergoes an isotonic contraction while the fruit is being lifted, but the contraction becomes isometric as the arm stops to hold the banana at the mouth.

Other Contractions An animal's body is not limited to pure isotonic and isometric contractions. Muscle length and tension frequently vary throughout a range of motion. Think about drawing a bow and arrow. The tension of the biceps muscle continuously increases to overcome the progressively increasing resistance as the bow is stretched further. At the same time, the muscle progressively shortens as the bow is drawn farther back. Such a contraction occurs at neither constant tension nor constant length.

Some skeletal muscles do not attach to bones at both ends but still produce movement. For example, the muscles of the tongue are not attached at the free end. Isotonic contractions of the tongue muscles maneuver the free, unattached portion of the tongue to facilitate eating (and speech in humans). The external eye muscles attach to the skull at their origin but to the eye at their insertion. Isotonic contractions of these muscles produce the eye movements that enable an animal to track moving objects, for example. A few skeletal muscles that are completely unattached to bone actually prevent movement. These are the rings of skeletal muscles known as **sphincters** that guard the exit of urine and feces from the body by isotonically contracting. (Sphincters are examples of muscles that are expanded by fluid pressure as an antagonistic mechanism.)

The velocity of shortening is related to the load

The load is also an important determinant of the velocity, or speed, of shortening (Figure 8-19). During a concentric contraction, the greater the load, the lower the velocity at which

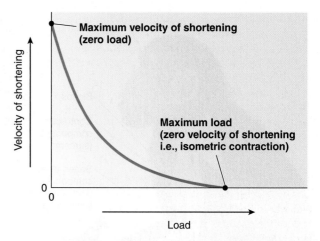

FIGURE 8-19 Load–velocity relationship in concentric contractions. The velocity of shortening decreases as the load increases.

© Cengage Learning, 2013

a single muscle fiber (or a constant number of contracting fibers within a muscle) shortens. The velocity of shortening is maximal when there is no external load, progressively decreases with an increasing load, and falls to zero (no shortening—isometric contraction) when the load cannot be overcome by maximal tetanic tension. You have frequently experienced this **load–velocity relationship**. You can quickly lift light objects requiring little muscle tension, whereas you can lift very heavy objects only slowly, if at all. This relationship between load and shortening velocity is a fundamental property of muscle, presumably because it takes the cross bridges longer to stroke against a greater load.

Whereas load and velocity for shortening are *inversely* related for *concentric* contractions, load and velocity for lengthening are *directly* related for *eccentric* contractions. The greater the external force (load) stretching a muscle that is contracting to resist the stretch, the greater the speed with which the muscle lengthens, likely because the load breaks some of the stroking cross-bridge attachments.

Skeletons include endoskeletons, exoskeletons, and hydrostatic skeletons

As you have seen, skeletal muscles typically create movement via skeletal elements. We have already mentioned the two major rigid skeletons: the *endoskeleton* of vertebrates and the *exoskeleton* of arthropods. However, there is a third kind common in the animal kingdom: the *hydrostatic* skeleton, which is based on the fact that water is essentially incompressible. A pressurized nonmoving fluid can produce rigid states, while a pressurized moving fluid can create large-scale movement and stretch out contracted muscles. In its simplest form, such as in a sea anemone (see Figure 5-2), a hydrostatic skeleton consists of a closed chamber of body fluid (in this case its digestive tract, sealed by a closed mouth) surrounded by two sets of muscles: *circular* muscles ringing the chamber and *longitudinal* muscles running the length of the chamber. In effect, these are antagonistic muscles but working through fluid pressure rather than a hard skeleton. When the circular muscles contract, the incompressible fluid builds up pressure against its walls, stiffening the body region around it. However, if the longitudinal muscles are relaxed, the chamber and body region will become narrower and longer due to pressurized fluid movement. This is how a sea anemone extends its body and then keeps it upright, and how an earthworm extends its anterior into soil for burrowing.

The fluid being pressurized and moved can be that of a body cavity, a circulatory system, or the interior of cells. The latter is the case for an octopus's arms: Its muscles can (1) contract to pull on other muscles and connective tissue, (2) pressurize their own intracellular fluids to become stiff skeletal elements for other muscles to work on, and (3) relax to act as flexible joints pulled on by other muscles. These versions of hydrostatic skeletons are called *muscular hydrostats*. Hydrostatic skeletons and hydrostats provide little protection for body parts compared to hard skeletons, but they have one major advantage over the latter: They can switch between a static state to maintain rigidity and a mobile state that is more flexible than any jointed hard skeleton. An octopus, for example, can reach its arm into twisting rock crevices to find prey in a way no rigid limb can. And a cat-sized octopus can squeeze its entire body out of a box with only a 2 cm hole!

Hydrostatic skeletons (including muscular hydrostats) are also found in many animals with hard skeletons. Mammalian ones include the tongue, the erectile genitalia (penis, clitoris), and—perhaps the most sophisticated example—the highly dexterous trunk of an elephant, which can reach around and through dead brush to grab a live branch for food, pick up objects as small as a coin, and even shuck an ear of corn. Finally, as you'll see in Chapter 9, many spiders extend their exoskeletal legs rapidly for jumping not by extensor muscles but by circulatory fluid squeezed into the leg by muscles in the main body.

Interactive units of skeletal muscles, tendons, skeletons, and joints form lever systems

Let's return to the mechanics of rigid skeletons. Most skeletal muscles are attached to bones or exoskeleton segments across **joints**, forming lever systems. A **lever** is a rigid structure capable of moving around a pivot point known as a **fulcrum**. In the vertebrate body, the bones function as levers, the joints serve as fulcrums, and the skeletal muscles provide the force to move the bones. The part of a lever between the fulcrum and the point where an upward force is applied is called the **power arm**; the part between the fulcrum and the downward force exerted by a load is the **load arm** (Figure 8-20a).

The most common type of lever system in the mammalian body is exemplified by flexion of the human elbow joint. Skeletal muscles, such as the biceps whose contraction flexes the elbow joint, consist of many parallel (side-by-side) tension-generating fibers that can exert a large force at their insertion but shorten only a small distance and at a relatively slow velocity. The lever system of the elbow joint amplifies the slow, short movements of the biceps to produce more rapid movements of the hand that cover a greater distance.

Consider how an object weighing 5 kg is lifted by the human hand (Figure 8-20b). When the biceps contracts, it exerts an upward force at the point where it inserts on the forearm bone about 5 cm away from the elbow joint, the fulcrum. Thus, the power arm of this lever system is 5 cm long. The length of the load arm, the distance from the elbow joint to the hand, averages 35 cm. In this case, the load arm is seven times longer than the power arm, which enables

The octopus *Octopus aculeatus* uses its hydrostat arms to walk along the seafloor and to mimic algae.

the load to be moved a distance seven times greater than the shortening distance of the muscle and at a velocity seven times greater (the hand moves 7 cm during the same length of time the biceps shortens 1 cm).

The disadvantage of this lever system is that at its insertion the muscle must exert a force seven times greater than the load. The product of the length of the power arm times the upward force applied must equal the product of the length of the load arm times the downward force exerted by the load. Because the load arm times the downward force is 35 cm × 5 kg, the power arm times the upward force must be 5 cm × 35 kg (the force that must be exerted by the muscle to be in mechanical equilibrium). Thus, some skeletal muscles typically work at a mechanical disadvantage in that they must exert a considerably greater force than the actual load to be moved. Nevertheless, the amplification of velocity and distance afforded by the lever arrangement enables muscles to move loads faster over greater distances than would otherwise be possible. This amplification provides valuable maneuverability and speed.

In contrast, some muscles are arranged to amplify load lifting, at the expense of speed. An example is the "calf" muscle or *gastrocnemius* in the hind limbs of terrestrial vertebrates (used for lifting or jumping the body up on the toes). This muscle pulls on the heel bone at a distance farther from the fulcrum—the toe joints—than the load being lifted. That load is the weight of the body being supported by the foot, usually centered over the middle of the foot. A modest force produced by the muscle can lift a body weight greater than that force, but as a trade-off, speed is de-amplified; that is, the lifting muscle moves a greater distance than the body being lifted.

Elastic storage structures can greatly increase the efficiency and speed of locomotion

Finally, recall that elastic "spring" elements can store muscle-generated energy, not only to help stretch out contracted muscles but also for rapid locomotion, such as the chameleon tongue's feeding strike, or a flea's rapid jump to distances

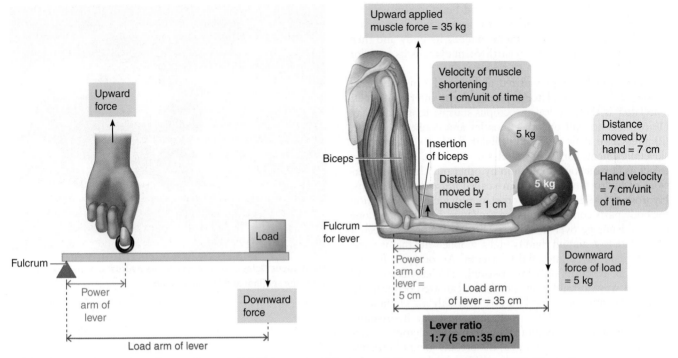

(a) Most common type of lever system in mammalian limbs

(b) Flexion of elbow joint as example of body lever action

FIGURE 8-20 Lever systems of muscles, bones, and joints. Note that the lever ratio (length of the power arm to length of the load arm) is 1:7 (5 cm:35 cm), which amplifies the distance and velocity of movement seven times (distance moved by the muscle [extent of shortening] = 1 cm, distance moved by the hand = 7 cm, velocity of muscle shortening = 1 cm/unit of time, hand velocity = 7 cm/unit of time), but at the expense of the muscle having to exert seven times the force of the load (muscle force = 35 kg, load = 5 kg).

© Cengage Learning, 2013

over 100 times its body length. Let's look at the latter more closely. Fleas are thought to jump by first using muscles to compress a pad with a highly elastic rubberlike protein called **resilin**. Once compressed, resilin releases almost all of its stored energy to extend the leg for a fast, powerful push-off. Elastic apodemes are similarly used by jumping locusts and grasshoppers, and resilin springs are also key to the rapid wing beats of some flying insects, as you will see later. Resilin contains coiled peptide chains in a three-dimensional network stabilized by tyrosines cross-linked to form *dityrosine* and *trityrosine*. The network stores and releases a higher percentage of energy (97%) than any other known elastic substance, and is currently being tested for uses in springy athletic shoes and restoring elasticity to damaged arteries.

Elastic springlike mechanisms also increase energy efficiency. For example, ligaments of a human's arched foot and the gastrocnemius tendon of most mammals are important energy components of running and jumping. Consider what happens when you run. First, you push with one leg to create thrust, but also have to use energy to lift it off the ground. That energy could be lost, but when gravity pulls the foot down to the ground, much of that energy is captured by ligaments in your foot arch (stretched as your foot flattens slightly) and your gastrocnemius (Achilles) tendon, stretched as your lower leg bends forward. Now, when that foot pushes off again with the gastrocnemius muscle, it is aided by its tendon and arch ligaments rebounding. This mechanism is perhaps most effectively used in hopping animals like kangaroos (see photograph at the chapter start). Their large tendons store large amounts of energy as they hit the ground, much more forcefully in a hop than during running. In fact the faster a kangaroo hops, the *less* energy it uses to cover a given distance! This is not the case during running.

Elastic storage and release are also found in fluid-based locomotory systems. This is how a jellyfish's bell and a squid's mantle reopen after muscles contract to create a jet of water.

We now shift our discussion from muscle mechanics to the metabolic means by which muscles power these movements.

check your understanding 8.4

A mussel is attempting to avoid being eaten by a starfish by tightly closing its shell. Is this an isometric contraction or an isotonic contraction? Explain.

A person eating a crab claw usually picks out only one—very large—muscle. Is it the flexor or the extensor, and why? And why is the other muscle—extensor or flexor—usually too small to bother eating?

8.5 Skeletal Muscle Metabolism and Fiber Types

Recall that three different steps in the contraction–relaxation process require ATP:

1. Splitting of ATP by myosin ATPase provides (indirectly) the energy for the power stroke of the cross bridge.
2. Binding (but not splitting) of a fresh molecule of ATP to myosin permits detachment of the bridge from the actin filament at the end of a power stroke so that the cycle can be repeated (see p. 346). This ATP is subsequently split to provide energy for the next stroke of the cross bridge (step 1).
3. The active transport of Ca^{2+} back into the sarcoplasmic reticulum during relaxation depends on energy derived from the breakdown of ATP.

Muscle fibers have alternate pathways for forming ATP

Because ATP is the only energy source that can be directly used for these activities, for contraction and relaxation to continue ATP must constantly be supplied. Only limited stores of ATP are immediately available in muscle tissue, but three pathways supply additional ATP as needed during muscle contraction: (1) transfer of a high-energy phosphate from a *phosphagen* to ADP, (2) *oxidative phosphorylation* (the citric acid cycle and electron transport system), and (3) *glycolysis*. We discussed these at length in Chapter 2 (pp. 48–53) and give only a brief overview here.

Phosphagens The phosphagens *creatine phosphate* and *arginine phosphate*, described on p. 55, are often the first energy storehouse tapped at the onset of contractile activity (Figure 8-21, step 3a). Like ATP, the phosphagens contain a high-energy phosphate group, which can be donated directly to ADP to form ATP. This reaction is reversible; energy and phosphate from ATP can be transferred to creatine or arginine to form the phosphate. In vertebrates, which use creatine phosphate, the reaction is catalyzed by the muscle cell enzyme creatine kinase:

$$\text{Creatine phosphate} + \text{ADP} \overset{\text{creatine}}{\underset{\text{kinase}}{\longleftrightarrow}} \text{creatine} + \text{ATP}$$

In accordance with the law of mass action (see p. 533), as the energy reserves are built up in a resting muscle, the increased concentration of ATP favors the transfer of the high-energy phosphate group to the phosphagen. By contrast, at the onset of contraction when myosin ATPase splits the meager reserves of ATP, the resultant fall in ATP favors transfer of the high-energy phosphate group from stored creatine phosphate to form more ATP. A rested vertebrate muscle contains about five times as much creatine phosphate as ATP. Thus, most energy is stored in vertebrate muscle in creatine phosphate pools. Because only one enzymatic reaction is involved in this energy transfer, ATP can be formed rapidly (within a fraction of a second) by using creatine phosphate.

Muscle ATP levels actually remain fairly constant early in contraction, but phosphagen stores become depleted. In fact, short bursts of high-intensity contractile effort are supported primarily by ATP derived at the expense of phosphagen. Other energy systems do not have a chance to become operable before the activity is over.

Oxidative Phosphorylation If the energy-dependent contractile activity is to be continued, most animal muscles shift to the pathways of oxidative phosphorylation and glycolysis to form ATP. These multistepped pathways require time to pick up their rates of ATP formation to match the increased demands for energy, time provided by the immediate supply of energy from the one-step phosphagen system.

If enough O_2 is present, oxidative phosphorylation (see p. 51) takes place within the muscle mitochondria. Oxygen is required to support the mitochondrial electron-transport system, which, together with chemiosmosis by ATP synthase, efficiently harnesses energy captured from the breakdown of nutrient molecules and uses it to generate ATP. This pathway is fueled by glucose or fatty acids, depending on the intensity and duration of the activity (Figure 8-21, step 3b). Although it provides a rich yield of ~30 ATP molecules for each glucose molecule processed, oxidative phosphorylation is relatively slow because of the number of steps involved, and it requires a constant supply of O_2 and nutrient fuel.

During light activity (such as walking) to moderate activity (such as flight or swimming), muscle cells can form sufficient amounts of ATP through oxidative phosphorylation to keep pace with the modest energy demands of the contractile machinery for prolonged periods of time. To sustain ongoing oxidative phosphorylation, the exercising muscles depend on delivery of adequate O_2 (such as by circulatory systems or tracheae) and nutrient supplies. Activity that can be supported in this way is known as **aerobic** ("with O_2") or **endurance-type** activity. The O_2 required for oxidative phosphorylation in vertebrates is primarily delivered by the blood. Increased O_2 is made available to vertebrate muscles during exercise by several mechanisms: Enhanced breathing brings in more O_2 (Chapter 11), the heart contracts more rapidly and forcefully to pump more oxygenated blood to the tissues, more blood is diverted to the exercising muscles by dilation of the blood vessels supplying them (Chapter 9), and the **hemoglobin** molecules that carry the O_2 in the blood release more O_2 in exercising muscles (Chapter 11). Furthermore, some types of muscle fibers have an abundance of **myoglobin (Mb)**, which is similar to hemoglobin. Myoglobin can store amounts of O_2, akin to a storage tank, but it also increases the rate of O_2 transfer from the blood into muscle fibers.

Glucose and fatty acids, ultimately derived from ingested food, are also delivered to the muscle cells by the circulatory fluid. Excess ingested nutrients not immediately used are stored (such as in vertebrate liver and adipose tissue, and fat bodies in insects; pp. 685–690). In addition, muscle cells can store limited quantities of glucose in the form of glycogen. Only a small volume of the muscle fiber is normally committed to fuel storage, with the combined volume of lipids and glycogen usually less than 3%, although storage of glycogen is an extraordinary 17% in the shaker muscle of the rattlesnake.

Insects have an advantage in that concentrations of **trehalose** (diglucose) in their hemolymph occur at concentrations around 2% (0.06 M), in contrast to mammals, where blood concentrations of glucose are only 0.1% (0.005 M). These high concentrations of disaccharide represent a substantial source of fuel for muscle contractions, particularly in flying insects. Trehalose has an advantage over glucose and other reducing sugars in being nonreducing and consequently is less toxic; it does not exhibit the damaging reactions that glucose has with tissue proteins in diabetes in mammals (see p. 564).

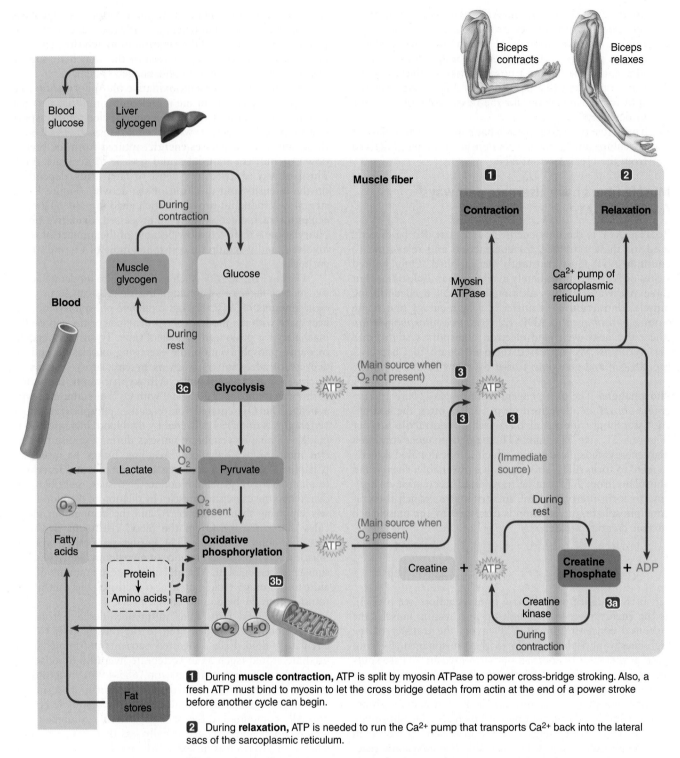

1. During **muscle contraction,** ATP is split by myosin ATPase to power cross-bridge stroking. Also, a fresh ATP must bind to myosin to let the cross bridge detach from actin at the end of a power stroke before another cycle can begin.

2. During **relaxation,** ATP is needed to run the Ca^{2+} pump that transports Ca^{2+} back into the lateral sacs of the sarcoplasmic reticulum.

3. The **metabolic pathways that supply the ATP** needed to accomplish contraction and relaxation are

3a. transfer of a high-energy phosphate from **creatine phosphate** to ADP (immediate source);

3b. **oxidative phosphorylation** (the main source when O_2 is present), fueled by glucose derived from muscle glycogen stores or by glucose and fatty acids delivered by the blood; and

3c. **glycolysis** (the main source when O_2 is not present). Pyruvate, the end product of glycolysis, is converted to lactate when lack of O_2 prevents the pyruvate from being further processed by the oxidative phosphorylation pathway.

FIGURE 8-21 Metabolic pathways producing ATP used during muscle contraction and relaxation.

Glycolysis There are respiratory and cardiovascular limits to the amount of O_2 that can be delivered to a vertebrate muscle. In near-maximal contractions, the blood vessels that course through the muscle are compressed almost closed by the powerful contraction, severely limiting the O_2 available to the muscle fibers. Furthermore, even when O_2 is available, the relatively slow oxidative-phosphorylation system may not be able to produce ATP rapidly enough to meet the muscle's needs during intense activity. A skeletal muscle's energy consumption may increase up to 100-fold when going from rest to high-intensity exercise. When O_2 delivery or oxidative phosphorylation cannot keep pace with the demand for ATP formation as the intensity of activity increases, the muscle fibers rely increasingly on glycolysis (Figure 8-21, step 3c) to generate ATP (see p. 48). Glycolysis alone has two advantages over the oxidative-phosphorylation pathway: (1) glycolysis can form ATP in the absence of O_2 (operating anaerobically, that is, "without O_2"), and (2) it can proceed much more rapidly than oxidative phosphorylation because it requires fewer steps and can work with only local cellular molecules (that is, no external oxygen is needed). Although glycolysis extracts considerably fewer ATP molecules from each nutrient molecule processed, it can proceed so much more rapidly that it can out produce oxidative phosphorylation in ATP yield over a given period of time if enough glucose is present. Activity that can be supported in this way is **anaerobic** or **high-intensity**.

However, there is a disadvantage of the low efficiency of glycolysis (recall that glycolysis yields a net of 2 ATP molecules for each glucose molecule degraded, whereas the oxidative-phosphorylation pathway can extract up to about 30 ATP molecules from each glucose; see p. 56). Muscle cells can store limited quantities of glucose in the form of glycogen, but anaerobic glycolysis rapidly depletes the muscle's glycogen supplies to produce ATP rapidly but with this low efficiency.

Another aspect of glycolysis is the end products it yields, which must be produced from pyruvic acid to regenerate NAD^+ (see p. 55). There are different end products in different animal groups (p. 56), but the best studied are the following:

1. *Lactate.* The end product of anaerobic glycolysis, pyruvic acid, is converted in most vertebrates to **lactate** (by the enzyme **lactate dehydrogenase**; p. 55):

$$\text{Pyruvic acid} + NADH \rightarrow \text{lactate} + NAD^+$$

Lactate leads to acidosis (low pH/high acidity). Lactate is also produced in many nonvertebrates such as insects.

2. *Octopine.* As we noted in Chapter 2, some nonvertebrates produce other glycolytic end products, some with less impact on pH than lactate. In many mollusks, **octopine** is produced by the enzyme octopine dehydrogenase, by combining an acid (pyruvic acid) and a basic amino acid (arginine):

$$\text{Pyruvic acid} + \text{arginine} + NADH \rightarrow \text{octopine} + H_2O + NAD^+$$

In the cuttlefish, *Sepia officinalis,* for example, one study found octopine to be the primary end product in mantle muscle (used for jetting) during hypoxia (low oxygen) and exhaustive swimming. As octopine concentration increased, there was a corresponding decrease in muscle glycogen and arginine phosphate. This reaction, in addition to regenerating NAD^+, detoxifies arginine, a solute that, at the high concentrations produced by phosphagen breakdown, can perturb structures of proteins. Octopine, in contrast, has little effect on protein structures and is known as a *compatible solute* (see Chapter 13).

Lactate picked up by the blood produces the metabolic acidosis accompanying intense activity. Therefore, anaerobic high-intensity activity can be sustained for only a short duration, in contrast to the body's prolonged ability to sustain aerobic endurance-type activities. Both depletion of energy reserves and the fall in muscle pH caused by lactate accumulation in vertebrates are believed to play a role in the onset of muscle fatigue, a topic to which we turn next.

Fatigue may be of muscle or central origin

Contractile activity in a particular skeletal muscle cannot be maintained at a given level indefinitely. Eventually, the tension in the muscle declines as fatigue sets in. There are two different types of fatigue: muscle fatigue and central fatigue.

Muscle fatigue occurs when an exercising muscle can no longer respond to stimulation with the same degree of contractile activity. Muscle fatigue is a defense mechanism that protects a muscle from reaching a point at which it can no longer produce ATP. An inability to produce ATP would result in rigor mortis (obviously not an acceptable outcome of exercise). The underlying causes of muscle fatigue are unclear. The primary implicated factors include the following:

1. The *local increase in ADP and inorganic phosphate* from ATP breakdown may directly interfere with cross-bridge cycling and/or block Ca^{2+} release and uptake by the sarcoplasmic reticulum.
2. *Accumulation of lactate* may inhibit key enzymes in the energy-producing pathways and/or excitation–contraction coupling process.
3. The *accumulation of extracellular K^+* that occurs in the muscle when the Na^+/K^+ pump cannot actively transport K^+ back into the muscle cells as rapidly as this ion leaves during the falling phase of repeated action potentials (see p. 120) causes a local reduction in membrane potential. This altered potential may decrease the release of Ca^{2+} intracellularly by impairing the coupling of voltage-gated dihydropyridine receptors in the T tubules and Ca^{2+}-release channels in the sarcoplasmic reticulum.
4. *Depletion of glycogen energy reserves* may lead to muscle fatigue in exhausting exercise.
5. The time of onset of fatigue varies with the *type of muscle fiber,* some fibers being more resistant to fatigue than others, and with the intensity of the exercise, more rapid onset of fatigue being associated with high-intensity activities.

Central fatigue occurs when the CNS no longer adequately activates the motor neurons supplying the working muscles. The animal slows down or stops exercising even though the muscles are still able to perform. The mechanisms involved in central fatigue are poorly understood. In some cases, central fatigue may be psychological or may stem from biochemical insufficiencies within the brain.

Increased oxygen consumption is necessary to recover from activity

An animal continues to breathe deeply and rapidly for a period of time after exhaustive activity. The necessity for the elevated O_2 uptake during recovery from exercise is due to a variety of factors. The best known is repayment of an **oxygen deficit** incurred during the activity, when ATP derived from nonoxidative sources such as creatine phosphate and anaerobic glycolysis was supporting contractile activity. During the activity, phosphagen stores of active muscles are reduced, lactate may accumulate, and glycogen stores may be tapped; the extent of these effects depends on the intensity and duration of the activity. Oxygen is needed for recovery of the energy systems. During the recovery period, fresh supplies of ATP are formed by oxidative phosphorylation using the newly acquired O_2, which is provided by the sustained increase in respiratory activity after the activity has stopped. Much of this ATP is used to resynthesize phosphagen to restore its reserves. This can be accomplished in a matter of a few minutes in mammals. Any accumulated lactate (or octopine) is converted back into pyruvic acid, part of which is used by the oxidative-phosphorylation system for ATP production (some of this takes place in other organs such as vertebrate liver and heart, which take up lactate from the bloodstream). The remainder of the pyruvic acid is converted back into glucose (primarily by the liver in vertebrates). Most of this glucose in turn is used to replenish the glycogen stores drained from the muscles and liver during exercise. These biochemical transformations involving pyruvic acid require O_2 and take several hours for completion. Thus, as the O_2 debt is repaid, the phosphagen system is restored, lactate (or octopine) is removed, and glycogen stores are at least partially replenished.

Unrelated to increased O_2 uptake is the need to restore nutrients after grueling activity, such as a fight between two males competing for females, or fleeing from a predator. In such cases, glycogen stores are severely depleted, and long-term recovery can take a day or more because the exhausted energy stores require nutrient intake for full replenishment. Therefore, depending on the type and duration of activity, recovery can be complete within a few minutes or can require more than a day.

Part of the extra O_2 uptake during recovery is not directly related to the repayment of energy stores but instead is the result of a general metabolic disturbance after exercise. For example, the local increase in muscle temperature arising from heat-generating contractile activity speeds up the rate of all chemical reactions in the muscle tissue, including those dependent on O_2. Likewise, the secretion of epinephrine, a vertebrate hormone that increases O_2 consumption, is elevated during moderate activity. Until the circulating level of epinephrine returns to its pre-exercise state, O_2 uptake is above normal.

Although muscles can accomplish work, much of the energy is converted to heat

One of the limits to muscle energy is its relatively low efficiency. In an isotonic contraction, it is only about 25%. That is, of the ATP consumed by the muscle during the contraction, 25% is realized as external work whereas the remaining 75% is converted to heat, in part due to friction.

Recall that in an isometric contraction, when no object is moved, external work is zero. All energy consumed by the muscle during contraction is converted to heat. However, this is not all wasted energy since holding a limb in place can obviously be very useful.

Muscle heat is not always wasted in another way, because it is important in some animals in maintaining elevated core-body temperatures (such as tunas, bumblebees, birds, and mammals). In fact, shivering, a form of involuntarily induced skeletal muscle contraction in mammals, birds, and some insects, is a well-known mechanism for increasing heat production on a cold day or to fight infections with a fever. Heavy exertion on a hot day, in contrast, may overheat the body, because the normal heat-loss mechanisms may be unable to compensate for this increase in heat production (see Chapter 15).

We have been looking at the contractile and metabolic activities of skeletal muscle in general. Yet not all skeletal muscle fibers use these mechanisms to the same extent. We are next going to examine the different types of muscle fibers, based on their speed of contraction and how they are metabolically equipped to generate ATP.

There are three types of skeletal muscle fibers, which differ in ATP hydrolysis and synthesis

No one muscle-fiber type can perform all the activities essential for an animal to survive. For this reason, muscle fibers show the greatest diversity of any tissue in an animal's body. Based on their biochemical capacities and physiological requirements, vertebrate skeletal muscle fibers have been classified into three major groups (Table 8-2):

1. Slow-oxidative (SO) fibers or "slow" fibers (Type I)
2. Fast-oxidative (FO) fibers or "FR": Fast Resistant fibers (Type IIa)
3. Fast-glycolytic (FG) fibers or "FF": Fast Fatigable fibers (Type IIb , IId or IIx) (In addition, there is a type IIc fiber, an undifferentiated form that can become other type II forms).

As their names imply, the two main differences between these fiber types are their speed of contraction (slow or fast) and the type of enzymatic machinery they primarily use for ATP formation (oxidative or glycolytic). There also exist a small number of muscle fibers that do not obviously belong to those groups listed above that are termed "Fast Intermediate." These fibers are characterized by a fast contraction time but are unable to maintain their initial force production in response repeated stimulation.

Fast versus Slow Fibers As is well known, compared with large animals, small ones have a much higher power output when normalized to body weight. Correspondingly, fast twitch muscles of these animals have a larger unloaded shortening velocity than their counterparts in very large animals. Fast fibers have higher myosin-ATPase (ATP-splitting) activity than slow fibers because of different myosin isoforms (see *Molecular Biology and Genomics: Myosin Family Genetics*). The higher the ATPase activity, the more rapidly ATP is split and the faster the rate at which energy is made available for cross-bridge cycling. The result is a fast twitch, compared to the slower twitches of those fibers that split ATP more slowly. On average, the time to peak twitch ten-

TABLE 8-2 Characteristics of Skeletal Muscle Fibers

	TYPE OF FIBER		
Characteristic	Slow-Oxidative (Type I)	Fast-Oxidative (Type IIa)	Fast-Glycolytic (Type IIx)
Myosin–ATPase Activity	Low	High	High
Speed of Contraction	Slow	Fast	Fast
Resistance to Fatigue	High	Intermediate	Low
Oxidative Phosphorylation Capacity	High	High	Low
Enzymes for Anaerobic Glycolysis	Low	Intermediate	High
Mitochondria	Many	Many	Few
Capillaries	Many	Many	Few
Myoglobin Content	High	High	Low
Color of Fiber	Red	Red	White
Glycogen Content	Low	Intermediate	High

© Cengage Learning, 2013

sion for fast fibers is 20–40 msec compared to 60–100 msec for slow fibers. Thus, two factors determine the speed with which a muscle contracts: the load (load–velocity relationship) and the myosin ATPase activity of the contracting fibers (fast or slow twitch).

Oxidative versus Glycolytic Fibers Fiber types also differ in ATP-synthesizing ability. Those with a greater capacity to form ATP are more resistant to fatigue. Some fibers are better equipped for oxidative phosphorylation, whereas others rely primarily on anaerobic glycolysis for synthesizing ATP. Because oxidative phosphorylation yields considerably more ATP from each nutrient molecule processed, it does not readily deplete energy stores. Furthermore, it does not result in lactate accumulation. Oxidative types of muscle fibers are therefore more resistant to fatigue than are glycolytic fibers.

Other related characteristics distinguishing these three fiber types are summarized in Table 8-2. As would be expected, the oxidative fibers, both slow and fast, contain an abundance of mitochondria, the organelles that house the enzymes involved in the oxidative-phosphorylation pathway. Mitochondrial volume in most skeletal-muscle fibers lies between 4 and 42%, with low values found in muscles relying on anaerobic metabolism (so-called white muscle), and high values in those relying on aerobic metabolism (so-called red muscle; we discuss the details of white and red muscle shortly). For example, mitochondrial volume of the

tail-shaker muscle in rattlesnakes (*Crotalus*) is 26%. However, there are extreme values, as low as 1% in some white fish muscle and as much as 50% in the flight muscle of locusts.

Oxidative fibers (type I) are richly supplied with capillaries and also have a high myoglobin (Mb) content. The highest known myoglobin contents are those in swimming muscles of diving penguins and mammals; for example, the pectoralis muscle (which operates the wings) of a king penguin has 4.3 g of Mb per 100 g of muscle, whereas a gull's pectoralis has only 0.55 g Mb/100g and a chicken's pectoralis only 0.03 g Mb/100g. Myoglobin not only helps support oxidative fibers' O_2 dependency but also imparts a red color to them, just as oxygenated hemoglobin is responsible for the red color of arterial blood. Accordingly, these muscle fibers are referred to as **red fibers**.

In contrast, the fast fibers specialized for glycolysis (IIb, IId, IIx) contain few mitochondria but have a high content of glycolytic enzymes instead. Also, to supply the large amounts of glucose necessary for glycolysis, they contain an abundance of stored glycogen. Because the glycolytic fibers need relatively less O_2 to function, they receive only a meager capillary supply compared with the oxidative fibers. The glycolytic fibers contain very little myoglobin and therefore are pale in color, so they are sometimes called **white fibers**. (The most readily observable comparison between red and white fibers is the dark and white meat in poultry, and the red meat of a tuna compared to the white meat of a halibut.) In addition to their higher myosin ATPase content, these fibers are also larger in diameter than the slow-oxidative fibers because of a greater abundance of actin and myosin filaments. With these additional power-generating filaments, the fast-glycolytic fibers are capable of rapidly producing large amounts of tension, but only for a short duration because of their dependence on glycolysis for energy release. For example, the explosive "burst" of flapping flight in many species of galliform birds (such as grouse) is used to escape predators but can be sustained for only a few minutes. Appropriately, these species of birds have relatively small heart masses and a reduced capacity to supply the flight muscles with oxygen. For these reasons, galliform birds are restricted to anaerobic flights of short duration and are adapted to a largely ground-based life.

Finally, fast-oxidative fibers share characteristics with each of the other two types. They have high ATPase activity like the fast-glycolytic fibers and high oxidative capacity like the slow-oxidative fibers. They contract more rapidly than the slow-oxidative fibers and can maintain the contraction for a longer period of time than the fast-glycolytic fibers. However, because their rate of ATP production by oxidative phosphorylation cannot keep pace with the high rate of ATP splitting, they rely partially on glycolysis and are more prone to fatigue than the slow-oxidative fibers. Thus, the flight muscles of migratory birds contain an abundance of fast-oxidative glycolytic fibers. Normally, the flight muscles of small birds are composed entirely of this muscle type, whereas larger birds have a mixture of different fiber types.

Consider as well the energy requirements of the muscle fibers in a horse as its gait changes from a walk into a canter. While the horse is walking, Type I fibers are recruited because the muscle contracts slowly and energy requirements are minimal. During this phase the muscle fibers rely primarily on fat as an energy source. However, as speed increases

Molecular Biology and Genomics
Myosin Family Genetics

As noted in the text, there are at least three different functional categories of muscle fibers: In humans, these are termed slow-oxidative or Type I, fast-oxidative or Type IIa, and fast-glycolytic or Type IIx. Scientists have known for some time that each of these expresses a different myosin protein from its own myosin gene; each myosin has catalytic properties suited to its muscle type. Thus, researchers long suspected that there are at least three different myosin genes. Genetic analysis has revealed that there is indeed a large variety of related myosins, with as many as 16 different myosin gene types in muscles among different mammals and among different muscle tissues. Key examples are designated as follows:

- Embryonic
- Neonatal
- Cardiac α
- Cardiac β or slow type I (in skeletal muscle)
- Fast type IIa
- Fast type IIx
- Fast type IId
- Fast type IIb
- Extraocular
- Mandibular or masticatory (m-MHC)
- Smooth muscle type (SMMHC)

All these are considered to be *isoforms* of myosin *class II*, the type in muscle (not to be confused with subtype IIa, and so on). In addition, there are numerous nonmuscle myosin classes such as myosin class VI, associated with movements of single cells. Mutations in the myosin VI gene in mice lead to deafness because the stereocilia used in hearing (p. 219) are nonfunctional. Including all classes, the human genome has a total of 39 myosin genes!

The class II muscle myosins have played important roles in evolutionary adaptations. For example, fast-glycolytic types termed IIb and IId are found in rodents, animals that scurry around at high limb velocities, and are faster than type IIx, the fastest type found in skeletal muscles of large mammals (including humans). (Type IIb gene is expressed in eye muscles in some large mammals, but in others such as horses it has become a pseudogene.) As another example, nonhuman primates have an active mandibular (m-MHC) gene (also called MYH16) that produces very strong jaw muscles. In humans, the gene is mutated, leading to our relatively small, weak jaws. It is speculated that this mutation allowed for finer jaw control needed for speech.

The expression of myosin genes changes with activity and other environmental factors related to muscle use, in concert with changes in muscle fiber types (see main text). In human muscles repeatedly loaded, as in weight training, the genes of fast IIx fibers stop transcribing the IIx myosin and begin transcribing the IIa myosin, as the fibers convert from IIx type to IIa type. (In rodents subjected to the equivalent of weight training, fibers convert from IIb to IIx.) In contrast, in sedentary people, much of the limb muscle converts to type IIx (which fits the way their muscles are used: quiet most of the time, with intermittent short bursts of usage).

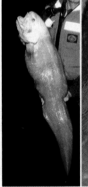
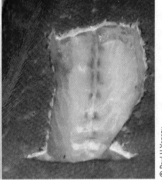

(a) Icelandic horses standing and moving. As these animals increase their activity from resting to higher and higher speeds, more motor units are activated and different fiber types are recruited. While standing or walking, Type I (aerobic) dominates, while in burst activity, such as rearing, Type IIx (anaerobic) dominates. Unlike most horses, which have four gaits, these have five: walk, trot, tölt, gallop, and a smooth flying pace.

(b) A deep-sea cusk eel (with muscles under the skin shown at the right). This fish cruises slowly in the deep using thin bands of red Type I (aerobic) muscle, then bursts occasionally to escape or catch prey using white Type IIx (anaerobic) muscle (seen here as light pink).

to a trot, Type I fibers can no longer contract rapidly enough to propel the horse. Type IIa fibers are now recruited. Because fat cannot be metabolized anaerobically, glycogen serves as the primary energy source for ATP production. As the speed of the horse increases to a gallop, Type IIx muscle fibers are recruited. ATP production is now almost exclusively generated through anaerobic pathways. However, although ATP is rapidly generated, it is associated with the generation of lactate, which results in a decline in muscle pH and the onset of fatigue.

Genetic Endowment of Muscle Fiber Types The muscles of most terrestrial vertebrates contain a mixture of all three fiber types because almost all must cope with gravity using red fibers, while also producing faster movements. However the type of activity for which the muscle is specialized largely determines the percentage of each type. Accordingly, a high proportion of slow-oxidative fibers are found in muscles specialized for maintaining low-intensity contractions for long periods of time without fatigue, such as the muscles of the back and legs that support an animal's weight against the force of gravity. The deep layers of the pectoralis muscles of birds adapted for soaring and gliding modes of flight contain more slow-oxidative fibers (red). Fast-glycolytic fibers are preponderant in the arm muscles of primates, adapted for rapid, forceful movements. In fish, which do not need to cope with gravity, muscle fibers are often completely separate. In species that are inactive much of the time (such as a halibut), almost all the muscle is fast glycolytic (white), used for burst activity, with red muscle forming a distinct, small band under the lateral line (p. 218).

Muscle fibers adapt considerably in response to the demands placed on them

Not only is the nerve supply to a skeletal muscle essential for initiating contraction, but the motor neurons supplying a skeletal muscle are also important in maintaining the muscle's integrity and chemical composition. Different types of activity produce different patterns of neuronal discharge to the muscle involved. Depending on the pattern of neural activity, long-term adaptive changes occur in the muscle fibers, enabling them to respond most efficiently to the types of demands placed on the muscle. Therefore, skeletal muscle has a high degree of *plasticity* (see p. 171). Three types of changes can be induced in muscle fibers: changes in their ATP-synthesizing capacity, in myosin isoforms and thus contractile speed and force, and in their diameter.

Improvement in Oxidative Capacity Regular endurance (aerobic) activities, such as flight or swimming, induce metabolic changes within the oxidative fibers, which are the ones primarily recruited during aerobic exercise. These changes enable the muscles to use O_2 more efficiently. For example, the number of mitochondria and the number of capillaries supplying blood to these fibers both increase. Muscles so adapted are better able to endure prolonged activity without fatiguing, but they do not enlarge. In the example of the horse whose gait is increased from a walk to a canter, sufficient endurance training ultimately enhances the aerobic ability of its muscle to use fuels through oxidation. Blind mole rats, with most of their muscle mass located in the anterior end of the body, find themselves digging in tunnels with scant oxygen supplies. Researchers found that, as an adaptation to these extreme conditions, capillary density was approximately 30% higher in mole rat muscles compared to white rat muscles. Although mole rats have less muscle mass than white rats, mitochondrial density is about 50% higher.

Muscle Hypertrophy The actual size of the muscles can also be increased by regular bouts of anaerobic, short-duration, high-intensity training. The resulting muscle enlargement comes primarily from an increase in diameter (hypertrophy)

of the fast-glycolytic fibers that are called into play during such powerful contractions. Most of the fiber thickening is a consequence of increased synthesis of myosin and actin filaments, which permits a greater opportunity for cross-bridge interaction and subsequently increases the muscle's contractile strength. The resultant bulging muscles are better adapted to activities that require intense strength for brief periods, but endurance has not improved.

Influence of Hormones Muscle fibers of many mammalian males are thicker, and accordingly, their muscles are larger and stronger than those of females because of the actions of testosterone, a steroid hormone secreted primarily in males. Testosterone promotes the synthesis and assembly of myosin and actin and is responsible for the naturally larger muscle mass of males. Consider the effect of the photoperiod-induced increase in testosterone on the development of muscle in the necks of male deer prior to the onset of the rutting season. Only the fittest males (those that dominate in battles that involve clashes with the head and antlers) have access to the harem and can reproduce. While testosterone increases male-specific muscle mass, IGF-1 production from the liver as induced by growth hormone (p. 292) increases muscle mass associated with normal growth in both males and females with overexpression of the IGF-1 gene in transgenic mice resulting in muscle hypertrophy. **Myostatin**, a growth factor secreted primarily by skeletal muscle, is a negative regulator of muscle growth. Mutations in the myostatin gene in mice, sheep, and cattle lead to gross muscle hypertrophy as a result of hyperplasia (increased cell number) and hypertrophy (increased cell size).

Interconversion between Fast Muscle-Fiber Types All the muscle fibers within a single vertebrate motor unit are of the same fiber type. This pattern usually is established early in life, but the two types of fast-twitch fibers are interconvertible, depending on usage. Regular endurance activities can convert fast-glycolytic fibers into fast-oxidative fibers, whereas fast-oxidative fibers can be shifted to fast-glycolytic fibers in response to power events such as weight training. Slow and fast fibers are not interconvertible, however, but the proportions can change to some extent with activity. Adaptive changes that take place in skeletal muscle gradually reverse to their original state over a period of months if the regular exercise program that induced these changes is discontinued. During these changes, the type of myosin isoform that is expressed also changes (see *Molecular Biology and Genomics: Myosin Family Genetics*).

Although training can induce changes in muscle fibers' metabolic support systems, whether a fiber is fast or slow twitch depends on the fiber's nerve supply. Slow-twitch fibers are supplied by motor neurons that exhibit a low-frequency pattern of electrical activity, whereas fast-twitch fibers are innervated by motor neurons that display intermittent rapid bursts of electrical activity. Experimental switching of motor neurons supplying slow muscle fibers with those supplying fast fibers gradually reverses the speed at which these fibers contract. Innervation of muscle fibers by motor neurons is also responsible for differentiation.

Muscle Atrophy At the other extreme, if a muscle is not used or an animal undergoes metamorphosis, its organelles, cytoplasm, and protein content decreases, its fibers become smaller, and the muscle accordingly **atrophies** (decreases in mass) and becomes weaker. Muscle atrophy can take place in two ways. **Disuse atrophy** occurs when a muscle experiences metamorphosis or if it is not used for a long period of time even though the nerve supply is intact. Sarcomeric proteins are also normally rapidly lost during periods of fasting. **Denervation atrophy** occurs after the nerve supply to a muscle is lost. If the muscle is stimulated electrically until innervation can be reestablished, such as during regeneration of a severed peripheral nerve, atrophy can be diminished but not entirely prevented. Contractile activity itself obviously plays an important role in preventing atrophy; however, poorly understood factors released from active nerve endings, perhaps packaged with the ACh vesicles, apparently contribute to the integrity and growth of muscle tissue.

In contrast, **aestivation** ("summer sleep") and **hibernation** ("winter sleep"), periods of an animal's life when it experiences prolonged inactivity as a result of inhospitable environmental conditions (e.g., aridity or excessive temperature), do not result in any significant muscle atrophy for reasons that include the marked reduction in the animal's metabolic rate. A well-studied example is the aestivation of the Australian green-striped burrowing frog (*Cyclorana alboguttata*), which escapes its cocoon of shed skin and mucus to greet the seasonally heavy summer rains with muscles capable of force production comparable to when they initially burrowed underground. Similarly, prolonged hibernation of black bears results in only marginal declines in muscle performance, which suggests that animals have evolved specialized mechanisms to reduce the effects of disuse atrophy. In the case of the black bear, researchers have suggested that these animals may use occasional bouts of shivering as an isometric exercise to maintain a low-level register of activity. These animals may also preserve skeletal muscle by recycling the nitrogen from *urea* (see Chapter 12) back into amino acids that can be used in protein synthesis.

As exemplified by metamorphosis in the gastropod mollusk, *Haliotis rufescens*, specific larval muscles can also undergo rearrangement and biogenesis of the new juvenile muscle system. The larval retractor muscle, by far the major muscle of the larva, reorganizes at metamorphosis, with the ventral cells of the muscle degenerating, while the dorsal cells become part of developing juvenile mantle musculature. Associated with these changes in myofibrillar structure, tropomyosin mRNA prevalence declines until becoming undetectable in the ventral cells, concurrent with an increase in the dorsal cells. Programmed cell death is a common fate of redundant larval tissues at metamorphosis in a wide range of animal taxa. Regulatory mechanisms operating at both transcriptional and posttranscriptional levels ultimately control the biogenesis and atrophy of the various larval and postlarval muscles during this process.

Limited Repair of Muscle When a muscle is damaged, limited repair is possible, even though muscle cells cannot divide mitotically to replace lost cells. Small populations of inactive muscle-specific stem cells called satellite cells are located close to the muscle surface. When a muscle fiber is damaged, locally released factors activate the satellite cells, which divide to give rise to **myoblasts**, the same undifferentiated cells that formed the muscle during embryonic development. A group of myoblasts fuse to form a large, multinucleated cell, which imme-

diately begins to synthesize and assemble the intracellular machinery characteristic of the muscle, ultimately differentiating completely into a mature muscle fiber. With extensive injury, this limited mechanism is not adequate to completely replace all the lost fibers. In that case, the remaining fibers often hypertrophy to compensate.

unanswered Questions | What signals trigger changes in muscle mass and fiber type?

As you have seen, muscle features change considerably under different patterns of neural stimulation and usage. But what are the actual signals involved? The answers are not complete, but neural stimulation of muscles causes changes in several candidate signal mechanisms: titin kinase (p. 339), neurotrophic growth factors, cAMP, calcium, and nitric oxide. *Nitric oxide,* a widespread paracrine (p. 93), is produced in skeletal muscle in relation to the load on the muscle, and may thus trigger changes involved in hypertrophy. Calcium, in contrast, builds up in chronically stimulated muscle cells and activates the protein *calcineurin.* Its role was suggested with mice that were genetically engineered to overexpress calcineurin. Researchers found that these mice were converting fast fibers to slow ones, with no hypertrophy. Calcineurin appears to lead to activation of a transcription factor that in turn activates genes involved in slow-muscle fiber development. Thus, two distinct signaling pathways may be involved in hypertrophy and fiber specification, respectively. <<

check your understanding 8.5

What is the first energy storehouse to be tapped at the onset of contractile activity?

Which energy storehouse is utilized in brief *anaerobic* activity versus ongoing *aerobic* activity? Explain why each fuel is appropriate for the particular metabolic needs.

8.6 Adaptations for Flight: Continuous High Power at High Contraction Frequencies

Consider the price a flying animal has to pay in order to fly. Bats and birds must be lightweight, and so have evolved very lightweight hollow bones. For this reason bats hang from their feet when not flying: Their leg bones are simply too thin to support their body weight. But even with this skeletal adaptation, it is not immediately clear how muscles are up to the task of flying. Recall that muscle used to generate high power is anaerobic (and thus is used intermittently) and is most effective at low-frequency output, whereas muscle that operate continuously at high frequency is aerobic, but generates low power output. Given this contradiction, that high-force output

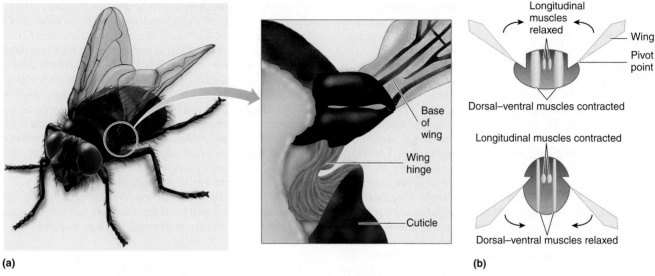

FIGURE 8-22 Mechanics of flight in the housefly. (a) The exoskeleton of insects relying on asynchronous flight muscles is hinged. Each wing is moved through an elastic, hingelike structure as a result of changes in the shape of the thorax. (b) Flight muscles do not attach directly to the hinges or to the wings but rather pull on the sides of the thorax to which they are attached. The two sets of flight muscles are arranged perpendicular to each other. Upward movement of the wings occurs when the vertical dorsoventral muscles contract and pull the flexible dorsal roof downward, flicking the wings up. The downstroke occurs when the longitudinal muscles contract, distorting the roof upward and forcing the wings down. Only the up or down positions are stable, and because of the elasticity of the hinge, much of the energy used to power muscle wing movement is recycled into the subsequent stroke. Smaller muscles acting directly on the hinge allow the wings to move at an angle, resulting in forward movement, so that a complete wing cycle actually involves the wingtip performing a figure-eight pattern.

© Cengage Learning, 2013

and high-frequency operation are mutually exclusive, it would seem impossible for an animal to have both attributes. Yet this is precisely what animals must have in order to fly. What evolutionary developments have arisen to solve this problem?

To provide for faster contraction rates, the mechanics of contraction need to speed up. One factor that increases reaction times is an increase in muscle temperature. Increasing body temperature from 30 to 40°C results in a 2.2-fold increase in the rate of ATP synthesis, equivalent to doubling the effectiveness of the mitochondria within each cell. Birds have higher body temperatures than mammals, a subtle but effective adaptation to increasing reaction times and an important adaptation for flight. Many insects cannot fly without warming up because they cannot supply ATP at a sufficient rate to power the flight muscles; thus, some (such as butterflies) bask in the morning sun before takeoff, whereas others (such as honey bees) shiver their muscles to warm up before flight. An increase in body temperature also permits Ca^{2+} pumps to work more efficiently, resulting in faster clearance of Ca^{2+} from the cytoplasm, which permits muscles to relax more rapidly. In birds and insects the structure of mitochondria is altered such that the inner mitochondrial membrane has twice the packing of cristae (folds of the inner membranes) compared to mammalian mitochondria. Thus, each mitochondrion consumes O_2 at approximately twice the rate (7 to 10 versus 4 to 5 mL O_2 per cm^{-3} per min) measured in mammals. Combined, each of these factors contributes to the ability of flight muscle to operate at higher frequencies. This is exemplified by the high wing-beat frequency of hummingbirds (70 wing beats per second) as well

as for large insects that use **direct** or **synchronous** muscle contractions to power flight muscle.

High-speed synchronous skeletal muscles are those in which each muscle action potential evoked by incoming neuronal activity results in a single mechanical contraction. For example, cicadas generate a calling song by vibrating a pair of resilin-based elastic cuticular plates termed **tymbals**. Contraction of the tymbal muscle bends the tymbals inward, which then recoil outward when the muscle relaxes. Each distortion cycle results in a burst of resonant sound vibration with a frequency of approximately 550 Hz.

Synchronous muscle arrangements are also found in butterflies, moths, locusts, and dragonflies and consist of attachment of the flight muscle directly to the wing. Each signal from a motor neuron triggers contraction of the flight muscle, so that the contractions that raise and lower the wing are synchronous. As has been discussed, muscle fibers are activated by the release of Ca^{2+} and deactivated by the reuptake of Ca^{2+}. However, animals cannot fly at frequencies of greater than 100 Hz with synchronous muscles. How have some animal groups found a solution to this problem?

Asynchronous muscle contractions are characterized by a nearly constant myoplasmic Ca^{2+} concentration

The most significant space- as well as energy-saving solution for flight is the development of **asynchronous** muscle contractions (Figure 8-22). For muscles that operate at extraordinarily high frequencies, their design results in the generation of more power, although capacity for fine neural control declines. Asynchronous muscles are believed to have evolved from synchronous ones and represent a design breakthrough.

Asynchronous or "indirect" muscles are associated with the flight muscle of approximately three quarters of known insect species including Hemiptera (true bugs, cicadas, aphids, and bedbugs), Diptera (flies), and Hymenoptera (bees and wasps). In these animals, wing beat frequency is too great to be controlled directly by neuronal activation. For example, wing beat frequencies greater than 1,000 per second have been recorded for midges, although for most species the number of muscle contractions ranges from 100 to 300 per second. Associated with the rapid wing-beat frequency is a lowered number of nerve impulses that are now asynchronous with muscle contraction. In these circumstances, a single Ca^{2+} pulse maintains the muscle in an activated state for numerous successive cycles. In addition, the flight muscles are attached to the walls of the thorax rather than directly to the wings.

Flight muscles consist of two sets of large antagonistic power muscles oriented more or less perpendicular to each other: the **dorsal ventral muscles** and the **dorsal longitudinal muscles**. Contraction of asynchronous flight muscle is normally triggered by stretch ("stretch activation") and deactivated by "shortening deactivation" both in the presence of elevated myoplasmic Ca^2+ concentrations. With the flight muscles antagonistically situated, a contraction or relaxation changes the shape of the thorax—which has *resilin*-based elasticity in some species (p. 358)—into one of two stable positions: one with the wings elevated and the thorax depressed and the other with the wings depressed and the thorax elevated. Let's examine the situation as the fruit fly (*Drosophila melanogaster*) prepares itself for flight. The fruit fly must beat each wing up to 240 times per second to achieve enough power for lift. Just before takeoff, intermittent nervous stimulation primes the two muscle sets with increased intracellular Ca^2+. The fly then leaps into the air and becomes airborne. Contraction of dorsal ventral muscles lengthens the dorsal longitudinal flight muscles simultaneously. These longitudinal muscles respond to the stretch with a delayed rise in tension (the stretch activation response), causing it to contract. This in turn causes an antagonistic reciprocal stretch and delayed contraction in the dorsal ventral muscles. As the two sets of muscles alternately contract and relax, the thorax rapidly oscillates between the two stable set points, causing deformations at the wing hinges that make the wings beat at the appropriate resonant frequency. The release of the catecholamine octopamine further increases both the force and efficiency of flight muscle contractions. At the same time, octopamine stimulates the oxidation of carbohydrate and fat in the flight muscle. Wing movements are ended when action potentials are no longer initiated, permitting muscle fiber repolarization and Ca^2+ uptake into the sarcoplasmic reticulum.

Insects having an asynchronous arrangement of flight muscle conserve cellular volume in two ways. The first is achieved by a marked reduction in the volume of the sarcoplasmic reticulum. As an example, the sound-producing synchronous muscles of cicadas have approximately 34% sarcoplasmic reticulum, whereas an asynchronous flight muscle has a sarcoplasmic volume less than 4%. Secondly, intracellular space is saved by the reduced requirement for ATP-producing mitochondria. Synchronous muscles require a substantial volume of mitochondria in order to supply ATP for the recycling of Ca^2+ into sarcoplasmic reticulum. Thus, the reduction in Ca^2+ cycling in asynchronous muscle reduces the requirement for mitochondria, which translates into a substantial space savings. Those mitochondria that are present are committed to supplying ATP for the cross bridges, whereas the force-generating myofibrils can instead use the volume not occupied by mitochondria. As a rule of thumb, muscle fibers in all flying animals contain a minimum of 50% myofibrils. Also, the reduction in amount of ATP used to power Ca^2+ cycling reduces energy costs, which lessens the direct expenses of flight, making the entire process more efficient.

For the remaining section on skeletal muscle, we will examine the central and local mechanisms involved in regulating the motor activity performed by these muscles.

check your understanding 8.6

How does asynchronous flight muscle differ from that of synchronous? Which is capable of faster wing-beat frequency and why?

Compare the composition of flight muscles with a powerful locomotive muscle in terms of mitochondria, myofibrils, or sarcoplasmic reticulum (assume that these three components make up 100% of the muscle volume). Explain your answer.

8.7 Control of Motor Movement

Particular patterns of motor unit output govern vertebrate motor activity, ranging from maintenance of posture and balance to stereotypical locomotory movements such as walking or swimming, to individual, highly skilled motor activity such as swinging and leaping through trees. Control of any motor movement, no matter what its level of complexity, depends on converging input to the motor neurons of specific motor units. The motor neurons in turn trigger contraction of the muscle fibers within their respective motor units by means of the events that occur at the neuromuscular junction.

Multiple motor inputs influence vertebrate motor unit output

Three levels of input control motor neuron output in vertebrates (Figure 8-23):

1. *Input from afferent neurons,* usually through intervening interneurons, at the level of the spinal cord—that is, spinal reflexes (see p. 182).

2. *Input from the primary motor cortex.* Fibers originating from neuronal cell bodies known as **pyramidal cells** within the primary motor cortex (see p. 189) descend directly without synaptic interruption to terminate on motor neurons (or on local interneurons that terminate on motor neurons) in the spinal cord. These fibers make up the **corticospinal** (or pyramidal) motor system.

3. *Input from the brain stem* as part of the multineuronal motor system. The pathways composing the **multi-**

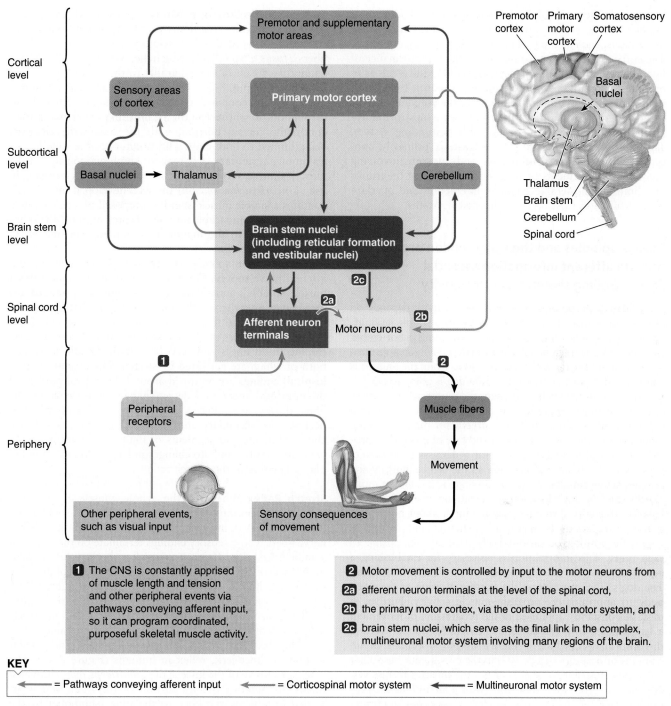

FIGURE 8-23 Motor control (shown for a human). Arrows imply influence, whether excitatory or inhibitory; connections are not necessarily direct but may involve interneurons.

© Cengage Learning, 2013

neuronal (or **extrapyramidal**) **motor system** include a number of synapses that involve many regions of the brain (*extra* means "outside of;" *pyramidal* refers to the pyramidal system). The final link in multineuronal pathways is the brain stem, especially the reticular formation, which in turn is influenced by motor regions of the cortex, the cerebellum, and the basal nuclei. In addition, the motor cortex itself is interconnected with the thalamus as well as with premotor and supplementary motor areas, all part of the multineuronal system.

The only brain regions that directly influence motor neurons are the primary motor cortex and brain stem; the other involved brain regions indirectly regulate motor activity by adjusting motor output from the motor cortex and brain stem. A number of complex interactions take place between these various brain regions; the most important are represented in Figure 8-23. (See Chapter 5 for further discussion of the specific roles and interactions of these brain regions.)

Spinal reflexes involving afferent neurons are important in maintaining posture and in executing basic protective movements, such as the withdrawal reflex (see Figure 5-22). The corticospinal system primarily mediates performance of fine, discrete movements of the hands and digits (such as in primates). Premotor and supplementary motor areas, with input from the cerebrocerebellum, plan the voluntary motor command that the primary motor cortex issues to the appropriate motor neurons through this descending system. The multineuronal system, in contrast, is primarily concerned with regulation of overall body posture involving involuntary movements of large muscle groups of the trunks and limbs. Considerable complex interaction and overlapping of function exist between these two systems.

Muscle spindles and the Golgi tendon organs provide afferent information essential for controlling skeletal muscle activity

Coordinated, purposeful skeletal muscle activity depends on afferent input from a variety of sources. At a simple level, afferent signals indicating that a foot is touching a hot ember (see Figure 5-21) triggers reflex contractile activity in appropriate limb muscles to withdraw the foot from the injurious stimulus. At a more complex level, if a dog is going to catch a stick thrown in the air by a human, the motor systems of the dog's brain must program sequential motor commands that will move and position its body correctly for the catch, using predictions of the stick's direction and rate of movement provided by visual input (see p. 185 for a discussion on predictions made by brain regions such as the cerebellum). Many muscles acting simultaneously or alternately at different joints rapidly shift the dog's location and position while maintaining balance. It is critical to have ongoing input about body position with respect to the surrounding environment, as well as about the position of various body parts in relationship to each other. This information is necessary for establishing a neuronal pattern of activity to perform the desired movement. The dog's CNS must know the starting position of the dog's body to appropriately program muscle activity. Further, it must be constantly appraised of the progression of movement it has initiated so it can adjust as needed. An animal receives this information, which is known as *proprioceptive input,* from **proprioceptors** (see p. 192) in the eyes, joints, vestibular apparatus, and skin, as well as from the muscles themselves. For example, you can demonstrate your joint and muscle proprioceptive receptors in action by closing your eyes and bringing the tips of your right and left index fingers together at any point in space. You can do so without seeing where your hands are, because afferent input from the joint and muscle receptors inform your brain at all times about the position of your hands and other body parts.

Two types of muscle proprioceptors—**muscle spindles** and **Golgi tendon organs**—monitor changes in muscle length and tension. This information is used in two ways: (1) to apprise motor areas of the brain about muscle length and tension and (2) to control muscle length and tension in negative-feedback fashion by means of local spinal reflexes. Muscle length is monitored by muscle spindles, whereas changes in muscle tension are detected by Golgi tendon organs. Both these receptor types are activated by muscle stretch, but they convey different types of information. Let's see how.

Muscle Spindle Structure **Muscle spindles**, which are distributed throughout the fleshy part of a skeletal muscle, are collections of specialized muscle fibers known as **intrafusal fibers** (*fusus,* "spindle"), which lie within spindle-shaped connective tissue capsules parallel to the "ordinary" **extrafusal fibers** (Figure 8-24). Unlike an ordinary skeletal muscle fiber, which contains contractile elements (myofibrils) throughout its entire length, an intrafusal fiber has a noncontractile central portion, with the contractile elements being limited to both ends. The structure of the spindle varies among vertebrate classes with Amphibia considered the most primitive vertebrate to posses this sensory organ. In fish (*Oncorhynchus masou*) the well-developed jaw muscle contains a simple muscle spindle composed of a single intrafusal fiber. A comparable **muscle receptor organ (MRO)** has been described in the mandible of a pierce-sucking insect *Oncopeltus fasciatus.*

Each muscle spindle has its own private efferent and afferent nerve supply. The efferent neuron that innervates a muscle spindle's intrafusal fibers is known as **gamma (γ) motor neuron**, whereas the motor neurons that supply the ordinary extrafusal fibers are designated as **alpha (α) motor neurons.** Two types of afferent sensory endings terminate on the intrafusal fibers and serve as muscle spindle receptors, both of which are activated by stretch. The **primary (annulospiral) endings** are wrapped around the central portion of the intrafusal fibers; they detect changes in the length of the fibers during stretching as well as the speed with which it occurs. The **secondary (flower-spray) endings,** which are clustered at the end segments of many of the intrafusal fibers, are sensitive only to changes in length. Muscle spindles play a key role in the stretch reflex.

Stretch Reflex Whenever the whole muscle is passively stretched, the intrafusal fibers within its muscle spindles are likewise stretched, increasing the rate of firing in the afferent nerve fibers whose sensory endings terminate on the stretched spindle fibers. The afferent neuron directly synapses on the α motor neuron that innervates the extrafusal fibers of the same muscle, resulting in contraction of that muscle (Figure 8-25a, pathway 1 → 2). This stretch reflex serves as a local negative-feedback mechanism to resist any passive changes in muscle length so that optimal resting length can be maintained.

The classic example of the stretch reflex is the **patellar tendon,** or knee-jerk, reflex in humans (Figure 8-26). The extensor muscle of the knee is the *quadriceps femoris,* which forms the anterior (front) portion of the thigh and is attached just below the knee to the tibia (shinbone) by the *patellar tendon.* Tapping this tendon with a rubber mallet passively stretches the quadriceps muscle, activating its spindle receptors. The resultant stretch reflex brings about contraction of this extensor muscle, causing the knee to extend and raise the foreleg in the well-known knee-jerk fashion. This test is routinely performed as a preliminary assessment of nervous system function.

The primary purpose of the stretch reflex is to resist the tendency for the passive stretch of extensor muscles caused by gravitational forces when a person is standing upright. Whenever the knee joint tends to buckle because of gravity, the quadriceps muscle is stretched. The resultant enhanced contraction of this extensor muscle brought about by the stretch reflex quickly straightens out the knee, holding the limb extended so that the person remains standing.

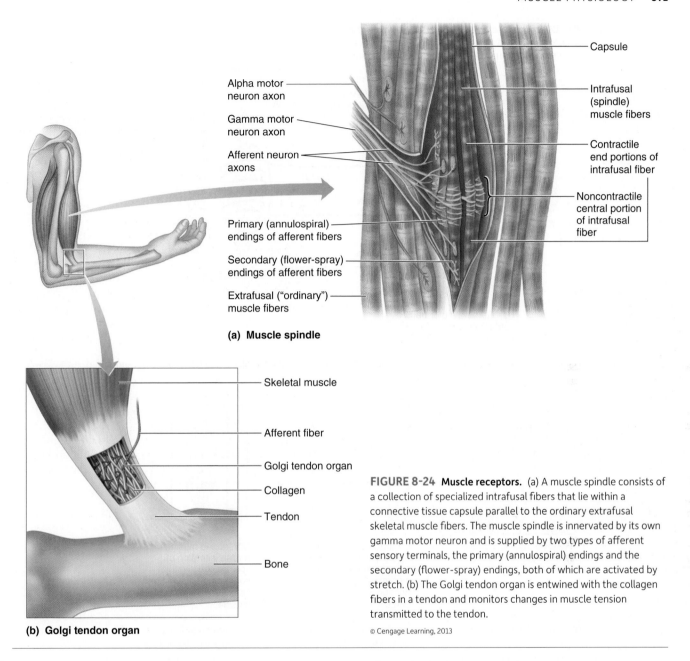

Capsule

Alpha motor
neuron axon

Gamma motor
neuron axon

Afferent neuron
axons

Intrafusal
(spindle)
muscle fibers

Contractile
end portions of
intrafusal fiber

Noncontractile
central portion
of intrafusal
fiber

Primary (annulospiral)
endings of afferent fibers

Secondary (flower-spray)
endings of afferent fibers

Extrafusal ("ordinary")
muscle fibers

(a) Muscle spindle

Skeletal muscle

Afferent fiber

Golgi tendon organ

Collagen

Tendon

Bone

(b) Golgi tendon organ

FIGURE 8-24 Muscle receptors. (a) A muscle spindle consists of a collection of specialized intrafusal fibers that lie within a connective tissue capsule parallel to the ordinary extrafusal skeletal muscle fibers. The muscle spindle is innervated by its own gamma motor neuron and is supplied by two types of afferent sensory terminals, the primary (annulospiral) endings and the secondary (flower-spray) endings, both of which are activated by stretch. (b) The Golgi tendon organ is entwined with the collagen fibers in a tendon and monitors changes in muscle tension transmitted to the tendon.

© Cengage Learning, 2013

Coactivation of γ and α Motor Neurons Gamma (γ) motor neurons initiate contraction of the muscular end regions of intrafusal fibers (Figure 8-25a, pathway 3). This contractile response is too weak to have any influence on whole-muscle tension, but it does have an important localized effect on the muscle spindle itself. If there were no compensating mechanisms, shortening of the whole muscle by α motor-neuron stimulation of extrafusal fibers would cause slack in the spindle fibers so that they would be less sensitive to stretch and therefore not as effective as muscle length detectors (Figures 8-25b and c). Co-activation of the γ-motor-neuron system along with the α-motor-neuron system during reflex and voluntary contractions (Figure 8-25, pathway 4) takes the slack out of the spindle fibers as the whole muscle shortens, permitting these receptor structures to maintain their high sensitivity to stretch over a wide range of muscle lengths. When γ motor-neuron stimulation triggers simultaneous contraction of both end muscular portions of an in-

trafusal fiber, the noncontractile central portion is pulled in opposite directions, tightening this region and taking out the slack (Figure 8-25d). Whereas the extent of α-motor-neuron activation depends on the intended strength of the motor response, the extent of simultaneous γ motor-neuron activity to the same muscle depends on the anticipated distance of shortening.

Golgi Tendon Organs In contrast to muscle spindles, which lie within the belly of the muscle, Golgi tendon organs are located in the tendons of the muscle, where they can respond to changes in the muscle's externally applied tension rather than to changes in its length. Because a number of factors determine the tension developed in the whole muscle during contraction (for example, frequency of stimulation or length of the muscle at the onset of contraction), it is essential that motor control systems be apprised of the tension actually achieved so that adjustments can be made if necessary.

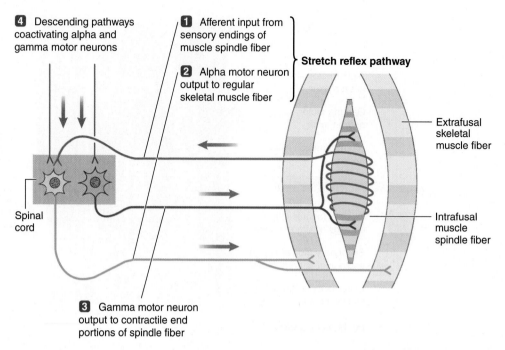

4 Descending pathways coactivating alpha and gamma motor neurons

1 Afferent input from sensory endings of muscle spindle fiber

Stretch reflex pathway

2 Alpha motor neuron output to regular skeletal muscle fiber

Extrafusal skeletal muscle fiber

Intrafusal muscle spindle fiber

Spinal cord

3 Gamma motor neuron output to contractile end portions of spindle fiber

(a) Pathways involved in monosynaptic stretch reflex and coactivation of alpha and gamma motor neurons

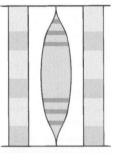
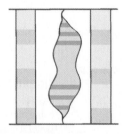
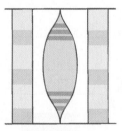

Relaxed muscle; spindle fiber sensitive to stretch of muscle

Contracted muscle in hypothetical situation of no spindle coactivation; slackened spindle fiber not sensitive to stretch of muscle

Contracted muscle in normal situation of spindle coactivation; contracted spindle fiber sensitive to stretch of muscle

(b) Relaxed muscle

(c) Contracted muscle with no spindle coactivation

(d) Contracted muscle with spindle coactivation

FIGURE 8-25 Muscle spindle function.

© Cengage Learning, 2013

Golgi tendon organs consist of endings of afferent fibers entwined within bundles of connective tissue fibers that make up the tendon. When the extrafusal muscle fibers contract, the resultant pull on the tendon tightens the connective tissue bundles, which in turn increase the tension exerted on the bone to which the tendon is attached. In the process, the entwined Golgi organ afferent-receptor endings are stretched, causing the afferent fibers to fire; the frequency of firing is directly related to the tension developed. This afferent information is sent to the brain for processing, much of which is used subconsciously for smoothly executing motor activity. Unlike afferent information from the muscle spindles, afferent information from the Golgi tendon organ reaches the level of conscious awareness, at least in humans. You are aware of the tension within a muscle, but you are not aware of its length.

Traditionally, physiologists thought the Golgi tendon organ triggered a protective spinal reflex that halted further contraction and brought about sudden reflex contraction when the muscle tension became great enough, thus helping prevent damage to the muscle or tendon from excessive, tension-developing muscle contraction. Scientists now believe, however, that this receptor is a pure sensor and does not initiate any reflexes. Other unknown mechanisms are apparently involved in inhibiting further contraction to prevent tension-induced damage.

check your understanding 8.7

How do muscle spindles differ from Golgi tendon organs?

Is the stretch reflex monosynaptic or polysynaptic? Explain.

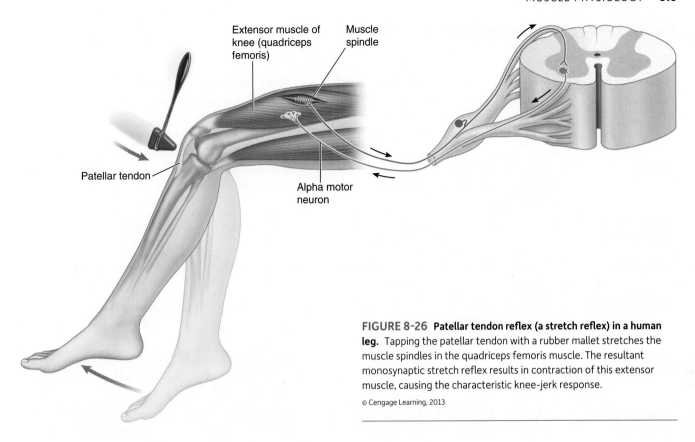

FIGURE 8-26 Patellar tendon reflex (a stretch reflex) in a human leg. Tapping the patellar tendon with a rubber mallet stretches the muscle spindles in the quadriceps femoris muscle. The resultant monosynaptic stretch reflex results in contraction of this extensor muscle, causing the characteristic knee-jerk response.

© Cengage Learning, 2013

8.8 Smooth Muscle and Cardiac Muscle

Having completed our discussion of skeletal muscle, we will finish with an examination the other two types.

Smooth and cardiac muscle share some basic properties with skeletal muscle

The other types of muscle—smooth muscle and cardiac muscle—share some basic properties with skeletal muscle, but each also displays unique characteristics (Table 8-3). For example, smooth muscle evolved independently from the skeletal and cardiac muscles and is found in all vertebrates as well as some nonvertebrates. However, the three muscle types do have several features in common. First, they all have a specialized contractile apparatus made up of thin actin filaments that slide relative to stationary thick myosin filaments in response to a rise in cytosolic Ca^{2+} to accomplish contraction. Second, they all directly use ATP as the energy source for cross-bridge cycling. However, the structure and organization of fibers within these different muscle types vary, as do their mechanisms of excitation and the means by which excitation and contraction are coupled. Furthermore, there are important distinctions in the contractile response itself. We spend the remainder of this chapter highlighting unique features of smooth and cardiac muscle as compared with skeletal muscle, reserving a more detailed discussion of their function for chapters devoted to organs containing these muscle types.

Smooth muscle cells are small and unstriated

The majority of smooth muscle cells in vertebrates lie in the walls of hollow organs and tubes. Their contraction exerts pressure on and regulates the forward movement of the contents of these structures.

Both smooth and skeletal muscle cells are elongated, but in contrast to their large, cylindrical, multinucleated skeletal muscle counterparts, smooth muscle cells are spindle-shaped, have a single nucleus, and are considerably smaller (2 to 10 μm in diameter and 50 to 400 μm long). Also unlike skeletal muscle cells, a single smooth-muscle cell does not extend the full length of a muscle. Instead, groups of smooth muscle cells are typically arranged in sheets (Figure 8-27a).

A smooth muscle cell has three types of filaments: (1) thick myosin filaments, which are longer than those found in skeletal muscle; (2) thin actin filaments, which contain tropomyosin but lack the regulatory protein troponin; and (3) filaments of intermediate size, which are unique to smooth muscle and which do not directly participate in the contractile process but serve as part of the cytoskeletal framework that supports the shape of the cell. Smooth muscle filaments do not form myofibrils and are not arranged in the sarcomere pattern found in skeletal muscle. Thus, smooth muscle cells do not display the banding or striation found in skeletal muscle, hence the term "smooth" for this muscle type.

Lacking sarcomeres, smooth muscle does not have Z lines as such, but **dense bodies** containing the same protein constituent found in Z lines are present (Figure 8-27b). Dense bodies are positioned throughout the smooth muscle

TABLE 8-3 Comparison of Muscle Types

	TYPE OF MUSCLE			
Characteristic	Skeletal	Multiunit Smooth	Single-Unit Smooth	Cardiac
Location	Attached to skeleton	Large blood vessels, eye, and hair follicles	Walls of hollow organs in digestive, reproductive, and urinary tracts and in small blood vessels	Heart only
Function	Movement of body in relation to external environment	Varies with structure involved	Movement of contents within hollow organs	Pumps blood out of heart
Mechanism of Contraction	Sliding filament mechanism	Sliding filament mechanism	Sliding filament mechanism	Sliding filament mechanism
Innervation	Somatic nervous system (alpha motor neurons)	Autonomic nervous system	Autonomic nervous system	Autonomic nervous system
Level of Control	Under voluntary control; also subject to subconscious regulation	Under involuntary control	Under involuntary control	Under involuntary control
Initiation of Contraction	Neurogenic	Neurogenic	Myogenic (pacemaker potentials and slow-wave potentials)	Myogenic (pacemaker potentials)
Role of Nervous Stimulation	Initiates contraction; accomplishes gradation	Initiates contraction; contributes to gradation	Modifies contraction; can excite or inhibit; contributes to gradation	Modifies contraction; can excite or inhibit; contributes to gradation
Modifying Effect of Hormones	No	Yes	Yes	Yes
Presence of Thick Myosin and Thin Actin Filaments	Yes	Yes	Yes	Yes
Striated by Orderly Longitudinal Arrangement of Filaments	Yes	No	No	Yes
Presence of Troponin and Tropomyosin	Yes	Tropomyosin only	Tropomyosin only	Yes
Presence of T Tubules	Yes	No	No	Yes
Level of Development of Sarcoplasmic Reticulum	Well developed	Poorly developed	Poorly developed	Moderately developed
Cross Bridges Turned on by Ca^{2+}	Yes	Yes	Yes	Yes
Source of Increased Cytosolic Ca^{2+}	Sarcoplasmic reticulum	ECF and sarcoplasmic reticulum	ECF and sarcoplasmic reticulum	ECF and sarcoplasmic reticulum
Site of Ca^{2+} Regulation	Troponin in thin filaments	Myosin in thick filaments	Myosin in thick filaments	Troponin in thin filaments
Mechanism of Ca^{2+} Action	Physically repositions troponin–tropomyosin complex to uncover actin cross-bridge binding sites	Chemically brings about phosphorylation of myosin cross bridges so they can bind with actin	Chemically brings about phosphorylation of myosin cross bridges so they can bind with actin	Physically repositions troponin–tropomyosin complex

TYPE OF MUSCLE

Characteristic	Skeletal	Multiunit Smooth	Single-Unit Smooth	Cardiac
Presence of Gap Junctions	No	Yes (very few)	Yes	Yes
ATP Used Directly by Contractile Apparatus	Yes	Yes	Yes	Yes
Myosin ATPase Activity; Speed of Contraction	Fast or slow, depending on type of fiber	Very slow	Very slow	Slow
Means by which Gradation Accomplished	Varying number of motor units contracting (motor unit recruitment) and frequency at which they are stimulated (twitch summation)	Varying number of muscle fibers contracting and varying cytosolic Ca^{2+} concentration in each fiber by autonomic and hormonal influences	Varying cytosolic Ca^{2+} concentration through myogenic activity and influences of the autonomic nervous system, hormones, mechanical stretch, and local metabolites	Varying length of fiber (depending on extent of filling of the heart chambers) and varying cytosolic Ca^{2+} concentration through autonomic, hormonal, and local metabolite influence
Presence of Tone in Absence of External Stimulation	No	No	Yes	No
Clear-Cut Length–Tension Relationship	Yes	No	No	Yes

© Cengage Learning, 2013

cell as well as being attached to the internal surface of the plasma membrane. The dense bodies are held in place by a scaffold of intermediate filaments. The actin filaments are anchored to the dense bodies. Considerably more actin is present in smooth muscle cells than in skeletal muscle cells, with 10 to 15 thin filaments for each thick myosin filament in smooth muscle and two thin filaments for each thick filament in skeletal muscle.

The thick- and thin-filament contractile units are oriented slightly diagonally from side to side within the smooth muscle cell in an elongated, diamond-shaped lattice, rather than running parallel with the long axis as myofibrils do in skeletal muscle (Figure 8-28a). Relative sliding of the thin filaments past the thick filaments during contraction causes the filament lattice to reduce in length and expand from side to side. As a result, the whole cell shortens and bulges out between the points where the thin filaments are attached to the inner surface of the plasma membrane (Figure 8-28b).

Unlike in skeletal muscle, myosin molecules are arranged in a smooth-muscle thick filament so that cross bridges are present along the entire filament length (that is, there is no bare portion in the center of a smooth-muscle thick filament). As a result, the surrounding thin filaments can be pulled along the thick filaments for longer distances than in skeletal muscle. Also dissimilar to skeletal muscle, the myosin proteins in smooth-muscle thick filaments are organized so that half of the surrounding thin filaments are pulled in one direction and the other half are pulled in the opposite direction (Figure 8-28b).

Smooth muscle cells are turned on by Ca^{2+}-dependent phosphorylation of myosin

The thin filaments of smooth muscle cells do not contain troponin, and tropomyosin does not block actin's cross-bridge binding sites. What, then, prevents actin and myosin from binding at the cross bridges in the resting state, and how is cross-bridge activity switched on in the excited state? Lightweight chains of proteins are attached to the heads of myosin molecules, near the "neck" region. These so-called light chains are only of secondary importance in skeletal muscle, but they have a crucial regulatory function in smooth muscle. The smooth muscle myosin heads can interact with actin only when the myosin light chain is *phosphorylated* (that is, has an inorganic phosphate from ATP attached to it). During excitation, the increased cytosolic Ca^{2+} acts as an intracellular messenger, initiating a chain of biochemical events that results in phosphorylation of the myosin light chain (Figure 8-29). Smooth muscle Ca^{2+} binds with calmodulin, an intracellular protein found in most cells that is structurally similar to troponin (see p. 339). This Ca^{2+}-calmodulin complex binds to and activates another protein, myosin light chain kinase (MLC kinase), which in turn phosphorylates the myosin light chain. Note that this inorganic phosphate on the myosin light chain is in addition to the inorganic phosphate accompanying ADP on the myosin cross-bridge ATPase site. The P_i on the ATPase site is part of the energy-supplying cycle that powers cross-bridge bending. The P_i on the light chain permits the myosin cross bridge to bind with actin so that cross-bridge cycling can

Smooth muscle cells   Nucleus

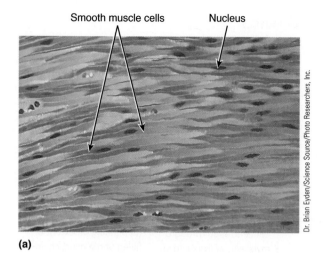

(a)

Smooth muscle cells   Dense bodies

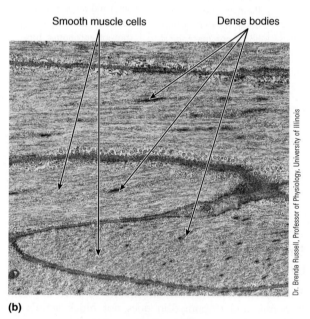

(b)

FIGURE 8-27 Microscopic view of smooth muscle cells. (a) Note the spindle shape and single, centrally located nucleus. (b) Note the presence of dense bodies and lack of banding.

begin. Therefore, smooth muscle is triggered to contract by a rise in cytosolic Ca^{2+} similar to what happens in skeletal muscle, but in smooth muscle, Ca^{2+} ultimately turns on the cross bridges by inducing a *chemical* change in myosin in the *thick* filaments, whereas in skeletal muscle it exerts its effects by invoking a *physical* change at the *thin* filaments (Figure 8-30). Recall that in skeletal muscle Ca^{2+} moves troponin and tropomyosin from their blocking position, so actin and myosin are free to bind with each other.

Phasic smooth muscle contracts in bursts of activity; tonic smooth muscle maintains an ongoing level of contraction

The means by which cytosolic Ca^{2+} concentration increases in smooth muscle cells to turn on the cross bridges also differs from that for skeletal muscle and even varies among the highly diversified smooth muscle found in different organs.

Smooth muscle can be classified in various ways, depending on the timing and means of increasing cytosolic Ca^{2+}: phasic or tonic, multiunit or single-unit, and neurogenic or myogenic smooth muscle. Each smooth muscle belongs to one class of each of these three categories. That is, the smooth muscle in one organ might be phasic, multiunit, and neurogenic, and the smooth muscle in another organ might be tonic, single-unit, and myogenic. Let's examine each of these categories.

Smooth muscle can be grouped into two categories depending on how its cytosolic Ca^{2+} concentration increases and its level of ongoing contractile activity: *phasic smooth muscle* and *tonic smooth muscle*. **Phasic smooth muscle** contracts in bursts, triggered by action potentials that lead to increased cytosolic Ca^{2+}. These bursts in contraction are characterized by pronounced increases in contractile activity. Phasic smooth muscle is most abundant in the walls of hollow organs that push contents through them, such as the digestive organs. Phasic digestive contractions mix the food with digestive juices and propel the mass forward for further processing. **Tonic smooth muscle** is usually partially contracted at all times. This state of partial contraction is called **tone**. Tone exists because this type of smooth muscle has a relatively low resting potential of -55 mV to -40 mV. Some surface-membrane voltage-gated Ca^{2+} channels are open at these potentials. The resultant Ca^{2+} entry maintains a state of partial contraction. Therefore, maintenance of tone in tonic smooth muscle does not depend on action potentials. Tonic smooth muscle does not display bursts of contractile activity but instead incrementally varies its extent of contraction above or below this tonic level in response to regulatory factors, which alter the cytosolic Ca^{2+} concentration. The smooth muscle in walls of arterioles is an example of tonic smooth muscle. The ongoing tonic contraction in these small blood vessels squeezes down on the blood flowing through them and is one of the major contributing factors to maintenance of blood pressure.

A smooth muscle cell has no T tubules and a poorly developed sarcoplasmic reticulum. In phasic smooth muscle, the increased cytosolic Ca^{2+} that triggers depolarization and contraction comes from two sources: Most Ca^{2+} enters from the ECF, but some is released intracellularly from the sparse sarcoplasmic reticulum stores. Unlike their role in skeletal muscle cells, C-type Ca^{2+} channels (voltage-gated dihydropyridine receptors) in the plasma membrane of smooth muscle cells function as Ca^{2+} channels. When these surface-membrane channels are opened in response to an action potential, Ca^{2+} enters down its concentration gradient from the ECF. The entering Ca^{2+} triggers the opening of Ca^{2+} channels in the sarcoplasmic reticulum so that small additional amounts of Ca^{2+} are released intracellularly from this meager source. Because smooth muscle cells are so much smaller in diameter than skeletal muscle fibers, most Ca^{2+} entering from the ECF can influence cross-bridge activity, even in the central portions of the cell, without requiring an elaborate T tubule–sarcoplasmic reticulum mechanism.

One of the major means of increasing cytosolic Ca^{2+} concentration and thus increasing contractile activity in tonic smooth muscle is binding of an extracellular chemical messenger, such as norepinephrine or various hormones, to a G-protein coupled receptor, which activates the IP_3/Ca^{2+} second-messenger pathway (see p. 98). The membrane of the sarcoplasmic reticulum in tonic smooth muscle has IP_3 receptors, which like ryanodine receptors, are Ca^{2+}-release channels. IP_3 binding leads to release of contractile-inducing

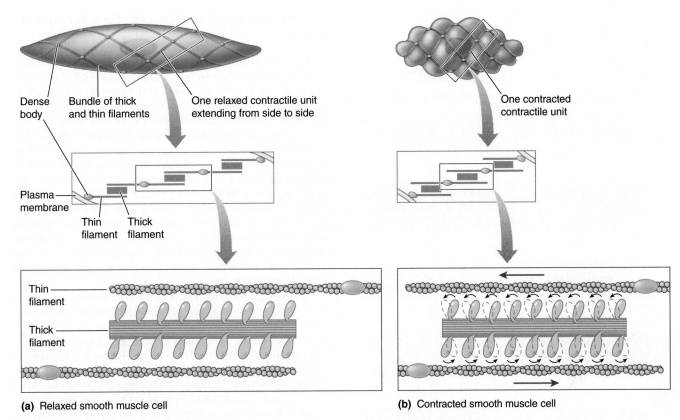

(a) Relaxed smooth muscle cell

(b) Contracted smooth muscle cell

FIGURE 8-28 **Arrangement of thick and thin filaments in a smooth muscle cell in relaxed and contracted states.**

© Cengage Learning, 2013

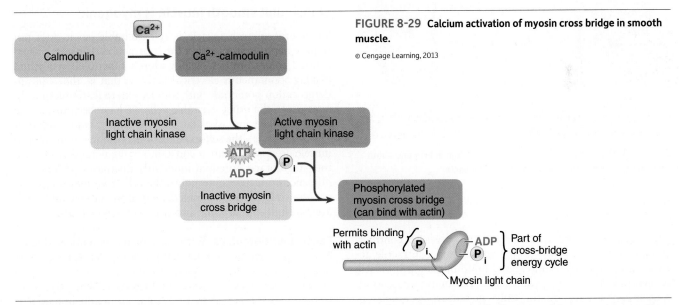

FIGURE 8-29 **Calcium activation of myosin cross bridge in smooth muscle.**

© Cengage Learning, 2013

Ca^{2+} from this intracellular store into the cytosol. This is how norepinephrine released from the sympathetic nerve endings acts on arterioles to increase blood pressure.

Relaxation is accomplished by removal of Ca^{2+} as it is actively transported out across the plasma membrane and back into the sarcoplasmic reticulum. When Ca^{2+} is removed, myosin is dephosphorylated (the phosphate is removed) and can no longer interact with actin, so the muscle relaxes.

We still have not addressed the question of how smooth muscle becomes excited to contract; that is, what opens the Ca^{2+} channels in the plasma membrane and sarcoplasmic reticulum? Smooth muscle is grouped into two categories—*multiunit* and *single-unit smooth muscle*—based on differences in how the muscle fibers become excited. Let's compare these two types of smooth muscle.

Multiunit smooth muscle is neurogenic

Multiunit smooth muscle exhibits properties partway between skeletal muscle and single-unit smooth muscle. As the name implies, a multiunit smooth muscle consists of multiple discrete units that function independently of each other

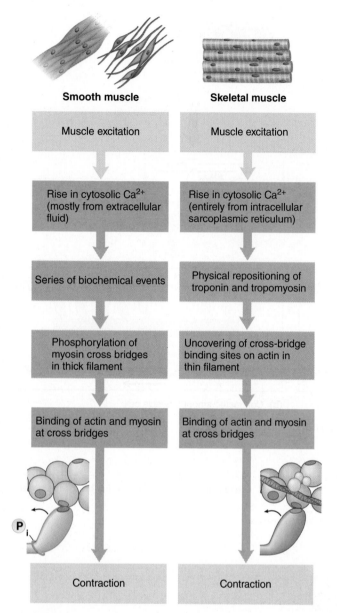

Smooth muscle	Skeletal muscle
Muscle excitation	Muscle excitation
Rise in cytosolic Ca^{2+} (mostly from extracellular fluid)	Rise in cytosolic Ca^{2+} (entirely from intracellular sarcoplasmic reticulum)
Series of biochemical events	Physical repositioning of troponin and tropomyosin
Phosphorylation of myosin cross bridges in thick filament	Uncovering of cross-bridge binding sites on actin in thin filament
Binding of actin and myosin at cross bridges	Binding of actin and myosin at cross bridges
Contraction	Contraction

FIGURE 8-30 Comparison of the role of calcium in bringing about contraction in smooth muscle and skeletal muscle.

© Cengage Learning, 2013

and must be separately stimulated by nerves to contract, similar to skeletal muscle motor units. Thus, contractile activity in both skeletal muscle and multiunit smooth muscle is **neurogenic** ("nerve-produced"). All multiunit smooth muscle is phasic, contracting only when neurally stimulated. Whereas skeletal muscle is innervated by the voluntary somatic nervous system (motor neurons), multiunit (as well as single-unit) smooth muscle is supplied by the involuntary autonomic nervous system.

Mammalian multiunit smooth muscle is found (1) in the walls of large blood vessels; (2) in large airways to the lungs; (3) in the muscle of the eye that adjusts the lens for near or far vision; (4) in the iris of the eye, which alters the size of the pupil to adjust the amount of light entering the eye; and (5) at the base of hair follicles, contraction of which causes raising of hairs in response to fear or cold.

Single-unit smooth muscle cells form functional syncytia

Most smooth muscle is **single-unit** muscle, alternatively called **visceral smooth muscle** because it is found in the walls of the hollow organs or viscera (for example, the digestive, reproductive, and urinary tracts and small blood vessels). The term *single-unit smooth muscle* derives from the fact that the muscle fibers that make up this type of muscle become excited and contract as a single unit. Gap junctions electrically link the muscle fibers in single-unit smooth muscle (see p. 66). When an action potential occurs anywhere within a sheet of single-unit smooth muscle, it is quickly propagated via these special points of electrical contact throughout the entire group of interconnected cells, which then contract as a single coordinated unit. This group of interconnected muscle cells that function electrically and mechanically as a unit is a *functional syncytium* (see p. 336 for skeletal muscle version).

Thinking about the role of the uterus during the process of labor in a mammal can help you appreciate the significance of this arrangement. Muscle cells composing the uterine wall act as a functional syncytium. They repetitively become excited and contract as a unit during labor, exerting a series of coordinated "pushes" that are eventually responsible for delivering the neonate. Independent, uncoordinated contractions of individual muscle cells in the uterine wall would not exert the uniformly applied pressure needed to expel the neonate. Single-unit smooth muscle elsewhere in the body is arranged in similar functional syncytia.

Single-unit smooth muscle is myogenic

Single-unit smooth muscle is **self-excitable**, so it does not require nervous stimulation for contraction. Clusters of specialized smooth-muscle cells within a functional syncytium display spontaneous electrical activity; that is, they can undergo action potentials without any external stimulation. In contrast to the other excitable cells we have been discussing (such as neurons, skeletal muscle fibers, and multiunit smooth muscle), the self-excitable cells of single-unit smooth muscle do not maintain a constant resting potential. Instead, their membrane potential inherently fluctuates without any influence by factors external to the cell. Two major types of spontaneous depolarizations displayed by self-excitable cells are *pacemaker potentials* and *slow-wave potentials*.

Pacemaker Potentials With pacemaker potentials (Figure 8-31a), the membrane potential gradually depolarizes on its own because of shifts in passive ionic fluxes accompanying automatic changes in channel permeability. When the membrane has depolarized to threshold, an action potential is initiated. After repolarizing, the membrane potential once again depolarizes to threshold, cyclically continuing in this manner to self-generate action potentials.

Slow-Wave Potentials **Slow-wave potentials** are gradually alternating hyperpolarizing and depolarizing swings in potential caused by automatic cyclic changes in the rate at which sodium ions are actively transported across the membrane (Figure 8-31b). The potential moves farther from threshold during each hyperpolarizing swing and closer to threshold during each depolarizing swing. If

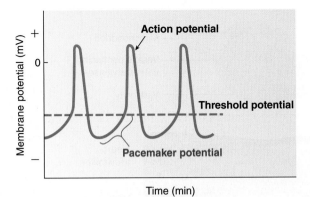

(a) Pacemaker potential

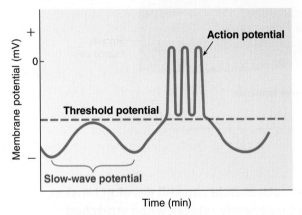

(b) Slow-wave potential

FIGURE 8-31 Self-generated electrical activity in smooth muscle.
(a) With pacemaker potentials, the membrane gradually depolarizes to threshold on a regular periodic basis without any nervous stimulation. These regular depolarizations cyclically trigger self-induced action potentials. (b) In slow-wave potentials, the membrane gradually undergoes self-induced hyperpolarizing and depolarizing swings in potential. A burst of action potentials occurs if a depolarizing swing brings the membrane to threshold.

© Cengage Learning, 2013

threshold is reached, a burst of action potentials occurs at the peak of a depolarizing swing. Threshold is not always reached, however, so the oscillating slow-wave potentials can continue without generating action potentials. Whether threshold is reached depends on the starting point of the membrane potential at the onset of its depolarizing swing. The starting point, in turn, is influenced by neural and local factors.

Myogenic Activity Self-excitable smooth muscle cells are specialized to initiate action potentials, but they are not equipped to contract. Only a very small proportion of the total number of cells in a functional syncytium are noncontractile, pacemaker cells. Pacemaker cells are typically clustered together in a specific location. The vast majority of smooth muscle cells in a functional syncytium are specialized to contract but cannot self-initiate action potentials. However, once an action potential is initiated by a self-initiating smooth muscle cell, it is conducted to the remaining contractile, nonpacemaker cells of the functional syncytium via gap junctions, so that the entire group

of cells contracts without any nervous input. Such nerve-independent contractile activity initiated by the muscle itself is called myogenic ("muscle-produced") activity, in contrast to the neurogenic activity of skeletal muscle and multiunit smooth muscle. Recall that tonic, single-unit smooth muscle cells have sufficient levels of cytosolic Ca^{2+} to maintain a low level of tension even in the absence of action potentials, so they too are myogenic. (Thus, multiunit smooth muscles are all neurogenic and phasic; single-unit smooth muscles are all myogenic and may be phasic or tonic.)

Gradation of single-unit smooth muscle contraction differs considerably from that of skeletal muscle

Single-unit smooth muscle differs from skeletal muscle in the way gradation of contraction is accomplished. Gradation of skeletal muscle contraction is entirely under neural control, primarily involving motor unit recruitment and twitch summation. In smooth muscle, the gap junctions ensure that an entire smooth-muscle mass contracts as a single unit, making it impossible to vary the number of muscle fibers contracting. Only the tension of the fibers can be modified to achieve varying strengths of contraction of the whole organ. The portion of cross bridges activated and the tension subsequently developed in single-unit smooth muscle can be graded by varying the cytosolic Ca^{2+} concentration. A single excitation in smooth muscle does not cause all the cross bridges to switch on, in contrast to skeletal muscles where a single action potential triggers the release of enough Ca^{2+} to permit all cross bridges to cycle. As Ca^{2+} concentration increases in smooth muscle, more cross bridges are brought into play, and greater tension develops.

Smooth Muscle Tone Many single-unit smooth muscle cells have sufficient cytosolic Ca^{2+} to maintain a low level of tension, or tone, even in the absence of action potentials. A sudden drastic change in Ca^{2+}, such as accompanies a myogenically induced action potential, brings about a contractile response superimposed on the ongoing tonic tension. Besides self-induced action potentials, a number of other factors, including autonomic neurotransmitters, can influence contractile activity and the development of tension in smooth muscle cells by altering their cytosolic Ca^{2+} concentration.

Smooth muscle activity can be modified by the autonomic nervous system

Both branches of the vertebrate autonomic nervous system typically innervate smooth muscle. In single-unit smooth muscle (both phasic and tonic), this nerve supply does not *initiate* contraction, but it can *modify* the rate and strength of contraction, either enhancing or retarding the inherent contractile activity of a given organ. Recall that the isolated motor end-plate region of a skeletal muscle fiber interacts with ACh released from a single axon terminal of a motor neuron. In contrast, the receptor proteins that bind with autonomic transmitters are dispersed throughout the entire surface membrane of a smooth muscle cell. Smooth muscle cells are sensitive to varying degrees and in

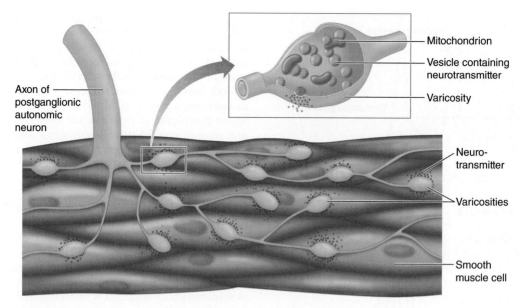

Axon of postganglionic autonomic neuron

Mitochondrion

Vesicle containing neurotransmitter

Varicosity

Neuro-transmitter

Varicosities

Smooth muscle cell

FIGURE 8-32 Innervation of smooth muscle by autonomic postganglionic nerve terminals.

© Cengage Learning, 2013

varying ways to autonomic transmitters, depending on their distribution of cholinergic and adrenergic receptors (see p. 164).

Each terminal branch of a postganglionic autonomic fiber travels across the surface of one or more smooth muscle cells, releasing transmitter from the vesicles within its multiple **varicosities** (bulges) as an action potential passes along the terminal (Figure 8-32). The neurotransmitter diffuses to the many receptor sites specific for it on the cells underlying the terminal. Thus, in contrast to the discrete one-to-one relationship at motor end plates, a given smooth muscle cell can be influenced by more than one type of neurotransmitter, and each autonomic terminal can influence more than one smooth muscle cell.

Other Factors Influencing Smooth Muscle Activity Other factors (besides autonomic neurotransmitters) can influence the rate and strength of both multiunit and single-unit smooth muscle contraction, including certain hormones, local metabolites, mechanical stretch, and specific drugs. Recent evidence suggests that the abundant caveolae found in smooth muscle may serve as sites of integration of the extracellular signals that modify smooth muscle contractility (see p. 89). All these factors ultimately act by modifying the permeability of Ca^{2+} channels in the plasma membrane, the sarcoplasmic reticulum, or both, through a variety of mechanisms. Thus, smooth muscle is subject to more external influences than is skeletal muscle, even though smooth muscle can contract on its own, whereas skeletal muscle cannot.

For now, we are going to consider only the effect of mechanical stretch on smooth muscle contractility in more detail as we look at the length–tension relationship in smooth muscle. We defer an examination of the extracellular chemical influences on smooth muscle contractility to other chapters, where we discuss regulation of the various organs that contain smooth muscle.

Smooth muscle can still develop tension, yet inherently relaxes when stretched

The relationship between the length of the muscle fibers before contraction and the tension that can be developed on a subsequent contraction is less closely linked in smooth muscle than in skeletal muscle. The range of lengths over which a smooth muscle fiber can develop near-maximal tension is much greater than for skeletal muscle. Smooth muscle can still develop considerable tension even when stretched up to 2.5 times its resting length, for two probable reasons. First, in contrast to skeletal muscle, in which the resting length is at l_o, in smooth muscle the resting (nonstretched) length is much shorter than the l_o. Therefore, smooth muscle can be stretched considerably before even reaching its optimal length. Second, the thin filaments still overlap the much longer thick filaments even in the stretched-out position, so that cross-bridge interaction and tension development can still take place. In contrast, the thick and thin filaments of skeletal muscle are completely pulled apart and no longer able to interact when the muscle is stretched only three fourths longer than its resting length.

The ability of a considerably stretched smooth-muscle fiber to still develop tension is important because the smooth muscle fibers within the wall of a hollow organ are progressively stretched as the volume of the organ's contents increases. Consider the urinary bladder as an example. Even though the muscle fibers in the urinary bladder are stretched as the bladder gradually fills with urine, they still maintain their tone and are even capable of developing further tension in response to inputs regulating bladder emptying. If considerable stretching prevented development of tension, as in skeletal muscle, a filled bladder could not contract to empty.

Stress Relaxation Response When a smooth muscle is suddenly stretched, it initially increases its tension, much like the tension in a stretched rubber band. The muscle quickly adjusts to this new length, however, and inherently relaxes to the tension level prior to the stretch, probably from rearrangement of cross-bridge attachments. Smooth muscle cross bridges detach comparatively slowly. Physiologists speculate that on sudden stretching, any attached cross bridges would strain against the stretch, contributing to a passive (not actively generated) increase in tension. As these cross bridges detach, the filaments would be permitted to slide into an unstrained stretched position, restoring the tension to its original level. This inherent property of smooth muscle is called the stress relaxation response.

Advantages of the Smooth-Muscle Length–Tension Relationship

These two responses of smooth muscle to being stretched—being able to develop tension even when considerably stretched and inherently relaxing when stretched—are highly advantageous. They enable smooth muscle to exist at a variety of lengths with little change in tension. As a result, a hollow organ enclosed by smooth muscle can accommodate variable volumes of contents with little change in the pressure exerted on the contents except when the contents are to be pushed out of the organ. At that time, the tension is deliberately increased by fiber shortening. It is possible for smooth muscle fibers to contract to half their normal length, enabling hollow organs to dramatically empty their contents on increased contractile activity; thus, smooth-muscled viscera can easily accommodate large volumes but can empty to practically zero volume. This length range in which smooth muscle normally functions (anywhere from 0.5 to 2.5 times the normal length) is considerably greater than the limited length range within which skeletal muscle remains functional.

Smooth muscle contains abundant connective tissue, which resists being stretched. Unlike skeletal muscle in which the skeletal attachments restrict how far the muscle can be stretched, this connective tissue prevents smooth muscle from being overstretched and thus places an upper limit on the volume capacity of a smooth-muscled hollow organ.

Smooth muscle is slow and economical, especially in latch and catch types

A smooth muscle contractile response proceeds at a more leisurely pace than does a skeletal muscle twitch. The rate of ATP splitting by myosin ATPase is much slower in smooth muscle, so cross-bridge activity and filament sliding occur about 10 times more slowly compared to striated muscle. A single smooth muscle contraction may last as long as 3 seconds (3,000 msec), compared to the maximum of 100 msec required for a single contractile response in skeletal muscle. Smooth muscle also relaxes more slowly because of a slower rate of Ca^{2+} removal. Slowness should not be equated with weakness, however. Smooth muscle can generate the same contractile tension per unit of cross-sectional area as skeletal muscle, but it does so more slowly and at considerably less energy expense. Because of the low rate of cross-bridge cycling, cross bridges are maintained in the attached state for a longer period of time during each cycle compared with skeletal muscle.

Some smooth muscles are specialized to carry this low-cycling attached state to an extreme known as a **latch state** in vertebrates, and a **catch state** in nonvertebrates (Figure 8-33). In these states, the cross bridges "latch onto" the thin fila-

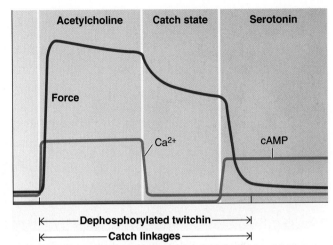

FIGURE 8-33 Model for the regulation of the catch state in the bysuss retractor muscle of the common mussel *Mytilus*. The energy-saving catch state occurs in certain specialized smooth muscles of mollusks after cessation of ACh stimulation. During catch, force is maintained for prolonged periods in the absence of elevated ICF Ca^{2+} concentrations. In this model, the chainlike protein twitchin (which resembles the giant vertebrate protein titin) is the key structure that maintains catch force during contraction due to its linkages between the thick and thin filaments. Dephosphorylation of twitchin induces catch, whereas phosphorylation of twitchin inhibits the catch state. Dephosphorylation of twitchin is induced by a Ca^{2+}-dependent phosphatase. Relaxation of contraction occurs in response to the release of serotonin from synapses of catch-terminating neurons. Serotonin increases intracellular cAMP concentrations, which activates a cAMP-dependent protein kinase (PKA). In turn, PKA phosphorylates the thick filament protein twitchin with the result that catch is now inhibited.

Source: Avrova, J. (2009). Twitchin of mollusc smooth muscles can induce "catch"-like properties in human skeletal muscle: support for the assumption that the **"catch"** state involves twitchin linkages between myofilaments. *Comp. Physiol.* B 179:945–950.

ments for minutes to hours, enabling the smooth muscle to maintain tension with very low ATP consumption (because each cross-bridge cycle uses up one molecule of ATP). The mechanisms for this are unclear, and may be different in different types of animals, but some evidence is emerging. Let's look at a bivalve mollusk's **adductor muscle** as an example. Bivalve shells have elastic elements that spring the shells open in the absence of muscle activity. The adductor muscle contains smooth muscle that must hold the two shells closed for long periods (such as when the tide is out, or a predator is attacking). The adductor also contains some white skeletal-type muscle to close the shell rapidly, as during an initial disturbance, or in the jet-propelled swimming of a scallop (which rapidly opens and closes its shell to draw in and squirt out water). Acetylcholine (ACh) stimulates smooth muscle contraction and the catch state, whereas serotonin triggers relaxation. The rate at which the myosin heads detach from the actin is greatly slowed by the presence of **twitchin**, a titin-like protein associated with the myosin. Phosphorylation of twitchin by protein kinase A, perhaps because of the actions of serotonin, inhibits the catch state by allowing the myosin heads to detach at a faster (noncatch) rate.

Because of its slowness and the less ordered arrangement of its filaments, smooth muscle has often been mistakenly viewed as a poorly developed version of skeletal

Courtesy of NOAA, photo by Dann Blackwood, USGS

Dann Blackwood

A scallop has an adductor muscle to close its shell. In a *catch* state, the shell can be closed for long periods of time. The muscle can also be contracted and relaxed rapidly for jet-propelled swimming.

muscle. Actually, smooth muscle is just as highly specialized for the demands placed on it—that is, being able to economically maintain tension for prolonged periods without fatigue and being able to accommodate considerable variations in the volume of contents it encloses with little change in tension. It is an extremely adaptive, efficient tissue.

Nutrient and O_2 delivery are generally adequate to support the smooth-muscle contractile process. Smooth muscle can use a wide variety of nutrient molecules for ATP production. There are no energy storage pools comparable to creatine phosphate in smooth muscle; they are not necessary. Oxygen delivery is usually adequate to keep pace with the low rate of oxidative phosphorylation needed to provide ATP for the energy-efficient smooth muscle. If necessary, anaerobic glycolysis can sustain adequate ATP production if O_2 supplies are diminished.

Cardiac muscle is striated but also has some smooth-muscle characteristics

Cardiac muscle, found only in the heart, shares structural and functional characteristics with both skeletal and single-unit smooth muscle. Like skeletal muscle, cardiac muscle is striated, with its thick and thin filaments highly organized into a regular banding pattern. Cardiac thin filaments contain troponin and tropomyosin, which constitute the site of Ca^{2+} action in turning on cross-bridge activity, as in skeletal muscle. Also similar to skeletal muscle, cardiac muscle has a clear length–tension relationship. Like the oxidative skeletal muscle fibers, cardiac muscle cells have an abundance of mitochondria and myoglobin. They also have T tubules and a moderately well-developed sarcoplasmic reticulum.

As in smooth muscle, Ca^{2+} enters the cytosol from both the ECF and the sarcoplasmic reticulum during cardiac excitation. Ca^{2+} entry from the ECF through voltage-gated dihydropyridine receptors, alias Ca^{2+} channels, in the T tubule membrane triggers the release of Ca^{2+} intracellularly from the sarcoplasmic reticulum. Like single-unit smooth muscle, the heart displays pacemaker (but not slow-wave) activity, initiating its own action potentials without any external influence. Cardiac cells are interconnected by gap junctions that enhance the spread of action potentials throughout the heart,

just as in single-unit smooth muscle. Also similarly, the heart is innervated by the autonomic nervous system, which, along with certain hormones and local factors, can modify the rate and strength of contraction. In contrast to skeletal muscle, the heart's output is graded by controlling contraction frequency and modulating mechanical output of each cell, not the number of activated cells.

Unique to cardiac muscle, the cardiac fibers are joined together in a branching network, and its action potentials have a much longer duration before repolarizing. Further details and the importance of cardiac muscle's features are addressed in the next chapter.

check your understanding 8.8

In a recent study, skinned human skeletal muscle fibers were exposed to a solution containing twitchin. What changes in stretch resistance (stiffness) might be observed? What would be the effect of adding protein kinase A to the medium?

Explain how more than one smooth muscle cell can be influenced by an action potential generated in an autonomic postganglionic nerve terminal.

making connections

How Muscle Physiology Contributes to the Body as a Whole

The skeletal muscles make up the muscular system itself. Cardiac and smooth muscle are part of the organs that comprise other body systems. Cardiac muscle is found only in the heart, which is part of the circulatory system. Smooth muscle is found in the walls of vertebrate hollow organs and tubes, including the blood vessels in the circulatory system, airways in the respiratory system, bladder in the urinary system, stomach and intestines in the digestive system, and uterus and ductus deferens (the duct that provides a route of exit for sperm from the testes) in the reproductive system.

Muscles contribute to both desirable changes as well as homeostasis. Contraction of skeletal muscles moves the body parts in relation to each other and moves the whole body in relation to the external environment. Thus, these muscles permit movement through and manipulation of the external environment. Some of these movements are aimed at maintaining homeostasis, such as moving the body toward a food source or away from predators, the chewing and swallowing of food for further breakdown in the digestive system into usable energy-producing nutrient molecules (the mouth and throat muscles are all skeletal muscles), and breathing to obtain O_2 and eliminate CO_2 (the gill and respiratory muscles are all skeletal muscles). In some species the generation of heat by contracting skeletal muscles also helps maintain body temperature.

All the other systems of the body, except the immune (defense) system, depend on their nonskeletal muscle components to enable them to accomplish their homeostatic functions. For example, contraction of cardiac muscle in the vertebrate heart pushes life-sustaining blood forward into the blood vessels, and contraction of smooth muscle in the stomach and intestines pushes the ingested food through the digestive tract at a rate appropriate for the digestive juices secreted along the route to break down the food into usable units.

Chapter Summary

- Whole muscles are groups of muscle fibers (cells) bundled together by connective tissue, often attached to skeletal elements in antagonistic pairs. Skeletal muscle consists of muscle fibers lying parallel to each other, while smooth muscle cells are small and unstriated.

- Each fiber is striated by a highly organized internal arrangement and consists of thick filaments, which are mostly myosin. Actin, along with tropomyosin and troponin, forms the thin filaments. Binding of calcium to troponins causes tropomyosin to move so that myosin "heads" can form cross bridges with actin.

- During contraction, cycles of cross-bridge binding and bending pull the thin filaments closer together between the stationary thick filaments, causing shortening of the sarcomeres, while complete shortening is accomplished by repeated cycles of cross-bridge binding and bending. ATP triggers the release of myosin and energizes it.

- Excitation of muscle begins with neurotransmitter at a synapse. The consequent depolarization is carried by T-tubules, which penetrate deep into the muscle fiber. Calcium is the link between excitation and contraction, and the removal of calcium is the key to muscle relaxation.

- The number of fibers contracting within a vertebrate muscle depends on the extent of motor unit recruitment. The frequency of stimulation can influence the tension developed by each vertebrate skeletal muscle fiber. Twitch summation results from a sustained elevation in cytosolic calcium.

- Arthropod muscle tension is controlled by gradation of contraction within a motor unit, with the strength of contraction dependent on the summation of postsynaptic potentials. Asynchronous muscle contractions are characterized by a nearly constant myoplasmic Ca^{2+} concentration.

- There is an optimal muscle length at which maximal tension can be developed on a subsequent contraction. The two primary types of contraction are isotonic and isometric.

- Muscle tension is transmitted to skeletal elements as the contractile component tightens the series-elastic component.

- There are three types of skeletal muscle fibers, which differ in ATP hydrolysis and synthesis and fatigue rate: fast glycolytic (for short but powerful bursts), slow oxidative (for long, endurance actions), and the fast oxidative (for rapid, sustained actions).

- Muscle spindles and the Golgi tendon organs provide afferent information essential for controlling skeletal muscle activity.

- Smooth muscle cells are turned on by Ca^{2+}-dependent phosphorylation of myosin. Phasic smooth muscle contracts in bursts of activity; tonic smooth muscle maintains an ongoing level of contraction.

- Smooth muscle is slow and economical, especially in latch and catch types. Its activity can be modified by the autonomic nervous system.

Review, Synthesize, and Analyze

1. Until the 1930s, it was hypothesized that an increase in lactate neutralized the negative charges inside the muscle fiber, which then resulted in shortening. Design a study to show that this was indeed not the case.

2. What advantages are there of setting the resting length of smooth muscle lower than l_o?

3. How do the structures associated with the series-elastic component of muscle contribute to the production of a tetanic force?

4. In most fish, the bulk of the trunk muscles are fast glycolytic, which are not used for steady cruising. Explain.

5. A rattlesnake spies you walking along the trail, which elicits contraction of the shaker muscle. List three roles for ATP in the shaker muscle.

6. Why can neither red nor white *swimming* muscles produce sound?

7. In your view, what is the best technology to increase the muscle mass of production farm animals? Explain.

8. An emerging concept in muscle biology is that signals dependent on muscle activity, specifically the muscle load, may arise in the sarcomere and be transmitted to the nucleus to affect gene expression. Explain the putative role of titin in this process.

9. How does the entry of Ca^{2+} into the inner mitochondrial matrix affect the mitochondrial membrane potential? How might this affect the energy available for the operation of the ATP synthase?

Suggested Readings

Alexander, R. M., & G. Goldspink, eds. (1977). *Mechanics and Energetics of Animal Locomotion*. London: Chapman and Hill. A classic analysis of locomotion.

Alexander, R. M. (1982). *Locomotion of Animals*. Glasgow: Blackie. Another classic.

Armstrong, C. F. (2007). ER-Mitochondria communication. How Privileged? *Physiology* 22:261–268. Describes the role of mitochondria in calcium regulation.

Baldwin, J., J-P. Jardel, T. Montague, & R. Tomkin. (1984). Energy metabolism in penguin swimming muscles. *Molecular Physiology* 6:33–42.

Butler, T. M., S. R. Narayan, S. U. Mooers, D. J. Hartshorne, & M. J. Siegman. (2001). The myosin cross-bridge cycle and its control by twitchin phosphorylation in catch muscle. *Biophysical Journal* 80:415–26.

Demirel, H. A., S. K. Powers, H. Naito, M. Hughes, & J. S. Coombes. (1999). Exercise-induced alterations in skeletal muscle myosin heavy chain phenotype: Dose–response relationship. *Journal of Applied Physiology* 86:1002–1008. How exercise alters muscles.

Granzier, H. L., & S. Labeit. (2006). The giant muscle protein titin is an adjustable molecular spring. *Exercise and Sport Science Review* 34:50–53

Huxley, H. E. (1969). The mechanism of muscular contraction. *Science* 164:1356–1365. A classic description of the sliding filament theory.

Nahirney, P. C., J. G. Forbes, H. D. Morris, S. C. Chock, & K. Wang. (2006). What the buzz was all about: Superfast song muscles rattle the tymbals of male periodicals cicadas. *FASEB Journal* 20:2017–2026.

Naya, F. J., B. Mercer, J. M. Shelton, J. A. Richardson, R. S. Williams, & E. N. Olson. (2000). Stimulation of slow skeletal muscle fiber gene expression by calcineurin *in vivo*. *Journal of Biological Chemistry* 275:4545–4548.

Rome, L. C., & S. Lindstedt. (1998). The quest for speed: Muscles built for high-frequency contractions. *News in Physiological Science* 13:261–268. Describes the toadfish song.

Shanshan Lv, Daniel M. Dudek, Yi Cao, M. M. Balamurali, John Gosline, Hongbin Li. (2010). Designed biomaterials to mimic the mechanical properties of muscles. *Nature* 465:69. Describes practical applications of titin's properties.

Suarez, R. K. (2000). Energy metabolism during insect flight: Biochemical design and physiological performance. *Physiological and Biochemical Zoology* 73:765–770.

Syme, D. A., & R. K. Josephson. (2002). How to build fast muscles: Synchronous and asynchronous designs. *Integrative and Comparative Biology* 42:762–770.

 CourseMate

Circulation serves as an animal's internal transport system. Moving fluid transports and distributes O$_2$, nutrients, wastes, hormones, and heat. The fluid is usually moved under pressure created by a heart or other pump. As an extreme example, consider a giraffe. Its heart must generate high pressures in order to overcome gravity to provide adequate blood flow to its brain. It also has tight-fitting skin on its legs that helps prevent blood pooling there, as occurs in humans who sit or stand immobile for extended periods.

Oleg Znamenskiy/Shutterstock.com

Circulatory Systems

9.1 Evolution of Circulation

Most multicellular organisms with specialized cells need some form of internal transport to move important molecules (and often cells) from one tissue to another in a reasonable amount of time. Transported molecules in animals include gases, nutrients, wastes, and hormones. Heat may also be usefully transported, especially in *endothermic* animals ("warm-blooded" ones such as birds and mammals that regulate body temperature via internal heat production), but also in some *ectothermic* animals ("cold-blooded" ones whose body temperatures depend on external sources). For example, many reptiles (ectotherms) speed up the heating process when they move into the sun, simultaneously increasing their heart rate and rapidly shunting the heat from the skin to the body core (see Chapter 15 for details on thermoregulation). In this situation, the increase in heart rate clearly costs energy. Why doesn't the animal just let heat conduct on its own throughout the body? Similarly, why not just let oxygen diffuse in through the skin rather than expend energy pumping it around the body?

Circulatory systems evolved to overcome the limits of diffusion

It is important to realize that diffusion of molecules, and to a lesser extent conduction of heat, are slow processes. For example, it takes a glucose molecule in calm water only about 0.06 microseconds (µsec) to diffuse across a membrane (about 10 nanometers, or 10^{-8} meters thick), but it takes about 20 *years* to diffuse 1 meter (3.3 feet)! (Recall Fick's law, p. 76, and the physics of diffusion in Chapter 3.) For single-celled organisms and very small animals such as rotifers, movement of molecules occurs primarily by diffusion, since distances are quite small and metabolic demands are not high. Similarly, flatworms (though they may be quite large in body length) are—as their name implies—thin enough to rely on diffusion, aided by an internal branching digestive tract that brings digested materials close to all cells.

However, as animals evolved thicker and larger bodies and higher metabolic rates, the need for internal circulatory systems became paramount because of diffusion limits. Briefly, circulatory systems overcome the slowness of diffusion by the much faster process of **bulk transport**: the movement of the medium that contains the molecules (and cells) of interest. We begin our examination of these systems with some general features and principles.

Circulatory systems have up to three distinct components: fluid, pump, and vessels

From a functional standpoint, a circulatory system can have the following components:

1. The *fluid* itself, which carries the transported molecules and cells; typically called *blood* or *hemolymph*. Terms associated with circulatory fluid often contain the root *hemo* or *emia* (from the Greek for blood).

FIGURE 9-1 Diagram of the circulatory systems of a sponge, Cnidarian, and nematode. (a) Circulation of external medium in a gastrovascular cavity of a sponge; (b) Circulation of external medium in a gastrovascular cavity of a cnidarian; (c) Circulation of fluid in the pseudocoelom of a nematode.

Source: P. Withers. (1992). *Comparative Animal Physiology*. Belmont, CA: Brooks/Cole, p. 667.

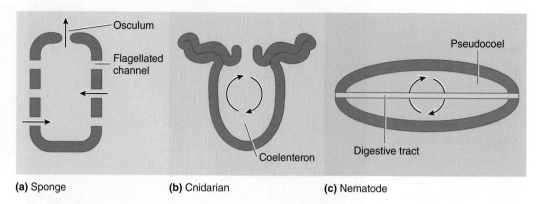

(a) Sponge — Osculum, Flagellated channel

(b) Cnidarian — Coelenteron

(c) Nematode — Pseudocoel, Digestive tract

2. A *pump* to move the fluid; dedicated pumps are usually called *hearts*, and terms associated with them often contain the root *cardio* (Greek for heart).
3. *Vessels* to carry the fluid between the pump and body tissues; these are called *vascular* (Latin, relating to small vessels) components.

A system with all three components should be quite familiar to you. For example, a city water supply typically has pumps (at reservoirs, wells, and water towers) that move a vital fluid, water, through pipes to your house and other buildings. However, not all three components are necessary for some types of circulation. In particular, completely enclosed vessels are absent in many animal groups. Also, biological pumps are not always distinct dedicated organs (hearts). The simplest example is that of the water canal system of a sponge (Figure 9-1a). A sponge has no true internal circulatory fluid; rather, environmental water is pumped through hundreds of pores and internal chambers by thousands of collar cells (*choanocytes*) with beating flagella. Thus, water with food items and oxygen is brought close to most cells, whereas water with wastes leaves a large exit pore. In large Cnidarians (jellies, sea anemones, and so on; Figure 9-1b), materials move by a combination of diffusion and muscle-driven movement of ambient water in and out of the digestive cavity (*coelenteron*) that takes up most of their interiors. Pumping as such is accomplished as part of the digestive process. The tissues are generally thin, and all cells are close enough to the ambient water (outside) and the digestive cavity to survive on diffusion.

Circulatory systems can be open or closed

Most other animals have an internal extracellular fluid (ECF) that is separate from the environment. Small animals with an internal fluid-filled body cavity, such as nematodes (roundworms), may move materials through body motions that simply move the internal fluid (Figure 9-1c). But more complex animals have dedicated pumps and vessels that actively circulate the fluid. Traditionally, internal circulation of fluids is divided into two broad categories:

- *Open* systems (Figure 9-2a and b), in which the fluid—called **hemolymph**—moves via pumping through vessels that open like garden hoses into extracellular spaces among the tissues, bathing them directly for molecular exchanges with cells. The entire space filled with hemolymph is called the **hemocoel**, which may be subdivided into smaller spaces called **sinuses**. The fluid may be moved by body movements, by cilia, or by hearts; the latter may have outflowing vessels called *arteries*, and intake pores called *ostia* or intake vessels called *veins*. This design is similar to your home water system: Water arrives in pipes; emerges into the open spaces in your shower, toilet, sinks, dishwasher, and so on; then re-enters pipes through drains (although the analogy fails at this point, because whereas the water does not return to its starting point, hemolymph does).

Various versions of open systems are found in many animal groups, including most mollusks and all arthropods (an example is shown in Figure 9-2b).

- *Closed* systems (Figure 9-2c and d), in which the fluid—called **blood**—exits a heart through vessels that are continuous all the way back to the intake side of the heart. The delivery (outgoing) vessels branch and become smaller and smaller until they become tiny leaky **capillaries**, where flow slows down and exchange of materials occurs with body cells (Figure 9-2c). These vessels then merge and become larger in size and fewer in number as blood is returned to the heart. Capillaries are widely regarded as the primary structure distinguishing a closed from an open system because, while some open systems have delivery vessels that branch into quite small ones, they are not contiguous with return vessels. An analogy is the cooling system in your refrigerator, in which a compressor pumps fluid out to coiled tubes and back again. (The open–closed distinction is not clear-cut: As you'll see, blood fluid may leak out of capillaries and bathe nearby cells much as hemolymph does before returning to nearby vessels.) Closed systems are found in several animal groups, including cephalopod mollusks, annelids, and vertebrates (e.g., Figure 9-2d).

Now let's examine each of the three circulatory system components individually. Then we will examine how pumps and vessels are integrated and regulated as one system.

check your understanding 9.1

How do circulatory systems overcome the limits of diffusion?

What are the differences between open and closed circulatory systems?

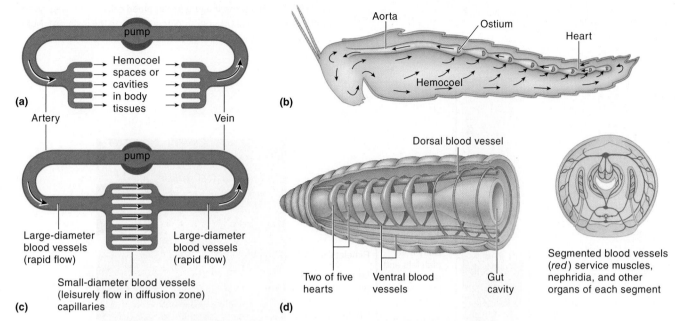

FIGURE 9-2 Flow through open and closed circulatory systems. (a, b) Open system of a grasshopper. A "heart" pumps blood through a vessel (aorta). Blood moves into tissue spaces and mingles with fluid bathing cells, then reenters the heart at opening (ostia) in the heart wall. (c, d) Closed system of an earthworm. Blood is confined within several pairs of muscular "hearts" near the head end and within blood vessels.

Source: C. Starr. (2000). *Biology: Concepts and Applications*, 4th ed. Belmont, CA: Brooks/Cole, Figure 33.3.

9.2 Circulatory Fluids and Cells

Circulatory fluids are traditionally divided into two components: a liquid called **plasma**, primarily water containing a variety of dissolved and dispersed molecules, and **cellular elements**, of which there may be several specialized kinds. In hemolymph, various cell types are called *hemocytes,* which are responsible for immune functions (Chapter 10), clotting (see below), and sometimes oxygen transport (Chapter 11). In vertebrates, these functions belong to three types of specialized cellular elements suspended in the plasma: (1) *erythrocytes* (or red blood cells, for oxygen transport; see below), (2) *leukocytes* (or white blood cells, for immunity; Chapter 10), and (3) *thrombocytes* or *platelets* (for clotting; see below).

Methodology: The Hematocrit If a sample of whole blood or hemolymph is placed in a test tube, treated to prevent clotting, and centrifuged, the heavier cellular elements move to the bottom and the lighter plasma rises to the top. This gives the **hematocrit**, or **packed cell volume** (Figure 9-3). Hematocrit values in vertebrates give a good indication of the oxygen delivery capacity of a species or individual; for example, well over 90% of the packed cells are erythrocytes in mammals. Stressed animals often have abnormal values. Vertebrates evolutionarily adapted or acclimatized to higher aerobic demands or low oxygen usually have higher values. Here are some examples:

- Human, *Homo sapiens:* 45% in males, 42% in females (average values; often higher in athletes), can increase dangerously by 20% in dengue fever
- Weddell seal, *Leptonychotes weddellii* (a deep-diving species): 46.5% in pups; 63.5% in adults

- Pekin duck, *Anas domesticus:* 45% at sea level, 56% after four weeks acclimation at 5,640 meters.
- Striped bass, *Morone saxatilis:* 39% if acclimated to 5°C, 53% if acclimated to 25°C (note that warm water cannot dissolve as much oxygen as cold water).
- Icefish of Antarctica (Channichthyidae family): <1% (see box on p. 532).
- Ark snails (Arcidae family): 6–7%.

Plasma accounts for the remaining volume. Let's first consider the properties of this component before turning to the cellular elements.

Plasma is an aqueous medium for inorganic ions, gases, and numerous organic solutes

Plasma typically consists of 90% or more water, which serves as a medium for a large number of organic and inorganic substances being carried in the fluid. In mammals, the most plentiful organic constituents are the *plasma proteins,* which compose 6 to 8% of plasma's total weight, compared to about 1% for inorganic constituents. The most abundant electrolytes (ions) in virtually all animal plasmas are Na^+ and Cl^- (see p. 103). There are smaller amounts of HCO_3^-, K^+, Ca^{2+}, and others. The most notable functions of these ECF ions are their roles in membrane potential, osmotic distribution of fluid between the ECF and cells, and buffering of pH; these functions are discussed elsewhere (Chapters 3, 4, and 13). The remaining small percentage of plasma is occupied by nutrients (for example, glucose, amino acids, lipids, and vitamins); waste products (such as creatinine, bilirubin, and urea in mammals); dissolved gases (O_2 and CO_2); and hormones. Most of these substances are merely being transported and have no functions in the plasma.

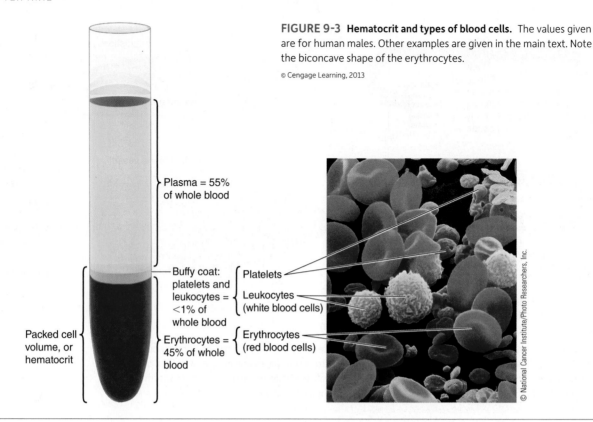

FIGURE 9-3 Hematocrit and types of blood cells. The values given are for human males. Other examples are given in the main text. Note the biconcave shape of the erythrocytes.

© Cengage Learning, 2013

Plasma = 55% of whole blood

Buffy coat: platelets and leukocytes = <1% of whole blood

Platelets

Leukocytes (white blood cells)

Erythrocytes = 45% of whole blood

Erythrocytes (red blood cells)

Packed cell volume, or hematocrit

© National Cancer Institute/Photo Researchers, Inc.

Plasma proteins carry out many of the functions of plasma

The **plasma proteins** are the one group of plasma constituents not present just for the ride. Here are the two of the most important nonspecific functions:

1. *Colloid osmotic pressure.* Unlike other plasma constituents that are dissolved in the plasma water, the plasma proteins exist in a colloidal dispersion. Furthermore, as largest in size of the plasma constituents, plasma proteins usually do not exit through the pores in capillary walls (of closed systems) to enter the interstitial fluid. Thus, plasma proteins establish an osmotic gradient between the blood (where they are present) and interstitial fluid (where they are absent). This **colloid osmotic pressure** is the primary force responsible for limiting the loss of plasma from the capillaries into the interstitial fluid, which helps maintain plasma volume (see p. 444). In open systems, plasma proteins create part of the osmotic pressure of the entire ECF but not the ICF.
2. *Buffering.* Plasma proteins are partially responsible for the plasma's capacity to buffer changes in pH (see Chapter 13, p. 641).

In addition to these nonspecific functions, each plasma protein has its own specific role(s).

Arthropods Insects, crustaceans, and arachnids also have a variety of plasma proteins in their hemolymph. There are clotting factors as well as proteins involved in exoskeleton production, because arthropods periodically molt their exoskeletons so that they can grow in size. Making new exoskeleton involves, in part, the polymerization of *phenols*

from the plasma. Three major protein classes found in arthropod plasma are as follows:

1. **Coagulogens** are key proteins in clotting (see p. 399).
2. **ProPOs** are inactive forms of **phenoloxidases (POs)** that are activated during exoskeletal synthesis (after molting, and during wound repair). POs in turn polymerize phenols. ProPOs are also involved in arthropod immune responses (p. 462).
3. **Hexamerins** and other hemolymph proteins in insects are thought to transport phenols to exoskeletal-producing epithelial cells.

Vertebrates These have three major groups of plasma proteins, which are classified according to their various physical and chemical properties:

1. **Fibrinogen,** like coagulogen, is a key factor in the blood-clotting process (see Figure 9-11, p. 397).
2. **Albumins,** the most abundant of mammalian plasma proteins, bind many substances (for example, bilirubin, bile salts, and fatty acids) for transport through the plasma and contribute most extensively to the colloid osmotic pressure by virtue of their numbers.
3. **Globulins** come in three forms—**alpha (α), beta (β),** and **gamma (γ)**—which act as:

    - *Transporters.* Specific α- and β-globulins bind and transport a number of substances in the plasma, such as thyroid hormone, cholesterol, and iron. The most common is **transferrin,** which binds and transports iron atoms.
    - *Clotting agents.* Many of the factors involved in the process of blood clotting, which will be described shortly, are α- or β-globulins.

- *Regulators.* Inactive proteins that are precursors of regulators such as hormones, and that are activated as needed by specific signals, belong to the α-globulin group (for example, the α-globulin angiotensinogen is activated to angiotensin, which plays an important role in regulating blood pressure and salt balance in the body; p. 587).
- *Immune effectors.* The γ-globulins are the immunoglobulins (antibodies), which are crucial to vertebrate defense mechanism.

These proteins are predominately synthesized by the liver. Some are produced by vascular endothelial (lining) cells. The functions of some of these proteins are elaborated on elsewhere in the text; for example, the γ-globulins in immune responses are covered in Chapter 10.

Lipoprotein complexes carry energy lipids and structural lipids for biosynthesis

Because lipids are hydrophobic, they are not very soluble in blood or hemolymph. For this reason, most lipids in circulation are converted into droplets attached to specific plasma-protein carriers called **apolipoproteins.** The resulting droplets called **lipoproteins** can disperse in circulatory fluids and then transport lipids between tissues.

Cells of most animals need such lipids for two purposes: (1) *energy* lipids, such as *triglycerides,* form large stores of chemical energy (p. 26), while (2) *structural* lipids are used for making membranes. There are two types of the latter: *phospholipids* for constructing the primary bilayer itself, and *cholesterol* for stabilizing the membrane (see Chapter 3). In addition, a few special cell types use cholesterol as a precursor for the synthesis of secretory products, such as steroid hormones and bile salts. Most cells cannot manufacture enough of these lipids and therefore must rely on the circulation to deliver supplemental lipid. There are two sources for that: *dietary* intake, with animal tissues being especially rich in cholesterol compared to plant tissues; and *biosynthesis* by other organs, for example, the liver in vertebrates.

Now let's look at key examples of lipoproteins:

Arthropods Except for Crustacea, arthropods have **lipophorins** in their hemolymph that deliver energy lipids, phospholipids, and cholesterol to all cells. These droplets have lipids with apolipoproteins called *apolipophorins.* The lipoproteins are named for their densities based on lipid-protein ratios. As shown in Figure 9-4a for an insect:

1. **High-density lipophorins (HDLp)** and the apolipophorin *apoLp-III* in the hemolymph pick up lipids, primarily energy-rich *diacylglycerides* from the gut (step 1a, DAG) and from triacylglyceride breakdown in the fat body (equivalent to a liver; step 1b, TAG and DAG). The apolipophorin helps keep the droplet dispersible in the hemolymph and can be recognized by muscle receptors.
2. The droplets are now **low-density lipophorins (LDLp)** because they contain more lipid (which is less dense than protein). LDLp delivers lipids primarily to muscles (step 2).
3. Muscles use lipids for energy storage or immediate usage as free fatty acids (step 3, FFA).

4. With many lipids removed, the droplets become HDLp again and release the apoliprotein; both recycle to start the transport over again (steps 4a, b).

Vertebrates These have four major lipoproteins also named by density properties:

- **Chylomicrons (CM),** produced by intestinal absorptive cells, which transport triglycerides, cholesterol, and phospholipids after a meal.
- **High-density lipoproteins (HDL),** which contain the most protein, some phospholipids, and least cholesterol.
- **Low-density lipoproteins (LDL),** which contain less protein, some phospholipids, and more cholesterol.
- **Very-low-density lipoproteins (VLDL),** which contain the least protein and most lipid, but the lipid they carry is triglyceride for energy.

Let's look at these in more detail; note that all the transport droplets have apolipoproteins (Figure 9-4b; see Chapter 14 for more on chylomicrons):

1. Lipids in the diet for energy (triglycerides) and structures (phospholipids, cholesterol) are packaged by the small intestine into *chylomicrons* that enter the lymph and blood (step 1).
2. Some tissues as adipose extract some of the triglycerides as free fatty acids (step 2a; FFA) to be reconverted into triglycerides for storage; then the leftovers—*chylomicron remnants*—go to the liver (step 2b).
3. The liver repackages the lipids into *VLDLs,* which are then transported into blood (step 3a); VLDLs deliver energy lipids to adipose and other body cells (step 3b).
4. The leftovers become intermediate-density lipoproteins (IDLs) and then *LDLs,* which are dominated by structural lipids (steps 4a, b).
5. LDLs deliver cholesterol and phospholipids to most body cells (step 5), including those lining the blood vessel walls. LDL is often called "bad" cholesterol because *plaques*—pathological accumulations of LDL cholesterol in blood vessel walls—are a leading cause of cardiovascular disease in humans, many domestic mammals such as pigs, and may possibly occur in some wild vertebrates such as salmon.
6. In contrast, cholesterol carried in HDL complexes is dubbed "good" cholesterol, because HDL removes excess cholesterol *from* the cells and transports it to the liver (step 6)
7. There the excess is partly eliminated as bile (step 7). Bile enters the intestinal tract, where bile salts participate in the digestive process (Chapter 14). Most of the secreted cholesterol and bile salts are later reabsorbed from the intestinal tract into the blood to be recycled to the liver, but the cholesterol molecules not reclaimed by absorption are eliminated in the feces.

Cells accomplish phospholipid and cholesterol uptake from the blood by receptor proteins in their plasma membranes specifically capable of binding the apolipoprotein of LDLs (Figure 9-4b). When an LDL particle binds to one of the membrane receptors, the cell engulfs the particle by endocytosis. Within the cell, lysosomal enzymes break down the LDLs to free the lipids, making them available to the cell for synthesis of new cellular membrane.

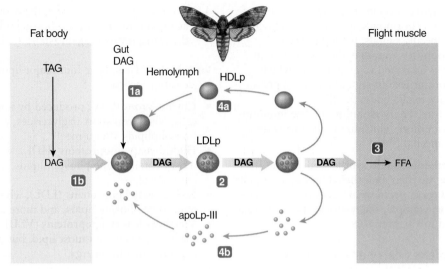

(a) Lipid transport in insect hemolymph

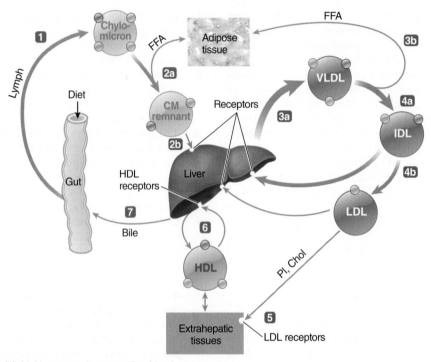

(b) Lipid transport in mammalian blood

FIGURE 9-4 Role of the circulating lipoproteins in lipid transport. (a) Insects. High-density lipophorins (*HDLp*) and low-density lipophorins (*LDLp*) complexed with an apolipoprotein *apoLp-III* and are involved in hemolymph transport of energy lipids (*DAG*—diacylglycerides) to muscles from the gut and fat body (converted there from *TAG*—triacylglycerides). See text for description of each step. (b) Mammals. Chylomicrons (CM), very-low-density lipoproteins (VLDL), low-density lipoproteins (LDL), and high-density lipoproteins (HDL) deliver lipids (such as FFA, free fatty acid; Pl, phospholipids; Chol, cholesterol) to various tissues. The small colored spheres are apolipoproteins, proteins that bind to the droplets to make them dispersible in the blood and recognizable by cell receptors. See text for description of each step.

Sources: (a) Modified from P.M. M. Weers & R. O. Ryan. (2006). Apolipophorin III: role model apolipoprotein. *Insect Biochemistry and Molecular Biology* 36:231–240. (b): Modified from R. E. Olson. (1998). Discovery of the lipoproteins, their role in fat transport and their significance as risk factors. *Journal of Nutrition* 128:439S–443S.

Respiratory pigments carry oxygen

In many animals, other crucial types of circulatory proteins function as **respiratory pigments**. These are oxygen-binding proteins critical for transporting O_2 for cellular respiration. Respiratory pigments in animals vary considerably by phylum and may exist is one or more isoforms in a particular species, as you will see in detail in Chapter 11; we give a brief summary here. The two most common ones are **hemoglobins**, proteins in vertebrates and many nonvertebrates that use an iron atom to bind an O_2 molecule; and **hemocyanins**, proteins in many mollusks and arthropods such as crustaceans, that use two copper atoms to bind one oxygen molecule. Most insects do not have a respiratory pigment, because their tracheal system (Chapter 11) carries O_2 directly to tissues. However, arachnids do have hemocyanin because some species lack tracheae (using book lungs, for example; see Chapter 11), whereas others have tracheae that do not extend into the tissues. Hemocyanins circulate in arachnid hemolymph to transfer O_2 between the lungs or tracheal termini to muscles and other organs.

The pigment proteins may be dissolved in the hemolymph or blood, which is the case for most annelids, arthropods, and some other nonvertebrates. In many animals, however, these proteins are not in the plasma itself but are contained in specialized cells. If they contain hemoglobin, which is reddish when oxygen is bound, these cells are called **erythrocytes** (*erythro*, "red"), to which we now turn our attention.

Erythrocytes serve primarily to transport oxygen

Each milliliter (mL) of vertebrate blood contains numerous erythrocytes (or *red blood cells*), for example, about 3 billion per mL in chickens, 7 billion per mL in cows and pigs, 10 billion per mL in horses, and 15 billion per mL in llamas (which live at high altitude in the Andes). The main function of erythrocytes is to transport O_2 from the lungs or gills to the tissues. In most vertebrates, these cells are oblong oval (ovoid) shapes (Figure 9-5), with amphibians and lungfish having the largest. In contrast, mammalian erythrocytes are flat, disc-shaped (biconcave) cells without nuclei, indented in the middle on both sides like a doughnut with a flattened center instead of a hole (Figure 9-3). See *Unanswered Questions: Why are erythrocytes in birds and mammals so different?* for details.

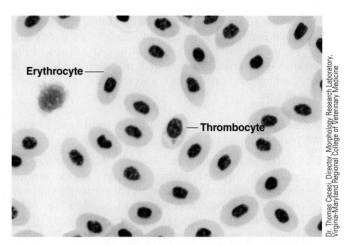

FIGURE 9-5 Anatomic characteristics of chicken erythrocytes (and thrombocytes). Compare these to the mammalian biconcave shape in Figure 9-3.

Dr. Thomas Caceci, Director, Morphology Research Laboratory, Virginia-Maryland Regional College of Veterinary Medicine

brates (this narrowness presumably evolved to support the high metabolism of mammals). The cells, each with a diameter of 6 μm, can deform amazingly without rupturing as they squeeze single file through capillaries as narrow as 3 μm in diameter.

In other vertebrates, erythrocytes are at lower densities, have a larger ovoid shape, and keep their nuclei, although in aging birds the red cells become progressively condensed (and unused). However, birds, which typically have higher oxygen requirements than mammals, have ovoid red cells twice the diameter of mammalian ones. Correspondingly, bird capillaries, though smaller than those of most vertebrates, are twice as wide as mammals'. Yet oxygen delivery obviously functions quite well in birds. Why these differences evolved is uncertain; might the faster circulation or more efficient lung of birds be the answer? See pp. 414, 515, and 527. <<

Vertebrate erythrocytes also play a role in returning carbon dioxide (CO_2) to the lungs or gills. As you will see shortly, both the cytoplasm and hemoglobin contribute to CO_2 transport.

Oxygen Transport To maximize its hemoglobin content, a mammalian erythrocyte is stuffed with several hundred million hemoglobin molecules, to the exclusion of almost everything else including nucleus, other organelles, or ribosomes. These structures are extruded during the cell's development to make room for more hemoglobin. Maturation of erythrocytes is aided by immune cells called *macrophages* (p. 465), which bind to the immature nucleated erythrocytes (erythroblasts) to form "islands" consisting of a central macrophage that extends cytoplasmic protrusions to a ring of maturing erythroblasts. The macrophages nurture red cell development and aid in the loss of their nuclei. Ultimately then, a mammalian red blood cell is mainly a plasma-membrane-enclosed sac full of hemoglobin. Ironically, these mammalian cells cannot use the O_2 they are carrying for energy production since they lack the mitochondria. Instead, they rely entirely on glycolysis for ATP formation (see p. 48).

unanswered Questions | Why are erythrocytes in birds and mammals so different?

Mammalian red cells are unique in having a biconcave shape and no nucleus; they are also the smallest of vertebrate erythrocytes and so more can be packed into a given blood volume. The mammalian red cell's small size and shape is thought to aid O_2 transport by (1) providing a larger surface area for diffusion of O_2 across the membrane than would a spherical cell of the same volume; (2) enabling O_2 to diffuse rapidly between the exterior and innermost regions of the cell due to its thinness; (3) having more room for hemoglobin due to the lack of a nucleus; and (4) allowing the cell to squeeze through mammalian capillaries, which are the narrowest among all verte-

Because of its iron content, hemoglobin appears reddish when combined with O_2 and bluish when deoxygenated, acidified, and combined with CO_2. Thus, fully oxygenated arterial blood is red in color, and venous blood, which has lost some O_2 and gained some CO_2, has a bluish cast.

Carbon Dioxide Transport Erythrocytes contribute to CO_2 transport in two ways:

1. *Bicarbonate production:* Though lacking most enzymes, the mature mammalian erythrocyte retains glycolytic enzymes and **carbonic anhydrase (CA)**. CA is crucial in CO_2 transport (see also Chapter 11, p. 541). It catalyzes a key reaction that ultimately leads to the conversion of metabolically produced CO_2 into **bicarbonate ion (HCO_3^-)**, which is the primary form in which CO_2 is transported in the blood. (This bicarbonate is a major pH buffer in the extracellular fluid—see p. 640)
2. *Hemoglobin* also contributes by binding CO_2, although it does not carry as much of this gas as it carries O_2. CO_2 binds to the protein itself (the globin), not the iron (p. 540).

Hemoglobins and erythrocytes have additional transport functions

In addition to carrying CO and O_2, hemoglobin and erythrocytes can also transport or regulate the following:

- *Bicarbonate (HCO_3^-).* In crocodiles, hemoglobin binds this ionic form of carbon dioxide, which in turn enhances oxygen release (Chapter 11).
- *H^+* from ionized carbonic acid, generated at the tissue level from CO_2. Hemoglobin binds to and thus buffers this proton so that it minimally alters the pH of the blood (Chapter 13).
- *Nitric oxide (NO).* In mammalian lungs, the vasodilator nitric oxide (see p. 93) binds to a sulfur atom within the hemoglobin molecule to form SNO. This nitric oxide is released at the tissues, where it relaxes and dilates the local arterioles. This vasodilation helps ensure that the O_2-rich blood can make its vital rounds and also helps stabilize blood pressure.
- *Hydrogen sulfide (H_2S).* Mammalian erythrocytes generate small amounts of this "rotten egg" gas that is toxic at high levels. H_2S, like NO, stimulates vessel dilation. This may be one reason garlic is thought to aid cardiovascular health in humans: it contains organic sulfur compounds that erythrocytes convert into H_2S. A very different animal, the giant annelid tubeworm *Riftia*, needs H_2S to live. This animal, found at deep-sea hydrothermal vents in the Pacific Ocean, lives exclusively off sulfide-oxidizing bacterial symbionts in its *trophosome* (a saclike structure that replaces the gut). *Riftia* has a hemoglobin that transports both H_2S (from the hot waters of the vents) and O_2 (from cold ocean water) to its symbionts, via a closed circulatory system with a heart. In the trophosome, the symbionts oxidize H_2S using O_2 to provide the energy for conversion of CO_2 into glucose, which then powers metabolism of both the bacteria and their animal host.

Hemopoietic tissues continuously replace worn-out erythrocytes

Avian and mammalian erythrocytes have a short life span, presumably because they have nonfunctioning nuclei (in birds) or no nuclei (in mammals). For example, they survive an average of only about 100 to 110 days in domestic mammals, and only 28 days in a parrot and 30 days in a chicken. During this short life span, the cell's plasma membrane, which cannot be repaired, becomes fragile as the cell repeatedly squeezes through capillaries.

One organ involved with erythrocyte regulation is the **spleen**. It has two important roles:

1. *Removal of old red cells.* Most old red-blood cells meet their final demise in the spleen, because this organ's narrow, winding, capillary network is a tight fit for these fragile cells. Macrophages engulf and destroy the dying cells.
2. *Red cell reservoir.* The spleen can store healthy erythrocytes in its pulpy interior; it also serves as a reservoir site for platelets and contains an abundance of lymphocytes, a type of white blood cell. In many mammals, the stored erythrocytes form a reserve of oxygen that can be released (by contraction of the organ) during locomotory activity. In horses, for example, splenic contraction during exercise can double the hematocrit. This also occurs in many marine mammals during a dive (p. 545).

Because erythrocytes cannot divide to replenish their own numbers, the old ruptured cells must be replaced by new cells produced in an erythrocyte factory, called **hemopoietic** tissues. The **kidney** and **spleen** produce these cells in bony fishes. In birds and mammals, it is the **bone marrow**—the soft, highly cellular tissue that fills the internal cavities of bones. Regions of bone marrow called *red bone marrow* normally generate new red blood cells, a process known as **erythropoiesis**, to keep pace with the demolition of old cells (at the amazing rate of 2 to 3 million per second in humans). In adult humans, red bone marrow is found in the sternum (breastbone), vertebrae (backbone), ribs, base of the skull, and upper ends of the long limb bones. Red marrow not only produces red blood cells but is the ultimate source for leukocytes and platelets as well. Undifferentiated **pluripotent stem cells** reside in the red marrow (or in blood of lower fishes, and the kidney and spleen in bony fishes), where they continuously divide and differentiate to give rise to each of the types of blood cells. Regulatory factors act on the *hemopoietic* ("blood-producing") marrow to govern the type and number of cells generated and discharged into the blood. Of the blood cells, the mechanism for regulating red blood cell production is the best understood. Let's consider it now for vertebrates.

Erythropoiesis in mammals and probably other vertebrates is controlled by erythropoietin from the kidneys

The number of circulating erythrocytes normally remains fairly constant in an animal living with consistent oxygen supply, indicating that erythropoiesis must be closely regulated in negative-feedback fashion. If O_2 delivery to the tissues is reduced (due to loss of erythrocytes or lower

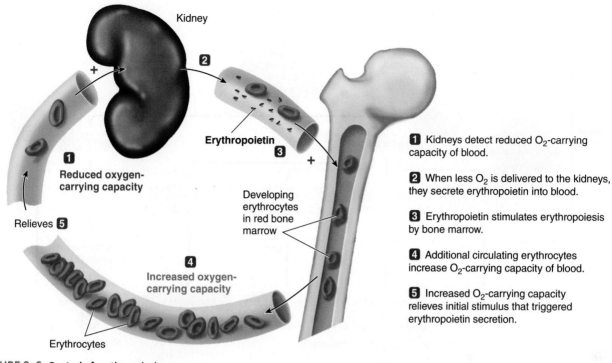

FIGURE 9-6 **Control of erythropoiesis.**

© Cengage Learning, 2013

Kidney

Erythropoietin

Reduced oxygen-carrying capacity

Relieves

Developing erythrocytes in red bone marrow

Increased oxygen-carrying capacity

Erythrocytes

1 Kidneys detect reduced O_2-carrying capacity of blood.

2 When less O_2 is delivered to the kidneys, they secrete erythropoietin into blood.

3 Erythropoietin stimulates erythropoiesis by bone marrow.

4 Additional circulating erythrocytes increase O_2-carrying capacity of blood.

5 Increased O_2-carrying capacity relieves initial stimulus that triggered erythropoietin secretion.

environmental O_2), increased erythropoiesis is indeed stimulated. However, low O_2 delivery does not cause this directly, bur rather it triggers the *kidneys* to secrete the hormone **erythropoietin (EPO)** into the blood. This hormone in turn stimulates erythropoiesis by the hemopoietic tissues (Figure 9-6). (Synthetic EPO was developed to treat humans with kidney failure and other causes of anemia, but you may be more familiar with EPO due to its illegal use by some human athletes; it is also sometimes illegally injected into racing animals such as horses.) EPO has been mostly studied in mammals but is found in other vertebrates including some bony fish; for example, in zebrafish, the gene is regulated by low-oxygen conditions just as it is in mammals. EPO acts on derivatives of undifferentiated stem cells that are already committed to becoming red blood cells, stimulating their proliferation and maturation into mature erythrocytes (Figure 9-7). This increased erythropoietic activity elevates the number of circulating red blood cells, thereby restoring O_2 delivery to the tissues to normal. Once normal O_2 delivery is achieved, EPO secretion is turned down until needed again. In this way, erythrocyte production is normally balanced against loss of these cells so that the O_2-carrying capacity in the blood remains fairly constant. In response to severe loss of erythrocytes, as in hemorrhage, or to prolonged impairment of O_2 delivery, such as livestock being taken to high altitude or a fish entering low-O_2 water, the rate of erythropoiesis is upregulated.

Leukocytes are key components of vertebrate immune systems

Leukocytes, or white blood cells, are the mobile units of vertebrate immune systems. We leave a more detailed discussion of their names and functions for Chapter 10. For now, note that all leukocytes ultimately originate from the same undifferentiated pluripotent stem cells that also give rise to erythrocytes and platelets (Figure 9-7). To direct the differentiation and proliferation of each cell type, specific hormones analogous to erythropoietin are required.

Thrombocytes and platelets function in clotting

The third type of blood cell or cell derivative is the **thrombocyte** (a living cell) or **platelet** (a cell fragment), which serves in the clotting mechanism. Thrombocytes are cells found in all vertebrates except mammals; these cells circulate in an inactive state, and when activated by an injury to nearby tissue, they begin to break up into platelet-like fragments. Mammals are different in that the precursor cells become platelets, which become the primary inactive form that circulates in the blood. Specifically, platelets are small cell fragments (about 2 to 4 μm in diameter) that are shed off the outer edges of extraordinarily large (up to 60 μm in diameter), bone-marrow-bound cells known as **megakaryocytes** (Figure 9-8), each of which typically produces about 1,000 platelets. Megakaryocytes are derived from the same stem cells that give rise to the erythrocytic and leukocytic cell lines (Figure 9-7). Platelets are essentially detached vesicles containing portions of cytoplasm wrapped in plasma membrane.

Platelets remain functional for less than two weeks, at which time they are removed from circulation by macrophages, especially those in the spleen and liver, and are replaced. The hormone **thrombopoietin,** made by hepatocytes (*hepato,* "liver"), increases the number of megakaryocytes in the bone marrow and stimulates each megakaryocyte to produce more platelets (Figure 9-7). Platelets lack nuclei, but they can produce secretory products from storage granules, and contain the contractile proteins actin and myosin (Chapters 2 and 8) to enable them to contract. These features are important in hemostasis, our next topic.

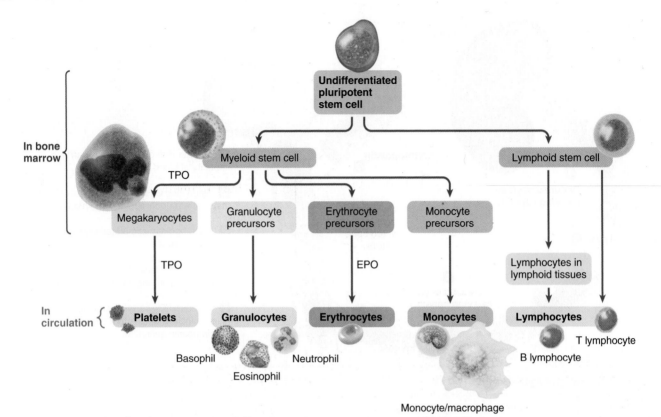

FIGURE 9-7 **Blood cell production (hemopoiesis) in a mammal.** All the blood cell types ultimately originate from the same undifferentiated pluripotent stem cells in the red bone marrow. The hormones EPO and TPO (thrombopoietin) stimulate cell differentiation and production where indicated; see main text for details.

© Cengage Learning, 2013

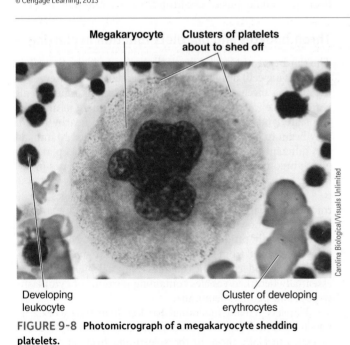

FIGURE 9-8 **Photomicrograph of a megakaryocyte shedding platelets.**

check your understanding 9.2

What is the basic composition of plasma?

Discuss the roles of HDLs and LDLs in mammals.

What are the major transport functions of vertebrate erythrocytes?

9.3 Circulatory Fluids: Hemostasis

Hemostasis is the arrest of bleeding. (Be sure not to confuse this with the term *homeostasis*.) Bleeding occurs when there is a rupture in a vessel or hemocoel wall, and the pressure inside is greater than the pressure outside. Hemostatic mechanisms in vertebrates normally are adequate to seal tears and stop loss of blood through small damaged capillaries, arterioles, and venules (pp. 429–448), small vessels that are frequently ruptured by minor traumas. The much rarer occurrence of bleeding from medium- to large-size vessels usually cannot be stopped by the hemostatic mechanisms alone. Bleeding from a severed artery is more profuse and therefore more dangerous than venous bleeding because the driving pressure is greater in the arteries than in the veins (as you will see later).

Here we focus primarily on the mammalian system, which involves three major steps: (1) *vascular spasm*, (2) *formation of a platelet plug*, and (3) *blood coagulation (clotting)*.

Vascular spasm reduces blood flow through an injured vessel

A cut or torn blood vessel in a vertebrate immediately constricts as a result of an inherent vascular response to injury, release of chemicals by platelets or thrombocytes (described shortly), and sympathetically induced vasoconstriction. This constriction slows blood flow through the defect and thus minimizes blood loss. Also, as the opposing endothelial (inner) surfaces of the vessel are pressed together by this initial **vascular spasm**, they become sticky and adhere to each other, further sealing off the damaged vessel. These physical measures are important in minimizing blood loss until the other hemostatic measures can actually plug up the defect.

Platelets aggregate to form a plug at a vessel defect by positive feedback

Platelets and thrombocytes normally do not adhere to the smooth endothelial surface of blood vessels, but when this lining is disrupted because of vessel injury, platelets in mammals—or, in other vertebrates, platelet-like fragments that begin to form from thrombocytes—can now attach to the exposed collagen, a fibrous protein present in the underlying connective tissue. Once platelets or fragments start aggregating at the site of the defect, they release several important chemicals from their storage granules (Figure 9-9). Among these chemicals is *adenosine diphosphate (ADP)*, which causes the surface of nearby circulating platelets to become sticky, so that they adhere to the first layer of aggregated platelets. These newly aggregated platelets release more ADP, which causes more platelets to pile on, and so on. Thus, a **platelet plug** is rapidly built up at the defect site in a *positive-feedback* fashion. The actin-myosin protein complex within the aggregated platelets contracts to compact and strengthen what was originally a fairly loose plug.

Given the self-perpetuating nature of platelet aggregation, why is the platelet plug limited to the site of vessel injury once it is initiated? (In other words, why doesn't the platelet plug continue to expand over the surface of the adjacent normal vessel lining?) A key reason why this does not happen is that ADP and other chemicals released by the activated platelets stimulate the release of *prostacyclin* and *nitric oxide (NO)* from the adjacent normal endothelium. Both chemicals profoundly inhibit platelet aggregation. Thus, the platelet plug is limited to the defect and does not spread to normal tissue (Figure 9-9).

The aggregated platelet plug not only physically seals the break in the vessel but also performs other important roles. (1) The chemicals released from the platelet plug include *serotonin, epinephrine,* and *thromboxane A_2*, which induce profound constriction of the affected vessel to reinforce the vascular spasm. (2) The aggregated platelets secrete a chemical that helps promote the invasion of fibroblasts ("fiber formers") from the surrounding connective tissue into the wounded area of the vessel to repair it, forming a scar at the vessel defect. (3) The platelet plug releases other chemicals that enhance blood coagulation, the next step of hemostasis.

A triggered chain reaction and positive feedback involving clotting factors in the plasma results in blood coagulation

Although the platelet-plugging mechanism alone is often sufficient to seal the myriad minute tears in capillaries that occur daily, larger holes in the vessels require the formation of a *clot* to completely stop the bleeding. **Blood coagulation**, or **clotting**, is the transformation of blood from a liquid into a solid gel. Formation of a clot on top of the platelet plug strengthens and supports the plug, reinforcing the seal over a break in a vessel. Furthermore, as blood in the vicinity of the vessel defect solidifies, it can no longer flow. Clotting is the most powerful hemostatic mechanism, and it works as follows.

Clot Formation The ultimate step in clot formation is the conversion of **fibrinogen**, a large, soluble plasma protein produced by the liver and normally always present in the plasma, into **fibrin**, an insoluble, threadlike molecule. This conversion into fibrin is catalyzed by the enzyme **thrombin** at the site of the injury. Fibrin molecules adhere to the damaged vessel surface, forming a loose, netlike meshwork that traps blood cells, including aggregating platelets. The resulting mass, or **clot**, typically appears red because of the abundance of trapped red cells, but the foundation of the clot is formed of fibrin (Figure 9-10).

The original fibrin web is rather weak because the fibrin strands are only loosely interlaced. However, chemical linkages rapidly form between adjacent strands to strengthen and stabilize the clot meshwork. This cross-linkage process is catalyzed by a clotting factor known as **factor XIII (fibrin-stabilizing factor)**, which normally is present in the plasma in inactive form.

Fibrin is one of the stretchiest proteins that scientists have ever studied. On average fibrin fibers can be passively stretched to 2.8 times their original length and still snap back to their starting length and can be stretched to 4.3 times their original length before they break. This highly elastic property accounts for the extraordinary stretchiness of blood clots.

Roles of Thrombin Thrombin, in addition to converting fibrinogen into fibrin (step 1a in Figure 9-11) also activates factor XIII to stabilize the resultant fibrin mesh (step 1b), acts in a *positive-feedback* fashion to facilitate its own formation (step 1c), and enhances platelet aggregation (step 1d), which in turn is essential to the clotting process (step 2).

Because thrombin converts the ever-present fibrinogen molecules in the plasma into a blood-stanching clot, thrombin must normally be absent from the plasma except in the vicinity of vessel damage. Otherwise, blood would always be coagulated—a situation incompatible with life. How can thrombin normally be absent from the plasma, yet be readily available to trigger fibrin formation when a vessel is injured? The solution lies in thrombin's existence in the plasma in the form of an inactive precursor called **prothrombin**. What converts prothrombin into thrombin when blood clotting is desirable? This conversion involves the clotting cascade.

The Clotting Cascade Yet another activated plasma clotting factor, **factor X**, converts prothrombin to thrombin; factor X itself is normally present in the blood in inactive form and

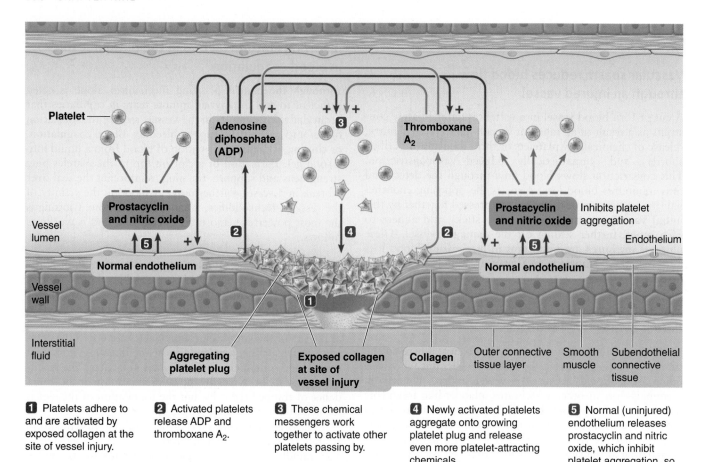

Platelet

Adenosine
diphosphate
(ADP)

3

Thromboxane
A₂

Prostacyclin
and nitric oxide

Prostacyclin
and nitric oxide

Inhibits platelet
aggregation

Vessel
lumen

5

2

4

2

5

Endothelium

Normal endothelium

Normal endothelium

Vessel
wall

1

Interstitial
fluid

**Aggregating
platelet plug**

**Exposed collagen
at site of
vessel injury**

Collagen

Outer connective
tissue layer

Smooth
muscle

Subendothelial
connective
tissue

1 Platelets adhere to and are activated by exposed collagen at the site of vessel injury.

2 Activated platelets release ADP and thromboxane A₂.

3 These chemical messengers work together to activate other platelets passing by.

4 Newly activated platelets aggregate onto growing platelet plug and release even more platelet-attracting chemicals.

5 Normal (uninjured) endothelium releases prostacyclin and nitric oxide, which inhibit platelet aggregation, so platelet plug is confined to site of injury.

FIGURE 9-9 Formation of a platelet plug. Platelets aggregate at a vessel defect through a positive-feedback mechanism involving the release of adenosine diphosphate (ADP) and thromboxane A₂ from platelets, which stick to exposed collagen at the site of the injury. Platelets are prevented from aggregating at the adjacent normal vessel lining by the release of prostacyclin and nitric oxide from the undamaged endothelial cells.

© Cengage Learning, 2013

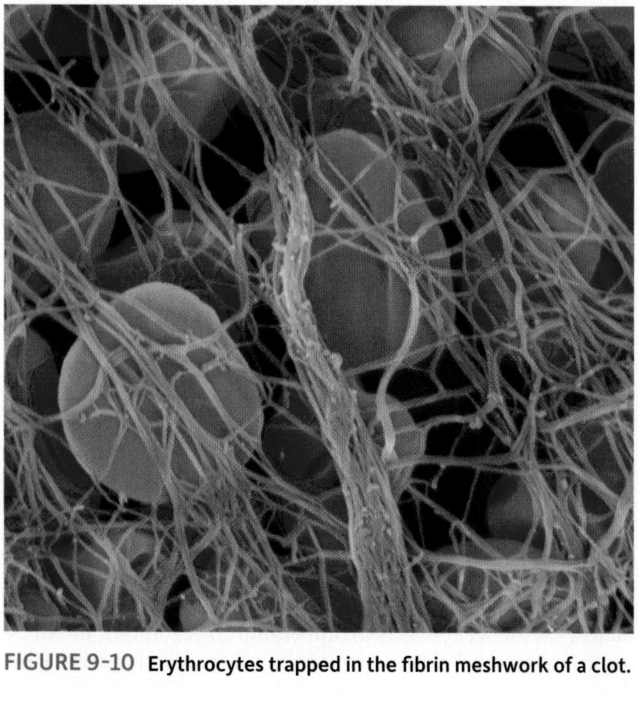

FIGURE 9-10 Erythrocytes trapped in the fibrin meshwork of a clot.

Dennis Kunkel Microscopy, Inc./Visuals Unlimited, Inc.

must be converted into its active form by still another activated factor, and so on. Altogether, 12 plasma-clotting factors participate in essential steps that lead to the final conversion of fibrinogen into a stabilized fibrin mesh (Figure 9-12). These factors are designated by Roman numerals in the order in which the factors were discovered, not the order in which they participate in the clotting process. (The term *factor VI* is no longer used, as it has now been determined to be an activated form of factor V.) Most of these clotting factors are plasma proteins synthesized by the liver. Normally, they are always present in the plasma in an inactive form, such as fibrinogen and prothrombin. In contrast to fibrinogen, which is converted into insoluble fibrin strands, prothrombin and the other precursors, when converted to their active form, act as *proteolytic* (protein-splitting) enzymes. These enzymes activate another specific factor in the clotting sequence. Once the first factor in the sequence is activated, it in turn activates the next factor, and so on, in a series of sequential reactions known as the **clotting cascade**, until thrombin catalyzes the final conversion of fibrinogen into fibrin. Several of these steps require the presence of plasma Ca²⁺ and *platelet factor 3 (PF3)*, a phospholipid secreted by the aggregated platelet plug. Thus, platelets also contribute to clot formation.

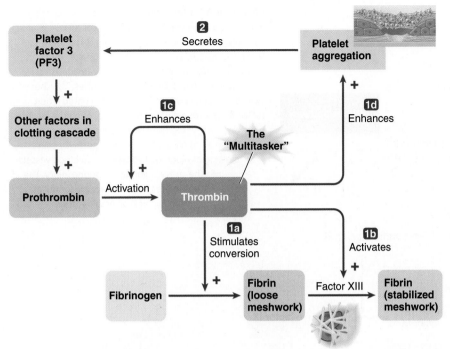

2 Thrombin, a component of the clotting cascade, plays multiple roles in hemostasis:

1a stimulates conversion of fibrinogen to fibrin

1b activates factor stabilizing fibrin meshwork of clot

1c enhances activation of more prothrombin into thrombin through positive feedback

1d enhances platelet aggregation

2 Through positive feedback, aggregated platelets secrete PF3, which stimulates clotting cascade that results in thrombin activation.

FIGURE 9-11 **Roles of thrombin in hemostasis.**

© Cengage Learning, 2013

The clotting cascade may be triggered by the intrinsic pathway or the extrinsic pathway

Two distinct clotting cascades called the *intrinsic* and *extrinsic* mechanisms usually operate simultaneously. When tissue injury involves rupture of vessels, the intrinsic mechanism stops blood in the injured vessel, whereas the extrinsic mechanism clots the blood that escaped into the tissue before the vessel was sealed off.

- The **intrinsic pathway** precipitates clotting within damaged vessels as well as clotting of blood samples in test tubes. This pathway, which involves seven separate steps (shown in blue in Figure 9-12), is set off when **factor XII** is activated by coming into contact with a negatively charged surface, especially exposed collagen in an injured vessel (or a foreign surface such as a glass test tube). Remember that exposed collagen also initiates platelet aggregation, so formation of a platelet plug and the chain reaction leading to clotting are simultaneously set in motion. Positive feedback further contributes: The aggregated platelets secrete PF3, which is essential for the clotting cascade that in turn enhances further platelet aggregation (Figures 9-11 and 9-13).
- The **extrinsic pathway** takes a shortcut and requires only four steps (shown in gray in Figure 9-12). This pathway, which requires contact with tissue factors external to the blood, initiates clotting of blood that has escaped into the tissues. When a tissue is traumatized, it releases a protein complex known as **tissue thromboplastin**. Tissue thromboplastin directly activates factor X, thereby bypassing all preceding steps of the intrinsic pathway. From this point on, the two pathways are identical. Indeed, it is not clear whether the intrinsic pathway with factor XII is essential. Non-avian reptiles have it, but birds have lost it. Some marine mammals (dolphins, whales) have lost it as well.

unanswered Questions | Why is mammalian clotting so complex?

The complex multistep clotting process in mammals clearly evolved from simpler beginnings, as revealed by genomic analysis. All vertebrates use the fibrinogen–fibrin mechanism, but fishes lack many of the mammalian clotting factors and steps. Some of these steps appeared later in the first land vertebrates, but only mammals have the full system described above. This addition of clotting steps in vertebrate evolution occurred in part by the process of **gene duplication** (in which an error in meiosis produces a second copy of a gene, which may mutate to yield new functions), as revealed by closely related gene sequences among many of the clotting factors.

Clotting clearly works in the simpler systems of nonmammals, so why did the complexity in mammals evolve? For one, *cascade amplification* is accomplished during many of the steps. One molecule of an activated factor can activate perhaps a hundred molecules of the next factor in the sequence, each of which can activate many more molecules of the next factor, and so on. This increases the speed of clotting, which may be crucial in the higher-pressure circulation that birds and mammals have compared to other vertebrates. (A similar amplification occurs in second-messenger systems, p. 99.) Also, Russell Doolittle (the biochemist who has documented the stepwise evolution of vertebrate clotting) speculates that mammalian complexity, in addition to boosting amplification, allows for more versatile regulation of clotting over a wider range of environments including a larger range of blood pressures, higher metabolic rates, and additional organs such as placentas. <<

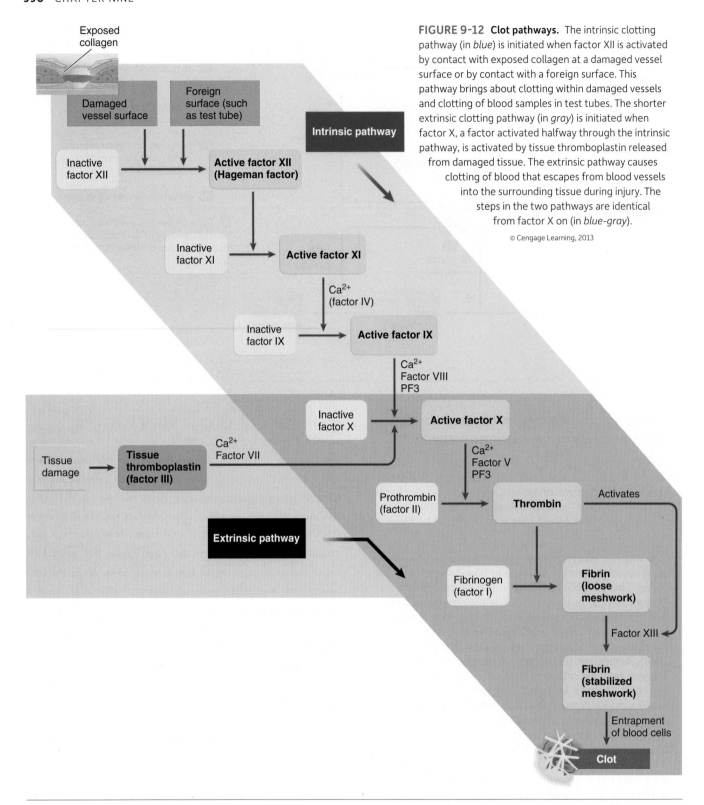

FIGURE 9-12 **Clot pathways.** The intrinsic clotting pathway (in *blue*) is initiated when factor XII is activated by contact with exposed collagen at a damaged vessel surface or by contact with a foreign surface. This pathway brings about clotting within damaged vessels and clotting of blood samples in test tubes. The shorter extrinsic clotting pathway (in *gray*) is initiated when factor X, a factor activated halfway through the intrinsic pathway, is activated by tissue thromboplastin released from damaged tissue. The extrinsic pathway causes clotting of blood that escapes from blood vessels into the surrounding tissue during injury. The steps in the two pathways are identical from factor X on (in *blue-gray*).

© Cengage Learning, 2013

Fibrinolytic plasmin dissolves clots and prevents inappropriate clot formation

A clot is not meant to be a permanent solution to vessel injury. It is a transient device to stop bleeding until the vessel can be repaired. Moreover, because positive feedback is involved, clotting could get out of control and spread beyond the wound. Thus, there are crucial anticlotting regulatory processes.

Clot Dissolution Simultaneous with the healing process, the clot, which is no longer needed to prevent hemorrhage, is slowly dissolved by a fibrinolytic (fibrin-splitting) enzyme called **plasmin**. If clots were not removed after they per-

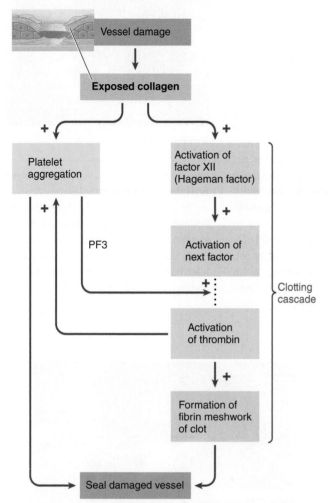

FIGURE 9-13 Concurrent platelet aggregation and clot formation.
Exposed collagen at the site of vessel damage simultaneously initiates platelet aggregation and the clotting cascade. These two hemostatic mechanisms positively reinforce each other as they seal the damaged vessel.

© Cengage Learning, 2013

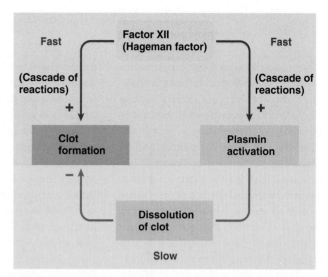

FIGURE 9-14 Role of factor XII in clot formation and dissolution.
Activation of factor XII (Hageman factor) simultaneously initiates a fast cascade of reactions that result in clot formation and a fast cascade of reactions that result in plasmin activation. Plasmin, which is trapped in the clot, subsequently slowly dissolves the clot. This action removes the clot when it is no longer needed after the vessel has been repaired.

© Cengage Learning, 2013

formed their hemostatic function, the vessels, especially the small ones that endure tiny ruptures on a regular basis, would eventually become obstructed by clots.

Plasmin, like the clotting factors, is a plasma protein produced by the liver and present in the blood in an inactive precursor form, **plasminogen**. Plasmin is activated in a fast cascade of reactions involving many factors, among them factor XII, which also triggers the chain reaction leading to clot formation (Figure 9-14). When a clot is rapidly being formed, activated plasmin becomes trapped in the clot and later dissolves it by slowly breaking down the fibrin meshwork. Phagocytic white blood cells gradually remove the products of clot dissolution.

Preventing Inappropriate Clot Formation In addition to removing clots that are no longer needed, plasmin functions continually to prevent clots from forming inappropriately. Throughout the vasculature, small amounts of fibrinogen are constantly being converted into fibrin, triggered by unknown mechanisms. Clots do not develop, however, because the fibrin is quickly disposed of by plasmin activated by **tis-**

sue plasminogen activator (tPA) from the tissues, especially the lungs. Normally, the low level of fibrin formation is counterbalanced by a low level of fibrinolytic activity, so inappropriate clotting does not occur. Only when a vessel is damaged do additional factors precipitate the explosive chain reaction that leads to more extensive fibrin formation and results in local clotting at the site of injury.

Inhibition of clotting is also a predatory adaptation in blood-sucking parasites. See *A Closer Look at Adaptation: Vampires and Medicine* (p. 400).

Hemocytes and hemolymph proteins provide hemostasis in arthropods

Hemostasis mechanisms also exist in animals with hemolymph systems, but these processes are not nearly as well understood. In arthropods such as insects and horseshoe crabs, wounds in the exoskeleton trigger coagulation of the hemolymph in part by the conversion of the hemolymph protein *coagulogen* into *coagulin* in a mechanism similar to fibrinogen–fibrin conversion. Hemolymphs contain a variety of cells collectively called **hemocytes** (about 30 to 50 million per milliliter). Some of these (called *hyaline hemocytes*), when triggered by a wound, release enzymes for coagulin formation and protein strands, which interact with a variety of dissolved hemolymph proteins to form insoluble clots at the wound site. Molecular analysis reveals that one of these proteins is related to mammalian clotting factor XIII (the cross-linking enzyme), but the other proteins are different. Remarkably, recent analysis of barnacle glue by Daniel Rittschof and colleagues has found that one of its key proteins is also very similar to the same cross-linking factor XIII. Barnacles apparently use a clotlike cross-linking of proteins to adhere the animals tightly to rocks, whale skin, and the hulls of ships.

The arthropod clot system can also be induced by bacterial invasion. Because of this, a horseshoe crab's hemolymph extract is required by U.S. law to be used by the pharmaceutical and medical device industries to test their products for bacterial contamination. Exposure of a contaminated drug, for example, to horseshoe crab extract will rapidly cause the extract to gel into a clot, making a testing process that is much faster and more reliable than other methods.

check your understanding 9.3

Describe the key events of clotting (without naming every step) and the role of positive feedback.

What mechanisms keep platelet plugs and clots from spreading inappropriately?

A grasshopper has a heart with an open circulatory system, but its main heart stops pumping when the animal is moving. This is because movements of the body wall and limbs are sufficient to pump the blood during locomotion.

9.4 Circulatory Pumps: Evolution

Circulating body fluids require some form of pump. Here we will look at different types.

Pumping mechanisms include flagella, extrinsic skeletal muscles, peristaltic muscular pumps, and hearts (chamber muscle pumps)

At least four different types of pumping mechanisms have evolved, as follows:

1. **Flagella:** Internal fluids may be moved slowly by beating flagella on epithelial cells. We have seen how flagella do this in sponges, where the fluid is environmental water (Figure 9-1a). In echinoderms such as sea urchins, which have an internal fluid in their main body cavity, the *coelom,* flagella of epithelial cells create slow currents.

2. **Extrinsic muscle or skeletal pumps:** Fluids may be moved by motions of muscles or skeletal elements that are not part of the circulatory system itself. Most often, extrinsic pumping occurs only during locomotion. For example, a seastar's body-wall muscles function mainly to move the flexible arms, but they also cause the coelomic fluid to move. Various body-wall and exoskeletal muscle movements in arthropods similarly enhance circulation of hemolymph. In fact, in grasshoppers the primary heart only beats when the animal is inactive! Apparently, it is not needed when extrinsic pumping is working. Active skeletal muscles of vertebrates can also push circulatory fluids in vessels adjacent to or within them (Figure 9-15a). This is an important mechanism in tails of hagfish and legs of tall animals such as humans, for example. In both, flow is aided by active skeletal muscles squeezing blood through one-way valves in

veins and by the action of the respiratory muscles in breathing. We discuss both processes later (p. 448).

A final example of extrinsic pumping involves skeletal and connective elements. Some mammals with hooves (*perissodactyls* such as rhinoceros and horse) have a cartilaginous plate, the "frog" or **cuneus ungulae**, at the base of the hoof that is flexed upward with each footstep (Figure 9-15b). This flexing pushes on an overlying elastic cushioning tissue that is compressed and moves outward, expanding nearby cartilage and the hoof walls. It also compresses nearby veins, sending blood into the leg above. When the foot is lifted, elastic rebound of these tissues draws blood into the local veins. This pumping action is particularly important in horses, with their long, thin legs, because they do not have enough skeletal muscle mass below the knee to act as extrinsic muscle pumps. Gravity provides the direct energy for flexing the plate, although this gravitational energy originated from upper limb muscles, which lift the limb off the ground.

3. **Peristaltic (tubular) muscle pumps:** Peristaltic pumping occurs when muscles in the walls of vessels contract in a moving wave that pushes fluid in front of it. By repeated cycles of this action, fluid is moved in one direction. These pumps are called "hearts" if they occur in specialized sections of vessels. This is the pumping method in many annelids such as earthworms, where regions of dorsal blood vessels are specialized for this purpose (Figure 9-15c). Many arthropods also have peristaltic hearts.

4. **Chamber muscle pumps:** These are the more familiar form of hearts, consisting of a chamber or chambers with muscles to squeeze the fluid within. Chamber hearts are the primary pumps in all vertebrates and many arthropods and mollusks. Unlike directional peristaltic pumps, chamber pumps usually need one-way valves to create flow in one direction to prevent backflow when the pump relaxes (Figure 9-15d). In some animals such as arthropods, the heart has a single chamber. But many chamber hearts such as those of most mollusks and all vertebrates have (at a minimum) two types of chambers: an **atrium**, which collects returning fluid, and a **ventricle** that provides the primary force for outgoing fluid. (As you will see, there are three or more chambers in more complex hearts.)

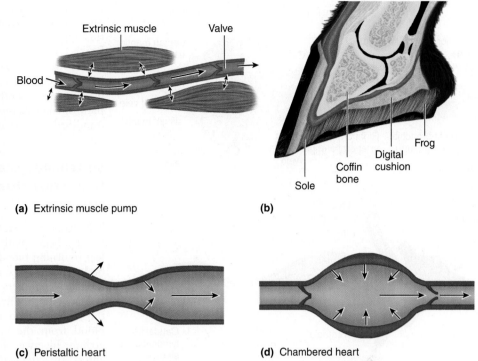

FIGURE 9-15 Types of pumps in animals. (a) External muscle pump expands and compresses a blood vessel (double-headed arrows) and forces blood to flow along the vessel (right-pointing arrows); valves maintain a unidirectional flow, e.g., in nematodes, scaphopod mollusks, and some leeches with no heart to propel blood flow, and the skeletal muscle pump of vertebrates. (b) Structure of the horse foot. The cartilaginous "frog" is pushed by the weight of the horse to aid pumping of the blood (see main text). (c) Peristaltic tubular heart forces blood along a vessel (which may have valves to ensure one-way flow) by peristaltic contractions, e.g., insect heart. (d) A chambered heart propels blood by a coordinated contraction of the muscular wall; valves ensure a one-way flow of blood, e.g., most mollusk and vertebrate hearts.

Source: P. Withers. (1992). *Comparative Animal Physiology.* Belmont, CA: Brooks/Cole, a, c, d, Figure 14-2, p. 668; b, Figure 14-8, p. 676).

Many animals have primary hearts aided by auxiliary pumps

Peristaltic or chamber hearts are found in most animal groups more advanced than flatworms (although there are major exceptions, such as nematodes and echinoderms, which have no hearts). Interestingly, most of these animals have both a primary (often called **systemic**) heart that drives the initial, usually oxygenated fluid to most organs, and one or more **auxiliary** pumps that aid flow returning to the primary heart or flow going to critical organs. We have already noted how extrinsic pumping by body walls, skeletal muscles, and skeletal elements can aid flow during locomotion. But some animals have true nonextrinsic auxiliary hearts:

- *Cephalopods,* the only mollusks with a closed system, have a systemic heart and two auxiliary hearts that boost flow to their gills (Figure 9-16a).
- Most *vertebrates* have two or more *lymph hearts* (Figure 9-16b), part of a separate circulatory pathway, the *lymphatic system,* which we discuss later (p. 444). *Hagfish* have several auxiliary hearts; the *caudal heart* in their tails has a chamber that is squeezed by a cartilaginous rod, which in turn is pushed by extrinsic skeletal muscles.
- *Insects* often have auxiliary hearts at the base of legs, wings, and antennae (Figure 9-16c).

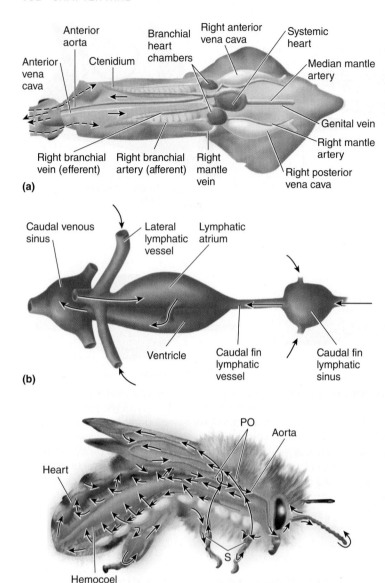

(a)

(b)

(c)

FIGURE 9-16 Examples of auxiliary hearts. (a) Diagram of the closed circulatory system of a cephalopod circulatory system. These animals have two branchial hearts that pump blood to the gills, in addition to a main systemic heart. (b) Diagram of a lymph heart in an eel. Muscles in the heart wall push lymph from the caudal lymphatic sinus into the caudal venous sinus. (c) Diagram of the open circulatory system of an insect. A main dorsal vessel (aorta) carries blood from the heart into side channels and open spaces. Flow to the wings is boosted by auxiliary pumps called *pulsatile organs* (PO). Septa (S) in the limbs channel flow through the spaces in specific directions.

Sources: (a) Modified from Sherman and Sherman. (1976). *The Invertebrates: Function and Form*, 2nd ed. New York: Macmillan. (b) From Withers, P. C. (1992). *Comparative Animal Physiology*. Fort Worth, TX: Saunders College Publishing as modified from O. F. Kampmeier. (1969). *Evolution and Comparative Morphology of the Lymphatic System*. Springfield, MA: C. C. Thomas. (c) Modified from Wigglesworth (1972).

Arthropod systemic hearts are dorsally located and have many valved openings

Crustaceans, insects, and arachnids (the major arthropod groups) have many similarities in their primary hearts. Each is located in a dorsal region of the body, and each has one or more porelike openings called **ostia** (singular, **ostium**), which allow hemolymph to reenter the heart after passing

through the body (Figure 9-2b). The ostia often have **one-way valves** to prevent backflow (you will see how this type of valve works in a mammalian example later).

However, there are some differences. Crustacean hearts, present mainly in larger species such as decapods (lobsters, crabs, and so on), are single-chamber pumps, whereas insect and arachnid hearts are tubular (modified vessels) and can have peristaltic pumping.

Vertebrate systemic hearts evolved from a two-chambered to a four-chambered structure

The primitive systemic heart of vertebrates is thought to have begun with two chambers, one atrium to collect returning blood and one ventricle to pump to the body. Fishes have this basic design, but one or two auxiliary chambers have evolved. Blood of all jawless and jawed fishes first enters a chambered extension of the atrium called the **sinus venosus**, which collects blood from the veins before entering the adjacent atrium. In jawed fishes, blood leaving the ventricles enters an extension forming a fourth chamber, the **conus arteriosus** (in cartilaginous fish) or **bulbus arteriosus** (in bony fish), which dampens the pulsatile pressure output of the ventricle (a function taken over by the *aorta* in reptiles (including birds) and mammals, in a process we describe on p. 430). Although these fish hearts have four chambers, they are considered to have two primary chambers with two auxiliary chambers (see Figure 9-17 for a cartilaginous fish).

A significant change in vertebrate hearts began when the first air-breathing fishes evolved. The crucial feature that drove this change was the evolution of a separate circuit, called the **pulmonary circulation**, from the heart to the newly evolved lungs and back. The two primary chambers began to subdivide with internal septa to support this separate circuit, with one side of the atrium collecting blood from the body and one side collecting from the lungs, and with one side of the ventricle pumping blood to the lungs and one side pumping to the body. This culminated in the hearts of birds and mammals, which have two completely separate atrium–ventricle pairs, which pump to the pulmonary circulation to the lungs and the **systemic circulation** to the rest of the body. We describe these evolutionary changes and the flow patterns in more detail in the next section of this chapter, on circulatory pathways and vessels.

Avian and mammalian hearts are dual pumps

Even though anatomically the heart is a single organ, in birds and mammals there are separate right and left sides that act as two separate pumps (we examine the hearts of other vertebrates later). Indeed they begin in embryos as *two separate hearts that fuse* together. These hearts have four chambers, an upper and a lower one, within each half (Figure 9-18a). The upper chambers, the **atria** (singular, **atrium**), receive blood returning to the heart and transfer it to the lower chambers, the **ventricles**, which pump the blood from

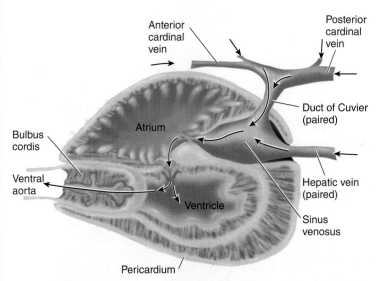

FIGURE 9-17 Diagram of an elasmobranch heart. Venous blood enters the sinus venosus (*upper right*) and is then pumped by the atrium, ventricle, and bulbus cordis into the aorta (*lower left*).

Source: P. Withers. (1992). *Comparative Animal Physiology*. Belmont, CA: Brooks/Cole; as modified from A. J. Waterman, B. E. Frye, K. Johansen, et al. (1971). *Chordate Structure and Function*. New York: Macmillan.

the heart. The vessels that return blood from the tissues to the atria are **veins**, and those that carry blood away from the ventricles to the tissues are **arteries**. The two halves of the heart are separated by the **septum**, a continuous muscular partition that prevents mixture of blood from the two sides of the heart. This separation is extremely important, because the right half of the heart is receiving and pumping O_2-depleted CO_2-rich blood whereas the left side of the heart receives and pumps O_2-rich CO_2-poor blood.

Let's examine how the heart functions as a dual pump by tracing a drop of blood through one complete circuit in a mammal (Figure 9-18b). Blood returning from the systemic circulation enters the right atrium via large veins known as the **venae cavae**. The drop of blood entering the right atrium has returned from the body tissues, where O_2 has been extracted from it and CO_2 has been added to it. This partially deoxygenated blood flows from the right atrium into the right ventricle, which pumps it out through the **pulmonary artery** to the lungs. Thus, *the right side of the heart pumps blood into the pulmonary circulation*. Within the lungs, the drop of blood loses its extra CO_2 and picks up a fresh supply of O_2 before being returned to the left atrium via the **pulmonary veins**. This O_2-rich blood returning to the left atrium subsequently flows into the left ventricle, the pumping chamber that propels the blood to all body systems except the pulmonary circulation of the lungs; that is, *the left side of the heart pumps blood into the systemic circulation*. The large artery carrying blood away from the left ventricle is the **aorta** (Figure 9-18a). Major arteries branch from the aorta to supply the various tissues of the body.

Both sides of the heart simultaneously pump equal amounts of blood. The pulmonary circulation is a low-pressure, low-resistance system, whereas the systemic circulation is a high-pressure, high-resistance system. Therefore, even though the right and left sides of the heart pump the same amount of blood, the left side performs more work

because it pumps an equal volume of blood at a higher pressure into a higher-resistance system. Accordingly, the heart muscle on the left side is much thicker than the muscle on the right side, making the left side a stronger pump (Figure 9-18c).

This left–right difference leads to difficulties in rapidly growing animals, such as broiler chickens. Oxygen usage by the growing tissues reduces oxygen in the blood. In response, the volume of blood pumped by the left ventricle increases to restore oxygen delivery. The right ventricle must in turn work harder, and it hypertrophies (grows larger). However, for unclear reasons, the right ventricle's growth does not keep pace with the growth of the body, so the ventricle becomes unable to maintain sufficient pressure to propel the entire blood flow. Eventually, the overstressed right ventricle is unable to take in and pump returning blood, resulting in life-threatening **edema** (fluid accumulation) in various organs, a condition termed **ascites syndrome**. Current research is focused on selecting strains of chickens resistant to developing this condition. A similar phenomenon, called **brisket disease**, occurs in cattle and horses that spend some time at high altitudes and then return to low altitude.

Heart valves ensure that the blood flows in the proper direction through the heart

As you have seen, blood flows through vertebrate hearts in one fixed direction through the heart. The presence of one-way valves ensures this unidirectional flow of blood. Heart valves are positioned so that they open and close passively because of pressure differences, similar to a one-way door (Figure 9-19). A forward pressure gradient (that is, a greater pressure behind the valve) forces the valve open, much as you open a door by pushing on one side of it, whereas a backward pressure gradient (that is, a greater pressure in front of the valve) forces the valve closed, just as you apply pressure to the opposite side of the door to close it. (Functionally similar valves are found in arthropod heart ostia, as you saw earlier.)

These valves are in all vertebrate hearts; two are shown in a cartilaginous fish in Figure 9-17. Let us examine these valves more closely in mammals, which have four (Figure 9-20). The **right** and **left atrioventricular (AV) valves** are positioned between the atrium and the ventricle on the right and left sides, respectively. These valves allow one-way flow from the atria into the ventricles during ventricular filling (when atrial pressure exceeds ventricular pressure). The other two valves, the **aortic** and **pulmonary valves**, are located at the juncture where the major arteries leave the ventricles. They are known as **semilunar** ("half-moon") **valves** because they consist of three cusps, each resembling a shallow half-moon–shaped pocket. These valves are forced open when the left and right ventricular pressures exceed the pressure in the aorta and pulmonary arteries, respectively, during ventricular contraction and emptying. Closure results when the ventricles relax and ventricular pressures fall below the aortic and pulmonary artery pressures.

Even though there are no valves between the atria and veins, backflow into the veins usually is not a significant problem, because (1) atrial pressures are not much higher than venous pressures, and (2) the junctions of the veins and atria are partially compressed during atrial contraction.

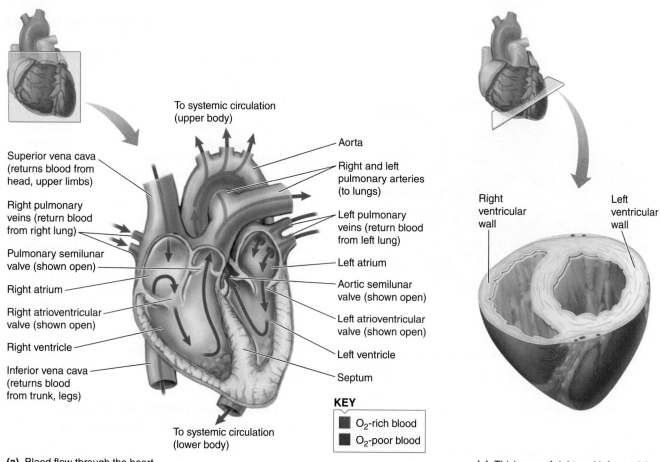

To systemic circulation (upper body)

Superior vena cava (returns blood from head, upper limbs)

Right pulmonary veins (return blood from right lung)

Pulmonary semilunar valve (shown open)

Right atrium

Right atrioventricular valve (shown open)

Right ventricle

Inferior vena cava (returns blood from trunk, legs)

To systemic circulation (lower body)

Aorta

Right and left pulmonary arteries (to lungs)

Left pulmonary veins (return blood from left lung)

Left atrium

Aortic semilunar valve (shown open)

Left atrioventricular valve (shown open)

Left ventricle

Septum

KEY
- O$_2$-rich blood
- O$_2$-poor blood

(a) Blood flow through the heart

Right ventricular wall

Left ventricular wall

(c) Thickness of right and left ventricles

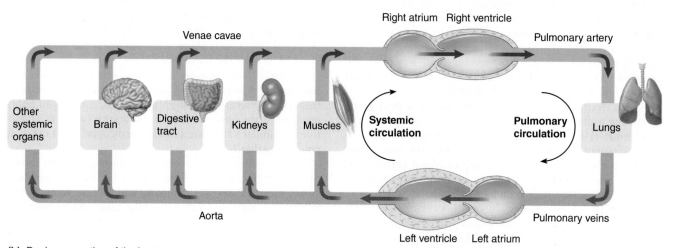

Venae cavae

Right atrium   Right ventricle

Pulmonary artery

Other systemic organs

Brain

Digestive tract

Kidneys

Muscles

Systemic circulation

Pulmonary circulation

Lungs

Aorta

Pulmonary veins

Left ventricle   Left atrium

(b) Dual pump action of the heart

FIGURE 9-18 Blood flow through and pump action of the mammalian heart. (a) The arrows indicate the direction of blood flow. To illustrate the direction of blood flow through the heart, all of the heart valves are shown open, which is never the case. The right side of the heart receives O$_2$-poor blood from the systemic circulation and pumps it into the pulmonary circulation. The left side of the heart receives O$_2$-rich blood from the pulmonary circulation and pumps it into the systemic circulation. (b) Note the parallel pathways of blood flow through the systemic organs. (The relative volume of blood flowing through each organ is not drawn to scale.) (c) Note that the left ventricular wall is much thicker than the right wall.

© Cengage Learning, 2013

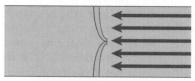

When pressure is greater behind the valve, it opens.

Valve opened

When pressure is greater in front of the valve, it closes. Note that when pressure is greater in front of the valve, it does not open in the opposite direction; that is, it is a one-way valve.

Valve closed; does not open in opposite direction

FIGURE 9-19 Mechanism of valve action.

© Cengage Learning, 2013

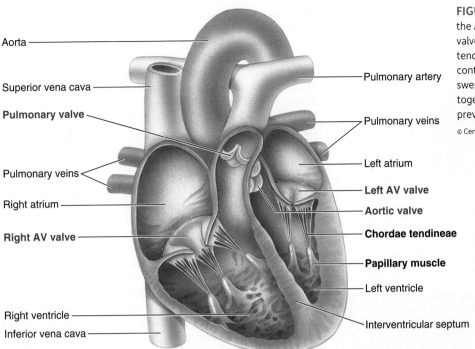

FIGURE 9-20 **Heart valves.** Eversion of the AV valves is prevented by tension on the valve leaflets exerted by the chordae tendineae when the papillary muscles contract. When the semilunar valves are swept closed, their upturned edges fit together in a deep, leakproof seam that prevents valve eversion.

© Cengage Learning, 2013

Aorta

Superior vena cava

Pulmonary valve

Pulmonary veins

Right atrium

Right AV valve

Right ventricle

Inferior vena cava

Pulmonary artery

Pulmonary veins

Left atrium

Left AV valve

Aortic valve

Chordae tendineae

Papillary muscle

Left ventricle

Interventricular septum

(a) Location of the heart valves in a longitudinal section of the heart

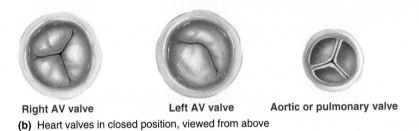

Right AV valve

Left AV valve

Aortic or pulmonary valve

(b) Heart valves in closed position, viewed from above

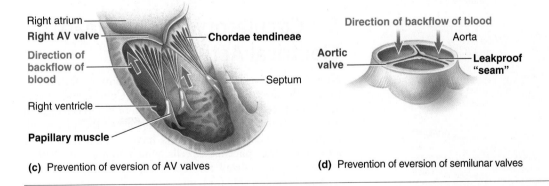

Right atrium

Right AV valve

Direction of backflow of blood

Right ventricle

Papillary muscle

Chordae tendineae

Septum

(c) Prevention of eversion of AV valves

Direction of backflow of blood

Aorta

Aortic valve

Leakproof "seam"

(d) Prevention of eversion of semilunar valves

Vertebrate heart walls are composed primarily of spirally arranged cardiac muscle fibers interconnected by intercalated discs

The bulk of the vertebrate heart wall is the **myocardium** (*myo,* "muscle"), a middle layer of cardiac muscle lying between an endothelial layer (the **endocardium**) and a thin outer sheath, the **epicardium**. In mammals, the myocardium consists of interlacing bundles of cardiac muscle fibers arranged spirally around the heart circumference. As a result of this arrangement, when the ventricular muscle contracts, the diameter of the ventricular chambers is reduced while

the apex is simultaneously pulled upward toward the top of the heart in a rotating manner. This "wringing" effect efficiently exerts pressure on the blood within the chambers and directs it upward to the arteries that exit the ventricles.

The individual cardiac muscle cells are interconnected to form branching fibers, with adjacent cells joined end-to-end at specialized structures known as **intercalated discs**. Within an intercalated disc are two types of membrane junctions we saw in Chapter 2: *desmosomes* and *gap junctions* (Figure 9-21). Recall that desmosomes are a type of adhering junction that mechanically holds cells together under high mechanical stress, whereas gap junctions are channels

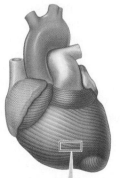

(a) Bundles of cardiac muscle are arranged spirally around the ventricle. When they contract, they "wring" blood from the apex to the base where the major arteries exit.

Intercalated discs

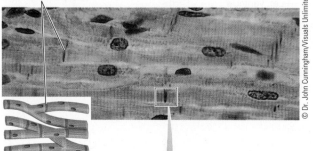

© Dr. John Cunningham/Visuals Unlimited

(b) Cardiac muscle fibers branch and are interconnected by intercalated discs.

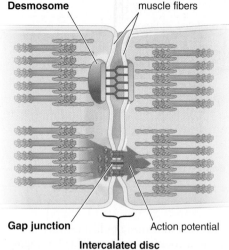

Plasma membranes of adjacent cardiac muscle fibers

Desmosome

Gap junction **Action potential**

Intercalated disc

(c) Intercalated discs contain two types of membrane junctions: mechanically important desmosomes and electrically important gap junctions.

FIGURE 9-21 Organization of mammalian cardiac muscle fibers. Bundles of cardiac muscle fibers are arranged spirally around the ventricle. Adjacent cardiac muscle cells are joined end to end by intercalated discs, which contain two types of specialized junctions: desmosomes, which act as spot rivets mechanically holding the cells together, and gap junctions, which permit action potentials to spread from one cell to adjacent cells.

© Cengage Learning, 2013

that allow action potentials to spread from one cardiac cell to the next (see pp. 65–67).

As we explore later, electrical impulses trigger heart contraction. Once initiated, an impulse spreads to all the other cells that are joined by gap junctions so that they contract as a single functional **syncytium** (a group of joined cells acting as one functional unit). Vertebrate atrial and ventricular muscles form separate syncytia; each contracts synchronously to produce the force necessary to eject the enclosed blood. However, an impulse starting in the right atrium will spread to the rest of the heart. Therefore, unlike skeletal muscle, where graded contractions can be produced by varying the number of muscle cells that are contracting within the muscle (recruitment of motor units), either all the cardiac muscle fibers contract or none do. A "half-hearted" contraction is not possible.

Cardiac muscle cells in most vertebrates have two features that indicate how much the energy for contraction relies on aerobic metabolism: (1) 30–50% by volume of mitochondria, the O_2-dependent energy organelles (see p. 47); and (2) an abundance of *myoglobin* (p. 533), which stores limited amounts of O_2 for immediate use, and also facilitates the diffusion of O_2 arriving at the cell to the mitochondria for rapid processing.

check your understanding 9.4

Describe four different pumping mechanisms for circulation that are not systemic hearts (include at least one nonvertebrate).

How do heart valves work?

What are intercalated discs and their functions?

9.5 Circulatory Pumps: Heart Electrical Activity

Now that you've learned the basic anatomy of hearts, we turn to the electrical control of beating. Contraction of cardiac muscle cells to bring about ejection of blood is triggered by *action potentials* (p. 116) sweeping across the muscle cell membranes. Many animal hearts have a small number of **autorhythmic** or **pacemaker cells**, which do not contract but instead are specialized for repeatedly initiating the action potentials responsible for contraction of the contractile cells. Hearts that have pacemaker cells are called **myogenic** ("generating from muscle"), and are found in many animals including all vertebrates, tunicates (Urochordates), insects, and some mollusks and crustaceans. They are presumably advantageous because they ensure blood or hemolymph flow even if there is severe neural trauma. (However, myogenic insect hearts appear to be largely under the control of external neural stimuli under normal circumstances.) Some crustaceans such as decapods (crabs, lobsters, and so on) have **neurogenic** hearts, which do not beat without an external neural stimulus. This type of regulation may allow for a heart to be stopped completely in animals that do not need continuous circulatory flow in some situations (see also the grasshopper on p. 400).

Pacemaker cells have channels that cyclically move them to "firing" threshold

Let's examine the role of the pacemaker cells in the origin and spread of a heartbeat. In contrast to most nerve and skeletal muscle cells, in which the membrane remains at constant resting potential unless the cell is stimulated (see p. 116), the cardiac pacemaker cells do not have a stable resting potential. Instead, their membrane potential slowly depolarizes, or drifts, between action potentials until threshold is reached, at which time the membrane "fires," that is, has an action potential (Figure 9-22). Through repeated cycles of drifting and firing, these autorhythmic cells cyclically initiate action potentials, which then spread throughout the heart to trigger rhythmic beating without any nervous stimulation.

Consider how this coordinates pumping in an insect (as in Figure 9-16c). Pacemaker activity dominates in the most posterior segment of the dorsal heart and spreads forward into the anterior aorta. This ensures that contraction moves in a wave from posterior to anterior for proper hemolymph flow.

Pacemaker Potential and Action Potential in Autorhythmic Cells Complex interactions of several different ionic mechanisms are responsible for the **pacemaker potential**, which is an autorhythmic cell membrane's slow drift to threshold. The most important changes in ion movement that give rise to the mammalian pacemaker potential are (1) an increased inward Na^+ current, (2) a decreased outward K^+ current, and (3) an increased inward Ca^{2+} current (Ca^{2+} influx is also essential to cardiac action potentials in insects). Let's examine the details:

1. The initial phase of the slow depolarization to threshold is caused by net Na^+ entry through a type of voltage-gated Na^+ channel—an **HCN** or **hyperpolarization-activated cyclic nucleotide-gated channel**—found only in cardiac pacemaker cells and some neurons. Most ion channels open when the membrane becomes less negative (depolarizes), but these unique channels open when the potential becomes more negative (hyperpolarizes) at the end of repolarization from the previous action potential. Because of their unusual behavior, currents through HCN channels are called **funny**, or I_f, currents (recall that "I" refers to current in Ohm's Law, p. 115). When one action potential ends and the HCN channels open, the resultant inward I_f Na^+ current starts immediately moving the pacemaker cell's membrane potential toward threshold once again.

2. At the same time, a progressive reduction occurs in the passive outward flux of K^+. Unlike in neurons and skeletal muscles, the cardiac K^+ channels that opened during the falling phase of the preceding action potential slowly close at negative potentials. This slow closure gradually diminishes the outflow of positive potassium ions down their concentration gradient. Thus, K^+ efflux slowly declines at the same time as the inward leak of Na^+, adding to the drift toward threshold.

3. Finally, increased Ca^{2+} entry occurs in the second half of the pacemaker potential, as the HCN channels close and *transient* Ca^{2+} channels (**T-type Ca^{2+} channels**) open before the membrane reaches threshold. The resultant brief influx of Ca^{2+} further depolarizes the membrane, bringing it to threshold, at which time these Ca^{2+} channels close. At threshold, the rapid rising phase of the

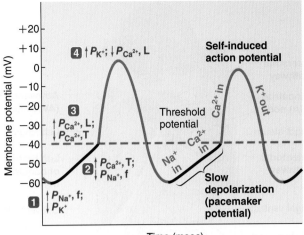

FIGURE 9-22 Pacemaker activity of mammalian cardiac autorhythmic cells. The first half of the pacemaker potential (1) is the result of simultaneous opening of the funny (f or HCN) channels, which permit inward Na^+ current, and closure of the K^+ channels, which reduce outward K^+ current. The second half of the pacemaker potential (2) is the result of opening of T-type Ca^{2+} channels. Once threshold is reached (*green dashed line*), the action potential rise (3) results from opening of L-type Ca^{2+} channels; the falling phase (4) results from the opening of K^+ channels.

© Cengage Learning, 2013

action potential occurs from voltage-activation of a second class of Ca^{2+} channel, the *long-lasting* (**L-type Ca^{2+} channel**) type, which produces a large influx of Ca^{2+} (in contrast to the rising phase resulting from Na^+ influx in neurons and skeletal muscles).

The falling phase is the result, as usual, of the K^+ efflux that occurs when K^+ permeability increases on activation of voltage-gated K^+ channels, coupled with closure of the L-type Ca^{2+} channels. (Inhibited function of these K^+ channels, resulting from a genetic mutation or drugs such as *erythromycin*, interferes with repolarization and has been shown to cause sudden cardiac arrest in horses and humans.)

The sinoatrial node is the normal pacemaker of the vertebrate heart

The specialized, noncontractile cardiac cells capable of autorhythmicity are found in several locations in vertebrate hearts (Figure 9-23a). The main sites are as follows:

- The **sinoatrial node (SA node)**, a small, specialized region in the right atrial wall in reptiles (including birds) and mammals. In fishes and amphibians, the equivalent of the SA node lies in the wall of the **sinus venosus.**

- The **atrioventricular node (AV node)**, a small bundle of specialized cardiac muscle cells located at the base of the right atrium near the septum, just above the junction of the atria and ventricles.

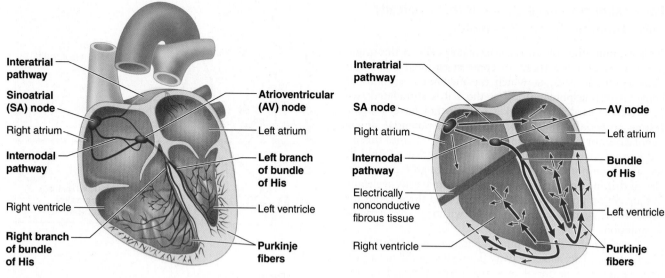

(a) Specialized conduction system of the heart

(b) Spread of cardiac excitation

FIGURE 9-23 **Specialized conduction system of the mammalian heart and spread of cardiac excitation.** An action potential initiated at the SA node first spreads throughout both atria. Its spread is facilitated by two specialized atrial conduction pathways, the interatrial and internodal pathways. The AV node is the only point where an action potential can spread from the atria to the ventricles. From the AV node, the action potential spreads rapidly throughout the ventricles, hastened by a specialized ventricular conduction system consisting of the bundle of His and Purkinje fibers.

© Cengage Learning, 2013

In all vertebrates, the **SA node** (or sinus venosus node) normally exhibits the fastest rate of autorhythmicity, inherently at 90 to 100 action potentials per minute in humans, and drives the rest of the heart at this rate. Thus, it is the main pacemaker of the heart, which excites the entire heart, triggering the contractile cells to contract (and thus the heart to beat) at the pace set by pacemaker autorhythmicity. The AV node does not assume its own naturally slower rate, because it is activated by action potentials generated from the SA node before it reaches threshold at their own slower rhythm.

The spread of cardiac excitation is coordinated to ensure efficient pumping

Once initiated in the SA node, an action potential spreads throughout the rest of the heart. For efficient cardiac function, the spread of excitation achieves the following:

1. *Atrial excitation and contraction is complete before the onset of ventricular contraction.* This ensures complete ventricular filling. During cardiac relaxation, the AV valves are open so that venous blood entering the atria continues to flow directly into the ventricles. Most ventricular filling occurs by this means prior to atrial contraction. When the atria do contract, additional blood is squeezed into the ventricles to complete ventricular filling. Only then does ventricular contraction eject blood into the arteries.

2. *Excitation of cardiac muscle fibers is coordinated to ensure that each heart chamber contracts as a unit to accomplish efficient pumping.* If the muscle fibers in a heart chamber were to become excited and contract randomly rather than contracting simultaneously in a coordinated fashion, they would be unable to eject blood. A smooth, uniform ventricular contraction is essential. In contrast, random, uncoordinated excitation and contraction is known as **fibrillation.** Ventricular fibrillation rapidly causes death because the heart cannot pump blood into the arteries.

3. *The pair of atria and pair of ventricles are coordinated so that both members of the pair contract simultaneously.* This permits synchronized pumping of blood into the pulmonary and systemic circulation.

The normal spread of cardiac excitation is carefully orchestrated to achieve these three events, as described next (Figure 9-23b):

1. *Atrial excitation:* An action potential originating in the SA node first spreads throughout both atria, primarily from cell to cell via gap junctions. The signal appears to spread symmetrically, ensuring synchronized atrial contraction.

2. *Transmission between the atria and the ventricles:* The AV node is the only point of electrical contact between the atria and ventricles; other connections consist of electrically nonconductive fibrous tissue. In the AV node, the impulse is delayed about 0.1 second (the **AV nodal delay**), which enables the atria to become completely depolarized and to contract, emptying their contents into the ventricles, before ventricular contraction occur.

3. *Ventricular excitation:* After the AV nodal delay, the impulse travels rapidly along two specialized tracts (Figure 9-23b):

   - The **bundle of His (atrioventricular bundle),** a tract of specialized autorhythmic cells that originates at the AV node and enters the interventricular septum, where it divides to form the right and left bundle branches that travel down the septum, curve around the tip of the ventricular chambers, and travel back toward the atria along the outer walls.

- **Purkinje fibers,** small terminal fibers that extend from the bundle of His and spread through some (in most mammals) or all (in ruminants) of the ventricular myocardium, much like small twigs of a tree branch.

The network of fibers in this ventricular conduction system is specialized for rapid propagation of action potentials to ensure that the ventricles contract as a unit. Although this system carries the action potential rapidly to a large number of cardiac muscle cells, it does not terminate on every cell. The impulse quickly spreads from the excited cells to the remainder of the ventricular muscle cells by means of gap junctions. Altogether, these conduction steps lead to almost simultaneous activation of the ventricular myocardial cells in both ventricular chambers, efficiently ejecting blood into the systemic and pulmonary circulations simultaneously.

The action potential of contractile cardiac muscle cells shows a characteristic plateau

The action potential in contractile cardiac muscle cells, although initiated by the nodal pacemaker cells, varies considerably in ionic mechanisms and shape from the SA node potential (compare Figures 9-22 and 9-24). Unlike autorhythmic cells, contractile cells remain essentially at rest until excited by electrical activity from the pacemaker. Once a myocardial contractile cell is excited, an action potential is generated by a complicated interplay of permeability changes and membrane potential changes as follows (Figure 9-24):

1. During the rising phase of the action potential, the membrane potential rapidly climbs to about +20 mV to +30 mV as a result of activation of voltage-gated Na^+ channels. At peak potential, the Na^+ permeability then rapidly plummets. This is essentially identical to events in neurons and skeletal muscle, using the same channels (see pp. 117–120).

2. At peak potential, another subclass of K^+ channels transiently opens. The resultant limited efflux of K^+ through these transient channels brings about a brief, small repolarization as the membrane becomes slightly less positive.

3. Unique to cardiac muscle cells, the membrane potential is maintained at this near peak positive level for several hundred msec (compared to 1 to 2 msec in neurons and skeletal muscles), producing a **plateau phase.** This plateau, which prolongs contraction for squeezing the heart chambers, is maintained by two voltage-dependent permeability changes triggered by the sudden voltage change during the rising phase of the action potential: activation of "slow" L-type Ca^{2+} channels and a marked decrease in K^+ permeability. Because Ca^{2+} is in greater concentration in the ECF, opening of the L-type Ca^{2+} channels results in a slow, inward diffusion of Ca^{2+} that prolongs the positivity inside the cell. This effect is enhanced by the concomitant decrease in K^+ permeability on closure of both the briefly opened transient K^+ channels and the leaky K^+ channels open at resting potential.

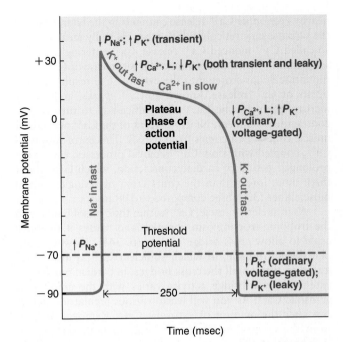

FIGURE 9-24 Action potential in mammalian contractile cardiac muscle cells. The action potential in cardiac contractile cells differs considerably from the action potential in cardiac autorhythmic cells (compare with Figure 9-22). The rapid rising phase of the action potential in contractile cells is the result of Na^+ entry on opening of fast Na^+ channels at threshold. The early, brief repolarization after the potential reaches its peak is because of limited K^+ efflux on opening of transient K^+ channels, coupled with inactivation of the Na^+ channels. The prolonged plateau phase is the result of slow Ca^{2+} entry on opening of L-type Ca^{2+} channels, coupled with reduced K^+ efflux on closure of several types of K^+ channels. The rapid falling phase is the result of K^+ efflux on opening of ordinary voltage-gated K^+ channels, as in other excitable cells. Resting potential is maintained by opening of leaky K^+ channels.

© Cengage Learning, 2013

4. The rapid falling phase of the action potential results from inactivation of the Ca^{2+} channels and delayed activation of "ordinary" voltage-gated K^+ channels, yet another subclass of K^+ channels identical to the ones in neurons and skeletal muscle cells. As the inward movement of Ca^{2+} stops, the sudden increase in K^+ permeability simultaneously promotes rapid outward movement of positive K^+. Thus, as in other excitable cells, the cell returns to resting potential as K^+ leaves the cell. At resting potential, the voltage-gated K^+ channels close and the leaky K^+ channels open once again.

Let's now see how this action potential brings about contraction.

Ca^{2+} entry from the ECF induces a much larger Ca^{2+} release from the sarcoplasmic reticulum

In cardiac contractile cells, the L-type Ca^{2+} channels lie primarily in the *T tubules* (p. 343), opening during a local action potential. Thus, unlike in skeletal muscle, Ca^{2+} moves into the cytosol from the ECF across the T tubule membrane in cardiac muscle. This entering Ca^{2+} triggers the opening of nearby *ryanodine* Ca^{2+}-release channels in the lateral sacs of the sarcoplasmic reticulum (SR; see p. 343). By means of this so-called Ca^{2+}-**induced** Ca^{2+} **release**, Ca^{2+} entering the cytosol from the ECF induces a much larger release of Ca^{2+} from the intracellular stores into the cytosol. The resultant local bursts of Ca^{2+} release, known as *Ca^{2+} sparks*, collectively increase the cytosolic Ca^{2+} pool sufficiently to turn on the contractile machinery. Ninety percent of the Ca^{2+} needed for muscle contraction comes from the SR. This extra supply of Ca^{2+}, coupled with slow Ca^{2+} removal processes, causes the prolonged period of cardiac contraction, which lasts about three times longer than the contraction of a single skeletal muscle fiber (300 msec compared to 100 msec).

As in skeletal muscle, Ca^{2+} within the cytosol binds with the troponin-tropomyosin complex and causes it to move aside to allow cross-bridge cycling (p. 344). However, unlike skeletal muscle, in which sufficient Ca^{2+} is always released to turn on all the cross bridges, in cardiac muscle the extent of cross-bridge activity varies with the amount of cytosolic Ca^{2+}. As you will learn, various regulatory factors can alter the amount of cytosolic Ca^{2+}. Removal of Ca^{2+} from the cytosol by active transport in both the plasma membrane and SR restores the blocking action of troponin and tropomyosin so that contraction ceases and the heart muscle relaxes.

Tetanus of cardiac muscle is prevented by a long refractory period

Like other excitable tissues, cardiac muscle has a *refractory* period. Recall that this occurs immediately after an action potential and makes it impossible or very difficult for another action potential to be generated (p. 124). In skeletal muscle, the refractory period is very short compared with the duration of the resultant contraction, so the fiber can be restimulated again before the first contraction is complete to produce summation of contractions, resulting in a sustained, maximal contraction known as *tetanus* (see Figure 8-16, p. 352).

In contrast, cardiac muscle has a long refractory period that lasts about 250 msec. This is almost as long as the period of contraction (Figure 9-25). Consequently, cardiac muscle cannot be restimulated until contraction is almost over, making summation of contractions and tetanus of cardiac muscle impossible. This is a valuable protective mechanism because a prolonged tetanic contraction would prove fatal. (Think about why.)

The chief factor responsible for the long refractory period is inactivation, during the plateau phase, of the Na^+ channels that were activated during the initial Na^+ influx of the rising phase. Not until the membrane recovers from this inactivation process (when the membrane has already repolarized to resting), can the Na^+ channels be activated once again to begin another action potential (compare to neurons, pp. 124–125).

Methodology: The Electrocardiogram (ECG) The electrical currents generated by cardiac muscle during depolarization and repolarization spread into the tissues surrounding the heart and are conducted through the body fluids. A small portion of this electrical activity reaches the body surface, where it can be detected using recording electrodes on the skin, generally on limbs (Figure 9-26a) and/or chest (al-

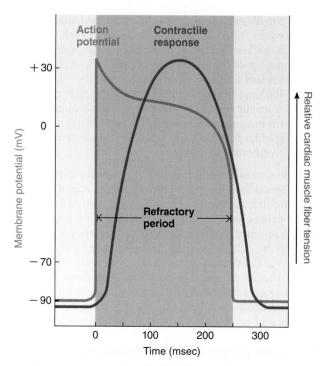

FIGURE 9-25 Relationship of an action potential and the refractory period to the duration of the contractile response in cardiac muscle.

© Cengage Learning, 2013

though on horses they may be placed on the neck and shoulder). It can also be detected with internally implanted devices. The record produced is an **electrocardiogram**, or **ECG**. (Alternatively, the term **EKG** is used, from the ancient Greek "kardia" instead of "cardia" for *heart*.) Two important points should be remembered when analyzing an ECG:

1. It is a recording of that portion of the electrical activity induced in the body fluids by the cardiac impulse that reaches the surface of the body, not a direct recording of the actual electrical activity of the heart.
2. It is a complex recording representing the sum of electrical activity spreading throughout all of the heart's muscle cells during depolarization and repolarization.

A normal vertebrate ECG exhibits three distinct waveforms (Figure 9-26b and d), as follows:

- The **P wave** represents atrial depolarization.
- The **QRS complex** represents ventricular depolarization.
- The **T wave** represents ventricular repolarization.

(These letters do not signify anything; the founder of the technique simply started in the middle of the alphabet when naming the waves.) The following important points about the ECG record should also be noted:

1. Firing of the SA node does not generate sufficient electrical activity to reach the surface of the body, so no wave is recorded for SA nodal depolarization. Therefore, the first recorded wave, the P wave, occurs when the impulse spreads across the atria.
2. No separate wave is detected for atrial repolarization, because that normally occurs simultaneously with ventricular depolarization and is masked by the QRS complex.

FIGURE 9-26 **Electrocardiograms (ECGs).** (a) Recording an ECG of a cow standing in the four saline-filled trays containing electrodes. The right leg tray was used to ground the animal. (b) Electrocardiogram wave forms. (c, d) Goldfish ECG sensor and ECG recording.

Source: L. A. Geddes. (2002). Electrocardiograms from the turtle to the elephant that illustrate interesting physiological phenomena. *Pacing and Clinical Electrophysiology* 25:1762–1770, Figure 12.

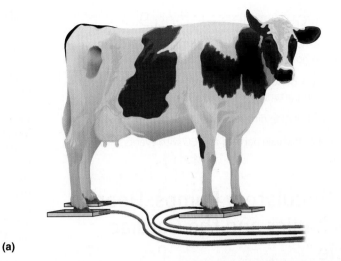

(a)

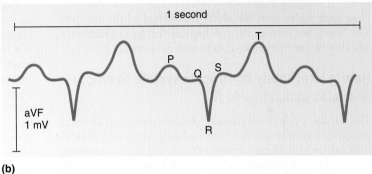

(b)

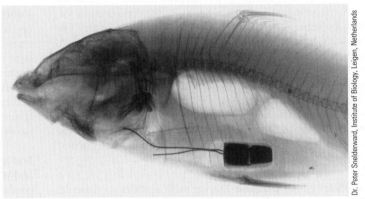

Dr. Peter Snelderward, Institute of Biology, Leigen, Netherlands

(c) Goldfish with implanted sensor for ECG

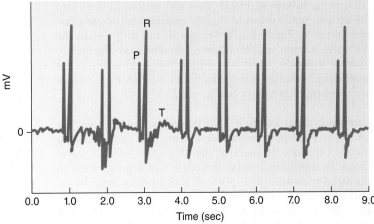

(d) Electrocardiogram of a goldfish

3. The P wave is much smaller than the QRS complex because the atria have a much smaller muscle mass than the ventricles and thus generate less electrical activity.

4. There are three times when no current is flowing in the heart musculature and the ECG remains at baseline:

   a. *During the AV nodal delay.* This delay is the interval of time between the end of the P wave and the onset of the QRS wave, and is termed the **PR segment.** (It is not called the PQ segment because the Q deflection is small and sometimes absent.) Current is flowing through the AV node, but it is too weak to be detected by the ECG electrodes.

   b. *When the ventricles are completely depolarized and the cardiac contractile cells are undergoing the plateau phase of their action potential before they repolarize.* This is represented by the **ST segment,** the interval between QRS and T; during this time, ventricular activation is complete and the ventricles are contracting and emptying.

   c. *When the heart muscle is completely at rest and ventricular filling is taking place, after the T wave and before the next P wave.* This is called the **TP interval.**

ECG measurements are conducted on humans and domestic animals such as horses to diagnose certain heart problems. For example, a P wave with no subsequent QRST events (a skipped beat) is called an "AV block," and may be due to damaged conduction fibers. Horses may exhibit such skipped beats occasionally during rest, but a healthy horse never has one during exercise. ECGs are also measured on a variety of other vertebrates as part of physiological monitoring during studies on behavior and effects of temperature, drugs, stress, exercise, and so on (see the example for a goldfish in Figure 9-26c and d).

ECGs have also been measured in arthropods, although the PQRST pattern is not present. One study, using electrodes under the cuticle, demonstrated cephalic control of the housefly's heart. The ECG of an anesthetized fly had rhythmic variations averaging 133 beats/min. However, when neural input to the heart was eliminated the pattern became highly regular, at a steady 120 beats/min. Thus, the brain was apparently modulating an intrinsic mechanism.

check your understanding 9.5

Describe how the vertebrate heart pacemaker fires rhythmically and how this controls heartbeat.

What is the plateau phase of cardiac action potentials and why is it important?

How is fibrillation different from tetanus?

Correlate the main parts of an ECG with events in the heart.

9.6 Circulatory Pumps: Heart Mechanics and the Cardiac Cycle

Now that we've seen the electrical control of the mammalian heart, we turn to the mechanical features, which (usually due to pacemaker activity) occur in a cycle.

Hearts alternately contract in systole to empty and relax in diastole to fill

The *cardiac cycle* in all chamber hearts consists of alternate periods of **systole** (contraction and emptying) and **diastole** (relaxation and filling). In vertebrates, the atria and ventricles go through separate cycles of systole and diastole. Contraction occurs as a result of the spread of excitation (started by the pacemaker) across the heart, whereas relaxation follows the subsequent repolarization of the musculature. The following discussion, again focusing on the mammalian heart, correlates various events that occur concurrently during the cardiac cycle, including pressure changes, volume changes, and valve activity. Look at Figure 9-27 to understand this discussion. Only the events on the left side of the heart are described, but keep in mind that identical events are occurring on the right side of the heart, except that the pressures are lower. Our discussion follows one full cardiac cycle.

Early Ventricular Diastole During early *ventricular* diastole, the atrium is also in diastole. This stage corresponds to the TP interval on the ECG—the resting stage. Because of the continuous inflow of blood from the venous system into the atrium, atrial pressure slightly exceeds ventricular pressure (point 1, Figure 9-27). Because of this differential, the AV valve is open, and blood flows directly from the atrium into the ventricle (heart *a* in Figure 9-27). As a result, the ventricular volume slowly continues to rise even before atrial contraction takes place (point 2).

Late Ventricular Diastole Late in ventricular diastole, the SA node reaches threshold and fires. The depolarization impulse spreads throughout the atria (the ECG P wave) (point 3), triggering atrial contraction, which causes a rise in the atrial pressure curve (point 4) and squeezes more blood into the ventricle. The corresponding rise in ventricular pressure (point 5) is due to this additional blood added from the atria (point 6) (heart *b*). Throughout atrial contraction, atrial pressure still slightly exceeds ventricular pressure, so the AV valve remains open.

End of Ventricular Diastole Ventricular diastole ends at the onset of ventricular contraction. By this time, atrial contraction and ventricular filling are completed. The volume of blood in the ventricle at the end of diastole (point 7) is known as the **end-diastolic volume,** which we will discuss in detail later.

Ventricular Excitation and Onset of Ventricular Systole Following atrial excitation, the impulse passes through the AV node and specialized tracts to excite the ventricle (the ECG QRS complex) (point 8). The ventricular pressure curve sharply increases shortly after the QRS complex, signaling the onset of ventricular systole (point 9). As ventricular contraction begins, ventricular pressure immediately exceeds atrial pressure. This backward pressure differential forces the AV valve closed (point 9).

Isovolumetric Ventricular Contraction After the AV valve has closed, the ventricular pressure must continue to increase before it can open the aortic valve. Therefore, there is a brief period of time when the ventricle remains a closed chamber (point 10). Thus, no blood can enter or leave the ventricle during this time. This interval is termed the period of **isovolumetric ventricular contraction** (*isovolumetric* means "constant volume") (heart *c*). The muscle fibers remain at constant length, similar to an isometric contraction in skeletal muscle. During this period, ventricular pressure continues to increase as volume remains constant (point 11).

Ventricular Ejection When ventricular pressure finally exceeds aortic pressure (point 12), the aortic valve is forced open and ejection of blood begins (heart *d*). The aortic pressure curve rises as blood is forced into the aorta from the ventricle faster than blood is draining off into the smaller vessels at the other end (point 13). The ventricular volume decreases substantially as blood is rapidly pumped out (point 14). Ventricular systole includes the periods of both isovolumetric contraction and ventricular ejection.

End of Ventricular Systole The ventricle does not empty completely during ejection. Normally, about half the blood contained within a ventricle at the end of diastole is pumped out during the subsequent systole. The amount of blood remaining in the ventricle at the end of systole is known as the **end-systolic volume** (point 15). This is the least amount of blood that the ventricle will contain during this cycle. The amount of blood pumped out of each ventricle with each contraction is known as the **stroke volume;** it is equal to the end-diastolic volume minus the end-systolic volume. As an example, in some horses the end-diastolic volume is 800 mL, the end-systolic volume is 350 mL, and the stroke volume is 450 mL.

Ventricular Repolarization and Onset of Ventricular Diastole The T wave signifies ventricular repolarization occurring at the end of ventricular systole (point 16). As the ventricle starts to relax on repolarization, ventricular pressure falls below aortic pressure and the aortic valve closes (point 17). Closure of the aortic valve produces a disturbance or notch on the aortic pressure curve known as the **dicrotic notch** (point 18). No more blood leaves the ventricle during this cycle.

FIGURE 9-27 Mammalian cardiac cycle. This graph depicts various events that occur concurrently during the cardiac cycle. Follow each horizontal strip across to see the changes that take place in the electrocardiogram; aortic, ventricular, and atrial pressures; ventricular volume; and heart sounds throughout the cycle. Late diastole, one full systole and diastole (one full cardiac cycle), and another systole are shown for the left side of the heart. Follow each vertical strip downward to see what happens simultaneously with each of these factors during each phase of the cardiac cycle. See the text (pp. 412 and 414) for a detailed explanation of the circled numbers. The sketches of the heart illustrate the flow of O₂-poor (*dark blue*) and O₂-rich (*bright red*) blood in and out of the ventricles during the cardiac cycle.

© Cengage Learning, 2013

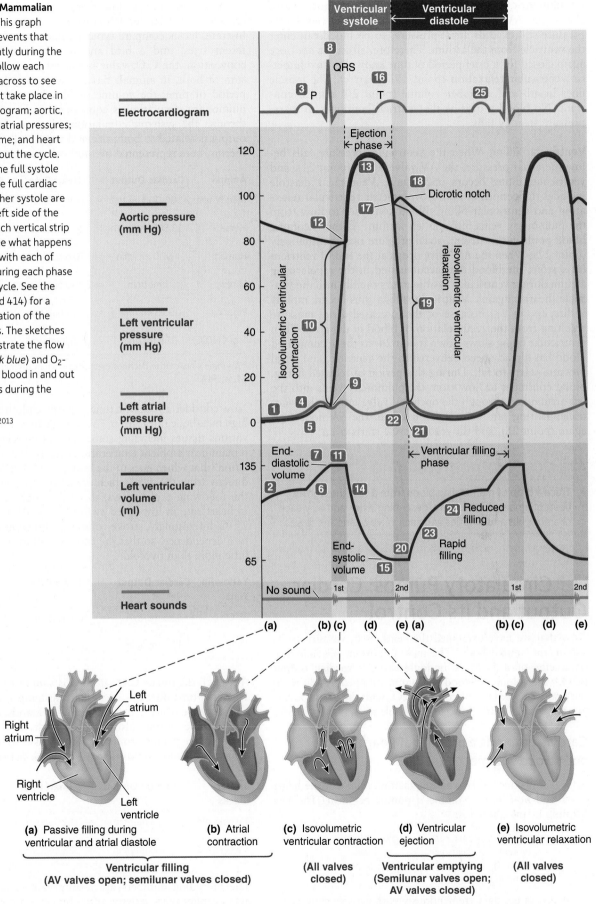

Isovolumetric Ventricular Relaxation When the aortic valve closes, the AV valve is not yet open, because ventricular pressure still exceeds atrial pressure, so no blood can enter the ventricle from the atrium. Therefore, all valves are once again closed for a brief period of time known as **isovolumetric ventricular relaxation** (point 19) (heart *e*). The muscle fiber length and chamber volume (point 20) remain constant. No blood moves as the ventricle continues to relax; pressure steadily falls.

Ventricular Filling When the ventricular pressure falls below the atrial pressure, the AV valve opens (point 21), and ventricular filling occurs once again. Ventricular diastole includes the periods of both isovolumetric ventricular relaxation and ventricular filling. Blood continues to flow from the pulmonary veins into the left atrium. As this incoming blood pools in the atrium, atrial pressure rises continuously (point 22); When the AV valve opens at the end of ventricular systole, the blood that accumulated in the repolarizing atrium during ventricular systole pours rapidly into the ventricle (heart *a* again). Ventricular filling thus occurs rapidly at first (point 23) because of the increased atrial pressure resulting from the accumulation of blood in the atria. Then ventricular filling slows down (point 24) as the accumulated blood has already been delivered to the ventricle, and atrial pressure starts to fall. During this period of reduced filling, blood continues to flow from the pulmonary veins into the left atrium and through the open AV valve into the left ventricle. During late ventricular diastole, the SA node fires again (point 25), and the cardiac cycle starts over.

check your understanding 9.6

Trace a drop of mammalian blood from a vein through all the heart chambers to the aorta, describing when and why each chamber fills and empties and each valve opens and closes.

9.7 Circulatory Pumps: Cardiac Output and Its Control

Now that we have examined the heart itself, we turn to its role in the animal body. The most important physiological parameter of a heart is undoubtedly the **cardiac output (C.O.)**: the volume of blood per minute pumped by a heart to the body. (*Note*: Do not confuse C.O. with the gas CO, carbon monoxide.) This flow is responsible for transport needs.

Cardiac output is a product of heart rate and stroke volume

C.O. is a product of two simple factors: *how fast* the heart beats and *how much* volume it pumps per beat. The key formula in any animal is:

$$\text{C.O.} = \text{heart rate} \times \text{stroke volume}$$

| (volume per minute) | (beats per minute) | (volume pumped per beat or stroke) |

C.O. values vary tremendously with activity state of an individual animal, and across the animal kingdom. Larger animals tend to have slower heart rates (see the discussion on scaling in Chapter 15) but larger stroke volumes. To illustrate, let us compare typical values for mammals of different sizes, and a bird and a fish of similar size. (By convention, the C.O. value is that of one ventricle, not the sum of both, in animals having two ventricles; during any period of time, the volume of blood flowing through the pulmonary circulation is equivalent to the volume flowing through the systemic circulation.) As you can see, cardiac output is related to body size and is generally lower in ectotherms, here represented by rainbow trout:

Animal	Cardiac Output	=	Heart Rate	×	Stroke Volume
Blue Whale:	2,100,000 mL/min	=	6 beats/min*	×	350,000 mL/beat
Horse:	13,500 mL/min	=	30 beats/min	×	450 mL/beat
Human:	4,900 mL/min	=	70 beats/min	×	70 mL/beat
Shrew:	1 mL/min	=	1,000 beats/min	×	0.001 mL/beat
Pigeon:	195.5 mL/min	=	115 beats/min	×	1.7 mL/beat
Trout (10°C):	17.4 mL/min	=	37.8 beats/min	×	0.46 mL/beat

*A whale's heart rate declines even lower when diving; see Chapter 11, p. 544.

Now consider a yellowfin tuna, a partly endothermic fish with high metabolic rates (see p. 750). While they are able to warm various tissues including some skeletal muscles, their hearts remain near ambient temperature because the *coronary* circulation (that which goes to the heart; see p. 418) receives blood directly from the gills, which are at ambient temperature. As the following table shows, this fish suffers a fivefold drop in heart rate in low-speed swimming at 10° C compared to 25° C. Interestingly, stroke volume shows an inverse temperature dependence, so that the cardiac output decreases only a little more than twofold:

Yellowfin Tuna	Cardiac Output	=	Heart Rate	×	Stroke Volume
25°C:	43.5 mL/min	=	106 beats/min	×	0.41 mL/beat
10°C:	19.8 mL/min	=	19.6 beats/min	×	1.01 mL/beat

Although the tuna's C.O. at 10°C is similar to the trout's at 10°C, the tuna does not do well at this temperature, presumably because of its higher metabolic demands.

Cardiac output also changes with development. For example, in young broiler chicks, stroke volume (but not heart rate) almost doubles in a two-week period:

Broiler Chicks	Cardiac Output	=	Heart Rate	×	Stroke Volume
4 weeks old:	253 mL/min	=	362 beats/min	×	0.70 mL/beat
6 weeks old:	434 mL/min	=	328 beats/min	×	1.33 mL/beat

The preceding numbers are resting values; cardiac output also changes with activity, often by large amounts (fivefold or more). The following are some values that have been

TABLE 9-1 Effects of the Autonomic Nervous System on the Heart and Structures That Influence the Heart

Area Affected	Effect of Parasympathetic Stimulation	Effect of Sympathetic Stimulation
SA Node	Decreases the rate of depolarization to threshold; decreases the heart rate	Increases the rate of depolarization to threshold; increases the heart rate
AV Node	Decreases excitability; increases the AV nodal delay	Increases excitability; decreases the AV nodal delay
Ventricular Conduction Pathway	No effect	Increases excitability; hastens conduction through the bundle of His and Purkinje cells
Atrial Muscle	Decreases contractility; weakens contraction	Increases contractility; strengthens contraction
Ventricular Muscle	No effect	Increases contractility; strengthens contraction
Adrenal Medulla (an Endocrine Gland)	No effect	Promotes adrenomedullary secretion of epinephrine, a hormone that augments the sympathetic nervous system's actions on the heart
Veins	No effect	Increases venous return, which increases the strength of cardiac contraction through the Frank-Starling mechanism

© Cengage Learning, 2013

measured at maximal activity levels (running, flying, swimming):

Thoroughbred Horse:	300,000 mL/min
Human (Untrained):	25,000 mL/min
Human (Athlete):	40,000 mL/min
Pigeon:	1,072 mL/min
Rainbow Trout (10°C):	53 mL/min
Yellowfin Tuna (10°C):	26 mL/min
Yellowfin Tuna (25°C):	99 mL/min

How can cardiac output vary so tremendously, depending on the demands of the animal body? The regulation of cardiac output depends on antagonistic control of both heart rate and stroke volume, topics that are discussed next.

Heart rate is determined primarily by antagonistic regulation of autonomic influences on the SA node

Most vertebrate hearts are innervated by both divisions of the autonomic nervous system, which can modify the rate (as well as the strength) of contraction in antagonistic fashion. Again, we focus on mammals. The "rest-and-digest" parasympathetic nerve to the mammalian heart, the *vagus nerve* (p. 183), primarily supplies the atrium, especially the SA and AV nodes, in order to reduce C.O. "Fight-or-flight" sympathetic nerves also supply the atria, including the SA and AV nodes, and—unlike the vagus nerve—richly innervate the ventricles as well in order to boost C.O.

Both the parasympathetic and sympathetic nervous system affect C.O. and stroke volume by altering the activity of

ion channels and the cyclic AMP second-messenger system in the innervated cardiac cells, as follows:

- The parasympathetic neurotransmitter **acetylcholine (ACh)** from the vagus nerve binds to *muscarinic* receptors which open up K^+ channels via a G protein, and to another inhibitory G protein which reduces activity of the cyclic AMP pathway (see pp. 97 and 164).
- The sympathetic neurotransmitter **norepinephrine (NE)** binds with a β1-adrenergic receptor that is coupled to a stimulatory G protein, which accelerates the cyclic AMP pathway.

Let's examine the specific effects that these neurotransmitters have on the heart.

Effect of Parasympathetic Stimulation on the Mammalian Heart Parasympathetic stimulation *reduces cardiac output* through these effects:

1. *ACh decreases heart rate* (Figure 9-28 and Table 9-1). Enhanced K^+ permeability because of ACh hyperpolarizes the SA node membrane (in most vertebrates), because more K^+ leaves than normal, making the inside even more negative. Inhibition of the cAMP pathway depresses both the inward movement of Na^+ and Ca^{2+} through the HCN-I_f and T-type channels, respectively, further slowing the depolarization to threshold. The "resting" potential starts even farther away from threshold, so it takes longer to reach threshold.
2. *ACh decreases excitability of the AV node*, prolonging transmission of impulses to the ventricles even longer than the usual AV nodal delay.
3. *Parasympathetic stimulation of the atrial contractile cells shortens the action potential*, reducing the slow inward current carried by Ca^{2+}; that is, the plateau phase is shortened.

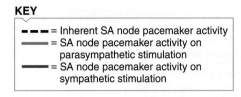

KEY

-·-·- = Inherent SA node pacemaker activity

⎯⎯ = SA node pacemaker activity on parasympathetic stimulation

⎯⎯ = SA node pacemaker activity on sympathetic stimulation

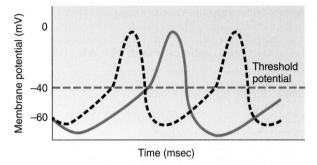

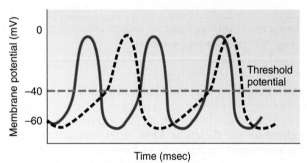

(a) Autonomic influence on SA node potential

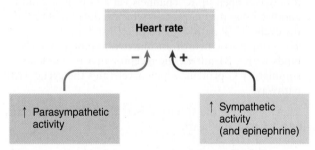

(b) Control of heart rate by autonomic nervous system

FIGURE 9-28 Autonomic control of SA node activity and heart rate.
(a) Parasympathetic stimulation decreases the rate of SA nodal depolarization so that the membrane reaches threshold more slowly and has fewer action potentials, whereas sympathetic stimulation increases the rate of depolarization of the SA node so that the membrane reaches threshold more rapidly and has more frequent action potentials.
(b) Because each SA node action potential ultimately leads to a heartbeat, increased parasympathetic activity decreases the heart rate, whereas increased sympathetic activity increases the heart rate.

© Cengage Learning, 2013

Thus, the heart is more "leisurely" under parasympathetic influence—it beats less rapidly, the time between atrial and ventricular contraction is stretched out, and atrial contraction is weaker. These actions are appropriate in relaxed situations when a high cardiac output is not needed, and in the *dive reflex* in air-breathing diving animals such as many seals and cetaceans (Chapter 11). These effects are similar in other vertebrates as well.

Effect of Sympathetic Stimulation on the Mammalian Heart
Through the following effects, *sympathetic stimulation increases C.O.* (Figure 9-28 and Table 9-1):

1. *NE increases heart rate* through its effect on cAMP, which increases the influx of Na^+ and Ca^{2+}. The resulting swifter drift to threshold permits a greater frequency of action potentials and a correspondingly more rapid heart rate.
2. *NE reduces the AV nodal delay* at the AV node by increasing conduction velocity, by enhancing the slow, inward Ca^{2+} current.
3. *NE speeds up the spread of the action potential* throughout the specialized conduction pathway.
4. *NE increases contractile strength* of the atrial and ventricular contractile cells, both of which have numerous sympathetic nerve endings, so that the heart beats more forcefully and squeezes out more blood (a higher stroke volume). This effect is produced by increasing inward Ca^{2+} movement through prolonged opening of L-type Ca^{2+} channels, which enhances the slow Ca^{2+} influx and intensifies Ca^{2+} participation in excitation–contraction coupling.

The overall effect of sympathetic stimulation on the heart, therefore, is to increase pumping by increasing the heart rate, decreasing the delay between atrial and ventricular contraction, decreasing conduction time throughout the heart, and increasing the force of contraction. These actions are appropriate for emergency or exercise situations.

Control of Heart Rate Thus, as is typical of the autonomic nervous system, parasympathetic and sympathetic effects on heart rate are an example of antagonistic regulation (p. 16). At any given moment, the heart rate is determined largely by the existing balance between the inhibitory effects of the vagus nerve and the stimulatory effects of the cardiac sympathetic nerves. The relative level of activity in these two branches in turn is primarily coordinated by the *cardiovascular control center* located in the brain stem (p. 437).

Although autonomic innervation is the primary means by which heart rate is regulated, other factors can affect it as well. The most important is *epinephrine*, a hormone that is secreted into the blood from the adrenal medulla (p. 308) on sympathetic stimulation and that acts in a manner similar to norepinephrine (NE) to increase the heart rate.

Stroke volume is determined by the extent of venous return and by sympathetic activity

Note that both autonomic systems affect heart rate, but the sympathetic also boosts stroke volume, the amount of blood pumped out by each ventricle during each beat. Actually, there are two types of controls that influence stroke volume: (1) *intrinsic control* related to the extent of *venous return*, which increases during activity and (2) *extrinsic control* from sympathetic stimulation. Recall that intrinsic/extrinsic control is part of the hierarchical regulation we described in Chapter 1 (p. 19). Both factors increase stroke volume by increasing the strength of contraction of the heart (Figure 9-29). Let's examine each of these factors' effects in detail.

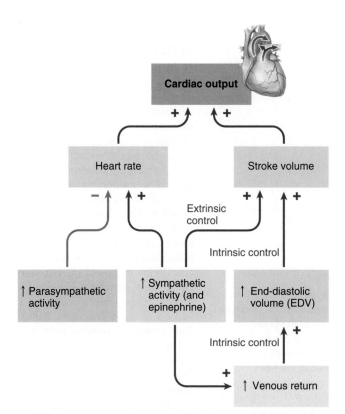

FIGURE 9-29 Control of cardiac output. Cardiac output equals heart rate times stroke volume. Heart rate in turn is increased by sympathetic activity and decreased by parasympathetic activity, while stroke volume is increased by sympathetic activity and higher venous return.

© Cengage Learning, 2013

Increased end-diastolic volume intrinsically results in increased stroke volume

Intrinsic control of stroke volume refers to the heart's inherent ability to vary the stroke volume, based on a direct correlation between end-diastolic volume (EDV) and stroke volume (SV). That is, *as more blood is returned to the vertebrate heart, the heart pumps out more blood.* However the relationship is not quite as simple as it appears, because the heart does not eject all the blood it contains. This control depends on the *length–tension* relationship of cardiac muscle, which is similar to that of skeletal muscle. For skeletal muscle, the resting muscle length is approximately the optimal length at which maximal tension can be developed during a subsequent contraction. When the skeletal muscle is longer or shorter than this optimal length, the subsequent contraction is weaker (see Figure 8-17, p. 354). For cardiac muscle, the resting cardiac muscle-fiber length is less than optimal. Therefore, an increase in cardiac muscle-fiber length increases the contractile tension of the heart on the following systole (Figure 9-30). Unlike in skeletal muscle, there is another more important effect: As a cardiac muscle fiber is stretched from greater ventricular filling, its myofilaments are pulled closer together. Because of this reduced distance between filaments, more cross-bridge interactions between myosin and actin can take place during contraction. Thus, the length–tension relationship in cardiac muscle depends not on muscle fiber length per se but on the resultant variations in the lateral spacing between the myosin and actin filaments.

Frank-Starling Law of the Heart What causes cardiac muscle fibers to vary in length before contraction without being attached to any bones? The main determinant of cardiac muscle-fiber length is the *degree of diastolic filling*. An analogy is a balloon filled with water—the more water you put in, the larger the balloon becomes, and the more it is stretched. Likewise, the greater the extent of diastolic filling, the larger the volume, the more the heart is stretched, and the longer the initial cardiac-fiber length is before contraction. The increased length results in a greater force on the subsequent cardiac contraction and, consequently, a greater SV. This intrinsic relationship between EDV and SV is known as the **Frank-Starling law of the heart**. Stated simply, the law says that *the greater the volume of blood entering the heart during diastole, the greater the volume of blood ejected during systole*. In Figure 9-30, assume that the EDV increases from point A to point B, due to increased venous return. You can see that this increase in EDV is accompanied by a corresponding increase in SV from point A^1 to point B^1. This effect is not unique to vertebrates; for example, mollusk hearts also respond in this way to increased filling (they also beat faster in response).

The built-in relationship matching SV with venous return has two important advantages. First, this intrinsic mechanism equalizes output between the right and left sides of the avian and mammalian hearts so that the blood pumped out by the heart is equally distributed between the pulmonary and systemic circulation. If, for example, the right side of the heart ejects a larger SV, more blood enters the pulmonary circulation, so venous return to the left side of the heart is increased accordingly. The increased EDV of the left side of the heart causes it to contract more forcefully, so it too pumps out a larger SV. In this way, equality of output of the two ventricular chambers is maintained.

The contractility of the heart and venous return are increased by sympathetic stimulation

In addition to intrinsic control, stroke volume is also subject to extrinsic control, the most important of which are actions of the cardiac sympathetic nerves and epinephrine. Sympathetic NE and epinephrine act in two ways:

1. As we noted earlier, both enhance the heart's **contractility**, which is the strength of contraction at any given EDV; that is, the heart contracts more forcefully and squeezes out a greater percentage of the blood it contains on sympathetic stimulation. This increased contractility is due to the increased Ca^{2+} influx triggered by NE and epinephrine. The extra cytosolic Ca^{2+} allows the myocardial fibers to generate more force through greater cross-bridge cycling than they would without sympathetic influence. Normally, the human EDV 135 mL and the end-systolic volume (ESV) is 65 mL for a SV of 70 mL (Figure 9-31a). Under sympathetic influence, for the same EDV the ESV might be 35 mL, yielding an SV of 100 mL (Figure 9-31b).

2. Sympathetic stimulation increases stroke volume also by constricting veins, squeezing more blood from the veins to the heart and thus *enhancing venous return* (Figure 9-31c). A value of about 140 mL may be achieved in humans. The resultant increase in EDV automatically increases SV correspondingly.

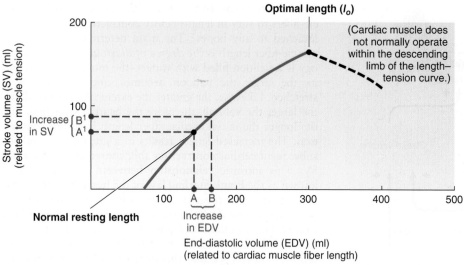

FIGURE 9-30 **Intrinsic control of stroke volume (Frank-Starling curve).** The cardiac muscle fiber's length, which is determined by the extent of venous filling, is normally less than the optimal length for developing maximal tension. Therefore, an increase in end-diastolic volume (that is, an increase in venous return), by moving the cardiac muscle fiber length closer to optimal length, increases the contractile tension of the fibers on the next systole. A stronger contraction squeezes out more blood. Thus, as more blood is returned to the heart and the end-diastolic volume increases, the heart automatically pumps out a correspondingly larger stroke volume.

© Cengage Learning, 2013

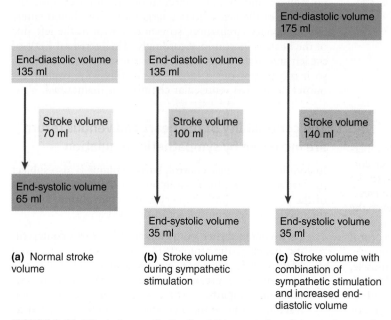

FIGURE 9-31 **Effect of sympathetic stimulation on stroke volume.** Values are for average human heart.

© Cengage Learning, 2013

The strength of cardiac muscle contraction and, accordingly, the stroke volume can thus be varied. Because activity also increases heart rate, these two factors act together to increase the C.O. so that more blood can be delivered to the active muscles. All the factors that determine the cardiac output by influencing the heart rate or stroke volume are summarized in Figure 9-29.

The heart receives most of its own blood supply through the coronary circulation

Although all the blood passes through the heart, the heart muscle cannot extract O_2 or nutrients from the blood within its chambers, in part because the walls are too thick to permit diffusion of O_2 and other supplies from the blood in the chamber to all the cardiac cells. Therefore, like most tissues, heart muscle must receive blood through blood vessels, specifically by means of the **coronary circulation**, which first evolved in active fishes (this circulation is lacking in most sedentary fishes). The coronary arteries branch in fishes from the branchial arteries leaving the gills, in mammals from the aorta just beyond the aortic valve.

During locomotory activity, the rate of coronary blood flow increases several-fold above its resting rate. This is accomplished primarily by vasodilation, or enlargement, of the coronary vessels, which allows more blood to flow through them. Coronary blood flow is adjusted primarily in response to changes in the heart's O_2 requirements. The major link that coordinates coronary blood flow with myocardial O_2 needs is *adenosine*, which is formed from adenosine triphosphate (ATP) during cardiac metabolic activity. Increased formation and release of adenosine from the cardiac cells occur (1) when there is a cardiac O_2 deficit or (2) when cardiac activity is increased and the heart accordingly requires more O_2 for ATP production. The adenosine induces dilation of the coronary blood vessels, allowing more O_2-rich blood to flow to the more active cardiac cells to meet their increased demand. This is a negative-feedback process, summarized as follows:

Reduced O_2 in cardiac myocytes → adenosine release → vasodilation of coronary vessels → increased blood flow and O_2 delivery to myocytes

This matching of O_2 delivery with O_2 needs is critical, because the heart muscle depends on oxidative processes, not anaerobic, to generate energy. The heart primarily uses free fatty acids and, to a lesser extent, glucose and lactate as fuel sources, depending on their availability, and it can shift metabolic pathways to use whatever nutrient is available. The primary danger of insufficient coronary blood flow is not fuel shortage but O_2 deficiency.

Obstruction of coronary arteries is a leading cause of death in humans. Hardening called **arteriosclerosis** occurs in the artery walls for a variety of reasons, such accumulation of cholesterol plaques (p. 389) following damage from high blood pressure (hypertension). Curiously, arteriosclerotic lesions in the coronary vessels of migratory salmonids (Pacific and Atlantic salmon and steelhead trout) are very common and increase with age, affecting up to 100% of adults. These lesions are absent or less severe in nonmigratory salmonids and other fish. It is not certain why this is; researchers speculate that the damage arises from various natural stresses, such as migrating, feeding, and avoiding predation. As in humans, stress leads to *hypertension* (p. 449), which may cause the initial damage. Another possible factor relates to the anatomy of the coronary artery: It lies on top of the bulbus arteriosus (recall its pressure storage-release function, p. 402) that overly expands during stress. This may cause damaging distortion of the coronary artery.

Now that we have concluded the discussion of circulatory pumps and circulatory fluids, we turn our attention to circulatory pathways and vessels.

check your understanding 9.7

What is the cardiac output equation and why is it important?

What is the Frank-Starling law of the heart and why is it important?

Discuss how the parasympathetic and sympathetic systems of vertebrates control the heart.

9.8 Circulatory Pathways and Vessels: Hemodynamics and Evolution

Now that we have discussed the fluids and pumps of circulatory systems, we turn to the details of the flow pathways. As we discussed at the beginning of this chapter, most animals have an *open* (hemolymph) or *closed* (blood) systems (or both). Before we examine these in detail, we begin with the physics of fluid flow.

Flow of fluid obeys certain physical laws called *hemodynamic laws*, which are important to understanding how flow is controlled. Here we focus on flow through vessels.

Blood flow varies directly with the pressure gradient and inversely with resistance

The **flow rate** of blood through a vessel (that is, the volume of blood passing through per unit of time) is expressed in the **hemodynamic flow law**, one of the key hemodynamic equations:

$$Q = \Delta P/R$$

where

Q = flow rate of fluid through a vessel (quantity per unit time)
ΔP = pressure gradient, or $P_1 - P_2$, where
P_1 = pressure at the inflow end of a vessel
P_2 = pressure at the outflow end of a vessel
R = resistance of blood vessels

This equation might look familiar; it is similar to *Ohm's law* for electrical current (current $\Delta I = \Delta V/R$; p. 115). In both,

A crocodile and impala both have four-chambered hearts. But their plumbing is different. The crocodile's heart and vessels allow flow to be diverted away from the lungs when the animal is underwater, while the impala's blood must always go to the lungs before going to the rest of the body.

flow changes in direct proportion to the driving force (pressure or voltage gradient) and inversely with resistance. Let's look at these factors.

Pressure Gradient The **pressure gradient**—the difference in pressure between the beginning and end of a vessel—is the main driving force for flow through the vessel; that is, blood flows from an area of higher pressure (P_1) to an area of lower pressure (P_2) down a pressure gradient (ΔP). Contraction of the heart imparts pressure to the blood, but because of frictional losses (resistance), the pressure decreases as blood flows through a vessel, so it is higher at the beginning than at the end. This maintains a positive ΔP for forward flow of blood through the vessel. The greater the pressure gradient forcing blood through a vessel, the greater the rate of flow through that vessel (Figure 9-32, vessels 1 versus 2). Think of a garden hose attached to a faucet. If you turn on the faucet slightly, a small stream of water flows out of the end of the hose because the pressure is only slightly greater at the beginning than at the end of the hose. If you open the faucet all the way, the pressure gradient increases tremendously so that the rate of water flow through the hose is much greater, and water spurts forth from the end of the hose. Note that the *difference* in pressure (hence the delta symbol in ΔP) between the two ends of a vessel, not the absolute pressures, determines flow rate (Figure 9-32, vessels 2 versus 3).

Gravity is another major factor in establishing the pressure gradient. This is particularly important in a tall terrestrial animal such as a human or a giraffe (Figure 9-33). Continuing with the garden hose analogy, if you point a spurting hose upward, the water will not travel as great a distance from the hose as it does if you point the hose horizontally. To push a column of water upward 1 m requires a pressure of about 70 mm Hg. In effect, gravity increases the P_2 (the pressure at the end of the flow being measured), so that a higher initial P_1 is needed to maintain a given flow. Thus, to maintain a reasonable pressure well above zero in the brain, the average driving pressure from the heart in

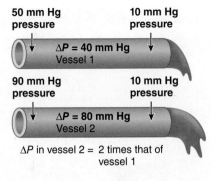

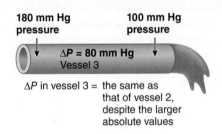

FIGURE 9-32 Relationship of flow to the pressure gradient in a vessel. As the difference in pressure (ΔP) between the two ends of a vessel increases, the flow rate increases proportionately (vessel 1 versus 2). Flow rate is determined by the *difference* in pressure between the two ends of a vessel, not the magnitude of the pressures at each end (vessel 2 versus 3).

© Cengage Learning, 2013

humans needs to be about 100 mm Hg, whereas in a giraffe with its brain about 2 m above its heart (Figure 9-33b), driving pressure needs to be about 200 mm Hg! In contrast, fishes (in which gravity effects are minimal) have lower pressures: 2 to 14 mm Hg in hagfish, and up to about 40 mm Hg in bony fish.

unanswered Questions | How could a brachiosaur pump blood to its brain?

The giraffe heart generates 200 mm Hg to ensure oxygen supply to its brain. But consider the large dinosaurs such as brachiosaurs and barosaurs, which had necks of 25 m or more in length! Pressures of 700 mm Hg or more would be needed from a heart in the chest to create sufficient flow to the brain. Is this likely to have been the case, or can you think of another mechanism to create adequate blood flow to the head? After generating your own hypotheses, read the article by Choy and Altman (1992) in the Suggested Readings. <<

Resistance The other factor influencing flow rate through a vessel is the **resistance** (*R*), which is a measure of the hindrance to blood flow through a vessel caused by friction between the moving fluid and the stationary vascular walls. As *R* increases, it is more difficult for blood to pass through

the vessel, so flow decreases (as long as the pressure gradient remains unchanged). When resistance increases, the pressure gradient ΔP must increase correspondingly to maintain the same flow rate. Accordingly, when the vessels offer more resistance to flow, the heart must work harder to maintain adequate circulation.

Resistance to blood flow depends on several factors. For an idealized smooth flow known as **laminar flow** (where layers of the fluid slide smoothly over each other), there are three key factors: (1) *viscosity* of the fluid, η; (2) vessel *length, L;* and (3) vessel *radius, r,* which is by far the most important. The exact relationship, for idealized laminar, nonpulsatile flow in a rigid tube, is

$$R = 8\eta L/\pi r^4$$

Viscosity (η) arises from friction developed between the molecules of a fluid as they slide over each other during flow of the fluid. Thus, higher viscosity creates higher resistance to flow. In general, the thicker a liquid, the more viscous it is. For example, molasses flows more slowly than water because molasses has greater viscosity. Viscosity of blood is determined by two factors: the concentration of plasma proteins and, more importantly, the number of red blood cells per unit volume. Normally, these two factors are relatively constant for a species and are not important in the control of resistance. Occasionally, however, blood viscosity and, accordingly, resistance are altered by an abnormal number of red blood cells. Recall that increased synthesis of EPO increases number of red blood cells (p. 393), which thus increases blood viscosity. Blood flow can be more sluggish than normal when excessive red blood cells are present; this is a consequence, for example, of a migratory bird or mammal acclimatizing to a high elevation (from a low elevation) or EPO injections.

Because blood "rubs" against the lining of the vessels as it flows past, the greater the vessel surface area in contact with the blood, the greater the resistance to flow. Surface area is determined by both the length (*L*) and radius (*r*) of the vessel. At a constant radius, the longer the vessel, the greater the surface area and the greater the resistance to flow. Because vessel length is typically constant in an adult body, it is not a variable in the control of vascular resistance. *Therefore, the major determinant of resistance to flow is the vessel's radius.* Fluid passes more readily through a large vessel than through a smaller vessel because a given volume of blood comes into contact with much more of the surface area of a small-radius vessel than of a larger-radius vessel, resulting in greater resistance (Figure 9-34a). Furthermore, a slight change in the radius of a vessel brings about a notable change in flow because the resistance is inversely proportional to the *fourth power* of the radius:

$$R \propto 1/r^4$$

Thus, doubling the radius (Figure 9-34b) decreases *R* 16-fold and therefore increases flow through the vessel 16-fold (at the same pressure gradient). The converse is also true. Only 1/16th as much blood flows through a vessel at the same driving pressure when its radius is halved. As you will see the radius of *arterioles* is subject to regulation and is the most important factor in the control of resistance.

The factors that affect the flow rate through a vessel are integrated in an idealized way in the **Poiseuille-Hagen law,** as follows:

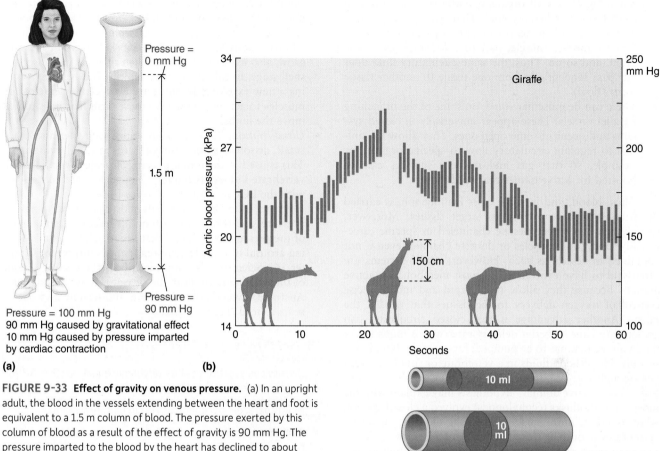

FIGURE 9-33 Effect of gravity on venous pressure. (a) In an upright adult, the blood in the vessels extending between the heart and foot is equivalent to a 1.5 m column of blood. The pressure exerted by this column of blood as a result of the effect of gravity is 90 mm Hg. The pressure imparted to the blood by the heart has declined to about 10 mm Hg in the lower-leg veins because of frictional losses in preceding vessels. The pressure caused by gravity (90 mm Hg) added to the pressure imparted by the heart (10 mm Hg) produces a venous pressure of 100 mm Hg in the ankle and foot veins. Similarly, the capillaries in the region are subjected to these same gravitational effects. (b) Effect on aortic blood pressure of raising a giraffe's head.

Source: (a) © Cengage Learning, 2013; (b) P. Withers. (1992). *Comparative Animal Physiology.* Belmont, CA: Brooks/Cole, p. 676, Figure 14-8.

$$Q = \pi\Delta Pr^4/ 8\ \eta L$$

Although blood vessels are not rigid tubes with nonpulsatile flow (to which this law applies), it works reasonably well for some blood vessels (small arteries and veins). But even where the law is not wholly accurate, the simplified version, $Q = \Delta P/R$, is correct. For the most part, we use this version, noting viscosity and radius effects where appropriate. These factors will come more into focus when we later embark on a voyage through animal circulation.

Circulatory fluids transport materials in a parallel manner, especially in closed systems

Now that we have seen the basic physics of flow, let's see how flow is directed and regulated. There are two important structural features in closed, and often in open, systems:

1. Closed systems always have initial *parallel* branching in arteries, and many open systems do as well (see Figure 9-2; you will see more detailed figures shortly). Parallel

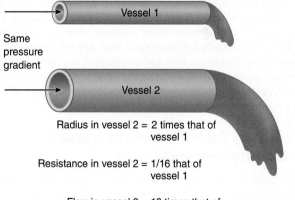

(a) Comparison of contact of a given volume of blood with the surface area of a small-radius vessel and a large-radius vessel

Radius in vessel 2 = 2 times that of vessel 1

Resistance in vessel 2 = 1/16 that of vessel 1

Flow in vessel 2 = 16 times that of vessel 1

Resistance ∝ 1/R⁴
Flow ∝ R⁴

(b) Influence of vessel radius on resistance and flow

FIGURE 9-34 Relationship of resistance and flow to the vessel radius. (a) The same volume of blood comes into contact with a greater surface area of a small-radius vessel compared to a larger-radius vessel. Accordingly, the smaller-radius vessel offers more resistance to blood flow, because the blood "rubs" against a larger surface area. (b) Doubling the radius decreases the resistance to 1/16 and increases the flow 16 times because the resistance is inversely proportional to the fourth power of the radius.

© Cengage Learning, 2013

plumbing allows all organs or body regions to obtain *fresh* blood (or hemolymph). That is, part of the fluid pumped out by the heart can go to a particular muscle, part to another muscle, part to a kidney, part to the brain, and so on. Thus, the same circulatory fluid does not pass sequentially from one tissue to another tissue (*series* flow).

2. There can be muscular *valves* on some of the branching parallel vessels. These appear universally in closed systems and occur in some open ones. They allow the animal to regulate circulatory flow to specific organs, for example, so that gas and nutrient supplies can be boosted for active tissues that need more.

Thus, blood, and in some cases hemolymph, is carried by arteries until very close to target tissues. Moreover, hemolymph can sometimes be channeled in specific directions by body wall muscles or discrete channels even if not contained in vessels as such. However, closed systems are often said to have evolved in the most metabolically active animals because they allow more rapid and more precise control of oxygen delivery to the tissues that need it the most. (Another advantage is that closed circulation can achieve the same or better delivery rates with a much lower fluid volume that has to be pumped; for example, blood may be only 5% of body fluids in a vertebrate or squid, whereas hemolymph can be 30 to 50% in a noncephalopod mollusk.) This "active animal" hypothesis may explain why the most active mollusks (cephalopods) evolved a closed system, whereas other, more sluggish mollusks did not. Similarly, vertebrates with their closed systems are generally more active than most nonvertebrates. (Highly active flying insects are an exception because they have a separate tracheal system for oxygen delivery.) This explanation for closed systems is compelling, but certain exceptions seem to contradict it; for example, nemerteans (ribbon worms) and most annelids (segmented worms) have closed systems but are not more active than many other animals with open systems (such as crustaceans). However, it does appear that the most highly active animals that use circulation for oxygen transport do have closed systems. (And as you will see, large active crustaceans have open systems that approach closed ones in terms of parallel flow.)

Circulatory fluids are driven by pressure and can transmit useful force

Directing the fluid flow in a parallel manner can serve another purpose—that of *force transmission* to specific organs. To move from one point to another, circulatory fluids must be pressurized, as we discussed earlier. In addition to creating circulation, pressure can be used to exert a force for other functions, including the following (Figure 9-35):

- *Movement.* For example, hemolymph pressure (rather than skeletal muscle) is used to extend the legs in arachnids such as spiders. Arteries branch to the legs and flow to them can be controlled. Hemolymph entering into a bent leg at high pressure (from an open artery) makes the leg straighten out (Figure 9-35a), much as a coiled water hose may straighten out when high-pressure water enters it. Cephalothorax muscles, not the spider's heart, create the pressure. The same process is used to extend the foot in mollusks, such as in a clam burrowing into

sand (Figure 9-35b). These animals have hemolymph sinuses with valves to control flow direction in the foot. Researchers have discovered that blue crabs, which like all arthropods must molt their exoskeletons in order to grow, also use fluid pressure when they are in the soft-shell stage. In this stage, after molting but before growing a new exoskeleton, there are no skeletal elements for muscles to act on, so muscles must push internal fluid to move the limbs.

- *Ultrafiltration.* Blood pressure can force water and small, dissolved solutes out of pores in capillary linings. This is used in the initial process of urine formation in vertebrate kidneys (Chapter 12), and for interactions with the ECF in many tissues (see p. 443).

- *Erection.* Blood can enter a flaccid organ under high pressure, and if the exiting blood is restricted, the force of the pressure will inflate that organ. This occurs during arousal of erectile genitalia (penis, clitoris; see Chapter 16). Scientists also think it inflates the sensitive snout of the echidna (a monotreme or egg-laying mammal of Australia), which pokes its snout into termite and ant nests to feed (Figure 9-35c).

check your understanding 9.8

Discuss each part of the hemodynamic equation ($Q = \Delta P/R$) and the major factors affecting each.

Why is parallel flow important in circulatory systems?

In what ways can circulatory fluid be used in force transmission?

9.9 Circulatory Pathways and Vessels: Open Circulation

Now that you've seen common principles and features of circulation, let's look at details in specific animals, starting with those having open circulation. There are many similarities but also significant differences among the various systems. For example, some nonvertebrates have two or more separate ECF compartments. In echinoderms, the ECF is divided into the main body cavity or *coelom*, a *water-vascular* system of canals to operate their unique tube feet, plus a *perihemal* system and a *hemal* system of connected spaces with uncertain functions. However, the most widely studied animals with open circulations are dominated by a single hemolymph space. Let's look at three examples—mollusks, insects, and crustaceans.

Mollusks (Noncephalopods) Most mollusks—snails, chitons, bivalves, and so on—have open circulations with myogenic chamber hearts (Figure 9-36a). (Recall that cephalopods have closed systems.) Although there are no capillaries in the non-cephalopods, hemolymph flow is not completely random. As you saw earlier, burrowing clams can control hemolymph for burrowing. Also, flow to individual organs in at least one species, the black abalone (*Haliotis cracheroidii*), matches each organ's metabolic demands, clearly indicating some directionality in hemolymph flow.

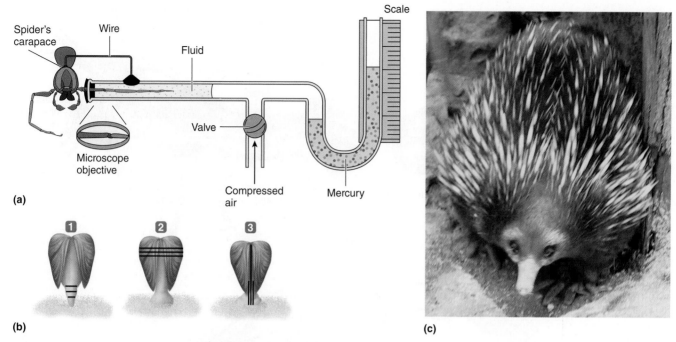

FIGURE 9-35 **Examples of the use of circulatory fluid pressure for force transmission.** (a) An experiment measuring hemolymph pressure in a spider's leg. Spiders use hemolymph pressure to extend their legs. In the experiment, a living spider (*Tegenaria*) is held with one leg sealed into a tube. While watching the turgid membrane of one joint of the leg, through a microscope, the experimenter slowly raises the pressure in the tube by admitting compressed air. When the external pressure in the tube just exceeds the internal pressure of the leg, the membrane is seen to collapse. Normal pressures in a quiet spider were approximately 5 cm mercury. When the spider was stimulated, pressures up to 45 cm mercury were recorded. Intermediate pressures extend the legs during normal walking. (b) A bivalve uses its foot to burrow into sand. Black lines represent muscles being used. In step 1, foot muscles squeeze the hemolymph to extend the foot, while the shells remain open to form an anchor. In step 2, adductor muscles contract and squeeze more hemolymph into the foot, which flares to form an anchor. In step 3, other foot muscles contract to pull the shell downwards. (c) The spiny echidna uses blood pressure to stiffen its snout for probing into the ground for insects.

Source: (a) V. J. Pearse, M. & R. Buchsbaum. (1987). *Living Invertebrates.* Palo Alto, CA: Blackwell Scientific, p. 543; (b) R. M. Alexander. (1982). *Locomotion of Animals.* London: Blackie, p. 115; (c) Photo by Paul Yancey.

Crustacea Larger crustaceans such as decapods (lobsters, crabs, crayfish, and so on) have a well-developed circulatory system with numerous parallel structures (Figure 9-36b). At least some decapods have fine, highly branched capillary-sized arteries within target tissues (see the remarkable crab system in Figure 9-36c). These capillary-like arteries open up into local **lacunae** (very small meshlike spaces), and the larger sinuses that collect the fluid for return to the heart are defined channels rather than random open spaces. Thus, the circulation in these animals blurs somewhat the distinction between open and closed systems. Flow can be characterized as (Figure 9-36b):

Dorsal heart → anterior and posterior aortas → arteries branching in parallel into capillary-like ends → tissue lacunae → sinuses → gills → dorsal heart via ostia

The various branching arteries have valves to control flow to different regions. Later you will see how such valves work for vertebrates. Note that all the hemolymph flows to the gills before returning to the heart. This is part of the flow *in series* rather than in parallel, the significance of which is explained shortly.

Insects Flow in insects is in some sense cruder than in decapods, with a tubular rather than chamber heart, and with much less parallel branching and control. Recall that insects do not rely on circulation for oxygen delivery, which probably explains why they have not evolved more complex, crustacean-like vessels. Flow can be characterized as (Figure 9-2b and 9-16c):

Posterior dorsal heart → dorsal aorta and arterial branches → head and other anterior regions → open spaces throughout body → abdomen → dorsal heart via ostia

Flow through the open spaces is not completely random. In the limbs, for example, longitudinal sheaths channel the hemolymph in specific directions. Thus, some parallel flow is evident.

check your understanding 9.9

Compare the circulatory systems of mollusks, crustaceans, and insects.

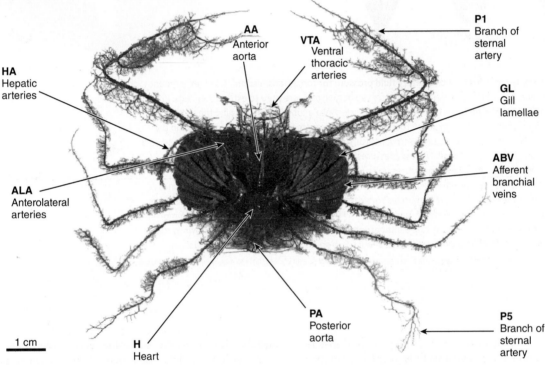

(a) Bivalve mollusk

(b) Crayfish

(c) Crab

I. J. McGaw & C. L. Reiber, (2002), Cardiovascular system of the blue crab *Callinectes sapidus*, *Journal of Morphology*, 252: 1–21. Reprinted by permission of Wiley-Liss, Inc., a subsidiary of John Wiley & Sons, Inc.

FIGURE 9-36 **Detailed examples of open circulations.** (a) A bivalve mollusk has a heart (ventricle) that pumps hemolymph into an aorta, which empties into sinuses in the foot; from there the fluid enters vessels that lead to the gills (ctenidium). The heart also pumps into a mantle vessel, which empties into the hemocoel to bathe most organs. (b) A crayfish has a dorsal heart that pumps hemolymph into arteries leading into the abdomen, where the fluid empties into sinuses. The fluid then enters vessels that travel to the anterior of the animal before returning to the heart. (c) A crab (*Callinectes sapidus*) has a heart that pumps hemolymph into a complex set of arteries that repeatedly branch into fine, capillary-like endings. This remarkable photograph shows those vessels in a resin cast that preserves their structure (shown in dorsal view). The heart (H) and pericardial sinus are situated slightly posterior to the midline of the animal. The anterior aorta (AA), anterolateral arteries (ALA), and hepatic arteries (HA) branch off of the anterior end of the heart, while the posterior aorta (PA) exits posteriorly. The ventral thoracic arteries (VTA) supply the mouthparts, and the pereiopod arteries (P1–P5) supply the limbs. The branchial veins (ABV) run between the gill lamellae (GL).

9.10 Circulatory Pathways and Vessels: Closed Circulation

We now examine closed circulation in detail, starting with an overview.

Parallel and series flow are both important for reconditioning the blood

Let's return to the concepts of parallel and series flow. Circulatory fluid is constantly "reconditioned" so that its composition remains relatively constant despite an ongoing drain of supplies to support metabolic activities and the continual addition of wastes from the tissues. The organs that recondition the fluid may receive substantially more blood than necessary to meet their basic metabolic needs so they can perform homeostatic adjustments. To maximize gas exchange with the environment, gills of most fishes and lungs of birds and mammals receive all of the blood flow, that is, in series. Similarly, all blood flows from the heart of a cephalopod to its two gills, then to its two auxiliary (branchial) hearts. For gas delivery, flow in all these animals then branches in parallel fashion, in part so that large percentages of the cardiac output may be distributed to a digestive tract (to pick up nutrient supplies), to kidneys (to eliminate metabolic wastes and adjust water and electrolyte composition), and to skin (to eliminate or pick up heat) (Figure 9-37). Flow to the other organs—heart, skeletal muscles, and so on—is solely supplying these tissues' metabolic needs and can be adjusted according to their level of activity. For example, during locomotory activity in a vertebrate additional blood is delivered to the active muscles to meet their increased metabolic needs.

Because reconditioning organs receive blood flow in excess of their own needs, they can withstand temporary reductions in blood flow much better than can organs that do not have this extra margin of blood supply. In particular, the brain and heart in many vertebrates can suffer irreparable damage when transiently deprived of blood supply. (A notable exception is the zebra fish heart, which is capable of regenerating itself after a major injury.) Unless an animal is preconditioned to *ischemia* (p. 544) or has evolved a mechanism to withstand *hypoxia* (low O_2; p. 543), permanent brain damage can occur after only a few minutes without O_2. (Brains in birds and mammals are particularly vulnerable because they are not enzymatically equipped to support their metabolic needs anaerobically.) In contrast, digestive organs, kidneys, and skin frequently undergo significant reductions in circulatory flow for a considerable length of time. For example, during locomotory activity in a vertebrate, some of the blood that normally flows through the digestive organs and kidneys is diverted instead to the skeletal muscles. Likewise, blood flow through mammalian skin is markedly restricted during exposure to cold to conserve body heat.

Later in this chapter, after we examine the evolution of circulation in vertebrates, you will see how the distribution of cardiac output is adjusted according to an animal body's changing needs as we examine the various vessels that make up the vascular system.

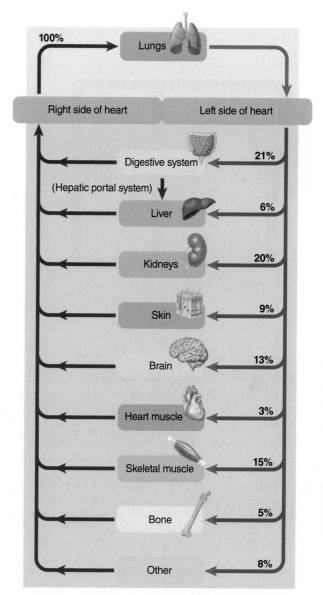

FIGURE 9-37 Distribution of cardiac output at rest. Values are for a typical human. The lungs receive all the blood pumped out by the right side of the heart, whereas the systemic organs each receive a portion of the blood pumped out by the left side of the heart. The percentage of pumped blood received by the various organs under resting conditions is indicated. This distribution of cardiac output can be adjusted as needed.

© Cengage Learning, 2013

The vertebrate vascular system evolved from one circuit to two separate circuits

In the vertebrates, the closed system presumably began as a simple loop:

Heart → artery → gills → aorta and branching parallel arteries → arterioles → capillaries in body organs → veins → heart

As we noted earlier, the flow to the gills is in series with the rest of the circulation, with the rest of the flow in a parallel system. (Recall that in series flow in crustaceans, gills receive

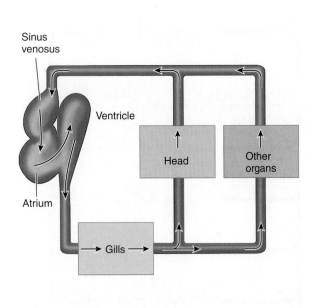

(a) Flow in a bony fish

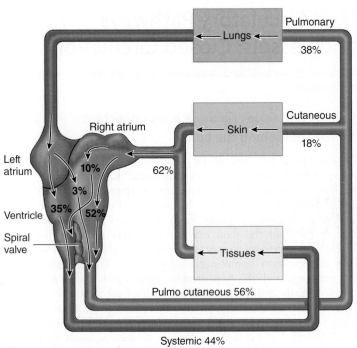

(b) Flow in a bullfrog

FIGURE 9-38 Circulatory flow in fish and amphibia. (a) A typical bony fish. (b) A bullfrog (*Rana catesbeiana*), showing distribution of total flow (20.5 mL/min) through the pulmonary, cutaneous and systemic systems, and the partial separation of oxygenated and deoxygenated blood in the heart.

Source: P. C. Withers. (1992). *Comparative Animal Physiology*. Belmont, CA: Brooks/Cole, p. 708, Figure 14-36; as modified from E. Rogers. (1986). *Looking at Vertebrates: a Practical Guide to Vertebrate Adaptations*. London, UK: Longman; and from H. Tazawa, M. Mochizuki, & J. Piper. (1979). Respiratory gas transport by the incompletely separated ventricle in the bullfrog, *Rana catesbeiana. Respiratory Physiology* 36:77–95.

hemolymph *before* the heart.) While useful for maximal gas exchange, series flow creates a problem because the tiny diameters of the capillaries in the gills create an enormous resistance to flow. Consequently, blood flow from the gills to the rest of the body tends to be somewhat sluggish. This means that whereas there is sufficient oxygen in the blood, it is not rapidly delivered to body muscles. As you will see, this situation does not work on land. In the transition to land, a separate loop began to evolve for the new respiratory organ, the lung. Let us now trace this evolutionary change.

Fishes Water-breathing jawed fishes have essentially the same one-circuit flow pattern just discussed (Figure 9-38a). Hagfish (jawless fishes) are similar but have partially open systems with large sinuses in the head and under the skin. They also have *auxiliary hearts* (as you saw earlier; p. 401) in veins of the head, abdomen, and tail to assist return flow. Hagfish flow can be summarized as follows (note we are omitting the vessels from now on, because they are the same):

Systemic heart → gills → body organs and sinuses →
(auxiliary hearts) → systemic heart

Air-breathing fishes such as lungfish evolved a new respiratory organ, the lung, in addition to gills. This required a separate circuit, because the lung is not always used by these fishes. This in itself would not be very efficient if the oxygenated blood returning from the lungs were to mix with the deoxygenated blood returning from the systemic veins, or if the oxygenated blood were to travel on to the gills. For this reason the lungfish heart evolved partial barriers or *septa* in the atrium and ventricle that subdivide those chambers into somewhat separate pumping functions on the right and left sides. The septa reduce mixing of oxygenated with deoxygenated blood.

In lungfish, the flow to the lungs occurs after the gills, where the blood can either go to the body (during water breathing), or be diverted to the lungs (during air breathing), and is summarized as follows:

Right side of heart → gills → body organs →
right side of heart (water breathing)
Right side of heart → gills → lungs → left side of heart →
body organs → right side of heart (air breathing)

Amphibians As the first amphibians evolved to live part time on land, gills were lost in adults of some species. Today such amphibians primarily use their skins to absorb O_2 when they are in water, and lungs when they are in air. Thus, the pulmonary circulation evolved into a completely separate circuit, going to the lungs and skin, with the systemic circulation going to the main body without passing through a respiratory organ. The atrium is split completely into two chambers, such that amphibians have essentially a three-chambered heart (although most retain a sinus venosus and conus arteriosus). The single ventricle is not subdivided well except for some modest folds, so that oxygenated and deoxygenated blood does mix somewhat in that chamber. Flow is summarized as follows (Figure 9-38b):

Right atrium → right side of ventricle (56%) → lung (or skin) → left or right atrium → left side of ventricle (44%)
→ body organs → right atrium

Some Reptiles Reptiles evolved into purely air-breathing animals, relying on lungs only. The heart lost the auxiliary chambers of fishes and amphibians but retained the two atria that amphibians have. However, the ventricle is different. In most reptiles, it is one large chamber but has two large subchambers partially divided by thick muscle rather than a true septum: the *cavum arteriosum* and *cavum pulmonale*. At the top of both (and connected to both) is a small third subchamber, the *cavum venosum*. (In turtles, a true but partial septum separates the two larger subchambers.) Flow through these subchambers can follow two different pathways if the reptile is a diver (such as a turtle or seasnake). Blood flow while breathing air is as follows (Figure 9-39):

Right atrium → c. venosum → c. pulmonale → lungs →left atrium → c. arteriosum → c. venosum → body organs → right atrium

During a dive, the lungs can be bypassed via the c. venosum, a useful adaptation when the animal is holding its breath for long periods:

Right atrium → c. venosum → body organs → right atrium

Crocodiles Crocodiles and their relatives have a complete four-chambered heart with two atria and two ventricles, with this basic flow (Figure 9-40a):

Right atrium → right ventricle → lungs → left atrium → left ventricle → body organs → right atrium

The right ventricle has an apparent oddity: In addition to a vessel leading to the lungs, it also has one leading to the main body. In addition, this vessel is connected to some of the flow from the left ventricle by a short vessel, the **foramen of Panizza**. Normally, flow through this from the left ventricle is at a higher pressure, such that it blocks any flow from the right ventricle. This oddity is actually a useful adaptation, again for prolonged breath holding during a dive. The flow to the pulmonary arteries is restricted during a dive, so that pressure of the right atrium is directed into the systemic circulation:

Right atrium → right ventricle → body organs → right atrium

In the estuarine crocodile *Crocodylus porosus* and perhaps other species, there are unique, coglike valves between the right ventricle and pulmonary arteries that control this diversion of blood flow (Figure 9-40b). Unlike the loose flaps that form heart valves in other vertebrates, these are stiffer with teethlike projections, which can close during a dive, diverting flow to the main body.

Birds and Mammals The hearts of birds and mammals have a complete four-chambered structure similar to crocodiles, but without the foramen or other shunt to divert flow in different ways. Recall that these four-chambered hearts are two separate pumps fused together. Thus, mammalian and avian blood travels with this pattern (Figure 9-41):

Right atrium → right ventricle → lungs → left atrium → left ventricle → body organs → right atrium

With the complete separation of pulmonary and systemic flow, all blood pumped by the right side of the heart passes through the lungs for O_2 pickup and CO_2 removal. Then the oxygenated blood pumped by the left side of the

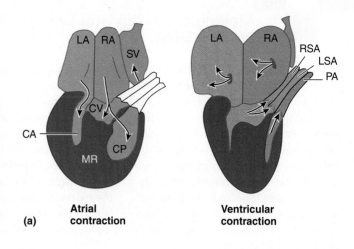

(a) Atrial contraction   Ventricular contraction

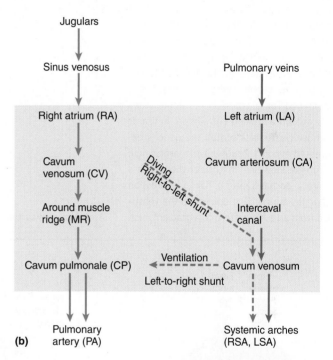

FIGURE 9-39 **Hearts of most reptiles.** (a) Flow through the heart of a varanid lizard during atrial contraction (*left*) and ventricular contraction (*right*). Arrows in red areas indicate the direction of oxygenated blood flow, and arrows in blue areas indicate the direction of deoxygenated blood flow. (b) Normal pattern of blood flow (*solid arrows*) through the reptile heart and pattern of blood flow (*dashed arrows*) for a left-to-right intracardiac shunt (e.g., during apnea, such as diving) and a right-to-left shunt (e.g., during pulmonary ventilation).

Source: P. Withers. (1992). *Comparative Animal Physiology*. Belmont, CA: Brooks/Cole, p. 709, Figure 14-37; as modified from Waterman et al. (1971). (See Figure 9-17); from A. S. Romer & T. S. Parsons. (1986). *The Vertebrate Body*. Philadelphia: Saunders; and from F. N. White. (1968). Functional anatomy of the heart of reptiles. *American Zoologist* 8:211–219.

heart is parceled out in various proportions to the systemic organs through the parallel vessels that branch from the aorta. Researchers think this complete separation evolved independently in birds and mammals and was necessary for the high endothermic metabolisms of these vertebrates. A gene associated with this evolutionary step has been found; see the box, *Molecular Biology and Genomics: From Three to Four*.

Molecular Biology and Genomics
From Three to Four

The evolution of a complete four-chambered heart, in which the ventricle became completely divided by a septum, was vital for the high-powered endothermic metabolisms of birds and mammals. How did this happen? Recent research led by Benoit Bruneau has found that a single regulatory gene may have played a major role. The gene, called *Tbx-5*, codes for a "master" transcription factor (p. 31) that regulates a variety of pattern-forming genes in vertebrate embryos. In avian and mammalian hearts, the gene is expressed in a distinct pattern: It is "on" in the left but not the right ventricle, and a septum forms precisely where Tbx-5 activity sharply changes. Mutations in this gene in humans lead to defects in the septum. When the researchers examined Tbx-5 expression in reptiles, they discovered a striking correlation. In lizards (whose hearts lack septa), the gene is expressed evenly throughout the large single ventricle. However, in turtles (whose hearts have a partial septum), Tbx-5 expression changes during development in the ventricle from an even pattern to a gradient, high on the left and low on the right. Correlating with this gradient formation (which lacks the sharp drop-off in endotherms), a partial septum grows down the middle of the ventricle. Thus, a relatively simple change in regulation of this "master" gene may have led to the four-chambered heart.

You might be wondering why the ectothermic crocodilians have a four-chambered heart. Fossil evidence suggests that their ancestors were fast land-based predators, perhaps even endothermic. Genomic evidence supports this idea. Mitochondrial DNA (mtDNA) in endotherms evolves at about twice the rate in ectotherms, a phenomenon attributed to the higher metabolic rates of endotherms. Researchers were surprised to find that alligator mtDNA has also been evolving at about the same rate as in mammals. This too hints at an endothermic ancestry. Imagine being chased by a fleet-footed crocodile with the metabolism of a lion!

An Amazonian turtle.

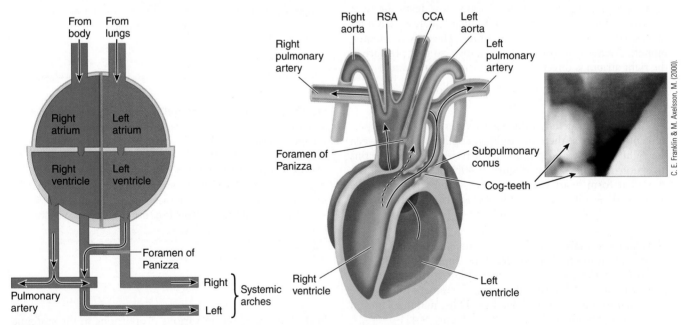

FIGURE 9-40 A crocodile heart. (a) Representation of the pattern of blood flow through the heart, showing the shunting of oxygenated blood from the right to the left systemic arch via the foramen of Panizza. (b) Diagram showing the flow in the heart and the location of special cog teeth, with a photograph showing two of these teeth (see text for description of the teeth's function). RSA, right systemic arch; CCA, common carotid artery.

Sources: (a) P. Withers. (1992). *Comparative Animal Physiology*. Belmont, CA: Brooks/Cole, Figure 14-38. (b) C. E. Franklin & M. Axelsson. (2000). An actively controlled heart valve, *Nature* 406:847–848, and A. Thomas. (2000). *Secret of the Crocodile Heart*, available online at www.abc.net.au/science/news/space/SpaceRepublish_167223.htm, accessed March 3, 2004.

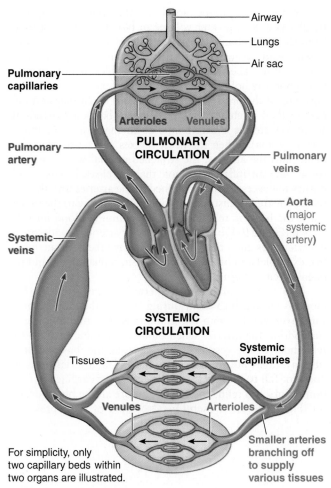

Pulmonary capillaries

Pulmonary artery

Systemic veins

Airway
Lungs
Air sac

Arterioles   Venules

PULMONARY CIRCULATION

Pulmonary veins

Aorta (major systemic artery)

SYSTEMIC CIRCULATION

Tissues

Venules   Arterioles

Systemic capillaries

Smaller arteries branching off to supply various tissues

For simplicity, only two capillary beds within two organs are illustrated.

FIGURE 9-41 Mammalian and avian circulation. Arteries progressively branch as they carry blood from the heart to the tissues. A separate small arterial branch delivers blood to each of the various organs. As a small artery enters the organ it is supplying, it branches into arterioles, which further branch into an extensive network of capillaries. The capillaries rejoin to form venules, which further unite to form small veins that leave the organ. The small veins progressively merge as they carry blood back to the heart.

© Cengage Learning, 2013

The fetus of a placental mammal is an exception. The fetus is not breathing, so the lungs are not functional. There are two bypasses in the fetal circulation: (1) the **foramen ovale**, an opening in the septum between the right and left atrium; and (2) the **ductus arteriosus**, a vessel connecting the pulmonary artery and aorta as they both leave the heart. These bypasses are similar to those we discussed for the crocodile heart, and have a similar function.

Now let's examine the individual vessels, focusing on mammals, with other vertebrates noted on occasion. Vessel types differ considerably in their structures and functions. We begin with an overview. *Arteries*, which carry blood from the heart to the tissues, branch into a "tree" of progressively smaller vessels, with the various parallel branches to different regions of the body. When a small artery reaches the organ it is supplying, it branches into numerous parallel *arterioles*, which provide flow control. Arterioles branch further within the organs into *capillaries*, the smallest of ves-

sels, across which all exchanges are made with surrounding cells and which is the primary goal of the entire circulation. Capillaries rejoin to form small *venules*, which further merge to form small *veins* that leave the organs. The small veins progressively unite to form larger veins that eventually empty into the heart. The arterioles, capillaries, and venules are collectively referred to as the **microcirculation** because they are only visible microscopically.

check your understanding 9.10

Discuss the major differences in circulation and heart structures among the major vertebrate groups.

9.11 Circulatory Vessels: Arteries

The consecutive segments of the vascular tree are specialized to perform specific tasks (Table 9-2). Arteries, the initial vessels, are specialized for two major functions:

1. A *rapid-transit passageways* for blood from the heart to the tissues; and
2. A *pressure reservoir* to provide the driving force for blood when the heart is relaxing (recall that the conus or bulbus arteriosus does this in fishes and amphibians; p. 402).

Arteries are wide with low total cross-sectional area and store and release pressure for high-velocity delivery

Because of their large radii, arteries offer little resistance to blood flow. However they also have another important geometrical feature: their *total cross-sectional area is lower* than those of the downstream arterioles and capillaries (Table 9-2). As we will explore in detail in the capillary discussion, this ensures that blood flows at a high velocity from heart to organs.

Now let's examine how arteries serve as a pressure reservoir. The heart alternately contracts to pump blood into the arteries and then relaxes to refill from the veins. No blood is pumped out when the heart is relaxing and refilling. However, capillary flow is continuous; that is, it does not fluctuate between systole and diastole. The driving force for the continued flow of blood to the tissues during cardiac relaxation is provided by the elastic properties of the arterial walls, whose pressures do fluctuate between systole and diastole. All vessels are lined with a layer of smooth, flattened endothelial cells (continuous with the heart's endocardial lining). Surrounding the arterial endothelial lining is a thick wall containing smooth muscle and two types of connective tissue fibers; *collagen fibers,* which provide tensile strength against the high driving pressure of blood ejected from the heart, and *elastin fibers,* which give the arterial walls elasticity so that they behave much like a balloon (Figure 9-42). Arteries of the giraffe, for example, are more richly endowed with these components in order to withstand the higher blood pressures generated by the heart.

Elastin fibers

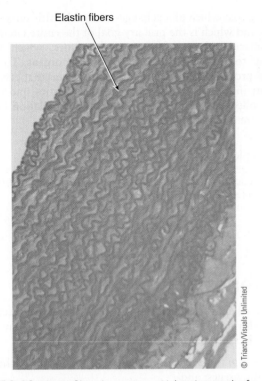

© Triarch/Visuals Unlimited

FIGURE 9-42 Elastin fibers in an artery. Light micrograph of a portion of the aorta wall in cross section, showing numerous wavy elastin fibers, common to all arteries.

As the heart pumps blood into the arteries during ventricular systole, a greater volume of blood enters the arteries from the heart than leaves them to flow into smaller vessels downstream because the smaller vessels have a greater resistance to flow. The arteries' elasticity enables them to expand to temporarily hold this excess volume of ejected blood, storing some of the pressure energy imparted by cardiac contraction in their stretched walls—just as a balloon expands to accommodate the extra volume of air that you blow into it (Figure 9-43a). When the heart relaxes and ceases pumping blood into the arteries, the stretched arterial walls passively recoil, like an inflated balloon that is released. This recoil pushes the excess blood contained in the arteries into the vessels downstream, ensuring continued blood flow to the tissues when the heart is relaxing and not pumping blood into the system (Figure 9-43b). This property of arteries is widespread in vertebrates and other animals that have them, occurring in (for example) amphibians, reptiles, fish, cephalopods, and the open systems of crustaceans.

The maximum pressure exerted in the arteries when blood is ejected into them during systole, the **systolic pressure,** averages 120 mm Hg in humans. The minimum pressure within the arteries when blood is draining off into the remainder of the vessels during diastole, the **diastolic pressure,** averages 80 mm Hg in humans. The arterial pressure does not fall to 0 mm Hg, because the next cardiac contraction occurs and refills the arteries before all the blood drains off (Figure 9-44). A useful way to think of these pressures

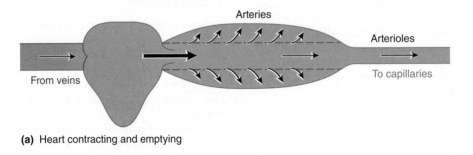

(a) Heart contracting and emptying

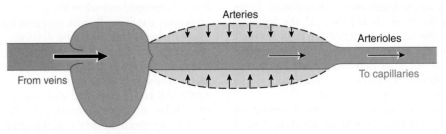

(b) Heart relaxing and filling

FIGURE 9-43 Arteries as a pressure reservoir. Because of their elasticity, arteries act as a pressure reservoir. (a) The elastic arteries distend during cardiac systole as more blood is ejected into them than drains off into the narrow, high-resistance arterioles downstream. (b) The elastic recoil of arteries during cardiac diastole continues driving the blood forward when the heart is not pumping.

TABLE 9-2 Features of Blood Vessels (with Human Values)

Feature	VESSEL TYPE			
	Arteries	Arterioles	Capillaries	Veins
Number	Several hundred*	Half a million	Ten billion	Several hundred*
Special Features	Thick, highly elastic, walls; large radii*	Highly muscular, well-innervated walls; small radii	Very thin walled; large to-tal cross-sectional area	Thin walled compared to arteries; highly distensible; large radii*
Functions	Passageway from heart to organs; serve as pres-sure reservoir	Primary resistance vessels; determine distribution of cardiac output	Site of exchange; deter-mine distribution of extra-cellular fluid between plasma and interstitial fluid	Passageway to heart from or-gans; serve as blood reservoir

Structure

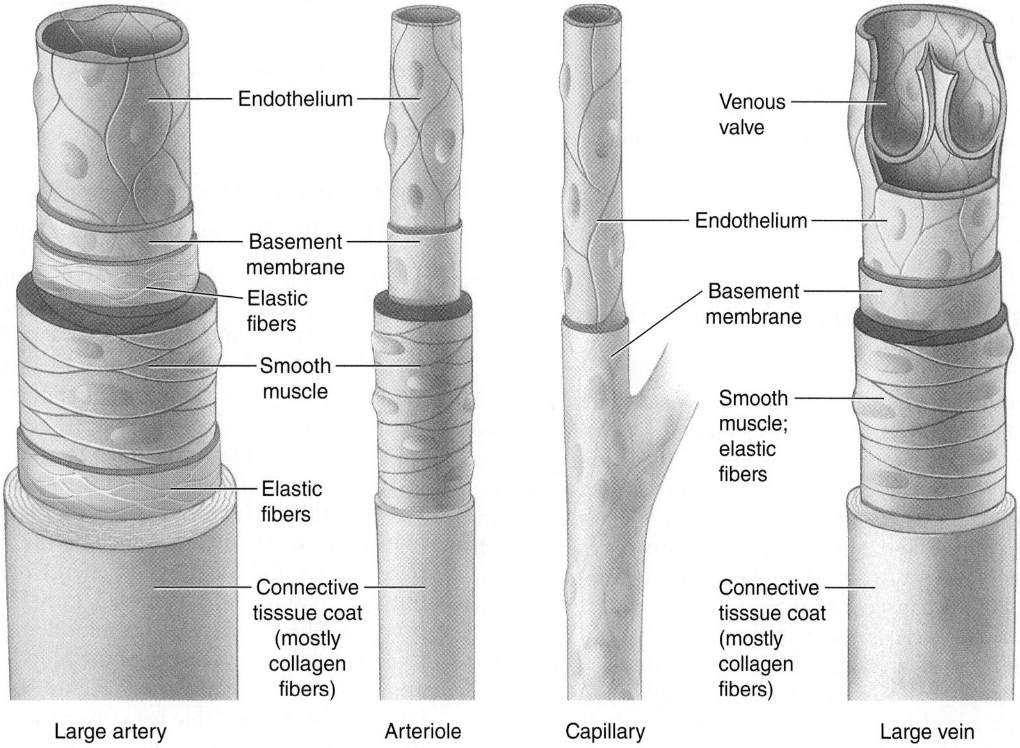

Large artery   Arteriole   Capillary   Large vein

Relative Thickness of Layers in Wall
Endothelium
Elastic fibers
Smooth muscle
Collagen fibers

*These numbers and special features refer to the large arteries and veins, not to the smaller arterial branches or venules.

© Cengage Learning, 2013

and the elastic recoil is to assume that the human left ven-tricle exerts about 200 mm Hg of systolic pressure. Of this, 120 mm Hg are used to push the blood immediately, and 80 mm Hg of pressure energy are stored in the elastic arter-ies. During diastole, when the ventricle pressure falls to about 0, the rebounding arteries release the 80 mm Hg as pressure to keep the blood moving.

The systolic–diastolic pressures are typically written as pairs, for example, 120/80 for humans. Values for other mammals are not very different, for example, 130/95 for horses, 110/80 for rabbits. Birds usually have higher values, such as 180/130 for starlings. Ectotherms have lower pres-sures in general, such as 31/21 for frogs, 43/33 for trout, and 35/21 for lobster (open circulation). There are significant exceptions to this. Systolic pressures in an octopus can reach 75 mm Hg, whereas in a jumping spider they can reach 400 mm Hg! Recall that spiders use pressure to extend their legs, such as for jumping (pp. 422–423).

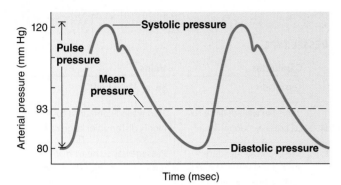

FIGURE 9-44 Arterial blood pressure. Values are for an average human. The systolic pressure is the peak pressure exerted in the arteries when blood is pumped into them during ventricular systole. The diastolic pressure is the lowest pressure exerted in the arteries when blood is draining off into the vessels downstream during ventricular diastole. The pulse pressure is the difference between systolic and diastolic pressure. The mean pressure is the average pressure throughout the cardiac cycle.

© Cengage Learning, 2013

Mean arterial pressure is the main driving force for blood flow

More important than the fluctuating systolic and diastolic pressures is the **mean arterial pressure,** which is the *average pressure* responsible for driving blood forward into the tissues throughout the cardiac cycle. Contrary to what you might expect, mean arterial pressure is not the halfway value between systolic and diastolic pressure (for example, with a human blood pressure of 120/80, mean pressure is *not* 100 mm Hg), because, at resting heart rate, about two thirds of the cardiac cycle is spent in diastole and only one third in systole. As an analogy, if a race car traveled 80 miles per hour (mph) for 40 minutes and 120 mph for 20 minutes, its average speed would be 93 mph, not the halfway value of 100 mph. Similarly, a good approximation of the mean arterial pressure can be determined using the following formula:

Mean arterial pressure = diastolic pressure +
1/3 (systolic pressure – diastolic pressure)

Here is a human example at 120/80 systolic/diastolic pressures:

Mean arterial pressure = 80 + (120–80)/3 = 93 mm Hg

For practice, you may want to apply this formula to another animal, using values just given for a horse, starling, and so on. It is this mean arterial pressure, not the systolic or diastolic pressures, that is generally regulated by blood pressure reflexes to be described in the final section of this chapter.

check your understanding 9.11

Explain how pressure fluctuations are dampened by arteries.

9.12 Circulatory Vessels: Arterioles

When an artery reaches the organ it is supplying, it branches into numerous *arterioles*. Despite their name ("little arteries"), they are very different from arteries. In brief, the radii (and, accordingly, the resistances) of arterioles supplying individual organs can be adjusted independently to accomplish two major functions:

1. To *variably distribute the cardiac output among the systemic organs,* depending on the body's momentary needs, and
2. To help *regulate bodywide arterial blood pressure.*

First we discuss how arterioles affect resistance, then we look at the mechanisms involved in adjusting their resistance. Finally we will consider how such adjustments are important in accomplishing these two functions.

Arterioles are the major resistance vessels and can be dilated or constricted

The radii of arterioles can be small enough to offer considerable resistance to flow, particularly when they are constricted by their muscular layers (see following section). In fact, the *arterioles are the major resistance vessels* in the vascular tree. (Even though the capillaries have smaller radii than the arterioles, you will see later how collectively the capillaries do not offer as much resistance to flow as the collective arterioles do.) In contrast to the low resistance of the arteries, the high degree of arteriolar resistance causes a marked drop in local pressure as the blood flows through these vessels and the pressure energy dissipates. On average, human pressure falls from 93 mm Hg, the mean arterial pressure (the pressure of the blood entering the arterioles), to 37 mm Hg, the pressure of the blood leaving the arterioles and entering the capillaries (Figure 9-45). Arteriolar resistance is also responsible for converting the pulsatile systolic-to-diastolic pressure swings in the arteries into the nonfluctuating pressure present in the capillaries.

Vasoconstriction and Vasodilation Unlike arteries, arteriolar walls contain very little elastic connective tissue. However, they do have a thick layer of smooth muscle that is richly innervated by sympathetic nerve fibers. The smooth muscle layer runs circularly around the arteriole (Figure 9-46a), so when it contracts the vessel's radius becomes smaller, increasing resistance and decreasing flow through that vessel. **Vasoconstriction** is the term applied to such narrowing (Figure 9-46c), while **vasodilation** refers to enlargement in the radius of a vessel as a result of relaxation of its smooth muscle layer (Figure 9-46d). Vasodilation leads to decreased resistance and increased flow through that vessel.

Vascular Tone Arteriolar smooth muscle normally displays a state of partial constriction known as **vascular tone,** which establishes a baseline of arteriolar resistance (Figure 9-46b). Two factors are responsible for vascular tone. First, arteriolar smooth muscle has considerable myogenic activity; that

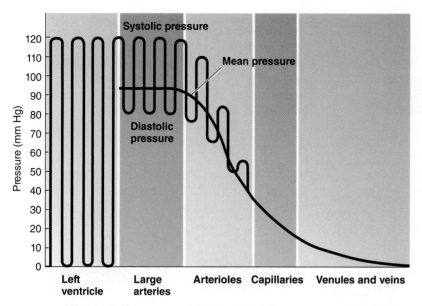

FIGURE 9-45 Pressures throughout the mammalian systemic circulation. Values shown are for an average human. Left ventricular pressure swings between a low pressure of 0 mm Hg during diastole to a high pressure of 120 mm Hg during systole. Arterial blood pressure, which fluctuates between a peak systolic pressure of 120 mm Hg and a low diastolic pressure of 80 mm Hg each cardiac cycle, is of the same magnitude throughout the large arteries. Because of the arterioles' high resistance, the pressure drops precipitously and the systolic-to-diastolic swings in pressure are converted to a nonpulsatile pressure when blood flows through the arterioles. The pressure continues to decline but at a slower rate as blood flows through the capillaries and venous system.

© Cengage Learning, 2013

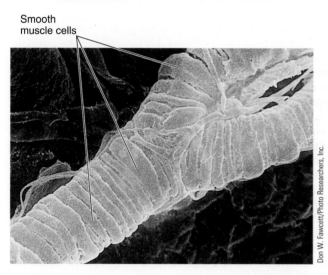

(a) Scanning electron micrograph of an arteriole showing how the smooth muscle cells run circularly around the vessel wall

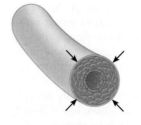

(b) Normal arteriolar tone

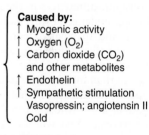

Caused by:
↑ Myogenic activity
↑ Oxygen (O_2)
↓ Carbon dioxide (CO_2) and other metabolites
↑ Endothelin
↑ Sympathetic stimulation
Vasopressin; angiotensin II
Cold

(c) Vasoconstriction (increased contraction of circular smooth muscle in the arteriolar wall, which leads to increased resistance and decreased flow through the vessel)

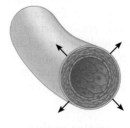

Caused by:
↓ Myogenic activity
↓ O_2
↑ CO_2 and other metabolites
↑ Nitric oxide
↓ Sympathetic stimulation
Histamine release
Heat

(d) Vasodilation (decreased contraction of circular smooth muscle in the arteriolar wall, which leads to decreased resistance and increased flow through the vessel)

FIGURE 9-46 Arteriolar vasoconstriction and vasodilation.
© Cengage Learning, 2013

is, its membrane potential fluctuates without any neural or hormonal influences, leading to self-induced contractile activity (see p. 378). Second, the sympathetic fibers supplying most arterioles continually release norepinephrine (NE, to be discussed later), which further enhances the vascular tone. This ongoing tonic activity makes it possible to either increase or decrease the level of contractile activity to accomplish vasoconstriction or vasodilation, respectively. Were it not for tone, it would be impossible to reduce the tension in an arteriolar wall to accomplish vasodilation.

Distribution of Blood Flow The amount of cardiac output received by each organ is determined by the number and caliber (diameter) of the arterioles supplying that area. Recall that flow rate $(Q) = \Delta P/R$ (p. 419). Because blood is delivered to all tissues at the same mean arterial pressure, the driving force ΔP for flow is identical for each organ. Therefore, differences in flow Q to various organs are completely determined by differences in the extent of

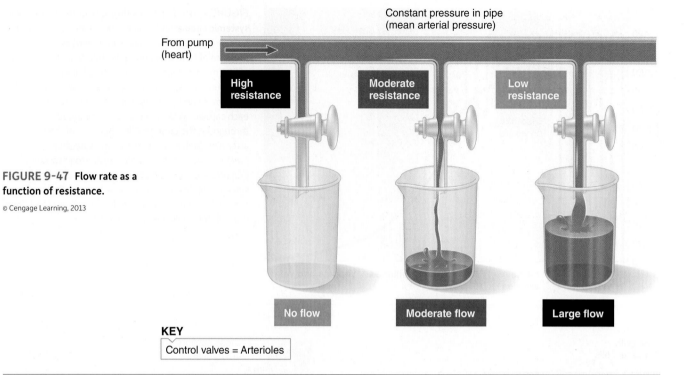

Constant pressure in pipe
(mean arterial pressure)

From pump
(heart)

High
resistance

Moderate
resistance

Low
resistance

No flow

Moderate flow

Large flow

FIGURE 9-47 Flow rate as a function of resistance.

© Cengage Learning, 2013

KEY

Control valves = Arterioles

vascularization (density of blood vessels) and by differences in resistance R offered by the arterioles supplying each organ. On a moment-to-moment basis, the distribution of cardiac output can be varied by differentially adjusting arteriolar R in the various vascular beds.

As an analogy, consider a pipe carrying water with a number of adjustable valves located throughout its length (Figure 9-47). Assuming that water pressure in the pipe is constant, differences in the amount of water flowing into a beaker under each valve depend entirely on which valves are open and to what extent. No water enters beakers under closed valves (high resistance), and more water flows into beakers under valves that are opened completely (low resistance) than into beakers under valves that are only partially opened (moderate resistance). Similarly, more blood flow Q goes to areas whose arterioles offer the least resistance R to its passage. You will see later how this works during elevated muscular activity, when skeletal muscles need more blood but many other organs do not.

A variety of factors can influence the level of contractile activity in arteriolar smooth muscle, thereby substantially changing resistance to flow in these vessels. These factors fall into the two hierarchy categories we saw for cardiac output: *local (intrinsic) controls,* which are important in matching blood flow to the metabolic needs of the specific tissues in which they occur; and *extrinsic controls,* which are important for diverting blood among organs to where it is needed and also for bodywide blood pressure regulation.

Local (intrinsic) influences on arteriolar radius help match blood flow with the tissues' "selfish" needs

Let's see how these controls work at the two levels. Local (intrinsic) controls are changes within a tissue that alter the radii of the vessels and hence adjust blood flow through the

tissue by directly affecting the smooth muscle of the tissue's arterioles. The goal is rapid regulation without major feedback delays: Local metabolic changes quickly alter arteriolar radius in order to match the blood flow through a tissue with the tissue's metabolic needs. Such "selfish" controls are especially important in skeletal muscle and the heart, the tissues whose metabolic activity and need for blood supplies normally vary most extensively, and in the brain, whose overall metabolic activity is relatively constant.

Local influences may be either chemical or physical in nature, including (1) local metabolic changes, (2) histamine release, (3) exposure to heat or cold, and (4) stretch. The chemical influences often work through local mediators. Let's examine the role and mechanism of each of these.

Local Metabolic Changes A variety of local chemical changes act together in a cooperative, redundant manner to bring about these "selfish" local adjustments in arteriolar caliber. Specifically, the following factors produce relaxation of arteriolar smooth muscles:

1. *Decreased O_2.* During increased activity, such as exercise when a skeletal muscle is contracting at an elevated rate, the local O_2 concentration drops.
2. *Increased CO_2.* More CO_2 is generated as a by-product during the stepped-up pace of aerobic metabolism that accompanies increased activity.
3. *Increased acid.* More carbonic acid is generated from the increased CO_2 produced as the metabolic activity of a cell increases. Also, lactate accumulates if the glycolytic pathway alone is used for ATP production (see p. 55).
4. *Increased K^+.* Repeated action potentials that outpace the ability of the Na^+/K^+ pump to restore the resting concentration gradients (see p. 120) result in an increase in K^+ in the tissue fluid of an actively contracting muscle or a more active region of the brain.

5. *Increased osmolarity.* Osmolarity (the concentration of osmotically active solutes) may increase during elevated cell metabolism because of increased formation of osmotically active particles.
6. *Adenosine release.* Especially in cardiac muscle, adenosine is released in response to increased metabolic activity or O_2 deprivation (see p. 418).

By triggering relaxation of the nearby arteriolar smooth muscle, all these factors increase blood flow to that particular area, a response called **active hyperemia**, to bring in more O_2 and nutrients and remove metabolic wastes at a higher rate. Conversely, when a tissue, such as a resting skeletal muscle, is less active metabolically and thus has reduced needs for blood delivery, many of the factors above change in the opposite direction, which triggers local arteriolar vasoconstriction and a subsequent reduction in blood flow to the area. Again, local metabolic changes can adjust blood flow as needed with little delay, as could occur with neural or hormonal regulation.

Histamine Histamine is not released in response to local metabolic changes, nor is derived from vessel cells. Rather, it is released by local immune cells in tissues that are injured or during allergic reactions. It triggers an increase in blood flow into the area that is a major feature in defensive inflammatory responses (see Chapter 10).

Local Heat or Cold Exposure to heat causes localized arteriolar vasodilation. Conversely, cold exposure usually causes vasoconstriction. These processes help maintain body temperature in endotherms and some ectotherms (see Chapter 15). Therapeutic application of heat or cold to an injury takes advantage of these responses; heating promotes blood flow, bringing in oxygen and nutrients to an injured site, whereas cold reduces inflammation triggered by histamine.

Stretch Many arteriole smooth muscles myogenically respond to stretch by vasoconstricting, and conversely respond to a reduction in stretch by vasodilating. This results in a phenomenon called **autoregulation** in some organs, in which local flow *Q* is kept homeostatic. This occurs when external factors lead to a sudden change in blood pressure in an arteriole. For example, if pressure rises and thus flow increases, the resulting stretch of the arteriole causes it to contract to increase *R*, and so flow *Q* is restored (and vice versa for a drop in pressure). You will see why this is important in kidneys in Chapter 12.

Local Vasoactive Mediators Most of the local chemical changes do not act directly on vascular smooth muscle to change its contractile state. Instead, they arise from **endothelial cells**, the single layer of specialized epithelial cells that line the lumen of all blood vessels and heart chambers. In addition to serving as a physical barrier between the blood and the remainder of the vessel wall, endothelial cells have these functions (some of which will be described in more detail elsewhere):

- Secrete substances that stimulate new vessel growth and proliferation of smooth muscle cells in vessel walls.
- Participate in the exchange of materials between the blood and surrounding tissue cells across capillaries through vesicular transport (see p. 440).

- Influence formation of platelet plugs, clotting, and clot dissolution (see pp. 398–399).
- Participate in the determination of capillary permeability by contracting to vary the size of the pores between adjacent endothelial cells.
- Secrete *vasoactive* ("acting on vessels") mediators in response to local chemical and physical changes (such as a reduction in O_2); these substances cause vasodilation or vasoconstriction of the underlying smooth muscle.

Let's look at the last function more closely. Local mediators include *endothelin*, *prostaglandins* (p. 779), *epoxyeicosatrienoic acids (EETs)*, and *nitric oxide (NO)* (a widespread paracrine that is produced in numerous other tissues besides endothelial cells; p. 93). NO is the best studied and has these (and probably other) roles in promoting blood flow:

- NO causes local arteriolar vasodilation by inducing relaxation of arteriolar smooth muscle cells in the vicinity. It does so by increasing the concentration of the intracellular second messenger cyclic GMP, which leads to activation of an enzyme that reduces phosphorylation of myosin. Remember that smooth-muscle myosin can bind with actin and initiate contraction only when myosin is phosphorylated (see p. 375). In this way, NO plays an important role in controlling blood flow through the tissues and in maintaining mean arterial blood pressure. Extrinsic regulation can activate NO production, for example, in dilating the arterioles of the penis and clitoris (p. 782).
- NO interferes with platelet function and blood clotting at sites of vessel damage (p. 395; see also its role in the blood-sucking kissing bug, p. 568).

In 1998 Robert Furchgott, Louis Ignarro, and Ferid Murad received the Nobel Prize for their discovery that nitric oxide signals blood vessels to dilate. Scientists were initially skeptical when in 1986 Furchgott and Ignarro first reported that nitric oxide, a gas and common air pollutant, could be linked to so many physiological processes in the mammalian body. It was Murad in 1977 who discovered that nitroglycerin—a common treatment for heart attacks that increases blood flow to the heart—acted by generating nitric oxide. Strangely enough, Alfred Nobel, the inventor of dynamite and who endowed the foundation for Nobel Prizes in his will, was prescribed nitroglycerin—one of the components in dynamite—for heart trouble.

Extrinsic sympathetic control of arteriolar radius is important in the regulation of both blood flow during activity and arterial blood pressure

Extrinsic control allows an animal's "higher" regulatory systems to control blood flow for the good of the entire body, for both functions of flow distribution and arterial blood-pressure regulation. Control of arteriolar radius includes both neural and hormonal influences, with the effects of the sympathetic nervous system (via NE) being dominant (as we noted earlier for vascular tone).

Sympathetic Nerve Influence The NE released from sympathetic nerve endings combines with α_1-adrenergic receptors (see p. 163) on arteriolar smooth muscle to bring about vasoconstriction. Cerebral (brain) and pulmonary alveolar (air sac) arterioles are the only ones that do not have α_1 receptors, so no vasoconstriction occurs in these areas. It is important that these arterioles not be reflexly constricted by neural influences, because lung and brain blood flow must remain consistent no matter what is going on elsewhere in the body. In fact, reflex vasoconstrictor activity in the rest of the cardiovascular system functions to maintain an adequate pressure head for blood flow to the vital brain and often the heart (see the box on diving mammals in Chapter 11, pp. 544–545).

Adrenal Hormone Influence The sympathetic system also works indirectly on arterioles via the "fight-or-flight" reflex through the adrenal gland medulla, which, under sympathetic stimulation, releases epinephrine and NE (p. 159). Adrenal medullary NE combines with the same α receptors as sympathetically released NE to produce generalized vasoconstriction. However, epinephrine, the more abundant of the adrenal medullary hormones, combines with both β_2 and α_1 receptors but has a much greater affinity for the β_2 receptors. Activation of β_2 receptors produces vasodilation, but not all tissues have β_2 receptors; they are most abundant in the arterioles of the heart and skeletal muscles. During sympathetic discharge, the released epinephrine combines with the β_2 receptors in the heart and skeletal muscle to reinforce local vasodilatory mechanisms in these tissues, increasing delivery of oxygen and nutrients to meet the increased energy demands. Arterioles in digestive organs and kidneys, in contrast, are equipped only with α receptors. Therefore, the arterioles of these organs undergo profound vasoconstriction (Table 9-3). Later you will see how this works during locomotory activity.

Parasympathetic Influence There is no significant level of parasympathetic innervation to arterioles in most vertebrates. Vasodilation is usually produced by decreasing sympathetic vasoconstrictor activity below its tonic level. There are exceptions. Some species have acetylcholine-releasing fibers in skeletal muscles that trigger vasodilation in anticipation of locomotor activity. And there is abundant parasympathetic vasodilator supply to the arterioles of the mammalian penis and clitoris. The rapid, profuse vasodilation induced by parasympathetic stimulation in these organs (by promoting release of NO) is largely responsible for accomplishing erection.

Extrinsic Controls and Redistribution of Blood during Muscular Activity As you have seen, intrinsic regulation can regulate blood flow to some extent during muscular activity, when skeletal muscles produce local changes that dilate their own arterioles (and *precapillary sphincters*, as you will see later). However, the extrinsic sympathetic system (both direct neural and indirect adrenal pathways) is necessary to coordinate bodywide needs. During locomotory activity, for example, not only does it boost cardiac output, but it can also vasodilate arterioles in skeletal muscle and heart in *anticipation* of activity, and, simultaneously, it can vasoconstrict arterioles in the digestive tract to divert more blood to skeletal muscles. This action can even override the gut's "selfish" needs such as its local demand for more blood during food processing. Again you will see this function in detail later.

TABLE 9-3 Arteriolar Smooth Muscle Adrenergic Receptors

Characteristic	RECEPTOR TYPE	
	α_1	β_2
Location of the Receptor	All arteriolar smooth muscle except in the brain	Arteriolar smooth muscle in the heart and skeletal muscles
Chemical Mediator	Norepinephrine from sympathetic fibers and the adrenal medulla Epinephrine from the adrenal medulla (less affinity for this receptor)	Epinephrine from the adrenal medulla (greater affinity for this receptor)
Arteriolar Smooth Muscle Response	Vasoconstriction	Vasodilation

© Cengage Learning, 2013

Extrinsic Control of Arterial Blood Pressure For the maintenance of proper arterial blood pressure, changes in arteriolar resistance are the main adjustments that extrinsic regulation can perform quickly. The formula $Q = \Delta P/R$ applies to the entire circulation as well as to a single vessel:

- *Q:* Looking at the circulatory system as a whole, *flow (Q) through all the vessels in either the systemic or pulmonary circulation is equal to the cardiac output, C.O.*
- ΔP: *The pressure gradient (ΔP) for the entire systemic circulation is the mean arterial pressure.* ΔP equals the difference in pressure between the beginning and the end of the systemic circulatory system. Because the beginning pressure is the mean arterial pressure as the blood leaves the left ventricle at an average of 93 mm Hg (in humans) and the end pressure in the right atrium is about 0 mm Hg, $\Delta P = 93$ mm Hg (that is, 93 minus 0). (For the pulmonary circulation, ΔP = mean pulmonary arterial pressure [15 mm Hg] minus the pressure in the left atrium [0 mm Hg] = 15 mm Hg.)
- *R:* By far the greatest percentage of the total resistance (*R*) offered by all the systemic peripheral vessels (**total peripheral resistance**) is due to arteriolar resistance.

Therefore, for the entire systemic circulation, rearranging $Q = \Delta P/R$ and using C.O. for flow Q gives us the following equation:

$$\Delta P = C.O. \times R$$
Mean arterial pressure = cardiac output ×
total peripheral resistance

Thus, *the extent of total peripheral resistance offered collectively by all the systemic arterioles influences the mean arterial blood pressure immensely.* A dam provides an analogy to this relationship. At the same time it restricts the flow of water downstream, a dam increases the pressure upstream by elevating the water level in the reservoir behind the dam. Similarly, generalized, sympathetically induced vasoconstriction reflexly reduces blood flow downstream to the tissue cells while elevating the upstream mean arterial pressure, thereby increasing the main driving force for blood flow to all the organs.

These effects seem counterproductive. Why increase the driving force for flow to the organs by increasing arterial blood pressure while reducing flow to the organs by narrowing the vessels supplying them? In effect, the sympathetically induced arteriolar tone helps maintain the appropriate driving pressure head. If all arterioles were dilated, blood pressure would fall substantially, so there would not be an adequate force for (1) overcoming gravity to the brain (in animals with heads above their main bodies), (2) proper flow to the heart, and (3) filtering pressure in the kidneys (see p. 579). Thus, tonic sympathetic activity constricts most vessels (with the exception of those in the brain), assuring an *adequate driving force for blood flow to the brain, heart, and kidneys* at the expense of organs and tissues that can better withstand reduced blood flow.

The medullary cardiovascular control center, other brain regions, and hormones regulate blood flow and pressure

The main region of the brain responsible for adjusting sympathetic output to the arterioles is the same as that for heart rate—the *cardiovascular control center* in the medulla of the brain stem. This is the integrating center for blood pressure and flow regulation. It in turn is under the influence of "higher" brain centers that can, among other actions, *anticipate* circulatory changes needed for activity. You'll see later how the all these systems work together when we look at cardiovascular integration at the chapter's end.

Several other brain regions and hormones also extrinsically influence arteriolar tone. Two major ones are *vasopressin* (also regulated by the hypothalamus and important in regulating body water balance) and *angiotensin II,* part of a hormonal pathway, the *renin–angiotensin–aldosterone pathway* (important in the regulation of the body's salt balance). Both also help maintain adequate pressure for filtration. The functions and control of these hormones are discussed in Chapters 12 and 13.

This completes our discussion of the various factors that affect total peripheral resistance, the most important of which are controlled adjustments in arteriolar radius. These factors are summarized in Figure 9-48.

check your understanding 9.12

What are the key factors that regulate dilation and constriction of arterioles, and how is such regulation useful?

9.13 Circulatory Vessels: Capillaries

Capillaries, which branch extensively to bring blood within the reach of every cell, are the ultimate functional units of circulation. As the sites for exchange of materials between the blood and tissues, they have one primary function: *to enhance diffusion*. With few exceptions, carrier-mediated transport systems do not occur across capillaries (see p. 439). Rather,

exchange of most materials across capillary walls is accomplished by the process of diffusion. Thus, evolutionary enhancements of diffusion are crucial, since diffusion is slow.

Capillary anatomy increases diffusion rates by minimizing distance ΔX while maximizing surface area A and the diffusion coefficient D of Fick's law

Recall that diffusion follows Fick's law (Table 3-1 and p. 76):

$$\text{Rate of diffusion } (Q) = \frac{\Delta C \times A \times D}{\Delta X}$$

Capillaries and regulation of their blood flow have evolved to maximize diffusion by enhancing every component of the law. Let's start with capillary anatomy, which directly minimizes diffusion distances ΔX while maximizing surface area A and the diffusion coefficient D as follows:

1. ΔX: Diffusing molecules have only a short distance (ΔX) to travel between the blood and surrounding cells because of the following factors:
   a. Capillary walls are very *thin* (1 μm in thickness; in comparison, the diameter of a human hair is 100 μm). Capillaries are composed of only a single layer of flattened endothelial cells—essentially the lining of the other vessel types. No smooth muscle or connective tissue is present (Figure 9-49a).
   b. Each capillary is so *narrow* that red blood cells have to squeeze through single file (p. 391 and Figure 9-49b). Consequently, oxygen does not have to travel far from hemoglobin inside those cells to leave the capillary (or vice versa).
   c. Because of extensive capillary *branching*, it is estimated that no cell is farther than 0.1 mm (4/1,000 inch) from a capillary.
2. A: Because capillaries are distributed in such incredible numbers (estimates range from 10 to 40 billion capillaries), a tremendous total surface area (A) is available for exchange (an estimated 600 m²). Generally, tissues that are more metabolically active have a greater density of capillaries (*vascularization*). Muscles, for example, have relatively more capillaries than their attached tendons. Despite this large number of capillaries, at any point in time they contain only 5% of the total blood volume. As a result, a small volume of blood is exposed to an extensive surface area. If all the capillary surfaces of a human were stretched out in a flat sheet and the volume of blood contained within the capillaries were spread over the top, this would be roughly equivalent to spreading a cup of paint over the floor of a high-school gymnasium. Imagine how thin the paint layer would be!
3. D: Diffusion across capillary walls also depends on the walls' permeability to the materials being exchanged (included in the D of Fick's law). The endothelial cells forming the capillary walls fit together in jigsaw-puzzle fashion. In most capillaries, narrow, water-filled pores are present at the junctions between the cells (Figure 9-50). These pores permit passage of water-soluble substances. (Lipid-soluble substances, such as O_2 and CO_2, can readily pass through the endothelial cells' lipid bilayer.)

FIGURE 9-48 **Factors affecting total peripheral resistance.** The primary determinant of total peripheral resistance is the adjustable arteriolar radius. Two major categories of factors influence arteriolar radius: (1) local (intrinsic) control, which is primarily important in matching blood flow through a tissue with the tissue's metabolic needs and is mediated by local factors acting on the arteriolar smooth muscle, and (2) extrinsic control, which is important in the regulation of blood pressure and is mediated primarily by sympathetic influence on arteriolar smooth muscle.

© Cengage Learning, 2013

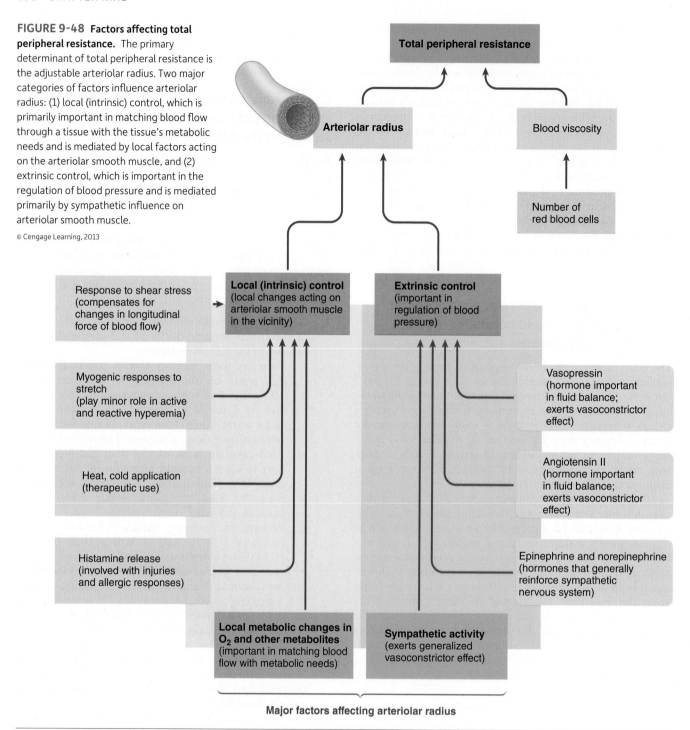

Major factors affecting arteriolar radius

The size of the capillary pores varies from organ to organ. At one extreme, the endothelial cells in brain capillaries are joined by tight junctions so that pores are nonexistent. These junctions prevent transcapillary passage of materials between the cells and thus constitute part of the protective *blood–brain barrier* (p. 169). In most tissues, small, water-soluble substances such as ions, glucose, and amino acids can readily pass through the water-filled clefts, but large, non-lipid-soluble materials such as plasma proteins are excluded from passage. At the other extreme, liver capillaries have such large pores that even proteins pass through readily. This is adaptive because the liver's functions include

synthesis of plasma proteins and the metabolism of protein-bound substances such as cholesterol (see HDLs, for example, p. 389). These proteins must all pass through the liver's capillary walls. The leakiness of capillary beds is therefore a function of how tightly the endothelial cells are joined, which varies according to the different organs' needs.

Vesicular transport also plays a limited role in the passage of materials across the capillary wall. Large non-lipid-soluble molecules such as proteinaceous hormones that must be exchanged between the blood and surrounding tissues are transported from one side of the capillary wall to the other in endocytotic–exocytotic vesicles (see p. 42).

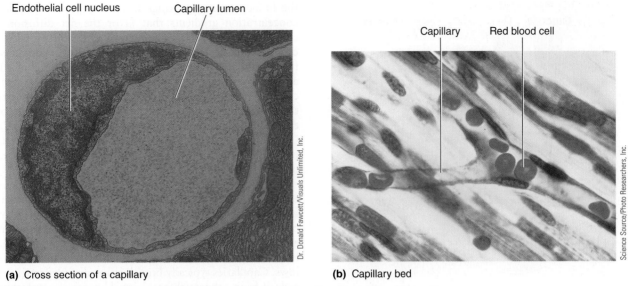

(a) Cross section of a capillary

(b) Capillary bed

FIGURE 9-49 Capillary anatomy. (a) Electron micrograph of a cross section of a capillary. The capillary wall consists of a single layer of endothelial cells. The nucleus of one of these cells is shown. (b) Photograph of a capillary bed. The capillaries are so narrow that the red blood cells must pass through single file.

Blood composition and perfusion regulation enhance capillary-tissue concentration gradients ∆C

The component of Fick's law we have not addressed is the concentration gradient ∆C between capillary blood and tissues. It too has can be enhanced by regulation of blood composition and **perfusion**—the volume per minute of blood entering a capillary bed. First, the chemical composition of arterial blood is carefully regulated to carry in oxygen, glucose, and other "desirable" solutes at concentrations usually higher than inside cells. Meanwhile, cells are constantly using up supplies and generating metabolic wastes. Diffusion of each solute continues independently as a result

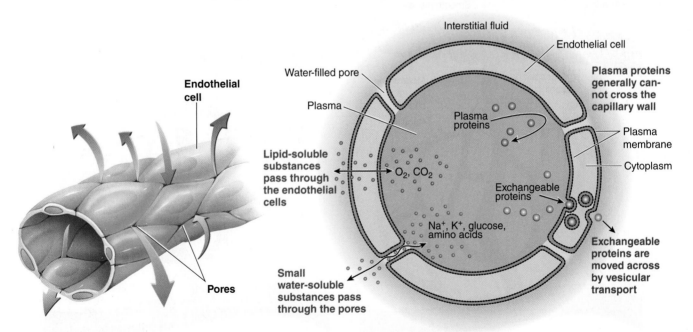

(a) Continuous capillary

(b) Transport across a continuous capillary wall

FIGURE 9-50 Exchanges across a continuous capillary wall, the most common type of capillary.
(a) Slitlike gaps between adjacent endothelial cells form pores within the capillary wall. (b) As depicted in this cross section of a capillary wall, small water-soluble substances are exchanged between the plasma and the interstitial fluid by passing through the water-filled pores, whereas lipid-soluble substances are exchanged across the capillary wall by passing through the endothelial cells. Proteins to be moved across are exchanged by vesicular transport. Plasma proteins generally cannot escape from the plasma across the capillary wall.

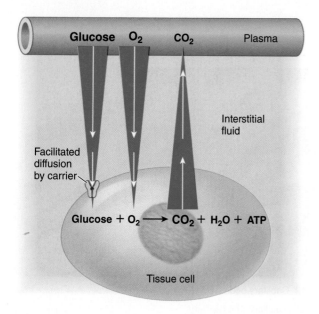

the blood constantly brings in fresh supplies, maintaining concentration gradients that favor the net diffusion into cells. Simultaneously, ongoing net diffusion of CO_2 and other metabolic wastes from cells to blood is maintained by the continual production of these wastes at the cellular level and their constant removal from the tissue level by the circulating blood. Also, as cells increase their level of activity, they use up more O_2 and produce more CO_2, among other things. This creates larger concentration gradients (ΔC) for O_2 and CO_2 between cells and blood, so more O_2 diffuses out of the blood into the cells and more CO_2 proceeds in the opposite direction.

Second, regulation of perfusion through capillaries also enhances concentration gradients. We have already discussed how this occurs with arterioles, which constrict or dilate to control capillary perfusion. This alters concentration gradients, for example, by bringing in more fresh, oxygenated blood. Capillaries beds can also influence perfusion, as follows. Capillaries typically branch either directly from an arteriole or from a thoroughfare channel known as a **metarteriole**, which runs between an arteriole and a venule (Figure 9-52). Unlike the true capillaries within a capillary bed, metarterioles are sparsely surrounded by wisps of spiraling smooth muscle cells to form **precapillary sphincters**, each consisting of a ring of smooth muscle around the entrance to a capillary as it arises from a metarteriole. Precapillary sphincters are not innervated, but they have a high degree of myogenic tone, such that most are closed in a resting tissue (e.g., 90% closed in resting muscle). Importantly, they are sensitive to local

of the concentration difference ΔC for that solute between the blood and surrounding cells (Figure 9-51). This process repeats itself continuously. As cells use up O_2 and glucose,

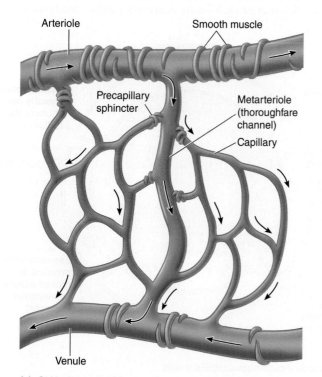

(a) Sphincters relaxed

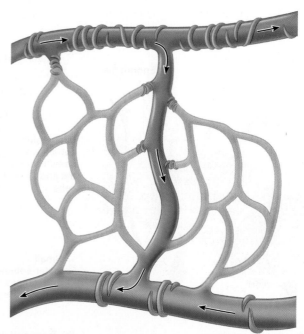

(b) Sphincters contracted

FIGURE 9-52 Capillary bed. Capillaries branch either directly from an arteriole or from a metarteriole, a thoroughfare channel between an arteriole and venule. Capillaries rejoin at either a venule or a metarteriole. Smooth muscle cells form precapillary sphincters that encircle capillaries as they arise from a metarteriole. (a) When the precapillary sphincters are relaxed, blood flows through the entire capillary bed. (b) When the precapillary sphincters are contracted, blood flows only through the metarteriole, bypassing the capillary bed.

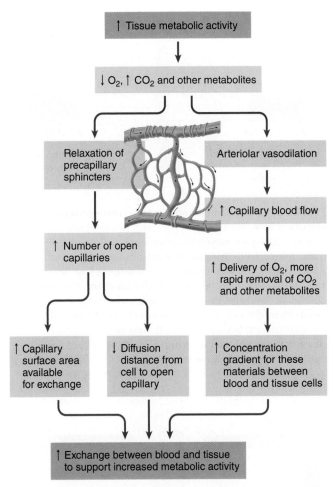

FIGURE 9-53 **Complementary action of precapillary sphincters and arterioles in adjusting blood flow through a tissue in response to changing metabolic needs.**

© Cengage Learning, 2013

metabolic changes, acting as stopcocks to control perfusion through the particular capillary that each one guards.

Thus, arterioles and precapillary sphincters both regulate perfusion, for example during activity. As chemical concentrations start to change during activity in a region of a muscle tissue supplied by closed-down capillaries, both precapillary sphincters and arterioles in the region relax. As a result of higher perfusion due to more open capillaries, concentration gradients increase. Other parts of Fick's law are also enhanced: As capillaries open, total volume and surface area available for exchange increase, and the diffusion distance between the cells and blood decreases (Figure 9-53). Restoration of the chemical concentrations to normal causes precapillary sphincters to close once again and the arterioles to return to normal tone.

Capillary anatomy increases time for diffusion by slowing blood flow due to a high total cross-sectional area

Despite this enhancement to Fick's law, diffusion is still slow. Capillaries thus have another important feature: *Blood flows much more slowly* in them than elsewhere in the circulatory system, allowing more time for diffusion to occur.

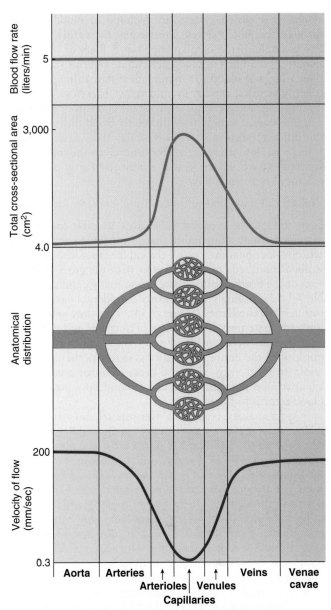

FIGURE 9-54 **Comparison of blood flow rate and velocity of flow in relation to total cross-sectional area.** The blood flow rate (*red curve*) is identical through all levels of the circulatory system and is equal to the cardiac output (5 liters/min at rest in humans). The velocity of flow (*purple curve*) varies throughout the vascular tree and is inversely proportional to the total cross-sectional area (*green curve*) of all the vessels at a given level. Note that the velocity of flow is slowest in the capillaries, which have the largest total cross-sectional area.

© Cengage Learning, 2013

The extensive capillary branching is responsible for this slow velocity of blood flow through the capillaries. Let's see how blood slows in the capillaries.

First let us clarify a potentially confusing point. The term *flow* can be used in two different contexts—the *flow rate*, which refers to the *volume* of blood flowing through a given segment of the circulatory system per unit of time (this is the flow we have been talking about in relation to the pressure gradient and resistance), and *velocity of flow*, which refers to the linear *speed* with which blood flows forward from one point to another. Because the circulatory system is a closed system, the volume of blood flowing through the whole system must equal the cardiac output. For example, if the heart pumps out 5 liters of blood per minute, and 5 liters of blood per minute return to the heart, then 5 liters of blood per minute must flow through the arteries, arterioles, capillaries, and veins. Therefore, the flow rate (*Q*) is the same at all levels of the circulatory system.

However, the *velocity* with which blood flows through the different segments of the vascular tree varies because velocity of flow is inversely proportional to the total cross-sectional area of all the vessels at any given level of the circulatory system as follows:

Velocity of flow (cm/sec) = flow rate (cm³/sec) / πr^2 (cm²)

Even though the cross-sectional area of each capillary is extremely small compared to that of the large aorta, the *total* cross-sectional area of all the capillaries added together is about hundreds of times greater than the cross-sectional area of the aorta because there are so many capillaries (Table 9-1). Accordingly, blood slows considerably as it passes through the capillaries (Figure 9-54). This slow velocity allows adequate time for exchange of nutrients and metabolic end products between blood and tissues, which is the sole purpose of the entire circulatory system. As the capillaries rejoin to form veins, the total cross-sectional area is once again reduced, and the velocity of blood flow increases as blood returns to the heart.

As an analogy, consider a river (the arterial system) that widens into a swamp with many small, parallel channels (the capillaries), then narrows into a river again (the venous system) (Figure 9-55). The flow rate is the same throughout the length of this body of water, that is, identical volumes of water are flowing past all the points along the bank of the river and through the whole swamp. However, the velocity of flow is slower in the wide swamp than in the narrow river because the identical volume of water, now spread out over a larger cross-sectional area, moves forward a much shorter distance in the wide swamp (even though it is in many channels) than in the narrow river during a given period of time. You could readily observe the forward movement of water in the swift-flowing river, but the forward motion of water in the swamp would be harder to detect. Notice that velocity increases in the outgoing river even though driving force may be less at this point, simply because of geometry.

Interstitial fluid is a passive intermediary between the blood and cells

Exchanges between blood and the tissue cells are not made directly. Interstitial fluid, the true internal environment in immediate contact with the cells, acts as the go-between (Figure 9-56). This fluid is in a sense the evolutionary descendant of an open circulatory system. In fact, only about 20% of mammalian ECF circulates as plasma, with the remaining 80% in the interstitial fluid. Cells exchange materials directly with this fluid, the type and extent of exchange being governed by the properties of the cellular plasma membranes. Exchange is usually so thorough that the interstitial fluid takes on essentially the same composition as the incoming arterial blood, with the exception of the large plasma proteins that usually do not escape from the blood. Therefore, when we speak of exchanges between blood and tissue cells, we tacitly include interstitial fluid as a passive intermediary.

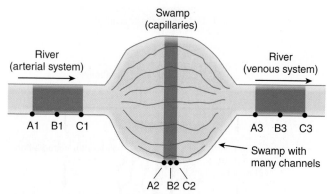

FIGURE 9-55 Relationship between total cross-sectional area and velocity of flow. The three dark blue areas represent equal volumes of water. During one minute, this volume of water moves forward from points A to points C. Therefore, an identical volume of water flows past points B1, B2, and B3 during this minute; that is, the flow rate is the same at all points along the length of this body of water. However, during that minute the identical volume of water moves forward a much shorter distance in the wide swamp (A2 to C2) than in the much narrower river (A1 to C1 and A3 to C3). Thus, velocity of flow is much slower in the swamp than in the river. Similarly, velocity of flow is much slower in the capillaries than in the arterial and venous systems.

© Cengage Learning, 2013

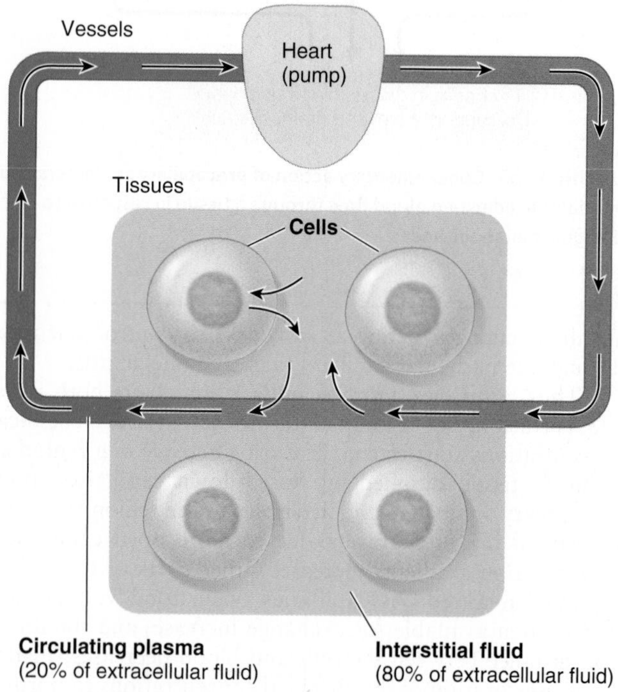

FIGURE 9-56 Interstitial fluid acting as an intermediary between blood and cells.

© Cengage Learning, 2013

Bulk flow across the capillary wall is important in extracellular fluid distribution

Although, as we noted earlier, passive diffusion down concentration gradients is the primary mechanism for exchange of individual solutes, *bulk flow* also occurs (recall that bulk flow or transport refers to constituents of the fluid moving

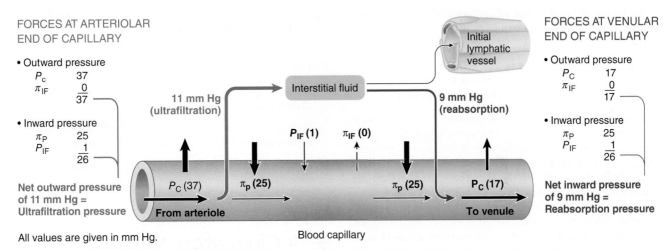

FORCES AT ARTERIOLAR
END OF CAPILLARY

• Outward pressure

P_C	37
π_{IF}	0
	37

• Inward pressure

π_P	25
P_{IF}	1
	26

Net outward pressure
of 11 mm Hg =
Ultrafiltration pressure

FORCES AT VENULAR
END OF CAPILLARY

• Outward pressure

P_C	17
π_{IF}	0
	17

• Inward pressure

π_P	25
P_{IF}	1
	26

Net inward pressure
of 9 mm Hg =
Reabsorption pressure

11 mm Hg (ultrafiltration) — Interstitial fluid — 9 mm Hg (reabsorption) — Initial lymphatic vessel

P_{IF} (1) π_{IF} (0)

P_C (37) π_p (25) π_p (25) P_C (17)

From arteriole   **To venule**

All values are given in mm Hg.   Blood capillary

FIGURE 9-57 Bulk flow across the capillary wall. Ultrafiltration occurs at the arteriolar end and reabsorption occurs at the venule end of the capillary as a result of imbalances in the physical forces acting across the capillary wall.

© Cengage Learning, 2013

together in bulk with the fluid itself). This process accomplishes the totally different function of determining the distribution of the ECF volume between the vascular and interstitial fluid compartments. Let's see how.

A volume of protein-free plasma actually filters out of the capillary, mixes with the surrounding interstitial fluid, and is subsequently reabsorbed. When pressure inside the capillary exceeds pressure on the outside, fluid is pushed out through the pores. The majority of the plasma proteins and blood cells are retained on the inside during this process, although a few do escape. This process is known as **ultrafiltration** (which we explore further in kidneys, Chapter 12). Because all other constituents in the plasma are carried along with the fluid leaving the capillary, the filtrate is essentially a protein-free plasma. When inward-driving pressures exceed outward pressures across the capillary wall, net inward movement of fluid from the interstitial fluid compartment into the capillaries takes place through the pores, a process known as **reabsorption.**

Factors Influencing Bulk Flow Bulk flow in this form essentially makes the so-called closed circulatory system into an open system, creating a cell-bathing fluid analogous to hemolymph. However, as you will see, its primary purpose is not molecular exchange.

Bulk flow occurs because of differences in the *hydrostatic* and *colloid osmotic pressures* between the plasma and interstitial fluid (Figure 9-57):

1. **Capillary blood pressure (P_C)** is the hydrostatic pressure exerted on the inside of the capillary walls by the blood. This pressure tends to force fluid *out of* the capillaries into the interstitial fluid. Mean blood pressure has dropped substantially by the level of the capillaries because of frictional losses in pressure in the high-resistance arterioles upstream. On the average, the hydrostatic pressure is 37 mm Hg at the arteriolar end of a tissue capillary and has declined even further to 17 mm Hg at the venular end.

2. **Plasma-colloid osmotic pressure (Π_p)** is a force caused by the colloidal dispersion of plasma proteins (p. 388); it arises because the plasma proteins remain in the plasma rather than entering the interstitial fluid. Accordingly, there is also an osmotic difference between these two regions, causing water movement from the interstitial fluid *into* the capillaries by *osmosis* (pp. 77–80). (The other plasma constituents are equal in the plasma and interstitial fluid.) The plasma-colloid osmotic pressure averages 25 mm Hg in humans and cats, 20 in dogs, 11 in chicks, 6 in turtles, and 5 in frogs.

3. **Interstitial-fluid hydrostatic pressure (P_{IF})** is the fluid pressure exerted on the outside of the capillary wall by the interstitial fluid. This pressure tends to force fluid *into* the capillaries. Because of the difficulties encountered in measuring interstitial-fluid hydrostatic pressure, the actual value of the pressure is a controversial issue. It is either at, slightly above, or slightly below atmospheric pressure. For purposes of illustration, we will say it is 1 mm Hg above atmospheric pressure.

4. **Interstitial fluid–colloid osmotic pressure (Π_{IF})** results from the small fraction of plasma proteins that leak across the capillary walls into the interstitial spaces (these are normally returned to the blood by means of the lymphatic system). This protein concentration is extremely low, so the interstitial fluid–colloid osmotic pressure is very close to zero. If plasma proteins pathologically leak into the interstitial fluid, however, as they do when *histamine* widens the intercellular clefts during tissue injury (p. 467), the leaked proteins exert an osmotic effect that tends to promote movement of fluid *out of* the capillaries into the interstitial fluid.

In summary, P_C and Π_{IF} cause movement out of the capillaries, while Π_p and P_{IF} do the opposite. Now let's analyze the fluid movement that occurs from these forces (Figure 9-57).

Net Exchange of Fluid across a Capillary Wall Net exchange at a given point across the capillary wall can be calculated as follows:

$$\text{Net exchange pressure} = \underset{\substack{\text{(outward} \\ \text{pressure)}}}{(P_C + \Pi_{IF})} - \underset{\substack{\text{(inward} \\ \text{pressure)}}}{(\Pi_P + P_{IF})}$$

A positive net pressure (outward pressure > inward pressure) represents an ultrafiltration pressure, which takes place at the beginning of the capillary as this outward pressure gradient forces a protein-free filtrate through the capillary pores, largely due to P_C. By the time the venular end of the capillary is reached, P_C has dropped, but the other pressures have remained essentially constant, so the outward pressure has fallen below that of the inward pressure. Reabsorption of fluid therefore takes place as the pressure gradient now forces fluid back into the capillary. Ultrafiltration and reabsorption, which collectively create bulk flow, are thus due to a shift in the balance between the physical forces acting across the capillary wall. No local energy expenditures are involved in this flow.

It is important to realize that we have been discussing "snapshots" at two points—at the beginning and at the end—in a hypothetical idealized capillary. The pressures used in Figure 9-57 are mammalian values, and controversial at that. In fact, a recent theory proposes that net ultrafiltration occurs throughout the length of all *open* capillaries (which would have high outward pressures), whereas net reabsorption occurs throughout the length of all *closed* capillaries (which would have low outward pressures).

Role of Bulk Flow in Fluid Balance Bulk flow plays only a minor role in the exchange of individual solutes between blood and tissues because the quantity of solutes moved across the capillary wall by bulk flow is small compared to the much larger transfer of solutes by direct diffusion. So, although this process is somewhat akin to hemolymph flow, its primary function is different. Essentially, it is extremely important in regulating the distribution of ECF between the plasma and interstitial fluid. Maintenance of proper arterial blood pressure depends in part on an appropriate volume of circulating blood. If plasma volume is reduced (for example, by hemorrhage), blood pressure falls. The resultant lowering of capillary blood pressure alters the balance of forces across the capillary walls. Because the net outward pressure is decreased while the net inward pressure remains unchanged, extra fluid is shifted from the interstitial compartment into the plasma as a result of reduced filtration and increased reabsorption. The extra fluid soaked up from the interstitial fluid provides additional fluid for the plasma, temporarily compensating for the loss of blood. Meanwhile, reflex mechanisms acting on the heart and blood vessels (to be described later) also come into play to help maintain blood pressure until long-term mechanisms, such as thirst and reduction of urinary output, can restore the fluid volume to compensate for the loss.

Conversely, if the plasma volume becomes overexpanded, as with excessive fluid intake, the resultant elevation in capillary blood pressure forces extra fluid from the capillaries into the interstitial fluid, temporarily relieving the expanded plasma volume until the excess fluid can be eliminated from the body by long-term measures, such as increased urinary output.

check your understanding 9.13

Explain how capillaries enhance all components of Fick's law of diffusion.

Distinguish between velocity and volume flow of blood, how capillaries affect velocity, and why that is important.

Explain how bulk flow occurs in and out of capillaries, and its importance.

9.14 Circulatory Vessels: Lymphatic System

Even under normal circumstances, slightly more fluid is filtered out of the capillaries into the interstitial fluid than is reabsorbed from the interstitial fluid back into the plasma. On average, the net ultrafiltration pressure at the beginning of the capillary is slightly higher than the net reabsorption pressure at the vessel's end. Because of this pressure differential, on average more fluid is filtered out of the first half of the capillary than is reabsorbed in its last half. The extra fluid filtered out as a result of this deficient reabsorption is picked up by the *lymphatic system*. This accessory "open" circulation has at least two additional functions, in immune defense and in fat transport.

The lymphatic system is an accessory route by which interstitial fluid can be returned to the blood

The **lymphatic system** consists of an extensive network of one-way vessels for returning the remaining interstitial fluid to the blood. It functions much like a storm sewer that picks up and carries away excess rainwater so that it does not accumulate and flood an area. Even though only a small fraction of the filtered fluid is not reabsorbed by the blood capillaries, the cumulative effect of this process being repeated with every heartbeat results in the equivalent of more than the entire plasma volume being left behind in the interstitial fluid each day. Obviously, this fluid must be returned to the circulating plasma, and lymph vessels accomplish this task. The system is thought to occur in all vertebrates, though lymphatic vessels have been documented in only one bony fish species, the zebra fish.

Pickup and Flow of Lymph Small, blind-ended terminal lymph vessels known as **initial lymphatics** permeate almost every tissue of the body (Figure 9-58a). The endothelial cells forming the walls of initial lymphatics slightly overlap like shingles on a roof, with their overlapping edges being free instead of attached to the surrounding cells. This arrangement creates one-way, valvelike openings in the vessel wall. Fluid pressure on the outside of the vessel pushes the innermost edge of a pair of overlapping edges inward, creating a gap between the edges (that is, opening the valve), thus permitting interstitial fluid to enter (Figure 9-58b). Once interstitial fluid enters a lymphatic vessel, it is called **lymph.** Fluid pressure on the inside forces the overlapping edges together, closing the valves so that lymph does not escape. These valvelike lymphatic openings are much larger than the pores in blood capillaries. Consequently, large particulates in the interstitial fluid, such as escaped plasma proteins and bacteria, can gain access to initial lymphatics but are excluded from blood capillaries.

Initial lymphatics converge to form larger and larger **lymph vessels,** which eventually empty into the venous system near the point where the blood enters the right atrium (Figure 9-59). In all amphibians and reptiles, as well as bird embryos, there are **lymph hearts** (usually in pairs on each side of the body) that create flow back to the blood. The flow of lymph through these hearts are assisted by movements of skeletal muscles and lungs.

Mammals and most adult birds (ostriches are an exception) do not have these distinct auxiliary hearts; flow is accomplished by two mechanisms. First, lymph vessels beyond the initial lymphatics are surrounded by smooth muscle, which contracts rhythmically as a result of myogenic initiation of action potentials. When this muscle is stretched because the vessel is distended with lymph, the muscle inherently contracts more forcefully, thereby pushing the lymph through the vessel. This intrinsic "lymph pump" is the major force for propelling lymph. Second, because lymph vessels lie between skeletal muscles, contraction of these extrinsic muscles squeezes the lymph out of the vessels. One-way valves spaced at intervals within the lymph vessels ensure the flow of lymph toward its venous outlet in the chest.

The lymphatic system is important for immune functions and fat transport

The lymph percolates through **lymph nodes** located en route within the lymphatic system. Passage of this fluid through the lymph nodes is an important aspect of immune defenses against disease. This is covered in Chapter 10. The system is also important in the absorption of fat from the digestive tract. As described in detail in Chapter 14, the end products of the digestion of dietary fats are packaged by cells lining the digestive tract into large particles (*chylomicrons,* pp. 389 and 700) that are too large to gain access to the blood capillaries but can easily enter the initial lymphatics.

check your understanding 9.14

How does lymph form? What are the functions of lymph and the lymphatic system?

9.15 Circulatory Vessels: Venules and Veins

The venous system completes the circulatory circuit. Recall that blood leaving the capillary beds enters into venules and then veins for transport back to the heart (Figure 9-41).

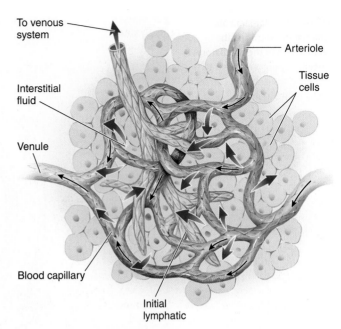

(a) Relationship between initial lymphatics and blood capillaries

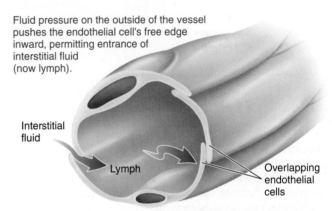

Fluid pressure on the outside of the vessel pushes the endothelial cell's free edge inward, permitting entrance of interstitial fluid (now lymph).

Fluid pressure on the inside of the vessel forces the overlapping edges together so that lymph cannot escape.

(b) Arrangement of endothelial cells in an initial lymphatic

FIGURE 9-58 Initial lymphatics. (a) Blind-ended initial lymphatics pick up excess fluid filtered by blood capillaries and return it to the venous system in the chest. (b) Note that the overlapping edges of the endothelial cells create valvelike openings in the vessel wall.

© Cengage Learning, 2013

Veins serve as a blood reservoir as well as passageways back to the heart

Veins have large radii, so they offer little resistance to flow. Furthermore, because the total cross-sectional area of the venous system gradually decreases as smaller veins converge into progressively fewer but larger vessels, the velocity of blood flow *increases* as the blood approaches the heart despite the lower pressure (to review why, see Figure 9-55).

In addition to serving as low-resistance passageways to return blood from the tissues to the heart, systemic veins

FIGURE 9-59 Mammalian lymphatic system and its relationship to the circulatory system. Lymph empties into the venous system near its entrance to the right atrium.

© Cengage Learning, 2013

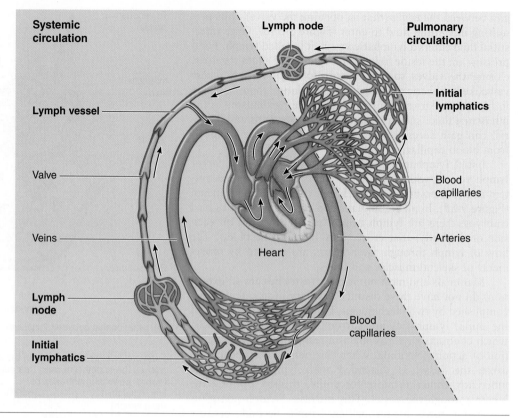

also serve as a *blood reservoir*. Because of their storage capacity, veins are often referred to as **capacitance vessels.** Veins have much thinner walls with less smooth muscle than do arteries. Because collagen fibers are considerably more abundant than elastin fibers in venous connective tissue, veins have very little elasticity, in contrast to arteries. Also, unlike arteriolar smooth muscle, venous smooth muscle has little inherent myogenic tone. Because of these features, veins are highly distensible, or stretchable, and have little elastic recoil. They easily distend to accommodate additional volumes of blood with only a small increase in venous pressure. Arteries stretched by an excess volume of blood recoil (p. 430), but veins containing an extra volume of blood simply stretch to accommodate the additional blood without tending to recoil. In this way veins serve as a blood reservoir; that is, when demands for blood are low, the veins can store extra blood in reserve because of their distensibility. (Note that blood in the veins is not normally stagnant, but is moving constantly.) Under resting conditions, mammalian veins contain more than 60% of the total blood volume. When the stored blood is needed, such as during exercise, extrinsic factors (soon to be described) drive the extra blood from the veins to the heart, inducing an increased cardiac stroke volume in accordance with the Frank-Starling law of the heart (see p. 417). If too much blood pools in the veins instead of being returned to the heart, cardiac output is abnormally diminished. Thus, a delicate balance exists between the capacity of the veins, the extent of venous return, and the cardiac output. Let's now examine the factors that affect venous capacity and contribute to venous return.

Venous return is enhanced by a number of extrinsic factors

Venous capacity (the volume of blood that the veins can accommodate) depends on the distensibility of the vein walls (how much they can stretch to hold blood) and the influence of any externally applied pressure squeezing inwardly on the veins. At a constant blood volume, as venous capacity increases, more blood remains in the veins instead of being returned to the heart. Such venous storage decreases the effective circulating volume. Conversely, when venous capacity decreases more blood is returned to the heart and continues circulating. Thus, changes in venous capacity directly influence the magnitude of venous return, which in turn is an important (although not the only) determinant of effective circulating blood volume. The magnitude of the total blood volume is also influenced on a short-term basis by passive shifts in bulk flow between the vascular and interstitial fluid compartments and on a long-term basis by factors that control total ECF volume, such as salt and water balance.

Recall that *venous return* refers to the volume of blood entering each atrium per minute from the veins (pp. 416–417), and that the magnitude of flow through a vessel is directly proportional to the pressure gradient ΔP. Much of the driving pressure imparted to the blood by cardiac contraction has been lost by the time the blood reaches the venous system because of frictional losses along the way, especially in the high-resistance arterioles. By the time the blood enters the venous system, mean pressure in humans averages only 17 mm Hg (Figure 9-45, p. 433). However, because atrial

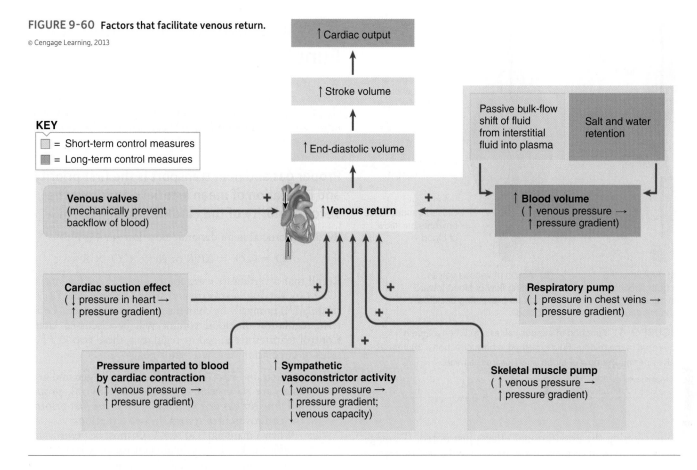

FIGURE 9-60 Factors that facilitate venous return.
© Cengage Learning, 2013

pressure is near 0 mm Hg, a small but adequate driving pressure still exists to promote the flow of blood through the large-radius, low-resistance veins.

In addition to the driving pressure imparted by cardiac contraction, five other factors enhance venous return: sympathetically induced venous vasoconstriction, skeletal muscle activity, the effect of venous valves, respiratory activity, and the effect of cardiac suction (Figure 9-60). Most of these secondary factors affect venous return by influencing the ∆P between the veins and the heart. We examine each in turn.

Effect of Sympathetic Activity on Venous Return Veins are not very muscular and have little inherent tone, but venous smooth muscle is abundantly supplied with sympathetic nerve fibers. Sympathetic stimulation produces vasoconstriction, which modestly elevates venous pressure; this, in turn, increases the pressure gradient to drive more of the stored blood from the veins into the right atrium. Even when constricted, the veins still have a relatively large diameter, thus minimizing any effect of resistance on flow.

It is important to recognize the different outcomes of vasoconstriction in arterioles and veins. Arteriolar vasoconstriction immediately *reduces* flow through these vessels because of their increased resistance (less blood can enter and flow through a narrowed arteriole), whereas venous vasoconstriction immediately *increases* flow through these vessels because of their decreased capacity (narrowing of veins

squeezes out more of the blood that is already present in the veins, thus increasing blood flow through these vessels).

Effect of Skeletal Muscle Activity on Venous Return Many of the large veins in the extremities lie between skeletal muscles, so when the muscles contract, the veins are compressed. This external venous compression decreases venous capacity and increases venous pressure, in effect squeezing fluid contained in the veins forward toward the heart. This pumping action, known as the **skeletal muscle pump** (p. 400), is one way extra blood stored in the veins is returned to the heart during locomotory activity. Increased sympathetic vasoconstriction also accompanies locomotory activity, further enhancing venous return. This also helps overcome the effects of gravity, which tends to cause blood to pool in the lower extremities of tall terrestrial animals.

Effect of Venous Valves on Venous Return Venous vasoconstriction and skeletal pumping both drive blood in the direction of the heart. Yet if you squeeze a fluid-filled tube in the middle, fluid is pushed in both directions from the point of constriction (Figure 9-61a). Why, then, is blood not driven backward as well as forward? The answer is that the large veins are equipped with *one-way valves* (spaced at 2- to 4-cm intervals in humans), which permit blood to move forward toward the heart but prevent it from moving back (Figure 9-61b). These valves also play a role in

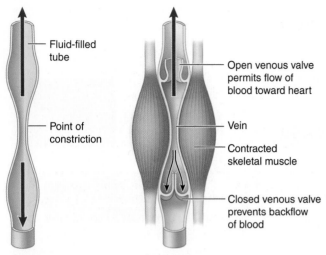

(a) **Fluid moving in both directions on squeezing a fluid-filled tube**

(b) **Action of venous valves, permitting flow of blood toward heart and preventing backflow of blood**

Fluid-filled tube

Point of constriction

Open venous valve permits flow of blood toward heart

Vein

Contracted skeletal muscle

Closed venous valve prevents backflow of blood

FIGURE 9-61 Function of venous valves. (a) When a tube is squeezed in the middle, fluid is pushed in both directions. (b) Venous valves permit the flow of blood only toward the heart.

© Cengage Learning, 2013

counteracting the gravitational effects just noted. (Damage to these valves in humans leads to so-called *varicose veins*.)

Effect of Respiratory Activity on Venous Return As a result of respiratory activity, the pressure within the mammalian chest cavity averages 5 mm Hg less than atmospheric pressure (see Chapter 11). As the venous system returns blood to the heart from the lower regions of the body, it travels through the chest cavity, where it is exposed to this sub-atmospheric pressure. Thus, a pressure gradient exists between the lower veins (which are at atmospheric pressure) and the chest veins (at 5 mm Hg less than atmospheric pressure). This difference enhances flow from the lower veins to the chest veins, promoting increased venous return. This facilitation of venous return is known as the **respiratory pump**. Increased respiratory activity as well as skeletal muscle pumping and venous vasoconstriction all enhance venous return during locomotory activity.

Effect of Cardiac Suction on Venous Return The extent of cardiac filling may not depend entirely on factors affecting the veins. The heart may play a role in its own filling, as we noted earlier, by a suction effect. This would also enhance venous return. Biologists have traditionally considered that suction is the dominant force in heart filling in fishes, whereas venous pressure is the primary force in mammals. However, recent studies show that venous return pressure is probably dominant in fishes and perhaps in all vertebrates, so that suction may be a minor effect.

check your understanding 9.15

Explain the ways that veins return blood to the heart, including external pumping and valves.

9.16 Integrated Cardiovascular Function

We have now examined the three major components—fluid, pumps, and vessels—of a circulatory system. We end this chapter by examining how the components are integrated into a smoothly functioning delivery system—the *cardiovascular* system.

Proper gas and heat transport is the first priority and regulation of mean arterial pressure is the second priority of the cardiovascular system

Recall the crucial hemodynamic flow law (p. 419):

$$Q = C.O. = \Delta P/R \text{ or } \Delta P = C.O. \times R$$

Recall that a vertebrate can rapidly alter (1) *cardiac output (C.O.)* by varying heart rate and stroke volume, and (2) *resistance (R)* primarily by changing arteriole diameter. (Long-term regulation of blood volume will be discussed later.) Control features are aimed at regulating these two key features of circulation, for two major goals:

1. *Proper gas and heat transport.* Because oxygen and heat balance are so crucial to short-term survival, *proper blood flow for gas and heat transport is the first priority* of the cardiovascular system, in two situations:
   a. When a mammal is at rest, C.O. must be *homeostatically* regulated to maintain consistent delivery of O_2 to key organs, especially brain, heart, and kidneys (and, CO_2 and acid must be removed properly).
   b. During activity, C.O. cannot be homeostatic but must be *reset* higher to boost flow to skeletal muscles, and often to the skin for heat removal.
2. *Arterial blood pressure.* This is somewhat homeostatically regulated, for three reasons, the first of which directly assists gas transport, the other two as secondary priorities:
   a. Pressure must be high enough to overcome gravity, friction, and other resistance factors; without proper pressure, the brain and other tissues would not receive adequate flow, no matter what local adjustments are made in the resistance of the arterioles supplying them. For example, if diastolic pressure drops to 50 mm Hg or less, blood cannot overcome gravity to reach the top of the human brain. Here, pressure is regulated to protect C.O. to the brain, the first priority of the system.
   b. Pressure must be high enough for ultrafiltration in the kidneys; in mammals, a minimum average pressure of about 80 mm Hg is needed for this. Here pressure itself, not flow, is the goal of regulation.
   c. Pressure must not be so high that it creates extra work for the heart and increases the risk of vascular damage and possible rupture of small blood vessels. Again, pressure rather than flow is the direct goal of regulation. Thus, during activity when C.O. must be reset higher, *total R in the body must be lowered* to keep the pressure from rising excessively. However, we stated that this is only somewhat homeostatic because pressures can vary without harm over a moderate range.

Overall *homeostasis of arterial pressure is the second regulatory priority*, with transport needs "trumping" pressure needs if there is a conflict. This can be seen in two situations:

- *Hypertension* (chronic high blood pressure), a common disorder of some mammals such as humans and cats. Excessive resistance R in these animals impedes flow and so ΔP is increased by the cardiovascular integrator in the medulla oblongata to restore proper C.O. In theory, the integrator could let C.O. drop to keep ΔP normal. But it does not. It allows ΔP to rise so that C.O. remains normal. (Examine the flow law to make sure you understand these physical interactions.) Presumably this response evolved because low C.O. (and thus low O_2 delivery) can be lethal within minutes, whereas high blood pressure may take years to kill.

  Factors causing hypertension are often uncertain, Experiments with high-salt diets result in elevated arterial pressures in trout and in some (but not all) humans, but the mechanism is unclear. In other cases, hypertension is a symptom of certain diseases. For example, kidney failure with resulting fluid retention is a major cause in cats. In addition, factors that harden and obstruct the arteries (p. 389) eventually impede blood flow with the consequence that blood pressure increases as a means to deliver sufficient O_2 to the brain.
- *Temperature regulation.* The hypothalamus will dilate cutaneous (skin) arterioles to eliminate excess heat from a mammal's body. As a result, blood pressure can fall even though the reflex responses (which we will examine shortly) are calling for cutaneous vasoconstriction to help maintain adequate total arteriolar resistance (see Chapter 15).

Gas transport and blood pressure are monitored by arterial sensors and regulated in part by the cardiovascular control center

As we noted earlier (pp. 416 and 437), the **cardiovascular control center** in the medulla can regulate cardiac output and arteriolar resistance. As an integrator, it receives input from sensors on the afferent side. Its efferent pathway is the autonomic nervous system. Let's see how the system works in negative feedback for the two priorities:

1. *Proper blood flow for gas transport.* Delivery of O_2 and removal of CO_2 are monitored by arterial **chemosensors** in the carotid and aortic arteries. They are sensitive to low O_2 or high acid levels in the blood. These chemoreceptors' main function is to reflexly increase respiratory activity to bring in more O_2 or to blow off more acid-forming CO_2, but they also reflexly increase blood flow by sending excitatory impulses to the cardiovascular center. Regulation using these chemosensors is also discussed in Chapter 11. Also, as you will see soon, during muscular activity the cardiovascular center can be activated in *anticipation* fashion by higher brain centers.
2. *Arterial pressure* is constantly monitored by **baroreceptors** (pressure sensors, a type of mechanoreceptor type; p. 209). The most important vertebrate baroreceptors involved in regulation of blood pressure are located in

the *aortic arch* (in amphibians, reptiles, birds, and mammals) and the *carotid sinus* (in mammals only). These are strategically located (Figure 9-62) to provide critical information about arterial blood pressure in (1) the vessels leading to the brain (the carotid sinus baroreceptor), for protecting blood flow to the brain, and in (2) the major arterial trunk before it gives off branches that supply the rest of the body (the aortic arch baroreceptor), particularly important for protecting blood flow to the heart. When deviations from normal are detected, reflex responses are initiated. Baroreceptors are sensitive to changes in both mean arterial pressure and pulse pressure. Their responsiveness to fluctuations in pulse pressure enhances their sensitivity as pressure sensors because small changes in systolic or diastolic pressure may alter the pulse pressure without changing the mean pressure.

Next let's look at those reflexes in more detail in a resting animal, then see how the system regulates gas transport and pressure during increasing muscular activity.

The baroreflex is the most important mechanism for short-term regulation of blood pressure during rest or low activity

Any change in mean blood pressure triggers an autonomically mediated **baroreflex** that influences the heart and blood vessels to adjust cardiac output and total peripheral resistance in an attempt to restore blood pressure to normal. Like any reflex, the baroreflex includes a receptor, an afferent pathway, an integrating center, an efferent pathway, and effector organs:

- The baroreceptors constantly provide information to the cardiovascular center about blood pressure by continuously generating action potentials in response to the ongoing pressure within the arteries. When arterial pressure (either mean or pulse pressure) increases, the receptor potential of these receptors increases, thus increasing the rate of firing in the corresponding afferent neurons. The converse happens when blood pressure decreases (Figure 9-63).
- The integrator, the cardiovascular control center, alters the ratio between sympathetic and parasympathetic activity to the effector organs (the heart and blood vessels). To show how autonomic changes alter arterial blood pressure, Figure 9-64 provides a review of the major effects of parasympathetic and sympathetic stimulation on the heart and blood vessels.

Let's fit all the pieces of the baroreflex together now by tracing the reflex activity that occurs to compensate for an elevation or fall in blood pressure by negative feedback. Imagine that for some reason arterial pressure rises above normal (Figure 9-65a); this could happen when a giraffe lowers its head to drink, as gravity now adds to the blood pressure going to the brain rather than reducing it. The carotid sinus baroreceptors increase the rate of firing in their respective afferent neurons. On being informed by increased afferent firing that arterial pressure has become too high, the cardiovascular control center responds by decreasing sympathetic and increasing parasympathetic activity to the car-

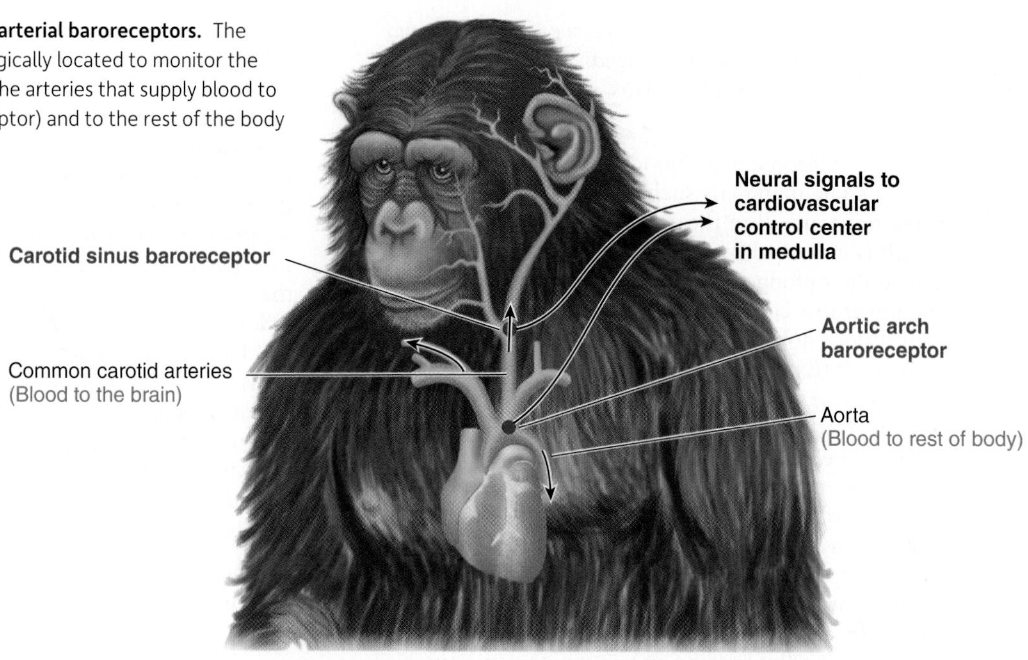

FIGURE 9-62 Location of the arterial baroreceptors. The arterial baroreceptors are strategically located to monitor the mean arterial blood pressure in the arteries that supply blood to the brain (carotid sinus baroreceptor) and to the rest of the body (aortic arch baroreceptor).

© Cengage Learning, 2013

Carotid sinus baroreceptor

Common carotid arteries
(Blood to the brain)

Neural signals to
cardiovascular
control center
in medulla

Aortic arch
baroreceptor

Aorta
(Blood to rest of body)

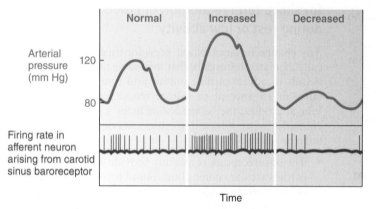

FIGURE 9-63 Firing rate in the afferent neuron from the carotid sinus baroreceptor in relation to the magnitude of mean arterial pressure.

© Cengage Learning, 2013

diovascular system. These efferent signals decrease heart rate and stroke volume, and produce arteriolar and venous vasodilation, which in turn lead to a decrease in cardiac output and a decrease in total peripheral resistance, with a subsequent decrease in blood pressure back toward normal.

Conversely, when blood pressure falls below normal (Figure 9-65b), for example by gravity when an animal stands up, baroreceptor activity decreases, inducing the cardiovascular center to increase sympathetic cardiac and vasoconstrictor nerve activity while decreasing its parasympathetic output. In a giraffe raising its head from a lowered position, constriction of vessels in the lower body raises the pressure and diverts blood to the brain. This efferent pattern of activity leads to an increase in heart rate and stroke volume coupled with arteriolar and venous vasoconstriction. These changes result in an increase in both cardiac output and total peripheral resistance, producing an elevation in blood pressure back toward normal.

Regulation of gas transport and blood pressure are coordinated during locomotory activity by higher brain centers

Now let's see how these two regulatory priorities, gas transport and ΔP, interact in a common event, that of increased locomotory activity (such as an animal running away from a predator). Increased gas transport for muscles is the first priority, but also dangerous pressures must not arise. How are both goals achieved? Two major cardiovascular changes must be activated simultaneously:

1. A substantial *increase in cardiac output* (C.O.) is triggered to increase oxygen delivery by increases in both heart rate and stroke volume (recall that C.O. = HR × SV); and

2. A large *increase in skeletal muscle, heart and skin blood flow* accompanied by a *decrease in overall resistance R* are achieved by dilation of arterioles in those organs. At the same time, constriction of gut arterioles diverts more blood to skeletal muscle, heart and skin. Figure 9-66 charts these changes. In summary, intense activity results in increased blood flow to those organs that most need it, while ΔP is prevented from rising too drastically as total R decreases (the increasing R in the gut is far outweighed decreasing R in the perfused organs). Indeed, there is only a modest increase in mean arterial pressure (Table 9-4).

Role of Intrinsic and Extrinsic Regulation Including Anticipation Increase in blood flow to skeletal and heart muscles occurs in part from the intrinsic mechanisms we discussed in detail earlier (pp. 416 and 434). Increasing CO_2 and acid levels, for example, dilate both local arterioles and precapillary sphincters. However, this will not replace the bodywide depletion of O_2 nor reduce bodywide acid buildup. Extrinsic regulators are needed for these adjustments. The simple

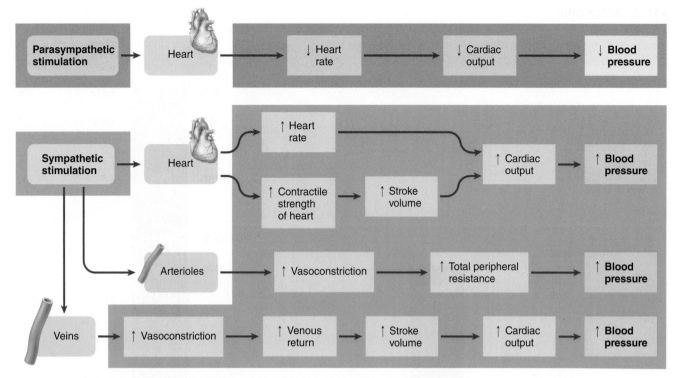

FIGURE 9-64 Summary of the effects of the parasympathetic and sympathetic nervous systems on factors that influence mean arterial blood pressure.

© Cengage Learning, 2013

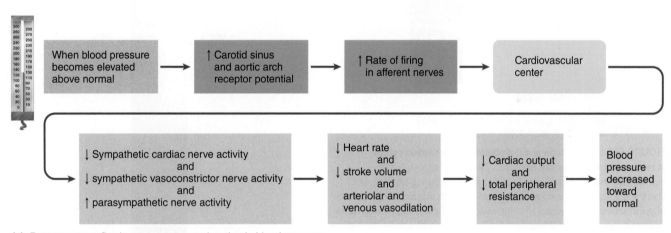

(a) Baroreceptor reflex in response to an elevation in blood pressure

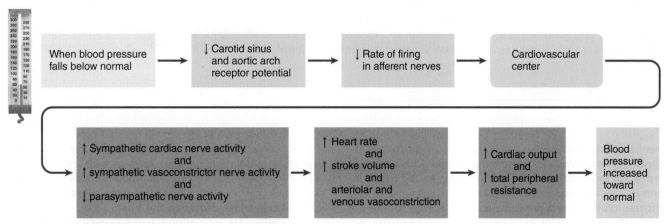

(b) Baroreceptor reflex in response to a fall in blood pressure

FIGURE 9-65 Baroreflexes to restore the blood pressure to normal.

© Cengage Learning, 2013

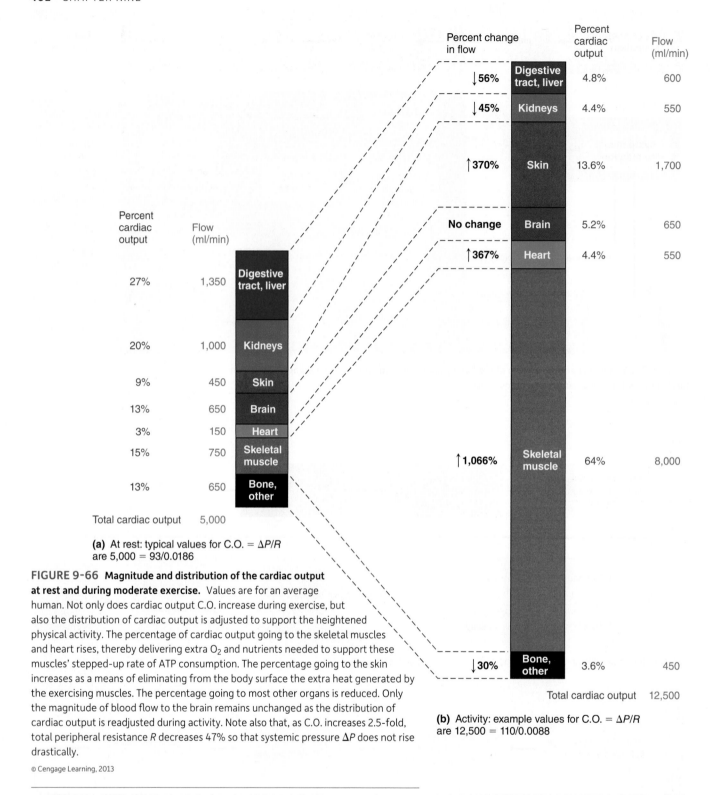

(a) At rest: typical values for C.O. = ΔP/R are 5,000 = 93/0.0186

FIGURE 9-66 Magnitude and distribution of the cardiac output at rest and during moderate exercise. Values are for an average human. Not only does cardiac output C.O. increase during exercise, but also the distribution of cardiac output is adjusted to support the heightened physical activity. The percentage of cardiac output going to the skeletal muscles and heart rises, thereby delivering extra O_2 and nutrients needed to support these muscles' stepped-up rate of ATP consumption. The percentage going to the skin increases as a means of eliminating from the body surface the extra heat generated by the exercising muscles. The percentage going to most other organs is reduced. Only the magnitude of blood flow to the brain remains unchanged as the distribution of cardiac output is readjusted during activity. Note also that, as C.O. increases 2.5-fold, total peripheral resistance R decreases 47% so that systemic pressure ΔP does not rise drastically.

© Cengage Learning, 2013

(b) Activity: example values for C.O. = ΔP/R are 12,500 = 110/0.0088

mechanism is negative feedback: Arterial chemosensors inform the cardiovascular center that blood CO_2 and acid levels are increasing and O_2 is decreasing, so the center increases cardiac output to compensate (at the same time that the respiratory centers are stimulating breathing). But notice that extrinsic mechanisms have a *delay* problem (p. 16). Only after the blood returns through the heart and lungs and back to the aortic sensors would the cardiovascular

center "know" to change C.O. and R, by which time the muscles would have gone many seconds with low O_2. However, it has long been known that blood gases do not exhibit significant changes at the onset of exercise—that is, there are no feedback delays. How can this be?

The answer is *anticipation*, which occurs in at least two ways (changes that occur even if the activity does not actually take place):

TABLE 9-4 Cardiovascular Changes during Strenuous Activity

Cardiovascular Variable	Change	Comment
Heart Rate	Increases	Occurs as a result of increased sympathetic and decreased parasympathetic activity to the SA node
Venous Return	Increases	Occurs as a result of sympathetically induced venous vasoconstriction and increased activity of the skeletal muscle pump and respiratory pump
Stroke Volume	Increases	Occurs both as a result of increased venous return by means of the Frank-Starling mechanism (unless diastolic filling time is significantly reduced by a high heart rate) and as a result of a sympathetically induced increase in myocardial contractility
Cardiac Output	Increases	Occurs as a result of increases in both heart rate and stroke volume
Blood Flow to Active Skeletal Muscles and Heart Muscle	Increases	Occurs as a result of locally controlled arteriolar vasodilation, which is reinforced by the vasodilatory effects of epinephrine and overpowers the weaker sympathetic vasoconstrictor effect
Blood Flow to the Brain	Unchanged	Occurs because sympathetic stimulation has no effect on brain arterioles; local control mechanisms maintain constant cerebral blood flow whatever the circumstances
Blood Flow to the Skin	Increases	Occurs because the hypothalamic temperature control center induces vasodilation of skin arterioles; increased skin blood flow brings heat produced by exercising muscles to the body surface where the heat can be lost to the external environment
Blood Flow to the Digestive System, Kidneys, and Other Organs	Decreases	Occurs as a result of generalized sympathetically induced arteriolar vasoconstriction
Total Peripheral Resistance	Decreases	Occurs because resistance in the skeletal muscles, heart, and skin decreases to a greater extent than resistance in the other organs increases
Mean Arterial Blood Pressure	Increases (modest)	Occurs because cardiac output increases more than total peripheral resistance decreases

© Cengage Learning, 2013

1. Evidence suggests that discrete locomotory centers—thought to be in the motor cortex and/or the midbrain—induce cardiac and vascular changes for locomotory activity in *anticipatory* fashion, boosting C.O. and reducing R *before* activity leads to disturbances.
2. Perception of stress can trigger (via the hypothalamus) the widespread changes in cardiovascular activity accompanying the *fight-or-flight* response even in the absence of actual activity. This too is an anticipatory mechanism that prepares the body for action (p. 308).

Blood pressure regulation must integrate both short-term and long-term influences

As we have seen, although gas (and heat) transport is the higher priority, pressure must be regulated within a safe range once transport demands are met. Elaborate mechanisms involving the integrated action of the various components of the circulatory system and other body systems are dedicated to regulating mean arterial pressure. So far we have focused primarily on short-term regulation. But there are other, long-term influences on pressure, the most important of these being *blood volume regulation*. As we mentioned earlier (e.g., Figure 9-48), water and salt balance regulators affect the plasma volume, which in turn alters blood pressure. This involves the hypothalamus, kidneys, and so on (see Chapters 12 and 13).

Let's review all the factors that affect mean arterial blood pressure by working our way through Figure 9-67. Even though we've covered all these factors before, it is useful to pull them all together. The circled numbers in the text correspond to the numbers in the figure and indicate the portion of the figure being discussed.

- Mean arterial pressure depends on cardiac output and total peripheral resistance ([1] on Figure 9-67); that is, $\Delta P = C.O. \times R$.
- Cardiac output depends on heart rate and stroke volume [2], both of which increase during locomotory activity.
- Heart rate depends on the relative balance of parasympathetic activity [3], which decreases heart rate, and sympathetic activity (tacitly including epinephrine throughout this discussion) [4], which increases heart rate.
- Stroke volume increases in response to sympathetic activity during locomotory activity [5] (extrinsic control of stroke volume).

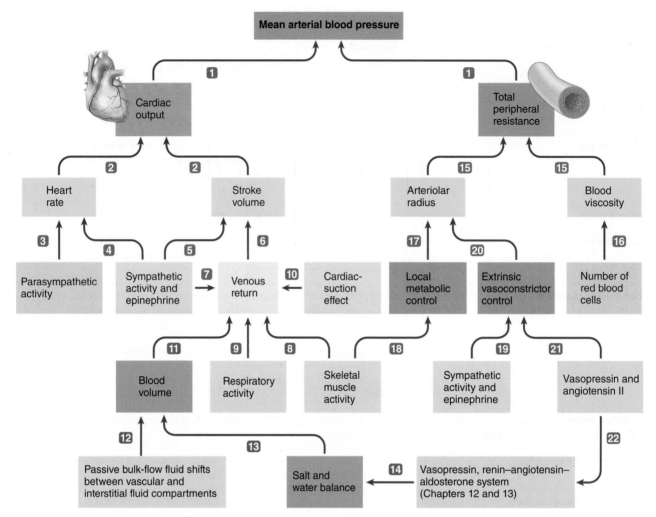

FIGURE 9-67 Determinants of mean arterial blood pressure. Note that this figure is basically a composite of Figure 9-29, p. 417, "Control of cardiac output"; Figure 9-48, p. 438, "Factors affecting total peripheral resistance"; and Figure 9-60, p. 447, "Factors that facilitate venous return." See the text for a discussion of the circled numbers.

© Cengage Learning, 2013

- Stroke volume also increases as venous return increases during locomotory activity [6] (intrinsic control of stroke volume by means of the Frank-Starling law of the heart).
- Venous return is enhanced during locomotory activity by sympathetically induced venous vasoconstriction [7], the skeletal-muscle pump [8], the respiratory pump [9], and cardiac suction [10].
- The effective circulating blood volume also influences how much blood is returned to the heart [11]. The blood volume depends in the short term on the magnitude of passive bulk-flow fluid shifts between the plasma and interstitial fluid across the capillary walls [12]. In the long term, the blood volume depends on salt and water balance [13], which are hormonally controlled by the renin–angiotensin–aldosterone system and vasopressin, respectively [14].
- The other major determinant of mean arterial blood pressure, total peripheral resistance, depends on the radius of all arterioles as well as blood viscosity [15]. The

major factor determining blood viscosity is the number of red blood cells [16]. However, arteriolar radius is the more important factor determining total peripheral resistance.
- Arteriolar radius is influenced by local (intrinsic) metabolic controls that match blood flow with metabolic needs [17]. For example, local changes that take place in active skeletal muscles cause local arteriolar vasodilation and increased blood flow to these muscles [18].
- Arteriolar radius is also influenced by sympathetic activity [19], an extrinsic control mechanism that causes arteriolar vasoconstriction [20] to increase total peripheral resistance and mean arterial blood pressure, but which also (via adrenal epinephrine) dilates skeletal muscle and heart arterioles during activity [20].
- Arteriolar radius is also extrinsically controlled by the hormones vasopressin and angiotensin II, which are potent vasoconstrictors [21] as well as being important in salt and water balance [22].

Altering any of the pertinent factors produces a change in blood pressure, unless a compensatory change in another variable keeps the blood pressure within the safe range. Blood flow to any given tissue depends on the driving force of the mean arterial pressure and on the degree of vasoconstriction of the tissue's arterioles. Because mean arterial pressure depends on the cardiac output and the degree of arteriolar vasoconstriction, if the arterioles in one tissue dilate, the arterioles in other tissues will have to constrict to maintain an adequate arterial blood pressure to provide a driving force to push blood not only to the vasodilated tissue but also to the brain, heart, and kidneys. Thus, the cardiovascular variables must be continuously juggled to maintain a consistent blood pressure despite tissues' varying needs for blood.

This concludes our examination of circulatory systems. However, circulation plays a vital role in all the following chapters because it links all body systems and carries cells and molecules involved in other major processes such as respiration. Also, as we have just discussed, other systems and processes affect circulation.

check your understanding 9.16

Explain why regulation of gas transport and blood pressure are important, and why the former is the top priority.

Discuss the changes that occur during locomotory activity and how they are regulated.

making connections

How Circulation Contributes to the Body as a Whole

Homeostasis depends on continual delivery of needed supplies to all the cells throughout the body and on ongoing removal of wastes generated by the cells. Regulated changes such as those needed during muscular activity are also dependent on circulatory delivery and removal. Thus, circulatory systems contribute to the whole animal by serving as a body's transport system, linking all organ systems together. Circulatory systems provide a means of rapidly moving heat and materials—oxygen, nutrients, hormones, wastes, and cells—from one part of the body to another, overcoming the limits of diffusion.

Blood or hemolymph has special transport capabilities that enable it to move its cargo efficiently throughout the body. For example, O_2 is poorly soluble in water, but circulatory fluids may be equipped with O_2-carrying proteins such as hemoglobin, often in erythrocytes (red blood cells). Likewise, homeostatically important water-insoluble hormonal messengers are shuttled in the blood by plasma protein carriers.

Specific components of blood or hemolymph perform the following additional homeostatic activities that are unrelated to transport function:

- Circulatory fluid helps maintain the proper pH in the internal environment by buffering changes in the acid–base load of the body.
- It helps maintain body temperature in endotherms by absorbing heat produced by heat-generating tissues and distributing it throughout the body, or carrying it to the body surface for elimination to the external environment.
- The electrolytes in the plasma are important in membrane excitability, which in turn forms the basis of nerve and muscle function.
- The electrolytes in the plasma are also important in the osmotic distribution of fluid between the extracellular and intracellular fluid, and the plasma proteins play a critical role in the distribution of extracellular fluid between the plasma and interstitial fluid.
- Through their hemostatic functions, clotting factors and cells minimize the loss of life-sustaining blood following vessel injury.
- Circulating immune cells, their secretory products, and certain types of plasma proteins, such as antibodies, constitute immune defense systems.

Circulation usually requires a pump such as a heart. Hearts often exhibit their own intrinsic homeostasis, such as the ability to beat continuously without external signals. Local mechanisms within the vertebrate heart ensure that blood flow to the cardiac muscle normally meets the heart's need for O_2. In addition, the heart has built-in capabilities to vary its strength of contraction, depending on the amount of blood returned to it. The vertebrate heart does not act entirely autonomously, however. It is innervated by the autonomic nervous system and is influenced by the hormone epinephrine, both of which can vary the rate and contractility of the heart, depending on the body's needs for blood delivery.

Chapter Summary

- Circulatory systems evolved to overcome diffusion, which is extremely slow over all but very small distances, by creating bulk transport. Circulatory systems have up to three distinct components: fluid, pump, and vessels.

- Circulatory fluids contain plasma and often cells. Plasma is a water medium for inorganic ions, gases, and numerous organic solutes. Many of the functions of plasma are carried out by plasma proteins, such as antibodies. Lipoprotein such as LDL and HDL complexes carry energy lipids and structural lipids for biosynthesis.

- Respiratory pigments such as hemoglobins and hemocyanins carry oxygen. Erythrocytes in some animals contain these pigments and so serve primarily to transport oxygen. Hemopoietic tissues such as bone marrow continuously replace worn-out erythrocytes. Erythropoiesis in mammals and probably other vertebrates is controlled by erythropoietin from the kidneys.

- Thrombocytes and platelets function in clotting. Hemostasis prevents blood loss from damaged small vessels. Hemocytes and hemolymph proteins provide hemostasis in arthropods. In mammals, first vascular spasm reduces blood flow through an injured vessel. Then, platelets aggregate to form a plug at a vessel defect by positive feedback. Finally, a triggered chain reaction and positive feedback involving clotting factors in the plasma results in blood coagulation. Plasmin dissolves clots and prevents inappropriate clot formation.

- Pumping mechanisms include flagella, extrinsic skeletal muscles, peristaltic muscular pumps, and hearts (chamber muscle pumps). Many animals have primary hearts aided by auxiliary pumps, such as two extra hearts in cephalopods to aid gill flow. Arthropod hearts are dorsally located and have many valved openings.

- Vertebrate hearts evolved from a two-chambered (in fish) into a four-chambered structure, that is, crocodilian, avian and mammalian hearts are dual pumps with two atria, two ventricles. Heart valves ensure that the blood flows in the proper direction.

- Vertebrate heart walls are composed of spirally arranged cardiac muscle fibers to create squeezing action, and are interconnected by intercalated discs with gap junctions for electrical connectivity. Pacemaker cells have channels that cyclically push them to "firing" threshold to keep myogenic hearts beating independently. The sinoatrial (SA) node in the right atrium is the normal pacemaker of the mammalian heart. The spread of cardiac excitation is coordinated to ensure efficient pumping, starting at the SA node, spreading through the atria, then to the atrioventricular node and tracts running through the ventricles. The electrocardiogram (ECG) records these electrical events.

- The action potential of contractile cardiac muscle cells shows a characteristic plateau to aid pumping. Ca^{2+} entry from the ECF induces a much larger Ca^{2+} release from the sarcoplasmic reticulum also to aid pumping. Tetanus of cardiac muscle is prevented by a long refractory period, so the muscle can contract cyclically.

- Hearts alternately contract in systole to empty and relax in diastole to fill.

- Cardiac output (C.O.) depends on the heart rate times the stroke volume. Heart rate is determined primarily by antagonistic regulation of autonomic influences on the SA node. Stroke volume is determined by the extent of venous return and by sympathetic activity. Increased end-diastolic volume, which occurs during exercise as skeletal muscles squeeze more blood into the heart, results in increased stroke volume.

- The heart receives most of its own blood supply through the coronary circulation.

- Blood flow Q through vessels depends on the pressure gradient ΔP and inversely on vascular resistance R: $Q = \Delta P/R$. Pressure is dependent on many factors such as gravity; resistance depends on many factors but blood vessel radius is the most important.

- Circulatory fluids transport materials in a parallel manner, especially in closed systems, so that each organ gets a fair share of fluid. However, flow to respiratory organs may be in series for maximal gas exchange. Circulatory fluids are driven by pressure and can transmit useful force, such as extending spider legs for jumping.

- The vascular system evolved from one circuit to two separate circuits—pulmonary to the lungs and systemic to the rest of the body—in vertebrates, along with a change from a two-chambered heart to a four-chambered one.

- Arteries store and release pressure to keep blood flowing by being elastic. Arterioles control blood distribution and bodywide blood pressure by having sphincter muscles, and are the major resistance vessels. Local (intrinsic) metabolic influences on arteriolar radius help match blood flow with the tissues' needs. Extrinsic sympathetic control of arteriolar radius (by nerves and by epinephrine) is important in the regulation of arterial blood pressure and in redistributing blood flow among organs during activity, in which blood flow is increased to skeletal muscles and reduced to the digestive tract.

- Capillary anatomy increases diffusion rates by minimizing distance ΔX while maximizing surface area A and the diffusion coefficient D of Fick's law. Blood composition and perfusion regulation enhance capillary-tissue concentration gradients ΔC. Capillary anatomy increases time for diffusion by slowing blood flow due to a high total cross-sectional area. Interstitial fluid is a passive intermediary between the blood and cells. Bulk flow across the capillary wall is important in extracellular fluid distribution.

- The lymphatic system is an accessory route by which interstitial fluid can be returned to the blood, and is important in immune responses and fat transport.

- Veins serve as a blood reservoir as well as passageways back to the heart. Venous return is enhanced by a number of extrinsic factors, such as skeletal muscle pumping and one-way valves.

- Regulation of gas transport and mean arterial blood pressure is accomplished by controlling cardiac output, total peripheral resistance, and blood volume. C.O. must increase in exercise but ΔP must not get too high; thus, overall R is reduced (mainly by dilation in skeletal muscles). The medullary cardiovascular center and several hormones regulate these factors. Gas transport and blood pressure are monitored by arterial baroreceptors and chemoreceptors signaling the cardiovascular center.

- The baroreflex is the most important mechanism for short-term regulation of blood pressure: dropping pressure to the brain signals the cardiovascular center to alter C.O. and R to compensate. Regulation of gas transport and blood pressure are coordinated during locomotory activity by feedback and by anticipatory regulation, increasing C.O. and reducing overall R before activity. Several other systems can influence regulation, for example, those involved in salt and water balance and temperature regulation.

Review, Synthesize, and Analyze

1. Compare lipid transport in insects and mammals.

2. What are the roles of clotting factor XIII in mammals and in arthropods?

3. Compare the electrical events in heart contraction, skeletal muscle contraction, and neuron signal transmission, including similarities and differences in ion channels, Ca^{2+} release, and length of each event.

4. Define the difference between *parallel* and *series* flow in circulation, then discuss where each occurs and why. Use examples from fish and mammals.

5. Arteries and veins are both large vessels that allow for fast blood flow. But in other regards, they are quite different. Discuss the differences and why they evolved.

6. Compare delivery of nutrients and gases, and removal of wastes, in a noncephalopod mollusk and a mammal, focusing on the presence or absence of capillaries.

7. In Chapter 1, we noted the lethal positive-feedback events of congestive heart failure: A weak mammalian heart results in a drop in blood pressure. Using Figure 9-63, discuss how the fluid-retention mechanisms lead to eventual heart failure.

8. EPO injection by human athletes or injected into racehorses can increase blood viscosity. Discuss why, and also why this is potentially dangerous.

9. Using the flow law, discuss what would happen in locomotory activity if C.O. increased fivefold, as it often does during fight or flight, with no change in *R*.

10. Use the following diagram, where the big arrows show *flow Q*, identical in both systems.

   a. CIRCLE the following statement(s) that is/are correct:

      i. Velocity is SLOWER in *Pipes D1–5* than in *Chamber X*

      ii. Velocity is FASTER in *Pipes D1–5* than in *Chamber X*

      iii. Velocity is SLOWER in *Pipes D1–5* than in *pipe B*

      iv. Velocity is FASTER in *Pipes D1–5* than in *pipe B*

   b. In the diagram, if Q = 10, what is the individual *flow* in each of *Pipes D1–5?* _____; what is the *velocity* in *Pipe C?* _____; and the *velocity* in *Pipes D1–5?* _____

   c. Which of these diagrams is like an insect system, and which like a vertebrate system? Explain.

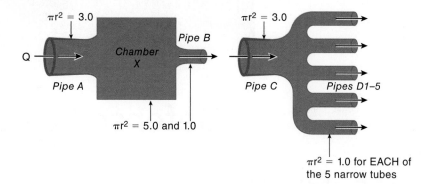

Suggested Readings

Choy, D. S. J., & P. Altman. (1992). The cardiovascular system of barosaurus: An educated guess. *Lancet* 340:534–536. A hypothesis regarding this long-necked animal.

Dukes, H. H., M. J. Swenson, & W. O. Reece. (1993). *Dukes' Physiology of Domestic Animals.* Comstock. Ithaca, NY: Cornell University Press. A classic text.

Franklin, C. E., & M. Axelsson. (2000). An actively controlled heart valve. *Nature* 406:847.

Geddes, L. A. (2002). Electrocardiograms from the turtle to the elephant that illustrate interesting physiological phenomena. *Pacing and Clinical Electrophysiology* 25:1762–1770.

Goetz, R. H., J. V. Warren, O. Gauer, et al. (1960). Circulation of the giraffe. *Circulation Research* 8:1049–57.

Kampmeier, O. F. (1969). *Evolution and Comparative Morphology of the Lymphatic System.* Springfield, MA: C. C. Thomas.

Liberatore, G. T., A. Samson, C. Bladin, W. D. Schleuning, & R. L. Medcalf. (2003). Vampire bat salivary plasminogen activator (desmoteplase): A unique fibrinolytic enzyme that does not promote neurodegeneration. *Stroke* 34:537–543.

McGaw, I. J., & C. L. Reiber. (2002). Cardiovascular system of the blue crab. *Callinectes sapidus. Journal of Morphology* 251:1–21.

Pearse, V. J., M. Buchsbaum, & R. Buchsbaum. (1987). *Living Invertebrates.* Palo Alto, CA: Blackwell Scientific. A classic text on nonvertebrate anatomy and physiology.

Pollitt, C. C. (1992). Clinical anatomy and physiology of the normal equine foot. *Equine Veterinary Education* 4:219–224.

Satchell, G. H. (1991). *Physiology and Form of Fish Circulation.* New York: Cambridge University Press.

Sherman, I. W., & V. G. Sherman. (1976). *The Invertebrates: Function and Form,* 2nd ed. New York: Macmillan. A classic lab manual on nonvertebrate anatomy and physiology.

Terwilliger, N. B. (1999). Hemolymph proteins and molting in crustaceans and insects. *American Zoologist* 39:589.

Vogel, S. (1993). *Vital Circuits: On Pumps, Pipes, and the Workings of Circulatory Systems.* Oxford, UK: Oxford University Press. A biomechanical analysis of circulation.

CourseMate

10

© Brandon Cole Marine Photography/Alamy

Hagfish (jawless vertebrates) have *innate* "ready-to-go" immunity similar to that found in all animals. For example, their skin produces copious amounts of mucus that, in addition to suffocating predators, contains *antimicrobial peptides* that quickly kill most bacteria. However, unlike most other vertebrates, hagfish lack an *acquired* immune system, which first "learns" about internal invaders such as pathogenic bacteria, then responds with highly specific defenses.

Defense Systems

10.1 Evolution of Defense Systems

Animals could not survive beyond early infancy were it not for defense mechanisms. Homeostasis can be optimally maintained, and thus life sustained, only if body cells are not physically injured or functionally disrupted by pathogenic microorganisms or are not replaced by abnormally functioning cells, such as traumatized cells or cancer cells. Attack by other organisms has undoubtedly been a major feature of life for billions of years (long before multicellular animals evolved). Various defense mechanisms have been evolving all that time in a process called **coevolution:** Adaptive changes in one organism favor selection for adaptations in a second organism, which in turn favor selection for a different adaptation in the first organism, and so on. In the context of defense, this is sometimes called an evolutionary "arms race." For example, an animal may evolve a defense mechanism against a predator or pathogen (an invader such as a disease-causing bacterium), but this provides only a temporary respite in evolutionary terms. The antagonistic species after many generations may evolve a way to overcome this defense. In turn, the first species may evolve yet another improvement to its defenses. Thus,

Paul Yancey

This poison dart frog has innate defenses such as skin mucus with antimicrobial peptides. The mucus has alkaloids toxic to other animals. Internally, the frog also has an acquired immune system.

over time, both species can evolve very sophisticated offenses and defenses.

Effective immune systems are so important that they can influence mate choice behavior in some animals. In many birds, for example, red and yellow pigment molecules called *carotenoids* (from plants) have two functions: They provide colors for male beaks

and feathers, which are used to attract females, and they benefit immunity of the birds because they function as antioxidants and suppress overactive inflammatory reactions. These are not unrelated functions, because research has revealed a linkage between mate attraction and immune function in male zebra finches. Males given extra dietary carotenoids had more brightly colored beaks and faster-responding immune systems compared to control males; moreover, females were more likely to mate with them than with control males with drabber beaks. Thus, a male's plumage signals to the female how healthy he is, clearly a major factor in producing healthy offspring. It is unlikely that the female consciously perceives the male's health; rather, females in the past who were predisposed to pick the most brightly colored males produced healthier offspring, spreading the genes for color preference.

Effective immunity is undoubtedly one of the most important physiological systems for survival, but it can be expensive. For example, one study in 2005 found that male crickets with stronger immune systems had weaker sperm compared to males with less robust immunity. This finding illustrates the concept of *trade-offs* in the evolution of adaptations (see p. 2).

Animal defenses target external as well as internal threats including pathogenic bacteria, viruses, parasites, and cancer

Animal defense systems fall into two broad categories. The first deals with external antagonists, primarily predators and competitors. These defenses may be *anatomic* features such as the spines of a porcupine, *chemical* agents such as the nerve inhibitor tetrodotoxin (p. 143) in the skin of a pufferfish, behavioral programs such as *escape responses* (for example, the rapid tail-flip of a lobster), and *reproductive* strategies such as the ability of coral to breed rapidly following decimation by attacking seastars. We will not consider this category of defenses any further.

The second type, and the focus of this chapter, is internal defense, called **immunity**, which is the ability to resist or eliminate potentially harmful foreign materials or abnormal cells inside or attached parasitically to the surface of an animal. The following activities are attributable to *immune systems*, which function to block, recognize, and either destroy or neutralize materials that are foreign to the "normal self."

1. Defense against invading *pathogens*.
2. Removal of "worn-out" cells (such as aged red blood cells) and tissue debris (for example, tissue damaged by trauma or disease), which is necessary before repair can occur.
3. Identification and destruction of abnormal or mutant cells that have originated in the body. This *immune surveillance* is the primary internal defense mechanism against cancer.

The primary foreign enemies against which the immune system defends are disease-causing biological invaders called **pathogens**. The disease-producing power of a pathogen is known as its virulence. Pathogenic bacteria that invade animal bodies induce tissue damage and produce disease largely by releasing enzymes or toxins that physically injure or functionally disrupt affected cells and organs. The disruption may release nutrients for bacterial growth. Internal **parasites** include larger eukaryotes such as protozoa, yeast, fungi and animals, which take up residence inside a larger animal (called the **host**) and use the latter's nutrients for growth and reproduction. External parasites such as leeches (annelids) and ticks (arachnids) attach to skin but exert their effects internally on the host by ingesting blood; and they may transmit microbial diseases to the host. For example, the most common tickborne disease in the Northern Hemisphere is Lyme disease. Deer ticks (*Ixodes scapularis*) infect mammals (including domestic ones) with bacteria belonging to the genus *Borrelia*. Tick saliva, which accompanies the bacteria into the wound site, contains substances that provide a protective environment for the bacteria to establish the infection where, if left untreated, can persist for years. For disease transmission to occur, the nymphal-stage tick must remain attached to the host for a period of at least 24 hours. The tick may also infect the host with *Babesia*, a protozoan that reproduces in and (and ruptures) erythrocytes in the disease *babesiosis*. Some internal parasites such as tapeworms may cause harm mostly through nutrient depletion or physical damage to host cells, but others also release damaging chemical agents. The distinction between bacterial and parasitic pathogens is thus somewhat arbitrary.

In contrast to these living cells, **viruses** are not self-sustaining cellular entities. They consist only of DNA or RNA enclosed by a protein coat. Because they lack cellular machinery for energy production and protein synthesis, viruses cannot carry out metabolism and reproduce unless they invade a **host cell** (a body cell of the infected individual) and take over the cellular biochemical facilities for their own purposes. The viral nucleic acids direct the host cell to synthesize proteins needed for viral replication, sapping the host cell's energy resources, synthesizing toxins that can lyse (burst) the cell, and/or transforming it into a cancer cell. However, if a virus becomes incorporated into a host cell, the host defense system may destroy the cell (as you will see), because it is no longer recognized as a "normal self" cell.

Immune responses can be either innate or acquired

Internal immunity is traditionally divided into two separate but interdependent components: the ancient *innate immune systems* (which has many separate aspects) and the more recently evolved *adaptive* or *acquired immune systems*. The responses of these two systems differ in their timing and in the degree of selectivity of the defense mechanisms employed:

- **Innate immune systems** encompass an animal's immune responses that come into play rapidly (even immediately) on exposure to any threatening agent and that *do not depend on prior exposure to that agent*. These responses are "ready-to-go" mechanisms that defend against foreign or abnormal material of almost type, including pathogens, chemical irritants, and tissue injury accompanying mechanical trauma and burns. All animal types including sponges have innate immunity mechanisms, such as *phagocytotic* cells that engulf and destroy foreign materials. Innate systems are often called *nonspecific*; however, this is misleading since, as you will see, they involve binding to specific features of pathogens, such as the carbohydrates typically attached to the bacterial cell walls.
- **Acquired or adaptive immune systems**, in contrast, rely on *learned* immune responses selectively targeted against a particular foreign material following the first exposure. In essence, this component of the immune system retains *memories* of attacking pathogens in the environment. Since it requires learning, acquired immunity takes considerably more time to be mounted and takes on one specific foe at a time. Probably the most familiar components of this system are *antibodies*, which may take days or weeks to build up after your first exposure to a disease (or vaccine). Until recently, this learning mechanism was thought to have evolved only in vertebrates with jaws, but at the end of the chapter we look at recent evidence for similar systems in lampreys (jawless fish) and perhaps some arthropods and mollusks. Although this defense is often called *adaptive* immunity, many biologists feel the term should be reserved for evolutionary changes over many generations. Indeed, learned defenses are similar to *acclimatization* (see p. 17)—useful prolonged changes in a body triggered in response to a new situation. Therefore, we will use the term *acquired*.

Immune responses use pattern-recognition receptors to distinguish "self" from "nonself"

Both innate and acquired immunity require that an organism recognizes its own tissues and can distinguish them from foreign materials. This *self-recognition* ability has ancient roots: Even cells of the most primitive metazoans, the sponges (Porifera), can distinguish "self" from foreign cells. Another ancient example, illustrated in Figure 10-1a, has been called the "anemone clone wars." Some species of these Cnidarians grow in distinct patterns: groups of densely packed sea anemones with barren areas between them. A researcher curious about this pattern discovered that each cluster is a clonal lineage; that is, all the anemones are genetically identical and arose from asexual budding starting with one individual. Furthermore, other clusters are also clones, each with their own unique genomes. The barren zones arise when one cluster grows close to another. By detecting unique molecules on the animals' surfaces, individuals on the edge of each clone recognize the others as "nonself," and they react violently. Using special stinging war tentacles called **acrorhagia**, the animals sting and try to drive off (or kill) the foreign anemones, leaving the rock barren in between. The recognition process is essentially the same used by internal immune systems to detect invaders (as you will see) and is also similar to the process that causes rejection of organ transplants in humans.

Recognition of self and nonself is accomplished by special **pattern-recognition receptors**, or **PRRs**, in plasma membranes, which one immunologist has called "the eyes of the innate immune system." A prime example are **Toll receptors** in the fruit fly *Drosophila*, which participate in pattern recognition in both immunity and development. The first Toll gene was discovered in flies as a mutated nonfunctional immune gene that resulted in flies highly prone to fungal infection. Later, many immune cells in mammals were found to have similar receptors and were named **Toll-like receptors** (**TLRs**). TLR and other PRRs act as sensors in the host, recognizing and binding with common foreign macromolecules, such as *lipopolysaccharides (LPS)* of the bacterial cell wall (Figure 10-1b). We will investigate PRR roles in detail later in the chapter.

Immune cells communicate with cytokines and by direct contact

A final universal aspect of immunity is the methods by which immune cells communicate during defensive reactions. The primary mechanism involves the secretion and reception of **cytokines** (*cyto*, cell, and *kinos*, movement), which are defined as *signal molecules produced by nongland cells*. These "immunohormones" are analogous to hormones, neurohormones, and neurotransmitters in the other major regulatory systems; some act locally in paracrine fashion, while others travel long distances through circulatory fluids like true hormones. In vertebrates, most cytokines are called **interleukins** (**IL**) (*inter*, between, and *leukin*, a protein of leukocytes or white blood cells). Vertebrate immune cells also communicate by direct contact with other cells via a so-called **immunosynapse**, in which two cells' membranes join either by receptor–receptor binding or by *nanotubes* (p. 91) to initiate signaling responses. Immunosynapses probably occur in nonvertebrates but have not been well studied.

(a)

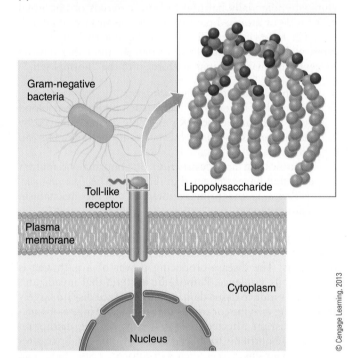

(b)

FIGURE 10-1 Recognizing self vs. nonself. (a) Anemone clone wars. Conflicts occur between two colonies of sea anemones formed by asexual budding (cloning). When one colony member contacts a member of the other colony, each recognizes the other as foreign and begins an attack. The result is an open patch of habitat between the two groups. (b) Toll-like receptors (TLRs). These proteins on some mammalian immune cells have binding sites that "recognize" common components of bacterial cell walls (such as lipopolysaccharides), flagella, etc.

Some animals behaviorally acquire defense components from other organisms

Before we examine innate and acquired immunity in depth, note that there is another form of immunity in some species. This is the "borrowing" of immune defenses from other organisms (an example of a behavioral effector, p. 16). Humans are of course renowned for this—virtually every human culture makes use of medicinal compounds from plants, fungi, and the like. In most cases, the compounds, which aid human health, are also defense compounds in the originating species. Aspirin, for example, was originally obtained from the willow tree and is widely used to treat fever, inflammation, and pain in humans. It occurs in a wide variety of plants, where it appears to be involved in activating a plant's defense mechanisms when an infection occurs.

Other primates can also take advantage of the remedies provided by nature. For example, chimpanzees in eastern Africa swallow whole leaves of the *Aspilia* plant (a relative of sunflowers). In contrast to leaves eaten for food, which are chewed first, these leaves are probably medicinal. They contain large amounts of *thiarubrine-A*, which readily kills parasitic fungi and worms. Similarly, capuchin monkeys in Venezuela find noxious millipedes with defensive compounds called *benzoquinones*. The monkeys smear mashed millipedes on their fur, where the noxious compounds repel biting mosquitoes.

Many birds also seem to make use of defense compounds from other organisms. Most birds that reuse their nests (which may easily become breeding grounds for parasites and bacteria) are seen to line the nests with branches and leaves from certain plants. Starlings, for example, carefully select specific (often rare) plants to line their nests during breeding season. Some of these plants have been tested and are found to contain volatile defense compounds that repel external parasites such as ticks.

We now turn to a detailed look at innate defenses, starting with nonvertebrates.

check your understanding 10.1

What are the major disturbances that internal immunity defends against?

What is the basic difference between innate and acquired immunity?

How do immune systems distinguish "self" from "nonself"?

How do immune cells communicate?

10.2 Innate Immune Systems in Nonvertebrates

Because all animals have innate immunity systems, these have been evolving for a very long time. Due to this, plus the fact that the kinds of invaders and toxins encountered by animals vary so widely among species and habitats, evolution has produced a wide diversity of innate defenses. Nevertheless, there are many similarities and common themes. In particular, most animals have (1) *barrier tissues*, (2) *phagocytic* cells that "eat" or consume invaders often in an *inflammatory* process, (3) *antimicrobial peptides* that destroy the invaders directly, and (4) *opsonin* proteins that "tag" invaders to aid recognition by phagocytic cells. Arthropods also have an encapsulation defense. Let's examine how each of these work.

Barrier tissues are the first line of defense, using passive and active mechanisms

Barrier tissues are linings that form impermeable walls to keep invaders out. As such, they are said to be the first lines of defense. The wall aspect is a passive one, but these tissues may also have glands and other active components which can aid in impeding an invasion. The most obvious barrier is the **integument** (*integere*, "to cover"), which covers the outside of an animal's body. Examples include skin *epidermis* in mammals and exoskeletal *epicuticle* in insects. The insect **epicuticle** (the outermost layer of the exoskeleton) has layers of dense lipoproteins, waxes and protein-polyphenol complexes that form not only a tight "fence" but also provide waterproofing to limit moisture loss. Integuments often have glands that secrete defensive proteins. For example, epithelial cells in damaged moth cuticles secrete *cecropin,* an antimicrobial peptide. You will see this and other examples in detail later.

Other organs systems exposed directly to the environment also form protective barriers. These include the *digestive, respiratory*, and *genitourinary* systems, all of which have ducts connected to the outside world and so are potentially vulnerable to invasion. As we'll see for mammals later, their epithelial linings have glands for defensive secretions, such as microbe-killing acid in some digestive systems.

Phagocytotic cells and inflammatory responses are the major form of cell-mediated second lines of defense

If a barrier tissue is breached by an invader, so-called second lines of innate defense take over. Two of these—*phagocytotic cells and inflammation*—often go together. Although most nonvertebrates do not have enclosed circulatory systems, all have some form of internal fluid cavity. In these fluids are typically found wandering immune cells called **amoebocytes**, which provide what is often called *cell-mediated* defense. These patrol the fluid spaces looking out for foreign invaders. They can often distinguish "self" and "nonself" using PRRs that evolved to bind common bacterial and fungal cell wall components (Figure 10-1b), and possibly viral components. To kill the invaders, they engulf them through **phagocytosis** ("cell eating"; Figure 10-2a), the same process used by amoebas to engulf food. This involves the capture of the foreign object in an internal membrane vesicle made by *endocytosis*, fusion of that vesicle with a *lysosome* (p. 44), and digestion of the object by the lysosome's hydrolytic enzymes. In some arthropods, immune cells have also been found to make *reactive oxides* that kill microbes, just as some mammalian phagocytes do (as you will see later).

Amoebocytes may be called by different names in different animals. For example, in sponges they are called **archaeocytes**. In echinoderms, which have a fluid-filled body cavity called the coelom, these cells are called **coelomocytes**. In arthropods, mollusks, and other animals having an open circulatory system filled with hemolymph (Chapter 9), they are called **hemocytes** (Figure 10-2b). As you will see later, vertebrates have multiple types.

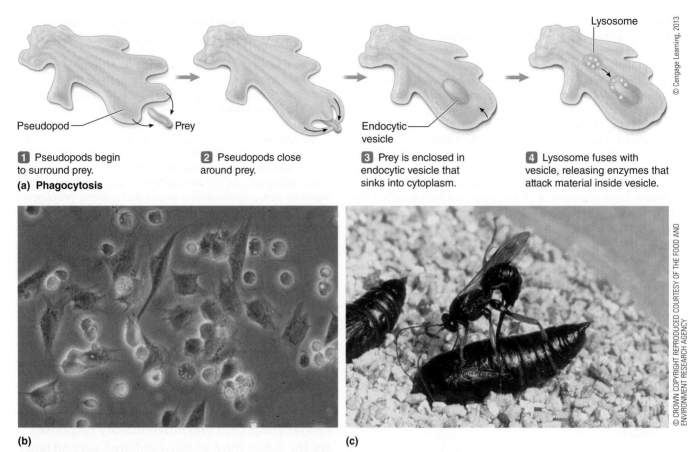

Lysosome

Pseudopod — → Prey

Endocytic — vesicle

1 Pseudopods begin to surround prey.

(a) Phagocytosis

2 Pseudopods close around prey.

3 Prey is enclosed in endocytic vesicle that sinks into cytoplasm.

4 Lysosome fuses with vesicle, releasing enzymes that attack material inside vesicle.

© Cengage Learning, 2013

(b)

(c)

© CROWN COPYRIGHT REPRODUCED COURTESY OF THE FOOD AND ENVIRONMENT RESEARCH AGENCY

FIGURE 10-2 Phagocytosis and Encapsulation. (a) White blood cells (in vertebrates) or hemocytes (in many nonvertebrates) internalize multimolecular particles such as bacteria or old red blood cells by extending pseudopods that wrap around and seal in the targeted material. A lysosome fuses with and degrades the vesicle contents. (b) Hemocytes (blood cells) from larval moth *Lacanobia oleracea*. Many are phagocytes; a subset called a plasmatocyte is involved in the encapsulation response. Some parasitoid venom components inhibit plasmatocyte development or activity, and thus prevent the normal immune response. (c) The ectoparasitic wasp *Eulophus pennicornis* laying eggs in a host pupa. This wasp injects a cocktail of venom components that induce changes in the host's immune system (such as blocking plasmatocytes), and prevents the process of metamorphosis by interfering with hormone levels.

Source for (b) and (c): www.csl.gov.uk/science/organ/environ/invertebrate/maff1.cfm

Inflammation-Like Responses Amoebocytes in many non-vertebrates participate in an inflammation-like response to wounds and invaders. **Inflammation** refers to immune-cell (and often fluid) buildup due to chemotaxis to the site of injury, where the cells are activated to engage in phagocytotic destruction of invaders. **Chemotaxis** is directed movement due to attraction to *chemotaxins* released at the site of damage by damaged cells and/or other immune cells. Binding of chemotaxins with protein receptors on the plasma membrane of a phagocytotic cell increases Ca^{2+} entry into it. Ca^{2+}, in turn, switches on the microfilament contractile apparatus that leads to amoeba-like crawling (p. 63). Because the concentration of chemotaxins progressively increases toward the site of injury, phagocytotic cells move toward this site along a chemotaxin concentration gradient.

In addition to creating a concentrated attack, inflammation helps keep the infection from spreading. Some of the cells also produce antimicrobial peptides (as you will see shortly). Here we find possible evolutionary precursors to the verte-

brate system. Once stimulated, phagocytotic cells (coelomocytes, etc.) in echinoderms, tunicates, worms, and insects secrete a cytokine very similar in structure to a major mammalian one called *interleukin-1* or *IL-1* (which you will encounter several times later). These IL-1 cytokines seem to coordinate the cell defenses by acting as chemotaxins to draw in phagocytotic cells and also trigger them to reproduce.

Encapsulation in Arthropods Another second line of defense found in arthropods is **encapsulation**. In this process a foreign invader that is too large for the phagocytes to engulf is walled off by a new barrier. It is commonly used by many insects against parasitic worms, fungi, and eggs laid by parasitic wasps within them (Figure 10-2c). Surface proteins on an invader trigger in certain hemocytes a cascade of enzymatic reactions, which ultimately converts **pro-phenoloxidase (proPO)**, an inactive enzyme that is always present in the cellular membrane, into the active form **phenoloxidase**. This in turn converts phenols into a variety of products, one of which forms a black polymer called **melanin** (the same molecule that gives human skin its pigment color). Factors that attract other hemocytes to the infection site are also produced, a *positive-feedback* process to accelerate the process. The invader is surrounded with a nodule or capsule several cell layers thick, and the cells closest to the pathogen secrete the melanin to form an impermeable wall. Research studies have demonstrated that if the melanization reaction is inhibited, a higher susceptibility to microbial infection occurs.

The vertebrate clotting system (Chapter 9, p. 395) also can wall off a site of invasion. We will look at this more closely when we examine mammalian inflammation later.

Antimicrobial and other defensive peptides of barrier tissues and immune cells are common noncellular second lines of defense

Another universal second-line defense involves various **antimicrobial peptides** produced by integument, gut and other barrier tissues, phagocytotic cells, and sometimes other tissues (for example, in insects, the fat body appears to be a major source of such peptides). These are small proteins, released by injured or activated cells or free-floating in body fluids, that can bind to and often kill microbes on their own. Many of these, such as **cecropins** in the hemolymph of many insects (first discovered in the silk moth *Hyalophora cecropia*), kill bacteria by a "hole-punch" process: The proteins insert into the bacterial membrane to form pores (Figure 10-3). This hole makes the membrane extremely leaky, resulting in an influx of ECF sodium. This disrupts ions gradients, and may also be followed by an osmotic flux of water into the victim cell, causing it to swell and lyse. Cecropins and similar pore-forming peptides subsequently found in a wide variety of animals and plants are often called **defensins**. Most defensins have numerous positive charges, which are attracted to bacterial membranes to form pores because that membrane is dominated by negatively charged phospholipids. Eukaryotic membranes do not have such strong charges and so are generally safe from attack by these peptides.

Since the discovery of moth cecropin in 1981, hundreds of defensins and other defensive peptides have been found in numerous taxonomic groups including some plants, fruit flies, horseshoe crabs, frogs, fishes, and mammals. Numerous names have been given to these amazing molecules, such as *magainins* in the skin of *Xenopus* toads, *piscidins* in mast cells (p. 465) of bony fishes, *squalamine* in shark livers. Some of these peptides do not form pores, and in most cases their mechanisms are not yet understood. Moreover, some defensive peptides have been found to have antifungal, antiparasitic, and even anticancer activities. The antifungal *drosomycin* in *Drosophila* is one example. Considerable research is now being devoted to these molecules in the hopes that some might be used to fight a variety of diseases in humans and other animals. For example, *brevinin* peptides found in skin of Rana frogs come in different forms, some of which kill bacteria via pores, while others show ability to kill human cancer cells by activating apoptosis (p. 101).

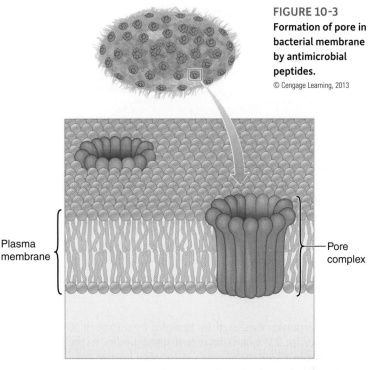

FIGURE 10-3
Formation of pore in bacterial membrane by antimicrobial peptides.
© Cengage Learning, 2013

Plasma membrane

Pore complex

Opsonin proteins "tag" invaders for destruction by phagocytotic cells

While phagocytotic cells can often recognize invaders with their PRRs (e.g., the Toll receptors in *Drosophila*), sometimes they need help from extracellular proteins called opsonins. An **opsonin** ("to prepare for eating") enhances phagocytosis by "tagging" an invader for easy recognition by phagocytotic cells (Figure 10-4). Typically, one portion of an opsonin binds to a surface molecule of an invading bacterium while another portion binds to receptor sites specific for it on the phagocytotic cell's plasma membrane. This link ensures that the bacterial victim does not have a chance to "get away" before the phagocyte can perform its lethal attack. Perhaps all animals have a spectrum of opsonins as innate defenses (and *antibodies*, which we'll see later, are "acquired" opsonins in vertebrates).

One common opsonin class, best studied in arthropods and vertebrates, are the **lectins**, proteins that bind to glycoproteins on foreign cell walls. For example, *carcinolectins* in horseshoe crabs come in several forms capable of binding glycoproteins of many viruses, bacteria, and fungi. The lectins also bind to each other, causing the foreign cells to clump, much as occurs with antibodies (as you will see), and makes them easy targets for phagocytes. Also, a number of defensive proteins in many nonvertebrates have been found to contain a so-called Ig-fold. This is a region in a protein with a structure similar to the fold found in vertebrate antibodies (also called *immunoglobins* or *Ig*) that bind to foreign proteins. Although Ig-fold proteins are not antibodies, they appear to bind proteins commonly found on microbes and to act as opsonins. Indeed, these may be the evolutionary precursors of antibodies. An example of an Ig-fold protein is **hemolin**, found in the hemolymph of moths and the tobacco hornworm (*Manduca sexta*). Suppression of

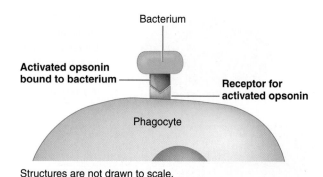

Structures are not drawn to scale.

FIGURE 10-4 Mechanism of opsonin action. An opsonin protein links a foreign cell, such as a bacterium, and a phagocytic cell by nonspecifically binding with the foreign cell and specifically binding with a receptor on the phagocyte. This link ensures that the intruder does not escape before it can be engulfed by the phagocyte.

© Cengage Learning, 2013

immune responses mediated by hemolin by means of RNA interference (p. 33) causes increased susceptibility to the insect pathogen *Photorhabdus* in the tobacco hornworm.

At least in the fruit fly *Drosophila*, opsonin tagging can also trigger release of antimicrobial peptides. Soluble PRRs circulating in the hemolymph first bind to invading bacteria. Subsequently, a PRR (activated by bacteria) binds to the Toll receptor on the fly's hemocyte immune cells, activating genes for production of antimicrobial peptides. Recent studies have found that this activation of genes uses transcription factors that are homologous between flies and mammals, suggesting an ancient history for this aspect of innate immunity.

check your understanding 10.2

What are the first line and three major second lines of immune defense, and how do they work?

What features of bacterial surfaces attract the attention of antimicrobial peptides?

10.3 Immune Systems in Vertebrates: Overview

Most vertebrates have not two but three lines of defenses—the two innate levels plus acquired immunity, which work in concert to contain, and then eliminate, harmful agents. As previously discussed the components of the innate system are always on guard as a rapid but limited, sometimes crude, repertoire of defense mechanisms that contain and limit the spread of infection. These ready-to-go responses work well enough for species survival most of the time (indeed, innate immunity of nonvertebrates may be superior to those in vertebrates with acquired immunity). However, in most vertebrates, the innate defenses are important for keeping highly invasive foes at bay until the acquired immune system, with its highly selective weapons, can be prepared to take over and set in motion strategies to completely eliminate the invader.

Active immune responses are regulated by feedback systems with sensors, integrators, and effectors that are primarily leukocytes and proteins

Although a few innate mechanisms are largely passive, such as the physical barrier formed by the skin, most innate and acquired mechanisms are active responses that we would expect to be regulated by feedback systems (as are almost all biological processes). Many of these feedback processes were believed to be localized to the specific tissues involved. But in recent years, immunologists have come to realize that some responses in vertebrates constitute sophisticated feedback loops coordinating many parts of the body. Indeed, some have proposed that the vertebrate immune system be considered a third "whole-body" regulatory system on a par with the neural and endocrine systems. Certainly the immune system has sensors, integrators, and effectors that communicate both locally and across the whole body. However, a single immune cell can act as more than one regulatory component (for example, as both an effector and sensor), and thus, the regulatory features are not as distinct as they are in neural and endocrine systems. Let's briefly examine these regulatory mechanisms before turning to details.

With a few exceptions, the active immune components are individual cells and extracellular proteins, rather than tissues and organs as in the neural and endocrine systems. As in all animals, vertebrate immune proteins include antimicrobial peptides and opsonins (including *antibodies*) as effectors. The cells responsible for the various immune defense strategies are the circulating **leukocytes** (white blood cells) and their noncirculating derivatives. They communicate through interleukin cytokines and through immunosynapses.

Leukocytes of innate and acquired immunity are found in circulation and localized in tissues

The key mammalian immune cells, which are found either in circulatory fluids (blood, lymph) or localized in barrier tissues, and which may participate in innate or acquired immunity or both, are briefly reviewed as follows (Figure 10-5).

Circulating Leukocytes: Sensors, Integrators, and Effectors

1. *Neutrophils* are highly mobile phagocytotic specialists that engulf and destroy unwanted materials. They are innate effectors in the sense that they correct the problem by eating disturbing agents, that is, "nonself" antigens. But because they can often recognize invaders (with PRRs) and destroy them without needing other regulatory commands, they are in some sense both sensors and effectors.

2. *Eosinophils* secrete chemicals that destroy parasitic worms and are involved in allergies.

3. *Basophils* release *histamine*, a trigger of inflammation, and *heparin,* a clotting inhibitor. They also are involved in allergic manifestations. Basophils can be considered sensor-integrators, in the same way that a home thermostat both senses temperature and regulates heating and cooling effectors (Chapter 1, p. 14), or a pancreatic islet cell senses blood glucose and releases insulin as a regulatory hormone (Chapter 7, p. 316).

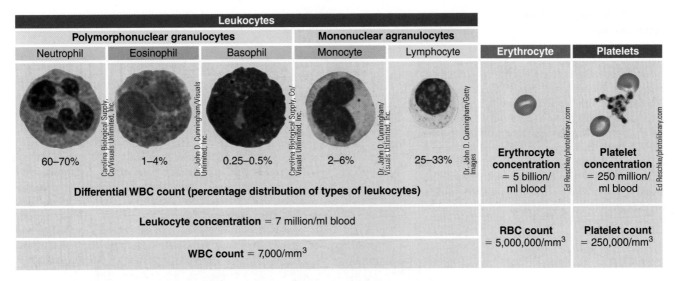

FIGURE 10-5 Normal blood cellular elements and typical human blood cell count. In the far right photograph, the larger cells are erythrocytes, while the smaller objects are platelets.

4. *Monocytes* are transformed into **macrophages,** which are large, tissue-bound phagocytotic specialists, acting as effectors or as sensor-integrator-effectors such as neutrophils. Macrophages can also act as sensors in acquired immunity by becoming *antigen-presenting* cells, or *APCs,* as you will see later. Recall that microglia, "cousins" of monocytes, are the immune defense cells of the central nervous system (p. 169).

5. *Natural killer (NK) cells,* are a special class of immune cells that destroy virus-infected host cells and cancer cells. These seem to act independently as innate sensor-integrator-effectors.

6. *Lymphocytes* are the primary cells of acquired immunity in jawed vertebrates, and are classified into two types, which we'll explore in detail later. Briefly, *B-lymphocytes* (also called simply *B cells*) secrete antibodies (effector proteins) that lead to the destruction of very specific foreign material. *T-lymphocytes (T cells)* come in three main types. One is an effector that directs destruction of virus-invaded and cancer cells, while the others are integrators that stimulate or suppress acquired immune effector responses.

Localized Leukocyte Derivatives: Sensors and Integrators

1. *Mast cells* reside in barrier tissues (integumentary, digestive, respiratory) and are analogous to sensing-integrating endocrine glands or a thermostat. They sense an invasion and secrete cytokines such as *histamine* to activate effectors in the local area under attack.

2. *Dendritic cells* also reside in barrier and many other tissues and are now thought to be the key sensors in the acquired defenses. They sense invaders via innate mechanisms but can also become antigen presenting cells *(APCs)* like macrophages to activate acquired immunity.

   Although most leukocytes are circulating cells, that does not mean they are primarily in the blood or lymph itself. Most are out in the tissues on defense missions, widely dispersed throughout the body to defend in any location. Al-

most all leukocytes originate from common precursor stem cells in the *bone marrow* (see Figure 9-7, p. 394). We first examine in more detail mammalian innate immunity before turning our attention to acquired immunity.

check your understanding 10.3

What are the major functions of neutrophils, monocytes, NK cells, lymphocytes, mast cells, and dendritic cells?

10.4 Innate Immunity in Mammals

Innate defenses in mammals are similar to those you saw for other animals, but much more detail is known about them. They include:

1. *Barrier tissues and glands.*
2. *Inflammation,* sensed and integrated by *mast cells,* with phagocytotic specialists—*neutrophils* and *macrophages*—as primary effectors. *Basophils* also trigger the inflammatory process responsible for allergies.
3. *The complement system,* a group of inactive plasma proteins that, when activated, act as antimicrobial peptides and opsonins.
4. *Interferon,* a family of effector proteins that nonspecifically defend against viral infection.
5. *Natural killer (NK) cells,* which destroy virus-infected host cells and cancer cells.
6. *Symbiotic bacteria.* In the digestive system, on the skin, and in the female reproductive tract reside countless numbers of resident bacteria. For the most part these do no detectable harm (although it has been shown that some bacteria may convert antioxidants into reactive oxidants) and are often beneficial (for example, some mammalian colon bacteria synthesize vitamin K; other

FIGURE 10-6 Anatomy of the mammalian skin. The skin consists of two layers, a keratinized outer epidermis and a richly vascularized inner connective tissue dermis. Special infoldings of the epidermis form the sweat glands, sebaceous glands, and hair follicles. The epidermis contains four types of cells: keratinocytes, melanocytes, Langerhans cells, and Granstein cells. The skin is anchored to underlying muscle or bone by the hypodermis, a loose, fat-containing layer of connective tissue.

© Cengage Learning, 2013

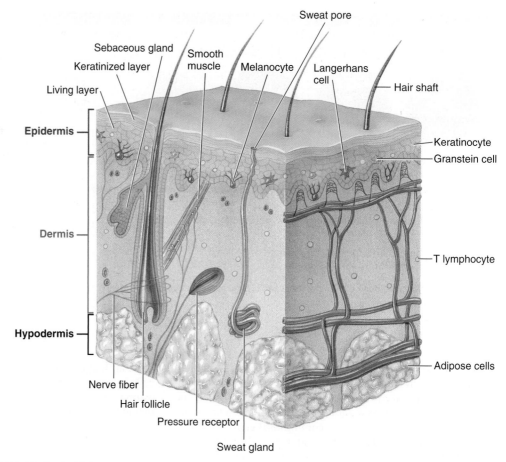

gut bacteria are crucial to digestion in ruminants; p. 708). An additional benefit is that they can inhibit the growth of invaders, either by outcompeting them or by secreting defense chemicals. For example, "friendly" bacteria that live on the back of the human tongue convert food-derived nitrate into nitrite, which is swallowed. Acidification of nitrite on reaching the highly acidic stomach generates nitric oxide, which is toxic to a variety of microorganisms.

We will now examine the first five of these defenses in detail.

Barrier tissues form passive walls and can have active mechanisms to kill or remove invaders

Recall that barrier tissues—the first lines of defense—are those exposed to the environment.

Integument The skin (Figure 10-6), typically the largest organ of an animal, not only serves as a mechanical barrier between the external environment and the underlying tissues but is dynamically involved in defense mechanisms (and other important functions). The vertebrate **epidermis** consists of numerous layers of epithelial cells without a direct blood supply. Its cells are nourished only by diffusion of nutrients from a rich vascular network in the underlying **dermis** (the connective tissue of the skin). Newly forming cells in the inner layers constantly push the older cells closer to the surface, farther and farther away from their nutrient supply. Epidermal cells are tightly bound together by spot

desmosomes which interconnect with intracellular *keratin* filaments (see p. 65) to form a strong, cohesive, keratinized layer. As the scales of the outermost keratinized layer slough or flake off through abrasion, they are continuously replaced by cell division in the deeper layers. The keratinized layer is airtight, fairly waterproof, and impervious to most substances. It impedes passage into the body of most materials that come into contact with the body surface, including bacteria and toxic chemicals.

Special infoldings of the epidermis into the dermis form the skin's *exocrine* glands—the sweat glands in some mammals, and sebaceous glands—as well as hair follicles. **Sweat glands** are crucial to temperature regulation (Chapter 15), but they can also produce antimicrobial peptides such as *dermcidin*, which not only help keep pathogens out of the glands but also skin wounds. **Sebaceous glands** produce the oily secretion **sebum** that is released into hair follicles. From there the sebum flows to the surface of the skin, oiling both the hairs and the outer keratinized layers of the skin to help waterproof them and prevent them from drying and cracking. Sebum may also help inhibit the growth of microorganisms.

The epidermis contains two distinct innate defensive cell types. **Melanocytes** produce the pigment **melanin**, which they disperse to surrounding skin cells. Melanin protects by absorbing harmful ultraviolet light rays. **Keratinocytes**, the most abundant epidermal cells, are specialists in keratin production. As they die, they form the outer protective keratinized layer. They are also responsible for generating hair, claws, talons, and nails. A surprising, recently discovered function is that keratinocytes are also important immunologically. For example, they make antimicrobial peptides called *cathelicidins*, which can protect skin wounds from streptococcal bacteria. (The latter function was shown with knockout mice lacking a gene for one cathelicidin; they could not fight off infection by certain streptococcal bacteria.) Cathelicidins in humans, pigs, and cows have other functions, such as attracting other immune cells into the site of invasion during inflammation, stimulating wound healing, and triggering *apoptosis* (p. 101) of damaged body cells in the wound site.

The epidermis also synthesizes a precursor of vitamin D, which is derived from a precursor molecule closely related to cholesterol, in the presence of sunlight. The biologically active vitamin D promotes the absorption of Ca^{2+} from the digestive tract into the blood (p. 329). Recently considerable evidence has accumulated showing that vitamin D boosts the actions of many immune cells. This precursor is made by an undetermined cell type in most mammals. In birds and mammals with thick fur, the precursor is produced in sebum.

Two other epidermal cell types play important roles in acquired immunity (Figure 10-6). A type of *dendritic* cells called **Langerhans cells** serve as sensors (APCs) to activate acquired immune responses (as you will see later). In contrast, **Granstein cells** appear to serve as a "brake" on skin-activated immune responses. Significantly, Langerhans cells are more susceptible to damage by UV radiation (as from the sun) than are Granstein cells. Loss of Langerhans cells due to UV radiation can lead to the braking signals dominating, leaving the skin more vulnerable to microbial invasion and cancer cells.

The various epidermal components of the immune system are collectively termed **skin-associated lymphoid tissue**, or **SALT**. SALT probably plays an even more elaborate role in active immune defense than described here. This is appropriate, because the skin serves as a major interface with the external environment.

Other Barrier Tissues Other barrier tissues also have active innate defenses in addition to being physical barriers and having symbiotic bacteria.

- The *digestive tract* has many defenses because it is frequently exposed to ingested pathogens. Saliva contains the enzyme *lysozyme* and various *defensins* such as *histatins*, which kill bacteria and fungi and promote wound healing (all benefits of licking wounds with the tongue). Furthermore, many surviving bacteria that are swallowed are killed by the strongly acidic fluids in the stomach. Farther down the tract, the intestinal lining has **GALT** (**gut-associated lymphoid tissue**). GALT makes antimicrobial peptides such as the *defensin* HD-5 in the small intestine. HD-5 kills an intestinal bacterium called *Salmonella typhimurium*, which is spread by contaminated food and causes severe diarrhea. The gene for HD-5 was genetically engineered into mice, which are normally unable to resist *S. typhimurium*. The engineered mice all survived an orally induced infection that killed all the control mice.
- Within the *genitourinary* (reproductive and urinary) system, would-be invaders encounter hostile conditions in acidic, often salty urine and acidic vaginal secretions with "territorial" symbiotic bacteria. The genitourinary

organs also produce a sticky mucus, which, like flypaper, entraps small invading particles. Later the particles are either engulfed by phagocytes or are swept out (for example, with urination or menstrual flow).

- The *respiratory system* in air breathers has several defense mechanisms against inhaled invaders and particulate matter. Larger airborne particles are filtered out of the inspired air by hairs at the nasal passage entrance. Later you will learn about *lymphoid tonsils* and *adenoids*, which protect the entrance to the respiratory system. Farther down in the airways, millions of tiny cilia (see Figure 2-28, p. 61) constantly beat in an outward direction. The airways are coated with a layer of thick, sticky mucus with *cathelicidins* secreted by epithelial cells. This mucus sheet, laden with any inspired particulate debris (such as dust) that adheres to it, is constantly moved upward to the throat by ciliary action. This moving mucus is known as the **mucus escalator**. The dirty mucus is either expectorated (spit out) or in most cases swallowed. In addition, an abundance of phagocytotic specialists called the alveolar macrophages scavenge debris within the air sacs (alveoli, p. 504) of the lungs. Further respiratory defenses include common reflex mechanisms that involve forceful outward expulsion of material in an attempt to remove irritants from the trachea by *coughing* or the nose by *sneezing*.

Inflammation in part is triggered by histamine released by mast cells

Recall that *inflammation* is an innate response to invasion and tissue injury that impedes the spread of damage and brings fluids, phagocytes, and defensive proteins to the invaded or injured area. The following sequence of events typically occurs during the mammalian inflammatory response in the case of bacterial entry into a break in the skin (Figure 10-7).

Defense by Resident Tissue Macrophages On bacterial invasion through a break in the external barrier of skin, the macrophages already present in the area immediately begin phagocytizing the foreign microbes (provided they can recognize them as nonself). Although the resident macrophages are usually not present in sufficient numbers to meet a severe challenge alone, they defend against infection during the first hour or so, before other mechanisms can be mobilized.

These phagocytotic cells are studded with *Toll-like receptors* (*TLRs*, Figure 10-1b). TLR binding to a pathogen triggers the phagocyte to engulf and destroy the invader. There are a number of TLRs, which recognize (bind) different antigen molecules. For example, TLR2 recognizes peptidoglycans (polymer of amino acids and sugars) from bacterial cell walls, TLR5 binds to flagellin in bacterial flagella, and TLR9 recognizes bacterial DNA. Another type of pattern-recognition receptor (PRR), **retinoic acid inducible gene I (RIG-I)**, has recently been found inside immune cells and recognizes viral RNA. Phagocytes are also activated by the release of intracellular chemicals from damaged body cells.

Integration by Mast Cells: Localized Capillary Vasodilation and Increased Permeability Almost immediately on microbial invasion, mast cells detect the injury and release the

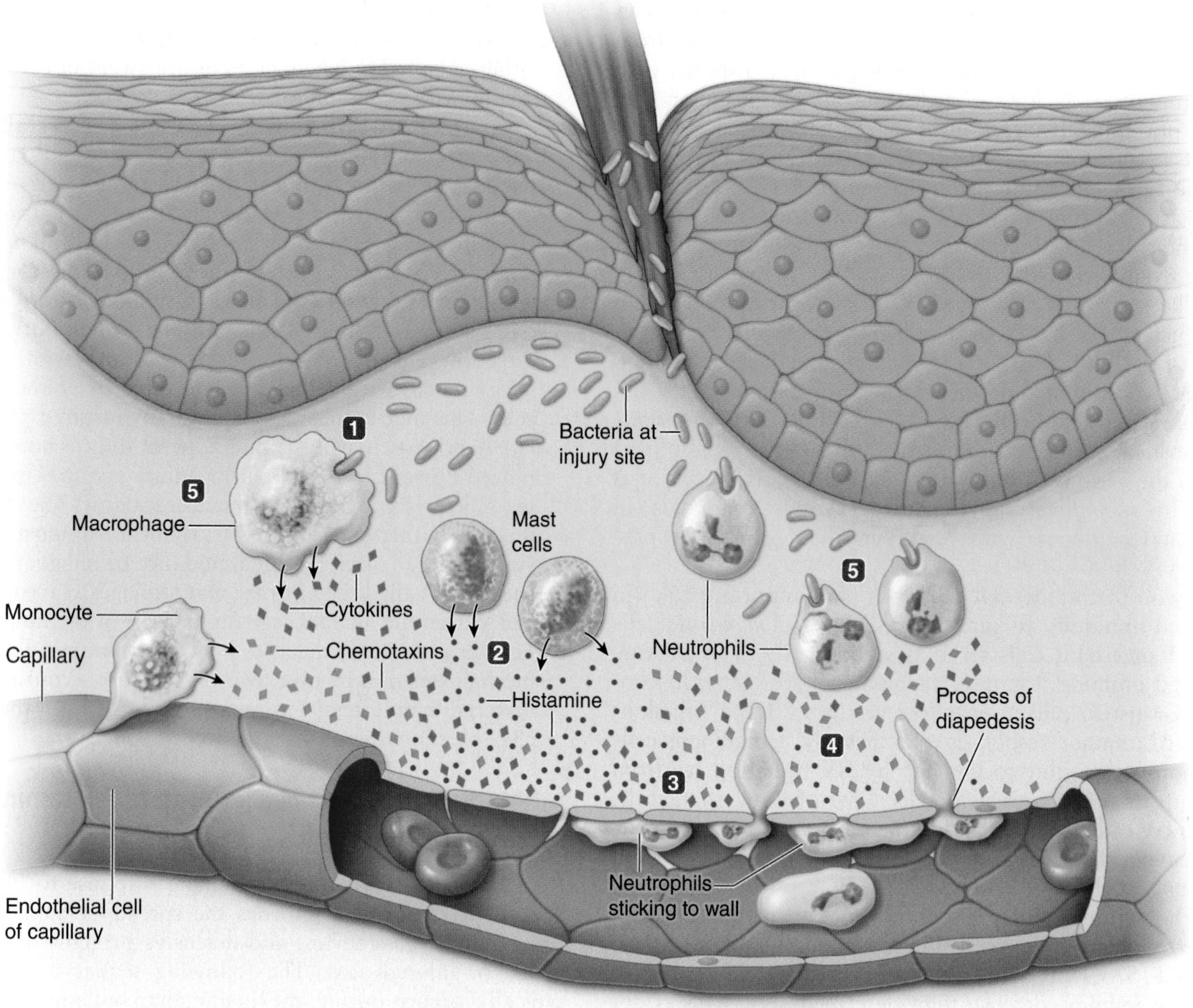

FIGURE 10-7 Steps producing inflammation. Chemotaxins released at the site of damage attract phagocytes to the scene. Note the leukocytes emigrating from the blood into the tissues by assuming amoeba-like behavior and squeezing through the capillary pores, a process known as *diapedesis*. Mast cells secrete vessel-dilating, pore-widening histamine. Macrophages secrete cytokines that exert multiple local and systemic effects.

© Cengage Learning, 2013

paracrine cytokine called *histamine*. (One signal that activates mast cells in skin is a type of cathelicidin released from resident macrophages.) This causes arterioles and capillaries within the area to dilate, increasing blood flow to the site of injury and thus aiding the delivery of more phagocytotic cells and plasma proteins such as *complement* (which we will describe shortly). Mast cells also release *chemotaxins* to attract phagocytes. Histamine also increases the capillaries' permeability by enlarging the capillary pores (the spaces between the endothelial cells; see p. 440) so that plasma proteins and cells that normally are prevented from leaving the blood can escape into the inflamed tissue. The familiar

swelling that accompanies inflammation is due to these histamine-induced vascular changes. Pain is caused both by local distension within the swollen tissue and by the direct effect of locally produced substances such as *prostaglandins* on the receptor dendrites of pain neurons.

Clotting On exposure to injured tissue and to specific chemicals secreted by phagocytes on the scene, *fibrinogen*, the final factor in the clotting system, is converted into *fibrin*. Fibrin forms interstitial fluid clots in the spaces around the bacterial invaders and damaged cells. This walling off of the injured region from the surrounding tissues delays the spread of bacterial invaders and their toxic products. (For details of clotting, see p. 395.)

Emigration of Leukocytes

Due to chemotaxis, within an hour after the injury, the involved area is teeming with leukocytes that have exited from the vessels. Neutrophils arrive first, followed by monocytes, which swell and mature into macrophages. Leukocyte emigration from the blood into the tissues uses an amoeba-like mechanism known as *diapedesis*. A leukocyte pushes a long, narrow projection through a capillary pore; then the remainder of the cell flows forward into the projection. In this way, the leukocyte can wriggle its way in about a minute through the capillary pore—even though it is much larger than the pore. Outside the vessel, the leukocyte moves in an amoeboid fashion toward the site of tissue damage and bacterial invasion.

Phagocytotic Destruction of Tagged and Untagged Bacteria

Distinguishing "self" from "nonself" is aided both by phagocytic TLRs as well as the foreign particles being coated with plasma opsonins (Figure 10-4). The most important opsonins are the innate proteins of the *complement* system (see below) and *antibodies* of the acquired system (p. 474). These increase phagocytosis by neutrophils and macrophages, which clear the inflamed area of infectious and toxic agents as well as tissue debris. Phagocytosis can engulf a bacterium in less than 0.01 second. Phagocytes eventually die because of the accumulation of toxic byproducts from foreign particle degradation or inadvertent release of destructive lysosomal chemicals into the cytosol. The pus that forms in an infected wound is a collection of these phagocytotic cells, both living and dead; necrotic (dead) tissue liquefied by lysosomal enzymes released from the phagocytes; and bacteria.

Nonphagocytotic Destruction of Tagged and Untagged Bacteria

Nonphagocytotic mechanisms also exist to kill invaders. For example, neutrophils also secrete cathelicidins (p. 467), and macrophages produce and release *nitric oxide (NO)*, a reactive chemical that is toxic to nearby microbes (see p. 93). They can also create other toxic microbe-destroying *reactive oxygen species (ROS*, p. 53), released in bursts in the vicinity of invaders. Perhaps most important, the complement system also kills by nonphagocytotic means. As a subtler means of destruction, neutrophils secrete *lactoferrin*, a protein that tightly binds with iron, making it unavailable for use by invading bacteria that depend on high concentrations of available iron for growth. Macrophages also secrete the cytokines **interleukin-1 (IL-1)** and **interleukin-6 (IL-6)**, which decrease the plasma concentration of iron by altering iron metabolism within the liver, spleen, and other tissues, inhibiting bacterial multiplication. Thus, for humans at least, taking iron supplements can sometimes defeat this strategy and the lactoferrin defense and make a disease (such as bubonic plague) actually worse.

Tissue Repair

The inflammatory process serves not just to isolate and destroy injurious agents, but also to clear the area for tissue repair. In some tissues (for example, skin, bone, and liver), the healthy organ-specific cells surrounding the injured area undergo cell division to replace the lost cells, often accomplishing complete regeneration of the damaged region. In nonregenerative tissues such as nerve and muscle in mammals, however, lost cells are replaced by scar tissue. Local fibroblasts, a type of connective tissue cell, start to divide rapidly in the vicinity and secrete large quantities of the protein collagen (see p. 65), which fills in the region vacated by the lost cells and forms scar tissue. Even in a tissue as readily replaceable as the skin, scar tissue sometimes forms when deep wounds permanently destroy complex underlying structures.

Mediation of Inflammation by Phagocyte-Secreted Chemicals

Microbe-stimulated phagocytes release many chemicals that function as mediators of the inflammatory response. These cytokines and effector proteins take part in a broad range of interrelated immune activities, varying from local responses to whole-body manifestations. In addition to killing invaders, these proteins stimulate mast cells (and thus boost inflammation), while others assist the clotting processes. One chemical secreted by neutrophils, **kallikrein**, converts specific plasma protein precursors produced by the liver into activated **kinins**. Activated kinins augment a variety of inflammatory events. For example, the end product of the kinin cascade, **bradykinin**, activates nearby pain receptors and thus partially produces the soreness associated with inflammation. In *positive-feedback* fashion, kinins also act as powerful chemotaxins to entice more neutrophils to join the battle.

Fever

IL-1, IL-6, and TNF also all function as **endogenous pyrogen (EP)**, which induces the development of fever (*pyro* means "fire"; *gen* means "production"). This response occurs when the invading organisms spread into the blood. EPs, in turn, cause the release within the hypothalamus of *prostaglandins*, locally acting chemical messengers that "turn up" the hypothalamic "thermostat" that regulates body temperature. The function of the resulting elevation in body temperature in fighting infection remains unclear. The fact that fever is such a common systemic manifestation of disease suggests it plays an important beneficial role in defense, as supported by recent evidence. For example, infected lizards will behaviorally choose a higher environmental temperature than normal (see also Chapter 15 on thermoregulation). In mammals, higher body temperatures augment phagocytosis and increase the rate of the many enzyme-dependent inflammatory activities. Furthermore, an elevated body temperature may interfere with bacterial multiplication by increasing the bacterial requirements for iron. Nevertheless, although a mild fever may possibly be beneficial, an extremely high fever can harm the central nervous system.

This list of events augmented by chemicals secreted by phagocytes is by no means complete, but it serves to illustrate the diversity and complexity of responses elicited by these mediators.

The complement system punches holes in microorganisms directly, acts as opsonins, works with antibodies, and augments inflammation

Several times now we have mentioned the **complement system**, another "second-line" defense mechanism based on free-floating proteins. These can be called into play by the presence of almost any foreign invader. The complement system is activated in two ways (Figure 10-8):

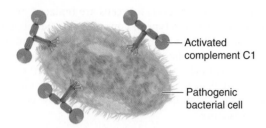

Alternate complement pathway: Binding directly to a foreign invader nonspecifically activates the complement cascade (an innate immune response).

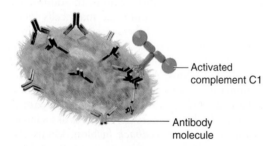

Classical complement pathway: Binding to antibodies (Y-shaped molecules) produced against and attached to a particular foreign invader specifically activates the complement cascade (an adaptive immune response).

FIGURE 10-8 **Activation of complement system via the alternate pathway and the classical pathway.**

© Cengage Learning, 2013

1. By exposure to particular carbohydrate chains expressed on the surface of microorganisms but not on body cells, an innate immune response called the **alternative pathway.**
2. By exposure to antibodies produced against a specific foreign invader, an acquired immune response called the **classical pathway.** This is the source of the system's name—the fact that it "complements" the action of antibodies. In fact it is the major mechanism activated by antibodies to kill foreign cells.

The fact that the complement system is involved in both innate and acquired defense mechanisms illustrates an important point. The various components of the immune system are highly interactive and interdependent, making the system highly sophisticated and effective but also complex and difficult to sort out. As we have done a few times already, we will continue to point out significant cooperative relationships known to exist among immune features as we examine various components separately.

The Membrane Attack Complex The complement system consists of plasma proteins designated C1 to C9 that are produced by the liver and circulate in the blood in an inactive form. Once the first component, C1, is activated, it activates the next component, C2, and so on, in a sequential cascade of activation reactions. The five final components,

C5 through C9, assemble into a large, doughnut-shaped, protein complex, the **membrane attack complex (MAC)**, which attacks the surface membrane of nearby microorganisms by imbedding itself so to create a pore, just as the ancient antimicrobial peptide system (Figure 10-3). In fact, many of the complement proteins are evolutionarily related to defensins of nonvertebrates. The pore's osmotically induced lysis of the intruder is the major means of directly killing microbes without phagocytizing them.

Augmentation of Inflammation Besides the direct destruction of foreign cells accomplished by the MAC, various other activated complement components augment the inflammatory process by:

- *Serving as chemotaxins* to attract phagocytes to the site of microbial invasion.
- *Acting as opsonins* by binding with microbes and thereby enhancing their phagocytosis.
- *Stimulating the release of histamine* from mast cells in the vicinity, which in turn enhances the local vascular changes characteristic of inflammation.
- *Activating kinins*, which further reinforce inflammatory reactions.

NSAIDs and glucocorticoid drugs suppress the inflammatory response

Many drugs can suppress the inflammatory process; the most effective are the *nonsteroidal anti-inflammatory drugs*, or *NSAIDs* (aspirin, ibuprofen, and related compounds) and *glucocorticoids* (drugs similar to the steroid hormone *cortisol*, which is secreted by the adrenal cortex; p. 305). For example, aspirin interferes with the inflammatory response by decreasing histamine release, thus reducing pain, swelling, and redness. Furthermore, aspirin inhibits clotting and reduces fever by inhibiting production of prostaglandins, the mediators of EP-induced fever. NSAIDs are of course widely used by humans, but they are also sometimes prescribed by veterinarians for domestic mammals to reduce inflammation.

Glucocorticoids suppress almost every aspect of the inflammatory response. In addition, they destroy lymphocytes in lymphoid tissue and reduce antibody production. These therapeutic agents, also used by humans and given to domestic mammals, are useful for treating undesirable immune responses, such as allergic reactions (for example, asthma or severe reaction to a bee sting). However, suppression of immune responses also reduces the body's ability to resist infection. Aspirin can also lead to dangerous bleeding. For these reasons, glucocorticoids and NSAIDs should be used discriminatingly.

Recent evidence suggests that cortisol, whose secretion is increased in response to any stressful situation, exerts anti-inflammatory activity even at normal physiological levels (see p. 306). According to this proposal, the anti-inflammatory effect of cortisol modulates stress-activated immune responses, preventing them from overshooting and thus protecting against damage by potentially overreactive defense mechanisms. This is probably important because there are positive-feedback loops in the inflammation response, as we saw earlier. Positive feedback can spiral out of control without some way to interrupt it.

FIGURE 10-9 Mechanism of action of interferon in preventing viral replication. Interferon, which is released from virus-infected cells, binds with other uninvaded host cells and induces these cells to produce inactive enzymes capable of blocking viral replication. The inactive enzymes are activated only if a virus subsequently invades one of these prepared cells.

© Cengage Learning, 2013

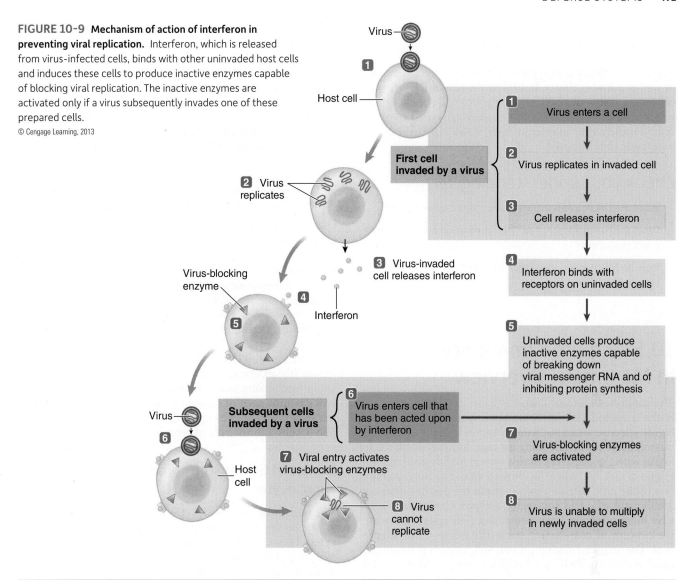

Natural killer cells destroy virus-infected cells and cancer cells on first exposure to them

Natural killer (NK) cells are naturally occurring, lymphocyte-like cells that nonspecifically destroy virus-infected cells and cancer cells by releasing chemicals that directly lyse (rupture) the membranes of such cells on first exposure to them. NK cells recognize general features of virus-infected cells and cancer cells by means largely unknown (they do not have TLRs like macrophages or specific receptors like effector T cells have). NK cells provide a rapid, nonspecific defense against virus-invaded cells and cancer cells before the more specific and abundant effector T cells become operational, which requires previous exposure and a prolonged maturation period. NK cells form an immunosynapse with their target cells in order to inject toxins into them, and they can also attach to escaping cancer cells with *nanotubes* (p. 91) that pull the target cell back like a bungee cord!

Interferon transiently inhibits multiplication of viruses in most cells

Besides the inflammatory response, another innate defense mechanism is the release of **interferon** from virus-infected cells. Interferon, a group of three related cytokines, briefly provides nonspecific resistance to viral infections by transiently interfering with replication of the same or unrelated viruses in other host cells. In fact, interferon was named for its ability to "interfere" with viral replication.

Antiviral Effect of Interferon When a virus invades a cell, the cell synthesizes and secretes interferon in response to being exposed to viral nucleic acid. Once released into the ECF, interferon binds with receptors on the plasma membranes of healthy neighboring cells or even distant cells via the blood, signaling these cells to prepare for possible viral attack. Interferon thus acts as a "whistle-blower," forewarning healthy cells of potential viral attack. Interferon triggers potential host cells to produce enzymes that can break down viral messenger RNA and inhibit protein synthesis. Both these processes are essential for viral replication (Figure 10-9). Because interferon activation takes place only during a limited time span, this is a short-term defense mechanism.

Anticancer Effects of Interferon Interferon exerts anticancer as well as antiviral effects. It markedly enhances the actions of cell-killing cells—the NK cells and effector T-lymphocytes—which attack and destroy both virus-infected cells and cancer cells. Furthermore, interferon itself slows cell division and suppresses tumor growth.

check your understanding 10.4

What are the active features of the major barrier tissues?

What are the major steps of (1) inflammation and (2) complement responses?

How do NK cells and interferon deal with cancerous and virus-infected cells?

10.5 Acquired Immunity in Mammals: Overview

The sophisticated third line of defense of jawed vertebrates is the acquired (adaptive) immune system, which directs highly selective attacks at foreign agents but only after previous exposure to the offending invader. Modern immunology began with the discovery of this system by Edward Jenner in England in 1796 when he showed that exposing a young boy to the mild disease cowpox (from the pus of blisters on a milkmaid who had contracted cowpox from a dairy cow) gave him protection against smallpox. The system illustrates the concept of *cost–benefit trade-offs* (see p. 2) in adaptation. The acquired system (as you will see) can produce millions of different PRRs (and soluble versions called antibodies) from a relatively few number of genes to bind to very specific foreign molecules. This ability allows an individual animal to "keep up" with rapidly evolving pathogens, such as new strains of influenza virus. This may be particularly important in long-lived animals with relatively few offspring, such as sharks (in which the vertebrate system may have first evolved, as you will see later). In contrast, short-lived animals with huge numbers of offspring, such as insects, may rely more on evolutionary selection for favorable immune gene mutations over many generations.

However, were all of these variants to be produced in large quantities at once, it would be far too costly, even dangerous, in terms of energy for biosynthesis and for the heart to pump blood made viscous by a high cell density. Instead, very few of each variant is produced at first (not enough for defense), but once one variant is activated by a particular invader, it is amplified to high concentrations to take on that specific invader.

Acquired immunity includes antibody-mediated and cell-mediated responses

There are two classes of acquired immune responses:

- **Antibody-mediated,** or **humoral immunity,** based on **B-lymphocytes** (also called simply **B cells**). B cells transform into *plasma cells*, which secrete *antibodies* (effector proteins) that directly or indirectly lead to the destruction of foreign material. B cells can be thought of as part of the effector response that specifically kills invaders. Because antibodies are bloodborne, antibody-mediated immunity is sometimes known as *humoral* immunity, in reference to the ancient Greek use of the term *humors* for the various body fluids.
- **Cell-mediated immunity with T-lymphocytes** (also called simply **T cells**): as noted earlier, some T cells are effectors—called *cytotoxic T cells*—that are responsible for cell-mediated immunity involving both direct destruction of virus-invaded cells and cancer cells.

Both responses depend on integrating signals from two other types of T cells: (1) **helper T cells** are the primary integrators of the acquired immune response, receiving input from sensory APC cells and then activating the effector cells; and (2) **T-regulatory cells** suppress immune responses; they are integrators that keep immune reactions in check.

Lymphocytes can specifically recognize and selectively respond to an almost limitless variety of foreign agents as well as cancer cells. The recognition and response processes are different in B cells and in T cells. In general, B cells recognize free-existing foreign invaders; effector T cells specialize in recognizing and destroying body cells gone awry. We will examine each of these processes in detail in the upcoming sections. First, we explore the different life histories of B and T cells and the role of *lymphoid tissues.*

Lymphoid tissues store, produce, and/or process lymphocytes

The term lymphoid tissues (Figure 10-10) refers collectively to the tissues that store, produce, and/or process lymphocytes. These include the *lymph nodes, spleen, thymus, tonsils, adenoids, bone marrow,* aggregates of lymphoid tissue in the lining of the digestive tract (**GALT,** p. 467), and the lymphoid system of the skin (**SALT,** p. 467). Lymphoid tissues of barrier tissues are strategically located to intercept invading microorganisms before they have a chance to spread very far. For example, the lymphocytes populating the *tonsils* and *adenoids* are situated to respond to inhaled microbes, whereas microorganisms invading through the digestive system immediately encounter lymphocytes in the *appendix* (in some primates including humans) and other GALT such as *Peyer's patches* (Figure 10-10). Potential pathogens that gain access to the lymph are filtered through *lymph nodes,* which have sticky chambers as well as many lymphocytes and macrophages. The *spleen,* the largest of the lymphoid tissues, performs immune functions on the blood similar to those the lymph nodes perform on the lymph. The spleen's macrophages and lymphocytes clear the blood that passes through it of microorganisms and other foreign matter and also removes worn-out red blood cells (see p. 392). Table 10-1 summarizes the major functions of the lymphoid tissues.

Origins of B and T Cells Both types of lymphocytes, like all blood cells, are derived from common stem cells in the bone marrow (see p. 394). Whether a lymphocyte and all its progeny are destined to be B or T cells depends on the site of final differentiation and maturation of the original cell in the

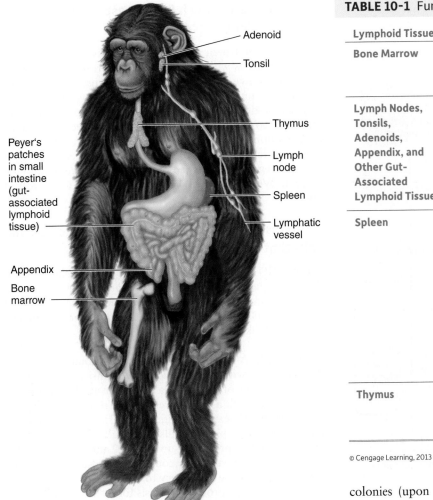

Peyer's patches in small intestine (gut-associated lymphoid tissue)

Adenoid

Tonsil

Thymus

Lymph node

Spleen

Lymphatic vessel

Appendix

Bone marrow

FIGURE 10-10 **Lymphoid tissues in a mammal.** The lymphoid tissues, which are dispersed throughout the body, produce, store, or process lymphocytes.

© Cengage Learning, 2013

TABLE 10-1 Functions of Lymphoid Tissues

Lymphoid Tissue	Functions
Bone Marrow	Origin of all blood cells Site of maturational processing for B lymphocytes
Lymph Nodes, Tonsils, Adenoids, Appendix, and Other Gut-Associated Lymphoid Tissue	Exchange lymphocytes with the lymph (remove, store, produce, and add them) Resident lymphocytes produce antibodies and activated T cells, which are released into the lymph Resident macrophages remove microbes and other particulate debris from the lymph
Spleen	Exchanges lymphocytes with the blood (removes, stores, produces, and adds them) Resident lymphocytes produce antibodies and activated T cells, which are released into the blood Resident macrophages remove microbes and other particulate debris, most notably worn-out red blood cells, from the blood Stores a small percentage of red blood cells, which can be added to the blood by splenic contraction as needed
Thymus	Site of maturational processing for T-lymphocytes Secretes the hormone thymosin

© Cengage Learning, 2013

colonies (upon appropriate stimulation) rather than from the bone marrow.

An antigen induces an immune response against itself

Both B and T cells must be able to specifically recognize "self" from "nonself." The presence of *antigens* enables them to make this distinction. An **antigen** is a large, complex, unique molecule that triggers a specific immune response against itself when it gains entry into the body. In general, the more complex a molecule, the greater its antigenicity, because it has multiple sites to which antibodies can be generated. Foreign proteins are the most common antigens because of their size and structural complexity, although other macromolecules, such as large polysaccharides, can also act as antigens. Antigens may exist as isolated molecules, such as bacterial toxins, or they may be an integral part of a multimolecular structure, such as being on the surface of an invading microbe.

We first examine the role of B cells, then T cells.

check your understanding 10.5

What are the basic differences between antibody- and cell-mediated responses?

Where do B and T cells originate and mature?

What is a lymphoid tissue? An antigen?

lineage (Figure 10-11). B cells differentiate and mature in the bone marrow in mammals, in the *bursa* (an offshoot of the cloaca) in birds. As for T cells, during fetal life and early childhood, some of the immature lymphocytes from the bone marrow migrate through the blood to the *thymus*, where they undergo further processing to become T-lymphocytes (named for their site of maturation). The **thymus** is a lymphoid tissue located midline within the chest cavity anterior to the heart in the space between the lungs (Figure 10-10). The thymus produces **thymosin**, a hormone that enhances proliferation of new T cells within the peripheral lymphoid tissues and augments the immune capabilities of existing T cells. Secretion of thymosin decreases in older mammals. This decline has been suggested as a contributing factor to the increased susceptibility to cancer and viruses with age.

On being released into the blood from either the bone marrow or the thymus, mature B and T cells take up residence and establish lymphocyte colonies in the peripheral lymphoid tissues. After the early developmental years of a mammal, most new lymphocytes are derived from these

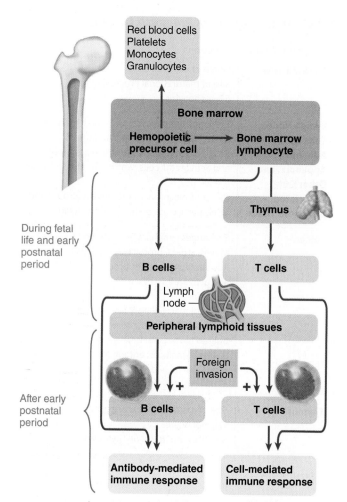

Red blood cells
Platelets
Monocytes
Granulocytes

Bone marrow

Hemopoietic precursor cell → **Bone marrow lymphocyte**

Thymus

During fetal life and early postnatal period

B cells   **T cells**

Lymph node

Peripheral lymphoid tissues

Foreign invasion

After early postnatal period

B cells +   + **T cells**

Antibody-mediated immune response   **Cell-mediated immune response**

FIGURE 10-11 Origins of B and T cells. B cells are derived from lymphocytes that matured and differentiated in the bone marrow, whereas T cells are derived from lymphocytes that originated in the bone marrow but matured and differentiated in the thymus. After early childhood, new B and T cells are produced primarily by colonies of B and T cells established in peripheral lymphoid tissues during fetal life and early childhood.

© Cengage Learning, 2013

10.6 Acquired Immunity in Mammals: B-Lymphocytes

The key to B- and T-cell specificity lies in their membrane PRRs that bind an antigen. We discuss how these receptors are generated later, but for now, keep in mind that B and T cells have unique PRRs—B-cell receptors (BCRs) and T-cell receptors (TCRs)—that can be thought of as the sensors of the acquired system; importantly, each receptor binds to only one particular type of the multitude of possible antigenic sites. This is in contrast to the TLRs (p. 460) of the innate system, which bind to numerous generic markers on many microorganisms.

The antigens to which B cells respond can be T-independent or T-dependent

B cells can bind with and be directly activated by polysaccharide antigens without any assistance from T cells. These antigens, known as **T-independent antigens**, stimulate production of antibody without any T cell involvement. By contrast, **T-dependent antigens**, which are typically protein antigens, do not directly stimulate the production of antibody without the assistance of a *helper T cell*. For now we will consider activation of B cells by binding with antigen without regard to whether or not helper T cells must be present. We will discuss how helper T cells are involved later.

Antigens stimulate B cells to convert into plasma cells that produce antibodies

When BCRs (Figure 10-12a) bind with antigen, signal transduction triggers the B cell to divide and differentiate into active *plasma cells* while others become dormant *memory cells*. We first examine plasma cells and their antibodies, then later look at memory cells.

Plasma Cells A **plasma cell** produces **antibodies** that can then combine with the specific type of antigen that stimulated activation of the plasma cell. During differentiation into a plasma cell, a B cell swells as the rough endoplasmic reticulum greatly expands (Figure 10-13). Because antibodies are proteins, plasma cells essentially become prolific protein factories, producing up to 2,000 antibody molecules per second. So great is the commitment of a plasma cell's protein-synthesizing machinery to antibody production that it cannot maintain protein synthesis for its own viability and growth. Consequently, it dies after a brief (five- to seven-day), highly productive life span.

Antibodies are secreted into the blood or lymph, depending on the location of the activated plasma cells, but all antibodies eventually gain access to the blood, where they are known as **gamma globulins**, or **immunoglobulins (Ig)**.

Antibodies are Y-shaped and classified according to properties of their tail portion

Antibodies are composed of four interlinked polypeptide chains—two long, *heavy chains* and two short, *light chains*—arranged in the shape of a Y (Figure 10-14). Characteristics of the arm regions of the Y determine the *specificity* of the antibody (that is, with what antigen the antibody can bind). Properties of the tail portion of the antibody determine the *functional properties* of the antibody (what the antibody does once it binds with an antigen).

An antibody has two identical antigen-binding sites, one at the tip of each arm. These **antigen-binding fragments (Fab)** are unique for each different antibody so that each antibody can interact only with an antigen that specifically matches it, much like a lock and key. The tremendous variation in Fabs of different antibodies leads to the extremely large number of unique antibodies that can bind specifically with millions of different antigens.

In contrast to these variable Fab regions at the arm tips, the tail portion of every antibody within each immunoglobulin subclass is identical. The tail, the antibody's so-

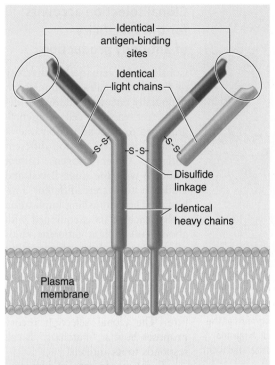

(a) B-cell receptor (BCR)

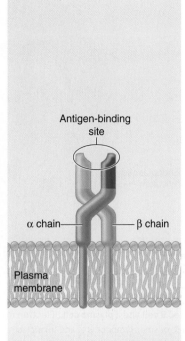

(b) T-cell receptor (TCR)

FIGURE 10-12 B-cell and T-cell receptors.

© Cengage Learning, 2013

called **constant region**, contains binding sites for particular mediators of antibody-induced activities, which vary among different subclasses, as follows.

Antibody Subclasses Antibodies are grouped into five subclasses; note that within each functional subclass are millions of different antibodies, each able to bind only with a specific antigen:

- **IgM** immunoglobulin serves as the B-cell receptor (BCR) for antigen attachment and is produced in the early stages of plasma cell response.
- **IgG**, the most abundant immunoglobulin, is produced copiously when the body is subsequently exposed to the same antigen. The constant tail region of IgG antibodies, when activated by antigen binding in the Fab region, binds with phagocytotic cells and serves as an opsonin to enhance phagocytosis. Together, IgM and IgG antibodies produce most specific immune responses against bacterial invaders and a few types of viruses.
- **IgE** immunoglobulins help protect against parasitic worms and is the antibody mediator for common allergic responses, such as hay fever, asthma, and hives. The constant tail region of IgE antibodies attaches to mast cells and basophils, even in the absence of antigen. Once the triggering antigen binds to the specific antibody (such as IgE in this case), this triggers the release of histamine from the affected mast cells and basophils. Histamine, in turn, induces the allergic manifestations that follow.
- **IgA** immunoglobulins are found in secretions of the digestive, respiratory, and genitourinary systems, as well as in milk and tears.

- **IgD** is present on the surface of many B cells, but its function is uncertain.

Antibodies largely amplify innate immune responses to promote antigen destruction

Antibodies generally do not destroy foreign organisms or other unwanted materials on binding with antigens on their surfaces (although there is some evidence that certain antibodies can catalyze the formation of ozone to kill microbes). Instead, they usually exert their protective influence by physically hindering antigens or, more commonly, by amplifying the innate immune responses (Figure 10-15).

Neutralization and Agglutination Antibodies physically hinder some antigens from exerting their detrimental effects in two ways:

- In **neutralization**, antibodies combine with bacterial toxins and some types of viruses, preventing these harmful chemicals from interacting with susceptible cells (Figure 10-15a).
- In **agglutination**, multiple antibody molecules cross-link numerous antigen molecules into chains or lattices of antigen–antibody complexes (Figure 10-15b). Through this means foreign cells, such as bacteria or mismatched transfused red blood cells, bind together in a nonmotile clump.

Amplification of Innate Immune Responses These physical hindrance mechanisms play only a minor protective role. The most important function of antibodies by far is to profoundly augment the innate immune responses already initiated by the invaders. Antibodies do so by the following methods:

1. *Activating the complement system.* When an appropriate antigen binds with an antibody, receptors on the tail portion of the antibody bind with and activate C1, the first component of the complement system. This sets off the "classic pathway" of events leading to formation of the MAC, which is specifically directed at the membrane of the invading cell that bears the antigen that initiated the activation process (Figure 10-15c). By tagging the invading bacterium, the antibody initiates the most important attack mechanism by which it exerts its (indirect) protective influence.
2. *Enhancing phagocytosis.* As we noted earlier, IgG acts as an opsonin (Figure 10-15d).
3. *Stimulating natural killer (NK) cells.* Binding of antibody to antigen also induces attack of the antigen-bearing target cell by NK cells. NK cells have receptors for the constant tail region of antibodies. In this case, when the target cell is coated with antibodies, the tail

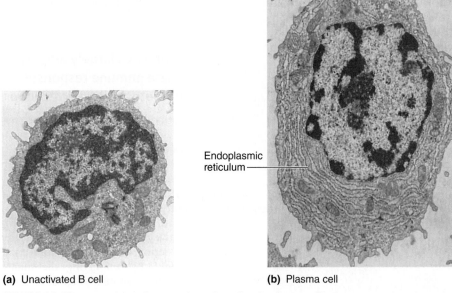

Endoplasmic reticulum

(a) Unactivated B cell

(b) Plasma cell

FIGURE 10-13 Comparison of an unactivated B cell and a plasma cell. Electron micrographs at the same magnification of (a) an unactivated B cell, or small lymphocyte, and (b) a plasma cell. A plasma cell is an activated B cell. It is filled with an abundance of rough endoplasmic reticulum distended with antibody molecules.

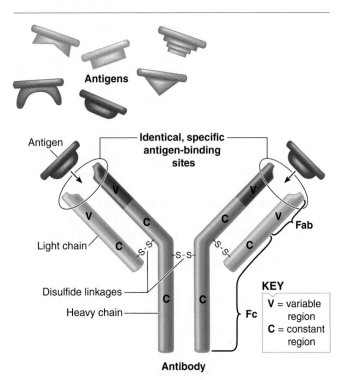

FIGURE 10-14 Antibody structure. An antibody is Y-shaped. It is able to bind only with the specific antigen that "fits" its antigen-binding sites (Fab) on the arm tips. The tail region (Fc) binds with particular mediators of antibody-induced activities.

© Cengage Learning, 2013

Clonal selection accounts for the specificity of antibody production

Consider the remarkable diversity of foreign molecules a mammal can potentially encounter during a lifetime. Yet each B cell is preprogrammed to respond to only one of perhaps over 100 million different antigens. Other antigens cannot combine with the same B cell and induce it to secrete different antibodies. The astonishing implication is that each of us is equipped with about 100 million different kinds of preformed B-lymphocytes, at least one for every possible antigen that we might ever encounter—including those specific for synthetic substances that do not exist in nature. The clonal selection theory proposes how a "matching" B cell responds to its antigen.

Early researchers in immunologic theory believed antibodies were "made to order" whenever a foreign antigen gained entry to the body. In contrast, the currently accepted **clonal selection theory,** first proposed by the Danish immunologist Niels Jerne (who received the Nobel Prize), states that diverse B-lymphocytes are produced during fetal development, each capable of synthesizing an antibody against a particular antigen before ever being exposed to it. All offspring of a particular ancestral B-lymphocyte form a family of identical cells, or a **clone,** that is committed to producing the same specific antibody. B cells remain dormant—known as **naive lymphocytes**—and low in number, not actually secreting their particular antibody product nor undergoing rapid division until (or unless) they come into contact with the appropriate antigen. When this antigen gains entry to the body, the particular clone of B cells that bear receptors (BCRs) on their surface uniquely specific for that antigen becomes activated or "selected" by antigen binding with the BCRs, hence the term *clonal selection theory* (Figure 10-16).

Selected clones differentiate into active plasma cells and dormant memory cells

Antigen also causes a small proportion of the activated B-cell clone to multiply and differentiate into another cell type—**B memory cells.** These do not participate in the current immune attack against the antigen but instead remain dormant. If the animal is ever reexposed to the same antigen, these memory cells remain primed and ready for immediate action.

Recently, evidence has accumulated for another type of B cell, a *B regulatory cell,* that helps suppress overactive immune responses. However, currently the *T-regulatory cell's* role is better known, and we will describe that later.

Primary and Secondary Responses During initial contact with a foreign antigen, the response is delayed for several days as plasma cells are formed and antibody production increases until reaching a peak over a period of weeks. This

portions link the target cell to NK cells, which destroy the target (Figure 10-15e).

Note that each of these innate mechanisms is originally activated by innate means, with their actions being further enhanced by the acquired system's antibodies.

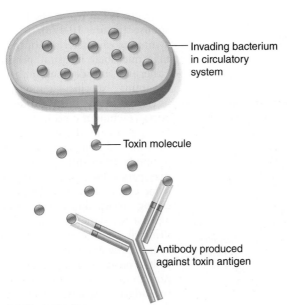

(a) **Neutralization**

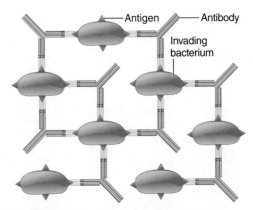

(b) **Agglutination** (clumping of antigenic cells) and **precipitation** (if soluble antigen–antibody complex is too large to stay in solution)

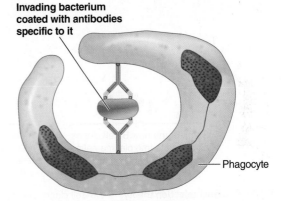

(d) **Enhancement of phagocytosis (opsonization)**

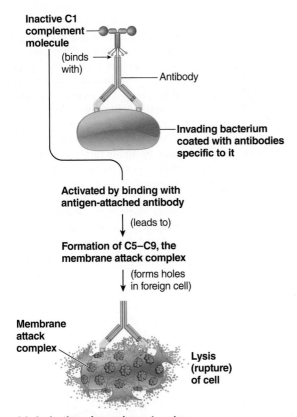

(c) **Activation of complement system**

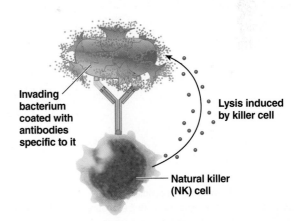

(e) **Stimulation of natural killer (NK) cells:** antibody-dependent cellular cytotoxicity

Structures are not drawn to scale.

FIGURE 10-15 **How antibodies help eliminate invading microbes.** Antibodies physically hinder antigens through (a) neutralization or (b) agglutination and precipitation. Antibodies amplify innate immune responses by (c) activating the complement system, (d) enhancing phagocytosis by acting as opsonins, and (e) stimulating natural killer cells.

Population of unactivated B cells, each a member of a different B-cell clone that makes a specific receptor, which is displayed on the membrane surface as a BCR

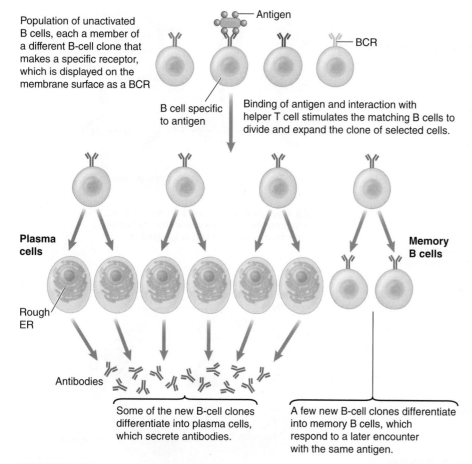

B cell specific to antigen

Binding of antigen and interaction with helper T cell stimulates the matching B cells to divide and expand the clone of selected cells.

Plasma cells

Memory B cells

Rough ER

Antibodies

Some of the new B-cell clones differentiate into plasma cells, which secrete antibodies.

A few new B-cell clones differentiate into memory B cells, which respond to a later encounter with the same antigen.

FIGURE 10-16 Clonal selection theory. The B-cell clone specific to the antigen proliferates and differentiates into plasma cells and memory cells. Plasma cells secrete antibodies that bind with free antigen not attached to B cells. Memory cells are primed and ready for subsequent exposure to the same antigen.

© Cengage Learning, 2013

response is known as the **primary response** (Figure 10-17a). Meanwhile, symptoms characteristic of the particular microbial invasion persist until either the invader succumbs to the mounting specific immune attack against it or the infected animal dies. After reaching the peak, the antibody levels gradually decline over a period of time, although some circulating antibody from this primary response may persist for a prolonged period, in some cases years after the initial triggering event. Long-term protection against the same antigen, however, is primarily attributable to the memory cells. If the same antigen ever reappears, memory cells launch a more rapid, more potent, and longer-lasting **secondary response** than occurred during the primary response (Figure 10-17b). This swifter, more powerful immune attack is frequently adequate to prevent or minimize overt infection on subsequent exposures to the same microbe, forming the basis of long-term immunity against a specific disease. **Vaccination** (**immunization**), discovered by Jenner and now used widely on humans and domestic animals, is a powerful tool to prevent disease. It takes advantage of this process by deliberately exposing an animal to a pathogen that has been stripped of its disease-inducing capability but that can still induce antibody formation against itself. The mechanisms of long-term immunity induced by disease and vaccines are summarized in Figure 10-18.

FIGURE 10-17 Primary and secondary immune responses. (a) Primary response on first exposure to a microbial antigen. (b) Secondary response on subsequent exposure to the same microbial antigen. The primary response does not peak for a couple of weeks, whereas the secondary response peaks in a week. The magnitude of the secondary response is 100 times that of the primary response. (The relative antibody response is in the logarithmic scale.)

© Cengage Learning, 2013

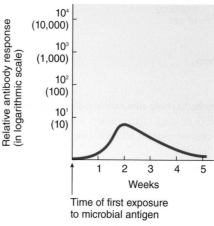

(a) **Primary immune response**

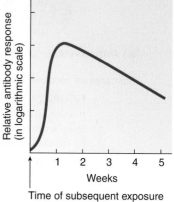

(b) **Secondary immune response**

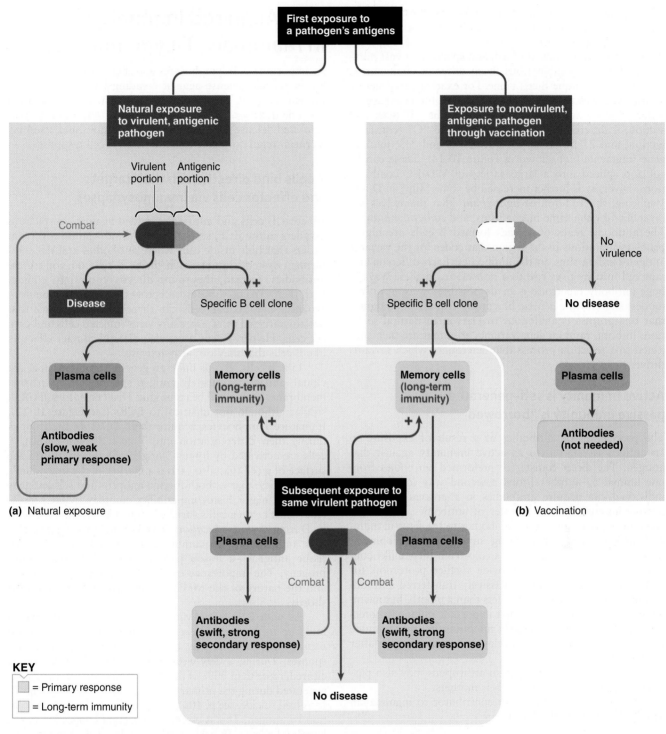

FIGURE 10-18 Means of acquiring long-term immunity. Long-term immunity against a pathogen can be acquired through having the disease or being vaccinated against it. (a) Exposure to a virulent (disease-producing) pathogen. (b) Vaccination with a modified pathogen that is no longer virulent (that is, can no longer produce disease) but is still antigenic. In both cases, long-term memory cells are produced that mount a swift, secondary response that prevents or minimizes symptoms on a subsequent natural exposure to the same virulent pathogen.

© Cengage Learning, 2013

The huge repertoire of B cells is built by recombination of a small set of gene fragments

Considering the millions of different antigens against which each mammal has the potential to actively produce antibodies, how is it possible for an individual to have such a tre-

mendous diversity of B cells and their corresponding antibodies? Antibodies as proteins require a nuclear DNA blueprint. Because all cells of the body, including the B cells, contain the same nuclear DNA, it is hard to imagine how enough DNA could be packaged within the nuclei of every cell to code for a 100 million different antibodies, along with all the other genetic instructions used by other cells. Actually, only a relatively small number of gene fragments code for antibody synthesis, but during B cell development these fragments are cut, shuffled, and spliced in a vast number of different combinations. Each different combination gives rise to a unique B-cell clone. For example, to generate various *heavy chains* of IgG (Figure 10-14), humans have 51 different gene fragments that can code for the "V" (variable) section, 4 different gene fragments for the "C" (constant) section, and 27 different "D" (diversity) and "J" (joining) gene fragments (not shown in Figure 10-14). These can be cut and spliced into a large variety of V-D-J-C combinations. Diversity is further increased by some "slip" in D and J splicing, the addition of *light chains* that themselves are diversified by the same mechanisms, and *somatic mutation*: The antibody genes of already-formed B cells are highly prone to mutations in the region that codes for the variable antigen-binding sites on the antibodies. Each different mutant cell in turn gives rise to a new clone. In these ways, a huge antibody repertoire is made possible by using only a modest share of the genetic blueprint. You might recognize that this process is analogous to Darwinian natural selection: In both, random gene variants arise that are then subjected to a selection process that determines which variants thrive.

Active immunity is self-generated; passive immunity is "borrowed"

The production of antibodies as a result of exposure to an antigen is referred to as **active immunity** against that antigen. The direct transfer of preformed antibodies from one animal to another forms a second way in which an individual can acquire antibodies in a process known as **passive immunity**. Such transfer of antibodies of the IgG class normally occurs from the mother to the fetus in mammals across the placenta during intrauterine development. In addition, a lactating female's colostrum (first milk) contains IgA antibodies that provide further protection for mammary-gland-fed infants. Passively transferred antibodies are usually broken down in less than a month, but meanwhile the newborn is provided important immune protection until it can begin actively mounting its own immune responses, which does not develop for about a month after birth in humans.

As you will see later, certain arthropods may also passively acquire immunity from their mothers.

We now turn to the other contributor to mammalian acquired immunity, the T cell.

check your understanding 10.6

How do B cells become plasma and memory cells, and what is their function?

What are antibodies, how do they work, and how are they genetically diversified?

How do specific B cells get selected to produce specific antibodies?

10.7 Acquired Immunity in Mammals: T-Lymphocytes

As important as B-lymphocytes and their antibody products are in specific defense against invading bacteria and other foreign material, they represent only half of the acquired immune defenses. The T cells are equally important in *cell-mediated* defense against most viral infections and also play a crucial regulatory role in both T and B cell responses.

T cells bind directly with their targets and effector cells via immunosynapses

Whereas B cells and antibodies defend against conspicuous invaders in the ECF, some T cells defend against covert invaders that hide inside cells where antibodies and the complement system cannot reach them. T cells do not secrete antibodies. Instead, they must directly come into contact with their targets, through receptors that form an immunosynapse. Cytotoxic T cells can then release chemicals that destroy targeted cells, especially virus-infected cells and cancer cells. Helper T cells must also directly contact effectors like B cells that they can also activate.

Like B cells, T cells undergo genetic recombination, are clonal, and are exquisitely antigen specific. On its plasma membrane, each T cell bears unique *T-cell receptors (TCRs)*, similar although not identical to BCRs (see Figure 10-12). Immature lymphocytes acquire their TCRs in the thymus during their differentiation into T cells. Unlike B cells, T cells are activated by foreign antigen only when it is on the surface of a cell that also carries a marker of the individual's own identity; that is, both foreign antigens and **self-antigens** known as **major histocompatibility complex (MHC) molecules** must be on a cell's surface before a T cell can bind with it. During *thymic education*, T cells learn to recognize foreign antigens only in combination with the animal's own tissue antigens—a lesson passed on to all T cells' future progeny. The importance of this dual antigen requirement and the nature of the MHC self-antigens will be described shortly.

As with B cells, activated T cells take several days to build up to therapeutic levels. Also like B cells, they form a *memory* pool and display both primary and secondary responses. During a few-week period after the infection is cleared, more than 90% of the huge number of effector T cells generated during the primary response die by means of *apoptosis* (cell suicide; see p. 101), triggered by the lack of stimulation by antigen. This is essential to prevent congestion in the lymphoid tissues. (B cells, recall, rapidly work themselves to death producing antibodies.) The remaining surviving effector T cells become long-lived memory T cells that migrate to all areas of the body, where they are poised for a swift secondary response to the same pathogen in the future.

The three types of T cells are cytotoxic T cells, helper T cells, and regulatory T cells

Recall that are three subpopulations of T cells, which we'll now examine in detail:

- **Cytotoxic T cells** destroy host cells harboring anything foreign and thus bearing foreign antigen, such as body cells invaded by viruses, cancer cells that have mutated proteins resulting from malignant transformations, and transplanted cells. The T-cell receptors (TCRs) for cytotoxic T cells are associated with coreceptors designated CD8, which are inserted into the plasma membrane as these cells pass through the thymus. Therefore, these cells are also known as **CD8+ T cells.**

- **Helper T cells** do not directly participate in immune destruction of invading pathogens. Instead, they modulate activities of other immune cells. Because of the important role they play in "turning on" the full power of all the other activated lymphocytes and macrophages, helper T cells constitute the immune system's "master integrator." Helper T cells are by far the most numerous T cells, making up 60 to 80% of circulating T cells. The T-cell receptors for helper T cells are associated with co-receptors designated CD4. Accordingly, helper T cells are also called **CD4+ T cells.**

- **Regulatory T cells** (T_{regs}) are a recently identified small subset of CD4+ cells. They have the same CD4 coreceptors as the helper T cells, but in addition they also have CD25, a component of a receptor for IL-2, which promotes T_{reg} activities. Thus, these cells are also referred to as **CD4+CD25+ T cells.** Regulatory T cells, which represent 5 to 10% of CD4 cells, suppress immune responses. T_{regs} are specialized to inhibit both innate and acquired immune responses in a check-and-balance fashion to minimize harmful immune pathology. Researchers hope that the ability of T_{regs} to put the brakes on helper T cells, B cells, NK cells, and dendritic (macrophage-like) cells can be used therapeutically to curb autoimmune diseases and prevent rejection of transplanted organs.

We now examine the functions of the two most abundant and best known T cells, cytotoxic T cells and helper T cells, in further detail.

Cytotoxic T cells secrete chemicals that destroy target cells

Cytotoxic T cells are microscopic "hit men." The most frequent targets of these destructive cells are host cells infected with viruses. When a virus invades a body cell, as it must to survive, the cell breaks down the envelope of proteins surrounding the virus and loads a fragment of this viral antigen piggyback onto a newly synthesized MHC self-antigen. This self-antigen and viral antigen complex is inserted into the host cell's surface membrane, where it serves as a red flag indicating the cell is harboring the invader (Figure 10-19, steps 1 and 2). Cytotoxic T cells of the clone specific for this particular virus recognize and bind to the viral-antigen and self-antigen on the surface of an infected cell (Figure 10-19, step 3—the immunosynapse). Thus, activated by viral antigen, a cytotoxic T cell can kill the infected cell and its internal virus by either direct or indirect means, depending on the type of lethal chemicals the activated T cell releases:

- An activated cytotoxic T cell may directly kill the victim cell by releasing **perforins** that lyse the attacked cell before viral replication can begin (Figure 10-19, step 4) by the same hole-punching technique used by antimicrobial peptides (Figure 10-3) and the complement MAC (membrane attack complex, p. 470).

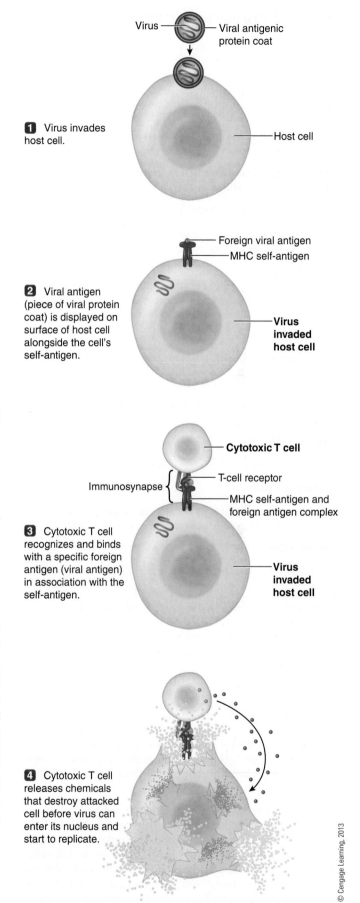

1 Virus invades host cell.

Virus — Viral antigenic protein coat

Host cell

2 Viral antigen (piece of viral protein coat) is displayed on surface of host cell alongside the cell's self-antigen.

Foreign viral antigen — MHC self-antigen

Virus invaded host cell

3 Cytotoxic T cell recognizes and binds with a specific foreign antigen (viral antigen) in association with the self-antigen.

Cytotoxic T cell

Immunosynapse { — T-cell receptor — MHC self-antigen and foreign antigen complex

Virus invaded host cell

4 Cytotoxic T cell releases chemicals that destroy attacked cell before virus can enter its nucleus and start to replicate.

© Cengage Learning, 2013

FIGURE 10-19 A cytotoxic T cell lysing a virus-invaded cell. The combination of the T cell receptors with the other cell's membrane proteins is called an *immunosynapse.*

TABLE 10-2 Defenses against Viral Invasion

When the virus is free in the ECF,

Macrophages

Destroy the free virus by phagocytosis

Process and present the viral antigen to helper T cells

Secrete interleukin-1, which activates B and T cell clones specific to the viral antigen

Plasma Cells Derived from B Cells Specific to the Viral Antigen Secrete Antibodies That

Neutralize the virus to prevent its entry into a host cell

Activate the complement cascade that directly destroys the free virus and enhances phagocytosis of the virus by acting as an opsonin

When the virus has entered a host cell (which it must do to survive and multiply, with the replicated viruses leaving the original host cell to enter the ECF in search of other host cells),

Interferon

Is secreted by virus-infected cells

Binds with and prevents viral replication in other host cells

Enhances the killing power of macrophages, natural killer cells, and cytotoxic T cells

Natural Killer Cells

Nonspecifically lyse virus-infected host cells

Cytotoxic T Cells

Are specifically activated by the viral antigen and lyse the infected host cells before the virus has a chance to replicate

Helper T Cells

Secrete cytokines, which enhance cytotoxic T-cell activity and B-cell antibody production

When a virus-infected cell is destroyed, the free virus is released into the ECF, where it is attacked directly by macrophages, antibodies, and the activated complement components.

- A cytotoxic T cell can also indirectly bring about death of an infected host cell by releasing **granzymes**, which are enzymes similar to digestive enzymes. Granzymes enter the target cell through the perforin channels. Once inside, these chemicals trigger the virus-infected cell to self-destruct through apoptosis.

The virus released on destruction of the host cell by either of these methods is directly destroyed in the ECF by phagocytic cells, antibodies, and the complement system. Meanwhile the cytotoxic T cell can move on to kill other infected host cells. Recall also that other nonspecific defense mechanisms also combat viral infections, most notably NK cells, interferon, macrophages, and the complement system. Thus, an intricate web of interplay exists among the defenses launched against viral invaders (Table 10-2).

Antiviral Defense in the Nervous System The previously described approach to destroying virus-infected host cells is not appropriate for the nervous system. If cytotoxic T cells destroyed virus-infected neurons, the lost cells could not be replaced, because neurons cannot reproduce. Fortunately, virus-infected neurons are spared from extermination by the immune system, but how, then, are neurons protected from viruses? Immunologists long thought that the only antiviral defenses for neurons were those aimed at free viruses in the ECF. Surprising new research has revealed, however, that antibodies not only target viruses in the ECF but can also eliminate viruses inside neurons. It is unclear whether antibodies can actually enter the neurons and interfere directly with viral replication (neurons have been shown to take up antibodies near their synaptic endings) or bind with the surface of nerve cells and trigger intracellular changes that stop viral replication.

Helper T cells secrete chemicals that amplify the activity of other immune cells

In contrast to cytotoxic T cells, helper T cells function as integrators, not effectors. They secrete cytokines that augment nearly all aspects of the immune response. B cells stimulated by T-dependent antigen cannot develop into antibody-secreting plasma cells without the help of a subset of helper T cells called **T helper 2** (T_H2) cells. Additionally, the **T helper 1** (T_H1) cells secrete cytokines that enhance actions of the cytotoxic T cells and activate macrophages. This is why **acquired immunodeficiency syndrome** (**AIDS**), caused by the **human immunodeficiency virus** (**HIV**), is so devastating to human immune defense system. The AIDS virus selectively invades helper T cells, destroying or incapacitating the cells that normally orchestrate much of the immune response.

The following are among the best known of helper T-cell cytokines, although the list is rapidly growing as researchers uncover more cytokines with other roles:

1. **B cell growth factors (IL-4, IL-5, and IL-6)** contribute to B cell function in concert with the IL-1 secreted by macrophages. Antibody production may be greatly reduced or absent without these.
2. **T cell growth factor (IL-2)** augments the activity of cytotoxic T cells and even of other helper T cells responsive to the invading antigen. In typical interplay fashion, IL-1 secreted by macrophages not only enhances the activity of both the appropriate B- and T-cell clones but also stimulates secretion of IL-2 by activated helper T cells.
3. **Macrophage-migration inhibition factor** keeps large phagocytotic cells in the affected region by inhibiting their outward migration. As a result, a great number of chemotactically attracted macrophages accumulate in the infected area. This factor also converts macrophages into **angry macrophages** that have more powerful phagocytotic ability.

We next examine how T cells are activated by antigen-presenting cells and the roles of MHC molecules.

T-lymphocytes respond only to antigens presented to them by antigen-presenting cells

Relevant T cells cannot recognize "raw" foreign antigens entering the body; before reacting to it, a T-cell clone must be formally "introduced" to the antigen. **Antigen-presenting**

FIGURE 10-20 **Dendritic cell.**

David Scharf/photolibrary.com

What triggers dendritic cells to begin their migration? There is evidence that, like macrophages, they use TLRs (p. 460) to sense unusual or foreign "patterns" (antigens). This concept of sensing nonself is sometimes called the "stranger" hypothesis. However, an alternative idea called the "danger" hypothesis proposes that immune sensing is based on chemicals released from damaged body cells. Recently, for example, researchers have found that *nucleotides* and crystals of *uric acid* (p. 558), released from damaged body cells, are potent activators of dendritic cells. Both "stranger" and "danger" signals may be at work in a complementary fashion; that is, tissue danger signals may prime dendritic cells so they more potently react to foreign materials.

How does a dendritic cell become an APC? As shown in Figure 10-21, invading bacteria are phagocytized by dendritic cells (step 1). Within the cell, the lysosome system enzymatically breaks down the bacterium's proteins into antigenic peptides (small protein fragments) (step 2). Each antigenic peptide then binds to an MHC molecule, which has been newly synthesized within the endoplasmic reticulum/Golgi complex (steps 3 and 4). Loading of the antigenic peptide onto an MHC molecule takes place on a specialized organelle within APCs, the **peptide-loading compartment**. An MHC molecule has a deep groove into which a variety of antigenic peptides can bind, depending on what the macrophage has engulfed. The MHC molecule then transports the bound antigen to the cell surface, where it is presented to passing T cells (step 5). The combined presentation of these self- and nonself antigens alerts the helper T cell—in particular, one with the appropriate receptor for the antigen—to the presence of an undesirable bacterium within the body.

Macrophages become APCs by a similar mechanism. B cells do not phagocytize antigenic particles like macrophages whereas dendritic cells do, but B cells can internalize antigens bound to their surface receptor by receptor-mediated endocytosis (see p. 44).

When the helper T cell binds to the APC, the APC secretes the cytokines IL-1 and TNF, which activate the associated T cell, among other effects. The activated T cell then secretes various cytokines, which act in an autocrine manner to stimulate that cell's clonal expansion and act in a paracrine manner on adjacent B cells, cytotoxic T cells, and NK cells to enhance their responses to the same antigen.

Next we will look at the role of MHCs in more detail.

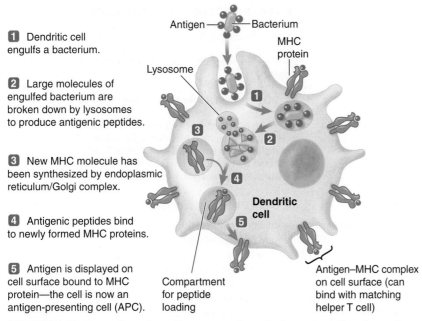

1 Dendritic cell engulfs a bacterium.

2 Large molecules of engulfed bacterium are broken down by lysosomes to produce antigenic peptides.

3 New MHC molecule has been synthesized by endoplasmic reticulum/Golgi complex.

4 Antigenic peptides bind to newly formed MHC proteins.

5 Antigen is displayed on cell surface bound to MHC protein—the cell is now an antigen-presenting cell (APC).

Antigen — Bacterium

MHC protein

Lysosome

Dendritic cell

Compartment for peptide loading

Antigen–MHC complex on cell surface (can bind with matching helper T cell)

FIGURE 10-21 **Generation of an antigen-presenting cell when a dendritic cell engulfs a bacterium.**

© Cengage Learning, 2013

cells (**APCs**), which we have mentioned several times, handle the formal introduction; they process and present antigen, complexed with MHC molecules, on their surface to the T cells via an immunosynapse. APCs include (1) macrophages of innate defenses, (2) the closely related dendritic cells, and (3) B cells. The helper T cells in turn enhance antibody-mediated immunity. Let's examine how this works.

You are already familiar with macrophages. **Dendritic cells** are specialized APCs that act as sentinels in almost every tissue, but especially the linings of the barrier tissues where invasion is most likely (e.g., skin with its *Langerhans cells*; Figure 10-6). They are so named because they have many surface projections that resemble the dendrites of neurons (*dendros* means "tree") (Figure 10-20). After exposure to the appropriate antigen, dendritic cells leave their tissue home and migrate through the lymphatic system to nearby lymph nodes, where they cluster and activate T cells.

The major histocompatibility complex (MHC) is the code for self-antigens

Self-antigens are plasma-membrane-bound glycoproteins arising from the MHC genes. This complex spans 4 million base pairs within the DNA molecule and contains 128 genes. The MHC genes are the most variable ones in mammals. For example, more than 100 different MHC molecules have been identified in human tissue, but each individual has

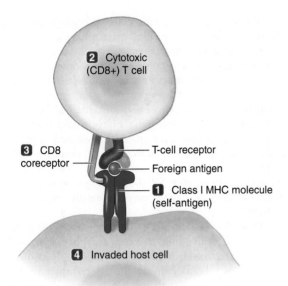

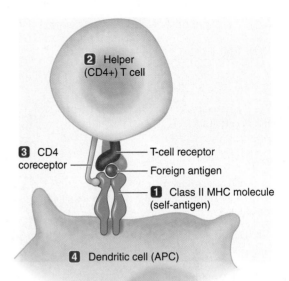

1 Class I MHC molecules are found on surface of all cells.

2 They are recognized only by cytotoxic (CD8+) T cells.

3 CD8 coreceptor links the two cells together.

4 Linked in this way, cytotoxic T cells can destroy body cells if invaded by foreign (viral) antigen.

1 Class II MHC molecules are found on the surface of immune cells with which helper T cells interact: dendritic cells, macrophages, and B cells.

2 They are recognized only by helper (CD4+) T cells.

3 CD4 coreceptor links the two cells together.

4 To be activated, helper T cells must bind with a class II MHC-bearing APC (dendritic cell or macrophage). To activate B cells, helper T cell must bind with a class II MHC-bearing B cell with displayed foreign antigen.

(a) Class I MHC self-antigens

(b) Class II MHC self-antigens

FIGURE 10-22 Distinctions between class I and class II major histocompatibility complex (MHC) glycoproteins. Specific binding requirements for the two types of T cells ensure that these cells bind only with the target cells with which they can interact. Cytotoxic (CD8+) T cells can recognize and bind with foreign antigen only when the antigen is in association with class I MHC glycoproteins, which are found on the surface of all body cells. This requirement is met when a virus invades a body cell, whereupon the cell is destroyed by the cytotoxic T cells. Helper (CD4+) T cells, which are activated by and/or enhance the activities of dendritic cells, macrophages, and B cells, can recognize and bind with foreign antigen only when it is in association with class II MHC glycoproteins, which are found only on the surface of these other immune cells. The CD8+ or CD4+ T-cell's coreceptor CD8 or CD4 links these cells to the target cell's class I or class II MHC molecules, respectively, in an immunosynapse.

a code for only three to six of these possible antigens. Because of the tremendous number of different combinations possible, the exact pattern of MHC molecules varies from one individual to another, much like a "biochemical fingerprint" or "molecular identification card," except in identical twins (triplets, and so on), who have the same MHC self-antigens.

The major histocompatibility (*histo*, "tissue") complex was so named because these genes and the self-antigens they encode were first discerned in relation to tissue typing (similar to blood typing), which is done to obtain the most compatible matches for tissue grafting and transplantation. However, the transfer of tissue from one individual to another does not normally occur in nature. The natural func-

tion of MHC antigens lies in their ability to direct the responses of T cells, not in their artificial role in rejecting transplanted tissue.

Each mammal has two main classes of MHC-encoded molecules that are differentially recognized by cytotoxic T and helper T cells—*class I* and *class II*, respectively (Figure 10-22). These markers serve as signposts to guide cytotoxic and helper T cells to the precise cellular locations where their immune capabilities are most effective. The coreceptors on the T cells bind with the MHC molecules on the target molecule, linking the two cells together in an immunosynapse. In addition to receptor–receptor binding shown in Figure 10-22, nanotubes (p. 91) may also join the two cells to increase the rate of communication signal transfer.

TABLE 10-3 Innate and Adaptive Immune Responses to Bacterial Invasion

Innate Immune Mechanisms	Adaptive Immune Mechanisms
Inflammation Resident tissue macrophages engulf invading bacteria.	B cells specific to T-independent antigen are activated on binding with the antigen.
Histamine-induced vascular responses increase blood flow to the area, bringing in additional immune-effector cells and plasma proteins.	B cells specific to T-dependent antigen present antigen to helper T cells. On binding with the B cells, helper T cells activate the B cells.
A fibrin clot walls off the invaded area.	The activated B-cell clone proliferates and differentiates into plasma cells and memory cells.
Neutrophils and monocytes/macrophages migrate from the blood to the area to engulf and destroy foreign invaders and to remove cell debris.	Plasma cells secrete customized antibodies, which specifically bind to invading bacteria. Plasma cell activity is enhanced by: Interleukin-1 secreted by macrophages; and Helper T cells, which have been activated by the same bacterial anti-gen processed and presented to them by macrophages or dendritic cells.
Phagocytic cells secrete cytokines, which enhance both innate and adaptive immune responses and induce local and systemic symptoms associated with an infection.	
Nonspecific Activation of the Complement System Complement components form a hole-punching membrane attack complex that lyses bacterial cells.	Antibodies bind to invading bacteria and enhance innate mechanisms that lead to the bacteria's destruction.
Complement components enhance many steps of inflammation.	Specifically, antibodies: Act as opsonins to enhance phagocytic activity, Activate the lethal complement system, and Stimulate natural killer cells, which directly lyse bacteria.
	Memory cells persist that are capable of responding more rapidly and more forcefully should the same bacteria be encountered again.

© Cengage Learning, 2013

Class I MHC Glycoproteins Cytotoxic T cells (CD8+ T cells) can respond to foreign antigens only in association with **class I MHC glycoproteins**, which are found on the surface of all nucleated body cells, because the T cell's co-receptor CD8 can interact only with MHC class I. To carry out their role of dealing with pathogens that have invaded host cells, it is appropriate that cytotoxic T cells bind only with body cells that viruses have infected—that is, with for-eign antigens in association with self-antigens. Furthermore, these deadly T cells can also link up with any cancerous body cell because class I MHC molecules may also display mutated (and thus appearing "foreign") cellular proteins characteristic of these abnormal cells. Because any nucleated body cell can be invaded by viruses or become cancerous, essentially all cells display class I MHC glycoproteins, en-abling cytotoxic T cells to attack any virus-invaded host cell or any cancer cell. Because cytotoxic T cells do not bind to MHC self-antigens in the absence of foreign antigen, normal body cells are protected from lethal immune attack.

Class II MHC Glycoproteins In contrast, **class II MHCs** are recognized by helper T cells (CD4+ T cells) and are restricted to the surface of APCs—macrophages, dendritic cells, and B cells. The CD4 coreceptor of helper cells can interact only with MHC class II glycoproteins. Thus, the specific binding requirements for cytotoxic T cells and helper T cells ensure the appropriate T-cell responses. For example, binding of the antigen-bearing B cell with the matching helper T cell causes the T cell to secrete cytokines that activate this specific B cell, leading to clonal expansion and conversion of this B cell clone into antibody-producing plasma cells and memory cells (Fig-

ure 10-23). This is the primary pathway by which the ac-quired immune system fights bacteria. Table 10-3 summarizes the innate and acquired immune strategies that defend against bacterial invasion.

You may have noticed a problem: Why doesn't the ac-quired immune system unleash its powerful defenses against an animal's own self-antigens? We examine this issue next.

The immune system is normally tolerant of self-antigens

The term **immune tolerance** refers to the phenomenon of preventing the immune system from attacking the animal's own tissues. During the genetic "cut, shuffle, and paste" pro-cess that goes on during lymphocyte development, some B and T cells are by chance formed that target the body's own antigens. If these lymphocyte clones were allowed to func-tion, they would destroy the individual's own body. Fortu-nately, this rarely happens. At least eight different mecha-nisms are involved in tolerance; we will examine six of them:

1. *Clonal deletion.* In response to continuous exposure to body antigens early in development, B and T cell clones capable of attacking these self-antigens in most cases are permanently destroyed within the thymus. This **clonal deletion** is accomplished by triggering apoptosis of im-mature cells that would react with the body's own pro-teins. This physical elimination is the major mechanism by which tolerance is developed.

2. *Clonal anergy.* Recall that a T-lymphocyte must receive two specific simultaneous signals to be activated, one

Activation of helper T cells by antigen presentation

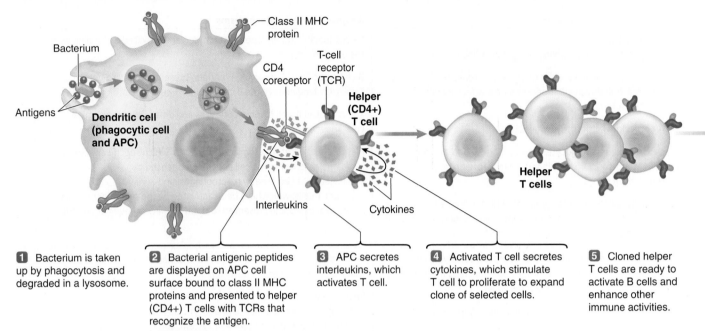

1 Bacterium is taken up by phagocytosis and degraded in a lysosome.

2 Bacterial antigenic peptides are displayed on APC cell surface bound to class II MHC proteins and presented to helper (CD4+) T cells with TCRs that recognize the antigen.

3 APC secretes interleukins, which activates T cell.

4 Activated T cell secretes cytokines, which stimulate T cell to proliferate to expand clone of selected cells.

5 Cloned helper T cells are ready to activate B cells and enhance other immune activities.

FIGURE 10-23 Interactions among large phagocytic cells (APCs), helper T cells, and B cells responsive to T-dependent antigen via immunosynapses.

© Cengage Learning, 2013

from its compatible antigen and a stimulatory cosignal found only on an APC. Both signals are present for foreign antigens introduced to T cells by APCs. In contrast, these dual signals are never present for self-antigens because these antigens are not handled by cosignal-bearing APCs. The first exposure to a single signal from a self-antigen turns *off* the compatible T cell, rendering the cell unresponsive to further exposure to the antigen, a reaction referred to as **clonal anergy** (*anergy* means "lack of energy").

3. *Receptor editing.* A newly identified means of ridding the body of self-reactive B cells is **receptor editing**. With this mechanism, once a B cell that bears a receptor for one of the body's own antigens encounters the self-antigen, the B cell escapes death by swiftly changing its antigen receptor to a nonself version, in effect "rehabilitating" the self-reactive B cell.

4. *Active suppression by regulatory T cells.* T_{reg} cells play a role in tolerance by inhibiting throughout life some lymphocyte clones specific for the body's own tissues; the mechanism by which it accomplishes this feat is being actively studied.

5. *Immunological ignorance.* Some intracellular self-molecules are normally hidden from the immune system, since they are never in the ECF.

6. *Immune privilege.* A few tissues, most notably the testes and the eyes, have **immune privilege** because they escape immune attack even when they are transplanted in an unrelated individual. Scientists recently discovered that the plasma membranes in these immune-privileged tissues have a specific molecule that triggers apoptosis of approaching activated lymphocytes that could attack the tissues.

Autoimmunity Occasionally the immune system fails to distinguish self from nonself and begins to attack the body's own tissues, resulting in *autoimmune diseases*. Over 80 such diseases are known, including multiple sclerosis and Type I diabetes mellitus (the destruction of the insulin-producing beta cells; p. 316) in humans, and rheumatoid arthritis and systemic lupus erythematosus in humans and some dog breeds. Autoimmune diseases may arise from (1) exposure during injury of previously sequestered antigens; (2) modification of self-antigens by external chemicals, viruses or mutation; and (3) exposure of T cells to a foreign antigen almost identical to a self-antigen. Such foreign antigens may be bacterial, but may also be on fetal cells that accidentally enter a mother's bloodstream during the trauma of labor and birth.

Let's now look in more detail at the role of T cells in defending against cancer.

Immune surveillance against cancer cells involves an interplay among immune cells and interferon

Besides destroying virus-infected host cells, another important function of the T-cell system is a process known as **immune surveillance**: the recognition and destruction of new, potentially cancerous tumor cells before they have a chance to multiply and metastasize (spread through the body). At least once a day, on average, the human immune system destroys such cells. Any normal cell in an animal may be transformed into a tumor cell if mutations occur within its genes that govern cell division and growth. Such mutations may occur by chance alone or, more frequently, by

Activation of B cells responsive to T-dependent antigen

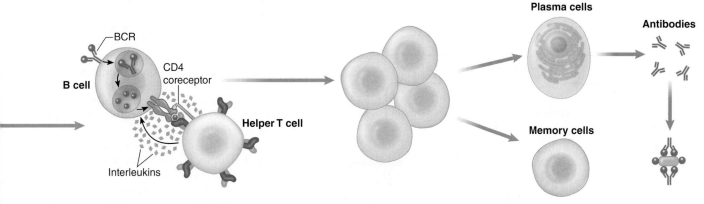

6 BCR binds to antigen. Antigen is internalized by receptor-mediated endocytosis and its macromolecules degraded. Antigenic peptides produced are displayed on cell surface bound to class II MHC proteins.

7 TCR of a helper T cell recognizes specific antigen on B cell, and CD4 coreceptor links the two cells together.

8 Helper T cell secretes interleukins, which stimulate B cell proliferation to produce clone of selected cells.

9 Some cloned B cells differentiate into plasma cells, which secrete antibodies specific for the antigen, while a few differentiate into memory B cells.

10 Antibodies bind with antigen, targeting antigenic invader for destruction by the innate immune system.

exposure to **carcinogenic** (cancer-causing) **factors** such as ionizing radiation, certain environmental chemicals, or physical irritants. Alternatively, a few cancers are caused by tumor viruses, which turn the cells they invade into cancer cells. The immune system recognizes cancer cells (Figure 10-24) because they bear new and different surface antigens alongside the cell's normal self-antigens, a consequence of either mutation or invasion by a tumor virus.

Fortunately, only a fraction of the mutations involve loss of control over the cell's growth and multiplication. A cell usually becomes cancerous only after an accumulation of multiple independent mutations. This requirement contributes at least in part to the much higher incidence of cancer in older individuals, in whom mutations have had more time to accumulate in a single cell lineage.

Effectors of Immune Surveillance Immune surveillance against cancer depends on an interplay among three types of cells—*cytotoxic T cells, macrophages, and NK cells*—as well as *interferon* (Figure 10-25). Not only can all three of these cell types attack and destroy cancer cells directly but all three also secrete interferon. Interferon, in turn, inhibits multiplication of cancer cells and increases the killing ability of the immune cells.

Because NK cells do not require prior exposure and activation in response to a cancer cell before being able to launch a lethal attack, they are the first line of defense against cancer. In addition, cytotoxic T cells take aim at cancer cells after being activated by mutated surface proteins alongside normal class I MHC molecules. On contacting a cancer cell, both these killer cells release perforin and other toxic chemicals that destroy the mutant cell (Figure 10-25). Macrophages can phagocytose cancer, as well as clear away debris.

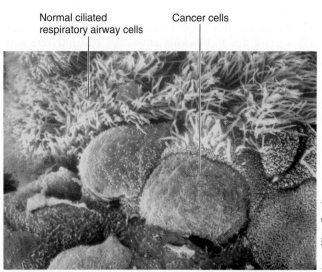

FIGURE 10-24 Comparison of normal and cancerous cells in the large respiratory airways. The normal cells display specialized cilia, which constantly contract in whiplike motion to sweep debris and microorganisms from the respiratory airways so they do not gain entrance to the deeper portions of the lungs. The cancerous cells are not ciliated, so they are unable to perform this specialized defense task.

The fact that cancer does sometimes occur means that cancer cells occasionally escape immune surveillance. Some cancer cells are believed to survive by evading immune detection, for example, by failing to display identifying antigens. This is thought to be the case in Devil facial tumour disease, a lethal facial cancer of Tasmanian devils that can

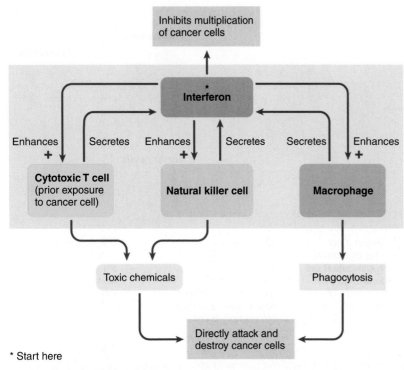

Inhibits multiplication of cancer cells

* Start here

FIGURE 10-25 Immune surveillance against cancer. Anticancer interactions of cytotoxic T cells, natural killer cells, macrophages, and interferon.

© Cengage Learning, 2013

actually spread from one animal to another! Some cancer cells survive due to **blocking antibodies** that interfere with T-cell function. For unknown reasons, B cells occasionally make antibodies to the antigenic sites on the cancer cells, but instead of triggering complement, they "hide" these sites from recognition by cytotoxic T cells. A new finding reveals that other successful cancer cells thwart immune attack by inducing the cytotoxic T cells that bind with them to commit suicide. In addition, some cancer cells secrete large amounts of a chemical messenger that recruits regulatory T cells and programs them to suppress cytotoxic T cells.

A regulatory loop links the immune system with the nervous and endocrine systems

As you have seen, factors regulating the immune system are complex. Until recently, the immune system was believed to function independently of other control systems in animals. Investigations now indicate that the immune system both influences and is influenced by the two major whole-body regulatory systems, the nervous and endocrine systems. For example, IL-1 can turn on the stress response by activating a sequence of nervous and endocrine events that result in the secretion of cortisol, one of the major hormones released during stress. This linkage between a mediator of the immune response and a mediator of the stress response is appropriate. Cortisol mobilizes the body's nutrient stores so that metabolic fuel is readily available to keep pace with the body's energy demands at a time when the person is sick and may not be eating enough (or, in the case of a nonhuman animal, may not be able to search for food). Furthermore, cortisol mobilizes amino acids, which serve as building

blocks to repair any tissue damage sustained during the encounter that triggered the immune response.

In the reverse direction, lymphocytes and macrophages are responsive to bloodborne signals from the nervous system and from certain endocrine glands. These important immune cells possess receptors for a wide variety of neurotransmitters, hormones, and other chemical mediators. For example, epinephrine released in the short-term stress response (p. 308) can boost the activity of lymphocytes. In contrast, cortisol and other chemical mediators of the stress response have a profound immunosuppressive effect, as we have described (p. 470). Thus, negative-feedback loops appear to exist between the immune system and the nervous and endocrine systems. In large part because long-term stress suppresses the immune system, stressful physical, psychological, and social life events are linked with increased susceptibility to infections and cancer. Thus, the body's resistance to disease can be influenced by an animal's mental state. For example, infant monkeys separated from their mothers exhibit markedly suppressed immune responses.

There are other important links between the immune and nervous systems in addition to the stress connection. For example, many immune system organs, such as the thymus, spleen, and lymph nodes, are innervated by the sympathetic nervous system, the branch of the nervous system called into play during stress-related "fight-or-flight" situations (see p. 308). Also, the vagus nerve, the main nerve of the parasympathetic nervous system (p. 183), innervates the spleen and other sites where macrophages are abundant. Macrophages, the orchestrators of many aspects of the inflammatory process, have receptors for acetylcholine (ACh). As a parasympathetic postganglionic neurotransmitter, ACh suppresses inflammation. In the reverse direction, immune cytokines act on the brain to produce fever and other general symptoms that accompany infections. Furthermore, immune cells secrete some traditional hormones thought to be produced only by the endocrine system. For example, many of the hormones secreted by the pituitary gland are produced by lymphocytes as well. Scientists are only beginning to sort out the mechanisms and implications of the many complex neuro–endocrine–immune interactions.

check your understanding 10.7

How does each of the three types of T cells work?

What is the role of MHCs in T-cell function?

How does the system avoid attacking self-antigens?

What are the major mechanisms of immune surveillance?

How is cortisol involved in neuro–endocrine–immune interactions?

10.8 Acquired Immunity in Other Animals

As we noted earlier, the acquired immune system has only been thoroughly documented in jawed vertebrates. In recent years, many interesting discoveries have been made in other animals.

unanswered Questions | How did the acquired immune system arise in jawed vertebrates?

The acquired immune system of higher vertebrates seems to have originated rather abruptly in evolutionary terms. How could this have happened by the known mechanisms of evolution? For a recent hypothesis, see the box, *Molecular Biology and Genomics: An Accidental Origin for Acquired Immunity?* <<

Sharks generate diverse B and T cells with a lower diversity but a greater number of gene fragments than mammals

Recent research on nonmammalian vertebrates and nonvertebrates has focused on understanding the genetic evolution of acquired immunity. Other jawed vertebrates appear to have B- and T-cell systems similar to mammals, most generating a comparable diversity of these cells in ways similar to mammals. But not all do. It is perhaps most instructive to compare mammals to chondrichthyes (sharks, skates, chimaeras, and rays), the most evolutionarily ancient jawed vertebrates to have acquired immunity. Like mammals, chondrichthyes have B and T cells and exhibit gene shuffling to make millions of different antibodies and receptors. However, shark antibodies differ in other ways from those of mammals. First, shark B cells make four different classes of antibodies, but only one of these—the IgM class—is the same as one of the five classes found in mammals. Second, they do not mutate to increase diversity with the result that they are theoretically limited in the scope of antigens that they can respond to (p. 480). Third, gene shuffling in sharks involves the recombination of a much lower diversity of gene fragments than in mammals. But shark genomes make up for this in ways not found in mammals. For example, the gene fragments that are recombined in mammals are found only in one cluster on one chromosome. Sharks, however, have hundreds of antibody-generating gene clusters on many chromosomes. Thus, although each cluster has a lower diversity of gene fragments to shuffle around, there are many more such clusters available. In addition, there are also hundreds of other gene clusters that already have the coding fragments joined together, so that they can rapidly produce antibodies without waiting for genetic recombination. Those gene combinations are fixed and heritable, so they do not generate more diversity than what the animal was born with. However, they work well (and much faster than mammalian counterparts) for common pathogens that sharks have encountered for millennia.

These studies on sharks reveal that the basic B- and T-cell system, with gene shuffling of receptor and antibody gene clusters to increase diversity, evolved early on in the jawed vertebrates. But since the origin of acquired immunity, evolution has led to different uses of these gene clusters and the coding fragments within them. We cannot say that one version is better than the other. After all, the shark system, although somewhat different, has helped these animals survive successfully far longer than mammals have been on Earth.

Lamprey lymphocytes shuffle ancient PRR genes to generate variable receptors and antibody-like proteins

Recently, lampreys have been found to have a unique system of acquired immunity based on gene recombination that probably evolved independently of the jawed vertebrate system. These jawless fish have lymphocytes, which in 2009 were shown to be comparable to B or T cells. It is not clear whether these are related to the true B and T cells of jawed vertebrates. In any case, they also generate diverse proteins from relatively few genes, but their mechanism is clearly different: The cells make **variable lymphocyte receptors (VLRs)** by shuffling the genes of a class of PRRs, an ancient type different from the antibody-type. The T-like cells use these as membrane receptors, while the B-like cells secrete soluble versions of VLRs that may act like antibodies. The lamprey studies suggest that there may be many ways to evolve immune systems based on genetic recombination.

Some arthropods and mollusks have immunity mechanisms resembling vertebrate acquired processes

We end this chapter with a look at recent studies suggesting that some nonvertebrates may have independently evolved defenses with some similarities to the vertebrate acquired system.

First, research on the water flea (*Daphnia magna*) found that offspring are more resistant to a particular pathogenic bacterium (but not other bacteria) if their mother was exposed to it before she lays her eggs. It appears that the mother's immune system "learns" that a particular pathogen is present and passes on specific protective mechanisms to her eggs. How she does this is not yet known.

In another study, *Macrocyclops albidus*, a species of copepod (one of the most common animal groups on Earth; p. 128), was exposed to parasitic tapeworms twice. It was better able to resist the tapeworms during the second infection if the subsequent exposure used tapeworms closely related to those used in the first infection. This suggests that the copepod immune system somehow "learned" the antigenic markers of an individual tapeworm and later recognized those markers on its relatives. How this takes place is not yet known. Similarly, researchers found that giving *Drosophila* a sublethal dose of *Streptococcus pneumoniae* bacteria induces changes that subsequently protect them from an otherwise lethal dose of the same (but not other) bacteria.

How might specific immunity work in nonvertebrates? One possibility involves diversification of ancient PRR genes, as in lampreys. Research on the freshwater snail *Biomphalaria glabrata* (a host of the human schistosomiasis

Molecular Biology and Genomics
An Accidental Origin for Acquired Immunity?

All animals have innate immunity, but (as far as we know) only vertebrates with jaws have the acquired system based on generating millions of different receptors and antibodies (in T and B cells). For example, lancelets (nonvertebrates closely related to the first vertebrates) have genomes with codes that are similar to the gene fragments that give rise to antibody diversity in mammals. However, the functions of these lancelet genes are unknown, and they do not undergo the gene shuffling of acquired immunity. Lampreys appear to have lymphocyte-like cells, but their acquired system appears to be independently evolved. Thus, the system in jawed vertebrates seems to have appeared suddenly in evolutionary terms, with few precursors. How could this complex immunity innovation evolve so quickly, and why only in jawed vertebrates? Recent genomic studies suggest that the initial event might have been an evolutionary accident involving a virus!

Recall that the acquired immune system can generate millions of different proteins by shuffling and joining a relatively small number of gene fragments. This process, called genetic recombination, is dependent on two genes called *RAG1* and *RAG2* ("recombinant-activating genes"). The genes code for a transposase, an enzyme that cuts out sections of DNA at one site in a chromosome, then inserts them at another location. Transposase activity has a partially random nature that is the key to generating millions of different B-cell antibodies and T-cell receptors.

Researchers used DNA probes to study other vertebrates and discovered a distinct divide: All jawed vertebrates have RAG1 and RAG2, but hagfish and lamprey do not. Nor do those ancient fishes have any genes with similar coding that might provide a precursor. There seems to have been a sudden evolutionary transition, estimated at 450 million years ago. Transposase events have been well documented in other organisms (p. 34), where the genes have "jumped" from one species to another, causing a sudden genetic leap. The jumping transposase genes were most likely carried by viruses. If this mechanism is indeed the origin of the RAG genes in higher vertebrates, we may owe our complex immune systems to an ancestral fish that survived a viral infection!

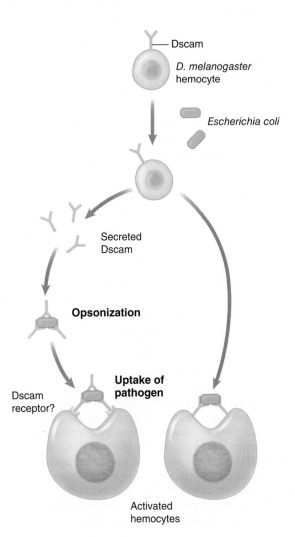

Activated hemocytes

parasite) found that these molluscs' circulating defense cells (hemocytes) generate a wide diversity of *lectin* defense proteins (p. 463). The diversity appears to arise from both increased mutation and gene shuffling (recombination).

Arthropods also appear to diversify their *Ig-like* PRRs (p. 463) called *Dscam* (Down syndrome cell-adhesion molecules) by *alternative splicing* (p. 32), which generates over 30,000 variants from very few genes. (In contrast, an antibody system might have evolved from gene recombination of ancient Ig-like PRRs.) Much like B cells, arthropod hemocytes use Dscam proteins as receptors and secrete smaller versions that act much like antibodies (Figure 10-26). A possible learning aspect of this system has been found in some insects. In mosquitoes, specific Dscam proteins were produced after infection that enhanced survival upon later infection. Moreover, when those particular Dscam's mRNAs were blocked by RNA interference (p. 33) to prevent translation, survival during a specific infection was impaired.

However, it is not clear if the snail or insect system has the ability to learn in the vertebrate sense, that is, to selectively amplify receptors that have encountered a specific antigen and to retain a lifelong memory. Still, this and the other nonvertebrate discoveries indicate a need for more study in all types of animals for possible acquired immunity.

FIGURE 10-26 Dscam (Down syndrome cell-adhesion molecule) receptors in *Drosophila melanogaster* (fruit fly) immunity. Dscams are pattern-recognition receptors that bind to foreign molecules on invaders. Hemocytes can make many diverse Dscams by alternative RNA splicing, and secrete them to act as opsonins in antibody-like fashion, enhancing phagocytosis, using Dscam receptors that have not been fully characterized.

Source: L. M. Stuart & R. A. Ezekowitz. (2008). Phagocytosis and comparative innate immunity: learning on the fly. *Nature Reviews Immunology* 8:131–141.

check your understanding 10.8

How do sharks and lampreys differ from other vertebrates in making diverse immune cells?

What is the evidence in *Drosophila* for immune features similar to those of vertebrate acquired immunity?

making connections

How Defense Systems Contribute to the Body as a Whole

Animals could not survive beyond early infancy were it not for defense mechanisms. Homeostasis can be optimally maintained only if body cells are not physically injured or functionally disrupted by pathogens or are not replaced by cancer cells. Immune defense systems—complex, multifaceted, interactive networks of barrier tissues, defense proteins, and specialized cells for sensing, integrating, and effecting defenses—keep cells alive so that they can perform their specialized activities to maintain a stable internal environment. Immune systems protect from foreign agents, eliminate newly arisen cancer cells, and clear away debris to pave the way for replacement with healthy new cells. These defenses involve both positive- and negative-feedback loops, interacting with neural and endocrine systems in ways that are still being explored.

Chapter Summary

- Animal survival depends on immune systems that fight external as well as internal threats including pathogenic bacteria, viruses, parasites, and cancerous cells. Immune responses can be either "ready-to-go" innate ones or "learning" acquired ones that must be first exposed to a pathogen. Immunity uses pattern-recognition receptors (PRRs) to distinguish "self" from "nonself" by having binding sites for foreign molecules. Immune cells coordinate their actions with external signals called cytokines.

- Innate barrier tissues—integumentary, digestive, respiratory, genitourinary—are the first line of defense. They form passive walls and have active glands to aid defense. If barriers are breached, the second lines of defense are (1) phagocytotic cells, with PRRs to sense common foreign molecules such as bacterial cell wall components, that engulf and digest invaders; (2) inflammatory signals to draw phagocytotic cells to an invasion site; (3) antimicrobial peptides, many of which punch holes in bacterial membranes; and (4) opsonin proteins, which bind to common foreign molecules and "tag" them for phagocytotic ingestion. Arthropods can also "wall off" large invaders with a polymerized capsule.

- Vertebrate active immune responses are regulated by feedback systems with sensors, integrators, and effectors that are primarily leukocytes and proteins. Circulating leukocytes include phagocytotic neutrophils, macrophages, and natural killer (NK) cells of innate immunity and lymphocytes of acquired immunity. Localized leukocytes include mast cells and dendritic cells in barrier tissues.

- Mammalian barrier tissues have active defenses; for example, skin glands release antimicrobial peptides, stomachs produce acid, respiratory systems use mucus, cilia, and sneezing and coughing to transport foreign materials out.

- In mammalian inflammation, mast cells detect damage and release histamine, which promotes fluid accumulation. Chemotaxins draw in phagocytes, which also release other cytokines that can promote fever. The complement protein system (among other things) punches holes in microbes and acts as opsonins to enhance phagocytosis.

- NK cells destroy virus-infected cells and cancer cells on first exposure to them by uncertain mechanisms. Interferon transiently inhibits multiplication of viruses in most cells and warns other cells of potential viral attack.

- Acquired immunity includes B-cell antibody-mediated and T-cell cell-mediated responses, both aided by integrating helper and regulatory T cells. Both cells are produced and mature in lymphoid tissues and organs including bone marrow and thymus. B cells react to antigens, foreign molecules that induce an immune response against itself.

- Antigens stimulate B cells to convert into plasma cells that produce antibodies, Y-shaped proteins classified according to properties of their tail portion. They amplify innate immunity (e.g., becoming opsonins) to promote antigen destruction. The huge variety of B cells results from shuffling a small set of gene fragments. Via clonal selection, only the B cell with a receptor for a particular antigen will multiply and produce specific antibodies. Some B cells become memory cells to mount a much faster defense upon second exposure to an antigen.

- T cells make diverse receptors by the same shuffling mechanism, and also form memory cells. To be activated, they must bind other body cells through immunosynapses. Cytotoxic T cells secrete chemicals that destroy virus-infected or cancerous body cells that "present" improper proteins on their major histocompatibility complexes (MHCs). Helper T cells secrete chemicals that amplify the activity of other immune cells, for example, B cells. Helper T cells are activated by antigen-presenting cells (APCs), for example, macrophages and dendritic cells that ingest and present antigens on their MHCs.

- The MHC is the code for self-antigens: MHC I is found on all cells and can present viral or cancerous proteins to T cells; MHC II is found on APCs and presents foreign antigens. The immune system is normally tolerant of self-antigens; among other mechanisms, B and T clones specifically capable of attacking self-antigens in most cases are permanently destroyed within the thymus.

- Immune surveillance, a constant search-and-destroy mission against cancer cells, involves an interplay among cytotoxic T cells, NK cells, and interferon.

- A regulatory loop links the immune system with the nervous and endocrine systems. Cortisol, for example, released in long-term stress, can suppress overactive immunity.

- Sharks generate diverse B and T cells with a lower diversity but a greater number of gene fragments than mammals. Lamprey lymphocytes shuffle PRR genes to generate variable receptors and antibody-like proteins. Some arthropods and mollusks have alternative-RNA splicing mechanisms to make diverse defense proteins from a limited gene set.

Review, Synthesize, and Analyze

1. In what ways is the innate immune system of mammals similar to that of insects?

2. Both mast and dendritic cells reside in mammalian barrier tissues. Compare their functions.

3. Dendritic cells, helper T cells, and B cells with their antibodies can be considered as sensors, integrators, and effectors, respectively, of a negative-feedback system. Explain.

4. Explain the events that occur in mammalian skin following a mosquito bite.

5. Explain how the mammalian body and its acquired immune system respond when the body is invaded by an influenza virus that has evolved new antigens.

6. Antimicrobial peptides, hemolin, complement, perforins, and antibodies are all extracellular defensive proteins. Compare how they work.

7. Cheetahs are highly susceptible to diseases such as feline infectious peritonitis (FIP), which kills less than 10% of other cats but 50% of cheetahs. Cheetahs are nearly identical genetically, apparently due to a crash in population numbers perhaps 10,000 years ago. Explain why low genetic diversity in these cheetahs and also in inbred domestic animals makes acquired immunity less effective.

8. Feline immunodeficiency virus (FIV) is a common and ultimately fatal disease of cats; much like HIV, FIV damages T cells. Explain all the possible problems that could result from this. Also, like HIV, FIV mutates rapidly into new strains. Veterinarians can give cats an FIV vaccine; explain how it might work to protect the cat and why it might not work.

Suggested Readings

Banchereau, J. (2002). The long arm of the immune system. *Scientific American* 287:52–59. An overview of dendritic cells.

Blount, J. D., N. B. Metcalfe, T. R. Birkhead, & P. F. Surai. (2003). Carotenoid modulation of immune function and sexual attractiveness in zebra finches. *Science* 300:125–127.

Jancin, B. (2002). Antimicrobial peptides are first line of defense against skin infections. *Skin & Allergy News* 33:22.

Kurtz, J., & K. Franz. (2003). Evidence for memory in invertebrate immunity. *Nature* 425:37. A test for acquired immunity in copepods.

Litman, G. (1996). Sharks and the origins of vertebrate immunity. *Scientific American* 275:67–71.

O'Neil, L. A. J. (2005). Immunity's early-warning system. *Scientific American* 292:38–45. A discussion of the surprisingly sophisticated innate defenses in mammals.

Stuart, L. M. & R. A. Ezekowitz. (2008). Phagocytosis and comparative innate immunity: Learning on the fly. *Nature Reviews Immunology* 8:131–141. On the evolution of phagocytosis as a universal innate defense.

Travis, J. (1998). The accidental immune system. *Science News* 154:302–303. How the acquired immune system may have arisen by viral invasion.

Zasloff, M. (2002). Antimicrobial peptides of multicellular organisms. *Nature* 415:389–396.

CourseMate

Access an interactive eBook, chapter-specific interactive learning tools, including flashcards, quizzes, videos and more in your Biology CourseMate, accessed through **CengageBrain.com**.

Elephants crossing the Zambezi River in Botswana must either hold their breaths for extended periods, or else use their trunks as snorkels (as the elephant on the left is doing).

Respiratory Systems

11.1 Gas Demands: General Problems and Evolutionary Solutions

Energy is essential for sustaining all life-supporting cellular activities. For most animals to survive for extended periods of time, **aerobic** (oxygen-using) **metabolism** (p. 47) is necessary. In fact, most (though not all) scientists believe that the first nonmicroscopic animals did not evolve until the O_2 content of Earth's atmosphere reached a critical point some 550–600 million years ago. In addition to obtaining O_2, animals must eliminate the CO_2 produced as a by-product of aerobic metabolism at the same rate it is produced to prevent dangerous fluctuations in pH (that is, to maintain the acid–base balance), because CO_2 generates carbonic acid. The term **respiration** refers to all of the processes of gas movement and metabolism, with two distinct components:

- The term **internal** or **cellular respiration** refers to the intracellular metabolic processes carried out within the mitochondria, which use O_2 and produce CO_2 during the derivation of en-

ergy from nutrient molecules (see pp. 47–53).
- The term **external respiration** refers to the sequence of events involved in the exchange of O_2 and CO_2 between the external environment and cellular mitochondria.

External, not cellular, respiration is the topic of this chapter. We begin with an overview.

External respiration involves up to four major steps of transport and exchange

For the most part, gases in the external-respiration sequence move by two methods:

1. *Exchange* (i.e., O_2 and CO_2 moving oppositely) across membranes by the relatively slow process of *diffusion,* and
2. *Bulk transport*—the movement of the medium that contains the O_2 and CO_2, which (as discussed in Chapter 9 for circulation):
   a. bypasses diffusion altogether to move gases much more quickly than diffusion; and

   b. enhances diffusion gradients (as we will explore in detail in this chapter)

These two processes can occur in up to four steps, summarized in Figure 11-1:

Step 1. *Ventilation:* bulk transport of external media across a gas exchange surface. This motion, called **ventilation,** arises from environmental movements (e.g., winds or currents), an animal's locomotion, and/or **breathing** if it is actively and specifically produced.

Step 2. *Respiratory exchange:* gas diffusion between the environmental medium and internal body fluids. This occurs at a *respiratory surface,* which may be a dedicated organ such as a lung or gill, and the internal fluid may be the ECF in a circulatory system. The gases move through membranes without the need of protein carriers; however, CO_2 in the form of bicarbonate (HCO_3^-) may also be moved via facilitated diffusion carriers (p. 84) into an aquatic environment. (In

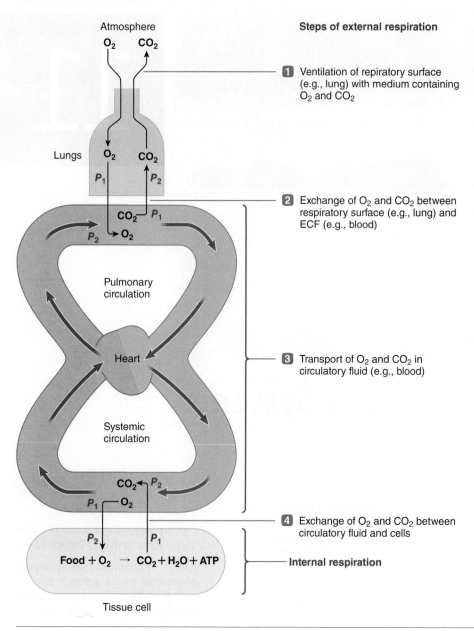

Atmosphere

O_2 CO_2

Lungs

O_2 CO_2

P_1 P_2

CO_2 P_1

P_2 O_2

Pulmonary circulation

Heart

Systemic circulation

CO_2 P_2

P_1 O_2

P_2 P_1

Food + O_2 → CO_2 + H_2O + ATP

Tissue cell

Steps of external respiration

1 Ventilation of repiratory surface (e.g., lung) with medium containing O_2 and CO_2

2 Exchange of O_2 and CO_2 between respiratory surface (e.g., lung) and ECF (e.g., blood)

3 Transport of O_2 and CO_2 in circulatory fluid (e.g., blood)

4 Exchange of O_2 and CO_2 between circulatory fluid and cells

Internal respiration

FIGURE 11-1 **The four steps of external respiration, shown for an air-breathing vertebrate.** External respiration encompasses the steps involved in the exchange of O_2 and CO_2 between the external environment and the tissue cells (steps 1 through 4). Gases flow from areas of high (P_1) to low (P_2) partial pressures (defined in Figure 11-2). Internal respiration, shown at the bottom, encompasses the intracellular metabolic reactions involving the use of O_2 to derive ATP from food, producing CO_2 as a by-product.

© Cengage Learning, 2013

---

general, gases are not moved by membrane active transport.)

Step 3. *Circulation:* bulk transport of the ECF (Chapter 9).

Step 4. *Cellular exchange:* gas diffusion between the outside (typically in the ECF, but sometimes the external environment) and the ICF. This is the ultimate goal of external respiration: exchange of gases between the cell's immediate surroundings and its mitochondria.

All animals have gas exchange with their cells (step 4) but not necessarily all other steps; for example, vertebrates have all four steps, but insects have steps 1, 2, and 4. When we refer to an animal's **respiratory system,** we are not talking about all these steps of external respiration, rather only steps 1 and 2. We will examine respiratory systems in detail later, but first, we begin with general problems and principles of gas exchange and transport.

Gas diffusion follows Fick's law for partial pressure gradients

Since two of the four steps involve diffusion, we begin by revisiting the physics of that process. In Chapter 3, you learned that diffusion follows **Fick's law,** in which solutes move down a concentration gradient. For gases, the term C (concentration) is replaced with the *partial pressure P* as follows (compare this version to that in Table 3-1, p. 76):

$$Rate\ of\ diffusion\ (Q) = \frac{\Delta P \cdot A \cdot D}{\Delta X}$$

where:

  D = the *diffusion coefficient* (which depends on, among other things, molecular weight, temperature, and the permeability of any barrier between two points)

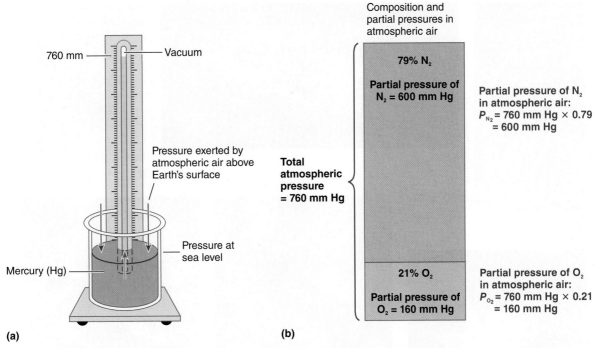

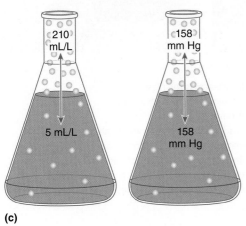

FIGURE 11-2 Gas Pressures. (a) Atmospheric pressure. The pressure exerted on objects by the atmospheric air above Earth's surface at sea level can push a column of mercury to a height of 760 mm. Therefore, atmospheric pressure at sea level is 760 mm Hg. The International System of Units (SI system) uses the pascal (Pa) unit, with 1 atm being equivalent to 101.3 kPa. (b) Concept of partial pressures. The partial pressure exerted by each gas in a mixture equals the total pressure times the fractional composition of the gas in the mixture. (c) A comparison of concentration and partial pressures between air and liquid. On the left side, air (sea level) has a concentration of 210 mL/L O_2, and due to poor solubility, only about 5 mL/L can dissolve in seawater at 25°C. Thus, the air and water here are in equilibrium; but if we plug the values into Fick's law for concentration, we would erroneously think there is a gradient of 205 (210 − 5) and, thus, that diffusion into the water is still occurring. To prevent this error, we convert concentrations to partial pressures (*right side*). Air has 158 mm Hg of O_2 (21% of 760 mm Hg). The O_2 concentration (5 mL/L) is converted by dividing it by its *solubility* (a physical constant). At 25°C in seawater, the solubility is 0.0316 mL/L per mm Hg air pressure. So, dividing 5 mL/L gives 158 mm Hg, showing there is an equilibrium.

© Cengage Learning, 2013

A = the *surface area* for gas exchange
ΔP = the gas *gradient* (in *partial pressure*) = $P_1 - P_2$, where P_1 is the partial pressure in one compartment and P_2 is the pressure in the other (do not confuse this with blood pressure)
ΔX = the *distance* the gas must cover

Partial Pressures Before we proceed, it is important to understand the term *partial pressure*. Atmospheric air is a mixture of gases that contains about 79% nitrogen (N_2) and 20.9% O_2 or 209 mL per liter of air, with very small percentages of CO_2, H_2O vapor, other gases, and pollutants in dry air. Altogether, these gases exert a total atmospheric pressure of 760 mm Hg at sea level (Figure 11-2a). This total pressure is equal to the sum of the pressures that each gas in the mixture partially contributes. The pressure exerted by a particular gas is directly proportional to the percentage of that gas in the total air mixture. Every gas molecule, no matter what its size, exerts the same amount of pressure; for example, an

N_2 molecule exerts the same pressure as an O_2 molecule. Because 79% of the air consists of N_2 molecules, 79% of the 760 mm Hg atmospheric pressure, or 600 mm Hg, is exerted by the N_2 molecules. Similarly, because O_2 represents about 21% of the atmosphere, 21% of the 760 mm Hg atmospheric pressure, or about 160 mm Hg, is exerted by O_2 (Figure 11-2b). The individual pressure exerted independently by a particular gas within a mixture of gases is known as its **partial pressure**, designated by P_{gas}. Thus, the partial pressure of O_2 in dry atmospheric air, P_{O_2}, is normally about 160 mm Hg. The atmospheric partial pressure of CO_2, P_{CO_2}, is negligible at 0.2 mm Hg.

Gases dissolved in water or blood are also considered to exert a partial pressure. Water that is in equilibrium with air will have the same gas partial pressures as in the air, *but the actual concentrations can be very different*. The amount of a gas that will dissolve in the blood depends on the gas's solubility in water and on the partial pressure of the gas in the environment (alveolar air or water) to which the blood is

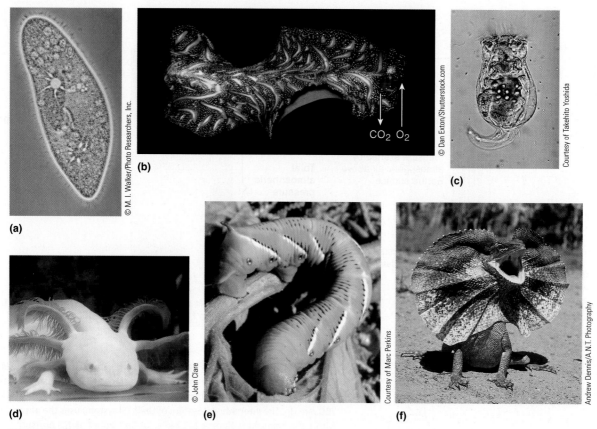

FIGURE 11-3 Diversity of gas exchange structures. (a) Respiratory gases diffuse across the plasma membrane of single-celled organisms, such as *Paramecium*. (b) In small multicellular animals, such as this flatworm, gas exchange occurs through the skin or integument; the distance from the integument to the tissues is small enough that gases can diffuse down concentration gradients in adequate time. (c) Some small aquatic animals, such as this rotifer, have cilia that create both feeding and breathing currents. (d) Most large aquatic animals, such as this axolotl, exchange gases at gills, evaginations of the respiratory surface. (e) The respiratory surface of terrestrial animals is invaginated, such as the tracheae of insects, with openings called spiracles (the brass-colored pores on this caterpillar). (f) Air-breathing vertebrates have lungs, with noses and mouths for air entry, as seen in this frilled lizard.

exposed. Because O_2 is nonpolar, it is very poorly soluble in water. So even if air and water have the same P_{O_2}, the actual concentration is much higher in air than water. You'll see the consequences of this later; for now, examine Figure 11-2c to make sure you understand why we rarely use gas concentrations.

In summary, keep in mind that *gas diffusion follows partial pressure ΔP gradients and changes linearly with surface area A, coefficient D, and inversely with distance ΔX.*

External respiratory processes must meet the demands of size, metabolism, and habitat

Recall that diffusion is quite slow except over very small distances (p. 385). Indeed, August Krogh (p. 3) calculated that diffusion can supply the energy demands of a typical cell only over distances of 1mm or less. For very small organisms, simple diffusion using steps 2 and 4, or step 4 only (Figure 11-1), can be sufficient. However, some larger organisms rely primarily on diffusion by having morphological adaptations that enhance it. As indicated by the ΔX component of Fick's law, flattened and thin shapes (which reduce distances compared to more spherical shapes) are the most efficient for exchange of O_2 and CO_2. This can be seen even in unicellular organisms: While very small ones are often spherical, larger unicellular protists are often elongated and narrow (such as *Paramecium*, Figure 11-3a), a shape that reduces ΔX between the environment and the mitochondria (step 4).

In animals, exchange occurs across **respiratory surfaces**— tissues that are exposed to the environment and across which O_2 and CO_2 diffuse, either directly to internal cells (step 4) or first into an ECF (step 2). The site of gas exchange in the first animals was presumably through the relatively unspecialized body surface or integument. This has persisted to the present in many small, thin, and/or sluggish animals such as Cnidaria (e.g., Figure 9-1b, p. 386), which have thin tissues close to external water, and thin-bodied flatworms (Figure 11-3b), which rely substantially on this **integumentary** (or **cutaneous**) **exchange**. Both can also exchange gases with digestive tract fluid, often highly branched to increase A.

Even in such organisms, there can be forms of bulk transport. For example, the cytoplasm of many protists such as *Paramecium* (Figure 11-3a) exhibits a cyclical cytoskeletal-powered movement called *cytoplasmic streaming*. Ventilation may also occur: The cilia of *Paramecium*, for example, produce not only locomotion but also a flow of fresh oxygenated water across the membrane. Many small animals such as rotifers and aquatic larvae have similar dual-purpose integumentary cilia (Figure 11-3c). In Cnidaria, ciliary movements that move fluid into the digestive tract bring in O_2, while some sea anemones have cilia on their tentacles that move water for both feeding and respiration.

However, three major factors in the evolution of many animals groups have favored enhancements in gas exchange—for every component of Fick's law—and more advanced forms of bulk transport:

1. Animal *size* has increased in many phyla since the first animals evolved. As this occurred, diffusion became too slow to cover internal distances in a reasonable time.

2. Animal *habitats* vary in their availability of O_2, from anoxic ("no oxygen") aquatic mud to sea level air. Table 11-1 shows the wide range of availability of O_2 among various habitats. The most distinctive difference is found when comparing aquatic to terrestrial habitats. Note in particular that in water, *gas solubility decreases as temperature increases*, such that oxygen can be much more available in a cold-water habitat. Note also that aquatic habitats, even at their highest O_2 content, hold only a fraction of the content of air. And levels can get much lower in mud (which impedes diffusion much more than sand) and in waters such as the Oxygen Minimum Zones found at intermediate depths in many areas of the oceans (Table 11-1). These occur where it is too dark for photosynthetic O_2 production; bacteria use up the available O_2, and there are no O_2-rich surface waters sinking down to freshen the zones. (The deepest waters of most oceans, however, are adequately oxygenated by the sinking of cold surface waters in polar regions.) The bottoms of some ponds, lakes, and ocean basins also have low concentrations of O_2 when there are no vertical currents to mix surface with deeper water.

   Animal lifestyles also come into play regarding habitat O_2; for example, air-breathing aquatic divers like sperm whales may need to survive over an hour without O_2 intake.

3. Animal *metabolisms* vary in their demand for O_2, with the highest demands arising in birds, mammals, some fishes, some cephalopods, and flying insects.

Let's now look at the general ways that external respiration has been enhanced in animals that are larger, live in lower-oxygen environments, and/or have more active metabolisms.

Gas exchange may be increased by respiratory organs, bulk transport, and diffusion-enhancing proteins

Diffusion at steps 2 and/or 4 can be enhanced, beyond the basic ways found in flatworms, by four broad adaptations at the cellular, organ, and system levels, as follows:

TABLE 11-1 Oxygen Concentration in Various Habitats, in mL per Liter of Medium

Air at sea level	209
Air at Mount Everest (8,848 m)	59
Fresh water at 0°C (maximum)	10.3
Fresh water at 30°C (maximum)	5.6
Seawater at 0°C (maximum)	8.0
Seawater at 30°C (maximum)	4.5
Seawater in Oxygen Minimum Zones, 200–1,000 m depths	0–0.7

© Cengage Learning, 2013

1. *Ventilation* (step 1). For diffusion between two nonmoving fluids, a gas gradient can quickly become equilibrated (thus, $\Delta P = 0$), and no further net gas movement can occur. Bulk transport delays this equilibration. In the case of ventilation, respiratory surfaces are provided with a continuous fresh supply of external medium with a high P_1 for O_2 (accompanied by a low P_2 for CO_2), keeping gas gradients high in the appropriate directions at step 2.

2. *Respiratory organs* (step 2). The integument may not provide enough surface area for gas exchange, and it may also have a low D and large diffusion ΔX, because of protective layers. Thus, many animal groups have evolved dedicated gas-exchange organs such as *gills, tracheae,* and *lungs* (Figure 11-3d, e, and f), which enhance step 2 as follows:

   a. A high D value due to high permeability; for example, a fish gill epithelium is much more permeable to gases than mucus- and scale-covered skin;

   b. A low ΔX (distance), using very thin epithelia;

   c. A high A value, through specialized *folds* (*evaginations*, or "outpockets," and *invaginations*, or "inpockets") that increase surface area for gas diffusion. Surface areas are generally higher in animals with higher metabolisms. For example, the lungs of some amphibians are relatively simple sacs, but in more active (non-avian) reptiles and all mammals, the tubes branch into numerous tiny blind-end sacs (*alveoli*, which we explore later) to provide a much higher surface area.

3. *Circulation* (step 3). Internal bulk transport aids diffusion in the same way as ventilation. At the respiratory surface (step 2), ventilation may be complemented by circulatory *perfusion*, which brings to the exchange surface a continuous supply (P_2) of low-O_2 body fluid (with a high P_1 for CO_2). At internal tissues (step 4), which are consuming O_2 and producing CO_2, the circulatory fluid—reoxygenated at the respiratory surface—provides a fresh supply (P_1) of high O_2 (accompanied by a low P_2 for CO_2).

   An internal circulatory system may also enhance diffusion at steps 2 and 4 by increasing D (e.g., permeable capillaries), increasing A (e.g., high vascularization), and reducing ΔX (e.g., narrow capillaries in closed circulations, or direct bathing of cells by circulatory fluids in open circulations).

4. *Proteins* in the circulatory fluid and inside some cells that convert gases between *diffusible* and *nondiffusible* forms (at steps 2 and 4). In its nondiffusible form, a molecule does not contribute to ΔP, and so the gas gradient ($P_1 - P_2$) is not equilibrated nearly as quickly. Think of the gas as being "trapped" so that it cannot diffuse backward once it has reached a certain location, keeping P_2 low. Conversely, when a "trapped" gas is released back into a diffusible form, P_1 increases. This greatly increases the amount of gas that the circulation can carry. For O_2, this mechanism typically involves special *respiratory pigments* (p. 391) such as *hemoglobin* and *myoglobin*. For CO_2, this involves both binding to hemoglobin and enzymatic conversion to and from *bicarbonate* HCO_3^- ion, which unlike the free gas cannot diffuse through membranes without a membrane transporter.

Breathing as a form of bulk transport can be tidal or flow-through

Now let's look at bulk transport itself for moving gases rapidly. As we noted earlier, ventilation may occur by environmental movements, for example, a water current flowing over the gills of the axolotl in Figure 11-3d. True breathing, however, requires energy-using devices to move the external medium. Typically, these devices are either flagella or cilia that line the surfaces themselves, or separate muscular pumps that create negative and positive pressures to move the medium (as with the respiratory muscles of the lizard in Figure 11-3f). You will later see how these work, but for now, understand that there are two distinct designs to consider in terms of the type of flow created by the pumps:

1. In **tidal breathing,** the external medium is moved in and out of the same opening, in two distinct steps termed **inhalation** (or **inspiration**) and **exhalation** (or **expiration**). It is relatively inefficient because fresh medium typically mixes with depleted medium, and because fresh medium can only be brought in about half the time.

2. In **flow-through breathing,** the external medium enters one opening and leaves through a separate opening. Fresh and depleted medium need not mix much in this process, and flow of fresh medium can be nearly or fully continuous. This also allows for *countercurrent flow,* a process we will explore later that increases gas exchange efficiency. Thus, flow-through systems can be much more effective at gas exchange than tidal ones.

We will examine how all these adaptations and processes work in detail in this chapter, except for some details of circulatory covered in Chapter 9. We first turn to specific respiratory systems (that is, steps 1 and 2 in Figure 11-1).

check your understanding 11.1

What are the four possible steps in external respiration?

How have body size, metabolism, and habitat affected the evolution of external respiration?

How do respiratory adaptations enhance the limits of diffusion?

Compare tidal and flow-through ventilation.

11.2 Water Respirers

Let us begin our examination of respiratory systems by focusing on the most distinctive adaptational contrasts: water respirers and air respirers.

Water is a more difficult medium than air for gas exchange

Though life began in water, it contains much less O_2 than air (Table 11-1). It is also problematic for gas exchange in other ways. A comparison of air and water reveals the following points:

1. *Viscosity:* Water has a higher viscosity than air (approximately 850-fold) that necessitates a higher energy output to maintain flow over the respiratory surface. The higher density of water (60-fold greater than air) also entails a higher cost of acceleration.

2. *Solubility:* CO_2 is and its product, bicarbonate, being polar, are in much higher concentrations than in air. However, as we noted earlier, O_2 concentration of water is considerably smaller than in the gas phase: 1 L of water at 15°C contains 7 mL of O_2, whereas 1 L of air contains 209 mL of O_2. For this reason approximately 30 times more water than air must be moved to achieve the same ventilatory transport of O_2.

3. *Diffusion:* The rate of diffusion of gases in water is about 10,000-fold slower than in air.

4. *Salinity:* The solubility of O_2 and CO_2 decrease with increases in salinity; no comparable effect occurs in air.

5. *Temperature:* Diffusion rates for gases increase with temperature in both air and water; however, solubility of O_2 and CO_2 are *decreased* as ambient temperature *increases*. No comparable effect occurs in air. These changes in solubility are critical; as temperature is raised from 0 to 35°C, O_2 concentration is reduced by 50% at constant P_{O_2}. Because an increase in temperature raises the metabolic rate in ectotherms (animals whose body temperatures depend on the environment; Chapter 15), the associated reduction in O_2 solubility in water exerts an ecological pressure for aquatic species to obtain access to air.

6. *Habitat variation:* As we noted earlier, the O_2 content of water is subject to greater variation than that of air. In addition to areas with permanently low O_2 (muddy sediments, warm areas, and Oxygen Minimum Zones; Table 11-1), bays, swamps, and intertidal pools can have large swings in P_{O_2}, which rise during daylight hours as a result of photosynthesis and decline at night as a result of biological O_2 demand. The P_{O_2} values vary from 0 to 450 mm Hg in these shallow habitats, whereas the surface waters of the open ocean have O_2 levels at close to that of air, about 155 mm Hg.

7. *Composition:* Unlike air, water can contain many life-sustaining components other than gases. These include dissolved ions and organic matter, and, of course, water itself (although air can have water vapor).

The limitations of diffusion in water are overcome with gills and other external thin, high-surface-area structures, internal circulation, and flow-through breathing

Enhancements to the components of the diffusion equation are critical for water respirers. Bulk transport is similarly important, with the more effective flow-through arrangements being more common than tidal ones. Earlier we saw how cell shape, cytoplasmic streaming, and cilia aid gas exchange in *Paramecium* (Figure 11-3a). In addition, cilia of *Paramecium* and many animals such as sea anemones often beat in synchrony in one direction, essentially creating a "flow-across" process akin to a flow-through organ system.

Integument We have already noted integumentary respiration in flatworms and Cnidaria. It also occurs to some extent in many larger and less thin aquatic animals and is sufficient to maintain a normal resting metabolic rate, even in some gill-bearing vertebrates such as certain eels and catfish. However, unlike flatworms and Cnidaria, internal circulation (step 3, Figure 11-1) enhances this process. For example, in the earthworm (an annelid) a single, thin layer of mucus-coated epidermal cells covers a dense network of capillaries. (O_2 and CO_2 can cross cell membranes only when they are dissolved in water, so this "terrestrial" animal is effectively a water respirer.)

Integumentary respiration is important in virtually all amphibians and aquatic reptiles and in most fishes. During hibernation, for example, frogs and turtles exchange all their respiratory gases through this route while submerged in ponds or buried beneath the bank of a stream. Eels normally exchange 60% of their CO_2 and O_2 through their highly vascular skins. Of particular note are the lungless salamanders of the family *Plethodontidae*. Because the skin of these animals is also important for protection against the environment, the efficiency of gas exchange is compromised.

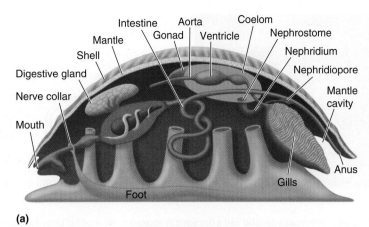

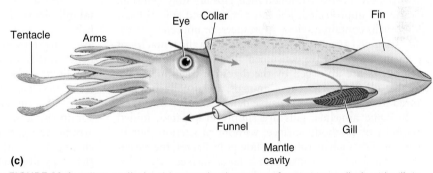

FIGURE 11-4 **Gills in mollusks.** (a) Generalized anatomy of a primitive mollusk, with gills in the posterior mantle cavity protected by a shell and ventilated by cilia on the gill. (b) A nudibranch (gastropod mollusk, *Chromodoris annae*); its gills are the tufts on the middle top. (c) A squid, showing muscle-driven respiratory currents flowing into the mantle cavity across the gill.

Sources: (a) Phillip C. Withers. (1992). *Comparative Animal Physiology*, 1st ed. Belmont, CA: Brooks/Cole, p. 577, Figure 12-9; (b) © Steve Norvich/Visuals Unlimited; (c) http://tolweb.org/accessory/Cephalopod_Jet_Propulsion?acc_id=2060.

Gills Nonintegumentary structures for gas exchange evolved in most water breathers with larger and/or less-thin bodies and higher metabolisms. Most of these specialized respiratory organs are called **gills** (Figure 11-3d and Figures 11-4 to 11-6), evaginations of tissue protruding into the external medium. Gills range from simple integumentary bulges, to branched filaments, to stacks of **lamellae** (flat platelets on filaments). Filaments and lamellae greatly increase surface area A.

Typically, gills are delicate structures because, as we noted earlier, they are composed of thin cell layers for a small ΔX as well as a high A. They are also highly perfused by a circulatory system and may have associated

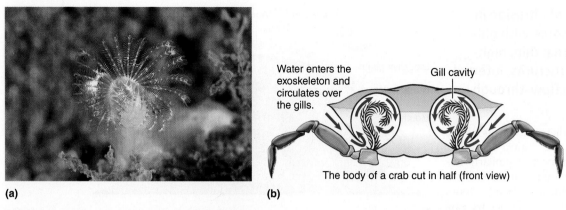

(a)   **(b)**

Water enters the
exoskeleton and
circulates over
the gills.

Gill cavity

The body of a crab cut in half (front view)

FIGURE 11-5 Gills of a polychaete annelid and a decapod crustacean. (a) A polychaete featherduster tubeworm (*Hydroides norvegica*), with its anterior gills (also used for feeding) emerging from a calcareous tube. (b) Cross section through the carapace of a crab, showing the gill extending as a branch from the basal segment of a limb into a protected branchial chamber.

Sources: (a) www.UWPhoto.no © Erling Svensen; (b) http://www.arthursclipart.org/biologya/biology/page_02.htm).

flow-through breathing mechanisms. For example, gill filaments of mollusks (Figure 11-4a) have cilia that create unidirectional respiratory currents across the filament surfaces.

Due to the fragile composition of gills, some form of protection is generally required. Often the organs are internal in the sense of being protected by a hard cover like a mollusk's shell (thus, the gills are not externally visible; Figure 11-4a). But even then, the respiratory surfaces evaginate to form delicate structures in direct contact with the external medium through openings to the environment. Protection of a different kind is found in the specialized elaborate filaments of shell-less sea slugs called *nudibranchs* ("naked gills") (Figure 11-4b). Although their gills are fully external, they are not unprotected. The integuments and gills of most nudibranchs contain noxious substances, or unfired nematocysts (stinging capsules, p. 42) "stolen" from the tentacles of sea anemones they have consumed! The (somehow) unfired nematocysts are absorbed by the digestive tract and move into the gill filaments. These animals are usually brightly colored to advertise their invisible defenses.

In some annelids, gills are simply evaginated, folded extensions of the body surface, without protection; but in other annelids such as tube-dwelling polychaetes, the extensions form dense filaments with a large surface area (also used for feeding; Figure 11-5a). Though unprotected when in use, these filaments can be quickly withdrawn into the animal's hard tube when threatened.

The gills of arthropods such as crabs and lobsters (Figure 11-5b) are protected by their exoskeltons. Those of fishes (Figure 11-6) are also protected with hard parts. Let's look at fish in more detail. The gill system of fish consists of **gill arches,** the tissue that provides skeletal support for a double series of gill filaments, each of which carries a row of lamellae on either side (Figure 11-6a–c). Normally there are five separate pairs of these arches, covered with a protective bony plate called the **operculum,** which opens on one side to the outside world. The basic unit of gas exchange in the gill is the lamella, not the arch or filament. Lamellae form a sievelike structure which minimizes diffusion distances for gas exchange between water and the extensive blood supply found in each lamella. The thickness of each lamella ranges from 10 to 25 µm. Each is comprised of two epithelial cells, separated by a series of *pillar cells* between which blood can flow. The degree of activity of a fish species is correlated with the amount of spacing between subsequent lamellae, which can vary from 10 to 60 per mm. Depending on the species, total lamellar numbers can also vary. Generally, to accommodate the metabolic requirements of fast-swimming species, the size of the lamella is larger and the interlamellar spacing is smaller. For example, in an inactive bottom dweller such as the toadfish the number of lamellae is approximately 0.5 million, whereas in a comparably sized active species such as the bluefin tuna, lamellar number is over 6 million.

Exposure to the environment has its risks. **Environmental gill disease (EGD)** is a syndrome that commonly affects fish reared using aquaculture technology. The initial stages are characterized by inflammation of the gill lamellae. It is believed that stress as well as environmental factors triggers the disease process. If exposure to the irritant (such as bacteria, amoebas, fungi) is short-lived, damage to the gill lamellae is minimal. However, if the irritant exposure continues, a serous (blood plasma) exudate accumulates around the capillary endothelium, which reduces gas exchange. Fish with *epithelial-capillary endothelial separation (ECS)* exhibit symptoms of respiratory distress that ultimately becomes fatal if left untreated.

Other Respiratory Tissues Integuments and gills are not the only respiratory surfaces in aquatic animals. For example, in water-breathing mollusks, the **mantle** (Figure 11-4a) as well as the gill is highly vascularized and can take up O_2 from ventilatory currents flowing over these surfaces. In shelled mollusks, the ventilatory current also flushes the **visceral mass** (main body) with oxygenated water. A comparatively large sinus lying beneath the visceral mass can actually be inflated, increasing the amount of surface area exposed to the ventilatory current.

Sea cucumbers (soft-bodied echinoderms) are animals with a unique respiratory structure. Echinoderms are not typically high-metabolism animals, and most (such as seastars and sea urchins) exchange gases through small body-wall evaginations and/or tube feet. But a sea cucumber

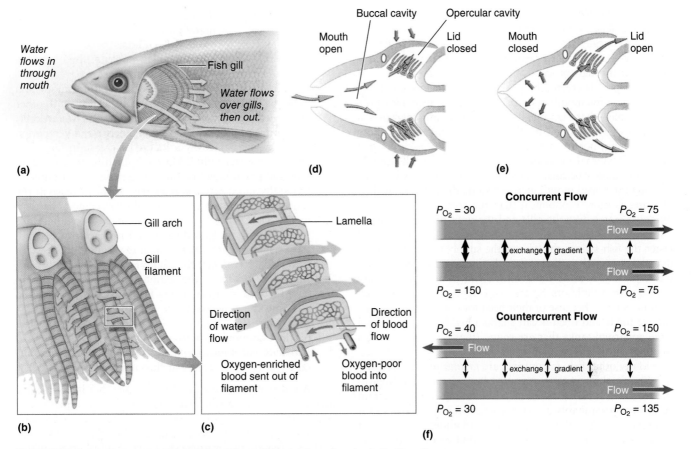

FIGURE 11-6 Respiratory system of a teleost (bony fish). (a) One of a pair of gills. The gill cover or operculum has been removed for this sketch. Water drawn into the fish's mouth is forced across the gills. (b) Each gill has filaments bearing lamellae, which are richly vascularized respiratory surfaces. (c) The flow of oxygen-rich water is countercurrent to the flow of oxygen-depleted blood entering the lamellae. (d) Opening the mouth and closing the lid forces water into the buccal cavity and across the gill. (e) Closing the mouth and opening the lid forces the water to exit through the opercular cavity. (f) In concurrent flow (*top*), gases reach a 50:50 equilibrium in the two tubes. However, in countercurrent flow (*bottom*) as in a gill, up to 90% of a gas can be transferred because blood moving through the one tube continually encounters water whose O_2 content is higher than what is present in the blood. Thus, there is a positive ΔP at every point, keeping gases diffusing into the blood.

Source: C. Starr & R. Taggart. (2004). *Biology: The Unity and Diversity of Life,* 10th ed. Belmont, CA: Brooks/Cole, p. 710, Figure 40.6, a, b, c, d, e.

draws in seawater through its anus into a branching structure called a **respiratory tree** that protrudes into its coelomic (body cavity) fluid, not into the environment. Pumping involves muscles in the tree, but it is relatively slow and involves a low-efficiency tidal flow, so it cannot be used to support a high metabolism. This outgrowth of the digestive tract is structurally more like a lung than a classic gill!

Breathing muscles in water provide rapid and often flow-through transport

Gas exchange in the most active aquatic species requires the aid of a dedicated muscle-driven form of breathing. The most active of the mollusks—cephalopods—exemplify this: They have abandoned ciliary-driven bulk transport on their gills and instead rely on a muscular pumping mechanism. This involves the muscular walls of their mantle cavity and funnel (Figure 11-4c). During inhalation, the funnel closes and the mantle cavity expands, drawing in large amounts of water to

ventilate the gills rapidly. To exhale, the mantle opening seals up and the funnel opens, whereupon the mantle contracts and squirts the water out the siphon. This can also be used as a means of rapid **jet propulsion** when the animal needs to move quickly. Note that, because the intake and exit paths are different, this is largely a flow-through process (Figure 11-4c).

Another flow-through type of muscular pump is readily seen in fishes (Figure 11-6d, e). Water is pumped over the gills of bony fish by skeletal muscle pumps in the **buccal** (mouth) and **opercular cavities.** Ventilation of the gills operates by a cycle of negative and positive pressure gradients. As the mouth opens, the opercula are shut, which leads to a negative pressure gradient in the buccal and opercular cavities, resulting in the inward movement of O_2-rich water. The mouth is then closed and the opercular cavity constricts, forcing water through the gills. At this point the water that is trapped in the gills is under higher pressure than outside. When the opercula open, the positive pressure gradient forces water outward across the gills and through the opercular exit. Note again that this is a flow-through system, whose efficiency we explore later.

Paradoxically (it seems at first), the most active fish of all—tuna—have lost their breathing muscles! They instead rely on a flow-through process called **ram ventilation,** in which bulk transport is created by the animal's own motion, along with keeping the mouth and opercula slightly open. However, the principle—that high metabolisms require muscular-driven bulk transport—remains the same, but the muscles are those of swimming, not the traditional respiratory muscles. Large masses of external medium are moved over the gills rapidly and continuously because these animals never stop swimming. Although the open mouth increases drag, the continuous flow of water is more economical than the costs associated with pumping water over the gills and the drag of the turbulence that would be caused by flapping opercula. In addition, the faster the fish swims, the greater the supply of O_2-rich water that flows over the gills to meet the increased tissue requirements for O_2. Because tunas must swim to breathe, they are *obligate* ram ventilators. So are many open-ocean sharks such as white sharks. Because of their need for constant swimming, captive tunas and open-ocean sharks must be kept in special large aquarium tanks that are round (that is, lacking corners to bump into) with circular water flow.

Many other fish, in contrast, are *facultative* ram ventilators, that is, they switch from active buccal-opercular breathing to ram ventilation when swimming above certain velocities, presumably to save energy. Rainbow trout, for example, switch as they reach a speed of about 24 cm/sec, at which point their total body O_2 consumption drops by 10%.

Countercurrent blood flow enhances gas pressure gradients in fishes

Another way to maximize gas exchange relies on the arrangement of blood flow. To be able to extract the maximum amount of O_2 from the water flowing through the gills, blood flows in a direction opposite or **countercurrent** to that of water flow (Figure 11-6c and f); that is, blood flow through the capillaries is opposite to the stream of water flowing through the gill lamellae, such that blood moving through the lamellae continually encounters water whose O_2 content is higher than what is present in the capillary. Thus, there is a positive ΔP at every point, keeping gases diffusing into the blood. This anatomical arrangement permits the O_2 content of arterial blood to exceed that in the respired water. As you will see later when we compare water and air respirers, this yields a much greater efficiency of gas exchange than other forms of flow.

Interestingly, some species of fish use *retrograde flow* of water from the opercular opening into the buccal chamber to ventilate their gills. Lampreys, for example, use this unusual adaptation because, while feeding, the lamprey's mouth remains attached to a host fish. Thus, the gills must be ventilated by this tidal flow of water generated by cyclic contraction of the muscular walls lining the gill pouches.

Aquatic respiratory systems can perform numerous nonrespiratory functions

In water respirers the respiratory system (such as gills) provides the greatest exposure between internal and external environments. Therefore, the system often performs several important nonrespiratory functions related to its high exchange capacity:

1. *Feeding*—many gills are also feeding structures, as in bivalve mollusks, many polychaetes (Figure 11-5a), brachiopods, and so on. These animals have ciliated filaments that create flow-through currents for obtaining plankton as well as for exchanging gases.
2. *Fluid and solute balance*—the regulation of osmotic and ionic balance. Gills in crustaceans, fishes, and other groups continually adjust wastes, ions, and water levels in blood (Chapter 13), and so any increase in ventilation can be accompanied by a marked increase in ionic loss (Na^+) and water influx. Marine bony fish, however, gain ions and potentially lose body water to the environment. Generally more active species rely on an increase in the functional surface area of the gills to accommodate an increase in O_2 needs, whereas sedentary species increase blood flow to the gills during increased activity. Because these fish rarely require maximal O_2 ventilation, they are better able to tolerate the resulting ion losses.
3. *Acid–base balance*—the maintenance of normal acid–base balance by transporting H^+ or HCO_3^- through the gills into the aqueous media (Chapter 13).
4. *Excretion*—for example, removing the toxic waste product ammonia (Chapter 12).
5. *Nutrient and mineral uptake*—for example, calcium intake from the environment through the gills.

Another process affected by gills, one that is not always adaptive, is heat transfer. The vast majority of aquatic animals have body temperatures within 1 to 2°C of the surrounding environment because of heat loss through the high surface area of respiratory structure. Endothermic water respirers such as tuna require special adaptations to minimize heat loss (Chapter 15).

check your understanding 11.2

What are some nongill methods of gas exchange in aquatic animals?

Describe three examples of gills in different nonvertebrate phyla.

How do fish create flow-through breathing, and how does gill countercurrent flow aid gas exchange?

11.3 Air Respirers: Overview and Nonvertebrates

The evolutionary transition to land involved new challenges for external respiration.

Air breathers may have less efficient exchange surfaces and breathing mechanisms, and their exchange surfaces must be protected from drying out

Firstly, recall that air is much less viscous than water and that it contains much more O_2 than any body of water (Table 11-1). These facts have two implications for diffusion and bulk transport:

- Surface area need not be as high as in water. For example, one remarkable animal, the land crab *Scopimera*, can take up O_2 from special "thigh windows," thinned areas of the cuticle located in its walking legs. These have much less surface area than gills of aquatic crabs but are just as effective in total gas exchange. Nevertheless, the most active air breathers do require high surface areas, as you will see.
- Efficient external bulk transport (which has a potentially significant metabolic cost) is less necessary for air respirers. For example, small insects rely partly on diffusion for gas movement in and out of their tracheae. Also, simple tidal breathing is more common among terrestrial animals than the (often more complex) more efficient flow-through systems. For example, mammals (which have among the highest known metabolic rates) have relatively inefficient tidal breathing. However, birds, with their even higher metabolic rates, have partial flow-through arrangements (as you will see later).

Secondly, because respiratory surfaces are typically thin for diffusion, they can potentially collapse in air under gravity and, more importantly, dry out rapidly if exposed to air. Therefore, in air respirers the respiratory surfaces must be kept moist to maintain the integrity of the respiratory surface. There are two broad ways to do so:

- *Remain in moist conditions.* Such land animals closely resemble small aquatic animals in respiratory features; for example, many rely on integumentary exchange. But because air with its high O_2 content is close at hand, this type of exchange can work well in larger animals. Earlier, you saw how the earthworm can be considered aquatic in this regard. Sometimes certain regions of the skin are enhanced for respiration. For example, the East African annelid swamp worm, *Alma emini*, spends the dry season burrowed in decomposing plant material, where it breathes moist air. During the rainy season, the local environment becomes anoxic, so the swamp worm extends its highly vascularized, flattened tail out of its burrow into the air. The tail is even more versatile than this, however. When the animal withdraws into its burrow (to avoid predation or drying out), its tail can roll up into a covered cylinder that takes a bubble of air down with it underground. The tail then slowly absorbs the O_2 from this "scuba tank" of gas!
- *Have covered or fully internal structures* for gas exchange, with secretions to moisten the surface. Structures include internal air tubes called **tracheae**, which take air directly to tissues (insects, some spiders), gill-like **book lungs** (scorpions and some spiders), **mantle cavities** (some snails), or vascularized sacs called **lungs,** which provide air for internal circulation (some snails, and land vertebrates, including those such as penguins and whales that have reevolved into aquatic habitats). Some of these form as invaginations rather than the evaginations; others form as evaginations off of digestive tracts, but are fully internal.

Let's now examine these internal structures in more detail.

Land slugs and snails use skin, mantle tissue, or lungs

Mollusks are primarily aquatic. Some species such as many bivalves are semiterrestrial in the intertidal zone, where they are exposed alternately to water and air. They generally use the same (gill) structures for gas exchange with air or water. However, these animals limit exposure of their epithelial surfaces to the atmosphere because of desiccation risk, so they typically close up to minimize water loss and then depend on anaerobic metabolism.

Only in certain groups of gastropods (snails, slugs) have any mollusks become terrestrial. Land slugs require moist habitats, relying on integumentary exchange through a moistened skin, whereas one group of snails (the *prosobranchs*) uses its mantle tissue under its shell. The most successful groups are the pulmonate ("lung-bearing") snails, some of which have adapted to deserts. They have an actual lung in which a well-perfused mantle tissue forms a complete chamber with a small opening to the outside to reduce water loss. Gases move primarily by diffusion, although some tidal transport may occur from body movements.

Arachnids use book lungs or tracheae

The first animals to venture onto land were probably arthropods, over 400 million years ago. Recent fossil evidence suggests that these were scorpion-like animals that lived in empty mollusk shells to keep their gills from drying out (much like modern hermit crabs do; see photograph below). The gills eventually became invaginated into an internal chamber to become the first lungs. We see these in modern scorpions and some spiders (arachnids) in the form of **book lungs,** called this because they consist of stacked lamellae appearing like pages of a book, with air spaces between the folds for gas exchange (Figure 11-7). These stacks of lamellae invaginate from the cuticle into the abdomen, forming numerous thin air chambers. In more advanced spiders, these have evolved into tubular extensions into the tissues and thus form a true *tracheal* system (which, however, evolved independently of the insect system, which we examine next).

A land hermit crab in Costa Rica. It protects its gills from drying out by entering the shell of a dead snail, much like the first land animals were thought to have done some 400 million years ago.

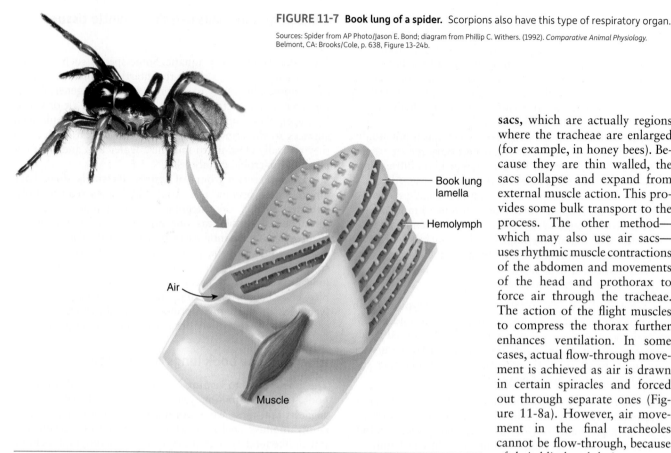

FIGURE 11-7 **Book lung of a spider.** Scorpions also have this type of respiratory organ.

Sources: Spider from AP Photo/Jason E. Bond; diagram from Phillip C. Withers. (1992). *Comparative Animal Physiology.* Belmont, CA: Brooks/Cole, p. 638, Figure 13-24b.

Book lung lamella

Hemolymph

Air

Muscle

sacs, which are actually regions where the tracheae are enlarged (for example, in honey bees). Because they are thin walled, the sacs collapse and expand from external muscle action. This provides some bulk transport to the process. The other method—which may also use air sacs—uses rhythmic muscle contractions of the abdomen and movements of the head and prothorax to force air through the tracheae. The action of the flight muscles to compress the thorax further enhances ventilation. In some cases, actual flow-through movement is achieved as air is drawn in certain spiracles and forced out through separate ones (Figure 11-8a). However, air movement in the final tracheoles cannot be flow-through, because of their blind-end design.

In insects, internal air-filled tubes—tracheae—supply oxygen directly to the tissues

In insects (as well as centipedes, millipedes, onychophorans, and most spiders), gas exchange occurs through a system of internal air-filled tubes, the tracheae (Latin *trachea*, "windpipe"), which directly supply O_2 to the tissues (Figure 11-8). It is not known how these evolved, but regulatory genes that induce their formation in insects are related to genes of crustacean gill development, so tracheae may have started as invaginated gill folds. Tracheae have been found in fossilized millipedes from about 400 million years ago.

Most insect tracheae are reinforced with rings of chitin. The circulatory system is thus important *only* for the transport of nutrients and not respiratory gases to the respiring cells. The tracheae connect to the outside through openings called **spiracles** (Latin *spiraculum*, "airhole") that are pores or openings through the exoskeleton (Figures 11-3e and 11-8). These can be closed by processes we will discuss later (see p. 546). Most species have three pairs of spiracles above the legs and on the first seven abdominal segments. The tracheae ultimately break up into finer branches, the **tracheoles,** which are about 0.2 μm in diameter. Tracheoles can actually indent the surface of target cells such as muscle fibers and are "functionally internalized," although they never become intracellular. Thus, insects use steps 2 and 4 of Figure 11-1, bypassing step 3. The distribution of tracheae reflects the O_2 demands of particular tissues. For this reason, the brain, sense organs, and flight muscles of insects contain an abundant supply.

Insects can also have step 1. Both larger and flying insects have long been known to rely on active tidal pumping of gas through the tracheal system. One method uses **air**

It used to be thought that simple diffusion was sufficient for gas movement in and out of tracheae in small insects. This is probably true in some cases. However, in 2003, Mark Westneat and colleagues observed tracheae alternately compressing and expanding in head and thorax regions of many insects, including very small ones such as carpenter ants. Compression, which forces air out of the tracheae, is thought to be due to body muscles contracting around the tracheae. Expansion, which creates suction to draw air into the tracheae (in tidal fashion), is due to elastic rings in the tracheae walls that expand like springs after being compressed.

check your understanding 11.3

Compare how ventilation and gas exchange work in a land snail, spider, and insect.

11.4 Air Respirers: Vertebrates

The respiratory system of vertebrate lung breathers includes the respiratory structures leading into the lungs, the lungs themselves, and the structures of the thorax (chest) involved in moving air through the airways into and out of the lungs. The **respiratory airways** are tubes that carry air between the atmosphere and the lung surfaces such as **alveoli** (small sacs in grapelike clusters). In some cases, these airways can also function in gas exchange. An unusual solution to gas exchange observed in submerged turtles are the well-vascularized membranes lining the mouth, which take up O_2 from water as well as from the air. The tortuous nasal passages of some birds also

Challenges and Controversies
Why Are There No Large Insects?

Watch a bee or dragonfly accelerate and zoom—those feats of flight rely on muscles that receive the most O_2 per second of any known animal muscle—up to 200 times faster than in mammals. How can they do this? As a result of pumping air directly to muscles, tracheae are actually much more effective at delivering O_2 than the vertebrate blood system, which requires many more stages and the relatively slow pumping of a liquid. Insect flight muscle can elevate its metabolism 100-fold, but mammals rarely increase muscle metabolism more than 20-fold. And yet, tracheae may explain why insects are so small. Careful *scaling* analysis of small to large beetles, done by Alexander

Kaiser and colleagues in 2006 with intense X-rays, found that tracheae take up a proportionally greater part of an insect body as body size increases. That is, the tubes must increase in width more than body size increases in order to deliver enough O_2, and the researchers calculate that at above a 15 cm body size—interestingly, the size of the largest insects (rhinoceros beetles)—spiracles and tubes begin to take up too much space, particularly in legs. This idea is supported by the fossil record. Some 300 million years ago, Earth's atmosphere was about 35% O_2 (instead of the 21% today), which would have boosted tracheal O_2 deliv-

ery. At that time, there were large insects including giant dragonflies with wingspans of almost a meter!

A rhinoceros beetle in Costa Rica.

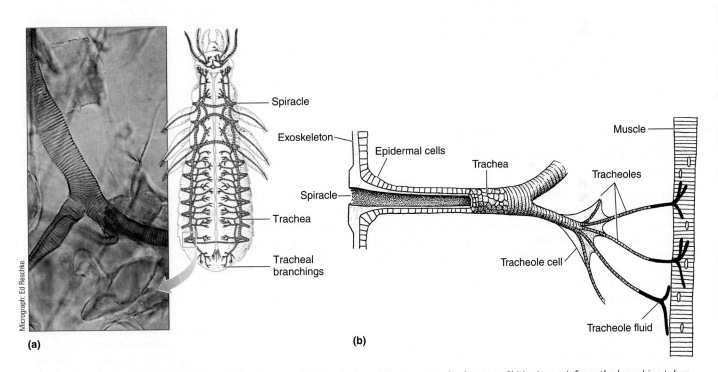

(a)

(b)

FIGURE 11-8 Respiratory system of an insect. (a) Generalized body plan of the insect tracheal system. Chitin rings reinforce the branching tubes of the trachea and prevent their collapse. (b) Oxygen entering the trachea moves through the tracheoles to fluid-filled tips located in specific target tissues. An increased number of tube tips are found on muscle and other tissues with high O_2 demands.

Sources: C. Starr & R. Taggart. (2004). *Biology: The Unity and Diversity of Life*, 10th ed., Belmont, CA: Brooks/Cole, p. 709 (Figure 40.5); (b) E. E. Ruppert, R. S. Fox, & R. D. Barnes. (2004). *Invertebrate Zoology: A Functional Evolutionary Approach*, 7th ed. Belmont, CA: Brooks/Cole, Figure 21-10c.

represent an additional site of O_2 pickup. Let's examine the evolution of these airways and lungs in more detail.

The first air-breathing vertebrates were bimodal

Vertebrate air breathing is believed to have been favored among those animals living in habitats similar to modern tropical lowlands, where warm water and seasonal drought

results in stagnation of ponds leading to aquatic *hypoxia* (low oxygen), and even desiccation. The first evolutionary transition from aquatic to terrestrial life produced distinct fishes with a unique set of adaptations to respiring in both media. Some had modified gills (such as air-breathing catfish, with more arch support to prevent gill collapse), but others were **bimodal breathers** with, in addition to gills in water, the ability to breathe air by integumentary exchange (such as some eels), parts of the digestive system including oral or stomach linings, or separate air sacs called *lungs.* Bimodal breathing is used by some fishes in waters that become anoxic (such as stagnant swamps) or in ponds that dry out seasonally. In fact, the mangrove rivulus (*Rivulus marmoratus*) a tropical killifish, can survive for over two months without water and food. When their habitat dries up, they seek refuge in logs hollowed out by insects as they await a subsequent rainfall. In addition, some have gills, lungs, *and* some integumentary (skin) gas exchange with the result that they are actually **trimodal breathers** (such as the reedfish *Calamoichthys calabaricus*).

The most successful vertebrate groups in terms of terrestrial adaptation are those that evolved true lungs. Lungs in fishes originally evolved as a simple ventral (internal) evagination of the pharynx, and in some groups evolved into a dorsal gas bladder (see p. 539 and Figure 11-32). Eventually the gas bladder transformed into a structure specialized for both sound reception and buoyancy control and in many cases lost its respiratory function. In at least one group of fishes, however, the air sac became paired and ventrally oriented, leading eventually to the lungs of terrestrial amphibians, reptiles, birds, and mammals. The tube connecting the pharynx to the lungs became the **trachea** (windpipe), splitting into two primary **bronchi** for the paired lungs (the anatomy found in most terrestrial vertebrates).

Frog lungs may be simple or moderately folded and are inflated by positive pressure

Amphibians, the first class of vertebrates to evolve from lung-bearing fishes, are noted for their bimodal lifestyles, living both in the water and air. Aquatic and terrestrial amphibians rely to varying degrees on integuments, gills, and lungs for gas exchange. For example, some species of salamanders are without lungs and are solely dependent on cutaneous gas exchange. In these animals the skin is moist and well vascularized. At the other extreme, one salamander (*Siren lacertina*) has trimodal breathing.

In frogs, most larval stages have gills, whereas most adults have lungs that are comparatively simple and noncompartmentalized (Figure 11-9a). Air enters these lungs from the pharynx via the trachea and bronchi; note that the pharynx also branches into the **esophagus,** the tube through which food passes to the stomach, reflecting the way that lungs first evolved. In larger species needing greater surface area because of scaling problems, the lung is folded into alveoli. However, alveoli are larger and less numerous than in vertebrates with higher metabolisms. Air is forced by *positive pressure* into their lungs by a **buccal pump** (originating in the mouth) that is similar in function to that used in fish (Figure 11-10a–d). Normally, several inspiratory oscillations are required to completely fill the lungs, whereas one long exhalation empties the lungs. These animals are bi-

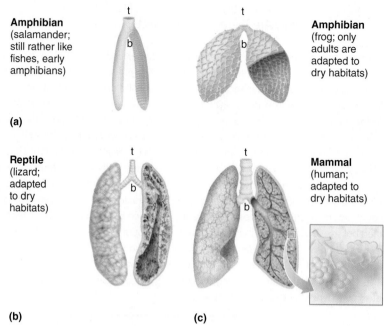

FIGURE 11-9 Comparative lung structure of amphibians, reptiles, and mammals. (a) Amphibian lungs range from comparatively simple sacs surrounded by few blood vessels to pouches divided into many large compartments with dense vascularization. The latter is helpful for terrestrial amphibians that spend more time on land to reduce the use of integumentary exchange. (b) Lungs of a typical reptile rely on a bellowslike pumping of posterior air sacs that help airflow into and out of the lungs. Gas exchange occurs in a forward, spongy region of each lung. (c) Highly compliant mammalian lungs rely on gas-exchanging alveoli (lower right-hand box). t = trachea, b = bronchus

Source: C. Starr & R. Taggart. (2004). *Biology: The Unity and Diversity of Life,* 10th ed., Belmont, CA: Brooks/Cole, p. 711, Figure 40.8.

modal skin-lung breathers (Figure 11-10e), but O_2 uptake is primarily through the lung, whereas CO_2 elimination is almost exclusively cutaneous. Terrestrial caecilians (tropical, legless amphibians), which spend most of their time in burrows, have single, elongate lungs that extend almost 70% of their body length. Structurally, caecilian lungs have a high degree of internal compartmentalization, forming numerous alveoli that markedly increase the available surface area for gas exchange.

In bimodal gill–lung breathers, branchial (gill) contractions alternate between periods of pumping water through the gills to pumping air into the lung. In the ancestors of the first exclusively air breathers, water pumping was completely abandoned with the buccal muscles exclusively functioning as a positive-pressure air pump to inflate the lungs. As the dependence on branchial O_2 uptake decreased, the gills and branchial arteries became reduced and internalized. The diffuse, externally oriented chemoreceptors for monitoring gas concentrations located in the gills evolved into discrete arterial chemoreceptors located near the carotid artery and in the aortic arch (see the section on breathing regulation later). Adaptations to air breathing also included a decrease in the affinity of *hemoglobin* for O_2 that reflected the much greater O_2 availability in air (see Section 11.9 later).

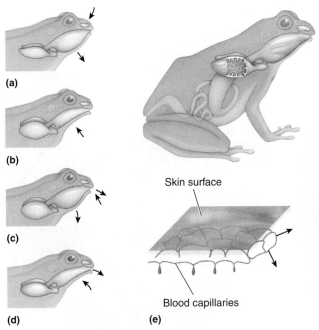

(a)

(b)

(c)

Skin surface

(d)

(e)

Blood capillaries

FIGURE 11-10 Respiration in a frog. (a) During inspiration, the frog lowers the floor of its mouth and inhales air through its nostrils into the buccal cavity. (b) The frog then closes its nostrils, opens the glottis, and raises the mouth's floor. This action forces air into the lungs. (c) Gas exchange occurs in the lungs. The nostrils open at the end of this phase. (d) Air is forced out of the lungs when muscles in the body wall above the lungs contract and the lungs elastically recoil. Lung filling generally requires several successive active inflation cycles, whereas lung deflation is usually accomplished by a single passive expiration. (e) Integumentary "open" exchange in a frog, showing blood flowing in capillaries underneath the skin.

Sources: (a to d): C. Starr & R. Taggart. (2004). *Biology: The Unity and Diversity of Life,* 10th ed. Belmont, CA: Brooks/Cole, p. 711, Figure 40.7); (e): Phillip C. Withers. (1992). *Comparative Animal Physiology.* Belmont, CA: Brooks/Cole, p. 574, Figure 12-6.

Reptiles and mammals have respiratory systems ranging from simple to elaborately folded chambers inflated by negative pressure

Compartmentalized lungs are an important adaptation in some reptiles as well as birds and mammals, but what separates these groups from all others is their mechanism of ventilation. Instead of forcing air from the buccal cavity into the lungs (amphibians and some reptiles), air flows into the lungs because of a *negative pressure*. Various muscles forcefully *expand* the lungs creating a subatmospheric pressure that draws air inward. You will later see how negative pressure and these muscles work, but note that all work indirectly, that is, they do not connect directly to the lungs themselves.

Most Reptiles Except for birds and crocodilians, reptile lungs are expandable tidally ventilated sacs, which may be simple or with several large sacs and even some alveoli (Figure 11-9b). The simplest reptilian lung is somewhat analogous to a single, oversized alveolus. Vascularized ingrowths, or **septae** (dividing walls or partitions), penetrate centrally from the lung's perimeter and subdivide the pulmonary lu-

men into a series of spatial units called **ediculae** (if they are as wide as they are deep, as in tortoises, chameleons, and geckos) or **faveoli** (if they are deeper than they are wide, as in snakes and iguanids). Gas exchange occurs principally on the septae, although they are poorly vascularized compared to alveoli. Unlike mammalian alveoli, ediculae are relatively passive participants during ventilation and do not contribute to the movement of air during inhalation and exhalation. Airflow during ventilation is tidal and bellowslike, as you will see for mammals later.

Lizards and snakes rely solely on **costal** (rib muscle) ventilation resulting from simple rib-cage expansion to create negative pressure for inhalation, and contraction for exhalation. They work indirectly in ways similar to the mammalian system, to be described shortly. Animals whose rib cage is fixed rely on alternative solutions to expand their lungs. Tortoises, for example, must extend their limbs to expand the thoracic cavity.

Crocodilians Crocodiles, caymans, and alligators were long thought to have a tidally ventilated lung, but a recent study by C. G. Farmer and K. Sanders reveals that alligators, at least, have a birdlike *flow-through* system! Inhaled air enters the lungs via the trachea and bronchi, but instead of becoming a dead-end sac, each bronchus branches into secondary bronchi and then into very small tertiary or *parabronchi*, where gas exchange takes place. Importantly, the air continues in one direction into other secondary bronchi that reenter the original bronchus for exhalation. Though tidal in the initial airways, flow is unidirectional at the gas-exchange sites. We will explore the similar but more elaborate flow-through parabronchi of birds later.

For ventilation in crocodilians, contraction of a **diaphragmaticus muscle** (attached between the liver and a distinctive elongated region of the pubis bone) pulls the liver away from the lungs in a pistonlike manner that results in expansion of the lungs. The crocodilian **diaphragm** consists of a sheet of nonmuscular connective tissue that adheres tightly to the dome-shaped anterior surface of the liver. Exhalation is accomplished by abdominal muscles, which push the liver against the lungs. You will see how similar muscles work in mammals shortly.

Birds and Mammals To meet the energy demands of birds and mammals with their increased metabolic rates, the surface area of the lung available for gas exchange had to markedly increase. The mammalian solution was the elaboration of numerous very small alveoli (Figures 11-9c and 11-11), whereas the avian respiratory system, which is the most complex of the vertebrates, evolved parabronchi with *air capillaries*. The cardiovascular system became increasingly modified to perfuse the gas exchange organ. Oxygenated blood from the lung achieved complete separation from systemic venous blood with a four-chambered heart (also found in crocodilians; Chapter 9). With the evolution of these two adaptations both groups were now able to exploit lifestyles that could make use of their high metabolic rates.

Most mammals are bimodal due to integumentary gas exchange, as exemplified by bats: in flight they are capable of eliminating almost 12% of their CO_2 across the surface of their wings. Approximately 1 to 2% of the total CO_2 and O_2 in many other mammals can be exchanged directly through the skin. This one signal by which a mosquito finds

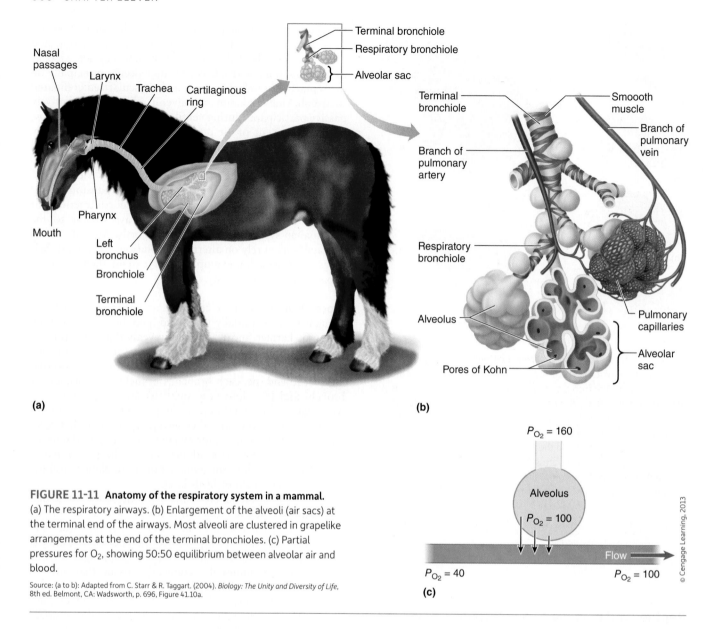

(a)

(b)

(c)

$P_{O_2} = 160$

Alveolus

$P_{O_2} = 100$

$P_{O_2} = 40$

Flow

$P_{O_2} = 100$

© Cengage Learning, 2013

FIGURE 11-11 Anatomy of the respiratory system in a mammal.
(a) The respiratory airways. (b) Enlargement of the alveoli (air sacs) at the terminal end of the airways. Most alveoli are clustered in grapelike arrangements at the end of the terminal bronchioles. (c) Partial pressures for O_2, showing 50:50 equilibrium between alveolar air and blood.

Source: (a to b): Adapted from C. Starr & R. Taggart. (2004). *Biology: The Unity and Diversity of Life,* 8th ed. Belmont, CA: Wadsworth, p. 696, Figure 41.10a.

its victims, and a tick detects a potential host, including humans. Carbon dioxide release serves as the primary attractant as the tick scuttles its way up the concentration gradient. A greater CO_2 production rate increases the likelihood of a particular individual becoming the tick's next meal!

Mammalian airways terminate in alveoli, which are involved in both ventilation and gas exchange

Mammalian airways begin with the **nasal passages** (in a **nose** or blowhole) (Figure 11-11a). Adaptations for running in the horse and cheetah include nostrils that are pliable and easily dilated to increase the volume of air inspired. Concomitantly with the evolution of endothermy, the evolution of **maxilloturbinals,** thin curls of bone deep within the nasal cavity, was favored to retain heat and water (see Chapter 15). The nasal passages open into the **pharynx** (throat), which serves as a common passageway for both the respiratory and digestive systems. Again, reflecting evolutionary

ancestry, two tubes lead from the pharynx—the trachea and the esophagus. Because the pharynx serves as a common passageway for food and air, reflex mechanisms exist to close off the trachea during swallowing so that food enters the esophagus and not the airways. The esophagus remains closed except during swallowing to prevent air from entering the stomach during breathing. The trachea is lined with two kinds of cells: mucus-secreting cells, which lubricate it, and ciliated cells lined with tiny hairs that continually beat upward, pushing impurities such as inhaled particles and dust, up and out of the trachea.

Again, the trachea divides into right and left bronchi, each entering one lung. Within each lung, the bronchus continues to branch into progressively narrower, shorter, and more numerous airways called **bronchioles,** much like the branching of a tree. Clustered at the ends of the terminal bronchioles in mammals are the alveoli (Figure 11-11b).

To permit airflow in and out of the gas-exchanging portions of the lungs, the continuum of conducting airways, from the entrance through the terminal bronchioles to the alveoli, must remain open most or all of the time. The trachea and larger bronchi in most mammals are fairly rigid, nonmuscular tubes encircled by a series of cartilaginous rings that prevent the tubes' compression. For example, the trachea of a giraffe is supported along its length by more than a hundred of these tracheal rings. The smaller bronchioles have no cartilage to hold them open. Their walls contain smooth muscle that is innervated by the autonomic nervous system and is sensitive to certain hormones and local chemicals. These factors, by varying the degree of contraction of bronchiolar smooth muscle and hence the caliber of these small terminal airways, are able to regulate the amount of air passing between the atmosphere and each cluster of alveoli. We'll examine their regulation later.

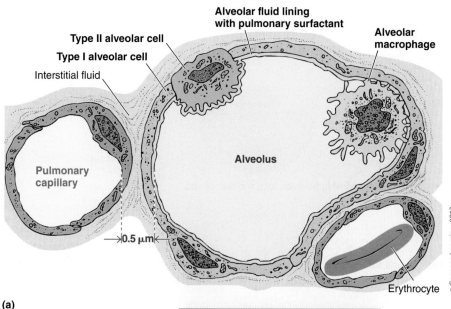

(a)

© Cengage Learning, 2013

Alveoli These tidally ventilated sacs cannot have countercurrent exchange with blood, so they can only achieve equal partial pressures of gases in air and blood P_{O_2} (compare Figure 11-11c to the gill countercurrent in Figure 11-6f, in which arterial gas pressure is *higher* than exhaled water's gas pressure). However, in other ways, alveoli are well-adapted for gas exchange. Recall that, according to Fick's law, enhancements arise by decreasing the distance for diffusion (ΔX), and increasing the surface area (A). The alveoli accomplish both of these enhancements well, as readily visible in Figure 11-12. The alveolar walls consist of a single layer of highly flattened **Type I alveolar cells** (Figure 11-12a). The linings of the dense network of pulmonary capillaries encircling each alveolus also consist of a single layer of very thin cells. The interstitial space between an alveolus and the surrounding capillary network is also an extremely thin barrier, with only 0.25 µm separating the air in the alveoli from the blood in the pulmonary capillaries. (A sheet of tracing paper is about 50 times thicker than this air-to-blood barrier.)

Furthermore, the alveolar air–blood interface presents a tremendous surface area for exchange. The lungs of a human contain about 300 million alveoli in the lungs, each about 300 µm (11 mm) in diameter. So dense are the pulmonary capillary networks that each alveolus is encircled by an almost continuous sheet of blood (Figure 11-12b). The total surface area thus exposed between alveolar air and pulmonary capillary blood is about 75 m² (about the size of a tennis court). In contrast, if the lungs consisted of a single, hollow chamber of the same dimensions instead of being divided into myriad alveolar units, the total surface area would be only about 0.01 m².

The alveolar surface area varies among mammals. One of the hallmarks of *scaling* (p. 10) in animals is that smaller species have higher basal metabolic rates, that is, they require more energy per unit mass. As you will see in Chapter 15, this phenomenon in birds and mammals is traditionally attributed to the higher surface-area-to-volume ratio of small species, with concomitant high rate of heat loss. To

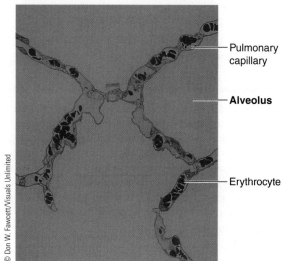

© Don W. Fawcett/Visuals Unlimited

(b) Transmission electron micrograph of several alveoli and surrounding pulmonary capillaries

FIGURE 11-12 Alveolus and associated pulmonary capillaries (mammalian lung). (a) A schematic representation of a detailed electron microscope view of an alveolus and surrounding capillaries. A single layer of flattened Type I alveolar cells forms the alveolar walls. Type II alveolar cells embedded within the alveolar wall secrete pulmonary surfactant. Wandering alveolar macrophages are found within the alveolar lumen. (The size of the cells and respiratory membrane is exaggerated compared to the size of the alveolar and pulmonary capillary lumens. The diameter of an alveolus is actually about 600 times larger than the intervening space between air and blood.) (b) Each alveolus is encircled with a dense network of pulmonary capillaries and therefore surrounded by an almost continuous sheet of blood.

compensate, smaller mammals can take up proportionately more O_2 per unit of body weight than large mammals by (1) reducing the size of each alveolus with a concomitant increase in the density (and thus total surface area) of alveoli, and (2) increasing the density of capillaries for gas exchange. You can see the result in *scaling relationships* that

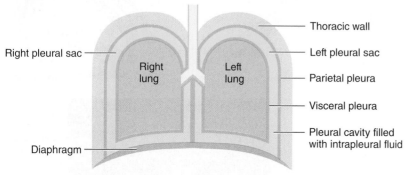

(a) Analogy of relationship between lung and pleural sac

(b) Relationship of lungs to pleural sacs, thoracic wall, and diaphragm

FIGURE 11-13 Pleural sac. (a) Pushing a lollipop into a water-filled balloon produces a relationship analogous to that between each double-walled, closed pleural sac and the lung that it surrounds and separates from the thoracic wall. (b) Schematic representation of the relationship of the pleural sac to the lungs and thorax. One layer of the pleural sac, the *visceral pleura,* closely adheres to the surface of the lung (*viscus* means "organ"), then reflects back on itself to form another layer, the *parietal pleura,* which lines the interior surface of the thoracic wall (*paries* means "wall"). The relative size of the pleural cavity between these two layers is grossly exaggerated for the purpose of visualization.

© Cengage Learning, 2013

have been determined by actual measurements of lungs. Among mammals, lung volume (V_l, in liters) is directly related to body weight by the equation

$$V_l = 0.035 m_b^{1.06}$$

where m_b = body mass (kilograms). The exponent 1.06 tells us that larger mammals actually have slightly *larger* lung volumes per unit mass than do smaller species. However, alveolar surface area (A_a, in cm^2) follows the following relationship:

$$A_a = 119 m_b^{0.75}$$

This exponent of 0.75 (3/4) tells us that larger animals have comparably reduced surface areas for gas exchange. This value is also essentially identical to the exponent for basal metabolic rate (as you will see in Chapter 15), showing that gas exchange ability precisely matches metabolic needs.

In addition to the thin, wall-forming Type I cells, the alveolar epithelium also contains **Type II alveolar cells** (Figure 11-12a), which secrete **pulmonary surfactant,** a phospholipoprotein complex that facilitates alveolar expansion (described later). Surfactant can be found in the lungs of amphibians, reptiles, birds, and mammals as well as the gas bladders of fish. Also present within the air sac lumen are defensive *macrophages* (p. 465).

Minute **pores of Kohn** (Figure 11-11b) present in the alveolar walls permit airflow between adjacent alveoli, a process known as **collateral ventilation.** These passageways are especially important in allowing fresh air to enter an alveolus whose terminal conducting airway is blocked because of disease.

A pleural sac separates each lung from the thoracic wall

Mammalian lungs occupy most of the volume of the **thoracic (chest) cavity,** the only other structures in the chest being the heart and associated vessels, the esophagus, the thymus, and some nerves. Separating each lung from the thoracic wall and other surrounding structures is a double-walled, closed sac called the **pleural sac** (Figure 11-13). The interior of the pleural sac is known as the **pleural cavity.** Its size is greatly exaggerated in the illustration to aid visualization; in reality the layers of the sac are in close contact with one another. The surfaces of the pleura secrete a thin **intrapleural fluid,** which lubricates the pleural surfaces as they slide past each other during respiratory movements. The pleural cavity plays a crucial role in the flow of air, as you will see shortly.

Surprisingly, elephants are without a pleural cavity and instead their lungs are surrounded by loose connective tissue. A clue to explain this mystery arises from the chapter opener photograph of the "snorkeling" elephant. Elephants are the only mammal that can remain underwater while breathing at depth, using their trunks as snorkels. The presence of a pleural cavity would create a significant pressure difference between the interior of the lungs and the blood vessels lining the cavity. Because the vessels would also be exposed to the higher pressures from the surrounding water, the pressure gradient could be sufficient to cause rupture of the vessels (at 2 meters under water, pressure would be about 20% greater than in the lungs). By replacing the pleural membranes with layers of connective tissue minus a vascular component, the loose tissue permits the lung to expand against the chest wall while swimming. Molecular evidence reveals that elephants are related to manatees, while paleontology studies suggest that elephants evolved from an aquatic ancestor to a land animal.

Mammalian lungs are inflated and deflated tidally by cyclical changes in intra-alveolar pressure generated indirectly by respiratory muscles

Now let's see how ventilatory bulk transport works in mammalian lungs. Alveoli have no muscles to help them inflate and deflate during the breathing process. Moreover, respiratory muscles that accomplish breathing do not act directly on the alveoli or even the lung organ to change their volume. Instead, these muscles change the volume of the thoracic cavity, causing a corresponding change in lung volume because the thoracic wall and lungs are linked together by the pleural sac.

Let's follow the changes that occur in a mammal during one respiratory cycle—that is, one breath in (*inspiration*) and out (*expiration*).

Inspiration: Contraction of Inspiratory Muscles Before the beginning of inspiration, the respiratory muscles are relaxed (Figure 11-14a), no air is flowing, and **intra-alveolar pressure** (pressure inside the alveoli) is equal to atmospheric pressure. At the onset of inspiration, these muscles are stimulated to contract, resulting in enlargement of the thoracic cavity. The major *inspiratory muscles* are the *diaphragm* and *external intercostal muscles* (Figure 11-14b):

1. The relaxed **diaphragm** assumes a dome shape that protrudes into the thoracic cavity. When the diaphragm contracts on stimulation by the phrenic nerve, it contracts downward, thereby enlarging the volume of the thoracic cavity (Figure 11-14b). Contraction of the muscular diaphragm can account for up to two thirds of the increase in pulmonary volume in mammals.

2. Whereas contraction of the diaphragm enlarges the thoracic cavity in the vertical dimension, contraction of the **external intercostal muscles,** whose fibers run downward and forward between adjacent ribs, enlarges the thoracic cavity in both the lateral (side-to-side) and anteroposterior (front-to-back) dimensions (Figure 11-14b). They do so by expanding the ribs and subsequently the sternum outward.

Exhalation: Elastic Recoil and Relaxation of Inspiratory Muscles At the end of inspiration (Figure 11-14c), the inspiratory muscles relax. The diaphragm rebounds to its original dome-shaped position as it relaxes; the expanded rib cage contracts when the external intercostals relax; and the chest wall and stretched lungs recoil to their pre-inspiratory size because of their elastic properties, much as a stretched balloon would on release (we will examine recoil properties on p. 520). As the lungs become smaller in volume, the intra-alveolar pressure rises because the greater number of air molecules contained within the larger lung volume at the end of inspiration are now compressed into a smaller volume. Air now leaves the lungs down its pressure gradient. Outward flow of air ceases when intra-alveolar pressure becomes equal to atmospheric pressure.

Exhalation: Contraction of Expiratory Muscles In most mammals, the inspiratory and expiratory phases are cyclical and smooth. Expiration in humans is normally a *passive* process during quiet breathing because it is accomplished by elastic recoil of the lungs on relaxation of the inspiratory muscles, with no muscular exertion or energy expenditure required. In contrast, inspiration is *active* because it is brought about by contraction of inspiratory muscles at the expense of energy use. To empty the lungs more completely and more rapidly than is accomplished during quiet breathing, as during the deeper breaths accompanying locomotory activity, expiration does become active in mammals. The intra-alveolar pressure must be increased even further above atmospheric pressure than can be accomplished by simple relaxation of the inspiratory muscles and elastic recoil of the lungs. To produce such a **forced,** or **active expiration,** two sets of **expiratory muscles** must contract to further reduce the volume of the thoracic cavity and lungs (Figure 11-14d):

1. The most important expiratory muscles are the **abdominal wall muscles.** As the abdominal muscles contract, the resultant increase in intra-abdominal pressure exerts a force on the diaphragm, pushing it further into the thoracic cavity than in its relaxed position, thus decreasing the vertical dimension of the thoracic cavity even more.

2. The other expiratory muscles are the **internal intercostal muscles,** whose contraction pulls the ribs inward, flattening the chest wall and further decreasing the size of the thoracic cavity; this action is antagonistic to that of the external intercostal muscles (Figure 11-14b, c). These muscles increase the differential between intra-alveolar and atmospheric pressure above that of passive expiration, so more air leaves down the pressure gradient before equilibration is achieved. In this way, the lungs are emptied more completely during forceful, active expiration than during quiet, passive expiration.

This general pattern of respiration is somewhat different in the horse, where both the inspiratory and expiratory phases have two separate components even at rest. Inspiration begins from passive recoil but rapidly becomes an active process with contributions from the diaphragm, external intercostals, and (again in horses) the *triangularis sterni* connected to the sternum. Similarly expiration begins passively by recoil and rapidly becomes active as the internal intercostals and abdominal muscles become involved. This pattern of respiration occurs because a portion of the energy spent during both active inspiration and expiration is stored and used as passive recoil energy to initiate the subsequent breathing cycle. Both inspiration and expiration thus have an active as well as a passive component.

Ventilation and gas exchange are separated in avian respiratory systems due to air sacs and parabronchi with air capillaries

Finally, let's see how the most complex vertebrate lung works. A greater efficiency is achieved in birds compared to mammals due to a complete *separation of ventilation and gas-exchange functions.* In contrast to those of mammals, the lungs of birds only perform gas exchange and are comparatively small and inelastic (and thus do not change in volume during the respiratory cycle). Instead, ventilation arises from expandable **air sacs** (Figure 11-15a, b) that perform a bellowslike tidal function comparable to the mammalian lung but without gas exchange. As you will see, this separation of function allows tidal flow to be converted into a flow-through pattern in the lungs, which results in an increase in exchange efficiency.

Again, air enters the nasal passages, the trachea, and bronchi. The tracheal volume of a bird is considerably larger than that of a comparably sized mammal. To perform some functions assigned to forelimbs, birds can have necks up to threefold longer than mammals of comparable body size.

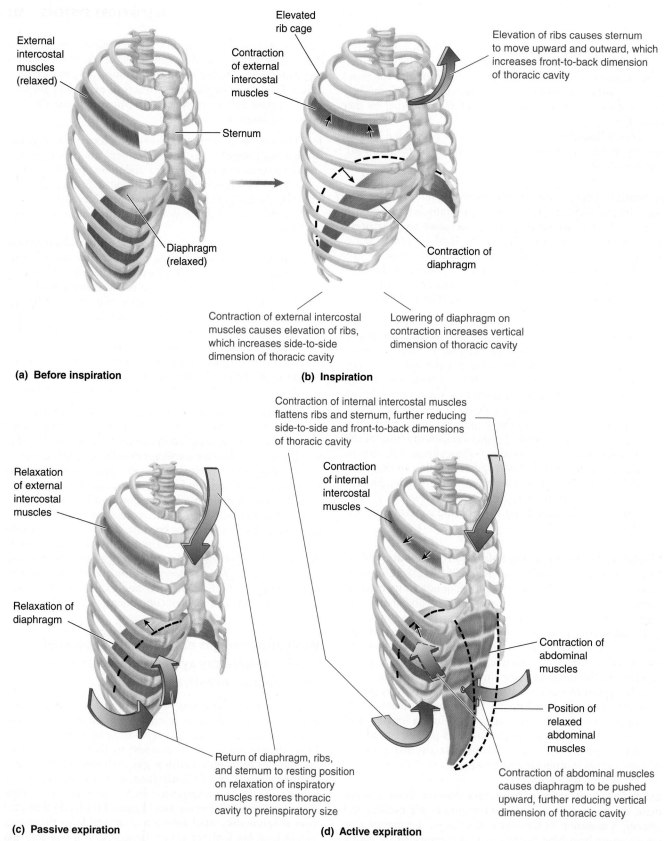

(a) Before inspiration

External intercostal muscles (relaxed)

Sternum

Diaphragm (relaxed)

(b) Inspiration

Elevated rib cage

Contraction of external intercostal muscles

Elevation of ribs causes sternum to move upward and outward, which increases front-to-back dimension of thoracic cavity

Contraction of diaphragm

Contraction of external intercostal muscles causes elevation of ribs, which increases side-to-side dimension of thoracic cavity

Lowering of diaphragm on contraction increases vertical dimension of thoracic cavity

(c) Passive expiration

Relaxation of external intercostal muscles

Relaxation of diaphragm

Return of diaphragm, ribs, and sternum to resting position on relaxation of inspiratory muscles restores thoracic cavity to preinspiratory size

(d) Active expiration

Contraction of internal intercostal muscles flattens ribs and sternum, further reducing side-to-side and front-to-back dimensions of thoracic cavity

Contraction of internal intercostal muscles

Contraction of abdominal muscles

Position of relaxed abdominal muscles

Contraction of abdominal muscles causes diaphragm to be pushed upward, further reducing vertical dimension of thoracic cavity

FIGURE 11-14 Respiratory muscle activity during inspiration and expiration in a human. (a) Before inspiration, all respiratory muscles are relaxed. (b) During *inspiration*, the diaphragm descends on contraction, increasing the vertical dimension of the thoracic cavity. Contraction of the external intercostal muscles elevates the ribs and subsequently the sternum to enlarge the thoracic cavity from front to back and from side to side. (c) During *quiet passive expiration*, the diaphragm relaxes, reducing the volume of the thoracic cavity from its peak inspiratory size. As the external intercostal muscles relax, the elevated rib cage falls because of the force of gravity. This also reduces the volume of the thoracic cavity. (d) During *active expiration*, contraction of the abdominal muscles increases the intra-abdominal pressure, exerting an upward force on the diaphragm. This reduces the vertical dimension of the thoracic cavity further than it is reduced during quiet passive expiration. Contraction of the internal intercostal muscles decreases the front-to-back and side-to-side dimensions by flattening the ribs and sternum.

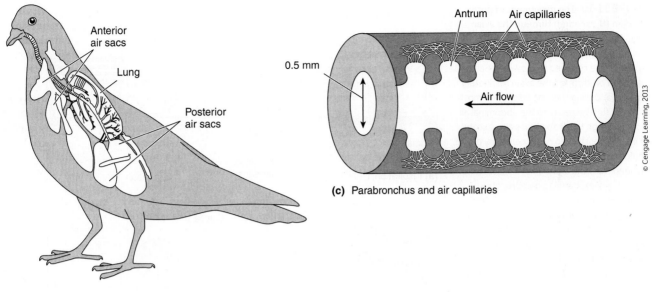

(c) Parabronchus and air capillaries

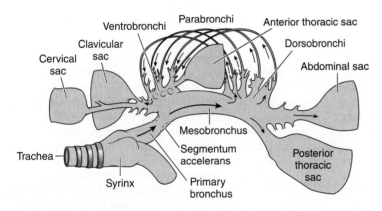

(a) Lungs and air sacs

(b) Schematic of air sacs and bronchi showing aerodynamic valve (segmentum accelerans; see text p. 515)

FIGURE 11-15 Respiratory system of a bird. (a) A lateral view of the major anterior (cranial) and posterior (caudal) air sacs. (b) A lateral view of major components including the parallel parabronchi. (c) A section through a parabronchus, showing the air capillaries branching off from side chambers called *antra* (*antrum*, singular).

Sources: (a) K. Liem, W. Bemis, W. F. Walker, & L. Grande. (2001). *Functional Anatomy of the Vertebrates: An Evolutionary Perspective*, 3rd ed. Belmont, CA: Brooks/Cole, Figure 18-17; (b) N. Wang, R. B. Banzett, C. S. Nations, & F. A. Jenkins, Jr. (1992). An aerodynamic valve in the avian primary bronchus. *J. Exp. Zool.* 262: 441–445.

The longer trachea increases the resistance to airflow, but this is compensated, in part, by an increase in the radius of the trachea. For this reason the resistance to tracheal airflow in birds is similar to a mammal of similar body weight.

Within each lung a bronchus gives rise to several sets of secondary bronchi, named according to the regions of the lung that they supply. The most important groups include the *ventrobronchi* and the *dorsobronchi*. As in alligators, the secondary bronchi are connected to each other by a number of *parabronchi* from which the **air capillaries** (gas exchange surface) originate (Figure 11-15c). Parabronchi interconnect freely with each other and tend to have a constant internal diameter, ranging from 0.5 mm in hummingbirds to approximately 2 mm in chickens. Parabronchi are cylindrical and run in parallel with each other (shown schematically in Figure 11-15b). Comparable to the bronchial muscles of mammals, smooth muscles lining the parabronchus are innervated and regulate the diameter of the parabronchi.

Together, the secondary bronchi and their interconnecting parabronchi form an integrated unit that enables flow-through air movement. Airflow is from the dorsobronchi, through the parabronchi and air capillaries, to the ventrobronchi. Recall that flow-through systems are more efficient; indeed, in birds air is not diluted with "old" air as in mammalian alveoli. Shortly you will see how the airflow is achieved.

Air Capillaries In birds, the air capillaries form an extensive network of air-carrying tubules jacketed with a profuse network of blood capillaries. A comparison of the gas exchange surfaces between birds (Figure 11-15c) and mammals (Figure 11-11a) reveals the following:

1. *Narrow size:* The diameter of an air capillary varies from only 3 μm in songbirds to 10 μm in swans and penguins, compared to the 35 μm of alveoli in the smallest mammals. The mean thickness of the epithelial cells lining the air capillaries is also lower than in alveoli. Thus, the diffusion distance ΔX is at least 30% less than in mammals.

2. *Flow-through:* Air capillaries are not blind ending; rather, they freely anastomose (connect) with each other. This allows air to flow through in one direction.

3. *Rigidity:* Unlike alveoli, the diameter of air capillaries does not change significantly during the respiratory cycle. This allows them to be rigid and thus to resist damage more effectively than alveoli can. Galloping horses,

FIGURE 11-16 The movement of pure nitrogen (N_2; shaded) through the avian lung. N_2 is used as an inert tracer. Two complete cycles of inspiration and expiration are required to move a specific volume of air through the avian respiratory system. (a) During the first inspiration most of the N_2 flows directly to the abdominal air sacs. Although the cranial sacs also expand on inhalation, they do not receive any of the inhaled N_2. (b) On the first exhalation, nitrogen from the abdominal air sacs flows into the gas-exchanging parabronchus. (c) On the subsequent inhalation, N_2 from the lung now moves into the cranial air sacs. (d) Finally, on the second exhalation, N_2 from the cranial air sacs flows through the trachea to the outside.

Source: K. Liem, W. Bemis, W. F. Walker, & L. Grande. (2001). *Functional Anatomy of the Vertebrates: An Evolutionary Perspective*, 3rd ed. Belmont, CA: Brooks/Cole, p. 594, Figure 18-18.

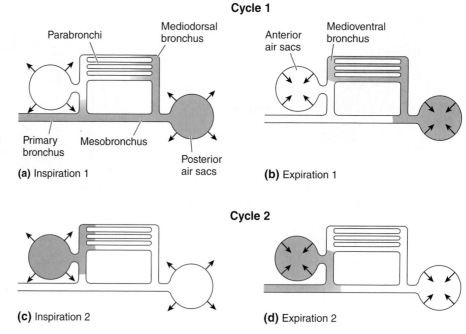

for example, often suffer ruptured alveolar blood capillaries due to the stretch-recoil action of alveoli, but birds do not have such problems.

4. *Greater blood volume:* The volume of the pulmonary-capillary blood per gram of body weight is approximately 20% greater in birds.

Air Sacs In most birds the air sacs arise in the embryo from six primordial pairs of sacs, two of which fuse at the time around hatch. Thus, there are eight air sacs in the adult domestic fowl: one cervical and one clavicular sac, two cranial thoracic and two caudal thoracic sacs in addition to the two abdominal sacs. Functionally, though, there are two groups of air sacs (Figure 11-15a, b), an *anterior* (or *cranial*) group and a *posterior* (or *caudal*) group. The anterior group consists of several smaller sacs that connect with the ventro-bronchi and receive expelled air from the lungs, whereas the posterior group, which includes the comparatively large abdominal sacs, connects to the caudal end of the main bronchus and receives relatively fresh air from the trachea. Because the blood supply to the air sacs is essentially negligible, no gas exchange occurs across this surface (as we noted earlier). Air sacs are held open by their attachment to the structures that surround them and occasionally even penetrate the medullary (inner portion) cavities of some bones. The air sacs are also a common site of disease in birds. Generally, inspired pathogens lodge in the posterior air sacs because of their role in receiving freshly inspired air.

Flow-through breathing plus crosscurrent blood flow enhances gas exchange in birds

Now let us see how these structures achieve efficient gas exchange. First, the sacs must inflate and deflate tidally, much like mammalian lungs. Birds do not have a muscular diaphragm but instead have a **membranous diaphragm** that is attached to the body wall by muscles. Birds rely primarily

on a highly modified costal (rib) movement in association with a well-developed sternum (the long, flat breastbone) to power air-sac ventilation. In addition, in some species of birds the joints of the ribs permits fore–aft movement of the rib cage during ventilation. As a result, avian ribs rotate such that the posterior end of the sternum is depressed on inhalation and generates negative intra-abdominal pressures that helps fill the comparatively large abdominal air sacs.

Although they do not need to be as efficient as fishes because air is a much better medium for oxygen uptake, birds do require greater efficiency than mammals because of their higher metabolisms. This occurs at three levels:

1. *Narrower diffusion distance between air and blood capillaries*, as described earlier.
2. *Flow-through movement in the parabronchi and air capillaries*. What is particularly intriguing about bird respiration is precisely how the ventilatory muscles and air sacs are coordinated in order to achieve this highly efficient means of gas exchange. To comprehend this process, let's follow the passage of an inert component such as N_2 through the avian respiratory system (Figure 11-16). The inhaled N_2 gas is initially mixed with gases present in the dead air space of the trachea and primary bronchus of each lung. Then the posterior air sacs inflate for the first inspiration, and the N_2 flows into these air sacs (Figure 11-16a). Gases in these air sacs are rich in O_2 and relatively low in CO_2. The posterior sacs initiate the first expiration, with valving that directs the flow with the N_2 into the gas-exchanging parabronchi region of the lung (Figure 11-16b). Here the composition of the respiratory gases is modified to suit the metabolic needs of the bird. Next, the anterior sacs inflate to create a second inspiration, with the N_2 gas now drawn into those air sacs (Figure 11-16c). Finally, with second expiration resulting from the anterior sacs and valving control, the gas is expelled to the outside atmosphere (Figure 11-16d). Thus,

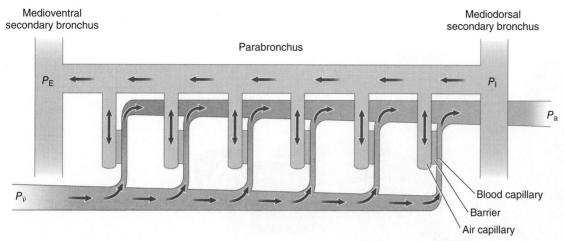

FIGURE 11-17   Crosscurrent flow in the avian parabronchus. Air (in *turquoise*) enters a parabronchus with a high P_{O_2} at the inhalation side on the right (P_I), and is exhaled with a low P_{O_2} (P_E). Blood enters on the left with low P_{O_2} (*blue* color at $P_{\bar{v}}$) then branches to run along air capillaries at a right angle to the parabronchus. Each successive branch picks up more gas because the air capillaries have successively increasing P_{O_2} levels, much as in the fish countercurrent system. Blood then leaves with a relatively high P_{O_2} (*red* color at P_a).

Source: Modified from P. Scheid. (1979). Mechanisms of gas exchange in bird lungs. *Rev. Physiol. Biochem. Pharmacol.* 86:137–186.

in birds, it requires two complete respiratory cycles to move air through the entire system. This remarkable pattern of gas flow through the avian lung is achieved through both aerodynamic mechanisms and valving properties of the air sacs. The end result is that gas flow is continuous and unidirectional from the mediodorsals through the parabronchi to the medioventrals.

The nature of the valving that ensures unidirectional flow is still not fully understood. There are no muscular sphincter valves that shut off flow. However there is evidence for *aerodynamic valving*, in which a bronchial constriction before a junction increases air velocity, much as narrowing a nozzle increases the speed of water out of a garden hose (see the *segmentum accelerans* (Figure 11-15b). The higher velocity air then jets past the inappropriate joining tubes without entering them.

3. *Crosscurrent blood flow at the parabronchus.* The anatomic arrangement of airflow through the gas-exchanging air capillaries in concert with the surrounding blood capillary network is such that blood flow approaches the parabronchus at approximately right angles, providing a **crosscurrent** system of gas exchange (Figure 11-17). Blood in the capillaries becomes oxygenated to varying degrees depending on how far along the parabronchus the blood capillary is located. As in fish gills (Figure 11-6f), and in contrast to alveoli (Figure 11-11c), the arterial blood P_{O_2} is actually *higher* than expired P_{O_2} because arterial blood P_{O_2} results from the mixing of the blood from the differently oxygenated blood capillaries (Figure 11-17).

unanswered Questions | Soaring birds from Saurischian dinosaurs?

The unique bird respiratory system has traditionally been thought to have evolved in concert with the energetic demands of flight. However, the recent finding of birdlike lungs in alligators (p. 507)

and recent fossil evidence suggest a different origin. Certain bones of flightless Saurischian dinosaurs such as *Haplocanthosaurus* had large holes very similar to the ones in bird bones that house their air sacs. Jonathan Codd, Phil Manning, and colleagues have also found that carnivorous Saurischians called *theropods*—from which birds are descended—had small birdlike bones that moved the ribs and sternum for breathing (an example was the feathered cheetah-like theropod *Velociraptor*).

Theropods also had a diaphragmaticus muscle for respiration like crocodiles. Moreover, Saurischians first evolved during a period (175–275 million years ago) when atmospheric O_2 levels were at about half of that found today. Paleontologist Peter Ward has hypothesized that Saurischians had birdlike flow-through ventilation, making them better adapted than mammals at gas exchange in that low-O_2 world. Perhaps birds owe their ability to soar over mountaintops in our current high O_2 era to these flightless progenitors who lived at sea level with O_2 levels equivalent to today's mountaintops. <<

Aerial respiratory systems can perform numerous nonrespiratory functions

As with gills, lungs and other respiratory structures can have nonrespiratory functions:

1. *Regulation of water loss and heat exchange.* Venous blood originating from the walls of the nasal passages can cool the arteries that supply the brain during intense activity. Inspired atmospheric air is humidified and warmed by the respiratory airways. Moistening of inspired air is essential to prevent the respiratory surfaces from drying out. O_2 and CO_2 cannot diffuse through dry membranes. The *maxilloturbinals* are especially effective for both purposes (Chapter 15, p. 745).

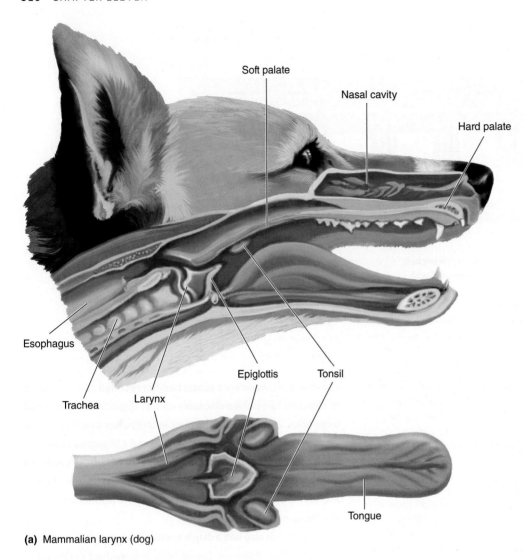

Soft palate

Nasal cavity

Hard palate

Esophagus

Trachea

Larynx

Epiglottis

Tonsil

Tongue

(a) Mammalian larynx (dog)

FIGURE 11-18 The mammalian larynx and avian syrinx. (a) The mammalian larynx sits at the junction of the pharynx and trachea, where it keeps food from entering the latter, and has folds for sound production. (b) The avian syrinx sits at the trachea–bronchi junction, and is also involved in sound production.

Source: www.biology.eku.edu/RITCHSO/ birdcommunication.html.

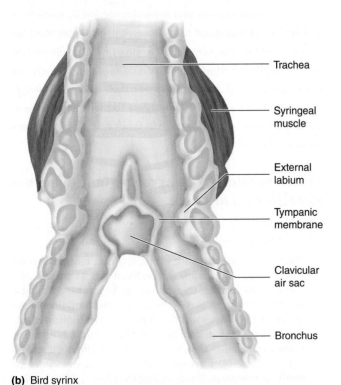

Trachea

Syringeal muscle

External labium

Tympanic membrane

Clavicular air sac

Bronchus

(b) Bird syrinx

2. *Circulation*—enhancement of venous return (see the "respiratory pump," p. 448).

3. *Acid–base balance*—the maintenance of normal acid–base balance by altering the amount of H^+-generating CO_2 exhaled (Chapter 13, p. 642).

4. *Defense* against inhaled foreign matter (Chapter 10, p. 467).

5. *Removal, modification, activation, or inactivation of various materials* passing through the pulmonary circulation. All blood returning to the heart from the tissues must pass through the avian and mammalian lungs before being returned to the systemic circulation. The lungs, therefore, are uniquely situated to partially or completely remove specific materials that have been added to the blood at the tissue level before they have a chance to reach other parts of the body. For example, *prostaglandins,* a collection of chemical messengers released in numerous tissues to mediate particular local responses, may spill into the blood, but they are inactivated during passage through the lungs so that they cannot exert systemic effects. In contrast, the lungs generate *angiotensin II,* a hormone that plays an important role in regulating both blood pressure as well as the concentration of Na^+ in the extracellular fluid (p. 587).

6. *Olfaction.* The nose has chemoreceptors for airborne odorants (Chapter 6, p. 236).

7. *Sounds: Vocalization, warning and mating calls, and so on.* In mammals, the **larynx** (or **voicebox**) at the tracheal entrance (Figure 11-18a) is a set of cartilage folds that helps prevent food from entering the trachea (though not as effectively in a few species, including many aquatic mammals, red deer, and adult humans, in which the larynx descends down the trachea). The larynx also has vibrating **vocal folds or cords** that allow for modulation of air flow for the production of sounds. The *laryngeal nerve,* which controls the larynx, illustrates the historical and often illogical nature of evolution. This nerve evolved as a branch from a fish vagus nerve running from the brain directly to the gills. As a branch evolved to the larynx in an early terrestrial vertebrate, it turned back up to the neck. In a giraffe, this illogical feature results in a looping 6-meter (20-foot) nerve from the brain down to near the lungs and then nearly all the way back up!

Unlike the larynx, the **syrinx** of birds lies at the exit of the trachea (Figure 11-15b) and generally consists of both tracheal and bronchial contributions (Figure 11-18b). Normally, three to six rings at the base of the trachea are enlarged and form a rigid structure termed the **tympanum.** The upper three rings of each bronchus are also enlarged, and together the composite structure is supplied with muscles, air sacs, and vibrating membranes. The entrance to each bronchus is constricted by a narrow slit, the so-called *tympanic membrane,* and the extent of opening is subject to variable muscle tension from the *syringeal muscles.* Air pressure generated from the lungs and air sacs sets the tympanic membrane in motion, and the syrinx operates as a valve that regulates both pitch and volume. The complexity of a bird's song is closely related to the number of syringeal muscles that connect with the various rings and assorted membranes. Several species of birds, including vultures, storks, and ostriches, lack any syringeal muscles, whereas true songbirds have around five pairs. Because each bronchus has its own tympanic membrane and the muscles are innervated separately, it is possible for some species, such as the wood thrush (*Hylocichla mustelina*), to simultaneously sing notes of separate frequencies. Contributing to vocalization in some species (swans, geese, and cranes) are tracheae of exceptional length. In the whooping crane (*Gus americana*), the trachea is almost as long as the bird itself and serves as a resonating tube to amplify its vocalization. More powerful vocalizations are evidence of "fitness," which increase the male's likelihood of attracting a mate and passing on its genes.

This completes our overview of respiratory systems. In the following sections, we delve into each of the steps of Figure 11-1 in greater detail, beginning with breathing.

check your understanding 11.4

What are examples of bimodal and trimodal breathing, and why did they evolve?

Compare respiratory systems of frogs and nonavian reptiles, including crocodilians.

Compare mammalian and bird respiratory systems (structures and methods of ventilation).

11.5 Breathing: Respiratory Mechanics in Mammals

In this section, we examine details of the mechanics involved in breathing (step 1, Figure 11-1) in more depth, focusing primarily on mammals.

Interrelationships among atmospheric, intra-alveolar, and intrapleural pressures are important in respiratory mechanics of mammals

Recall that air flows in and out of mammalian lungs during the act of breathing by moving down alternately reversing pressure gradients, established between the lungs and the atmosphere by cyclical respiratory-muscle activity. Let's look at this more closely. Three different pressure considerations are important in ventilation (Figure 11-19a):

1. **Atmospheric (barometric) pressure** is the pressure exerted by the weight of the air in the atmosphere on objects on Earth's surface. At sea level it equals 760 mm Hg or 1 atmosphere (atm) or 101.3 kilopascals (Figure 11-2a). Atmospheric pressure diminishes with increasing altitude above sea level (Table 11-1), and minor fluctuations occur at any height because of changing weather conditions (as you can see using a barometer).

2. **Intra-alveolar pressure** was defined earlier (p. 511). Because the lung communicates with the atmosphere through the conducting airways, air quickly flows down its pressure gradient any time this pressure differs from atmospheric pressure; the airflow continues until the two pressures equilibrate.

3. **Intrapleural pressure** is the pressure exerted outside the lungs within the thoracic cavity, that is, within the pleural sac. This pressure is usually less than atmospheric pressure, averaging 756 mm Hg at rest (4 mm Hg less). Intrapleural pressure does not equilibrate with atmospheric or intra-alveolar pressure, because there is no direct communication between the pleural cavity and either the atmosphere or the lungs.

The lungs are normally stretched to fill the larger thorax

The thoracic cavity is larger than the unstretched lungs because the thoracic wall grows more rapidly than the lungs during development. However, two forces—the intrapleural fluid's cohesiveness and the *transmural pressure gradient*—hold the thoracic wall and lungs in close apposition, stretching the lungs to fill the larger thoracic cavity:

1. *Intrapleural fluid's cohesiveness:* The polar water molecules in the intrapleural fluid resist being pulled apart because of their attraction to each other. The resultant cohesiveness of the intrapleural fluid tends to hold the pleural surfaces together. Thus, the intrapleural fluid can be considered very loosely as a "glue" between the lining of the thoracic wall and the lung. If you ever tried to pull apart two glass slides held together by a thin layer of liquid, you know that the two surfaces act as if the thin layer of water sticks them together. Even though you can easily slip the slides back and forth relative to each other

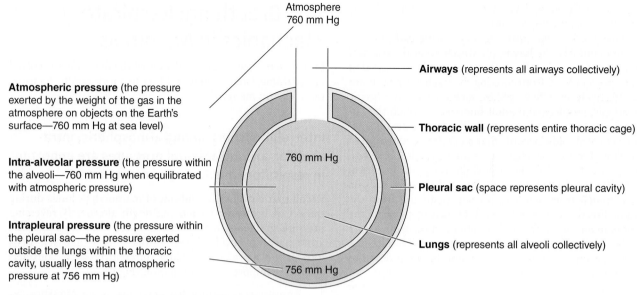

Atmosphere
760 mm Hg

Airways (represents all airways collectively)

Thoracic wall (represents entire thoracic cage)

760 mm Hg

Pleural sac (space represents pleural cavity)

Lungs (represents all alveoli collectively)

756 mm Hg

Atmospheric pressure (the pressure exerted by the weight of the gas in the atmosphere on objects on the Earth's surface—760 mm Hg at sea level)

Intra-alveolar pressure (the pressure within the alveoli—760 mm Hg when equilibrated with atmospheric pressure)

Intrapleural pressure (the pressure within the pleural sac—the pressure exerted outside the lungs within the thoracic cavity, usually less than atmospheric pressure at 756 mm Hg)

(a) Pressures important in mamalian lung ventilation

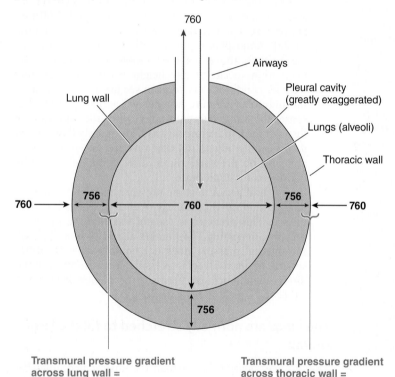

760

Airways

Lung wall

Pleural cavity
(greatly exaggerated)

Lungs (alveoli)

Thoracic wall

760 → | 756 | ← 760
760
760 → | 756 | ← 760

756

Transmural pressure gradient across lung wall = intra-alveolar pressure minus intrapleural pressure

Transmural pressure gradient across thoracic wall = atmospheric pressure minus intrapleural pressure

Numbers are mm Hg pressure.

(b) Transmural pressure gradient

FIGURE 11-19 Pressure and mammalian lung inflation. (a) Atmospheric, intra-alveolar and intrapleural pressures. (b) Across the lung wall, the intra-alveolar pressure of 760 mm Hg pushes outward, while the intrapleural pressure of 756 mm Hg pushes inward. This 4-mm Hg difference in pressure constitutes a transmural pressure gradient that pushes out on the lungs, stretching them to fill the larger thoracic cavity. Across the thoracic wall, the atmospheric pressure of 760 mm Hg pushes inward, while the intrapleural pressure of 756 mm Hg pushes outward. This 4-mm Hg difference in pressure constitutes a transmural pressure gradient that pushes inward and compresses the thoracic wall.

© Cengage Learning, 2013

(just as the intrapleural fluid facilitates movement of the lungs against the interior surface of the chest wall), you can pull the slides apart only with great difficulty because the molecules within the intervening liquid resist being separated. This relationship is partly responsible for the fact that changes in thoracic dimension are always accompanied by corresponding changes in lung dimension; that is, when the thorax expands, the lungs, being stuck

to the thoracic wall by virtue of the intrapleural fluid's cohesiveness, do likewise.

2. *Transmural pressure gradient:* An even more important reason that the lungs follow the movements of the chest wall are the **transmural pressure gradient** that exists across the lung and thoracic walls (*trans*, "across"; *mural*, "wall") (Figure 11-19b). First, intra-alveolar pressure at 760 mm Hg is greater than the intrapleural

FIGURE 11-20 Boyle's law. Each container has the same number of gas molecules. Given the random motion of gas molecules, the likelihood of a gas molecule striking the interior wall of the container and exerting pressure varies inversely with the volume of the container at any constant temperature. The gas in container B exerts more pressure than the same gas in larger container C but less pressure than the same gas in smaller container A. This relationship is stated as Boyle's law: $P_1V_1 = P_2V_2$. As the volume of gas increases, the pressure of the gas decreases proportionately; conversely, the pressure increases proportionately as the volume decreases.

© Cengage Learning, 2013

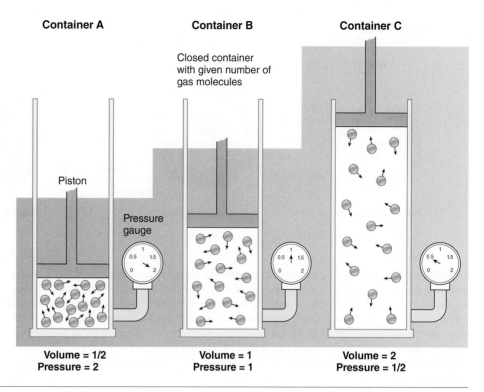

Container A | Container B | Container C

Closed container with given number of gas molecules

Piston

Pressure gauge

Volume = 1/2
Pressure = 2

Volume = 1
Pressure = 1

Volume = 2
Pressure = 1/2

pressure of 756 mm Hg, so a greater pressure is pushing outward than is pushing inward across the lung wall. This net outward pressure differential, the lung-wall transmural pressure gradient, pushes out on the lungs, stretching or distending them (Figure 11-19b). Because of this pressure gradient, the lungs are always forced to expand to fill the thoracic cavity.

Second, external atmospheric pressure pushing inward on the thoracic wall is again greater than the intrapleural pressure of 756 mm Hg, creating a thoracic-wall transmural pressure that tend to "squeeze in" or compress the chest wall compared to what it would be in an unrestricted state. The effect of the lung-wall transmural pressure gradient is much more pronounced, however, because the highly distensible lungs are influenced by this modest pressure differential to a much greater extent than is the more rigid thoracic wall.

Because neither the lungs nor the thoracic wall are in their natural position when they are held in apposition to each other, they constantly try to assume their own inherent dimensions. The stretched lungs have a tendency to pull inward away from the thoracic wall, whereas the compressed thoracic wall tends to move outward away from the lungs. The transmural pressure gradient and intrapleural fluid's cohesiveness, however, prevent these structures from pulling away from each other except to the slightest degree. Even so, the resultant ever-so-slight expansion of the pleural cavity is sufficient to drop the pressure in this cavity by 4 mm Hg to the subatmospheric value of 756 mm Hg.

In summary, the lungs are stretched and the thorax is compressed because a transmural pressure gradient exists across their walls as a result of the presence of a subatmospheric intrapleural pressure. The intrapleural pressure, in turn, is subatmospheric because the stretched lungs and compressed thorax tend to pull away from each other, slightly expanding the pleural cavity.

Intrapleural and lung pressures decrease during inhalation and increase during exhalation

Because air flows down a pressure gradient, for air to flow into the lungs during inspiration the intra-alveolar pressure must be less than atmospheric pressure. Similarly, the pressure gradient must reverse during expiration. Altering the volume of the lungs, in accordance with Boyle's law, can change intra-alveolar pressure. **Boyle's law** states that at any constant temperature, the *pressure* exerted by a gas varies inversely with the *volume* of the gas (Figure 11-20).

Just before inspiration, intra-alveolar pressure is equal to atmospheric pressure so no air is flowing into or out of the lungs (Figure 11-21a). As the thoracic cavity expands, the intrapleural pressure falls from 756 to 754 mm Hg because the same number of air molecules now occupy a larger lung volume. Thus, the lungs are forced to expand to fill the larger thoracic cavity, typically causing the intra-alveolar pressure to drop 1 mm Hg to 759 mm Hg (Figure 11-21b). Because this is now less than atmospheric pressure, air flows into the lungs until intra-alveolar pressure equals atmospheric pressure.

In a resting expiration, as the lungs and diaphragm recoil, the intra-alveolar pressure increases about 1 mm Hg above atmospheric level to 761 mm Hg (Figure 11-21c). Now air flows outwards from the lungs to the environment.

Respiratory diseases often increase airway resistance

In Chapter 9, we discussed how fluid flow depends on both the pressure gradient and resistance (flow = $\Delta P/R$; see p. 419). This is true of airflow as well as blood flow. In a

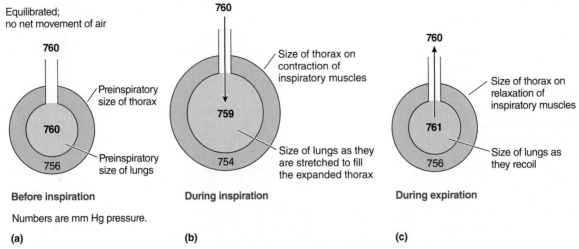

FIGURE 11-21 Changes in lung volume and intra-alveolar pressure during inspiration and expiration (mammalian lung). (a) Before inspiration, at the end of the preceding expiration. Intra-alveolar pressure is equilibrated with atmospheric pressure and no air is flowing. (b) Inspiration. As the lungs increase in volume during inspiration, the intra-alveolar pressure decreases, establishing a pressure gradient that favors the flow of air into the alveoli from the atmosphere; that is, an inspiration occurs. (c) Expiration. As the lungs recoil to their pre-inspiratory size on relaxation of the inspiratory muscles, the intra-alveolar pressure increases, establishing a pressure gradient that favors the flow of air out of the alveoli into the atmosphere; that is, an expiration occurs.

© Cengage Learning, 2013

healthy respiratory system, the radius of the conducting system is sufficiently large that resistance remains extremely low. Therefore, the pressure gradient between the alveoli and the atmosphere is usually the primary factor determining the airflow rate. Indeed, the airways normally offer such low resistance that only very small pressure gradients of 1 to 2 mm Hg need be created to achieve adequate rates of airflow in and out of the lungs. However, various respiratory diseases can narrow the airways such that resistance becomes a limiting factor. For example, **chronic obstructive pulmonary disease (COPD)** is a group of human lung diseases (including *chronic bronchitis, asthma,* and *emphysema*) characterized by increased airway resistance resulting from the narrowing of the lumen of the lower airways. Other mammals can develop similar diseases, such as *equine restrictive lung diseases* in horses. When airway resistance increases, a larger pressure gradient must be established to maintain even a normal airflow rate. Accordingly, mammals with such diseases must work harder to breathe.

Elasticity in mammalian lungs depends on connective tissue and alveolar surface tension, which is reduced by surfactant

The term **elastic recoil** refers to how readily the lungs rebound after having been stretched. It is responsible for the lungs returning to their pre-inspiratory volume when the inspiratory muscles relax at the end of inspiration. Elastic recoil depends mainly on two factors: highly elastic connective tissue in the lungs and alveolar surface tension:

1. *Elastin:* Pulmonary connective tissue contains large quantities of *elastin* fibers (see p. 65). Not only do these fibers exhibit elastic properties themselves, but they are

arranged into a meshwork that amplifies their elastic behavior, much like the threads in a piece of stretch-knit fabric. These rebound to a smaller length after being stretched.

2. *Alveolar surface tension:* An even more important factor influencing elastic behavior of the lungs is the **alveolar surface tension** displayed by the thin liquid film that lines each alveolus. At an air–water interface, the water molecules at the surface are more strongly attracted to other surrounding water molecules than to the air above the surface. This unequal attraction produces a force known as *surface tension* at the surface of the liquid. Surface tension is responsible for a twofold effect. First, the liquid layer resists any force that increases its surface area; that is, it opposes expansion of the alveolus because the surface water molecules oppose being pulled apart. Accordingly, the greater the surface tension, the less compliant the lungs. Second, the liquid surface area tends to become as small as possible because the surface water molecules, being preferentially attracted to each other, try to get as close together as possible. Thus, the surface tension of the liquid lining an alveolus tends to reduce the size of the alveolus, squeezing in on the air within it (Figure 11-22). This property, along with the rebound of the stretched elastin fibers, is responsible for the lungs' elastic recoil back to their pre-inspiratory size after inspiration.

However, there is a problem. The cohesive forces between water molecules are so strong that if the alveoli were lined with water alone, the surface tension would be so great that the lungs would collapse; the recoil force attributable to the elastin fibers and high surface tension would exceed the opposing stretching force of the transmural pressure gradient. Furthermore, the lungs would be very poorly compliant, so exhausting

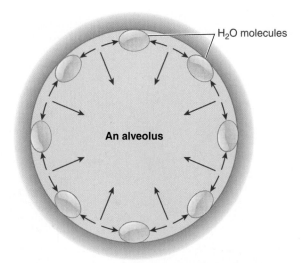

H₂O molecules

An alveolus

FIGURE 11-22 Alveolar surface tension (mammalian lung). The attractive forces between the water (H₂O) molecules in the liquid film that lines the alveolus are responsible for surface tension. Because of its surface tension, an alveolus (1) resists being stretched, (2) tends to be reduced in surface area or size, and (3) tends to recoil after being stretched.

© Cengage Learning, 2013

muscular efforts would be required to accomplish stretching and inflation of the alveoli.

Pulmonary Surfactant The tremendous surface tension of pure water is normally counteracted by pulmonary surfactant, a detergent-like mixture of lipids and proteins secreted by the Type II alveolar cells (Figure 11-12a) in response to mechanical stimulation and autonomic nerves. Pulmonary surfactant intersperses between the water molecules in the fluid lining the alveoli and lowers the alveolar surface tension because the cohesive force between a water molecule and an adjacent pulmonary surfactant molecule is very low. This provides two important benefits: (1) It increases pulmonary compliance, thus reducing the work of inflating the lungs; and (2) it reduces (but does not eliminate) the lungs' tendency to recoil so that they do not collapse as readily.

Pulmonary surfactant's role in reducing the alveoli's tendency to recoil, thereby discouraging alveolar collapse, is important in helping maintain lung stability. The division of the lung into a myriad of tiny air sacs provides the advantage of a tremendously increased surface area for the exchange of O_2 and CO_2, but it also presents the problem of maintaining the stability of all these alveoli. Recall that the pressure generated by alveolar surface tension is directed inward, squeezing in on the air in the alveoli. If the alveoli are visualized as spherical bubbles, according to **LaPlace's law,** the magnitude of the inward-directed collapsing pressure is directly proportional to the surface tension and inversely proportional to the radius of the bubble:

$$P = 2T/r$$

where P is the inward-directed collapsing pressure; T is the surface tension; and r is the radius of bubble (alveolus) (Figure 11-23). Because the collapsing pressure is inversely pro-

portional to the radius, the smaller the alveolus, the greater its tendency to collapse at a given T. Accordingly, if two alveoli of unequal size but the same T are connected by the same terminal airway, the smaller alveolus—because it generates a larger collapsing pressure—has a tendency to collapse and empty its air into the larger alveolus (Figure 11-23a). This does not normally happen, however, because pulmonary surfactant reduces the surface tension of small alveoli more than it reduces it in larger ones. This is because the surfactant molecules are more densely packed together in the smaller alveoli. The surfactant-induced lower T of small alveoli offsets the effect of their smaller radii in determining the inward-directed pressure. Therefore, the presence of surfactant causes the collapsing pressure of small alveoli to become comparable to that of larger alveoli and minimizes the tendency for small alveoli to collapse (Figure 11-23b).

In nonmammals surfactant functions as an "antiglue" that prevents adhesion of adjacent respiratory surfaces close to each other. Complete collapse of the lung occurs at the end of expiration in a variety of species including sea snakes and some small frogs. Without surfactant, inspiration after lung collapse would be costly in terms of energy expenditure. Moreover, as body temperature decreases, lung volumes in a lizard are reduced and contact of the respiratory surfaces within the lung are increased. Surfactant is thus essential in this situation to prevent adhesion of the surfaces. The cooler ambient temperatures trigger the release of surfactant via cholinergic stimulation of the Type II cells.

check your understanding 11.5

Compare atmospheric, intra-alveolar, and intrapleural pressures and their roles.

Explain pulmonary elastic recoil in terms of surface tension and elastic fibers.

State the source and function of pulmonary surfactant.

11.6 Breathing: Lung Volumes in Mammals

Next we turn to details of lung dynamics and volume changes that occur in mammalian respiratory cycles.

Changes in lung volumes are important for analyzing different respiratory efforts, disease states, aerobic activity capacities, and species differences

Let's begin with key volume measurements that are often made for a respiratory cycle.

Methodology: The Spirometer Lung volumes have traditionally been measured with a **spirometer.** Spirometric measurements can be obtained from trained or anesthetized animals wearing a mouthpiece appropriate to their anatomy. Basically, a spirometer consists of an air-filled inverted drum

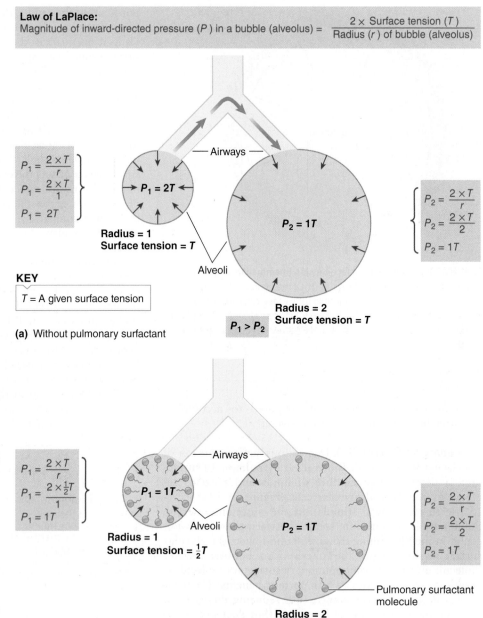

Law of LaPlace:
Magnitude of inward-directed pressure (P) in a bubble (alveolus) = $\dfrac{2 \times \text{Surface tension } (T)}{\text{Radius } (r) \text{ of bubble (alveolus)}}$

FIGURE 11-23 **Role of pulmonary surfactant in counteracting the tendency for small alveoli to collapse into larger alveoli.** (a) According to the law of LaPlace, if two alveoli of unequal size but the same surface tension are connected by the same terminal airway, the smaller alveolus—because it generates a larger inward-directed collapsing pressure—has a tendency (without pulmonary surfactant) to collapse and empty its air into the larger alveolus. (b) Pulmonary surfactant reduces the surface tension of a smaller alveolus more than that of a larger alveolus. This reduction in surface tension offsets the effect of the smaller radius in determining the inward-directed pressure. Consequently, the collapsing pressures of the small and large alveoli are comparable. Therefore, in the presence of pulmonary surfactant, a small alveolus does not collapse and empty its air into the larger alveolus.

© Cengage Learning, 2013

floating in a water-filled chamber. As the animal breathes air in and out of the drum through a tube connecting the mouth to the air chamber, the drum rises and falls in the water chamber (Figure 11-24). This rise and fall can be recorded as a **spirogram,** which is calibrated to volume changes (typically analysed by a computer).

Spirograms yield several lung volumes and **capacities** (sums of two or more lung volumes), as shown in Figure 11-25 and defined below. Figure 11-25a shows total capacity and resting volume for an average human male (generally the values are lower for females). Let's now look at the significances of these and other volumes and capacities.

Volumes and Capacities Most mammals have similar respiratory cycles, although the volumes and capacities can differ greatly among species. Figure 11-25b is a hypothetical example of a spirogram and a table of values for a healthy young human compared a healthy young horse. Let us compare them:

1. **Total lung capacity, or TLC**—the maximum amount of air that the lungs can hold. On average, in healthy young horses this is about 42 L, whereas it is only about 5.7 L in humans. Within a species, anatomic build, age, distensibility of the lungs, and presence or absence of

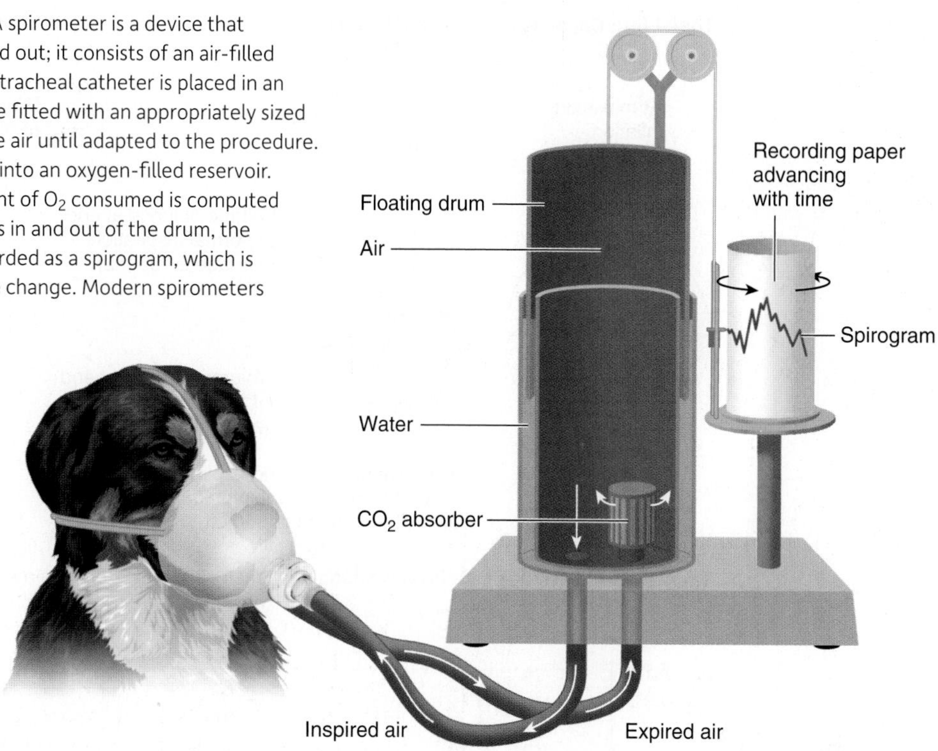

FIGURE 11-24 Spirometry in animals. A spirometer is a device that measures the volume of air breathed in and out; it consists of an air-filled drum floating in a water-filled chamber. A tracheal catheter is placed in an anesthetized animal (or trained animals are fitted with an appropriately sized mask). Initially, the animal breathes outside air until adapted to the procedure. A valve is closed, and breathing continues into an oxygen-filled reservoir. Carbon dioxide is absorbed, and the amount of O_2 consumed is computed from the recording. As the animal breathes in and out of the drum, the resultant rise and fall of the drum are recorded as a spirogram, which is calibrated to the magnitude of the volume change. Modern spirometers record the output on a computer.

© Cengage Learning, 2013

respiratory disease all affect TLC. However, not all this air can be exchanged.

2. **Tidal volume, or TV**—the volume of air entering or leaving the lungs during a single breath. The resting TV (**rTV**) is a common measurement. During each typical breath at rest, a horse inspires about 4 to 6 L (0.4 to 0.5 L in humans) of air, and the same quantity is expired, so during quiet breathing the lung volume varies between 24 L (2.2 L in humans) at the end of expiration to 30 L (2.7 L in humans) at the end of inspiration. In most mammals, rTV is about 10 to 14% of TLC.

3. **Functional residual reserve, or FRC**—the volume of air in the lungs at the end of a normal passive expiration. Normally, during quiet breathing, the lungs are not close to maximal inflation, nor are they deflated to their minimum volume. Thus, the lungs normally remain moderately inflated throughout the respiratory cycle. At the end of a normal quiet expiration, the lungs still contain about 24 L (2.2 L in humans) of air.

4. **Residual volume, or RV**—the minimum volume of air remaining in the lungs even after a maximal expiration. During maximal expiration, lung volume can be decreased to 12 L (1.2 L in humans), but the lungs can never be completely deflated because the small airways collapse during forced expirations at low lung volumes, blocking further outflow of air. This results in:

5. **Vital capacity, or VC**—the maximum volume of air that can be moved out during a single breath following a maximal inspiration. It is rarely used because the maximal muscle contractions involved become exhausting, but it is useful in ascertaining the functional capacity of the lungs. For the average horse it is 30 L (4.5 L in hu-

mans). VC declines in many lung diseases such as asthma.

The inability to empty the lungs completely is a drawback because fresh inspired air is mixed with the large volume of old air remaining in the lungs (FRC during rest, RV during maximal activity), reducing the O_2 partial pressure. However, as we will explain later (p. 528), this does have the benefit of preventing wide fluctuations in gas exchange. Furthermore, recall that it takes less effort to inflate a partially inflated alveolus than a totally collapsed one.

Alveolar ventilation is less than pulmonary ventilation because of the presence of dead space

Various changes in lung volume represent only one factor in the determination of **pulmonary** or **minute ventilation,** which is the volume of air breathed in and out in one minute. The other important factor is **respiratory rate,** which averages 12 breaths per minute in both humans and horses. The *pulmonary ventilation equation* parallels that of cardiac output (cardiac output = stroke volume × beat rate; p. 414):

Pulmonary ventilation = tidal volume × respiratory rate
    (L/breath)   (L/min)   (breaths/min)

Pulmonary ventilation *scales* with mammalian body size. Tidal volume does not; that is, it is almost exactly proportional to body mass (m_b in kg):

$$TV = 0.0062 m_b^{1.01}$$

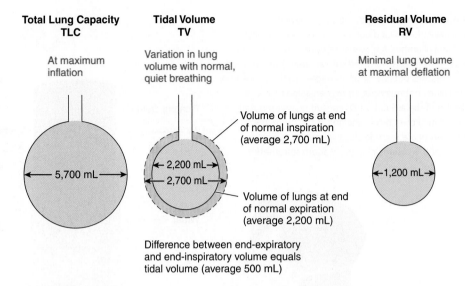

Total Lung Capacity TLC	Tidal Volume TV	Residual Volume RV
At maximum inflation	Variation in lung volume with normal, quiet breathing	Minimal lung volume at maximal deflation

Volume of lungs at end of normal inspiration (average 2,700 mL)

←2,200 mL→
←2,700 mL→

←5,700 mL→

←1,200 mL→

Volume of lungs at end of normal expiration (average 2,200 mL)

Difference between end-expiratory and end-inspiratory volume equals tidal volume (average 500 mL)

(a) Values are average for a healthy young adult male; values for females are somewhat lower.

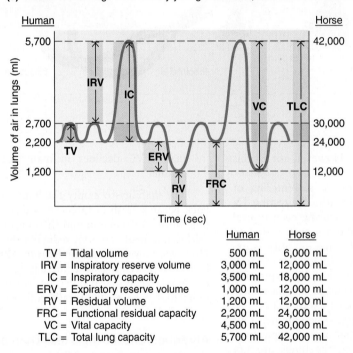

	Human	Horse
TV = Tidal volume	500 mL	6,000 mL
IRV = Inspiratory reserve volume	3,000 mL	12,000 mL
IC = Inspiratory capacity	3,500 mL	18,000 mL
ERV = Expiratory reserve volume	1,000 mL	12,000 mL
RV = Residual volume	1,200 mL	12,000 mL
FRC = Functional residual capacity	2,200 mL	24,000 mL
VC = Vital capacity	4,500 mL	30,000 mL
TLC = Total lung capacity	5,700 mL	42,000 mL

(b) Spirogram and table of values for adult male human and horse

FIGURE 11-25 Variations in lung volume. (a) Normal range and extremes of lung volume in a human (healthy young adult male). (b) Normal spirogram of a human (healthy young adult male) compared to a horse. See main text for details on the various volumes (such as tidal volume, TV) and capacities (such as vital capacity, VC). (The residual volume cannot be measured with a spirometer but must be determined by another means.)

© Cengage Learning, 2013

However, resting respiratory rate RR follows the equation:

$$RR = 53.5 m_b^{-0.26}$$

The negative exponent, –0.26, means that larger animals have slower rates. (Recall that alveolar area scaled to 0.75 or 3/4 power (p. 510), while here we see an approximate negative 1/4 power; the significance of these will be explored in Chapter 15.) On the next page are some average examples for resting and active mammals (differing slightly for humans compared to Figure 11-25b due to different studies). Among the resting animals, note that the giraffe's respiratory rate of 9/min is 20% less than the scaling relationship predicts (11/min), perhaps because of the high friction in its long trachea. Also, compare the resting and active human and horse. Note that both can increase ventilation over 20-fold during high activity, and that TV increases about as much as respiratory rate. We explore why this is important next.

Animal at Rest	Pulmonary Ventilation (PV)	=	Resting Tidal Volume (rTV)	×	Respiratory Rate
Rat (0.22 kg)	0.12 L/min	=	0.001 L/breath	×	120 breaths/min
Human (70 kg)	6.0 L/min	=	0.4 L/breath	×	15 breaths/min
Giraffe (400 kg)	30 L/min	=	3.3 L/breath	×	9 breaths/min
Horse (450 kg)	72 L/min	=	6 L/breath	×	12 breaths/min
Elephant (7,000 kg)	235 L/min	=	47 L/breath	×	5 breaths/min
Active Animals			**Active TV**		
Human, fast run:	150 L/min	=	2.5 L/breath	×	60 breaths/min
Horse, fast trot:	1,500 L/min	=	25 L/breath	×	60 breaths/min

Anatomic Dead Space Pulmonary ventilation does not give a proper estimate of gas delivery. When increasing pulmonary ventilation, *it is usually more advantageous to have a greater increase in tidal volume* than in respiratory rate because of the presence of **anatomic dead space,** which results from the fact that not all the inspired air gets down to the site of gas exchange in the alveoli. Part of it remains in the conducting airways, where it is not available for gas exchange. The volume of the conducting passages in an adult human averages about 0.15 L, whereas it is about 1.8 L in the horse. This volume is considered anatomic dead space (Figure 11-26) because air within these conducting airways is useless for exchange purposes. Anatomic dead space has a pronounced effect on the efficiency of pulmonary ventilation. In horses, even if 6 L of air are moved in and out with each breath, only 4.2 L are actually exchanged between the atmosphere and alveoli because of the 1.8 L occupying the anatomic dead space.

Alveolar Ventilation and Breathing Patterns Because the amount of atmospheric air that reaches the alveoli and is actually available for exchange with the blood is more important than the total amount breathed in and out, **alveolar ventilation (AV)**—the volume of air exchanged between the atmosphere and alveoli per minute—is more important than pulmonary ventilation (PV). In determining AV, the amount of wasted air in the anatomic dead space must be taken into account, as follows:

AV = (TV – dead-space volume/breath) × respiratory rate

If we use this for average resting values for a horse, we get:

$$AV = (6 \text{ L/breath} - 1.8 \text{ L/breath})$$
$$\times 12 \text{ breaths/min} = 50.4 \text{ L/min}$$

Since pulmonary ventilation is *72 L/min,* this is a loss of 21.6 L/min (30%).

To understand how important dead-space volume is in determining the magnitude of AV, let's examine the effect of various breathing patterns on human AV when pulmonary ventilation (PV) is constant at 6.0 L/min. The values for normal resting yield an AV of 4.2 L/min, also a 30% loss:

$$AV, \text{ normal rest} = (0.5 \text{ TV} - 0.15 \text{ dead-space L/breath})$$
$$\times 12 \text{ breaths/min} = 4.2 \text{ L/min}$$

Now suppose a human deliberately breathes deeply (for example, a TV of 1.2 L) and slowly (e.g., at 5 breaths/min). PV is still 6.0 L/min, but AV *increases:*

$$AV, \text{ deep/slow} = (1.2 \text{ TV} - 0.15 \text{ dead-space L/breath})$$
$$\times 5 \text{ breaths/min} = 5.25 \text{ L/min}$$

In contrast, if a person deliberately breathes shallowly (for example, a TV of 0.15 L) and rapidly (at 40 breaths/min), PV is still be 6.0 L/min; however, AV would be *zero!*

$$AV, \text{ shallow/rapid} = (0.15 \text{ TV} - 0.15$$
$$\text{dead-space L/breath}) \times 40 \text{ breaths/min} = 0 \text{ L/min}$$

In effect, the person would only be drawing air in and out of the anatomic dead space without any atmospheric air being exchanged with the alveoli, where it could be useful. The individual could voluntarily maintain such a breathing pattern for only a few minutes before losing consciousness, at which time normal breathing would resume.

The value of reflexively bringing about a large increase in depth of breathing (TV) as well as in rate of breathing when PV is increased during activity should now be apparent. When TV is increased, the entire increase goes toward elevating AV, whereas an increase in respiratory rate does not entirely do so. When respiratory rate is increased, the frequency with which air is wasted in the dead space is also increased because a portion of *each* breath must move in and out of the dead space. Here are human values for high activity yielding a PV of 150 mL/min; note that AV is only 7% less:

$$AV, \text{ high activity} = (2.5 \text{ TV} - 0.15 \text{ dead-space L/breath})$$
$$\times 60 \text{ breaths/min} = 141 \text{ L/min}$$

In trotter horses at a maximal trot with a PV of 1,500 L/min, we get a similar 8% drop:

$$AV, \text{ maximal trot} = (25 \text{ TV} - 1.8 \text{ dead-space L/breath})$$
$$\times 60 \text{ breaths/min} = 1,392 \text{ L/min}$$

However, trotters forced to gallop at maximum are only able to manage a TV of 12 L/min, so are forced to breathe at a much faster rate of 140! This yields an impressive PV of 1,680 L/min, but the AV is *15%* less in this case:

$$AV, \text{ maximal trot} = (12 \text{ TV} - 1.8 \text{ dead-space L/breath})$$
$$\times 140 \text{ breaths/min} = 1,428 \text{ L/min}$$

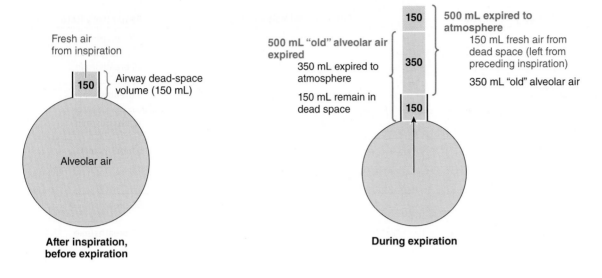

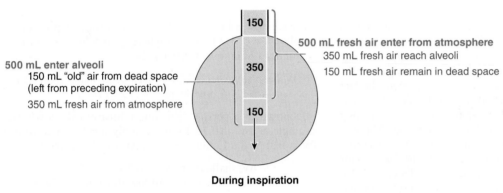

The numbers in the figure represent ml of air.

KEY

"Old" alveolar air that has exchanged O_2 and CO_2 with the blood

Fresh atmospheric air that has not exchanged O_2 and CO_2 with the blood

FIGURE 11-26 **Effect of dead-space volume on exchange of tidal volume between the atmosphere and the alveoli for humans.** Even though 500 mL of air move in and out between the atmosphere and the respiratory system and 500 mL move in and out of the alveoli with each breath, only 350 mL are actually exchanged between the atmosphere and the alveoli because of the anatomic dead space (the volume of air in the respiratory airways).

© Cengage Learning, 2013

---

check your understanding 11.6

Define the various lung volumes and capacities and their significance.

Compare pulmonary ventilation and alveolar ventilation. What is the consequence of anatomic and alveolar dead space? Give some examples with actual animal data.

11.7 Breathing: Flow-Through versus Tidal Respirers

We temporarily end our examination of ventilation and breathing (step 1, Figure 11-1) with comparisons between flow-through and tidal respirers. (We will return to breathing at the end of the chapter when we examine neural, hormonal and local regulation.)

Dead-space volumes are lower in fishes but higher in birds compared to mammals

The dead-space problem we just discussed also occurs in animals with gills. Not all the water entering the buccal cavity perfuses through the gill lamellae and is, instead, lost through the spaces between the gill arches. This fraction of the water unavailable for gas exchange is also termed the **anatomical dead space.** However, flow-through systems have much lower dead-space volumes because of unidirectional flow.

However, the same does not hold true for the partial flow-through systems of birds, but this is because their tracheae are longer and wider than mammals'. In fact, bird tracheae—which have tidal, not flow-through, ventilation—have an average 4.5 greater dead-space volume than mammals (when comparing animals of the same size). To compensate, a bird has a much higher TV and a much lower respiratory rate than a mammal of the same size. For example, compare a bird and a mammal, both weighing 2.6 kg:

Mammal's PV = 0.02 TV L/breath × 42 breaths/min
= 0. 84 L/min
Duck's PV = 0.07 TV L/breath × 16 breaths/sec
= 1.12 L/min

The work of normal breathing requires as little as 2% of total energy expenditure in some mammals to as much as 50% in some fishes

During normal quiet breathing, the respiratory muscles must work during inspiration to expand the lungs against their elastic forces and to overcome airway resistance, whereas expiration is a passive process. Normally, the mammalian lung is highly compliant and airway resistance is low, so only about 2% of the total energy expended by the body is used to accomplish quiet breathing.

During strenuous activity, the amount of energy required to power pulmonary ventilation may increase up to 25-fold. However, because total energy expenditure by the body is increased up to 15- to 20-fold during heavy activity, the energy used to accomplish the increased ventilation still represents only about 5% of total energy expended. There is an exception to this: Animals with chronic lung disease require considerably more energy for respiration, which may increase to 30% of total energy expenditure.

In contrast, respiration in water assumes a greater metabolic cost because of the fluid properties of water, which resist movement as compared to air. Estimates of the metabolic cost of respiration in a resting fish vary from 5 to 50% (normally approximately 20%) of the resting metabolic rate. However, were it not for the efficiencies of flow-through ventilation—including the switch from active breathing to ram ventilation at high speeds—the costs would presumably be even higher.

Flow-through ventilation yields a greater oxygen uptake than tidal breathing

Recall that flow-through ventilation allows for counter- or crosscurrent blood flow. Consequently, the P_{O_2} in outgoing blood of fish and birds is distinctly higher than in the expired medium from the respiratory organ (Figures 11-6f and 11-17), unlike the case for alveoli. However, as we noted earlier, the tidal lung of mammals can only yield an outgoing blood P_{O_2} that is equal to that of expired air (Figure 11-11c). In terms of the total O_2 extracted from the incoming medium, efficiency from the water can be as high as 90% in some species of fish, crustaceans, and amphibians. In birds, because they are partially tidal with dead spaces, efficiency of extraction from air ranges from only 30 to 40%. But in mammals, it averages a mere 25%. (In taking to the air, bats—while retaining the basic mammalian tidal system—

do somewhat better, having evolved the highest capacity gas exchange system among mammals with relatively larger lungs and thinner alveolar barriers to diffusion. A flying bat can maintain metabolic rates 2.5 to 3 times the maximal rates attainable by comparably sized nonflying rodents.) Recall, however, that, the low O_2 content of water relative to air requires a greater expenditure of energy for water breathers compared to air breathers with similar metabolisms.

Oxygen uptake—actually, O_2/CO_2 exchange—at respiratory surfaces is step 2 of external respiration, to which we now turn our focus.

check your understanding 11.7

Compare vertebrate tidal and flow-through breathers in terms of dead space, energy costs, and oxygen uptake efficiency.

11.8 Gas Exchange at Vertebrate Respiratory Organs and Body Tissues

Having examined the details of ventilatory bulk transport (step 1, Figure 11-1), we now turn to gas exchange in depth, focusing on vertebrates. The ultimate purpose of step 1 is to provide a continual supply of fresh O_2 for pickup by the blood and to constantly remove CO_2 unloaded from the circulation. This is the gas exchange step 2 of Figure 11-1. In turn, circulation serves the same purpose at the blood-tissue exchange step 4. Since both of these involve Fick's law and its gas gradients (ΔP), will examine them together, starting with step 2.

Lung air has a lower partial pressure of oxygen than atmospheric air

Let's begin with the gas gradients in air breathers. In tidal systems, there is a drawback: Generally, lung air (alveolar air in mammals) is not of the same composition as inspired atmospheric air, for two reasons. First, as soon as atmospheric air enters the respiratory passages, it becomes saturated with H_2O by exposure to the moist airways. Water vapor exerts a partial pressure just like any other gas. At mammalian body temperature, the partial pressure of H_2O vapor is 47 mm Hg. Humidification of inspired air in effect "dilutes" the partial pressure of the inspired gases by 47 mm Hg because the sum of the partial pressures approximates the atmospheric pressure of 760 mm Hg. In moist air, P_{H_2O} = 47 mm Hg, P_{N_2} = 563 mm Hg, P_{O_2} = 150 mm Hg, and P_{CO_2} = 0.2 mm Hg.

Second, compared to air, alveolar P_{O_2} is lower and P_{CO_2} is much higher because fresh inspired air is mixed with the large volume of old air that remained in the lungs and dead space at the end of the preceding expiration (the FRC—functional residual capacity, Figure 11-25b). In a human, only about one seventh of the total alveolar air is replaced by fresh atmospheric air with each normal breath. Thus, at the end of inspiration, less than 15% of the air in the alveoli is

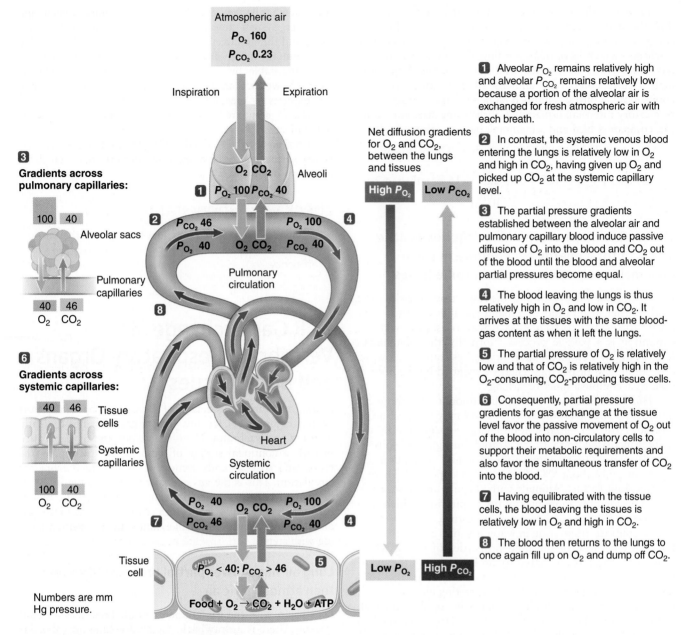

Atmospheric air

P_{O_2} 160

P_{CO_2} 0.23

Inspiration   Expiration

Net diffusion gradients for O_2 and CO_2, between the lungs and tissues

High P_{O_2} Low P_{CO_2}

3

Gradients across pulmonary capillaries:

100   40

Alveolar sacs

40   46
O_2 CO_2

Pulmonary capillaries

O_2 CO_2 Alveoli

1 P_{O_2} 100 P_{CO_2} 40

2 P_{CO_2} 46 P_{O_2} 100 **4**

P_{O_2} 40 O_2 CO_2 P_{CO_2} 40

Pulmonary circulation

8

6

Gradients across systemic capillaries:

40   46   Tissue cells

Systemic capillaries

100   40
O_2 CO_2

7 P_{O_2} 40 O_2 CO_2 P_{O_2} 100 **4**

P_{CO_2} 46 P_{CO_2} 40

Heart

Systemic circulation

Tissue cell

5 $P_{O_2} < 40$; $P_{CO_2} > 46$

Food + O_2 → CO_2 + H_2O + ATP

Low P_{O_2} High P_{CO_2}

Numbers are mm Hg pressure.

1 Alveolar P_{O_2} remains relatively high and alveolar P_{CO_2} remains relatively low because a portion of the alveolar air is exchanged for fresh atmospheric air with each breath.

2 In contrast, the systemic venous blood entering the lungs is relatively low in O_2 and high in CO_2, having given up O_2 and picked up CO_2 at the systemic capillary level.

3 The partial pressure gradients established between the alveolar air and pulmonary capillary blood induce passive diffusion of O_2 into the blood and CO_2 out of the blood until the blood and alveolar partial pressures become equal.

4 The blood leaving the lungs is thus relatively high in O_2 and low in CO_2. It arrives at the tissues with the same blood-gas content as when it left the lungs.

5 The partial pressure of O_2 is relatively low and that of CO_2 is relatively high in the O_2-consuming, CO_2-producing tissue cells.

6 Consequently, partial pressure gradients for gas exchange at the tissue level favor the passive movement of O_2 out of the blood into non-circulatory cells to support their metabolic requirements and also favor the simultaneous transfer of CO_2 into the blood.

7 Having equilibrated with the tissue cells, the blood leaving the tissues is relatively low in O_2 and high in CO_2.

8 The blood then returns to the lungs to once again fill up on O_2 and dump off CO_2.

FIGURE 11-27 O_2 and CO_2 exchange across pulmonary and systemic capillaries caused by partial pressure gradients (shown for a mammal).

© Cengage Learning, 2013

fresh air. As a result of humidification and the small turnover of alveolar air, *the average alveolar P_{O_2} is 100 mm Hg*, compared to (nonmoist) atmospheric P_{O_2} of about 160 mm Hg (Figures 11-11c and 11-27). P_{CO_2} is even worse at 40 mm Hg in the lungs compared to atmospheric level of 0.2 mm Hg (Figure 11-27).

It is logical to think that alveolar P_{O_2} and P_{CO_2} would change during inspiration with the arrival of fresh air during expiration. Only small fluctuations of a few mm Hg occur, however, for two reasons. First, because only a small proportion of the total alveolar air is exchanged with each breath, the relatively small volume of high-P_{O_2} low-P_{CO_2} air that is inspired is quickly mixed with the much larger volume of retained alveolar air, which has a lower P_{O_2} and

higher P_{CO_2}. Thus, the gases in the inspired air can only slightly change the total alveolar gas pressures. Even a potentially small elevation of P_{O_2} is diminished for another reason. Oxygen is continually moving by passive diffusion down its partial pressure gradient from the alveoli into the blood. The O_2 arriving in the alveoli in the newly inspired air simply replaces the O_2 diffusing out of the alveoli into the pulmonary capillaries. Therefore, the alveolar P_{O_2} remains relatively constant at about 100 mm Hg throughout the respiratory cycle. Because the pulmonary blood P_{O_2} equilibrates with the alveolar P_{O_2}, the P_{O_2} of the blood leaving the alveoli likewise remains fairly constant at this same value. Similar events occur for CO_2 but in reverse (Figure 11-27).

O₂ enters and CO₂/HCO₃⁻ leaves the capillaries in the lungs or gills passively down partial pressure or concentration gradients

Now let's look at the specific gradients, starting with mammals. The blood entering the pulmonary capillaries is systemic venous blood pumped to the lungs through the pulmonary arteries (Chapter 9). This blood, having just returned from the body tissues, is relatively low in O_2, with a $P_{2_{O_2}}$ of 40 mm Hg, and is relatively high in CO_2, with a $P_{1_{CO_2}}$ of 46 mm Hg. As this blood flows through the pulmonary capillaries, it is exposed to alveolar air with a higher $P_{1_{O_2}}$ at 100 mm Hg (Figure 11-27). Therefore O_2 diffuses down its partial pressure gradient ($\Delta P = 60$ mm Hg) into the blood until no further gradient exists. As the blood leaves the pulmonary capillaries, it has a P_{O_2} equal to alveolar P_{O_2} at 100 mm Hg. The partial pressure gradient for CO_2 is in the opposite direction. Alveolar $P_{2_{CO_2}}$ is only 40 mm Hg, 6 mm Hg less than the blood. Thus, CO_2 diffuses from the blood into the alveoli until blood P_{CO_2} equilibrates with alveolar P_{CO_2}, so the blood leaving the pulmonary capillaries has a P_{CO_2} of 40 mm Hg. After leaving the lungs, the blood is returned to the heart to be subsequently pumped out to the body tissues as systemic arterial blood (Chapter 9).

Note that blood returning to the lungs from the tissues still contains O_2 at an average P_{O_2} of 40 mm Hg. However, that value varies. The amount of O_2 picked up in the lungs matches the amount extracted and used by the tissues. When the tissues metabolize more actively (for example, during strenuous activity), more O_2 is extracted from the blood at the tissue level, reducing the systemic venous P_{O_2} even lower—for example, to a $P_{2_{O_2}}$ of 30 mm Hg. At the lungs, a larger-than-normal P_{O_2} gradient exists, now at 70 mm Hg (100 mm Hg vs 30 mm Hg), compared to the normal gradient of 60 mm Hg. Therefore, more O_2 diffuses from the alveoli into the blood down the larger partial-pressure gradient before blood P_{O_2} equals alveolar P_{O_2}. This additional transfer of O_2 into the blood replaces the increased amount of O_2 consumed, so O_2 uptake matches O_2 use even when O_2 consumption increases. At the same time, ventilation is stimulated so that O_2 enters the alveoli more rapidly from the atmosphere to replace the O_2 diffusing into the blood.

Similarly, blood leaving the lungs still contains CO_2 at an average P_{CO_2} of 40 mm Hg). The CO_2 remaining in the blood even after passage through the lungs plays an important role in the acid–base balance of the body, because CO_2 generates carbonic acid (Chapter 13). Furthermore, arterial P_{CO_2} is important in driving respiration. This mechanism is described later. As with O_2, the amount of CO_2 given up to the alveoli from the blood matches the amount of CO_2 picked up at the tissues. Once again, an increase in ventilation associated with increased activity assures that increased amounts of CO_2 delivered to the alveoli are blown off to the atmosphere.

Water breathers are exposed to very low concentrations of CO_2, with in-current water having only about 0.2 mm Hg, the same as atmospheric P_{CO_2} (in contrast to the 30 to 40 mm Hg in most air breathing vertebrates). This is because flow-through ventilation prevents buildup of the gas in the gill chamber. The P_{CO_2} in expired water and in arterial blood of fish is also remarkably low, around 1 to 4 mm Hg, but higher than the incoming water. The low arterial level arises because much of the gas in the blood is actually in the

form of *bicarbonate*, HCO_3^-, which is highly soluble and higher than in most external waters:

Venous blood (tuna) $HCO_3^- = 12$ mmol/L;
seawater $HCO_3^- = 2$ mmol/L

Thus, the ion can readily diffuse (through facilitated transporters) out of the gills into the external water (down a concentration rather than a partial pressure gradient). Air breathers, in contrast, cannot excrete bicarbonate into the air. Conversion of CO_2 into HCO_3^- is primarily due to the enzyme *carbonic anhydrase* in erythrocytes. We'll explore how this enzyme works in mammalian erythrocytes later.

Factors other than the partial pressure gradient influence the rate of gas transfer

We have been discussing diffusion of O_2 and CO_2 between the blood and the environment in terms of the gases' partial pressure (ΔP) gradients. Recall that, according to Fick's law of diffusion, the rate of diffusion of a gas through a sheet of tissue also depends on the surface area (A) and thickness of the membrane (ΔX) through which the gas is diffusing and on the diffusion coefficient (D). Let's examine these factors more closely.

In an individual animal, D is not typically alterable, except that it increases somewhat as temperature increases. However, during intense activity, A and ΔX can be physiologically altered to enhance the rate of gas transfer. During resting conditions in mammals, some of the pulmonary capillaries are typically closed because the normally low pressure of the pulmonary circulation is inadequate to keep all the capillaries open. During activity, when the pulmonary blood pressure is raised as a result of increased cardiac output, many of the previously closed pulmonary capillaries are forced open. This increases the surface area of blood available for exchange. Furthermore, the alveolar membranes are stretched further than normal during activity because of the larger tidal volumes (deeper breathing). Such stretching increases the alveolar surface area and decreases the thickness (ΔX) of the alveolar membrane.

Similarly, in resting fishes the functional surface area is the total surface area available for gas exchange. For example, during quiet breathing of water in trout only 60% of the available lamellae are perfused with water. As the level of activity increases, more water channels are recruited for ventilation.

Inadequate gas exchange can occur when the thickness of the barrier separating the air or water from the blood is pathologically increased. As the thickness increases, the rate of gas transfer decreases because a gas takes longer to diffuse through the greater thickness. Thickness increases because of (1) *low or high environmental pH* in water breathers: Acute exposure of freshwater fish to pH values between 4.0 to 5.5 or between 8.5 to 10.5 results in mucification and inflammatory swelling of the gill epithelium; (2) *pulmonary edema* in air breathers, an excess accumulation of interstitial fluid between the alveoli and pulmonary capillaries, caused by pulmonary inflammation or left-sided congestive heart failure; (3) *pulmonary fibrosis* in air breathers involving replacement of delicate lung tissue with thick fibrous tissue in response to certain chronic irritants; and (4) *pneumonia* in air breathers, which is characterized by inflammatory fluid accumulation within or around the alveoli. Most commonly, pneumonia is due to bacterial or viral infection of the lungs, but it may also arise from accidental aspiration (breathing in) of food, vomitus, or chemical agents.

The preceding discussion focused on individual animals. As we have touched on earlier, all the components of Fick's law can be evolutionarily adapted over generations if not always within an individual. For example, the area of the gill surface available for gas exchange and the diffusion distance have evolved to match the lifestyles of different fish species. At one extreme are the air-breathing fish that have comparatively small gill surface areas and thick lamellae, so much so that in some cases the gills are virtually functionless in terms of gas exchange. Alternatively, fast-swimming fishes such as tuna and mackerel have particularly thin lamellae and a comparatively large surface area available for gas exchange.

Gas exchange across the systemic capillaries also occurs down partial pressure gradients

Just as they do at the pulmonary and branchial (gill) capillaries, O_2 and CO_2 move between the systemic capillary blood and the tissue cells by simple passive diffusion down partial pressure gradients. Refer again to Figure 11-27 for a mammal. The arterial blood that reaches the systemic capillaries is essentially the same blood that left the lungs by means of the pulmonary veins—that is, arterial P_{O_2} is 100 mm Hg and P_{CO_2} is 40 mm Hg—because the only two places in the entire circulatory system at which gas exchange can take place are the pulmonary capillaries and the systemic capillaries.

Cells constantly consume O_2 and produce CO_2 through oxidative metabolism (Chapter 2). Cellular P_{O_2} in a mammal averages about 40 mm Hg and P_{CO_2} about 46 mm Hg, although these values are highly variable, depending on the level of cellular metabolic activity. Oxygen moves by diffusion down its partial pressure gradient from the entering systemic capillary blood ($P_{1_{O_2}}$ = 100 mm Hg) into the adjacent cells ($P_{2_{O_2}}$ = 40 mm Hg) until equilibrium is reached with the P_{O_2} of venous blood leaving the systemic capillaries equal to the tissue P_{O_2} at an average of 40 mm Hg. The reverse situation exists for CO_2. Carbon dioxide rapidly diffuses out of the cells ($P_{1_{CO_2}}$ = 46 mm Hg) into the entering capillary blood ($P_{2_{CO_2}}$ = 40 mm Hg) down the gradient created by the ongoing production of CO_2, until blood P_{CO_2} equilibrates with tissue P_{CO_2}. Accordingly, the blood leaving the systemic capillaries has an average P_{CO_2} of 46 mm Hg. This systemic venous blood returns to the heart and is subsequently pumped to the lungs as the cycle repeats itself (Figure 11-27).

The more actively a tissue is metabolizing, the lower the cellular P_{O_2} falls and the higher the cellular P_{CO_2} rises. As a consequence of the larger blood-to-cell partial pressure gradients, more O_2 diffuses from the blood into the cells and more CO_2 moves in the opposite direction before blood P_{O_2} and P_{CO_2} achieve equilibrium with the surrounding cells. Thus, the amount of O_2 transferred to the cells and the amount of CO_2 carried away from the cells depend on the rate of cellular metabolism.

This ends our examination of the detailed gas-exchange gradients at steps 2 and 4 of Figure 11-1. Next we look in depth at the effects of the bulk transport step 3.

check your understanding 11.8

Describe how all the gas gradients of Figure 11-27 arise; apply them to fish.

11.9 Circulatory Transport and Gas Exchange

Oxygen picked up by the circulation at the respiratory organs must be transported to the tissues for cell use. Conversely, CO_2 produced at the cellular level must be transported to the respiratory organs for elimination. In theory, this could be done by simply dissolving the gases in the circulatory water—and in some animals that is the case. However, in most circulatory systems, there is another way.

Most O_2 in many animal circulatory systems is transported bound to hemoglobins, hemocyanins, and other metal-containing respiratory pigments

Oxygen can present in the blood in two forms: physically dissolved as *free gas*, and chemically bound to a *respiratory protein or pigment*. As you saw in Chapter 9 (p. 391), for vertebrate *hemoglobin*, the pigment may be carried in circulating cells such as erythrocytes. However in some animals, the pigments are extracellular. There is a wide diversity of these pigments, but all use a *metal ion* to bind and release O_2:

1. *Hemoglobins (Hb)*, which have two parts: (1) the **globin** portion, a protein made up of a highly folded polypeptide chain, and (2) one iron-containing, nonprotein group known as **heme** bound to each of the polypeptides (Figure 11-28a, b). It appears blue when deoxygenated and red when oxygenated. Hemoglobins are the most common respiratory pigments, found not only in vertebrates but in many annelids, and some mollusks and crustaceans. They are usually intracellular. Most consist of two or more subunits; for example, mammalian Hb is called a **tetramer** because it has four subunits (Figure 11-28a).

    Hemoglobins are found in some species of plants, fungi, protists, and even bacteria; and related proteins called *protoglobins* are found in some archaea. The protein–heme–iron complex may have originally evolved to detoxify toxic gases like H_2S, only later becoming adapted for O_2 transport.

2. *Hemocyanins (Hc)*, which use a pair of *copper* ions directly bound to amino acid side chains to bind O_2. As a result of the copper atoms, the blood of these animals appears blue (colorless when deoxygenated). Hemocyanins are the second most common type of respiratory pigment. They are usually extracellular and are therefore very large proteins that will not leak out in excretory organs, and fall into two groups with very different structures:

    (i) Arthropods (in the Ecdysozoa branch of the animal kingdom, Figure 1-1) have **hexameric** (six-unit) Hc complexes akin to hemoglobin, but several hexamers can join together. For example, spider Hc forms

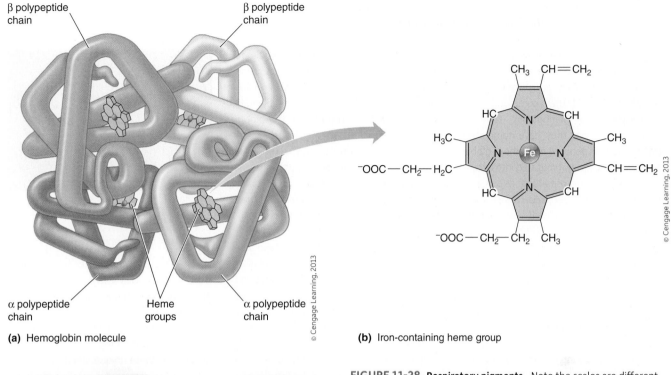

β polypeptide chain

β polypeptide chain

α polypeptide chain

Heme groups

α polypeptide chain

© Cengage Learning, 2013

(a) Hemoglobin molecule

© Cengage Learning, 2013

(b) Iron-containing heme group

(c)

(d)

FIGURE 11-28 Respiratory pigments. Note the scales are different for each figure. (a, b) Hemoglobin molecule from a mammal. This tetrameric hemoglobin molecule consists of (a) four highly folded polypeptide chains (the globin portion) and (b) four iron-containing heme groups. (c) Hemocyanin from a tarantula, made from four hexameric complexes; each polypeptide chain has copper-containing heme groups. (d) Hemocyanin from a keyhole limpet, assembled from decamers that form a cylinder of 160 total polypeptide chains and 160 copper-containing heme groups.

Sources: (c) Voit, R. et al. Complete Sequence of the 24-mer Hemocyanin of the Tarantula Eurypelma californicum. Published, JBC Papers in Press, August 28, 2000, DOI 10.1074/jbc.M005442200. (d) Reprinted from Journal of Molecular Biology, 385/3, Christos Gatsogiannis and Jürgen Markl, Keyhole Limpet Hemocyanin: 9-Å CryoEM Structure and Molecular Model of the KLH1 Didecamer Reveal the Interfaces and Intricate Topology of the 160 Functional Units, 963-983, 2009, with permission from Elsevier.

by four hexamers joining (24 total proteins) (Figure 11-28c).

(ii) Annelids and mollusks (in the Lophotrochozoan branch, Figure 1-1) have very different hemocyanins based on **decamers** (10-protein) clusters that form hollow rings. Two or more decamers, in turn, may assemble into huge cylinders (Figure 11-28d) containing up to 160 copper ions, each of which can bind one O_2!

3. *Hemerythrin* also uses iron and is red, but the iron is not in a heme complex. This pigment is found intracellularly in certain groups of marine brachiopods (lamp shells) and some worm groups such as sipunculids (peanut worms) and one known annelid.

4. *Chlorocruorin and erthyrocruorin* are pigments found only in some annelids. Both are iron/heme proteins like hemoglobin, but are much larger and found extracellularly. Chlorocruorin appears green to red when oxygenated, while erythrocruorin is red.

Annelids are the perhaps most remarkable group: In this one phylum can be found *five* different respiratory pigments: hemoglobin, hemocyanin, chlorocruorin, erthyrocruorin, and hemerythrin.

Insects, as air breathers with direct tracheal input to their tissues, normally lack a respiratory pigment. However, in the aquatic larvae of the caddis fly (*Chironomus*) and the water bugs *Anisops* and *Buenoa*, hemocyanin plays a role in O_2 uptake. Recall that O_2 uptake is more difficult in water than in air (p. 498).

For the rest of this section, we focus primarily on vertebrate respiratory pigments. In a mammal, very little O_2 is physically dissolved in plasma because O_2 is poorly soluble in warm fluids (Table 1-1). The amount dissolved is directly proportional to the P_{O_2} of the blood; the higher the P_{O_2}, the more O_2 dissolved. At a normal arterial P_{O_2} of 100 mm Hg and body temperature of 37°C, only 3 mL of O_2 can dissolve in 1 liter of blood. Thus, in a human, only 15 mL of O_2/min can be dissolved in the normal pulmonary blood flow of 5 L/min (the resting cardiac output in a human). Even under resting conditions, human cells consume 250 mL of O_2/min in total, and this may increase up to 25-fold during strenuous activity. To deliver the O_2 required by the tissues

A Closer Look at Adaptation
Life in the Cold: The Hemoglobinless Icefish

Antarctic icefish (Chaenichthyidae) are noted for having a low metabolic rate and sluggish lifestyle. They are unique among vertebrates in that they completely lack erythrocytes and hemoglobin. Genome analysis shows that they have completely lost the β-globin gene, whereas the α-globin gene is still present but mutated into a transcriptionally inactive pseudogene (p. 33). In the absence of red blood cells, these fish appear whitish and translucent—hence their common name. Icefish are sizable: They can weigh over 2 kg and reach a length of approximately 0.6 m.

Icefish are ectotherms with body temperatures essentially equal to that of the environment. A lack of hemoglobin in this species represents an adaptation to a stable environmental condition of cold (−2°C) as well as water with a relatively high O_2 content (recall that cold water dissolves gases better than warm water). In the low ambient temperatures, the viscosity of the blood is increased. Elimination of red blood cells helps to offset the decrease in blood flow. However, in the absence of hemoglobin, only O_2, which is physically dissolved in solution, can be transported in the blood. Because icefish live in cold, well-oxygenated waters, the arterial P_{O_2} is a relatively high 110 mm Hg. Cardiac output is also remarkably elevated in icefish, which helps ensure that sufficient quantities of O_2 can be taken up at the gill and delivered to the tissues despite a low arterial–venous O_2 difference. Relative heart size and blood volume are also increased in these animals. The increased cardiac output is maintained by an increase in stroke volume and not heart rate, which is a relatively slow 14 beats per minute. Peripheral resistance to blood flow is also reduced in icefish, which ensures that the work actually performed by the heart is kept to a minimum despite the large volume of blood being pumped. Compared to most fishes, the gill area of

An Antarctic icefish appears pale in part due to lack of hemoglobin.

icefish is reduced and the secondary lamellae are thick. Gas exchange does occur across the gills, although up to 40% of their O_2 requirement can be obtained through the skin. An increased density of capillaries in the skin ensures adequate gas exchange. Because of these unique adaptations, Antarctic icefish can survive without the presence of Hb in their blood, but they are limited to well-oxygenated, icy cold waters.

---

even at rest, the cardiac output would have to be 83.3 L/min if O_2 could only be transported in dissolved form! Obviously, there must be an additional mechanism for transporting O_2 to the tissues.

In vertebrates, this mechanism takes the form of hemoglobin (Hb). Jawless fishes (lampreys and hagfish) have one to three of these polypeptides making up their hemoglobins. Other vertebrates have a tetrameric Hb (Figure 11-28a). In adult mammals, there is one type of tetramer designated $\alpha_2\beta_2$, made from two proteins, α- and β-globin. Many fish, however, have multiple variants (isomers) of α- and β-globin such that there are different tetramer combinations. Each of the four iron atoms in the tetrameric form can combine reversibly with one molecule of O_2; thus, each hemoglobin molecule in most vertebrates can pick up four O_2 passengers in the lungs or gills. Only 1.5% of the O_2 in the blood of mammals is free (dissolved); the remaining 98.5% is transported in combination with Hb. In fishes the amount of free O_2 is somewhat higher, at about 5% of the total O_2 content at physiological P_{O_2} levels. One notable exception is the Antarctic icefish, unique among vertebrates because it lacks Hb (see *A Closer Look at Adaptation: Life in the Cold: The Hemoglobinless Icefish*).

Recall that O_2 *bound to Hb does not contribute to the* P_{O_2} *of the blood;* thus, blood P_{O_2} is a measure not of the total O_2 content of the blood but only of the portion of O_2 that is dissolved. As we noted earlier (p. 498), bound O_2 does not factor into the diffusion equation (Fick's law, p. 494) and by doing so, helps to maximize diffusion and thus gas exchange. Later we will show how this works in more detail.

Hemoglobin has the ability to form a loose, easily reversible combination between Fe and O_2. When not combined with O_2, Hb is referred to as **reduced** or **deoxyhemoglobin;** when combined with O_2 (thereby releasing protons), it is called **oxyhemoglobin (HbO$_2$)**:

$$HbH_4^+ + 4O_2 \leftrightarrow Hb(O_2) + 4H^+$$
$$\text{deoxyhemoglobin} \quad \text{oxyhemoglobin}$$

The concentration of Hb in an animal is not fixed. For example, hemoglobin increases in the plasma of mammals can occur with:

1. *Seasonal changes.* During cold, acclimatization concentrations of Hb increase (such as in the horseshoe rabbit).
2. *Locomotory activity.* This stimulates some animals to increase the release of erythrocytes from the spleen (such as horse, diving mammals, fish), and prolonged aerobic activity triggers hormonal stimulation (EPO, p. 393) of the bone marrow to produce more erythrocytes.
3. *High altitude.* This and other prolonged exposure to low ambient O_2 can also trigger the production of more erythrocytes (via EPO).
4. *Stress and disease.* For example, *anemia*—a state in which erythrocyte levels are abnormally low—can be triggered by stress hormones, immune cytokines (p. 469) and many diseases.

Myoglobin stores oxygen in aerobic muscle and may facilitate diffusion from the blood to the mitochondria

Vertebrates also have other respiratory pigments, including one called **myoglobin (Mb)**. It is a monomer (single protein unit) with one heme that serves to store O_2 in muscle, especially Type I (red aerobic) skeletal fibers (p. 362) and heart (contributing to these muscles' red color). Myoglobin's function as a reserve for O_2 is well characterized; for example, because contraction of skeletal muscle and heart reduces blood flow by squeezing actions, Mb can release O_2 to compensate. But a longstanding hypothesis predicts that Mb will provide an alternate pathway for O_2 diffusion, with Mb picking up O_2 near the membrane (where P_{O_2} is higher) and diffusing to the mitochondria where the O_2 would be released (where P_{O_2} is lower). It has proved quite difficult to test this hypothesis in the past. The technique of "knockout" genes (Chapter 2) has allowed researchers to make mice completely lacking myoglobin! The heart and muscles of these mice are pale pink rather than the normal robust red color. Amazingly, physiological studies showed no impairment in muscular or cardiovascular activity in these animals. Does this mean that myoglobin is a useless protein? No, because one study found that the hearts of these mice have a much higher density of capillaries, which delivers enough O_2 to make up for the lack of Mb storage. By unknown signals during development, low supplies of O_2 to heart fibers presumably triggered this compensatory angiogenesis. These experiments in fact show that Mb is useful and that its absence requires compensation to maintain homeostasis; but they did not clarify whether Mb's role is simply for storage or to enhance diffusion. As we noted in Chapter 2, the technique of knocking out genes does not always yield simple results and answers, because of the complex feedback processes in living systems (p. 35).

Neurons contain neuroglobin, while fibroblasts contain cytoglobin

By searching genome data for globin-type codes, researchers have found two more classes of vertebrate globin, named **neuroglobin (Nb)** and **cytoglobin (Cb)**. Neuroglobin is found in neurons including the retina, as well as some endocrine cells. This oxygen-binding protein is thought to enhance neural O_2 supply in a way similar to myoglobin. Researchers have found that Nb can protect brains of mice from stroke damage resulting from insufficient O_2. The mice were injected with a virus that was engineered to produce high levels of Nb. Recently, Nb was also found to block a key step in apoptosis, preventing neuronal cell death following injury.

Cytoglobin has been found in fibroblasts (p. 469) and a few other cell types. Its function is uncertain, but there is evidence that it can protect the cell from free-radical damage.

The P_{O_2} is the primary factor determining the percent hemoglobin saturation

Let's return to hemoglobin. We need to answer several important questions about the role of respiratory pigments in O_2 transport. For example, what determines whether O_2 and Hb are combined or dissociated (separated)? Why does Hb combine with O_2 in the lungs and release O_2 at the tissues? How can a variable amount of O_2 be released at the tissue level, depending on the level of tissue activity? How can we talk about O_2 transfer between blood and surrounding tissues in terms of O_2 partial pressure gradients when most of the O_2 is bound to Hb and thus does not contribute to the P_{O_2} of the blood at all?

Tetrameric hemoglobin is considered *fully saturated* when all the Hb present is carrying its maximum O_2 load of four O_2s each. The **percent hemoglobin (% Hb) saturation**, a measure of the extent to which the Hb present is combined with O_2, can vary from 0 to 100%. The most important factor determining the % Hb saturation is the P_{O_2} of the blood, which, in turn, is related to the concentration of O_2 physically dissolved in the blood (P_{O_2}). According to the **law of mass action,** if the concentration of one of the substances involved in a reversible reaction is increased, the reaction is driven toward the opposite side. Conversely, if the concentration of one of the substances is decreased, the reaction is driven toward that side. Applying this law to the reversible reaction involving Hb and O_2 (Hb + O_2 ↔ HbO_2), when the blood P_{O_2} is increased, as it is in the pulmonary capillaries, the reaction is driven toward the right side of the equation, resulting in increased formation of HbO_2 (increased % Hb saturation). When the blood P_{O_2} is decreased, as it is in the systemic capillaries, the reaction is driven toward the left side of the equation and O_2 is released from Hb as HbO_2 dissociates (decreased % Hb saturation). Thus, because of the difference in P_{O_2} at the lungs and other tissues, Hb automatically "loads up" on O_2 in the lungs, where fresh supplies of O_2 are continually being provided by ventilation, and "unloads" it in the tissues, which are constantly using up O_2. Let us examine how this works.

Cooperativity among hemoglobin subunits allows for effective delivery of oxygen

Binding of ligands to most proteins (such as substrates to enzymes) follows a *hyperbolic* curve, not a linear relationship, as shown in Figure 11-29a (orange curve) for myoglobin (Mb) and O_2. This type of curve results from *saturation* of the protein; that is, after a certain concentration of ligand is reached, no more bound complex can form because all the proteins are occupied by ligands. The relationship between blood P_{O_2} and % Hb saturation is not hyperbolic, however, a point that is very important physiologically. The relationship between these variables follows an S-shaped *sigmoidal* curve, known as the **O_2–Hb dissociation** (or **saturation**) **curve** (hemoglobin curves of Figures 11-29a and b). With the exception of species that contain monomeric oxygenated Hb (such as agnathans), the curves are sigmoidal in shape because hemoglobins exhibit **cooperativity** of binding; that is, the loading or unloading of the first O_2 on Hb makes it sequentially easier for subsequent O_2s to load or unload. Cooperativity arises from conformational changes that occur in the globins, which oscillate between one of two states: a high-affinity or "relaxed" form called the **R-state,** and a low-affinity or "tense" form called the **T-state:** The R-state is favored at high P_{O_2} levels (thus, "R" is easier to remember as the "red" form), whereas the T-state is favored at low P_{O_2} levels. To see how this works, examine Figure 11-29b.

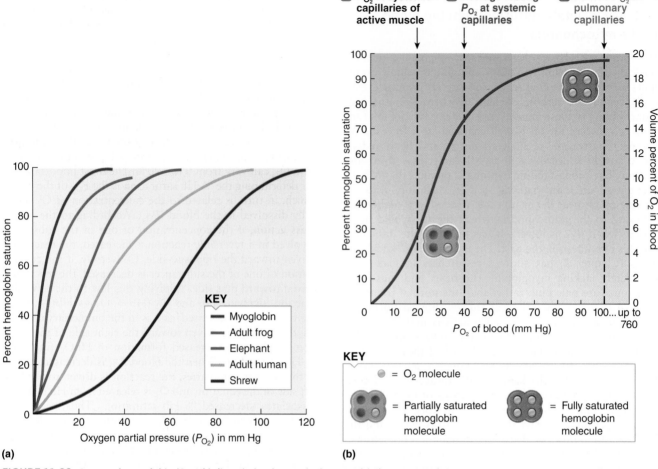

FIGURE 11-29 Oxygen–hemoglobin (O₂–Hb) dissociation (saturation) curve. (a) The percent Hb saturation (the scale on the left side of the graph) depends on the P_{O_2} of the blood. The relationship between these two variables is depicted by a curve running from zero to high P_{O_2}, where the protein is saturated. Shown here are the curves for the myoglobin pigment, and hemoglobins from a frog, an elephant, a human, and a shrew. The myoglobin has a hyperbolic curve, while the hemoglobins have an S-shaped sigmoidal curve, with a plateau region between a blood P_{O_2} of 60 and 100 mm Hg and a steep portion between 0 and 60 mm Hg. (b) O₂–Hb curve for a human. Points 1, 2, and 3 show the binding at the partial pressures in the lungs, average tissues, and active muscles, respectively (see text for details). Another way of expressing the effect of blood P_{O_2} on the amount of O₂ bound with Hb is the volume percent of O₂ in the blood (mL of O₂ bound with hemoglobin in each 100 mL of blood). That relationship is represented by the scale on the right side of the graph.

© Cengage Learning, 2013

Unloading Oxygen At the lungs (point 1 of Figure 11-29b), the high P_{O_2} of the alveolar input favors an all R (R₄) state, so the Hb is nearly 100% saturated. Now, as the blood proceeds into systemic circulation to capillaries in a typical tissue (point 2) with an intermediate P_{O_2} level of 40 mm Hg, the T-state becomes increasingly favorable, and the Hb is in an average R₃T₁ conformation, releasing about 25% of its O₂. Moreover, the globin that releases the first O₂ not only shifts to the T-state but will also affect its neighboring globin units, making them more likely to shift to the T-state and thus releasing their oxygens. This is co-called cooperativity makes the curve become steeper at this point.

The value of this steepness is apparent when the blood instead goes to an active muscle (point 3) with a P_{O_2} of 20 mm Hg. As the plunging curve indicates, even more O₂'s are rapidly released, with the Hb now in an average R₁T₃ state (75% unloaded). Thus, in the range of tissue P_{O_2}

(about 0–60 mm Hg), a modest drop in systemic capillary P_{O_2} can result in large amounts of O₂ immediately available to meet the O₂ needs of the more actively metabolizing tissues. As much as 85% of the Hb may give up its O₂ to actively metabolizing cells during strenuous activity. In addition to this more thorough withdrawal of O₂ from the blood, even more O₂ is made available to actively metabolizing cells, such as exercising muscles, by circulatory and respiratory adjustments that increase the flow rate of oxygenated blood through the active tissues.

Thus, for O₂ delivery, the T-state is of particular importance. Iron has an inherently high, nearly irreversible affinity for O₂ (we know the basic Fe–O₂ reaction as rust!). In the R-state, the iron does have a very high affinity. However, the protein in the T-state essentially alters the iron atom's electrons in the heme to reduce its affinity for the gas. This is why the globin is necessary; without it, the iron-heme com-

TABLE 11-2 Hemoglobin P_{50} Values for Various Mammals and Birds

Animal	Habitat; Body Size	Hemoglobin P_{50} (mm Hg)
Elephant (African)	Low altitude; 5,000 kg	20
Human	Low altitude; 70 kg	27
Deer (sika)	Low altitude; 40 kg	28
Vicuña (Andes)	3,500–5,500 m; 40 kg	21
Rodents (rat, shrew, etc.)	Low altitude; 0.005–0.8 kg	36 or higher
Long-tailed Andean chinchilla	Up to 3,000 m; 0.55 kg	27
Short-tailed Andean chinchilla	3,000–5,000 m; 0.5 kg	23
Mole (European)	Subterranean; 0.05 kg	21
Greylag goose (India)	Low altitude; 3.5 kg	39.5
Andean goose	5–6,000 m; 3 kg	34
Bar-headed goose (Himalayas)	9,000 m; 3 kg	30

© Cengage Learning, 2013

plex would readily bind O_2 at the lungs, but would then not release it unless O_2 were extremely low (as seen for myoglobin in Figure 11-29a).

Loading Oxygen The reverse sigmoidal process occurs as an unloaded hemoglobin enters an area of high P_{O_2} level (i.e., as it returns to the lungs) for rapid reloading of O_2's. This takes place on the right side of the graph where saturation climbs steeply, then reaches a plateau. Like the steep part of the curve, the plateau also has physiological significance. Recall that mammalian systemic arterial blood leaving the lungs normally has a P_{O_2} of 100 mm Hg. Looking at the O_2–Hb curve, note that at a blood P_{O_2} of 100 mm Hg, Hb is 97.5% saturated. Therefore, Hb in systemic arterial blood normally is almost fully saturated.

If arterial P_{O_2} falls below normal, there is little reduction in the total amount of O_2 transported by the blood until the P_{O_2} falls below 60 mm Hg (in the human example) because of the plateau. If the arterial P_{O_2} falls 40%, from 100 to 60 mm Hg, the % Hb saturation is still remarkably high at 90%. Accordingly, the total O_2 content of the blood is only slightly decreased despite the 40% reduction in P_{O_2} because Hb is still carrying an almost-full load of O_2. However, even if the blood P_{O_2} is greatly increased, say, to 600 mm Hg by breathing pure O_2, very little additional O_2 is added to the blood. Thus, the plateau portion of the O_2–Hb curve provides a good margin of safety in O_2-carrying capacity of the blood. A P_{O_2} of 60 mm Hg is about what human alveolar air would be at 2,400 m (8,000 ft) elevation; the drop in % Hb saturation to 90% is clearly noticeable to a person arriving suddenly from sea level, but the drop is generally not life-threatening in a healthy individual. However, on Mt. Everest, arterial P_{O_2} values in climbers have been measured at 28 mM Hg!

Arterial P_{O_2} may be reduced because of pulmonary or gill diseases accompanied by inadequate ventilation or defective gas exchange or by circulatory disorders that result in inadequate blood flow to the respiratory surfaces. How-

ever, unless the arterial P_{O_2} becomes markedly reduced in either pathological conditions or abnormal environmental circumstances, near-normal amounts of O_2 can still be carried to the tissues.

Evolutionary adaptation results in different hemoglobin P_{50} values

So far we've looked at the human curve in detail, but as Figure 11-29a shows, other hemoglobins can have different binding curves. Although all have a sigmoidal shape, they differ in their relative affinity of Hb for O_2. Affinity is characterized by the P_{50} value—the partial pressure of O_2 at which 50% saturation of the blood is reached. For adult human hemoglobin (Figure 11-29b), the P_{50} is 27 mm Hg. If the affinity for O_2 is greater (that is, Hb gives up O_2 less readily), then the dissociation curve is located further to the *left* of the human one; that is, it has a *lower* P_{50} value. For example, fetal humans have a Hb made of α and γ subunits with a P_{50} value of 19 mm Hg, so it can readily take up O_2 from the mother's Hb. Consequently, the plateau region is shifted to the left, meaning the Hb can pick up O_2 at the respiratory surface at lower alveolar P_{O_2} levels than the adult human Hb. If the affinity for O_2 is lower, and so Hb gives up O_2 more easily, the curve is shifted to the *right*; that is, it has a *higher* P_{50} value. A higher P_{50} value does not indicate a lower carrying capacity for O_2; it simply means that O_2 is more easily unloaded at the tissues.

Let's look at the evolution of P_{50} values in terms habitats and gas gradients. Note that that there is (1) a positive correlation between Hb O_2 affinity and body size, and (2) an inverse correlation between O_2 affinity and environmental O_2 (summarized in Table 11-2).

P_{50} **and Body Size** In mammals, O_2 affinity increases (thus, P_{50} decreases) with body size. For example, as seen in Figure 11-29a and Table 11-2, elephant Hb has a P_{50} around 20 mm Hg, while shrews (Figure 11-29a) as well as mice and rats have P_{50} values of 36 or higher. This is a *scaling* phenomenon, relating to the fact that metabolic rate per unit mass increases as body size decreases (see Chapter 15). For the small rodents, at the typical alveolar P_{O_2} of 100 mm Hg, Hb saturates well, but in their tissues, Hb unloads O_2 more readily than does human or elephant Hb to support their very high metabolic rates. Conversely, the elephant's relative low metabolic rate does not require as effective off-loading of O_2, but its Hb is better able to pick up O_2 at the lungs if environmental P_{O_2} drops.

P_{50} and Environmental P_{O_2} Not surprisingly, high affinity is also common in animals adapted to high altitudes, where a low P_{50} allows Hb to saturate at the lungs even if alveolar P_{O_2} drops below 60 mm Hg. For example, compare a lowland sika deer and a high-altitude Andean vicuña (with similar body sizes) in Table 11-2. Small mammals at high altitude also show the same trend; for example, in Table 11-2 compare low-altitude rodents to the long-tailed chinchilla of the Andes, which lives on the mountain slopes up to 3,000 m, and to the congeneric short-tailed chinchilla from 3,000–5,000 m.

Animals living underground may frequently experience low ambient O_2 levels. Correspondingly, the subterranean mole *Tulpa europea* has a low P_{50} (Table 11-2).

Birds show similar trends (Table 11-2). The Andean goose from 5,000–6,000 m has a Hb with a P_{50} about 6 mm Hg lower than low-altitude greylag geese. And the bar-headed goose, which has been observed flying over Mt. Everest at over 9,000 m, has a P_{50} about 10 mm Hg lower. Molecular analysis shows that only one or two amino acid differences in this bird's Hb can account for this adaptation. We will return to this remarkable bird later.

Similarly, we find low-P_{50} (high-affinity) hemoglobins in aquatic animals living in low-oxygen waters. Sluggish fish, for example, such as the plaice (a type of bottom-dwelling flatfish) and carp, rely on Hb with a low P_{50} (for example, 18 mM Hg in the European plaice) to work in an environmental P_{O_2} which may be 35 mm Hg (versus a venous P_{O_2} of 11 mm Hg; hematocrit 14%). Under these circumstances there is still a large partial pressure gradient between the water and the gill capillaries, although the gradient between the blood and the tissues is reduced. These species have thus optimized O_2 extraction from the water while maintaining a sedentary lifestyle that rarely requires strenuous activity. Recall also that many fish have multiple Hb variants. Some of these have slightly different temperature responses to aid in thermal acclimatization, but in some they have different P_{50} values. For example, the turbot (*Scophthalmus maximus*, a flatfish), has three different Hb tetramers, one of which has a higher O_2 affinity than the others.

One such hypoxia-adapted fish, the Namibian bearded goby (*Sufflogobius bibarbatus*), is being hailed as an "environmental savior"! This fish spends its days hiding in the highly anoxic seafloor, eating anaerobic bacteria. At night, it swims to the surface to reoxygenate. Overfishing off the coast of Namibia has led to a collapse in the once-dominant fish, sardines, and an explosion in jellyfish. The bearded goby, however, has begun to eat the jellyfish (the only predator to do so), preventing the ecosystem shift from getting worse.

Active aquatic species in well-oxygenated waters such as trout have a comparatively low-affinity Hb (high P_{50}),

befitting their high arterial P_{O_2} (113 mm Hg versus a venous P_{O_2} of 32 mm Hg, hematocrit 40%). Trout ($P_{50} = 24$ mM Hg) have a large diffusion gradient between the blood and the metabolizing tissues. One disadvantage of this strategy is the comparatively low efficiency of O_2 extraction from the water should the environmental P_{O_2} drop. However, active species normally inhabit well-aerated waters. Moreover, during periods of intense activity, the O_2-carrying capacity of their blood can be rapidly increased by the release of erythrocytes from their storage site in the spleen (p. 545).

By acting as a storage depot, respiratory pigments promote the net transfer of O_2 from the respiratory surfaces to the circulatory fluid

Let's return to the role of respiratory pigments in gas exchange. Recall our earlier discussion of O_2 being driven from the alveoli to the blood by the ΔP_{O_2} between the alveoli, P_1, and the blood, P_2 (Figure 11-28). As we have noted (p. 530), respiratory pigments play a crucial role in permitting the transfer of large quantities of O_2 before blood P_{O_2} equilibrates with the surrounding tissues (Figure 11-30). It does so by keeping P_2 low through acting as a "storage depot" for O_2, removing the O_2 from solution by "trapping" it in a nondiffusible form as soon as it enters the blood from the alveoli. Because only diffusible (dissolved) O_2 contributes to the P_{O_2}, the O_2 bound to Hb cannot contribute to blood P_{O_2}. When systemic venous blood enters the pulmonary capillaries, its P_{O_2} is considerably lower than the alveolar P_{O_2}, so O_2 immediately diffuses into the blood, raising the blood P_{O_2}. As soon as the P_{O_2} of the blood increases, the percentage of Hb that can bind with O_2 likewise increases, as indicated by the O_2–Hb curve. Consequently, most of the O_2 that has diffused into the blood combines with Hb. As O_2 is removed from solution by combining with Hb, blood P_{O_2} falls to about the same level it was when the blood entered the lungs, although the total quantity of O_2 in the blood actually has increased. Because the blood P_{O_2} is once again below alveolar P_{O_2}, more O_2 diffuses from the alveoli into the blood, only to be soaked up by Hb again. Essentially, this maximizes the ΔP of Fick's law by keeping P_2 low.

Net diffusion of O_2 from alveoli or water to blood occurs continuously until Hb becomes saturated with O_2 as completely as it can be at that particular environmental P_{O_2}. Not until Hb is maximally saturated for that P_{O_2} does all the O_2 transferred into the blood remain dissolved and directly contribute to the P_{O_2}. Only at this time does the blood P_{O_2} rapidly equilibrate with the alveolar P_{O_2} and halt further O_2 transfer.

The reverse situation occurs at the tissue level. Because the P_{O_2} of blood entering the systemic capillaries is considerably higher than the P_{O_2} of the surrounding tissue, the gradient favors the immediate diffusion of O_2 from the blood into the tissues, lowering the blood P_{O_2}. When blood P_{O_2} falls, Hb is forced to unload some of its stored O_2. As the released O_2 dissolves in the blood, blood P_{O_2} increases and is once again above the P_{O_2} of the surrounding tissues. This favors further movement of O_2 out of the blood by keeping the ΔP component positive, although the total quantity of O_2 in the blood has already been reduced. Only when Hb can no longer release any more O_2 into solution can blood P_{O_2} become as

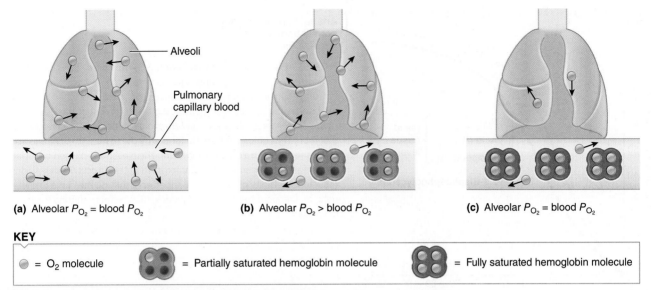

(a) Alveolar P_{O_2} = blood P_{O_2} **(b)** Alveolar P_{O_2} > blood P_{O_2} **(c)** Alveolar P_{O_2} = blood P_{O_2}

KEY

= O_2 molecule   = Partially saturated hemoglobin molecule   = Fully saturated hemoglobin molecule

FIGURE 11-30 Hemoglobin (Hb) facilitating a large net transfer of O_2 by acting as a storage depot to keep P_{O_2} low. (a) In the hypothetical situation in which no Hb is present in the blood, the alveolar P_{O_2} and the pulmonary capillary blood P_{O_2} are at equilibrium. (b) Hemoglobin has been added to the pulmonary capillary blood. As the Hb starts to bind with O_2, it removes O_2 from solution. Because only dissolved O_2 contributes to blood P_{O_2}, the blood P_{O_2} falls below that of the alveoli, even though the same number of O_2 molecules are present in the blood as in part (a). By "soaking up" some of the dissolved O_2, Hb favors the net diffusion of more O_2 down its partial pressure gradient from the alveoli to the blood. (c) Hemoglobin is fully saturated with O_2, and the alveolar and blood P_{O_2} are at equilibrium again. The blood P_{O_2} resulting from dissolved O_2 is equal to the alveolar P_{O_2}, despite the fact that the total O_2 content in the blood is much greater than in part (a) when blood P_{O_2} was equal to alveolar P_{O_2} in the absence of Hb.

© Cengage Learning, 2013

low as in the surrounding tissue. At this time, further transfer of O_2 ceases.

Hemoglobin thus permits the transfer of tremendously more O_2 from the blood into the cells than would be possible in its absence. Note that this means that measurements of blood P_{O_2} do not tell us the total O_2 being transported. If Hb levels are reduced to half of normal, as in a severely flea-bitten puppy, the O_2-carrying capacity of the blood is reduced by 49% even though the arterial P_{O_2} is the normal 100 mm Hg with normal Hb saturation.

Increased CO_2, acidity, temperature, and organic phosphates shift the O_2–Hb dissociation curve to the right, favoring unloading

Even though the primary factor determining the % Hb saturation is the P_{O_2} of the blood, other factors can affect the affinity, or bond strength, between Hb and O_2 and, accordingly, can shift the O_2–Hb curve (that is, change the % Hb saturation at a given P_{O_2}). These other factors are CO_2, acidity, temperature, and organic phosphates, which we examine separately. The O_2–Hb dissociation curves with which you are already familiar (Figure 11-29) are typical curves at normal arterial CO_2 and acidity levels, normal body temperatures, and normal organic phosphate concentrations.

Effect of P_{CO_2} An increase in P_{CO_2} shifts the O_2–Hb curve to the right (Figure 11-31a). The % Hb saturation still depends on the P_{O_2}, but for any given P_{O_2}, the amount of O_2 and Hb

that can be combined is reduced. This effect is important, because the P_{CO_2} of the blood increases in the systemic capillaries as CO_2 diffuses down its gradient from the cells into the blood. This additional CO_2 in the blood in effect decreases the affinity of Hb for O_2, so Hb unloads even more O_2 at the tissue level than it would if the reduction in P_{O_2} in the systemic capillaries were the only factor affecting % Hb saturation.

Effect of pH: The Bohr and Root Effects An increase in acidity also shifts the curve to the right (Figure 11-31a). Because CO_2 generates carbonic acid (H_2CO_3), the blood becomes more acidic at the systemic capillary level as it picks up CO_2 from the tissues. The resultant reduction in Hb affinity for O_2 in the presence of increased acidity aids in releasing even more O_2 at the tissue level for a given P_{O_2}. In actively metabolizing cells, such as exercising muscles, not only is more carbonic acid–generating CO_2 produced, but lactate also may be produced (see p. 55). The resultant local elevation of acid in the working muscles facilitates further unloading of O_2 in the very tissues that have the greatest O_2 need. The same effect occurs with arthropod hemocyanins.

The influence acid on the release of O_2 is known as the **Bohr effect,** discovered in 1904 by the Danish physiologist Christian Bohr. Both CO_2 and the hydrogen ion (H^+) component of acids can combine reversibly with Hb at sites other than the O_2-binding sites. The result is an alteration in the molecular structure of Hb that reduces its affinity for O_2. However, the magnitude of the Bohr effect is not the same for all animals with Hb. Generally it depends on overall

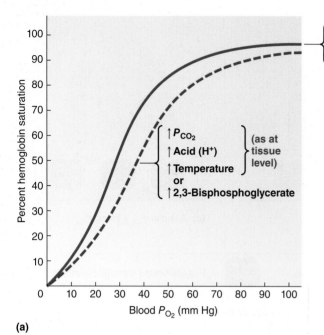

(a)

Source: (b) D. H. Evans. (1997). *The Physiology of Fishes*, 2nd ed. Boca Raton, FL: CRC Press, p. 103, Figure 1. Used by permission.

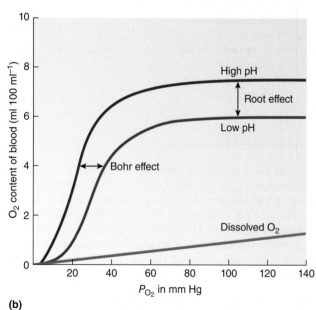

(b)

FIGURE 11-31 Key factors that alter the O_2–Hb curve. (a) Effect of increased P_{CO_2}, H^+, temperature, and 2,3-bisphosphoglycerate on the O_2–Hb curve (human). Increased P_{CO_2}, acid, temperature, and 2,3-bisphosphoglycerate, as found at the tissue level, shift the O_2–Hb curve to the right. As a result, less O_2 and Hb can be combined at a given P_{O_2} so that more O_2 is unloaded from Hb for use by the tissues. (b) Bohr and Root effects for Hb from a teleost (bony fish). The plot shows oxygen dissociation curves for teleost hemoglobins demonstrating the effects of acid. Acidosis reduces the affinity of Hb for O_2 (Bohr effect), shifting the curve to the right, and reduces the capacity of binding (Root effect), shifting the curve downward.

sitivity to pH and nucleoside triphosphates. The differences between the isoforms become evident under conditions where the acidity of the blood is increased. In these circumstances the maximal O_2-binding capacity of one form is reduced. The effect of acidosis on lowering blood O_2 content is termed the **Root effect** (Figure 11-31b). Hydrogen ion binds to this isoform of Hb and drastically reduces its total capacity (not just its affinity, which is the Bohr effect) for O_2. The Root effect is important when establishing the elevated P_{O_2} values in the gas bladder (see below) and may also play a role during periods of lactate accumulation associated with strenuous activity. The presence of an acid-insensitive isoform of Hb helps minimize disturbances in blood O_2 transport. However, in some species with multiple forms of Hb, such as the carp, each form displays a pronounced Root effect.

Effect of Temperature In a similar manner, an elevation in temperature shifts the O_2–Hb curve to the right (Figure 11-31a), unloading more O_2 at a given P_{O_2}. This is of particular importance to ectotherms because many routinely undergo increases in body temperature and thus increases in metabolic rate. The increased O_2 demands at the higher temperature can be met by increased unloading of O_2 from Hb. In endotherms, a working muscle or other actively metabolizing cell also produces heat. The resultant local elevation in temperature enhances O_2 release from Hb for use by the more active tissues.

Recently, Kevin Campbell and colleagues were able to obtain DNA from 43,000-year-old wooly mammoth remains, and to clone the Hb gene from this extinct relative of elephants that lived in cold habitats during the last Ice Ages. Unlike other hemoglobins, the mammoth Hb's P_{50} was unaffected by temperature. Since cold temperature reduces the release of O_2 by other Hbs, the researchers propose that this insensitivity allowed mammoth Hb to release O_2 normally in the animal's extremities. Like many extant polar endotherms, the mammoth's limbs, trunk, tail and ears may have been very cold to prevent heat loss from the body core (due to countercurrent heat exchangers; see Chapter 15).

Overall, increases in CO_2, acidity, and temperature (in most cases) at the tissue level, all of which are associated with increased cellular metabolism and increased O_2 consumption, enhance the effect of a drop in P_{O_2} in facilitating

body size. For example, small mammals, with a higher metabolic rate and O_2 requirement, have Hb that is more sensitive to acid, which thus aids in delivery of O_2 to the tissues. Turtles, lizards, and snakes have smaller Bohr effects than mammals of comparable body size because of a reduced affinity of hemoglobin for O_2. Alternatively, the large Bohr effect observed in crocodilians is considered an adaptation associated with the extended periods they need to remain submerged in order to drown their prey. In fish, the Bohr effect is observed after an increase in H^+ ion but not CO_2. Recall that the concentration of CO_2 in water breathers is remarkably low and has little influence on O_2 binding.

In most species of fish and reptiles, there are multiple forms of Hb. Some forms are characterized by a low affinity for O_2 and a heightened sensitivity to pH (and to *nucleoside triphosphates*, which we describe shortly), whereas other forms are characterized by a high affinity and relative insen-

the release of O_2 from Hb. These effects are largely reversed at the pulmonary level, where the extra acid-forming CO_2 is blown off and the local environment is cooler. Appropriately, therefore, Hb has a higher affinity for O_2 in the pulmonary capillary environment, thus enhancing the effect of the elevation in P_{O_2} in loading O_2 onto Hb.

Effect of Organic Phosphates The preceding changes take place in the *environment* of the red blood cells, but a factor *inside* the red blood cells of some species can also affect the degree of O_2–Hb binding: **organic phosphates.** Specific types are (1) *2,3-diphosphoglycerate (DPG)* in most mammals (such as horses, humans, dogs, and rats), (2) *inositol pentaphosphate (IPP)* in birds, and (3) *nucleoside triphosphates (NTPs)* in fishes. *Adenosine triphosphate* (ATP) is the predominant NTP of salmonoids, sharks, and rays, whereas *guanosine triphosphate* (GTP) is found in eels, carp, and goldfish. These erythrocyte constituents, which are produced during red blood cell metabolism, bind reversibly with Hb and reduce its affinity for O_2. Thus, an increased level of organic phosphate shifts the O_2–Hb curve to the right, just as with CO_2 and H^+ (Figure 11-31), enhancing O_2 unloading as the blood flows through the tissues. However, unlike the other factors (which normally are present only in the systemic capillaries, where the right-shift is advantageous in unloading O_2), DPG is present in the red blood cells throughout the circulatory system and, accordingly, shifts the curve to the right to the same degree in both the tissues and the lungs. As a result, DPG decreases the ability to load O_2 at the pulmonary level. Accordingly, in order to favor O_2 loading in respiratory organs, the response of water-breathing vertebrates to chronic hypoxia is a *decline* in NTP to increase in O_2 affinity. Similarly, mammals like llamas that have evolved to live at high altitude have less DPG than sealevel species.

Interestingly, many artiodactyls (ruminants, giraffes, deer) and cats have inherently low-affinity hemoglobins that are relatively insensitive to the effects of DPG. In contrast, in most other sealevel mammals (including humans), have high-affinity hemoglobins with DPG sensitivity. Moreover, DPG production by red blood cells gradually *increases* whenever O_2 delivery is chronically below normal, as occurs at high altitudes or in certain types of circulatory or respiratory diseases. An increase in DPG may help maintain O_2 availability when there is a decreased arterial O_2 supply by promoting O_2 release at the tissue level (Hb right shift). However, at high altitude at least, it may also simply counteract a Hb left shift ensuing as O_2-deprived animals hyperventilate (which offloads more CO_2 than normal and so raises blood pH).

Filling of the gas bladder is aided by the Root effect

The ability of an animal to achieve buoyancy or weightlessness in water is of obvious benefit because it greatly decreases the amount of energy required to maintain itself at any particular depth. A fish whose total body density is equal to that of the surrounding environment would thus be effectively weightless. One means of reducing density is to create a specific compartment containing low-density material. For example, all cartilaginous fishes have relatively light, nonmineralized bones. And some deep-sea sharks decrease their density by storing large amounts of lipid (squalene) in their livers. Consequently, these animals have

exceptionally large livers. But even these features do not make these animals neutrally buoyant, so they still tend to sink if they stop swimming. An alternative solution is to store gases—far more buoyant than lipid—within a particular compartment.

The **gas bladder** (or swimbladder) is a large, centrally located sac ventral to the spinal column in the abdominal cavity of teleost fish (Figure 11-32a). Whether a particular species of fish has a gas bladder is related to its physiological needs. For example, some bottom-dwelling and some deep-sea species lack gas bladders, and some deep-sea species have fat-filled rather than air-filled bladders because of the difficulty of inflating the bladder at high pressures. The extent to which the gas bladder must be filled to achieve buoyancy at any particular depth depends on whether the fish is a freshwater or marine type. Fresh water is less dense than seawater and so provides less buoyancy, necessitating a larger gas bladder (7 to 11% of body volume) than in marine fishes (4 to 6% of body volume), to prevent sinking.

The key to the success of the gas bladder is its ability to vary the total quantity of gas as a function of its depth in the water. If the gas bladder were to contain an unchanging concentration of gas, the particular fish would be weightless at only one depth. Because pressure increases with depth, the gas in the bladder is compressed, so decreasing the bladder's volume and increasing the relative density of the fish. To keep its increasingly denser body from sinking, the fish would have to expend energy to maintain its depth. Conversely, when a fish rises, the volume of the bladder increases and the fish becomes too light. In situations where a fish ascends from extreme depths to the surface, the bladder can actually burst. In some species, such as the carp, a duct connects the gas bladder to the esophagus, so gas can then be expelled through the mouth and gill cavities as it rises. In these species, gas may be added to the bladder by swallowing air at the surface. However, for most fishes, rising to the surface to gulp air prior to swimming at depths is impractical, and in these species the gas bladder is completely enclosed from the outside. The actual gases contained within the bladder are the same as those in water but are present in varying proportions. In most fish the gas bladder contains 80 to 95% O_2, but in some species, such as the whitefish (*Coregonus albus*), it is almost pure N_2.

The ability to adjust the quantity of gas within the bladder arises from several sources:

1. A **gas gland** within the wall of the gas bladder. The gland operates anaerobically to secrete lactate into the blood, which lowers the pH of the blood leaving the gland.
2. An unusual blood supply called the **rete mirabile** ("wonderful net"), which perfuses the gas gland. The rete mirabile is a dense bundle of capillaries arranged side by side in *countercurrent* fashion.
3. Hemoglobins exhibiting the **Root effect.**

Using these features, gas can be transferred from the blood vessels lining the gas bladder into the gas bladder. Inflating the gas bladder is an active process, because it is accomplished against a pressure gradient within the bladder, as follows (Figure 11-32b):

1. The blood arriving at the gas bladder is carrying gas at a pressure equal to that of the water passing through the gills, but with most of the O_2 in nondiffusible form as Hb–O_2.

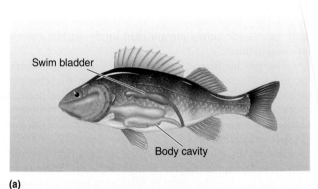

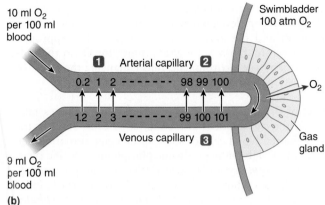

FIGURE 11-32 The gas bladder of a fish. (a) A lateral view of a teleost showing the relative location of the gas or swimbladder in the abdominal cavity. (b) Countercurrent multiplier system for filling the gas bladder. Production of lactate from the gas gland increases O_2 tension in the venous capillary (values are shown in atmospheres or atm). As a result, O_2 diffuses down its concentration gradient into the arterial capillary, thus remaining within the rete. However, the O_2 content of the venous blood exiting the rete is now lower than the incoming arterial blood with the difference deposited in the gas gland.

Sources: (a) K. Liem, W. Bemis, W. F. Walker, & L. Grande. (2001). *Functional Anatomy of the Vertebrates: An Evolutionary Perspective*, 3rd ed. Belmont, CA: Brooks/Cole, p. 358, Figure 11-6); (b): K. Schmidt Nielsen, (1997), "*Animal Physiology: Adaptation and Environment*", 5th ed. (Cambridge, UK: Cambridge University Press), p. 447, Figure 10-33. Copyright © 1997 Cambridge University Press. Reprinted with permission of Cambridge University Press.

2. The gas gland secretes lactate into the incoming capillaries, triggering the release of O_2 from isoforms of Hb having both the Bohr and Root effects. (Lactate also forces nitrogen out of solution.) The local partial pressure thus increases in the capillary, favoring diffusion of gas into the bladder. However, the gas exits the area by the outgoing capillary, and the partial pressure of this gas is equal to that of the gas within the gas bladder, much higher than that in the incoming capillary.

3. Because of the countercurrent blood flow between the opposing vessels, gas diffuses from the outgoing blood into the incoming blood that is supplying the gas gland. The dissolved gas increases in the vessel and so returns to the bladder. This process is repeated continuously, with higher and higher concentrations of gas collecting at the junction point of the incoming and outgoing capillaries within the gas gland.

It is the pH-dependent release of O_2 from Hb via the Root effect that ultimately accounts for the ability of fish to force O_2 into the gas bladder. Diffusion of gases stored in the bladder is restricted by the presence of multiple sheets of crystalline guanine, which line the wall of the gas bladder. The countercurrent rete prevents dissipation of the gradient.

Deflating the gas bladder as a fish decreases its depth is a passive process (favored when the gas gland is less active). Gases move down their partial pressure gradients into the surrounding capillaries, and are then transported to the gills.

Most CO_2 is transported in the blood as bicarbonate

When arterial blood flows through the tissue capillaries, CO_2 diffuses down its partial pressure gradient from the tissue cells into the blood. Carbon dioxide in vertebrates is transported in the blood in three ways (Figure 11-33):

1. *Physically dissolved.* As with dissolved O_2, the amount of free CO_2 *physically dissolved* in the blood depends on the P_{CO_2}. Because CO_2 is more soluble than O_2 in the blood, a greater proportion of the total CO_2 in the blood is physically dissolved compared to O_2. Even so, only about 5 to 10% of an air breather's total blood CO_2 is carried this way at the normal systemic venous P_{CO_2} level. This amount is further reduced in fish and is estimated to be less than 5% of the total CO_2 carried in the blood.

2. *Bound to hemoglobin.* Another 25 to 30% (in mammals) of the CO_2 combines with Hb to form **carbamino hemoglobin (HbCO$_2$)**. Carbon dioxide binds with the globin (protein) portion of Hb, reacting with the nitrogen of the last amino acid of the protein chain:

$$CO_2 + Hb-NH_2 \rightarrow Hb-NH-COOH$$

This is in contrast to O_2, which combines with the heme iron. However, $HbCO_2$ is unimportant in fish, because their Hb has been modified (acetylated), preventing binding.

3. *As bicarbonate.* By far, the most important means of CO_2 transport is as **bicarbonate (HCO$_3^-$)** ion. In fish, HCO_3^- generally constitutes approximately 90 to 95% of the total CO_2 in the blood, whereas in mammals it is about 60 to 70%. The CO_2 is converted into HCO_3^- in part by the following chemical reaction:

$$CO_2 + H_2O \leftrightarrow H_2CO_3 \leftrightarrow H^+ + HCO_3^-$$

In the first step, CO_2 combines with H_2O to form **carbonic acid (H$_2$CO$_3$)**. As is characteristic of acids, some of the carbonic acid molecules spontaneously dissociate into hydro-

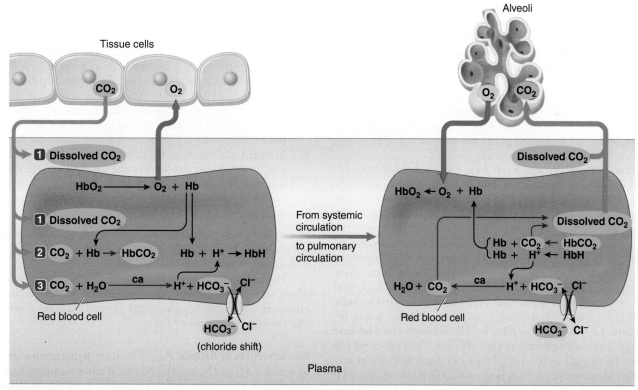

ca = Carbonic anhydrase

FIGURE 11-33 Gas transport in the blood (vertebrate). *Left:* Carbon dioxide (CO_2) picked up at the tissue level is transported in the blood to the lungs in three ways: (1) physically dissolved, (2) bound to hemoglobin (Hb), and (3) as bicarbonate ion (HCO_3^-). Oxygen enters the cells primarily from Hb, which is present only in the red blood cells, as is carbonic anhydrase, the enzyme that catalyzes the production of HCO_3^-. The H^+ generated during the production of HCO_3^- also binds to Hb. Bicarbonate moves by facilitated diffusion down its concentration gradient out of the red blood cell into the plasma, and chloride (Cl^-) moves by means of the same passive carrier into the red blood cell down the electrical gradient created by the outward diffusion of HCO_3^-. *Right:* The reactions that occur at the tissue level are reversed at the pulmonary level, where CO_2 diffuses out of the blood to enter the alveoli, and O_2 enters where most of it binds to Hb.

© Cengage Learning, 2013

gen ions (H^+) and bicarbonate ions (HCO_3^-). The one carbon and two oxygen atoms of the original CO_2 molecule are thus present in the blood as an integral part of HCO_3^-. This is beneficial because HCO_3^- (1) is more soluble in the blood than CO_2 is, and (2) cannot diffuse through membranes on its own, making it more controllable than the free gas.

This reaction takes place slowly in the plasma, but it proceeds swiftly within the red blood cells because of the presence of the erythrocyte enzyme **carbonic anhydrase,** which catalyzes (speeds up) the reaction. In fact, under the influence of carbonic anhydrase, the reaction proceeds directly from CO_2 to HCO_3^- without the intervening H_2CO_3 step:

$$CO_2 + OH^- \xleftrightarrow{CA} HCO_3^-$$

OH^- ion is derived from the ionization of water (below). As the OH^- ion is consumed by CA, the reaction is driven to the right, resulting in the generation of H^+ ion.

$$H_2O \leftrightarrow H^+ + OH^-$$

Carbonic anhydrase is also found in lungs and kidneys of mammals, lungs of amphibians, and in gills of fishes. Small mammals contain higher concentrations of this enzyme because of the ability of H^+ ion to increase the rate of O_2 delivery to the tissue (Bohr effect).

Unlike hemoglobins found in most animals, crocodile Hb has evolved the unique ability to bind to HCO_3^-. Under conditions where the crocodile remains submerged for prolonged periods, the binding of HCO_3^- ions on Hb results in the progressive unloading of O_2 from its binding sites on Hb, permitting this vast reservoir of O_2 to be available to the respiring tissues.

Chloride Shift As this reaction proceeds, HCO_3^- and H^+ start to accumulate within the red blood cells in the systemic capillaries. The red-cell membrane has a HCO_3^-/Cl^- *antiport carrier* (p. 85) that passively facilitates the diffusion of these ions in opposite directions across the membrane (Figure 11-33). Thus, HCO_3^- diffuses down its concentration gradient out of the erythrocytes into the plasma, where it can be transported to the gills or lungs. Because HCO_3^- is a negatively charged ion, the efflux of HCO_3^- unaccompanied by a comparable outward movement of positively charged ions creates an electrical gradient. Chloride ions (Cl^-), the dominant plasma anions, conduct into the red blood cells down this electrical gradient to restore electric neutrality. This inward shift of Cl^- in exchange for the outflux of HCO_3^- is known as the **chloride (Cl^-) shift.**

The membrane is relatively impermeable to H^+, where the vast majority generated by the dissociation of H_2CO_3 becomes bound to Hb (exerting the Bohr effect). Because only free dissolved H^+ contributes to the acidity of a solution, the venous blood would be considerably more acidic than the arterial blood if it were not for Hb mopping up most of the H^+ generated at the tissue level.

Haldane Effect As you have seen, the binding of both H^+ and CO_2 to oxyhemoglobin (HbO_2) favors release of O_2 (the Bohr effect). Conversely, deoxy-Hb has a greater affinity for both CO_2 and H^+ than does HbO_2. The unloading of O_2 from Hb in the tissue capillaries therefore facilitates the picking up of CO_2 and H^+ by Hb. The ability of deoxygenated Hb to pick up CO_2 and CO_2-generated H^+ is known as the **Haldane effect.** The Haldane effect and Bohr effect work in synchrony to facilitate gas exchange: Increased CO_2 and H^+ cause increased O_2 release from Hb by means of the Bohr effect; increased O_2 release from Hb, in turn, causes increased CO_2 and H^+ uptake by Hb through the Haldane effect. The entire process is very efficient. Deoxygenated Hb must be carried back to the lungs to reload with O_2 in any event. Meanwhile, after O_2 is released, Hb picks up new passengers—CO_2 and H^+—that are going in the same direction to the lungs.

The reactions that occur at the tissue level as CO_2 enters the blood from the tissues are reversed due to mass action once the blood reaches the lungs or gills and CO_2 leaves the blood to enter the alveoli (Figure 11-33) or ventilatory water outside the gill epithelium.

Various respiratory states are characterized by abnormal blood gas levels

Both gases can enter in abnormal states, although some of these states occur within the normal lifestyles of some animals.

Abnormalities in Arterial P_{O_2} The term **hypoxia** refers to insufficient O_2 at the cellular level. There are four general categories of hypoxia:

1. *Hypoxic hypoxia* is characterized by a low arterial blood P_{O_2} accompanied by inadequate Hb and blood saturation. It is caused by (a) a respiratory malfunction involving inadequate gas exchange, typified by a normal alveolar (or gill chamber) P_{O_2} but a reduced arterial P_{O_2}, or (b) exposure to an environment where environmental P_{O_2} is reduced, so that alveolar (or gill chamber) and arterial P_{O_2} are likewise reduced. This occurs in low-oxygen habitats for water breathers, at high altitudes, or during diving for air breathers (see *A Closer Look at Adaptation: High Fliers and Deep Divers* on p. 544).

2. *Anemic hypoxia* refers to a reduced O_2-carrying capacity of the blood. It can be brought about by (a) a decrease in circulating red blood cells, (b) an inadequate amount of Hb within the red blood cells, or (c) Hb poisoning (such as by *carbon monoxide,* a gas that binds more tightly to heme than does O_2). In all cases of anemic hypoxia, the arterial P_{O_2} is at a normal level, but the O_2 content of arterial blood is lower than normal because of the reduction in available Hb.

3. *Circulatory hypoxia* arises when too little oxygenated blood is delivered to the tissues. Circulatory hypoxia can be restricted to a limited area as a result of a local vascular spasm or blockage. In contrast, widespread circulatory hypoxia can result from congestive heart failure or circulatory shock. The arterial P_{O_2} and O_2 content may be normal, but too little oxygenated blood reaches the cells.

4. In *histotoxic hypoxia*, O_2 delivery to the tissues is normal, but the cells cannot use the O_2 available to them. The classic example is *cyanide poisoning*. Cyanide blocks cellular enzymes of the mitochondrial electron-transport chain (p. 51).

Hypoxia triggers a number of genetic and cellular processes; see the box, *Molecular Biology and Genomics: HIF to the Rescue* on p. 549.

Hyperoxia, an above-normal arterial P_{O_2}, cannot occur when an animal is breathing atmospheric air at sea level. However, breathing supplemental O_2 can increase alveolar and consequently arterial P_{O_2}.

Abnormalities in Arterial P_{CO_2} The term **hypercapnia** refers to excess CO_2 in the arterial blood; it can be caused by **hypoventilation** (ventilation inadequate to meet the metabolic needs for O_2 delivery and CO_2 removal) or by exposure to high environmental CO_2.

Hypocapnia, below-normal arterial P_{CO_2} levels, is brought about by **hyperventilation;** this occurs when an animal "overbreathes," that is, when the rate of ventilation is in excess of the body's metabolic needs for CO_2 removal so that CO_2 is blown off to the atmosphere more rapidly than it is produced in the tissues and arterial P_{CO_2} falls. Hyperventilation can be triggered by anxiety states and by fever. Alveolar P_{O_2} increases during hyperventilation as more fresh O_2 is delivered to the alveoli from the atmosphere than is extracted from the alveoli by the blood for tissue consumption, and arterial P_{O_2} increases correspondingly (Figure 11-34). However, because Hb is almost fully saturated at the normal arterial P_{O_2}, very little additional O_2 is added to the blood. Except for the small extra amount of dissolved O_2, blood O_2 content remains essentially unchanged during hyperventilation.

Increased ventilation is not synonymous with hyperventilation. Increased ventilation that matches an increased metabolic demand, such as the increased need for O_2 delivery and CO_2 elimination during activity, is termed **hyperpnea.** During activity, alveolar P_{O_2} and P_{CO_2} remain constant, with the increased atmospheric exchange just keeping pace with the increased O_2 consumption and CO_2 production.

Consequences of Abnormal Blood Gas Levels The consequences of reduced O_2 availability to the tissues during hypoxia are apparent. The cells need adequate O_2 supplies to sustain their energy-generating metabolic activities. The consequences of abnormal blood CO_2 levels are less obvious. *Changes in blood CO_2 concentration primarily affect acid–base balance.* Hypercapnia results in an elevated production of carbonic acid. The subsequent generation of excess H^+ produces an acidic condition termed **respiratory acidosis.** Conversely, less-than-normal amounts of H^+ are generated

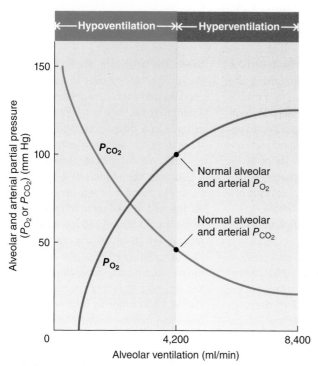

FIGURE 11-34 Effects of hyperventilation and hypoventilation on arterial P_{O_2} and P_{CO_2}.

© Cengage Learning, 2013

through carbonic acid formation in conjunction with hypocapnia. The resultant alkalotic (less acidic than normal) condition is called **respiratory alkalosis** (see Chapter 13).

Other unusual respiratory states such as apnea are not necessarily pathological

Not all unusual respiratory conditions are abnormal.

Apnea One of these is **apnea,** a temporary cessation of breathing. It is a normal characteristic of hibernating mammals and of reptiles, and of diving air breathers. In general, terrestrial reptiles have comparatively brief periods of apnea relative to their aquatic counterparts. This may be due, in part, to the energy costs associated with surfacing to breathe. Aquatic turtles and snakes exhibit periods of apnea of several minutes, whereas in crocodilians and some diving mammals, breathing may cease for periods over an hour. Periods of *sleep apnea* in elephant seals can range from 10 to 25 minutes and are followed by several minutes of regular breathing. Sleep apnea is considered similar to the diving response (see *A Closer Look at Adaptation: High Fliers and Deep Divers*) because both heart rate and metabolism decline. Sleep apnea does occur as a disorder in some humans, and is a threat to health.

Long-Term Hypoxia and Anoxia Some animals have adapted to very long periods of extreme hypoxia, or **anoxia**—the complete lack of oxygen. In 2011, researchers reported a nematode worm (*Halicephalobus mephisto*) living over 1 km beneath the Earth's surface in warm, almost anoxic waters, eating bacteria which were previously thought to be the only life this deep. Recall that certain deep-sea Loriciferans may go without O_2 their entire lives (p. 57). Other animals are thought to require O_2 for at least some periods, but some do

show a most remarkable resistance to anoxia. Goldfish (*Carassius auratus*) and their close relatives, the Crucian carp (*Carassius carassius*), are well-known examples. Indeed, this is why they became favorite pets in simple bowls in households and display fish in artificial ponds, for aeration is not required for them to survive. Unlike other pond animals such as turtles—which eventually become comatose in low O_2—goldfish and carp remain active for one to two days with no detectable O_2 in the water. And they can survive over winter in the bottom of an oxygen-less pond sealed with ice. How can they do this?

The most critical organ in anoxia is the brain, which needs a constant high supply of ATP to maintain ion balance via its Na^+/K^+ ATPase "pumps" (p. 84). Recall that, without O_2, only 2 ATPs are produced for each glucose molecule (compared to perhaps 30 in the presence of O_2) (p. 53). The carp and goldfish solve this low-ATP problem in part by reducing metabolism during anoxia. They also have large stores of glycogen in the liver, which is the main source of glucose for anaerobic metabolism. But in addition to energy problems, an animal in anoxia must avoid the lactate poisoning that results from normal anaerobic metabolism (Chapter 2). Remarkably, these fish (in their muscles primarily) make **ethanol**—the intoxicating ingredient of fermented beverages—as an end product, just as yeast do (p. 56). Ethanol (unlike lactate) has the advantages of being nonacidic and being able to readily diffuse out of the gills. Using this strategy, fish avoid the acid–base problems that freshwater turtles must endure (see Chapter 13). Pond-living carp employ two additional strategies to cope with the anoxic water conditions found in winter. During the fall, in response to the decline in water temperatures, these fish store large amounts of glycogen in their brain, an approximately 15-fold increase compared to that measured during the summer months. In addition, the activity of the brain's Na^+/K^+ ATPase "pumps" decreases approximately 10-fold, which dramatically lowers its energy requirements. Thus, during the winter when energy demands are low, the glycogen level that supports brain function in the carp for 16 hours would provide the same support for only eight minutes during the summer.

Equally as astounding as the goldfish's ability to withstand anoxia is the ability of the epaulette shark to survive periods of up to three hours with complete oxygen deprivation. Occasionally this reef inhabitant is left stranded in the shallows during low tide, where O_2 concentrations rapidly fall to levels that result in permanent brain damage for other fish species. How can the epaulette shark master this feat? Studies have shown that a series of biochemical events unfold as the duration of anoxia increases:

1. As P_{O_2} in the tidal pool falls, the epaulette shark reduces both its rate of respiration and heartbeat, and decreases arteriolar resistance to the brain.
2. GABA—a depressive neurotransmitter (p. 133)—is released in order to "turn off" nonessential neurological regions of the brain.
3. The animal greatly reduces its neural electrical activities, at least in the eye. Researchers found that retinas of quiescent epaulette sharks were completely unresponsive to light. They speculate that other neural circuits are also suppressed.
4. Most critically, perhaps, the anoxic shark can "turn down" the activity of its mitochondria. Normally, in most

A Closer Look at Adaptation
High Fliers and Deep Divers

As you have seen, some animals have respiratory systems that work at incredible extremes. For high altitude, the champion is probably the bar-headed goose (*Anser indicus*). It migrates biannually above 9,000 m over the Himalayas, often fighting fierce headwinds in air with O_2 less than 30% of that at sea level. Earlier, we noted that it has (1) a Hb with a higher affinity than those of other birds. However, this is not enough to explain its amazing ability. In addition, it has (2) unusually large wings to provide lift in the thin atmosphere and make use of updrafts efficiently, (3) unusually dense muscle capillary networks for enhanced O_2 delivery, (4) a much greater tidal volume than other birds (which enhances parabronchial ventilation more than increased respiratory rate would), (5) muscle mitochondria placed much closer to the fiber membranes (and thus closer to capillaries), and (6) higher aerobic capacity in terms of mitochondrial enzymes.

Now let's look at another extreme habitat, in some ways the opposite of the Himalayas: the deep sea. Perhaps most astounding are those diving mammals that are capable of taking a breath at the surface, diving to extraordinary depths in a comparatively short period of time, and then resurfacing. Let's start with the situation as a deep-sea diver descends underwater. The body is immediately exposed to greater than atmospheric pressure. Pressure increases with depth as a result of the weight of the water, increasing by about 1 atm per each 10 m deeper. High pressure has many direct effects on biological structures, including reduced membrane fluidity and perturbed protein folding. We explore adaptations to these elsewhere (pp. 145 and 625). In humans, who are not adapted to high pressure, a malady called **high-pressure neurological syndrome (HPNS)** occurs in divers at depths below 200 m, manifested as tremors and convulsions. Pressure decreases rate of action potential propagation and excitatory synaptic transmission, apparently as the result of a decrease in neurotransmitter release.

What about pressure and gases? Recall that (1) the amount of a gas in solution is directly proportional to the partial pressure of the gas and that (2) air has 79% N_2. Nitrogen is poorly soluble in body tissues, but the high pressure that occurs during deep-sea diving causes more of this gas than normal to dissolve in the body tissues. The small amount of N_2 dissolved in the tissues at sea level has no known effect, but as more N_2 dissolves at greater depths, **nitrogen narcosis,** or **raptures of the deep,** ensues in humans. Nitrogen narcosis is believed to result from a reduction in the excitability of neurons because of the highly lipid-soluble N_2 dissolving in their lipid membranes.

Another problem associated with deep-sea diving occurs during ascent. If a diver who has been submerged long enough for a significant amount of N_2 to become dissolved in the tissues suddenly ascends to the surface, the rapid reduction in pressure causes N_2 to quickly come out of solution and form bubbles of gaseous N_2 in the blood, blocking capillaries. The consequences depend on the amount and location of the bubble formation. This condition is called **decompression sickness** or **the bends,** the latter term arising because the victim often bends over in pain. Decompression sickness can be prevented by slow ascent to the surface or by gradual decompression in a decompression tank so that the excess N_2 can slowly escape through the lungs without bubble formation.

Contrast the abilities of a human with those of a diving marine mammal. Foraging escapades can take some of these animals to depths of 1,500–2,000 m below the surface in less than 20 minutes. At these depths, pressure of 150–200 atm (over a ton per square inch) forms, and yet Weddell and elephant seals and sperm whales can stay at 1,500 m for an hour or more! The champion may be Cuvier's beaked whale, recently tracked in a dive to about 2,000 m for 85 minutes, presumably to hunt for giant squid found at these remarkable depths. These animals' ability to accomplish these feats seems to defy all known physiological limits, yet they can repeat this feat again and again over a 24-hour period. How is this possible? Several adaptations have been found over the years:

- *Dive reflex:* First, early studies of seals discovered an adaptation called the **dive reflex:**
    - (i) Elimination of blood flow to most organs (other than the heart, brain, and adrenal gland), saving O_2 by reducing metabolic demands. Interestingly, maintenance of blood flow to the adrenal gland may be important for normal release of cortisol during

species of animals, as O_2 is restored to the anoxic tissues (reperfusion), the mitochondria mysteriously generate free radicals (p. 53) from whatever O_2 becomes available. This results in brain injury and apoptosis (cell death, p. 101). Because the epaulette shark's mitochondria are functioning at a lower level, the return of O_2 to the brain cells does not result in the generation of free radicals.

As a consequence the shark becomes limp and flaccid; but should a predator or potential prey move into the vicinity, its apparently unsuppressed electroreceptors (p. 261) cause the seemingly dead animal to revert back to life!

Ischemic Preconditioning Recent studies from the laboratory of Roberto Bolli have revealed the protective mechanisms of *ischemic preconditioning* in heart tissue. In this procedure brief periods of **ischemia** (lack of blood flow to an organ), in this case vascular occlusion for 3 to 5 minutes, results in the heart developing resistance to the potential damage resulting from later ischemia, as might occur in a heart attack. For this protection to occur three cardioprotective proteins must be induced in response to the oxidative stress:

- *iNOS* (Nitric oxide synthase) which increases nitric oxide (NO). NO reacts with the superoxide radicals generated in response to the oxidative stress to generate the harmful nitrogen radical peroxynitrite ($ONOO^-$). NO may suppress mitochondrial respiration, and it also induces:
- *HO-1* (heme oxygenase) which produces carbon monoxide (CO) (along with biliverdin; p. 688). In turn, CO induces:

a dive. Evidence suggests cortisol stabilizes neurons, thus preventing the onset of HPNS.

(ii) Suppression of shivering and reduction of body temperature by up to 3˚C.

(iii) Apnea accompanied by a decrease in heart rate to as little as 2–6 beats/min at depth. This *bradycardia* ("slow heart") decreases the work load of the heart and thus its O_2 requirement.

Phylogenetic studies indicate that this diving reflex is ancestral in all mammals (including humans), so is not special to diving mammals except that it is more effective than in nondivers. But other adaptations are wholly unique to divers:

• *Lungs:* The effects of N_2 are mostly avoided with distinct lung features. Shallow-diving birds and turtles rely on O_2 from their lungs to sustain them during their dives. However, the lungs of the (deep-diving) seals actually collapse by approximately 40 m below the surface. Seals exhale prior to diving, which significantly lowers gas volumes within the lungs. Considering the animals' body size, the lungs are proportionately small, further reducing their capacity for the uptake of N_2. High pressure squeezes the remaining gas out of the lungs into the bronchial airduct system. Both the bronchi and bronchioles are supported by rings of cartilage, which resist compression at depth, with the result that these air-

ways store the remaining gases and do not introduce them into the circulation. Another advantage of exhaling before diving is to reduce the buoyancy of the animal, which aids it in the descent. However, recent studies have demonstrated that at least sperm whales suffer some complications from decompression illness. When the dissolved nitrogen reverts to a gas and forms bubbles in the bone and cartilage of these animals, the bone dies and is not repaired. In time, the injuries form deep gaps in the bone, a condition termed **osteonecrosis** in humans. Characteristically, the older the sperm whale at the time of death, the greater the amount of bone damage sustained from the bends.

• *Oxygen storage* has been greatly enhanced by adaptations involving respiratory pigments. The blood of divers has up to two times greater hemoglobin concentration than does blood of nondivers, with the hematocrit of diving mammals being 60% in the Weddell seal versus 35 to 40% in humans. Similarly, red muscle has a greater myoglobin content. Most recently, diving mammals, as well as very active shallow mammals, were found to have high neuronal levels of *neuroglobin* and *cytoglobin* (p. 533). Diving mammals can also squeeze out O_2-rich erythrocytes from their abnormally enlarged spleens (as do the spleens of horses during intense activity). Blood constitutes approxi-

mately 20% of the body weight of the elephant seal, compared to only 7% in humans.

• *Buffering:* Anaerobic acidosis is prevented during the dive by confining anaerobic metabolism to the skeletal muscles, which have extremely high buffer content in the form of histidine dipeptides (p. 641). Because blood flow to this tissue is cut off, lactate cannot be released into the blood until the animal resurfaces. Only then can the liver and kidneys metabolize this by-product.

• *Gliding:* Some diving mammals use a gliding technique to save energy during the dive. They stop all locomotory movements, and let gravity pull them down once their lungs collapse so that they are no longer buoyant. This behavioral effector saves oxygen.

A bottlenose dolphin (*Tursiops*) with a video camera. The camera was used to document gliding behavior during a dive. Physiological instruments to record heart rate, blood oxygen levels, body temperatures, and so forth are also attached to diving animals.

Source: http://www.sciencenews.org/20000408/fob4.asp.

• *ecSOD* (extracellular superoxide dismutase) which inactivates the superoxide radical and ONOO⁻ and induces more iNOS.

If any of these three proteins is inhibited from expression the protective effect is gone, while upregulating any one of these three proteins increases protection because each is linked to the other.

This finishes our detailed examination of circulatory bulk transport (step 3 of Figure 11-1) as a gas transport mechanism, and its effects on gas exchange. We end the chapter with a look at mechanisms for regulating ventilation and coordinating it with circulation.

check your understanding 11.9

Compare hemoglobins with hemocyanins.

List the methods of O_2 and CO_2 transport in the blood.

What are the significances of the plateau and the steep portions of the O_2–Hb dissociation curve?

Explain the ways that hemoglobin P_{50} has evolved in vertebrates, with examples.

Explain the Bohr, Root, and Haldane effects and their role in oxygen delivery and in fish gas-bladder filling.

Define the following: *hypoxic hypoxia, anemic hypoxia, histotoxic hypoxia, hypercapnia, hypocapnia, hyperventilation, hypoventilation, apnea.*

11.10 Control of Respiration

Like the heartbeat and circulation, ventilation must occur frequently if not continuously to sustain life processes in many animals. Both systems (cardiac, respiratory) must also be altered coordinately during elevated activity. The underlying controls of these two systems are similar in some ways, but remarkably different in others. Both, however—in typical hierarchical control (p. 19)—are under *intrinsic* (local) and *extrinsic* (neural, hormonal) regulation. Recall that intrinsic control allows for local needs to be met with little feedback delay, while extrinsic controls can coordinate actions for bodywide needs, as well as *anticipate* actions shortly to be needed.

Spiracles that control tracheal ventilation in insects are regulated intrinsically and extrinsically for both gas exchange and evaporation control

Consider tracheal ventilation in terrestrial insects, which is often continuous during high activity, and discontinuous (intermittent) during rest. Discontinuous breathing may reduce (although the evidence is debated) a major problem of the tracheal system, that of water loss. Alternatively or in addition, it has been proposed that the purpose of discontinuous breathing is to maintain low concentrations of O_2 in the trachea. Because the tracheal system is very efficient at obtaining O_2, and too much O_2 is damaging (see p. 53), then during periods of inactivity the priority may be to limit O_2 intake. Indeed, studies conducted on the moth *Attacus atlas* over a range of ambient O_2 concentrations found that tracheal O_2 concentrations were consistently low.

Ventilation is in part controlled by valvelike spiracles (Figures 11-3e and 11-8), which in turn are regulated by both local and neural mechanisms. For example, locust spiracles will relax if exposed to CO_2, which directly reduces electrical activity at the muscle itself. This response increases ventilation rapidly as CO_2 builds up. Extrinsically, however, spiracles are tightly regulated by motor output originating in the **metathoracic ganglia**. An increase in the firing rates stimulates *closer muscles* to reduce the diameter of the elastic spiracles, which then spring open when the muscles relax. However, in some insects such as firebrats (relatives of silverfish), opening involves higher firing rates and thus contraction of the closer muscles, which cause a flap to pull back from the opening; when the muscles relax, the flap springs back to cover the opening. In either case, the resistance to gas flow as well as the total ventilation volume can respond to chemoreceptors that detect changes in CO_2 (and possibly O_2 levels) in the hemolymph. Recent studies show that in the fruit fly *Drosophila*, spiracles are opened and closed during flight such that O_2 needs are precisely met, while at the same time water loss is reduced by having the spiracles closed periodically.

Extrinsic and intrinsic controls act on the smooth muscle of mammalian airways and arterioles to maximally match ventilation to perfusion

Now let's look at ventilation in mammals, starting with the airways themselves. In Chapter 9 you learned that blood flow is adjusted to match both local and bodywide needs during varying levels of activity. This is accomplished primarily at the *arterioles,* which have muscular valves controlled by both extrinsic and intrinsic mechanisms (pp. 432–437). For example, during elevated muscular activity, arterioles in skeletal muscles are *vasodilated* by both extrinsic epinephrine and local chemical changes (such as decreasing O_2), resulting in increased volume of blood to those muscles. This is important for matching blood supply (perfusion) to local metabolic needs. At the same time, arterioles going to the gut are *vasoconstricted* by sympathetic nerves to divert blood flow elsewhere.

Adjustments in airway size can be accomplished similarly and for comparable reasons. In particular, this is to ensure that *ventilation rates match perfusion rates* for the most effective gas exchange. If, for example, perfusion of the alveoli were to increase during activity without an increase in ventilation, the blood would not be oxygenated sufficiently to meet the increased bodywide demand. Let's see how this matching occurs.

Extrinsic Regulation The autonomic nervous system can adjust airflow to suit the animal's respiratory needs. Parasympathetic stimulation, which occurs in quiet, relaxed situations when the demand for airflow is not high, promotes **bronchoconstriction** from smooth-muscle contraction, increasing airway resistance. In contrast, sympathetic stimulation and to a greater extent epinephrine, bring about **bronchodilation** and thus decreased airway resistance by promoting bronchiolar smooth-muscle relaxation. Since sympathetic domination predominates when increased demands for O_2 uptake are anticipated or actually occurring, bronchodilation ensures that the pressure gradients in the airways can achieve maximum airflow with minimum resistance. Importantly, activation of the sympathetic system simultaneously increases blood flow (via cardiac output and vasodilation), matching perfusion with ventilation.

These processes are relevant only to the contribution of the overall resistance of all the airways collectively. However, the resistance of individual airways supplying specific alveoli can be adjusted independently in response to changes in the airways' local environment.

Intrinsic Regulation: Effect of CO_2 on Bronchiolar Smooth Muscle Like arteriolar smooth muscle, bronchiolar smooth muscle is sensitive to local changes within its immediate environment, particularly to local CO_2 levels. If an alveolus is receiving too little ventilation in comparison to its perfusion, CO_2 levels will increase in the alveolus and surrounding tissue as more CO_2 is dropped off by the blood than is being exhaled into the atmosphere. This local increase in CO_2 acts directly on the bronchiolar smooth muscle to induce the airway supplying the under-aerated alveolus to relax. The resultant decrease in airway resistance leads to an increased airflow to the involved alveolus, so its airflow now matches its blood supply (Figure 11-35). The converse is also true. A localized decrease in CO_2 associated with an alveolus that is receiving too much air for its blood supply directly increases contractile activity of the airway smooth muscle involved, causing the airway supplying this over-aerated alveolus to constrict. The result is a reduction in airflow to the overaerated alveolus.

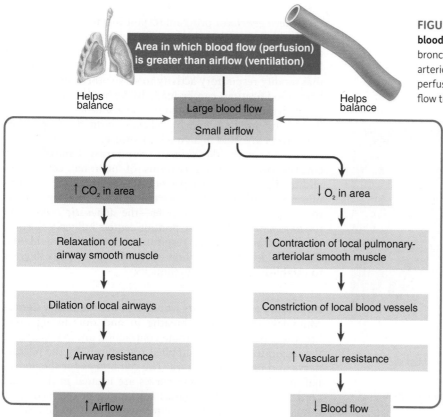

FIGURE 11-35 Local controls to match airflow and blood flow to an area of the lung. CO_2 acts locally on bronchiolar smooth muscle and O_2 acts locally on arteriolar smooth muscle to adjust ventilation and perfusion, respectively, to match airflow and blood flow to an area of the lung.

© Cengage Learning, 2013

(a) Local controls to adjust ventilation and perfusion to lung area with large blood flow and small airflow

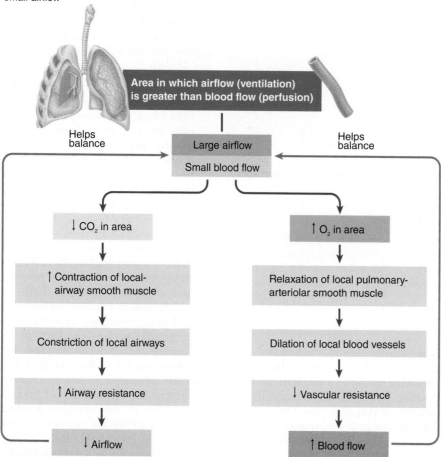

(b) Local controls to adjust ventilation and perfusion to a lung area with large airflow and small blood flow

Intrinsic Regulation: Effect of O_2 on Pulmonary Arteriole Smooth Muscle

A similar locally induced effect on pulmonary vascular smooth muscle also takes place simultaneously to maximally match blood flow to airflow. Just as in the systemic circulation, distribution of the cardiac output to different alveolar capillary networks can be controlled by adjusting the resistance to blood flow through specific pulmonary arterioles. If the blood flow is greater than the airflow to a given alveolus, the O_2 level in the alveolus and surrounding tissues will fall below normal as more O_2 than usual is extracted from the alveolus by the overabundance of blood. The local decrease in O_2 concentration causes vasoconstriction of the pulmonary arteriole supplying this particular capillary bed; the result is a reduction in blood flow to match the smaller airflow. Conversely, an increase in alveolar O_2 concentration caused by a mismatched large airflow and small blood flow brings about pulmonary vasodilation, which increases blood flow to match the larger airflow. Note that the local effect of O_2 on pulmonary arteriolar smooth muscle is appropriately just the opposite of its effect on systemic arteriolar smooth muscle:

- Systemic arteries: vasodilate with decreased O_2; vasoconstrict with increased O_2
- Pulmonary arteries: vasoconstrict with decreased O_2; vasodilate with increased O_2

Respiratory centers in the vertebrate brain stem establish a rhythmic breathing pattern

Unlike the heartbeat in myogenic hearts, which is rhythmically triggered by an intrinsic pacemaker (p. 406), vertebrate breathing is regulated by a rhythmic neural program in the **medullar respiratory center** in the brain stem (Figure 11-36). This neural integrator has inputs from chemosensors

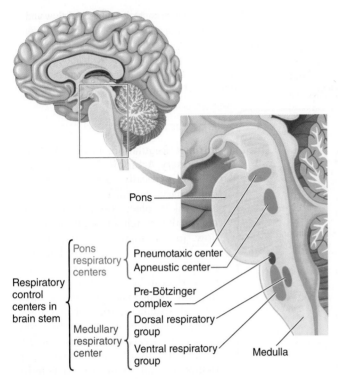

FIGURE 11-36 **Respiratory control centers in the brain stem (human).**

© Cengage Learning, 2013

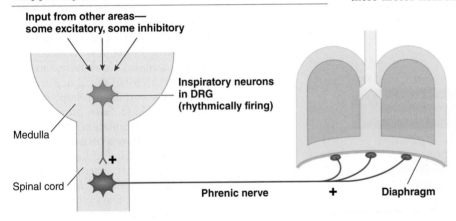

Not shown are intercostal nerves to external intercostal muscles.

FIGURE 11-37 **Schematic representation of medullary dorsal respiratory group (DRG) control of inspiration.** Inspiration takes place when the inspiratory neurons are firing and activating the motor neurons that supply the inspiratory muscles. Expiration takes place when the inspiratory neurons cease firing, so that the motor neurons supplying the inspiratory muscles are no longer activated.

© Cengage Learning, 2013

ral *pattern generator* program responsible for the alternating inspiration/expiration rhythm, (2) the factors that regulate the magnitude of ventilation (that is, the rate and depth of breathing) to match physiological needs, and (3) the factors that modify respiratory activity to serve other purposes. The latter modifications can be made by higher brain centers, as in the breath control required for vocalization or for nonreflex breath holding, and by reflexes, as in the respiratory maneuvers involved in a cough or sneeze.

In mammals, the primary respiratory control center consists of several aggregations of neuronal cell bodies within the medulla that provide output to the respiratory muscles. In addition, there are two other medullary respiratory centers higher in the pons—the *apneustic center* and *pneumotaxic center*. These pontine centers influence the output from the medullary respiratory center (Figure 11-36). Here is a description of how these various regions interact to establish respiratory rhythmicity.

Inspiratory and Expiratory Neurons in the Medullary Center Mammals rhythmically breathe in and out during quiet breathing because of alternate contraction and relaxation of the inspiratory muscles supplied respectively by the phrenic nerve and intercostal nerves. The cell bodies for the neuronal fibers composing these nerves are located in the spinal cord (Figure 11-37). Impulses originating in the medullary center terminate on these motor-neuron cell bodies. When these motor neurons stimulate the inspiratory muscles, this leads to inspiration; when these neurons are not firing, the inspiratory muscles relax and expiration takes place.

The medullary respiratory center consists of two neuronal clusters (Figure 11-36).

1. The **dorsal respiratory group (DRG)** consists mostly of *inspiratory neurons* whose descending fibers terminate on the motor neurons that supply the inspiratory muscles. When the DRG inspiratory neurons fire, inspiration takes place; when they cease firing, expiration occurs. Expiration is brought to an end as the inspiratory neurons once again reach threshold and fire.

2. The **ventral respiratory group (VRG)**, connected to the DRG, is composed of *inspiratory neurons* and *expiratory neurons*, both of which remain inactive during normal quiet breathing. This region is called into play by the DRG as an "overdrive" mechanism during periods when demands for ventilation are increased. It is especially important in active expiration. No impulses are generated in the descending pathways from the expiratory neurons during quiet breathing. Only during active expiration do the expiratory neurons stimulate the motor neurons supplying the expiratory muscles (the abdominal and internal intercostal muscles). Furthermore, the VRG inspiratory neurons, when stimulated by the DRG, rev up inspiratory activity when demands for ventilation are high.

that measure blood gas levels, and outputs to breathing effectors (muscles) in negative-feedback fashion. The lack of intrinsic regulation is presumably because breathing requires coordination among multiple organs. However, feedback delays are minimized by having a regulatory center that is close to the sensors and effectors and that can operate independently of higher brain centers.

We focus primarily on mammalian breathing, but note important features of other vertebrates. Neural control of respiration involves three distinct components: (1) the neu-

When an air breather accustomed to low altitude travels to high altitude, or a water breather accustomed to well-oxygenated waters encounters hypoxic water, internal hypoxia ensues from low oxygen availability. This is not an uncommon phenomenon in some species: consider human mountaineers or cattle migrating to high summer pastures. The body's initial reaction is an increase in ventilation and circulation, but over prolonged time at high altitude, acclimatization changes occur, including (1) an *increase in red blood cell numbers* triggered by the hormone *EPO* from the kidney (p. 393); and (2) *angiogenesis*, the growth of new capillaries triggered by the paracrine *vascular endothelial growth factor (VEGF)*. For decades, physiologists have sought the sensory mechanisms by which cells producing these hormones detect low O_2. The key appears to be regulation of **hypoxia-inducible factor (HIF)**, a transcription factor that binds to response elements of genes (see p. 31) for proteins that are needed for hypoxia acclimatization, including genes for (1) and (2) above, for—in most cells—genes for (3) increased glucose transport and metabolism and (4) a subunit of a cytochrome in mitochondrial electron transport that enhances the efficiency of oxygen use. HIF, which is found in many different phyla, consists of two subunits, HIFα and HIFβ. In *normoxia* (normal oxygen levels), HIFα is hydroxylated by **proline hydroxylase (PH)**, converting HIFα into a form that then binds to **von Hippel-Lindau protein (pVHL)**. Together, the HIFα–pVHL complex activates **ubiquitin ligase (UL),** an enzyme complex that adds *ubiquitin* to HIFα. As you saw in Chapter 2 (p. 46), ubiquitin is a "tag" that targets a protein for destruction by proteasomes.

Thus, in normoxia, the HIF system is relatively inactive. However, during hypoxia, PH no longer targets HIFα for destruction; therefore, HIFα combines with HIFβ to form a functional transcription factor. In turn, the acclimatization genes are activated. It appears that PH requires O_2 to function, and thus it becomes less active when O_2 levels are low.

Interestingly, native Tibetan humans living permanently at high altitudes in the Himalayas have Hb and red-cell concentrations similar to humans at sea level! (In contrast, humans native to the high Andes do have elevated Hb and red-cell levels). Tibetans make up for low P_{O_2} with unusually large ventilatory volumes, no pulmonary va-soconstriction in hypoxia (p. 547), and higher levels of nitric oxide (p. 435) to promote vasodilation and thus blood flow to tissues body-wide. Recent genomic analysis indicates that genes in their HIF system have mutated to become less sensitive to hypoxia, thus blunting the EPO response. This adaptation may be beneficial because the high red-cell count increases blood viscosity (p. 420), putting a strain on the heart, especially during pregnancy and fetal development.

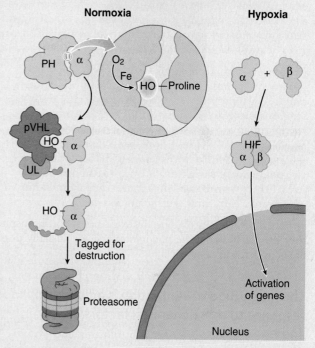

Source: M. K. Campbell & S. O. Farrell. (2003). *Biochemistry*, 4th ed. Belmont, CA: Brooks/Cole. Modified from H. Hsu & H. F. Bunn. (2001). How do cells sense oxygen? *Science* 292:449–451.

Fish brainstem medullas have two centers, one active during opercular opening, and one active during closing. Primitive air breathing in vertebrates evolved as a new motor pattern, possibly arising from modified coughing and suction feeding movements. The central rhythm generator in the brain stem of higher vertebrates may have evolved through an amalgamation of these two preexisting neural generators.

Generation and Fine Tuning of Respiratory Rhythm The generation of a respiratory rhythm in mammals is now widely believed to lie, not in the DRG, but in the **pre-Botzinger complex,** a region located near the upper (head) end of the medullary respiratory center (Figure 11-36). A network of neurons in this region display pacemaker activity, undergoing self-induced action potentials similar to those of the SA node of the heart. Scientists believe the rate at which the DRG inspiratory neurons rhythmically fire is driven by synaptic input from this complex. Thus, while both heartbeat and breathing involve pacemakers, the former's is intrinsic while the latter's is extrinsic.

Recently, researchers have found that a single gene, called *Teashirt-3* (Tshz3), may be a master regulator of the pacemaker complex's formation. In mice with this gene knocked out, fetuses failed to start breathing at birth. Autopsies showed that there was a pacemaker complex in the fetuses' brainstems, but that they had failed to develop properly.

The respiratory centers in the pons exert "fine-tuning" influences over the medullary center to help produce normal, smooth inspirations and expirations. The **pneumotaxic center** sends impulses to the DRG that help "switch off" the inspiratory neurons, limiting the duration of inspiration. Without the pneumotaxic brakes, the breathing pattern consists of deep inspiratory gasps at a low rate. In contrast, the **apneustic center** prevents the inspiratory neurons from being

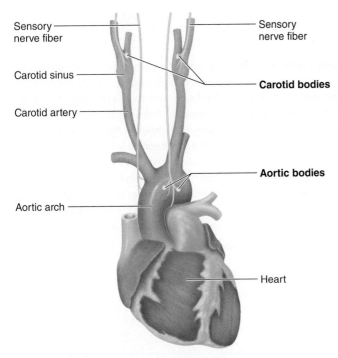

FIGURE 11-38 Location of the peripheral chemoreceptors (mammal). The carotid bodies are located in the carotid sinus, and the aortic bodies are located in the aortic arch.

© Cengage Learning, 2013

switched off, thus providing an extra boost to the inspiratory drive. However, the pneumotaxic center usually dominates over the apneustic center.

The magnitude of ventilation is adjusted in response to three chemical factors: P_{O_2}, P_{CO_2}, and H^+

No matter how much O_2 is extracted from the blood or how much CO_2 is added to it at the tissue level, the P_{O_2} and P_{CO_2} of the systemic arterial blood leaving the lungs are held remarkably constant, indicating that arterial blood-gas content is precisely regulated. Arterial blood gases are maintained within the normal range by varying the magnitude of ventilation to match the animal's needs for O_2 uptake and CO_2 removal. If more O_2 is extracted from the alveoli and more CO_2 is dropped off by the blood because the tissues are metabolizing more actively, ventilation increases correspondingly to bring in more fresh O_2 and blow off more CO_2.

The medullary respiratory center receives inputs that provide information about the body's needs for gas exchange. It responds by sending appropriate signals to the motor neurons supplying the respiratory muscles (Figure 11-37), to adjust the rate and depth of ventilation to meet those needs. The two most obvious signals to increase ventilation are a decreased arterial P_{O_2} or an increased arterial P_{CO_2}. Intuitively, you would suspect that if O_2 levels in the arterial blood declined or if CO_2 accumulated, ventilation would be stimulated to obtain more O_2 or to eliminate excess CO_2. These two factors do indeed influence the magnitude of ventilation, but not to the same degree nor through the same pathway. Also, a third chemical factor, H^+, notably

influences the level of respiratory activity. We next examine the role of each of these chemical factors in the control of ventilation.

Feedback monitoring of blood gases and H^+ occurs both peripherally and centrally

In mammals arterial P_{O_2} is monitored by *peripheral* chemoreceptors known as the **carotid bodies** and **aortic bodies,** which are positioned at the bifurcation of the common carotid arteries and in the arch of the aorta, respectively so that they detect hypoxia long before it reaches the central nervous system (Figure 11-38). In fishes, chemoreceptors resembling those of mammals are located on the arterial side downstream from the gills.

As you will see, peripheral receptors also react to H^+. In addition, there are **central chemoreceptors** in mammals, located in the medulla in the vicinity of the respiratory center. These respond most strongly to H^+, but also to O_2 (Table 11-3). These peripheral and central chemoreceptors, which respond to specific changes in the chemical content of the arterial blood that bathes them, are distinctly different from the baroreceptors located in the same vicinity as the peripheral chemoreceptors. Recall that the baroreceptors inform the brainstem's cardiovascular center about blood pressure (p. 449). However, these two sets of sensors help coordinate ventilation and perfusion through the two closely located and closely coordinated brainstem regulators.

Decreased arterial P_{O_2} increases ventilation more strongly in fish than in air breathers

In water-breathing vertebrates, P_{O_2} in the blood is the primary homeostatic variable. In fish, a reduction in blood P_{O_2} detected by the peripheral chemoreceptors leads to a marked increase in the amplitude and frequency of gill ventilation. P_{CO_2} is not a major signal, because the high solubility of CO_2 in the external water results in only small percentage changes in this gas in the blood at the gills.

Chemoreceptors of air breathers are moderately sensitive to reductions in arterial P_{O_2}. A decrease from 100 mm Hg to 80 mm Hg is sufficient to evoke afferent nerve activation in the carotid body. However, a more severe hypoxia with arterial P_{O_2} falling below 40 mm Hg triggers appropriate autonomic changes that include stimulation of breathing as well as an increase in blood pressure. Low P_{O_2} inhibits O_2-sensitive K^+ channels, which results in depolarization of the sensory cells and the subsequent release of transmitters. Little or no adaptation of the carotid body sensors occurs during sustained hypoxia in most species, while in some such as goats, sensory activity may increase. The signal leads to reflex stimulation of respiration that serves as an important emergency mechanism in dangerously low arterial P_{O_2} states. Indeed, this reflex mechanism is a lifesaver because a low arterial P_{O_2} tends to directly depress the respiratory center via its central chemoreceptors, as it does all the rest of the brain (Figure 11-39; Table 11-3). Except for the peripheral chemoreceptors, the level of activity in all nervous tissue becomes reduced in the face of O_2 deprivation. Were it not for stimulatory intervention of the peripheral chemoreceptors when the arterial P_{O_2} falls threateningly low, a vicious cycle ending in cessation of breathing would ensue. Direct depression of the respira-

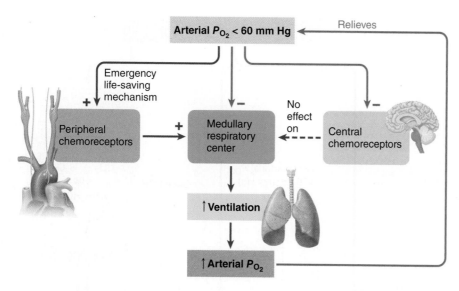

FIGURE 11-39 **Effect of threateningly low arterial P_{O_2} (<60 mm Hg) on ventilation (mammal).**

© Cengage Learning, 2013

TABLE 11-3 Influence of Chemical Factors on Mammalian Respiration

Chemical Factor	Effect on the Peripheral Chemoreceptors	Effect on the Central Chemoreceptors
P_{O_2} in the Arterial Blood	Stimulates only when arterial P_{O_2} has fallen below 80 mm Hg and maximally at life-threatening levels (40–60 mm Hg); an emergency mechanism	Directly depresses the central chemoreceptors and the respiratory center itself when <60 mm Hg
P_{CO_2} in the Arterial Blood (H+ in the Brain ECF)	Weakly stimulates	Strongly stimulates; is the dominant control of ventilation (Levels >70–80 mm Hg directly depress the respiratory center and central chemoreceptors)
H+ in the Arterial Blood	Stimulates; important in acid–base balance	Does not affect; cannot penetrate the blood–brain barrier

© Cengage Learning, 2013

tory center by the markedly low arterial P_{O_2} would further reduce ventilation, leading to an even greater fall in arterial P_{O_2}, which would even further depress the respiratory center until ventilation ceased and death occurred. Of particular note, the carotid body, despite its small size of a few micrograms, expresses as many types of neurotransmitters as the brain, some being excitatory while others are inhibitory! Discerning the function of this array of neurotransmitters remains an active area of study.

Because the peripheral chemoreceptors respond to the P_{O_2} of the blood, *not* the total O_2 content of the blood, O_2 content in the arterial blood can fall to dangerously low or even fatal levels without the peripheral chemoreceptors adequately responding to reflexly stimulate respiration. Remember that only physically dissolved O_2 contributes to blood P_{O_2}. The total O_2 content in the arterial blood can be reduced in anemic states, in which O_2-carrying Hb is reduced, or in carbon monoxide (CO) poisoning, when the Hb is preferentially bound to this molecule rather than to O_2. In both cases, arterial P_{O_2} is normal, so respiration is not stimulated, even though O_2 delivery to the tissues may be so reduced that the animal dies from cellular O_2 deprivation.

Carbon dioxide–generated H+ in the brain is normally the main regulator of ventilation in air breathers

In contrast to arterial P_{O_2}, arterial P_{CO_2} is the most important input regulating the magnitude of ventilation under resting conditions. This role is appropriate, because changes in alveolar ventilation have an immediate and pronounced effect on arterial P_{CO_2}, whereas changes in ventilation have little effect on % Hb saturation until the arterial P_{O_2} falls by more than 40%. Even slight deviations from normal in arterial P_{CO_2} induce a significant reflex effect on ventilation:

- An increase in arterial P_{CO_2} reflexly stimulates the respiratory center, with the resultant increase in ventilation promoting elimination of the excess CO_2 to the atmosphere.
- A fall in arterial P_{CO_2} reflexly reduces the respiratory drive. The subsequent decrease in ventilation allows metabolically produced CO_2 to accumulate so that P_{CO_2} can be returned to normal.

Most fish also respond to an increase in P_{CO_2} and acidosis by increasing ventilation. Researchers believe that the response to CO_2 is mediated by their branchial (gill) chemoreceptors. Further, mechanoreceptors located along the respiratory surfaces of fish also play a role in respiration; they are important in regulating the position of the gill during the respiratory cycle and they interact with chemoreceptors to control the transition from active breathing to ram ventilation.

Perhaps surprisingly, given the key role of arterial P_{CO_2} in regulating respiration in air breathers, most species have no important receptors that monitor arterial P_{CO_2} per se. The primary signal, as we shall see, is acid (H^+). The carotid and aortic bodies are only weakly responsive to changes in arterial P_{CO_2} (Table 11-3), so they play only a minor role in reflexively stimulating ventilation in response to an elevation in arterial P_{CO_2}. In reptiles (including birds), however, intrapulmonary chemoreceptors sense P_{CO_2}, although they are unusual in that their discharge rate is inversely proportional to pulmonary P_{CO_2} concentrations.

More important in linking changes in arterial P_{CO_2} to compensatory adjustments in ventilation are the central chemoreceptors. These central chemoreceptors are most sensitive to changes in, not CO_2, but CO_2-induced H^+ concentration in the brain extracellular fluid (ECF) that bathes them. Movement of materials across the brain capillaries is restricted by the blood–brain barrier (see p. 169). Because this barrier is readily permeable to CO_2, any increase in arterial P_{CO_2} causes a similar rise in brain ECF P_{CO_2} as CO_2 diffuses down its pressure gradient from the cerebral blood vessels into the brain ECF. The increased P_{CO_2} within the brain ECF causes a corresponding increase in the concentration of H^+ according to the law of mass action as it applies to the reaction we described earlier: $CO_2 + H_2O \leftrightarrow H_2CO_3 \leftrightarrow H^+ + HCO_3^-$. An elevation in H^+ concentration in the brain ECF directly stimulates the central chemoreceptors, which in turn increase ventilation by stimulating the respiratory center through synaptic connections (Figure 11-40). As the excess CO_2 is subsequently blown off, the arterial P_{CO_2} and the P_{CO_2} and H^+ concentration of the brain ECF are returned to normal. Conversely, a decline in arterial P_{CO_2} below normal is paralleled by a fall in P_{CO_2} and H^+ in the brain ECF, the result of which is a central-chemoreceptor-mediated decrease in ventilation. As CO_2 produced by cellular metabolism is consequently allowed to accumulate, arterial P_{CO_2} and P_{CO_2} and H^+ of the brain ECF are restored toward normal.

Unlike CO_2, H^+ cannot readily permeate the blood–brain barrier, so H^+ present in the plasma cannot gain access to the central chemoreceptors. Accordingly, the central chemoreceptors are responsive only to H^+ generated within the

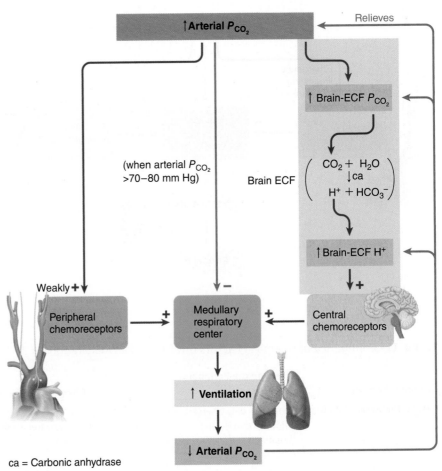

ca = Carbonic anhydrase

FIGURE 11-40   Effect of increased arterial P_{CO_2} on ventilation (mammal).

© Cengage Learning, 2013

brain ECF itself as a result of CO_2 entry. Thus, the major mechanism controlling ventilation under resting conditions is specifically aimed at regulating the brain ECF H^+ concentration, which in turn is a direct reflection of the arterial P_{CO_2}. However, little evidence suggests that fish have any central receptors and so researchers believe their response to P_{CO_2} is mediated solely by the peripheral chemoreceptors.

The powerful influence of the central chemoreceptors on the respiratory center is responsible for most humans' inability to deliberately hold their breath for more than a minute or so. While breath-holding, metabolically produced CO_2 continues to accumulate in the blood and subsequently to build up the H^+ concentration in the brain ECF. Finally, the increased P_{CO_2}-H^+ stimulant to respiration becomes so powerful that central chemoreceptor excitatory input overrides voluntary inhibitory input to respiration, so breathing resumes despite deliberate attempts to prevent it. Breathing resumes long before arterial P_{O_2} falls to the threateningly low levels that trigger the peripheral chemoreceptors. Therefore, you cannot deliberately hold your breath long enough to create a dangerously high level of CO_2 or low level of O_2 in the arterial blood.

Direct Effect of a Large Increase in P_{CO_2}   In contrast to the normal reflex stimulatory effect of the increased P_{CO_2}-H^+ mechanism on respiratory activity, very high levels of CO_2,

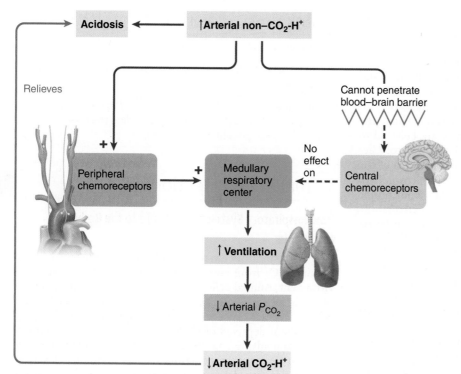

FIGURE 11-41 Effect of increased arterial non-carbonic-acid-generated hydrogen ion (non–CO_2-H^+) on ventilation (mammal).

© Cengage Learning, 2013

as may occur when an animal is buried in an avalanche, directly depress the entire brain, including the respiratory center, just as very low levels of O_2 do. Up to a P_{CO_2} of 70 to 80 mm Hg, progressively higher P_{CO_2} levels induce correspondingly more vigorous respiratory efforts in an attempt to blow off the excess CO_2. A further increase in P_{CO_2} beyond 70 to 80 mm Hg, however, does not further increase ventilation but actually depresses the respiratory neurons. For this reason, CO_2 must be chemically removed and O_2 supplied in closed environments such as closed-system anesthesia machines, submarines, or space capsules, and must not be allowed to accumulate in animal burrows. Otherwise, CO_2 could reach lethal levels, not only because of its depressant effect on respiration but also because of the resultant severe state of respiratory acidosis.

Adjustments in ventilation in response to changes in arterial H^+ are important in acid–base balance

The aortic and carotid body peripheral chemoreceptors are also highly responsive to fluctuations in arterial H^+ concentration, in contrast to their weak sensitivity to deviations in arterial P_{CO_2} and also arterial P_{O_2} until it falls 40% below normal. The result is the same as for central chemoreceptors. However, these changes in ventilation mediated peripherally are far less important than the powerful central mechanism in adjusting ventilation in response to changes in CO_2-generated H^+ concentration.

The peripheral chemoreceptors do play a major role in adjusting ventilation in response to alterations in arterial H^+ concentration unrelated to fluctuations in P_{CO_2}. In many situations, even though P_{CO_2} is normal, arterial H^+ concentration is changed by the addition or loss of non-CO_2-generated acid from the body. For example, arterial H^+ concentration increases from lactate during activity and also from diabetes mellitus because excess H^+-generating keto acids (p. 316) are abnormally produced and added to the blood. Changes in ventilation by this mechanism are extremely important in regulating an animal's acid–base balance (Figure 11-41; see also Chapter 13 for acid–base details).

Activity profoundly increases ventilation, possibly due to anticipatory mechanisms

Alveolar ventilation may increase up to 25-fold during heavy locomotory activity to keep pace with the increased demand for O_2 uptake and CO_2 output. The cause of increased ventilation during activity such as exercise is still largely speculative. It would seem logical that changes in the "big three" chemical factors—decreased P_{O_2}, increased P_{CO_2}, and increased H^+—could account for the increase in ventilation. This does not appear to be the case, however.

1. As you saw in Chapter 9 (p. 452), despite the marked increase in O_2 use during activity, arterial P_{O_2} does not decrease but remains normal or may actually increase slightly because the increase in alveolar ventilation and circulatory flow both keep pace with or even slightly exceeds the stepped-up rate of O_2 consumption.
2. Likewise, despite the marked increase in CO_2 production during activity, arterial P_{CO_2} does not increase but remains normal or decreases slightly because the extra CO_2 is removed as rapidly or even more rapidly than it is produced by the increase in ventilation.
3. During mild or moderate activity, H^+ concentration does not increase, because H^+-generating CO_2 is held constant. During heavy activity, H^+ concentration does increase somewhat from release of H^+-generating lactate into the blood by anaerobic metabolism in the exercising muscles. Even so, the elevation in H^+ concentration resulting from lactic acid formation is not enough to account for the large increase in ventilation accompanying activity.
4. Ventilation increases abruptly at the onset of activity (within seconds), long before changes in arterial blood gases could become important influences on the respiratory center (which requires a matter of minutes). Simple feedback from the chemosensors cannot explain this, since that has a delay (as explained on p. 452).

Researchers have suggested that a number of factors besides blood gases and acid, including the following, play a role in the ventilatory response to activity such as exercise:

1. *Reflexes originating from body movements.* Joint and muscle receptors excited during muscle contraction reflexly stimulate the respiratory center, abruptly increasing ventilation. Even passive movement of the limbs (for example, someone else alternately flexing and extending a person's knee) may increase ventilation several-fold through activation of these receptors, although no actual exercise is occurring. Thus, the mechanical events of muscular activity are believed to play an important role in coordinating respiratory activity with the increased metabolic requirements of the active muscles.

2. *Anticipatory activation by epinephrine.* Epinephrine also stimulates ventilation as well as blood circulation, as we have described (pp. 546 and 417). It can be released by the sympathetic nervous system under stress conditions in advance of physical activity.

3. *Anticipatory activation by the cerebral cortex.* Especially at the onset of activity, the motor areas of the cerebral cortex are believed to simultaneously stimulate both the medullary respiratory and cardiovascular (p. 449) centers at the same time it activates the motor neurons of the exercising muscles. In this way, the motor cortex calls forth increased ventilatory and circulatory responses to support the increased physical activity it is about to orchestrate. These anticipatory adjustments occur *before* any homeostatic factors actually change.

None of these factors or combinations of factors is fully satisfactory in explaining the abrupt and profound effect activity has on ventilation, nor can they completely account for the high degree of correlation between respiratory activity and an animal's needs for gas exchange during activity. Much remains to be learned about external respiration.

check your understanding 11.10

How are insect spiracles regulated and why?

How are ventilation and perfusion matched at airways and pulmonary vessels?

Describe the respiratory brain centers of mammals and how they work by negative feedback, including the roles of P_{O_2}, P_{CO_2}, H^+, and (possibly) anticipation.

making connections

How Respiratory Systems Contribute to the Body as a Whole

Animal respiratory systems contribute to homeostasis by obtaining O_2 from and eliminating CO_2 to the external environment. Animal cells ultimately need an adequate supply of O_2 to use in oxidizing nutrients to generate ATP. Brain cells, which are especially dependent on a continual supply of O_2, die if deprived of O_2 for more than a few minutes in most mammals. Even cells that can resort to anaerobic metabolism for energy production, such as strenuously exercising muscles, can do so only transiently by incurring an O_2 debt that ultimately must be repaid.

As a result of oxygen-using metabolism, large quantities of CO_2 are produced that must be eliminated from internal fluids. Because CO_2 and H_2O form carbonic acid, adjustments in the rate of CO_2 elimination by a respiratory system are important in the regulation of internal acid–base balance (Chapter 13). Cells can survive only within a narrow pH range.

Respiratory systems can also play a role in other physiological processes, such as fluid and solute balances, feeding, heat transfer, sound production, and immunity.

Chapter Summary

- External respiration—events in the exchange of O_2 and CO_2 between the external environment and cellular mitochondria—involves up to four major steps of bulk transport and exchange by diffusion. Step 1. Bulk transport of external media, called ventilation or breathing if it is active. Step 2. Exchange by diffusion between the environmental medium and cells or ECF at a respiratory surface or organ. Step 3. Bulk transport of ECF in a circulatory system. Step 4. Exchange by diffusion between the environment or ECF and the ICF. Not all animals have all steps.

- Gas diffusion follows Fick's law for partial pressure, not concentration, gradients. The individual pressure exerted independently by a particular gas within a mixture of gases is known as its partial pressure, designated by P_{gas}. Diffusion rate changes linearly with surface area A, coefficient D, and inversely with distance ΔX.

- External respiratory processes must meet the demands of size, metabolism, and habitat. Small, thin, and/or sluggish animals (e.g., flatworms) utilize simple integumentary exchange (steps 2 and 4 above). But larger, more active animals and those in O_2-poor habitats (e.g., high altitude, warm waters) need enhanced bulk transport and exchange.

- Gas exchange (steps 2, 4) may be increased by (1) respiratory organs with folds (and so on) having high A, low ΔX; bulk transport by (2) ventilation/breathing and (3) circulation, which enhance ΔP; and (4) diffusion-enhancing proteins such as hemoglobins, which enhance ΔP by converting cases to and from nondiffusible forms, and which also boost transport capacity of circulation.

- Breathing as a form of bulk transport can be tidal (medium moving in and out of the same opening) or the more efficient flow-through, which ventilates more continuously and with little mixing of stale and fresh medium.

- Water breathers have a harder time breathing than air breathers due to water's greater viscosity, lower diffusion rates, decreased solubility with temperature, and wider habitat variation. The limitations of diffusion in water are overcome with gills (often highly branched, e.g., with pagelike lamellae) and other thin, high-surface-area structures, internal circulation, and flow-through breathing (as in fish buccal–opercular pumps). Countercurrent flow enhances gas pressure gradients in fishes because blood moving through gill lamellae continually encounters flow-through water with an O_2 content that is higher than what is present in the capillary.

- Air breathers may not need highly efficient exchange surfaces or breathing mechanisms, but their exchange surfaces need protection from drying out. Land slugs and snails use skin in moist habitats, or mantle tissue or lungs under a shell. Arachnids use internalized gill-like book lungs and/or tracheae, invaginated tubes going directly to internal organs. Insects also use tracheae, which may be actively ventilated with air sacs and body movements.

- The first air-breathing vertebrates were bimodal or trimodal, with gills and air-breathing device(s)—skin and/or lungs. All lungs except those of birds and crocodilians have tidal ventilation. Frog lungs may be simple or moderately folded and are inflated by positive pressure. Reptiles as well as birds and mammals have respiratory systems ranging from simple to elaborately folded, inflated by negative pressure.

- Mammalian airways terminate in alveoli, which are involved in both ventilation and gas exchange. Mammalian lungs are inflated and deflated tidally by cyclical changes in intra-alveolar pressure generated indirectly by respiratory muscles (ribs, diaphragm).

- Ventilation and gas exchange are separated in crocodilian and avian respiratory systems due to air sacs and parabronchi with air capillaries. Flow-through breathing due to air sacs, plus crosscurrent flow in the parabronchi, enhances gas exchange in crocodiles and birds.

- Atmospheric, intra-alveolar, and intrapleural pressures are important in respiratory mechanics of mammals, whose lungs are normally stretched to fill the thorax. Intrapleural and lung pressures decrease during inhalation and increase during exhalation, which in part depends on elasticity. Elasticity depends on connective tissue and alveolar surface tension, which is kept collapsing the alveoli by surfactant.

- Changes in certain lung volumes such as tidal volumes, measured with a spirometer, are important for analyzing disease states, aerobic activity capacities, and species differences. Pulmonary ventilation (tidal volume $\times$ respiratory rate) scales with mammalian body mass. Alveolar ventilation is less than pulmonary ventilation because of the presence of dead space. It is important to increase tidal volume during activity since that benefits alveolar ventilation more than increased respiratory rate.

- Fish have lower dead-space problems but higher respiratory energy costs than mammals. Fish flow-through ventilation yields up to 90% O_2 extraction, while bird lungs achieve up to 40% and mammalian ones only 25%.

- Lung air has a lower partial pressure of O_2 than does atmospheric air due to humidification and dead-space mixing. Gases move to/from alveoli to blood and blood to/from body cells down partial pressure gradients maintained by breathing and circulation to be in the proper directions.

- Most O_2 in many circulatory systems is transported bound to hemoglobins (Hb), hemocyanins, and other metal-containing respiratory pigments. These trap O_2, so they cannot contribute to P_{O_2}, thus favoring gas diffusion. Most Hbs are tetramers of globin subunits each with an iron-containing heme group.

- The P_{O_2} is the primary factor determining the percent hemoglobin saturation. O_2 binding by Hb follows a sigmoidal curve with a steep rise at intermediate P_{O_2} levels and a plateau at high levels. This arises from cooperativity, in which subunits switch between high- and low-affinity states and affect their neighbors' states. At high P_{O_2} in the lungs, Hb is nearly 100% saturated. As it enters low-P_{O_2} tissues, cooperativity ensures proper release of O_2.

- Acid (the Bohr effect), temperature, and certain organic phosphates enhance O_2 release. O_2 affinity, measured by P_{50} values, has evolved to enhance gas delivery in different species; for example, larger mammals and those at high altitudes have higher affinities. Total saturation declines with acidity in the Root effect in fish Hbs, aiding the filling of gas bladders with acid secretion and countercurrent trapping.

- Most CO_2 is transported in vertebrates as bicarbonate due to carbonic anhydrase in erythrocytes. This reaction is also critical to acid–base balance.

- Ventilation of airways is regulated to match perfusion by pulmonary vessels by local and extrinsic mechanisms. Breathing is regulated in vertebrates by a negative-feedback system—which may be activated in anticipation of activity—of peripheral chemosensors in the arteries and brainstem medulla, multiple regulatory centers in the medulla, and effector breathing muscles. P_{O_2} is an important homeostasis signal in water breathers; P_{CO_2} and the H^+ it generates are more important in air breathers.

Review, Synthesize, and Analyze

1. Compare the respiratory mechanisms and events in a flying insect and a flying bird.

2. Compare and explain the gas exchange efficiencies of fish, birds, and mammals.

3. If an animal lives 1,600 m above sea level, where the atmospheric pressure is 630 mm Hg, what would the P_{O_2} of the inspired air be once it is humidified in the respiratory airways before it reaches the alveoli?

4. Suppose a medium-size mammal has a tidal volume of 350 mL. While breathing at 12 breaths/min, its alveolar ventilation is 80% of the pulmonary ventilation. What is its anatomic dead-space volume?

5. Compare the respiratory and blood-transport features of a bar-headed goose and an acclimatized human climber on Mt. Everest at 8,848 m altitude.

6. In mammals, fetal Hb has a lower P_{50} than adult Hb. Explain how this favors O_2 transfer across the placenta.

7. Even when partial pressure of O_2 in air is approximately the same as in water, the concentration of O_2 is much greater. Explain.

8. What mechanisms have animals come up with to survive extended periods of anoxia?

9. An increase in respiratory rate along with inappropriate voiding of your pet dog's bladder have brought you for a consultation with your veterinarian. What diagnosis do you think the practitioner will come up with and why?

Suggested Readings

Art, T., W. Bayly, & P. Lekeux. (2002). Pulmonary function in the exercising horse. In P. Lekeux, ed., *Equine Respiratory Diseases.* Ithaca, NY: International Veterinary Information Service.

Campbell, K.L., J. E. E. Roberts, L. N. Watson, et al. (2010). Substitutions in woolly mammoth hemoglobin confer biochemical properties adaptive for cold tolerance. *Nature Genetics* 42:536–540.

Daniels, C. B., S. Orgeig, P. G. Wood, L. C. Sullivan, O. V. Lopatko, & A. W. Smits. (1998). The changing state of surfactant lipids: New insights from ancient animals. *American Zoologist* 38:305–320.

Evans, D. H., J. B. Claiborne, eds. (2006). *The Physiology of Fishes,* 3rd ed. New York: CRC Marine Science Series. A classic book on this topic, updated with molecular advances.

Farmer, C. G., & K. Sanders. (2010). Unidirectional airflow in the lungs of alligators. *Science* 327:338–340. The discovery of bird-like lungs in alligators.

Gill, F. B. (2007). *Ornithology*, 3rd ed. New York: W. H. Freeman and Company. A leading text on ornithology for undergraduate students.

Hochachka, P. W. (1980). *Living without Oxygen.* Cambridge, MA: Harvard University Press. A classic book on animal life with low or no oxygen.

Prabhakar, N. R. (2006). O_2 sensing at the mammalian carotid body: Why multiple O_2 sensors and multiple transmitters? *Experimental Physiology* 91:17–23.

Routly, M., G. Nilsson, & G. Renshaw. (2002). Exposure to hypoxia primes the respiratory and metabolic responses of the epaulette shark to progressive hypoxia. *Comparative Biochemistry and Physiology* 131A:313.

Wedel, M. J. (2009). Evidence for bird-like air sacs in saurischian dinosaurs. *Journal of Experimental Biology A* 311:611–628.

West, J. B., R. R. Watson, & Z. Fu. (2007). The human lung: Did evolution get it wrong? *European Respiratory Journal* 29:11–17. Compares mammalian and avian respiration.

CourseMate

Cormorants have a kidney that excretes nitrogenous waste as uric acid, a metabolically costly but nontoxic compound that requires little water to excrete and has antioxidant properties. They imbibe seawater (which is saltier than their body fluids), and have special nasal salt glands to excrete excess NaCl. The harbor seal also must excrete excess NaCl and nitrogenous waste (primarily urea). Urea is a semitoxic compound that requires some water to excrete. The seal has very efficient kidneys for concentrating salt and urea.

Paul Yancey

Excretory Systems

12.1 Evolution of Excretory Systems

One of the most crucial aspects of homeostasis is the maintenance of an internal aqueous environment that is conducive to life. This "inland sea" is regulated within narrow limits to maintain a beneficial composition of water and important solutes and a low level of wastes. Dietary intake, metabolic products, and losses or excess intake of water and ions frequently create imbalances in the internal environment. Consider a female mosquito that has just sucked blood from your skin. The animal is bloated with a relatively huge amount of water and NaCl (the major components of your blood), making it more difficult to fly and potentially disturbing internal fluid movements. The insect also digests your nutritious blood proteins (especially hemoglobin), gaining useful products but producing harmful nitrogenous wastes. Thus, she must somehow remove excess water, salt, and wastes from her body. Or consider a seal or a cormorant (see opening photograph). Because seawater is saltier than body fluids of these vertebrates, it tends to dehydrate cells by osmosis. Thus, these animals may need to excrete excess salt (as well as wastes) while conserving body water.

Selective excretion is crucial to internal fluid homeostasis and involves several systems

In these and other animals, regulation of salt and water balance and excretion of wastes depend on mechanisms of selective retention and removal of molecules. These mechanisms are generally the purpose of *excretory systems*. Note that the term "excretory" is a bit misleading, because these systems often perform selective retention as well; indeed, *retention always accompanies excretion*. In particular, the primary functions of these organs for fluid homeostasis are as follows:

1. *Maintenance of proper internal levels of inorganic solutes* (Na^+, K^+, Cl^-, H^+, CO_2, and so forth). Even minor fluctuations in concentrations of some of these electrolytes in the ECF (extracellular fluid, p. 27) can have profound influences on metabolism. For example, changes in ECF concentration of K^+ can potentially lead to fatal cardiac dysfunction.

2. *Maintenance of proper plasma water volume*, important for proper circulatory-fluid pressure and general state of tissue hydration.

3. *Removal of nonnutritive and harmful substances resulting from metabolism* (ammonia, urea, bilirubin, and so forth) or *ingestion* (such as plant alkaloids, drugs), and *removal of hormones* after they have had their desired effect. This must be done without losing useful organic molecules such as nonhormone proteins and glucose.

4. *Maintenance of osmotic balance* (which results from a combination of water and dissolved solutes) by selective retention and excretion of water and ions.

Simple aquatic animals (sponges, Cnidarians, and possibly echinoderms) do not have specialized organs for fluid-regulating processes. They rely on diffusion and membrane transporters for nonsolid wastes. (Individual cells of freshwater sponges also have *contractile vacuoles*, specialized vesicles that remove and eject excess water out of the cell.) As aquatic animals evolved larger sizes and greater complexity, diffusion and membrane transport became inadequate for removing wastes from internal body fluids. Thus, the evolution of specialized excretory tissues was favored with **transport epithelia**: linings with specific membrane mechanisms to remove nonfecal wastes and to regulate other solutes and water. In some cases, epithelial surfaces of existing organs such as skin, gills, and hindguts evolved such mechanisms. However, in many animals these exchange surfaces became inadequate to process all body fluids or molecules for two reasons. First, as body size increased, these surfaces were too small and too distant from many body fluids. Second, evolution onto land created additional problems. Because of the lack of an aquatic environment to directly provide water for terrestrial animals and absorb their wastes, external epithelia were unable to regulate most solutes or water. To be free to move about in an often dry and ever-changing external environment, these animals had to evolve more elaborate mechanisms to maintain internal fluid homeostasis. Thus, larger and more complex aquatic animals, and all terrestrial ones, evolved specialized *tubules* lined with transport epithelia dedicated to excreting and selectively retaining key molecules of the body.

In the animal kingdom, four organ systems can be involved in excretion and retention:

1. *Respiratory systems (gills and lungs)*, which help regulate CO_2 (although we do not often think of it in these terms, the act of exhaling CO_2 is a form of waste excretion). Gills can also help remove other solutes such as ammonia and HCO_3^-. Some of these functions are covered in Chapter 11 (on respiratory physiology) and Chapter 13 (on fluid and acid–base balance).
2. *Digestive systems*, which remove not only undigested food but also some end products of internal metabolism such as biliverdin from the vertebrate liver (p. 688). Indeed, the liver in this case is acting as an excretory organ. Intestinal tract epithelia may also help regulate ions and water, and in some animals the gut receives the output of renal organs (see 4 below). Some of these functions are covered in Chapter 14 (on digestion).
3. *Integument (skin) and glands,* some of which can excrete organic wastes as secondary functions (such as sweat glands in humans), and others that evolved primarily to excrete excess inorganic ions (**salt glands** of brine shrimp and marine reptiles). The skin of many amphibians also regulates salt and water uptake.
4. *Renal organs (renalis*, Latin for "kidney") with tubules that *filter* body fluids and regulate water, ions, and many organic substances, and then *selectively reabsorb or secrete* these molecules. We will examine several examples such as insect Malpighian tubules and vertebrate kidneys. The output of renal organs is called **urine.** These organs, together with ductwork and any **bladders** (urine-storage chambers), compose the **urinary system.**

In the next chapter, we focus on how these and other systems work in an integrated way to maintain water and solute homeostasis. In this chapter, we focus primarily on renal organs and, to a lesser extent, skin and its glands: how they process water, ions, and wastes, and how they are regulated. We begin by examining a common problem of animals: how to deal with waste produced by nitrogen metabolism.

Nitrogen metabolism creates special stresses and produces three major end products: ammonia, urea, and uric acid

Metabolism of all foodstuffs results in CO_2 as a waste product, but protein and nucleic acid metabolism produces various **nitrogenous wastes** as well. Breakdown of proteins yields amino acids (Figure 12-1, top), which may be used (1) to make new proteins for growth and maintenance, and (2) in many marine animals as *osmolytes* (see p. 617) (Figure 12-1, lobster). Excess amino acids are routinely broken down to produce ATP, yielding CO_2, water, and **ammonia** (NH_3) (Figure 12-1, right), a potentially toxic strong base that can potentially alter acid–base balance as it binds protons and becomes **ammonium ion** (NH_4^+). The ammonium ion is also toxic; among other effects, it interferes with the Na^+/K^+ ATPase transporters of cell membranes by substituting for K^+. This toxicity manifests in many ways but is particularly evident in neurons, which suffer morphological changes, interrupted ion conduction, and disrupted neurotransmitter metabolism. Just 0.05 mM NH_4^+ in mammals and 2 mM in fishes severely impairs neural function. Thus, NH_3 and NH_4^+ must either be highly diluted and rapidly excreted, or be converted into a less toxic form. The most common of these forms are **urea** and **uric acid** (Figures 12-1 and 12-2). The soluble form of uric acid in the blood is actually **urate**, although we will use the term "uric acid" in the text. Let's examine these waste products in detail.

1. *Ammonia.* Most aquatic animals that breathe water—including most bony fishes, larval amphibians, and most nonvertebrates—rely on ammonia excretion, usually via gills. This feature is called **ammonotely** (Figure 12-1, fish). Water is sufficiently available to dilute the ammonia, which readily leaves the body down a favorable diffusion gradient. Thus, it is relatively easy to keep it below its toxic level. (NH_3, not NH_4^+, is the primary form that leaves an aquatic animal's body.)

    Interestingly, many deep-sea squids retain high levels of NH_4Cl in some body fluids because it is very buoyant (Figure 12-1, squid). It is not certain how toxicity of NH_4^+ is avoided, but it may be due to sequestration of the ions in special fluid chambers.
2. *Urea.* Most terrestrial animals do not have the option of greatly diluting ammonia and instead must spend metabolic energy to convert ammonia into a less toxic form such as urea or uric acid, usually in the liver, and to transport the product to the excretory organs. Urea excretion is termed **ureotely.** This concept is exemplified in amphibians, whose aquatic larval stages use ammonia excretion via their gills, whereas semiaquatic adults switch to urea production and excretion using their livers and kidneys. Urea is the primary nitrogenous waste not only of most adult amphibians but also mammals (Figure 12-1, bear) and some reptiles such as sea turtles. Urea is produced, in most vertebrates that make it, from two ammonium ions (one via aspartic acid) and a bicarbonate ion, using ATP, in the **ornithine–urea cycle,** or OUC (Figure 12-2a). (Some bony fishes that produce urea use the **uricolytic pathway;** Figure 12-2b.) Urea is the primary nitrogenous waste of marine chondrichthyan fishes (Figure 12-1, shark) and coelacanths (meaning "hollow spine," referring to hollow fin rays), "living

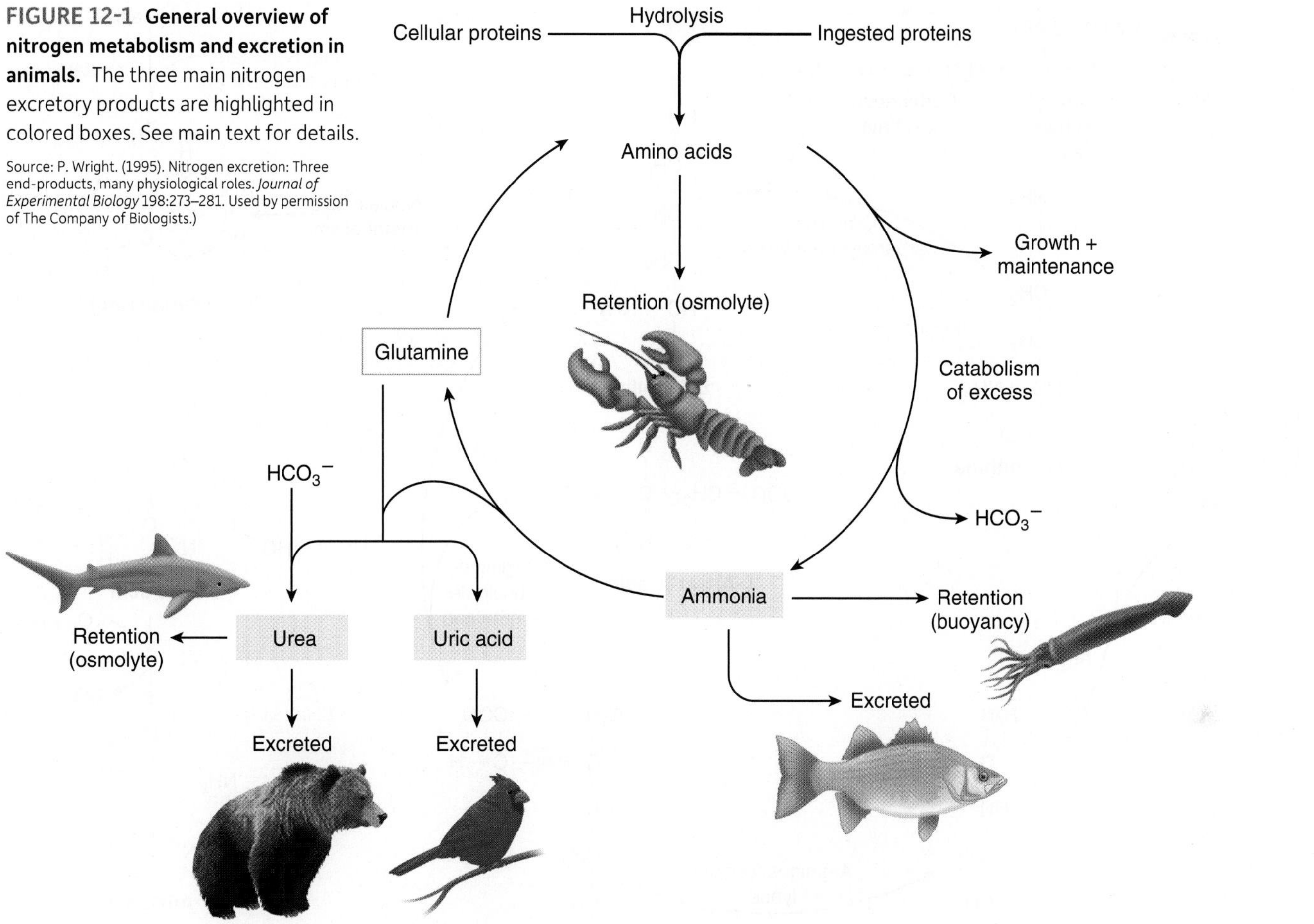

FIGURE 12-1 **General overview of nitrogen metabolism and excretion in animals.** The three main nitrogen excretory products are highlighted in colored boxes. See main text for details.

Source: P. Wright. (1995). Nitrogen excretion: Three end-products, many physiological roles. *Journal of Experimental Biology* 198:273–281. Used by permission of The Company of Biologists.)

fossil" lobe-finned fishes once thought to be extinct. In these fishes, urea is accumulated as their major osmolyte (see p. 621). Urea is 10 to 100 times less toxic than ammonia and thus can be accumulated to much higher concentrations, and has the added benefit of removing two nitrogens per molecule. Thus, it takes about 10 times less water to excrete a given amount of nitrogen as urea than as ammonia. Urea is, however, toxic at concentrations greater than 100 mM or so because it binds to proteins and destabilizes their structures and interferes with their ability to bind ligands. Thus, it must be kept at moderate levels, or where it is highly concentrated in chondrichthyan fishes, coelacanths, and the mammalian inner kidney, it must be *counteracted* with a protein stabilizer (p. 619; see also *Molecular Biology and Genomics: Surviving Salt and Urea*, p. 606).

In ruminants, urea has another function. After synthesis in the liver, it can enter the rumen via the salivary glands, where symbiotic microbes ferment the food; there, microbes convert urea back to ammonia, using the enzyme *urease*. The ammonia can then be used directly by these beneficial microbes for their own protein synthesis (Chapter 14).

3. *Uric acid.* In other terrestrial animals—insects, most reptiles including birds, and even a tree frog and a desert toad—uric acid is usually the primary nitrogenous waste (although some insects excrete some urea and ammonia). This is called **uricotely** (Figure 12-1, bird). Uric acid

is the end product of amino acid and purine (nucleic acid) metabolism (Figure 12-2b) and requires even more ATP than urea (Figure 12-2a) to produce. It is less toxic because it is highly insoluble, has the added benefit of removing four nitrogens per molecule, and is excreted in a semisolid form, generally into the hindgut. (In the physiological pH range, 99% of uric acid exists as urate, the monovalent anionic form.) It takes about 50 to 100 times less water to excrete a given amount of nitrogen as uric acid than as ammonia. Compared to urea, which is essentially infinitely soluble in solution, uric acid is highly insoluble (precipitates at concentrations greater than about 0.4 mM). Uric acid also serves as a protective antioxidant (see *A Closer Look at Adaptation: Avian Longevity: Unraveling the Mystery*, p. 564).

An interesting question arises from these patterns: If uric acid is less toxic and more favorable to water conservation than urea, why do not all terrestrial animals use it as the primary nitrogenous waste? One possibility lies in the high metabolic cost of synthesizing uric acid compared to urea (12 or more ATPs versus 4 ATPs/mole, respectively), and different modes of reproduction. Birds, reptiles, and many insects lay sealed yolky eggs, often in fairly dry areas. Without a way to remove nitrogen wastes produced from metabolism of yolk proteins, such eggs may require the least toxic (but most expensive) form, which also exerts no osmotic disturbances in its crystallized state. Adult insects, because of their

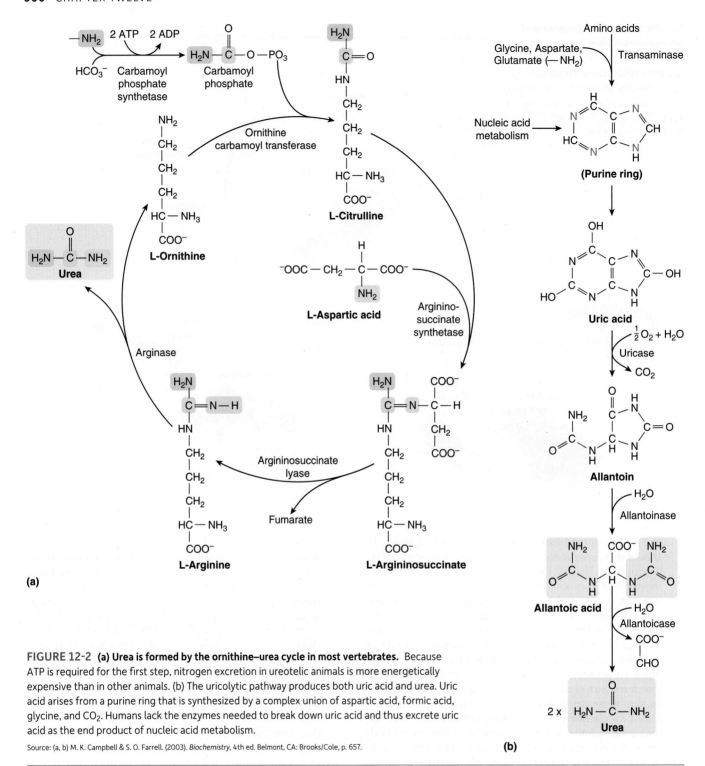

FIGURE 12-2 (a) Urea is formed by the ornithine–urea cycle in most vertebrates. Because ATP is required for the first step, nitrogen excretion in ureotelic animals is more energetically expensive than in other animals. (b) The uricolytic pathway produces both uric acid and urea. Uric acid arises from a purine ring that is synthesized by a complex union of aspartic acid, formic acid, glycine, and CO_2. Humans lack the enzymes needed to break down uric acid and thus excrete uric acid as the end product of nucleic acid metabolism.

Source: (a, b) M. K. Campbell & S. O. Farrell. (2003). *Biochemistry*, 4th ed. Belmont, CA: Brooks/Cole, p. 657.

small sizes and thus high surface-area-to-volume ratios (p. 10), may also need the most effective water conservation mechanisms. In contrast, amphibian adults (which usually enter water frequently) and prenatal stages of all types of mammals (whose wastes are removed by the mother's circulation) can more easily dispose of their nitrogenous wastes. Thus, using urea as a primary waste saves considerably on metabolic energy of synthesis. In any case, mammals have not evolved mechanisms to handle high concentrations of the highly insoluble uric acid, and excessive levels can lead to

arthritic **gout** and **kidney stones** because of crystals of this highly insoluble molecule forming in inappropriate places (the joints and kidney tubules, respectively). Mammals do make uric acid, but most only synthesize small amounts that are removed by the kidneys (though long-lived primates lack the degradative enzyme **uricase** and may accumulate modest amounts of uric acid). See *A Closer Look at Adaptation: Avian Longevity: Unraveling the Mystery*, p. 564.

Overall, it appears that *the use of ammonia, urea, or uric acid as the primary nitrogen waste correlates strongly with water availability*. As we have seen, most aquatic animals use ammonia, whereas most animals with terrestrial eggs use uric acid. As another example, aquatic birds such as ducks excrete an evenly balanced mixture of uric acid and ammonia, whereas terrestrial birds excrete primarily uric acid. But there are exceptions, including the following. (1) The tilapia *Alcolapia grahamithe* in the highly alkaline Lake Magadi in Africa is mainly ureotelic, using the OUC (absent in most fish) (Figure 12-2a). This presumably evolved because high pH impairs ammonia excretion. (2) The Gulf toadfish *Opsanus beta* has an OUC and switches from ammonotelic to ureotely when stressed, for example, by air exposure, crowding, or confinement, releasing urea in pulses through membrane urea transporters in the gills. (3) Some marine snails use uric acid. (4) Monotremes (egg-laying mammals—echidna [p. 423] and platypus) use primarily urea, but their embryonic development relies very little on yolk; most development occurs before the (porous) egg is laid, during which time nutrients and wastes are exchanged between the embryo and mother (p. 801). (5) The most water-stressed mammals, desert rodents such as kangaroo rats (which may never drink water in their entire lives), use urea (see p. 602). "Physiological logic" would suggest that these rodents would benefit from uric acid as their major nitrogenous waste. Perhaps evolutionary constraints (p. 2) have resulted in mammals being "locked in" to using urea because they lack the hindgut system that seems to be essential for uric acid processing. Similarly, (6) nectar-feeding birds like hummingbirds deal with an excess of liquid intake, and perhaps could benefit energetically by making urea rather uric acid for nitrogen waste; but they do not. Perhaps birds too are "locked in."

There are a few other forms of nitrogenous wastes. Three of these are *guanine*, the primary nitrogenous waste of arachnids; *allantoin* (Figure 12-2b), produced from uric acid in some insects; and *creatinine*, a by-product of vertebrate muscle metabolism that is constantly produced, transported, and removed by the kidneys. As discussed later, (p. 593) renal physiologists and physicians use this process to measure key renal functions.

Excretory organs have transport epithelia, which typically use Na+/K+ ATPases

As we noted earlier, excretory and related processes in complex animals rely on specialized transport epithelia. Recall that such transport may be *passive* and *active*. In passive types, no steps in the transport of a substance across an epithelium use direct metabolic energy, because energy is provided by electrochemical or osmotic gradients (see p. 75). In contrast, a substance is said to be actively transported if any one of the steps requires direct use of energy, even if other steps are passive. In active transport, net movement of the substance occurs against an electrochemical gradient. Though transport tissues vary in their membrane transporters, most involve two mechanisms: (1) the energetic cost of transport typically lies in an ion *ATPase "pump,"* which not only results in ion transport but also in an ion gradient that can be used to move other solutes in symport or antiport fashion (see Figure 3-15); and (2) *no active transporters for water molecules are known to exist*, so if water movement is involved, epithelia cells must first transport solute(s), which then build up osmotic pressure and draw water by osmosis.

Figure 12-3a and b show two common examples of active epithelial transport that use the common Na+/K+ ATPase

(p. 85). Note the orientation of the cells: The *basolateral* side typically faces interstitial fluid, while the *apical* side typically faces the environment or a tubular **lumen** (the cavity inside a biological tube).

Salt Transport Only Transporting salt without water involves two steps:

1. The Na+/K+ ATPase "pump" uses ATP to increase the Na+ concentration outside the cell's basolateral side, and decrease it inside the cell; some Na+ then moves passively into the cell from the apical side through an *ENaC* channel (p. 235).
2. The build-up of positively charged Na+ in the basolateral fluid passively draws Cl− ions by charge attraction. Cl− moves through either or both of the following:
   a. A **ClC** (**chloride-activated chloride**) channel. ClC channels are an ancient, large family of chloride channels found in all domains and kingdoms of life. As gated channels, many ClC channels open in response to increasing Cl− concentration, hence their name.
   b. A CFTR (**cystic fibrosis transmembrane-conductance regulator**) chloride channel, so named because its mutated, nonworking form is responsible for the defective salt transport characteristic of *cystic fibrosis*. It is regulated by cAMP and phosphorylation. Humans with cystic fibrosis have nonfunctioning CFTRs, and cannot properly transport salt and water through membranes of several epithelia such as linings of airways, sweat glands, and intestine. In airways, for example, harmful thick, sticky mucus builds up.

Overall, then, NaCl is moved across the tissue through the cells that is, **transcellularly**. The tissue in Figure 12-3a is fairly waterproof, so little or no water moves.

Salt and Water Transport In the second tissue, Figure 12-3b, the first two steps are the same, but there are two more:

3. The osmotic pressure of transported NaCl builds up within *lateral spaces* between the cells, where it induces osmosis of H_2O from the cells. The accumulation of H_2O in those lateral spaces results in a buildup of hydrostatic pressure, flushing H_2O out of the spaces into the interstitial fluid (and, in animals with blood vessels, into nearby capillaries).
4. In some cases, some water and ions are thought to move through spaces between the cells, that is **paracellularly**, as shown here for H_2O.

We now turn to renal organs, which rely heavily on such transport epithelia.

check your understanding 12.1

What are the four major functions of excretory systems, and what are the four organ systems that can serve this role?

Discuss the usefulness and evolutionary patterns of ammonia, urea, and uric acid as nitrogenous wastes.

How do transport epithelia move NaCl alone or NaCl and water?

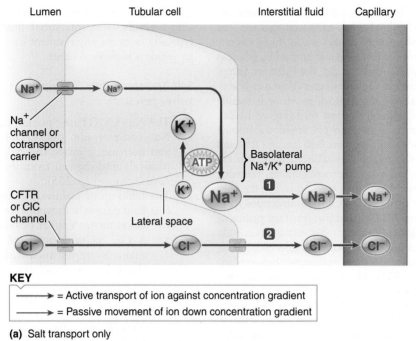

Lumen   Tubular cell   Interstitial fluid   Capillary

Na$^+$
channel or
cotransport
carrier

CFTR
or ClC
channel

Basolateral
Na$^+$/K$^+$ pump

Lateral space

KEY

⟶ = Active transport of ion against concentration gradient

⟶ = Passive movement of ion down concentration gradient

(a) Salt transport only

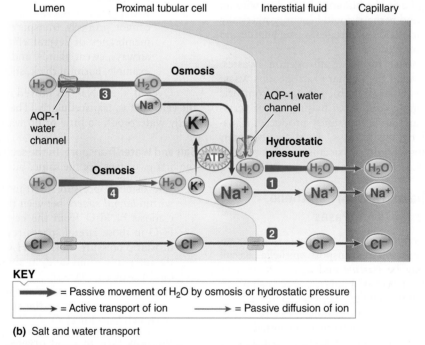

Lumen   Proximal tubular cell   Interstitial fluid   Capillary

Osmosis

AQP-1
water
channel

AQP-1 water
channel

**Hydrostatic
pressure**

Osmosis

KEY

➡ = Passive movement of H$_2$O by osmosis or hydrostatic pressure

⟶ = Active transport of ion   ⟶ = Passive diffusion of ion

(b) Salt and water transport

FIGURE 12-3 Properties of transport epithelia and tubular excretory organs. (a) Salt absorption by a waterproof epithelium. (1) Basolateral Na$^+$/K$^+$ ATPase pumps increase Na$^+$ externally, which (2) draws Cl$^-$ by electric charge through ClC and/or CFTR channels. (b) Fluid absorption by a water-permeable epithelium in a well-supported model first proposed by Peter Curran and John Macintosh in 1962. This has the same steps 1 and 2, but also (3) H$_2$O is drawn across osmotically through aquaporins by the NaCl buildup in lateral spaces between cells. H$_2$O is not drawn from the capillary because it is permeable to NaCl (recall that osmosis requires nonpermeable solutes). (4) H$_2$O and Cl$^-$ may also move paracellularly. See main text for other details.

Sources: (a, b): J. J. Wine, (1999), The genesis of cystic fibrosis lung disease, 'Journal of Clinical Investigation', 103:309. Reproduced with permission via Copyright Clearance Center. (c) P. C. Withers. (1992). *Comparative Animal Physiology* Belmont, CA: Brooks/Cole, p. 841, Figure 17-9.

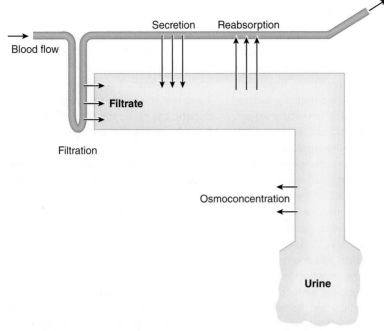

(c) The four basic renal processes (shown for a vertebrate): filtration, secretion, reabsorption, and osmoconcentration

FIGURE 12-3, CONT'D **Properties of transport epithelia and tubular excretory organs.** (c) Schematic of the principal functions of a tubular renal organ, such as the vertebrate nephron: filtration, secretion, and reabsorption. Insect, avian, and mammalian tubules may also osmoconcentrate urine.

12.2 Renal Excretory Organs: Overview

Renal organs have tubules that filter solutes and water from body fluids and subsequently modify the fluid in the tubular lumen using transport epithelia, specifically for water and solute homeostasis and waste removal. Let's look at the key processes.

Renal tubules produce urine using the processes of filtration, secretion, reabsorption, and osmoconcentration

Renal tubules operate by two or more of four basic renal processes (Figure 12-3c):

1. *Filtration,* in which solutes are selectively separated from a solution by passing through a boundary. Often this is a mostly nonselective process, in which some water with all its small solutes passes through the barrier, except that cells and large molecules such as proteins (and some water) remain behind, a process called **ultrafiltration.** Ultrafiltration is typically driven by hydrostatic pressure. Another type of filtration relies on specific transporters and is thus more selective.

2. *Secretion,* in which transport epithelia move (often actively) specific solutes into the tubule lumen for excretion.

3. *Reabsorption,* in which transport epithelia move (often actively) specific solutes (and often water) back into the body from the lumen.

4. *Osmoconcentration,* in which water is removed from the lumen fluid while leaving solutes behind, producing an excretory fluid more concentrated (*hyperosmotic*) than body fluids and thus conserving water. Some reabsorption processes are tied to this mechanism.

Methodology Numerous techniques are used to explore the functions of renal excretory systems. Here we describe two that have contributed greatly to the understanding of renal functions, which we will explore shortly:

- For studying the tubules we have just discussed, perhaps the most important method is that of the **isolated perfused tubule** (Figure 12-4). A researcher begins by carefully dissecting an intact tubule out of the animal's renal system and placing it in an appropriate bathing solution. Two holding pipettes are applied, one to each open end of the tubule. Within one pipette, a much finer pipette (the perfusion pipette) is inserted to allow injection of defined solutions into the tubule. A fine pipette within the other holding pipette is used to collect fluid exiting the tubule. By measuring solutes in this collected fluid and in the bathing solution, researchers can determine transport and permeability features. Other devices may also be used on these tubules, such as patch clamps (p. 112) and microelectrodes for measuring potentials.

- In studying the effect of the whole kidney on retention and removal, the **plasma clearance** method has been important. To understand this method, you first need to understand certain tubule functions; we return to plasma clearance later.

The major renal organs are protonephridia, meso- and metanephridia, and Malpighian tubules

Several different kinds of renal organs have evolved, and there are several ways to classify these. One useful scheme recognizes three common types—*protonephridia, meso- and metanephridia, Malpighian tubules*—based on which body fluids are processed and how:

- **Protonephridia** (bundles of *flame cells*) use ultrafiltration driven by cilia, with secretion and reabsorption. These are the types of specialized filtration system thought to have evolved first in freshwater nonvertebrates. These are generally blind-end ducts that project into the body cavity (of which there is only one: a general interstitial

A Closer Look at Adaptation:
Avian Longevity: Unraveling the Mystery

Have you ever wondered why we live for so much longer than our companion animals? If our pet dog or cat lives to be 15 or more years of age, we regard this as exceptional. What makes this even more intriguing is that birds live considerably longer than mammals of the same body size. For example, a hummingbird lives for approximately 9 years, whereas a comparably sized mouse may live for only 1.5 years. This mystery has intrigued the scientific community for as long as it has been aware of the paradox. Given certain inherently avian characteristics, prevailing theories suggest that birds should be more susceptible than mammals to the degenerative process of aging. These include the following: Birds have plasma glucose concentrations typically 2 to 6 times higher than those of mammals, metabolic rates as much as 2 to 2.5 times higher than those of similar-sized mammals, and body temperatures approximately 3°C higher. The higher metabolic rates, coupled with the extreme longevity of many avian species, result in a much greater lifetime energy expenditure per unit mass by birds than mammals. This is important because, within a class of animals, longevity is inversely correlated with metabolic rate (see scaling discussion in Chapter 15, p. 720). With mammals, physiologists assume that lifetime energy expenditure correlates with an organism's cumulative exposure to potentially damaging free radicals (and other reactive oxygen species, or *ROS*) produced as normal by-products of mitochondrial metabolism (p. 53). In addition, immune cells generate ROS with which they destroy pathogens (p. 469). Normally this is beneficial, but any abnormal activation of the immune cells, as found in certain disease states, can lead to the constant generation of ROS, resulting in chronic inflammation and accelerated tissue aging. Thus, the increased avian metabolic rate should expose birds to a higher concentration of ROS and, consequently, accelerated tissue damage.

In addition, the high glucose levels (along with high body temperature) should promote accelerated tissue aging. Consider the situation in hummingbirds. They have blood sugar levels much higher than do people with Type II (noninsulin-dependent) *diabetes mellitus* (p. 323). The tissues of human and domestic dogs and cats, as a consequence of the elevated blood-sugar levels and oxidative stress, are susceptible to tissue-specific complications that manifest over time. Diabetes is the primary cause of blindness, kidney failure, and limb amputation in America and contributes to cardiovascular disease. It has been well established that the higher the blood sugar in a diabetic, the earlier the onset of complications. Do birds suffer a similar fate to that found for diabetics?

First, how does glucose damage the body's tissues and organs? One mechanism begins with the formation of advanced **Maillard products:** Glucose covalently binds (nonenzymatically) to proteins. This *glycosylation* process may be involved in normal aging-related tissue degeneration and is accelerated in diabetes. (Interestingly, glucose is among the least reactive of the monosaccharides, which may be one reason why animals rely on this particular carbohydrate for its energy needs.) Maillard products are further modified by free radicals in an animal's body, which contribute to the formation of irreversible cross-links between adjacent proteins. This impairs protein functions; for example, elasticity of the lung and blood vessels diminishes, wrinkles appear, and the basement membranes in the kidney thickens, which reduces substrate filtration. Research has shown the diabetic population has an elevated concentration of free radi-

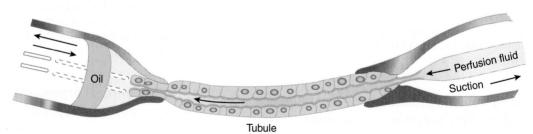

FIGURE 12-4 Perfusion of an isolated segment of a renal tubule. A specific perfusion fluid is injected on the right, and a sample of the outgoing fluid is taken on the left. Any modifications to the fluid made by the tubule cells can then be determined.

Source: Diagram courtesy of Maurice Burg, National Institutes of Health.

space, a pseudocoelom, or a coelom). They have low (fluid) hydrostatic pressure inside them relative to the body cavity, most often because of the beating of cilia in the duct, which resembles a flame (giving the cell its name), move fluid outward. This relatively negative pressure draws water and small solutes nonselectively in through a specialized cell with filtering slits or pores (ultrafiltration) from the body cavity. The duct after the filter may further modify the filtrate by reabsorbing desired molecules and secreting undesired ones. (Details of similar mechanisms of filtration, secretion, and reabsorption are covered for the mammalian kidney later.) The action of the cilia eventually moves this modified fluid or urine out through a pore in the epidermis into the external environment. Protonephridia are generally found in more simple animals having one internal fluid compartment, such as rotifers, flatworms, larval annelids, and larval mollusks; they are also found in larval fishes and amphibians. An example is shown in Figure 12-5a.

cals and oxidative stress as a result of an immune system triggered into action by glycosylated tissue proteins.

What about the hummingbirds? With their elevated metabolism and higher body temperatures, these birds actually generate substantially more free radicals than a comparably sized mammal, yet remarkably, the concentration of glucose-derived cross-links is considerably lower than in mammals. How can this be?

Longevity appears to be a balance between free radical production and quenching of the free radicals by a combination of both endogenous (internally produced) and exogenous (dietary) *antioxidants*. Perhaps, then, birds have in their anti-aging repertoire some powerful antioxidants. A clue arises from the dual role of **uric acid,** a nitrogenous waste product (p. 559) that may also play a role in the body's antioxidant system. As you have seen, uric acid is the major nitrogenous waste in birds (and most other reptiles, some amphibians, and insects). These animals are uric acid "factories." However, uric acid is not just a waste product: It has been found to be important in combating oxidative stress and functions something like a catcher's mitt that can "pick off" any stray free radicals missed by the other antioxidant systems (see p. 53). A reduction in uric acid concentrations in chickens dramatically increases oxidative

stress and accelerates the rate of tissue aging, whereas the opposite effect is observed if uric acid levels are increased. Of interest, shorter-lived animals posses the enzyme **uricase,** which degrades uric acid into allantion. It is an established scientific finding that long-lived species fail to express this enzyme and so enjoy an extended lifetime due in part to greater antioxidant protection.

What happens if uric acid concentrations rise too high? This results in a condition called **gout,** which can occur in humans, some other mammals, and some reptiles. Uric acid is highly insoluble and so when its concentrations increase above a certain threshold, it begins to come out of solution and crystallize in various tissues. For example, uric acid crystals forming in kidney *nephrons* can permanently impair kidney function. For unknown reasons, this increases plasma *renin* concentrations, which ultimately raises blood pressure (p. 587). Colder temperatures enhance crystallization; thus, uric acid tends to form painful urate crystals in the toes, fingers, and edge of the ears in primates. However, individuals with gout are protected against certain neurodegenerative diseases that afflict the general population. How then can birds tolerate much higher concentrations of uric acid in their tissues and blood? (In fact, both domestic and wild birds can

get gout, but not at rates higher than mammals.) One obvious answer is the much higher body temperature of birds, which increases the solubility of uric acid. The second is the unusual reaction that occurs in the nephron of the avian kidney. As soon as uric acid begins to crystallize in the avian kidney, the microvilli in the *proximal tubule* (p. 571) encapsulate the crystal with a protein coat, forming a *urate ball,* a highly soluble colloidal particle. These spheres are not physically destructive and serve to transport the uric acid crystal safely through the nephron. The colloidal suspension then passes through the ureters into the cloaca of the bird. Here the spheres are refluxed back into the ceca, pouchlike attachments off the large intestine, and microorganisms recover the protein. Most of the uric acid is also degraded in the ceca. Specific transporters located in the epithelial cells of the large intestine take up any free amino acids and sugars.

Because birds have a high concentration of blood glucose, it has been suggested that they evolved a mechanism to limit the associated damage. This mechanism includes a more efficient antioxidant system that takes the form of an apparently simple waste product: uric acid. This discovery may have important clinical applications for treating conditions associated with inflammatory stress.

- **Mesonephridia** and **metanephridia** use ultrafiltration typically driven by fluid pressure, with secretion, reabsorption, and occasionally osmoconcentration. These tubules are in animals with two or more major fluid spaces—a coelom and a circulatory fluid at a minimum. The tubules begin with a filtering capsule or funnel-like opening of some kind that ultrafilter the circulatory fluid. This is followed by tubule segments that perform selective secretion and reabsorption (and osmoconcentration in some groups) and may empty via a connecting canal or **ureter** into a hindgut or a bladder.

  In vertebrates, the renal tubules are called **nephrons** (Greek *nephros,* for "kidney"), and they go through significant developmental changes in most groups. In all groups, the first renal structure is a ciliated *pronephric* one, which becomes functional only in larval fishes, amphibians, adult hagfish, and lampreys. In all other adult fishes and amphibians, the pronephric tubules degenerate except the duct, which becomes the foundation of new pressure-driven *mesonephric* tubules (the kidneys of these aquatic animals). Finally, in terrestrial vertebrates, both the pro- and mesonephric tubules form in embryos but are replaced

by the *metanephric* tubules (more adapted for water-limited life on land). The mesonephric tubules, however, become the ducts of reproductive organs in males.

  Metanephric-type tubules have evolved independently several times—in adult mollusks, in annelids (segmented worms), in some arthropods (such as crustaceans), and in vertebrates. In some cases, a single filtering tubule forms the renal organ, as in the (misnamed) *crustacean antennal gland* (Figure 12-5b). Each single tubule (one on each side of the head) begins with a capsule (end sac) that filters hemolymph; enters a *labyrinth* of epithelia and fluid spaces followed by a coiled *nephridial canal,* where selective absorption and secretion take place; and then empties into a bladder. In other animals, many meso- and metanephridial tubules are assembled together with connective tissue into large organs called **kidneys,** as in the case in "higher" vertebrates.

- **Malpighian tubules** begin filtration driven by active secretion of ions from an arthropod's hemolymph (Chapter 9), and then, in conjunction with the hindgut, use reabsorption and sometimes osmoconcentration. We begin with these.

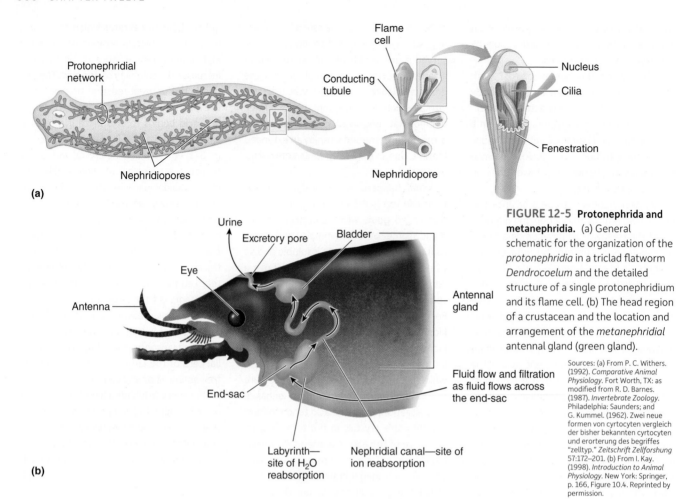

(a)

(b)

FIGURE 12-5 **Protonephrida and metanephridia.** (a) General schematic for the organization of the *protonephridia* in a triclad flatworm *Dendrocoelum* and the detailed structure of a single protonephridium and its flame cell. (b) The head region of a crustacean and the location and arrangement of the *metanephridial* antennal gland (green gland).

Sources: (a) From P. C. Withers. (1992). *Comparative Animal Physiology*. Fort Worth, TX: as modified from R. D. Barnes. (1987). *Invertebrate Zoology*. Philadelphia: Saunders; and G. Kummel. (1962). Zwei neue formen von cyrtocyten vergleich der bisher bekannten cyrtocyten und erorterung des begriffes "zelltyp." *Zeitschrift Zellforshung* 57:172–201. (b) From I. Kay. (1998). *Introduction to Animal Physiology*. New York: Springer, p. 166, Figure 10.4. Reprinted by permission.

check your understanding 12.2

Describe the four major renal processes.

How do protonephridia, mesonephridia, and metanephridia differ?

12.3 Insect Malpighian Tubules

Let's first examine in detail the excretory system of the most successful group of animals, the insects, with Malpighian tubules for excretion. These were named after their discoverer, Marcelo Malpighi, who—in Italy in the 1600s—pioneered microscopic study of anatomy (also discovering insect tracheae). Malpighian tubules are blind-end epithelial ducts, one-cell-layer thick, and are not organized into a distinct organ. The tubules project into the hemolymph from the hindgut (Figure 12-6a) in varying numbers; for example, there are four tubules in two pairs in *Drosophila melanogaster*, the common laboratory fruit fly (mosquitoes, in contrast, have five tubules; large insects like locusts may have hundreds). Some insects have small muscles that wave the tubules in the hemolymph, perhaps to create circulation. Recent genomic and proteomic studies show that the tubules have innate immune functions (p. 459) as well, at least in fruit flies.

Based on both classical microscopy studies and functional properties, the tubules can be divided into distinct regions. In fruit flies, one pair of tubules, which project anteriorly, begin with large *initial segments*, followed by *transitional segments, main segments,* a *lower segment,* and a junction between the two forming a common *ureter* that joins the hindgut. The other pair, which project posteriorly, have only main segments and a ureter junction. Other insects have different patterns of tubular differentiation; for example, the house cricket's tubules have three regions called *proximal, mid,* and *distal.*

Malpighian tubules initiate excretion by ion secretion, which causes osmosis

Malpighian tubules filter the hemolymph not by pressure, as in most other renal organs, but by secreting K^+ (and often Na^+) into tubule lumen (Figure 12-6a). This occurs in the main tubule segment in fruit flies and mid segment in crickets. Rather than using the Na^+/K^+ ATPase, this process uses a **V-ATPase**, one of the widespread types of proton (H^+) pumps we noted in Chapter 3 (p. 85). The V-ATPase is thought to set up a proton gradient (with H^+ higher in the lumen), which can be used to pull in K^+ and Na^+ against a gradient. This is thought to occur through *CPA (cation–proton antiport)* transporters, found in large so-called *principal cells* in fruit flies and mosquitoes (Diptera). As you

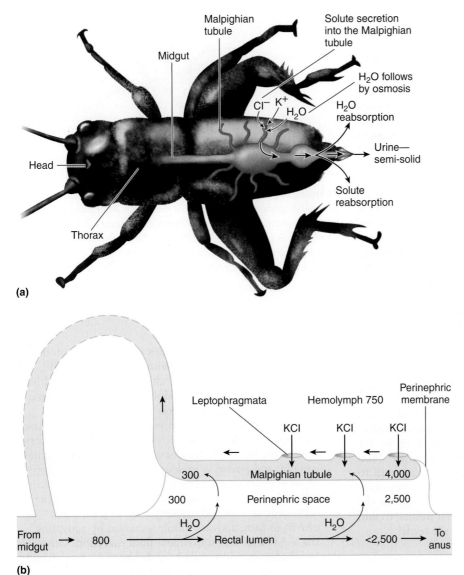

(a)

(b)

FIGURE 12-6 Malpighian tubules. (a) The arrangement of these tubules in a typical insect. First, K⁺ ions are transported into the proximal end of the tubule, drawing Cl⁻ in by charge attraction and then water by osmosis. The tubule fluid is further modified as it travels distally in the tubule; in the rectum, large amounts of water are reabsorbed. (b) Countercurrent flow between the Malpighian tubule and hindgut in some desert insects, with osmolarities shown in mOsm. In these animals, the tubules bend posteriorly to lie against the rectum. Unlike in other insects, the side of the tubule facing the hemolymph is waterproof so that K⁺ and Cl⁻ transported into the tubule without water following osmotically can accumulate to high levels (e.g., 4,000 mOsm). This in turn draws water osmotically from the nonwaterproof rectum, concentrating the fluid leaving the anus.

Sources: (a) From I. Kay. (1998). *Introduction to Animal Physiology.* New York: Springer, Figure 10.3. Reprinted by permission. (b) P. Willmer, G. Stone, & I. Johnston, (2000), "Environmental Physiology of Animals", p. 110, Figure 5.20. Copyright © 2000 Blackwell Publishing. Reproduced with permission of Blackwell Publishing Ltd.

electrical gradient (as in Figure 12-3a, b). Cl⁻ is thought to move transcellularly through ClC channels (p. 561) in smaller *stellate cells* in fruit flies, and paracellularly in mosquitos. The net accumulation of KCl (and often NaCl) in turn makes the tubule fluid more concentrated, so water from the hemolymph moves in by osmosis through aquaporins in stellate cells and also by paracellular pathways (Figures 12-3b and 12-6a). The joint movement of salts and water results in a filtrate that is *isosmotic* with the hemolymph.

The tubules and hindgut modify the lumen fluid by specific secretion and reabsorption

The osmotic movement of water creates a bulk flow down the tubule, whose epithelium further modifies the lumen fluid through specific secretion and reabsorption. In particular, organic wastes such as uric acid are secreted into the tubule by transporters. Some solutes are moved by active transport, and others diffuse passively down concentration gradients through channel proteins because the lumen fluid, created by KCl, NaCl, and water initially, begins with no organic solutes, whereas the hemolymph has some. How many transporters and channels are in the tubules is not known, but the fly genome contains numerous genes with codes closely related to organic-solute transporters in mammals (see p. 592). Moreover, the tubule cells are enriched in *detoxification* enzymes related to those in vertebrate livers, which may break down toxic molecules such as insecticides (natural or human-made).

Antidiuresis and Diuresis The tubule empties into the gut, excreting a fluid with essentially the same osmotic pressure as body fluids (that is, *isosmotic*), but with a very different solute concentration. Once in the hindgut, specialized epithelial cells of the rectum can further modify the urine. Some insects face frequent dehydrating conditions and must undergo **antidiuresis**, that is, excretion of a low-volume *hyperosmotic* (p. 78) urine, which conserves water in the body. This *osmoconcentration* process occurs by various mechanisms in different species but typically involves complex tissue anatomy that uses active transport of ions (such as KCl), followed by water osmosis. Uric acid precipitates as crystals as water is removed.

In the most efficient osmoconcentrating insects, such as mealworm beetle larvae (*Tenebrio molitor*)—which consume dry food such as flour—a *countercurrent* arrangement of the gut and Malpighian tubules is thought to be used (Figure 12-6b). Basically, the tubule runs parallel to the rectum and fluid moves in opposite directions in the two tubes.

learned for sodium and glucose in Chapter 3 (Figure 3-15, p. 88), this *secondary active transport* can move substances against a gradient using the energy of another substance with a downhill gradient.

The accumulation of K⁺ and Na⁺ in the lumen makes the fluid more positive, which then draws in Cl⁻ ions via the

It is thought that specialized cells called **leptophragmata**, found in waterproof domelike swellings in the initial segment, transport KCl into the upper portions of the tubule. Since water does not follow (as in Figure 12-3a), the KCl increases the lumen osmotic pressure to very high levels (e.g., 4,000 mOsm). This then draws water out of the end of the rectum through nonwaterproof membranes (Figure 12-6b), with the countercurrent arrangement enhancing the concentrating process in a manner similar to the mammalian nephron, which you will learn about later. These animal larvae appear to survive exclusively on *metabolic water* (p. 602) when consuming food with no water.

Other insects must cope with excessive loads of water. Freshwater insects are obvious examples. Another example is a nectar-feeding insect such as the butterfly, which imbibes large amounts of sugary nectar that can have a lower osmotic pressure than the insect's body fluids, thus potentially causing detrimental dilution of those fluids. These insects' tubules undergo **diuresis**—a high rate of urine production—which in this case is *hypo-osmotic*. Here, the hindgut recovers ions without concomitant water uptake.

Finally, other insects face excessive loads of both water and salt. Recall the mosquito we discussed at the beginning of the chapter, or consider the South American blood-sucking kissing bug, *Rhodnius prolixus* (see photograph)—both using piercing mouthparts to obtain huge loads of salty liquid in the form of blood from mammalian skin. The mosquito and bug rapidly bloat up from their blood meals, sometimes by a factor of 10 in mass! In both cases, the Malpighian tubules rapidly undergo diuresis, but with an approximately *isosmotic* urine: They increase their fluid excretion rates (by a factor of 1,000 in the kissing bug) to remove the excess water and salt from the body. In mosquitoes, this increase in fluid excretion occurs even before the animal has finished sucking in its meal. That means, to add insult to injury, she is urinating on you as she feeds!

A kissing bug, *Rhodnius*. The saliva of these animals contains a protein called *nitrophorin,* which binds nitric oxide (NO). After biting a human, the bug injects its saliva and the nitrophorin releases NO, which (as seen in Chapter 9, p. 392) dilates blood vessels. This action also prevents clotting. The bug subsequently takes in an enormous water and salt load (human blood), with which its excretory system must deal.

unanswered Questions | Why are there virtually no marine insects?

Insects are the most successful animal group in terms of species, but curiously, they have not invaded Earth's largest habitat, the oceans. Although several hypotheses have been proposed for this, none has been well tested and found valid yet. One of these hypotheses focuses on the excretion systems, proposing that Malpighian tubules and hindguts in insects are inherently incapable of coping with seawater. How might this hypothesis be tested? «

Insect excretion is regulated by diuretic and antidiuretic hormones

Much less is known about the regulation of excretion in insects than in vertebrates. However, it is known that insects have both **diuretic hormones,** which promote water loss, and **antidiuretic hormones,** which promote water conservation.

- *Diuretic hormones*: In the kissing bugs and mosquitoes, which must excrete considerable fluid loads, both *serotonin* and several neuropeptides stimulate excretion. One of these is *mosquito natriuretic peptide* (MNP), which selectively increases transcellular secretion of Na^+ and water after a blood meal. MNP and other diuretic hormones are structurally related to *corticotropin-releasing hormone* (CRH) in vertebrates (p. 289). These hormones are thought to be released by neurosecretory cells in the brain (*corpora cardiaca*, p. 280), abdominal ganglia, and posterior midgut after a blood meal. Serotonin appears to stimulate Ca^{2+} influx into the principal cells, whereas MNP triggers cAMP production in the same cells as a second messenger (p. 97). Ca^{2+} and cAMP, in turn, activate tubule secretion by activating the V-ATPases and subsequent fluid excretion. Other small neuropeptides called *myokinins* (e.g., *leukokinin-VII* in cockroaches) have been found in some insects that stimulate the stellate cells. The hormones act synergistically; that is, secretion is faster with them together than predicted by adding the effects of each hormone individually.

- *Antidiuretic hormones*: As antagonists, peptide hormones such as *cardioacceleratory peptide 2b* (*CAP2b*) in kissing bugs and *Tenmo-ADFa* in mealworm beetle larvae (*Tenebrio molitor*) trigger cGMP production in principal cells, which (in opposition to cAMP) reduces secretion and conserves water. Injection of Tenmo-ADFa into mosquitoes acts comparably. Curiously, CAP2b is a diuretic hormone in fruit flies. And in "same-key different-locks" fashion, CAP2b also has unrelated actions such as heartbeat acceleration (thus its name).

We now turn to vertebrate kidneys and their nephrons.

check your understanding 12.3

Describe how initial salt and water movements are created in Malpighian tubules, and how the fluid is modified by the tubules and hindgut in osmoconcentrating insects.

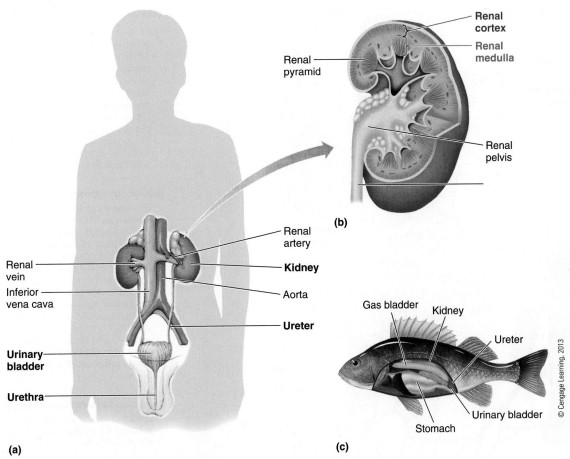

FIGURE 12-7 Urinary system components in two vertebrates. (a) Location of the components of the human male urinary system. The pair of kidneys form the urine, which the ureters carry to the urinary bladder. Urine is stored in the bladder and periodically emptied to the exterior through the urethra. (b) Longitudinal section of a human kidney. The kidney consists of an outer, granular-appearing renal cortex and an inner, striated renal medulla. The renal pelvis at the medial inner core of the kidney collects urine after it is formed. (c) The long, thin mesonephric kidneys of a fish lie under the spine.

Sources: (a, b) Adapted from Ann Stalheim-Smith & Greg K. Fitch. *Understanding Human Anatomy and Physiology,* p. 888, Figure 23.4. Copyright © 1993 West Publishing Company

12.4 Vertebrate Urinary Systems and Extrarenal Organs

Vertebrate kidneys come in a variety of morphologies and functional capacities. In addition, as we noted earlier, animals may also use respiratory, digestive, and integumentary systems for nonfecal excretion and internal fluid homeostasis. Some of these *extrarenal* excretory mechanisms can be important for salt regulation in all vertebrate classes except mammals and nonmarine birds, the groups that have evolved the most effective osmoconcentrating kidneys. Here we take a look at kidneys and extrarenal organs in the major vertebrate classes. In the next chapter, you will see how each type of vertebrate integrates its mechanisms for solute and water regulation.

Vertebrate kidneys form the urine; the remainder of the urinary system is ductwork that carries the urine to the bladder or hindgut

The **urinary system** consists of the urine-forming organs—the kidneys—and the structures that carry the urine from the kidneys to the outside for elimination from the body (Figure 12-7a, b). Kidneys are paired organs that lie in the dorsal side of the abdominal cavity, one on each side of the vertebral column. Only the mammalian kidney has a "kidney-bean" shape; see Figure 12-7c for elongated mesonephric kidneys of bony fish. The kidneys of birds are divided into cranial, middle, and caudal subdivisions, with the surface of the kidney covered in rounded projections, the **renal lobules.** Each kidney is supplied by a *renal artery* and a *renal vein.* Urine drains into the two ureters (ducts walled with smooth

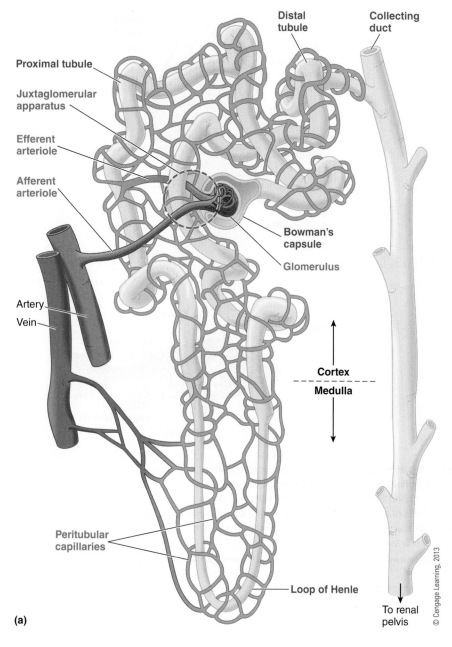

(a)

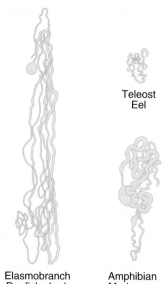

(b)

Overview of Functions of Parts of a Nephron

Vascular component
- Afferent arteriole—carries blood to the glomerulus
- Glomerulus—a tuft of capillaries that filters a protein-free plasma into the tubular component
- Efferent arteriole—carries blood from the glomerulus
- Peritubular capillaries—supply the renal tissue; involved in exchanges with the fluid in the tubular lumen

Tubular component
- Bowman's capsule—collects the glomerular filtrate
- Proximal tubule—uncontrolled reabsorption and secretion of selected substances occur here
- Loop of Henle—establishes an osmotic gradient in the renal medulla that is important in the kidney's ability to produce urine of varying concentration
- Distal tubule and collecting duct—variable, controlled reabsorption of Na^+ and H_2O and secretion of K^+ and H^+ occur here; fluid leaving the collecting duct is urine, which enters the renal pelvis

Combined vascular/tubular component
- Juxtaglomerular apparatus—produces substances involved in the control of kidney function

FIGURE 12-8 Nephrons in vertebrate kidneys. (a) A schematic representation of a nephron in a mammal, with both tubular and vascular components. (b) Schematic representations of the tubular components of nephrons from a shark (dogfish *Squalus acanthias*), a teleost or bony fish (eel *Anguilla anguilla*), and an amphibian (mudpuppy *Necturus maculosus*).

Source: (b) W. H. Dantzler. (1989). *Comparative Physiology of the Vertebrate Kidney*. Berlin: Springer.

muscle), one carrying urine from each kidney to a bladder (in fish, amphibians, and mammals) or hindgut (in reptiles including birds), which temporarily stores urine (and in some animals may modify it). Periodically urine is emptied to the outside; in vertebrates with bladders, this is through another tube, the **urethra** (which in male mammals serves both as a route for eliminating urine and as a passageway for semen from the testes; Figure 12-7a). In monotreme mammals, the urethra empties into the hindgut as in reptiles (hence their name: *mono*, "one"; *treme*, "hole").

The nephron is the functional unit of the vertebrate kidney, using three or four of the basic renal processes

A nephron is the smallest functional unit capable of the kidney's basic functions. (Recall that a functional unit is the smallest structure within an organ capable of performing all of that organ's functions, although the mammalian renal pelvis [p. 603] may modify the urine after it leaves all the nephrons.) Nephrons are found in all vertebrate kidneys, and each kidney consists of numerous nephrons (for example, about 1.25 million in pigs, 1 million in humans, 0.5 million in dogs, 0.25 million in cats) bound together by connective tissue. Each nephron consists of the tubules and an associated *vascular* (blood vessel) component, both of which are intimately related structurally and functionally (shown for a mammal in Figure 12-8a). The arrangement of nephrons within mammalian kidneys gives rise to two distinct regions—an outer, granular-appearing region, the renal **cortex,** and an inner, striated region called the renal **medulla** (Figure 12-7b). Some mammals (generally small ones such as rodents and rabbits) have a **unipapillary** medulla, a single, striated, cone-shaped body (the papilla). Larger mammals (such as dogs and humans) have a compound medulla, containing several striated-appearing triangles, the renal **pyramids** (Figure 12-7b).

Let's first look at the associated vascular system (Figure 12-8a). On entering the kidney, the renal artery systematically subdivides to form many small vessels known as **afferent arterioles,** one of which supplies each nephron. The afferent arteriole delivers blood to the **glomerulus,** a ball-like knot of capillaries (in the cortex in mammals) through which part of the water and solutes are filtered from the blood passing through. The glomerular capillaries rejoin to form another arteriole, the **efferent arteriole,** through which blood that was not filtered into the tubular component leaves the glomerulus. This pattern is unusual (typically, arterioles break up into capillaries that rejoin to form venules) but similar to the arterioles of the hypothalomo-hypophyseal portal system (p. 288). At the glomerular capillaries, no O_2 or nutrients are extracted from the blood for use by the kidney tissues, nor are waste products picked up from the surrounding tissue. The efferent arteriole then subdivides into the **peritubular capillaries,** which supply the renal tissue with blood and are important in exchanges between the tubular system and blood during conversion of the filtered fluid into urine. These capillaries, as their name implies (*peri,* "around"), are intertwined around the tubular system. The capillaries rejoin to form venules that ultimately drain into the renal vein, by which blood leaves the kidney.

Returning to the glomerulus, filtered fluid leaving the capillaries passes through the tubules of the nephron. The tubules are hollow, fluid-filled tubes formed by a single layer of epithelial cells. Even though each tubule is continuous, it is specialized into various segments with different structures and functions along its length. The first three basic renal processes discussed previously—filtration, reabsorption, secretion (Figure 12-3c)—are involved in the formation of urine in all vertebrates, with the fourth (osmoconcentration) found in birds and mammals. Let's examine these functions along a mammalian nephron in Figure 12-8a:

- *Bowman's (or glomerular) capsule and filtration*:
  Around a glomerular capillary bed is an expanded, double-walled cup-shaped invagination called **Bowman's (or glomerular) capsule** that collects the fluid filtered from the capillaries there. This is known as **glomerular filtration:** As blood flows through the capillaries, blood pressure forces protein-free plasma into Bowman's capsule. This ultrafiltration process is the first step in urine formation in most vertebrates and is normally continuous. In fact, in humans the kidneys filter the entire plasma volume about 65 times per day. If everything filtered were to pass out in the urine, the total plasma volume would be urinated out in less than half an hour! This does not happen, however, because the kidney tubules and peritubular capillaries are intimately related throughout their lengths (Figure 12-8a), so that materials can be transferred between the fluid inside the tubules and the blood within the peritubular capillaries. This filtration process is nondiscriminating; that is, with the exception of blood cells and most plasma proteins, all constituents within the blood—H_2O, nutrients, electrolytes, wastes, dietary toxins, small proteins, peptides such as hormones, and so on—are nonselectively filtered.

- *Proximal tubule and reabsorption/secretion:* From this capsule, the filtered fluid passes into the **proximal tubule,** which lies within the cortex and is highly convoluted throughout much of its course. Here, substances of value such as glucose, amino acids, and much of the water are returned to the peritubular capillary plasma (by mechanisms we will examine in detail later). This selective movement of substances from inside the tubule (the tubular lumen) into the blood uses energy and is called **tubular reabsorption.** Reabsorbed substances are not lost from the body in the urine but are carried instead by the peritubular capillaries to the venous system and then to the heart to be recirculated. Of the volume of plasma filtered per day, generally over 90% of the water and 100% of the glucose and amino acids are reabsorbed in mammals, with the remaining fluid eliminated as urine. In general, substances that need to be conserved are selectively reabsorbed, whereas unwanted substances that need to be eliminated remain in the urine.

  The proximal tubule also performs **tubular secretion,** the selective transfer of substances from the peritubular capillary blood into the tubular lumens, a second route for substances to enter the renal tubules from the blood. Glomerular filtration (the first route) filters only about 20% of the plasma, with the remaining 80% flowing on into the peritubular capillaries. Some substances such as organic ions can be selectively transferred as well, by tubular secretion from the plasma in the peritubular capillaries into the ECF and then into the tubular lumen. Tubular secretion provides a mechanism for more rapidly eliminating selected substances from the plasma by extracting an additional quantity of a particular substance from the (unfiltered) plasma in the peritubular capillaries and adding it to the quantity of the substance already present in the tubule as a result of filtration.

  Note that this multistage system starts with indiscriminate filtration followed by selective adjustment mechanisms. Because there are mechanisms for eliminating unwanted solutes by specific secretion, you might wonder why the apparently wasteful step of glomerular filtration exists. After all, it also loses useful molecules initially, and the nephron (unlike a Malpighian tubule) has to expend energy to recover them in the reabsorption process. One possible benefit is that indiscriminant filtration helps remove *all* small harmful solutes from the body, including all those for which no specific transporters have evolved, such as novel plant toxins that an animal has not previously encountered.

- *Loop of Henle (or nephron loop) and osmoconcentration.* Next, the tubule in birds and mammals forms a sharp U-shaped or hairpin loop that dips into the renal medulla. The **descending limb** of this **loop of Henle** plunges from the cortex deep into the medulla; the **ascending limb** traverses back up into the cortex, running *countercurrent* to the descending limb. This loop is a major adaptive feature of avian and mammalian nephrons, resulting in the medulla region to become highly concentrated in NaCl (and urea in mammals); that is, the medullary ECF becomes a brine bath. As you will see later, this *osmoconcentration* process allows avian and mammalian kidneys to recover and conserve more water if necessary, further concentrating the waste products in the urine. This process is often classified as a type of tubular reabsorption, but as you will see later, it is conceptually useful to think of this as a process separate from the other three.

  No loop is present in other vertebrates, which instead may have a short narrow **intermediate segment,** often ciliated and possibly involved in selective reabsorption and/or secretion. In at least one reptile, the American alligator, it makes a U-shape turn, perhaps similar to the first step that led to the evolution of the loop in birds and mammals.

- *Distal tubule and reabsorption/secretion and osmoconcentration.* The tubule once again becomes highly coiled to form the **distal tubule,** which also lies entirely within the cortex in mammals. (This tubule is absent in marine bony fishes.) It too performs selective reabsorption and/or secretion, and participates in osmoconcentration (see p. 595).

- *Collecting duct and osmoconcentration.* The distal tubule empties into a **collecting duct,** each of which in mammals drains fluid from up to eight separate nephrons. Each collecting duct releases its fluid contents (which have now been converted into urine) into ureters in most vertebrates, though in mammals, each duct empties into the renal *pelvis,* a cavity at the inner core of each kidney (Figure 12-7b). In birds and mammals, this duct is also important for osmoconcentration, as we will explain later.

This ends our survey of the main tubule; however, there is one other feature (Figure 12-8a):

- *Juxtaglomerular apparatus and regulation.* The ascending limb returns to the glomerular region of its own nephron, where it passes through the fork formed by the afferent and efferent arterioles. Here, both the tubular and vascular cells are specialized to form the **juxtaglomerular apparatus,** or JGA (*juxta,* "next to"), a structure that lies next to the glomerulus and plays an important role in sensing blood osmolarity and pressure and regulating kidney function. It or similar structures are found in most vertebrate kidneys.

Now let's examine different vertebrates, which differ in their nephron morphologies—see Figures 12-8b and schematics of 12-9—as well as in extrarenal organs.

Elasmobranch fishes retain urea and trimethylamine oxide, and use gills, kidneys, and rectal glands for excretion and retention

Sharks and their relatives (*elasmobranchs*) are *isosmotic* or slightly *hyperosmotic* relative to seawater (recall that seawater's osmotic pressure is about 1,000 milliosmoles/liter or mOsm; see p. 79), but not because internal NaCl is equal to that of seawater. Rather, they retain **urea** and **trimethylamine oxide (TMAO)** as major **osmolytes** (solutes used to regulate osmotic balance; see p. 617 for details). Thus, they face the problem of excess NaCl entering through the gills and diet, but they do not have a problem with osmotic water loss. The gills are relatively impermeable to urea and TMAO, in part because of high cholesterol content. Also, the nephrons reabsorb most of these osmolytes from the filtrate, excreting only what is necessary for nitrogen balance. Although the nephrons have only *glomeruli, proximal,* and *distal tubules* (see schematic in Figure 12-9a), they are among the most complex known, apparently exhibiting multiple *countercurrent* arrangements that are probably related to this reabsorption ability (Figure 12-8b, shark). (We will examine countercurrent function in mammalian nephrons later.) Beyond this, however, extrarenal organs perform most other excretion and osmoregulatory roles. *Gills* are thought to remove some of the excess NaCl, but in particular, these fish have a specialized hindgut organ called the **rectal gland,** which has numerous blind-end tubules that excrete a hypertonic fluid high in NaCl. It actively transports salt out of the blood, but it is not considered a renal organ, because no filtration or waste excretion is involved (see p. 558).

Marine bony fishes use gills for most excretion and retention processes

In contrast to elasmobranchs, most marine bony fishes are highly *hypo-osmotic* (about 350 to 400 mOsm in most) compared to seawater (about 1,000 mOsm), so they face constant influx of excess salt through the gills and diet, and constant water loss through the gills (in Chapter 13, we will speculate on how hypo-osmotic regulation evolved). The gills of these fishes are responsible for most of the functions we typically associate with kidneys. To reverse the water loss, these fish must drink seawater, a process known to be stimulated by *angiotensin II* (which we will examine later in mammals, p. 588). Initially, this also creates an excess of salt in the blood. The gills compensate for this excess, using specialized epithelial *chloride cells,* which actively transport NaCl outward faster than water leaks out (see Chapter 13, p. 625).

The kidneys play a minor role in correcting salt imbalances: The vast majority of marine fishes studied cannot produce a sufficiently hyperosmotic urine that would remove NaCl and conserve water. Kidneys in most species have small *glomeruli* and *proximal tubules* but no distal tubules (which would unnecessarily create a hypo-osmotic urine) (Figures 12-8b and 12-9b); they function primarily to

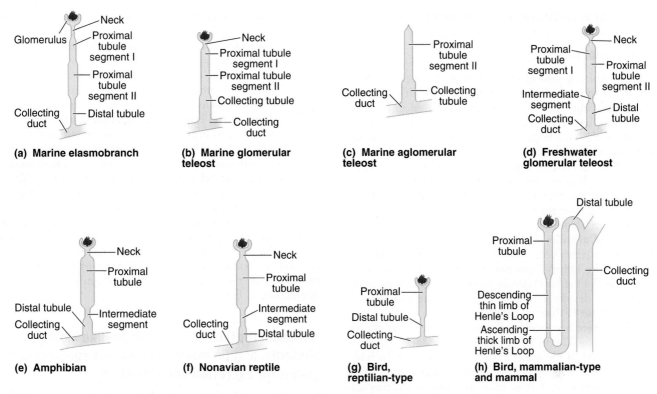

FIGURE 12-9 Schematic representations of nephrons from the major classes of vertebrates showing the major tubule segments. See main text for details.

Source: W. H. Dantzler. (1989). *Comparative Physiology of the Vertebrate Kidney.* Berlin: Springer.

remove excess divalent ions (such as Mg^{2+}) in a low-volume, isosmotic urine.

Kidneys also have little role in nitrogenous waste removal. Instead, the gills excrete NH_3 and NH_4^+, the primary nitrogenous wastes, probably by mechanisms better understood in freshwater fish (which we will discuss shortly). Thus, the gills serve as extrarenal osmoregulatory and excretory organs (see Chapter 13).

A few **aglomerular** species (Figure 12-9c), including many Antarctic fish, lack glomeruli and so do not use ultrafiltration. These species, such as toadfish (*Opsanus tau*), may have lost their glomeruli through evolution because they (and all marine bony fishes) produce very little urine. What urine they do produce begins initially by salt and water secretion into the proximal tubule lumen, much like Malpighian tubules (p. 566). Furthermore, the lack of glomeruli in some Antarctic fish may help them retain the small *antifreeze glycoproteins* they have in their blood (Chapter 15).

Freshwater bony fishes excrete water with their kidneys, and their gills excrete wastes and take up salts

Freshwater bony fishes must maintain an internal osmotic pressure far above that of the environment—that is, they are *hyperosmotic*—and are faced with constant influx of water through the gills and mouth (from eating, because they rarely drink). Accordingly, their kidneys have evolved to excrete a

voluminous, highly dilute (hypo-osmotic) urine. Kidneys have large *glomeruli* with nephrons having *proximal tubules*, sometimes ciliated *intermediate tubules, distal tubules,* and *collecting ducts* (Figure 12-9d). Most of these segments reabsorb nutrients and minerals; the *distal tubule* helps create the hypo-osmotic urine by removing ions from the lumen (a process you will learn about for the mammalian loop later).

Dietary intake and active transport by the gills brings in NaCl and other ions to replace those lost in the urine. As in marine fish, the gills are primarily responsible for excreting NH_3 and NH_4^+. The charged form NH_4^+, which does not pass through lipid bilayers, is thought to be transported by Na^+/K^+ symporters and ATPases, which can transport NH_4^+ in place of K^+. Although the gas form NH_3 can diffuse passively through membrane bilayers, its permeability is now thought to be greatly increased by **Rh (rhesus) channels** in gill-cell membranes. The Rh family of proteins was first discovered on erythrocyte membranes in humans and rhesus monkeys as critical factors in blood transfusions (because erythrocytes of people having a certain form of Rh protein trigger immune rejection in other people without that form). But only recently have scientists discovered one of their functions, namely as NH_3 channels. In models developed by Patricia Wright, Chris Wood, and others, gills of freshwater fish have Rh channels that not only allow NH_3 to diffuse rapidly out to the environment but also transport H^+ out at the same time. The H^+ ions react with the external NH_3 to convert it to NH_4^+, which cannot diffuse back into the cell through the Rh channel.

Terrestrial vertebrates evolved new renal and extrarenal mechanisms to maintain an "internal sea"

As lungs replaced gills in the evolution of vertebrates onto land, renal organs evolved new functions (with the loss of mesonephric and gain of metanephric tubules), in part because lungs, unlike gills, can neither excrete NH_3/NH_4^+ nor regulate NaCl. To a large extent, vertebrates that live on dry land independent of the sea succeed because of their kidneys, which, in concert with hormonal and neural extrinsic control, have become crucial regulators of ECF volume and composition. In effect, they maintain a consistent "internal sea" in the ECF for the benefit of cells. In mammals, kidneys are the most important organs for such maintenance. By adjusting the quantity of water and various plasma constituents that are either conserved for the body or eliminated in the urine, the kidneys can maintain water and electrolyte balance within the very narrow range compatible with mammalian cells, despite wide variations in intake and losses of these constituents through other avenues.

When there is a surplus of water or a particular electrolyte such as NaCl in the ECF, mammalian kidneys can eliminate the excess in the urine. If there is a deficit in the body, kidneys cannot actually provide additional quantities of the depleted constituent, but in birds and mammals, they can limit (though not completely halt) the urinary losses of the material in short supply and thus conserve it until more of the depleted substance can be ingested. Accordingly, kidneys can compensate more efficiently for excesses than for deficits.

Terrestrial kidneys are also the primary route for eliminating potentially toxic metabolic wastes and foreign compounds from the body. Most wastes usually cannot be eliminated in solid form; they must be excreted in solution (especially in mammals). Because the H_2O eliminated in the urine is derived from the plasma, animals unable to obtain H_2O eventually urinate themselves to death by depleting the plasma volume to a fatal level. However, except under such extreme circumstances, the kidneys can maintain stability in the internal fluid environment despite the usual variations in intake of fluids and electrolytes.

Not only can kidneys adjust for wide variations in ingestion of H_2O, salt, and other electrolytes, but they also compensate for abnormal losses from thermoregulatory evaporation (sweating, panting), vomiting, diarrhea, or hemorrhage. Thus, urine composition varies widely as the kidneys adjust for differences in intake as well as losses of various substances in an attempt to maintain the ECF within the narrow limits required by most terrestrial vertebrates' cells.

Amphibians use kidneys and bladders for excretion and retention

Amphibians demonstrate the transition to a life on land. Metanephric nephrons in adult amphibians are similar to the mesonephric ones of freshwater fish (Figures 12-8b and 12-9e), with similar functions (to excrete water and reabsorb ions and nutrients) but with the added function of urea excretion. Also unlike fish, terrestrial frogs can shut down glomerular filtration to almost zero when faced with dehydration (a similar shutdown occurs in mammals under acute stress and severe dehydration).

Moreover, the urinary bladder in terrestrial amphibians has a secondary role, that of a temporary water reservoir. After the salty urine reaches the bladder, active transporters move the salt back to the blood, leaving highly purified water behind (much like the functioning of the thick ascending loop of Henle in mammals; see p. 595). Certain frogs such as the Australian water-holding frog (*Cyclorana platycephala*) that burrow and **estivate** (dry-weather equivalent of hibernation) store this nearly pure water to be returned to the blood when needed. The native human inhabitants of Australia have long made use of this as a source of safe drinking water in the desert, released by squeezing the animal! When these and other terrestrial amphibians suffer dehydration stress, the hypothalamic hormone *arginine vasotocin (AVT)* triggers water uptake into the main body from the wall of the bladder via *aquaporins* (p. 81). (The closely related hormone *arginine vasopressin AVP* in mammals works similarly, as you will see later; p. 598.) Uptake of water through the skin may also occur, driven by sodium uptake by the mechanism of Figure 12-3b.

Nonavian reptiles use amphibian-like kidneys, hindguts, and (in marine and desert species) salt glands for excretion and retention

Nonavian reptiles have nephrons similar to aquatic vertebrates in having *capsules, proximal tubules,* ciliated *intermediate tubules, distal tubules,* and *collecting ducts* (Figure 12-9f). The ureters carry urine in a liquid or semisolid form into the **cloaca**, the final chamber of the hindgut (Figure 12-10). In order to remove excess water from the body, freshwater reptiles have larger glomeruli compared to terrestrial species. For the latter, without a loop of Henle, the nephrons cannot produce a significantly hyperosmotic urine to conserve body water. However, they can conserve water in three ways. (1) Uric acid as the primary nitrogenous waste conserves water, as we have seen. (2) The *cloaca* or *lower intestine* can reabsorb some water by first transporting salt (as in Figure 12-3b). This precipitates the uric acid left behind even further, creating a more solid urine, although it usually does not make a hyperosmotic urine. (3) Marine and some desert reptiles (nonavian) have an extrarenal organ, the *nasal salt gland* (see Figure 12-11 for the avian equivalent) that is dedicated to excreting a highly salty fluid by the same mechanisms found in birds (see the next section). For marine species, NaCl is the primary salt, but glands of herbivorous lizards also excrete KCl to compensate for the high potassium content of ingested plant food.

Birds have kidneys with reptile- and mammal-like nephrons, hindguts, and (in marine species) salt glands for excretion and retention

Since birds are now classified as reptiles, it is not surprising that their urinary system also involves a liquid or semisolid urine, and a cloaca and lower intestine (Figure 12-10) that can recover some water. Bird kidneys are also usually dominated by so-called *reptilian type* nephrons (Figure 12-9g), which have the same segments as (nonavian) reptile nephrons. However birds also have some so-called *mammalian type* nephrons with loops of Henle that can form a concen-

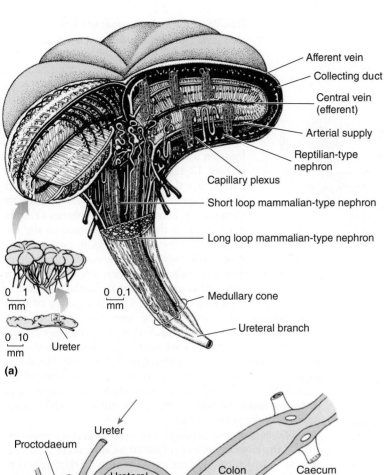

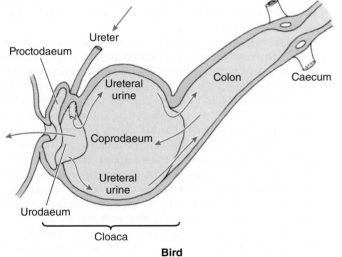

FIGURE 12-10 Avian urinary systems. (a) The kidney, showing its organization into lobes with short "reptilian"-type and long "mammalian"-type nephrons. (b) The hindgut, showing retrograde flow of urine from the ureter into the coprodaeum and colon.

Sources: (a) From P. C. Withers. (1992). *Comparative Animal Physiology*. Fort Worth, TX: Saunders, as modified from E. J. Braun, & W. H. Danztler. (1972). Function of mammalian-type and reptilian-type nephrons in the kidney of the quail. *American Journal of Physiology* 222:617–629. (b) From P. C. Withers. (1992). *Comparative Animal Physiology*. Fort Worth, TX: Saunders, as modified from I. Chosniak, B. G. Munck, & E. Skadhauge. (1977). Sodium chloride transport across the chicken coprodeum: Basic characteristics and dependence on the sodium chloride intake. *Journal of Physiology* 271:489–504.

trated urine (Figure 12-9h). In turn, there are two mammalian types called *cortical* and *juxtamedullary,* which we will explore in detail for mammals in the next section. For now, note that the juxtamedullary nephrons are responsible for making a concentrated urine, important for water

conservation. As an example, G. Casotti and K. C. Richardson found that arid-zone white-fronted honeyeaters (*Phylidonyris albifrons*) in Australia have a higher percentage of juxtamedullary nephrons in their kidneys that do wet-zone New Holland honeyeaters (*Phylidonyris novaehollandiae*) of similar size.

Most birds form only modestly hyperosmotic urines, but (as we noted earlier) much of their primary nitrogenous waste is uric acid. (Uric acid is transported into the renal tubules by the organic ion mechanism, p. 592.) Most of this uric acid precipitates as crystals (as uric acid is inherently insoluble), which exert no osmotic pressure but can potentially damage the nephron because of their "spiky" structure. For this reason the precipitated uric crystals are then covered by a protein coat (albumin) to form **urate balls.** These are sufficiently small enough to be filtered through the pores into the capsule. (Protein concentration in mammalian urine is normally less than 0.1 mg/mL, while in a chicken's urine it may be 3–5 mg/mL.) Because of this consolidation of uric acid we should not expect very high osmotic pressures in the urines of birds. Nevertheless some species of songbirds, which have no salt glands, produce urine osmolalities up to 2,000 mOsm if they live in arid or high-salt habitats. For example, the Chilean Seaside Cinclodes (*Cinclodes nigrofumosus*), which eats marine nonvertebrates, needs to produce a urine over 1,000 mOsm, because its prey have extracellular fluid concentrations of NaCl equal to seawater (see Chapter 13, p. 617). In contrast, the White-Winged Cinclodes (*C. atacamensis)* of river and grassland habitats produces urines of 765 mOsm or less.

Salt Glands Most marine birds, which must drink seawater, also have a **nasal salt gland** located near the eyes, with ducts leading to the nasal passages (Figure 12-11). These glands contain blind-end tubules lined with active salt-secreting cells that apparently transport NaCl out of the blood without concomitant osmotic water movement (as in Figure 12-3a). Unlike kidneys but like shark rectal glands, these extrarenal cells perform no filtration or waste excretion; moreover, they are not constantly active but rather work only when the bird is dehydrated or overloaded with salt.

The NaCl concentration excreted by the salt gland depends on what a particular species eats. A cormorant (see opening figure of this chapter) eats bony fish, which (as we noted earlier) are very hypo-osmotic compared to seawater (p. 572). Cormorant salt glands accordingly excrete a fluid with about 500 mM NaCl (930 mOsm), osmotically more concentrated than its body fluids, but not quite equal to seawater. In contrast, a petrel that eats marine nonvertebrates secretes a fluid that is 1,100 mM NaCl (over 2,000 mOsm).

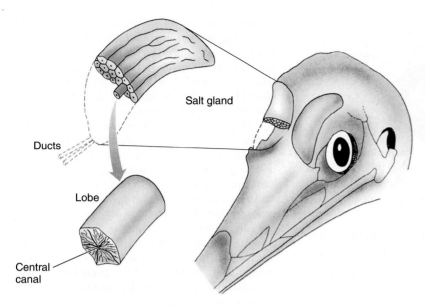

Salt gland

Ducts

Lobe

Central
canal

FIGURE 12-11 The salt glands of a gull.

Source: J. C. Welty & L. Baptista. (1990). *The Life of Birds.* Fort Worth, TX: Saunders, as modified from K. Schmidt-Nielsen. (1960). *Animal Physiology.* Englewood Cliffs, NJ: Prentice Hall.

check your understanding 12.4

Describe a vertebrate nephron and the major roles of the capsule + glomerulus, proximal and distal tubules, loop of Henle, and collecting duct.

Compare excretory processes of marine elasmobranchs, marine bony fish, freshwater bony fish, and amphibians.

Compare excretory processes of nonavian reptiles with birds.

12.5 Mammalian Urinary System: Overview and Glomerular Filtration

Mammalian kidneys, as well as those of many of the other vertebrates we just surveyed, are effectors for the four major functions noted at the beginning of the chapter (p. 557): regulating (1) major inorganic solutes (including ions involved in acid–base balance), (2) plasma volume, (3) harmful or unneeded organic molecules, and (4) osmotic balance. In mammals, kidneys have the following additional functions (the first two of which are not effector but rather integrator functions):

5. *Secretion of erythropoietin,* a hormone that stimulates red blood cell production (p. 393).
6. *Secretion of renin,* a hormone important for salt conservation (as you will see later).
7. *Conversion of vitamin D into its active form* (p. 329).
8. *Excretion of pheromones* for sexual signaling, marking territories, and so forth.

Mammalian (and avian) kidneys have cortical and juxtamedullary nephrons

Let's begin with an in-depth examination of nephron function. Knowing the structural arrangement of an individual nephron that we described earlier (Figure 12-8a) is essential for understanding the distinction between the *cortical* and *medullary*

regions of the kidney and, more importantly, for understanding detailed renal function. In addition, the medulla is divided into *outer* and *inner* regions. As we just noted for birds, there are two types of nephrons distinguished by the location and length of some of their structures (Figure 12-12). All nephrons originate in the cortex, but the glomeruli of **cortical nephrons** lie in the outer layer of the cortex, whereas the glomeruli of **juxtamedullary nephrons** lie in the inner layer of the cortex, adjacent to the medulla. The presence of all glomeruli and their Bowman's capsules in the cortex give this region a granular appearance. These two nephron types differ most markedly in the structure of their loops of Henle. The hairpin loop of a cortical nephron dips only slightly into the outer medulla, whereas the loop of a juxtamedullary nephron plunges through into the inner medulla. This relatively simple dichotomy is more complex in animals with relatively large renal medullas, such as desert rodents (see p. 601), in which the loops of juxtamedullary nephrons vary considerably: some barely penetrate the inner medulla, some reach midway, while others descend to the innermost tip.

The peritubular capillaries of juxtamedullary nephrons form hairpin vascular loops known as **vasa recta** ("straight vessels"), which run in close association with the loops (Figures 12-8a and 12-12). As they course through the medulla, the collecting ducts of both cortical and juxtamedullary nephrons run parallel to the ascending and descending limbs of the juxtamedullary nephrons' loops of Henle and vasa recta. This parallel arrangement of tubules and vessels creates the medullary tissue's striated appearance in the papilla or pyramids (see Figure 12-7b). More importantly, as you will see, this arrangement plays a key role in the kidneys' ability to produce urine of varying concentrations, and in the evolution of mammals to different habitats.

All plasma constituents that gain access to the tubules—that is, are filtered or secreted—but are not reabsorbed, remain in the tubules and pass into the renal pelvis to be excreted as urine and eliminated from the body. (Do not confuse *excretion* with *secretion.*) Conversely, anything filtered and subsequently reabsorbed or not filtered at all enters the venous blood and thus is conserved for the body, despite passage through the kidneys. The kidneys act only on the plasma. However, because of the free exchange between plasma and interstitial fluid across the capillary walls (with the exception of plasma proteins), interstitial fluid composition reflects the composition of plasma. Thus, by performing their regulatory and excretory roles on the plasma, the kidneys maintain the proper interstitial fluid environment for optimal cell function. Now we will discuss how the renal processes are accomplished and regulated to help maintain homeostasis. (The processes described are thought to be the same in all mammals, but the particular concentrations and rates given are mainly from humans and laboratory mammals.)

Recall that the process begins with *glomerular filtration.* Let's now examine how these remarkable cleansing systems work.

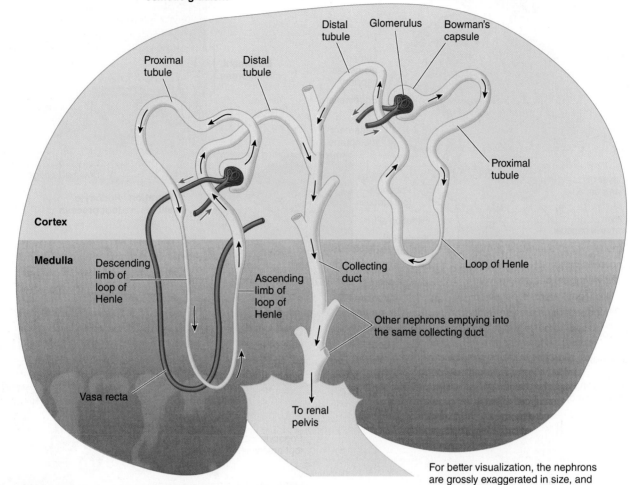

**Juxtamedullary nephron:
long-looped nephron important in
establishing the medullary vertical
osmotic gradient**

Cortical nephron

Distal tubule

Glomerulus

Bowman's capsule

Proximal tubule

Distal tubule

Proximal tubule

Cortex

Medulla

Descending limb of loop of Henle

Ascending limb of loop of Henle

Collecting duct

Loop of Henle

Other nephrons emptying into the same collecting duct

Vasa recta

To renal pelvis

For better visualization, the nephrons are grossly exaggerated in size, and the peritubular capillaries have been omitted, except for the vasa recta.

FIGURE 12-12 Comparison of juxtamedullary and cortical nephrons (mammalian kidney). The glomeruli of cortical nephrons lie in the outer cortex, whereas the glomeruli of juxtamedullary nephrons lie in the inner part of the cortex next to the medulla. The loops of Henle of cortical nephrons dip only slightly into the medulla, but the juxtamedullary nephrons have long loops of Henle that plunge deep into the medulla. The juxtamedullary nephrons' peritubular capillaries form hairpin loops known as *vasa recta*.

© Cengage Learning, 2013

The glomerular membrane has three layers forming a fine molecular sieve

Fluid filtered from a glomerulus into Bowman's capsule must pass through the three layers that make up the glomerular membrane. Collectively, these layers function as a fine molecular sieve that retains the blood cells and most

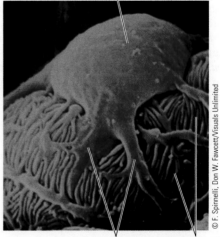

Cell body of podocyte

Foot processes   Filtration slits

© F. Spinnelli, Don W. Fawcett/Visuals Unlimited

FIGURE 12-13 Bowman's capsule podocytes with foot processes and filtration slits (mammalian kidney). Note the filtration slits between adjacent foot processes on this scanning electron micrograph. The podocytes and their foot processes encircle the glomerular capillaries.

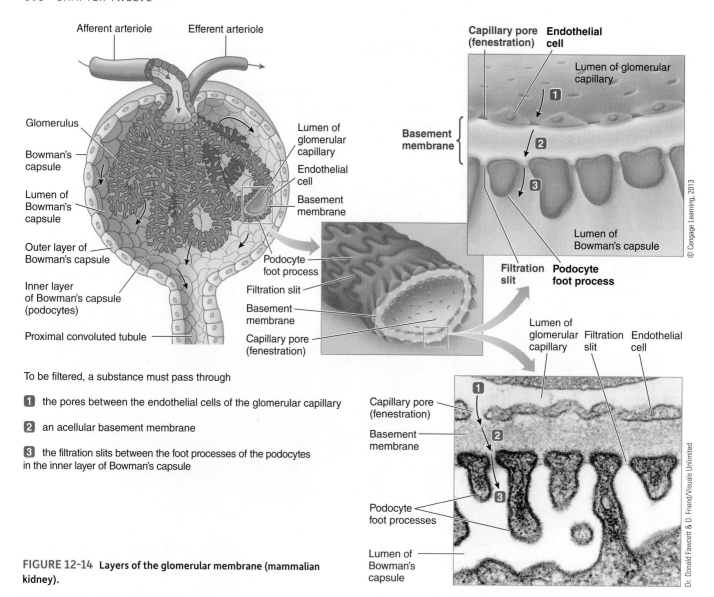

FIGURE 12-14 Layers of the glomerular membrane (mammalian kidney).

plasma proteins but permits H_2O and solutes of small molecular size to filter through (Figures 12-13 and 12-14):

1. The *glomerular capillary wall* consists of a single layer of flattened endothelial cells. It is perforated by many pores (*fenestrations*) that make it more than 100 times more permeable to H_2O and solutes than capillaries elsewhere in the body. The pores, though large, are too small for most plasma proteins to fit through.

2. The *basement membrane* is not a phospholipid bilayer but rather a noncellular gelatinous layer composed of collagen and glycoproteins that is sandwiched between the glomerulus and Bowman's capsule. The collagen provides structural strength, and the glycoproteins discourage the filtration of small plasma proteins. The wall pores are just barely large enough to permit passage of albumin, the smallest of plasma proteins. However, because the glycoproteins are negatively charged, they repel albumin and other plasma proteins, which are also negatively charged. Therefore, less than 1% of the albumin molecules escape into the capsule. In mammals, unlike in reptiles (including birds), the small proteins

that do slip into the filtrate are picked up by the proximal tubule by endocytosis, then degraded into constituent amino acids that are returned to the blood. Thus, urine is considered protein free. This process of protein exclusion from filtrate can be disrupted during certain disease conditions, such as diabetes.

3. The final layer is the *inner layer of Bowman's capsule*. It consists of **podocytes** (*podo*, "foot"), octopus-like cells that encircle the glomerular tuft. Each podocyte bears many elongated foot processes (meaning projections) that interdigitate with foot processes of adjacent podocytes, much as you interlace your fingers between each other when you cup your hands around a ball (Figure 12-13). The narrow slits between adjacent foot processes, known as **filtration slits,** provide a pathway through which fluid leaving the glomerular capillaries can enter the lumen of Bowman's capsule.

Thus, the route that filtered substances take across the glomerular membrane is completely extracellular—first through capillary pores, then through the acellular basement membrane, and finally through capsular filtration slits (Figure 12-14).

The glomerular capillary blood pressure is the major force that causes glomerular ultrafiltration

To accomplish glomerular filtration, a force—pressure—is required to drive a portion of the plasma out of the glomerular capillaries and into Bowman's capsule. (There is no active transport as in Malpighian tubules.) Because the glomerulus is a tuft of capillaries, the same principles of fluid dynamics that are responsible for ultrafiltration across other capillaries apply (see p. 443), except for two important differences: (1) The glomerular capillaries are much more permeable than capillaries elsewhere, so more fluid is filtered for a given filtration pressure, and (2) the balance of forces across the glomerular membrane is such that filtration occurs throughout the entire length of the capillaries, in contrast to the reabsorption that occurs toward the ends of other capillaries (see p. 444).

Three physical forces are involved in glomerular filtration (Table 12-1); the first favors filtration while the second two oppose it:

1. The *glomerular capillary blood pressure* is the fluid pressure exerted by the blood within the glomerular capillaries. It ultimately depends on contraction of the heart to create blood pressure and the resistance to blood flow offered by the afferent and efferent arterioles. The glomerular capillary blood pressure, at an estimated average value of 55 mm Hg in humans, is higher than capillary blood pressure elsewhere, because the diameter of the afferent arteriole is larger than that of the efferent arteriole. Because blood can more readily enter the glomerulus through the wide afferent arteriole than it can leave through the narrower efferent arteriole, glomerular capillary blood pressure is elevated by blood "damming up" in the glomerular capillaries. This high glomerular blood pressure tends to push fluid out of the glomerulus into Bowman's capsule along the glomerular capillaries' entire length, and it is the major force responsible for producing glomerular filtration.

2. *Plasma-colloid osmotic pressure* is caused by the unequal distribution of plasma proteins across the glomerular membrane. Recall that plasma proteins are in the glomerular capillaries but not in Bowman's capsule. Thus, the concentration of H_2O is higher in the capsule than in the glomerular capillaries, so H_2O is drawn osmotically back into the glomerulus. This movement opposes glomerular filtration, with an osmotic force averaging 30 mm Hg.

3. *Bowman's capsule hydrostatic pressure*, the pressure exerted by the fluid in this initial part of the tubule, is estimated to be about 15 mm Hg. This pressure opposes the movement filtration of fluid out of the glomerular capillaries.

Glomerular Filtration Rate As can be seen in Table 12-1, the forces acting across the glomerular membrane are not in balance. The glomerular capillary blood pressure at 55 mm Hg is opposed by the total of the other two pressures at about 45 mm Hg. The difference (10 mm Hg) is called the **net filtration pressure**. This modest pressure forces large volumes of fluid from the blood through the highly permeable glomerular membrane. The actual rate of filtration, the **glomerular filtration rate (GFR)**, depends not only on the net filtration pressure but also on how much glomerular surface area is available for penetration and how permeable the glomerular membrane is (that is, how "holey" it is). These properties of the glomerular membrane are collectively referred to as the **filtration coefficient** (K_f). Accordingly,

$$GFR = K_f \times \text{net filtration pressure}$$

Normally, in humans about 20% of the plasma that enters the glomerulus is filtered at the net filtration pressure of 10 mm Hg. This produces 180 liters of glomerular filtrate each day for an average GFR of 125 mL/min in male humans (versus an average GFR of 115 mL/min in female humans). GFR is lower in the kidneys of most newborn animals compared to mature animals (correlating with low arterial pressures in newborns), pony foals being an exception where GFR remains relatively constant throughout the postnatal period. To compare species, GFR is expressed in terms of body weight, revealing that GFR *scales* (p. 10) with body weight such that GFR per unit mass decreases with body size, just as does metabolic rate (see Chapter 15, p. 270). For example, in a 70-kg male human the GFR is about 1.8 mL/min/kg, whereas in dogs, it is approximately 4 mL/min/kg.

Changes in the GFR result primarily from changes in glomerular capillary blood pressure

Because the net filtration pressure is simply the result of an imbalance of opposing physical forces between the glomerular capillary and capsule fluids, alterations in any of these physical forces can affect the GFR. Let's examine important alterations.

Unregulated Influences on the GFR Plasma-colloid osmotic pressure and the capsule hydrostatic pressure are not subject to regulation and normally do not vary much. However,

TABLE 12-1 Forces Involved in Glomerular Filtration

Force	Effect	Magnitude (mm Hg)
Glomerular Capillary Blood Pressure	Favors filtration	55
Plasma-Colloid Osmotic Pressure	Opposes filtration	30
Bowman's Capsule Hydrostatic Pressure	Opposes filtration	15
Net Filtration Pressure (Difference between Force Favoring Filtration and Forces Opposing Filtration)	Favors filtration	10 $55 - (30 + 15) = 10$

© Cengage Learning, 2013

they can change pathologically and thus inadvertently affect GFR. Because plasma-colloid osmotic pressure opposes filtration, a decrease in plasma protein concentration leads to an increase in the GFR. Such a reduction might occur, for example, in severely burned victims who lose a large quantity of protein-rich, plasma-derived fluid through the exposed burned surface of their skin. Conversely, in situations in which the plasma protein concentration is elevated, such as in cases of dehydrating diarrhea, the GFR is reduced.

Bowman's capsule hydrostatic pressure can become uncontrollably elevated, and filtration can subsequently decrease. This is a common occurrence, if a urinary tract obstruction, such as a kidney stone (a problem in humans and certain cat and dog breeds) or enlarged prostate in males, is present. The damming up of tubular fluid behind the obstruction elevates capsular hydrostatic pressure.

Controlled Adjustments in the GFR Unlike plasma-colloid osmotic and capsule hydrostatic pressures, glomerular capillary blood pressure can be controlled to adjust the GFR to suit the body's needs. Assuming that all other factors stay constant, as the glomerular capillary blood pressure goes up, the net filtration pressure (and therefore GFR) increases. The magnitude of the glomerular capillary blood pressure depends on the rate of blood flow within each glomerulus. The amount of blood flowing into a glomerulus per minute is determined largely by the magnitude of the mean systemic arterial blood pressure and the resistance offered by the afferent arterioles. If *resistance increases* in the *afferent* arteriole, *less blood* flows into the glomerulus, decreasing the GFR. Conversely, if afferent arteriolar *resistance is reduced, more blood* flows into the glomerulus and the GFR increases. Two major control mechanisms regulate glomerular blood flow by regulating the radius and thus the resistance of the afferent arteriole: (1) *autoregulation,* intrinsic mechanisms aimed at preventing spontaneous changes in GFR; and (2) *extrinsic sympathetic* control, which is aimed at long-term regulation of arterial blood pressure.

Autoregulation of GFR Because arterial blood pressure drives blood into the glomerulus, the glomerular capillary blood pressure and, accordingly, the GFR would increase in direct proportion to an increase in arterial pressure if everything else remained constant (Figure 12-15a). Similarly, a fall in arterial blood pressure would be accompanied by a decline in GFR. Such spontaneous, inadvertent changes in GFR are largely prevented by regulatory mechanisms initiated by the kidneys themselves, an example of intrinsic regulation called **auto-regulation** (see p. 435). The kidneys can, within limits, maintain a constant blood flow into the glomerular capillaries (and thus a stable GFR) despite changes in the driving arterial pressure. They do so primarily by altering af-

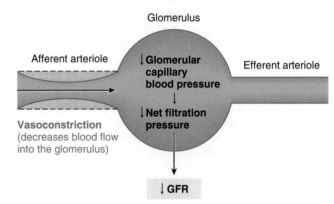

(a) Direct effect of aterial blood pressure on the glomerular filtration rate (GFR).

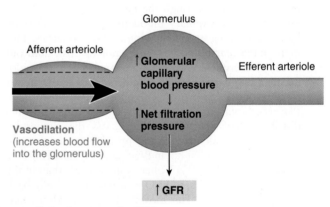

(b) Arteriolar vasoconstriction decreases the GFR

(c) Arteriolar vasodilation increases the GFR

FIGURE 12-15 Alteration mechanisms for glomerular filtration rate (GFR; mammalian kidney). (a) Direct effect of arterial blood pressure on the GFR. (b) Arteriolar adjustment to reduce the GFR. (c) Arteriolar adjustment to increase the GFR.

© Cengage Learning, 2013

ferent arteriolar caliber, thereby adjusting resistance to flow through these vessels. For example, if the GFR increases as a result of a rise in arterial pressure, the net filtration pressure and GFR can be reduced to normal by constriction of the afferent arteriole to decrease the flow of blood into the glomerulus (Figure 12-15b). This local adjustment lowers the GFR to normal. The opposite effects occur if there is a decline in arterial pressure (Figure 12-15c), which leads to vasodilation. The resulting increase in glomerular blood volume increases local blood pressure and thus GFR.

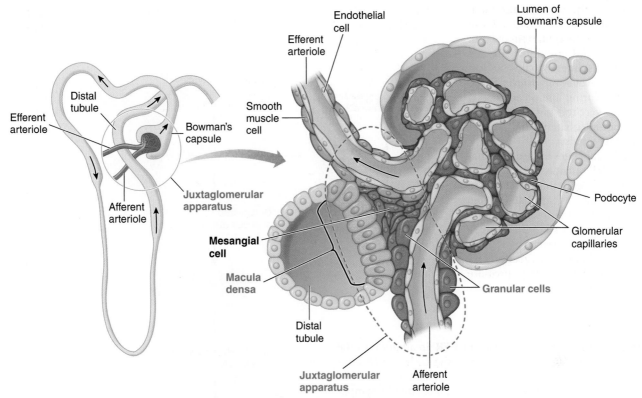

FIGURE 12-16 **The juxtaglomerular apparatus (mammalian kidney).** The juxtaglomerular apparatus consists of specialized vascular cells (the granular cells) and specialized tubular cells (the macula densa) at a point where the distal tubule passes through the fork formed by the afferent and efferent arterioles of the same nephron.

© Cengage Learning, 2013

Two intrarenal mechanisms contribute to autoregulation:

1. The **myogenic** ("muscle produced") mechanism is a common property of arteriolar vascular smooth muscle, which contracts inherently in response to the stretch accompanying increased pressure within the vessel (see p. 435). Accordingly, the afferent arteriole automatically does this when it is stretched because of an increased arterial driving pressure, limiting blood flow into the glomerulus despite the elevated arterial pressure. Conversely, inherent relaxation of an unstretched afferent arteriole when pressure within the vessel is reduced increases blood flow into the glomerulus despite the fall in arterial pressure.

2. The **tubuloglomerular feedback (TGF)** mechanism involves the *juxtaglomerular apparatus (JGA),* the specialized structure near the glomerulus (Figures 12-8a and 12-16). Unlike the rest of the kidney, which is an effector organ, the JGA is an integrator in a negative-feedback sense (p. 14). The smooth muscle cells within the wall of the afferent arteriole in the JGA are specialized to form glandlike **granular cells,** so called because they contain many secretory granules. Specialized tubular cells in the JGA are collectively known as the **macula densa,** which detect changes in the salt level of the fluid flowing past them through the tubule.

If the GFR is increased due to an elevation in arterial pressure, more fluid than normal is filtered and flows through the distal tubule. In response to the resultant rise in salt delivery to the distal tubule, the macula densa cells release ATP and *adenosine.* These are paracrines (p. 91), which act on the adjacent afferent arteriole, causing it to constrict and thus reducing glomerular blood flow and returning GFR to normal. In the opposite situation, when less salt is delivered to the distal tubule because of a spontaneous decline in GFR accompanying a fall in arterial pressure, the macula densa releases less paracrines. The resultant afferent arteriolar vasodilation increases the glomerular flow rate, restoring the GFR to normal. To exert even more exquisite control over TGF, the macula densa cells also secrete the vasodilator *nitric oxide* (p. 435), which puts the brakes on the action of ATP and adenosine at the afferent arteriole. By means of the TGF mechanism, the tubule of a nephron is able to monitor the salt level in the fluid flowing through it and adjust its own GFR accordingly to maintain fluid and salt delivery in the distal tubule.

The myogenic and TGF mechanisms work in unison to autoregulate the GFR during the transient changes in blood pressure that accompany daily activities unrelated to the need for the kidneys to regulate H_2O and salt excretion, such as the normal elevation in blood pressure accompanying exercise (p. 452). Autoregulation is important because unintentional shifts in GFR could lead to dangerous imbalances of fluid, electrolytes, and wastes. Since a given portion of the filtered fluid is always excreted, the amount of fluid excreted in the urine is automatically increased as the GFR increases. If autoregulation did not occur, the GFR could increase and H_2O and solutes would be lost needlessly whenever arterial pressure rises during intense activity. If the GFR were too low, the kidneys could not adequately eliminate unneeded materials.

When changes in mean arterial pressure fall outside the range over which autoregulation works, 80 to 180 mm Hg in humans, these mechanisms cannot compensate. Therefore, dramatic changes in mean arterial pressure (<80 mm Hg or >180 mm Hg) directly cause the glomerular capillary pressure and, accordingly, the GFR to decrease or increase in proportion to the change in arterial pressure.

Extrinsic Sympathetic Control of the GFR In addition to the intrinsic autoregulatory mechanisms, the GFR can be controlled by extrinsic control mechanisms that override the autoregulatory responses. This control, mediated by sympathetic nervous system, is aimed at regulating overall body arterial blood pressure.

If plasma volume is decreased—for example, by hemorrhage—the resulting fall in arterial blood pressure is detected by the arterial carotid sinus and aortic arch baroreceptors, which initiate the neural *baroreceptor reflex* to raise blood pressure toward normal (Figure 12-17). Recall from Chapter 9 (p. 449) that this reflex response is coordinated by the cardiovascular control center in the brainstem and is mediated primarily through increased sympathetic activity to the heart and blood vessels. Although the resulting increase in both cardiac output and total peripheral resistance helps raise blood pressure toward normal, plasma volume is still reduced by the disturbance and must be restored to normal. One compensation for a depleted plasma volume is reduced urinary output to conserve fluid, which is partially accomplished by reducing the GFR.

Recall that during the baroreceptor reflex, vasoconstriction occurs in most arterioles throughout the body as a compensatory mechanism to increase total peripheral resistance. The glomerular afferent arterioles have α1 adrenergic receptors (see p. 163) and are heavily innervated with sympathetic vasoconstrictor fibers compared to the efferent arterioles. When the glomerular afferent arterioles constrict from increased sympathetic activity, less blood flows into the glomeruli and GFR is reduced (Figure 12-15b). Thus, less urine forms and some of the H_2O and salt that would otherwise have been lost are saved for the body, helping to restore plasma volume to normal. Other mechanisms, such as increased tubular reabsorption of H_2O and salt as well as increased thirst (described more thoroughly elsewhere), also contribute to long-term maintenance of blood pressure by helping to restore plasma volume.

Conversely, if blood pressure is elevated (for example, because of an expansion of plasma volume following ingestion of excessive fluid), the opposite responses occur. When

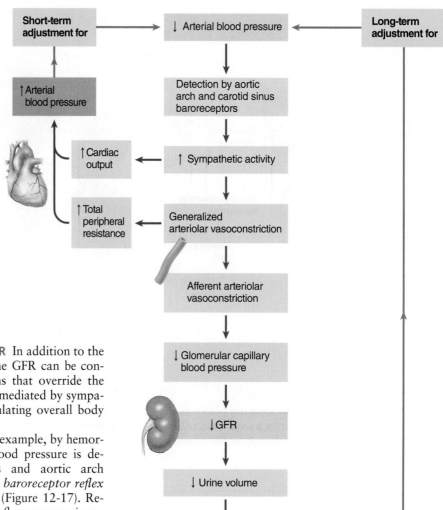

FIGURE 12-17 Baroreceptor reflex influence on the GFR in long-term regulation of arterial blood pressure (mammal).

© Cengage Learning, 2013

the baroreceptors detect a rise in blood pressure, sympathetic vasoconstrictor activity to the arterioles is reflexly reduced, allowing afferent arteriolar vasodilation to occur. As more blood enters the glomeruli, glomerular capillary blood pressure rises, and increases the GFR (Figure 12-15c) and urinary output. A hormonally adjusted reduction in the tubular reabsorption of H_2O and salt also contributes to the increase in urine volume. These two renal mechanisms are key in removing excess fluid from the body.

The GFR can be influenced by changes in the filtration coefficient

Thus far, we have discussed changes in the GFR as a result of changes in net filtration pressure. The rate of glomerular filtration, however, also depends on the filtration coefficient

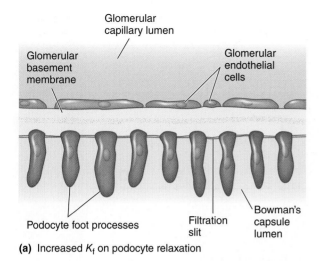

(a) Increased K_f on podocyte relaxation

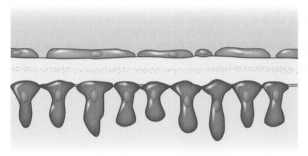

(b) Decreased K_f on podocyte contraction

FIGURE 12-18 Change in the number of open filtration slits caused by podocyte relaxation and contraction (mammalian kidney).
(a) Podocyte relaxation narrows the bases of the foot processes, increasing the number of fully open intervening filtration slits spanning a given area. (b) Podocyte contraction flattens the foot processes and thus decreases the number of intervening filtration slits.

Source: Adapted from *Proceedings of the Federation of American Societies for Experimental Biology*, vol. 42, pp. 3046–3052, 1983. Reprinted by permission.

(K_f). For years K_f was considered a constant, except in disease situations in which the glomerular membrane becomes leakier than usual. However, research indicates that K_f is subject to physiological control. Both factors on which K_f depends—the surface area and the permeability of the glomerular membrane—can be modified by contractile activity within the membrane.

The surface area available for filtration within the glomerulus is represented by the extent of the capillary surface that faces the fluid in Bowman's capsule. Each tuft of glomerular capillaries is held together by **mesangial cells,** which form rings around the capillaries (Figure 12-16). These cells contain actinlike contractile elements. Like vascular sphincters, contraction of these mesangial cells reduces the radius of the filtering capillaries, which reduces the surface area available for filtration within the glomerular tuft. When the net filtration pressure remains unchanged, this reduction in K_f decreases GFR. Sympathetic stimulation causes the mesangial cells to contract, thus providing a second mechanism (besides afferent arteriolar vasoconstriction) by which sympathetic activity can decrease the GFR.

Podocytes also possess actinlike contractile filaments, whose contraction or relaxation can, respectively, decrease or increase the number of filtration slits open in the inner membrane of Bowman's capsule by changing the shapes and proximities of the foot processes (Figure 12-18). The more open slits, the greater the permeability. Contractile activity of the podocytes, which in turn affects permeability and the K_f, is under physiologic control by mechanisms incompletely understood.

check your understanding 12.5

How do cortical and juxtamedullary nephrons differ?

Describe the three layers of the glomerular membrane, their roles in the filtration process, and the roles of mesangial cells and podocytes.

Describe the pressure forces in glomerular filtration and how both autoregulation and extrinsic sympathetic regulation come into play.

12.6 Mammalian Kidneys: Tubular Reabsorption

As we noted earlier, filtration through the glomerular capillaries is largely indiscriminant. Filtered constituents include not only wastes but also nutrients, electrolytes, and other substances the body cannot afford to lose in urine. Indeed, through ongoing glomerular filtration greater quantities of these materials are filtered per day than are even present in the entire mammalian body. It is important that these essential materials be returned to the blood by the process of *tubular reabsorption* (p. 571). As we noted earlier, reabsorption of most substances occurs in the proximal tubule.

Tubular reabsorption involves passive and active transepithelial transport and is tremendous, highly selective, and variable

Tubular reabsorption is a highly selective process. All constituents except plasma proteins (and protein-bound solutes such as iron) are at the same concentration in the glomerular filtrate as in the plasma. In most cases, the quantity of each material reabsorbed is the amount required to maintain the proper composition and volume of the internal fluid environment. In general, the tubules have a high reabsorptive capacity for substances needed by the body and little or no reabsorptive capacity for substances of no value (Table 12-2). Accordingly, only a small percentage, if any, of filtered plasma constituents that are useful to the body are present in the urine. Only excess amounts of essential materials such as electrolytes are excreted in the urine. As H_2O and valuable solutes are reabsorbed, the waste products remaining in the tubular fluid become highly concentrated. Considering the magnitude of glomerular filtration, the extent of tubular reabsorption is tremendous: Mammalian tubules typically reabsorb 99% of the filtered H_2O and salt and 100% of the filtered glucose and amino acids (Table 12-2).

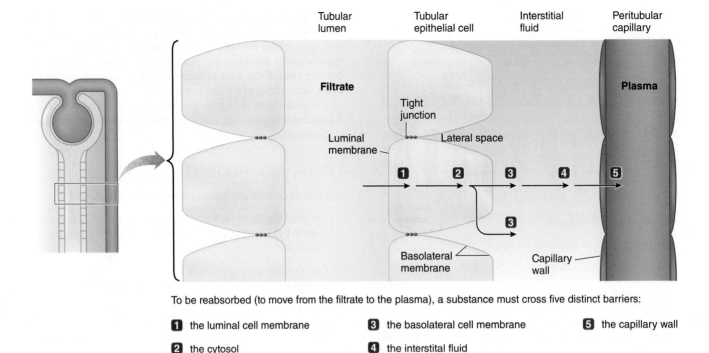

To be reabsorbed (to move from the filtrate to the plasma), a substance must cross five distinct barriers:

1 the luminal cell membrane   **3** the basolateral cell membrane   **5** the capillary wall

2 the cytosol   **4** the interstital fluid

FIGURE 12-19 **Steps of transepithelial transport (mammalian kidney).**

© Cengage Learning, 2013

Recall that, throughout its entire length, the tubule is in close proximity to a surrounding peritubular capillary (Figure 12-8a) and it is also only one-cell thick (Figure 12-19). Adjacent tubular cells touch each other only where they are joined by tight junctions at their lateral edges near their apical membranes, which face the tubular lumen. Interstitial fluid lies in the gaps between adjacent cells—the **lateral spaces**—as well as between the tubules and capillaries. The basolateral membrane faces the interstitial fluid at the base and lateral edges of the cell. The tight junctions largely prevent substances except H_2O from moving between the cells, so materials must move transcellularly to leave the tubular lumen and gain entry to the blood. To be reabsorbed, a substance must traverse five distinct barriers (Figure 12-19). It must (1) leave the tubular fluid by crossing the apical membrane of the tubular cell, (2) pass through the cytosol from one side of the tubular cell to the other, (3) traverse the basolateral membrane of the tubular cell to enter the interstitial fluid, (4) diffuse through the interstitial fluid, and (5) penetrate the capillary wall to enter the blood plasma. This entire sequence is known as **transepithelial transport** and relies on transport epithelia that we discussed in general terms earlier (e.g., Figure 12-3).

Recall that there can be both *passive* and *active* transport (p. 84). Substances that are actively reabsorbed against a gradient and are of particular importance to the body include organic nutrients such as glucose and amino acids, and electrolytes such as Na^+ and PO_4^{3-}. Rather than specifically describing the reabsorptive process for each of the many filtered substances that are returned to the plasma, we provide illustrative examples of the general mechanisms involved after first highlighting the unique and important case of Na^+ reabsorption.

TABLE 12-2 Fate of Various Substances Filtered by Mammalian Kidneys

Substance	AVERAGE PERCENTAGE OF FILTERED SUBSTANCE	
	Reabsorbed	*Excreted*
Water	99*	1
Sodium	99.5	0.5
Glucose	100	0
Urea (a waste product)	50	50
Phenol (a waste product)	0	100
Amino acids	100	0

*Value for average humans; may be less in some situations and species

© Cengage Learning, 2013

Sodium reabsorption is essential for the reabsorption of nutrients, Cl⁻, and H₂O, and for the regulation of ECF osmolality and volume

Sodium reabsorption is unique and complex—passive in some sections of the tubule but active in most sections. In fact, about 80% of the total energy requirement of the kidneys is estimated to be used for Na^+ transport, indicating the importance of this process. Of the Na^+ filtered in a laboratory rodent, 99% is normally reabsorbed, of which on average 67% is reabsorbed in the proximal tubule, 25% in the loop of Henle, and 8% in the distal tubules and collecting ducts. Sodium reabsorption plays different important roles in each of these segments, as will become apparent as our discussion continues.

- Na$^+$ reabsorption in the proximal tubule plays a pivotal role in the reabsorption of glucose, amino acids, Cl$^-$, and urea, as well as H$_2$O for the regulation of ECF volume.
- Na$^+$ reabsorption in the loop of Henle, along with Cl$^-$ reabsorption, plays a critical role in the kidneys' ability to adjust ECF osmolality by producing urine of varying concentrations and volumes, depending on the need to conserve or eliminate salt and/or H$_2$O.
- Na$^+$ reabsorption in the distal portion of the nephron is important in the regulation of ECF volume and in secretion of K$^+$ and H$^+$.

Sodium is reabsorbed throughout the tubule with the exception of the descending limb of the loop of Henle. You will learn about the significance of this exception later. Throughout all Na$^+$-reabsorbing segments of the tubule, the active step in Na$^+$ reabsorption involves the energy-dependent Na$^+$/K$^+$ ATPase pump located in the tubular cell's basolateral membrane (Figure 12-3a, b). The apical Na$^+$ channels and/or transport carriers that permit movement of Na$^+$ from the lumen into the cell vary for different parts of the tubule, but in each case, movement of Na$^+$ across the apical membrane is always a passive step. For example, in the proximal tubule, Na$^+$ crosses the apical border by means of a cotransport carrier that simultaneously moves Na$^+$ and an organic nutrient such as glucose from the lumen into the cell (see p. 87). By contrast, in the collecting duct, Na$^+$ crosses the apical border through an Na$^+$ leak channel. Na$^+$ continues to move down its electrochemical gradient from its high concentration in the lateral space into the surrounding interstitial fluid and finally into the peritubular capillary blood. Thus, net transport of most of the Na$^+$ from the tubular lumen into the blood occurs at the expense of energy. This obviously helps conserve NaCl, but the movement of Na$^+$ can also help transport other molecules, as we show next.

Water Reabsorption Water is passively reabsorbed by osmosis in several regions of the tubule. Of the H$_2$O filtered, 45 to 65% is reabsorbed in the proximal tubules and about 15% in the descending loops of Henle (discussed later), osmotically following solute reabsorption (Figure 12-3b). This amount is not regulated extensively. The remaining 20 to 25% is reabsorbed in the distal tubules and collecting ducts, and is subject to hormonal control, depending on the body's state of hydration (see the section on osmoconcentration later).

During reabsorption, H$_2$O passes primarily through water channels or *aquaporins* (AQPs; p. 81). Different AQPs are selectively expressed in the various parts of the nephron. For example, the water channels in the proximal tubule, AQP-1, are always open, accounting for the high H$_2$O permeability of this region. However, the AQP-2 channels in the distal tubule and collecting duct are regulated by the hormone *vasopressin*, which accounts for the variable H$_2$O reabsorption in this region.

This return of filtered H$_2$O to the plasma is enhanced by the fact that the plasma-colloid osmotic pressure is greater in the peritubular capillaries than elsewhere because of their elevated concentration of plasma proteins (p. 579). The resulting plasma-colloid osmotic pressure tends to "pull" H$_2$O into the peritubular capillaries, simultaneous with the "push" of the hydrostatic pressure in the lateral spaces that drives H$_2$O toward the capillaries. Note that ATP is not directly utilized for this tremendous reabsorption of H$_2$O.

Chloride Reabsorption The negatively charged chloride ions are passively reabsorbed down the electrical gradient created by the active reabsorption of the positively charged sodium ions. CFTR Cl$^-$ channels are found here (Figure 12-3a). The amount of Cl$^-$ reabsorbed is determined by the rate of active Na$^+$ reabsorption instead of being directly controlled by the kidneys.

Glucose and amino acids are reabsorbed by Na$^+$-dependent secondary active transport

Large quantities of nutritionally important organic molecules such as glucose and amino acids are filtered each day, but normally none of these materials is excreted in the urine. This rapid and thorough reabsorption occurs early, in the proximal tubules, where glucose and amino acids are actively moved uphill against their concentration gradients from the tubular lumen into the blood. Again, no ATP is directly used to operate the glucose or amino acid carriers. These nutrients are transferred by means of secondary active transport utilizing a specialized symporter that simultaneously transfers both Na$^+$ (down it gradient) and the specific organic molecule (up its gradient) from the lumen into the cell (such as the sodium–glucose cotransporter *SGLT* in Figure 3-15, p. 88), just as the actively generated H$^+$ gradient moves K$^+$ or Na$^+$ in the Malpighian tubule. Again, the lumen-to-cell Na$^+$ electrochemical gradient maintained by the energy-consuming basolateral Na$^+$/K$^+$ ATPase pump drives this cotransport system. Because the overall process of glucose and amino acid reabsorption depends on the use of ATP, these organic molecules are considered actively reabsorbed, even though ATP is not used directly to transport them across the membrane.

Once transported into the tubular cells, glucose and amino acids passively diffuse down their concentration gradients across the basolateral membrane into the plasma, facilitated by a non-energy-dependent carrier (such as the glucose transporter *GLUT* in Figure 3-15).

With the exception of Na$^+$, actively reabsorbed substances exhibit a tubular maximum

All actively reabsorbed substances bind with plasma membrane carriers that transfer them across the membrane against a concentration gradient. Each carrier is specific for the types of substances it can transport. As you learned in Chapter 3 (Figure 3-12, p. 83), because a limited number of each carrier type are present in the cells lining the tubules, there is an upper limit on the quantity of a particular substance that can be actively transported from the tubular fluid in a given period of time. The maximum reabsorption rate is reached when all the carriers specific for a particular substance are fully "occupied" or saturated. This transport maximum, designated as the **tubular** (or **transport**) **maximum,** or T_m, in the kidney tubules is the maximum amount of a substance that the tubular cells can actively transport within a given time period. With the exception of Na^+, all actively reabsorbed substances display a T_m. (Sodium does not display a T_m, because *aldosterone*, which we discuss later, promotes the synthesis of more active Na^+/K^+ ATPase carriers as needed.) Any quantity of a substance filtered beyond its T_m fails to be reabsorbed, and escapes instead into the urine.

The plasma concentrations of some but not all substances that display carrier-limited reabsorption are regulated by the kidneys. To illustrate, we compare glucose, a substance that has a T_m but is not regulated by the kidneys, with phosphate, a T_m-limited substance that is regulated by the kidneys.

Glucose reabsorption is not regulated by the kidneys and has a T_m that is exceeded in diabetes

Because glucose is freely filterable at the glomerulus, it passes into Bowman's capsule at the same concentration it has in the plasma, which is about 100 mg/100 mL in humans and many other mammals. Ordinarily, no glucose appears in the urine because all the filtered glucose is reabsorbed. Not until the filtered load of glucose exceeds the T_m does glucose start spilling into the urine. The plasma concentration at which the T_m of a particular substance is reached and the substance first starts appearing in the urine is known as the **renal threshold**. This value is about 180 mg/100 mL in humans and dogs, and 240–290 mg/100 mL in domestic cats, well above the filtered value of 100. As glucose concentration in the filtrate increases above the threshold, reabsorption begins to taper off and urine levels begin to climb, until about 300 mg/100 mL, where the T_m is reached. Above that, reabsorption remains constant at its maximum rate (Figure 12-20), with the excess filtered glucose remaining in the filtrate to be excreted. As you saw in Chapter 7 (p. 322), the plasma glucose concentration can become extremely high in diabetes mellitus. Glucose in the urine was used as the original diagnosis of diabetes mellitus (*mellitus* is Latin for "honey-sweet"), which the Chinese called *sweet urine disease*.

With the T_m for glucose well above the normal filtered load, the kidneys normally transport all the luminal glucose, thereby protecting against the loss of this important nutrient. However, the kidneys do not regulate glucose per se, because they do not maintain glucose at some specific plasma concentration; instead, this concentration is normally regulated by endocrine and liver mechanisms, with the kidneys merely maintaining whatever plasma glucose concentration is set by these other mechanisms (except at excessively high levels). The same general principle holds true for other organic plasma nutrients, such as amino acids and water-soluble vitamins.

PO_4^{3-} (phosphate) reabsorption is regulated by the kidneys

The kidneys do directly contribute to the regulation of many electrolytes, such as phosphate (PO_4^{3-}) and calcium (Ca^{2+}), because the renal thresholds of these inorganic ions equal their normal plasma concentrations. The transport carriers for these electrolytes are located in the proximal tubule. We use PO_4^{3-} as an example. Some animal diets are rich in PO_4^{3-}, but because the tubules can reabsorb up to the nor-

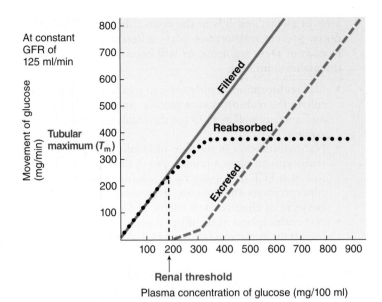

FIGURE 12-20 Renal handling of glucose as a function of plasma glucose concentration. At a constant GFR, the quantity of glucose filtered per minute is directly proportional to the plasma concentration of glucose. All the filtered glucose can be reabsorbed up to the renal threshold, the plasma concentration at which glucose first starts appearing in the urine. If the amount of glucose filtered per minute exceeds the tubular maximum T_m, no more glucose is reabsorbed and the rest stays in the filtrate to be excreted in urine.

© Cengage Learning, 2013

mal plasma concentration's worth of PO_4^{3-} and no more, the excess ingested PO_4^{3-} is quickly spilled into the urine, restoring the plasma concentration to normal.

However, unlike the reabsorption of organic nutrients, the reabsorption of PO_4^{3-} and Ca^{2+} is also subject to hormonal control. *Parathyroid hormone* can alter the renal thresholds for these ions, thus adjusting the quantity of these electrolytes conserved, depending on the body's momentary needs (p. 324).

In general, unwanted waste products, except urea, are not reabsorbed

Because of extensive reabsorption of H_2O in the proximal tubule, the original volume of filtrate is progressively reduced. Substances that have been filtered but not reabsorbed become progressively more concentrated in the tubular fluid as H_2O is reabsorbed. Waste products from metabolism and ingested toxins become concentrated for later excretion and are not subject to physiological control. When renal function is normal, these uncontrolled excretory processes proceed at a satisfactory rate.

Urea molecules are an important exception because they are the only wastes to be passively reabsorbed as a result of this concentrating effect. Urea's initial concentration at the glomerulus is identical to its concentration in the plasma entering the peritubular capillaries. However, like other wastes, it is concentrated several-fold at the end of the proximal tubule. A concentration gradient for urea is thus created, which results in urea's passive diffusion from the tubular lumen into the peritubular capillary plasma (thought to leak through paracellular pathways due to its small molecular size). Even so, the walls of the proximal tubules are only moderately permeable to urea, so only about 40% of the filtered urea is passively reabsorbed by this means. (As you will see later, urea permeability depends on urea transporters in other nephron segments.) Even though only 60% of the filtered urea is eliminated from the plasma with each pass through the nephrons, this rate is adequate.

The renin–angiotensin–aldosterone system stimulates Na$^+$ reabsorption in distal tubules and collecting ducts, elevates blood pressure, and stimulates thirst and salt hunger

Reabsorption of filtered Na$^+$ is subject to hormonal control in the nephron. This occurs primarily in the distal tubule and collecting ducts, and to a lesser extent in the proximal tubule as well. The extent of this controlled reabsorption is inversely related to the magnitude of the Na$^+$ load in the body. If there is too much Na$^+$, little Na$^+$ reabsorption occurs and is instead lost in the urine. In contrast, if Na$^+$ is depleted, most or all Na$^+$ is reabsorbed and conserved for the body.

The Na$^+$ load in the body is reflected by the ECF volume. Sodium and chloride account for more than 90% of the ECF's osmotic pressure (p. 103). When the Na$^+$ load is above normal, making the ECF hyperosmotic, osmosis of H$_2$O out of cells expands the ECF volume. Conversely, when the Na$^+$ load is below normal, thereby making the ECF hypo-osmotic, less H$_2$O than normal stays in the ECF, so its volume shrinks. Because plasma is a component of the ECF, the most important consequence of a change in ECF volume is the corresponding change in blood pressure accompanying expansion (increased blood pressure) or reduction (decreased blood pressure) of the plasma volume (see Figure 9-67, p. 454). Thus, long-term control of arterial blood pressure ultimately depends on Na-regulating mechanisms. We now turn our attention to these mechanisms.

Cardiotonic Steroids A new class of hormones called *cardiotonic steroids* (CTS) has been found that are related to exogenous compounds like *ouabain* (p. 87) that affect the Na$^+$/K$^+$ ATPase. Moreover, that pump has been found to be a receptor in complexes residing in *caveolae* (p. 89), tethering and regulating tyrosine kinases. Recent studies show that high-Na$^+$ or low-K$^+$ diets increase CTS levels (from the adrenal gland, probably stimulated by the hypothalamus). CTSs in turn bind the ATPase-receptor complexes in the proximal tubule. The resulting kinase signal cascade causes down-regulation of the ATPases, so less Na$^+$ is reabsorbed, enhancing **natriuresis** (urinary sodium excretion).

Activation of the Renin–Angiotensin–Aldosterone System The most important and best-known hormonal system involved in regulating Na$^+$ is the **renin–angiotensin–aldosterone system (RAAS)**. It begins with integrating granular cells of the JGA (Figure 12-16) secreting a hormone, **renin**, into the blood in response to a fall in NaCl/ECF volume/blood pressure. This function is in addition to the role the JGA plays in autoregulation. Specifically, the following three inputs to the granular cells increase renin secretion:

1. The granular cells themselves function as *intrarenal baroreceptors*. They are sensitive to pressure changes within the afferent arteriole. When the granular cells detect a fall in blood pressure, they secrete more renin.
2. The macula densa cells in the tubular portion of the juxtaglomerular apparatus are sensitive to the NaCl moving past them through the tubular lumen. In response to a fall in NaCl, the macula densa cells trigger the granular cells to secrete more renin.
3. The granular cells are innervated by the sympathetic nervous system. When blood pressure falls below normal, the baroreceptor reflex increases sympathetic activity, which stimulates the granular cells to secrete more renin.

These interrelated signals for increased renin secretion function together to increase salt in the ECF and to expand the plasma volume to increase the arterial pressure to normal on a long-term basis. Through the complex RAAS, increased renin secretion brings about these changes in several ways, as we will examine next (Figure 12-21).

Effects of the RAAS Once secreted into the blood, renin functions as an enzyme to convert **angiotensinogen** into **angiotensin I**. Angiotensinogen is a plasma protein synthesized by the liver and always present in the plasma in high concentration. On passing through the lungs via the pulmonary circulation, angiotensin I is converted into **angiotensin II** by **angiotensin-converting enzyme (ACE)**, which is abundant in the pulmonary capillaries. Angiotensin II is the primary stimulus for the secretion of the hormone **aldosterone** from the adrenal cortex (p. 304). These do the following:

1. *Na$^+$ reabsorption:* Among its actions, aldosterone increases Na$^+$ reabsorption by the distal tubules and collecting ducts. Specifically, it works on *principal* cells, one of two major cell types making up these tubules (the other type, *intercalated* cells, are involved in acid–base balance which is covered in Chapter 13.) Aldosterone is a steroid hormone, which exerts its effects by binding to an internal receptor and regulating gene transcription (Chapters 2 and 7). For aldosterone, gene regulation leads to the insertion of additional ENaC Na$^+$ channels into the apical membranes and Na$^+$/K$^+$ ATPase pumps into the basolateral membranes (Figure 12-3a) of the distal-tubule and collecting-duct cells. (Angiotensin II itself also stimulates Na$^+$ reabsorption in the proximal and distal tubules and the collecting ducts.) The net result is a greater passive inward flux of Na$^+$ into the tubular cells from the lumen and increased active pumping of Na$^+$ out of the cells into the plasma, with Cl$^-$ following passively and H$_2$O following by osmosis. The RAAS thus promotes salt retention resulting in H$_2$O retention and elevation of arterial blood pressure.
2. *Water retention:* Angiotensin II stimulates hypothalamic release of *vasopressin*, a hormone that increases H$_2$O retention by the kidneys (p. 597), elevating blood pressure.
3. *Vasoconstriction:* Angiotensin II is so named (*angeion*, Greek for blood vessel, plus "tense") because it is also a potent constrictor of the systemic arterioles, thereby directly and rapidly increasing blood pressure by increasing total peripheral resistance (see Figure 9-48, p. 438). Furthermore, stimulation of vasopressin (*vas*, Latin for blood vessel, plus "pressure") adds to this effect since it is also a vasoconstrictor.

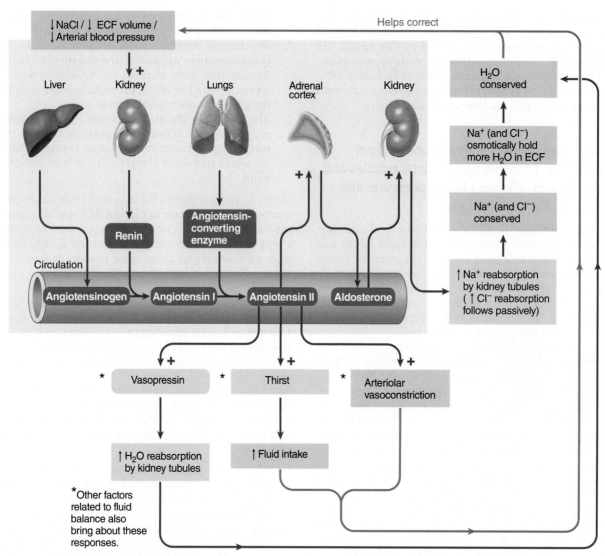

FIGURE 12-21 Renin–angiotensin–aldosterone system (RAAS) in a mammal. The kidneys secrete the enzymatic hormone renin in response to reduced NaCl, ECF volume, and arterial blood pressure. Renin activates angiotensinogen, a plasma protein produced by the liver, into angiotensin I. Angiotensin I is converted into angiotensin II by angiotensin-converting enzyme (ACE) produced in the lungs. Angiotensin II stimulates the adrenal cortex to secrete the hormone aldosterone, which stimulates Na^+ reabsorption by the kidneys. The resulting retention of Na^+ exerts an osmotic effect that holds more H_2O in the ECF. Together, the conserved Na^+ and H_2O help correct the original stimuli that activated this hormonal system. Angiotensin II also exerts other effects that help rectify the original stimuli, such as by promoting arteriolar vasoconstriction.

© Cengage Learning, 2013

4. *Thirst and salt hunger:* Angiotensin II stimulates *thirst.* It also stimulates **salt hunger** in the brain, that is, it triggers a craving for dietary salt intake. Aldosterone appears to do this as well, but angiotensin is the most important stimulant. Salt hunger and thirst trigger corrective behaviors that fall into the category of *behavioral effector processes* (p. 16). Thirst results in increased water consumption, while salt hunger favors the consumption of salty foods or licking salt deposits. Both actions help correct the imbalances.

Thus, acting in negative-feedback fashion, activation of the RAAS alleviates the factors that triggered the initial release of renin—namely, salt depletion, plasma volume reduction, and decreased arterial blood pressure (Figure 12-21).

Suppression of the RAAS The opposite situation exists when the Na^+ load, ECF and plasma volume, and arterial blood pressure are above normal. Under these circumstances, renin secretion is inhibited, which prevents the angiotensin II–induced release of aldosterone (and vasopressin). Without aldosterone, the modest aldosterone-dependent (regulated) portion of Na^+ reabsorption in the proximal and distal tubules does not occur. Instead, this unreabsorbed Na^+, and the Cl^- and water that would have come with it, are lost in urine. This ongoing loss can rapidly remove excess NaCl (and water) from the body.

In the complete absence of aldosterone, a human may excrete 20 g of salt per day. With maximum aldosterone secretion, almost all the filtered Na^+ (and Cl^-) is reabsorbed, so salt excretion in the urine is very low (although probably never zero). The amount of aldosterone secreted, and consequently the relative amount of salt conserved versus salt excreted varies according to intake. Herbivores, for example, have a high-potassium, low-sodium diet because of the ion composition of plants, while carnivores may have a substantial sodium intake. With the renin mechanism, the kidneys maintain the salt load and ECF volume/arterial blood pressure at a relatively constant level despite wide variations in salt consumption and abnormal losses of salt-laden fluid (hemorrhage).

This system is responsible in part for the fluid retention and edema accompanying congestive heart failure, a condition suffered by many aging mammals. When a fall in blood pressure occurs due to a failing heart, the salt- and fluid-retaining reflexes triggered by the low blood pressure are inappropriate. Excess sodium in the ECF and the resulting volume expansion exacerbates the problem because the weakened heart cannot pump the additional plasma volume. As you saw in Chapter 1, this is an example of a *positive-feedback process* that is not beneficial (p. 18). Treatment normally involves the blockage of aldosterone function in afflicted animals, usually by administering an **inhibitor of ACE** (angiotensin converting enzyme in the lung) to block the conversion of angiotensin I into angiotensin II.

Atrial and brain natriuretic peptides antagonize the RAAS, inhibiting Na⁺ reabsorption and reducing blood pressure

Whereas RAAS exerts the most powerful influence on the renal handling of Na^+, this Na^+-retaining, blood pressure–raising system is opposed by an Na^+-losing, blood pressure–lowering system that involves the hormones **atrial natriuretic peptide (ANP)** and **brain natriuretic peptide (BNP)** (*natriuretic* means "inducing excretion of sodium in the urine"). The heart, in addition to its pump action, has endocrine cells that release ANP and BNP. As its name implies, ANP is produced in the atrial cardiac muscle cells. BNP was first discovered in the brain but is produced primarily in the ventricular cardiac muscle cells. ANP and BNP are stored in granules and released when the heart muscle cells are mechanically stretched by an expansion of the circulating plasma volume when the ECF volume is increased. This expansion, which occurs as a result of Na^+ and H_2O retention, increases arterial blood pressure. In turn, both ANP and BNP promote natriuresis and accompanying *diuresis*, decreasing the plasma volume, and also directly lower the blood pressure, as follows (Figure 12-22):

1. *Natriuresis and diuresis:* The main function of ANP and BNP is to directly inhibit Na^+ reabsorption in the distal parts of the nephron, thus increasing Na^+ and accompanying osmotic H_2O excretion in the urine. They further increase Na^+ excretion in the urine by inhibiting two steps of the Na^+-conserving RAAS. The NPs inhibit renin secretion by the kidneys and act on the adrenal cortex to inhibit aldosterone secretion. In addition, they inhibit the secretion and actions of vasopressin, the H_2O-conserving hormone. ANP and BNP dilate the afferent arterioles and constrict the efferent arterioles, thus raising glomerular capillary blood pressure and increasing the GFR. They further increase the GFR by relaxing the glomerular mesangial cells, leading to an increase in K_f. As more salt and water are filtered, more salt and water are excreted in the urine.

2. *Reduced blood pressure:* Besides indirectly lowering blood pressure by reducing the Na^+ load and hence the fluid load in the body, ANP and BNP directly lower blood pressure by decreasing the cardiac output and reducing peripheral vascular resistance by inhibiting sympathetic nervous activity to the heart and blood vessels.

The relative contributions of ANP and BNP in maintaining salt and H_2O balance and blood pressure regulation are being intensively investigated.

check your understanding 12.6

Compare the reabsorption processes in the proximal tubule for Na^+, H_2O, and glucose.

Describe how the RAAS works and is opposed by the ANP/BNP system.

12.7 Mammalian Kidneys: Tubular Secretion

Next we turn to the counterpart of reabsorption, which also involves transepithelial transport, but now the steps are reversed. By providing a second route of entry into the tubules for selected substances, **tubular secretion**—the discrete transfer of substances from the peritubular capillaries into the tubular lumen—is a supplemental mechanism that hastens elimination of selected compounds from the body. The most important substances secreted by the tubules are *hydrogen ion (H^+), potassium (K^+)*, and *organic anions and cations*, many of which are foreign to the body.

Hydrogen ion secretion is important in acid–base balance

Renal H^+ secretion is extremely important in the regulation of the acid–base balance in the body. Hydrogen ion can be added to the filtered fluid by being secreted by the proximal, distal, and collecting segments, using proton-pumping *V-ATPases* (see p. 85 and Malpighian tubules, p. 566). The extent of H^+ secretion depends on the acidity of the body fluids. When the body fluids are too acidic, H^+ secretion increases. Conversely, H^+ secretion is reduced when the H^+ concentration in the body fluids is too low. (See Chapter 13 for further details.)

Potassium secretion is controlled by aldosterone

Potassium is one of the most abundant cations in the body, but about 98% of the K^+ is in the ICF because the Na^+/K^+ pump actively transports K^+ into the cells. Because only a relatively small amount of K^+ is in the ECF, even slight changes in the ECF K^+ load can have a pronounced effect on the plasma K^+ concentration. Changes in the plasma K^+ concentration can be dangerous, as we will describe.

© Cengage Learning, 2013

FIGURE 12-22 Atrial and brain natriuretic peptides (mammal). The cardiac atria secrete the hormone atrial natriuretic peptide (ANP) and the cardiac ventricles secrete brain natriuretic peptide (BNP) in response to being stretched by Na^+ retention, expansion of the ECF volume, and increase in arterial blood pressure. ANP and BNP, in turn, promote natriuretic, diuretic, and hypotensive effects to help correct the original stimuli that resulted in their release.

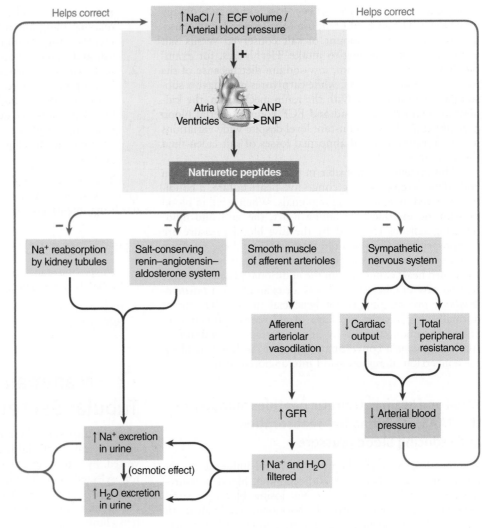

Therefore, plasma K^+ concentrations are tightly controlled, primarily by the kidneys.

Potassium ion is selectively moved in opposite directions in different parts of the tubule. It is actively reabsorbed in the proximal tubule, but it can also actively secreted by principal cells in the distal and collecting tubules, where, moreover, one type of intercalated cell actively secretes K^+ and another type actively reabsorbs K^+ in conjunction with H^+ transport (see Chapter 13). Early in the tubule K^+ reabsorption is unregulated, whereas K^+ secretion by the principal cells in the distal tubule is variable and subject to regulation. Because the filtered K^+ is almost completely reabsorbed in the proximal tubule, most K^+ in the urine is derived from controlled K^+ secretion in the distal parts of the nephron.

During periods of K^+ depletion, K^+ secretion in the distal portions of the nephron is reduced to a minimum, so only the small percentage of filtered K^+ that escapes reabsorption in the proximal tubule is excreted in the urine. In this way, K^+ that normally would have been lost in the urine is conserved for the body. In contrast, when plasma K^+ levels are elevated, K^+ secretion is adjusted so that just enough K^+ is added to the filtrate for elimination to reduce the plasma K^+ concentration to normal. Thus, K^+ secretion, not the filtration or reabsorption of K^+, is varied in a controlled fashion to regulate the rate of K^+ excretion and maintain the desired plasma K^+ concentration.

Mechanism and Control of K^+ Secretion Potassium ion secretion in the distal tubules and collecting ducts is coupled to Na^+ reabsorption by means of the basolateral Na^+/K^+ pump (Figure 12-23). This pump not only moves Na^+ into the lateral space but also transports K^+ into the tubular cells. The resulting high intracellular K^+ concentration favors the net diffusion of K^+ from the cells into the tubular lumen. Movement across the apical membrane occurs passively through the large number of K^+ leak channels present in this barrier. By keeping the interstitial fluid concentration of K^+ low as it transports K^+ into the tubular cells from the surrounding interstitial fluid, the basolateral pump encourages the passive diffusion of K^+ out of the peritubular capillary plasma into the interstitial fluid. Potassium exiting the plasma in this manner is subsequently pumped into the cells, from which it diffuses into the lumen. In this way, the basolateral pump actively induces the net secretion of K^+ from the peritubular capillary plasma into the tubular lumen.

Note that the different pathways for K^+ movement are dependent on the location of the passive K^+ channels. In the distal tubules and collecting ducts, these channels are found mainly in the apical membrane, providing a route for K^+

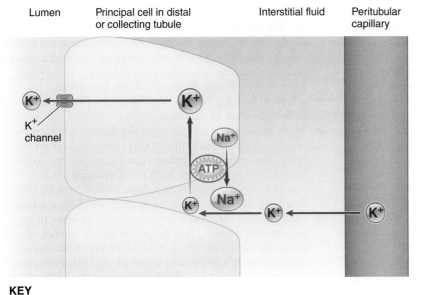

Lumen | Principal cell in distal or collecting tubule | Interstitial fluid | Peritubular capillary

KEY

⟶ = Active transport
⟶ = Passive diffusion

FIGURE 12-23 Potassium secretion (mammalian kidney). The basolateral pump simultaneously transports Na^+ into the lateral space and K^+ into the tubular cell. In the parts of the tubule that secrete K^+, this ion leaves the cell through channels located in the apical border, thus being secreted. (In the parts of the tubule that do not secrete K^+, the K^+ pumped into the cell during the process of Na^+ reabsorption leaves the cell through channels located in the basolateral border, thus being retained in the body.)

© Cengage Learning, 2013

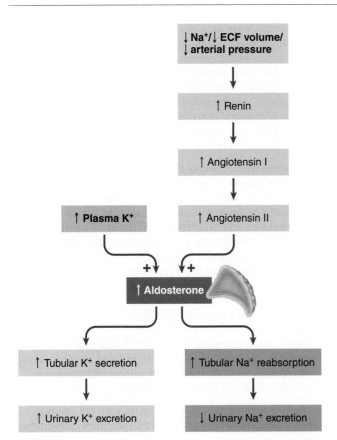

FIGURE 12-24 Dual control of aldosterone secretion of K^+ and Na^+ (mammal).

© Cengage Learning, 2013

cling permits the ongoing operation of the Na^+/K^+ pump to accomplish Na^+ reabsorption with no local net effect on K^+.

Several factors can alter the rate of K^+ secretion, the most important being *aldosterone,* which stimulates K^+ secretion by the tubular cells late in the nephron simultaneous to enhancing these cells' reabsorption of Na^+. An elevation in plasma K^+ concentration directly (independently of the RAAS) stimulates the adrenal cortex to increase output of aldosterone, which in turn promotes urinary excretion and elimination of the excess K^+. Conversely, a decline in plasma K^+ concentration reduces aldosterone secretion, which correspondingly reduces aldosterone-stimulated renal K^+ secretion.

Note that a rise in plasma K^+ directly stimulates aldosterone secretion, whereas a fall in plasma Na^+ concentration stimulates aldosterone secretion via the RAAS (Figure 12-24). Thus, aldosterone secretion can be stimulated by two separate pathways. No matter what the stimulus, however, increased aldosterone secretion always promotes simultaneous Na^+ reabsorption and K^+ secretion. For this reason, K^+ secretion can be inadvertently stimulated as a result of increased aldosterone activity brought about by Na^+ depletion, ECF volume reduction, or a fall in arterial blood pressure. The resulting inappropriate loss of K^+ can lead to K^+ deficiency.

Because this process has the net effect of swapping K^+ for Na^+ in the urine, and because it can be regulated, the amount of KCl and NaCl excreted in the final urine can vary widely among animals and among diets. Herbivores, for example, eat food often with a high potassium and low sodium content (most plants) and therefore excrete more KCl than NaCl. Indeed, such mammals exhibit salt hunger (stimulated by angiotensin II and aldosterone, p. 589), including a strong behavioral attraction to salt licks (natural or human-made deposits of NaCl and other minerals).

Normally the kidneys exert a fine degree of control over plasma K^+ concentration. This is extremely important because even minor fluctuations in plasma K^+ concentration can

pumped into the cell to exit into the lumen. In the other tubular segments, the K^+ channels lie primarily in the basolateral membrane. As a result, K^+ pumped into the cell from the lateral space by the Na^+/K^+ pump simply diffuses back out into the lateral space through these channels. This K^+ recy-

have life-threatening consequences. Potassium ion plays a key role in the membrane potential and electrical activity of excitable tissues (Chapters 3 and 4). Either increases or decreases in the plasma (ECF) K^+ concentration can alter the intracellular-to-extracellular K^+ concentration gradient. A rise leads to a reduction in resting potential (bringing the cell closer to threshold potential) and a subsequent increase in excitability, especially in cardiac and neural tissue. This cardiac overexcitability can lead to a rapid heart rate and even fatal cardiac arrhythmias, although pharmacological solutions of K^+ are used to slow the heart during surgical procedures. Conversely, a fall in ECF K^+ concentration results in hyperpolarization of nerve and muscle cell membranes, which reduces their excitability. The manifestations of ECF K^+ depletion are skeletal muscle weakness, diarrhea and abdominal distention caused by smooth muscle dysfunction, and abnormalities in cardiac rhythm and impulse conduction.

As we noted earlier, inadvertent loss of K^+ because of Na^+ depletion can occur. This has important implications for domestic herbivores such as cattle because they are sometimes fed herbage with extremely low NaCl and high KCl content. In one early study, cattle were fed such a diet and monitored carefully. After two months, they began licking walls and dirt, chewing the bark of trees, eating manure, and drinking the urine of other cows, driven by an intense salt hunger. The ion content of their urines was dominated by KCl with virtually no NaCl. After 10 months, the animals began to lose weight, while some developed muscle tremors and eventually died. Autopsies revealed hypertrophied (excessively enlarged) adrenal cortexes, probably because of a high rate of aldosterone production.

Another factor that can inadvertently alter the magnitude of K^+ secretion is the acid–base status of the body. An antiport transporter in the distal tubules' intercalated cells can secrete either K^+ or H^+ in exchange for reabsorbed Na^+. An increased rate of secretion of either K^+ or H^+ is accompanied by a decreased rate of secretion of the other ion. Normally the kidneys secrete a preponderance of K^+, but when the body fluids are too acidic and H^+ secretion is increased as a compensatory measure, K^+ secretion is correspondingly reduced. This reduced secretion leads to inadvertent K^+ retention in the body fluids.

Organic anion and cation secretion helps eliminate foreign compounds from the body

The proximal tubule contains two distinct types of secretory carriers, one for organic anions and another for organic cations. These carriers serve several important functions:

- By adding more of a particular type of organic ion to the quantity that has already gained entry to the tubular fluid by means of glomerular filtration, these organic secretory pathways facilitate excretion of these substances. Included among these organic ions are certain bloodborne chemical messengers such as prostaglandins, histamine, and norepinephrine that, having served their purpose, must be rapidly removed from the blood so that their biological activity is not unduly prolonged.
- In some important instances, organic ions are poorly soluble in water. To be transported in blood, they are extensively but not irreversibly bound to plasma proteins. Because they are attached to plasma proteins,

these substances cannot be filtered through the glomeruli. Tubular secretion facilitates elimination of these nonfilterable organic ions in urine. Even though a given organic ion is largely bound to plasma proteins, a small percentage of the ion always exists in free or unbound form in the plasma. Removal of this free organic ion by secretion permits "unloading" of some of the bound ion, which can then be secreted.
- Most important is the ability of the organic ion secretory systems to eliminate many foreign compounds from the body. The organic ion systems can secrete a large number of different organic ions, both those produced endogenously (within the body) and those foreign organic ions that have gained access to the body fluids. This nonselectivity permits these organic ion secretory systems to hasten removal of many (but by no means all) foreign organic chemicals, including food additives, environmental pollutants (such as pesticides), drugs, and other nonnutritive organic substances that have entered the body.

Even though these mechanisms help rid the body of unwanted compounds, they do not appear subject to physiologic adjustments. The carriers cannot pick up their secretory pace when confronting an elevated load of these organic ions.

The vertebrate liver plays an important role in this process. Many foreign organic compounds are not ionic in their original form, so they cannot be secreted by the organic ion systems. The liver converts these foreign substances into an anionic form that facilitates their secretion by the organic anion system, which thus accelerates their removal.

This completes our discussion of the reabsorptive and secretory processes that occur across the proximal and distal portions of the nephron. To generalize:

- The *proximal tubule* does most of the reabsorbing. This "mass reabsorber" transfers much of the filtered water and needed solutes back into the blood in unregulated fashion. The proximal tubule is also the major site of secretion (with the exception of K^+).
- The *distal tubules* and *collecting ducts* then determine the final amounts of H_2O, Na^+, K^+, and H^+ that are excreted in urine. They do so by "fine-tuning" the amount of Na and H_2O reabsorbed and the amount of K^+ and H^+ secreted.
- Many of these processes are subject to control, especially in the distal part of the nephron, depending on the body's momentary needs. The unwanted filtered waste products are left behind to be eliminated in urine, along with excess amounts of filtered or secreted nonwaste products that fail to be reabsorbed.

Before turning to the last function of the mammalian nephron—osmoconcentration—let's see how researchers measure the combined actions of filtration, reabsorption, and secretion for individual substances by a parameter called **plasma clearance**.

Plasma clearance is the volume of plasma cleared of a particular substance per minute

Compared to the plasma entering the kidneys through the renal arteries, the plasma leaving the kidneys through the renal veins lacks the materials that were left behind to be eliminated in the urine. By excreting substances in the urine, the kidneys clean or "clear" the plasma flowing through them of these substances. For any substance, its **plasma clearance** is defined as the volume of plasma that is completely cleared of that substance by the kidneys per minute. Plasma clearance is actually a more useful measure than urine excretion; it is more important to know what effect urine excretion has on removing materials from body fluids than to know the volume and composition of the urine. Plasma clearance expresses the kidneys' effectiveness in removing various substances from the internal fluid environment.

Plasma clearance can be calculated for any plasma constituent as follows:

$$
\begin{array}{l}
\text{Clearance rate} \\
\text{of a substance} \\
\text{(mL/min)}
\end{array}
=
\dfrac{
\begin{array}{l}
\text{urine concentration} \\
\text{of the substance} \quad \times \quad \begin{array}{l}\text{urine flow} \\ \text{rate (mL/min)}\end{array} \\
\text{(quantity/mL urine)}
\end{array}
}{
\text{plasma concentration of the substance}
}
$$

The plasma clearance rate varies for different substances, depending on how the kidneys handle each substance, as follows.

1. *Plasma clearance rate equals the GFR if a substance is freely filtered but not reabsorbed or secreted.* Assume that a plasma constituent, substance X, has these properties. If, for example, 125 mL/min of plasma are filtered and none of X is subsequently reabsorbed or secreted, all of substance X originally contained within the 125 mL is left behind in the tubules to be excreted. Thus, 125 mL of plasma are cleared of substance X each minute.

   There is no endogenous chemical with these characteristics: All substances normally present in the plasma are reabsorbed or secreted to some extent. However, **inulin** (do not confuse with insulin), a harmless carbohydrate produced by Jerusalem artichokes (which are not artichokes, but the tubers of a sunflower species), is freely filtered and not reabsorbed or secreted—an ideal substance X. Inulin can be injected and its plasma clearance determined as a clinical means of ascertaining the GFR. Because all glomerular filtrate formed is cleared of inulin, the volume of plasma cleared of inulin per minute equals the volume of plasma filtered per minute—the GFR.

   Although the determination of inulin plasma clearance is accurate, it is not very convenient because inulin must be infused continuously during the analysis. Therefore, the plasma clearance of *creatinine*—an end product of muscle metabolism (p. 561)—is often used instead to give a rough estimate of the GFR. Creatinine is produced at a relatively constant rate, is freely filtered and not reabsorbed but is only slightly secreted. Accordingly, creatinine clearance is not a completely accurate reflection of the GFR, but it does provide a close approximation and can be more readily determined than inulin clearance.

2. *If a substance is filtered and reabsorbed but not secreted, its plasma clearance rate is always less than the GFR.* Because less than the filtered volume of plasma will have been cleared of the substance, the plasma clearance rate of a reabsorbable substance is always less than the GFR. For example, the plasma clearance for glucose is normally zero since the filtered glucose is reabsorbed.

For a substance that is partially reabsorbed, such as urea, only part of the filtered plasma is cleared of that substance. With about 40% of the filtered urea being passively reabsorbed, only half of the filtered plasma, or 62.5 mL, is cleared of urea each minute.

3. *If a substance is filtered and secreted but not reabsorbed, its plasma clearance rate is always greater than the GFR.* Tubular secretion allows the kidneys to clear certain materials from the plasma more efficiently. Only 20% of the plasma entering the kidneys is filtered. The remaining 80% passes unfiltered into the peritubular capillaries. The only means by which this unfiltered plasma can be cleared of any substance during this trip through the kidneys before being returned to the general circulation is by the process of secretion. An example is H^+. Not only will the plasma that is filtered be cleared of nonreabsorbable H^+, but the plasma from which H^+ is secreted will also be cleared of H^+. For example, at the normal GFR of 125 mL/min, if the quantity of H^+ that is secreted is equivalent to the quantity of H^+ present in 25 mL of plasma, the clearance rate for H^+ will be 150 mL/min (125 mL through the process of filtration and failure of reabsorption, and 25 more mL through the process of secretion).

check your understanding 12.7

Describe mechanisms of K^+ secretion, and its regulation and importance.

Describe the three different types of plasma clearance in relation to GFR.

12.8 Mammalian Kidneys: Osmoconcentration

Having considered how the glomerulus and proximal and distal tubules operate, we now focus on renal handling of NaCl and H_2O in the loop of Henle (in concert with the distal tubule and collecting duct) in a process we call *osmoconcentration*. Although these segments are engaged in reabsorption of NaCl and water, we consider this process of osmoconcentration a distinct renal function because it occurs only in birds and mammals (among vertebrates) and because NaCl reabsorption from the tubules does not necessarily return NaCl to the body.

The ability to excrete urine of varying concentrations varies among species and depends on hydration state

The ECF osmolarity (solute concentration) depends on the relative amount of H_2O compared to solute, which is primarily NaCl. At normal fluid balance and solute concentration, body fluids are said to be *isotonic* or *isosmotic* at an osmolarity of about 300 milliosmoles/liter (mOsm) in most terrestrial vertebrates (Chapter 3, p. 79, and Chapter 13, p. 625). Recall that, if there is too much H_2O relative to the solute load, body fluids are *hypotonic* or *hypo-osmotic* (osmolarity <300 mOsm). In contrast, if an H_2O deficit exists relative to the solute load, body fluids are too concentrated, that is, are *hypertonic* or *hyperosmotic* (osmolarity >300 mOsm).

Generally speaking, the osmolarity of the ECF is uniform throughout the body (except, as you will see, in the renal medulla, and also intestinal villi exposed to swallowed water). Knowing that the driving force for H_2O reabsorption throughout the entire length of the tubules is an osmotic gradient between the tubular lumen and surrounding interstitial fluid, you might expect, based on osmotic considerations, that the kidneys could not excrete urine of different osmolarity than the body fluids. Indeed, this would be the case if the interstitial fluid surrounding the tubules in the kidneys were identical in osmolarity to the remaining body fluids. Water reabsorption would proceed only until the tubular fluid equilibrated osmotically with the interstitial fluid, and there would be no way to eliminate excess H_2O when the body fluids were hypotonic or to conserve H_2O in the presence of hypertonicity.

However, a large *vertical osmotic gradient* is uniquely generated in the interstitial fluid of the medulla of avian and mammalian kidneys. The concentration of the interstitial fluid progressively increases from the cortical boundary down through the depth of the renal medulla until it reaches a maximum, which is at the junction with the renal pelvis in mammals (Figure 12-25). This vertical osmotic gradient remains there (although it may change in magnitude) regardless of the fluid balance of the body.

The presence of this gradient enables the kidneys to produce urine that ranges in concentration from less than 100 mOsm to highly hypertonic, depending on the species and a body's state of hydration. Examples of maximal concentrating ability will be discussed later, but in brief, these run from a modest 520 mOsm in the beaver (a freshwater animal with plentiful water supply) to a reported 9,400 mOsm in one desert rodent! As a benchmark value, recall that seawater is about 1,000 mOsm, mostly NaCl. As shown in Figure 12-25, humans can osmoconcentrate up to 1,200–1,400 mOsm, but some of this is due to urea. Human kidneys cannot concentrate NaCl to 1,000 mOsm, so humans cannot safely drink seawater, because it causes a net accumulation of salt in the body. Desert rodents, marine mammals, and some marine birds, in contrast, have no trouble drinking highly saline water (and eating salty food) and excreting excess salt, with a net gain of water to the body.

Antidiuresis and Diuresis When a mammal is in ideal fluid balance, isotonic urine is produced at a moderate rate (1 mL/min in humans). This can be altered between two extremes of antidiuresis and diuresis. When the body is dehydrated, the kidneys can put out a small volume of concentrated urine (hypertonic; down to 0.3 mL/min in humans), thus conserving H_2O (antidiuresis). Conversely, when the body is overhydrated (too much H_2O), the kidneys can produce a large volume of dilute urine (hypotonic at 100 mOsm or less; up to 25 mL/min in humans), thus eliminating the excess H_2O in the urine (diuresis).

Unique anatomic arrangements and complex functional interactions between the various nephron components in the renal medulla are responsible for the establishment and use of the vertical osmotic gradient. Recall that the hairpin loop of Henle in the *juxtamedullary nephrons* plunges through

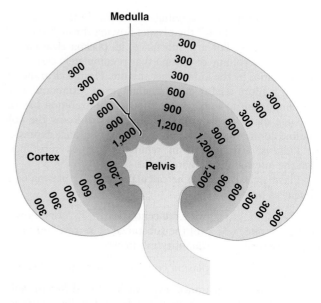

FIGURE 12-25 Vertical osmotic gradient in the human renal medulla. Schematic representation of the kidney rotated 90° from its normal position in an upright human for better visualization of the vertical osmotic gradient in the renal medulla. All values are in mOsm. The osmolarity of the interstitial fluid throughout the renal cortex is isotonic at 300 mOsm, but the osmolarity of the interstitial fluid in the renal medulla increases progressively from 300 mOsm at the boundary with the cortex to a maximum of 1,200 mOsm at the junction with the renal pelvis.

© Cengage Learning, 2013

the entire depth of the medulla so that the tip of the loop lies near the renal pelvis (Figure 12-12). Also, the vasa recta of the juxtamedullary nephrons have a similar deep hairpin loop. Flow in the loops of Henle and the vasa recta is considered **countercurrent,** because the flow in the two closely adjacent limbs of the loop and blood vessel is in opposite directions. Also running through the medulla, in the descending direction only, are the collecting ducts that serve both types of nephrons on their way to the renal pelvis. This arrangement, coupled with the permeability and transport characteristics of these tubular segments, plays a key role in the kidneys' ability to provide for water conservation or elimination. Briefly, the juxtamedullary nephrons' long loops of Henle establish the vertical osmotic gradient; their vasa recta prevent the dissolution of this gradient while providing blood to the renal medulla, and the collecting ducts use the gradient, in conjunction with the hormone vasopressin, to produce urine ranging from diuretic to antidiuretic. Collectively, this entire functional organization is known as the medullary **countercurrent-multiplier** system. We examine each of its facets in greater detail.

The medullary vertical osmotic gradient is established by countercurrent multiplication

We will follow the filtrate through a juxtamedullary nephron that lies primarily in the outer medulla to see how this structure establishes a vertical osmotic gradient. Immediately after the filtrate is formed, uncontrolled osmotic reabsorption of filtered H_2O occurs in the proximal tubule secondary to active Na^+ reabsorption. As a result, by the end of the proximal tubule, 45 to 65% of the filtrate has been reabsorbed, but the 35 to 55% remaining in the tubular lumen still has the same osmolarity (300 mOsm) as the body fluids. That is, it is still isotonic. An additional 15% of the filtered H_2O is obligatorily reabsorbed from the loop of Henle during the establishment and maintenance of the vertical osmotic gradient, with the osmolarity of the tubular fluid being altered greatly in the process.

Properties of the Limbs of a Long Henle's Loop The following functional distinctions among the *descending limb* of a long Henle's loop (which carries fluid from the proximal tubule down into the depths of the medulla), the *hairpin turn,* and the *ascending limb* (which carries fluid up and out of the medulla into the distal tubule) are crucial for establishing the incremental osmotic gradient in the medullary interstitial fluid. The ascending limb is further subdivided into a *thin segment* in the inner medulla (for longer nephrons) and a *thick segment* in the outer medulla (see Figure 12-8a and Figure 12-26, initial scene).

The *descending limb (primarily the upper part):*

1. is highly permeable to H_2O (via abundant, always-open AQP-1 water channels).
2. does not actively transport Na^+.

The *hairpin turn* region, including the lower descending and the *thin ascending limb* in the inner medulla:

1. has low water permeability (due to lack of aquaporins).
2. has no active salt transport, but in the loop and thin ascending segments is permeable to NaCl due to leak channels.

The *thick ascending limb:*

1. actively transports NaCl out of the tubular lumen into the surrounding interstitial fluid.
2. is always impermeable to H_2O, so salt leaves the tubular fluid without H_2O osmotically following along.

Mechanism of Countercurrent Multiplication The close proximity and countercurrent flow of the two limbs allow important interactions between them. Even though the flow of fluids is continuous through the loop of Henle, we will visualize what happens step by step in Figure 12-26 for the outer medulla, much like an animated film run so slowly that each frame can be viewed:

- *Initial scene:* if no vertical osmotic gradient is established, the medullary interstitial and tubular fluid concentrations would be uniformly 300 mOsm, isosmotic with other body fluids.
- *Step 1.* The active salt pump in the waterproof *thick ascending limb* transports Na^+ followed by Cl^- (via ClC channels; see Figure 12-3a) out of the lumen until the surrounding interstitial fluid is 200 mOsm more concentrated than (hyperosmotic to) the tubular fluid in this limb. Water cannot follow osmotically from the ascending limb, but it does osmose from the *upper descending limb* through AQP-1 into the (hypertonic) interstitial fluid. This passive movement of H_2O out of the descending limb continues until the osmolarities of the fluid in the descending limb and interstitial fluid become equili-

brated. Thus, the isosmotic tubular fluid entering the loop of Henle immediately starts to become more concentrated as it loses H_2O. At equilibrium, the osmolarity of the ascending limb fluid is 200 mOsm and the osmolarities of the interstitial fluid and descending limb fluid are equal at 400 mOsm in this hypothetical situation.

- *Step 2.* If we now advance the entire column of fluid in the loop of Henle several frames, a mass of 200 mOsm fluid exits from the top of the ascending limb into the distal tubule, and a new mass of isotonic fluid at 300 mOsm enters the top of the descending limb from the proximal tubule. At the bottom of the loop, a comparable mass of 400 mOsm fluid from the descending limb moves forward around the tip into the ascending limb, placing it opposite a 400 mOsm region in the descending limb. In the *hairpin turn* and *thin* ascending limb, some NaCl escapes through leak channels (including ClC types) into the medulla, raising its osmolarity (not shown). Note that the 200 mOsm concentration difference has been lost at both the top and the bottom of the loop.
- *Step 3.* The thick ascending limb pump again transports NaCl out, while H_2O passively leaves the descending limb until a 200 mOsm difference is reestablished between the ascending limb and both the interstitial fluid and descending limb at each horizontal level. Note, however, that the concentration of tubular fluid is progressively increasing in the descending limb and progressively decreasing in the ascending limb.
- *Steps 4 and 5.* As the tubular fluid is advanced still further, the 200 mOsm concentration gradient is disrupted once again at all horizontal levels. Again, active extrusion of NaCl from the ascending limb, coupled with the net diffusion of H_2O out of the descending limb, reestablishes the 200 mOsm gradient at each horizontal level.
- *Step 6 and so on.* As the fluid flows slightly forward again and this stepwise process continues, the fluid in the descending limb becomes progressively more hypertonic until it reaches a maximum concentration of 1,200 mOsm (in humans) at the bottom of the loop, four times the normal concentration of body fluids. Because the interstitial fluid always achieves equilibrium with the descending limb, an incremental vertical concentration gradient ranging from 300 to 1,200 mOsm is likewise established in the medullary interstitial fluid. In contrast, the concentration of the tubular fluid progressively decreases in the ascending limb as salt is pumped out, but H_2O is unable to follow. In fact, the tubular fluid even becomes hypotonic before leaving the ascending limb to enter the distal tubule at a concentration of 100 mOsm or less, one-third the normal concentration of body fluids.

Note that although a gradient of only 200 mOsm exists between the ascending limb and the surrounding fluids at each medullary horizontal level, a much larger vertical gradient exists from the top to the bottom of the medulla. Even though the ascending limb pump can generate a gradient of only 200 mOsm, this effect is multiplied into a large vertical gradient because of the countercurrent flow within the loop. This is the concentrating mechanism accomplished by the loop of Henle known as **countercurrent multiplication.** This is also termed a **single-effect** process since Na^+ pumping drives all the major events.

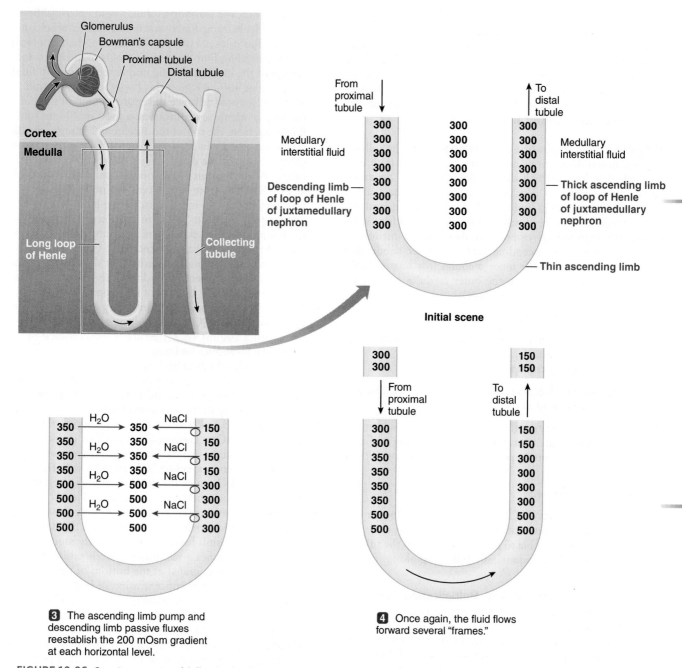

FIGURE 12-26 Countercurrent multiplication by the "single effect" in the mammalian renal medulla. All values are in mOsm.

© Cengage Learning, 2013

We have artificially described single-effect countercurrent multiplication in a stop-and-flow, stepwise fashion to facilitate understanding of this process. It is important to realize that once the incremental medullary gradient is established, it stays relatively constant because of the continuous flow of fluid coupled with the ongoing ascending limb active transport and accompanying descending limb passive fluxes.

Benefits of Countercurrent Multiplication If you consider only what happens to the tubular fluid as it flows through the loop of Henle, the whole process seems an exercise in futility. The isotonic fluid that enters the loop becomes progressively more concentrated as it flows down the descending limb, achieving a maximum concentration of about 1,200 mOsm in humans, only to become progressively more dilute as it flows up the ascending limb, finally leaving the loop at a minimum concentration of 100 mOsm. What is the point of concentrating the fluid fourfold and then turning around and diluting it until it leaves at one-third the concentration at which it entered? Such a mechanism offers two benefits:

1. It establishes the vertical osmotic gradient in the medullary interstitial fluid; this gradient, in turn, is used by the collecting ducts to concentrate the tubular fluid so that a low-volume urine *more concentrated* than normal body fluids can be excreted (antidiuresis).

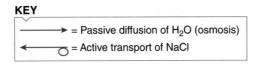

KEY

⟶ = Passive diffusion of H₂O (osmosis)

⟵○ = Active transport of NaCl

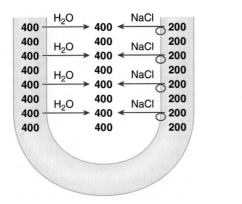

1 The active salt pump in the ascending limb establishes a 200 mOsm gradient at each horizontal level.

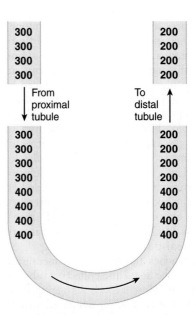

2 As the fluid flows forward several "frames," a mass of 200 mOsm fluid exits into the distal tubule and a new mass of 300 mOsm fluid enters from the proximal tubule.

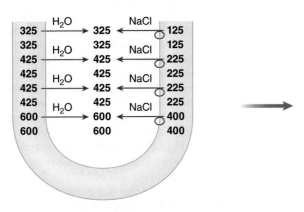

5 The 200 mOsm gradient at each horizontal level is established once again.

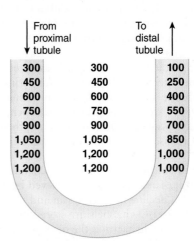

6 The final vertical osmotic gradient is established and maintained by the ongoing countercurrent multiplication of the long loops of Henle.

2. Conversely, the fact that the fluid is hypotonic as it enters the distal parts of the tubule enables the kidneys to excrete a voluminous urine *more dilute* than normal body fluids (diuresis). Let's see how.

Vasopressin-controlled, variable H₂O reabsorption occurs in the final tubular segments

After obligatory H₂O reabsorption from the proximal tubule and loop of Henle, the remaining filtered H₂O remains in the lumen to enter the distal and collecting tubules for variable reabsorption that is under hormonal control. The fluid leaving the loop of Henle enters the distal tubule at 100 mOsm, so it is hypotonic to the surrounding isotonic (300 mOsm) interstitial fluid of the renal cortex through which the distal tubule passes. The distal tubule then empties into the collecting duct, which is bathed by progressively increasing concentrations of surrounding interstitial fluid as it descends through the medulla.

Role of Vasopressin For H₂O absorption to occur across a segment of the tubule, two criteria must be met: (1) an osmotic gradient must exist across the tubule, and (2) the tubular segment must be permeable to H₂O. The distal and collecting tubules are *impermeable* to H₂O except in the presence of **vasopressin,** or **antidiuretic hormone** (ADH), which increases their permeability to H_2O. In most mammals including humans, it is a version called **arginine vasopressin (AVP)** to distinguish it from other forms such as *lysine vasopressin* in pigs. Vasopressin is produced by several specific neuronal cell bodies in the *hypothalamus,* then stored in the *posterior pituitary gland* (p. 284), where its release into the blood is regulated by the hypothalamus. Osmoreceptors in the hypothalamus are sensitive to changes in osmotic pressure so that, in negative-feedback fashion, vasopressin secretion is triggered by an H_2O deficit when the ECF is too concentrated (that is, hypertonic) and H_2O must be conserved for the body. Conversely, vasopressin secretion is inhibited by an excess of H_2O in the ECF (that is, hypotonic) and surplus H_2O must be eliminated in urine. Low blood pressure can also stimulate its release.

Vasopressin reaches the basolateral membrane of the principal cells lining the distal and collecting tubules through the circulatory system. Here it binds with specific G-protein-coupled receptors (Figure 12-27), which activate the cyclic AMP (cAMP) second-messenger system within the tubular cells (p. 97). This leads to increased permeability of the opposite apical membrane to H_2O by promoting exocytotic docking and fusion of vesicles laden with aquaporins (specifically, *AQP-2)* on this membrane. Without AQPs, the apical membrane is impermeable to H_2O. Once H_2O enters the tubular cells from the filtrate through these vasopressin-regulated apical water channels, it passively leaves the cells down the osmotic gradient across the cells' basolateral membrane to enter the interstitial fluid. The aquaporins in the basolateral membrane of the distal and collecting tubule *(AQP-3* and *AQP-4)* are always present and open, so this membrane is always permeable to H_2O. Overall, then, there is increased H_2O reabsorption from the filtrate into the interstitial fluid.

The increase in apical membrane water channels is not permanent, however. The channels are retrieved by endocytosis in response to a decline in vasopressin and thus cAMP concentrations. These H_2O channels are stored in the internalized vesicles ready for reinsertion the next time vasopressin secretion increases. This shuttling of AQP-2 into and out of the apical membrane provides a means of rapidly controlling H_2O permeability of the distal and collecting tubules, responding in turn to the degree of hydration in the body.

Vasopressin influences H_2O permeability only in the distal part of the nephron, especially the collecting ducts. Recall that the ascending limb of Henle's loop is always impermeable to H_2O, even in the presence of vasopressin.

This system is crucial to two important situations:

1. *H_2O deficit (dehydration):* When vasopressin secretion increases, the permeability of the distal and collecting tubules to H_2O accordingly increases. As the hypotonic tubular fluid (about 100 mOsm) enters the distal part of the nephron, progressively more H_2O enters the interstitial fluid due to the osmotic pressure generated by the increasing osmolarity of the medullary interstitial fluid (Figure 12-28a). Consequently, the tubular fluid loses progressively more H_2O by osmosis and becomes increasingly concentrated, reaching about the maximum found in the interstitium (e.g., about 1,200 mOsm in humans).

The result of the vasopressin-induced reabsorption of H_2O in the late segments of the tubule is antidiuresis (a small volume of concentrated urine excreted). The reabsorbed H_2O entering the medullary interstitial fluid is picked up by the peritubular capillaries and returned to the general circulation, thus being conserved for the body.

Although vasopressin promotes H_2O conservation by the body, it cannot completely halt urine production, because a minimum volume of H_2O must be excreted with the solute wastes. Collectively, the waste products and other constituents eliminated in human urine average 600 milliosmoles each day. Because the maximum urine concentration is 1,200 mOsm, the minimum volume of urine required to excrete these wastes is 500 mL/day (600 milliosmoles of wastes/day ÷ 1,200 milliosmols/liter of urine = 0.5 liter, or 500 mL/day, or 0.3 mL/min). Thus, under maximal vasopressin influence, 99.7% of the 180 liters of human plasma H_2O filtered per day is returned to the blood, with an obligatory H_2O loss of 0.5 liter.

The kidneys' response to dehydration by tremendously concentrating urine to minimize H_2O loss when necessary is possible only because of the presence of the vertical osmotic gradient in the medulla. If this gradient did not exist, the kidneys could not produce a urine more concentrated than the body fluids no matter how much vasopressin was secreted, because the only driving force for H_2O reabsorption is a concentration differential between the tubular fluid and the interstitial fluid.

2. *H_2O excess (overhydration):* Conversely, when a mammal consumes large quantities of H_2O, the excess H_2O must be removed from the body without simultaneously losing solutes that are critical for maintaining homeostasis. Under these circumstances, no vasopressin is secreted, so the distal and collecting tubules remain impermeable to H_2O. The tubular fluid entering the distal tubule is hypotonic (100 mOsm), having lost salt without an accompanying loss of H_2O in the ascending limb of Henle's loop. As this hypotonic fluid passes through the distal and collecting tubules (Figure 12-28b), the medullary osmotic gradient cannot exert any influence because of the late tubular segments' impermeability to H_2O. Thus, no water in the filtered fluid that reaches the distal tubule is reabsorbed. Meanwhile, excretion of wastes and other urinary solutes remains constant. The net result is diuresis: a large volume of dilute urine, which helps rid the body of excess H_2O. Urine osmolarity may be as low as 100 mOsm, the same as in the fluid entering the distal tubule. Human urine flow may be increased up to 25 mL/min in the absence of vasopressin, compared to the normal urine production of 1 mL/min.

The ability to produce urine less concentrated than the body fluids depends on the fact that the tubular fluid is hypotonic as it enters the distal part of the nephron, due to the ascending limb's active NaCl transport in the absence of H_2O movement. Therefore, the loop of Henle, by simultaneously establishing the medullary osmotic gradient and diluting the tubular fluid before it enters the distal segments, plays a key role in allowing the kidneys to excrete urine over a wide range of volumes and osmolarities.

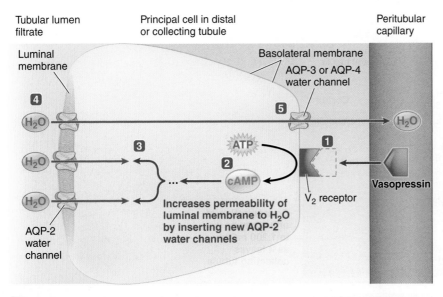

Tubular lumen filtrate

Principal cell in distal or collecting tubule

Peritubular capillary

Luminal membrane

Basolateral membrane

AQP-3 or AQP-4 water channel

4

H_2O

ATP

5

H_2O

H_2O

3

2

cAMP

1

H_2O

Vasopressin

H_2O

...

V_2 receptor

AQP-2 water channel

Increases permeability of luminal membrane to H_2O by inserting new AQP-2 water channels

1 Bloodborne vasopressin binds with its receptor sites on the basolateral membrane of a principal cell in the distal or collecting tubule.

2 This binding activates the cyclic AMP (cAMP) second-messenger system within the cell.

3 Cyclic AMP increases the opposite luminal membrane's permeability to H_2O by promoting the insertion of AQP-2 water channels into the membrane. This membrane is impermeable to water in the absence of vasopressin.

4 Water enters the tubular cell from the tubular lumen through the inserted water channels.

5 Water exits the cell through a different water channel (either AQP-3 or AQP-4) permanently positioned at the basolateral border, and then enters the blood, in this way being reabsorbed.

FIGURE 12-27 Mechanism of action of vasopressin in a mammal.

© Cengage Learning, 2013

Countercurrent exchange within the vasa recta conserves the medullary vertical osmotic gradient

The renal medulla must be supplied with blood to nourish the tissues in this area as well as to transport water that is reabsorbed by the loops of Henle and collecting ducts back to the general circulation. In doing so, however, it is important that circulation of blood through the medulla does not disturb the vertical gradient of hypertonicity established by the loops of Henle. Consider the situation if blood were to flow straight through from the cortex to the inner medulla and then directly into the renal vein (Figure 12-29a). Because capillaries are freely permeable to NaCl and H_2O, the blood would progressively pick up salt and lose H_2O through passive fluxes down concentration and osmotic gradients as it flowed through the depths of the medulla. Isotonic blood entering the medulla, upon equilibrating with each medullary level, would leave the medulla as a very hypertonic solution. It would be impossible to establish and maintain the medullary hypertonic gradient because the NaCl pumped into the medullary interstitial fluid would continuously be carried away by the circulation.

This dilemma is avoided by the hairpin construction of the vasa recta, which, by looping back through the concentration gradient in reverse, allows the blood to leave the medulla and enter the renal vein essentially isotonic to in-coming arterial blood (Figure 12-29b). As blood passes down the descending limb of the vasa recta, equilibrating with the progressively increasing concentration of the surrounding interstitial fluid, it picks up salt and loses H_2O until it is very hypertonic by the bottom of the loop. Then, as blood flows up the ascending limb, salt diffuses back out into the interstitium, and H_2O reenters the vasa recta as progressively decreasing osmolarities are encountered in the surrounding interstitial fluid. This passive exchange of solutes and H_2O between the two limbs of the vasa recta and the interstitial fluid is known as **countercurrent exchange**. Unlike countercurrent multiplication, it does not *establish* the concentration gradient. Rather, it *preserves (prevents the dissolution of)* the gradient. Because blood enters and leaves the medulla at the same osmolarity as a result of countercurrent exchange, the medullary tissue is nourished with blood, yet the gradient of hypertonicity in the medulla is preserved.

Passive urea recycling in the renal medulla contributes to medullary hypertonicity

In addition to NaCl, accumulation of *urea* in the medullary interstitial fluid also adds to its hypertonicity, especially in the inner medulla (as we will explain). For example, the fluid in the inner medulla of a laboratory rat on normal water and protein intake is 800 mOsm from salts and 900 mOsm from urea. The latter can rise dramatically on high protein diets and drop to very low levels on low-protein diets (Table 12-3). However, unlike NaCl, urea accumulation appears to be entirely passive and due to a process called *urea recycling*.

Like the thick ascending limb of the loop, the top and middle portion of the collecting duct is impermeable to urea (because of lack of urea transporters; Figure 12-30). Consequently, urea becomes progressively more concentrated inside this segment as H_2O is reabsorbed in the presence of vasopressin. By the time the fluid reaches the innermost portion of the collecting duct, its urea concentration is greater than in the surrounding interstitial fluid. This last section of the duct contains many urea transporters called *UT-As* in its membranes (which are upregulated by vasopressin). These transporters are closely related to the ones in toadfish that we noted on p. 561. The concentration difference favors the diffusion of urea out of the duct into the interstitial fluid, creating a high urea concentration in the inner medullary interstitial fluid. Moreover, some urea enters the bottom of the long loops of Henle (via UT-A transporters) and gets carried back up to the distal tubule, to undergo the same process again. This is the recycling aspect (Figure 12-30).

Urea movement out of the collecting duct has at least one major function: prevention of urea-induced water loss

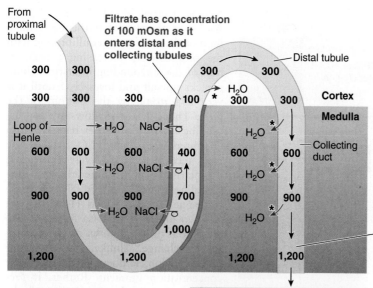

From proximal tubule

Filtrate has concentration of 100 mOsm as it enters distal and collecting tubules

Distal tubule

Cortex

Medulla

Loop of Henle

Collecting duct

Concentration of urine may be up to 1,200 mOsm as it leaves collecting tubule

Vasopressin present: distal and collecting tubules permeable to H₂O

Small volume of concentrated urine excreted; reabsorbed H₂O picked up by peritubular capillaries and conserved for body

(a) In the face of a water deficit

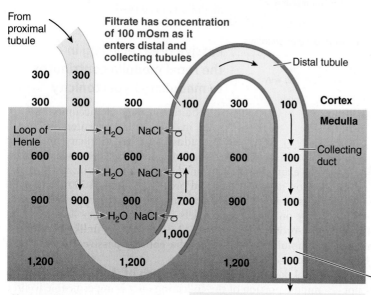

From proximal tubule

Filtrate has concentration of 100 mOsm as it enters distal and collecting tubules

Distal tubule

Cortex

Medulla

Loop of Henle

Collecting duct

Concentration of urine may be as low as 100 mOsm as it leaves collecting tubule

No vasopressin present: distal and collecting tubules impermeable to H₂O

Large volume of dilute urine; no H₂O reabsorbed in distal portion of nephron; excess H₂O eliminated

(b) In the face of a water excess

KEY

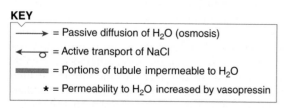

= Passive diffusion of H₂O (osmosis)

= Active transport of NaCl

= Portions of tubule impermeable to H₂O

★ = Permeability to H₂O increased by vasopressin

FIGURE 12-28 Excretion of urine of varying concentrations depending on the mammalian body's needs (values shown for humans). All values are in mOsm.

© Cengage Learning, 2013

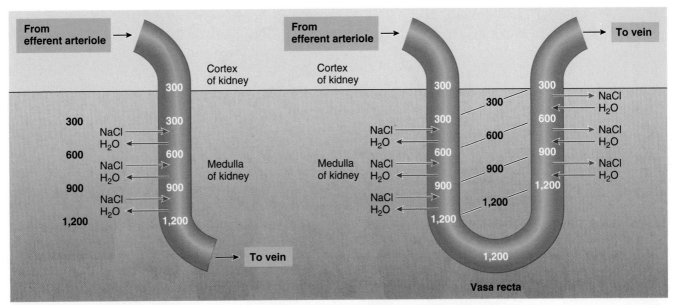

(a) Hypothetical pattern of blood flow

(b) Actual pattern of blood flow

KEY

← = Passive diffusion of H₂O (osmosis)   → = Passive diffusion of NaCl

FIGURE 12-29 Countercurrent exchange in the mammalian renal medulla. All values are in mOsm. (a) If the blood supply to the renal medulla flowed straight through from the cortex to the inner medulla, the blood would be isotonic on entering but very hypertonic on leaving, having picked up salt and lost H_2O as it equilibrated with the surrounding interstitial fluid at each incremental horizontal level. It would be impossible to maintain the vertical osmotic gradient, because the salt pumped out by the ascending limb of Henle's loop would be continuously flushed away by blood flowing through the medulla. (b) Blood equilibrates with the interstitial fluid at each incremental horizontal level in both the descending limb and the ascending limb of the vasa recta, so blood is isotonic as it enters and leaves the medulla. This countercurrent exchange prevents dissolution of the medullary osmotic gradient while providing blood to the renal medulla.

© Cengage Learning, 2013

associated with antidiuresis. As urea becomes progressively concentrated as it moves down the collecting ducts, its osmotic pressure could begin to oppose the reabsorption of water from those ducts. Allowing it to leave the innermost duct region prevents this from happening. However, urea recycling may have another role: enhancing the process of osmoconcentration, which is not fully understood for the inner medulla (Figure 12-26 is primarily for the outer medulla). See *Challenges and Controversies: The Inner Mystery* (p. 603).

Different osmoconcentrating abilities among species depend on nephron anatomy and metabolic rates

For osmoconcentration, we have been using mainly human values, with a 1,200–1,400 mOsm maximum. However, ability to osmoconcentrate the urine varies enormously among mammals, as shown in Table 12-3. Note in particular the desert rodents. How do they manage such high levels? Two features are thought to explain this:

1. *Relative loop length and medullary area.* Weak osmoconcentrators such as the beaver have primarily cortical nephrons; in contrast, the best osmoconcentrators (such as desert rodents) usually have all juxtamedullary nephrons with very long loops within an elongated medulla.

This anatomy provides a relatively long length over which the countercurrent effect can operate. Indeed, it was observations of these differences that first led physiologists to suspect that the loop was responsible for osmoconcentration. Related to this, concentrating ability correlates with *relative medullary area* (Figure 12-31), that is, with the proportion of the kidney devoted to the medulla.

2. *Metabolic rate.* The correlation between concentrating ability and both absolute medullary length and relative area cannot fully explain the amazing osmoconcentrating ability of some desert rodents (Table 12-3). Whereas these animals do have much longer loops relative to rodents from moist habitats, the loops are much shorter on an absolute scale than those of a large mammal that is a poor or modest osmoconcentrator (such as a pig). Another factor is necessary, and may be the metabolic rate of the animal, which could enhance the multiplier aspect of the countercurrent. As you will see in Chapter 15, smaller animals have higher metabolic rates per unit mass than larger animals (a *scaling* phenomenon; see p. 10). In the thick ascending limb, the cells have a higher density of mitochondria and active transport proteins and thus can move NaCl out much faster than the limbs of a larger mammal. Thus, the multiplier effect of the loop that depends on active metabolism can be much higher in a small mammal.

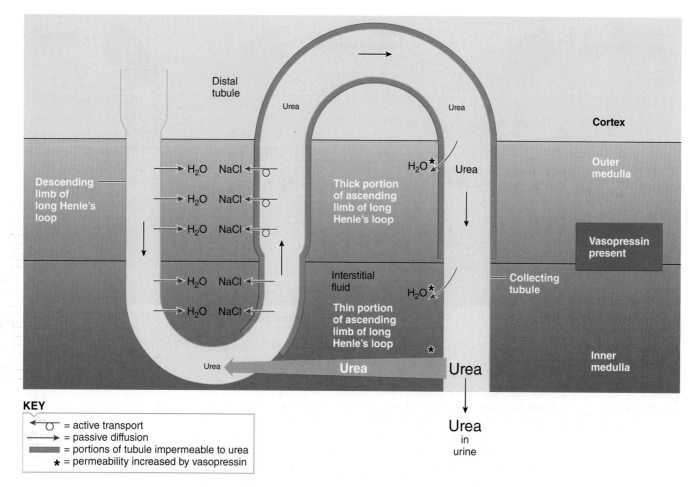

FIGURE 12-30 Urea recycling in the mammalian renal medulla. Urea becomes more concentrated in the fluid in the early portion of the collecting tubules as H_2O is osmotically reabsorbed in the presence of vasopressin; the urea cannot move out down its concentration gradient because this segment is impermeable to urea. Urea does diffuse out via UT-A transporters of the late portion of the collecting tubule down its concentration gradient into the surrounding interstitial fluid and into the urea-permeable bottoms of the long Henle's loop. Inside the loop fluid, urea is carried back up through urea-impermeable segments to start the process all over again; thus, the process is called *urea recycling*. The buildup of urea in interstitial fluid contributes to hypertonicity in the inner medulla, preventing urea from drawing water into the collecting duct and possibly enhancing NaCl buildup (see main text).

© Cengage Learning, 2013

Desert rodents are particularly remarkable in their urine-concentrating abilities. Some species live on relatively dry plant materials in a hot, dehydrating environment without drinking any free water for their entire lives. Kangaroo rats of western North American deserts have been closely studied (by the Schmidt-Nielsens, using the Krogh principle, p. 3; see also Chapter 13, p. 630) to determine how they accomplish this amazing feat. First, like any animal, kangaroo rats have an internal source of water called **metabolic water**. Recall that basic aerobic metabolism (p. 53) produces water as an end product ($C_6H_{12}O_6 + 6\ O_2 \rightarrow 6\ CO_2 + 6\ H_2O$ for glucose). Kangaroo rats get about 90% of their water needs from metabolic water and 10% from water in its food. (Humans can use metabolic water, of course, but only about 12% of water needs can be met with this.) Beyond use of metabolic water, adaptation in the kangaroo rat involves both behavior and internal physiology. *Avoidance* (p. 13), in the form of nocturnal foraging above ground and daytime residence in an insulating burrow, reduces both heat exposure and water loss (behavior effectors). Physiological adaptations include a nasal passage that traps moisture (pp. 515 and 745), and reduced excretory loss. The lower intestine removes almost all the fecal water, so that feces consist of nearly dry pellets. And, perhaps most importantly, their kidneys can produce a urine at up to 5,500 mOsm (Table 12-3). This saves water, while aldosterone regulation also saves sodium, which is low in this animal's seed-based diet: The urine osmolality is primarily K-salts and urea rather than Na-salts.

Another issue involving osmoconcentration is the extreme osmotic stress faced by renal cells, particularly in the medulla. This is an area of active research; see *Molecular Biology and Genomics: Surviving Salt and Urea: Organic Osmolytes and Gene Regulation* (p. 606).

We have now finished our examination of the mammalian kidney's mechanisms. In summary, mammalian kidney

Challenges and Controversies
The Inner Mystery

Thus far, you might conclude that the NaCl and H_2O exchanges between the two limbs of the long loops of Henle and the interstitial fluid, driven by the thick ascending limb's "single-effect" NaCl pumping, are solely responsible for the vertical osmotic gradient in the medullary interstitial fluid. Indeed, countercurrent processes are thought to be the major factors contributing to medullary hypertonicity and the resulting ability of the kidneys to concentrate urine. However, while the model of Figure 12-26 works well for the outer medulla, it does not fully account for the inner medulla where concentrations are highest, especially in the best osmoconcentrators like desert rodents, which can generate urine of up to 9,400 mOsm (Table 12-3). Mathematical modeling of the countercurrent process suggests that the basic countercurrent model does not fully explain this ability, particularly since no active NaCl transport occurs in the inner medulla. Let's look at some hypotheses.

- *Urea recycling or "solute-separation/ solute-mixing":* One major hypothesis is that NaCl passively leaves the *thin* ascending limb driven by a concentration gradient created by urea recycling. Recall that the thin ascending limbs, which lie in the inner medulla, can passively "leak" NaCl but do not actively extrude NaCl; only the thick ascending limb does this. In this hypothesis, high osmotic pressure of urea in the inner medulla pulls H_2O out of the *descending* limbs of the long loops as they plunge through the inner medulla. Osmotic removal of H_2O from the descending limbs increases the NaCl concentration inside them. As this NaCl-rich fluid enters into the NaCl-permeable, H_2O-impermeable loop and thin ascending limbs, NaCl would passively diffuse down its concentration gradient into the interstitial fluid. Anita Layton and colleagues have termed this the "solute-separation/solute-mixing" mechanism, in which NaCl and urea are *separated* in the thick ascending limb of the outer medulla, then *mixed* in the inner medullary interstitium.

This model for the enhancing effect of urea on NaCl concentration, though widely accepted, has several problems. First, mathematical models using known urea and NaCl permeabilities of key nephron segments do not completely match the model. Second, the desert sand rat (*Psammomys obesus*), which eats salty dry seeds, can concentrate urine NaCl more effectively than most rodents but is relatively poor at concentrating urea. Similarly, laboratory rats fed a high-salt, low-protein diet excrete very salty urines with very low urea in either the urine or medulla. Third, mice with UT-A genes knocked out had lower urea but the same NaCl concentrations compared to controls in their inner medullas (although whether there was still a salt gradient within that region was not determined). Finally, many birds can concentrate NaCl quite effectively with no urea at all; for example, the salt-marsh savannah sparrow (*Passerculus sandwichensis beldingi*) can produce a urine of about 2,000 mOsm (primarily due to salt) in the laboratory, an ability reflecting its diet of salty marine organisms.

- *Pelvic pumping:* Bodil Schmidt-Nielsen (p. 4), Mark Knepper, and co-workers have proposed that the renal pelvis uses *hydrostatic* pressure created by muscles to increase *osmotic* pressure. The pelvis has been observed to contract and relax rhythmically, moving the urine into and out of the papilla. This process may set up pressure gradients that help draw water out of the collecting ducts, enhancing the concentration process begun in the countercurrent loop. In support of this, osmotic pressure of the inner medulla decreases in the absence of pelvic pumping.

- *External osmolytes:* Some researchers have proposed that the generation of organic osmolytes in the inner medulla could enhance the osmoconcentration process if they were transported into the interstitial fluid. However, this hypothesis has not yet been tested.

- *Three countercurrent systems:* Several researchers have noted that the three-

dimensional configurations of the loops, collecting ducts, and vasa recta may be crucial and, moreover, are different in the upper and lower parts of the inner medulla. In this model, tubules are arranged around tightly packed vascular bundles, with the descending limbs positioned closer to the bundles. As a consequence, interactions between the tubules and vas recta are enhanced, leading to a more efficient countercurrent exchange, urea recycling, and inner medullary urea accumulation. Anita and Harold Layton, William Dantzler, and Thomas Pannabecker have proposed that three countercurrent systems (two in the upper inner medulla and one in the lower inner medulla) are required for the complete concentrating process. In the lower inner medulla, loops having wide lateral bends wrapped around collecting ducts greatly increase the surface area for NaCl release in close proximity to the ducts.

For further details on this and the other hypotheses, see the following:

Dwyer, T. M., & B. Schmidt-Nielsen. (2003). The renal pelvis: Machinery that concentrates urine in the papilla. *News in Physiological Sciences* 18:1–6.

Jen, J. F. & J. L. Stephenson. (1994). Externally driven counter-current multiplication in a mathematical model of the urinary concentrating mechanism of the renal inner medulla. *Bulletin of Mathematical Biology* 56:491–514.

Knepper, M. A., G. M. Saidel, V. C. Hascall, & T. Dwyer. (2003). Concentration of solutes in the renal inner medulla: interstitial hyaluronan as a mechano-osmotic transducer. *American Journal of Physiology* 284:F433–F446.

Layton, A. T., H. E. Layton, W. H. Dantzler, & T. L. Pannabecker. (2009). The mammalian urine concentrating mechanism: hypotheses and uncertainties. *Physiology* 24:250–256.

TABLE 12-3 Urine-Concentrating Abilities of the Kidney of Selected Mammals, and Urine Concentrations of Urea of Rats on Different Protein Diets

Mammal	Maximum Urine Osmolarity (mOsm)
Beaver	520
Pig	1,100
Human	1,400
White rat	2,900
Cat (domestic)	3,100
Kangaroo rat (North American deserts)	5,500
Hopping mouse (Australian desert)	9,400
Rabbit, normal water intake	400
Rabbit, no water for 2 days	1,300
	Urea Concentration in Urine (mM)
White rat, low-protein diet	26
White rat, high-protein diet	1,400

Sources: Data are from R. E. MacMillen, & A. K. Lee. (1967). *Science* 158:383–385; D. P. Peterson, K. M. Murphy, R. Ursino, K. Streeter, & P. H. Yancey. (1992). *American Journal of Physiology* 263:F594–F600; B. Schmidt-Nielsen & R. O'Dell. (1961). *American Journal of Physiology* 200:1119–1124; K. Schmidt-Nielsen. (1964). *Desert Animals: Physiological Problems of Heat and Water,* Oxford, UK: Clarendon Press; K. Schmidt-Neilsen. (1997). *Animal Physiology,* Cambridge, UK: Cambridge University Press; P. H. Yancey, & M. B. Burg. (1989). *American Journal of Physiology* 257:F602–F607.

A kangaroo rat, *Dipodomys ordii,* of western North American deserts. Its efficient kidneys and other adaptations allow it to live without ever drinking free water.

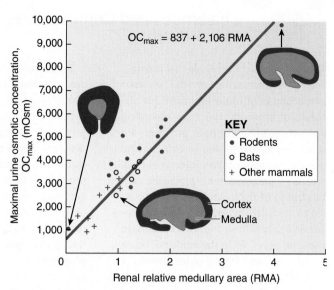

FIGURE 12-31 The relationship between relative medullary area in the mammalian kidney (taken at the midline in sagittal section) and the maximal urine concentration that can be produced. In general, proportionally larger medullas produce a more concentrated product; note that the relatively largest medullas and very concentrated urines mostly occur in rodents, which are of small body size and common in arid habitats. Insets show diagrammatic cross sections for three kidneys (not to scale).

Source: P. C. Withers. (1992). *Comparative Animal Physiology.* Fort Worth, TX: Saunders, p. 873, Figure 17.33.

nephrons use four processes—glomerular filtration, tubular reabsorption, tubular secretion, and osmoconcentration—to produce a urine of selective composition. Table 12-4 summarizes the key movements of sodium and water in these processes. Hormones—primarily aldosterone, angiotensin II, ANP, and vasopressin—regulate some of these processes, so that the final composition that is excreted helps maintain homeostasis of internal body fluids. Then there are two final stages: storage of urine in the bladder, and removal from the body.

check your understanding 12.8

Describe how countercurrent multiplication in the loop of Henle results in the osmotic gradient of the kidney medulla, and how the vasa recta prevents its dissipation.

Describe how vasopressin regulation works in dehydration and overhydration.

How does urea recycle in the kidneys, and why might this be important?

Describe how renal concentration abilities differ between a beaver and a desert rat.

TABLE 12-4 Handling of Sodium and Water by Various Nephron Segments

Tubular Segment	Na⁺ Reabsorption — *Distinguishing Features*	H₂O Reabsorption — *Distinguishing Features*
Proximal Tubule	Active; plays a pivotal role in the reabsorption of glucose, amino acids, Cl^-, H_2O, and urea	Obligatory osmotic reabsorption following active Na^+ reabsorption
Loop of Henle	Active Na^+ and Cl^- reabsorption from the thick ascending limb, along with countercurrent flow, establish the medullary interstitial vertical osmotic gradient; important in the kidneys' ability to produce urine of varying concentrations and volumes	Obligatory osmotic reabsorption from the descending limb as the thick ascending limb extrudes NaCl into the interstitial fluid (that is, reabsorbs NaCl)
Distal and Collecting Tubules	Active; variable and subject to aldosterone control; important in the regulation of ECF volume; linked to K^+ secretion and H^+ secretion; subject to angiotensin II control	Variable quantities of "free" H_2O reabsorption subject to vasopressin control; driving force is the vertical osmotic gradient in the medullary interstitial fluid established by the long loops of Henle; important in the regulation of ECF osmolarity

© Cengage Learning, 2013

12.9 Mammalian Urinary System: Bladder Storage and Micturition

In mammals, amphibians, and fishes, urine formed in the kidneys is carried through the ureters to the urinary bladder.

Urine is temporarily stored in the bladder, which is emptied by micturition

The mammalian bladder wall consists of smooth muscle lined by transitional epithelium with special cells called *umbrella cell*. Umbrella cells are joined by tight junctions (p. 66) that form an impermeable barrier. Physiologists once assumed the bladder was an inert sac. However, both its epithelium and smooth muscle actively participate in its ability to accommodate large fluctuations in urine volume. The umbrella cells can increase and decrease their surface area by the orderly process of membrane recycling as the bladder alternately fills and empties. Membrane-enclosed cytoplasmic vesicles are inserted into the plasma membrane by means of exocytosis into the surface area during bladder filling; then the vesicles are withdrawn by endocytosis to shrink the surface area after emptying. As is characteristic of smooth muscle, bladder muscle can stretch tremendously without a buildup in bladder wall tension. In addition, the highly folded bladder wall flattens out during filling, to increase bladder storage capacity.

The bladder smooth muscle is richly supplied by parasympathetic fibers, stimulation of which causes bladder contraction. If the passageway through the urethra to the outside is open, bladder contraction empties urine from the bladder. The exit from the bladder, however, is guarded by two sphincters, the *internal urethral sphincter* and the *external urethral sphincter* (Figure 12-32). The **internal urethral sphincter**—which consists of smooth muscle and is not un-

der cortical control—is not really a separate muscle but instead consists of the last portion of the bladder. Although not a true sphincter, it performs the same function. When the bladder is relaxed, the arrangement of the internal urethral sphincter region closes the outlet of the bladder. The **external sphincter** is skeletal muscle and is under cortical control via a motor neuron, which is normally active to keep the sphincter closed.

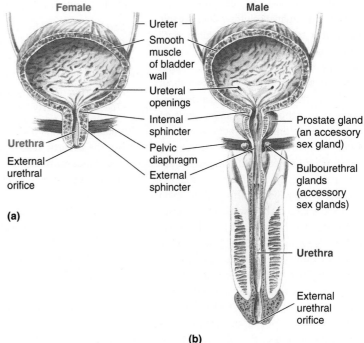

FIGURE 12-32 Comparison of the urethra in human females and males. (a) In females, the urethra is straight and short. (b) In males, the urethra, which is much longer, passes through the prostate gland and penis.

© Cengage Learning, 2013

Molecular Biology and Genomics
Surviving Salt and Urea: Organic Osmolytes and Gene Regulation

As we discussed in Chapter 3, osmosis and cell volume regulation are among the most crucial processes facing cells in general. Consider the external osmotic environment of a nephron cell in the mammalian renal medulla. The countercurrent mechanism, in a dehydrated animal, produces very high levels of NaCl and urea, threatening the cells with both osmotic shrinkage from the salt and protein damage from urea. How do the cells cope with this? It was not until after the publication of research on *osmolytes* in marine organisms that this medically important question was answered; moreover, it led to a possible treatment for cystic fibrosis.

As you will see in detail in Chapter 13, cells of most organisms subjected to high osmotic stress—such as seawater—accumulate organic solutes called **organic osmolytes** to maintain cell volume. NaCl cannot be used at high levels inside cells because it disrupts proteins and damages DNA. Medical researchers who read about organic osmolytes in marine organisms discovered that the mammalian kidney is no exception—to prevent osmotic water loss from the high NaCl in the medulla, renal cells exclude excess NaCl and instead accumulate high levels of the sugar alcohols *sorbitol* and *myo-inositol*, the sulfur amino acid *taurine*, and the methylamines *glycerophosphorylcholine (GPC)* and *glycine betaine (GB)*. These are the same chemical types of osmolytes used by marine algae and animals to balance the high osmotic pressure of seawater (see p. 617). These osmolytes have been identified in the renal medullas of all mammalian species examined (humans, dogs, rats, rabbits, and desert rodents). In all cases, the total concentrations of these osmolytes correlates with osmotic pressure of medulla fluids; that is, animals with a low water intake have increased NaCl levels in the ECF and organic osmolytes in the ICF (compared to well-hydrated animals). Notably, GPC is the only osmolyte that increases in response to an increase in medullary levels of both salt and urea; the others rise primarily in response to increased salt concentration.

Why are sugar alcohols, taurine, and methylamines used as osmolytes? Again, as you will see in Chapter 13 in detail, organic osmolytes are generally *compatible* with cellular macromolecules, that is—unlike salt ions—they do not perturb protein functions and are therefore safe to accumulate in cells over a wide concentration range. The methylamine osmolytes such as GPC provide another benefit: They strongly *stabilize* proteins, by maintaing the protein in its folded state. Urea is a noncompatible protein denaturant that unfolds proteins, and it penetrates most medullary cells. Rats on a high-protein diet, whose kidneys must concentrate very high levels of urea (Table 12-3), have a correspondingly high level of GPC in their medullary cells. Why might this be? The first clue came from the discovery that the methylated amine *TMAO* in sharks and their relatives (pp. 572 and 622) has protein-stabilizing properties that *counteract* the deleterious effects of urea (their main osmolyte) on proteins (p. 559). GPC has since been shown to act similarly.

Interestingly, kidneys do not malfunction in cystic fibrosis patients despite the presence of CFTR-type chloride channels (p. 561) in nephrons. Some researchers have speculated that organic osmolytes enable the mutant CFTR proteins to fold properly. Indeed, Marybeth Howard, William Welch, and colleagues showed that addition of renal osmolytes to CFTR cells in culture restores normal chloride transport.

How do renal cells exposed to high osmotic stress regulate the accumulation of osmolytes? Salt stress leads to the activation of genes for key enymes and transporters:

- Sorbitol increases by conversion from glucose due to the enzyme *aldose reductase (AR)*.
- GPC increases due to an enzyme that synthesizes it and by simultaneous inhibition (by both NaCl and urea) of an enzyme that breaks it down.
- GB, myo-inositol, and taurine are accumulated from external fluids via *betaine/GABA transporters (BGT1), sodium/myo-inositol transporters (SMIT),* and *taurine transporters (TauT)*, respectively.

Physiologists now know these genes share a highly similar regulatory promoter (Chapter 2) called the **osmotic response element (ORE)** or **tonicity enhancer (TonE)**. Moreover, the genes for aquaporin *AQP-2*, urea transporter *UT-A*, and *vasopressin* have TonE sequences.

Here are some actual enhancer codes "upstream" from the protein-coding genes:

Rabbit AR ORE	*CGGAAAATCAC* . . . [coding gene for AR]
Canine BGT1 TonE	*TGGAAAAGTCC* . . . [coding gene for BGT1]
Human SMIT TonE	*TGGAAAACTAC* . . . [coding gene for SMIT]
Rat UT-A TonE	*TGGAAAACTCC* . . . [coding gene for UT-A]
Mouse AQP-2 TonE	*TGGAAATTTCTT* . . . [coding gene for AQP-2]

Studies have revealed an **ORE-binding** or **TonE-binding protein** (OREBP or **TonEBP**), which functions as a *transcription factor* (Chapter 2). These regulatory proteins bind to the OREs/TonEs in a renal cell's DNA and, in antidiuretic states, activate transcription of all the osmolyte (and AQP-2 and UT-A) genes simultaneously (and presumably the vasopressin gene in the hypothalamus). Knockout of TonEBP/OREBP leads to embryonic death in mice, and the few survivors have severe atrophy of the inner medulla and cannot regulate osmolyte genes in antidiuresis. How is osmotic stress sensed and transduced into signal(s) that activate TonEBP? So far, researchers have found that high NaCl in renal cells (1) stabilizes TonEBP mRNA, leading to more protein synthesis; (2) triggers dimerization (two proteins joined) and phosphorylation of TonEBPs, making them active; and (3) activates proteins that rapidly import TonEBPs into the nucleus from the cytoplasm (where they reside at low osmotic stress). The exact processes by which NaCl does all this are actively being studied.

Micturition, or urination, the process of bladder emptying, is governed by two mechanisms: the micturition spinal reflex and cortical control. The micturition reflex is initiated when stretch receptors within the bladder wall are stimulated (Figure 12-33). The greater the distension beyond a certain minimum—250 to 400 mL in a human—the greater the extent of receptor activation. Afferent fibers from the stretch receptors carry impulses into the spinal cord and eventually, via interneurons, stimulate the parasympathetic supply to the bladder and inhibit the motor neuron supply to the external sphincter. Parasympathetic stimulation of the bladder causes it to contract. No special mechanism is required to open the internal sphincter; changes in the shape of the bladder during contraction mechanically pull the internal sphincter open. Simultaneously, the external sphincter relaxes as its motor neuron supply is inhibited. Now both sphincters are open, and urine is expelled through the urethra by the force of bladder contraction.

In addition to triggering the micturition reflex, bladder filling also sends signals to the brain cortex, giving rise to the urge to urinate. Control of micturition by the brain can override the micturition reflex so that bladder emptying can take place at the animal's convenience rather than at the time bladder filling first reaches the point of activating the stretch receptors. See p. 138 for the mechanism involved. Micturition is also an animal's normal response to an extremely stressful situation, because emptying the bladder reduces body weight.

Urination cannot be delayed indefinitely. As the bladder continues to fill, reflex input from the stretch receptors increases with time. Finally, reflex inhibitory input to the external-sphincter motor neuron becomes so powerful that it can no longer be overridden by cortical excitatory input, so the sphincter relaxes and the bladder uncontrollably empties.

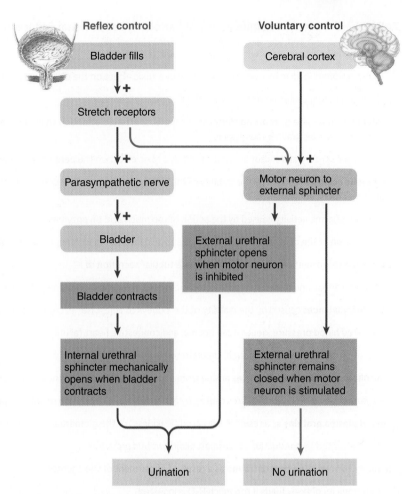

FIGURE 12-33 Reflex and higher-cortical control of micturition (mammal).
© Cengage Learning, 2013

check your understanding 12.9

Describe how the urine fills and then empties, including external regulatory features.

12.10 Renal Diseases

Urine excretion and the resulting clearance of wastes and excess electrolytes from the plasma are crucial for maintaining homeostasis. Thus, kidney pathologies are dangerous and are unfortunately relatively common ailments of aging humans and domestic cats and small dogs.

Renal diseases are numerous and have wide-ranging consequences

Renal problems have a variety of causes, some of which begin elsewhere in the body and affect renal function secondarily. Among the causes are (1) *cellular damage* resulting from excessive glucose in diabetes, or infectious organisms, either bloodborne or gaining entrance to the urinary tract through the urethra; (2) *toxic agents,* such as lead, arsenic, pesticides, or even long-term exposure to high doses of aspirin or ibuprofen in humans; (3) *inappropriate immune responses,* such as glomerulonephritis, which occasionally follows certain infections as antigen–antibody complexes are deposited in the glomeruli; (4) *obstruction* of urine flow because of the presence of infectious agents, kidney stones, tumors, cysts, or enlargement of the prostate gland, with the back pressure reducing GFR as well as damaging renal tissue; and (5) an *insufficient renal blood supply* that leads to inadequate filtration pressure. The latter can occur secondarily to diabetes or circulatory disorders such as heart failure, hemorrhage, shock, or narrowing of the renal arteries as a result of atherosclerosis.

In aging domestic cats, for example, renal disease can be due, among other things, to diabetes, parasites, bacteria, feline leukemia virus—which can trigger glomerular nephritis and tumor formation—and ingested toxins such as antifreeze from car radiators. Most antifreezes contain ethylene glycol, which is converted in the liver and kidney to a toxic product that changes the pH of the bloodstream and destroys the kidneys by depositing calcium oxalate crystals in the renal tubules. In some cat breeds, *polycystic kidney disease (PKD)* is also a common genetic problem, in which damaging cysts slowly form in the kidney with age.

TABLE 12-5 Potential Ramifications of Various Types of Renal Failure in a Mammal

Uremic toxicity caused by retention of waste products

Nausea, vomiting, diarrhea, and ulcers caused by a toxic effect on the digestive system

Bleeding tendency arising from a toxic effect on platelet function

Mental changes—e.g., reduced alertness, insomnia, shortened attention span, progressing to convulsions and coma—caused by toxic effects on the central nervous system

Abnormal sensory and motor activity caused by a toxic effect on the peripheral nerves

Metabolic acidosis* caused by the inability of the kidneys to adequately secrete H^+ that is continually being added to body fluids as a result of metabolic activity

Altered enzyme activity caused by the action of too much acid on enzymes

Depression of the central nervous system due to acid interfering with neuronal excitability

Potassium retention* resulting from inadequate tubular secretion of K^+

Altered cardiac and neural excitability due to changes in resting membrane potential

Sodium imbalances caused by the inability of the kidneys to adjust Na^+ excretion to balance the changes in Na^+ consumption

Elevated blood pressure, generalized edema, and congestive heart failure

Hypotension and, if severe enough, circulatory shock if too little Na^+ is consumed

Phosphate and calcium imbalances arising from impaired reabsorption of these electrolytes

Disturbances in skeletal structures caused by abnormalities in deposition of calcium phosphate crystals, which harden bone

Loss of plasma proteins as a result of increased "leakiness" of the glomerular membrane

Edema caused by a reduction of plasma-colloid osmotic pressure

Inability to vary urine concentration as a result of impairment of the countercurrent system

Hypotonicity of body fluids if too much H_2O is ingested

Hypertonicity of body fluids if too little H_2O is ingested

Hypertension arising from the combined effects of salt and fluid retention and vasoconstrictor action of excess angiotensin II

Anemia caused by inadequate erythropoietin production

Depression of the immune system most likely caused by toxic levels of wastes and acids

Increased susceptibility to infections

*Among the most life-threatening consequences of renal failure.

Domestic male cats are also very prone to **feline urologic syndrome (FUS)**: reduced or blocked urination because of infections, magnesium ammonium phosphate crystals, and other damage to the bladder or urethra. Crystal blockage illustrates the importance of understanding evolutionary adaptation: Cats are essentially desert animals that normally obtain most or all of their water from moist prey meat and have kidneys capable of producing urine about 10 times more hypertonic than plasma (Table 12-3). Highly concentrated urine can lead to mineral crystallization, causing FUS. However, wild cats, which as pure carnivores eat a high-meat diet, have an acidic urine that prevents crystallization. Problems occur in domestic cats that are fed foodstuffs not formulated to match natural diets (those containing an unnaturally high content of fat, nonmuscle protein, and/or carbohydrate), leading to a high-pH urine that enhances crystallization. Similar problems are present in laboratory rats, where minute fluctuations in dietary mineral, protein, and lipid content can initiate the development of renal disease.

When the functions of both kidneys are disrupted to the point that they cannot perform their regulatory and excretory functions sufficiently to maintain homeostasis, **renal failure** is said to exist. An animal may die from acute renal failure, or the condition may be reversible and lead to full recovery. Chronic renal failure, in contrast, is not reversible. Gradual, permanent destruction of renal tissue eventually proves fatal.

We will not sort out the different stages and symptoms associated with various renal disorders, but Table 12-5, which summarizes the potential consequences of renal failure, can give you an idea of the broad effects that kidney impairment can have. The extent of these effects should not be surprising, considering the central role the kidneys play in maintaining homeostasis. By the time end-stage renal failure occurs, literally every body system has become impaired to some extent.

check your understanding 12.10

Describe four general causes of renal problems, with one specific example of each.

making connections

How Excretory Systems Contribute to the Body as a Whole

Excretory organs in more complex animals contribute to internal fluid homeostasis more extensively than most other single organs. Major examples are gills in fishes and kidneys in mammals. Both help maintain the plasma constituents they regulate within the narrow range compatible with life, despite wide variations in intake and losses of these substances through other avenues. Excretory organs can sometimes serve as sensors and integrators that contribute to regulating the excretory processes by negative feedback.

Here are the specific ways in which mammalian kidneys contribute to homeostasis:

1. *Effector functions* include:
   - Regulating the quantity and concentration of most ECF electrolytes, important in osmotic balance and membrane potentials.
   - Maintaining proper plasma volume, which is important in the long-term regulation of arterial blood pressure, by controlling the salt and water balance.
   - Maintaining water balance for maintaining proper ECF osmolarity (concentration of solutes), to prevent cells from swelling or shrinking by osmosis.
   - Excreting the often-toxic end products of metabolism (such as ammonia, urea, uric acid), as well as many foreign compounds that gain entrance to the body.
   - Conserving useful substances, such as glucose, and water in animals subject to dehydration.
   - Contributing to maintenance of proper pH (Chapter 13).
2. *Integrator/hormonal functions* include:
   - Secretion of erythropoietin (EPO) to stimulate red blood cell production (p. 393).
   - Secretion of renin to initiate the renin–angiotensin–aldosterone pathway for controlling renal tubular Na^+ reabsorption.
3. *Metabolic functions:*
   - The kidneys help convert vitamin D into its active form. Vitamin D is essential for Ca^{2+} absorption from the digestive tract.

Chapter Summary

- Selective excretion is crucial to internal fluid homeostasis and involves respiratory, digestive, integumentary, and renal systems.

- Nitrogen metabolism creates special stresses and produces three major end products: ammonia (metabolically "cheap" but toxic; mainly in aquatic organisms), urea (moderately costly and toxic; in elasmobranchs, adult amphibians, and mammals), and uric acid (nontoxic but costly; in insects, reptiles including birds, and mammals).

- Excretory organs have transport epithelia, which typically use Na^+/K^+ ATPases to set up cation gradients, which can be used to move other substances secondarily.

- Renal tubules produce urine using the processes of filtration, secretion, reabsorption, and osmoconcentration. The major renal organs are protonephridia, meso- and metanephridia, and Malpighian tubules.

- Malpighian tubules in insects initiate excretion by ion secretion, which causes osmosis. The tubules and hindgut modify the lumen fluid by specific secretion and reabsorption. Insect excretion is regulated by diuretic and antidiuretic hormones.

- Vertebrate kidneys form the urine; the remainder of the urinary system is ductwork that carries the urine to the bladder or hindgut. The nephron is the functional unit of the vertebrate kidney, using the basic renal processes. The glomerulus and Bowman's capsule perform ultrafiltration; the proximal and distal tubules perform specific reabsorption and secretion; and the loop of Henle and collecting duct perform osmoconcentration (only in birds and mammals).

- Elasmobranch fishes retain urea and trimethylamine oxide (TMAO) as osmolytes, and use gills, kidneys, and rectal glands for excretion and retention. Marine bony fishes use gills for most excretion and retention processes. Freshwater bony fishes excrete water with their kidneys, and their gills excrete wastes and take up salts.

- Terrestrial vertebrates evolved new renal and extrarenal mechanisms to maintain an "internal ocean." Amphibians use kidneys and bladders for excretion and retention. Nonavian reptiles use amphibian-like kidneys, hindguts, and (in marine and desert species) salt glands for excretion and retention. Birds have kidneys with reptile- and mammal-like nephrons, hindguts, and (in marine species) salt glands for excretion and retention. Mammalian (and avian) kidneys have cortical and juxtamedullary nephrons.

- For glomerular filtration, the glomerular membrane has three layers forming a fine molecular sieve. The glomerular capillary blood pressure is the major force that causes glomerular ultrafiltration. Changes in the glomerular filtration rate or GFR result primarily from changes in glomerular capillary blood pressure.

- Tubular reabsorption involves passive and active transepithelial transport and is tremendous, highly selective, and variable. Sodium reabsorption is essential for the reabsorption of nutrients, Cl^-, and H_2O, and for the regulation of ECF osmolality and volume. Glucose and amino acids are reabsorbed by Na^+-dependent secondary active transport. Most actively reabsorbed substances exhibit a tubular maximum. Glucose reabsorption is not regulated by the kidneys and has a maximum important in diabetes. PO_4^{3-} (phosphate) reabsorption, in contrast, is regulated. In general, unwanted waste products, except some urea, are not reabsorbed.

- The renin–angiotensin–aldosterone system (RAAS) stimulates Na^+ reabsorption in distal tubules and collecting ducts, elevates blood pressure, and stimulates thirst and salt hunger. Atrial and brain natriuretic peptides (ANP, BNP) antagonize the RAAS, inhibiting Na^+ reabsorption and reducing blood pressure.

- In tubular secretion, H^+ secretion is important in acid–base balance. K^+ secretion is controlled by aldosterone. Organic anion and cation secretion helps eliminate foreign compounds from the body. Plasma clearance, the volume of plasma cleared of a particular substance per minute, is used to characterize secretion.

- In osmoconcentration, the ability to excrete urine of varying concentrations varies among species and depends on hydration state. A medullary vertical osmotic gradient is established by countercurrent multiplication with the thick ascending limb of the loop pumping out NaCl without water following. The interstitial NaCl buildup draws water out of the descending limb. In dehydration, vasopressin from the hypothalamus–pituitary triggers insertion of aquaporins in the collecting duct. Water is then drawn out by the medullary gradient to conserve water (antidiuresis). In overhydration, no vasopressin is released, so the collecting duct retains its water, excreting excess from the body (diuresis).

- Countercurrent exchange within the vasa recta conserves the medullary vertical osmotic gradient. Passive urea recycling in the renal medulla contributes to medullary hypertonicity and prevents urea diuresis. Different osmoconcentrating abilities among species depend on nephron anatomy and metabolic rates, with animals having long loops and high metabolic rates capable of the highest levels of osmoconcentration.

- Urine is temporarily stored in the bladder, which is emptied by micturition.

- Renal diseases such as feline urologic syndrome are numerous and have wide-ranging consequences.

Review, Synthesize, and Analyze

1. Distinguish between *secretion* and *excretion*.

2. Compare and contrast insect Malpighian tubules and mammalian nephrons. In doing so, describe all tubular transport processes that are linked to the basolateral Na^+/K^+ ATPase in nephrons and the V-ATPase in Malpighian tubules.

3. The Gulf toadfish *Opsanus beta* (p. 561) switches from ammonotely to ureotely when crowded or confined, releasing urea in pulses. Hypothesize why it might do this.

4. Explain *tubular maximum* (T_m) and *renal threshold*. Compare two substances that display a T_m: one substance that *is* and one that *is not* regulated by the kidneys.

5. Calculate an animal's rate of urine production, given an inulin clearance of 75 mL/min and urine and plasma concentrations of inulin of 300 mg/liter and 3 mg/liter, respectively.

6. If the urine concentration of a substance is 7.5 mg/mL of urine, its plasma concentration is 0.2 mg/mL of plasma, and the urine flow rate is 2 mL/min, what is the clearance rate of the substance? Is the substance being reabsorbed or secreted by the kidneys?

7. If the plasma concentration of substance X is 200 mg/100 mL and the GFR is 125 mL/min, what is the filtered load of this substance? If the T_m for substance X is 200 mg/min, how much of the substance will be reabsorbed at a plasma concentration of 200 mg/100 mL and a GFR of 125 mL/min? How much of substance X will be excreted?

8. *Conn's syndrome* is an endocrine disorder in some older domestic cats brought about by excessive growth of the adrenal cortex (sometimes due to a tumor). The result is secretion of unregulated excessive aldosterone. Given what you know about the functions of aldosterone, describe what the most prominent features of this condition would be.

9. Because of a mutation, an animal was born with an ascending limb of Henle that was water permeable. What would be the minimum/maximum urine osmolarities (in units of mOsm) the animal could produce?

10. A dog suffers permanent damage of the lower spinal cord in an accident and is paralyzed from the waist down. Describe what governs bladder emptying in this animal.

Suggested Readings

Beyenbach, K. W., H. Skaer, & J. A. T. Dow. (2010). The developmental, molecular, and transport biology of Malpighian tubules. *Annual Review of Entomology* 55: 351–374.

Burg, M. B., J. D. Ferraris, & N. I. Dmitrieva. (2007). Cellular response to hyperosmotic stresses. *Physiological Reviews* 87: 1441–1474. A review of genes and osmolytes.

Fenton, R. A., & M. A. Knepper. (2007). Urea and renal function in the 21st century: Insights from knockout mice. *Journal of the American Society of Nephrology* 18: 679–688.

Hazard, L. C. (2001). Ion secretion by salt glands of desert iguanas (*Dipsosaurus dorsalis*). *Physiological and Biochemical Zoology* 74:22–31.

Jeon, U. S., J.-A. Kim, M. R. Sheen, & H. M. Kwon. (2006). How tonicity regulates genes: Story of TonEBP transcriptional activator. *Acta Physiologica* 187:241–247.

Ruppert, E. E., & P. R. Smith. (1988). The functional organization of filtration nephridia. *Biological Reviews* 63:231–258. A review of the evolution of filtration tubules.

Schmidt-Nielsen, K. (1964). *Desert Animals: Physiological Problems of Heat and Water*. Oxford, UK: Clarendon Press. A classic text on desert animal physiology.

Simoyi, M. F., K. Van Dyke, & H. Klandorf. (2002). Manipulation of plasma uric acid in broiler chicks and its effect on leukocyte oxidative activity. *American Journal of Physiology* 282:R791–R796. On the relationship between concentrations of uric acid and oxidative stress in birds.

Wright, P., & P. Anderson, eds. (2001). *Fish Physiology*. Vol. 20: *Nitrogen Excretion*. New York: Academic Press.

Wright, P. (1995). Nitrogen excretion: Three end products, many physiological roles. *Journal of Experimental Biology* 198:273–281.

CourseMate

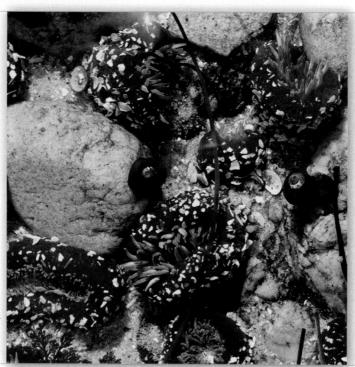

Paul Yancey

Intertidal anemones and snails. Both animals and their relatives (marine Cnidarians and mollusks, respectively) are osmoconformers; that is, the osmotic pressure of their body fluids is equal to that of the environment. The snails have $CaCO_3$ shells that form well at the naturally basic pH of seawater, but their shells may now be weakening from increasing acidity resulting from anthropogenic CO_2.

Fluid and Acid–Base Balance

13.1 Introduction

The concept of homeostasis as first delineated by Claude Bernard and Walter Cannon (p. 11) focused on the *milieu interieur*, that is, the *extracellular fluid (ECF)* that bathes all cells of an animal. The idea was that individual cells could be spared the cost of struggling against disturbances to their *intracellular fluids (ICF)* if the ECF is regulated on their behalf. The composition of the ICF and ECF results from a variety of key molecular components: the solvent water and its solutes—salts, nutrients such as glucose, proteins, and other organic molecules, ions such as calcium for shells and skeletons, and hydrogen ions. In accordance with the original homeostasis concept, many animals can regulate these constituents at the level of the ECF. The *total concentration* of all solutes, which determines osmotic pressure; the *concentration of individual solutes;* and the total fluid *volume* are the key components that need to be regulated.

As you have seen in Chapter 12, excretory organs play a key role in such regulation. However, single excretory organs such as kidneys do not regulate every aspect of the ECF. Other organs contribute (CO_2, for example, is mainly regulated by respiratory

systems), and the ECF may vary in composition in some types of animals, such that individual cells must cope on their own. Overall, though, critical features of cellular homeostasis are still achieved, as you will see. Therefore, the regulation of the ECF and ICF composition is not attributable to one single physiological system. Rather, it is an inherently integrative process involving individual cells and more than one organ system. In Chapter 12, we focused primarily on the excretion and retention of specific solutes (salts, wastes) and water. In this chapter, we look at fluid-related regulation at a more integrative level by focusing on three aspects of the ECF and ICF that are crucial to the whole animal and to its individual cells: (1) the *osmotic balance,* determined primarily by the total concentration of major solutes; (2) the total *fluid volume* of the body, which directly affects circulatory pressure; and (3) the *acid–base balance.* Hydrogen ions alter the acid–base status (measured as pH) of body fluids, a factor crucial to the functioning of proteins.

Before we examine these, however, let's look at how the composition of the ECF can change in basic terms.

If balance is to be maintained, input must equal output

The quantity of any particular substance in the ECF is considered a readily available internal pool. The amount of the substance in the pool may be increased either by transferring more in from the external environment (such as via the gut or the integument) or by metabolically producing it within the body (Figure 13-1). Substances may be removed from the body by loss to the outside (through excretion, sweating, and so forth) or by being used up in a metabolic reaction. If the quantity of a substance is to remain stable within the body, its input must be balanced by an equal output by means of excretion or metabolic consumption. This relationship, known as the **balance concept,** is extremely important in maintaining homeostasis. Not all input and output pathways are applicable for every body fluid constituent. For example, salt is not synthesized or altered metabolically by organisms, so the stability of salt concentration in

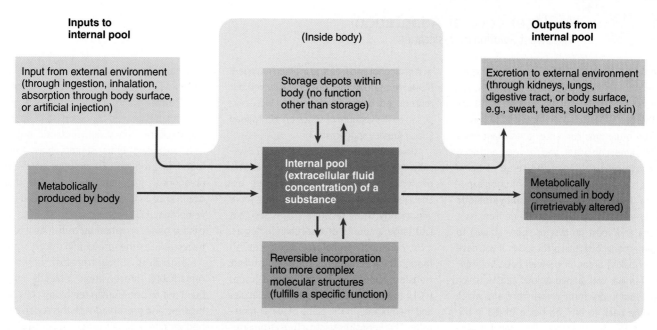

FIGURE 13-1 Inputs to, and outputs from, the internal pool of a body constituent.
© Cengage Learning, 2013

body fluids depends entirely on a balance between salt intake and salt excretion.

The ECF pool can further be altered by transferring a particular ECF constituent into storage within cells or certain extracellular features (such as shells, bladder fluid). If a body as a whole has a surplus or deficit of a particular stored substance, the storage site can be expanded or partially depleted to maintain the ECF concentration of the substance within homeostatically prescribed limits. For example, some amphibians use their bladders for temporary water storage (see p. 574). Some storage involves chemical conversions; for instance, after a mammal has absorbed a meal, when more glucose is entering the plasma than is being consumed by the cells, the excess glucose can be temporarily stored in liver cells in the form of glycogen. This storage depot can then be tapped between meals as necessary to maintain the plasma glucose level when no new nutrients are being added to the blood by eating. It is important to recognize, however, that internal storage capacity is limited. Although an internal exchange between the ECF and a storage depot can temporarily restore the plasma concentration of a particular substance to normal, in the long run any excess or deficit of that constituent must be compensated for by appropriate adjustments in total body input or output.

Some exchanges between the pool and the remainder of the body involve structures with functions beyond mere storage. Consider calcium: It is an essential component of an animal's skeleton or shell for support and protection, but it is also is necessary as a cellular messenger and trigger of muscle contraction. If plasma and cell calcium fall too low, calcium may be taken out of the shell or skeleton. This process differs from simple storage, which serves no additional purpose.

Some body constituents may be poorly controlled. For example, the plasma volume and composition of many *osmoconformers* (animals whose internal osmotic pressure is equal to that of the environment; see p. 616) passively changes with the external environment, forcing individual cells to adjust at least in part on their own. Some constituents undergo frequent and/or large disturbances. Salt and H_2O can be lost to the terrestrial environments to varying degrees through the digestive tract (vomiting, diarrhea), skin (sweating), lungs (panting), and elsewhere without regard for salt or H_2O balance in the body. Hydrogen ions are uncontrollably generated internally (especially during muscle activity) and added to body fluids. If possible, compensatory adjustments must be made for these uncontrolled changes.

Body water is distributed between the intracellular and extracellular fluid compartments

Water is by far the most abundant component of most animals, though the proportion varies considerably. For example, water is about 60% of mammalian body weight on average, and it can be more than 90% in some soft aquatic animals. Yet water content may drop well below 10% in certain **anhydrobiotic** animals, such as encysted tardigrades, rotifers, and brine shrimp embryos that have entered a dormant state. However, anhydrobiotic states exhibit no detectable metabolism, illustrating that normal metabolic functions are associated with high water content. (Note that some anhydrobiotic stages of organisms can remain viable under harsh conditions for years, even many millenia in the case of some prokaryotes. It is not fully understood how they accomplish this. See *A Closer Look at Adaptation: The Sweet Solution to Desiccation*.)

Among the major body components, plasma, as you might suspect, is the most watery, typically more than 90% H_2O. But even soft tissues such as skin, muscles, and internal organs consist of 70 to 80% H_2O. Drier structural materials such as shells and skeletons may be only about 20% H_2O. Fat and external coverings such as hair are the driest tissues of all, having only 10% H_2O content or less. These drier examples include many metabolically inactive components; in contrast, the active cytosols and organelles of the cells within shells, skeletons, and fat tissue have high water

A Closer Look at Adaptation
The Sweet Solution to Desiccation

Biologists routinely talk about life being incompatible with the absence of water. Although this is basically correct (at least for life on Earth), a remarkable group of unrelated organisms can enter a dormant state with almost no water, and for years and even decades can survive in the sense of being able to restore normal functions upon exposure to free water. These **anhydrobiotic** ("no-water life") organisms range from bacteria and yeast (in cyst or spore stages) to nematodes to certain arthropods. If you have ever baked bread or brewed beer, you may be familiar with anhydrobiotic yeasts, which are sold in dry form in the store and which "come back to life" on being mixed in the moist dough or beer mash. Less familiar, perhaps, are the brine shrimp *Artemia,* sold as a novelty item often called "sea monkeys." As adults, these inhabit highly saline lakes such as the Great Salt Lake in Utah. However, they and their eggs can dry out if washed ashore or if a desert pond they inhabit dries out. This kills the adults, but an early embryonic stage can encyst and survive for years with less than 2% water internally. The Arctic springtail (*Megaphorura arctica*) also survives winter by drying up to prevent freezing. More remarkably, **tardigrades** (see photograph), small, segmented, exoskeletal animals found in ponds and forest litter, can encyst as adults with less than 1% water and survive for years and under harsh treatments such as immersion in alcohol and freezing in liquid nitrogen! Moreover, tardigrades not only survived but reproduced normally after they were exposed to the vacuum of space on a European Space Agency spacecraft.

How can these organisms do this? Typically, they have an impermeable outer cover of dense proteins or wax to reduce exchanges with the environment. But more importantly, these resistant stages accumulate large amounts of disaccharide sugars, most commonly **trehalose** (diglucose). Although the mechanism is still being studied, trehalose appears to use its hydroxyl groups (—OH) to hydrogen bond to macromolecules and membranes in the place of water molecules. This keeps these structures from irreversible damage (although they are nonfunctional until water returns). The sugar not only stabilizes proteins and membranes, but also turns into a glasslike material that holds the cytoplasm and all its constituents in a smooth, undamaged condition (see Chapter 15). Recently, many anhydrobiotic organisms, including brine shrimp, have been found to produce *LEA* proteins (late embryogenesis abundant), named after their occurrence in late stages of plant seed development. LEA proteins are thought to facilitate glass formation by trehalose and to help stabilize other proteins in the dry state.

Regardless of how trehalose works, its remarkable preservation abilities have been put to use: Human red blood cells can be dried and preserved with this sugar, carried around in a lightweight form (such as by a soldier), and then rehydrated into usable form when needed for transfusions.

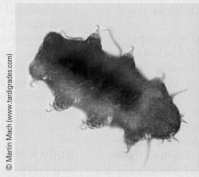 

An active tardigrade (*left*) and a dormant encysted one (*right*).

Source: M. Mach, *Tardigrades*, www.tardigrades.com, accessed March 19, 2004.

---

contents, but those cell spaces are a minor component of those tissues.

Major Fluid Compartments Body H_2O is distributed between two major fluid compartments: the ICF and the ECF (Figure 13-2). Table 13-1 shows values for a mammal (human). The ICF compartment comprises about two thirds of total body H_2O in insects and vertebrates, but only about one third to one half in marine mollusks and crustaceans. Even though each contains its own unique mixture of constituents, these numerous minute fluid compartments are similar enough to be considered collectively as one large fluid compartment.

FIGURE 13-2 A model of the major body compartments in an animal and exchange routes (arrows) between them. ECF, extracellular fluid; ICF, intracellular fluid.

Source: P. Willmer, G. Stone, & I. Johnson, (2000), "Environmental Physiology of Animals", Figure 1.8. Copyright © 2000 Blackwell Publishing. Reproduced with permission of Blackwell Publishing Ltd.

TABLE 13-1 Classification of Body Fluid for a Human

Compartment	Volume of Fluid (in liters)	Percentage of Body Fluid	Percentage of Body Weight
Total Body Fluid	42	100%	60%
Intracellular Fluid (ICF)	28	67	40
Extracellular Fluid (ECF)	14	33	20
Plasma	2.8	6.6 (20% of ECF)	4
Interstitial fluid	11.2	26.4 (80% of ECF)	16
Lymph	Negligible	Negligible	Negligible
Transcellular fluid	Negligible	Negligible	Negligible

© Cengage Learning, 2013

The remaining body H_2O—in the ECF compartment—is primarily the hemolymph in animals with open circulatory systems (Chapter 9). In animals with closed circulation, such as cephalopods and vertebrates, the ECF is subdivided primarily into plasma and interstitial fluid. Vertebrate plasma, which makes up about 15% (in bony fishes) to 40% (in birds) of the ECF volume, is the fluid portion of the blood. The interstitial fluid, which represents most of the remaining ECF compartment, constitutes the true internal environment that bathes the tissue cells.

Minor ECF Compartments Some animals have two other relatively minor ECF fluids. *Lymph*, found in most vertebrates, is fluid being returned from the interstitial fluid to the plasma via the lymphatic system (Chapter 9). **Transcellular** fluids consist of a number of small specialized fluid volumes that are secreted by specific cells into a particular body cavity to perform some specialized function. In vertebrates, these include *cerebrospinal fluid* (p. 169); *intraocular fluid* (maintaining the shape of and nourishing the eye); *synovial fluid* (lubricating the joints); and *pericardial, intrapleural,* and *peritoneal* fluids (lubricating movements of the heart, lungs, and intestines, respectively). Although important functionally, these fluids usually represent an insignificant fraction of the total body H_2O (Table 13-1). Furthermore, many transcellular fluids do not reflect changes in the body's fluid balance. For example, the mammalian cerebrospinal fluid does not decrease in volume when the body as a whole is experiencing a negative H_2O balance. Therefore, the transcellular compartment can usually be ignored when analyzing fluid balance.

Finally, a fluid compartment found in all animals is that which is produced by the digestive tract. Technically this is part of the external environment because the digestive tract connects to the outside world. However, this fluid contains components from the body (such as digestive enzymes, and sometimes water) and can be particularly dynamic in ruminant and postgastric fermenters (see Chapter 14). It is usually a small percentage of the ECF but cannot always be ignored, because it potentially can be major source of gains and losses of water and solutes.

In vertebrates, the plasma and interstitial fluid are similar in composition, but the ECF and ICF are markedly different in all animals

Several barriers separate vertebrate body fluid compartments, limiting the movement of H_2O and solutes between the various compartments to differing degrees. The two components of the ECF—plasma and interstitial fluid—are separated by the walls of the blood vessels. However, H_2O and all plasma constituents with the exception of plasma proteins are continuously and freely exchanged between the plasma and the interstitial fluid by passive means across capillary walls (p. 440). Accordingly, the two fluids are nearly identical in composition, except that interstitial fluid lacks plasma proteins. Any change in one of these ECF compartments is quickly reflected in the other compartment.

In contrast to the close similarity between plasma and interstitial fluid, the composition of the ECF as a whole differs considerably from that of the ICF (Figure 13-3) in all animals. Each cell is surrounded by a highly selective plasma membrane that permits passage of certain materials while excluding others. Movement through the membrane barrier occurs by both passive and active means and may be highly discriminating. Among the major differences between the ECF and ICF are:

1. *Cellular proteins* that cannot permeate the plasma membranes to leave the cells.
2. *Cellular organic osmolytes* (see p. 617; typically higher in the ICF than in the ECF).
3. *Na^+ and K^+* and their attendant anions: In most organisms, Na^+ is the primary ECF cation, and K^+ is the primarily ICF cation. This is due in part to action of the membrane-bound Na^+/K^+ ATPase pump that is present in all cells (although this is not the only factor; see p. 103). The unequal distribution of Na^+ and K^+, coupled with differences in membrane permeability to these ions, is responsible for the electrical properties of cells, including the action potentials in excitable tissues (see Chapters 3 and 4).

Except for the extremely small electrical imbalance in the intracellular and extracellular ions involved in membrane potential, the majority of the ECF and ICF ions are electrically balanced. In the ECF, Na^+ is accompanied primarily by the anion Cl^- and to a lesser extent by HCO_3^- (bicarbonate). In the ICF, K^+ is accompanied primarily by the anion PO_4^{3-} (phosphate) and by the negatively charged proteins trapped within the cell.

Although all cells' plasma membranes display selective permeability, cells are typically permeable to H_2O. The movement of H_2O between the plasma and the interstitial fluid across capillary walls is governed by relative imbalances between capillary blood pressure (a fluid, or hydrostatic, pressure) and colloid osmotic pressure (see p. 443). In contrast, the net transfer of H_2O between the interstitial fluid and the ICF across plasma membranes occurs as a

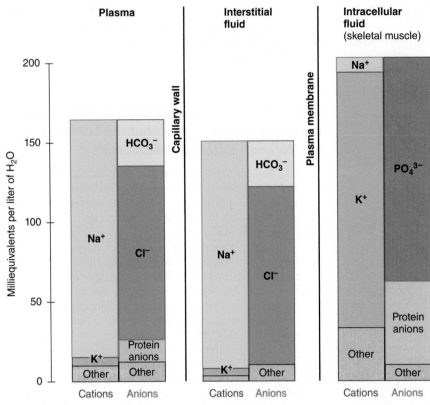

FIGURE 13-3 Ionic composition of the major body fluid compartments in a mammal.

© Cengage Learning, 2013

Osmotic problems that threaten cells and animals include salinity, evaporation, ingestion and excretion, freezing, and certain pathologies

One problem that all animal cells face is a potential swelling arising from the unequal Na^+ distribution in the ECF and ICF. Because ECF Na^+ has both an inward electrical and a chemical gradient (Chapter 4), Na^+ leaking into cells raises the solute concentration and thus can elevate osmotic pressure, which in turn would cause water to enter the cell by osmosis. The primary, and nearly universal, mechanism to correct this situation is the Na^+/K^+ ATPase "pump." By exporting more Na^+ than the K^+ it imports (3 Na^+ out for each 2 K^+ in), the pump not only maintains ion gradients but also reduces total cellular osmotic pressure.

Beyond this basic problem of cells, internal osmotic balance can be disturbed in several major ways. The ECF can become too concentrated or too dilute because of the following:

result of osmotic effects alone because these fluids have very low hydrostatic pressures.

check your understanding 13.1

Explain the balance concept.

Outline the distribution of animal body H_2O in the various fluid compartments.

Compare the ionic composition of plasma, interstitial fluid, and intracellular fluid.

13.2 Osmotic and Volume Balance: Overview and Osmoconformers

As you saw in Chapter 3, osmotic forces are among the crucial factors with which cells must cope. Lacking true cell walls as found in bacteria, plants, algae, and fungi, most cells of animals become osmotically equalized with the ECF that bathes them. Thus, the osmotic composition of the ECF directly affects cell volume homeostasis. Recall that water molecules alone apparently cannot be moved directly by ATPase pumps (p. 561), but rather must passively follow the movement of solutes. Inorganic ions, especially NaCl, are dominant in this process (see Figure 12-3).

1. *Salinity* of the environment, probably the first osmotic stress faced by the earliest life. Marine organisms must prevent dehydration and limit excess salt influx in high salinity, while freshwater organisms have the opposite problems. Organisms in estuaries and aquatic migrators such as salmon (which move between freshwater and the ocean) face both situations.

2. *Evaporation* of body water into air: The evolution of terrestrial organisms created this problem, sometimes accompanied by salt loss, through processes such as breathing, sweating, panting, transpiration, and general loss through outer layers.

3. *Ingestion and excretion*: Intake of water and solutes through the digestive tract may not be of the correct amounts and types for an animal body, while excretion of waste products often requires water, a problem more critical for terrestrial than for aquatic organisms.

4. *Freezing*, which locks up water in ice crystals and concentrates ions in the residual unfrozen water.

5. *Pathologies* such as diabetes in which ECF osmolarity increases because of excessive glucose levels, cystic fibrosis in which salt transport is impaired, and *hyponatremia* (low blood sodium) due to excessive water drinking.

Changes in the ECF osmolarity in turn can cause cells to swell or shrink. Let's use environmental salinity to illustrate how disturbances can occur. The universal dissolved constituents of cells—macromolecules, metabolites, K^+, and so forth—typically yield an osmotic concentration of roughly 300–400 milliosmolar (mOsm) in most organisms (this does not include nonuniversal *organic osmolytes*, which we discuss shortly). Recall that open ocean contains about 3.5%

salt, yielding about 1,000 mOsm (see p. 79). *Brackish* waters have less than 3.0% but greater than 0.5%. The lowest salinities are considered freshwater, approximating as low as 0 mOsm; in contrast, some brine lakes such as the Dead Sea in Israel can reach over 20% salt at 6,000 mOsm! In any environment greater than 300–400 mOsm, organisms must have adaptations to maintain cell volume in the face of this potentially dehydrating force. In contrast, lower osmolarities threaten to overhydrate organisms.

Animals have evolved two strategies to cope with osmotic challenges: osmoconforming and osmoregulating

How do animals maintain a proper balance of ICF and ECF volumes and compositions in the face of different environments? Adaptations include a cell membrane (or the extracellular layer next to it) that is impermeable to water, an option that turns out to be extremely limiting because it reduces the ability of a cell to exchange vital solutes. This solution has evolved in some epithelial layers exposed to osmotically challenging environments and is usually achieved with extracellular waterproofing layers on one side of a cell layer. For example, epidermal cells of insects secrete a noncellular cuticle of waterproof *waxes,* whereas the epidermis of terrestrial vertebrates produces a *keratinized* layer with low permeability (p. 466). Another solution is found in some freshwater sponges, which have special water-pumping organelles called *contractile vacuoles* (p. 557). However, most animal cells, lacking cell walls, impermeable membranes, and contractile vacuoles, typically must have the total concentration of molecules in their ICF equal to that of the ECF. Thus, in the long run there are no osmotic gradients between most animal cells' ICF and the nearby ECF.

What solutions have animals evolved to limit swelling or shrinking of their cells, given that the external environment may have quite a different osmotic concentration and that perturbing changes in the environment may alter the osmotic state of the ECF (and thus alter the ICF)? There are two very different strategies of maintaining osmotic balance (Figure 13-4a):

- **Osmoconformers:** Body fluids and cells are generally equal in osmotic pressure to the environment (most commonly salty waters), with adjustments to both ECF and ICF made using small molecules called *osmolytes.*
- **Osmoregulators:** The osmotic pressure of body fluids is homeostatically regulated and typically very different from that of the external environment.

As shown in Figure 13-4b and c, there are varying degrees of osmoconforming and osmoregulating. Let's look at osmoconformers first.

Osmoconformers use compatible and counteracting organic osmolytes in their cells

Osmoconformers have little tendency to gain or lose water, except when the environment changes. The main examples of osmoconformers are found in the oceans, where osmolarity averages 1,000 mOsm. Most marine organisms—especially soft-bodied ones, which have no impermeable coverings to reduce osmotic forces—are also about 1,000 mOsm internally. In these animals, the ECF is very similar to seawater. Na^+ and Cl^- are critical to maintaining osmotic balance and are considered **inorganic osmolytes** as well as electrolytes. What about the ICF? Following small, short-term disturbances to the ECF, cells often do alter their internal K^+ and other ion levels to prevent swelling or shrinking. But for larger and long-term osmotic challenges, the story is quite different. Instead, ICF osmotic pressures are raised predominantly with certain organic solutes called *compatible* and *counteracting* **organic osmolytes.** Thus, a marine osmoconformer has:

- An ECF at the same osmotic pressure as seawater (e.g., 1,000 mOsm), dominated by NaCl.
- An ICF with the same osmotic pressure, but due to roughly 400 mOsm of universal solutes (K^+, macromolecules, and so forth) and about 600 mOsm of organic osmolytes.

Structures of common organic osmolytes are shown in Figure 13-5, and distributions in the animal kingdom are listed in Table 13-2. Note that organic osmolytes fall into a few broad categories of chemical types:

- *Carbohydrates* such as polyols and sugars (e.g., myoinositol in mammalian kidneys).
- *Free amino acids* such as glycine and taurine (an unusual, sulfur-based amino acid found in many marine animals and also mammalian kidney and brain).
- *Methylamine* solutes. A common example in marine animals is TMAO (**trimethylamine oxide**). Whether you know it or not, you are probably quite familiar with this compound: Its breakdown is the primary cause of "fishy" smell in many marine animals. On the animal's death, bacteria convert TMAO into a volatile gas, *TMA (trimethylamine),* which has a strong odor of rotten fish; and minor amounts of TMA are present in live animals. Indeed, the marine food industry monitors TMA in fish flesh as an indicator of spoilage.
- *Urea (and other nitrogenous wastes).* As we noted in Chapter 12, urea is the major organic osmolytes of Chondrichthyes (p. 572). Mammalian kidneys also concentrate it for excretion and may use it as an osmolyte to help the concentrating mechanism (p. 603).
- *Methylsulfonium* solutes. The main example is DMSP (dimethylsulfoniopropionate), a major osmolyte in many marine algae and also found in some marine animals such as corals (Cnidaria) with symbiotic unicellular algae. DMSP is also familiar to most people who have been to an ocean beach: Its breakdown product is a gas DMS (dimethyl sulfide), which is the sweet sulfurous odor of decaying seaweed often called "the smell of the sea." DMS and DMSP are also major components of the Gaia hypothesis for a homeostatic planet Earth (see *Challenges and Controversies: Can a Planet Have Physiology?* p. 14, Chapter 1).

Why do osmoconformers use these particular organic solutes as osmolytes, which cost metabolic energy, rather than the readily available inorganic ions? At first glance, it might seem that simply moving the high external NaCl of the ECF into the cell would easily eliminate osmotic imbalances. However, this does not appear feasible, for two reasons.

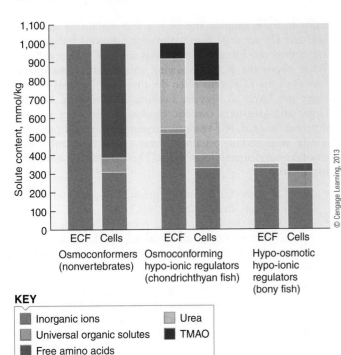

FIGURE 13-4 **Osmotic adaptations of marine animals.** (a) Generalized osmotic contents of three different types of adaptation in marine animals. TMAO is trimethylamine oxide. Osmoconformers use primarily free amino acids as organic osmolytes in their cells, while osmoconforming hypo-ionic regulators use primarily urea and TMAO. In contrast, hypo-osmotic hypo-ionic regulators have very low levels of organic osmolytes. (b) General categories of responses of animal body fluids to variations in external concentrations, with (c) examples.

Sources: (b, c) P. Willmer, G. Stone, & I. Johnson, (2000), "Environmental Physiology of Animals", Figure 5-2. Copyright © 2000 Blackwell Publishing. Reproduced with permission of Blackwell Publishing Ltd.

© Cengage Learning, 2013

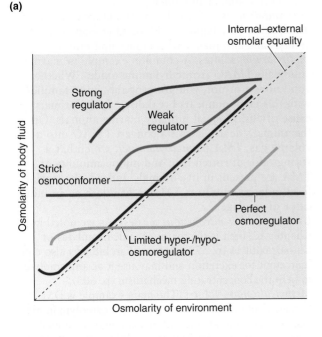

(b) Principles

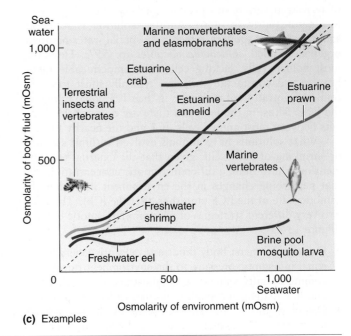

(c) Examples

1. Many cells use Na^+ gradients for useful, basic functions such as coupled transport and signal conduction, whereas movement of Cl^- ions into the cell is unfavorable because of the cell's negative charge (p. 108).

2. More universally, inorganic ions (especially Na^+ and Cl^-) at high concentrations can disrupt the structure and function of macromolecules, in part because of charge interactions (Figure 13-6). Were NaCl to accumulate in cells to the level of the ECF and seawater, the conformation of many proteins would become abnormal. Moreover, exposure of a variety of eukaryotic cells to elevated NaCl leads to double-strand breaks in DNA.

Of the common inorganic ions, K^+ disrupts macromolecules least, perhaps explaining why it is the universal cellular cation; but it too is disruptive at high concentrations (Figure 13-6).

Organic osmolytes differ from inorganic ions in two important ways:

- *Compatibility:* Most organic osmolytes do not disturb macromolecules even at high concentrations, and so are termed **compatible solutes** (Figure 13-6). (A major exception is urea, which we discuss later.) In general, compatible solutes can be safely up-regulated or down-

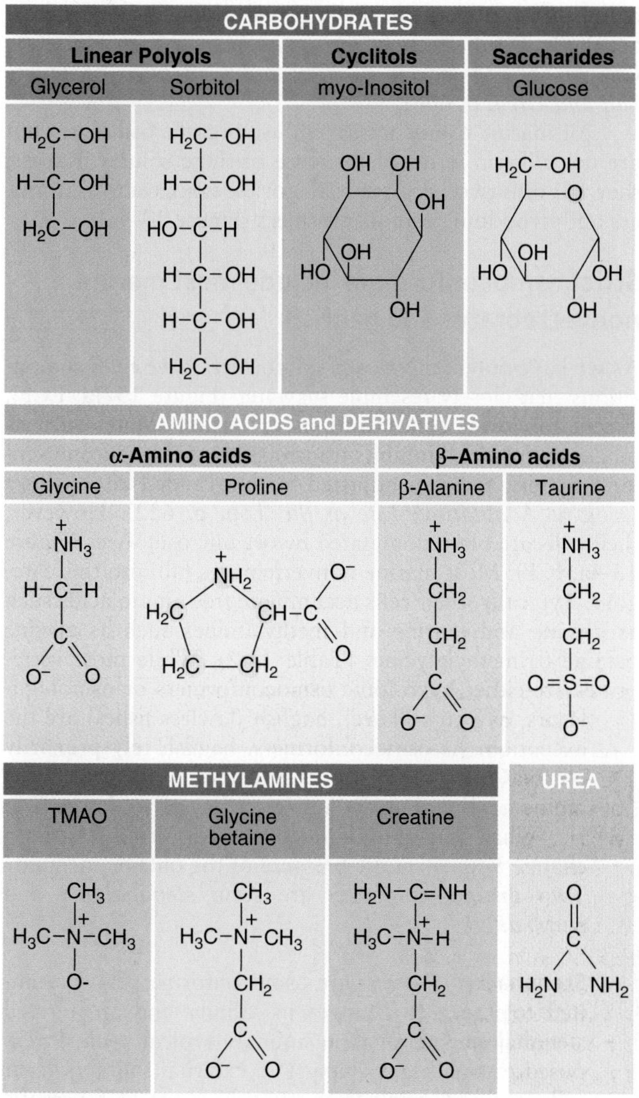

CARBOHYDRATES

AMINO ACIDS and DERIVATIVES

METHYLAMINES / UREA

FIGURE 13-5 Examples of major organic solutes used as osmolytes by animals. Not shown are methylsulfonium solutes such as *dimethylsulfoniopropionate* (DMSP), which has the formula $(CH_3)_2–S–CH_2–CH_2–COOH$.

© Cengage Learning, 2013

TABLE 13-2 Major Classes of Organic Osmolytes and Their Distribution among Animals

Elevated concentrations of osmolyte-type solutes for cryoprotection (freezing avoidance) are included.

Animal and Water Stress	Organic Solute
1. Polyhydric Alcohols (Polyols) and Sugars	
Insects—freezing	Glycerol, sorbitol, erythritol, sucrose
Insects—salinity	Trehalose
Bony fish (Arctic)—freezing	Glycerol
Amphibia—freezing	Glucose, glycerol
Mammals—kidney salinity	Sorbitol, myo-inositol
Mammals—brain salinity	Myo-inositol
2. Amino Acids and Amino Acid Derivatives	
Nonvertebrates—salinity	Glycine, alanine, proline, serine, taurine, etc.
Insects, aquatic—salinity	Proline, serine
Cyclostomes—salinity	Glycine, alanine, proline
Chondrichthyes—salinity	Taurine, glycine, β-alanine
Osteichthyes—salinity	Taurine, myo-inositol
Amphibians—salinity	Various α-amino acids
Mammal kidney, brain—salinity	Taurine, glutamine
Bird erythrocytes—salinity	Taurine
3. Methylamine Compounds	
Nonvertebrates—salinity	Glycine betaine, TMAO, proline betaine
Cyclostomes—salinity	TMAO
Chondrichthyes—salinity	TMAO, glycine betaine, sarcosine
Coelacanths—salinity	TMAO, glycine betaine
Amphibia, marine—salinity	Glycerophosphorylcholine (GPC)
Mammal kidney, brain—salinity	GPC, glycine betaine
4. Methylsulfonium Compound	
Cnidaria—salinity	Dimethylsulfoniopropionate (DMSP)
5. Urea (and Other Nitrogenous Wastes)	
Gastropods, terrestrial—estivation	Urea, uric acid, guanine
Chondrichthyes—salinity	Urea with methylamines
Coelacanth—salinity	Urea with methylamines
Lungfish—estivation	Urea only?
Amphibians, marine—salinity	Urea with methylamines and amino acids
Amphibians, terrestrial—estivation; hibernation	Urea with methylamines and amino acids
Mammal kidney—salinity	Urea with methylamines

Source: Modified from G.N. Somero & P.H. Yancey, (1997), Osmolytes and cell volume regulation: Physiological and evolutionary principles, in J. F. Hoffman, J. D. Jamieson, eds., "Handbook of Physiology", Sec. 14, (Oxford, UK: Oxford University Press). Used by permission of Oxford University Press, Inc.

regulated to keep an osmoconforming cell's osmotic pressure equal to that of the ECF. Cell volume changes are therefore small. Importantly, some compatible solutes protect cells through chemical reactions, for example, by acting as antioxidants.

- *Counteraction:* Some organic osmolytes are not simply compatible, but rather have the ability to stabilize macromolecules against denaturing forces, such as high temperature and perturbing solutes such as NaCl and urea. When stabilizing solutes are used in cells for both osmotic balance and to offset destabilizing forces, they are termed **counteracting solutes**. As you will see, the methylamine osmolytes such as TMAO (Figure 13-4) have been found to be the strongest stabilizers in many situations (Figure 13-7). Unlike compatible solutes that undergo protective reactions, counteracting solutes do not change, but rather are thought to affect the hydrogen-bonded network of water in ways that favor protein folding.

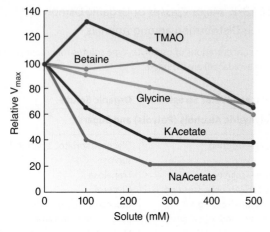

FIGURE 13-6 Effect of various solutes on the activity of an enzyme, lactate dehydrogenase, from a marine polychaete worm. Note that salts of Na and K are quite inhibitory, whereas the compatible organic osmolytes (glycine, betaine) are much less so, while the counteracting osmolyte TMAO is stimulatory except at high concentrations.

Source: M E. Clark & M. Zounes, 1977, The effects of selected cell osmolytes on the activity of lactate dehydrogenase from the euryhaline polychaete Nereis succinea, 'Biological Bulletin', 153:468–484. Reproduced with permission via Copyright Clearance Center.

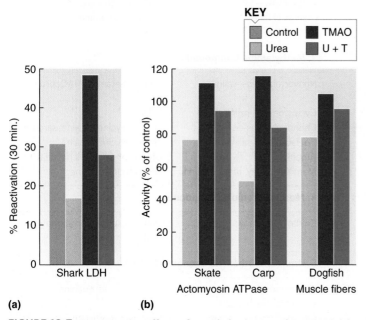

FIGURE 13-7 Counteracting effects of a methylamine osmolyte, TMAO (T), on urea (U) inhibition of protein functions. (a) Refolding of a denatured enzyme, lactate dehydrogenase, from a shark (which uses urea and TMAO as its major cellular osmolytes). (b) Activity of a muscle protein complex, actomyosin ATPase, from a skate (cartilaginous fish) and a carp (bony fish), and contractile force of whole muscle fibers from a shark.

Source: Modified from P. H. Yancey, & G. N. Somero. (1979). Counteraction of urea destabilization of protein structure by methylamine osmoregulatory compounds of elasmobranch fishes. *Biochemical Journal* 182:317–323; and P. H. Yancey. (1985). Organic osmotic effectors in cartilaginous fishes. In R. Gilles & M. Gilles-Ballien, eds., *Transport Processes, Iono- and Osmoregulation*. Berlin: Springer.

The use of organic osmolytes is widespread in nature and not just in marine organisms. Organisms subjected to freezing, for example, often use them for both cell volume maintenance and lowering freezing points (see Chapter 15). Mammalian kidney cells use them to osmoconform to the

high osmotic pressure in the kidney medulla and to counteract the harmful effects of urea, a major waste product (see Chapter 12, *Molecular Biology and Genomics: Surviving Salt and Urea*, p. 606).

All marine osmoconformers use organic osmolytes but are not alike in terms of their use of these solutes. Rather, they fall into two broad subcategories: strict osmoconformers and hypo-ionic osmoconformers (Figure 13-4a).

Strict osmoconformers include most marine nonvertebrates and hagfish

As we have noted, most osmoconformers have ECF compositions that closely resemble seawater (Figure 13-4a, ECF), except for lower concentrations of certain solutes such as magnesium (which inhibits transmission at the neuromuscular junction) and ions adjusted for buoyancy (see *A Closer Look at Adaptation: Life at the Top*, p. 622). However, their cells are often dominated by organic osmolytes (Figure 13-4a, ICF). Most marine nonvertebrates fall into this category. Typically, their cells accumulate free amino acids such as glycine and taurine and methylamines such as glycine betaine (trimethylglycine) (Table 13-2). While most vertebrates are either hypo-ionic osmoconformers or osmoregulators (as you will see), hagfish (jawless fishes) are the exception. As osmoconformers, hagfish rely primarily on NaCl to maintain ECF osmotic pressure, with amino acids and methylamines as osmolytes in the ICF.

What happens if an osmoconformer faces a change in the osmotic pressure of the outside medium? Two distinct responses are seen: *stenohalinity* and *euryhalinity*.

Stenohalinity Some pure osmoconformers have a limited tolerance to changes in salinity and are termed **stenohaline**. Stenohaline animals typically die if exposed to significant changes in external salinity. Their cells swell or shrink from osmosis as the ECF osmotic state changes because they cannot regulate their osmolytes to compensate. To survive, their primary strategy is *avoidance* (p. 13). Most marine Cnidarians (such as jellies, corals, and sea anemones) and all echinoderms (such as seastars, sea urchins, and sea cucumbers) are examples of animals restricted to life within a narrow range of salinity, and they avoid disturbances by remaining in the ocean (however, if their larvae are swept by currents into a low-salinity environment, they will likely die). Indeed, to our knowledge no echinoderm has managed to adapt to freshwater, whereas some Cnidarians (such as hydra) have. Hagfish are also stenohaline; because they live primarily in deep water, they probably never face changing salinities.

Euryhalinity In contrast, some osmoconformers are relatively tolerant of changes in the salinity of the surrounding medium and can survive substantial dilution or concentration of their ECFs; these animals are termed **euryhaline**. While many aquatic habitats from freshwater to brine are relatively stable in salinity, some habitats can change significantly. Animals living in the intertidal zone live a particularly precarious existence. During parts of the day they are covered by seawater. However, when the tide is out they may face increased salinities because of evaporation of water from a tide pool, or evaporation directly from their bodies if the animals are on a rock face. Or they may face a marked decline in salinity because of heavy rainfall. Consider also animals in brackish estuaries, where freshwater streams and rivers mix with ocean water. In both intertidal and estuary environments, salinity and hence the osmotic pressure change with the tides, and do so unpredictably, because of variations in river flow, precipitation, and tidal height. Other organisms are euryhaline over their life cycle when migration between freshwater and marine environments is involved (as is the situation for some species of eels and salmon).

Euryhaline osmoconformers can successfully adapt to variable salinities by regulating the organic osmolytes in their cells, thereby keeping the ICF in balance with the ECF. For an example, let us consider the Eastern oyster (*Crassostrea virginica*). Some populations are resident of the estuaries bordering the east coast of North America. Oysters are not mobile: They attach themselves to river bottoms and form vast oyster beds. During high tides saltwater is forced inland such that during a 24-hour period animals anchored in this environment are alternately exposed to both freshwater and saltwater. Cell volume in these bivalves is ultimately regulated by changing the concentration of intracellular compatible amino acids in response to changes in extracellular osmotic pressure. During the transition to saline water, cellular mitochondria produce amino acids, which are then transported to the cytosol. The process is thought to be triggered by the increasing levels of cellular Na⁺. Conversely, while in brackish water these organic osmolytes are reduced to low concentrations by being transported out of the cells (Figure 13-8). Osmotic neutrality is thus maintained under both conditions, and the cells undergo limited volume changes as a result. (What occurs in freshwater exposure is discussed later.) Studies have shown that the acute response to osmotic stress involves an increase in the amino acid *alanine*, whereas chronic (several days) exposure to saltwater results in the successive replacement of alanine with *glycine* and *proline*. Physiologists do not know why the amino acids change in this way.

These estuarine oysters cannot adapt to long-term exposure to full-strength seawater. In contrast, populations of the same species on the open Atlantic coast live well in seawater. Interestingly, their tissues use methylamine *glycine betaine* and the nonprotein sulfonic amino acid *taurine* (Figure 13-5) as osmolytes. The latter can make up as much as 70% of the cytosolic amino acid pool. Whether these particular osmolytes are better suited for adaptation to full seawater than are typical amino acids remains unknown. The reasons for taurine (a derivative of the amino acid cysteine) are particularly unclear. In mammals, it serves not only an osmolyte (p. 606), but for unknown reasons, it is also essential for brain development and may be an antioxidant and modulator of Ca²⁺. However, these roles in mammals are not fully understood (for example, taurine is a major ingredient in many sports drinks for no clear reasons), and similar roles (if any) have not been well studied in nonvertebrates.

In fact, the composition of osmolytes used by marine nonvertebrates varies, with some relying primarily on protein-building amino acids such as glycine (in sponges, for example), whereas others (such as many mollusks) use a mixture of taurine, glycine betaine, and glycine. Researchers are not certain why this might be. However, they have re-

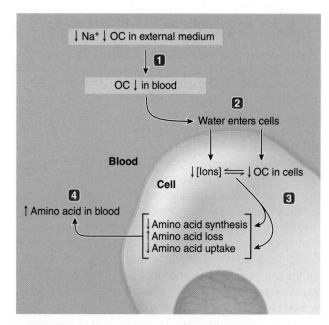

FIGURE 13-8 Adjustments to changes in Na⁺ levels in a brackish-water nonvertebrate. A reduction in external Na⁺ and osmotic concentration (OC) causes a reduction in the blood (step 1), leading to osmosis of water into cells (step 2). This in turn triggers a reduction in cellular amino acid osmolytes (step 3), which enter the blood (step 4) to be broken down.

Source: P. Willmer, G. Stone, & I. Johnston, (2000), "Environmental Physiology of Animals", Figure 10.18. Copyright © 2000 Blackwell Publishing. Reproduced with permission of Blackwell Publishing Ltd.

cently discovered one pattern related to the environment. Cells of shallow-water marine shrimp use mainly glycine, with small amounts of the methylamines glycine betaine and TMAO. However, in deep-sea shrimp (examined at 2,000 and 3,000 m), TMAO levels increase (and glycine levels decrease) with depth. Why might this be? We will return to this pattern later when we discuss deep-sea fish (p. 625).

Because ECF and ICF osmotic pressures vary considerably in euryhaline osmoconformers, they do not exhibit homeostasis of body fluids in the classic sense (of Bernard and Cannon) first applied to mammals. But they do exhibit the crucial homeostasis of cell volume regulation by regulating osmolytes. This form of homeostasis is called *enantiostasis* (Chapter 1).

Although evidence shows that rising cellular Na⁺ levels triggers osmolyte accumulation, in most cases the "sensors" and genes involved in adapting to higher osmolarity are poorly characterized. See *Molecular Biology and Genomics: A Worm in Saltwater*, p. 624.

Hypo-ionic osmoconformers include marine chondrichthyans and coelacanths and some arthropods

The other subcategory of osmoconformation, which also involves cellular osmolytes, is termed *osmoconforming hypo-ionic regulation* (Figure 13-4a). These animals, exemplified by the chondrichthyan fishes (sharks, skates, rays, chimaeras), also have ECF and ICF osmolarities equal to or slightly higher than the environment. Thus, again they show

A Closer Look at Adaptation
Life at the Top

An important adaptation closely related to water and solute regulation is that of buoyancy. Pelagic aquatic animals—those that swim or drift above the bottom—often benefit from mechanisms that help them float. This is a problem because some biological materials—proteins, shells, and skeletons—are heavier (denser) than water. A major example of an adaptation is the gas bladder of bony fish (see Chapter 11, p. 539). Another example is found in a variety of zooplankton (animal drifters). Jellies (scyphozoan Cnidarians), for example, float with the currents and catch prey using an array of stinging tentacles extending below the main body, or bell. Jellies can stay up in the water by pulsing their bells, but many must halt this pulsing when they extend their tentacles in a cylindrical or complex trap, something like a spider's web. While "fishing" this way, it is important for the jelly to remain neutrally buoyant because any change in the configuration of the extended web reduces both the effectiveness of the web and the volume of water that it controls. Furthermore, any active swimming movements made to maintain its position in the water might alert potential prey of its presence. How then can the jelly maintain buoyancy? Measurement of concentrations of solutes found in the body fluids of the jelly provided a clue. Jellies are osmoconformers. However, examination of the ion profile in the jelly revealed some intriguing differences. In the jelly *Aurelia*, the concentration of sulfate ion (SO_4^{2-}) in the body fluids is maintained at around 16 mM, whereas in the surrounding seawater, it is approximately 29 mM. Chloride ion in seawater is 560 mM, whereas in the body fluids of *Aurelia,* the concentration is maintained at approximately 580 mM. Why then might the jelly benefit from excluding SO_4^{2-} and replacing it with Cl^-? Consider that the molecular weight of SO_4^{2-} is a comparatively hefty 98, while Cl^- is a mere

A marine jelly, which removes SO_4^{2-} ions from its body fluids to increase buoyancy.

35. Exclusion of SO_4^{2-} coupled with replacement with Cl^- is a solution numerous marine nonvertebrates use to achieve buoyancy. Squid and larval tunicates use a similar solution as an aid to buoyancy but instead accumulate NH_4^+ at the expense of heavier cations such as Ca^{2+}, Mg^{2+}, and Na^+. Some vertebrates may use similar strategies: The osmolytes urea and TMAO of chondrichthyan fishes are more buoyant than seawater, and unusual buoyant gelatinous layers with low ion contents have been found in some deep-sea fishes.

---

no tendency to lose water and may even benefit from a modest gain. However, in contrast to pure osmoconformers, they also actively regulate their extracellular fluids to have considerably lower salt concentrations than the environment. They achieve osmotic parity with seawater by producing the organic solutes *urea* and *TMAO,* found throughout their bodies in both ECF and ICF. In both fluid compartments, urea is at 300–400 mOsm, whereas TMAO is typically 40–70 mOsm in the ECF and 150–200 mOsm in the ICF. These animals are called **ureosmotic** conformers because urea is typically the dominant osmolyte. Some species are stenohaline, unable to regulate their osmolytes in the face of environmental changes. In euryhaline species such as the Nicaraguan shark that migrates from the ocean to freshwater lakes, urea and TMAO are down-regulated (to a greater extent than inorganic ions) for osmotic adjustments.

In contrast to most organic osmolytes, which are compatible with cellular macromolecules, urea at the concentrations in the fluids of these animals would be inhibitory or even fatal to most animals because of the destabilizing effects it has on protein structure and function. As you saw in Chapter 12, urea is the major nitrogenous waste of mammals, which must not be allowed to build up in the blood (think about that the next time you eat shark meat). The reasons why chondrichthyan fishes can thrive with elevated urea concentrations in their fluids is threefold:

1. *Urea resistance:* Some proteins (such as in the eye lens) in these animals have evolved structures resistant to the denaturing effects of urea.

2. *Urea requirement:* Some proteins (such as the enzyme lactate dehydrogenase, p. 55) have evolved overly stable features and require urea to "loosen" them up for proper function.

3. *Counteraction:* TMAO, when present at concentrations about half that of urea, counteracts the destabilizing effects of urea. An example is shown in Figure 13-7. Keeping a 2:1 ratio of urea to TMAO is thought to be the most common adaptation in these fishes.

A summary of body water and solutes in marine chondrichthyan fishes is shown in Figure 13-9a. Regulation of urea, TMAO, and water involves the gills, digestive tract, kidneys, and the *rectal gland*. See Chapter 12 (p. 572) for more information.

The coelacanth fish (p. 558) also uses urea and TMAO as its major organic osmolytes. Another vertebrate using this strategy of osmotic adaptation is the crab-eating frog (*Rana cancrivora*) of Southeast Asia. When this frog moves from land to brackish estuarine water, it accumulates urea in all body fluids (along with amino acids in its cells), rather than salt, to osmoconform to the environment.

Finally, a few arthropods are hypo-ionic osmoconformers. In particular, larvae of some species of *Culex* mosquitoes develop in salt ponds that have a salinity about 70% that of seawater. They regulate their hemolymph ions to stay relatively constant and, like chondrichthyan fishes, build up additional osmotic pressure with organic osmolytes in their hemolymph and cells. These osmolytes include *proline* and *trehalose.*

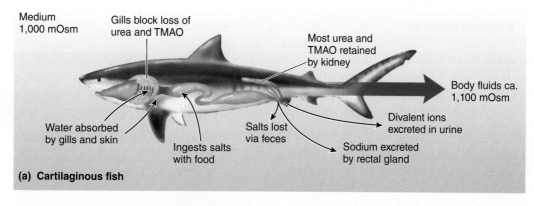

Medium 1,000 mOsm

Gills block loss of urea and TMAO

Most urea and TMAO retained by kidney

Body fluids ca. 1,100 mOsm

Divalent ions excreted in urine

Water absorbed by gills and skin

Ingests salts with food

Salts lost via feces

Sodium excreted by rectal gland

(a) Cartilaginous fish

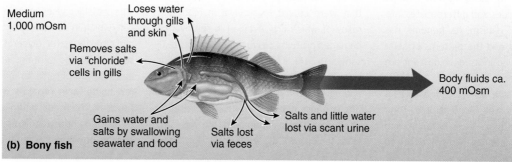

Medium 1,000 mOsm

Loses water through gills and skin

Removes salts via "chloride" cells in gills

Body fluids ca. 400 mOsm

Gains water and salts by swallowing seawater and food

Salts lost via feces

Salts and little water lost via scant urine

(b) Bony fish

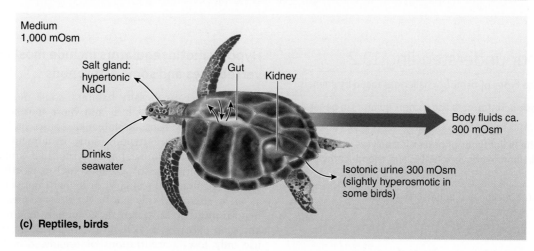

Medium 1,000 mOsm

Salt gland: hypertonic NaCl

Gut

Kidney

Body fluids ca. 300 mOsm

Drinks seawater

Isotonic urine 300 mOsm (slightly hyperosmotic in some birds)

(c) Reptiles, birds

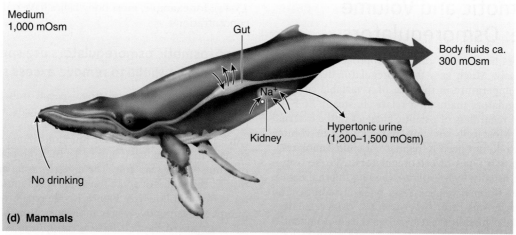

Medium 1,000 mOsm

Gut

Body fluids ca. 300 mOsm

Na^+

Hypertonic urine (1,200–1,500 mOsm)

Kidney

No drinking

(d) Mammals

FIGURE 13-9 Diagrams of major osmotic gains, losses, regulation, and body fluid concentrations in marine vertebrates.

Source: (a, b) C. E. Bond. (1996). *Biology of Fishes,* 2nd ed. Belmont, CA: Brooks/Cole Thomson Learning, p. 407, Figure 24-5, and p. 411, Figure 24-6. (c, d) P. Willmer, G. Stone, & I. Johnston. (2000). *Environmental Physiology of Animals.* Oxford, UK: Blackwell Science, p. 299, Figure 9.39.

Molecular Biology and Genomics
A Worm in Saltwater

In the last couple of decades, a soil-dwelling nematode worm, *Caenorhabditis elegans,* has emerged as a important model animal in the study of development and genetics. Like other nematodes, every individual of this remarkable species has an exact number of cells after it reaches maturity—959 cells in this species, of which 81 cells are muscles, 302 are neurons, and so on. Thus, it is a model organism for studying how a single fertilized egg divides and differentiates into a multicellular organism. Its genome has been sequenced, and the patterns of gene regulation involved in its development are being investigated.

Physiologists are using *C. elegans* as a model animal as well. For example, Todd Lamitina, Keith Chloe, Kevin Strange, and colleagues examined this worm's ability to adapt to salty conditions in conjunction with gene expression. They found that the worms could acclimate to 400 mM NaCl, equivalent to about 75% seawater. In doing so, they accumulate large amounts of the compatible osmolyte *glycerol*. The researchers also analyzed the worm's genetic responses to high osmolarity using RNA interference (RNAi, p. 33) to inhibit expression of 19,000 genes individually. They found 49 genes that were crucial to survival in high salt, many of which are involved in removal of damaged proteins by lyosomes and proteasomes (p. 44). This is probably important because hypertonicity leads to damaged and aggregated proteins in the cell. Inhibition of these genes, and others that regulate protein translation and folding, also leads to constant glycerol production. Their

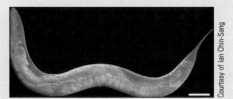

C. elegans, a nemotode worm that is used as a model animal for many biological studies.

experiments suggest that hypertonic exposure initially causes protein damage in the cell, which in some way acts as a signal to activate genes for glycerol production. As some damaged proteins are cleared out, and as glycerol replaces inorganic ions, the remaining cell proteins resume normal function. This model worm is one of the first to provide understanding of genetic mechanisms for osmotic adaptation.

---

check your understanding 13.2

Describe the types of osmotic challenges faced by animals.

Compare osmoconformers and osmoregulators in general terms.

What are the major types of organic osmolytes, and how are they used in marine nonvertebrates and chondrichthyan fishes?

Compare stenohaline and euryhaline animals.

13.3 Osmotic and Volume Balance: Osmoregulators

Osmoregulators have a very different approach to cell volume homeostasis. Much as a thermoregulator maintains a steady body temperature in the face of a variable environment, osmoregulators maintain a relatively steady ECF osmotic pressure regardless of osmotic changes in the external environment. Most cells are thus not exposed to volume changes and so do not need to regulate organic osmolytes.

There are two types of osmoregulators. Marine osmoregulators are usually *hypo-osmotic* (having internal osmolalities less than the environment's; see p. 78). In contrast, all freshwater organisms are *hyperosmotic* osmoregulators (having internal osmolalities greater than the environment's) by necessity because to osmoconform would mean to have virtually *no* cellular solutes, a condition clearly incompatible with life. Let's examine each of these.

Hypo-osmotic regulators include most marine vertebrates and some arthropods

Some crustaceans in salty habitats, some mosquito larvae in saline ponds and lakes, and most marine vertebrates (lampreys, bony fish, amphibians, reptiles including birds, and mammals) are hypo-osmotic osmoregulators. These vertebrates, for example, have internal ECF and cellular osmotic concentrations of 250–400 mOsm whether the environment is brackish (estuary) or higher than seawater (such as an evaporating tide pool). As a consequence of being hypo-osmotic to seawater, these animals tend to lose water and gain excess salt. Cells in these marine animals usually use only low concentrations of organic osmolyte (Figure 13-4a); for example, most bony fishes have only low TMAO concentrations.

Hypo-osmotic osmoregulators use special transport epithelia to remove excess salt

To achieve osmotic regulation, special transport epithelia (p. 561), plus an impermeable skin, are generally needed. In Chapter 12, we discussed the contributions of excretory organs. Let's look at how their functions integrate with other systems in marine fish (Figure 13-9b). To replace lost water, they must drink seawater. The digestive tract epithelium can restrict the influx of some ions, but NaCl must be imported for water to follow (see Figure 12-3b; recall that no ATPase pumps actively move water molecules alone). Interestingly, Cl^- is imported by a Cl^-/HCO_3^- antiporter, which also has the role of stopping uptake of excess Ca^{2+}: The HCO_3^- transported into the gut reacts with Ca^{2+} imbibed in seawater to form $CaCO_3$ precipitates, which are then excreted. Scientists now think that this process is so prolific that it is responsible for 15% or more of the marine carbon cycle!

The gut's import process creates an excess of salt in the blood. The gills then use specialized epithelial **chloride cells** (Figure 13-10) to actively transport NaCl outward with little simultaneous water movement. Unlike the basic salt-transport model of Figure 12-3a (p. 562), fish (as well as crab) gills use *NKCC* symporters that move 1 Na⁺, 1 K⁺, and 2 Cl⁻ ions (Figure 13-10); but the principles are the same as in Figure 12-3.

Removal of excess salt from the blood requires other organs in marine air breathers. Marine birds and reptiles use special *salt glands* (Figure 12-11, p. 576). Marine mammals have efficient *kidneys* with long nephron loops for the same purpose. Water and solute movements in these marine vertebrates are summarized in Figure 13-9c and d.

Hypo-osmotic regulation occurs in some arthropods in briny habitats. For example, the brine shrimp *Artemia* (see *A Closer Look at Adaptation: The Sweet Solution to Desiccation*, p. 614) can live in desert brine lakes in the Great Basin of the United States. In broad terms, their adaptations are similar to fishes, although some details are different: These crustaceans must drink the briny water (as much as 8% of their body weight per day) to gain needed water and unwanted ions, retain water with a highly impermeable exoskeleton, and then excrete the excess ions with special salt glands on their gills. Larvae of some *Aedes* mosquito species are also strong hypo-osmotic regulators, developing in brine ponds up to three times more saline than the ocean, while maintaining a lower and constant internal ion level. These larvae have evolved an extra segment on their rectums, which actively transport ions out of the body (see the mechanisms of insect excretion in Chapter 12, p. 566).

Some osmoregulating vertebrates have high levels of organic osmolytes

Some exceptions to the basic osmotic pattern of osmoregulating vertebrates have been found.

Polar and Deep-Sea Fishes Some marine fishes in the Arctic accumulate the compatible osmolyte *glycerol* to the point that they become isosmotic with seawater at 1,000 mOsm! Glycerol functions as an antifreeze (see Chapter 15, p. 739). Some polar marine fishes and in the deep sea have high TMAO content in their cells and high NaCl in their blood, with osmotic concentrations at 600 mOsm or more. These fish are still hypo-osmotic to seawater, but with higher concentrations than typical bony fish. TMAO accumulation in polar fishes is not consistent: One species may have high levels, whereas a related species may not. However, the pattern is clearer for the deep sea: *TMAO content in deep-sea fish increases with depth* in the ocean. This is the same pattern seen in osmoconforming deep-sea shrimp (p. 621). Why do osmoregulating fish need to retain such large amounts of TMAO? Possibly reducing the water gradient between the internal and environmental fluids lowers the energetic costs of hypo-osmoregulation in these environments, where cold temperatures and (in the deep sea) low food supply may make energy more limiting. Alternatively, recall that TMAO is a *counteracting* osmolyte that can stabilize proteins against urea in Chondrichthyes (p. 622). Experiments in the

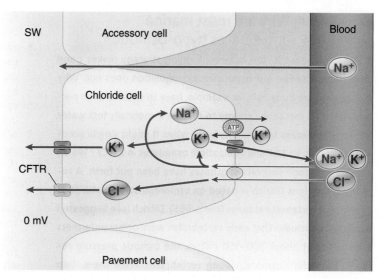

FIGURE 13-10 **Model for ion transport mechanisms in seawater teleost chloride cell.** The process starts with active transport of Na⁺ and K⁺ ("pump" indicated by "ATP"). The resulting Na⁺ electrochemical gradient is used by an *NKCC* cotransporter (shown as the small blue membrane protein) to move one Na⁺, one K⁺, and two Cl⁻ ions into the cell. The resulting high electrochemical gradient for Cl⁻ (primarily due to the cell's negative charge) allows those ions to exit into seawater via a CFTR channel. This creates a negative charge that helps draw Na⁺ across the tight junctions between cells (paracellular pathway, top green arrow) as well as some K⁺ from the cell.

Source: Modified from W. S. Marshall. (1995). *Cellular Approaches to Fish Ionic Regulation*, ed. C. M. Wood & T. J. Shuttleworth. San Diego: Academic Press.

laboratory have shown that TMAO can also counteract the effects of hydrostatic pressure, which tends to inhibit folding and activity of proteins by trapping a dense layer of water molecules around them. Moreover, in Chondrichthyes, not only does the stabilizer TMAO increase with depth, but the destabilizer urea decreases with depth.

Estivating Animals Some transitional vertebrates (living in water and on land), although they are ion regulators, use organic osmolytes in dormancy. Lungfish are found in Africa, South America, and Australia; the former two enter a state of **estivation** (summer hibernation) when their habitats become dry. They burrow and seal themselves in a mucus cocoon and become largely dormant. During estivation, they build up large amounts of urea (reportedly up to 400 mM), presumably as a less toxic alternative to ammonia for nitrogen waste storage during estivation (Table 13-2). By elevating osmotic pressure, the urea also helps retain water and may draw in water from the soil by osmosis. Similarly, some amphibians (such as the American spadefoot toad and wood frog) and Australian desert frogs estivate during dry periods and accumulate urea. (Some terrestrial snails also do the same.) These estivators accumulate only small amounts of counteracting osmolytes, perhaps because the counteraction is unnecessary: The inhibitory effects of urea may actually be useful in this situation where suppression of metabolism is important for survival. For example, Timothy Muir, Jon Costanzo, and Richard Lee found that addition of urea to organs isolated from three species of urea-using estivators markedly inhibited metabolism.

unanswered
Questions | Why are most marine vertebrates hypo-osmotic?

If you think about it, the osmoconforming strategy makes good physiological sense, but hypo-osmotic regulation does not. Why should a marine bony fish, for example, have to spend a large percentage of its metabolic energy to replace osmotically lost water and remove excess salt from its body, when it might simply accumulate NaCl in its ECF and compatible osmolytes in its ICF? We do not know, though several hypotheses have been put forth. A recent one by Hans Ditrich is based on osmoregulatory features of developing vertebrate kidneys (see p. 565). Ditrich (see Suggested Readings) concludes that early vertebrates were osmoconformers in estuaries at about 300–350 mOsm, the osmotic pressure retained in modern osmoregulating vertebrates. From there, two different paths into the ocean were taken. Hagfish and chondrichthyans stayed as osmoconformers at higher salinity by evolving mechanisms to regulate osmolytes, while the ancestors of lampreys and bony fishes evolved homeostatic osmoregulation to maintain 300–400 mOsm internally even in full seawater. Unless definitive fossil evidence is found, we may never know if this hypothesis is correct. <<

Hyperosmotic regulation is found in all freshwater animals

Across the planet, fewer species of life reside in freshwater than in the oceans. This may indicate that life did not evolve first in freshwater, and it also indicates that this habitat is not easily adapted to. The primary reason is osmotic stress. In effect, organisms in freshwater are constantly taking in unwanted water that they must somehow "bale" out. In addition, valuable solutes are constantly lost through gills, feces, and urine. Gills are particularly problematic because of their comparatively large surface areas and the necessity of minimizing protective coverings to maximize gas exchange (see p. 499). What mechanisms have evolved to enable an animal to survive in a low salinity or freshwater environment? A summary of water and solute movements in freshwater fish is shown in Figure 13-11; details for these and some nonvertebrates are as follows:

- *Active uptake of ions:* To combat the loss of salts from its body fluids, freshwater animals must expend energy (ATP) in transport epithelia to move ions from the environment into the body. Crustacea may have feathery accessory gills in the form of modified *epipodites* (outer branch of some crustacean legs) that do this. Fish gills have *chloride cells* (different from those of marine fishes) that take up Cl^- and Ca^{2+} and *pavement cells* that take up Na^+ via an electrical potential generated by H^+ V-ATPases (p. 85). Extrusion of H^+ makes the cell more negative, which increases the electrical gradient for by Na^+ to enter the cell via sodium channels.

  For freshwater frogs and worms, the skin is an important site of ion uptake. Frog skin, for example, also uses H^+ V-ATPases in the same way that fish gills do. Indeed, it was August Krogh (p. 3) over 70 years ago who

first showed that frogs can take up salt from water with NaCl as low as 0.01 millimolar!

- *Hypotonic fluid excretion:* To combat excess water gained from osmosis and eating, most freshwater animals remove it by excreting a voluminous fluid that is hypo-osmotic to body fluids. This requires (1) transport epithelia that secrete ions followed by water into an excretory space, (2) other epithelia to recover the ions while leaving the water behind, and (3) excretion of the subsequent hypo-osmotic fluid. In freshwater hydra (Cnidarians), cells lining the coelenteron (primitive gut) are thought to do this, beginning with secretion into special intercellular vacuoles. In more complex animals, kidneys provide these functions. To produce diuresis, freshwater fish tend to have more *glomeruli* (p. 571), the filtering structures of renal nephrons, than do marine fish. Conversely, freshwater Prussian carp acclimated to 40% seawater lost most of their glomeruli.

- *Lower internal osmolarities:* To help reduce water influx, solute concentrations in freshwater animals are maintained at lower levels in comparison to their marine relatives. For example, freshwater teleost fish have internal osmolalities in the range of 250–300 mOsm, slightly lower than the 300–400 mOsm of most marine species. With this strategy the animal expends less energy to maintain homeostasis. Similarly, most freshwater crustaceans have osmolalities of 300–500 mOsm, in contrast to about 1,000 mOsm in their marine counterparts. The difference in freshwater animals lies primarily in less NaCl in the ECF and little or no amino acid osmolytes in the ICF.

- *Low permeability of integument:* To reduce efflux of solutes and influx of water in freshwater, many animals have reduced permeability of the outer body surface. For example, freshwater decapod crustaceans (crabs, crayfish) are approximately 10-fold less permeable to Na^+ and Cl^- than their marine counterparts. Problems arise in those species (such as mussels that open their shells to feed) that expose sizable portions of their unprotected body surfaces to freshwater. Uncontrolled water entry and loss of valuable ions are the price these animals must pay. Amazingly, solute concentrations in the freshwater mussel *Anodonta* are reported to be only about 66 mOsm, probably the lowest osmotic concentration recorded in a living organism and deserving of further study.

Some animals such as salmon alternate between modes of osmotic adaptation

Osmoconforming is the most common adaptation in the oceans, but it has a limit: It cannot be used to adapt fully to freshwater (although some osmoconformers such as estuarine bivalves may avoid damage by "clamming up," that is, avoiding the problem temporarily). Thus, euryhaline animals that adapt to both habitats must have at least some osmoregulatory capacity. Some of the best animals at doing so are killifish in tide pools. Killifish were noted in Chapter 1 (p. 16) to illustrate the use of both physiological and behavioral processes to cope with salinity changes. Fishes that spawn in either fresh or seawater and migrate to the other habitat to grow are also highly adaptable. Here we look at salmon. These fish are particularly interesting because, as

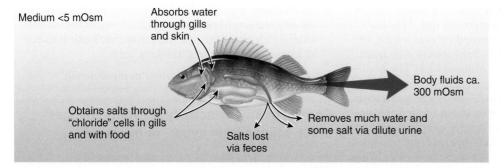

Medium <5 mOsm

Absorbs water through gills and skin

Body fluids ca. 300 mOsm

Obtains salts through "chloride" cells in gills and with food

Salts lost via feces

Removes much water and some salt via dilute urine

FIGURE 13-11 Diagram of major osmotic gains, losses, regulation, and body fluid concentration in a freshwater bony fish.

Source: From C. E. Bond. (1996). *Biology of Fishes*, 2nd ed. Belmont, CA: Brooks/Cole Thomson Learning, p. 405, Figure 24-4.

excellent osmoregulators, they are hypo-osmotic in the oceans but hyperosmotic in rivers. The changes these fish make are classic examples of *acclimatization* regulation (Chapter 1) coupled with an *anticipation* mechanism that reduces delays in adaptation. When they hatch in freshwater and begin migrating to the sea, *growth hormone,* which is controlling their general maturation, and the stress hormone *cortisol* (see Chapter 7) begin to alter the gill epithelial cells even before seawater exposure (an anticipation mechanism). As they enter estuaries, increased internal Na^+ triggers more cortisol production. Together, the hormones trigger the growth of the seawater-type chloride cells, reversing the direction of ion transport, as well as increasing Na^+/K^+ ATPase activity for removing excess salt from the blood.

After maturing for a few years at sea, the fish are ready to return to their home streams to spawn, so the gills must be altered yet again with freshwater chloride cells. The hormone *prolactin* (p. 288) has been found to induce this change in salmon. In freshwater fishes in general, prolactin maintains proper ion and water permeabilities in the epithelial cells of osmoregulatory organs, including the gill, intestine, kidney, and bladder. Hypophysectomy (removal of the pituitary gland) of freshwater fish results in a loss of ions and eventually death unless prolactin is administered to the surgically manipulated animals.

Some animals switch between osmoconforming and osmoregulating (e.g., crab in Figure 13-4c). An example is the green shore crab (*Carcinus maenas*); it is native to the Atlantic coast of Western Europe, but due to in part to its euryhaline adapations, it has invaded bays and estuaries worldwide (probably by its larvae being carried in ballast water of ships) and is having disruptive effects on ecosystems. This animal and other euryhaline crabs are osmoconformers at high salinities but osmoregulate at low salinities. To do so, they up-regulate transcription of Na^+/K^+ ATPase in their gills to increase salt uptake from the environment. Using DNA microarrays (p. 35), David Towle and colleagues found that reduced salinity alters transcription of a variety of other genes, including an increase in mitochondrial genes involved in ATP production. This may be necessary to power the increasing activity of Na^+/K^+ ATPases as environmental NaCl declines.

Many terrestrial animals osmoregulate in the face of both low water and salt availability

The problems of osmotic balance are different on land compared to water. For animals in the ocean, salt is in excess, whereas water is limiting; in freshwater, the problems are

Terrestrial animals

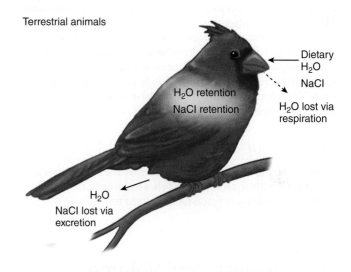

Dietary H_2O

NaCl

H_2O retention

NaCl retention

H_2O lost via respiration

H_2O

NaCl lost via excretion

FIGURE 13-12 Diagram of major osmotic gains, losses, regulation, and body fluid concentration in a terrestrial vertebrate.

Source: Modified from Y. Takei. (2000). Comparative physiology of body fluid regulation in vertebrates with special reference to thirst regulation, *Japanese Journal of Physiology* 50:171–186.

reversed. But on land, both water and salt may limiting factors. Losses of water and ions occur via evaporation and through the excretory and digestive systems, and in most cases, water and solutes must be obtained via dietary intake (Figure 13-12). Terrestrial animals are by necessity osmoregulators to some extent because the environment (air) has no solutes (exceptions are animals such as earthworms in the soil, but because they live only in moist conditions they are effectively aquatic). Excretory and digestive systems are also the most common sites of regulation (as you saw in Chapter 12), but other mechanisms come into play. In addition to physiological mechanisms, behaviors as effectors are also important. For example, most amphibians venturing onto land seek moist, cool habitats.

The most successful fully terrestrial animals are arthropods and vertebrates. Reflecting their ancestry, terrestrial vertebrates typically have osmotic concentrations at about 280–350 mOsm, with the ECF dominated by NaCl. Insect fluids can have even higher osmotic concentrations in some species, in part because of high concentrations of amino acids in the hemolymph; and whereas the dominant ECF cation is usually Na^+, some herbivorous insects (such as some Coleoptera) have higher K^+ than Na^+, although individual neurons are loosely wrapped in a neural sheath that maintains elevated Na+ concentrations (pp. 126–127).

Although water is typically restored by oral intake, interesting exceptions are found in certain insects and arachnids, which can extract water out of humid air! The house dust mite *Dermatophagoides,* for example, uses its salivary glands to secrete a fluid droplet that is so concentrated with inorganic and organic osmolytes that it can draw water vapor into it. A beetle (*Onymacris*) in the coastal desert of Namibia has a bumpy cuticle, with hydrophobic troughs and hydrophilic peaks that can attract water molecules from fog that forms from the ocean. Other arthropods use the excretory systems; for example, the rat flea *Xenopsylla* draws moist air into its rectum and shows a net water gain. How it accomplishes this is uncertain.

As for most physiological processes, research on mammals has provided the most details of volume, solute, and water regulation, including hormonal control. As you will see, osmotic regulation is also closely tied to volume regulation. Therefore, we now turn to these animals to illustrate fluid homeostasis in detail.

check your understanding 13.3

Compare the osmoregulatory problems and adaptations of a hypo-osmotic shallow marine fish, a hyperosmotic freshwater fish, a deep-sea fish, a salmon, and a terrestrial vertebrate.

13.4 Osmotic and Volume Balance in Mammals

Recall that cells in osmoregulators usually do not experience any net gain (swelling) or loss (shrinking) of volume, because the concentration of nonpenetrating solutes in the ECF is carefully regulated. In mammals, this is done primarily by the kidneys, which work to maintain the same osmolarity (about 300 mOsm) as found within most body cells.

Fluid balance is maintained by regulating ECF volume and osmolarity

Plasma is the only fluid that can be directly acted on to control its volume and composition. However, as we noted earlier, changes in the volume and composition of the plasma also directly change the interstitial fluid bathing the cells. Thus, any control mechanism that operates on the plasma in effect regulates almost the entire ECF. The ICF in turn is influenced by changes in the ECF to the extent permitted by the permeability of membranes surrounding the cells.

Two factors are regulated to maintain **fluid balance** in the mammalian body: ECF volume and ECF osmolarity. Although regulation of these two factors is closely interrelated, both being dependent on the relative NaCl and H_2O load in the body, the reasons they are closely controlled differ significantly:

- *ECF osmolarity* must be closely regulated to prevent swelling or shrinking of the cells, as discussed. Maintenance of both water and salt balance are important for this process.

- *ECF volume* must be closely regulated to help maintain *blood pressure*. Maintenance of both water and salt balance is important for this regulation, with salt balance being more important in the long term.

Let's examine each of these factors in more detail.

Control of ECF osmolarity prevents changes in ICF volume from hyper- or hypotonicity

As you have seen, normally the osmolarities of the ECF and ICF are the same because the total concentration of K+ and other effectively nonpenetrating solutes inside the cells is equal to the total concentration of Na+ and other effectively nonpenetrating solutes in the ECF. It is the number (not the nature) of the particles per volume that determines the fluid's osmolarity.

However, any circumstance that results in a loss or gain of free H_2O that is not accompanied by comparable solute deficit or excess leads to changes in ECF osmolarity. If there is a deficit of free H_2O in the ECF, the solutes become too concentrated, making the osmolarity abnormally high (that is, *hypertonic;* p. 80). If there is excess free H_2O in the ECF, the solutes become too dilute, so the osmolarity becomes abnormally low (that is, *hypotonic*). When the ECF osmolarity changes with respect to the ICF osmolarity, osmosis takes place, with H_2O either leaving or entering the cells. The osmolarity of the ECF must therefore be regulated to prevent these undesirable shifts of H_2O into or out of the cells.

First we examine the causes of hypertonic or hypotonic ECF. Then we consider how water balance and subsequently ECF osmolarity are normally maintained to minimize detrimental changes in cell volume. Although this section focuses on mammals, we note a few examples of other animals for comparison.

ECF Hypertonicity Hypertonicity of the ECF is usually associated with dehydration, or a negative free H_2O balance. Dehydration with accompanying hypertonicity can be brought about in several ways:

1. *Insufficient H_2O intake,* such as might occur during a drought or in a desert habitat.
2. *Excessive H_2O loss,* such as might occur in a mammal from heavy sweating or panting, vomiting, diarrhea, or diabetes (even though both H_2O and solutes can be lost during these conditions, relatively more H_2O is usually lost, so the remaining solutes become more concentrated). Also, an air-breathing animal exercising heavily in cold, dry air can suffer excessive water loss, as we discuss later. Finally, a stenohaline freshwater animal exposed to saltwater would also lose too much water and gain too much salt.
3. *Drinking hypertonic saline water,* as occurs in marine mammals, reptiles, and birds. Here, although water is entering the body, the salts are excessively high and may dehydrate cells.
4. *Alcohol* inhibits vasopressin secretion and can lead to ECF hypertonicity by promoting excessive free H_2O excretion.

On rare occasions, the ECF becomes hypertonic in the absence of dehydration because of the abnormal accumulation of osmotically active solutes that do not normally contribute significantly to ECF osmotic activity. This occurs, for example, with the high blood glucose in diabetes and with the high blood urea levels in uremia (from kidney failure; see p. 607).

Whenever the ECF compartment becomes hypertonic, H_2O moves out of the cells by osmosis, and so the cells shrink. Of particular concern is the fact that considerable shrinking of neurons causes disturbances in brain function, which can bring about convulsions or coma in more severe hypertonic conditions.

Although shrinkage does occur in such situations, many cells exhibit a process called **regulatory volume increase (RVI)** to compensate. In osmoregulators such as mammals, RVI occurs by transport of ions (Na^+, K^+, Cl^- serving as inorganic osmolytes) into the cell to draw water in osmotically, but there are limits to this strategy because those ions cannot be allowed to build up to levels that inhibit proteins and disrupt membrane potential. Thus, either shrinkage cannot be stopped beyond a certain point, or cells must act like euryhaline osmoconformers and import or synthesize cellular organic osmolytes. These compatible solutes can be accumulated to levels that prevent shrinkage altogether.

Interestingly, a few types of mammalian cells can use organic osmolytes for RVI. In Chapter 12 we described how cells of the inner kidney accumulate organic osmolytes to prevent shrinkage (see *Molecular Biology and Genomics: Surviving Salt and Urea*, p. 606). To some extent neurons, glial cells, and cardiac cells act similarly, accumulating compatible organic osmolytes such as *taurine* to stop shrinkage in severe dehydration. But these cells can adapt to only modest changes in osmolarity. (It is important that attempts to reduce the osmolarity of severely dehydrated animals including humans be done *slowly*. If it is done rapidly, the neurons do not have time to excrete their organic osmolytes and can suffer damage from cell swelling.)

Some mammals can tolerate whole-body dehydration and hypertonicity to an extraordinary degree. One is the camel (such as the dromedary), which can survive up to two weeks without drinking and tolerate about 30% loss of body water (compared to most mammals, which do not survive more than about 15% loss). To survive severe dehydration, the camel slows its metabolism, maintains plasma volume at the expense of other fluid compartments (which shrink), and extracts most of the water from feces in the colon and much of the water from the filtrate in the kidneys. On finding water after prolonged dehydration, the camel can also restore water balance via rapid drinking in a few minutes, a process that would also be lethal to other mammals because of the osmotic shock to cells. Camels survive because water is very slowly absorbed from their gut and because their red blood cells are more tolerant of swelling than are those of other mammals. These cells are oval rather than disc shaped as in most mammals (p. 391), have a hemoglobin that holds onto more water molecules than do other mammalian hemoglobins, and have membranes that have densely packed proteins that seem to form a "skeleton" to prevent rupture.

ECF Hypotonicity Hypotonicity of the ECF is usually associated with overhydration; that is, excess free H_2O is present, making the ECF is more dilute than normal. Because it is generally NaCl that is too low, this condition may be called **hyponatremia** ("low sodium"). Usually, any surplus free H_2O is promptly excreted in the urine, so hypotonicity/hyponatremia is rare. However, hypotonicity can arise in two ways:

1. *Intake of relatively more H_2O than solutes* more rapidly than the kidneys can compensate. This would occur if, for example, a stenohaline marine fish were exposed to freshwater (and possibly if a euryhaline fish is exposed too quickly). Also, many human athletes have suffered (and some have died) from drinking too much water without sufficient salt intake after heavy sweating.

2. *Retention of excess H_2O without solute* as a result of inappropriate secretion of vasopressin. Vasopressin in mammals (discussed in Chapter 12, p. 598, and below) is normally secreted in response to an H_2O deficit, which is relieved by increasing H_2O reabsorption in the kidney. However, vasopressin secretion can be increased in response to pain, acute infections, trauma, and other stressful situations, even when there is no H_2O deficit. The resultant H_2O retention may be appropriate in anticipation of potential blood loss in the stressful situation: The extra retained H_2O could minimize the effect a loss of blood volume would have on blood pressure. However, if there is no such injury, excess retention can occur.

Whichever way it is brought about, a hypotonic ECF causes cells to swell as H_2O moves into them osmotically. Like shrinking, pronounced swelling of brain cells also leads to brain dysfunction. Symptoms include irritability, lethargy, vomiting, drowsiness, and in severe cases, convulsions, coma, and death. Nonneural symptoms of overhydration include weakness caused by the swelling of muscle cells and circulatory disturbances, including hypertension and edema, caused by expansion of the plasma volume.

As with shrinkage, swelling in many cells triggers a compensating process called **regulatory volume decrease (RVD)**. In osmoregulators, RVD occurs through the transport of osmolytes (KCl for most cells and organic osmolytes for those cells that use them) out of the cell, followed by water. However, once again there are limits to this strategy because the cell must have a certain level of ions for macromolecular functions and membrane potential.

Isotonic Fluid Gain or Loss Let's contrast the situations of hypertonicity and hypotonicity with what happens as a result of isotonic fluid gain or loss. The major problem created by such changes is a *blood volume* disturbance, which we consider shortly. Isotonic gains are a rare occurrence in the animal world. Examples include drinking of brackish water at about 300 mOsm, or the therapeutic intravenous administration of an isotonic solution. When the isotonic fluid is injected into the ECF compartment, the ECF volume increases, but the concentration of ECF solutes remains unchanged (isotonic). The ECF compartment has increased in volume without causing a shift of H_2O into the cells.

Isotonic loss is more common because it occurs in hemorrhage. The loss is confined to the ECF with no corresponding loss of fluid from the ICF because there is no osmotic gradient to draw H_2O out of the cells. Of course, many other mechanisms come into play to counteract the loss of blood, but the ICF compartment is not directly affected by the loss.

Control of ECF volume is important in the long-term regulation of blood pressure

Now let's turn to the second major component of body fluid homeostasis—the ECF volume. A reduction in ECF volume lowers arterial blood pressure by decreasing plasma volume. Conversely, a rise in ECF volume increases the arterial blood pressure by expanding the plasma volume. It is not volume itself is but rather its effect on blood pressure that is sensed by regulatory mechanisms. Two compensatory measures come into play to transiently adjust blood pressure until the ECF volume can be restored to normal:

1. *Baroreceptor reflex* mechanisms alter both cardiac output and total peripheral resistance through autonomic nervous system effects on the heart and blood vessels (see p. 449). These rapid cardiovascular responses minimize the effect that any deviation in circulating volume has on blood pressure.

2. *Fluid shifts* occur temporarily and automatically between the plasma and interstitial fluid. A reduction in plasma volume is partially compensated for by a shift of fluid out of the interstitial compartment into the blood vessels, expanding the circulating plasma volume at the expense of the interstitial compartment. Conversely, when plasma volume is too large, much of the excess fluid is shifted into the interstitial compartment. These shifts occur immediately and automatically as a result of changes in the balance of hydrostatic and osmotic forces acting across the capillary walls that arise when plasma volume deviates from normal (see p. 443).

These two measures provide temporary relief to help keep blood pressure fairly constant, but they are not long-term solutions. Furthermore, these short-term compensatory measures have a limited ability to minimize a change in blood pressure. If the plasma volume is too low, blood pressure also remains too low no matter how vigorous the pump action of the heart, how constricted the resistance vessels, or what proportion of interstitial fluid shifts into the blood vessels. Conversely, if the plasma volume is greatly over-expanded, blood pressure cannot be restored to normal even with maximum dilation of the resistance vessels and other short-term measures.

Thus, for disturbances to both ECF osmolarity and volume, long-term regulatory mechanisms are important. These involve regulation of both water and solutes, mainly NaCl. Let's now look at those mechanisms, starting with water.

Water balance depends on input from food, drink, and metabolism, and output via urine, feces, and insensible and sensible cutaneous and respiratory losses

Control of H_2O balance is crucial for maintaining ECF osmolarity and also affects ECF volume. To maintain a stable H_2O balance, H_2O input must equal H_2O output (the balance concept). Table 13-3 shows data for a human and a kangaroo rat as examples. Let's first examine input, which has several routes:

1. *Food consumption:* Perhaps surprisingly, a significant amount of water is obtained in most animals just from

TABLE 13-3 Typical Daily Water Balance in Two Mammals

1. Human

	WATER INPUT		WATER OUTPUT	
Avenue	Quantity (mL/day)		Avenue	Quantity (mL/day)
Fluid intake	1,250		Insensible loss (from lungs and non-sweating skin)	900
H_2O in food intake	1,000			
Metabolically produced H_2O	350		Sweat	100
			Feces	100
			Urine	1,500
Total input	2,600		**Total output**	2,600

2. Kangaroo Rat (in Laboratory Fed on Dry Barley; 25°C, 33% Relative Humidity)*

Avenue	Quantity (g H_2O/ 100 g barley)		Avenue	Quantity (mL/day)
Fluid intake	0		Insensible loss	43.9
H_2O in food intake	6		Feces	2.6
Metabolically produced H_2O	54		Urine	13.5
Total input	60		**Total output**	60

*Source: B. Schmidt-Nielsen & K. Schmidt-Nielsen. (1951). A complete account of the water metabolism in kangaroo rats and an experimental verification. *Journal of Cellular and Comparative Physiology* 38:165–181.

eating solid food. Recall that muscles consist of about 75% H_2O; meat is therefore 75% H_2O because it is animal muscle. Likewise, fruits and vegetables consist of 60 to 90% H_2O. Eaters of dry food like seeds (such as some birds and rodents), in contrast, receive much less water this way.

2. *Drinking* external water: This is triggered by thirst sensations in the hypothalamus.

3. *Metabolically produced H_2O:* Recall that aerobic metabolism converts food and O_2 into energy stores, producing CO_2 and H_2O in the process. Amazingly, some desert mammals can survive primarily on this water source (see Chapter 12, p. 602, and Table 13-3).

As you saw earlier, water vapor in humid air may be taken up by some insects and arachnids. But there is no evidence that mammals can do this.

On the output side of the H_2O balance tally, there are several pathways of loss:

1. *Urine excretion.* This is the most important output mechanism in most mammals.

2. *Insensible cutaneous and respiratory loss.* During the process of respiration, inspired air becomes saturated with H_2O within the airways. This H_2O is lost when the moistened air is subsequently expired. This **insensible** (that is, unregulated) loss can be recognized on cold days, when the H_2O vapor condenses (because cold air cannot hold as much vapor as warm body air) and exhaled air is "visible." Cold dry air is particularly dehydrating. (Birds and mammals have folds in their nasal passages to reduce water and heat loss; see Chapter 15, p. 745.) The other insensible loss is the continual loss of H_2O from ordinary skin even in the absence of sweating. Water molecules can move through the cells of ordinary skin and evaporate. Mammalian and avian skin is fairly waterproof because of its keratinized exterior layer, which protects against a great loss of H_2O by this avenue (although some species use cutaneous evaporation in thermoregulation; see p. 747). And some mammals, such as marine mammals with their blubber layers, have skins that are effectively fully waterproof. Reptile skin is also quite waterproof.

3. *Sensible cutaneous and respiratory loss.* H_2O may be lost through sweating and panting, primarily for the purpose of thermoregulation. Regulated loss of water this way (called **sensible** loss) can vary substantially, of course, depending on the environmental temperature and humidity and the degree of physical activity and stress levels.

4. *Feces loss.* This is normally a minor pathway. During the process of fecal formation in the large intestine, most of the H_2O is absorbed out of the tract lumen into the blood (p. 703). Additional losses of H_2O can occur from the tract through vomiting or diarrhea.

Water balance is regulated primarily by the kidneys and thirst

Of the many sources of H_2O input and output, only two are regulated significantly to maintain H_2O balance:

1. On the intake side, *thirst* influences the amount of fluid ingested.
2. On the output side, the *kidneys* can adjust the amount of urine formed. This is the more important mechanism in mammals.

Some of the other factors are regulated but not for maintaining H_2O balance. Food intake is regulated to maintain energy balance, whereas control of sweating and panting is important in maintaining body temperature. Metabolic H_2O production and insensible losses are generally unregulated (although there are exceptions; see p. 747).

Control of Water Output in the Urine by Vasopressin In mammals, fluctuations in ECF osmolarity caused by imbalances between H_2O input and output are quickly compensated for by adjusting the urinary excretion of H_2O without changing the usual excretion of salt; that is, H_2O reabsorption and excretion are partially dissociated from solute reabsorption and excretion, so the amount of free H_2O retained or eliminated can be varied to quickly restore ECF osmolarity to normal. Free H_2O reabsorption and excretion are regulated through changes in *vasopressin* secretion (as discussed in Chapter 12, p. 598).

Control of Water Input by Thirst and Drinking Thirst is the internal craving that drives an animal to ingest H_2O (although only humans can report the subjective perception of thirst). A *thirst center* in the hypothalamus lies in close proximity to the vasopressin-secreting cells. Drinking increases H_2O input as a behavioral effector.

Simultaneous Regulation of Vasopressin and Thirst The hypothalamic control centers that regulate vasopressin secretion (and thus urinary output) and thirst (and thus drinking input) act in concert. Vasopressin secretion and thirst are both stimulated by a free-H_2O deficit (hypertonic blood) and suppressed by a free-H_2O excess (hypotonic blood). Thus, appropriately, the same circumstances that call for reduced urinary output to conserve body H_2O also give rise to the sensation of thirst to replenish body H_2O. As we noted in Chapter 12 (p. 598), the predominant excitatory input for both vasopressin secretion and thirst comes from hypothalamic *osmoreceptors* located near the vasopressin-secreting cells and thirst center. These osmoreceptors monitor the osmolarity of the fluid surrounding them, which in turn reflects the osmolarity of the entire ECF. They are thought to use TRP receptor proteins closely related to those found in many sensor cells (p. 217). As the osmolarity increases, vasopressin secretion and thirst are both stimulated (Figure 13-13).

In addition to an increase in ECF osmolarity, the vasopressin-secreting cells and thirst center are both influenced to a moderate extent by changes in ECF volume mediated by input from the *left atrial volume receptors*. Located in the left atrium, these volume receptors monitor blood pressure, which reflects the ECF volume. In response to a major reduction in ECF volume and arterial pressure, as during hemorrhage, the left atrial volume receptors reflexly stimulate both thirst and vasopressin secretion. Vasopressin in the circulation exerts a potent vasoconstrictor effect on arterioles (hence its name), in addition to affecting the kidney tubules. By increasing total peripheral resistance, vasopressin helps relieve the low blood pressure that elicited vasopressin secretion (Figure 13-13).

Conversely, vasopressin and thirst are both inhibited when ECF/plasma volume and arterial blood pressure are raised. The resultant suppression of H_2O intake, coupled with elimination of excess ECF/plasma volume in urine, helps restore blood pressure to normal.

As we noted in Chapters 7 and 12, vasopressins are mammalian hormones, the most common being *arginine vasopression (AVP)*, but closely related hypothalamic peptide hormones are found in all vertebrates. Reptiles, including birds, fishes, and some amphibians, have *arginine vasotocin (AVT)*, which differs from AVP by one amino acid in its peptide chain. It too promotes water conservation. For example, AVT enhances water uptake by the skin in toads (p. 547).

One other hormone, *angiotensin II*, also leads to increases in ECF volume by altering water balance, as you saw in Chapter 12 (see p. 587). In addition, thirst has its own regulatory process: There is evidence of some kind of "oral H_2O metering" that constitutes an example of *anticipation* control of basic homeostasis (p. 16). At least in some mammals, a thirsty animal will rapidly drink only enough H_2O to satisfy its H_2O deficit. It stops drinking before the ingested H_2O has had time to be absorbed from the digestive tract and actually return the ECF compartment to normal.

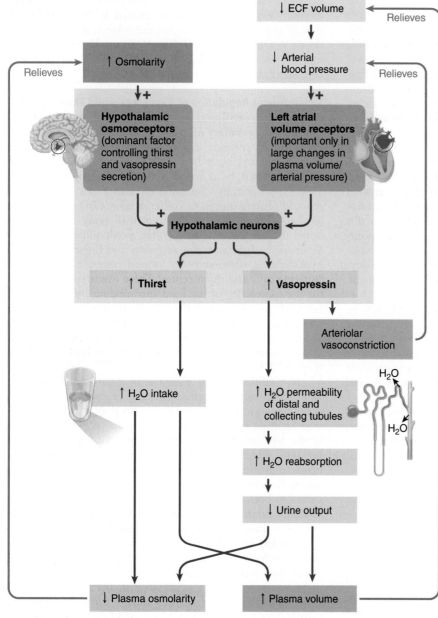

FIGURE 13-13 Control of increased vasopressin secretion and thirst during an H_2O deficit (shown for a human).

© Cengage Learning, 2013

ECF therefore determines the ECF volume, and, appropriately, long-term regulation of ECF volume depends on controlling salt balance even more than on water balance.

To achieve balance, salt input must equal salt output. The only avenue for salt input is ingestion, which can exceed some mammals' need to replace obligatory salt losses. Carnivores (meat eaters), which naturally get sufficient NaCl in fresh meat (because meat contains an abundance of salt-rich ECF) and marine mammals (as well as other marine vertebrates), normally do not show a physiological appetite to seek additional salt. In contrast, terrestrial vertebrate herbivores, which have low NaCl in their plant diets, develop *salt hunger* due to angiotensin II. As a result, mammalian herbivores will travel miles to a salt lick (as we noted in Chapter 12, p. 591).

The excess salt ingested by carnivores, omnivores, and marine mammals must be excreted to maintain salt balance. The three avenues for salt output are obligatory loss of salt in sweat and feces and controlled excretion of salt in the urine (see Table 13-4 for human data as an example). Fecal and excretory losses are most common because few mammals sweat. The total amount of sweat produced is unrelated to salt balance, being determined instead by factors that control body temperature. However, *aldosterone* (p. 591) can reduce the sweat's salt content, thus helping conserve salt in a hot environment. The small salt loss in the feces is thought not to be regulated, and it can be exacerbated by diseases. In mammals, the most important regulation is again accomplished by the kidneys, which precisely excrete excess salt in urine to maintain salt balance. By regulating the rate of urinary salt excretion, kidneys normally keep the total Na^+ mass in the ECF constant despite any notable changes in dietary intake of salt or unusual losses through sweating, diarrhea, or other means. By keeping the total Na^+ mass in the ECF constant, ECF volume in turn is maintained within the narrowly prescribed limits essential for normal circulatory function. (Recall that other osmoregulating marine vertebrates have other excretory organs for salt, such as salt glands in marine reptiles including birds, and gills in bony fishes; see pp. 572–576.)

In contrast, mammalian herbivores typically face an excess of potassium rather than sodium from eating plants. Their kidneys compensate for this by swapping Na^+ for K^+ in the nephron's distal tubule (regulated by aldosterone; p. 591). For example, the omnivorous laboratory rat has about equal concentrations of Na^+ and K^+ in its urine, but a

Exactly what factors are involved in signaling that enough H_2O has been consumed is still uncertain. It may be a learned anticipatory response based on past experience.

Now let's turn to the other major regulated component, salt.

Salt balance is regulated primarily by the kidneys and salt hunger

Because sodium and its attendant anions account for more than 90% of the ECF solutes in a mammal, it is the main contributor to ECF osmolarity. Moreover, the total Na^+ load (the total quantity of NaCl, not its concentration) in the ECF determines the total amount of H_2O that is osmotically retained in the ECF. The total mass of Na^+ salts in the

TABLE 13-4 Daily Salt Balance in a Human

SALT INPUT		SALT OUTPUT	
Avenue	Amount (g/day)	Avenue	Amount (g/day)
Ingestion	10.5	Obligatory loss in sweat and feces	0.5
		Controlled excretion in urine	10.0
Total input	10.5	**Total output**	10.5

© Cengage Learning, 2013

desert pocket mouse (which eats primarily seeds containing high levels of K^+) has about twice as much K^+ as Na^+ in its urine.

Deviations in ECF volume accompanying changes in salt load are responsible for triggering renal compensatory responses that quickly bring the Na^+ load and ECF volume back into line. Sodium is freely filtered at the glomerulus and actively reabsorbed, but it is not secreted by the tubules, so the amount of Na^+ excreted in urine represents the amount of Na^+ that is filtered but is not subsequently reabsorbed:

$$Na^+ \text{ excreted} = Na^+ \text{ filtered} - Na^+ \text{ reabsorbed}$$

The kidneys accordingly adjust the amount of salt excreted by controlling two processes: (1) glomerular filtration rate (GFR) and (2) more importantly, tubular reabsorption of Na^+.

Control of the Amount of Na⁺ Filtered through Regulation of the GFR The GFR is deliberately changed to alter the amount of salt and fluid filtered as part of the general baroreceptor reflex response to a change in blood pressure (see Chapter 12, p. 582). Because changes in plasma Na^+ can alter blood pressure via its effects on ECF volume, it is fitting that baroreceptors monitoring fluctuations in blood pressure are responsible for adjusting the amounts of Na^+ filtered and eventually excreted. (Interestingly, recent studies suggest that total body sodium is itself a regulated state, but how this might be sensed is unknown.)

Control of the Amount of Na⁺ Reabsorbed through the Renin–Angiotensin–Aldosterone and ANP Systems The amount of Na^+ reabsorbed also depends on regulatory systems that play an important role in controlling blood pressure. As you saw in Chapter 12 (p. 587), the main factor controlling the extent of Na^+ reabsorption in the mammalian nephron is the powerful *renin–angiotensin–aldosterone system (RAAS)*, which promotes Na^+ reabsorption and thereby Na^+ retention. Sodium retention in turn promotes os-

motic retention of H_2O and the subsequent expansion of plasma volume and elevation of arterial blood pressure. Appropriately, this Na^+-conserving system is activated by a reduction in ECF NaCl and volume and arterial blood pressure.

Clearly, control of GFR and Na^+ reabsorption are highly interrelated and intimately tied in with long-term regulation of ECF volume as reflected by blood pressure. Specifically, a fall in arterial blood pressure brings about a twofold response in the renal handling of Na^+ (Figure 13-14): (1) a reflex reduction in the GFR to decrease the amount of Na^+ filtered and (2) an RAA-adjusted increase in the amount of Na^+ reabsorbed. Together, these effects reduce the amount of Na^+ excreted, thereby conserving the Na^+ and accompanying H_2O necessary to compensate for the fall in arterial pressure.

Conversely, a rise in arterial blood pressure brings about (1) increases in the amount of Na^+ filtered and (2) a reduction in RAA activity, which decreases salt (and fluid) reabsorption. This is partly controlled by the *ANP* (*atrial natriuretic peptide*) hormone system from the right atrium, which as you saw in Chapter 12 is a feedback antagonist to the RAAS (p. 589). Together, these actions increase salt (and fluid) excretion, eliminating the extra fluid that was expanding plasma volume and raising arterial pressure.

The RAAS is the most important factor in regulating mammalian ECF volume and blood pressure, with the vasopressin and thirst mechanism playing only a supportive role. In addition, as we noted in Chapter 12 (p. 587), angiotensin II acts directly on the brain to stimulate both thirst and va-

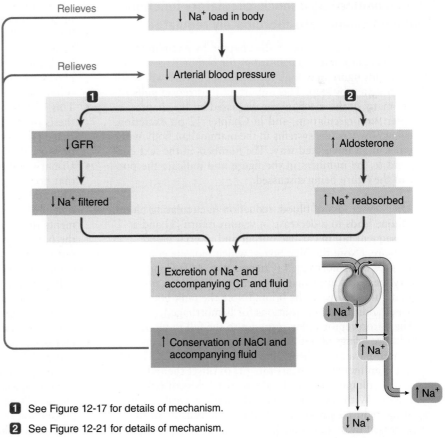

❶ See Figure 12-17 for details of mechanism.

❷ See Figure 12-21 for details of mechanism.

FIGURE 13-14 Dual effect of a fall in arterial blood pressure on renal handling of Na⁺. See main text for details.

© Cengage Learning, 2013

TABLE 13-5 Summary of the Regulation of ECF Volume and Osmolarity

Regulated Variable	Need to Regulate the Variable	Outcomes if the Variable Is Not Normal	Mechanism for Regulating the Variable
ECF Volume	Important in the long-term control of arterial blood pressure	↓ ECF volume → ↓ arterial blood pressure ↓ ECF volume → ↓ arterial blood pressure	Maintenance of salt balance; salt osmotically "holds" H_2O, so the Na^+ load determines the ECF volume. Accomplished primarily by aldosterone-controlled adjustments in urinary Na^+ excretion
ECF Osmolarity	Important to prevent detrimental osmotic movement of H_2O between the ECF and ICF	↓ ECF osmolarity (hypotonicity) → H_2O enters the cells → cells swell ↑ ECF osmolarity (hypertonicity) → H_2O leaves the cells → cells shrink	Maintenance of free H_2O balance. Accomplished primarily by vasopressin-controlled adjustments in excretion of H_2O in the urine

© Cengage Learning, 2013

sopressin release to enhance renal H_2O reabsorption. The resultant increased H_2O intake and decreased urinary output help correct the reduction in ECF volume that triggered the renin–angiotensin–aldosterone system. (Angiotensin II is an ancient vertebrate hormone; for example, it stimulates drinking in fishes. The hormone induces drinking in most birds but not desert birds, which drink whenever possible, nor in carnivorous and succulent-plant-eating birds, which obtain water from each meal.)

During hemorrhagic shock, circulatory functions and fluid balance are coordinately regulated

We conclude this section of the chapter by examining the consequences of and compensations for hemorrhage (Figure 13-15). This figure may look intimidating, but we will work through it step by step. This important illustration pulls together many of the mechanisms discussed in this chapter, in Chapter 9 on circulation, and in Chapter 12 on excretion, showing how various systems in the mammalian body work together in an integrated way. The numbers in the text correspond to the numbers in the figure and indicate the portion of the figure being discussed.

- After severe loss of blood, reduction in circulating blood volume leads to a decrease in venous return [1] and a subsequent fall in cardiac output and arterial blood pressure. (Note the blue boxes, which indicate consequences of hemorrhage.) (Chapter 9, p. 447) Compensatory measures immediately attempt to maintain adequate blood flow to the brain. (Note the pink boxes, which indicate compensations for hemorrhage.)
- The baroreceptor reflex response to the fall in blood pressure brings about increased sympathetic and decreased parasympathetic activity to the heart [2]. The result is an increase in heart rate [3] to offset the reduced stroke volume [4] brought about by the loss of blood volume. With severe fluid loss, the pulse is weak because of the reduced stroke volume but rapid because of the increased heart rate. (Chapter 9, p. 451)
- As a result of increased sympathetic activity to the veins, generalized venous vasoconstriction occurs [5], increasing venous return by means of the Frank-Starling mechanism [6]. (Chapter 9, p. 417)

- Simultaneously, sympathetic stimulation of the heart increases the heart's contractility [7] so that it beats more forcefully and ejects a greater volume of blood, also increasing the stroke volume. (Chapter 9, p. 417)
- The increase in heart rate and increase in stroke volume collectively lead to an increase in cardiac output [8].
- Sympathetically induced generalized arteriolar vasoconstriction [9] leads to an increase in total peripheral resistance [10]. (Chapter 9, p. 435)
- Together, the increase in cardiac output and total peripheral resistance bring about a compensatory increase in arterial pressure [11]. (Chapter 9, p. 436)
- The original fall in arterial pressure is also accompanied by a fall in capillary blood pressure [12], which results in fluid shifts from the interstitial fluid into the capillaries to expand the plasma volume [13]. This response is sometimes referred to as *autotransfusion*. (Chapter 9, p. 444)
- This ECF fluid shift is enhanced by plasma protein synthesis by the liver during the next few days after hemorrhage [14]. The plasma proteins exert a colloid osmotic pressure that helps retain extra fluid in the plasma.
- Urinary output is reduced, thereby conserving water that normally would have been lost from the body [15]. This additional fluid retention helps restore plasma volume [16]. Expansion of plasma volume further augments the increase in cardiac output brought about by the baroreceptor reflex [17]. Reduction in urinary output results from decreased renal blood flow caused by compensatory renal arteriolar vasoconstriction [18] (Chapter 12, p. 582). The reduced plasma volume also triggers increased secretion of vasopressin and activation of the salt- and water-conserving RAAS, which further reduces urinary output [19] (pp. 587 and 633).
- Increased thirst is also stimulated by a fall in plasma volume [20]. The resultant increased fluid intake helps restore plasma volume.
- Over a longer course of time (a week or more), lost red blood cells are replaced through increased red blood cell production triggered by reduced O_2 delivery to the kidneys [21]. (Chapter 9, p. 393)

This completes our examination of fluid balance. Table 13-5 summarizes mammalian regulation of ECF volume and osmolarity.

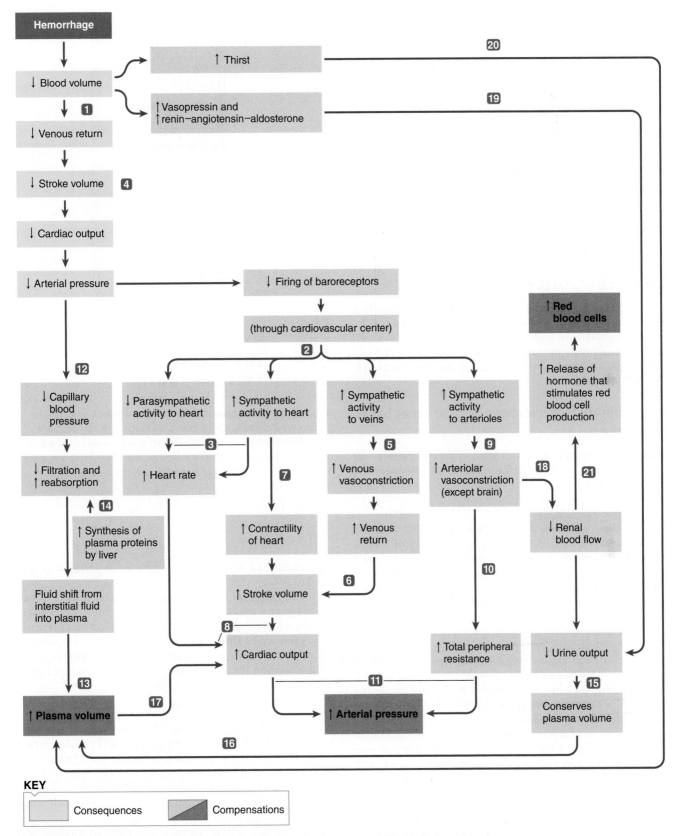

FIGURE 13-15 Consequences and compensations of hemorrhage in a mammal. The reduction of blood volume resulting from hemorrhage leads to a fall in arterial pressure. (Note the *blue* boxes, representing consequences of hemorrhage.) A series of compensations ensue (*light pink* boxes) that ultimately restore plasma volume, arterial pressure, and the number of red blood cells toward normal (*dark pink* boxes). Refer to the text (on the previous page) for an explanation of the circled numbers and a detailed discussion of the compensations.

check your understanding 13.4

Why is the regulation of both ECF osmolarity and ECF volume important?

What are the causes and consequences of ECF hypertonicity and ECF hypotonicity?

Outline the sources of input and output in a daily salt balance and a daily H_2O balance in a typical mammal. Which are subject to control to maintain the body's fluid balance?

Discuss the roles of vasopressin, thirst, and the RAA hormone system in fluid balance.

13.5 Acid–Base Balance: General Concepts

The final aspect of internal fluid and solute regulation that we consider is that of **acid–base balance**—the precise regulation of free hydrogen ions (H^+) in body fluids. This is crucial to the entire organism because protein structure and function is typically highly dependent on the concentration of H^+ ions around them.

Acids liberate free hydrogen ions, whereas bases accept them

Water itself normally exists in an equilibrium between its uncharged form and two charged products: *hydronium* and *hydroxide* ions:

$$2 \, H_2O \leftrightarrow H_3O^+ + OH^-$$

In pure water, which is 55.6 M, this equilibrium is far to the left, with 10^{-7} M of the ions at 25°C (a pH of 7.0, as we explain shortly). Technically, acid–base balance correlates with hydronium ions, but because acids and bases react with hydrogen ions, from which hydronium ions are formed, hydronium ions are treated as hydrogen ions by convention.

Acids are a special group of hydrogen-containing substances that dissociate, or separate, when in solution to liberate free H^+ and anions (negatively charged ions). Many other substances (for example, carbohydrates) also contain hydrogen, but they are not classified as acids, because the hydrogen is tightly bound within their molecular structure and is never liberated as free H^+.

A strong acid has a greater tendency to dissociate in solution than does a weak acid; that is, a greater percentage of a strong acid's molecules separate into free H^+ and anions. Hydrochloric acid (HCl) is an example of a strong acid; virtually every HCl molecule dissociates into free H^+ and Cl^- (chloride) when dissolved in H_2O. With a weaker acid such as *carbonic acid* (H_2CO_3), only a portion of the molecules dissociate in solution into H^+ and HCO_3^- (*bicarbonate* anions). The remaining H_2CO_3 molecules remain intact. Only the free hydrogen ions contribute to the acidity of a solution, so H_2CO_3 is a weaker acid than HCl because H_2CO_3 does not yield as many free hydrogen ions

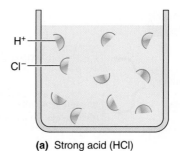

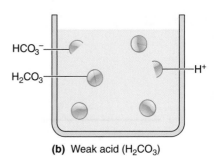

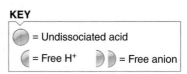

KEY

= Undissociated acid

= Free H^+ = Free anion

FIGURE 13-16 Comparison of a strong and a weak acid. (a) Five molecules of a strong acid. A strong acid such as HCl (hydrochloric acid) completely dissociates into free H^+ and anions in solution. (b) Five molecules of a weak acid. A weak acid such as H_2CO_3 (carbonic acid) only partially dissociates into free H^+ and anions in solution.

© Cengage Learning, 2013

per number of acid molecules present in solution (Figure 13-16).

The extent of dissociation for a given acid is always constant at a constant temperature and ionic composition; that is, when in solution, the same proportion of a particular acid's molecules always separate to liberate free H^+, with the other portion always remaining intact. The constant degree of dissociation for a particular acid "A" is expressed by its *dissociation constant*, K_a, as follows:

$$[H^+] \, [A^-]/[HA] = K_a$$

To indicate the concentration of a chemical, its symbol is enclosed in brackets; thus, $[H^+]$ designates H^+ concentration. Thus, $[H^+] \, [HA^-]$ represents the concentration of ions resulting from acid dissociation, whereas $[HA]$ represents the concentration of intact (undissociated) acid. Note that dissociation constants vary among different acids, with stronger acids having lower values.

Conversely, a **base** is a substance that can combine with a free H^+ and thus remove it from solution. A strong base can bind H^+ more readily than a weak base can. A base "B" is characterized by an *association constant*, K_b, as follows:

$$[HB^+] \, [OH^-]/[B] = K_b$$

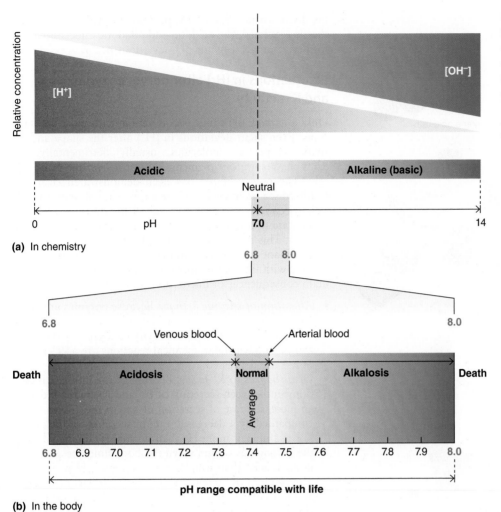

FIGURE 13-17 pH considerations in chemistry and physiology.
(a) Relationship of pH to the relative concentrations of H⁺ and base (OH⁻) under chemically neutral, acidic, and alkaline conditions. (b) Blood pH range under normal, acidotic, and alkalotic conditions.
© Cengage Learning, 2013

The pH designation is used to express hydrogen ion concentration

The $[H^+]$ in a typical mammalian ECF is about 4.3×10^{-8} or 0.00000004 equivalents per liter. The concept of pH has been developed to express $[H^+]$ more conveniently. Specifically, pH equals the logarithm (log) to the base 10 of the reciprocal of the hydrogen ion concentration:

$$pH = \log 1/[H^+]$$

Thus, the pH of typical mammalian ECF is 7.4. Three important points should be noted about this formula:

1. Because $[H^+]$ is in the denominator, *a high $[H+]$ corresponds to a low pH, and a low $[H+]$ corresponds to a high pH.*
2. *Every unit change in pH actually represents a 10-fold change in $[H+]$* because of the logarithmic relationship. For example, the log of 10 = 1, whereas the log of 100 = 2. A solution with a pH of 7 has a $[H^+]$ 10 times less than that of a solution with a pH of 6 (1 pH-unit difference) and 100 times less than that of a solution with a pH of 5 (2 pH-unit difference).

The pH of pure H_2O is 7.0 at 25°C, and 6.81 at 37°C (typical mammalian body temperature); either value is con-

sidered chemically neutral. Because OH⁻ can bind with H⁺ to once again form an H_2O molecule, it is considered basic. Because an equal number of acidic hydrogen ions (actually hydronium) and basic hydroxyl ions are formed, H_2O is neutral, being neither acidic nor basic. Solutions having a pH less than 7.0 contain a higher $[H^+]$ than pure H_2O and are considered acidic. Conversely, solutions having a pH value greater than 7.0 have a lower $[H^+]$ and are considered basic or alkaline (Figure 13-17a). Figure 13-18 compares the pH values of common solutions.

Animal Blood pH The pH of arterial blood of a typical mammal is normally 7.45 and the pH of venous blood is 7.35, for an average blood pH "set point" of 7.4. The pH of venous blood is slightly lower (more acidic) than that of arterial blood because of H⁺ generated by the formation of H_2CO_3 from CO_2 picked up at the tissue capillaries. **Acidosis** exists whenever the blood pH falls below 7.35, whereas **alkalosis** occurs when the blood pH is above 7.45 (Figure 13-17b). Note that the reference point for determining the body's acid–base status is not neutral pH (6.81 at 37°C), but the normal plasma pH of 7.4. Thus, a plasma pH of 7.2 is considered acidotic even though in chemistry a pH of 7.2 is considered basic. Death can occur if arterial pH falls outside the range of 6.8 to 8.0 for more than a few seconds because

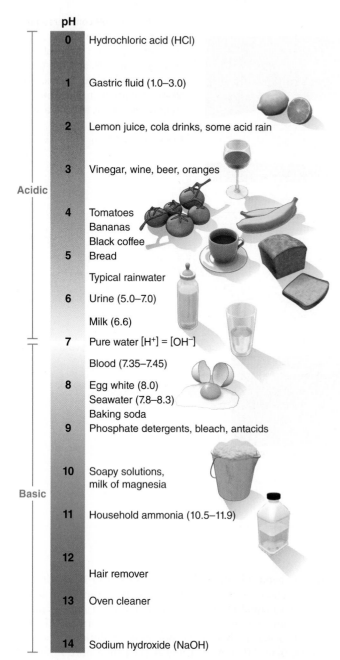

pH

pH	
0	Hydrochloric acid (HCl)
1	Gastric fluid (1.0–3.0)
2	Lemon juice, cola drinks, some acid rain
3	Vinegar, wine, beer, oranges
4	Tomatoes Bananas Black coffee
5	Bread Typical rainwater
6	Urine (5.0–7.0) Milk (6.6)
7	Pure water $[H^+] = [OH^-]$ Blood (7.35–7.45)
8	Egg white (8.0) Seawater (7.8–8.3) Baking soda
9	Phosphate detergents, bleach, antacids
10	Soapy solutions, milk of magnesia
11	Household ammonia (10.5–11.9)
12	Hair remover
13	Oven cleaner
14	Sodium hydroxide (NaOH)

Acidic

Basic

FIGURE 13-18 Comparison of pH values of common solutions.
© Cengage Learning, 2013

(as discussed in Chapter 15, p. 738). Regardless of the pH that is regulated, the acidotic and alkalotic effects do apply to these animals if major pH deviations occur.

Fluctuations in $[H^+]$ alter nerve, enzyme, and K^+ activity

Only a narrow pH range is compatible with most life functions. Even slight deviations in $[H^+]$ alter the shape and activity of protein molecules, usually detrimentally; for example, respiratory pigments do not deliver oxygen properly. The power of pH has been demonstrated clearly in fertilized sea urchin eggs, where an increase of only 0.13 pH units is enough to activate metabolism from a nearly dormant state to a level that supports the new embryo's development. This represents a useful pH change, but in normal, nondormant multicellular animals, changes are not desirable (except with internal temperature changes; see p. 738). The main consequences of fluctuations in $[H^+]$ are as follows:

- *Alteration of enzyme activity.* Because enzymes are proteins, a shift in the body's acid–base balance disturbs the normal pattern of cellular activity catalyzed by these enzymes. Some cellular chemical reactions are accelerated; others are depressed.
- *Disturbances to K^+ levels.* Because of the intimate relationship between secretion of H^+ and K^+ by the kidneys, an increased rate of secretion of one of these ions is accompanied by a decreased rate of secretion of the other. For example, if more H^+ than normal is eliminated by the kidneys, as occurs when the body fluids become acidotic, less K^+ than usual can be excreted (see p. 592). The resultant K^+ retention can affect cardiac, neural, and muscular functions, in addition to having other detrimental consequences.
- *Changes in excitability of nerve and muscle cells.* The major effect of increased $[H^+]$ (acidosis) at the whole-body level, presumably because of effects on proteins and K^+, is depression of the central nervous system. Acidotic vertebrates, for example, become disoriented and, in more severe cases, eventually die in a coma. In contrast, the major effect of decreased $[H^+]$ (alkalosis) is overexcitability of the nervous system, first the peripheral and later the central nervous system. Peripheral nerves become so excitable that they fire even in the absence of normal stimuli, bringing about muscle twitches and, in more pronounced cases, severe muscle spasms. In extreme alkalosis death may occur because spasm of the respiratory muscles seriously impairs breathing. In addition, severely alkalotic animals may die of convulsions resulting from overexcitability of the central nervous system. In less serious situations, CNS overexcitability is manifested as extreme nervousness in mammals.

Changes in body fluid pH can arise from both internal and external disturbances, as we examine next.

Hydrogen ions are continually being added to body fluids from metabolic activities

As with any other constituent, to maintain a constant $[H^+]$ in body fluids, input of hydrogen ions must be balanced by an equal output. On the input side, only a small amount of acid capable of dissociating to release H^+ is taken in with

an arterial pH of less than 6.8 or greater than 8.0 is not compatible with many cellular functions. Obviously, therefore, $[H^+]$ in body fluids must be carefully regulated.

These particular pH values do not necessarily apply to other animals. The reason is that pH set points in animals vary, and also change with temperature. A *poikilotherm* (animal whose body temperature varies with that of the environment; p. 733) from a warm habitat similar to that of mammalian body temperatures may also have a blood pH of about 7.4. But a poikilotherm in a cold habitat (such as a polar fish) typically has a higher internal pH (such as 8.0), though may reduce its pH if it is warmed up. The reasons for this are uncertain, although it has been suggested to be a mechanism for stabilizing proteins with temperature changes

food, such as the weak citric acid found in oranges. Most H^+ in the body fluids is generated internally from metabolic activities.

Internal Sources of H^+ Normally, H^+ is continually being added to body fluids from the following three sources:

1. *Carbonic acid produced in metabolism.* The major source of H^+ is through H_2CO_3 formation from metabolically produced CO_2. Recall that cellular oxidation of nutrients yields CO_2 and H_2O as end products (p. 53). CO_2 and H_2O spontaneously react to form H_2CO_3, although rather slowly. Once formed, H_2CO_3 mostly dissociates to liberate free H^+ and HCO_3^- (bicarbonate):

$$CO_2 + H_2O \leftrightarrow H_2CO_3 \leftrightarrow H^+ + HCO_3^-$$

Recall that in some cell types such as erythrocytes, the reaction is rapidly catalyzed by the enzyme *carbonic anhydrase* (CA, p. 392), although in an indirect way:

$$CO_2 + OH^- \overset{CA}{\leftrightarrow} HCO_3^-$$

The reduction in OH^- in turn causes a change in H^+ formation from water:

$$H_2O \rightarrow H^+ + OH^-$$

These reactions are readily reversible, proceeding in either direction depending on the concentrations of the substances involved as dictated by the law of mass action, as we described in Chapter 11 (p. 541). In brief, while CO_2 production by respiring cells increases acidity, exhalation of CO_2 at the lungs reduces it. When the respiratory system can keep pace with the rate of metabolism, there is no net gain or loss of H^+ in the body fluids from metabolically produced CO_2. This is a major regulatory process we will discuss later (p. 642). If the rate of CO_2 removal by the lungs does not match the rate of CO_2 production at the tissue level, however, the resultant accumulation or deficit of CO_2 in the body leads to an excess or shortage, respectively, of free H^+ in the body fluids.

2. *Inorganic acids produced during catabolism.* Dietary proteins and other ingested nutrient molecules that are found abundantly in meat contain a large quantity of sulfur and phosphorus. When these molecules are broken down, sulfuric acid and phosphoric acid are produced as by-products. Being moderately strong acids, these two inorganic acids dissociate to a large extent, liberating free H^+ into the body fluids. In contrast, catabolism of plants produces bases that, to some extent, neutralize the acids derived from protein metabolism. Generally, however, more acids than bases are produced during the breakdown of ingested food, leading to an excess of these acids.

3. *Organic acids resulting from metabolism.* Numerous organic acids are produced during normal metabolism (see pp. 55 and 316). For example, fatty acids are produced during fat metabolism, and vertebrate muscles produce lactate during strenuous activity. These acids partially dissociate to yield free H^+.

Hydrogen ion generation therefore normally goes on continuously as a result of ongoing metabolic activities. Furthermore, in certain disease states additional acids may be produced that further contribute to the total body pool of H^+. One of these, diabetes mellitus, will be discussed later (p. 649).

External Sources of H^+ Finally, inputs from outside the body may also add to the total load of H^+ that must be handled by the body. Examples include types of acid-producing medications and carbonic acid arising from excess atmospheric CO_2 from human activities such as fossil-fuel burning. As you have seen, CO_2 acidifies water when it dissolves. We will examine this issue in more depth later (p. 650).

Three Lines of Defense against $[H^+]$ Changes Generation of H^+ is unceasing, highly variable, and essentially unregulated in an animal. The key to H^+ balance is maintaining the normal alkalinity of the ECF (such as pH 7.4 in most mammals) despite this constant onslaught of acid. The generated free H^+ must be largely removed from solution while in the body, and ultimately must be eliminated so that the pH of body fluids can remain within the normal range. Mechanisms must also exist to compensate rapidly for the occasional situation in which the ECF becomes too alkaline.

Three lines of defense against changes in $[H^+]$ operate to maintain the $[H^+]$ of body fluids at a nearly constant level despite unregulated input: (1) chemical buffer systems, (2) respiratory mechanisms of pH control, and (3) excretory mechanisms of pH control. We look at each of these methods, primarily in vertebrates.

check your understanding 13.5

Distinguish acids from bases, and define pH.

Why is pH important to animals, and what are the sources of excess H^+?

13.6 pH Regulation: Buffers

When pH begins to change, the first corrective response to act is that of *buffering*.

Chemical buffer systems act as the first line of defense against changes in $[H^+]$ by binding with or releasing free H^+

A **chemical buffer system** is a mixture in a solution of two (or perhaps three) chemical compounds that minimize pH changes when either an acid or a base is added to or removed from the solution. A buffer system consists of a pair of substances involved in a reversible reaction—one substance that can yield free H^+ as the $[H^+]$ starts to fall and another that can bind with free H^+ (thus removing it from solution) when $[H^+]$ starts to rise. An important example of such a buffer system is the carbon dioxide–bicarbonate buffer system, which is involved in the following reversible reaction that we noted earlier:

$$H^+ + HCO_3^- \leftrightarrow H_2CO_3 \leftrightarrow H_2O + CO_2$$

When a strong acid such as HCl is added to an unbuffered solution, all the dissociated H^+ remains free in the solution (Figure 13-19a). In contrast, when HCl is added to a

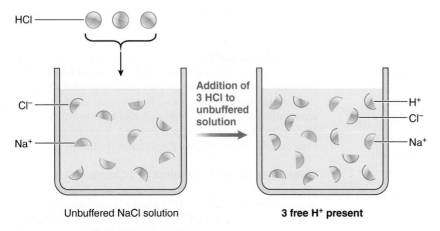

(a) Addition of HCl to an unbuffered solution

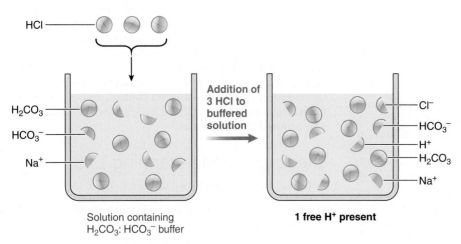

(b) Addition of HCl to a buffered solution

FIGURE 13-19 Action of chemical buffers. (a) Addition of HCl to an unbuffered solution. All the added hydrogen ions (H⁺) remain free and contribute to the acidity of the solution. (b) Addition of HCl to a buffered solution. Bicarbonate ions (HCO₃⁻), the basic member of the buffer pair, bind with some of the added H⁺ and remove them from solution so that they do not contribute to its acidity.

© Cengage Learning, 2013

3. *Hemoglobin buffer system:* primary erythrocyte buffer for carbonic acid changes.
4. *Phosphate buffer system:* secondary ICF and primary urinary buffer.

Each serves a different important role, as we will now discuss.

The CO₂–HCO₃⁻ buffer system is the primary ECF buffer for noncarbonic acids

The CO₂–HCO₃⁻ buffer system dominates in the ECF, where it is very effective for two reasons:

1. HCO₃⁻ is abundant in the ECF, so this system is readily available to absorb excess protons, and conversely there is sufficient CO₂ and H₂CO₃ to produce protons when they decrease below normal concentration. If, for example, proton concentrations increase through lactate released into the ECF from exercising muscles, the buffer reaction is driven toward the right side of the equation, so the rise in [H⁺] is abated. In the converse situation, when the plasma [H⁺] occasionally falls below normal for some reason other than a change in CO₂ (such as the loss of plasma-derived HCl in the gastric juices during vomiting), the reaction is driven toward the left side of the equation.
2. More importantly, key components of this buffer pair are closely regulated. The kidneys regulate H⁺ and HCO₃⁻, and the respiratory system regulates CO₂, which generates H₂CO₃.

solution containing the CO₂–HCO₃⁻ buffer system (Figure 13-19b), the HCO₃⁻ immediately binds with the free H⁺ to form H₂CO₃ and thus H₂O + CO₂. In contrast, when the pH of the solution starts to rise because of the addition of base or loss of acid, the H⁺-yielding member of the buffer system, H₂CO₃, releases H⁺ to minimize the rise in pH (reducing H₂O + CO₂).

All chemical buffer systems act immediately, within fractions of a second, to minimize changes in pH. When [H⁺] is altered, the buffer systems' reversible chemical reactions shift at once in favor of compensating for the change in [H⁺]. Accordingly, the buffer systems are the first line of defense against changes in [H⁺] because they are the first mechanism to respond. There are four buffer systems in vertebrates:

1. *CO₂–HCO₃⁻ buffer system:* primary ECF buffer for noncarbonic acids.
2. *Peptide and protein buffer system:* primary ICF buffer.

This buffer system cannot buffer changes in pH induced by fluctuations in H₂CO₃. A buffer system cannot buffer itself. Consider, for example, the situation in which the plasma [H⁺] is elevated because of CO₂ retention associated with a respiratory problem. The rise in CO₂ drives the reaction to the left according to the law of mass action, resulting in an elevation in [H⁺]. The increase in [H⁺] occurs because of an increase in CO₂, so the elevated [H⁺] cannot drive the reaction to the right to buffer the increase in [H⁺]. Only if the increase in [H⁺] is brought about by some mechanism other than CO₂ accumulation can this buffer system be shifted to the CO₂ side of the equation and be effective in reducing the [H⁺]. Likewise, in the opposite situation, the CO₂–HCO₃⁻ buffer system cannot compensate for a reduction in [H⁺] caused by a deficit of CO₂ by generating more H⁺-yielding H₂CO₃ when the problem in the first place is a shortage of H₂CO₃–forming CO₂. Other mechanisms, described shortly, are available for resisting fluctuations in pH caused by changes in CO₂ levels.

Henderson-Hasselbalch Equation The relationship between [H⁺] and the members of a buffer pair can be expressed according to the **Henderson-Hasselbalch equation**, which, for the CO_2–HCO_3^- buffer system, is as follows:

$$pH = pK_a + \log [HCO_3^-]/[CO_2]$$

Although you do not need to know the mathematical manipulations involved, it is helpful to understand how this formula is derived. Recall the dissociation constant K_a formula for acids (p. 636); for H_2CO_3, it is

$$[H^+][HCO_3^-]/[H_2CO_3] = K_a$$

Practically speaking, $[H_2CO_3]$ is a direct reflection of the concentration of dissolved CO_2, henceforth referred to as $[CO_2]$, because most of the CO_2 in the plasma is converted into H_2CO_3. Thus, we substitute H_2CO_3 with CO_2 and then solve the formula for $[H^+]$ (that is, $[H^+] = K_a \times [CO_2]/[HCO_3^-]$). Finally, by converting this to the logarithmic form, we come up with the Henderson-Hasselbalch equation (recall that pH is the logarithm of $1/[H^+]$; similarly, pK_a is the logarithm of $1/K_a$).

For H_2CO_3 derived from CO_2, the pK_a is 6.1 under mammalian conditions. Because the pK_a is, like K_a, a constant at a given temperature and ionic strength, changes in pH are associated with changes in the ratio between $[HCO_3^-]$ and $[CO_2]$. Normally, the ratio between $[HCO_3^-]$ and $[CO_2]$ in the ECF is 20 to 1; that is, there is 20 times more HCO_3^- than CO_2. Plugging this ratio into our formula:

$$pH = pK_a + \log [HCO_3^-]/[CO_2]$$
$$= 6.1 + \log 20:1$$

The log of 20 is 1.3. Therefore pH = 6.1 + 1.3 = 7.4, which is the normal pH of mammalian plasma.

When the ratio of $[HCO_3^-]$ to $[CO_2]$ increases above 20:1, pH increases. Accordingly, either a rise in $[HCO_3^-]$ or a fall in $[CO_2]$, both of which increase the $[HCO_3^-]/[CO_2]$ ratio if the other component remains constant, shifts the acid–base balance toward the alkaline side. In contrast, when the $[HCO_3^-]/[CO_2]$ ratio decreases below 20:1, pH decreases toward the acid side. This can occur either if the $[HCO_3^-]$ decreases or if the $[CO_2]$ increases while the other component remains constant. Because $[CO_2]$ is regulated by the lungs and gills, and both $[H^+]$ and $[HCO_3^-]$ are regulated by the kidneys and gills, the pH of the plasma can be shifted up and down by renal and respiratory contributions. Also because of the Henderson-Hasselbalch relationship, renal or respiratory dysfunction can induce acid–base imbalances by altering the $[HCO_3^-]/[CO_2]$ ratio.

Before we leave the topic of CO_2–HCO_3^- buffer systems, note that other forms of carbonate can participate in buffering in some animals, such as those that must cope with much greater acid loads than others. One example is the painted turtle *Chrysemys picta*, found in ponds in the northern United States and southern Canada. In the winter, it can survive in iced-over waters with no oxygen for months. In part, it survives by having large energy stores (glycogen in the liver, muscle, and heart) for anaerobic metabolism, coupled with a greatly reduced metabolic rate in the cold (p. 751). Nevertheless, months of metabolism without O_2 produces large quantities of lactate, as it would in most vertebrates. The turtles do have extremely high bicarbonate levels to help absorb protons released by lactate. But this is not nearly enough. How can they cope? The extraordinary

A red-eared slider turtle *(Trachemys scripta)*, which (like the painted turtle) uses its shell as an aid in acid–base regulation.

answer lies in their massive shells, which make up about a third of their body weight and are the major repository of the animals' minerals. The shell helps deal with acid load in two ways. First, acid buildup begins to demineralize the shell and appears primarily to cause the release of calcium and magnesium carbonate (CO_3^-). The latter anion is thought to buffer the protons (binding them). Second, lactate along with its acidic proton accumulate in the shell itself, perhaps bound to calcium and carbonate, respectively. This effectively removes as much as half of the acid threat from the blood. Weeks later, the turtle emerges into the spring sunshine to begin breathing and metabolizing normally again, and the lactate can be safely released and metabolized. In this way, the shell itself serves as a buffer!

The peptide and protein buffer system, including hemoglobin in erythrocytes, is primarily important intracellularly

The most plentiful buffers of the ICF are the cell proteins. A more limited number of plasma proteins reinforce the CO_2–HCO_3^- system in ECF buffering. Proteins are excellent buffers because they contain both acidic and basic groups that can bind or release H^+. Quantitatively, the protein system is most important in buffering changes in $[H^+]$ in the ICF because of the sheer abundance of the intracellular proteins. The most important buffering amino acid is *histidine* because it is the only one with a pK_a close to 7. In addition, muscles of some fishes, birds, and mammals have large concentrations of **dipeptides** (chains of two amino acids) containing histidine. These serve as a reserve buffer system in muscles that produce a substantial load of acid during activity. For example, the highest known concentration of histidine dipeptides has been found in diving mammals, whose muscles must work using anaerobic metabolism for considerable periods (p. 545).

Hemoglobin (Hb) in vertebrate erythrocytes is a key example of a protein buffer. It buffers the H^+ generated from metabolically produced CO_2 in transit between the tissues and the lungs or gills. At the systemic capillary level, CO_2 continuously diffuses into the blood from the tissue cells where it is being produced. Recall that most of this CO_2 forms H^+ and HCO_3^-. Simultaneously, some of the oxyhemoglobin (HbO_4) releases O_2, which diffuses into the tissues (Figure 11-33, p. 541). Reduced (unoxygenated) Hb has a greater affinity for H^+ than HbO_2 does. Therefore, most of the H^+ generated from CO_2 at the tissue level becomes bound to histidines of reduced Hb and no longer contributes to the acidity of the body fluids:

$$H^+ + Hb \leftrightarrow HHb^+$$

At the respiratory organ the reactions are reversed. As Hb picks up O_2 diffusing from the lungs or gills into the red blood cells, the affinity of Hb for H^+ is decreased, so H^+ is released from histidines. This liberated H^+ combines with HCO_3^- to yield H_2CO_3, which in turn produces CO_2, which is removed from the body. Meanwhile, the hydrogen ion has been reincorporated into neutral H_2O molecules. Were it not for Hb, the blood would become much too acidic after picking up CO_2 at the tissues. Because of the tremendous buffering capacity of the Hb system, venous blood is only slightly more acidic than arterial blood despite the large volume of H^+-generating CO_2 carried in the venous blood.

The phosphate buffer system is important in the ICF and urine

The phosphate buffer system consists of an acid phosphate salt (NaH_2PO_4) that can donate a free H^+ when the $[H^+]$ falls and a basic phosphate salt (Na_2HPO_4) that can accept a free H^+ when the $[H^+]$ rises. Basically, this buffer pair can alternately switch an H^+ for an Na^+ as demanded by the $[H^+]$:

$$Na_2HPO_4 + H^+ \leftrightarrow NaH_2PO_4 + Na^+$$

The phosphate pair in the ECF is rather low, so it is not very important as an ECF buffer. Because phosphates are more abundant within the cells, this system contributes significantly to intracellular buffering, being rivaled only by the more plentiful intracellular proteins. Even more importantly, the phosphate system serves as an excellent urinary buffer. When an animal consumes more phosphate than is needed, the excess phosphate filtered through the kidneys is not reabsorbed but remains in the tubular fluid to be excreted (because the renal threshold for phosphate is exceeded; see p. 586). This excreted phosphate buffers the urine as it is being formed by removing from solution the H^+ secreted into the tubular fluid. None of the other body fluid buffer systems are present in the tubular fluid to play a role in buffering urine during its formation. Most or all of the filtered HCO_3^- and CO_2 are reabsorbed, whereas Hb and most plasma proteins are not even filtered.

Buffers are a temporary mechanism because they do not eliminate excess [H⁺]

Through the mechanism of buffering, most hydrogen ions seem to "disappear" from the body fluids between the time of their generation and their elimination. It must be emphasized, however, that none of the chemical buffer systems actually eliminates H^+ from the body by themselves. Protons are merely removed from solution by being incorporated

within one of the members of the buffer pair, thus preventing the hydrogen ions from contributing to the body fluids' acidity. Because each buffer system has a limited capacity to "soak up" H^+, the H^+ that is unceasingly produced must ultimately be removed from the body. If H^+ were not eventually eliminated, soon all the body fluid buffers would already be bound with H^+ and there would be no further buffering ability.

As we have mentioned, actual removal of excess H^+ occurs in excretory and respiratory systems (for example, excretion of NaH_2PO_4 in urine, and exhaling of CO_2 by lungs). However, these systems respond more slowly than the chemical buffer systems. We now turn our attention to these other defenses against changes in acid–base balance.

check your understanding 13.6

Define buffer and the Henderson-Hasselbalch equation for the bicarbonate system.

Discuss the primary buffers of the ECF and the ICF (including hemoglobin).

13.7 pH Regulation: Respiration and Excretion

If a deviation in $[H^+]$ is not swiftly and completely corrected by the buffer systems, respiratory systems come into action a few minutes later.

Respiratory systems as a second line of defense regulate [H⁺] through adjustments in ventilation

Respiratory systems play an important role in acid–base balance through their ability to alter ventilation to control exchange H^+-generating CO_2 in gills and lungs, or to alter transport of H^+ and HCO_3^- in gills. These responses are slower than buffering because of the time it takes to transport all the affected blood past the respiratory surfaces, plus the time it takes to alter ventilation rates. Thus, they can be considered as the second line of defense against changes in $[H^+]$. (An exception to this delay occurs in exercise and stress, in which breathing is activated in *anticipation* of both the oxygen needs and pH regulation necessary during heavy muscle use; see p. 554.) Let's compare lungs and gills.

Air Breathers Every day, lungs remove from the body fluids manyfold more H^+ derived from carbonic acid than the kidneys can eliminate from noncarbonic-acid sources. Furthermore, the respiratory system, through its ability to regulate arterial $[CO_2]$, can adjust the amount of H^+ added to body fluids from this source as needed to restore pH toward normal when fluctuations in $[H^+]$ from noncarbonic-acid sources occur. (Noncarbonic acids include keto acids; p. 316.) Because of this ability, it is not surprising that the level of respiratory activity is governed at least in part by the arterial $[H^+]$, as you saw in Chapter 11. Let's review this in terms of pH regulation:

- When arterial $[H^+]$ increases (pH decreases), the respiratory center in the terrestrial-vertebrate brain stem is reflexly stimulated (see p. 552) to increase ventilation (the rate at which gas is exchanged between the respiratory organ and the atmosphere). As the rate and depth of breathing increase, more CO_2 than usual is blown off, so less H_2CO_3 than normal is added to the body fluids.

- Conversely, when arterial $[H^+]$ falls (pH increases), ventilation is reduced. As a result of slower, shallower breathing, metabolically produced CO_2 diffuses from the cells into the blood faster than it is removed from the blood, so higher-than-usual amounts of acid-forming CO_2 accumulate in the blood, restoring $[H^+]$ toward normal.

Interestingly, insects regulate pH in very similar ways. The opening and closing of spiracles and abdominally driven breathing are both controlled by the pH of the hemolymph. Insects may even have feedforward regulation of this in anticipation of high activity.

Water Breathers In fishes, kidneys play a minor role in acid–base adjustments, leaving the gills with the primary responsibility. Acid loads are primarily handled by membrane transporters in the gills rather than by alterations in breathing (ventilation). Since these are essentially excretory mechanisms, they are discussed later.

Respiratory Diseases When changes in $[H^+]$ occur because of fluctuations in $[CO_2]$ that arise from respiratory abnormalities, the respiratory mechanism cannot contribute at all to the control of pH; for example, if acidosis exists because of CO_2 accumulation caused by gill or lung disease, the impaired organs are clearly not compensating properly. The buffer systems (other than the CO_2–HCO_3^- pair) plus renal regulation are the only mechanisms available for defending against respiratory-induced acid–base abnormalities. Let's now turn to renal regulation.

Excretory systems are a powerful third line of defense that aid acid–base balance by controlling both $[H^+]$ and $[HCO_3^-]$ in the ECF

Excretory organs are the third line of defense against changes in $[H^+]$ in body fluids, although they require hours to days to compensate for changes in body fluid pH. Note that gills in fishes and many crustaceans serve both as excretory and respiratory organs. Gill transporters serving as excretory mechanisms include *H⁺ V-ATPases* (p. 85) to remove protons, an *Na⁺/H⁺ antiport* (which takes up one sodium ion for each proton excreted), and a *Cl⁻/HCO₃⁻ antiport* that may help remove excess base. Skin breathers such as frogs may also have similar transporters.

In some animals, nonrenal excretory organs contribute significantly to acid–base balance. The bladder fluid of reptiles can be acidified by an H^+ ATPase that is activated in response to high CO_2 levels. In insects, the hindgut is the key excretory organ for acid–base regulation. For example, the mosquito *Aedes dorsalis* can live in alkaline brine lakes in Africa at pH 10.5 while maintaining a hemolymph pH of 7.6. Bicarbonate is actively excreted by the hindgut in this process.

Here we focus on mammals, in which the kidneys are the most potent acid–base regulatory mechanism. Not only can they vary removal of H^+ from any source, but they can

also variably conserve or eliminate HCO_3^-, depending on the acid–base status of the body. For example, during renal compensation for acidosis for each H^+ excreted in the urine, a new HCO_3^- is added to the plasma. By simultaneously removing acid (H^+) from and adding base (HCO_3^-) to the body fluids, the kidneys can restore the pH toward normal more effectively than the lungs, which can adjust only the amount of H^+-forming CO_2 in the body. Indeed, kidneys can return the pH almost exactly to normal. In contrast, when some nonrespiratory abnormality has altered the $[H^+]$, the mammalian respiratory system alone can return the pH to only 50 to 75% of the way toward normal; this is because the driving force governing the compensatory ventilatory response is diminished as the pH moves toward normal. In comparison, the kidneys continue to respond to a change in pH until compensation is essentially complete.

The kidneys control the pH of the body fluids by adjusting three interrelated factors: (1) H^+ excretion, (2) HCO_3^- excretion, and (3) ammonia (NH_3) secretion.

Hydrogen Ion Excretion Since lungs can adjust pH only by eliminating CO_2, the task of eliminating H^+ derived from sulfuric, phosphoric, lactic, and other acids rests with the kidneys in mammals.

Almost all the excreted H^+ enters the urine by means of secretion. Recall that the filtration rate of H^+ equals plasma $[H^+]$ times GFR (p. 593). Because plasma $[H^+]$ is quite low (less than in pure H_2O except during extreme acidosis), the filtration rate of H^+ is likewise extremely low. The majority of excreted H^+ gains entry into the tubular fluid by being secreted by the proximal and distal tubules and collecting ducts (p. 571). Because the kidneys normally excrete H^+, urine is usually acidic, having an average pH of 6.0.

The H^+ secretory process begins in the tubular cells with CO_2 arising from three sources: (1) diffusing in from plasma, (2) diffusing in from tubular fluid, or (3) metabolically produced within the tubular cells, where carbonic anhydrase (CA) catalyzes the conversion of CO_2 and H_2O into H^+ and HCO_3^- (p. 639). An energy-dependent carrier in the apical (luminal) membrane then transports H^+ out of the cell into the tubular lumen. The types of carriers differ in different parts of the nephron as follows:

- *Proximal tubule:* In this tubule, H^+ is secreted by primary active transport via H^+ ATPase pumps and by secondary active transport via Na⁺/H⁺ antiporters as in fish gills (see p. 643).
- *Distal and collecting tubules:* Recall that two types of cells are located in the distal and collecting tubules, *principal cells* and *intercalated cells* (p. 587). Principal cells are the ones crucial for Na⁺, Cl⁻, and K⁺ balance under the influence of aldosterone and maintenance of H_2O balance under the influence of vasopressin. Intercalated cells, which are interspersed among the principal cells, are involved in fine regulation of acid–base balance. There are two types of intercalated cells, Type A and Type B:
  1. **Type A intercalated cells** are H^+-secreting, HCO_3^--reabsorbing, K⁺-reabsorbing cells. They actively secrete H^+ into the tubular lumen via H^+/K^+ ATPase (which secrete H^+ in exchange for the uptake of K⁺) and H^+ ATPase pumps. Both of these carriers are in the apical membrane (Figure 13-20a). The HCO_3^-

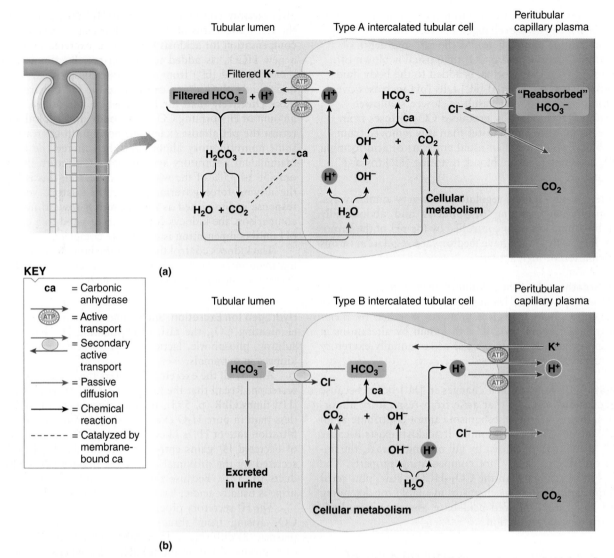

FIGURE 13-20 Renal ion secretion. (a) Hydrogen ion secretion coupled with bicarbonate reabsorption in a Type A intercalated cell. The H⁺-secreting pumps are located at the luminal membrane, and the HCO₃⁻-reabsorbing antiporters are located at the basolateral membrane. Because the disappearance of a filtered HCO₃⁻ from the tubular fluid is coupled with the appearance of another HCO₃⁻ in the plasma, HCO₃⁻ is considered to have been "reabsorbed." (b) Bicarbonate secretion coupled with hydrogen ion reabsorption in a Type B intercalated cell. The HCO₃⁻-secreting antiporters are located at the luminal membrane and the H⁺-reabsorbing pumps are located at the basolateral membrane.

© Cengage Learning, 2013

generated in the process of H⁺ formation from CO_2 under the influence of carbonic anhydrase enters the blood in exchange for Cl⁻ at the basolateral membrane via Cl⁻/HCO₃⁻ antiporters.

2. **Type B intercalated cells** are HCO₃⁻-secreting, H⁺-reabsorbing, K⁺-secreting cells, just the opposite of the Type A cells. In the Type B cells, the active H⁺ ATPase pumps and H⁺/K⁺ ATPase pumps are in the basolateral membrane and the Cl⁻/HCO₃⁻ antiporters are in the apical membrane. In this case, when H⁺ and HCO₃⁻ are generated from the hydration of CO_2 under the influence of carbonic anhydrase, HCO₃⁻ is secreted into the tubular lumen in exchange for Cl⁻, and H⁺ is reabsorbed into the plasma in exchange for K⁺ (Figure 13-20b). (Even though K⁺ is actively secreted by the Type B cells, quantitatively much more

K⁺ is actively secreted by the principal cells under the control of aldosterone.)

Type A intercalated cells are more active than Type B intercalated cells under normal circumstances, and their activity increases even more during acidosis. Type B intercalated cells become more active during alkalosis.

Bicarbonate Excretion Before being eliminated by the kidneys, the H⁺ generated from noncarbonic acids is buffered to a large extent by plasma HCO₃⁻. Appropriately, therefore, renal handling of acid–base balance also involves the adjustment of HCO₃⁻ excretion, depending on the H⁺ load in the plasma.

The kidneys regulate plasma [HCO₃⁻] by three interrelated mechanisms: (1) variable reabsorption of filtered HCO₃⁻

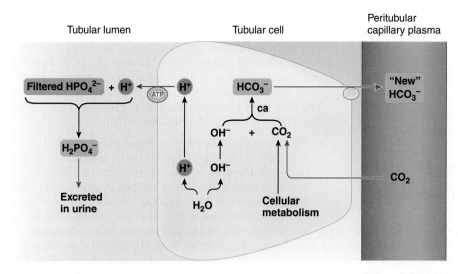

FIGURE 13-21 **Hydrogen ion secretion and excretion coupled with the addition of new HCO$_3^-$ to the plasma.** Secreted H$^+$ does not combine with filtered HPO$_4^{2-}$ and is not subsequently excreted until all the filtered HCO$_3^-$ has been "reabsorbed," as depicted in Figure 13-20a. Once all the filtered HCO$_3^-$ has combined with secreted H$^+$, further secreted H$^+$ is excreted in the urine, primarily in association with urinary buffers such as basic phosphate. Excretion of H$^+$ is coupled with the appearance of new HCO$_3^-$ in the plasma. The "new" HCO$_3^-$ represents a net gain rather than merely a replacement for filtered HCO$_3^-$.
© Cengage Learning, 2013

back into the plasma in conjunction with H$^+$ secretion, (2) variable addition of new HCO$_3^-$ to the plasma in conjunction with H$^+$ secretion, and (3) variable secretion of HCO$_3^-$ in conjunction with H$^+$ reabsorption. The first two mechanisms of renal handling of HCO$_3^-$ are inextricably linked with H$^+$ secretion, primarily by proximal tubular cells and to a lesser extent by Type A intercalated cells. Every time an H$^+$ is secreted into the tubular fluid, an HCO$_3^-$ is simultaneously transferred into the peritubular capillary plasma. Whether a filtered HCO$_3^-$ is reabsorbed or a new HCO$_3^-$ is added to the plasma in accompaniment with H$^+$ secretion depends on whether filtered HCO$_3^-$ is present in the tubular fluid to react with the secreted H$^+$, as follows.

Bicarbonate is freely filtered, but because the apical membranes of tubular cells are impermeable to filtered HCO$_3^-$, it cannot diffuse into these cells. Therefore, reabsorption of HCO$_3^-$ must occur indirectly. We will use the Type A intercalated cell as an example (Figure 13-20a). A hydrogen ion secreted into the tubular fluid combines with filtered HCO$_3^-$ to form H$_2$CO$_3$. Under the influence of carbonic anhydrase, which is present on the surface of the apical membrane, H$_2$CO$_3$ becomes CO$_2$ and H$_2$O within the filtrate. Unlike HCO$_3^-$, CO$_2$ can easily penetrate tubular cell membranes. Within the cells, CO$_2$ and H$_2$O, under the influence of intracellular carbonic anhydrase, form H$^+$ and HCO$_3^-$. Because HCO$_3^-$ can permeate these tubular cells' basolateral membrane by means of the Cl$^-$/HCO$_3^-$ antiporter, it diffuses out of the cells and into the peritubular capillary plasma. Meanwhile, the generated H$^+$ is actively secreted (Figure 13-21). Because the disappearance of an HCO$_3^-$ from the tubular fluid is coupled with the appearance of another HCO$_3^-$ in the plasma, an HCO$_3^-$ has, in effect, been "reabsorbed." Even though the HCO$_3^-$ entering the plasma is "new," that is, not the same ion that was filtered, the net result is the same as if HCO$_3^-$ were directly reabsorbed. Meanwhile, the secreted H$^+$ combines with urinary buffers such as phosphate (HPO$_4^{2-}$) and is excreted (Figure 13-21).

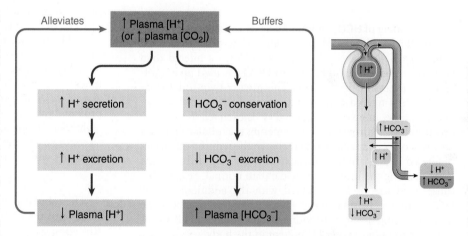

FIGURE 13-22 **Control of the rate of tubular H$^+$ secretion and HCO$_3^-$ reabsorption (mammalian kidney).**
© Cengage Learning, 2013

The same steps are involved in HCO$_3^-$ reabsorption in the proximal tubular cells, except in addition to having basolateral Cl$^-$/HCO$_3^-$ antiporters, these cells also have more abundant basolateral Na$^+$/HCO$_3^-$ symporters that simultaneously reabsorb Na$^+$ and HCO$_3^-$.

Renal Handling of H$^+$ during Acidosis and Alkalosis The kidneys are able to exert a fine degree of control over body pH. Renal handling of H$^+$ and HCO$_3^-$ depends primarily on a direct effect of the plasma's acid–base status on the kidney's tubular cells. Under normal circumstances, the proximal tubular cells and Type A intercalated cells are predominantly active, promoting net H$^+$ secretion and HCO$_3^-$ reabsorption. This pattern of activity is adjusted when pH deviates from the set point. Let's look first at the influence of acidosis and alkalosis on H$^+$ secretion (Figure 13-22):

- When the [H$^+$] of the plasma passing through the peritubular capillaries is elevated above normal, the proximal tubular cells and Type A intercalated cells respond by secreting greater-than-usual amounts of H$^+$ from the plasma into the tubular fluid to be excreted.

- Conversely, when plasma [H^+] is lower than normal, the kidneys conserve H^+ by reducing its secretion by proximal cells and Type A intercalated cells. Also, Type B intercalated cells become more active to compensate for alkalosis by increasing H^+ reabsorption. Together, these actions decrease H^+ excretion.

Because chemical reactions for H^+ secretion begin with CO_2, the rate at which they proceed is also influenced by [CO_2].

- When plasma [CO_2] increases, the rate of H^+ secretion speeds up (Figure 13-22).
- Conversely, the rate of H^+ secretion slows when plasma [CO_2] falls below normal.

These responses are especially important in renal compensations for acid–base abnormalities involving a change in H_2CO_3 caused by respiratory dysfunction. Thus, the kidneys can adjust H^+ excretion to compensate for changes in both carbonic and noncarbonic acids.

Renal Handling of HCO_3^- during Acidosis and Alkalosis When plasma [H^+] is elevated during acidosis, more H^+ is secreted than normal. At the same time, less HCO_3^- is filtered than normal because more of the plasma HCO_3^- is used up in buffering the excess H^+ in the ECF. This greater-than-usual inequity between filtered HCO_3^- and secreted H^+ has two consequences. First, more of the secreted H^+ is excreted in the urine because more hydrogen ions are entering the tubular fluid at a time when fewer are needed to reabsorb the reduced quantities of filtered HCO_3^-. In this way, extra H^+ is eliminated from the body, making the urine more acidic than normal. Second, because excretion of H^+ is linked with the addition of new HCO_3^- to the plasma, more HCO_3^- than usual enters the plasma passing through the kidneys. This additional HCO_3^- is available to buffer excess H^+ present in the body.

In the opposite situation of alkalosis, the rate of H^+ secretion diminishes, while the rate of HCO_3^- filtration increases compared to normal. When plasma [H^+] is below normal, a smaller proportion of the HCO_3^- pool is tied up buffering H^+, so plasma [HCO_3^-] is elevated above normal. As a result, the rate of HCO_3^- filtration correspondingly increases. Not all the filtered HCO_3^- is reabsorbed, because HCO_3^- ions are in excess of secreted H^+ ions in the tubular fluid and HCO_3^- cannot be reabsorbed without first reacting with H^+. Excess HCO_3^- is left in the tubular fluid to be excreted in the urine, thus reducing plasma [HCO_3^-] while making the urine alkaline. Furthermore, Type B intercalated cells come into play during alkalosis, further decreasing the excess HCO_3^- load in the body by secreting HCO_3^- into the urine.

Ammonia Secretion The energy-dependent H^+ carriers in the tubular cells can secrete H^+ against a concentration gradient until the tubular fluid (urine) becomes 800 times more acidic than the plasma. At this point, further H^+ secretion ceases because the gradient becomes too great for the secretory process to continue. It is impossible for kidneys of most mammals to acidify the urine beyond a gradient-limited urinary pH of 4.5. If left unbuffered as free H^+, only about 1% of the excess H^+ typically excreted daily would produce a urinary pH of this magnitude at normal urine flow rates, and elimination of the other 99% of the usually secreted H^+ load would be prevented, a situation that would be intolerable. For H^+ secretion to proceed, the majority of secreted H^+ must be buffered in the tubular fluid so that it does not exist as free H^+ and, accordingly, does not contribute to the tubular acidity.

Bicarbonate cannot buffer urinary H^+ as it does the ECF, because HCO_3^- is not excreted in the urine simultaneously with H^+. There are, however, two critical urinary buffers: (1) filtered phosphate buffers (p. 586) and (2) (the more important one) secreted ammonia (NH_3):

- *Phosphate as urinary buffer:* Normally, secreted H^+ is first buffered by the phosphate buffer system, which is in the tubular fluid because excess ingested phosphate has been filtered but not reabsorbed (Figure 13-21). The basic form of phosphate, HPO_4^{2-}, binds with secreted H^+. Basic phosphate is present in the tubular fluid by virtue of dietary excess, not because of any deliberate mechanism to buffer secreted H^+. When H^+ secretion is high, the buffering capacity of urinary phosphates is exceeded, but the kidneys cannot respond by excreting more basic phosphate. As soon as all the basic phosphate ions that are coincidentally excreted have "soaked up" H^+, the acidity of the tubular fluid quickly rises as more H^+ ions are secreted. Without additional buffering capacity from another source, H^+ secretion would soon be halted abruptly as the free [H^+] in the tubular fluid quickly increased to a critical limiting threshold.
- *Ammonia as urinary buffer:* The enzyme *glutaminase* in proximal tubular cells generates NH_3 from the amino acid *glutamine*. When acidosis exists, NH_3 enables the kidneys to continue secreting additional H^+ ions because NH_3 combines with free H^+ to form ammonium ion (NH_4^+):

$$NH_3 + H^+ \rightarrow NH_4^+$$

The tubular membranes transport both NH_3 and NH_4^+ into the lumen by membrane proteins (including *Rh* types; see p. 573). Since membranes are not very permeable to NH_4^+, these ions remain in the tubular fluid and are lost in the urine, each one taking an H^+ with it. Thus, NH_3 during acidosis serves to buffer excess H^+ in the tubular fluid, so that large amounts of H^+ can be secreted into the urine before the pH falls to the limiting value of 4.5. The rate of NH_3 secretion is related to the amount of excess H^+ to be transported in the urine. When an individual mammal has been acidotic for more than two or three days, the rate of NH_3 production increases substantially due to increased protein breakdown (to yield amino acids), glutamine uptake, and glutaminase activity (the regulatory signals for this are not fully known). This extra NH_3 provides additional buffering capacity to allow H^+ secretion to continue after the normal buffering capacity of phosphate is overwhelmed during renal compensation for acidosis.

check your understanding 13.7

Discuss the role of lungs in acid–base balance.

Compare the means by which H^+ and HCO_3^- are handled in the proximal tubules and in the Type A and Type B intercalated cells of the distal and collecting tubule.

What is the role of ammonia in renal regulation of acid–base balance?

13.8 Acid–Base Imbalances: Respiratory, Environmental, and Metabolic Disturbances

So far, we have mentioned some disturbances that alter body pH. Let's now look at these systematically. Deviations from normal acid–base status are divided into six general categories, depending on the source and direction of the abnormal change in [H$^+$]. These categories are *respiratory, metabolic,* and *environmental acidosis* and *respiratory, metabolic,* and *environmental alkalosis.*

Recall that changes in [H$^+$] are reflected by changes in the ratio of [HCO$_3^-$] to [CO$_2$]. Recall that the normal ratio is 20:1 (pH = 7.4; note these are mammalian values). Determinations of [HCO$_3^-$] and [CO$_2$] provide more meaningful information about the underlying factors responsible for a particular acid–base status than do direct measurements of [H$^+$] alone. The following rules of thumb apply when examining acid–base imbalances *before any compensation take place:*

1. A change in pH that has a respiratory cause is associated with an abnormal [CO$_2$], giving rise to a change in carbonic acid–generated H$^+$. Most environmental causes also involve abnormal [CO$_2$]. In contrast, a pH deviation of metabolic origin is associated with an abnormal [HCO$_3^-$] as a result of the participation of HCO$_3^-$ in buffering abnormal amounts of H$^+$ generated from noncarbonic acids.
2. Anytime the [HCO$_3^-$]/[CO$_2$] ratio falls below 20:1, an acidosis exists. The log of any number lower than 20 is less than 1.3 and, when added to the pK of 6.1, yields an acidotic pH below 7.4. Anytime the ratio exceeds 20:1, an alkalosis exists. The log of any number greater than 20 is more than 1.3 and, when added to the pK of 6.1, yield an alkalotic pH above 7.4. (Note these are mammalian values.)

We next examine each of the six imbalances separately in more detail in terms of possible causes and the compensations that occur. The "balance beam" concept, presented in Figure 13-23 in conjunction with the Henderson-Hasselbalch equation, may help you better visualize the contributions of lungs and kidneys to the causes and compensations of various acid–base disorders. The normal situation is represented in Figure 13-23a.

Respiratory acidosis arises from an increase in [CO$_2$]

Respiratory acidosis is the result of abnormal CO$_2$ buildup in body fluids, typically arising from hypoventilation (see p. 542). As less-than-normal amounts of CO$_2$ are lost through the lungs (or gills), the resultant increase in H$_2$CO$_3$ leads to an elevated [H$^+$]. Possible causes include lung or gill disease, depression of the respiratory center by drugs or disease, nerve or muscle disorders that reduce respiratory muscle ability, or (transiently) even the simple act of holding one's breath.

In uncompensated respiratory acidosis (Figure 13-23b), [CO$_2$] is elevated (in our example, it is doubled) while [HCO$_3^-$] is normal, so the ratio is 20:2 (10:1) and pH is reduced. Let us clarify a potentially confusing point. You

might wonder why when [CO$_2$] is elevated and drives the reaction CO$_2$ + H$_2$O ↔ H$_2$CO$_3$ ↔ H$^+$ + HCO$_3^-$ to the right, we say that [H$^+$] becomes elevated but [HCO$_3^-$] remains normal, although the same quantities of H$^+$ and HCO$_3^-$ are produced when CO$_2$-generated H$_2$CO$_3$ dissociates. The answer lies in the fact that normally the [HCO$_3^-$] is 600,000 times the [H$^+$] in a mammal. For every one hydrogen ion and 600,000 bicarbonate ions present in the ECF, generating one additional H$^+$ and one HCO$_3^-$ doubles the [H$^+$] (a 100% increase) but only increases the [HCO$_3^-$] 0.00017% (from 600,000 to 600,001 ions). Therefore, an elevation in [CO$_2$] brings about a pronounced increase in [H$^+$], but [HCO$_3^-$] remains essentially normal.

Compensatory measures act to restore pH to normal:

- The chemical buffers immediately take up additional H$^+$, but the respiratory mechanism usually cannot respond with compensatory increased ventilation, because an impairment in respiratory activity is the problem in the first place. Also, if high environmental CO$_2$ is the cause, increasing ventilation only makes the problem worse because the gradient favors inward diffusion of the gas.
- Thus, in mammals, the kidneys are most important in compensating for respiratory acidosis. They conserve all the filtered HCO$_3^-$ and add new HCO$_3^-$ to the plasma while simultaneously secreting and, accordingly, excreting more H$^+$. As a result, HCO$_3^-$ stores become elevated. In our example (Figure 13-23b), the plasma [HCO$_3^-$] is doubled, so the [HCO$_3^-$]–[CO$_2$] ratio is 40:2 rather than 20:2 as it was in the uncompensated state. A ratio of 40:2 is equivalent to a normal 20:1 ratio, so the pH is once again the normal 7.4. Enhanced renal conservation of HCO$_3^-$ has fully compensated for CO$_2$ accumulation, restoring the pH to normal, although both the [CO$_2$] and the [HCO$_3^-$] are now distorted.

Note that maintenance of normal pH depends on preserving a normal ratio between [HCO$_3^-$] and [CO$_2$], no matter what the absolute values of each of these buffer components are. (Compensation is never fully complete because the pH can be restored close to but not precisely to normal. In our examples, however, we assume full compensation for ease in mathematical calculations. Also bear in mind that the values used are only representative. Deviations in pH actually occur over a range, and the degree to which compensation can be accomplished varies.)

Respiratory alkalosis usually arises from a decrease in [CO$_2$]

The primary defect in respiratory alkalosis is excessive loss of CO$_2$ from the body as a result of hyperventilation. When pulmonary ventilation increases out of proportion to the rate of CO$_2$ production, too much CO$_2$ is blown off. Consequently, less H$_2$CO$_3$ is formed and [H$^+$] decreases. Possible causes of respiratory alkalosis include fever, anxiety, and ingested toxins, all of which excessively stimulate ventilation without regard to the status of O$_2$, CO$_2$, or H$^+$ in the body fluids. Respiratory alkalosis also occurs as a result of physiological mechanisms at high altitude. When the low concentration of O$_2$ in the arterial blood reflexly stimulates ventilation in an attempt to obtain more O$_2$ for the body, too much CO$_2$ is blown off in the process, inadvertently

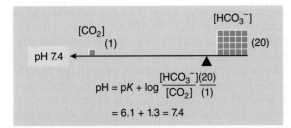

$$pH = pK + \log \frac{[HCO_3^-]}{[CO_2]} \frac{(20)}{(1)}$$
$$= 6.1 + 1.3 = 7.4$$

(a) Normal acid–base balance

Uncompensated acid–base disorders   Compensated acid–base disorders

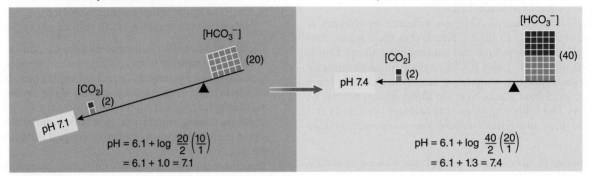

$$pH = 6.1 + \log \frac{20}{2} \left(\frac{10}{1}\right)$$
$$= 6.1 + 1.0 = 7.1$$

$$pH = 6.1 + \log \frac{40}{2} \left(\frac{20}{1}\right)$$
$$= 6.1 + 1.3 = 7.4$$

(b) Respiratory acidosis

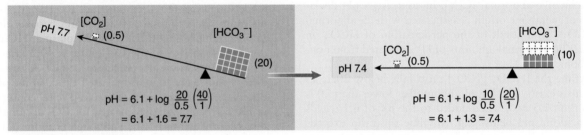

$$pH = 6.1 + \log \frac{20}{0.5} \left(\frac{40}{1}\right)$$
$$= 6.1 + 1.6 = 7.7$$

$$pH = 6.1 + \log \frac{10}{0.5} \left(\frac{20}{1}\right)$$
$$= 6.1 + 1.3 = 7.4$$

(c) Respiratory alkalosis

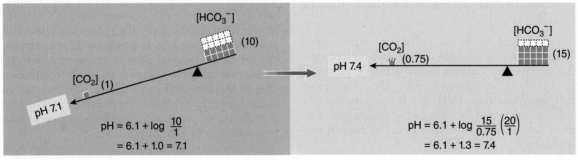

$$pH = 6.1 + \log \frac{10}{1}$$
$$= 6.1 + 1.0 = 7.1$$

$$pH = 6.1 + \log \frac{15}{0.75} \left(\frac{20}{1}\right)$$
$$= 6.1 + 1.3 = 7.4$$

(d) Metabolic acidosis

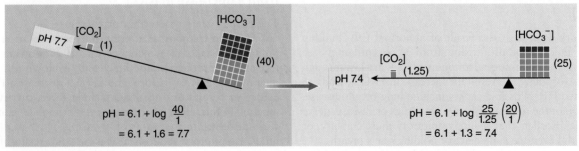

$$pH = 6.1 + \log \frac{40}{1}$$
$$= 6.1 + 1.6 = 7.7$$

$$pH = 6.1 + \log \frac{25}{1.25} \left(\frac{20}{1}\right)$$
$$= 6.1 + 1.3 = 7.4$$

(e) Metabolic alkalosis

FIGURE 13-23 **Relationship of [HCO$_3^-$] and [CO$_2$] to pH in various acid–base statuses, shown visually as a balance beam and mathematically as a solution to the Henderson-Hasselbalch equation.** Note that the lengths of the arms of the balance beams are not to scale. (a) When acid–base balance is normal, the [HCO$_3^-$]/[CO$_2$] ratio is 20:1. Each of the four types of acid–base disorders has an uncompensated state and a compensated state. (b) In uncompensated respiratory acidosis, the [HCO$_3^-$]/[CO$_2$] ratio is reduced (20:2) because CO$_2$ has accumulated. In compensated respiratory acidosis, HCO$_3^-$ is retained to balance the CO$_2$ accumulation, which restores the [HCO$_3^-$]/[CO$_2$] ratio to a normal equivalent (40:2). (c) In uncompensated respiratory alkalosis, the [HCO$_3^-$]/[CO$_2$] ratio is increased (20:0.5) by a reduction in CO$_2$. In compensated respiratory alkalosis, HCO$_3^-$ is eliminated to balance the CO$_2$ deficit, which restores the [HCO$_3^-$]/[CO$_2$] ratio to a normal equivalent (10:0.5). (d) In uncompensated metabolic acidosis, the [HCO$_3^-$]/[CO$_2$] ratio is reduced (10:1) by an HCO$_3^-$ deficit. In compensated metabolic acidosis, HCO$_3^-$ is conserved, which partially makes up for the HCO$_3^-$ deficit, and CO$_2$ is reduced; these changes restore the [HCO$_3^-$]/[CO$_2$] ratio to a normal equivalent (15:0.75). (e) In uncompensated metabolic alkalosis, the [HCO$_3^-$]/[CO$_2$] ratio is increased (40:1) by excess HCO$_3^-$. In compensated metabolic alkalosis, some of the extra HCO$_3^-$ is eliminated, and CO$_2$ is increased; these changes restore the [HCO$_3^-$]/[CO$_2$] ratio to a normal equivalent (25:1.25).

© Cengage Learning, 2013

leading to an alkalotic state (see p. 542), which causes constriction of blood vessels going to the brain. This can cause fainting or even death. Interestingly, blood flow to the brain does not drop in birds hyperventilating. The mechanisms are not known but are essential for flight at high altitudes.

If we look at the biochemical abnormalities in uncompensated respiratory alkalosis (Figure 13-23c), the increase in pH reflects a reduction in [CO$_2$] (half the normal value in our example) while the [HCO$_3^-$] remains normal. This yields an alkalotic ratio of 20:0.5, which is comparable to 40:1.

Compensatory measures in mammals act to shift the pH back toward normal:

- The chemical buffer systems liberate H$^+$ to diminish the severity of the alkalosis.
- As plasma [CO$_2$] and [H$^+$] fall below normal because of excessive ventilation, two of the normally potent stimuli for driving ventilation are removed. This effect tends to "put brakes" on the extent to which some non-respiration-related factors such as fever or anxiety can overdrive ventilation. In those cases, hyperventilation does not continue unabated.
- If the situation continues for a few days, the kidneys compensate by conserving H$^+$ and excreting more HCO$_3^-$. If, as in our example (Figure 13-23c), the HCO$_3^-$ stores are reduced by half by loss of HCO$_3^-$ in the urine, the [HCO$_3^-$]/[CO$_2$] ratio becomes 10:0.5, equivalent to the normal 20:1. Therefore, the pH is restored to normal by reducing the HCO$_3^-$ load to compensate for the CO$_2$ loss.

Metabolic acidosis is associated with a reduction in plasma [HCO$_3^-$]

Metabolic acidosis encompasses all types of acidosis besides that caused by excess CO$_2$ in the body fluids. In the uncompensated state (Figure 13-23d), metabolic acidosis is always characterized by a reduction in plasma [HCO$_3^-$] (in our ex-

ample it is halved), while [CO$_2$] remains normal, producing an acidotic ratio of 10:1. The problem may arise from the body's excessive loss of fluids rich in HCO$_3^-$ or from an accumulation of noncarbonic acids. In the latter case, plasma HCO$_3^-$ is used up in the process of buffering the additional H$^+$.

This type of acid–base disorder is the one more frequently encountered. The following are the most common causes in mammals:

1. *Severe diarrhea.* During digestion, a bicarbonate-rich digestive juice is normally secreted into the digestive tract from the pancreas and is subsequently reabsorbed back into the plasma when digestion is completed. During diarrhea, this HCO$_3^-$ is lost from the body rather than being reabsorbed. The reduction in plasma [HCO$_3^-$] without a corresponding reduction in [CO$_2$] lowers the pH. That is, loss of HCO$_3^-$ shifts the H$^+$ + HCO$_3^-$ ↔ CO$_2$ + H$_2$O reaction to the left to compensate for the HCO$_3^-$ deficit, increasing [H$^+$] above normal in the process.
2. *Diabetes mellitus ketoacidosis.* Abnormal fat metabolism resulting from the inability of cells to preferentially use glucose in the absence of insulin results in the formation of excess *keto acids* whose dissociation causes an increase in plasma [H$^+$].
3. *Strenuous activity.* When muscles resort to anaerobic glycolysis during strenuous activity (see p. 55), excess lactate is produced, leading to a rise in plasma [H$^+$]. In exercising alligators, for example, plasma pH may drop from 7.4 to 7.0, then take two hours to recover.
4. *Uremic acidosis.* In severe renal failure (uremia), the kidneys cannot rid the body of even the normal amounts of H$^+$ generated from the noncarbonic acids formed by the body's ongoing metabolic processes, so H$^+$ starts to accumulate in the body fluids. Also, the kidneys cannot conserve an adequate amount of HCO$_3^-$ to be used for buffering the normal acid load. The resultant fall in [HCO$_3^-$] without a concomitant reduction in [CO$_2$] is correlated with a decline in pH to an acidotic level.

Except in uremic acidosis, metabolic acidosis is compensated for by both respiratory and renal mechanisms as well as by chemical buffers:

- The buffers take up extra H$^+$.
- The lungs blow off additional H$^+$-generating CO$_2$.
- The kidneys excrete more H$^+$ and conserve more HCO$_3^-$.

In our example (Figure 13-23d), these compensatory measures restore the ratio to normal by reducing [CO$_2$] to 75% of normal and by raising [HCO$_3^-$] halfway back toward normal (up from 50 to 75% of the normal value). This brings the ratio to 15:0.75 (equivalent to 20:1).

Note that in compensating for metabolic acidosis, the lungs deliberately displace [CO$_2$] from normal in an attempt to restore [H$^+$] toward normal. Whereas in respiration-induced acid–base disorders, an abnormal [CO$_2$] is the cause of the [H$^+$] imbalance, in metabolic acid–base disorders, [CO$_2$] is intentionally shifted from normal as an important compensation for the [H$^+$] imbalance.

When kidney disease is the cause of metabolic acidosis, complete compensation is not possible, because the renal mechanism is not available for pH regulation. Recall that the respiratory system can compensate only up to 75% of the

TABLE 13-6 Summary of [CO_2], [HCO_3^-], and pH in Uncompensated and Compensated Respiratory and Metabolic Acid–Base Abnormalities

Acid–Base Status	pH	[CO_2] (compared to normal)	[HCO_3^-] (compared to normal)	[HCO_3^-]/[CO_2]
Normal	Normal	Normal	Normal	20/1
Uncompensated Respiratory Acidosis	Decreased	Increased	Normal	20/2 (10/1)
Compensated Respiratory Acidosis	Normal	Increased	Increased	40/2 (20/1)
Uncompensated Respiratory Alkalosis	Increased	Decreased	Normal	20/0.5 (40/1)
Compensated Respiratory Alkalosis	Normal	Decreased	Decreased	10/0.5 (20/1)
Uncompensated Metabolic Acidosis	Decreased	Normal	Decreased	10/1
Compensated Metabolic Acidosis	Normal	Decreased	Decreased	15/0.75 (20/1)
Uncompensated Metabolic Alkalosis	Increased	Normal	Increased	40/1
Compensated Metabolic Alkalosis	Normal	Increased	Increased	25/1.25 (20/1)

way toward normal. Uremic acidosis is very serious, because the kidneys cannot restore the pH all the way to normal.

Metabolic alkalosis is associated with an elevation in [HCO_3^-]

Metabolic alkalosis is a reduction in plasma [H^+] caused by a relative deficiency of noncarbonic acids. This acid–base disturbance is associated with an increase in [HCO_3^-], which, in the uncompensated state, is not accompanied by a change in [CO_2]. In our example (Figure 13-23e), [HCO_3^-] is doubled, producing an alkalotic ratio of 40:1. This condition arises most commonly in mammals from *vomiting*, which results in abnormal loss of H^+ from the body as a result of loss of acidic gastric juices. Hydrochloric acid is secreted into the stomach lumen during the process of digestion. Bicarbonate is added to the plasma during gastric HCl secretion. This HCO_3^- is neutralized by H^+ as the gastric secretions are eventually reabsorbed back into the plasma, so normally there is no net addition of HCO_3^- to the plasma from this source. However, when this acid is lost from the body during vomiting, not only is plasma [H^+] decreased, but also reabsorbed H^+ is no longer available to neutralize the extra HCO_3^- added to the plasma during gastric HCl secretion. Thus, loss of HCl in effect increases plasma [HCO_3^-].

In metabolic alkalosis:

- The chemical buffer systems immediately liberate H^+.
- Ventilation is reduced so that extra H^+-generating CO_2 is retained in the body fluids.
- If the condition persists for several days, the kidneys conserve H^+ and excrete the excess HCO_3^- in the urine.

The resultant compensatory increase in [CO_2] (up 25% in our example—Figure 13-23e) and the partial reduction in [HCO_3^-] (75% of the way back down toward normal in our example) together restore the [HCO_3^-]/[CO_2] ratio back to the equivalent of 20:1 at 25:1.25.

Overview of Compensated Acid–Base Disorders Clearly, an animal's acid–base status cannot be assessed on the basis of

pH alone. Uncompensated acid–base abnormalities can readily be distinguished on the basis of deviations of either [CO_2] or [HCO_3^-] from normal (Table 13-6). When compensation has been accomplished and the pH is essentially normal, determinations of [CO_2] and [HCO_3^-] can reveal the presence of an acid–base disorder. Note, however, that in both compensated respiratory acidosis and compensated metabolic alkalosis, [CO_2] and [HCO_3^-] are both above normal. With respiratory acidosis, the original problem is an abnormal increase in [CO_2], and a compensatory increase in [HCO_3^-] restores the [HCO_3^-]/[CO_2] ratio to 20:1. Metabolic alkalosis, in contrast, is characterized by an abnormal increase in [HCO_3^-] in the first place; then [CO_2] is deliberately increased to restore the ratio to normal. Similarly, compensated respiratory alkalosis and compensated metabolic acidosis share similar patterns of [CO_2] and [HCO_3^-]. Respiratory alkalosis starts out with a reduction in [CO_2], which is compensated by a reduction in [HCO_3^-]. With metabolic acidosis, [HCO_3^-] falls below normal, followed by a compensatory decrease in [CO_2].

Environmental acidosis arises from natural and anthropogenic increases in environmental [CO_2] and sulfuric acid

Finally, let's turn to environmental disturbances. Acidosis can arise from CO_2 and acids in the environment. One example would be a human or pet exposed to output from a CO_2-based fire extinguisher. Other environmental acids include sulfuric acid derived from sulfur compounds in emissions from volcanoes and coal-burning power plants. Sulfuric acid is a major cause of *acid rain* that can fall downwind of some power plants or volcanoes and damage aquatic ecosystems. And there is anthropogenic (human-made) atmospheric CO_2.

Ocean Acidification In terms of environmental acidosis, the greatest concern is the increase of CO_2 in Earth's atmosphere as a result of human burning of fossil fuels. The gas has increased in the air from 280 ppm (parts per million) to nearly 387 ppm (as of 2009) since the Industrial Revolution began.

In addition to the "greenhouse" warming effect of CO_2 (see Chapter 15), up to half of the excess is dissolving in the oceans and thus, through the bicarbonate reaction, acidifying them. The oceans have had a consistent pH of about 8.2 for millennia, but by 2008, they had dropped below 8.1. Since pH is logarithmic, this is a 25% increase in $[H^+]$!

The most direct impact of this *ocean acidification* is a weakening of shells and coral skeletons, whose $CaCO_3$ matrix cannot form properly if water is too acidic. Recent studies show that some animals are already being affected by this, including:

- *Corals:* Glenn De'ath and colleagues reported in 2008 that growth of some corals on the Great Barrier Reef has declined by 13% since 1990, in correlation with rising acidity. Another study in 2011 on Indonesian coral reefs near natural CO_2 seeps reveals that coral diversity drops as pH falls from 8.1 to 7.8 and that coral growth stops at pH 7.7.
- *Bivalves:* Stephanie Talmage and Christopher Gobler reported in 2010 that larvae of two commercially important bivalves grew faster and developed thicker shells when grown at pre-industrial CO_2/pH levels than when grown under current conditions. And at an oyster hatchery in Oregon, acidic ocean water entering the tanks in 2008 and 2009 killed most of the oyster larvae.
- *Fish:* Fisheries biologists of the National Oceanic and Atmospheric Administration tested pH effects on young pollock (a commercially important bony fish) in Oregon. The fish were found to compensate for increased acidity by boosting bicarbonate buffer in their blood, but this could harm their growth since that process presumably costs energy.

One possible result of such effects is rather dramatic: The density of marine jellies is increasing worldwide, choking the waters in some places like the Mediterranean and Japan Seas. At the same time, other organisms such as fishes are declining. Increasing acidity is thought to be one cause because—unlike the declining organisms—jellies so far appear to thrive in these changing ocean conditions that harm other animals. However, the role of pH in this trend remains controversial.

Some people have proposed that excess atmospheric CO_2 be dumped into the deep sea, where low temperatures and high pressures convert the gas into a liquid. Unfortunately, fish placed in cages next to experimental liquid CO_2 dumps died from acidosis as the liquid gas slowly dissolved into seawater.

Corals in Hawai'i. These corals are being threatened worldwide due to ocean acidification.

Environmental alkalosis arises from basic minerals

Environmental alkalosis is not common but can arise from basic minerals like sodium carbonate in water. In Chapter 12, you learned about the excretory adaptations of tilapia living in the highly alkaline Lake Magadi in Africa (p. 561). Another example, Pyramid Lake in Nevada, has been slowly drying up for many years, and the resulting concentration of sodium carbonate is continuously increasing the lake's alkalinity (now at pH 9.4). There is concern that this threatens the resident cutthroat trout. In laboratory studies, trout exposed to water at pH 10 suffered a 0.25-pH-unit increase in blood pH, and over half of the fish died after 72 hours. The fish had excessively high levels of plasma ammonia and low levels of plasma Na^+ and Cl^-, suggesting that excretion and ion regulation were inhibited.

check your understanding 13.8

What are the causes of the six categories of acid–base imbalances?

Why is uremic acidosis so serious?

making connections

How Fluid and Acid–Base Regulation Contribute to the Body as a Whole

Homeostasis depends on maintaining a balance between the input and output of all constituents present in the internal fluid environment. Regulation of fluid balance involves two separate components: control of salt balance and control of H_2O balance. Control of H_2O balance is important in preventing changes in both ECF volume and osmolarity. Volume changes affect blood pressure, while osmolarity changes can induce detrimental osmotic shifts of H_2O between the cells and the ECF. Such shifts of H_2O into or out of the cells would cause the cells to swell or shrink, respectively. Cells, especially brain neurons, do not function normally when swollen or shrunken. In marine osmoconformers, the ICF has organic osmolytes to control osmotic pressure, and sometimes to protect proteins from destabilizing effects of urea and other stressors. In osmoregulators such as bony fishes and mammals, ECF salt, osmolarity, and volume are kept relatively constant.

A balance between input and output of H^+ is also critical to maintaining a body's acid–base balance within the narrow limits compatible with life. Deviations in the internal fluid environment's pH lead to altered neuromuscular excitability, to changes in enzymatically controlled metabolic activity, and to K^+ imbalances, which can cause cardiac arrhythmias. Hydrogen ions are uncontrollably and continually being added to body fluids as a result of metabolic activities, yet the ECF's pH must be kept relatively constant (at a slightly alkaline level of 7.4 for optimal function in mammals). In vertebrates, control of ECF pH by the kidneys is the main effector for achieving H^+ balance. Assisting the kidneys are the lungs, and buffers, which provide temporary pH stability in both ECF and ICF. Buffering mechanisms take up or liberate H^+, thereby transiently keeping its concentration constant within the body until its output can be brought into line with its input.

Chapter Summary

- If water–solute balance is to be maintained, input must equal output. Body water is distributed between the intracellular and extracellular fluid compartments (plasma, lymph, and interstitial and transcellular fluids). In vertebrates, the plasma and interstitial fluid are similar in composition, but the ECF and ICF are markedly different in all animals, with NaCl dominating in the ECF and K^+ and organic molecules in the ICF.

- Osmotic problems that threaten cells and animals include salinity, evaporation, ingestion and excretion, freezing, and certain pathologies. Animals have evolved two strategies to cope with osmotic challenges: osmoconforming and osmoregulating.

- Osmoconformers, which have internal osmolarities equal to that of the environment, use compatible and counteracting organic osmolytes in their cells. These include certain small carbohydrates, free amino acids, methylamines, methylsulfonium solutes, and urea. Compatible osmolytes do not perturb membranes and macromolecules, unlike high salt levels. Counteracting solutes offset factors that perturb macromolecules and membranes.

- Strict osmoconformers include most marine nonvertebrates and hagfish, which have ECF compositions similar to that of seawater but ICFs dominated by compatible organic osmolytes such as glycine and taurine. Stenohaline species cannot tolerate major changes in external salinity, while euryhaline ones can, by regulating cellular osmolytes to match ECF and external osmolarity. Hypoionic osmoconformers, with ECF salt levels less than seawater's, include marine chondrichthyans and coelacanths, which use the protein-destabilizing osmolyte urea counteracted by the protein-stabilizing osmolyte trimethylamine oxide (TMAO).

- Osmoregulators, which maintain a consistent internal osmolarity, rely on special transport epithelia. Hypo-osmotic regulators include most marine vertebrates and some arthropods, which use ion-excretion tissues such as gills. Some osmoregulating vertebrates have high levels of organic osmolytes, such as TMAO in deep-sea fishes and urea in estivating lungfish and frogs.

- Hyperosmotic regulation is found in all freshwater animals, and involves ion-uptake tissues such as gills, hypotonic urines, and impermeable integuments. Some animals such as salmon alternate between hyper- and hypo-osmotic regulation.

- Many terrestrial animals osmoregulate in the face of both low water and salt availability, using a variety of physiological (e.g., renal) and behavioral mechanisms.

- Fluid balance in mammals involves regulation of ECF volume and osmolarity. Control of ECF osmolarity prevents changes in ICF volume from hyper- or hypotonicity. Control of ECF volume is important in the long-term regulation of blood pressure and is adjusted in the short term through the baroreceptor reflex and plasma–interstitial fluid shifts. Both osmolarity and volume regulation are based on water and salt balances.

- Water balance depends on input from food, drink, and metabolism, and output via urine, feces, and insensible and sensible cutaneous and respiratory losses. Control of water balance by means of vasopressin and thirst is of primary importance in regulating ECF osmolarity.

- Long-term salt balances, which are more critical for volume regulation than water balance, depend on GFR changes and the renin–angiotensin–aldosterone system. During hemorrhagic shock, circulatory functions and fluid balance are coordinately regulated by many of these and other systems.

- Acid–base balance—the regulation of free hydrogen ions (H^+) in body fluids—is crucial to survival. Acids liberate free hydrogen ions, whereas bases accept them. The pH designation is used to express hydrogen ion concentration. Fluctuations in $[H^+]$ alter nerve, enzyme, and K^+ activity. Hydrogen ions are continually being added to body fluids from metabolic activities.

- Chemical buffer systems act as the first line of defense against changes in $[H^+]$ by binding with or releasing free H^+. Buffering is described by the Henderson-Hasselbalch equation; for example, $pH = pK + \log [HCO_3^-]/[CO_2]$

- The CO_2–HCO_3^- buffer system is the primary ECF buffer for noncarbonic acids. The peptide and protein buffer system, including hemoglobin in erythrocytes, is primarily important intracellularly. Buffers are a temporary mechanism because they do not eliminate excess $[H^+]$, necessitating the second and third lines of defense.

- Respiratory systems as a second line of defense regulate $[H^+]$ through adjustments in ventilation. Increased breathing removes CO_2, while decreased breathing retains it.

- Excretory systems are a third line of defense that aid acid–base balance by controlling both $[H^+]$ and $[HCO_3^-]$ in the ECF. Mammalian kidneys, for example, have cells that can secrete H^+ and reabsorb HCO_3^-, and other cells to do the reverse. In acidosis, some cells secrete NH_3, which can trap H^+ as NH_4^+.

- Respiratory acidosis arises from an increase in $[CO_2]$, typically from hypoventilation. Respiratory alkalosis usually arises from a decrease in $[CO_2]$, such as in hyperventilation.

- Metabolic acidosis is associated with a reduction in plasma $[HCO_3^-]$, caused by severe diarrhea, diabetes, strenuous activity, and uremia. Respiratory alkalosis usually arises from a decrease in $[CO_2]$, such as in hyperventilation. Metabolic alkalosis is associated with an elevation in $[HCO_3^-]$, and can result from vomiting.

- Environmental acidosis arises from natural and anthropogenic increases in environmental $[CO_2]$ and sulfuric acid. Anthropogenic CO_2 is already acidifying the oceans and harming some organisms, especially those with $CaCO_3$ shells and skeletons. Environmental alkalosis arises from basic minerals, as in some alkaline lakes.

Review, Synthesize, and Analyze

1. What is *anhydrobiosis?* What are some animal examples, and how do they survive?

2. Compare the use of *counteracting* organic osmolytes in marine Chondrichthyes, deep-sea fishes, and (from Chapter 12) the mammalian kidney.

3. If a horse loses 10 liters of salt-rich sweat and drinks 7 liters of water during the same time period, what will happen to vasopressin and aldosterone secretion? Why is it important to replace both the water and the salt?

4. If a solute that can penetrate the plasma membrane, such as dextrose (a type of sugar), is dissolved in sterile water at a concentration equal to that of normal body fluids and then is injected intravenously, what is the impact on the body's fluid balance?

5. Describe the effects of *hemorrhage* on fluid balance and the ensuing compensatory responses.

6. A pet cat has had pronounced diarrhea for more than a week as a result of having acquired a bacterial intestinal infection. What impact has this prolonged diarrhea had on its fluid and acid—base balance? In what ways has the cat's body been trying to compensate for these imbalances?

7. Describe the three lines of defense against changes in $[H^+]$ in terms of their mechanisms and speed of action.

8. Given that mammalian plasma pH = 7.4, arterial P_{CO_2} = 40 mm Hg, and each mm Hg partial pressure of CO_2 is equivalent to a plasma $[CO_2]$ of 0.03 mM, what is the value of plasma $[HCO_3^-]$?

9. Death of most mammals occurs if plasma pH falls outside the range of 6.8 to 8.0 for an extended time. What is the concentration range of H^+ represented by this pH range?

Suggested Readings

Claiborne, J. B., S. L. Edward, & A. I. Morrison-Shetlar. (2002). Acid—base regulation in fishes: Cellular and molecular mechanisms. *Journal of Experimental Zoology* 293:302–319.

Bradley, T. (2009). Animal osmoregulation. Oxford, UK: Oxford University Press. A current and thorough review of this topic.

Clegg, J. S. (2001). Cryptobiosis—A peculiar state of biological organization. *Comparative Biochemistry and Physiology* 128B:613–624. A review of mechanisms of dormancy.

Crowe, J. H., L. M. Crowe, & D. Chapman. (1984). Preservation of membranes in anhydrobiotic organisms: The role of trehalose. *Science* 223:701.

Ditrich, H. (2007). The origin of vertebrates: a hypothesis based on kidney development. *Zoological Journal of the Linnean Society* 150: 435–441.

Harrison, J. F. (2001). Insect acid—base physiology. *Annual Review of Entomology* 46:221–250.

Hochachka, P., & G. N. Somero. (2002). *Biochemical Adaptation: Mechanism and Process in Physiological Evolution.* Oxford, UK: Oxford University Press. A classic text.

Hoffmann, E. K., I. H. Lambert, & S. F. Pedersen. (2007). Physiology of cell volume regulation in vertebrates. *Physiological Reviews* 89: 193–277.

Jackson, D. C. (2000). How a turtle's shell helps it survive prolonged anoxic acidosis. *News in Physiological Sciences* 15:181–185.

Koeppen, B. M. (2009). The kidney and acid—base regulation. *Advances in Physiological Education* 33: 275–281.

Parker, A. R., & C. R. Lawrence. (2001). Water capture by a desert beetle. *Nature* 414:33–34.

Yancey, P. H. (2005). Organic osmolytes as compatible, metabolic, and counteracting cytoprotectants in high osmolarity and other stresses. *Journal of Experimental Biology* 208:2819–2830.

Yancey, P. H., M. E. Clark, S. C. Hand, R. D. Bowlus, & G. N. Somero. (1982). Living with water stress: Evolution of osmolyte systems. *Science* 217:1214–1222.

CourseMate

The rumen of a Scottish Highland cow enables it to survive under conditions of low food quality and availability.

Digestive Systems

14.1 Introduction: Feeding Strategies and Evolution

To maintain homeostasis, nutrient molecules are needed for two purposes. First, some are used as "fuel" for energy production and must continually be replaced by new, energy-rich nutrients. Second, nutrient molecules such as amino acids are needed as "building blocks" for ongoing synthesis of new cells and cell components in the course of tissue turnover and growth. Similarly, water and electrolytes must be replenished regularly. The digestive system contributes to homeostasis by transferring nutrients, water, and electrolytes from the external environment to the internal environment. Because living cells depend on a continuing flow of nutrients to sustain their metabolic needs, most animals commit considerable time and energy to acquiring and digesting food. Carbohydrates, fats, and proteins represent the main fuels; they are broken down into absorbable units (specifically simple sugars, fatty acids, and amino acids), which an animal's individual cells can use to produce ATP to carry out their particular energy-dependent activi-

ties, such as active transport, contraction, synthesis, and secretion. The units can also be used as building blocks. However, the digestive system does not directly regulate the concentration of any of these constituents in the internal environment; rather, it optimizes conditions for ingesting needed materials (food, water, and so forth), then digests and absorbs what is ingested.

We first provide an overview of digestive systems, examining the common features of the various components, before we begin a detailed tour of the tract from beginning to end.

Animal digestion evolved from an intracellular to an extracellular process in a specialized sac or tube connected to the environment

Digestion of food inside the body but outside the cell is accomplished by a digestive tract. Intracellular digestion is limited because only food items small enough to be internalized by the cells can be digested, whereas extracel-

lular digestion permits an animal to store and break down larger varieties and quantities of food items. The first digestive systems probably evolved as an infolding of the epidermis into an interior sac to trap food, providing the ability to capture much larger quantities of food. This is seen in Cnidaria (*knidos*, "stinging nettle" in Greek; jellies, corals, and their relatives), which are composed of an outer epidermis and an inner endoderm or *gastrodermis* ("stomach skin") lining a blind-end sac. In later animal groups, the sac fused with the opposite end of the body to form a complete tube with an entrance (mouth) and exit (anus). Cells lining this digestive tube could evolve specialized functions; for example, cells responsible for acidic digestion could be compartmentalized and separated from those involved in alkaline digestion. Other parts of the tube could be specialized for receiving and storing food and processing wastes (Figure 14-1). Indigestible items could remain outside and not take up space within cells.

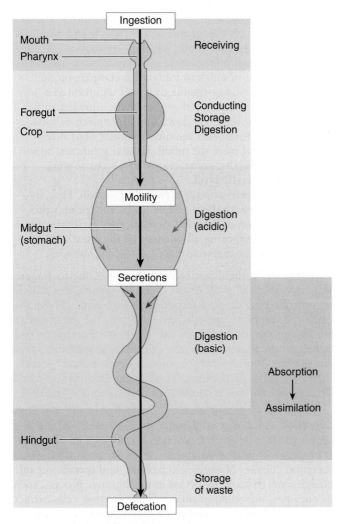

FIGURE 14-1 A generalized digestive tract. A digestive tract with one-way passage of food allows simultaneous operation of sequential stages in the processing of food and reduces mixing of digested and undigested matter. The crop is a storage region found in some animals.

© Cengage Learning, 2013

Thus, an animal's digestive tract can be thought of as a hollow sac or a tube with an entrance and exit, lying within the body and specialized for the initial processing of food items. At the openings, the tract is in direct contact with the outside world, so the lumen of the tract, although within the animal's body, is also technically part of the external environment. Until nutrients are actually absorbed across this tube, they cannot be considered "inside" the body. Conventionally, the gut tube in most animals can be divided into three specific regions: the **foregut, midgut,** and the **hindgut** (Figure 14-1).

Animals can be classified according to their primary feeding methods

Numerous methods for obtaining food have evolved. These include:

- **Filter or suspension feeders,** which range from sponges to baleen whales, trap organic material suspended in water (which may include dead particles, bacteria, algae,

protozoa, and small animals). An example of a filtering device can be seen in Figure 11-5, a polychaete's ciliated tentacles (the baleen mechanism will be described later).
- **Detritivores,** or **deposit feeders,** such as earthworms, many sea cucumbers and mites, feed on dead and living material in sediments.
- **Fluid feeders,** such as mosquitoes and vampire bats, which suck or lick fluids of a larger animal without ingesting the whole animal. Some fluid feeders are termed **parasites** if they remain attached to or within a host animal for long periods. A key example is the tapeworm, which has no digestive tract and lives within host intestines, where it absorbs small organic molecules through its body wall.
- **Carnivores** primarily eat other animals. They specialize in capturing live prey or rely on their abilities to locate carrion, and typically eat a high-energy, high-protein diet because of the high percentage of muscle tissue ("meat") in most animals.
- **Herbivores** depend on the intake of algal or plant materials, such as kelp blades and plant leaves, fruits, or seeds. Some of these materials have comparatively little directly available energy, and herbivores eating them have two options. The first, as seen in the dairy cow and the deer, uses **pregastric fermentation** (the processing of foodstuffs by microbes) as a means to break down plant material into absorbable units. In these animals the foregut is modified such that microorganisms can proliferate in a fermentation vat, the *rumen.* In a second type of herbivore, those with simple stomachs such as the horse and rabbit, microbial digestion occurs in the *hindgut,* which includes the *colon* and *cecum.* Herbivores either can be nonselective in their choice of food items or, as in the case of the panda, with its dependence on bamboo, can be highly selective. Herbivorous fish may feed on the fruits, seeds, flowers, and leaves of trees, whereas others may feed on algae.
- **Omnivores** such as humans, monkeys, and pigs can eat both animals and other organisms such as plants or fungi, and have digestive tracts more similar to carnivore tracts that rely on enzymatic digestion of foodstuffs. Because of seasonal changes in food availability, gut size adapts to low-quality feeds by increasing in length. For example, the starling is primarily a carnivore during the warmer months when insects and worms are available, but during the harsh winter months it is forced to become a herbivore and consume low-quality foods (buds, and so forth). In response, gut length increases a remarkable 450%.
- **Symbiotic autotroph-bearing** animals in the oceans require little or no external organic food (autotrophs are primary producers, namely photosynthesizers and chemosynthesizers). Key examples are (1) reef-building corals (Cnidarians) with photosynthesizing dinoflagellate symbionts that provide food molecules in addition to those ingested; and (2) vestimentiferan annelid worms living at hydrothermal vents (p. 741) and hydrocarbon (petroleum and natural gas) seeps. These worms use food molecules solely produced by symbiotic chemosynthesizing bacteria.

Depending on the life cycle of some animals, feeding specializations can be related to a particular stage of development (as the vestimentiferan worm illustrates). Insect larvae normally have completely different food habits and digestive systems than adults. Tadpoles of some frog species are herbivorous and have a longer and more complex intestine than does the adult, which is carnivorous. And young mammals shift from milk to an adult diet.

External food supply ultimately represents one of the key factors that determines animal survivability, the others being predation, parasitism, disease, competition, and beneficial symbioses (such as microbes in ruminants' digestive tracts and algae in coral Cnidarians). The reproductive success of many species can also be linked to food availability. Rainfall, light, and temperature determine plant and algal growth and consequently the amount of food available to herbivores (and therefore the rest of the food chain). Plant material also varies in composition during the course of a year. Immature grasses and leaves are characterized by a higher content of protein compared with other nutrients than later in the season. In the oceans, the amount of particulate food generally decreases with depth and also varies according to location and season. In environments characterized by seasonal variation in climate, natural selection favors reproductive cycles that can best exploit these limiting resources.

check your understanding 14.1

Describe the different modes of animal feeding.

14.2 General Aspects of Digestion

Now let's turn to general features of digestive systems common to most animals.

Digestive systems perform four basic digestive processes

There are four basic digestive processes: *motility, secretion, digestion,* and *absorption.*

Motility The term **motility** refers to the muscular contractions within the gut tube that mix and move forward the contents of the digestive tract. However, some worm-shaped animals rely on muscular contractions of the body to indirectly force food through the gut tube. Although the smooth muscle in the walls of digestive tracts is typically *phasic* smooth muscle (p. 376) that displays action potential–induced bursts of contraction, it also maintains a constant low level of contraction known as **tone.** Tone is important in maintaining a steady pressure on the digestive tract contents as well as in preventing the walls of the digestive tract from remaining permanently stretched following distension.

Two basic types of digestive motility are superimposed on this ongoing tonic activity:

- *Propulsive movements* propel, or push, the contents through the digestive tract at varying speeds, with the rate of propulsion dependent on the functions accomplished by the different regions; that is, food is moved forward in a given segment at an appropriate velocity to allow that segment to "do its job." For example, transit of food through the foregut of arthropods and esophagus of vertebrates is rapid, which is appropriate because these structures merely serve as a passageway from the mouth to the stomach or midgut. In comparison, in the small intestine—the major site of digestion and absorption—the contents move slowly, allowing sufficient time for the breakdown and absorption of food. Propulsive movements are often thought of as being unidirectional from the mouth to anal sphincter; however, antiperistaltic contractions also occur in some species, especially birds.

- *Mixing movements* serve a twofold function. First, by mixing food with the digestive juices, movements promote digestion of the food. Second, they facilitate absorption by exposing all portions of the intestinal contents to the digestive tract's absorbing surfaces.

Motility in the digestive tract is mainly due to smooth muscle within the walls of the digestive organs; this muscle is controlled by complex autonomic ("involuntary") mechanisms. Exceptions include the ends of the mammalian tract—the mouth and some or all of the esophagus (see p. 662) at the beginning and the external anal sphincter at the end—where motility may involve skeletal muscle. Accordingly, the acts of chewing, swallowing (rumination), and defecation have motor cortex control ("voluntary," p. 192).

Secretion A number of digestive juices are secreted into the digestive tract lumen by exocrine glands located along the route, each with its own specific secretory product(s). Each secretion consists of water, electrolytes, and specific organic constituents that are important in the digestive process, such as enzymes, bile salts, or mucus. The secretory cells extract large volumes of water and those raw materials necessary to produce their particular secretion from blood. Secretion of all digestive juices requires energy, both for active transport of some of the raw materials into the cell (others diffuse in passively) and for synthesis of secretory products by the endoplasmic reticulum. On appropriate neural or hormonal stimulation, the secretions are released into the digestive tract lumen. Normally, the digestive secretions are reabsorbed in one form or another (back into the blood in vertebrates) after they participate in digestion. Failure to do so (because of vomiting or diarrhea, for example) results in loss of this fluid that has been "borrowed" from the plasma.

Digestion Digestion is the process whereby the structurally complex foodstuffs of the diet are broken down into smaller, absorbable units by enzymes produced within the digestive system. Complex carbohydrates, proteins, and fats are large molecules unable to cross plasma membranes intact. So, to be absorbed from the lumen of the digestive tract into the blood or lymph, they must be broken down into smaller nutrient molecules (although some animals use endocytosis, for example, of whole proteins in neonatal mammals). Digestion begins in the anterior regions of animal tracts and is accomplished for all nutrients with enzymatic **hydrolysis** ("breakdown by water"). Hydrolytic enzymes (**hydrolases**) add H_2O at the bond site, thereby breaking down the bonds that hold the small molecular subunits within the nutrient molecules together, thus setting the small molecules free (Figure 14-2a). These small subunits were originally joined

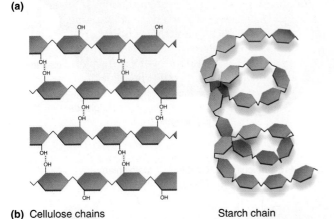

(a)

(b) Cellulose chains   Starch chain

FIGURE 14-2 Structure and hydrolysis of common dietary carbohydrates. (a) An example of hydrolysis. In this example, the disaccharide maltose (the intermediate breakdown product of polysaccharides) is broken down into two glucose molecules by the addition of H_2O at the bond site. (b) Hydrogen bonding relationships for glucose units in cellulose and starch. In cellulose, hydrogen bonds form between glucose chains. This pattern stabilizes the chains and permits them to become tightly bundled. In amylose, a form of starch, the bonds form between units within a chain, which permits the glucose units to coil.

Source: C. Starr. (2000). *Biology: Concepts and Applications*, 4th ed., Belmont, CA: Brooks/Cole, p. 39, Figure 3.7.

to form nutrient molecules by the removal of H_2O at the bond sites. All digestive enzymes are hydrolytic, and the remarkable similarity of enzymes used by different animal species attests to their appearance early in evolution. Digestive enzymes are specific in the bonds they can hydrolyze. As food moves through the tract, it is subjected to various enzymes, each of which breaks down the food molecules even further. In this way, large food molecules are converted to simple absorbable units in a progressive, stepwise fashion as the tract contents are propelled forward. The major molecules are as follows:

1. **Carbohydrates** include the following:
    a. **Monosaccharides** ("one-sugar" molecules), such as **glucose, fructose,** and **galactose,** very few of which are normally found in most diets. They are the absorbable units.
    b. **Disaccharides** ("two-sugar" molecules) are another source of dietary carbohydrate and include **sucrose** (table sugar, which consists of one glucose and one fructose), **trehalose** (the most important carbohydrate for flight in insects, made up of two glucose molecules in opposite orientations), **maltose** (also made of two glucose molecules but in the same orientation), and **lactose** (milk sugar, made up of

one glucose and one galactose). They are hydrolyzed into monosaccharides in digestive tracts synthesizing the appropriate enzymes.

    c. **Polysaccharides** ("many-sugar" molecules), which consist of chains of interconnected monosaccharide molecules, are the most common carbohydrates of animal diets. The most common polysaccharides that are derived from plant sources (Figure 14-2b) are **starch** (which has α-bonds between the glucose units) and **cellulose** (which has β-bonds). To be usable, they also must be hydrolyzed into monosaccharides. Cellulose is the most abundant organic molecule in the biosphere and is the major component of plant material, comprising over half of the plant cell wall. However, few animals can directly exploit this energy source. Given the fibrous "staircase" structure of cellulose resulting from its β-bond linkages (Figure 14-2b), it is extremely difficult to hydrolyze. In animals that use cellulose for fuel, cellulose digestion is accomplished by symbiotic microorganisms that proliferate in the digestive tract and that have the necessary enzymes—**cellulases**—to hydrolyze cellulose. These microbes may be housed in a specific area of the gut or in special organs connected to the gut, as in the case of termites and many wood-boring beetles. Some termites and wood wasps actually culture fungi that produce cellulase. **Chitin** is the principal component of the hard exoskeleton of arthropods and, like cellulose, is indigestible by most vertebrates. **Glycogen** represents the polysaccharide storage form of glucose in animals. After digestion of a meal, unused glucose is stored as glycogen (in liver and muscle in vertebrates; p. 26). *Cecropia,* a bamboolike tree, is the only plant known to synthesize glycogen. A particularly aggressive ant, *Azteca,* consumes the glycogen and, in turn, protects the tree from potential agents that could damage it.

2. **Proteins** are the second category of foodstuffs, which consist of various combinations of **amino acids** held together by peptide bonds. Through the process of digestion, proteins are degraded primarily into their constituent amino acids as well as a few **polypeptides** (several amino acids linked together by peptide bonds), both of which are the primary absorbable units for digested protein.

3. **Fats** represent the third category of foodstuffs. In most terrestrial food chains, dietary fat is in the form of **triglycerides,** which are neutral fats, each consisting of a combination of glycerol with three **fatty acid** molecules attached. During digestion, two of the fatty acid molecules split off, leaving a **monoglyceride,** a glycerol molecule with one fatty acid molecule attached. Thus, the end products of fat digestion are monoglycerides and free fatty acids, which are the absorbable units of fat. However, in ruminant species the triglycerides are degraded mostly to free fatty acids and glycerol, with the glycerol being fermented. For this reason glycerol synthesis must occur to a greater extent than in other species. In marine food chains, **waxes** (a fatty alcohol and fatty acid linked together) are common lipids. Waxes are hydrolyzed by the activity of esterases synthesized in microorganisms.

4. **Nucleic acids** can also form a significant portion of ingested nitrogen. Pancreatic nucleases, *RNase* and *DNase,* split the nucleic acids into their component nucleotides. The *duodenum* of ruminants is particularly effective in digesting nucleic acids because approximately 20% of the nitrogen entering the *abomasum* (acid-secreting portion of the stomach) originates from the microorganisms spilled from the rumen.

Absorption After digestion is completed, the process of absorption occurs in the middle to posterior sections of most digestive systems (such as the small intestine in vertebrates). Here the small, absorbable units that result from digestion, along with water, vitamins, and electrolytes, are transferred from the digestive tract lumen into the blood or hemolymph or body cavity. Specialized transporters in the epithelial cells carry the hydrophilic units through the membrane. The uptake of amino acids and/or glucose depends on cotransport with Na$^+$. To enhance this process (in accordance with Fick's law of diffusion, p. 76), *surface area* is often greatly increased by folds. For example, in annelids (such as earthworms) and mollusks, long folds forming ridges called **typhlosoles** line parts of the tract (Figure 14-3a). You will see later how *villi* increase surface area in vertebrate intestines.

The ocean contains large amounts of dissolved organic matter, including sugars and amino acids. Some aquatic species can also absorb these nutrients through their gills (as in the mussel *Mytilus*), or through the epidermis (as in many marine nonvertebrates including annelids and echinoderms).

The digestive tract and accessory digestive organs make up digestive systems

The digestive system of an animal consists of the digestive tract plus the accessory digestive organs. In addition to its role in the digestive process, additional functions of the tract include osmoregulation, endocrine secretions, immune function, and the elimination of toxins.

In annelids and arthropods the foregut is referred to as the **stomodaeum,** the midgut as the **mesenteron,** and the hindgut as the **proctodaeum** (Figure 14-3b). Insect stomodaea and proctodaea are lined internally with a thin layer of cuticle, the **intima,** which is shed along with the outer skeleton during molt. The stomodaeum, which is specialized for the reception of food, consists of the *pharynx, esophagus, crop,* and *proventriculus*. The **stomodael valve** regulates the passage of food between the foregut and the midgut. The midgut is an elongated tube and generally consists of two or more sections. The midgut of most insects, for example, contains **diverticula** (blind tubules), which arise from the main passage and contain the **gastric caecae** (saclike structures, open at only one end) near the anterior end. The

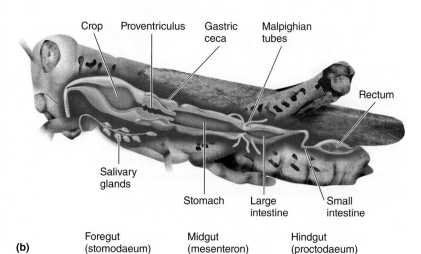

(a)

(b)

Foregut (stomodaeum)   Midgut (mesenteron)   Hindgut (proctodaeum)

FIGURE 14-3 Comparison of mollusk and insect digestive tracts. (a) A bivalve mollusk tract. Many bivalve and some gastropod mollusks have a crystalline style, which is rotated by cilia and the head of which grinds food against a hard shield in the stomach. The style also contains enzymes that are released by grinding. (b) Tract of a typical insect.

Sources: (a) V. Pearse, J. Pearse, M. Buchsbaum, & R. Buchsbaum. (1987). *Living Invertebrates*. Palo Alto, CA: Blackwell Scientific, p. 352, top left. (b) Modified from R. L. Patton. (1963). *Introductory Insect Physiology*. Philadelphia: Saunders, 1963, p. 31, Figure 3-1.

midgut is specialized for storing and digesting food items and subsequent midgut absorption of nutrients. In some cases, the same cell can carry out secretion and absorption. The hindgut extends from the **pyloric valve,** which separates the midgut from the hindgut, to the anus. The hindgut consists of two sections, the anterior *intestine* and the posterior *rectum*. Recall that the Malpighian tubules, which function as excretory organs (p. 563), empty their contents into the anterior end of the hindgut.

The digestive (or *gastrointestinal*) tract of most vertebrates includes the following organs (Figure 14-4): *mouth; pharynx* (throat); *esophagus; stomach* (and *rumen* in ruminants), or *proventriculus-gizzard complex* in birds; *small intestine; large intestine;* and *anus*. Note that these organs are continuous with each other and are discussed as separate entities only because of their regional modifications, which allow them to specialize in particular digestive activities.

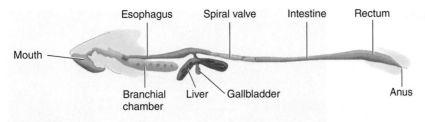

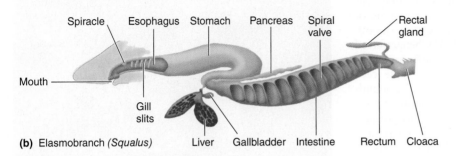

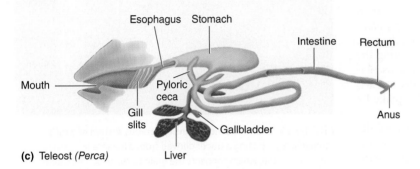

(a) Cyclostome *(Petromyzon)*

(b) Elasmobranch *(Squalus)*

(c) Teleost *(Perca)*

FIGURE 14-4 Vertebrate digestive tracts showing progressive anatomic specialization for digestion (stomach and accessory digestive glands such as pancreas and liver) and absorption surface area of small intestine.

Source: D. C. Withers. (1992). *Comparative Animal Physiology.* Fort Worth, TX: Saunders, p. 925, Figure 18-28.

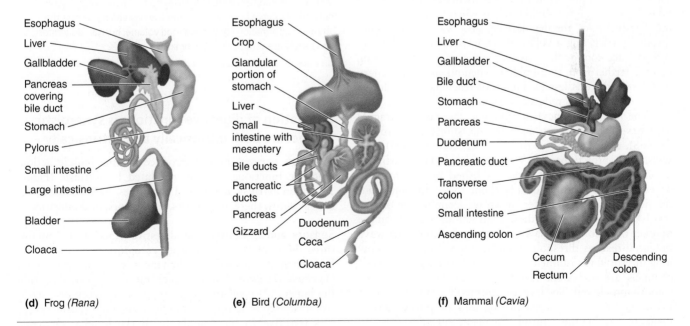

(d) Frog *(Rana)* **(e)** Bird *(Columba)* **(f)** Mammal *(Cavia)*

In addition to the specialized gut tubes, many phyla have **accessory organs** that aid digestion. The accessory digestive organs of most vertebrates include the *salivary glands,* the *exocrine pancreas,* and the *biliary system,* which consists of the *liver* and, in some species, the *gallbladder* (Figure 14-4). (Most metazoans have salivary glands; see Figure 14-3). These exocrine organs are located outside of the wall of the digestive tract and empty their secretions through ducts into the digestive tract lumen. They develop from outpocketings of the embryonic digestive tube and maintain their connection with the digestive tract through the ducts that are formed.

Some smooth muscles of digestive tracts have autonomous activity

Digestive motility and secretion are carefully regulated to maximize digestion and absorption of the ingested food. First let's examine local action of the smooth muscles.

Autonomous Smooth-Muscle Function As we discussed in Chapter 8, some smooth muscle cells do not have a constant resting potential but rather display rhythmic, spontaneous variations in membrane potential. The prominent type of self-induced electrical activity in vertebrate digestive smooth muscle are *slow-wave* potentials (Figure 8-31), here referred to as the **basic electrical rhythm (BER)**, The BER is initiated by "pacesetter" cells, musclelike but noncontractile cells known as the **interstitial cells of Cajal**, located at the boundary between the *longitudinal* and *circular* smooth muscle layers (described on p. 671). These cells periodically produce electrical signals that spread to the adjacent contractile smooth-muscle cells as the BER. Recall that sheets of smooth muscle cells are connected by gap junctions; thus, the whole muscle sheet behaves like a functional *syncytium* (p. 378). Recall also that slow waves are not action potentials and do not directly induce muscle contraction; they are fluctuations in membrane potential that cyclically bring the membrane closer to or farther from threshold. These oscillations are believed to be due to cyclical variations in Ca^{2+} release from the endoplasmic reticulum and Ca^{2+} uptake by the mitochondria of the pacesetter cell.

Whether threshold for an action potential is reached depends on the effect of various mechanical, nervous, and hormonal factors that influence the "resting" potential, or the starting point around which the BER oscillates. If the starting point is nearer the threshold level, as when food is present in the digestive tract, the depolarizing slow-wave peak reaches threshold, triggering a volley of action potentials and its accompanying contractile activity. Conversely, if the starting point is farther from threshold, as when an animal is hungry or starved, there is less likelihood of reaching threshold, so contractile activity is reduced or eliminated.

The *rate* (frequency) of rhythmic digestive contractile activities—such as *peristalsis* in the stomach or rumen (p. 668), *segmentation* in the small intestine (p. 693), and *haustral* contractions (p. 703) in the large intestine—depends on the inherent rate established by the involved pacesetter cells. Specific details about these rhythmic contractions are discussed later, when we examine the organs involved. The *intensity* of these contractions depends on the number of action potentials that occur when the slow-wave potential reaches threshold, which in turn depends on how long threshold is sustained. At threshold, voltage-gated Ca^{2+} channels are activated, resulting in Ca^{2+} influx into the smooth muscle cell. The level of contractility can range from low-level tone to vigorous mixing and propulsive movements by varying the cytosolic Ca^{2+} concentration.

Furthermore, the intensity of the contraction also depends on the number of muscle fibers in the organ. Species vary tremendously in muscle mass within the walls of digestive tract organs. For example, the thick and thin muscle pairs lining the *gizzard* (organ of mastication) (p. 681) of a ruffed grouse can crack the shell of an acorn, whereas these muscle pairs are absent in most carnivorous birds, whose stomach is more similar to that of mammalian carnivores.

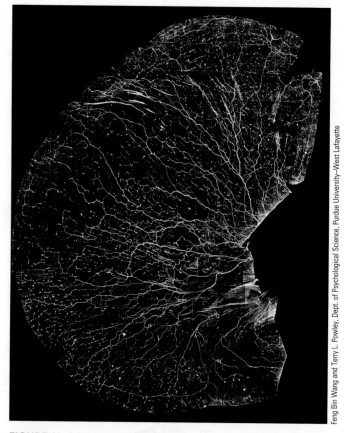

FIGURE 14-5 The enteric (intrinsic) nervous system of a rat's stomach. By injecting a tracer derived from a horseradish enzyme into the vagus nerve, which connects the brain to the esophagus and stomach, researchers were able to reveal the extent of the nerve network. As nerve fibers fray out into the tiny endings across the stomach, information concerning food volume, hunger, discomfort, and satiety are sent back to the brain.

Feng Bin Wang and Terry L. Powley, Dept. of Psychological Science, Purdue University–West Lafayette

Regulation of digestive functions is synergistic and occurs both intrinsically and extrinsically

Although pacesetters may produce routine muscle action, digestive functions are also regulated by intrinsic (Figure 14-5) and extrinsic nerves and hormones. As in most regulatory systems, there are sensors, integrators, and effectors.

Sensors and Effectors The digestive tract wall contains three different types of sensory receptors that respond to local chemical or mechanical changes in the digestive tract: (1) *chemoreceptors* sensitive to chemical components within the lumen, (2) *mechanoreceptors* (pressure receptors) sensitive to stretch or tension within the wall, and (3) *osmoreceptors* sensitive to the osmolarity of the luminal contents. Stimulation of these receptors elicits neural reflexes or secretion of hormones by endocrine cells, both of which alter the level of activity in the digestive system's effector cells. These effector cells are smooth muscle cells (for modifying motility) and exocrine gland cells (for controlling secretion of digestive juices) (Figure 14-6).

Now let's look at the integrators and their reflex pathways.

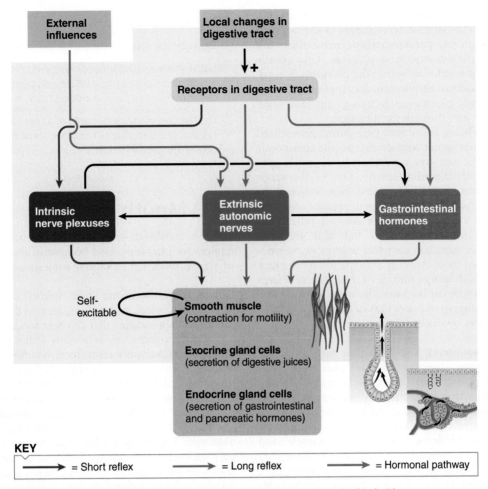

FIGURE 14-6 Summary of pathways controlling digestive system activities. Receptor proteins bind with and respond to gastrointestinal hormones, neurotransmitters, and local chemical mediators.

© Cengage Learning, 2013

Intrinsic Nerve Plexuses In vertebrates, the **intrinsic nerve plexuses** (*plexus,* "network") are the two major networks of nerve fibers—the **myenteric plexus** and the **submucous plexus**—located entirely within the digestive tract wall and running its entire length. A nerve plexus is a type of nerve net (p. 151). Thus, unlike any other organ system, the digestive tract has its own intramural ("within wall") nervous system, which contains as many neurons as the spinal cord and endows the tract with a considerable degree of self-regulation (a prime example of hierarchic distributed regulation, p. 19). In vertebrates, the two plexuses together are often termed the **enteric nervous system** (Figure 14-5). Insects have an analogous system, the **stomatogastric nervous system,** a network of peripheral ganglia along the gut.

The intrinsic plexuses influence all facets of digestive tract activity. Various types of neurons are present in the intrinsic plexuses. Some are afferent sensory neurons, which have receptors that respond to specific local stimuli in the digestive tract. Efferent local neurons innervate the smooth muscle cells and exocrine and endocrine cells of the digestive tract to directly affect digestive tract motility, secretion of digestive juices, and secretion of gastrointestinal hormones. As with the central nervous system, these input and output neurons of the enteric nervous system are linked by integrating interneurons. Some of the output neurons are excitatory, and some are inhibitory. For example, vertebrate enteric neurons that release *acetylcholine* (*ACh*) as a neurotransmitter promote contraction of digestive-tract smooth muscle, whereas the neuromodulators *nitric oxide* and *vasoactive intestinal peptide* act in concert to cause its relaxation. These intrinsic nerve networks are primarily responsible for coordinating local activity within the digestive tract. To illustrate, if a large chunk of food gets stuck in the esophagus, local contractile responses coordinated by the intrinsic plexuses are initiated to push the food forward.

Extrinsic Nerves In hierarchical fashion, extrinsic neural and hormonal regulation can modulate the nerve plexuses, or, in some cases, can act directly on the smooth muscle and glands. Extrinsic nerves in vertebrates are the nerve fibers from both branches of the autonomic system that originate outside the digestive tract and innervate the various digestive organs. Recall that in general, the sympathetic and parasympathetic nerves supplying any given tissue exert opposing actions on that tissue (p. 159): The "fight-or-flight" sympathetic system inhibits or slows down digestive tract blood flow and motility, while the "rest-and-digest" parasympathetic system dominates in quiet, relaxed situations, when general maintenance activities such as digestion can proceed optimally. Accordingly, the parasympathetic nerve fibers supplying the digestive tract, which arrive primarily by way of the *vagus nerve,* increase smooth muscle motility and promote secretion of digestive enzymes and hormones. Unique to the parasympathetic nerve supply to the digestive tract, the postganglionic parasympathetic nerve fibers are actually a part of the intrinsic nerve plexuses. They are the ACh-secreting output neurons within the plexuses. Accordingly, ACh is released in response to local reflexes coordinated by the intrinsic plexuses as well as to extrinsic vagal stimulation, which acts through the plexuses.

In addition to being called into play during generalized sympathetic or parasympathetic discharge, the autonomic nerves supplying the digestive system can be discretely activated to modify only digestive activity. One of the major purposes of specific activation of extrinsic innervation is the coordination of activity between different regions of the digestive system; for example, the act of chewing reflexly increases not only salivary secretion but also stomach, pancreatic, and liver secretion via vagal reflexes in *anticipation* of the arrival of food. Another anticipation mechanism is a pathway by which factors outside of the digestive system can influence digestion, as, for example, with a vagally mediated increase in digestive juices that occurs before eating when an animal sees or smells food.

Gastrointestinal Hormones Tucked within the mucosa of certain regions of the digestive tract are endocrine gland cells that, on appropriate stimulation, release hormones into the blood. These **gastrointestinal hormones,** some of which we will discuss later, are carried through the blood to other areas of the digestive tract, where they exert either excitatory or inhibitory influences on smooth muscle and exocrine gland cells. Interestingly, many of these same hormones are released from neurons in the brain, where they act as neurotransmitters and neuromodulators. During embryonic development, certain cells of the developing neural tissue migrate to the digestive system, where they become endocrine cells.

Reflex Pathways Now let's summarize intrinsic and extrinsic regulatory actions. Receptor activation may bring about two types of neural reflexes—*short* reflexes and *long* reflexes. When the intrinsic plexuses respond to local stimulation and alter effectors, all elements of the reflex are located within the wall of the digestive tract itself; that is, a **short reflex** takes place. Extrinsic nervous activities, which can be superimposed on the local controls to modify smooth muscle and glandular responses for coordination and anticipation, are known as **long reflexes** because they involve the long pathways between the central nervous system and digestive system.

In addition to the sensory receptors within the digestive tract wall that monitor luminal content and wall tension, the plasma membranes of the digestive system's effector cells have receptor proteins that bind with and respond to gastrointestinal hormones, neurotransmitters, and local chemical mediators.

From this overview, it is clear that regulation of gastrointestinal function is very complex, being influenced by many synergistic, interrelated pathways that evolved to ensure that the appropriate responses occur to digest and absorb the ingested food. Nowhere else is such a level of overlapping control exercised.

We are now going to take a "tour" of the digestive tract, beginning with the mouth and ending with the rectum, focusing primarily on vertebrates and insects.

check your understanding 14.2

Describe the four basic digestive processes.

List the three categories of energy-rich foodstuffs and the absorbable units of each. Which polysaccharide is the most common storage form in animals? Why is this so?

List the common components of a representative digestive system, and describe its cross-sectional anatomy and its neural–hormonal features.

14.3 Mouth

There is considerable diversity in the mechanisms and behaviors for obtaining food because of the enormous variety of animal diets. Let's examine some examples.

Obtaining and Receiving Food Animals may use shredding and cutting devices, such as bony *teeth* in many vertebrates, or specialized appendages that can tear food items into smaller pieces prior to ingestion, as in many arthropods. For example, many insects (such as locusts) have *mandibles* adapted for cutting and grinding. Other mouthparts associated with reducing food size include chisels and chitinous teeth. Liquid food such as nectar does not require reduction and is typically obtained by suction. External parasites have modified mouthparts adapted to feeding on the blood or other body fluids of a suitable host. The mosquito, for example, pierces the skin of its prey with an array of six needlelike mouthparts. One of these injects an anticoagulant (p. 400), which leads to the uncomfortable itch that follows the bite. It is through this route that foreign organisms such as those causing malaria, encephalitis, and yellow fever gain entry to the body. Only the egg-laying females (who have a greater protein requirement) are the nuisances, because the males dine exclusively on nectar.

Other body parts can aid in food acquisition, the hands of primates being obvious examples. As another example, Lophiiformes—a marine group that includes goosefishes, frogfishes, and anglerfishes—use modifications in their first dorsal fin to function as a lure to attract potential prey. The fin is used to dangle a lure that may resemble a source of live food immediately above their upwardly directed mouths. In deep-sea anglers, the lure contains symbiotic bioluminescent bacteria, whose glow attracts prey animals to investigate.

The **mouth** (or **oral cavity**) itself, which initially receives the food entering the gut, may also be specialized. The mouth of a snake, for example, requires unique adaptations to swallow a meal up to 1.5 times the snake's own body weight. A snake's jaw can open to an angle of up to 130 degrees, whereas a human's jaw can be opened only 30 degrees. Snakes have a hinged jaw as opposed to the single bone found in mammals. The snake's lower jaw is also split into two parts, with the halves connected by an elastic ligament, which permits the snake to completely stretch its head around its victim's body. Powerful muscles in the cheek and throat steadily push its prey head first toward the esophagus helped in part by the backward curvature of the teeth. Among modern-day sharks, such as the white-tipped reef shark, courtship rituals include the male biting the female's back, neck, and then the pectoral fin, a move that enables him to hold onto the female while copulating. Based on this observa-

A male Atlantic puffin demonstrates its hunting prowess to female puffins. The specialized beak can hold a dozen or more fishes inside. It has backward-pointing spikes in its interior, which hold fish in place while other fish are being hunted.

tion, John Long speculated that jaws may have first evolved to improve mating success and not for chewing food.

Although ancient birds had teeth (found as fossils), modern birds lack them, but their beaks are often specialized for teethlike functions. The beak represents one of the more evolutionarily plastic components of the avian digestive system, having been molded to conform to particular feeding habits. In herbivorous birds, the beak is an enlarged, crushing forceps, whereas in carnivorous birds the beak is a sharp-edged tearing instrument. The beaks (bills) of some birds, including ducks and geese, are richly innervated with sensory endings. The density of mechanoreceptors along the edges and at the tip of the bill is actually greater than in the fingertips of a human's index finger. Interestingly, as sophisticated beaks evolved, key genes controlling teeth development became pseudogenes (p. 33), although genes for constructing teeth remain intact in avian genomes. Given exposure to a tooth-inducing tissue from a mouse embryo, a chicken embryo's beak tissue will produce crocodilian-like teeth, and the Talpid2 mutant strain of chicken is an example of a bird that makes such teeth.

In mammals, the mouth is ringed by the muscular **lips,** which aid in **prehension,** the seizing and conveying of food to the mouth. For example, the horse uses its lips to move foodstuffs into the mouth. The lips also serve nondigestive functions. They are important in communication; for example, facial expression and the articulation of many sounds depend on a particular lip formation.

The **palate,** which forms the arched roof of the oral cavity, separates the mouth from the nasal passages. Its presence allows breathing and chewing or sucking to take place simultaneously. Toward the front of the mouth, the palate is made of bone, forming what is known as the **hard palate,** There is no bone in the portion of the palate toward the rear of the mouth; this region is called the **soft palate,** However, in most avian species there is no soft palate, but the hard palate is separated by a median slit that permits direct communication with the nasal cavities. The floor of the avian mouth is membranous and, in some species such as pelicans, serves a storage

capacity. In some primates, including humans, hanging down from the soft palate at the rear of the throat is a dangling projection, the **uvula,** which plays an important role in sealing off the nasal passages during swallowing (p. 667).

The **tongue,** which forms the floor of the oral cavity, consists of voluntarily controlled skeletal muscles that form a *muscular hydrostat* (see p. 357). Methods of obtaining food and drink vary considerably; for example, the ballistic tongue of most lizards relies on simple surface tension (stickiness created when a wet tongue contacts dry prey) to catch prey. However, the tongue pouch of the chameleon functions more like a suction cup as it is capable of consuming prey much larger than what simple surface tension could handle. In ruminant mammals, the tongue is the organ of prehension because the lips have limited motility and the upper incisor teeth are absent. A giraffe can wrap its 46-cm (18-inch) tongue around a leafy branch and tear the leaves off into its mouth. Tongues can also be used for drinking. Cats and dogs, for example, curl their tongues downward into water, then pull it up rapidly to create a rising column of water. The jaws then snap over the column to engulf it before it falls back. A domestic cat can do this four times per second!

Movements of the tongue are important in guiding food within the mouth during chewing and swallowing, and also play an important role in vocalization. Embedded within the tongue are the *taste buds* (p. 234), which are also dispersed in the soft palate, throat, and linings of the cheeks. The tongue also continually synthesizes an *antimicrobial peptide* (p. 467). For this reason, tongues are rarely infected and heal quickly after injury.

Mastication The first step in the digestive process is **mastication** (chewing), the motility of the mouth that involves slicing, tearing, grinding, and mixing ingested foods by specialized mouthparts such as teeth. The purposes of mastication are (1) to grind and break food up into smaller pieces to facilitate swallowing and to increase the food surface area on which salivary enzymes will act, (2) to mix food with saliva, and (3) to stimulate the taste buds, which reflexly increases salivary, gastric, pancreatic, and bile secretion to prepare for the arrival of food in the lower digestive tract. However, carnivores normally swallow food without chewing and mixing the contents with saliva. Ruminants chew the feed only briefly before swallowing, sufficiently grinding the forage into particles just small enough to be formed into a food bolus. Only when ruminants regurgitate (movement of a food bolus from the stomach back to the mouth) is the food then thoroughly masticated. In contrast, the horse thoroughly masticates its food prior to swallowing.

The act of chewing in vertebrates is normally regulated by a chewing center in the medulla oblongata. The presence of food in the oral cavity activates mechanoreceptors that, via afferent pathways to this center, activate motor neurons that open the oral cavity while inhibiting the neurons that cause the closure of the cavity.

Let's look at functions of teeth, which are firmly embedded in and protrude from the jawbones. The exposed part of a tooth is covered by **enamel,** the hardest structure in most vertebrates. Enamel forms before the tooth's eruption by special cells that are lost as the tooth erupts. Teeth in fish, amphibians, and reptiles are specialized for holding and tearing food. **Polyphydontia,** the replacement of teeth throughout an animal's lifetime, occurs in dinosaurs, sharks, and nonavian reptiles. Tooth replacement varies considerably among species and age. For example, it occurs every two weeks in most sharks, but about every two years in an adult crocodile.

Mammals have **incisors** (cutting teeth) in front to grab and hold food and flat **molars** in the back for grinding fibrous material. Carnivorous mammals are noted for sharp **canines** (next to incisors) for seizing, piercing, and tearing prey. Finally there are **premolars** between canines and molars, with intermediate properties. As an example, all Old World primates (including humans) have 8 incisors, 4 canines, 8 premolars, and 12 molars. The extent to which molars are used varies between species and is related to the composition of the diet. Simple up-and-down movements of molars in carnivores and omnivores are inadequate to grind tough, coarse feeds, so herbivores make lateral grinding movements of the jaw. The upper jaw is slightly wider than the lower jaw, and the animal masticates on one side of the mouth at a time. Because of the lateral jaw movement, premolars may develop sharp edges that interfere with chewing or may injure the tongue or cheeks. This problem in domestic herbivores can be corrected by "floating the teeth," that is, filing the projections down. Herbivores also have a long **diastema,** or gap, between the front teeth and the premolars on each side that allow ready manipulation of food by the tongue. Ruminants have a unique dentition; the upper incisors and canines are absent, while the lower incisors bite against a hard gum or dental pad in the upper jaw. They manipulate forage into the mouth with the tongue and lips and then cut or tear it with their lower incisors and dental pads. The teeth of rabbits, and the lower incisors of possums, koalas and rodents, grow continually throughout their lifetime, a consequence of abrasive high fiber diets, which continually wear down the enamel coating.

Teeth can evolve other roles, such as prey-paralyzing venomous fangs in snakes. Recent research has revealed that the first snake fangs evolved about 60 million years ago from the posterior tooth-forming layer in the upper jaw. At some point that layer became detached from the rest of the jaw and so became free to evolve independently of other teeth.

Not all vertebrates rely on teeth. Teeth in embryonic baleen whales (Mysticeti) are vestigial and are reabsorbed and replaced with long plates of **baleen** (parallel, fused filaments of *keratin,* the protein of hair) that hang from the jaws in place of teeth. These are used as straining devices for filter feeding, primarily on small marine animals such as krill (crustaceans) and fish. Such filtering is highly efficient: Recent energy estimates suggest that baleen whales gain at least 90 times more energy than they expend when filtering schools of prey such as krill. This may explain why blue whales can achieve body masses of 100,000 kg or more. In contrast, energy calculations predict that carnivores such as polar bears, which target large single prey, cannot obtain enough energy if they exceed 1,100 kg in size. Indeed, no such carnivores (living or extinct) exceed that limit.

Saliva aids in mastication but plays a more important role in lubricating food boluses before swallowing

Saliva, the secretion associated with the mouth, is produced by salivary glands, which are located outside the oral cavity and discharge saliva through short ducts into the mouth. The functions of saliva include the following:

1. *Moistening:* Saliva facilitates swallowing by moistening food particles, thereby holding them together, and (in vertebrates but not insects) by providing lubrication through the presence of mucus. Saliva keeps the oral cavity moist (and aids in vocalization by facilitating movements of the lips and tongue). In insects saliva provides lubrication of the mouthparts.

2. *Digestion:* In most animals (excluding ruminants), saliva begins digestion of starch in the mouth through action of **salivary amylase,** an enzyme that breaks down starches into maltose, a disaccharide consisting of two glucose molecules. In insects both the salivary gland and midgut cells can secrete amylases and various proteases. In some species of bees, the salivary gland secretes **invertase** (an enzyme that breaks down sucrose), which is taken into the body with the ingested nectar. Some insect and mammalian saliva also contains the lipid-splitting enzyme *lingual lipase.*

3. *Defense:* Saliva in many vertebrates exerts some antibacterial action by means of a twofold effect—first by *lysozyme,* an enzyme that lyses certain bacteria by breaking down their cell walls; second, by *salivary agglutin,* a glycoprotein that complexes with IgA antibodies (p. 475) and then binds to bacteria; third, by *lactoferrin,* which tightly binds to iron needed for bacterial multiplication (p. 469); and fourth, by rinsing away material that may serve as a food source for bacteria.

4. *Taste:* Saliva is a solvent for molecules that stimulate the taste buds. Only molecules in solution can react with taste bud receptors. You can demonstrate this for yourself: Dry your tongue and then drop some sugar on it; you cannot taste the sugar until it is moistened.

5. *Neutralization:* Saliva is often rich in bicarbonate buffers, which neutralize acids in food as well as acids produced by bacteria in the mouth. In ruminants, salivary bicarbonate is responsible for neutralizing the pH of rumen fluid and is an important buffering system for maintaining the acid–base equilibrium of the rumen contents.

6. *Thermoregulation:* Saliva can also be used as a means of temperature control in animals without sweat glands. In fact, most mammals and birds have no sweat glands and rely on panting as a means of evaporative cooling (Chapter 15). Kangaroos actually spread saliva over their bodies when outside temperatures approach lethal limits.

7. *Poisons:* Venomous reptiles have special oral glands (associated with fangs) that produce highly toxic secretions, which, once injected into prey, exert **hemotoxic** (affecting the circulatory system), **neurotoxic** (affecting the nervous system and brain), or **myotoxic** (muscle necrosis) effects. Venoms are approximately 90% proteins (dry weight) and most are enzymes. These components act rapidly to prevent the prey from escaping. Some nonvertebrates such as spiders also have saliva with poisons for immobilizing or killing prey.

8. *Anticoagulation:* recall the secretion of anticoagulants in blood-sucking vampire bats, insects, and leeches (pp. 400 and 459). Mosquito saliva glands secrete **apyrase,** which breaks down ADP, and so prevents platelet aggregation (see p. 395).

9. *Silk:* In Lepidoptera, Hymenoptera, and Trichoptera, the glands secrete silk proteins for constructing cocoons and shelters, and net-spinning caddis flies use silk for food gathering.

10. *Pheromones:* The salivary glands of many insects and some vertebrates release these external hormones. For example, male boars stimulated by a receptive female will produce copious amounts of saliva laden with androsterones. These stimulate the female to take a mating posture and may warn off other boars. (You will see examples in insects shortly.)

Let's now look more closely at saliva in insects and vertebrates.

Insects   In insects there are up to four pairs of salivary glands associated with the oral cavity, each of which is named after its associated mouthpart: the *mandibular, maxillary, hypopharyngeal,* and *labial* glands (Figure 14-3b). Of note, a particular gland may be present in only one of its life stages. For example, in Lepidoptera (moths and butterflies) the mandibular glands are present in the larval stage but not in the adult. In young worker honey bees, the mandibular gland produces a pheromonal secretion called *royal jelly* that is fed to larvae to control their development, whereas in older workers, the mandibular gland produces an alarm pheromone. In insects, ducts from paired salivary glands extend forward and unite into a common duct that opens near the base of the **labium** or **hypopharynx** mouthparts. Generally, these salivary glands have a secretory region that produces the primary saliva and a reabsorptive region that reabsorbs ions from the saliva. Both exchange regions are characterized by deep infoldings to increase surface area and numerous mitochondria to fuel the active transport of these ions. During feeding, a valve on the hypopharynx is opened as the hypopharynx is raised, thus permitting the release of saliva. However, when the insect is not feeding, the hypopharynx remains in a lowered position, which shuts off the valve and permits the saliva to be stored in the respective glands. Regulation of these glands is not as well known as in vertebrates, but in the American cockroach, serotonin induces the secretion of proteinaceous saliva, whereas dopamine induces the release of protein-free saliva.

Vertebrates   Salivary glands in many vertebrates combine two cell types: One type produces a wetting *serous* (watery) fluid that is high in enzyme content, whereas the other produces *mucus,* which is thick and slippery for lubrication. Fish and chickens do not have serous cells, so their total daily output of saliva is comparatively small, and it lubricates the food in preparation for swallowing; wetting occurs later, in the crop. The salivary glands of many swifts and swallows can actually increase in size during nest-building season. For example, the glands of the edible-nest swiftlet, *Collocalia fuciphaga,* enlarge approximately 50-fold because their nests consist entirely of cemented saliva, the primary ingredient of "bird's nest soup," a Chinese delicacy.

In mammals saliva consists of about 99.5% H_2O and 0.5% electrolytes and protein. There are three major glands: the *parotid* for serous secretions, the *sublingual* for mucus secretions, and the *submandibular,* which makes both types. The salivary NaCl concentration is only one seventh of that in the plasma, which is important in perceiving salty tastes. Similarly, glucose is absent from saliva so that discrimina-tion of sweet tastes is not impaired. Depending on the species of animal, the osmolality of saliva compared to blood plasma can range from hypotonic to hypertonic. However, saliva is always isotonic in ruminants and generally hypotonic in dogs and cats.

The salivary glands of ruminants produce copious amounts of alkaline saliva. When swallowed, this saliva enables selective compartments of its stomach to house a population of microbes capable of digesting dietary cellulose. Salivary bicarbonates and phosphates buffer the acids produced during fermentation and maintain the pH of the *ruminoreticulum* (compartment of a ruminant's stomach; see p. 707) within a narrow range. Urea is also recycled via the salivary glands back to the rumen microbes for protein synthesis. In nonruminant mammals, saliva is ultimately not as critical for digesting and absorbing foods, because enzymes produced by the pancreas and small intestine can digest food even in the absence of salivary and gastric secretion.

The continuous low level of salivary secretion can be increased by simple and conditioned reflexes

Salivary secretion in vertebrates is the only digestive secretion entirely under neural control. Both nervous system reflexes and hormones regulate all other secretions. One exception is the ruminants, where the parotid gland secretes saliva spontaneously in the absence of any neural stimuli.

Saliva in mammals is continuously secreted, even in the absence of apparent stimuli, because of constant low-level stimulation by the parasympathetic nerve endings that terminate in the salivary glands. This basal secretion is important in keeping the mouth and throat moist at all times. The basal rate of salivary secretion in humans is about 0.5 mL/min, which can increase to 5 mL/min in response to a potent stimulus such as sucking on a lemon. Remarkably, salivary secretion in ruminants is of such a magnitude that it can be regarded as a second circulatory system for body fluids and electrolytes. Cattle produce approximately 140 liters of saliva per day, whereas sheep and ponies secrete around 10 liters per day. Almost 50% of the total extracellular sodium in the animal may be found in the rumen contents at any one time, whereas half of the total body water passes through the salivary glands into the rumen every day.

In addition to a continuous low-level secretion, salivary secretion may be enhanced by two different types of salivary reflexes (Figure 14-7):

1. The **simple,** or **unconditioned, salivary reflex** occurs when chemoreceptors and pressure receptors within the oral cavity respond to the presence of food. On activation, these receptors initiate impulses in afferent nerve fibers that carry the information to the **salivary center** located in the medulla oblongata, as are all the brain centers that control digestive activities. The salivary center, in turn, sends impulses via the extrinsic autonomic nerves to the salivary glands to promote increased salivation.

2. The **acquired,** or **conditioned, salivary reflex** occurs without oral stimulation. A zoo mammal hearing the sound of a food cart, or a dog watching the preparation of a meal, initiates salivation through this reflex. All of us have experienced such "mouth watering" in anticipa-

tion of something to eat. This reflex is a learned response based on previous experience and is an example of feed-forward or anticipation regulation. Inputs that arise outside the mouth and are mentally associated with eating act through the cerebral cortex to stimulate the medullary salivary center.

Autonomic Influence on Salivary Secretion

Unlike the autonomic nervous system elsewhere in the vertebrate body, sympathetic and parasympathetic controls from the salivary center to the salivary glands are not antagonistic. Instead, both act to *increase* salivary secretion, but the quantity and consistency of saliva change with the needs of the animal. The rate of saliva secretion depends on the quantity of food in the mouth, whereas its consistency depends on the composition of the food consumed. Both the amount and quality of saliva are commensurate with the need for sensory evaluation of the food as well as the need for lubrication of dry foodstuffs. Parasympathetic stimulation, which exerts the dominant role, increases blood flow through the glands and is accompanied by a prompt and abundant flow of serous saliva. Normally, sympathetic stimulation reduces blood flow but allows a continued saliva output of smaller volume of mucus saliva. Because sympathetic stimulation in humans elicits a smaller volume of saliva, the mouth is drier than usual during circumstances when the sympathetic system is dominant, such as stress situations. Thus, people experience a dry feeling in the mouth when they are nervous about giving a speech. Cats, in contrast, secrete voluminous watery saliva in response to sympathetic stimulation.

Digestion in the mouth is minimal

Most digestion of foodstuffs in vertebrates is accomplished farther down the tract after the swallowing. Stomach acid inactivates amylase, but in the center of the food mass, where acid has not yet reached, salivary amylase can continue to function for several hours. **Lingual lipase** (a fat-hydrolyzing salivary enzyme) is, however, activated by the low pH in the stomach.

Cats have a strong aversion to diets containing medium-chain triglycerides or hydrogenated coconut oil. Presumably lingual lipase breaks down some triglycerides into fatty acids in their mouths, which then activates specific taste receptors and produces the finicky response.

check your understanding 14.3

Discuss the key features of mastication.

What is the role of saliva in digestion?

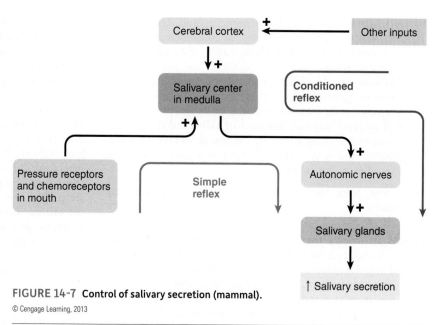

FIGURE 14-7 **Control of salivary secretion (mammal).**
© Cengage Learning, 2013

14.4 Pharynx, Esophagus, and Crop

The **pharynx** (throat) is the cavity at the rear of the oral cavity. In vertebrates, it acts as a common passageway for both the digestive system (by serving as the link for passing food between the mouth and esophagus) and the respiratory system (as a path for water moving across gills, or by providing access between the nasal passages and trachea for air). This arrangement necessitates mechanisms (described shortly) to guide food and air into the proper passageways beyond the pharynx. Some species of insects have *pharyngeal* and *cibarial* pumps that suction up their food and help it move into the esophagus.

Swallowing in vertebrates is a sequentially programmed all-or-none reflex

The motility associated with the pharynx and esophagus is **swallowing,** or **deglutition,** Most of us think of swallowing as the act of moving food out of the mouth and into the esophagus. However, swallowing actually is the entire process of moving food from the mouth through the esophagus into the stomach.

Swallowing in a vertebrate is initiated when a **bolus,** or ball of food, is forced by the tongue to the rear of the mouth into the pharynx. The pressure of the bolus in the pharynx stimulates pharyngeal pressure receptors, which send afferent impulses to the **swallowing center** located in the medulla. This center then reflexly activates, in the appropriate sequence, the muscles that are involved in swallowing. Swallowing is an example of a sequentially programmed all-or-none reflex in which multiple responses are triggered in a specific timed sequence; that is, a number of highly coordinated activities are initiated in a regular pattern over a period of time to accomplish the act of swallowing. Once initiated, swallowing cannot be stopped.

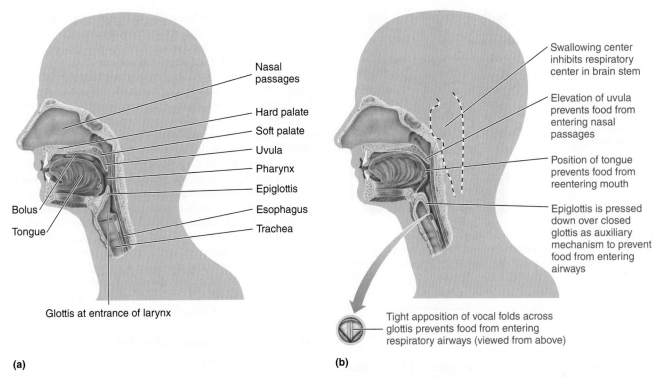

FIGURE 14-8 Oropharyngeal stage of swallowing in humans. (a) Position of the oropharyngeal structures at rest. (b) Changes that occur during the oropharyngeal stage of swallowing to prevent the bolus of food from entering the wrong passageways.

© Cengage Learning, 2013

During the oropharyngeal stage of swallowing, food is directed into the esophagus and prevented from entering the wrong passageways

Swallowing is arbitrarily divided into two stages: the *oropharyngeal stage* and the *esophageal stage*. The **oropharyngeal stage** consists of moving the bolus from the mouth through the pharynx and into the esophagus. When the bolus enters the pharynx during swallowing, it must be directed into the esophagus and prevented from entering the other openings that communicate with the pharynx. In other words, food must be prevented from reentering the mouth, from entering the nasal passages, and from entering the trachea (where it causes choking). All this is accomplished by the following coordinated activities (Figure 14-8 for a human):

- The position of the tongue against the hard palate prevents food from reentering the mouth during swallowing.
- The *uvula* is elevated and lodges against the back of the throat, sealing off the nasal passage from the pharynx so that food does not enter the nose.
- Food is prevented from entering the trachea primarily by elevation of the larynx and tight closure of the vocal folds across the laryngeal opening, or **glottis,** The first portion of the mammalian trachea is the *larynx* (see Figure 11-18, p. 518), across which are stretched the *vocal folds.* During swallowing, the vocal folds serve a purpose unrelated to vocalization. Contraction of laryngeal muscles aligns the vocal cords in tight apposition to each other, sealing the glottis entrance. Also, the bolus tilts a small flap of cartilaginous tissue, the **epiglottis** (*epi*

means "upon"), backward down over the closed glottis as further protection from food entering the respiratory airways. In some animals, such as horses, the nasal and tracheal tubes line up so precisely that they can breathe and drink at the same time, with the water passing to the left and right of the respiratory connection.
- Because the respiratory passages must be temporarily sealed off during swallowing, respiration in humans is briefly inhibited so that futile respiratory efforts are not attempted.
- With the larynx and trachea sealed off, pharyngeal muscles contract to force the bolus into the esophagus.

That the air and food passageways cross over in air-breathing vertebrates, necessitating these elaborate mechanisms to prevent choking, seems to be an evolutionary accident. Air breathing first evolved in some fishes as ventral lungs (filled by gulping) that developed in embryos as evaginations off the digestive tract. A dorsal tube from the same tract, with nostrils, evolved later from the tract upward for obtaining air independently of the mouth, creating the crossover.

The esophagus is a muscular tube guarded by sphincters at both ends

The **esophagus** is a fairly straight, muscular tube that extends between the pharynx and stomach (Figures 14-1, 14-3, and 14-4). In ruminants and dogs the entire musculature of the esophagus is skeletal, whereas in birds and humans it is entirely smooth muscle. In the horse and cat, the bottom third of the esophagus is smooth muscle. Lying for

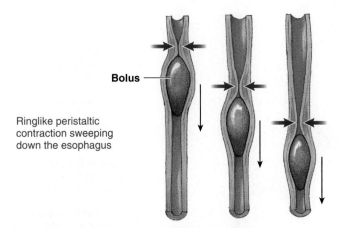

Bolus

Ringlike peristaltic
contraction sweeping
down the esophagus

FIGURE 14-9 **Peristalsis in the esophagus.** As the wave of peristaltic contraction sweeps down the esophagus, it pushes the bolus into the relaxed area ahead of it, propelling the bolus toward the stomach.

© Cengage Learning, 2013

the most part in the thoracic cavity, the esophagus joins the stomach in the abdominal cavity.

The esophagus is guarded at the upper end by the **pharyngoesophageal sphincter** and at the lower end by the **gastroesophageal sphincter** (or cardiac sphincter). These sphincter muscles, when closed, prevent passage through the tube.

Because the esophagus in most air-breathing vertebrates is exposed to subatmospheric intrapleural pressure as a result of respiratory activity (see p. 517), a pressure gradient exists between the atmosphere and the esophagus. Except during a swallow, the pharyngoesophageal sphincter remains closed as a result of neurally induced tonic contraction of the sphincter's circular skeletal muscle. This keeps the entrance to the esophagus closed to prevent large volumes of air from entering the esophagus and stomach during breathing. Otherwise, the digestive tract would be subjected to large volumes of gas, which would lead to excessive **eructation** (burping). During swallowing, this sphincter opens and allows the bolus to pass into the esophagus. Once the bolus has entered the esophagus, the pharyngoesophageal sphincter closes, the respiratory airways are opened, and breathing resumes.

Ruminants have another valve. On average, rumen gases consist of approximately 65% CO_2, 27% methane, 7% nitrogen, 0.5% oxygen, and 0.2% hydrogen, in addition to a trace amount of hydrogen sulfide. Most of these gases are diverted from the rumen into the trachea by the closing of the **nasopharyngeal sphincter,** the junction between the nasal cavity and the throat. Most of the gases are ultimately expelled with normal respiratory activity. (Gases produced by ruminants are currently of global concern: See *Challenges and Controversies: Global Warming and the Rumen.*)

Peristaltic waves push the food through the esophagus

After the oropharyngeal stage, the **esophageal stage** of the swallow begins once the food bolus contacts the pharyngeal mucosa. The swallowing center initiates a **primary peristaltic** wave that sweeps from the beginning to the end of the

esophagus, forcing the bolus ahead of it through the esophagus to the stomach. The term **peristalsis** refers to ringlike contractions of the circular smooth muscle that move progressively forward, pushing the bolus ahead of the contraction (Figure 14-9). Food and liquid can be propelled to the stomach even when an animal's head is lower than its body. The duration of the peristaltic wave depends on the length of the esophagus, and in humans requires about 5 to 9 seconds to reach the lower end. Progression of the wave is controlled by the swallowing center via the vagus nerve.

If a large or sticky swallowed bolus fails to be carried along to the stomach by the primary wave of peristalsis, the lodged bolus distends the esophagus, stimulating pressure receptors within its walls and thus initiating a second, more forceful peristaltic wave that is mediated by the intrinsic nerve plexuses at the site. These **secondary peristaltic** waves do not involve the swallowing center. Distension of the esophagus also reflexly increases salivary secretion. The trapped bolus is eventually dislodged and moved forward through the combination of lubrication by the extra swallowed saliva and the forceful secondary waves.

The gastroesophageal sphincter prevents reflux of gastric contents

Except during swallowing, the gastroesophageal sphincter remains contracted, to maintain a barrier between the stomach and esophagus, reducing the possibility of reflux of acidic gastric contents into the esophagus. If gastric contents do flow back into the esophagus despite the sphincter, the acidity of these contents irritates the esophagus, causing the esophageal discomfort known in humans as **heartburn.** (The heart itself is not involved at all.) The gastroesophageal sphincter relaxes reflexly as the peristaltic wave sweeps down the esophagus so that the bolus can pass into the stomach. After the bolus has entered the stomach, the gastroesophageal sphincter again contracts.

Achalasia is a condition that occurs commonly in dogs (and in humans) in which the lower esophageal sphincter fails to relax during swallowing but instead contracts more vigorously. Food accumulates in the esophagus, which enormously distends the esophagus when the food's passage into the stomach is greatly delayed. The underlying defect is apparently the result of damage to the *myenteric* nerve plexus (p. 661) in the region of the gastroesophageal sphincter.

Esophageal secretion is entirely protective

In most vertebrates, mucus is secreted throughout the length of the entire digestive tract. This lubricates the food, lessens the likelihood that the esophagus will be damaged by any sharp edges in the newly entering food, and protects the esophageal wall from acid and enzymes in gastric juice if gastric reflux occurs.

The crop is a modified section of the esophagus and functions mainly as a storage organ

The **crop,** a saclike outpouching of the esophagus, is commonly found in a number of animals including snails, insects (Figure 14-3b), and annelids; in vertebrates only some species of birds (and dinosaurs) have this structure (Figure 14-4e). The crop mainly serves a storage function that is par-

Challenges and Controversies
Global Warming and the Rumen

In ruminants, eructation helps relieve the pressure of the gases generated during the fermentation process. For example, in a 24-hour period, a cow belches (ejecting gas spasmodically from the rumen) approximately 500 liters of methane and 1,050 liters of CO_2. Approximately 2 to 12% of the ingested energy consumed by cattle is thus lost as methane. This represents an important loss of an energy source, as well as a factor that contributes to global warming. Global warming results from gases (primarily carbon dioxide and methane) in the upper atmosphere that trap heat (infrared) radiation that is leaving Earth. Methane affects climate directly through its interaction with long-wave infrared energy and indirectly through atmospheric oxidation reactions that produce CO_2. Atmospheric concentrations of methane had remained fairly stable at 750 ppm until approximately 100 years ago, when concentrations began to rise to their present levels of 1,800 ppm. (Similarly, atmospheric CO_2 has risen from about 280 ppm to about 390 ppm.) More than 500 million metric tons of methane enter the atmosphere annually, which exceeds the capacity of the atmosphere to oxidize it. At this rate, methane will contribute 14 to 17% of the factors leading to global warming in the next 50 years.

There are two significant sources of methane production, those of fossil origin, which contributes 20 to 30%, and those that yield contemporary carbon and produce 70 to 80%. Examples of the former include gas drilling, mining, and wetland emissions that contain methane that has been stored for thousands of years, whereas enteric fermentation from ruminants and insects (termites in particular), natural wetlands, biomass burning, oceans and lakes, and waste treatment are examples of the latter. Cattle are expected to contribute about 2% to global warming over the next 50 years.

Strategies to reduce methane emissions by livestock are thus of international importance. Scientists know that the type of carbohydrate fermented determines the level of methane production, both by affecting rumen pH in addition to the ruminal population. For example, the fermentation of cell wall fiber in coarse diets results in higher methane production than do concentrate diets high in soluble carbohydrates, where methane losses may drop to as little as 3% of the ingested energy. Grinding and pelleting forages can also decrease methane production, but doing so decreases ruminal pH, which can lead to ruminal acidosis. Beef cattle are increasingly reared in feedlots and given methane-inhibiting substances (such as ionophores and dicarboxylic acids). It is thus possible to assume modest reductions in methane emissions of livestock while enhancing productivity by modifying diet quality. Although methane production is largely dependent on diet quality and feed intake, differences in methane emissions (per kg dry matter intake between animals) of 40 to 50% have been reported. This suggests that methane production is heritable, and current research (mainly in Australia and New Zealand) is focused on genetic selection for low-methane-emitting animals. Future technologies may develop methods to alter the microbial population in ways that would benefit the animal without negatively impacting the environment.

---

ticularly important in fluid-feeding insects, which have relatively large crops for storing blood or nectar. In some insects, the crop is a simple enlargement of the foregut, or, as in mosquitoes and Lepidoptera, a lateral diverticulum off the digestive tract. The crop in domestic fowl is structurally identical to that of the esophagus except for a scarcity of mucus glands. Mucus glands are also relatively sparse in the region of the esophagus below the crop because movement of the bolus largely involves prewetted material from the crop. Granivorous ("grain-eating") and fish-eating birds have comparatively large crops for food storage, whereas carnivorous and insectivorous birds have limited need for a crop, which therefore are considerably reduced in size or are absent.

A food bolus moving down the esophagus in crop-bearing animals has two possibilities. It can either continue on to the *proventriculus-gizzard* organ(s) (in oligochaete worms, numerous arthropods, and most birds; Figures 14-3b and 14-4e), or it can enter the crop for temporary storage. The stored food is moistened by water as the animal drinks, but essentially no digestion occurs here. While empty, the crop of birds contracts once per minute, but if food is present in the proventriculus, then there is a complete inhibition of this contractile activity and the food remains within. The motility of the crop is regulated by vagal impulses. Motility and emptying are coordinated so as to release ingesta at a rate matching the emptying rate of the proventriculus and gizzard. If the food bolus is too large for the gizzard, contractions subside and return the excess portion to the crop. We'll examine details of avian proventriculus and gizzard later (p. 681).

The epithelial cells in the crop of some birds, including pigeons and doves, are sensitive to the hormone *prolactin,* the same hormone that promotes milk synthesis in mammals (p. 296). In both males and females, the increase in prolactin secretion during incubation of the eggs results in a marked proliferation of this tissue. Crop epithelia synthesize large stores of lipid material. Once the eggs hatch, these cells are sloughed off and mixed with food already present in the adult crop. This nutritive secretion, known as **crop milk,** is regurgitated into the esophagus of the squabs (chicks). In another species of bird, the **hoatzin** (*Opisthocomus hoazin*), the crop and esophagus function as a modified rumen (see p. 705), because this storage vat contains bacteria capable of digesting cellulose and detoxifying harmful plant alkaloids. Leafy plant material ferments in a large *double crop,* producing volatile fatty acids as smelly fermentation by-products, which gives the hoatzin a musky odor. The anterior sternum is markedly reduced in size to accommodate the voluminous fermentation structures, thus considerably reducing the area for flight muscle attachment. Nestlings can also be fed this regurgitated material. The highly developed digestive strategy of the hoatzin may have arisen from an evolutionary trade-off between detoxification of plant chemical defenses and enhanced use of cellulotic leaf-cell wall as a nutritional resource.

Bees, too, can regurgitate the contents of their crops. As in birds, the proventriculus controls the movement of food from the crop but can also selectively remove pollen from a nectar suspension in the crop. The lips of the proventriculus are modified so that they can strain out the pollen grains for digestion while leaving the nectar in the crop for eventual regurgitation and processing into honey.

check your understanding 14.4

Describe how swallowing works.

Describe the functions of the esophagus and its modified section, the crop.

14.5 Stomach or Midgut

In most animals, the midgut or **stomach** provides for the initial digestion (beyond the minor salivary contribution) and storage of ingested food items. We will examine these roles primarily in insects and vertebrates.

Digestion in many insect midguts is aided by peritrophic membranes and filter chambers

In arthropods, the midgut is also referred to as the **mesenteron,** In many species of insects, the midgut is mildly acidic in the anterior two thirds and more alkaline in the posterior third. For example, the anterior midgut of the American cockroach (*Periplaneta americana*) is acidic (pH ~5.7), whereas the posterior midgut is slightly less (pH ~6.1). (Note that insect midguts are not nearly as acidic as vertebrate stomachs, and thus cannot use acid-activated pepsinlike enzymes.)

In numerous species of insects, a **peritrophic membrane** separates the midgut epithelium from the food. Structurally it is a permeable (porous) network of chitin fibrils set in a matrix of protein and mucopolysaccharides (proteins with attached chains of repeating sugar units). The role of the peritrophic membrane has not been established but is believed to include the following:

- *Protection:* It may protect the epithelial cells of the midgut from abrasion by the food.
- *Motility:* It may help move food through this region of the gut tube.
- *Defense:* It may inhibit the movement of some pathogens from the food into the insect's tissues. Although the pore size is normally smaller than the size of most bacteria, many bacterial enzymes and toxins can traverse the membrane and damage the midgut. However, large-molecular-weight tannins, which are toxic, cannot penetrate the pores of the peritrophic membrane.
- *Digestion:* It compartmentalizes the midgut lumen into an **ectoperitrophic** (*ecto,* "outer") and **endoperitrophic** (*endo,* "inner") space within which specialized digestive functions can occur. Trypsin and amylase are localized to the endoperitrophic space, where they begin the initial digestion of food. Aminopeptidase and trehalase are localized to the ectoperitrophic space, where they complete the process of digestion.

The **filter chamber** represents an additional variation in midgut structure. In Homoptera—insects like aphids that feed on large quantities of plant juices—the filter chamber permits water from the ingested sap to pass directly from the anterior region of the midgut to the hindgut. In this manner the sap is concentrated before being digested in the posterior region of the midgut. The excess fluid is released from the hindgut as **honeydew,** which in turn is a food for other insects. To produce this secretion, the midgut actually loops back onto itself and becomes internalized along with the Malpighian tubules.

Diverticula (small outpocketings or folds) often supplement or replace the midgut of many other nonvertebrates (including tunicates, echinoderms, mollusks, and crustacean) and function in enzymatic digestion and absorption.

The vertebrate stomach stores food, begins protein digestion, and forms chyme

In vertebrates, only some species of bony fish, lampreys, hagfishes, and larval toads completely lack a stomach, with the esophagus opening directly into the intestine. For example, parrot fish (*Scaridae*) grind their algal food between heavy pharyngeal plates to disrupt the algal cell walls, the finely ground material then passing directly into the moderately alkaline intestine.

In monogastric mammals, the stomach is a simple, muscular, saclike chamber lying between the esophagus and small intestine. Generally the stomach of amphibians, fish, reptiles, and most mammals is a simple tubular expansion of the digestive tract. In humans it is arbitrarily divided into three sections based on anatomic, histologic, and functional distinctions (Figure 14-10). The dorsal region, or the **fundus,** is responsible for the storage of ingested food and adaptation to changes in volume. The middle, or main, part of the stomach is the body or **corpus,** Here the digesta is mixed with gastric secretions. The smooth muscle layers in the fundus and corpus are relatively thin, but the lower portion of the stomach, the **antrum,** has a much heavier musculature that is used to regulate expulsion of food into the small intestine as well as mix the contents. The terminal portion of the stomach consists of the **pyloric sphincter,** which acts as a barrier between the stomach and the upper part of the small intestine, the **duodenum.**

Variations between different animal groups with simple stomachs are based on the distribution and composition of the epithelium lining of the stomach. There are also glandular differences in the mucosa of these regions, as described later. Some species may have a more complex, multichambered stomach. The stomach of rats, for example, is divided into separate compartments. Voluminous chambers that are compartmentalized and perform separate functions characterize the digastric stomach of ruminants. As we noted earlier, some species of fish, such as the carp, completely lack a stomach, and food enters the small intestine directly.

The stomach performs three main functions:

1. *Storage:* The most important function is to store ingested food until it can be emptied into the small intestine at a rate appropriate for optimal digestion and absorption. In only a matter of minutes carnivores can consume a meal that then takes hours to digest and absorb. Because the small intestine is the primary site for this digestion and absorption, it is important that the stomach store the

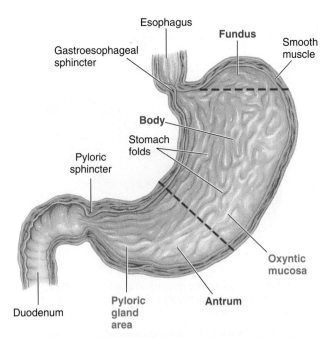

FIGURE 14-10 Anatomy of the stomach (mammalian; shown for a human). The stomach is divided into three sections based on structural and functional distinctions—the fundus, body, and antrum. The mucosal lining of the stomach is divided into the oxyntic mucosa and the pyloric gland area based on differences in glandular secretion.

© Cengage Learning, 2013

food and meter it into the duodenum at a rate that does not exceed the small intestine's capacities.

2. *Digestion:* The stomach secretes hydrochloric acid (HCl) and enzymes that begin protein digestion.

3. *Chyme formation:* Through mixing movements, the ingested food is pulverized and mixed with gastric secretions to produce a thick, liquid mixture known as **chyme,** The stomach contents must be converted to chyme before they can be emptied into the duodenum.

We now discuss how the stomach accomplishes these functions as we examine the four basic digestive processes—motility, secretion, digestion, and absorption—as they relate to the stomach. The details are primarily for mammals, but other animals are noted where comparison is illustrative. Starting with motility, gastric motility is complex and subject to multiple regulatory inputs. The four aspects of gastric motility are (1) gastric filling, (2) gastric storage, (3) gastric mixing, and (4) gastric emptying. We begin with gastric filling.

Gastric filling involves receptive relaxation

Stomachs can have a remarkable ability to accommodate significant changes in volume. Consider the blood-sucking leech, which feeds only once during its life and ingests the equivalent of nine times its body weight in about 30 minutes. Distension of the body eventually causes the animal to cease feeding. By comparison, the empty human stomach has a volume of about 50 mL, but it can expand to a capacity of about 1 liter (1,000 mL) during a meal. The mammalian stomach can accommodate such a 20-fold change in volume with little tension in its walls and little rise in intragastric pressure through the following mechanism. The interior of the stomach is thrown into deep folds. During a meal, the folds

get smaller and flatten out as the stomach relaxes slightly with each mouthful. This reflex relaxation of the stomach as it is receiving food is called **receptive relaxation;** it enables the stomach to accommodate the extra volume of food with little rise in stomach pressure. Receptive relaxation is triggered by the act of eating and is mediated by the vagus nerve.

Gastric storage takes place in the body of the stomach

A group of pacesetter cells (interstitial cells of Cajal) located in the upper fundus region of the mammalian stomach generate BERs (p. 660) that sweep down the length of the stomach toward the pyloric sphincter at a rate of a few per minute. This occurs continuously and may or may not be accompanied by contraction of the stomach's circular smooth muscle layer. Depending on the level of excitability in the smooth muscle, it may be brought to threshold by this flow of current and undergo action potentials, which in turn initiate peristaltic waves that sweep over the stomach in pace with the BER. Once initiated, the peristaltic wave spreads over the fundus and body to the antrum and pyloric sphincter. Because the muscle layers are thin in the fundus and body, the peristaltic contractions in this region are weak. When the waves reach the antrum, they become much stronger and more vigorous because the muscle there is much thicker.

Because only feeble mixing movements occur in the body and fundus, food emptied into the stomach from the esophagus is stored in the relatively quiet body without being mixed. The fundic area usually does not store food but contains only a pocket of gas. Food is gradually fed from the body into the antrum, where mixing does take place.

Gastric mixing takes place in the antrum of the stomach

The strong antral peristaltic contractions are responsible for mixing food with gastric secretions to produce chyme. Each antral peristaltic wave propels chyme forward toward the pyloric sphincter. Tonic contraction of the pyloric sphincter normally keeps it almost, but not completely, closed. The opening is large enough for water and other fluids to pass through with ease but too small for the thicker chyme to pass through except when a strong antral peristaltic contraction pushes it through. Even then, usually only a few milliliters of antral contents are forced into the duodenum with each peristaltic wave. Before more chyme can be squeezed out, the peristaltic wave reaches the pyloric sphincter and causes it to contract more forcefully, sealing off the exit and blocking further passage into the duodenum. The bulk of the antral chyme that was being propelled forward but failed to be pushed into the duodenum is abruptly halted at the closed sphincter and tumbled back into the antrum, only to be propelled forward and tumbled back again as the new peristaltic wave advances (Figure 14-11). This tossing back and forth, called **retropulsion,** thoroughly mixes the chyme in the antrum.

Gastric emptying is largely controlled by factors in the duodenum

In addition to accomplishing gastric mixing, the antral peristaltic contractions provide the driving force for gastric emptying. The amount of chyme that escapes into the

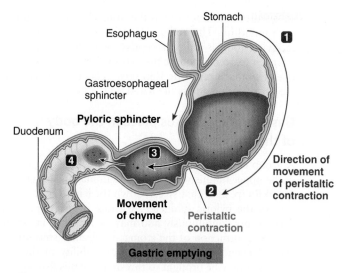

1. A peristaltic contraction originates in the upper fundus and sweeps down toward the pyloric sphincter.

2. The contraction becomes more vigorous as it reaches the thick-muscled antrum.

3. The strong antral peristaltic contraction propels the chyme forward.

4. A small portion of chyme is pushed through the partially open sphincter into the duodenum. The stronger the antral contraction, the more chyme is emptied with each contractile wave.

5. When the peristaltic contraction reaches the pyloric sphincter, the sphincter is tightly closed and no further emptying takes place.

6. When chyme that was being propelled forward hits the closed sphincter, it is tossed back into the antrum. Mixing of chyme is accomplished as chyme is propelled forward and tossed back into the antrum with each peristaltic contraction.

FIGURE 14-11 Gastric emptying and mixing as a result of antral peristaltic contractions (mammalian stomach, shown for a human).

© Cengage Learning, 2013

duodenum with each peristaltic wave before the pyloric sphincter closes tightly depends largely on the strength of peristalsis and on the particle size of the ingested material. During the digestive phase of gastric motility, only particles less than 2 mm in diameter escape through the pyloric sphincter. The intensity of antral peristalsis can vary markedly under the influence of different signals from both the stomach and the duodenum; thus, gastric emptying is regulated by both gastric and duodenal factors. These factors influence the stomach's excitability by slightly depolarizing or hyperpolarizing the gastric smooth muscle. This excitability in turn is a determinant of the degree of antral peristaltic activity. The greater the excitability, the more frequently the BER generates action potentials, the greater the degree of peristaltic activity in the antrum, and the faster the rate of gastric emptying (Table 14-1).

Factors in the Stomach that Influence the Rate of Gastric Emptying The main gastric factor that influences the strength of contraction is the amount of chyme, as well as the chyme's fluidity, in the stomach. Other things being equal, the stomach empties at a rate proportional to the volume of chyme in it at any given time. Distension of the stomach triggers increased gastric motility through a direct effect of stretch on the smooth muscle as well as through involvement of the intrinsic plexuses, the vagus nerve, and the stomach hormone *gastrin*. (The control and other functions of this hormone, which is secreted by special endocrine cells in the antrum, are described later.)

Factors in the Duodenum that Influence the Rate of Gastric Emptying Despite these gastric influences, factors in the duodenum are of primary importance in controlling the rate of gastric emptying. The duodenum must be ready to receive the chyme and can act to delay gastric emptying by reducing peristaltic activity in the stomach until the duodenum is ready to accommodate more chyme. Even if the stomach is distended and its contents are in a liquid form, it cannot empty until the duodenum is ready to process the chyme.

The four most important factors in the duodenum that influence gastric emptying are *fat, acid, hypertonicity,* and *distension*. The presence of one or more of these stimuli in the duodenum activates appropriate duodenal receptors, triggering either a neural or hormonal response that slow gastric motility by reducing the excitability of the gastric smooth muscle. The subsequent reduction in antral peristaltic activity slows down the rate of gastric emptying.

- The *neural response* is mediated through both the intrinsic nerve plexuses (short reflex) and the autonomic nerves (long reflex). Collectively, these reflexes are called the **enterogastric reflex** (*entero* means "intestine").
- The *hormonal response* involves the release from the duodenal mucosa of several hormones collectively known as **enterogastrones,** The blood carries these hormones to the stomach, where they inhibit antral contractions to reduce gastric emptying. Three of these enterogastrones have been positively identified: *secretin, cholecystokinin (CCK),* and *gastric inhibitory peptide*

TABLE 14-1 Factors Regulating Gastric Motility and Emptying (Vertebrate)

Factors	Mode of Regulation	Effects on Gastric Motility and Emptying
Within the Stomach		
Volume of chyme	Distension has a direct effect on gastric smooth muscle excitability, as well as acting through the intrinsic plexuses, the vagus nerve, and gastrin	Increased volume stimulates motility and emptying
Degree of fluidity	Direct effect; contents must be in a fluid form to be evacuated	Increased fluidity allows for more rapid emptying
Within the Duodenum		
Presence of fat, acid, hyperto-nicity, or distension	Initiates the enterogastric reflex or triggers the release of enterogastrones (secretin, cholecystokinin)	These factors in the duodenum inhibit further gastric motility and emptying until the duodenum has coped with factors already present
Outside the Digestive System		
Emotion	Alters autonomic balance	Stimulates or inhibits motility and emptying
Intense pain	Increases sympathetic activity	Inhibits motility and emptying
Decreased glucose use in the hypothalamus	Increases vagal activity	Stimulates motility; accompanied by hunger pangs

© Cengage Learning, 2013

(or *glucose-dependent insulinotrophic peptide*). **Secretin** is produced by endocrine cells known as **S cells** and CCK by endocrine cells known as **I cells** in the duodenal and jejunal mucosa. Secretin was the first hormone discovered, by Bayliss and Starling in 1902. Because it was a secretory product that entered the blood in response to an increase in acidity in the duodenum, it was termed *secretin*. The name **cholecystokinin** derives from the fact that this same hormone also governs contraction of the bile-containing gallbladder (*chole* means "bile," *cysto* means "bladder," and *kinin* means "contraction"). The name **gastric inhibitory peptide** is self-explanatory; it is a peptide hormone that inhibits the stomach (but see the hormone summary at the end of this chapter for an explanation of its new name, *glucose-dependent insulino-trophic peptide*). A possible fourth intestinal hormone, **neurotensin**, inhibits both gastric acid secretion as well as motility of the small intestine. Birds have a unique hormone **avian pancreatic polypeptide (APP)** that also is effective in reducing gastric motility.

Let's examine why it is important that each of these stimuli in the duodenum (fat, acid, hypertonicity, and distension) delays gastric emptying (acting through the enterogastric reflex or one of the enterogastrones).

- *Fat:* Fat is digested and absorbed more slowly than the other nutrients. Furthermore, fat digestion and absorption take place only within the small-intestine lumen. Therefore, when fat is already present in the duodenum, further gastric emptying of more fatty stomach contents into the duodenum is prevented until the small intestine has processed the fat already there. In fact, fat is the most potent stimulus for inhibition of gastric motility. This is evident in humans when you compare the rate of emptying of a high-fat meal (after six hours, some of a

bacon-and-eggs meal may still be in the stomach) to that of a protein-and-carbohydrate meal (lean meat and potatoes may require only three hours to empty).

- *Acid:* Because the stomach secretes hydrochloric acid (HCl), highly acidic chyme is emptied into the duodenum, where it is neutralized by sodium bicarbonate ($NaHCO_3$) secreted into the duodenal lumen from the pancreas. Unneutralized acid irritates the duodenal mucosa and inactivates the pancreatic digestive enzymes that are secreted into the duodenal lumen. Appropriately, therefore, unneutralized acid in the duodenum inhibits further emptying of acidic gastric contents until complete neutralization is accomplished.

- *Hypertonicity:* As molecules of protein and starch are digested in the duodenal lumen, large numbers of amino acid and glucose molecules are released. If absorption of these amino acid and glucose molecules does not keep pace with the rate at which protein and carbohydrate digestion proceeds, these large numbers of molecules remain in the chyme and increase the osmolarity of the duodenal contents. Osmolarity depends on the number of molecules present, not on their size (recall colligative properties, p. 79), and one protein molecule may be split into several hundred amino acid molecules, each of which has nearly the same osmotic activity as the original protein molecule. The same holds true for one large starch molecule, which yields many smaller but equally osmotically active glucose molecules. Because water is freely diffusible across the duodenal wall, water enters the duodenal lumen from the plasma as the duodenal osmolarity rises. Large volumes of water entering the intestine from the plasma lead to intestinal distension, and, more importantly, circulatory disturbances ensue because of the reduction in plasma volume. To prevent these effects, gastric emptying is reflexly inhibited when the osmolarity of the duodenal contents starts to rise. Thus, the amount of food entering the duodenum for further digestion into a multitude of additional osmotically active particles is reduced until absorption processes have had an opportunity to catch up.

Emptying and stress can influence gastric motility

In conjunction with the sensation of hunger, peristaltic contractions begin some time after the stomach empties, sweeping over the nearly empty antrum. This arousal of stomach motility appears to be mediated by increased parasympathetic activity. An animal may experience the sensation of hunger pangs when these peristaltic contractions are occurring, but the contractions themselves are not responsible for the sensation.

Other factors unrelated to digestion, such as stress, also can alter gastric motility by acting through the autonomic nerves to the smooth muscles. Even though the effect of emotions on gastric motility varies from one animal to another and is not always predictable, fear generally tends to decrease motility, whereas aggression tends to increase it. In addition to these influences, intense pain from any part of the body tends to inhibit motility, not just in the stomach but also throughout the digestive tract. Most of these responses are brought about by increased sympathetic activity and a corresponding decrease in parasympathetic activity.

The act of vomiting is powered by respiratory muscles, not the stomach

Vomiting, or **emesis,** the forceful expulsion of gastric contents out through the mouth, is generally perceived as being caused by abnormal gastric motility. However, animals with simple stomachs do not vomit by reverse peristalsis, as might be predicted. Actually, the stomach itself does not actively participate in the act of vomiting. The stomach, the esophagus, the esophageal sphincters, and the pyloric sphincter are all relaxed during vomiting. The major force for expulsion comes, perhaps surprisingly, from contraction of the respiratory muscles—namely, the diaphragm (the major inspiratory muscle) and the abdominal muscles (the muscles of active expiration) (see Figure 11-14, p. 511). Vomiting, however, is unusual in horses because the tone of the gastroesophageal sphincter is so great that it limits regurgitation of materials into the esophagus. Rats, too, cannot vomit and are careful to taste novel foods in small quantities.

Some species of birds, particularly carnivores, can "vomit" or **egest** indigestible foodstuffs. Owls, for example, egest the bones, feathers, and fur of their prey in the form of pellets. The presence of nutrients in the digesta inhibits egestion, but once digestion of the meal is complete, the rate and frequency of gastric contractions then increase. This compacts and shapes the pellet. Continued contractions move the pellet into the lower esophagus, where it is egested by rapid reverse peristalsis. Other birds such as penguins egest food to feed their young, while birds such as the fulmar and petrel can spew oil by-products at potential predators.

Vomiting in mammals begins with a deep inspiration and closure of the glottis. The contracting diaphragm descends downward on the stomach while simultaneous con-

traction of the abdominal muscles compresses the abdominal cavity, increasing the intra-abdominal pressure and forcing the abdominal viscera upward. As the flaccid stomach is squeezed between the diaphragm from above and the compressed abdominal cavity from below, the gastric contents are forced into the esophagus and out through the mouth. The glottis and uvula (in some primates) move to their blocking positions to prevent vomited material from entering the respiratory airways.

This complex act of vomiting is coordinated by a **vomiting center** in the medulla of the brain stem. Vomiting can be initiated by afferent input to the vomiting center from a number of receptors throughout the body. The causes of vomiting include (1) irritation or distension of the stomach and duodenum, and (2) chemical agents, including drugs or noxious substances that initiate vomiting (that is, **emetics**) either by acting in the upper portions of the gastrointestinal tract, or by stimulating chemoreceptors in a specialized **chemoreceptor trigger zone** in the medulla. Activation of the trigger zone initiates the vomiting reflex.

With excessive vomiting an animal experiences considerable losses of secreted fluids and acids that normally would be reabsorbed. The resultant reduction in plasma volume can lead to dehydration and circulatory problems, whereas loss of acid from the stomach can lead to metabolic alkalosis (see p. 650). However, limited vomiting brought about by irritation of the digestive tract can provide a useful service in removing noxious material from the stomach. Snakes, for example, that swallow too large a prey item are sometimes forced to vomit the remains before they putrefy inside their guts.

Gastric digestive juice is secreted by glands located at the base of gastric pits

The rate and the amount of gastric juice secretion are dependent on the frequency of feeding. For example, gastric juice production is continuous in ruminants and unpredictable in reptiles (carnivorous reptiles are dependent on prey availability and environmental temperature for digestion to occur). Once under way, gastric secretion in a snake carries on for up to six days to completely digest a rat, whereas in hibernators digestive functions virtually come to a standstill during hibernation.

The cells responsible for gastric secretion in a mammal are located in the lining of the stomach, the gastric mucosa, which is divided into two distinct areas whose distribution varies between species: (1) the **oxyntic mucosa,** which lines the body and fundus, and (2) the **pyloric gland area (PGA),** which lines the antrum. The luminal surface of the stomach is pitted with deep pockets formed by infoldings of the gastric mucosa. The invaginations in this first portion are called **gastric pits,** at the base of which lie the **gastric glands,** A variety of secretory cells line these invaginations, some exocrine and some endocrine or paracrine (Table 14-2). Two types of glands normally line the proventriculus of birds: simple mucosal glands that secrete mucus in addition to compound submucosal glands that secrete HCl and *pepsinogen* (see p. 677). Interestingly, unlike in mammals, both HCl and pepsinogen are synthesized within the same cell—the chief or **oxynticopeptic cell,** Oxynticopeptic cells also line the stomach of fish.

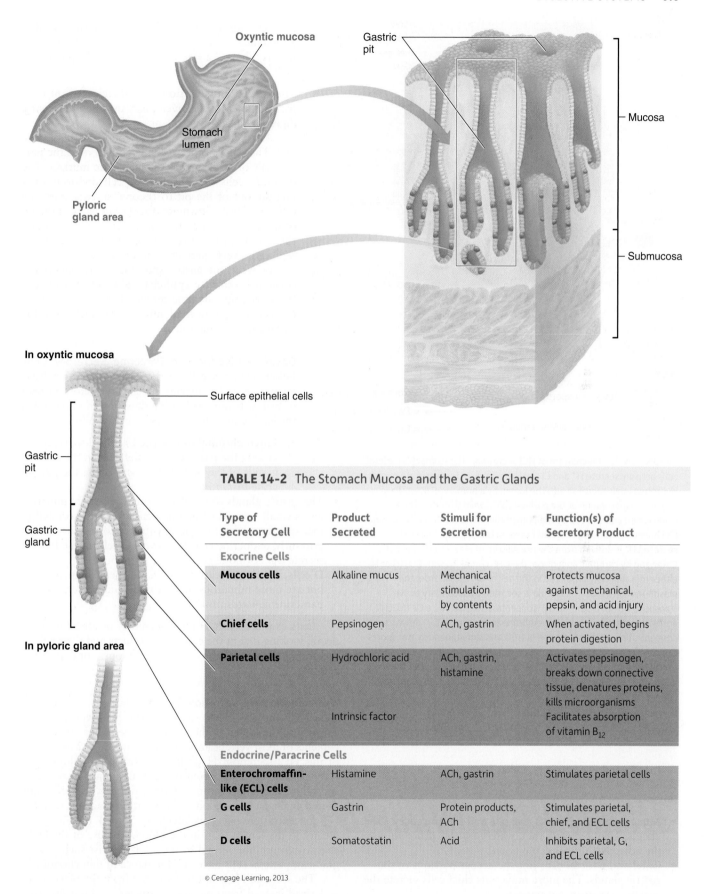

Oxyntic mucosa

Gastric pit

Mucosa

Stomach lumen

Submucosa

Pyloric gland area

In oxyntic mucosa

Surface epithelial cells

Gastric pit

Gastric gland

In pyloric gland area

TABLE 14-2 The Stomach Mucosa and the Gastric Glands

Type of Secretory Cell	Product Secreted	Stimuli for Secretion	Function(s) of Secretory Product
Exocrine Cells			
Mucous cells	Alkaline mucus	Mechanical stimulation by contents	Protects mucosa against mechanical, pepsin, and acid injury
Chief cells	Pepsinogen	ACh, gastrin	When activated, begins protein digestion
Parietal cells	Hydrochloric acid	ACh, gastrin, histamine	Activates pepsinogen, breaks down connective tissue, denatures proteins, kills microorganisms
	Intrinsic factor		Facilitates absorption of vitamin B_{12}
Endocrine/Paracrine Cells			
Enterochromaffin-like (ECL) cells	Histamine	ACh, gastrin	Stimulates parietal cells
G cells	Gastrin	Protein products, ACh	Stimulates parietal, chief, and ECL cells
D cells	Somatostatin	Acid	Inhibits parietal, G, and ECL cells

© Cengage Learning, 2013

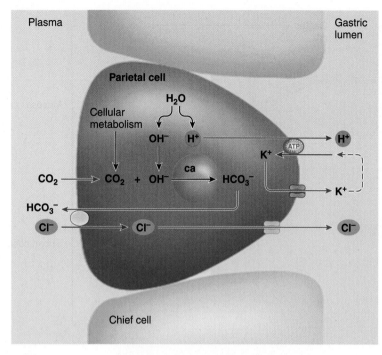

Plasma ... Gastric lumen

Parietal cell

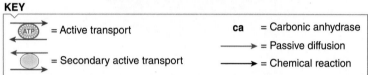

Chief cell

KEY

ATP	= Active transport	**ca**	= Carbonic anhydrase
		→	= Passive diffusion
	= Secondary active transport	→	= Chemical reaction

FIGURE 14-12 Mechanism of HCl secretion. The stomach's parietal cells actively secrete H^+ and Cl^- by the actions of two separate pumps. Hydrogen ion is secreted into the lumen by a primary H^+/K^+ ATPase active transport pump at the parietal cell's luminal border. The K^+ transported into the cell by the pump promptly exits through a luminal K^+ channel, thus being recycled between the cell and lumen. The secreted H^+ is derived from the breakdown of H_2O into H^+ and OH^-. Catalyzed by carbonic anhydrase, the OH^- combines with CO_2 (that is either metabolically produced in the cell or diffuses in from the plasma) to form HCO_3^-. Chloride is secreted by secondary active transport. Driven by the HCO_3^- concentration gradient, a Cl^-/HCO_3^- antiporter in the basolateral membrane transports HCO_3^- down its concentration gradient into the plasma and simultaneously transports Cl^- into the parietal cell against its concentration gradient. Chloride secretion is completed as the Cl^- that entered from the plasma moves out of the cell down its electrochemical gradient through a luminal Cl^- channel into the lumen.

© Cengage Learning, 2013

Gastric Exocrine Secretory Cells Three types of secretory cells are found in the walls of the pits in the oxyntic mucosa:

- **Mucous cells** line the gastric pits and the entrance of the glands. They secrete a thin, watery *mucus*. (*Mucous* is the adjective; *mucus* is the noun.)
- Chief and parietal cells line the deeper portions of the gastric glands. The more numerous **chief cells** secrete the enzyme precursor *pepsinogen*.
- The **parietal** (or **oxyntic**) **cells,** which secrete *HCl* and *gastric intrinsic factor.* The parietal cells are located on the outer wall of the gastric pits and do not come into

contact with the pit lumen. (*Parietal* means "wall," a reference to the location of these cells. *Oxyntic* means "sharp," a reference to these cells' potent HCl secretory product.)

These exocrine secretions are all released into the gastric lumen. Collectively, they make up the gastric digestive juice.

A few stem cells are also found in the gastric pits. These cells rapidly divide and serve as the parent cells of all new cells of the gastric mucosa. The daughter cells that result from cell division either migrate out of the pit to become surface epithelial cells or migrate down into the deeper parts of the pit to differentiate into chief or parietal cells. Through this activity, the entire stomach mucosa is regularly replaced (every three days in humans).

Between the gastric pits, the gastric mucosa is covered by **surface epithelial cells,** which secrete a thick, viscous, alkaline mucus that forms a visible protective layer several millimeters thick over the surface of the mucosa.

Gastric Endocrine and Paracrine Secretory Cells Other secretory cells in the gastric mucosa release endocrine and paracrine regulatory factors instead of products involved in the digestion of nutrients in the gastric juice lumen. These are as follows:

- **Enterochromaffin-like (ECL) cells** dispersed among the parietal and chief cells in the gastric glands of the oxyntic mucosa secrete the paracrine *histamine.*
- The gastric glands of the PGA primarily secrete mucus and a small amount of pepsinogen; no acid is secreted in this area, in contrast to the oxyntic mucosa. More importantly, endocrine cells known as **G cells** in the PGA secrete the hormone *gastrin* into the blood.
- **D cells,** which are scattered in glands near the pylorus but are more numerous in the duodenum, secrete the paracrine *somatostatin.*

Let's examine each of these secretory products in more detail, looking first at the exocrine products and their role in digestion, then at the exocrine secretions, among other actions.

Hydrochloric Acid Secretion The parietal cells actively secrete HCl into the lumen of the gastric pits, which in turn empty into the lumen of the stomach. The pH of the luminal contents of most vertebrates may fall to as low as 2 as a result of this HCl secretion. Hydrogen ion (H^+) and chloride ion (Cl^-) are actively transported by separate pumps in the parietal cell's plasma membrane. Hydrogen ion is actively transported against a tremendous concentration gradient, with the H^+ concentration being as much as 3 to 4 million times greater in the lumen than in the blood. Because of the high-energy expenditure needed to move H^+ against such a large gradient, parietal cells have an abundance of mitochondria.

The secreted H^+ is not transported from the plasma but is derived instead from metabolic processes within the parietal cell (Figure 14-12). Specifically the H^+ to be secreted is derived from the breakdown of H_2O molecules into H^+ and OH^- within the parietal cells. This H^+ is secreted into the stomach lumen by H^+/K^+ ATPase in the parietal cell's luminal membrane. This primary active transport carrier also pumps K^+ into the cell from the lumen. The transported K^+ then passively leaks back into the lumen through luminal K^+ channels, thus leaving K^+ levels unchanged by the process of H^+ secretion.

Chloride is also actively secreted but against a much smaller concentration gradient of only 1.5 times. The process begins with the enzyme *carbonic anhydrase* (*ca* in Figure 14-12; see also p. 541), which parietal cells contain in abundance. In the presence of carbonic anhydrase, OH^- readily combines with CO_2, which either has been produced within the parietal cell by metabolic processes or has diffused in from the blood. The combination of OH^- and CO_2 form HCO_3^-. The reduction in OH^- causes H_2O to split into more H^+ and OH^-. The generated H^+ in essence replaces the one secreted.

The generated HCO_3^- is moved into the plasma down its electrochemical gradient by the Cl^-/HCO_3^- antiporter in the basolateral membrane of the parietal cells and simultaneously transports Cl^- from the plasma into the parietal cell against its electrochemical gradient. By building up Cl^- inside the parietal cell, the Cl^-/HCO_3^- antiporter establishes a Cl^- concentration gradient between the parietal cell and gastric lumen. Because of this concentration gradient and because the cell interior is negative compared to the luminal contents, the negatively charged Cl^- now moves out of the cell down its electrochemical gradient through Cl^- channels in the luminal membrane into the gastric lumen, completing the Cl^- secretory process. In the meantime, the blood leaving the stomach is alkaline, the so-called **alkaline tide,** because HCO_3^- has been added to it. The alkaline tide is greatest in the alligator (*Alligator mississippiensis*) in which the pH increases from 7.4 to 7.6, a change of 0.2 pH units. Comparably the change in pH units in *Rana catasbeina* is 0.1; *Bufo marinus,* 0.05; and the python (*Molurus*), 0.02.

Although HCl does not actually digest anything and is not absolutely essential to gastrointestinal function, it does perform several functions. Specifically, HCl:

- Activates the enzyme precursor pepsinogen to an active enzyme, pepsin, and provides an acid medium that is optimal for pepsin activity.
- Aids in the breakdown of connective tissue and muscle fibers, thereby reducing large food particles into smaller particles.
- Denatures protein; that is, it uncoils proteins from their tertiary structure, thus exposing more of the peptide bonds for enzymatic attack.
- Along with salivary lysozyme, kills most microorganisms ingested with the food (although some resistant ones may escape and continue to grow and multiply in the intestine).

Pepsinogen, once activated, begins protein digestion

The major digestive constituent of gastric secretion is **pepsinogen,** an inactive enzymatic molecule produced in all vertebrates having stomachs except cyclostomes. (In stomachless pufferfish, the gene is not expressed in the gut, though it is in the skin.) Moreover, in animals lacking highly acidic digestive regions (such as insects), no equivalent en-

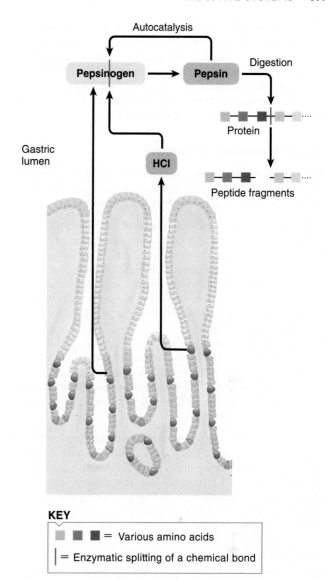

KEY

⬜ ⬛ ⬛ = Various amino acids

| = Enzymatic splitting of a chemical bond

FIGURE 14-13 Pepsinogen activation in the stomach lumen (vertebrate). In the lumen, hydrochloric acid (HCl) activates pepsinogen to its active form, pepsin, by cleaving off a small fragment. Once activated, pepsin autocatalytically activates more pepsinogen and begins protein digestion. Secretion of pepsinogen in the inactive form prevents it from digesting the protein structures of the cells in which it is produced.

© Cengage Learning, 2013

zyme has evolved. Pepsinogen is stored in the chief cell's cytoplasm within secretory vesicles known as **zymogen granules,** from which it is released by exocytosis (p. 41) on appropriate stimulation. When pepsinogen is secreted into the gastric lumen, HCl cleaves off a small fragment of the molecule, converting it to the active form of the enzyme, **pepsin** (Figure 14-13). Once formed, pepsin acts on other pepsinogen molecules to produce more pepsin. A mechanism such as this, whereby an active form of an enzyme activates other molecules of the same enzyme, is referred to as an **autocatalytic** ("self-activating") **process** and is another example of positive feedback. Only when there is no more protein in the stomach does the secretion of pepsinogen cease.

Pepsin initiates protein digestion by splitting certain amino acid linkages in proteins to yield peptide fragments (small amino acid chains); it works most effectively in a low-pH environment. Because pepsin can digest protein, it must be stored and secreted in an inactive form, so it does not digest proteins such as cytoskeletons in the cells in which it is formed. While inside the cell, therefore, pepsin is maintained in the inactive form of pepsinogen until it reaches the gastric lumen, where it is activated by HCl.

Stomach mucus is protective

The surface of the gastric mucosa is covered by a layer of mucus, which serves as a protective barrier against several forms of potential injury to the gastric mucosa:

- By virtue of its lubricating properties, mucus protects the gastric mucosa against mechanical injury.
- Mucus helps protect the stomach wall from self-digestion because pepsin is inhibited when it comes in contact with the mucus coating the stomach lining. (However, mucus does not affect pepsin activity in the lumen, where digestion of dietary protein proceeds without interference.)
- Being alkaline, mucus helps protect against acid injury by neutralizing HCl in the vicinity of the gastric lining, but it does not interfere with the function of HCl in the lumen. Whereas the pH in the lumen may be as low as 2, the pH in the mucus layer adjacent to the mucosal cell surface is about 7.

Four chemical messengers primarily influence the secretion of gastric digestive juices. Parietal cells have separate receptors for each of these messengers. Three of them—acetylcholine (ACh), gastrin, and histamine—are stimulatory. They bring about increased secretion of HCl by promoting the insertion of additional H⁺/K⁺ ATPases into the parietal cell's plasma membrane. The fourth regulatory agent—somatostatin—inhibits acid secretion. Gastrin and ACh also increase pepsinogen secretion through their stimulatory effect on the chief cells. We now consider each of these chemical messengers in further detail:

- *Acetylcholine* is a neurotransmitter released from the intrinsic nerve plexus in response to vagus stimulation. ACh stimulates the parietal and chief cells as well as the G cells and ECL cells. It is the most important trigger of HCl secretion in fishes.
- The G cells secrete the hormone **gastrin** into the blood in response to protein products in the stomach lumen as well in response to ACh. Like secretin and CCK, gastrin is a major gastrointestinal hormone. After being carried by the blood back to the body and fundus of the stomach, gastrin stimulates (that is, it is *tropic*) the parietal and chief cells, promoting secretion of a highly acidic gastric juice. In addition to directly stimulating the parietal cells, gastrin also indirectly promotes HCl secretion by stimulating the ECL cells to release histamine. Gastrin is the main factor responsible for increasing HCl secretion during digestion of a meal. Gastrin is also *tropic* (growth promoting) to the mucosa of the stomach and small intestine, thereby maintaining their secretory capabilities.
- *Histamine*, a paracrine, is released from the ECL cells in response to gastrin and ACh. Histamine acts locally on nearby parietal cells to increase the rate of HCl secre-

tion and potentiates the actions of ACh and gastrin. Histamine and gastrin are the most important triggers of HCl secretion in frogs and mammals.
- *Somatostatin* is released from the D cells in response to acid. It acts locally as a paracrine, in negative-feedback fashion, to inhibit secretion by the parietal cells, G cells, and ECL cells, thus turning off the HCl-secreting cells and their most potent stimulatory pathway.

ACh and gastrin both operate through IP₃/Ca²⁺ second-messenger pathways; histamine activates cAMP to bring about its effects. These messengers all bring about increased secretion of HCl by promoting the insertion of additional H⁺/K⁺ ATPases into the parietal cells' plasma membrane. A pool of these pumps is stored within the parietal cell in intracellular vesicles, which fuse with the luminal membrane via exocytosis to add more of these active carriers to the membrane as needed to increase HCl secretion (recall that renal aquaporins are similarly regulated; p. 585).

From this list, it is obvious not only that multiple chemical messengers influence the parietal and chief cells but also that these chemicals also influence each other. Next, as we examine the phases of gastric secretion, you will see under what circumstances each of these regulatory agents is released.

Control of gastric secretion involves three phases

The rate of gastric secretion can be increased by (1) factors arising before food ever reaches the stomach, (2) factors resulting from the presence of food in the stomach, and (3) factors originating in the duodenum after food has left the stomach. Accordingly, gastric secretion is divided into three phases—the cephalic, gastric, and intestinal phases (Table 14-3).

Cephalic Phase The **cephalic phase** of gastric secretion refers to the increased secretion of HCl and pepsinogen that occurs in feedforward (anticipatory) fashion in response to stimuli acting in the head even before food reaches the stomach (*cephalic* means "head"). As we noted earlier, response to food-related stimuli (anticipating a meal, odor, chewing and swallowing) gastric secretion is increased by the vagus nerve. Recall that vagal stimulation results in two effects: (1) It promotes increased secretion of ACh, by intrinsic plexuses, which in turn leads to increased secretion of HCl and pepsinogen by the secretory cells. (2) It triggers G cells to release of gastrin, which in turn further enhances secretion of HCl and pepsinogen, with the effect on HCl being potentiated by gastrin promoting the release of histamine. The secretion of gastric juice in response to cephalic stimuli does not occur in ruminants because of the lack of association between the ingestion of food and the appearance of the ingesta in the compartmentalized ruminant stomach.

Gastric Phase The **gastric phase** of gastric secretion begins when food actually reaches the stomach. Stimuli acting in the stomach—namely *protein*, especially peptide fragments; *distension; caffeine;* and *alcohol*—increase gastric secretion by overlapping efferent pathways. For example, protein in the stomach, the most potent stimulus, stimulates chemoreceptors that activate the intrinsic nerve plexuses, which in

The photo above is of Stella, the successor to Ellie, the first cow at the Cornell College of Veterinary Medicine, to be fitted with a fistula (portal in the side) that functioned as a "window" into ruminant digestion. Ellie, born in 1988, died of old age in 2002; she outlived most of the other Holsteins. The removable cap on the fistula fitting allowed veterinary students to sample her digestive fluids and determine the rate of digestion of various foodstuffs. This fistula also allowed her to donate rumen fluid to other, less healthy cows, whose digestive systems needed a boost.

turn stimulate the secretory cells. Furthermore, protein brings about activation of the extrinsic vagal fibers to the stomach, enhancing the events of the cephalic phase. Through these synergistic and overlapping pathways, protein induces the secretion of a highly acidic, pepsin-rich gastric juice, which continues the digestion of the protein that first initiated the process.

Cephalic- and gastric-phase events have been tested as follows. First, if an animal (non ruminant) is fitted with an **esophageal fistula** (tube connecting the esophagus with the outside world) such that any ingested food is diverted from the stomach into a collecting port, there is an increase in secretion of HCl and pepsinogen in the animal's stomach when the animal is eating due to cephalic stimulation. Alternatively, if a region of the stomach is experimentally denervated and pinched off with an opening to the outside world (a **Heidenhain pouch**) such that acid secretion can be measured, there is still acid secretion in the pouch as soon as food is put into it. Because the animal did not "see" the food, no cephalic phase is triggering the stomach. Rather, when food is placed in the stomach, gastric secretion is increased by the processes of the gastric phase.

Pioneering studies of gastric function were performed by William Beaumont, the "father of gastric physiology," on a patient who retained a hole in his stomach after an accident. Beaumont would tie a piece of food to a string, insert it through the hole into the stomach, and at various intervals would remove the string and observe how much food remained. This led to the observation that it was primarily stomach secretions and not the mixing process that resulted in the digestion of the food item. Thus, digestion was understood to be primarily a chemical process and not simply a mechanical one.

Intestinal Phase The **intestinal phase** of gastric secretion encompasses the factors originating in the small intestine that influence gastric secretion, as with gastric emptying. Even if the stomach is completely isolated from the intestinal tract, placement of food into the upper part of the duodenum can still induce acid secretion in the stomach. Several important factors that originate in the small intestine influence gastric secretion. The intestinal phase has both an excitatory and an inhibitory component.

To a limited extent, the presence of the products of protein digestion in the duodenum stimulates further gastric secretion by triggering the release of **intestinal gastrin** that is carried by the blood to the stomach. This is the *excitatory component* of the intestinal phase of gastric secretion. It is as if the small intestine, on noting the arrival of protein fragments from the stomach, offers the stomach a "helping hand" in digesting the protein by enhancing gastric secretion.

The *inhibitory* component of the intestinal phase of gastric secretion is dominant over the excitatory component, as we will examine next.

Gastric secretion gradually decreases as food empties from the stomach into the intestine

You now know what factors turn on gastric secretion before and during a meal, but how is the flow of gastric juices shut off as chyme begins to be emptied from the stomach into the small intestine? Gastric secretion is gradually reduced by three different means as the stomach empties (Table 14-4):

- *Emptying:* As food is gradually emptied into the duodenum, the major stimulus for enhanced gastric secretion—the presence of protein in the stomach—is withdrawn.
- *Somatostatin:* After foods leave the stomach and gastric juices accumulate to such an extent that gastric pH falls very low (pH less than 3), somatostatin is released, inhibiting gastric secretion.
- *The inhibitory component of the intestinal phase of gastric secretion:* The same stimuli that inhibit gastric motility (fat, acid, hypertonicity, or distension in the duodenum brought about by stomach emptying) inhibit gastric secretion as well. The enterogastric reflex and the enterogastrones suppress the gastric secretory cells, while they simultaneously reduce the excitability of the gastric smooth muscle cells. For example, when the pH of the intestine becomes too low, *enteroglucagon* is released into the circulation and "shuts off" the flow of gastric juices.

The stomach lining is protected from gastric secretions by the gastric mucosal barrier

How can the stomach contain strong acid contents and proteolytic enzymes without destroying itself? We have already learned that mucus provides a protective coating. Furthermore, the surface mucus-secreting cells secrete HCO_3^- that is trapped in the mucus and neutralizes acid in the vicinity. In addition, other barriers to acid damage of the mucosa are provided by the mucosal lining itself. First, the luminal membranes of the gastric mucosal cells are almost impermeable to H^+, so acid cannot penetrate *into* the cells and thus cause cell damage. Furthermore, the lateral edges of these

TABLE 14-3 Stimulation of Gastric Secretion (Mammal)

Phase	Stimuli	Excitatory Mechanism for Enhancing Gastric Secretion
Cephalic Phase of Gastric Secretion	Stimuli in the head—seeing, smelling, tasting, chewing, swallowing food	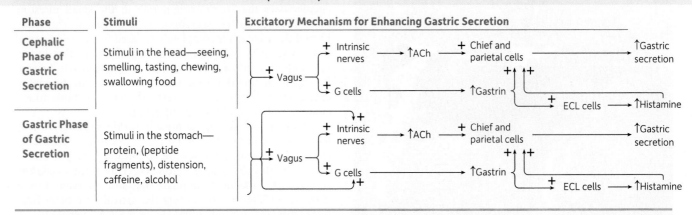
Gastric Phase of Gastric Secretion	Stimuli in the stomach—protein, (peptide fragments), distension, caffeine, alcohol	

TABLE 14-4 Inhibition of Gastric Secretion (Mammal)

Region	Stimuli	Inhibitory Mechanism for Gastric Secretion
Body and Antrum	Removal of protein and distension as the stomach empties	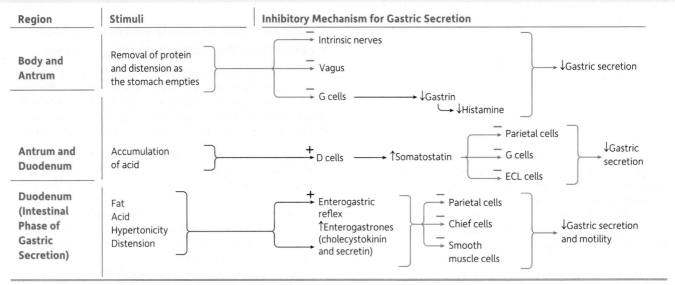
Antrum and Duodenum	Accumulation of acid	
Duodenum (Intestinal Phase of Gastric Secretion)	Fat Acid Hypertonicity Distension	

cells are joined together near their luminal borders by tight junctions (see p. 66), so acid cannot diffuse or conduct *between* the cells from the lumen into the underlying submucosa. Cardiac glands in some mammals (such as swine) secrete bicarbonate ion into the overlying mucus to protect the epithelial cells from H^+. The properties of the gastric mucosa that enable the stomach to contain acid without injuring itself constitute the **gastric mucosal barrier** (Figure 14-14). These protective mechanisms are further enhanced by the fact that the entire stomach lining is replaced (every three days in humans) before they are irreparably damaged by the wear and tear of harsh gastric conditions.

Despite these protections (Figure 14-14), the barrier occasionally is broken so that the gastric wall is injured by its acidic and enzymatic contents. When this occurs, an erosion, or **peptic ulcer,** of the stomach wall results. A major contributor in humans is *Helicobacter pylori,* a bacterium that tolerates stomach. (Excessive gastric reflux into the esophagus and dumping of excessive acidic gastric contents into the duodenum, as well as *H. pylori,* can lead to peptic ulcers in

these locations as well.) Pigs are notable for their susceptibility to stomach ulcers, particularly in the esophageal area. Generally, there is a higher incidence in barrows (male pigs that have been castrated before reaching puberty) than gilts (female pigs that have not given birth), with both crowding and heat stress increasing the severity of the lesion.

Carbohydrate digestion continues in the body of the stomach, whereas protein digestion begins in the antrum

Two separate digestive processes take place within the stomach. Food in the body of the stomach remains a semisolid mass because peristaltic contractions in this region are too weak for mixing to occur. Because food is not mixed with gastric secretions in the body of the stomach, very little protein digestion occurs here. Acid and pepsin can attack only the surface of the food mass. Carbohydrate digestion, however, continues in the interior of the mass under the influ-

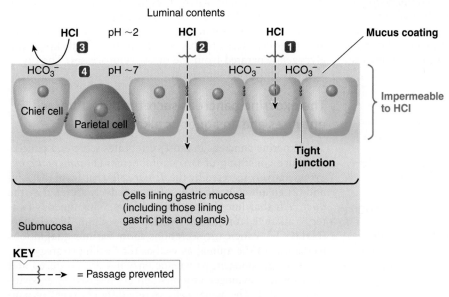

Luminal contents

The components of the gastric mucosal barrier enable the stomach to contain acid without injuring itself:

1 The luminal membranes of the gastric mucosal cells are impermeable to H^+ so that HCl cannot penetrate into the cells.

2 The cells are joined by tight junctions that prevent HCl from penetrating between them.

3 A mucus coating over the gastric mucosa serves as a physical barrier to acid penetration.

4 The HCO_3^--rich mucus also serves as a chemical barrier that neutralizes acid in the vicinity of the mucosa. Even when luminal pH is 2, the mucus pH is 7.

KEY

= Passage prevented

FIGURE 14-14 Gastric mucosal barrier.

ence of salivary amylase. Even though acid inactivates salivary amylase, the unmixed interior of the food mass is free of acid.

Digestion by the gastric juice itself is accomplished in the antrum of the stomach, where the food is thoroughly mixed with HCl and pepsin, initiating protein digestion.

In most vertebrates, at this point, fat digestion has not even begun. Although it is possible for fat molecules, which are lipid soluble, to pass through the lipid portions of the cell membranes that line the stomach, this does not occur to any appreciable extent, because dietary fat in the gastric lumen separates into large fat droplets that float in the chyme. However, in some species lingual lipase (p. 664) has begun the process of lipid digestion.

The proventriculus and gizzard begin the process of digestion in birds and insects

Animals that possess gizzards include birds, which have no stomach per se, but rather an organ is divided into two components we noted earlier (p. 660): the **proventriculus**, which serves as the "glandular stomach," and a separate **gizzard**, which serves as the "muscular stomach" to macerate and grind the ingesta. The proventriculus is comparatively small in herbivorous birds and is considerably larger in aquatic carnivores, which also use it as a storage space for captured prey. Motility functions of the proventriculus are to expel ingesta and digestive secretions into the gizzard.

Two pairs of smooth muscle groups, the *thin (musculi intermedii)* and *thick (musculi laterales)* encircle the gizzard and act in opposition to each other. Most birds cannot chew their food, so the gizzard, which contains **grit** (swallowed pebbles and small stones), is responsible for the mechanical breakdown of ingesta. The mucosal lining of the gizzard is covered by a coating, **koilin**, a tough protein-polysaccharide complex that has an amino acid composition similar to that of feather protein (keratin). Structurally, koilin is a composite of filaments formed in tubular glands lining the gizzard. This

hard filament eventually forms a rod that protrudes through the luminal surface of the mucosa and forms the *gizzard teeth* (the abrasive lining of the gizzard's surface).

The gizzard is a comparatively large organ evolved to develop high pressures for particle size reduction. In birds the myoglobin content of the gizzard is approximately 100-fold greater than in the breast muscle, and mitochondrial numbers are also elevated, indicating a high aerobic capacity. Digesta from the gizzard is refluxed back into the proventriculus for further digestion and mixed with gastric secretions. Movement of digesta from the gizzard back into the proventriculus allows feed to be exposed to digestive enzymes and low pH for extended times, consequently improving digestive efficiency and increasing gut health. Gut health improvements are also associated with the acidic destruction of pathogenic bacteria that are ingested with feed.

In domestic poultry, gastroduodenal motility is tightly coordinated and begins with contraction of the thin, circular muscles lining the gizzard, followed by several waves of peristalsis in the duodenum. A powerful contraction of the thick muscles then precedes peristalsis of the proventriculus. Contractions of the proventriculus and duodenum are generated by intrinsic neural connections originating from the gizzard and are not significantly affected by loss of extrinsic innervation. At the end of the contraction of the thin muscles, ingesta flows from the gizzard into the duodenum, whereas after contraction of the thick muscles, the ingesta flows into the proventriculus. Contraction of the proventriculus then forces material back into the gizzard. In poultry, these cycles are normally repeated two to three times per minute, with food deprivation decreasing the frequency of contractions.

Gizzards are also found in alligators, crocodiles, and in several species of fish, including the mullet, which is found in estuarine waters worldwide. The gizzard also occurred in herbivorous dinosaurs. Many annelids, mollusks, and arthropods also have (nonhomologous) gizzards: In cockroaches, for example, six cutting edges or plates are arranged radially around the gizzard such that when compressed together, they form a powerful grinding tool to macerate digesta. Only when food particles are sufficiently ground can they continue their passage from the foregut into the midgut. This function helps limit the amount of food leaving the crop and entering the midgut.

check your understanding 14.5

Describe the three phases of gastric secretion and the factors that affect each.

What enzyme is produced in the stomach, and how is it activated?

How does the parietal cell accomplish the production of HCl?

What are the functions of the proventriculus and gizzard?

14.6 Digestive Accessory Organs: Pancreas, Liver, Gallbladder, Fat Body

When gastric contents are emptied into the vertebrate small intestine, they are mixed not only with juice secreted by the small-intestine mucosa but also with the secretions of the exocrine pancreas and liver that are emptied into the duodenal lumen. We discuss the roles of each of these accessory digestive organs before we examine the small intestine itself.

The (hepato)pancreas is a mixture of exocrine and endocrine tissue

The **pancreas** in most vertebrates is a mixed gland that contains both exocrine and endocrine tissue (Figure 14-15). Approximately 98% of the vertebrate pancreas is committed to its exocrine function. The predominant exocrine portion consists of grapelike clusters of secretory cells that form sacs known as **acini,** which connect to ducts that eventually empty into the duodenum. The smaller endocrine portion consists of isolated islands of endocrine tissue, the **islets of Langerhans,** which are dispersed throughout the pancreas. The most important hormones secreted by the islet cells are *insulin* and *glucagon* (p. 316). In most species the islets are unequally distributed throughout the gland. For example, the splenic lobe of the avian pancreas is approximately 4 to 5% endocrine in function, compared to less than 0.5% for the remaining three lobes. The exocrine and endocrine tissues of the pancreas are derived from different tissues during embryonic development and share only their location in common. Although both are involved with the metabolism of nutrient molecules, they have different functions under the control of different regulatory mechanisms.

The **hepatopancreas** of crustaceans, gastropods, and fishes contains cells with functions comparable to those found in the mammalian pancreas and liver. Structurally the crustacean hepatopancreas consists of numerous blind ended tubules lined with epithelium that secrete various digestive enzymes. Ducts from the hepatopancreas are connected to

the anterior midgut where they are refluxed back to the proventriculus. Alternatively chyme from the midgut enters the hepatopancreas where it is further digested and eventually absorbed. Cells in this organ can store both lipids and glycogen generated from the digestive processes. The organ also has endocrine functions.

The exocrine pancreas secretes digestive enzymes and an aqueous alkaline fluid

The **exocrine pancreas** in most vertebrates secretes a pancreatic juice consisting of two components: (1) *pancreatic enzymes* actively secreted by the *acinar cells* that form the acini and (2) an *aqueous alkaline solution* actively secreted by the *duct cells* that line the pancreatic ducts. The aqueous (watery) alkaline component is rich in sodium bicarbonate ($NaHCO_3$). Pancreatic secretions show various adaptations to the diet of the animal as well as the feeding strategy.

Like pepsinogen, pancreatic enzymes are produced and stored within zymogen granules and are released by exocytosis as needed. The acinar cells secrete different types of pancreatic enzymes that can digest foodstuffs. In nonruminants, these pancreatic enzymes are important because they can almost completely digest food in the absence of all other digestive secretions. The principal types of vertebrate pancreatic enzymes are (1) *proteolytic enzymes,* which are involved in protein digestion; (2) *pancreatic amylase* and *chitinase* (in some vertebrates), which contributes to carbohydrate digestion; and (3) *pancreatic lipase,* for fat digestion.

Pancreatic Proteolytic Enzymes The three major proteolytic enzymes secreted by the vertebrate pancreas are *trypsinogen, chymotrypsinogen,* and *procarboxypeptidase,* each of which is secreted in an inactive form. When **trypsinogen** is secreted into the duodenal lumen, it is activated to its active enzyme form, **trypsin,** by **enterokinase,** an enzyme embedded in the luminal border of the cells that line the duodenal mucosa. Trypsin then autocatalytically activates more trypsinogen. Like pepsinogen in the chief cells of the stomach, trypsinogen must remain inactive within the pancreas to prevent this proteolytic enzyme from digesting the cells in which it is formed. As further protection, the pancreatic tissue also produces a chemical known as **trypsin inhibitor,** which blocks trypsin's actions if activation of trypsinogen inadvertently occurs within the pancreas.

Chymotrypsinogen and **procarboxypeptidase,** the other pancreatic proteolytic enzymes, are converted by trypsin to their active forms, **chymotrypsin** and **carboxypeptidase,** respectively, within the duodenal lumen. Thus, once enterokinase has activated some of the trypsin, trypsin then governs the rest of the activation process.

Each of these proteolytic enzymes attacks different peptide linkages. The end products that result from this action are a mixture of amino acids and small peptide chains. Mucus secreted by the intestinal cells protects against digestion of the small-intestine wall by the activated proteolytic enzymes.

Pancreatic Amylase Like salivary amylase, pancreatic amylase plays an important role in carbohydrate digestion by converting polysaccharides into disaccharides. Amylase is secreted in the pancreatic juice in an active form because active amylase does not present a danger to the secretory cells. These cells do not contain any polysaccharides.

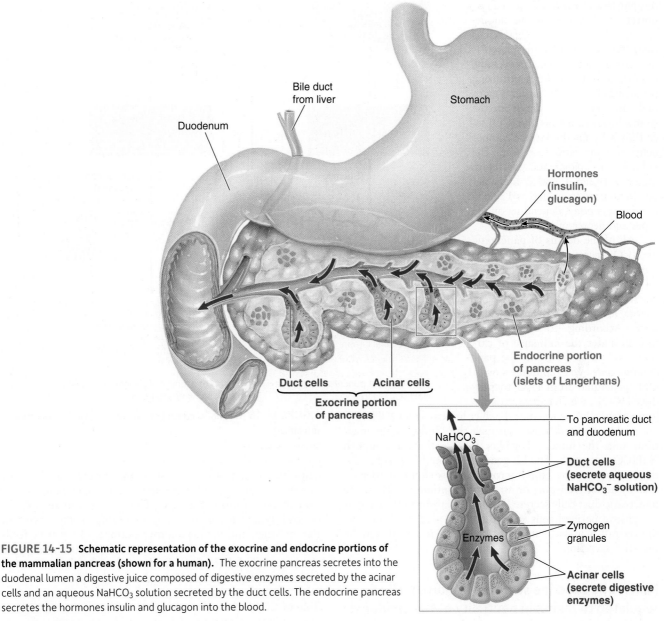

FIGURE 14-15 Schematic representation of the exocrine and endocrine portions of the mammalian pancreas (shown for a human). The exocrine pancreas secretes into the duodenal lumen a digestive juice composed of digestive enzymes secreted by the acinar cells and an aqueous NaHCO₃ solution secreted by the duct cells. The endocrine pancreas secretes the hormones insulin and glucagon into the blood.

Pancreatic Chitinase If cellulose is the most common biological polysaccharide, then what is the second? It is *chitin,* a compound found in the cuticle of arthropods and in cell walls of fungi where it serves a structural function comparable to that of cellulose. **Chitin** consists of chains of glucose molecules, each linked by β-bonds, and each of which has an acetylamino group (–N–CO–CH₃) in place of a hydroxyl (–OH) group on the second carbon. Compared to cellulose, this is a more digestible nutrient for vertebrates, although only fish and some species of marine birds have a pancreatic chitinase that breaks down chitin into N-acetyl-glucosamine.

Pancreatic Lipase In vertebrates pancreatic lipase is extremely important because it is the principal enzyme that can digest fat. Pancreatic lipase (like salivary lipase) hydrolyzes dietary triglycerides into monoglycerides and free fatty acids, which are the absorbable units of fat. Like amylase, lipase is secreted in its active form because there is no risk of

pancreatic self-digestion by lipase. When pancreatic enzymes are deficient, digestion of food is incomplete. Because the pancreas is the only significant source of lipase, pancreatic enzyme deficiency results in serious maldigestion of fats. The principal clinical manifestation of pancreatic exocrine insufficiency in mammals is **steatorrhea,** or excessive undigested fat in the feces. Up to 60 to 70% of the ingested fat may be excreted in the feces. Digestion of protein and carbohydrates is less impaired because salivary, gastric, and small-intestinal enzymes contribute to the digestion of these two foodstuffs.

Pancreatic Aqueous Alkaline Secretion Pancreatic enzymes function best in a neutral or slightly alkaline environment, yet the highly acidic gastric contents are emptied into the duodenal lumen in the vicinity of pancreatic enzyme entry into the duodenum. This acidic chyme must be quickly neutralized in the duodenal lumen, not only to allow optimal functioning of the pancreatic enzymes but also to prevent acid damage to the duodenal mucosa. The alkaline (NaHCO$_3$-rich) fluid secreted by the pancreas into the duodenal lumen serves the important function of neutralizing the acidic chyme as the latter is emptied into the duodenum from the stomach. This aqueous NaHCO$_3$ secretion is by far the largest component of pancreatic secretion. The volume of pancreatic secretion ranges between 1 and 2 liters per day in humans to as high as 12 liters per day in a pony, depending on the types and degree of stimulation.

All of the details of pancreatic NaHCO$_3$ secretion are not known, but carbonic anhydrase (see Figure 14-12, p. 676) is involved and the Na$^+$/K$^+$ pump provides the driving energy for secondary active transport. According to the current model, under the influence of carbonic anhydrase, CO$_2$ in the pancreatic duct cell combines with OH$^-$ generated from H$_2$O to produce HCO$_3^-$, which exits across the luminal membrane to enter the duct lumen via a Cl$^-$/HCO$_3^-$ antiporter. Sodium moves through the "leaky" tight junctions into the lumen. Together these actions accomplish NaHCO$_3$ secretion. The H$^+$ generated from H$_2$O within the duct cell enters the blood across the basolateral border either by active transport or an Na$^+$/H$^+$ antiporter. Thus, the pancreatic duct cells secrete HCO$_3^-$ and absorb H$^+$, whereas the gastric parietal cells secrete H$^+$ and absorb HCO$_3^-$, so the overall acid–base status of the body is not altered by digestive secretion.

Pancreatic exocrine secretion is hormonally regulated to maintain neutrality of the duodenal contents and to optimize digestion

Pancreatic exocrine secretion is regulated primarily by hormonal mechanisms. A small amount of cholinergic-induced (parasympathetic) pancreatic secretion occurs during the cephalic phase of digestion, with a further token increase occurring during the gastric phase in response to gastrin. However, the predominant stimulation of pancreatic secretion occurs during the intestinal phase of digestion, when chyme is in the small intestine. The release of the two major enterogastrones, secretin and cholecystokinin (CCK), in response to chyme in the duodenum plays the central role in the control of pancreatic secretion (Figure 14-16).

Role of Secretin in Pancreatic Secretion Of the factors that stimulate enterogastrone release (fat, acid, hypertonicity, and distension), the primary stimulus for secretin release is acid in the duodenum. Secretin, in turn, is carried by the blood to the pancreas, where it stimulates the duct cells to markedly increase their secretion of a NaHCO$_3$-rich aqueous fluid into the duodenum. Even though other stimuli may

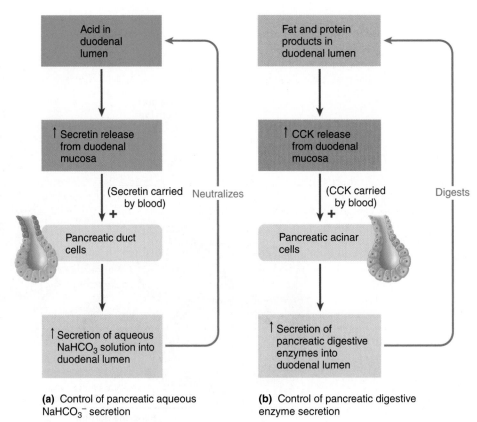

(a) Control of pancreatic aqueous NaHCO$_3^-$ secretion

(b) Control of pancreatic digestive enzyme secretion

FIGURE 14-16 Hormonal control of pancreatic exocrine secretion (mammal).
© Cengage Learning, 2013

cause the release of secretin, it is appropriate that the most potent stimulus is acid, since the result is pancreatic secretion that neutralizes the acid. This mechanism provides a control system for maintaining neutrality of the chyme in the intestine, and is thus another example of negative feedback. The amount of secretin released is proportional to the amount of acid that enters the duodenum, so the amount of NaHCO$_3$ secreted parallels the duodenal acidity.

Role of CCK in Pancreatic Digestion Cholecystokinin is important in the regulation of pancreatic digestive enzyme secretion. The main stimulus for release of CCK from the duodenal mucosa into the blood is the presence of fat in the lumen and, to a lesser extent, protein products. The circulatory system transports CCK to the pancreas, where it stimulates the pancreatic acinar cells to increase digestive enzyme secretion.

All three types of pancreatic enzymes are packaged together in the zymogen granules, so all the pancreatic enzymes are released together on exocytosis of the granules. Therefore, even though the total amount of enzymes released varies depending on the type of meal consumed (the most being secreted in response to fat), the proportion of enzymes released does not vary on a meal-to-meal basis. That is, a high-protein meal does not cause the release of a greater proportion of proteolytic enzymes. Evidence suggests, however, that long-term adjustments in the proportion of the types of enzymes produced may occur as an adaptive response to a prolonged or seasonal change in diet. For example, a greater proportion of proteolytic enzymes are produced with a long-term switch to a high-protein feed. CCK may play a role in pancreatic enzyme adaptation to changes in diet.

Just as gastrin is trophic ("nourishing") to the stomach and small intestine, CCK and secretin exert trophic effects on the exocrine pancreas to maintain its integrity.

We now look at the contributions of the remaining vertebrate accessory digestive units, the liver and gallbladder.

The vertebrate liver performs various important functions, including bile production, nutrient processing, and detoxification

Besides pancreatic juice, the other secretory product that is emptied into the duodenal lumen is **bile,** The **biliary system** includes the *liver,* the *gallbladder,* and associated ducts. Bile is continuously secreted in ruminants, horses, and pigs, whereas in most animals that feed at infrequent intervals, such as the dog and cat, bile is stored in the gallbladder until periods of feeding and digestion.

Liver Functions The **liver** is the largest and most important metabolic organ in the vertebrate body; it can be viewed as the body's major "biochemical factory." The liver performs a wide variety of functions, including the following:

- *Secretion of bile salts* for the digestive tract.
- *Metabolic processing of the major categories of nutrients* (carbohydrates, proteins, and lipids) after their absorption from the digestive tract, including *gluconeogenesis*—the process of converting noncarbohydrate nutrients into glucose.
- *Detoxification or degradation* of body wastes and hormones as well as drugs and other foreign compounds.
- *Synthesis of plasma (lipo)proteins,* including those necessary for the clotting of blood, and those that transport steroid and thyroid hormones, and phospholipids and cholesterol in the blood (see Figure 9-4).
- *Storage* of glycogen, fats, iron, copper, and many vitamins.
- *Activation of vitamin D,* which the liver accomplishes in conjunction with the kidneys.
- *Removal of bacteria and worn-out red blood cells,* thanks to its resident macrophages known as **Kupffer cells.**
- *Secretion of the hormones thrombopoietin* (stimulates platelet production; p. 393), *hepcidin* (inhibits iron uptake from the intestine), *insulin-like growth factor-I* (stimulates growth; p. 292), and *angiotensinogen* (important in salt regulation; p. 587).
- *Excretion of cholesterol, biliverdin, and bilirubin* (p. 688), the latter two being breakdown products derived from the destruction of worn-out red blood cells.
- *Synthesis of ascorbic acid* (vitamin C) in some species of mammals (excluding primates and guinea pigs) and birds.
- *Buoyancy* in sharks because of high amounts of low-density lipids.

Given this wide range of complex functions, there is amazingly little specialization of cells within the liver. Each liver cell, or **hepatocyte** (*hepato,* "liver") performs the same wide variety of metabolic and secretory tasks. The specialization comes from the highly developed organelles within each hepatocyte. The only liver function not accomplished by the hepatocytes is the phagocytic activity carried out by Kupffer cells.

In some species the liver performs additional functions during particular stages of development. For example, the size of the liver is increased in some animals engaged in reproduction. Synthesis of egg constituents (lipogenesis for yolk and some albumin synthesis) is carried out in the avian liver. As a consequence, fatty livers are a common problem in laying hens. Fat content of the liver can increase from 5 to 50% and can markedly increase mortality rate. **Fatty liver syndrome** arises when carbohydrate intake exceeds the liver's ability to synthesize lipoproteins for secretion into the plasma. However, there is not a corresponding increase in fat content of mammalian livers, because both the fetus and the mammary gland contribute to the metabolic demands associated with carbohydrate intake.

Liver Blood Flow To carry out these wide-ranging tasks, the anatomic organization of the liver permits each hepatocyte to be in direct contact with blood from two sources: venous blood coming directly from the digestive tract and arterial blood coming from the aorta. Venous blood enters the liver by means of a unique and complex vascular connection between the digestive tract and the liver, known as the **hepatic portal system** (Figure 14-17). The veins draining the digestive tract do not directly join the inferior vena cava, the large vein that returns blood to the heart. Instead, the veins from the stomach and intestine enter the hepatic portal vein, which carries the products absorbed from the digestive tract directly to the liver for processing, storage, or detoxification before they gain access to the general circulation. Within the liver, the portal vein once again breaks up into a capillary network called **sinusoids** to permit exchange between the blood and hepatocytes before draining into the hepatic vein, which joins the inferior vena cava. The hepatocytes also are provided with fresh arterial blood, which supplies their oxygen and delivers bloodborne metabolites and toxins for hepatic processing.

The liver lobules are delineated by vascular and bile channels

The liver is organized into functional units known as **lobules,** which are hexagonal arrangements of tissue surrounding a central vein (Figure 14-18a). At the outer edge of each "slice" of the lobule are three vessels: a branch of the hepatic artery, a branch of the portal vein, and a bile duct. Blood from the branches of both the hepatic artery and the portal vein flows from the periphery of the lobule into large, expanded sinusoids, which run between rows of liver cells to the central vein like spokes on a bicycle wheel (Figure 14-18b). The Kupffer cells line the sinusoids and engulf and destroy old red blood cells and bacteria that pass through in the blood. The hepatocytes are arranged between the sinusoids in plates two cell layers thick, so that each lateral edge faces a sinusoidal pool of blood. The central veins of all the liver lobules converge to form the hepatic vein, which carries the blood away from the liver. A thin, bile-carrying channel, a **bile canaliculus,** runs between the cells within each hepatic plate. Hepatocytes continuously secrete bile into these thin channels, which carry the bile to a bile duct at the periphery of the lobule. The bile ducts from the various lobules converge to eventually form the hepatic duct, which transports the bile from the liver to the duodenum. Each hepatocyte is in contact with a sinusoid on one side and a canaliculus on the other side.

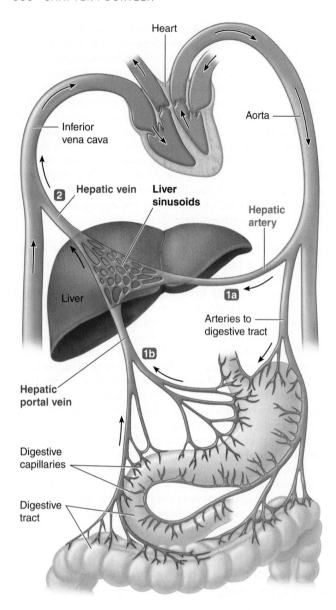

The liver receives blood from two sources:

1a Arterial blood, which provides the liver's O_2 supply and carries bloodborne metabolites for hepatic processing, is delivered by the **hepatic artery**.

1b Venous blood draining the digestive tract is carried by the **hepatic portal vein** to the liver for processing and storage of newly absorbed nutrients.

2 Blood leaves the liver via the **hepatic vein**.

FIGURE 14-17 Schematic representation of liver blood flow (human).

© Cengage Learning, 2013

Bile is continuously secreted by the vertebrate liver and is diverted to the gallbladder between meals

The opening of the bile duct into the duodenum is guarded by the **sphincter of Oddi,** which prevents bile from entering the duodenum except during digestion of meals (Figure 14-19). When this sphincter is closed, most of the bile secreted by the liver is diverted back up into the **gallbladder,** a small, saclike structure tucked beneath the liver. Those vertebrates with a gallbladder can concentrate the bile salts approximately 10- to 20-fold by selectively removing electrolytes and water. Bile is subsequently stored and concentrated in the gallbladder between meals. After a meal, bile enters the duodenum as a result of the combined effects of gallbladder emptying and increased bile secretion by the liver. The amount of bile secreted per day in humans ranges from 250 mL to 1 liter, depending on the degree of stimulation. In birds, two bile ducts exit the liver, one from each lobe. The cystic duct is either connected to the gallbladder or directly to the small intestine, whereas the hepatic duct directly communicates with the small intestine. Animals without a gallbladder include the rat, horse, pigeon, deer, giraffe, elephant, and camel.

Bile salts are recycled through the enterohepatic circulation

Bile contains several organic constituents, namely *bile salts, cholesterol, lecithin,* and *bilirubin* or *biliverdin,* all derived from hepatocyte activity. These are released in an aqueous alkaline fluid (added by the duct cells) similar to the pancreatic $NaHCO_3$ secretion. Even though bile does not contain any digestive enzymes, it is important for the digestion and absorption of fats, as we will describe shortly.

Bile salts are derivatives of cholesterol. To reduce their toxicity, mammals conjugate bile acids with either *taurine* or *glycine,* whereas birds and fish form only the taurine conjugate. (Steroid bile salts have not been identified in any nonvertebrates.) Bile salts are actively secreted into the bile and eventually enter the duodenum along with the other biliary constituents. Following their participation in fat digestion and absorption, most bile salts are reabsorbed back into the blood by special active-transport mechanisms located in the mammalian terminal ileum, the last portion of the small intestine. (In birds, the majority of absorption takes place in the jejunum, the second intestinal segment.) From here the bile salts are returned via the hepatic portal system to the liver, which resecretes them into the bile. This recycling of bile salts (and some of the other biliary constituents) between the small intestine and liver is referred to as the **enterohepatic circulation** (Figure 14-19).

On average, bile salts cycle between the liver and small intestine twice during the digestion of a typical meal. Approximately 5% of the secreted bile salts ultimately escape into the feces daily. These lost bile salts are replaced by new bile salts synthesized by the liver; thus, the size of the pool of bile salts is kept constant.

Bile salts aid fat digestion and absorption through their detergent action and micellar formation, respectively

Bile salts aid fat digestion through their detergent action (*emulsification*) and facilitate fat absorption through their participation in the formation of *micelles.* Both functions are related to the structure of bile salts. Let's see how.

Emulsification by Bile Salts **Emulsification** refers to bile salts' ability to convert large fat globules into a **lipid emulsion** that consists of many small fat droplets suspended in

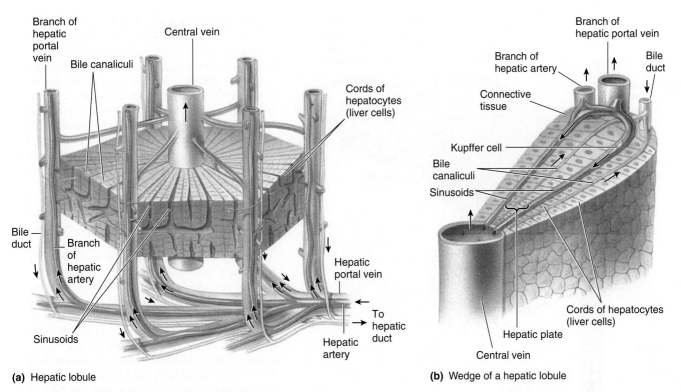

(a) Hepatic lobule

(b) Wedge of a hepatic lobule

FIGURE 14-18 Anatomy of the mammalian liver.

© Cengage Learning, 2013

the aqueous chyme, thus increasing the surface area available for attack by pancreatic lipase. Fat globules, no matter their size, are made up primarily of undigested triglyceride molecules. For lipase to digest fat, it must come into direct contact with the triglyceride molecules. Because fat molecules are not soluble in water, they tend to aggregate into large droplets in the aqueous environment of the small-intestine lumen. If bile salts did not emulsify these large droplets, lipase could act on the lipids only at the surface of the large droplets, and triglyceride digestion would be greatly prolonged.

Bile salts act like the detergent you use to break up grease when you wash dishes. Bile salt molecules and detergents are **amphipathic:** Each contains a hydrophobic (lipid-soluble) portion (a steroid derived from cholesterol) plus a negatively charged, hydrophilic (water-soluble) portion (from taurine or glycine). Bile salts *adsorb* on the surface of a fat droplet; that is, the lipid-soluble portion of the bile salt interacts with the fat droplet, leaving the charged portion projecting from the surface of the droplet to dissolve in water (Figure 14-20a). Intestinal mixing movements break up large fat droplets into smaller ones. These small droplets would quickly recoalesce were it not for adsorbed bile salts creating a "shell" of water-soluble negative charges on the surface of each little droplet. Because like charges repel, these negatively charged groups on the droplet surfaces cause the fat droplets to repel each other (Figure 14-20b). This prevents the small droplets from recoalescing into large fat droplets, producing a lipid emulsion that increases the surface area available for lipase action. The small emulsified fat droplets range in diameter from 200 to 5,000 nm.

Although bile salts increase the surface area available for attack by pancreatic lipase, lipase alone cannot penetrate

the layer of bile salts adsorbed on the surface of the small emulsified fat droplets. To solve this dilemma, the pancreas secretes the polypeptide **colipase** along with lipase. Like bile salts, colipase is amphipathic; it displaces some bile salts and lodges at the surface of the fat droplets, where it binds to lipase, thus anchoring this enzyme to its site of action amid the bile-salt coating.

During digestion of a meal, when chyme reaches the small intestine, the presence of food, especially fat products, in the duodenal lumen triggers the release of CCK. This hormone stimulates contraction of the gallbladder and relaxation of the sphincter of Oddi, so bile is discharged into the duodenum, where it appropriately aids in the digestion and absorption of the fat that initiated the release of CCK. A similar process happens in the small intestine of fish with a gastrin/CCK-related peptide that stimulates gallbladder contraction.

Micellar Formation Bile salts—along with cholesterol and lecithin, which are also constituents of bile—play an important role in facilitating fat absorption through micellar formation. Like bile salts, **lecithin** (a phospholipid used in nonstick cooking sprays) is amphipathic, whereas cholesterol is almost totally insoluble in water. In a **micelle,** the bile salts and lecithin aggregate in small clusters with their fat-soluble portions huddled together in the middle to form a hydrophobic core, while their water-soluble portions form an outer hydrophilic shell (Figure 14-21). A micelle is 3 to 10 nm in diameter, compared to an average diameter of 1,000 nm for an emulsified lipid droplet. Micelles are water soluble by virtue of their hydrophilic shells, but they can dissolve water-insoluble (and hence lipid-soluble) substances in their lipid-soluble cores. Micelles thus provide a handy

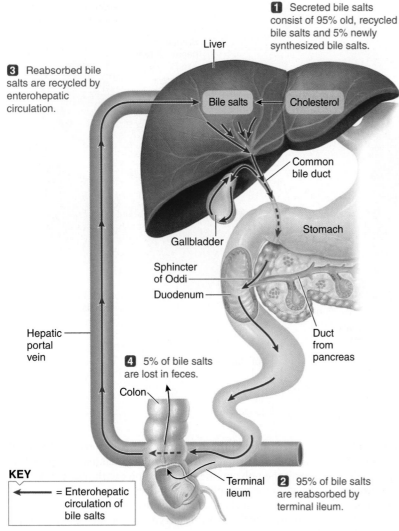

1 Secreted bile salts consist of 95% old, recycled bile salts and 5% newly synthesized bile salts.

3 Reabsorbed bile salts are recycled by enterohepatic circulation.

Liver

Bile salts ← Cholesterol

Common bile duct

Gallbladder

Stomach

Sphincter of Oddi

Duodenum

Duct from pancreas

Hepatic portal vein

4 5% of bile salts are lost in feces.

Colon

Terminal ileum

2 95% of bile salts are reabsorbed by terminal ileum.

KEY

← = Enterohepatic circulation of bile salts

FIGURE 14-19 Enterohepatic circulation of bile salts (human). The majority of bile salts are recycled between the liver and small intestine through the enterohepatic circulation, designated by the blue arrows. After participating in fat digestion and absorption, most bile salts are reabsorbed by active transport in the terminal ileum and returned through the hepatic portal vein to the liver, which resecretes them in the bile.
© Cengage Learning, 2013

vehicle for carrying water-insoluble substances through the watery luminal contents. The most important lipid-soluble substances thus carried are the products of fat digestion (monoglycerides and free fatty acids) as well as fat-soluble vitamins, which are all transported to their sites of absorption by means of the micelles. If they did not hitch a ride in the water-soluble micelles, these nutrients would float on the surface of the aqueous chyme (just as oil floats on top of water), never reaching the absorptive surfaces of the small intestine.

Biliverdin and bilirubin are waste products of hemoglobin excreted in the bile

Biliverdin (in most vertebrates) and *bilirubin* (in mammals), the other major constituents of bile, do not play a role in digestion at all but instead are two of the few waste products excreted in the bile. Worn-out red blood cells are removed from the blood by macrophages in the spleen (p. 392) and

other areas in the body. Kupffer cells degrade these worn-out red blood cells to yield heme (iron-containing portion of the hemoglobin). Heme, which is toxic, is converted to **biliverdin** (green in color), releasing *carbon monoxide* (CO) in the process. (CO, although toxic at high levels, may have a physiological role at low concentrations as a paracrine vasodilator to boost local blood flow.) Amphibians and reptiles (including birds) excrete biliverdin as the major means of pigment elimination, with the result that the color of their bile is dark green. The excretion rate of bile pigment is greater in birds than mammals because of the much shorter life of avian red blood cells (see p. 392).

In mammals, biliverdin is transformed by a uniquely mammalian enzyme into bilirubin (yellow in color) and then released into the lobule sinusoid. Bilirubin synthesis may have evolved because the molecule is a potent antioxidant, reverting to biliverdin as it scavenges free radicals (perhaps replacing uric acid, p. 565). Excess bilirubin is toxic, however, and much is extracted from the blood by the hepatocytes and actively excreted into the bile. A small amount of bilirubin is normally reabsorbed by the small intestine back into the blood, and when it is eventually excreted in the urine, it is largely responsible for the urine's yellow color. If bilirubin is formed more rapidly than it can be excreted, it accumulates in the body and causes **jaundice,** Mammals with this condition appear yellowish, with this color being most apparent in the whites of their eyes. Human babies with jaundice are exposed to fluorescent light, which helps break down bilirubin in the skin.

The fat body of insects plays a major role in intermediary metabolism and is the central storage depot of nutrients and energy reserves

The insect **fat body** houses the analogues of the vertebrate liver, hematopoietic and immune systems, and adipose tissue in a single functional unit. The fat body is analogous to the vertebrate liver because it plays a role in intermediary metabolism, is regulated by hormones, and serves as a storage place for food reserves. Because lipids can be accumulated in such massive quantities in the fat body, it appears as a diffuse organ of fat located throughout the **hemocoel** (body cavity filled with blood) and abdomen of insects. Structurally, the fat body is arranged in thin lobes that are bathed by the hemolymph. The vast majority of the cells in the fat body that perform metabolic functions are termed **adipocytes,** which (like vertebrate adipose cells) are characterized by the presence of numerous lipid droplets. Another fat body cell type includes **urate cells,** which are involved in storing uric acid crystals in species lacking Malpighian tubules (see p. 566). The fat body is structurally organized to maximize exchange with the hemolymph, which is the

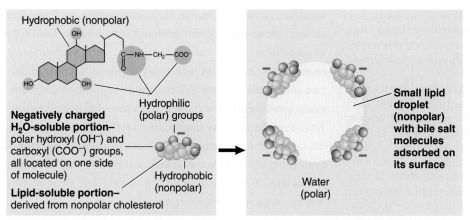

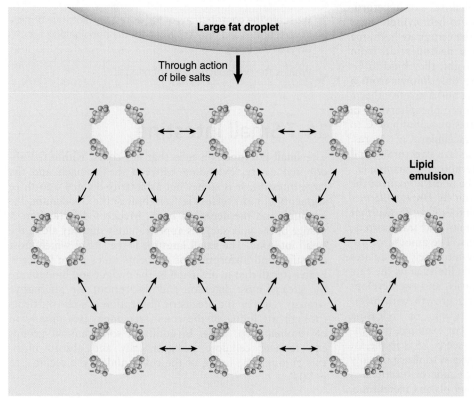

(a) Structure of bile salts and their adsorption on the surface of a small lipid droplet

(b) Formation of a lipid emulsion through the action of bile salts

FIGURE 14-20 Schematic structure and function of bile salts. (a) A bile salt consists of a lipid-soluble part that dissolves in the fat droplet and a negatively charged, water-soluble part that projects from the surface of the droplet. (b) When a large fat droplet is broken up into smaller fat droplets by intestinal contractions, bile salts adsorb on the surface of the small droplets, creating shells of negatively charged, water-soluble bile salt components that cause the fat droplets to repel one another. This emulsifying action holds the fat droplets apart and prevents them from recoalescing, increasing the surface area of exposed fat available for digestion by pancreatic lipase.

© Cengage Learning, 2013

FIGURE 14-21 A micelle. Bile constituents (bile salts, lecithin, and cholesterol) aggregate to form micelles that consist of a hydrophilic (water-soluble) shell and a hydrophobic (lipid-soluble) core. Because the outer shell of a micelle is water soluble, the products of fat digestion, which are not water soluble, can be carried through the watery luminal contents to the absorptive surface of the small intestine by dissolving in the micelle's lipid-soluble core. This figure is not drawn to scale compared to the lipid emulsion droplets in Figure 14-20b. An emulsified fat droplet ranges in diameter from 200 to 5,000 nm (average 1,000 nm) compared to a micelle, which is 3 to 10 nm in diameter.

© Cengage Learning, 2013

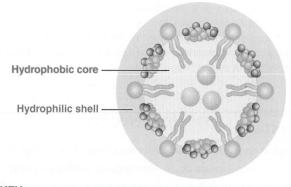

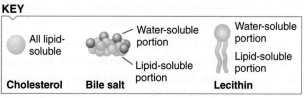

carrier for products taken up or released by the fat body. For example, during feeding cycles, the fat body synthesizes lipids, carbohydrates, and proteins, whereas during strenuous activity, it mobilizes the release of lipids and carbohydrates (see Figure 9-4, p. 390).

Locomotion and contraction of the flight muscles in insects depends on the oxidation of substrates, mainly carbohydrates and lipids, although in tsetse flies and certain beetles, the amino acid proline can be metabolized. During a short flight, proline is utilized as an energy source, whereas during a long, flight triglycerides are mobilized. These substrates are mobilized from their storage sites and provided to the flight muscles by the action of small peptide hormones synthesized in the insect's *corpora cardiaca* (CC; p. 280). These hormones are generically known as **adipokinetic** (**AKH;** "lipid-mobilizing"), **hypertrehalosemic (HrTHs)** ("increasing the hemolymph sugar trehalose"), or **hyperprolinemic** hormones. Their contribution to hemolymph sugar homeostasis is comparable to that of the vertebrate hormone *glucagon* (p. 320). After release of these neuropeptides from the CC into the hemolymph during flight, they bind to G-protein coupled receptors, which are only known from a very limited number of insects to date, and activate either a triacylglycerol (TAG) lipase or a glycogen phosphorylase in the fat body. In addition, AKH activates phospholipase C to induce the release of Ca^{2+} from intracellular Ca^{2+} stores. Interestingly, the insect AKH receptors are structurally and evolutionarily related to the GnRH receptors of mammals.

Nutrient sensing also appears to be in the domain of the fat body. Studies of the laboratory fruit fly *Drosophila melanogaster* and, more recently, mosquitoes have shown that the fat body can express specific amino acid transporters, which can function as nutrient sensors. The amount of nutrient reserves stored in the fat body ultimately modulates several important events in the insect's life such as the rate of growth, the timing of metamorphosis, and egg development. The fat body coordinates insect growth with metamorphosis or reproduction by storing or releasing components essential to these events. For example, the synthesis of the egg protein **vitellogenin** (see p. 765) in the fat body of *Aedes aegypti* female mosquitoes is transcriptionally up-regulated after a blood meal. A cascade of reactions is then initiated that begin at the fat body plasma membrane, where specific amino acids present in the hemolymph are "sensed" by amino acid transporters. This signal activates an evolutionarily conserved nutritional signaling cascade that results in the translation of a transcriptional activator of vitellogenin gene expression. In response, the synthesis of vitellogenin is stimulated, which reaches a peak ~30 hours after ingestion of the blood meal.

unanswered Questions | A solar-powered hornet?

Biological dogma states that plants obtain metabolic energy from the sun, while animals do so from organic molecules (food). In 2008, that dichotomy was partly demolished by the discovery of a seaslug that has obtained photosynthesis genes from algae (p. 34). More recently, a hornet that may also utilize solar energy for specific metabolic functions has been described. The oriental hornet *Vespa orientalis* has a cuticle with a bright yellow abdomi-

nal stripe. Jacob Ishay, Marian Plotkin, and colleagues found that shining light on the abdomen generates several hundred millivolts of potential across the cuticle. Analysis of cuticle microstructure revealed that the cuticles reflect very little light, but instead capture and channel light waves into a yellow pigment, *xanthopterin*, which generates electric currents when activated by light energy. What does the wasp use this electricity for? We do not know, but intriguingly, the researchers found that the yellow stripe has enzymes normally found in the fat body, and that shining light on the stripes alters the activities of those enzymes. Thus, these hornets may have a solar-powered metabolic tissue! **<<**

check your understanding 14.6

What are the contributions of the accessory digestive organs in vertebrates and insects? What are the nondigestive functions of the liver?

What is the principal function of bile salts?

14.7 Small Intestine

The **small intestine** is a tube that lies coiled within the abdominal cavity, extending between the stomach and the large intestine. It is somewhat arbitrarily divided into three segments in most terrestrial vertebrates: the **duodenum,** the **jejunum,** and the **ileum,** In birds, **Meckel's diverticulum,** a vestige of the yolk sac, lies approximately midway along the small intestine. The small intestine is the site at which most digestion and absorption of nutrients takes place in vertebrates (recall that in arthropods, the midgut and hindgut are the sites of most digestion and absorption). An alternative strategy relies on the pregastric absorption of some nutrients through the lining of the ruminoreticulum (see discussion of ruminants, p. 705). In addition, some animal groups can ferment cellulose postgastrically and absorb nutrients through the lining of the colon and/or the cecum (see p. 704).

The digestive tract wall has four layers: mucosa, submucosa, muscularis externa, and serosa

The digestive tract wall has the same general structure throughout most of its length from the esophagus to the anus, with some local variations characteristic for each region. A cross section of the small intestine (Figure 14-22) reveals four major tissue layers. From the innermost layer of the tract outward, they are the *mucosa,* the *submucosa,* the *muscularis externa,* and the *serosa:*

1. The **mucosa** lines the luminal surface of the digestive tract. It is divided into three layers:
   a. A **mucous membrane,** an inner epithelial layer that serves as a protective surface as well as being modified in particular areas for secretion and absorption. It contains *exocrine cells* for secreting digestive juices, *endocrine cells* for secreting gastrointestinal hormones, and *epithelial cells* specialized for absorbing digested nutrients.

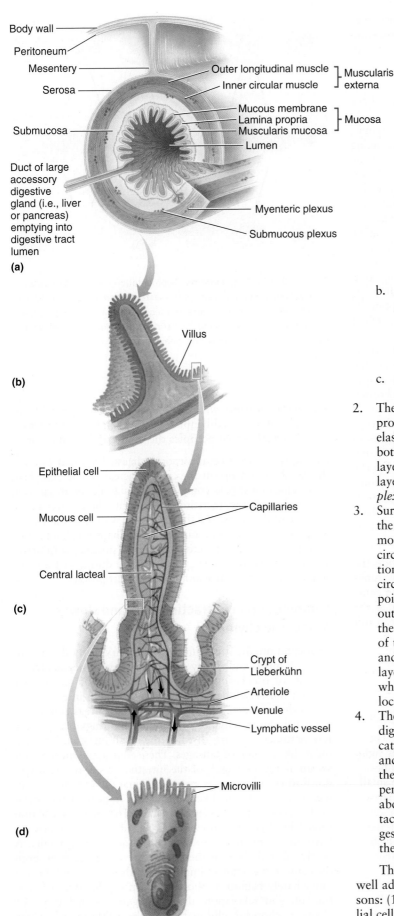

Body wall
Peritoneum
Mesentery
Serosa
Submucosa

Outer longitudinal muscle ⎤ Muscularis
Inner circular muscle ⎦ externa

Mucous membrane ⎤
Lamina propria ⎬ Mucosa
Muscularis mucosa ⎦
Lumen

Duct of large
accessory
digestive
gland (i.e., liver
or pancreas)
emptying into
digestive tract
lumen

Myenteric plexus
Submucous plexus

(a)

Villus

(b)

Epithelial cell
Mucous cell
Central lacteal

Capillaries

(c)

Crypt of
Lieberkühn
Arteriole
Venule
Lymphatic vessel

Microvilli

(d)

FIGURE 14-22 Tissue structure of the mammalian digestive tract. (a) Generalized layers of the digestive tract wall, consisting of four major layers: from innermost out, the mucosa, submucosa, muscularis externa, and serosa. (b–d) The mammalian small intestine: (b) One of the circular folds of the small-intestine mucosa, which collectively increase the absorptive surface area threefold. (c) Microscopic fingerlike projection known as a *villus*. Collectively, the villi increase the surface area another 10-fold. (d) A villus epithelial cell, depicting the presence of microvilli on its luminal border to form a *brush border;* the microvilli increase the surface area another 20-fold. Altogether, these surface modifications increase the small intestine's absorptive surface area 600-fold.

© Cengage Learning, 2013

b. The **lamina propria,** a thin middle layer of connective tissue on which the epithelium rests. Small blood vessels, lymph vessels, and nerve fibers pass through the lamina propria, and it houses the *gut-associated lymphoid tissue (GALT),* which is important in the defense against intestinal bacteria (p. 467).

c. The **muscularis mucosa,** a sparse outer layer of smooth muscle that lies adjacent to the submucosa.

2. The **submucosa** is a thick layer of connective tissue that provides the digestive tract with its distensibility and elasticity. It contains the larger blood and lymph vessels, both of which send branches inward to the mucosal layer and outward to the surrounding thick muscle layer. Also, a nerve network known as the *submucosal plexus* lies within the submucosa.

3. Surrounding the submucosa is the **muscularis externa,** the major smooth-muscle coat of the digestive tube. In most parts of the tract, it consists of two layers: an inner circular layer and an outer longitudinal layer. Contraction of the inner circular fibers, which run around the circumference of the tube, constricts the lumen at the point of contraction. Contraction of the fibers in the outer layer, which run longitudinally along the length of the tube, shortens the tube. Together, contractile activity of these smooth-muscle layers produces the propulsive and mixing movements. Lying between the two muscle layers is another nerve network, the *myenteric plexus,* which, along with the submucous plexus, helps regulate local gut activity.

4. The **serosa** is the outer connective tissue covering of the digestive tract, which secretes a serous fluid that lubricates and prevents friction between the digestive organs and surrounding viscera. Throughout much of the tract, the serosa is continuous with the **mesentery,** which suspends the digestive organs from the inner wall of the abdominal cavity like a sling (Figure 14-22a). This attachment provides relative fixation, supporting the digestive organs in proper position, while still allowing them freedom for mixing and propulsive movements.

The mucous lining of the small intestine is remarkably well adapted to its special absorptive function, for two reasons: (1) It has a very *large surface area,* and (2) the epithelial cells in this lining have a variety of *specialized transport mechanisms.*

Adaptations that Increase the Small Intestine's Surface Area
The following three modifications of the small-intestine mucosa greatly increase the surface area available for absorption (Figure 14-22b, c, d):

1. *Folds:* The inner surface of the small intestine is arranged in circular folds that are visible to the naked eye and that increase the surface area threefold.

2. *Villi:* Projecting from this folded surface are microscopic, fingerlike projections known as **villi,** which give the lining a velvety appearance and increase the surface area by another 10 times (Figure 14-22b, c). Epithelial cells cover the surface of each villus. Villi are dynamic and readily change dimensions in response to numerous dietary factors. For example, the mass of a python's small intestine increases two- to threefold overnight after swallowing its prey, associated with a 60-fold increase in enzyme content. Each epithelial cell lining the small intestine increases in size and develops longer projections, adaptations that enable the ingesta to be digested more quickly. Interestingly, the energetic cost to the python for adapting its small intestine to the arrival of food is approximately a third to a half of the energy derived from its prey.

3. *Microvilli:* Even smaller, hairlike projections known as **microvilli** arise from the luminal surface of these epithelial cells (Figure 14-22d), increasing the surface area another 20-fold in a formation called the **brush border,** Each epithelial cell has as many as 3,000 to 6,000 of these microvilli, which are visible only with an electron microscope. Generally, villi are longest in the jejunum and progressively decrease in length through the ileum. It is within the membrane of this brush border that the enzymes of the small intestine perform their functions. Arising from the surface of the microvilli is the **glycocalyx,** a meshwork of carbohydrate-rich filaments. These filaments are components of the digestive enzymes that project into the lumen of the small intestine.

Altogether the folds, villi, and microvilli provide the small intestine with a luminal surface area about 600 times greater than it would have if it were a tube of the same length and diameter lined by a flat surface. In fact, if the surface area of the human small intestine were spread out flat, it would cover an entire tennis court.

Structure of a Villus Absorption across the digestive tract wall involves transepithelial transport similar to movement of material across the kidney tubules (see p. 583). Each villus has the following major components (Figure 14-22c):

- *Epithelial cells that cover the surface of the villus.* The epithelial cells are joined at their lateral borders by tight junctions, which limit passage of luminal contents between the cells, although the tight junctions in the small intestine are "leakier" than those in the stomach. These epithelial cells have, within their luminal brush borders, carriers for absorption of specific nutrients and electrolytes from the lumen as well as the brush-border digestive enzymes that complete carbohydrate and protein digestion.

- *A capillary network.* Each villus is supplied by an arteriole that breaks up into a capillary network within the villus. The capillaries rejoin to form a venule that drains away from the villus.

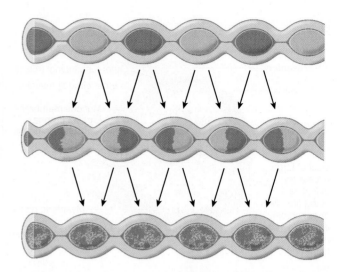

FIGURE 14-23 Segmentation. Segmentation consists of ringlike contractions along the length of the vertebrate small intestine. Within a matter of seconds, the contracted segments relax and the previously relaxed areas contract. These oscillating contractions thoroughly mix the chyme within the small-intestine lumen.

© Cengage Learning, 2013

- *A terminal lymphatic vessel.* Each villus is supplied by a single, blind-ended lymphatic vessel known as the **central lacteal,** which occupies the center of the villus core.

During the process of absorption, digested substances enter the capillary network or the central lacteal. To be absorbed, a substance must pass completely through the epithelial cell, diffuse through the interstitial fluid within the connective tissue core of the villus, and then cross the wall of a capillary or lymph vessel. Like renal transport (see Chapter 12), intestinal absorption may be an active or passive process, with active absorption involving energy expenditure during at least one of the steps in the transepithelial transport process.

Segmentation contractions mix and slowly propel the chyme

Segmentation, the small intestine's primary method of motility, both mixes and slowly propels the chyme. Segmentation consists of oscillating, ringlike contractions of the circular smooth muscle along the length of the small intestine; between the contracted segments are relaxed areas containing a small bolus of chyme. The contractile rings occur every few centimeters apart, dividing the small intestine into segments like a chain of sausages. These contractile rings do not sweep along the length of the intestine as peristaltic waves do. Rather, after a brief period of time the contracted segments relax, and ringlike contractions appear in the previously relaxed areas (Figure 14-23). The new contraction forces the chyme in a previously relaxed segment to move in both directions into the now relaxed adjacent segments. A newly relaxed segment therefore receives chyme from both the contracting segment immediately ahead of it and the one immediately behind it. Shortly thereafter, the areas of contraction and relaxation alternate again. In this way, the chyme is chopped, churned, and thoroughly mixed. These contractions can be compared to squeezing a pastry tube with your hands to mix the contents. This mixing serves the dual functions of mixing the chyme with the digestive juices secreted into the small-intestine lumen and exposing all the chyme to the absorptive surfaces of the small intestine mucosa.

Initiation and Control of Segmentation Segmentation contractions are initiated by the small intestine's pacesetter cells, which produce a basic electrical rhythm (BER; p. 660) similar to the gastric BER responsible for peristalsis in the stomach. If the small-intestine BER brings the circular smooth-muscle layer to threshold, segmentation contractions are induced, with a frequency following the frequency of the BER.

The circular smooth muscle's degree of responsiveness and thus the intensity of segmentation contractions can be influenced by distension of the intestine, by the hormone gastrin, and by extrinsic nerve activity. All these factors influence the excitability of the small-intestine smooth-muscle cells by moving the "resting" potential around which the BER oscillates closer to or farther from threshold. Segmentation is slight or absent between meals but becomes very vigorous immediately after feeding. Both the duodenum and ileum start to segment simultaneously when the digesta first enters the small intestine. The duodenum starts to segment primarily in response to local distension caused by the presence of chyme. Segmentation of the empty ileum, in contrast, appears to be brought about by gastrin, which is secreted in response to the presence of chyme in the stomach, a mechanism known as the **gastroileal reflex.** Extrinsic nerves can modify the strength of these contractions. Parasympathetic stimulation enhances segmentation, whereas sympathetic stimulation depresses segmental activity.

Segmentation not only accomplishes mixing but also is the primary factor responsible for slowly moving chyme through the small intestine. How can this be, when each segmental contraction propels chyme both forward and backward? The chyme slowly progresses forward because the frequency of segmentation declines along the length of the small intestine. The pacesetter cells in the duodenum spontaneously depolarize at a faster rate than those farther down the tract, with the pacesetter cells in the terminal ileum exhibiting the slowest rate of spontaneous depolarization. For example, the rate of segmentation in the human duodenum is 12/min, compared to a rate of only 9/min in the terminal ileum. As a result, more chyme on average is pushed forward than is pushed backward, so chyme is moved very slowly down the small intestine while being shuffled back and forth to accomplish thorough mixing and absorption. This slow movement is advantageous because it allows ample time for the digestive and absorptive processes to take place. The contents usually take three to five hours to move through the human small intestine.

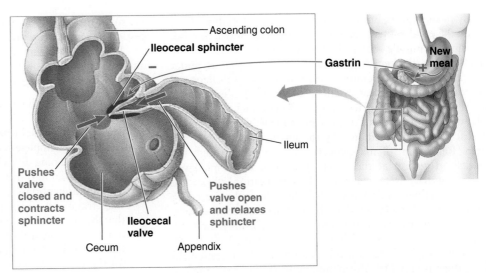

FIGURE 14-24 Control of the ileocecal valve/sphincter (shown for a human). The juncture between the ileum and large intestine consists of the ileocecal valve, which is surrounded by thickened smooth muscle, the ileocecal sphincter. Pressure on the cecal side pushes the valve closed and contracts the sphincter, preventing the bacteria-laden colonic contents from contaminating the nutrient-rich small intestine. The valve/sphincter opens and allows ileal contents to enter the large intestine in response to pressure on the ileal side of the valve and to the hormone gastrin secreted as a new meal enters the stomach.

© Cengage Learning, 2013

The migrating motility complex sweeps the intestine clean between meals

When most of the meal has been absorbed, segmentation contractions cease and are replaced by the **migrating motility complex,** or "intestinal housekeeper," This between-meal motility consists of weak, repetitive peristaltic waves that move a short distance down the intestine before dying out. These short waves start at the stomach and take about 100 to 140 minutes in humans to gradually migrate from the stomach to the end of the small intestine, with each contraction "sweeping" any remnants of the preceding meal plus mucosal debris and bacteria forward toward the colon, just like a good housekeeper. After the end of the small intestine is reached, the cycle begins again and continues to repeat itself until the next meal. The hormone *motilin* (p. 712), released by intestinal *crypt* cells, plays a role in stimulating this migrating motility complex. When the next meal arrives, segmental activity is triggered again, and the migrating motility complex ceases.

The ileocecal juncture prevents contamination of the small intestine by colonic bacteria

At the juncture between the small and large intestines, the last portion of the ileum empties into the *cecum* (Figure 14-24). Two factors contribute to this region's ability to act as a barrier between the small and large intestines. First, the anatomic arrangement is such that valvelike folds of tissue protrude from the ileum into the lumen of the cecum. When the ileal contents are pushed forward, this **ileocecal valve** is easily pushed open, but the folds of tissue are forcibly closed when the cecal contents attempt to move backward. Second, the smooth muscle within the last several centimeters of the ileal wall is thickened, forming a sphincter that is under neural and hormonal control. Most of the time this **ileocecal sphincter** remains at least mildly constricted. Pressure on the cecal side of the sphincter causes it to contract more forcibly; distension of the ileal side causes the sphincter to relax, a reaction that is mediated by the intrinsic plexuses in the area. Relaxation of the sphincter is enhanced through the release of gastrin at the onset of a meal, when increased gastric activity is taking place. This relaxation allows the undigested fibers and unabsorbed solutes from the preceding meal to be moved forward as the new meal enters the tract.

The digestive tract of numerous carnivorous mammals, including some species of bats and marsupials, lack a constriction or valve between the small and large intestine. In fact, there is very little distinction between the small and the large intestinal tracts, and generally these animals also lack a cecum. For example, the short terminal segment of the large intestine of a mink is slightly larger than the small intestine and lacks a sphincter or valve at the junction between the two segments. In the horse, there is a sphincter at both the *ileocecal* and the *cecocolic* junction that ensures that the intestinal contents enter the cecum before passage through the colon. The ceca are paired in birds, although they are not present in all avian species.

Small-intestine secretions do not contain any digestive enzymes

Each day the exocrine glands located in the small-intestine mucosa secrete into the lumen large volumes of an aqueous salt and mucus solution known as the **succus entericus** (*succus*, "juice"; *entericus*, "of the intestine"; about 1.5 liters in humans). However, no digestive enzymes are secreted into this intestinal juice. The small intestine synthesizes digestive enzymes, but they function within the borders of the epithelial cells that line the lumen instead of being secreted directly into the lumen. As in other parts of the tract, mucus in the secretion provides protection and lubrication. Furthermore, this aqueous secretion provides an abundance of H_2O to participate in the enzymatic digestion of food, which proceeds most efficiently when all the reactants are in solution.

The regulation of small-intestine secretion is not clearly understood. As well, the factors that specifically alert the intestine of the upcoming meal are not completely known. However, the secretion of succus entericus does increase after a meal. The most effective stimulus for secretion appears to be local stimulation of the mucosa by the presence of chyme.

Small-intestine enzymes complete the process of digestion at the brush border membrane

Digestion within the small-intestine lumen is accomplished by the pancreatic enzymes, with fat digestion being enhanced by bile secretion. As a result of those enzymes' actions, fats are completely reduced to their absorbable units of monoglycerides and free fatty acids, proteins are broken down into small peptide fragments and some amino acids, and carbohydrates are reduced to disaccharides and some monosaccharides. Thus, fat digestion is completed within

the small-intestine lumen, but carbohydrate and protein digestion have not been brought to completion.

Recall that the luminal surface of the mammalian small-intestine epithelial cells has a *brush border* (see p. 692). This contains different categories of enzymes:

1. **Enterokinase**, which activates the pancreatic enzyme trypsinogen.
2. The **disaccharidases** (**maltase, sucrase, lactase,** and **trehalase**), which complete carbohydrate digestion by hydrolyzing the remaining disaccharides (*maltose, sucrose, lactose,* and *trehalose*) into their constituent monosaccharides.
3. The **aminopeptidases**, which hydrolyze the small peptide fragments into their amino acid components, thereby completing protein digestion. Thus, carbohydrate and protein digestion are completed at the brush border. (Table 14-5 summarizes the digestive processes for the three major categories of nutrients.)

Not all vertebrate groups express each of these enzymes in their brush border. For example, birds have no need of lactase but do have an intestinal amylase. Many passerine songbirds lack sucrase, whereas hummingbirds, with a sucrose-rich nectar diet, express sufficient sucrase activity in their intestines. Because the microbes in the reticulorumen of a ruminant scavenge most of the carbohydrates, the concentrations of disaccharidases in the small intestine are low, with sucrase not synthesized at all. Vertebrates whose diet normally contains chitin have evolved a chitinase enzyme. Bats, rodents, turtles, and insectivorous lizards synthesize substantial quantities of chitinase in their gastric mucosa. Similarly, many species of fish and birds produce chitinase in their pancreas, whereas other species rely on intestinal bacteria to synthesize this enzyme. Gut microflora of some species may contain chitinase as well as cellulase.

The small intestine is highly adaptable for its primary role in absorption

Normally, all products of carbohydrate, protein, and fat digestion, as well as most of the ingested electrolytes, vitamins, and water, are absorbed by the small intestine indiscriminately. Usually only the absorption of calcium and iron is adjusted to an individual animal's needs in the short run. Thus, the more food that is consumed, the more that is digested and absorbed. However, this does not mean food absorption capacity is static. In many vertebrates, the length of the intestine and the density of its transporters can be altered. You have already seen how a python's small intestine can change radically and quickly in mass. In other animals, changes can occur during long-term dietary changes. For example, in omnivores the glucose transporter is up-regulated on a prolonged high-carbohydrate diet (a form of acclimatization; p. 16).

The lengths of the small intestine and the types and densities of its transporters are remarkably matched to typical diets of species. For instance, herbivores typically have a higher density of intestinal glucose transporters than do carnivores. Cats require taurine and arginine in their diet because they cannot synthesize them and can suffer from even short-term deficits in these amino acids. Taurine is essential for neural development, osmotic balance of neural and renal tissues, and other functions that are poorly understood. Arginine is used in the production of urea (see Figure 12-2), which carnivores produce in high amounts because of

TABLE 14-5 Digestive Processes for the Three Major Categories of Nutrients (Mammals)

Nutrients	Enzymes for Digesting Nutrient	Source of Enzymes	Site of Action of Enzymes	Action of Enzymes	Absorbable Units of Nutrients
Mammals					
Carbohydrate	Amylase	Salivary glands	Mouth and body of stomach	Hydrolyzes polysaccharides to disaccharides	
		Exocrine pancreas	Small-intestine lumen		
	Disaccharidases (maltase, sucrase, lactase)	Small-intestine epithelial cells	Small-intestine brush border	Hydrolyze disaccharides to monosaccharides	Monosaccharides, especially glucose
Protein	Pepsin	Stomach chief cells	Stomach antrum	Hydrolyzes protein to peptide fragments	
	Trypsin, chymotrypsin, carboxypeptidase	Exocrine pancreas	Small-intestine lumen	Attack different peptide fragments	
	Aminopeptidases	Small-intestine epithelial cells	Small-intestine brush border	Hydrolyze peptide fragments to amino acids	Amino acids and a few small peptides
Fat	Lipase	Exocrine pancreas	Small-intestine lumen	Hydrolyzes triglycerides to fatty acids and monoglycerides	Fatty acids and monoglycerides
	Bile salts (not an enzyme)	Liver	Small-intestine lumen	Emulsify large fat globules for attack by pancreatic lipase	
Insects (Locust)	Amylase	Salivary gland, hindgut	Foregut, hindgut	Hydrolyzes polysaccharides to disaccharides	
	Maltase, invertase, trehalase	Foregut lining	Foregut, ceca	Hydrolyze disaccharides to monosaccharides	Monosaccharides
	Lipase, protease	Ceca, midgut, hindgut	Ceca, midgut, hindgut	Hydrolyze triglycerides to fatty acids and proteins to peptides and amino acids	Fatty acids, amino acids, peptides

© Cengage Learning, 2013

their high protein intake. To ensure adequate intake, cats have exceptionally high levels of intestinal taurine and arginine transporters.

Most absorption occurs in the duodenum and jejunum; very little occurs in the ileum, not because the ileum does not have absorptive capacity but because most absorption has already been accomplished before the intestinal contents reach the ileum. The small intestine has an abundant reserve absorptive capacity. In fact, about 50% of the small intestine can be removed with little interference to absorption—with at least two exceptions. If the terminal portion of the ileum is removed, vitamin B_{12} and bile salts are not properly absorbed, because the specialized transport mechanisms for these two substances are located only in this region. All other substances can be absorbed throughout the length of the small intestine.

The mucosal lining experiences rapid turnover

Dipping down into the mucosal surface between the villi are shallow invaginations known as the **crypts of Lieberkühn** (Figure 14-22c). Unlike the gastric pits, these intestinal crypts do not secrete digestive enzymes, but they do secrete water and electrolytes, which, along with the mucus secreted

by the cells on the villus surface, constitute the succus entericus. They also secrete the hormone motilin (p. 712).

Furthermore, the crypts function as stem cell "nurseries." The epithelial cells lining the small intestine slough off and are replaced at a rapid rate as a result of high mitotic activity in the crypts. New cells that are continually being produced at the bottom of the crypts migrate up the villi and, in the process, push off the older cells at the tips of the villi into the lumen. In this manner, more than 100 million intestinal cells are shed per minute. In humans, the entire trip from crypt to tip averages about three days, so the epithelial lining of the small intestine is replaced approximately every three days. Because of this rapid growth, John Gurdon in 1962 used nuclei from adult frog intestines in the first successful animal cloning experiment, in which adult nuclei were transplanted into an egg that had its own nucleus destroyed.

The new cells undergo several changes as they migrate up the villus. The concentration of brush border enzymes increases and the capacity for absorption improves, so the cells at the tip of the villus have the greatest digestive and absorptive capability. Just at their peak, these cells are pushed off by the newly migrating cells. Thus, the luminal contents are constantly exposed to cells that are optimally equipped to complete the digestive and absorptive functions.

Furthermore, just as in the stomach, the rapid turnover of cells in the small intestine is essential because of the harsh luminal conditions. Dogs infected with canine *parvovirus* exhibit a high mortality because the virus attacks and destroys the rapidly dividing cells found in the crypts. Bacteria residing in the intestine can then cross into the blood stream, causing **sepsis,** a systemic inflammatory state.

In addition to the stem cells, **Paneth cells** are found in the crypts. They serve a defensive function, safeguarding the stem cells. They produce two chemicals that thwart bacteria: (1) *lysozyme,* the bacterial-lysing enzyme also found in saliva, and (2) *defensins,* small antimicrobial proteins (p. 467).

Let's now turn our attention to the mechanisms through which the specific dietary constituents are normally absorbed.

Energy-dependent Na^+ absorption drives passive H_2O absorption

Sodium may be absorbed both passively and actively. When the electrochemical gradient favors movement of Na^+ from the lumen to the blood, passive diffusion of Na^+ can occur *between* the intestinal epithelial cells through the "leaky" tight junctions into the interstitial fluid within the villus. Movement of Na^+ *through* the cells is energy dependent and involves at least two different carriers, similar to the process of Na^+ reabsorption across the kidney tubules (see p. 562). Sodium enters the epithelial cells across the luminal border either by itself passively through Na^+ channels or in the company of another ion or a nutrient molecule by secondary active transport via three different carriers: Na^+/Cl^- symporter, Na^+/H^+ antiporter, or Na^+–glucose (or amino acid) symporter. Sodium is actively pumped out of the cell by the Na^+/K^+ pump at the basolateral membrane into the interstitial fluid in the lateral spaces between the cells where they are not joined by tight junctions.

As with the renal tubules in the early portion of the nephron, the absorption of Cl^-, H_2O, glucose, and amino acids from the small intestine is linked to this energy-dependent Na^+ absorption. Chloride passively follows down the electrical gradient (through *CFTR* channels; p. 561) created by Na^+ absorption, resulting in a concentrated area of high osmotic pressure in a localized region between the cells, similar to the situation in the kidneys (see Figure 12-3, p. 562). As a result, H_2O is drawn osmotically from the lumen into the lateral space where it is picked up by the capillary network. Meanwhile, more Na^+ is pumped into the lateral space to encourage more H_2O absorption.

Carbohydrate and protein are both absorbed by secondary active transport and enter the blood

Absorption of the digestion end products of both carbohydrates and proteins involves special carrier-mediated transport systems that require energy expenditure and Na^+ cotransport, and both categories of end products are absorbed into the blood.

Carbohydrate Absorption Dietary carbohydrate is presented to the small intestine for absorption mainly in the forms of the disaccharides maltose (the product of polysaccharide digestion), sucrose, and lactose (Figure 14-25a). The disaccharidases located in the brush borders of the small-intestine cells further reduce these disaccharides into the absorbable monosaccharide units of glucose, galactose, and fructose.

Glucose and galactose are both absorbed by secondary active transport, in which symport carriers, such as the *sodium–glucose cotransporter (SGLT;* see Figure 14-25b) on the luminal membrane transport both the monosaccharide and Na^+ from the lumen into the interior of the intestinal cell. Recall that the operation of these carriers, which do not directly use energy themselves, depends on the Na^+ concentration gradient established by the energy-consuming basolateral Na^+/K^+ pump (see p. 84). Glucose (or galactose), having been concentrated in the cell by these symporters, leaves the cell down its concentration gradient by facilitated diffusion (passive carrier-mediated transport; see p. 318) via the *glucose transporter GLUT-2* in the basal border to enter the blood within the villus. In addition to glucose being absorbed through the cells by means of the symporter, recent evidence suggests that a significant amount of glucose crosses the epithelial barrier through the leaky tight junctions between the epithelial cells. *Fructose* is absorbed into the blood solely by facilitated diffusion. It enters the epithelial cells from the lumen via GLUT-5 and exits these cells to enter the blood via GLUT-2 (Figure 14-25b).

Protein Absorption The protein presented to the small intestine for absorption is primarily in the form of amino acids and a few small peptide fragments (Figure 14-26a). Amino acids are absorbed across the intestinal cells by symporters, similar to glucose and galactose absorption (Figure 14-26b). The amino acid symporters are selective for different amino acids. Small peptides gain entry by means of yet another Na^+-dependent carrier in a process known as **tertiary active transport** (*tertiary,* "third"). In this case, the symporter simultaneously transports both H^+ and the peptide from the lumen into the cell, driven by H^+ moving down its concentration gradient and the peptide moving against its concentration gradient (Figure 14-26b). The H^+ gradient is established by an antiporter in the luminal membrane that is driven by Na^+ moving into the cell down its concentration gradient and H^+ moving out of the cell against its concentration gradient. The Na^+ concentration gradient that drives the antiporter in turn is established by the energy-dependent Na^+/K^+ pump at the basolateral membrane. Thus, glucose, galactose, amino acids, and small peptides all get a "free ride" on the energy expended for Na^+ transport. The small peptides are broken down into their constituent amino acids by the aminopeptidases in the brush border membrane or by intracellular peptidases (Figure 14-25a). Like monosaccharides, amino acids enter the capillary network within the villus.

In addition to dietary proteins, in nonruminants some endogenous proteins that have entered the intestinal lumen from the three following sources are digested and absorbed as well:

1. Digestive enzymes, all of which are proteins that have been secreted into the lumen.
2. Proteins within the cells that are pushed off from the villi into the lumen during the process of mucosal turnover.
3. Small amounts of plasma proteins that normally leak from the capillaries into the digestive tract lumen.

About 20 to 40 g of endogenous proteins enter the lumen of humans each day from these three sources. All

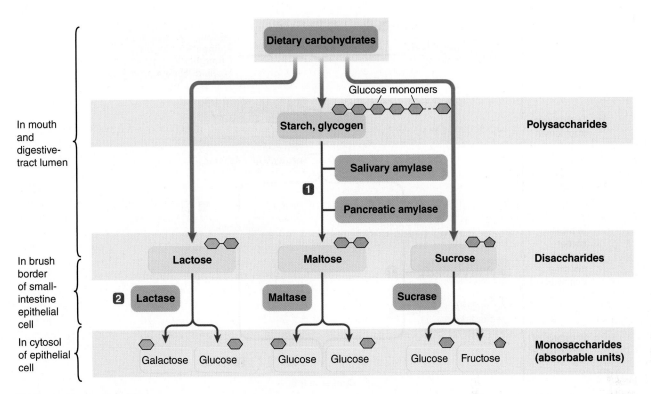

(a) Carbohydrate digestion

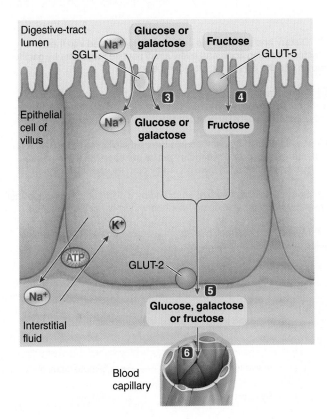

(b) Carbohydrate absorption

1 The dietary polysaccharides starch and glycogen are converted into the disaccharide maltose through the action of salivary and pancreatic amylase.

2 Maltose and the dietary disaccharides lactose and sucrose are converted to their respective monosaccharides by the disaccharidases (maltase, lactase, and sucrase) located in the brush borders of the small-intestine epithelial cells.

3 The monosaccharides glucose and galactose are absorbed into the epithelial cells by Na$^+$- and energy-dependent secondary active transport (via the symporter SGLT) located at the luminal membrane.

4 The monosaccharide fructose enters the cell by passive facilitated diffusion via GLUT-5.

5 Glucose, galactose, and fructose exit the cell at the basal membrane by passive facilitated diffusion via GLUT-2.

6 These monosaccharides enter the blood by simple diffusion.

KEY

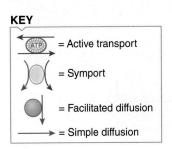

= Active transport

= Symport

= Facilitated diffusion

= Simple diffusion

FIGURE 14-25 **Carbohydrate digestion (a) and absorption (b), for vertebrate small intestine.**

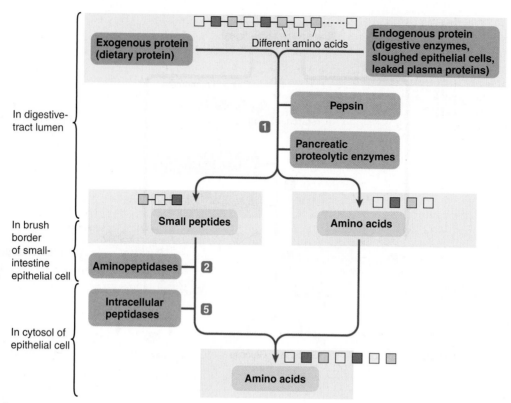

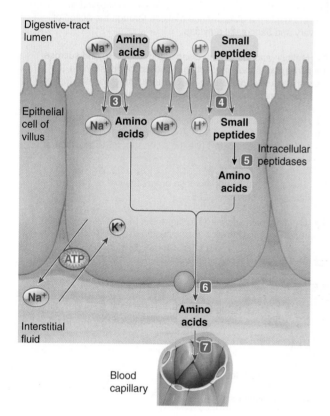

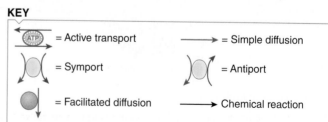

(a) Protein digestion

(b) Protein absorption

① Dietary and endogenous proteins are hydrolyzed into their constituent amino acids and a few small peptide fragments by gastric pepsin and the pancreatic proteolytic enzymes.

② Many small peptides are converted into their respective amino acids by the aminopeptidases located in the brush borders of the small-intestine epithelial cells.

③ Amino acids are absorbed into the epithelial cells by means of Na^+- and energy-dependent secondary active transport via a symporter. Various amino acids are transported by carriers specific for them.

④ Some small peptides are absorbed by a different type of symporter driven by H^+, Na^+-, and energy-dependent tertiary active transport.

⑤ Most absorbed small peptides are broken down into their amino acids by intracellular peptidases.

⑥ Amino acids exit the cell at the basal membrane via various passive carriers.

⑦ Amino acids enter the blood by simple diffusion. (A small percentage of di- and tripeptides also enter the blood intact.)

KEY

(ATP) = Active transport	⟶ = Simple diffusion
= Symport	= Antiport
= Facilitated diffusion	⟶ Chemical reaction

FIGURE 14-26 **Protein digestion (a) and absorption (b), for vertebrate small intestine.**

endogenous proteins must be digested and absorbed along with the dietary proteins to prevent depletion of the body's protein stores. The amino acids absorbed from both food and the endogenous protein are used primarily to synthesize new protein in the body.

An exception to the need to digest proteins is found in neonatal mammals. Their intestines can accomplish limited uptake of whole proteins via the active process of endocytosis (see p. 44). This is primarily used to absorb whole antibodies of the IgA class (see p. 475) from maternal milk, and helps protect the young animal from pathogens while its immune system is still developing. There is also some evidence that large protein molecules are absorbed intact into an animal's bloodstream through a process termed *persorption*. This form of passage occurs between the cells and is possible wherever single-layered epithelium covers the mucosa of the gastrointestinal tract. Researchers have shown that ingested proteins, which were radioactively labeled, reach peak concentrations in the blood of a rabbit within 2 hours. Absorption rates ranged from 1 to 10% of the ingested protein.

Fat Absorption Fat absorption is quite different from carbohydrate and protein absorption because the insolubility of fat in water presents a special problem. Fat must be transferred from the watery chyme through the watery body fluids even though it is not water soluble. Therefore, fat must undergo a series of transformations to circumvent this problem during its digestion and absorption (Figure 14-27).

When the stomach contents are emptied into the duodenum, the ingested fat is aggregated into large, oily, triglyceride droplets that float in the chyme. Recall that through the bile salts' detergent action in the small intestine, the large droplets are dispersed into a lipid emulsification of small droplets, thereby exposing a much greater surface area of fat for digestion by pancreatic lipase. The products of lipase digestion (monoglycerides and free fatty acids) are also not very water soluble, so very little of these end products of fat digestion can diffuse through the aqueous chyme to reach the absorptive lining. However, biliary components facilitate absorption of these fatty end products by forming micelles, as you learned earlier (p. 686). Once these micelles reach the luminal membranes of the epithelial cells, the monoglycerides and free fatty acids passively diffuse from the micelles through the lipid component of the epithelial cell membranes to enter the interior of these cells. As these fat products leave the micelles and are absorbed across the epithelial cell membranes, the micelles can pick up more monoglycerides and free fatty acids, which have been produced from digestion of other triglyceride molecules in the fat emulsion.

Bile salts continuously repeat their fat-solubilizing function down the length of the small intestine until all the fat is absorbed. Then, the bile salts themselves are reabsorbed in the terminal ileum by special active transport. This is an efficient process because relatively small amounts of bile salts can facilitate digestion and absorption of large amounts of fat, with each bile salt performing its ferrying function repeatedly before it is reabsorbed.

Once within the interior of the epithelial cells, the monoglycerides and free fatty acids are resynthesized into triglycerides. These triglycerides conglomerate into droplets and are coated with a layer of lipoprotein (synthesized by the endoplasmic reticulum of the epithelial cell), which renders the fat droplets water soluble. The largest of the lipoproteins, known as **chylomicrons** (p. 389), are extruded by exocytosis from the epithelial cells into the interstitial fluid within the villus. In mammals the triglyceride-rich chylomicrons are secreted into the central lacteals and enter lymphatic vessels, whereas in birds they are absorbed into capillaries of the villi. Fatty acids with short- or medium-length carbon chains also enter the blood.

The actual transfer of monoglycerides and free fatty acids from the chyme across the apical membranes of the intestinal epithelial cells is a passive process because the lipid-soluble fatty end products merely dissolve in and pass through the lipid portions of the membrane. Fat absorption is therefore considered to be a passive process. However, the overall sequence of events necessary for fat absorption does require energy. For example, bile salts are actively secreted by the liver, the resynthesis of triglycerides and formation of chylomicrons within the epithelial cells are active processes, and the exocytosis of chylomicrons requires energy.

Vitamin absorption is largely passive

Water-soluble vitamins (B complex vitamins and vitamin C) are primarily absorbed passively with water, whereas fat-soluble vitamins (vitamins A, D, E, and K) are carried in the micelles and absorbed passively with the end products of fat digestion. Carriers, if necessary, can also accomplish absorption of some of the vitamins. Vitamin B_{12} is unique in that it must be combined with *gastric intrinsic factor* from stomach parietal cells (p. 676 and Table 14-2) for absorption by special transport in the terminal ileum.

Most absorbed nutrients immediately pass through the liver for processing

Recall that the venules that leave the small-intestine villi, along with those from the remainder of the digestive tract, empty into the portal vein (Figure 14-17), which carries the blood to the liver. Consequently, anything absorbed into the digestive capillaries first must pass through the hepatic biochemical factory before entering the general circulation. Thus, the products of carbohydrate and protein digestion are channeled into the liver, where many of these energy-rich products are subjected to immediate metabolic processing. Furthermore, the liver detoxifies harmful substances that may have been absorbed from the gut. Only after this processing does the venous blood that began in the intestine rejoin the general circulation.

Recall that fat in the form of chylomicrons is picked up by the central lacteal and enters the lymphatic system instead, thereby bypassing the hepatic portal system. Contractions of the villi, accomplished by the muscularis mucosa, periodically compress the central lacteal and "milk" the lymph out of this vessel. The lymph vessels eventually converge to form the *thoracic duct*, a large lymph vessel that empties into the venous system within the chest. By this means, fat ultimately gains access to the circulatory system. Therefore, the liver does have a chance to act on the digested fat, but not until the fat has been diluted by the blood in the general circulatory system and has been somewhat reduced by uptake into adipose cells (see p. 389). This dilution of fat presumably protects the liver from being inundated with more fat than it can handle at one time.

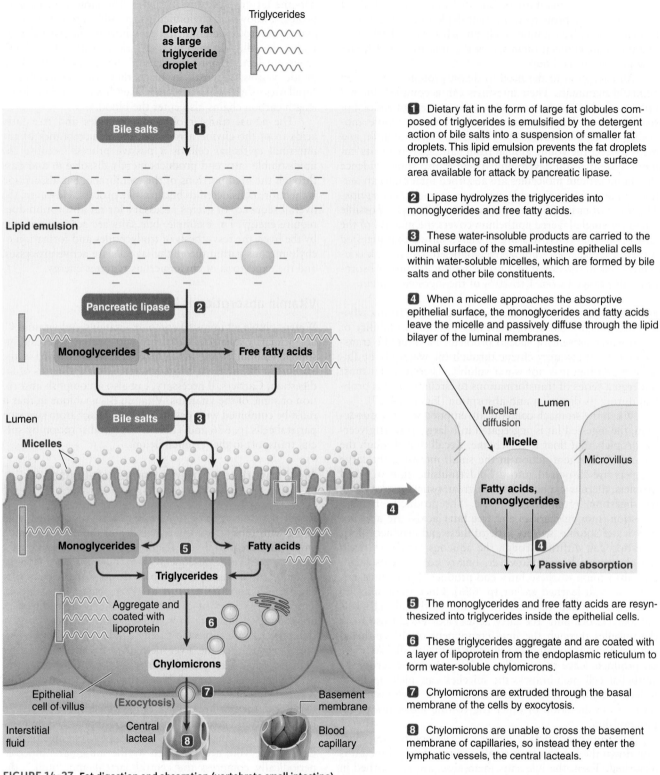

FIGURE 14-27 Fat digestion and absorption (vertebrate small intestine).
© Cengage Learning, 2013

1 Dietary fat in the form of large fat globules composed of triglycerides is emulsified by the detergent action of bile salts into a suspension of smaller fat droplets. This lipid emulsion prevents the fat droplets from coalescing and thereby increases the surface area available for attack by pancreatic lipase.

2 Lipase hydrolyzes the triglycerides into monoglycerides and free fatty acids.

3 These water-insoluble products are carried to the luminal surface of the small-intestine epithelial cells within water-soluble micelles, which are formed by bile salts and other bile constituents.

4 When a micelle approaches the absorptive epithelial surface, the monoglycerides and fatty acids leave the micelle and passively diffuse through the lipid bilayer of the luminal membranes.

5 The monoglycerides and free fatty acids are resynthesized into triglycerides inside the epithelial cells.

6 These triglycerides aggregate and are coated with a layer of lipoprotein from the endoplasmic reticulum to form water-soluble chylomicrons.

7 Chylomicrons are extruded through the basal membrane of the cells by exocytosis.

8 Chylomicrons are unable to cross the basement membrane of capillaries, so instead they enter the lymphatic vessels, the central lacteals.

The liver plays an important role in lipid transport by synthesizing three types of plasma lipoproteins. The first studies to characterize these plasma lipoproteins were conducted in the 1920s on horse serum. Liver-produced lipoproteins were named on the basis of their density of protein as compared to lipid. High-density lipoprotein (HDL) was isolated first. With the advent of analytical high-speed ultracentrifuges, subse-quent studies identified low-density lipoprotein (LDL), which contains less protein and more cholesterol, and very-low-density lipoprotein (VLDL), which contains the least protein and most lipid. As you saw in Chapter 9 (p. 389), the LDLs and HDLs transport primarily cholesterol (and phospholipid) for cell membrane production. VLDLs are rich in triglycerides, which are used for energy storage in adipose tissue.

Extensive absorption by the small intestine keeps pace with secretion

The small intestine of humans normally absorbs about 9,000 mL of water per day, which contains the absorbable units of nutrients, vitamins, and electrolytes. How can that be, when humans normally ingest only about 1,250 mL of fluid and 1,250 g of solid food (80% of which is H_2O) per day? Table 14-6 illustrates the tremendous daily absorptive accomplishments performed by the small intestine. Each day, about 9,500 mL of H_2O and solutes enter the small intestine. Note that of this 9,500 mL, only 2,500 mL are ingested from the external environment. The remaining 7,000 mL of fluid consist of digestive juices that are essentially derived from plasma. Considering that the entire plasma volume is only about 2,750 mL, it is obvious that absorption must closely parallel secretion to prevent the plasma volume from falling sharply. Of the 9,500 mL of fluid entering the small-intestine lumen per day, about 95%, or 9,000 mL of fluid, is normally absorbed by the small intestine into the plasma, with only 500 mL of the small-intestine contents passing on into the colon. Thus, most digestive juices are not lost from the body. After the constituents of the juices are secreted into the digestive tract lumen and perform their function, they are returned to the plasma. The only secretory product that escapes from the body is bilirubin (p. 688).

Biochemical balance among the stomach, pancreas, and small intestine is normally maintained

Because the secreted juices are normally absorbed back into the plasma, the acid–base balance of the body is not altered by digestive processes. Within the small-intestine lumen, the HCl secreted by the parietal cells of the stomach is neutralized by the $NaHCO_3$ secreted by the pancreatic duct cells:

$$HCl + NaHCO_3 \rightarrow NaCl + H_2CO_3$$

The resultant H_2CO_3 decomposes into $CO_2 + H_2O$:

$$H_2CO_3 \rightarrow CO_2 + H_2O$$

The end products of these reactions—Na^+, Cl^-, CO_2, and H_2O—are all absorbed by the intestinal epithelium into the blood. Thus, through these interactions, the body normally does not experience a net gain or loss of acid or base during digestion.

Diarrhea results in loss of fluid and electrolytes

When vomiting or diarrhea occurs, these normal neutralization processes cannot take place. We have already described vomiting and the subsequent loss of acidic gastric contents leading to metabolic alkalosis in the section on gastric motility. The other common digestive disturbance that can lead to a loss of fluid and an acid–base imbalance is **diarrhea,** This condition is characterized by passage of a highly fluid fecal matter, often with increased frequency of defecation. Just as with vomiting, the effects of diarrhea can be either beneficial or harmful. Diarrhea is beneficial when rapid emptying of the intestine hastens the elimination of harmful material from the animal's body, which happens frequently in domestic dogs, which are prone to eat such materials. However, not only are

TABLE 14-6 Volumes Absorbed by the Small and Large Intestine per Day (Human)

Volume entering the small intestine per day			
Sources	Ingested	Food eaten	1,250 g[a]
		Fluid drunk	1,250 mL
	Secreted from the plasma	Saliva	1,500 mL
		Gastric juice	2,000 mL
		Pancreatic juice	1,500 mL
		Bile	500 mL
		Intestinal juice	1,500 mL
			9,500 mL
Volume absorbed by the small intestine per day			9,000 mL
Volume entering the colon from the small intestine per day			500 mL
Volume absorbed by the colon per day			350 mL
Volume of feces eliminated from the colon per day			150 g[a]

*One milliliter of H_2O weighs 1 g. Therefore, because a high percentage of food and feces is H_2O, we can roughly equate grams of food or feces with milliliters of fluid.

some of the ingested materials lost, but some of the secreted materials that normally would have been reabsorbed are lost as well. Excessive loss of intestinal contents causes dehydration, loss of nutrient material, and metabolic acidosis resulting from the loss of HCO_3^- (see p. 649). The abnormal fluidity of the feces in diarrhea usually occurs because the intestine cannot absorb fluid as extensively as normal. This extra unabsorbed fluid passes out in the feces. Diarrhea is the leading cause of death of human children worldwide.

The most common cause of diarrhea is excessive intestinal motility, which arises either from local irritation of the gut wall caused by bacterial or viral infection of the intestine or from emotional stress. Rapid transit of the intestinal contents does not allow sufficient time for adequate absorption of fluid to occur.

check your understanding 14.7

How is the biochemical balance in pH between the stomach and small intestine maintained?

Describe how the three major categories of nutrients are digested in the small intestine.

14.8 Hindgut/Large Intestine

We now turn to the final segments of the digestive tract.

The hindgut of insects actively transports nutrients and may house symbionts

The hindgut of insects is divided into two predominant regions: the intestine, or **ileum** (anterior region), and the **rectum** (posterior region). In contrast to vertebrates, the hindgut of

Many insects are similar to ruminant mammals in that they rely on symbiotic bacteria in their digestive tracts (p. 707). One well-studied model animal is the pea aphid *Acyrthosiphon pisum,* which feeds on plant sap that is poor in several nutrients. In special cells around the aphid's digestive tract, symbiotic bacteria (*Buchnera*) manufacture key nutrients, especially several amino acids that the aphid cannot make (known as *essential amino acids*). Until recently, it has been difficult to study these endosymbionts, because they do not grow in the laboratory. However, genetic analysis is shedding some light on the relationship and its evolution. By comparing the genome of free-living *Buchnera* with that of the symbiont, researchers discovered that the latter have lost most of the genes necessary for the microbes to live on their own. For example, genes for defensive cell wall proteins are absent. Furthermore, the genes for making essential amino acids for the host aphid have been duplicated manyfold and are located in *plasmids* (small circlets of

Pea aphid, *Acyrthosiphon pisum*.

DNA separate from the main chromosome). Plasmids serve as dedicated factories producing RNA for the enzymes to make large quantities of the amino acids, which are exported to the host.

These studies reveal much about the genetic processes of evolution. First, because they have no selective benefit, genes that are no longer used mutate into unusable forms and may shrink as pieces are accidentally lost during DNA replication. Second, intimate symbioses such as these may be irreversible, with each partner no longer capable of living without the other. Finally, the process suggests how mitochondria and chloroplasts, which also have greatly reduced genomes, may have evolved from symbiotic microbes.

insects constitutes the reabsorptive portion of both the digestive and excretory systems as ion and nutrient transport occurs in both regions. For example, the rectum in the hindgut of locusts has been demonstrated to transport some amino acids via an Na^+-independent process. The rectum works in concert with the Malpighian tubules (see p. 566), which are excretory in function, as part of the excretory system.

As you will see for vertebrates shortly, hindguts often house beneficial symbiotic microbes. Cellulose-digesting anaerobic microorganisms are housed in special organs connected to the hindgut of wood-eating termites and the wood roach *Cryptocercus.* (See also *Molecular Biology and Genomics: Big Bugs Have Little Bugs…*)

The vertebrate hindgut or large intestine consists of the colon, cecum, and rectum or cloaca

The hindgut of vertebrates consists of the *colon, cecum, and rectum/cloaca* (Figure 14-28); however, in amphibians the hindgut is difficult to distinguish from the midgut, although the diameter is somewhat larger. In most species a valve or sphincter separates the two compartments. The cecum was described earlier (p. 693). The small, fingerlike projection at the bottom of the cecum in humans and some apes is the **vermiform appendix,** a lymphoid tissue that stores lymphocytes but that has no direct digestive function. It is thought to be a remnant of a larger cecum in ancestral primates and has long been thought to be vestigial. Recent studies indicate that it may serve as a refuge for symbiotic bacteria, which may repopulate the main tract if symbionts there have been killed off by diseases or toxins. However, since the appendix is prone to obstruction from trauma, infections, and foreign bodies—which can lead to lethal swelling and rupture—it is not an optimal adaptation (see p. 2).

The **colon,** which makes up most of the mammalian large intestine, varies considerably between animals in structure depending on diet (as you will see later). Generally the colon of mammals shows a remarkable range of structural variation and tends to be proportionately longer than that of other vertebrates. In humans, there is a distinct *ascending colon, transverse colon,* and *descending colon.* The ascending colon of ruminants, pigs, and horses are dramatically modified in both size and structure. For example, in ruminants and pigs it is much elongated and organized into coils. In the horse it is markedly enlarged and occupies a considerable portion of the abdominal cavity. The transverse colon is comparatively short, whereas the descending colon is continuous with the **rectum** (*rectum* means "straight"). In amphibians, reptiles (including birds), and some mammals the hindgut terminates, along with the renal and reproductive systems, in a **cloaca,** whereas in fish the ducts from the urinary and genital tracts exit from the body separate from that of the digestive tract. In reptiles including birds, the cloaca is divided by ridges into three regions (Figure 12-10, p. 575). The anterior **coprodaeum** receives the excreta from the intestines; the middle **urodaeum** receives the fluid from the kidneys through the ureters, as well as the material from the oviduct; and the **proctodaeum** stores the excreta. The proctodaeum opens externally into the muscular anus. Lying on the dorsal wall of the urodaeum is the **bursa of Fabricius,** a lymphoid organ (p. 473) important in newly hatched birds.

The large intestine serves as a temporary storage site for excreta, functions in electrolyte and fluid balance, and harbors microbes for fermentation and production of volatile fatty acids

The colon of a human normally receives about 500 mL of chyme from the small intestine each day. Because most digestion and absorption have been accomplished in the small intestine, the contents delivered to the colon consist of indigestible food residues (such as cellulose), unabsorbed biliary components, and the remaining fluid. The colon extracts more H_2O and salt from the contents, so it is important in fluid-electrolyte balance. Sodium is actively absorbed, Cl^- follows passively down the electrical gradient, and H_2O follows osmotically. What remains to be eliminated is known as *excreta* or **feces,** The primary function of the large intestine is to store this fecal material before defecation. Cellulose and other indigestible substances in the diet provide bulk and help maintain regular bowel movements by contributing to the volume of the colonic contents.

Symbiotic microbes live in the cecum and colon in most if not all species, where they synthesize some vitamins such as vitamin K (p. 710), although in nonruminants only vitamin K is produced in significant quantities. Microbes also make **volatile fatty acids (VFAs;** fatty acids with 6 or fewer carbons). For example, acetate, butyrate, and propionate are produced by symbiotic bacteria in broiler chickens, in which these VFAs may provide usable nutrients to the chicken and inhibit growth of pathogenic bacteria.

Diet composition governs variation in structure of the large intestine

Carnivores (such as cats and dogs) tend to have the simplest digestive tracts, with comparatively short, unstructured colons. For example, carnivorous fish have gut lengths normally slightly greater than the length of their own body. In these animals there is little distinction between the small and large intestine and the cecum is either absent or marginal in its development. Thus, in carnivores, the major function of the colon is the absorption of electrolytes, water, and other materials that escaped absorption in the small intestine.

In omnivores and particularly in herbivores that ingest considerable amounts of complex polysaccharides, notably rabbits and horses, the structure of the large intestine tends to be more complex. Generally the ceca and/or the colon are **sacculated** (having expandable side sacs) and voluminous, providing a site for microbial digestion. Intestinal length is also longer in omnivorous fishes compared to carnivorous fishes. In herbivorous fishes, gut length can exceed 20 times body length. Exceptions include the kangaroo and sheep, whose cecum and colon are neither sacculated nor particularly enlarged. Sacculations are formed when the outer longitudinal smooth-muscle layer of the colon does not completely surround the large intestine. In these animals, it consists of separate, conspicuous, longitudinal bands of muscle, the **taeniae coli,** which run the length of the cecum and large intestine. Most animals with taeniae have three, but a horse has four. Approximately half of the proximal coiled region of the pig colon has two taeniae, whereas the remainder is without. These taeniae coli are actually shorter than the underlying circular smooth-muscle and mucosal layers would be if the latter were stretched out flat. Because of this, the underlying layers are gathered into pouches or sacs called **haustra,** The haustra are not merely passive gath-

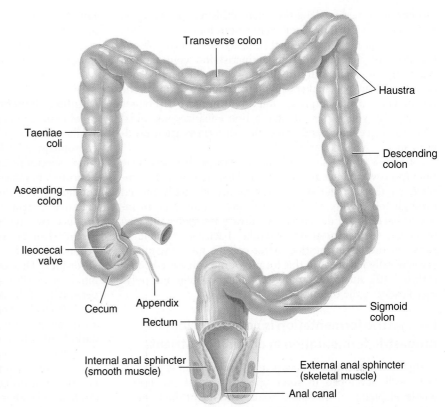

FIGURE 14-28 **Anatomy of the large intestine (mammal, shown for a human).**
© Cengage Learning, 2013

ers, however; they actively change location as a result of contraction of the circular smooth-muscle layer.

Haustral contractions prolong the retention of digesta

Most of the time, movements of the large intestine are slow and nonpropulsive, as is appropriate for its absorptive and storage functions. The colon's primary method of motility are **haustral contractions** initiated by the autonomous rhythmicity of colonic smooth muscle cells. These contractions, which throw the large intestine into haustra, are similar to small-intestine segmentations but occur much less frequently. Whereas human small-intestine segmentation occurs at rates of between 9 and 12 contractions per minute, there may be 30 minutes between haustral contractions. The location of the haustral sacs gradually changes as a relaxed segment that has formed a sac slowly contracts while a previously contracted area simultaneously relaxes to form a new sac. These movements are nonpropulsive; they slowly shuffle the contents in a back-and-forth mixing movement that exposes the colonic contents to the absorptive mucosa. By prolonging the retention of the colonic contents, more time is made available for microbial digestion. This is particularly important in the pig and horse because they lack a rumen in which to digest dietary cellulose. Locally mediated reflexes involving the intrinsic plexuses largely control haustral contractions.

Peristaltic contractions also propel the flow of intestinal and cecal contents toward the rectum. A pacemaker in the midcolon region generates slow waves, which travel both directions along the colon, which contributes to the net movement of ingesta along the colon. Birds rely primarily on peristaltic rather than haustral contractions for colon motility.

In addition to the haustral contractions, which impede the flow of ingesta, are the **antiperistaltic contractions,** which function primarily to fill the cecum. Generally they originate from pacemakers in the proximal region of the colon. These contractions are important in most herbivores and pigs, in which their mixing action enhances microbial digestion of cellulose and absorption of fermentation products, such as volatile fatty acids.

Because of this slow colonic movement, bacteria have time to grow and accumulate in the large intestine. In contrast, contents in the small intestine are normally moved through too rapidly for bacteria to proliferate. Not all ingested bacteria are destroyed by salivary lysozyme and gastric HCl, so the surviving anaerobic bacteria continue to thrive in the large intestine. Most colonic microorganisms are generally harmless in this location (and some are beneficial; p. 710), surviving on any remaining protein and polysaccharides in the digesta.

Postgastric fermentation is not as efficient as pregastric fermentation in obtaining nutrients

Postgastric fermentation occurs in all animals but is particularly well developed in some species such as the horse, whale, elephant, possum, and koala. The site of the fermentation chamber is the combined colon and upper cecum. By positioning the fermentation chamber after the stomach, the host animal has the first chance at the available carbohydrates and protein in the food but loses the opportunity to exploit the high-quality microbially synthesized protein. Further, the digestive capabilities of the stomach and small intestine cannot contribute to the digestion of microbial products. However, VFAs and some vitamins synthesized by the microbes in the cecum and colon are absorbed through the mucosa of the large intestine. In many species the cecal wall is arranged in spiral ridges, which increases the surface area available for absorption. Because the cecum and colon do not absorb nutrients as efficiently as the small intestine, the quality of these animals' diet is more important in sustaining normal metabolic activity. Recycling of nitrogen (see p. 709) is also less important in mammalian postgastric fermenters. Thus, when dietary sources become scarce, ruminants have a survival edge over nonruminant herbivores.

The environment within the colon/cecum of postgastric fermenters must be able to support a substantial microbe population in a buffered environment. In the horse, large quantities of bicarbonate and phosphate buffers are secreted by the ileum and transferred to the cecum, an action comparable to the role of the salivary gland in ruminants. The animal must also be able to absorb the products of fermentation as well as to reclaim as much fluid as possible from the digesta before it is excreted.

Herbivorous avian species such as the galliformes (chicken, turkeys, and so forth), ostriches, ducks, and geese have enlarged ceca. It has been shown in the galliformes that uric acid and urea from their urine passes down the ureters into the cloaca, then reflux antiperistaltically into the colon and ceca. Microbes decompose the urea and uric acid into ammonia, which may then be used by bacteria to synthesize amino acids. The microbes themselves primarily use these amino acids, but some are absorbed into the blood of the host bird to meet its amino acid requirements. Ammonia is also absorbed directly from the ceca and is used to synthesize amino acids in the liver. Ammonia "recycling" is an important adaptation that permits the galliformes to exploit habitats otherwise unavailable to most animals.

Nutritional Benefit from the Reingestion of Feces A solution to obtaining nutritional benefit from protein and vitamin synthesis by microbes in the large intestine is the practice of **coprophagy** (Greek *copros* or "feces" and *phagein* "to eat"). Many species practice coprophagy for various reasons, some as normal practice (e.g., stink bugs) and others under specific conditions. For example, some species that lack a pregastric digestive system, such as rabbits, hamsters, and capybara, produce two types of fecal material. The soft fecal pellets produced in the nighttime are reingested, thus making the products of cecal fermentation available for digestion and absorption. Delivery of the soft feces is accomplished by inhibition of motility in the proximal colon associated with hyperactivity of the distal region close to the anus. In the rabbit, the soft fecal pellets contain over 50% bacteria and provide adequate intake of vitamin B$_{12}$. The second type of feces is a dark, hard pellet, which is not reingested.

Many hindgut digestors practice coprophagy when very young. Colts, young elephants, pandas, koalas, and hippos practice it in order to establish a microbe population in the hindgut, then abandon this practice as adults. Because chickens normally practice coprophagy, outbreaks of *coccidiosis* are relatively common in birds raised on litter. Coccidiosis is caused by single-celled protozoa (*Eimeria*) that colonize the large intestine. Affected animals may develop symptoms that include diarrhea, bloody droppings, and an abrupt decrease in food intake.

Mass movements propel colonic contents long distances

Generally after meals, a marked increase in motility takes place during which large segments of the colon contract simultaneously, driving the feces one third to three fourths of the length of the colon in a few seconds. These massive contractions, appropriately called **mass movements,** drive the colonic contents into the distal portion of the large intestine, where material is stored until defecation occurs. In the dog, mass movements originate near the ileocecal valve and can evacuate the ingesta the entire length of the colon, whereas in the cat the movements are generated lower in the colon.

When food enters the stomach, mass movements occur in the colon primarily by means of the **gastrocolic reflex,** which is mediated from the stomach to the colon by gastrin and by the extrinsic autonomic nerves. Thus, when a new meal enters the digestive tract, reflexes are initiated to move the existing contents farther along down the tract to make way for the incoming food. The gastroileal reflex moves the remaining small-intestine contents into the large intestine, and the gastrocolic reflex pushes the colonic contents into the rectum, triggering the defecation reflex.

Feces are eliminated by the defecation reflex

When mass movements of the colon move fecal material into the rectum, the resultant distension of the rectum stimulates stretch receptors in the rectal wall, thus initiating the

defecation reflex. This reflex causes the **internal anal sphincter** (which consists of smooth muscle) to relax and the rectum and sigmoid colon to contract more vigorously. If the **external anal sphincter** (which consists of skeletal muscle) is also relaxed, defecation occurs. Being skeletal muscle, the external anal sphincter is under voluntary control. The initial distension of the rectal wall is accompanied by the conscious urge to defecate. If circumstances are unfavorable for defecation, voluntary tightening of the external anal sphincter can prevent defecation despite the defecation reflex. Ground-nesting animals generally defecate at considerable distances from their young to decrease the likelihood of detection by a predator. If defecation is delayed, the distended rectal wall gradually relaxes, and the urge to defecate subsides until the next mass movement propels more feces into the rectum, once again distending the rectum and triggering the defecation reflex. During periods of inactivity, both anal sphincters remain contracted to ensure fecal continence.

When defecation does occur, voluntary straining movements that involve simultaneous contraction of the abdominal muscles and a forcible expiration against a closed glottis usually assist it. This maneuver brings about a large increase in intra-abdominal pressure, which helps eliminate the feces. Defecation is also a normal response to fear, presumably in response to activation of neural centers in the brain.

Large-intestine secretion is protective in nature

The large intestine does not secrete any digestive enzymes. Colonic buffers consist of an alkaline (HCO_3^- and PO_4^{3-}) mucous solution, whose function is to protect the large-intestine mucosa from mechanical and chemical injury. In the horse and pig, pancreatic HCO_3^- is of sufficient volume so as to contribute to the buffering capability in the colon. In ruminants, considerable amounts of PO_4^{3-} are secreted in the saliva, whereas in nonruminants, the source of PO_4^{3-} is the diet. The uptake of PO_4^{3-} is low in the intestine, with the result that its concentration increases as the colon absorbs fluid. The mucus provides lubrication to facilitate passage of the feces, whereas the buffers neutralize acids produced by local bacterial fermentation.

The large intestine converts the luminal contents into feces

Fecal material normally consists of undigested cellulose and other unabsorbed food residues, bilirubin (or biliverdin), small amounts of salt and water, and bacteria. Some of these items were never actually a part of the body. As we noted earlier, some absorption takes place within the vertebrate colon, but not to the same extent as in the small intestine. Because the luminal surface of the colon is fairly smooth, it has considerably less absorptive surface area than the small intestine. In many vertebrates, specialized transport mechanisms are not present in the colonic mucosa for absorption of glucose or amino acids, as there are in the small intestine. A notable exception is the colon of birds, where both glucose and amino acids are absorbed by secondary active transport. Thus, the glucose and amino acids not reabsorbed in the kidney can either be reabsorbed into the blood or used by the microorganisms in the ceca. When excessive small-intestine motility delivers the contents to the colon before absorption of nutrients has been completed,

the colon cannot absorb these materials, and they are lost in diarrhea.

14.9 Ruminant Digestion

Ruminants are a group of animals that have achieved the highest degree of specialization in fermenting plant material. Animals having a rumen can exploit the vast supplies of structural carbohydrates found in the plant cell wall and convert them into a source of nutrients. Ruminants, so named because they **ruminate** (chew the cud), also voluntarily regurgitate partially digested food back into the mouth to complete the mechanical grinding of the ingested plant material. This has often been thought of as an evolutionary development that has allowed ruminants (often prey animals) to consume feed in haste and later mechanically process their food during leisure times of safety. Ruminant species have evolved a stomach of sufficient size and motility to house a population of microbes able to break down cellulose and other complex polysaccharides into end products of fermentation, which meet the nutritional requirements of the host animal. Consequently, ruminants are among the most widely distributed groups of mammals on Earth, having adapted to Arctic, desert, and tropical environments.

The rumen is divided into separate compartments

The stomach of true ruminants (*Ruminantia*), which includes cattle, sheep, goats, deer, giraffe, and antelope, is divided into four compartments, which occupy approximately three quarters of the abdominal cavity. The **forestomach** (the pregastric region) consists of three chambers: the *rumen, reticulum,* and *omasum,* each of which is involved in the storage and passage of ingested food (Figure 14-29). *Pseudoruminants* (Tylpoda), including llamas and camels, have a three-compartment stomach, the omasum being absent.

The **rumen** and its continuation, the **reticulum,** also have an important role in the absorption of nutrients and simple molecules. It is in these two compartments that the bulk of anaerobic fermentation (see p. 47) of plant material occurs and cellulose is catabolized into digestible units. Energy is obtained through fermentation that would otherwise be lost to the host animal. However, the fermentative process requires specific temperatures, pH, motility, and secretions to maintain the microbial populations. In all ruminant animals, the rumen is the most spacious structure, followed by the abomasum. In a majority of ruminant species, the reticulum is greater in size than the omasum, which is the smallest of the four compartments. The variability in size of the latter two compartments is ultimately determined by the

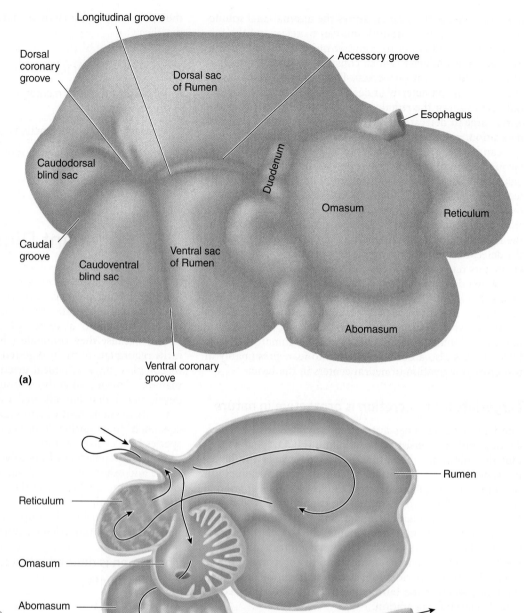

Longitudinal groove

Dorsal
coronary
groove

Dorsal sac
of Rumen

Accessory groove

Esophagus

Caudodorsal
blind sac

Duodenum

Omasum

Reticulum

Caudal
groove

Ventral sac
of Rumen

Caudoventral
blind sac

Abomasum

Ventral coronary
groove

(a)

Reticulum

Omasum

Abomasum

Rumen

To small
intestine

(b)

FIGURE 14-29 Comparison of rumen structure and function in a selective versus a nonselective feeder.

(a) Stomach of a nonselective feeder, the ox. (b) Rumen function in an antelope stomach. Note that the relative size of the rumen is considerably smaller than in the ox. Food moves from the rumen and reticulum into the omasum for sorting. The abomasum is the acid-secreting region.

Sources: (a) R. D. Frandson & T. L. Spurgeon (1992). *Anatomy and Physiology of Farm Animals*, 5th ed. Philadelphia: Lippincott Williams & Wilkins, p. 317. (b) C. Starr. (2004). *Biology: Unity and Diversity of Life*, 10th ed. Belmont, CA: Brooks/Cole, p. 727, Figure 41.4b.

feeding strategies of the particular animal group as well as by seasonal changes in food availability.

Surrounding the ruminant stomach is an outer muscular layer, which consists of a thin sheet of longitudinally directed smooth muscle, whereas the inner layer is thick and circular in structure. The muscular layers mix the ingesta and provide for the progressive movement of this material through the stomach toward the small intestine. Rumen mixing also promotes the turnover of indigestible particles that would quickly clog the rumen if allowed to accumulate.

The ruminal cavity is divided into internal compartments by the presence of **pillars.** Longitudinal pillars partially divide the rumen into dorsal and ventral sacs. When contracted, these muscular folds enable the mixing and movement of large volumes of fluid. These pillars also help

stabilize the fluid contents of the ruminoreticulum and limit the movement of the digesta toward the reticulum. In deer, for example, rapid movements of the body would potentially result in a considerable displacement of ruminal fluid were it not for the limitations imposed by these structures. Because this volume may equal as much as 14% of an animal's body weight, it is vital that it be stabilized. Lining the rumen are fingerlike projections, the **papillae,** that increase the surface area available for absorption of nutrients. Feeding *high-concentrate diets* (replacing the roughage component of the diet with higher-energy grains) to cattle results in the gradual degradation of these structures. Starvation also decreases the size of the papillae.

The ruminal compartment is separated from the reticulum by the **ruminoreticular fold,** However, the separation does not inhibit the free exchange of digesta between the two compartments. Within these two sacs occurs most fermentation and absorption of nutrients. Hardware and other foreign objects accumulate in the reticulum because of its position relative to the esophagus. **Hardware disease** occurs in ruminants when a foreign object, usually a piece of metal that has been swallowed, penetrates the reticulum. The condition is particularly common in cattle because they do not separate foreign objects—nails, wire, and so forth—from feed, and swallow everything before it is completely chewed. Penetration of the reticulum wall permits bacteria and partially digested food to spill out into the abdominal cavity, resulting in *peritonitis*.

The esophagus terminates at the junction between the reticulum and the rumen, the *cardia,* and opens directly over the cavity of the reticulum. It is encircled by strands of smooth muscle that are organized to form a *reticular groove.* The reticular grove extends from the opening of the esophagus to the *reticulo-omasal orifice.* The luminal surface of the reticulum is lined with ridges, which are thought to be involved in the sorting of particles that pass near the reticulo-omasal orifice. Ingesta ultimately entering the omasum are identical in consistency to that found in the ruminoreticulum.

The omasum provides a channel for the passage of ingesta from the reticulum into the glandular stomach, the abomasum. The luminal surface is characterized by the presence of leaflike structures that absorb water and nutrients as well as limit the passage of large particles. At the junction with the abomasum is the *omasoabomasal orifice,* a well-developed structure that lacks a sphincter to limit backflow. It also regulates the movement of materials from the omasum into the abomasum during digestion and inhibits the movement of ingesta from the omasum when the abomasum is full.

The abomasum is similar in function to the stomachs of nonruminants

The acid-secreting region of the stomach, the **abomasum,** functions in the digestion of protein and lysis of rumen microbes. It is structurally similar to that of stomachs of nonruminants except for the presence of relatively large spiral folds. It has a much larger fundic area and a larger pyloric region at the junction of the abomasum and small intestine. In ruminants that are fed coarse plant material, gastric juice secretion is continuous, whereas it is cyclic in animals provided concentrate diets at intervals during the day. **Displaced abomasum syndrome** results from feeding high-concentrate diets to dairy cattle as a consequence of the abomasum becoming enlarged with fluid and gas, which results in a shift of the abomasum from the right half of the abdominal cavity to either the left or the right.

Motility of the ruminant stomach is predominantly regulated by central nervous system reflex mechanisms

Vagal and splanchnic nerves innervate the ruminant stomach. Vagal motor nerve fibers originating in the gastric centers of the medulla oblongata in the brain stem can increase the rate at which the ruminoreticulum contracts. For example, excitatory input generated from the sight of food, the process of chewing or rumination itself, or an increase in the distension of the ruminoreticulum all increase rumen motility. In contrast, input from the splanchnic nerve is inhibitory. Distension of the abomasum, for example, decreases the rate at which the ruminoreticulum contracts.

Rumination is the regurgitation, remastication, and reswallowing of ingesta

Rumination cycles (or "cud chewing") are closely linked with ruminoreticular motility and last approximately a minute. Regurgitation is associated with an extra contraction of the reticulum followed by a primary contraction. The extra contraction is associated with a transient negative pressure created by closure of the glottis prior to an inspiratory cycle. The influx of digesta into the esophagus occurs simultaneously with the opening of the lower esophageal sphincter and a relaxing of the upper esophageal sphincter. The semiliquid digesta is propelled up the esophagus by reverse peristalsis. Once food enters the mouth the liquid portion is immediately reswallowed, but the particulate material remains in the mouth for maceration by the teeth and *reinsalivation*, mainly from the parotid gland on the side of chewing. Each rumination cycle is divided into a prolonged chewing phase followed by a short redeglutition phase. The food bolus is reformed and reswallowed. The amount of time spent ruminating depends on the species of animal as well as the composition of the food. In general, the greater the roughage content of the diet, the more time that is spent ruminating the digesta. Forages and hay require more chewing than do feed concentrates.

The environment of the rumen fosters the growth of anaerobic microbes

As we noted earlier, enhanced fermentation of plant material is the key to ruminants' success. Microorganisms contained in the rumen break down cellulose into short-chain fatty acids that are absorbed through the rumen wall. Most of these microorganisms are attached to partially digested feed particles (approximately 75%), whereas the remaining either adhere to the rumen wall or are present in the rumen fluid. The rumen is normally well buffered by the copious flow of saliva (which contains both bicarbonates and phosphates), production of short-chain organic acids, as well as the buffering capacity of the animal's diet. Further, the presence of ammonia in the rumen can limit decreases in rumen pH, which is optimally around 6.5 to 7.0, but due to the generation of VFAs is normally somewhat lower.

The microbial populations found in the rumen include anaerobic species of *bacteria, archaea, protozoa,* and *fungi.* Of these, bacteria are the greatest in number and diversity of species. Over 200 species of bacteria have been identified in concentrations of 10^9 to 10^{10} cells per gram of rumen contents. The predominant bacterial species found in rumen fluid from wild ruminants such as deer and moose are similar to those species identified in domesticated ruminants. Bacteria can be either liquid-associated and feed on soluble carbohydrates or proteins, or be solid-associated (adherent bacteria) and bound to food particles. The latter bacteria digest the insoluble polysaccharides as well as the less soluble proteins. Immediately after the animal's feeding, some of the adherent bacteria detach from the particulate matter and colonize the new feed particles. Soluble nutrients leaching from the cut ends of the plant material attract both microbes and protozoa by chemotaxis. A third group of bacteria *(epimural bacteria: epi-,* "on"; *mural,* "wall") is associated with the epithelial cells lining the rumen wall. These organisms use the O_2 that diffuses from the blood through the epithelial cells to the rumen fluid. In this manner, they help maintain the anaerobic environment of the rumen. These cells are also important in converting the urea that diffuses through the rumen wall into ammonia (via the enzyme *urease*) because very few microbe populations synthesize ureases. Epimural bacteria also digest the cells that are sloughed from the rumen wall. Were it not for these microbes, these highly keratinized cells (fibrous proteins) would be indigestible.

Protozoa numbers vary from 10^5 to 10^6 per gram of contents and include about 60 species in domestic animals. While the diversity of protozoa in wild ruminants is considerably less than in domestic ruminants, compartmentalization of protozoa is similar to that of bacteria in that there are liquid-associated, solids-associated, and epimural populations. Protozoa face a particular problem in that the time required to reproduce is longer than the time the ingesta remains in the rumen. For this reason protozoa must adhere to large feed particles or the rumen wall to avoid being washed out with the other digesta. Protozoa are particularly sensitive to the composition of an animal's diet. For example, in dairy cows the number of protozoa increases in proportion to the amount of grain fed. Protozoa feed on ruminal bacteria, fungal zoospores, and other readily digestible materials, including plant starch granules and some polyunsaturated fatty acids. Interestingly, certain species of methanogenic bacteria (those converting CO_2 to CH_4) colonize the cilia of protozoa. Thus, many feed-grade antibiotics such as *monensin* or *lasalocid* that target protozoal species often decrease methane production by reducing the methanogenic host.

Archaea contribute from 0.3 to 3.3% of the total ruminal microbe population. Ruminal archaea are strictly anaerobic methanogens, which survive by utilizing the electrons derived from H_2 or formate as energy sources to reduce CO_2 to methane.

At least 14 species of fungi have been identified, but a reliable estimate of their numbers has not been made. The vegetative stage of fungi is termed a *sporangium*. Sporangia generate flagellate, motile *zoospores* that remain in the fluid compartment until their attachment to a food particle. However, once they are attached to the digesta it is difficult to quantify population numbers. Fungal hyphae (filaments) penetrate deep into the plant cell walls and thus gain access to a pool of cellulose unavailable to protozoa and bacteria. Their weakening of the cell wall enables bacteria to eventually gain access and increase the rate of digestion of these insoluble plant fibers. Chemotaxis also plays a role in the location and colonization of plant material. Fungi secrete a more soluble form of cellulase than bacteria and are more successful in fermenting coarse particles, although their overall rate of digestion is slower compared to bacteria. Because of their attachment to large particles, both protozoa and fungi are cleared from the ruminoreticulum at relatively low rates, compared to bacteria.

Nutrients for the host ruminant are generated by anaerobic microbes

Fermentation is the most important process to occur in the rumen. It is also the first step in the generation of nutrients suitable for the host ruminant. Because the diet of ruminant animals contains varying proportions of insoluble celluloses,

hemicelluloses (structural component of plant cells), pectins (polysaccharide isolated from fruits), and starch (polysaccharide that stores energy in plants), these complex polysaccharides need to be converted into simpler forms to be metabolized by the ruminant microbes. Complicating this digestive process are the seasonal changes in the proportion of cellulose and hemicellulose as the herbage matures. Aging of plant walls is associated with an increase in the structural component, **lignin,** a noncarbohydrate polymer, which effectively resists degradation by most rumen microbes. Most bacteria attach to feed particles and digest the structurally complex polysaccharides via cellulase enzyme complexes tightly bound to the surface of the bacteria. Similar cellulase complexes are used by protozoa to digest engulfed feed particles, whereas fungi can digest polysaccharides by bound enzymes or by releasing the cellulases into the surrounding fluid. Starch and cellulose are degraded into glucose, whereas hemicellulose and pectin are converted into xylose (monosaccharide). Sugars and pectins are the most rapidly fermented, followed by starches and fiber. Most simple sugars do not accumulate in the rumen liquor, because they are taken up and metabolized by the microbes populating the forestomach, regardless of whether they participated in the initial hydrolysis of the complex polysaccharides. However, depending on the rate and extent of digestion, some carbohydrates escape the rumen. For example, most fiber, considerable starch, and some simple sugars escape the rumen environment.

Generation of Volatile Fatty Acids by Microbes The metabolism of glucose and xylose by the microbes is similar to that of cells in the animal (Figure 14-30). These simple sugars are anaerobically metabolized through the glycolytic pathway into the intermediate *phosphoenolpyruvate (PEP)*. Methane, CO_2, acetate, and some butyrate are generated from PEP. Alternatively, PEP can be further metabolized into pyruvate, which ultimately leads to generation of the short-chain fatty acids propionate and butyrate. Under normal conditions, the rumen fluid content is approximately 60 to 70% acetate, 14 to 20% propionate, and 10 to 14% butyric acid. If the diet contains high concentrations of soluble carbohydrates or starch, the concentration of propionic and lactic acids increase, whereas high-fiber diets cause increases in acetic acid and declines in propionate and lactate.

The VFAs acetate, propionate, and butyrate are the most important product of the microbial fermentation process because they can be directly used as a source of energy by the host animal. Propionate is particularly important because it is the only VFA that can be used to synthesize glucose. Under normal conditions, approximately 70% of the ruminant's glucose and glycogen is generated from propionate, whereas the catabolism of protein contributes only about 20%. The other VFAs provide energy by entering the TCA cycle as acetyl CoA, with the excess energy stored as body fat in the tissues. These end products become milk fat and lactose in a lactating ruminant. Acetate, in particular, is important in the synthesis of milk fat in the udder and body fat. Other aromatic compounds can enter the vascular system and reach the mammary gland. For example, digestion of onions, garlic, and leek in the rumen generates VFAs that eventually flavor the milk.

VFAs produced by the rumen microbes are passively absorbed through the rumen wall. The rate of absorption is dependent on chain length, pH, concentration, and osmolarity. For example, as the osmotic pressure in the rumen in-

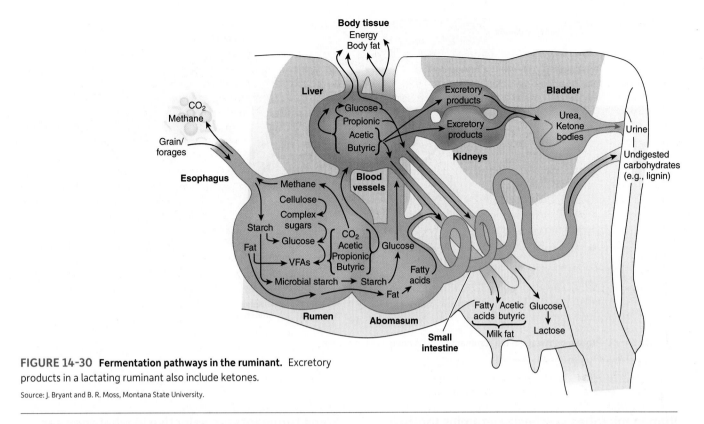

FIGURE 14-30 Fermentation pathways in the ruminant. Excretory products in a lactating ruminant also include ketones.

Source: J. Bryant and B. R. Moss, Montana State University.

creases, absorption of VFAs from the rumen decreases. Osmotic pressure in the rumen averages 280 mOsm, but pressures above 350 mOsm can completely stop rumination. Between the pHs of 4.5 and 6.5, the rates of VFA absorption are butyrate > propionate > acetate, whereas the rates are similar when rumen pH is above 6.5. Low pHs are also associated with a diminished attachment of rumen microbes to fiber particles, which reduces the microbial degradation of cellulose. At pH levels lower than 5.0, there is a complete cessation of fiber digestion in addition to a decline in the number of cellulolytic (cellulose-digesting) microbes.

Protein Metabolism by Microbes Each of the microbial populations in the rumen can hydrolyze protein into constituent polypeptides and amino acids and use these products as a source of nitrogen for their own growth (Figure 14-31). Although most bacterial proteases (protein-digesting enzymes) remain bound to the cell, some species liberate their proteases into the rumen fluid to generate polypeptides. Polypeptides up to six amino acids long can be absorbed by the rumen bacteria and further hydrolyzed to end products such as amino acids, or are deaminated (the nitrogenous amine group is removed) to produce ammonia. Ammonia as well as amino acids are readily absorbed through the lining of the ruminoreticulum and omasum. Deamination of some amino acids such as valine and leucine is responsible for the generation of **isoacids**, or branched-chain fatty acids. These isoacids are essential in small amounts for the growth of cellulolytic microbes. Peak ammonia production occurs about one to two hours after feeding, although the appearance of free amino acids and ammonia occurs within minutes after feeding. Rumen microbes use the ammonia along with some amino acids and polypeptides to synthesize their own microbial protein. Once these bacteria are swept into the abomasum, their cell proteins are digested and eventually absorbed through the small intestine. The synthesis of microbial protein also ensures that the host is

supplied with essential as well as nonessential amino acids. Approximately 30 to 40% of the dietary protein entering the rumen is not degraded and is classified as **rumen-undegradable protein.** However, the gastrointestinal proteolytic enzymes readily hydrolyze these proteins.

Ammonia generated in the ruminoreticulum is either converted into microbial protein or converted to urea in the liver. Some of this urea is returned to the forestomach directly through the wall of the ruminoreticulum or through the salivary glands. High urease (urea-digesting) activity is found along the ruminal wall and is responsible for the rapid degradation of urea to ammonia for metabolism into microbial protein. Recall that presence of urease in the rumen permits the incorporation of urea in the diet of ruminants. However, feeding ruminants a diet too high in urea or rapidly hydrolyzed protein can lead to ammonia toxicity.

Wild herbivores experience marked changes in food quality and quantity throughout the year. When environmental conditions become sufficiently harsh and the only herbage available is deficient in protein, the concentration of ammonia in the rumen is low, as are the number of rumen microbes, thus slowing the breakdown of cellulose. However, the total amount of nitrogen returned to the rumen as urea exceeds that absorbed from the rumen as ammonia. Nitrogen recovered from this process is converted to microbial protein. Ultimately, the total amount of protein arriving in the intestine can be greater than that in the original food. A wild ruminant can conserve an important nitrogen source by returning to the rumen urea that would otherwise have been excreted in its urine. Alternatively, under conditions when protein degradation is greater than its synthesis, such as in animals fed protein-rich concentrates, ammonia can potentially accumulate in the rumen fluid. Under these circumstances, some of the ammonia that is converted to urea in the liver is excreted in the urine and thus wasted.

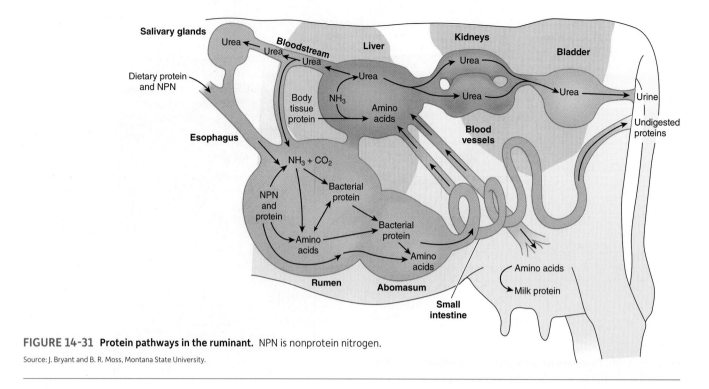

FIGURE 14-31 Protein pathways in the ruminant. NPN is nonprotein nitrogen.

Source: J. Bryant and B. R. Moss, Montana State University.

Rumen microbes synthesize vitamins for their host and detoxify some ingested toxins

One important by-product of fermentation is the synthesis of all B vitamins. Rumen microorganisms, if enough cobalt is available, even synthesize the vitamin B_{12} complex. The microbes responsible for this process require many of the B vitamins. Thus, ruminants do not require a dietary source of these vitamins.

Regardless of the location of microbes in an animal's digestive tract, these microorganisms can also protect the host animal from some toxins that are ingested in the diet. Microbes are thus sometimes considered the *first line of defense* against dietary toxins and nutritional limiting compounds. Livestock are more likely to consume poisonous plants when pasture is short, a consequence of restricted grazing practices, drought, or late-season grazing. Most poisonous plants are unpalatable and, if there is something better to eat, are usually not eaten in quantities sufficient to become toxic. Consumption of various poisonous molecules, such as oxalates (in rhubarb leaves), can be neutralized by rumen microbes into harmless by-products. Protozoa and bacteria can convert another poison, nitrate, in foodstuffs. Often some plants store minerals such as calcium and phosphorous in complexes known as **phytate**. This phytate complex is resistant to normal mammalian digestion and often additional dietary phosphorous is necessary to meet a monogastric's (swine and poultry) phosphorous requirements. Ruminants, however, due to their microbial inhabitants, are able to utilize these minerals because of microbial degradation of the phytate complex. Some species of baleen whales can survive the ingestion of pollutant-laden krill (a tiny crustacean). Anaerobic microorganisms found in the ceca of bowhead whales (though not ruminants) can degrade PCBs, anthracene, and naphthalene, major constituents of oil spills.

Some ruminants are selective in what they eat, whereas others simply graze the available forage

Cattle are representative of a group of herbivores that are relatively nonselective in their consumption of forage. They intensively consume forage at limited periods during the day, followed by intense periods of rumination. In general, the larger the animal, the more likely it can be classified within this grouping. Ruminants with a larger forestomach have a greater digestive capacity in which to ferment lower-quality plant material. The ability to digest fiber is also associated with an increase in total intestinal length and a reduction in transit time through the intestine. A slower rate of passage improves nutrient absorption by increasing time of contact with absorptive cells and increases digestibility of dietary fiber by allowing more time for microbial fermentation. However, management practices of the beef cattle industry include the feeding of concentrate diets to enhance growth rate. Because these animals are slaughtered for meat, they normally do not live long enough to develop diet-related problems associated with this feeding practice.

At the other end of the spectrum are the *selective feeders* that consume only the most nutritious parts of the herbage. This activity normally requires more effort and time and so tends to be employed by smaller ruminants that need less total energy consumption. Associated with the high-quality diet is an increased rate of VFA production. Their rumens are more highly folded, which increases the available surface area for absorption of amino acids and VFAs. Selectors tend to have smaller rumens relative to body size as well as shorter total intestinal lengths. The pillars and ruminal structure have evolved to increase the passage rate of digesta. Antelopes are examples of ruminants that require frequent meals and whose forestomachs are comparatively less developed. The turnover of material in the rumen of these animals is faster than in their nonselective counterparts. Concentrate selectors such as whitetail and roe deer can exploit the seasonal variability in plant growth and show marked increases in food intake during the summer months, when food sources are plentiful. The increase in turnover of rumen contents at this time permits them to store the excess energy as fat reserves to help them through the winter season. Rumen capacity in deer is also increased by 50% during the summer months, which lets the animal accommodate a greater digesta load. Goats and caribou are examples of intermediate feeders and show a less marked increase in food intake and rumen size during the summer.

check your understanding 14.9

Describe the processes that distinguish ruminants from non-ruminants, and the advantages those give the former over the latter.

What compound(s) can be absorbed from the rumen?

14.10 Overview of the Gastrointestinal Hormones

Throughout our discussion of digestion, we have repeatedly mentioned different functions of the several major gastrointestinal hormones in vertebrates, especially gastrin, secretin, and cholecystokinin (CCK). Let's now fit all these functions together so that you can appreciate the overall adaptive importance of these interactions. Furthermore, we now introduce more recently identified hormones as gastrointestinal hormones. All of these are small peptides, and most perform their functions by binding to G-protein coupled receptors on their target cells to bring about the desired responses (via second messengers; see Figure 3-17c, p. 95). This list is far from complete; at least 20 possible signal peptides have been isolated from the mammalian digestive tract, and the functions of many of these remain uncertain.

Gastrin Protein in the stomach stimulates the release of **gastrin,** which has the following functions:

1. It stimulates parietal and chief cells to increase secretion of HCl and pepsinogen, two substances of primary importance in initiating digestion of the protein that promoted their secretion.
2. It enhances gastric motility, stimulates ileal motility, relaxes the ileocecal sphincter, and induces mass movements in the colon—functions that are all aimed at keeping the contents moving through the tract on the arrival of a new meal.
3. It is trophic (see p. 272) not only to the stomach mucosa but also to the small-intestine mucosa, helping maintain a well-developed, functionally viable digestive tract lining.

Predictably, gastrin secretion is inhibited by an accumulation of acid in the stomach and by the presence in the duodenal lumen of acid and other constituents that necessitate a delay in gastric secretion.

Secretin As the stomach empties into the duodenum, the presence of acid in the duodenum stimulates the release of secretin into the blood, which performs the following inter-related functions:

1. It inhibits gastric emptying to prevent further acid from entering the duodenum until the acid that is already present is neutralized.
2. It inhibits gastric secretion to reduce the amount of acid being produced.
3. It stimulates the pancreatic duct cells to produce a large volume of aqueous $NaHCO_3$ secretion, which is emptied into the duodenum to neutralize the acid.
4. It stimulates secretion by the liver of a $NaHCO_3$-rich bile, which likewise is emptied into the duodenum to assist in neutralization of the acidic chyme. This helps prevent damage to the duodenal walls and provides a suitable environment for the optimal functioning of the pancreatic digestive enzymes, which are inhibited by acid.
5. Secretin and CCK are both trophic to the exocrine pancreas.

CCK As chyme is emptied from the stomach, fat and other nutrients enter the duodenum. These nutrients, especially fat and to a lesser extent protein products, cause the release of **cholecystokinin (CCK)** which performs the following inter-related functions:

1. It inhibits gastric motility and secretion, allowing adequate time for the nutrients already in the duodenum to be digested and absorbed.
2. It stimulates the pancreatic acinar cells to increase secretion of pancreatic enzymes, which continue the digestion of these nutrients in the duodenum. (This action is especially important for fat digestion because pancreatic lipase is the only enzyme that digests fat.)
3. It causes contraction of the gallbladder and relaxation of the sphincter of Oddi so that bile is emptied into the duodenum to aid fat digestion and absorption. Bile salts' detergent action is particularly important in enabling pancreatic lipase to perform its digestive task. Once again, the multiple effects of CCK are remarkably well adapted to dealing with the fat and other nutrients whose presence in the duodenum triggered this hormone's release.
4. CCK has also been implicated in long-term adaptive changes in the proportion of pancreatic enzymes produced in response to seasonal changes in diet.
5. Besides facilitating the digestion of ingested nutrients, CCK is an important regulator of food intake. It plays a key role in satiety or fullness (see Chapter 15, p. 727).

GIP A more recently recognized hormone released by the duodenum, **GIP,** helps promote metabolic processing of the nutrients once they are absorbed. This hormone was originally named *gastric inhibitory peptide (GIP)* for its presumed role as an enterogastrone. It was believed to inhibit gastric motility and secretion, similar to secretin and CCK. GIP does this, but only weakly; we now know that this hormone's main role is to stimulate insulin release by the pancreas. So it is now called **glucose-dependent insulinotropic peptide** (once again, GIP). Again, this is remarkably adaptive in anticipation or "feedforward" fashion (p. 16). As soon as the meal is absorbed, the body has to shift its metabolic gears to use and store the newly arriving nutrients. The metabolic activities of this postabsorptive phase are largely under the control of insulin. Stimulated by the presence of a meal, especially glucose, in the digestive tract, GIP initiates the release of insulin *before* the absorption of the meal. Insulin is especially important in promoting the uptake and storage of glucose in the muscle and liver.

GIP structure has evolved interesting differences. Consider the Gila monster, a venomous lizard of the Sonoran desert, which eats only a few times a year. Its salivary glands and digestive tract release a very potent form of GIP that stimulates insulin rapidly during one of these rare large meals to both slow intestinal absorption (to prevent a blood glucose spike) and stimulate insulin production for better glucose postabsorptive processing. Gila monster GIP is now being used by some Type II diabetic humans (p. 323) whose pancreatic endocrine cells only weakly make insulin after a meal.

Motilin The hormone **motilin,** secreted by gland cells in the crypt of the small intestine, triggers peristaltic pumping in the mammalian intestine between meals. In support of its housecleaning function, motilin stimulates the contraction of the antrum and fundus of the stomach and accelerates gastric emptying.

Ghrelin and Obestatin The peptide hormone **ghrelin** (the "hunger hormone") is secreted by epithelial cells lining the stomach fundus, primarily during fasting periods just before normal mealtimes (in rats and humans). It stimulates release of growth hormone from the anterior pituitary, and—perhaps most importantly—greatly increases appetite. The gene that encodes grehlin also expresses the opposing hormone **obestatin** (the "fullness hormone"), whose function may be to decrease appetite. The roles of these two hormones in food intake and energy balance are further discussed in Chapter 15.

PYY$_{3-36}$ The recently discovered peptide **PYY$_{3-36}$** is secreted by cells lining the ileum and colon. It is released in proportion to the amount of food inside it. Locally, like CCK, it inhibits gastric motility and stimulates bile and pancreas secretion. It also travels to the brain (hypothalamus), where it suppresses appetite. Its role in food intake and energy balance is further discussed in Chapter 15.

Nefstatin-1 The peptide **nefstatin-1** was discovered in 2006 as an appetite suppressant in fish and mammals. More recently it has been found in gastric mucosa and pancreas, although its function there is unclear. Its role in the brain is covered in Chapter 15.

This overview of the multiple, integrated, adaptive functions of the gastrointestinal hormones provides an excellent example of the remarkable efficiency of an animal's body, and the remarkable nature of the vertebrate's "second brain" and its associated endocrine system located in the gut. Much remains for science to learn about these regulatory functions.

check your understanding 14.10

Summarize the functions of the gastrointestinal hormones gastrin, CCK, secretin, and GIP.

making connections

How Digestive Systems Contribute to the Body as a Whole

To maintain constancy in the internal environment, materials that are used up in the body (such as nutrients and O_2) or uncontrollably lost from the body (such as evaporative H_2O loss from the airways or salt loss in sweat) must constantly be replaced by new supplies of these materials from the external environment. All these replacement supplies except O_2 are acquired through the digestive system in most animals. The large, complex food that is ingested is broken down by the digestive system into small units, which are then absorbed. If excess nutrients are ingested and absorbed, the extra is placed in storage, such as in the insect fat body and vertebrate adipose tissue (fat), for later use; or it is metabolized, so that the blood level of nutrient molecules is kept relatively constant. In a sense, *the entire digestive system is an effector for energy, mineral, and water homeostasis.*

Unlike most body systems, regulation of digestive-system activities themselves is not aimed at maintaining homeostasis. The quantity of nutrients and H_2O ingested is subject to control, but the quantity of ingested materials absorbed by the digestive tract is not subject to control, with few exceptions. The hunger mechanism governs food intake to help maintain energy balance (Chapter 15), and the thirst mechanism controls H_2O intake to help maintain H_2O balance (Chapter 13). Once these materials are in the digestive tract, the digestive system does not vary its rate of nutrient, H_2O, or electrolyte uptake according to body needs; rather, it optimizes conditions for digesting and absorbing what is ingested. Truly, except for indigestible and other unabsorbable items, what an animal consumes is what it gets.

Chapter Summary

- To maintain homeostasis, nutrient molecules used for energy production must continually be replaced by new, energy-rich nutrients. Thus, the digestive system contributes to homeostasis by transferring nutrients, water, and electrolytes from the external environment to the internal environment.

- Carbohydrates, fats, and proteins represent the main fuels; they are broken down into absorbable units (specifically simple sugars, fatty acids, and amino acids), which an animal's individual cells can use as fuel or as "building blocks" to make their own, larger molecules.

- Animal digestion evolved from an intracellular to an extracellular process in a specialized sac or tube connected to the environment. Conventionally, the gut tube in most animals can be divided into three specific regions: the foregut, midgut, and hindgut.

- Animals can be classified according to their primary feeding methods, including filter-feeding, deposit feeding, herbivory, carnivory, and so forth.

- Digestive systems perform four basic digestive processes: motility, secretion, digestion, and absorption.

- The prominent type of self-induced electrical activity in digestive smooth muscle are slow-wave potentials: basic electrical rhythms (BER) or pacesetter potentials.

- In vertebrates, the intrinsic nerve plexuses are the two major networks of nerve fibers—the myenteric plexus and the submucous plexus—that are located entirely within the digestive tract wall and run its entire length.

- Saliva aids in mastication but plays a more important role in lubricating food boluses before swallowing.

- The esophagus is a fairly straight, muscular tube that extends between the pharynx and stomach and is guarded at both ends by sphincters. The crop is a modified section of the esophagus and functions mainly as a storage organ.

- In most animals, the midgut or stomach provides for the initial digestion (beyond the minor salivary contribution) and storage of food items. Strong antral peristaltic contractions are responsible for mixing food with gastric secretions to produce chyme. Before emptying, the stomach contents must be converted into a finely divided, thick liquid form.

- The four most important factors in the vertebrate duodenum that influence gastric emptying are fat, acid, hypertonicity, and distension.

- Gastric (stomach) digestive juice in mammals is secreted by glands located at the base of gastric pits. The major digestive constituent of gastric secretion is pepsinogen, an inactive enzymatic molecule produced by the stomach of all vertebrates.

- Mammalian parietal cells actively secrete HCl into the lumen of the gastric pits, which in turn empty into the lumen of the stomach. Intrinsic factor, a secretory product of the parietal cells (also secreted from the pylorus and beginning of the duodenum in pigs) is important in absorption of vitamin B_{12}. Gastric secretion gradually decreases as food empties from the stomach into the intestine

- The vertebrate exocrine pancreas secretes a pancreatic juice consisting of two components: (1) pancreatic enzymes actively secreted by the acinar cells that form the acini and (2) an aqueous alkaline solution actively secreted by the duct cells that line the pancreatic ducts. Pancreatic exocrine secretion is hormonally regulated to maintain neutrality of the duodenal contents and to optimize digestion.

- The vertebrate liver performs various important biochemical functions, including bile production. Bile is continuously secreted by the vertebrate liver and is diverted to the gallbladder between meals. Bile salts aid fat digestion and absorption through their detergent action and micellar formation, respectively.

- The fat body of insects plays a major role in intermediary metabolism, and it is the central storage depot of nutrients and energy reserves. Nutrient sensing also appears to be in the domain of the fat body.

- The small intestine is the site at which most digestion and absorption of nutrients takes place in vertebrates. (In arthropods, the midgut and hindgut are the sites of most digestion and absorption.) Small-intestine enzymes complete the process of digestion at the brush border membrane. Carbohydrate and protein are both absorbed by secondary active transport and enter the blood. Extensive absorption by the small intestine keeps pace with secretion.

- The hindgut or large intestine functions in electrolyte and fluid balance, harbors microbes for fermentation and VFA production, and serves as a temporary storage site for excreta. Diet composition governs the variation in structure of the large intestine. The large intestine absorbs primarily salt and water, converting the luminal contents into feces.

- Ruminant species have evolved a stomach of sufficient size and motility to house a population of microbes able to break down cellulose and other complex polysaccharides into end products of fermentation, which meet the nutritional requirements of the host animal. Pregastric fermenters have a survival edge over monogastrics and hindgut fermenters in nutrient-limiting situations.

- Acetate, propionate, and butyrate are the most important volatile fatty acids of the microbial fermentation process because they can be directly used as a source of energy by the host animal. VFAs produced by the rumen microbes are passively absorbed through the rumen wall.

- There are numerous gastrointestinal hormones including gastrin, secretin, CCK, ghrelin, and GIP, each having important roles in regulating digestive-related processes.

Review, Synthesize, and Analyze

1. What four general factors are involved in regulating digestive system function? What is the role of each?

2. Describe the types of motility in each component of the digestive tract. What factors control each type of motility?

3. State the composition of the digestive juice secreted by each component of the digestive system. Describe the factors that control each digestive secretion.

4. List the enzymes involved in digesting each category of foodstuff. Indicate the source and control of secretion of each of the enzymes.

5. Describe the process of mucosal turnover in the stomach and small intestine.

6. Why are some digestive enzymes secreted in inactive form? How are they activated?

7. Speculate on why insects do not have a pepsin-type enzyme released from their gastric mucosa.

8. What absorption processes take place within each component of the digestive tract? What special adaptations of the small intestine enhance its absorptive capacity?

9. Describe the absorptive mechanisms for salt, water, carbohydrate, protein, and fat.

10. The number of immune cells in the *gut-associated lymphoid tissue (GALT)* housed in the mucosa is estimated to be equal to the total number of these defense cells in the rest of the body. Speculate on the adaptive significance of this extensive defense capability of the digestive system.

11. After bilirubin is extracted from the blood by the liver, it is conjugated (combined) with glucuronic acid by the enzyme glucuronyl transferase within the liver. Only when conjugated can bilirubin be actively excreted into the bile. For the first few days of life, the liver does not make adequate quantities of glucuronyl transferase. Explain how this transient enzyme deficiency leads to the common condition of jaundice in newborns.

12. Why might coprophagy be a better solution to the digestion of cellulose in rabbits than pregastric fermentation?

Suggested Readings

Batterham, R. L., M. A. Cowley, C. J. Small, et al. (2002). Gut hormone PYY$_{3-36}$ physiologically inhibits food intake. *Nature* 418:650–654.

Chivers, D. J., & P. Langer, eds. (2005). *The Digestive System in Mammals: Food Form and Function.* Cambridge, UK. Cambridge University Press.

Church, D. C. (1988). *The Ruminant Animal. Digestive Physiology and Nutrition.* Englewood Cliffs, NJ: Prentice Hall.

Gershon, M. D. (1998). *The Second Brain.* New York: Harper Perennial. A detailed look at the enteric neuroendocrine system.

Hillson, S. (2005). *Teeth,* 2nd ed. Cambridge: Manuals in Archaeology. Cambridge University Press.

Moran, N. A., & P. Baumann. (2000). Bacterial endosymbionts in animals. *Current Opinions in Microbiology* 3:270–275.

Prosser, C. L., & E. J. DeVillez. (1991). Feeding and digestion. In C. L. Prosser, ed., *Environmental and Metabolic Animal Physiology: Comparative Animal Physiology,* 4th ed. New York: Wiley-Liss.

Stevens, C. E., & I. D. Hume. (2004). *Comparative Physiology of the Vertebrate Digestive System,* 2nd ed. Cambridge, UK: Cambridge University Press.

Swenson, M. J., & W. O. Reese, eds. (1993). *Duke's Physiology of Domestic Animals,* 11th ed. Ithaca, NY: Cornell University Press.

Taylor, R.E., & Field T. G. (2012). *Scientific Farm Animal Production,* 10th ed. Upper Saddle River, NJ: Prentice Hall.

Triplehorn, C. A., & N. F. Johnson. (1989). *Borror's Introduction to the Study of Insects,* 7th ed. Belmont, CA: Brooks/Cole.

CourseMate

A butterfly basks in the sun to heat its flight muscles before takeoff. Butterflies are partly ectotherms, dependent on the environment for their internal temperature. However, during flight, the butterfly becomes partly endothermic as its muscles generate heat.

Paul Yancey

15

Energy Balance and Thermal Physiology

15.1 Introduction: Thermodynamics and Life

Each cell needs energy to perform the functions essential for its own survival and to carry out its specialized contribution toward maintaining homeostasis. All energy used by animal cells is ultimately provided by food (exogenous energy-rich nutrients), intake of which is regulated to maintain **energy balance,** the internal homeostasis of nutrients available for the cells. In this chapter, we show how this balance occurs. We also consider temperature, one of the key components of energy equations. Temperature profoundly affects the rate of chemical reactions within cells, and animals have evolved various strategies to cope with these effects. Compare, for example, the butterfly above with the hummingbird on the next page. Moreover, solid evidence shows that Earth's atmosphere and oceans have been warming at least in part due to anthropogenic CO_2, and effects of this rise on organisms are already being documented. Before we examine energy balance and temperature, however, let us set the stage by examining the key energy equations.

Life follows the laws of thermodynamics

Perhaps the most distinctive characteristic of life that delineates it from nonliving systems is the way it uses energy. Whereas nonliving energetic processes—volcanoes, earthquakes, storms, and so forth—create disorder, living systems somehow produce their own order unique to each individual. Yet all these processes (indeed, all known processes in the universe) obey fundamental laws regarding energy—the **laws of thermodynamics.** There are three of these laws, with two that are relevant to physiology:

- The *First Law of Thermodynamics* is that energy can be neither created nor destroyed. Therefore, energy is subject to the same kind of input–output balance as are the chemical components of life such as H_2O and salt (see p. 612). The First Law of Thermodynamics suggests that life might operate forever on its internal energy content, simply converting one form to another as necessary. However, the Second Law says this is not possible.

- The *Second Law of Thermodynamics* is that the **entropy** (a measure of disorder or randomness) of a system *plus its surroundings* increases over time as the energy content degrades to unusable heat. This is expressed in the following equation, where Δ indicates change and S is entropy:

$$\Delta S_{net} = \Delta S_{surroundings} + \Delta S_{system}$$
$$>0 \text{ for spontaneous reactions}$$

The Second Law of Thermodynamics is in many ways the most important physical law regarding life in the universe. Most people are aware that the universe seems to run down over time, and only by expending energy can we reverse this trend. But life itself seems to defy this decay (until death, at least), and indeed some people have suggested that life violates the Second Law. However, they are forgetting that the law covers a system *plus its surroundings,* not just a system

715

Paul Yancey

A hummingbird is endothermic: It regulates its internal body temperature primarily by internal metabolism. However, it is also heterothermic: At night, its metabolism and body temperature may decrease by many degrees to save energy.

by itself. The system called "life" does *not* violate the law, because the order it creates can arise only at the expense of its environment. That is, as life becomes more organized, the surrounding environment becomes even more disorganized. The net result—increased order in an organism, increased disorder in the surroundings—is *always* an increase in net disorder (entropy). Plants, for example, create order at the expense of the sun, which is creating enormous disorder as it radiates out light and particles. Animals create order at the expense of the local environment: They eat ordered molecules (food such as glucose) and release less ordered waste molecules (such as CO_2) and—perhaps most importantly—heat, which directly increases disorder. Thus, life must continuously extract energy from the environment to maintain itself, whereas the rest of the universe degrades even faster than it would on its own. In short, the "metabolic cost of living" is a constant struggle against entropy (a struggle that is ultimately lost through aging and death).

To compensate for entropy, animals require input of food energy, most of which is ultimately converted to heat output

Since energy cannot be created or destroyed, it is subject to the same kind of *input–output balance* we described for other body components such as water and salt (p. 612).

Energy Input *Input* energy is that in ingested foodstuffs for most animals. As you have just seen, nondormant organisms need ongoing energy input from the environment. Each cell in an animal needs energy to perform functions essential for its own survival (for example, reversing the effects of entropy by restoring ion gradients and repairing damaged structures) and to carry out its specialized contributions to-

ward maintenance of the whole animal. This requires food (from the environment, or in some animals, from internal symbionts). Chemical energy locked in the bonds that hold atoms together in food molecules is released when these molecules are broken down in the body. However (as a consequence of the Second Law of Thermodynamics), 100% of the food energy cannot be used usefully, because some energy is always lost to entropy.

Energy Output Energy *output* or *expenditure* falls into two categories (Figure 15-1). **External work** is the energy expended when skeletal muscles are contracted to move external objects or to move the body in relation to the environment. **Internal work** constitutes all other forms of biological energy expenditure that do not accomplish mechanical work outside the body. That is, it encompasses all the internal energy-expending activities that must go on all the time just to sustain life, such as skeletal muscle contractions associated with postural maintenance and shivering, the work of pumping blood and breathing, the energy required for active transport of critical materials across plasma membranes, and the energy used during synthetic reactions essential for the maintenance, repair, and growth of cell structures.

Although energy cannot be created or destroyed, it can be converted from one form to another. For example, **chemical energy** in ATP is converted into **kinetic energy** of locomotion by muscle contractile proteins. But not all energy in nutrient molecules can be harnessed to perform useful biological work like this (in accordance with the Second Law). The energy in nutrient molecules that is not used for work is transformed into **thermal energy,** or **heat.** During biochemical processing, only about 50% of the energy in nutrient molecules is transferred to ATP; the other 50% of nutrient energy is immediately lost as heat. During ATP expenditure by the cells, another 25% of the energy derived from ingested food becomes heat. Because animal bodies are not heat engines, they cannot convert heat into work. Therefore, not more than 25% of nutrient energy is available to accomplish work, whether external or internal. The remaining 75% is lost as heat during the sequential transfer of energy from nutrient molecules to ATP to cellular systems.

Furthermore, of the energy actually captured for use by animals, *almost all expended energy eventually becomes heat*. To exemplify, energy expended by a heart to pump blood is gradually changed into heat by friction as blood flows through vessels. Likewise, energy used in the synthesis of cellular structural protein initially reduces entropy, but it eventually appears as heat when that protein is degraded during the normal course of turnover of bodily constituents. Even in the performance of external work, skeletal muscles convert chemical energy inefficiently, with as much as 75% of the expended energy being lost as heat, and the 25% converted into kinetic energy being dissipated as heat when an animal stops moving. Thus, all energy that is liberated from ingested food but not used for net growth eventually becomes heat. This heat is not necessarily wasted energy, however, because it is used to maintain body temperature in some animals (which affects the rate at which chemical reactions can proceed), but not to do work.

Because most animal energy expenditure eventually appears as heat, energy is normally expressed in terms heat energy, the basic units of which are as follows:

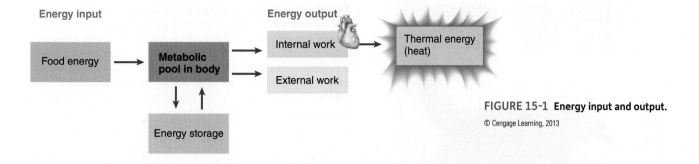

FIGURE 15-1 Energy input and output.

© Cengage Learning, 2013

1. The **calorie** is the amount of heat required to raise the temperature of 1 g of H_2O by 1°C (specifically, from 14.5 to 15.5°C). This unit is too small to be convenient when discussing animals because of the magnitude of heat involved, so the **kilocalorie (kcal)** or **Calorie,** equal to 1,000 calories, is used. When nutritionists speak of "calories" in quantifying the energy content of food, they are actually referring to kilocalories or Calories.

2. The **joule,** the international unit for energy of any kind, is equal to 0.239 calories. The term **kilojoule (kJ)** (1,000 joules) is commonly used, again because of the magnitude of animal energies. *Most scientific research now uses the joule or kilojoule rather than the calorie,* but in the United States the Calorie is still used in human nutrition.

As a common example, 4.1 kcal or 17.1 kJ of heat energy are released when 1 g of glucose is oxidized or "burned," whether the oxidation takes place inside or outside a body.

check your understanding 15.1

Describe the laws of thermodynamics and how they apply to life.

Define *internal* and *external work* in an animal and why heat is the primary output.

15.2 Energy Balance: General Principles

Now let's examine how animals achieve energy balance as limited by thermodynamics.

Energy input must equal energy output, which is measured as metabolic rates of basal metabolism, activity, diet-induced thermogenesis, and production

Because energy cannot be created or destroyed, energy input *must* equal energy output (heat plus storage), as represented by the following equation (where E = energy):

$$E_{input} = E_{output}$$

Metabolic Rate and the "Animal Energy Equation" In the balance equation above, *output* energy (expenditures used for external and internal work) is generally measured as the *rate* at which energy is expended. This is broadly termed the **metabolic rate:**

$$\text{Metabolic rate} = \frac{\text{energy expenditure}}{\text{unit of time}}$$

Rather than simply using internal and external work, physiologists break this rate into specific major expenditures, in what might be called "*the animal energy equation*":

$$E_{input} - E_{loss} = E_{SMR/BMR} + E_{activity} + E_{DIT} + E_{production}$$

where *input* is the total food energy obtained and *loss* is that portion lost per unit time via feces, urine, skin sloughing, and so forth. Thus. the left side of the equation is *net energy intake.* The other four right-side components—the output expenditures—are as follows:

1. *SMR and BMR* are **standard** and **basal metabolic rates.** These internal-work expenditures are the "idling speed": *the minimal amount of energy needed per unit time to sustain waking (nonsleep) life under optimal conditions.*

2. *Activity* energy is the cost per unit time of neuromuscular efforts above the SMR or BMR level. This involves both internal and external work.

3. *DIT* is **diet-induced thermogenesis** (*thermo-,* "heat"; *genesis,* "production"), also called **specific dynamic action (SDA),** is an increase in metabolic rate above the SMR or BMR level that occurs as a consequence of food intake (internal work).

4. *Production* refers to the rate of energy storage, such as adipose deposition, net growth during development, and reproduction (all internal work). This is the only component that does not rapidly become heat (increasing entropy), but rather decreases entropy as an increase in body mass. However, it can also be negative when a fasting animal taps into its reserves for energy, and so this component effectively becomes energy input for the animal.

We will examine each of these expenditure components in detail, but first, let's see how metabolic rates are measured.

Methodology The rate of energy use can be measured directly or indirectly.

1. **Direct calorimetry** assumes all energy ends up as heat and involves the cumbersome procedure of placing the subject in an insulated chamber with H_2O circulating through the walls. The difference in the temperature of the H_2O entering and leaving the chamber reflects the amount of heat liberated by the subject and picked up

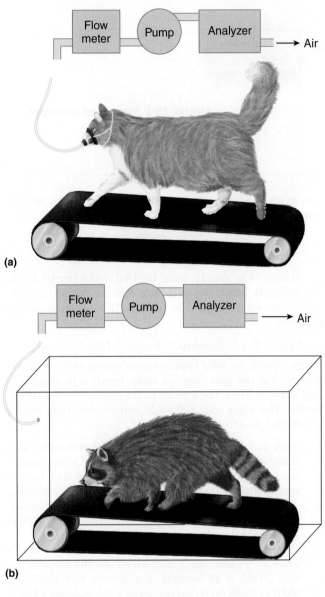

(a)

(b)

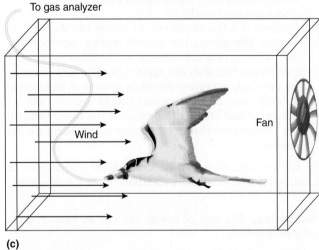

To gas analyzer

Wind

Fan

(c)

FIGURE 15-2 Measuring oxygen consumption by animals.
(a, b) Animals on treadmills, which force an animal to run at controlled speeds. The animal may be fitted with a mask for measuring gases in exhaled breath, or may be in an enclosed chamber from which gases are sampled. (c) A bird flying in a wind tunnel, which forces the animal to fly at controlled speeds. A face mask samples gases in exhaled breath.

© Cengage Learning, 2013

calorimetry for determining metabolic rates were developed for widespread use.

2. In the indirect **respirometry method,** only the subject's O_2 uptake per unit of time (called the V_{O_2}) is measured, which is a straightforward task in most cases (Figure 15-2). Recall that food energy is liberated with the use of oxygen during aerobic metabolism. Accordingly, a direct relationship exists between the volume of O_2 used and the quantity of heat produced. This relationship also depends on the type of food being oxidized, with the following average values, known as the *physiological fuel values:*

Carbohydrates: 4.1 kcal/g = 17.1 kJ
Proteins: 4.3 kcal/g = 17.9 kJ
Fats: 9.3 kcal/g = 38.9 kJ

An estimate, the **energy equivalent of O_2,** can be made of the quantity of heat produced per liter of O_2 consumed, using the animal's typical diet. For a typical human in North America, for example, the average is 4.8 kcal (20.0 kJ) per liter of O_2 consumed. Thus, in determining metabolic rate, a simple measurement of O_2 consumption can be used to reasonably approximate heat production by multiplying by this factor. However, for nonhuman animals the composition of the diet may not be easily obtained; thus, (1) metabolic rates may be reported simply as liters of oxygen consumed per unit time, or (2) an estimate may be made using the **respiratory quotient** (*RQ*), the ratio of CO_2 produced to O_2 consumed. This requires measurements of carbon dioxide production (V_{CO_2}), an increasingly common technique. The *RQ* varies depending on the foodstuff consumed. When carbohydrate is being used, the *RQ* is 1; that is, for every molecule of O_2 consumed, one molecule of CO_2 is produced:

$$C_6H_{12}O_6 + 6\ O_2 \rightarrow 6\ CO_2 + 6\ H_2O + 30\text{–}32\ ATP + heat$$

For fat use, the *RQ* is 0.7; for protein, it is 0.8. Using the *RQ* and the physiological fuel values, an energy equivalent of O_2 can be closely estimated. (Note that a slight correction is necessary to convert gas volumes to moles: 1 mole of O_2 = 22.39 L, whereas 1 mole of CO_2 = 22.26 L.)

On a diet consisting of a mixture of these three nutrients, resting O_2 consumption in a human averages about 250 mL/min, and CO_2 production averages about 200 mL/min, for an average *RQ* of 0.8:

$$RQ = \frac{CO_2 \text{ produced}}{O_2 \text{ consumed}} = \frac{200 \text{ mL/min}}{250 \text{ mL/min}} = 0.8$$

Measurement of *RQ* in a newly hatched or migrating bird or in a diving seal reveals a value closer to 0.7, indicating their reliance on fats during these energy-

by the H_2O as it passes through the chamber. Even though this method provides a direct measurement of heat production, it is not practical, because a calorimeter chamber can be costly, particularly for large animals. Therefore, other more practical methods of **indirect**

intensive activities. Conversely, because carbohydrates yield more energy per O_2 consumed, it is thought that high-altitude animals rely more on carbohydrate that fat compared to low-altitude relatives. Marie-Pierre Schippers and colleagues recently verified this for the Andean leaf-eared mouse.

3. The **doubly labeled water** method is another form of indirect calorimetry, in which animals (often captured in the wild) are given $D_2^{18}O$, water made from heavy hydrogen (deuterium D) and the oxygen isotope ^{18}O. Deuterium is eliminated from the body as water (D_2O), but ^{18}O is eliminated as both water ($D_2^{18}O$) and $C^{18}O_2$. Researchers measure the difference between the elimination rates of each, calculate CO_2 production rate, and convert it into energy expenditure using the *RQ*. This is the method used to measure **field metabolic rate (FMR)**, an average metabolic expenditure over the time of the analysis.

Now let's see how respirometry is used to measure minimal metabolic rates. An animal's metabolic rate varies, depending on a variety of factors, such as genetics, external temperature, exercise, food intake, shivering, and anxiety. To determine the minimal energy requirements for survival, metabolic rate must be measured under standardized conditions established to control the variables that can alter rate. It is measured under the following specified conditions:

1. To eliminate any nonresting muscular exertion, the animal should be quiescent and should have refrained from muscular activity for some time. (Note that metabolic rate during sleep is lower than the minimal rate as defined here.)
2. The animal should have minimal anxiety or stress levels. Stress usually increases metabolic rates in vertebrates; this is due to secretion of epinephrine, which increases heart rate (p. 416). For many animals this necessitates an acclimation period to laboratory conditions prior to the actual measurement. This factor has not been well studied.
3. The animal should not have eaten any food within an appropriate amount of time before the rate determination, to avoid diet-induced thermogenesis (DIT).
4. There should be no elevated energy costs of reproduction, such as pregnancy, egg brooding, lactation, or seasonal gamete production.
5. The measurement should be performed at an optimal temperature for that organism.

Regarding point 5, recall that animals fall into two broad categories of thermal adaptation: **ectotherms** (such as amphibians), which depend on external heat for their body temperatures, and **endotherms** (such as mammals), which use internal heat to regulate body temperatures (these are discussed in detail later in this chapter). Because of this difference, there are two types of this minimal metabolic rate:

- **Standard metabolic rate (SMR)** is an ectotherm's resting metabolic rate at a particular temperature, which may be the average temperature experienced naturally, the temperature the animal prefers if given a choice, or any temperature within its normal range. SMR values usually change dramatically with temperature, and thus the temperature of the test conditions must be reported.
- **Basal metabolic rate (BMR)** is an endotherm's resting metabolic rate in its **thermal neutral zone (TNZ)**, a range of external temperatures that do not induce thermoregulatory processes by the animal. For example, the TNZ for lightly clothed humans lies around so-called room temperature. In the TNZ, there is no shivering, significant sweating, or other thermoregulatory processes that would raise the metabolic rate. (See TNZ later, p. 748.)

Measurements of BMR/SMR are difficult to make, and any exceptions to the conditions just listed must be reported. Another term is used if measurements are made under less-controlled conditions: the **resting metabolic rate (RMR)**, the rate of a quiescent animal in which all the strict conditions are not met. RMR values are typically 10 to 25% higher than the BMR but are often used because measurements in a controlled laboratory with all of the strict conditions are not always practical.

Let's know look at the four expenditure components in detail, beginning with $E_{production}$.

Energy of production determines net energy balance

The production component ($E_{production}$) leads to the concept of energy balance, the factor that determines whether an animal will be at a stable weight, losing weight, or gaining weight, and whether that individual is maintaining a mass* that is optimal or is over- or underweight. For example, gaining significant mass is normal in juvenile growth and in hibernators laying down internal stores for the winter, but in other cases it leads to **obesity**, defined as having a body mass that is high enough to impair health. Conversely, starvation in famines, as may occur in overpopulated deer, may lead to health-threatening low body mass. Thus, understanding of energy balance is of great importance not only for human health but also for health of all animals, domestic or wild. There are three possible states of balance:

- *Neutral energy balance.* If the net amount of energy in food intake (input minus losses) exactly equals the amount of energy expended (output) other than production, then body mass remains constant ($E_{production} = 0$).
- *Positive energy balance.* If the net energy intake exceeds the amount of nonproduction energy expended, and the extra energy taken in is not used immediately, but is stored (a for example, fat stored in adipose tissue or a growing fetus), then body mass increases ($E_{production}$ is positive).
- *Negative energy balance.* Conversely, if the net energy intake is less than the body's immediate nonproduction energy requirements, the animal must use its stored energy to supply energy needs, and, accordingly, body mass decreases ($E_{production}$ is negative).

Thus, the other expenditures—BMR/SMR, activity, and DIT—determine whether there is sufficient energy left over from net intake for production. Let's look at BMR/SMR next.

---

*Note that *mass* is a measure of the amount of material in an object, with units in grams. *Weight* is popularly used in the same way, but actually, it is the gravitational force acting on a mass, in units of newtons. We will generally use *mass* except for terms like *overweight*.

BMR and SMR are regulated in part by thyroid hormones

The BMR or SMR is not a constant for a species; for example, it can vary between individuals, life stages, and sexes. We do not know all of the factors that regulate the rate of BMR and SMR, but in vertebrates, thyroid hormones (p. 301) are the primary (though not sole) determinant: As thyroid hormones increase, the SMR or BMR increases correspondingly. For example, in an experiment on Western fence lizards (*Sceloporus occidentalis*), SMR fell 31% in animals with their thyroids removed, and SMR was restored to normal by injections of thyroid hormone (T_4) (p. 298). Other factors include stress production of epinephrine, which also increases metabolic rate.

Recent research has found that *bile acids*, which are released during fat digestion (p. 686), bind to receptors on adipose cells in mice and skeletal muscle in humans. The receptors activate a gene for an enzyme (deiodinase) that converts the inactive T_4 into the biologically active T_3 (p. 301). In this way, consuming excess fat may temporarily lead to increased basal metabolic rates.

BMR and SMR scale with body mass

BMR and SMR can also differ considerably among species; in particular, endotherms have much higher BMRs than the SMRs of ectotherms. We will examine this later in the section titled Thermal Physiology. Here we examine the phenomenon of *scaling*. As discussed in previous chapters (e.g., pp. 10, 155, 414, 505, and 601), some physiological features depend on the size of organisms because of scaled factors such as surface area and volume. Perhaps the most consistent, yet controversial, scaling relationship is that between BMR or SMR and body mass. Numerous studies have found that although larger animals certainly consume more energy than smaller animals (Figure 15-3a), they have lower metabolic rates *per unit mass* than smaller ones (Figure 15-3b). Basically, a gram of elephant costs much less energy to maintain than a gram of shrew!

The actual scaling factor (the rate at which metabolism changes with body size) is found by plotting the BMR values for whole animals against their body masses on a log–log plot (Figure 15-3a). A straight line with a positive slope is obtained, with the following formula:

$$M = aW^b$$

where *M* is metabolic rate, *a* is the metabolic rate per unit mass (usually rate per kg), *W* is the body mass, and *b* is the slope. Again, this plot shows that larger animals do use more energy. But the slope (*b*) is always less than 1, which signifies the lower metabolic rates (per unit mass) of larger animals (if *b* were 1, then all animals would use energy at the same rate per unit mass). In fact, for mammals, *a consistent slope (b) of about 0.75 (3/4)* has been obtained in numerous studies. Similar scaling is found for birds, again with a slope between 0.7 and 0.8. Other physiologic and anatomic features also scale to body mass with slopes of about 0.75, such as alveolar surface area (see p. 509).

Scaling and Thermoregulation Following the discovery of the 0.75 scaling by Max Kleiber in 1932, a widely accepted hypothesis was generated: Scaling reflects properties of the ratio between surface area and volume with respect to temperature regulation by endotherms. The argument goes as follows: (1) Surface area increases as the square of the radius, whereas volume increases as the cube; therefore, smaller organisms have a higher ratio of surface area to volume (see Figure 1-4). Due to the square-cube difference, area/volume ratio scales by a power of 2/3 (0.67), at least for spheres. (2) Endotherms produce heat by their volumes (mass) and lose it from their surface areas. (3) Therefore, a smaller endotherm must have a higher metabolic rate per unit mass to compensate for its higher rate of heat loss (via surface area). In fact, as this hypothesis predicts, metabolic rates within a single mammalian species do often scale with size by a factor of 0.67.

This hypothesis makes perfect physical sense, but it has two major flaws. First, the hypothesis predicts that BMR will scale with a slope of about 0.67, but instead a slope of 0.75 has been repeatedly found among species. Second, other studies have reported that ectotherms such as non-avian reptiles and fishes (and even unicellular organisms) have SMRs that scale by factors less than 1.0 (see Figure 15-3c for insects, showing a slope of 0.825). The hypothesis is irrelevant to most of these organisms, because most do not generate (and retain) body heat for the purposes of thermoregulation. So why does metabolism scale with an apparent 3/4 factor among endotherms, and by slopes of less than 1 in animals in general? Neither of these flaws has been satisfactorily explained. See *Challenges and Controversies: A Universal Scale of Life?* (p. 722) for recent hypotheses.

Increased muscle activity is the factor that can most increase metabolic rate, as indicated by metabolic scope

The next expenditure category is neuromuscular *activity*. Expressed primarily as locomotion powered by skeletal muscles (external work) and breathing by respiratory muscles and circulatory flow from the heart (internal work), this is potentially the greatest energy usage for an animal, at least for short periods of time. Even slight increases in muscle tone notably elevate metabolic rate, and metabolic rates can increase manyfold during extreme activity. Table 15-1 shows examples for a human and a hummingbird. Note that the rates for the bird are much smaller than for the human, as befitting the bird's much smaller size. However, if we compare these animals on a per-kg basis, a very different perspective is found. The human marathon runner expends about 77 kJ/hr per unit kg, but the hovering hummingbird expends about 870! This is thought to be the maximum possible for vertebrate muscle (see the Suggested Reading by Suarez). And yet, the metabolic rate of a flying honeybee can be at least three times that of the hummingbird.

Metabolic Scope A convenient indicator of an animal's activity capacity is the ratio of metabolic rate at its highest level to the BMR or SMR—the **metabolic scope**. For example, the ratio for a human marathoner is about 17 (1,300/77 in kcal/hr, Table 15-1). **Aerobic scope** values are derived from activity measured using oxygen consumption. This has been done for many animals, using laboratory devices such as treadmills, wind tunnels, and water flumes (Figure 15-2). Aerobic scope values for terrestrial vertebrates (including

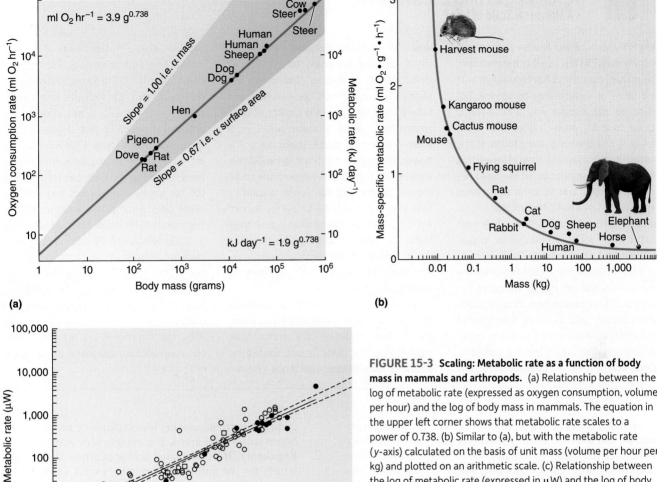

(a)

(b)

(c)

FIGURE 15-3 Scaling: Metabolic rate as a function of body mass in mammals and arthropods. (a) Relationship between the log of metabolic rate (expressed as oxygen consumption, volume per hour) and the log of body mass in mammals. The equation in the upper left corner shows that metabolic rate scales to a power of 0.738. (b) Similar to (a), but with the metabolic rate (y-axis) calculated on the basis of unit mass (volume per hour per kg) and plotted on an arithmetic scale. (c) Relationship between the log of metabolic rate (expressed in μW) and the log of body mass in ants (*open squares*), beetles (*closed circles*) and spiders (*open circles*). The metabolic rate here scales to a power of 0.825.

Sources: (a) From P. C. Withers. (1992). *Comparative Animal Physiology*. Fort Worth, TX: Saunders, using data from M. Kleiber. (1932). Body size and animal metabolism. *Hilgardia* 6:315–353. (b) From K. Schmidt-Nielsen. (1960). *Animal Physiology*. Englewood Cliffs, NJ: Prentice-Hall, p. 193, Figure 5.9. (c) Lighton, J. R. B. and L. J. Fielden, (1995), Mass scaling of standard metabolism in ticks: A valid case of low metabolic rates in sit-and-wait strategists, 'Physiological Zoology', 68:43-62. Copyright © 1995 University of Chicago Press.

both ectotherms and endotherms) fall in the range of about 5 to 40, following a scaling pattern. That is, whole-animal aerobic metabolic rate can be elevated above the basal level by a factor of about 5 to 10 for a small vertebrate like a hummingbird, up to 24-fold in human medium-distance runners and cross-country skiers, and 40 or more in large mammalian runners like horses. In contrast, for some flying insects, the aerobic scope factor can be over 100!

Why is aerobic scope limited to the 5 to 40 range for so many vertebrates? The limit appears to be the delivery of oxygen by the circulatory system. There seems to be a strict limit to the rate of pumping through a network of narrow vessels. The values for some insects show that the insect respiratory system, with trachea that carry oxygen directly to muscles, may be much more effective (although limited by body size; see p. 505).

Locomotory Costs and Speed Metabolism during locomotion generally increases with speed, as you would expect,

but there are some interesting exceptions. For running mammals, the increase in metabolism is roughly linear with velocity (Figure 15-4a). For a swimming fish, the cost increases exponentially as drag from the environment begins to have more and more effect (Figure 15-4b) because water is so much more dense than air. Flight energetics show varying patterns: In some species, metabolic rate is independent of speed at medium to high levels, whereas in others metabolism actually *decreases* at medium speeds (Figure 15-4c). This is because at low speeds much energy is needed to overcome gravity when a bird is first ascending from a perch. Once airborne at medium speeds, a bird can save energy by *gliding*, using its forward momentum to generate lift with its wings to stay aloft. (Bird wings have an *aerofoil* shape: flat or slightly concave on the bottom and curved on the top so that air moving over the wing must travel a greater distance across the top than it does across the bottom. This creates negative pressure (lift) on the top relative to the bottom, drawing the wing up.)

Why is metabolism not linearly proportional to body mass? As discussed in the main text, the primary hypothesis has focused on heat and ratios of surface area to volume. But these do not explain why a relationship scaled to a 3/4 power rather than a 2/3 power is found among many forms of life. Indeed, some scientists have claimed that the 3/4 power applies to all life from microbes to trees, and in some sense must represent a universal phenomenon of life. Because the heat loss hypothesis (see main text) cannot apply to ectothermic organisms, several other hypotheses have been generated. One of these might be called the *four-dimensional* or *fractal transport* hypothesis. Three researchers (Brian Enquist, James Brown, and Geoffrey West) noted that all organisms require a means of transporting and distributing internal nutrients, gases, and wastes. In multicellular organisms, this usually involves networks of increasingly finer branching tubes, such as blood vessels, tracheae, and xylem (in plants). The researchers showed that any network that maximizes the packing of two-dimensional tubular surface areas in a three-dimensional space scales to a 3/4 power. After joining the team, Jamie Gillooly showed that factoring in temperature could account for the fact that colder organisms have scaling lines that fall below the lines of warmer organisms. These ideas lead to their Metabolic Theory of Ecology, which states that temperature and metabolic rate govern most interactions and patterns in ecosystems.

This is the first hypothesis that seems to apply to many different kinds of organisms, and is thus attractive. But critics have pointed out that BMR is not limited by transport systems, because those systems are only pushed to their limits at maximal metabolism. Also, some researchers are not convinced that all life forms follow the 3/4 scaling pattern; for example, some ectotherms show scaling slopes considerably different from 0.75 (see Figure 15-3c). Moreover, recent studies have found that the slope varies among taxonomic groups from 0.4 to 1.0; the slope tends to be near 0.6 for smaller vertebrates and 0.87 for larger ones. Another study suggested that the true slope for mammals is actually close to 0.67, with the 0.75 value arising from large herbivores whose prolonged digestive processes (DIT) yield laboratory measurements of MR that are not true BMRs. Regardless, it does appear that larger organisms are more efficient than smaller ones. But the causes are likely to remain controversial for many years.

Finally, changes in **gait** (modes of limb use during locomotion) can alter the linear pattern. As you learned in Chapter 8, most terrestrial vertebrates have elastic tendons in their legs that can store energy each time a leg hits the ground. This energy can be released on the next step, so that much of the energy required to lift the limbs is not wasted. Horses also have tendons in their backs, which come into play during a gallop gait. The ultimate example is found in hopping kangaroos and wallabies, which have been found to have no increase in metabolism as they increase hopping velocity (although they do show the expected linear increase while walking). See Figure 15-4d, and p. 358 for details.

Actual metabolic rates during locomotion (as well as scope) are scaled with body size, so that larger animals use less energy per unit body mass to cover a given distance than do smaller animals (Figures 15-4a and b). That is, it costs much less to move a gram of dog than a gram of mouse the same distance. This appears to be true of insects as well as vertebrates. It is not readily apparent why this should be, although one of the current hypotheses to explain scaling suggests that it may reflect the scaling of oxygen delivery (see *Challenges and Controversies: A Universal Scale of Life?*).

Diet-induced thermogenesis occurs after eating in most animals

The third energy-expenditure category is *diet-induced thermogenesis* (DIT; or *specific dynamic action*, SDA), which manifests as a rise in metabolism and heat production for several hours after eating. DIT has two components.

1. **Obligatory DIT** is the result of the increased metabolic activity associated with the costs of obtaining and processing food, that is, gut motility, production of gut secretions, nutrient uptake, biosynthesis of proteins, glycogen stores, and lipids, and excretion of wastes.

2. **Regulatory DIT** is increased heat production after a meal, largely for the purpose of removing excess nutrients, for example, "burning off" excess calories ingested (p. 729).

Obligatory DIT has been found in all phyla of animals tested; regulatory DIT has historically been studied in endotherms but may also occur in many types of animals. We examine the role of regulatory DIT in body mass control shortly.

Although the exact costs involved with DIT are not all certain, there are clear patterns. Within an individual, large and/or hard-to-digest meals induce a greater DIT than small and/or readily digested meals. Species differences are diverse; for example, DIT reaches a peak within a few hours in most endotherms, but it takes up to several days in sluggish ectotherms such as seastars. The extent of DIT also varies; for instance, humans show an increase of up to 50% in metabolism after a large meal, whereas other mammals, ectothermic vertebrates, and insects may exhibit 100% to 800% increases or more. The record may be held by the Burmese python, which shows a 4,400% increase in metabolism after swallowing a rodent! This animal (and other snakes that eat large, infrequent meals) actually undergoes *atrophy* of its digestive organs in between meals, presumably to save energy. This down-regulation is manifested as lower liver and intestinal masses and lower density of nutrient transporters in the intestine (see p. 692). Then, on ingesting its large but infrequent prey item, the snake restores all these atrophied features in concert with the movement of the load through the respective regions of the digestive tract. This up-regulation accounts for the enormous DIT.

Now that we have examined the details of energy balance and its components, we turn to its regulation.

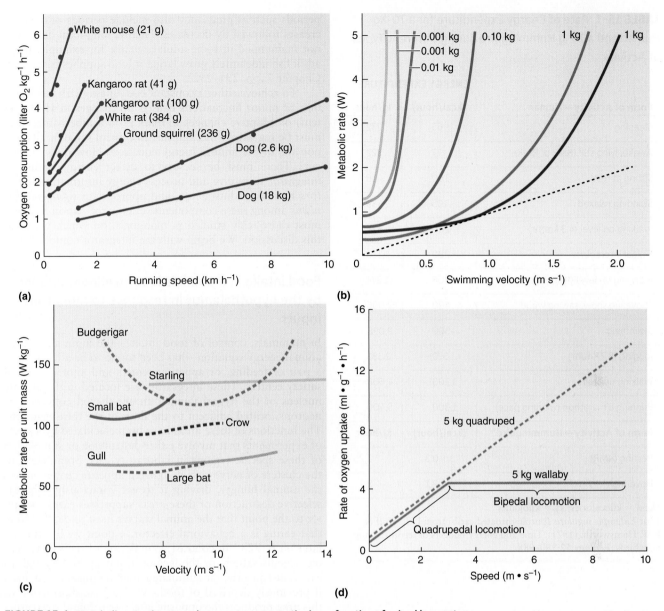

FIGURE 15-4 **Metabolic rates (measured as oxygen consumption) as a function of animal locomotory speed.** (a) Metabolic rates (per body mass) for running terrestrial mammals. Rate increases linearly with speed, with costs being higher in smaller mammals. (b) Metabolic rate (per whole animal) for a swimming fish. Rate tends to increase exponentially with speed. (c) Metabolic rate (per body mass) for flying birds and bats. A decrease in rate is seen at intermediate speeds due to gliding in some species. (d) Metabolic rate (per body mass) for a wallaby. When the animal uses all four limbs (quadrapedal locomotion), the rate increases linearly, as in part (a). But after the animal switches to bipedal hopping at higher speeds, rate becomes independent of speed.

Sources: (a) From P. Willmer, G. Stone, & I. Johnston. (2000). *Environmental Physiology of Animals*. Oxford, UK: Blackwell Science, as modified from C. R. Taylor, K. Schmidt-Nielsen, & J. L. Raab. (1970). Scaling of the energetic cost of running to body size in mammals. *American Journal of Physiology* 219:1104–1107. (b) From P. Willmer, G. Stone, & I. Johnston. (2000). *Environmental Physiology of Animals*. Oxford, UK: Blackwell Science, as modified from R. H. Peters. (1983). *The Ecological Implications of Body Size*. Cambridge UK: Cambridge University Press. (c) P. Willmer, G. Stone, & I. Johnston. (2000). *Environmental Physiology of Animals*. Oxford, UK: Blackwell Science, as modified from R. M. Alexander. (1999). *Energy for Animal Life*. Oxford, UK: Oxford University Press. (d) Modified from R. V. Baudinette, G. K. Snyder, & P. B. Frappell. (1992). Energetic cost of locomotion in the tammar wallaby. *American Journal of Physiology* 262:R771–R778.

check your understanding 15.2

Define *metabolic rate, standard* and *basal metabolic rate (SMR/BMR)*, and *dietary-induced thermogenesis (DIT)*. Explain how these are measured by indirect calorimetry.

Discuss scaling features and hypotheses regarding BMR/SMR.

Describe the three states of energy balance.

Discuss aerobic scope and the costs of different modes of locomotion.

TABLE 15-1 Rate of Energy Expenditure for a 70-kg Human and a 10-g Hummingbird during Different Types of Activity

Form of Activity—Human	ENERGY EXPENDITURE*	
	(kcal/hour)	kJ/hour
Sleeping	65	272
Awake, lying still (BMR or RMR)	77	322
Sitting at rest	100	418
Standing relaxed	105	439
Walking on level (4.3 km/hr)	200	836
Sexual intercourse	280	1,170
Bicycling on level (9 km/hr)	304	1,271
Shoveling snow, sawing wood	480	2,006
Swimming	500	2,090
Jogging (8.7 km/hr)	570	2,383
Walking up stairs	1,100	4,600
Running a marathon (winning pace)	1,300	5,400
Form of Activity—Hummingbird[†]	(kcal/hour)	kJ/hour
Resting (RMR)	0.3	1.25
Hovering	2.1	8.7

*kcal = kilocalories; kJ = kilojoules
†For *Eulampis jugularis* (purple-throated carib) from L. L. Wolf & F. W. Hainsworth. (1971). Time and energy budgets of territorial hummingbirds. *Ecology* 52:980–988.

15.3 Energy Balance: Regulation

Some adult animals maintain long-term neutral energy balance, while others undergo regulated periods of positive or negative energy balance

Physiologists have observed that many adult mammals, including humans, maintain a fairly constant mass (weight) over long periods of time. This implies that homeostatic mechanisms (with a set point) exist to maintain a long-term balance between energy intake and energy expenditure. Moreover, individuals may have different set points for body masses (you may have noticed that some humans stay lean no matter how much they eat, whereas other, overweight humans have great difficulty losing weight).

Conversely, periodic changes in body mass are adaptive in some animals. For example, hibernators (such as a marmot, Figure 15-5a) typically increase their body mass greatly during active (feeding) seasons and decrease during hibernation as they live off of their internal energy stores. Migratory birds may similarly gain and lose mass during premigration feeding and the migration itself, respectively. Reproductive periods such as pregnancy also include periods of mass increases followed by decreases. Finally, long-term balance is not maintained in some adult animals; for example, many adult fish indefinitely grow larger if food supplies permit (see Chapter 7, pp. 291–292).

To achieve either a constant body mass (with the exception of minor fluctuations caused by changes in H_2O content) or adaptive changes in mass as in a hibernator, there must be regulatory mechanisms to control some or all components of the animal energy equation. How does this occur? There must be *sensors* to detect energy status, an *integrator* to receive and process energy information, *effectors* to make adjustments, and appropriate *signal mechanisms* among these components. Energy regulation has been most thoroughly studied in mammals, on which we focus this discussion. We begin with the integrator component.

Food intake in mammals is controlled primarily by the hypothalamus in response to numerous inputs

In mammals, control of food intake—the input side of the animal energy equation—has been found to be a function of a pair of **feeding,** or **appetite, centers** and another pair of **satiety centers.** These integrators are located in the **arcuate nucleus** of the hypothalamus, an arc-shaped collection of neurons located adjacent to the floor of the third ventricle. The functions of these areas have been elucidated by a series of experiments that involve either destruction or stimulation of these specific regions in laboratory rats. Stimulation of the clusters of nerve cells designated appetite centers makes the animal hungry, driving it to eat voraciously, whereas selective destruction of these areas suppresses eating behavior to the point that the animal starves itself to death. (Note that eating is a behavioral effector, a point we'll return to later on p. 728.) In contrast, stimulation of the satiety centers signals satiety, or the feeling of having had enough to eat. Consequently, the stimulated animal refuses to eat, even if previously deprived of food. As expected, destruction of this area produces the opposite effect—profound overeating and obesity—because the animal never achieves a feeling of being full (Figure 15-5b). Thus, the feeding centers tell animals when to eat, whereas the satiety centers tell them when they have had enough.

Although it is convenient to consider these specific areas as direct regulators of feeding behavior, this approach is too simplistic. Physiologists now know that complex systems and numerous signals provide information about the body's energy status. Recent studies have uncovered multiple, highly integrated, redundant pathways, crisscrossing into and out of the hypothalamus, that are involved in controlling food intake and maintaining energy balance. Integration of multiple molecular signals ensures that effector feeding behavior is synchronized with the body's immediate and long-term energy needs. Let's examine these, beginning with a closer look at the hypothalamic integrator.

Arcuate Nucleus: The Integrator The arcuate nucleus has two subsets of neurons that function in an opposing manner. One subset releases *neuropeptide Y,* and the other releases *melanocyte-stimulating hormones (MSHs)* derived from *pro-opiomelanocortin (POMC),* a precursor molecule

(a)

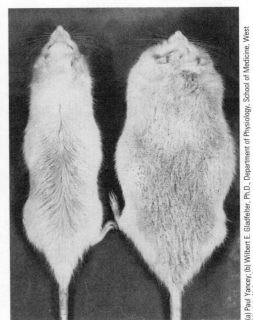

(b)

FIGURE 15-5 **"Fat" mammals.** (a) The hoary marmot (*Marmota caligata*) of western North America. This heterotherm must build up large deposits of fat before winter, when it hibernates. (b) Comparison of a normal rat with a rat whose satiety center (in the hypothalamus) has been destroyed. Several months after destruction of the classical satiety center in the ventromedial area of the hypothalamus, the rat on the right had gained considerable mass as a result of overeating compared to its normal littermate on the left. Rats sustaining lesions in this area also display less grooming behavior, accounting for the soiled appearance of the fat rat.

that can be cleaved in different ways to produce several different hormones (see p. 271). **Neuropeptide Y (NPY),** one of the most potent appetite stimulators ever found, leads to increased food intake, thus promoting mass gain. MSHs, a group of hormones known to be important in regulating skin color in some species (see p. 284), have been shown to exert an unexpected role in energy homeostasis in mammals. Most notably **α melanocyte-stimulating hormone (α-MSH)** suppresses appetite, thus leading to reduced food intake and mass loss.

Beyond the Arcuate Nucleus: Orexins and Others NPY and α-MSH are not the final signals in appetite control. These arcuate-nucleus chemical messengers, in turn, influence the release of neuropeptides in other parts of the brain that exert more direct control over food intake. Scientists are currently trying to unravel the other factors and regions that act upstream and downstream from arcuate nucleus to regulate appetite. Two such regions in the hypothalamus have been found to be richly supplied by axons from the NPY- and MSH-secreting neurons of the arcuate nucleus. These areas are the **lateral hypothalamic area (LHA)** and **paraventricular nucleus (PVN).** In a recently proposed model, the LHA and PVN release chemical messengers in response to input from the arcuate nucleus neurons. These messengers act downstream from the NPY and α-MSH signals to regulate appetite. The LHA produces two closely related neuropeptides known as **orexins** A and B, which are potent stimulators of food intake (*orexis* means "appetite"). NPY stimu-

lates and α-MSH inhibit the release of orexins, thus regulating appetite and food intake. However, orexins (which are produced in a circadian rhythm) also increase wakefulness and physical activity, so are not committed to mass gain. By contrast, the PVN releases neuromodulators, for example, **corticotropin-releasing hormone (CRH),** that decrease appetite and food intake. (As its name implies, CRH is better known for its role as a hormone; see p. 289.)

The release of all of these brain regulators requires sensory input that "informs" the hypothalamus of the body's energy status for both long-term maintenance of energy balance and the short-term control of food intake. Let's next look at the key input signals based on current research (Figure 15-6).

Leptin and insulin are signals of long-term energy balance

Scientists' notion of fat cells (**adipocytes**) in adipose tissue as merely storage space for triglyceride fat underwent a dramatic change late in the last century with the discovery of their active role in energy homeostasis. Adipocytes secrete several hormones, collectively termed **adipokines,** that play important roles in energy balance and metabolism. Thus, adipose tissue is now considered an endocrine gland. For example, **adiponectin** increases sensitivity to insulin (which helps protect against Type II diabetes mellitus), decreases body mass, and exerts anti-inflammatory actions. Unfortunately, obesity suppresses adiponectin secretion. Here we

Molecular Biology and Genomics
Discovering the Obesity Gene

The discovery and characterization of the leptin gene in 1994 by Jeffrey Friedman illustrates the importance of genetics to physiology. Physiologists had known for decades that obesity has a genetic component because different animal individuals on the same diets could have very different—and stable—body masses, with an apparent set point that could vary widely among individuals. Furthermore, the propensity for obesity could be passed from one generation to the next. Manipulations of the hypothalamus showed appetite was controlled in that brain region. Researchers widely suspected that mammalian bodies must have feedback signals that would tell the hypothalamus about the energy status of the body.

But physiologists had no luck finding such a signal. Enter the geneticists in the 1980s. By then, several strains of obese mice were established, including the *ob* strain, weighing about three times more than other mice. After eight years of genetic breeding and cloning experiments, Friedman and his colleagues showed that obesity was due to the functional absence of a single gene, dubbed the *ob* gene. They isolated and cloned the normal *ob* gene (in non-*ob* mice), and subsequent work showed that it codes for the protein hormone leptin, which has turned out to be one of the key energy signals in mammals. Without modern genetic methods, this might never have been discovered.

Using genome analysis and antibody detection, researchers have found leptin in most other vertebrates (reptiles including birds, and fishes). Leptin in broiler chickens is produced not only by adipose but also by liver tissue. The discovery of chicken leptin is being used to investigate methods for obesity control in these animals! Broiler chickens, as a source of human food, have been bred for maximal growth, but this has produced obese parent birds with low reproductive rates. For this reason, these birds are placed on restricted diets from an early age: Even so, both egg production and fertility dramatically decline with age. Researchers hope to manipulate leptin concentrations to reduce adipose stores while enhancing lean meat production.

Leptin-related genes and proteins have been studied in a reptile (the fence lizard *Scoleporus undulatus*) and a fish (the green sunfish *Lepomis cyanellus*). As in mammals, fasting sunfish have lower levels of plasma leptin than fed sunfish. Injecting mouse leptin into these animals induced effects similar to that seen in mammals: The injected lizard raised its body temperature slightly and its RMR by 250%, and showed a reduced appetite, whereas the injected sunfish began to metabolize its adipose stores. Thus, leptin is an ancient vertebrate signal of energy balance.

---

will focus on **leptin** (*leptin*, "thin"), an adipocyte hormone essential for normal body-mass regulation in mammals. As a sensory process, the amount of leptin in the blood informs the brain of the total amount of triglyceride fat stored in adipose tissue: That is, the larger the fat stores, the more leptin released into the blood. This bloodborne signal, discovered in the mid-1990s by genetic analysis of obese laboratory mice, was the first molecular satiety signal found by genetics (see *Molecular Biology and Genomics: Discovering the Obesity Gene*).

The arcuate nucleus is the major site for leptin action. Acting in negative-feedback fashion, increased leptin from burgeoning fat stores serves as a "trim-down" signal. Leptin suppresses appetite, thus decreasing food consumption and promoting mass loss, by inhibiting hypothalamic output of appetite-stimulating NPY and stimulating output of appetite-suppressing α-MSH. Conversely, a decrease in fat stores and the resultant decline in leptin secretion bring about an increase in appetite, leading to mass gain. The leptin signal is generally considered the dominant factor responsible for the long-term matching of food intake to energy expenditure so that total body energy content remains balanced and body mass remains constant.

Interestingly, leptin has recently been shown to also be important in reproduction, for example, as one of the triggers for the onset of puberty through its effects on *kisspeptin* neurons (see p. 798). The importance of its role in energy balance, and how that affects survival, was shown in recent experiments on Siberian hamsters. Susannah French and colleagues used a pump to infuse leptin into pregnant hamsters to keep the hormone artificially high. These mothers had significantly more offspring than mothers with a placebo treatment. Apparently, the treatment tricked the hypothalamus into "thinking" that energy was in excess, so the mother's body diverted more energy to the growing embryos. But, following the principles of energy balance, this came at a cost: The treated mothers had weakened immune responses and were more prone to bacterial infections.

Leptin may also be involved in the annual cycle of hibernating mammals such as Arctic ground squirrels, animals that do not maintain constant mass year-round because they must store up large adipose deposits in the productive seasons and survive on them in the winter. Some studies suggest that the brain ignores leptin signals in the autumn, allowing the animals to gain mass, whereas during the winter hibernation, the high leptin output of the stored fat effectively suppresses appetite.

Another bloodborne signal besides leptin that plays an important role in long-term control of body mass is **insulin,** a hormone secreted by the pancreas in response to a rise in the concentration of glucose and other nutrients in the blood following a meal (p. 316). It stimulates cellular uptake, use, and storage of these nutrients. Thus, the increase in insulin secretion that accompanies nutrient abundance, use, and storage appropriately inhibits the NPY-secreting cells of the arcuate nucleus, thus suppressing further food intake.

Gastrointestinal Hormones: Signals of Short-Term Energy Balance In addition to insulin's and leptin's importance in the long-term regulation of body energy and mass, other factors are believed to play a role in controlling the timing and size of meals. Early proposals suggested that cues of

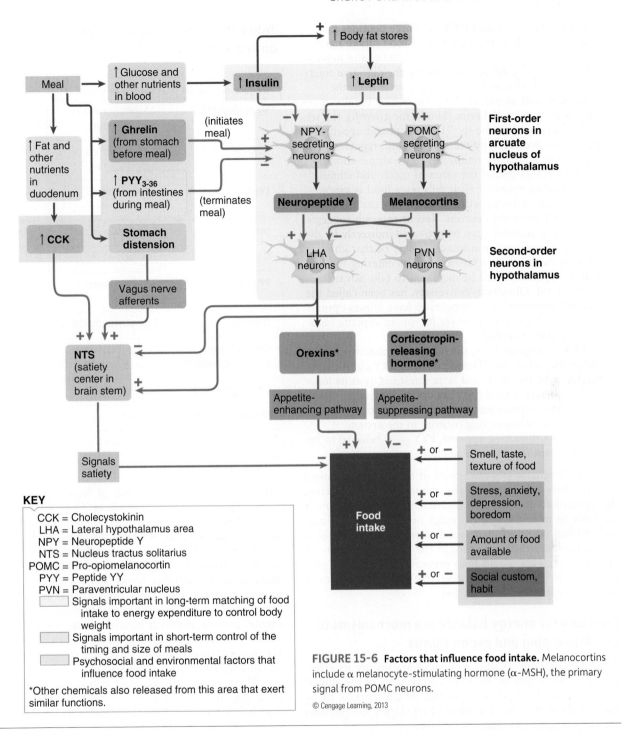

FIGURE 15-6 Factors that influence food intake. Melanocortins include α melanocyte-stimulating hormone (α-MSH), the primary signal from POMC neurons.

© Cengage Learning, 2013

emptiness or fullness of the digestive tract signaled hunger or satiety, respectively. For example, stimulation of gastric stretch receptors has been shown to suppress food intake. However, neural input arising from stomach distension plays a more important role in controlling the rate of gastric emptying than in signaling satiety (see p. 672). Recent studies have found that the digestive tract has food-sensing receptors similar or identical to those in the tongue's taste buds. There is now good evidence that the following internal bloodborne signals reflecting the depletion or availability of energy-producing substances, sensed by the gut, are important in the initiation and cessation of eating.

- **Cholecystokinin (CCK):** One of the gastrointestinal hormones released from the duodenal mucosa during digestion of a meal, CCK is an important satiety signal for regulating the size of meals. CCK is secreted in response to the presence of nutrients in the small intestine. Through multiple effects on the digestive system, CCK facilitates the digestion and absorption of these nutrients (see p. 684). It is appropriate that this bloodborne signal, whose rate of secretion is correlated with the amount of nutrients ingested, also contributes to the sense of being filled after a meal has been consumed but before it has actually been digested and absorbed.

- **Ghrelin, Obestatin, and PYY$_{3-36}$:** Three peptides involved in the short-term control of food intake have recently been identified: *ghrelin, obestatin,* and *peptide YY$_{3-36}$ (PYY$_{3-36}$).* All are secreted by the digestive tract; moreover, ghrelin and obestatin are arise from the same gene and its encoded protein, which breaks in half to form them. **Ghrelin** (Hindu for growth), the so-called hunger hormone (p. 712), is a potent appetite stimulator produced by the stomach and regulated by the feeding status. Secretion of this mealtime stimulator rises before and at the onset of meals and stimulates appetite behavior, then falls once food is eaten. Ghrelin stimulates appetite by activating the hypothalamic NPY-secreting neurons. Recent research suggests that ghrelin secretion can be activated indirectly by dietary fats entering the stomach. It may be that ghrelin informs the brain that fat-rich food is entering the gut and so appetite should be increased to take advantage of that food. **Obestatin,** conversely, has been called the "satiety hormone" as it appears to have effects opposite those of ghrelin. However its role in appetite regulation is still unsettled.

PYY$_{3-36}$ appears to be the main antagonist to ghrelin rather than obestatin. The secretion of PYY$_{3-36}$, which is produced by the small and large intestines, is at its lowest level before a meal but rises during meals and signals satiety. This peptide acts by inhibiting the appetite-stimulating NPY-secreting neurons in the arcuate nucleus. By thwarting appetite, PYY$_{3-36}$ is believed to be an important mealtime terminator.

Table 15-2 summarizes the effects of these gastrointestinal signals. These could explain why animals stop eating before the ingested food is actually digested, absorbed, and made available to meet the body's energy needs. An animal may feel satisfied when adequate food to replenish the stores is in the digestive tract, even though the body's energy stores are still low. This would help prevent the overshoot phenomenon (in this case, overeating) that characterizes simple negative-feedback systems (p. 16).

Effectors for energy balance are mechanisms to regulate eating and expenditures

The final components of a regulatory system are its effectors. Potentially any or all of the components of the animal energy equation could be regulated. Much remains to be learned about this, but a number of regulatory mechanisms are known. Let's examine some of these.

Regulation of Food Intake (Eating) On the input side of the equations, you have already seen how eating, the behavioral effector controlled in part by the hypothalamus, is affected by sensory signals such as leptin and ghrelin. Over a 24-hour period the energy in ingested food rarely matches energy expenditure for that day. The correlation between total caloric intake and total nonproduction energy output is strong, however, over long periods of time for animals with long-term neutral energy balance.

However, that balance can be disrupted by psychosocial and other environmental influences. For example, the amount of pleasure derived from eating can reinforce feeding behavior. This has been demonstrated in an experiment in which

TABLE 15-2 Effect of Involuntary Regulatory Signals on Appetite

Regulatory Signal	Source of Signal	Effect of Signal on Appetite
Neuropeptide Y	Arcuate nucleus of hypothalamus	Increases
α-MSH	Arcuate nucleus of hypothalamus	Decreases
Leptin	Adipose tissue	Decreases
Insulin	Endocrine pancreas	Decreases
Ghrelin	Stomach	Increases
PPY$_{3-36}$	Small and large intestines	Decreases
Orexins	Lateral hypothalamus	Increases*
Corticotropin-Releasing Hormone	Paraventricular nucleus of hypothalamus	Decreases
Cholecystokinin	Small intestine	Decreases
Stomach Distension	Stomach	Decreases

*Also increases in energy expenditures and wakefulness.

© Cengage Learning, 2013

rats were offered their choice of highly palatable human foods. They overate by as much as 70 to 80% and became obese. When the rats returned to eating their regular monotonous but nutritionally balanced rat chow, their obesity was rapidly reversed as their food intake was controlled once again by physiological drives rather than by urges for the apparently tastier offerings. Stress, anxiety, depression, and apparent boredom have also been shown to alter feeding behavior in ways that are unrelated to energy needs in experimental animals as well as humans. Some types of stress, particularly involving *cortisol* (p. 286), lead to overeating (as if the hypothalamus "thinks" it needs to prepare for future starvation), while other types lead to undereating.

There is no question that regulation of eating is the most important factor in the long-term maintenance of energy balance and body mass. However, in laboratory studies on different strains of rats, genetically lean mice eat as much or more than genetically obese mice while remaining much leaner. Thus, regulation of expenditures (output) must also be important.

Compensatory Changes in Expenditures in Response to Underfeeding Some studies suggest that after several weeks of eating less than normal (e.g., as a result of starvation, or deliberate "dieting" in a human), a mammal may initiate small counteracting decreases in BMR, possibly by reductions in thyroid hormones (p. 301), and/or decreases in muscular activity by uncertain mechanisms. This energy-saving effect partially explains why some humans or laboratory animals on a diet become stuck at a plateau after having shed the first grams easily.

Chronic underfeeding can also lead to suppression or even reversal of production output. For example, starvation (as well as some other stresses) in mammals can lead to lower gonadotropin output and thus reduced spermatogenesis and oogenesis. This may occur in part because declining leptin levels in starvation cause a decrease in the release of *kisspeptin*, a regulator of reproductive hormones (p. 798). Many vertebrates such as rodents and ovoviviparous fishes actually reabsorb their internal embryos or fetuses. Some vertebrates will even eat developing eggs (oviparous animal) or newborns. Nonvertebrates also exhibit similar energy-saving reproductive adjustments. Starving aphids, for example, reabsorb their developing oocytes, while starving great pond snails (*Lymnaea stagnalis*) stop producing egg-laying hormone due to reduced neurosecretory output from the brain.

Compensatory Increases in Expenditures in Response to Overfeeding Conversely, physiological processes that "burn off" excess calories account in part for the inability of some mammals (including some humans) to gain much mass despite excessive food intake. These involve regulatory DIT and muscular activity.

1. *Increased regulatory DIT: Uncoupling proteins and brown adipose tissue.* Humans and laboratory mammals have been found to release "unwanted" calories by activating **uncoupling proteins (UCP)** in some tissues. UCPs are channels for protons (H^+) that let the H^+ gradients generated from food energy in mitochondria dissipate, with no ATP synthesis. A major site of UCPs is **brown adipose tissue (BAT)**, which probably evolved as a thermoregulatory organ in small and newborn mammals (as we will discuss later; p. 746). However, BAT can also be activated after a large meal (as regulatory DIT). BAT does not store fats but rather converts available energy in lipids into heat using **UCP-1,** and is now recognized as a major contributor to energy balance. Larger mammals were thought to lose their BAT as they aged, but new research shows that some lean humans have some BAT, while obese humans have little or none. White adipose tissue and skeletal muscle, containing UCP-2 and UCP-3, respectively, appear to be similarly used, and again, lean humans have higher densities of both UCPs than do obese individuals.

   BAT is activated by the sympathetic nervous system. Interestingly, the ghrelin system may suppress BAT. Yuxiang Sun and colleagues created two groups of knock-out mice, one lacking ghrelin and the other missing the ghrelin receptor. The latter group, in contrast to the ghrelin knock-out or control animals, did not gain weight on a high-fat diet. The researchers found that BAT, lacking the ghrelin receptor, increased its levels of UCP-1 and thus its ability to "burn off" excess fat. Thus, the normal ghrelin receptor may favor fat storage.

2. *Increased exercise and non-exercise activity thermogenesis.* Muscular activity levels may also be regulated for disposal of excess energy. In humans, involuntary movements called "fidgeting," like foot tapping, have been found to be more frequent in lean than in heavy individual humans. Lean rodents on a high-fat diet also exhibit purposeless repetitive movements. Such behavior is called **non-exercise activity thermogenesis (NEAT)**

and is regulated by the hypothalamus. Laboratory rodents injected with leptin into blood, or orexin-A into the PVN, show a spontaneous increase in muscular (including NEAT) activity, and either lose more mass or gain less than control animals. In humans, voluntary activities such as exercise are of course well known to contribute to energy balance.

Different energy-balance set points among species result from evolutionary adaptations to different food supplies

Clearly, even though many mammals maintain a relatively constant body mass, species and individuals vary greatly in the mass they actually maintain (as we noted earlier). This has led to the proposal that the hypothalamic centers determining long-term satiety can evolve different *set points* for energy balance, possibly for adipose storage in particular. The two extreme set points are often termed "lean" and "obese" (or "heavy"). Although humans in modern Western societies tend to attach positive and negative connotations to lean and obese states, respectively, physiologists view these in a different light. Both extremes (and set points in between) can be evolutionarily advantageous depending on the environment:

- Lean animals do better in habitats where food is plentiful year-round so that storing a high amount of body fat is unnecessary. The advantage here is that such animals are more agile than heavy ones, better able to avoid predators and (in some species) to obtain food in the first place. Thus, carnivores in the tropics (such as tigers) are lean. (Although the leanness may be attributed to high activity levels, many carnivores in fact are sedentary for much of the time.) The trade-off between storage and agility was revealed in zebra finches by researchers in Scotland, who videotaped and weighed numerous birds on a daily basis. Some finches had access to excessive calories from human sources. A mere 7% gain in body mass led to a 33% loss in speed and maneuverability, making these overfed birds easier prey for cats and sparrow hawks. Thus, an obese or heavy set point has negative consequences in a food-plentiful habitat, including increased cardiovascular and oxidative stresses causing internal health problems, as well as reduced maneuverability. Household pets, as well as humans themselves, often suffer from the consequences of obesity.

- Conversely, a high adipose set point has very clear advantages in "feast or famine" habitats, where food is only abundant at limited times of the year. A clear example is the mammalian hibernator (see Figure 15-5a). An Arctic ground squirrel must reach an "obese" state (up to 50% body fat) before winter sets in or it will not survive the eight months during which it hibernates in its burrow. Desert habitats that have abundant food only during short rainy seasons also select for this adaptation. Overall, it is now recognized that having a propensity toward being "overweight" is not necessarily an abnormal physiological state but may be an evolutionary adaptation for a particular environment.

Researchers do not yet fully understand what determines these different set points in the neurological structure of the brain. But, as discussed earlier in this chapter, fat storage levels and the associated set point may be closely tied to leptin and how the hypothalamus reacts to it. The hypothalamus of a lean-adapted mammal reacts strongly to a modest rise in leptin, shutting off appetite and acting as if it had a low set point. The hypothalamus of "obese" mammals, conversely, reacts only to very high levels of leptin. Also, humans and laboratory animals appear to differ in the thermogenic activity of tissues with uncoupling proteins, correlating with lean or obese set points. Research is continuing on these numerous fronts.

check your understanding 15.3

Describe the role of the hypothalamus in energy balance regulation.

Describe the source and role of the following in regulation of energy balance: leptin, insulin, ghrelin, PYY_{3-36}, and cholecystokinin (CCK).

Describe the effectors that regulate energy input and energy expenditures.

Discuss evolutionary differences in body-mass "set points."

15.4 Thermal Physiology: General Principles

We now turn to a major habitat factor intimately involved with energy. When physiological ecologists examine the distribution of species around the planet, noting that most species live in rather restricted ranges, they ask what factors are responsible for these restrictions. One of the most pervasive factors is temperature, which has direct and profound effects on biological functions.

Temperature alters rates of chemical reactions and denatures macromolecules

Temperature is a measure of heat energy, which in turn is manifested primarily in kinetic energy (movement) of molecules. The primary effects of this are shown in Figure 15-7, which shows a typical optimal curve of biological function and temperature. Basically, two different temperature effects create this curve:

1. *Kinetic energy of reactants:* On the left (up-trending) side of the plot, reaction rates increase with temperature simply from the increased kinetic energy of molecules. Reacting molecules (such as substrates and enzymes, for example) encounter each other more frequently. It has become customary to characterize this temperature response with a value called the Q_{10}, which is the ratio of a reaction rate (or metabolic process) at one temperature to the rate at 10°C cooler:

$$Q_{10} = \frac{\text{rate at temperature } T}{\text{rate at temperature } T_{-10}}$$

Q_{10} values are useful measures of the temperature sensitivity of a biological process (from single reactions to a

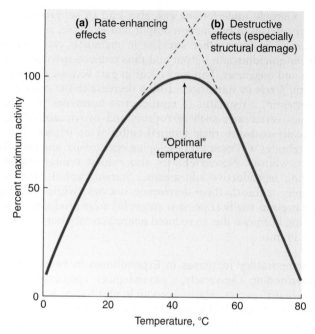

FIGURE 15-7 **The relative activity of an enzymatic reaction as a function of temperature.** The decrease in activity below 30°C is primarily due to reduced kinetic energy, while the decrease in activity above 50°C is due to disturbed protein structures, including denaturation.

Source: Modified from M. K. Campbell & S. Farrell. (2003). *Biochemistry*, 4th ed. Belmont, CA: Thomson, p. 137, Figure 5.2.

whole organism's metabolism). In general, researchers have found that many reactions tend to double or triple for each 10°C rise, at least over a moderate range. However, at some point with rising temperature, the curve begins to flatten, reaches an optimum, and then declines.

2. *Denaturation of macromolecules.* Above the optimum (right side of the plot), rates of biological processes begin to decrease with temperature for one basic reason: The macromolecules (such as enzymes) responsible for the process begin to denature (that is, lose their tertiary levels of structure). As we discussed in Chapter 2 (p. 27), weak bonds stabilize many of these structures so that they can be flexible, but this makes them susceptible to loss of function at higher kinetic energies.

In general, this curve means that organisms cannot operate over a wide range of body temperatures. For animals, active metabolism seems to be restricted to a range from about –2°C (the freezing point of seawater) to 55 to 60°C. (A penguin living actively at –50°C air temperature in Antarctica does not violate this restriction, because its cells are warmed to about 40°C from internal heat production and retention.) This is a relatively narrow range, given that a temperature of –89.2°C (–128.5°F) has been recorded (in Antarctica) and hydrothermal vent temperatures may reach about 400°C (750°F)! Some animals have evolved ways to cope with body temperatures colder than freezing (as you will see), but metabolism in these animals is generally in a dormant state. At the other extreme, there is no convincing evidence of any animal tolerating over 55°C for anything but very brief periods (on some desert sands and at some

FIGURE 15-8 Schematic diagrams of membrane structures and their relationship to temperature. (a) Saturation level of membrane phospholipids changes with temperature adaptation. The *left* figure represents a warm-adapted membrane: The fatty-acid "tails" are highly saturated, allowing for close stacking of neighboring tails. This enhances thermostability at higher temperatures. The *right* figure represents a cold-adapted membrane: The fatty-acid "tails" are more unsaturated, creating double bonds that interfere with stacking of neighboring tails. This enhances flexibility at colder temperatures. (b) Stiffening of the membrane with cholesterol. Cholesterol reduces fluidity by stabilizing extended forms of the phospholipid tails. (c) The relationship between adaptation temperature and ratio of unsaturated phospholipid tails (phosphatidylethanolamine, PE) to saturated ones (phosphatidylcholine, PC) in gill membranes of rainbow trout transferred to different temperatures as shown.

Sources: (a) © Cengage Learning, 2013 (b) Modified from M. K. Campbell & S. O. Farrell. (2003). *Biochemistry*, 4th ed. Thompson, p. 202, Figure 7.14. (c) Hochachka, P. W. and G. N. Somero, (2002), Biochemical Adaptation, Oxford: Oxford University Press, as modified from J. A. Logue, A. L. DeVries, E. Fodor, & A. R. Cossins, (2000), Lipid compositional correlates of temperature-adaptive interspecific differences in membrane physical structure, 'Journal of Experimental Biology', 203:2105-2115. Reproduced with permission via Copyright Clearance Center.

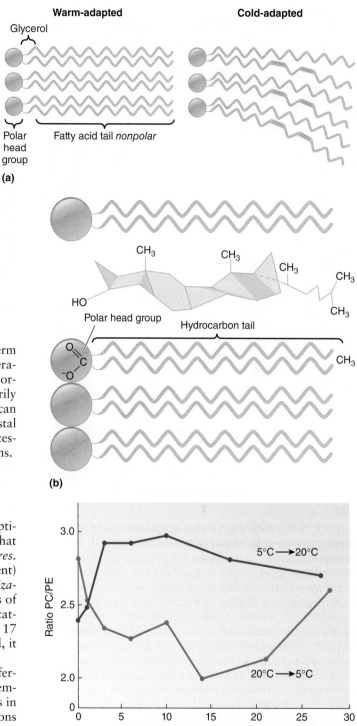

(a) (b) (c)

marine hydrothermal vents; see p. 741). Ignoring long-term acclimatization for the moment, excessively high temperatures are usually more dangerous than cold ones. Many organisms can survive lower body temperatures temporarily because reactions merely slow down (although freezing can set a lower limit because of the lysis of cells from ice crystal formation in organisms that cannot prevent that), but excessive heat can irreversibly damage membranes and proteins.

Biomolecules can be altered to work optimally at different temperatures

For the most part, biological systems are subject to the optimal curve of Figure 15-7, but it is important to realize that *the optimum itself can change with different temperatures.* Changes can occur over two possible (and very different) time courses: within an organism's lifetime in *acclimatization* processes, and over many generations by the process of mutation and natural selection, that is, *evolution.* Both categories are sometimes loosely called "adaptation" (see p. 17 for more details on terminology). If "adaptation" is used, it is crucial to carefully explain the time course.

Not all organisms can successfully acclimatize to different temperatures, nor can all species evolve to survive temperatures outside of their normal ranges. Indeed, changes in habitat temperatures may be a major cause of extinctions over geological time; moreover, the current rise in global air and water temperatures is already affecting organismal distributions and physiologies (as we will discuss later). Regardless, optima clearly differ among species and within some species experiencing different thermal habitats, in part because of changes in two critical classes of biomolecules— membranes and proteins.

Membranes and the Homeoviscous Adaptation Biological membranes are held weakly together by hydrophobic interactions between the fatty acid chains of its phospholipids (p. 71), giving them a certain *viscosity* or fluidity that is vital to functions such as transport and diffusion. However, the weak structure makes them very susceptible to temperature

changes. At low temperatures, viscosity increases and the membrane becomes too rigid for optimal function. At higher temperatures, membranes lose viscosity and become too fluid. If you have used butter from a refrigerator and also from a kitchen counter on a hot day, you are aware of these effects. As you saw in Chapter 3 (p. 72), a membrane can be changed to counter these effects by the length of the fatty acids and by the degree of *saturation* of its fatty acids (and, to some extent, by altering its cholesterol content). A hypothetical example is shown in Figures 15-8a and b, with a real

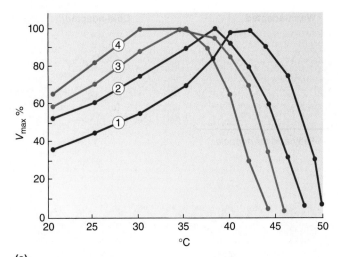

(a)

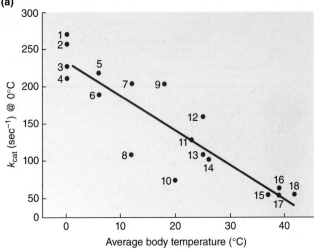

(b)

FIGURE 15-9 **Relationship between enzyme activity and temperature in animals adapted to different temperatures.** (a) The relative activity max (V_{max}) of myosin ATPase of four lizards, showing that animals adapted to higher temperatures have enzymes with higher thermal optima. 1. *Dipsosaurus dorsalis,* which has a preferred body temperature of 38.8°C; 2. *Uma notata,* preferred body temperature of 37.5°C; 3. *Scleroporus undulatus,* preferred body temperature of 36.3°C; 4. *Gerrhonotus multicarinatus,* preferred body temperature of 30.0°C. (b) The enzyme catalytic rate constant (k_{cat}) for *lactate dehydrogenases* (LDH, A_4 type) from various vertebrates adapted to different temperatures, showing that, in general, animals adapted to higher temperatures have enzymes with lower catalytic rates. Antarctic notothenioid fishes are 1. *Lepidonotothen nudifrons;* 2. *Parachaenichthys charcoti;* 3. *Champsocephalus gunnari;* 4. *Harpagifer antarcticus.* South American notothenioid fishes are 5. *Patagonotothen tessellata;* 6. *Eleginops maclovinus.* Temperate and subtropical fishes are 7. *Sebastes argentea* (rockfish); 8. *Hippoglossus stenolepis* (halibut); 9. *Sphyraena argentea* (barracuda); 10. *Squalus acanthias* (dogfish shark); 11. *Sphyraena lucasana* (barracuda); 12. *Gillichthys mirabilis* (goby); 13. *Thunnus thynnus* (bluefin tuna). A tropical fish is 14. *Sphyraena ensis* (barracuda). Terrestrial animals are 15. *Bos taurus* (cow); 16. *Gallus gallus* (chicken); 17. *Meleagris gallapavo* (turkey); and 18. *Dipsosaurus dorsalis* (desert iguana).

Sources: (a) Modified from P. Licht. (1967). Thermal adaptation in the enzymes of lizards in relation to preferred body temperature. In *Molecular Mechanisms of Temperature Adaptation,* ed. C. L. Prosser. Washington: American Association for the Advancement of Science. (b) P. W. Hochachka & G. N. Somero. (2002). *Biochemical Adaptation.* Oxford, UK: Oxford University Press; as modified from P. A. Fields & G. N. Somero. (1998). Hot spots in cold adaptation: Localized increases in conformational flexibility in lactate dehydrogenases A_4 orthologs of Antarctic notothenioid fishes. *Proceedings of the National Academy of Sciences USA* 95:11476–11481.

example in Figure 15-8c. First, consider a warm-adapted organism. Its membranes typically consist of *saturated fatty acids* (SFAs), in which the carbons in the "tails" are fully hydrogenated. This allows the tails to stack together tightly, giving the whole membrane enough viscosity to work properly at warm temperatures (Figure 15-8a, left). Cholesterol may be high as well, to further reduce fluidity (Figure 15-8b). However, these membranes become too rigid to work in the cold (as butter does in a refrigerator). Conversely, a cold-adapted organism typically has a higher proportion of *polyunsaturated fatty acids* (PUFAs), in which some carbons have fewer hydrogens and thus form double bonds with neighbor carbons. This puts "kinks" in the tails, making tight stacking impossible (Figure 15-8a, right). This makes the whole membrane less viscous and thus more fluid in the cold. But these membranes become too fluid to work in warm conditions. Thus, there is a trade-off between thermal stability and fluidity. Both membranes have approximately the same viscosity *when they are each measured at their adaptational temperatures.* Hence, the term **homeoviscous adaptation** was coined to refer to these saturation changes that preserve an optimal viscosity of the cell membrane.

Saturation levels are catalyzed by enzymes called **desaturases** (used in the cold to remove hydrogens) and **saturases** (used in warm conditions to add hydrogens). As noted, the time course of changes is critical when discussing this adaptation. In general, organisms from different thermal habitats have lipid compositions that follow the saturation principles just described. However, only some organisms can acclimatize (from winter to summer, for example) by regulating the saturases and desaturases and thus their membranes. In addition to the fish described in Chapter 3 (p. 72), an example is shown in Figure 15-8c.

Proteins and the "Homeoflexibility Adaptation" A similar type of change can occur with proteins, although generally only over evolutionary time. In this case, the trade-off is between *thermostability and protein flexibility* rather than fluidity. Recall from Chapter 2 (e.g., Figure 2-1) that many proteins need to be flexible to "do their job." As you saw earlier, increased temperature can cause these flexible structures to denature (Figure 15-7), but lower temperatures can make an enzyme too rigid to function well. As shown in Figure 15-9a for myosin (the "motor" protein of muscle; see Chapter 8), optimal temperatures for functionality do vary among species. How does this occur? A major clue was found in studies of *catalytic rate constants* (k_{cat}) for some enzymes. These values indicate how fast an enzyme can catalyze a particular reaction. The studies revealed an initially surprising result: Enzymes of animals with warm bodies, such as mammals, have relatively inefficient (low k_{cat}) enzymes, whereas cold-bodied animals have faster (high k_{cat}) enzymes. As with membranes, there appears to be a trade-off between the flexibility needed for fast catalysis and the thermostability needed to resist heat.

One particular enzyme, **lactate dehydrogenase** (LDH), has been analyzed in some detail. LDH is the final enzyme of anaerobic glycolysis, catalyzing the conversion of pyruvate and NADH to lactate and NAD^+ (p. 55), and its action is paramount to survival in anaerobic burst activity, such as pouncing on food or escaping a predator. As shown in Figure 15-9b, k_{cat} of LDHs is inversely related to an animal's average body temperature. In its tertiary structure LDH has

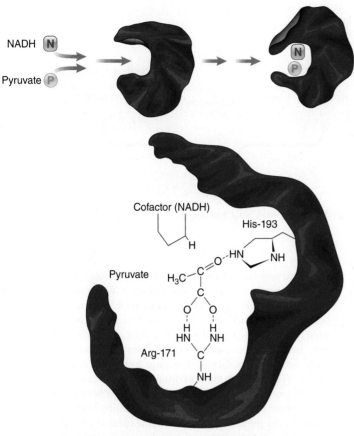

FIGURE 15-10 **Schematic diagram of the enzyme lactate dehydrogenase (LDH), which binds NADH and pyruvate and catalyzes their conversion into NAD$^+$ and lactate.** The *upper* diagram shows the conformational changes that occur upon binding, as a flexible loop of the enzyme closes over the active site. The flexibility of this loop evolves to match a species' thermal habitat (greater flexibility in the cold). The *lower* diagram shows the active site itself and the amino acids (Histidine-193 and Arginine-171) that bind pyruvate. The binding ability of the histidine may be protected from thermal disturbances by an animal decreasing its fluid pH with rising temperature (see main text, p. 738).

Source: P. W. Hochachka & G. N. Somero. (2002). *Biochemical Adaptation*. Oxford, UK: Oxford University Press; as modified from P. A. Fields & G. N. Somero. (1998). Hot spots in cold adaptation: Localized increases in conformational flexibility in lactate dehydrogenases A$_4$ orthologs of Antarctic nototheniold fishes. *Proceedings of the National Academy of Sciences USA* 95:11476–11481.

a flexible loop, which opens and closes during its reaction (Figure 15-10). Warm-adapted LDHs have relatively stiff loops, rigid enough to resist denaturation at higher temperatures. But they are too inflexible to work at cold temperatures. Conversely, cold-adapted LDHs have relatively loose loops, able to open and close well in the cold. But they lose their structure more easily at higher temperatures.

Such differences in thermostability may evolve fairly straightforwardly, as revealed in recent studies on malate dehydrogenases (MDH, an enzyme of the citric-acid cycle, p. 50) in intertidal limpets of the western United States. Yunwei Dong and George Somero compared MDHs from *Lottia digitalis* of northern habitats and its close relative *L. austrodigitalis* of southern habitats. The MDH from the southern *L. austrodigitalis* is considerably more thermosta-

ble and can bind substrate more effectively at higher temperatures than the MDH of *L. digitalis* (northern). The differences are the result just one amino acid substitution. In the active site, MDH of *L. austrodigitalis* has a serine where *L. digitalis* MDH has a glycine. Importantly, serine allows for additional hydrogen bonding, improving binding at higher temperatures. Knowing such differences in thermal adaptation are critical to understanding the effects of global warming on species' distribution and survival.

The thermal adaptation strategies of animals depend on their primary source of heat

Let's now turn to the ways in which animals cope with environmental temperature changes that threaten their biomolecules. Recall that physiologists generally classify animals into two broad categories: **ectotherms** (*ecto*, "external")—those dependent on external sources for body heat—and **endotherms** (*endo*, "internal")—those more dependent on internal sources. Physiologists still frequently use a pair of older terms: **poikilotherms** (*poikilos*, "changeable")—animals whose body temperatures vary with the environment—and **homeotherms**—animals with narrowly varying body temperatures (recall that *homeo* means "similar," not "constant"). These terms often lead to confusion. Because we humans are mammals (endotherms) with fairly constant body temperatures (homeotherms), some people mistakenly believe the two terms are synonymous. But they are not, as we show in detail later. For now, consider an ectothermic lizard that can use sun and shade to keep its body temperature fairly constant (that is, it is homeothermic some of the time). Also consider a hibernating endotherm such as a marmot (p. 725): It is not homeothermic when considered on an annual basis. Rather it is an example of a **heterotherm,** an animal that has endothermy but is not fully homeothermic.

Note that there is no clear dividing line between a heterotherm and a homeotherm. For example, a camel in the desert absorbs a large heat load during the day, increasing its body temperature by several degrees (for example, from as low as 34°C to as high as 41°C) to avoid losing water by evaporative cooling. At night, it releases the heat and its body temperature drops. Does this level of change constitute homeothermy, or heterothermy? Some physiologists suggest that homeothermy be defined as internal temperature variations of no more than plus or minus 2°C, but in some ways the definitions do not matter. Rather, it is important to realize that there is a continuum of temperature regulation ranges and abilities, and that no endotherm has an absolutely constant internal temperature (hence the prefix *homeo* for "similar" rather than *homo* for "same"). Even your body temperature drops when you sleep and increases with heavy exercise and fever.

The internal body temperatures of both ectotherms and endotherms depend simply on the difference between heat input and heat output (Figure 15-11). *Heat input* occurs by gain from the external environment (which dominates in ectotherms) and from internal heat production (the most important source for endotherms). Conversely, *heat output* occurs by way of heat loss from exposed body surfaces to the external environment (in both ectotherms and endotherms). Let's first examine the inputs and outputs involving the environment.

Heat exchange between the body and the environment takes place by radiation, conduction, convection, and evaporation

Endotherms as well as ectotherms are influenced by heat exchanges with their environments, so it is important to understand how such exchanges occur. All heat loss or heat gain between the body and the external environment must take place between the body surface and its surroundings. The same physical laws of nature that govern heat transfer between inanimate objects also control the transfer of heat between the body surface and the environment. The temperature of an object may be thought of as a measure of the concentration of heat within the object. Accordingly, heat always moves down its "concentration" gradient, that is, down a **thermal gradient** from a warmer to a cooler region.

Organisms are subject to, and make use of, four mechanisms of heat transfer: *radiation, conduction, convection,* and *evaporation.*

1. **Radiation** refers to light energy, which exists as electromagnetic waves (and photons) that travel through space (Figure 15-12, step 1). Objects can both emit (source of heat loss) and absorb (source of heat gain) radiant energy. Heat is emitted from all objects (if not at absolute zero, –273°C) in the form of **infrared radiation** (wavelengths longer than red light). In the reverse direction, when radiant energy (especially infrared, but also visible light) strikes an object and is absorbed, the energy of the wave motion is transformed into heat within the object (manifested as increased vibrational energy within molecules). Whether a body loses or gains heat by radiation depends on the difference in temperature between the skin surface and the surfaces of various other objects in the body's environment, and on the amount of direct sunlight striking the skin. Because net transfer of heat by radiation is always from warmer objects to cooler ones, a body gains heat by radiation from objects warmer than the skin surface, such as the sun, rocks, or soil in the sun, or burning wood. In contrast, an animal loses heat by radiation to objects in its environment whose surfaces are cooler than the surface of the skin, such as trees, soil in the shade, and so forth.

2. **Conduction** is the transfer of heat between objects of differing temperatures that are in direct contact with each other (Figure 15-12, step 2). Heat moves down its thermal gradient from the warmer to the cooler object by being transferred from molecule to molecule. Except at absolute zero, all molecules are constantly vibrating, with warmer molecules moving faster than cooler ones. When molecules of differing heat content touch each other, the faster-moving, warmer molecule agitates the cooler molecule into more rapid motion, "warming up" the cooler molecule. During this process, the original warmer molecule loses some of its thermal energy as it slows down and cools off a bit. Therefore, given enough time, the temperature of the two touching objects eventually equalizes. The rate of heat transfer by conduction depends on the *temperature difference* between the touching objects and the *thermal conductivity* of the substances involved (that is, how easily heat is conducted by the molecules of the substances).

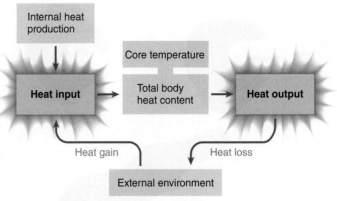

FIGURE 15-11 **Heat input and output.**

© Cengage Learning, 2013

Heat can be lost or gained by conduction when the skin is in contact with a good conductor. When standing on snow, for example, an animal's foot becomes cold because heat conducts from the foot to the snow. Conversely, when an animal sits on a rock warmed by the sun, heat is transferred directly from the rock to its body. Similarly, bodies lose or gain heat by conduction to the layer of air in direct contact with the body. Only a small percentage of total heat exchange between the skin and air takes place by conduction alone because air is not a very good conductor of heat. Heat is conducted much more rapidly between a body surface and water, a better conductor than air.

3. **Convection** refers to the transfer of heat energy by *currents* of air or water (the medium). Let's follow an example for air next to skin, air that is warming from body heat. Because warm air is lighter (less dense) than cool air, the warmed air rises, and cooler air moves in next to the skin to replace the vacating warm air. The process is then repeated (Figure 15-12, step 3). Such movements in the external medium, known as *convection currents,* can carry heat away from a body. If it were not for convection currents, no further heat could be dissipated from the skin by conduction once the temperature of the layer of the medium immediately around the body equilibrated with skin temperature. The combined conduction–convection process of removing heat from a body is enhanced by forced movement of the medium across the body surface, either by external movements, such as those caused by wind and water currents, or by movement of the body through the medium, such as during locomotion.

4. **Evaporation** is the final method of heat transfer used by a body, but only in air. When water evaporates from the skin surface, such as that of a frog in air, the heat required to transform water from a liquid to a gaseous state is absorbed from the skin, thereby cooling the animal (Figure 15-12, step 4). Evaporative heat loss can occur from the linings of the respiratory airways as well as from the skin (*cutaneous* loss). Evaporation is particularly important for regulation if the air temperature rises above the temperature of skin. In that situation, the temperature gradient reverses itself so that heat is gained from the environment. Evaporation is the only means of heat loss under these conditions, and the pro-

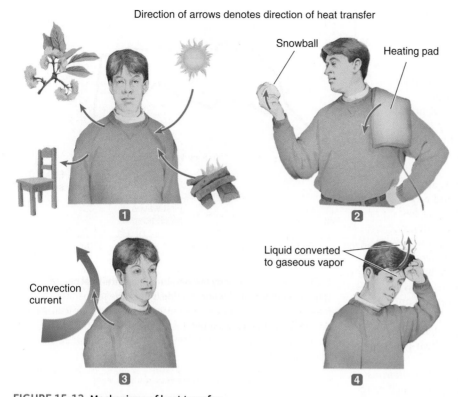

Direction of arrows denotes direction of heat transfer

Snowball

Heating pad

Convection current

Liquid converted to gaseous vapor

1 **Radiation**—the transfer of heat energy from a warmer object to a cooler object in the form of electromagnetic waves ("heat waves"), which travel through space.

2 **Conduction**—the transfer of heat from a warmer to a cooler object that is in direct contact with the warmer one. The heat is transferred through the movement of thermal energy from molecule to adjacent molecule.

3 **Convection**—the transfer of heat energy by air currents. Cool air warmed by the body through conduction rises and is replaced by more cool air. This process is enhanced by the forced movement of air across the body surface.

4 **Evaporation**—conversion of a liquid such as sweat into a gaseous vapor, a process that requires heat (the heat of vaporization), which is absorbed from the skin.

FIGURE 15-12 **Mechanisms of heat transfer.**

© Cengage Learning, 2013

cess can be understood in terms of the fundamental entropy equation given earlier. Recall that

$$\Delta S_{net} = \Delta S_{surroundings} + \Delta S_{system}$$
$$>0 \text{ for spontaneous reactions}$$

Here the system is the water droplet on the skin, and the surroundings are the body being cooled. When the air is hotter than the body, and yet the latter is being cooled, the $\Delta S_{surroundings}$ is thermodynamically unfavorable because entropy is actually decreasing during cooling against a heat gradient. However, the unfavorable factor is more than "paid for" by a **change of state** in water. During evaporation, water molecules go from a moderately ordered liquid state to a highly disordered vapor state, giving an enormously favorable ΔS_{system}. Essentially, the heat energy of the body is absorbed in the breaking of hydrogen bonds between water molecules as water vaporizes. This change of state, which is very effective, is exactly the process used in refrigerators and air conditioners, where liquid refrigerants vaporize in cooling coils, removing heat from the desired space. Energy in the form of a compressor motor must be used to reliquefy the refrigerant, recycling this costly resource, whereas in organisms, the cooling fluid—water—is lost and must be replaced.

Evaporation cannot succeed, however, if the air is saturated with water vapor, that is, at 100% relative humidity (RH). At that point, the rate of water vapor reforming hydrogen bonds and thus condensing into liquid is equal to the reverse process. *If the air is hotter than an animal's body at 100% RH, then there is no physiological means for cooling off.* Animals must allow

their body temperatures to rise or must use behavioral mechanisms to avoid the situation.

Evaporation often occurs in a maladaptive way, that is, not for thermoregulation. A frog in a warm, dry area is in danger of dying from dehydration because it cannot stop cutaneous water loss. Inadvertent respiratory losses of water can also be dangerous; for example, exercising in cold, dry winter air can cause serious dehydration in humans.

Heat gain versus heat loss determines core body temperature, with mechanisms to gain external heat, retain internal heat, generate more internal heat, and lose excess heat

From a thermoregulatory viewpoint, an animal body may conveniently be viewed as a *central core* surrounded by an *outer shell.* The temperature within the inner core, which in a vertebrate consists of the abdominal and thoracic organs, the central nervous system, and the skeletal muscles, is the subject of regulation in animals that can thermoregulate, ectotherm and endotherm alike. That is, this internal **core temperature** remains fairly constant in homeotherms and may also vary less than environmental temperature in poikilotherms. The skin and subcutaneous fat constitute the outer shell. Temperatures within the shell may vary considerably more than in the core and are most often cooler than the core in endotherms.

The core temperature is a result of the difference between heat input and heat output (Figure 15-11). The means by which heat gains and heat losses can be balanced to regulate core body temperature generally fall into four broad categories—in short, *gain, retain, generate,* and *lose* heat:

- *Gain external heat/avoid loss to cold environs* by using solar radiation, conduction from a warm surface or other source of environmental heat, and by avoiding cold areas. The heat gains are **ectothermic sources,** made use of primarily through specific thermoregulatory *behaviors* (such as solar basking), aided by *anatomic* features such as dark surfaces to absorb solar radiation effectively.
- *Retain internal heat.* Because of entropy, all metabolism produces some heat. This and any externally gained heat can be retained using *behavior* (such as entering an insulated burrow), *insulation* (such as hair), *reduced blood flow* to the integument to reduce losses, *countercurrent exchangers* in circulation that retain core heat, and *larger body sizes.* All these reduce heat loss via conduction, radiation, and sometimes convection and evaporation.
- *Generate more internal heat.* This is the definitive feature of **endothermy,** requiring significant heat-generating tissues.
- *Lose excess internal heat/avoid gains from hot environs* by *behavior* (such as seeking shade), *anatomy* (such as reflective white layers on the shell of an intertidal snail, and long, thin legs of desert insects which hold their bodies well above the hot ground), enhancing transfer via the integument through *increased blood flow,* and increasing *evaporation* (panting, sweating, cutaneous loss).

This list is only a brief overview; we now turn to the details for ectotherms.

check your understanding 15.4

Describe the basic effects of temperature on biological systems.

Discuss the evolutionary and functional aspects of homeoviscous adaptations of membranes and stability versus flexibility of proteins.

Define *endothermy, ectothermy, poikilothermy,* and *homeothermy.*

Discuss the features of the four heat transfer mechanisms: radiation, conduction, convection, and evaporation.

Discuss body heat balance and key features of "gain, retain, generate, and lose heat."

15.5 Ectothermy

Body temperatures of ectotherms may follow the environment, or may be regulated by heat gains, retention, and losses

What happens to an ectotherm when the environmental temperature is not near the optimum for its molecular structures? There are two broad categories of response: *poikilothermy* and *ectothermic regulation.*

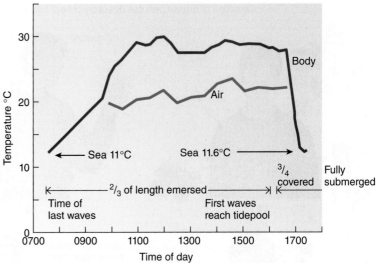

FIGURE 15-13 The body temperature of an intertidal mussel (*Mytilus californianus*) when the tide is out. The temperature is measured with a thermistor inserted inside the shell (*blue* line). As the tide goes out, the body temperature soars more than the air temperature (*green* line). Upon reemersion, the temperature drops rapidly.

Source: Modified from T. Carefoot. (1977). *Pacific Seashores.* Seattle, WA: University of Washington Press by arrangement with J. J. Douglas, Vancouver Canada. p. 68, Figure 67.

Poikilothermy Poikilothermic ectotherms that live in thermally variable environments undergo variations in body temperatures as well. Metabolic rates in such animals typically decrease in the cold (thus saving energy) and speed up as the environment warms (thus becoming better able to obtain food, up to a point). Survival may depend on using behavior to avoid extreme temperatures and/or limited physiological heat-loss mechanisms such as evaporation. If extreme temperatures cannot be avoided, poikilotherms may enter dormancy during the extremes or compensate through changes in internal biochemical optima (as you will see later). Some poikilotherms have body temperatures essentially identical to that of the environment—especially aquatic animals, because of the high heat conductivity and specific heat of water. These animals are sometimes called pure *thermoconformers*. But some have body temperatures that, although not nearly constant, vary less than the environment. This is particularly true of many ectotherms in terrestrial habitats because air transfers heat much less effectively than water.

Most life on this planet is poikilothermic. An extreme example is given in Figure 15-13, showing a mussel on an intertidal rock on the West Coast of North America. When the tide is in, its body temperature will be the same as the ocean, which is usually cold (about 11°C in this case). When the tide is out on a hot summer day, the mussel's body temperature soars and may reach 30°C or more. Some evaporative cooling may keep its temperature lower than it would be otherwise, but nevertheless it undergoes a large change in internal temperature.

Ectothermic Regulation An ectotherm in a constant environment, such as the cold deep sea or some polar ocean regions, is even more homeothermic (almost "homothermic") than any mammal. However, these animals typically cannot cope with any significant temperature changes. For example,

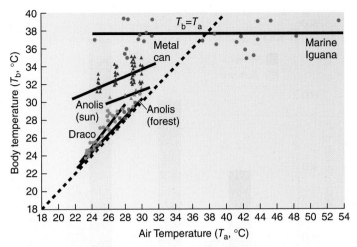

FIGURE 15-14 **Relationship between body temperature and air temperature for lizards.** The dotted black line is the relationship for pure poikilothermy (equal body [T_b] and air [T_a] temperature). Thermoconforming lizards *Draco* and *Anolis* in shaded forest habitat show a similar pattern (*red squares, orange circles,* respectively), while thermoregulating lizards marine iguana and *Anolis* in open sun have flatter lines (*green circles, purples triangles,* respectively). Also shown is the relationship for a water-filled metal can in the sun (*blue triangles*).

Source: From P. C. Withers. (1992). *Comparative Animal Physiology.* Fort Worth, TX: Saunders, p. 140, Figure 5-13.

Antarctic fishes can suffer heat-associated mortality at 5°C! Of more interest are ectotherms that actively use external heat exchanges to maintain their body temperatures at or near an optimum. Well-studied examples include lizards and butterflies (see opening photograph of this chapter). Let's examine the four strategies of heat regulation in a lizard to illustrate (see Figure 15-14):

- *Gain external heat/avoid loss to cold environs.* On cold mornings, the lizard basks in sunlight, soaking up infrared radiation and heat via conduction from warmed surfaces. (It also basks for the purpose of vitamin D production; see p. 329.) Once warmed up to an optimal level, it becomes active, seeking food while using both behavioral and physiological mechanisms to maintain a consistent body temperature.
- *Retain internal heat.* The lizard may *vasoconstrict* blood vessels going to its skin to reduce heat loss. In very large ectotherms, a phenomenon called **gigantothermy** may occur, in which an animal like a large turtle may absorb heat and then easily retain it for some time in a cooler environment due to their comparatively low surface-area-to-volume ratio.
- *Generate more internal heat.* This does not occur significantly in pure ectotherms. However, some very large (varanid) lizards, such as the Komodo dragon, warm up temporarily from high locomotory activity, retaining heat again because of their low surface area to volume ratio.
- *Lose excess internal heat/avoid gains from hot environs.* Blood vessels in the lizard's skin may dilate to increase heat loss. Evaporation from the mouth also provides cooling (one lizard, the Gila monster of southwestern United States, evaporates water from its cloaca rather than its mouth). When it gets too hot, the animal seeks

shade. As another example, Sylvain Pincebourde and colleagues have recently shown that the intertidal ochre seastar (*Pisaster ochraceus* of the North American Pacific Northwest), instead of being a strict poikilotherm as long thought, can pump cold seawater into its coelom to lower its body temperature when overheating due to exposure to the sun at low tide. In order for certain animals to go outdoors during the colder winter temperatures, some zoos heat up the animal during the night so that they have sufficient heat to radiate during the daylight hours.

Using such mechanisms, some ectotherms can regulate body temperature as effectively as a mammal can, at least for some periods.

On a seasonal basis, some ectotherms can *migrate* to help minimize changes in body temperature. The monarch butterfly (p. 14), for example, spends summers in northern North America (such as Alaska and Canada) but joins a massive migration in the autumn to specific southern sites in California and Mexico for the winter months (as you will see later, such migrations are being altered by global warming). Some migratory fishes have similar behaviors; for example, the dogfish shark of the western Atlantic migrates between Florida in the winter and Maine in the summer.

Although these two categories of poikilothermy and ectothermic regulation are convenient, it is important to note that there is no distinct boundary between them. All ectothermic regulators that live in thermally variable habitats are also poikilotherms part of the time. For example, when there is no sun, warm burrow, or other heat source, a lizard cannot regulate its body temperature. At night, it usually cools off to approximately the air temperature. During this period, it is purely poikilothermic.

Some ectotherms can compensate biochemically for changes in body temperatures

Ectotherms in many habitats are subject to temperature changes that impair metabolism because they cannot behaviorally or physiologically adjust body temperature. The onset of winter and of summer are the most common examples. When it is too cold, a poikilotherm's metabolism may slow down to the point that it cannot obtain any food. Freezing may also kill it. If it is too hot, its membranes, proteins, and nucleic acids may suffer irreversible damage. To avoid problems, many poikilotherms undergo adaptive physiological or biochemical changes when exposed to temperature changes. Some of these changes occur on a daily basis, whereas others are seasonally induced. Some involve dormancy, whereas other changes allow for continued activity even with no mechanism for regulating body temperature.

In some cases, acclimatization to temperature changes allows an animal to maintain a useful level activity in a process called **metabolic compensation.** In this form of homeostasis, an animal's metabolic reactions in cold temperatures are increased to a level that is closer to that of warm-acclimated animals *even though their body temperatures are that of the environment's.* The reverse occurs with warm acclimatization. How is metabolic compensation achieved? In many cases, the mechanisms are not known, but several have been revealed in various animals, all involving specific biochemical changes:

1. *Enzyme concentration changes.* In cold acclimation and acclimatization (see definitions, p. 17), many ectotherms such as some fishes, frogs, and reptiles will increase their levels of metabolic enzymes, particularly those involved in aerobic pathways. An example is shown in Figure 15-15 for alligators, comparing animals acclimatized to winter and summer seasons. This particular study examined the activity levels of key enzymes in muscle tissue. Note that for LDH (structure in Figure 15-10), measured in the laboratory at 15, 22.5, and 30°C, the winter-acclimated animals have considerably higher enzyme activities at all measurement temperatures. This biochemical adjustment allows the cold-acclimated alligator to move better in the cold. Note also that the LDH level of the winter animals at 15°C is about the same as the level of the summer animals at 30°C, so this is a form of homeostasis. However, a cold-acclimated alligator will die of heat-related problems at warm temperatures at which the warm-acclimated animals will thrive. Thus, there appears to be a trade-off between metabolic rate and heat tolerance. Moreover, it takes energy to produce more enzymes, and because energy input is usually reduced in the cold, this strategy has limits.

2. *Homeoviscous membrane adaptation.* As we discussed previously, restructuring of membranes to maintain proper fluidity is a common strategy. The common carp, for example, uses this adaptation.

3. *pH regulation.* Most ectotherms in the cold have higher internal pH values than do warmer animals; that is, warmer animals are more acidic. To some extent, pH changes with temperature automatically because the neutral pH of water and the acid constants (pK_a) of protein buffers (p. 641) decrease with heating and increase with cooling. However, some buffer systems such as phosphate and bicarbonate do not have this property, so apparently animals must actively regulate pH to some extent. The primary hypothesis to explain this observation focuses on the amino acid *histidine*. Of the 20 amino acids used to build proteins, histidine is the most important in acting as a buffer (p. 641), because its pK_a is close to physiological pH, so it is nearly ideal for releasing or binding hydrogen ions over the normal physiological pH range. Binding and release of hydrogen ions by histidines are also crucial in many enzyme active sites for catalyzing reactions, as seen in LDH in Figure 15-10, where histidine with a bound H^+ binds to pyruvate, then donates that H^+ to pyruvate. Importantly, histidine's affinity for H^+ decreases with increasing temperature, so it binds H^+ too weakly to serve effectively as a buffer or catalyst if fluid pH is constant:

$$\text{Histidine} - H^+ \underset{\text{colder}}{\overset{\text{warmer}}{\rightleftharpoons}} \text{histidine} + H^+$$

Indeed, studies on LDHs from a variety of animals show that binding of pyruvate weakens greatly as temperature rises. The converse is also true: Binding becomes too tight in the cold. This problem can be offset if [H^+] in body fluids is regulated. Take the case of increasing heat. If an animal regulates its body fluids to become more acidic, the higher [H^+] via mass action increases the concentration of histidines in the charged

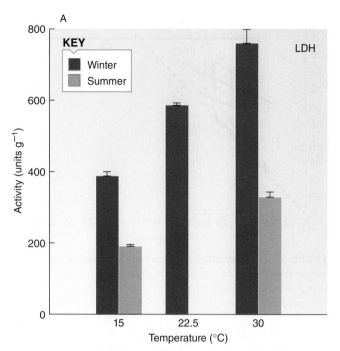

FIGURE 15-15 Activity of lactate dehydrogenase in muscle from alligators acclimatized to winter and summer conditions. Activity increases in the winter to make up for lower thermal energy. The x-axis indicates temperatures at which the activities were measured in the laboratory.

Source: From F. Seebacher, H. Guderley, R. M. Elsey, & P. L. Trosclair III. (2003). Seasonal acclimatization of muscle metabolic enzymes in a reptile (*Alligator mississippiensis*). *Journal of Experimental Biology* 206:1193–1200.

form (histidine-H^+) to the level found at colder temperatures, and so restores binding affinity.

4. *Isoform regulation.* A final strategy requires that different forms of the same protein—called *isoforms*—be used at different temperatures. That is, there may be a set of winter proteins and another set of summer proteins, with the genes for these being activated only under appropriate environmental conditions. For example, the common carp expresses distinctly different forms of its myosin and myosin light-chain protein (MLC) in the summer and in the winter. Myosin is the main contractile protein of muscle (p. 337), whereas MLC binds to myosin and modifies its activity. Together, the isoforms expressed in the cold give the fish greater speed and force in the cold. However, the cold-adapted muscle then does not work well at higher temperatures.

Isoform regulation appears to be uncommon, and many of the mechanisms of metabolic compensation are still unknown. Currently, DNA microarrays (p. 35) are being used to look for major changes in gene expression that occur during acclimation. A study by Andrew Gracey and colleagues of the common carp, which acclimatizes annually to winter temperatures, found that about 252 genes (out of 13,400 measured on a microarray) exhibit significantly upregulated expression in all tissues during cold exposure. The genes are involved in RNA processing and translation, mitochondrial metabolism, proteasome function, and restructuring of lipid membranes. Such studies are revealing previously unsuspected mechanisms of temperature acclimatization.

Ectotherms survive extreme cold by metabolic dormancy and by either freeze avoidance or freeze tolerance

Ectotherms generally cannot stay active in extreme cold. Many can survive low temperatures by dramatically lowering their metabolic rates, thereby reducing the rate at which ATP is consumed. Although they may be technically starving, they have sufficient reserves to survive the season. Frequently, metabolic rate can be reduced to as low as 1 to 10% of the normal resting rate: A 10-fold reduction in energy expenditure gains the animal a 10-fold extension of the time that fuel reserves will last. For example, in the Crucian carp the glycogen level that supports brain function for 16 hours in the anoxic winter ponds when energy demand is low, would support the fish for only eight minutes in the summer with the same amount of oxygen. Some species of insects completely arrest their development in a state of *diapause*. Turtles hibernate at depths below the frost line along the banks of streams and at the bottom of ponds. These reptiles survive the entire winter without breathing (p. 641) and with a heart rate of only one beat per minute.

Special adaptations are needed for ectothermic animals whose body temperatures drop below freezing. There are two broad categories of these adaptations: *freeze tolerance* and *freeze avoidance*.

Freeze Tolerance Some ectotherms actually survive freezing of some body fluids (with up to 80% of body water being frozen in such animals). Woolly bear caterpillars spend up to 10 months frozen in the high Arctic, barnacles and mussels freeze when exposed to subfreezing temperatures at low tide, and even some reptiles and amphibians freeze for variable periods of time. For example, spring peeper frogs (*Pseudacris crucifer*) and wood frogs (*Rana sylvatica*) can revive after having as much as 65% of body water converted to ice. During the frozen state, ATP use virtually ceases and neurological activity is minimally detectable. Energy needs of a tissue are normally supplied by nutrients delivered by the circulatory system. However, in a thawing animal the tissues are completely reliant on immediately available energy sources.

What happens to the tissues that actually freeze? Consider a key problem of freezing—the osmotic stress placed on an animal cell as it enters the frozen state. Formation of ice crystals in the ECF (extracellular fluid) immediately changes the osmotic balance with the ICF (intracellular fluid). Because ice excludes solutes from its structure, during the freezing process the composition of the ECF becomes progressively more concentrated as the concentration of water declines. In response, water moves out of the cell. Only when the concentration of solute is great enough can the transformation of water into ice be retarded. The majority of freeze-tolerant animals reach the critical minimum cell volume when approximately 65% of the total body water is converted into the frozen state.

To understand how some animals survive the stresses of freezing on metabolism and cell volume, let's look at what happens to a peeper or wood frog, starting when it is initially exposed to freezing temperatures. When water freezes on the frog's outer skin, a signal sent to the liver triggers a massive breakdown of glycogen in the liver. Consequently, a flood of *glucose* molecules enters the circulatory system, raising blood sugar concentrations over 450-fold (over 10-fold greater than a diabetic human with moderately uncontrolled blood-sugar concentrations). This serves as an *antifreeze* by lowering the freezing point through simple colligative properties (p. 79). In less than eight hours, the frog's organs become loaded with this nutrient, which has now assumed an additional role as a **compatible cryoprotectant**. The cryoprotectant acts as an *osmolyte* (p. 617) to help keep the cells in osmotic balance with the increasing osmotic pressure in the ECF. Cryoprotectants must also be *compatible* with macromolecules (p. 617). Thus, inorganic ions cannot be used: They are just as effective in lowering freezing points by colligative mechanisms (and that is why humans dump salt on icy roads in many wintry locales), but unlike the compatible solutes such as glucose, they disrupt macromolecules. (Compatibility is the reason humans use small carbohydrates such as ethylene glycol rather than corrosion-promoting salts in automobile radiators.) Wood frogs, at least, also accumulate *urea*, which at modest levels appears to reduce freezing damage to cells and may also help suppress metabolism. In addition, during thawing—the time when cells have no delivery of oxygen—glucose is immediately available as a fuel to generate ATP. Studies have shown that a frozen frog ventricle regains its ability to contract if it is thawed in the presence of glucose but not in the presence of other cryoprotectants such as glycerol. Using this strategy, the frog is ready to resume its active life.

Other animals are also freeze tolerant. For example, Arctic earthworms accumulate glucose in the winter, which serves as both a cellular antifreeze and fuel source. Many insects employ similar mechanisms, often using other carbohydrates such as *trehalose* (p. 614) as compatible cryoprotectants and antifreezes. In addition, many of these animals actually enhance the freezing of their ECFs with **ice-nucleating agents,** small proteins that trigger the formation of small ice crystals. Apparently these agents ensure that ice only forms in the ECF and that ice forms in an ordered fashion rather than in the "normal" way, which would create large crystals that could rupture membranes.

Freeze Avoidance Most overwintering animals are considered intolerant of internal freezing yet may still have body temperatures well below the freezing point. Again, many such animals have antifreeze compounds, but they are used somewhat differently. First, some overwintering animals use compatible cryoprotectants such as sorbitol and glycerol, but throughout the ECF and ICF rather than just the latter. Thus, the ECF is also protected from freezing. Examples include many insects, arachnids, and some Arctic fishes such as the rainbow smelt. However, other insects and polar fishes have special **antifreeze proteins,** which do not work by classic colligative mechanisms. These proteins are effective at much lower concentrations than the compatible cryoprotectants. They generally contain very hydrophilic amino acids and sugar side chains that are thought to bind to growing ice crystals and thus prevent their growth.

In addition to, or instead of using antifreezes, some overwintering animals appear to use the phenomenon of **supercooling.** This is a state of water in which the temperature is well below the freezing point but there is no trigger or nucleation site to begin ice formation. Some polar fishes and arachnids, such as the Antarctic mite, employ this strategy. Painted turtles (*Chrysemys picta,* common in wetlands across eastern North America) also supercool in the winter, as does at least one mammal, the Arctic ground squirrel (see p. 751). Supercooling is a dangerous strategy, however, because any encounter with external ice can trigger a catastrophically rapid crystallization of body fluids.

Finally, some polar animals such as the Arctic springtail (*Megaphorura arctica*) avoid freezing through *cryoprotective dehydration*. While supercooling, these small insects lose water from their bodies to the point that they cannot freeze. At the same time, they build up *trehalose,* the protective solute we described in Chapter 13 (p. 614) that builds up in many dehydrated organisms in nonfreezing situations.

Ectotherms may survive temporary extreme heat with the heat shock response

Finally, let's examine what ectotherms do when exposed to excess heat. Given enough time, ectotherms may evolve (over many generations) thermostable proteins and membranes. These adaptations appear to have a thermal limit in eukaryotes well below that for prokaryotes; see *Challenges and Controversies: What Is the Maximum Temperature for Life?*

But what happens in the short term? If an organism is exposed to a sudden jump of about 5°C or more in cell temperature, an ancient mechanism is induced: the *heat shock response*. This occurs in all types of organisms from archaea to humans, and involves the activation of genes for **heat shock proteins (HSPs)**, also known as **stress proteins** because other stresses (such as osmotic and chemical shocks) can induce them as well. As shown in Figure 15-16a, these small, thermostable, hydrophobic proteins bind to larger, unfolded proteins and assist their folding into functional conformations. They are attracted to hydrophobic amino acids that are normally in the interior of folded proteins but become exposed on denaturation (unfolding). Some HSPs are always present in cells, where they assist in the folding of newly formed polypeptides coming off ribosomes (hence HSPs are also called *molecular chaperones*). But during heat shock, other HSPs are rapidly induced at the gene transcription level, protecting the cell from heat death. This occurs as follows: In an unstressed cell, there are small amounts of HSPs such as **hsp70**, the most common form (whose weight is about 70,000 daltons). The hsp70s are bound in an inactive state to a protein called **HSF-1** (heat-shock factor 1). Under a sudden heat stress, the hsp70s detach from HSF-1 and bind to and stabilize unfolded proteins. The release also activates HSF-1, which is a *transcription factor* (p. 31). When active, HSF-1 binds to two other HSF-1 proteins to form a trimer (protein complex made of three separate proteins). The trimer enters the nucleus and binds to a *response element* in the DNA (p. 31) called **HSE (heat shock element)**; this action in turn activates the transcription of genes for heat shock proteins. (This mechanism of transcription factors and response elements is a typical process of gene regulation in eukaryotes, as we discussed in Chapter 2.)

As noted, HSPs have been found in all forms of life. Mammals, for example, make them during a fever. An example for a poikilotherm is shown in Figure 15-16b, focusing on hsp70. Two different species of limpets (*Collisella* spp.) live at different levels on a rocky intertidal habitat. The species that lives higher up—and thus is more exposed to excessive heat from the sun when the tide is out—has higher levels of hsp70 and can make more hsp70 than can species living lower in the intertidal zone.

Although HSPs are thought to be universal, heat shock induction can apparently be lost in some species over evolutionary time. Nototheniid fishes, for example, appear to have no heat shock response whatsoever. These Antarctic fish live in a nearly constant temperature environment, about –2°C year-round. In the laboratory, they begin to die of heat stress at 5°C and above! Yet even at 12°C, they cannot make additional HSPs above the basal levels.

check your understanding 15.5

Describe the different responses in pure poikilothermy and ectothermic regulation.

Discuss the biochemical adjustments that account for metabolic compensation in ectotherms.

Discuss the differences in freeze tolerance and avoidance, with examples.

Discuss how heat shock proteins aid survival at extreme temperatures.

15.6 Endothermy and Homeothermy

Endothermy has several benefits. The ectotherm's problem of either finding external heat sources or suffering a slowing metabolism from cooling, can be avoided. Since endotherms usually have body temperatures warmer than their habitats, they have consistently faster biochemical processes than do most ectotherms in the same habitat. This can provide a competitive advantage: expansion of daily activity into nighttime, higher digestion rates, sustained locomotion, habitat expansion, and regulated incubation temperature. Moreover, some endotherms achieve homeothermy—a consistently warm body temperature that allows proteins and membranes to always be at their optima.

But there is a large metabolic cost: Endotherms must generate heat internally to raise body temperature and must do so continuously if they are homeothermic. Heat production ultimately depends on the oxidation of metabolic fuel derived from food, and thus endotherms consume much more energy (5 to 20 times) than do ectotherms of the same mass. Conversely, a high metabolism increases the risk of overheating. So endotherms must also have mechanisms to remove excess heat that are more effective than those of ectotherms. In light of these advantages and disadvantages, we still do not fully understand the evolution of endothermy (see *Unanswered Questions: How, when, and why did endothermy evolve?*).

Although hydrothermal vents can reach 400°C, no life has been found that can tolerate anything close to that level. In 2003, the highest temperature for life yet recorded was demonstrated to be 130°C (266°F), achieved by a hydrothermal vent archaean. What about animals at the vents? A polychaete annelid, *Alvinella*, which lives at deep-sea hydrothermal vents, has been reported to live at temperatures as high as 80°C. However, many researchers believe the body temperature measurements were flawed (they are difficult to make using a submersible vehicle on an animal that retreats inside a small hard calcium-carbonate tube). Nevertheless, one worm wrapped around the submersible's temperature probe reportedly survived a brief exposure to 105°C! In the laboratory, however, *Alvinella*'s proteins do not work above about the 40 to 45°C range and its collagen (a major connective-tissue protein; p. 65) is stable up to 65°C, suggesting the field data are mis-

leading. In work on a related species, *Paralvinella sulfincola,* Peter Girguis and Raymond Lee placed worms in high-pressure chambers with a regulated temperature gradient ranging from 20°C at one end to 61°C at the other. When kept at their natural habitat pressure, the worms crawled about and settled around the area at 50°C, where they appeared to behave normally. In other experiments, Lee and Christian Rinke found that some individuals could survive short periods at 60°C, and that the species has key enzymes stable up to 50 to 60°C.

On land, two animals rival the vent worm. Saharan desert ants (*Cataglyphis*) and Namibian desert pseudoscorpions (*Eremogaryus perfectus*) have been found to tolerate 55°C body temperatures for a few minutes, and the pseudoscorpion can tolerate up to 65°C in the laboratory! The ant experiences 55°C while running across hot sand to forage at midday. (This behavior gives them a competitive advantage because other animals

Paralvinella sulfincola.

Paul Yancey

are hiding in burrows to avoid the midday heat.) The ants survive in part by loading up on heat shock proteins before foraging. What actually sets the upper limit for these African ants, hydrothermal vent worms, arthropods, and other eukaryotes is not known, but it is suspected to be fundamental features of mitochondrial respiration and membrane stability, gene transcription, and/or translation that cannot be stabilized beyond a certain temperature.

How, when, and why did endothermy evolve?

Despite the common view that endothermy and homeothermy have great advantages, scientists are not certain how, when, or even why they evolved. Consider the "when" question: While birds are endothermic, we are still not sure if their dinosaur ancestors were also. Because of their enormous sizes, some dinosaur species would have exhibited gigantothermy (p. 736) but not necessarily true endothermy. But what about smaller dinosaurs? Some fossil evidence suggests that at least some were endothermic—for example, some dinosaur bones have capillary densities similar to those of mammals and higher than in modern (non-avian) reptiles. Recent fossil discoveries of *theropod* dinosaurs (the direct ancestors of birds) have found remnants of feathers and birdlike air sacs (p. 513), consistent with endothermy, as are the birdlike bipedal stances of theropods like *Velociraptor.* Moreover, some fossil embryos and juveniles reveal a rapid growth rate as found in today's endotherms but not ectotherms. However, no dinosaur fossils show evidence of maxilloturbinals, the nasal heat exchangers (p. 745) found in living endotherms. Such exchangers do appear in the late Permian era in some fossils of therapsids, advanced reptiles that are the ancestors of mammals.

The "how" and "why" questions are also perplexing. In shivering mammals, flying insects, and swimming tunas, aerobic loco-

motory muscles are a major or sole source of extra heat. Thus, some biologists have proposed that enhanced aerobic locomotion evolved first and that only later was the heat output used for thermoregulation. Others propose the converse: that heat generation evolved first and that later benefited locomotory ability. Another hypothesis is that high body temperature initially arose because it greatly restricts infections by fungi. Yet other researchers suggest that the ability to keep embryos at an optimal developmental temperature was the first selective advantage of endothermy in land animals. Finally, the latest hypothesis is that endothermy began in herbivores. Eating plant matter yields a lot of carbon but very little nitrogen. Endothermy allows not only a way to eat more food to obtain that nitrogen but also the ability to "burn off" the excess carbon. Definitive evidence to distinguish among these competing ideas is still lacking. <<

Birds and mammals maintain a consistent internal core temperature (homeothermy)

Many organisms have evolved endothermy: birds, mammals, some fishes and non-avian reptiles, some insects, and even some plants! In all cases, these organisms rely on high levels of *aerobic metabolism* for sustainable heat production. High aerobic ability has another function: Most endotherms rely on aerobic metabolism for locomotion more than do ectotherms (with few exceptions) (see *Unanswered*

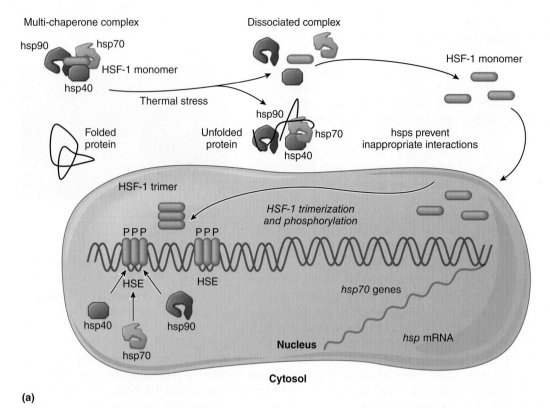

(a)

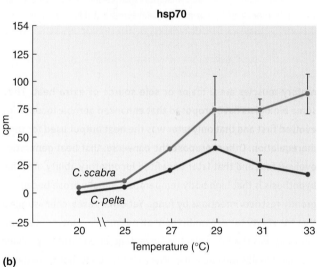

(b)

FIGURE 15-16 Role and regulation of heat shock proteins.
(a) Model for regulation of hsp70, a major heat shock protein whose activity is activated by sudden temperature increases and other stresses. Under nonstress conditions, hsp70 (along with hsp40 and 90) is bound in an inactive state to heat-shock factor 1 (HSF-1) in the cytoplasm. Under temperature stress, this complex dissociates, and the hsp proteins bind to other, unfolded cell proteins, preventing aggregations and assisting in their refolding. The unbound HSF-1 moves to the nucleus, forms a trimer (three HSF-1 proteins), and binds to a promoter element in the DNA called the heat shock enhancer (HSE). This in turn activates the gene for hsp70, leading to synthesis of hsp70 mRNA and thus more hsp70 proteins to aid in the stress response. (b) *Collisella* limpets and their production of hsp70 at various temperatures. *C. scabra* lives higher in the rocky intertidal zone than *C. pelta,* and thus is exposed to higher temperatures more often. Hsp70 production is quantified by the incorporation of radioactive amino acids, determined by radioactive counts per minute (cpm).

Sources: (a) From Hochachka, P. W., and G. N. Somero, (2002), "Biochemical Adaptation", Oxford: Oxford University Press, Fig. 7.14, as modified from R. I. Morimoto and M. G. Santoro, (1998), Stress-inducible responses and heat shock proteins: New pharmacologic targets for cytoprotection, 'Nature Biotechnology', 16:833–838. Reproduced with permission of Nature Publishing Group via Copyright Clearance Center. (b) From B. M. Sanders, C. Hope, V. M. Pascoe, & L. S. Martin. (1991). Characterization of the stress protein response in two species of *Collisella* limpets with different temperature tolerances. *Physiological Zoology* 64:1471–1489.

Questions: How, when, and why did endothermy evolve?). Thus, a lizard (ectotherm) may have tremendous burst (anaerobic) capacity that allows it to outrun a mammal of the same size over short distances. However, in a long-distance chase the mammal's aerobic capacity allows it to outrun the lizard, which fatigues quickly.

Among all the different groups of endotherms, only birds and mammals exhibit homeothermy in all of their core organs. We consider these first. The other animals, which are more variable, are discussed later as *heterotherms.*

It is important to recognize that core body temperatures vary among species of birds and mammals (Table 15-3).

Monotreme mammals (platypuses, echidnas), have lower set points than most placental mammals, which in turn have lower set points than most birds. Why these different set points have evolved is not clear. However, one consideration is that it is physiologically easier to warm up than to cool off. Heat is produced by metabolism (following the Second Law of Thermodynamics). But cooling off in an environment hotter than an animal's body requires evaporation of water—often a precious resource, and one that works poorly or not at all at high humidity (p. 735). Thus, it has been proposed that endotherms have evolved optima that are slightly above the average high environmental temperature of their ancestral habitats.

TABLE 15-3 Core Body Temperatures (T_b) Characteristic of Endothermic Vertebrates

Taxon	Common Names	T_b (°C)
Mammals		
Monotremes	Echidna, platypus	30
Edentates	Anteaters, etc.	33–34
Marsupials	Possums, kangaroos, etc.	36
Insectivores	Hedgehogs, moles, etc.	36
	Shrews	37–38
Chiropterans	Bats	37
Cetaceans	Whales, etc.	
Pinnipeds	Seals, etc.	
Rodents	Mice, rats, etc.	37–38
Perissodactyls	Tapir, rhinoceros, horses	
Primates	Monkeys, humans, etc.	
Carnivora	Dogs, cats, etc.	
Artiodactyls	Cows, camels, pigs, etc.	38–39
Lagomorphs	Rabbits	
Birds		
	Penguins	38
	Ostrich, petrels, etc.	39–40
	Pelicans, parrots, ducks, gamebirds	41–42
	Passerine songbirds	42

Source: Willmer, P., G. Stone and I. Johnston, (2000), "Environmental Physiology of Animals", p. 223, Table 8.11. Copyright © 2000 Blackwell Publishing. Reproduced with permission of Blackwell Publishing Ltd.

A roadrunner ruffles it feathers to expose its dark skin, which absorbs sunlight to help heat the animal's blood.

Body temperatures also vary among individuals, and vary throughout the day within individuals. For diurnal mammals (including humans), body temperatures typically fall 1 to 2° C at night in response to a biological clock (p. 276). (This is one reason why most humans feel particularly sluggish if awakened before dawn.) The camel, which we described earlier (p. 733), varies even more. Furthermore, there is no one body temperature, because temperature can vary from organ to organ, even in the core; for example, active skeletal muscles are warmer than most other organs.

To maintain a stable core temperature, heat gains must balance heat loss, with mechanisms to gain, retain, generate, and lose excess heat

To maintain a constant total heat content and thus a stable core temperature, an animal body must balance heat input and output (Figure 15-11). We now elaborate on the means by which birds and mammals can adjust heat gains and losses to regulate core body temperature. Recall that the mechanisms fall into broad categories of *gain, retain, generate,* and *lose* (p. 736).

Gaining External Heat/Avoiding Loss to Cold Environs Birds and mammals evolved from ectothermic ancestors and thus have inherited many of the thermoregulatory behaviors and physiology of ectotherms. These include the following:

1. *Ectothermic behavior.* A cold bird or mammal may seek out sunshine or a warm surface, whereas a warm endotherm avoids excessive environmental cold, just as a lizard does. Thus, basking behaviors (serving as effectors) are common in many birds and mammals, especially smaller ones with their high ratios of surface area to volume, and especially in cold weather. House cats, for example, are well known to bask in patches of sunlight. Larger mammals may also exhibit such behaviors: A human basking on the beach or soaking in a hot tub is behaving like a lizard! Long-distance migrations may also serve in part as a thermoregulatory adaptation. For example, one reason why humpback and gray whales migrate from polar feeding waters in the summer to tropical birthing areas in the winter may be to provide a warm habitat for newborns.

2. *Anatomic features such as dark skin that help absorb solar radiation.* Insulation, such as feathers, actually interferes with this absorption. For this reason hummingbirds and roadrunners, which undergo *torpor* (reduced metabolism and body temperatures, p. 750) at night and bask in the sun in the morning to warm up, erect the feathers of their upper backs. Unlike other skin (which is pink), their skin here is black, to more efficiently absorb the solar radiation.

Retaining Internal Heat Retaining internal heat is accomplished by several mechanisms. Some of these also occur in some ectotherms, but in general these mechanisms are more prominent in endotherms:

1. *Vasoconstriction.* The insulative capacity of skin can be varied by controlling the amount of blood flowing through it. Blood flow to the skin serves two functions. First, it provides a nutritive blood supply to the skin. Second, because blood has been heated in the central core, it carries this heat to the skin. But an animal

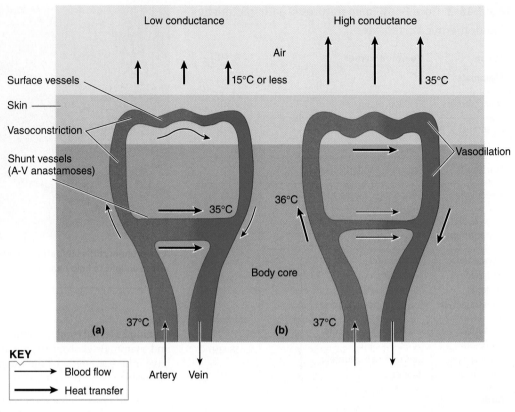

Source: Willmer, P., G. Stone and I. Johnston, (2000), "Environmental Physiology of Animals", p. 226, Fig. 8.36. Copyright © 2000 Blackwell Publishing. Reproduced with permission of Blackwell Publishing Ltd.

FIGURE 15-17 Changes in blood flow to the skin for thermoregulation. (a) In a cold environment, blood can be diverted through shunts (anastomoses) deep under the skin to reduce heat loss. (b) If body heat needs to be released to the environment, or if external heat needs to be absorbed from a warm environment, vessels just under the skin surface can be opened.

suffering excessive heat loss can vasoconstrict the skin vessels (Figure 15-17a). This reduces blood flow through the skin, decreasing heat loss by keeping the warm blood closer to the core and thus more insulated from the external environment. As you will see later, the hypothalamus regulates this blood flow in vertebrates. Some ectotherms, such as lizards, can use this process to some extent.

2. *Anatomic insulation.* Birds and mammals have evolved integumentary structures that help trap heat in their bodies. These *insulating* (low-heat-conducting) structures include feathers, hair, and subcutaneous adipose layers (including blubber). Such insulation may appear to be static anatomy, but there are some forms of regulation. Many mammals grow more hair and/or adipose tissue in the autumn. For example, the fur of white-tailed deer is thin in the summer and thick and long in the winter. Furthermore, the winter hairs are hollow, allowing them to trap heat inside the shaft. Birds such as goldfinches increase their feather masses up to 50% by wintertime. In animals with dense fur or feathers, contracting the tiny muscles at the base of the hair or feather shafts lifts the hair or feathers off the skin surface (the *piloerection* reflex, which humans retain vestigially). This puffing up traps a layer of poorly conductive air between the skin surface and the environment, increasing the insulating barrier between the core and the cold air and thus reducing heat loss.

3. *Behavioral insulation.* Additional insulation may result from behaviors such as nest building, postural changes, burrowing, and huddling. Certain postural changes, such as a cat curling up in a ball, reduce the exposed surface area from which heat can escape. Emperor penguins on the Antarctic ice often huddle in the thousands, a behavior estimated to save 25 to 50% energy. Another thermoregulatory behavior in many birds is perching on one leg, tucking the other leg against the body feathers (the leg is a major site of heat loss in birds.) Consider also the sea otter, which grooms its hair and coats it with an oily secretion to help trap a layer of insulating air.

4. *Larger body size in colder climates.* A trend in avian and mammalian species is the evolution of larger average body sizes (such as penguins) in cold-climate species compared to warm-climate relatives. This trend, known as *Bergman's rule*, is predictable based on the properties of ratios between surface area and volume that we discussed earlier (p. 720).

5. *Countercurrent exchangers.* Blood flow can be used to retain core heat in another way known as a **countercurrent exchanger**, found between the core body and many exposed peripheral organs that could lose heat rapidly. (Recall examples of countercurrent flow in fish gills, fish swimbladder systems, and mammalian nephrons.) The exchanger consists of a set of veins and arteries (or venules and arterioles) placed closely together in a dense array known as a **rete mirabile** (Figure 15-18a). Because the vessels are so closely packed, they are nearly in thermal equilibrium. Crucial to the rete's function is that blood flows in opposite directions (countercurrent) in the two types of vessels. Heat moves in the following way: Warm core blood moves out the arteries toward the cold peripheral tissue. In the rete, it encounters cold blood from that periphery. By conduction, the heat moves into

the cold vein and thus returns to the core. The venous blood leaving the rete is thus nearly at the temperature of the periphery so that little core heat is lost.

Retes are found in many endotherms and heterotherms (Figures 15-18b, c, and d). In mammals, they are often in the limbs, such as a dolphin's fluke and flippers. In birds, they are often in the legs to limit heat loss from the feet. You will see later how heterotherms such as tunas use retes.

A different type of countercurrent exchange is found in the nasal passages of birds and mammals. One drawback of endothermy is the potential for high water loss, and not just from thermoregulatory evaporation. Because the lungs must be kept moist for respiratory gases to dissolve into the cells lining the capillaries, lung air in endotherms is always warm and at 100% relative humidity. There is a danger of losing this body water and its heat content to the external environment. Instead, some of the moisture contained in the air from the lungs condenses onto the comparatively cool surface of the **maxilloturbinals** (folds in the nasal cavity; see p. 508), which can be very elaborate. For example, the camel's nose has an enormous surface area in these folds. The subsequent inhalation of drier air from the external environment evaporates this water and cools the surface of the maxilloturbinals in preparation for the next cycle. This type of countercurrent process is called a *temporal countercurrent exchanger,* where the opposing flows of a fluid are separated in time (using one tube) rather than in space (using two separate tubes). Mammals on average reclaim as much as 45% of the water (and its heat content) from the exhaled air, whereas kangaroo rats recycle up to 88% of their body water in this fashion. Birds also have such structures. However, ectothermic vertebrates, including non-avian reptiles, do not.

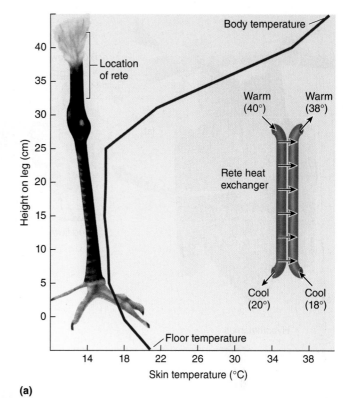

(a)

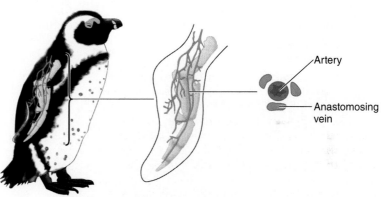

(b) Venae comitantes

FIGURE 15-18 Countercurrent heat exchangers (retes) in endotherms. (a) A rete exchanger in the leg of a stork at an air temperature of 12°C and floor temperature of 20°C, showing body temperature (*plotted line*) along the leg. The inset shows the principle of countercurrent exchange in a rete. (b) A *venae comitantes* rete, with two or more anastomosing veins surrounding a central artery (shown for a penguin flipper). (c) A centralized rete, with one central large artery surrounded by many separate, small veins (shown for fluke of a whale). (d) An artery–vein network rete, with many small arteries and veins together (shown for the limb of a loris).

Source: (a) From P. C. Withers. (1992). *Comparative Animal Physiology.* Fort Worth, TX: Saunders, as modified from M. P. Kahl. (1963). Thermoregulation in the wood stork, with special reference to the role of the legs. *Physiological Zoology* 36:141–151; (b) P. G. H. Frost, W. R. Siegfried, & P. J. Greenwood. (1975). Arterio-venous heat exchange systems in the jackass penguin *Spheniscus demersus. Journal of Zoology* 175:231–241; (c) P. F. Scholander & W. E. Schevill. (1955). Countercurrent vascular heat exchange in the fins of whales. *Journal of Applied Physiology* 8:279–282; (d) P. F. Scholander. (1957). The wonderful net. *Scientific American* 196:96–107.

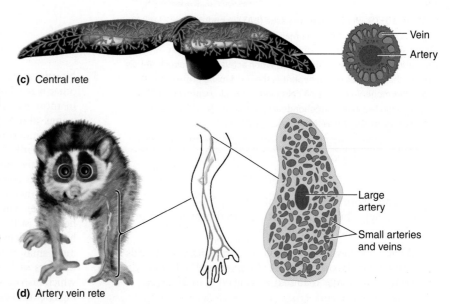

(c) Central rete

(d) Artery vein rete

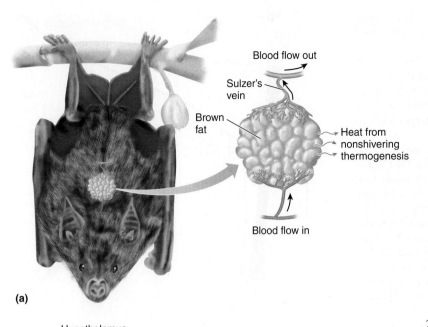

(a)

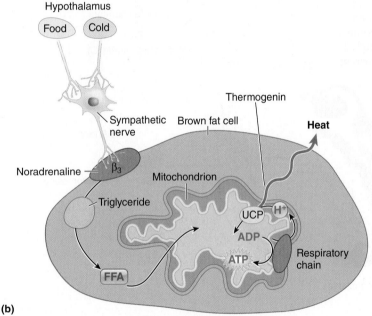

(b)

FIGURE 15-19 Brown adipose tissue (BAT) in endotherms. (a) BAT deposits in bats and other mammals are found between the shoulder blades and other locations not shown. (b) Schematic diagram illustrating how BAT generates heat. When activated by cold or diet via a nerve that releases norepinephrine, the uncoupling protein (UCP) called thermogenin in the BAT cells opens and allows protons (H+) to pass through the inner mitochondrial membrane. This dissipates the proton gradient as heat without making ATP.

Source: (a) Modified from D. Randall, W. Burggren, & K. French. (2002). *Eckert Animal Physiology.* New York: W. H. Freeman and Company, Figure 17-22. (b) Modified from A. G. Dulloo. (1999). UCP2 and UCP3: genes for thermogenesis or lipid handling? *Obesity Matters* 2:5–8.

Generating More Internal Heat The hallmark of endotherms is their ability to produce internal heat for the purpose of thermoregulation, as follows:

1. *BMR.* One key adaptation was a large increase in birds' and mammals' overall BMR ("idling speed"). Indeed, the BMR of a bird or mammal is typically 5 to 20 times

greater than the SMR of an ectotherm of the same mass. This appears to be due to "leaky" cell membranes. In one study comparing the liver, kidney, and brain of a mammal (laboratory rat) and a reptile (bearded dragon) of the same mass and body temperature, the endotherm (rat) tissues consumed O_2 two to four times faster, in part because the rat Na^+/K^+ ATPase pump used ATP three to six times faster. The pump runs faster because the rat's membranes are leakier to these ions, so the pump must work harder to restore the gradients.

Thyroid hormones (p. 301) regulate the number of active Na^+/K^+ ATPase pump units in vertebrate cell membranes, with basal levels of these hormones being considerably higher in endothermic than in ectothermic vertebrates. In a resting mammal, most body heat is produced by the thoracic and abdominal organs.

2. *Shivering and other muscular activity.* The rate of heat production can be variably increased above the "idling" BMR level in the cold through skeletal muscle activity. Muscles constitute the largest organ system in the avian and mammalian body and are a ready source of heat from the contractile process. Any type of increased muscular activity (such as exercise and NEAT, p. 729) will increase heat production, but there are also specific thermoregulatory mechanisms. In response to a fall in core temperature caused by exposure to cold, skeletal muscle *tone* gradually increases. (Muscle tone is the constant level of tension within the muscles.) This produces some heat. Soon shivering begins. **Shivering** consists of rhythmic, oscillating skeletal-muscle contractions that occur at a rapid rate of 10 to 20 per second. This mechanism is very effective in increasing heat production; all the energy liberated during these muscle tremors is converted to heat because no external work is accomplished. Within a matter of seconds to minutes, internal heat production may increase two- to fivefold as a result of shivering.

3. *Nonshivering (chemical) thermogenesis.* There is one other way to boost heat production above the BMR level. In most experimental mammals, chronic cold exposure brings about an increase in metabolic heat production that is independent of muscle contraction, appearing instead to involve changes in heat-generating chemical activity. This **nonshivering thermogenesis** is mediated by the hormones epinephrine and thyroid hormone and by sympathetic nerves (which use the neurotransmitter norepinephrine). The cellular mechanisms triggered to produce heat are still not fully understood. One mechanism that has been delineated is found in newborn mammals and most small mammals (especially hibernators). These have deposits of *brown adipose tissue* (BAT) (Figure 15-19a), which (as you saw earlier) is especially capable of converting chemical energy into heat. BAT cells, which are devel-

opmentally derived from muscle rather than adipose cells, contain deposits of triglycerides packed with specialized mitochondria (the combination of fat and mitochondria gives this tissue its color). In their inner membranes, these mitochondria contain the *uncoupling protein* UCP-1 (p. 729), a gated proton (H^+) channel also called **thermogenin** (Figure 15-19b). Recall that normal mitochondria generate ATP by using a proton gradient (see Figure 2-20, p. 52). Normally, the protons move down their gradient through an ATP-synthesizing enzyme complex. But when thermogenin channels open, the two processes (H^+ flow and ATP synthesis) become uncoupled and the protons simply flow down their gradient (through thermogenin) without generating ATP, with the energy from this H^+ flow being completely dissipated as heat.

BAT is particularly important in newborn mammals, which have a higher ratio of surface area to volume than do adults, and in hibernators, where it is used to warm the mammal up rapidly as hibernation ends. Also, as we mentioned earlier (p. 729), there is some evidence that BAT may help lean animals "burn off" excess calories.

Losing Excess Heat/Avoiding Gains from Hot Environs A rise in core body temperature is potentially more dangerous than a decline, because of irreversible denaturation of proteins. Thus, several mechanisms exist to remove excess body heat, which can occur not only from hot weather but also from high muscular activity. These mechanisms are as follows:

1. *Reduced insulation.* Desert endotherms may have inherently low insulation; for example, the camel has virtually no subcutaneous fat, storing its adipose instead in its characteristic hump. As we noted earlier, layers of insulation may be regulated. The shedding of hair by a house cat or dog in the springtime is a familiar example. A bird in hot weather may loosely ruffle its feathers to convect heat from the skin.

2. *Enhanced radiation and vasodilation.* In the process of thermoregulation, skin blood flow can vary tremendously. The more blood that reaches the skin from the warm core, the closer the skin's temperature is to the core temperature. The skin's blood vessels diminish the effectiveness of the skin as an insulator by carrying heat to the surface, where it can be lost from the body by radiation, conduction, and convection. Accordingly, vasodilation of the skin vessels (Figure 15-17b), which permits increased flow of heated blood through the skin, increases heat loss. Some external structures with large surface areas help with heat removal from the blood, such as elephant ears and the large beak of the toucan.

3. *Enhanced evaporation.* As you saw earlier, evaporation is the only way to cool off if the environment is hotter than the skin. Recall the importance of the relative humidity (RH, p. 735) of the surrounding air. When the RH is high, the air is nearly or fully saturated with H_2O, so it has limited or no ability to take up additional moisture from the animal. Thus, little or no heat loss can occur on hot, humid days.

Endotherms have at least two evaporation mechanisms specifically adapted for thermoregulation: **respiratory panting** and **"insensible" cutaneous loss** through the skin (see p. 631). Cutaneous loss is an important route, especially in smaller endotherms. For example, it ac-

counts for well over half of evaporative loss in the desert kangaroo rat (p. 630). Panting involves a shallow, rapid, breathing pattern that permits large volumes of air to move over the hot, moist tongue and respiratory airways. The resultant increase in evaporative heat loss from the respiratory tract provides a cooling effect. Numerous species of birds supplement panting with a rapid fluttering of the well-vascularized esophageal region. This rhythmic inflation of the hyoid apparatus is termed **gular fluttering.** Because air movement in panting and gular fluttering is associated with surfaces that do not participate in gas exchange, problems with imbalances in the blood concentration of O_2 and CO_2 (hypocapnia and alkalosis; see pp. 542 and 647) are avoided. In many bird species, cutaneous evaporation accounts for about half of water loss at moderate temperatures, whereas at high temperatures most birds rely on respiratory evaporation. For example, the desert verdin (*Auriparus flaviceps*) at 50°C uses respiratory evaporation for 85% of total water loss. For unknown reasons the response of some species differs; for example, Spinifex pigeons increase their rate of cutaneous evaporation as temperatures rise.

A few mammals have a third mechanism: **Sweating** is an active evaporative heat-loss process from specialized integumentary glands under sympathetic nervous control. Sweat is a dilute salt solution actively extruded to the surface of the skin by the glands and dispersed. For heat loss to occur, sweat must be evaporated from the skin. Humans and horses sweat over much of their bodies, but most mammals either lack or have limited distribution of sweat glands; for example, dogs have sweat glands only on their paw pads.

4. *Countercurrent exchange.* The rete mechanism discussed earlier for heat retention can also be used to prevent overheating of certain organs. In particular, some fast-running animals such as gazelles have a rete between their brains and the core body, in which warm blood in the carotid artery passes by cool venous blood from the nose, sinuses, and facial skin. This protects the brain from the high heat produced from the skeletal muscles when the animal is running from, say, an attacking cheetah (which also has a similar rete). Horses may protect their brains from overheating in a different way: In addition to cooling sinuses, they have unusual **guttural pouches** (sacs extending from the auditory tubes) surrounding the carotid arteries, which may also help cool the blood.

5. *Avoidance behavior.* Again, like their ectothermic ancestors, birds and mammals seek cooler environs when overheating. This can be a local change, such as moving under a shady tree, or a long-distance process, such as a seasonal migration.

6. *Anatomic reduction of heat gain.* Light-colored surfaces, such as skin, fur, or feathers, reflect sunlight. Camels, for example, have wool in the winter and shiny, reflecting hair in the summer (tropical camels maintain this reflecting hair year-round).

As we noted earlier, evaporative cooling must be balanced with the body's need for water; the camel is an excellent example of this trade-off in a hot climate (p. 733). Some desert birds use the same strategy; that is, they let their body temperatures rise during the day to reduce evaporative cooling, thus saving up to 50% of body water. Another mechanism involves physiological regulation of the skin barrier. For example, zebra finches deprived of water add layers of lipids to their skin, reducing cutaneous water loss by 75% or more.

The hypothalamus and/or spine integrates a multitude of thermosensory inputs from both the core and the surface of the body

In vertebrates, the hypothalamus controls most mechanisms of thermoregulation. In birds, the hypothalamus appears to control metabolic rate, panting, vasodilation, and shivering, but a reflex integrator in the spine has also been identified that regulates vasodilation and constriction, panting, shivering, and erection of plumage. The hypothalamus (plus the spinal integrator in birds) acts like a "thermostat" in negative-feedback fashion. As we discussed in Chapter 1 (Figure 1-8), a house thermostat keeps track of the temperature in a room and triggers a heating mechanism (the furnace) or a cooling mechanism (the air conditioner) as necessary to maintain the room temperature at the indicated setting. Similarly, the hypothalamus receives afferent information about body temperature from various regions of the body and (particularly in mammals) initiates coordinated adjustments in heat gain and heat loss mechanisms as necessary to correct any deviations in core temperature from the set point. The hypothalamic thermostat is far more sensitive than your house thermostat. The hypothalamus in endotherms can respond to changes in blood temperature as small as 0.01°C.

To make the appropriate adjustments in the delicate balance between the heat loss mechanisms and the opposing heat-producing and heat-conserving mechanisms, the hypothalamus must be continuously appraised of both the skin temperature and the core temperature by means of specialized temperature-sensitive *thermoreceptors*. The core temperature is monitored by *central thermoreceptors*, which are located in the hypothalamus itself as well as elsewhere in the central nervous system and the abdominal organs. These sensors monitor the core temperature that is actually being defended homeostatically. *Peripheral thermoreceptors* monitor skin temperature throughout the body and transmit information about changes in surface temperature to the hypothalamus. To some extent, these sensors serve in *anticipation* fashion. That is, changes in skin temperature give advance warning of potential environmental threats to the core temperature, allowing corrections to be initiated before the core temperature is actually disturbed.

Two centers for temperature regulation are in the hypothalamus. The *posterior region*, activated by cold, triggers reflexes that mediate heat production and heat conservation. The *anterior region*, activated by warmth, initiates reflexes that mediate heat loss. Together, these work to maintain a remarkably consistent core temperature (which, however, oscillates slightly because of feedback delays; see p. 16). Let's see how.

To regulate core temperature homeostatically, the hypothalamus simultaneously coordinates heat production, heat loss, and heat conservation mechanisms

Let's now pull together the coordinated adjustments in heat production as well as heat loss and heat conservation in response to exposure to either a cold or a hot environment for a mammal (Figure 15-20). In response to cold exposure, the posterior region of the hypothalamus directs increased heat production, such as increased muscle tone and shivering and BAT activity, while simultaneously decreasing heat loss (that is, conserving heat) by skin vasoconstriction and other measures. Because there is a limit to the ability to reduce skin temperature through vasoconstriction, when the external temperature falls too low, even maximum vasoconstriction is not sufficient to prevent excessive heat loss. Accordingly, other measures must be instituted to further reduce heat loss, such as puffing up of hair. After maximum skin adjustments have been achieved physiologically, further heat loss can be prevented only by behavioral adaptations (as we described on p. 736).

Under the opposite circumstance—heat exposure—the anterior part of the hypothalamus reduces heat production by decreasing skeletal muscle activity and promoting increased heat loss by inducing skin vasodilation. When even maximal skin vasodilation is inadequate to rid the body of excess heat, sweating or panting is brought into play to accomplish further heat loss through evaporation.

Thermal Neutral Zone The consequences of thermoregulatory actions on metabolism are illustrated in Figure 15-21, which shows the metabolic rate as a function of environmental temperature. In the middle of this diagram lies the **thermal neutral zone** (**TNZ**), a range of environmental temperatures in which the animal does not need to expend significant energy for thermoregulation. In this zone, mechanisms involving insulation, blood vessels, and low-energy behaviors are sufficient. Below the neutral zone, at a point called the **lower critical temperature**, metabolic rate increases as the special heat-generating mechanisms are activated. If these mechanisms are insufficient, the animal may suffer from **hypothermia** (dangerously low body temperature). Above the neutral zone, at a point called the **upper critical temperature**, metabolism increases because of panting (which uses rapid muscle activity) or heavy sweating (which uses ion transport processes). These activities counterproductively but unavoidably generate even more heat to deal with, leading to the potential for **hyperthermia** (dangerously high body temperatures).

The width of the TNZ and the critical temperatures depend on the effectiveness of the thermoregulatory adaptations that do not require significant energy. Insulation is often the major factor. For example, the lower critical temperature for the lightly insulated cardinal (*Cardinal cardinalis*) is 18°C, whereas for the heavily insulated emperor penguin (*Aptenodytes forsteri*) it is −10°C. In tropical hummingbirds (such as *Colibri delphinae*), there are no critical temperatures; metabolic rate decreases linearly with environmental temperature as the latter is increased from 4 to 40°C. Seasonal acclimatization can alter the TNZ. For example, with a summer hair coat the critical lower temperature for a beef cow averages 15°C (59°F), whereas with a heavy winter coat wind chills of −8°C (18°F) are comfortably tolerated. Fermentation in rumens also produces heat as a by-product, contributing to the comparatively low critical lower temperatures observed in ruminants.

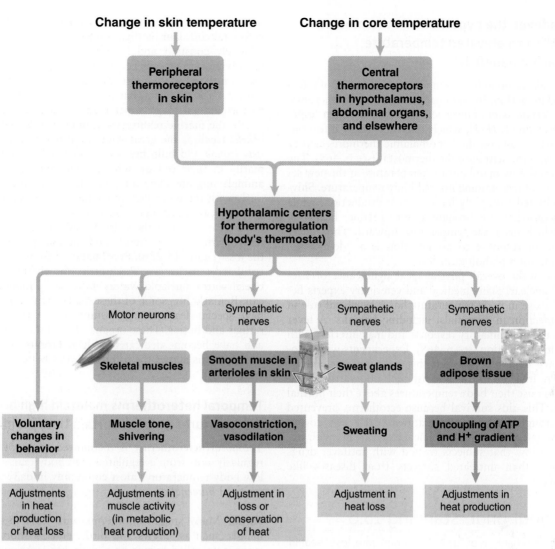

FIGURE 15-20 Major thermoregulatory pathways in a mammal.

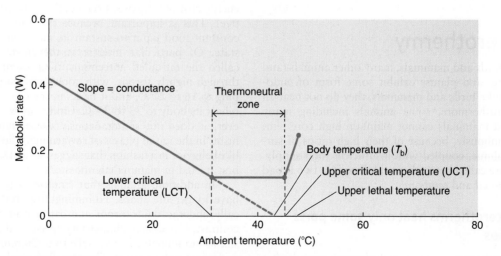

FIGURE 15-21 The influence of ambient temperature on the metabolic rate (MR) of a small temperate endotherm, showing the thermoneutral zone where MR is constant.

Source: Willmer, P., G. Stone and I. Johnston, (2000), "Environmental Physiology of Animals", Fig. 8.33. Copyright © 2000 Blackwell Publishing. Reproduced with permission of Blackwell Publishing Ltd.

During a fever, the hypothalamic thermostat is "reset" to an elevated temperature, which can be beneficial

Fever is an elevation in body temperature as a result of infection or inflammation. In mammals, in response to microbial invasion, certain white blood cells release a **pyrogen** (especially *interleukin-1, IL-1*), which among its many infection-fighting effects, acts on the hypothalamic thermoregulatory center to raise the setting of the thermostat (see p. 469). The hypothalamus now maintains the temperature at the new set level instead of maintaining normal body temperature. Shivering is initiated to rapidly increase heat production, while skin vasoconstriction is brought about to reduce heat loss, both of which drive the temperature upward. Thus, fever production in response to an infection is a "deliberate" mechanism, not a pathological one.

Although the overall physiological significance of a fever is still unclear, many medical and veterinary experts believe that a rise in body temperature has a beneficial role in fighting infection in mammals, including humans. A fever augments the inflammatory response and may interfere with bacterial multiplication (see p. 469). Experiments with ectotherms have demonstrated that this is an ancient defense mechanism. Infected lizards and fish seek out warmer habitat areas to raise their body temperatures above their normal optimum. This aids survival because ectotherms prevented from elevating their body temperatures in this way do not survive the infection as well. Medical experiments with humans also show that subjects treated with antifever drugs suffer more than untreated subjects from diseases like pneumonia.

check your understanding 15.6

Describe mechanisms of "gain, retain, generate, lose" heat in achieving homeothermy in birds and mammals.

Discuss how the hypothalamus maintains homeothermy with respect to temperatures out of the thermal neutral zone, and when and why it elevates temperature to create a fever.

15.7 Heterothermy

In addition to birds and mammals, many other animals (and even some fungi and plants) exhibit some form of endothermy, but unlike birds and mammals, they do not heat all their cores. Furthermore, some animals including many small birds and mammals cannot maintain high core temperatures continuously, because of their high ratios of surface area to volume, coupled with insufficient food supply. Such animals are called **heterotherms**, falling into two broad categories: *regional* and *temporal*.

Regional heterotherms heat only some parts of their bodies

Regional heterothermy occurs in endothermic animals that heat only certain organs or body regions, not their entire bodies. For example, flying insects such as bees and moths heat their thoraxes by the activity of flight muscles. As a preflight warm-up, some species shiver their flight muscles before takeoff. During flight they can keep their thoraxes at relatively constant and high temperatures by controlling hemolymph flow between the thorax and abdomen, which are connected by countercurrent flow channels (Figure 15-22a). The abdomen has a thin ventral surface, the *thermal window*, through which excessive heat can be lost.

In the marine realm, two groups of fishes—the *lamnid sharks* (such as the great white) and many *scombroid teleosts* (tunas and billfishes)—have evolved endothermy, primarily in their red or aerobic swimming muscles. These animals migrate over long distances, swimming continuously with these muscles and generating constant heat output. But it is not enough to heat the entire body, because too much heat is lost at the gills. A countercurrent rete system between the aerobic muscle and the gills prevents most of the loss (Figure 15-22b). Presumably, these warm, thermally regulated muscles give these fishes the ability to move between warm and cold waters more easily than ectotherms can. Remarkably, some of these fishes such as the swordfish have special heater organs behind their eyes to heat their retinas and brains. These organs evolved out of eye muscles and have become dedicated to heat production only! Presumably, this allows the fish to see prey better in cold, dark waters.

Temporal heterotherms maintain high body temperatures only for certain time periods

Temporal heterothermy is manifested in endotherms that regularly shift from a regulated high body temperature to a low body temperature, most commonly in daily *torpor* and seasonal *hibernation*.

Torpor Many small endotherms with high metabolic demands, such as shrews and birds, enter a short-term dormant state called **torpor** on a daily basis. At night, for example, a deer mouse's body temperature (Figure 15-23a) drops from around 35°C to as low as 15 to 20°C (depending on environmental temperatures), saving a large amount of energy. (Note that these animals usually do not let body temperatures drop close to freezing, possibly because the energy cost of rewarming on a daily basis would be prohibitive). This is important because this mouse needs almost constant food input to sustain its metabolism in the active state. Of particular interest is the Australian marsupial called the fat-tailed antechinus. This small mammal goes through nightly torpor, with core body temperatures dropping to 16 to 27°C, and it basks in the sun each morning to reheat its body to 33 to 37°C. Unlike other mammals, however, it does not simultaneously use endothermic mechanisms in the initial phases of rewarming. Thus, it uses lizard-like behavior in a fashion that suggests how thermoregulation first started in mammalian ancestors.

In addition, animals that employ daily torpor tend to have longer life spans. Hummingbirds, for example, typically undergo daily torpor and can live up to 10 years. In contrast, Norwegian shrews (six species, all about the same size as hummingbirds) live only 16 to 20 months and do not undergo regular torpor. The so-called southern African elephant shrew (*Elephantus myurus*), which is not a true shrew but is of similar size, does undergo torpor, and it lives four to six years. Thus, reductions in metabolism on a regular

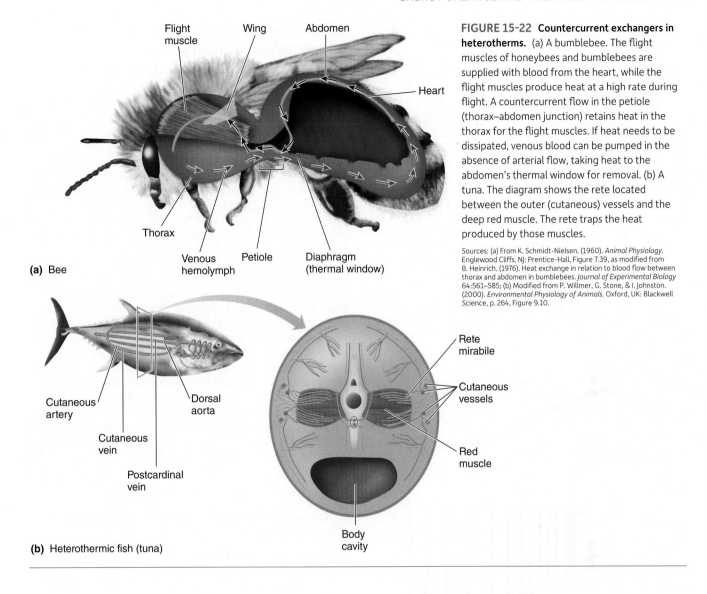

(a) Bee

Flight muscle — Wing — Abdomen — Heart — Thorax — Venous hemolymph — Petiole — Diaphragm (thermal window)

(b) Heterothermic fish (tuna)

Cutaneous artery — Dorsal aorta — Cutaneous vein — Postcardinal vein — Body cavity — Rete mirabile — Cutaneous vessels — Red muscle

FIGURE 15-22 Countercurrent exchangers in heterotherms. (a) A bumblebee. The flight muscles of honeybees and bumblebees are supplied with blood from the heart, while the flight muscles produce heat at a high rate during flight. A countercurrent flow in the petiole (thorax–abdomen junction) retains heat in the thorax for the flight muscles. If heat needs to be dissipated, venous blood can be pumped in the absence of arterial flow, taking heat to the abdomen's thermal window for removal. (b) A tuna. The diagram shows the rete located between the outer (cutaneous) vessels and the deep red muscle. The rete traps the heat produced by those muscles.

Sources: (a) From K. Schmidt-Nielsen. (1960). *Animal Physiology.* Englewood Cliffs, NJ: Prentice-Hall, Figure 7.39, as modified from B. Heinrich. (1976). Heat exchange in relation to blood flow between thorax and abdomen in bumblebees. *Journal of Experimental Biology* 64:561–585; (b) Modified from P. Willmer, G. Stone, & I. Johnston. (2000). *Environmental Physiology of Animals.* Oxford, UK: Blackwell Science, p. 264, Figure 9.10.

basis may slow the manifestation of aging, much of which is thought to be caused by reactive oxygen species (ROS, p. 53) generated in proportion to metabolic rate.

Hibernation Many small endotherms become nearly or fully poikilothermic as winter approaches, entering a long-term dormant state called **hibernation,** in which body temperatures may drop to near ambient (Figure 15-23b). Remarkably, the Arctic ground squirrel even supercools slightly, with body temperatures of –3°C. Hibernators must store up large amounts of unsaturated fats (which do not turn hard like butter at cold body temperatures) to serve as energy reserves. However, these animals are usually not fully ectothermic. Mammalian hibernators still have a functioning hypothalamic thermostat, but with a greatly lowered set point. The hypothalamus has been shown to regulate a lowering of body temperature and also to regulate heating tissues such as BAT to prevent internal freezing. Many also go through periodic "bouts" of rewarming (Figure 15-23b) and then recooling every few days or weeks. It is not known why they do this, but one hypothesis is that they must warm up in order to void wastes and sometimes to eat cached food, with the latter serving in particular to restore glycogen stores in their brains. Another hypothesis is that they do this periodic awakening to reactivate their immune systems temporarily in case they have been infected.

Contrary to popular belief, only mammals about marmot size (3 kg) or smaller can truly hibernate, because of the principles of ratios between surface area and volume, discussed earlier. A bear is simply too large to cool off greatly and then reheat in a reasonable time in the springtime, so it essentially sleeps through much of the winter with a body temperature only about 5°C lower than normal. This may be more properly called *winter sleep,* although the term *hibernation* is still often used.

Though hibernation is usually associated with high latitudes, at least one tropical mammal, the Madagascan fat-tailed dwarf lemur (*Cheirogaleus medius*), hibernates. Its body temperature can be purely ectothermic at times, fluctuating by up to 20°C in a day.

Regulation of hibernation is still incompletely understood. In a startling discovery, Mark Roth and colleagues reported in 2005 that mice breathing moderate concentrations of H_2S (the "rotten-egg" gas that is toxic at high levels) went into a reversible, deep hibernation-like state. This is remarkable since mice do not hibernate naturally. How H_2S does

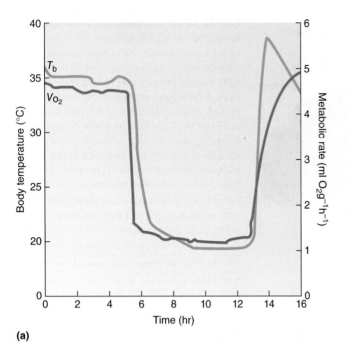

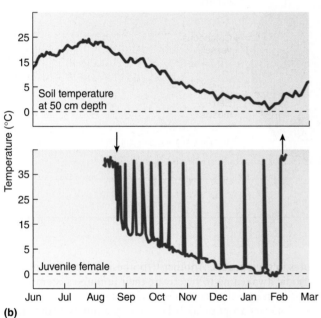

(b)

FIGURE 15-23 **Temperatures of mammals in torpor and hibernation.** (a) Metabolic rate (V_{O_2}) and body temperature (T_b) of a deer mouse during daily torpor. (b) Soil temperature and body temperature in a Richardson's ground squirrel (*Spermophilus richardsonii*) hibernating through the winter, showing periodic "bouts" of rewarming.

Sources: (a) From P. C. Withers. (1992). *Comparative Animal Physiology*. Fort Worth, TX: Saunders, using data from J. R. Nestler. (1990). Relationships between respiratory quotient and metabolic rate during entry to and arousal from daily torpor in deer mice (*Peromyscus maniculatus*). *Physiological Zoology* 63:504–515; (b) From G. R. Michener. (1998). Sexual differences in reproductive effort of Richardson's ground squirrels. *Journal of Mammalogy*, 79:1–19.

this, and whether natural hibernators use this gas, is not certain. Some evidence shows that the gas inhibits mitochondrial respiration, but it is also known to be a body regulatory signal (much like NO; p. 93) at minute concentrations.

Some colonial heterotherms create homeothermic "superorganisms"

Finally, spectacular examples of homeothermy can be found in certain animal colonies in which the individual animals are heterothermic or poikilothermic. Several social insects form large colonies that are thermoregulated by the collective behavior of the group. These "superorganisms" include many beehives and termite mounds, in which groups of animals construct ventilation tubes and use their wing beats to heat up or convectively cool the colony. Honeybees also carry water into their hives for evaporative cooling. Thus, beehives can maintain temperatures that are much less variable than the environment. In tropical habitats where the environmental temperature does not vary extensively, some termite mounds are homeothermic all year round.

Thermoregulated hives have recently been shown to be crucial to larval bee development. Pupated larvae were raised at 32, 34, and 36°C in the laboratory. Those raised at the lower temperatures had abnormal neural development. Amazingly, when the colony is infected with a pathogenic fungus, the bees can even create a feverlike state. The fungus, which can kill bee larvae, is greatly inhibited by the elevated temperature. Swarms of honeybees have also been documented using rapid wing beats to create heat intense enough (over 50°C) to kill parasitic wasps attempting to invade the hive.

One amazing group of mammals, the naked mole rats, has evolved a hivelike system that resembles these insect colonies. See *A Closer Look at Adaptation: The Naked Mole Rat—Mammalian Hive Ectotherm.*

check your understanding 15.7

Discuss regional heterothermy and the mechanisms involved in a tuna and a bee.

Discuss the features and adaptive value of hibernation and torpor.

15.8 Thermal Physiology and Climate Change

By the first decade of the 21st century, science had firmly established that the atmosphere and oceans are warming up. Indeed, the years 2005 and 2010 were the hottest years on record. Average ocean surface temperature in July 2009 was the warmest ever recorded. This **global warming** is clearly due, at least in part, to anthropogenic (man-made) "greenhouse" gases, mainly CO_2 from fossil-fuel burning and methane from cattle (p. 669), rice paddies, and other sources. (Recall from Chapter 13 that CO_2 is also acidifying the oceans.) These gases, which prevent infrared (heat) radiation from leaving the Earth, have increased dramatically in the last century or so in parallel with temperature increases.

Arctic regions have warmed the most, and climate is becoming more extreme

The warming trend is not by any means uniform. In particular, in the last 30 years, southern temperate regions have not experienced net warming, while the Arctic has experienced

A Closer Look at Adaptation
The Naked Mole Rat—Mammalian Hive Ectotherm

Although traditional physiology states that all mammals are endotherms, there is at least one remarkable exception: the naked mole rat. Twelve known species of naked mole rats live in underground tunnel complexes underneath savanna and grasslands of equatorial Africa. *Heterocephalus glaber* has been the most studied species. These mole rats have almost no body fur (although they do have fine hairs) and no subcutaneous fat, and thus have no effective insulation. Most amazingly, they cannot individually regulate their body temperatures via endothermy. Thus, their body temperatures do fluctuate somewhat. But a degree of homeothermy around 31°C is still achieved using mechanisms like those of social insects. First, they occasionally bask in the sun at their tunnel entrances. Second, the animals do produce some heat, and by huddling together while sleeping (along with some insulation provided by the soil around the tunnel), they can trap much of this heat. Altogether, their thermoregulatory mechanisms are very similar to those of termites in their mounds.

Thermoregulatory physiology of these mammals is not the only parallel to insect physiology. Both species have a social structure (termed *eusocial*) that is very much like the hive societies of social insects. There is a single queen mole rat, for example, that mates with a few dominant males; there are many workers that dig the tunnels and gather food; and there are a few soldiers that do not work but protect the colony from predators such as snakes. Thus, cooperation within the colony is essential to survival because a pair of mole rats on their own would not be able to perform all the functions needed to cope with environmental challenges.

A naked mole rat.

These animals are becoming laboratory models for medical research because they have other amazing physiological features relating to many other chapters in this book. For example, they have no functional nociceptors for noxious chemicals (Chapter 6, p. 260), allowing them to tolerate the buildup of ammonia in their burrows. They are incredibly tolerant of hypoxia (their brains surviving after 30 minutes without O_2), an obvious adaptation to burrow life (Chapter 11, p. 542). Their two large teeth evolved from fixed chewing structures (Chapter 14, p. 663) into independently moveable digging and manipulating devices, with about 30% of the animal's somatosensory cortex (Chapter 5, p. 192) devoted to those teeth. Perhaps most remarkably, their lifespan is about 30 years, compared to 3 years or less for a similar sized rat (lifespan normally scales inversely with body size; Chapter 1, p. 10). They do not get osteoporosis (Chapter 7, p. 328) and appear to avoid cancer as well, perhaps due to better immune-surveillance mechanisms (Chapter 10, p. 486), and they have proteins highly resistant to reactive oxygen species (ROS; Chapter 2, p. 53).

the greatest warming (Figure 15-24a) including increasing loss of the Arctic ice cap. Climate appears to be getting more chaotic, and the term **climate change** is increasingly used more often than global warming. Earth's climate has many (poorly understood) feedback components ("geophysiology," p. 14) and may exhibit the "flaws" inherent in feedback systems (p. 16) such as oscillating under- and overshoots (e.g., hotter summers but colder winters). Uncertain is whether climate change is subject to negative or positive feedbacks or both. For example, warmer oceans lead to more evaporation, which has increased rainfall and (in winter) snowfall in some areas; likewise, evaporation along with more DMS (p. 14) may also trigger more cloud formation that would block sunlight, thus cooling the planet (negative feedback). Conversely, increased melting of sea ice and snow (which reflect sunlight) exposes more water and dark land to the rays of the sun. The water and land in turn absorb more solar energy, which could accelerate global warming and melting, and so on (positive feedback).

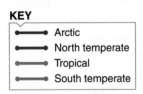

KEY

- Arctic
- North temperate
- Tropical
- South temperate

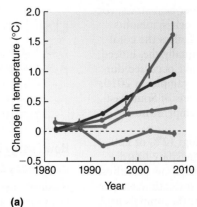

(a)

(b)

FIGURE 15-24 Global temperature changes and their effects on animal metabolic rates. (a) Recent changes in average temperatures for various geographical regions; (b) Predicted effects of those changes on mass-specific metabolic rates for ectotherms in those regions.

Source: M. E. Dillon, G. Wang & R. B. Huey, (2010), Global metabolic impacts of recent climate warming, 'Nature', 467:704–706. Used by permission from Macmillan Publishers Ltd, copyright 2010.

Animal physiology, behavior, and genes are already being affected by warming trends

As you might expect, due to the pervasive importance of temperature, habitat temperature changes are expected to impact all types of organisms. Animal physiologists are increasingly studying such impacts, at the theoretical level using physiological principles, in the laboratory to test potential effects, and in the natural world to test for effects already occurring. At the theoretical level, a study by Michael Dillon, George Wang, and Raymond Huey in 2010 yielded a perhaps unexpected prediction. While the Arctic is warming faster than the tropics (Figure 15-24a), metabolic rates (MR) of tropical ectotherms will likely rise faster than those of Arctic ectotherms (Figure 15-24b). This is because of the Q_{10} effect we described earlier (p. 730), in which MR doubles (or more) with each 10°C rise in temperature. This is an exponentially rather than linear effect. For example, an MR of 1 kJ/hr at 0°C might rise to 2 at 10°C, but an MR of 6 kJ/hr at 25°C could rise to 12 at 35°C. Thus, the absolute change is much greater in the tropics.

Laboratory studies are also revealing. For example, Hans Pörtner and colleagues have shown that, in fish at least, both cooling and warming reduce an animal's *aerobic scope* (p. 720) and that this might be the first major effect of climate change.

What about real-world effects? Considerable evidence has accumulated showing they are already occurring. Let's look at some of the studies.

Heat Shock Although global averages have risen only a few degrees, temporary but severe heat waves have greatly increased. For example, 30,000 flying foxes died of heat stress in Australia between 1994 and 2010 during unprecedented heat waves reaching 43°C or more. Tropical species are predicted to be hardest hit, not only because of the Q_{10} effect, but also because their physiologies are adapted to fairly consistent temperatures rather than seasonally changing ones.

Energy Balance Effects Habitat warming is affecting the "animal energy equation" for many animals, such as the following:

- *Tropical corals* that build reefs rely heavily on symbiotic algae for sugars and amino acids. Under heat stress, corals will "bleach"—that is, they expel their symbionts (with their photosynthetic pigments). Unless the coral can obtain more symbionts, they eventually die. Indeed, large-scale deaths of bleached coral have occurred during unprecedented temperatures. For example, in 2010, 50% of coral colonies died in many reefs of Southeast Asian and Indian Ocean after water temperature increased by 4°C above normal.
- *Tropical lizards* of Mexico are suffering: 12% of 200 populations of *Sceloporus serrifer* (blue spiny lizards) have died out in recent decades. Studies show that those lost populations were in areas where springtime temperatures have increased to the point that the animals probably had to reduce foraging time, staying in burrows to avoid overheating.
- *Yellow-bellied marmots* in Europe are getting fatter (and perhaps more successful) with warming due to longer foraging seasons, but *Arctic foxes* in Iceland have shrunk in size, perhaps as an adaptive way to cool off more easily, or due to decreased food supply.

Habitat Change Some animals have shifted habitats; for example, in 2003 Terry Root and colleagues gathered data from 143 studies and showed that hundreds of Northern Hemisphere species including mollusks, amphibians, birds, and mammals, as well as plants, have shifted north in correlation with warming. Other shifts include these:

- The two *limpet* species we discussed earlier (p. 733) have shifted ranges, with the southern species displacing the northern species in Monterey Bay, which has warmed up.
- The number of continental *butterfly* species in Britain has increased as they are now migrating north, presumably to find cooler temperatures.
- *Birds* in the University of California's Deep Canyon Reserve have moved upslope by 117 m in the last 30 years, correlating with an increase of 5°C in average temperature in the breeding season and a decline in rainfall by 44%.

Reproduction Finally, reproductive effects are also occurring, some of them genetic.

- The *grapevine moth* of Spain now breeds 12 days earlier than a century ago. The change correlates with a 3°C rise in springtime temperatures between 1984 and 2006. This pest, which eats grapevines, may increase its damage to vineyards.
- *Red squirrels* of the Yukon are breeding 18 days earlier than in the past, as springtime temperatures and food supply are increasing. This appears to be a genetic, evolved change.
- *Pitcher-plant mosquitoes* go into winter dormancy based on photoperiod, with northern populations entering dormancy earlier than southern ones. Now, in concert with warming, the northern groups are entering dormancy later, a change that is genetic (evolved).

As you can see, some species may be adapting to warming effects (some behaviorally, some by evolutionary genetic changes), while some are affected negatively and others positively. The long-term effects on ecological interactions are still uncertain. Only time will tell.

check your understanding 15.8

Discuss four findings that show effects of recent global warming on animal physiology.

making connections

How Energy Balance and Thermal Adaptations Contribute to the Body as a Whole

Because energy can be neither created nor destroyed, for body mass and body temperature, respectively, to remain constant, input must equal output in both an endotherm's total energy balance and its heat–energy balance. If total energy input exceeds total energy output, the extra energy is stored in the body, and body mass increases. Similarly, if the input of heat energy exceeds its output, body temperature increases. Conversely, if output exceeds input, body mass decreases or body temperature falls. The hypothalamus is the major integrating center in mammals for maintaining both a constant total energy balance (and thus a constant body mass) and a constant heat–energy balance (and thus a constant body temperature). However, the effectors for temperature and energy regulation include components of many other systems covered in previous chapters. Energy balance, for example, obviously requires the digestive system and endocrine as well as hormonal regulation (such as insulin) and neural regulation (including behaviors such as eating). Temperature regulation involves (among other things) the skin and its blood vessels (and sweat glands in some mammals), respiration (panting), and skeletal muscles (shivering, or behavioral avoidance or seeking).

Body temperature is one of the most pervasive influences on body functions, which slow down if too cold, and suffer from denaturing macromolecules if too warm. Some ectotherms achieve partial homeostasis by controlling external heat exchanges (mainly through behavior), whereas others are not homeostatic but must adapt through dormancy or internal biochemical adjustments. Endotherms achieve homeostasis (homeothermy) much more consistently with physiological mechanisms.

Chapter Summary

- Life follows the laws of thermodynamics. Energy cannot be created or destroyed (First Law), and useful energy is continually lost to entropy (Second Law). To compensate for entropy, animals require input of food energy, most of which is converted to heat output.

- Energy input must equal energy output, which is measured as metabolic rates in the form of basal metabolism, activity, diet-induced thermogenesis (DIT), and production. Indirect calorimetry through respirometry of O_2 consumption is a common measurement method.

- Energy of production determines net energy balance, which can be positive, negative, or neutral. Basal and standard metabolic rates (BMR, SMR) are the "idling costs" for life, regulated in part by thyroid hormones. BMR and SMR are scaled to body mass: Smaller animals need more energy per unit mass than larger ones, for unclear reasons. Increased muscle activity is the factor that can most increase metabolic rate, as indicated by metabolic scope (ratio of maximal activity to BMR or SMR). DIT occurs after eating in most animals, and involves the costs of processing food.

- Some adult animals maintain long-term neutral energy balance, while others undergo regulated periods of positive or negative energy balance. Food intake in mammals is controlled primarily by the hypothalamus in response to numerous inputs, with internal signals such as orexins. Leptin from adipose and insulin from the pancreas are signals of long-term energy balance, while others such as ghrelin from the stomach and CCK from the intestine signal short-term energy deficiency or sufficiency. Effectors for energy balance are mechanisms to regulate eating and expenditures via regulated changes in BMR, DIT, activity, and production. Brown adipose tissue (BAT), for example, can burn off excess food energy.

- Different energy-balance set points among species result from evolutionary adaptations to different food supplies. Plentiful food favors "lean" set points to maintain agility, while intermittent food favors "heavy" set points for storing up reserves for famine periods.

- Temperature alters rates of chemical reactions and denatures macromolecules. Biomolecules can be altered to work optimally at different temperatures: Membrane fluidity is kept optimal by changes in saturation (homeoviscous adaptation), while proteins evolve to balance thermostability with flexibility needed for activity.

- The thermal adaptation strategies of animals depend on their primary source of heat: Endotherms rely on internal heat, ectotherms on external heat; poikilotherms have variable body temperatures, homeotherms have nearly constant body temperatures.

- Heat exchange between the body and the environment takes place by radiation, conduction, convection, and evaporation. Heat gain versus heat loss determines core body temperature. Animals gain heat from the environment while avoiding losses, retain body heat, generate more body heat, and lose excess heat.

- Body temperatures of ectotherms may follow the environment (poikilothermy), or may be regulated by external exchanges such as solar basking, shade seeking (ectothermic regulation via gain, retain, lose). Some ectotherms can compensate biochemically for changes in body temperatures with internal adjustments in enzyme concentration, pH, membrane saturation, and protein isoforms.

- Ectotherms survive extreme cold by metabolic dormancy and by either freeze tolerance (often with antifreeze carbohydrates in cells to prevent ICF but not ECF freezing), or freeze avoidance with antifreeze proteins, supercooling, or dehydration. Ectotherms may survive temporary extreme heat with heat shock proteins that protect other proteins.

- Birds and mammals maintain a consistent internal core temperature (homeothermy). To do so, heat gain must balance heat loss, using numerous mechanisms to gain (e.g., ectothermic behavior), retain (e.g., insulation, vasoconstriction, countercurrent exchangers), generate (e.g., high BMR, shivering, BAT), and lose excess (e.g., sweating, panting, vasodilation) heat.

- The hypothalamus and/or spine integrates a multitude of thermosensory inputs from both the core and the surface of the body. To regulate core temperature homeostatically, the hypothalamus simultaneously coordinates heat production, heat loss, and heat conservation mechanisms. During a fever, the hypothalamic thermostat is "reset" to an elevated temperature, which can be beneficial.

- Regional heterotherms heat only some parts of their bodies. Temporal heterotherms maintain high body temperatures only for certain time periods. Some colonial heterotherms create homeothermic "superorganisms" such as termite mounds and beehives.

- Climate change, particularly warming, is having documented effects on animal migration, metabolism, habitat location, and reproduction.

Review, Synthesize, and Analyze

1. If more food energy is consumed than is expended, explain the possible fates of the excess energy.

2. Explain how drugs that selectively block key appetite signals would affect feeding behavior.

3. The basal metabolic rate (BMR) is about 72 kcal/hr (301 kJ/hr) in an average human, with nearly all the energy converted to heat. If we could not lose this heat, our temperature would rise until we boiled (although we would die long before reaching that temperature). How long would that take? If an amount of energy ΔU is put into a liquid of mass m, the temperature change ΔT (in °C) is given by the following formula:

$$\Delta T = \Delta U/m \times C$$

Here, C is the specific heat of the liquid. For water, $C = 1.0$ kcal/kg-°C. Calculate how long it would take for BMR heat to boil your body fluids (assume 42 liters of water in your body and a starting body temperature at 37°C).

4. Why is it dangerous to engage in heavy exercise on a hot, humid day?

5. Discuss the advantages and disadvantages of ectothermy and endothermy, and why the latter might have evolved in some groups.

6. Different groups of homeotherms have different body temperatures. Why might that be?

7. Do you think fish run a fever when they have a systemic infection? Why or why not?

8. Compare the role of BAT in body-mass regulation and body-temperature regulation.

9. Small endotherms such as hummingbirds are temporal heterotherms, saving energy at night by undergoing torpor. Why don't large animals such as humans do this as well?

10. The discovery that H_2S triggers hibernation in a nonhibernating rodent has led to proposals for human applications. What might those be, and why?

Suggested Readings

Alexander, R. M. (1982). *Locomotion of Animals.* Glasgow, UK: Blackie. A classic text on mechanisms and energetics of animal locomotion.

Block, B. A., J. Finnerty, A. F. R. Stewart, & J. A. Kidd. (1993). Evolution of endothermy in fish: Mapping physiological traits on a molecular phylogeny. *Science* 260:210–214.

Burton, R. F. (2002). Temperature and acid–base balance in ectothermic vertebrates: The imidazole alphastat hypotheses and beyond. *Journal of Experimental Biology* 205:3587–3600. A review of the hypothesis on pH regulation as an adaptation to temperature.

Cannon, B., & J. Nedergaard. (2004). Brown adipose tissue: Function and physiological significance. *Physiological Reviews* 84:277–359.

Clarke, A., & H.-O. Pörtner. (2010). Temperature, metabolic power and the evolution of endothermy. *Biological Reviews* 85:703–727. A hypothesis on how endothermy evolved.

Dunnan, J. G. (2001). Antifreeze and ice nucleator proteins in terrestrial arthropods. *Annual Review of Physiology* 63:327–357.

Feder, M. E., & G. E. Hofmann. (1999). Heat-shock proteins, molecular chaperones, and the stress response: Evolutionary and ecological physiology. *Annual Review of Physiology* 61:243–282.

Gillooly, J. F., J. H. Brown, G. B. West, V. M. Savage, & E. L. Charnov. (2001). Effect of size and temperature on metabolic rate. *Science* 293:2248–2251. On the possible universal nature of scaling and its implications.

Goodman, B. (1998). Where ice isn't nice: How the icefish got its antifreeze and other tales of molecular evolution. *BioScience* 48:586–591.

Heinrich, B. (1996). *The Thermal Warriors. Strategies of Insect Survival.* Cambridge, MA: Harvard University Press.

Hochachka, P., & G. N. Somero. (2002). *Biochemical Adaptation: Mechanism and Process in Physiological Evolution.* Oxford, UK: Oxford University Press. A classic text.

Isaac, N. J. B., & C. Carbone. (2010). Why are metabolic scaling exponents so controversial? Quantifying variance and testing hypotheses. *Ecology Letters* 13:728–735.

Pörtner, H. O., & A. P. Farrell. (2008). Physiology and climate change. *Science* 322:690–692. On effects of climate change on aerobic scope of animals.

Schmidt-Nielsen, K. (1998). *The Camel's Nose, Memoirs of a Curious Scientist.* Washington, DC: Shearwater Books. A lively, engaging discussion of a prominent comparative physiologist's pioneering research.

Secor, S. M., & J. M. Diamond. (2000). Evolution of regulatory responses to feeding in snakes. *Physiological and Biochemical Zoology* 73:123–141.

Storey, K. B. (1997). Organic solutes in freezing tolerance. *Comparative Biochemistry and Physiology* 117A:319–326.

Suarez, R. K. (1992). Hummingbird flight: Sustaining the highest mass-specific metabolic rates among vertebrates. *Experientia* 48:565–570.

Watanabe, M. E. (2005). Generating heat: New twists in the evolution of endothermy. *BioScience* 55:470–475. Another hypothesis on how endothermy evolved.

Wehner, R., A. C. Marsh, & S. Wehner. (1992). Desert ants on a thermal tightrope. *Nature* 357:586–587. About an ant that may be the most thermotolerant animal.

Weibel, E. R. (2002). The pitfalls of power laws. *Nature* 417:131–132.

West, G. B., J. H. Brown, & B. J. Enquist. (1999). The fourth dimension of life: Fractal geometry and allometric scaling of organisms. *Science* 284:1677–1679.

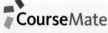
CourseMate

Access an interactive eBook, chapter-specific interactive learning tools, including flashcards, quizzes, videos and more in your Biology CourseMate, accessed through **CengageBrain.com**.

Kittiwake (*Rissa tridactyla*) with chicks (Bass Rock, Scotland).

Reproductive Systems

16.1 Introduction: Reproductive Processes

The central theme of this book has been the physiological processes aimed at maintaining homeostasis to ensure survival of the individual animal. However, as we noted in Chapter 1 and in several subsequent chapters, some processes, such as growth and development, are not aimed toward homeostasis, but rather constitute regulated changes. We end this book with a discussion of reproductive systems, which also are not aimed toward homeostasis and, moreover, are not necessary for survival of an individual. Instead, they are essential for perpetuation of species. Ever since the first life on Earth evolved, only cells give rise to new cells; that is, each species survives beyond the lives of its individual members only through reproduction using a complex genotypic "recipe." Our understanding of the origin of self-reproducing life is highly incomplete, but progress is continuing. Craig Venter in 2010 created the first synthetic genome: a putative "Frankenstein" microbe (*Mycoplasma laboraorium*). His team synthesized a DNA molecule and placed it in a bacterial cell, which had its DNA removed, thus instructing the cell to operate and reproduce with an entirely new blueprint. Theoretically, this feat may yield important insight into the initial evolution of life; it may also open the door to the creation of novel genomes, which could be designed to perform specific reactions, leading to advances in environmental and energy applications.

Animals may reproduce asexually or sexually

Those first cells on Earth, and indeed all extant prokaryotes, reproduce by mitosis: a *cloning* process in which one cell divides to give rise to two genetically identical cells. Eukaryotic cells also reproduce this way, but eukaryotes also evolved an innovation called *sex*. Let's look at animal reproduction, which can occur in three ways (with more than one occurring in some species):

1. **Budding** or **fission,** in which an animal produces a "copy" of itself without embryonic development; for example, some sea anemones split in half to produce two adults from one.

2. **Parthenogenesis,** in which an unfertilized egg develops into an embryo (see p. 767).

3. **Sexual reproduction,** which depends on the union of male and female **gametes** (reproductive, or **germ,** cells), each with a single set of chromosomes (*haploid*), to form a new individual with a unique *diploid* (that is, twice haploid) set of chromosomes (Figure 16-1).

Perhaps surprisingly, biologists since Darwin have struggled to understand why sexual reproduction is so widespread among eukaryotes but not prokaryotes. The first two methods above are called *asexual* cloning, a much faster way to reproduce that seemingly should allow rapid takeover of a habitat. However, even sea anemones regularly switch from cloning to sex. Why? Three hypotheses have found experimental support in recent years. (1) Sex allows different beneficial mutations arising in different individuals to come together, enhancing adaptation to changing environments. (2) Cloning passes on all harmful mutations,

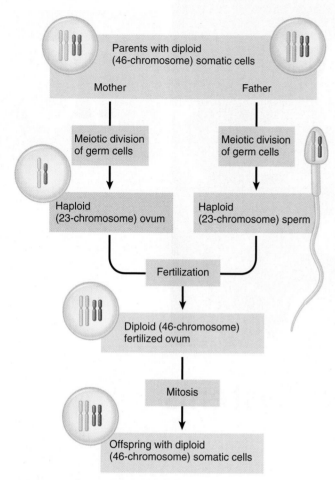

FIGURE 16-1 Chromosomal distribution in sexual reproduction.

© Cengage Learning, 2013

whereas sex allows an individual to pick up a healthy gene from a mate for its offspring. Indeed, water fleas (*Daphnia*) that reproduce asexually have more harmful mutations than sexually reproducing populations. (3) By increasing genetic diversity, sex reduces the effectiveness of parasites that have evolved to attack a specific genome. In support of this, Curt Lively and colleagues have found that a group of asexual New Zealand snails died off due to evolving parasites (and were eventually replaced by a rare, different strain of asexual snails); however, a nearby group of sexual snails have had much more steady population levels.

In this chapter we will focus primarily on sexual reproduction. Let's begin with major variations of that process.

Animals may reproduce by *r*-selected, *K*-selected, or intermediate ecological strategies

Animals employ a variety of strategies to ensure that gamete union occurs successfully and sufficient offspring are produced. Conventionally, sexual reproductive processes are classified according to their ecological and physiological adaptations. The ecological approach considers parents to have limited energy resources for producing offspring. Use of parental energy is classified according to two opposite strategies termed *r-selected* and *K-selected* reproduction (*r* is the rate of population growth, and *K* is the carrying capacity of the environment). A highly *r*-selected species places virtu-

ally all its reproductive resources into producing thousands (or more) of offspring, with no parental care and minimal nourishment. Most of the offspring do not survive, but the overwhelming numbers usually ensure that a few survive to perpetuate the species. *K*-selected species, in contrast, produce relatively few offspring, with most of the parental reproductive energy being used for nourishing and/or protecting the offspring. With this strategy, a much greater percentage of offspring survive. Any given species may be fully *r*-selected, such as a seastar that releases millions of eggs with very little yolk (nutritional stores) into the ocean; fully *K*-selected, such as a gorilla that gives birth to a few live offspring that are subsequently nourished and protected; or intermediate, such as a female garibaldi fish that produces hundreds of eggs with yolk for nutrition while the male provides some parental protection of developing eggs.

Animals may reproduce physiologically by oviparity, ovoviviparity, or viviparity

Physiological classification of sexual reproduction uses three categories (Figure 16-2):

- **Oviparous,** meaning that young develop in (and hatch from) eggs after expulsion from the mother's body (*ovi-*, "egg"; *parous,* "produce") (most nonvertebrates, fishes, amphibians, nonavian reptiles).
- **Ovoviviparous,** the production of eggs (not necessarily with shells) that use *yolk* to develop partly or fully *within* the body of the parent and may also hatch there to emerge "live" (*vivi,* "live") (some marine nonvertebrates, a few insects, some fishes and nonavian reptiles, all birds, and many pelagic sharks such as sand tiger and great white sharks).
- **Viviparous,** the production of young within the parent's body that use maternal blood nutrients (usually via a *placenta*) to develop and that emerge "live" (almost all mammals and some pelagic sharks such as the hammerhead).

Note the differences in nourishment for developing young: yolk or other forms of nutrients within the egg in the case of oviparous and ovoviviparous and placenta in the case of viviparous.

Rattlesnakes, some sharks, and guppies are examples of ovoviviparity: The female develops eggs in utero, but because no nutrients are taken from the female during development, these animals are not considered viviparous. As exemplified in the sand tiger shark (*Carcharias taurus*) the yolk is rapidly absorbed and the embryos then feed on unfertilized eggs or other embryos in the uterus. In such species only a single pup may be born from each uterus.

Similarly, because copulation does not occur in true oviparous species (insemination as well as fertilization is external to the genital tract), birds are also classified as ovoviviparous. This classification illuminates some of the reproductive similarities between birds and mammals as well as distinguishing birds from true oviparous species. For example, in true oviparous species the membranes surrounding the egg must be penetrable by the sperm. Some fish and arthropods have **micropiles,** small pores in the shell to allow sperm entrance. Further, in oviparous species embryonic development is external to the maternal genital tract, and thus the maternal contribution to the offspring arises solely

(a)

(b)

(c)

(d)

FIGURE 16-2 Reproductive strategies in animals. (a) Snails are oviparous, meaning they release eggs from which the young later hatch. (b) Many snakes and lizards, and some fishes, are ovoviviparous. Their fertilized eggs develop inside the mother, and then offspring are born alive. Yolk reserves sustain development in the egg. In this figure, the live born copperhead offspring are still encased in their egg sacs. (c) Birds are ovoviviparous because there is some internal development in the female. Their fertilized eggs contain large yolk reserves, and they develop and hatch outside the mother's body. (d) Most mammals are viviparous; their young are born live. The young of the opossum complete development in a pouch on the mother's ventral surface. The juvenile stages continue to draw nourishment from mammary glands in the mother's pouch.

from the nucleus and (typically yolky) cytoplasm of the egg. In contrast, insemination and fertilization are inside the maternal genital tract of ovoviviparous animals, including birds, before addition of the shell membrane. Thus, with the earliest embryonic development occurring within the genital tract, the maternal parent has some opportunity to affect embryonic development in addition to providing the nucleus and cytoplasm of the egg.

Viviparous animals have internal insemination and fertilization, with embryonic development occurring entirely within the maternal genital tract using maternal blood nutrients rather than yolk. Offspring from viviparous species are comparatively fully developed, and are to some extent free-living animals, although there are some variations. For example, rat pups are born **altricial,** meaning they are absolutely dependent on continued parental care, whereas guinea pig piglets are born **precocial** and so require minimal parental care. However, gestation is 64 days in guinea pigs and only 21 in the rat. Rat pups are born blind, hairless, and immobile, whereas the guinea pig piglets are fully furred; the eyes are open, and the young begin foraging soon after birth. The maternal parent may thus affect the developing offspring by providing the egg itself, that is, the nucleus and cytoplasm, the environment for fertilization, and subsequently the environment (including nutrients) for gestation of the developing fetus.

Unlike most reptiles, viviparous plesiosaurs, giant sea creatures living about 75 million years ago, gave birth to their young in the water rather than laying eggs on land. All mammals are viviparous except monotremes, the egg-laying duck-billed platypus (*Ornithorhynchus*) and the spiny anteater (*Echidna*; p. 423). These mammals represent the evolutionary transition between ovoviviparity and viviparity: The porous egg remains developing in the female tract for a considerable time, receiving nutrients from the mother's fluids without a placenta, before being laid with a small amount of yolk (p. 801). The transition is further seen in yolk genes: Whereas birds have three or more copies of a gene for a major yolk protein, *vitellogenin* (p. 795), platypuses have only one (which became a pseudogene in other mammals). Marsupials (*marsupium*, "pouch"), in contrast, are viviparous with a placenta, but their newborn young are highly altricial: They leave the uterus at a comparatively undeveloped stage and climb (using limbs that develop more rapidly than in any other vertebrate) into the pouch of the female, where they attach to a *teat* (p. 810).

Animals use various strategies to ensure breeding at the appropriate time

To optimize the number of offspring that can be raised to maturity, animals need to reproduce at an appropriate time. Let's examine the factors and adaptations involved.

Seasonal Breeding One major strategy for reproductive success is to breed seasonally. Seasonal **anestrus** (reproductive quiescence) evolved as a means of preventing females from giving birth during periods of the year when survival of the offspring is unlikely. Most nontropical seasonal breeders give birth in spring, when environmental and nutritional conditions support growth of the offspring. In birds, for example, seasonal breeding is particularly highly evolved, in part because the young depend on fresh food from the day of hatching. The energy requirements of the parent increase between 20 and 70% above normal because of the requirements associated with yolk synthesis and deposition of yolk in the maturing oocytes; thus, food availability is crucial for the laying female as well as for the hatchling. *Circannual* (biological) clocks in the brain (Chapter 7, p. 277) are partly in control of this seasonal reproduction, but the clocks are set or *entrained* by environmental signals (see Figure 7-8). Photoperiod, the amount of time during the day when there is light, is by far the most important factor determining the onset of the breeding season. The unique regularity of the annual change in day length makes it ideal for regulating seasonal breeding, with the result that most groups of organisms have evolved the ability to use photoperiod as the primary factor regulating reproduction. Thus, in temperate latitudes, virtually all bird species breed in the spring. However, seasonal conditions for all photoperiodic species are not identical each year, so other environmental cues are used for fine-tuning the reproductive processes set in motion by changes in day length. These modifying factors can, within specified limits, accelerate or retard photoperiod-induced gonadal growth. These factors include but are not limited to temperature, presence of the male, and the availability of a particular food source. The great tit, for example, is completely dependent on the availability of caterpillars and thus on the ambient temperature, which affects caterpillar popu-

lations in the spring and alters the onset of their breeding cycle. Similarly, the crossbill reproduces very early in the spring when pine seeds become available.

Photoperiod is not the deciding factor for many tropical and desert species, and breeding cycles may correlate with rainy seasons or may be irregular. In these species, changes in temperature or rainfall determine the timing of gonadal development. For example, the zebra finch (*Poephila gutta*) is an "opportunistic" breeder that lives in the desert and ovulates within a day of rainfall, the increase in humidity apparently triggering the reproductive cycle. If favorable conditions persist, the zebra finch lays a succession of clutches. The fact that the zebra finch remains in a state of readiness is somewhat comparable to the strategy used by *induced ovulators* (see p. 786). Within tropical regions, where there are pronounced wet and dry seasons, reproduction tends to occur in association with the decrease in light intensity associated with the rainy season. However, the environmental triggers are not clear in all cases. Recent studies on tropical (Panamanian) antbirds have shown that even slight changes in photoperiod (from 12 hours to 13 hours daylight) can trigger gonadal growth leading to breeding (only at the equator is daylight time 12 hours year round). In Panama, the longer daylight months are the rainy season.

The time of egg hatching in some species such as reptiles (for example, snapping turtles), arachnids, and insects also normally depends on environmental conditions. For example, eggs that have overwintered or **diapaused** (entered dormancy) rely on seasonal increases in temperature to initiate the hatching process. Normally, a cumulative period above a critical temperature, rather than a simple temperature threshold, is required. Alternatively, the eggs of species such as mosquitoes and dragonflies rely on decreasing oxygen concentrations in water, as well as on increasing temperatures, to cue the time of hatch.

Synchronization and Mating Behaviors A final set of reproductive strategies ensures that both sexes are ready to breed at the same time. Optimal conditions include both environmental (such as springtime) and internal (such as when eggs are ripe and ready for fertilization). Reproduction, especially for females, is energetically costly, and success of a species can depend on efficient use of this energy. We have already discussed seasonal cues that trigger reproduction at an optimal time, but these cues are not very precise, and could, for example, result in some individuals being ready to breed a few days ahead of others. Thus, a variety of other cues and triggers have evolved to synchronize reproduction between the sexes.

For animals with external fertilization and without complex behavior, other environmental cues and pheromones or other biologically produced agents (such as light in fireflies, p. 96) serve as synchronizing signals. A spectacular example is found in the annelid palolo worm (*Leodice viridis*) of South Pacific coral reefs. In the springtime (September to November in the southern hemisphere), the posterior segments of these worms become ripe with gametes. During the last quarter moon of October and November, typically at low tide, this part of the worm detaches (becoming an *epitoke*) and floats to the surface, where it releases its gametes. Because all mature worms in the vicinity do this at the same time, external fertilization has a high probability of success. Female epitokes also emit a pheromone that attracts wriggling male epitokes and stimulates release of sperm.

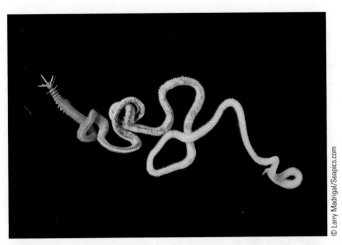

© Larry Madrigal/Seapics.com

Palolo worms, *Palola siciliensis* (entire worm with palolo or epitoke attached).

(Many marine annelids reproduce with epitokes.) Similarly, many reef corals spawn synchronously just after full moon in summertime. Although corals lack eyes, recent studies show they have genes for *cryptochrome* proteins, which react to blue light (in this case in moonlight) and were first discovered in biological clocks of more advanced animals (pp. 264 and 276).

Animals with complex behavioral capacities may use pheromones, typically produced by females when they are ready for fertilization, but they often perform elaborate reproductive behaviors called **courtship displays.** These are most commonly ritualistic patterns of movement and/or songs that allow males and females of the same species to recognize and attract each other and to signal and synchronize mating readiness. Courtship behaviors are thought to advertise the genetic fitness of the potential mate, as another way to ensure efficient use of reproductive resources. These behaviors occur in the greatest complexity in vertebrates. It usually (but not always) the male that displays to the female. The most elaborate vertebrate courtships are among the bowerbirds of New Guinea and nearby islands. The male bowerbirds of many species construct elaborate *bowers,* a patch of ground partly enclosed with woven twigs, decorated with small objects such as leaves, shells, dead insects, flowers, paper, and glass, creating a splash of color that stands out in the forest (see Figure 6-1c, p. 211). Females that are ready to mate begin visiting bowers and appear to be attracted to the more elaborate and colorful ones. But that is not enough. The male, on a female's arrival, also performs a ritualistic strutting dance (with sound and puffing of feathers) that needs to be pleasing to the female, or she leaves for another bower. Recent studies using electronic robots made to mimic female bowerbirds show that the males adjust the intensity of their displays according to whether the female appears to be gaining or losing interest. Thus, feedback synchronization occurs between the sexes.

Many arthropods also have complex courtship displays. For many male spiders, these displays are a matter of life or death. A female spider is almost always ready to eat any small animal; thus, the male (who is generally smaller than the female) must signal her in such a manner that suppresses her appetite and increases her readiness to mate. For example, a male wolf spider waves his distinctly marked front legs and **pedipalps** (limbs modified to carry sperm) in ritualistic

patterns in front of the female prior to inserting his pedipalp with sperm into her vaginal opening.

Reproductive capability ultimately depends on intricate relationships among the environment, the nervous and endocrine systems, reproductive organs, and target cells of sex hormones. In this chapter, we focus primarily on the basic sexual and reproductive functions that are under nervous and hormonal control.

check your understanding 16.1

List specific differences between oviparous, ovoviviparous, and viviparous as reproductive strategies.

What strategies are employed to ensure that both sexes are ready to breed at the same time?

16.2 Reproductive Systems and Genetics

Unlike the other body systems, which are essentially identical in the two sexes, the reproductive systems of males and females are remarkably sexually dimorphic, befitting their different roles in the reproductive process.

The reproductive system of vertebrates includes the hypothalamus, gonads, and reproductive tract

The primary reproductive organs, or **gonads,** of most vertebrates consist of *testes* in the male and *ovaries* in the female. In both sexes in vertebrates, the mature gonads are found in pairs and perform the dual function of (1) producing gametes (**gametogenesis**), that is, **spermatozoa (sperm)** in the male and **ova (eggs)** in the female, and (2) secreting sex hormones, primarily *testosterone* in males and *estrogens* (the major form being *estradiol*) and *progesterone* in females (see Figure 7-3, p. 272). However, as discussed later, males also make estrogens and females make testosterone with crucial roles—for example, males need estrogens for bone maintenance and sperm production; females need testosterone for *libido*.

In addition to the gonads, the reproductive system in each sex includes a **reproductive tract** encompassing a system of ducts that are specialized to transport or house the gametes after they are released, plus **accessory sex glands** that empty their supportive secretions into these passageways. The accessory glands of insects are paired tubular secretory structures that release their secretions into the genital tract. *Mammary glands* (from Latin *mamma* for "breast," the origin of the term "mammal") in mammalian females and *prostate glands* in mammalian males also are considered accessory reproductive organs. The externally visible portions of the reproductive system are known as **external genitalia** as opposed to those internally housed portions, termed the **internal genitalia.**

The anatomy of the reproductive organs is related to the reproductive strategies characteristic for a particular species. Female reproductive organs differ considerably between those animals that require internal fertilization and those that use external fertilization. In addition, female anatomy differs between those species that store large volumes of yolk in the egg and species that store a relatively small volume.

In some species, evolution of **secondary sexual characteristics,** the many external characteristics of vertebrates that are not directly involved in reproduction itself, can be used to distinguish between males and females. These features include body configuration, hair distribution, feather coloration, and sexual weaponry. Often the secondary sexual characteristics are of great importance in courting and mating behavior: The rooster's comb attracts the hen's attention, and the stag's antlers are useful to ward off other stags and establish dominance in the herd. For the most part, testosterone in the male and estrogens in the female govern the development and maintenance of these characteristics. While the evolution of weaponry in females is infrequently observed, the available examples support their importance. For example, in the African gemsbok antelope (*Oryx gazella*) defense against predators likely drove the evolution of female horns. And the female dung beetle (*Onthophagus sagittaris*) with the heftier horns controls more dung and is able to lay more dung-encased eggs.

Overview of Male Reproductive Functions and Organs The essential reproductive functions of the male are as follows:

1. Production of sperm (*spermatogenesis*) continuously in huge numbers, since only a small percentage survive the hazardous journey to the site of fertilization, and many spermatozoa may be needed to break down the barriers surrounding the female gamete (oocyte).
2. Delivery of sperm to the female.

In most mammals the sperm-producing organs, the **testes,** are suspended outside the abdominal cavity in a skin-covered sac, the **scrotum,** which lies within the angle between the posterior (inferior) appendages. In fishes the paired testes are suspended by membranes from the dorsal part of the body cavity. Mammals have a testicular cooling mechanism, which we consider later in this chapter.

The male reproductive system is designed to deliver sperm to the female reproductive tract in a liquid vehicle, **semen,** which is conducive to sperm viability. In mammals, secretions from the major male accessory sex glands, the **seminal vesicles, prostate gland,** and **bulbourethral glands** (Figure 16-3) provide the bulk of the seminal fluid. The **penis,** or **hemipenis** (in avian drakes), is the organ used to deposit semen in the female. Sperm exit the mammalian testes through the **vasa efferentia, epididymis, ductus (vas) deferens, ejaculatory duct,** and **urethra,** the latter being a canal that runs the length of the penis. Instead of an epididymis, in birds the *vasa efferentia* conduct sperm from the testis to a short epididymal duct that is continued as the vas deferens, where it eventually opens into an enlarged area prior to entering the **cloaca.** Both the vas deferens and enlarged region serve as sperm storage sites. Lacking the accessory glands of

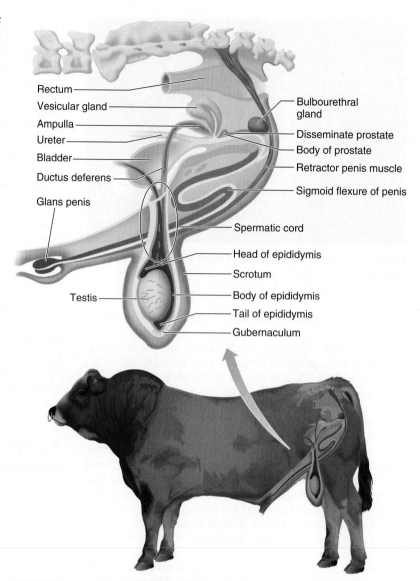

FIGURE 16-3 Reproductive tract of a bull. Sagittal view of the bull reproductive tract.

Source: P. L. Senger. (2003). *Pathways to Pregnancy and Parturition,* 2nd ed. Current Conceptions Inc., Washington State University Research & Technology Park, pp. 48, 49.

mammals, birds derive seminal fluids from the *seminiferous tubules* and the *vasa efferentia.*

Overview of Female Reproductive Functions and Organs The female's role in reproduction is often more complicated than the male's, especially if internal fertilization is involved. The major structures of the female reproductive tract in reptiles including birds and mammals include the **vagina, cervix, uterus, oviducts,** and external genitalia with internal extensions. These and the **ovaries** lie within the pelvic and abdominal cavity (Figure 16-4a). The essential female reproductive functions include:

1. Production of ova (**oogenesis**) and ovulation in the **ovaries** (although management of the ovulated oocyte depends on species).
2. Reception of sperm.
3. Transport of the sperm and ovum to a common site for *fertilization* (or *conception*).
4. Giving birth to the young (**parturition**) in a viviparous animal, or laying eggs in an oviparous animal.

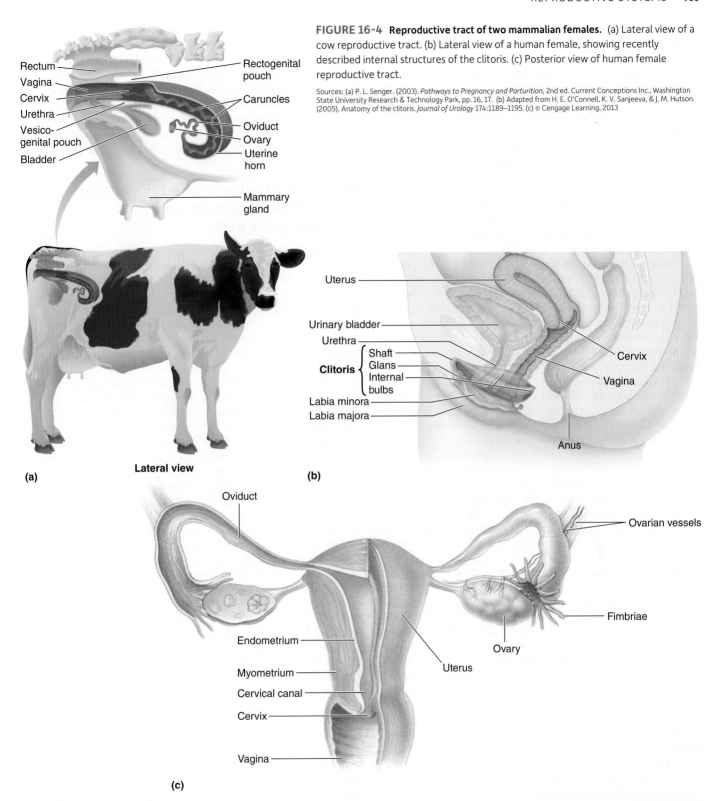

FIGURE 16-4 Reproductive tract of two mammalian females. (a) Lateral view of a cow reproductive tract. (b) Lateral view of a human female, showing recently described internal structures of the clitoris. (c) Posterior view of human female reproductive tract.

Sources: (a) P. L. Senger. (2003). *Pathways to Pregnancy and Parturition,* 2nd ed. Current Conceptions Inc., Washington State University Research & Technology Park, pp. 16, 17. (b) Adapted from H. E. O'Connell, K. V. Sanjeeva, & J. M. Hutson. (2005). Anatomy of the clitoris. *Journal of Urology* 174:1189–1195. (c) © Cengage Learning, 2013

5. Nourishing the offspring by milk production (**lactation**) in mammals. Placental and marsupial mammals have an additional role: nourishment of the developing fetus internally until it can survive in the outside world (**gestation,** or **pregnancy**) via a **placenta** (structurally different in the two mammalian groups; see p. 803), a vascular structure that supplies the fetus with nutrients in exchange for waste products generated by the fetus.

The mammalian female **oviducts** (**Fallopian tubes** in mammals) capture ova on ovulation (Figures 16-4a and 16-4c) and serve as the site of fertilization if fertilization occurs internally. In some vertebrates, the oviduct is lined with secretory and ciliated cells, which provide a suitable environment for the ova and help transport the spermatozoa. The portion of each mammalian tube adjacent to its respective ovary expands to form the **infundibulum,** a thin, funnel-shaped structure that captures the egg after release from the ovary. The "catching" of the egg involves active participation on the part of the infundibulum and is not simply a passive process by which the egg is funneled into the oviduct. The infundibulum is covered in fingerlike projections called **fimbriae** (Figure 16-4c), which function to cause the infundibulum to slip over the ovarian surface at ovulation, increasing the probability of "catching" the oocyte.

The product of fertilization is known as an **embryo** (**zygote** before cell division) during early development when tissue differentiation is taking place. The best demarcation is the establishment of organ systems and a functioning placenta in placental mammals. Beyond this time, features of the developing mammal are discernible and the embryo, now termed a **fetus,** is connected to the mother through the placenta for the duration of gestation.

Also in mammals, the thick-walled, hollow **uterus** serves as a protective and nutritive structure for maintaining the fetus. Considerable variation exists among species with regard to the anatomic organization of oviducts and uterus. Among mammals there are three distinct anatomical uterine structures: the **simplex uterus** (primates) with a single uterine body, the **bicornuate uterus** (dog, mare, cow) with a small uterine body and two uterine horns or **cornua,** and the **duplex uterus** (marsupials, rabbits), with two cervical canals that separate each uterine horn into distinct compartments. The corpus is comparatively large in the mare, less extensive in the cow and sheep, and small in the dog and pig while in some species, such as the rat and rabbit, no uterine body is present.

In oviparous and ovoviviparous vertebrates, the oviducts are the site for deposition of nonembryonic egg materials (albumin). In birds and reptiles, the uterus serves as a **shell gland** for calcification of the egg's outer layer.

The caudal portion of the uterus is the **cervix,** which projects into the vagina. The **cervical canal,** a single small opening, serves as the site of semen deposition in some species or as a pathway for sperm deposited in the anterior vagina to pass into the uterus in others. During mammalian pregnancy the cervix effectively closes the external opening of the uterus, producing a viscous mucus that prevents the entry of foreign material into the uterus. The cervix becomes greatly dilated during parturition, as it serves as the passageway for delivery of the fetus from the uterus to the vagina.

The **vagina,** a muscular, expansible tube connecting the uterus to the external environment (Figure 16-4), is primarily an organ of copulation and serves as a receptacle for sperm. The female external genitalia are collectively referred to as the **vulva** in mammals. In humans and Old World monkeys, two pairs of skin folds, the **labia minora** and **labia majora,** surround the vaginal and urethral openings laterally. The minora are homologous to the male prepuce, whereas the majora are a homolog of the male scrotum (see Figure 16-7 later). In mammals, the **clitoris,** an erectile structure composed of tissue comparable to the glans penis, lies at the anterior end of the folds of the labia minora. Recent work by Helen O'Connell and colleagues has shown that the human clitoris is much larger than is traditionally shown, because much of its mass remains hidden internally, consisting of large, highly vascular internal bulbs that surround the urethra and vagina (Figure 16-4b). These bulbs engorge during intercourse, perhaps to squeeze the urethra closed to prevent infections, to support the vaginal wall during penile penetration, and/or to increase pleasure signaling. These

structures were missed by the prudish Victorian anatomists who, until very recently, provided the diagrams widely used. Interestingly, this situation becomes exaggerated in the female hyena, because the external size of her clitoris is similar to the penis of the male, and she has a false scrotum, making it particularly difficult to distinguish between the two sexes.

The ventral portion of the cloaca of birds is comparable to the mammalian vulva. The cloaca serves as a common pathway for urinary, fecal, and reproductive products.

Reproductive systems in insects include the neuroendocrine organs, gonads, and reproductive tract

The reproductive system of male insects consists of a pair of testes, seminal vesicles, and accessory glands (Figure 16-5a). Most insects have one pair of accessory glands but the desert locust can have up to 15 pairs. Secretions from the accessory glands may include a supply of nutrients to the female and seminal fluid that serves as a carrier for the spermatozoa or that hardens about them to form a sperm-containing capsule, the **spermatophore.** The vas efferens and vas deferens connect the **sperm tube** (site of spermatogenesis) to the ejaculatory duct. As you will soon see, the process of spermatogenesis in each sperm tube is comparable to that in other animals.

In many species such as bumblebees, the formation of a temporary **mating** or **copulatory plug** (also found in some vertebrates, p. 779) after fertilization prevents both loss of semen from the female tract and entry of sperm into the vagina (genital chamber) from a competing male. Contraction of the female ducts then transports the spermatozoa to the **spermatheca,** the organ that stores the sperm prior to entry into the vagina (Figure 16-5b).

The reproductive system of a female insect consists of a pair of ovaries, a system of ducts through which the eggs navigate to the outside, the accessory glands, and the spermatheca (Figure 16-5b). There are three types of insect ovaries: the ancestral form, or **panoistic,** those without nurse cells (cells that provide nutrients to the oocyte in the early stages of development by cytoplasmic processes called **nutritive chords**); and **meroistic,** those ovaries with nurse cells. There are two subdivisions of meroistic ovaries: **telotrophic,** in which the nurse cells are found in the germarium, and **polytrophic,** where the nurse cells are enclosed in the follicle. The **germarium** forms the anterior portion of the ovariole and it is in this egg-producing portion of the ovary that cystocyte divisions occur and the clusters of cystocytes become enveloped by follicle cells. Externally, female insects have an **ovipositor,** an extension of the abdomen, used to lay eggs, for example, by insertion into the ground or into a host animal (as is done by parasitic wasps; see Figure 10-2c, p. 462). The ovipositor is modified in many species for other functions. Considered one of the most feared defenses among animals, the hymenopteran *stinger* is a modified ovipositor. In some species such as infertile honey bees, this sexual structure has completely lost its reproductive function. In such a bee, an entire section of the animal's abdomen is torn off when it stings its victim, killing the bee in the process.

Recall that in insects it is the *quantity of juvenile hormone* (*JH*) released at each molt that determines the *quality* of the molt, that is, the time to maturity (p. 281). As the concentration of JH falls, each molting episode triggered by *ecdysone*

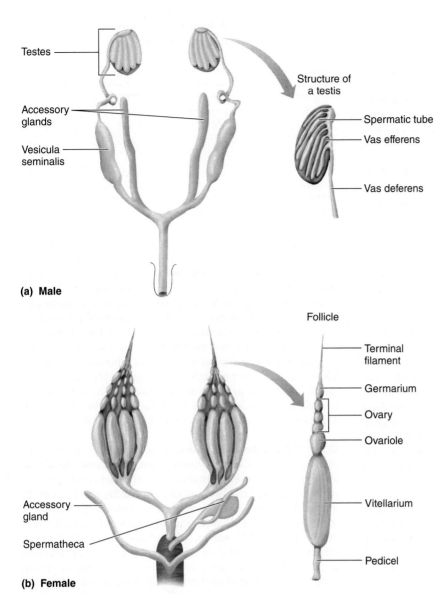

(a) Male

(b) Female

FIGURE 16-5 Reproductive systems of insects. (a) Male reproductive system. (b) Female reproductive system.

Source: C. A. Triplehorn, D. J. Borrer, & N. F. Johnson. (1989). *Introduction to the Study of Insects*, 6th ed. Belmont, CA: Thomson/Brooks/Cole, p. 57, Figure 3-31.

put such as ingestion of a blood meal in mosquitoes. Ecdysone, in turn, initiates vitellogenesis and egg maturation in concert with secretions from the ovarioles. Ovulation is believed to be a hormonally controlled event involving the release of a brain peptide, whereas **oviposition** (egg lay) also requires the contribution of the nervous system because the female can carefully select the time and the site for oviposition. Secretions from the accessory glands are species dependent and include such functions as encasing the eggs in frothy or gelatinous secretions, storing them in protective casings (**oothecae**) or simply gluing them to a particular surface, normally the food plant consumed during juvenile stage of development.

Reproductive cells each contain a half set of chromosomes

The DNA molecules that carry the cell's genetic code are not randomly crammed into the nucleus but are precisely organized into **chromosomes.** Each chromosome consists of a different DNA molecule that contains a unique set of genes. Somatic (body) cells contain matching pairs of chromosomes (the **diploid number**), which can be sorted on the basis of various distinguishing features. Chromosomes composing a matched pair are termed **homologous chromosomes,** one member of each pair having been derived from the individual's maternal parent and the other from the paternal parent. Gametes (that is, sperm and eggs) contain only one member of each homologous pair (the **haploid number**).

Gametogenesis occurs by meiosis, resulting in genetically unique sperm and ova

Most cells in an animal's body can reproduce themselves, a process important in growth, replacement, and repair of tissues. Cell division in somatic cells is accomplished by **mitosis.** In mitosis, the chromosomes (which contain the DNA) replicate (make duplicate copies of themselves), then the identical chromosomes are separated so that a complete set of genotypic information (that is, a diploid number of chromosomes) is distributed to each of the two new daughter cells. Nuclear division in the specialized case of gametes is accomplished by **meiosis,** in which only one of the two sets of homologous chromosomes (that is, a haploid number of chromosomes) is distributed to each of four new daughter cells. Importantly, not all the maternally derived chromosomes go to one daughter cell and the paternally derived chromosomes to the other cell. Because of this process, each sperm and oocyte has a unique complement of chromosomes (Table 16-1). When fertilization takes place, a sperm and ovum fuse to form a new individual with one member of each chromosomal pair having been inherited from the maternal parent and the other member from the paternal parent (see Figure 16-1).

results in a progressively less juvenile morphology until finally the adult genes are completely expressed and the adult form emerges. The gonads and their accessory glands either develop slowly during the molting cycles or not until the final molt to the adult stage, when the inhibitory effects of JH are absent. Most adult male insects emerge with fully formed reproductive systems, including a supply of mature spermatozoa.

JH plays a role in oogenesis in some species, primarily to initiate **vitellogenesis** (deposition of yolk in an oocyte, including the protein *vitellogenin* made in the fat body; see p. 281), which leads to egg development in the ovarioles (tubular structures forming the ovary). In other species, egg development is delayed until the adult stage, when the release of another brain hormone stimulates ecdysone release from the prothoracic gland in response to environmental cues such as temperature, photoperiod, or even specific in-

The sex of an individual is determined by the combination of sex chromosomes or by environmental stimuli

In some vertebrates, including all mammals and birds, individuals are destined to become males or females by **genetic sex determination (GSD)**, in which a sex is determined by the *sex chromosomes* they inherit. Recall that, as the chromosome pairs are separated during meiosis, each sperm or ovum receives only one member of each chromosome pair. Of the chromosome pairs, **autosomal chromosomes** include all the chromosomes that are shared by the males and females of a species. The remaining pair of chromosomes are the **sex chromosomes** and represent the single pair of chromosomes whose makeup is different for males and females, of which there are two genotypically different types. In contrast, many other vertebrates exhibit **environmental sex determination (ESD)**, in which external factors like temperature determine an embryo's sex. Let's look at the details in major groups.

Mammals: XX/XY Mammals have GSD using a larger **X chromosome** and a smaller **Y chromosome**. Sex depends on the combination of these: **Genotypic males** have both an X and a Y chromosome and are called **heterogametic; genotypic females** have two X chromosomes and are called **homogametic**. Thus, in mammals, the genotypic difference responsible for all the anatomic and functional distinctions between normal males and females is the single Y chromosome. Males have it; females do not. When the XY homologs separate during sperm formation, half the sperm receive an X chromosome and the other half a Y chromosome. In contrast, during oogenesis every ovum receives an X chromosome because separation of the XX sex chromosome pair yields only X chromosomes. During fertilization, combination of an X-bearing sperm with an X-bearing ovum produces a genotypic female, XX, whereas union of a Y-bearing sperm with an X-bearing ovum results in a genotypic male, XY. Thus, genotypic sex is determined at the time of conception and depends on which type of sex chromosome is contained within the fertilizing sperm.

Amazingly, the duck-billed platypus (a monotreme) has five Y and five X chromosomes in males and ten X chromosomes in females! One of the X chromosomes has the *DMRT1* gene found in the bird Z chromosome, to which we now turn.

Birds: ZZ/ZW Birds use the ZZ/ZW GSD system, with the heterozygote ZW being the female and the homozygote ZZ being the male. The Z and W terms are used to avoid confusion with species in which the female is homogametic. After meiosis, all the sperm cells carry a large Z chromosome, whereas only half of the egg cells carry a Z chromosome and the other half carry a small W chromosome. The mechanism of sex determination is not fully understood but involves differential steroidogenesis under the control of the Z-chromosome *DMRT1* gene, which activates male development. Later you will see how a sex-regulating gene works in mammals.

Reptiles (Nonavian) Many other vertebrates exhibit *ESD*. In all crocodilians, many turtles and lizards, and a few snakes, sex determination depends on the temperature at which eggs are incubated. In some turtles, embryos become males when

TABLE 16-1 Haploid Number of Chromosomes in the Gametes of Some Animal Species

Cat	19	Mouse	20
Chicken	39	Pigeon	31
Cow	30	Pig	19
Dog	39	Rat	21
Goat	30	Sheep	27
Horse	32	Turkey	41
Mink	15		

© Cengage Learning, 2013

they are incubated at low temperatures and into females when incubated at high temperatures. In crocodillians, females develop at high and low temperatures, and males at intermediate ones. In many species, the female can control the sex ratio of her offspring by choosing a particular thermal habitat for her eggs. In species that exhibit ESD, the mechanism is still genetic (a result of the particular genes inherited). For example, researchers have recently cloned a gene, *Sf1* (*stereodogenesis factor 1*) that is key to sexual development in turtles. Although this gene is present in all turtles of that species, its expression changes with environmental temperature.

Other Vertebrates Most urodele amphibians (newts, salamanders) have a ZZ/ZW system, whereas anurans (frogs) have an XX/XY system. Fish are even more complex, exhibiting one of the most diverse collections of reproductive strategies in the animal kingdom. Some have diffuse, indiscernible chromosomes, making sex determination cytologically difficult. Some exhibit the XX/XY system, others have the ZZ/ZW system, others have ESD based on temperature or social interactions, and some rely on complex combinations of genes; moreover, some are **hermaphroditic** (having both sexes in one individual). The latter come in two forms: **simultaneous** hermaphrodism, in which both sets of reproductive systems are active at the same time, and **sequential** hermaphrodism, in which an individual fish changes gender during its life cycle. The simultaneous form is rare; an example is the mangrove killifish *Rivulus marmoratus*. To complicate things even more, the sequential form can be protandry, protogyny, or both ways. **Protandry,** in which a male becomes a female, is less common; an example is the anemone (clown) fish *Amphiprion*. **Protogyny,** in which a female becomes a male, is common in many coral-reef fishes such as wrasses. Both-ways switching, in which a fish can change between genders in both directions, is rare, but has been seen in some reef gobies. The triggers for switching sexes are not known for all species, but for some the trigger is age (body size), whereas in others it is social interactions such as male–male interactions. In anemone fish, if a female is lost, a male changes into a female. Occasionally some species of fish express multiple forms of one sex. Some mature parrot fish, for example, come in one female and two distinct male forms: a small one (nonmating) and a large "supermale" form (which controls a harem of females and has exclusive mating rights).

Insects In most insects the male has one X (sex) chromosome, whereas the female has two. Males are referred to as XO (if only one chromosome in the pair is present) or XY (heterogametic with the Y chromosome being different in size from the X chromosome), whereas the female is XX (homogametic). One exception to this pattern is found in the Lepidoptera (butterflies and so forth), where the female is heterogametic.

An unfertilized egg can develop into a male or female in parthenogenesis

Parthenogenesis (p. 757) also occurs in many species including water fleas, some insects, a few reptiles (such as whiptail lizards), and some fishes (including occasionally in hammerhead sharks). Parthenogenetic development that produces females implies that either the eggs failed to undergo meiosis or that the two cleavage nuclei fused to restore the diploid condition, whereas production of a male involves the loss of an X chromosome. Depending on the time of year, gall wasps and aphids produce either males or females parthenogenetically.

At least one group, the bdelloid rotifers, living in an all-female world, reproduce only parthenogenetically (spawning cloned daughters instead of using sex). Recently it has been found that they may avoid the low genetic diversity of asexual reproduction by promiscuous gene transfer: Their genomes are riddled with genes of other animals, fungi, plants, and even bacteria!

Sex differentiation in mammals depends on the presence or absence of Y-chromosome masculinizing determinants during critical periods of embryonic development

Differences between mammalian males and females exist at three levels: genotypic, gonadal, and phenotypic (anatomic and/or physiologic) sex (Figure 16-6).

Genotypic and Gonadal Sex Genotypic sex, which depends on the combination of sex chromosomes at the time of conception, in turn determines **gonadal sex,** that is, whether testes or ovaries develop. The presence or absence of a Y chromosome determines gonadal differentiation in mammals. All mammalian embryos have the potential to differentiate along either male or female lines because the developing reproductive tissues include the precursors of both sexes. Gonadal specificity in the human appears during the seventh week of intrauterine life, when the indifferent gonadal tissue of a genotypic male begins to differentiate into testes under the influence of the **sex-determining region** of the Y chromosome (**SRY**), the single gene that is responsible for sex determination. This gene triggers a chain of reactions that leads to physical development of a male. The sex-determining region of the Y chromosome "masculinizes" the gonads (induces their development into testes) by stimulating production of **testis-determining factor (TDF)** by primordial gonadal cells. TDF, which is a specific plasma membrane protein found only in males, directs differentiation of the gonads into testes. Because genotypic females lack the SRY gene and so do not produce TDF, their gonadal cells never receive a signal for testicular formation, so the undifferentiated gonadal tissue starts developing during the ninth week into ovaries instead.

Because of this pattern, female sex development has been called the "default" state: If the SRY region of the Y chromosome is present, the embryo becomes a male; if not, it "automatically" becomes a female. This must involve regulatory genes, but while the SRY/TDF "master regulator" for male development has been known for some time, genes for female development are not fully known. One gene, *Wnt-4,* was discovered first as a key regulator of kidney development. Yet it also orchestrates female sexual development: It suppresses the production of testosterone and triggers the development of the *Müllerian duct,* the embryonic duct that develops into the female reproductive tract (see below). In mice with a mutant *Wnt-4* gene, the female does not develop a normal oviduct, uterus, or vagina; instead, she develops internal male structures. Other candidate genes such as a transcription factor called *FoxL2* are being analyzed; mice with nonfunctioning *FoxL2* do not develop ovarian granulosa cells. How *Wnt-4, FoxL2,* and other regulatory genes exert their effects is not yet known.

Phenotypic Sex **Phenotypic sex,** the apparent anatomic sex of an individual, depends on the genotypically determined gonadal sex. **Sexual differentiation** concerns the embryonic development of the external genitalia and reproductive tract along either male or female lines. As with the undifferentiated gonads, embryos of both sexes have the potential to develop either male or female reproductive tracts and external genitalia. Differentiation into a male-type reproductive system is induced by **androgens,** which are masculinizing hormones secreted by the developing testes. Testosterone is the most potent androgen. The absence of these testicular hormones in female fetuses results in the development of a female-type reproductive system. Let's examine the internal tract and external genitalia separately.

- *Reproductive tract:* Two primitive duct systems— the **mesonephric ducts,** or **Wolffian ducts,** and the **paramesonephric,** or **Müllerian ducts**—develop in all mammalian embryos. In males, the reproductive tract develops from the Wolffian ducts and the Müllerian ducts degenerate, whereas in females the Müllerian ducts differentiate into the reproductive tract and the Wolffian ducts regress (Figure 16-6). Because both duct systems are present before sexual differentiation occurs, the early embryo has the potential to develop either a male or a female reproductive tract. Differentiation into a male-type reproductive system is induced by (1) androgens from the newly developed Leydig cells, with **dihydrotestosterone (DHT),** the reduced form of testosterone, being the most potent one; and (2) **anti-Müllerian hormone (AMH)** from the early Sertoli cells, which trigger the destruction of the precursor of the female genital reproductive tract (Figure 16-6). A hormone from the placenta, *chorionic gonadotropin* (p. 805), activates this early testicular secretion. Testosterone induces development of the Wolffian ducts into the male tract (epididymis, ductus deferens, ejaculatory duct, and seminal vesicles). This hormone, after being converted into DHT, is also responsible for differentiating the external genitalia into the penis and scrotum. Meanwhile, AMH causes regression of the Müllerian ducts. In the absence of testosterone and AMH in females, the Wolffian ducts regress, the Müllerian ducts develop into the female tract (oviducts, uterus, and vagina), and the genitalia differentiate into the clitoris and labia.

- *External genitalia:* Unlike the tract, male and female external genitalia develop from the same embryonic

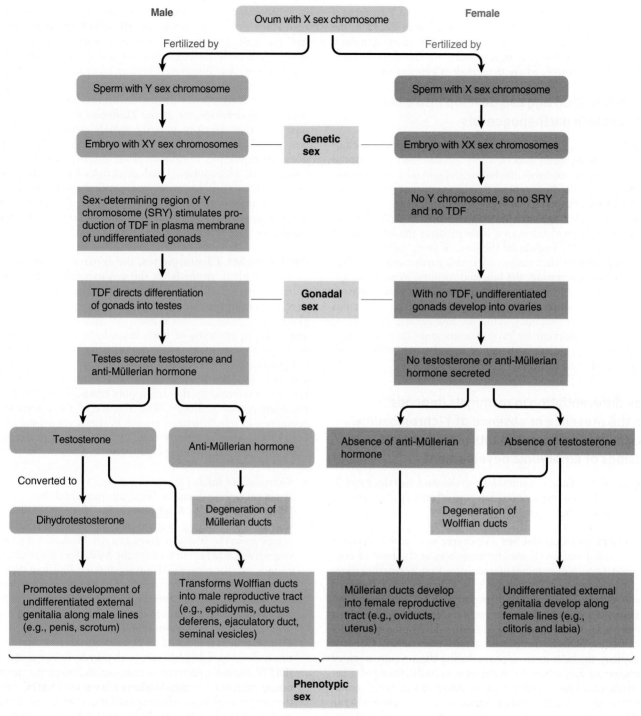

FIGURE 16-6 Sexual differentiation in mammals.

© Cengage Learning, 2013

tissue in mammals. In both sexes in mammals, the undifferentiated external genitals consist of a *genital tubercle*, paired *urethral folds* surrounding a urethral groove, and, more laterally, *genital (labioscrotal) swellings* (Figure 16-7). The **genital tubercle** gives rise to highly sensitive tissue—in males the *glans penis* (the distal end of the penis) and in females the *clitoris*. The major distinctions between the glans penis and clitoris are the smaller size of the external parts of the clitoris, the lack of spongiosum, and the urethral opening at the end of the glans

penis. The urethra is the tube through which urine is transported from the bladder to the outside and also serves in males as a passageway for exit of semen through the penis to the outside. In males, the **urethral folds** fuse around the urethral groove to form the penis, which encircles the urethra. The **genital swellings** similarly fuse to form the scrotum and **prepuce,** a fold of skin (a double fold in the stallion) that extends over the end of the penis and more or less completely covers the glans penis.

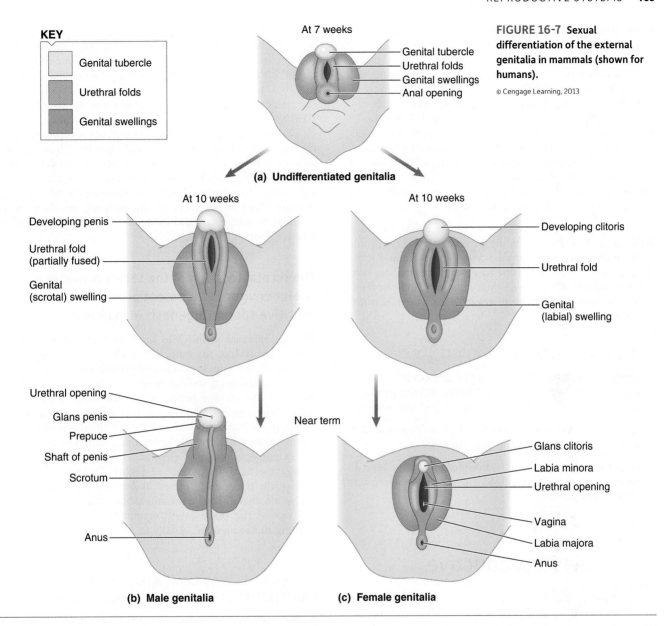

KEY

- Genital tubercle
- Urethral folds
- Genital swellings

At 7 weeks

- Genital tubercle
- Urethral folds
- Genital swellings
- Anal opening

(a) Undifferentiated genitalia

At 10 weeks

- Developing penis
- Urethral fold (partially fused)
- Genital (scrotal) swelling

At 10 weeks

- Developing clitoris
- Urethral fold
- Genital (labial) swelling

Near term

- Urethral opening
- Glans penis
- Prepuce
- Shaft of penis
- Scrotum
- Anus

- Glans clitoris
- Labia minora
- Urethral opening
- Vagina
- Labia majora
- Anus

(b) Male genitalia

(c) Female genitalia

FIGURE 16-7 Sexual differentiation of the external genitalia in mammals (shown for humans).

© Cengage Learning, 2013

Note that the indifferent embryonic reproductive tissue develops into a female structure unless *actively* suppressed by masculinizing factors. Although considered a "default" process, in fact female differentiation requires an equally well-orchestrated (though poorly understood) developmental program. However it is *not due to estrogens* from ovaries. In the absence of male testicular hormones, a female reproductive tract and external genitalia develop regardless of the genotypic sex of the individual. Ovaries do not need to be present for feminization of the fetal genital tissue. Such a control pattern for determining sex differentiation is appropriate, considering that fetuses of both sexes are exposed to high concentrations of female sex hormones throughout gestation. If female sex hormones exerted influence over the development of the reproductive tract and external genitalia, all fetuses would be feminized. It is worth noting that masculinization of the brain at least in part is the result of "female" estrogens, which are generated by an enzyme located in the neural tissue, *P450 aromatase* (see p. 772).

Errors in Sexual Differentiation In the usual case, genotypic sex and phenotypic sex are compatible; that is, a genotypic male appears to be a male anatomically and functions as a male, and the same compatibility holds true for females. Occasionally, however, discrepancies occur between genotypic and phenotypic sexes because of errors in sexual differentiation, as the following examples illustrate:

- In cattle carrying twin fetuses, one of each sex with co-circulation of blood between the two, exposure of the female co-twin to anti-Müllerian hormone regresses the genital reproductive tract (that is, Müllerian duct derivatives), but not the urogenital (that is, posterior vagina and external genitalia), with the consequence that these animals are infertile. Such heifers are termed **freemartins.**
- If testes in a genotypic male fail to properly differentiate and secrete hormones, the result is the development of an apparent anatomic female in a genotypic male, which, of course, will be sterile.

- Because testosterone acts on the Wolffian ducts to convert them into a male reproductive tract but the testosterone derivative DHT masculinizes the external genitalia, a genotypic deficiency of the enzyme that converts testosterone into DHT results in a genotype with male abdominal (that is, undescended) testes and a male reproductive tract but with female external genitalia.
- The adrenal gland normally secretes a weak androgen, *dehydroepiandrosterone*, in insufficient quantities to masculinize females. However, pathologically excessive secretion of this hormone in a genotypically female fetus during critical developmental stages imposes differentiation of the genitalia along male lines.

Ultimately current research studies suggest that a cell's genetic identity "trumps" any hormonal instructions. Studies in **gynandormorphs,** an animal with both male and female characteristics resulting from early-stage anomalies, reveal that individual cells maintain their individual male or female identities during development regardless of their hormonal environment.

check your understanding 16.2

Compare the primary reproductive organs, sex hormones, reproductive tract, accessory sex glands, external genitalia, and secondary sexual characteristics in males and females.

Discuss the differences between males and females with regard to genetic, gonadal, and phenotypic sex.

What parts of the male and female reproductive systems develop from the Wolffian and Müllerian ducts? What regulates these developments?

16.3 Male Reproductive Physiology

Testes in most mammals descend into the scrotum to become external

In the mammalian embryo, the testes develop from the gonadal ridge located at the rear of the abdominal cavity. In the last phase of fetal life (dependent on the species), they begin a slow descent, passing out of the abdominal cavity through an opening in the abdominal wall called the **inguinal canal** into the scrotum, one testis dropping into each pocket of the scrotal sac. Testosterone from the fetal testes is responsible for inducing descent of the testes into the scrotum. In most instances in which the testes are retained within the inguinal canal at birth, descent occurs naturally before sexual maturity or can be encouraged with administration of testosterone. Rarely, a testis remains undescended into adulthood, a condition known as **cryptorchidism** ("hidden testis"). Subnormal amounts of androgen are synthesized in the cryptorchid testes, and the elevated core body temperature interferes with mitosis of spermatogonia, so spermatozoa are not normally produced (see next section). This condition is more prevalent in humans, horses, and pigs. In a bull with a **short scrotum,** the testes can be artificially forced into the dorsal region of the scrotum by placing a large rubber band around the lower portion of the scrotum. Within four weeks the lower scrotum sloughs off at the juncture of the rubber band because of the impaired circulation. Although such a bull is infertile, the marked reduction in scrotal size does not affect testosterone concentrations, with the result that animals maintain a high rate of growth and at slaughter have leaner (reduced fat content) carcasses.

Following descent of the testes into the scrotum through the inguinal canal, the abdominal opening closes snugly around the vas deferens, cremaster muscle, blood vessels, and nerves that traverse between each testis and the abdominal cavity. Incomplete closure or rupture of this opening permits abdominal viscera to slip through, resulting in an **inguinal hernia,** a condition most frequently observed in humans, horses, and swine.

The scrotal location of the testes provides a cooler environment essential for temperature-sensitive spermatogenesis in mammals

The temperature within the scrotum averages several degrees Celsius less than normal body (core) temperature. Descent of the testes into this cooler environment is essential because spermatogenesis in most mammals is temperature sensitive and cannot occur at normal body temperature. This is why a cryptorchid is unable to produce viable sperm.

In those mammals without a scrotum, alternative mechanisms associated with temperature regulation are found. For example, whales and elephants have countercurrent exchange mechanisms (retes; see Figure 15-18a, p. 745) for cooling the testes despite their abdominal location. In some mammals, including the rat and rabbit, the testes move into and out of the body cavity seemingly at will through an opening in the inguinal canal.

unanswered Questions | Why can't sperm develop at core body temperatures?

On the face of it, that mammalian sperm cannot develop at core body temperatures (37 to 39°C) makes no sense. After all, other tissues in mammals have membranes and macromolecules adapted to work at normal body temperatures. Moreover, measurements in domestic chicken's testes (which are internal) show that this bird, at least, has normal spermatogenesis at core temperatures of 40–41°C. (It was once thought that air sacs cooled the testes in birds, but evidence is lacking.) Why mammalian sperm require temperatures of 33 to 35°C is not known. One suggestion is that mutation rate increases with temperature, something that might be detrimental to cells that reproduce so rapidly and in such high numbers. Another idea is that testicles first became external as a sexual display in some mammals, and sperm later lost their tolerance for normal body temperatures. However, we do not know the answer. <<

The position of an external scrotum in relation to the abdominal cavity can be varied by a spinal reflex mechanism that plays an important role in regulating testicular temperature. Reflex contraction of the *external cremaster muscle* on

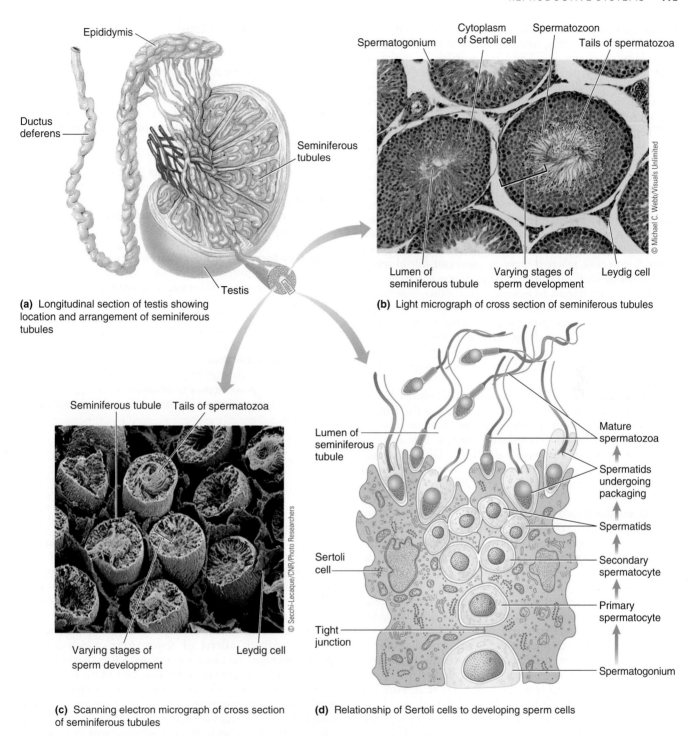

(a) Longitudinal section of testis showing location and arrangement of seminiferous tubules

(b) Light micrograph of cross section of seminiferous tubules

(c) Scanning electron micrograph of cross section of seminiferous tubules

(d) Relationship of Sertoli cells to developing sperm cells

FIGURE 16-8 Anatomy of testis depicting the site of spermatogenesis. (a) The seminiferous tubules are the sperm-producing portion of the testis. (b) The undifferentiated germ cells (the spermatogonia) lie in the periphery of the tubule, and the differentiated spermatozoa are in the lumen, with the various stages of sperm development in between. (c) Note the presence of the highly differentiated spermatozoa (recognizable by their tails) in the lumen of the seminiferous tubules. (d) Relationship of the Sertoli cells to the developing sperm cell.

(a, d) © Cengage Learning, 2013

exposure to a cold environment raises the testicles and the contraction of the *tunica dartos muscle* decreases the scrotal surface area to bring the testes closer to the warmer abdomen. Conversely, on exposure to heat, muscle relaxation permits the scrotal sac to become more pendulous, moving the testes farther from the warm core of the body.

The testicular Leydig cells secrete masculinizing testosterone during the breeding season

The majority of vertebrate testicular mass consists of highly coiled **seminiferous tubules** within which spermatogenesis takes place (Figure 16-8). The endocrine cells that produce testosterone—the **Leydig,** or **interstitial, cells**—are located in the connective tissue (interstitial tissue) between the seminiferous tubules. Thus, the portions of the testes that produce sperm and secrete testosterone are structurally and functionally distinct. During the breeding season, the testes perform the dual function of producing sperm and secreting testosterone.

Testosterone is a steroid hormone derived from cholesterol, as are the female sex hormones, estrogen and progesterone (see Figure 7-3, p. 272). The Leydig cells contain a high concentration of the enzymes required to direct cholesterol through the testosterone-yielding pathway. Once produced, some of the testosterone is secreted into the blood and transported, primarily bound to plasma proteins, to its target sites of action. A substantial portion of the newly synthesized testosterone is bound by androgen-binding protein in the lumen of the Sertoli cells, where it plays an important role in sperm production by locally increasing androgen concentrations.

To exert its effects, testosterone (and other androgens) bind with androgen receptors in the cytoplasm of target cells. The androgen-receptor complex moves to the nucleus, where it binds with the androgen response element on DNA, leading to transcription of genes that direct synthesis of new proteins that carry out the desired cellular response (see Figure 7-4, p. 274).

Most but not all of testosterone's actions ultimately are directed toward ensuring delivery of sperm to the female. The effects of testosterone can be grouped into five categories: (1) effects on the reproductive system before birth, (2) effects on sex-specific tissues after birth, (3) other reproduction-related effects, (4) effects on secondary sexual characteristics, and (5) nonreproductive actions (Table 16-2).

Effects on the Reproductive System before Birth

Before birth of a mammal, testosterone secretion by the fetal testes masculinizes the reproductive tract and external genitalia and promotes descent of the testes into the scrotum, as already described. Testosterone can cross the blood–brain barrier and does so in the male; in the brain, it is then converted to the estrogen *estradiol* by the enzyme **P450 aromatase,** which defeminizes the hypothalamus by eliminating the GnRH surge center so that a male pattern of hormone secretion occurs. In order to prevent the estradiol produced from the fetal ovaries from defeminizing the female hypothalamus, **alpha-fetoprotein,** a glycoprotein synthesized in the embryonic yolk sac and later the fetal liver, binds estradiol and prevents it from crossing the blood–brain barrier.

After birth, testosterone secretion is markedly reduced, and then slowly increases as the individual animal reaches sexual maturity or *puberty,* supporting the slow, but timely development of the testes and remainder of the reproductive system so it is ready to function at puberty. For example, in the colt, FSH and LH concentrations both peak at around 40 weeks of age. FSH remains elevated until the time of puberty (~83 weeks), whereas LH concentrations return to baseline. Interestingly, testosterone concentrations prior to puberty remain low and do not increase at 40 weeks of age in conjunction with the increase in FSH and LH, but rapidly peak at the time of puberty.

Effects on Sex-Specific Tissues after Birth

Puberty is considered a gradual process, not a single event, and is recognized as the attainment of sexual maturity with the ability to

---

TABLE 16-2 Effects of Testosterone

Effects before Birth
Masculinizes the reproductive tract and external genitalia
Promotes descent of the testes into the scrotum of most mammals

Effects on Sex-Specific Tissues
Promotes growth and maturation of the reproductive system at puberty
Essential for spermatogenesis
Maintains the reproductive tract throughout adulthood

Other Reproduction-Related Effects
Develops the sex drive at puberty (sexual maturity)
Controls gonadotropic hormone secretion

Effects on Secondary Sexual Characteristics
Induces the male pattern of hair or feather growth
Causes the voice to deepen because of thickening of the vocal cords
Promotes muscle growth responsible for the male body configuration

Nonreproductive Actions
Exerts a protein anabolic effect
Promotes bone growth at puberty (sexual maturity)
Closes the epiphyseal plates after being converted to estrogen by aromatase
Induces aggressive behavior

© Cengage Learning, 2013

---

reproduce. Puberty is denoted by the first ovulation in females or the presence of active sperm in the ejaculate in males. At puberty in males, the secretion of GnRH is at the appropriate frequency and quantity to stimulate gonadotropin (LH, FSH; p. 290) release by the anterior pituitary (adenohypophysis). In turn, the gonadotropins promote the Leydig cells to secrete sufficient testosterone as a result of the decreasing sensitivity of the hypothalamus to the negative-feedback effects of testosterone, which initiates spermatogenesis. Testosterone, usually acting in the dihydro form, governs the growth and maturation of the entire male reproductive system. Ongoing testosterone secretion is essential for spermatogenesis and for maintaining a mature male reproductive tract throughout adulthood. Studies using a knock-out mouse with the androgen receptor gene deleted demonstrate an alteration in the expression of several key steroidogenic enzymes in Leydig cells, indicating that testosterone is also an autocrine factor regulating its own production. In this same model there was also observed an arrest of spermatogenesis at the spermatid stage of development. Following castration (surgical removal of the testes) or testicular failure caused by disease, the sex organs regress in size and function.

In contrast, it is believed that female mammals must develop a minimum body fat content before reproductive cycles can be initiated. The link between metabolic status and gonadotropin release, which may involve the peptide signals *leptin* and *kisspeptin,* is not fully understood, as we will discuss later (p. 798).

TABLE 16-3 Average Age (Range) of Puberty in the Male and Female of Various Mammals

Species	Male	Female
Cat	9 months (8–10)	8 months (4–12)
Cow	11 months (7–18)	11 months (9–24)
Camel	3–5 years	3 years
Dog	9 months (5–12)	12 months (6–24)
Sheep	7 months (6–9)	7 months (6–16)
Swine	7 months (5–8)	6 months (5–7)
Horse	14 months (10–24)	18 months (12–19)
Elephant	13–15 years	11–14 years

© Cengage Learning, 2013

The onset of puberty varies considerably among and within species (Table 16-3). For example, most birds reproduce in the year after hatch, but some, notably seabirds, hawks, and ratites, exhibit deferred sexual maturity and breed only when three or more years of age. Large albatrosses and condors reach sexual maturity only between the ages of 9 and 11 years. Even within a particular species there is considerable diversity in the age at which an animal first breeds.

Other Reproduction-Related Effects Testosterone governs development of sexual **libido** (drive or desire for sexual activity) in advance of puberty and helps maintain that drive in the adult male. That drive stimulated by testosterone is important for facilitating mating. Once libido has developed, testosterone is no longer absolutely required for its maintenance. Castrated animals sometimes remain sexually active but at a reduced level.

The testes of the boar secrete large quantities of compounds known as C-16 unsaturated androgens. These androgens function as pheromones when secreted in the **preputial glands** of the boar. The prepuce surrounds the penis and contains a diverticulum (pouch). Release of the pheromone stimulates a sow in heat to assume the immobile mating stance.

Effects on Secondary Sexual Characteristics: Antlers and Their Regeneration All male secondary sexual characteristics depend on testosterone for their development and maintenance (see the box, *A Closer Look at Adaption: A Rut for One Season*, p. 795). These male characteristics induced by testosterone are important for mate acquisition and/or retention and include (1) the male pattern of hair, antler, or feather growth; (2) thick skin; and (3) the male body configuration (for example, the heavy musculature associated with the necks of deer in rut) as a result of protein deposition. In human males, (4) a deep voice is caused by enlargement of the larynx and thickening of the vocal cords. A male castrated before puberty (human: eunuch; sheep: wether; pig: barrow; cattle: steer; and so forth) does not mature sexually or develop secondary sexual characteristics.

Antlers are a characteristic feature of male deer and function as weapons and status symbols when rival males compete over females, food, and territory. Antlers are composed of skin, bone, nerves, blood vessels, fibrous tissue, and cartilage and should not be confused with horns, which are a keratinized tissue that grow from their base under the control of underlying mesenchymal cells and remain a permanent fixture for the duration of the animals' life. Testosterone controls the initial development of the antlers at puberty, normally in the spring of the animal's second year of life, as well as the seasonal cycle of casting and regrowth of the antlers in the adult. Antlers are actually bony structures that develop from pedicles (bony protuberances) on the frontal bones of the skull. They are cast and regrown each year in association with the breeding season. The high blood concentrations of testosterone in the autumn (1) trigger a growth phase, during which the antlers are covered in a velvety hair that is endowed with an exceptionally rich blood supply, and (2) prevent the process of rejection of this tissue for the duration of the breeding season.

Once the antlers become fully calcified, the overlying skin peels off; the mature antlers are now dead, insensitive structures that function as weapons for the rutting season. At the end of the breeding season, as the testes regress and concentrations of testosterone decline, the dead antlers are cast, and new ones are regenerated from the living pedicles.

Nonreproductive Actions Testosterone exerts several important effects not directly related to reproduction: induction of aggressive behavior and a general protein anabolic (synthesis) effect that promotes greater muscling and bone growth. These effects play an indirect role in reproduction by enhancing the ability of males to compete for females. Ironically, testosterone not only stimulates bone growth but eventually prevents further growth by sealing the growing ends of the long bones (that is, ossifying, or "closing," the epiphyseal plates—see p. 293).

Spermatogenesis yields an abundance of highly specialized, mobile sperm

The basic pattern of spermatogenesis is functionally similar among all animals. In fact, many of the original descriptions of spermatogenesis were obtained from insect models. Two functionally important cell types are present in the mammalian sperm-producing seminiferous tubules (Figure 16-8a): **germ cells,** progenitors that are mostly in various stages of sperm development, and **Sertoli cells,** which provide crucial support for spermatogenesis (Figure 16-8b, c, and d). **Spermatogenesis** is a complex process by which relatively undifferentiated germ cells, the **A-type spermatogonia** (each of which contains a diploid complement of 2N chromosomes), proliferate and are converted into extremely specialized, motile spermatozoa (sperm), each bearing a randomly distributed haploid set of 1N chromosomes.

Microscopic examination of a seminiferous tubule reveals layers of germ cells in an anatomic progression of sperm development, starting with the least differentiated in the outer layer and moving inward through various stages of division to the lumen, where the highly differentiated sperm are ready for exit from the seminiferous epithelium (*spermiation*) (Figure 16-8b, c, and d). In bulls, spermatogenesis takes 64 days for development from a spermatogonium to a mature sperm. In normal male animals, a large number of spermatozoa are produced each day: approximately 6.0×10^9 in the bull and 16.5×10^9 in the boar.

Stages of
spermatogenesis

Chromosomes
in each cell

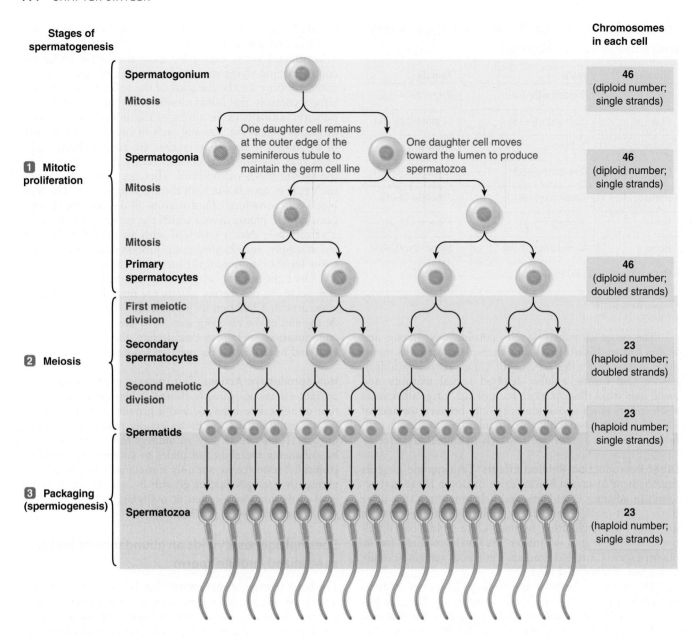

FIGURE 16-9 **Spermatogenesis.**

© Cengage Learning, 2013

Comparably, recall from Figure 16-5a the sperm tube of a male insect: Each tube has a **germarium** (site where spermatogenesis is initiated) and three zones. Zone 1, or the region of growth at the periphery of the sperm tube, is associated with the maturation of spermatogonia to spematocytes. Zone 2 is the site of maturation and reduction, and Zone 3 is the site of the transformation of spermatids into spermatozoa.

Spermatogenesis encompasses three major stages: *mitotic proliferation (spermatocytogenesis)*, *meiosis*, and *packaging (spermiogenesis)* (Figure 16-9).

Mitotic Proliferation Spermatogonia located in the basal (outermost) layer of the tubule continuously divide mitotically, with all new cells bearing the full complement of 2*n* chromosomes that are identical to those of the parent cell. Such proliferation provides a continual supply of new germ

cells. Following mitotic division of a spermatogonium, one of the daughter cells remains at the outer edge of the tubule as an undifferentiated spermatogonium, thus maintaining the germ cell line. The other daughter cell starts moving toward the lumen while undergoing the various steps required to form sperm, which are released into the lumen. The sperm-forming daughter cell divides mitotically several times more to form four identical **primary spermatocytes.** After the last mitotic division, the primary spermatocytes enter a resting phase during which the chromosomes are duplicated and the doubled strands remain together in preparation for the first meiotic division.

Meiosis During meiosis, each primary spermatocyte (with a diploid number of 2*N*, 4*n* doubled chromosomes with pairing of the homologous chromosomes) forms two **secondary sper-**

matocytes (each with a haploid number of 1*N*, 2*n* doubled chromosomes), referred to as the *reductional division* because the number of sets of homologous chromosomes has been reduced, even though there are still two copies of each. The secondary spermatocytes then divide during the second meiotic division separating the two copies of each chromosome into daughter cells during the first meiotic division, yielding four **spermatids** (each with 1*N*, 1*n* haploid complement).

No further division takes place beyond this stage of spermatogenesis. Each spermatid is remodeled into a single spermatozoon. Because each sperm-producing spermatogonium mitotically produces four primary spermatocytes and each primary spermatocyte meiotically yields four spermatids (spermatozoa-to-be), the spermatogenic sequence in mammals can theoretically produce 16 spermatozoa each time a spermatogonium initiates this process. Usually, however, some cells are lost at various stages, so the efficiency of spermatogenesis is rarely this high.

Packaging After meiosis, spermatids still resemble undifferentiated spermatogonia structurally, except for their 1*N* complement of chromosomes. Production of extremely specialized, motile spermatozoa from spermatids requires extensive remodeling of cellular elements, a process known as **spermiogenesis.** Sperm are essentially "stripped down" cells in which most of the cytosol and any organelles not needed for the task of delivering the sperm's genotypic information to an ovum have been extruded. Thus, sperm travel lightly, taking only the bare essentials to accomplish fertilization with them.

A **spermatozoon** has three parts (Figure 16-10): a head capped with an acrosome, a midpiece, and a tail. The **head** consists primarily of the nucleus, which contains the sperm's complement of genotypic information. The **acrosome,** an enzyme-filled vesicle that caps the tip of the head, is used as an "enzymatic drill" for penetrating the ovum. It is formed by aggregation of vesicles comprising the endoplasmic reticulum–Golgi complex before these organelles are discarded. The acrosomal enzymes remain inactive until the sperm contacts an egg, at which time the enzymes are released. A long, whiplike tail that grows out of one of the centrioles provides motility for the spermatozoon. Motility is powered by energy generated by the mitochondria concentrated within the **midpiece** of the sperm. (Insect sperm lack a midpiece, and the mitochondria are in the tail.)

Until sperm maturation is complete, the developing germ cells arising from a single primary spermatocyte remain joined by cytoplasmic bridges. These connections, which result from incomplete cytoplasmic division, permit cytoplasm to be exchanged among the four developing sperm. This linkage is important in mammals because the X chromosome, but not the Y chromosome, contains genes that code for cellular products essential for development of the sperm. (Whereas the large X chromosome contains several thousand genes, the small Y chromosome has only a few dozen, the most important of which are the *SRY* gene and others that play critical roles in male fertility.) Recall that half the sperm receive an X and the other half a Y chromosome. If it were not for the sharing of cytoplasm so that all the haploid cells are provided with the products coded for by X chromosomes until sperm development is complete, the Y-bearing, male-producing sperm would not be able to develop and survive.

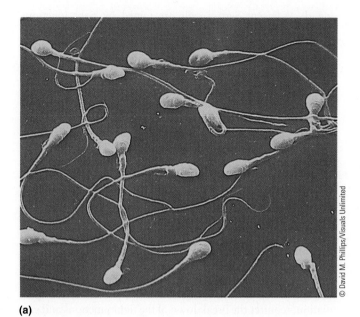

(a)

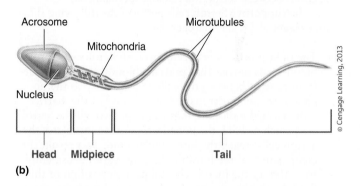

Acrosome   Microtubules

Mitochondria

Nucleus

Head   Midpiece   Tail

(b)

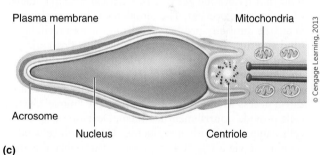

Plasma membrane   Mitochondria

Acrosome

Nucleus   Centriole

(c)

FIGURE 16-10 Anatomy of a spermatozoon (mammal). (a) A phase-contrast photomicrograph of human spermatozoa. (b) Schematic representation of a spermatozoon in "frontal" view. (c) Longitudinal section of the head portion of a spermatozoon in "side" view.

Throughout their development, vertebrate sperm remain intimately associated with Sertoli cells

The seminiferous tubules house the **Sertoli cells** in addition to the spermatogonia and developing sperm cells. The Sertoli cells, which are epithelial cells, lie side by side and form a ring that extends from the outer basement membrane to the lumen of the tubule. Each Sertoli cell spans the entire distance from the outer membrane to the fluid-filled lumen (Figure 16-8b and d). Adjacent Sertoli cells are joined by tight junctions (p. 66) at a point slightly beneath the outer membrane. Developing sperm cells are tucked between adjacent Sertoli cells, with spermatogonia lying at the outer perimeter of the tubule, outside the tight junction. During spermatogenesis, developing sperm cells arising from spermatogonial mitotic activity pass through the tight junctions, which transiently separate to make a path for them, then migrate toward the lumen in intimate association with the adjacent Sertoli cells, undergoing their further divisions during this migration. The cytoplasm of the Sertoli cells envelops the migrating germ cells, which remain buried within these cytoplasmic recesses throughout their development. Sertoli cells form tight junctions and gap junctions with the developing sperm cells. Recall that gap junctions allow for small molecules to pass directly from one cell's cytoplasm to another's (p. 67). At all stages of spermatogenic maturation, the developing sperm and Sertoli cells exchange small molecules and communicate with one another by means of this direct cell-to-cell binding. Final release of a mature spermatozoa from the Sertoli cell, a process called **spermiation,** requires the breakdown of the tight junctions and gap junctions between the Sertoli cell and spermatozoa.

The supportive Sertoli cells perform the following functions essential for spermatogenesis:

- The tight junctions between adjacent Sertoli cells form a **blood–testis barrier** that prevents bloodborne substances from passing between the cells to gain entry to the lumen of the seminiferous tubule. Because of this barrier, only selected molecules that can pass through the Sertoli cells (such as FSH, the gonadotropin that stimulates the Sertoli cells) reach the intratubular fluid. As a result, the composition of the intratubular fluid varies considerably from that of the blood. The unique composition of this fluid that bathes the germ cells is critical for later stages of sperm development. The blood–testis barrier also prevents the antibody-producing cells in the extracellular fluid entering the seminiferous tubule, thus preventing the formation of antibodies against the highly differentiated spermatozoa.
- Because the secluded developing sperm cells do not have direct access to bloodborne nutrients, the "nurse" Sertoli cells provide nourishment for them. Developing sperm cells cannot efficiently use glucose. The Sertoli cells take up glucose via a GLUT-1 symporter, metabolize the glucose to lactate, then transfer the lactate to the sperm cells, which can use lactate as an energy source.
- The Sertoli cells have an important phagocytic function. They engulf the cytoplasm extruded from the spermatids during their remodeling and destroy defective germ cells that fail to successfully complete all stages of spermatogenesis.
- The Sertoli cells secrete into the lumen **seminiferous tubule fluid,** which "flushes" the released sperm from the tubule into the epididymis for storage and further processing.
- An important component of Sertoli secretion is **androgen-binding protein.** As the name implies, this protein binds androgens (that is, testosterone), thus maintaining a very high level of this hormone within the lumen of the seminiferous tubules. This high local concentration of testosterone is essential for sustaining sperm production. Androgen-binding protein is necessary to retain testosterone within the lumen because this steroid hormone is lipid soluble and could easily diffuse across the plasma membranes and leave the lumen. Testosterone itself stimulates production of androgen-binding protein.
- The Sertoli cells are the site of action for control of spermatogenesis by both testosterone and follicle-stimulating hormone (FSH). IGF-1 also stimulates the proliferation of Sertoli cells and spermatogenesis. The Sertoli cells themselves release another hormone, the peptide *inhibin*, which negatively regulates FSH secretion.
- During fetal development, Sertoli cells also secrete anti-Müllerian factor (AMH).

LH and FSH from the anterior pituitary control testosterone secretion and spermatogenesis

The vertebrate testes are controlled by the two gonadotropic hormones secreted by the anterior pituitary, **luteinizing hormone (LH)** and **follicle-stimulating hormone (FSH)**, both of which are produced by the same cell type, the gonadotrope. Each hormone in each sex acts on the gonads by activating cAMP.

Feedback Regulation of Testicular Function LH and FSH act on separate components of the testes (Figure 16-11). Luteinizing hormone acts on the Leydig (interstitial) cells and regulates testosterone secretion. Follicle-stimulating hormone acts on the Sertoli cells to enhance spermatogenesis. Secretion of both LH and FSH from the anterior pituitary is stimulated in turn by a single hypothalamic hormone, **gonadotropin-releasing hormone (GnRH)** or inhibited by a **gonadotropin-inhibiting hormone (GnIH)**. GnIH was first localized in the hypothalamic paraventricular nucleus of birds, while in mammals, GnIH cell bodies were found clustered in the mediobasal hypothalamus with pronounced projections and terminals throughout the central nervous system.

Recent studies have shown that GnIH also has direct inhibitory actions on GnRH neurons. GnIH reduces rates of firing and induces hyperpolarization in substantial portions of the total population of GnRH neurons. Moreover, since GnIH neurons are not neurosecretory, it appears that GnIH may have largely a hypothalamic site of action on gonadotropin secretion. It is well known that stress inhibits reproduction, in part due to an increase in GnIH production during exposure to acute stressors. Studies in male rats reveal that GnIH production also reduces the expression of sexual behaviors.

GnRH is released from the hypothalamus in secretory bursts (pulses). In seasonally breeding vertebrates, the rate at which GnRH is secreted decreases during the nonbreeding season, as does the amount associated with each burst (see Figure 7-8). Because GnRH stimulates the gonadotropic hormone secretory cells in the anterior pituitary, this pulsatile pattern of hypothalamic secretion results in similar episodic bursts in LH and FSH secretion. Unexpectedly it was determined that GnRH neurons do not contain steroid receptors and so would not be directly responsive to seasonal changes in day length. These being the case, gonadal steroids were suggested to regulate GnRH release through an alternative mechanism. The answer was eventually found in neurons that express steroid receptors, specifically those located in the *periventricular nucleus* and the *arcuate nucleus* (pp. 282 and 724) among others. Projections from these cell bodies are sent to the preoptic area of the hypothalamus, where there

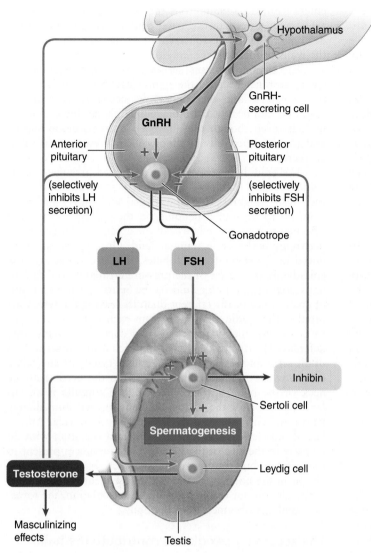

FIGURE 16-11 Control of testicular function.

© Cengage Learning, 2013

responsiveness of the LH secretory cells to GnRH. The latter action explains why testosterone exerts a greater inhibitory effect on LH secretion than on FSH secretion. The testicular inhibitory signal specifically directed at controlling FSH secretion is inhibin from the Sertoli cells. **Inhibin** acts directly on the anterior pituitary to inhibit FSH secretion. This feedback inhibition of FSH by a Sertoli cell product is appropriate because FSH stimulates spermatogenesis by acting on the Sertoli cells.

Roles of Testosterone and FSH in Spermatogenesis

Both testosterone and FSH play critical roles in controlling spermatogenesis, each exerting its effect by acting on the Sertoli cells. Testosterone is essential for both mitosis and meiosis of the germ cells, whereas FSH is required to initiate spermatogenesis before puberty as well as for spermatid remodeling. Testosterone concentration is much higher in the testes than in the blood because a substantial portion of this hormone produced locally by the Leydig cells is retained in the intratubular fluid, complexed with androgen-binding protein secreted by the Sertoli cells. Only this high concentration of testicular testosterone is adequate to sustain sperm production.

Conversion of Testosterone to Estrogen in Males

Although testosterone is classically considered to be the male sex hormone and estrogens to be female sex hormones, the distinctions are not as clear-cut as once thought. In addition to the small amount of estrogens produced by the adrenal cortex (see p. 305), a portion of the testosterone secreted by the testes is converted to estradiol outside of the testes by the enzyme *aromatase* which is widely distributed (e.g., the brain, p. 772) but most abundant in adipose tissue. The boar and the stallion testicle both have significant aromatase activity in the Sertoli cells and as a result secrete substantial quantities of estradiol. Because of this conversion, it is sometimes difficult to distinguish effects of testosterone itself and testosterone-turned-estradiol inside cells. For example, scientists recently learned that closure of the epiphyseal plates in males is induced not by testosterone per se but by testosterone converted into estradiol by aromatization. Estrogen receptors have been identified in the mammalian male brain, testes, prostate, bone, and elsewhere. Recent findings indicate that estrogens play an essential role in male reproductive health, including being important in spermatogenesis and contributing to normal sexuality. Also, it probably contributes to bone homeostasis. The depth, breadth, and mechanisms of action of estrogens in males are only beginning to be explored. (Likewise, in addition to the weak androgenic hormone DHEA produced by the adrenal cortex in both sexes, the ovaries in females also secrete a small amount of testosterone, which is important to libido but may have other functions that remain unclear.)

Gonadotropin-releasing hormone activity increases at puberty

Even though the fetal testes secrete testosterone, which directs masculine development of the reproductive system, after birth the testes become quiescent. Subsequently, in

is an abundance of GnRH cell bodies. The stimulatory peptide released from these neurons was named **kisspeptin,** a reference to its discovery in Hershey, Pennsylvania, the home of the American chocolate "Hershey's kisses."

Even though GnRH controls the relative secretion of both LH and FSH, the blood concentrations of these two gonadotropic hormones do not always parallel each other for two reasons. (1) LH is removed from the blood more rapidly between the secretory bursts than is the more slowly metabolized FSH, so the pulsatile variations in blood concentrations of LH are much more pronounced than are those of FSH. (2) Two other regulatory factors besides GnRH—*testosterone* and *inhibin*—differentially influence the release of LH and FSH.

Testosterone, the product of LH stimulation of the Leydig cells, acts in negative-feedback fashion to inhibit LH secretion in two ways. The predominant negative-feedback effect of testosterone is to decrease the episodes of GnRH release by acting on the hypothalamus, thus indirectly decreasing both LH and FSH release by the anterior pituitary. In addition, testosterone acts directly on the anterior pituitary to reduce the

response to the increasing secretion of gonadotropins, there is a gradual increase in the secretion of testosterone until puberty. The prepubertal delay in the onset of reproductive capability allows time for the animal to mature physically enough to handle offspring, for example, mastering the skills necessary to obtain sufficient food items.

Early in the prepubertal period, GnRH secretory pulses occur (at about four-hour intervals in humans), causing brief increases in LH secretion and accordingly, testosterone secretion. The frequency of episodic GnRH secretion gradually increases until the adult pattern of GnRH, FSH, LH, and testosterone secretion is established. Under the influence of the rising testosterone concentrations, particularly in the peripubertal period, the physical changes that encompass the secondary sexual characteristics and reproductive maturation become evident.

The low frequency of GnRH activity during the prepubertal period appears due to active inhibition of GnRH release by both hormonal and neural mechanisms. Before puberty, the hypothalamus is extremely sensitive to the negative-feedback actions of testosterone, so the very small amounts of testosterone produced by the prepubertal testes can inhibit GnRH release. During the prepubertal period the hypothalamus becomes less sensitive to feedback inhibition by testosterone. Because the low concentrations of testosterone no longer suppress the hypothalamus, GnRH and gonadotropic hormone concentrations rise until sufficient to induce spermatogenesis and therefore puberty.

The ducts of the reproductive tract store and concentrate sperm and increase their fertility

The remainder of the male reproductive system (besides the testes) is designed to deliver sperm to the female reproductive tract. Essentially, it consists of (1) a tortuous pathway of tubes that transport sperm from the testes to the outside of the body; (2) several glands, which contribute secretions that are important to the viability and motility of the sperm; and (3) the penis or its equivalent, which is designed to penetrate and deposit the sperm within the vagina (or in the uterus, as in the stallion and boar). We next examine each of these duct parts in greater detail, beginning with the reproductive tract.

Functions of the Epididymis and Ductus Deferens
The epididymis and ductus deferens perform several important functions. They serve as the sperm's exit route from the testis. As they leave the testis, the sperm can neither move nor fertilize. They gain both capabilities during their passage through the epididymis. This maturational process is stimulated by the testosterone retained within the tubular fluid bound to androgen-binding protein. Sperm's capacity to fertilize is enhanced even further by exposure to secretions of the female reproductive tract, which remove seminal plasma proteins from the surface of the sperm cell. This enhancement of sperm's capacity as a result of the removal of seminal plasma proteins in the female reproductive tracts is known as **capacitation.** Scientists believe that a type of *defensin* (p. 463), a protein secreted by the epididymis that defends sperm from microorganisms, may serve a second role by boosting sperm's motility. The epididymis concentrates the sperm 100-fold by absorbing most of the fluid that

enters from the seminiferous tubules. The maturing sperm are slowly moved through the epididymis into the ductus deferens by rhythmic contractions of the smooth muscle in the walls of these tubes.

The ductus deferens serves as an important site for sperm storage. Because the tightly packed sperm are relatively inactive and their metabolic needs are accordingly low, they can be stored in the epididymis tail for prolonged periods, even though they have no nutrient blood supply and are nourished only by simple sugars present in the tubular secretions.

The epididymis, because of its folded arrangement, is comma-shaped and loosely attached to the caudal surface of each testis (Figures 16-3 and 16-8a) and ranges in length from 30 to 60 mm depending on the species. During the breeding season the cells within the epididymis undergo hypertrophy and secrete seminal fluid. Once sperm are produced in the seminiferous tubules, they are swept into the epididymis as a result of the pressure created by the continual secretion of tubular fluid by the Sertoli cells. As a result of coalescence of the efferent ductules leaving the rete testis tubules, the epididymal ducts from each testis converge to form a large, thick-walled, muscular duct—the **ductus (vas) deferens.** The ductus deferens from each testis passes up out of the scrotum and through the inguinal canal into the pelvic cavity, where it eventually empties into the urethra at the neck of the bladder (Figure 16-3). The **ampulla** is an enlarged portion of the ductus deferens located immediately before the entrance into the urethra or the cloaca. An enlarged muscle layer characterizes it, in association with an increase in the size of the lumen. The ampulla is present in the primate, bull, ram, stallion, and cock, but is reduced or absent in the boar, dog, and fox. Subsequently, the urethra carries sperm out of the penis during **ejaculation,** the forceful expulsion of semen from the body.

The accessory sex glands contribute the bulk of the semen

Several accessory sex glands—especially the seminal vesicles and prostate—empty their secretions into the duct system before it joins the urethra (Figure 16-3). A pair of saclike *seminal vesicles* empty into the last portion of the two ductus deferens, one on each side. The short segment of duct that passes beyond the entry point of the seminal vesicle to join the urethra constitutes the *ejaculatory duct*. The *prostate* is a large single gland that completely surrounds the ejaculatory ducts and urethra (some species have only a small *bulb*, as in humans, whereas others have both a bulb and a longer *disseminate* prostrate that surrounds a longer portion of the pelvic urethra). Another pair of accessory sex glands, the *bulbourethral glands,* drains into the urethra after it has passed through the prostate, just before it enters the penis. Numerous mucus-secreting glands are located along the length of the urethra.

Semen
During ejaculation, the accessory sex glands contribute secretions that provide support for the continuing viability of the sperm inside the female reproductive tract (in animals with internal fertilization) or in water (such as fish semen). These secretions constitute the bulk of the **semen,** which consists of a mixture of accessory sex-gland secretions, sperm, and mucus. Sperm make up only a small percentage of the total ejaculated fluid. In some mammals, such

as rodents and the boar and stallion, the seminal fluid has special coagulation properties that plug the female reproductive tract (as in some insects; p, 764) and minimize loss of spermatozoa after ejaculation. In rodents, this solidified *copulatory plug* is later lost from the vagina, signaling that successful copulation has occurred.

The Male Accessory Glands Although the accessory sex gland secretions are not absolutely essential for fertilization, they do make species-specific contributions that greatly facilitate the fertilization process in mammals:

- The **seminal vesicles** or **vesicular glands** are paired glands that empty into the pelvic urethra and (1) supply fructose, which serves as the primary energy source for ejaculated sperm; (2) secrete *prostaglandins*, which stimulate contractions of the smooth muscle in both the male and female reproductive tracts, helping to transport sperm from their storage site in the male to the site of fertilization in the female oviduct; (3) provide more than half the semen volume, which helps wash the sperm into the urethra and also dilutes the thick mass of sperm, thus enabling them to develop motility; and (4) secrete fibrinogen, a precursor of fibrin, which forms the meshwork of a clot (see Figure 9-11, p. 397). Seminal vesicles vary significantly between species but are absent in the dog, fox, and birds.
- The **prostate gland** (1) secretes an alkaline fluid that neutralizes the acidic vaginal secretions, an important function because sperm are more viable in a slightly alkaline environment; (2) provides clotting enzymes; and (3) releases **prostate-specific antigen (PSA)**. The prostate clotting enzymes act on fibrinogen from the seminal vesicles to produce fibrin, which "clots" the semen, thus helping to keep the ejaculated sperm in the female reproductive tract during withdrawal of the penis. Shortly thereafter, the seminal clot is broken down by prostate-specific antigen, a fibrin-degrading enzyme from the prostate, releasing motile sperm within the female tract. The prostate varies in structure from branched tubular to tubuloalveolar. The prostate gland is large in the boar, comparatively smaller in the bull, and absent in the smaller ruminants. It is absent in the cock, but is extremely well developed in the dog and fox.
- During sexual arousal, the **bulbourethral glands** secrete a mucuslike substance that neutralizes the acidic pH in the urethra, provides lubrication, and causes the seminal plasma to coagulate following ejaculation. These glands are particularly enlarged in the boar, but are absent in the fox, dog, and birds.

Before turning our attention to the act of delivering sperm to the female, we are briefly going to digress and discuss the diverse roles of *prostaglandins,* which were first discovered in semen but are abundant throughout the body.

Prostaglandins are ubiquitous, locally acting chemical messengers

Although **prostaglandins** were first identified in the semen and were believed to be of prostate gland origin (hence their name, even though the seminal vesicles are now known to produce more prostaglandins than the prostate), their production and actions are by no means limited to the reproductive system. These 20-carbon fatty acid derivatives are among the most ubiquitous and physiologically active substances in vertebrates. They are produced in virtually all tissues from arachidonic acid, a fatty-acid constituent of the phospholipids within the plasma membrane. Prostaglandins and other closely related arachidonic-acid derivatives, namely, *prostacyclins, thromboxanes,* and *leukotrienes,* are collectively known as **eicosanoids,** and are among the most biologically active compounds known. On appropriate stimulation, arachidonic acid is split from the plasma membrane by a cytosolic enzyme and then is converted into the appropriate eicosanoid, which acts as a paracrine locally within or near its site of production. After eicosanoids act, they are inactivated rapidly by local enzymes before they gain access to the blood, or if they do reach the circulatory system, they are swiftly degraded on their first pass through the lungs so that large quantities are not dispersed through the systemic arterial system.

Prostaglandins are designated as belonging to one of nine groups—PGA, PGB, PGC, PGD, PGE, PGF, PGG, PGH, or PGI (prostacyclin)—according to structural variations in the five-carbon ring that they contain at one end (Figure 16-12). Within each group, prostaglandins are further identified by the number of double bonds present in the two side chains that project from the ring structure (for example, PGE_1 has one double bond and PGE_2 has two double bonds).

Prostaglandins exert a bewildering variety of effects. Not only are slight variations in prostaglandin structure accompanied by profound differences in biological action, but the same prostaglandin molecule may even exert opposite effects in different tissues. Besides enhancing sperm transport in semen, these abundant chemical messengers are known or suspected to exert other actions in the female reproductive system and in the respiratory, urinary, digestive, nervous, and endocrine systems, in addition to affecting platelet aggregation, fat metabolism, and inflammation and fever (p. 469). Salicylate drugs such as aspirin and other nonsteroidal anti-inflammatory drugs inhibit the production of prostaglandins (see p. 461).

Next, before considering the female in greater detail, we examine the means by which males and females come together to accomplish reproduction.

The penis in many vertebrates is both erectile and extendible

The mammalian penis consists of three parts, the **root, body** or **shaft,** and **glans.** The body normally constitutes the major portion of the organ. In some mammals (such as mink, raccoons, bats, and moles, but not humans, horses, bulls, rams, boars, bucks, and most primates) the penis contains a penile bone called the **os penis** or **baculum** (Figure 16-13). The free end of the penis, the glans, is richly supplied with nerves. In some species the penis is highly erectile. In the boar the penis has the unique feature of existing as a left-handed corkscrew, which actually penetrates, with rotary motion, into the cervix and locks in place, providing the pressure needed for the stimulation of ejaculation.

Some mammals (boars, bulls, and rams) have a penis containing an increased quantity of connective tissue relative to the amount of erectile tissue. The penis of these animals contains a **sigmoid flexure,** an S-shaped configuration along the shaft of the penis (Figure 16-13). The straightening of

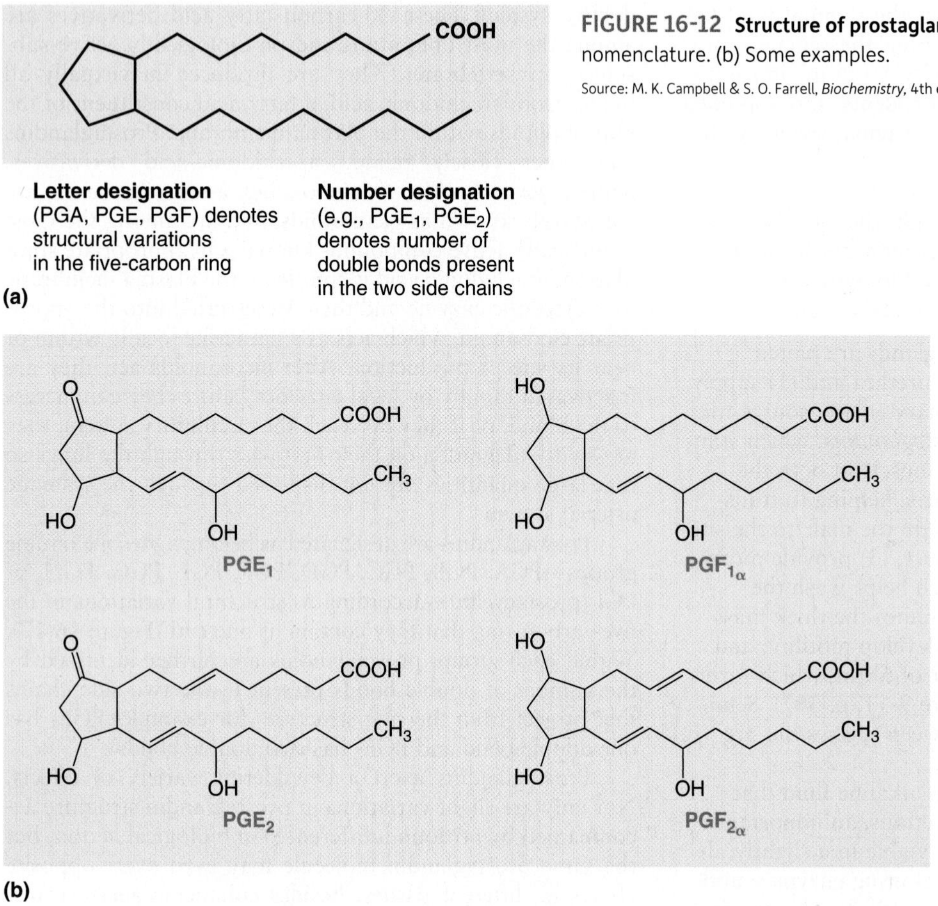

FIGURE 16-12 Structure of prostaglandins. (a) General features and nomenclature. (b) Some examples.

Source: M. K. Campbell & S. O. Farrell, *Biochemistry*, 4th ed. Belmont, CA: Brooks/Cole, p. 219, Figure 7.33.

the curve of the penis results in extension of the penis from the sheath. Consequently, during erection there is little change in the diameter of the penis. The **retractor penis muscle** is a paired muscle located on the ventral face of the penis. It originates at the root of the penis and shunts across the S-shaped curve. These muscles control the degree of extension of the penis by their action on the sigmoid flexure and retract the penis into the sheath after extension.

In lizards and snakes, the penis is paired and represents an extension of the cloaca. During copulation this **hemipenis** is everted into the cloaca of the female. To sustain intromission in the female, the hemipenis contains spiny ridges. The body of the normally flaccid penis can become erect as a consequence of testicular vasocongestion (engorgement with blood). In birds the pelvic urethra ends in either a cloaca or a hemipenis. In some species of sharks, skates, and other chondrichthyan fishes, males transfer sperm to females through a groove in their **claspers,** a pair of modified pectoral fins.

check your understanding 16.3

Describe the three major stages of spermatogenesis. Discuss the functions of each part of a spermatozoon. What are the roles of Sertoli cells?

Discuss the control of testicular function.

Discuss the source and functions of testosterone.

16.4 Mating

Ultimately, union of male and female gametes often requires *mating* behavior to accomplish delivery of sperm-laden semen to mature ova. In species with internal fertilization, sperm enter the female vagina through the **sex act,** also called **sexual intercourse, coitus,** or **copulation.** Extinct fish called placoderms (armored or "plated skin") that lived approximately 375 million years ago were likely the first backboned creatures to copulate and give birth to live young. In most vertebrates, mating occurs at the peak of female fertility, which is *estrus* in most mammals (see p. 784). A few species, including humans, bonobos, chimpanzees, and dolphins, engage in sex acts (including same-sex interactions) for pleasure in addition to reproduction, in part as a means of reinforcing social bonds and status.

The male sex act is characterized by erection and ejaculation

After a suitable partner has been located, species-specific courtship behavior (p. 761) is initiated in many species. In many mammals, a mating posture assumed by the female (**lordosis**) serves as a powerful trigger for sexual stimulation in the male. Arousal of the male stimulates the erection and protrusion of the penis. The *male sex act* involves two components: (1) **erection,** the straightening or hardening of the penis to permit its entry into the vagina, and (2) **ejaculation,**

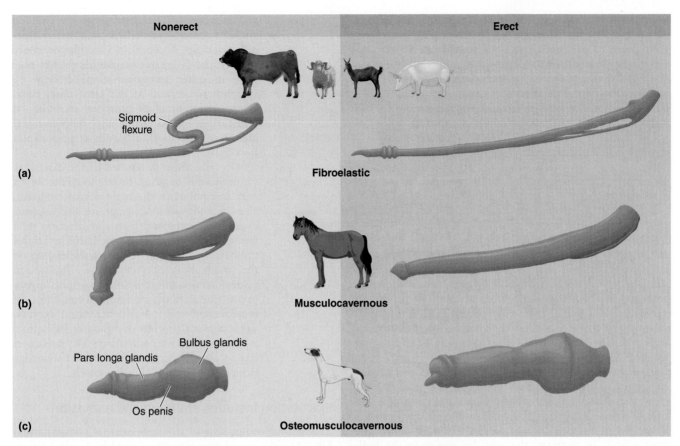

FIGURE 16-13 Classification of penile structures. (a) The penises of rams, goats, and boars are fibroeleastic. Erection results in a straightening of the sigmoid flexure. Engorgement of the corpus cavernosum with blood distends and elongates this class of penis. (b) The penis of cats and horses is musculocavernous. Erection results in a more complex change in the position, length, and stiffness of the penis compared to fibroelastic forms. (c) The musculocavernous penis of dogs contains a central bony portion (*os penis*) that aids in intromission into the female's vagina with only a partial erection. Erection is completed in the female's vagina. The local engorgement of the *bulbus glandis* is prolonged by occlusion of the penile venous return by the vaginal constrictor muscles. This contraction maintains and lengthens the duration of the intravaginal erection. Intravaginal engorgement associated with dilation of the penile bulb prevents withdrawal of the penis from the vagina, leading to the so-called tie between the two.

Source: Y. Ruckebusch, L.-P. Phaneuf, & R. Dunlop. (1991). *Physiology of Small and Large Animals.* Philadelphia: B. C. Decker, p. 582, Figure 55-1.

or forceful expulsion of semen into the urethra and out of the penis. In addition to these strictly reproduction-related components, the **sexual response cycle** also encompasses broader physiological responses that can be divided into four phases:

1. The *excitement phase,* which includes erection (and heightened sexual awareness in humans, at least).
2. The *plateau phase,* which is characterized by intensification of these responses, plus more generalized body responses, such as steadily increasing heart rate, blood pressure, respiratory rate, and muscle tension.
3. The *orgasmic phase,* which includes ejaculation as well as other responses that culminate in a burst of stimulating neural signals in pleasure centers of the brain.
4. The *resolution phase,* which returns genitalia and body systems to their prearousal state.

Erection is accomplished by penis vasocongestion

In those species in which the penis consists almost entirely of **erectile tissue,** the body consists of three columns of spongelike vascular spaces extending the length of the organ. In the absence of sexual excitation, the erectile tissues contain little blood, because the arterioles that supply these vascular chambers are constricted and venous drainage is maximal. As a result, the penis remains small and flaccid. During sexual arousal, these arterioles reflexly dilate and the erectile tissue fills with blood, causing the penis to enlarge both in length and width and to become more rigid. In the stallion, for example, these vascular spaces are particularly enlarged such that, during erection, considerable increases in the penal size are achieved as a result of accumulation of blood in

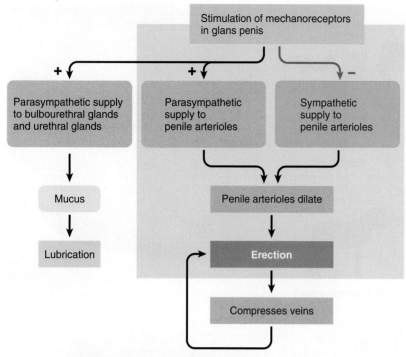

FIGURE 16-14 Erection reflex in a male mammal.

© Cengage Learning, 2013

these spaces. In contrast, the vascular spaces in the bull, ram, and boar are comparatively small except in the distal bend of the sigmoid flexure. Additional buildup of blood and further enhancement of erection are achieved by a reduction in venous outflow. The veins that drain the erectile tissue are compressed as a result of engorgement and expansion of the vascular spaces brought about by increased arterial inflow.

Erection Reflex The erection reflex is a spinal reflex triggered by stimulation of highly sensitive mechanoreceptors located in the glans penis. The actual **erection-generating center** lies in the sacral segments of the spinal cord. Tactile stimulation of the glans reflexly triggers, by means of this center, increased parasympathetic vasodilator activity and decreased sympathetic vasoconstrictor activity to the penile arterioles. The result is rapid, pronounced vasodilation of these arterioles and an ensuing erection (Figure 16-14).

This parasympathetically induced vasodilation is the major instance of direct parasympathetic control over blood vessel caliber (internal diameter) in the body. Parasympathetic stimulation brings about relaxation of penile arteriolar smooth muscle by means of *nitric oxide* (NO), which causes arteriolar vasodilation in response to local tissue changes elsewhere in the body (see p. 392). Specifically, NO activates a membrane-bound enzyme, guanylate cyclase, within nearby arteriolar smooth muscle cells. This enzyme activates cyclic guanosine monophosphate (cGMP), an intracellular second messenger similar to cAMP (see p. 435). Cyclic GMP in turn leads to relaxation of the penile arteriolar smooth muscle, bringing about a pronounced local vasodilation. Arterioles are typically supplied only by sympathetic nerves, with increased sympathetic activity producing vasoconstriction and decreased sympathetic activity resulting in

vasodilation. Concurrent parasympathetic stimulation and sympathetic inhibition of penile arterioles accomplish vasodilation more rapidly and in greater magnitude than is possible in other arterioles supplied only by sympathetic nerves. At the same time, parasympathetic impulses promote secretion of lubricating mucus from the bulbourethral glands and the urethral glands in preparation for coitus.

Recent research has led to the discovery of numerous regions throughout the brain that can influence the male sexual response. The erection-influencing brain sites appear extensively interconnected and function as a unified network to either facilitate or inhibit the basic spinal erection reflex, depending on the momentary circumstances. Failure to achieve an erection may occur despite appropriate stimulation as a result of inhibition of the erection reflex by higher brain centers. Treatment of erectile dysfunction in human males by drugs that stimulate NO production has become a major application of knowledge gained from physiological research.

Ejaculation includes emission and expulsion

The second component of the male sex act is ejaculation. Like erection, ejaculation is accomplished by a spinal reflex. The same types of tactile and psychic stimuli that induce erection cause ejaculation when the level of excitation intensifies to a critical peak. The ejaculatory response occurs in two phases: emission and expulsion. First, sympathetic impulses cause sequential contraction of smooth muscles in the prostate, reproductive ducts, and seminal vesicles. This contractile activity delivers prostatic fluid, then sperm, and finally seminal vesicle fluid (collectively, *semen*) into the urethra. This phase of the ejaculatory reflex is known as **emission.** During this time, the sphincter at the neck of the bladder is tightly closed to prevent semen from entering the bladder and urine from being expelled along with the ejaculate through the urethra. Second, the filling of the urethra with semen triggers nerve impulses that activate a series of skeletal muscles at the penis base. Rhythmic contractions of these muscles increase the pressure within the penis, forcibly expelling the semen through the urethra to the exterior. This is the **expulsion** phase of ejaculation.

Orgasm in turn, refers to the pelvic and overall systemic responses that culminate the sex act. These include the rhythmic contractions that occur during semen expulsion and the involuntary rhythmic throbbing of pelvic muscles associated with the peak intensity of the overall body responses that were manifested during the earlier phases. While only humans can report the intense pleasure associated with orgasm, observations of chimpanzees and bonobos suggest that they also experience this reinforcing phenomenon.

During the resolution phase following orgasm, sympathetic vasoconstrictor impulses slow the inflow of blood into the penis, causing the erection to subside. Muscle tone returns to normal while the cardiovascular and respiratory systems return to their prearousal level of activity. Once ejaculation has occurred, a temporary refractory period of variable duration ensues before sexual stimulation can trigger another erection.

Volume and Sperm Content of the Ejaculate The volume and sperm content of the ejaculate depend on the length of time between ejaculations. Production of spermatozoa in mammals ranges from <1 to 25 billion spermatozoa per day for both testes. In humans, an average ejaculate contains about 180 million sperm (66 million/mL), but some ejaculates contain as many as 400 million sperm. In birds, the volume of the ejaculate is comparatively small (200–800 μL in a turkey), but the density of spermatozoa (4,000–7,000 per μL) is usually higher. Because the increased body size of domesticated male turkeys (toms) relative to the hens makes normal mating extremely difficult, the poultry industry uses artificial insemination extensively in turkeys.

Even though only one spermatozoon actually fertilizes the ovum, large numbers of accompanying sperm are often needed to provide sufficient acrosomal enzymes (the acrosome is the membrane-bound organelle of the spermatozoon that contains proteolytic enzymes) to break down the barriers surrounding the ovum (zona pellucida) until the fertilizing sperm is engulfed into the ovum's cytoplasm. The quality of sperm must be taken into account when assessing the fertility potential of a semen sample. The presence of substantial numbers of sperm with abnormal motility or structure, such as sperm with distorted tails, reduces the chances of fertilization.

The female sex act is similar to the male sex act

In humans at least, both sexes experience the same four phases of the sexual cycle—excitement, plateau, orgasm, and resolution. Furthermore, the physiologic mechanisms responsible for orgasm are fundamentally the same in males and females. The excitement phase in human females can be initiated by either physical or psychological stimuli. Tactile stimulation of the externally exposed portion of the glans clitoris and surrounding perineal area is an especially powerful sexual stimulus. These stimuli trigger spinal reflexes that bring about parasympathetically induced vasodilation of arterioles throughout the vagina and external genitalia, especially the clitoris. The latter—like its male homolog, the penis—is composed largely of erectile tissue (recall that it is much larger than traditionally reported; Figure 16-4b).

Vasocongestion of the vaginal capillaries forces fluid out of the vessels into the vaginal lumen. This fluid, which is the first positive indication of sexual arousal, serves as the primary lubricant for intercourse. Additional lubrication is provided by the mucus secretions from the male and by mucus released during sexual arousal from glands located at the outer opening of the vagina.

During the plateau phase, the changes initiated during the excitement phase intensify, while systemic responses similar to those in the male (such as increased heart rate, blood pressure, respiratory rate, and muscle tension) occur. If erotic stimulation continues, the sexual response culminates in orgasm as sympathetic impulses trigger rhythmic contractions of the pelvic musculature as in males. Systemic responses identical to those of the male orgasm also occur. Again, observations of chimpanzee and bonobo females suggest that they experience intense pleasure as do human fe-

males. During resolution, pelvic vasocongestion and the systemic manifestations gradually subside.

Oxytocin has several roles in mating and other social interactions

Oxytocin is a pituitary hormone most closely associated with labor, lactation, and other female reproductive processes that you will see in detail later. But it is also found in males. Why? Oxytocin appears to play roles in mating-related behaviors in both sexes. It is released during arousal and orgasm, and may stimulate orgasmic muscle contractions in both sexes. Oxytocin has been localized in the testes and semen of several mammalian species and may be a paracrine factor inducing contraction of the seminiferous tubules during ejaculation. Most notably, oxytocin in females, as well as vasopressin in males (as in the prairie vole, p. 286), reinforce pair-bonding between mates. Oxytocin's actions go beyond mating: Human males who sniffed an experimental oxytocin spray were much more likely to trust other males in their own group, while simultaneously becoming more antagonistic toward strangers. Though oxytocin is popularly dubbed the "trust hormone," its affects are thus more complex.

check your understanding 16.4

How do penile arterioles accomplish vasodilation more rapidly and in greater magnitude than is possible in other arterioles?

Compare the stages of the sex act and roles of oxytocin in males and females.

16.5 Female Reproductive Physiology

In some vertebrates, female reproductive physiology is more complex than the male's.

Complex cycling characterizes female reproductive physiology in many vertebrates, including estrous cycles in most mammals

As in the vertebrates male, female reproduction involves interactions of internal and environmental factors mediated by the hypothalamic–pituitary–gonadal axis. However, unlike the continuous sperm production and relatively constant testosterone secretion in males, the release of ova or deposition of eggs is intermittent, and secretion of female sex hormones displays wide cyclical swings. In mammals, there are two separate but coordinated cycles:

1. The **ovarian cycle,** in which an oocyte matures in the ovary and is *ovulated* to travel to the uterus. As the primary female reproductive organs, ovaries perform the dual function of producing ova (*oogenesis*) and secreting the female sex hormones estrogens and progesterone.
2. The **uterine cycle,** in which the uterine lining is prepared for a fertilized egg to implant (primarily due to the ovarian hormones).

TABLE 16-4 Reproductive Cycles of Various Mammals

Genus species[a]	Cycle Type[b]	Cycle Length[c]	Duration of Estrus	Ovulation Type[d]	Time	Luteal Phase in Absence of Coitus
Felis domesticus (cat)	S, P	14	4 days	I	24–30 hr after coitus	No
Bos taurus (cow)	C, P	17–27	13–14 hr	Sp	12–16 hr after end of estrus	Yes
Canus familiaris (dog)	S, M	60	7–9 days	Sp	1–3 days after onset of estrus	Yes
Ovis aries (sheep)	S, P	15–18	30–36 hr	Sp	12–14 hr before the end of estrus	Yes
Mustelo fero (ferret)	S	—	Continuous	I	30 hr after coitus	No
Vulpes vulpes (red fox)	S, M	90	1–5 days Seasonally	Sp	1–2 days after onset of estrus	Yes
Capra hirca (goat)	S, P	20–21	39 hr	Sp	30–36 hr after onset of estrus	Yes—shorter cycles in some breeds
Cavia porcellus (guinea pig)	C, P	16	6–11 hr	Sp	10 hr after onset of estrus	Yes
Mesocricetus auratus (hamster)	C, P	4	20 hr	Sp	8–12 hr after onset of estrus	No
Homo sapiens (human)	C, Mn	28	None	Sp	14 days before the onset of menses	Yes—extremely variable cycle
Equus caballus (horse)	S, P	19–23	4–7 days	Sp	5–6 hr after the onset of estrus	Yes
Macaca mulatta (rhesus monkey)	C, Mn	28	None	Sp	11–14 days after the onset of menses	Yes
Mus musculus (mouse)	C, P	4	10 hr	Sp	2–3 hr after onset of estrus	No
Oryctolagus cuniculus (rabbit)	C	—	Continuous	I	10–12 hr after coitus	No
Rattus norvegicus (rat)	C, P	4–5	13–15 hr	Sp	8–10 hr after onset of estrus	No
Sus scrofa (pig)	C, P	18–23	2–3 days	Sp	36 hr after onset of estrus	Yes

[a]Domesticated or in captivity.
[b]C = continuous; S = seasonal; P = polyestrous; M = monoestrous; Mn = menstrual.
[c]Length of cycle in days.
[d]I = Induced (reflex) ovulators; Sp = spontaneous ovulators (male not necessary).

Source: H. E. Kidder and R. Cochrane, West Virginia University.

If fertilization does not occur, the cycles repeat. If fertilization does occur, the cycles are interrupted, while the female system adapts to nurture and protect the newly conceived embryo until it has developed the ability to live outside the maternal environment. Furthermore, the mammalian female continues her reproductive duties after birth by producing milk (lactation) for the young's nourishment. Thus, the female reproductive system is characterized by complex cycles that are interrupted only by more complex changes should pregnancy ensue.

The ovarian and uterine cycles together constitute an **estrous cycle** in most mammals, defined as the time from one period of sexual receptivity to the next. Sexual receptivity is called **estrus** (a noun, not to be confused with the adjective *estrous*) or **heat,** which occurs periodically under certain specified conditions (often at the time of ovulation).

Ovarian cycles are very similar among mammals, but uterine cycles differ in two versions. In the *estrous* type, if there is not a pregnancy, the uterine lining that has been prepared for an embryo is reabsorbed. In contrast, some primates, including humans, have a **menstrual cycle,** in which uterine lining is sloughed off periodically (*menstruation*). The cycle begins at the end of each menstruation and occurs uniformly throughout the year. These primates also usually have distinct *estrus* (receptivity) periods, although human females are considered to be continuously receptive.

The duration of estrous and menstrual cycles is species specific but variable within a species (Table 16-4). In general, the principles governing reproduction are the same for the different species, but there are notable differences. For example, the frequency of estrous cycles differs between

For the past 60 years, humans have been polluting the environment with synthetic endocrine-disrupting chemicals (**EDCs,** p. 275) as an unintended side effect of industrialization. As you saw in Chapter 7, these hormonelike pollutants bind with the receptor sites normally reserved for the naturally occurring hormones, either mimicking or blocking normal hormonal activity. Many EDCs have been found to alter sex steroid functions. Some are *estrogenic* and exert feminizing effects as estrogen agonists. Others (androgen antagonists) block a male's own androgens from accomplishing their intended masculinizing effects. For example, studies suggest that bacteria in wastewater from pulp mills can convert the sterols in pine pulp into androgens. By contrast, anti-androgen compounds have been found in the fungicides commonly sprayed on vegetable and fruit crops. Anti-androgenic action demasculinizes males, producing an outcome similar to that brought about by estrogenic EDCs. Yet another cause for concern are the androgens used by the livestock industry to enhance the production of muscle (that is, meat) in feedlot cattle. (Androgens have a protein anabolic effect.) These drugs do not end up in the meat, but they can get into drinking water and other food as hormone-laden feces contaminate rivers and streams.

Estrogenic pollutants are everywhere in our environment. They contaminate food, drinking water, and air. Proven feminizing synthetic compounds include (1) certain weed killers and insecticides, (2) some detergent breakdown products, (3) petroleum by-products found in car exhaust, (4) common food preservatives used to retard rancidity, and (5) softeners that make plastics flexible. These plastic softeners are commonly found in food packaging and can readily leach into food with which they come in contact, especially during heating. Plastic softeners are among the most plentiful industrial contaminants in our environment. An estimated 87,000 synthetic chemicals are already in our environment. Scientists suspect that the estrogen-mimicking chemicals among these may underlie a spectrum of reproductive disorders that have been rising in the past 60 years—the same time period during which large amounts of these pollutants have been introduced into the environment.

Evidence of Gender Bending in Animals

Some fish and wild animal populations that have been severely exposed to environmental EDCs—such as those living in or near water heavily polluted with chemical wastes—display a high rate of grossly impaired reproductive systems. Examples include male fish that are hermaphrodites (having both male and female reproductive parts) and male alligators with abnormally small penises. Similar reproductive abnormalities have been identified in land mammals. Presumably, excessive exposure to estrogen agonists is emasculating these populations.

Environmental estrogens are implicated in the rising incidence of breast cancer in female humans. Breast gland cancer is 25 to 30% more prevalent now than in the 1940s. Many of the established risk factors for mammary gland cancer, such as starting to menstruate earlier than usual and undergoing menopause later than usual, are associated with an elevation in the total lifetime exposure to estrogen. Because increased exposure to natural estrogens bumps up the risk for mammary gland cancer, prolonged exposure to environmental estrogens may be contributing to the rising prevalence of this malignancy among women (and men as well).

Some environment-disrupting chemicals are the natural estrogen-like compounds called **phytoestrogens,** synthesized by some plant species. Usually the ingestion of these compounds is disadvantageous to the reproductive success of an animal, but in some situations can be of benefit. The California quail (*Lophortyx californicus*) is a photoperiodic species that relies on day length to regulate the time of the breeding season. In the more arid regions of its range, it breeds irregularly, depending on the amount of rainfall in the previous winter. During dry years, high concentrations of phytoestrogens are produced in the desert plants on which the quail feed. As with endocrine-disrupting chemicals, these phytoestrogens suppress reproductive development, in this situation by feeding back onto the hypothalamic–pituitary axis. By suppressing reproductive activity in these lean years, the animal does not use scarce body reserves on a clutch of eggs that will not likely reach sexual maturity.

---

species: Cattle, swine, and rodents are **continuously polyestrous** (cycles occuring uniformly throughout the year), while others are **seasonally polyestrous** (estrous cycles are restricted to a particular time of year). Deer, sheep, and goats are examples of **short-day breeders** because they exhibit estrous cycles as day length decreases, whereas bears, hamsters, and horses are **long-day breeders;** that is, they come into estrous as day length increases. **Seasonally monoestrous** females (most carnivores, including bears, dogs, foxes, and wolves) are characterized as having a single estrous event followed by a long period of **anestrus** (period of time without regular cyclic activity). For example, most female dogs have about two estrus periods in a year. In these animals the period of estrus is prolonged and lasts for several days, thus increasing the probability of securing a mate.

Reproductive capability begins at puberty in mammalian females, but reproductive potential of many primates, some rodents, whales, dogs, rabbits, elephants, and domestic livestock, ceases gradually during middle age (unlike males, who typically have reproductive potential through the remainder of life, although it gradually declines). In primates with menstrual cycles, this cessation is called **menopause** due to the complete cessation of **menses** (the outflow uterine materials during menstruation).

An understanding of the hormonal interactions governing follicular development and ovulation has permitted the manipulation of ovulation of some domestic species for management purposes. Thus, domesticated animals may have regular, repeated cycles during the year, whereas animals in the wild may have more sporadic estrous cycles because they are either pregnant or lactating. For this reason, it may be more appropriate to consider **calving intervals,** literally from one birth to the next. Considering the juvenile period and the times of gestation and lactation, plus any decline in reproductive capacity with age, *anestrus* may constitute the more common condition of a normal female animal.

Although ovarian cycles are similar in many ways among vertebrates, differences can be found in the type of ovulation. Most animals are **spontaneous ovulators** that ovulate with a regular frequency and do not require copulation. Some species of birds, for example, ovulate daily for extended periods of time without any contact with the male. **Reflex (induced) ovulation** has been described for a large number of mammals including the cat, mink, ferret, rabbit, and camel. In this strategy ovulation is induced by stimulation of sensory receptors in the vagina and cervix during coitus, either mechanically or by semen components (camelids). Although the strategy is considered a more primitive condition than spontaneous ovulation, very few examples have been described in any taxa other than mammals. Exceptions include some reptiles including the whiptail lizard (*Cnemidophorus inornatus*), redsided garter snake (*Thamnophis sirtalis parietalis*), and loggerhead sea turtle (*Caretta caretta*).

Functions of Estrogens and Progesterone As we noted earlier, the ovaries synthesize estrogens and progesterone. These sex hormones act together to promote the opportunity for fertilization of the ovum and, in mammals, to prepare the female reproductive system for pregnancy. Estrogens in the female govern many functions similar to those carried out by testosterone in the male, such as maturation and maintenance of the entire female reproductive system and establishment of female secondary sexual characteristics. In general, the actions of estrogens are important to preconception events. Estrogens are essential for the female to become sexually receptive and permit copulation, ova maturation and release, and transport of sperm from the vagina to the site of fertilization in the oviduct. Two different nuclear estrogen receptors can be identified in estrogen-responsive tissues—$ER\alpha$ and $ER\beta$. These receptors differ in their distribution among estrogens target cells and in the responses they trigger when bound with an estrogen. Some target cells have only $ER\alpha$ receptors, some only $ER\beta$ receptors, and some both. For example, both ER receptors are localized in Leydig cells, whereas $ER\beta$ is mostly confined to Sertoli cells and developing germ cells. These findings may help explain the complex effects of *environmental estrogens* (see *Concepts and Controversies: Environmental Estrogens: Bad News for Reproduction*).

Estrogens in mammals contribute to mammary gland development in anticipation of lactation. The other ovarian steroid, progesterone, is important in preparing a suitable environment for nourishing a developing embryo/fetus and for contributing to the mammary glands ability to produce milk (see p. 810).

The steps of gametogenesis are the same in both sexes, but the timing and outcome differ sharply

Oogenesis contrasts sharply with spermatogenesis in several important aspects, even though the identical steps of chromosome replication and division take place during gamete production in both sexes. The undifferentiated primordial germ cells in the fetal ovaries, the **oogonia** (comparable to the spermatogonia), divide mitotically during gestation and/ or neonatally, after which time mitotic proliferation ceases.

Formation of Primary Oocytes and Primary Follicles During the last part of fetal life, the oogonia begin the early steps of the first meiotic division but do not complete it. Known now as **primary oocytes,** they contain the diploid number of 2N replicated chromosomes, which are gathered into homologous pairs but do not separate. The primary oocytes remain in this state of meiotic arrest until they are prepared for ovulation. A presumed purpose of this nuclear arrest is to inactivate the DNA in the female gamete in order to reduce the possibility of biochemical insult during the lifetime of the female.

Before birth, each primary oocyte is surrounded by a single layer of flattened **granulosa cells.** Together, an oocyte and surrounding granulosa cells make up a **primordial follicle.** Oocytes that fail to be incorporated into follicles self-destruct by apoptosis (cell suicide). At birth only a fraction of the original primordial follicles remain, each containing a single primary oocyte capable of producing a single ovum. In most mammals, no new oocytes or follicles appear after birth; the follicles already present in the ovaries at birth serve as a reservoir from which all ova throughout the reproductive life of a female must be drawn. Of these follicles, only a small percentage (about 400 in humans) mature and release ova.

The pool of primary follicles present at birth gives rise to an ongoing trickle of developing follicles. Once it starts to develop, a follicle is destined for one of two fates: It reaches maturity and ovulates, or it degenerates, a process known as **atresia.** Until puberty, all the follicles that start to develop undergo atresia in the early stages without ever ovulating. Even after puberty, many of the follicles are **anovulatory** (that is, no ovum is released). Of the initial pool of follicles, most (for example, 99.98% in humans) never ovulate but instead undergo atresia at some stage in development. At the end of the female's reproductive life, only a few follicles remain in the ovaries, and soon even these succumb to atresia.

This limited gamete potential, which is thought to be determined at birth in mammalian females, is in sharp contrast to the continual process of spermatogenesis in males. (Although in rabbits at least, there is evidence of some new oocytes forming after puberty.) Furthermore, there is considerable chromosome wastage in oogenesis compared with spermatogenesis. Let's see how.

Formation of Secondary Oocytes and Secondary Follicles The primary oocyte within a primary follicle is still a diploid cell that contains 2N doubled chromosomes. A portion of the resting pool of follicles starts developing into **secondary (antral) follicles.** The number of developing follicles at any time is roughly proportional to the size of the pool, but the mechanisms that determine which follicles in the pool develop during a given cycle are unknown. Development of a secondary follicle is characterized by growth of the primary oocyte and by expansion (proliferation) and differentiation of the surrounding cell layers. The extent of oocyte enlargement is species dependent. Oocyte enlargement is due to a buildup of cytoplasmic and/or yolk materials that will be needed by the early embryo. The follicle next acquires a dependence on gonadotropins, in part as a result of the dif-

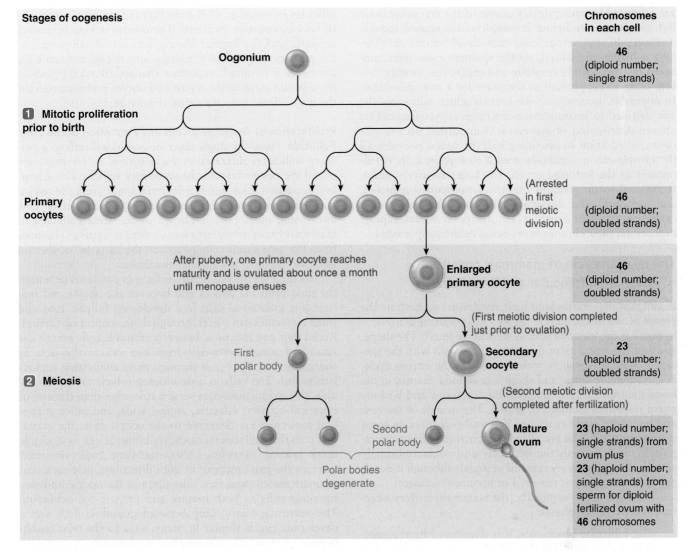

FIGURE 16-15 Oogenesis. Compare with Figure 16-9, p. 774, spermatogenesis.

© Cengage Learning, 2013

ferentiation of the cells surrounding the granulosa into theca cells, which markedly increases estrogen production by the follicular cells.

Just before ovulation, the primary oocyte, whose nucleus has been in meiotic arrest, often for years, completes its first meiotic division, except in the horse and the carnivores, which ovulate a primary oocyte in association with the extended period of estrus. This division yields two daughter cells, each receiving a haploid set of *n* doubled chromosomes, analogous to the formation of secondary spermatocytes (Figure 16-15). However, almost all the cytoplasm remains with one of the daughter cells, now called the **secondary oocyte,** which is destined to become the ovum. The chromosomes of the other daughter cell together with a small share of cytoplasm form the **first polar body.** In this way, the ovum-to-be loses half of its chromosomes to form a haploid gamete but retains all its nutrient-rich cytoplasm. The nutrient-poor polar body soon degenerates.

Formation of a Mature Ovum Actually, the secondary oocyte, and not the mature ovum, is ovulated and fertilized, but common usage refers to the developing female gamete as

an *ovum* even in its primary and secondary oocyte stages. Sperm entry into the secondary oocyte is needed to trigger the second meiotic division. Secondary ocytes that are not fertilized never complete this final division. During this division, a half set of chromosomes along with a thin layer of cytoplasm is extruded as the **second polar body.** The other half set of *n* unpaired chromosomes remains behind in what is now the **mature ovum** (sometimes called the *ootid,* comparable to the spermatid, until the polar bodies disintegrate and the mature ovum alone remains). These 1*N* maternal chromosomes unite with the 1*N* paternal chromosomes of the penetrating sperm to complete fertilization. If the first polar body has not already degenerated, it too undergoes the second meiotic division at the same time the fertilized secondary oocyte is dividing its chromosomes.

Comparison of Steps in Oogenesis and Spermatogenesis
The steps involved in chromosome distribution during oogenesis parallel those of spermatogenesis, except that the cytoplasmic distribution and time span for completion sharply differ. Just as four haploid spermatids are produced by each primary spermatocyte, four haploid daughter cells

are produced by each primary oocyte (if the first polar body did not degenerate before it completes the second meiotic division). In spermatogenesis, each daughter cell develops into a highly specialized, motile spermatozoon unencumbered by unessential cytoplasm and organelles, its only destiny being to supply half of the genes for a new individual. In oogenesis, however, of the four daughter cells, only the one destined to become the ovum receives cytoplasm. This uneven distribution of cytoplasm is important because the ovum, in addition to providing half the genes, provides all the cytoplasmic components needed to support early development of the fertilized ovum. The large, relatively undifferentiated ovum contains numerous nutrients, organelles, and structural and enzymatic proteins. The three other cytoplasm-deficient daughter cells, the polar bodies, rapidly degenerate, their chromosomes being deliberately wasted.

The ovarian cycle of mammals consists of alternating follicular and luteal phases

Recall that the ovaries (and their regulators) orchestrate the events of the ovarian cycle (as well as producing hormones to regulate the uterine cycle, as we will see later). The shape, size and functionality of the ovaries vary both with the species of animal as well as with the stage of the estrous cycle. For example, in cattle and sheep it is almond shaped, in the horse the ovary is bean shaped, and in the sow and bird the ovary resembles a cluster of grapes. The ovaries of the cow and horse are easily examined by rectal palpation. In some species ovaries can be visualized by *transrectal ultrasonography.* In chickens, only the left ovary and oviduct normally mature. The right ovary remains vestigial, although it develops if the left ovary is removed or becomes diseased.

After the onset of puberty, the mammalian ovary alternates between two phases:

1. The **follicular phase,** which is dominated by the presence of maturing follicles, which produce a mature egg ready for ovulation and the steroids responsible for maturing the oocyte. The follicular hormones also govern the development of the reproductive tract and facilitate sexual receptivity.
2. The **luteal phase,** which is characterized by the presence of the *corpus luteum* (to be described shortly), the dominant ovarian structure during this phase. It secretes progesterone and, in so doing, prepares the female reproductive tract for pregnancy if the released egg is fertilized.

In estrous mammals, the follicular phase is comparatively reduced, usually encompassing no more than 20% of the estrous cycle, with the luteal phase encompassing the remaining 80%. In primates with menstrual cycles, the phases are more equal. We will now examine each of these phases in detail.

The follicular phase is characterized by the development of maturing follicles

At any given time throughout an estrous or menstrual cycle, portions of the primary follicles are starting to develop. In monovulators the follicle that will become the ovulatory follicle leaves the resting pool many cycles prior to the one in which it will become the ovulatory follicle. Estradiol and in-

hibin (as in males, p. 777) from that follicle reduce secretion of FSH by negative feedback. The others, lacking hormonal support, undergo atresia. During follicular development, as the primary oocyte is synthesizing and storing materials for future use if fertilized, important changes are taking place in the cells surrounding the reactivated oocyte in preparation for the egg's release from the ovary (Figure 16-16).

Proliferation of Granulosal Cells and Formation of the Zona Pellucida First, the single layer of *granulosal cells* in a primary follicle proliferates to form several layers that surround the oocyte termed the secondary follicle. The granulosa secretes a thick, gel-like material that covers the oocyte and separates it from the surrounding granulosal cells. This intervening membrane is known as the **zona pellucida** in mammals (most animal ova have a similar coating). Tendrils from the granulosal cells penetrate the zona pellucida and surround the oocyte's surface membrane.

As with sperm and Sertoli cells, gap junctions penetrate the zona pellucida and extend between the oocyte and surrounding granulosal cells in a developing follicle. Ions and small molecules can travel through these connecting tunnels. Recall that gap junctions between excitable cells permit the spread of action potentials from one cell to the next as charge-carrying ions pass through these connecting tunnels (see p. 66). The cells in a developing follicle are not excitable, so gap junctions here serve a role other than transfer of electrical activity. Glucose, amino acids, and other important molecules are delivered to the oocyte from the granulosa cells through these tunnels, enabling the egg to stockpile these critical nutrients. Also, signaling molecules pass through the gap junctions in both directions, helping to coordinate the changes that take place in the oocyte and surrounding cells as both mature and prepare for ovulation. The nurturing relationship between granulosa cells and a developing egg is similar in many ways to the relationship between Sertoli cells and developing sperm.

Proliferation of Thecal Cells; Estrogen Secretion At the same time the oocyte is enlarging and the granulosal cells are proliferating, specialized ovarian connective tissue cells in contact with the expanding granulosa proliferate and differentiate to form an outer layer of **thecal** cells in response to paracrines secreted by the granulosa cells. The thecal and granulosa cells, collectively known as **follicular cells,** function as a unit to secrete estrogen. Of the three physiologically important estrogens—estradiol, estrone, and estriol (the last being important only in the human)—the principal ovarian estrogen is usually estradiol (although in the sow it is estrone).

Formation of the Antrum The early stages of follicular development that occur without gonadotropin influence are not part of the follicular phase of the ovarian cycle. Only preantral follicles that have developed sufficiently to respond to FSH stimulation are "recruited" at the beginning of the follicular phase when FSH levels rise. The hormonal environment that exists during the follicular phase promotes enlargement and development of the follicular cells' secretory capacity, converting the primary follicle into a tertiary, or **antral, follicle** capable of estrogen secretion. This stage of follicular development is characterized by the formation of a fluid-filled **antrum** surrounded by the granulosal cells (Fig-

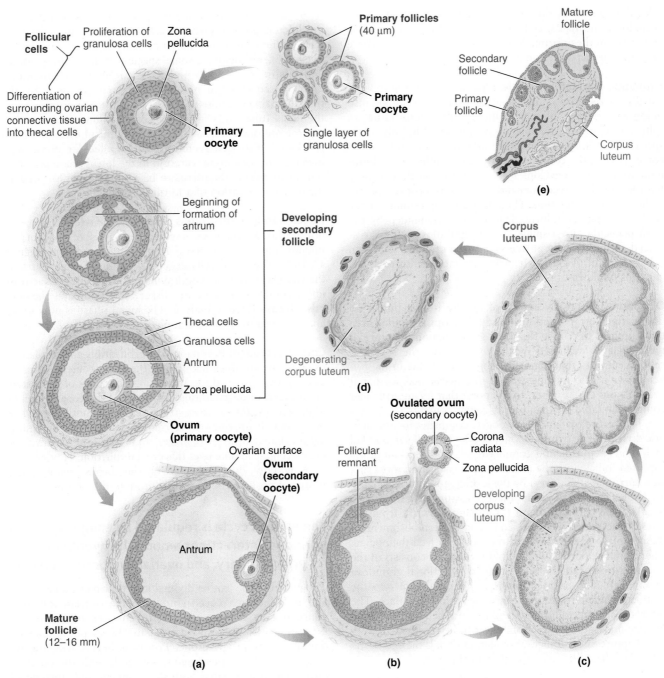

FIGURE 16-16 Development of the follicle, ovulation, and formation of the corpus luteum in a mammal.
(a) Stages in follicular development from a primary follicle through a mature follicle. (b) Rupture of a mature follicle and release of an ovum (secondary oocyte) at ovulation. (c) Formation of a corpus luteum from the old follicular cells after ovulation. (d) Degeneration of the corpus luteum if the released ovum is not fertilized. (e) Ovary (actual size in a human female), showing development of a follicle, ovulation, and formation and degeneration of a corpus luteum.

© Cengage Learning, 2013

ure 16-16). The follicular fluid originates partially from transudation (passage through capillary pores) of plasma and partially from follicular cell secretions. As the follicular cells start producing estrogen, some of this hormone is secreted into the blood for distribution throughout the body. However, a portion of the estrogen collects in the hormone-rich antral fluid.

The oocyte has nearly reached full size by the time the antrum begins to form. The shift to an antral follicle initiates a period of rapid follicular growth. During this time, the follicle increases in size from a diameter of less than 1 mm to 12 to 16 mm, depending on the species of mammal (30 to 50 mm in equids). In contrast, in the chicken there is not an antrum but an oocyte of nearly 3 to 4 cm. Part of the follicular growth is due to continued proliferation of the granulosal and thecal cells, but most is due to a dramatic expansion of the antrum. As the follicle grows, estrogens are produced in increasing quantities.

Formation of a Mature Follicle One or more of the follicles usually grows more rapidly than the others, developing into **dominant** (or **preovulatory**) follicle(s). (The term **Graafian** follicle is often used for this in humans only.) In polyovular species (that is, most mammals) more than one follicle becomes part of the cohort of preovulatory follicles. A dominant follicle that develops into a mature follicle generally has the most FSH receptors and therefore is most responsive to hormonal stimulation. The antrum occupies most of the space in a mature follicle. The oocyte, surrounded by the zona pellucida and a single layer of granulosa cells, is displaced asymmetrically at one side of the growing follicle in a little mound of cells that protrudes into the antrum, the **cumulus oophorus.**

Ovulation The greatly expanded mature follicle bulges on the ovarian surface, and ultimately develops a thinned area that ruptures to release the oocyte at **ovulation.** Rupture of the follicle is facilitated by the release from the follicular cells of enzymes that digest the connective tissue in the follicular wall. The bulging wall is thus weakened so that it balloons out even further, to the point that it can no longer contain the rapidly expanding follicular contents.

Just before ovulation, the oocyte completes its first meiotic division. The ovum (secondary oocyte), still surrounded by its zona pellucida and granulosal cells, is swept out of the ruptured follicle into the infundibulum (which, in most species, surrounds the ovary) by the leaking antral fluid (Figure 16-16b). The layer of granulosa cells surrounding the oocyte is called the **corona radiata** ("radiating crown") from the time of the secretion of the zona and the establishment of the cytoplasmic extensions of these cells to feed the oocyte. The cumulus oocyte complex that is ovulated consists of the oocyte, corona radiata, and a mass of additional granulosa (nonmural). The released ovum is quickly drawn into the oviduct, where fertilization may or may not take place. The other developing follicles that failed to reach maturity and ovulate undergo degeneration, never to be reactivated.

Rupture of the follicle at ovulation signals the end of the follicular phase and ushers in the luteal phase.

The luteal phase is characterized by the presence of a corpus luteum and secretion of progesterone

The ruptured follicle that is left behind in the ovary following release of the ovum undergoes a rapid change. The granulosal and thecal cells remaining in the remnant follicle first collapse into the emptied antral space that has been partially filled by clotted blood.

Formation of the Corpus Luteum; Estrogen and Progesterone Secretion These old follicular cells soon undergo a dramatic structural transformation to form the **corpus luteum** in a process called **luteinization** (Figure 16-16c). The follicular-turned-luteal cells hypertrophy and are converted into very active steroidogenic (steroid hormone–producing)

tissue. Beta-carotene, stored in lipid droplets within the corpus luteum, gives this tissue a yellowish appearance, particularly in the cow, hence its name (*corpus,* "body"; *luteum,* "yellow"), although in some species, such as sheep, pigs, and rats, it is a reddish pink. The corpus luteum becomes highly vascularized as blood vessels from the theca invade the luteinizing granulosa. These changes are appropriate for the luteal function, which is to secrete abundant quantities of progesterone along with lesser amounts of estrogens into the blood. Estrogen secretion in the follicular phase followed by progesterone secretion in the luteal phase is essential for preparing the uterus to be a suitable site for attachment or implantation of a fertilized ovum.

Degeneration of the Corpus Luteum If the released ovum is not fertilized and does not implant, lysis of the corpus luteum is initiated by $PGF_{2\alpha}$ from the endometrium (Figure 16-16d). The luteal cells degenerate and are phagocytized, the vascular supply is withdrawn, and connective tissue rapidly fills in to form a fibrous tissue mass known as the **corpus albicans** ("white body"). The luteal phase is now over, and one ovarian cycle is complete.

Corpus Luteum of Pregnancy If fertilization and implantation/attachment does take place, the corpus luteum continues to grow and produces increasing quantities of progesterone instead of degenerating. Now called the **corpus luteum of pregnancy,** this ovarian structure persists until the end of pregnancy. It provides the hormones essential for maintaining pregnancy in women, mares, and ewes until the developing placenta can take over this crucial function. However, in the sow, cow, doe, goat, and rat the corpus luteum is the source of progesterone for the duration of gestation.

The estrous cycle is regulated by complex hormonal interactions among the hypothalamus, anterior pituitary, and ovarian endocrine units

The estrous cycle can be divided into four phases: *proestrus* and *estrus,* which correspond with the ovarian follicular phase and are dominated by estrogens, and *metestrus* and *diestrus,* which correspond with the luteal phase and are dominated by progesterone. **Proestrus** begins with the regression of the corpus luteum and ends with the onset of estrus and is associated with development and ripening of the follicles. **Estrus** is the time of sexual receptivity and usually terminates with ovulation. **Metestrus,** the early postovulatory period, is associated with the development of the corpus luteum. **Diestrus** is associated with a functioning corpus luteum and begins after ovulation and ends with regression of the corpus luteum. If conception occurs, the female enters a period of *anestrus* during pregnancy, which ultimately ends with the positive-feedback cycle of *parturition* or the process of giving birth to the offspring.

As you have seen, the ovary has two related endocrine units: the estrogen-secreting follicle and the corpus luteum, which secretes progesterone. These units are sequentially triggered by complex cyclical hormonal relationships among the hypothalamus, anterior pituitary, and these two ovarian endocrine units. As in the male, gonadal function in the female is controlled directly by the anterior-pituitary gonadotropic hormones, follicle-stimulating hormone (FSH), and

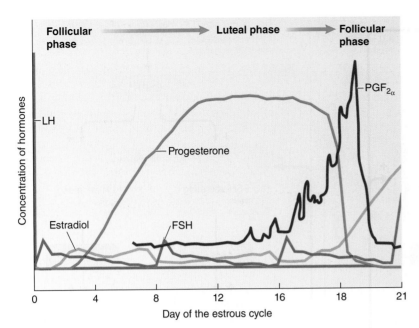

FIGURE 16-17 **Correlation between hormonal levels and the follicular and luteal phases of the bovine estrous cycle.** The cycle begins with a surge in LH (day 0) in mid-follicular phase, which initiates the transition to the luteal phase. The upper grey arrows indicate the luteal phase only.

Source: Courtesy Keith Inskeep, West Virginia University.

luteinizing hormone (LH). These hormones, in turn, are regulated by pulsatile episodes of hypothalamic gonadotropin-releasing hormone (GnRH) and feedback actions of gonadal hormones.

Differing from the male, however, control of the female gonads is complicated by the cyclical nature of ovarian function. For example, the effects of FSH and LH on the ovaries depend on the stage of the estrous cycle. Also in contrast to the male, FSH is not strictly responsible for gametogenesis, nor is LH solely responsible for gonadal hormone secretion. Using the bovine (cow) estrous cycle as an example, we consider control of follicular function, ovulation, and the corpus luteum, noting that the processes described are similar to those in other mammals (Table 16-4). We will follow Figures 16-17 and 16-18 to integrate the various concurrent and sequential activities that take place throughout the cycle. The cycle length is marked by behavioral estrus, which occurs, on the average, at intervals of 21 to 22 days, although estrous cycles of 17 to 24 days are considered normal.

Ovulation by Positive Feedback After regression of the corpus luteum in the cow, the rapidly increasing secretion of estradiol by the dominant follicle initiates a **luteinizing hormone (LH) surge** (day 0; Figures 16-17 and 16-18a). Ovulation of the dominant follicle occurs 24 to 27 hours after the LH surge. In the primate menstrual cycle, the LH surge occurs in midcycle, but in estrous mammals, the LH surge occurs at or very close to the onset of estrus at the beginning of the cycle, as shown in Figure 16-17. (Note that Figure 16-18 also applies to the menstrual cycle, which however begins with part c.) The surge brings about four major changes in the follicle:

- It halts estrogen synthesis by the follicular cell.
- It reinitiates meiosis in the oocyte of the developing follicle by blocking secretion of an *oocyte maturation–inhibiting substance* produced by the granulosal cells. This substance is believed to be responsible for arresting meiosis in the primary oocytes once they are wrapped within granulosal cells in the fetal ovary.

- It triggers production of locally acting prostaglandin E_2, which supports ovulation by promoting vascular changes that cause rapid swelling of the follicle while inducing enzymatic digestion of the follicular wall, the latter aided by a preovulatory increase in the intrafollicular secretion of progesterone. Prostaglandin $F_{2\alpha}$ ($PGF_{2\alpha}$) then stimulates contractions of myoepithelial or smooth-muscle-type cells in the theca externa at the base of the follicle. Together these actions lead to rupture of the weakened wall that covers the bulging follicle.
- It causes *differentiation of follicular cells into luteal cells.* Because the LH surge triggers both ovulation and luteinization, formation of the corpus luteum automatically follows ovulation.

Thus, the surge in LH secretion is a dramatic point in the cycle; it terminates the follicular phase and initiates the luteal phase.

Two different modes of LH secretion—tonic secretion of LH responsible for promoting follicular development and hormone secretion and the LH surge that causes ovulation—not only occur at different times and produce different effects on the ovaries, but also are controlled by different mechanisms. Tonic LH secretion is reduced to low frequency by the increasing concentrations of progesterone during the luteal phase (Figure 16-18b) and is briefly suppressed by the inhibitory action of the low concentrations of estrogen during the follicular phase (Figure 16-18c). Because tonic LH secretion stimulates both estrogens and progesterone secretion, this is a typical negative-feedback control system.

In contrast, the LH surge is triggered by a *positive-feedback* effect of estrogen. Whereas the low, rising concentrations of estrogens early in the follicular phase *inhibit* LH secretion, the high concentration of estrogens that occur during peak estrogen secretion late in the follicular phase *stimulates* LH secretion and initiates the LH surge (Figure 16-18a). Thus, LH enhances estrogen production by the follicle, and the resultant peak estrogen concentration stimulates LH secretion. The high plasma concentration of estrogens acts

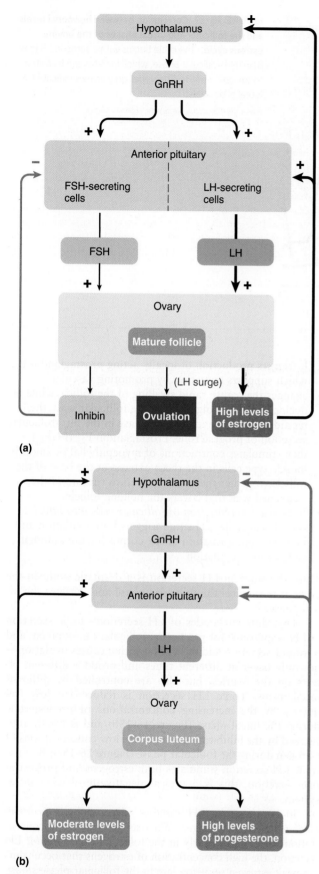

(a)

(b)

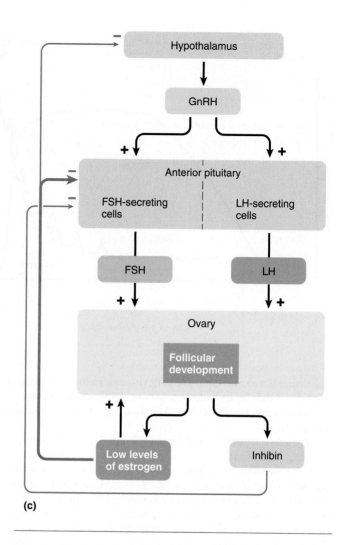

(c)

FIGURE 16-18 Reproductive hormones in a female mammal.
(a) Control of the LH surge at ovulation. (b) Feedback control during the luteal phase. (c) Feedback control of FSH and tonic LH secretion during the follicular phase.

directly on the hypothalamus to increase the frequency of GnRH pulses, thereby increasing both LH and FSH secretion. It also acts directly on the anterior pituitary to specifically increase the sensitivity of LH-secreting cells to GnRH. The latter effect accounts in large part for the much greater surge in LH secretion compared to FSH secretion at estrus. Also, continued inhibin secretion by the follicular cells preferentially inhibits the FSH-secreting cells, keeping the FSH concentrations from rising as high as the LH concentrations. There is no known requirement for the modest surge in FSH that accompanies the pronounced and pivotal LH surge; however, low doses of FSH injected directly into sheep follicles cause ovulation. Because only a mature, preovulatory follicle (not follicles in earlier stages of development) can secrete sufficient amounts of estrogens to trigger the LH surge, ovulation is not induced until a follicle has reached the proper size and degree of maturation. In a way, then, the follicle lets the hypothalamus know when it is ready to be stimulated to ovulate. The LH surge lasts for only one to two days (prior to ovulation in both the horse and pig), although the surge duration is shorter in the cow and ewe, while in the mare the surge continues after ovulation.

Luteal Phase The luteal phase (Figure 16-18b), dominated by the corpus luteum (as you have seen), is measured from the time of ovulation until regression of the corpus luteum (luteolysis). It is the longest portion of the estrous cycle. Following ovulation, the corpus luteum begins to form at the site of ovulation and can be seen from ovulation until day 3 as a **corpus hemorrhagicum**. At about day 3 to 5, the corpus luteum loses its bloody appearance and begins to increase in size and to produce sufficient progesterone such that the increase in hormone secretion is detectable in the peripheral circulation. By approximately midcycle, the corpus luteum reaches its maximum size and production of progesterone (Figure 16-17).

LH "maintains" the corpus luteum (hence its name); that is, after triggering development of the corpus luteum, LH stimulates ongoing steroid hormone secretion by this ovarian structure, with progesterone being its most abundant hormonal product (estrogen is not produced by the corpus luteum of cows and ewes). Progesterone has two major regulatory roles:

1. It inhibits the release of LH. High concentrations of progesterone reduce the frequency of secretion of pulses of gonadotropin-releasing hormone (GnRH) from the hypothalamus and thus the frequency of secretion of pulses of LH from the anterior pituitary gland, and prevent the surge of LH and subsequent ovulation (Figure 16-18b).

2. It directs the endometrial secretion of prostaglandin $F_{2\alpha}$ and controls the timing of secretion of $PGF_{2\alpha}$ for luteolysis by regulating the timing of the initial increase in secretion of $PGF_{2\alpha}$, which usually occurs around day 12 of the estrous cycle in the pig and sheep and day 14 in the cow (Figure 16-17). In ruminants, progesterone modulates the secretion of $PGF_{2\alpha}$, keeping it at midrange episodic values until luteal regression has begun. As concentrations of progesterone decline, greater peaks of secretion of $PGF_{2\alpha}$, produced by the uterus and the corpus luteum, complete luteal regression and end the luteal phase. Notice in Figure 16-17 that the maximal secretion of $PGF_{2\alpha}$ occurs after luteal function (secretion of progesterone) has already ceased.

Follicular Phase The follicular phase is initiated only after luteolysis is under way. As luteolysis proceeds, the decreasing concentrations of progesterone reduce the negative feedback on pulsatile secretion of LH by the anterior pituitary. The increased frequency of pulses of LH promotes the final maturation of the large antral follicle. During the follicular phase, LH reaches higher concentrations than in the luteal phase, because negative feedback from progesterone on the hypothalamus is removed (Figure 16-18c).

Dominant follicles, as they continue to grow, secrete increasing concentrations of estradiol and inhibin so that FSH declines during the follicular phase but estradiol does not limit pulse frequency of LH. When the dominant follicle reaches adequate size (10 mm in diameter is approximately the minimum required for ovulation in the cow, as compared to 7 mm in a pig) and secretion of estradiol from the follicles reaches threshold, a preovulatory surge of LH and FSH occurs (Figure 16-17). Immediately after the preovulatory surge of LH, concentrations of estradiol abruptly decline and a secondary smaller pulse of FSH occurs on the day of ovulation.

While early stages of ovarian follicular development are not dependent on gonadotropins, pulses of FSH recruit a cohort of secondary follicles and is required for their growth from approximately 4 to 9 mm in diameter. Thus, the process of follicular growth and degeneration occurs continuously throughout the entire estrous cycle. However, more frequent pulses of LH are necessary for development of the follicle beyond about 9 mm. FSH induces antrum formation, further follicular development, and estrogen secretion. As the follicles grow, they secrete increasing amounts of estradiol and inhibin. Each of these hormones exerts negative-feedback effects on the secretion of FSH (estradiol limits secretion of GnRH by the hypothalamus, and inhibin reduces responsiveness of the anterior pituitary to GnRH; Figure 16-18c). In the pig, in which concentrations of progesterone are an order of magnitude greater than in the cow and ewe, no large antral follicles develop during the luteal phase.

FSH and estrogens stimulate proliferation of the granulosal cells, whereas LH and FSH are required for synthesis and secretion of estrogens by the follicle, although these hormones act on different cells and at different steps in the estrogen production pathway. Both granulosa and thecal cells participate in estrogen production. The conversion of cholesterol into an estrogen requires a number of sequential steps, the last of which is conversion of androgens into estrogens (see Figure 7-3). Thecal cells readily produce androgens but have limited capacity to convert them into estrogens except in the pig, whose thecal cells have copious aromatase activity. Granulosal cells, in contrast, can readily convert androgens into estrogens but cannot produce androgens in the first place. Luteinizing hormone acts on the thecal cells to stimulate androgen production, whereas FSH acts on the granulosa cells to promote the conversion of thecal androgens (which diffuse into the granulosal cells from the thecal cells) into estrogens. Because low basal concentrations of FSH are sufficient to promote this final conversion to estrogen, the rate of estrogen secretion by the follicle primarily depends on the circulating concentration of LH, which continues to rise. Furthermore, as the follicle continues to grow, more estrogen is produced simply because more estrogen-producing follicular cells are present.

This differential sensitivity of FSH- and LH-producing cells to negative feedback by estrogens is at least in part responsible for the fact that the serum FSH concentration, unlike the serum LH concentration, declines as the estrogen concentrations rise. Another contributing factor to the fall in FSH is the secretion of inhibin by the follicular cells, which preferentially inhibits FSH secretion at the anterior pituitary, just as it does in the male. The decline in FSH secretion brings about atresia of all but the single most mature of the developing follicles. As FSH concentrations decline, the dominant follicle becomes more dependent on LH. Continued pulses of LH maintain the growth of the dominant follicle until it is either ovulated or becomes **atresic** (degenerated and reabsorbed).

In contrast to FSH, LH secretion continues to rise slowly despite inhibition of GnRH (and thus indirectly, LH) secretion. This seeming paradox is due to the fact that estrogens alone cannot completely suppress tonic (low-level, ongoing) LH secretion; progesterone is required to more fully curtail the frequency of tonic LH secretion. Because progesterone does not appear until the luteal phase of the cycle, the basal level of circulating LH slowly increases during the follicular phase under incomplete inhibition by estrogens alone.

Subnormal Luteal Phase A **subnormal luteal phase** is defined by low secretion of progesterone or can occur during treatment with other *progestogens* (progesterone-related steroids) for synchronization of estrus. Lower circulating concentrations of progestogen lead to an *increased* frequency of release of pulses of LH from the anterior pituitary gland. Increased LH stimulates continued growth of the largest ovarian follicle, which is followed by increased secretion of estradiol by the **persistent largest follicle;** that is, the largest dominant follicle continues to increase in size and becomes *persistent.* Low dosages of administered progestogens, native or synthetic, result in reduced fertility in association with follicular persistence. In contrast, a low frequency of pulses of LH fails to support continued follicular growth and allows earlier degeneration or more frequent replacement of the largest follicle. Treatment with higher dosages of progesterone or a progestogen can be used to cause atresia of the largest follicle and initiation of growth of a new cohort of follicles.

The uterine changes that occur during the estrous cycle reflect hormonal changes during the ovarian cycle

The fluctuations in circulating levels of estrogens and progesterone that occur during the ovarian cycle induce profound cyclic changes in the uterus in the *uterine cycle.* In estrous mammals, the complete removal of the uterus, **hysterectomy,** after ovulation results in the persistence of the corpus luteum as if the female were pregnant. Because it reflects hormonal changes that occur during the ovarian cycle, the uterine cycle averages 17 to 24 days (in cattle), as does the ovarian cycle, although there is considerable variation from this mean even in normal animals. This variability is primarily a reflection of differing lengths of the follicular phase; the duration of the luteal phase is fairly constant within a species. However, across species the duration of the luteal phase varies considerably. Consistent changes that take place throughout the cycle include the preparation of the uterus for implantation should a released ovum be fertilized, which involves a thickening of the uterine lining (*endometrium*) that is reduced at the end of the luteal phase in the absence of conception (recall that it is reabsorbed in estrous species and sloughed off in menstrual species).

We briefly examine the influences of estrogens and progesterone on the uterus and then consider the effects of cyclic fluctuations of these hormones on uterine structure and function.

Influence of Estrogen and Progesterone on the Uterus The uterus consists of two main layers: the **myometrium,** the outer smooth-muscle layer; and the **endometrium,** the inner lining that contains numerous blood vessels and glands. Estrogens stimulate growth of both the myometrium and the endometrium. It also induces the synthesis of progesterone receptors in the endometrium. Thus, progesterone can exert an effect on the endometrium only after it has been "primed" by estrogen. Progesterone acts on the estrogen-primed endometrium to convert it into a hospitable and nutritious lining

suitable for implantation of a fertilized ovum. Under the influence of progesterone, the endometrial connective tissue becomes loose and edematous as a result of an accumulation of electrolytes and water, which facilitates implantation of the fertilized ovum. Progesterone further prepares the endometrium to sustain an early-developing embryo by inducing the endometrial glands to secrete and store large quantities of glycogen and by causing tremendous growth of the endometrial blood vessels. Progesterone also reduces the contractility of the uterus to provide a "quiet" environment for implantation and embryonic growth in the cow, pig, and ewe but not in the mare. In this species, contractions of the myometrium serve to transport the embryo throughout the uterine lumen so as to reduce secretion of $PGF_{2\alpha}$ and prevent luteolysis.

Fluctuating concentrations of estrogens and progesterone produce cyclic changes in cervical mucus

Hormonally induced changes in the cervix occur during the ovarian cycle as well. Under the influence of estrogens during the follicular phase, the mucus secreted by the cervix becomes abundant, clear, and thin. This change, which is most pronounced when estrogens are at their peak and ovulation is approaching, facilitates passage of sperm through the cervical canal. After ovulation, under the influence of progesterone from the corpus luteum, the mucus becomes thick and sticky, essentially forming a plug across the cervical opening. This plug constitutes an important defense mechanism by preventing bacteria that might threaten a pregnancy (should conception have occurred) from entering the uterus from the vagina. Sperm cannot penetrate this thick mucus barrier.

The reproductive cycles of nonviviparous vertebrates are fundamentally similar but have some unique differences

The details of female reproductive features we have been discussing are primarily those of mammals. How does the reproductive cycle of a viviparous cow compare to the ovulatory cycle of an ovoviviparous bird or oviparous fish? Each reproductive strategy represents separate evolutionary trends, with the result that there are as many similarities between these strategies as there are differences. For example, as in mammalian species, the hormonal pathway of reproduction in birds and fishes revolves around the hypothalamic–pituitary–gonadal axis. The hypothalamus is activated by specific environmental cues: day length in some species of birds and pheromones in some species of fish. Following this activation, kisspeptin stimulates GnRH synthesis and release into the hypothalamic–hypophyseal portal vasculature. However, different bird and fish species have two to three forms of GnRH, the number depending on the species, although it is believed that only one form of GnRH governs the release of LH. We initially focus on the ovulatory cycle of the domestic chicken.

Ovarian Follicular Growth and the Ovulatory Cycle in the Domestic Hen At the time of hatching, the chick embryo contains approximately a million oocytes, many of which

A Closer Look at Adaptation
A Rut for One Season

As day length declines in the fall, thick-necked bucks start shredding trees, making "scrapes" (areas scraped by the bucks' front hooves), focusing on does, and fighting other bucks. Basically the culmination of the rut, which translates to successfully breeding as many does as possible, involves all the activity that leads to this event. During this period, deer release various pheromones as they primarily use scent for communication at this time. Studies suggest that each reproductively active deer exudes a unique odor that communicates its identity as well as its sexual status.

Glands that are used for communication purposes include:

- *Tarsal gland:* Early in the rut, this gland on the inside of the back knee joint of dominant bucks turns dark brown in color. Urine deposited on the gland reacts with bacteria that live on the surface, which then steadily darkens with time. As the time nears to when the fe-

males ovulate, these glands are almost black in color. Scent is subsequently released onto the surrounding vegetation as the buck moves through his territory. Females also urinate over the tarsal gland (rub urination), and when it eventually turns black, they are in estrus, meaning that they are about to ovulate.

- *Metatarsal gland:* Approximately two-thirds of the way down the leg between the knee joint and the hoof is the metatarsal gland, one whose function has not been elucidated, although it has been suggested to produce an "alarm scent."
- *Forehead gland:* Located between the base of the antlers and the eyes, this gland is easily identified in adult bucks because the hair in this region is particularly dark and brown. Prior to the rut, the bucks rub trees and overhanging branches with the base of their antlers, thus exposing the moist undersurface of the trees to the secretions from the fore-

head gland. In this fashion, bucks are advertising their territory and relative dominance, for the most dominant bucks have the most active forehead glands.

- *Preorbital gland:* Located on the inside corner of the eye, this gland advertises the presence of the buck and warns off potential intruders. Scents from the preorbital gland generally are positioned on branches that overhang the scrape.
- *Interdigital gland:* Located between the hooves, this gland enables the deer to follow other deer by smell.
- *Vomeronasal gland:* This gland is located in the nasal cavity (p. 238). When a doe is close to estrus, bucks will approach her when she urinates, occasionally licking the urine and curling their lip in the Flehman response. The VNO detects when the does are beginning to go into estrus. Secretions from the uterus in association with behavioral changes in the doe signal she is ready to mate.

---

become atresic in the first weeks. However, the remaining follicles enter a stage of slow growth to 1 to 2 mm in size that continues for months or even years. During this time, neutral lipids are deposited as yolk in the oocyte. In the weeks before the onset of lay (egg production), which in the domestic hen begins at approximately 18 weeks of age, a small proportion of these oocytes begins to enlarge. This stage of follicular development lasts for approximately eight to nine weeks, with recruitment of oocytes continuing throughout the female's reproductive life. As a consequence of the deposition of a white, proteinaceous, primordial yolk, these oocytes increase in size to about 2 to 6 mm in diameter. When the follicles reach 5 to 8 mm in diameter, they enter the final rapid-growth phase, which is characterized by the rapid deposition of yellow yolk. During this stage, the follicles grow to more than 35 mm in diameter in as few as 7 to 11 days.

One of the most obvious differences in follicular development between birds and mammals now becomes evident, because in the bird's ovary the follicles do not mature synchronously. The yellow yolky follicles form an orderly hierarchy of different sizes so that at the peak of egg production, the largest (**F1 follicle**) of about 40 to 45 mm in diameter is the one destined to be ovulated next, the second largest follicle (**F2**) the following day, the third largest (**F3**) two days later, and so on. Normally, the ovary of the hen contains up to 10 yellow yolky follicles, together with a greater number of small yellow follicles and numerous small white follicles awaiting recruitment or atresia (Figure 16-19). The actual number of rapidly growing follicles generally remains stable in an individual hen, with just one

follicle from the intermediate phase recruited into the final growth phase once the largest follicle is ovulated. However, it is not known how new follicles are recruited into the intermediate phase.

In response to the increase in estrogen concentrations prior to the onset of lay, very-low-density lipoprotein (VLDL, p. 389) and vitellogenin (similar forms of which are found in most yolk-producing animals; see pp. 765 and 798) are synthesized in the liver. These are the major precursors of yellow yolk and are transported to the ovary, where they are deposited in the yolk. Researchers have studied the growth of yellow yolky follicles by feeding the hen differently colored lipid-soluble dyes (Figure 16-19a). The dye is laid down in the yolk as concentric rings marking the day of feeding. Using this technique, researchers showed that follicles require from 11 to 13 days to mature from an intermediate follicle to one of ovulable size.

Because yolk is an energy-expensive material to produce, the number of follicles in the hierarchy, their size relative to body weight, and the interval between successive ovulations differ greatly among different species of birds. Most birds lay an egg each day until the **clutch** is complete. A longer interval between egg lay has energy advantages but creates difficulties in synchronizing the timing of hatch and increases the likelihood of predation. Normally clutch size is smaller, and the interval between eggs is longer, in species producing larger eggs. Many species of birds are **determinate** layers; that is, they lay a fixed number of eggs in a clutch, whereas others are **indeterminate** layers and can continue to lay eggs for extended periods of time, particularly if the eggs are removed from the nest as they are laid. With the onset

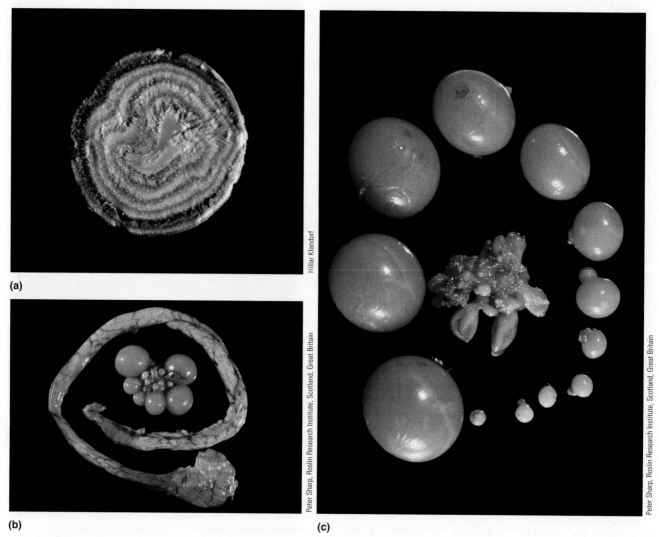

(a)

(b)

(c)

Hillar Klandorf

Peter Sharp, Roslin Research Institute, Scotland, Great Britain

Peter Sharp, Roslin Research Institute, Scotland, Great Britain

FIGURE 16-19 Ovary of a laying hen. (a) Midsagittal section through the follicle of an egg. Xanthophylls (yellow to orange carotenoid pigments) from the diet are transported rapidly from the bloodstream to the developing yolk; consequently, more pigment is deposited when the hen is eating. This process gives rise to 7 to 11 dark (xanthophyll-rich) and light concentric yolk layers. (b) Ovary, follicles, and reproductive tract of a laying hen. (c) Follicular hierarchy. Follicles are arranged in order from the largest (F1) to the smallest. F1 also denotes the follicle next in line to be ovulated. The F1 follicle is the principal source of progesterone, the steroid hormone that induces the preovulatory surge of LH from the anterior pituitary.

of agriculture and domestication of species over 4,000 years ago, humans have exploited this property of indeterminate layers and developed strains of chickens, ducks, and quail that can lay over 300 eggs each year.

Eggs laid on successive days are known as a **sequence,** whereas days on which no eggs are laid are termed **pause** days. The first egg of a long sequence may be laid earlier in the day than the first egg of a hen producing a short sequence. The general rule is that ovulation occurs 6 to 8 hours after the ovulatory surge of LH, with the egg spending about 24 hours in the oviduct before it is laid (**oviposited**). There is a close relationship between ovulation and oviposition, with ovulation occurring 15 to 75 minutes after oviposition, except if it is the last egg of a sequence. Because of this relationship, the time of lay becomes a practical guide to the time of ovulation in a bird. To achieve sequences of 50 to 100 eggs, ovulation occurs every day at approximately

the same time, and the egg from that ovulation is subsequently laid 24 hours later. Between sequences, there are one or more pause days. Chickens held under normal light–dark cycles of 14L:10D lay their eggs in the first half of the light period, whereas under the same lighting conditions, quail lay their eggs late in the day and early into the dark period. In both situations, an ovulatory surge of LH occurs six hours before oviposition, which means that the ovulatory surge in chickens occurred during the night, whereas in quail, it occurred during the morning. Egg laying is thus *entrained* by the light–dark cycle and can readily be affected by any changes in the photoperiod.

Two additional features pertinent to the ovulatory cycle need to be pointed out. The first is that each ovulation in a hen occurs slightly later on subsequent days. The second is that the preovulatory release of LH is restricted to a 10- to 11-hour period each day, the so-called **open period.** The

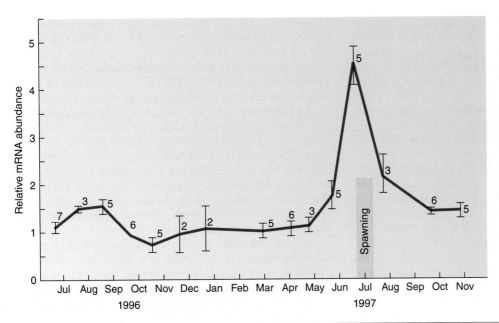

FIGURE 16-20 Expression of the LH receptor (LHR) gene in the channel catfish ovary. Cyclic changes in the LH receptor gene through a complete spawning cycle reveal an increase in expression associated with ovulation. The LHR binds to the LH-like gonadotropin (GTH-2) of fish.

Source: R. S. Kumara, S. Ijiri, & J. M. Trant, (2001), Molecular biology of channel catfish gonadotropic receptors: I. Cloning of a functional, luteinizing hormone receptor and preovulatory induction of gene expression. 'Biology of Reproduction', 64:1016. Reproduced with permission via Copyright Clearance Center.

preovulatory release of LH can be initiated only from the onset of darkness to about one hour after the onset of the light period.

With age there is a decline in the rate of recruitment of follicles into the hierarchy and in the maturation of these follicles, and an increase in the rate of atresia of small follicles. If we induce a cessation in lay (**forced molt**), older hens lay eggs more frequently and of a greater quality (thicker shells) when production is resumed.

Unlike the situation in female mammals, ovulation of the F1 (largest) follicle in birds is preceded by an increase in the concentrations of progesterone and LH. The preovulatory surge of LH is caused by a positive-feedback interaction, comparable to the effect of estrogens at the hypothalamus in mammals, but it is progesterone that stimulates the release of GnRH and hence LH. LH, in turn, stimulates the secretion of more progesterone from the granulosa of the mature preovulatory follicle. Thus, in birds, it is the granulosa that produce progesterone, most of which diffuses to the thecal cells, where it is converted to estrogen. As the follicles mature, the enzyme activity diminishes within the theca but increases in the granulosa, resulting in the enhanced production of progesterone at the time of ovulation. There are two possible reasons for this switch in the pattern of steroid synthesis by the maturing follicle. First, birds require estrogens for synthesis of Ca^{2+}-ATPase in the shell gland as well as for yolk synthesis by the liver, and estrogens must be available in sufficient concentrations well before ovulation. Secondly, progesterone is the hormone responsible for the positive-feedback cycle, generating both the surge in LH and ovulation of the F1 follicle, and so the synthesis of this steroid needs to be restricted to the few mature follicles. The preovulatory releases of progesterone and LH are initiated 10 to 12 hours before ovulation, with LH peaking 6 to 8 hours before ovulation and progesterone peaking 2 to 4 hours before ovulation.

Ultimately, the expulsion of the egg from the uterus (shell gland) into the vagina is mediated by an increase in the release of arginine vasotocin (AVT), a neurohypophyseal hormone (p. 285). Once the egg enters the vagina, rhythmic contractions of abdominal and vaginal muscles force expulsion of the egg. This event, termed **bearing down**, results from a nervous reflex generated by the egg stretching the vaginal lumen. Vaginal muscles contract in response to the increased concentrations of AVT.

Spawning Cycles of Fish Most information on the spawning cycles of viviparous species of fish has been obtained from studies in teleosts. Three modes of oocyte development are recognized, based on the recruitment of new oocytes: **synchronous**, **group synchronous**, and **asynchronous**. In those species showing synchronous development, all the oocytes mature and ovulate at the same time. This strategy is used by some species of salmon and eels that spawn once and die. In species employing group synchronous development, oocytes are partitioned into groups that mature at different rates and are then ovulated periodically throughout the breeding season. This group is categorized as being multiple spawners with comparatively short spawning seasons. The third strategy is the most complex, because oocytes are found at all stages of development throughout the spawning cycle. Oocytes are recruited continuously into the pool of maturing oocytes. As with group synchronous fish, asynchronous fish are multiple spawners but they have a prolonged spawning season.

In the hypothalamic–pituitary–gonadal axis in fishes, the pituitary gonadotropins *GTH-1* has FSH-like activity and *GTH-2* has LH-like activity, and each functions comparably to their mammalian counterparts. However, neuroendocrine regulation of GTH-2 in teleost fish is under dual hypothalamic control. GTH-2 release is inhibited by dopamine, which functions as a gonadotropin release–inhibiting hormone. Dopamine acts directly at the pituitary to modulate the actions of GnRH as well as the release of GTH-2.

GTH-1 is important for vitellogenesis and early gonadal development, and is secreted for an extended period of time. For example, concentrations of LH receptor gene and GTH-1 in coho salmon and catfish steadily increase from early in May until they reach peak concentrations in September and then decline through December (Figure 16-20).

In response to GTH-1, the ovarian granulosal cells synthesize estradiol, which stimulates the production of *vitellogenin* in the liver. GTH-1 is responsible for generating GTH-2 in the pituitary gland. Once vitellogenesis is completed, a preovulatory surge of GTH-2 is generated, which initiates final oocyte maturation and ovulation of the mature oocytes. The surge in GTH-2 is comparatively rapid (several days), although the actual profile is species dependent.

Normally the surge in GTH-2 is initiated by external environmental cues such as photoperiod, water temperature, and adequate water velocities along with conditions suitable for the survival of the larvae. Other factors that can elicit the GTH-2 surge include pheromones, sound production by male fishes, and the courtship behavior of males, factors that reliably ensure that the eggs are fertilized once ovulated. A dopamine antagonist can artificially induce ovulation; in cobia (*Rachycentron canadum*), a migratory pelagic species of fish, human chorionic gonadotropin (p. 806) can induce ovulation.

In addition to GTH-2, final oocyte maturation and ovulation in fish require an additional factor produced by the follicular cells. *Maturation-inducing hormone (MIH)* is a progestogen synthesized via a two-cell mechanism comparable to that by which estradiol is synthesized. In this model, there are receptors for GTH-2 on both the thecal and granulosal cells, unlike the mechanism for estradiol synthesis.

Among the various fish groups the path that sperm and eggs take to get to the outside differs considerably. Oocytes can be ovulated either into the coelom of the fish and moved into the oviduct and the cloaca by ciliated epithelium lining the peritoneal cavity or into an ovarian cavity and moved down the oviduct to the cloaca without entering the peritoneal cavity itself.

Sexual maturity events in mammalian females are similar to those in males

As with males, sexual maturation (puberty) in mammalian females is not a binary switch but rather a long progression of greater and greater function of the hypothalamic–hypophyseal–gonadal axis until the time when the hypothalamus has become sufficiently mature that it allows the next follicle to produce enough estrogens to elicit the first LH surge. Unlike the fetal testes, the fetal ovaries do not need to be functional, because feminization of the female reproductive system automatically takes place in the absence of fetal testosterone secretion without the presence of female sex hormones. As the prepubertal female matures, an increase in the frequency of GnRH pulses is associated with a decrease in the sensitivity of the hypothalamus to estrogen. Puberty occurs at about three to four months of age in rabbits. However, it can vary from 6 to 16 months in sheep, and goats depending on the season of birth (spring or fall). Puberty is from 6 to 7 months in pigs, 10 to 15 months in cattle, and 15 to 18 months in horses. Domestication of the wolf, ancestor of dogs, has reduced the age of puberty from 2 years to between 6 and 12 months in modern dog breeds.

Puberty can be best considered as the age at which a female can support pregnancy. In most female mammals, a requisite body size is required before the onset of puberty can occur, and there is evidence that metabolic signals can affect the development of the hypothalamic neurons and the rate of GnRH release. Reproductive function is influenced by body weight and nutritional status, for gonadotropin secretion is reduced in animals and humans that are nutritionally restricted. GnRH begins stimulating the release of the anterior-pituitary gonadotropic hormones, which in turn stimulate ovarian activity. The resultant secretion of estrogens by the activated ovaries induces growth and maturation of the female reproductive tract as well as development of the female secondary sexual characteristics. Estrogen's prominent action in the latter regard is to promote fat deposition. Enlargement of the mammary glands at puberty is due primarily to fat deposition in the mammary gland tissue and not to functional development of the mammary glands. The pubertal rise in estrogens does close the epiphyseal plates, however, halting further growth in height, similar to the effect of testosterone in males.

Leptin, a peptide produced by adipocytes and a potent satiety factor (see p. 725), may be a signal that informs the hypothalamus that metabolic stores are sufficient for initiating reproductive function. In this model, leptin-responsive neurons in the arcuate nucleus synthesize kisspeptin, whose release is influenced by nutritional status. Food deprivation in prepubertal rats decreased the expression of kisspeptin mRNA, while administration of kisspeptin to chronically undernourished prepubertal rats increased the suppressed levels of LH and estrogen. In some species leptin may play only a permissive function, functioning as a "barometer" of body function, and not necessarily an initiator of puberty, because leptin concentrations in monkeys do not change around the onset of puberty. The feeding of high-protein feedstuffs (flushing) to animals causes an acute response in terms of gonadotropin release to improve ovulation rates; this method has long been practiced in agriculture. Whether leptin is involved in these effects has not been determined. Other factors can modulate the onset of puberty, including the photoperiod during the peripubertal period (sheep), season of birth (sheep), as well as the presence or absence of siblings during the peripubertal period (cattle and swine).

Seasonal Regression and Recrudescence of Ovarian Function After puberty in most seasonally breeding animals, other factors, in turn, influence the timing of GnRH release and the recrudescence (reawakening) of ovarian function (see Figure 7-8). As we have discussed previously, photoperiod at temperate latitudes is the most important factor determining the onset and termination of the breeding cycle. Light is initially perceived in the retina (although in birds the eyes are not required for the stimulation of reproduction) and transferred by the optic nerve to a specific area of the hypothalamus known as the *suprachiasmatic (SCN) nucleus* (*mediobasal hypothalamus* in birds; see p. 282). From the SCN, a nerve tract travels to the *superior cervical ganglion (SCG)*. From here, a synapse is made with cells in the pineal gland. **Pinealocytes** in turn cease the secretion of the hormone *melatonin*, which is secreted during the period of dark (**scotophase**). During the daylight hours (**photophase**), the light detected by the retinal cells of the eye, through the SCG, activates neurons that limit the release of melatonin from the pineal gland (again, in birds, a definitive role for the pineal has not been conclusively established). However, during the dark period, this inhibitory pathway is shut down by a reduction in the firing rate of nerves in the light-sensitive areas of the retina. The amount of melatonin released from the pinealocytes is a function of the duration of the period of darkness, although data in experimental animals indicate that receiving only brief flashes of light, separated by a given number of hours of dark (skeleton photoperiod), results in similar quantities of melatonin released. Under normal conditions in short-day breeders, sufficient quantities of melatonin are released only when a critical length of the dark period is reached, which then triggers an increase in GnRH release from the hypothalamus. In seasonal long-day breeders, termination of the breeding season is associated with the development of **photorefractoriness,** the inability of the long day lengths to sustain the breeding cycle. In birds this is believed to be the result of an increase in GnIH associated with a decline in the release of GnRH. GnIH is also elevated in birds incubating an egg and is believed to be responsible for suppressing gonadotropic function at this time.

check your understanding 16.5

Compare oogenesis with spermatogenesis.

Describe the events of the follicular and luteal phases of the ovarian cycle.

Compare an estrous cycle of a mammal with an ovulatory cycle of a bird. What key similarities and differences are most evident?

16.6 Fertilization and Implantation

You have now learned about the events that take place if fertilization does not occur in a mammal. Because the primary function of the reproductive system is, of course, reproduction, we next turn our attention to the sequence of events that ensue when this function is accomplished.

The oviduct is the site of fertilization

Fertilization, the union of male and female gametes, in most mammals occurs in the **ampullary–isthmic junction,** the midpoint of the oviduct. Thus, both the ovum and the sperm must be transported from their gonadal sites of production to the ampullary–isthmic junction in the female.

Ovum Transport to the Oviduct At ovulation the ovum is released into the abdominal cavity, but the oviduct picks it up quickly. The fimbriae (Figure 16-4c) contract in a sweeping motion to guide the released ovum into the oviduct. Furthermore, the fimbriae are lined by cilia that beat in waves toward the ostium (interior) of the oviduct, further assuring the ovum's passage into the oviduct. Within the oviduct, the ovum is propelled rapidly by peristaltic contractions and ciliary action to the ampullary–isthmic junction.

Conception can take place during a limited time span in each cycle, the **fertile period.** If not fertilized, the ovum begins to disintegrate within 4 hours in the rabbit and 24 hours in the guinea pig and human. In most mammalian species studied, the survival of the egg is so short as to preclude fertilization in the uterus. Cells that line the reproductive tract are responsible for phagocytizing the ovum.

Fertilization must therefore occur in the brief time when the ovum is still viable. In contrast, the ability of sperm to survive in the female reproductive tract is highly variable among species. For example, in some species of bats sperm survive from the time of mating in the fall until the time of ovulation in the spring because of the presence of specific anatomic features involved in sperm storage. In farm and laboratory animals the survivability of sperm is considerably shorter and ranges from 30 to 35 days in the hen (which also has a sperm storage organ), 5 days in the mare and human, and 26 to 30 hours in the rabbit. (Contrast these values with those in a queen bee, who, after a single insemination, can continue to lay fertilized eggs for as long as four years.)

Sperm Transport to the Oviduct In some mammals sperm are deposited in the vagina at ejaculation (primates, cows, sheep, rabbits), whereas in others the sperm are delivered to the cervix or are deposited in the uterus (horses and pigs). These sperm must travel through the uterus and then up to the egg into the upper third of the oviduct. The first sperm arrive in the oviduct within a half hour after ejaculation, but these may not be capacitated. To accomplish this formidable journey, sperm need the help of the female reproductive tract. The first hurdle is passage through the cervical canal. Throughout most of the cycle, because of high progesterone or low estrogen, the cervical mucus is too thick to permit sperm penetration. The cervical mucus become thin and watery enough to permit sperm to penetrate only when estrogen is high, as occurs in the presence of a mature follicle about to ovulate. Sperm migrate up the cervical canal under their own power. The canal remains penetrable for only two or three days during each cycle, around the time of ovulation. In species in which the sperm are deposited in the vagina, the process of capacitation (p. 778) is initiated as they pass through the cervix. However, in those species where the sperm are deposited directly in the uterus, capacitation is initiated here and is completed in the oviduct.

Once the sperm have entered the uterus, contractions of the myometrium churn them around in "washing-machine" fashion. This action quickly disperses the sperm throughout the uterine cavity. When sperm reach the utero-tubal junction, they are moved up the oviduct toward the site of fertilization with the help of smooth muscle contractions and cilia that move in the direction of the ampullary–isthmic junction (region of the oviduct where the isthmus makes an anatomic transition into the ampulla) under the influence of high concentrations of estrogens present near ovulation. Furthermore, new research indicates that ova are not passive partners in conception. Mature eggs have been shown to release **allurin,** a recently identified chemical that attracts sperm and causes them to propel themselves toward the waiting female gamete, although this likely only influences cells that are very close (<1 cm) to the oocyte. Interestingly, the allurin receptor, called *hOR17-4,* is an olfactory receptor (OR) similar to those found in the nose for smell perception (see p. 237). Therefore, sperm are believed to "smell" the egg. According to current thinking, activation of the hOR17-4 receptor on binding with a chemoattractant from the egg triggers a second-messenger pathway in sperm that brings about intracellular Ca^{2+} release. This Ca^{2+} turns on the microtubule sliding that brings about tail movement and sperm swimming in the direction of the chemical signal.

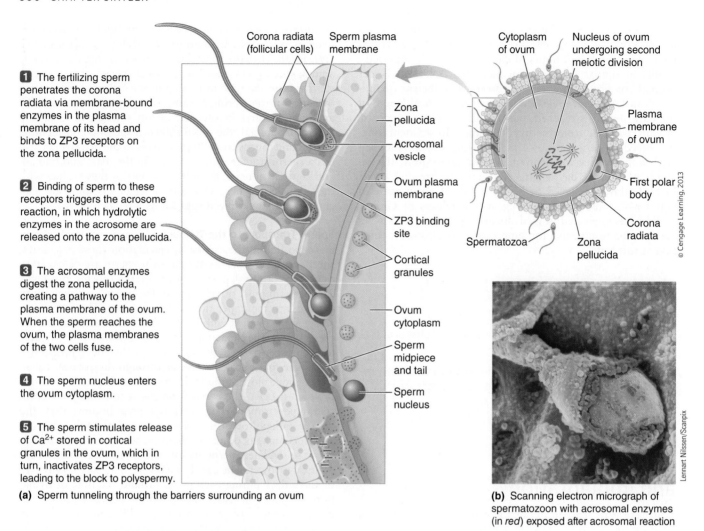

1 The fertilizing sperm penetrates the corona radiata via membrane-bound enzymes in the plasma membrane of its head and binds to ZP3 receptors on the zona pellucida.

2 Binding of sperm to these receptors triggers the acrosome reaction, in which hydrolytic enzymes in the acrosome are released onto the zona pellucida.

3 The acrosomal enzymes digest the zona pellucida, creating a pathway to the plasma membrane of the ovum. When the sperm reaches the ovum, the plasma membranes of the two cells fuse.

4 The sperm nucleus enters the ovum cytoplasm.

5 The sperm stimulates release of Ca²⁺ stored in cortical granules in the ovum, which in turn, inactivates ZP3 receptors, leading to the block to polyspermy.

Corona radiata (follicular cells)
Sperm plasma membrane
Zona pellucida
Acrosomal vesicle
Ovum plasma membrane
ZP3 binding site
Cortical granules
Ovum cytoplasm
Sperm midpiece and tail
Sperm nucleus

Cytoplasm of ovum
Nucleus of ovum undergoing second meiotic division
Plasma membrane of ovum
First polar body
Corona radiata
Zona pellucida
Spermatozoa

© Cengage Learning, 2013
Lennart Nilssen/Scanpix

(a) Sperm tunneling through the barriers surrounding an ovum

(b) Scanning electron micrograph of spermatozoon with acrosomal enzymes (in *red*) exposed after acrosomal reaction

FIGURE 16-21 Process of fertilization (mammal).

Even around ovulation time, when sperm can penetrate the cervical canal, of the several hundred million sperm deposited in a single ejaculate only a few thousand make it to the oviduct. The fact that only a very small percentage of the deposited sperm ever reach their destination is one reason why sperm concentration must be so high. Sperm, like ova, have limited life spans. Another reason so many sperm are needed is that the acrosomal enzymes assist in digesting the zona pellucida, allowing sperm to penetrate the zona and arrive in the perivitelline space (Figure 16-21). In addition, in some species a high percentage of sperm have nonfertilizing roles, as you will see shortly.

Fertilization Although the tail of the sperm is used to maneuver for final penetration of the ovum, motility ceases once contact with the ovum is made. To fertilize an ovum, a sperm must pass through the corona radiata and zona pellucida surrounding it (or equivalent jelly coating in nonmammalian animals). Contact of the sperm with the cumulus-oocyte complex results in hyperactivation of the sperm cell, which is responsible in part for denuding the oocyte of the cumulus cells. Sperm can penetrate the zona pellucida only after binding with specific receptor sites on the surface of this layer. The binding partners between the sperm and ovum have recently been identified. **Fertilin,** a protein found on the plasma

membrane of the sperm, binds with glycoproteins known as **ZP3** in the outer layer of the zona pellucida. Only sperm of the same species can bind to these zona pellucida sites and pass through. Binding of sperm triggers the **acrosome reaction,** in which the acrosomal membrane disrupts and the acrosomal enzymes are released. The acrosomal enzymes digest the zona pellucida, enabling the sperm, with its tail still beating, to tunnel a path through this protective barrier.

The first sperm to reach the ovum itself fuses with the plasma membrane of the ovum (actually a secondary oocyte), and its head enters the ovum's cytoplasm. The sperm's tail is frequently lost in this process, but the head carries the crucial genetic information. Sperm–egg fusion triggers a chemical change in the ovum's surrounding membrane that makes this outer layer impenetrable to the entry of any more sperm. This phenomenon is known as **block to polyspermy** ("many sperm"). Specifically, fertilization-induced release of intracellular Ca²⁺ into the cytosol triggers the exocytosis of enzyme-filled **cortical granules** that are located in the outermost, or cortical, region of the egg into the space between the egg membrane and the zona pellucida. These enzymes diffuse into the zona pellucida, where they inactivate the ZP3 receptors so that other sperm reaching the zona pellucida cannot bind with it. The enzymes also crosslink molecules in the zona pellucida, hardening it and sealing off

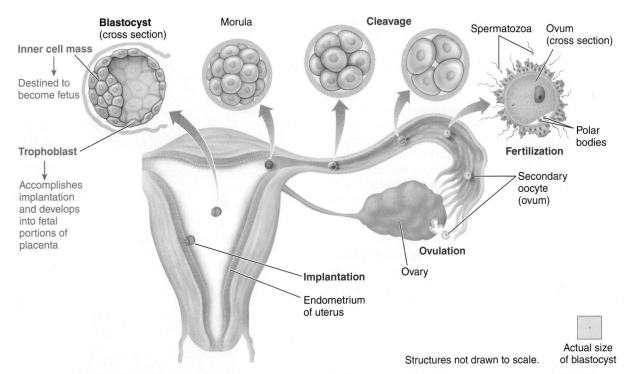

FIGURE 16-22 **Early stages of development from fertilization to implantation (mammal; shown for a human).** Note that the fertilized ovum progressively divides and differentiates into a blastocyst as it moves from the site of fertilization in the upper oviduct to the site of implantation in the uterus.

© Cengage Learning, 2013

tunnels in progress to keep other penetrating sperm from advancing. Furthermore, the released Ca^{2+} triggers the second meiotic division of the egg, which is now ready to unite with the sperm to complete the fertilization process.

Within an hour, the sperm and egg pronuclei fuse, thanks to a centrosome (microtubule organizing center) provided by the sperm that forms microtubules to bring the male and female chromosome sets together for uniting. In addition to contributing its half of the chromosomes to the fertilized ovum, now called a **zygote,** the fertilizing sperm also activates ovum enzymes that are essential for the early embryonic developmental program.

We have been describing these events for mammals, but they are broadly similar in many phyla. Indeed, many of the steps of fertilization—penetration of the egg jelly coat, block to polyspermy, role of calcium—were first delineated in sea urchins (echinoderms).

Sperm may compete among males

In mammals with females that routinely copulate with many males (including many primates such as chimpanzees), males are said to engage in **sperm competition.** They typically have testes much larger than the average mammalian species in order to produce large quantities of sperm to outcompete, and possibly to block and kill, rival sperm in the female tract. (In contrast, primates such as the gorilla in which one dominant male mates exclusively with females, the testes are comparatively smaller in size.) Recent work on honey bees and leafcutter ants by Den Boer and Boris Baer has shown that seminal fluid contains chemicals that enhance survival of a male's own sperm but which damage sperm from other males.

The mammalian blastocyst implants in the endometrium through the action of its trophoblastic enzymes

During the first few days following fertilization, the zygote usually remains within the ampulla because the high, decreasing concentrations of estrogens are still directing contractions and ciliary motion toward the site of fertilization. As the concentration of estrogens continues to decline, contractions and ciliary movement in the isthmic portion of the oviduct reverse, transporting the zygote to the uterus.

The Beginning Steps in the Ampulla The zygote is not idle during this time, however. It rapidly undergoes a number of mitotic cell divisions to form a solid ball of cells called the **morula** (Figure 16-22). Meanwhile, the rising concentrations of progesterone from the newly developing corpus luteum that formed after ovulation has a quieting influence on the uterus and stimulates the release of a nutrient medium (glycogen) from the endometrium into the reproductive tract lumen for use as energy by the early embryo. The nutrients stored in the cytoplasm of the mammalian ovum can sustain the product of conception for less than a day. The concentration of secreted nutrients increases more rapidly in the small confines of the ampulla than in the uterine lumen.

Descent of the Morula to the Uterus Several days after ovulation, progesterone is being produced in sufficient quantities to relax the oviduct constriction, thus permitting the morula to be propelled into the uterus by oviductal peristaltic contractions and ciliary activity. The temporary delay before the developing embryo passes into the uterus allows sufficient nutrients to accumulate in the uterine lumen to support the embryo until implantation can take place. If the morula arrives prematurely, it dies.

When the morula descends into the uterus, it floats freely within the uterine cavity for another three to four days, living on the endometrial secretions and continuing to divide. During the first six to seven days after ovulation, whereas the developing embryo is in transit in the oviduct and floating in the uterine lumen, the uterine lining is simultaneously being prepared for implantation or attachment under the influence of luteal-phase progesterone. During this time, the uterus is in its secretory, or progestational phase, storing up glycogen and becoming richly vascularized.

Implantation of the Blastocyst in the Prepared Endometrium In the usual case, by the time the endometrium is suitable for implantation (about a week after ovulation), the morula has descended to the uterus and continued to proliferate and differentiate into a blastocyst capable of implantation. The week's delay after fertilization and before implantation allows time for both the endometrium and the developing embryo to prepare for implantation.

A **blastocyst** is a single-layered sphere of cells encircling a fluid-filled cavity, with a dense mass of cells grouped together at one side (Figure 16-22). This dense mass, called the **inner cell mass,** is destined to become the embryo itself. The remainder of the blastocyst will never be incorporated into the fetus but serves a supportive role during intrauterine life. The thin outermost layer, the **trophectoderm,** is responsible for accomplishing implantation, after which it develops into the fetal portion of the placenta. The **amnion** forms either by cavitation, or opening up of a cavity (in rodents and humans), or folding (in noninvasive species).

When the blastocyst is ready to implant, its surface becomes sticky. By this time the endometrium is ready to accept the early embryo and it too has become more adhesive through increased formation of cell adhesion molecules (CAMs, p. 65) that help "Velcro" the blastocyst when it first contacts the uterine lining. Implantation occurs within two to five weeks after fertilization, the interval as short as two weeks in the cat and as long as five weeks in cattle and horses. The blastocyst adheres to the uterine lining on the side of its inner-cell mass (Figure 16-23).

During the preattachment phase, the blastocyst experiences phenomenal growth. For example, blastocysts of the pig increase from 2 mm in size on day 10 to 200 mm on day 12 and 1,000 mm by day 16, mainly due to development of the extraembryonic membranes, or **placentation.**

Placentation is the vascularization of the chorionic epithelium and is not the same thing as elongation. Elongation in noninvasive species allows for maximum interaction with as much uterine surface as possible to ensure **maternal recognition of pregnancy** is accomplished and that surface area for attachment is sufficient to provide nutrients to the developing conceptus in a situation where intimate interaction with the maternal vasculature does not exist.

Preventing Rejection of the Embryo/Fetus What prevents the mother from immunologically rejecting the embryo/fetus, which is actually a "foreigner" to the mother's immune system, being half-derived from genetically different

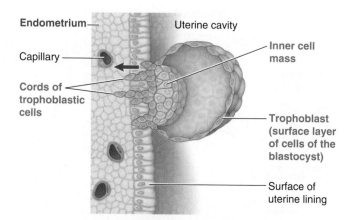

1 When the free-floating blastocyst adheres to the endometrial lining, cords of trophoblastic cells begin to penetrate the endometrium.

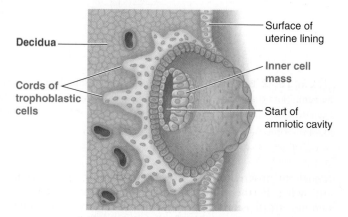

2 Advancing cords of trophoblastic cells tunnel deeper into the endometrium, carving out a hole for the blastocyst. The boundaries between the cells in the advancing trophoblastic tissue disintegrate.

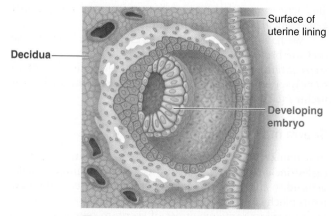

3 When implantation is finished, the blastocyst is completely buried in the endometrium.

FIGURE 16-23 Implantation of the blastocyst (primate).

© Cengage Learning, 2013

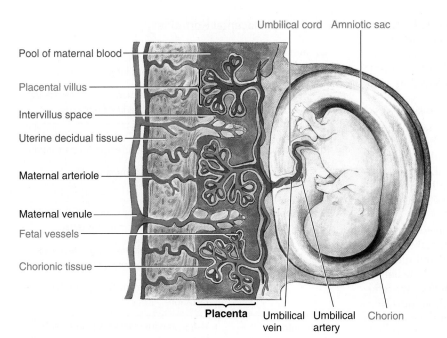

Pool of maternal blood

Placental villus

Intervillus space

Uterine decidual tissue

Maternal arteriole

Maternal venule

Fetal vessels

Chorionic tissue

Umbilical cord   Amniotic sac

Placenta   Umbilical   Umbilical   Chorion
vein   artery

FIGURE 16-24 Schematic representation of interlocking maternal and fetal structures that form the placenta in a human. Fingerlike projections of chorionic (fetal) tissue form the placental villi, which protrude into a pool of maternal blood. Decidual (maternal) capillary walls are broken down by the expanding chorion so that maternal blood oozes through the spaces between the placental villi. Fetal placental capillaries branch off of the umbilical artery and project into the placental villi. Fetal blood flowing through these vessels is separated from the maternal blood by only the thin chorionic layer that forms the placental villi. Maternal blood enters through the maternal arterioles, then percolates through the pool of blood in the intervillus spaces. Here, exchanges are made between the fetal and maternal blood before the fetal blood leaves through the umbilical vein and maternal blood exits through the maternal venules.

Source: Adapted from C. Starr, *Biology: Concepts and Applications*, 4th ed., Fig. 38.25b, p. 655. Copyright © 2000 Brooks/Cole.

paternal chromosomes? Following are several proposals under investigation. New evidence indicates that the trophoblasts produce **Fas ligand,** which binds with **Fas,** a specialized receptor on the surface of approaching activated maternal cytotoxic T cells. Cytotoxic T cells are the immune cells that carry out the job of destroying foreign cells (see p. 480). This binding triggers the immune cells that are targeted to destroy the developing foreigner to undergo apoptosis, sparing the embryo/fetus from immune rejection. Other researchers have found that the fetal portion of the placenta, which is derived from trophoblasts, produces an enzyme, **indoleamine 2,3-dioxygenase (IDO),** which destroys tryptophan. Tryptophan, an amino acid, is a critical factor in the activation of maternal cytotoxic T cells. Thus, the embryo/fetus, through its trophoblast connection, is believed to defend itself against rejection by shutting down the activity of the mother's cytotoxic T cells within the placenta that would otherwise attack the developing foreign tissues. Furthermore, recent studies demonstrate that production of **regulatory T cells** is doubled or tripled in pregnant experimental animals. Regulatory T cells suppress maternal cytotoxic T cells that might target the fetus (see p. 480).

The placenta is the organ of exchange between maternal and fetal blood

The glycogen stores (in rats and humans) in the endometrium are sufficient to nourish the embryo only during its first few weeks. To sustain the growing embryo for the duration of its intrauterine life, the *placenta,* a specialized organ of exchange between the maternal and fetal blood, is rapidly developed (Figure 16-24). The placenta is derived from both trophoblastic and decidual tissue. In noninvasive species, **histotroph,** which consists of proteins (some of which are growth factors), prostaglandins, ions, carbohydrates, and so on, is produced by the endometrium and subsequently absorbed by the embryo and digested in the yolk sac.

Modes of placentation differ among species

By the time the embryo has attached to the uterine luminal epithelium, the trophoblastic layer is two cell layers thick and is called the **chorion.** As the chorion continues to expand, it fuses (in placental mammals) with a structure called the **allantois** (an outgrowth of the hindgut) forming the vascularized *chorioallantoic* membrane. (In marsupials, the placenta is *choriovitellic,* forming from the chorion and yolk-sac or vitelline lining.) Placentas can be classified according to their degree of invasiveness and the anatomy of those regions of greatest maternal–fetal interaction (Figure 16-24). If we describe the potential placental layers available and then begin to remove layers as we describe progressively more invasive placentation, you will be able to see the interrelationship of the different modes of placentation. On the maternal side the maximum number of layers could include endothelial cells lining the blood vessels, a basement membrane, endometrial stromal tissue, another basement membrane, and the uterine luminal epithelium. On the fetal side we could potentially have chorionic epithelium, a basement membrane, chorioallantoic stromal tissue, another basement membrane, and endothelial cells lining the blood vessels. Thus, six potential tissue layers separate maternal and fetal blood.

- In the pig, cow, and ewe, all six tissue layers are present, and the type of placentation is referred to as **epithelial-** (the maternal component) **chorial-** (the fetal component). In the pig, chorion is in contact with the uterine luminal epithelium throughout the surface of the placenta, so this arrangement is called **diffuse.**
- In ruminants, the greatest contact of the chorion with the uterine luminal epithelium occurs at discreet sites at which fetal membranes interact with maternal *caruncles* (round thickenings) such that chorionic villi project into the caruncles forming **placentomes,** and so we call this **placentomal.**
- In the carnivores, the conceptus erodes the uterine luminal epithelium and underlying stroma at sites of greatest

attachment such that chorion is in contact with maternal endothelium, an arrangement called **endothelial-chorial** (four tissue layers). The erosion of maternal tissue occurs only in regions around a band in the central portion of the placenta, or zone, and is referred to as **zonary attachment.**

- In the human and higher primates, the conceptus erodes maternal uterine luminal epithelium, stroma, and the underlying endothelial cells, allowing the chorionic epithelium to have direct access to maternal blood, a type of placentation referred to as **hemo-chorial.** Because this is a very invasive placentation, only a small site on the placental surface is involved, and the shape of placentation is referred to as **discoid.**

- In the most invasive placentation, found in rodents, not only are all three tissue layers on the maternal side eroded, but the chorioallantoic epithelium and underlying stroma are eroded and the conceptus endothelial cells are in direct contact with maternal blood; we refer to this as **hemo-endothelial placentation.** It is found in the shape of a disk of interaction between the conceptus and maternal system.

Throughout gestation, fetal blood continuously travels between the placental villi and the fetal circulatory system via the **umbilical artery** and **umbilical vein,** which are wrapped within the **umbilical cord,** a lifeline between the primate fetus and the placenta (Figure 16-24). The maternal blood within the placenta is continuously replaced as fresh blood enters through the uterine arterioles; percolates through the intervillus spaces, where it exchanges substances with fetal blood in the surrounding villi; then exits through the uterine vein.

During intrauterine life, the placenta performs the functions of the digestive system and the respiratory system for the "parasitic" fetus. This is not to say that the fetus does not have these organ systems, but rather that they cannot (and do not need to) function within the uterine environment. Intrauterine urination is common and contributes to the volume of amniotic fluid. Nutrients and O_2 diffuse from the maternal blood across the thin placental barrier into the fetal blood, whereas CO_2 and other metabolic wastes simultaneously diffuse from the fetal blood into the maternal blood. The nutrients and O_2 brought to the fetus in the maternal blood are acquired by the maternal digestive and respiratory systems, and the maternal lungs and kidneys eliminate the CO_2 and wastes transferred into the maternal blood, respectively. Thus, the maternal digestive tract, respiratory system, and kidneys serve the fetus's needs as well as the mother's.

Transport of Substances across the Placental Barrier The means by which materials move across the placenta depends on the substance. Some substances that can permeate the placental membrane, such as O_2, CO_2, water, and electrolytes, cross by simple diffusion. Some traverse the placental barrier by special mediated-transport systems in the placental membranes, such as glucose by facilitated diffusion and amino acids by secondary active transport. Other substances such as cholesterol in the form of LDL move across by receptor-mediated endocytosis. Unfortunately, many drugs, environmental pollutants, other chemical agents, and microorganisms in the female's bloodstream can cross the placental barrier, some of which are harmful to the developing fetus.

TABLE 16-5 Placental Hormones

Hormone	Function
Chorionic Gonadotropin	Maintains corpus luteum of pregnancy
	Stimulates secretion of testosterone by developing testes in XY embryos
Estrogen *(also secreted by corpus luteum of pregnancy)*	Stimulates growth of myometrium, increasing uterine strength for parturition
	Helps prepare mammary glands for lactation
Progesterone *(also secreted by corpus luteum of pregnancy)*	Suppresses uterine contractions to provide quiet environment for fetus
	Promotes formation of cervical mucus plug to prevent uterine contamination
	Helps prepare mammary glands for lactation
Chorionic Somatomammotropin *(also called* placental lactogen; *has a structure similar to both prolactin and growth hormone)*	Helps prepare mammary glands for lactation (similar to prolactin)
	Believed to reduce maternal use of glucose and to promote breakdown of stored fat (similar to growth hormone) so that greater quantities of glucose and free fatty acids can be shunted to the fetus
Relaxin *(also secreted by corpus luteum of pregnancy)*	Softens the cervix in preparation for cervical dilation at parturition
	Loosens connective tissue between pelvic bones in preparation for parturition
Placental PTHrp *(parathyroid hormone-related peptide)*	Increases maternal plasma Ca^{2+} levels for use in calcifying fetal bones; if necessary, promotes localized dissolution of maternal bone, mobilizing mother's Ca^{2+} stores for use by developing fetus

© Cengage Learning, 2013

The placenta assumes yet another important responsibility—it becomes a temporary endocrine organ during pregnancy, a topic to which we now turn.

Hormones secreted by the corpus luteum and placenta play a critical role in maintaining pregnancy

The placenta has the remarkable capacity to secrete a number of peptide and steroid hormones essential for maintaining pregnancy. The most important are *chorionic gonadotropin, estrogen,* and *progesterone* (Table 16-5). Serving as the major endocrine organ of pregnancy, the placenta is unique among endocrine tissues in two regards. First, it is a transient tissue. Second, secretion of its hormones is not subject to extrinsic control, in contrast to the stringent, often-complex mechanisms that regulate the secretion of other hormones. Instead, the type and rate of placental hormone secretion depend primarily on the species of animal and stage of pregnancy.

In all mammals, the corpus luteum is important in the early stages of pregnancy for secreting progesterone, but, as exemplified in the human, ewe, and mare, its contribution declines with time as the placental contribution of progesterone increases. For example, the corpus luteum is not required in women and ewes after approximately 50 days and in the mare after 100 days of pregnancy, whereas the corpus luteum is required for the duration of pregnancy in the pig, cow, goat, and deer.

One of the first events in primates is the secretion by the developing chorion of *chorionic gonadotropin (CG)*, a polypeptide hormone that prolongs the life span of the corpus luteum. In the mare, *equine chorionic gonadotropin (eCG)* is also called *pregnant mare's serum gonadotropin (PMSG)*. The placenta of the mare is notable for unique structures termed **endometrial cups,** temporary endocrine units ranging in size from millimeters to several centimeters that produce eCG. After approximately 60 days of gestation, these cups are sloughed into the uterine lumen. Recall that during the menstrual cycle in humans (and some other primates), the corpus luteum degenerates and the highly prepared, luteal-dependent uterine lining sloughs if fertilization and implantation have not occurred. When fertilization does occur, the implanted human blastocyst saves itself by producing human *CG (hCG)*. This hormone, which is functionally similar to LH, stimulates and maintains the corpus luteum so that it does not degenerate. Now called the *corpus luteum of pregnancy*, this ovarian endocrine unit grows even larger and produces increasingly greater amounts of progesterone until the placenta takes over secretion of these steroid hormones (see p. 790). Because of the persistence of circulating estrogens and progesterone, the thickened endometrial tissue of the human is maintained instead of sloughing. In menstrual primates, accordingly, menstruation ceases during pregnancy. The eCG of the mare is produced around 60 days of gestation, and like the hCG, eCG primarily exhibits LH-like activity, although if administered to other species eCG exhibits FSH-like activity. This is important for the mare because the corpora lutea do not persist beyond about day 70 of gestation and are supplemented by the growth and luteinization of new follicles under the influence of eCG to save the pregnancy. These accessory corpora lutea only maintain pregnancy for another short period until the placenta finally takes over progesterone production at around 100 days of gestation.

Maternal body systems respond to the increased demands of gestation

The period of gestation (pregnancy) is about 1 month in the rabbit, 3.8 months in the sow, 5 months in the ewe and goat, 9 months in the cow and woman, 15 to 16 months in the sperm whale, and 1.8 years in African forest and savanna elephants. During gestation, the fetus continues to grow and develop to the point of being able to leave its maternal life support system. Meanwhile, a number of changes take place within the female to accommodate the demands of the pregnancy. The most obvious is uterine enlargement. The uterus expands and increases in weight more than 16 times in women, exclusive of its contents. The mammary glands enlarge and develop the capability to produce milk.

Body systems other than the reproductive system make needed adjustments. The volume of blood increases by 30%, and the cardiovascular system responds to the increasing demands of the growing placental mass. The weight gain experienced during pregnancy is due only in part to the weight of the fetus. The remainder is primarily caused by the growth of the uterus, including the placenta, and the increased blood volume. Respiratory activity is increased by about 16% to handle the additional fetal requirements for O_2 use and CO_2 removal. Urinary output increases, and the kidneys excrete the additional wastes from the fetus.

The increased metabolic demands of the growing fetus increase nutritional requirements for the female. In general, the fetus takes what it needs from the female, even if this leaves the female with a nutritional deficit. For example, the placental hormone *chorionic somatomammotropin (CS)*, also known as *placental lactogen (PL)*, is thought to be responsible for the decreased use of glucose by the female and the mobilization of free fatty acids from maternal adipose stores, similar to the actions of growth hormone (see p. 292). (In fact, CS has a structure similar to both growth hormone and prolactin. In the mammary gland, CS helps prepare these glands for lactation, similar to prolactin's effect, although much weaker.) As a result of the CS-induced metabolic changes in the female, greater quantities of glucose and fatty acids are available for shunting to the fetus. The extent to which CS exerts somatotropic versus lactogenic effects is ultimately species dependent. For example, in the ewe CS is more effective in promoting lactation. Also, if the female does not consume sufficient Ca^{2+} in her diet, placental *parathyroid-hormone-related peptide* mobilizes Ca^{2+} from the maternal bones to ensure adequate calcification of the fetal bones.

check your understanding 16.6

How are the ovum and spermatozoa transported to the site of fertilization? Describe the process of fertilization.

Describe the process of implantation and placenta formation.

What are the functions of the placenta? What hormones does the placenta secrete?

What is the role of chorionic gonadotropin?

16.7 Parturition and Lactation

Parturition (labor, delivery, or birth) is the process by which the uterus expels the fetus and placenta from the female. Parturition is divided into three stages: (1) dilation of the cervical canal to accommodate passage of the fetus from the uterus through the vagina and to the outside, (2) contractions of the uterine myometrium that are sufficiently strong to expel the fetus, and (3) expulsion of the placenta.

Changes during late gestation prepare for parturition

Several events take place near the end of pregnancy in preparation for parturition. Throughout gestation, the cervix remains sealed. As parturition approaches, the cervix begins to soften (or "ripen") as a result of the dissociation of its tough connective tissue (collagen) fibers. This softening is believed to be caused by **relaxin,** a peptide hormone produced in most mammals by the corpus luteum of pregnancy and by the placenta. However, in species such as the rabbit, relaxin is produced entirely in the placenta. Relaxin also "relaxes" the pelvic bones. Furthermore, $PGF_{2\alpha}$ released by the placental membranes overlying the cervix stimulates the synthesis of relaxin and promotes the production of cervical enzymes that degrade the collagen fibers and help soften the cervix. $PGF_{2\alpha}$ initiates regular, nondirectional uterine contractions and regression of the CL. These small contractions (sometimes called *contractures*) help orient the fetus(es) so that the head and front appendages are in contact with the cervix in preparation for exiting through the birth canal. If the fetus is not positioned correctly, a difficult birth (*dystocia*), results.

Rhythmic, coordinated contractions, usually painless at first, begin at the onset of true labor in response to pituitary *oxytocin* (see p. 286). As labor progresses, the contractions occur with increasing frequency and intensity and are accompanied by increasing discomfort. These strong, rhythmic, and directional contractions force the fetus against the cervix, resulting in dilation of the cervix. Then, after having dilated the cervix sufficiently for passage of the fetus, these contractions force the fetus out through the birth canal.

The factors that trigger the onset of parturition are only partially understood

The exact factors responsible for triggering this change in uterine contractility and thus initiating parturition are not fully established, although considerable progress has been made in recent years in unraveling the sequence of events. During early gestation, maternal estrogen levels are relatively low, but, as gestation proceeds, placental estrogen secretion continues to rise, associated with the increase in concentrations of $PGF_{2\alpha}$. However, the timing and magnitude of the increase in estrogens vary among species. In the immediate days before the onset of parturition, soaring levels of estrogens bring about changes in the uterus and cervix to prepare them for labor and delivery (Figure 16-25). First, high levels of estrogens promote the synthesis of connexons within the uterine smooth-muscle cells. These myometrial cells are not functionally linked to any extent throughout most of gestation. The newly manufactured connexons are inserted in the myometrial plasma membranes to form gap junctions (see p. 66) that electrically link the uterine smooth-muscle cells so that they become able to contract as a coordinated unit.

Simultaneously, the high levels of estrogens cause a dramatic, progressive increase in the concentration of myometrial receptors for oxytocin. Together, these myometrial changes collectively bring about the increased uterine responsiveness to oxytocin that ultimately initiates labor. In addition to preparing the uterus for labor, the increasing levels of estrogen promote production of local prostaglandins that contribute to cervical ripening by stimulating cervical enzymes that locally degrade collagen fibers. Furthermore, these prostaglandins themselves increase uterine responsiveness to oxytocin.

Role of Oxytocin Oxytocin is a peptide hormone that is produced by the hypothalamus, stored in the posterior pituitary, and released into the blood from the posterior pitu-

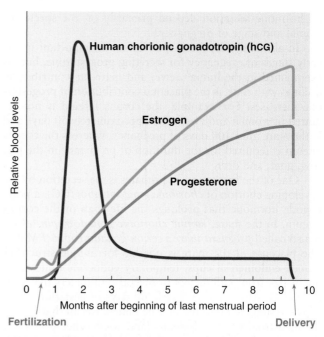

FIGURE 16-25 Secretion rates of human placental hormones.
© Cengage Learning, 2013

itary on nervous stimulation by the hypothalamus (see p. 286). Oxytocin exerts its effects via the $IP_3/Ca^{2+}/DAG$ pathway. A powerful uterine muscle stimulant, oxytocin plays the key role in the progression of labor. However, this hormone was discounted as serving as the trigger to parturition because the circulating concentrations of oxytocin remain constant prior to the onset of labor. The discovery that uterine responsiveness to oxytocin is greater at term than in nonpregnant females (because of the increased concentration of myometrial oxytocin receptors) led to the now widely accepted conclusion that labor begins when the oxytocin receptor concentration reaches a critical threshold that permits the onset of strong, coordinated contractions in response to ordinary concentrations of circulating oxytocin.

In addition to preparing the uterus for labor, the increasing levels of estrogens promote production of the local prostaglandins that contribute to cervical ripening. Oxytocin has other roles, including the triggering of uterine contractions during orgasm, milk ejection during lactation, and maternal behavior related to bonding with the father and the offspring (see p. 286). *Pitocin* is the synthetic form of oxytocin used to induce labor if it is not progressing well in humans and many domestic mammals (it is also used to treat birds and reptiles suffering from egg binding, in which eggs have become trapped in the reproductive tract).

Role of Corticotropin-Releasing Hormone Until recently, scientists were baffled by the factors responsible for the rising levels of placental estrogen secretion. Recent research has shed new light on the probable mechanism. Evidence suggests that *corticotropin-releasing hormone* (CRH) secreted by the fetal portion of the placenta into both the maternal and fetal circulation not only drives the synthesis of placental estrogen, thus ultimately dictating the timing of the onset of labor, but also promotes changes in the fetal lungs needed for breathing air; see Figure 16-27. Recall that

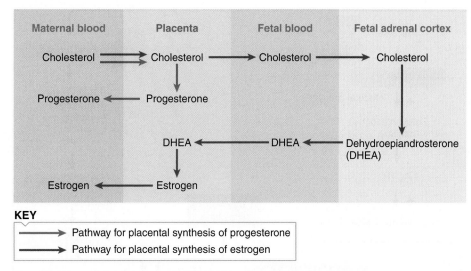

FIGURE 16-26 Secretion of estrogen and progesterone by the human placenta. The placenta secretes increasing quantities of progesterone and estrogen into the maternal blood after the first trimester. The placenta itself can convert cholesterol into progesterone (*green* pathway) but lacks some of the enzymes necessary to convert cholesterol into estrogen. However, the placenta can convert DHEA derived from cholesterol in the fetal adrenal cortex into estrogen when DHEA reaches the placenta by means of the fetal blood (*blue* pathway).

© Cengage Learning, 2013

CRH is normally secreted by the hypothalamus and regulates the output of ACTH by the anterior pituitary. In turn, ACTH stimulates production of both cortisol and DHEA by the adrenal cortex. In the fetus, much of the CRH comes from the placenta rather than solely from the fetal hypothalamus. The additional cortisol secretion summoned by the extra CRH promotes fetal lung maturation. Specifically, cortisol stimulates the synthesis of pulmonary surfactant, which facilitates lung expansion and reduces the work of breathing (see p. 521).

The bumped-up rate of DHEA secretion by the adrenal cortex in response to placental CRH leads to the rising concentrations of placental estrogen secretion. Recall that the human placenta converts DHEA from the fetal adrenal gland into estrogen, which enters the maternal bloodstream (Figure 16-26). In most other species, the placenta has the ability to synthesize estrogens *de novo*.

When sufficiently high, this estrogen combined with lowered progesterone sets in motion the events that initiate labor. Thus, pregnancy duration and delivery timing are determined largely by the placenta's rate of CRH production. That is, a *"placental clock"* ticks out a predetermined length of time until parturition. The timing of parturition is established early in pregnancy, with delivery at the end point of a maturational process that extends throughout the entire gestation. The ticking of the placental clock is measured by the rate of placental secretion. As the pregnancy progresses, CRH concentrations in maternal plasma rise. Researchers can accurately predict the timing of parturition by measuring the maternal plasma concentrations of CRH. Higher-than-normal concentrations are associated with premature deliveries, whereas lower-than-normal concentrations indicate late deliveries. These and other data suggest that a critical level of maternal CRH of placental origin may directly trigger parturition. Placental CRH ensures that when labor

begins, the infant is ready for life outside the womb. It does so by concurrently increasing the fetal cortisol needed for lung maturation and the estrogens needed for the uterine changes that bring on labor. The remaining unanswered question regarding the placental clock is: What controls placental secretion of CRH?

Role of Inflammation and Lung Maturation Surprisingly, recent research suggests that *inflammation* plays a central role in the labor process, in the onset of both full-term labor and premature labor. Less surprisingly, recent research also indicates that the maturing *lungs* are involved. The inflammatory mechanism begins with activation of *nuclear factor κB (NF-κB)* in the uterus. NF-κB boosts production of inflammatory cytokines such as *interleukin-8 (IL-8)* (see p. 469) and prostaglandins that increase the sensitivity of the uterus to contraction-inducing chemical messengers and help soften the cervix. What activates NF-κB, setting off an inflammatory cascade that helps prompt labor? Various factors associated with the onset of full-term labor and premature labor can cause an upsurge in NF-κB. These include stretching of the uterine muscle and the presence of the pulmonary *surfactant protein SP-A* (stimulated by the action of CRH on the fetal lungs) in the amniotic fluid from the fetus. SP-A promotes the migration of fetal macrophages to the uterus (Figure 16-27). These macrophages, in turn, produce the inflammatory cytokine *interleukin-1β (IL-1β)* that activates NF-κB. In this way, fetal lung maturation contributes to the onset of labor, just the time at which the lungs are ready for the transition to air breathing.

Bacterial infections and allergic reactions can lead to premature labor by activating NF-κB. Also, multiple-fetus pregnancies are at risk for premature labor, likely because the increased uterine stretching triggers earlier activation of NF-κB.

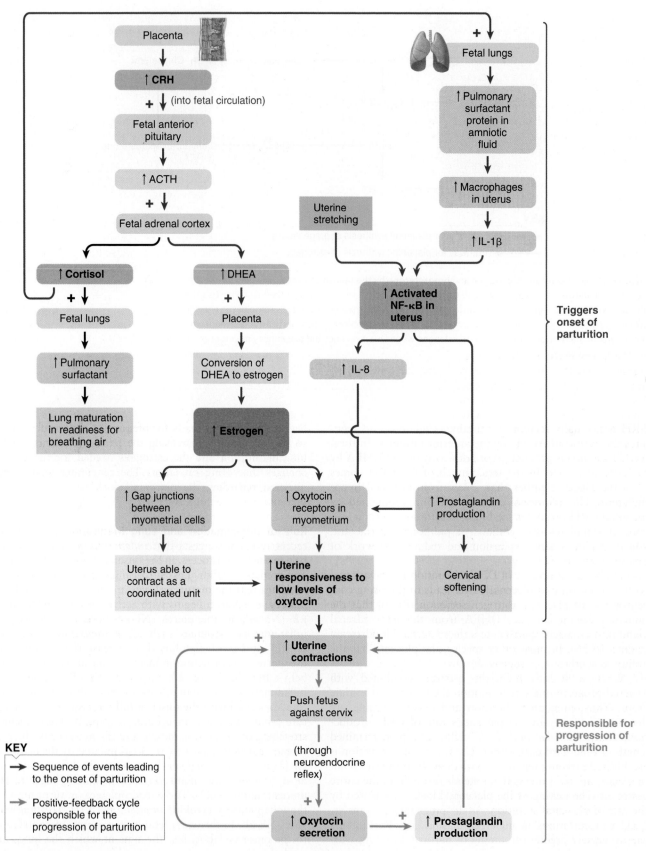

FIGURE 16-27 Initiation and progression of parturition (mammal).

© Cengage Learning, 2013

Parturition is accomplished by a positive-feedback cycle

Once the high concentrations of estrogens have increased uterine responsiveness to oxytocin to a critical level and regular uterine contractions have begun, myometrial contractions progressively increase in frequency, strength, and duration throughout labor until the uterine contents are expelled. This progressive increase in contractions is a classical example of a positive-feedback cycle, which in this case involves *cervical stretch, oxytocin,* and *prostaglandin* (Figure 16-27; see also Figure 1-10, p. 18). Each uterine contraction begins at the oviductal end of the uterus and sweeps downward, forcing the fetus(es) toward the cervix. Pressure of the fetus against the cervix accomplishes two things. First, the fetal head pushing against the softened cervix acts as a wedge to dilate the cervical canal. Second, cervical stretch in response to fetal pressure stimulates stretch receptors that produce a reflex neural signal, which travels up the spinal cord to the hypothalamus. This in turn triggers oxytocin release from the posterior pituitary. Additional oxytocin promotes more powerful uterine contractions. As a result, the fetus is pushed more forcefully against the cervix, increasing stretch and so stimulating the release of even more oxytocin, and so on. This cycle is reinforced as oxytocin stimulates prostaglandin production by the decidua. As a powerful myometrial stimulant, prostaglandin further enhances uterine contractions. Oxytocin secretion, $PGF_{2\alpha}$ production, and uterine contractions continue to increase in this positive-feedback fashion throughout labor until the pressure on the cervix is relieved by delivery.

Stages of Labor Recall that labor has three stages: (1) cervical dilation, (2) delivery of the young, and (3) delivery of the placenta. At the onset of labor or sometime during the first stage, the membranes surrounding the amniotic sac rupture. The fluid surrounding the amnion is *chorioallantoic fluid.* The amnion does not rupture until much later during parturition, so it can protect the fetus from physical damage during the early stages of parturition. As the chorioallantoic fluid escapes out of the vagina, it helps lubricate the birth canal. During the first stage, the cervix of ungulates (hoofed mammals) is forced to dilate to accommodate the diameter of the young's head and front feet. This stage is the longest, lasting from 1 to 12 hours in the sow, 2 to 6 hours in the ewe and cow, and several hours to as long as 24 hours in humans. If another part of the fetus's body other than the head is oriented against the cervix, it is generally less effective than the head as a wedge, although in the sow, half of the fetuses come headfirst and half rear-end first. (Also, whales are born at sea tail first to prevent drowning.) The head has the largest diameter of the fetus's body. In some species, if the fetus approaches the birth canal feet first, the cervix may not be dilated sufficiently by the feet to permit passage of the head. Without medical intervention in such a case, the fetus's head would remain stuck behind the too-narrow cervical opening.

The second stage of labor, the actual birth, begins once cervical dilation is complete. When the offspring begins to move through the cervix and vagina, stretch receptors in the vagina activate a neural reflex that triggers contractions of the abdominal wall in synchrony with the uterine contractions. These abdominal contractions greatly increase the force pushing the fetus through the birth canal. Stage 2 is usually much shorter than the first stage and lasts 30 to 90 minutes. The fetus is still attached to the placenta by the umbilical cord at birth, and this cord breaks as the newborn animal begins to move about following delivery, with the stump shriveling up in a few days to form the umbilicus (navel).

Shortly after delivery of the fetus, a second series of uterine contractions causes the placenta to separate from the endometrium and be expelled through the vagina. Delivery of the placenta, or **afterbirth,** constitutes the third stage of labor, which is typically the shortest stage, being completed within 15 to 30 minutes in humans, 1 hour in the mare, 5 to 8 hours in the ewe, and as long as 12 hours in the cow, after the young is born. After the placenta is expelled, continued contractions of the myometrium constrict the uterine blood vessels supplying the site of placental attachment to prevent hemorrhage. In sows, cows, ewes, mares, and so on, there is no erosion of maternal tissue and therefore no danger of hemorrhage. Most female terrestrial mammals then eat the placenta (**placentophagy**), which contain high levels of prostaglandins that may foster *involution* (see next section). The placenta also contains **placental opioid-enhancing factor (POEF),** which is thought to trigger pain-inhiting opioids (p. 310) in the mother's brain.

Uterine Involution After delivery, the uterus shrinks to its pregestational size, a process known as **involution,** which requires 4 weeks in the ewe and sow, 12 weeks in the dog, and 4 to 6 weeks in the human. During involution, the remaining endometrial tissue not expelled with the placenta gradually disintegrates and sloughs off, producing a vaginal discharge called **lochia** that continues for three to six weeks following parturition. After this period, the endometrium is restored to its nonpregnant state. Involution occurs largely because of the precipitous fall in circulating estrogens and progesterone when the placental source of these steroids is lost at delivery. The process is facilitated by suckling the infant(s) because of the oxytocin released in response to mammary gland stimulation. In addition to playing an important role in lactation, this periodic nursing-induced postpartum release of oxytocin promotes myometrial contractions that stimulate uterine muscle tone, thus enhancing involution.

Lactation requires multiple hormonal inputs

The female reproductive system of mammals supports the new being from the moment of its conception through gestation and continues to nourish it during its early life outside the supportive uterine environment. However, only two species of mammals, whales and bears, can fast during the energetically challenging period of lactation. This strategy permits baleen whales to feed in the seasonally productive polar regions of the world's oceans and retain the advantage of breeding in the warmer, tropical regions of the world.

Because the evolution of the mammary gland preceded that of placental gestation, even egg-laying mammals (monotremes) produce milk for their young (hence the name "mammal"). **Lactogenesis** is the process by which mammary alveolar cells acquire the ability to secrete milk (or its equivalent), for milk is essential for survival of the newborn, and, interestingly, the basic structure of the mammary gland is remarkably consistent across the class Mammalia. Accordingly, development begun during puberty is completed during gestation as the mammary glands are prepared for lactation (milk production).

The mammary gland is a modified sweat gland, the basic structure and location established during embryonic development. The glands develop from the so-called mammary or **milk line,** two rows that extend from the thoracic region, parallel to the midline of the abdomen that appears in the early embryo. Subsequently, the milk lines differentiate into mammary buds, the number and location dependent on the species of mammal. For example, the mammary glands of some mammals (deer, cattle, goats, camels, horses, and sheep) have an inguinal (located near the groin) location and in nonpregnant females consist mostly of adipose tissue and a rudimentary duct system. In primates and elephants, the mammary glands develop in the pectoral region, whereas in litter-bearing species, such as rabbits, dogs, cats, and pigs, mammary glands develop along the entire length of the milk line. Pigs normally have seven pairs of mammary glands, whereas dogs and cats usually have five pairs of glands. The sea mammals (whales, dolphins, seals, and manatees) develop mammary buds in either the pectoral or inguinal region. Structures associated with the mammary gland include the **teat** or **nipple,** the part of the mammary gland from which the young suckle milk (Figure 16-28). At its distal end, the teat has fine openings that permit expulsion of the milk. The number of openings can be as few as one, such as in cattle, which have a collecting **cistern,** which permits pooling of milk from several ducts before exiting the gland, as compared to the nipple, which has multiple lactiferous ducts.

The mammary gland (udder) of the cow has distinct right and left halves, and each has a front and hindquarter. Each half of the mammary gland is independent with regard to its blood supply, innervation, and lymphatic drainage. Under the hormonal environment present during pregnancy, the mammary glands develop the internal glandular structure and function necessary for milk production. A mammary gland capable of lactating consists of a network of progressively smaller ducts that branch out from the teat and terminate in lobules (Figure 16-28). Each lobule is made up of a cluster of saclike epithelial-lined, milk-producing glands known as **alveoli** that constitute the milk-producing glands. Milk is synthesized by the epithelial cells and then secreted into the alveolar lumen, which is drained by a milk-collecting duct that transports the milk to the teat cistern. Ejection of milk from the teat occurs via the teat canal, which is normally kept closed by a muscle sphincter (circular muscle whose contraction closes the canal). **Mastitis,** infection of the mammary gland by microorganisms, can occur if the sphincter is not kept tightly closed.

Prevention of Lactation during Gestation Most mammary gland growth occurs during pregnancy, when the high concentration of estrogens promotes extensive duct development and the high level of progesterone stimulates abundant alveolar-lobular formation. The adipose tissue is steadily consumed and replaced by ducts, lobular alveoli, blood and lymph vessels, and the necessary connective tissue structures associated with the suspensory apparatus. Elevated concentrations of **prolactin** (an anterior pituitary hormone stimulated by the rising levels of estrogen) and **chorionic somato-**

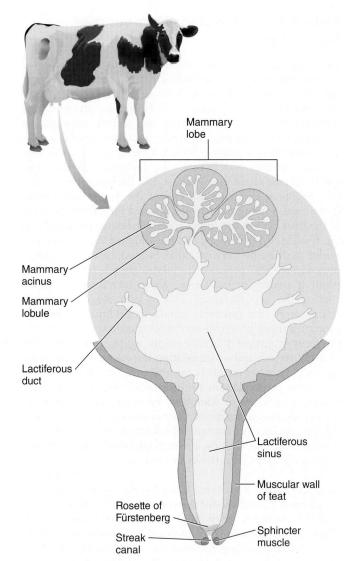

FIGURE 16-28 **A cow mammary gland or teat.**

Source: Y. Ruckebusch, L.-P. Phaneuf, & R. Dunlop. (1991). *Physiology of Small and Large Animals.* Philadelphia: B. C. Decker, p. 620, Figure 59-4.

mammotropin (a peptide hormone produced by the placenta that has a structure similar to both prolactin and growth hormone) contribute to mammary gland development by inducing the synthesis of enzymes needed for milk production.

By the middle of pregnancy, the mammary glands are fully capable of producing milk; however, milk secretion does not occur until parturition, even though prolactin—the primary stimulant of milk secretion—is present. The high estrogen and progesterone concentrations during the last half of pregnancy prevent lactation by blocking prolactin's stimulatory action on milk secretion. Thus, even though the high levels of placental steroids induce the development of the milk-producing machinery in the mammary glands, they prevent these glands from becoming operational until the young is born and milk is needed. The abrupt decline in estrogens and progesterone that occurs with loss of the placenta at parturition initiates lactation. (The functions of estrogens and progesterone during gestation and lactation as well as throughout the reproductive life of females are summarized in Table 16-6.)

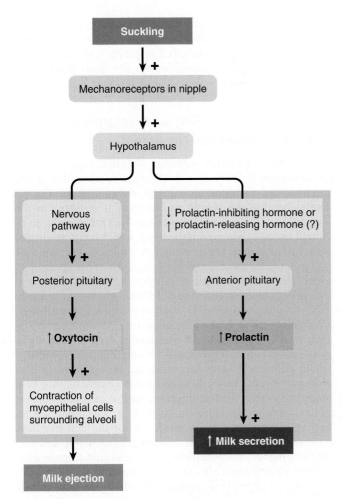

FIGURE 16-29 Suckling reflexes (mammal).

© Cengage Learning, 2013

TABLE 16-6 Actions of Estrogens and Progesterone (in Mammals)

ESTROGENS

Effects on Sex-Specific Tissues

Essential for egg maturation and release

Stimulates growth and maintenance of entire female reproductive tract

Stimulates granulosa cell proliferation, which leads to follicle maturation

Thins cervical mucus to permit sperm penetration

Enhances transport of sperm to oviduct by stimulating upward contractions of uterus and oviduct

Stimulates growth of endometrium and myometrium

Induces synthesis of progesterone receptors in endometrium

Triggers onset of parturition by increasing uterine responsiveness to oxytocin during late gestation through a twofold effect: by inducing synthesis of myometrial oxytocin receptors and by increasing myometrial gap junctions so that the uterus can contract as a coordinated unit in response to oxytocin

Other Reproductive Effects

Promotes development of secondary sexual characteristics

Controls GnRH and gonadotropin secretion

Low levels inhibit secretion

High levels responsible for triggering LH surge

Stimulates duct development in mammary glands during gestation

Inhibits milk-secreting actions of prolactin during gestation

Nonreproductive Effects

Promotes fat deposition

Increases bone density; closes epiphyseal plates

Promotes vasodilation by increasing nitric oxide production in arterioles (cardioprotective)

PROGESTERONE

Prepares a suitable environment for nourishment of a developing embryo/fetus

Promotes formation of a thick mucus plug in the cervical canal

Inhibits hypothalamic GnRH and gonadotropin secretion

Stimulates alveolar development in the mammary gland during gestation

Inhibits the milk-secreting actions of prolactin during gestation

Inhibits uterine contractions during gestation

© Cengage Learning, 2013

Once milk production begins after delivery, two hormones maintain lactation: (1) prolactin, which acts on the alveolar epithelium to promote secretion of milk, and (2) oxytocin, which produces **milk ejection.** The latter refers to the forced expulsion of milk from the lumen of the alveoli out through the ducts. Release of both of these hormones is stimulated by a neuroendocrine reflex triggered by suckling (Figure 16-29). Let's examine each of these hormones and their roles in detail.

Oxytocin Release and Milk Ejection The neonate cannot directly suck milk out of the alveolar lumen. Instead, milk must be actively squeezed out of the alveoli into the ducts and hence toward the nipple by contraction of specialized myoepithelial cells (musclelike epithelial cells) that surround each alveolus. Suckling of the mammary gland by the neonate stimulates sensory nerve endings in the nipple, initiating action potentials that travel up the spinal cord to the hypothalamus. Thus activated, the hypothalamus triggers a burst of oxytocin release from the posterior pituitary. Oxytocin in turn stimulates contraction of the myoepithelial cells in the mammary glands to bring about milk ejection, or "milk letdown." Cetacean (whale) young lack proper lips, so gripping the nipple is difficult at best. For this reason the female actually squirts the milk into the mouth of the young, using contractions of both the myoepithelial cells as well as surrounding cutaneous muscles. Milk letdown continues only as long as the infant continues to nurse. In this way, the milk ejection reflex ensures that milk exits the mammary glands only in the amount needed by the young. Even though the alveoli may be full of milk, the milk cannot be released without oxytocin. Often the presence of the calf or other conditioned reflexes can generate a release of oxytocin, whereas milk letdown is inhibited in situations in which the female is stressed.

Prolactin Release and Milk Ejection Suckling not only triggers oxytocin release but also stimulates prolactin secretion. Prolactin output by the anterior pituitary is controlled by two hypothalamic secretions: *prolactin-inhibiting hormone* (*PIH*) and *prolactin-releasing hormone* (*PRH*). PIH has been shown to be dopamine, which also serves as a neurotransmitter in the brain. The chemical nature of PRH has not been identified with certainty, but scientists suspect that in most mammals (though not in the cow) PRH is oxytocin secreted by the hypothalamus into the hypothalamic–hypophyseal portal system to stimulate prolactin secretion by the anterior pituitary (see p. 288). This role of oxytocin is distinct from the roles of oxytocin produced by the hypothalamus and stored in the posterior pituitary. The oxytocin released by the posterior pituitary plays a key role in parturition by causing powerful uterine muscle contractions and also brings about milk ejection by stimulating contraction of the myoepithelial cells of lactating mammary glands.

Throughout most of the female's life, PIH is the dominant influence, so prolactin concentrations normally remain low. During lactation, a burst in prolactin secretion occurs each time the infant stimulates the udder and/or teats. Afferent impulses initiated in the nipple on suckling are carried by the spinal cord to the hypothalamus. This reflex ultimately leads to prolactin release by the anterior pituitary, although it is unclear whether this outcome is accomplished by inhibition of PIH or stimulation of PRH secretion or both. Prolactin then acts on the alveolar epithelium to promote secretion of milk to replenish that lost from the alveoli by milk ejection. In some species of mammals (cows and goats) once lactation has been established, the basal concentration of prolactin as well as the induced release of prolactin at milking decline without affecting milk yield.

Suckling concurrently stimulates milk ejection and milk production, therefore ensuring that the rate of milk synthesis keeps pace with the young's needs for milk. The more the infant nurses, the more milk is removed by letdown and the more milk is produced for the next suckling. Thus, twins elicit more milk production than do single offspring. In addition to prolactin, which is the most important factor controlling synthesis of milk, at least four other hormones are essential for their permissive role in milk production: cortisol, insulin, parathyroid hormone, and growth hormone (see Chapter 7). Among its actions, growth hormone enhances milk production by increasing the mammary gland's uptake from the blood of nutrients needed for the synthesis of milk (see p. 292).

Prolactin fittingly appears to facilitate bonding, or attachment, between a mother and the newborn in birds and mammals. Prolactin (as well as oxytocin, p. 296) is thought to trigger maternal behaviors such as denning in wolves, incubation behavior in birds, bonding of mothers to offspring, and suppression of sex drive during nursing periods. For more on prolactin, see the box, *Molecular Biology and Genomics: Prolactin: A Key with Many Locks in Females and Males.*

Broody behavior in commercial turkeys involves in a pause or cessation in egg lay correlating with a surge in prolactin. Because of the decreased production, research efforts have investigated means to lower concentrations of prolactin in affected birds. Seasonal increases in the concentrations of plasma prolactin are believed to play a role in determining the end of the breeding season for birds breeding at temperate latitudes.

Mammary gland feeding is advantageous to both the infant and the parent

Milk consists of variable amounts of water, triglyceride fat, the carbohydrate lactose (milk sugar), a number of proteins, vitamins, and the minerals calcium and phosphate. For example, because the period of lactation is so short in the Weddell seal, the percentage of energy-rich fat in milk can be as high as 42%, whereas in the mare, which has a prolonged period of lactation, it can be as low as 1.3%. Human milk contains significant amounts of *omega-3* fatty acids (p. 72), which may promote brain development. Protein content of milk is high in the Weddell seal (15.8%) and low in primates (0.8% in humans and 1.5% in other primates). Remarkably, given the differing ages of the neonates undergoing development in the brood patch of marsupials, each of four mammary glands can produce a milk of differing composition!

In addition to supplying nutrients, milk contains a host of immune cells, antibodies, and other chemicals that help protect the infant against infection until it can mount an effective immune response on its own a few months after birth. **Colostrum,** the milk produced for the first five days postpartum, contains lower concentrations of fat and lactose but higher concentrations of protein and immunoprotective components. In many mammals some passive immunity is transferred from the female to the fetus during gestation through the passage of antibodies across the placenta (see p. 480). These antibodies are short-lived, however, and often do not persist until the infant can fend for itself immunologically. Mammary-gland-fed young gain additional protection during this vulnerable period through a variety of mechanisms:

- Mammary gland milk—especially in colostrum—contains an abundance of immune cells—both B and T lymphocytes, macrophages, and neutrophils (see p. 480)—that produce antibodies and destroy pathogenic microorganisms outright.
- **Secretory IgA,** a special type of antibody, is present in great amounts in mammary gland milk. Secretory IgA consists of two IgA antibody molecules joined together with a so-called secretory component that helps protect the antibodies from destruction by the infant's acidic gastric juice and digestive enzymes. The collection of IgA antibodies ingested by a mammary-gland-fed young is specifically targeted against the particular pathogens in the environment of the female—and, accordingly, of the infant as well.
- Some components in milk, such as mucus, adhere to potentially harmful microorganisms, thus preventing them from attaching to and crossing the intestinal mucosa.
- **Lactoferrin** is a mammary-gland milk constituent that thwarts growth of harmful bacteria by decreasing the availability of iron, a mineral needed for multiplication of these pathogens (see p. 469).
- **Bifidus factor** in mammary gland milk, in contrast, promotes multiplication of the nonpathogenic microorgan-

As you have seen, prolactin is classically known for its role in milk production (hence its name). But it has many other roles, some only recently discovered. Recall that a particular chemical messenger can have multiple, sometimes unrelated roles because of different receptor systems on different target cells ("same key, different locks," p. 93). Prolactin is a prime example because its roles have diversified during vertebrate evolution. First, it is an ancient hormone found in all vertebrates, and its genetic codes show that it probably evolved along with growth hormone (GH) from an even more ancient ancestral gene. In fishes and amphibians, prolactin is involved in regulating osmotic balance, particularly in retaining salt in these animals' adaptation to freshwater (p. 627). This is a function that some physiologists suspect occurs in mammals: It reduces Na$^+$ and K$^+$ output in the urine and content in sweat, and increases uptake of these ions from the intestines.

As you have seen for birds and mammals in this chapter, prolactin signaling evolved another set of functions involved in reproduction. These functions are not only those for milk and parental behavior but also are for early stages of reproduction. Female knock-out mice lacking prolactin receptors cannot reproduce, because the uterus does not allow embryonic implantation. The reason is that the corpus luteum undergoes apoptosis (programmed death) in the absence of prolactin stimulation; in turn, the corpus luteum does not make enough progesterone to prepare the uterus.

Males also use prolactin in reproduction. It may trigger paternal behavior: For example, prolactin is high during brooding in male birds that help care for eggs and hatchlings; and California mice (*Peromyscus californicus*), unusual rodents because males help care for the young, have high prolactin levels and more prolactin receptors in their hypothalamus than do other rodents. Also, Tillmann Krüger and colleagues have found that prolactin levels surge during orgasms in both human sexes and that prolactin is associated with feelings of relaxation and satisfaction. In males, it inhibits erection in postcoitus recovery.

Prolactin also has developmental roles in all vertebrates. Its relation to GH is seen in fishes and amphibians that undergo metamorphoses: Prolactin appears to oppose metamorphosis and enhance growth of the larval stage. Yet another developmental role is found in fishes, adult amphibians, reptiles, birds, and some mammals: Prolactin stimulates various aspects of skin development (e.g., molting in reptiles and the defeathering of bird brood patches). Furthermore, prolactin receptors have been found in numerous other nonreproductive cells in mammals, where their role is often uncertain. One example is the mammalian lymphocyte system: T and B cells (p. 374) have prolactin receptors and can also produce prolactin. Mice in which prolactin binding is blocked have weak immune responses and are thus more susceptible to disease.

Prolactin also violates the dogma for peptide-hormone signal transduction. Like all peptide hormones, prolactin binds to target cell receptors on the plasma membrane to exert its effects (a *tyrosine kinase* system; see p. 95). The dogma is that such lipophobic hormones always work this way, whereas lipophilic hormones such as steroids work by binding to internal receptors that directly regulate gene expression (p. 274). But now many exceptions to these "rules" have been found. Progesterone, for example, is a steroid that has been found to bind to both membrane and internal receptors. Prolactin has been found to bind to a protein called *cyclophilin B* (*CypB*) at the membrane of breast cancer cells, but then moves to the nucleus with CypB to activate genes. Knock-down of CYPB with RNA interference (p. 33) has shown that these genes are involved in cell growth, motility and tumorogenesis. In this way, prolactin may be a major culprit in the spread of breast cancer cells in human females.

---

ism *Lactobacillus bifidus* in the infant's digestive tract. Growth of this harmless bacterium helps crowd out potentially harmful bacteria.
- Other components in mammary gland milk promote maturation of the young's digestive system so that it is less vulnerable to diarrhea-causing bacteria and viruses.
- Still other factors in mammary gland milk hasten the development of the infant's own immune capabilities.

Toothed whales (*Odontoceti*) wean their young any time from 2 to 13 years, much later than other marine and terrestrial mammals. It has been observed that reproductively senescent pilot whales, whose weaning period lasts four to five years, are still lactating. In many species a resumption of estrus does not occur until after the young are weaned. Suckling suppresses the estrous cycle by inhibiting LH and FSH secretion, probably through inhibition of GnRH, thus permitting all the female's resources to be directed toward the newborn instead of being shared with a new embryo. This becomes a very important mechanism for spacing births.

Cessation of Milk Production at Weaning After weaning, two mechanisms contribute to the cessation of milk production. First, without suckling, prolactin secretion is not stimulated, thus removing the primary stimulus for continued milk synthesis and secretion. Also because of the lack of suckling, milk letdown does not occur in the absence of oxytocin release. Because milk production does not immediately shut down, milk accumulates in the alveoli, engorging the mammary glands. The resultant pressure buildup acts directly on the alveolar epithelial cells to suppress further milk production. Cessation of lactation at weaning therefore results from a lack of suckling-induced stimulation of both prolactin and oxytocin secretion.

The end is a new beginning

Reproduction is an appropriate way to end our discussion of physiology from genes to organisms. The single cell resulting from the union of male and female gametes divides mitotically and differentiates into a multicellular individual made

Paul Yancey

A newborn animal represents a new beginning for all the processes of physiology.

up of a number of different organ systems that interact cooperatively to maintain homeostasis and regulate useful changes. All these life-supporting processes discussed throughout this book begin all over again at the start of a new life and serve to perpetuate a particular species.

check your understanding 16.7

What factors contribute to the initiation of parturition? What is the role of oxytocin?

Describe the hormonal factors that play a role in lactation.

making connections

How Reproductive Systems Contribute to the Body as a Whole

The reproductive system is unique in that it is not essential for individual homeostasis or for survival of the individual, but it is essential for sustaining the thread of life from generation to generation. Reproduction depends on the union of male and female gametes (reproductive cells), each with a half set of chromosomes, to form a new individual with a full, unique set of chromosomes. Unlike the other body systems, which are essentially identical in the two sexes, the reproductive systems of males and females are remarkably different, befitting their different roles in the reproductive process.

In addition to basic reproduction, sexual recombination appears to be critical to the long-term survival of most eukaryotic species. The reasons are not fully understood, but probably involve purging of harmful mutations, faster adaptation to changing environments, and/or warding off parasites that have evolved to attack a specific genome.

Chapter Summary

- Reproduction depends on the union of male and female gametes (reproductive, or germ, cells), each with a single set of chromosomes (haploid), to form a new individual with a diploid (that is, twice haploid), unique set of chromosomes.

- Physiological classification of reproductive processes uses three categories: oviparous, meaning the parents release eggs from which the young hatch after expulsion from the body; ovoviviparous, the production of eggs (not necessarily with shells) that hatch within the body of the parent; and viviparous, the production of living young within the body.

- The primary reproductive organs, or gonads, of most vertebrates consist of testes in the male and ovaries in the female. In both sexes in vertebrates, the mature gonads are found in pairs and perform the dual function of (1) producing gametes (gametogenesis), that is, spermatozoa (sperm) in the male and ova (eggs) in the female, and (2) secreting sex hormones, primarily testosterone in males and estrogens (the major form being estradiol) and progesterone in females.

- The male reproductive system is designed to deliver sperm to the female reproductive tract in a liquid vehicle, semen, which is conducive to sperm viability. In contrast, the female's role in reproduction is often more complicated, especially if internal fertilization is involved.

- Nuclear division in the specialized case of gametes is accomplished by meiosis, in which only one of the two sets of homologous chromosomes (that is, a haploid number of chromosomes) is distributed to each of four new daughter cells. Each reproductive cell contains a half set of chromosomes.

- Sex differentiation in mammals depends on the presence or absence of masculinizing determinants during critical periods of embryonic development.

- Puberty in both sexes is a gradual process and not a single event and is recognized as the attainment of sexual maturity and the ability to reproduce.

- Testosterone, usually acting in the dihydro form, governs the growth and maturation of the entire male reproductive system. Ongoing testosterone secretion is essential for spermatogenesis and for maintaining a mature male reproductive tract throughout adulthood.

- A spermatozoon has three parts: a head capped with an acrosome, a midpiece, and a tail. Throughout their development, vertebrate sperm remain intimately associated with Sertoli cells.

- Secretion of both LH and FSH from the anterior pituitary is stimulated in turn by gonadotropin-releasing hormone (GnRH) or inhibited by gonadotropin-inhibiting hormone (GnIH).

- The ducts of the reproductive tract store and concentrate sperm and increase their fertility. During ejaculation, the accessory sex glands contribute secretions that provide support for the continuing viability of the sperm inside the female reproductive tract (in animals with internal fertilization) or in water (such as fish semen).

- In those species in which the penis consists almost entirely of erectile tissue, the body consists of three columns of spongelike vascular spaces extending the length of the organ.

- Female mammals have two interlinked cycles: ovarian and uterine. Together, they form an estrous cycle in most mammals, the time from one period of sexual receptivity to the next. Receptivity is called estrus, or heat, occurring under certain specified conditions. Some primates have a menstrual cycle in which the uterine lining is periodically shed rather than reabsorbed as in estrous. Most animals are spontaneous ovulators that ovulate with a regular frequency and do not require copulation.

- Oogenesis contrasts sharply with spermatogenesis in several important aspects, even though the identical steps of chromosome replication and division take place during gamete production in both sexes. The undifferentiated primordial germ cells in the fetal ovaries, the oogonia (comparable to the spermatogonia), divide mitotically during gestation and/or neonatally, after which time mitotic proliferation ceases.

- After the onset of puberty, the ovary alternates between two phases: the follicular phase, which is dominated by the presence of maturing follicles, and the luteal phase, which is characterized by the presence of the corpus luteum.

- Just before ovulation, the oocyte completes its first meiotic division. The ovum (secondary oocyte), still surrounded by its zona pellucida and granulosa cells, is swept out of the ruptured follicle into the infundibulum (which, in most species, surrounds the ovary) by the leaking antral fluid.

- Rupture of the follicle at ovulation signals the end of the follicular phase and ushers in the luteal phase. Old follicular cells soon undergo a dramatic structural transformation to form the corpus luteum, in a process called luteinization. The follicular-turned-luteal cells hypertrophy and are converted into very active steroidogenic tissue.

- As with males, sexual maturation (puberty) in mammalian females is a long progression leading to the time when the hypothalamus is sufficiently mature to allow the next follicle to produce enough estrogens to elicit the first LH surge.

- Fertilization is the union of male and female gametes. Conception can take place during a limited time span in each cycle, the fertile period. In mammals, copulation involves stimulation of erectile tissues (penis, clitoris), delivery of sperm into the vagina by ejaculation, and orgasm in both sexes.

- The mammalian blastocyst implants in the endometrium through trophoblastic enzymes. To sustain the growing embryo for its intrauterine life, the placenta, a specialized organ of exchange between the maternal and fetal blood, is rapidly developed. The placenta secretes a number of peptide and steroid hormones essential for maintaining pregnancy. The most important are chorionic gonadotropin, estrogen, and progesterone.

- Parturition (labor, delivery, or birth) involves the uterus expelling the fetus and placenta. Pregnancy duration and delivery timing are determined largely by the placenta's rate of CRH production and lung maturation. Labor is divided into three stages: (1) cervical dilation, (2) delivery of the young, and (3) delivery of the placenta.

- The mammalian female continues to nourish the newborn. Suckling concurrently stimulates milk ejection and milk production, therefore ensuring that the rate of milk synthesis keeps pace with the young's needs for milk. Milk consists of variable amounts of water, triglyceride fat, the carbohydrate lactose (milk sugar), a number of proteins, vitamins, and the minerals calcium and phosphate.

Review, Synthesize, and Analyze

1. What feature of the reproductive system makes it unique from each of the other body systems? (*Hint:* Consider the survival of an individual versus a species.)

2. Of what functional significance is the scrotal location of the testes?

3. Occasionally, testicular tumors composed of interstitial cells of Leydig may secrete up to 100 times the normal amount of testosterone. When such a tumor develops in young animals, they grow up much shorter than their genetic potential. Explain why. What other symptoms would be present?

4. The hypothalamus releases GnRH in pulsatile bursts once every two to three hours, with no secretion occurring in between. The blood concentration of GnRH depends on the frequency of these bursts of secretion. A promising line of research for a new method of contraception involves administration of GnRH-like drugs. In what way could such drugs act as contraceptives when GnRH is the hypothalamic hormone that triggers the chain of events leading to ovulation? (*Hint:* The anterior pituitary is "programmed" to respond only to the normal pulsatile pattern of GnRH.)

5. How are the ovum and spermatozoa transported to the site of fertilization? Describe the process of fertilization.

6. Compare oogenesis with spermatogenesis.

7. Describe the events of the follicular and luteal phases of the ovarian cycle. Correlate the phases of the uterine cycle with those of the ovarian cycle.

8. Explain the physiologic basis for administering a posterior pituitary extract to induce or facilitate labor.

9. Discuss why sexual reproduction is more widespread than asexual reproduction among animals.

Suggested Readings

Bole-Feysot, C., V. Goffin, M. Edery, N. Binart, & P. A. Kelly. (1998). Prolactin (PRL) and its receptor: Actions, signal transduction pathways and phenotypes observed in PRL receptor knockout mice. *Endocrine Reviews* 19:225–268.

Griffiths, R. (2000). Sex identification using DNA markers. In A. J. Baker, ed., *Molecular Methods in Ecology.* London: Blackwell Science.

Gubernick, D. J. (1990). Prolactin and paternal behavior in the biparental California mouse, *Peromyscus californicus. Hormones and Behavior* 23:203–210.

Hau, M. (2001). Timing of breeding in variable environments: Tropical birds as model systems. *Hormones and Behavior* 40:281–290.

Johnson, A. L. (2000). Reproduction in the female. *Sturkie's Avian Physiology,* 5th ed., ed. G. C. Whittlow. San Diego: Academic Press.

Knobil, E., & J. D. Neill, eds. (2006). *Physiology of Reproduction,* 3rd ed. New York: Academic Press.

Lincoln, G. A., & R. V. Short. (1980). Seasonal breeding: Nature's contraceptive. *Recent Progress in Hormone Research.* 36:1–52.

Packer, C., M. Tatar, & A. Collins. (1998). Reproductive cessation in female mammals. *Nature* 392:807.

Patricelli, G. L., J. A. C. Uy, G. Walsh, & G. Borgia. (2002). Sexual selection: Males adjust displays in response to female signals. *Nature* 415:279–280.

Reece, W. O. (1997). *Physiology of Domestic Animals,* 2nd ed. Baltimore: Williams and Wilkins.

Ross, R. M. (1990). The evolution of sex-change mechanisms in fishes. *Environmental Biology of Fishes* 29:81–93.

Senger, P. L. (2003). *Pathways to Pregnancy and Parturition,* 2nd ed. Current Conceptions, Inc. Washington State University Research & Technology Park. Pullman, WA. Covers the principles of reproduction in food-producing animals.

Wilhelm, D., S. Palmer, & P. Koopman. (2007). Sex determination and gonadal development in mammals. *Physiological Reviews* 87:1–28.

Zimmer, C. (2009). On the origin of sexual reproduction. *Science* 324:1254–1256.

Access an interactive eBook, chapter-specific interactive learning tools, including flashcards, quizzes, videos and more in your Biology CourseMate, accessed through **CengageBrain.com**.

Answers to Check Your Understanding Questions

check your understanding 1.1

- Proximate explanations in physiology answer the question, "How does it work?" Evolutionary ones answer the question, "How did it get to be this way?" through history. For example, a proximate description is that the human spine is a movable support device for the upper body and a protective conduit for the major nerve cords. The evolutionary explanation addresses our spine's flaws by analyzing its ancestral origin as a horizontal flexible swimming device in fishes.
- Adaptations are beneficial features that enhance overall survival of the species. They are not always optimal because they are constrained by evolutionary history.
- "For a large number of problems there will be some animal of choice, or a few such animals, on which it can be most conveniently studied." For example, to study basic neuron functions, John Zachary Young chose the giant squid axon because it was much easier to pierce with electrodes than the much smaller mammalian neurons.

check your understanding 1.2

- (1) Ask a question about nature. (2) Propose alternative hypotheses to explain the phenomenon. (3) Design experiments or observations that test the hypotheses by making testable predictions. (4) Conduct the observations or experiments. (5) Using the outcome of these tests, refine the earlier questions and hypotheses and design new tests.

check your understanding 1.3

- The basic function of life include self-organization, self-regulation, self-support and movement, and self-reproduction (see details on p. 7).
- The four major types of tissue are (1) epithelial—linings and endocrine and exocrine glands; (2) connective tissue—loose, tendons, bone, blood or hemolymph; (3) muscular—skeletal, cardiac, smooth; (4) nervous.
- Organs form from two or more tissues, e.g., an epithelium forming lining with glands, and an underlying connective tissue "glue." Organs may also have muscles for movement with nerves to control them and the glands.

check your understanding 1.4

- Surface area increases as the square of the radius, whereas volume increases as the cube. Thus, to obtain nutrients and get rid of wastes through the surface, a larger organism is at a disadvantage because of its smaller surface-area-to-volume ratio. But to retain heat, the larger organism has an advantage.

check your understanding 1.5

- *Homeostasis* means "similar state," the maintenance of relatively consistent internal conditions. This allows internal structures and functions to operate at their optima.

- Negative feedback occurs when a change in a controlled variable triggers a response that opposes the change, driving the variable in the opposite direction of the initial change. A referenced system has a sensor to measure the variable, an integrator to make decisions using a set point for the desired state, and effectors to make corrections. An example is the thermostat system for home heating/cooling. An antagonistic system has opposing effectors, such as a furnace and air conditioner, for better control.
- Anticipation activates corrective responses of negative feedback before a disturbance occurs, so there is no delay in the correction. Acclimatization improves a component of the system so that it works better under new conditions.

check your understanding 1.6

- Reset systems raise or lower the set point of a negative-feedback system to a new level, which can be useful for a new situation such as a fever. Positive-feedback systems rapidly accentuate changes rather than opposing them; this can be useful when a rapid response is needed, such as blood clotting.

check your understanding 1.7

- Intrinsic controls are those regulated by a single tissue or organ on its own; extrinsic controls are regulatory mechanisms initiated outside an organ to alter its activity. For example, a mammalian heart beats regularly due to its own internal pacemaker, but nerves and hormones can speed it up or slow it down to meet demands of the whole body.
- Many simple local needs of organs, such as blood flow, can be more rapidly and precisely adjusted if they do not have to wait for signals to and from distant regulators.
- The functional systems are based on organ contributions to the entire organism, based on specific body needs such as circulation, respiration, etc. Thus, the systems approach allows us to study how the involved organs all work together and what processes they have in common.

check your understanding 2.1

- A eukaryotic cell's three major subdivisions are the (1) Plasma membrane, (2) cytoplasm, and (3) nucleus.
- Flexibility is necessary for cell shape changes (e.g., passage of a red blood cell through a capillary) and for many transport proteins to move molecules through membranes.

check your understanding 2.2

- In the process of transcription, the gene is transcribed into a pre-messenger RNA molecule by an enzyme called RNA polymerase. This pre-mRNA is a complementary copy of the gene that contains the coding and noncoding sequences. The pre-mRNA, in turn, is processed into a final messenger RNA by removal of the dispersed introns and ligation of the exon. This step is followed by the addition of noncoding signal sequences to the leading and trailing ends of the molecule. The mature mRNA can now exit the nucleus through its membrane pores.
- Different cell types transcribe different sets of genes and thus synthesize different proteins for their specialized roles. In addition, many genes are active only under certain conditions or life stages.
- Differential gene expression is accomplished through two levels of control: (1) regulation of individual genes with promoters and transcription factors and (2) regulation of transcription factors in different tissues and at different stages. See also the discussion on pp. 31–32.

check your understanding 2.3

- Gene therapy could potentially be used to treat a defect arising from a mutated gene by inserting a normal gene.
- The discovery of iPSCs demonstrates the ability to rejuvenate adult cells by the introduction of various genes that are normally active during the embryonic stage of development.
- Cloning refers to the process by which the nucleus of a fertilized egg is removed and destroyed and a carefully selected nucleus from an adult cell is injected in its place. In both approaches, factors synthesized within the egg somehow rewind the internal clock of the adult genetic material and restore it to the pluripotent state, much like the response of an adult cell to the introduction of embryonic genes or gene products.

check your understanding 2.4

- The ER is an elaborate, fluid-filled membranous system distributed extensively throughout the cytosol, where it is primarily a protein-manufacturing factory. The rough ER consists of stacks of relatively flattened interconnected sacs, whereas the smooth ER is a meshwork of tiny interconnected tubules (Figure 2-9). The rough ER and its ribosomes synthesize and release a variety of new proteins into the ER lumen, whereas the smooth ER serves primarily as a central packaging and discharge site for molecules that are to be transported from the ER. Newly synthesized proteins and lipids pass from the rough ER to gather in the smooth ER, where they are packaged for transport out of the cell.

check your understanding 2.5

- Finished proteins are pinched off the Golgi complex to form a membrane-enclosed vesicle. Each distinct type of vesicle takes up a specific product before budding off. Vesicles with their selected cargo destined for different intracellular sites are wrapped in membranes containing distinctly different surface proteins. For extracellular transport, numerous large secretory vesicles, which contain proteins to be secreted, bud off from the Golgi stacks and, on the appropriate signal, move to the cell's periphery, fuse with the plasma membrane, and empty the contents to the outside.

- Before budding off from the outermost Golgi sac, the portion of the Golgi membrane that will be used to enclose the secretory vesicle becomes "coated" with a layer of specific proteins from the cytosol. Coat proteins, in turn, bind with another specific protein facing the outer surface of the membrane. The linking of these coat proteins causes the surface membrane of the Golgi sac to curve and form a dome-shaped bud around the captured cargo. Eventually, the surface membrane closes and pinches off the vesicle. Golgi vesicles with their selected cargo destined for different sites also contain distinctly different surface proteins or docking markers. These vesicles can then "dock" and "unload" selected cargo only at the appropriate docking-marker acceptor, a protein located only at the proper destination within the cell.

check your understanding 2.6, 2.7

- Lysosomes remove worn out organelles as well as fuse with aged or damaged organelles to remove parts of the cell. This selective self-digestion makes way for new replacement parts. Lysosomes also play an important role in tissue regression, such as during the normal reduction in the uterine lining following pregnancy. Another important intracellular digestive apparatus is the proteasome, which has the job of taking in internal cell proteins and chopping them up into reusable amino acids, notably those proteins that are no longer useful. Peroxisomes are similar to lysosomes in that they are membrane-enclosed sacs containing enzymes, but unlike the lysosomes, which contain hydrolytic enzymes, peroxisomes house several powerful oxidative enzymes and contain most of the cell's catalase.
- Antioxidants inactivate potentially harmful free radicals before they can damage the cell or its functions.

check your understanding 2.8

- The density of mitochondria in muscle fiber varies greatly, depending on energy needs. The greater the energy expenditure, the greater the mitochondrial content. For example, animals that are continually active (e.g., flying animals) contain muscle fibers with high densities of mitochondria, whereas those that are mostly sessile have low mitochondrial content.
- End products from mitochondrial metabolism include H_2O and free radicals (reactive oxygen species). Free radicals react readily with other molecules, either acquiring or giving up an electron to achieve stability.
- Mitochondria are enclosed by a double membrane—a smooth outer membrane that surrounds the mitochondrion itself and an inner membrane that forms a series of infoldings called cristae, which project into an inner cavity filled with a gel-like solution known as the matrix (Figure 2-16). Cristae contain crucial proteins that ultimately are responsible for converting much of the energy in food into a usable form (the electron transport proteins). The folds of the inner membrane greatly increase the surface area available for housing these important proteins. The matrix consists of a concentrated mixture of hundreds of dif-

ferent dissolved enzymes (the citric acid cycle enzymes) that are important in respiration.
- Oxygen deficiency forces cells to rely on glycolysis and other anaerobic reactions. Most animals can rely on anaerobic metabolism for only relatively brief periods. However, some species of organisms—the facultative anaerobes—can adapt to anaerobic conditions for periods ranging from days to months by substituting other electron acceptors for oxygen or by entering dormant states.

check your understanding 2.9, 2.10

- Vaults are noted for their octagonal shape that fit precisely into the nuclear pores. Because they are hollow, it is believed that they transport specific molecules such as mRNA or ribosome subunits synthesized in the nucleus and transport them to other sites in the cell.
- In vertebrate adipose tissue, the stored triglycerides can occupy almost the entire cytosol, coalescing to form one large fat droplet as compared with other cells in which the fat appears as small droplets known as inclusions. Fats and sugars are stored as triglycerides and glycogen, respectively.

check your understanding 2.11, 2.12

- Microtubules are one of the components of the cytoskeleton; they anchor many of the membranous organelles and serve as "highways" along which vesicles are transported within the cell by "molecular motors." They are also essential for maintaining an asymmetric shape, such as that of a nerve cell. During cell division, the centrioles form a mitotic spindle out of microtubules to direct movement of chromosomes. Microfilaments serve at least two functions: (1) They play a vital role in various cellular contractile systems, and (2) they act as mechanical stiffeners for several specific cellular projections.
- Microtubules are essential for maintaining an asymmetric cell shape, such as that of a nerve cell, whose elongated axon may extend 1 meter in length or more in a large vertebrate, from the origin of the cell body in the spinal cord to the termination of the axon at a muscle.

check your understanding 2.13

- Cells subjected to considerable stress rely on (1) desmosomes, which are composed of dense, buttonlike cytoplasmic thickenings known as plaque located on the inner surface of each of the two adjacent cells, and (2) strong glycoprotein filaments containing cadherins (a type of CAM) that extend across the space between the two cells and attach to the plaque on both sides.
- Electrical activity in one cell can be transmitted to another through gap junctions: small connecting tunnels known as connexons.

check your understanding 3.1

- Phospholipids have a polar head containing a negatively charged phosphate group and two nonpolar fatty acid tails. When in contact with water (Figure 3-2b), these two-sided molecules self-assemble into a lipid bilayer, a double layer of lipid molecules. The hydrophobic tails of phospholipids bury themselves in the center away from the water, whereas the hydrophilic heads line up on both sides in contact with the water. The membrane's hydrophobic interior limits the passage of polar molecules either from the ECF or the ICF.
- A certain degree of membrane fluidity is required, for example, for the proper action of transport and channel proteins and for cell shape changes. Cholesterol contributes to both

the fluidity and the stability of the membrane. By being tucked in between the phospholipid molecules, cholesterol molecules prevent the fatty acid chains from packing together and crystallizing, a process that would drastically reduce membrane fluidity. Cholesterol provides some rigidity to the membrane and can counter some of the detrimental effects of elevated temperatures. Second, the addition of double bonds into the fatty acid molecule results in chains that do not stack together well and, therefore, do not become too rigid in the cold, thus offsetting the effects of low temperatures on fluidity.
- A variety of different proteins within the plasma membrane serve the following specialized functions: channels, carriers, receptors, docking-marker acceptors, enzymes, CAMs, and self-identity markers.
- The sugar chains on the outer-membrane surface serve as self-identity markers that enable cells to identify and interact with each other in the following ways: (1) The unique combination of sugar chains projecting from the surface membrane serves as the "trademark" of a particular cell type, enabling a cell to recognize others of its own kind; and (2) carbohydrate-containing surface markers are also involved in tissue growth, which is normally held within certain limits of cell density.

check your understanding 3.2

- Diffusion occurs if a substance can move from a region of higher concentration to a region of lower concentration. Diffusion assumes that each molecule moves separately and randomly in any direction. Osmosis is the net movement of water through a membrane down its concentration gradient.
- The movement of ions (electrically charged particles that have either lost or gained an electron) down an electrical gradient is not an example of diffusion because they are not randomly moving in any direction but are rather conducted along an electrical gradient.
- If ECF solute concentrations increase, water moves osmotically down its concentration into the cell, causing it to swell. If ECF solute concentrations decrease, water moves down its concentration out of the cell, causing it to shrink.
- Factors in addition to the concentration gradient influence the rate of net diffusion across a membrane. See Table 3-1 and the discussion on pp. 76–77.

check your understanding 3.3

- Aquaporins are channel proteins evolved specifically for the passage of water molecules. They are found in all membranes where rapid water movement is necessary.
- Facilitated diffusion uses a carrier to facilitate the transfer of a particular substance across the membrane "downhill" from high to low concentration. This process is passive and does not require energy because movement occurs naturally down a concentration gradient. Active transport, in contrast, requires the carrier to expend energy to transfer its passenger "uphill" against a concentration gradient, from an area of lower concentration to an area of higher concentration. Both forms of transport do not directly require energy; although with secondary active transport, energy is required in the entire process but is not directly required to run the pump. See Figures 3-11 and 3-15 and the discussion on p. 84.
- The Na^+/K^+ ATPase pump transports Na^+ out of the cell, concentrating it in the ECF of multicellular organisms, and picks up K^+ from the

outside, concentrating it in the ICF (Figure 3-14). The Na$^+$/K$^+$ pump maintains Na$^+$ and K$^+$ concentration gradients across the plasma membrane of all cells. It helps regulate cell volume, and the energy used to run the Na$^+$/K$^+$ pump also indirectly serves as the energy source for the cotransport of glucose and amino acids across intestinal and kidney cells.

- Vesicular transport requires energy expenditure by the cell, so this is an active method of membrane transport. Energy is needed to accomplish vesicle formation and vesicle movement within the cell.

check your understanding 3.4

- Signal transduction is the process by which instructions from extracellular chemical messengers are conveyed to the target cell's interior for execution. Lipophilic extracellular messengers gain entry into the cell by dissolving in and passing through the lipid bilayer of the target cell's plasma membrane, whereas lipophobic messengers first bind with surface membrane receptors specific for that given messenger. The desired intracellular response is achieved by only three general means: (1) by opening or closing specific channels in the membrane to regulate the movement of particular ions into or out of the cell, (2) by activating an enzyme that phosphorylates a cell protein, or (3) by transferring the signal to an intracellular chemical messenger (the second messenger), which, in turn, triggers a preprogrammed series of biochemical events within the cell. See also Figure 3-17 and the general discussion on pp. 93–95.
- Deliberate programmed cell death is termed *apoptosis*. A cell signaled to commit suicide detaches itself from its neighbors and decreases in size. The suicidal cell activates a cascade of normally inactive intracellular protein-cutting enzymes, the caspases, which destroy the cell from within. Once unleashed, the caspases act like molecular scissors to systematically dismantle the cell. Snipping protein after protein, they chop up the nucleus, disassembling its life-essential DNA, then break down the internal shape-holding cytoskeleton, and finally fragment the cell itself into disposable membrane-enclosed packets.

check your understanding 3.5

- The simultaneous existence of an electrical gradient and concentration (chemical) gradient for a particular ion is called an electrochemical gradient, which contributes to the electrical properties of the plasma membrane as a result of the unequal separation of both ions and charges.
- The force that keeps K$^+$ high inside the cell is due to the activity of the Na$^+$/K$^+$ pump. This active transport mechanism counterbalances the rate of leakage (Figure 3-26). At resting potential, the pump transports K$^+$ back into the cell, essentially the same number of K$^+$ ions that leaked out.
- For the same reason, the force that keeps Na$^+$ low inside the cell is the result of the activity of the Na$^+$/K$^+$ pump. The pump effectively transports to the outside the Na$^+$ ions that leaked in. In addition, although Na$^+$ continually leaks inward down its electrochemical gradient, it does so only slowly because of its low permeability, that is, because of the scarcity of Na$^+$ leak channels.

check your understanding 4.1

- Two types of cells, neurons and muscle cells, can undergo transient, rapid changes in their membrane potentials. These fluctuations in potential serve as electrical signals.

- *Polarization:* Charges are separated across the plasma membrane, so the membrane has potential to do work. Any time the value of the membrane potential is not 0 mV, in either the positive or negative direction, the membrane is in a state of polarization. *Depolarization:* A change in potential that makes the membrane less polarized than at resting potential. *Hyperpolarization:* A change in potential that makes the membrane more polarized (more negative) than at resting potential. *Repolarization:* The membrane returns to resting potential after having been depolarized. *Resting potential:* The constant membrane potential present when a cell is electrically at rest, that is, not producing electrical signals.
- Ions responsible for carrying charge can cross the membrane only through channels specific to them. There are two types of channels: (1) leak channels, or nongated channels, which are open all the time, and (2) gated channels, which can be opened or closed in response to specific triggering events.

check your understanding 4.2

- Graded potentials are produced by a triggering event that causes gated Na$^+$ ion channels to open in a specialized region of the excitable cell membrane. The magnitude of the graded potential decreases the farther it moves away from the initial active area.
- The passive electrotonic flow of current along a membrane does not require the opening of channels and is therefore faster than current flow generated by action potentials, which require the opening of Na$^+$ voltage-gated channels. The passive movement of the electrical depolarization can be described by Ohm's law, where ΔV_m is the actual voltage change across the membrane; ΔI, the amount of current (amperes); and R, the resistance encountered by the current passing along the membrane (ohms).

check your understanding 4.3

- *Threshold potential:* The critical potential that must be reached before an action potential is initiated in an excitable cell. *Action potential:* A brief, rapid, large change in membrane potential that serves as a long-distance electrical signal in an excitable cell. *Refractory period:* The time period when a recently activated patch of membrane is refractory to further stimulation, which prevents the action potential from spreading backward into the area through which it has just passed and ensures the unidirectional propagation of the action potential away from the initial site of activation. *All-or-none law:* An excitable membrane either responds to a stimulus with a maximal action potential that spreads nondecrementally throughout the membrane or does not respond with an action potential at all.
- A neuron consists of three basic parts: the cell body, dendrites, and axon. The dendrites and cell body are the neuron's input zone, because these components receive and integrate incoming signals. Action potentials are then conducted along the axon from the axon hillock to the typically highly branching endings at the axon terminals. Functionally, therefore, the axon is the conducting zone of the neuron, whereas the axon terminals constitute the output zone.
- Saltatory conduction propagates action potentials more rapidly than does conduction by local current flow, because the action potential leaps over myelinated sections, whereas it must be regenerated within every section of an un-

myelinated axonal membrane from beginning to end. Contiguous conduction thus involves the spread of the action potential along every patch of membrane down the length of the axon. An increase in fiber diameter increases the speed of propagation in both unmyelinated and myelinated fibers.

check your understanding 4.4

- Neurotransmitters are very short-range chemical messengers, which diffuse from their site of release across a narrow extracellular space to act locally on only an adjoining target cell, which is usually another neuron, muscle, or gland. Neurotransmitters come in several chemical classes of signal molecules, but the most common ones are amines.
- Synapses typically operate in one direction only so that false messages do not inadvertently travel from effectors to integrators; that is, the presynaptic neuron brings about changes in membrane potential of the postsynaptic (target) membrane, but the postsynaptic membrane does not influence the potential of the presynaptic neuron.
- Chemical synapses involve the release of neurotransmitters followed by the opening and closing of channels on the postsynaptic membrane, which requires time, whereas action potential waveforms are essentially transmitted across electrical synapses unperturbed, moving along the axon as if the synapse wasn't present.

check your understanding 4.5

- See Figure 4-16 and the discussion on pp. 130–132.
- Neurotransmitters usually function by changing the conformation of chemically gated channels in so-called fast synapses. Another mode of synaptic transmission used by some neurotransmitters, such as serotonin, involves the activation of intracellular second messengers, such as cAMP, within the postsynaptic neuron. Synapses that lead to responses mediated by second messengers are known as slow synapses, because these responses take longer and often persist for longer periods (because of second-messenger cascades) than those accomplished by fast synapses.
- At an excitatory synapse, the response to the neurotransmitter–receptor combination is the opening of nonspecific cation channels within the subsynaptic membrane that permit simultaneous passage of Na$^+$ and K$^+$ through them. In contrast, at an inhibitory synapse, the combination of a chemical messenger with its receptor site increases the permeability of the subsynaptic membrane to either K$^+$ or Cl$^-$ by altering these ions' respective channel conformations.

check your understanding 4.6

- See Figure 4-18 and the discussion on p. 134.
- The muscle cells' electrical response is turned off by an enzyme present in the motor end-plate membrane, acetylcholinesterase, which inactivates ACh. This occurs because ACh binding to the receptor is reversible.
- Unlike most vertebrate neuromuscular systems, neurons that innervate arthropod muscles are both inhibitory as well as excitatory. For example, the neuromuscular systems regulating claw opening and closing in decapod crustaceans (such as crabs) contain both excitatory and inhibitory motor input. Inhibitory motor neurons also have endings on the presynaptic terminals of the excitatory motor neurons.

check your understanding 4.7

- Threshold potential is lowest at the axon hillock because this region has a much greater density of voltage-gated Na+ channels than anywhere else in the neuron. When summation of EPSPs takes place, the lower threshold of the axon hillock is reached first, so an action potential originating here is propagated to the end of the axon.
- A presynaptic axon terminal may itself be innervated by another axon terminal (Figure 4-21). Neurotransmitter released from a modulatory terminal binds with receptor sites on the axon terminal. This binding alters the amount of transmitter released in response to action potentials. If the amount of transmitter released from the axon terminal is reduced, the phenomenon is known as presynaptic inhibition. If the release of transmitter is enhanced, the effect is presynaptic facilitation.
- A neuron can influence the level of activity in many other cells by divergence of output. The term *divergence* refers to the branching of axon terminals so that a single cell synapses with and influences many other cells. A neuron may also have many other neurons synapsing on it. Such a relationship is known as convergence. Through this converging input, a single cell is influenced by thousands of other cells.

check your understanding 4.8

- Temperature and pressure affect action potential propagation acutely in unadapted animals. Temperature directly affects the rates of all reactions, including channel opening and closing, whereas high pressure inhibits the proper folding of proteins, reduces the movement of proteins during functional shape changes, and reduces membrane fluidity. The consequences of high pressure include a reduction in the activity of the Na+/K+ pump, a decrease in the rate at which an action potential is propagated, a decrease in the rate of axonal repolarization, an increase in axon potential duration, and an inhibition of excitatory synaptic transmission because of reduced neurotransmitter release from the presynaptic terminal and a reduced binding of ACh to its receptor on the muscle membrane. High temperature can reverse some of the effects of high pressure; that is, a moderate increase in temperature restores fluidity to the membrane and increases the activity of the Na+/K+ pump.
- The mechanisms by which toxins modify synaptic transmission between neurons include (1) altering the synthesis, axonal transport, storage, or release of a neurotransmitter; (2) modifying neurotransmitter interaction with the postsynaptic receptor; (3) influencing neurotransmitter reuptake or destruction; and (4) replacing a deficient neurotransmitter with a substitute transmitter.

check your understanding 5.1

- The minimum components of a reflex arc include a sensory neuron that controls an effector cell directly. Most reflex arcs have two or more neurons involved, with information being transmitted through one or more intermediate synapses (Figure 5-1b, c).
- Over evolutionary time in many animal groups, the interneuronal component progressively expanded and formed more complex interconnections, especially at the head end of the nervous system, forming a brain (cephalization). From the brain, many neural control pathways became routed through the centralized nerve cords (centralization). Associated with this evolution was the acquisition of more specialized

receptors for sensing the external environment and the routing of their information through the nerve cord to the brain for processing.
- Complex ganglionic nervous systems are characteristic of higher nonvertebrates because a centralized brain is not always necessary and may even be detrimental for some kinds of neural regulation. For this reason, ganglia in each segment of an arthropod have become specialized in coordinating the regional function of each body segment, that is, legs, wings, tail or abdomen, and structures on the head.
- The logarithm of brain weight as a function of body weight is typically linear for a variety of vertebrates, with the positive slope indicating that the overall size of an animal is related to its relative brain size (Figure 5-4). A larger brain is required of animals that must use their intellect to locate ripened fruits, compared with those that simply find leaves. Carnivores (who must hunt wary prey) also tend to have larger brains than herbivores.

check your understanding 5.2

- In vertebrates, the CNS consists of the brain and spinal cord, and the PNS consists of nerve fibers that carry information between the CNS and other parts of the body (the periphery) (Figure 5-5). The PNS also includes ganglia (also in a segmented pattern) that contain the cell bodies of most sensory neurons and some nonmotor effector neurons. The PNS is further subdivided into afferent and efferent divisions.
- Because each autonomic neurotransmitter and medullary hormone stimulate activity in some tissues but inhibit activity in others, the particular responses depend on the specific receptor types available for each neurotransmitter. For example, two types of acetylcholine (cholinergic) receptors—nicotinic and muscarinic—have been identified on the basis of their response to particular drugs (Table 5-4). There are also two major classes of adrenergic receptors for norepinephrine and epinephrine.

check your understanding 5.3

- Four major features help protect the vertebrate CNS from injury: (1) It is enclosed by hard, bony structures: the cranium (skull) encases the brain, and the vertebral column surrounds the spinal cord. (2) Three protective and nourishing membranes, the meninges, lie between the bony covering and the nervous tissue. From the outermost to the innermost layer, these are the dura mater, the arachnoid mater, and the pia mater (*mater,* "mother"). (3) The brain "floats" in a special cushioning fluid, the cerebrospinal fluid (CSF). (4) A highly selective blood–brain barrier limits access of bloodborne materials into the vulnerable brain tissue.
- Systemic capillaries are characterized by holes or pores between the cells making up the capillary wall, which permits free exchange of all plasma components, except the large plasma proteins, with the surrounding interstitial fluid. In the brain capillaries, the endothelial cells are joined by tight junctions, which completely seal the capillary wall so that nothing can be exchanged across the wall by passing between the cells (Figure 5-14).
- *Neural plasticity* refers to the ability of the nervous system to generate new synaptic connections, axonal and dendritic branches, and neurons. There are two main sites of neurogenesis in the adult brain. The first is the hippocampus (a key memory center), in addition to the subventricular zone (SVZ) associated with

the lateral ventricles of the brain. Cell division also can occur in the hypothalamus in response to seasonal and other environmental signals.

check your understanding 5.4

- The brainstem is the smallest and least changed region of the vertebrate brain (Table 5-6). It controls many of the life-sustaining processes, such as breathing and circulation—processes that have remained essentially unchanged during vertebrate evolution.
- The forebrain has changed the most over evolutionary time, in parallel with the dramatic increase in the number of interconnections, a reflection of this structure being the highest, most complex integrating area of the brain.

check your understanding 5.5

- A nerve is a bundle of peripheral nerve axons, some afferent and some efferent, enclosed by a connective tissue covering and following the same pathway, whereas a neuron is a nerve cell specialized to initiate, propagate, and transmit electrical signal, typically consisting of a cell body, dendrites, and axon.
- A reflex that is integrated at the spinal cord without involvement of the brain is termed a spinal reflex. Often neuronal connections that produce patterned or stereotyped movements find their origins in the spinal cord (e.g., rudimentary walking or swimming motion). In contrast, a fixed action pattern represents a short sequence of behavior that constitutes a unique part of the behavioral makeup of a species (e.g., courtship or threat display).
- Built into the withdrawal reflex is inhibition of the muscle that antagonizes (opposes) the desired response. This type of neuronal connection, involving stimulation of the nerve supply to one muscle and simultaneous inhibition of the nerves to its antagonistic muscle, is known as reciprocal innervation.

check your understanding 5.6

- Key functions of the brainstem include (1) the receipt and integration of all synaptic input via the reticular formation; (2) sensation input and motor output in the head and neck via cranial nerves; (3) reflex control of the heart, blood vessels, respiration, and digestion; (4) modulation of the sense of pain; and (5) housing of the centers responsible for sleep.
- The cerebellum compares the "intentions" or "orders" of the higher centers (especially the motor cortex) with the "performance" of the muscles and then corrects any "errors" or deviations from the intended movement even in the absence of feedback from sensors (Figure 5-25). The cerebellum even appears able to predict the position of a body part in the next fraction of a second and to make adjustments accordingly. These ongoing anticipatory adjustments are especially important for rapidly changing activities such as climbing, swinging through trees, and running, in which delays in sensory feedback would make the action clumsy.

check your understanding 5.7

- The basal nuclei's role in motor control include (1) inhibiting muscle tone throughout the body (proper muscle tone is normally maintained by a balance of excitatory and inhibitory inputs to the neurons that innervate skeletal muscles); (2) selecting and maintaining purposeful motor activity while suppressing useless or unwanted patterns of movement; and (3) helping monitor and coordinate slow, sustained contractions, especially those related to posture and support.

- The hypothalamus is the area of the brain most notably involved in regulating the direct homeostasis of the internal environment. The hypothalamus is an integrating center for many important homeostatic functions, including (1) controlling body temperature (although this ability varies considerably among vertebrates); (2) controlling thirst and urine output; (3) controlling food intake; (4) controlling anterior pituitary hormone secretion; (5) synthesizing posterior pituitary hormones; (6) controlling uterine contractions and milk ejection; (7) serving as a major autonomic nervous system coordinating center, which, in turn, affects all smooth muscle, cardiac muscle, and exocrine glands; (8) playing a role in emotional and behavioral patterns; and (9) participating in the sleep–wake cycle.

- The amygdala of the mammalian limbic system processes inputs that give rise to the sensation of fear, a sense that normally helps animals avoid danger. In this region, a link between an unconditioned stimulus and a conditioned stimulus is formed, which leads to a strengthening of synaptic connections in this region, which, once laid down, are retained for the lifetime of the animal. Basic innate behavioral patterns are also controlled in part by the limbic system and include those directed toward perpetuation of the species (sociosexual behaviors conducive to mating).

check your understanding 5.8

- Functions associated with the four divisions of the mammalian cerebrum are as follows: The occipital lobes are a sensory and association area that carries out the initial perception and processing of visual input in the cortex. Similarly, auditory (sound) sensation is initially received and processed by the temporal lobes. The parietal lobes are primarily responsible for receiving and processing body sensory input (e.g., touch, temperature, pain). The frontal lobes contain motor and higher association areas and are responsible for four main functions: (1) nonreflex motor activity; (2) vocal ability (in mammals with this capability); (3) some types of long-term memory; and (4) higher mental functions, such as planning, involving multiple sensory inputs and memory.

- The sensory cortex deals with sensory perception; the motor cortex initiates nonreflex or learned movements; and the association cortexes are the sites of complex memory, integration, and planning, as well as (in some species) self-awareness, language, and personality traits.

- The greater the sensory input density from a specific area of the body, the greater the area of the cortex devoted to processing this information. For each species, the extent of each body part is indicative of the relative proportion and location of the somatosensory cortex devoted to that area (Figure 5-31).

check your understanding 5.9

- *Habituation* represents a decreased responsiveness to repetitive presentations of an indifferent stimulus, that is, one that is neither rewarding nor punishing; *sensitization* refers to increased responsiveness to mild stimuli following a strong or noxious stimulus. Habituation is the most common form of learning; by learning to ignore indifferent stimuli, the animal is free to focus on other more important stimuli.

- Long-term memory storage requires the activation of specific genes that control protein synthesis. These proteins, in turn, are needed for the formation of new synaptic connections. Thus, long-term memory storage involves permanent physical changes in the brain.

- Formation of enduring memories involves not only the activation of positive regulatory factors (CREB-1) that favor memory storage but also the turning off of inhibitory, constraining factors (CREB-2) that prevent memory storage. In life-threatening situations, sufficient MAP kinase molecules are presumably sent to the nucleus rapidly enough to inactivate the CREB-2 molecules so that protein kinase A activates sufficient CREB molecules to ensure that the animal recalls the details of this experience for future reference.

check your understanding 6.1

- Modalities include mechanoreception for physical stimuli, chemoreception for chemicals, thermoreception for hot or cold, photoreception for light, electroreception for electric fields, and magnetoreception for magnetic fields; chemoreceptive and mechanoreceptive nociceptors for pain are sometimes listed separately. Their roles are exteroreception for external stimuli, interoreception for internal states, or proprioception for motion and position.

- The most ancient, found in all life, were probably mechanically gated channels, which open in response to physical changes, such as swelling, and chemically gated channels and receptors that signal in response to specific chemicals. Light sensing begins with receptors. Animals later evolved voltage-gated channels to transmit long-range signals and thermally gated channels to sense specific temperatures.

check your understanding 6.2

- Tonic receptors fire continuously as long as they are stimulated—for example, joint proprioceptors often do this since continual information on limb position is crucial. Phasic receptors shut off after a while under continuous stimulus—for example, olfactory receptors do this since paying attention to an ongoing unimportant odor diverts brain power.

- For touch, each sensory neuron responds to stimuli only within a circumscribed region of the skin surface surrounding it; this region is its receptive field. Smaller fields increase acuity or discriminating ability, as does lateral inhibition. See Figure 6-7.

- An adequate stimulus is one that a receptor most readily responds to; it alters the receptor's permeability, leading to a local graded voltage change (i.e., one proportional in strength to the stimulus). This is the receptor potential; it may initiate action potentials in an afferent neuron for long-distance signaling, but only if it reaches threshold.

check your understanding 6.3

- Proprioceptors inform the CNS about body and limb position and movement, which are essential for coordinated motions.

- Hair cells have stereocilia, which depending on which way they bend due to external movements such as current or sound waves, open and close ion channels, as shown in Figure 6-11. In the fish lateral line, these sense water pressure and flow changes.

- All are proprioceptors for position and movement. Statocysts (Figure 6-9) are gravity sensors in many animal phyla, using signals made by mineralized moveable grains called statoliths. The utricle (for vertical motion) and saccule (for horizontal motion) of vertebrate heads (Figure 6-14) work similarly with otoliths. The vertebrate semicircular canals (Figure 6-13) detect rotational acceleration or deceleration of the head in three dimensions due to relative fluid movement within them.

check your understanding 6.4

- Intensity or loudness depends on a sound wave amplitude, in decibels (dB); hearing ranges from a leaf rustling (about 10 dB) to a waterfall roaring (perhaps 100 dB, a 100 million times louder than the leaf rustling) or higher. Pitch is determined by frequency. Infrasound attenuates only slightly in the environment and so can function in long-distance communication (e.g., by whales and elephants). Ultrasound is useful because its numerous small waves can reflect well off very small features, making it ideal for echolocation in bats and toothed whales. Timbre depends on overtones (other frequencies superimposed on the basic pitch); echolocating dolphins use it to discriminate objects, while many birds and mammals use timbre to distinguish voices.

- The pinna is a skin-covered flap of cartilage that collects sound waves and channels them down the external ear canal to the tympanic membrane, a taut drumlike structure that vibrates, pushing on the three ossicle bones, which act as levers to push on the oval window of the cochlea (Figure 6-17). The window vibrates fluid in the cochlea, which then transmits the vibrations to the basilar membrane. This discriminates sounds as shown in Figures 6-19 and 6-20. Figure 6-21 summarizes the events.

check your understanding 6.5

- See Figure 6-23 for (1) salt, which aids in acquiring electrolytes; (2) bitter, which helps avoid plant toxins; (3) sweet, for energy-rich carbohydrates; (4) acid, for avoiding rotting food; and (5) umami, for proteins.

- The bulb for general odor detection in mammals is on the roof of the nasal cavity (Figure 6-24); the VNO for pheromones lies near the nasal entrance. Receptors are activated by binding of specific chemicals, with each receptor binding several types of molecules weakly to strongly. By comparing strengths of binding signals from multiple receptors, the brain can distinguish far more odors than there are receptors.

check your understanding 6.6

- The answer begins with retinal in the opsin protein and should include all steps of Figure 6-36 to the bipolar and ganglion cells, which work as shown in Figure 6-39.

- Rods have rhodopsin, in which retinal can absorb most wavelengths of light and therefore allow animals to see shades of gray. Cones have various scotopsins, which absorb at different wavelengths, and therefore allow animals to see specific colors. L types are for red; M, for green; S, for blue; and UV, for ultraviolet. There are various combinations among animals—for example, shallow fish have L, M, and S, but deep ones have lost L types due to lack of red light. Many birds see UV, which aids locating the sun on cloudy days.

- Vertebrate eyes are camera types as shown in Figure 6-28 with ciliary receptor cells such as rods. Phototransduction was described earlier. Insect eyes are compound types with numerous rhabdomeric receptor cells (Figure 6-44).

Images are made in the brain as if each cell were a pixel in a newspaper photo. Transduction is similar to vertebrates, but the events lead to channel opening rather than closing.

check your understanding 6.7

- In mammalian skin, one receptor responds to increased temperature (warmth sensors) above the skin's temperature, while the other receptor responds to decreasing temperature (cold sensors) below skin temperature. These help the brain anticipate threats to core temperature. Infrared receptors in pit vipers help them detect body heat of prey mammals and perhaps also habitat temperatures for behavioral thermoregulation. All involve TRP-type channel proteins, which appear to be thermally gated.
- This depends on the type of animal and its habitat. In general, increased heat is more likely to be immediately harmful than a comparable decrease in temperature.

check your understanding 6.8

- Fast pain is a quick response to direct mechanical or heat damage (which may subside) that is transmitted over large, myelinated A-delta fibers. It is followed by a slower, prolonged chemical response to inflammation and cell contents released by damage that is transmitted by small, unmyelinated C fibers.
- See Figure 6-46.

check your understanding 6.9

- Passive electroreception serves to detect extraneous electric fields, not an animal's own electrical currents, and is used for electrolocation of other animals, as in prey location by sharks. Sharks and their relatives have ampullary receptors filled with a clear, salty gel (a glycoprotein complex) that readily transmits minute electrical currents. Active electroreception resembles echolocation in that the animal assesses its environment by actively emitting signals with electric organs (Figure 6-47) and receiving the feedback signal. Magnetorecep-tion detects the Earth's magnetic field and is used as a compass sense for migration, as in many birds. The mechanism is unclear, but three hypotheses are given on p. 264.

check your understanding 7.1

- Hormones are chemically classified into three categories: peptides and proteins, amines, and steroids. The peptide and protein hormones consist of specific amino acids arranged in a chain of varying length. Amines are derived from the amino acid tyrosine. Steroids are neutral lipids derived from cholesterol. All peptides and catecholamines are hydrophilic, whereas all steroid and thyroid hormones are lipophilic.
- Hormones produce their effects by altering intracellular proteins through ion fluxes, second messengers, and transcription factors.
- The plasma concentration of a hormone depends on the hormone's rate of secretion into the blood by the endocrine gland: for a few hormones, the rate of metabolic activation; for lipophilic hormones, the extent of binding to plasma proteins; and finally the rate of removal from the circulation by metabolic inactivation and excretion in urine.

check your understanding 7.2

- See Figure 7-7b and the discussion on p. 280. The quantity of JH released from the corpora allata determines whether metamorphosis will occur.

check your understanding 7.3

- Short-day breeders rely on steadily increasing concentrations of melatonin (Figure 7-8), whereas long-day breeders rely on the progressively decreasing concentrations of melatonin to trigger the onset of the breeding season.
- Once synchronized with a regular environmental cycle such as day–night or summer–winter, clocks permit organisms to prepare for these cyclical changes in advance of their actual occurrence. Neural clocks regulate many neuromuscular functions, such as feeding and other behaviors, as well as hormonal secretion. Clock genes interact with each other through both transcription and negative-feedback loops to produce a cycle that lasts approximately a day. The mechanism is detailed in the box, *Molecular Biology and Genomics: Clocks and Genes*, on p. 284.
- Anterior pituitary hormone release is regulated by peptides released from the hypothalamus through a unique vascular link, whereas the posterior pituitary hormones are synthesized in the hypothalamus and transported to the posterior pituitary by a neural pathway.
- MSH plays a vital role in the camouflage of selected species. α-MSH controls skin coloration via the dispersion of storage granules containing the pigment melanin.

check your understanding 7.4

- GH's indirect growth-promoting actions are mediated by IGFs, which act on the target cells to cause growth of both soft tissues and bones, whereas GH can directly bind with receptors on its target organs, namely adipose tissue, skeletal muscles, and liver, to exert its metabolic effects.
- Administering GH to swine dramatically increases protein accretion and muscle mass, in addition to reducing adipose tissue growth.
- Brood patch development is related to both the gender that incubates and the form of steroid that is most effective, in combination with prolactin, to produce a full brood patch. In those species in which the female alone incubates, it is estrogen; in those species in which the male incubates, it is androgen.

check your understanding 7.5

- Hypothalamic TRH triggers the release of TSH secretion by the anterior pituitary, whereas T_3, in negative-feedback fashion, "turns off" TSH secretion by inhibiting the anterior pituitary and hypothalamus.
- A primary role of thyroid hormone in endotherms is to regulate metabolic heat production, whereas in ectotherms, thyroid hormone may be elevated during periods of metabolically demanding activity. Some ectotherms (e.g., tadpoles) rely on a surge of thyroid hormone to induce metamorphosis.
- Thyroglobulin incorporates the thyroid hormones in their various stages of synthesis and then stores them until needed. Once freed from thyroglobulin, the hydrophilic thyroid hormones simply diffuse away into the circulation.

check your understanding 7.6

- See Figure 7-22, Table 7-3, and the discussion on p. 310.
- The catecholamine epinephrine strengthens sympathetic responses and reaches places not innervated by the sympathetic system to perform additional functions, such as mobilizing carbohydrate and fat stores.

check your understanding 7.7

- The pancreatic hormones insulin and glucagon shift the metabolic pathways back and forth from net anabolism to net catabolism and glucose sparing, depending on whether the body is feasting or fasting, respectively. See Figure 7-27.
- The fed state lowers blood levels of glucose, fatty acids, and amino acids and promotes their storage. As these nutrient molecules enter the blood during the absorptive state, insulin promotes their cellular uptake and conversion into glycogen, triglycerides, and protein, respectively. Insulin facilitates glucose transport into most cells, stimulates glycogenesis while decreasing glycogenolysis, and inhibits gluconeogenesis in the liver. During periods of fasting, glucagon increases hepatic glucose production and release and thus raises blood glucose levels. Glucagon exerts its hyperglycemic effects by decreasing glycogen synthesis, promoting glycogenolysis, and stimulating gluconeogenesis. Several other hormones also play a secondary role in fuel metabolism (Table 7-7). For example, in long-term starvation, cortisol helps maintain blood glucose concentrations by promoting gluconeogenesis through protein catabolism.

check your understanding 7.8

- Minor variations in the concentration of free ECF Ca^{2+} can profoundly and immediately affect the sensitivity of excitable tissues. For example, a fall in free Ca^{2+} results in overexcitability of nerves and muscles; conversely, a rise in free Ca^{2+} depresses neuromuscular excitability. Fluctuations in free ECF Ca^{2+} can also affect excitation–contraction coupling in cardiac and smooth muscle, stimulus–secretion coupling, maintenance of tight junctions between cells, and clotting of blood.
- The overall effect of PTH is to increase the Ca^{2+} concentration of plasma (and, accordingly, of the entire ECF), thereby preventing hypocalcemia. By its combined actions on bone, kidneys, and intestine (effectors), PTH raises the plasma Ca^{2+} level when it starts to fall, so hypocalcemia and its effects are normally avoided. Vitamin D is a steroidlike compound that is essential for Ca^{2+} absorption in the intestine.

check your understanding 8.1

- Almost all living cells have rudimentary intracellular machinery for producing movement. White blood cells and amoebas use intracellular contractile microfilaments to propel themselves through their environment with pseudopodia. Other cells such as sperm and paramecia move with microtubule-based propulsive structures called cilia and flagella. As the first animals evolved, larger-scale movements evolved from these intracellular ones. For example, Porifera (sponges) create large-scale water currents using numerous flagellated cells. Many aquatic larvae, flatworms, and comb jellies move slowly through the environment by the beating of coordinated cilia. However, other (especially faster) locomotory movements of animals required the evolution of a highly organized microfilament-based structure called a muscle: a specialized cell that contracts to pull an external structure or squeeze a fluid.

check your understanding 8.2

- The area between two Z lines is called a sarcomere, which is the functional unit of skeletal muscle. A functional unit of any organ is the smallest component that can perform all the

- functions of that organ. Accordingly, a sarcomere is the smallest component of a muscle fiber that is capable of contraction.
- Muscle contraction is regulated by intracellular Ca^{2+} concentrations. When the troponin is not bound to Ca^{2+}, this protein complex stabilizes tropomyosin in its blocking position over actin's cross-bridge binding sites (Figure 8-6a). When Ca^{2+} binds to troponin, the shape of this complex is changed in such a way that tropomyosin is allowed to slide away from its blocking position (Figure 8-6b). With tropomyosin out of the way, the myosin head can bind to actin to form cross bridges, resulting in muscle contraction.

check your understanding 8.3

- During contraction in mammalian striated muscle, the thin filaments on each side of a sarcomere slide inward toward the A band's center (Figure 8-7). As they slide inward, the thin filaments pull closer together the Z discs to which they are attached, so the sarcomere shortens. Because all the sarcomeres throughout the muscle fiber's length shorten simultaneously, the entire fiber becomes shorter.
- An action potential on the outside of a muscle fiber triggers the contractile process on the inside by travelling down a T tubule, which triggers the release of Ca^{2+} from the sarcoplasmic reticulum into the cytosol to trigger contraction. A T tubule runs perpendicularly from the surface of the muscle cell membrane into the central portions of the muscle fiber (Figure 8-9).
- Foot proteins, known as ryanodine receptors, extend from the sarcoplasmic reticulum and span the gap between the lateral sac and T tubule and serve as Ca^{2+}-release channels. T tubule channels (dihydropyridine receptors) are located like mirror images in contact with every other foot protein protruding from the sarcoplasmic reticulum (Figure 8-10b, c). These Ca^{2+} channels are voltage-gated sensors. When an action potential is propagated down the T tubule, the local depolarization activates the voltage-gated dihydropyridine receptors. These activated T tubule channels, in turn, trigger the opening of the directly abutting foot proteins in the adjacent lateral sacs of the sarcoplasmic reticulum. Opening of half of these Ca^{2+}-release channels in direct contact with the dihydropyridine channels triggers the opening of the other half that are not in direct contact.
- Although mitochondria are positioned close to the Ca^{2+} release sites in the sarcoplasmic reticulum, they are not designed to respond to an action potential with a release of Ca^{2+}. With only a few specific exceptions, mitochondria keep up with the cytoplasmic Ca^{2+} transients with only a few milliseconds delay. An exception would be the repeated stimulation of a muscle fiber, which results in tetanus.

check your understanding 8.4

- This is an isometric contraction—the muscle is prevented from shortening because the shell is already closed, so tension develops at constant muscle length.
- A crab claw (a modified limb) is opened with an extensor and closed with a flexor. The claw opener muscle requires limited muscle effort (as with the opening of an alligator's mouth), but the claw closing muscle is used for fighting and is therefore comparatively much larger in size (as are the muscles that close an alligator's mouth).

check your understanding 8.5

- ATP is the only energy source that can be directly used for contractile activity: For contraction and relaxation to continue, ATP must constantly be supplied. The phosphagens creatine phosphate and arginine phosphate are often the first energy storehouse tapped at the onset of contractile activity.
- If the energy-dependent contractile activity is to be continued, most animal muscles shift from the phosphagens (brief anaerobic activity) to the pathway of oxidative phosphorylation to form ATP. Oxidative phosphorylation takes place within the muscle mitochondria; this multistepped pathway requires time to pick up its rate of ATP formation to match the increased demands for energy, time provided by the immediate supply of energy from the one-step phosphagen system. Oxygen is required to support the mitochondrial electron transport system, which together with chemiosmosis by ATP synthase, efficiently harnesses energy captured from the breakdown of nutrient molecules and uses it to generate ATP. This pathway is fueled by glucose or fatty acids, depending on the intensity and duration of the activity.

check your understanding 8.6

- High-speed synchronous skeletal muscles are those in which each muscle action potential evoked by incoming neuronal activity results in a single mechanical contraction. The development of asynchronous muscle contractions (Figure 8-21) represents a solution for muscles that operate at extraordinarily high frequencies: Their design results in the generation of more power, although capacity for fine neural control declines. Associated with the rapid wing-beat frequency is a lowered number of nerve impulses that are now asynchronous with muscle contraction. In these circumstances, a single Ca^{2+} pulse maintains the muscle in an activated state for numerous successive cycles.
- Myofibrils generate the force, mitochondria are committed to supplying ATP for the cross bridges, while the sarcoplasmic reticulum is the "on–off" switch. In a powerful locomotive muscle, the myofibrils would be most represented.

check your understanding 8.7

- Muscle spindles, which are distributed throughout the fleshy part of a skeletal muscle, are collections of specialized muscle fibers known as intrafusal fibers. Whenever the whole muscle is passively stretched, the intrafusal fibers within its muscle spindles are likewise stretched, increasing the rate of firing in the afferent nerve fibers whose sensory endings terminate on the stretched spindle fibers. This stretch reflex serves as a local negative-feedback mechanism to resist any passive changes in muscle length so that optimal resting length can be maintained. Golgi tendon organs are located in the tendons of the muscle, where they can respond to changes in the muscle's externally applied tension rather than to changes in its length.
- The classic example of the stretch reflex is the patellar tendon, a monosynaptic reflex (no intermediate interneurons) in humans (Figure 8-26). The extensor muscle of the knee is the quadriceps femoris, which forms the anterior (front) portion of the thigh and is attached just below the knee to the tibia (shinbone) by the patellar tendon. Tapping this tendon with a rubber mallet passively stretches the quadriceps muscle. The resultant stretch reflex brings about contraction of this extensor muscle, causing the knee to extend and raise the foreleg in the well-known knee-jerk fashion.

check your understanding 8.8

- The exposure of skinned human skeletal muscle fibers to twitchin induces a catchlike state in the skeletal muscle fibers. The rate at which the myosin heads detach from the actin is greatly slowed by the presence of twitchin, a titinlike protein associated with the myosin. Phosphorylation of twitchin by protein kinase A, perhaps because of the actions of serotonin, inhibits the catch state by allowing the myosin heads to detach at a faster (noncatch) rate.
- Single-unit smooth muscle is supplied by the involuntary autonomic nervous system: Muscle fibers that make up this type of muscle become excited and contract as a single unit. Gap junctions electrically link the muscle fibers in single-unit smooth muscle. When an action potential occurs anywhere within a sheet of single-unit smooth muscle, it is quickly propagated via these special points of electrical contact throughout the entire group of interconnected cells, which then contract as a single coordinated unit.

check your understanding 9.1

- Circulatory systems overcome the slowness of diffusion by the much faster process of bulk transport: the movement of the medium that contains the molecules (and cells) of interest.
- In open systems (Figure 9-2), fluid called hemolymph moves via pumping through arteries (outgoing vessels) that open like garden hoses into extracellular spaces among the tissues, bathing them directly for molecular exchanges with cells. Veins (return vessels) may pick up the fluid. In closed systems, fluid called blood exits a heart through vessels that are continuous all the way back to the intake side of the heart, with capillaries joining arteries and veins.

check your understanding 9.2

- Plasma is typically 90% or more water, with various proteins, electrolytes (mainly NaCl), nutrients, waste products, and hormones.
- Low-density lipoproteins (LDLs) are blood complexes that deliver cholesterol and phospholipids to most body cells (e.g., for membrane synthesis). Pathological accumulations of LDL cholesterol in blood vessel walls are a leading cause of cardiovascular disease. High-density lipoproteins (HDLs), in contrast, remove excess cholesterol from cells and transport it to the liver for removal.
- Erythrocytes primarily transport gases for respiration: oxygen bound to hemoglobin; carbon dioxide primarily as bicarbonate ions (formed by carbonic anhydrase in the cells); and carbon dioxide bound to hemoglobin. The cells also transport NO and H_2S, which vasodilate vessels.

check your understanding 9.3

- Clotting transforms blood into a solid gel on top of a platelet plug. As shown in Figures 9-11 and 9-12, tissue damage, exposed collagen, and platelet aggregation activate protein factors in the blood, which ultimately convert prothrombin to thrombin. Thrombin, in turn, converts fibrinogen into fibrin, which forms a mesh (the clot). In positive-feedback fashion, thrombin also facilitates its own formation from prothrombin and enhances platelet aggregation.
- (1) ADP and other chemicals released by activated platelets stimulate the release of prostacyclin and NO from the adjacent normal endothelium. Both inhibit platelet aggregation, so a platelet plug is limited to the defect and does not spread. (2) Clotting factors convert

plasminogen into active plasmin, which slowly dissolves a clot's fibrin. At nonwound sites, inappropriate fibrin formation is stopped by plasmin activated by tissue plasminogen activator (tPA) from the tissues, especially the lungs.

check your understanding 9.4

- (1) Flagella, as in sponges; (2) extrinsic pumping of fluid by skeletal muscles, as in mammalian leg veins; (3) peristaltic pumping of vessel walls, as in annelids; (4) auxiliary hearts for blood, as in cephalopods for pumping blood to the gills
- Valves consist of flaps of connective tissue in a vessel that can open in one direction only, due to fluid pressure. A backward pressure causes the flaps to close together to prevent backflow.
- Individual cardiac muscle cells are interconnected as branching fibers, with adjacent cells joined end-to-end by these discs. Within a disc are two types of membrane junctions: (1) Desmosomes are a type of adhering junction that mechanically holds cells together under high mechanical stress, whereas (2) gap junctions are channels that allow action potentials to spread from one cardiac cell to the next, ensuring the cells beat in synchrony.

check your understanding 9.5

- The pacemaker fires repeatedly because its membrane potential slowly depolarizes between action potentials until threshold is reached, at which time the membrane "fires." This, in turn, triggers the heartbeat so that the heart beats regularly without needing extrinsic triggers. The slow depolarization involves Na^+ and Ca^{2+} influx through channels as described on p. 407.
- As the cardiac membrane potential reaches its peak positive level, it remains there for several hundred msec in a plateau phase. It is maintained by activation of "slow" Ca^{2+} channels and a marked decrease in K^+ permeability. This plateau prolongs contraction for squeezing the heart chambers properly.
- Tetanus is a sustained full contraction of a muscle. Fibrillation is random, uncoordinated contraction–relaxation cycles.
- The P wave represents atrial depolarization. The QRS complex represents ventricular depolarization. The T wave represents ventricular repolarization.

check your understanding 9.6

- See Figure 9-27 and the discussion on p. 412.

check your understanding 9.7

- C.O. = heart rate × stroke volume; the volume of blood per minute pumped by a heart to the body. This flow is critical because it is responsible for all blood or hemolymph transport functions.
- The Frank-Starling law says that the greater the volume of blood entering the heart during diastole, the greater the volume of blood ejected during systole. (1) This provides equalization of output between the right and left sides of the avian and mammalian hearts, so the blood pumped is equally distributed between the pulmonary and systemic circulation. (2) This response allows cardiac output to increase intrinsically when venous return increases during exercise.
- The parasympathetic system releases ACh onto muscarinic receptors, which favor resting conditions, while the sympathetic releases NE to α_1-adrenergic receptors, which stimulate changes for action. See also Table 9-1.

check your understanding 9.8

- Q = flow rate of fluid through a vessel (quantity per unit time). It is the outcome of ΔP, the pressure gradient, divided by R, the resistance of blood vessels. The pressure gradient is established by the force of the heart pumping and can be altered by gravity. The resistance is mainly determined by the radii and length of blood vessels and the viscosity of the blood.
- Parallel flow ensures that each organ/tissue receives an appropriate share of oxygenated blood rather than depleted blood that has already passed through other organs/tissues.
- Circulatory fluid force can create movement, as in extension of spider legs, pressure for ultrafiltration in kidneys, and inflation of erectile tissue.

check your understanding 9.9

- Most mollusks (except cephalopods) and insects have open circulation with large open spaces, with body muscles providing some directional flow, but large crustacea have many branching vessels leading to small local open spaces.

check your understanding 9.10

- The answer should involve the evolution from a two-chambered heart with one circulatory circuit (water-breathing fishes) to a four-chambered heart with separate pulmonary and systemic circuits (crocodilians, birds, mammals), as found in Figures 9-37 to 9-41. The special shunt of reptiles and fetal mammals should be included.

check your understanding 9.11

- The elasticity of arteries enables them to expand temporarily during systole, storing some of the pressure energy in their stretched walls (Figure 9-43). When the heart relaxes and ceases pumping blood into the arteries, the stretched arterial walls passively recoil, like an inflated balloon that is released. This recoil pushes the excess blood contained in the arteries into the vessels downstream, ensuring continued blood flow to the tissues when the heart is relaxing.

check your understanding 9.12

- Local myogenic activity and sympathetic stimulation cause partial contraction of arteriole muscles to give vascular tone, helping maintain normal systemic pressure. Tone can be altered intrinsically to match blood flow with local needs through (1) local metabolic changes, (2) histamine release, (3) exposure to heat or cold, and (4) stretch. For example, decreased O_2 reduces tone, so more blood flows to the tissue. Extrinsic regulation also works to adjust tone for whole-body needs. For example, during exercise, sympathetic nerves constrict arterioles to the gut more, while epinephrine dilates those to the skeletal muscles, diverting blood to the latter.

check your understanding 9.13

- Capillaries decrease ΔX by being very thin and narrow and numerous; increase D by greater permeability through pores; increase A by being numerous; and increase ΔC by increased perfusion of incoming blood having high levels of oxygen, glucose, etc.
- *Flow rate* refers to the volume of blood flowing through a given segment of the circulatory system per unit of time, but *velocity of flow* refers to the linear speed with which blood flows forward through a given segment of the circula-

tory system. Velocity slows as cross-sectional area of a vessel network increases. Because that area is much higher in a capillary bed compared with the incoming arteriole, blood slows in the bed. This is important to allow time for diffusion to occur between blood and tissues.
- The factors affecting bulk flow are described in Figure 9-57. This flow exiting the capillaries gives the fluid closer contact with cells, enhancing glucose and waste transport and pushing out pathogens into the lymph. The flow reentering capillaries prevents fluid buildup in tissues, although a small amount remains to create the lymph.

check your understanding 9.14

- Lymph forms from the small amount of exiting capillary bulk flow that is not reabsorbed by capillaries. The system is important for immune responses and fat transport from the small intestine.

check your understanding 9.15

- Veins return blood to the heart by (1) weak systemic pressure; (2) systemic constriction of vein walls; and (3) extrinsic pumping by skeletal muscles, respiratory pressure changes, and cardiac suction. One-way valves inside the veins ensure that flow does not reverse.

check your understanding 9.16

- Gas transport is essential for minute-to-minute survival of tissues and must increase when body activity increases. Blood pressure must not get too high; otherwise, vessel damage can occur and the heart can be stressed. Likewise, it must not get too low, because flow may be unable to overcome gravity and other resistance factors and because insufficient filtration pressure in the kidneys may result. Pressure is a lower priority since changes may be damaging over the long term, but rarely minute-to-minute.
- A substantial increase in cardiac output (C.O.) is triggered to increase oxygen delivery, and a large increase in skeletal muscle, heart, and skin blood flow accompanied by a decrease in blood flow to the digestive system are achieved by dilation of skeletal muscle, skin and heart arterioles, and constriction of gut arterioles. Changes occur extrinsically by sympathetic nerves and epinephrine, regulated anticipatory brain centers via the cardiovascular center of the brainstem. Intrinsic factors discussed earlier also alter blood flow locally.

check your understanding 10.1

- Immune systems defend against invading pathogens (e.g., viruses, bacteria, parasites), worn-out body tissue, and cancer.
- Innate immune systems are "ready-to-go" responses that come into play on exposure to any threatening agent and do not depend on prior exposure to that agent. Acquired systems rely on learned immune responses selectively targeted against a particular foreign material following the first exposure. Learning takes considerably more time to be mounted.
- Recognition of nonself is accomplished by pattern recognition receptors, or PRRs, in plasma membranes. These sensor proteins recognize and bind with common macromolecules on pathogens, such as lipopolysaccharides (LPS) of the bacterial cell wall (Figure 10-1b).
- Immune cells communicate with cytokines (hormonelike secretions) and by direct contact through immunosynapses, both involving receptor proteins.

check your understanding 10.2

- Barrier tissues such as skin and respiratory linings are the first lines of defense; they physically block invaders and may have active defenses such as coughing. Second lines of defense are (1) phagocytotic cells that use PRRs to recognize and consume invaders, often in an inflammatory process; (2) antimicrobial peptides that destroy the invaders directly (e.g., by forming leaky pores); and (3) opsonin proteins that "tag" invaders to aid recognition by phagocytotic cells.
- Some antimicrobial peptides are attracted to particular glycoproteins found on bacterial surfaces, whereas others are attracted to the strong electrical charge that characterizes many bacterial membranes.

check your understanding 10.3

- The major functions of neutrophils, monocytes, NK cells, lymphocytes, mast cells, and dendritic cells are listed on pp. 464–465. Briefly, the first four are in the circulation, and the last two reside in barrier tissues (although dendritic cells become mobile upon activation).

check your understanding 10.4

- Active features of major barrier tissues include the following: (1) in skin, sweat glands and keratinocytes that make antimicrobial peptides, melanocytes to absorb UV, dendritic cells as sensors of acquired immunity; (2) in digestive tracts, enzyme and acidic secretions, symbiotic bacteria, and lymphoid tissues that make antimicrobial peptides; (3) in genitourinary tracts, acidic secretions, sticky mucus, and symbiotic bacteria; and (4) in respiratory tracts, antimicrobial peptides and cilia-driven mucus elevator with coughing, sneezing.
- The major steps of (1) inflammation and (2) complement responses are shown in Figures 10-7 and 10-8, respectively.
- NK cells destroy virus-infected cells and cancer cells by releasing chemicals that directly lyse the membranes of such cells on first exposure to them, by means largely unknown. Interferon transiently interferes with replication of viruses (Figure 10-9) and enhances action of cancer-killing NK and T cells.

check your understanding 10.5

- In antibody-mediated immunity, activated B cells recognizing a specific antigen secrete proteins called antibodies to attach to antigens to target them for destruction as opsonins, whereas in cell-mediated immunity, activated cytotoxic T cells recognize and destroy virus-invaded and cancerous body cells.
- B cells differentiate and mature in the bone marrow in mammals and in the bursa in birds. T cells do so primarily in the thymus.
- Lymphoid tissues (lymph nodes, spleen, thymus, tonsils, adenoids, bone marrow) store, produce, and/or process lymphocytes. Antigens are foreign molecules that induce an immune response against themselves.

check your understanding 10.6

- B cells become plasma either by direct activation by a specific antigen binding to its specific receptor or by activation by a helper T cell with the same receptor. Plasma cells make antibodies. Some activated B cells become dormant memory cells, becoming active if there is a later invasion by the same antigen.
- Antibodies are Y-shaped proteins, each arm having a tip with a binding site for a specific antigen. They bind to antigen to neutralize their interactions, agglutinate them for mass engulfment by phagocytes, and amplify innate responses. They are genetically diversified into millions of variants from relatively few genetic sequences by gene recombination and mutation as described on pp. 479–480.
- Specific B cells get selected to produce specific antibodies most likely by the clonal selection mechanism, shown in Figure 10-16, in which unactivated B cells remain low in number while ones activated by specific antigens or helper T cells rapidly clone themselves into large numbers.

check your understanding 10.7

- (1) Cytotoxic T cells make destructive proteins such as hole-punching perforins to destroy body cells bearing foreign antigen, such as cells invaded by viruses, cancer cells with mutated proteins, and transplanted cells. (2) Helper T cells are integrators that recognize specific antigens and then activate appropriate effector B and cytotoxic T cells. (3) Regulatory T cells suppress both innate and acquired immune responses.
- MHCs (self-antigens) are necessary to trigger T-cell function. (1) Antigen-presenting cells (APCs) such as dendritic cells phagocytize invaders and place foreign protein fragments (antigens) on MHC class II in their membranes, which helper T cells with the right receptor can recognize. Compromised body cells may present viral or mutated self proteins on MHC class II to similarly activate the right cytotoxic T cells.
- T cells avoid attacking self-antigens through (1) clonal deletion, (2) clonal anergy, (3) receptor editing, (4) suppression by regulatory T cells, (5) immunological ignorance, and (6) immune privilege, as detailed on pp. 485–486.
- Immune surveillance relies on cytotoxic T cells, macrophages, NK cells, and interferon. The cells recognize newly arisen, potentially cancerous tumor cells and destroy them and/or release interferon before the cancer cells have a chance to multiply and spread.
- In neuro–endocrine–immune interactions, interleukins can activate the hypothalamus to trigger the release of adrenal cortisol, which mobilizes the body's nutrient stores for fuel and amino acids for repair during injury and illness. In turn, cortisol can suppress immune reactions, perhaps to dampen overreactions.

check your understanding 10.8

- Sharks have B and T cells with gene shuffling to make millions of different antibodies and receptors, but unlike other jawed vertebrates, they do not have increased mutation and they have far fewer gene segments to shuffle. However, they have many more gene clusters on many chromosomes, some of which are already joined to code for antibodies. Lampreys have lymphocytes that generate diverse defense proteins from relatively few genes, but those are from a class of PRR genes not closely related to antibody genes.
- In *Drosophila*, evidence of acquired immunity is shown by a sublethal dose of bacteria inducing changes that subsequently protect the flies from an otherwise-lethal dose of the same (but not other) bacteria. This may involve PRR-type genes called Dscam (Down syndrome cell-adhesion molecules), which can generate over 30,000 variants of receptors and antibody-like proteins from a few genes by alternative splicing.

check your understanding 11.1

- The four possible steps in external respiration are (1) bulk transport of external media—ventilation; (2) exchange by diffusion between the environmental medium and internal body fluids, at a respiratory surface; (3) bulk transport of ECF in a circulatory system; and (4) exchange by diffusion between the ECF and ICF.
- As body size and metabolic rates increase in evolution, diffusion becomes increasingly too slow for O_2 delivery; therefore, better respiratory mechanisms are needed. Animal habitats vary in their O_2 availability of oxygen, from anoxic ("no oxygen") aquatic mud to sea level air, again necessitating various respiratory adaptations.
- The limits of diffusion are overcome by ventilatory and circulatory bulk transport, which bypass diffusion altogether and also enhance gas diffusion gradients; respiratory organs that have a high permeability D, low distance ΔX, high surface area A; and special proteins that convert gases between diffusible and nondiffusible forms.
- Tidal ventilation involves breathing in and out of the same opening, so fresh medium is available only about half the time and may mix with stale medium. In contrast, in flow-through ventilation, the medium enters and exits through separate openings, so those problems are avoided.

check your understanding 11.2

- Some nongill methods of gas exchange in aquatic animals include integument (e.g., flatworms), mantle of mollusks, and respiratory tree of sea cucumbers.
- Three examples of gills in nonvertebrate phyla are external folded projections of sea slugs, dense external filaments of some annelids, and the folded external projections of crustacea that are protected inside an exoskeletal chamber.
- Fish create flow-through breathing by the mechanism of Figure 11-6. Countercurrent flow aids gas exchange because blood flow through lamellar capillaries is opposite to the stream of water, such that blood moving through the lamellae continually encounters water whose O_2 content is higher than what is present in the capillary. Thus, there is a positive gas gradient at every point, keeping gases diffusing into the blood.

check your understanding 11.3

- Fully terrestrial pulmonate snails have a lung as a well-perfused mantle tissue in a chamber with a small opening to the outside to reduce water loss. Gases move primarily by diffusion. Spiders may have book lungs, invaginated gill-like stacked lamellae appearing like pages of a book, with air spaces between the folds for gas exchange. Other spiders have internal air tubes called tracheae, as do insects. These start at openings called spiracles and branch to reach most internal cells. Book lungs and tracheae may be ventilated by diffusion or by active pumping.

check your understanding 11.4

- Bimodal breathers have gills and air-breathing systems and include certain fishes and most amphibians. Trimodal breathers such as redfish have gills and two air-breathing systems such as skin and lungs. These modes evolved in the transition from water to land, particularly in aquatic animals whose habitats regularly dried up.

- To compare respiratory systems of frogs and nonavian reptiles excluding crocodilians, follow Figure 11-9 and the positive-pressure steps of Figure 11-10 for frogs. For nonavian reptiles, refer to the role of negative pressure as described on p. 507.
- Mammalian respiratory systems use tidal flow to alveoli as shown in Figure 11-11 for structure and in Figure 11-14 for respiratory muscles that create negative pressure for inhalation. Crocodilian and avian respiratory systems are partly flow-through systems with air capillaries, as explained for birds in Figures 11-15 and 11-16.

check your understanding 11.5
- To compare atmospheric, intra-alveolar, and intrapleural pressures and their roles, follow Figure 11-19.
- Lungs rebound (recoil) after they are stretched by inhalation due to (1) elastic fibers made of the protein elastin and (2) surface tension, in which water molecules attracted to each tend to draw stretched, wet surfaces (the alevolar linings) into smaller surface areas.
- Pulmonary surfactant is a detergent-like water–lipid–protein secretion from Type II alveolar cells; surfactant lowers the alveolar surface tension because the cohesive force between a water molecule and an adjacent pulmonary surfactant molecule is very low. By doing so, surfactant (1) increases pulmonary compliance, thus reducing the work of inflating the lungs, and (2) reduces the lungs' tendency to recoil so that they do not collapse as readily.

check your understanding 11.6
- The various lung volumes and capacities (TLC, TV, FRC, RV, VC) and their significances are listed on pp. 522–523 and in Figure 11-25.
- Pulmonary ventilation = tidal volume × respiratory rate; alveolar ventilation, which takes into account the effect of dead space, = (tidal volume – dead-space volume/breath) × respiratory rate. This is important because dead space does not exchange stale with fresh air, so it detracts from gas exchange. Some examples with actual animal data are listed on p. 526.

check your understanding 11.7
- Dead space is less in fishes compared with mammals due to flow-through mechanism, but dead space in birds is higher due to their long tracheae. Energy costs for air breathers (up to 5% of metabolic energy) are less than that in fishes (up to 50%) due to difficulty of pumping water. O_2 uptake efficiency is much higher in flow-through systems: up to 90% in fishes, 40% in birds, and 25% in mammals.

check your understanding 11.8
- Describe the gas gradients of Figure 11-27 by following the discussion on pp. 528–529. To apply them to fish, use the gas concentrations given on p. 529 for CO_2 and Figure 11-6f for O_2.

check your understanding 11.9
- Hemoglobins (Hb) use iron to bind O_2 and are usually modest-size proteins within circulatory cells. Mammalian Hb is a tetramer (4 unit proteins together). Hemocyanins (Hc) use copper

and are often gigantic protein complexes in ECFs. Arthropod Hc's are hexamers that may assemble into large groups; annelid and mollusk Hc's are decamers that may assemble into large cylinders.
- O_2 is transported in blood in most vertebrates by iron within hemoglobins, with a small amount as dissolved gas. CO_2 is transported primarily as bicarbonate ion, less as bound to the end of hemoglobin proteins, and even less as dissolved gas.
- The plateau part of the O_2–Hb dissociation curve is important if arterial P_{O_2} falls below normal: Within the plateau, there is little reduction in the total amount of O_2 transported by the blood until the P_{O_2} falls below 60 mm Hg (human example). In contrast, the steep portion of the curve aids gas delivery: In the range of tissue P_{O_2} (about 0–60 mm Hg), a modest drop in systemic capillary P_{O_2} can result in large amounts of O_2 immediately available to meet the O_2 needs of the actively metabolizing tissues.
- The hemoglobin P_{50} has evolved in vertebrates (Table 11-2) in two ways: (1) P_{50} increases and thus, affinity decreases as size decreases to better offload O_2 for high metabolic rates, and (2) P_{50} decreases and thus, affinity increases as environmental O_2 decreases.
- In the Bohr effect, acidity decreases affinity of Hb for O_2 in all vertebrates, boosting gas release at active tissues generating acid. In the Root effect for some fish Hb's, acidity also reduces the total Hb capacity, even further enhancing gas release to help fill swim bladders (Figure 11-32). In the Haldane effect, deoxy-Hb has a greater affinity for both CO_2 and H^+ than does HbO_2, so O_2 release at active tissue facilitates binding and transport of CO_2 and acid.
- The terms *hypoxic hypoxia*, *anemic hypoxia*, *histotoxic hypoxia*, *hypercapnia*, *hypocapnia*, *hyperventilation*, *hypoventilation*, and *apnea* are defined on pp. 542 and 543.

check your understanding 11.10
- Insect spiracles are regulated to be open during high activity but discontinuously at rest, perhaps to reduce water loss or buildup of ROS. Local and neural influences are involved.
- So that ventilation and perfusion are matched, the autonomic nervous system (from brainstem control centers) adjusts simultaneously both blood flow and airflow to suit the animal's current or anticipated respiratory needs. Intrinsic factors such as local decreases in blood and airway O_2 may also work at the same time to increase gas delivery.
- The respiratory brain centers of mammals are in the brainstem. They work by negative feedback using brainstem and arterial chemoreceptors for P_{O_2}, P_{CO_2}, and H^+ and by controlling airway and breathing muscles as effectors to adjust gas delivery. Table 11-3 shows the effects of these variables on the receptors. The centers' responses may be activated in anticipation of activity by higher centers such as the motor cortex to prevent delays in gas delivery.

check your understanding 12.1
- The major functions of excretory systems are maintenance of proper internal levels of inorganic solutes, plasma water volume, and osmotic balance, as well as removal of hormones and nonnutritive and harmful substances resulting from metabolism. The four organ systems that can serve this role are the respiratory

and digestive systems, integuments, and renal organs.
- The evolutionary patterns of ammonia, urea, and uric acid as nitrogenous (N) wastes are shown in Figure 12-1. Use tends to follow water availability, although with exceptions (see p. 561). Ammonia, the most toxic, is the primary N waste of most aquatic animals, since it is inexpensive in terms of ATP and can be diluted easily. Urea has intermediate toxicity and cost and is the major N waste in mammals. Uric acid is the least toxic but most costly, and it dominates in terrestrial insects and most reptiles, including all birds, possibly in part to reduce toxicity in eggs.
- The mechanisms by which transport epithelia move NaCl alone or NaCl and water are shown in Figure 12-3. In brief, Na is transported first, which electrically draws Cl; then in nonwaterproof cells, water follows the NaCl by osmosis.

check your understanding 12.2
- The four major renal processes are (1) filtration, in which solutes are selectively separated from a solution by passing through a boundary; (2) secretion, in which transport epithelia move specific solutes into a renal tubule lumen for excretion; (3) reabsorption, in which transport epithelia move specific solutes (and often water) back into the body from the lumen; and (4) osmoconcentration, in which water is removed from the lumen fluid while leaving solutes behind, producing an excretory fluid more concentrated than body fluids to conserve water.
- Protonephridia use ultrafiltration driven by cilia, with secretion and reabsorption. These are generally blind-end ducts that project into the body cavity. Mesonephridia and metanephridia, in contrast, have ultrafiltration typically driven by fluid pressure; the tubules begin with a filtering capsule or funnel-like opening of some kind that filters the circulatory fluid. This is followed by tubule segments that perform selective secretion and reabsorption (and osmoconcentration in some groups).

check your understanding 12.3
- Initial salt and water movements are created in Malpighian tubules secreting K^+ (via a proton gradient) into tubule lumen (Figure 12-6a), creating osmotic pressure to draw water. The fluid is modified by the tubules and hindgut in osmoconcentrating insects by various mechanisms in different species but typically involves complex tissue anatomy that uses active transport of ions (such as KCl), followed by water osmosis. Uric acid precipitates as crystals as water is removed. A detailed example is described in Figure 12-6b.

check your understanding 12.4
- A vertebrate nephron is the smallest functional unit capable of the kidney's basic functions, as shown in Figure 12-8. The capsule and glomerulus are responsible for ultrafiltration; proximal and distal tubules, for selective reabsorption (of desired molecules) and secretion (of wastes); and the loop of Henle and collecting ducts, for osmoconcentration.
- Marine elasmobranchs' kidneys retain urea and trimethylamine oxide as osmolytes to osmoconform to seawater; their gills, kidneys, and rectal glands excrete excess salts. Marine bony fish are very hypoosmotic; they drink seawater, and their gills remove excess salts. Freshwater bony fish are hyperosmotic; they

do not drink, their gills take up salts, and their kidneys excrete excess water. Amphibian adult kidneys conserve ions and excrete excess water and urea; in some, the bladder is a temporary water reservoir.

- Nonavian reptiles use amphibian-like kidneys, hindguts, and (in marine and desert species) salt glands for excretion and retention. Birds have kidneys with reptile- and mammal-like nephrons, hindguts, and (in marine species) salt glands for excretion and retention. See pp. 574–575 for details.

check your understanding 12.5

- The glomeruli of cortical nephrons lie in the outer layer of the cortex, whereas the glomeruli of juxtamedullary nephrons lie in the inner layer of the cortex. The hairpin loop of cortical nephrons dips only slightly into the outer medulla, whereas the loop of juxtamedullary nephrons plunges through into the inner medulla.
- The three layers of the glomerular membrane are (1) the glomerular capillary wall of a single layer of flattened endothelial cells, perforated by many pores that let out water and small solutes but not most proteins; (2) the basement membrane, an acellular gelatinous layer that provides structural strength and discourages the filtration of small plasma proteins; and (3) the inner layer of Bowman's capsule of podocytes, octopus-like cells that encircle the glomerular tuft to form filtration slits. Mesangial cells encircle glomerular capillaries and can constrict them to reduce filtration. Podocytes also can contract or relax to decrease or increase the number of filtration slits open for filtration control.
- The pressure forces in glomerular filtration are delineated on p. 579 and in Table 12-1. In autoregulation, spontaneous, inadvertent changes in GFR are largely prevented by intrinsic regulatory mechanisms of the kidneys themselves—the myogenic and tubuloglomerular feedback mechanisms described on p. 581. Extrinsic sympathetic regulation can override autoregulation to regulate overall body arterial pressure; see Figure 12-17, for an example.

check your understanding 12.6

- In reabsorption in the proximal tubule, Na^+ is moved by Na^+/K^+ ATPase pumps in the basolateral membrane but crosses the apical membrane passively. H_2O moves passively through aquaporins, drawn by Na^+ and plasma colloidal pressure. Glucose is transferred by secondary active transport utilizing a specialized cotransport carrier that simultaneously transfers both Na^+ (down its gradient) and glucose (against its gradient).
- The RAA system stimulates Na^+ reabsorption in distal tubules and collecting ducts, elevates blood pressure, and stimulates thirst and salt hunger (Figure 12-21). It is opposed by the ANP/BNP system, which inhibits Na^+ reabsorption and lowers blood pressure (Figure 12-22).

check your understanding 12.7

- K^+ secretion is regulated by aldosterone in the distal tubules and collecting ducts; it is coupled to Na^+ reabsorption by the basolateral Na^+/K^+ pump (Figure 12-23). That moves Na^+ into the lateral space while also transporting K^+ into the tubular cells. Regulation is important because K^+ plays a key role in the membrane potential and electrical activity of excitable tissues.
- Plasma clearance rate varies as follows: (1) It equals the GFR if a substance is freely filtered but not reabsorbed or secreted. (2) If a sub-

stance is filtered and reabsorbed but not secreted, its plasma clearance rate is always less than the GFR. (3) If a substance is filtered and secreted but not reabsorbed, its plasma clearance rate is always greater than the GFR.

check your understanding 12.8

- How countercurrent multiplication in the loop of Henle results in the osmotic gradient of the kidney medulla is described in detail in Figure 12-26. How the vasa recta prevents its dissipation is shown in Figure 12-29.
- How vasopressin regulation works in dehydration and overhydration is delineated in Figure 12-28.
- How urea recycles in the kidneys is described in Figure 12-30. This is important for prevention of urea-induced water loss associated with antidiuresis. It may also aid in osmoconcentration as discussed in the box on p. 603.
- Weak osmoconcentrators such as the beaver have primarily cortical nephrons, which produce fairly dilute urines only. Desert rodents usually have all juxtamedullary nephrons with very long loops within an elongated medulla. This anatomy, coupled with a high metabolic rate, provides a relatively long length over which the countercurrent effect can operate.

check your understanding 12.9

- How bladder urine fills and then empties, including external regulatory features, is summarized in Figure 12-33.

check your understanding 12.10

- General causes of renal problems include (1) cellular damage, for example, resulting from excessive glucose in diabetes; (2) toxic agents, such as long-term exposure to high doses of aspirin; (3) inappropriate immune responses, such as glomerulonephritis; (4) obstruction of urine flow because of kidney stones, for example; and (5) an insufficient renal blood supply that leads to inadequate filtration pressure, due for example to heart failure.

check your understanding 13.1

- The balance concept states that, if the quantity of a substance is to remain stable within the body, its input must be balanced by an equal output by means of excretion or metabolic consumption (Figure 13-1).
- The distribution of animal body H_2O in various fluid compartments is shown in Figure 13-2; that for a human is shown in Table 13-1.
- The ionic composition of plasma and interstitial fluid are similar (NaCl dominating), but intracellular fluid differs considerably, in particular due to K dominating (Figure 13-3).

check your understanding 13.2

- The types of osmotic challenges faced by animals are (1) salinity of aquatic environments, (2) evaporation of body water into air, (3) ingestion and excretion, (4) freezing, and (5) pathologies such as diabetes.
- Osmoconformers have body fluids and cells generally equal in osmotic pressure to the environment, with adjustments to ECF and ICF made using small molecules called osmolytes. Osmoregulators have internal osmotic pressures homeostatically regulated and often very different from that of the external environment.
- The major types of organic osmolytes are small carbohydrates, free amino acids, methylamines, methylsulfonium solutes, and urea. In marine nonvertebrates, free amino acids such as glycine

and taurine are used as ICF osmolytes because they are compatible with cellular functions (unlike salts). In chondrichthyan fishes, urea (protein destabilizer) and methylamines (protein stabilizer) are used in ICFs to counteract each other's effects on cellular functions.

- Stenohaline animals such as Cnidaria have a limited tolerance to changes in salinity; they may die if exposed to significant changes in external salinity because theirs cells swell or shrink from osmosis. Euryhaline animals such as many intertidal crustacean are relatively tolerant of changes in external salinity and can survive substantial dilution or concentration of their ECFs.

check your understanding 13.3

- The osmotic problems and solutions for marine, freshwater, and terrestrial vertebrates are summarized in Figures 13-9, 13-11, and 13-12. A deep-sea fish faces high hydrostatic pressure and has higher levels of trimethylamine oxide than shallow fish, possibly to stabilize proteins against pressure. A salmon alternates between hypo-osmotic and hyperosmotic conditions; they restructure their gills to reverse direction of salt transport.

check your understanding 13.4

- ECF osmolarity must be closely regulated to prevent swelling or shrinking of the cells. ECF volume must be closely regulated to help maintain blood pressure.
- The causes of ECF hypertonicity, which can cause cells to shrink, are insufficient water intake, excessive water loss, consumption of hypersaline water, and alcohol consumption. The causes of ECF hypotonicity, which can cause cells to swell, are intake of more water than solutes and retention of excess water.
- The sources of input and output in daily H_2O balance in two mammals are shown in Table 13-3. Sources for daily salt balance for a human are shown in Table 13-4. Only oral intake and urine output can be regulated for water–solute balance of the body.
- The roles of vasopressin and thirst in fluid balance are shown in Figure 13-13. The role of the RAA hormone system, the major regulator of ECF volume and blood pressure, is to stimulate Na retention, thirst, salt hunger, and vasoconstriction.

check your understanding 13.5

- Acids are hydrogen-containing substances that dissociate when in solution to liberate free H^+ and anions; bases, in contrast, have a strong affinity for H^+. The term *pH* refers to the negative log of the hydrogen ion concentration (e.g., pH 7 indicates $[H^+]$ is 10^{-7} M).
- pH is important to animals because changes alter neuromuscular excitability, enzymatic activity, and K^+ levels. The sources of excess H^+ are carbonic acid from CO_2, inorganic acids from nutrient breakdown, and organic acids such as lactate from metabolism.

check your understanding 13.6

- A buffer is a mixture in a solution of two chemical compounds that minimize pH changes from added acid or base: One substance reversibly yields free H^+ as the $[H^+]$ starts to fall, and the other reversibly binds with free H^+ when $[H^+]$ starts to rise. The Henderson-Hasselbalch equation for the bicarbonate system is $pH = pK + \log [HCO_3^-]/[CO_2]$.

- Animal buffers are (1) CO_2–HCO_3^-, primary ECF buffer for noncarbonic acids; (2) peptides and proteins, primary ICF buffer; (3) hemoglobin, primary erythrocyte buffer for carbonic acid changes; and (4) phosphate system, secondary ICF and primary urinary buffer.

check your understanding 13.7

- The role of lungs in acid–base balance is to alter ventilation to control exchange of H^+-generating CO_2 in gills and lungs or to alter transport of H^+ and HCO_3^- in gills.
- In the proximal tubule, H^+ is secreted by both H^+ ATPase pumps and by Na^+/H^+ antiporters. The means by which H^+ and HCO_3^- are handled in the Type A and Type B intercalated cells of the distal and collecting tubule are summarized in Figure 13-20.
- Kidneys of most mammals cannot acidify the urine beyond a gradient-limited urinary pH of 4.5. Ammonia production in proximal tubule cells helps this because the NH_3 binds a proton. Since membranes are not very permeable to NH_4^+, these ions remain in the tubular fluid and are lost in the urine, each one taking an H^+ with it. Thus, NH_3 secretion allows large amounts of H^+ to be secreted into the urine before the pH falls to the limit of 4.5

check your understanding 13.8

- The six categories of acid–base imbalances are respiratory, environmental, and metabolic acidosis, plus the same three for alkalosis.
- Uremic acidosis is very serious because it results from severe renal failure, so even normal daily pH imbalances cannot be corrected.

check your understanding 14.1

- The different modes of animal feeding include filter or suspension feeders, detritivores or deposit feeders, fluid feeders, carnivores, herbivores, omnivores, and symbiotic autotroph-bearing animals. See also p. 655.

check your understanding 14.2

- The four basic digestive processes include motility, secretion, digestion, and absorption. *Motility* refers to the muscular contractions within the gut tube that mix and move forward the contents of the digestive tract. Digestive juices are *secreted* into the digestive tract lumen by exocrine glands located along the route, each with its own specific secretory product(s). *Digestion* is the process whereby the structurally complex foodstuffs of the diet are broken down into smaller, absorbable units by enzymes produced within the digestive system. *Absorption* refers to the transfer of nutrients from the digestive tract lumen into the blood or hemolymph or body cavity.
- Carbohydrates can be digested into monosaccharides such as glucose, fructose, and galactose, which are the absorbable units. Proteins—the second category of foodstuffs—are degraded primarily into their constituent amino acids as well as a few polypeptides, both of which are the primary absorbable units for digested protein. Fats are the third category of foodstuffs; the end products of fat digestion are monoglycerides and free fatty acids, which are the absorbable units of fat. Glycogen represents the polysaccharide storage form of glucose in animals.
- The gut tube in most animals can be divided into three specific regions: the foregut, midgut, and hindgut. See also Figure 14-1 and the discussion on p. 655.

check your understanding 14.3

- *Mastication* is the motility (chewing) of the mouth that involves slicing, tearing, grinding, and mixing ingested foods by specialized mouthparts such as teeth. The purposes of mastication are (1) to grind and break food up into smaller pieces to facilitate swallowing and increase the food surface area on which salivary enzymes will act, (2) to mix food with saliva, and (3) to stimulate the taste buds, which reflexly increases salivary, gastric, pancreatic, and bile secretion to prepare for the arrival of food in the lower digestive tract.
- The functions of saliva include the moistening of food items, digestion of starch through the action of salivary amylase, defense against some bacteria, solubilization of molecules that stimulate the taste bud, and the neutralization of acids. In some species, saliva can be used as a means of temperature control, whereas in others, saliva can be used as venom to immobilize prey. Other uses may include anticoagulation, production of silk, and release of pheromones.

check your understanding 14.4

- See Figure 14-8 and the discussion on p. 667.
- The esophagus serves as a passageway for the rapid transit of a bolus of food from the mouth to the stomach (or the reverse in ruminants). The crop mainly serves a storage function that is particularly important in fluid-feeding insects, which have relatively large crops for storing blood or nectar.

check your understanding 14.5

- See Table 14-4 and the discussion on pp. 678–679.
- The enzyme pepsin is synthesized in the chief cells lining the stomach and is maintained in the inactive form of pepsinogen until it reaches the gastric lumen, where it is activated by HCl.
- See Figure 14-12 and the discussion on pp. 676–677.
- The proventriculus serves as the "glandular stomach," whereas the separate gizzard serves as the "muscular stomach" to macerate and grind the ingesta.

check your understanding 14.6

- See the discussion in Section 14.6 for the contributions of the pancreas, liver, gallbladder, and fat body to digestion. Nondigestive functions of the liver include detoxification or degradation of body wastes and hormones, as well as drugs and other foreign compounds; synthesis of plasma proteins; activation of vitamin D; removal of bacteria and worn-out red blood cells; secretion of the hormones thrombopoietin, hepcidin, and insulin-like growth factor-I; synthesis of ascorbic acid in some species; and buoyancy in sharks.
- Bile salts aid fat digestion through their detergent action and facilitate fat absorption through their participation in the formation of micelles.

check your understanding 14.7

- See Figure 14-28 and the discussion on pp. 701–703.
- See the discussion on how the three major categories of nutrients (carbohydrates, proteins, and fats) are digested on pp. 694–699. See also Table 14-5 and Figures 14-25, 14-26, and 14-27.

check your understanding 14.8

- By positioning the fermentation chamber after the stomach (postgastric fermentation), the host animal loses the opportunity to exploit the high-quality microbially synthesized protein produced in the rumen (pregastric fermentation). Further, the digestive capabilities of the stomach and small intestine cannot contribute to the digestion of microbial products.
- The large intestine serves as a temporary storage site for excreta, functions in electrolyte and fluid balance, and harbors microbes for fermentation and production of volatile fatty acids.
- In carnivores, the major function of the colon is the absorption of electrolytes, water, and other materials that escaped absorption in the small intestine. In omnivores, and particularly in herbivores that ingest considerable amounts of complex polysaccharides, the structure of the large intestine tends to be voluminous, providing a site for microbial digestion. The environment within the colon/cecum of postgastric fermenters must also be able to support a substantial microbe population in a buffered environment. In many vertebrates, specialized transport mechanisms are not present in the colonic mucosa for absorption of glucose or amino acids, as there are in the small intestine. A notable exception is the colon of birds, where both glucose and amino acids are absorbed by secondary active transport.

check your understanding 14.9

- Ruminant species have evolved a stomach to house a population of microbes able to break down cellulose and other complex polysaccharides into end products of fermentation (VFAs), which contribute to the nutritional requirements of the host animal. VFAs provide energy by entering the TCA cycle as acetyl CoA, with the excess energy stored as body fat in the tissues. These microorganisms can also protect the host animal from some toxins that are ingested in the diet. Nonruminants are unable to exploit this food source, which restricts their habitat range.
- VFAs produced by the rumen microbes are passively absorbed through the rumen wall. Ammonia as well as amino acids also are readily absorbed through the lining of the ruminoreticulum and omasum.

check your understanding 14.10

- See Section 14.10 and the discussion on pp. 711–712.

check your understanding 15.1

- The First Law of Thermodynamics is that energy can be neither created nor destroyed. The Second Law is that the entropy (a measure of disorder or randomness) of a system plus its surroundings increases over time as the energy content degrades to unusable heat. The consequence is that life must continuously extract energy from the environment to maintain itself, whereas the rest of the universe degrades even faster than it would on its own.
- External work is the energy expended when skeletal muscles contract to move external objects or to move the body through the environment. Internal work constitutes all other forms of energy expenditure: skeletal muscle activity used for nonexternal purposes (e.g., breathing); pumping of blood; active transport across plasma membranes; and synthetic reactions

for maintenance, repair, and growth. Heat is the primary output because entropy increases for all processes except net growth (order).

check your understanding 15.2

- *Metabolic rate* (MR) is energy expenditure per unit time. *Standard* and *basal MR* (SMR, BMR) are the minimal amounts of energy needed per unit time to sustain waking (non-sleep) life under optimal conditions. BMR is for endotherms in thermal neutral zones; SMR is for ectotherms at a defined natural temperature, such as that the animal prefers. Diet-induced thermogenesis is an increase in MR above the SMR or BMR due to food processing. Indirect calorimetry by respirometry measures an animal's O_2 uptake per unit time, then makes an estimate of MR using physiological fuel values for a typical diet.
- Some physiological features depend on the organism size, because of scaled factors, such as surface area (increases with radius squared) and volume (increases with radius cubed). Although larger animals consume more energy than smaller animals, they have lower metabolic rates per unit mass than smaller ones. BMR scaling with body mass, with an exponent of 0.75, has been attributed to heat retention because heat is lost via surface area but produced by volume. Problems with this and alternative hypotheses are discussed in the box on p. 722.
- The three states of energy balance are (1) neutral (net food intake equals energy output), (2) positive (net food intake exceeds output, so body mass increases), and (3) negative (net food intake is less than output, so body mass decreases).
- Aerobic scope is the ratio of MR measured by O_2 consumption at its highest level to the BMR or SMR; it is larger in larger animals and may be limited by O_2 delivery. With increasing speed, running MR increases linearly, swimming MR exponentially, and flying MR in varying ways—for example, gliding saves energy after costly take-off. Hopping MR may not increase at all with speed.

check your understanding 15.3

- The arcuate nucleus in the hypothalamus has a pair of appetite centers and another pair of satiety centers. These monitor various signals (see next answer) and adjust eating behavior and energy expenditures to keep body energy levels at a set point.
- The roles in energy balance (along with sources) for leptin, insulin, ghrelin, PYY_{3-36}, and cholecystokinin (CCK) are given in Table 15-2.
- The effector that regulates energy input is primarily eating behavior. Effector actions (which may be increased or decreased) for regulating energy expenditures are BMR, gametogenesis, DIT resulting in part from brown adipose tissue, and exercise and nonexercise activity thermogenesis.
- The evolutionary differences in body-mass "set points" have two extremes. Lean animals evolve where food is plentiful much of the year; they are agile but can starve easily. Heavy animals evolve in "feast or famine" habitats; they readily store energy for famine times but are less agile.

check your understanding 15.4

- The basic effects of temperature on biological systems (Figure 15-7) are (1) increasing the rate of reactions with temperature rise due to kinetic energy, and (2) at relatively high temperatures, causing macromolecules to begin denaturing.
- In homeoviscous adaptation, cold-adapted membranes have more unsaturated fatty acids to give them proper fluidity in the cold, but they are too fluid as temperature rises. Warm-adapted membranes have more saturated lipids to resist that problem, but they are too stiff to work in the cold. Similarly, cold-adapted proteins have fewer stabilizing bonds between folds to make them properly flexible in the cold, but they are too flexible as temperature rises. Warm-adapted proteins have more stabilizing bonds to resist that problem, but they are too stiff to work in the cold.
- *Endothermy, ectothermy, poikilothermy,* and *homeothermy* are defined on p. 733.
- Figure 15-12 summarizes heat transfers by four methods. Briefly, (1) radiation is light energy, particularly infrared, which transfers heat; (2) conduction is transfer of heat between objects of differing temperatures that are in direct contact with each other; (3) convection refers to the transfer of heat energy by currents of air or water; and (4) evaporation involves the heat required to transform water from a liquid to a gaseous state (and can be absorbed from the skin).
- Body heat balance depends on the sum of inputs and outputs (Figure 15-11). Animals may (1) gain external heat from the sun, warm rocks, etc.; (2) retain internal heat with insulation, behavior, reduced skin blood flow, etc.; (3) generate more heat with heat-generating tissues; and (4) lose excess heat by behavior, increased skin blood flow, evaporation (sweating, panting), etc.

check your understanding 15.5

- Pure poikilotherms' body temperatures follow the environment, and they must avoid extremes, enter dormant states, and/or compensate biochemically (see next answer). Ectothermic regulation involves regulation (mostly by behavior) of "gain, retain, and lose" mechanisms noted on pp. 736–737.
- Biochemical adjustments that account for metabolic compensation in the face of changing body temperatures include (1) increased enzyme concentrations in the cold; (2) homeoviscous membrane adaptation; (3) regulation of internal pH to a higher level—$[H^+]$ lower—in the cold to compensate for tighter H^+ binding by proteins; and (4) isoform regulation (activating different genes for similar proteins at different temperatures).
- Freeze tolerance involves actual freezing of some body fluids, but generally only ECF, as in the peeper frog whose fluids are flooded with glucose so as an antifreeze, enough to keep ICFs unfrozen. Freeze avoidance involves compatible antifreeze carbohydrates to prevent all internal freezing without disrupting macromocules, antifreeze proteins to block ice-crystal growth, supercooling, and cryprotective dehydration. An example is the Arctic springtail, which dehydrates while building up the sugar trehalose.
- Heat shock proteins (HSPs) are small, thermostable, hydrophobic proteins that bind to larger, unfolded proteins and assist their folding into functional conformations. HSP genes are induced by sudden rises in cell temperatures.

check your understanding 15.6

- To achieve homeothermy, birds and mammals may (1) gain external heat through behavior, dark skin, etc.; (2) retain internal heat by vaso-constriction, insulation, behavior, larger body size, countercurrent exchangers; (3) generate more heat with high BMR, muscles shivering, and brown adipose with uncoupling proteins; and (4) lose excess heat by reducing insulation, vasodilation, sweating and panting, exchangers, behavior, light-colored skin, etc.
- The hypothalamus maintains homeothermy using negative feedback (Figure 15-20). It elevates temperature to create a fever during infections when leukocytes release a pyrogen. There is some evidence that fever inhibits growth of pathogens and augments inflammation.

check your understanding 15.7

- In regional heterothermy, a tuna and a bee heat only parts of their bodies. In tunas, red swimming muscles constantly produce considerable heat, whereas in bees, flight muscles do so during flight. Both retain heat with countercurrent exchanges between them and sites of heat loss (gills in tunas, the abdomen in bees).
- Hibernation and torpor involve endotherms becoming nearly poikilothermic in winter and daily, respectively. Most still thermoregulate, but at much lower body temperatures. This saves energy, especially in small endotherms with their high surface-area-to-volume ratios.

check your understanding 15.8

- Findings that show effects of recent global warming on animal physiology include (1) theoretical calculations showing that effects on metabolic rates of ectotherms will be worst in the tropics; (2) corals bleaching and dying from unprecedented warming of seawater; (3) dying out of populations of Mexican lizards in areas with increasing springtime temperatures; (4) temperate species of limpets and butterflies moving northwards; and (5) earlier breeding in springtime in grapevine moths and red squirrels in correlation with warmer springs.

check your understanding 16.1

- *Oviparous* refers to a reproductive strategy in which the parents release eggs from which the young hatch after expulsion from the body; *ovoviviparous* refers to the strategy of producing eggs (not necessarily with shells) that hatch within the body of the parent; and *viviparous* is the production of living young within the body.
- Strategies used to ensure that both sexes are ready to breed at the same time include photoperiod (day length), temperature, presence of the male, and availability of a particular food source.

check your understanding 16.2

- See Figures 16-3 and 16-4 and the discussion on pp. 762–764.
- Genotypic sex, which depends on the combination of sex chromosomes at the time of conception, determines gonadal sex, that is, whether testes or ovaries develop. The presence or absence of a Y chromosome determines gonadal differentiation in mammals. Phenotypic sex, the apparent anatomic sex of an individual, depends on the genotypically determined gonadal sex.
- Two primitive duct systems—the mesonephric ducts, or Wolffian ducts, and the paramesonephric, or Müllerian ducts—develop in all mammalian embryos. In males, the reproductive tract develops from the Wolffian ducts and

the Müllerian ducts degenerate, whereas in females, the Müllerian ducts differentiate into the reproductive tract and the Wolffian ducts regress (Figure 16-6).

check your understanding 16.3

- Spermatogenesis encompasses three major stages: mitotic proliferation (spermatocytogenesis), meiosis, and packaging (spermiogenesis). See also Figure 16-9 and the discussion on pp. 773–775.
- LH and FSH act on separate components of the testes (Figure 16-11). LH acts on the Leydig (interstitial) cells and regulates testosterone secretion. FSH acts on the Sertoli cells to enhance spermatogenesis. Secretion of both LH and FSH from the anterior pituitary is stimulated, in turn, by a single hypothalamic hormone, GnRH, or inhibited by a GnIH.
- Testosterone is synthesized in the Leydig cells of the testes. Its concentration is much higher in the testes than in the blood, because a substantial portion is retained in the intratubular fluid, complexed with androgen-binding protein secreted by the Sertoli cells. Only this high concentration of testicular testosterone is adequate to sustain sperm production. In some species, evolution of secondary sexual characteristics, the many external characteristics of vertebrates that are not directly involved in reproduction itself, are dependent on secretion of the sex steroid.

check your understanding 16.4

- Arterioles are typically supplied by sympathetic nerves, with increased sympathetic activity producing vasoconstriction and decreased sympathetic activity resulting in vasodilation. Concurrent parasympathetic stimulation and sympathetic inhibition of penile arterioles accomplish vasodilation more rapidly and in greater magnitude than is possible in other arterioles supplied only by sympathetic nerves.
- Both sexes experience the same four phases of the sexual cycle—excitement, plateau, orgasm, and resolution. The excitement phase can be initiated by either physical or psychological stimuli. During the plateau phase, the changes initiated during the excitement phase intensify (such as increased heart rate, blood pressure, respiratory rate, and muscle tension). If erotic stimulation continues, the sexual response culminates in orgasm as sympathetic impulses trigger rhythmic contractions of the pelvic musculature. During the resolution phase following orgasm, sympathetic vasoconstrictor impulses slow the inflow of blood into the penis, causing the erection to subside. Oxytocin appears to play roles in mating-related behaviors in both sexes. It is released during arousal and orgasm and may stimulate orgasmic muscle contractions in both sexes.

check your understanding 16.5

- See Figures 16-9 and 16-15 and the discussion on pp. 786–788.
- After the onset of puberty, the ovary alternates between two phases: the follicular phase, which is dominated by the presence of maturing follicles, and the luteal phase, which is characterized by the presence of the corpus luteum. See also Figure 16-16 and the discussion on pp. 788–790.
- See Figure 16-17 and the discussion on pp. 791 and 794. Key similarities include the hormonal pathway of reproduction. In birds and mammals, the pathway revolves around the hypothalamic–pituitary–gonadal axis. The hypothalamus is activated by specific environmental cues, including day length as a trigger for onset of gonadal development. One of the most obvious differences in follicular development between birds and mammals is that in the bird's ovary the follicles do not mature synchronously. Unlike the situation in female mammals, ovulation of the largest follicle in birds is preceded by an increase in the concentrations of progesterone and LH. LH, in turn, stimulates the secretion of more progesterone from the granulosa of the mature preovulatory follicle. Thus, in birds, it is the granulosa that produce progesterone, most of which diffuses to the thecal cells, where it is converted to estrogen.

check your understanding 16.6

- At ovulation the ovum is released into the abdominal cavity, but the oviduct picks it up quickly. The fimbriae (Figure 16-4c) contract in a sweeping motion to guide the released ovum into the oviduct. Sperm migrate up the cervical canal under their own power. Once the sperm have entered the uterus, contractions of the myometrium disperse the sperm throughout the uterine cavity. When sperm reach the uterotubal junction, they are moved up the oviduct toward the site of fertilization with the help of smooth muscle contractions and cilia that move in the direction of the ampullary–isthmic junction. See also Figure 16-21 and the discussion on p. 799.
- See Figures 16-22 and 16-23 and the discussion on p. 801. When the blastocyst is ready to implant, its surface becomes sticky. By this time, the endometrium is ready to accept the early embryo, and it too has become more adhesive through increased formation of cell adhesion molecules that help attach the blastocyst when it first contacts the uterine lining. By the time the embryo has attached to the uterine luminal epithelium, the trophoblastic layer is two cell layers thick and is called the chorion. As the chorion continues to expand, it fuses with a structure called the allantois (an outgrowth of the hindgut), forming the vascularized chorioallantoic membrane. Placentas can be classified according to their

degree of invasiveness and the anatomy of those regions of greatest maternal–fetal interaction (Figure 16-24).
- To sustain the growing embryo for the duration of its intrauterine life, the placenta, a specialized organ of exchange between the maternal and fetal blood, is rapidly developed (Figure 16-24). The most important hormones secreted by the placenta include chorionic gonadotropin, estrogen, and progesterone (Table 16-5).
- Chorionic gonadotropin is a polypeptide hormone that prolongs the life span of the corpus luteum and its secretion of progesterone until the placenta takes over.

check your understanding 16.7

- Corticotropin-releasing hormone (CRH) secreted by the fetal portion of the placenta into both the maternal and fetal circulation drives the synthesis of placental estrogen, ultimately dictating the timing of the onset of labor. Thus, pregnancy duration and delivery timing are determined largely by the placenta's rate of CRH production. In the immediate days before the onset of parturition, soaring levels of estrogens bring about changes in the uterus and cervix to prepare them for labor and delivery (Figure16-25). First, high levels of estrogens promote the synthesis of connexons within the uterine smooth muscle cells. The newly manufactured connexons are inserted in the myometrial plasma membranes to form gap junctions so that the uterine smooth muscle cells become able to contract as a coordinated unit. Second, high levels of estrogens prior to parturition cause a progressive increase in the concentration of myometrial receptors for oxytocin. Together, these myometrial changes bring about the increased uterine responsiveness to oxytocin that ultimately initiates labor. A powerful uterine muscle stimulant, oxytocin plays the key role in the progression of labor. Labor begins when the oxytocin receptor concentration reaches a critical threshold that permits the onset of strong, coordinated contractions in response to ordinary concentrations of circulating oxytocin.
- Hormonal factors that play a role in lactation include the high concentration of estrogens, which promote extensive duct development in the mammary gland, as well as the high level of progesterone that stimulates abundant alveolar-lobular formation. Elevated concentrations of prolactin and chorionic somatomammotropin contribute to mammary gland development by inducing the synthesis of enzymes needed for milk production. Once milk production begins after delivery, two hormones maintain lactation: (1) prolactin, which acts on the alveolar epithelium to promote secretion of milk, and (2) oxytocin, which produces milk ejection.

Glossary

A band One of the dark bands that alternate with light (I) bands to create a striated appearance in skeletal or cardiac muscle fibers when these fibers are viewed with a light microscope

abomasum The fourth compartment of the ruminant stomach, which has functions similar to the glandular stomach of nonruminants

absorptive state The metabolic state following a meal when nutrients are being absorbed and stored; fed state

accessory digestive organs Exocrine organs outside the wall of the digestive tract that empty their secretions through ducts into the digestive tract lumen

accessory sex glands Glands that empty their secretions into the reproductive tract

acclimation A laboratory phenomenon in which the chronic response of an animal to a change in environment is measured; normally the old and new environments differ in one or two highly specific ways

acclimatization The (usually slow) process of changing physiological processes to function more optimally under new conditions

accommodation In the eye, the ability to adjust the strength of the lens so that both near and far sources can be focused on the retina

acetylcholine (ACh) (as′-uh-teal-KŌ-lēn) The neurotransmitter released from all autonomic preganglionic fibers, parasympathetic postganglionic fibers, and motor neurons

acetylcholinesterase (AChE) (as′uh-teal-kō-luh-NES-tuh-rās) An enzyme present in the motor end-plate membrane of a skeletal muscle fiber that inactivates acetylcholine

ACh See *acetylcholine*

achalasia A condition that occurs commonly in dogs (and in humans) in which the lower esophageal sphincter fails to relax during swallowing but instead contracts more vigorously

AChE See *acetylcholinesterase*

acid A hydrogen-containing substance that on dissociation yields a free hydrogen ion

acidosis (as-i-DŌ-sus) Blood pH below the normal range (<7.35 in most mammals)

acini (ĀS-i-nī) The secretory component of saclike exocrine glands, such as enzyme-producing pancreatic glands or milk-producing mammary glands

acquired immune responses Responses (also called *adaptive*) that are selectively targeted against particular foreign material to which the body has previously been exposed; see also *antibody-mediated immunity* and *cell-mediated immunity*

acrorhagia Special stinging "battle" tentacles in some Anthozoan Cnidarians (anemones, corals)

ACTH See *adrenocorticotropic hormone*

actin The contractile protein that forms the backbone of the thin filaments in muscle fibers; also contributes to the motility of other kinds of cells

active expiration Emptying of the lungs more completely than when at rest by contracting the expiratory muscles; also called *forced expiration*

active reabsorption The condition when any step in the transepithelial transport of a substance requires energy expenditure

active transport Active carrier-mediated transport involving transport of a substance against its concentration gradient across the plasma membrane

acuity Discriminative ability; the ability to discern two different points of stimulation

adaptation (1) A reduction in receptor potential despite sustained stimulation of the same magnitude; (2) any feature of an organism that enhances evolutionary fitness

adenohypophysis See *anterior pituitary gland*

adenosine diphosphate (ADP) (uh-DEN-uh-sēn) The two phosphate products formed from the splitting of ATP to yield energy for the cell's use

adenosine triphosphate (ATP) The common energy "currency" of all life, which consists of an adenosine with three phosphate groups attached; splitting of the high-energy, terminal phosphate bond provides energy to power cell activities

adenylyl cyclase (ah-DEN-il-il sī-klās) The membrane-bound enzyme that is activated by a G-protein intermediary in response to binding of an extracellular messenger with a surface membrane receptor and that, in turn, activates cyclic AMP, an intracellular second messenger

adequate stimulus The main type of stimulus to which a particular receptor (sensory cell) is specialized to respond

ADH See *vasopressin*

adiponectin A hormone produced by adipose tissue in nonobese mammals; may serve in regulating energy balance

adipose tissue The tissue specialized for storage of triglyceride fat; found under the skin in the hypodermis

ADP See *adenosine diphosphate*

adrenal cortex (uh-DRĒ-nul) The outer portion of the vertebrate adrenal gland; secretes three classes of steroid hormones: glucocorticoids, mineralocorticoids, and sex hormones

adrenal medulla (muh-DUL-uh) The inner portion of the vertebrate adrenal gland; an endocrine gland that is a modified sympathetic ganglion that secretes the hormones epinephrine and norepinephrine into the blood in response to sympathetic stimulation

adrenergic fibers (ad′-ruh-NUR-jik) Nerve fibers that release norepinephrine as their neurotransmitter

adrenocorticotropic hormone (ACTH) (ad-rē′-nō-kor′-tuh-kō-TRŌP-ik) An anterior pituitary hormone that stimulates cortisol secretion by the adrenal cortex and promotes growth of the adrenal cortex

aerobes Organisms that rely on oxygen-based metabolism

aerobic Referring to a condition in which oxygen is available; **aerobic metabolism** is the process in which ATP formation is accomplished through oxidative phosphorylation

afferent arteriole (AF-er-ent ar-TIR-ē-ōl) The vessel that carries blood into the glomerulus of the mammalian kidney's nephron

afferent division The portion of the peripheral nervous system that carries information from the periphery to the central nervous system; made of **afferent neurons**, which have sensory receptors at their peripheral endings

after hyperpolarization (hī′-pur-pō-luh-ruh-ZA-shun) A slight, transient hyperpolarization that sometimes occurs at the end of an action potential

agonist A nonnative signal molecule (e.g., a drug or toxin) that mimics the effects of a native signal molecule (for opposing term, see *antagonist*)

agranulocytes (ā-GRAN-yuh-lō-sīs′) Leukocytes that do not contain granules, including lymphocytes and monocytes

air capillary Small airways that branch off from the principal respiratory tubules, the parabronchi, within the avian lung; these are the site of gas exchange between the air and blood

air sac A thin-walled, air-filled component of the avian respiratory system with little musculature and few blood vessels

albumin (al-BEW-min) The smallest and most abundant of the plasma proteins; binds and transports many water-insoluble substances in the blood; contributes extensively to plasma-colloid osmotic pressure

aldosterone (al-dō-steer-OWN or al-DOS-tuh-rōn) The adrenocortical hormone that stimulates Na^+ reabsorption by the distal and collecting tubules of the kidney's nephron during urine formation

alkalosis (al′-kuh-LŌ-sus) Blood pH above the normal range (>7.45 in most mammals)

all-or-none law An excitable membrane either responds to a stimulus with a maximal action potential that spreads nondecrementally throughout the membrane or does not respond with an action potential at all

allosteric ("other site") A noncatalytic binding site on an enzyme that binds regulatory molecules

alpha (α) cells The endocrine pancreatic islet cells that secrete the hormone glucagon

alpha (α) motor neuron A motor neuron that innervates ordinary skeletal muscle fibers

altricial Referring to young that are absolutely dependent on continued parental care

alveolar surface tension (al-VĒ-ō-lur) The surface tension of the fluid lining the alveoli in the lungs; see *surface tension*

alveolar ventilation Volume of air exchanged between the atmosphere and alveoli per minute; equals (tidal volume – dead-space volume) × respiratory rate

alveoli (al-VĒ-ō-lī) The air sacs across which O_2 and CO_2 are exchanged between the blood and air in mammalian (and some other vertebrate) lungs

amines (ah-means) Hormones derived from amino acids, especially tyrosine; include thyroid hormone and catecholamines

ammonotelic Having ammonia as a primary nitrogenous waste product

amoebocyte Wandering immune cell in nonvertebrates

amoeboid movement (uh-MĒ-boid) "Crawling" movement of white blood cells, similar to the means by which amoebas move

ampullary electroreceptors Found in almost all nonteleost fishes and in some teleosts as well as in several amphibian species; respond to low-frequency electric signals that are typical of electrical output from animal nerves and hearts

amygdala Structure of mammalian forebrain (limbic system) located on the interior underside of the temporal lobe; believed to be the homologue of the reptilian and avian paleostriatum; appears to store memories of highly emotional events

anabolism (ah-NAB-ō-li-zum) The buildup, or synthesis, of larger organic molecules from the small organic molecular subunits

anaerobes Organisms that can survive on metabolism that does not require oxygen

anaerobic (an'-uh-RŌ-bik) Referring to a condition in which oxygen is not present; in **anaerobic metabolism**, ATP formation is accomplished by anaerobic glycolysis, usually only for brief periods of time when O_2 delivery is inadequate to support oxidative phosphorylation

analgesic (an-al-JEE-zic) Pain relieving

anatomy The study of body structure

androgen A vertebrate masculinizing "male" sex hormone; includes testosterone from the testes and dehydroepiandrosterone from the adrenal cortex

anemia A reduction below normal in the O_2-carrying capacity of the blood

anestrus Condition where a female mammal does not cycle into reproductive readiness, because of insufficient hormonal stimuli

angiotensin-converting enzyme (ACE) An enzyme in the lung that converts inactive angiotensin I to angiotensin II

anhydrobiotic Literally, life without water; referring to dormant stages of organisms with very little water content

anion (AN-ī-on) Negatively charged ion that has gained one or more electrons in its outer shell

anoxia The complete lack of oxygen

ANP See *atrial natriuretic peptide*

antagonism A process opposing another; e.g., when one hormone causes the loss of another hormone's receptors, reducing the effectiveness of the second hormone

antagonist A process opposing another; e.g., (1) a nonnative signal molecule (such as a drug or toxin) blocking the effects of a native signal molecule (for opposite term, see *agonist*); (2) a muscle that moves in opposition to another

anterior pituitary The nonneural endocrine portion of the vertebrate pituitary gland, having three parts with variable representation in various species: pars tuberalis, pars intermedia, and pars distalis; stores and secretes six different hormones: GH, TSH, ACTH, FSH, LH, and prolactin

antibody A vertebrate immunoglobulin produced by a specific activated B lymphocyte (plasma cell) against a particular antigen; binds with the specific antigen against which it is produced and promotes the antigenic invader's destruction by augmenting nonspecific immune responses already initiated against the antigen

antibody-mediated immunity A specific immune response accomplished by antibody production by B cells

antidiuresis A state of low urine output

antidiuretic hormone (an'-ti-dī'-yū-RET-ik) See *vasopressin*

antigen A large complex molecule that triggers a specific immune response against itself when it gains entry into a vertebrate body

antigen-presenting cell (APC) A vertebrate immune cell that can ingest and digest antigens, then "present" them to helper T cells to activate the acquired immune response

antimicrobial peptides Small defensive proteins produced by barrier tissues and some immune cells in many, if not all animals; many kill microbes by creating pores in their membranes

antioxidant A substance that helps inactivate biologically damaging free radicals

antiporter A transporter protein in a membrane that moves two (or more) molecules or ions in the opposite direction

antrum (of ovary) The fluid-filled cavity formed within a developing ovarian follicle

antrum (of stomach) The lower portion of the vertebrate stomach

aorta (a-OR-tah) The large vessel that carries blood from the vertebrate heart to the body

aortic valve A one-way valve that permits the flow of blood from the mammalian left ventricle into the aorta during ventricular emptying but prevents the backflow of blood from the aorta into the left ventricle during ventricular relaxation

apnea The absence of breathing

apodemes In arthropods, ridges that project from the inner face of the exoskeleton for the attachment of muscles

apoptosis (ā-pop-TŌ-sis) Programmed cell death; deliberate self-destruction of a cell (as opposed to necrosis, unintended cell death)

appetite centers Neuronal clusters in the lateral regions of the vertebrate hypothalamus that drive the animal to eat

aquaporin A protein that forms a channel in a membrane that allows water to diffuse through it

aqueous humor (Ā-kwē-us) The clear watery fluid in the anterior chamber of the vertebrate eye; provides nourishment for the cornea and lens

archaea Ancient prokaryotes forming one of three distinct forms of life (along with bacteria and eukarya), often found in extreme habitats

such as hot springs, salt brines; thought to be related to the ancestor of the main eukaryotic cell

aromatase The enzyme that converts androgens (e.g., testosterone) into estrogens (estradiol)

arterioles (ar-TIR-ē-ōlz) The highly muscular, high-resistance vessels of vertebrates, the caliber of which can be changed subject to control to determine how much of the cardiac output is distributed to each of the various tissues

artery A vessel that carries blood away from the heart

ascending tract A bundle of nerve fibers of similar function that travels up the vertebrate spinal cord to transmit signals derived from afferent input to the brain

astrocyte A type of glial cell in the vertebrate brain; major functions include holding the neurons together in proper spatial relationship and inducing the brain capillaries to form tight junctions important in the blood–brain barrier

asynchronous Out of step or phase control of flight muscles; a single Ca^{2+} release "turns on" the flight muscle; when in this state, the muscle becomes activated by stretch and deactivated by shortening

atherosclerosis (ath-uh-rō-skluh-RŌ-sus) A progressive, degenerative arterial disease that leads to gradual blockage of vertebrate vessels, reducing blood flow through them

atmospheric pressure The pressure exerted by the weight of the air in the atmosphere on objects on Earth's surface; equals 760 mm Hg at sea level

ATP See *adenosine triphosphate*

ATPase An enzyme that hydrolyses ATP into ADP and energy

ATP synthase (SIN-thās) The enzyme within the mitochondrial inner membrane that phosphorylates ADP to ATP

atrial natriuretic peptide (ANP) (Ā-trē-al NĀ-tree-ur-eh'tik) A peptide hormone released from the vertebrate cardiac atria that promotes urinary loss of Na^+ in mammals

atrioventricular (AV) node (ā'-trē-ō-ven-TRIK-yuh-lur) A small bundle of specialized cardiac cells located at the junction of the atria and ventricles that serves as the only site of electrical contact between the atria and ventricles in vertebrates

atrioventricular (AV) valve A one-way valve that permits the flow of blood from the atrium to the ventricle during filling of the mammalian heart but prevents the backflow of blood from the ventricle to the atrium during emptying of the heart

atrium (pl., **atria**) (Ā-tree-um) A chamber of the heart that receives blood from the veins or hemolymph and transfers it to a ventricle

atrophy (AH-truh-fē) Decrease in mass of an organ by loss of cells

autocrine A locally acting signal molecule regulating a cellular process of the cell that secreted it ("self-stimulation")

autoimmune disease Disease characterized by erroneous production of antibodies against one of the body's own tissues

autonomic nervous system The portion of the efferent division of the vertebrate peripheral nervous system that innervates smooth and cardiac muscle and exocrine glands; composed of two subdivisions, the sympathetic nervous system and the parasympathetic nervous system

autorhythmicity The ability of an excitable cell to rhythmically initiate its own action potentials

AV valve See *atrioventricular valve*

avoiders Organisms that reduce disturbances to a particular physiological state by behaviorally avoiding environmental changes

axon A single, elongated tubular extension of a neuron that conducts action potentials away from the cell body; also known as a *nerve fiber*

axon hillock The first portion of a neuronal axon plus the region of the cell body from which the axon leaves; the site of action potential initiation in most neurons

axon terminals The branched endings of a neuronal axon, which release a neurotransmitter that influences target cells in close association with the axon terminals

B lymphocytes (B cells) Vertebrate white blood cells that produce antibodies against specific *antigens* to which they have been exposed

bacteria Prokaryotes forming one of three distinct forms of life (along with archaea and eukarya); included are those related to the ancestors of the eukaryotic mitochondria and chloroplast

baculum Penis bone in some mammals

bag cells Two clusters of cells in crustacean connective tissue above the abdominal ganglion; produce egg-laying hormone (ELH)

baleen Parallel fused filaments of keratin, the same protein of hair fibers that hang from the upper jaws of baleen whales in place of teeth

baroreceptor reflex An autonomically mediated reflex response that influences the vertebrate heart and blood vessels to oppose a change in mean arterial blood pressure

baroreceptors Receptors located within the vertebrate circulatory system that monitor blood pressure

basal metabolic rate (BMR) (BĀ-sul) The minimal rate of internal energy expenditure; an endotherm body's "idling speed"

basal nuclei Several masses of gray matter located deep within the white matter of the cerebrum of the vertebrate brain; play an important inhibitory role in motor control

basal transcription complex An assembly of proteins that initiate gene transcription in eukaryotes

base A substance that can combine with a free hydrogen ion and remove it from solution

basic electrical rhythm (BER) Self-induced electrical activity of the digestive tract smooth muscle

basilar membrane (BAS-ih-lar) The membrane that forms the floor of the middle compartment of the vertebrate cochlea and bears the organ of Corti, the sense organ for hearing

basophils (BAY-so-fills) White blood cells of vertebrates that synthesize, store, and release histamine, which is important in allergic responses

BER See *basic electrical rhythm*

beta (β) cells The endocrine pancreatic cells that secrete the hormone insulin

bicarbonate (HCO₃⁻) The anion resulting from dissociation of carbonic acid, H_2CO_3

bile salts Cholesterol derivatives secreted in vertebrate bile that facilitate fat digestion through their detergent action and facilitate fat absorption through their micellar formation

biliary system (BIL-ē-air′-ē) The bile-producing system, consisting of the liver, gallbladder, and associated ducts in vertebrates

bilirubin (bill-eh-RŪ-bin) A bile pigment antioxidant and waste product derived from biliverdin

biliverdin A bile pigment that is a waste product derived from the degradation of hemoglobin during the breakdown of old red blood cells

bimodal Having two respiratory exchange surfaces, including the gills and either cutaneous exchange or lungs

blastocyst The developmental stage of the fertilized mammalian ovum by the time it is ready to implant; consists of a single-layered sphere of cells encircling a fluid-filled cavity

blood–brain barrier Special structural and functional features of vertebrate brain capillaries that limit access of materials from the blood into the brain tissue

body system A collection of organs that perform related functions and interact to accomplish a common activity that is essential for survival of the whole body; for example, the digestive system

bone marrow The soft, highly cellular tissue that fills the internal cavities of vertebrate bones and is the source of most blood cells in terrestrial vertebrates

book lung Respiratory organ in many arachnids consisting of numerous membranous folds arranged like the pages of a book with sheetlike air spaces

Bowman's capsule The beginning of the tubular component of the mammalian kidney's nephron that cups around the glomerulus and collects the glomerular filtrate as it is formed

Boyle's law (boilz) At any constant temperature, the pressure exerted by a gas varies inversely with the volume of the gas

brain The most anterior, most highly developed portion of a central nervous system

brainstem The portion of the vertebrate brain that is continuous with the spinal cord, serves as an integrating link between the spinal cord and higher brain levels, and controls many life-sustaining processes, such as breathing, circulation, and digestion

bronchioles (BRONG-kē-ōlz) The small branching airways within vertebrate lungs

bronchoconstriction Narrowing of the respiratory airways

bronchodilation Widening of the respiratory airways

brush border The collection of microvilli projecting from the luminal border of epithelial cells lining the digestive tract and kidney tubules

buffer system A mixture in a solution of two or more chemical compounds that minimize pH changes when either an acid or a base is added to or removed from the solution

bulbourethral glands (bul-bo-you-RĒTH-ral) Male accessory sex glands that secrete mucus for lubrication (in mammals)

bulbus arteriosus In bony fish, a chambered extension of the ventricle through which blood exits the heart

bulk flow Movement in bulk of a protein-free plasma across the capillary walls between the blood and surrounding interstitial fluid; encompasses ultrafiltration and reabsorption

bulk transport Active movement of the medium (gas or liquid)

bundle of His (hiss) A tract of specialized cardiac cells that rapidly transmits an action potential down the interventricular septum of the avian and mammalian heart

bursa of Fabricius A gut-related lymphoid tissue unique to birds; site of B-cell maturation

C cells The thyroid cells that secrete calcitonin

calcitonin (kal′-suh-TŌ-nun) A vertebrate hormone secreted by the thyroid C cells that lowers plasma Ca^{2+} levels

calcium homeostasis Maintenance of a constant free plasma Ca^{2+} concentration

calmodulin (kal′-MA-jew-lin) An intracellular Ca^{2+}-binding protein that, on activation by Ca^{2+}, induces a change in structure and function of another intracellular protein; especially important in smooth muscle excitation–contraction coupling

CAMs See *cell adhesion molecules*

capacitance The ability of an insulating material, such as the cell membrane, to store an electrical charge

capacitation The stage in which mammalian spermatozoa become capable of fertilization

capillaries The thin-walled, pore-lined, smallest blood vessels, across which exchange between the blood and surrounding tissues takes place

capsaicin The "hot"-tasting chemical of chili peppers

carbonic anhydrase (an-HĪ-drās) The enzyme that catalyzes the conversion of CO_2 and OH^- into HCO_3^-

cardiac cycle One period of systole and diastole

cardiac muscle The specialized muscle found only in the heart

cardiac output (C.O.) The volume of blood pumped by each ventricle each minute; equals stroke volume × heart rate

cardiovascular control center The integrating center located in the medulla of the vertebrate brainstem that controls mean arterial blood pressure

carnivore Animal that captures and consumes live prey or that locates and consumes carrion

carrier molecules Membrane proteins that, by undergoing reversible changes in shape so that specific binding sites are alternately exposed at either side of the membrane, can bind with and transfer particular substances unable to cross the plasma membrane on their own

carrier-mediated transport Transport of a substance across the plasma membrane facilitated by a carrier molecule

cascade A series of sequential reactions that culminates in a final product, such as a clot

catabolism (kuh-TAB-ō-li-zum) The breakdown, or degradation, of large energy-rich molecules within cells

catalase (KAT-ah-lās) An antioxidant enzyme found in peroxisomes that decomposes potent hydrogen peroxide into harmless H_2O and O_2

catch state In some nonvertebrates such as bivalve mollusks, a state of continuous contraction of smooth muscle that maintains tension with little metabolic cost

catecholamines (kat′-uh-KŌ-luh-means) The chemical classification of the vertebrates adrenomedullary hormones

cations (KAT-ī-onz) Positively charged ions that have lost one or more electrons from their outer shell

caveolae (kā-vē-Ō-lē) Cavelike indentations in the outer surface of the plasma membrane that contain an abundance of membrane receptors and serve as important sites for signal transduction

cDNA DNA code that is synthesized in the laboratory as a complementary copy of an RNA isolated from a tissue

cell The smallest unit capable of carrying out the processes associated with life; the basic unit of both structure and function of living organisms

cell adhesion molecules (CAMs) Proteins that protrude from the surface of the plasma membrane and form loops or other appendages that the cells use to grip each other and the surrounding connective tissue fibers

cell body The portion of a neuron that houses the nucleus and organelles

cell-mediated immunity A specific immune response accomplished by activated T lymphocytes, which directly attack unwanted cells

center A functional collection of cell bodies within the central nervous system

central chemoreceptors (kē-mō-rē-SEP-turz) Receptors located in the vertebrate medulla near the respiratory center that respond to changes in ECF H^+ concentration resulting from changes in arterial P_{CO_2} and adjust respiration accordingly

central lacteal (LAK-tē-ul) The initial lymphatic vessel that supplies each of the vertebrate small intestinal villi

central nervous system (CNS) The brain and longitudinal nerve cord(s), which integrate input from sensory neurons and output to effectors

central sulcus (SUL-kus) A deep infolding of the mammalian brain surface that runs roughly down the middle of the lateral surface of each cerebral hemisphere and separates the parietal and frontal lobes

centrioles (SEN-tree-ōls) A pair of short, cylindrical structures within a cell that form the mitotic spindle during cell division

cerebellum (ser′-uh-BEL-um) The portion of the vertebrate brain attached to the brainstem and concerned with maintaining proper position of the body in space and subconscious coordination of motor activity

cerebral cortex The outer shell of gray matter in the vertebrate cerebrum; site of initiation of all voluntary motor output and final perceptual processing of all sensory input as well as integration of most higher neural activity

cerebral hemispheres The cerebrum's two halves, which are connected by a thick band of neuronal axons

cerebrospinal fluid (CSF) (ser′-uh-brō-SPĪ-nul or sah-REE-brō-SPĪ-nul) A special cushioning fluid that is produced by, surrounds, and flows through the central nervous system of vertebrates

cerebrum (SER-uh-brum or sah-REE-brum) The division of the vertebrate brain that consists of the basal nuclei and cerebral cortex

channel Small water-filled passageway through the plasma membrane; formed by membrane proteins that span the membrane and provide a highly selective passage for small water-soluble substances such as ions

chemically gated Referring to channels in the plasma membrane that open or close in response to the binding of a specific chemical messenger with a membrane receptor site that is in close association with the channel

chemoreceptor (kē-mo-rē-sep′-tur) A sensory receptor sensitive to specific chemicals

chemotaxin (kē-mō-TAK-sin) A chemical released at an inflammatory site that attracts phagocytes to the area

chemotaxis Movement of an organism toward or away from a chemical substance

chief cells The vertebrate stomach cells that secrete pepsinogen

chitin A polymer of repeating N-acetylglucosamine, serving as the primary structural molecule of arthropod exoskeletons

chloride cells Specialized cells, which transport NaCl, in the gill epithelia of fishes

cholecystokinin (CCK) (kō′-luh-sis-tuh-kī-nun) A hormone released from the vertebrate duodenal mucosa primarily in response to the presence of fat; inhibits gastric motility and secretion, stimulates pancreatic enzyme secretion, and stimulates gallbladder contraction

cholesterol A type of lipid molecule that serves as a precursor for steroid hormones and bile salts and is a stabilizing component of the plasma membrane

cholinergic fibers (kō′-lin-ER-jik) Nerve fibers that release acetylcholine as their neurotransmitter

chorionic gonadotropin (CG) (kō-rē-ON-ik gō-nad′-uh-TRŌ-pin) A hormone secreted by the early mammalian embryo and developing placenta that stimulates and maintains the corpus luteum of pregnancy

chromatophores Pigment-containing cells that are found in the integument of some animals and are responsible for physiological color change; the color granules are dispersed or concentrated depending on the animal's perception of its surroundings

chronic obstructive pulmonary disease A group of lung diseases characterized by increased airway resistance resulting from narrowing of the lumen of the lower airways; includes asthma, chronic bronchitis, and emphysema

chyme (kīm) A thick liquid mixture of food and digestive juices

cilia (SILL-ee-ah) Motile hairlike protrusions from the surface of many cells, such as in ciliate protozoa, flatworm integument, and linings of mammalian respiratory airways and oviducts

ciliary body The portion of the vertebrate eye that produces aqueous humor and contains the ciliary muscle

ciliary muscle A circular ring of smooth muscle, within the vertebrate eye, whose contraction increases the strength of the lens to accommodate for near vision

circadian rhythm (sir-KĀ-dē-un) Repetitive oscillations in the set point of various body activities, such as hormone levels and body temperature, that are very regular and have an approximate frequency of 24 hours, usually linked to light–dark cycles; diurnal rhythm; biological rhythm

circannual Biological rhythm that fluctuates yearly

cistern Structure associated with the mammalian teat; permits pooling of milk from several ducts before exiting the gland

citric acid cycle A cyclical series of biochemical reactions that involves the further processing of intermediate breakdown products of nutrient molecules, resulting in the generation of carbon dioxide and the preparation of hydrogen carrier molecules for entry into the high-energy–yielding electron transport chain

cloaca The final chamber of the hindgut of birds and reptiles

clone An exact genetic copy of a gene, or an organism arising from an exact copy of a genome

clutch The number of eggs incubated at one time

CNS See central nervous system

cochlea (KOK-lē-uh) The snail-shaped portion of the vertebrate inner ear that houses the receptors for sound

coevolution The process in which adaptive changes in one organism favor selection for adaptations in a second organism, which, in turn, favor selection for a different adaptation in the first organism, and so on

collecting tubule The last portion of tubule in the mammalian kidney's nephron that empties into the renal pelvis

colligative properties Solution properties of an idealized solute at 1 mole in 1 kg of water (defined as 1 molal), which has an osmotic pressure of 22.4 atmospheres (atm), raises boiling point by 0.54°C, depresses freezing point by 1.86°C, and also reduces vapor pressure

colloid (KOL-oid) The thyroglobulin-containing substance enclosed within the thyroid follicles

colloid osmotic pressure The difference in osmotic pressure that exists between the ECF and ICF due to differences in their concentrations of nonpermeating proteins

compatible osmolyte An organic osmolyte that elevates the osmotic pressure of a body fluid without perturbing cell functions

complement system A collection of vertebrate plasma proteins that are activated in cascade fashion on exposure to invading microorganisms, ultimately producing a membrane attack complex that destroys the invaders

compliance The distensibility of a hollow, elastic structure, such as a blood vessel or the lungs; a measure of how easily the structure can be stretched

compound eye Multifaceted arthropod eye composed of numerous optical units called ommatidia

concentration gradient A difference in concentration of a particular substance between two adjacent areas

conduction Transfer of heat between objects of differing temperatures that are in direct contact with each other

cones The vertebrate eye's photoreceptors used for color vision in the light

conformers Organisms in which a particular physiological state matches that of the environment

congestive heart failure The inability of the cardiac output to keep pace with the body's needs for blood delivery, with blood damming up in the veins behind the failing heart

connective tissue Tissue that serves to connect, support, and anchor various body parts; distinguished by relatively few cells dispersed within an abundance of extracellular material

contiguous conduction The means by which an action potential is propagated throughout a nonmyelinated nerve fiber; local current flow between an active and adjacent inactive area brings the inactive area to threshold, triggering an action potential in a previously inactive area

contractile proteins Myosin and actin, whose interaction brings about shortening (contraction) of a muscle fiber

conus arteriosus In cartilaginous fish, a chambered extension of the ventricle through which blood exits the heart

convection Transfer of heat energy by air or water currents

convergence (1) The converging of many presynaptic terminals from thousands of other neurons on a single neuronal cell body and its dendrites so that activity in the single neuron is influenced by the activity from many other neurons; (2) the independent evolution of similar structures in different species without a common ancestral structure

convex Curved out, as a surface in a lens that converges light rays

cooperativity The binding of a ligand, such as O_2, to its binding site on a multisubunit protein, such as hemoglobin, subsequently affecting the ability of other binding sites on the same protein to associate with additional ligand

coprodaeum Anterior portion of the avian cloaca, receives the excreta from the digestive tract

coprophagy Reingestion of feces

copulatory plug Temporary mating plug that forms after fertilization, which prevents both the loss of semen from the female tract as well as entry of sperm into the vagina (genital chamber) from a competing male

core temperature The temperature within the inner core of a body (abdominal and thoracic organs, central nervous system, and skeletal muscles)

cornea (KOR-nee-ah) The clear, most anterior, outer layer of an eye, through which light rays pass to the interior of the eye

coronary circulation The blood vessels that supply the vertebrate heart muscle

corpus luteum (LOO-tē-um) The ovarian structure that develops from a ruptured follicle following ovulation in a mammal

cortisol (KORT-uh-sol) The vertebrate adrenocortical hormone that plays an important role in carbohydrate, protein, and fat metabolism and helps the body resist stress; functionally similar to corticosterone

cotransport See *symporter* and *antiporter*

counteracting osmolyte An organic osmolyte that elevates the osmotic pressure of a body fluid and that counteracts the effects of a perturbant of cellular macromolecules

countercurrent A design in which fluid in two juxtaposed tubes flows in opposite directions (alternatively, in which flow in a single tube alternates between inflow and outflow)

cranial nerves The 12 pairs of vertebrate peripheral nerves, the majority of which arise from the brainstem

crop An enlarged posterior portion of the foregut, just anterior to the esophagus

crop milk Lipid material synthesized by epithelial cells of the crop of adult pigeons and doves; these cells are sloughed off and mixed with food already present in the crop and fed by regurgitation to juveniles

cross bridges The myosin molecules' globular heads that protrude from a thick filament within a muscle fiber and interact with the actin molecules in the thin filaments to shorten the muscle fiber during contraction

crosscurrent exchange Airflow and blood flow in the parabronchi of the avian lung that occur perpendicular to one another

cryoprotectant A molecule that is used to prevent freezing by lowering the freezing point in a body fluid

cyclic adenosine monophosphate (cyclic AMP or cAMP) An intracellular second messenger derived from adenosine triphosphate (ATP)

cytokines Signal molecules not produced by classic or distinct endocrine glands

cytoplasm (SĪ-tō-plaz′-um) The portion of the cell interior not occupied by the nucleus

cytoskeleton A complex intracellular protein network that acts as the "bone and muscle" of the cell

cytosol (SĪ-tuh-sol′) The semiliquid portion of the cytoplasm not occupied by organelles

cytotoxic T cells (sī-tō-TOK-sik) The population of T cells (T lymphocytes) that destroys host cells bearing foreign antigen, such as body cells invaded by viruses or cancer cells

dead-space volume The volume of medium that occupies the respiratory pathways as air or water is moved in and out and that is not available to participate in exchange of O_2 and CO_2 between the respiratory surfaces and external medium

decompression sickness, or "the bends" Pathology caused by the presence of nitrogen bubbles in the tissues, inducing a state of profound stupor, unconsciousness, or arrested activity

defensin A general term for antimicrobial peptides in all animals, with numerous positive charges that are attracted to negatively charged bacterial membranes to form pores

dehydration A water deficit in the body

dehydroepiandrosterone (DHEA) (dē-HĪ-drō-ep-i-and-row-steer-own) The androgen (masculinizing hormone) secreted by the vertebrate adrenal cortex in both sexes

dendrites Projections from the surface of a neuron's cell body that carry signals toward the cell body

dendritic cell A major antigen-presenting cell in vertebrate skin and other barrier tissues

deoxyhemoglobin Hemoglobin that is not combined with O_2

deoxyribonucleic acid (DNA) (dē-OK-sē-rī-bō-new-klā-ik) The cell's genetic material, which is found within the nucleus, mitochondria, and chloroplasts in eukarya, provides codes for protein synthesis, and serves as a blueprint for cell replication

depolarization (de′-pō-luh-ruh-ZĀ-shun) A reduction in membrane potential from resting potential; movement of the potential from resting toward 0 mV

dermis The connective tissue layer that lies under the epidermis in the skin; in vertebrates, it contains the skin's blood vessels and nerves

descending tract A bundle of nerve fibers of similar function that travels down the vertebrate spinal cord to relay messages from the brain to efferent neurons

desmosome (dez′-muh-sōm) An adhering junction between two adjacent but nontouching cells formed by the extension of filaments between the cells' plasma membranes; most abundant in tissues that are subject to considerable stretching

detritivore Animal that consumes dead and living organic material in sediments

DHEA See *dehydroepiandrosterone*

diabetes mellitus (muh-LĪ-tus) An endocrine disorder characterized by inadequate insulin action

diaphragm (DIE-uh-fram) A dome-shaped sheet of skeletal muscle that forms the posterior end of the thoracic cavity in many air-breathing vertebrates; a major inspiratory muscle

diaphragmaticus muscle A muscle attached between the liver and a distinctive elongated region of the pubis bone in crocodiles; pulls the liver away from the lungs in a pistonlike manner that results in expansion of the lungs

diastema Gap that extends between the front teeth and the cheek teeth (premolars and molars) of ruminants

diastole (dī-AS-tō-lē) The period of cardiac relaxation and filling

diencephalon (dī′-un-SEF-uh-lan) The division of the vertebrate brain that consists of the thalamus and hypothalamus

diestrus Associated with a functioning corpus luteum (dominance of progesterone) in a mammal and begins about four days after ovulation and ends with regression of the corpus luteum

diet-induced thermogenesis The increase in metabolism that follows ingestion of a meal; obligatory form is associated with the costs of processing the food, whereas the regulatory form is activated for the purpose of removing excess calories as heat; also called *specific dynamic action (SDA)*

diffusion Random collisions and intermingling of molecules as a result of their continuous, thermally induced random motion

digestion The conversion process whereby the structurally complex foodstuffs of the diet are broken down into smaller absorbable units by the enzymes produced within the digestive system

2,3-diphosphoglycerate (DPG) An organic phosphate synthesized in erythrocytes that reduces the affinity of most vertebrate hemoglobins for O_2

diploid number (DIP-loid) A complete set of chromosomes (e.g., 23 pairs in humans), as found in all somatic cells

distal tubule A highly convoluted tubule that extends between the loop of Henle and the collecting duct in the mammalian kidney's nephron

diuresis A state of high urine output

divergence The diverging, or branching, of a neuron's axon terminals so that activity in this single neuron influences the many other cells with which its terminals synapse

diverticula Blind tubules that arise from the main passage of the midgut in insects and contain the gastric caecae

DMS (dimethyl sulfide) A gas produced by the breakdown of DMSP; triggers cloud formation and so is important in climate regulation

DMSP (dimethylsulfoniopropionate) A major compatible osmolyte in many marine algae

DNA See *deoxyribonucleic acid*

dormancy A temporary state of very low metabolic rate

dorsal root ganglion A cluster of afferent neuronal cell bodies located adjacent to the vertebrate spinal cord

dorsobronchi Caudal group of secondary bronchi in birds that branch to form a fan-shaped covering of the mediodorsal surface of the avian lung

downregulation A decrease in the capacity of a cellular or organ function

DPG See *2,3-diphosphoglycerate (DPG)*

ecdysis The shedding of the old exoskeleton

ecdysone Hormone secreted by the prothoracic gland of insects

ECF See *extracellular fluid*

ECG See *electrocardiogram*

ectotherm An organism dependent on external heat for body temperature

edema (i-DĒ-muh) Swelling of tissues as a result of excess interstitial fluid

EDV See *end-diastolic volume*

EEG See *electroencephalogram*

effector organs The organs, especially muscles or glands, that are controlled by a regulatory system and that carry out that system's orders to bring about a desired effect, such as a particular movement or secretion

efferent division (EF-er-ent) The portion of the peripheral nervous system that carries instructions from the central nervous system to effector organs

efferent neuron Neuron that carries information from the central nervous system to an effector organ

efflux (Ē-flux) Movement out of a cell

egest Discharge of indigestible matter from the digestive tract

eicosanoid A class of signal molecules derived from lipids such as arachidonic acid

elastic recoil Rebound of the lungs after having been stretched

electrical gradient A difference in charge between two adjacent areas

electrocardiogram (ECG) The graphic record of the electrical activity that reaches the surface of the body as a result of cardiac depolarization and repolarization

electrochemical gradient The simultaneous existence of an electrical gradient and concentration (chemical) gradient for a particular ion

electrocyte Modified neurons of electric fish that, when linked in series, are capable of generating a substantial electrical discharge

electrolytes Solutes that form ions in solution and conduct electricity

enantiostasis A form of regulation in which changes in one physiological state help offset a disturbance to another physiological state

end-diastolic volume (EDV) The volume of blood in the ventricle at the end of diastole, when filling is complete

endocrine glands Ductless glands that secrete hormones into the blood or hemolymph

endocrine-disrupting chemicals (EDCs) Human-made substances that are released into the environment (or generated in sewage) and that interfere with the endocrine functions of animals

endocytosis (en'-dō-sī-TŌ-sis) Internalization of extracellular material within a cell as a result of the plasma membrane forming a pouch that contains the extracellular material, then sealing at the surface of the pouch to form a small, intracellular, membrane-enclosed vesicle with the contents of the pouch trapped inside

endogenous opiates (en-DAJ'-eh-nus ō'-pē-ātz) Endorphins and enkephalins, which bind with opiate receptors and are important in the vertebrate body's natural analgesic system

endogenous pyrogen (pī'-ruh-jun) A chemical released from vertebrate macrophages during inflammation that acts by means of local prostaglandins to raise the set point of the hypothalamic thermostat to produce a fever

endometrial cup Temporary endocrine units in the uterine horn of the mare (placental origin), ranging in size from millimeters to several centimeters that produce eCG (equine chorionic gonadotropin)

endometrium (en'-dō-MĒ-trē-um) The lining of the mammalian uterus

endoplasmic reticulum (en'-dō-PLAZ-mik ri-TIK-yuh-lum) An organelle consisting of a continuous membranous network of fluid-filled tubules and flattened sacs, partially studded with ribosomes; synthesizes proteins and lipids for formation of new cell membrane and other cell components and manufactures products for secretion

endothelium (en'-dō-THĒ-lē-um) The thin, single-celled layer of epithelial cells that lines the entire vertebrate circulatory system

endotherm An organism that uses internal heat to regulate body temperature

end-plate potential (EPP) The graded receptor potential that occurs at the motor end plate of a skeletal muscle fiber in response to binding with acetylcholine

end-systolic volume (ESV) The volume of blood in the ventricle at the end of systole, when emptying is complete

enhancer A promoter DNA sequence that is specific for a particular gene, helping regulate that gene's transcription (by binding of transcription factors) in a tissue-specific manner

enterogastrones (ent'-uh-rō-GAS-trōnz) Hormones secreted by the vertebrate duodenal mucosa that inhibit gastric motility and secretion; include secretin, cholecystokinin, and gastric inhibitory peptide

enterohepatic circulation (en'-tur-ō-hi-PAT-ik) The recycling of bile salts and other bile constituents between the vertebrate small intestine and liver by means of the hepatic portal vein

entropy A measure of disorder in a system

enzyme A protein molecule that speeds up (catalyzes) a particular chemical reaction in an organism

eosinophils (ē'-uh-SIN-uh-fils) White blood cells that are important in allergic responses and in combating internal parasite infestations

epidermis (ep'-uh-DER-mus) The outer layer of the skin, consisting of numerous layers of epithelial cells

epinephrine (ep'-uh-NEF-rin) The primary hormone secreted by the vertebrate adrenal medulla; important in preparing the body for "fight-or-flight" responses and in regulation of arterial blood pressure; adrenaline

epiphyseal plate (eh-pif-i-SEE-al) A layer of cartilage that separates the diaphysis (shaft) of a vertebrate long bone from the epiphysis (flared end); the site of growth of bones in length before the cartilage ossifies (turns into bone)

epithelial tissue (ep'-uh-THĒ-lē-ul) A functional grouping of cells specialized in the exchange of materials between the cell and its environment; lines and covers various body surfaces and cavities and forms secretory glands

EPSP See *excitatory postsynaptic potential*

equilibrium potential The potential that exists when the concentration gradient and opposing electrical gradient for a given ion exactly counterbalance each other so that there is no net movement of the ion

equine chorionic gonadotropin (eCG) A peptide hormone secreted by the developing chorion of the mare; see *chorionic gonadotropin*

eructation Belching

erythrocyte (i-RITH-ruh-sīt) A red blood cell, which is a plasma membrane–enclosed bag of hemoglobin that transports O_2 and to a lesser extent CO_2 and H^+ in the blood

erythropoiesis (i-rith'-rō-poi-Ē-sus) Erythrocyte production by the bone marrow

erythropoietin (EPO) The hormone released from vertebrate kidneys in response to a reduction in O_2 delivery to the kidneys; stimulates the bone marrow to increase erythrocyte production

esophagus (i-SOF-uh-gus) A straight, muscular tube that extends between the pharynx and stomach

estivation A dormant or low-metabolism state of an animal, usually during dry summer months

estrogens A group of feminizing "female" sex hormones such as **estradiol**

estrous cycle The time from one period of sexual receptivity to the next

estrus The period of sexual receptivity in a mammalian female

ESV See *end-systolic volume*

eukarya One of three distinct forms of life (along with archaea and bacteria); eukaryotic cells have organelles including a nucleus, mitochondria, and endoplasmic reticulum; Protista, Plantae, Animalia, and Fungi

euryhaline Capable of surviving over a wide range of external salinities

evolutionary explanation In biology, the evolutionary reason for the existence of a structure or process ("why it arose this way"), as opposed to the *mechanistic explanation*

excitable tissue Tissue capable of producing electrical signals when excited; includes nervous and muscle tissue

excitation–contraction coupling The series of events linking muscle excitation (the presence of an action potential) to muscle contraction (filament sliding and sarcomere shortening)

excitatory postsynaptic potential (EPSP) (pōst'-si-NAP-tik) A small depolarization of the postsynaptic membrane in response to neurotransmitter binding, thereby bringing the membrane closer to threshold

excitatory synapse (SIN-aps') Synapse in which the postsynaptic neuron's response to neurotransmitter release is a small depolarization of the postsynaptic membrane, bringing the membrane closer to threshold

exocrine glands Glands that secrete through ducts to the outside of the body or into a cavity that communicates with the outside

exocytosis (eks'-ō-sī-TŌ-sis) Fusion of a membrane-enclosed intracellular vesicle with the plasma membrane, followed by the opening of the vesicle and the emptying of its contents to the outside

expiration A breath out (exhalation)

expiratory muscles The skeletal muscles whose contraction moves respiratory air or water out of a respiratory chamber

extensors Skeletal muscles that straighten a limb out

external intercostal muscles Inspiratory muscles whose contraction expands the vertebrate rib cage, thereby enlarging the thoracic cavity (in most air-breathing vertebrates)

external work Energy expended by contracting skeletal muscles to move external objects or to move the body in relation to the environment

exteroreceptors The "classic" senses that detect external stimuli such as light, chemicals, touch, temperature, and sound

extracellular fluid All the body's fluid found outside the cells; consists of interstitial fluid and plasma; ECF

extracellular matrix An intricate meshwork of fibrous proteins embedded in a watery, gel-like substance; secreted by local cells

extrafusal muscle fiber A vertebrate muscle fiber not associated with a muscle-spindle stretch receptor

extrinsic controls Regulatory mechanisms initiated outside of an organ that alter the activity of the organ; accomplished by the nervous and endocrine systems

facilitated diffusion Passive carrier-mediated transport involving transport of a substance down its concentration gradient across the plasma membrane

fat body The insect equivalent of a liver; it stores large amounts of triglycerides and carbohydrates

fatigue Inability to maintain muscle tension at a given level despite sustained stimulation

feedforward mechanism A response designed to prevent an anticipated harmful change, or activate an anticipated useful change, in a controlled variable

fibrinogen (fī-BRIN-uh-jun) A large soluble plasma protein that is converted into an insoluble, threadlike molecule that forms the meshwork of a clot during blood coagulation

Fick's law of diffusion The rate of net diffusion of a substance across a membrane is directly proportional to the substance's concentration gradient, the membrane's permeability to the substance, and the surface area of the membrane and inversely proportional to the substance's molecular weight and the diffusion distance

fight-or-flight response The changes in activity of the various organs innervated by the vertebrate autonomic nervous system in response to sympathetic stimulation, which collectively prepare the body for strenuous physical activity in the face of an emergency or stressful situation, such as a physical threat from the outside environment

filter chamber A modification of the digestive tract of Homoptera, insects that feed on large quantities of plant juices; the filter chamber permits water from the ingested sap to pass directly from the anterior region of the midgut to the hindgut

filter feeding Trapping of dead and/or living material suspended in water by straining through specialized entrapment devices; also called suspension feeding

firing The event when an excitable cell undergoes an action potential

first messenger An extracellular messenger, such as a hormone, that binds with a surface membrane receptor and activates an intracellular second messenger to carry out the desired cellular response

flagellum (fluh-JEL-um) The single, long, whip-like appendage of some cells, serving, for example, as the tail of a spermatozoon or the current producer of a sponge cell

flexors Skeletal muscles that bend a limb

flow-through breathing Movement of air or water in and out of a respiratory chamber through separate openings, for the purposes of gas exchange

follicle (of ovary) A developing ovum and the surrounding specialized cells

follicle-stimulating hormone (FSH) An anterior pituitary hormone that stimulates ovarian follicular development and estrogen secretion in females and stimulates sperm production in males

follicular cells (of ovary) (fah-LIK-you-lar) Collectively, the granulosa and thecal cells

follicular cells (of thyroid gland) The cells that form the walls of the colloid-filled follicles in the thyroid gland and secrete thyroid hormone

follicular phase The phase of the mammalian ovarian cycle dominated by the presence of maturing follicles prior to ovulation

forestomach Ruminoreticular region of the ruminant stomach; contains the populations of microbes involved in the fermentation of foodstuffs

fovea A small depression or pit in the vertebrate retina containing densely packed cones; the region of the eye with the greatest visual resolution

Frank-Starling law of the heart Intrinsic control of the heart, such that increased venous return resulting in increased end-diastolic volume leads to an increased strength of contraction and increased stroke volume; that is, the heart normally pumps out all the blood returned to it

free radicals Very unstable electron-deficient particles that are highly reactive and destructive

freemartin Sterile female calf twin born with a male

frontal lobes The lobes of the vertebrate cerebral cortex that lie at the top of the brain in front of the central sulcus and that, in mammals, are responsible for nonreflexive motor output and complex planning

FSH See *follicle-stimulating hormone*

functional syncytium (sin-sish'-ē-um) A group of cells that are interconnected by gap junctions and function electrically and mechanically as a single unit

functional unit The smallest component of an organ that can perform all the functions of the organ

G protein A membrane-bound intermediary, which, when activated on binding of an extracel-

lular first messenger to a surface receptor, activates the enzyme adenylyl cyclase on the intracellular side of the membrane in the cAMP second-messenger system

gametes (GAM-ētz) Reproductive or germ cells, each containing a haploid set of chromosomes; sperm and ova

gamma motor neuron A motor neuron that innervates the fibers of a muscle spindle receptor

ganglion (pl., **ganglia**) (1) A group or cluster of nerve cell bodies, often with related functions; (2) in the vertebrate eye, the nerve cells in the outermost layer of the retina and whose axons form the optic nerve

gap junction A communicating junction formed between adjacent cells by small connecting tunnels that permit passage of charge-carrying ions between the cells so that electrical activity in one cell is spread to the adjacent cell

gas bladder (swim bladder) A large centrally located sac ventral to the spinal column, in the abdominal cavity of teleost fish, used for buoyancy control

gastrin A hormone secreted by the pyloric gland area of the vertebrate stomach that stimulates the parietal and chief cells to secrete a highly acidic gastric juice

genome The complete set of genetic codes in a particular species

genomics The study of the structure, regulation, and information content of genes

gestation Pregnancy

gill Evaginated respiratory organ of water-breathing animals

gill arch Fish tissue that provides skeletal support for a double series of gill filaments, each of which carries a row of secondary lamellae on either side

gizzard Muscular portion of the stomach of most birds where food (e.g., grains, seeds) is ground up with the aid of ingested pebbles

gland Epithelial tissue derivative specialized for secretion

glial cells (glē-ul) Cells that serve as the connective tissue of the vertebrate CNS and help support the neurons both physically and metabolically; include astrocytes, oligodendrocytes, ependymal cells, and microglia

glomerular filtration (glow-MER-yū-lur) Filtration of a protein-free plasma from the glomerular capillaries into the tubular component of the vertebrate kidney's nephron as the first step in urine formation

glomerular filtration rate (GFR) The rate at which glomerular filtrate is formed

glomerulus (glow-MER-yū-lus) (1) A ball-like tuft of capillaries in the vertebrate kidney's nephron that filters water and solute from the blood as the first step in urine formation; (2) small ball-like neural junction in the vertebrate olfactory bulb

glucagon (GLOO-kuh-gon) The vertebrate pancreatic hormone that raises blood glucose and blood fatty-acid levels

glucocorticoids (gloo'-kō-KOR-ti-koidz) The adrenocortical hormones that are important in intermediary metabolism and in helping the body resist stress; primarily cortisol

gluconeogenesis (gloo'-kō-nē-ō-JEN-uh-sus) The "new formation of glucose"; the conversion of pyruvate or amino acids into glucose, running glycolysis essentially in reverse

glycation A reaction in which glucose and other reducing sugars spontaneously bind with proteins such as collagen and hemoglobin, forming a covalent bond that can alter its structure and function

glycogen (GLĪ-kō-jen) The storage form of glucose in the liver and muscle

glycogenesis (glī′-kō-JEN-i-sus) The conversion of glucose into glycogen

glycogenolysis (glī′-kō-juh-NOL-i-sus) The conversion of glycogen to glucose

glycolysis (glī-KOL-uh-sus) A biochemical process that takes place in the cell's cytosol and involves the breakdown of glucose into two pyruvic acid molecules

glycoprotein A protein that has sugars covalently bound to it; common on outer sides of cell membranes

GnRH See *gonadotropin-releasing hormone*

Golgi complex (GOL-jē) An organelle consisting of sets of stacked, flattened membranous sacs; processes raw materials transported to it from the endoplasmic reticulum into finished products and sorts and directs the finished products to their final destination

gonad The primary reproductive organ, which produces the gametes; testes and ovaries

gonadotropin-releasing hormone (GnRH) (gō-nad′-uh-TRŌ-pin) The hypothalamic hormone that stimulates the release of FSH and LH from the vertebrate anterior pituitary

gonadotropins FSH and LH; vertebrate hormones that are tropic to the gonads

gradation of contraction Variable magnitudes of tension produced in a single whole muscle

graded potential A local change in membrane potential that occurs in varying grades of magnitude; serves as a short-distance signal in excitable tissues

granulocyte (gran′-yuh-lō-sīt) Vertebrate leukocyte that contains granules, including neutrophils, eosinophils, and basophils

granulosa cell (gran′-yuh-LŌ-suh) A cell immediately surrounding a developing oocyte within a vertebrate ovarian follicle

gray matter The portion of the vertebrate central nervous system composed primarily of densely packaged neuronal cell bodies and dendrites

grit Hard granules, such as sand and pebbles found in the gizzard of birds

growth hormone (GH) An anterior pituitary hormone that is primarily responsible for regulating overall body growth and is also important in intermediary metabolism; somatotropin

gular fluttering Panting with a rapid fluttering of the well-vascularized esophageal region of the hyoid apparatus, in birds

gustation Taste, or detection of molecules in objects in contact with the body

H⁺ See *hydrogen ion*

habituation Decreased responsiveness to repetitive presentation of an indifferent stimulus (neither rewarding nor punishing), resulting in a decrease in synaptic activity

haploid number (HAP-loid) The number of chromosomes found in gametes; a half set of chromosomes, one member of each pair

Hb See *hemoglobin*

hCG Human version of *chorionic gonadotropin*

heat shock proteins (HSPs) See *stress proteins*

helper T cells The population of T cells that enhances the activity of other immune-response effector cells

hematocrit (hi-mat′-uh-krit) The percentage of blood volume occupied by erythrocytes as they are packed down in a centrifuged blood sample

hemerythrin Iron-containing respiratory pigment found in several nonvertebrate phyla, including annelid worms

hemipenis Organ used by the drake to deposit semen in the female duck

hemocoel The body cavity of an animal with a circulating hemolymph

hemocyanin An oxygen-transporting protein in many mollusks and arthropods such as crustaceans that uses two copper atoms to bind one oxygen molecule

hemodynamic Pertaining to the physics of blood flow

hemoglobin (Hb) (HĒ-muh-glō′-bun) A large iron-bearing protein molecule that binds with and transports most O_2 in blood; in vertebrates, it is contained within erythrocytes, and it also carries some of the CO_2 and H^+

hemolymph The circulatory fluid in an animal that does not have capillaries

hemolysis (hē-MOL-uh-sus) Rupture of red blood cells

hemostasis (hē′-mō-STĀ-sus) The stopping of bleeding from an injured vessel

hepatic portal system (hi-PAT-ik) A complex vascular connection between the vertebrate digestive tract and liver such that venous blood from the digestive system drains into the liver for processing of absorbed nutrients before being returned to the heart

hepatopancreas An organ in crustaceans that secretes digestive enzymes, stores glycogen and lipids, and absorbs nutrient molecules

herbivore Animal that depends on the intake of algal or plant materials

hermaphrodism State in which an animal has both functional male and female reproductive tracts, although not necessarily at the same time

heterogametic Producing two or more different kinds of gametes that differ in their sex chromosome content

heterotherm An animal that has some degree of endothermy but that is not fully homeothermic

hibernation A state of dormancy in an animal usually associated with winter months; involves lower body temperatures in endotherms

hippocampus (hip-oh-CAM-pus) The elongated, medial portion of the temporal lobe that is a part of the vertebrate limbic system and is especially crucial for forming long-term memories

histamine A chemical released from vertebrate mast cells or basophils that brings about vasodilation and increased capillary permeability; important in allergic responses

homeostasis (hō′-mē-ō-STĀ-sus) Maintenance by coordinated, regulated actions of body systems of relatively stable (though rarely constant) chemical and physical conditions in the internal fluid environment and in other body states

homeotherm Animal with narrowly varying body temperatures

homeoviscous Having the same fluidity or viscosity; referring to cell membranes being altered

at different temperatures to maintain an optimal fluidity

homogametic Producing one type of gamete with respect to sex chromosome content

homunculus Orderly distribution of cortical sensory processing in the vertebrate brain; for each species of vertebrate, the extent of each body part is indicative of the relative proportion and location of the somatosensory cortex devoted to that area

honeydew Nutritious liquid discharged from the anus of some Homoptera

hormone A long-distance chemical mediator that is secreted by an endocrine gland into the blood or hemolymph, which transports it to its target cells

host cell A body cell infected by a virus

HRE See *response element*

hydrogen ion (H^+) The cationic portion of a dissociated acid; a proton without an electron

hydrolysis (hī-DROL-uh-sis) The digestion of a nutrient molecule by the addition of water at a bond site

hydrostatic pressure (hī-dro-STAT-ik) The pressure exerted by a fluid on the walls that contain it

hyperglycemia (hī′-pur-glī-SĒ-mē-uh) Elevated blood glucose concentration

hyperplasia (hī-pur-PLĀ-zē-uh) An increase in the number of cells

hyperpolarization An increase in membrane potential from resting potential; potential becomes even more negative than at resting potential

hypertension (hī′-pur-TEN-shun) Sustained, above-normal mean arterial blood pressure

hyperthermia Abnormally high body temperature

hypertonic (hī′-pur-TON-ik) Having an osmolarity greater than normal body fluids; more concentrated than normal

hypertrophy (hī-PUR-truh-fē) Increase in the size of an organ

hyperventilation Overbreathing; when the rate of ventilation is in excess of the body's metabolic needs for CO_2 removal

hypomagnesemic tetany, or grass tetany A pathological state in herbivores such as cattle, characterized by low magnesium concentrations in the blood and cerebrospinal fluid; affected animals appear nervous and show muscular twitching around the face and ears

hyponatremia A pathological low level of sodium in the blood

hypophysiotropic hormones (hi-PŌ-fiz-ē-oh-TRŌ-pik) Vertebrate hormones secreted by the hypothalamus that regulate the secretion of anterior pituitary hormones; see also *releasing hormone*

hypotension (hi-po-TEN-chun) Sustained, below-normal mean arterial blood pressure

hypothalamic–hypophyseal portal system (hī-pō-thuh-LAM-ik hī-pō-FIZ-ē-ul) The vascular connection between the hypothalamus and anterior pituitary gland used for the pickup and delivery of hypophysiotropic hormones

hypothalamus (hī′-pō-THAL-uh-mus) The vertebrate brain region located beneath the thalamus that is concerned with regulating many aspects of the internal fluid environment, such as water and salt balance and food intake; serves as an important link between the autonomic nervous system and endocrine system

hypothermia Abnormally low body temperature

hypothetico-deductive method The most widely accepted version of "the scientific method"; data are gathered about an aspect of nature in the "discovery" phase, then alternative hypotheses are made to explain them; testable deductions must be made from the hypotheses, with experiments and observations designed to test them, leading to support, refinement, or rejection of hypotheses

hypotonic (hī′-pō-TON-ik) Having an osmolarity less than normal body fluids

hypoventilation Underbreathing; ventilation inadequate to meet the metabolic needs for O_2 delivery and CO_2 removal

hypoxia (hī-POK-sē-uh) Insufficient O_2 at the cellular level

I band One of the light bands that alternate with dark (A) bands to create a striated appearance in skeletal or cardiac muscle fibers when these fibers are viewed with a light microscope

ice-nucleating agent Small proteins in some animals that trigger the formation of small ice crystals once temperatures have dropped to the freezing point

ICF See *intracellular fluid*

IGF See *insulin-like growth factor*

immune surveillance Recognition and destruction of newly arisen cancer cells by the immune system

immunity Ability to resist or eliminate potentially harmful foreign materials or abnormal cells

immunoglobulins (im′-ū-nō-GLOB-yū-lunz) Antibodies; gamma globulins

impermeable Prohibiting passage of a particular substance through the plasma membrane

implantation The burrowing of a mammalian blastocyst into the endometrial lining

inflammation An innate, nonspecific series of highly interrelated events in vertebrates, especially involving neutrophils, macrophages, and local vascular changes, that are set into motion in response to foreign invasion or tissue damage

influx Movement into a cell

inhibin (in-HIB-un) A vertebrate hormone secreted by the Sertoli cells of the testes or by the ovarian follicles that inhibits FSH secretion

inhibitory postsynaptic potential (IPSP) (pōst′-si-NAP-tik) A small hyperpolarization of the postsynaptic membrane in response to neurotransmitter binding, thereby moving the membrane farther from threshold

inhibitory synapse (SIN-aps′) Synapse in which the postsynaptic neuron's response to neurotransmitter release is a small hyperpolarization of the postsynaptic membrane, moving the membrane farther from threshold

innate immune responses Inherent defense responses that nonselectively defend against foreign or abnormal material, even on initial exposure to it; see also *inflammation, interferon, natural killer cells*

inorganic Referring to substances that do not contain carbon (and usually hydrogen)

insensible Referring to body water loss that is not sensed or regulated by an animal

inspiration An inward breath (inhalation)

inspiratory muscles The vertebrate skeletal muscles whose contraction enlarges the thoracic cavity, bringing about lung expansion and movement of air into the lungs from the atmosphere

instar The period between molts, or ecdysis, in arthropod development

insulin (IN-suh-lin) The vertebrate pancreatic hormone that lowers blood levels of glucose, fatty acids, and amino acids and promotes their storage

insulin-like growth factor (IGF) See *somatomedins*

integrating center A region that determines efferent output based on processing of afferent input

integument (in-TEG-yuh-munt) The skin and underlying connective tissue

intercostal muscles (int-ur-KOS-tul) The muscles that lie between the vertebrate ribs; see also *external intercostal muscles* and *internal intercostal muscles*

interferon (in′-tur-FĒR-on) A chemical released from virus-invaded vertebrate cells that provides nonspecific resistance to viral infections by transiently interfering with replication of the same or unrelated viruses in other host cells

interleukins (int-ur-LOO-kins) Chemical mediators released from immune cells that enhance immune responses; for example, interleukin-1 (IL-1) in vertebrates is made by macrophages and enhances B cells, whereas interleukin-2 (IL-2) is secreted by helper T cells and augments the activity of all T cells

intermediary metabolism The collective set of intracellular chemical reactions that involve the degradation, synthesis, and transformation of small nutrient molecules

intermediate filaments Threadlike cytoskeletal elements that play a structural role in parts of the cell subject to mechanical stress

internal intercostal muscles Expiratory muscles whose contraction pulls the ribs inward, thereby reducing the size of the thoracic cavity (in most air-breathing vertebrates)

internal respiration The intracellular metabolic processes carried out within the mitochondria that use O_2 and produce CO_2 during the derivation of energy from nutrient molecules

internal work All forms of biological energy expenditure that do not accomplish mechanical work outside of the body

interneuron Neuron that lies entirely within the central nervous system and is important for integrating responses to peripheral information

interoreceptors Sensors that detect information about the internal body fluids usually crucial to homeostasis, such as blood pressure and O_2 concentration inhalation

interstitial fluid (in′-tur-STISH-ul) The portion of the extracellular fluid that surrounds and bathes all the body's cells

intima The cuticular lining of the insect foregut, tracheae, and hindgut

intra-alveolar pressure (in′-truh-al-VĒ-uh-lur) The pressure within alveoli (in mammals, some reptiles)

intracellular fluid The fluid collectively contained within all the body's cells; ICF

intrafusal muscle fiber A specialized muscle fiber associated with a muscle-spindle stretch re-

ceptor that can be activated by a gamma motor neuron

intrapleural pressure (in′-truh-PLOOR-ul) The pressure within the pleural sac (in mammalian lungs)

intrinsic controls Local control mechanisms inherent to an organ

intrinsic factor A special substance secreted by the parietal cells of the vertebrate stomach that must be combined with vitamin B_{12} for this vitamin to be absorbed by the intestine; deficiency produces pernicious anemia

intrinsic nerve plexuses Interconnecting networks of nerve fibers within the vertebrate digestive-tract wall

inulin A naturally occurring polysaccharide used to determine the glomerular filtration rate, as it is neither secreted nor absorbed

ion An atom that has gained or lost one or more of its electrons, so it is not electrically balanced

IPSP See *inhibitory postsynaptic potential*

iris A pigmented smooth muscle that forms the colored portion of the vertebrate eye and controls pupillary size

islets of Langerhans (LAHNG-er-honz) The endocrine portion of the vertebrate pancreas that secretes the hormones insulin and glucagon into the blood

isoacids Branched-chain fatty acids generated by the deamination of some amino acids such as valine and leucine

isometric contraction (ī′-sō-MET-rik) A muscle contraction in which the development of tension occurs at constant muscle length

isotonic (ī′-sō-TON-ik) Having an osmolarity equal to normal body fluids

isotonic contraction A muscle contraction in which muscle tension remains constant as the muscle fiber changes length

joule The international unit for energy of any kind, equal to 0.239 calorie

juvenile hormone Insect hormone secreted by the corpora allata gland that stimulates its target tissues to retain immature characteristics

juxtaganglionar organ A crustacean organ composed of scattered cells in the connective tissue around the cerebral ganglion; the cells release peptides of uncertain function into the hemolymph during egg laying

juxtaglomerular apparatus (juks′-tuh-glō-MER-yū-lur) A cluster of specialized vascular and tubular cells at a point where the ascending limb of the loop of Henle passes through the fork formed by the afferent and efferent arterioles of the same nephron in the mammalian kidney

keratin (CARE-uh-tin) The protein found in the intermediate filaments in vertebrate skin cells that gives the skin strength and helps form a waterproof outer layer

ketone bodies A group of fatty-acid derivatives produced by the vertebrate liver during glucose sparing

kinase An enzyme that phosphorylates (adds a phosphate group to) other proteins, thereby increasing or decreasing the activity of those proteins

kinesin (kī-NĒ′-sin) The transport or motor protein that transports secretory vesicles along the microtubular highway within neuronal axons by "walking" along the microtubule

kinocilium A type of cilium that extends from the apex of all sensory hair cells in all vertebrates except mammals

knockout genetics A procedure in which a specific gene is removed from an organism's genome, usually in the fertilized egg stage; a means of testing a gene's and its coded protein's function by the consequences of its absence

koilin Mucosal lining of the gizzard; a tough protein–polysaccharide complex that has an amino acid composition similar to that of feather protein

Krogh principle "For a large number of problems there will be some animal of choice, or a few such animals, on which it can be most conveniently studied"; formulated by German physiologist Hans Kreb in 1975

K-selected Referring to species that produce relatively few offspring, with most of the parental reproductive energy used for nourishing and/or protecting the offspring

lactate An end product formed from pyruvic acid during the anaerobic process of glycolysis

lactation Milk production by the mammary glands

larynx (LARE-inks) The "voicebox" at the entrance of the mammalian trachea

latch state In vertebrates, a state of continuous contraction of smooth muscle that maintains tension with little metabolic cost

lateral geniculate nucleus The first stop in the mammalian brain for information in the visual pathway; in the thalamus

lateral inhibition The phenomenon in which the most strongly activated signal pathway originating from the center of a stimulus area inhibits the less excited pathways from the fringe areas by means of lateral inhibitory connections within sensory pathways

lateral sacs The expanded saclike regions of a muscle fiber's sarcoplasmic reticulum; store and release calcium, which plays a key role in triggering muscle contraction

law of mass action If the concentration of one of the substances involved in a reversible reaction is increased, the reaction is driven toward the opposite side, and if the concentration of one of the substances is decreased, the reaction is driven toward that side

left ventricle The heart chamber that pumps blood into the systemic circulation (in reptiles, birds, mammals)

length–tension relationship The relationship between the length of a muscle fiber at the onset of contraction and the tension the fiber can achieve on a subsequent tetanic contraction

lens A transparent, biconvex structure of a camera-type eye that refracts (bends) light rays and (in some animals) whose strength can be adjusted to accommodate for vision at different distances

leptin An endocrine signal originating from mammalian adipose tissue that informs the hypothalamus that metabolic stores are sufficient for normal energy homeostasis or for the initiation of reproduction at puberty

leukocyte endogenous mediator (LEM) (LOO-kō-sīt en-DAJ-eh-nus MĒ-de-ā-tor) A chemical mediator secreted by vertebrate macrophages that is identical to endogenous pyrogen and exerts a wide array of effects associated with inflammation

leukocytes (LOO-kuh-sīts) White blood cells of vertebrates; the immune system's mobile defense units

Leydig cells (LĪ-dig) The interstitial cells of vertebrate testes that secrete testosterone

LH See *luteinizing hormone*

LH surge The burst in LH secretion that occurs at midcycle of the mammalian ovarian cycle and triggers ovulation

lignin Rigid polymer derived from phenylalanine and tyrosine; with cellulose, forms the woody cell walls of plants and cements them together

limbic system (LIM-bik) A functionally interconnected ring of vertebrate forebrain structures that surrounds the brainstem and is concerned with emotions, basic survival, and sociosexual behavioral patterns, motivation, and learning

lipid emulsion A suspension of small fat droplets held apart as a result of adsorption of bile salts on their surface

long-term potentiation A long-lasting amplification in signal transmission between two neurons that results from stimulating them synchronously

loop of Henle (HEN-lē) A hairpin loop that extends between the proximal and distal tubule of the mammalian kidney's nephron

lordosis Mating posture assumed by a female

lower critical temperature In an endotherm, the temperature at which internal heat-generating mechanisms are activated to maintain thermal homeostasis

lumen (LOO-men) The interior space of a hollow organ or tube

luteal phase (LOO-tē-ul) The phase of the mammalian ovarian cycle dominated by the presence of a corpus luteum

luteinization (loot'-ē-un-uh-ZĀ-shun) Formation of a postovulatory corpus luteum in the mammalian ovary

luteinizing hormone (LH) An anterior pituitary hormone of vertebrates that stimulates ovulation, luteinization, and secretion of estrogen and progesterone in females and stimulates testosterone secretion in males

lymph Interstitial fluid that is picked up by vertebrate lymphatic vessels and returned to the venous system, meanwhile passing through the lymph nodes for defense purposes

lymphocyte Vertebrate white blood cell that provides immune defense against targets for which it is specifically programmed; B and T cells

lymphoid tissue Tissue that produces and stores lymphocytes, such as lymph nodes and tonsils (in vertebrates)

lysosomes (LĪ-sō-sōmz) Organelles consisting of membrane-enclosed sacs containing powerful hydrolytic enzymes that destroy unwanted material within the cell, such as internalized foreign material or cellular debris

macrophage (MAK-ruh-fāj) Large tissue-bound phagocyte

magnetosome Magnetic crystals arranged in a chain

Maillard product Glucose covalently bound to a protein (browning reaction)

Malpighian tubules Filtering excretory tubules that begin filtration driven by secretion of ions from an arthropod's hemolymph

mantle External body wall lining the shell of mollusks, or the mantle chamber in shell-less cephalopods

mast cells Cells located within vertebrate connective tissue that synthesize, store, and release histamine, as during allergic responses

Mauthner neurons Paired giant interneurons found in some teleost fish and salamanders that elicit the startle response; activity generated in one neuron simultaneously inhibits the activity in the other

maxilloturbinal Thin curl of bone deep within the avian and mammalian nasal cavity, used to retain heat and water

mean arterial blood pressure The average pressure responsible for driving blood forward through the arteries into the tissues throughout the cardiac cycle; equals cardiac output × total peripheral resistance

mechanically gated Referring to a channel that opens or closes in response to stretching or other mechanical deformation

mechanistic (or proximate) explanation Explanation of body functions in terms of mechanisms of action, that is, the "how" of events that occur in the body

mechanoreceptor (meh-CAN-oh-rē-SEP-tur or mek'-uh-nō-rē-SEP-tur) A sensory receptor sensitive to mechanical energy, such as stretching or bending

Meckel's diverticulum Vestige of the yolk sac; forms a tube connected to the lower ileum of birds

medullary respiratory center (MED-you-LAIR-ē) Several aggregations of neuronal cell bodies within the vertebrate medulla oblongata that provide output to the respiratory muscles and receive input important for regulating the magnitude of ventilation

meiosis (mī-ō-sis) Cell division in which the chromosomes replicate followed by two nuclear divisions so that only a half set of chromosomes is distributed to each of four new daughter cells

melanocyte-stimulating hormone (MSH) (mel-AH-nō-sīt) A vertebrate hormone produced by the anterior pituitary in humans and by the intermediate lobe of the pituitary in lower vertebrates; regulates skin coloration by controlling the dispersion of melanin granules in lower vertebrates; involved with control of food intake and possibly memory and learning in humans

melatonin (mel-uh-TŌ-nin) A vertebrate hormone secreted by the pineal gland during darkness that helps entrain the body's biological rhythms with the external light/dark cues

membrane attack complex A collection of the five final activated components of the vertebrate complement system that aggregate to form a porelike channel in the plasma membrane of an invading microorganism, with the resultant leakage leading to destruction of the invader

membrane potential A separation of charges across the membrane; a slight excess of negative charges lined up along the inside of the plasma membrane and separated from a slight excess of positive charges on the outside

memory cells B or T cells of vertebrates that are newly produced in response to a microbial invader but that do not participate in the current immune response against the invader; instead they remain dormant, ready to launch a swift, powerful attack should the same microorganism invade again in the future

menstrual cycle (men'-stroo-ul) The cyclical changes in the uterus that accompany the hormonal changes in the ovarian cycle of humans and some other primates

mesenteron The midgut, or middle portion of the insect digestive tract

mesotocin (MT) A hormone that influences the blood flow to some organs in nonmammalian vertebrates

messenger RNA (mRNA) Carries the transcribed genetic blueprint for synthesis of a particular protein from nuclear DNA to the cytoplasmic ribosomes, where the protein synthesis takes place

metabolic acidosis (met-uh-bol´-ik) Acidosis resulting from any cause other than excess accumulation of carbonic acid in the body

metabolic alkalosis (al´-kuh-LŌ-sus) Alkalosis caused by a relative deficiency of noncarbonic acid

metabolic rate Energy expenditure per unit of time

metabolic scope The degree to which an animal's metabolism can be activated above the resting level, as in maximal muscular activity in an emergency

metanephridia Filtering excretory tubules that begin with a filtering capsule or funnel-like opening of some kind, followed by tubules that perform selective secretion and reabsorption; generally found in larger, more complex animals including adult mollusks, annelids (segmented worms), some arthropods (e.g., crustaceans), and vertebrates

metathoracic ganglia Ganglia located in the posterior segment of the insect thorax

metestrus Stage of the mammalian estrous cycle between ovulation and formation of a functional corpus luteum

methanogen A microbe that produces methane from degrading organic material; usually archaea; found in many anaerobic habitats, including ruminant stomachs

micelle (mī-SEL) A water-soluble aggregation of bile salts, lecithin, and cholesterol that has a hydrophilic shell and a hydrophobic core; carries the water-insoluble products of fat digestion to their site of absorption

microarray A glass slide on which has been fixed, in a grid, a set of cDNAs from an organism

microfilaments Cytoskeletal elements made of actin molecules (as well as myosin molecules in muscle cells); play a major role in various cellular contractile systems and serve as a mechanical stiffener for microvilli

microtubules Cytoskeletal elements made of tubulin molecules arranged into long, slender, unbranched tubes that help maintain asymmetric cell shapes and coordinate complex cell movements

microvilli (mī´-krō-VIL-ī) Actin-stiffened, nonmotile, hairlike projections from the luminal surface of epithelial cells lining the digestive tract and kidney tubules; tremendously increase the surface area of the cell exposed to the lumen

micturition (mik-too-RISH-un or mik-chuh-RISH-un) The process of bladder emptying; urination

mineralocorticoids (min-uh-rul-ō-KOR-ti-koidz) The vertebrate adrenocortical hormones that are important in Na^+ and K^+ balance; primarily aldosterone

mitochondria (mī-tō-KON-drē-uh) The energy organelles of eukaryotic cells, which contain the enzymes for oxidative phosphorylation

mitosis (mī-TŌ-sis) Cell division in which the chromosomes replicate before nuclear division, so each of the two daughter cells receives a full set of chromosomes

mitotic spindle The system of microtubules assembled during mitosis along which the replicated chromosomes are directed away from each other toward opposite sides of the cell prior to cell division

modality The kind of stimulus a sensor cell or organ reacts to

molting The process of replacing an old exoskeleton with a new one (see *ecdysis*)

monocyte (MAH-nō-sīt) Vertebrate white blood cell that emigrates from the blood, enlarges, and becomes a macrophage

monoestrous Displaying one period of sexual receptivity (estrus) during a year in a mammal

monomer A unit molecule used to make a larger molecule called a *polymer*

monosaccharide (mah´-nō-SAK-uh-rīd) Simple sugar, such as glucose; the absorbable unit of digested carbohydrates

motor activity Movement of the body accomplished by contraction of skeletal muscles

motor end plate The specialized portion of a skeletal muscle fiber that lies immediately underneath the terminal button of the motor neuron and has receptor sites for binding acetylcholine released from the terminal button

motor neuron A neuron that innervates skeletal muscle and whose axons constitute the somatic nervous system

motor unit One motor neuron plus all the muscle fibers it innervates

motor unit recruitment The progressive activation of a muscle fiber's motor units to accomplish increasing gradations of contractile strength

mucosa (mew-KŌ-sah) The innermost layer of the digestive tract that lines the lumen

multiunit smooth muscle A smooth muscle mass that consists of multiple discrete units that function independently of each other and that must be separately stimulated by autonomic nerves to contract

muscarinic receptor Type of cholinergic receptor found at the effector organs of all parasympathetic postganglionic fibers

muscle fiber A relatively long and cylindrical single muscle cell

muscle tension See *tension*

muscle tissue A functional grouping of cells specialized for contraction and force generation

myelin (MĪ-uh-lun) An insulative lipid covering that surrounds myelinated nerve fibers at regular intervals along the axon's length (in vertebrates and in a few nonvertebrates)

myelinated fiber Neuronal axon covered at regular intervals with insulative myelin

myocardium (mī´-ō-KAR-dē-um) The cardiac muscle layer within heart walls

myofibril (mī´-ō-FĪB-rul) A specialized intracellular structure of muscle cells that contains the contractile apparatus

myogenic Muscle contraction that is initiated by the muscle cell itself

myometrium (mī´-ō-mē-TRĒ-um) The smooth muscle layer of the mammalian uterus

myosin (MĪ-uh-sun) The contractile protein that forms the thick filaments in muscle fibers

Na^+/K^+ pump (ATPase) A carrier that actively transports Na^+ out of the cell and K^+ into the cell, using ATP for energy

nanotube Long tubes with an internal actin-filament support surrounded by a plasma membrane that permits contiguous contact between two cells

nasal fossae Uppermost tract of the respiratory airways

natriuretic Loss of sodium in the urine

natural killer (NK) cells Naturally occurring, vertebrate lymphocyte-like cells that nonspecifically destroy virus-infected cells and cancer cells by directly lysing their membranes on first exposure to them

natural selection The process in which members of a species having genetic characteristics that enhance survival are able to produce more surviving offspring than other members not having those characteristics; the primary mechanism of Darwinian evolution

negative balance Situation in which the losses for a substance exceed its gains so that the total amount of the substance in the body decreases

negative feedback A regulatory mechanism in which a change in a controlled variable triggers a response that opposes the change, thus maintaining a relatively steady set point for the regulated factor

nephron (NEF-ron´) The functional unit of the vertebrate kidney; consisting of an interrelated vascular and tubular component; the smallest unit that can form urine

nerve A bundle of peripheral neuronal axons, some afferent and some efferent, enclosed by a connective tissue covering and following the same pathway

nerve net A network of interconnected neurons that are distributed throughout a body or organ; characterized by a diffuse spread of excitation; occurs in most lower organisms, including Cnidarians, and in visceral organs of most higher phyla

nerve plexus Network of interwoven nerves

nervous system One of the two major regulatory systems of an animal body; in general, coordinates rapid activities of the body, especially those involving interactions with the external environment

nervous tissue One of four animal tissue types (along with epithelial, connective, and muscular); a functional grouping of cells specialized for initiation and transmission of electrical signals

net filtration pressure The net difference in the hydrostatic and osmotic forces acting across the vertebrate glomerular membrane that favors the filtration of a protein-free plasma into Bowman's capsule

neuroendocrinology The study of the interaction between the nervous and endocrine systems

neurogenic Producing a pacemaking electrical signal from neural tissue

neuroglia See *glial cells*

neuroglobin Oxygen-binding protein found in vertebrate neurons

neurohormones Hormones released into the blood by neurosecretory neurons

neuromodulator Chemical messenger that binds to neuronal receptors at nonsynaptic sites and brings about long-term changes that subtly depress or enhance synaptic effectiveness

neuromuscular junction The juncture between a motor neuron and a skeletal muscle fiber; a highly specialized synapse

neuron (NER-on) A nerve cell, typically consisting of a cell body, dendrites, and an axon and specialized to initiate, propagate, and transmit electrical signals

neuropeptides Large slow-acting peptide molecules released from axon terminals along with classical neurotransmitters; most neuropeptides function as neuromodulators

neuropil Region of the nervous system where the synapses between axons and dendrites are concentrated

neurotransmitter The chemical messenger that is released from the axon terminal of a neuron in response to an action potential and influences another neuron or an effector with which the neuron is anatomically linked

neutrophils (new′-truh-filz) Vertebrate white blood cells that are phagocytic specialists and important in inflammatory responses and defense against bacterial invasion

nicotinic receptor (nick′-o-TIN-ik) Type of cholinergic receptor found at all vertebrate autonomic ganglia and the motor end plates of skeletal muscle fibers

nitric oxide A local chemical mediator released from endothelial cells and other tissues; exerts a wide array of effects, ranging from causing local arteriolar vasodilation to acting as a toxic agent against foreign invaders to serving as a unique type of neurotransmitter

nitrogen narcosis Syndrome of reduced neural excitability in human divers with increasing depth, caused by the highly lipid-soluble N_2 dissolving in neural membranes

nociceptor (nō-sē-SEP-tur) A pain receptor, sensitive to tissue damage

node of Ranvier (RAN-vē-ā) The portions of a myelinated neuronal axon between the segments of insulative myelin; the axonal regions where the axonal membrane is exposed to the ECF and membrane potential exists

nontropic hormone A hormone that exerts its effects on nonendocrine target tissues

norepinephrine (nor′-ep-uh-NEF-run) The neurotransmitter released from vertebrate sympathetic postganglionic fibers; noradrenaline

nucleus (of brain) (NŪ-klē-us) A functional aggregation of neuronal cell bodies within the brain

nucleus (of cells) A distinct spherical or oval structure that is usually located near the center of a cell and that contains the cell's genetic material, deoxyribonucleic acid (DNA)

O_2–Hb dissociation curve A plot of the relationship between arterial P_{O_2} and percent hemoglobin (Hb) saturation

occipital lobes (ok-SIP′-ut-ul) The lobes of the vertebrate cerebral cortex that are located posteriorly and are responsible for initially processing visual input

olfaction Smell, or detection of molecules released from a distant object

oligodendrocytes (ol-i-gō′-DEN-drō-sitz) The myelin-forming cells of the vertebrate central nervous system

omasum The third chamber of the ruminant stomach

ommatidium The structural unit of the nonvertebrate compound eye, composed of a cornea, a focusing cone and several receptor cells connected to the optic nerve

omnivore Animal that can eat both animals and other organisms such as plants or fungi

oncotic pressure Form of osmotic pressure exerted by proteins in the blood that attracts water into the circulatory system

oogenesis (ō′-ō-JEN-uh-sus) Egg production

ootheca The covering or case of an egg mass

operculum (1) Bony plate that covers fish gill arches; (2) hardened protein disc on the foot of a gastropod that seals up the shell opening when the animal is withdrawn

opsonin (OP′-suh-nun) Immune chemicals that "tag" bacteria for destruction, making the bacteria more susceptible to phagocytosis

optic chiasm Located underneath the vertebrate hypothalamus where the two optic nerves meet; depending on the species, some axons cross the midline at this point and project to the contralateral side of the brain

optic nerve The bundle of nerve fibers that leaves the retina, relaying information about visual input

optimal length The length before the onset of contraction of a muscle fiber at which maximal force can be developed on a subsequent tetanic contraction

organ A distinct structural unit composed of two or more types of primary tissue organized to perform one or more particular functions; for example, the stomach

organ of Corti (KOR-tē) The sense organ of hearing within the vertebrate inner ear that contains hair cells whose hairs are bent in response to sound waves, setting up action potentials in the auditory nerve

organelle (or′-gan-EL) Distinct, highly organized, membrane-bound intracellular compartment, containing a specific set of chemicals for carrying out a particular cellular function

organic Referring to substances that contain carbon (and usually hydrogen)

organism A living entity, either single-celled (unicellular) or made up of many cells (multicellular)

osmolarity (oz′-mō-LAR-ut-ē) A measure of the concentration of solute molecules in a solution; also known as **osmotic pressure**

osmolyte A molecule used to elevate a cell's or body fluid's osmotic pressure in order to balance a high external osmolarity; can be inorganic or organic

osmosis (os-MŌ-sis) Movement of water across a membrane down its own concentration gradient toward the area of higher solute concentration, exerting an **osmotic pressure** in doing so

osteoblasts (OS-tē-ō-blasts′) Bone cells that produce the organic matrix of bone

osteoclasts Bone cells that dissolve bone in their vicinity

ostium (pl., **ostia**) Large porelike opening, such as that which allows hemolymph to reenter the arthropod heart after passing through the body

otolith organs (ŌT′-ul-ith) Sense organs in the vertebrate inner ear that provide information

about rotational changes in head movement; include the utricle and saccule

oval window The membrane-covered opening that separates the air-filled vertebrate middle ear from the upper compartment of the fluid-filled cochlea in the inner ear

ovarioles Tubular structures forming the insect ovary

overhydration Water excess in the body

oviparous Referring to females that release eggs from which the young hatch after expulsion from the body

oviposition In birds, the act of laying an egg

ovipositor An extension of the female insect abdomen used to lay eggs, for example, by insertion into the ground or into a host animal

ovoviviparous Referring to production of eggs that hatch within the body of the parent

ovulation (ov′-yuh-LĀ-shun) Release of an ovum from a mature ovarian follicle

oxidative phosphorylation (fos′-fōr-i-LĀ-shun) The entire sequence of mitochondrial biochemical reactions that uses oxygen to extract energy from the nutrients in food and transforms it into ATP, producing CO_2 and H_2O in the process

oxygen debt Condition in which oxygen demand is greater than oxygen supply to a tissue

oxyhemoglobin (ok-si-HĒ-muh-glō-bun) Hemoglobin combined with O_2

oxyntic mucosa (ok-SIN-tic) The mucosa that lines the body and fundus of the vertebrate stomach; contains gastric pits lined by mucous neck cells, parietal cells, and chief cells

oxytocin (ok′-sē-TŌ-sun) A hypothalamic neurohormone that is stored in the posterior pituitary and, in mammals, stimulates uterine contraction and milk ejection

pacemaker activity Self-excitable activity of an excitable cell in which its membrane potential gradually depolarizes to threshold on its own

Pacinian corpuscle (pa-SIN-ē-un) A rapidly adapting skin receptor in mammals that detects pressure and vibration

papilla A long, cone-shaped inner medulla in the kidneys of some mammals, such as rodents

parabronchi Open-ended, small parallel tubes in bird lungs, each several millimeters long and 0.5 mm wide

paracrine (PEAR-uh-krin) A local chemical messenger whose effect is exerted only on neighboring cells in the immediate vicinity of its site of secretion

parasympathetic nervous system (pear′-uh-sim-puh-THET-ik) The subdivision of the vertebrate autonomic nervous system that dominates in quiet, relaxed situations and promotes body maintenance activities such as digestion and emptying of the urinary bladder

parathyroid glands (pear′-uh-THĪ-roid) Small glands, located on or near the posterior surface of the vertebrate thyroid gland, that secrete parathyroid hormone

parathyroid hormone (PTH) A vertebrate hormone that raises plasma Ca^{2+} levels

parietal cells (puh-RĪ-ut-ul) The vertebrate stomach cells that secrete hydrochloric acid and intrinsic factor

parietal lobes The lobes of the vertebrate cerebral cortex that lie at the top of the brain behind the central sulcus and contain the somatosensory cortex

parthenogenesis Sexual reproduction in which an egg develops without entry of a sperm; common among ants, bees, wasps, and aphids

partial pressure The individual pressure exerted independently by a particular gas within a mixture of gases

partial pressure gradient A difference in the partial pressure of a gas between two regions that promotes the movement of the gas from the region of higher partial pressure to the region of lower partial pressure

parturition (par′-too-RISH-un) Delivery of a newborn mammal

passive expiration Expiration accomplished during quiet breathing as a result of elastic recoil of the lungs on relaxation of the inspiratory muscles, with no energy expenditure required

passive force A force that does not require expenditure of cellular energy to accomplish transport of a substance across the plasma membrane

passive reabsorption When none of the steps in the transepithelial transport of a substance reabsorbed requires energy expenditure

patch clamping A process in which a tiny pipette is attached to a plasma membrane with gentle suction, forming a tight seal around one patch of a membrane; with a fine enough pipette, a single ion channel or receptor protein can be isolated

pathogens (PATH-uh-junz) Disease-causing microorganisms, such as bacteria or viruses

pathophysiology (path′-ō-fiz-ē-OL-ō-gē) Abnormal functioning of the body associated with disease

pecten Thin, comblike projections from the choroid into the vitreous humor (avian retina)

pepsinogen (pep-SIN-uh-jun) An enzyme secreted in inactive form by the stomach that, once activated, becomes **pepsin,** which begins protein digestion

peptide hormones Hormones that consist of a chain of specific amino acids of varying length

percent hemoglobin saturation A measure of the extent to which the hemoglobin present is combined with O_2

perception The brain's interpretation of the external world as created from a pattern of nerve impulses delivered to it

perfusion The pumping of a fluid through a tissue or organ by means of a circulatory vessel

peripheral chemoreceptors (kē′-mō-rē-SEP-turz) The carotid and aortic bodies, which respond to changes in arterial P_{O_2}, P_{CO_2}, and H^+ and that adjust respiration accordingly (in vertebrates)

peripheral nervous system (PNS) Nerve fibers that carry information between the central nervous system and other parts of the body

peristalsis (per′-uh-STOL-sus) Ringlike contractions of the circular smooth muscle of a tubular organ that move progressively forward with a stripping motion, pushing the contents of the organ ahead of the contraction

peritrophic membrane Layer that separates the midgut epithelium from the food in insects

peritubular capillaries (per′-i-TŪ-bū-lur) Capillaries that intertwine around the tubules of the mammalian kidney's nephron; they supply the renal tissue and participate in exchanges between the tubular fluid and blood during the formation of urine

permeable Permitting passage of a particular substance

permissiveness The condition when one hormone must be present in adequate amounts for the full exertion of another hormone's effect

peroxisomes (puh-ROK′-suh-sōmz) Organelles consisting of membrane-bound sacs that contain powerful oxidative enzymes that detoxify various wastes produced within the cell or foreign compounds that have entered the cell

pH The logarithm to the base 10 of the reciprocal of the hydrogen ion concentration; pH log $1/[H^+]$ or pH log$[H^+]$

phagocytosis (fag′-oh-sī-TŌ-sus) A type of endocytosis in which large multimolecular solid particles are engulfed by a cell

pharynx (FARE-inks) The throat, which in vertebrates serves as a common passageway for the digestive and respiratory systems

phasic receptors Sensory receptors that produce action potentials at the onset and offset of a sustained stimulus

pheromone Chemical signal released into the environment, usually by glands, which travels through the air or water to sensory cells in another animal; used for sexual activity (e.g., signaling of readiness to mate), marking of territories, and other behaviors related to interactions among individuals of a species

phosphagen An energy-storing cellular molecule with a phosphate attached by a high-energy bond, such as creatine phosphate and arginine phosphate; its phosphate can be transferred to ADP to make ATP

phosphorylation (fos′-fōr-i-LĀ-shun) Addition of a phosphate group to a molecule

photoperiod The amount of time during the day when there is light

photoreceptor A sensory receptor responsive to light

phototransduction The mechanism of converting light stimuli into electrical activity by the rods and cones of the eye

physiology (fiz-ē-OL-ō-gē) The study of organismal functions

phytoestrogen Natural estrogenic-like compounds synthesized by some plant species

pineal gland (PIN-ē-ul) A small endocrine gland located in the center of the vertebrate brain that secretes the hormone melatonin

pinna A skin-covered flap of cartilage around the entrance to the mammalian ear, collects sound

pinocytosis (pin-oh-cī-TŌ-sus) Type of endocytosis in which the cell internalizes fluid

pitch The tone of a sound, determined by the frequency of vibrations (i.e., whether a sound is a C or G note)

pituitary gland (pih-TWO-ih-tair-ee) A small vertebrate endocrine gland connected by a stalk to the hypothalamus; consists of the anterior pituitary and posterior pituitary

placenta (plah-SEN-tah) The organ of exchange between the maternal and fetal blood in placental mammals and some sharks

plasma The liquid portion of the blood or hemolymph

plasma cell An antibody-producing derivative of an activated vertebrate B lymphocyte

plasma clearance The volume of plasma that is completely cleared of a given substance by the kidneys per minute

plasma membrane A protein-studded lipid bilayer that encloses each cell, separating it from the extracellular fluid

plasma proteins The proteins that remain within the plasma, where they perform a number of important functions; include albumins, globulins, and fibrinogen in vertebrates

plasma-colloid osmotic pressure (KOL-oid os-MOT-ik) The force caused by the unequal distribution of plasma proteins between the blood and surrounding fluid that encourages fluid movement into the capillaries

plasticity (plas-TIS-uh-tē) The ability of portions of the nervous system to assume new responsibilities in response to the demands placed on it

platelet (PLĀT-let) Specialized cell fragment in mammalian blood that participates in hemostasis by forming a plug at a vessel defect

pleural sac (PLOOR-ul) A double-walled, closed sac that separates each lung from the thoracic wall in air-breathing vertebrates

pluripotent stem cell Precursor cell that resides in certain adult tissues (e.g., bone marrow in mammals) and continuously divides and differentiates to give rise to specialized cells (e.g., blood cells in mammals)

poikilotherm Animal whose body temperature varies with environmental temperature

polarity Unequal sharing of electrons between atoms

polarized light Light in which the waves are all oriented at the same angle

polyestrous Mammalian estrous cycles occurring uniformly throughout the year without seasonal influence

polymer A chain of monomers linked together

polysaccharide (pol′-i-SAK-uh-rīd) Complex carbohydrate consisting of chains of interconnected glucose molecules (starch, glycogen, cellulose)

polyunsaturated fatty acid (PUFA) A fatty acid in which more than one of the bonds between carbon atoms in the carbon chain backbone of the molecules are double bonds

positive balance Situation in which the gains via input for a substance exceed its losses via output, so the total amount of the substance in the body increases

positive feedback A regulatory mechanism in which the input and the output in a control system continue to enhance each other so that the controlled variable is progressively moved further from a steady state

postabsorptive state The metabolic state after a meal is absorbed during which endogenous energy stores must be mobilized (and glucose must be spared for the glucose-dependent brain in vertebrates); fasting state

posterior pituitary The neural portion of the pituitary that stores and releases into the blood on hypothalamic stimulation two hormones produced by the hypothalamus, vasopressin and oxytocin (in vertebrates)

postganglionic fiber (pōst′-gan-glē-ON-ik) The second neuron in the two-neuron autonomic nerve pathway; originates in an autonomic ganglion and terminates on an effector organ (in vertebrates)

postsynaptic neuron (pōst'-si-NAP-tik) The neuron that conducts its action potentials away from a synapse

precocial Referring to young (hatchlings) with well-developed legs, open eyes, and alertness; able to leave the nest and feed independently

preganglionic fiber The first neuron in the two-neuron autonomic nerve pathway; originates in the central nervous system and terminates on an autonomic ganglion (in vertebrates)

pregastric fermentation Positioning the microbe-containing fermentation chamber of ruminants before the acid-secreting stomach

preprohormones Large precursor proteins that undergo a series of posttranslational modifications to yield the biologically active hormone end product

pressure gradient A difference in pressure between two regions that drives the movement of blood or air from the region of higher pressure to the region of lower pressure

presynaptic facilitation Enhanced release of neurotransmitter from a presynaptic axon terminal as a result of excitation of another neuron that terminates on the axon terminal

presynaptic inhibition A reduction in the release of neurotransmitter from a presynaptic axon terminal as a result of excitation of another neuron that terminates on the axon terminal

presynaptic neuron (prē-si-NAP-tik) The neuron that conducts its action potentials toward a synapse

primary active transport A carrier-mediated transport system in which energy is directly required to operate the carrier and move the transported substance against its concentration gradient; see *secondary active transport*

primary follicle A primary oocyte surrounded by a single layer of granulosa cells in the ovary (in vertebrates)

primary motor cortex The portion of the vertebrate cerebral cortex that lies anterior to the central sulcus and is responsible for voluntary motor output

proctodaeum In birds, the hind portion of the cloaca, stores the excreta; in insects, the hindgut from the Malpighian tubules to the anus

proestrus The stage of the estrous cycle between the regression of the corpus luteum and the onset of estrus

prokaryote A microbial cell that has no classic organelles such as a nucleus; bacteria and archaea

prolactin (PRL) (prō-LAK-tun) An anterior pituitary hormone with numerous roles; for example, stimulates mammary gland development and milk production in female mammals, and regulates osmoregulation in fishes (among other effects)

promoter A regulatory or "switch" section of DNA that can be bound by specific transcription-regulating proteins, thereby regulating the transcription of a nearby RNA-coding gene

pro-opiomelanocortin (POMC) (prō-op'Ē-ō-ma-LAN-oh-kor'-tin) A large precursor molecule produced by the anterior pituitary that is cleaved into adrenocorticotropic hormone, melanocyte-stimulating hormone, and endorphin (in vertebrates)

proPOs Inactive forms of POs (phenoloxidases) that are activated during exoskeletal synthesis and immune responses in arthropods

proprioception (prō'-prē-ō-SEP-shun) Sensing of position of body parts in relation to each other and to surroundings

proprioceptors Sensors that send information about movement and position of an animal's body or certain parts such as limbs

prostaglandins (pros'-tuh-GLAN-dins) Local chemical mediators (paracrines) that are derived from a component of the plasma membrane, arachidonic acid

prostate gland A vertebrate male accessory sex gland that secretes an alkaline fluid, which neutralizes acidic vaginal secretions in internal fertilizers

protandry Process in which a male becomes a female

proteasome An organelle that destroys unwanted cell proteins that have been tagged by ubiquitin

protein kinase (KĪ-nase) An enzyme that phosphorylates and thereby induces a change in the shape and function of a particular intracellular protein

proteolytic enzymes (prōt'-ē-uh-LIT-ik) Enzymes that digest protein

proteomics (proteonomics) The study of the structure, regulation, and functions of proteins

protogyny Process in which a female becomes a male

protonephridia Blind-end ducts that project into the body cavity and that filter wastes using ultrafiltration driven by cilia, followed by secretion, and reabsorption; generally found in more primitive animals having one internal fluid compartment, such as rotifers, flatworms, larval annelids, and larval mollusks

proventriculus (1) True stomach of a bird; (2) in some nonvertebrates, a dilation of the foregut

proximal tubule A highly convoluted tubule that extends between Bowman's capsule and the loop of Henle in the mammalian kidney's nephron

proximate explanation See *mechanistic explanation*

pseudogene A gene sequence that once coded for a protein but no longer does so

PTH See *parathyroid hormone*

pubertin See *leptin*

pulmonary artery (PULL-mah-nair-ē) The large vessel that carries blood from the heart to the lungs (in air-breathing vertebrates)

pulmonary circulation The loop of blood vessels carrying blood between the heart and lungs (in air-breathing vertebrates)

pulmonary surfactant (sur-FAK-tunt) A phospholipoprotein complex secreted by the Type II alveolar cells that intersperses between the water molecules that line the alveoli, thereby lowering the surface tension within the lungs (in mammals)

pulmonary valve A one-way valve that permits the flow of blood from the right ventricle into the pulmonary artery during ventricular emptying but prevents the backflow of blood from the pulmonary artery into the right ventricle during ventricular relaxation (in mammals, birds)

pulmonary vein The large vessel that carries blood from the lungs to the heart (in air-breathing vertebrates)

pulmonary ventilation The volume of air breathed in and out of the lungs in one minute; equals tidal volume × respiratory rate

pupil A round opening in the center of the iris through which light passes to the interior portions of the eye (adjustable in some animals)

Purkinje fibers (pur-KIN-jē) Small terminal fibers that extend from the bundle of His and rapidly transmit an action potential throughout the mammalian ventricular myocardium

pyloric gland area (PGA) (pī-LŌR-ik) The specialized region of the mucosa in the antrum of the mammalian stomach that secretes gastrin

pyloric sphincter (pī-lōr'-ik SFINGK-tur) The juncture between the stomach and duodenum (in vertebrates)

radiation Emission of heat energy from the surface of a warm body in the form of electromagnetic waves

reabsorption The net movement of interstitial fluid into the capillary

receptor See *sensory receptor* or *receptor site*

receptor potential The graded potential change that occurs in a sensory receptor in response to a stimulus; generates action potentials in the afferent neuron fiber

receptor site Membrane protein that binds with a specific extracellular chemical messenger, thereby bringing about a series of membrane and intracellular events that alter the activity of the particular cell

rectal gland A specialized hindgut organ of cartilaginous fishes, which has numerous blind-end tubules that remove and excrete NaCl using active transport

reflex Any response that occurs automatically without modification by "higher" control centers; the components of a reflex arc include a receptor, afferent pathway, integrating center, efferent pathway, and effector

reflex arc A neuronal pathway that controls the reflex motor response to a specific sensory stimulus

refraction Bending of a light ray

refractory period (rē-FRAK-tuh-rē) The time period when a recently activated patch of membrane is refractory (unresponsive) to further stimulation, preventing the action potential from spreading backward into the area through which it has just passed, thereby ensuring the unidirectional propagation of the action potential away from the initial site of activation

regulators Organisms in which a particular physiological state is kept relatively constant in the face of a changing environment

regulatory T cell A type of T cell that suppresses the actions of other T cells, keeping them in check

regulatory volume decrease (RVD) A process in which cells subjected to osmotic swelling decrease their internal osmotic pressure by removing osmolytes in order to restore cell volume

regulatory volume increase (RVI) A process in which cells subjected to osmotic shrinkage increase their internal osmotic pressure by accumulating osmolytes in order to restore cell volume

releasing hormone A hypothalamic hormone that stimulates the secretion of a particular anterior pituitary hormone (in vertebrates)

renal cortex An outer, granular-appearing region of the mammalian and avian kidney

renal medulla (RĒ-nul muh-DUL-uh) An inner, striated-appearing region of the mammalian and avian kidney

renal threshold The plasma concentration at which the T_m (tubular maximum) of a particular substance is reached and the substance first starts appearing in the urine

renin (RĒ-nin) An enzymatic hormone released from mammalian kidneys in response to a decrease in NaCl/ECF volume/arterial blood pressure; activates angiotensinogen

renin–angiotensin–aldosterone system (RAAS) (an′jē-ō-TEN-sun al-dō-steer-OWN) The salt-conserving system triggered by the release of renin from mammalian kidneys, which activates angiotensin, which stimulates aldosterone secretion, which stimulates Na^+ reabsorption by the kidney tubules

repolarization (rē′-pō-luh-ruh-ZĀ-shun) Return of membrane potential to resting potential following a depolarization

reset In physiological negative-feedback regulation, the process of changing the set point from one level to another

residual volume The minimum volume of air remaining in respiratory pathways even after a maximal expiration

resistance Hindrance of flow of liquid or air through a passageway (e.g., a blood vessel or respiratory airway)

respiration The sum of processes that accomplish ongoing movement of O_2 from the atmosphere to the tissues, as well as the continual movement of metabolically produced CO_2 from the tissues to the atmosphere

respiratory acidosis (as-i-DŌ-sus) Acidosis resulting from abnormal retention of CO_2 arising from hypoventilation

respiratory airways The system of tubes that conducts air between the atmosphere and the alveoli of the lungs

respiratory alkalosis (al′-kuh-LŌ-sus) Alkalosis caused by excessive loss of CO_2 from the body as a result of hyperventilation

respiratory rate Breaths per minute

response element A promoter DNA sequence that is the target of action of signal molecules such as hormones; binds to a hormone and its nuclear receptor; also termed hormone response element (HRE)

resting membrane potential The membrane potential that exists when an excitable cell is not displaying an electrical signal

rete mirabile Extensive countercurrent network of arterial and venous capillaries

reticular activating system (RAS) (ri-TIK-ū-lur) Ascending fibers that originate in the reticular formation and carry signals upward to arouse and activate the cerebral cortex (in vertebrates)

reticular formation A network of interconnected neurons that runs throughout the vertebrate brainstem and initially receives and integrates all synaptic input to the brain

reticulum The second stomach of ruminants (see *rumen*)

retina The innermost layer in the posterior region of an eye that contains the eye's photoreceptors, the rods and cones

rhabdomere Specialized region of the retinular cell of a nonvertebrate compound eye

rheostasis Regulated change of a physiological state

rhopalium Cnidarian medusa sense organ; has ocelli (simple eye structures) and statocysts

ribonucleic acid (RNA) (rī-bō-new-KLĀ-ik) A nucleic acid that exists in three major forms (messenger RNA, ribosomal RNA, and transfer RNA), which participate in gene transcription and protein synthesis (other forms have also been recently discovered, such as small nuclear and microRNAs)

ribosomes (RĪ-bō-sōms) Special ribosomal RNA–protein complexes that synthesize proteins under the direction of nuclear DNA

right atrium (Ā′-trē-um) The heart chamber that receives venous blood from the systemic circulation (in reptiles, birds, mammals)

right ventricle The heart chamber that pumps blood into the pulmonary circulation (in reptiles, birds, mammals)

rigor mortis "Stiffness of death"; a generalized locking-in-place of the skeletal muscles that begins sometime after death

RNA See *ribonucleic acid*

rods The eye's photoreceptors used for night vision

root effect A decrease in the O_2-carrying capacity of hemoglobin in response to a decrease in pH

round window In higher vertebrates, the membrane-covered opening that separates the lower chamber of the cochlea in the inner ear from the middle ear

r-selected Referring to species that place virtually all reproductive resources into producing hundreds to thousands (or more) of offspring, with no parental care and minimal nourishment

rumen The first compartment of the ruminant stomach; microbial populations found in the rumen include anaerobic species of bacteria, protozoa, and fungi

ruminant Various hoofed mammals that chew food already swallowed and that have a stomach divided into four (or three) compartments

ruminoreticulum The rumen and its continuation, the reticulum; two compartments of the ruminant stomach

SA node See *sinoatrial node*

salivary amylase (AM-uh-lās′) An enzyme produced by salivary glands that begins carbohydrate digestion in the mouth and continues in the digestive tract after the food and saliva have been swallowed

salt gland A specialized gland associated with the eyes or nasal cavities of marine birds and reptiles; it removes excess NaCl from the blood

saltatory conduction (SAL-tuh-tōr′-ē) The means by which an action potential is propagated throughout a myelinated fiber, with the impulse jumping over the myelinated regions from one node of Ranvier to the next (in vertebrates)

sarcomere (SAR-kō-mir) The functional unit of skeletal muscle; the area between two Z lines within a myofibril

sarcoplasmic reticulum (ri-TIK-yuh-lum) A fine meshwork of interconnected tubules that surrounds a muscle fiber's myofibrils; contains expanded lateral sacs, which store calcium that is released into the cytosol in response to a local action potential

satiety centers (suh-TĪ-ut-ē) Neuronal clusters in the ventromedial region of the hypothalamus that inhibit feeding behavior (in vertebrates)

saturation The condition when all the binding sites on a carrier molecule are occupied

scaling The phenomenon that many anatomical and physiological parameters vary with other parameters (e.g., body mass) in nonproportional relationships

Schwann cells (shwah′-n) The myelin-forming cells of the vertebrate peripheral nervous system

second messenger An intracellular chemical that is activated by binding of an extracellular first messenger to a surface receptor site and that triggers a preprogrammed series of biochemical events, which result in altered activity of intracellular proteins to control a particular cell activity

secondary active transport A transport mechanism in which a carrier molecule for glucose or an amino acid is driven by an Na^+ concentration gradient established by the energy-dependent Na^+ pump to transfer the glucose or amino acid uphill without directly expending energy to operate the carrier

secondary follicle A developing ovarian follicle that is secreting estrogen and forming an antrum (in vertebrates)

secondary sexual characteristics The many external characteristics that are not directly involved in reproduction but that distinguish males and females

secretin (si-KRĒT-′n) A hormone released from the vertebrate duodenal mucosa primarily in response to the presence of acid; inhibits gastric motility and secretion and stimulates secretion of an $NaHCO_3$ solution from the pancreas

secretion Release to a cell's exterior, on appropriate stimulation, of substances that have been produced by the cell

secretory vesicles (VES-i-kuls) Membrane-enclosed sacs containing proteins that have been synthesized and processed by the endoplasmic reticulum/Golgi complex of the cell and that will be released to the cell's exterior by exocytosis on appropriate stimulation

segmentation The vertebrate small intestine's primary method of motility; consists of oscillating, ringlike contractions of the circular smooth muscle along the small intestine's length

self-antigens Antigens that are characteristic of an animal's own cells

semen (SĒ-men) A mixture of accessory sex-gland secretions and sperm

semicircular canal Sense organ in the inner ear that detects rotational or angular acceleration or deceleration of the head (in vertebrates)

semilunar valves (sem′-ī-LEW-nur) The aortic and pulmonary valves (in the avian and mammalian heart)

seminal vesicles (VES-i-kuls) Male accessory sex glands that supply fructose to ejaculated sperm and secrete prostaglandins (in vertebrates)

seminiferous tubules (sem′-uh-NIF-uh-rus) The highly coiled tubules within vertebrate testes that produce spermatozoa

sensilla Specialized projections from insect cuticles for tasting

sensory afferent Pathway coming into the central nervous system that carries information

sensory input Includes somatic sensation and special senses

sensory receptor An afferent neuron's peripheral ending, which is specialized to respond to a particular stimulus in its environment

septum (pl., **septa**) Dividing wall or partition

series-elastic component The noncontractile portions of a skeletal muscle fiber, including the connective tissue and sarcoplasmic reticulum

Sertoli cell (sur-TŌ-lē) Cell located in the seminiferous tubules that supports spermatozoa during their development (in vertebrates)

set point The desired level at which homeostatic control mechanisms maintain a controlled variable

signal transduction The sequence of events in which incoming signals (instructions from extracellular chemical messengers such as hormones) are conveyed to the cell's interior for execution

single-unit smooth muscle The most abundant type of smooth muscle; made up of muscle fibers that are interconnected by gap junctions so that they become excited and contract as a unit; also known as *visceral smooth muscle*

sinoatrial (SA) node (sī-nō-Ā-trē-ul) A small specialized autorhythmic region in the right atrial wall of the heart that has the fastest rate of spontaneous depolarizations and serves as the normal pacemaker of the heart (in reptiles, birds, mammals)

sinus venosus In all fishes, a chambered extension of the atrium into which blood first enters the heart

skeletal muscle Striated muscle, which is attached to the endoskeleton or exoskeleton and is responsible for movement of the parts of the skeleton in purposeful relation to one another; innervated by the somatic nervous system and under both reflex and nonreflex control

slow-wave potentials Self-excitable activity of an excitable cell in which its membrane potential undergoes gradually alternating depolarizing and hyperpolarizing swings

Smith predictor A control device that predicts the next step in a regulated process and triggers it to begin before feedback information is received about the immediately preceding step; thought to be the mechanism by which the mammalian cerebellum controls skilled locomotory movements

smooth muscle Muscle without organized sarcomeres; found in the walls of hollow organs and tubes; innervated by the autonomic nervous system in vertebrates

solutes Molecules dissolved in liquid (e.g., water)

somatic cells (sō-MAT-ik) Body cells, as contrasted with reproductive cells

somatic nervous system The portion of the efferent division of the vertebrate peripheral nervous system that innervates skeletal muscles; consists of the axonal fibers of the alpha motor neurons

somatic sensation Sensory information arising from the body surface or interior, including somesthetic sensation and proprioception

somatomedins (sō′-mat-uh-MĒ-dinz) Hormones secreted by the vertebrate liver or other tissues, in response to growth hormone, that act directly on the target cells to promote growth

somatosensory cortex The region of the vertebrate parietal lobe immediately behind the central sulcus; the site of initial processing of somesthetic and proprioceptive input

somesthetic sensation Sensory information arising mainly from the body surface (touch, pressure, temperature, pain, and electrical fields)

sound waves Traveling vibrations of air that consist of regions of high pressure caused by compression of air molecules alternating with regions of low pressure caused by rarefaction of the molecules

spatial summation The summing of several postsynaptic potentials arising from the simultaneous activation of several excitatory (or several inhibitory) synapses

specific dynamic action See *diet-induced thermogenesis*

specificity Ability of carrier molecules to transport only specific substances across the plasma membrane

spermatheca Organ that stores the sperm prior to entry into the vagina

spermatogenesis (spur′-mat-uh-JEN-uh-sus) Sperm production

spermatophore Sperm-containing capsule formed by hardened seminal fluid

sphincter (sfink-tur) A ring of muscle that controls passage of contents through an opening into or out of a hollow organ or tube

spiracle Surface opening of the tracheal system in insects and certain arachnids

spleen A vertebrate lymphoid tissue in the upper left part of the abdomen that stores lymphocytes and platelets and destroys old red blood cells

stanniocalcin A hormone regulating calcium homeostasis in bony fish (formerly called *hypocalcin*); synthesized in small organs termed the corporacles of Stannius located within the kidney

state of equilibrium A condition in which no net change in a system is occurring

statolith Small dense particle, that readily moves within the fluid content of a statocyst, to provide directional information

stem cell Relatively undifferentiated precursor cell that gives rise to highly differentiated, specialized cells

stenohaline Having a limited tolerance to changes in salinity

stereocilia Nonmotile, ciliary projections of a hair cell

steroid (STEER-oid) Hormone derived from cholesterol

stimulus A detectable physical or chemical change in the environment of a sensory receptor

stomodaeum The foregut of an insect

stress The generalized, nonspecific response of the body to any factor that overwhelms, or threatens to overwhelm, the body's compensatory abilities to maintain homeostasis

stress proteins Low-molecular-weight proteins synthesized in many types of organisms after a sudden increase in a stress factor such as temperature, pollutants, or oxygen deprivation; help properly fold other proteins perturbed by the stress factor; if associated with heat stress, these are called *heat shock proteins* (HSPs)

stretch reflex A monosynaptic reflex in which an afferent neuron originating at a stretch-detecting receptor in a skeletal muscle terminates directly on the efferent neuron supplying the same muscle to cause it to contract and counteract the stretch

stroke volume (SV) The volume of blood pumped out of a ventricle with each contraction, or beat, of the heart

subcortical regions The vertebrate brain regions that lie under the cerebral cortex, including the basal nuclei, thalamus, and hypothalamus

submucosa The connective tissue layer of the digestive tract that lies under the mucosa and (in vertebrates) contains the larger blood and lymph vessels and a nerve network

substance P The neurotransmitter released from mammalian pain fibers

subsynaptic membrane (sub-sih-NAP-tik) The portion of the postsynaptic cell membrane that lies immediately underneath a synapse and contains receptor sites for the synapse's neurotransmitter

supercooling A state of water in which the temperature is below the freezing point, but with no trigger or nucleation site to begin ice formation

suprachiasmatic nucleus (SCN) (soup′-ra-kī-as-MAT-ik) A cluster of nerve cell bodies in the hypothalamus that serves as the master biological clock, acting as the pacemaker that establishes many of the mammalian body's circadian rhythms

surface tension The force at the liquid surface of an air–water interface resulting from the greater attraction of water molecules to the surrounding water molecules than to the air above the surface; a force that tends to decrease the area of a liquid surface and resists stretching of the surface

sympathetic nervous system The subdivision of the vertebrate autonomic nervous system that dominates in emergency ("fight-or-flight") or stressful situations and prepares the body for strenuous physical activity

symporter A transporter protein in a membrane that moves two (or more) molecules or ions in the same direction

synapse (SIN-aps′) The specialized junction between two neurons where an action potential in the presynaptic neuron influences the membrane potential of the postsynaptic neuron by means of the release of a chemical messenger that diffuses across the small cleft that separates the two neurons

synchronous In step or in phase control of flight muscles, which are activated by the release of Ca^{2+} and deactivated by Ca^{2+} reuptake

synergism (SIN-er-jiz′-um) When several actions are complementary, causing their combined effect to be greater than the sum of their separate effects

syrinx Vocal organ of birds, located at the lower end of the trachea, where the two bronchi join; acts as a sound-producing valve in which air forced by pressure from the lungs sets the tympanic membranes in motion

systemic circulation (sis-TEM-ik) The closed loop of vertebrate blood vessels carrying blood between the heart and body systems

systole (SIS-tō-lē) The period of cardiac contraction and emptying

T lymphocytes (T cells) Vertebrate white blood cells that accomplish cell-mediated immune responses against targets to which they have been previously exposed; see also *cytotoxic T cells* and *helper T cells*

T tubule See *transverse tubule*

T₃ See *tri-iodothyronine*

T₄ See *thyroxine*

taeniae coli Three separate, conspicuous, longitudinal bands of muscle lining the large intestine

tapetum lucidum Reflecting structure found in the choroid coat of some nocturnal animals

target cell receptors Receptors located on a target cell that are specific for a particular chemical mediator

target cells The cells that a particular extracellular chemical messenger, such as a hormone or a neurotransmitter, influences

TATA box A promoter found next to almost all eukaryotic RNA-coding genes; site for the initial binding of RNA polymerase, an enzyme that makes an RNA copy of the coding gene sequence

teleological approach (tē′-lē-ō-LA-ji-kul) Explanation of body functions in terms of their particular purpose in fulfilling a bodily need, that is, the "why" of body processes (in contrast to *mechanistic* and *evolutionary explanations*)

telomere The end section of eukaryotic chromosomes; pieces of it are lost during cell replication

temporal lobes The lobes of the mammalian cerebral cortex that are located laterally and that are responsible for initially processing auditory input

temporal summation The summing of several postsynaptic potentials occurring very close together in time because of successive firing of a single presynaptic neuron

tension In muscles, the force produced during muscle contraction by shortening of the sarcomeres, resulting in stretching and tightening of the muscle's elastic connective tissue and tendon, which transmit the tension to the bone to which the muscle is attached

terminal button A vertebrate motor neuron's enlarged knoblike ending that terminates near a skeletal muscle fiber and releases acetylcholine in response to an action potential in the neuron

testosterone (tes-TOS-tuh-rōn) The vertebrate male sex hormone, secreted by the Leydig cells of the testes

tetanus (TET′-n-us) A smooth, maximal muscle contraction that occurs when the fiber is stimulated so rapidly that it does not have a chance to relax at all between stimuli

thalamus (THAL-uh-mus) The vertebrate brain region that serves as a synaptic integrating center for preliminary processing of all sensory input on its way to the cerebral cortex

thecal cells (THAY-kel) The outer layer of specialized ovarian connective tissue cells in a maturing follicle (in vertebrates)

thermal neutral zone (TNZ) A range of environmental temperatures in which an animal does not need to expend significant energy for thermoregulation

thermogenin An uncoupling protein in brown adipose tissue of mammals

thermoreceptor (thur′-mō-rē-SEP-tur) A sensory receptor sensitive to heat and cold

thick filaments Specialized cytoskeletal structures within skeletal muscle that are made up of myosin molecules and interact with the thin filaments to accomplish shortening of the fiber during muscle contraction

thin filaments Specialized cytoskeletal structures within skeletal muscle that are made up of actin, tropomyosin, and troponin molecules and interact with the thick filaments to accomplish shortening of the fiber during muscle contraction

thoracic cavity (thō-RAS-ik) Chest cavity

threshold potential The critical potential that must be reached before an action potential is initiated in an excitable cell

thrombocyte A blood cell in nonmammalian vertebrates that participates in clotting

thrombus An abnormal clot attached to the inner lining of a blood vessel

thymus (THIGH-mus) A vertebrate lymphoid gland located midline in the chest cavity that processes T lymphocytes and produces the hormone thymosin, which maintains the T-cell lineage

thyroglobulin (thī′-rō-GLOB-yuh-lun) A large complex molecule on which all steps of thyroid hormone synthesis and storage take place

thyroid gland An endocrine gland that traps and stores iodide from the blood of vertebrates and that secretes thyroxine and tri-iodothyronine, hormones that regulate vertebrate development and metabolism

thyroid hormone Collectively, the vertebrate hormones secreted by the thyroid follicular cells, namely, thyroxine and tri-iodothyronine

thyroid-stimulating hormone (TSH) An anterior pituitary hormone that stimulates secretion of thyroid hormone and promotes growth of the thyroid gland; thyrotropin

thyroxine (thī-ROCKS-in) The most abundant hormone secreted by the vertebrate thyroid gland; important in the regulation of overall metabolic rate; also known as tetraiodothyronine, or T_4

tidal breathing Movement of air or water in and out of a respiratory chamber via the same opening, in two distinct steps termed *inspiration* and *expiration*, for the purposes of gas exchange

tidal volume The volume of air entering or leaving tidal-type lungs during a single breath

tight junction An impermeable junction between two adjacent epithelial cells formed by the sealing together of the cells' lateral edges near their luminal borders; prevents passage of substances between the cells

tissue (1) A functional aggregation of cells of a single specialized type, such as nerve cells forming nervous tissue; (2) the aggregate of various cellular and extracellular components that make up a particular organ, such as lung tissue

titin Cytoskeletal elastic protein in the myofibril; forms a flexible filamentous network that can stretch under tension; may act as a locomotory spring that assists in the return of the sarcomere to its relaxed conformation

T_m See *transport maximum* and *tubular maximum*

TMAO (trimethylamine oxide) A counteracting osmolyte in many marine organisms, capable of offsetting the perturbing effects of urea, pressure, and other stressors on proteins

tone The ongoing baseline of activity in a given system or structure, as in muscle tone, sympathetic tone, or vascular tone

tonic receptors Sensory receptors that continue to generate action potentials throughout the duration of a stimulus; directly convey information about the duration of the stimulus

torpor A state of reduced metabolic rate in an animal occurring on a daily basis; involves lower body temperatures in endotherms

total peripheral resistance The resistance offered by all the peripheral blood vessels, with arteriolar resistance contributing most extensively

trachea (TRA-kē-uh) (1) The "windpipe"; the conducting airway that extends from the pharynx and branches into two bronchi, each entering a lung (in air-breathing vertebrates); (2) an air tube

in an air-breathing arthropod that leads directly to a tissue requiring oxygen

tracheole System of air-filled tubules that exchange respiratory gases from the outside world with insect tissues

tract A bundle of nerve fibers (axons of long interneurons) with a similar function within a long nerve cord (e.g., the spinal cord)

transducin G protein that links the excitation of rhodopsin molecules by light with a change in the current flowing across the membrane of photoreceptors

transduction Conversion of stimuli into action potentials by sensory receptors

transepithelial transport (tranz-ep-i-THĒ-lē-al) The entire sequence of steps involved in the transfer of a substance across the epithelium between either the renal tubular lumen or digestive tract lumen and the blood

transgenic organism An organism that has had genes from another species incorporated into its genome

transmural pressure gradient The pressure difference across the lung wall (intra-alveolar pressure is greater than intrapleural pressure) that stretches the lungs to fill the thoracic cavity, which is larger than the unstretched lungs (in air-breathing vertebrates)

transport maximum (T_m) The maximum rate of a substance's carrier-mediated transport across the membrane when the carrier is saturated; in the kidney tubules, known as *tubular maximum*

transporter recruitment The phenomenon of inserting additional transporters (carriers) for a particular substance into the plasma membrane, thereby increasing membrane permeability to the substance, in response to an appropriate stimulus

transverse tubule (T tubule) A perpendicular infolding of the surface membrane of a muscle fiber; rapidly spreads surface electric activity into the central portions of the muscle fiber

trehalose Diglucose (a disaccharide); in insects, synthesized in the fat body as a metabolic substrate for flight

triglyceride (trī-GLIS-uh-rīd) Neutral fat composed of one glycerol molecule with three fatty-acid molecules attached

tri-iodothyronine (T_3) (trī-ī-ō-dō-THĪ-rō-nēn) The most potent hormone secreted by the vertebrate thyroid follicular cells; important in the regulation of overall metabolic rate in birds and mammals

trimodal Three respiratory exchange surfaces, including the gills, lungs, and some cutaneous surfaces

trophic hormone "Nourishing" hormone involved in triggering cell growth and development; not to be confused with *tropic* hormone

trophoblast (TRŌ-fuh-blast′) The outer layer of cells in a mammalian blastocyst, responsible for accomplishing implantation and developing the fetal portion of the placenta

trophocytes Cells in the fat body of insects that perform a role in intermediary metabolism

tropic hormone (TRŌ-pik) A hormone that regulates the secretion of another hormone; not to be confused with *trophic* hormone

tropomyosin (trōp′-uh-MĪ-uh-sun) One of the regulatory proteins found in the thin filaments of muscle fibers

troponin (tro-PŌ-nun) One of the regulatory proteins found in the thin filaments of muscle fibers

TSH See *thyroid-stimulating hormone*

tuberous electroreceptors Found on the anterior body surface in the lateral line system, these receptors respond to the high-frequency signals from the fish's own electric organ discharge

tubular maximum (T_m) The maximum amount of a substance that renal tubular cells can actively transport within a given time period; the kidney cells' equivalent of transport maximum

tubular reabsorption The selective transfer of substances from the tubular fluid into the peritubular capillaries during the formation of urine

tubular secretion The selective transfer of substances from the peritubular capillaries into the tubular lumen during the formation of urine

twitch A brief, weak contraction that occurs in response to a single action potential in a muscle fiber

twitch summation The addition of two or more muscle twitches as a result of rapidly repetitive stimulation, resulting in greater tension in the fiber than that produced by a single action potential

twitchin A titinlike protein associated with the myosin for regulating catch states

tympanum A vibrating membrane; (1) an auditory membrane of certain insects, or eardrum of the vertebrate middle ear; (2) a rigid structure composed of three to six enlarged rings at the base of the trachea; used for song production in some species of birds

Type I alveolar cells (al-VĒ-ō-lur) The single layer of flattened epithelial cells that forms the wall of the alveoli within the mammalian lungs

Type II alveolar cells The cells within the mammalian alveolar walls that secrete pulmonary surfactant

ubiquitin A molecular tag that is added to cell proteins that are no longer needed, targeting them for destruction in proteasomes

ultrafiltration The net movement of a protein-free fluid across a tissue boundary

umami One of the five taste qualities in mammals; savoriness taste stimulated by amino acids

uncoupling protein A gated channel for hydrogen ions found in the inner membrane of mitochondria in some tissues in mammals; serves to dissipate the mitochondrial H^+ gradient into heat

upper critical temperature In an endotherm, the temperature at which internal heat-removing (active) mechanisms are activated to maintain thermal homeostasis

upregulation An increase in the capacity of a cellular or organ function

ureotelic Having urea as a primary nitrogenous waste product

ureter (yū-RĒ-tur) A duct that transmits urine from the kidney to the bladder

urethra (yū-RĒ-thruh) A tube that carries urine from the bladder to outside the body

uricotelic Having uric acid as a primary nitrogenous waste product

urinary excretion The elimination of substances from the body in the urine; anything filtered or secreted and not reabsorbed is excreted

urodaeum Middle portion of the avian cloaca; receives the fluid from the kidneys through the ureters in addition to material from the oviduct

vagus nerve (VĀ-gus) The vertebrate 10th cranial nerve, which serves as the major parasympathetic nerve

vasoconstriction (vā′-zō-kun-STRIK-shun) The narrowing of a blood vessel lumen as a result of contraction of the vascular circular smooth muscle

vasodilation The enlargement of a blood vessel lumen as a result of relaxation of the vascular circular smooth muscle

vasopressin (vā-zō-PRES-sin) A vertebrate hormone secreted by the hypothalamus, then stored and released from the posterior pituitary; increases the permeability of the distal and collecting tubules of mammalian kidneys to water and promotes arteriolar vasoconstriction; also known as *antidiuretic hormone (ADH)*

vasotocin A water-conserving hormone in many vertebrates, closely related to vasopressin in mammals

vault Organelle shaped like octagonal barrel; believed to serve as transporters for messenger RNA and/or the ribosomal subunits from the nucleus to sites of protein synthesis

vein A vessel that carries blood toward the heart

venous return (VĒ-nus) The volume of blood returned to each atrium per minute from the veins

ventilation The movement of air in and out of the lungs, or water across gills; called *breathing* if actively produced by the animal

ventricle (VEN-tri-kul) A lower chamber of the heart that pumps circulatory fluid into the arteries

ventrobronchi Cranial group of secondary bronchi in birds, that branch to form a fan-shaped covering of the mediodorsal surface of the avian lung

vesicle (VES-i-kul) A small intracellular, fluid-filled, membrane-enclosed sac

vesicular transport Movement of large molecules or multimolecular materials into or out of the cell by means of being enclosed in a vesicle, as in endocytosis or exocytosis

vestibular apparatus (veh-STIB-yuh-lur) The component of the vertebrate's inner ear that provides information essential for the sense of equilibrium and for coordinating head movements with eye and postural movements; consists of the semicircular canals, utricle, and saccule

vestigial Having no (or very minimal) useful function; evolved from useful features in ancestors but not yet eliminated by natural selection

villus (pl., **villi**) (VIL-us) Microscopic fingerlike projection from the inner surface of the small intestine

virulence (VIR-you-lentz) The disease-producing power of a pathogen

visceral afferent A pathway coming into the central nervous system that carries information derived from the internal viscera

visceral smooth muscle (VIS-uh-rul) See *single-unit smooth muscle*

viscosity (vis-KOS-i-tē) The friction developed between molecules of a fluid as they slide over each other during flow of the fluid; the greater the viscosity, the greater the resistance to flow

visual field Field of view that can be seen without moving the head

vital capacity The maximum volume of air that can be moved out during a single breath following a maximal inspiration

vitellogenesis Deposition of yolk in an oocyte

viviparous Referring to production and nourishment of living young within the body

voltage clamp A procedure in which the potential across a membrane is held at a constant value by an electronic circuit

voltage-gated channel Channel in the plasma membrane that opens or closes in response to changes in membrane potential

vomeronasal organ (**VNO**) Accessory olfactory organ in some mammals that detects pheromones; connected to two small openings in the anterior of the mouth behind the upper lip

white matter The portion of the central nervous system composed of myelinated nerve fibers

X-organs Organs in crustacean eyestalks that send axons into sinus glands (also in eyestalks) to release several neurohormones

Y-organs Organs in the crustacean head that make crustecydsone to promote molting

Z line A flattened, disclike cytoskeletal protein that connects the thin filaments of two adjoining sarcomeres

zeitgeber Any environmental factor that entrains a biological rhythm

zona fasciculata (zō-nah fa-SIK-ū-lah-ta) The middle and largest layer of the adrenal cortex; major source of glucocorticoids

zona glomerulosa (glō-MER-yū-lō-sah) The outermost layer of the adrenal cortex; sole source of aldosterone

zona reticularis (ri-TIK-yuh-lair-us) The innermost layer of the adrenal cortex; produces the sex steroid DHEA

ANATOMICAL TERMS USED TO INDICATE DIRECTION AND ORIENTATION

Anterior	situated in front of or in the front part of
Posterior	situated behind or toward the rear
Ventral	toward the belly or front surface of the body; synonymous with *anterior*
Dorsal	toward the back surface of the body; synonymous with *posterior*
Medial	denoting a position nearer the midline of the body or a body structure
Lateral	denoting a position toward the side or farther from the midline of the body or a body structure
Superior	toward the head
Inferior	away from the head
Proximal	closer to a reference point
Distal	farther from a reference point
Sagittal section	a vertical plane that divides the body or a body structure into right and left sides
Longitudinal section	a plane that lies parallel to the length of the body or a body structure
Cross section	a plane that runs perpendicular to the length of the body or a body structure
Frontal or coronal section	a plane parallel to and facing the front part of the body
Apical	surface of the plasma membrane that faces the lumen
Basolateral	surface of the plasma membrane that faces toward the interstitium

WORD DERIVATIVES COMMONLY USED IN PHYSIOLOGY

a; an-	absence or lack	epi-	above; over	osteo-	bone
ad-; af-	toward	erythro-	red	oto-	ear
adeno-	glandular	gastr-	stomach	para-	near
angi-	vessel	-gen; -genic	produce	pariet-	wall
anti-	against	gluc-; glyc-	sweet	peri-	around
archi-	old	hemi-	half	phago-	eat
-ase	splitter	hemo-	blood	-pod	footlike
auto-	self	hepat-	liver	-poiesis	formation
bi-	two; double	homeo-	similar	poly-	many
-blast	former	homo-	same	post-	behind; after
brady-	slow	hyper-	above; excess	pre-	ahead of; before
cardi-	heart	hypo-	below; deficient	pro-	before
cephal-	head	inter-	between	pseudo-	false
cerebr-	brain	intra-	within	pulmon-	lung
chondr-	cartilage	kal-	potassium	rect-	straight
-cide	kill; destroy	leuko-	white	ren-	kidney
contra-	against	lip-	fat	reticul-	network
cost-	rib	macro-	large	retro-	backward
crani-	skull	mamm-	breast	acchar-	sugar
-crine	secretion	mening-	membrane	sarc-	muscle
crypt-	hidden	micro-	small	semi-	half
cutan-	skin	mono-	single	-some	body
-cyte	cell	multi-	many	sub-	under
de-	lack of	myo-	muscle	supra-	upon; above
di-	two; double	natr-	sodium	tachy-	rapid
dys-	difficult; faulty	neo-	new	therm-	temperature
ecto-; exo-; extra-	outside; away from	nephr-	kidney	-tion	act or process of
ef-	away from	neuro-	nerve	trans-	across
-elle	tiny; miniature	oculo-	eye	tri-	three
-emia	blood	-oid	resembling	vaso-	vessel
encephalo-	brain	ophthalmo-	eye	-uria	urine
endo-	within; inside	oral-	mouth		